GATE

COMPUTER SCIENCE AND INFORMATION TECHNOLOGY

THE COMPLETE REFERENCE

GATE

COMPUTER SCIENCE AND INFORMATION TECHNOLOGY

THE COMPLETE REFERENCE

By

E. Srinivasa Reddy, B.Tech., M.Tech., Ph.D
Vice-principal and Professor,
University College of Engineering,
Acharya Nagarjuna University, Guntur.

J.V.N. Ramesh, M.Tech., Ph.D
Associate Professor
Department of Computer Science and Engineering
Narasaraopeta Institute of Technology, Narasaraopeta, Guntur.

BSP **BS Publications**
A unit of **BSP Books Pvt. Ltd.**
4-4-309, Giriraj Lane, Sultan Bazar,
Hyderabad - 500 095 - A.P.
Phone : 040 - 23445605, 23445688

Published by :

BSP BS Publications
A unit of **BSP Books Pvt. Ltd.**
4-4-309, Giriraj Lane, Sultan Bazar,
Hyderabad - 500 095 - A.P.
Phone : 040 - 23445605, 23445688
e-mail : contactus@bspublications.net
website: www.bspublications.net

Printed at

Adithya Offset Process (I) Pvt. Ltd.
Hyderabad.

Price : Rs. 895.00
ISBN : 978-93-81075-61-6

CONTENTS

UNIT - XVI NUMERICAL REASONING

UNIT - XVII GATE PAPERS

SYLLABUS - 2011
COMPUTER SCIENCE AND INFORMATION TECHNOLOGY

GENERAL APTITUDE(GA)

Verbal Ability : English grammar, sentence completion, verbal analogies, word groups, instructions, critical reasoning and verbal deduction

Numerical Ability : Numerical computation, numerical estimation, numerical reasoning and data interpretation.

ENGINEERING MATHEMATICS

Mathematical Logic : Propositional Logic; First Order Logic.

Probability: Conditional Probability; Mean, Median, Mode and Standard Deviation; Random Variables; Distributions, uniform, normal, exponential, Poisson, Binomial.

Set Theory & Algebra: Sets; Relations; Functions; Groups; Partial Orders; Lattices; Boolean Algebra.

Combinatorics : Permutations; Combinations; Counting; Summation; generating functions; recurrence relations; asymptotics.

Graph Theory: Connectivity; spanning trees; Cut vertices & edges; covering; matching; independent sets; Colouring; Planarity; Isomorphism.

Linear algebra: Algebra of matrices, determinants, systems of linear equations, Eigen values and Eigen vectors.

Numerical Methods: LU decomposition for systems of linear equations; numerical solutions of non linear algebraic equations by Secant, Bisection and Newton-Raphson Methods; Numerical integration by trapezoidal and Simpson's rules.

Calculus : Limit, Continuity & differentiability, Mean value Theorems, Theorems of integral calculus, evaluation of definite & improper integrals, Partial derivatives, Total derivatives, maxima & minima.

COMPUTER SCIENCE AND INFORMATION TECHNOLOGY

Digital Logic: Logic functions, Minimization, Design and synthesis of Combinational and Sequential circuits; Number representation and Computer Arithmetic (fixed and floating point);

Computer Organization and Architecture : Machine instructions and addressing modes, ALU and Data-path, CPU control design, Memory interface, I/O interface (Interrupt and DMA mode), Instruction pipelining, Cache and main memory, Secondary storage. Programming and Data Structures : Programming IN C; Functions, Recursion, Parameter passing, Scope, Binding : Abstract data types, Arrays, Stacks, Queues, Linked Lists, Trees, Binary search threes, Binary heaps.

Algorithms : Analysis, Asymptotic notation, Notions of space and time complexity, Worst and average case analysis; Design : Greedy approach, Dynamic programming, Divide - and - conquer; Tree and graph traversals, Connected components, Spanning trees, Shortest paths; Hashing, Sorting, Searching. Asymptotic analysis (best, worst, average cases) of time and space, upper and lower bounds, Basic concepts of complexity classes -P , NP, NP - hard, NP-complete.

Theory of Computation : Regular languages and finite automata, Context free languages and Push-down automata, Recursively enumerable sets and Turing machines, Un-decidability;

PROGRAMMING AND DATA STRUCTURES

Operating System : Processes, Threads, Inter - process communication, Compiler Design, Concurrency, Synchronization, Deadlock, CPU scheduling , Memory management and virtual memory, File systems, I/O systems, Protection and security.

Data bases: ER-model , Relational model (relational algebra, tuple calculus), Database design (integrity constraints, normal forms), Query languages (SQL), File structures (sequential files, indexing , B and B+ trees). Transactions and concurrency control

Information Systems and Software Engineering : information gathering, requirement and feasibility analysis, data flow diagrams, process specifications, input / output design, process life cycle, planning and managing the project, design , coding, testing, implementation, maintenance.

Computer Networks : ISO/OSI stack, LAN technologies (Ethernet, Token ring), Flow and error control techniques, Routing algorithms, Congestion control, TCP/UDP and sockets, IP (v4), Application layer protocols (icmp, dns, smtp, pop, ftp, http); Basic concepts of hubs, switches, gateways, and routers. Network security - basic concepts of public key and private key cryptography, digital signature, firewalls.

Web technologies : HTML, XML, basic concepts of client - server computing.

Unit - I
Engineering Mathematics

UNIT - I ENGINEERING MATHEMATICS

I. PERMUTATIONS AND COMBINATIONS

- **Fundamental Principle**
- **Types of Permutations**
- **Combinations**
- **Counting**

1. Introduction

Objects (or things) can be arranged in many ways. Suppose there are 3 objects marked a, b, c on a table. From these two objects can be selected at a time in three different ways as ab, ac, bc. In this way selection of two objects from 3 objects in three ways is called Combination.

The above selections ab, ac, bc can also be arranged as ba, ca, cb. So, We can understand that 2 objects can be selected from 3 objects and arranged in 6 ways. These arrangements are called Permutations.

2. Fundamental Principle

If A is a finite set, then the number of different elements in A is denoted by n(A). Note that here n(A) is a non-negative integer.

Example:

1. If A = {2,5,7} then n (A) = 3
2. If $A = \{a, \frac{2}{3}, 5, 0, \frac{1}{3}\}$ then n(A) = 5
3. If C = ϕ (empty set), then n(C) = 0.

Note : Let us assume that there are three routes, say a_1, a_2, a_3 from Hyderabad (H) to Vijayawada (V) and there are two routes, say b_1, b_2 from Vijayawada to Tirupathi (T). If they are written as $A = \{a_1, a_2, a_3\}$, $B = \{b_1, b_2\}$ then n(A)=3 and n(B)=2. Now we can match the route a_1 from H to V with the two routes b_1, b_2 from V to T. That is (a_1, b_1), (a_1, b_2). Similarly by the remaining routes can be written as (a_2, b_1), (a_2, b_2), (a_3, b_1), (a_3, b_2). So to travel from H to V via V, there are 6 different routes (a_1, b_1), (a_1, b_2), (a_2, b_1), (a_2, b_2), (a_3, b_1), (a_3, b_2).

In this arrangement the a's in A should come in the first place and b's in B come the next. So in the above 6 pairs importance is given for the places of the elements. Here one can understand that these 6 ways are nothing but the elements of the cartesian product of the two sets A and B.

That is $A \times B = \{(a_1, b_1), (a_1, b_2), (a_2, b_1), (a_2, b_2), (a_3, b_1), (a_3, b_2)\}$

$$n(A \times B) = 6 = 3 \times 2 = n(A).\ n(B)$$

3. Fundamental Principle of Counting

If a man can do a work in p different ways and another work in q different ways, then he can do the two works one after the other in pq ways. Now treating one way of doing a work as one element, we form the two sets A and B :

We denote the set of ways of doing the first work by A and the set of ways of doing the second work by B. Then we have n(A)=p and n(B)=q. The number of ways one can do the two works in an order is $n(A \times B) = n(A) \times n(B) = pq$. If the same principle is extended to 'm' works then we can write $n(A_1 \times A_2 \times \ldots \times A_m) =$

$$n(A_1)\, n(A_2) \ldots\ldots n(A_m).$$

4. Permutations

I. Linear Permutations

From the elements of a finite set A (In general distinct elements) taking some or all the elements and arranging them linearly is called a linear permutation.

II. Similar Permutations

If, in a given set of things, some are alike, permutations formed from them are known as similar permutations. In this case the number of permutations will change because we can not identify the change among the like things.

III. Circular Permutations

If 'n' distinct things are arranged in a circular manner giving preference only to relative positions then we call it a circular permutation.

Note : In a circular permutation if the places are changed circularly, the permutations will not change.

Ex : If A = {4, 5, 6} then the permutations obtained by arranging all elements in A in an order are : 456, 465, 546, 564, 645, 654.

Note : From a set A of n elements, the number of permutations taken $r(\leq n)$ at a time is denoted by ${}^nP_r$, or P(n, r).

Notation : The product of first 'n' natural numbers is called factorial n and is denoted by n!. We define 0!=1 and we observe that 1! = 1.

For example : $6! = 6 \times 5 \times 4 \times 3 \times 2 \times 1 = 720$.

Example : Taking all the letters at a time from A, A, B, the possible number of permutations is 3. They are AAB, ABA, BAA.

Here we can not find difference between A's but if you find it then the number of permutations is $3p_3 = 3! = 6$.

Theorem : The number of permutations of 'n' dissimilar things taken $r(r \leq n)$ at a time is

$${}^nP_r = n(n-1)\,(n-2) \ldots\ldots (n-r+1)$$

Note : A finite bijective mapping is a permutation. If S = {a, b, c, d, e} and defining a permutation 'f' as f(a) = b; f(b)=d; f(c)=a; f(d)=e; f(e)=c then we denote the permutation f as $f = \begin{pmatrix} a & b & c & d & e \\ b & d & a & e & c \end{pmatrix}$

Note : ${}^nP_n = n(n-1)\,(n-2) \ldots\ldots (n-r+1) =$

$$n(n-1)(n-2) \ldots\ldots 2.1 = 1.2.3 \ldots\ldots (n-2)(n-1)n = n!$$

That is, the number of permutations containing all objects from 'n' distinct things is $^np_n = n!$

$$\therefore {}^nP_r = \frac{n!}{(n-r)!}$$

SOLVED EXAMPLES

Ex 1 : If ${}^nP_r = 3024$, find n and r ?

Sol : Since ${}^nP_r$ is the product of consecutive positive integers we have

${}^nP_r = 3024 = 9 \times 8 \times 7 \times 6$. $\therefore$ n = 9 and r = 4.

Ex 2: If ${}^nP_r = 30\ {}^nP_4$ find the value of n?

Sol : We have ${}^nP_6$ = n(n-1) (n-2) (n-3) (n-4) (n-5)

${}^nP_4$ = n(n-1) (n-2) (n-3)

$\therefore$ From the hypothesis n (n-1) (n-2) (n-3) (n-4) (n-5) = 30n (n-1) (n-2)(n-3)

$\Rightarrow$ (n-4)(n-5) = 30 $\Rightarrow n^2$-9n-10 = 0 $\Rightarrow$ (n+1)(n-10) = 0

$\Rightarrow$ n=-1 or n=10

Since n is a positive integer, n=10.

Ex 3 : In an Assembly elections if 7 members are contesting for a single Assembly constituency, then in how many ways can their names be printed on the ballot paper?

Sol : The required number of ways of adjusting the names of the 7 members in 7 places is ${}^7P_7 = 7! = 7 \times 6 \times 5 \times 4 \times 3 \times 2 \times 1 = 5040$

$\therefore$ The number of ways of printing the names of 7 members on ballot paper is 5040.

Ex 4 : How many ways can 5 men and 3 woman be seated in a row so that two woman cannot sit side by side ?

Sol : Since there is no restriction on men, they can sit in a row in 5! = 120 ways.

Since woman cannot sit side by side they can sit in between two men or in the first place or in the last place as shown below.

X [M] X [M] X [M] X [M] X [M] X

If the sign 'X' shows the woman place, there are 6 vacancies for woman to sit. In these 6 vacancies 3 woman can sit in ${}^6P_3$ = 120 ways.

$\therefore$ The number of ways where 5 men and 3 woman can sit with the given condition is 120 $\times$ 120 = 14400.

Theorem : ${}^nP_r = (n-1)_{P_r} + r\,(n-1)_{P_{(r-1)}}$

To find the number of Circular Permutations

Circular permutations are of two types, one is clockwise and the other is anti-clockwise.

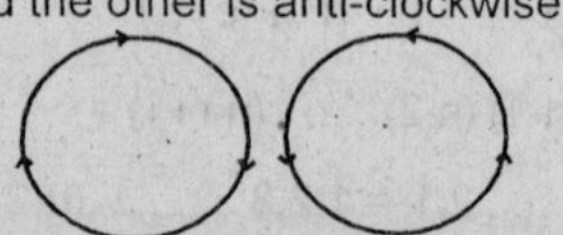

clockwise Anticlock-wise

Note : Considering the direction, the above two permutations are different otherwise they are the same.

From the required 'n' objects, the object 'a' can be put in any one of 'n' places. After filling a place by 'a' the remaining (n-1) vacancies can be filled by (n-1) objects in (n-1)! ways. So the number of circular permutations formed with 'n' objects is (n-1)!.

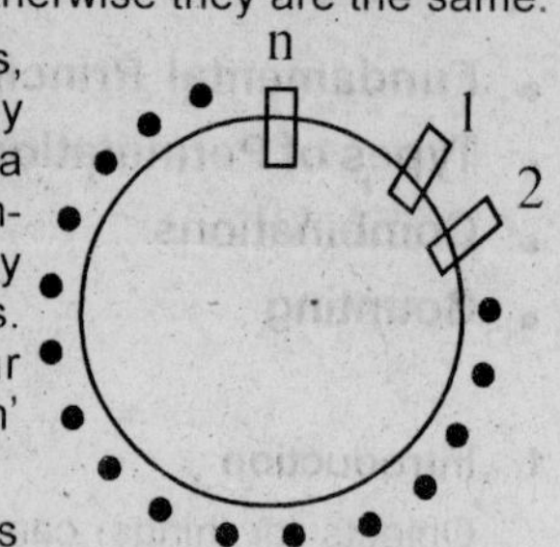

Note : In circular permutations the direction of the arrangements of the objects is also important. The above mentioned circular permutations should be considered as total number of clock-wise and anti-clockwise permutations. Since the number of permutations are equal in both the directions and if the direction is not considered then the number of circular permutations is $\frac{(n-1)!}{2}$

Theorem : The number of circular permutations of n different things taken all at a time is (n-1)!

SOLVED EXAMPLES

Ex 1 : In how many ways can 6 people sit on a round table ?

Sol : From the above Theorem, the number of permutations of sitting is (6-1)! = 5! = 120.

If the clock-wise and anti-clockwise directions are not considered, the number of permutations is $\frac{120}{2} = 60$

Ex 2 : Find the number of ways that 5 boys and 5 girls can sit circularly such that a boy should sit between two girls.

Sol : 5 girls can sit circularly in (5-1)! = 4! = 24 ways

As shown in the figure 5 boys can sit in five vacant places between 5 girls. After the girls are made to sit, we should consider in which place should the first boy sit. If we consider 5 boys should sit in a row they can sit in 5 vacancies in 5! = 120 ways.

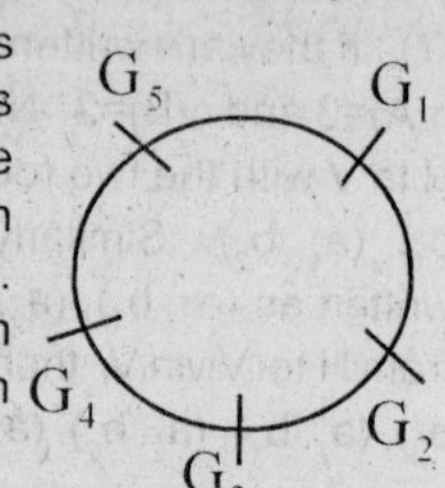

Therefore 5 boys and 5 girls can sit circularly with the given condition in 24 $\times$ 120 = 2880 ways.

Theorem : The number of permutations of 'n' dissimilar things, taken 'r' at a time, when the repetition of things is allowed any number of times, is n^r.

Ex 3 : There are 9 objects and 9 boxes. Out of 9, 5 objects cannot fit to be put in 3 small boxes. How many arrangements can be made, such that each object can be put in one box only ?

Sol : Since 5 objects cannot fit to be put 3 small boxes, these 5 objects can be put in the remaining 6 boxes in ${}^6P_5$ ways. The remaining 4 objects can be put in ${}^4P_4$ ways.

∴ The total number of arrangements = $6P_5 \times 4P_4$
$= (6 \times 5 \times 4 \times 3 \times 2) \times (4 \times 3 \times 2 \times 1) = 720 \times 24 = 17280.$

Theorem : There are 'n' objects. Out of which m(< n) objects are identical and the remaining are dissimilar. Using all objects, the number of permutations from 'n' objects is $\frac{n!}{m!}$

Note : In 'n' objects, if 'm' objects are of one type, 'p' objects are of another type and 't' objects are of different type etc., then, the number of permutations formed from n = (m+p+....+t) objects is $\frac{n!}{m!p!........t!}$

Note : In 'n' objects, if 'm' objects are of one type and (n-m) objects are of another type, then the number of permutations using all objects at a time is $\frac{n!}{m!\,(n-m)!}$

Note : In 'n' objects, if 'm' objects are of one type, q objects are of another type and (n-p-q) objects are of different type then using all the objects, the number of permutations formed is $\frac{n!}{p!\,q!\,(n-p-q)!}$

Ex 4: In the word 'MANORAMA', find

I. The number of permutations formed taking all letters.

II. Out of these the number of permutations with all A's together.

III. The number of permutations which start with A and end with 'M'.

Sol : I. In the given word there are 8 letters. Out of which 2 'M's, 3 'A's, 1 'N', 1 'O' and 1 'R'.

The number of permutations taking all letters

$= \frac{8!}{2!\,3!\,1!\,1!\,1!} = \frac{8 \times 7 \times 6 \times 5 \times 4 \times 3 \times 2 \times 1}{(2 \times 1)(3 \times 2 \times 1)} = 3360$

II. In the given word there are 3 'A's. Keeping them as one letter, there will be 6 letters. Out of these, there are 2 'M's and the remaining are different.

The number of permutations formed with these letters

$= \frac{6!}{2!} = \frac{6 \times 5 \times 4 \times 3 \times 2 \times 1}{2 \times 1} = 360$

III. To find the words which start with A and end with M, in the remaining 6 letters there 2 'A's, 1 'M', 1 'N', 1 'O' and 1 'R'.

∴ The required number of permutations $= \frac{6!}{2!1!1!1!1!}$

$= \frac{6 \times 5 \times 4 \times 3 \times 2 \times 1}{2 \times 1} = 360$

5. Combinations

Definition : n, r are integers. If n≥r, for the set of n elements a subset of r elements is called a combination.Therefore a combination is an unordered selection of r elements from a set of n elements. The number of combinations of r elements selected from n elements is denoted by nc_r or $\binom{n}{r}$ or C(n,r)

Example : Out of 40 students in a class any 4 students can sit on a bench without considering the order in $40c_4$ ways.

Theorem : To find the number of combinations of 'n' dissimilar things taken 'r' at a time is $nc_r = \frac{n!}{r!(n-r)!}$

Note : There is no importance for the order of arrangements in combinations.

Theorem : $nc_r = nc_{n-r}$

Note : I. ${}^nC_n = \frac{n!}{n!\,(n-n)!} = \frac{n!}{n!\,0!} = 1\ (\because 0! = 1)$

II. ${}^nC_0 = \frac{n!}{0!(n-0)!} = \frac{n!}{n!} = 1$

Theorem : If ${}^nC_r = {}^nC_s$ then neither r = s or r + s = n.

Theorem : $(n-1)_{C_{(r-1)}} + (n-1)_{C_r} = n_{C_r}$

Theorem : If n, r are positive integers with 1≤ r ≤ n, then $n_{C_r} + n_{C_{(r-1)}} = (n+1)_{C_r}$

Note : $(n+1)c_r - nc_r = nc_{(r-1)}$ or $(n+1)c_r - nc_{(r-1)} = nc_r$.

Corollary : If m and n are positive integers with m ≠ n, then the total m+n distinct objects can be divided into two groups of m objects and n objects in $\frac{(m+m)!}{m!n!}$ ways.

Note : Let 2n objects are to be divided into 2 equal groups. Here applying the above corollary, there will not be any change in the division, if the bags A and B are filled with the objects n and n instead of m and n objects. But the interchange can be done in 2! ways. So, the number of dividing 2n objects into n and n objects is $\frac{(2n)!}{2!n!n!} = \frac{(2n)!}{2!(n!)^2}$

Note : (m+n+p) objects can be divided into 3 groups of m objects, n objects and p objects in $\frac{(m+n+p)!}{m!n!p!}$ ways

Note : 3n objects can be divided into 3 equal groups in $\frac{(3n)!}{3!(n!)^3}$ ways.

Corollary : Suppose that there are p objects of one type and q objects are of another type, then the number of combinations selected from the total (p+q) objects is [(p+1)(q+1) - 1].

Note : If there are p objects of one type q objects of 2nd type and so on....and t objects of another type, then the total number of combinations formed from (p+q+.....+t) objects is {[(p+1)(q+1) ....(t+1)]-1}.

In the above note if p =q =......=t = 1, and they are 'n' in number, then the total number of combinations formed

$= (1+1)\,(1+1)\, \ldots\ldots\, (1+1) - 1$
(n times)
$= (2.\,2.\, \ldots\ldots\ldots\, 2) - 1 = 2^n - 1$
(n times)
$= 2^n - 1.$

Important Note : Remember that the number of subsets in a power set formed from the 'n' elements is 2^n-1 (except the empty set).

SOLVED EXAMPLES

Ex 1 : If $nc_6 = nc_8$, then what is nc_9 ?

Sol : Since $nc_6 = nc_8$, we have n = 6+8 = 14

$$\therefore\ 14c_9 = 14c_5 = \frac{14.13.12.11.10}{5.4.3.2.1} = 2002$$

Ex 2 : In how many ways a team of 3 boys and 3 girls can be selected from 6 boys and 4 girls ?

Sol : 3 boys can select from 6 boys in

$$6c_3 = \frac{6!}{3!3!} = \frac{6\times5\times4}{3\times2\times1} = 20 \text{ ways.}$$

3 girls can be selected from 4 girls in $4c_3 = 4c_1 = 4$ ways
∴ From the fundamental principle, the required number of ways of selection is 20 × 4= 80.

Note : The number of straight lines drawn through 'n' points on a circle is $nC_2 = \frac{n(n-1)}{2\times1} = \frac{n(n-1)}{2}$

Ex 3 : How many straight lines can be drawn through 10 points on a circle ?

Sol : Since two points are necessary to draw a straight line, the number of straight lines drawn through 10 points is $10c_2 = \frac{10\times9}{2\times1} = 45.$

Ex 4 : Find the diagonals of a polygon of n sides. Find the number of triangles formed by joining the vertices of the polygon.

Sol : The number of straight lines formed by joining the 2 vertices from 'n' vertices of the polygon is equal to the number of ways of selecting 2 vertices from 'n' vertices.

Therefore they are $nc_2 = \frac{n(n-1)}{2\times1}$ Out of these straight lines, 'n' lines are the sides of polygon, so the number of diagonals $= \frac{n(n-1)}{2\times1} - n$

$$= \frac{n^2-n-2n}{2} = \frac{n^2-3n}{2} = \frac{n(n-3)}{2}$$

Since a triangle is formed with any three diagonals of the polygon the number of triangles formed with 3 vertices.

$$= nc_3 = \frac{n(n-1)\,(n-2)}{3\times2\times1} = \frac{n(n-1)\,(n-2)}{6}$$

Ex 5 : There are 3 questions in 1st section, 3 questions in the 2nd section and 2 questions in the 3rd section in a question paper of an exam. In how many ways can a student who appeared for the exam select to answer any 5 questions of the paper choosing at least 1 question from each section.

Sol : Out of 8 questions, the student can select 5 questions in following ways.

The number of ways of selection in 1st way

$$= 3c_3 \times 3c_1 \times 2c_1 = 6$$

The number of ways of selection in 2nd way

$$= 3c_2 \times 3c_2 \times 2c_1 = 18$$

The number of ways of selection in 3rd way

$$= 3c_2 \times 3c_1 \times 2c_2 = 9$$

The number of ways of selection in 4th way

$$= 3c_1 \times 3c_2 \times 2c_2 = 9$$

The total number of ways of selection

$$= 6 + 18 + 9 + 9 = 42$$

6. Counting : Counting problems are encountered through out Mathematics and Computer Science. As an example, to determine the probabilites of occurrence of discrete events, one has to count the successful outcomes of experiments and all the possible outcomes of these experiments.

Basic Counting rules : There are two basic counting rules which can be used to solve many different counting problems. They are the sum rule and the product rule.

The Sum rule : If a first task can be done in n_1 ways and a second task in n_2 ways, and if these tasks are such that they cannot be done at the same time then there are $n_1 + n_2$ ways to do either task.

The Product rule : Whenever a procedure can be broken down into two tasks and then there are n_1 ways to do the first task and n_2 ways to do the next task after the first task has been done, then there are $n_1 n_2$ ways to do the procedure.

SOLVED EXAMPLES

Ex 1 : In a shop, you have the right to choose from three almirahs, articles of your choice. These almirahs contain 27, 29 and 32 articles respectively. How many possible articles are there to choose from ?

Sol : You can choose an article from the first almirah in 27 ways, from the second almirah in 29 ways and from the third almirah in 32 ways. Hence, applying the sum rule, there are 27+29+32 = 88 articles to choose from.

Ex 2 : Threre are 32 rows of chairs in an auditorium. Each row has 20 chairs. How many different chairs to a row in the auditorium are there ?

Sol : The choice of chair has two tasks. The first task is to select the row and then picking up a chair in the row. Since there are 32 rows, there are 32 ways to choose a row i.e., the first task. After completing it, there are 20 ways to choose a chair, no matter which row has been selected. Hence, applying product rule, there are 32 × 20 = 640 chairs.

Ex 3 : How many different bit strings are there of length nine ?

Sol : Each of the nine bits can be chosen in two ways. Therefore, the product rule shows that there are a total of $2^9 = 512$ different bit strings of length nine.

Ex 4 : How many one-one functions are there from a set with m elements to one with n elements ?

Sol : Whenever $m > n$ there are no one-one functions from a set with m elements to a set with n elements. Let us say $m \leq n$. If the elements in the domain are assumed to be $a_1, a_2, ..., a_n$, then there are n ways to choose the value of the function at a_1.

Since the function is one-one there are only (n-1) ways to pick up for a_2, as the value of the function. In general the value of the function can be chosen in n-k+1 ways for a_k. Applying the product rule, there are n (n-1) ..... (n- k+1) one-one functions from a set with m elements to one with n elements.

Ex 5 : How many bit strings of length nine either start with 1 bit or end with the two bits 00 ?

Sol : Procedure can be divided into two tasks. The first task is constructing a bit string of length nine beginning with a 1 bit. This can be done in 2^8 ways = 256 ways.

The second task of constructing a bit string of length nine ending with two bits 00, can be done in 2^7 = 128 ways.

From this we have to remove the number of ways of choosing 1 in the first bit and corresponding 00 in the last two bits. This can be done in 2^6 ways = 64 ways.

Therefore, the total number of bit strings of length nine that can be constructed , that begin with 1 and end with a 00 is equal to 256 + 128 -64 = 320 ways.

Ex 6 : How many bit strings of length four do not have two consecutive 1's?

Sol : Let us be solve it by drawing a tree diagram.

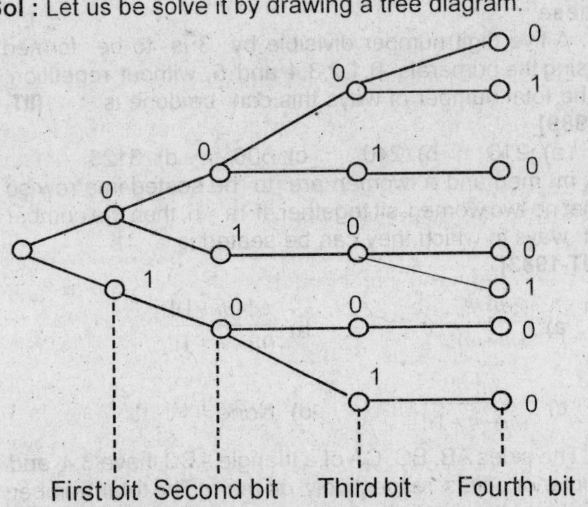

Fig : Tree diagram for bit strings of length four without two consecutive 1's

From Figure, by inspection it can be found out that there are eight strings of length four without two consecutive 1's.

OBJECTIVE QUESTIONS - I

1. The number of different permutations of the word BANANA is **[IIT- 1960]**

a) 720 b) 60 c) 120 d) 360

2. Seven women and seven men are to sit round a circular table such that there is a man on either side of every women; the number of seating arrangements is **[JNTU- 1990]**

a) $(7!)^2$ b) $(6!)^2$ c) $6! \times 7!$ d) $7!$

3. 12 persons are to be arranged to a round table. If two particular persons among them are not to be side by side, the total number of arrangements is **[JNTU- 1994]**

a) 9 (10 !) b) 2 (10 !) c) 45 (8 !) d) 10 !

4. The number of ways in which a team of eleven players can be selected from 22 players including 2 of them and excluding 4 of them is **[JNTU- 1995]**

a) ${}^{16}C_{11}$ b) ${}^{16}C_5$ c) ${}^{16}C_9$ d) ${}^{20}C_9$

5. In how many ways can 5 read and 4 white balls be drawn from a bag containing 10 red and 8 white balls **[JNTU- 1991]**

a) ${}^{8}C_5 \times {}^{10}C_4$ b) ${}^{10}C_5 \times {}^{8}C_4$

c) ${}^{18}C_9$ d) None

6. How many 10 digit numbers can be written by using the digits 1 and 2 **[IIT- 1992]**

a) ${}^{10}C_1 + {}^{9}C_2$ b) 2^{10}

c) ${}^{10}C_2$ d) 10 !

7. The number of diagonals that can be drawn by joining the vertices of an octagon is **[JNTU- 1993]**

a) 28 b) 48 c) 20 d) None

8. In an examination there are three multiple choice Questions and each question has 4 choices. Number of ways in which a student can fail to get all answers correct is **[JNTU- 1990]**

a) 11 b) 12 c) 27 d) 63

9. Ramesh has 6 friends. In how many ways can he invite one or more of them at a dinner ? **[JNTU- 1994]**

a) 61 b) 62 c) 63 d) 64

10. If ${}^{n}C_{12} = {}^{n}C_8$, then n = **[JNTU- 1979]**

a) 20 b) 12 c) 6 d) 30

11. Total number of words formed by 2 vowels and 3 consonants taken from 4 vowels and 5 consonants is equal to **[JNTU- 1990]**

a) 60 b) 120 a) 7200 d) None

12. The number of words that can be formed out of the letters of the word 'COMMITTEE' is **[JNTU- 1986]**

a) $\frac{9!}{(2!)^3}$ b) $\frac{9!}{(2!)^2}$ c) $\frac{9!}{2!}$ d) 9 !

13. In how many ways can 5 prizes be distributed among 4 boys when every boy can take one or more prizes ? **[JNTU- 1990]**

a) 1024 b) 625 c) 120 d) 600

14. The number of different signals which can be given from 6 flags of different colours taking one or more at a time, is **[JNTU- 1990]**

a) 1958 b) 1956 c) 16 d) 64

15. How many different committees of 5 can be formed from 6 men and 4 women on which exact 3 men and 2 women serve ? **[JNTU- 1995]**

a) 6 b) 20 c) 60 d) 120

16. The number of ways to arrange the letters of the word CHEESE are **[JNTU-1995]**

a) 120 b) 240 c) 720 d) 6

ANSWER KEY - 1				
1. b	**2.** c	**3.** a	**4.** c	**5.** b
6. b	**7.** c	**8.** d	**9.** c	**10.** a
11. c	**12.** a	**13.** a	**14.** b	**15.** d
16. a				

EXPLANATORY NOTES - I

1. There are 6 letters in the given word, of which 2 are N's, 3 are A's. So total no. of words in the number of arrangements of 6 things of which 3 are similar kind, 2 are similar kind. i.e., $\frac{6!}{3!2!}=\frac{720}{3\times2\times2}=60$

2. The seven women can be seated with a circle table such that there is a vacant seat between every pair of women is $(7-1)!=6!$ ways. Hence, the required number of seating arrangements $=6!\times7!$.

3. To find the number of ways in which 2 particular persons never sit together. We subtract the number of ways in which they sit together from the total number of seating arrangements of 12 persons. Clearly 12 persons can seated round a circle in $(12-1)!=11!$ ways. Hence the required number of seating arrangements $=11!-2\times10!=9(10!)$.

4. The number of ways in which a team of eleven players can be selected from 22 players including 2 of them and excluding 4 of them is

$$^{22-4-2}C_{22-4-2-11}={}^{16}C_5$$

5. The number of ways can 5 red and 4 white balls be drawn from a bag containing 10 red balls and 8 white balls is $^{10}C_5\times{}^8C_4$

6. The number of ways of 10 digits numbers can be written by using the 1 and 2 digits is 2^{10}.

7. We know that the formula is number of diagonals of n sided polygon is $\frac{n(n-3)}{2}$. Given octagon contains 8 sides.

$\therefore \frac{8(5)}{2}=20$

9. Ramesh has invites all the friends in $^6C_6$ ways. He invite 5 friends in $^6C_5$ ways. He invites 4 friends in $^6C_4$ ways. He invites 3 friends in $^6C_3$. He invites 2 friends in $^6C_2$ ways. He invites one friend in $^6C_1$ ways. The number of ways he can invite one or more of them is $^6C_6+{}^6C_5+{}^6C_4+{}^6C_3+{}^6C_2+{}^6C_1=63$.

10. If $^nC_{12}={}^nC_8$ then

$12+8=n \quad \therefore n=20$

11. 2 vowels out of 4 and 3 consonants out of 5 can be chosen $^4C_2\times{}^5C_3$ ways. Thus there are $^4C_2\times{}^5C_3$ groups, each group contains 5 letters, which can be arranged among themselves in 5 ! ways. Hence, the required number of words

$$\left(^4C_2\times{}^5C_3\right)\times5!=\frac{4\times3}{1\times2}\times\frac{5\times4\times3}{2\times3}\times120=7200$$

12. There are 9 letters in the given word of which 2 are M's, 2 are T's, 2 are E's. So total number of arrangements of 9 things of which 2 are same kind, 2 are same kind, 2 are same kind. i.e., $\frac{9!}{2!2!2!}=\frac{9!}{(2!)^3}$

13. The first prize can be given away in 4 ways as it may be given to any of the 4 boys. The second prize can also be given away in 4 ways. The third prize can also be given away in 4 ways. The forth prize can also be given in 4 ways. The fifth prize can be given away in 4 ways. Hence the number of ways in which all prizes can be given away $=4\times4\times4\times4\times4=1024$

14. The signals can be made by using at a time one or more flags. The total number of signals when r flags are used at a time from 6 flags is equal to the number of arrangements of 6, taking r at a time i.e., $^6C_r\times r!$. Since r can take values 1, 2, 3, 4, 5, 6. Hence the number of different signals that can be generated is

$$\sum_{r=1}^{6}{}^6C_r\times r!=6+30+120+360+720+720=1956$$

15. Selecting 2 women out of 4 and 2 men out of 6. This can be done in $^4C_2\times{}^6C_3$ ways.

$$\therefore \frac{4\times3}{1\times2}\times\frac{6\times5\times4}{1\times2\times3}=120$$

16. There are 6 letters in the given word of which 3 are E's. So total no. of words in the number of arrangements of 6 things of which 3 are similar.

i.e., $\frac{6!}{3!}=\frac{6\times5!}{3\times2}=120$

OBJECTIVE QUESTIONS - II

1. Every body in a room shakes hands with every body else. The total number of hand shakes is 66. The total number of persons in the room is **[JNTU-1991]**

a) 11 b) 12 c) 13 d) 14

2. Ten different letters of an alphabet are given. Words with five letters are formed from these given letters. Then the number of words which have at least one letter repeated is **[IIT-1980]**

a) 69760 b) 30240 c) 99748 d) None of these

3. A five digit number divisible by 3 is to be formed using the numerals 0,1,2,3,4 and 5, without repetition. The total number of ways this can be done is **[IIT-1989]**

a) 216 b) 240 c) 600 d) 3125

4. m men and n women are to be seated in a row so that no two women sit together. If m > n, then the number of ways in which they can be seated is **[IIT-1983]**

a) $\frac{m!\,n!}{(m+n)!}$ b) $\frac{m!\,(m+1)!}{(m-n+1)!}$

c) $\frac{m!\,n!}{(m-n+1)!}$ d) None

5. The sides AB, BC, CA of a triangle ABC have 3,4 and 5 interior points respectively on them. The total number of triangles that can be constructed by using these points as vertices is **[IIT-1984]**

a) 220 b) 204 c) 205 d) 195

6. There are (n + 1) white and (n + 1) black balls each set numbered 1 to n + 1. The number of ways in which the balls can be arranged in a row so that the adjacent balls are of different colours is **[JNTU-1991]**

a) $(2n+2)!$ b) $(2n+2)! \times 2$

c) $(n+1)! \times 2$ d) $2\{(n+1)!\}^2$

7. In a football champion ship, 153 matches were played. Every team played one match with each other. The number of teams participating in the championship is **[JNTU-1992]**

a) 17 b) 18 c) 9 d) None

8. There are 10 lamps in a hall. Each one of them can be switched on independently. The number of ways in which the hall can be illuminated is **[IIT - 1990]**

a) 10^2 b) 1023 c) 2^{10} d) 10 !

9. Three persons enter a railway compartment. If there are 5 seats vacant, in how many ways can they take these seats?

[JNTU-1985]

a) 60 b) 20 c) 15 d) 125

10. The number of ways in which 10 persons can sit around a circular table so that none of them has the same neighbours in any two arrangements, is **[JNTU-1986]**

a) 9 ! b) $\frac{1}{2}(9!)$ c) 10 ! d) $\frac{1}{2}(10!)$

11. There are 13 players of cricket, out of which 4 are bowlers. In how many ways a team of eleven be selected from them so as to include at least two bowlers ?

[JNTU-1988]

a) 72 b) 78 c) 42 d) None

12. The number of words from the letters of the word 'BHARAT' in which B and H will never come together, is **[IIT-1997]**

a) 360 b) 240 c) 120 d) None

13. Six teachers and six students have to sit round a circular table such that there is a teacher between any two students. The number of ways in which they can sit is

a) $6!\times 6!$ b) $5!\times 6!$ c) $5!\times 5!$ d) None

14. In how many ways can 10 people be seated in a row so that a given pair is not next to each other ? [Hyd.Univ-2001]

a) 10 ! - 9 ! b) 10 ! - 2 (9 !) c) 9 ! d) None

15. The number of ways in which three girls and nine boys can be seated in two vans each having numbered seats with four in the back and three in the front ? **[Hyd. Univ-2001]**

a) 7 ! b) 2 (7 !) c) 7 (13!) d) 12 !

16. The number of ways to cut a six sided convex polygon whose vertices are labeled into four triangles using diagonal lines that do not cross is **[Hyd. Univ-1999]**

a) 13 b) 14 c) 12 d) 11

17. There are three identical red balls and four identical blue balls in a bag. Three balls are drawn. The number of different colour combinations is **[Hyd. Univ-1998]**

a) 20 b) 35 c) 40 d) 30

ANSWER KEY - II

1. b	**2.** a	**3.** a	**4.** b	**5.** c
6. d	**7.** b	**8.** b	**9.** a	**10.** b
11. b	**12.** b	**13.** b	**14.** b	**15.** c
16. c	**17.** d			

EXPLANATORY NOTES - II

1. It is note that, when 2 persons shake hands, it is counted as one hand shake, not two. The total number of hand shakes is same as the number of ways of selecting 2 persons among 12 persons $= {}^{12}C_2 = \frac{12!}{10!2!} = 66$

2. The number of 5 letter words can be formed from 10 letters when one or more of its letters are repeated $= 10\times10\times10\times10\times10 = 10^5$. The number of 5 letter words which can be formed when none of their letters is repeated.

= Number of arrangements of 10 letters by taking 5 at a time $= {}^{10}P_5 = 30240$

4. M men and n women are seated in a row so that no two women sit together. If m > n then the number of ways in which they can be seated is $\frac{m!(m+1)!}{(m-n+1)!}$

6. We know that there are n+1 while and n+1 black balls each set numbered 1 to n+1. The number of ways in which the balls can be arranged in a row. So that the adjacent balls are of different colour is $2\{(n+1)!\}^2$

7. It is note that, when 2 teams play a game. It is counted as one match not a two. Total number of matches is same as the no. of ways of selecting 2 teams among 18 teams is $= {}^{18}C_2 = \frac{18\times17}{1\times2} = 153$

8. There are 10 lamps in a hall. Each one of them can switched on independently. The number of ways hall is illuminated is $2^{10}-1=1023$.

9. Three persons are taken 5 seats is ${}^5P_3 = 5\times4\times3 = 60$.

10. The number of ways in which 10 persons can sit around a circle is $(10-1)! = 9!$. But none of them has same neighbours is $\frac{1}{2}\times 9!$

13. 6 teachers can be sit in a round circle table is $(6-1)!$ ways. In between the teachers the students can be sit together is 6 !. So six teachers and six students have sit round circle table such that there is teacher between any 2 students. The number ways in which they can sit is $5!\times6!$.

14. Let pair is xy. The pair xy is not next to each other = total number of arrangement - total number of arrangements in which pair xy or yx are always together. Total number of arrangement is 10 People be seated 10 Places is ${}^{10}P_{10} = 10!$

Total number of arrangements in which pair xy or yx are always together is $9!+9!$.

$\therefore$ The Pair xy is not next to each other $=10!-2(9!)$.

15. total no. of Persons = 3 girls + 9 boys = 12

total no. of seats = $=2\times3+4\times2=14$

no. of ways in which 12 Persons can be seated 14 seats = number of arrangements or 14 seats by taking 12 at time ${}^{14}P_{12} = \frac{14!}{2!} = \frac{14(13!)}{2} = 7(13!)$.

16. There are 6 vertices of polygon one triangle is foamed by selecting a group of 3 vertices from the given 6 vertices. This can be done in ${}^6C_3$ ways.i.e., $\frac{6!}{3!3!} = 20$

But we can cross the diagonals are 8. So, we remove that diagonals = 20-8 = 12.

17. So we can draw 1 Red ball and 2 blue balls or we can draw 2 red balls and 1 blue ball is

$= {}^3C_1\times{}^4C_2 + {}^3C_2\times{}^4C_1 = 3\times6+3\times4 = 30$.

II. ELEMENTS OF PROBABILITY

- **Random Experiment**
- **Addtion Theorem of Probability**
- **Conditional Probability**
- **Multiplication Theorem of Probability**
- **Independent Random Variables**

1. Random Experiment

Definition 1 : An experiment which can be repeated any number of times under essentially identical conditions and which is associated with a set of known results, is called a random experiment or trial if the result of any single repetition of experiment is not certain and is any one of the associated set.

Ex 1: " Tossing a coin" is a trial

Ex 2: " Rolling a die" is a trial

Ex 3: " Drawing a card from a pack" is a trial

Definition 2: The result of any single repetition of a random experiment is called a random event or elementary or simple event.

Ex 1: In the trial 'Tossing a coin', the elementary events are " getting head (H), getting tail (T)".

Ex 2: In the trial "Rolling die",the elementary events are getting the number 1,2,3,4,5,6.

Definition 3: The list of all elementary events in a trail is called list of exhaustive events.

Ex : The list of exhaustive events in the trial

I. Tossing a coin is { H,T }

II. Rolling a die is {1,2,3,4,5,6}

Definition 4: Elementary events are said to be equally likely if they have the same chance of happening.

Ex : In the trial "Tossing a coin ", the elementary events, "getting head, getting tail" are equally likely.

Definition 5: A combination of two or more elementary events in a trial is called an event.

Ex : In the trial "Rolling a die", getting an even number is an event.

Definition 6 : The favourable cases to a particular event of an experiment are called successes and the ramaining cases are called failures with respect to that event.

2. Classical Definition Of Probability

Definition 7: If there are n exhaustive equally likely elementary events in a trial and m of them are favourable to an event A,then $\frac{m}{n}$ is called the probability of A. It is denoted by P(A).

Ex: The probability of getting an even number when a die is rolled is $\frac{3}{6}=\frac{1}{2}$

$$P(A)=\frac{\text{Number of favourable cases (out comes) with respect A}}{\text{Number of all cases (out comes) of the experiment}}$$

3. Von Mises Statistical Definition Of Probability

Definition 8 : If a trial is conducted n times and m of them are favourable to an event A, then m/n is called relative frequency of A and is denoted by R(A). If $\underset{n\to\infty}{Lt}$ R(A) exists, then the limit is called probability of A.

4. Mathematical Definition Of Probability

Definition 9: The set of all possible outcomes (results) in a trial is called sample space for the trial. It is denoted by S. The elements of S are called sample points.

Ex: The sample space for the trial

I. Tossing a coins is { H,T }

II. Tossing two coins is { HH,HT,TH,TT }

III. Rolling a die is { 1,2,3,4,5,6 }

Note: We shall consider the sample spaces containing the sample points which are equally likely only.

Definition 10: Let S be a sample space of a random experiment. Every subset of S is called an event.

Ex 1: A ={1,3,5} is an event in the sample space

S = {1,2,3,4,5,6}

Ex 2: A = { HH,TT } is an event in the sample space

S ={ HH,HT,TH,TT }

Note: If S is a sample space, then $\phi\subseteq S$, $S\subseteq S$ and hence ϕ ,S are two events in S.

Definition 11: Let S be a sample space. The event ϕ is called impossible event and the event S is called certain event in S.

Definition 12: Let S be a sample space. An event E in S is said to be a simple event if E contains only one sample point.

Note : Simple events in a sample space are precisely the elementary events in the random experiment.

Definition 13: Two events A,B in a sample space S are said to be disjoint or mutually exclusive if $A\cap B=\phi$

Definition 14: The events $A_1, A_2, \ldots, A_n$ in a sample space S are said to be mutually exclusive or pairwise disjoint if every pair of the events $A_1, A_2, \ldots, A_n$ are disjoint

Ex: The events {1,2}, {3,5} are disjoint in the sample space S ={1,2,3,4,5,6}.

Definition 15: Two events A,B in a sample space S are said to be exhaustive if $A\cup B=S$

Definition16: The events $A_1, A_2, \ldots, A_n$ in a sample space S are said to be exhaustive if $A_1\cup A_2\cup\ldots\cup A_n=S$.

Ex : The events {1,2,3,5}, {2,4,6} are exhaustive in the sample space S={1,2,3,4,5,6}.

Definition 17: Two events A ,B in a sample space S are said to be complementary if $A\cup B = S$, $A\cap B=\phi$

Ex: {1,3,5},{2,4,6} are complementary in the sample space S={1,2,3,4,5,6}.

Definition 18 : Let A be an event in a sample space S. An event B in S is said to be complement of A if A,B are complementary in S. The complement B of A is denoted by $\overline{A}$.

Ex : The complement of {1,2} in {1,2,3,4,5,6} is {3,4,5,6}

Note: The complement of an event A in a sample space S is unique.If $\overline{A}$ is the complement of A then

$$A \cup \overline{A} = S,\ A \cap \overline{A} = \phi \ and \left(\overline{\overline{A}}\right) = A$$

Definition 19: let S be a finite sample space. A real valued function P: $p(S) \to R$ is said to be a probability function on S if I. $P(A) \geq 0\ \forall A \in p(S)$ II. $P(S) = 1$

III. $A, B \in p(S),\ A \cap B = \phi \Rightarrow P(A \cup B) = P(A) + P(B)$

Definition 20: let S be a finite sample space and P be a probability function on S. If A is an event in S then P(A), the image of A, is called probability of A.

Theorem 1: If $A_1, A_2, \ldots A_n$ are n mutually exclusive events in a sample space S then

$P(A_1 \cup A_2 \cup \ldots \cup A_n)\ P(A_1) + P(A_2) + \ldots + P(A_n)$

Theorem 2 : $P(\phi) = 0$

Theorem 3: If A is an event in a sample space S, then $P(\overline{A}) = 1 - P(A)$

Theorem 4: Let A,B be two events in a sample space S. If $A \subseteq B$ then $P(A) \leq P(B)$

Theorem 5: If A is an event in a sample space S, then $0 \leq P(A) \leq 1$,

Theorem 6: Let S be sample space containing n sample points. If E is an elementary event in S, then $P(E) = 1/n$

Theorem 7: let S be a sample space containing n sample points.

If A is an event in S containing m sample points, then

$$P(A) = \frac{m}{n}$$

Note : If A is an event in a sample space S then $P(A) = \frac{n(A)}{n(S)}$ where n(A) is the number of sample points in A and n(S) is the number of sample points in S.

Definition 21: If A is an event in a sample space S,then the ratio $P(A) : P(\overline{A})$ is called the odds favour to A and $P(\overline{A}) : P(A)$ is called odds against to A.

SOLVED EXAMPLES

Ex 1 : Find the probability of getting two heads when tossing seven coins.

Sol: Let A be the event of getting two heads and S be the sample space when tossing seven coins.

$$\therefore n(s) = 2^7 = 128\ n(A) = {}^7C_2 = \frac{7 \times 6}{1 \times 2} = 21$$

$$\therefore P(A) = \frac{n(A)}{n(S)} = \frac{21}{128}$$

Ex 2 : When two dice are thrown, find the probability of obtaining total score seven.

Sol: Let A be the event of getting total score seven and S be the sample space when two dice are thrown.

$\therefore$ n (S) = 6^2 = 36,

A = {1+ 6, 2 + 5, 3+ 4, 4 + 3, 5 + 2, 6 + 1}

$$\therefore \text{Thus } \therefore n(A) = 6, \quad \therefore P(A) = \frac{n(A)}{n(S)} = \frac{6}{36} = \frac{1}{6}$$

Ex 3 : Find the probability that a leap year contains 53 Sundays.

Sol : Let A be the event of containing 53 Sundays in leap year and S be the sample space. A leap year contains 366 days i.e., 52 weeks and 2 days extra. The extra two days may be (Sun, Mon)or (Mon, Tues) or (Tues, Wednes) or (Wednes, Thurs) or (Thurs, Fri) or (Fri, Satur) or (Satur, Sun).

$\therefore$ n (S)=7, n (A) =2. $\therefore$ P (A) = 2 / 7.

Ex 4 : Three dice are rolled simultaneously. Find the probability that the numbers on them is 8.

Sol : Let A be the event of getting the sum 8 when three dice are rolled and S be sample space. Now n(S) = 6^3 =216.

A = {6+1+1, 5+1+2, 5+2+1, 4+1+3, 4+2+2, 4+3+1, 3+1+4, 3+2+3, 3+3+2, 3+4+1, 2+1+5, 2+2+4, 2+3+3, 2+4+2, 2+5+1, 1+1+6, 1+2+5, 1+3+4, 1+4+3, 1+5+2, 1+6+1}

$$\therefore \text{n (A) =21. Thus P (A) = } \frac{n(A)}{n(S)} = \frac{21}{216} = \frac{7}{72}$$

Ex 5 : If three people are chosen at random then find the probability that no two of them were born in the same date of the month of september.

Sol: Let A be the event that no two of three persons chosen were born in the same date of the month of september and S be the sample space. Since a person can be born any date of september, he can born in 30 ways. Thus n(S) =30^3. If no two persons can born on the same date,the first person can born in 30 ways, second person in 29 ways and the third person in 28 ways. Thus n(A) = 30 × 29 × 28.

$$\therefore P(A) = \frac{n(A)}{n(S)} = \frac{30 \times 29 \times 28}{30^3} = \frac{203}{225}$$

Ex 6: There are n letters and n addresssed envolopes. If the letters are put at random in the envolopes, find the probability that atleast one letter may be placed in wrongly addressed envolope.

Sol: Let A be the event that atleast one letter may be placed in wrongly addressed envolope and S be the sample space n letters can be placed in n envelopes in n! ways. Thus n(S) =n! Now $\overline{A}$ denotes the event of placing the n letters in the n envelopes correctly. Now $n\left(\overline{A}\right) = 1$

$$\therefore P\left(\overline{A}\right) = \frac{1}{n!} \qquad \therefore P(A) = 1 - P(\overline{A}) = 1 - \frac{1}{n!}$$

5. Addition Theorem Of Probability

Theorem 1: If A, B are two events in a sample space S then $P(A \cup B) = P(A) + P(B) - P(A \cap B)$

Note 1: If A,B are two disjoint events in a sample space S, then $A \cap B = \phi$ and hence $P(A \cup B) = P(A) + P(B)$

Note 2 : If A,B are two events ina sample space S, then

$P(B - A) = P(B) - P(A \cap B)$ and

$$P(A - B) = P(A) - P(A \cap B)$$

Corollary : If A, B, C are three events in a sample space S, then $P(A \cup B \cup C)$

$= P(A) + P(B) + P(C) - P(A \cap B) - P(B \cap C) - P(C \cap A) + P(A \cap B \cap C)$

SOLVED EXAMPLES

Ex 1 : If $P(A) = 3/4$, $P(B) = 2/5$ and $P(A \cap B) = 1/4$ then find $P(A \cup B)$

Sol : $P(A \cup B) = P(A) + P(B) - P(A \cap B)$

$$= \frac{3}{4} + \frac{2}{5} - \frac{1}{4} = \frac{15+8-5}{20} = \frac{18}{20} = \frac{9}{10}$$

Ex 2 : When two dice are rolled, find the probability of getting the sum 7 or 8.

Sol : Let A be the event of getting the sum 7,B be the event of getting the sum 8 when two dice are rolled and S be the sample space.

Now n(S) = 6^2 =36, A = {1+6, 2+5, 3+4, 4+3, 5+2, 6+1}, B = {2+6, 3+5, 4+4,5+3, 6+2}.

Thus n(A) = 6, n(B) = 5 and $A \cap B = \phi$

$$\therefore P(A) = \frac{n(A)}{n(S)} = \frac{6}{36},\ P(B) = \frac{n(B)}{n(S)} = \frac{5}{36}$$

$$P(A \cup B)\ P(A) + P(B)(A \cap) = \frac{6}{36} + \frac{5}{36} - 0 = \frac{11}{36}$$

Ex 3 : When two dice are rolled simultaneously, find the probability that their sum is a prime number.

Sol : When two dice are rolled, the sum of the numbers on the dice may be a number from 2 to 12. The prime numbers from 2 to12 are 2,3,5,7,11. Let A,B,C,D,E be the events of getting the sums 2,3,5,7,11respectively and S be the sample space. Now A,B,C,D ,E are mutually disjoint and n(S) = 6^2 =36.

A = {1+1}, B ={1+2,2+1}, C = {1+4, 2+3, 3+2, 4+1}, D = {1+6, 2+5, 3+4, 4+3, 5+2, 6+1}, E= {6+5, 5+6}.

∴ n (A) = 1, n (B) =2, n (C) =4, n (D) =6; n(E) =2

$$\therefore P(A) = \frac{1}{36},\ P(B) = \frac{2}{36},\ P(C) = \frac{4}{36},\ P(D) = \frac{6}{36},\ P(E) = \frac{2}{36}$$

$P(A \cup B \cup C \cup D \cup E) = P(A) + P(B) + P(C) + P(D) + P(E)$

$$= \frac{1}{36} + \frac{2}{36} + \frac{4}{36} + \frac{6}{36} + \frac{2}{36} = \frac{15}{36} = \frac{5}{12}$$

Ex 4 : A bag contains 4 red balls,5 white balls, 6 blue balls. A ball is drawn at random. Find the probability that it is a red ball or a white ball.

Sol: Let A be the event of getting a red ball, B be the event of getting a white ball when a ball is drawn from the bag containing 4 red balls, 5 white balls, 6 blue balls and S be the sample space.

Now n(S) = ${}^{15}C_1 = 15. n(A) = {}^4C_1 = 4. n(B)\ {}^5C_1 = 5$,,

A and B are disjoint.

$$\therefore P(A) = \frac{4}{15},\ P(B) = \frac{5}{15},\ P(A \cup B) = P(A) + P(B)$$

$$= \frac{4}{15} + \frac{5}{15} = \frac{9}{15} = \frac{3}{5}$$

Ex 5: Three horses A,B and C run a race.if the probability of A's win is twice that of B and the probability of B's winis twice that of C then find the probability of their winning. What is the probability that A losing the race?

Sol: Let A,B,C be the events that the horses A,B, C win the race respectively.

Given P (A) = 2 P(B), P (B) = 2 P(C) .

P (A) = 2 P(B) = 2x2 P (C) =4 P (C)

Since the horses A,B and C run the race.

$A \cup B \cup C$ =S and A,B,C are mutually disjoint.

$\therefore P(A \cup B \cup C) = P(A) + P(B) + P(C)$

$\Rightarrow$ P (S) =4P (C) +2P (C) + P(C)

$\Rightarrow$ 1= 7 P(C) $\Rightarrow$ P (C) = 1/ 7

P (A) = 4P (C) = 4/ 7, P (B) = 2P (C) = 2/ 7

The probability that A winning the race =

$$P(\bar{A}) = 1 - P(A) = 1 - 4/7 = 3/7$$

Ex 6 : Out of 7 gentleman and 4 ladies a commitee of 5 is to be formed. Find the probability that the commitee consists of atleast 2 ladies.

Sol: Since commitee contains atleast 2 ladies the memebers of the commitee may be of the following three types.

(I) 3 gents, 2 ladies (II) 2 gents, 3 ladies (III) 1 gent, 4 ladies.

Let A,B,C be the events that the commitee consisting of the members in the above types respectively. Now A,B,C are mutually disjoint and

$$n(S) = {}^{11}C_5 = \frac{11 \times 10 \times 9 \times 8 \times 7}{1 \times 2 \times 3 \times 4 \times 5} = 462$$

$$n(A) = {}^7C_3 \times {}^4C_2 = 35 \times 6 = 210$$

$$n(B) = {}^7C_2 \times {}^4C_3 = 21 \times 4 = 84$$

$$n(C) = {}^7C_1 \times {}^4C_4 = 7 \times 1 = 7$$

$$\therefore P(A) = \frac{210}{462},\ P(B) = \frac{84}{462},\ P(C) = \frac{7}{462}$$

$$P(A \cup B \cup C) = P(A) + P(B) + P(C) = \frac{210}{462} + \frac{84}{462} + \frac{7}{462} = \frac{301}{462} = \frac{43}{66}$$

Ex 7 : Three electric lamps are fitted in a room. Three bulbs are chosen at random from 10 bulbs having 6 good bulbs. Find the probability that the room is lighted.

Sol : The room is lighted if the selection of bulbs contains atleast one good bulb. Thus the number of bulbs in a favourable selection is one of the following types . (I) one good bulb and 2 bad bulbs (II) two good bulbs and 1 bad bulb (III) three good bulbs and no bad bulb.

Let A,B,C be the events of selecting bulbs in the above types respectively and S be the sample space.

Now $n(S) = {}^{10}C_3 = \dfrac{10 \times 9 \times 8}{1 \times 2 \times 3} = 120$

$n(A) = {}^6C_1 \times {}^4C_2 = 6 \times 6 = 36,\ n(B) = {}^6C_2 \times {}^4C_1$

$= 15 \times 4 = 60,\ n(C) = {}^6C_3 \times {}^4C_0 = 20$

and A,B,C are disjoint

$\therefore P(A) = \dfrac{36}{120},\ P(B) = \dfrac{60}{120},\ P(C) = \dfrac{20}{120}$

$$P(A \cup B \cup C) = P(A) + P(B) + P(C) = \frac{36}{120} + \frac{60}{120} + \frac{20}{120} = \frac{116}{120} = \frac{29}{30}$$

Second method : Let A be the event that the room is unlighted when three bulbs are selected at random from 10 bulbs having 6 good bulbs and S be the sample space.

Now n (S) = ${}^{10}C_3$ =120 and n (A) = ${}^4C_3$ =4.

$\therefore P(A) = \dfrac{1}{120} = \dfrac{1}{30}$

The probability that the room is lighted

$= P(A) = 1 - P(A) - \dfrac{1}{30} = \dfrac{29}{30}$.

6. Conditional Probability

Definition 1: If A,B are two events in a sample space then the event of happening B after the event A happing is called conditional event. It is denoted by $B|A$

Definition 2: If A,B are two events in a sample space S and $P(A) \neq 0$ then the probability of B after the event A has occured is called conditional probability of B given A. It is denoted by $P(B \mid A)$

Note : If A,B are two events in a sample space S such that $P(A) \neq 0$ then $P(B \mid A) = \dfrac{n(A \cap B)}{n(A)}$

Ex : If two dice are thrown simultaneously,then the sum of the numbers on them is 6. Find the probability that 2 is on any one of them.

Sol : Let A be the event that the sum is 6 and B be the event that 2 is on any one of the dice and S be the sample space when two dice are thrown simultaneously.

n(A) = n {1+5, 2+4, 3+3, 4+2, 5+1} =5

$n(A \cap B) = n\{2+4+4+2\} = 2$

$\therefore P(B \mid A) = \dfrac{n(A \cap B)}{n(A)} = \dfrac{2}{5}$

$\therefore$ The probability that 2 is on any oneof the dice $= \dfrac{2}{5}$.

7. Multiplication Theorem Of Probability Or Theorem Of Compound Probability

Theorem 1: Let A,B be two events in a sample S such that $P(A) \neq 0,\ P(B) \neq 0$

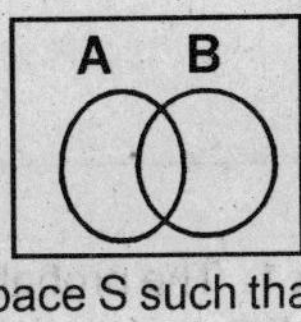

Then I. $P(A \cap B) = P(A)\ P(B \mid A)$

II. $P(A \cap B) = P(B)\ P(A \mid B)$

Ex : If A,B are two events in a sample space S such that $P(A) = 2/5$

$P(A \cap B) = 3/10$ then find P (B/A)

Sol : $P(B/A) = \dfrac{P(A \cap B)}{P(A)} = \dfrac{3/10}{2/5} = \dfrac{3}{4}$

Definition 3: Two events A,B in a sample space S are said to be independent if $P(B|A)\ P(B)$

Corollary : Two events A,B in a sample space S are independent iff $P(A \cap B) = P(A)\ P(B)$

Note : The events $A_1, A_2, \ldots, A_n$ are independent iff

$P(A_1 \cap A_2 \cap \ldots \cap A_n) = P(A_1)\ P(A_1) \ldots P(A_n)$

Ex : If P (A) =3/4 , P(B)=3/5 , A and B are independent, then find $P(A \cup B)$

Sol : A,B are independent

$\Rightarrow P(A \cap B) = P(A)\ P(B) = \dfrac{3}{4} \cdot \dfrac{3}{5} = \dfrac{9}{20}$

$\therefore P(A \cup B) = P(A) + P(B) - P(A \cap B) = \dfrac{3}{4} + \dfrac{3}{5} - \dfrac{9}{20}$

$= \dfrac{15 + 12 - 9}{20} = \dfrac{18}{20} = \dfrac{9}{10}$

Theorem 2: If A,B are two independent events in a sample space S. Then

I. $\overline{A}$,B are independent

II. A, $\overline{B}$ are independent

III. $\overline{A}, \overline{B}$ are independent.

Note: If $A_1, A_2, \ldots, A_n$ are independent events in a sample space S then $\overline{A}_1, \overline{A}_2, \ldots, \overline{A}_n$ are also independent events.

Theorem 3: If $A_1, A_2, \ldots, A_n$ are mutually exclusive and exhaustive events in a sample space S such that $P(A_i) > 0$ for i = 1,2,......,n and E is any event then

$P(E) = \sum_{i=1}^{n} P(A_i)\ P(E \mid A_i)$.

Note 1 : If A_1, A_2 are two mutually exclusive and exhaustive events and E is any event then

$P(E) = P(A_1)\ P(E|A_1) + P(A_2)\ P(E|A_2)$

Note 2 : If A_1, A_2, A_3, are three mutually exclusive and exhaustive events and E is any event then

$P(E) = P(A_1)\ P(E|A_1) + P(A_2)\ P(E|A_2) + P(A_3) P(E|A_3)$

8. Bayes' Theorem

Theorem 4: If $A_1, A_2, \ldots, A_n$ are mutually exclusive and exhaustive events in a sample space S such that $P(A_i) > 0$ for i = 1,2,....n and E is any event with

$P(E) > 0$ then

$$P(A_K | E) = \frac{P(A_K)\,P(E | A_K)}{\sum_{i-1}^{n} P(A_i)\,P(E | A_i)} \quad \text{for } k = 1,2,....n$$

SOLVED EXAMPLES

Ex 1: The probability that a boy will pass the examination is 3/5 and that for girl is 2/5. What is the probability that atleast one of them passes the examination.

Sol : If A is the event that the boy will pass the examination and B is the event that the girl will pass the examination, then P(A) = 3/5, P (B) = 2/5 and A,B are independent.

$$P(A \cup B) = P(A) + P(B) - P(\cap B) = P(A) + P(B) - P(A)\,P(B)$$

$$= \frac{3}{5} + \frac{2}{5} - \frac{3}{5}.\frac{2}{5} = \frac{15+10-6}{25} = \frac{19}{25}.$$

Ex 2 : The probability of a problem being solved by three students are 1/3, 1/4, 1/6. Find the probability that the problem being solved.

Sol: Let A,B,C be the events that the problem being solved by the three students respectively.

Then P (A) =1/3, P (B) = 1/4, P (C) = 1/6 and A,B,C are independent.

$$\therefore\ P(A \cup B \cup C)$$

$$= P(A) + P(B) + P(C) - P(A \cap B) - P(B \cap C) - P(C \cap A) + P(A \cap B \cap C)$$

$$= P(A) + P(B) + P(C) - P(A)P(B) - P(B)P(C)$$

$$- P(C) + P(A) + P(A)\,P(B)P(C)$$

$$= \frac{1}{3} + \frac{1}{4} + \frac{1}{6} - \frac{1}{3}.\frac{1}{4} - \frac{1}{4}.\frac{1}{6} - \frac{1}{6}.\frac{1}{3} + \frac{1}{3}.\frac{1}{4}.\frac{1}{6}$$

$$= \frac{24+18+12-6-3-4+1}{72} = \frac{42}{72} = \frac{7}{12}$$

Ex 3 : An antiaircraft gun can take a maximum of three shots at an enemy plane moving away from it. The probabilities of hitting the plane at the first, second, third shot are 0.5, 0.4, 0.3 respectively. Find the probability that the gun hits the plane.

Sol : Let A,B,C be the events that the gun hits the plane in first, second and third shot respectively. Then P (A) = 0.5, P (B)= 0.4, P(C) = 0.3 and A,B,C are independent.

$\therefore P(\overline{A}) = 0.5\ P(\overline{B}) = 0.6,\ P(\overline{C}) = 0.7$ and

$\overline{A}, \overline{B}, \overline{C}$ are independent.

The probability that the gun hits the plane = 1- The probability that the gun does not hit the plane in all the

shots $= 1 = -P\left(\overline{A} \cap \overline{B} \cap \overline{C}\right) = 1 - P(\overline{A})\,P(\overline{B})\,P(\overline{C})$

$$= 1 - (0.5)\,(0.6)\,(0.7) = 1 - 0.21 = 0.79$$

Ex 4 : The bag A contains 3 white and 2 black balls and the bag B contains 2 white and 4 black balls. One of the bags is chosen at random and a ball is drawn from it. Find the probability that the ball is white.

Sol : Let A be the event of selecting bag A,B be the event of selecting bag B and E be the event of drawing a white ball. Then

$$P(A) = \frac{1}{2},\ P(B) = \frac{1}{2},\ P(E / A) = \frac{3}{5},\ P(E / B) = \frac{2}{6} = \frac{1}{3}$$

OBJECTIVE QUESTIONS - I

1. Four fair coins are tossed simultaneously. The probability that at least one head and one tail turn up is **[GATE- 2002]**

a) 1/16 b) 1/8 c) 7/8 d) 15/16

2. How many four digit EVEN numbers have all 4 digits distinct? **[GATE-2001]**

a) 2240 b) 2296 c) 2620 d) 4536

3. Seven (distinct) car accidents occurred in a week. What is the probability that they all occurred on the same day? **[GATE-2001]**

a) $\frac{1}{7^7}$ b) $\frac{1}{7^6}$ c) $\frac{1}{2^7}$ d) $\frac{7}{2^7}$

4. The minimum number of cards to be dealt from an arbitrarily shuffled deck of 52 cards to guarantee that three cards are from same suit is **[GATE-2000]**

a) 3 b) 8 c) 9 d) 12

5. A die is rolled three times. The probability that exactly one ODD number turns up among the three outcomes is **[GATE-1998]**

a) 1/6 b) 3/8 c) 1/8 d) 1/2

6. The probability of getting atleast two heads when tossing a coin three times is **[JNTU-1988]**

a) 1/8 b) 3/8 c) 1/2 d) 5/8

7. The probability of getting exactly 2 tails in 6 tosses of a fair coin is **[JNTU-1996]**

a) 3/8 b) 1/4 c) 15/64 d) 49/64

8. Six coins are tossed simultaneously. The probability of getting atleast 4 heads is **[JNTU-1995]**

a) 11/64 b) 11/32 c) 15/44 d) 21/32

9. The probability of getting a total of 10 in a single throw of two dice is **[JNTU-1994]**

a) 1/9 b) 1/12 c) 1/6 d) 5/36

10. When two dices are thrown, the probability of getting equal numbers is **[JNTU-1991]**

a) 1/6 b) 1/12 c) 1/18 d) 5/36

11. Two dice thrown simultaneously. The probability of getting even numbers on both the dice is **[JNTU-1991]**

a) 1/3 b) 1/4 c) 1/6 d) 1/12

12. Three identical dice are rolled. The probability that the same number will appear on each of them is **[IIT-1984]**

a) 1/6 b) 1/36 c) 1/18 d) 3/28

13. Three dice are rolled simultaneously. The probability that the sum of the numbers on them is 7 is

a) 7/216 b) 5/216 c) 7/72 d) 5/72

14. When two balls are drawn from a bag containing 2 White, 4 Red and 6 Black Balls, the chance for both of them to be Red is **[JNTU-1991]**

a) 1/11 b) 6/11 c) 3/11 d) 4/12

15. The probability that in a family of 4 children there will be atleast one Boy is

a) 1/16 b) 3/16 c) 5/16 d) 15/16

16. The probability that a leap year have 53 Sundays is **[JNTU-1987]**

a) 1/7 b) 2/7 c) 5/7 d) 6/7

17. Card is drawn at random from a pocket of 100 cards numbered 1 to 100. The probability of drawing a number which is a square is **[JNTU-1989]**

a) 1/100 b) 2/100 c) 1/10 d) 3/10

18. A card is taken out of a pack of 52 cards numbered 2 to 53. The probability that the numbers on the card is a prime number less than 20 is **[JNTU-1990]**

a) 1/13 b) 2/13 c) 3/13 d) 4/13

19. The probability of getting a number between 1 and 100. Which is divisible by one and itself only is **[JNTU-1988]**

a) 27/185 b) 23/97 c) 25/98 d) None

20. If A and B are two elements such that $P(A)=2/3$, $P(B)=3/4, P(A\cap B)=1/2$ then $P(A\cup B)$=?

a) 8/9 b) 11/12 c) 17/18 d) 7/12

21. If A and B are two events such that $P(A)=1/4$, $P(B)=1/2$, $P(A\cup B)=5/8$ then $P(A\cap B)$=? **[JNTU-1995]**

a) 3/8 b) 1/8 c) 2/8 d) 5/8

22. If A and B are two events such that $P(A\cup B)=0.65$, $P(A\cap b)=0.15$ then $P(\bar{A})+P(\bar{B})$ =? **[JNTU-1988]**

a) 0.6 b) 0.8 c) 1.2 d) 1.4

23. The probability that at least one of the events, A, B occurs is 0.6. If A and B occur simultaneously with Probability 0.2 then $P(\bar{A})+P(\bar{B})$ =? **[IIT-1987]**

a) 0.4 b) 0.8 c) 1.2 d) 1.4

24. When two dice are thrown, the probability of getting the sum 10 or 11 is **[JNTU-1996]**

a) 7/36 b) 5/36 c) 5/18 d) 7/18

25. If A, B, C are three events, then $P[A\cap(B\cup C)]$= **[JNTU-1994]**

a) $P(A)+P(B)+P(C)-P(A\cap B)-P(A\cap C)$

b) $P(A)+P(B)+P(C)-P(B)P(C)$

c) $P(A\cap B)+P(A\cap C)-P(A\cap B\cap C)$

d) $P(B\cap C)+P(A\cap B)-P(A\cap B\cap C)$

26. If A and B are two arbitrary events, then $P(A\cap B)$ cannot be

a) Not less than $P(A)+P(B)-1$

b) Not greater than $P(A)+P(B)$

c) Equal to $P(A)+P(B)-P(A\cup B)$

d) Equal to $P(A)+P(B)+P(A\cup B)$

27. If M and N are two events, then the probability that exactly one of them occurs is not equal to **[IIT-1984]**

a) $P(M)+P(N)-2P(M\cap N)$ b) $P(M)+P(N)-P(M\cap N)$

c) $P(\bar{M})+P(\bar{N})-2P(\bar{M}\cap\bar{N})$ d) $P(M\cap\bar{N})+P(\bar{M}\cap N)$

28. If A,B,C are three mutually exclusive and exhaustive events of a trail such that $P(A)=2P(B)=3P(C)$ then $P(A)$=? **[JNTU-1986]**

a) 1/3 b) 2/5 c) 3/8 d) 6/11

29. If two dice are rolled then the probability that their sum is a prime number is **[JNTU-1996]**

a) 5/12 b) 5/6 c) 5/36 d) 2/5

30. A card is drawn at random from a normal pack of cards. The probability that it is either a Spade or a Queen is **[JNTU-1987]**

a) 1/13 b) 4/14 c) 1/4 d) 13/51

31. The probability of drawing a card which is atleast a Spade or a King from well shuffled pack of cards is **[JNTU-1985]**

a) 1/13 b) 4/13 c) 1/4 d) 13/51

32. A bag contains 3 Red, 4 White and 5 Blue balls. 2 balls are drawn at random. The probability that they are of different colours is **[JNTU-1987]**

a) 47/66 b) 10/33 c) 5/22 d) 2/11

33. A box contains 10 Mangoes out of which 4 are rotten. Two Mangoes are taken out together. If one of them is found to be good, then the probability that the other is also good is **[JNTU-1992]**

a) 1/3 b) 8/15 c) 5/13 d) 2/3

34. If $P(A)=0.4$, $P(A\cup B)=0.7$ and A,B are independent then $P(B)$=? **[JNTU-1992]**

a) 0.2 b) 0.3 c) 0.5 d) 0.6

35. The probabilities of a problem being solved by two students are 1/2 and 1/3. The probability of the problem being solved is **[JNTU-1991]**

a) 2/3 b) 4/3 c) 1/3 d) 1

36. If A and B are any two events then which one of the following pairs of events are mutually exclusive? **[R.E.C.Trichy-1995]**

a) $A\cup B$ and $A\cap\bar{B}$ b) $A\cup B$ and $\bar{A}\cap B$

c) $A\cup B$ and $\bar{A}\cap\bar{B}$ d) $A\cup B$ and $A\cap B$

37. A bag contains 3 Red balls, 4 White balls and 7 Black balls. The probability of drawing a Red or Black ball is **[R.E.C.Trichy-1995]**

a) 5/7 b) 2/7 c) 3/7 d) 4/7

38. $P(A\cup B)=0.65$, $P(A\cap B)=0.15$, then $P(\bar{A})+P(\bar{B})$= **[R.E.C.Trichy-1995]**

a) 0.6 b) 0.8 c) 1.4 d) 1.2

39. What is the probability of getting HEAD at least once if we toss a coin four times **[UGC NET-1996]**

a) 1/4 b) 1/2 c) 3/4 d) 15/16

40. Probability that two randomly selected cards, from a set of two red cards and two black cards, are of the same colour is two black cards, are of the same colour, is **[UGC NET-1995]**

a) 1/2 b) 1/3 c) 2/3 d) None

ANSWER KEY - I

1. c	**2.** b	**3.** b	**4.** c	**5.** c
6. c	**7.** c	**8.** b	**9.** b	**10.** a
11. b	**12.** b	**13.** d	**14.** a	**15.** d
16. d	**17.** c	**18.** b	**19.** c	**20.** b
21. b	**22.** c	**23.** c	**24.** b	**25.** c
26. c	**27.** b	**28.** d	**29.** a	**30.** b
31. b	**32.** a	**33.** c	**34.** c	**35.** a
36. c	**37.** a	**38.** d	**39.** d	**40.** c

EXPLANATORY NOTES - I

1. (c) **Explanation :** Let A be event of getting atleast one head and one tail turn up when tossing 4 coins.

$n(S)=2^4=16$, $n(A)=14$

The 14 situations are (T H H H, H T H H, H H T H, H H H T, H T T T, T H T T, T T H T, T T T H, T T H H, H H T T, H T T H, T H H T, T H T H, H T H T)

$$p(A)=\frac{\text{favourable no. of cases}}{\text{exhaustive no. of cases}}=\frac{14}{16}=\frac{7}{8}$$

2. (d) **Explanation :** We can't have 0 at the thousands place. So thousands place can be filled with 9 digits 1, 2, 3, 4 ......9. So there are 9 ways of filling the thousands place. Now 9 digits are left including 0. So hundreds place can be filled with any of the remaining 9 digits in 9 ways. Now the tenth palce can be filled with any of the remaining 8 digits. Now units palce can be filled with any of the remaining 7 digits. Hence total no. of required $= 9\times9\times8\times7=4536$.

3. (b) **Explanation :** Let A be a event cars accidents occured on the same day when 7 car acidents occured in a week. S is a sample space $n(S)=7^7$. 7 car accidents in one day so $n(A)={}^7C_1$

$$P(A)=\frac{\text{favourable no. of cases}}{\text{exhaustive no. of cases}}=\frac{7}{7^7}=\frac{1}{7^6}$$

5. (d) **Explanation :** A die is rolled 3 times. So total no. of cases is $3\times6=18$. A die consists 3 odd no's and 3 even no's exactly one odd no. and 2 even no's is ${}^3c_1\times{}^3c_2$

$$p(A)=\frac{\text{favourable no. of cases}}{\text{exhaustive no. of cases}}=\frac{3\times3}{18}=\frac{1}{2}$$

7. (c) **Explanation :** Let A be the event of getting exactly 2 tails and S be the sample space when coin tosses 6 times. $n(S)=2^6=64$ $n(A)={}^6C_2=\frac{6\times5}{1\times2}=15$

$$P(A)=\frac{\text{favourable no. of cases}}{\text{exhaustive no. of cases}}=\frac{15}{64}$$

8. (b) **Explanation :** Let A be the event of getting at least 4 heads. S be the sample space when 6 coins tosses simultaneously. $n(S)=2^6=64$ $n(A)=22$

$$P(A)=\frac{\text{favourable no. of cases}}{\text{exhaustive no. of cases}}=\frac{22}{64}=\frac{11}{32}$$

9. (b) **Explanation :** Let A be event of getting of sum 10 when 2 dice are rolled. And S be sample space Now

$n(S)=6^2=36$ $A=\{6+4, 5+5, 4+6\}$

$$P(A)=\frac{\text{favourable no. of cases}}{\text{exhaustive no. of cases}}=\frac{3}{36}=\frac{1}{12}$$

10. (a) **Explanation :** Let A be event of getting the equal numbers when 2 dice are rolled. And S be sample space now $n(S)=6^2=36$ $A=\{(1,1)(2,2)(3,3)(4,4)(5,5)(6,6)\}$

$$P(A)=\frac{\text{favourable no. of cases}}{\text{exhaustive no. of cases}}=\frac{6}{36}=\frac{1}{6}$$

11. (b) **Explanation :** Let A be event of getting no. on both dice and S be sample space now $n(S)=6^2=36$

$A=\{(2,2)(2,4)(2,6)(4,2)(4,4)(4,6)(6,2)(6,4)(6,6)\}$

$$P(A)=\frac{\text{favourable no. of cases}}{\text{exhaustive no. of cases}}=\frac{9}{36}=\frac{1}{4}$$

12. (b) **Explanation :** Let A be the event of getting the same number will appear on each of them when tossing 3 dice. Let S be a sample space $n(S)=6^3=216$

$A=\{(1,1,1)(2,2,2)(3,3,3)(4,4,4)(5,5,5)(6,6,6)\}$

$$P(A)=\frac{\text{favourable no. of cases}}{\text{exhaustive no. of cases}}=\frac{6}{216}=\frac{1}{36}$$

13. (d) **Explanation :** Let A be the event of getting the sum of the no's on them is 7 when tossing 3 dice. Let S be a sample space $n(S)=6^3=216$

A = {5+1+1, 4+2+1, 3+3+1, 2+4+1, 1+5+1, 3+2+2, 2+4+1, 1+4+2, 1+1+5, 1+3+3, 1+3+1, 1+2+4, 2+1+4, 4+2+1, 2+3+2, 2+2+3}

$$P(A)=\frac{\text{favourable no. of cases}}{\text{exhaustive no. of cases}}=\frac{15}{16}=\frac{5}{72}$$

14. (a) **Explanation :** Let A be the event of getting a red ball when two balls are drawn at random from the bag and S be the sample space. Now when 2 balls

$n(S)={}^{12}C_2=\frac{12\times11}{1\times2}=6\times11$, $n(A)={}^{11}c_2=\frac{4\times3}{1\times2}=6$

$$P(A)=\frac{\text{favourable no. of cases}}{\text{exhaustive no. of cases}}=\frac{6}{6\times11}=\frac{1}{11}$$

15. (d) **Explanation :** Let A be the event of getting atleast one boy. S be the sample space. $n(S)=2^4=16$

Sample space = { B B B B, B G G G, G B G G, G G B G, G G G B, G B B B, B G B B, B B G B, B B B G, B B G G, G G B B, B G B G, G B G B, G B B G, B G G B, G G G G}

$$P(A)=\frac{\text{favourable no. of cases}}{\text{exhaustive no. of cases}}=\frac{15}{16}$$

17. (c) **Explanation :** Let A be the event of drawn a card from a pocket of 100 cards. S be a sample space.

$n(S)=100$, $P(A)=\frac{\text{favourable no. of cases}}{\text{exhaustive no. of cases}}=\frac{10}{100}=\frac{1}{10}$

18. (b) **Explanation :** Let A be the event of drawn a card fromapackof 52cards numbered 2 to 53. $n(S)={}^{52}C_1=52$,

The prime numbers lessthan 20 = {2,3,5,7,11,13,17,19}

$$P(A)=\frac{\text{favourable no. of cases}}{\text{exhaustive no. of cases}}=\frac{8}{52}=\frac{2}{13}$$

19. (c) **Explanation :** Let A be the event of drawn a no. between 1 and 100 of the prime numbers. S be a sample space. $n(S)=98$. The prime no's between 1 and 100 = {2, 3, 5, 7, 11, 13, 17, 19, 23, 29, 31, 37, 41, 43, 47, 53, 59, 61, 67, 71, 73, 79, 83, 89, 97}

$$n(A)=25,\ P(A)=\frac{\text{favourable no. of cases}}{\text{exhaustive no. of cases}}=\frac{25}{98}$$

20. (b) **Explanation :** $P(A\cap B)=P(A)+P(B)-P(A\cup B)$

$$=\frac{2}{3}+\frac{3}{4}-\frac{1}{2}=\frac{8+9-6}{12}=\frac{11}{12}$$

21. (b) **Explanation :** $P(A\cap B)=P(A)+P(B)-P(A\cup B)$

$$=\frac{1}{4}+\frac{1}{2}-\frac{5}{8}=\frac{2+4-5}{8}=\frac{1}{8}$$

22. (c) **Explanation :** $P(A\cap B)=P(A)+P(B)-P(A\cup B)$

$P(A)+P(B)=P(A\cap B)+P(A\cup B)$ ...... (1)

$P(\overline{A})+P(\overline{B})=1-P(A)+1-P(B)$

$=2-(P(A)+P(B))\ =2-[P(A\cap B)+P(A\cup B)]$ (from 1)

$=2-0.65+0.15\ =2-0.8\ =1.2$

24. (b) **Explanation :** Let A be event of getting the sum 10 or 11 when dies are thrown. S be the sample space.

$$A=\{6+4,5+5,4+6,5+6,6+5\}$$

$$P(A)=\frac{\text{favourable no. of cases}}{\text{exhaustive no. of cases}}=\frac{5}{36}$$

29. (a) **Explanation :** Let A be event of getting the sum is prime no. when 2 dice are rolled. S be a sample space. Now $n(S)=6^2=36$ the prime no are { 2, 3, 5, 7, 11}.

A = { 1+1, 1+2, 2+1, 1+4, 4+1, 2+3, 3+2, 1+6, 6+1, 2+5, 5+2, 3+4, 4+3, 6+5, 5+6 }

$$P(A)=\frac{\text{favourable no. of cases}}{\text{exhaustive no. of cases}}=\frac{15}{36}=\frac{5}{12}$$

30. (b) **Explanation :** Let A be event of drawn either spade or queen. When a card is drawn and S be sample space.

$$P(A)=\frac{\text{favourable no. of cases}}{\text{exhaustive no. of cases}}=\frac{^{13}C_1+^{3}C_1}{^{52}C_1}=\frac{16}{52}=\frac{4}{13}$$

31. (b) **Explanation :** Let A be event of getting at least a spade or king from well shuffled pack of cards. S be a sample space.

$$P(A)=\frac{\text{favourable no. of cases}}{\text{exhaustive no. of cases}}=\frac{^{13}C_1+^{3}C_1}{^{52}C_1}=\frac{16}{52}=\frac{4}{13}$$

37. (a) **Explanation :** Let A be ball is drawn from a bag that is a red ball or black ball is $n(A)={}^3C_1+{}^7C_1$.

The total no. of balls are 14 and we can draw a one ball is $n(S)={}^{14}C_1$.

$$P(A)=\frac{\text{favourable no. of cases}}{\text{exhaustive no. of cases}}=\frac{10}{14}=\frac{5}{7}$$

38. (d) **Explanation :** We know that

$P(A\cup B)=P(A)+P(B)-P(A\cap B)$

$=1-P(\overline{A})+1-P(\overline{B})-P(A\cap B)$

$=P(\overline{A})+P(\overline{B})=2-P(A\cap B)-P(A\cup B)$

$=2-0.65-0.15\ =1.2$

40. (c) **Explanation :** Let A be the event of getting 2 randomly selected cards from a set of 2 red cards and 2 black cards are of same color is two black cards are

$n(A)={}^2C_1\times{}^2C_1 \qquad n(S)={}^4C_2$

$$P(A)=\frac{\text{favourable no. of cases}}{\text{exhaustive no. of cases}}=\frac{^2C_1\times{}^2C_1}{^4C_2}=\frac{4}{6}=\frac{2}{3}$$

OBJECTIVE QUESTIONS - II

1. E1 and E2 are events in a probability space satisfying the following constraints:

* Pr (E1) = P_r (E2) * Pr(E1 ∪ E2) = 1

* E1 and E2 are independent **[GATE-2000]**

The value of P_r (E1), the probability of the event E1, is

a) 0 b) 1/4 c) 1/2 d) 1

2. Consider two events E_1 and E_2 such that probability of E_1, $P_r(E_1)=1/2$, probability of E_2, $Pr(E_2)=1/3$, and probability of E_1 and E_2, Pr (E_1 and E_2) =1/5. Which of the following statement(s) is/are TRUE? **[GATE-1999]**

a) Pr (E_1 or E_2) is 2/3

b) Events E_1 and E_2 and E_3 are independent

c) Events E_1 and E_2 are not independent

d) Pr (E_1 / E_2) =4/5

3. Two girls have picked 10 Roses, 15 Sunflowers and 14 Daffodils. What is the number of ways they can divide the flowers amongest themselves? **[GATE-1999]**

a) 1638 b)2100 c) 2640 d) None

4. The probability that it will rain today is 0.5. The probability that it will rain tomorrow is 0.6. The probability that it will rain either today or tomorrow is 0.7, What is the probability that it will rain today and tomorrow? **[GATE-1997]**

a) 0.3 b) 0.25 c) 0.35 d) 0.4

5. The probability that top and bottom cards of a randomly shuffled deck are both aces is **[GATE-1996]**

a) $\frac{4}{52}\times\frac{4}{52}$ b) $\frac{4}{52}\times\frac{3}{52}$ c) $\frac{4}{52}\times\frac{3}{51}$ d) $\frac{4}{52}\times\frac{4}{51}$

6. The probability that a number selected at random between 100 and 999 (both inclusive) will not contain the digit 7 is **[GATE-1995]**

a) $\frac{16}{25}$ b) $\left(\frac{9}{10}\right)^3$ c) $\frac{27}{75}$ d) $\frac{18}{25}$

7. A bag contains 10 white balls and 15 black balls. Two balls are drawn in succession. The probability that one of them is black and the other is white is **[GATE-1995]**

a) 2/3 b) 4/5 c) 1/2 d) 4/6

8. The probability of getting atleast two heads when tossing a coin three times is **[JNTU-1988]**

a) 1/8 b) 3/8 c) 1/2 d) 5/8

9. Three balls are drawn at random from a collection of 7 White, 12 Green and 4 Red balls. The probability that each is of different colour is [JNTU-1989]

a) 48/253 b) 64/253 c) 23/253 d) 56/ 253

10. Four tickets marked 00, 01, 10, 11 respectively are placed in a bag. A ticket is drawn at random five times being replaced each time. The probability that the sum of the numbers on the ticket is 27 is

a) 2/7 b) 25/256 c) 231/256 d) 0

11. 5 Engineering, 4 Mathematics and 2 Chemistry books are placed on a shelf at random. The probability that the books of each kind are all together is

a) 1/155 b) 2/255 c) 3/1255 d) 1/1155

12. A determinant is chosen at random from the set of all determinants of order 2 with elements 0 or 1 only. The probability that the value of the determinant chosen is positive is [IIT-1982]

a) 3/8 b) 3/16 c) 1/8 d) 1/16

13. In a multiple choice questions there are four alternative answers, of which one or more than one is correct. A candidate will get marks on the question only if he ticks all the correct answers. The candidate decides to tick answers at random. If he is allowed up to three chances to answer the question, the probability that he will get marks on it is [IIT-85]

a) 1/3 b) 1/5 c) 3/5 d) 1/15

14. A box contains 2 fifty-paise coins, 5 twentyfive-paise coins and a certain number n($\geq$,2) of ten and five Paise coins. Five coins are taken out of the box at random. The probability that the total value of these 5 coins is less than one rupee and fifty paise is [IIT-88]

a) $\frac{10(n+2)}{(n+7)c_5}$ b) $1-\frac{10(n+2)}{(n+7)c_5}$

c) $\frac{5(n+2)}{(n+2)c_5}$ d) $1-\frac{5(n+2)}{nc_5}$

15. Two events A and B have probabilities 0.25 and 0.5 respectively. The probability that both A and B occur simultaneously is 0.14. Then the probability that neither A nor B occurs is [IIT-1980]

a) 0.39 b) 0.25 c) 0.11 d) 0.65

16. In a class of 125 students, 70 passed in Mathematics, 55 in Statistics and 30 in both. The probability that a student selected at random from the class, has passed in only one subject is [JNTU -1993]

a) 13/25 b) 3/25 c) 17/25 d) 8/25

17. A man and a women appear in an interview for vacancies in the same post. The probability of man's selection is 1/4 and that of the women's selection is 1/3. The probability that none of them will be selected is [MNR-1988]

a) 1/2 b) 1/3 c) 1/4 d) 1/5

18. A and B are independent events. The probability that both A and B occur is 1/6 and the probability that neither of them occur is 1/3. The probability of occurrence of A is [IIT-1984]

a) 1/6 b) 1/5 c) 1/2 or 1/3 d) 2/5

19. The probabilities of solving a problem by three students A, B, C independently are1/3,1/4,1/5. The probability that the problem will be solved as [JNTU-1993]

a) 1/60 b) 36/60 c) 48/60 d) 57/60

20. From a pack of 52 cards, two cards are drawn, the first being replaced before the second is drawn. The probability that the first is a Diamond and the second is a King is [MNR-1979]

a) 1/4 b) 4/13 c) 1/52 d) 1/104

21. A speaks truth in 75% of the cases and B in 80% of the cases.The percentage of the cases of which they likely to contradict each other in stating the same fact is[IIT-75]

a) 20% b) 25% c) 30% d) 35%

22. Three groups of children contain 3 girls and one boy; 2 girls and 2 boys; one girl and 3 boys. One child is selected at random from each group. The probability that the three selected consist of 1 girl and 2 boys is [IIT-85]

a) 13/32 b) 19/32 c) 13/19 d) 6/19

23. A purse contains 4 copper coins, 3 silver coins; the second purse contains 6 copper coins and 2 silver coins. A coin is taken out of any purse. The probability that it is a copper coin is [MNR-85]

a) 4/7 b) 3/4 c) 3/7 d) 37/56

24. Bag A contains 2 White and 3 Red balls a and bag B contains 4 White and 5 Red balls. One ball is drawn at random from one of the bags and is found to be Red. The probability that it was drawn from bag B is [IIT-76]

a) 23/54 b) 25/52 c) 25/52 d) 27/55

25. Hyderabad sub-urban railway has 21 stations. Designating a ticket from one station to another station as distinct type of ticket, calculate the number of types of tickets required for the railway service is [Hyd.Univ-1995]

a) 441 b) 421 c) 420 d) 210

26. A computer science class consists of 13 females and 12 males. Six class members are to be chosen at random to plan a picnic. What is the probability that exactly 4 females and 2 males are chosen? [Hyd.Univ-1990]

a) 0.1 b) 0.3 c) 0.4 d) 0.2

27. A student appears for 4 interviews for jobs. He feels his chances are 50:50 for succeeding in any one interview. The probability of his getting two job's is [Hyd.Univ-1990]

a) 2/16 b) 4/16 c) 6/16 d) 8/16

28. From 7 Foreigners and 4 Indians a committee of 6 is to be formed; In how many ways can this be done so that committee contains atleast two Indians.[Hyd.Univ-1988]

a) 210 b) 41 c) 317 d) 371

29. Two numbers are drawn from 1to 100. The probability that their sum is divisibleby 5 is [Hyd.Univ- 1998]

a) 13/165 b) 197/495 c) 1/ 5 d) 14/165

30. Two unbiased dice are tossed together. The probability that the total score is a perfect square is [Hyd.Univ- 1996]

a) 1/3 b) 7/36 c) 1/6 d) 5/36

31. A box has 7 balls with numbers 1-7 written them. The number of ways of drawing first any two balls then any three balls and finally the last two balls without replacement is **[Hyd.Univ- 1996]**

a) 210 b) 120 c) 102 d) 480

32. Box 1 contains 4 defective and 16 non-defective light bulbs. Box 2 contains 1 defective and 1nondefective light bulb. Quality control manager tosses a coin and select a bulb form box one if it comes up heads otherwise he selects a bulb form box two.The probability that a selected bulb is defective is **[Hyd.Univ- 1996]**

a) 2/5 b) 2/7 c) 7/15 d) 7/20

33. The president of computer forum did not like the student's committee constituted earlier. The committee consisted of 2 students from M. Tech 1, 3 fromM.Tech 2 and 4 from M.Tech-3. The president wants to reduce the committee by 3. What is the probability that the three choosen remove at random are from different classes? **[Hyd. Univ - 1994]**

a) 1/24 b) 1/8 c) 1/4 d) None

ANSWER KEY - II

1. d	**2.** c	**3.** c	**4.** d	**5.** c
6. d	**7.** c	**8.** c	**9.** a	**10.** d
11. d	**12.** b	**13.** b	**14.** b	**15.** a
16. a	**17.** a	**18.** c	**19.** b	**20.** c
21. d	**22.** a	**23.** d	**24.** c	**25.** d
26. c	**27.** c	**28.** d	**29.** c	**30.** b
31. a	**32.** d	**33.** d		

EXPLANATORY NOTES - II

3. (c) **Explanation :** The no. of ways in which 10 Roses, 15 Sunflowers and 14 Daffodils can be divided the 2 girls is ${}^{10+2-1}C_{2-1} \times {}^{15+2-1}C_{2-1} \times {}^{14+2-1}C_{2-1}$

$$= {}^{11}C_1 \times {}^{16}C_1 \times {}^{15}C_1 = 11 \times 16 \times 15 = 2640$$

4. (d) **Explanation :** Let today is A and tomorrow is B

$$P(A)=0.5 \quad P(B)=0.6 \quad P(A\cup B)=0.7$$

$$P(A\cap B)=P(A)+P(B)-P(A\cup B) = 0.5+0.6-0.7=0.4$$

5. (c) **Explanation :** Probability of select the top card Aces is $\frac{{}^4C_1}{{}^{52}C_1}$. Next Probability of select bottom card Aces is. $\frac{{}^3C_1}{{}^{51}C_1}$ because one card will be draw 3 Aces are available and 51 cards are available. The probability that top and bottom cards of randomly shuffled deck are both aces is

$$\frac{{}^4C_1}{{}^{52}C_1} \times \frac{{}^3C_1}{{}^{51}C_1} = \frac{4}{52} \times \frac{3}{51}$$

6. (d) **Explanation :** We can't have 0, 7 at the hundreds place. So hundreds place can be filled with 8 digits. 1, 2, 3, 4, 5, 6, 7, 8, 9. So there are 8 ways of filling the hundreds place. Now 8 digits are left including 0. So ten's place can be filled with any of the remaining 8 digits in 8 ways. Now units place can be filled with in any of remaining 7 digits. Hence total no of required is $= 8 \times 8 \times 7 = 648$.

The probability that a number selected at random 100 and 999 will not contain the digit 7 is $= \frac{648}{900} = \frac{112}{225}$.

7. (c) **Explanation :** Let A be ball is drawn from a bag that is one of them is black and other is white

$$n(A) = {}^{10}C_1 \times {}^{15}C_1$$

The total no. of balls are 25 and we can draw 2 balls

$$n(S) = {}^{25}C_2 = \frac{25 \times 24}{2} = 25 \times 12$$

$$P(A) = \frac{\text{favourable no. of cases}}{\text{exhaustive no. of cases}} = \frac{10 \times 15}{25 \times 12} = \frac{1}{2}$$

8. (c) **Explanation :** Let A be event of getting at least 2 heads and S be the sample space when coin tosses 3 times. $n(S) = 2^3 = 8$

Sample space S = {T T T, T H H, H T H, H H T, H T T, T H T, T T H, H H H}

$$P(A) = \frac{\text{favourable no. of cases}}{\text{exhaustive no. of cases}} = \frac{4}{8} = \frac{1}{2}$$

9. (a) **Explanation :** Let A be event of drawing 3 balls from collection of 7 white balls, 12 Green and 4 Red balls.

Let S be a sample space. $P(S) = {}^{23}C_3 = \frac{23 \times 22 \times 21}{1 \times 2 \times 3}$

$$P(A) = \frac{\text{favourable no. of cases}}{\text{exhaustive no. of cases}} = \frac{{}^7C_1 \times {}^{12}C_1 \times {}^4C_1}{{}^{23}C_3}$$

$$\frac{7 \times 12 \times 4}{23 \times 11 \times 7} = \frac{48}{253}$$

14. (b) **Explanation :** We know that A box contains 2 fifty paise coins, 5 twenty five paise coins and certain number $n(\geq 2)$ of ten and five paise coins. Five coins are taken out of the box at random. The probability that the total value of these 5 coins is less than one rupee and fifty paise is $1 - \frac{10(n+2)}{{}^{(n+7)}C_3}$

16. (a) **Explanation :** The following shows vendiagram.

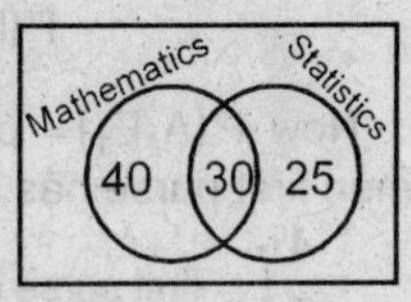

Let A be the event of student passed one subject and S be a sample space. $n(S)=125$

$$P(A) = \frac{\text{favourable no. of cases}}{\text{exhaustive no. of cases}} = \frac{65}{125} = \frac{13}{25}$$

20. (c) **Explanation :** Let A be the event of getting 2 cards of one is diamond and second is king. S be a sample space.

$$P(A) = \frac{\text{favourable no. of cases}}{\text{exhaustive no. of cases}} = \frac{{}^{13}C_1 \times {}^4C_1}{{}^{52}C_2} = \frac{13 \times 4}{\frac{52 \times 51}{1 \times 2}} = \frac{2}{51}$$

22. (a) **Explanation :** Consider the following events.

E_1 = selected child from first group

E_2 = selected child from second group.

E_3 = selected child from third group

$$P(E_1) = \frac{3}{4} \quad P(E_2) = \frac{2}{4} \quad P(E_3) = \frac{1}{4}$$

$\therefore$ $P(\bar{E}_1)$ = selected child from first group is boy.

$$= 1 - P(E_1) = 1 - \frac{3}{4} = \frac{1}{4}$$

$P(\bar{E}_2)$ = selected child from second group is boy.

$$=1-P(E_2)=1-\frac{2}{4}=\frac{2}{4}=\frac{1}{2}$$

$P(\overline{E}_3)=$ selected child from third group is boy.

$$=1-P(E_3)=1-\frac{1}{4}=\frac{3}{4}$$

now 1 girl and 2 boys can be selected in a mutually exclusive ways.

(I) girl from group 1, boy from group 2, boy from group 3 i.e., $E_1 \cap E_2 \cap \overline{E}_3$

(II) boy from group 1, girl from group 2, girl from group 3 i.e., $E_1 \cap \overline{E}_2 \cap E_3$

(III) boy from group 1, girl from group 2, girl from group 3. i.e., $\overline{E}_1 \cap E_2 \cap E_3$

$\therefore$ Required probability $=P(I)+P(II)+P(III)$

$$=P(E_1 \cap E_2 \cap \overline{E}_3)+P(E_1 \cap \overline{E}_2 \cap E_3)+P(\overline{E}_1 \cap E_2 \cap E_3)$$

$[\because E_1, E_2, E_3$ are independent events$]$

$$=\frac{3}{4}\times\frac{2}{4}\times\frac{3}{4}+\frac{3}{4}\times\frac{1}{2}\times\frac{1}{4}+\frac{1}{4}\times\frac{2}{4}\times\frac{1}{4}$$

$$=\frac{18}{64}+\frac{3}{32}+\frac{2}{64}=\frac{18+6+2}{64}=\frac{26}{64}=\frac{13}{32}$$

23. (d) **Explanation :** A copper coil can be drawn a mutually exclusive ways.

(I) selecting a purse 1 and then drawing a copper coil from it.

(II) Selecting a purse 2 and then drawing a copper coil from it.

$E_1=$ selecting purse 1, $E_2=$ selecting purse 2

A = drawing a copper coil

Since one of the 2 purses is selected randomly

$$\therefore P(E_1)=\frac{1}{2} \text{ and } P(E_2)=\frac{1}{2}$$

Now $P(A/E_1)=$ probability of drawing a copper coil when first purse has been chosen.

$=\frac{4}{7}$ [$\because$ first purse contains 4 copper coil and 3 silver coil]

and $P(A/E_2)=$ Probility of drawing a copper coil when the second purse has selected.

$=\frac{6}{8}$ [$\because$ the second purse contains 6 copper coil and 2 silver coil]

using the law of total probability we have

$$P(\text{redball})=P(A)=P(E_1)P(A/E_1)+P(E_2)P(A/E_2)$$

$$=\frac{1}{2}\times\frac{4}{7}+\frac{1}{2}\times\frac{6}{8}=\frac{2}{7}+\frac{3}{8}=\frac{16+21}{56}=\frac{37}{56}$$

25. (c) **Explanation :** In 1 railway station to 20 railway stations we can design a 20 tickets.

In 21 railway stations to design a ticket is 21 $\times$ 20 ways = 420 ways.

26. (d) **Explanation :** Let A be the exactly 4 females and 2 males are chosen of the 13 females and 12 males.

$$n(A)={}^{13}c_4\times{}^{12}c_2 \qquad n(S)={}^{25}c_6$$

$$P(A)=\frac{\text{favourable no. of cases}}{\text{exhaustive no. of cases}}=\frac{{}^{13}c_4\times{}^{12}c_2}{{}^{25}c_6}=\frac{858}{3220}=0.2$$

27. (c) **Explanation :** A student appears for 4 interviews for job. The probability of getting 2 job's is $=\frac{{}^4c_2}{16}=\frac{6}{16}$

28. (d) **Explanation :** The commitee can be formed in the following ways.

(1) By selecting 2 Indians and 4 foreigners is ${}^4c_2\times{}^7c_4$

(2) By selecting 3 Indians and 3 foreigners is ${}^4c_3\times{}^7c_3$

(3) By selecting 4 Indians and 2 foreigners is ${}^4c_4\times{}^7c_2$

Total no of ways of forming the commitee

$${}^4c_2\times{}^7c_4+{}^4c_3\times{}^7c_3+{}^4c_4\times{}^7c_2=371$$

29. (c) **Explanation :** First we can draw 1 then next we can draw 4, 9, 14, 29, 34, 39, 44, 49, 54, 59, 64, 69, 74, 79, 84, 89, 94, 99.

We can draw any no. The 20 types of no's sum will be divisible by 5.

We can draw 100 members then 2000 types of no's sum will be divisible by 5.

$$P(A)=\frac{\text{favourable no. of cases}}{\text{exhaustive no. of cases}}=\frac{2000}{100\times100}=\frac{1}{5}$$

30. (b) **Explanation :** Let A be the 2 unbiased dice are tossed together that the score is perfect square is 7 types these are (1, 3), (3, 1), (4, 5), (5, 4), (6, 3), (3, 6). $n(S)=36$

$$P(A)=\frac{\text{favourable no. of cases}}{\text{exhaustive no. of cases}}=\frac{7}{36}$$

31. (a) **Explanation :** A box has 7 balls. The no. of ways of drawing first any 2 balls then any 3 balls and finally the last 2 balls without replacement is $={}^7c_2\times{}^5c_3\times{}^2c_2$

$$=21\times10\times1=210$$

33. (d) **Explanation :** Let A be the 3 students chosen from different classes. Given the committee consists of 2 students from M.Tech 1, 3 students from M. Tech 2, 4 students from M.Tech 3.

$$n(A)={}^2c_1\times{}^3c_1\times{}^4c_1 \qquad n(S)={}^9c_3=84$$

$$P(A)=\frac{\text{favourable no. of cases}}{\text{exhaustive no. of cases}}=\frac{2\times3\times4}{84}=\frac{2}{7}$$

III. STATISTICS

- **Mean**
- **Median**
- **Mode**

1. Measures of Central Tendency : An average or a central value of a statistical series is the value of the variable which describe the characteristic of the entire distribution. In other words an average of a distribution is the value of the variable which is representative of the entire distribution.

The following are the five measures of central tendency :

(i) Arithmetic Mean (ii) Geometric Mean

(iii) Harmonic Mean (iv) Median and

(v) Mode.

2. Arithmetic Mean : If $x_1, x_2, \ldots, x_n$ are n values of variable X, then the arithmetic mean of these values is given by

$$\overline{X} = \frac{1}{n}(x_1 + x_2 + \ldots x_n) = \frac{1}{n}\left(\sum_{i=1}^{n} x_i\right)$$

In case of a frequency distribution x_i/f_i, i = 1,2,.....,n where f_i is the frequency of the variable x_i,

$$\overline{X} = \frac{f_1x_1 + f_2x_2 + \ldots + f_nx_n}{f_1 + f_2 + \ldots + f_n} = \frac{1}{N}\left(\sum_{i=1}^{n} f_i x_i\right) \quad \ldots (1)$$

where $N = f_1 + f_2 + \ldots + f_n = \sum_{i=1}^{n} f_i$

If $w_1, w_2, w_3 \ldots w_n$ be the weights assigned to the n values $x_1, x_2, \ldots, x_n$ respectively of a variable X, then the weighted AM is given by

$$\overline{X} = \frac{w_1x_1 + w_2x_2 + \ldots + w_nx_n}{w_1 + w_2 + \ldots + w_n} = \frac{\sum_{i=1}^{n} w_i x_i}{\sum_{i=1}^{n} w_i} \text{ or } \overline{X} = \frac{\Sigma wx}{\Sigma w}$$

If the values of x or (and) f are large, the calculation of AM by formula (i) is quite tedious and time consuming. In such a case we take deviations from an arbitrary point 'A'.

Let $d_i = x_i - A, i = 1,2,\ldots, n$. Then

$$f_i d_i = f_i(x_i - A);\ i = 1,2,\ldots,n$$

$$\Rightarrow \sum_{i=1}^{n} f_i d_i = \sum_{i=1}^{n} f_i x_i - A\sum_{i=1}^{n} f_i$$

$$\Rightarrow \frac{1}{N}\left(\sum_{i=1}^{n} f_i d_i\right) = \frac{1}{N}\left(\sum_{i=1}^{n} f_i x_i\right) - A$$

$$\Rightarrow \frac{1}{N}\sum_{i=1}^{n} f_i d_i = \overline{X} - A$$

$$\Rightarrow \overline{X} = A + \frac{1}{N}\sum_{i=1}^{n} f_i d_i \quad \ldots (ii)$$

In case of a grouped or continuous frequency distribution, x is taken to be the mid-point of the class-interval.

Sometimes the values of d_i are divisible by a common number h (say). In such a case the arithmetic is reduced to a greater extent by taking

$$u_i = \frac{x_i - A}{h}, i = 1,2,\ldots,n$$

$$\Rightarrow \quad hu_i = x_i - A,\ i = 1,2,\ldots,n$$

$$\Rightarrow \quad hf_i u_i = f_i x_i - A f_i,\ i = 1,2,\ldots,n$$

$$\Rightarrow \quad h\sum_{i=1}^{n} f_i u_i = \sum_{i=1}^{n} f_i x_i - A\sum_{i=1}^{n} f_i$$

$$\Rightarrow \quad h\left(\frac{1}{N}\sum_{i=1}^{n} f_i u_i\right) = \frac{1}{N}\sum_{i=1}^{n} f_i x_i - A$$

$$\Rightarrow \quad h\left(\frac{1}{N}\sum_{i=1}^{n} f_i u_i\right) = \overline{X} - A$$

$$\Rightarrow \quad \overline{X} = A + h\left(\frac{1}{N}\sum_{i=1}^{n} f_i u_i\right)$$

Properties of Arithmetic Mean :

(i) Algebraic sum of he deviations of a set of values from their arithmetic mean is zero

(ii) The sum of the squares of the deviations of a set of values is minimum when taken about mean.

(iii) Mean of the composite series : If $\overline{X}_1, \overline{X}_2, \ldots \overline{X}_k$ are the means of k series of sizes $n_1, n_2, \ldots n_k$ respectively then the mean $\overline{X}$ of the composite series is given by

$$\overline{X} = \frac{n_1\overline{X}_1 + n_2\overline{X}_2 + \ldots + n_k\overline{X}_k}{n_1 + n_2 + \ldots + n_k}$$

3 Geometric Mean : Geometric mean of a set of n observations is the nth root of their product. Thus, the geometric mean G of n non-zero observations $x_1, x_2, \ldots, x_n$ is

$$G = (x_1 . x_2 . x_3 \ldots x_n)^{1/n} \quad \ldots (i)$$

$$\Rightarrow \quad \log G = \frac{1}{n}(\log x_1 + \log x_2 + \ldots + \log x_n)$$

$$\Rightarrow \quad \log G = \frac{1}{n}\sum_{i=1}^{n} \log x_i$$

$$\Rightarrow \quad G = \text{antilog}\left(\frac{1}{n}\sum_{i=1}^{n} \log x_i\right) \quad \ldots (ii)$$

In a case of frequency distribution x_i/f_i; i = 1,2,...,n, geometric mean G is given by

$$\Rightarrow \quad G = \left(x_1^{f_1} . x_2^{f_2} . x_3^{f_3} \ldots x_n^{f_n}\right)^{1/N}, \text{ where } N = \sum_{i=1}^{n} f_i$$

$$\Rightarrow \quad \log G = \frac{1}{N}(f_1 \log x_1 + f_2 \log x_2 + \ldots + f_n \log x_n)$$

$$\Rightarrow \quad \log G = \frac{1}{N}\sum_{i=1}^{n} f_i \log x_i$$

$$\Rightarrow \quad G = \text{antilog}\left(\frac{1}{N}\sum_{i=1}^{n} f_i \log x_i\right) \quad \text{.... (iii)}$$

In case of a grouped or continuous frequency distribution, x is taken as the mid point of class intervals.

Properties of Geometric Mean : If G_1 and G_2 are the geometric means of two series of sizes n_1 and n_2 respectively, then the geometric mean G of the combined series is given by

$$\log G = \frac{n_1 \log G_1 + n_2 \log G_2}{n_1 + n_2} \quad \text{...... (iv)}$$

4. Harmonic Mean : Harmonic mean of a number of non-zero observations is the reciprocal of the AM of the reciprocals of the given values. Thus, harmonic mean H of n non-zero observations $x_1, x_2, \ldots\ldots, x_n$ is

$$H = \frac{1}{\frac{1}{n}\sum_{i=1}^{n}\left(\frac{1}{x_i}\right)} \quad \text{...... (i)}$$

In case of a frequency distribution, $x_i/f_i\,;\, i = 1,2,\ldots.,n$

$$H = \frac{1}{\frac{1}{N}\sum_{i=1}^{n}\left(\frac{f_i}{x_i}\right)} \quad \text{...... (ii)}$$

5. Median and other Partition Values : Median of a distribution is the value of the variable which divides it into two equal parts. It is the value such that number of observations above it is equal to the number of observations below it.

In case of individual observations $x_1, x_2, \ldots\ldots, x_n$ if the number of observations is odd, then median is the value of $\left(\frac{n+1}{2}\right)^{th}$ observation after the observations have been arranged in ascending or descending order of magnitude. In case of even number of observations median is the AM of the values of $\left(\frac{n}{2}\right)^{th}$ and $\left(\frac{n}{2}+1\right)^{th}$ observations.

In case of discrete frequency distribution $x_i/f_i\,;\, i = 1,2,\ldots,n$, the median is calculated by following algorithm.

Algorithm :

STEP I Find cumulative frequencies (c.f.)

STEP II Find $\frac{N}{2}$, where $N = \sum_{i=1}^{n} f_i$

STEP III See the cumulative frequency (c.f.) just greater than $\frac{N}{2}$

STEP IV The corresponding value of x is the median

In case of a grouped or continuous frequency distribution, the class corresponding to the c.f. just greater than $\frac{N}{2}$ is called the median class and the value of the median is calculated by the formula :

$$\text{Median} = l + \left(\frac{\frac{N}{2} - F}{f}\right) h$$

where l = lower limit of the median class.

f = frequency of the median class

h = size (width) of the median class

F = c.f. of the class preceding the median class

$$N = \sum_{i=1}^{n} f_i$$

As median divides a distribution into two equal parts, similarly the quartiles, quantiles, deciles and percentiles divide the distribution respectively into 4, 5, 10 and 100 equal parts. the ith quartile is given by

$$Q_i = l + \left(\frac{\frac{iN}{4} - F}{f}\right) h; \; i = 1,2,3$$

Q_1 is the lower quartile, Q_2 is the median and Q_3 is called the upper quartile.

The ith decile is given by

$$D_i = l + \left(\frac{\frac{iN}{10} - F}{f}\right) h; \; i = 1,2,\ldots.,9.$$

D_5 is the median

6. Mode :

The mode or modal value of a distribution is that value of the variable for which the frequency is maximum.

In case of a discrete frequency distribution the value of mode is determined by the method of grouping.

In case of a grouped or continuous frequency distribution, mode is given by the formula.

$$\text{Mode} = l + \frac{f - f_1}{2f - f_1 - f_2} \times h$$

where l = lower limit of the modal class.

h = width of the modal class

f_1 = frequency of the class preceeding the modal class

f_2 = frequency of the class following the modal class

f = frequency of the modal class

7. Symmetric Distribution

A distribution is a symmetric distribution if the values of mean, mode and median coincide. In a symmetric distribution frequencies are symmetrically distributed on bothsides of the centre point of the frequency curve.

A distribution which is not symmetric is called a skewed distribution. In a moderately asymmetric the intervel between the mean and the median is approximately one-third of the interval between the mean and the mode i.e. we have the following empirical relation between them:

Mean - Mode = 3 (Mean - Median)

$\Rightarrow$ Mode = 3 Median - 2 Mode

8. Measures of Dispersion

Measures of central tendency or averages gives us one single figure that repsents the entire data i.e. they give us one single figure around which the observations are concentrated. But the averages are inadequate to give us a complete idea of the distribution as they do not tell us the extent to which the observations vary from some central value. There can be two or more distributions having the same central value but still there can be wide disparities in the formulation of the distribution. Consider three distributions (i) 1, 5, 9, 13, 17, (ii) 3, 6, 9, 12, 15 (iii) 7, 8, 9, 10, 11. In all these distributions we have the same number of observations and the same mean $\overline{X}=9$. If we are given that the mean of 5 observations is 9, we are unable to say whether it is the average of first distribution or second distribution or third distribution. In first distribution the various about the central value are more in comparison to second and third distribution while their distribution has minimum variations. Dispersion is the measure of the variations. It measures the degree of scatteredness of the observations in a distribution around the central value.

The measures of dispersion commonly used are: (i) Range (ii) Quartile deviation or the semi-interquartile range (iii) Mean deviation and (iv) Standard deviation.

Range

The range is the difference between the greatest and the least observations of the distribution. Thus, if A and B are the greatest and the smallest observations respectively in a distribution, then its range = A - B.

The coefficient of range (or scatter) $= \dfrac{A-B}{A+B}$.

Quartile Deviation or Semi-Interquartile Range (Q.D.)

If Q_1 and Q_3 be the lower and upper quartiles then quartile deviation or semi-interquartile range Q is given by

$$Q = \frac{1}{2}(Q_3 - Q_1)$$

Coefficient of quartile deviation

$$= \frac{(Q_3-Q_1)/2}{(Q_3+Q_1)/2} = \frac{Q_3-Q_1}{Q_3+Q_1}$$

Mean Deviation (M.D.)

If $x_i/f_i; i=1,2,.....n$ is the frequency distribution, then mean deviation from an average A (median, mode or AM) is given by

$$\text{M.D.} = \frac{1}{N}\sum_{i=1}^{n} f_i|x_i - A|, \text{ where } \sum_{i=1}^{n} f_i = N$$

Mean coefficient of dispersion

$$= \frac{\text{Mean deviation from the mean}}{\text{Mean}}$$

Median coefficient of dispersion

$$= \frac{\text{Mean deviation from the median}}{\text{Median}}$$

Mode coefficient of dispersion

$$= \frac{\text{Mean deviation from the mode}}{\text{Mode}}$$

Standard Deviation (S.D.) and Variance

The standard deviation of variate X is the square root of the AM of the squares of all deviations of X from the AM of the observations and it is denoted by the Greek letter small sigma.

Thus, if x_i/f_i, $i = 1,2,.....,n$ is a frequency distribution, then

$$\sigma = \sqrt{\frac{1}{N}\sum_{i=1}^{n} f_i(x_i - \overline{X})^2} \quad \text{..... (i)}$$

Where $\overline{X}$ is the AM of the distribution and $N = \sum_{i=1}^{n} f_i$

The square of the standard deviation is called the variance and is given by

$$\sigma^2 = \frac{1}{N}\sum_{i=1}^{n} f_i(x_i - \overline{X})^2 \quad \text{..... (ii)}$$

Coefficient of dispersion $= \dfrac{\sigma}{\overline{X}}$

Coefficient of variation $= \dfrac{\sigma}{\overline{X}} \times 100$

For comparing the variability of two series, we calculate the coeff. of variations for each series. The series having greater C.V. is said to be more variable than the other and the series having lesser C.V. is said to be more consistent than the other.

If deviations of X are measured from an assumed mean A, then root-mean square deviation of X is denoted by s, and is given by

$$s = \sqrt{\frac{1}{N}\sum_{i=1}^{n} f_i(x_i - A)^2} \quad \text{..... (iii)}$$

s^2 is called mean square deviation.

Some Results :

(i) For a symmetrical distribution, the following area reltion-ship holds good :

$\overline{X} \pm \sigma$ covers 68.27% items

$\overline{X} \pm 2\sigma$ covers 95.45% items

$\overline{X} \pm 3\sigma$ covers 99.74% items

(ii) Relationship between different measures of dispersion

a) $\text{Q.D.} = \frac{2}{3}\sigma$ (approx.) b) $\text{M.D.} = \frac{4}{5}\sigma$ (approx.)

c) $\text{S.D.} = \frac{3}{2}\text{Q.D}$ (approx.) d) $\text{Q.D.} = \frac{5}{6}\text{M.D.}$ (approx.)

e) $\text{M.D.} = \frac{6}{5}\text{Q.D.}$ (approx.)

OBJECTIVE QUESTIONS

1. If the mean of a set of observatiions $x_1, x_2, ..., x_n$ is $\overline{X}$, then the mean of the observations $x_i + 2i$; $i = 1,2,....,n$ is

a) $\overline{X}+2$ b) $\overline{X}+2n$ c) $\overline{X}+(n+1)$ d) $\overline{X}+n$

2. If a variate X is expressed as a linear function of two variates U and V in the form $X = aU + bV$, then mean $\overline{X}$ of X is

a) $a\overline{U} + b\overline{V}$ b) $\overline{U} + \overline{V}$ c) $b\overline{U} + a\overline{U}$ d) none of these

3. The AM of n numbers of a series is $\overline{X}$. If the sum of first (n - 1) terms is k, then the nth number is

a) $\overline{X}-k$ b) $n\overline{X}-k$ c) $\overline{X}-nk$ d) $n\overline{X}-nk$

4. The means of a set of numbers is $\overline{X}$. If each number is divided by 3, then the new mean is

a) $\overline{X}$ b) $\overline{X}+3$ c) $3\overline{X}$ d) $\frac{\overline{X}}{3}$

5. The weighted AM of first n natural numbers whose weights are equal to the corresponding numbers is equal to

a) 2n+1 b) $\frac{1}{2}(2n+1)$ c) $\frac{1}{3}(2n+1)$ d) $\frac{2n+1}{6}$

6. The AM of the series 1, 2, 4, 8, 16,......, 2^n is

a) $\frac{2^n-1}{n}$ b) $\frac{2^{n+1}-1}{n+1}$ c) $\frac{2^n+1}{n}$ d) $\frac{2^n-1}{n+1}$

7. If $\overline{X}$ is the mean of $x_1, x_2, x_3....x_n$. Then the algebraic sum of the deviations about mean $\overline{X}$ is

a) 0 b) $\frac{\overline{X}}{n}$ c) $n\overline{X}$ d) none of these

8. The most stable measure of central tendency is

a) the mean b) the median

c) the mode d) none of these

9. The mean of the distribution, in which the vaues of X are 1,2,...,n, the frequency of each being unity is:

a) $\frac{n(n+1)}{2}$ b) $\frac{n}{2}$ c) $\frac{n+1}{2}$ d) none of these

10. 10 is the mean of a set of 7 observations and 5 is the mean of a set of 3 observations. The mean of the combined set is given by

a) 15 b) 10 c) 8.5 d) 7.5

11. A statistical measure which cannot be determined graphically is

a) median b) mode

c) harmonic mean d) mean

12. The measure which takes into account all the data items is

a) mean b) median c) mode d) none of these

13. The GM of the series 1, 2, 4, 8,16,...., 2^n is

a) $2^{n+1/2}$ b) 2^{n+1} c) $2^{n/2}$ d) 2^n

14. If G_1, G_2 are the geometric means of two series of observations and G is the GM of the ratios of the corresponding observations then G is equal to

a) $\frac{G_1}{G_2}$ b) $\log G_1 - \log G_2$

c) $\frac{\log G_1}{\log G_2}$ d) $\log(G_1, G_2)$

15. If G is the GM of the product of r sets of observations with geometric means $G_1, G_2, ... G_r$ respectively, then G is equal to

a) $\log G_1 + \log G_2 + ... + \log G_n$

b) $G_1.G_2......G_n$ c) $\log G_1.\log G_2......\log G_n$

d) none of these

16. A group of 10 items has arithmetic mean 6. If the arithmetic mean of 4 of these items is 7.5, then the mean of the remaining item is

a) 6.5 b) 5.5 c) 4.5 d) 50.

17. The arithmetic mean of a set of observations is $\overline{X}$. If each observation is divided by α and then is increased by 10, then the mean of the new series is

a) $\frac{\overline{X}}{\alpha}$ b) $\frac{\overline{X}+10}{\alpha}$ c) $\frac{\overline{X}+10\alpha}{\alpha}$ d) $\alpha\overline{X}+10$

18. The weighted mean of first n natural numbers whose weights are equal to the squares of corresponding numbers is

a) $\frac{n+1}{2}$ b) $\frac{3n(n+1)}{2(2n+1)}$ c) $\frac{(n+1)(2n+1)}{6}$ d) $\frac{n(n+1)}{2}$

19. If a variable takes values 0,1,2,....,n. with frequencies 1, ${}^nC_1$, ${}^nC_2$,......, ${}^nC_n$ then the AM is

a) n b) $\frac{2^n}{n}$ c) n + 1 d) $\frac{n}{2}$

20. The mean income of a group of workers is $\overline{X}$ and that of another group is $\overline{Y}$. If the number of workers in the second group is 10 times the number of workers in the first group, then the mean income of the combined group is

a) $\frac{\overline{X}+10\overline{Y}}{3}$ b) $\frac{\overline{X}+10\overline{Y}}{11}$ c) $\frac{10\overline{X}+\overline{Y}}{Y}$ d) $\frac{\overline{X}+10\overline{Y}}{9}$

21. The sum of the squares of deviations of a set of values is minimum when taken about

a) AM b) GM c) HM d) median

22. The mean age of a combined group of men and women is 25 yrs. If the mean age of the group of men is 26 and that of the group of women is 21, then the percentage of men and women in the group is

a) 60,40 b) 80,20 c) 20,80 d) 40,60

23. In a moderately skewed distribution the values of mean and median are 5 and 6 respectively. The value of mode in such a situation is approximately equal to

a) 8 b) 11 c) 16 d) none of these

24. If a variable takes discrete values $x+4, x-\frac{7}{2}, x-\frac{5}{2}, x-3, x-2, x+\frac{1}{2}, x-\frac{1}{2}, x+5$ $(x > 0)$ then the median is

a) $x-\frac{5}{4}$ b) $x-\frac{1}{2}$ c) $x-2$ d) $x+\frac{5}{4}$

25. If $y = f(x)$ be a monotonically increasing or decreasing function of x and M is the median of variable x, then the median of y is

a) f(M) b) $\frac{M}{2}$ c) $f^{-1}(M)$ d) none of these

26. In any discrete series (when all values are not same) the relationship between M.D. about mean and S.D. is

a) M.D. = S.D. b) M.D. $\geq$ S.D.

c) M.D. < S.D. d) M.D. $\leq$ S.D.

ANSWER KEY

1. c	2. a	3. b	4. d	5. c
6. b	7. a	8. a	9. c	10. c
11. c	12. a	13. a	14. a	15. b
16. d	17. c	18. b	19. d	20. b
21. a	22. b	23. a	24. a	25. a
26. d				

EXPLANATORY NOTES

1. It is given that $\overline{X} = \frac{x_1 + x_2 + ... + x_n}{n}$. Therefore $n\overline{X} = x_1 + x_2 + ... + x_n$. Let $\overline{Y}$ be the mean of observations $x_i + 2i; i = 1, 2, ..., n$ then

$$\overline{Y} = \frac{(x_1 + 2.1) + (x_2 + 2.2) + (x_3 + 2.3) + ... + (x_n + 2.n)}{n}$$

$$= \frac{\sum_{i=1}^{n} x_i + 2(1 + 2 + 3 + ... + n)}{n}$$

$$= \frac{1}{n}\sum_{i=1}^{n} x_i + \frac{2n(n+1)}{2n} = \overline{X} + (n+1)$$

2. We have : $\Sigma X = a\Sigma u + b\Sigma v$. Therefore,

$$\overline{X} = \frac{1}{n}\Sigma x = a.\frac{1}{n}\Sigma u + b\frac{1}{n}\Sigma v = a\overline{u} + b\overline{v}$$

3. Let the n-numbers be $x_1, x_2, ... x_n$. Then,

$$\overline{X} = \frac{1}{n}\sum_{i=1}^{n} x_i$$

$$\Rightarrow \overline{X} = \frac{x_1 + x_2 + ... + x_{n-1} + x_n}{n}$$

$$\Rightarrow \overline{X} = \frac{k + x_n}{n} \qquad [\because x_1 + x_2 + ... x_{n-1} = k]$$

$$\Rightarrow x_n = n\overline{X} - k$$

4. Let $x_1, x_n, ..., x_n$ be n numbers. Then $\overline{X} = \frac{1}{n}\sum_{i=1}^{n} x_i$. If each number is divided by 3, then the new mean $\overline{Y}$ is given by

$$\overline{Y} = \frac{1}{n}\sum_{i=1}^{n}\left(\frac{x_i}{3}\right) = \frac{1}{3}\left(\frac{1}{n}\sum_{i=1}^{n} x_i\right) = \frac{\overline{X}}{3}$$

5. The required weighted mean is

$$\overline{X} = \frac{1.1 + 2.2 + 3.3 + ... + n.n}{1 + 2 + 3 + ... + n} = \frac{\frac{n(n+1)(2n+1)}{6}}{\frac{n(n+1)}{2}} = \frac{2n+1}{3}$$

6. The required AM is

$$\overline{X} = \frac{1 + 2 + 2^2 + 2^3 + ... + 2^n}{n+1} = \frac{(2^{n+1} - 1)}{(n+1)(2-1)} = \frac{2^{n+1} - 1}{n+1}$$

10. We have, $n_1 = 7, \overline{X}_1 = 10; n_2 = 3, \overline{X}_2 = 5$

So, Combined mean $= \frac{n_1\overline{X}_1 + n_2\overline{X}_2}{n_1 + n_2} = \frac{85}{10} = 8.5$

13. $GM = (1.2.4.8.....2^n)^{1/n} = (1.2^1.2^2.2^3....2^n)^{1/n}$

$$= \left(2^{\frac{n(n+1)}{2}}\right)^{1/n} = 2^{\frac{n+1}{2}}$$

14. Let $x_1, x_2, ..., x_n$ and $y_1, y_2, ... y_n$ be two series of observations with geometric means G_1 and G_2 respectively. Then, $G_1 = (x_1.x_2.....x_n)^{1/n}$ and $G_2 = (y_1.y_2....y_n)^{1/n}$. Since G is the geometric mean of the ratios of the corresponding observations, therefore

$$G = \left(\frac{x_1}{y_1}.\frac{x_2}{y_2}......\frac{x_n}{y_n}\right)^{1/n} = \frac{(x_1x_2....x_n)^{1/n}}{(y_1.y_2.....y_n)^{1/n}} = \frac{G_1}{G_2}$$

15. Taking X as the product of variates $X_1, X_2, ..., X_r$ corresponding to r sets of observations i.e. $X = X_1X_2....X_r$, we have

$$\log X = \log X_1 + \log X_2 + ...\log X_r$$

$$\Rightarrow \Sigma \log X = \Sigma \log X_1 + \Sigma \log X_2 + ... + \Sigma \log X_r$$

$$\Rightarrow \frac{1}{n}\Sigma \log X = \frac{1}{n}\Sigma \log X_1 + \frac{1}{n}\Sigma \log X_2 + ... + \frac{1}{n}\Sigma \log X_r$$

$$\Rightarrow \log G = \log G_1 + \log G_2 + ... + \log G_r \Rightarrow G = G_1G_2....G_r$$

17. Let $x_1, x_2, ... x_n$ be n observatioins. Then,

$\overline{X} = \frac{1}{n}\Sigma x_i$. Let $y_1 = \frac{x_i}{\alpha} + 10$. Then,

$$\frac{1}{n}\sum_{i=1}^{n} y_i = \frac{1}{\alpha}\left(\frac{1}{n}\Sigma x_i\right) + \frac{1}{n}(10n)$$

$$\Rightarrow \overline{Y} = \frac{1}{\alpha}\overline{x} + 10 = \frac{\overline{x} + 10\alpha}{\alpha}$$

19. The required mean is

$$\overline{X} = \frac{0.1 + 1.{}^nC_1 + 2.{}^nC_2 + 3.{}^nC_3 + ... + n.{}^nC_n}{1 + {}^nC_1 + {}^nC_2 + ... + {}^nC_n}$$

$$= \frac{\sum_{r=0}^{n} r.{}^nC_r}{\sum_{r=0}^{n} {}^nC_r} = \frac{\sum_{r=1}^{n} r.\frac{n}{r}{}^{n-1}C_{r-1}}{\sum_{r=0}^{n} {}^nC_r} = \frac{n\sum_{r=1}^{n} {}^{n-1}C_{r-1}}{\sum_{r=0}^{n} {}^nC_r}$$

$$= \frac{n.2^{n-1}}{2^n} = \frac{n}{2}$$

22. The formula for combined mean is $\overline{X} = \frac{n_1\overline{X}_1 + n_2\overline{X}_2}{n_1 + n_2}$

We are given $\overline{X} = 25, \overline{X}_1 = 26, \overline{X}_2 = 21$. Let $n_1 + n_2 = 100$ and n_1 denotes men and n_2 women. For this $n_2 = 100 - n_1$

$$25 = \frac{26n_1 + 21(100 - n_1)}{100} \Rightarrow n_1 = 80. \text{ So, } n_2 = 20$$

Hence, the percentage of men and women is 80 and 20 respectively.

23. For moderately skewed distribution, we have

$$\text{Mode} = 3\text{Median} - 2\text{Mean} \Rightarrow \text{Mode} = 3(6) - 2(5) = 8$$

24. Arranging the given values in ascendnig order of magnitude, we get

$$x - \frac{7}{2}, x - 3, x - \frac{5}{2}, x - 2, x - \frac{1}{2}, x + \frac{1}{2}, x + 4, x + 5.$$

There are 8 observations in this series, therefore Median = AM of 4th and 5th observation

$$= \text{AM of } (x - 2) \text{ and } (x - 1/2) = \frac{x - 2 + x - \frac{1}{2}}{2} = x - \frac{5}{4}$$

IV. THEORETICAL DISTRIBUTIONS

- **Random Variables and Expectation**
- **Binomial Distribution**
- **Poisson Distribution**
- **Uniform Distribution**
- **Normal Distribution**
- **Exponential Distribution**

1. Random Variable

Definition :

1. Quantities which vary with some probability are called random variables.

2. By a random variable we mean a real number associated with the outcome of a random experiment.

Ex1 : Suppose two coins are tossed simultaneously then the sample space is S = {HH, HT, TH, TT}

Let X denote the number of heads. Then if X = 0 then the outcome is {T T} and $P(X=0) = \frac{1}{4}$

If X takes the value 1, then outcome is {HT, TH} and $P(X=1) = \frac{2}{4}$. Next if X takes the value 2 then the outcome is {HH} and $P(X=2) = \frac{1}{4}$.

The probability distribution of this random variable X is given by the following table :

X = x	0	1	2	Total
P (X = x)	$\frac{1}{4}$	$\frac{2}{4}$	$\frac{1}{4}$	1

Ex 2 : Out of 24 mangoes 6 are rotten, 2 mangoes are drawn. Obtain the probability distribution of the number of rotten mangoes that can be drawn.

Let X denote the number of rotten mangoes drawn then X can take values 0, 1, 2.

$$P(X=0) = \frac{18\,C_2}{24\,C_2} = \frac{18.17}{24.33} = \frac{52}{92}$$

$$P(X=1) = \frac{18\,C_1.6C_2}{24\,C_2} = \frac{18.6}{\frac{24.23}{1.2}} \doteq \frac{9}{23}$$

$$P(X=2) = \frac{6C_2}{24\,C_2} = \frac{6.5}{24.23} = \frac{5}{92}$$

The probability distribution of X is

X = x	0	1	2	Total
P (X = x)	$\frac{51}{92}$	$\frac{9}{23}$	$\frac{5}{92}$	1

2. Types Of Random Variables

There are two types of random variables :

I. Discrete random variables

II. Continuous random variables.

Discrete Random Variables : Quantities which are capable of taking only integral values are called discrete random variables.

Ex: 1 The number of childern in the families of a colony.

Ex: 2 The number of rooms in the houses of a township.

Continuous Random Variables : Quantities which are capable of taking all possible values in a certain range are called continous random variables.

Ex : 1 The weights of the students of a class.

Ex : 2 The height of a child as it grows.

3. Probability Mass Function (Probability Distribution)

Let X be a discrete random variable taking value x, x = 0, 1, 2.... then P (X=x) is called the probability mass function of X and it satisfies the following .

I. $P(X=x) \geq 0$ II. $\sum_{x=0} P(X=x) = 1$

Ex : A random variable X has the following probability function.

X = x	0	1	2	3	4	5	6	7
P (X = x)	0	k	2k	2k	3K	k^2	$2k^2$	$7k^2 + k$

I. Find the value of k, II. $P(X \leq 5)$ and $P(X > 5)$

III. $P(0 < X < 6)$

Sol : I. Since P (X = x) is a probability function

$\sum_x P(X=x) = 1$ i. e., $10k^2 + 9k = 1$

$10k^2 + 9k - 1 = 0 \quad \therefore \quad (k+1)(10K-1) = 0$

$\therefore \quad k = -1,\ k = \frac{1}{10}$ Since $P(X=1) = k$

$\therefore \quad k \neq -1 \Rightarrow k = \frac{1}{10}$

II. $P(X \leq 5) = P(X=0) + P(X=1) + P(x=2) + P(X=3) + P(X=4) + P(X=5)$

$$= 8k + k^2 = \frac{8}{10} + \frac{1}{100} = 0.81$$

$$P(X > 5) = 1 - P(X \leq 5) = 1 - 0.81 = 0.19$$

III. $P(0 < X < 6) = P(X=1) + P(X=2) + P(X=3) + P(X=4) + P(X=5)$

$$= 8k + k^2 = 0.81$$

4. Probability Density Function (p. d. f)

Let X be a continuous random variable taking value x, $a \leq x \leq b$ then $f(x) = P(X = x)$ is defined as the p. d. f of X and satisfies the following

I. $f(x) \geq 0$ **II.** $\int_a^b f(x)\,dx = 1$

Ex : A continuous random variable X has the p. d. f given by

$$f(x) = \begin{cases} kx, & 0 \le x \le 1 \\ k, & 1 \le x \le 2 \\ -x+3k, & 2 \le x \le 3 \\ 0 & \text{other wise} \end{cases}$$

Find the value of k. Also calculate P (X ≤ 1.5).

Sol : Since f (x) is the p. d. f

$$\therefore \int_{-\infty}^{\infty} f(x)\,dx = 1 \text{ i.e.,}$$

$$\int_{-\infty}^{0} f(x)\,dx + 1\int_{0}^{1} f(x)dx + \int_{1}^{2} f(x)dx + \int_{2}^{3} f(x)dx + \int_{3}^{\infty} f(x)\,dx = 1$$

Since f (x) = 0 from $-\infty < x < 0$ and f (x) = 0 from $3 < x < \infty$.

$$\therefore \int_0^1 kx\,dx + \int_1^2 k dx + \int_2^3 (-kx + 3x)\,dx = 1$$

$$k\left[\int_0^1 k dx + \int_1^2 dx + \int_2^3 (3-x)\,dx\right] = 1$$

$$k\left[\frac{1}{2} + (2-1) - \frac{1}{2}(0-1)\right] = 1 \quad \therefore \quad 2k = 1 \Rightarrow k = \frac{1}{2}$$

Now P (X ≤ 1.5) = P (0 ≤ X ≤ 1) + P (1 ≤ X ≤ 1.5)

$$= \int_0^1 kx\,dx \int_1^{1.5} k\,dx = k\left[\left.\frac{x^2}{2}\right|_0^1 + \left. x \right|_1^{1.5}\right] = \frac{1}{2}\left[\frac{1}{2}+\frac{1}{2}\right] = \frac{1}{2}(1) = \frac{1}{2}$$

5. Mathematical Expectation :

Let X be a discrete random variable taking value x, x = 0,1, 2 .... then the mathematical expectation of X is denoted by E(X) (read as expected value of X) and is defined as $E(X) = \sum_x x\,P(X = x)$

Similarly if k is a positive integer then $E(X^k) = \sum_x x^k P(X = x)$

Similarly if X is contiuous random variable taking value x, $-\infty < x < \infty$ with f (x) as the probability density function then $E(X) = \int_{-\infty}^{\infty} x f(x)\,dx$, $E(X^k) = \int_{-\infty}^{\infty} x^k\, f(x)\,dx$

Notation : E (X) is also called the mean or arithmetic mean of X denoted by μ. $\mu = E(X)$

Variance : If X is a random variable then the variance of X is denoted by V (X) and is defind as $V(X) = E\left[(X - E(X))^2\right]$

This can be simpified as $V(X) = E(X^2)\,[E(X)]^2$

Notation : The variance is denoted by σ^2

$$\therefore \quad \sigma^2 = V(X)$$

Standard Deviation : The positive square root of variance is defined as standard deviation and is denoted by .

$$\therefore \quad \sigma = +\sqrt{V(X)}$$

Properties of Mathematical Expectation and Variance

1. If a is a constant then E (a) = a
2. If a is a constant and X is a random variable then E (aX) = a E (X).
3. If X and Y are two random variables and a and b are constants then E (aX ± bY) = aE(X) ± bE (Y)
4. If a and b are constants and X is a random variable then E (a X + b) = aE(X) +b.
5. If a is a constant then V (a) =0.
6. If a and b are constants and X is a random variable then V (aX ± b) = a^2 V (X).
7. If X and Y are independent random variables then E (XY) = E (X) E (Y).

Ex 1 : Two coins are tossed simultaneously. Let X denote the number of heads. Find E (X) and V (X).

X = x	0	1	2	Total
P (X = x)	$\frac{1}{4}$	$\frac{2}{4}$	$\frac{1}{4}$	1

Mean : $\mu = E(X) = 0.\frac{1}{4} + 1\frac{2}{4} + 2\frac{1}{4} = 1$

Variance : $\sigma^2 = V(X) = E(X^2) - [E(X)]^2$

$$= 0^2.\frac{1}{4} + 1^2.\frac{2}{4} + 2^2\,\frac{1}{4} - (1)^2 \quad = \frac{2}{4} + 1 - 1 = \frac{1}{2}$$

Ex 2 : If it rains,a dealer in rain coats earns Rs.500/- per day and if it is fair, he loses Rs. 50/- per day. If the probability of a rainy day is 0.4, find his average daily income.

Sol : Let X denote his daily income. The distribution of X is

X =x	500	-50	Total
P (X =x)	0.4	0.6	0.1

Average = E (X) = 500 (0.4) + (-50) (0.6)
= 200 - 30 = Rs. 170.

6. Binomial Distribution

This distribution was discovered by james Bernoulli. This is a discrete distribution. It occurs in cases of repeated trials such as students writing an examination, births in a hospital etc. Here all the trials are assumed to be independent and each trial has only two outcomes namely success and failure.

Let an experiment consists of 'n' independent trials. Let it succeed 'x' times. Let 'p' be the probability of success and 'q' be the probability of failure in each trial.

$$\therefore \quad p + q = 1$$

The probability of getting x successes

= p.p...p (x times) = p^x

The probability of getting (n-x) failures

= q.q .....q ((n-x) times) = q^{n-x}

∴ From multiplication theorem, the probability of getting x successes and (n -x) failures is $p^x q^{n-x}$.

This is the probability of getting x successes in one combination. There are such $\binom{n}{x}$ mutually exclusive combinations each with probability $p^x q^{n-x}$.

$\therefore$ From addition theorem the probability of getting x success in $\binom{n}{x}$ $p^x q^{n-x}$.

Notation : b (x; n, p) denotes a binomial distribution with x successes, n trials and with p as the probability of success.

$\therefore$ b (x; n,p) = $\binom{n}{x} p^x q^{n-x}$, x = 0, 1,2,...n.

Parameters Of Binomial Distribution

In b (x; n,p) there are 3 constants namely n,p and q. Since q = 1-p, hence there are only 2 independent constants namely n and p. These are called the parameters of binomial distribution

Note : Since b (x;n,p) is same as the $(x+1)^{th}$ term in the binomial expansion $(q+p)^n$, hence this distribution is called the binomial distribution.

EX 1 : If a coin is tossed 5 times then the number X of heads getting is a random variable following binomial distribution with paramters n = 5, p =1/2

Ex 2 : If X is the number of heads getting when10 coins are tossed then the random variable X follows binomial distribution with paramters n =10,p =1/2

Theorem : If the random variable X follows a binomial distribution with parameters n and p then mean of X is np and the variance is npq where q =1-p.

SOLVED EXAMPLES

Ex 1 : Find the probability of getting 2 heads when 7 coins are tossed simultaneoulsy.

Sol : Let X be the number of heads getting when tossing 7 coins. If a coin is tossed, then probability of getting head is 1/2 and the probability of getting tail is 1/2.

$\therefore$ p = 1/2, q= 1/2

The number of trials n =7

Now X follows binomoial distribution with parameters n = 7, p = 1/2

The probability of getting 2 heads = P (X=2)

$$= {}^7C_2\,(1/2)^2\,(1/2)^{7-2} = 21 \times \frac{1}{2^7} = \frac{21}{128}$$

Ex 2: If X follows a binomial distribution with parameters 12 and 1/3 then find, mean and variance of X.

Sol : Given n = 12, p = 1/3.

$\therefore$ q = 1-p = 2/ 3 Mean = np = 12 (1/3) =4

Variance = npq = 12 (1/3) (2/3) = 8/3

Ex 3: If for a binomial distribution the mean is 20 and the variance is 15, then find p(X>5).

Sol : Mean np = 20, variance npq = 15

$$q = \frac{npq}{np} = \frac{15}{20} = \frac{3}{4} \qquad \therefore p = 1-q = 1-\frac{3}{4} = \frac{1}{4},$$

$$n = \frac{20}{p} = \frac{20}{1/4} = 80$$

$$P(X>5) = \sum_{r=6}^{80} P(X=r) = \sum_{r=6}^{80} {}^{80}C_r\, p^r q^{80-r} = \sum_{r=6}^{80} {}^{80}C_r (1/4)^r (3/4)^{80-r}$$

Ex 4: Five percent of objects prepared by a machine are defective what is the probability that in a sample of 20 objects, 4 will be defective ?

Sol : Let X be the number of defective objects in 20 objects. Probability of an item to be defective is

$$p = \frac{5}{100} = \frac{1}{20}$$

Now X follows binomial distribution with parameters

$$n = 20,\ p = \frac{1}{20}$$

$$\therefore P(X=4) = {}^{20}C_4\, p^4 q^{20-4} = {}^{20}C_4 (1/20)^4 (19/20)^{16}$$

$$= {}^{20}C_4 \frac{19^{16}}{20^{20}}$$

Ex 5 : The probability that a student is not a swimmer is 1/5. Out of five students, find the probability that

I. Four are swimmers and II. Atleast four are swimmers.

Sol : Let X be the number of swimmers, out of the five students.

The probability that a student is not a swimmer is q = 1/5.

The probability that a student is a swimmer is p = 4/5

Now X follows binomial distribution with parameters n = 5, p = 4 / 5

I. The probability that four are swimmers = P (X =4)

$$= {}^5C_4 (4/5)^4 (1/5)^1 = \frac{5 \times 4^4}{5^5} = \frac{4^4}{5^4} = \frac{256}{625}$$

II. The probability that atleast four are swimmers

$$= P(X \ge 4) = P(X=4) + P(X=5) = {}^5C_4 (4/5)^4 (1/5)^1 + {}^5C_5 (4/5)^5 (1/5)^0$$

$$= \frac{5 \times 4^4 + 4^5}{5^5} = \frac{2304}{3125}$$

7. Poisson Distribution

Definition : A random variable X is said to follow Poisson distribution if its probability mass function is given by

$$f(x, \lambda) = \frac{\lambda^x e^{-\lambda}}{x!}, \quad x = 0, 1, 2 \ldots\ldots\ldots$$

$$= 0 \quad \text{otherwise}$$

Note : 1. $e^{\lambda} = \sum_{x=0}^{\infty} \frac{\lambda^x}{x!} = \sum_{x=1}^{\infty} \frac{\lambda^{x-1}}{(x-1)!} = \sum_{x=2}^{\infty} \frac{\lambda^{x-2}}{(x-2)!}$

2. Show that $\sum_{x=0}^{\infty} f(x, \lambda) = 1$

$$\sum_{x=0}^{\infty} f(x, \lambda) = \sum_{x=0}^{\infty} \frac{e^{-\lambda} \lambda^x}{x!} = e^{-\lambda} \sum_{x=0}^{\infty} \frac{\lambda^x}{x!}$$

3. $\lambda > 0$ is called the parameter of the Poisson distribution.

4. $P(X=0) = \frac{e^{-\lambda} \lambda^0}{0!} = e^{-\lambda}$

Applications of Poisson Distribution:

Poisson distribution is applicable when n is very larege and p is very small. Hence some of the applications of Poisson distribution are as follows:

1. Number of faulty blades produced by a reputed firm.
2. Number of deaths from a disease such as heart attack or cancer.
3. Number of telephone calls received at a particular telephone exchange.
4. Number of cars passing a crossing per minute.
5. Number of printing mistakes in a page of a book.

Theorem : If a random variable X follows poisson distribution with parameter λ, then mean of X is λ and variance of X is λ.

SOLVED EXAMPLES

Ex 1 : If a random variable X follows poisson distribution such that P (X = 1) = P (X=2) then find P (X = 0)

Sol : Let λ be the parameter. Then $P(X=r) = \frac{e^{-\lambda}\lambda^r}{r!}$, for r = 0, 1, 2.....

$$P(X=1) = P(X=2) \Rightarrow \frac{e^{-\lambda}\lambda^1}{1!} = \frac{e^{-\lambda}\lambda^2}{2!} \Rightarrow \lambda\frac{\lambda^2}{2} \Rightarrow \lambda = 2$$

$$\therefore P(X=0) = \frac{e^{-2}2^0}{0!} = \frac{1}{e^2}$$

Ex 2 : A book contains 100 misprints distributed randomly trhroughout its 100 pages. Find the probability that a page observed at random contains atleast two misprints.

Sol : Let X represents the number of misprints in a page.

$$\text{Average of misprints in a page} = \frac{100}{100} = 1$$

Assume that X follows Poisson distribution with parameter λ. Now λ = mean = 1

$$\therefore P(X=r) = \frac{e^{-\lambda}\lambda^r}{r!} = \frac{e^{-1}}{r!}$$

Probability that a page contains atleast two misprints is

$$P(X \geq 2) = 1 - [P(X=0) + P(X=1)] = 1 - [\frac{e^{-1}}{0!} + \frac{e^{-1}}{1!}]$$

$$= 1 - 2e^{-1} = 1 - 2/e$$

Ex 3 : Cycle tyres are supplied in lots of 10 and there is a chance of 1 in 500 tyres to be defective. Using Poisson's distribution find the approximate number of lots containing no defective tyres in a consignment of 10000 lots.

Sol : The average of defective tyres in a lot

$$= \frac{1}{500} \times 10 = 0.02 \quad \lambda \Rightarrow \text{mean} = 0.02$$

The probability that no defective tyre in a lot

$$= \frac{e^{-\lambda}\lambda^0}{0!} = e^{-0.02} = 1 - \frac{0.02}{1!} + \frac{(0.02)^2}{2!} \text{........}$$

= 1-0.02 + 0.0002 -···· = 0.9802

The approximate number of lots containing no defective tyre in 10000 lots = 10000 × 0.9802 =9802

8. Normal Distribution

This is a continuous distribution. A random variable X is said to follow normal distribution with mean μ and variance σ^2 if its probability density function is given by

$$f(x) = \frac{1}{\sigma\sqrt{2\pi}} e^{-\frac{(x-\mu)^2}{2\sigma^2}}, \ -\infty < x < \infty, -\infty < \mu < \infty; \sigma > 0$$

$$= 0 \qquad \text{otherwise.}$$

The corresponding distribution function is

$$f(x) = \int_{-\infty}^{x} \frac{e^{-(t-\mu)^2/2\sigma^2}}{\sigma\sqrt{2\pi}} dt \quad \text{Let} \qquad Z = \frac{X-\mu}{\sigma}$$

Then the mean of Z is 0 and the variance is 1.

The correspoding probability density function is

$$\phi(z) = \frac{1}{\sqrt{2\pi}} e^{-z^2/2} \qquad -\infty < Z < \infty$$

Z is called standard normal variate.

Notation :

1. $X \sim N(\mu, \sigma^2)$ denotes that X is a normal variate with mean μ and variance σ^2
2. Z ~ N (0,1).

Graph of a Normal Curve :

The graph of f(x) is a bell shaped curve extending from $-\infty$ to ∞ with its peak at μ.

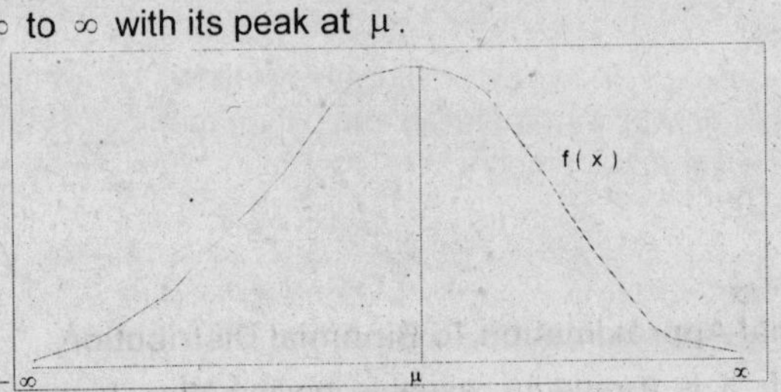

Graph of ϕ (Z):

The graph of ϕ (Z) is also a bell shapped curve extending from $-\infty$ to ∞ with its peak at the origin.

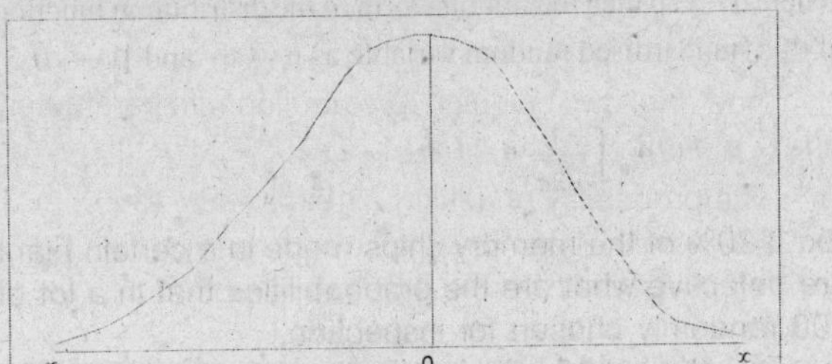

Note : 1. The mode of normal distribution is μ.

2. The median of normal distribution is also μ

Hence for a normal distribution the mean, median and mode coincide.

Some Important Observations

1. $\int_{-\infty}^{\infty} \phi(z)\, dz = 1$ 2. $\int_{-\infty}^{0} \phi(z)\, dz = \int_{0}^{\infty} \phi(z)\, dz = 0.5$

3. $\int_{-a}^{a}\phi(z)\,dz = 2\int_{0}^{a}\phi(z)\,dz$ 4. $\int_{-a}^{0}\phi(z)\,dz = \int_{0}^{a}\phi(z)\,dz$

Ex : X is normally distributed with mean 12 and S.D = 4. Find (a) $P(0 \le X \le 12)$ (b) $P(X \le 20)$

Sol : Given $\mu = 12,\ \sigma = 4$ $z = \dfrac{X-\mu}{\sigma}$

When X = 12, $z = \dfrac{12-12}{4} = 0$

When X = 0, $z = \dfrac{0-12}{4} = -3$

When X = 20, $z = \dfrac{20-12}{4} = 2$

(a) $P(0 < X < 12) = P(-3 < z < 0)$

$$= \int_{-3}^{0}\phi(z)\,dz = \int_{0}^{3}\phi(z)\,dz = 0.4986$$

(b) $P(X \le 20) = P(Z \le 2)$

$$= \int_{-\infty}^{2}\phi(z)\,dz = \int_{-\infty}^{0}\phi(z)\,dz + \int_{0}^{2}\phi(z)\,dz$$

$$= 0.5 + 0.4772 = 0.9772$$

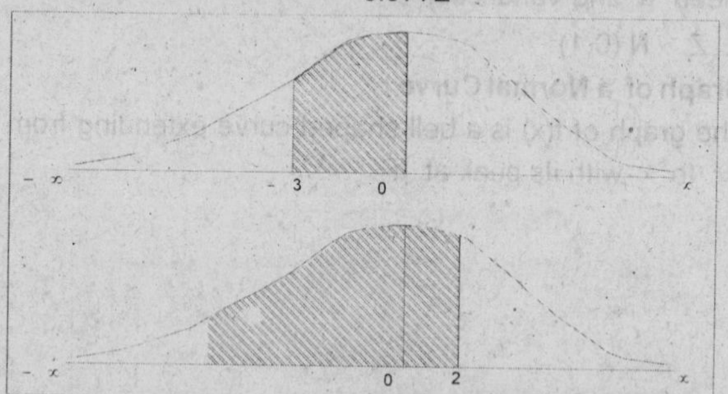

Normal Approximation To Binomial Distribution

If X is a random variable having the binomial distribution with parameters n and p and if $Z = \dfrac{X-np}{\sqrt{npq}}$

Where q = 1 - p then the limiting form of the distribution function of this standardised random variable as $n \to \infty$ and $p \to 0.5$

$$F(Z) = \int_{-\infty}^{z}\frac{1}{\sqrt{2\pi}}e^{-t^2/2}\,dt,\ -\infty < z < \infty$$

Ex: If 20% of the memory chips made in a certain plant are defective what are the praobabilities that in a lot of 100 randomly chosen for inspection,

(a) at most 15 will be defective

(b) exactly 15 will be defective.

Sol : Here n = 100, p = 0.20, q = 0.80

$\mu = np = 100\,(0.20) = 20$ $\sigma = \sqrt{npq} = \sqrt{100(0.20)\,(0.80)} = 4$

Treating the data as continuous, 15 can be taken as 15.5

$$\therefore\quad Z = \frac{155-20}{4} = -1.13$$

$$\therefore\quad P(X \le 15) = \int_{-\infty}^{-1.13}\phi(z)\,dz = \int_{-\infty}^{0}\phi(z)\,dz - \int_{-1.13}^{0}\phi(z)\,dz$$

$$= \int^{0}\phi(z)\,dz - \int_{0}^{1.13}\phi(z)\,dz = 0.5 - 0.3708 = 0.1292\ P(X = 15)$$

$$= \int^{15.5} f(x)\,dx - \int^{14.5}\phi(x)\,dx\ \int_{0}^{-1.135}\phi(x)\,dz - \int_{0}^{-1.38}\phi(z)\,dz$$

$$= \int_{-1.38}^{1-1.13}\phi(z)\,dz - \int_{-1.18}^{0}\phi(z)\,dz - \int_{-1.13}^{0}\phi(z)\,dz$$

$$= \int_{0}^{-1.18}\phi(z)\,dz - \int_{0}^{1.13}\phi(z)\,dz = 0.4162 - 0.3708 = 0.0454.$$

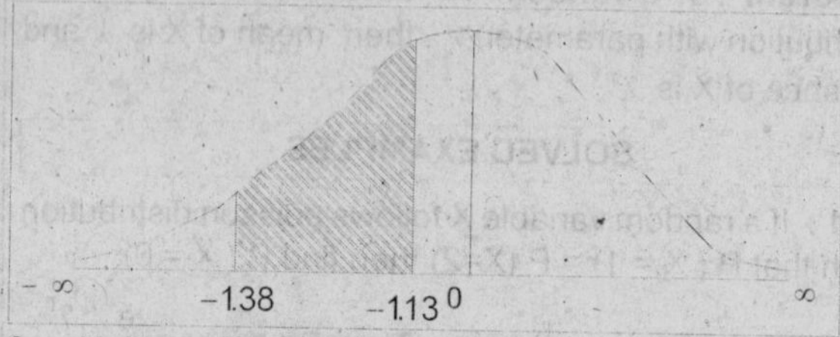

9. Uniform Distribution:

A random variable X is said to have a continuous uniform distribution over an interval (a,b) if its probability density function is constant = k (say), over the entire range of X, i.e.,

$$f(x) = \begin{cases} k, & a < x < b \\ 0, & \text{otherwise} \end{cases}$$

Since total probability is always unity, we have

$\int_a^b f(x)dx = 1 \Rightarrow k\int_a^b dx = 1$ i. e., $k = \dfrac{1}{b-a}$ $\therefore$ $f(x) = \begin{cases} \dfrac{1}{b-a}, & a < x < b \\ 0, & \text{otherwise} \end{cases}$

Remarks. 1. a and b, (a < b) are the two parameteres of the uniform distribution on (a,b).

2. The distribution is also known as rectangular distribution, since the curve y = f (x) describes a rectangle over the x-aixs and between the ordinates at x = a and x = b.

3. The distribution function F (x) is given by

$$F(x) = \begin{cases} 0, & \text{if } -\infty < x < a \\ \dfrac{x-a}{b-a}, & a \le x \le b \\ 1, & b < x < \infty \end{cases}$$

Since F (x) is not continuous at x = a and x = b, it is not differentiable at these points. Thus $\dfrac{d}{dx}F(x) = f(x) = \dfrac{1}{b-a} \ne 0$,exists every where except at the points x = a and x = b and consequently p.d.f. f(x) is given by (3.1).

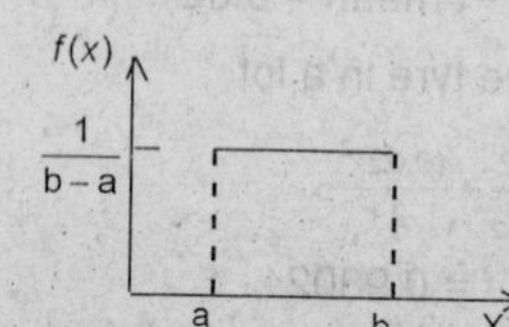

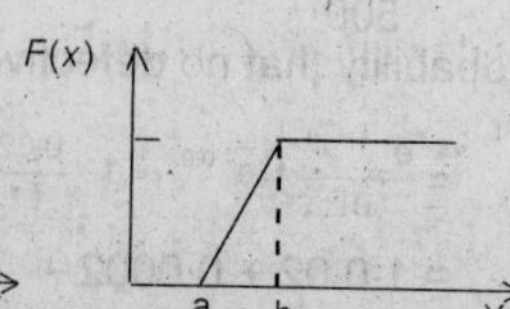

4. The graphs of uniform p.d.f. f(x) and the corresponding distribution function F (x) are given above.

5. For a rectangular or uniform variate X in (-a, a), p.d.f. is given by

$$f(x) = \begin{cases} \frac{1}{2a}, & -a < x < a \\ 0, & \text{otherwise} \end{cases}$$

Moments of Uniform Distribution.

$$\mu_r' = \int_a^b x^r f(x)\, dx = \frac{1}{(b-a)} \int_a^b x^r\, dx = \frac{1}{(b-a)}\left[\frac{b^{r+1}-a^{r+1}}{r+1}\right]$$

In particular Mean = $\mu_1' = \frac{1}{(b-a)}\left[\frac{b^2-a^2}{2}\right] = \frac{b+a}{2}$

and $\mu_2' = \frac{1}{(b-a)}\left[\frac{b^3-a^3}{3}\right] = \frac{1}{3}(b^2+ab+a^2)$

$$\therefore \mu_2 = \mu_2' - \mu_1'^2 = \frac{1}{3}(b^2+ab+a^2) - \left(\frac{b+a}{2}\right)^2 = \frac{1}{12}(b-a)^2$$

Ex 1: If X is uniformly distributed with mean 1 and variance 4/3. find P (X<0).

Sol : Let $X \sim U[a, b]$. so that $P(x) = \frac{1}{b-a}$: a<x<b. We are

given Mean $= \frac{b+a}{2} = 1 \quad \Rightarrow \quad b+a = 2$

Var (X) = $\frac{1}{12}(b-a)^2 = \frac{4}{3} \Rightarrow (b-a)^2 = 16 \Rightarrow b-a = \pm 4$

Solving, we get : a = -1 and b = 3; (a < b)

$$\therefore \quad p(x) = \frac{1}{4};\ -1 < x < 3 \quad P(X<0) = \int_{-1}^{0} p(x)dx = \frac{1}{4}|x|_{-1}^{0} = \frac{1}{4}$$

Ex 2 : Subway trains on a certain line run every half hour between mid-night and six in the morning. What is the probability that a man entering the station at a random time during this period will have to wait at least twenty minutes ?

Sol : Let the r.v. X denote the waiting time (in minutes) for the next train. Under the assumption that a man arrives at the station at random, X is distributed uniformly on

(0,30), with p.d.f., $f(x) = \begin{cases} \frac{1}{30}, & 0 < x < 30 \\ 0, & \text{otherwise} \end{cases}$

The probability that he has to wait at least 20 minutes is

$$P(X \ge 20) = \int_{20}^{30} f(x)\,dx = \frac{1}{30}\int_{20}^{30} 1.dx = \frac{1}{30}(30-20) = \frac{1}{3}$$

10. The Exponential Distribution:

A continuous random variable X assuming non-negative values is said to have an exponential distribution with parameter θ > 0, if its p.d.f. is given by

$$f(x) = \begin{cases} \theta e^{-\theta x}, & x \ge 0 \\ 0, & \text{otherwise} \end{cases}$$

The cumulative distribution function F (x) is given by

$$F(x) = \int_0^x f(u)\, du = \theta \int_0^x \exp(-\theta u)\, du$$

$$F(x) = \begin{cases} 1-\exp(-\theta x), & x \ge 0 \\ 0, & \text{otherwise} \end{cases}$$

Moment Generating Function of Exponential Distribution

$$M_x(t) = E(e^{tx}) = \theta \int_0^\infty e^{tx}\, e^{-\theta x}\, dx$$

$$= \theta \int_0^\infty \exp\{-(\theta - t)x\}\, dx = \frac{\theta}{(\theta - t)}, \theta > t$$

$$= \left(1 - \frac{t}{\theta}\right)^{-1} = \sum_{r=0}^{\infty}\left(\frac{t}{\theta}\right)^r$$

$$\therefore \quad \mu_r' = E(X^r) = \text{Coefficient of } \frac{t^r}{r!} \text{ in } M_x(t)$$

$$= \frac{r!}{\theta^r};\ r = 1, 2 \ldots \quad \therefore \quad \text{Mean} = \mu_1' = \frac{1}{\theta}$$

$$\text{Variance} = \mu_2 = \mu_2' - \mu_1'^2 = \frac{2}{\theta^2} - \frac{1}{\theta^2} = \frac{1}{\theta^2}$$

Ex : If X has exponential distribution with mean 2, find P(x < 1) | P (X < 2)

Sol: $P(x<1)\,|\,P(X<2) = \frac{(1-e^{-\theta})}{(1-e^{-2\theta})}$ where θ = 1/2

OBJECTIVE QUESTIONS - I

1. A random variable X has the following distribution

x	1	2	3	4
P (X=x)	c	2c	3c	4c

The value of c is **[JNTU- 1999]**

a) 0.1 b) 0.2 c) 10 d) 20

2. The probabillity distribution of random variable X is given below. Then k= **[JNTU- 2000]**

$X=x_i$	1	2	3	4
$P(X=x_i)$	2k	4k	3k	k

a) 0.1 b) 0.2 c) 0.3 c) 0.4

3. A random variable X has the follwoing distribution

$X=x_i$	1	2	3	4
$P(X=x_i)$	k	2k	3k	4k

The value of k and P(X<3) are equal to **[JNTU- 1998]**

a) k=1/10, P (X < 3) = 3/5 b) k= 1/10, P (X< 3) = 3/10
c) k=3/10, P (X< 3) =1/10 d) k= 1/24, P (X < 3) = 5/12

4. X is a random variable with distribution given below.

x :	0	1	2	3
P (X=x)	k	3k	3k	k

The value of k and its variance are **[JNTU- 1995]**

a) 1/8, 22/27 b) 1/8, 23/27 c)1/8, 24/27 d) 1/8, 27/36

5. In a binomial distribution n = 20, q =0.75. then mean = **[JNTU- 1983]**

a) 5 b) 15 c) 10 d) 7.5

6. For a binomial distribution n =10, q = 0.4. Then mean is **[JNTU- 2000]**

a) 1 b) 4 c) 6 d) 10

7. If the mean and variance of a binomial distribution are 15/4 and 15/16 then the number of trials is **[JNTU- 1991]**

a) 5 b) 2 c) 4 d) 6

8. In a binomial distribution n = 400, p = 1/5. It's standard deviation is **[JNTU- 1994]**

a) $10\sqrt{2}$ b) 1/800 c) 4 d) 8

9. If for a binomial distribution the mean is 6 and the standard deviation is $\sqrt{2}$, then P (X=r) =**[JNTU- 1984]**

a) ${}^9C_r(2/3)^r(1/3)^{9-r}$ b) ${}^9C_r(1/3)^r(2/3)^{9-r}$

c) ${}^{12}C_r(2/3)^r(1/3)^{12-r}$ d) ${}^{12}C_r(1/3)^r\ (2/3)^{9-r}$

10. For a binomial distribution $\bar{x} = 4, \sigma = \sqrt{3}$ Then P (X=r) = **[JNTU- 1990]**

a) ${}^{16}C_r(1/4)^r\ (3/4)^{16-r}$ b) ${}^{12}C_r(1/4)^r\ (3/4)^{12-r}$

c) ${}^{16}C_r(3/4)^r\ (1/4)^{16-r}$ d) ${}^{12}C_r(3/4)^r\ (1/4)^{12-r}$

11. Let X be a binomially distributed variable with mean 10 and variance 5. Then P (X>10)= **[JNTU- 1994]**

a) $\frac{1}{2^{20}}\sum_{11}^{20}{}^{20}C_k$ b) $\frac{1}{2^{20}}\sum_{1}^{11}{}^{20}C_k$

c) $\frac{1}{2^{20}}\sum_{1}^{20}{}^{10}C_k$ d) $\sum_{11}^{20}{}^{20}C_k\,\frac{1}{2^k}\left(\frac{2}{3}\right)30-k$

12. If the mean of a binomial distribution with 9 trials is 6, then its variance is **[JNTU- 1985]**

a) 2 b) 3 c) 4 d) $\sqrt{2}$

13. If the mean of the binomial distribution $(q+p)^{16}$ is 2, then mean is **[JNTU- 1986]**

a) 16 b) 8 c) 4 d) 6

14. If the mean of the binomial distribution is 25. Then the standard deviation lies in the interval**[JNTU- 1992]**

a) [0,5) b) (0,5] c) [0,25) d) (0,25]

15. X follows a binomial distribution with parameters n = 6 and p. If 4 P (X=4) = P (X = 2), then p=

a) 1/2 b) 1/4 c) 1/6 d) 1/3

16. The probability of a man hitting a target is 1/4. If he fires 7 times, the probability of hitting the target at least twice is **[JNTU- 1998]**

a) $1-\frac{5}{2}\left(\frac{3}{4}\right)^6$ b) $1-\frac{15}{2}\left(\frac{3}{4}\right)^6$ c) $1-\frac{5\times3^5}{6}$ d) $1-\left(\frac{3}{9}\right)^6$

17. In a Poisson distribution if P (X=0) = P (X=1), = k, the value of k is **[JNTU- 1999]**

a) 1 b) 1/e c) e d) $\sqrt{e}$

18. If X is a Poisson variate with P (X=0) = P (X=1), then P (X=2)= **[JNTU- 2000]**

a) e / 2 b) e / 6 c) 1/ (6e) d) 1/(2e)

19. If a random variable X has a Poisson distribution such that P (X=1) = P (X=2) then its mean and variance are **[JNTU- 2000]**

a) 1,1 b) 2,2 c) 2, $\sqrt{3}$ d) 2,4

20. X is a Poisson variate and P (X=1) = P (X=2). Then P (X=0) = **[JNTU- 1993]**

a) 1 b) 2 c) 1/e d) $1/e^2$

21. If X is a random Poisson variate such that α = P (X=1) = P (X=2) then P (X=4)= **[JNTU- 1994]**

a) 2α b) $\alpha/3$ c) $\alpha\ e^{-2}$ d) $\alpha\ e^2$

22. In a Poisson distribution P (X=0) is twice than P (X=1). The mean is **[JNTU- 1987]**

a) 1 b) 2 c) 1/2 d) 3/2

23. In a Poisson variate X, if P (X=0) = 0.2 then the variance of the distribution is **[JNTU- 1988]**

a) 2 b) 1 c) e d) log 5

24. Given a normal distribution with $\mu = \mathbf{200}$ and $\sigma = \mathbf{25}$, find the probability that X assumes a value greater than 240.

a) 0.0548 b) 0.0546 c) 0.0544 d) 0.0537

ANSWER KEY - I

1. a	**2.** a	**3.** b	**4.** d	**5.** a
6. c	**7.** a	**8.** d	**9.** a	**10.** a
11. a	**12.**a	**13.** b	**14.** a	**15.** d
16. a	**17.** b	**18.** d	**19.** b	**20.** d
21. b	**22.** c	**23.** d	**24.** a	

EXPLANATORY NOTES - I

1. (a) Explanation : $C+2C+3C+4C=1 \Rightarrow C=1/10$

3. (b) Explanation : $K+2K+3K+4K=1 \Rightarrow K=1/10$

$P(X<3)=P(X=1)+P(X=2)=3K=3/10$

4. (d) Explanation :

$K+3K+3K+K=1 \Rightarrow 8K=1 \Rightarrow K=1/8$

Mean : $\mu = 0\times K+1\times 3K+2\times 3K+3\times K = 12\times 1/8 = 3/2$

Variance $=\left(\sigma^2.K+1^2.3K+2^2.3K+3^2.K\right)-\mu^2$

$3K+12K+9K-9/4 = 24\times 1/8-9/4 = 3/4$

7. (a) Explanation : Mean = np = 15/4

Variance = npq = 15/16, p+q = 1

$\Rightarrow npq/np = 15/16\times 4/15 \Rightarrow q = 1/4$

$\therefore p=3/4 \quad \therefore n.3/4=15/4 \Rightarrow n=5$

8. (d) Explanation : n = 400, P = 1/5 $\therefore q=1-1/5=4/5$

$S.D.=\sqrt{npq}=\sqrt{400\times 1/5\times 4/5}=8$

10. (a) Explanation : $x=np=4, \sigma=\sqrt{npq}=\sqrt{3}$

$\Rightarrow npq=3, p+q=1, \therefore 4q=3 \Rightarrow 3/4, p=1/4, n=16$

$p(r)={}^{16}c_r.q^{16-r}.P^r={}^{16}c_r.(3/4)^{16-r}.(1/4)^r$

11. (a) Explanation :

$np=10, npq=5 \Rightarrow q=p=1/2 \ \therefore n=20$

$\therefore P(X>10)=\sum_{11}^{20}{}^{20}c_k.\frac{1}{2^{20}}=\frac{1}{2^{20}}\sum_{11}^{20}{}^{20}C_k$

13. (b) Explanation : $S.D=\sqrt{npq}=2 \Rightarrow npq=4.$

Distribution is $(q+P)^{16} \Rightarrow n=16, P+q=1.$

$\therefore 16P(1-P)=4 \Rightarrow 4P-4P^2=1 \Rightarrow 4P^2-4P+1=0$

$\Rightarrow (2P-1)^2=0 \quad \Rightarrow P=1/2 \ \therefore \text{Mean}=np=16\times 1/2=8$

14. (a) **Explanation :**

Mean = np = 25, Variance = $\sigma^2 = npq, P + q = 1$

$\sigma^2 = 25q$ and $0 \le 25q < 25$ ($\because P \ne 0$)

$0 \le \sigma^2 < 25 \Rightarrow 0 \le \sigma < 5 \Rightarrow \sigma \in [0.5]$

15. (d) **Explanation :** n = 6 and 4. p (x=4) = p (x=2)

$\Rightarrow 4.6c_4 q^2 P^4 = 6c_2 q^4 P^2 \Rightarrow 4P^2 = q^2 \Rightarrow 2P = q$

Now, $p+q=1 \Rightarrow 3p=1 \Rightarrow p=1/3$.

16. (a) **Explanation :** $p=1/4, q=3/4, n=7$

Required Probability $=1-[P(X=0)+P(X=1)]$

$=1-[7c_0 q^7 p^0 + 7c_1 q^6 P^1] = 1-\left[\left(\frac{3}{4}\right)^7 + 7\left(\frac{3}{4}\right)^6 \cdot \frac{1}{4}\right]$

$=1-\left\{\frac{3}{7}+\frac{7}{4}\right\}\left(\frac{3}{4}\right)^6 = 1-\frac{5}{2}\left(\frac{3}{4}\right)^6$

17. (b) **Explanation :** $P(X=0)=P(X=1)$

$\Rightarrow \frac{\lambda^0 . e^{-\lambda}}{0!} = \frac{\lambda^0 . e^{-\lambda}}{1!} \Rightarrow \lambda = 1 \therefore K = P(X=0) = \frac{1^0 e^{-1}}{0!} = 1/e$

19. (b) **Explanation :** For a poisson distribution,

$P(X=1)=P(X=2) \Rightarrow \frac{e^{-\lambda}.\lambda^1}{1!} = \frac{e^{-\lambda}.\lambda^2}{2!} \Rightarrow \lambda = 2$.

$\therefore$ Mean = 2, Variance = 2.

20.(d) **Explanation :** $P(X=1)=P(X=2) \Rightarrow \frac{e^{-\lambda}.\lambda^0}{1!} = e^{-2}$

21. (b) **Explanation :** $\alpha = P(X=1) = P(X=2)$

$\Rightarrow \frac{e^{-\lambda}.\lambda}{1!} = \frac{e^{-\lambda}.\lambda^2}{2!} \Rightarrow \lambda = 2 \therefore \alpha = P(X=1) = \frac{e^{-2}.2}{1!} = 2e^{-2}$

$\therefore P(X=4) = \frac{e^{-2}.2^4}{4!} = \frac{\alpha}{2}.\frac{16}{24} = \frac{\alpha}{3}$

22. (c) **Explanation :** $P(X=0)=2P(X=1)$

$\Rightarrow \frac{e^{-\lambda}.\lambda^0}{0!} = 2.\frac{e^{-\lambda}.\lambda^1}{0!} \Rightarrow \lambda = \frac{1}{2}$

23. (d) **Explanation :** $P(X=0) = 0.2 \Rightarrow \frac{e^{-\lambda}.\lambda^0}{0!} = 0.2 = 1/5$

$\Rightarrow e^{\lambda} = 5 \Rightarrow \lambda = \log_e^5 \therefore$ Variance $= \log_e^5$

OBJECTIVE QUESTIONS - II

1. A random variable X follows the following distribution

$X = x_i$:	0	1	2	3
$P(X=x_i)$:	2/6	3/6	0/6	1/6

The mean and the variance are **[JNTU- 1998]**

a) 1,1 b) 1,2 c) 2,1 d) 2,2

2. A coin is tossed 3 times. The probability of getting head once and tail two times is **[JNTU- 1998]**

a) 1/3 b) 1/4 c) 3/8 d) 1/2

3. In a binomial distribution, mean is 5 and the variance is 4. Then the number of trials is **[JNTU- 1988]**

a) 9 b) 20 c) 25 d) 125

4. If the difference between the mean and variance of a binomial distribution for 5 trials is 5/9 then the distribution is **[JNTU- 1985]**

a) $(2/5+3/5)^5$ b) $(2/3+1/3)^5$

c) $(1/3+2/3)^5$ d) $(3/4+1/4)^5$

5. Twenty identical coins each with probability P of showing heads are tossed. The probability of heads showing on 10 coins is same as that of heads showing on 11 coins. Then p = **[IIT- 1992]**

a) 10/11 b) 10/21 c) 11/21 d) 11/20

6. One hundred identical coins each with probability p showing up heads are tossed once. If 0 < p < 1 and the probability of heads showing on 50 coins is equal to that of heads showing on 51 coins, then the value of p is **[JNTU- 1988]**

a) 1/2 b) 49/101 c) 50/101 d) 51/101

7. The probability that a candidate secure a seat in Engineering through GATE is 1/10. Seven candidates are selected at random from a centre. The probability that exactly two will get seats is **[JNTU- 1995]**

a) $15(.1)^2(.9)^5$ b) $20(.1)^2(.9)^5$

c) $21(.1)^2(.9)^5$ d) $23(.1)^2(.9)^2$

8. In an experiment the success is twice that of failure. If the experiment is repeated 6 times, the probability that atleast 4 times favourable is **[JNTU- 1995]**

a) 64/729 b) 192/729 c) 240/729 d) 496/729

9. Out of 10,000 families with 4 childern each, the probability number of families all of whose childern are daughters is **[JNTU- 1989]**

a) 625 b) 1250 c) 2500 d) 9375

10. The probability that a bomb dropped from a plane strikes the target is 1/5. The probability that out of six bombs dropped at least 2 bombs strike the target is . **[JNTU- 1989]**

a) 0.345 b) 0.246 c) 0.543 d) 0.426

11. If the mean and the variance of a binomial variate X are 2 and 1 respectively, then the probability that X takes a value greater than one is equal to **[JNTU- 1991]**

a) 1/16 b) 5/16 c) 11/16 d) 15/16

12. If 3% of electric bulbs manufactured by a company are defective, then the probability that a sample of 100 bulbs has no defective is **[JNTU- 1990]**

a) 0 b) e^{-3} c) $1-e^{-3}$ d) $3e^{-3}$

13. Cycle tyres are supplied in lots of 10 and there is a chance of 1 in 500 tyres to be defective. Using Poisson's distribution the approximate number of lots containing no defective tyres in a consignment of 10,000 lots is **[JNTU- 1989]**

a) 9980 b) 9998 c) 9802 d) 9982

14. The mean weight of 500 boys at a certain college is 75 kg and the standard deviation is 7 kg. Assuming that the weights are normally distributed, find how many students weigh between 60 and 78 kg

a) 339 b) 400 c) 449 d) None

15. In a test on electric bulbs it was found that the life in hours of a particular make was normally distributed with an average ife of 2000 hours and a standard deviation of 60 hours. What of proportion of the bulbds can be expected to burn for more than 2100 hours.

a) 0.0334 b) 0.0485 c) 0.0045 d) None

16. In a sample of 120 workers in a factroy the mean and standard deviation of wages were Rs. 11.35 and Rs. 3.03 respectively. Find the percentage of workers getting wages between Rs. 9 and Rs. 17 in the whole factory assuming that the wages are normally distributed.

a) 75.12 % b) 75. 09 % c) 75.22 % d) 76.20 %

17. In a particular examination, an examinee can get marks ranging from 0 to 100. Last year, 1,00,000 students took this examination. The marks obtained by them had a normal distribution. What is the probability that the marks obtained by a student chosen at random would be exactly 63 ?

a) one b) two c) Zero d) three

18. Mean annual sales of a firm is worth Rs. 150 with a standard deviation of Rs. 20. For how many days in a year of 365 days his sales are expected to be worth lessthan Rs. 100 ?

a) 2.25 days b) 2.24 days c) 2.27 days d) 2.26 days

19. In a sample of 1,000 items, the mean weight is 45 kg. With a standard deviation of 15 kg. Assuming the normality of the distribution, find the number of items weighing between 40 kgs and 60 kgs.

a) 470 b) 471 c) 472 d) None

20. The marks obtained by the students in an examination are known to be normally distributed. If 10% of the students got less than 40 marks while 15% got over 80, what are the mean and standard deviation of marks ?

a) Mean = 62.00 SD = 17.20

b) Mean = 62 . 50 SD = 17.25

c) Mean = 62.07 SD = 17.24 d) None

ANSWER KEY - II

1. a	2. c	3. c	4. b	5. c
6. d	7. c	8. d	9. a	10. a
11. c	12. b	13. c	14. a	15. b
16. b	17. c	18. d	19. b	20. c

EXPLANATORY NOTES - II

1. (a) Explanation :

Mean of $X=0.\frac{2}{6}+1.\frac{3}{6}+2.\frac{0}{6}+3.\frac{1}{6}=1,$

Variance of $X=0^2.\frac{2}{6}+1^2\frac{3}{6}+2^2\frac{0}{6}+3^2\frac{1}{6}-1^2=1$

2. (c) Explanation : Total number of ways of getting heads or tails on 3 coins $=2\times2\times2$

One head and 2 tails on them can occur in $\frac{3C_1\times 2C_2}{2\times2\times2}=\frac{3}{8}$

3. (c) Explanation : $np=5, npq=4 \Rightarrow q=\frac{4}{5}, p=\frac{1}{5}$

$$n\left(\frac{1}{5}\right)=5\Rightarrow n=25.$$

4. (b) Explanation : $np-npq=\frac{5}{9},\ n=5,\ p+q=1$

$\Rightarrow p(1-q)=\frac{1}{9} \Rightarrow p^2=\frac{1}{9}\Rightarrow p=\frac{1}{3}, q=\frac{2}{3}$

$\therefore$ Distribution is $\left(\frac{2}{3}+\frac{1}{3}\right)^5$

7. (c) Explanation : $p=\frac{1}{10}=0.1, q=1-p, n=7$

prob. that exactly 2 will get seats

$$=7C_2.q^{7-2}.p^2=21\times(0.9)^5\times(0.1)^2$$

8. (d) Explanation :

p = prob. of successes, q = prob. of failures.

$\therefore p=2q$. But $p+q=1\Rightarrow 3q=1\Rightarrow q=\frac{1}{3}$

$\therefore$ prob. that atleast 4 times favourable

$=6C_4.q^2.p^4+6c_5qp^5+6c_6p^6$ since n = 6

$$=15.\frac{1}{9}.\frac{16}{81}+6.\frac{1}{3}.\frac{32}{243}+\frac{64}{729}=\frac{496}{729}$$

9. (a) Explanation : prob. for a family with 4 children to have all daughters $=\frac{1}{2}.\frac{1}{2}.\frac{1}{2}.\frac{1}{2}=\frac{1}{16}$

Out of 10, 000 families, number of families having all 4 as daughters $=10,000\times\frac{1}{16}=625.$

10. (a) Explanation :

prob. for a bomb to hit the target $=p=\frac{1}{5}$

prob. for a bomb not hitting the target $=q=1-p=\frac{4}{5}$

Here n = 6. If X is the number of bombs that hit the target, then $P(X\ge 2)=1-P(X=0)-P(X=1)$

$$=1-6C_0.q^6-6C_1q^5p^1=1-\left(\frac{4}{5}\right)^6-6.\left(\frac{4}{5}\right)^5.\left(\frac{1}{5}\right)$$

$$=1-\left(\frac{4}{5}\right)^5.\left(\frac{4+6}{5}\right)=0.345$$

13. (c) Explanation : $\lambda=np=10\times\frac{1}{500}=0.02,$

Approximate number of lots containing

no. of defective tyres $=10,000\times\frac{e^{-\lambda}.\lambda^0}{0!}=10,000\times e^{-0.02}$

$$=10,000\times 0.9802=9802.$$

14. (a) Explanation :

Mean$(\mu)=75$ Standard Deviation $\sigma=7$ $p(60<x<78)$

Let x = 60 $z=\frac{x-u}{\sigma}=\frac{60-75}{7}=-2.1$

Let x = 78 $z=\frac{x-\mu}{\sigma}=\frac{78-75}{7}=0.4$

$p(60<x<78)=p(-2.1<z<0.4)$

$=p(-2.1<z<0)+P(0<z<0.4)$

$=P(0<Z<2.1)+0.1554$ ($\therefore$ from table)

$=0.4821+0.1554=0.6375$

Therefore out of 500 boys the no. of student between weight 600 and 78 kg $0.6375 \times 500 = 338.75 = 339$

15. (b) **Explanation :** Mean$(\mu) = 2000$ S.D$(\sigma) = 60$

$P(x > 2100)$, Let $x = 2100$; $z = \frac{2100-2000}{60} = \frac{100}{60} = 1.6$

$P(z > 1.6) = 0.5 - P(0 \le z \le 1.6) = 0.5 - 0.4452 = 0.0548$

16. (d) **Explanation :** mean$(\mu) = 11.35$ S.D.$(\sigma) = 3.03$

$P(9 < x < 17)$, Let $x = 9$; $Z = \frac{x-\mu}{\sigma} = \frac{9-11.35}{3.03} \cdot = \frac{-2.35}{3.03}$

Let $x = 17$; $Z = \frac{x-\mu}{\sigma} = \frac{1.7-11.35}{3.03} = \frac{5.65}{3.03} = 1.8$

$P(-0.7 < Z < 1.8) = P(-0.7 < Z < 0) + P(0 < Z < 1.8)$

$= P(0 < Z < 0.7) + 0.4641 = 0.2580 + 0.4641 = 0.7221$

Out of 120 workers $= 0.7221 \times 120 = 8665.20$

The percentage this value = 86.65

18. (d) **Explanation :** mean $\mu = 150$, S.D.$(\sigma) = 20$

$P(x < 100)$, $Z = \frac{x-\mu}{\sigma} = \frac{100-150}{20} = -2.5$

$P(Z < -2.5) = P(Z > 2.5) = 0.5 - P(0 \le Z \le 2.5)$

$= 0.5 - 0.4938$ ($\because$ from tables) $= 0.0062$

out of 365 days his sales less than Rs. 100

$= 0.0062 \times 365 = 2.263$

20. (d) **Explanation :** $P(x < 40) = 0.1$, $P(x > 80) = 0.15$

The points x = 40 and x = 80 are located as shown in figure.

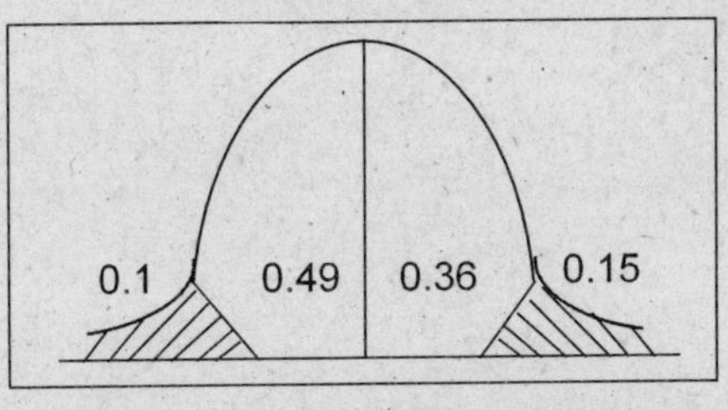

since the value x = 35 is located to the left of the ordinate at $x = \mu$. When x = 40 ; $Z = \frac{40-\mu}{\sigma} = -Z_1$

When x = 63 ; $Z = \frac{63-\mu}{\sigma} = Z_2$

$P(0 < Z < Z_2) = 0.35$; $P(0 < Z < Z_1) = 0.49$

Hence normal table we have $Z_2 = 1$ $Z_1 = 2.3$

$\frac{40-\mu}{\sigma} = -2.3$ ...... (1), $\frac{63-\mu}{\sigma} = 1$ ..... (2)

from (1) - (2) $\frac{13}{\sigma} = 3.3$, $\sigma = \frac{130}{3.3} = 3.9$

σ will be substituted from (2)

$63 - \mu = 3.9 \Rightarrow \mu = 63 - 3.9 = 59.1$

OBJECTIVE QUESTIONS - III

1. In a cetain test there are n questions. In this test 2^{n-k} students gave wrong answers to at least questions, where k = 1, 2, 3, ... n. If the total number of wrong answers given is 2047, then it is equal to

a) 10 b) 11 c) 12 d) 13

2. The number of ways in which 10 candidates $A_1, A_2, ... A_{10}$ can be ranked so that A_1 is always above A_2 is

a) 10 ! b) $\frac{10!}{2}$ c) 9 ! d) none of these

3. The number of integral solutions of x+y+z=0 with $x \ge -5, y \ge -5, Z \ge -5$ is

a) 135 b) 136 c) 455 d) 105

4. A man invites a party to $(m+n)$ friends to dinner and places m at one round table and n at another. The number of ways of arranging the guests is

a) $\frac{(m+n)!}{m!n!}$ b) $\frac{(m+n)!}{(m-1)!(n-1)!}$

c) $(m-1)!(n-1)!$ d) none of these

5. There are m copies of each of n different books in a university library. The number of ways in which one or more than one book can be selected is

a) $m^n + 1$ b) $(m+1)^n - 1$ c) $(n+1)^n - m$ d) m

6. Between two junction stations A and B there are 12 intermediate stations. The number of ways in which a train can be made to stop at 4 of these stations so that no two of these halting stations are consective is

a) ${}^8C_4$ b) ${}^9C_4$ c) ${}^{12}C_4$ d) none of these

7. In a football championship, 153 matches were played, Every team played one match with each other. The number of teams participating in the championship is

a) 17 b) 18 c) 9 d) none of these

8. There are 10 lamps in a hall. Each one of them can be switched on independently. The number of ways in which the hall can be illuminated is

a) 10^2 b) 1023 c) 2^{10} d) 10 !

9. The number of parallelograms that can be formed from a set of four parallel lines intersecting another set of three parallel lines is

a) 6 b) 18 c) 12 d) 9

10. There are 10 points in a plane, out of these 6 are collinear. The number of triangles formed by joining these points is

a) 100 b) 120 c) 150 d) none of these

11. In a college examinations, a candidate is required to answer 6 out of 10 questions which are divided into two sections each containing 5 questions. Further the candidate is not permited to attempt more than 4 questions from either of the section. The number of ways in which he can make up a choice of 6 questions, is

a) 200 b) 150 c) 100 d) 50

EXPLANATORY NOTES - III

1. (b) **Explanation :** The number of students answering exactly $k(1 \le k \le n-1)$ questions wrongly is $2^{n-k} - 2^{n-k-1}$. The number of students answering all questions wrongly is 2^0. Thus, the total number of wrong answers $= 1.(2^{n-1} - 2^{n-2}) + 2.(2^{n-2} - 2^{n-3})$

$$+ .... + (n-1)(2^1 - 2^0) + n.2^0$$

$$= 2^{n-1} + 2^{n-2} + 2^{n-3} + ... + 2^0 = 2^n - 1$$

Thus $2^n - 1 = 2047 \Rightarrow 2^n = 2048 \Rightarrow 2^n = 2^{11} \Rightarrow n = 11$

2. (b) **Explanation :** Ten candidates can be ranked in 10! ways. In half of these ways A_1 is above A_2 and in another half A_2 is above A_1. So, required number of ways is $\frac{10!}{2}$

3. (b) **Explanation :** Let $u = x+5$, $v = y+5$ and $w = x+5$. Then $u, v, w \ge 0$ and $x + y + z = 0 \Rightarrow u + v + w = 15$.

The number of solutions of this equation is

$^{15+3-1}C_{3-1} = {}^{17}C_2 = 136$

4. (d) **Explanation :** Required number of ways

$$= {}^{m+n}C_m \times (m-1)! \times (n-1)! = \frac{(m+n)!}{mn}$$

5. (b) **Explanation :** There are (m+1) choices for each of n different books. So, the total number of choices is $(m+1)^n$ including one choice in which we do not select any book. Hence, the required number of ways is $(m+1)^n - 1$.

6. (b) **Explanation :** Let x_1 be the number of stations before the first halting station, x_2 between first and second, x_3 between second and third, x_4 between third and fourth and x_5 on the right of the 4th stations. Then,

$x_1 \ge 0, x_5 \ge 0, x_2, x_3, x_4 \ge 1$ such that

$x_1 + x_2 + x_3 + x_4 + x_5 = 8$

The total number of ways is the number of solutions of the above equation.

Let $y = x_2 - 1, y_3 = x_3 - 1, y_4 = x_4 - 1$. Then, it reduces to $x_1 + y_2 + y_3 + y_4 + x_5 = 5$, where $y_2, y_3, y_4 \ge 0$. The number of solutions of this equation is $^{5+5-1}C_{5-1} = {}^9C_4$.

7. (b) **Explanation :** Let n be the number of participants. Then

$$^nC_2 = 153 \Rightarrow \frac{n(n-1)}{2} = 153 \Rightarrow n^2 - n - 306 = 0 \Rightarrow n = 18$$

8. (b) **Explanation :** Since each bulb has two choices, either switched on or off,

$\therefore$ required number $= 2^{10} - 1 = 1023$

9. (b)

10. (a) **Explanation :** A triangle is obtained by joining three non-colinear points. So number of triangles on joining 3 points out of 10 points $= {}^{10}C_3$. But 6 points are collinear and on joining any three out of these 6, we do not obtain a triangle. Hence, the required number of triangles $= {}^{10}C_3 - {}^6C_3 = 120 - 20 = 100$

11. (a) **Explanation :** The required number of ways

$$= {}^5C_4.{}^5C_2 + {}^5C_3.{}^5C_3 + {}^5C_2.{}^5C_4 = 50 + 100 + 50 = 200$$

V. MATRIX ALGEBRA

- Introduction
- Types of Matrices
- Determinant of Matrix
- Inverse of Matrix
- Linear Equations
- Rank of Matrix
- Eigen values of Matrix

1. Matrix

Definition 1 : If the complex numbers are arranged in the form of a rectangular array consisting of the complex numbers in horizontal and vertical lines then that arrangement is called a matrix. The horizontal lines of numbers in a matrix are called rows and the vertical lines of numbers in a matrix are called columns. The numbers in a matrix are called elements or entries.

Note 1 : The arrangement of elements of a matrix is enclosed by the brackets ( ) or [ ].

Note 2 : The rows of a matrix are counted from top to bottom and the columns are counted from left to right.

Eg : $\begin{bmatrix} 1 & 2 & 3 & 5 \\ 2 & -1 & 5 & 6 \\ 3 & 4 & 1 & 2 \end{bmatrix}, \begin{bmatrix} 2+3i & 3-2i & 5 \\ 6-2i & 7 & i-1 \end{bmatrix}$

Definition 2 : A matrix A is said to be of type or order or size m × n (read as m by n) if the matrix A has m rows and n columns.

Eg : If $A = \begin{bmatrix} 2 & 3 & -1 \\ 1 & -2 & 5 \end{bmatrix}$ then the order of A is 2×3.

Note : An m × n matrix A is usually written as

$$A = \begin{bmatrix} a_{11} & a_{12} & \dots & a_{1n} \\ a_{21} & a_{22} & \dots & a_{2n} \\ \dots & \dots & \dots & \dots \\ a_{n1} & a_{n2} & \dots & a_{nn} \end{bmatrix} \text{ or } A = (a_{ij})_{m \times n} \quad \begin{matrix} 1 \le i \le m \\ 1 \le j \le n \end{matrix}$$

2. Types Of Matrices

Definition 1: A matrix A is said to be a square matrix if the number of rows in A is equal to the number of columns in A.

Eg : $\begin{bmatrix} 2 & 3 \\ 5 & 4 \end{bmatrix}, \begin{bmatrix} 2 & 3 & -1 \\ 5 & 4 & 2 \\ 1 & 2 & 6 \end{bmatrix}$

Note 1: An n × n square matrix is called a square matrix of type n.

Note 2 : If A is a square matrix then the diagonal in A from the first element of the first row to the last element of the last row is called the principal diagonal of A.

Note 3 : If A is a square matrix then the sum of elements in the principal diagonal of A is called trace of A. It is denoted by tra of A.

Definition 2 : A matrix A is said to be a rectangular matrix if the number of rows in A is not equal to the number of columns in A.

Note 1: A is called a rectangular matrix if A is not a square matrix.

Eg : $\begin{bmatrix} 2 & 3 & 1 \\ 5 & 4 & 6 \end{bmatrix}, \begin{bmatrix} 3 & 5 \\ 1 & -2 \\ 7 & 2 \end{bmatrix}, \begin{bmatrix} 2 & 3 & 5 & 7 \\ 2 & 4 & 1 & 6 \\ -1 & 7 & 2 & -5 \end{bmatrix}$

Definition 3: A matrix A is said to be a zero matrix or null matrix if every element of A is equal to zero. An m × n zero matrix is denoted by O_{mxn} or O.

Eg : $\begin{bmatrix} 0 & 0 \\ 0 & 0 \end{bmatrix}, \begin{bmatrix} 0 & 0 & 0 \\ 0 & 0 & 0 \end{bmatrix}, \begin{bmatrix} 0 & 0 & 0 \\ 0 & 0 & 0 \\ 0 & 0 & 0 \end{bmatrix}$

Definition 4 : A matrix A is said to be a row matrix if A contains only one row.

Eg: [2 3], [2 1 -3]

Definition 5 : A matrix A is said to be a column matrix if A contains only one column.

Eg : $\begin{bmatrix} 2 \\ 3 \end{bmatrix}, \begin{bmatrix} 2 \\ -3 \\ 1 \end{bmatrix}$

Definition 6 : A square matrix $A = (a_{ij})_{n \times n}$ is said to be an upper triangular matrix if $a_{ij} = 0$ whenever $i > j$.

Eg : $\begin{bmatrix} 2 & -1 & 5 \\ 0 & 3 & 6 \\ 0 & 0 & 1 \end{bmatrix}, \begin{bmatrix} 1 & 5 & 0 \\ 0 & 6 & -2 \\ 0 & 0 & 7 \end{bmatrix}$

Definition 7 : A square matrix $A = (a_{ij})_{n \times n}$ is said to be lower triangular matrix if $a_{ij} = 0$ whenever $i < j$.

Eg : $\begin{bmatrix} 2 & 0 & 0 \\ 1 & 3 & 0 \\ 5 & 4 & 6 \end{bmatrix}, \begin{bmatrix} -1 & 0 & 0 \\ 0 & 3 & 0 \\ 7 & 5 & 2 \end{bmatrix}$

Definition 8 : A square matrix A is said to be a triangular matrix if A is either an upper triangular matrix or a lower triangular matrix.

Definition 9 : A square matrix A is said to be a diagonal matrix if A is both an upper triangular matrix and a lower triangular matrix.

Eg : $\begin{bmatrix} 2 & 0 & 0 \\ 0 & 3 & 0 \\ 0 & 0 & -1 \end{bmatrix}, \begin{bmatrix} 5 & 0 & 0 & 0 \\ 0 & 3 & 0 & 0 \\ 0 & 0 & 0 & 0 \\ 0 & 0 & 0 & 6 \end{bmatrix}$

Note 1 : A square matrix in which every element is equal to 0, except those of principal diagonal of the matrix is a diagonal matrix.

Note 2 : A matrix $A = (a_{ij})_{m \times n}$ where $a_{ij} = 0$ for $i \neq j$, is called a diagonal matrix. It is denoted by diag [a_{ij}, i=1, 2, ... n] or diag [a_{11} a_{22} ..... a_{mn}].

Definition 10 : A diagonal matrix A is said to be a scalar matrix if all elements in the principal diagonal are equal.

Eg : $\begin{bmatrix}2&0\\0&2\end{bmatrix}, \begin{bmatrix}3&0&0\\0&3&0\\0&0&3\end{bmatrix}, \begin{bmatrix}5&0&0&0\\0&5&0&0\\0&0&5&0\\0&0&0&5\end{bmatrix}$

Definition 11 : A diagonal matrix is said to be a unit matrix if every element in the principal diagonal is equal to unity. Unit matrix of order n is denoted by I_n.

$$\text{Eg} = I_1 = [1]\ I_2 = \begin{bmatrix}1&0\\0&1\end{bmatrix}\ I_3 = \begin{bmatrix}1&0&0\\0&1&0\\0&0&1\end{bmatrix}, I_4 = \begin{bmatrix}1&0&0&0\\0&1&0&0\\0&0&1&0\\0&0&0&1\end{bmatrix}$$

$I_n = (x_{ij})_{n\times n}$ where $x_{ij} = 1$ if $i=j$ $= 0$ if $i \neq j$

3. Equality Of Matrices

Definition : Two matrices A and B are said to be equal if

I. A, B are of same type and

II. the corresponding elements in A and B are equal. It is denoted by A = B.

Ex : If A = Eg :If $A = \begin{bmatrix}a&b&c\\x&y&z\end{bmatrix}, B = \begin{bmatrix}1&2&3\\4&5&6\end{bmatrix}$, then A = B

$\Leftrightarrow a = 1, b = 2, c = 3, x = 4, y = 5, z = 6.$

4. Addition Of Matrices (Matrix Addition)

Definition : If A and B are two matrices of same type then their sum is defined to be the matrix obtained by adding the corresponding elements of A and B. It is denoted by A+B. If $A=(a_{ij})_{m\times n}$ $B=(b_{ij})_{m\times n}$ then $A+B = (c_{ij})_{m\times n}$ where $c_{ij} = a_{ij} + b_{ij}$.

Eg :If $A = \begin{bmatrix}3&2&-1\\1&2&5\end{bmatrix}, B = \begin{bmatrix}4&-3&1\\3&2&-2\end{bmatrix}$, then $A+B = \begin{bmatrix}7&-1&0\\4&4&3\end{bmatrix}$

5. Properties Of Matrix Addition

Theorem 1 : Matrix addition is commutative, i.e..If A and B are two matrices of same type then A+B = B+A.

Theorem 2 : Matrix addition is associative, i.e., If A, B and C are three matrices of same type then (A+B)+C = A+(B+C)

Theorem 3 : If A is an m × n matrix then A+O = O+A = A.

Note : The matrix O is called additive identity matrix.

Theorem 4 : If A is an m x n matrix then there exists an m x n matrix B such that A+B = B+A = O

Definition 1 : If A is an mxn matrix then the matrix B such that A+B = B+A = O is called additive inverse matrix of A. It is denoted by - A.

Note : If A is an m x n matrix then -A is also an m x n matrix and A+(-A) = (-A)+A = O

Definition 2 : If A, B are two matrices of same type then A+(-B) is called the difference of B in A. It is denoted by A-B.

Ex: If $A = \begin{bmatrix}5&4&3\\2&-3&6\end{bmatrix}, B = \begin{bmatrix}2&-1&3\\-3&-2&5\end{bmatrix}$, then $A-B = \begin{bmatrix}3&5&0\\5&-1&1\end{bmatrix}$

6. Multiplication Of a Matrix by a Scalar (Scalar Multiplication)

Definition : If A is a matrix of type m × n and k is a complex number then their product is defined to be the matrix obtained by multiplying every element of the matrix A by the number k. It is denoted by kA.

Ex: If $A = \begin{bmatrix}2&-3&1\\1&3&-2\end{bmatrix}$, then $3A = \begin{bmatrix}6&-9&3\\3&9&-6\end{bmatrix}$

Theorem 1 : If A, B are two matrices of same type and k,l are complex numbers, then

I. k(A+B) = kA + kB, II. (k+l) A = kA + lA

III. (kl)A = k(lA) IV. 0A = 0 V. k0 = 0.

7. Multiplication of Two Matrices (Matrix Multiplication)

Definition 1 : Two matrices A and B are said to be conformable for multiplication if the number of columns in A is equal to the number of rows in B. If $A = (a_{ij})_{m\times n}$, $B = (b_{ij})_{n\times p}$ are two matrices (which are conformable for multiplication) then their product is defined to be the matrix $(c_{ik})_{m\times p}$ where $c_{ik} = \sum_{j=1}^{n} a_{ij} b_{jk}$. It is denoted by AB

Ex: $A = \begin{bmatrix}a_{11}&a_{12}&a_{13}\\a_{21}&a_{22}&a_{23}\end{bmatrix}, B = \begin{bmatrix}b_{11}&b_{12}\\b_{21}&b_{22}\\b_{31}&b_{32}\end{bmatrix}$ then

$$AB = \begin{bmatrix}a_{11}b_{11}+a_{12}b_{21}+a_{13}b_{31} & a_{11}b_{12}+a_{12}b_{22}+a_{13}b_{32}\\a_{21}b_{11}+a_{22}b_{21}+a_{23}b_{31} & a_{21}b_{12}+a_{22}b_{22}+a_{23}b_{32}\end{bmatrix}$$

Note : If the product AB exists then it is not necessary that the product BA will also exist.

Note : Matrix multiplication is not commutative. Even if AB and BA exist, they need not be equal.

Ex : If $A = \begin{bmatrix}2&0\\3&0\end{bmatrix}, B = \begin{bmatrix}0&0\\1&2\end{bmatrix}, C = \begin{bmatrix}0&0\\3&4\end{bmatrix}$, **show that AB=AC=0**

Sol : $AB = \begin{bmatrix}2&0\\3&0\end{bmatrix}\begin{bmatrix}0&0\\1&2\end{bmatrix} = \begin{bmatrix}0&0\\0&0\end{bmatrix}, AC = \begin{bmatrix}2&0\\3&0\end{bmatrix}\begin{bmatrix}0&0\\3&4\end{bmatrix} = \begin{bmatrix}0&0\\0&0\end{bmatrix}$

Note 1: If AB = 0, then either A or B need not be equal to 0.

Note 2 : If AB=AC then B need not be equal to C even if A ≠ 0.

Theorem 1 : Matrix multiplication is associative. i.e., if conformability is assured for the matrices A, B and C, then (AB)C = A(BC).

Note : The products (AB)C, A(BC) both can be written as ABC.

Theorem 2 : Matrix multiplication is distributive over matrix addition i.e., If conformability is assured for the matrices A, B and C, then

I. A(B+C) = AB+BC

II. (B+C)A = BA + CA.

Theorem 3 : If A is a matrix of type m × n then

$AI_n = I_m A = A.$

Note : If A is a square matrix, then AI = IA = A. The matrix I is called multiplicative identity matrix.

Definition 2 : If A is a square matrix and n is a positive integer, then A^n is defined as follows.

I. $A^1 = A$ II. $A^{n+1} = A^n A$

Note : If A is a square matrix, then $A^n = AAA\ldots\ldots A$ (n times)

Theorem 4 : If A is a square matrix, then $A^m A^n = A^{m+n} = A^n A^m$.

8. Idempotent Matrix

Definition : A square matrix A is said to be an idempotent matrix if $A^2 = A$.

Ex : Show that $A = \begin{bmatrix} 2 & -2 & -4 \\ -1 & 3 & 4 \\ 1 & -2 & -3 \end{bmatrix}$ is an idempotent matrix

Sol:

$$A^2 = A\ A = \begin{bmatrix} 2 & -2 & -4 \\ -1 & 3 & 4 \\ 1 & -2 & -3 \end{bmatrix}\begin{bmatrix} 2 & -2 & 4 \\ -1 & 3 & 4 \\ 1 & -2 & -3 \end{bmatrix} = \begin{bmatrix} 2 & -2 & 4 \\ -1 & 3 & 4 \\ 1 & -2 & -3 \end{bmatrix} = A$$

∴ A is an idempotent matrix.

9. Involutory Matrix

Definition : A square matrix A is said to be an involutory matrix if $A^2 = I$.

Ex : If $A = \begin{bmatrix} 0 & 1 \\ 1 & 0 \end{bmatrix}, B = \begin{bmatrix} 0 & -i \\ i & 0 \end{bmatrix}, C = \begin{bmatrix} 1 & 0 \\ 0 & -1 \end{bmatrix}$ show that A,B,C are involutory matrices

Sol: $A^2 = A.A = \begin{bmatrix} 0 & 1 \\ 1 & 0 \end{bmatrix}\begin{bmatrix} 0 & 1 \\ 1 & 0 \end{bmatrix} = \begin{bmatrix} 0+1 & 0+0 \\ 0+0 & 1+0 \end{bmatrix} = \begin{bmatrix} 0 & 1 \\ 1 & 0 \end{bmatrix} = I$

∴ A is involutory.

Similarly it can be verified that B and C are also involutory matrices.

10. Nilpotent Matrix

Definition : A square matrix A is said to be a nilpotent matrix if there exists a positive integer n such that $A^n=0$. If n is the least positive integer such that $A^n = 0$, then n is called the index of the nilpotent matrix A

Ex : Show that $A = \begin{bmatrix} 1 & -3 & -4 \\ -1 & 3 & 4 \\ 1 & -3 & -4 \end{bmatrix}$ is a nilpotent matrix

Sol : $A^2 = A.A = \begin{bmatrix} 1 & -3 & -4 \\ -1 & 3 & 4 \\ 1 & -3 & -4 \end{bmatrix}\begin{bmatrix} 1 & -3 & -4 \\ -1 & 3 & 4 \\ 1 & -3 & -4 \end{bmatrix}$

$$= \begin{bmatrix} 1+3-4 & -3-9+12 & -4-12+16 \\ -1-3+4 & 3+9-12 & 4+12-16 \\ 1+3-4 & -3-9+12 & -4-12+16 \end{bmatrix} = \begin{bmatrix} 0 & 0 & 0 \\ 0 & 0 & 0 \\ 0 & 0 & 0 \end{bmatrix} = 0$$

∴ A is a nilpotent matrix of index 2.

11. Transpose Of a Matrix

Definition : The matrix obtained by changing the rows of a given matrix A into columns is called transpose of A. It is denoted by A^T or A'. If $(a_{ij})_{m\times n}$ then $A^T = (a'_{ij})_{n\times m}$ where $a'_{ji} = a_{ij}$.

Ex : If $A = \begin{bmatrix} 2 & 3 & -1 \\ 1 & 2 & 3 \end{bmatrix}$ then $A^T = \begin{bmatrix} 2 & 1 \\ 3 & 2 \\ -1 & 3 \end{bmatrix}$

12. Properties Of Transpose

Theorem 1 : If A is any matrix, then $(A^T)^T = A$.

Theorem 2 : If A and B are two matrices of same type, then $(A+B)^T = A^T + B^T$.

Note : If A is any matrix and k is any complex number then $(kA)^T = kA^T$.

Theorem 3 : If A and B are two matrices for which conformability for multiplication is assured, then

$$(AB)^T = B^T A^T.$$

Corollary 1 : If $A_1, A_2, \ldots\ldots A_n$ are n matrices which are conformable for multiplication then $(A_1, A_2, \ldots\ldots A_n)^T = A_n^T A_{n-1}^T \ldots\ldots\ldots A_1^T$.

Note : If A is any matrix then AA^T and A^TA are defined. They are of square matrices. But their order may be different.

Ex :

If $A = \begin{bmatrix} 1 & 2 & 0 \\ -2 & 0 & 3 \end{bmatrix}$ then $A^T = \begin{bmatrix} 1 & -2 \\ 2 & 0 \\ 0 & 3 \end{bmatrix}$ and $A\ A^T = \begin{bmatrix} 5 & -2 \\ -2 & 13 \end{bmatrix}$

$$A^T\ A = \begin{bmatrix} 5 & 2 & -6 \\ 2 & 4 & 0 \\ -6 & 0 & 9 \end{bmatrix}$$

13. Symmetric Matrix

Definition : A square matrix A is said to be a symmetric matrix if $A^T = A$.

Ex : $\begin{bmatrix} 2 & 3 & 1 \\ 3 & 4 & 5 \\ 1 & 5 & 7 \end{bmatrix}, \begin{bmatrix} 1 & -1 \\ -1 & 2 \end{bmatrix}$ are symmetric matrices

14. Skew Symmetric Matrix

Definition : A square matrix A is said to be a skew symmetric matrix if $A^T = -A$.

Ex : $\begin{bmatrix} 0 & 1 & -2 \\ -1 & 0 & 3 \\ 2 & -3 & 0 \end{bmatrix}, \begin{bmatrix} 0 & 5 \\ -5 & 0 \end{bmatrix}$ are skew symmetric matrices.

Theorem : Every square matrix can be uniquely expressed as a sum of a symmetric matrix and a skew symmetric matrix.

15. Determinant Of 2 × 2 Matrix

Definition : If $A = \begin{bmatrix} a & b \\ c & d \end{bmatrix}$ then the number ad - bc is called determinant of A. It is denoted by det A or $|A|$ or $\begin{vmatrix} a & b \\ c & d \end{vmatrix}$.

Ex 1 : $\begin{vmatrix} 3 & 5 \\ 2 & 4 \end{vmatrix} = 12 - 10 = 2$

Ex 2 : $\begin{vmatrix} a_{11} & a_{12} \\ a_{21} & a_{22} \end{vmatrix} = a_{11}a_{22} - a_{21}a_{12}$

Note : If A is a square matrix of order 1, then the determinant of A is defined as the element of A.

16. Minor

Definition : If a_{ij} is an element which is in the ith row and jth column of a square matrix A, then the determinant of the matrix obtained by deleting the ith row and jth column of A is called minor of a_{ij}. It is denoted by M_{ij}.

Note: If $A = \begin{bmatrix} a_{11} & a_{12} & a_{13} \\ a_{21} & a_{22} & a_{23} \\ a_{31} & a_{32} & a_{33} \end{bmatrix}$

$M_{11} = \text{Minor of } a_{11} = \begin{vmatrix} a_{22} & a_{23} \\ a_{32} & a_{33} \end{vmatrix} = a_{22}a_{33} - a_{32}a_{23}$

$M_{12} = \text{Minor of } a_{12} = \begin{vmatrix} a_{21} & a_{23} \\ a_{31} & a_{33} \end{vmatrix} = a_{21}a_{33} - a_{31}a_{23}$

and so on.

17. Cofactor

Definition : If a_{ij} is an element which is in the ith row and jth column of a square matrix A, then the product of $(-1)^{i+j}$ and the minor of a_{ij} is called cofactor of a_{ij}. It is denoted by A_{ij}.

Note : If $A = \begin{bmatrix} a_{11} & a_{12} & a_{13} \\ a_{21} & a_{22} & a_{23} \\ a_{31} & a_{32} & a_{33} \end{bmatrix}$ then

$A_{11} = \text{Cofactor } a_{11} = (-1)^{1+1} M_{11} = +1 \begin{vmatrix} a_{22} & a_{33} \\ a_{32} & a_{33} \end{vmatrix} = a_{22}a_{33} - a_{32}a_{23}$

$A_{12} = \text{Cofactor of } a_{12} = \begin{vmatrix} a_{21} & a_{33} \\ a_{31} & a_{33} \end{vmatrix} = a_{21}a_{33} - a_{31}a_{23}$

and so on.

Ex : If $A = \begin{bmatrix} 1 & 2 & 3 \\ 4 & 5 & 6 \\ 7 & 8 & 9 \end{bmatrix}$ then find all cofactors of element of A.

Sol : $A_{11} = (-1)^{1+1} \begin{vmatrix} 5 & 6 \\ 8 & 9 \end{vmatrix} = +(45-48) = -3$

$A_{12} = (-1)^{1+2} \begin{vmatrix} 4 & 6 \\ 7 & 9 \end{vmatrix} = -(36-42) = 6$

$A_{13} = (-1)^{1+3} \begin{vmatrix} 4 & 5 \\ 7 & 8 \end{vmatrix} = +(32-35) = -3$

$A_{21} = (-1)^{2+1} \begin{vmatrix} 2 & 3 \\ 8 & 9 \end{vmatrix} = -(18-24) = 6$

$A_{22} = (-1)^{2+2} \begin{vmatrix} 1 & 3 \\ 7 & 9 \end{vmatrix} = +(9-21) = -12$

$A_{23} = (-1)^{2+3} \begin{vmatrix} 1 & 2 \\ 7 & 8 \end{vmatrix} = -(8-14) = 6$

$A_{31} = (-1)^{3+1} \begin{vmatrix} 2 & 3 \\ 5 & 6 \end{vmatrix} = +(12-15) = -3$

$A_{32} = (-1)^{3+2} \begin{vmatrix} 1 & 3 \\ 4 & 6 \end{vmatrix} = -(6-12) = 6$

$A_{33} = (-1)^{3+3} \begin{vmatrix} 1 & 2 \\ 4 & 5 \end{vmatrix} = +(5-8) = -3$

18. Determinant Of a Square Matrix

Definition : The sum of the products of the elements of the first row of a square matrix A with their corresponding cofactors is called the determinant of the matrix A.

Note 1 : The determinant of a square matrix A is equal to the sum of the products of the elements of a row (or column) of A with their corresponding cofactors.

Note 2 : If every element of a row (or column) of a square matrix is 0, then the determinant of the matrix is 0.

Ex : Find the determinant of the matrix $\begin{bmatrix} 1 & 4 & 2 \\ 2 & -1 & 4 \\ -3 & 7 & -6 \end{bmatrix}$

Sol : $\begin{vmatrix} 1 & 4 & 2 \\ 2 & -1 & 4 \\ -3 & 7 & -6 \end{vmatrix} = 1\begin{vmatrix} -1 & 4 \\ 7 & -6 \end{vmatrix} - 4\begin{vmatrix} 2 & 4 \\ -3 & -6 \end{vmatrix} + 2\begin{vmatrix} 2 & -1 \\ -3 & 7 \end{vmatrix}$

$= 1(16-28) - 4(-12+12) + 2(14-3) = -22 - 0 + 22 = 0$

19. Singular and Nonsingular Matrices

Definition : A square matrix A is said to be a

I. singular matrix if det A=0.

II. nonsingular matrix if det A $\neq$ 0.

Ex : Show that $\begin{bmatrix} 2 & 5 \\ 4 & 10 \end{bmatrix}$ is a singular matrix and $\begin{bmatrix} 3 & 2 \\ 5 & 1 \end{bmatrix}$ is a nonsingular matrix

Sol : $\begin{vmatrix} 2 & 5 \\ 4 & 10 \end{vmatrix} = 20 - 20 = 0 \quad \therefore \begin{bmatrix} 2 & 5 \\ 4 & 10 \end{bmatrix}$ is a singular matrix

$\begin{vmatrix} 3 & 2 \\ 5 & 1 \end{vmatrix} = 3 - 10 = -7 \neq 0 \quad \therefore \begin{bmatrix} 3 & 2 \\ 5 & 1 \end{bmatrix}$ is a nonsingualr matrix

20. Properties Of Determinants

Theorem 1 : If A is a square matrix, then det A = det A^T.

Proof : Let $A = \begin{bmatrix} a_1 & b_1 & c_1 \\ a_2 & b_2 & c_2 \\ a_3 & b_3 & c_3 \end{bmatrix}$

Now det $A = a_1A_1 + a_2A_2 + a_3A_3 = \det A^T$.

Theorem 2 : The sign of the determinant of a square matrix changes if any two rows (or columns) in the matrix are interchanged.

Theorem 3: If two rows (or columns) of a square matrix are identical, the value of the determinant of the matrix is zero.

Theorem 4: If all the elements of a row (or column) of a square matrix are multiplied by a number k then the value of the determinant of the matrix obtained is k times the determinant of the given matrix.

Note : If the elements of a row (or column) of a square matrix are k times the elements of another row (or column) then the value of the determinant of the matrix is zero.

Theorem 5 : If each element in a row (or column),of a square matrix is the sum of two numbers, then its determinant can be expressed as the sum of the determinants of two square matrices.

Theorem 6: If the elements of a row (or column) of a square matrix are added with k times the corresponding elements of another row (or column), then the value of the determinant of the matrix obtained is equal to the value of the determinant of the given matrix.

Note : The addition to the elements of i th row, the corresponding elements of j th row multiplied by a number k will be denoted by $R_i + k R_j$ [For columns $C_i + kC_j$]

Theorem 7 : The sum of the products of the elements of any row (or column) of a square matrix with the cofactors of the corresponding elements of another row (or column) of the matrix is zero.

Theorem 8: If the elements of a square matrix are polynomials in x and two rows (or columns) are identical when x =a, then (x-a) is a factor of the deteminant of the matrix.)

Theorem 9: If A and B are two square matrices of same type, then det (AB) = det A. det B.

21. Adjoint Matrix

Definition 1 : The transpose of the matrix obtained by replacing the elements of a square matrix A by the corresponding cofactors is called the adjoint matrix of A. It is denoted by Adj A or adj A.

Note 1 : If $A = \begin{bmatrix} a_{11} & a_{12} & a_{13} \\ a_{21} & a_{22} & a_{23} \\ a_{31} & a_{32} & a_{33} \end{bmatrix}$ then Adj $A = \begin{bmatrix} A_{11} & A_{21} & A_{31} \\ A_{12} & A_{22} & A_{32} \\ A_{13} & A_{23} & A_{33} \end{bmatrix}$

Ex : Find the adjoint matrix of the matrix $A = \begin{bmatrix} 1 & 1 & 1 \\ 1 & 2 & -3 \\ 2 & -1 & 3 \end{bmatrix}$

Sol : The cofactors of elements of A are

$A_{11} = +\begin{vmatrix} 2 & -3 \\ -1 & 3 \end{vmatrix} = 6-3, A_{12} = -\begin{vmatrix} 1 & -3 \\ 2 & 3 \end{vmatrix} = -(3+6) = -9$

$A_{13} = +\begin{vmatrix} 1 & 2 \\ 2 & -1 \end{vmatrix} = -1-4 = 15, A_{21} = -\begin{vmatrix} 1 & 1 \\ -1 & 3 \end{vmatrix} = -(3+1) = -4$

$A_{22} = +\begin{vmatrix} 1 & 1 \\ 2 & 3 \end{vmatrix} = 3-2 = 1, A_{23} = -\begin{vmatrix} 1 & 1 \\ 2 & -1 \end{vmatrix} = -(-1-2) = 3$

$A_{31} = +\begin{vmatrix} 1 & 1 \\ 2 & -3 \end{vmatrix} = -3-2 = -5, A_{32} = -\begin{vmatrix} 1 & 1 \\ 1 & -3 \end{vmatrix} = -(-3-1) = 4$

$A_{33} = +\begin{vmatrix} 1 & 1 \\ 1 & 2 \end{vmatrix} = 2-1 = 1$

$\therefore \text{Adj } A = \begin{bmatrix} A_{11} & A_{21} & A_{31} \\ A_{12} & A_{22} & A_{32} \\ A_{13} & A_{23} & A_{33} \end{bmatrix} = \begin{bmatrix} 3 & -4 & -5 \\ -9 & 1 & 4 \\ -5 & 3 & 1 \end{bmatrix}$

22. Inverse Of a Square Matrix

Definition : A square matrix A is said to be an invertible matrix if there exists a square matrix B such that AB= BA= I. The matrix B is called inverse of A.

Note 1 : A rectangular matrix can not be invertible.

Note 2 : Every square matrix need not be invertible.

Theorem 1 : An invertible matrix has unique inverse.

Note 1 : If A is an invertible matrix then its inverse is denoted by A^{-1} Now $A A^{-1} = A^{-1} A = I$

Note 2 : If I is a unit matrix then I is invertible and I^{-1}.

Theorem 2 : If A is an invertible matrix then A^{-1} is also invertible and $(A^{-1})^{-1} = A$

Theorem 3 : If A and B are two invertible matrices of same type then AB is also invertible and $(AB)^{-1} = B^{-1} A^{-1}$.

Corollary 1 : If $A_1, A_2, \ldots A_n$ are invertible matrices of same type then $A_1 A_2 \ldots A_n$ is also invertible and $(A_1 A_2 \ldots A_n)^{-1} = A_n^{-1} A_{n-1}^{-1} \ldots A_1^{-1}$.

Note : If A is an invertible matrix and k is a nonzero complex number then kA is invertible and $(kA)^{-1} = k^{-1} A^{-1}$.

Theorem 4 : If A is an invertible matrix then A^T is also invertible and $(A^T)^{-1} = (A^{-1})^T$

Theorem 5 : If A is a nonsingular matrix then A is invertible and $A^{-1} = \frac{\text{Adj } A}{\det A}$.

Note : If A is a square matrix, then A (Adj A) = (det A)I

Theorem 6 : If A is an invertible matrix then A is nonsingular.

Note : If A is an invertible matrix then $\det(A^{-1}) = \frac{1}{\det A}$

Theorem 7 : A square matrix A is non singular if A is invertible.

Corollary 2 : If $A = \begin{bmatrix} a & b \\ c & d \end{bmatrix}$ and $ad \neq bc$ then

$$A^{-1} = \frac{1}{ad-bc}\begin{bmatrix} d & -b \\ -c & a \end{bmatrix}$$

Ex : If $A = \begin{bmatrix} 2 & 3 \\ 5 & 7 \end{bmatrix}$ then find A^{-1}

Sol : $A^{-1} = \frac{1}{14-15}\begin{bmatrix} 7 & -3 \\ -5 & 2 \end{bmatrix} = -1\begin{bmatrix} 7 & -3 \\ -5 & 2 \end{bmatrix} = \begin{bmatrix} 7 & -3 \\ -5 & 2 \end{bmatrix}$

Theorem 8 : Let A, B, C be three square matrices of same type and A be a non-singular matrix. Then

I. $AB = AC \Rightarrow B = C$

II. $BA = CA \Rightarrow B = C$.

Theorem 9 : If A and B are nonsingular matrices of same type then Adj(AB) = (Adj B) (Adj A)

Theorem 10 : If A is a square matrix of type n then det (Adj A) = $(\det A)^{n-1}$

Ex : If A is a square matrix of type 3 and det A = 10 then find (Adj A).

Sol : det (Adj A) = $(\det A)^{3-1} = 10^2 = 100$

23. Linear Equations

The simultaneous linear equations of two or three variables are solved by using matrices Consider the system of simultaneous linear equations in two variables,

$a_1x + b_1y = c_1,\ a_2x + b_2y = c_2$ .... (1)

Let $A = \begin{bmatrix} a_1 & b_1 \\ a_2 & b_2 \end{bmatrix},\ X = \begin{bmatrix} x \\ y \end{bmatrix},\ B = \begin{bmatrix} C_1 \\ C_2 \end{bmatrix}$

The system of equations (1) can be represented as a matrix equation AX = B

Definition 1 : If x = a, y = b where a, b are complex numbers, satisfy the equations of system (1), then x=a, y = b is called a solution of (1)

Note : $x=\alpha, y=\beta$ is a solution of (1) $\Leftrightarrow x = X_1$

Where $X_1 = \begin{bmatrix} \alpha \\ \beta \end{bmatrix}$ is a solution of AX = B

Consider the system of simultaneous linear equations in three variables,

$a_1x + b_1y + c_1z = d_1$
$a_2x + b_2y + c_2z = d_2$ .... (2)
$a_3x + b_3y + c_3z = d_3$

Let $A = \begin{bmatrix} a_1 & b_1 & c_1 \\ a_2 & b_2 & c_2 \\ a_3 & b_3 & b_3 \end{bmatrix},\ X = \begin{bmatrix} x \\ y \\ z \end{bmatrix},\ B = \begin{bmatrix} d_1 \\ d_2 \\ d_3 \end{bmatrix}$

The system of equations (2) can be represented as a matrix equation AX = B

Definition 2 : If x = a, y = b, z = g where a, b,g are complex numbers, satisfy the equations of system (2) then x = a, y = b, z = g is called a solution of system (2).

Note : $x=\alpha,\ y=\beta,\ z=\gamma$ is a solution of (2) $\Leftrightarrow x = X_1$

where $X_1 = \begin{bmatrix} \alpha \\ \beta \\ \gamma \end{bmatrix}$ is a solution of AX = B

Definition 3 : The system of equations AX = B is said to be in consistent if AX = B has a solution.

Definition 4 : The system of equations AX = B is said to be in consistent if AX = B has no solution.

Eg 1 : x+y = 2, x-y =1 are consistent,

Eg 2 : 2x+y = 3, 4x+2y = 6 are consistent,

Eg 3 : x+y = 1, x+y = 2 are not consistent,

24. Matrix Inversion Method

Theorem 1 : If A is a nonsingular matrix then A^{-1} B is a solution AX = B.

Proof : A is nonsingular $\Rightarrow A^{-1}$ exists and $AA^{-1} = A^{-1}A = I$

$A(A^{-1}B) = (AA^{-1})B = IB = B \therefore X = A^{-1}B$ is a solution of AX = B.

Theorem 2 : If A is a nonsingular matrix then AX = B has a unique solution.

Theorem 3 : If A is a nonsingular matrix then $X = A^{-1}$ B is the unique solution of AX = B.

Note : The method of finding the solution $X = A^{-1}$ B of AX = B is called matrix inversion method.

Ex : solve the equations by matrix inversion method x + y + z = 6, 2x - y + z = 3, x + 2y - z = 2.

Sol : Let $A = \begin{bmatrix} 1 & 1 & 1 \\ 2 & -1 & 1 \\ 1 & 2 & -1 \end{bmatrix},\ X = \begin{bmatrix} x \\ y \\ z \end{bmatrix},\ B = \begin{bmatrix} 6 \\ 3 \\ 2 \end{bmatrix}$

The given equations can be written as AX = B

By matrix inversion method $X = A^{-1}B$ is the solution.

det A = 1(1-2) - 1(-2-1) + 1(4+1) = -1+3+5 = 7.

The cofactors of elements of A are

$A_{11} = +(1-2) = -1,\ A_{12} = -(-2-1) = 3,\ A_{13} = +(4+1) = 5$
$A_{21} = -(-1-2) = 3,\ A_{22} = +(-1-1) = -2,\ A_{23} = -(2-1) = -1$
$A_{31} = +(1+1) = 2,\ A_{32} = -(1-2) = 1,\ A_{33} = +(-1-2) = -3$

$$\text{Adj } A = \begin{bmatrix} A_{11} & A_{21} & A_{31} \\ A_{12} & A_{22} & A_{32} \\ A_{13} & A_{23} & A_{33} \end{bmatrix} = \begin{bmatrix} -1 & 3 & 2 \\ 3 & -2 & 1 \\ 5 & -1 & -3 \end{bmatrix}$$

$$A^{-1} = \frac{\text{Adj } A}{\det A} = \frac{1}{7}\begin{bmatrix} -1 & 3 & 2 \\ 3 & -2 & 1 \\ 5 & -1 & -3 \end{bmatrix}$$

$$A^{-1} = \frac{\text{Adj } A}{\det A} = \frac{1}{7}\begin{bmatrix} -1 & 3 & 2 \\ 3 & -2 & 1 \\ 5 & -1 & -3 \end{bmatrix}\begin{bmatrix} 6 \\ 3 \\ 2 \end{bmatrix} = \frac{1}{7}\begin{bmatrix} -6 & +9 & +4 \\ 18 & -6 & +2 \\ 30 & -3 & -6 \end{bmatrix} = \frac{1}{7}\begin{bmatrix} 7 \\ 14 \\ 21 \end{bmatrix} = \begin{bmatrix} 1 \\ 2 \\ 3 \end{bmatrix}$$

$\therefore$ x = 1, y = 2, z = 3 is the solution

25. Cramer's Rule

Theorem : Let $a_1x+b_1y+c_1z = d_1,\ a_2x+b_2y+c_2z = d_2,\ a_3x+b_3y+c_3z = d_3$-----> (1) be a system of linear equations.

If $\Delta = \begin{vmatrix} a_1 & b_1 & c_1 \\ a_2 & b_2 & c_2 \\ a_3 & b_3 & c_3 \end{vmatrix} \neq 0,\ \Delta_1 = \begin{vmatrix} d_1 & b_1 & c_1 \\ d_2 & b_2 & c_2 \\ d_3 & b_3 & c_3 \end{vmatrix},\ \Delta_2 = \begin{vmatrix} a_1 & d_1 & c_1 \\ a_2 & d_2 & c_2 \\ a_3 & d_3 & c_3 \end{vmatrix}$

$\Delta_3 = \begin{vmatrix} a_1 & b_1 & c_1 \\ a_2 & b_2 & c_2 \\ a_3 & b_3 & c_3 \end{vmatrix}$ then $x = \frac{\Delta_1}{\Delta}, y = \frac{\Delta_2}{\Delta}, z = \frac{\Delta_3}{\Delta}$

is the solution of (1)

Note 1 : $\Delta_1, \Delta_2, \Delta_3$ are also denoted as $\Delta_x, \Delta_y, \Delta_z$ respectively.

Note 2: $\frac{x}{\Delta_1} = \frac{y}{\Delta_2} = \frac{z}{\Delta_3} = \frac{1}{\Delta}$

Ex 2 : Solve the following equations by Cramer's rule. x+y+z=6, 2x-y+z=3, x+2y-z=2.

Sol : $\Delta = \begin{vmatrix} 1 & 1 & 1 \\ 2 & -1 & 1 \\ 1 & 2 & -1 \end{vmatrix} = 1(1-2)-1(-2-1)+1(4+1) = -1+3+5 = 7$

$\Delta_1 = \begin{vmatrix} 6 & 1 & 1 \\ 3 & -1 & 1 \\ 2 & 2 & -1 \end{vmatrix} = 6(1-2)-1(-3-2)+1(6+2) = -6+5+8 = 7$

$\Delta_2 = \begin{vmatrix} 1 & 6 & 1 \\ 2 & 3 & 1 \\ 1 & 2 & -1 \end{vmatrix} = 1(-3-2)-6(-2-1)+1(4-3) = -5+18+1 = 14$

By Cramer's rule $\dfrac{x}{\Delta_1}=\dfrac{y}{\Delta_2}=\dfrac{z}{\Delta_3}=\dfrac{1}{\Delta}$

$\therefore \dfrac{x}{7}=\dfrac{y}{14}=\dfrac{z}{21}=\dfrac{1}{7}$ $\quad\therefore x=1, y=2, z=3$ is the solution

Definition : Let $a_1x+b_1y+c_1z=d_1$, $a_2x+b_2y+c_2z=d_2$, $a_3x+b_3y+c_3z=d_3$ .... (1)

be a system of linear equations.

The matrix $A=\begin{bmatrix} a_1 & b_1 & c_1 \\ a_2 & b_2 & c_2 \\ a_3 & b_3 & c_3 \end{bmatrix}$ is called coefficient matrix and

the matrix $K=\begin{bmatrix} a_1 & b_1 & c_1 & d_1 \\ a_2 & b_2 & c_2 & d_2 \\ a_3 & b_3 & c_3 & d_3 \end{bmatrix}$ is called augmented matrix.

26. Elementary Row Transformations

Definition 1 : An elementary row transformation is an operation of any one of the following types.

1. The interchange of any two rows.
2. The multiplication of the elements of any row by a nonzero number.
3. The addition to the elements of row, the corresponding elements of any other row multiplied by a number.

Notation : The following notation will be used to denote the elementary transformations.

1. The interchange of i th and j th rows will be denoted by $R_i \leftrightarrow R_j$.
2. The multiplication of the elements of i th row by a nonzero number k will be denoted by $R_i \rightarrow k\,R_i$ or simply kR_i.
3. The addition to the elements of i th row, the corresponding elements of j th row multiplied by a number k will be denoted by $R_i \rightarrow R_i+kR_j$ or simply R_i+kR_j.

Definition 2 : Two matrices A and B are said to be equivalent if B is obtained from A by a finite number of elementary transformations on A. It is denoted by A ~ B.

Note : The relation, equivalent of matrices (~) is an equivalence relation on the set of all matrices of type m x n.

27. Gauss Jordan Method

Theorem : Let $a_1x+b_1y+c_1z=d_1$, $a_2x+b_2y+c_2z=d_2$, $a_3x+b_3y+c_3z=d_3$ ...... (1) be a system of linear equations.

If the augmented matrix $\begin{bmatrix} a_1 & b_1 & c_1 & d_1 \\ a_2 & b_2 & c_2 & d_2 \\ a_3 & b_3 & c_3 & d_3 \end{bmatrix}$ can be reduced to

the form $\begin{bmatrix} 1 & 0 & 0 & \alpha \\ 0 & 1 & 0 & \beta \\ 0 & 0 & 1 & \gamma \end{bmatrix}$ by using elementary row transformations then x=a, y=b, z=g is the solution of (1).

Ex : Solve the equations by Gauss - Jordan Method. x+y+z=6, 2x-y+z=3, x+2y-z=2.

Sol : The augmented matrix of the given equation is

$$\begin{bmatrix} 1 & 1 & 1 & 6 \\ 2 & -1 & 1 & 3 \\ 1 & 2 & -1 & 2 \end{bmatrix} \begin{matrix} R_2 & -2R_1 \\ R_3 & -R_1 \end{matrix} \begin{bmatrix} 1 & 1 & 1 & 6 \\ 0 & -3 & -1 & -9 \\ 0 & - & -2 & -4 \end{bmatrix} \begin{matrix} 3R_1 & +R_2 \\ 3R_3 & +R_2 \end{matrix} \begin{bmatrix} 3 & 0 & 2 & 9 \\ 0 & -3 & -1 & -9 \\ 0 & 0 & -7 & -21 \end{bmatrix}$$

$$\frac{-1}{7}R_3 \begin{bmatrix} 3 & 0 & 2 & 9 \\ 0 & -3 & -1 & -9 \\ 0 & 0 & -7 & -21 \end{bmatrix} \begin{matrix} R_1-2R_3 \\ R_2+R_3 \\ \sim \end{matrix} \begin{bmatrix} 3 & 0 & 0 & 3 \\ 0 & -3 & 0 & -6 \\ 0 & 0 & 1 & 3 \end{bmatrix}$$

$$\begin{matrix} \sim \\ \frac{1}{3}R_1 \\ \frac{-1}{3}R_2 \\ \sim \end{matrix} \begin{bmatrix} 1 & 0 & 0 & 1 \\ 0 & 1 & 0 & 2 \\ 0 & 0 & 1 & 3 \end{bmatrix} \therefore x=1, y=2, z=3 \text{ is the solution.}$$

28. Submatrix

Definition 1 : A matrix obtained by deleting some rows or columns (or both) of a matrix is called a submatrix.

Ex : If $A=\begin{bmatrix} 1 & 2 & 3 \\ 2 & 3 & 1 \\ -1 & 2 & 0 \end{bmatrix}$ and $B=\begin{bmatrix} 1 & 2 \\ -1 & 2 \end{bmatrix}$ then B is a submatrix of A.

Note 1 : Every element of a matrix is a submatrix of order 1.

Note 2 : Every matrix is a submatrix of itself.

Definition 2 : If B is a square submatrix of order r of a matrix A then det B is called an r-rowed minor of A.

Ex : If $A=\begin{bmatrix} 1 & 2 & 3 & 4 \\ 2 & 5 & 6 & -1 \\ -2 & 3 & 1 & 7 \end{bmatrix}$ then $\begin{vmatrix} 2 & 3 \\ 3 & 1 \end{vmatrix}$ is a 2-rowed minor of A.

29. Rank of a Matrix:

If A is a null matrix, then the rank of A is defined as zero.

Definition : If A is not a null matrix ; then the rank of the matrix A is the positive integer, if it possesses the following two properties:

I. Each minor of order n + 1 of the matrix A vanishes (i. e., is zero).

II. There exists at least one minor of order n of the matrix A which is not zero.

Some Properties of a Rank of Matrix:

1. Rank of A is denoted by p (A) and is UNIQUE.
2. Every matrix will have a rank
3. If A is a matrix of order m x n, then p(A) ≤ m or n (smaller of the two).
4. If p (A) = n, then every minor of order n + 1, n + 2, etc. is zero.
5. A is a matrix of order n × n. A is non- singualr ⇔ p (A) = n.
6. Rank of I_n = n.
7. A is a matrix of order m x n. If every k^{th} order minor (k < m, k < n) is zero, then P (A) < k.
8. A is a matrix of order m x n. If there is a minor of order k (k < m, k < n) which is not zero, then p(A) ≥ K .
9. The rank of a zero matrix is defined to be zero. The rank of a unit matrix is defined to be one.
10. If x is a non-zero column matrix and y is a non-zero row matrix, then P (x y) = 1.
11. The rank of a matrix is greater than or equal to (not lessthan) the rank of every sub-matrix there of.
12. If A is any n-rowed Square matrix of rank, n -1, then adj A ≠ 0.

13. The rank of the transpose of a matrix is equal to the rank of the original matrix.

14. The rank of a matrix does not change by pre-multiplication or post-multiplication with a non-singular matrix.

15. If A ~ B, then P (A) = P (B).

16. The rank of a product of two matrices cannot exceed the rank of either matrix.

17. The rank of the sum of two matrices cannot exceed the sum of their ranks.

18. Elementary transformations do not change the rank of a matrix.

Ex : Obtain the rank of the matrix $\begin{bmatrix} -1 & 2 & 0 \\ 3 & 7 & 1 \\ 5 & 9 & 3 \end{bmatrix}$

Sol : Let $A = \begin{bmatrix} -1 & 2 & 0 \\ 3 & 7 & 1 \\ 5 & 9 & 3 \end{bmatrix}$

$\therefore |A| = -1(21-9) - 2(9-5) + 0 = -20 \neq 0$

$\therefore$ A is non - singular and hence P (A) = 3

30.Characteristic Equation

For every square matrix A of order n, we can form a matrix $A - \lambda I$ where I is a unit matrix of order n and λ being a constant the determinant of this matrix equated to zero

namely $|A - \lambda I| = 0$ or $\begin{vmatrix} a_{11}-\lambda & a_{12} \ldots\ldots & a_{1n} \\ a_{21} & a_{22}-\lambda \ldots & a_{2n} \\ \ldots\ldots & \ldots\ldots & \ldots\ldots \\ a_{n1} & a_{n2} \ldots\ldots & a_{nn}-\lambda \end{vmatrix} = 0$

is called the characteristic equation of A. On expanding the determinant, we can write this equation as

$$(-1)^n \left[\lambda^n + P_1 \lambda^{n-1} + P_2 \lambda^{n-2} + \ldots + P_n\right] = 0$$

where $P_1, P_2 \ldots\ldots$, can be expressed in terms of the element a_{rs}

The roots of characterstic equation are called latent roots or Eigen values of matrix A.A matrix of order n will have n latent roots.

Example. Determine the eigen values of the matrix $\begin{bmatrix} 2 & -2 & 2 \\ 1 & 1 & 1 \\ 1 & 3 & -1 \end{bmatrix}$

The characterstic equations is $\begin{vmatrix} 2-\lambda & -2 & 2 \\ 1 & 1-\lambda & 1 \\ 1 & 3 & -1-\lambda \end{vmatrix} = 0$

$$(2-\lambda)\begin{vmatrix} 1-\lambda & 1 \\ 3 & -1-\lambda \end{vmatrix} + 2\begin{vmatrix} 1 & 1 \\ 1 & -1-\lambda \end{vmatrix} + 2 \geq \begin{vmatrix} 1 & 1-\lambda \\ 1 & 3 \end{vmatrix} = 0$$

$$(2-\lambda)(\lambda^2 - 1 - 3) + 2(-1-\lambda-1) + 2(3-1+\lambda) = 0$$

$$(\lambda-2)(\lambda^2-4) - 2(\lambda+2) + 2(\lambda+2) = 0$$

$$(\lambda-2)(\lambda^2-4) = 0 \text{ or } (\lambda-2)^2 (\lambda+2) = 0$$

$\therefore$ The latent roots (eigen values) are 2,2 and - 2.

31.Eigen Vectors

Consider the linear transformation $\lambda = AX$ .... (1)

which transforms a column vector X into another column vector I by means of the square matrix A. In physical applications we are often interested in finding the particular vectors which transforms into themselves or to a scalar multiple of themselves.

Let X be such a vector which transforms to it's multiple λX , by the transformation (1). Then

$$\lambda X = AX \text{ or } AX - \lambda IX = 0 , (A-\lambda!)X = 0$$

written in full, the above matrix equation gives n homogeneous equations in n unknowns, X_1, X_2, X_n. These equations will have a non-zero solution only when the coefficient matrix is singular i.e. when

$|A - \lambda I| = 0$ .... (2)

This is known as characteristic equation of transformation, and is same as the characteristic equation of matrix A. This has n roots. Corresponding to each root, equation (2) will have a non-zero solution.

$X = \begin{bmatrix} x_1 \\ x_2 \\ \ldots \\ x_n \end{bmatrix}$ which is known as an eigen vector.

We see from (2) that if X is a solution for root λ then so is CX, where C is an arbitrary constant. Thus the eigen vector corresponding to a root is not unique, but may be any one of the vectors CX.

Corresponding to n distinct latent root we generate n independent eigen vectors. When two (or more) roots are equal it may still be possible to obtain linearly independent eigen vectors corresponding to the repeated roots. (See Ex. below). In some cases however, it is not possible to obtain n independent eigen vectors.

Example. Find eigen values and eigen vectors of $\begin{bmatrix} -2 & 2 & -3 \\ 2 & 1 & -6 \\ -1 & -2 & 0 \end{bmatrix}$

The equation $(a-\lambda!)X = 0$ for the given matrix is

$$\begin{bmatrix} -2-\lambda & 2 & -3 \\ 2 & 1-\lambda & -6 \\ -1 & -2 & 0-\lambda \end{bmatrix} \begin{bmatrix} x \\ y \\ z \end{bmatrix} = 0 \quad \ldots\ldots\ldots (3)$$

For the non-zero solution, the determinant of the matrix should be zero i.e.

$$(-2-\lambda)\{(1-\lambda)(-\lambda) - 12\} - 2\{2(-\lambda) - 6\} - 3\{-4 + (1-\lambda)\} = 0$$

$$(\lambda+2)\{\lambda^2 - \lambda - 12\} - 4\lambda - 12 - 3\lambda - 9 = 0$$

$$\lambda^3 + \lambda^2 - 21\lambda - 45 = 0.$$

This is the characteristic equation. We find by trial that $\lambda = 5$ is one root. So we write

$$\lambda^2(\lambda-5)+6\lambda(\lambda-5)+9(\lambda-5)=0$$

$$(\lambda-5)\{\lambda^2+6\lambda+9\}=0 \quad or \quad (\lambda-5)(\lambda+3)^2=0$$

Hence the eigen values are 5, -3, -3,

Putting $\lambda = 5$ in (3) we see that the corresponding eigen vectors is given by the equations.

$-7x+2y-3z=0, 2x-4y-6z=0, -x-2y-5z=0$

Since these equations are linearly dependent, we can omit one of them solving the last two we get

$$\frac{x}{20-12}=\frac{-y}{-10-6}=\frac{z}{-4-4} \text{ or } \frac{x}{1}=\frac{y}{2}=\frac{z}{-1}$$

Similarly for the eigen vector corresponding to $\lambda = -3$ equation (3) gives

$x+2y-3z=0, 2x+4y-6z=0, -x-2y+3z=0$

Only one equation is independent in this case, the other two being multiples of this. So a number of different eigen vectors can be obtained. Two which are linearly independent are (2, - 1, 0) and (3, 0, 1).

These are obtained by respectively putting z and y equal to zero. Any other eigen vector corresponding to $\lambda = -3$ will be a linear combination of these two.

Hence three eigen vectors may be taken as (1, 2, -1), (2, -1, 0), (3, 0, 1).

OBJECTIVE QUESTIONS - I

1. The rank of the matrix $\begin{bmatrix} 1 & 1 \\ 0 & 0 \end{bmatrix}$ is **[GATE-2002]**

a) 4 b) 2 c) 1 d) 0

2. The determinant of the matrix $\begin{bmatrix} 6 & -8 & 1 & 1 \\ 0 & 2 & 4 & 6 \\ 0 & 0 & 4 & 8 \\ 0 & 0 & 0 & -1 \end{bmatrix}$ is **[GATE-1997]**

a) 11 b) -48 c) 0 d) -24

3. The matrices $\begin{bmatrix} \cos\theta & -\sin\theta \\ \sin\theta & \cos\theta \end{bmatrix}$ *and* $\begin{bmatrix} a & 0 \\ 0 & b \end{bmatrix}$ Commute under multiplication **[GATE-1996]**

a) If a =b or $\theta = n\pi$, n is an integer b) Always

c) Never d) If $a\cos\theta \neq b\sin\theta$

4. The rank of the following (n +1) × (n+1) matrix, where a is a real number is

$$\begin{bmatrix} 1 & a & a^2 & \cdots & a^n \\ 1 & a & a^2 & & a^n \\ \vdots & & & & \vdots \\ 1 & a & a2 & & a^n \end{bmatrix}$$

[GATE-1995]

a) 1 b) 2

c) n d) Depends upon the value of A

5. Let A and B be real symmetric matrices of size n × n, Then which of the following is TRUE? **[GATE-1994]**

a) $AA^T=1$ b) $A=A^{-1}$ c) AB = BA d) $(AB)^T=BA$

6. The rank of the matrix $\begin{bmatrix} 0 & 0 & -3 \\ 9 & 3 & 5 \\ 3 & 1 & 1 \end{bmatrix}$ is **[GATE-1994]**

a) 0 b) 1 c) 2 d) 3

7. If the two matrices $\begin{bmatrix} 1 & 0 & x \\ 0 & x & 1 \\ 0 & 1 & x \end{bmatrix}$ and $\begin{bmatrix} x & 1 & 0 \\ x & 0 & 1 \\ 0 & x & 1 \end{bmatrix}$ have the same determinant, Then The value of x is **[Hyd.Univ-1991]**

a) $\frac{1}{2}$ b) $\sqrt{2}$ c) $\pm\frac{1}{2}$ d) $\pm\frac{1}{\sqrt{2}}$

8. Find the characterstic roots of the matrix : $\begin{bmatrix} 2 & 3 \\ 3 & 2 \end{bmatrix}$ **[Hyd.Univ-1990]**

a) 2, 2 b) 2, 3 c) –1, 5 d) –5, 1

9. Find the cofactor of the element 7 in the matrix $\begin{bmatrix} 1 & 2 & 3 \\ 2 & 1 & 3 \\ 5 & 7 & 4 \end{bmatrix}$ is **[Hyd.Univ-1990]**

a) –3 b) 3 c) –11 d) 11

10. If A is a skew-symmetric matrix, A^T has the value **[Hyd.Univ-1990]**

a) - A b) A^{-1} c) - A^T d) – A^{-1}

11. The value of k for which the matrix $\begin{bmatrix} 3 & -1 & 2 \\ 2 & 1 & 3 \\ 1 & -3 & k \end{bmatrix}$ is singular, is **[Hyd.Univ-1989]**

a) 2 b) 1 c) –2 d) –1

12. The general expression for the n th power of A = $\begin{bmatrix} a & 1 \\ 0 & a \end{bmatrix}$ is **[Hyd.Univ-1989]**

a) $\begin{bmatrix} a^n & 1 \\ 0 & a^n \end{bmatrix}$ b) $\begin{bmatrix} a^n & n \\ 0 & a^n \end{bmatrix}$

c) $\begin{bmatrix} a^n & na^{n-1} \\ 0 & a^n \end{bmatrix}$ d) $\begin{bmatrix} a^n & na^{n-1} \\ -n & a^n \end{bmatrix}$

13. Eigen vectors of $\begin{bmatrix} 1 & \cos\theta \\ \cos\theta & 1 \end{bmatrix}$ are **[Hyd.Univ-1989]**

a) $\begin{bmatrix} \cos\theta \\ 1 \end{bmatrix}, \begin{bmatrix} 1 \\ \cos\theta \end{bmatrix}$ b) $\begin{bmatrix} 1 \\ 1 \end{bmatrix}, \begin{bmatrix} 1 \\ -1 \end{bmatrix}$

c) $\begin{bmatrix} \sin\theta \\ 1 \end{bmatrix}, \begin{bmatrix} 1 \\ \sin\theta \end{bmatrix}$ d) $\begin{bmatrix} \cos\theta \\ \sin\theta \end{bmatrix}, \begin{bmatrix} \cos\theta \\ -\sin\theta \end{bmatrix}$

14. The matrix $\begin{bmatrix} 0 & 0 & 1 \\ 0 & 1 & 0 \\ 1 & 0 & 0 \end{bmatrix}$ is **[Hyd.Univ-1994]**

a) Unit matrix b) Diagonal matrix

c) Square matrix d) Skew symmetric matrix

15. If the rank of a square matrix of order n is r, then how many eigen values will be zeroes. **[R.E.C. Trichy-1991]**

a) n-r b) n c) r d) None

16. If two rows or columns in a determinant are interchanged, then its value **[R.E.C. Trichy-1991]**

a) Remains same b) Becomes zero
c) Order change d) None

17. For what values of x the matrix $\begin{bmatrix} 9 & x-6 \\ 3 & x \end{bmatrix}$ has no inverse **[R.E.C. Trichy-1991]**

a) 0 b) –2 c) –3 d) 6

18. The rank of the matrix in which every element is 1 is : **[R.E.C. Trichy-1992]**

a) 0 b) 1 c) n d) n-1

19. If A and B are square matrices of order n and rank (A) = rank(B), then

a) rank(A^2) = rank(B^2) b) rank(AB) = rank(BA)
c) rank(A^T) = rank(B^T) d) rank(A^{-1}) = rank(B^{-1})

20. Which of the following is an eigen value of $\begin{bmatrix} 0 & 1 \\ 0 & 0 \end{bmatrix}$ is: **[R.E.C. Trichy-1993]**

a) 0 b) 1 c) 2 d) 3

21. Which of the following is a hermitian matrix : **[R.E.C. Trichy-1993]**

a) $\begin{bmatrix} 2 & 3+i \\ 3-i & 1 \end{bmatrix}$ b) $\begin{bmatrix} 2 & 3+i \\ 3+i & 1 \end{bmatrix}$

c) $\begin{bmatrix} 3+i & 2 \\ 1 & 3-i \end{bmatrix}$ d) $\begin{bmatrix} 3+i & 2 \\ 1 & -3+i \end{bmatrix}$

22. If a square matrix A satisfies A^TA =I then the matrix A is **[R.E.C. Trichy-1994]**

a) Idempotent b) Symmetric
c) Orthogonal d) Hermitian

23. If $\begin{bmatrix} 5 & k+2 \\ k+1 & -2 \end{bmatrix} = \begin{bmatrix} k+2 & 4 \\ 3 & -k \end{bmatrix}$ then k = **[R.E.C. Trichy-1994]**

a) 0 b) –2 c) 2 d) 1

24. The value of x in $\begin{bmatrix} x & 3 \\ 2 & -3 \end{bmatrix}\begin{bmatrix} 6 \\ 3 \end{bmatrix} = \begin{bmatrix} -3 \\ 3 \end{bmatrix}$ must be **[R.E.C. Trichy-1995]**

a) 2 b) -3 c) 3 d) -2

25. If A is a n×n matrix and If Rank (A)=n then A is **[R.E.C. Trichy-1997]**

a) Singular b) Non-singular
c) Not inversible d) None

26. If $\begin{vmatrix} 2+x & 3 & -4 \\ 2 & 3+x & -4 \\ 2 & 3 & -4+x \end{vmatrix}$ = 0 then the values of x are **[R.E.C. Trichy-1997]**

a) 0,0,1 b) 0,0,2 c) 0,0,-1 d) 0,0,-2

27. If the eigen values of a square matrix are all equal to unity, then **[R.E.C. Trichy-1997]**

a) Inverse exists and is identity matrix
b) Inverse exists and is diagonal matrix
c) Inverse does not exists
d) None

28. The matrix $\begin{bmatrix} \cos\theta & \sin\theta \\ -\sin\theta & \cos\theta \end{bmatrix}$ is a **[R.E.C. Trichy-1997]**

a) Diagonal matrix b) Nilpotent matrix
c) Scalar matrix d) Idempotent matrix

ANSWER KEY - I

1. c	2. b	3. a	4. a	5. d
6. c	7. a	8. c	9. b	10. a
11. c	12. c	13. b	14. c	15. a
16. d	17. c	18. b	19. c	20. a
21. a	22. c	23. c	24. d	25. b
26. b	27. a	28. b		

EXPLANATORY NOTES - I

1. (c) Explanation: Order of the given matrix A is 2 × 2

det A = 0 So the rank P < 2 ----------(1)

A is not a null ; so rank $\neq 0$-------(2)

from (1) & (2) $P = 1$

2. (b) Explanation: The determinant of the given matrix

$$= 6\begin{vmatrix} 2 & 4 & 6 \\ 0 & 4 & 8 \\ 0 & 0 & -1 \end{vmatrix} + 8\begin{vmatrix} 0 & 4 & 6 \\ 0 & 4 & 8 \\ 0 & 0 & -1 \end{vmatrix} - 1\begin{vmatrix} 0 & 2 & 6 \\ 0 & 0 & 8 \\ 0 & 0 & -1 \end{vmatrix} + 1\begin{vmatrix} 0 & 2 & 4 \\ 0 & 0 & 4 \\ 0 & 0 & 0 \end{vmatrix},$$

$= 6[2(-4) - 4(0) + 6(0) + 8(0) - 1(0) + 1(0)$

$= 6(-8 - 0 + 0)$

$= -48$

3. Explanation:

$$\begin{bmatrix} \cos\theta & -\sin\theta \\ \sin\theta & \cos\theta \end{bmatrix}\begin{bmatrix} a & 0 \\ 0 & b \end{bmatrix}$$

$$= \begin{bmatrix} a\cos\theta & -b\sin\theta \\ a\sin\theta & b\cos\theta \end{bmatrix}$$

If $a = b$ or $\theta = n\pi$, n is integer

6. (c) Explanation:

det $A = 0(3-5) - 0(9-8) - (9-9) = 0$

$\rho(A) < 3$------(1)

minor of A_{11} is $\begin{vmatrix} 3 & 5 \\ 1 & 1 \end{vmatrix} = 3 - 5 = -2 \neq 0$

$\rho(A) \geq 2$-------(2)

from (1) & (2) p(A) = 2

7. (a) **Explanation :**

det A = $1(x^2-1)-0(0)+x(0)=x^2-1$

det B = $x(-x)-1(+x)+0(x^2)=x^2-x$

$$\therefore x^2-1=-x^2-x\ ;\ 2x^2+x-1=0$$

$$2x^2+2x-x-1=0\ ;\ 2x(x+1)-1(x+1)=0$$

$$(x+1)\ (2x+1)=0;\ x=-1,\ 1/2$$

8. (c) **Explanation :** $\begin{vmatrix} 2-\lambda & 3 \\ 3 & 2-\lambda \end{vmatrix}=(2-\lambda)^2-9=0$

$$\lambda^2+4-4\lambda-9=0 \Rightarrow \lambda^2-4\lambda-5=0$$

$$\Rightarrow \lambda^2-5\lambda+\lambda-5=0 \Rightarrow \lambda(\lambda-5)+1(\lambda-5)=0$$

$$\lambda=5,-1$$

10. (a) **Explanation :** A is skew symetric matrix

$\therefore A^T=-A \Rightarrow$ Let A = $\begin{bmatrix} 0 & 5 \\ -5 & 0 \end{bmatrix}$ is skew symetric matrix

$$-A=\begin{bmatrix} 0 & -5 \\ 5 & 0 \end{bmatrix},\quad A^T=\begin{bmatrix} 0 & -5 \\ 5 & 0 \end{bmatrix}\ \therefore A^T=-A$$

11. (c) **Explanation :** Matrix is singular $\therefore$ det A =0

$$3(K+9)+1(2K-3)+2(-6-1)=0$$

$$3K+27+2K-3-12-2=0 \Rightarrow 5K+10=0$$

$$\Rightarrow 5K=-10 \Rightarrow K=-10/5 \Rightarrow K=-2$$

12. (c) **Explanation :**

$$A^2=\begin{bmatrix} a & 1 \\ 0 & a \end{bmatrix}\begin{bmatrix} a & 1 \\ 0 & a \end{bmatrix}=\begin{bmatrix} a^2 & 2a \\ 0 & a^2 \end{bmatrix}$$

$$A^3=\begin{bmatrix} a^2 & 2a \\ 0 & a^2 \end{bmatrix}\begin{bmatrix} a & 1 \\ 0 & a \end{bmatrix}=\begin{bmatrix} a^3 & 3a^2 \\ 0 & a^3 \end{bmatrix}$$

$$A^n=\begin{bmatrix} a^{n-1} & n-1a^{n-2} \\ 0 & a^{n-1} \end{bmatrix}\begin{bmatrix} a & 1 \\ 0 & a \end{bmatrix}=\begin{bmatrix} a^n & na^{n-1} \\ 0 & a^n \end{bmatrix}$$

14. (c) **Explanation :**

a. 2×2 unit matrix is $\begin{bmatrix} 1 & 0 \\ 0 & 1 \end{bmatrix}$

b. 2×2 diagonal matrix is $\begin{bmatrix} x & 0 \\ 0 & x \end{bmatrix}$

c. If m × n is square matrix then m=n

d. 2×2 skew symmetric matrix is $\begin{bmatrix} 0 & -5 \\ 5 & 0 \end{bmatrix}$

Then given matrix 3×3. So this is square matrix

16. (d) **Explanation :** Let A = $\begin{bmatrix} -1 & 2 & 0 \\ 3 & 7 & 1 \\ 5 & 9 & 3 \end{bmatrix}$

$$|A|=-1(21-9)-2(9-5)+0=-20$$

$$R_1 \leftrightarrow R_2\ \therefore A=\begin{bmatrix} 3 & 7 & 1 \\ -1 & 2 & 0 \\ 5 & 9 & 3 \end{bmatrix}$$

$|A|$ = 3(6)-7(-3)+1(-9-10) = -22

a. So not a same   b. Becomes not a zero

c. Not or dechange

17. (c) **Explanation :** Matrix has no inverse

$\therefore$ det A = 0 $\quad 9x-3x+18=0$

$$6x=-18$$

$$x=-3$$

23. (c) **Explanation :** Corresponding elements are equal

K+3=5 $\Rightarrow$ K = 2

24. (d) **Explanation :**

$$6x+9=-3 \Rightarrow 6x=-12 \Rightarrow x=-2$$

26. (c) **Explanation :**

$$2+x\left[(3+x)(x-4)+12\right]-3\left[2(x-4)+8\right]-4(6-6-x)$$

$$2+x\left[(3-12+x^2-4x+12\right]-3\left[2x-8+8\right]+4(2x)$$

$$6x+2x^2+x^3+3x^2-6x+8x-8x-4x^2$$

$$x^3+x^2=0 \Rightarrow x^2(x+1)=0 \Rightarrow x=0,0,-1$$

28. (d) **Explanation :**

a. Example of diagonal matrix $\begin{bmatrix} x & 0 & 0 \\ 0 & y & 0 \\ 0 & 0 & z \end{bmatrix}$

b. Example of nil potent matrix is A . A =0

$$A^2=\begin{bmatrix} 1 & -3 & -4 \\ -1 & 3 & 4 \\ 1 & -3 & -4 \end{bmatrix}\begin{bmatrix} 1 & -3 & -4 \\ -1 & 3 & 4 \\ 1 & -3 & -4 \end{bmatrix}\begin{bmatrix} 0 & 0 & 0 \\ 0 & 0 & 0 \\ 0 & 0 & 0 \end{bmatrix}$$

c. example of the scalar matrix is $\begin{bmatrix} 2 & 0 & 0 \\ 0 & 2 & 0 \\ 0 & 0 & 2 \end{bmatrix}$

So answer is none.

OBJECTIVE QUESTIONS - II

1. Consider the following statements:

S1: The sum of two singular n × n matrices may be non-singular

S2: The sum of two non-singular n × n matrices may be singular which ofthe following statements is correct?**[GATE-2000]**

a) S_1 and S_2 are both true b) S_1 is true, S_2 is false
c) S_1 is false, S_2 is true d) S_1 and S_2 are both false

2. The determinant of the matrix $\begin{bmatrix} 2 & 0 & 0 & 0 \\ 8 & 1 & 7 & 2 \\ 2 & 0 & 2 & 0 \\ 9 & 0 & 6 & 1 \end{bmatrix}$ is **[GATE-2000]**

a) 4 b) 0 c) 15 d) 20

3. The rank of the matrix $\begin{bmatrix} 1 & 4 & 8 & 7 \\ 0 & 0 & 3 & 0 \\ 4 & 2 & 3 & 1 \\ 3 & 12 & 24 & 2 \end{bmatrix}$ is**[GATE-1998]**

a) 3 b) 1 c) 2 d) 4

4. Consider the following determinant $\Delta = \begin{vmatrix} 1 & a & bc \\ 1 & b & ca \\ 1 & c & ba \end{vmatrix}$

Which of the following is a factor of Δ ? **[GATE-1998]**
a) a + b b) a – b c) a + b + c d) abc

5. Let A =(a_{ij}) be an n-rowed square matrix and I_{12} be the matrix obtained by interchanging the first and second rows of the n-rowed identity matrix. Then AI_{12} is such that its first **[GATE-1997]**
a) Row is the same as its second row
b) Row is the same as the second row of A
c) Column is the same as the second column of A
d) Row is all zero

6. Let Ax =b be a system of linear equations where A is an m×n matrix and b is a m×1 column vector and x is a n×1 column vector of unknowns. Which of the following is FALSE? **[GATE-1996]**
a) The system has a solution if and only if, both A and the augmented matrix [A b] have the same rank.
b) If m < n and b is the zero vector , then the system has infinitely many solutions
c) If m = n and b is a non zero vector, then the system has a unique solution.
d) The system will have only a trivial solution when m = n, b is the zero vector and rank(A)=n.

7. The eigen vector(s) of the matrix $\begin{bmatrix} 0 & 0 & \alpha \\ 0 & 0 & 0 \\ 0 & 0 & 0 \end{bmatrix}$, $\alpha \neq 0$ is(are): **[GATE-1993]**
a) $(0, 0, \alpha)$ b) $(\alpha, 0, 0)$ c) (0, 0, 1) d) $(0, \alpha, 0)$

8. If $A = \begin{bmatrix} 3 & -4 \\ 1 & -1 \end{bmatrix}$ then A^n = **[GATE-1991]**

a) $\begin{bmatrix} 3n & -4n \\ n & -n \end{bmatrix}$ b) $\begin{bmatrix} 2+n & 5-n \\ n & -n \end{bmatrix}$

c) $\begin{bmatrix} 3^n & (-4)^n \\ 1^n & (-1)^n \end{bmatrix}$ d) $\begin{bmatrix} 1+2n & -4n \\ n & 1-2n \end{bmatrix}$

9. The characterstic equation of the following matrix is $\begin{bmatrix} 2 & -1 & 1 \\ -1 & 2 & 1 \\ 1 & -1 & 2 \end{bmatrix}$ **[Hyd.Univ-1992]**
a) $8x^3 + 6x^2 - 9x + 81 = 0$ b) $-x^3 + 6x^2 - 9x + 4 = 0$
c) $x^3 + 6x^2 + 9x + 4 = 0$ d) $8x^3 - 6x^2 + 9x - 81 = 0$

10. P, Q, are two unimodular matrices. i.e., |P|=1, |Q|=1. If A is adjoint of B then which of the following are TRUE? **[Hyd. Univ-1992]**

I. PA is adjoint of $P^{-1}B$ II. BQ is adjoint of $Q^{-1}B$
III.PAQ is adjoint of $Q^{-1}BP^{-1}$
a) III only b) I, II and III c) I, II and not III d) None.

11. If U= $\begin{bmatrix} 0 & 1 & 0 & 0 \\ 0 & 0 & 1 & 0 \\ 0 & 0 & 0 & 1 \\ 0 & 0 & 0 & 0 \end{bmatrix}$ The rank of U^4 is **[Hyd.Univ-1992]**

a) 1 b) 2 c) 3 d) 4

12. The simultaneous equations **[Hyd.Univ-1992]**
2x + 6y + 11 = 0, 6x + 20y – 6z + 3 =0, 6y – 18z + 1 =0 have
a) A unique solution b) No solution
c) An infinite number of solutions
d) Consistent and finite number of solutions

13. If a, b, c are all different from zero and $\begin{vmatrix} 1+a & 1 & 1 \\ 1 & 1+b & 1 \\ 1 & 1 & 1+c \end{vmatrix} = 0$

then the value of 1 + 1/a + 1/b + 1/c is given by**[Hyd.Univ-1991]**
a) abc b) $\frac{1}{abc}$ c) $\frac{1}{a}+\frac{1}{b}+\frac{1}{c}$ d) ab + bc + ac

14. The determinant $\begin{vmatrix} a & b & ax+b \\ b & c & bx+c \\ ax+b & bx+c & 0 \end{vmatrix}$ is equal to zero if **[Hyd.Univ-1991]**
a) a, b, c are in arithmetic progression
b) a, b, c are in geometric progression
c) a, b, c are in harmonic progression
d) x is not a root of the equation $ax^2 + 2bx + c = 0$

15. Which of the following statements are TRUE concerning the solution of m simultaneous linear equations in n unknowns written in the form of Ax =b? **[Hyd.Univ-1990]**
I. If there are infinitely many solutions, then n>m
II. If the solution is unique, then n =m
III.If n >m, then there can not be a unique solution
IV.If b =0, then there is always atleast one solution
a) I and III only b) I and II only
c) III and IV only d) All

16. Which of the following statements are FALSE concerning the properties of deteriminants? **[Hyd.Univ-1990]**

I. The interchange of rows and columns multiplies the value of the determinant by –1

II. If each element is multiplied by a factor k, the value of the determinant is also multiplied by k.

III. If a determinant has two columns or two rows that are identical or are multiplies of each other, its value is zero.

a) I and II only b) III only c) II and III only d) II only

17. $\begin{vmatrix} 2 & 2\omega & 2\omega^2 \\ \omega & 1 & \omega^2 \\ \omega^2 & 1 & \omega \end{vmatrix}$ has the value : **[Hyd.Univ-1989]**

a) 1 b) 2 c) 0 d) $1+\omega+\omega^3=0$

18. If A, B, C etc are n × n square matrices. All of which are non- singular. Then the inverse of the product (A,B,C,....) is: **[Hyd.Univ-1989]**

a) Is the product of the inverses in the same order

b) Is the product of the inverses in reverse order

c) Is either (a) or (b) d) Can not be determined

19. The eigen values of the matrix $\begin{bmatrix} 4 & -3 & -2 \\ 2 & -1 & -1 \\ 3 & -3 & -1 \end{bmatrix}$ are : **[Hyd.Univ-1994]**

a) 1, -1, 2 b) 1, 1, 2 c) 1, 2, -2 d) 1, -1, -2

20. The system of equations

x + y + z = 3, x + 2y + 3z = 4, 2x + 3y + 4z = 7 has **[R.E.C. Trichy-1992]**

a) A unique solution

b) Three solutions with exactly one of x, y, z equal to zero in each solution

c) Infinite number of solutions d) No solution.

21. For what value of k the rank of the matrix $\begin{bmatrix} -1 & k & 1 \\ 1 & 1 & k \end{bmatrix}$ is equal to 1? **[R.E.C. Trichy-1992]**

a) –1 b) 1 c) –2 d) 2

22. If A ≠ 0 and B ≠ 0 are nxn matrices such that AB = 0 then **[R.E.C. Trichy-1994]**

a) rank(A) = rank(B) = 0

b) rank of one of A, B is equal to n and the rank of the other is less than n.

c) ranks of both A, B are equal to n

d) ranks of both A, B are less than n

23. If $f(\theta)=\begin{bmatrix} \cos\theta & \sin\theta \\ -\sin\theta & \cos\theta \end{bmatrix}$ then $f(\alpha)f(\beta)=$ **[R.E.C. Trichy-1994]**

a) $f(\alpha\beta)$ b) $f(\alpha+\beta)$ c) $f(\alpha-\beta)$ d) 2×2 zero matrix

24. If $A=\begin{bmatrix} 1 & 1 \\ 0 & 1 \end{bmatrix}$ and n is any positive integer then A^n is **[R.E.C. Trichy-1995]**

a) $\begin{bmatrix} n & n \\ 0 & 1 \end{bmatrix}$ b) $\begin{bmatrix} n & 1 \\ 0 & n \end{bmatrix}$ c) $\begin{bmatrix} 1 & n \\ 0 & 1 \end{bmatrix}$ d) $\begin{bmatrix} n & n \\ 0 & n \end{bmatrix}$

25. If two matrices A and B are such that $A+B=\begin{bmatrix} 1 & -2 \\ 2 & 4 \end{bmatrix}$ and $A-B=\begin{bmatrix} 3 & 2 \\ -1 & 0 \end{bmatrix}$ Then A **[R.E.C. Trichy-1995]**

a) $\begin{bmatrix} 1 & -4 \\ 3 & 2 \end{bmatrix}$ b) $\begin{bmatrix} -1 & 4 \\ 3 & 2 \end{bmatrix}$ c) $\begin{bmatrix} 1 & 4 \\ -3 & 2 \end{bmatrix}$ d) None

26. If $A=\begin{bmatrix} 3 & 2 \\ 6 & -5 \end{bmatrix}$ then the equation to determine eigen values is **[R.E.C. Trichy-1995]**

a) $\lambda^2+2\lambda-27=0$ b) $\lambda^2+4\lambda-12=0$

c) $\lambda^2+27\lambda=0$ d) None

27. Let P and Q be two 2×2 matrices, consider the following statements:

I. PQ = 0 implies P=0 or Q=0 or both

II. PQ = I^2 implies $P=Q^{-1}$

III. $(P-Q)^2=P^2-2PQ+Q^2$ **[R.E.C. Trichy-1997]**

a) All are false b) I and III are false,II is true

c) II and III are false, I is true d) None

28. If the rank of the matrix, $A=\begin{bmatrix} -1 & \lambda & 1 \\ 1 & 1 & \lambda \end{bmatrix}$ is 1, then the value of λ is **[R.E.C. Trichy-1997]**

a) 1 b) -1 c) ± 1 d) None

29. If $A=\begin{pmatrix} -3 & 0 & 0 & 0 \\ 6 & 2 & 0 & 0 \\ -1 & 2 & 1 & 0 \\ 8 & 10 & -11 & -1 \end{pmatrix}$ then $|3A|$ **[Hyd. Univ - 2001]**

a) 18 b) 6 c) 0 d) 466

30. Suppose A and B are two n-square symmetric matrices. The memory space required for efficient storage of A and B is ? **[Hyd. Univ - 2000]**

a) n^4 units b) n^2 units

c) $(n+1)\times(n+1)$ units d) $n\times(n+1)$ units

31. What is det (2A) if $A=\begin{pmatrix} 2 & 0 & 0 & 0 & 0 \\ 10 & 3 & 0 & 0 & 0 \\ -19 & 8 & -1 & 0 & 0 \\ 6 & -\frac{1}{5} & -9 & \frac{1}{2} & 0 \\ \frac{1}{6} & \frac{1}{8} & -2 & 14 & -\frac{1}{3} \end{pmatrix}$ **[Hyd. Univ - 2000]**

a) 32 b) $20\frac{3}{8}$ c) 1 d) 2

32. What is $3A^5+2A^4$ if $A=\begin{pmatrix} 3 & -2 \\ 4 & -3 \end{pmatrix}$ **[Hyd. Univ - 2000]**

a) $\begin{pmatrix} 11 & -10 \\ 20 & -15 \end{pmatrix}$ b) $\begin{pmatrix} 11 & 0 \\ 0 & -15 \end{pmatrix}$ c) $\begin{pmatrix} 11 & -6 \\ 12 & -7 \end{pmatrix}$ d) $\begin{pmatrix} 3 & -2 \\ 4 & -3 \end{pmatrix}$

33. $\begin{vmatrix} 1 & 1 & 1 \\ a^2 & b^2 & c^2 \\ a^3 & b^3 & c^3 \end{vmatrix}=$ **[Hyd. Univ - 1999]**

a) $(b-a)(c-b)(a-c)(a^2+b^2+c^2)$

b) $(a-b)(b-c)(c-a)(bc+ca+ab)$

c) $(a-b)(b-c)(c-a)(b-a)(c-b)(a-c)$

d) $2(a-b)(b-c)(c-a)(bc+ca+ab)$

34. Find the nth power of the square matrix M defined by $M=\begin{pmatrix}x\\y\\z\end{pmatrix}(a\ b\ c)$ **[Hyd. Univ - 1999]**

a) $\sum_{i=1}^{n}\binom{n}{i}(ax+by+cz)^{n-i}$ b) $(ax+by+cz)^{n(n-1)/2}$

c) $M^{n-1}K$ where $K = ax + by + cz$

d) $K^{n-1}M$ where $K = ax + by + cz$

35. The polynomial of minimal degree satisfied by $\begin{pmatrix}1&1&0&0&0\\0&1&1&0&0\\0&0&1&0&0\\0&0&0&1&1\\0&0&0&0&1\end{pmatrix}$ Is **[Hyd. Univ - 1998]**

a) x^3-1 b) $(x-1)^3$ c) x^2-1 d) x^2+1

36. If x,y and z are distinct and $\begin{vmatrix}1+x^3&x&x^2\\1+y^3&y&y^2\\1+z^3&z&z^2\end{vmatrix}=0$, then **[Hyd. Univ - 1998]**

a) xyz = 0 b) 1 - xyz = 0 c) 1 + xyz = 0 d) x + y + z = 0

37. For positive numbers x,y,z the numerical value of the determinant $\begin{vmatrix}1&\log_x y&\log_x z\\\log_y x&1&\log_y z\\\log_z x&\log_z y&1\end{vmatrix}$ is given by : **[Hyd. Univ - 1997]**

a) log x log y log z b) 0

c) (1 / log x) (1 / log y) (1/ log z) d) 1

38. The nullity of the following matrix is

$$\begin{pmatrix}2&-4&3&1&0\\1&-2&1&-4&2\\0&1&-1&3&1\\4&-7&4&-4&5\end{pmatrix}$$ **[Hyd. Univ - 1997]**

a) 0 b) 1 c) 2 d) 3

39. A matrix is said to be idempotent if $A^2 = A$. The eigen values of an indempotent matrix are **[Hyd. Univ - 1996]**

a) 0 and -1 only b) 0 only

c) 0 and 1 only d) -1 and 1 only

40. Given three matrices A, B and C as follows

$$A=\begin{pmatrix}1&1&-1\\2&-3&4\\3&-2&3\end{pmatrix}\ B=\begin{pmatrix}-1&-2&-1\\6&12&6\\3&10&5\end{pmatrix}\ C=\begin{pmatrix}-1&-1&1\\2&2&-2\\-3&-3&3\end{pmatrix}$$

which of the following is not TRUE ? **[Hyd. Univ - 1996]**

a) $AB\neq 0$ b) BA = 0 c) $AC\neq 0$ d) $CA\neq 0$

41. Consider the following determinants

$$\alpha=\begin{vmatrix}a+b&b+c&c+a\\b+c&c+a&a+b\\c+a&a+b&b+c\end{vmatrix}\quad \beta=\begin{vmatrix}a&b&c\\b&c&a\\c&a&b\end{vmatrix}$$

which of these is correct? **[Hyd. Univ - 1995]**

a) $\alpha=2\beta$ b) $\beta=2\alpha$ c) $\alpha=\beta$ d) $\alpha\neq\beta$

42. Which of the following is true for matrices ? For every 2 × 2 matrix A there exists a 2 × 1 matrix B and a 1 × 2 matrix C such that **[Hyd. Univ - 1995]**

a) A = BC b) B = AC c) A = B d) None of the above

43. The solution of the equation $\begin{vmatrix}1&4&20\\1&-2&5\\1&2x&5x^2\end{vmatrix}=0$ is **[Hyd. Univ - 1995]**

a) {2,0} b) {0,0} c) {1,0} d) {-1, 2}

44. If the matrix A is given as below what is the A^{101} ?

$$A=\begin{pmatrix}1&-1&2\\1&0&-1\\0&1&-1\end{pmatrix}$$ **[Hyd. Univ - 1995]**

a) A b) A^{-1} c) Indentity d) None of the above

45. If A is $\begin{pmatrix}1&1&0&0&0\\0&0&1&0&0\\0&0&1&0&0\\0&0&0&0&1\\0&0&0&1&0\end{pmatrix}$ then 4th row 2nd column of A^{25} is **[Hyd. Univ - 1994]**

a) 23 b) 6 c) 3 d) 0

46. Let A be an n × n matrix. If all the elements of a row are multiplied by 3 and added to the corresponding elements in another row, the ratio of the determinants of the original and new matrices are **[Hyd. Univ - 1994]**

a) 1/9 b) 1/3 c) 9 d) 3

47. Let A be an n × n skew symmetric matrix. Then $A + A^T$ =

a) 0 b) 2A

c) 2D where D is diagonal matrix formed out of the diagonal elements of A d) -2D

48. The rank of the matrix $\begin{pmatrix}7&6&5\\1&2&1\\3&-2&1\end{pmatrix}$ Is **[Hyd. Univ - 1994]**

a) 3 b) 2 c) 1 d) 0

49. The Eigen values of $\begin{pmatrix}0&0&1\\0&1&0\\1&0&0\end{pmatrix}$ are **[Hyd. Univ - 1994]**

a) 0,-1, 1 b) 0,0,-1 c) 0, 0, 1 d) 1, 1, -1

50. The rank of a 3 × 5matrix in which one row is a constant multiple of the other is less than or equal to **[Hyd. Univ - 1994]**

a) 3 b) 5 c) 2 d) 4

51. The value of $\begin{vmatrix} 1 & b & c+a \\ 1 & a & b+c \\ 1 & c & a+b \end{vmatrix}$ Is **[Hyd. Univ - 1994]**

a) 1 b) 0

c) a + b+ c + ab + bc+ ca

d) a +b + c + ab + bc+ ca + $2a^2 + 2b^2 + 2c^2$

52. Which of the following statements about a matrix A is not TRUE ? **[Hyd. Univ - 1993]**

a) If a row consists of zeroes, then det A = 0

b) det A^T = det A

c) If two rows are identical then det A= 0

d) If A is triangular then det A = $a_{11} a_{22} ..... a_{nn}$

53. The eigan values for the following matrix are

$\begin{pmatrix} 2 & -1 & 0 \\ -1 & 2 & -1 \\ 0 & -1 & 2 \end{pmatrix}$ **[Hyd. Univ - 1993]**

a) $2, 2\pm\sqrt{2}$ b) $-2, 2\pm\sqrt{2}$ c) $2, -2\pm\sqrt{2}$ d) $-2, -2\pm\sqrt{2}$

ANSWER KEY - II

1. a	**2.** a	**3.** a	**4.** b	**5.** c
6. c	**7.** b,d	**8.** ?	**9.** b	**10.** a
11. a	**12.** b	**13.** a	**14.** b	**15.** a
16. d	**17.** c	**18.** b	**19.** a	**20.** c
21. a	**22.** d	**23.** b	**24.** c	**25.** d
26. a	**27.** b	**28.** b	**29.** a	**30.** d
31. d	**32.** c	**33.** b	**34.** d	**35.** b
36. c	**37.** b	**38.** b	**39.** c	**40.** b
41. a	**42.** d	**43.** d	**44.** d	**45.** d
46. b	**47.** a	**48.** b	**49.** d	**50.** c
51. b	**52.** a	**53.** a		

EXPLANATORY NOTES - II

1. (a) **Explanation** : S_1 : Sum of 2 singular matrices are

$$\begin{bmatrix} 6 & 4 \\ 6 & 4 \end{bmatrix} + \begin{bmatrix} -3 & -7 \\ 3 & 7 \end{bmatrix} = \begin{bmatrix} 3 & -3 \\ 9 & 11 \end{bmatrix} = 33 + 27 = 60 \neq 0$$

is non singular. S_2: sum of 2 non singular matrices

$$\begin{bmatrix} 4 & 11 \\ 2 & 7 \end{bmatrix} + \begin{bmatrix} 6 & 9 \\ 4 & 5 \end{bmatrix} = \begin{bmatrix} 10 & 20 \\ 6 & 12 \end{bmatrix} = 120 - 120 = 0$$

is singular.

2. (a) **Explanation** : $2\begin{vmatrix} 1 & 7 & 2 \\ 0 & 2 & 0 \\ 0 & 6 & 1 \end{vmatrix} + 0 + 0 + 0$

$2[(2) - 7(0) + 2(0)] = 4$

3. (d) **Explanation** :

$$R_1 \leftrightarrow R_2 \begin{vmatrix} 0 & 0 & 3 & 0 \\ 1 & 4 & 8 & 7 \\ 4 & 2 & 3 & 1 \\ 3 & 12 & 24 & 2 \end{vmatrix} = -3 \begin{vmatrix} 1 & 4 & 7 \\ 4 & 2 & 1 \\ 3 & 12 & 2 \end{vmatrix}$$

$= -3[1(4-12) - 4(8-3) + 7(48-6)]$

$= -3[-8 - 4(5) + 7(42)] = -3[266] \neq 0$ rank =4.

4. (b) **Explanation** :

$1(b^2 a - c^2 a) - a(ba - ca) + bc(c-b)$

$a(b^2 - c^2) - a^2(c-b) + bc(c-b)$

$c - b\left[a^2 - a(b+c) + bc\right]$

$c - b\left[a(a-b) + c(a-b)\right]$

$(c-b)(a-b)(a+c)$

5. (c) **Explanation** : Let n = 2 ; Let A = $\begin{bmatrix} x & y \\ z & p \end{bmatrix}$

I be the identity matrix obtained by interchanging the first and second rows

So I = $\begin{bmatrix} 0 & 1 \\ 1 & 0 \end{bmatrix}$; $AI_{12} = \begin{bmatrix} x & y \\ z & p \end{bmatrix}\begin{bmatrix} 0 & 1 \\ 1 & 0 \end{bmatrix} = \begin{bmatrix} y & x \\ p & z \end{bmatrix}$

So Column is the same as second column of A.

6. (c) **Explanation** : A is an order of m×n

b is an order of m × 1, x is an order of n × 1

Ax = B is given, We know that

if m = n and b is a non zero then the system not has a unique solution.

9. (b) **Explanation** : $\begin{bmatrix} 2-x & -1 & 1 \\ -1 & 2-x & 1 \\ 1 & -1 & 2-x \end{bmatrix}$

$2 - x\left[(2-x)^2 + 1\right] + 1\left[x - 2 - 1\right] + 1\left[1 - 2 + x\right]$

$10 + 2\lambda^2 - 8\lambda - 5\lambda - \lambda^3 + 4\lambda^2 + 2\lambda - 4 = 0$

11. (c) **Explanation** : U is the identity matrix so U^4 is also identity matrix

$$U^4 = \begin{bmatrix} 0 & 1 & 0 & 0 \\ 0 & 0 & 1 & 0 \\ 0 & 0 & 0 & 1 \\ 0 & 0 & 0 & 0 \end{bmatrix}, |A| = 0, P(A) < 4$$

but a minor order of 3 of A is $\begin{vmatrix} 1 & 0 & 0 \\ 0 & 1 & 0 \\ 0 & 0 & 1 \end{vmatrix} \neq 0$

$\therefore P(A) = 3$

13. (b) **Explanation** :

$1 + a\left[(1+b)(1+c) - 1\right] - 1\left[(1+c) - 1\right] + 1\left[1 - 1 - b\right] = 1$

$1 + a\left[1 + b + c + bc - 1\right] - c - b = 1$

$b + c + bc + ab + ac + abc - c - b = 1$

$ab + bc + ac + abc = 1$

We can find out

$$1+1/a+1/b+\frac{1}{c}=\frac{abc+bc+ac+ab}{abc}=\frac{1}{abc}$$

14. (b) **Explanation :**

$$-b(bx+c)'+b(ax+b)(bx+c)+ax+b[b(bx+c)-c(ax+b)]$$

$$=-a(b^2x^2+c^2+2bcx)+b(abx^2+acx+b^2x+bc)$$

$$+b(abx^2+acx+b^2x+bc)-c(a^2x^2+abx+abx+b^2)=$$

$$-ab^2x^2-ac^2-2abcx+2ab^2x^2+2abcx+2b^3x$$

$$+2b^2c-ca^2x^2-abcx-abcx-b^2c=0$$

$$=ax^2(b^2-ac)-2bx(b^2-ac)+c(b^2-ac)=0$$

$$=(b^2-ac)(ax^2-2bx+c)=0$$

$$b^2-ac=0 \Rightarrow b^2=ac$$

$\therefore$ a,b,c are in geometric progression.

15. (a) **Explanation :** The solution of M simultaneous linear equations in n unknows written in the form of

AX = b, We know that if there are infinitely many solutions then n>m and if n>m then there can't be unique solution.

16. (d) **Explanation :** Let $A=\begin{bmatrix}1&2\\3&4\end{bmatrix} |A|=4-6=-2$

$$KA=\begin{bmatrix}K&2K\\3K&4K\end{bmatrix} |KA|=4K^2-6K^2=2K^2$$

The value of determent is also multiplied by K

$2K^3\neq-2$ So this is false.

17. (c) **Explanation :**

$$2(w-w^2)-2w(w^2-w^4)+2w^2(w-w^2)$$

$$2w-2w^2-2w^3+2w^5+2w^3-2w^4$$

$$2\omega-2\omega^2-2+2\omega^2+2-2\omega=0$$

19. (a) **Explanation :**

$$\begin{vmatrix}4-\lambda&-3&-2\\2&-1-\lambda&-2\\3&-3&-1-\lambda\end{vmatrix}=0$$

$$4-\lambda\left[(-1-\lambda)^2-6\right]+3(-2-2\lambda+6)-2(-6+3+3\lambda)$$

$$4-\lambda(\lambda^2+2\lambda-5)+3(-2\lambda+4)-2(-3+3\lambda)$$

$$4\lambda^2+8\lambda-20-\lambda^3-2\lambda^2+5\lambda-6\lambda+12+6-6\lambda=0$$

$$-\lambda^3+2\lambda^2+\lambda-2=0$$

$$(\lambda-1)(\lambda+1)(\lambda-2)=0 \Rightarrow \lambda=1,-1,2$$

22. (d) **Explanation :**

Let $A=\begin{bmatrix}1&0\\0&0\end{bmatrix}$. $B=\begin{bmatrix}0&0\\0&1\end{bmatrix}$

The order of A,B are 2×2 $AB=\begin{bmatrix}0&0\\0&0\end{bmatrix}$

AB rank is 0 ; rank of both A,B are less than n

23. (b) **Explanation :**

$$f(\alpha)\,f(\beta)=\begin{bmatrix}\cos\alpha&\sin\alpha\\\sin\alpha&\cos\alpha\end{bmatrix}\begin{bmatrix}\cos\beta&\sin\beta\\\sin\beta&\cos\beta\end{bmatrix}$$

$$\begin{bmatrix}\cos\alpha\cos\beta+\sin\alpha\sin\beta&\cos\alpha\sin\beta+\sin\alpha\cos\beta\\\sin\alpha\cos\beta+\cos\alpha\sin\beta&\sin\alpha\sin\beta+\cos\alpha\cos\beta\end{bmatrix}$$

$$\begin{bmatrix}\cos(\alpha+\beta)&\sin(\beta+\alpha)\\\sin(\beta+\alpha)&\cos(\alpha+\beta)\end{bmatrix}=f(\alpha+\beta)$$

26. (a) **Explanation :** $\begin{vmatrix}3-\lambda&2\\6&5-\lambda\end{vmatrix}$

$$(3-\lambda)(5-\lambda)-12=0$$

$$-15-3\lambda+5\lambda^2-12=0$$

$$\lambda^2+2\lambda-27=0$$

$$\lambda^2+2\lambda-27=0$$

29. (a) **Explanation :**

$$|A|=-3\begin{vmatrix}2&0&0\\2&1&0\\10&-11&-1\end{vmatrix}$$

$$=-3[2(-1)]=6$$

$$|3A|=18$$

30. (a) **Explanation :**

A matrix order is n × n, B matrix order is n × n

So A,B matrix storage space is n^2, n^2 units.

The efficent storage of memory space of A is n^2 then completed the work A then the same space is used to B.

32. (c) **Explanation :**

$$A^2=\begin{bmatrix}3&-2\\4&-3\end{bmatrix}\begin{bmatrix}3&-2\\4&-3\end{bmatrix}=\begin{bmatrix}9-8&-6+6\\12-12&-8+9\end{bmatrix}=\begin{bmatrix}1&0\\0&1\end{bmatrix}=I$$

$$A^4=I.I=I$$

$$A^5=A^4.A=I.A=A=\begin{bmatrix}3&-2\\4&-3\end{bmatrix}$$

$$3A^5+2A^4=\begin{bmatrix}9&-6\\12&-9\end{bmatrix}+\begin{bmatrix}2&0\\0&2\end{bmatrix}=\begin{bmatrix}11&-6\\12&-7\end{bmatrix}$$

36. (c) **Explanation :**

$$=\begin{vmatrix}1 & x & x^2\\ 1 & Y & Y^2\\ 1 & Z & Z^2\end{vmatrix}+\begin{vmatrix}X^3 & X & X^2\\ Y^3 & Y & Y^2\\ Z^3 & Z & Z^2\end{vmatrix}=0$$

In the second determent c_1 is modified to c_3

$$=\begin{vmatrix}1 & x & x^2\\ 1 & y & y^2\\ 1 & z & z^2\end{vmatrix}+xyz\begin{vmatrix}1 & x & x^2\\ 1 & y & y^2\\ 1 & z & z^2\end{vmatrix}=0 \Rightarrow \quad 1+xyz=0$$

37. (b) **Explanation :**

$$=\log_y^z \log_z^y - 1 - \log_x^y\left(\log_y^x - \log_y^z \log_z^x\right)$$

$$+\log_x^z\left(\log_y^x - \log_z^x\right)$$

$$=1-1-1+\log_x^y \log_y^z \log_z^x + \log_x^z \log_y^x \log_z^y - \log_x^z \log_z^x$$

= 1-1-1+1+1-1 = 0

38. (b) **Explanation :**

Null space and nullity of a matrix

Thus for a matrix A

rank + row nullity = no. of rows

$\therefore$ row nullity = no. of rows - rank

no of rows in the matrix 4 and rank is 3

row nullity = 4-3 = 1

41. (a) **Explanation :**

$$\alpha=\begin{vmatrix}a+b & b+c & c+a\\ b+c & c+a & a+b\\ c+a & a+b & b+c\end{vmatrix}=\begin{vmatrix}a & b & c\\ b & c & a\\ c & a & b\end{vmatrix}+\begin{vmatrix}b & c & a\\ c & a & b\\ a & b & c\end{vmatrix}$$

In the second matrics c_1 will be transfer to c_3

$$=\begin{vmatrix}a & b & c\\ b & c & a\\ c & a & b\end{vmatrix}+\begin{vmatrix}a & b & c\\ b & c & a\\ c & a & b\end{vmatrix}=\beta+\beta=2\beta$$

44. (d) **Explanation :**

$$A^2=\begin{bmatrix}1 & -1 & 2\\ 1 & 0 & -1\\ 0 & 1 & -1\end{bmatrix}\begin{bmatrix}1 & -1 & 2\\ 1 & 0 & -1\\ 0 & 1 & -1\end{bmatrix}=\begin{bmatrix}1-1 & -1+2 & 2+1-2\\ 1 & -1-1 & 2+1\\ 1 & -1 & -1+1\end{bmatrix}$$

$$=\begin{bmatrix}0 & 1 & 1\\ 1 & -2 & 3\\ 1 & -1 & 0\end{bmatrix}$$

A^2 is not A, A^2 is not A^{-1}, A^2 is not identity.

So A^{101} is not A, A^{101} is nct A^{-1}, A^{101} is not identity.

45. (d) **Explanation :**

$$A^2=\begin{bmatrix}1 & 1 & 0 & 0 & 0\\ 0 & 0 & 1 & 0 & 0\\ 0 & 0 & 1 & 0 & 0\\ 0 & 0 & 0 & 0 & 1\\ 0 & 0 & 0 & 1 & 0\end{bmatrix}\begin{bmatrix}1 & 1 & 0 & 0 & 0\\ 0 & 0 & 1 & 0 & 0\\ 0 & 0 & 1 & 0 & 0\\ 0 & 0 & 0 & 0 & 1\\ 0 & 0 & 0 & 1 & 0\end{bmatrix}\begin{bmatrix}1 & 1 & 1 & 0 & 0\\ 0 & 0 & 1 & 0 & 0\\ 0 & 0 & 1 & 0 & 0\\ 0 & 0 & 0 & 1 & 0\\ 0 & 0 & 0 & 0 & 1\end{bmatrix}$$

Clearly we can't modify A^{25} in the location of 4th row and 2nd column.

46. (b) **Explanation :**

$$Let\ A=\begin{vmatrix}1 & 4 & 20\\ 1 & -2 & 5\\ 1 & 5 & 20\end{vmatrix}$$

det A = - 40 - 20 - 4 (20 - 5) + 20 (7) = - 65 - 60 + 140= 15

All the elements of second row of A are multiplied by 3 and added to the corresponding elements of R_1 then

$$\begin{vmatrix}1 & 4 & 20\\ 4 & -1 & 35\\ 1 & 5 & 20\end{vmatrix}$$

determinant = (-20 -175) -4 (80 -35) + 20 (21)

= -195 -180 + 420 = -375 + 420 = 45

The ratio of the 2 determinants = 15 : 45 = 1 : 3 = 1/3

49. (d) **Explanation :**

$$\begin{bmatrix}0 & 0 & 1\\ 0 & 1 & 0\\ 1 & 0 & 0\end{bmatrix}=\begin{vmatrix}-\lambda & 0 & 1\\ 0 & 1-\lambda & 0\\ 1 & 0 & -\lambda\end{vmatrix}$$

$$\lambda^2(1-\lambda)+1(1-\lambda)=0$$

$$(\lambda^2+1)(1-\lambda)=0 \Rightarrow \lambda=\pm 1, 1$$

51. (b) **Explanation :**

$$1(a^2+ab)-bc-c^2-b[a+b-b-c]+c+a(c-a)$$

$$a^2+ab-bc-c^2-ab-b^2+b^2+bc+c^2-ac+ac-a^2=0$$

53. (a) **Explanation :**

$$\begin{vmatrix}2-\lambda & -1 & 0\\ -1 & 2-\lambda & -1\\ 0 & -1 & 2-\lambda\end{vmatrix}=2-\lambda\left[(2-\lambda)^2-1\right]+1\left[-1(2-\lambda)\right]=0$$

$$= 2-\lambda\left[4+\lambda^2-4\lambda-1\right]+\lambda-2=0$$

$$=8+2\lambda^2-8\lambda-2-4\lambda=\lambda^3+4\lambda^2+\lambda+\lambda-2=0$$

$$=-\lambda^3+6\lambda^2-10\lambda+4=0$$

$$(\lambda-2)(\lambda^2-4\lambda+2)=0$$

$$\frac{+4\pm\sqrt{16-8}}{2}=2\pm 2$$

$$2,\ 2+\sqrt{2},\ 2-\sqrt{2}$$

VI. NUMERICAL METHODS

- **Solution of Non-Linear Algebraic Equations**
- **Solutions of Simultaneous Linear Algebraic Equations**
- **Numerical Integration**

1. Solution of Transcendental and Polynomial Equations in one variable : In this article we will discuss numerical solution of equations of the form f(x)=0, where f(x) is a function of x.

If f(x) is an algebraic function or a polynomial of degree $n(n \in N)$ then it is expressible in the form $f(x) = a_0x^n + a_1x^{n-1} + a_2x^{n-2} + ... a_{n-1}x + a_n \ (a_0 \neq 0)$ and $f(x) = 0$ is called an algebraic equation or polynomial equation.

If f(x) contains algebraic and non-algebraic functions such as exponential, logarithmic and trigonometric, then $f(x) = 0$ is called a transcendental equation. For example, $7x^3 + 2e^x + 5\sin x = 0$, $\tan x - x = 0$ etc. are transcendental equations.

Any value of x which makes f(x) equal to zero, is called a zero or a root of the equation f(x) = 0. Geometrically $x = \alpha$ is root of f(x)=0 if the graph of $y = f(x)$ crosses the x-axis at the point $x = \alpha$. Thus $x = \alpha$ is a root of the equation $f(x) = 0$ iff $f(\alpha) = 0$.

We now mention some important properties and theorems of algebraic (polynomial) and transcendental equations.

Theorem 1. (Fundamental Theorem of Algebra): Every algebraic equation has at least one root.

Theorem 2. If f(x) is a polynomial of degree n, then f(x)=0 has exactly n roots.

Note that a transcendental equation have either no root or one or more roots. For example, $x - e^x = 0$ has no root, $x - e^{x} = 0$ has a root lying between 0 and 1 and the equation $\sin x - 2\cos x = 0$ has infinitely many roots.

Theorem 3. In any algebraic equation f(x) = 0

(i) the number of positive roots cannot exceed the number of changes of sign of the coefficients in the polynomial f(x).

(ii) the number of negative roots cannot exceed the number of changes of sign of the coefficients in f(-x)

Theorem 4. If f(x) is continuous on [a, b] and f(a), f(b) are of opposite signs, then there is at least one number $\alpha \in (a,b)$ such that $f(\alpha) = 0$.

According to this theorem, if $f(x) = 0$ is a polynomial equation and if we have two real numbers a and b such that f(a) and f(b) are of opposite signs, then there is at least one root of the equation f(x) = 0 in (a, b).

Theorem 5. If f(x) is continuous on [a, b] and a, b are two real numbers such that

(i) f(a) and f(b) are of opposite signs, then there are odd number of real roots of the equation f(x) = 0 in (a, b).

(ii) f(a) and f(b) are of the same sign, then either there is no real root or there are even number of real roots of the equation f(x) = 0 in (a, b).

Theorem 6. Every polynomial equation of even degree whose last term is negative and the coefficient of its first term being positive, has at least two real roots one positive and other negative.

Theorem 7. Every polynomial equation of an odd degree must have at least one real root whose sign is opposite to that of its last term, the coefficient of first term being positive.

Theorem 8. (Largest and Smallest Roots)

Let $f(x) = a_0x^n + a_1x^{n-1} + a_2x^{n-2} + ... + a_{n-1}x + a_n$. Then the largest root of polynomial f(x) = 0 may be approximated by the root of the linear equation $a_0x + a_1 = 0$ or by the root of the quadratic equation $a_0x^2 + a_1x + a_2 = 0$, larger in absolute value.

Similarly the smallest root of a polynomial equation f (x) = 0 may be approximated by the root of the equation $a_{n-1}x + a_n = 0$ or by the root of the equation $a_{n-2}x^2 + a_{n-1}x + a_n = 0$ smaller in absolute value.

These approximations are valid only when the largest root of equation $f(x) = 0$ is much larger in absolute value than all the remaining roots.

Theorem 9. (Position of Real Roots) If $f(x) = 0$ be a polynomial equation and $x_1, x_2, ... x_k$ are the consecutive real roots of $f'(x) = 0$, then positive or negative sign of the values of $f(-\infty), f(x_1), ..., f(x_k), f(\infty)$ will determine the intervals in which the roots of f(x) =0 will lie. Whenever there is a change of sign from $f(x_r)$ to $f(x_{r+1})$ the root lies in the interval $[x_r, x_{r+1}]$.

Ex 1. Separate the real roots of the equaion $x^5 - 80x + 6 = 0$

Sol : Let $f(x) = x^5 - 80x + 6$. Then $f'(x) = 5x^4 - 80$.

$\therefore \quad f'(x) = 0 \Rightarrow x = 2, -2$.

Signs of f(x) at $f(x) = -\infty, -2, 2, \infty$ are :

x:	$-\infty$	-2	2	∞
f (x):	-	+	-	+

This shows that there are three real roots of f(x) = 0 which lie in the intervals $(-\infty, -2)$; $(-2, 2)$ and $(2, \infty)$

Ex 2 : If all roots of the equation $x^3 - 3x + k$ are real, then find the range of values of k.

Sol : Let $f(x) = x^3 - 3x + k$. Then $f'(x) = 3x^2 - 3$ and so $f'(x) = 0 \Rightarrow x = \pm 1$. The values of f(x) at $x = -\infty, -1, 1$ and ∞ are :

x:	$-\infty$	-1	1	∞
f (x):	$-\infty$	k + 2	k - 2	∞

If all roots of the given equation are real, then

$k + 2 > 0$ and $k - 2 < 0. \Rightarrow -2 < k < 2$

Hence the range of k is (-2, 2).

Ex 3 : Show that the equation $x^5 - 5x + 2 = 0$ has three real roots. Also find their intervals in which they lie.

Sol : Let $f(x) = x^5 - 5x + 2$. Then $f'(x) = 5(x^4 - 1)$

$\therefore f'(x) = 0 \Rightarrow x = \pm 1$.

Signs of f(x) at $x = -\infty, -1, 1, \infty$ are :

x :	$-\infty$	-1	1	∞
f (x) :	$-\infty$	+	-	+

This shows that there are three real roots of $f(x) = 0$ which lie in the interval $(-\infty, -1)$, $(-1, 1)$ and $(1, \infty)$.

2. Successive Bijection Method : This method is a based on Theorem 4. Let f(x) = 0 be an equation. Following are the steps involved to solve the equation $f(x) = 0$ by successive bisectioin method :

STEP I : Find an interval $[x_0, x_1]$ such that $f(x_0)$ and $f(x_1)$ are of opposite signs. Then by theorem 4 a root of the equation $f(x) = 0$ lies between x_0 and x_1. Let $f(x_0) < 0$ and $f(x_1) > 0$.

STEP II : Let $x_2 = \frac{x_0 + x_1}{2}$. Then $f(x_2)$ may be positive, negative or zero. If $f(x_2) = 0$, then x_2 is the required root. If $f(x_2) > 0$, then according to theorem 4, the root lies between x_0 and x_2. If $f(x_2) < 0$, then root lies between x_2 and x_1.

Let us assume that $f(x_2) > 0$. Denote this x_2 by x_1.

STEP III : We have $f(x_0) < 0$. Therefore root of the equation $f(x) = 0$ lies between x_0 and x_1.

Let $x_2 = \frac{x_0 + x_2}{2}$. Then $f(x_2)$ may be positive, negative or zero. If $f(x_2) = 0$. Then $x = x_2$ is the required root.

Otherwise we repeat the process given in step I till we obtain the root correct upto the desired level of accuracy.

REMARK 1. If we want to find the root (s) correct upto n places of decimal, then in intermediate steps we retain the digits upto n+2 places of decimals and in the final answer we retain the digits upto n places of decimal and chop off the other digits.

3. Regula Falsi Method (Method of False Position or Method of Interpolation) : Let $f(x) = 0$ be an equation and let $[x_0, x_1]$ be an interval such that f(x) is continuous on $[x_0, x_1]$ and $f(x_0)$ and $f(x_1)$ are of opposite signs i.e. $f(x_0) f(x_1) < 0$. Then by theorem 4, a root of α of the equation $f(x) = 0$ lies between x_0 and x_1 and so the graph of the function $y = f(x)$ must cross the x-axis between x_0 and x_1.

The basic idea in this method is that the graph of the functiion in the small intera! $[x_0, x_1]$ is considered as a straight line as in the case of interpolation and hence the x-coordinate of the point of intersection of the straight line joining $(x_1, f(x_1))$ and $(x_0, f(x_0))$ gives approximate value of the desired root. In Fig.1 it is obvious that $x = OP$ is the actual value of the root and $x = OQ$ is the first approximation of the desired root.

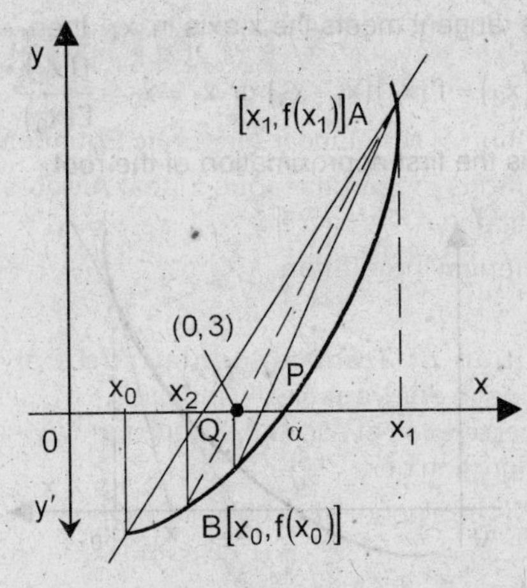

Figure. 1

Let the line through $A(x_1, f(x_1))$ and $B(x_0, f(x_0))$ intersects the x-axis in x_2. The equation of the line AB is

$$y - f(x_0) = \frac{f(x_1) - f(x_0)}{x_1 - x_0}(x - x_0)$$

This meets x-axis at a point given by

$$0 - f(x_0) = \frac{f(x_1) - f(x_0)}{x_1 - x_0}(x - x_0)$$

$$\Rightarrow \quad x = x_0 - \frac{f(x_0)}{f(x_1) - f(x_0)}(x_1 - x_0)$$

$$\text{Let } x_2 = x_0 - \frac{f(x_0)}{f(x_1) - f(x_0)}(x_1 - x_0) = \frac{x_0 f(x_1) - x_1 f(x_0)}{f(x_1) - f(x_0)} \quad ...(i)$$

If $f(x_2) = 0$, then $x = x_2$ is one of the roots. If $f(x_2) \neq 0$, then either $f(x_2) > 0$ or $f(x_2) < 0$.

If $f(x_2) > 0$, then a root lies between x_0 and x_2. If $f(x_2) < 0$, then a root lies between x_2 and x_1.

Let $f(x_2) < 0$, then the root lies in the interval $[x_2, x_1]$. Repeating the process of obtaining equation (i), we get the next approximation of the root as

$$x_3 = \frac{x_1 f(x_2) - x_2 f(x_1)}{x_2 - x_1} \quad ..... (ii)$$

Continuing this process we may write the iterative formula of Regula Falsi method as

$$x_{n+1} = \frac{x_{n-1} f(x_n) - x_n f(x_{n-1})}{x_n - x_{n-1}} \quad ..... (iii)$$

or

$$x_{n+1} = x_n - \frac{(x_n - x_{n-1}) f(x_n)}{f(x_n) - f(x_{n-1})} \quad ..... (iv)$$

provided that at each step $f(x_{n-1}) f(x_n) < 0$.

4. Newton - Raphson Method : Let $f(x) = 0$ be the equation whose solution is required. Let x_0 be a point near to a root of this equation. Consider the tangent to the curve $y = f(x)$ at $(x_0, f(x_0))$ (Fig. 2)

The equation of the tangent at x_0 is given by

$$y - f(x_0) = f'(x_0)(x - x_0)$$

If this tangent meets the x-axis in x_1, then

$$0 - f(x_0) = f'(x_0)(x_1 - x_0) \text{ or } x_1 = x_0 - \frac{f(x_0)}{f'(x_0)}$$

This is the first approximation of the root.

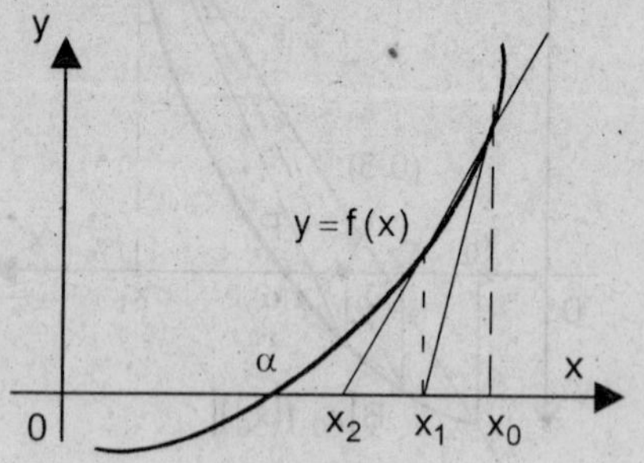

Figure. 2

If $f(x_1) = 0$, then $x = x_1$ is a root of the equation. Otherwise by taking x_1 as starting point the tangent at $P_1(x_1, f(x_1))$ gives $x_2 = x_1 - \frac{f(x_1)}{f'(x_1)}$

which is the second approximate value of the root. This process is continued further till we get a value for the root upto the desired level of accuracy i.e. till we get the answer to the specified number of decimal places.

$$x_{n+1} = x_n - \frac{f(x_n)}{f'(x_n)}, f'(x_n) \neq 0$$

Ex : By using Newton-Raphson's method, find the root of $x^4 - x - 10 = 0$ which is near to x = 2 correct to three places of decimal.

Sol : The approximate value of the root is given as x = 2.

Here $f(x) = x^4 - x - 10,\ f'(x) = 4x^3 - 1$

By the Newton-Raphson formula

$$x_{n+1} = x_n - \frac{f(x_n)}{f'(x_n)} = x_n - \frac{x_n^4 - x_n - 10}{4x_n^3 - 1} = \frac{3x_n^4 + 10}{4x_n^3 - 1}$$

Since $x_0 = 2$ (given), therefore the first approximation is

$$x_1 = \frac{3x_0^2 + 10}{4x_0^3 - 1} = \frac{3(2)^4 + 10}{4(2)^3 - 1} = 1.87096$$

Since, we have to find the value of root correct up to three places of decimal, so we take 5 digits after decimal.

The second approximation is

$$x_2 = \frac{3x_1^4 + 10}{4x_1^3 - 1} = \frac{3(1.87096)^4 + 10}{4(1.87096)^3 - 1} = 1.85689$$

The third approximation is

$$x_3 = \frac{3x_2^4 + 10}{4x_2^3 - 1} = \frac{3(1.85689)^4 + 10}{4(1.85689)^3 - 1} = 1.85643$$

Hence the required root is 1.856.

5. Rate of Convergence : Let α be the exact value of a root of the equation $f(x) = 0$ and let $x_1, x_2, x_3, \ldots x_n \ldots$ be successive approximations of α in first, second, third,....,nth,......iterations. Then a positive real number k is called the order of convergence of a method if

$$|x_{n+1} - \alpha| \le c\,|x_n - \alpha|^k, n \ge 0$$

and $C > 0$ is a finite constant.

or $\qquad \in_{n+1} \le C \in_n^k$

where $\in_n$ is the error in nth iteration,

In particular, if k = 1 and 0 < C < 1 then the convergence is said to be linear or first order. The constant C is called the rate of linear convergence. If k = 2, then the convergence is quadratic. The convergence is said to be rapid or slow according as 'c' is nearer to 0 or 1.

In Regular-Falsi method $\in_{n+1} = \left(\frac{f''(\alpha)}{2f'(\alpha)}\right)^{0.618} \in_n^{1.618}$

Therefore, the order of convergence of Regula-Falsi method is 1.618

In Newton-Raphson Method $\in_{n+1} = \left(\frac{1f''(\alpha)}{2f'(\alpha)}\right)\in_n^2$

Thus the condition for convergence is therefore $\frac{1}{2}\frac{f''(\alpha)}{f'(\alpha)} < 1$ and the order of convergence is quadratic which means that the number of correct decimal places at each iteration is almost doubled.

6. Solution of Simultaneous Linear Algebraic Equations: Gauss Elimination Method : This is an elementary elimination method in which we reduce the system of equations to an equivalent upper triangular system which can be easily solved by back substitution.

Consider the followin consistent system of n equation in n unknowns :

$$\left.\begin{aligned} a_{11}x_1 + a_{12}x_2 + a_{13}x_3 + \ldots\ldots + a_{1n}x_n &= b_1 \\ a_{21}x_1 + a_{22}x_2 + a_{23}x_3 + \ldots\ldots + a_{2n}x_n &= b_2 \\ \ldots \quad \ldots \quad \ldots \quad & \\ a_{n1}x_1 + a_{n2}x_2 + a_{n3}x_3 + \ldots\ldots + a_{nn}x_n &= b_n \end{aligned}\right\} \ldots\ldots (i)$$

Let us assume that $a_{11} \neq 0$, for if $a_{11} = 0$, the equations can be arranged in a suitable way such that $a_{11} \neq 0$.

We first multiply the first equation of (i) by $\frac{a_{21}}{a_{11}}$ and then we subtract it from the second equation. You can easily see that x_1 term vanishes in second equation of (i). In a similar manner, we remove x from third, fourth,........,nth equations of (i).

Now x_2 is eliminated from the third, fourth, ...nth equations by the same procedure. We continue this procedure till the system of equations takes the following form

$$\left.\begin{aligned} C_{11}x_2 + C_{12}x_2 + \ldots\ldots + C_{1n}x_n &= d_1 \\ C_{22}x_2 + \ldots\ldots + C_{2n}x_n &= d_2 \\ \ldots \quad \ldots \quad & \\ C_{nn}x_n &= d_n \end{aligned}\right\} \ldots\ldots (ii)$$

The last equation in (ii) gives the value of x_n. Substituting this value in (n - 1)th equation we get the value x_{n-1}. Continuing in this manner we can determine the values of $x_1, x_2, \ldots x_n$.

The following example illustrates the procedure.

Ex : Solve the following system of equations by Gauss elimination method.

$$x + 4y + 9z = 16$$
$$2x + y + z = 10$$
$$3x + 2y + 3z = 18$$

Sol : First, we eliminate x from second and third equations, using first equation. Adding (-2) times the first equation to second equation, (-3) times the first equation to third equation, we get the new system as

$$x + 4y + 9z = 16$$
$$-7y - 17z = -22$$
$$-10y - 24z = -30$$

Eliminating y from the third equation, using second equation, we get

$$x + 4y + 9z = 16$$
$$-7y - 17z = -22$$
$$\frac{2}{7}z = \frac{10}{7}$$

From the third equation, we get z = 5. Putting z = 5 in the second equation, we get y = -9. Substituting y and z in the first equation, we get x = 7. Thus, the solution is x = 7, y = -9, z = 5.

7. Gauss-Seidal Iterative Method : As we have seen, Gauss Elimination method is a direct method to solve a system of simultaneous linear equations. There are iterative or indirect methods also to solve a system of simultaneous linear equations. In these methods we start with an approximation to the true solution and by repeating the cycle of computations we obtain the desired level of accuracy. Thus in an iterative method the amount of computation depends on the accuracy required, while in a direct method the amount of computation is fixed.

The Gauss-Seidal Iterative Method is explained below :

Let
$$a_1x + b_1y + c_1z = d_1$$
$$a_2x + b_2y + c_2z = d_2$$
$$a_3x + b_3y + c_3z = d_3$$

be a consistent of system of equations having unique solution.

Let a_1, b_2, c_3 i.e. the leading diagonal elements (called pivotal elements) of the coefficient matrix are sufficiently large as compared to other coefficients. If not, interchange the equations in such a manner that the coefficient of x in first, coefficient of y in second and coefficient of z in the third equation is the largest.

The above system of equations can now be written as

$$x = \frac{1}{a_1}(d_1 - b_1y - c_1z) \quad \text{..... (i)}$$

$$y = \frac{1}{b_1}(d_2 - a_2x - c_2z) \quad \text{..... (ii)}$$

$$z = \frac{1}{c_1}(d_3 - a_3x - b_3y) \quad \text{..... (iii)}$$

Initialisation : We start with the initial solution y = 0, z = 0.

First Iteration : Putting y = 0, z = 0 in (i), we get $x = x_1$. Substituting $x = x_1, z = 0$ in (ii) we get $y = y_1$. Now putting $x = x_1$, $y = y_1$ in (iii), we get $z = z_1$.

So at the end of first iteration, we have the solution

$$x = x_1,\ y = y_1,\ z = z_1$$

Second Iteration : Putting $y = y_1$, $z = z_1$ in (i) we get $x = x_2$. Substituting $x = x_2$, $z = z_1$ in (ii), we obtain $y = y_2$. Now putting $x = x_2$, $y = y_2$ in (iii), we obtain $z = z_2$.

So at the end of second iteration, we have the solution

$$x = x_2,\ y = y_2,\ z = z_2$$

Continue the above process till the values of x, y, z are obtained to the desired level of accuray.

8. Approximation of Functions : Sometimes approximate values of a function f(x) can be calculated for a given values of x, upto a desired level of accuracy, with the help of its power series expansion.

Some of the commonly used power series expansions of some standard functions are given below:

(i) $(1+x)^n = 1 + nx + \frac{n(n-1)}{2!}x^2 + \frac{n(n-1)(n-2)}{3!}x^3 + \ldots + \frac{n(n-1)(n-2)\ldots(n(r-1))}{r!}x^r + \ldots$, where $|x| < 1$

(ii) $e^x = 1 + x + \frac{x^2}{2!} + \frac{x^3}{3!} + \ldots + \frac{x^n}{n!} + \ldots$ for all x

(iii) $a^x = 1 + (x\log a) + \frac{(x\log a)^2}{2!} + \frac{(x\log a)^3}{3!} + \ldots + \frac{(x\log a)^n}{n!} + \ldots$

(iv) $\log(1+x) = x - \frac{x^2}{2} + \frac{x^3}{3} - \frac{x^4}{4} + \ldots$ for $|x| \le 1$

(v) $\log(1-x) = -\left(x + \frac{x^2}{2} + \frac{x^3}{3} + \frac{x^4}{4} + \ldots\right)$ for $|x| < 1$

(vi) $\log\left(\frac{1+x}{1-x}\right) = 2\left(x + \frac{x^3}{3} + \frac{x^5}{5} + \ldots\right)$ for $|x| < 1$

(vii) $\sin x = x - \frac{x^3}{3!} + \frac{x^5}{5!} + \frac{x^7}{7!} + \ldots$ for all x

(viii) $\cos x = 1 - \frac{x^2}{2!} + \frac{x^4}{4!} - \frac{x^6}{6!} + \ldots$ for all x

(ix) $\tan x = x - \frac{x^3}{3} + \frac{2}{15}x^5 + \ldots$ for all ; $x \ne (2n+1)\frac{\pi}{2}$, $n \in Z$

To find the approximate value of given function we adopt the following procedure :

Step 1 : Write the power seires expansion of the given function.

Step 2 : Write the nth and (n+1)th terms denoted by T_n and T_{n+1} respectively.

Step 3 : Determine the value of T_1 for given value of x from the power series expansioin of the given function.

Step 4 : Determine the ratio $R = \frac{T_{n+1}}{T_n}$ and use the recurrence $T_{n+1} = RT_n$ to find a term when the preceding term is given.

Step 5 : Let S_n denotes the sum upto n terms of the given series. Then use the relation

$$S_{n+1} = S_n + T_{n+1} \text{ and take } S_0 = 0 \text{ and } S_1 = T_1$$

Step 6 : Repeat steps IV and V till the contribution of the following terms to the sum upto the desired level accuracy become zero.

Step 7 : The desired value of the function is the final value of S_{n+1}. The followed example illustrates the procedure :

Ex : Using power series expansion of e^x find an approximate value of e^2 upto two decimal places.

Sol : Step 1 : We have

$$e^x = 1 + x + \frac{x^2}{2!} + \frac{x^3}{3!} + ... + \frac{x^{n-1}}{(n-1)!} + ...$$

Step 2 : $T_n = \frac{x^{n-1}}{(n-1)!}$ and $T_{n+1} = \frac{x^n}{n!}$

Step 3 : Clearly $T_1 = 1$

Step 4 : We have $\frac{T_{n+1}}{T_n} = \frac{x^n}{n!}\frac{(n-1)}{x^n - 1} - \frac{x}{n}$

$$T_{n+1} = \frac{x}{n} T_n \Rightarrow T_{n+1} = \frac{2}{n} T_n \qquad [\because x = 2]$$

Step 5 : Let S_n denote the sum of n terms. Then

$S_0 = T_0 = 0$, $S_1 = S_0 + T_1 = 1$ and $S_{n+1} = S_n + T_{n+1}$ for $n \geq 1$

Now various calculations can be put in the form of a table as given below.

n	T_n	$T_{n+1} = \frac{2}{n} T_n$	$S_{n+1} = S_n + T_{n+1}$
-	-	$T_1 = 1$	$S_1 = 1$
1.	1	$T_2 = \frac{2}{1} \times 1 = 2$	$S_2 = 1 + 2 = 3$
2.	2	$T_3 = \frac{2}{2} \times 2 = 2$	$S_3 = 3 + 2 = 5$
3.	2	$T_4 = \frac{2}{3} \times 2 = 1.3333$	$S_4 = 6.3333$
4.	1.3333	$T_5 = \frac{2}{4} \times 1.3333 = 0.6666$	$S_5 = 6.9999$
5.	0.6666	$T_6 = \frac{2}{5} \times 0.6666 = 0.2666$	$S_6 = 7.2665$
6.	0.26666	$T_7 = \frac{2}{6} \times 0.2666 = 0.0888$	$S_7 = 7.3553$
7.	0.0888	$T_8 = \frac{2}{7} \times 0.0888 = 0.0253$	$S_8 = 7.3806$
8.	0.0253	$T_9 = \frac{2}{8} \times 0.0253 = 0.0063$	$S_9 = 7.3870$
9.	0.0063	$T_{10} = \frac{2}{9} \times 0.0063 = 0.0014$	$S_{10} = 7.3884$
10.	0.0014	$T_{11} = \frac{2}{10} \times 0.0014 = 0.0002$	$S_{11} = 7.3887$

Since the contribution of the following terms to the sum upto two places of decimal will be zero, therefore we stop the process.

$\therefore$ the value of $e^2 = 7.38$ (by chopping off)

9. Numerical Integration : If $\phi(x)$ is primitive of $f(x)$ defined on $[a, b]$, then $\int_a^b f(x)\,dx = \phi(b) - \phi(a)$

In most of the practical problems we are given a set of numerical values of the function f(x) corresponding to some values of x and in some cases either the primitive does not exist or it cannot be easily determined by elementary means; as a result the computation of the definite integral by the above formula may not be possible. In such circumstances, the numerical methods for integration are needed.

Numerical integration is the process of computing the value of a definite integral when we are given a set of numerical values of inte-grand f(x) corresponding to some values of the independent variable x.

The Trapezoidal Rule : Let y = f(x) be a function defined on [a, b] which is divided into n equal-sub-intervals each of width h so that b - a = nh

Let the values of f(x) for (n+1) equidistant arguments $x_0 = a$, $x_1 = x_0 + h$, $x_2 = x_0 + 2h, ..., x_n = x_0 + nh = b$ be $y_0, y_1, y_2, ..., y_n$ respectively. Then

$$\int_a^b f(x)dx = \int_{x_0}^{x_0+nh} ydx = h[\frac{1}{2}(y_0 + y_n) + (y_1 + y_2 + y_3) + ... + y_{n-1})]$$

This rule is knows as the Trapezoidal rule.

In the derivation of this formula it is assumed that the y is a linear function of x i.e. the equation of the curve is one of the form y = a + bx.

Simpson's One - Third Rule : Let y = f(x) be a function defined on [a, b] which is divided into n (an even number) equal parts each of width h so that b - a = nh.

Suppose the function y = f(x) attains values $y_0, y_1, y_2, ..., y_n$ at n+1 equidistant points $x_0 = a, x_1 = x_0 + h$, $x_2 = x_0 + 2h, ..., x_n = x_0 + nh = b$ respectively. Then

$$\int_a^b f(x)dx = \int_{x_0}^{x_0+nh} ydx = \frac{h}{3}[(y_0 + y_n) + 4(y_1 + y_3 + y_5 + ... + y_{n-1}) + 2(y_2 + y_4 + ... + y_{n-2})]$$

= (one-third of the distance between two consecutive ordinates) $\times$ [(sum of the extreme ordinates) + 4 (sum of odd ordinates) + 2 (sum of even ordinates)]

This formula is knows as Simpson's one-third rule. Its geometric significance is that we replace the graph of the given function by $\frac{n}{2}$ arcs of second degree polynomials, or parabolas with vertical axes. It is to note here that the interval [a, b] is divided into an even number of subinterval of equal width.

10. Comparison of Iterative Methods

The main basis of comparison of the various iterative methods presented in this chapter are:

I. The rate of convergence

II. The amount of computational effort per iteration

III. Sensitivity of the method to the starting value and the intermediate values.

Of the methods presented the Newton-Raphson method has the fastest convergence. It is an excellent method if one is already near the root. It is, however, quite sensitive as it may diverge if $f'(x)$ is near zero anytime during the iterative cycle. Besides this the amount of computational effort required in computing the values of f(x) and $f'(x)$ is considerable and the overall computing time may be large in spite of the inherently high rate of convergence.

The bisection method converges slowly. It is, however, the simplest iterative method and is guaranteed to converge. Convergence is slow and steady.

The false position method is guaranteed to converge. It is, however, slow as it is first order convergent. The amount of computational effort per iteration is one function evaluation. It is not necessarily a monotonic convergence to the root but most often it will be superior to bisection method.

The secant method is not guaranteed to converge. It,however, converges faster than the false position method.

The successive approximation method is suitable in some special cases. It is easy to program and is first order convergent. It is not guaranteed to converge.

Table 6.1 gives a summary of the methods.

Table 6.1. Comparison of Iterative Methods

Method	Iterative formulae	order of convergence	Evaluation of functions	Reliability of convergence
1. Bisection	$x_{i+1}=\frac{(x_i+x_{i-1})}{2}$ x_i and x_{i-1} enclose root	Gain of 1 one bit per iteration	1	Guaranteed convergence
2. False position	$x_{i+1}=\frac{(x_{i-1}f_i - x_i f_{i-1})}{(f_i - f_{i-1})}$ x_i and x_{i-1} enclose root	1	1	Guaranteed convergence
3. Newton Raphson	$x_{i+1}=x_i-\frac{f(x_i)}{f(x_i)}$	2	2	Sensitive to starting value.
4. Secant	$x_{i+1}=\frac{x_{i-1}f_i - x_i f_{i-1}}{(f_i - f_{i-1})}$ x_i and x_{i-1} enclose root	1.62	1	Convergence fast if starting point near the root No guarantee of convergence if not near root.
5. Successive	$x_i = g(x_{i-1})$	1	1	Easy to program. No guarantee of convergence

From this table it may be seen that a good practical way to solve nonlinear equation is to follow the steps below.

I. Plot f(x) and find the approximate location of root.

II. Use two initial guesses to enclose the required root.

III. Start iterations using bisection procedure.

IV. When a root is neared, switch to Newton-Raphson method, if the formula for $f'(x)$ is not complicated. If $f'(x)$ is difficult to compute switch to secant method.

OBJECTIVE QUESTIONS - I

1. The Newton-Raphson iteration $x_{n+1} = (x_n/2) + 3/(2x_n)$ can be used to slove the equation **[GATE- 2002]**

a) $x^2 = 3$ b) $x^3 = 3$ c) $x^2 = 2$ d) $x^3 = 2$

2. Which of the following statements applies to the bisection method used for finding roots of functions **[GATE-1998]**

a) Converges with in a few iterations

b) Guaranteed to work for all continous functions

c) Is faster than the Newton-Raphson method

d) Requires that there be no error in determining the sign of the function.

3. The Newton-Raphson method is used to find the root of the equation $x^2 - 2 = 0$. If the iterations are started from -1, the iterations will **[GATE-1997]**

a) Converge to -1 b) Converge to $\sqrt{2}$

c) Converge to - $\sqrt{2}$ d) Not converge

4. Which of the following statements is TRUE in respect of the convergence of the Newton-Raphson Procedure? **[GATE-1987]**

a) It converges always under all circumstances.

b) It does not converge to a tool where the second differential coefficient changes sign.

c) It does not converge to a root where the second differential coefficient vanishes.

d) None of these

5. The Newton's-Raphson iterative formula for finding $f(x) = x^2 - 1$ **[AUCET-2000]**

a) $x_{i+1}=\frac{x_i^2-1}{2x_i}$ b) $x_{i+1}=\frac{x_i^2+1}{2x_i}$

c) $x_{i+1}=\frac{2x_i^2+1}{2x_i}$ d) $x_{i+1}=\frac{2x_i}{2x_i^2+1}$

6. The Convergence of the following method is sensitive to starting value **[AUCET-2000]**

a) False position b) Gauss seidal method
c) Newton-Raphson method d) All of the above

7. For Regula-Falsi method x_{n+1}= **[AUCET-1999]**

a) x_n b) $\frac{x_{n-1} - f(x_{n-1})}{f(x_{n-1})}$

c) $x_n - \frac{x_n - x_{n-1}}{f(n_n) - f(x_{n-1})} f(x_n)$ d) $\frac{x_n f(b) - bf(x_n)}{f(b) - f(x_n)}$

8. Newton-Raphson method of solution of numerical equation is not preferred when **[AUCET-1998]**

a) When graph of A(B) is vertical
b) When graph of x(y) is not parallel
c) The graph of f(x) is nearly horizontal where it crosses the x-axis
d) None of these

9. Errors may occur in performing numerical computation on the computer due to: **[AUCET-2001]**

a) rounding errors b) power fluctuation
c) operator fatigue d) All the above

10. Regula falsi is better than bisection method because of its **[AUCET-2001]**

a) simplicity b) Accuracy
c) faster convergence d) none

11. In which of the following methods proper choice of initial value is very important **[AUCET-1999]**

a) Bisection method b) False position
c) Newton-Raphson d) Bairsto method

12. Newton-Raphson method is applicable to the solution of : **[AUCET-1997]**

a) Both algebraic and transcendental Equations
b) Both algebraic and transcendental and also used when the roots are complex
c) Algebraic Equations only
d) Transcendental Equations only

13. Newton-Raphson iteration formula for finding $\sqrt[3]{c}$, where $c > 0$ is, **[GATE-1996]**

a) $x_{n+1} = \frac{2x^3_n + \sqrt[3]{c}}{3x^2_n}$ b) $x_{n+1} = \frac{2x^3_n - \sqrt[3]{c}}{3x^2_n}$

c) $x_{n+1} = \frac{2x^2_n + c}{3x^2_n}$ d) $x_{n+1} = \frac{2x^3_n - c}{3x^2_n}$

14. The iteration formula to find the square root of a positive real number b using the Newton-Raphson method is **[GATE-1995]**

a) $x_{k+1} = \frac{3(x_{k+b})}{2x_k}$ b) $x_{k+1} = \frac{(x^2k + b)}{2x_k}$

c) $x_{k+1} = \frac{x_k - 2x_k}{(x^2k + b)}$ d) None

15. Match the following **[GATE-1994]**

I. Newton-Raphson P. Integration
II. Runge-kutta Q. Root finding
III. Gauss-seidel R. Ordinary Differential Equations
IV. Simpson's Rule S. Solution of system of Linear Equations

a) I- Q, II-R, III- S, IV-P b) I-R, II-Q, III-P, IV-S
c) I- P, II-S, III- Q, IV- R d) None of these

16. The root of x^3-2x-5=0 correct to three decimal places by using Newton-Raphson method is

a) 2.0946 b) 1.0404 c) 1.7321 d) 0.7011

17. Using Newton-Raphson method, find a root correct to three decimal places of the equation $e^x = 1 + 2x$

a) 1.256 b) 1.255 c) 1.286 d) None

18. Using Newton-Raphson method, find a root correct to three decimal places of the equation $\sin x = 1 - x$

a) 0.511 b) 0.500 c) 0.555 d) None

ANSWER KEY - I				
1. a	2. b	3. c	4. ?	5. b
6. c	7. c	8. c	9. a	10. c
11. c	12. a	13. c	14. b	15. a
16. a	17. a	18. a		

OBJECTIVE QUESTIONS - II

1. The maximum number of real roots of the equation $x^{2n} - 1 = 0$ is

a) 2 b) 3 c) n d) 2n

2. The number of real roots of the equation $x^2 - 5x + 2 = 0$ is

a) 3 b) 4 c) 5 d) none of these

3. If the equation $x^3 - 3x + k = 0$ has all real roots, then

a) $-2 < k < 2$ b) $-1 < k < 1$
c) $0 < k < \infty$ d) none of these

4. Let f(x) = 0 be an equation and x_1, x_2 be two real numbers, such that $f(x_1)f(x_2) < 0$, then

a) atleast one root of the equation lies in the interval (x_1, x_2)
b) no root of the equation lies in the interval (x_1, x_2)
c) either no root or more than one root of the equation lies in the interval (x_1, x_2)
d) none of these

5. Let f(x) =0 be an equation and let x_1, x_2 be two real numbers such that $f(x_1)\, f(x_2) > 0$. Then

a) at least one root of the equation lies in (x_1, x_2)
b) no root of the equation lies in (x_1, x_2)
c) either no root or an even number of roots lie in (x_1, x_2)
d) none of these.

6. The Newton-Raphson method converges fast if $f'(\alpha)$ is (α is the exact value of root)

a) small b) large c) 0 d) none of these

7. The order of convergence of the Newton-Raphson method is

a) linear b) quadratic c) cubic d) none

8. The order of convergence of Regula-Falsi method is

a) 1 b) 2 c) 1.618 d) none of these

9. The condition for convergence of the Newton-Raphson method to a root α is

a) $\frac{1}{2}\frac{f'(\alpha)}{f''(\alpha)} < 1$ b) $\frac{f'(\alpha)}{f''(\alpha)} < 1$ c) $\frac{1}{2}\frac{f'(\alpha)}{f''(\alpha)} > 1$ d) none

10. The Newton's iterative formula for finding the root of an equation f(x) = 0, is

a) $x_{n+1} = x_n - \frac{f'(x_n)}{f(x_n)}$ b) $x_{n+1} = x_n + \frac{f'(x_n)}{f(x_n)}$

c) $x_{n+1} = x_n - \frac{f(x_n)}{f'(x_n)}$ d) none of these

11. The number of positive roots of the equation $x^3 - 3x + 5 = 0$ is

a) 1 b) 2 c) 3 d) none

12. One root of the equation $x^3 - 5x + 1 = 0$ must lie in the interval

a) (0, 1) b) (1, 2) c) (-1, 0) d) none

13. The root of the equation f(x) = 0 lying in the interval (a, b) is given by

a) $\frac{af(b) - bf(a)}{b - a}$ b) $\frac{bf(a) - af(b)}{b - a}$

c) $\frac{af(b) - bf(a)}{f(b) - f(a)}$ d) none of these

14. If at the end of an iteration $f(a) < 0$ and $f(b) > 0$, then by the method of linear interpolation the value of the next approximation of the root is given by

a) $b - \frac{f(b)}{f(a) - f(b)}(a - b)$ b) $a - \frac{f(a)}{f(b) - f(a)}(b - a)$

c) $a - \frac{f(b)}{f(b) - f(a)}(b - a)$ d) none of these

15. the Gauss-Seidal method gives result faster when the pivotal elements are

a) smaller than other coefficients
b) larger than other coefficients
c) equal to other coefficients d) none of these

16. Newton-Raphson method is applicable only when

a) $f(x) \neq 0$ in the nbd of actual root $x = \alpha$
b) $f'(x) \neq 0$ in the nbd of actual root $x = \alpha$
c) $f''(x) \neq 0$ in the nbd of actual root $x = \alpha$ d) none

17. Simpson's one-third rule for evaluation of $\int_a^b f(x)\,dx$ requires the interval [a, b] to be divided into

a) an even number of sub intervals of equal width
b) an odd number of sub-intervals of equal width
c) any number of sub-intervals of equal width
d) any number of sub-intervals

18. In Simpson's one-third rule the curve y = f(x) is assumed to be a

a) circle b) parabola c) hyperbola d) none of these

19. Trapezoidal rule for evaluation of $\int_a^b f(x)dx$ requires the intervals (a, b) to be divided into

a) 2n sub-intervals of equal width
b) 2n + 1 sub-intervals of equal width
c) any number of sub-intervals of equal width
d) 3n sub-intervals of equal width.

20. If a and a + h are the two consecutive approximate roots of the equation f(x) = 0 as obtained by Newtons method, then h is equal to

a) $\frac{f(a)}{f'(a)}$ b) $\frac{f'(a)}{f(a)}$ c) $-\frac{f'(a)}{f(a)}$ d) $-\frac{f(a)}{f'(a)}$

21. Let $f(0) = 1, f(1) = 2.72$, then the trapezoidal rule gives approximate value of $\int_0^1 f(x)dx$

a) 3.72 b) 1.86 c) 1.72 d) 0.86

22. The value of f(x) given only at $x = 0, 1/3, 2/3, 1$. Which of the following can be used to evaluate $\int_0^1 f(x)\,dx$ apprximately :

a) Trapezoidal rule b) Simpson's rule
c) Trapezoidal as well as Simpson's Rule d) none

23. The equation $x^3 - 3x + 4 = 0$ has only one real root. What is its first approximate value as obtained by the method of false position in (-3, -2) ?

a) -2.125 b) 2.125 c) -2.812 d) 2.812

ANSWER KEY - II

1. a	2. a	3. b	4. a	5. c
6. b	7. b	8. c	9. c	10. c
11. d	12. a	13. c	14. b	15. b
16. b	17. a	18. b	19. c	20. d
21. b	22. c	23. b		

EXPLANATORY NOTES - II

1. Let $f(x) = x^{2n} - 1$, then $f(-x) = x^{2n} - 1$. Obviously there is only one variation of sign each in f(x) and f(-x). Therefore the equation has atmost two real roots one positive and one negative. Clearly, these roots are ± 1

6. The rate of convergence of Newton Raphson Methods is $\frac{1}{2}\frac{f''(\alpha)}{f'(\alpha)}$ which will be small if $f'(\alpha)$ is large.

11. Let $f(x) = x^3 - 3x + 5$. Since there are two changes of signs in f(x), therefore f(x) has atmost two positive roots. We have $f'(x) = 3x^2 - 3$. Therefore $f'(x) = 0 \Rightarrow x = \pm 1$

The signs of f(x) at $x = -\infty, -1, 1, \infty$ are :

x:	$-\infty$	-1	1	∞
f(x):	-	+	+	+

This shows that there is only are real root of f(x) which lies in the interval $(-\infty, -1)$ and it has no positive root.

12. Let $f(x) = x^3 - 5x + 1$. Then $f(0) = 1 > 0$ and $f(1) = -3 < 0$

Therefore one root lies between 0 and 1

20. We have,

$$a + h = a - \frac{f(a)}{f'(a)} \Rightarrow h = -\frac{f(a)}{f'(a)}$$

21. By trapezoidal rule, we have

$$\int_0^1 f(x)\,dx = \frac{1}{2}[f(0) + f(1)] = \frac{3.72}{2} = 1.86$$

23. We have, $x_2 = x_0 - \frac{x_1 - x_0}{f(x_1) - f(x_0)}$

Here, $x_0 = -3, x_1 = -2, f(x) = x^3 - 3x + 4$

$\therefore f(x_0) = -27 + 9 + 4 = -14, f(x_1) = -8 + 6 + 4 = 2$

So, $x_2 = -3 - \frac{1}{16} \times -14 = -3 + \frac{7}{8} = -\frac{17}{8}$

VII. CALCULUS

- **Limit, Continuity and Differentiability**
- **Mean Value Theorems**
- **Theorems of Integral Calculus.**
- **Definite Integrals**
- **Partial Derivatives**
- **Maxima and Minima**

I. Limits :

1. For all real numbers of n, $\lim\limits_{x\to a}\dfrac{x^n-a^n}{x-a}=n.a^{n-1}$

2. $\lim\limits_{x\to a}\dfrac{x^n-a^n}{x^m-a^m}=\dfrac{n}{m}a^{n-m}$

3. If $0<|x|<\dfrac{\pi}{2}$ and x is measured in radians.

$\lim\limits_{x\to 0}\dfrac{\sin x}{x}=1$ and $\lim\limits_{x\to 0}\dfrac{\tan x}{x}=1$

4. $\lim\limits_{x\to 0}\dfrac{\sin px}{x}=\lim\limits_{x\to 0}\dfrac{\tan px}{x}=p$

5. $\lim\limits_{x\to 0}\dfrac{x}{\sin px}=\lim\limits_{x\to 0}\dfrac{x}{\tan px}=\dfrac{1}{p}$

6. $\lim\limits_{x\to 0}\dfrac{\sin px}{\sin qx}=\lim\limits_{x\to 0}\dfrac{\sin px}{\tan qx}=\lim\limits_{x\to 0}\dfrac{\tan px}{\sin qx}=\lim\limits_{x\to 0}\dfrac{\tan px}{tax\ qx}=\dfrac{p}{q}$

7. If x is mesured in degrees

$\lim\limits_{x\to 0}\dfrac{\sin x^0}{x^0}=1$ and $\lim\limits_{x\to 0}\dfrac{\tan x^0}{x^0}=1$

8. $\lim\limits_{x\to\infty}\dfrac{\sin x}{x}=0$ and $\lim\limits_{x\to\infty}\dfrac{\cos x}{x}=0$

9. $\lim\limits_{x\to 0}\dfrac{\sin^{-1}x}{x}=1$ and $\lim\limits_{x\to 0}\dfrac{\tan^{-1}x}{x}=1$

10. $\lim\limits_{x\to 0}\dfrac{\sin hx}{x}=1$ and $\lim\limits_{x\to 0}\dfrac{\tan hx}{x}=1$

11. $\lim\limits_{x\to 0}\dfrac{\sin h^{-1}x}{x}=1$ and $\lim\limits_{x\to 0}\dfrac{\tan h^{-1}x}{x}=1$

12. $\lim\limits_{x\to 0}(1+x)^{1/x}=e$

13. $\lim\limits_{x\to 0}(1+ax)^{1/x}=e^a$

14. $\lim\limits_{x\to 0}\left(1+\dfrac{1}{x}\right)^x=e$

15. $\lim\limits_{x\to 0}\left(1+\dfrac{a}{x}\right)^x=e^a$ and $\lim\limits_{x\to 0}\left(1+\dfrac{1}{ax}\right)^x=e^{1/a}$

16. $\lim\limits_{x\to 0}\dfrac{e^x-1}{x}=1$

17. $\lim\limits_{x\to 0}\dfrac{a^x-1}{x}=\log_e{}^a\,(a>0)$

18. $\lim\limits_{x\to 0}\dfrac{a^x-b^x}{x}=\log_e(a/b)$

19. $\lim\limits_{x\to 0}\dfrac{a^x-1}{b^x-1}=\log_b a$

20. **Indeterminate forms :** The values of limits is in the form $\dfrac{0}{0},\dfrac{\infty}{\infty},\infty-\infty,0\times\infty,0^0,\infty^0,1^\infty,0^\infty$ etc. are indeterminate forms.

21. **L'Hospital's Rule :**

If $\lim\limits_{x\to a}\dfrac{f(x)}{g(x)}=$ indeterminate then $\lim\limits_{x\to a}\dfrac{f(x)}{g(x)}=\lim\limits_{x\to a}\dfrac{f'(x)}{g'(x)}$

If $\lim\limits_{x\to a}\dfrac{f'(x)}{g'(x)}=$ indeterminate then $\lim\limits_{x\to a}\dfrac{f(x)}{g(x)}=\lim\limits_{x\to a}\dfrac{f''(x)}{g''(x)}$

If $\lim\limits_{x\to a}\dfrac{f''(x)}{g''(x)}=$ indeterminate, then proceed as above till a finite limit is obtained.

22. Let $s=\{s,\sin x,\tan x,\sin hx,\tan hx,\sin^{-1}x,\tan^{-1}x,\sin h^{-1}x,\tan h^{-1}x\}$

If $f_1(x), f_2(x), g_1(x), g_2(x)\in s$ then

$\lim\limits_{x\to 0}\dfrac{f_1(mx)\pm f_2(nx)}{g_1(px)\pm g_2(qx)}=\dfrac{m\pm n}{p\pm q}$

23. If $f_1(x), f_2(x), g_1(x), g_2(x)\in s$ and $m+n=p+q$ then $\lim\limits_{x\to 0}\dfrac{f_1^m(ax)\,f_2^n(bx)}{g_1^p(cx)\,g_2^q(dx)}=\dfrac{a^m\,b^n}{c^p\,d^q}$

24. If $f_1(x), f_2(x)\in s$ then $\lim\limits_{x\to 0}\dfrac{1-\cos ax}{f_1(px)\,f_2(qx)}=\dfrac{a^2}{2pq}$

25. If $f_1(x), f_2(x)\in s$ then $\lim\limits_{x\to 0}\dfrac{1-\cos^n(ax)}{f_1(px)\,f_2(qx)}=\dfrac{na^2}{2pq}$

26. If $f_1(x), f_2(x).....f_{2n}(x)\in S$ then

$\lim\limits_{x\to 0}\dfrac{1-\cos\left(ax^n\right)}{f_1(p_1x)\,f_2(p_2x).....f_{2n}(p_{2n}x)}=\dfrac{a^2}{2p_1p_2....p_{2n}}$

27. If $f_1(x), f_2(x)\in s$ then $\lim\limits_{x\to 0}\dfrac{\cos ax-\cos bx}{f_1(cx).f_2(dx)}=\dfrac{b^2-a^2}{2cd}$

28. If $f_1(x), f_2(x)\in s$ then, $\lim\limits_{x\to 0}\dfrac{\cos^n ax-\cos^n bx}{f_1(cx).f_2(dx)}=\dfrac{n(b^2-a^2)}{2cd}$

29. If $f_1(x), f_2(x)....f_{2n}(x)\in s$ then,

$\lim\limits_{x\to 0}\dfrac{\cos\left(ax^n\right)-\cos\left(bx^n\right)}{f_1(p_1x)f_2(p_2x).....f_{2n}(p_{2n}x)}=\dfrac{b^2-a^2}{2p_1p_2....p_{2n}}$

30. If $f_1(x)\in s$ then $\lim\limits_{x\to 0}\dfrac{\tan^n(ax)-\sin^n(ax)}{\left[f_1(x)\right]^{n+2}}=\dfrac{n.a^{n+2}}{2}$

31. $\lim_{x\to 0} \dfrac{\sqrt{1+x^n}-\sqrt{1-x^n}}{x^n} = 1$

32. $\lim_{x\to 0} \dfrac{\sqrt[n]{a+x^m}-\sqrt[n]{a-x^m}}{x^m} = \dfrac{2}{n} a^{\frac{1}{n}-1}$

33. If $f(x) \in s$, then $\lim_{x\to 0} \dfrac{\sqrt{a+x}-\sqrt{a-x}}{f(x)} = \dfrac{1}{\sqrt{a}}$

34. If $f(x) \in s$, then $\lim_{x\to 0} \dfrac{\sqrt{a+x}-\sqrt{a}}{f(x)} = \dfrac{1}{2\sqrt{a}}$

35. $\lim_{x\to 0} \dfrac{xa^{\alpha x}-x}{1-\cos(px)} = \dfrac{2\alpha}{p^2}\log a$

36. $\lim_{x\to\infty} \dfrac{a^{x+\alpha}+b}{a^{x+\beta}+c} = a^{\alpha-\beta}$

37. $\lim_{x\to\infty} \left\{\sqrt{x^2+ax+b}-x\right\} = \dfrac{a}{2}$

38. $\lim_{n\to\infty} a^n = \begin{cases} \infty \text{ if } a>1 \\ 0 \text{ if } a<1 \end{cases}$ i.e. $a^\infty = \infty$, if $a > 1$ and $a^\infty = 0$ if $a < 1$

39. Some times following expansions are very useful to evaluate limits.

a) $(1+x)^n = 1+nx+\dfrac{(n-1)}{2!}x^2+\dfrac{n(n-1)(n-2)}{3!}x^3+...$

b) $e^x = 1+x+\dfrac{x^2}{2!}+\dfrac{x^3}{3!}+.....$

c) $a^x = 1+x\left(\log_e^a\right)+\dfrac{x^2}{2!}\left(\log_e^a\right)^2+.....$

d) $\log(1+x) = x-\dfrac{x^2}{2}+\dfrac{x^3}{3}-\dfrac{x^4}{4}+.....(|x|<1)$

e) $\log(1-x) = -x-\dfrac{x^2}{2}-\dfrac{x^3}{3}-\dfrac{x^4}{4}-.....(|x|<1)$

f) $\sin x = x-\dfrac{x^3}{3!}+\dfrac{x^5}{5!}$ ....

g) $\cos x = 1-\dfrac{x^2}{2!}+\dfrac{x^4}{4!}$ ....

h) $\tan x = x+\dfrac{x^3}{3}+\dfrac{2}{15}x^5+......$

i) $\sin^{-1}x = x+\dfrac{1}{2}\cdot\dfrac{x^3}{3}+\dfrac{1}{2}\cdot\dfrac{3}{4}\cdot\dfrac{x^5}{5}+\dfrac{1}{2}\cdot\dfrac{3}{4}\cdot\dfrac{5}{6}\cdot\dfrac{x^7}{7}+...(-1<x<1)$

j) $\tan^{-1}x = x-\dfrac{1}{3}x^3+\dfrac{1}{5}x^5....(-1<x<1)$

k) $\sec x = 1+\dfrac{x^2}{2!}+5.\dfrac{x^4}{4!}+.......$

40. Sandwich Theorem : If f, g and h are functions having a common domain D and defined near a. Suppose that $h(x) \le f(x) \le g(x)$ for $x \in D$ and $\lim_{x\to a} h(x) = \lim_{x\to a} g(x) = l$ then $\lim_{x\to a} f(x) = l$

41. $\lim_{x\to\infty} \dfrac{a_0x^p+a_1x^{p-1}+....+a_{p-1}x+a_p}{b_0x^p+b_1x^{q-1}+....b_{q-1}x+bq} = \dfrac{a_0}{b_0}$ if $p = q = 0$ if $p < q = \infty$ if $p > q$

42. If $\lim_{x\to a} f(x) = A > 0$ and $\lim_{x\to a} g(x) = B$, then $\lim_{x\to a}\left[f(x)\right]^{g(x)} = A^B$

43. If $\lim_{x\to a} f(x) = 1$ and $\lim_{x\to a} g(x) = \infty$, then $\lim_{x\to a}\left[f(x)\right]^{g(x)} = e^{\lim_{x\to a}[f(x)-1]}$

II. Continuity

1. Continuity of a function at a point : A function f(x) is said to be continuous at an interior point x = a of its domain if $\lim_{x\to a} f(x) = f(a)$

2. Continuity of a function on an open interval : A function f(x) is said to be continuous on an open internal (a, b) if it is continuous at each point of (a, b)

3. Continuity of a function on an closed interval : A function f(x) is said to be continuous on a closed interval [a, b] if

i) f(x) is continuous from right at x = a

i.e., $\lim_{h\to 0} f(a+h) = f(a)$

ii) f(x) is continuous from left at x = b

i.e., $\lim_{h\to 0} f(b-h) = f(b)$

iii) f(x) is continuous at each point of the open interval (a, b)

4. Discontinuity of a function : A function f(x) which is not continuous at a point x = a, is said to be discontinuous at that point.

The discontinuity may arise due to any of the following situations.

i) $\lim_{h\to 0} f(a-h) \ne \lim_{h\to 0} f(a+h)$ i.e. L.H.L. and R.H.L. exist, but are not equal.

ii) $\lim_{h\to 0} f(a-h) = \lim_{h\to 0} f(a+h) \ne f(a)$ i.e. L.H.L. and R.H.L. exist and are equal but are different from f(a).

iii) f(a) is not defined.

iv) At least one of the limits $\lim_{x\to a} f(a-h)$ or $\lim_{x\to a} f(a+h)$ does not exist or at least one of these limits is ∞ or $-\infty$

5. Geometrical meaning of continuity :

i) A function f(x) will be continuous at a point x = a if there is no break or cut or hole or gap in the graph of the function y = f(x) at the point, (a, f(a)) otherwise it is discontinuous at that point.

ii) A function f(x) will be continuous on the closed interval [a, b] if the graph of the function y = f(x) is an unbroken line (curved or straight) from the point (a, f(a)) to the point (b, f(b))

6. If 'f' and 'g' are continuous at x = a then

i) f + g is continuous at x = a

ii) f - g is continuous at x = a

iii) fg is continuous at x = a

iv) f/g is continuous at x = a provided $g(a) \neq 0$

v) kf is continuous at x = a where 'k' is any real constant.

vi) If 'f' is continuous at 'a' and 'g' is continuous at f(a) then gof is continuous at 'a'.

7. Every constant function is continuous in R.

8. Every polynomial function is continuous on R

9. The identify function is continuous on R.

10. The functions sinx and cosx are continuous on R.

11. The functions tanx and secx are continuous on $R-\left\{(2n+1)\frac{\pi}{2}\right\}$ where 'n' is any integer.

12. The functions cotx and cosecx are continuous on $R-\{n\pi\}$ where 'n' is any integer.

13. The function $f(x)=|x|$ is continuous on R.

14. The function $f(x)=e^x$ and $f(x)=a^x (a>0)$ are continuous on R.

15. The functions f(x) = [x] is continuous at all non integral values and discontinuous at all integral values.

III. Differentiability

1. If y = f(x) is differentiable at any point 'x' then its derivative is $f'(x)=\underset{x\to 0}{\text{Lt}}\frac{f(x+h)-f(x)}{h}$ and it is called the first derivative of f(x). A function 'f' is said to be derivable at x = a if $\underset{x\to a}{\text{Lt}}\frac{f(x)-f(a)}{x-a}$ exists. The limit is called the derivative of 'f' at x = a and is denoted by f '(a). i.e. $\underset{x\to a+}{\text{Lt}}\frac{f(x)-f(a)}{x-a}=\underset{x\to a-}{\text{Lt}}\frac{f(x)-f(a)}{x-a}$ then the limit value denoted by f '(a) is called the derivative of f(x) at x = a.

2. Let 'f' be a function defined on [a, b]. Then 'f' is said to be differentiable on [a, b], if

i) ' f ' is differentiable at 'c' i.e., $\underset{x\to c}{\text{Lt}}\frac{f(x)-f(c)}{x-c}$ where $c\in(a,b)$

ii) 'f' is right differentiable at 'a' i.e., $\underset{x\to a+}{\text{Lt}}\frac{f(x)-f(a)}{x-a}$

iii) 'f' is left differentiable at 'b' i.e., $\underset{x\to b-}{\text{Lt}}\frac{f(x)-f(b)}{x-b}$

3. If f(x), g(x) are two differentiable functions of 'x' then the linear combination of the two functions is also differentiable function of x.

i.e., $\frac{d}{dx}[c_1 f(x)+c_2 g(x)]=c_1\frac{d}{dx}f(x)+c_2\frac{d}{dx}g(x)$

where c_1, c_2 are constants.

IV. Mean Value Theorems :

1. **Rolle's theorem :** If $f:[a,b]\to R$ is a function such that i) f is continuous on [a, b], (ii) f is derivable on (a, b) and (iii) f (a) = f (b) then there exist $c\in(a, b) \ni f'(c)=0$.

2. **Geometrical interpretion of Rolle's theorem :** Let $f:[a, b]\to R$ be a function satisfying the three conditions of Rolle's Theorem. Then the graph of f is continuous from the point $A(a, f(a))$ to the point $B(b, f(b))$ and f has a tangent at every point on the graph. By Rolle's Theorem $\exists\, c\in(a, b)\ni f'(c)=0$. Thus there exists a point $C(c, f(c))$ on the curve between A, B at which the tangent to the curve is parallel to x-axis.

3. **Another form of Rolle's theorem :** If $f:[a, a+h]\to R$ is such that (i) f is continuous on [a, a + h] (ii) f is derivable on (a, a + h) and (iii) f(a) = f(a+h) then there exists $\theta\in(0, 1)$ such that $f'(a+\theta h)=0$

4. **Lagrange's Mean Value Theorem Or First Mean Value Theorem :** If a function $f:[a, b]\to R$ is i) continuous on [a, b] ii) derivable on (a, b) then $\exists\, c\in(a, b)\ni f'(c)=\frac{f(b)-f(a)}{b-a}$

5. **Another form of Lagrange's theorem :** If $f:[a, a+h]\to R$ is such that (i) f is continuous on [a, a+h] (ii) f is derivable on (a, a+h) then there exists $\theta\in(0, 1)$ such that $f(a+h)=f(a)+h\,f'(a+\theta h)$

6. **Geometrical interpretation of Lagrange's theorem:** Let $f:[a, b]\to R$ be a function satisfying the two conditions of Lagrange's mean value theorem. Then the graph of f is continuous from the point $A(a, f(a))$ to the point $B(b, f(b))$ and f has a tangent at every point on the graph between A and B. By Lagrange's mean value theorem $\exists\, c\in(a, b)\ni f'(c)=\frac{f(b)-f(a)}{b-a}$. Since $\frac{f(b)-f(a)}{b-a}$ = slope of the chord $\overleftrightarrow{AB}$.

There exists a point C (c, f(c)) on the curve between A, B at which the tangent to the curve is parallel to the chord $\overleftrightarrow{AB}$.

7. **Cauchy's theorem :** If $f:[a, b]\to R$, $g:[a, b]\to R$ such that (i) f, g are continuous on [a, b], (ii) f, g are derivable on (a, b) and (iii) $g'(x)\neq 0,\ \forall\, x\in(a, b)$ then there exists $c\in(a, b)$ such that $\frac{f(b)-f(a)}{g(b)-g(a)}=\frac{f'(c)}{g'(c)}$

8. **Another form of Cauchy's mean value theorem :** If $f:[a, a+h]\to R$, $g:[a, a+h]\to R$ are such that (i) f, g are continuous on [a, a + h], (ii) f, g are derivable on (a, a+h) and (iii) $g'(x)\neq 0,\ \forall\, x\in(a, a+h)$ then there exists $\theta\in(0, 1)$ such that $\frac{f(a+h)-f(a)}{g(a+h)-g(a)}=\frac{f'(a+\theta h)}{g'(a+\theta h)}$.

V. Integration :

1. A function F(x) is called an antiderivative or indefinite integral of a function f(x) if $\frac{d}{dx}[F(x)]=f(x)$

2. **Indefinite integrals:**

Let f(x) be a function. Then the collection of all its derivatives is called the indefinite integral of f(x) and it is denoted by $\int f(x)\,dx$

Thus $\frac{d}{dx}[F(x)+c]=f(x) \Leftrightarrow \int f(x)dx=F(x)dx$

Where F(x) is the antiderivative and c is the arbitrary constant known as the constant of integration.

3. $\int k\,dx = kx + c$

4. $\int x^n\,dx = \frac{x^{n+1}}{n+1} + c, n \neq -1$

5. $\int (ax+b)^n\,dx = \frac{(ax+b)^{n+1}}{a(n+1)} + c$

6. $\int \frac{1}{x}\,dx = \log|x| + c$

7. $\int e^x\,dx = e^x + c$

8. $\int a^x\,dx = \frac{a^x}{\log a} + c$

9. $\int \cos x\,dx = \sin x + c$

10. $\int \sin x\,dx = -\cos x + c$

11. $\int \tan x\,dx = -\log|\cos x| + c$ (or) $= \log|\sec x| + c$

12. $\int \cot x\,dx = \log|\sin x| + c = -\log|\operatorname{cosec} x| + c$

13. $\int \sec x\,dx = \log|\sec x + \tan x| + c = \log\left|\tan\left(\frac{\pi}{4} + \frac{x}{2}\right)\right| + c$

14. $\int \operatorname{cosec} x\,dx = \log|\operatorname{cosec} x - \cot x| + c = \log\left|\tan\frac{x}{2}\right| + c$

15. $\int \sec^2 x\,dx = \tan x + c$

16. $\int \operatorname{cosec}^2 x\,dx = -\cot x + c$

17. $\int \sec x \tan x\,dx = \sec x + c$

18. $\int \operatorname{cosec} x \cot x\,dx = -\operatorname{cosec} x + c$

19. $\int \cosh x\,dx = \sinh x + c$

20. $\int \sinh x\,dx = \cosh x + c$

21. $\int \tanh x\,dx = \log|\cosh x| + c$

22. $\int \coth x\,dx = \log|\sinh x| + c$

23. $\int \operatorname{sech} x\,dx = 2\operatorname{Tan}^{-1}(e^x) + c$

24. $\int \operatorname{cosech} x\,dx = \operatorname{Cos}^{-1}(e^x) + c$

25. $\int \operatorname{sech}^2 x\,dx = \tanh x + c$

26. $\int \operatorname{cosech}^2 x\,dx = -\coth x + c$

27. $\int \operatorname{sech} x \tanh x\,dx = -\operatorname{sech} x + c$

28. $\int \operatorname{cosech} x \coth x\,dx = -\operatorname{cosech} x + c$

29. $\int \frac{1}{\sqrt{1-x^2}}\,dx = \operatorname{Sin}^{-1} x + c\,(\text{or}) - \operatorname{Cos}^{-1} x + c$

30. $\int \frac{1}{\sqrt{x^2-1}}\,dx = \operatorname{Cosh}^{-1} x + c$

31. $\int \frac{1}{\sqrt{x^2+1}}\,dx = \operatorname{Sinh}^{-1} x + c$

32. $\int \frac{1}{1+x^2}\,dx = \operatorname{Tan}^{-1} x + c\,(\text{or}) - \operatorname{Cot}^{-1} x + c$

33. $\int \frac{1}{1-x^2}\,dx = \frac{1}{2}\log\left|\frac{1+x}{1-x}\right| + c$

34. $\int \frac{1}{x^2-1}\,dx = \frac{1}{2}\log\left|\frac{x-1}{x+1}\right| + c$

35. $\int \frac{1}{x\sqrt{x^2-1}}\,dx = \operatorname{Sec}^{-1} x\,(\text{or}) - \operatorname{Cosec}^{-1} x + c$

36. $\int \frac{1}{\sqrt{a^2-x^2}}\,dx = \operatorname{Sin}^{-1}\left(\frac{x}{a}\right) + c\,(\text{or}) - \operatorname{Cos}^{-1}\left(\frac{x}{a}\right) + c$

37. $\int \frac{1}{\sqrt{a^2-x^2}}\,dx = \frac{1}{a}\operatorname{Tan}^{-1}\left(\frac{x}{a}\right) + c\,(\text{or}) - \frac{1}{a}\operatorname{Cot}^{-1}\left(\frac{x}{a}\right) + c$

38. $\int \frac{1}{\sqrt{a^2+x^2}}\,dx = \operatorname{Sinh}^{-1}\left(\frac{x}{a}\right) + c$

39. $\int \frac{1}{\sqrt{x^2-a^2}}\,dx = \operatorname{Cosh}^{-1}\left(\frac{x}{a}\right) + c$

40. $\int \frac{1}{a^2-x^2}\,dx = \frac{1}{2a}\log\left|\frac{a+x}{a-x}\right| + c$

41. $\int \frac{1}{\sqrt{x^2-a^2}}\,dx = \frac{1}{2a}\log\left|\frac{x-a}{x+a}\right| + c$

42. $\int \sqrt{a^2-x^2}\,dx = \frac{x}{2}\sqrt{a^2-x^2} + \frac{a^2}{2}\operatorname{Sin}^{-1}\left(\frac{x}{a}\right) + c$

43. $\int \sqrt{a^2+x^2}\,dx = \frac{x}{2}\sqrt{a^2+x^2} + \frac{a^2}{2}\operatorname{Sinh}^{-1}\left(\frac{x}{a}\right) + c$

44. $\int \sqrt{x^2-a^2}\,dx = \frac{x}{2}\sqrt{x^2-a^2} - \frac{a^2}{2}\operatorname{Cos h}^{-1}\left(\frac{x}{a}\right) + c$

45. $\frac{d}{dx}\left[\int f(x)\,dx\right] = f(x)$

46. $\int k\,f(x)\,dx = k.\int f(x)\,dx$

47. $\int f(x) \pm g(x)\,dx = \int f(x)\,dx \pm \int g(x)\,dx$

48. $\int \frac{f'(x)}{f(x)}\,dx = \log|f(x)| + c$

49. $\int \frac{f'(x)}{\sqrt{f(x)}}\,dx = 2\sqrt{f(x)} + c$

50. $\int [f(x)]^n f'(x)\,dx = \frac{f^{n+1}(x)}{n+1} + c;\ n \neq -1$

51. $\int f(x^n)\,x^{n-1}\,dx = \frac{1}{n}\int f(t)\,dt$ where $x^n = t$

52. $\int \frac{f'(x)}{1+[f(x)]^2}\,dx = \operatorname{Tan}^{-1} f(x) + c$

Evaluation of Integrals of various types by using Standard Results

Type f Integral

53. $\int \frac{1}{ax^2+bx+c}\,dx$ (or) $\int \frac{1}{\sqrt{ax^2+bx+c}}\,dx$

(or) $\int \sqrt{ax^2+bx+c}\,dx$

Technique (or) Substitution

Express ax^2+bx+c as the sum or difference of two squares.

May be reduced to one of the forms like $\int\sqrt{a^2+x^2}\,dx$ (or) $\int\sqrt{a^2-x^2}\,dx$ or $\int\sqrt{x^2-a^2}\,dx$ or

$\int\frac{1}{\sqrt{a^2-x^2}}dx$ or $\int\frac{1}{\sqrt{x^2-a^2}}dx$ or $\int\frac{1}{\sqrt{a^2-x^2}}dx$

Type of Integral

54. $\int\frac{px+q}{ax^2+bx+c}dx$ (or) $\int\frac{px+q}{\sqrt{ax^2+bx+c}}dx$ (or)

$\int(px+q)\sqrt{ax^2+bx+c}\,dx$

Technique (or) Substitution

Write $px+q=A\frac{d}{dx}\left(ax^2+bx+c\right)+B$ where A,B are constants to be determined by equating the coefficients of similar terms on both sides.

Type of Integral Technique (or) Substitution

55. $\int\frac{1}{(px+q)\sqrt{ax^2+bx+c}}dx$ put $px+q=\frac{1}{t}$

56. $\int\frac{1}{\left(Ax^2+B\right)\sqrt{Cx^2+D}}dx$ Put $x=\frac{1}{t}$

57. $\int\sqrt{(x-\alpha)(\beta-x)}\,dx$ (or) Put $x=\alpha\cos^2\theta+\beta\sin^2\theta$

calculate $(x-\alpha)$, $(\beta-x)$ and dx

$\int\sqrt{\frac{x-\alpha}{\beta-x}}dx$ (or) $\int\frac{1}{\sqrt{(x-\alpha)(\beta-x)}}dx$

58. Type of Integral

$\int\frac{1}{a+b\sin x}dx$ (or) $\int\frac{1}{a+b\cos x}dx$ (or)

$\int\frac{1}{a\sin x+b\cos x+c}dx$

Technique (or) Substitution

Put $t=\tan\frac{x}{2}$

$\sin x=\frac{2\tan x/2}{1+\tan^2 x/2}$; $\cos x=\frac{1-\tan^2 x/2}{1+\tan^2 x/2}$

$dx=\frac{2}{1+\tan^2 x/2}dt$

59. Type of Integral

$\int\frac{1}{a+b\sin^2 x}dx$ (or) $\int\frac{1}{a+b\cos^2 x}dx$ (or)

$\int\frac{1}{a\sin^2 x+b\cos^2 x+c\cos x\sin x}dx$

Technique (or) Substitution

First multiply the numerator and denominator with $\sec^2 x$ and put $t=\tan x$

60. Type of Integral

$\int\frac{a\cos x+b\sin x}{c\cos x+d\sin x}dx$

Technique (or) Substitution

numerator = A (denominator) +

$B\left[\frac{d}{dx}(\text{denominator})\right]$

61. Type of Integral

$\int\frac{a\sin x+b\cos x+c}{p\sin x+q\cos x+r}dx$

Technique (or) Substitution

numerator = A (denominator) +

$B\frac{d}{dx}(\text{denominator})+\lambda$

where A, B and λ are constants.

Integration by Parts :

If U and V are two functions of x, then

$$\int UV\,dx=U\int V\,dx-\int\left(\frac{dU}{dx}\int V\,dx\right)dx$$

EXTENSION RULE :

$$\int UV\,dx=UV_1-U'V_2+U''V_3-U'''V_4+.....+U$$

where U', U'', U''' etc. are the successive derivatives of U and V_1, V_2, V_3 etc., are successive inegrals of V.

Note : This rule is very useful if one of the integrand is an algebraic function.

Formulae :

63. $\int e^{ax}\sin bx\,dx=\frac{e^{ax}}{a^2+b^2}[a\sin bx-b\cos bx]+c$

64. $\int e^{ax}\cos bx\,dx=\frac{e^{ax}}{a^2+b^2}[a\cos bx+b\sin bx]+c$

65. $\int e^x[f(x)+f'(x)]dx=e^x f(x)+c$

66. $\int e^{-x}[f(x)-f'(x)]dx=-e^{-x}f(x)+c$

67. $\int e^{ax}[af(x)+f'(x)]dx=e^{ax}f(x)+c$

Reduction Formulae :

68. If $I_n=\int x^n e^{ax}dx$ then $I_n=\frac{x^n e^{ax}}{a}-\frac{n}{a}I_{n-1}$

69. If $I_n=\int\sin^n x\,dx$ then $I_n=\frac{-\sin^{n-1}x\cos x}{n}+\frac{n-1}{n}I_{n-2}$

70. If $I_n=\int\cos^n x\,dx$ then $I_n=\frac{\cos^{n-1}x\sin x}{n}+\frac{n-1}{n}I_{n-2}$

71. If $I_n=\int\tan^n x\,dx$ then $I_n=\frac{\tan^{n-1}x}{n-1}-I_{n-2}$

72. If $I_n=\int\cot^n x\,dx$ then $I_n=\frac{-\cot^{n-1}x}{n-1}-I_{n-2}$

73. If $I_n=\int\sec^n x\,dx$ then $I_n=\int\frac{\sec^{n-2}x\tan x}{n-1}+\frac{n-2}{n-1}I_{n-2}$

74. If $I_n=\int\operatorname{cosec}^n x\,dx$ then

$$I_n = \frac{-\operatorname{cosec}^{n-2} x \cot x}{n-1} + \frac{n-2}{n-1} I_{n-2}$$

75. If $I_n = \int (\log x)^n \, dx$ then $I_n = x(\log x)^n - nI_{n-1}$

VI. Definite Integrals : If F(x) is an indefinite integral (or) anti derivative of a function f(x) defined on [a,b] then

$$\int_a^b f(x)\,dx = F(b) - F(a) \qquad ......\ (i)$$

The symbol $\int_a^b f(x)\,dx$ is called the definite integral of f(x) from x = a to x = b. The numbers 'a' and 'b' are called the limits of integration, 'a' is called the lower and 'b' is called the upper limit.

If we use the notation $[F(x)]_a^b$ to denote F(b) - F(a), then from (i)

$$\int_a^b f(x)\,dx = [(F(x)]_a^b = [(F(x) \text{ at } x=b) - (F(x) \text{ at } x=a)]$$

Fundamental Properties of Definite Integrals

Property : 1 : $\int_a^a f(x)\,dx = 0$

Property : 2 : $\int_a^b f(x)\,dx = \int_a^b f(t)\,dt$

Property : 3 : $\int_a^b f(x)\,dx = -\int_b^a f(x)\,dx$

Property : 4 : $\int_a^b f(x)\,dx = \int_a^c f(x)\,dx + \int_c^b f(x)\,dx$ if $a < c < b$

Property : 5 : $\int_0^a f(x)\,dx = \int_0^a f(a-x)\,dx$

Property : 6 : $\int_{-a}^a f(x)\,dx = \begin{cases} 2\int_0^a f(x)\,dx & f(x) \text{ is even} \\ 0 & f(x) \text{ is odd} \end{cases}$

Property : 7 : $\int_0^{2a} f(x)\,dx = \begin{cases} 2\int_0^a f(x)\,dx & \text{iff } (2a-x) = f(x) \\ 0 & \text{iff } (2a-x) = -f(x) \end{cases}$

Property : 8 : $\int_a^b f(x)\,dx = \int_a^b f(a+b-x)\,dx$

Property : 9 : $\int_0^{na} f(x)\,dx = n\int_0^a f(x)\,dx$ if $f(a+x) = f(x)$

Property : 10 : $\int_0^\infty f(x)\,dx = \underset{t\to\infty}{\text{Lt}} \int_0^t f(x)\,dx$

and $\int_{-\infty}^a f(x)\,dx = \underset{t\to-\infty}{\text{Lim}} \int_t^a f(x)\,dx$

and $\int_{-\infty}^\infty f(x)\,dx = \underset{t\to\infty}{\text{Lim}} \int_a^t f(x)\,dx + \underset{t\to-\infty}{\text{Lim}} \int_t^a f(x)\,dx$

Property : 11 : $\left|\int_a^b f(x)\,dx\right| \le \int_a^b |f(x)|\,dx$

Property : 12 : $f(x) \ge 0, \forall x \in [a,b]$, then $\int_a^b f(x)\,dx \ge 0$

and if $f(x) \ge g(x), \forall x \in [a,b]$, $\int_a^b f(x)\,dx \ge \int_a^b g(x)\,dx$

Reduction Formulae

(i) If $I_n = \int_0^{\pi/2} \sin^n x\,dx$ then $I_n = \frac{n-1}{n} I_{n-2}$

(ii) $I_n = \int_0^{\pi/2} \cos^n x\,dx$ then $I_n = \frac{n-1}{n} I_{n-2}$

(iii) $I_n = \int_0^{\pi/4} \tan^n x\,dx$ then $I_n = \frac{1}{n-1} - I_{n-2}$

(iv) $I_n = \int_0^{\pi/4} \sec^n x\,dx$ then $I_n = \frac{(\sqrt{2})^{n-2}}{n-1} + \frac{n-2}{n-1} I_{n-2}$

(v) If $I_n = \int_{\pi/4}^{\pi/2} \cot^n x\,dx$ then $I_n = \frac{1}{1-n} - I_{n-2}$

(vi) If $I_n = \int_{\pi/4}^{\pi/2} \operatorname{cosec}^n x\,dx$ then $I_n = \frac{-(\sqrt{2})^{n-2}}{n-1} + \frac{n-2}{n-1} I_{n-2}$

(vii) If $I_{m,n} = \int_0^{\pi/2} \sin^m x \cos^n x\,dx$ then $I_{m,n} = \frac{n-1}{m+n} I_{m,n-2}$

Definite Integral as the Limit of a Sum

If f(x) is an integrable function over [a,b] then

$$\int_a^b f(x)\,dx = \underset{h\to 0}{\text{Lim}}\, h[f(a) + f(a+h) + f(a+2h) + ...$$
$$+ f(a+(n-1)h)] \text{ where } h = \frac{b-a}{n}$$

(or) $\int_a^b f(x)\,dx = \lim_{h\to 0} h\sum_{r=0}^{n-1} f(a+rh)$

Summation of Series by Definite Integrals

The definition of the definite integral as the limit of a sum enables us to evaluate the sum upto infinity, of certain series by identify them with same definite integrals.

By definition of the definite integral as the limit of a sum we have

$$\int_a^b f(x)\,dx = \lim_{h\to 0}\sum_{r=0}^{n-1} f(a+rh) \quad \text{where } nh = b-a$$

by putting a = 0, b = 1, the above becomes

$$\int_0^1 f(x)\,dx = \lim_{n\to\infty}\sum_{r=0}^{n-1} f\left(\frac{r}{n}\right)$$

Working Rule

If a given series is expressible in the form of $\frac{1}{n}\Sigma f\left(\frac{r}{n}\right)$, then its limit as $n\to\infty$ can expressed as a definite integral by writing 'x' for $\frac{r}{n}$, dx for $\frac{1}{n}$ and integral sign $\int$ for Σ. The lower and upper limits of integration are the values of $\lim_{n\to\infty}\left(\frac{r}{n}\right)$ for least and greatest values of 'r' respectively.

VII. Partial Derivatives :

1. Let 'f' be a function of x and y.

Then $\frac{\partial y}{\partial x}$ = partial derivative of 'f' w.r.t. x

$$= \underset{h\to 0}{\text{Lt}} \frac{f(x+h, y) - f(x, y)}{h}$$

$\Rightarrow \frac{\partial y}{\partial x}$ = ordinary derivative of 'f' w.r.t. to x treating y as constant.

2. Let 'f' be a function of two variables x and y

1st order partial derivatives $\rightarrow \frac{\partial f}{\partial x} = f_x,\ \frac{\partial f}{\partial y} = f_y$

2nd order partial derivatives.

3. $f_{x^2} = f_{xx} = \frac{\partial^2 f}{\partial x^2} = \frac{\partial}{\partial x}\left(\frac{\partial f}{\partial x}\right)$

4. $f_{yx} = \frac{\partial}{\partial y}\left(\frac{\partial f}{\partial x}\right) = \frac{\partial^2 f}{\partial y\partial x}$

5. $f_{xy} = \frac{\partial}{\partial x}\left(\frac{\partial f}{\partial y}\right) = \frac{\partial^2 f}{\partial x\partial y}$

6. $f_{y^2} = f_{yy} = \frac{\partial}{\partial y}\left(\frac{\partial f}{\partial y}\right) = \frac{\partial^2 f}{\partial y^2}$

7. In general, $f_{xy} = f_{yx}$

8. If $u = f(v)$ and $v = g(x, y)$, then

$\frac{\partial u}{\partial x} = \frac{du}{dv}\cdot\frac{\partial v}{\partial x}$ and $\frac{\partial u}{\partial y} = \frac{du}{dv}\cdot\frac{\partial v}{\partial y}$

9. If $u = f(x, y)$ and $x = g(t),\ y = h(t)$, then

$$\frac{du}{dt} = \frac{\partial u}{\partial x}\cdot\frac{dx}{dt} + \frac{\partial u}{\partial y}\cdot\frac{dy}{dt}$$

10. If $u = f(x, y)$ then $du = \frac{\partial u}{\partial x}dx + \frac{\partial u}{\partial y}.dy$ and du is called as the total differential of 'u'.

11. If $u = f(x, y) = c$ where 'c' is a constant, then

$$\frac{dy}{dx} = \frac{-\left(\frac{\partial f}{\partial x}\right)}{\left(\frac{\partial f}{\partial y}\right)}$$

12. If $z = f(u, v)$, $u = g(x, y)$ and $v = h(x, y)$ then

$$\frac{\partial z}{\partial x} = \frac{\partial z}{\partial u}\cdot\frac{\partial u}{\partial x} + \frac{\partial z}{\partial v}\cdot\frac{\partial v}{\partial x},\ \frac{\partial z}{\partial y} = \frac{\partial z}{\partial u}\cdot\frac{\partial u}{\partial y} + \frac{\partial z}{\partial v}\cdot\frac{\partial v}{\partial y}$$

13. f(x, y) is said to be a homogenous function of degree 'n' if

i) $f(tx, ty) = t^n f(x, y)$ ii) $f(x, y) = x^n g\left(\frac{y}{x}\right)$

iii) $f(x, y) = y^n h\left(\frac{x}{y}\right)$

14. Euler's theorem :

If u = f(x, y) is a homogenous function of degree 'n', then

i) $x\frac{\partial u}{\partial x} + y\frac{\partial u}{\partial y} = nu$

ii) $x^2\frac{\partial^2 u}{\partial x^2} + 2xy\frac{\partial^2 u}{\partial x\partial y} + y\frac{\partial^2 u}{\partial y^2} = n(n-1)u$

iii) $x\frac{\partial^2 u}{\partial x^2} + y\frac{\partial^2 u}{\partial x\partial y} = (n-1)\frac{\partial u}{\partial x}$

iv) $x\frac{\partial^2 u}{\partial x\partial y} + y\frac{\partial^2 u}{\partial y^2} = (n-1)\frac{\partial u}{\partial y}$

VIII. Maxima and Minima :

1. **Increasing function :** A function f (x) is said to be an increasing function over the interval [a, b] if for every pair x_1 and x_2 in [a, b]

$x_1 < x_2 \Rightarrow f(x_1) \le f(x_2)$

2. **Decresing function :** A function f(x) is said to be a decreasing function over the interval [a, b] if for every pair x_1 and x_2 in [a, b] $x_1 < x_2 \rightarrow f(x_1) \ge f(x_2)$

Note : 1. If $x_1 < x_2 \Rightarrow f(x_1) < f(x_2)$, then 'f' is said to be strictly increasing.

2. If $x_1 < x_2 \Rightarrow f(x_1) > f(x_2)$ then 'f' is said to be strictly decreasing.

3. Let the function f(x) be continuous on the interval [a,b] and differentiable in (a, b) then

i) $f'(x) > 0\ \forall$ x in (a, b) $\Rightarrow$ f(x) is increasing in (a, b)

ii) $f'(x) < 0\ \forall$ x in (a, b) $\Rightarrow$ f(x) is decreasing in (a, b)

iii) $f'(x) = 0\ \forall$ x in (a, b) $\Rightarrow$ f(x) is constant in (a, b)

4. If the above function is differentiable at $x = x_0 \in [a, b]$, then

i) $f'(x_0) > 0 \Rightarrow$ 'f' is increasing at x_0

ii) $f'(x_0) < 0 \Rightarrow$ 'f' is decreasing at x_0

iii) $f'(x_0) = 0 \Rightarrow$ 'f' is neither increasing nor decreasing x at x_0.

5. **Stationary points :** If $f'(x) = 0$ at x_1, x_2, x_3........ then $f(x)$ is said to be stationary at $x = x_1, x_2, x_3$..... and $f(x_1)$ is called the stationary value at 'x_1' and $(x_1, f(x_1))$ is called as a stationary point.

6. **Critical points :** The values of $x(x_1, x_2, x_3.......)$ for which 'f' is stationary are called critical numbers of the function and corresponding points of the graph are called critical points of the curve.

7. **Non-decreasing (or a non increasing) function :**

If $f'(x_0) \geq 0$ [or $f'(x_0) \leq 0$] for all $x \in [a, b]$ then 'f' is called a non-decreasing (or a non-increasing) function on [a, b]

8. **Monotonic function :** A function which is either increasing or decreasing is called a monotonic (monotone) function.

9. **Maximum and minimum values :**

Let y = f(x) be any function of 'x' supposed to be continuous in the neighbourhood of x = a. Then the value of f(a) is said to be a maximum or minimum value of f(x) according as f (a+h) is less or greater than f(a), whatever be the value of the increment 'h', positive or negative, provided it is taken sufficiently small and is not equal to zero.

10. **Properties of maxima and minima :**

i) At least one maximum or minimum must lie between two equal values of a function.

ii) Maximum and minimum values must occur alternatively.

iii) There may be several maximum or minimum values of the same function.

iv) A function y = f(x) is maximum at x = a, if $\frac{dy}{dx}$ changes sign from +ve to -ve as x passes through 'a'.

v) A function y = f(x) is minimum at x = a, if $\frac{dy}{dx}$ changes sign from -ve to +ve as 'x' passes through 'a'.

vi) If the sign of $\frac{dy}{dx}$ does not change while 'x' passes through 'a', then 'y' is neither maximum nor minimum at x = a.

11. **Largest and smallest values :**

i) The largest (or greatest value) of a function f(x) in an interval [a, b] is either a maximum value of f(x) at a point inside the interval or the value of f(x) at an end point of the interval (i.e. at x = a or b), whichever is greatest of all of them.

ii) The smallest (or least) value of f(x) in an interval [a, b] is either a minimum value of f(x) at a point inside the interval or the value of f(x) at an end point of the interval (i.e. at x = a or x = b), whichever is smallest of all of them.

12. **Necessary and sufficient condition for maximum or minimum values :** A necessary condition for f(x) to be a maximum or minimum at x = a is that f '(a) = 0

13. **Working rule for maxima and minima of f(x) :**

y = f(x)

↓

$\frac{dy}{dx} = f'(x)$ equated to zero

↓

$f'(x) = 0 \Rightarrow x = x_1, x_2, x_3$ ....

↓

$f''(x) = ...$

$f''(x_1) < 0$	$f''(x_2) > 0$	$f''(x_3) = 0$
$\Rightarrow f(x)$ is maximum at $x = x_1$	$\Rightarrow f(x)$ is minimum at $x = x_2$	$\Rightarrow f(x)$ is neither maximum not minimum at $x = x_3$

14. If a > 0, the extreme value is the minimum value, and if a < 0, the extreme value is the maximum value.

Let $f'(a) = f''(a) = ... = f^{n-1}(a) = 0$ and $f^n(a) \neq 0$

a) If 'n' is an even integer and $f^n(a) > 0$, then 'f' has a minimum value at x = a.

b) If 'n' is an even integer and $f^n(a) < 0$, then 'f' has a maximum at x = a.

c) If 'n' is an odd integer and $f^n(a) > 0$, then 'f' is increasing at x = a.

d) If 'n' is an odd integer and $f^n(a) < 0$, then 'f' is decreasing at x = a.

15. **Absolute maxima and minima :**

Consider a function f(x) defined on the interval [a, b]. Let $c \in [a, b]$ then

i) f(c) is called an absolute maximum value of the function if and only if $f(c) \geq f(x)$ for all 'x' in [a, b]

ii) f(c) is called an absolute minimum value of the function if and only if $f(c) \leq f(x) \forall$ x in [a, b]

OBJECTIVE QUESTIONS

1. Evaluate $\lim_{x \to 0} \frac{|x|}{x}$

a) 1 b) -1 c) 0 d) None

2. Evaluate $\lim_{x \to 1} \frac{\sqrt{1 - \cos 2(x-1)}}{x - 1}$

a) 1 b) -1

c) Does not exist d) None of these

3. Evaluate $\lim_{n \to \infty} \left[\frac{1}{1-n^2} + \frac{2}{1-n^2} + \frac{3}{1-n^2} + ... + \frac{n}{1-n^2}\right]$

a) $\frac{1}{2}$ b) $-\frac{1}{2}$ c) 1 d) None of these

4. Evaluate $\lim_{x \to -\infty} \frac{x^2 - 3x + 2}{2x^2 + x - 3}$

a) $\frac{1}{2}$ b) 1 $-\frac{1}{2}$ d) -1

5. Evaluate $\lim_{n \to \infty} \left(\frac{1}{1.3} + \frac{1}{3.5} + ... + \frac{1}{(2n+1)(2n+3)}\right)$

a) 1 b) 2 3) $\frac{1}{2}$ d) None

6. Evaluate $\lim_{x\to a}\frac{\sqrt{x-b}-\sqrt{a-b}}{x^2-a^2}\ (a>b)$ is

a) $\frac{1}{a\sqrt{a-b}}$ b) $\frac{1}{4a\sqrt{a-b}}$ c) $4a\sqrt{a-b}$ d) None

7. Evaluate $\lim_{x\to 0} x^x$

a) 1 b) 0 c) ∞ d) None of these

8. Evaluate $\lim_{n\to\infty}\frac{2.3^{n+1}-3.5^{n+1}}{2.3^n+3.5^n}$

a) 5 b) -5 c) 1 d) -1

9. Evaluate $\lim_{x\to\infty}\frac{[x]}{x}$

a) 1 b) 0 c) ∞ d) None of these

10. Evaluate $\lim_{n\to\infty} 1+\frac{1}{2}+\frac{1}{2^2}+....+\frac{1}{2n}$

a) 2 b) $\frac{1}{2}$ c) 1 d) ∞

11. Evaluate : $\lim_{n\to\infty}\frac{1}{n^4}\left(1^2+\left(1^2+2^2\right)+....+\left(1^2+2^2+n^2\right)\right)$

a) $\frac{1}{6}$ b) $\frac{1}{16}$ c) $\frac{1}{12}$ d) None of these

12. Evaluate $\lim_{x\to 0}\frac{1}{x}$

a) 0 2. ∞ c) 1 d) None of these

13. $f(x)=|x|+|x-1|$ then f(x) at x = 1 is

a) Continuous b) Discontinuous

c) Can not be determined d) None of these

14. The function f(x) = x - [x] at x = 5 is

a) Continuous b) Discontinuous

c) Can not be determined d) None of these

15. The function $f(x)=\begin{cases}|x-3| & x\ge 1\\ \frac{x^2}{4}-\frac{3x}{2}+\frac{13}{4} & x<1\end{cases}$ is

a) Continuous at x = 1 b) Discontinuous at x = 1

c) Can not be determined d) None of these

16. If $f:R\to R$ is continuous such that $f(x+y)=f(x)+f(y)$, and $\forall\ x,y\in R$ and $f(1)=2$ then f (100)

a) 100 b) 200 c) 10 d) 20

17. If $f(x)=\begin{cases}[x]+[-x] & x\ne 2\\ k & x=2\end{cases}$ is continuous at x = 2, then k =

a) 1 b) -1 c) 0 d) None of these

18. Let $f(x)=\frac{x^3+x^2-16x+20}{(x-2)^2}$ if $x\ne 2$

$= k$ if $x = 2$

if f(x) is continuous for all x; then 'k' equal to

a) -7 b) 10 c) 7 d) None of these

19. If y = f(x) is an odd differentiable function defined on $(-\infty,\infty)$ such that $f'(3)=-2$, then $f'(-3)$ equals

a) 0 b) 2 c) -2 d) 4

20. For a differentiable function 'f', the value of $\underset{n\to 0}{\text{Lt}}\frac{[f(x+n)]^2-[f(x)]^2}{2n}$ is equal to

a) $[f'(x)]^2$ b) $f(x).f'(x)$

c) $\frac{1}{2}[f(x)]^2$ d) $\frac{1}{2}[f(x)]^2-[f(x)]^2$

21. At the point x = 1, the function f(x)

$= x^3-1, 1<x<\infty = x-1, -\infty<x\le 1$, is

a) Continuous and differentiable

b) Continuous and not differentiable

c) Discontinuous and differentiable

d) Discontinuous and not differentiable

22. Let $f(x)=\frac{\sqrt{x+1}-1}{\sqrt{x}}, x\ne 0, f(0)=0$ then the set of all points where f(x) is derivable is

a) $(1,\infty)$ b) $(-1,\infty)$ c) $[0,\infty)$ d) $(0,\infty)$

23. If $g(x)=(x^2+2x+3)\ f(x)$ and $f(x)=5$ and $\underset{x\to 0}{\text{Lt}}\frac{f(x)-5}{x}=4$, then $g(0)=$

a) 20 b) 14 c) 18 d) 12

24. If $f(x)=e^x+a^x, x<0$

$=b(x-1)^2, x\ge 0$ is differentiable at x = 0, then (a, b) is

a) (-3, -1) b) (-3, 1) c) (3, 1) d) (3, -1)

25. Let $f(x)=x-[x]$, then $f'(x)=1$ for

a) All $x\in R$ b) For all $x\in R-\{0\}$

c) All $x\in R-I$ d) For all $x\in I$

26. If $f(x)=x^3\ \text{sgn}(x)$, then

a) 'f' is derivable at x = 0

b) 'f' is continuous but not derivable at x = 0

c) f' (0 +) = 2 d) f'(0 -) = 1

27. The constant 'c' of Rolle's theorem for the function $f(x)=\sin x$ in $[0,2\pi]$ is

a) $\pi/6$ b) $\pi/3$ c) $\pi/4$ d) $\pi/2$

28. The constant 'c' of the Lagrange's mean value theorem for the function $f(x)=1+x^2$ on [1, 2] is

a) 5/4 b) 3/2 c) 7/4 d) 9/8

29. The constant c of Cauchy's mean value theorem for the functions $f(x)=\sqrt{x}, g(x)=1/\sqrt{x}$ in [1, 2] is

a) $\sqrt{2}$ b) $\sqrt{3}$ c) 3/2 d) 5/4

30. $\int\frac{1+\sin^2 x}{1-\cos 2x}dx$

a) $-\cot\frac{x}{2}+\frac{x}{2}+c$ b) $\cot\frac{x}{2}+\frac{x}{2}+c$

c) $\frac{\cot x}{2}-\frac{x}{2}+c$ d) None

31. $\int x\sqrt{x^2-a^2}\ dx=$

a) $(x^2-a^2)^{3/2}+c$ b) $(a^2-x^2)^{2/3}+c$

c) $\frac{1}{3}(x^2-a^2)^{3/2}+c$ d) None

32. $\int \frac{1}{4x^2-4x+3}dx =$

a) $\frac{1}{2}Tan^{-1}\left(\frac{2x-1}{\sqrt{2}}\right)+c$ b) $\frac{1}{\sqrt{2}}Tan^{-1}\left(\frac{2x-1}{\sqrt{2}}\right)+c$

c) $\frac{1}{2\sqrt{2}}Tan^{-1}\left(\frac{2x-1}{\sqrt{2}}\right)+c$ d) None

33. $\int \frac{1}{\sqrt{2x-x^2}}dx =$

a) $Sin^{-1}(x-1)+c$ b) $Sinh^{-1}(x-1)+c$

c) $Cosh^{-1}(x-1)+c$ d) $Cos^{-1}(x-1)+c$

34. $\int \frac{x}{x^2+x+1}dx =$

a) $\log(x^2+x+1)-\frac{1}{\sqrt{3}}Tan^{-1}\left(\frac{2x+1}{\sqrt{3}}\right)+c$

b) $\frac{1}{2}\log(x^2+x+1)+\frac{1}{\sqrt{3}}Tan^{-1}\left(\frac{2x-1}{\sqrt{3}}\right)+c$

c) $\frac{1}{2}\log(x^2+x+1)-\frac{1}{\sqrt{3}}Tan^{-1}\left(\frac{2x+1}{\sqrt{3}}\right)+c$ d) None

35. $\int x^2 e^x dx$

a) $e^x\left[x^2+2x+2\right]+c$ b) $e^x\left[x^2+2x-2\right]+c$

c) $e^x\left[x^2-2x+2\right]+c$ d) None

36. $\int \sin^3\theta\, d\theta =$

a) $\frac{-\sin^2\theta\cos\theta}{3}-\frac{2}{3}\cos\theta+c$

b) $\frac{1}{3}\sin^2\theta\cos\theta-\frac{2}{3}\cos\theta+c\theta$

c) $\frac{1}{3}\sin^2\theta\cos\theta+\frac{2}{3}\cos\theta+c$ d) None

37. $\int \frac{(x+1)}{x(1+xe^x)^2}dx =$

a) $ln\left|\frac{xe^x}{1+xe^x}\right|+1+xe^x+c$ b) $ln\left|\frac{xe^x}{1+xe^x}\right|-\frac{1}{1+xe^x}+c$

c) $ln\left|\frac{xe^x}{1+xe^x}\right|+\frac{1}{1+xe^x}+c$ d) None

38. $\int_0^a \frac{x-a}{x+a}dx =$

a) $a+2a\log 2$ b) $a-2a\log 2$

c) $2a\log 2-a$ d) $2a\log 2$

39. $\int_0^{\pi^2/4}\cos\sqrt{x}\,dx =$

a) $\pi-2$ b) $\pi+2$ c) $\frac{\pi}{2}-1$ d) None

40. $\int_0^{\pi/2}\frac{x+\sin x}{1+\cos x}dx =$

a) $\pi/2$ b) π c) 2π d) $\frac{\pi^2}{2}$

41. $\int_0^{\pi/2}\sin^4 x\,dx =$

a) $\frac{\pi}{16}$ b) $\frac{\pi}{8}$ c) $\frac{3\pi}{16}$ d) None

42. The value of $\int_1^{e^{37}}\frac{\pi\sin(\pi\log x)}{x}dx =$

a) 1 b) 2 c) 1/2 d) None

43. If $I_n = \int_0^{\pi}\frac{\sin 2nx}{\sin x}dx$ then I_n (if n is an integer) =

a) 0 b) $\pi/4$ c) $3\pi/4$ d) None

44. $\underset{n\to\infty}{Lt}\left\{\frac{1}{n+1}+\frac{1}{n+2}+...+\frac{1}{6n}\right\} =$

a) log 4 b) log 5 c) log 6 d) None

45. Let $\frac{d}{dx}F(x)=\frac{e^{\sin x}}{x}$, $x>0$

If $\int_1^4 \frac{2e^{\sin x^2}}{x}dx = F(K)-F(1)$ then one of the possible values of K is

a) 16 b) 20 c) 4 d) 2

46. If $u=\sqrt{\sin^2 x+\sin^2 y+\sin^2 z}$, then $\frac{\partial u}{\partial z}$ at $\left(0,0,\frac{\pi}{4}\right)=$

a) 0 b) $\frac{1}{\sqrt{2}}$ c) $\frac{1}{2}$ d) None of these

47. If $u=e^{\frac{x^2+y^2}{2y}}$ then $x\frac{\partial u}{\partial x}+y\frac{\partial u}{\partial y}=$

a) 0 b) 3 c) $\frac{4}{2}$ d) log u

48. If $u=\frac{xy}{x+y}$ then $x\frac{\partial^2 u}{\partial x^2}+y\frac{\partial^2 u}{\partial x\partial y}=$

a) u b) 2u c) 3u d) None of these

49. If $u=\log(\tan x+\tan y)$ then $\sin 2x\frac{\partial u}{\partial x}+\sin 2y\frac{\partial u}{\partial y}=$

a) 1 b) 2 c) 3 d) None of these

50. If $v=\left(\frac{x}{z}\right)^2+\left(\frac{y}{z}\right)^2+\frac{x}{y}$ then $\Sigma x\frac{\partial v}{\partial x}=$

a) 0 b) 1 c) v d) None of these

51. If $u=(x^2+y^2+z^2)^{3/2}$ then $\left(\frac{\partial u}{\partial x}\right)^2+\left(\frac{\partial u}{\partial y}\right)^2+\left(\frac{\partial u}{\partial z}\right)^2=$

a) 9u b) $9u^{4/3}$ b) $9u^2$ d) $u^{4/3}$

52. The function $f(x) = 1 - x^3$

a) Increases every where b) Decreases every where

c) Increases in $(0, \infty)$ d) None of these

53. The function $f(x) = \frac{2x}{\ln x}$ increases in

a) (0, 1) b) (1, e) c) (e, ∞) d) $(-\infty, e)$

54. $f(x) = x + \cos x - a$ increases if

a) 'a' is natural number b) 'a' is rational number
c) 'a' is real number d) None of these

55. Maximum value of $f(x) = \frac{x}{(x+a)(x+b)}$ occurs when x=

a) A.M. of a, b b) G.M. of a, b
c) H.M. of a, b d) None of these

56. The maximum value of $y = \frac{x^2 - x + 1}{x^2 + x + 1}$ is

a) 1 b) 3 c) $\frac{1}{3}$ d) None of these

57. If $f(x) = \frac{a \sin x + b \cos x}{c \sin x + d \cos x}$ is strictly decreasing for all $x \in$ domain of $f(x)$, then

a) ad > bc b) ad < bc c) ad = bc d) None of these

ANSWER KEY

1. d	2. c	3. b	4. a	5. c
6. b	7. a	8. b	9. a	10. a
11. c	12. d	13. a	14. a	15. a
16. b	17. b	18. c	19. c	20. b
21. b	22. d	23. b	24. b	25. c
26. a	27. d	28. b	29. a	30. a
31. c	32. c	33. a	34. c	35. c
36. a	37. c	38. b	39. a	40. a
41. c	42. b	43. a	44. c	45. a
46. b	47. a	48. d	49. b	50. a
51. b	52. b	53. c	54. c	55. b
56. b	57. b			

EXPLANATORY NOTES

1. L.H.L R.H.L.

$= \lim_{x \to 0^-} \frac{|x|}{x}$ $= \lim_{x \to 0^+} \frac{|x|}{x}$

$= \lim_{x \to 0^-} \frac{-x}{x}$ $= \lim_{x \to 0^+} \frac{x}{x}$

(bY def. of $|x|$) (by def. of $|x|$)

$= \lim_{x \to 0^-} -1$ $= \lim_{x \to 0^+} 1$

= -1 = 1

$\therefore$ L.H.L. $\neq$ R.H.L.

$\therefore \lim_{x \to 0} \frac{|x|}{x}$ does not exists.

2. $\lim_{x \to 1} \frac{\sqrt{1 - \cos 2(x-1)}}{x-1}$; $\lim_{x \to 1} \frac{\sqrt{2 \sin^2 (x-1)}}{x-1}$

$\lim_{x \to 1} \sqrt{2} \frac{|\sin (x-1)|}{x-1} = \sqrt{2} \lim_{x \to 1} \frac{|\sin (x-1)|}{x-1}$ put y = x - 1

$x \to 1 \Rightarrow y \to 0 = \sqrt{2} \lim_{y \to 0} \frac{|\sin y|}{y}$

(we know that $\lim_{x \to 0} \frac{|x|}{x}$ does not exists by ex.1)

$= \sqrt{2}$ does not exists = does not exists

4. Consider $\lim_{x \to -\infty} \frac{x^2 - 3x + 2}{2x^2 + x - 3}$

Put x = - y; y = -x

as $x \to -\infty$; $y \to +\infty$

$\lim_{y \to \infty} \frac{y^2 + 3y + 2}{2y^2 - y - 3}$ dividing by y^2.

$= \lim_{y \to \infty} \frac{1 + \frac{3}{y} + \frac{2}{y^2}}{2 - \frac{1}{y} - \frac{3}{y^2}}$ as $y \to \infty$; $\frac{1}{y} \to 0 = \frac{1+0+0}{2-0-0} = \frac{1}{2}$

6. $\lim_{x \to a} \frac{\sqrt{x-b} - \sqrt{a-b}}{x^2 - a^2}$

Apply L' Hospital rule

$= \lim_{x \to a} \frac{\frac{1}{2\sqrt{x-b}}}{2x} = \frac{1}{4a\sqrt{a-b}}$

(by writing the limit)

7. Let $y = \lim_{x \to 0} x^x$

Apply logarithem

$\log_e^y = \lim_{x \to 0} x \log_e^x = \lim_{x \to 0} \frac{\log_e^x}{\frac{1}{x}} = \lim_{x \to 0} \frac{\frac{1}{x}}{\frac{-1}{x^2}}$

(applying derivative)

$= \lim_{x \to 0} -x$

$\therefore \log_e^y = 0$ $y = e^0 = 1$

$\therefore \lim_{x \to 0} x^x = 1$

8. Consider $\lim_{n \to \infty} \frac{2.3^{n+1} - 3.5^{n+1}}{2.3^n + 3.5^n}$

$= \lim_{n \to \infty} \frac{2\left(\frac{3}{5}\right)^n .3 - 3.5}{2\left(\frac{3}{5}\right)^n + 3} = \frac{-3.5}{3} = -5$

9. By definition of [x] ; we have $x - 1 < [x] \le x$

$= \frac{x-1}{x} \le \frac{[x]}{x} \le \frac{x}{x} = 1 - \frac{1}{x} \le \frac{[x]}{x} \le 1$

$= \lim_{x \to \infty} \left(1 - \frac{1}{x}\right) \le \lim_{x \to \infty} \frac{[x]}{x} \le 1$

$= 1 \le \lim_{x\to\infty} \frac{[x]}{x} \le 1$

by sandwitch theorem

$\therefore \lim_{x\to\infty} \frac{[x]}{x} = 1$

10. For a G.P. with r < 1 $\lim_{n\to\infty} 1+\frac{1}{2}+....+\frac{1}{2^n}$

$$= \lim_{n\to\infty} \frac{1\left(1-\frac{1}{2^n}\right)}{\left(1-\frac{1}{2}\right)} = \lim_{n\to\infty} \frac{1-0}{1-\frac{1}{2}} = \frac{1}{\frac{1}{2}} = 2.$$

11. We know that $\Sigma\left(\Sigma n^2\right) = \frac{n(n+1)^2(n+2)}{12}$

$$\therefore \text{L.H.S.} = \lim_{n\to\infty} \frac{1}{n^4}\left(\frac{n(n+1)^2(n+2)}{12}\right)$$

$$= \frac{1}{12}\lim_{n\to\infty}\left(1.\left(1+\frac{1}{n}\right)^2\left(1+\frac{2}{n}\right)\right) = \frac{1}{12}$$

13. L.H.L. $\lim_{x\to1^-} f(x) = \lim_{x\to1^-} x-x+1 = 1$

R.H.L. $\lim_{x\to1^+} f(x) = \lim_{x\to1^+} x+x-1 = \lim_{x\to1^+} 2x-1$

= 2 - 1 = 1

$f(1) = |1| + |1-1| = 1$

$\therefore$ L.H.L. = R.H.L. = f(1)

$\therefore$ f(x) is continuous at x = 1

14. L.H.L. $\lim_{x\to5^-} f(x) = \lim_{x\to5^+} x-[x]$ $\quad x \to 5^-$

$= \lim_{x\to5^+} x-4$ $\quad x < 5$

= 5 - 4 = 1 $\quad [x] = 4$

R.H.L. $\lim_{x\to5^+} f(x) = \lim_{x\to5^+} x-[x]$ $\quad x\to5^+$

$x > 5$

$[x] = 5$

$= \lim_{x\to5^+} x-5 = 5-5 = 0$

$f(5) = (x-[x])_{x=5} = 5-[5] = 0$

$\therefore$ L.H.L. = R.H.L. = f(5)

$\therefore$ f(x) is continuous at x = 5

16. $f(x+4) = f(x)+f(y)$

$f(1) = 2$

$f(2) = f(1+1) = f(1)+f(1) = 2+2 = 4 = 2.2$

$f(3) = f(2+1) = f(2)+f(1) = 4+2 = 6 = 2.3$

Similarly in general $f(x) = 2x$

$\therefore f(100) = 2.100 = 200$

17. f(x) is continuous at x = 2

L.H.L. = f(2)

$\lim_{x\to2^-} f(x) = k$

$x+2^- \Rightarrow x<2 \Rightarrow [x] = 1, [-x] = -2$

$\Rightarrow \lim_{x\to2^-} [x]+[-x] = k \Rightarrow 1-2 = k \Rightarrow k = -1$

18. f(x) is continuous at x = 2

$= \lim_{x\to2} f(x) = f(2)$

$$\Rightarrow k = \lim_{x\to2} \frac{x^3+x^2-16x+20}{(x-2)^2}\left(\frac{0}{0}\right)$$

Applying derivative

$$= \lim_{x\to2} \frac{3x^2+2x-16}{2(x-2)}\left(\frac{0}{0}\right)$$

Applying derivative

$$k = \lim_{x\to2} \frac{6x+2}{2} = \frac{14}{2} = 7 \therefore k = 7$$

19. since f(x) is odd $\therefore f(-x) = -f(x)$

$\therefore f'(-x)(-1) = -f'(x)$ i.e., $f'(-x) = f'(x)$

$\therefore f'(-3) = f'(3) = -2; \Rightarrow f'(-3) = -2$

20. By definition : the given limit

$$= \frac{1}{2}\frac{d}{dx}(f(x))^2 = \frac{1}{2}2f(x).f'(x) = f(x).f'(x)$$

22. $$f'(0) = \frac{f(x)-f(0)}{x-0} = \underset{x\to0}{\text{Lt}} \frac{\sqrt{x+1}-1}{x\sqrt{x}}$$

$$= \underset{x\to0}{\text{Lt}} \frac{x+1-1}{x\sqrt{x}\left[\sqrt{x+1}+1\right]} = \underset{x\to0}{\text{Lt}} \frac{1}{\sqrt{x}\left(1+\sqrt{1+x}\right)}$$

Which does not exist, $\therefore$ f'(0) does not exist.

If $x \ne 0$, f'(x)

$$= \frac{\sqrt{x}\frac{1}{2\sqrt{x+1}} - \left(\sqrt{x+1}-1\right)\frac{1}{2\sqrt{x}}}{x} = \frac{\sqrt{x+1}-1}{2x^{3/2}\sqrt{x+1}}$$

Which exists if x > -1 and x > 0

$\therefore$ f'(x) exists if x > 0 i.e., in $(0, \infty)$.

23. $g(x) = (x^2+2x+3)f(x)$

$g(0) = 3f(0) = 3(5) = 15$

$$g'(0) = \underset{x\to0}{\text{Lt}} \frac{g(x)-g(0)}{x-0} = \underset{x\to0}{\text{Lt}} \frac{(x^2+2x+3)f(x)-15}{x}$$

$$= \underset{x\to0}{\text{Lt}}\left[x+2+3\left(\frac{f(x)-5}{x}\right)\right]$$

$$= 0+2+3\underset{x\to0}{\text{Lt}}\frac{f(x)-5}{x} = 2+3(4) = 14$$

25. $f'(x) = 1 \;\forall\, x \in R-I$

($\because [x]$ is differentiable for all $x \in R-I$)

26. $f(x) = x^3 \operatorname{sgn}(x)$

$= x^3$ if $x > 0$ $\qquad$ clearly 'f' is continuous at

= 0 if x = 0 $\qquad$ x = 0

$= -x^3$ if $x < 0$ $\qquad$ $Lf'(0) = \underset{x\to0^-}{\text{Lt}} \frac{f(x)-f(0)}{x-0}$

$$= \underset{x\to0^-}{\text{Lt}} \frac{-x^3-0}{x} = 0$$

$$Rf'(0) = \underset{x\to0^+}{\text{Lt}} \frac{f(x)-f(0)}{x-0} = \underset{x\to0^+}{\text{Lt}} \frac{x^3-0}{x} = 0$$

$\therefore f'(0) = 0$

$\therefore$ 'f' is derivable at x = 0

27. f (x) = sinx is continuous on $[0, 2\pi]$ and derivable on $(0, 2\pi)$. Now $f'(x) = \cos x,\ \forall\ x \in (0, 2\pi)$.

$f(0) = \sin 0 = 0,\quad f(2\pi) = \sin 2\pi = 0 \Rightarrow f(0) = f(2\pi)$

By Rolle's theorem :

$\exists\ c \in (0, 2\pi) \ni f'(c) = 0 \Rightarrow \cos(c) = 0 \Rightarrow c = \pi/2$ or $3\pi/2$

28. Clearly f(x) is continuous on [1, 2] and differentiable on (1, 2). Now $f'(x) = 2x$.

By Lagrange's mean value theorem

$\exists\ c \in (1, 2) \ni f'(c) = \frac{f(2)-f(1)}{2-1} \Rightarrow 2c = \frac{5-2}{2-1} \Rightarrow 2c = 3 \Rightarrow c = 3/2$

29. Clearly $f(x) = \sqrt{x}$ is continuous on [1, 2], derivable on (1, 2) and $g(x) = 1/\sqrt{x}$ is continuous on [1, 2] derivable on (1, 2)

Now $f'(x) = \frac{1}{2\sqrt{x}},\ g'(x) = \frac{-1}{2x\sqrt{x}}\ \forall\ x \in (1, 2)$.

Further $g'(x) \neq 0,\ \forall\ x \in (1, 2)$

$\therefore$ By Cauchy's mean value theorem

$$\exists\ c \in (1, 2) \ni \frac{f'(c)}{g'(c)} = \frac{f(2)-f(1)}{g(2)-g(1)}$$

$$\Rightarrow \frac{1/(2\sqrt{c})}{-1/(2c\sqrt{c})} = \frac{\sqrt{2}-1}{1/\sqrt{2}-1/1} \Rightarrow -c = \frac{2-\sqrt{2}}{1-\sqrt{2}}$$

$$\Rightarrow -c = -\sqrt{2}\frac{(1-\sqrt{2})}{1-\sqrt{2}} \Rightarrow c = \sqrt{2} \in (1, 2)$$

30. $$\int \frac{1+\sin^2 x}{1-\cos 2x}dx = \int \frac{1+\sin^2 x}{2\sin^2 x}\,dx$$

$$= \frac{1}{2}\left[\int \frac{1}{\sin^2 x} + \int \frac{\sin^2 x}{\sin^2 x}dx\right]$$

$$= \frac{1}{2}\left[\int \operatorname{cosec}^2 x\,dx + \int 1\,dx\right] = \frac{1}{2}(x - \cot x) + c$$

31. Put $t = x^2 - a^2$, $dt = 2x\,dx$

$1/2\ dt = x\,dx$

$$\therefore \int x\sqrt{x^2-a^2}\ dx = \int \frac{1}{2}t^{1/2}\,dt = \frac{1}{2}t^{3/2}\frac{2}{3} + c$$

$$= \frac{1}{3}t^{3/2} + c = \frac{1}{3}(x^2-a^2)^{3/2} + c$$

32. $$\int \frac{1}{4x^2-4x+3}dx = \frac{1}{4}\int \frac{1}{x^2-x+3/4}dx$$

$$= \frac{1}{4}\int \frac{1}{\left(x-\frac{1}{2}\right)^2 + \left(\frac{1}{\sqrt{2}}\right)^2}dx$$

$$= \frac{1}{2\sqrt{2}}\operatorname{Tan}^{-1}\left(\frac{x-\frac{1}{2}}{\frac{1}{\sqrt{2}}}\right) + c = \frac{1}{2\sqrt{2}}\operatorname{Tan}^{-1}\left(\frac{2x-1}{\sqrt{2}}\right) + c$$

35. We know $\int UV\,dx = UV_1 - U'V_2 + U''V_3 - \ldots$

Hence $\int x^2 e^x\,dx = x^2e^x - 2xe^x + 2e^x + c$

$= e^x\left[x^2 - 2x + 2\right] + c$

36. We know $I_n = \int \sin^n\theta\ d\theta$

then $I_n = \frac{-\sin^{n-1}\theta}{n} + \frac{n-1}{n}I_{n-2}$

By applying this formula we get

$$I_3 = \frac{-\sin^2\theta\cos\theta}{3} - \frac{2}{3}\cos\theta + c$$

38. $$\int_0^a \frac{x-a}{x+a}dx = \int_0^a \frac{x+a-2a}{x+a}dx$$

$$= \int_0^a\left(1 - \frac{2a}{x+a}\right)dx = \int_0^a 1\,dx - 2a\int_0^a \frac{1}{x+a}dx$$

$$= [x]_0^a - 2a\log[x+a]_0^a$$

$$= a - 0 - 2a[\log 2a - \log a] = a - 2a\log 2$$

39. Put $t = \sqrt{x}$ so that $t^2 = x$

Now $2t\,dt = dx$

When x = 0, t = 0 and when $x = \frac{\pi^2}{4}$, $t = \frac{\pi}{2}$

$$\therefore \int_0^{\pi^2/4} \cos\sqrt{x}\,dx = \int_0^{\pi/2} 2t\cos t\,dt$$

$$= 2[t\sin t + \cos t]_0^{\pi/2} = \pi - 2$$

40. $$\int_0^{\pi/2} \frac{x+\sin x}{1+\cos x}dx = \int_0^{\pi/2} x\frac{1}{2}\sec^2\frac{x}{2}dx + \int_0^{\pi/2}\tan\frac{x}{2}dx$$

$$= \left[x\tan\frac{x}{2}\right]_0^{\pi/2} - \int_0^{\pi/2}\tan\frac{x}{2}dx + \int_0^{\pi/2}\tan\frac{x}{2}dx = \frac{\pi}{2}$$

41. We know if $I_n = \int_0^{\pi/2}\sin^n x\,dx$ then $I_n = \frac{n-1}{n}I_{n-2}$

$$\therefore I_4 = \frac{3}{4}\times\frac{1}{2}\times\frac{\pi}{2} = \frac{3\pi}{16}$$

42. Put $t = \log x \Rightarrow dt = \frac{1}{x}dx$

$$\therefore \int_1^{e^{37}} \frac{\pi\sin(\pi\log x)}{x} = \int_0^{37}\pi\sin(\pi t)dt$$

$$= [-\cos\pi t]_0^{37} = 1 - \cos 37\pi = 1 - (-1) = 2$$

43. $$I_n = \int_0^\pi \frac{\sin 2nx}{\sin x}dx = \int_0^\pi \frac{\sin 2n(\pi - x)}{\sin(\pi - x)}dx$$

But $\sin 2n(\pi - x) = \sin(n2\pi - 2nx) = -\sin 2nx\ \forall\ n \in Z$ and $\sin(\pi - x) = \sin x$

$$\therefore I_n = -\int_0^\pi \frac{\sin 2nx}{\sin x}dx = -I_n \Rightarrow 2I_n = 0 \Rightarrow I_n = 0$$

44. $\underset{n\to\infty}{Lt}\sum_{r=1}^{5n}\frac{1}{n+r}=\underset{n\to\infty}{Lt}\frac{1}{n}\sum_{r=1}^{5n}\frac{1}{1+\frac{r}{n}}$

Put $x=\frac{r}{n}$ when $r=1\Rightarrow x=\frac{1}{n}$ as $n\to\infty, x\to 0$

$r=5n\Rightarrow x=5$ as $n\to\infty$, $x\to 5$

$\therefore \underset{n\to\infty}{Lt}\frac{1}{n}\sum_{r=1}^{5n}\frac{1}{1+\frac{r}{n}}=\int_0^5\frac{1}{1+x}dx=[\log(1+x)]_0^5=\log 6$

45. $I=\int_1^4\frac{2e^{\sin x^2}}{x}dx=\int_1^4\frac{2xe^{\sin x^2}}{x^2}dx$

Put $x^2=t$, then $2x\,dx=dt$

When x = 1, t = 1, and x = 4, t = 16

Thus $I=\int_1^{16}\frac{e^{\sin t}}{t}dt=[F(t)]_1^{16}$

$F(16)-F(1)\Rightarrow K=16$

46. $\frac{\partial u}{\partial z}=\frac{1}{2\sqrt{u}}(2\sin z\cos z)\Rightarrow\left(\frac{\partial u}{\partial z}\right)_{\left(0,0,\frac{\pi}{4}\right)}=\frac{1}{\sqrt{2}}$

47. $\log u=\frac{x^2+y^2}{2y}$ is a homogeneous functiion of degree 0.

$\Rightarrow x\frac{\partial}{\partial x}(\log u)+y.\frac{\partial}{\partial y}(\log u)=0(\log u)\Rightarrow x\frac{\partial u}{\partial x}+y\frac{\partial u}{\partial y}=0$

48. 'u' is a homogeneous function of degree 1.

$\therefore x\frac{\partial^2 u}{\partial x^2}+y\frac{\partial^2 u}{\partial x\partial y}=(n-1)u=(1-1)u=0$

49. $e^u\frac{\partial u}{\partial x}=\sec^2 x$, $e^u\frac{\partial u}{\partial y}=\sec^2 y$

$\therefore (\sin 2x)\frac{\partial u}{\partial x}+(\sin 2y)\frac{\partial u}{\partial y}=\frac{2\tan x+2\tan y}{e^u}=2$

50. 'v' is a homogeneous function of degree 0

$\therefore \Sigma x\frac{\partial v}{\partial x}=nv=0.v.=0$

51. $\frac{\partial u}{\partial x}=\frac{3}{2}\left(\Sigma x^2\right)^{1/2}(2x)=3x\left(\Sigma x^2\right)^{1/2}$

$\therefore \left(\frac{\partial u}{\partial x}\right)^2+\left(\frac{\partial u}{\partial y}\right)^2+\left(\frac{\partial u}{\partial z}\right)^2=9\left[\Sigma x^2\right]^2=9\left(u^{2/3}\right)^2=9.u^{4/3}$

52. $f'(x)=-3x^2<0, \forall x\in R$

53. $f'(x)=\frac{2\ln x-2}{(\ln x)^2}>0\Rightarrow \ln x>1\Rightarrow x>e$

54. f ' (x) does not depend on 'a'.

55. Let log [f(x)] = y

$\Rightarrow\frac{dy}{dx}=\frac{1}{x}-\frac{1}{x+a}-\frac{1}{x+b}=0$

$\Rightarrow\frac{a}{x(x+a)}=\frac{1}{x+b}\Rightarrow ba=x^2$

56. $x^2(y-1)+x(y+1)+(y-1)=0$

x is real $\Rightarrow (y+1)^2-4(y-1)^2\geq 0$

$\Rightarrow\left(y-\frac{1}{3}\right)(y-3)\leq 0$

$\Rightarrow y\in\left[\frac{1}{3}, 3\right]$

VIII. PRACTICE QUESTIONS

SET - I

1. If repetitions are not allowed the number of two digit numbers, can be formed from the numbers 2,3,5,6 are
(a) 12 (b) 21 (c) 13 (d) 16
2. The distinct permutations that can be formed from all the letters of the word 'that'
(a) 17 (b) 14 (c) 11 (d) 12
3. There are 6 balls in a box. The number of ways two balls can be drawn with replacement =
(a) 32 (b) 34 (c) 36 (d) 33
4. In how many ways a student can answer 3 questions out of 5 questions in an examination
(a) 8 (b) 10 (c) 9 (d) 7
5. A coin is tossed 3 times. The number of elements in sample space
(a) 8 (b) 6 (c) 9 (d) 10
6. The probability of having at least one head if two coins are tossed
(a) 4/3 (b) 3/4 (c) 1/3 (d) 1/4
7. If two dice are thrown. Sample space contains .... number of elements.
(a) 34 (b) 33 (c) 36 (d) 35
8. The probability of getting a sum of 6 if two dice are thrown is ......
(a) 4/36 (b) 5/36 (c) 7/36 (d) 6/36
9. The probability of getting two variables un equal when we throw two dice is ....
(a) 1/6 (b) 6/5 (c) 4/6 (d) 5/6
10. If $P(A)=\frac{1}{4}$, $P(B)=\frac{1}{3}$, and $P(A\cap B)=\frac{1}{5}$, then $P(A\cup B)$ =.......
(a) 23/60 (b) 27/60 (c) 21/60 (d) 29/60
11. $P(A)=\frac{2}{3}$ and $P(A\cap B)=\frac{1}{4}$ then $P(A\cap B^c)$ ......
(a) 1/8 (b) 3/8 (c) 5/8 (d) 7/8
12. If A and B are mutually exclusive events then $P(A\cup B)=$.............
(a) P(A) - P (B) (b) P (A) + P (B)
(c) P(A) P(B) (d) P(B) / P(A)
13. If A and B are independent events and $P(A)=\frac{2}{3}$ and $P(B)=\frac{3}{5}$ then $P(A\cup B^c)$ =.............
(a) 3/5 (b) 4/5 (c) 2/5 (d) 1/5
14. If A and B are mutually exclusive events then $P(A^c\cup B^c)$ =............
(a) 4 (b) 2
(c) 3 (d) 1
15. Probability of getting one Ace and one 10 when we draw 2 cards with replacement from a pack of 52 cards =..........
(a) 1/169 (b) 2/169
(c) 5/169 (d) 7/169
16. There are 5 cards numbered 1 to 5. Two cards are drawn together. Then the probability of getting a sum is odd is ....
(a) 3/25 (b) 9/25
(c) 6/25 (d) 7/25
17. In a bag there are 3 Red balls 2 Blue balls. One ball is drawn. The probability of getting a Red ball. =.......
(a) 1/5 (b) 3/5
(c) 4/5 (d) 2/5
18. If a die is thrown, the probability of getting 5 or 6 is ....
(a) 2 (b) 3
(c) 2/3 (d) 1/3
19. There are 3 boys and 2 girls in a class, students are selected at random one after the other such that they are alternate. The probability of selecting is ....
(a) 2/5 (b) 5
(c) 1/5 (d) 1/6
20. If $P(A)=\frac{1}{3}$, $P(B)=\frac{1}{4}$, $P(A\cup B)=\frac{1}{3}$ then $P(B/A)=$..........
(a) 5/4 (b) 3/4
(c) 4/3 (d) 1/4
21. Box I contains 3 Red balls 2 White balls, box II contains 2 Red balls, 2 White balls One box is selected and one ball is selected. The probability that it is Red =................
(a) 20/11 (b) 9/20
(c) 20/9 (d) 11/20
22. In the above problem probability that if ball is Red, that is from box I =...............
(a) 6/11 (b) 5/11
(c) 7/11 (d) 8/11
23. The probabilities of two students to solve the problem 'independently are $\frac{1}{2}$ and $\frac{1}{3}$. The probability that the problem can be solved =........
(a) 2/3 (b) 3/2
(c) 1/3 (d) 8/3
24. If two dice are thrown the probability that at least one of the faces is 2 is ....
(a) 11/36 (b) 12/11
(c) 36/11 (d) 11/12
25. If a dice is thrown three times the sample space contains .... elements.
(a) 216 (b) 212 (c) 214 (d) 210
26. If three coins are tossed the probability of getting two heads is ....
(a) 3/8 (b) 4/5 (c) 5/8 (d) 1/8

27. If a coin weighed such that probability of getting head is $\frac{2}{3}$, Then probability of getting one head when two coins of same type are tossed is..........
(a) 5/9 (b) 4/9
(c) 3/5 (d) 2/9

28. The probability of failing in English is $\frac{1}{3}$. The probability of failing in Telugu is $\frac{1}{5}$. The probability of failing in both the subjects $\frac{2}{15}$ then if a student is selected at random, if he failed in Telugu what is the probability that he failed in English is ....
(a) 2/15 (b) 1/15 (c) 1/3 (d) 2/3

29. In the above problem of the student selected is failed in English, the probability that he is failed in Telugu is ....
(a) 4/5 (b) 3/5
(c) 2/5 (d) 1/5

30. If $P(A)=\frac{1}{3}$, $P(B)=\frac{1}{6}$, $P(A\cup B)=\frac{1}{4}$ then $P(B/A)=$ ....
(a) 4/5 (b) 1/4
(c) 3/4 (d) 5/4

31. There are 5 Red balls, 3 Black balls in a box two balls are selected probability that one is Red and one is Black is ....
(a) 17/64 (b) 15/64
(c) 19/64 (d) 11/64

32. If a die is thrown the probability that the face is 2,3 and 5 is..........
(a) 1/2 (b) 2/3
(c) 3/2 (d) 2

33. There are 25 cards numbered 1 to 25. Two cards are chosen. The probability that the number on the card is a multiple of 3 is ....
(a) 7/25 (b) 4/25
(c) 9/25 (d) 8/25

34. If A and B are independent events $P(A)=\frac{1}{4}$ and $P(A\cup B)=\frac{2}{3}$ then $P(B)=$ ....
(a) 7/9 (b) 1/9
(c) 5/9 (d) 2/9

35. If $P(A\cup B)=\frac{3}{4}$, $P(A^c)=\frac{2}{3}$, given that A and B are mutually exclusive. Then $P(B)=$ ....
(a) 3/2 (b) 1/3
(c) 2/3 (d) 9/2

36. $P(A/B)=\frac{1}{10}$, $P(B)=\frac{1}{5}$ then $P(A^c\cup B^c)=$ ....
(a) 49/50 (b) 47/50
(c) 1 (d) 43/50

37. Probability of getting a multiple of 2 is $\frac{1}{2}$ and probability of getting a multiple of 3 is $\frac{1}{3}$ in throwing a die, then $P(A\cup B)=$..............
(a) 4/3 (b) 2/3 (c) 5/3 (d) 1/3

38. In a factory there are 3 machines. The products 40%, 30% and 30% produced from the three machines. Among the items produced from the three machines 3%, 2% and 1% are defective. One item is selected. The probability that it is defective is
(a) 23/1000 (b) 19/1000
(c) 18/1000 (d) 21/1000

39. 60% of the students in the class are boys. 3% of the boys and 2% of the girls wear spectacles. If a student selected wears spectacles the probability that the student a boy is ....
(a) 9/13 (b) 11/3 (c) 2/11 (d) 11/2

40. There are two coins one is fair and the other is so weighed that the probability of turning up head is $\frac{2}{3}$. Probability of getting head when one coin selected and tossed is ....
(a) 1/12 (b) 5/12
(c) 7 /12 (d) 11/12

41. Two persons are playing a game. Player A toss a coin. If head turns up he get one Rupee if tail turns up he loses 1 Rupee his expected gain is...........
(a) 0 (b) 1 (c) 3 (d) 4

42. Two cards are drawn from a pack of 52 cards. The probability of getting one spade and one Red card is ....
(a) 5/8 (b) 3/8 (c) 7/8 (d) 1/8

43. Two dice are thrown. The probability of getting at least one six is ....
(a) 11/36 (b) 13/36 (c) 11 (d) 36

44. In a box there are Ten cards numbered 1 to 10. Two cards are drawn. The probability that both are even is ....
(a) 4/9 (b) 2/9 (c) 7/9 (d) 5/9

45. In a box there are 50 cards numbered 1 to 50. If a card is drawn. The probability that it is a multiple of 3 and 5 is ....
(a) 3/50 (b) 7/50
(c) 1/50 (d) 1/25

46. The event $A=$ The sum is odd in throwing two dice. $B=$ Event that sum is 6. Then $P(A\cap B)=$ ....
(a) 7/72 (b) 1/9
(c) 11/72 (d) 5/72

47. In tossing three coins. The probability of getting at least one head is =..................

(a) 5/8 (b) 6/7 (c) 7/8 (d) 5/7

48. Suppose $P(A) = P_1$, $P(B) = P_2$ then A and B are independent, then $P((A \cup B)^c) =$

(a) $(1+P_1)(1+P_2)$

(b) $(1-P_1)(1-P_2)$

(c) $(1+P_1)(1-P_2)$

(d) $P_1(1-P_2)$

49. Box 1 contains 2 Red balls – 3 White balls.
Box 2 contains 4 Red balls – 5 White balls.
One box is chosen and one ball is drawn. If the ball is white the probability that it is from box 2 is ....

(a) 22/43 (b) 26/43
(c) 27/43 (d) 25/43

ANSWER KEY

1. (a)	2. (d)	3. (c)	4. (b)	5. (a)
6. (b)	7. (c)	8. (b)	9. (d)	10. (a)
11. (c)	12. (b)	13. (a)	14. (d)	15. (a)
16. (c)	17. (b) 1	8. (d)	19. (c)	20. (b)
21. (d)	22. (a)	23. (a)	24. (a)	25. (a)
26. (a)	27. (b)	28. (d)	29. (c)	30. (c)
31. (b)	32. (a)	33. (d)	34. (c)	35. (c)
36. (a)	37. (b)	38. (d)	39. (a)	40. (c)
41. (a)	42. (d)	43. (a)	44. (b)	45. (a)
46. (d)	47. (c)	48. (b)	49. (d)	

SET - II

1. The mean of the probability distribution of the number of heads obtained in two flips of a balanced coin is ....

(a) 1 (b) 4
(c) 2 (d) 3

2. The mean of the probability distribution of the numbers on face of die in throwing a die is ....

(a) 2/7 (b) 7/2
(c) 3/7 (d) 5/7

3. If the probability distribution is

x	0	1	2	3	4
$p(x)$	.5	k	-1	3k	k

of a random variable X. Then k = ...

(a) 2 (b) 3 (c) 1 (d) 4

4. If the probability distribution of the random variable X is

x	0	1	2	3	4	5
$p(x)$	$\frac{1}{25}$	$\frac{2}{25}$	$\frac{3}{25}$	$\frac{1}{5}$	$\frac{1}{5}$	$\frac{9}{25}$

$p(1 < x < 4) =$

(a) 8/25 (b) 7/25
(c) 6/25 (d) 4/25

5. If a coin so weighed that probability of getting head is $\frac{2}{3}$. X is the random variable assignment to the no of head in tossing a coin. Mean is ....

(a) 2 (b) 2/3 (c) 3/2 (d) 1/3

6. Variance of getting head in tossing a coin for (8) = ...

(a) 0 (b) 2 (c) 1 (d) 3

7. Probability of getting 2 heads in tossing 5 coins is

(a) 3/256 (b) 4/256
(c) 6/256 (d) 5/256

8. If probability of getting 1 or 6 is a success. The probability of getting atleast once 1 or 6 if we throw a dice four times is ....

(a) 55/81 (b) 75/81
(c) 65/81 (d) 85/81

9. Two dice are thrown; getting doubled is a success. The probability of getting doublet in throwing two dice, three times is ....

(a) 45/72 (b) 25/72
(c) 35/72 (d) 26/72

10. Two items are selected from 5 items of which 3 are defective. The expected number of defected items is ....

(a) 6/5 (b) 5/6
(c) 3/5 (d) 2/5

11. A sample of 3 items is selected from a box having 6 items of which 3 are defective variance of the distribution of defective items is ....

(a) 9/20 (b) 1/20
(c) 3/20 (d) 7/20

12. The probability of getting 3 heads in tossing 5 coins is ....
(a) 7/16 (b) 3/16
(c) 5/16 (d) 3/5

13. The mean of the distribution of getting heads in tossing a coin 5 times is.................
(a) 1/5 (b) 1/2
(c) 5/2 (d) 2/5

14. The probability of getting 1 or 3 or 5 twice in throwing a die four times is ....
(a) 1/8 (b) 3/8
(c) 5/8 (d) 7/8

15. The probability of getting a spade in drawing three cards with replacement is ....
(a) 11/64 (b) 7/64
(c) 5/64 (d) 3/64

16. Mean of Bionomial distribution is 4 and variance is 2 then, p = ...
(a) 3 (b) 1
(c) 1/2 (d) 2

17. Mean of a Binomial distribution is 6 and variance is 3, then n = ...
(a) 6 (b) 8
(c) 10 (d) 12

18. Mean of Binomial distribution is 6 and variance is 2, mode = ...
(a) 12 (b) 3
(c) 6 (d) 18

19. A coin is tossed 7 times. The mode of distribution of getting head is ....
(a) 2 (b) 4
(c) 7 (d) 8

20. The probability of getting one boy in a family of 4 children is ....
(a) 1/4 (b) 4 (c) 3/4 (d) 8

21. In a village three are 96 families of 4 children each, the number of families of getting atleast one boy are ....
(a) 70 (b) 60
(c) 80 (d) 90

22. Mean of binomial distribution = 8, variance = 6, then the mode of the distribution is
(a) 6 (b) 2 (c) 8 (d) 4

23. A Poisson distribution is such that probability at $x = 1$ = probability at $x = 2$ then mean is ....
(a) 2 (b) 1
(c) 3 (d) 4

24. If $\mu = 1$, the probability that $p(x = 1)$ if x is poisson variate is ....
(a) e (b) 1/e
(c) e - 1 (d) 1-e

25. X is a poisson variate such that $p(x = 3) = p(x = 5)$ then μ = ...
(a) $\sqrt{24}$ (b) $\sqrt{15}$
(c) $\sqrt{5/3}$ (d) $\sqrt{3/5}$

26. X is a poisson variate such that $\frac{5}{6}p(x = 4) = p(x = 6)$ then μ = ...
(a) 4 (b) 6 (c) 5 (d) 7

27. If X is a poisson variate which is a bimodal at $x=3$ and at $x=4$. Then μ = ...
(a) 4 (b) 6 (c) 3 (d) 2

28. If X is a poisson variate such that $p(x=1)=2$ and $p(x=2)=1$ then μ = ...
(a) 2 (b) 4 (c) 1 (d) 6

29. If the variance of a poisson distribution is 2 then $p(x = 0)$ = ...
(a) .135 (b) .153
(c) .351 (d) .251

30. Among the items manufactured in a factory 5% are defective. The probability of getting one defective blade in a pack of five blades is ....
(a) .0244 (b) .344
(c) .002 (d) .0344

31. A manufacturer knows that the condesers he makes contain an average of 1% defective. He packs them in boxes of 100. The probability that a box packed contain no defective items.
(a) .4 (b) .476
(c) .376 (d) .367

32. If X is a poisson variate such that $\mu = 2$, $p(x = 1)$ = ...
(a) .27 (b) .9 (c) 1 (d) .4

33. If X is a poisson variate such that $\mu = 3$ then $p(x \geq 1)$ = ...
(a) .4233 (b) .58
(c) .6422 (d) .5432

34. If $f(x) = cxe^{-x}$ $if\ 0<x<\infty = 0$ else where is a a.p.d.f of a continuous random variable then c = ...
(a) 5 (b) 2 (c) 1 (d) 4

35. If p.d.f = $f(x) = \frac{x^2}{9}\ in\ 0<x\leq 3$, mean = ...
(a) 9/4 (b) 4/9 (c) 2/3 (d) 3

36. If p.d.f = $k(2x+3)$ in $0 < x < 2$ then k = ...
(a) 1/10 (b) 3/10
(c) 5/10 (d) 7/10

37. p.d.f = $\frac{1}{2}(2-x)$ in $0\leq x\leq 2$, $p(0 \leq x \leq a) = \frac{3}{4}$ then a = ...
(a) 1 (b) 3/4
(c) 2 (d) 1/2

38. If X is a random variable assigned to the number of heads obtained in tossing two coins and $p(|X-1|) \geq 2 \leq 2$ ....

(a) 4 (b) 3/8
(c) 8 (d) 1/8

39. A coin is tossed 4 times $p(1) = .25$, then $p(x=2) = 1$ ....

(a) 4/25 (b) 3/8
(c) 5/8 (d) 1/8

40. A coin is tossed n times the probability at $x = 2 = \frac{3}{8}$ and probability at $x = 3 = \frac{1}{4}$ then n = ...

(a) 2 (b) 4
(c) 6 (d) 8

41. A die is thrown 5 times getting 1 or 6 is a success. Probability of getting two successes is $\frac{80}{243}$. The probability of getting 3 successes is ....

(a) 79/243 (b) 82/243 (c) 81/243
(d) 80/243

42. If the mean of the Poisson distribution is 1. $p(0) = .51$ then $p(+1) =$ ....

(a) .61 (b) .16
(c) .15 (d) .51

43. If the X is poisson variate $p(x=1) = .3679$, $p(x=2) = .184$ then $\mu =$ ....

(a) 4 (b) 3
(c) 1 (d) 2

44. If x is a standard normal variate $p(0 \leq z \leq 1.2) =$ ....

(a) .3849 (b) .4498
(c) 1 (d) .6999

45. Area under the normal curve between $z=0$ and $z=.68$ = ...

(a) .2666 (b) .3224
(c) .3289 (d) .2518

46. If $p(0 \leq z \leq z_1) = .3770$ then z_1 = ...

(a) ± 1.16 (b) ± 1.97
(c) ± 2.4 (d) ± 2.26

47. The probability of getting no head in ten tosses of a fair coin is ....

(a) .01 (b) .02 (c) .03 (d) .04

48. If mean = 5 variance = $\frac{10}{3}$ of a binomial distribution, then n = ...

(a) 6 (b) 5
(c) 3 (d) 2

49. 10% of tools produced in a certain manufacturing process terms out to be defective. The probability that in a sample of 10 tools chosen at random exactly 2 will be defective is ....

(a) .09 (b) .234
(c) .184 (d) .2

50. One of the modes of poisson distribution is 2 and $p(x=2) = .612$ then $\mu =$ ....

(a) 3 (b) 4
(c) 2 (d) 5

51. If X is a poisson variate such that $p(x=2)=.27$ and $p(x=3)=.18$ then $\mu =$ ....

(a) 4 (b) 3
(c) 18 (d) 2

52. If a binomial distribution is bimodel at $x = 4$ and 5 and $p = q$ then $n=$ ....

(a) 20 (b) 9
(c) 4 (d) 5

53. Six coins are tossed, probability of getting 2 to 4 heads using normal distribution is

(a) .985 (b) .878
(c) .785 (d) .9

ANSWER KEY

1. (a)	2. (b)	3. (c)	4. (a)	5. (b)
6. (a)	7. (d)	8. (c)	9. (b)	10. (a)
11. (a)	12. (c)	13. (c)	14. (b)	15. (d)
16. (c)	17. (d)	18. (c)	19. (b)	20. (a)
21. (d)	22. (c)	23. (a)	24. (b)	25. (a)
26. (c)	27. (b)	28. (c)	29. (a)	30. (a)
31. (d)	32. (a)	33. (a)	34. (c)	35. (a)
36. (a)	37. (a)	38. (d)	39. (b)	40. (b)
41. (d)	42. (d)	43. (c)	44. (a)	45. (d)
46. (a)	47. (a)	48. (b)	49. (c)	50. (a)
51. (d)	52. (b)	53. (c)		

SET - III

1. If the size of sample is 200 and of population is 2000, then correction factor is
   (a) 7 (b) 1
   (c) 9 (d) 5
2. If the size of the sample is 5 and size of the population is 2000. The correction factor is ....
   (a) 999 (b) 2
   (c) 587 (d) 768
3. If size of the sample 25 and standard deviation of the population is .1. Then standard error of the mean of sampling distribution is ....
   (a) 07 (b) 03
   (c) 02 (d) 05
4. If size of the sample is 121 and standard deviation of the population is 2. Then standard error of the mean of sampling distribution is...................
   (a) 123 (b) 182
   (c) 312 (d) 24
5. If a sample is taken from In infinite population and sample size is increased from 25 to 100. The effect on standard error is ....
   (a) divided by 5 (b) divided by 3
   (c) divided by 4 (d) divided by 2
6. If a sample is taken from an infinite population and if sample size is increased from 400 to 900. The effect on standard error is ....
   (a) multiplied 3
   (b) multiplied by 2
   (c) multiplied by 3/2
   (d) multiplied by 2/3
7. If a sample is taken from an infinite population and if the sample size is decreased from 800 to 200. The effect on standard error is ....
   (a) multiplied by 2
   (b) multiplied by 4
   (c) multiplied by 3
   (d) multiplied by 5
8. If a sample is taken from an infinite population and the sample size is decreased from 1000 to 160. The effect on standard error is ....
   (a) multiplied by 2/5
   (b) multiplied by 5/2
   (c) multiplied by 3/5
   (d) multiplied by 4/5
9. If a sample of size 64 is taken from a population whose standard deviation is .4. Then probable error is ....
   (a) 637 (b) 537
   (c) 337 (d) 437
10. A random sample of size 64 is taken from an infinite population having mean 50 and the variance 25. The probability that $\bar{x}$ is less than 51.5 is ....
    (a) 1.899 (b) 9918
    (c) 2 (d) 2
11. The size of the sample is 121. Mean and standard deviation of the population are 70 and 10. The probability that $\bar{x}$ is greater than 72 is ....
    (a) 9 (b) 139
    (c) 0139 (d) 001
12. $\mu = 30.5$, $n = 100$, $\bar{x} = 28.8$, $\sigma = 8.35$.
    Then $|z|$ = ...
    (a) 3.39 (b) 2.68
    (c) 4.28 (d) 3.86
13. The sample of size 4 has values 25, 28, 26, 25 then variance of the sample is
    (a) 2 (b) 3
    (c) 4 (d) 5
14. If the population is 2,4,6,8,10. If samples of size 2 are taken with replacement. Then the mean of the means of sample distribution is ....
    (a) 5 (b) 6
    (c) 7 (d) 10
15. For the same population in problem (14), the samples of size 2 are taken without replacement then mean of means of sampling distribution is ....
    (a) 6 (b) 14
    (c) 7 (d) 12
16. Samples of size 2 taken from the population 62, 57, 63, 58 variance of the mean of sampling distribution is ....
    (a) 1.17 (b) 2
    (c) 2.17 (d) 3.7
17. The marks of 5 students in one subject are 45, 47, 49, 61, 48. and mean of the population is 52 then t = ...
    (a) 1.7 (b) .7
    (c) 2.17 (d) 1

ANSWER KEY

1. (c)	2. (a)	3. (c)	4. (b)	5. (d)
6. (c)	7. (a)	8. (b)	9. (c)	10. (b)
11. (c)	12. (b)	13. (a)	14. (b)	15. (a)
16. (c)	17. (b)			

TEST YOURSELF - I

1. Newton - Raphson method is used to compute a root of the equation $x^2 - 13 = 0$ with 3.5 as the initial value. The approximation after one iteration is **[GATE - 10]**

a) 3.676 b) 3.575 c) 3.607 d) 3.667

2. What is the value of $\lim_{n\to\infty}\left(1-\frac{1}{n}\right)^{2n}$? **[GATE - 10]**

a) 0 b) e^{-2} c) $e^{-1/2}$ d) 1

3. Consider a company that assembles computers. The probability of a faulty assembly of any computer is *p*. The company therefore subjects each computer to a testing process. This testing process gives the correct result for any computer with a probability of q. What is the probability of a computer being declared faulty ? **[GATE - 10]**

a) pq + (1 - p)(1 - q) b) (1-q)p

c) (1 - p) q d) pq

4. What is the probability that divisor of 10^{99} is a multiple of 10^{96}? **[GATE - 10]**

a) 1/625 b) 4/625 c) 12/625 d) 16/625

5. Consider the following matrix

$$A = \begin{bmatrix} 2 & 3 \\ x & y \end{bmatrix}$$

If the eigenvalues of A are 4 and 8, then **[GATE - 10]**

a) $x = 4, y = 10$ b) $x = 5, y = 8$

c) $x = -3, y = 9$ d) $x = -4, y = 10$

6. An unbalanced dice (with 6 faces, numbered from 1 to 6) is thrown. The probability that the face value is odd is 90% of the probability that the face value is even. The probability of getting any even numbered face is the same.

If the probability that the face is even given that it is greater than 3 is 0.75, which one of the following options is closest to the probability that the face value exceeds 3? **[GATE - 09]**

a) 0.453 b) 0.468 c) 0.485 d) 0.492

7. $\int_0^{\pi/4}(1-\tan x)/(1+\tan x)dx$ evaluates to **[GATE - 09]**

a) 0 b) 1 c) ln 2 d) ½ ln 2

8. $\lim_{x\to\infty}\frac{x-\sin x}{x+\cos x}$ equals **[GATE - 08]**

(a) 1 (b) -1

(c) ∞ (d) $-\infty$

9. The following system of equations

$$x_1 + x_2 + 2x_3 = 1$$

$$x_1 + 2x_2 + 3x_3 = 2$$

$$x_1 + 4x_2 + \alpha x_3 = 4$$

has a unique solution. The only possible value (s) for α is/are **[GATE - 08]**

(a) 0 (b) either 0 or 1

(c) one of 0, 1or -1 (d) any real number

10. The minimum number of equal length subintervals needed to approximate $\int_1^2 xe^x\,dx$ to an accuracy of at least $\frac{1}{3}\times 10^{-6}$ using the trapezoidal rule is **[GATE - 08]**

(a) 1000 e (b) 1000

(c) 100 e (d) 100

11. The Newton-Raphson iteration $x_{n+1} = \frac{1}{2}\left(x_n + \frac{R}{x_n}\right)$ can be used to compute the **[GATE - 08]**

(a) square of R (b) reciprocal of R

(c) square root of R (d) logarithm of R

12. Let $P = \sum_{\substack{1\le i\le 2k \\ i\ odd}} i$ and $Q = \sum_{\substack{1\le i\le 2k \\ i\ even}} i$, where k is a positive integer. Then **[GATE - 08]**

(a) P = Q-K (b) P=Q+K

(c) P=Q (d) P=Q+2K

13. A point on a curve is said to be an extremum if it is a local minimum or a local maximum. The number of distinct extrema for the curve $3x^4 - 16x^3 + 24x^2 + 37$ is **[GATE - 08]**

(a) 0 (b) 1

(c) 2 (d) 3

14. Aishwarya studies either computer science or mathematics everyday. If she studies computer science on a day, then the probability that she studies mathematics the next day is 0.6. If she studies mathematics on a day, then the probability that she studies computer science the next day is 0.4. Given that Aishwarya studies computer science on Monday, what is the probability that she studies computer science on Wednesday? **[GATE - 08]**

(a) 0.24 (b) 0.36 (c) 0.4 (d) 0.6

15. How many of the following matrices have an eigenvalue 1?

$$\begin{bmatrix}1 & 0\\0 & 0\end{bmatrix},\begin{bmatrix}0 & 1\\0 & 0\end{bmatrix},\begin{bmatrix}1 & -1\\1 & 1\end{bmatrix} \text{ and } \begin{bmatrix}-1 & 0\\1 & -1\end{bmatrix}$$

[GATE - 08]

(a) one (b) two
(c) three (d) four

16. Let **X** be a random variable following normal distribution with mean +1 and variance 4. Let **Y** be another normal variable with mean - 1 and variance unknown. If $P(X \le -1) = P(P \ge 2)$. the standard deviation of **Y** is **[GATE - 08]**

(a) 3 (b) 2
(c) $\sqrt{2}$ (d) 1

17. A Sample space has two events A and B such that probabilities $P(A \cap B) = 1/2, P(\overline{A}) = 1/3, P(\overline{B}) = 1/3$

What is $P(A \cup B)$? **[GATE - 08]**

(a) 11/12 (b) 10/12
(c) 9/12 (d) 8/12

18. What is the probability that in a randomly chosen group of r people at least three people have the same birthday? **[GATE - 08]**

(a) $1 - \dfrac{365.364...(365 - r + 1)}{365^r}$

(b) $\dfrac{365.364...(365 - r + 1)}{365^r} + {}^rC_2.365.\dfrac{364.363...(364 - (r - 2) + 1)}{364^{r-2}}$

(c) $1 - \dfrac{365.364...(365 - r + 1)}{365^r} - {}^rC_2.365.\dfrac{364.363...(364 - (r - 2) + 1)}{364^{r-2}}$

(d) $\dfrac{365.364...(365 - r + 1)}{365^r}$

19. The exponent of 11 in the prime factorization of 300! is **[GATE - 08]**

(a) 27 (b) 28
(c) 29 (d) 30

20. In how many ways can b blue balls and r red balls be distributed in n distinct boxes? **[GATE - 08]**

(a) $\dfrac{(n + b - 1)!(n + r - 1)!}{(n - 1)!b!(n - 1)!r!}$

(b) $\dfrac{(n + (b + r) - 1)!}{(n - 1)!(n - 1)!(b + r)!}$

(c) $\dfrac{n!}{b!r!}$

(d) $\dfrac{(n + (b + r) - 1)!}{n!(b + r - 1)!}$

21. Consider the field C of complex numbers with addition and multiplication. Which of the following form(s) a subfield of C with addition and multiplication?

(S1) the set of real numbers

(S2) {a+ib) | a and b are rational numbers }

(S3) {a + ib} | $(a^2 + b^2) \le 1$}

(S4) {ia | a is real } **[GATE - 08]**

(a) only SI (b) S1 and S3
(c) S2 and S3 (d) S1 and S2

22. If M is a square matrix with a zero determinant, which of the following assertion (s) is (are) correct?

(S1) each row of M can be represented as a linear combination of the other rows

(S2) Each column of M can be represented as a linear combination of the other columns

(S3) MX=0 has a nontrivial solution

(S4) M has an inverse **[GATE - 08]**

(a) S3 and S2 (b) S1 and S4
(c) S1 and S3 (d) S1,S2 and S3

23. Consider the function $f(x) = x^2 - 2x - 1$. Suppose an execution of the Newton-Raphson method to find a zero of $f(x)$ starts with an approximation $x_o = 2$ of x. What is the value of x_2, the approximation of x that algorithm produces after two iterations, rounded to three decimal places? **[GATE - 08]**

(a) 2.417 (b) 2.419
(c) 2.423 (d) 2.425

24. If $f(x)$ is defined as follows, what is the minimum value of $f(x)$ for $x \in (0,2)$?

$$f(x)=\begin{cases} \frac{25}{8x} & when\, x \le \frac{3}{2} \\ x+\frac{1}{x} & otherwise \end{cases}$$ [GATE - 08]

(a) 2 (b) $2\frac{1}{12}$

(c) $2\frac{1}{6}$ (d) $2\frac{1}{2}$

25. Consider the following two statements about the function a $f(x)=|x|$ $P. f(x)$ is continuous for all real values of x

$Q. f(x)$ is differentiable for all real values of x

Which of the following is TRUE? [GATE - 07]

(a) P is true and Q is false
(b) P is false and Q is true
(c) Both P and Q are true
(d) Both P and Q are false

26. Let A be a 4×4 matrix with eigenvalues -5, -2, 1, 4. Which of the following is an eigenvalue of $\begin{bmatrix} A & I \\ I & A \end{bmatrix}$, where I is the 4×4 identity matrix? [GATE - 07]

(a) - 5 (b) - 7 (c) 2
(d) 1

27. Consider the set of (column) vectors defined by $X = \{X \in R^3 \mid x_1 + x_2 + x_3 = 0$. where $x^T = [x_1, x_2, x_3]^T\}$. Which of the following is TRUE? [GATE - 07]

(a) $\{[1,-1,0]^T, [1,0,-1]^T\}$ is a basis for the subspace X.

(b) $\{[1,-1,0]^T, [1,0,-1]^T\}$ is a linearly independent set, but it does not span X and therefore is not a basis of X

(c) X is not a subspace of R^3

(d) None of the above

28. Consider the series $x_{n+1} = \frac{x_n}{2} + \frac{9}{8x_n}, x_0 = 0.5$ obtained from the Newton - Raphson method. The series converges to [GATE - 07]

(a) 1.5 (b) $\sqrt{2}$
(c) 1.6 (d) 1.4

Statement for Linked Answer Questions 29 and 30 :
Suppose the letters a,b,c,d,e,f have probabilities $\frac{1}{2}, \frac{1}{4}, \frac{1}{8}, \frac{1}{16}, \frac{1}{32}, \frac{1}{32}$, respectively

29. Which of the following is the Huffman code for the letters a,b,c,d,e,f ? [GATE - 07]

(a) 0, 10, 110, 1110, 11110, 11111
(b) 11, 10, 011, 010, 001, 000
(c) 11, 10, 01, 001, 0001, 0000
(d) 110, 100, 010, 000, 001, 111

30. What is the average length of the correct answer to Q.29 ? [GATE - 07]

(a) 3 (b) 2.1875
(c) 2.25 (d) 1.9375

Statement for Linked Answer Questions 31 and 32 :
Suppose that a robot is placed on the cartesian plane. At each step it is allowed to move either one unit up or one unit right, i.e., if it is at (i,j) then it can move to either $(i+1, j)$ or $(i, j+1)$.

31. How many distinct paths are there for the robot to reach the point (10, 10) starting from the initial position (0, 0) ? [GATE - 07]

(a) $\binom{20}{10}$ (b) 2^{20}

(c) 2^{10} (d) None of the above

32. Suppose that the robot is not allowed to traverse the line segment from $(4,4)$ to $(5,4)$. With this constraint, how many distinct paths are there for the robot to reach $(10,10)$ starting from $(0,0)$? [GATE - 07]

(a) 2^9 (b) 2^{19}

(c) $\binom{8}{4} \times \binom{11}{5}$ (d) $\binom{20}{10} - \binom{8}{4} \times \binom{11}{5}$

33. Suppose there are two coins. The first coin gives heads with probability 5/8 when tossed, while the second coin gives heads with probability 1/4. One of the two coins is picked up at random with equal probability and tossed. What is the probability of obtaining heads ? [GATE - 07]

(a) $7/8$ (b) $1/2$
(c) $7/16$ (d) $5/32$

34. Let A be the matrix $\begin{bmatrix} 3 & 1 \\ 1 & 2 \end{bmatrix}$. What is the maximum value of $x^T Ax$ where the maximum is taken over all x that are the unit eigenvectors of A ? [GATE - 07]

(a) 5 (b) $\frac{5+\sqrt{5}}{2}$
(c) 3 (d) $\frac{5-\sqrt{5}}{2}$

35. The trapezoidal method is used to evaluate the numerical value of $\int_0^1 e^x\, dx$. Consider the following values for the step size h.

(i) 10^{-2} (ii) 10^{-3}
(iii) 10^{-4} (iv) 10^{-5}

For which of these values of the step size h, is the computed value guaranteed to be correct to seven decimal places. Assume that there are no round - off errors in the computation. [GATE - 07]

(a) (iv) only (b) (iii) and (iv) only
(c) (ii), (iii) and (iv) only (d) (i), (ii), (iii) and (iv)

36. We are given a set $X = \{x_1, \ldots\ldots, x_n\}$ where $x_i = 2^i$. A sample $S \subseteq X$ is drawn by selecting each x_i independently with probability $p_i = \frac{1}{2}$. The expected value of the smallest number in sample S is [GATE - 06]

(A) $\frac{1}{n}$ (B) 2 (C) $\sqrt{n}$ D) n

37. For each element in a set of size 2n, an unbiased coin is tossed. The 2n coin tosses are independent. An element is chosen if the corresponding coin toss were head. The probability that exactly n elements are chosen is [GATE - 06]

(A) $\frac{\binom{2n}{n}}{4^n}$ (B) $\frac{\binom{2n}{n}}{2^n}$

(C) $\frac{1}{\binom{2n}{n}}$ (D) $\frac{1}{2}$

38. Let E,F and G be finite sets.

Let $X = (E \cap F) - (F \cap G)$ and $Y = (E - (E \cap G)) - (E - F)$. Which one of the following is true ? [GATE - 06]

(A) $X \subset Y$
(B) $X \supset Y$
(C) $X = Y$
(D) $X - Y \neq \varnothing$ and $Y - X \neq \varnothing$

39. F is an $n \times n$ real matrix. b is an $n \times 1$ real vector. Suppose there are two $n \times 1$ vectors, u and v such that $u \neq v$, and $Fu = b, Fv = b$. Which one of the following statements is false ? [GATE - 06]

(A) Determinant of F is zero
(B) There are an infinite number of solutions to $Fx = b$
(C) There is an $x \neq 0$ such that $Fx = 0$
(D) F must have two identical rows

40. Given a set of elements $N = \{1, 2, \ldots\ldots\ldots, n\}$ and two arbitrary subsets $A \subseteq N$ and $B \subseteq N$, how many of the n! permutations π from N to N satisfy min $(\pi(A))$=min $(\pi(B))$, where $\min(S)$ is the smallest integer in the set of integers S, and $\pi(S)$ is the set of integers obtained by applying permutation π to each element of S ? [GATE - 06]

(A) $(n-|A\cup B|)|A||B|$

(B) $(|A|^2+|B|^2)n^2$

(C) $n!\frac{|A\cap B|}{|A\cup B|}$ (D) $\frac{|A\cap B|^2}{\binom{n}{|A\cup B|}}$

41. In a certain town, the probability that it will rain in the afternoon is known to be 0.6. Moreover, meteorological data indicates that if the temperature at noon is less than or equal to $25°C$, the probability that it will rain in the afternoon is 0.4. The temperature at noon is equally likely to be above $25°C$, ot at/below $25°C$. What is the probability that it will rain in the afternoon on a day when the temperature at noon is above $25°C$? [GATE - 06]

a) 0.4 b) 0.6

c) 0.8 d) 0.9

42. When a coin is tossed, the probability of getting a Head is $p, 0<p<1$. Let N be the random variable denoting the number of tosses till the first Head appears, including the toss where the Head appears. Assuming that successive tosses are independent, the expected value of N is [GATE - 06]

a) $\frac{1}{P}$ b) $\frac{1}{(1-P)}$

c) $\frac{1}{P^2}$ d) $\frac{1}{(1-P)^2}$

43. What are the eigenvalues of the matrix P given below

$$P=\begin{pmatrix} a & 1 & 0 \\ 1 & a & 1 \\ 0 & 1 & a \end{pmatrix}$$ **[GATE - 06]**

a) $a, a-\sqrt{2}, a+\sqrt{2}$ b) a, a, a

c) $0, a, 2a$ d) $-a, 2a, 2a$

44. Match the following iterative methods for solving algebraic equations and their orders of convergence.

Method	Order of convergence
(I) Bisection	(P) 2 or more
(II) Newton-Raphson	(Q) 1.62
(III) Secant	(R) 1
(IV) Regula falsi	(S) 1 bit per iteration

a) $I-R, II-S, III-P, IV-Q$ **[GATE - 06]**

b) $I-S, II-R, III-Q, IV-P$

c) $I-S, II-Q, III-R, IV-P$

d) $I-S, II-P, III-Q, IV-R$

45. The following definite integral evaluates to

$$\int_{-\infty}^{0} e^{-\left(\frac{x^2}{20}\right)} dx$$ **[GATE - 06]**

a) $\frac{1}{2}$ b) $\pi\sqrt{10}$ c) $\sqrt{10}$ d) π

<u>**Statement for linked Answer Questions 46 & 47:**</u>

Consider the following set of linear equations :

$x + y/2 = 9$

$3x + y = 10$

46. The value of the Frobenius norm for the above system of equations is [GATE - 06]

a) 0.5 b) 0.75

c) 1.5 d) 2.0

47. What can be said about the Gauss-Siedel iterative method for solving the above set of linear equations ? [GATE - 06]

a) It will converge

b) It will diverse

c) It will neither converge nor diverse

d) It is not applicable

ANSWER KEY				
1. c	2. b	3. a	4. a	5. d
6. b	7. d	8. a	9. d	10. a
11. c	12. a	13. b	14. c	15. a
16. a	17. b	18. c	19. a	20. a
21. d	22. d	23. a	24. a	25. a
26. d	27. b	28. a	29. a	30. d
31. a	32. d	33. c	34. b	35. d
36. b	37. a	38. c	39. d	40. d
41. c	42. a	43. a	44. d	45.
46. a	47. d			

EXPLANATIONS

1. $X_{n+1} = X_0 - \frac{(x_0)}{f'(x_0)}$

$n = 0$

$x_1 = x_0 - \frac{f(x_0)}{f^1(x_0)}$

given $x_0 = 3.5,\ f(x) = x^2 - 13$

$f(3.5) = (3.5)^2 - 13 = -0.75$

f '(x) = 2x

$f'(3.5) = 2 \times 3.5 = 7$

So after one iteration

$$x_1 = x_0 - \frac{f(x_0)}{f_1(x_0)} = 3.5 - \frac{(-0.75)}{7} = 3.607$$

2. $\underset{n\to\infty}{Lim}\left(1-\frac{1}{n}\right)^{2n} = \left(\underset{n\to\infty}{Lim}\left(1-\frac{1}{n}\right)^{-n}\right)^{-2} = e^{-2}$

3. Given, the probability of faulty assembly of the computers = p

∴ The probability of non faulty assembly of the computers = 1 - p

The probability of correct result for any computer in testing process = q

∴ The probability of correct result for any computer in testing process = 1-q

The probability of a computer being declared faulty = probability of being faulty and probability of being declared or probability of not being faulty and probability of declaring incorrectly. $= pq + (1-p)(1-q)$

4. The divisors of 10^{96} which are multiples of 10^{99} are

$10^{96}, 2\times10^{96}, 4\times10^{96}, 5\times10^{96}, 8\times10^{96}, 10\times10^{96},$

$20\times10^{96}, 25\times10^{96}, 40\times10^{96}, 50\times10^{96}, 100\times10^{96},$

$125\times1096, 200\times1096, 250\times1096,$

$500\times1096, 1000\times10^{96}$

That is a total of 16

The total number of divisors of

$10^{99} = 2^{99}\times5^{99}(99+1)(99+1) = 10000$

The required probability = 16/10000=1/625

5. $A = \begin{bmatrix} 2 & 3 \\ x & 4 \end{bmatrix}$

$[A - \lambda I] = 0$

$$\begin{bmatrix} 2-\lambda & 3 \\ x & y-\lambda \end{bmatrix} = 0$$

$(2-\lambda)(y-\lambda) - 3x = 0$

given eigen values : $\lambda = 4,8$

if we use $\lambda = 4$ then equation , - 2 (y-4)-3x= 0 ....(i)

if we use $\lambda = 8$ then equation, -6(y-8)-3x=0 ....(ii)

Solving equations (i) & (ii), we get, x = -4, y = 10

6. Given : P(ODD)=0.9*P(EVEN) # P(2)=P(4) = P(6) # P(EVEN | Exceeds 3)=0.75

Now as we know

P(ODD)=1-P(EVEN)

or0.9*P(EVEN)=1-P(EVEN)

{ Given P(ODD)=0.9*P(EVEN) }

or P(EVEN)=1/1.9

Because of independent events

$P(2)=P(4)=P(6)=P(2\cup4\cup6)/3=P(EVEN)/3=1/(3*1.9)$

Now P(EVEN|Exceeds 3)=P(EVEN ∩ Exceeds3)/ P(Exceeds3)

P(Exceeds3) =P(EVEN ∩ Exceeds 3) /P(EVEN / Exceeds3)

P(Exceeds3) = $P(4\cup6)/0.75$

P(Exceeds3) = (P(4)+P(6))/0.75

P(Exceeds3) = (1/(3*1.9)+1/(1.9*3))/0.75

P(Exceeds3) = 0.468

7. $\frac{1-\tan x}{1+\tan x}dx = \frac{\cos x - \sin x}{\cos x + \sin x}dx$

Let $\cos x + \sin x = t$

$\therefore\ (-\sin x + \cos x)dx = dt$

So (1/t) dt = lnt = ln (sinx + cosx) = ln $\left(\sin\frac{\pi}{4} + \cos\frac{\pi}{4}\right)$

= ln $\left(\frac{1}{\sqrt{2}} + \frac{1}{\sqrt{2}}\right) = \frac{1}{2}\ ln_2$

8. $\lim\limits_{x\to\infty}\frac{x-\sin x}{x+\cos x} = \lim\limits_{x\to\infty}\frac{x(1-\sin x/x)}{x(1+\cos nx/x)}$

$$= \lim_{x\to\infty}\frac{1-\sin x/x}{1+\cos x/x} = \frac{1-\lim\limits_{x\to\infty}\frac{\sin x}{x}}{1+\lim\limits_{x\to\infty}\frac{\cos x}{x}} = \frac{1-0}{1+0} = 1$$

9. For unique solution, $\Delta \neq 0$

$\Delta_1 = \Delta_2 = \Delta_3 = 0;\qquad \Delta = \begin{bmatrix} 1 & 1 & 2 \\ 1 & 2 & 3 \\ 1 & 4 & \alpha \end{bmatrix} \neq 0$

$$\Delta_1 = \begin{vmatrix} 1 & 1 & 2 \\ 2 & 2 & 3 \\ 4 & 4 & \alpha \end{vmatrix} = 0$$

$\Rightarrow$ 0 = 0 (columns same)

so, α can't be determined by Δ_1

$$\Delta_2 = \begin{vmatrix} 1 & 1 & 2 \\ 1 & 2 & 3 \\ 1 & 4 & \alpha \end{vmatrix} = \Delta$$, so can't be

equated to zero otherwise

otherwise 0 = 0

$$\Delta_3 = \begin{vmatrix} 1 & 1 & 1 \\ 1 & 2 & 2 \\ 1 & 4 & 4 \end{vmatrix} = 0 \quad \text{No } \alpha$$

α can be any real number as any value of α will be Δ_1 and $\Delta_3 = 0$

10. $\int_1^2 x e^x\, dx = xe^x - e^x = [e^x(x-1)]_1^2$

Using trapezoidal rule we get 1000 e.

11. According to Newton - Raphson iteration,

$$x_{n+1} = x_n - \frac{f(x_n)}{f'(x_n)}$$

Let us convert the given expression to above form

$$x_{n+1} = \frac{x_n}{2} + \frac{R}{2x_n}$$

$$= x_n - \frac{x_n}{2} + \frac{R}{2x_n}$$

$$= x_n - \left(\frac{x_n}{2} - \frac{R}{2x_n}\right)$$

$$= x_n - \left(\frac{x_n^2 - R}{2x_n}\right)$$

$\therefore$ $f(x) = x^2 - R$ and $f(x) = 2x$

$\Rightarrow$ $x^2 = R$

$\Rightarrow$ $x = \sqrt{R}$

12. P = 1+ 3 + 5.... (2k-1) = k/2 (2(k-1)2)

Q =2 + 4 + 6..... k = k/2 x 2k =k^2

where k is (+) ve integer

$$Q = k^2 + k$$

So $P = Q - k$

Alternately

$$P = \Sigma i = 1 + 3 + ... (2k-1)$$

$$i \le \underset{i=odd}{i} \le 2k \qquad = \frac{k}{2}(1 + 2k - 1) = k^2$$

$$Q = \Sigma i = 2 + 4 + ...2k$$

$$i \le \underset{i=even}{i} \le 2k$$

$$= \frac{k}{2}(2 + 2k)$$

$$= k(k+1) = k^2 + k$$

$\therefore P = k^2$ and $Q = k^2 + k$

13. Given : $y = 3x^4 - 16x^3 + 24x^2 + 37$

For maximum value or minimum

$$\frac{dy}{dx} = 12x^3 - 18x^2 + 43x = 0$$

$\Rightarrow$ $x(x^2 - 4x + 4) = 0$

$\Rightarrow$ $x(x-2)^2 = 0$

$\Rightarrow$ $x = 0, 2, 2$

$$\frac{d^2y}{dx^2} = 36x^2 - 96x + 48$$

At $x = 0$, $\frac{d^2y}{dx^2} = +48$

At $x = 2$,

$$\frac{d^2y}{dx^2} = 36 \times 1 - 96 \times 2 + 48$$

= 0

so only 1 local minima.

either x=0 or complex root

so only value =0

14. Combinations are CMC, CMM, CCC, CCM

Probability of CMC = 0.6 X 0.4 =0.24 -Favourable

CMM = 0.6 X 0.6 =0.36

CCC = 0.4 X 0.4 =0.16-Favourable

CCM = 0.4 X 0.6 =0.24

Total Probability =1

Favourable =0.24 + 0.16 =0.4

so P = 0.4/1

15 (i) $\begin{bmatrix} 1 & 0 \\ 0 & 0 \end{bmatrix}$

$$\therefore \quad \begin{bmatrix} 1-\lambda & 0 \\ 0 & -\lambda \end{bmatrix} = 0$$

$$\Rightarrow \quad (1-\lambda)(-\lambda) = 0$$

Hence eigen values are $\lambda = 0, 1$

(ii) $\begin{bmatrix} 0 & 1 \\ 0 & 0 \end{bmatrix}$

$$\therefore \quad \begin{bmatrix} 0-\lambda & 1 \\ 0 & -\lambda \end{bmatrix} = 0$$

$$\Rightarrow \quad (-\lambda)(-\lambda) = 0$$

Hence eigen values are $\lambda = 0, 0$

(iii) $\begin{bmatrix} 1 & -1 \\ 1 & 1 \end{bmatrix}$

$$\therefore \quad \begin{bmatrix} 1-\lambda & -1 \\ 1 & 1-\lambda \end{bmatrix} = 0$$

$$\Rightarrow \quad (1-\lambda)^2 + 1 = 0$$

$$\lambda^2 - 2\lambda + 2 = 0$$

$$\Rightarrow \quad \lambda \neq 1$$

(iv) $\begin{bmatrix} -1 & 0 \\ 1 & -1 \end{bmatrix}$

$$\therefore \quad \begin{bmatrix} -1-\lambda & 0 \\ 1 & -1-\lambda \end{bmatrix} = 0$$

$$\Rightarrow \quad (-1-\lambda)(-1-\lambda) - 1 = 0$$

$$\Rightarrow \quad (1+\lambda)^2 - 1 = 0$$

$$\Rightarrow \quad \lambda^2 + 2\lambda + = 0$$

Hence eigen values are $\lambda = 0, -2$

16. Here standard deviation is 3.

17. $P(A \cap B) = \frac{1}{2}; P(\bar{A}) = \frac{1}{3}; P(\bar{B}) = \frac{1}{3}$

Now $P(A) = 1 - P(\bar{A}) = 1 - \frac{1}{3} = \frac{2}{3}$

$$P(B) = 1 - P(\bar{B}) = 1 - \frac{1}{3} = \frac{2}{3}$$

$$\therefore \quad P(A \cup B) = P(A) + P(B) - P(A \cap B)$$

$$= \frac{2}{3} + \frac{2}{3} - \frac{1}{2} = \frac{4}{3} - \frac{1}{2} = \frac{8-3}{6}$$

$$= \frac{5}{6} = \frac{5}{6} \times \frac{2}{2} = \frac{10}{12}$$

18. Probability that from a group of r people, at least 3 have the same birthday. Therefore

Probability of 1st man's birthday = $\frac{365}{365}$

Probability of 2nd man's birthday = $\frac{364}{365}$

Probability of 3rd man's birthday = $\frac{363}{365}$

Probability of rth man's birthday = $\frac{365 - r + 1}{365}$

Probability that all have birthday an different days

$$= \frac{365.364.363....(365 - r + 1)}{365^r}$$

Probability that at least 3 have same birthday =

$$1 - \frac{365.364.....(365 - r + 1)}{365^r}$$

19. 300! = 300.299.298 .....2.1

From 1 to 100, we have 9 multiples of 11, i.e 11, 22, 33, 44, 55, 66, 77, 88,99.

From 101 to 200 , we have 9 multiples of 11, i.e 110, 121, 132, 143, 154, 165, 176, 187,198.

Similarly from 201 to 300 , we have 9 multiples of 11.

So, total = 9 + 9 + 9 = 27

20. By direct formula, we have

$$\frac{(n+b-1)!(n+r-1)!}{(n-1)!b!(n-1)!r!}$$

ways in which b blue balls and r red balls be distributed in n distinct boxes

21. Complex number c is in the form a+ib, where a and b are real numbers a and b can be rational i.e in the form of $\frac{p}{q}$

22. Square Matrix = M

$$|M|=0$$

Now, by linear combination, we can form a linear equation over a one degree equation with non identified constants.

Every matrix can be diagonalised according to extreme elements of each row and each column and it can combine in the form of 2 or 3 variables so according to their linear combination, we can represent each row and each column as a linear combination of other rows and columns respectively.

MX = 0 has nontrivial solution

Here, $\frac{tx}{A} \neq \frac{ty}{B} \neq \frac{tz}{C}$

where A,B,C are co-efficients of linear dependent variable

And inverse, $M^{-1} = \frac{adj\,M}{|M|}$

so,if $|M|=0,$ it can't have inverse

23.

$$f(x)=x^2-2x-1$$

$$x_o=2$$

$$f(x)=2x-2$$

$$x_1=x_o-\frac{f(x_o)}{f'(x_o)}$$

$$=2-\frac{(-1)}{2}=2+\frac{1}{2}$$

$$=\frac{5}{2}$$

$$x_2=x_1-\frac{f(x_1)}{f'(x_1)}$$

$$=\frac{5}{2}-\frac{1/4}{3}$$

$$=\frac{5}{2}-\frac{1}{12}=\frac{30-1}{12}$$

$$=\frac{29}{12}$$

$$=2.14666$$

24. $f(x)=\begin{cases}\frac{25}{8x}, & x\le\frac{3}{2}\\ x+\frac{1}{x} & otherwise\end{cases}$

Now putting $x=\frac{3}{2}$

$$f(x)=\frac{25}{8\times\frac{3}{2}}=\frac{25}{12}$$

$$=2\frac{1}{12}$$

When x will be maximum f(x) will be minimum if $x\le\frac{3}{2}$

And, we know that minimum value of $x+\frac{1}{x}$ is 2

$\therefore$ minimum of $\left(2\frac{1}{12},2\right)$ is 2

25. $f(x) = |x|$

$$= \begin{cases} x & x \geq 0 \\ -x & x < 0 \end{cases}$$

Graph of $f(x)$ will be

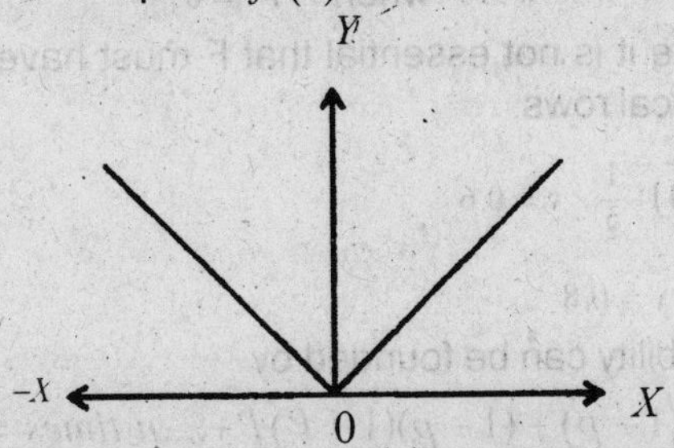

From above graph for $f(x)$, we can easily see that $f(x)$ is continuous for all real values of x.

But $f(x)$ is not defferentiable for all real values of x. For example at $x = 0$

Left hand derivative $= \lim_{h \to 0} \frac{f(0-h) - f(0)}{-h}$

$$= \lim_{h \to 0} \frac{-(0-h)}{-h} = -1$$

Right hand derivative $= \lim_{h \to 0} \frac{f(0+h) - (0)}{h}$

$$= \lim_{h \to 0} \frac{h}{h} = 1$$

Since, Left hand derivative $\neq$ Right hand derivative.

Therefore, $f(x)$ is not differentiable at $x = 0$, hence not differentiable for all real values of x.

28. The above series converges to 1.5.

29. Characters having higher probability will be needed lesses bits so number of bits needed are 1, 2, 3, 4, 5, 5

30. Average length = Σ bits required x probability

$$= \frac{1}{2} \times 1 + 2 \times \frac{1}{4} + 3 \times \frac{1}{8} + 4 \times \frac{1}{16} + 5 \times \frac{1}{32} + 5 \times \frac{1}{32}$$

$$= 1.9375$$

33. Probability of selecting one of two coins = $\frac{1}{2}$.

Probability of being head of the first coin = $\frac{5}{8}$

Probability of being head of second coin = $\frac{1}{4}$

So the Probability of obtaining heads =

$$\frac{1}{2}\left[\frac{5}{8} + \frac{1}{4}\right] = \frac{7}{16}$$

34. $A - \lambda I = 0$

$$\Rightarrow \begin{vmatrix} 3-\lambda & 1 \\ 1 & 2-\lambda \end{vmatrix} = 0$$

$$\Rightarrow (3-\lambda)(2-\lambda) - 1 = 0$$

$$\Rightarrow 6 - 3\lambda - 2\lambda + \lambda^2 - 1 = 0$$

$$6 - 5\lambda + \lambda^2 - 1 = 0$$

$$\Rightarrow \lambda^2 - 5\lambda + 5 = 0$$

$$\therefore \lambda = \frac{5 \pm \sqrt{25 - 4 \times 5 \times 1}}{2} = \frac{5 \pm \sqrt{5}}{2}$$

We take +ve sign because of maximum value of XT_{AX}. Hence answer is $\frac{5+\sqrt{5}}{2}$

36. Since the sample $S \subseteq X$, each element of S is an integral power of 2. The only possible value of the expected least number of S is 2 out of the options given.

37. Required favourable cases

= Number of ways of selecting the n elements out of the set of size 2n

$$= \binom{2n}{n}$$

Since each toss may result in either Head or tail (2 possibilities), hence for 2n tosses, total combinations $= 2^{2n} = 4^n$

$$\therefore \text{Required probability} = \frac{\text{Favourable cases}}{\text{Total cases}}$$

$$= \frac{\binom{2n}{n}}{4^n}$$

38. X and Y sets can be figured out using Venn Diagrams.

denotes $E \cap F$

denotes $F \cap G$

denotes $X = (E \cap F) - (F \cap G)$

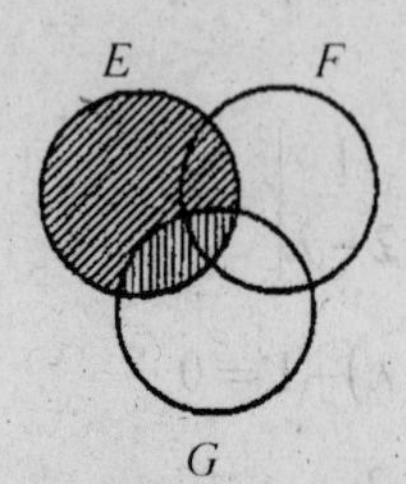

denotes $E \cap F$

denotes $E - (E \cap G)$

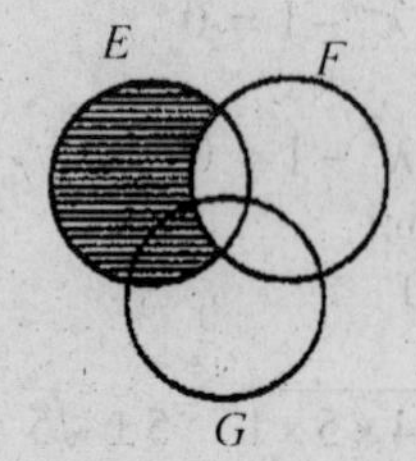

denotes $E - F$

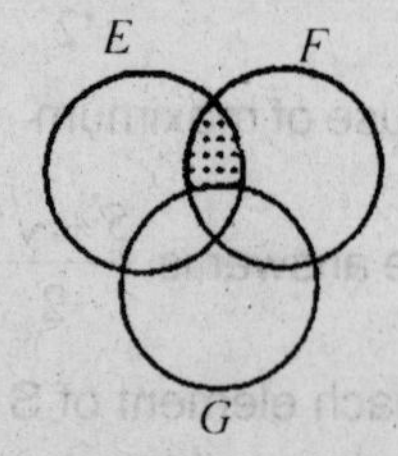

denotes $Y = (E - (E \cap G) - (E - F)$

$\therefore X = Y$

39. Lets take an example of F such that

$|F| = 0,\ n = 2$

$$\begin{bmatrix} 1 & 2 \\ 2 & 4 \end{bmatrix}\begin{bmatrix} x \\ y \end{bmatrix} = b \quad \text{.....(i)}$$

$$\begin{bmatrix} 1 & 2 \\ 2 & 4 \end{bmatrix}\begin{bmatrix} p \\ q \end{bmatrix} = b \quad \text{......(ii)}$$

where, $u = \begin{bmatrix} x \\ y \end{bmatrix},\ v = \begin{bmatrix} p \\ q \end{bmatrix}$

From equations (i) and (ii), we get

$$x + 2y = p + 2q$$

$$2x + 4y = 2p + 4q$$

The second equation reduces to the first equation such that we are left with a single condition for u and v to satisfy

$$x + 2y = p + 2q$$

This has a number of solutions.

e.g. $u = \begin{bmatrix} 2 \\ 3 \end{bmatrix}$ and $v = \begin{bmatrix} 4 \\ 2 \end{bmatrix}$

$\therefore \quad u \neq v$ when $|F| = 0$

Hence it is not essential that F must have two identical rows.

41. $\frac{1}{2} \times (0.4) + \frac{1}{2} \times x = 0.6$

$\therefore \quad x = 0.8$

42. Probability can be founded by

$$P + P(1-p) + (1-p)(1-P)P + \ldots .n \text{ times} = P$$

This is arithmetic Geometric progression on solving, we get the result.

43. Given $P = \begin{bmatrix} a & 1 & 0 \\ 1 & a & 1 \\ 0 & 1 & a \end{bmatrix}$

$$\therefore \quad P - \lambda I = \begin{bmatrix} a-\lambda & 1 & 0 \\ 1 & a-\lambda & 1 \\ 0 & 1 & a-\lambda \end{bmatrix}$$

$$\Rightarrow \quad (a-\lambda)\left[(a-\lambda)^2 - 1\right] - 1[a-\lambda] = 0$$

$$\Rightarrow \quad (a-\lambda)\left[a^2 + \lambda^2 - 2a\lambda - 1\right] - [a-\lambda] = 0$$

$$\Rightarrow \quad (a-\lambda)\left[a^2 + \lambda^2 - 2a\lambda - 1 - 1\right] = 0$$

$$\Rightarrow \quad (a-\lambda)\left[(a-\lambda)^2 - 2\right] = 0$$

$a = \lambda$ or $(a-\lambda)^2 = 2 \quad \Rightarrow (a-\lambda) = \pm\sqrt{2}$

$\therefore \quad \lambda = a,\ a - \sqrt{2},\ a + \sqrt{2}$

45. Given, $\int_{-\infty}^{0} e^{\left(\frac{x^2}{20}\right)} dx = \int_{0}^{\infty} e^{\left(\frac{-x^2}{20}\right)} dx$

$$\left[\because \int_{0}^{\infty} e^{-x^2} dx = \frac{\sqrt{\pi}}{2}\right]$$

Substituting $\frac{x^2}{20} = t^2 \Rightarrow x^2 = 20t^2$

$$\Rightarrow \quad x = \sqrt{20}\, t \Rightarrow dx = \sqrt{20}\, dt.$$

$$\therefore \quad \sqrt{20}\int_{0}^{\infty} e^{-t^2} dt = \sqrt{20} \times \frac{\sqrt{\pi}}{2} = 2\sqrt{5} \times \frac{\sqrt{\pi}}{2} = \sqrt{5\pi}$$

46. Frobenius norm $= \sqrt{trace\ of\left(AA^{H}\right)}$

A^{H} = Conjugate matrix or transpose matrix.

$$A=\begin{bmatrix}1 & 1/2\\ 3 & 1\end{bmatrix}$$

$$A.\,A^{H}=\begin{bmatrix}1+1/4 & 3+1/2\\ 3+1/2 & 9+1\end{bmatrix}=\begin{bmatrix}5/4 & 7/2\\ 7/2 & 10\end{bmatrix}$$

$$\therefore\ \text{Trace of}\begin{bmatrix}5/4 & 7/2\\ 7/2 & 10\end{bmatrix}=\frac{50}{4}-\frac{49}{4}=\frac{1}{4}$$

$$\sqrt{\frac{1}{4}}=0.5$$

TESTYOURSELF - II

1. Let P(E) denote the probability of the event E. Given P(A) = 1, P(B) = ½, the values of P(A/B) and P(B/A) respectively are **[GATE - 03]**

a) ¼, ½ b) ½, ¼ c) ½, 1 d) 1, ½

2. Let A be a sequence of 8 distinct integers sorted in ascending order. How many distinct pairs of sequences, B and C are there such that (i) each is sorted in ascending order, (ii) B has 5 and C has 3 elements, and (iii) the result of merging B and C gives A ? **[GATE - 03]**

a) 2 b) 30 c) 56 d) 256

3. n couples are invited to a party with the condition that every husband should be accompanied by his wife. However, a wife need not be accompanied by her husband. The number of different gatherings possible at the party is **[GATE - 03]**

a) $\binom{2n}{n} * 2^n$ b) $\frac{(2n)!}{2^n}$ c) 3^n d) $\binom{2n}{n}$

4. m identical balls are to be placed in n distinct bags. You are given that $m \geq kn$, where k is a natural number ≥ 1. In how many ways can the balls be placed in the bags if each bag must contain atleast k balls ? **[GATE - 03]**

a) $\binom{m-k}{n-1}$ b) $\binom{m-kn+n-1}{n-1}$

c) $\binom{m-1}{n-k}$ d) $\binom{m-kn+n+k-2}{n-k}$

5. Consider the following system of linear equations

$$\begin{bmatrix} 2 & 1 & -4 \\ 4 & 3 & -12 \\ 1 & 2 & -8 \end{bmatrix} \begin{bmatrix} x \\ y \\ z \end{bmatrix} = \begin{bmatrix} \alpha \\ 5 \\ 7 \end{bmatrix}$$

Notice that the second and the third columns of the coefficient matrix are linearly dependent. For how many values of α, does this system of equations have infinitely many solutions ? **[GATE - 03]**

a) 0 b) 1 c) 2 d) infinitely many

6. A piecewise linear function f(x) is plotted using thick solid lines in the figure below (the plot is drawn to scale)

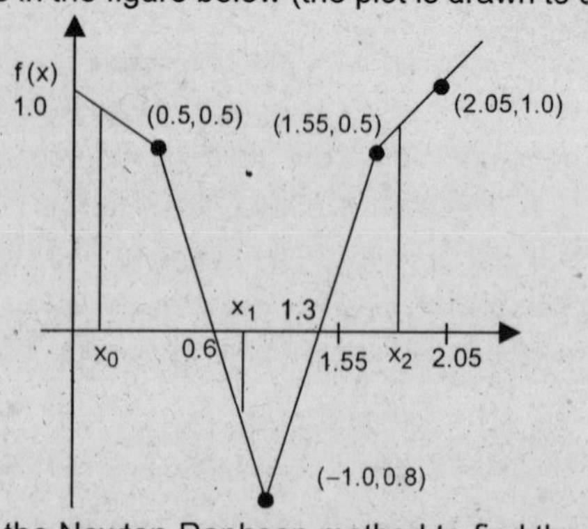

If we use the Newton-Raphson method to find the roots of f(x) = 0 using x_0, x_1 and x_2 respectively as initial guesses, the roots obtained would be **[GATE - 03]**

a) 1.3, 0.6 and 0.6 b) 0.6, 0.6 and 1.3

c) 1.3, 1.3, and 0.6 d) 1.3, 0.6 and 1.3

7. A program consists of two modules executed sequentially. Let $f_1(t)$ and $f_2(t)$ respectively denote the probability density functions of time taken to execute the two modules. The probability density function of the overall time taken to execute the program is given by **[GATE - 03]**

a) $f_1(t) + f_2(t)$ b) $\int_0^t f_1(x) f_2(x) dx$

c) $\int_0^t f_1(x) f_2(t-x) dx$ d) max $\{f_1(t), f_2(t)\}$

8. If a fair coin is tossed four times. What is the probability that two heads and two tails will result ? **[GATE - 04]**

a) 3/8 b) 1/2 c) 5/8 d) 3/4

9. The number of different n × n symmetric matrices with each element being either 0 or 1 is : (Note power (2, x) is same as 2^x) **[GATE - 04]**

a) power (2, n) b) power (2, n^2)

c) power (2, $(n^2+n)/2$) d) power(2, $(n^2-n)/2$)

10. Let A, B, C, D be n × n matrices, each with non-zero determinant. If ABCD = I then B^{-1} is **[GATE - 04]**

a) $D^{-1}C^{-1}A^{-1}$ b) CDA

c) ADC d) Does not necessarily exist

11. How many solutions does the following system of linear equations have ? **[GATE - 04]**

$$-x + 5y = -1$$
$$x - y = 2$$
$$x + 3y = 3$$

a) infinitely many b) two distinct solutions

c) unique d) none

12. An examination paper has 150 multiple-choice questions of one mark each, with each question having four choices. Each incorrect answer fetches -0.25 mark. Suppose 1000 students choose all their answers randomly with uniform probablity. The sum total of the expected marks obtained by all these students is **[GATE - 04]**

a) 0 b) 2550 c) 7525 d) 9375

13. Mala has a colouring book in which each English letter is drawn two times. She wants to paint each of these 52 prints with one of k colours, such that the colour-pairs used to colour any two letters are different. Both prints of a letter can also be coloured with the same colour. What is the minimum value of k that satisfies this requirement? **[GATE - 04]**

a) 9 b) 8 c) 7 d) 6

14. In an $M \times N$ matrix such that all non-zero entries are covered in a rows and b columns. Then the maximum number of non-zero entries, such that no two are on the same row or column, is **[GATE - 04]**

a) $\leq a+b$ b) $\leq \max\{a,b\}$

c) $\leq \min\{M-a, N-b\}$ d) $\leq \min\{a,b\}$

15. Two n bit binary strings, S_1 and S_2 are chosen randomly with uniform probability. The probability that the Hamming distance between these strings (the number of bit positions where the two strings differ) is equal to d is

a) ${}^nC_d/2^n$ b) ${}^nC_d/2^d$ **[GATE - 04]**
c) $d/2^n$ d) $1/2^d$

16. A point is randomly selected with uniform probability in the X-Y plane within the rectangle with corners at $(0, 0)$, $(1, 0)$, $(1, 2)$ and $(0, 2)$. If p is the length of the position vector of the point, the expected value of p^2 is **[GATE - 04]**

a) $2/3$ b) 1 c) $4/3$ d) $5/3$

17. In a population of N families, 50% of the families have three children 30% of the families have two children and the remaining families have one child. What is the probability that a randomly picked child belongs to a family with two children ? **[GATE - 04]**

a) $3/23$ b) $6/23$ c) $3/10$ d) $3/5$

18. In a class of 200 students, 125 students have taken Programming Language course, 85 students have taken Data Structures course, 65 students have taken Computer Organization course; 50 students have taken both Programming Language and Data Structures, 35 students have taken both Data Structures and Computer Organization; 30 students have taken both Data Structures and Computer Organization, 15 students have taken all the three course. How many students have not taken any of the three courses ? **[GATE - 04]**

a) 15 b) 20 c) 25 d) 35

19. What values of x, y and z satisfy the following system of linear equations ? **[GATE - 04]**

$$\begin{bmatrix} 1 & 2 & 3 \\ 1 & 3 & 4 \\ 2 & 2 & 3 \end{bmatrix}\begin{bmatrix} x \\ y \\ z \end{bmatrix} = \begin{bmatrix} 6 \\ 8 \\ 12 \end{bmatrix}$$

a) $x = 6, y = 3, z = 2$ b) $x = 12, y = 3, z = -4$
c) $x = 6, y = 6, z = -4$ d) $x = 12, y = -3, z = 0$

20. Let A be an $n \times n$ matrix of the following form.

$$A = \begin{bmatrix} 3 & 1 & 0 & 0 & 0 & \cdots & 0 & 0 & 0 \\ 1 & 3 & 1 & 0 & 0 & \cdots & 0 & 0 & 0 \\ 0 & 1 & 3 & 1 & 0 & \cdots & 0 & 0 & 0 \\ 0 & 0 & 1 & 3 & 1 & \cdots & 0 & 0 & 0 \\ \cdots\cdots \\ \cdots\cdots \\ 0 & 0 & 0 & 0 & 0 & \cdots & 1 & 3 & 1 \\ 0 & 0 & 0 & 0 & 0 & \cdots & 0 & 1 & 3 \end{bmatrix}_{n\times n}$$

What is the value of the determinant of A ? **[GATE - 04]**

a) $\left(\frac{5+\sqrt{3}}{2}\right)^{n-1}\left(\frac{5\sqrt{3}+7}{2\sqrt{3}}\right)+\left(\frac{5-\sqrt{3}}{2}\right)^{n-1}\left(\frac{5\sqrt{3}-7}{2\sqrt{3}}\right)$

b) $\left(\frac{7+\sqrt{5}}{2}\right)^{n-1}\left(\frac{7\sqrt{5}+3}{2\sqrt{5}}\right)+\left(\frac{7-\sqrt{5}}{2}\right)^{n-1}\left(\frac{7\sqrt{5}-3}{2\sqrt{5}}\right)$

c) $\left(\frac{3+\sqrt{7}}{2}\right)^{n-1}\left(\frac{3\sqrt{7}+5}{2\sqrt{7}}\right)+\left(\frac{3-\sqrt{7}}{2}\right)^{n-1}\left(\frac{3\sqrt{7}-5}{2\sqrt{7}}\right)$

d) $\left(\frac{3+\sqrt{5}}{2}\right)^{n-1}\left(\frac{3\sqrt{5}+7}{2\sqrt{5}}\right)+\left(\frac{3-\sqrt{5}}{2}\right)^{n-1}\left(\frac{3\sqrt{5}-7}{2\sqrt{5}}\right)$

21. Let X and Y be two exponentially distributed and independent random variables with mean α and β, respectively. If $Z=\min(X, Y)$, then the mean of Z is given by **[GATE - 04]**

a) $\frac{1}{\alpha+\beta}$ b) $\min(\alpha,\beta)$ c) $\frac{\alpha\beta}{\alpha+\beta}$ d) $\alpha\beta$

22. Let H_1, H_2, H_3........ be harmonic numbers. Then, for $n \in Z^+, \sum_{j=1}^{n} H_j$ can be expressed as **[GATE - 04]**

a) $nH_{n+1}-(n+1)$ b) $(n+1)H_n - n$
c) $nH_n - n$ d) $(n+1)H_{n+1}-(n+1)$

23. In how many ways can we distribute 5 distinct balls, $B_1, B_2 ..... B_5$ in 5 distinct cells, $C_1, C_2, ... C_5$ such that Ball B_i is not in cell $C_i, \forall i = 1,2 ..... 5$ and each cell contains exactly one ball ? **[GATE - 04]**

a) 44 b) 96 c) 120 d) 3125

24. If matrix $X = \begin{bmatrix} a & 1 \\ -a^2+a-1 & 1-a \end{bmatrix}$ and $X^2 - X + I = 0$ (I is the identity matrix and 0 is the zero matrix), then the inverse of X is **[GATE - 04]**

a) $\begin{bmatrix} 1-a & -1 \\ a^2 & a \end{bmatrix}$ b) $\begin{bmatrix} 1-a & -1 \\ a^2-a+1 & a \end{bmatrix}$

c) $\begin{bmatrix} -a & 1 \\ -a^2+a-1 & a-1 \end{bmatrix}$ d) $\begin{bmatrix} a^2-a+1 & a \\ 1 & 1-a \end{bmatrix}$

25. If $f(1) = 2$, $f(2) = 4$ and $f(4) = 16$, what is the value of $f(3)$ using Lagrange's interpolation formula ? **[GATE - 04]**

a) 8 b) $8\frac{1}{3}$ c) $8\frac{2}{3}$ d) 9

26. Consider the following iterative root finding methods and convergence properties : **[GATE - 04]**

Iterative root finding methods	Convergence properties
Q. False Position	I. Order of convergence = 1.62
R. Newton Raphson	II. Order of convergence = 2
S. Secant	III. Order of convergence = 1 with guarantee of convergence
T. Successive Approximation	IV. Order of convergence = 1 with no guarantee of convergence

The correct matching of the methods and properties is

a) Q-II, R-IV, S-III, T-I b) Q-III, R-II, S-I, T-IV
c) Q-II, R-I, S-IV, T-III d) Q-I, R-IV, S-II, T-III

27. Let $f(x)$ be the continuous probability density function of a random variable X. The probability that a $a < X \le b$ **[GATE - 05]**

a) $f(b-a)$ b) $f(b)-f(a)$ c) $\int_a^b f(x)\,dx$ d) $\int_a^b xf(x)\,dx$

28. Consider the following system of equations in three real variables x_1, x_2 and x_3 **[GATE - 05]**

$$2x_1 - x_2 + 3x_3 = 1$$
$$3x_1 - 2x_2 + 5x_3 = 2$$
$$-x_1 + 4x_2 + x_3 = 3$$

This system of equations has

a) no solution b) a unique solution
c) more than one but a finite number of solutions
d) an infinite number of solutions

29. What are the eigen values of the following 2×2 matrix? $\begin{bmatrix} 2 & -1 \\ -4 & 5 \end{bmatrix}$ **[GATE - 05]**

a) -1 and 1 b) 1 and 6
c) 2 and 5 d) 4 and -1

30. Let $G(x)=1/(1-x)^2=\sum_{i=0}^{\infty} g(i)x^i$ where $|x|<1$. What is g(i) ? **[GATE - 05]**

a) 1 b) i + 1 c) 2i d) 2^i

31. Box P has 2 red balls and 3 blue balls and box Q has 3 red balls and 1 blue ball. A ball is selected as follows :

(i) select a box

(ii) choose a ball from the selected box such that each ball in the box is equally likely to be chosen. The probabilities of selecting boxes P and Q are $\frac{1}{3}$ and $\frac{2}{3}$, respectively. Given that a bll selected in the above process is a red ball, the probability that it came from the box P is **[GATE - 05]**

a) $\frac{4}{19}$ b) $\frac{5}{19}$ c) $\frac{2}{9}$ d) $\frac{19}{30}$

32. A random bit string of length n is constructed by tossing a fair coin n times and setting a bit to 0 or 1 depending on outcomes head and tail, respectively. The probability that two such randomly generated strings are not identical is **[GATE - 05]**

a) $\frac{1}{2^n}$ b) $1-\frac{1}{n}$ c) $\frac{1}{n}!$ d) $1-\frac{1}{2^n}$

33. A bag contains 10 blue marbles, 20 green marbles and 30 red marbles. A marble is drawn from the bag, its colour recorded and it is put back in the bag. This process is repeated 3 times. The probability that no two of the marbles drawn have the same colour is **[GATE - 05]**

a) $\frac{1}{36}$ b) $\frac{1}{6}$ c) $\frac{1}{4}$ d) $\frac{1}{3}$

34. If the trapezoidal method is used to evaluate the integral $\int_0^1 x^2dx$, then the value obtained **[GATE - 05]**

a) is always $>\frac{1}{3}$ b) is always $<\frac{1}{3}$

c) is always $=\frac{1}{3}$ d) may be greater or lesser than $\frac{1}{3}$

35. The determinant of the matrix given below is

$$\begin{bmatrix} 0 & 1 & 0 & 2 \\ -1 & 1 & 1 & 3 \\ 0 & 0 & 0 & 1 \\ 1 & -2 & 0 & 1 \end{bmatrix}$$ **[GATE - 05]**

a) -1 b) 0 c) 1 d) 2

36. An unbiased coin is tossed repeatedly until the outcome of two successive tosses is the same. Assuming that the trails are independent, the expected number of tosses is **[GATE - 05]**

a) 3 b) 4 c) 5 d) 6

37. Let $n=p^2q$ whre p and q are distinct prime numbers. Howmany numbers m satisfy $1\le m\le n$ and $\gcd(m,n)=1$? Note that $\gcd(m,n)$ is the greatest common divisor of m and n. **[GATE - 05]**

a) $p(q-1)$ b) pq
c) $(p^2-1)(q-1)$ d) $p(p-1)(q-1)$

38. What is the value of $\int_0^{2\pi}(x-\pi)^3(\sin x)dx$ **[GATE - 05]**

a) -1 b) 0 c) 1 d) π

ANSWER KEY

1. d	2. d	3. a	4. b	5. b
6. b	7. c	8. a	9. d	10. b
11. c	12. d	13. a	14. a	15. a
16. d	17. b	18. c	19. c	20.
21. c	22. b	23. b	24. b	25. c
26.	27. c	28.	29.	30.
31.	32.	33. b	34. a	35. a
36. a	37. d	38.		

EXPLANATORY NOTES

2. Distinct pairs of sequence B and C = $2^8 = 256$

3. n couples means, number of persons in party = 2n. If wife need not accompanied by her husband, then, we choose, ${}^{2n}P_n\times 2^n$ way gatherings possible at the party.

4. Let n = 3; and k = 1; m = 3

Given that, n is number of distinct bags

○ ○ ○

Number of balls = 3

then number of ways balls be placed in 2 ways

then, number of ways = $3\times 2=6$

=< n ways

If K = 2 ○ ○ ○ (2, 2, 2)

Number of balls = nk = 6

We have choosen atleast k balls

Number of ways = one way.

5. $\begin{bmatrix} 2 & 1 & -4 \\ 4 & 3 & -12 \\ 1 & 2 & -8 \end{bmatrix}\begin{bmatrix} x \\ y \\ z \end{bmatrix}=\begin{bmatrix} \alpha \\ 5 \\ 7 \end{bmatrix}$

For this Linear equation, we write

$2x + y - 4z = \alpha$

$4x + 3y - 12z = 5$

$x + 2y - 8z = 7$

For infinitely solutions, D = 0

$$\begin{vmatrix} 2 & 1 & -4 \\ 4 & 3 & -12 \\ 1 & 2 & -8 \end{vmatrix} = 0$$

because 2nd and 3rd columns linearly dependent.
Now, we find the value of x, y, z
For x

$$D = \begin{vmatrix} \alpha & 1 & -4 \\ 5 & 3 & -12 \\ 7 & 2 & -8 \end{vmatrix} = 0$$

because, 2nd and 3rd columns are linearly dependent.

for y
$$D = \begin{vmatrix} \alpha & 1 & -4 \\ 5 & 5 & -12 \\ 1 & 7 & -8 \end{vmatrix} = 0$$

$$\Rightarrow 2(-40+84) - \alpha(-32+12) - 4(28-5) = 0$$

$$\Rightarrow 88 + 20\alpha - 92 = 0$$

$$\Rightarrow \alpha = \frac{4}{20} = \frac{1}{5} \quad \text{...... (i)}$$

for z
$$D = \begin{vmatrix} 2 & 1 & \alpha \\ 4 & 3 & 5 \\ 1 & 2 & 7 \end{vmatrix} = 0$$

$$\Rightarrow 2(21-10) - 1(28-5) + \alpha(8-3) = 0$$

$$\Rightarrow 22 - 23 + 5\alpha = 0$$

$$\Rightarrow \alpha = \frac{1}{5} \quad \text{.........(ii)}$$

We find from equation, α have only one value for infinitely solutions.

7. Probability density function of $f_1(t)$ and $f_2(t)$ is

$$= \int_0^t f_1(x) f_2(t-x)$$

8. Required probability = ${}^4C_2\left(\frac{1}{2}\right)^2\left(\frac{1}{2}\right)^2$

$$= \frac{3\times4}{2\times2} \times \frac{1}{4} \times \frac{1}{4} = \frac{3}{8}$$

10. $ABCD = I$

$$\Rightarrow AB = D^{-1}C^{-1}$$

$$(AB)^{-1} = \left(D^{-1}C^{-1}\right)^{-1} = B^{-1}A^{-1} = CD$$

$$\Rightarrow B^{-1} = CDA$$

11. $\frac{a_1}{a_2} \neq \frac{b_1}{b_2} \neq \frac{c_1}{c_2}$ Hence the linear equation has unique solution.

13. Want to paint 52 with k-color no adjacent letter will have the same colour. This problem is 9 colourable.
So, k = 9.

14. Given matrix is $M \times N$ (dimention) has
a - rows, and b - columns

Then maximum number of non-zero entries such that no two are on the same row and column is less than the sum of number of rows and column.
sum = a + b Therefore, number of zero entries $\leq a+b$

15. Simple space = 2^n

d-bit are choosen from all n-bit

so, favourable condition = ${}^nC_d$,

$$P = \frac{\text{favourable condition}}{\text{sample space}} = \frac{{}^nC_d}{2^n}$$

16. The expecation is $\frac{5}{3}$

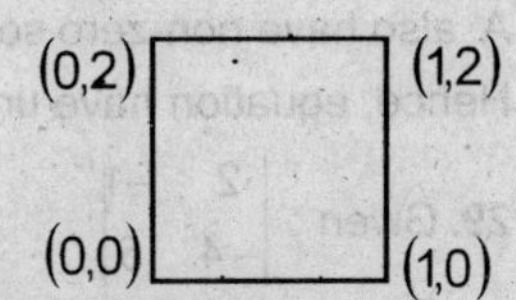

17. Required probability

$$= \frac{\frac{3N}{10}\times 2}{\frac{5N\times3}{10} + \frac{3N}{10}\times 2 + \frac{2N}{10}} = \frac{6N}{15N+6N+2N} = \frac{6}{23}$$

18. Total number student who have taken one or more courses = (125 + 85 + 65) - (50 + 35 + 30) + 15
= 275 - 115 + 15 = 175

22. Let H_1, H_2, H_3 be harmonic numbers, then

$$\sum_{j=1}^{n} H_j = (n+1)H_n - n.$$

23. Number of distribution

$$= 120 - 4\times3\times2\times1 = 120 - 24 = 96$$

24. $X = \begin{bmatrix} a & -1 \\ -a^2+a-1 & 1-a \end{bmatrix}$, Given : $X^2 - X + 1 = 0$

Multiplying both side by X^{-1}, we get

$$X - I + X^{-1} = 0 \Rightarrow X^{-1} = I - X$$

$$= \begin{bmatrix} 1 & 0 \\ 0 & 1 \end{bmatrix} - \begin{bmatrix} a & 1 \\ -a^2+a-1 & 1-a \end{bmatrix} = \begin{bmatrix} 1-a & -1 \\ a^2-a+1 & a \end{bmatrix}$$

25. $f(1) = 2$; $f(2) = 4$; $f(4) = 16$ $f(3) = ?$

Considering $X_0 = 1$; $X_1 = 2$, $X_2 = 4$, $X = 3$

$$f(3) = \frac{(3-2)(3-4)}{(1-2)(1-4)}\times 2 + \frac{(3-1)(3-3)}{(2-1)(2-4)}\times 4 + \frac{(3-1)(3-2)}{(4-1)(4-2)}\times 16$$

$$= \frac{1\times0}{(-1)(-2)}\times 2 + \frac{2\times0}{(1\times-1)}\times 4 + \frac{2\times1}{(3)(-2)}\times 16$$

$$= 0 + 0 + \frac{2}{6}\times 25 = \frac{25}{3} = 8\frac{1}{3}$$

28. Given :
$$2x_1 - x_2 + 3x_3 = 1$$
$$3x_1 + 2x_2 + 5x_3 = 2$$
$$-x_1 + 4x_2 + x_3 = 3$$

$$\therefore \quad A = \begin{vmatrix} 2 & -1 & 3 \\ 3 & 2 & 5 \\ -1 & 4 & 1 \end{vmatrix}$$

This is (3×3) matrix. To solve this equation first we find the rank of matrix. For any given equation if, $A' = A$ exist, then equation have unique solution.

Here, $A = 2[2-20]+1[3+8]+3[12+2]$

$= -36 + 42 + 8 = 14$

$A' = 3[-13]-10+35 = -49+35 = -14$

Hence, equation have non-zero solution. So, the rank of matrix is 3.

$$A' = \begin{vmatrix} 2 & 3 & -1 \\ -1 & 2 & 4 \\ 3 & 5 & 1 \end{vmatrix} = 2[2-20]-3[-1-12]-1[-5-6]$$

$= -36+39-11 = -8$

A' also have non-zero solution. So; rank of matrix is 3. Hence, equation have unique solution.

29. Given : $\begin{vmatrix} 2 & -1 \\ -4 & 5 \end{vmatrix}$

To solve this let

$\lambda I - A = 0$

Here, λ is identity matrix

$$\therefore \begin{vmatrix} \lambda & 0 \\ 0 & \lambda \end{vmatrix} - \begin{vmatrix} 2 & -1 \\ -4 & 5 \end{vmatrix} = 0$$

$$\Rightarrow \begin{vmatrix} (\lambda-2) & 1 \\ 4 & (\lambda-5) \end{vmatrix} = 0 \Rightarrow (\lambda-2)(\lambda-5)-4 = 0$$

$\Rightarrow \lambda^2 - 7\lambda + 10 - 4 = 0 \Rightarrow \lambda^2 - 7\lambda + 6 = 0$

$\Rightarrow \lambda^2 - 6\lambda - \lambda + 6 = 0 \Rightarrow \lambda(\lambda-6) - 1(\lambda-6) = 0$

$\Rightarrow (\lambda-6)(\lambda-1) = 0 \Rightarrow \lambda = 1, 6$

32. Probability for tossing a fair coin = n times

Hence, we select the number randomly 0 or 1.

Hence for random number, probability combination $= 2^n$

Here, we can't take same number 0 0 and 1 1

Hence, probability $= 1 - \frac{1}{2^n}$

1. Required Probability $= \left(\frac{10}{60} \times \frac{20}{60} \times \frac{30}{60}\right).3! = \frac{1}{6}$

2. $\int_0^1 x^2 dx = \frac{1}{2} n[(a_0 + a_n) + 2(a_1 + a_2 + a_3 + \ldots + a_{n-1})]$

where, n = 0.50 - 0.25 = 0.25

x	x^2
0.25	$0.0625 \to a_0$
0.50	$0.25 \to a_1$
0.75	$0.5625 \to a_2$
1.00	$1 \to a_n$

$\int_0^1 x^2 dx = \frac{1}{2} \times 0.25[1.0625 + 2(0.25 + 0.5625)]$

$= \frac{0.25}{2}[1.0625 + 2(0.8125)]$

$= \frac{0.25}{2}[1.0625 + 1.6250]$

$= \frac{0.25}{2}[2.6875] = 0.3359$

35. $$\begin{vmatrix} 0 & 1 & 0 & 2 \\ -1 & 1 & 1 & 3 \\ 0 & 0 & 0 & 1 \\ 1 & -2 & 0 & 1 \end{vmatrix} = -1\begin{vmatrix} -1 & 1 & 3 \\ 0 & 0 & 1 \\ 1 & 0 & 1 \end{vmatrix} = -1[-1(1)] = -1$$

37. Consider here example of 12

$12 = 2^2.3 \Rightarrow p = 2$ and $q = 3$ as $n = p^2q$

If we put in option (d), we get

$2(2-1)(3-1)$

$2.2 = 4$

Number which are co-prime with 12 are 1, 5, 7, 11

g c d (12, 1) = 1

g c d (12, 5) = 1

g c d (7, 12) = 1

g c d (7, 11) = 1

Unit - II
Discrete Mathematics

UNIT - II DISCRETE MATHEMATICS

I. SETS, RELATIONS AND FUNCTIONS

- Introduction
- Types of Sets
- Operations on Sets
- Applications of Set Theory
- Types of Relations
- Types of Functions

1. Set Theory and Its Basic Concepts

'A set is any collection of distinct and distinguishable objects around us'.

By the form 'distinct', we mean that no object is repeated and some lack the term 'distinguishable' we mean that whether that object is in our collection or not. The objects belonging to a set are called as elements or members of that set.

For example, say A is a set of stationary used by any student i.e.,

A = {Pen,Pencil,Eraser,Paper},

B = { Lucknow, Patna, Bhopal}

Here B is also a set of Capitals of State.

A set is represented by using all its elements between braces { } and by separating them from each other by commas (if there are more than one element).

As you have been sets are denoted by capital letters of English alphabet while the elements are divided in general, but small letters. If x is an element of a set A, we write $x \in A$ (read as 'x belongs to A'). If x is not an element of A, we write $x \notin A$ (read as x does not belong to A);

Let us see here two examples:

I. Let A = {4,2,8,2,6}. The elements of this collection are distinguishable but not distinct. Hence A is not a set.

II. Let B={a,e,i,o,u} i.e, B is set of vowels in English. Here elements of B are distinguishable as well as distinct. Hence B is a set.

2. Two Forms Of Representation Of a Set

1. 'Set-builders form' representation of set, and
2. 'Tabular form' or 'Roaster form' representation of set.

In 'Set-builder form' of representation of set, we write between the braces {} a variable x which stands for each of the elements of the set, then we state the properties possessed by x. We denote this property by p(x) and separate x and p(x) by a symbol: or (read as 'such that')

A = {x:p(x)}, A = {x : x is capital of States}

or A = {x : x is a natural number and 2<x<11}

In 'Tabular Form' or 'Roaster Form', the elements of a set are listed one by one within braces { } and one separated by each other by commas.

B = {Luknow, Patna,Bhopal} or B = { 3,4,5,6,7,8,9,10}

3. Types of Sets

1. Null set (or Empty Set or Void Set): A set having no elements is called as an empty set or void set. It is denoted by ϕ or { }

Ex: A = {x:x is an even number not divisible by 2 }

B = {the set of immortal men}

2. Singleton Set : A set having single element is called a singleton set.

Ex: A={x:x is present Prime Minister of India}, B={2}

3. Pair Set : A set having two elements is called a pair set.

Ex : {1,2},{0,3},{4,9} etc.

4. Finite Set: A set having a finite number of elements i.e., a set, where counting of elements is possible is called as a finite set.

Ex: A={1,2,4,6} is a finite set because it has four elements.

B=a null set ϕ is also a finite set because it has zero number of elements.

5. Infinite Set : A set having infinite number of elements i.e., a set where counting of elements is impossible, is called an infinite set.

Ex: A={x:x is a set of all natural numbers}

B= The set of all strengthness in a given plane.

4. Subsets, Super sets and Proper Subsets

Set A is said to be a subset of a set B if each element of set A is also an element of set B. If A is a subset of set B, we represent it as $A \subseteq B$. So, if $A \subseteq B \Leftrightarrow [x \in A \Rightarrow x \in B]$

Ex : Let A = {4,5,6} and B = {4,5,7,8,6} then we write $A \not\subseteq B$.

Let's see another example

$A=\{1,2,3\}, B=\{2,3,1\} \Rightarrow A \subseteq B$ also $B \subseteq A$

Here $A \subseteq B$ can also be expressed equivalently by writing $B \supseteq A$, read as B is superset of A.

So, a set A is said to be supeset of another set B, if and only if set B contains all the elements of set A.

Proper Subset : Set A is said to be a proper subset of a set B if

a) every element of set A is an element of set B, and

b) set B has at least one element which is not an element of set A.

This is expressed by writing $A \subset B$ or $B \subset A$ and read as A is a proper subset of B or B is a proper subset of A, if A is not a proper subset of B then we write it as $A \not\subset B$.

Example :

I. Let A = {4,5,6} and B = {4,5,6,7,8} so, $A \subset B$ and $B \supset A$

II. Let A = {1,2,3} and B = {3,2,9} so, $A \not\subset B$ and $B \not\supset A$

Comparability Of Sets : Two sets A and B are said to be comparable if either of these happens.

i) $A \subset B$ ii) $B \subset A$ iii) $A = B$

Similarly if neither of these above three exist i.e. $A \not\subset B$, $B \not\supset A$ and A'B, then let A and B are said to be incomparable.

Ex: A={1,2,3} and B={1,2}

Hence set A and B are comparable

But A={1,2,3} and B={2,3,6,7} are incomparable

Universal Set : Any set which is super set of all the sets under consideration is known as the universal set and is either denoted by W or S or U.

It is important to note that universal set can be chosen aribitarily for discussion, but once choosen, it is fixed for the discussion.

Ex : Let A={1,2,3}, B={3,4,6,9} and C={0,1}

We can take S={0,1,2,3,4,5,6,7,8,9} as Universal Set.

Power Set : The set or family of all the subsets of a given set A is said to be the power set of A and is expressed by P(A).

Mathematically, $P(A)=\{x : x \subseteq A\}$

So, $x \in P(A) \Rightarrow x \subseteq A$ It is to note that

i) $\phi \in P(A)$ and $A \in P(A)$ for all sets A.

ii) The elements of P(A) are the subsets of A.

For example : If, A={1} then P(A) = {ϕ ,{1}}

If A = {1,2}, then P(A) = {ϕ ,{1},{2}, {1,2}}

Similarly if A = {1,2,3}, then P(A)={ϕ ,{1},{2},{3}, {1,2},{1,3},{2,3},{1,2,3}}

So trends, show that if A has n elements then P(A) has 2^n elements. Further to note that it is not only true for n=1,2,3 but it is true of all n.

5. Some Operations On Sets

Union of Sets: The union of two sets A and B is the set of all those elements which are either in A or in B or in both. This set is denoted by AUB and read as 'A union B'.

Symbolically, $A \cup B = \{x:x \in A \text{ or } X \in B\}$

or $a \cup b = \{x:x \in AUx \in B\}$

For example: A = {4,5,6}, and B = {2,1,3,8}

$A \cup B = \{1,2,3,4,5,6,8\}$,

Intersection of Sets: The intersection of two sets A and B is the set of all the elements, which are common in A and B . This set is denoted by $A \cap B$ and read as 'A intersection of B'. i.e.,

$A \cap B = \{x:x \in A \text{ and } x \in B\}$ or $A \cap B = \{x:x \in A \cap x \in B\}$

For example: A = {1,2,3}, and B = {2,1,5,6} then $A \cap B = \{1,2\}$.

Difference of Sets: The difference of two sets A and B, is the set of all those elements of A which are not elements of B. Sometimes, we call difference of sets as the relative components of B in A . It is denoted by $A \cap B$ i.e.,

$A\text{-}B = \{x:x \in A \text{ and } x \notin B\}$

Similarly$B\text{-}A = \{x:x \in B \text{ and } x \notin A\}$

For example : A = {4,5,6,7,8,9}, and B = {3,5,2,7} then A-B = {4,6,8,9}

Similarly, B-A = {3,2}.

No need to mention that $A\text{-}B \neq B\text{-}A$

So, difference of two sets is not commutative.

Some important laws and Theorems

1. $A \cup A=A$, $A \cap A=A$. 2. $A \cup S=S$, $A \cap S=A$.

3. $A \cup \phi=A, A \cap \phi=\phi$ 4. $A\text{-}A=\phi, \phi\text{-}A=\phi, A\text{-}\phi=A$.

Theorem 1: For any two sets A and B,

a) $A \cup B=B \cup A$. b) $A \cap B=B \cap A$.

These two laws are known as commutative law of union and commutative law of intersection. This law holds even for three sets i.e.

i) $A \cup (B \cup C) = (A \cup B) \cup C$

ii) $A \cap (B \cap C)=(A \cap B) \cap C$.

Theorem 2: For any three sets A,B and C , the following distributive laws hold:

1) $A \cup (B \cap C)=(A \cup B) \cap (A \cup C)$

2) $A \cap (B \cup C)=(A \cap B) \quad (A \cap C)$.

Complements of Set: The complement of a set A, also known as ' absolute complement' of A is the sets of all those elements of the universal sets which are not element of A. It is denoted by A^c or A^1.

Infact A' or A^c = S-A. or $A'=\{x:x \in S \text{ and } x \in A\}$.

DeMorgan's Laws : For any two sets A and B the following laws known as De Morgan's Laws.

1. $(A \cup B)' = A' \cap B'$ and 2. $(A \cap B)' = A' \cup B'$.

Demorgan's Law can also be expressed as

(i) $A - (B \cup C) = (A - B) \cap (A - C)$

(ii) $A - (B \cap C) = (A - B) \cup (A - C)$

For any three sets A, B and C.

6. Venn Diagram and Some Of The Applications Of Set Theory

Here we will learn the operations on sets and its applications with the help of pictorial repersentation of the sets. The diagram formed by these sets is said to be the Venn Diagram of that statement.

A set is represented by circles or a closed geometrical figure inside the universal set. The Universal Set S, is represented by a rectangular region. First of all we will represent the set or a statement regarding sets with the help of the diagram or Venn Diagram. The shaded area represents the set written.

(a) Subset:

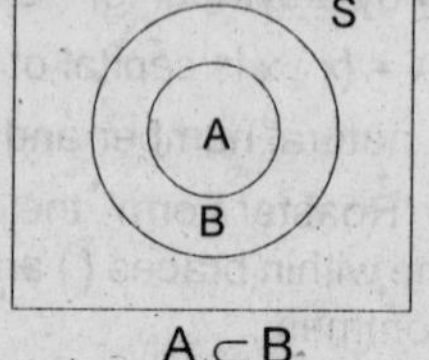

$A \subset B$

(b) Union of Sets : Let $A \cup B = B$. Here, whole area represented by B represents $A \cup B$.

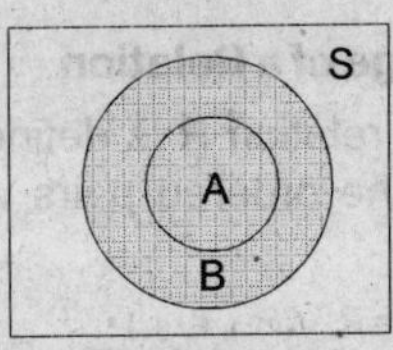

$A \subset B$ when $A \subset B$.

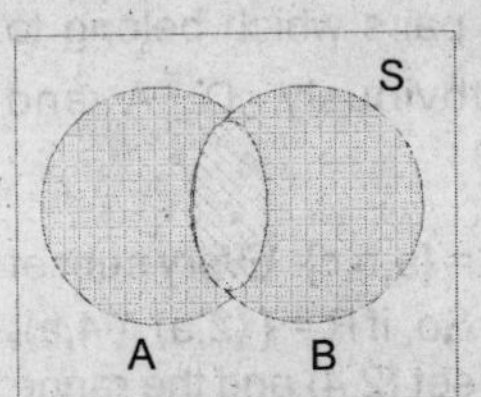

$A \subset B$ when neither $A \subset B$ nor $B \subset A$

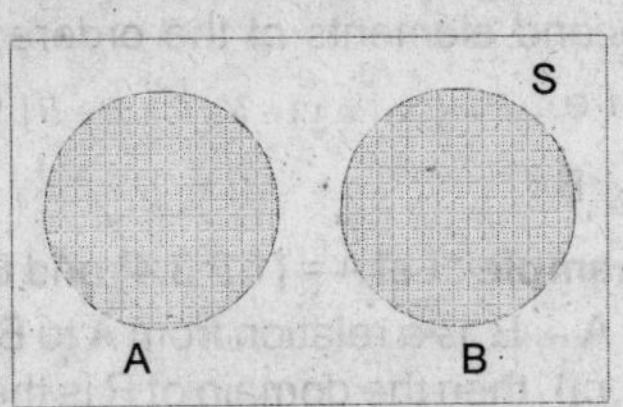

$A \cup B$ when A and B are disjoint events

(c) Intersection of Sets: $A \cap B$ represents the common area of A and B

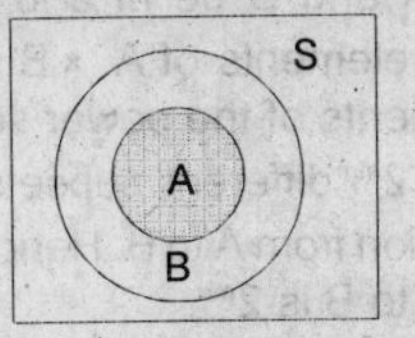

$(A \cap B)$ when $A \subset B$

$A \cap B = A$

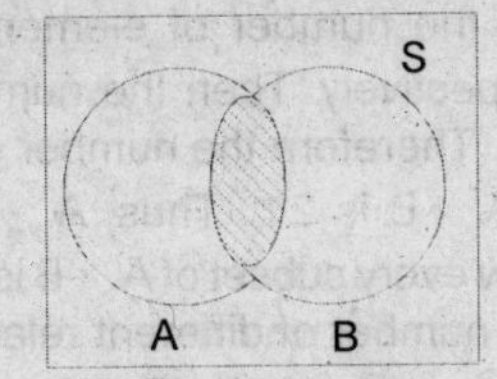

$A \cap B$ when neither $A \subset B$ nor $B \subset A$

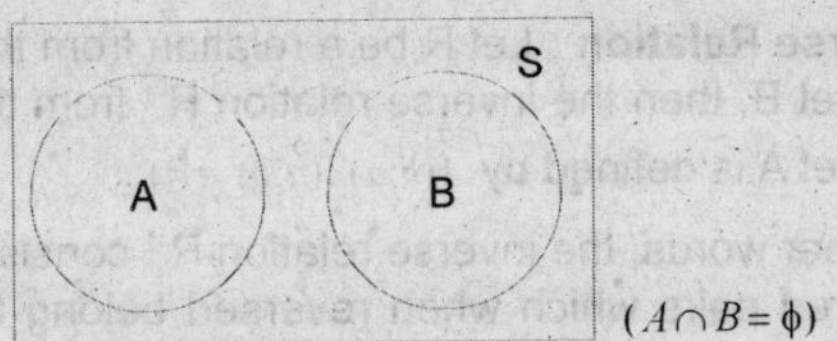

$(A \cap B = \phi)$

(d) Difference of Sets :(A - B).

A - B represents the area of A that is not in B.

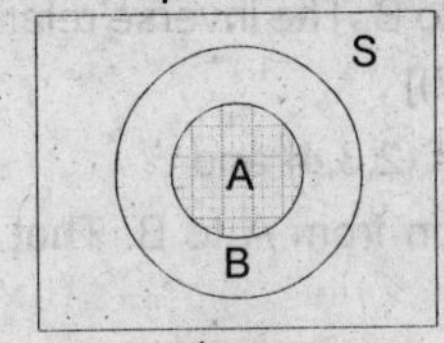

A - B when $A \subset B$ $(A - B = \phi)$

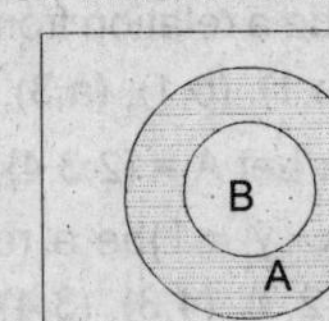

A - B when $B \subset A$

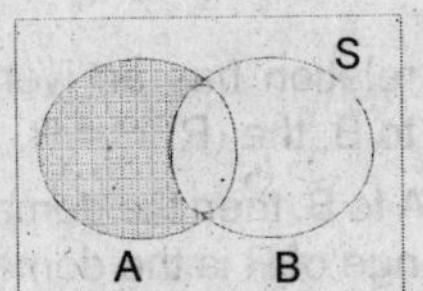

A-B when neither $A \subset B$ nor $B \subset A$

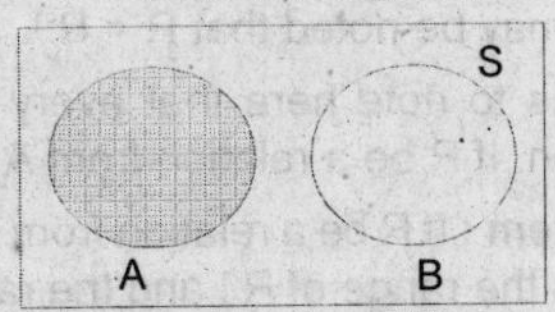

$A - B$ when A and B are disjoint sets $(A - B \subset A)$

(e) Complement of Sets : A' or A^c is the set of those points of Universal Set S which are not in A.

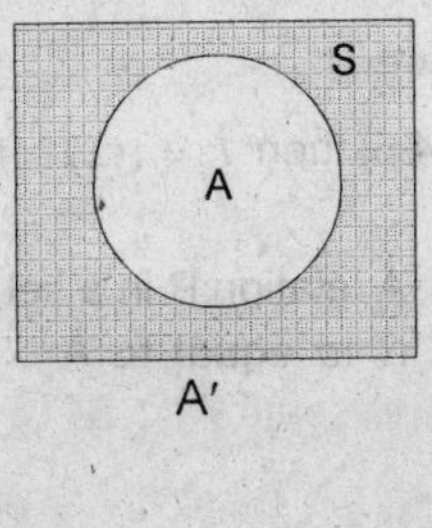

A'

7. Application of Set Theory

Why we study set theory ? Here the application of the same will make us clear about it.

The number of distinct elements of a finite set A is denoted by n (A).Let n (A),n (B) and $n(A \cap B)$,where A and B are non-empty sets. Then $n(A \cup B) = n(A) + n(b) - n(A \cap B)$

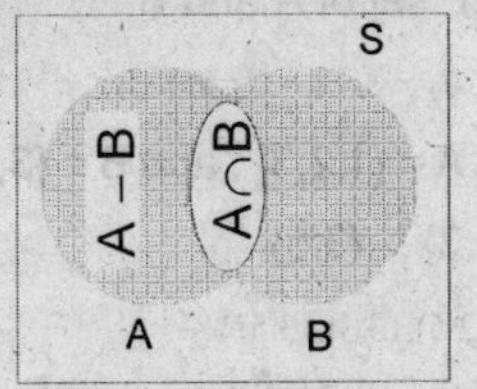

In case A and B are disjoint sets, so we have $A \cap B = \phi$

$\therefore \quad n(A \cup B) = n(A) + n(B)$

From the above Venn-Diagram, the following results appears, immediately. It will help in the application of set theory.

a) $n(A) = n(A - B) + n(A \cap B)$

b) $n(B) = n(B - A) + n(A \cap B)$

c) $n(A \cup B) = n(A - B) + n(B - A) + n(A \cap B)$

The result, $n(A \cap B) = n(A) + n(B) - n(A \cap B)$ can be generalised as,

$$n(A \cup B \cup C) = n(A) + n(B) + n(C) - n(A \cap B) - n(B \cap C)$$
$$- n(A \cap C) + n(A \cap B \cap C)$$

Worked out examples on the above result :

Ex 1 : If 63% of persons like oranges where 76% like apples. What can be said about the percentage of persons who like both oranges and apple?

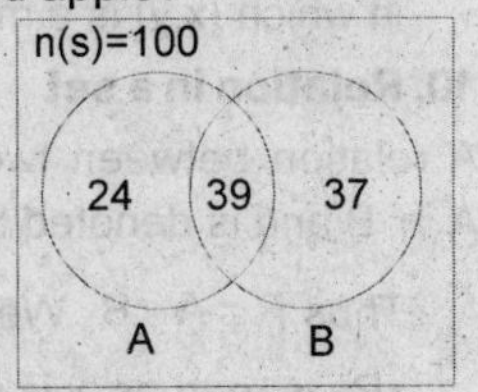

Sol : Let n(s)

= total number of persons = 100

A = {x : x likes oranges}

B = {x : x likes apples}

n (A) = 63, n (B) = 76

$A \cap B$ = {x:x likes oranges and apples both} Now ,

$\therefore \quad n(A \cup B) = n(A) + n(B) - n(A \cap B)$

$n(A \cap B) = n(A) + n(B) - n(A \cup B) = 63 + 76 - 100 = 39.$

$\therefore \quad n(A \cap B) = 39$

Hence, 39% people like both oranges and apples.

8. Cartesian Product Of Sets

Here before proceeding to learn about cartesian product of sets, we will first learn wht we mean by an ordered pair.

Ordered Pair : An ordered pair, usually denoted by (x,y) is a pair of elements x and y of some sets, which is ordered in the sense that $(x,y) \neq (y, x)$ whenever $x \neq y$. Here x is known as the first co-ordinate and y, as the second co-ordinate of the ordered pair (x,y).

Let us understand it by an example: The ordered pairs (1, 2) and (2, 1) though consists of the same elements

1 and 2, are different because they represent different points in the co-ordinate plane.

Cartesian Product : The cartesion product of two sets A and B is the set of all those pairs whose first co-ordinate is an element of A and the second co-ordinate is an element B. The set is denoted by $A \times B$ and is read as 'A cross B' or ' product set of A and B'.

i.e. $A \times B = \{(x,y) : x \in A \wedge y \in B\}$

For example : Let A = {1,2,3} , and B = {3,5}

$\therefore$ $A \times B = \{1,2,3\} \times \{3,5\}$

$\{(1,3),(1,5),(2,3),(2,5),(3,3),(3,5)\}$

and, $B \times A = \{(3,1),(1,2),(3,3),(5,1),(3,2),(5,3)\}$

So it is very clear that $A \times B \neq B \times A$

9. Relations

We often speak of relations which hold between two or more objects,e.g., 5 divides 15, Nehru wrote Freedom of India, 15 is less than 20 and 25, etc. These are the relations in every day situation. In these examples some sort of connections between pairs of objects are shown and hence they express some relations between pairs of objects. Relations may exist between several objects also.

Let us consider a set A = {1,2,3}, so that the cartesian product set is

$A \times A = \{(1,1),(1,2),(1,3),(2,1),(2,2),(2,3),(3,1)\ (3,2)\ (3,3)\}$

Consider a subset from this set in such a way that the first coordinate is 'less than' the second coordinate. If we denote this relation by R, then R is described by

$R = \{(x,y) \in A \times A : x < y\}$

in which (x,y) is a member of the relation R.

10. Relation in a set

A relation between two sets A and B is a subset of $A \times B$ and is denoted by R.

Thus $R \subseteq A \times B$. We write xRy, if and only if $(x,y) \in R$.

xRy is read as 'x is related to y'

If A = B then we say that R is a relation in the set A. Let us try to understand it by an example.

Example : Consider the relation

$R = \{(x,y) \in N \times N : x > y$, N is a set of natural numbers.}

Obviously, $R \subseteq N \times N$ and so R is relation N.

11. Binary Relation in a Set

A binary relation R is said to be defined in a set A if for any ordered pair $(x,y) \in A \times A$, it is meaningful to say that xRy is true or false.

In other words, R = { $(x,y) \in A \times A$: xRy is true}.

That is, a relation R in a set A is a subset of A × A. The meaning of binary is two. So, the binary relation is a relation between two sets, these sets may be different or identical, For the sake of convenience a binary relation will be written as only as relation.

12. Domain and Range of a Relation

The domain D of the relation R is defined as the set of all first elements of the ordered pairs which belong to R,i.e.,

$$D = \{x : (x,y) \in R,\ for\ x \in A\}$$

The range E of the relation R is defined as the set of all second elements of the ordered pairs which belong to R,i.e., $E = \{y : (x,y) \in R,\ for\ y \in B\}$ Obviously, $D \subseteq A$ and $E \subseteq B$.

Example : Let A = {1,2,3,4} and B = {a,b,c}. Every subset of $A \times B$ is a relation from A to B. So, if R = { (2,a), (4,a), (4,c)}, then the domain of R is the set {2,4} and the range of R is the set {a,c}.

13. Total number of distinct relations from a set A to set B

Let the number of elements of A and B be m and n respectively. Then the number of elements of $A \times B$ is mn. Therefore the number of elements of the power set of $A \times B$ is 2^{mn}. Thus, $A \times B$ has 2^{mn} different subsets. Now every subset of $A \times B$ is a relation from A to B. Hence the number of different relations A to B is 2^{mn}.

14. Types of Relations in a Set

We consider some special types of relations in a set A.

Inverse Relation : Let R be a relation from the set A to the set B, then the inverse relation R^{-1} from the set B to the set A is defined by $\{(b,a) : (a,b) \in R\}$.

In other words, the inverse relation R^{-1} consists of those ordered pairs which when reversed belong to R. Thus every relation R from the set A to the set B has an inverse relation R^{-1} from B to A.

Ex 1 : Let A = {1,2,3}, B = {a,b} and R = { (1,a), (1,b), (3,a), (2,b)} be a relation from A to B. The inverse relation R is R^{-1} = {(a,1), (b,1), (a,3), (b,2)}

Example 2 : Let A = {2,3,4}, B = {2,3,4} and R = {(x,y): $|x - y| = 1$}be a relation from A to B. That is, R = { (3,2), (2,3), (4,3), (3,4)}.

The inverse relation of R is R^{-1} = { (3,2), (2,3), (4,3), (3,4)}.

It may be noted that $R = R^{-1}$.

It is to note here that every relation has an iverse relation. If R be a relation from A to B, the $(R^{-1})^{-1} = R$.

Theorem : If R be a relation from A to B, then the domain of R is the range of R^{-1} and the range of R is the domain of R^{-1}.

Indentity Relation : A relation R in a set A is said to be identity relation, generally denoted by I_A, if

$$I_A = \{(x,x) : x \in A\}$$

Example : Let A = {2,4,6} then $I_A = \{(2,2),(4,4),(6,6)\}$ is an identity relation in A.

Universal Relation : A relation R in a set A is said to be universal relation if R is equal to $A \times A$. That is, if $R = A \times A$.

Example : Let A = {1,2,3} then R = A × A

= {(1,1), (1,2), (1,3), (2,1), (2,2), (3,1), (3,2), (3,3)}

is a universal relation in A.

Void Relation : A relation R in a set A is said to be a void relation if R is a null set i.e., if R = ϕ

Example : Let A = {2,3,7} and let R be difined as ' aRb if and only if a divides b' then we observe that R = $\phi \subset A \times A$ is a void relation.

15. Properties of Relations in a Set

Reflexive Relations : Let R be a relation in a set A. Then R is called a reflexive relation if (a,a) $\in$ R, for all a $\in$ A.

In other words, R is reflexive if every element in A is related to itself.

Thus, R is reflexive if aRa holds for all a $\in$ A.

a relation R in a set A is not reflexive if there is at least one element a $\in$ A, such that (a,a) $\notin$ R.

Example : Let A = { 1,2,3,4}. Then the relation R_1 = = {(1,1), (2,4), (3,3), (4,1), (4,4)} in A is not reflexive since 2 $\in$ A but (2,2) $\notin R_1$.

Anti-Reflexive Relations : Let R be a relation in a set A. Then R is called an anti-reflexive relation if (a,a) $\notin$ R for every a $\in$ A, i.e., a / Ra for all a $\in$ A.

Example : The relation in the set of natural numbers N defined by

I. ' x < y' is anti - reflexive, since a is not less than a for any natural number a,

II. Let A = {1,2,3,4}. Then the relation R = {(1,1), (2,3), (3,4)} is not anti-reflexive since, (1,1) $\in$ R.

Symmetric Relations : Let R be a relation in a set A. Then R is said to be symmetric relation if (a,b) $\in$ R $\Rightarrow$ (b,a) $\in$ R.

Thus R is symmetric if bRa holds whenever aRb holds.

A relation R in set A is not symmetric if there exist two distinct elements a,b $\in$ A, such that aRb, but b / Ra.

Example 1: Let L be the set of all straight lines in a plane. The relation R in L defined by ' x is parallel to y' is symmetric, since if a straight line a is parallel to a straight line b, then b is also parallel to a. Thus (a,b) $\in$ R $\Rightarrow$ (b,a) $\in$ R.

Example 2: Let R be relation in the natural numbers N which is defined by ' x-y > 0'. Then R is not symmetric since, 4-2 > 0 but 2 - 4 $\not>$ 0.

Thus , (4,2) $\in$ R but (2,4) $\notin$ R.

Remarks : Since (a,b) $\in$ R $\Rightarrow$ (b,a) $\in R^{-1}$, thus R is a symmetric relation if and only if R = R^{-1}.

Anti- Symmetric Relations : Let R be a relation in a set A, i.e., let R be a subset of A × A. Then R is said to be an anti-symmetric relation if (a,b) $\in$ R and (b,a) $\in$ R $\Rightarrow$ a = b.

Thus R is anti-symmetric if a $\neq$ b then possibly (a,b) $\in$ R or possibly (b,a) $\in$ R, but never both.

A relation R in a set A is not anti-symmetric if there exist elements a,b $\in$ A, a $\neq$ b such that (a,b) $\in$ R and(b,a) $\notin$ R.

Example : Let A be a family of sets and let R be the relation in A defined by ' x is a subset of y'. Then R is anti-symmetric since $A \subseteq B$ and $B \subseteq A \Rightarrow A = B$.

Transitive Relations : Let R be a relation in a set, i.e., let R be a subset of A × A. Then R is said to be a transitive relation if (a,b) $\in$ R and (b,c) $\in$ R $\Rightarrow$ (a,c) $\in$ R.

A relation R in a set A is not transitive if there exist elements a,b,c $\in$ A, not necessarily distinct, such that (a,b) $\in$ R, (b,c) $\in$ R but (a,c) $\notin$ R.

Example 1 : Let L be the set of all straight lines in a plane and R be the relation in L defined by ' x is parallel to y'. If a is parallel to b and b is parallel to c then obviously a is parallel to C .Thus (a,b) $\in$ R and

(b,c) $\in$ R $\Rightarrow$ (a,c) $\in$ R. Hence R is transitive.

Example 2 : Let A be the set of all Indians. Let R be the relation in A defined by ' x loves y' If a loves b and b loves c, it does not necessarily follow tht a loves c, So, R is not a transitive relation.

Equivalence Relations : Let R be a relation in a set A. Then R is an equivalence relation in A if and only if

(I) R is reflexive, i.e., for all a $\in$ R, (a,a) $\in$ R.

(II) R is symmetric, i.e., (a,b) $\in$ R $\Rightarrow$ (b,a) $\in$ R, for all a,b $\in$ A.

(III) R is transitive, i.e., (a,b) $\in$ R and (b,c) $\in$ R $\Rightarrow$ (a,c) $\in$ R for all a,b,c $\in$ A.

Example 1 : The most trivial example of an equivalence relation is that of ' equality' for any elements in any set

I. a = a, i.e., reflexive.

II. a = b $\Rightarrow$ b = a i.e., symmetric

III. a = b and b = c $\Rightarrow$ a = c, i.e., transitive.

Example 2 : Let R be the relation in the real numbers difined by ' x $\leq$ y'. Then

I. x $\leq$ x i.e., (x,x) $\in$ R, i.e., R is reflexive.

II. Let x $\leq$ y but y $\not\leq$ x i.e., (x,y) $\in$ R but not necessarily, (y,x) $\in$ R i.e., R is not symmetric Thus R is not an equivalence relation.

Theorem : If R be an equivalence relation in a set A, then R^{-1} is also an equivalence relation in A.

16. Function

Let A and B be two non-empty sets and f be a relation from A to B.If for each element a $\in$ A, there exists a unique b $\in$ B (denoted by f (a)) such that (a,b) $\in$ f, then f is called a function (a mapping) from A to B. It is denoted by f : A $\rightarrow$ B

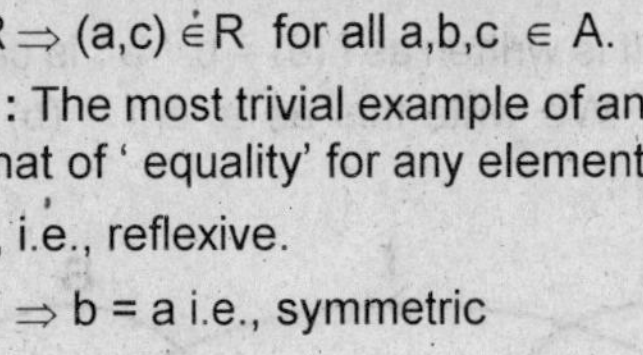

I. Take any element a $\in$ A. Since f : A$\to$ B is a function there exists a unique element b $\in$ B, such that f (a) = b. It may be wrtten as follows.

II. a $\in$A. f : A$\to$ B $\Rightarrow f(a)=b \in B$ or

III. f : A$\to$ B , a $\in$ A$\Rightarrow\exists$ unique (a,b) $\in$f.

Note : .

1. For all elements of A, images should exist in B.
2. All elements in B may or may not have pre-images in A
3. There may be one or more pre-images in A for elements in B.
4. $f(A)\subseteq B$ and f$\subseteq$A$\times$B .
5. Every function from A to B is a relation from A to B but every relation from A to B need not be a function from A to B.
6. If an element in A has more than one image in B, f is not a function.

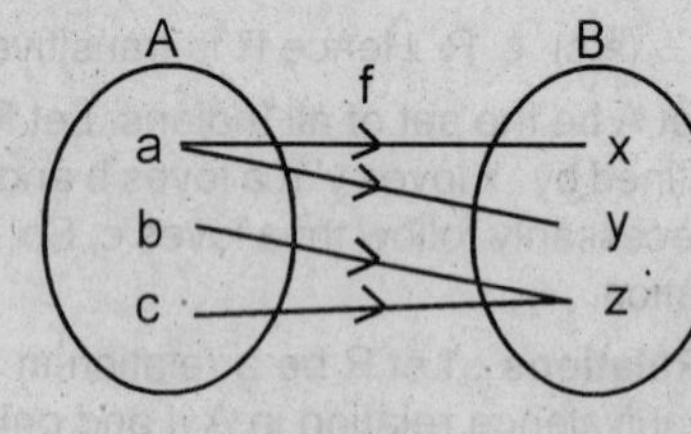

Fig Not a function

7. If n (A) = m,n (B) = k then the number of functions that can be defined from A to B is k^m.

17. Image, Pre- image

If f : A$\to$ B is a function and (a,b) $\in$f then b is called f -image of a. It is written as f (a) = b. " a " is called a pre-image of " b". We write this as a $\in$ f^{-1} (b) : f^{-1} (b) = $\{ x \in A | f(x) = t \}$.

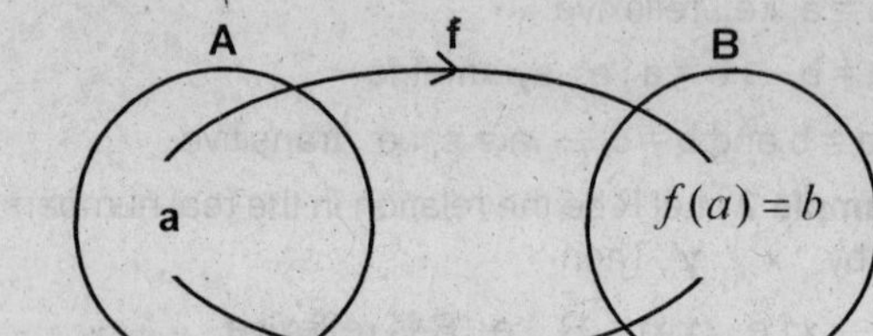

Every element of A has one and only one image in B. Every element of B may or may not have a pre-image in A. But one or more elements of A, may have the same image in B.

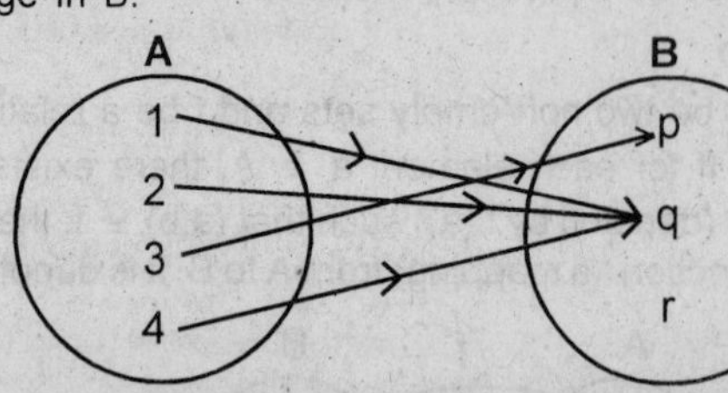

Note : In above figure there are more than one pre-images for q$\in$B . In fact, $f^{-1}(q)=\{a \in A | f(a)=q\}=\{1,2,4\}$

Example : R is the set of real numbers. If $f = \{(x,x^2)|x \in R\}$ then f(4) = 16. f (-4) = 16. Hence $f^{-1}(16)=\{4,-4\}$.

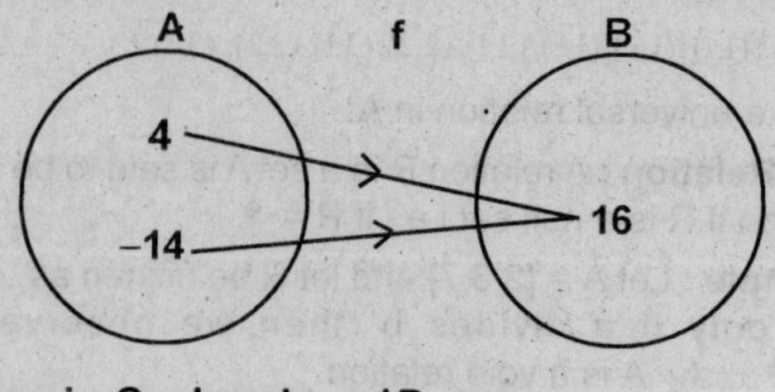

18. Domain, Co-domain and Range

If f : A$\to$ B is a function, then A is called the domain, B is called the co-domain of f.

Range : If f : A$\to$ B is a function, the set of all images of the elements of A is called the range of f. It is denoted by f (A) = $\{ f(a) / a \in A\}$. Range is a subset of the co-domain. That is $f(A)\subseteq B$.

Example: If A = {1,2,3,4}, B = {p,q,r} and f ={(1,q), (2,q), (3,q), (4,q)} then f is a function.

Domain = A, Co-domain = B.

Range of f = f (A) = $\{ f(1), f(2), f(3), f(4)\}$

$=\{q,q,p,q\}=\{p,q\}\quad \therefore f(A)\subseteq B$

There is no pre-image for the element r $\in$B. Even then f : A$\to$ B is a function. (hence for every element there may not be pre-images).

Example : Let A = {1,2,3,4}, B = {p,q,r} are $g=\{(1,p),(2,q),(3,p),(4,p),(1,r)\}$. The element 1 $\in$A has two images p and r in B,so g : A$\to$ B is not a function but it is only a relation.

Example : If N is the set of natural numbers and $f=\{(a,b)|a,b \in N, b=2a\}$ then f : N$\to$N is a function.

Example : If f (x) = |x|, $\forall$ x$\in$R then f :R$\to$ R is a function.

that is $f=\{(x|x|)\,|x \in R\},\ f(x)=x \ \ if \ \ x \geq 0$

$f(x)=-x \ \ if \ \ x<0$

19. Equality Of Two Functions

If f and g are functions defined on the same domain A and if f (a) = g (a) for every a $\in$ A, then the functions f and g are equal, that is, f = g.

Example 1: Let f(x) = x^2 + 5, where x is a real number and g (x) = x^2 + 5, where x is a complex number. Then the function f is not equal to the function g since they have different domains although f (x) = x^2 + 5 = g (x).

Example 2 : Let A = {1,2} and B = {1,2,3,4} and f = $\{(1,1),(2,4),\}$ Let a function g be defined by the formula g (x) = x^2 where the domain of g is the set {1,2}. Then f = g since they both have the same domain and since f and g assign the same image to each element in the domain.

20. Types Of Functions

1. Into Mappings : A mapping f : A$\to$ B is said to be an into mapping if f (A) is a proper subset of B. In this case we say that f maps A into B.

Example : Let $f : Z \to Z$ (the set of integers) be defined by f(x) = 2x, $x \in Z$. Then f is an into ma pping, because f (Z) (the set of all even integers) is a proper subset of the codomain set Z.

2. Onto Mappings : A mapping $f: A \to B$ is said to be an onto mapping if f (A) = B. In this case we say that f maps onto B.

Example : Let $f: Z \to Z$ be defined by f (x) = x + 1, $x \in Z$. Then every element y in the codomain set Z has a pre-image y-1 in the domain set Z. Therefore, f (z) = Z and f is an onto mapping.

3. Injective Mapping : A mapping $f: A \to B$ is said to be injective (or one - to- one or injection) if for each pair of distinct elements of A,their f-images are distinct.

Thus $f : A \to B$ is injective if $x_1 \neq x_2$ in A implies

$f(x_1) \neq (x_1)$ in B.

Example : Let $f:R \to R$ be defined by f (x) = 2x+1, $x \in R$ .Then for $x_1, x_2 \in R(x_1 \neq x_2)$

we have $f(x_1) \neq f(x_2)$.

4. Surjective Mapping : A mapping $f: A \to B$ is said to be surjective (or onto or surjection) if f (A) = B.

5. Bijective Mapping : A mapping $f: A \to B$ is said to be bijective (or bijection) if f is both injective and surjective.

Example : Let $f: Z \to Z$ be defined by f (x) = x+2, $x \in 1 \in Z$. The mapping f is both injective and surjective. So it is bijective.

Note : If f is injective, each element of B has at most one pre-image. If f is surjective, each element of B has atleast one pre- image.

6. Many one Mapping : A mapping $f : A \to A$ is said to be many - one if two or more distinct elements in A have the same image. That is if f (x) = f(x)' implies $x \neq x'$.

7. Constant Mapping : A function f of A into B is called a constant mapping if the same element $b \in B$ is assigned to every element in A. In other words, $f: A \to B$ is a constant mapping if the range of f consists of only one element.

Example : Let $f : R \to R$ be defined by the formula f (x) = 50. Then f is a constant function since 50 is assigned to every element.

8. Identity Mapping : Let A be a set,Let the function $f: A \to A$ be defined by the formula f (x) = x. Then f is called the identity function or the identity mapping or the identity transformation on A.

9. Inverse Mapping : Let $f: A \to B$ be a one-one onto mapping. Then the mapping $f^{-1}:B \to A$ which associates to each element $b \in B$ the element $a \in A$, such that f (a) = b, is called the inverse mapping of the mapping $f : A \to B$.

Hence if $f : A \to B$ be a one-one on to mapping, then $f^{-1}: B \to A$. The mapping f^{-1} is called the inverse mapping of the mapping f.

Example : Let $A = R - \{3\}$ and $B = R - \{1\}$ and the function $f : A \to B$ be defined by $f(x) = \frac{x-2}{x-3}$. Then the inverse function $f^{-1}: B \to A$ is given by $y = \frac{x-2}{x-3}$ or, $x = \frac{2-3y}{1-y}$.

Hence $f^{-1}(x) = \frac{2-3x}{1-x}$

21. Some of the Important Theorems

Theorem 1 : Only bijective mapping posses inverse mappings.

Theorem 2 : If $f: A \to B$ is bijective the $f^{-1}:B \to A$ is also bijective.

Theorem 3 : If $f: A \to B$ be bijective, then the inverse mapping of f is unique .

22. Product Or Compositions Of Mappings

Let :$f: A \to B$ and $g: B \to C$. Then the composite of the functions f and g denoted by (fog) or fg is a mapping of $A \to C$ given by (fog): $A \to C$ such that

(gof) (x) = $g\ f(x) = g[f(x)] \ = \forall x \in A$,

It may be noted that domain of g is equal to codomain of f.

Example : Let f (x) = 2x+1 and g(x) = 3x be two functions. Therefore, $(gof)(x) = g[f(x)] = g(2x+1) = g(y)$ (where y = 2x+1) = 3y [since g(x) = 3x]

= 3 (2x+1) = 6x+3.

Note : If $f: A \to A$ and $g: A \to A$, then we can find both the composite mapping go f and fog. But in general $(gof) \neq (fog)$.

For example : Let f(x) = 2x-1, g (x) = 5x.

Therefore, (fog)(x) = f [g (x)] = f (5x) = 2 (5x) -1 = 10x-1 and (gof) (x) = g [f (x)] = g (2x-1) = 5 (2x-1) = 10x-5. That is, fog $\neq$ gof.

Some Properties of Composites of Mappings

1. If $f: A \to B$ be a bijective mapping, then of $fof^{-1} = I_B$ and $f^{-1}of = I_A$, where I_A and I_B are the identity mappings of the sets A

2. If $f: A \to B$, then I_B of = f and fo I_A = f.

3. If $f: A \to B$, $g: B \to C$, $h: C \to D$ be three mappings. Then ho(gof) = (hog) of.

4. If $f: A \to B$ and $g: B \to C$ be both injective mappings then the composite mapping gof : $A \to C$ is injective.

5. If $f: A \to B$ and $g: B \to C$ be two mappings such that go f: $A \to C$ is injective then f is injective.

6. If $f: A \to B$ and $g: B \to C$ be both surjective then the composite mapping gof : $A \to C$ is surjective.

7. If $f: A \to B$ and $g: B \to C$ be two mappings such that gof : $A \to C$ is surjective then g is surjective.

8. If $f: A \to B$ and $g: B \to C$ be both bijective then the composite mapping go f : $A \to C$ is bijective.

23. Real Valued Functions

R is the set of real numbers. If $A \subseteq R$, then a function, $f: A \to R$ is called a real valued function.

Ex : Find the domain of the real valued function $f(x)=\sqrt{49-x^2}$

Sol : $\sqrt{49-x^2} \in R \Rightarrow 49-x^2 \geq 0$

$\Rightarrow x^2 \leq 49 \Rightarrow |x| \leq 7 \Rightarrow -7 \leq x \leq 7$

$\therefore$ Domain of f is $|-7,7|$

OBJECTIVE QUESTIONS - I

1. The binary relation S=ø (empty set) on set A={1, 2, 3} is **[GATE-2002]**

a) Neither reflexive nor Symmetric

b) Symmetric and Reflexive.

c) Transitive and Reflexive

d) Transitive and Symmetric.

2. A Relation R is defined on the set of integers as xRy iff (x+y) is even. Which of the following statements is TRUE? **[GATE–2000]**

a) R is not an equivalence relation.

b) R is an equivalence relation having one equivalence class.

c) R is an equivalence relation having two equivalence classes.

d) R is an equivalence relation having three equivalence classes.

3. Let P(S) denote the power set of set S. which of the following is always TRUE? **[GATE-2000]**

a) P(P(S))=P(S) b) P(S) ⋂S=P(S)

c) P(S) ⋂P(P(S))={ø} d) S ∉ P(S)

4. The number of binary relations on a set with n elements is: **[GATE-1999]**

a) n^2 b) 2^n c) $_2n^2$ d) None

5. Suppose A is a finite set with n elements. The number of elements in the largest equivalence relation of A is: **[GATE-1998]**

a) 1 b) n c) n+1 d) n^2

6. The number of functions from an m element set to an n element set is: **[GATE-1998]**

a) m + n b) m^n c) n^m d) m* n

7. The number of equivalence relations of the set { 1, 2, 3, 4} is **[GATE-1997]**

a) 4 b) 15 c) 16 d) 24

8. Let A and B be sets and A^c and B^c denote the complements of the sets A and B. The set (A-B) ⋃(B-A) ⋃ (A⋂B) is equal to **[GATE-1996]**

a) A⋃ B b) A^c ⋃B^c c) A⋂B d) A^c ⋂B^c

9. Let R be a non-empty relation on a collection of sets defined by A R B if and only if A ⋂B = ϕ. Then, (pick the TRUE statement) **[GATE-1996]**

a) R is reflexive and transitive.

b) R is symmetric and not transitive.

c) R is an equivalence relation.

d) R is not reflexive and not symmetric.

10. Let R be a symmetric and transitive relation on a set A. then **[GATE-1995]**

a) R is reflexive and hence an equivalence relation.

b) R is reflexive and hence a partial order.

c) R is not reflexive and hence not an equivalence relation.

d) None of these

11. The number of elements in the Power set P(S) of the set S={{ ϕ} , 1, {2,3}} **[GATE-1995]**

a) 2 b) 4 c) 8 d) None.

12. Let S be an infinite set and S1, S2, S3, ……, Sn be sets such that S1 ⋃S2⋃ S3⋃……Sn = S. Then **[GATE-1993]**

a) Atleast one of the sets S_i is a finite set.

b) Not more than one of the sets S_i can be finite.

c) Atleast one of the sets S_i is an infinite set.

d) None of these

13. Let A be a finite set of size n. the number of elements in the power set of A×A is **[GATE-1993]**

a) $_2 2^n$ b) $_2 n^2$ c) $(_2 n)^2$ d) None of these

14. Let A and B be sets with cardinalities m and n respectively. The number of one-one mappings(injections) from A to B, when m<n, is **[GATE-1993]**

a) m^n b) ^nP_m c) ^nC_m d) None

15. Which of the following sets are null sets?

a) {0} b) {ø} c) { } d) ø

16. Which of the following set(s) are empty?

a) {x: x = x} b) {x: x ≠ x} c) {x: x = x^2} d) {x: x ≠ x^2}

17. If A={x, y}, then the power set of A is

a) {{x}, {y}} b) {{ø},{x, y}} c) {ø,{x},{y}} d) None

18. If A={x, y}, then the proper subsets of A are

a) {x},{y} b) ø,{x, y} c) ø,{x},{y} d) None.

19. The number of subsets of a set of order three is

a) 3 b) 6 c) 8 d) 9

20. The number of proper subsets of a set of order three is

a) 3 b) 6 c) 8 d) 9

21. If A and B are sets and A⋃B= A ⋂B, then

a) A = ø b) B = ø c) A = B d) None

22. If X, Y, Z are sets such that X ⊂Y and Y ⊂ Z, then

a) X ⋃Y=Z b) X ⋂Y=Z c) X ⋂Y ⋂Z=Z d) None

23. If X and Y are two sets, then X ⋂ (Y ⋃ X $)^c$ equals **[IIT-1979]**

a) X b) Y c) ø d) None.

24. P, Q, R are subsets of a set A. R × (P^c ⋃ $Q^c$$)^c$ = (R×P) ⋂ (R×Q) **[IIT-1980]**

a) True b) False

25. For any two subsets X and Y of a set A, define X o Y = (X^c⋂ Y) ⋃ (X⋂Y^c). Then for any three subsets X, Y, Z of the set A is X o (Y o Z)=(X o Y) o Z **[IIT-1980]**

a) True b) False

26. The relation R defined on the set A={1, 2, 3, 4}, by R={ (1,1), (2,2), (3,3)} is

a) Reflexive b) Symmetric
c) Transitive d) None

27. The relation R is defined on N by R={ (a, b): a divides b; a, b $\in$ N }. Then R is

a) Reflexive b) Symmetric
c) Transitive d) None

28. The Relation R defined on a set N by R={ (a, b): |a-b| is divisible by 5 }, is

a) Reflexive b) Symmetric
c) Transitive d) Equivalence

29. The Relation R is defined on the set N as {(a, b): a, b are both odd }, is

a) Reflexive b) Symmetric
c) Transitive d) None

30. If R is an equivalence relation on a set A, then R^{-1} is

a) Reflexive b) Symmetric
c) Transitive d) None

31. "n/m" means that n is a factor of m, then the relation '/' is **[M.N.R-1985]**

a) Reflexive and symmetric
b) Transitive and symmetric
c) Reflexive, transitive and symmetric
d) Reflexive, transitive and not symmetric

32. The function f : N $\to$ N defined by f(n)=2n+3 is **[M.N.R-1987]**

a) Surjective b) Not surjective
c) Injective d) None

33. Given A={x, y, z}; B={u, v, w} the function $f : A \to B$ defined by $f(x)=u$, $f(y)=v$, $f(z)=w$ is **[M.N.R-1988]**

a) Surjective b) Bijective
c) Injective d) None

34. If $f : X \to Y$ and a, b$\subseteq$X then $f(a \cap b)$ is equal to **[Hyd.Univ-1995]**

a) $f(a)-f(b)$ b) $f(a) \cap f(b)$
c) A proper subset of $f(a) \cap f(b)$ d) $f(b)-f(a)$

35. The number of distinct relations on a set of 3 elements is **[Hyd.Univ-1993]**

a) 8 b) 9 c) 18 d) 512

36. If A and B are sets. Which of the following is FALSE? **[Hyd.Univ-1995]**

a) A-B$'$ = A$\cap$B b) A$\subset$B$\to$ B$' \subset$ A$'$
c) A-(A-B) = A $\cap$B d) None

37. A={string of 0's and 1's}. Relation d on A is defined as xdy if x is a sub string of y(Ex. 01 d 101). Then d is **[Hyd.Univ-1993]**

a) Symmetric b) Anti symmetric
c) Equivalence d) Reflexive and Symmetric

38. If the relation R over {a, b, c} is given by R={(a,a), (a,b), (b,a), (b,b), (c,c)}, which of the following properties does R have ? **[Hyd.Univ- 1990]**

I. Symmetry II. Anti symmetry.
III. Reflexive IV. Transitive.

a) I, III and IV only b) II and IV only
c) II and III only d) II, III and IV only

39. The number of elements in the power set of the set {{a, b},{c}} is **[Hyd.Univ-1989]**

a) 8 b) 4 c) 3 d) 7

40. If R={(a, b)/a, b are positive integers such that a-b is an ODD positive integer}. The relation R is **[Hyd.Univ-1988]**

a) Reflexive b) Symmetric
c) Transitive d) Equivalence

41. A-(B $\cup$ C) is **[Hyd.Univ-1988]**

a) (A-B)$\cup$ (A-C) b) A-B-C
c) (A-B) $\cap$ (A-C) d) $A-\left(\overline{B\cap C}\right)$

42. A $\cup$ B is NOT equivalent to **[R.E.C.Trichy - 1992]**

a) $\left(\overline{\overline{A}\cap\overline{B}}\right)$ b) $\left(\overline{\overline{A}-B}\right)$ c) $\left(\overline{B-A}\right)$ d) $\left(\overline{A-B}\right)$

43. A $\cup$B = A$\cap$B If and only if **[R.E.C.Trichy - 1992]**

a) A is empty set b) B is empty set.
c) A and B are non-empty sets d) A and B are empty sets.

44. Let A={x/-1<x<1}=B. The function $f(x)=x/2$ from A to B is **[R.E.C.Trichy - 1992]**

a) Injective b) Surjective
c) Both injective and surjective
d) Neither injective nor surjective.

45. The domain and range are same for **[R.E.C.Trichy - 1994]**

a) Constant function b) Identity function.
c) Absolute value function.
d) Greatest integer function.

46. The function f : IR $\to$ IR given by $f(x)=x^2$ is **[R.E.C.Trichy - 1994]**

a) One-one b) Onto
c) One-one and onto d) None.

47. The number of functions from an m element set to an n element set is **[Hyd. Univ-2001]**

a) m + n b) m^n c) n^m d) mn

48. Which of the following relations may be characterised as a function defined on the set I = {1, 2, 3, 4, 5} ? **[Hyd. Univ-2001]**

a) {(x, y) | x, y $\in$I, x < y} b) {(x, y) | x, y $\in$ I, x >y}
c) {(x, y) | x, y $\in$I, x = y/2} d) None

49. Which one of the following is a partition of the set S = {1, 2, 3, 4, 5, 6}? **[Hyd. Univ-2001]**

a) [{1, 3, 5}, {2, 4}, {3, 6}] b) [{1, 5}, {2}, {3, 6}]
c) [{1, 5}, {2}, {4}, {1, 5}, {3, 6}] d) [{1, 2, 3, 4, 5, 6}]

50. Find r^3 when A ={1, 2, 3, 4, 5} and r defined on A by x r y iff x + 1 = y. **[Hyd. Univ-2001]**

a) {(1, 3)(2, 4)} b) {(1, 3)(2, 5)}
c) {(1, 4)(2, 5)} d) {(1, 4)(4, 5)}

Note : Answer below three questions by using the following information. In a language survey of students it is found that 80 students know English, 60 know French, 50 know German, 30 know English and French, 20 know French and German, 15 know English and German and 10 students know all the three languages. **[Hyd. Univ-2001]**

51. How many students know at least one language?

a) 135 b) 30 c) 10 d) 45

52. How many students know at least two languages?

a) 135 b) 30 c) 10 d) 45

53. How many students know French and one but not both out of English and German?

a) 135 b) 30 c) 10 d) 45

Note : Answer the below 2 questions using the following information.

Let A = {1, 2, 3, 4} and B = {0, 1, -1, 2} and several sets of ordered pairs be specified as :

I. {(1, 0), (2, 1), (3, -1), (4, 2)}

II. {(1, 0), (2, -1),(3, 2)}

III.{(4, 1), (3, 2), (2, 3), (1, 4)}

IV.{(2, 1), (3, 0), (1, -1), (3, 1), (4, 2)}

V. {(2, 0), (1, -1), (3, 2), (4, 1)}

54. Which of the sets of ordered pairs are graphs of a function from A $\rightarrow$ B? **[Hyd. Univ-2000]**

a) II and III b) III and IV c) I and V d) IV and V

55. Which of the sets of ordered pairs are onto functions from A $\rightarrow$ B? **[Hyd. Univ-2000]**

a) II and III b) III and IV c) I and V d) IV and V

56. Let R ={(1, 1), (3, 1), (2, 3), (4, 2)}; which of the following represents R^2, where R^2 is R composite R? **[Hyd. Univ-2000]**

a) {(1, 1), (3, 1), (2, 3), (4, 2)}

b) {(1, 1), (9, 1), (4, 9), (16, 4)}

c) {(1, 3), (3, 3), (3, 4), (3, 2)}

d) {(1, 1), (2, 1), (4, 3), (3, 1)}

57. If I is set of positive integers and the relation R is defined over the set I by xRy if $x^y = y^x$. Then R is **[Hyd. Univ-2000]**

a) Reflexive but neither Symmetric nor Transitive

b) Reflexive and Transitive but not Symmetric

c) Reflexive and Symmetric but not Transitive

d) Reflexive, Symmetric and Transitive

58. The 'Subset' relation on a set of sets is **[Hyd. Univ-1999]**

a) A partial Ordering b) An Equivalence Relation

c) Transitive and Symmetric only

d) Transitive and Anti-Symmetric only

59. Given relation R = { (x, y)| x, y $\in$ N, x - y is divisible by 5} then R is **[Hyd. Univ-1999]**

a) Only Reflexive b) Only Reflexive and Transitive

c) Only Symmetric d) Equivalence Relation

60. Find the power set 2^s of the set S = {3, {1, 4}, 5} **[Hyd. Univ-1996]**

a) {S, 3, 1, 4, {1, 3, 5},{1,4,5} {3, 4}, ϕ }

b) {S, 3, {1, 4}, 5}

c) {S, {3}, {3, {1,4}}, {3, 5}, ϕ } d) None

61. Given a relation R on a set A such that R o R is a subset of R. Then **[Hyd. Univ-1995]**

a) R is transitive b) R is reflexive

c) R is symmetric d) None of the above

62. The function f :R – {2} $\rightarrow$ R defined by f (x) = $(x^2 + 2x) / (x - 2)$ is **[Hyd. Univ-1995]**

a) one-one and onto b) one-one but not onto

c) neither one-one nor onto d) not one-one but onto

63. The number of distinct reflexive and symmetric relations that can be defined on a 3 element set is **[Hyd. Univ-1994]**

a) 3 b) 4 c) 6 d) 8

64. Let P(A) be the collection of all subsets of A = {a, b, c}. Let R be a relation defined as "x is disjoint from y" over P(A). Then the number of elements in R is **[Hyd. Univ-1994]**

a) 8 b) 9 c) 17 d) 11

65. Given set S = {x | x is a student in your class} and relation (x R y $\leftrightarrow$ x sits in the same row as y). Which of the following statements about R is TRUE? **[Hyd. Univ-1993]**

a) reflexive only b) transitive only

c) reflexive, transitive and a symmetric

d) reflexive, transitive and symmetric

66. Set A has n elements. The number of functions that can be defined from A into A is **[IIT-1985]**

a) n^2 b) n ! c) n^n d) n

67. Set A has 3 elements and set B has 4 elements. The number of injections that can be defined from A into B is **[JNTU-1992]**

a) 144 b) 12 c) 24 d) 64

68. If $n \geq 2$ then the number of surjections that can be defined from {1,2,3,....n} onto {1,2} is **[JNTU-1992]**

a) 2n b) $^nP_2$ c) 2^n d) $2^n - 2$

69. The number of bijective functions from set A to itself when A contains 106 elements is **[JNTU- 1994]**

a) 106 b) 106^2 c) 106 ! d) 2^{106}

70. Let $f{:}Z \rightarrow Z$ be defined as $f(x) = x^2, x \in z$. The function f is **[JNTU-1995]**

a) Bijection b) Injection c) Surjection d) None

71. If $f{:}A \rightarrow B$ is one one, $g{:}B \rightarrow C$ is onto, then $g o f{:}A \rightarrow C$ is ... bijection.

a) a b) not a c) need not a d) none

ANSWER KEY - I

1. d	2. c	3. c	4. c	5. d
6. c	7. b	8. a	9. b	10. d
11. c	12. c	13. b	14. b	15. b
16. b	17. d	18. a	19. c	20. b
21. c	22. c	23. c	24. a	25. a
26. b	27. a,c	28. All	29. b,d	30. a,b,c
31. d	32. b,c	33. d	34. c	35. d
36. d	37. b	38. a	39. b	40. b
41. c	42. d	43. d	44. a	45. b
46. b	47. c	48. d	49. c	50. c
51. a	52. d	53. b	54. c	55. c
56. d	57. d	58. a	59. d	60. d
61. a	62. c	63. d	64. d	65. d
66. c	67. c	68. d	69. c	70. d
71. c				

EXPLANATORY NOTES - I

1. (d) **Explanation:** The binary relation $S=\phi$ (empty set) on set $A=\{1,2,3\}$ is transitive and symmetric.

2. (c) **Explanation:** We know that A Relation R is defined on the set of integers as $x\ R\ y$ iff $(x+y)$ is even. Then R is an equivalence relation having two equivalence classes.

3. (c) **Explanation** : Let $S=\{1,2\}$

Power Set of $P(s)=\{\{1\},\{2\},\{1,2\},\phi\}$

$P(p(s))=\{\{\{1\}\},\{\{2\}\},\{\{1\},\{2\}\},\ \phi,$

$\{\phi\},\{\{1\},\{1,2\}\},\{2\},\{1,2\}\}$

$P(s)\cap P(P(s))=\phi$

7. (b) **Explanation** : The no. of elements of the set is 4 number of equivalence relations is $2^4-1=15$

8. (a) **Explanation** :Let

$A=\{1,2,5\}\quad B=\{3,4,5\}\quad \mu=\{1,2,3,4,5,6\}$

$A^c=\{3,4,6\}\qquad B^c=\{1,2,6\}$

$(A\_B)\cup(B-A)\cup(A\cap B)=\{1,2\}\cup\{3,4\}\cup\{5\}$

$=\{1,2,3,4,5\}\qquad A\cup B=\{1,2,3,4,5\}$

$\therefore (A-B)\cup(B-A)\cup(A\cap B)=A\cup B$

10. (d) **Explanation** : Let R be a symmetric and transitive relation on a set then we can't know R is a reflexive or not

17. (c) **Explanation** : The set of all subsets of a set A is called power set of A.

$\therefore A=\{\phi,\{x\},\{y\}\}$

18. (c) **Explanation** : All other subsets of A are called Proper subsets.

$\therefore A=\phi,\{x\},\{y\}$

19. (c) **Explanation** : By known theorem A and B are sets and $A\cup B = A\cap B$ then A = B

20. (c) **Explanation** : Given $x<y$ then

Let $x=\{1,2,3\}\quad y=\{1,2,3,4\}$

given $y<z$ Let $Z=\{1,2,3,4,5\}$

$x\cap y\cap z=\{1,2,3\}=x$

21. (c) **Explanation** :Let

$x=\{1,2,3\},\ U=\{1,2,3,4,5,6\},\qquad y=\{3,4,5\}$

$y\cup x=\{1,2,3,4,5\},\qquad (y\cup x)^c=\{6\}$

$x\cap(y\cup x)^C=\phi$

22. (a) **Explanation** :

$R\times(P^c\cup Q^c)^C=R\times(P\cap Q)=(R\times P)\cap(R\times Q)$

23. (b) **Explanation** : A relation R in a set A is said to be symmetric if $(x,y)\in R\Rightarrow(y,x)\in R$

24. (a, c) **Explanation** :

i) for any $a\in X$, a divides a hence aRa $\therefore$ R is reflexive.

ii) for any $a,b\in X$ if 4 divides 2 then 2 not divides 4 i.e. $aRb\Rightarrow b\,Ra$. Thus R is not reflexive.

iii) for $a,b,c\in X$ if aRb, bRc then both a divides b and b divides c so that a divides c so R is Reflexive and Transitive.

25. (d) **Explanation** :

i) for any $a\in X, |a-a|$ is divisible 5 hence aRa

$\therefore$ R is reflexive.

ii) for any $a,b\in X$ *if* $|a-b|$ is divisible by 5 then $|b-a|$ is also divisible by 5. i.e. $aRb\Rightarrow bRa$. Thus R is symmetric.

iii) for $a,b,c\in X$, if aRb and bRc then both $|a-b|$ and $|b-c|$ are divisible by 3 so that $|a-c|=|a-b|+|b-c|$ is also divisible by 3 and hence aRc.

Thus R is transitive $\therefore$ R is an equivalence relation.

26. (d) **Explanation** :

i) for any $a\in X$, a divides a hence aRa $\therefore$ R is reflexive.

ii) for any $a,b\in X$ if 4 divides 2 then 2 not divides 4. i.e. $aRb\Rightarrow b\,Ra$ thus R is not reflexive.

iii) for $a,b,c\in X$ if aRb, bRc then both a divides b and b divides c so that a divides c So R is Reflexive and Transitive.

27. (b, c) **Explanation** : Since there is no X in N such that $2n+3=1$, f is not surjective and hence it is not bijective. If $f(x)=f(y)$ then $2x+3=2y+3\Rightarrow x=y$

$\therefore$ f is one one i.e. f is injective.

28. (d) **Explanation** :

$f=\{(x,u),(y,v)(z,v)\}\ f(y)=f(z)=V$

f is many one and range of $f=\{u,v\}\neq B$ so it is into

29. (c) **Explanation** :

If $f:x\rightarrow y$ and $a,b\subseteq X$ then $f(a\cap b)$ is equal to a proper subset of $f(a)\cap f(b)$

30. (b) **Explanation** : Let $A=\{1,2,3\}$

$A\times A=\{(1,1)(1,2)(1,3)(2,1)(2,2)(2,3)(3,1)(3,2)(3,3)\}$

31. (b) **Explanation** : The sets consists 2 elements So power set consists $2^2 = 4$

32. (d) **Explanation** : Let

$A = \{1,2,3\}$ $B = \{3,4,5\}$, $\mu = \{1,2,3,4,5,6\}$

$(A - B) = \{1, 2\}$

$\mu - (A - B) = \{3,4,5,6\} = \overline{A - B}$

$A \cup B = \{1,2,3,4,5\}$ $\therefore A \cup B \neq \overline{A - B}$

33. (c) The Vendiagram is

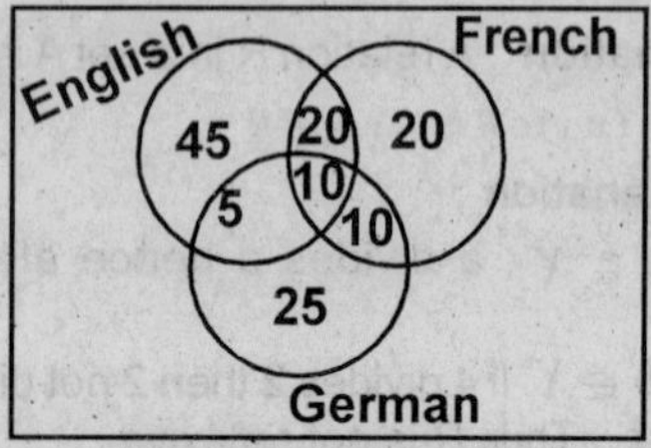

Explanation : We know that if I is set of positive integers and the relation R is defined over the set I by xRy if $x^y = y^x$ then R is Reflexive , Symmetric and transitive.

36. Explanation : Given set

S = { x/y is a student in you class }

and relation *(x Ry ↔ x sits in the same row as y)* then reflexive, transitive, symmetric.

OBJECTIVE QUESTIONS - II

1. Consider the following relations: **[GATE-2001]**

R1(a,b) iff (a+b) is even over the set of integers.

R2(a,b) iff (a+b) is odd over the set of integers.

R3(a,b) iff a.b>0 over the set of non zero rational numbers.

R4(a,b) iff |a-b|<=2 over the set of natural numbers.

Which of the following statements is correct?

a) R1 and R2 are equivalence relations, R3 and R4 are not.
b) R1 and R3 are equivalence relations, R2 and R4 are not.
c) R1 and R4 are equivalence relations, R2 and R3 are not.
d) R1, R2, R3 and R4 are all equivalence relations.

2. Consider the following statements: **[GATE-2001]**

S1: There exist infinite sets A, B, C such that $A \cap (B \cup C)$ is finite.

S2: There exist two irrational numbers x and y such that (x+y) is rational.

Which of the following is TRUE about S1 and S2?

a) Only S1 is correct. b) Only S2 is correct.
c) Both S1 and S2 are correct.
d) None of S1 and S2 is correct.

3. Let f : A → B be a function, and let E and F be subsets of A. Consider the following statements about images.

S1: $f(E \cup F) = f(E) \cup f(F)$, S2: $f(E \cap F) = f(E) \cap f(F)$

Which of the following is TRUE about S1 and S2? **[GATE-2001]**

a) Only S1 is correct b) Only S2 is correct.
c) Both S1 and S2 are correct
d) None of S1 and S2 are correct.

4. Let R1 and R2 be two equivalence relations on a set. Consider the following assertions: **[GATE-1998]**

I. R1 $\cup$ R2 is an equivalence relation.
II. R1 $\cap$ R2 is an equivalence relation.

Which of the following is correct?

a) Both assertions are true.
b) Assertion I is true but assertion II is not true.
c) Assertion II is true but assertion I is not true.
d) Neither I nor II is true.

5. The binary relation R={(1,1), (2,1), (2,2), (2,3), (2,4), (3,1), (3,2), (3,3), (3,4)} on the set A={1, 2, 3, 4} is **[GATE-1998]**

a) Reflexive, symmetric and transitive.
b) Neither reflexive, nor irreflexive but transitive.
c) Irreflexive, symmetric and transitive.
d) Irreflexive and antisymmetric.

6. In a room containing 28 people, there are 18 people who speak English, 15 people who speak Hindi and 22 people who speak Kannada. 9 persons speak both English and Hindi, 11 persons speak both Hindi and Kannada where as 13 persons speak both kannada and English. How many people speak all the three languages? **[GATE-1998]**

a) 6 b) 7 c) 8 d) 9

7. Suppose x and y are sets and |x| and |y| are their respective cardinalities. It is given that there are exactly 97 functions from x to y. From this one can conclude that **[GATE-1996]**

a) |x|=1, |y|=97 b) |x|=97, |y|=1
c) |x|=97, |y|=97 d) None

8. Let R denote the set of real numbers. Let f : R×R → R×R be a bijective function defined by f (x, y) =(x+y, x-y). The inverse function of f is given by **[GATE-1996]**

a) $f^{-1}(x,y) = \left[\frac{1}{(x-y)}, \frac{1}{(x-y)}\right]$

b) $f^{-1}(x,y) = (x-y, x+y)$

c) $f^{-1}(x,y) = \left[\frac{x+y}{2}, \frac{x-y}{2}\right]$

d) f^{-1} (x, y)=[2(x-y), 2(x+y)]

9. A relation over the set S ={x, y, z} is defined by: {(x,x), (x,y), (y,x), (x,z), (y,z), (y,y),(z,z)} what properties hold for this relation? **[GRE-CS]**

I. Symmetric, II. Reflexive,
III. Antisymmetric IV. Irreflexive

a) I only b) II only
c) I and II only d) I and IV only

10. A relation on the integers 0 through 4 is defined by :

R={(x,y): x+y ≤ 2x} Which of the properties listed below applies to this relation ? **[GRE-CS]**

I. Transivity. II. Symmetry. III. Reflexivity.

a) I only b) III only c) I and III d) II and III

11. The symmetric difference of sets A and B is

a) (A -B) $\cup$ (B-A) b) (A $\cup$ B) - (A $\cap$ B)
c) (A -B) $\cap$ (B-A) d) (A $\cap$ B) - (A $\cup$ B)

12. The order of the power set of a set of order n is

a) n b) 2n c) n^2 d) 2^n

13. If x, y are real numbers such that the ordered pairs (x+y, x-y) and (2x+3y,3x-2y) are equal, then (x, y) is equal to

a) (1,1) b) (2,3) c) (3,-2) d) (0,0)

14. Sets A and B have 3 and 6 elements each. What can be the minimum number of elements in A $\cup$ B is : **[IIT-1980]**

a) 3 b) 6 c) 9 d) 18

15. Set A has 3 elements and set B has 4 elements. The number of injections that can be defined from A to B is

a) 144 b) 12 c) 24 d) 64

16. If O(A) =m and O(B)=n then the number of possible relations R: A $\rightarrow$ B is

a) m n b) $(m\,n)^2$ c) 2^{m+n} d) $2^{m\,n}$

17. Assume R and S are (non empty) relations in a set A. Which of the following given relations is FALSE? **[M.N.R-1984]**

a) If R and S are transitive, then R $\cup$ S is transitive
b) If R and S are transitive, then R $\cap$ S is transitive
c) If R and S are symmetric, then R $\cup$ S is symmetric
d) If R and S are reflexive, then R $\cup$ S is reflexive.

18. If R be a relation < from A={1, 2, 3, 4} to B={1, 3, 5}. (i.e), (a, b) $\in$ R iff a<b,then Ro R^{-1} is: **[M.N.R-1984]**

a) { (1,3), (1,5), (2,3), (2,5), (3,5), (4,5)}
b) { (3,1), (5,1), (3,2), (5,2), (5,3), (5,4)}
c) { (3,3), (3,5), (5,3), (5,5)}
d) { (3,3), (3,4), (4,5)}

19. If f : R $\rightarrow$ R defined by $f(x)=x^2+1$ then the values of $f^{-1}(17)$ and $f^{-1}(-3)$ are respectively **[M.N.R-1986]**

a) {f },{4,-4} b) {3,-3},{ f }
c) {f },{3,-3} d) {4,-4},{ f }

20. If the function f : R $\rightarrow$ R is given by $f(x)=x^2+2x-3$ and the function g : R $\rightarrow$ R is given by g(x)=3x-4, then (go$f$)(x) is given by **[Hyd.Univ-1992]**

a) $9x^2+18x+5$ b) $3x^2+6x-13$
c) x^2+x-7 d) x^2-5x-1

21. With respect to the relation 'x divides y', which of the following sets are totally ordered? **[Hyd.Univ-1992]**

I. {36, 3, 9} II. {7, 77, 11}
III. {3, 6, 24, 12} IV. {1, 2, 3, ...}

a) I, II and III only b) II and III only
c) I and III only d) All

22. If Q be the set of non-zero rational numbers and the relation R be defined over the set Q by xRy if x=1/y, x,y $\in$ Q, which of the following are TRUE? **[Hyd.Univ-1991]**

I. R is an equivalence relation.
II. R is reflexive. III. R is symmetric.
IV. R is transitive.

a) I only b) II and III only
c) III and IV only d) III only

23. Let n (A) denotes the number of elements in set A. If n(A) =p and n(B)=q, how many ordered pairs (a, b) are there with a $\in$ A and b $\in$ B ? **[Hyd.Univ-1990]**

a) p^2 b) p ×q c) p + q d) 2pq

24. Determine which of the following relations are functions: **[Hyd.Univ- 1990]**

I. N={(x, y)/y= x^2, x $\in$ {-1, 0, 1, 2, 3}}
II. P={(x, y)/ y^2=x, x $\in$ {4, 9, 16}}
III. Q={(x, y)/y=4 x^2-10, x $\in$ {-1, 1, 2, 3}}

a) I only b) I and II only
c) I and III only d) III only

25. In a group of 72 students , 47 have background is Electronics, 59 have background in Mathematics and 42 have background in both the subjects. How many students do not have background in any of the subjects? **[R.E.C.Trichy - 1993]**

a) 8 b) 13 c) 25 d) 34

26. If [x] denotes the integer part of the real number, then the function $f(x)=x-[x]$ is **[R.E.C.Trichy - 1993]**

a) An even function b) An odd function
c) A periodic function d)A constant

27. In a beauty contest, half the number of experts voted for Mr. A and two thirds voted for Mr. B. 10 voted for both and 6 did not vote for either. How many experts were there in all? **[R.E.C.Trichy - 1995]**

a) 18 b) 36 c) 24 d) None.

28. Let R be a set of real numbers and f : R $\rightarrow$ R be a function defined by $f(x)=x^2$ then f=0 is **[R.E.C.Trichy - 1995]**

a) Injective but not surjective.
b) Surjective but not Injective.
c) Bijective. d) None.

29. In a classroom containing 28 people, there are 18 people who speak English, 15 people who speak Hindi and 22 people who speak Telugu. 9 people speak both English and Hindi, while 11 people speak both Hindi and Telugu. Further 13 persons speak both Telugu and English. How many people speak all three languages? **[Hyd. Univ-2001]**

a) 9 b) 8 c) 7 d) 6

30. Consider the sets **[Hyd. Univ-1999]**

I. A = {x | x = 2^n, n $\in$ N}
II. B = {x | x = 3n, n $\in$ N}
III. C = {x | x = 3n, n $\in$ Z}

Where N and Z are sets of all natural numbers and integers Which of the above set(s) is(are) closed under addition, subtraction and multiplication.

a) I b) II c) III d) I, III

31. The set of all Equivalence Classes of a set A of cardinality C **[Hyd. Univ-1999]**

a) has the same cardinality as A

b) forms a partition of A

c) is of cardinality 2^C d) is of cardinality C^2

32. Which of the following sets is a null set? **[Hyd. Univ-1999]**

I. X = {x | x = 9, 2x = 4} II. Y = {x | x = x}

III. Z = {x | x - 8 = 4}

a) I and II only b) I, II and III

c) I and III only d) II and III only

33. Let R be a relation "(x – y) is divisible by m", where x, y, m are integers and m >1, then R is **[Hyd. Univ-1999]**

a) symmetric but not transitive b) partial order

c) equivalence relation

d) anti symmetric and not transitive

34. Let s(w) denote the set of all the letters in w where w is an English word. Let us denote set equality, subset and union relations by =, ⊂ and ∪ respectively. Which of the following is NOT true? **[Hyd. Univ-1994]**

a) s(ten) ⊂ s(twenty) b) s(stored) = s(sorted)

c) s(sixty) ⊂ (s(six) ∪ s(twenty)) d) None

35. A relation R is defined on the set of positive integers as xRy if 2x + y ≤ 5. The relation R is **[Hyd. Univ-1994]**

I. Reflexive II. Symmetric III. Transitive

a) I only b) II only c) III only d) I and II only

36. Let R be an equivalence relation on the set {1, 2, 3, 4, 5, 6} given by {(1, 1), (1, 5), (2, 2), (2, 3), (2, 6), (3, 2), (3, 3), (3, 6), (4, 4), (5, 1), (5, 5), (6, 2), (6, 3), (6, 6)} The partition induced by R is **[Hyd. Univ-1999]**

a) {1, 2, 3, 4, 5, 6} b) {{1, 3, 5, 6}, {2, 4}}

c) {{1, 5}, {2, 3, 6}, {4}} d) {{1, 2, 3, 4}, {5, 6}}

37. Let **f:R→R** be defined by $f(x)=\begin{cases} x+2 & (x\le -1) \\ x^2 & (-1\le x\le 1) \\ 2-x & (x\ge 1\end{cases}$

Then the value of $f(-1.75)+f(0.5)+f(1.5)$ is **[JNTU- 2001]**

a) 0 b) 2 c) 1 d) -1

38. Let Z denote the set of all integers Define **f:Z→Z** by $f(x)=\begin{cases} x/2 & (x\ is\ even) \\ 0 & (x\ is\ odd)\end{cases}$ then f is **[JNTU- 2001]**

a) onto but not one-one b) one-one but not onto

c) one-one and onto d) neither one-one nor onto

39. If **f:A→B** is a bijective function then f^{-1}**of** of = **[JNTU- 1996]**

a) fof^{-1} b) f c) f^{-1}

d) I_A (Identity map of the set A)

40. To have inverse for the function f, f is **[JNTU-1988]**

a) one one b) onto

c) one one onto d) identity function

41. f: R→R is a function defined by f (x) = 10x - 7. If g = f^{-1} then g (x)= **[JNTU-1993]**

a) $\frac{1}{10x-7}$ b) $\frac{1}{10x+7}$ c) $\frac{x+7}{10}$ d) $\frac{x-7}{10}$

42. If f (x) and g (x) are defined on domains A,B respectively then the domain of f (x) + g (x) is **[IIT- 1988]**

a) A∪B b) A∩B c) AΔB d) A–B

43. The correspondence

f: N →N is such that f(x) = y and for $x = p_1^{e1}, p_2^{e2}, ...p_k^{ek}$ where P_i are distinct primes and integers $e_i \ge 1$ we have $y = e_1 + e_2 + ... + e_k$ then **[UGC NET-2000]**

a) f is not a function b) f is an onto function

c) f is an 1-1 function

d) f is neither 1 – 1 nor onto function

44. The number of functions f: (1, 2, 3, ..., n), → {1,2,.......m}, which are one-to-one is **[UGC NET-1996]**

a) m^n b) m(m -1)(m - 2) .... (m- n +1)

c) n^m d) n(n -1)(n - 2) ....(n- m +1)

ANSWER KEY - II

1. b	**2.** c	**3.** a	**4.** c	**5.** b
6. a	**7.** a	**8.** c	**9.** b	**10.** d
11. a	**12.** d	**13.** d	**14.** b	**15.** c
16. d	**17.** a	**18.** c	**19.** d	**20.** b
21. c	**22.** b	**23.** b	**24.** c	**25.** a
26. c	**27.** c	**28.** b	**29.** d	**30.** c
31. b	**32.** a	**33.** c	**34.** d	**35.** c
36. c	**37.** c	**38.** a	**39.** d	**40.** c
41. c	**42.** b	**43.** b	**44.** a	

EXPLANATORY NOTES - II

1. (b) **Explanation :** By known theorms $R_1(a-b)$ if $(a+b)$ is even over the set of integers is satisfies equivalence relations $R_3(a,b)$ iff $a.b>0$ over the set of non zero rational numbers is also satisfies equivalence relation and R_2 , R_4 are not equivalence relation.

2. (c) **Explanation :** By known theorms there exist infinite sets A,B,C such that $A\cap(B\cup C)$ is finite. There exist two irrational numbers x and y such that $(x+y)$ is rational. So Both S_1 and S_2 are correct.

3. (a) **Explanation :** $f:A\to B$ be a function and let E and F be subsets of A

S_1: $f(E\cup F)=f(E)\cup f(F)$ and

S_2: $f(E\cap F)\ne f(E)\cap f(F)$

So only S_1 is correct.

4. (c) **Explanation :** By known theorm R_1, R_2 be two equivalence relations on a set then $R_1\cap R_2$ is an equivalence relation and $R_1\cup R_2$ is not a equivalence relation.

5. (b) **Explanation** : R is not a reflexive because (4,4) is not available. R is not a irreflexive because (1,1) (2,2) (3,3) are available. But R is a transitive because $(2,3)\ (3,1) \rightarrow (2,1)$ is also available. This model of relations all satisfied.

6. (a) **Explanation** : The following is vendiagram of 28 people.

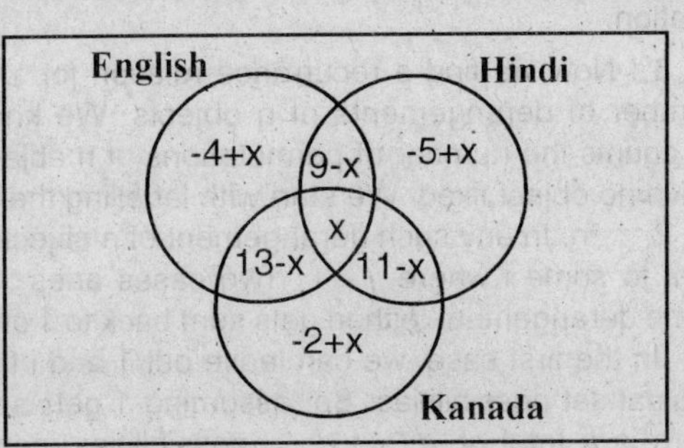

The room containing 28 people

$$-4+x-5+x-2+x+9-x+13-x+11-x+x=28$$

$$22+x=28,\ x=28-22=6,\ x=6$$

7. (a) **Explanation** : We know that x any y sets and $|x|\ and\ |y|$ are their respective cardinalities it is given that there are exactly 97 functions from x to y. This one is conclude that $|x|=1,\ |y|=97$

8. (b) **Explanation** :

$$x+y=z,\ x-y=z,\ x=z-y,\ -y=z-x,$$

$$f^{-1}(x)=x-y,\ y=x-z,\ f^{-1}(y)=x-y$$

$$f^{-1}(x,y)=(x-y,\ x-y)$$

9. (a) **Explanation** : $(x,x),\ (y,y),\ (z,z)$ are all in the set so this reflexive.

13. (d) **Explanation** : We can substitute (0,0) on $(x+y,\ x-y)$ and $(2x+3y,\ 3x-2y)$ then 2 values are equal.

19. (d) **Explanation** :

$$x^2+1=y$$

$$x^2=y-1$$

$$x=\sqrt{y-1}$$

$$f^{-1}(x)=\sqrt{x-1}$$

$$f^{-1}(17)=\sqrt{16}=\pm 4=\{4i,-4\}$$

$$f^{-1}(-3)=\sqrt{-4}=\pm 2i=\{\phi\}$$

20. (b) **Explanation** :

$$gof(x)=g[f(x)]=g[x^2+2x+3]=3(x^2+2x+3)-4$$

$$=3x^2+6x+9-4$$

$$=3x^2+6x+5$$

21. (c) **Explanation** :

Clearly x divides y in totally ordered sets are

{36,3,9}, {3,6,24,12}

24. (c) **Explanation:** Clearly

$y=x^2,\ x\in\{-1,0,1,2,3\}$ is function

clrearly

$y=4x^2-10,\ x\in\{-1,1,2,3\}$ is also function

Because Let $R\subseteq A\times B$ (i.e) relation from A to B. Then R is said to be a function if

(i) domain of R = A

(ii) (a, b), (a, c) $\in R\Rightarrow b=c$

25. (a) **Explanation:** Students don't background in subjects = group of students electronics students + Mathematics Students + both subjects students.

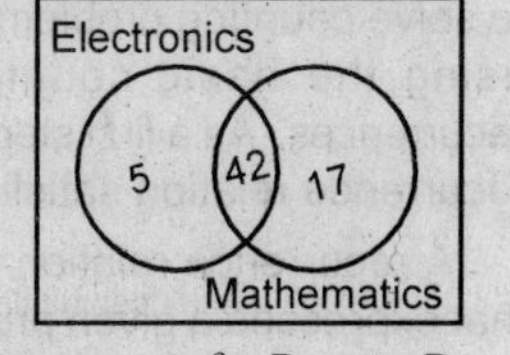

= 72- (42+17+5)

= 8

28. (b) **Explanation:**

Let R be a set of real numbers and $f: R\rightarrow R$

be a function defind by $f(x)=x^2$ Then $f=0$ is onto but not one - one. So $f=0$ is surjective but not a injective

32. (a) **Explanation:** We can't proved

$x=9\ and\ x=2\ \therefore$ I is null set

We can't proved $x\pm x\ \therefore$ II is also null set.

33. (c) **Explanation:**

(i) for any a $\in x, a-a$ is divisible by m hence a Ra

$\therefore$ R is reflexive

(i) for any $a,b,\in x$ if a-b is divisible by m then b-a is also divisible by m (i.e). is also divisible by m

Thus R is transitive $\therefore$ R is an equivalence relation.

36. (c) **Explanation:** $(1,1)(1,5)\rightarrow\{1,5\}$

$(2,2)\ (2,3)\ (3,6)\rightarrow\{2,3,6\}$ the remaining element

$(4,4)\rightarrow\{4\}\quad \therefore\{\{1,5\},\{2,3,6\},\{4\}\}$

37. (c) **Explanation:** f (-1.7) + f (0.5) + f (1.5)

$$x\leq -1+-1\leq x\leq 1+x\geq 1$$

x + 2 + x² + 2 - x, x² + 4.

41. (c) **Explanation:**

10x + 7 = y, 10x = y - 7

$$x=\frac{y-7}{10},\ f^{-1}(x)=\frac{x-7}{10}$$

but g(x) = f¹ (x) $\quad\therefore g(x)=\frac{x-7}{10}$

II. RECURRENCE RELATIONS

- Introduction
- Solving Recurrence Relations
- Fibonacci Relation
- Solving Recurrence Relations by substitution Method
- Using Generating Functions to Solve Recurrence Relations

1. Recurrence Relation

There are many kinds of problems that have to do with counting which can not be talked only with product rule or sum rule or permutations and combinations. For example look at the problem of counting the number of ways of filling up n - boxes labelled 1,2,3 .... n, with 0's and 1's in such a way that no two adjacent boxes have '0'. To solve this, and several such problems, we need to use the idea of recurrence relations. The basic method to solve counting problems that seen to reisist a solution using the basic counting tools is the method of recurrences. As a first step, we need to set up the required recurrence relation satisfied by the problem.

A recurrence relation or recurrences, is an equation that expresses a given problem for n - objects in terms of the same problem posed for less than n objectives. In this section we discuss in detail several problems that lead to recurrence relations, and give you an indication of how they can be solved.

Definition 1 : A recurrence relation for the sequence $\{a_n\}$ is an equation that expresses a_n interms of one or more of the previous terms of the sequence namely $a_0, a_1 .... a_{n-1}$ for all integers with $n \geq$ no where no is a non negative integer. A sequence is called a solution of a recurrence relation if its terms satisfy the recurrence relation.

Ex 1 : *Find the first 5 terms of the sequence defined by* $a_n = 6\ a_{n-1}, a_0 = 2$

sol: $a_1 = 6\ a_0 \Rightarrow a_1 = 6 \times 2 = 12$

$a_2 = 6a_{2-1} = 6a_1$, $a_2 = 6.(12) = 72$

$a_3 = 6a_2 = 6 \times 72 = 432$, $a_4 = 6a_3 = 6(432) = 2592$

$a_5 = 6a_4 = 6(2592)$

Ex 2 : Consider the set of all subsets of any non - empty set S, called its power set, and is denoted by $P(S)$. Let used determine a recurrence relation satisfied by $S_n = |P(S)|$ where $|S| = n$

Let $S = \{1,2,3......n\}$.

Now, any subset, A of S either contains the number *n* or does not. Let us consider these two mutually exclusive cases separately and count the number of such subsets.

If $n \in A$ Then $A = A' \cup \{n\}$. where A' is a subset of $\{1, 2....n-1\}$.

So, there are as many subsets A as there are subsets A'. Since A' is a subset of $\{1, 2, 3....n-1\}$. there are S_{n-1} such subsets on A. Also if $n \notin A$ then A is a subset of $\{1, 2....n-1\}$ and there are S_{n-1} of these too combining these we get $S_n = S_{n-1} + S_{n-1}$

$\Rightarrow S_n = 2S_{n-1} n \geq 1$ with $S_o = 1$ which is a recurrence relation.

Ex 3 : Now we find a recurrence relation for d_n, then number of derangements of n objects. We know that d_n counts the number of permutations of n objects that leave no object fixed. We start with labelling the objects : 1, 2, ... n. In any such derangement of n objects, 1 will sent to some i where $i \neq 1$. Two cases aries : for the same derangement, eitheri gets sent back to 1 or it does not. In the first case, we can leave out 1 and i from the original set possibilities. So, assuming 1 gets sent to i, there is a total of $d_{n-1} + d_{n-2}$ possibilities. now since $2 \leq i \leq n$ We have $d_n = (n-1)(d_{n-1} + d_{n-2})$ for $n \geq 3$. with $d_1 = 0, d_2 = 1$.

Ex 4 : Find arecurrence relation for the determinant of the $n \times n$ matrix with 1 along the main diagonal and with 1 one either side of the main diagonal in each row and zero else where.

Sol : Let Δ_n denote the required $n \times n$ determinant. Expanding about the first row, we get Δ_{n-1} minus the determinant which when expanded about its row. i.e., Δ_{n-2}. So the corresponding recurrence relation is $\Delta_n = \Delta_{n-1} - \Delta_{n-2}$ with $\Delta_1 = 1\ \ \Delta_2 = 0$.

Ex5 : Find a recurrence relation for the n - digit sequences of numbers using only the integers {0,1,2,3} having an even number of zeros.

Sol: Let a_n denoted the number of n - digit sequences containing an even number of zeros. Then there are $a_{n-1}(n-1)$ digit sequences that have an even number of 0's and $4^{n-1} - a_{n-1}(n-1)$ - digit sequences that have an odd number of zeros. To each of the a_{n-1} sequences that have an even number of zeros, the digit 1,2, or 3 can be appended to yield sequences of length n that contains an even number of 0's To each of the $4^{n-1} - a_{n-1}$ sequences that have an odd number of 0's the digit 0 must be appended to yield sequences of length n that contain an even number of 0's therefore for $n \geq 2$.

$a_n = 3a_{n-1} + 4^{n-1} - a_{n-1} \ = 2a_{n-1} + 4^{n-1}$ with $a_1 = 3$.

Ex 6 : Find a recurrence relation for the number of ways to arrange flags on a flag pole n feet tall using 4 types of flags : red flags 2 feet high, or white, blue and yellow flags each 1 foot high.

Sol : Let C_n be the number of ways to arrange flags on the flag pole n - feet tall. If the first flag is 1 foot high, then there are C_{n-1} ways to arrange the flags on the other n-1 feet of the pole. Since there are 3 colors for the first flag, in this case, there are 3 C_{n-1} ways to arrange the pole. If the first flag is red and is of 2 feet high, the there are C_{n-2} other $n-2$ feet of the pole. Hence the recurrence relation under given condition is

$$C_n = 3C_{n-1} + C_{n-2}$$

Ex 7 : Find a recurrence relation for the number of ways to make a pile of n chips using garnet, gold, red, white and blue chips such that no two gold chips are together.

Sol : If the first chip is gold there are 4 colors for the second chip and then there are a_{n-2} ways to make a pile of the other $n-2$ chips. Also there are 4 colors for the first chip to be other than a gold one and then a_{n-1} ways to make a pile of the othe n-1 chips.

There fore $a_n = 4a_{n-1} + 4a_{n-2}$

Ex 8 : Suppose n unit circles are drawn on the plane such that each circle intersects any of the other circles at exactly two points and no three circles meet at the same point. Derive a recurrence relation for the number r_n of regions into which the plane is divided by n circles.

Sol : Clearly we have $r_1 = 2,\ r_2 = 4,\ r_3 = 8,$ and $r_4 = 14$. (i.e., if we draw circles according to given condition).

Suppose that we have drawn n-1 unit circles that divide the plane in to $r_n - 1$ regions. The nth circle intersects these n-1 circles at $2(n-1)$ points i.e., the nth circle will be divided into $2(n-1)$ arcs. Since each of these arcs will divide one of the $r_n - 1$ regions into two we have the recurrence $r_n = r_{n-1} + 2(n-1)$ for $n \geq 2$.

Ex 9 : Suppose that a person deposits Rs. 10,000/- in a savings account at a bank yielding 12% per year with interest compounded annually. How much will be in the account after 30 years?

Sol : Let p_n denote the amount in the account after n years. Since the amount in the account after n years. equals to the amount in the account after n-1 years plus interest for the nth year, we have

$$p_n = p_{n-1} + 0.12\,p_{n-1} = (1.12)\,p_{n-1}$$

The initial conditions is $p_0 = 10{,}000$. Now we will find a formula for p_n. $p_1 = (1.12)\,p_0$,

$$p_2 = (1.12)\ p_1 = (1.12)\ (1.12)\ p_0 = (1.12)^2\ p_0$$

$$p_3 = (1.12)\ p_2 = (1.12)^3\ p_0$$

$$p_n (1.12) p_{n-1} = (1.12)^n\ p_0$$

Since $p_0 = 10{,}000$ we have $p_n = (1.12)^n 10.000$

If $n = 30$ then $p_{30} = (1.12)^{30}\ (10{,}000)$

Ex 10 : Find a recurrence relation for the number of bit strings of length n that do not have two consecutive 0s. How many such bit strings are there of length five?

Sol : Let a_n denote the number of bit strings of length n that do not have two consecutive zeros. By sum rule, the number of bit strings of length n that do not have two consecutive zeros equals the number of such bit strings ending with a zero plus the number of such bits strings ending with a 1. We will assume that $n \geq 3$, so that the bit string has at least three bits.

The bit strings of length n ending with a one that do not have two consecutive zeros are precisely the bit strings of length $n-1$ with no two consecutive zeros with a 1 added at the end. Consequently, there are a_{n-1} such bit strings.

Bit strings of length n ending with a zero that do not have two consecutive zeros must have 1 as their $(n-1)^{st}$ bit otherwise they would end with a pair of zeros. It follows that the bit strings of length n ending with a zero that do not have no two consecutive zeros are precisely the bit strings of length n-2 with no two consecutive zeros with 10 added at the end. Hence there are a_{n-2} such bit strings. Hence the required recurrence relation is $a_n = a_{n-1} + a_{n-2}$ Where $n \geq 3$.

The initial conditions are $a_1 = 2$ since both bit strings of length one, 0 and 1 donot have consecutive zeros and $a_2 = 3$. Since the valid bit strings of length two are 01, 10 and 11. Now

$$a_3 = a_2 + a_1 = 3 + 2 = 5,\ a_4 = a_3 + a_2 = 5 + 3 = 8$$

$$a_5 = a_4 + a_3 = 8 + 5 = 13$$

Def 2 : The order of the recurrence relation defined by $a_n = F(a_{n-1}, a_{n-2}, \ldots a_{n-k}, n)$ where F is a function of some of the variables $a_{n-1}, a_{n-2} \ldots\ldots a_0, n;$ *is k*, where a_n depends on one or more of the previous k-terms and k is the smallest such integer.

Def 3 : The degree of the recurrence relation is the degree of F, treated as a polynomial in its variables excluding n. If F is not a polynomial in its variables, no degree is assigned to the recurrence relation. A recurrence relation of degree one is called linear, degree two is called quadratic and so on.

Def 4 : A recurrence relation is called homogeneous if it contains no term that depends only on the variable n. If it is not homogeneous, then we say that it is non-homo geneous.

Examples :

1. $a_n = 3a_{n-1} + n^2$ is non homogeneous of order 1 and degree 1.
2. $a_n = n a_{n-2} + 2^n$ is non homogeneous of order 2 and degree 1.
3. $a_n = \sqrt{a_{n-1}} + a_{n-2}^2$ is homogeneous of order 2, but has no degree.
4. $a_n = a_{n-1} + a_{n-2} + \ldots + a_0$ is homogeneous has, nor order, but has degree.
5. $a_n = a_{n-1}^2 + a_{n-2}\,a_{n-3} \cdot a_{n-4}$ is homogeneous of order 4 and deree 3.
6. $a_n = \sin a_{n-1} + \cos a_{n-2} + \sin a_{n-3} + \ldots\ldots + e^n$ is non homogeneous, has no order and no degree.
7. $a_n = f_1(n)\ a_{n-1} + f_n(n)\ a_{n-2} + \ldots\ldots + f_{n-k}(a_n)\ a_{n-k} + g(n)$ represents the general form of a linear kth order recurrence relation. It is homogenous if g(n)=0 for each n, and non homogeneours otherwise.
8. $a_n = a_0, a_{n-1} + a_1 a_{n-2} + \ldots + a_{n-1}\ a_0$. with $a_0 = 0$, and $a_1 = 1$ is a non linear recurrence relation.

9. $a_n = a_{n/2} + 1$ with $a_1 = 0$ is a non linear recurrence relation.

2. SOLVING RECURRENCE RELATIONS

Solving Linear Homogeneous Recurrence Relations with Constant Coefficients :

The process of determining a closed form expression for the terms of a sequence from its recurrence relation is called solving the relation.

Definition 1 : nth Order Linear Relation : Let S be a sequence of numbers with domain $k \geq 0$. An nth order linear recurrence relation on S with constant coefficients is a recurrence relation that can be written in the form

$S(k) + c_1 S(k-1) + c_2 S(k-2) + ... + c_n S(k-n) = f(k)\ (k \geq n)$

Where $c_1, c_2 .... c_n$ are numbers and f is a numeric function that is defined for $k \geq n$.

Definition 2 : Homogeneous Recurrence Relation : An nth order linear relation is a homogenous recurrence relation if $f(k) = 0$ for all k. For each recurrence relation,

$S(k) + c_1 S(k-2) + ... + C_n S(k-n) = f(k)$ associated homogeneous relation is

$S(k) + c_1 S(k-1) + ... + C_n S(k-n) = 0$.

Ex 1 : Consider the second order homogenous relation $S(k) - 7S(k-1) + 12S(k-2) = 0$ together with the initial conditions $S(0) = 4$ and $S(1) = 4$. We can predict that the solution to this relation involves terms of the form ba_k, where b and a are non zero constants that must be determined. If the solutions were to equal to this quantity exactly, then

$S(k) = ba^k . S(k-1) = ba^{k-1}, S(k-2) = ba^{k-2}$

If we substitute the exponential expressions into the original equation, we get $ba^k - 7ba^{k-1} + 12ba^{k-2} = 0$

Each term on the left-hand side has a factor of ba^{k-2}, which is non zero. Dividing through by this common factor we get, $a^2 - 7a + 12 = (a-3)(a-4) = 0$

This equation is called the characteristic equation of the recurrence relation. The only possible values of a are 3 and 4. Sop general solution of the given equation will be of the form. $S(K) = b_1 e^K + b_2 4^K$ where b_1 and b_2 are real numbers.

This set of sequences is called the general solution of the recurrence relation.If we didn't have initial conditions for S, we would stop here. The initial conditions make it possible for us to obtain definite values for b_1 and b_2

$$\left.\begin{array}{l} S(0) = 4 \\ S(1) = 4 \end{array}\right\} \Rightarrow \begin{array}{l} b_1 3^0 + b_2 4^0 = 4 \\ b_1 3^1 + b_2 4^1 = 4 \end{array}$$

$b_1 + b_2 = 4 \quad \Rightarrow 3b_2 + 4b_2 = 4$

Solving we get $b_1 = 12$ and $b_2 = -8$ the solution is $S(k) = 12(3^k) - 8(4^k)$

Definition 3: Characteristic Equation : The characteristic equation of the homogeneous n^{th} order linear relation $S(k) + c_1 S(k-1) + c_1 S(k-1) + ...c_n S(k-n) = 0$ is the nth degree polynomial equation.

$$a^n + c_1 a^{n-1} + ... + c_{n-1} a + c_n = 0$$

The left hand side of this equation is called the characteristic polynomial.

Ex :1 The Characteristic equation of $F(k) - F(k-1) - F(k-2) = 0$ is $a^2 - a - 1 = 0$

Ex : 2 The characteristic equation of $Q(k) + 2Q(k-1) - 3Q(k-2) - 6Q(k-4) = 0$ is $a^4 + 2a^3 - 3a^2 - 6 = 0$

Working Rule for solving Homogeneous n[th] Order Linear Relations :

1. Write the characteristic equation of the relation $S(k) + c_1 S(k-1) + ... + c_n S(k-n) = 0$ which is $a^n + c_1 a^{n-1} + ... + c_{n-1} a + c_n = 0$
2. Find all roots of the characteristic equation called characteristic roots.
3. If there are n distinct characteristic roots, $a_1, a_2 ..... a_n$, then the general solution of the recurrence relation is

$$S(k) = b_1 a_1^k + b_2 a_2^k + ... + b_n a_n^k.$$

If there are fewer than n characteristic roots, then at least one root is a multiple root If a_j is a double root, then $b_j a_j^k$ is replaced with $(b_{j0} + b_{j1} k) a_j^k$.

In general, if a_j is a root of multiplicity p then $b_j a_j^k$ is replaced with $\left(b_{j0} + b_{j1} k + ... + b_{j(p-1)} k^{p-1}\right) a_j^k$.

4. If n initial conditions are given. Obtain n linear equations in n unknowns by substitution. Solve these equations.

Ex 1 : Solve $D(k) - 8D(k-1) + 16D(k-2) = 0$ where $D(2) = 16$ and $D(3) = 80$.

Sol : Consider the recurrence

$$D(k) - 8D(k-1) + 16D(k-2) = 0.$$

The characteristic equation is

$$a^2 - 8a + 16 = 0 \Rightarrow (a-4)^2 = 0$$

Therefore, there is a double characteristic root, 4.

The general solution is $(b_0 + b_1 k)4^k$

$D(2) = 16 \Rightarrow (b_0 + b_1 .2)16 = 16 \Rightarrow 16b_0 + 32b_1 = 16$

$D(3) = 80 \Rightarrow (b_0 + 3b_1)64 = 80 \Rightarrow 64b_0 + 192b_1 = 80$

Solving we get $b_0 = \frac{1}{2}$ and $b_1 = \frac{1}{4}$

Therefore $D(k) = \left(\frac{1}{2} + \frac{1}{4}k\right)4^k = (2+k)4^{k-1}$

Working Rule for Solving Non-homogeneous Finite Order Relations :

To solve the recurrence relation

$$S(k) + c_1 S(k-1) + ... + c_n S(k-n) = f(k)$$

1. Write the associated homogenous relation and find its general solution call this the homogeneous solution, denote this by $S^h(k)$
2. Find Particular solution with the help of the following table.

S.No.	Right Hand Side	Form of Particular Solution
1.	A constant, q	A constant d
2.	A linear function q_0+q_1k	A linear function, d_0+d_1k
3.	An mth degree polynomial $q_0+q_1k+....+q_mk^n$	An mth degree polynomial $d_0+d_1k+....+d_mk^m$
4.	An exponential function $q.a^k$	An exponential function $d.a^k$

3. Substitute the required form of particular solution in the recurrence relation and find the unknown co-efficients.

4. The general solution of the recurrence relation is the sum of the homogenous and particular solutions. If no initial conditions are given. Then $S(k)=S^h(k)+S^p(k)$, If n initial conditions are given, obtain 'n' linear equations in n unknowns and solve the system to get a complete solution.

Ex 1 : Solve the recurrence $S(k)+5S(k-1)=9,\ S(0)=6$

Sol : The characteristic equation is $a+5=0\ \therefore\ a=-5$

The homogeneous solution is $S^h(k)=b_1(-5)^k$

Since the right hand side is constant, the particular solution will be constant say d.

If we substitive $S^p(k)$ in the recurrence relation we get

$d+5d=9 \quad \Rightarrow 6d=9 \Rightarrow d=9/6=1.5$

The general solution is $s(k)=s^n(k)+s^p(k)$

$$=b_1(-5)^k+1.5$$

$$s(0)=6 \Rightarrow b_1(-5)^0+1.5=6$$

$$\Rightarrow b_1+1.5=6$$

$$\Rightarrow b=4.5$$

$$and S(k)=4.5(-5)^k+1.5$$

Ex 2 : Solve the recurrence relation

$T(k)-7T(k-1)+10T(k-2)=6+8k$ with

$T(0)=1$ and $T(1)=2$

Sol : The characteristic equation is $a^2-7a+10=0$

$\Rightarrow (a-2)(a-5)=0 \Rightarrow a=2,\ 5$

Therefore, the homogeneous solution is

$T^h(k)=b_1 2^k+b_2 5^k$

Since the right hand side is a linear polynomial, $T^p(k)$ will be of the form d_0+d_1k.

Substituting

$T^p(k)=d_0+d_1k$ in $T(k)-7T(k-1)+10T(k-2)=6+8k$

we get

$d_0 d_1 k-7(d_0+d_1(k-1))+10(d_0+d_1(k-2))=6+8k$

$\Rightarrow (4d_0-13d_1)+4d_1k=6+8k$

Two polynomials are equal if and only if their coefficients are equal. Therefore we have

$$4d_0-13d_1=6,\ 4d_1=8$$

Solving these two equations we get $d_1=2,\quad d_0=8$

$T^p(k)=8+2k$ Therefore the general solution

$T(k)=b_1 2^k+b_2 5^k+8+2k$

$T(0)=1 \quad \Rightarrow b_1+b_2+8=1$

$T(1)=2 \quad \Rightarrow 2b_1+5b_2+10=2$

$\Rightarrow b_1+b_2=-7,\ 2b_1+5b_2=-8$

Solving (3) and (4) we get $b_1=-9,\ b_2=2$.

$\therefore\ T(k)=-9(2^k)+2(5^k)+8+2k$

Ex 3 : Solve the recurrence relation

$$nS(n)-S(n-1)=0,\ S(0)=1$$

Sol : In general, the relation $P(n)=f(n)P(n-1),\ n\geq 0$ with $P(0)=f(0)$ where f is function that is defined for all $n\geq 0$ has the solution.

$$P(n)=f(0).f(1)f(2)....f(n)\prod_{K=0}^{n} f(x), n\geq 0$$

The given equation $\quad nS(n)-S(n-1)=0$

$$S(n)=\frac{1}{n}S(n-1) \quad and \quad S(0)=1$$

Here $f(n)=\frac{1}{n}$ and $f(0)=1=S(0)$

$f(1)=1,\ f(2)=\frac{1}{2}....$ The solution is

$$S(n)=1.1.\frac{1}{2}.\frac{1}{3}....\frac{1}{n}=\pi_{K=0}^{n}\frac{1}{n}=\frac{1}{n!}$$

3. Fibonacci Relation

Consider the problem how many pairs or rabbits are produced after n months if we start with one pair of rabbits exactly one month old, and if every month, to each pair of rabbits more than one month old and new pair is born?

Let F_n denoted the number of pairs of rabbits after n months. Then $F_1=1$. Since the pair does not breed in the second month $F_2=1$. To find the number of pairs after n months, we must add the number of pairs after n-1 months to the number of pairs born in the nth month. But the new borns come from pairs at least two months old i.e., from the pairs that already existed after $n-2$ months, there are F_{n-2} of these. Therefore, the sequence $\{F_n : n\geq 1\}$ meets the condition $F_n=F_{n-1}+F_{n-2}$ if $n\geq 3$ together with $F_1=1=F_2$ This sequence is called the Fibonacci sequence and the F_n are called Fibonacci numbers.

Ex : Find an explicit formula for Fibonacci numbers.

Sol : The recurrence relation for Fibonacci number is

$F_n=F_{n-1}+F_{n-2}$ with $F_0=F_1=1$

The characteristic equation is $r^2-r-1=0$

roots are $\frac{1+\sqrt{5}}{2}$ and $\frac{1-\sqrt{5}}{2}$

Therefore the solution is given by

$$f_n=\alpha_1\left(\frac{1+\sqrt{5}}{2}\right)^n+\alpha_2\left(\frac{1-\sqrt{5}}{2}\right)^n$$

for some constants α_1 and α_2

$F_0=0 \ \Rightarrow \alpha_1+\alpha_2=0$

$$F_1 = 1 \Rightarrow \alpha_1\left(\frac{1+\sqrt{5}}{2}\right)^n + \alpha_2\left(\frac{1-\sqrt{5}}{2}\right)^n$$

$$\Rightarrow \alpha_1 = \frac{1}{\sqrt{5}}, \alpha_2 = -\frac{1}{\sqrt{5}}$$

Therefore the fibonacci numbers are given by

$$F_n = \frac{1}{\sqrt{5}}\left(\frac{1+\sqrt{5}}{2}\right)^n - \frac{1}{\sqrt{5}}\left(\frac{1-\sqrt{5}}{2}\right)^n$$

4. Linear Non - Homogenous Recurrence Relations

A recurrence relation of the form

$a_n = c_1 a_{n-1} + c_2 a_{n-2} + \ldots + c_k a_{n-k} + F(n)$ where $c_1, c_2 \ldots c_k$ are real numbers and $F(n)$ is a function not identically zero depending only on n. The recurrence relation.

Result : If $\{a_n^{(P)}\}$ is a particular solution of the non-homogeneous linear recurrence relation with constant coefficients

$$a_n = c_1 a_{n-1} + c_2 a_{n-1} + \ldots + c_k a_{n-k} + F(n).$$

Then every solution is of the form $\{a_n^{(P)} + a_n^{(h)}\}$ where $\{a_n^{(h)}\}$ is a solution of the associated homogeneous recurrence relation.

$$a_n = c_1 a_{n-1} + c_2 a_{n-2} + \ldots + c_k a_{n-k}$$

Ex 1 : Find the complete solution of the recurrence

$$a_n = 3a_{n-1} - 4n, \; n \geq 1$$

Sol : The associated linear homogenous recurrence relation is $a_n = 3a_{n-1}$.

Characteristic equation is $t = 3$

Solution is $a_n = \alpha 1 . 3^n$

Now consider $a_n = 3a_{n-1} - 4^n$

Let us see if a_n can be of the form $A_n + B$ where A,B are constants.

$$A_n + B = 3[A(n-1) + B] - 4n$$
$$= n(3A - 4) - 3A + 3B$$

Comparing the coefficients of n we have $A = 3A - 4$ and $B = 3B - 3A$.

i.e., $A = 2$ and $B = 3$ i.e., An+B is a solution if and only if $A = 2$ and $B = 3$.

So $a_n = 2n + 3$ is a particular solution of the given recurrence. Therefore total solution of the recurrence will be $a_n = \alpha 1.3^n + 2n + 3$

5. Divide and Conquer Relations

This is a decomposition algorithm that solves a problem of size $n \in z^+$ by

1. Breaking it up into a number of instances of the same kind of problem with smaller input parameter.
2. Solving these sub problems and
3. Use their solutions to construct a solution for the original problem of size n.

Ex 1 : In a tennis tournament, each entrant plays a match in the first round. Next all winners from the first round play a second round match. Winners continue to move on to the next round, until finally only one player is left as the tournament winner. Assuming that tournament always involve $n = 2^k$ players, for some k, find the recurrence relation for the number of rounds in a tournament of n players.

Sol : The recurrence relation for a_n, the number of rounds is $a_n = a_{n/2} + 1$. Since after $a_{n/2}$ rounds there remains only two players, the winner of the sub tournament of the first $n/2$ players and the sub tournament of the second $n/2$ players. One more rounds picks the tournament winner from the two remaining players. Here $a_1 = 0$ since with one player, no tournament.

Ex 2 : Solve the divide and conquer relation $a_n - 7a_n/3 = 2n$ where $n = 3^k$ for $k \geq 1$ and $a_1 = 5/2$

Sol : We employ the change of variables $b_k = a_n = a_3 k$. Then the transformed relation is $b_k - 7b_{k-1} = 2.3^k$ and $a_1 = b_0 = 5/2$. The corresponding linear recurrence relation is $b_k - 7b_{k-1} = 0$ characteristic equation is $t - 7 = 0$ its root is 7. Therefore $b_k(h) = B.7^k$ for some constant B. Moreover, a particular solution of the non homogenous relation takes of the from $b_k^{(P)} = A.3^k$

By substituting we get the value of $A = -\frac{3}{2}$ so that

$$b_k = -\frac{3}{2}(3^k) + B.7^k, \; b_0 = 5/2 \Rightarrow B = 4$$

Thus $b_k = -\frac{3}{2}(3^k) + 4(7^k)$ Now

$n = 3^k \Rightarrow k = \log_3^n$ so $b_k = a_n = \left(-\frac{3}{2}\right)n + 4.7^{\log_3^n}$

Since $7^{\log_3^n} = n^{\log_3^7}$ Therefore $a_n = \left(-\frac{3}{2}\right)n + (4)n^{\log_3^7}$.

6. Solving Recurrence Relations by Substitution method

If c is a constant then we solve the recurrence relation $a_n = ca_{n-1} + f(n)$ for $n \geq 1$ by substitution method in the following way :

$$a_n = ca_{n-1} + f(n), a_1 = ca_0 + f(1), a_2 = ca_1 + f(2)$$

$$= c[ca_0 + f(1)] + f(2) = c^2 a_0 + cf(1) + f(2)$$

$$a_3 = ca_2 + f(3) = c(c^2 a_0 + cf(1)) + f(2) + f(3)$$

$$= c^3 a_0 = c^2 f(1)) + cf(2) + f(3), \; a_n = ca_{n-1} + f(n)$$

$$c^n a_0 + c^{n-1} f(1)) + c^{n-2} f(2) + \ldots c\, f(n-1) + f(n)$$ or

$$a_n = c^n a_0 + \sum_{k=1}^{n} c^{n-k} f(k)$$

Ex 1 : Solve the recurrence relation $a_n = a_{n-1} + f(n)$ for $n \geq 1$ by substitution

Sol : $a_1 = a_0 f(1), a_2 = a_1 + f(2) = a_0 + f(1) + f(2)$

$a_3 = a_2 + f(3) = a_0 f(1) + f(3)$ ....

$$a_n = a_0 + \sum_{k=1}^{n} f(k)$$

Ex 2 : Solve the recurrence relation $a_n = a_{n-1} + n$ where $a_0 = 2$ by substitution method.

Sol : Consider $a_n = a_{n-1} + n$, $a_1 = a_0 + 1$, $a_2 = a_1 + 2$
$= a_0 + 1 + 2$, $a_3 = a_2 + 3 = a_0 + 1 + 2 + 3$ ....

$$a_n = a_0 + 1 + 2 + 3 + ... + n = a_0 + \frac{n(n+1)}{2} = 2 + \frac{n(n+1)}{2}$$

Ex 3 : Solve the recurrence relation $a_n = a_{n-1} + 3^n$ where $a_0 = 1$ by substitution method.

Sol : $a_n = a_{n-1} + 3^n$, $a_1 = a_0 + 3 = a_0 + 3$

$a_2 = a_1 + 3^2$, $= (a_0 + 3) + 3^2$, $a_3 = a_2 + 3^3$

$= a_0 + 3 + 3^2 + ... + 3^n$ ..... $a_n = a_0 + 3 + 3^2 + ... + 3^n$

$= 1 + 3 + 3^2 + ... + 3^n$ Which is a geometric series with common ratio 3. Therefore $a_n = \frac{3^{n+1} - 1}{2}$

Ex 4 : Solve the recurrence relation

$$a_n = a_{n-1} + \frac{1}{n(n+1)} \text{ where } a_0 = 1$$

Sol : $a_n = a_{n-1} + \frac{1}{n(n+1)}$, $a_1 = a_0 + \frac{1}{1.2}$

$$a_2 = a_1 + \frac{1}{2.3} = a_0 + \frac{1}{1.2} + \frac{1}{2.3} \quad ......$$

$$a_n = a_0 + \frac{1}{1.2} + \frac{1}{2.3} + ...+ \frac{1}{n(n-1)} = 1 + \sum_{k=1}^{n} \left(\frac{1}{k} - \frac{1}{k+1} \right)$$

$$= 1 + \frac{1}{1} - \frac{1}{2} + \frac{1}{2} - \frac{1}{3} + ... + \frac{1}{n} - \frac{1}{n+1}$$

$$= 1 + 1 - \frac{1}{n+1} = 2 - \frac{1}{n+1} = \frac{2n+1}{n+1} \quad \therefore a_n = \frac{2n+1}{n+1}$$

7. Using Generating Functions to Solve Recurrence

Definition : The generating function for the sequence $a_0, a_1, a_2 ..... a_k ....$ of real numbers is the infinite series.

$$G(x) = a_0 + a_1 x + ... + a_k x^k + ... = \sum_{k=0}^{\infty} a_k x^k$$

Ex 1 : Solve the recurrence relation

$a_n - 7a_{n-1} + 10a_{n-2} = 0$ *for* $n \geq 2$

Sol: We number the steps of procedure.

1. Let $G(x) = \sum_{n=0}^{\infty} a_n x^n$

2. Multiply each term in the recurrence relation by x^n and sum from 2 to ∞ and simplify.

$$\sum_{n=2}^{\infty} a_n x^n - 7 \sum_{n=2}^{\infty} a_{n-1} x^n + 10 \sum_{n=2}^{\infty} a_{n-2} x^n = 0$$

$$\Rightarrow [G(x) - a_0 - a_1 x] - 7x[G(x) - a_0] + 10^2(G(x)) = 0$$

$$\Rightarrow G(x)[1 - 7x + 10^2] = a_0 + a_1 x - 7a_0 x$$

$$\Rightarrow G(x) \frac{a_0 + (a_1 - 7a_0)x}{1 - 7x + 10^2} = \frac{a_0 + (a_1 - 7a_0)x}{(1-2x)(1-5x)}$$

3. Decompose $G(x)$ as sum of partial fractions.

$$G(x) = \frac{c_1}{1-2x} + \frac{c_2}{1-5x} = c_1 \sum_{n=0}^{\infty} e^n x^n + c_2 \sum_{x=0}^{\infty} 5^n x^n$$

4. Express an as the coefficient of x^n in $G(x)$ and in the sum of the other series $a_n = c_1 2^n + c_2 5^n$ which is the solution of the given recurrence relation.

OBJECTIVE QUESTIONS - I

1. Find a recurrence relation for a_n the number of different ways to distribute either a \$1 bill, a \$ 2 bill, a \$ 5 bill, or a \$ 10 dollar bill on successive days until a total of n dollars has been distributed ?

a) $a_n = a_n + a_{n-1} + a_{n-2} + a_{n-3}$

b) $a_n = a_{n-2} + a_{n-3} + a_{n-4} + a_{n-5}$

c) $a_n = a_n + 2a_{n-1} + 3a_{n-2}$

d) $a_n = a_{n-1} + a_{n-2} + a_{n-5} + a_{n-10}$

2. Suppose that a school principal decides to give a prize away each day. Suppose further that the principal has 3 different kinds of prizes worth \$1 each and 5 different kinds of prizes worth \$4 each. Find a recurrence relation for a_n = the number of different ways to distribute prizes worth n dollars.

a) $a_n = 3a_{n-1} + 5a_{n-4}$ b) $a_n = a_{n-1} + a_{n-4}$

c) $a_n = 2a_{n-1} + 4a_{n-4}$ d) None

3. Consider a 1 x n chess board. Suppos we can color each square of the chess board either red or white. Let a_n = the number of ways of coloring the chess board in which no 2 red squares are adjacent. Find a recurrence relation that a_n satisfies.

a) $a_n = a_n + a_{n-1}$ b) $a_n = a_{n-1} + a_{n-2}$

c) $a_n = a_{n-1} + a_{n-2}$ d) None

4. Find a recurrence relation for the number of ways to arrange flags on a flag pole n feet tall using 4 types of flags : red flags 2 feet high, or white blue, and yellow flags each 1 foot high.

a) $a_n = 2a_{n-1} + 3a_{n-2}$ b) $a_n = a_{n-1} + a_{n-2}$

c) $a_n = 3a_{n-1} + a_{n-2}$ d) None

5. Let P_n be the number of permutations of m letters taken n at a time with repetitions but no 3 consecutive letters being the same. Derive a recurrence relation connecting P_n, P_{n-1} and P_{n-2}.

a) $P_n = (m-1)[P_{n-1} + 2P_{n-1}]$

b) $P_n = (m-1)[P_{n-1} + 2P_{n-2}]$

c) $P_n = (m-1)[P_n + 2P_{n-1}]$

d) $P_n = (m-1)[P_{n-1} + P_{n-2}]$

6. Find a recurrence relation for the number of n - digit ternary sequences that have an even number of 0's

a) $a_n = a_{n-1} + 3^{n-1}$ b) $a_n = a_{n-1} + 3^n$

c) $a_n = a_n + 3^n$ d) None

7. Suppose a coin is flipped until 2 heads appear and then the experiment stops. Find a recurrence relation for the number of experiments that end on the nth flip or sooner.

a) $a_n = a_{n-1} + 2(n-1)$ b) $a_n = a_{n-1} + (n-1)$

c) $a_n = a_{n-1} + n$ d) None

ANSWER KEY - I

1. d	2. a	3. b	4. c	5. d
6. a	7. b			

OBJECTIVE QUESTIONS - II

1. Suppose that the number of bacteria in a colony triples every hour. Then the recurrence relation for the number of bacterial after n hours have elapsed is

a) $a_n = a_{3n}$ b) $a_n = 3a_{n-1}$ c) $a_n = 3a_{n-1} - 1$ d) none

2. The recurrence relation for the number of bit strings of length n that contain a pair of consecutive zeros is

a) $a_n = a_{n-1} + a_{n-2}$ for $n \ge 2$

b) $a_n = a_{n-1} + a_{n-2} + 2^{n-1}$ for $n \ge 2$

c) $a_n = a_{n-1} + a_{n-2} + a_{n-3}$ for $n \ge 3$ d) none

3. The recurrence relation for the number of bit strings of length n that do not contain three consecutive zeros is

a) $a_n = a_{n-1} + 3a_{n-2}$ $n \ge 2$

b) $a_n = a_{n-1} + a_{n-2} + 3^{n-2}$ $n \ge 2$

c) $a_n = a_{n-1} + a_{n-2}, n \ge 2$ d) none

4. A string that contains only 0s, 1s and 2s is called a ternary string. Find a recurrence relation for the number of ternary strings that do not contain two consecutive 0's or two consecutive 1s

a) $a_n = a_{n-1} + a_{n-2}, n \ge 2$ b) $a_n = 2a_{n-1} + a_{n-2}, n \ge 2$

c) $a_n = a_{n-1} + 2a_{n-2}, n \ge 2$ d) $a_n = 2a_{n-1}$

5. The recurrence relation for R_n where R_n is the number of regions that a plane is divided into by n lines, if no two of the lines are parallel and no three of the lines go through the same point is

a) $R_n = nR_{n-1}, R_0 = 1$ b) $R_n = n + R_{n-1} + R_0 = 0$

c) $R_n = n + R_{n-1}, R_0 = 1$ d) none

6. The solution of the recurrence relation $a_n = a_{n-1} + 2a_{n-2}$ with $a_0 = 2$ and $a_1 = 7$ is

a) $3.2^n - (-1)^n$ b) $2^n - (-1)^n$

c) $3.2^n + 2^n$ d) none

7. The values of A and B so that $a_n = A_n + B$ is a solution of the recurrence relation $a_n = 2a_{n-1} + n + 5$ are

a) A = 1, B = 7 b) A = 7, B = 1

c) A = -1, B = -7 d) A = -7, B = -1

8. The general form of the particular solution of the linear non homogenous recurrence relation

$$a_n = 8a_{n-2} - 16a_{n-4} + F(n) \text{ if } F(n) = n^2 4^n$$

a) $(an^2 + bn + c)4^n$ b) $dn^2 4^n$

c) $dn\,4^n$ d) none

9. Suppose that $f(n) = f(n/5) + 3n^2$ where n is divisible by 5 and $f(t) = 4$, then $f(125) =$

a) 79 b) 48, 829 c) 89 d) 40

10. A group of 10 people being a chainletter with each person sending the letter to 4 other people. Each of these people sends the letter to 4 additional people. The number of letters sent at the nth stage is

a) $A_n = 4A_{n-1}$ b) $A_n = A\,4_{n-1}$

c) $A_n = 10.4^n$ d) none

11. The recurrence relation that arises in relation with the complexity of binary search is

a) $T(n) = T\left(\frac{n}{2}\right) + K$, K constant

b) $T(n) = 2T\left(\frac{n}{2}\right) + K$, K constant

c) $T(n) = T\left(\frac{n}{2}\right) + \log n$ d) $T(n) = T\left(\frac{n}{2}\right) + n$

12. The characterstic equation of the recurrence relation $a_n - a_{n-1} - 6a_{n-2} = 0$ for $n \ge 2$ is

a) $t^2 - t - 6 = 0$ b) $t^2 + t + 6 = 0$ c) $t - 6 = 0$ d) none

ANSWER KEY - II

1. b	2. c	3. a	4. b	5. c
6. a	7. c	8. a	9. b	10. c
11. a	12. a			

III. MATHEMATICAL LOGIC

- Introduction
- Connectives
- Well-formed formulas
- Normal Forms
- Propositional Calculus
- Predicate Calculus

1. Statements

A statement is a declarative sentence which has one and only one of two possible values, called ' truth values'. These two truth values are ' true' and 'false' denoted by the symbol T and F respectively; sometimes it is also denoted by the symbols 1 and 0. We assume that object language contains such declarative sentences. Since we allow only two possible truth values in our study, this logic is called two-valued logic. The sentence

This statement is false

is not a statement because we cannot assign a definite truth value to it. If we assign the value T, then the sentence says that the ' statement is false'; if we assign F, then the ' statement is true'; This is an example of what is called a semantic paradox. The sentence like 101+1= 110 is a statement. It has a truth value, but the truth value depends on the context; if we take decimal system, the statement is false, but in binary system the statement is true. Even then we consider such sentences as statements.

2. Connectives (Propositional Connectives Or Logical Connectives)

Just as we form new sentences from given sentences using 'and', 'but', 'if ', etc., We can get new propositions from given propositions using 'connectives'. But the new sentence obtained from the given propositions using connectives will be a proposition only when the new sentence has a truth value T or F (but not both). The truth value of the new sentence depends on the (logical) connetives and the truth value of the given propositions. Let us define the following connectives :

I. Negation (NOT) II. Conjunction (AND)
III. Disjunction (OR) IV. Implication (IF .... THEN ...)
V. If and only if .

I. Negation (NOT): If P is a proposition then negation P or NOT P (denoted by $\neg P$) is a proposition whose truth value is T if P has truth value F,and F if P has truth value T. Usually, the truth values of the proposition defined using a connective is given by a table called truth table for that connective (Table 3.1)

Table 3.1 Truth Table for Negation

P	$\neg P$
T	F
F	T

II. Conjunction (AND): If P and Q are two propositions, then the conjunction of P and Q (read as '' P and Q'') is a proposition whose truth values are given in the Truth Table 3.2 and denoted by $P \wedge Q$.

Table 3.2 Truth Table for Conjunction

P	Q	$P \wedge Q$
T	T	T
T	F	F
F	T	F
F	F	F

III. Disjunction (OR) : If P and Q are two propositions, then the disjunction of P and Q (read as P or Q) is a proposition whose truth values are given in the truth table (Table 3.3). and denoted by $P \vee Q$

Is should be noted that $P \vee Q$ is true if P is true or Q is true or both are true.

Table 3.3 Truth Table for Disjunction

P	Q	$P \vee Q$
T	T	T
T	F	T
F	T	T
F	F	F

Ex 1 : If P represents " This book is nice" and Q represents " This book is cheap", write the following sentences in symbolic form:

(a) This book is good and cheap.
(b) This book is not good but cheap.
(c) This book is costly but good.
(d) This book is neither good nor cheap.
(e) This book is good or cheap.

Sol : (a) is $P \wedge Q$. (b) is $(\neg P) \wedge Q$. (c) is $(\neg Q) \wedge P$.

(d) is $(\neg P) \wedge (\neg Q)$.(e) is $P \vee Q$.

IV. Implication (IF....THEN ....): If P and Q are two propositions, then " IF P THEN Q" is a proposition whose truth values are given in the truth table (Table 3.4) and is denoted by $P \Rightarrow Q$. We also read $P \Rightarrow Q$ as " P implies Q".

Table 3.4 Truth Table for Implication

P	Q	$P \Rightarrow Q$
T	T	T
T	F	F
F	T	T
F	F	T

We can note that $P \Rightarrow Q$ assumes the truth value F only if P has truth value T and Q has truth value F. In all the other cases, $P \Rightarrow Q$ assumes the truth value T. In the case of natural languages, we are concerned about the truth values of the sentence " IF P THEN Q" only when P is true. When P is false, we are not concerned about the truth value of " IF P THEN Q". But in the case of mathematical logic, we have to definitely specify the truth value of $P \Rightarrow Q$ in all cases. So the truth value of $P \Rightarrow Q$ is defined as T when P has truth value F (irrespective of the truth value of Q).

Ex 2: Find the truth values of the following propositions:

1. If 2 is not an integer, then 1/2 is an integer.
2. If 2 is an integer then 1/2 is an integer.

Sol : Let P and Q be " 2 is an integer", " 1/2 is an integer", respectively. Then proposition 1 is true (as P is false and Q is false) and proposition 2 is false (as P is true and Q is false). The above example illustrates the following. " We can prove anything if we start with a false assumption ." $P \Rightarrow Q$ whennever we want to ' translate' any one of the following: " P only Q", " P is a sufficient condition for Q", " Q is a necessary condition for p". " Q follows from P", " Q whenever P ", " Q provided P ".

V. If and Only If : If P and Q are two statements, then P if and only if Q (denoted by $P \Leftrightarrow Q$) is a statement whose truth value is T when the truth values of P and Q are the same and whose truth value is F when the statements differ. The truth tabel is given in Table 3.5

Table 3.5 Truth Table for If and Only If

P	Q	$P \Leftrightarrow Q$
T	T	T
T	F	F
F	T	F
F	F	T

Table 3.6 summarises the representation and meaning of logical connectives :

Table 3.6 Logical Connectives

Connective	Resulting proposition	Read as
Negation $\neg$	$\neg P$	NOT P
Conjunction $\wedge$	$P \wedge Q$	P AND Q
Disjunction $\vee$	$P \vee Q$	P OR Q (or both)
Implication $\Rightarrow$	$P \Rightarrow Q$	IF P THEN Q (P IMPLIES Q)
IF AND ONLY IF $\Leftrightarrow$	$P \Leftrightarrow Q$	P IF AND ONLY IF Q

Ex 3: Translate the following sentences into propositional forms:

(a) If it is not raining and I have time then I will go to a movie.

(b) It is raining and I will not go to a movie.

(c) It is not raining. (d) I will not go to a movie.

(e) I will go to movie only if it is not raining.

Sol : Let P be proposition " It is raining".

Let Q be the proposition " I have time".

Let R be proposition " I will go to a movie".

Then (a) is $\neg(P \wedge Q) \Rightarrow R$ (b) is $P \wedge \neg R$

(c) is .(d) is $\neg R$ (e) $R \Rightarrow \neg P$

3. Well- Formed Formulas

Consider $P \wedge Q$ and $Q \wedge P$, where P and Q are any two propositions. The truth tables of these two propositions are identical. This happens when we have any proposition in place of P and any proposition in place of Q. So we can develop the concept of a propositional variable (corresponding to propositions) and well formed formulas (corresponding to propositions involving connectives).

Definition 1 : A propositional variable is a symbol representing any proposition. We note that usually a real variable is represented by the symbol x. This means that x is not a real number but can take a real value.Similarly, a propositional variable is not a proposition but can be replaced by a proposition.

Definition 2 : A well - formed formula (wff) is defined recursively as follows :

I. If P is a propositional variable then it is a wff.

II. If α is a wff, then α is a wff.

III. If α and β are well-formed formulas, then $(\alpha \vee \beta)$, $(\alpha \wedge \beta)$, $(\alpha \Rightarrow \beta)$, $(\alpha \Leftrightarrow \beta)$ are well-formed formulas.

IV. A string of symbols is a wff if and only if it is obtained by finitely many applications of (I) - (III).

Note : A wff is not a proposition, but if we substitute the proposition in place of propositional variable, we get a proposition.For example,

I. $((P \vee Q) \wedge (Q \wedge R) \Rightarrow Q)$ is a wff.

II. $(P \wedge Q) \Leftrightarrow Q$ is a wff.

Note : We can drop parentheses when there is no ambiguity. For example, in propositions we can remove the outermost parentheses. We can also specify the hierarchy of connectives and avoid parentheses.

For the sake of convenience, we can refer to a wff as a formula.

4. Truth - Table For a Well-Formed Formula

If we replace the propositional variables in a formula α by propositions, we get a proposition involving connectives. The table giving the truth value of such proposition obtained by replacing the propositional variables by arbitrary propositions is called the truth tabe of α . If α involves n propositional constants, we have 2^n possible combinations of truth values of propositions replacing the variabels.

Ex 5 : Obtain the truth table for

$$\alpha = (P \vee Q) \wedge (P \Rightarrow Q) \wedge (Q \Rightarrow P)$$

Sol : The truth table of the given wff is given in Table 3.7.

Tabel 3.7 Truth Table of Example 5

P	Q	$P \vee Q$	$P \Rightarrow Q$	$(P \vee Q) \wedge (P \Rightarrow Q)$	$(Q \Rightarrow P)$	α
T	T	T	T	T	T	T
T	F	T	F	F	T	F
F	T	T	T	T	F	F
F	F	F	T	F	T	F

Ex 6 : Construct the truth table for

$$\alpha = (P \vee Q) \Rightarrow ((P \vee R) \Rightarrow (R \vee Q))$$

Sol : The truth table is given in Table 3.8.

Some formulas have the truth value T for all possible assignments of truth values to the propositional variables. For example $P \vee \rceil P$ has the truth value T irrespective of the truth value of P. such formulas are called tautologies.

Definition 3 : A tautology or a universally true formula is a well-formed formula whose truth value is T for all possible assignments of truth values to the propositional variables. For example, $P \vee \rceil P$, $(P \wedge Q) \Rightarrow P$ and $((P \Rightarrow Q) \wedge (Q \Rightarrow R)) \Rightarrow (P \Rightarrow R)$ are tautologies.

Note : When it is not clear whether a given formula is a tautology, we can construct the truth table and verify that the truth value is T for all combinations of truth values of the propositional variables appearing in the given formula.

Table 3.8 Truth Table of Example 6

P	Q	R	$P \vee R$	$R \vee Q$	$(P \vee R) \Rightarrow (R \vee Q)$	$(P \vee Q)$	α
T	T	T	T	T	T	T	T
T	T	F	T	T	T	T	T
T	F	T	T	T	T	T	T
T	F	F	T	F	F	T	F
F	T	T	T	T	T	T	T
F	T	F	F	T	T	T	T
F	F	T	T	T	T	F	T
F	F	F	F	F	T	F	T

Ex 7: Show that

$\alpha = (P \Rightarrow (Q \Rightarrow R)) \Rightarrow ((P \Rightarrow Q) \Rightarrow (P \Rightarrow R))$ is a tautology.

Sol : We give the truth tabel for α in Table 3.9

Table 3.9 Truth Table of Example 7

P	Q	R	$P \Rightarrow (Q \Rightarrow R)$	$P \Rightarrow Q$	$P \Rightarrow R$	$(P \Rightarrow Q) \Rightarrow (P \Rightarrow R)$	α
T	T	T	T	T	T	T	T
T	T	F	F	F	T	F	F
T	F	T	T	T	F	T	T
T	F	F	T	T	F	F	T
F	T	T	T	T	T	T	T
F	T	F	F	T	T	T	T
F	F	T	T	T	T	T	T
F	F	F	T	T	T	T	T

Definition 4 : A contradiction (or absurdity) is a wff whose truth value is F for all possible assignments fo truth values to the propositional variables. $P \wedge \rceil P$ and $(P \wedge Q) \wedge \rceil Q$ are, for example, contradictions.

Note : α is a contradiction if and only if $\rceil \alpha$ is a tautology.

5. Equivalence Of Well-Formed Formulas

Definition 5 : Two wffs α and β in propositional variables $P_1, P_2, \ldots\ldots P_n$ are equivalent (or logically equivalent) if the formula $\alpha \Leftrightarrow \beta$ is a tautology. When α and β are equivalent, we write $\alpha \equiv \beta$.

Note : a and b are equivalent if the truth tables for a and b are the same. For example, $P \wedge Q \equiv Q \wedge P$ and $P \wedge P \equiv P$.

Note : It is important to note the difference between $\alpha \Leftrightarrow \beta$ and $\alpha \equiv \beta$. $\alpha \Leftrightarrow \beta$ is a formula, whereas $\alpha \equiv \beta$ is not a formula, but it denotes the relation between α and β.

Ex 8 : Show that $(P \Rightarrow (Q \vee R)) \equiv ((P \Rightarrow Q) \vee (P \Rightarrow R))$

Sol : let $\alpha = (P \Rightarrow Q \vee R)$ and $\beta = ((P \Rightarrow Q) \vee (P \Rightarrow R)$. We construct the truth values of α and β for all assignments of truth values to the variables P, Q and R. These are given in Table 3.10.

Table 3.10 Truth Table of Example 8

P	Q	R	$Q \vee R$	$P \Rightarrow Q \vee R$	$P \Rightarrow Q$	$P \Rightarrow R$	$(P \Rightarrow Q) \vee (P \Rightarrow R)$
T	T	T	T	T	T	T	T
T	T	F	T	T	T	F	T
T	F	T	T	T	F	T	T
T	F	F	F	F	F	F	F
F	T	T	T	T	T	T	T
F	T	F	T	T	T	T	T
F	F	T	T	T	T	T	T
F	F	F	F	T	T	T	T

As the columns corresponding to α and β coincide, $\alpha \equiv \beta$.

As the truth value of a tautology is T, irrespective of the truth values of propositional variables, we denote any tautology by T. Similarly, we denote any contradiction by F.

6. Logical Identities

Some equivalences are useful for deducing other equivalences. we call them identities and give a list of such identities in Table 3.11.

The identities $I_1 - I_{12}$ can be used to simplify fromulas. If a formula β is part of another formula α, and β, is equivalent to β' then we can replace β by β' in α and the resulting wff is equivalent to α.

Ex 9 : Show that $(P \wedge Q) \vee (P \wedge \rceil Q) \equiv P$

Sol : L.H.S = $(P \wedge Q) \vee (P \wedge \rceil$

$\equiv P \wedge (Q \vee \rceil Q)$ by using Distributive law (i.e. I_4)

$\equiv P \wedge T$ by using I_8

$\equiv P$ by using I_8

Ex10 : Show that $(P \Rightarrow Q) \wedge (R \Rightarrow Q) \equiv (P \vee R) \Rightarrow Q$

Sol : L.H.S = $(P \Rightarrow Q) \wedge (R \Rightarrow Q)$

$\equiv (\rceil P \vee Q) \wedge (\rceil R \vee Q)$ by using I_{12}

$\equiv (Q \vee \rceil P) \wedge (Q \vee \rceil R)$ by using Commutative law

$\equiv Q \vee (\rceil P \wedge \rceil R)$ by using Distributive law

$\equiv Q \vee (\rceil (P \vee R))$ by using De Morgan's law

$\equiv (\rceil (P \vee R)) \vee Q$ by using Commutative law

$\equiv (P \vee R) \Rightarrow Q$ by using I_{12}

= R.H.S.

Table 3.11 Logical Identities

I_1	Idempotent laws : $P \vee P \equiv P$, $P \wedge P \equiv P$
I_2	Commutative laws : $P \vee Q \equiv Q \vee P$, $P \wedge Q \equiv Q \wedge P$
I_3	Associative laws : $P \vee (Q \vee R) \equiv (P \vee Q) \vee R$ $P \wedge (Q \wedge R) \equiv (P \wedge Q) \wedge R$
I_4	Distributive laws : $P \vee (Q \wedge R) \equiv (P \vee Q) \wedge (P \vee R)$, $P(Q \vee R) \equiv (P \wedge Q) \vee (P \wedge R)$
I_5	Absorption laws : $P \vee (P \wedge Q) \equiv P$. $P \wedge (P \vee Q) \equiv P$
I_6	De Morgan's laws : $\urcorner(P \vee Q) \equiv \urcorner P \wedge \urcorner Q$, $\urcorner(P \wedge Q) \equiv \urcorner P \vee \urcorner Q$
I_7	Double negation : $P \equiv \urcorner(\urcorner P)$.
I_8	$P \vee \urcorner P \equiv T$ $P \wedge \urcorner P \equiv F$
I_9	$P \vee T \equiv T$, $P \wedge T \equiv P$, $P \vee F \equiv P$, $P \wedge F \equiv F$
I_{10}	$(P \Rightarrow Q) \wedge (P \Rightarrow \urcorner Q) \equiv \urcorner P$
I_{11}	Contrapostitive : $P \Rightarrow Q \equiv \urcorner Q \Rightarrow \urcorner P$
I_{12}	$P \Rightarrow Q \equiv (\urcorner P \vee Q)$

7. Normal Forms Of Well-Formed Formulas

We have seen various well-formed formulas in terms of two propositional variables, say P and Q. We also know that two such formulas are equivalent if and only if they have the same truth table. The number of distinct truth tables for formulas in P and Q is 2^4 (As the possible combinations of truth values of P and Q are TT, TF, FT, FF, the truth table of any formula in P and Q has 4 rows. So the number of distinct truth tables is 2^4). Thus there are only16 distinct (nonequi valent)formulas, and any formula in P and Q is equivalent to one of these 16 formulas. In this section we give a method of reducing a given formula to an equivalent form called a ' normal form'. We also use 'sum' for disjunction. ' product' for conjunction, and 'literal' either for P or for $\urcorner P$, where P is any propositional variable.

Definition 6 : An elementary product is a product of literals. An elementary sum is a sum of literals. For example, $P \wedge \urcorner Q$, $\urcorner P \wedge \urcorner Q$, $P \wedge Q$, $\urcorner P \wedge Q$ are elementary products. $P \vee \urcorner Q$, $P \vee \urcorner R$ are elementary sums.

Definition 7 : A formula is in disjunctive normal form if it is a sum of elementary products. For example , $P \vee (Q \wedge R)$ and $P \vee (\urcorner Q \wedge R)$ are in disjunctive normal form. $P \wedge (Q \vee R)$ is not in disjunctive normal form.

8. Construction To obtain a Disjunctive Normal Form Of a Given Formula

Step 1 Eliminate $\Rightarrow$ and $\Leftrightarrow$ using logical identities. (We can use I_{12}, i.e., $\urcorner$.)

Step 2 Use De Morgan's laws (I_6) to eliminate $\urcorner$ before sums or products. The resulting formula has $\urcorner$ only before propositional variables, i.e., it involves sum, product and literals.

Step 3 Apply distributive laws (I_4) repeatedly to eliminate product of sums. The resulting formula will be a sum of products of literals i.e., sum of elementary products.

Ex 11 : Obtain a disjunctive normal form of $P \vee (\urcorner P \Rightarrow (Q \vee (Q \Rightarrow \urcorner R)))$

Sol : $P \vee (\urcorner P \Rightarrow (Q \vee (Q \Rightarrow \urcorner R)))$

$\equiv P \vee (\urcorner P \Rightarrow (Q \vee (\urcorner Q \vee \urcorner R)))$ (step 1 using I_{12})

$P \vee (P \vee (Q \vee (\urcorner Q \vee \urcorner R)))$ (step 1 using I_{12} and I_7)

$\equiv P \vee P \vee Q \vee \urcorner Q \vee \urcorner R$ by using I_3

$\equiv P \vee Q \vee \urcorner Q \vee \urcorner R$ by using I_1

Thus, $P \vee Q \urcorner Q \vee \urcorner R$ is a disjunctive normal form of the given formula.

Ex 12 : Obtain the disjunctive normal form of $(P \wedge \urcorner(Q \wedge R)) \vee (P \Rightarrow Q)$

Sol : $(P \wedge \urcorner(Q \wedge R)) \vee (P \Rightarrow Q)$

$\equiv (P \wedge \urcorner(Q \wedge R)) \vee (\urcorner P \vee Q)$ (step 1 using I_{12})

$\equiv (P \wedge (\urcorner Q \vee R)) \vee (\urcorner P \vee Q)$ (step 2 using I_7)

$\equiv (P \wedge (\urcorner Q) \vee (P \wedge \urcorner R) \vee \urcorner P \vee Q$ (step 3 using I_4 and I_3)

Therefore, $(P \wedge Q) \vee (P \wedge R) \vee P \vee Q$ is a disjunctive normal form of the given formula.

For the same formula, we may get different disjunctive normal forms. For example, $(P \wedge Q \wedge R) \vee (P \wedge Q \wedge R)$ and $P \wedge Q$ are disjunctive normal forms of $P \wedge Q$. So, we introduce one more normal form, called the principal disjunctive normal form or sum-of-products canonical form in the next definition. The advantages of constructing principal disjunctive normal forms are :

I. For a given formula, its prinicipal disjunctive normal form is unique.

II. Two formulas are equivalent if and only if their principal disjunctive normal forms coincide.

Definition 8 : A min term in n propositional variables $P_1, \ldots P_n$ is $Q_1 \wedge Q_2 \ldots \wedge Q_n$, where each Q_i is either P_i or $P_i \urcorner$ For example, the min terms in P_1 and P_2 are $P_1 \wedge P_2$, $P_1 \wedge P_2$, , $P_1 \wedge P_2$, $P_1 \wedge P_2$. The number of min terms in n variables is 2^n.

Definition 9: A formula α is in prinicipal disjunctive normal form if α is a sum of min terms.

9. Construction To Obtain The Principal Disjunctive Normal Form Of a Given Formula

Step 1 Obtain a disjunctive normal form.

Step 2 Drop elementary products which are contradictions (such s $P \wedge \urcorner P$).

Step 3 If P_i and $\urcorner P_i$ are missing in an elementary product α repalace α by $(\alpha \wedge P_i) \vee (\alpha \wedge \urcorner P_i)$.

Step 4 Repeat step 3 until all elementary products are reduced to sum of min terms. Use idempotent laws to avoid repetition of min terms.

Ex 13 : Obtain the canonical sum-of-products form (i.e., principal disjunctive normal form) of $\alpha = P \vee (\neg P \wedge \neg Q \wedge R)$.

Sol : α is already in disjunctive normal form. There are no contradictions. So we have to introduce missing variables (step3). $\neg P \wedge \neg Q \wedge R$ in α is already a min term.

Now, $P \equiv (P \wedge Q) \wedge (P \wedge \neg Q)$

$\equiv ((P \wedge Q \wedge R) \vee (P \wedge Q \wedge \neg R)) \vee (P \wedge \neg Q \wedge R) \vee (P \wedge \neg Q \wedge R)$

$\equiv ((P \wedge Q \wedge R) \vee (P \wedge Q \wedge \neg R)) \vee (P \wedge \neg Q \wedge R) \vee (P \wedge \neg Q \wedge \neg R)$

Therefore, the canonical sum-of- products form of α is

$(P \wedge Q \wedge R) \vee (P \wedge Q \wedge \neg R) \vee (P \wedge \neg Q \wedge R)$

$\vee (P \wedge \neg Q \wedge \neg R) \vee (\neg P \wedge \neg Q \wedge R)$

Ex14 : Obtain the principal disjunctive normal form of $\alpha = (\neg P \vee \neg Q) \Rightarrow (\neg P \wedge R)$

Sol : $\alpha = (P \vee Q) \Rightarrow (P \wedge R)$

$\equiv (\neg(\neg P \vee \neg Q)) \vee (\neg P \wedge R)$ by using I_{12}

$\equiv ((P \wedge Q) \vee (\neg P \wedge R))$ by using De Morgan's law

$\equiv ((P \wedge Q \wedge R) \vee (P \wedge Q \wedge \neg R)) \vee ((\neg P \wedge R \wedge Q) \vee (\neg P \wedge R \wedge \neg Q))$

$\equiv (P \wedge Q \wedge R) \vee (P \wedge Q \wedge \neg R) \vee (\neg P \wedge Q \wedge R) \vee (\neg P \wedge \neg Q \wedge \neg R)$

So the principal disjunctive normal form of α is

$(P \wedge Q \wedge R) \vee (P \wedge Q \wedge \neg R) \vee (\neg P \wedge Q \wedge R) \vee (\neg P \wedge \neg Q \wedge R)$

A min term of the form $Q_1 \wedge Q_2 \ldots \wedge Q_n$ can berepresented by $a_1 a_2 \ldots a_n$, where $a_i = 0$ if $Q_i = P_i$ and $a_i = 1$ if $Q_i = P_i$. So the principal disjunctive normal form can be represented by a 'sum' of binary strings. For example $(P \wedge Q \wedge R) \vee (P \wedge Q \wedge \neg R) \vee (\neg P \wedge \neg Q \wedge R)$ is represented by $111 \vee 110 \vee 001$.

The min terms in two variables P and Q are 00, 01, 10, and 11. Each wff is equivalent to its principal disjunctive normal form. Every principal disjunctive normal form corresponds to the min terms in it, and hence to a subset of {00, 01, 10, 11}. As the number of subsets is 2^4, the number of distinct formulas is 16.

The truth table and the principal disjunctive normal form of α are closely related. Each min term corresponds to a particular assignment of truth values to the variables yielding truth value To to α. For example, $P \wedge Q \wedge R$ corresponds to the assignment of T,T,F to P, Q and R, respectively. So, if the truth table of α is given, then the min terms are those corresponding to assignmens yielding truth value T to α.

Ex 15 : For a given formula α, the truth table is given in Table 3.12. Find the principal disjunctive normal form.

Table 3.12 Truth Table of Example 15

P	Q	R	α
T	T	T	T
T	T	F	F
T	F	T	F
T	F	F	T
F	T	T	T
F	T	F	F
F	F	T	F
F	F	F	T

Sol : We have T in the α - column corresponding to the rows 1, 4, 5 and 8. The min term corresponding to the first row is $P \wedge Q \wedge R$.

Similarly, the min terms corresponding to rows 4,5 and 8 are $P \wedge \neg Q \wedge \neg R$, $\neg P \wedge Q \wedge R$ and $P \wedge Q \wedge R$. Therefore, the principal disjunctive normal form of α is

$(P \wedge Q \wedge R) \vee (P \wedge \neg Q \wedge \neg R) \vee (\neg P \wedge Q \wedge R) \vee (\neg P \wedge \neg Q \wedge \neg R)$

We can form the ' dual ' of disjunctive normal form which is termed as conjunctive normal form.

Definition 10 : A formula is in conjunctive normal form if it is a product of elementary sums.

It α is in disjunctive normal form, then $\neg\alpha$ is in conjunctive normal form. (This can be seen by applying De Morgan's laws.) so to obtain the conjunctive normal form of α, we construct the disjunctive normal of $\neg\alpha$ and use negation.

Definition 11 : A max term in n propositional variables $P_1, P_2, \ldots, P_n$ is $Q_1 \vee Q_2 \ldots \vee Q_n$, where each Q_i is either P_i or $\neg P_i$.

Definition 12: A formula α is in principal conjunctive normal form if α is a product of max terms. For obtaining the principal conjunctive normal form of α, we can construct the principal disjunctive normal form of α and apply negation ($\neg$).

Ex 16 : Find the principal conjuctive normal form of $\alpha = P \vee (Q \Rightarrow R)$.

Sol : $\neg\alpha = \neg(P \vee (Q \Rightarrow R))$

$\equiv \neg(P \vee (\neg Q \Rightarrow R))$ by using I_{12}

$\equiv \neg(P \wedge (\neg(\neg Q \vee R))$ by using De Morgan's law

$\equiv \neg(P \wedge (Q \wedge \neg R)$ by using De Morgan's law and I_7

$\neg P \wedge Q \wedge \neg R$ is the principal disjunctive normal form of $\neg\alpha$.

Hence, the principal conjunctive normal form of α is

$\neg(\neg P \wedge Q \wedge \neg R) = P \vee \neg Q \vee R$

The logical identities given in Table 3.11 and the normal forms of well-formed formulas bear a close resemblance to identities in Boolean algebras and normal forms of Boolean functions. Actually, the propositions under $\vee, \wedge$ and $\neg$ form a Boolean algebra if equivalent propositions are identified.T and F act as bounds(i.e., 0 and 1of a Boolean algebra).Also, the statement formulas form a Boolean algebra under $\vee, \wedge$ and $\neg$ if equivalent formulas are identified. The normal forms of well-formed formulas correspond to normal forms of Boolean functions and we can'minimise' a formula in a similar manner.

10. Rules Of Inference For Propositional Calculus (Statement Calculus)

In a logical reasoning, a certain number of propositions are assumed to be true, and based on that assumption some other propositions are derived (deduced or inferred).

In this section we give some important rules of logical reasoning or rules of inference. The propositions that are assumed to be true are called hypotheses or premises. The proposition derived by using the rules of inference iscalled a conclusion. The process of deriving conclusions based on the assumption of premises is called a valid agrument. So in a valid argument we are concerned with the process of arriving at the conclusion rather than obtaining the conclusion.

The rules of inference are simply tautologies in the form of implication (i.e. $P \Rightarrow Q$). For example, $P \Rightarrow (P \vee Q)$ is such a tautology, and it is a rule of inference. We write this in the form $\dfrac{P}{\therefore P \vee Q}$. Here P denotes a premise. The proposition below the line, i.e., $P \vee Q$, is the conclusion. We give in Table 3.13 some of the important rules of inference. Of course we can derive more rules of inference and use them in valid arguments.

For valid arguments, we can use the rules of inference given in Table 3.13.As logical identities given in Table 3.11 are two - way implications,we can also use them as rules of inference.

Ex 17 : Can we conclude S from the following premises?

I. $P \Rightarrow Q$ II. $P \Rightarrow R$ III. $\rceil(Q \wedge R)$ IV. $S \vee P$

Sol : The valid argument for deducing S from the given four premises is given as a sequence. On the left, the well-formed formulas are given. On the right we indicate whether the proposition is a premise (hypothesis) or a conclusion. If it is a conclusion, we indicate the premises and the rules of inference or logical identities used for deriving the conclusion.

1. $P \Rightarrow Q$ Premise I
2. $P \Rightarrow R$ Premise II
3. $(P \Rightarrow Q) \wedge (P \Rightarrow R)$ Lines 1,2 and RI_2
4. $\rceil (Q \wedge R)$ Premise III
5. $\rceil Q \vee \rceil R$ Line 4 and De Morgan's law (I_6)
6. $\rceil P \vee \rceil P$ Line 3,5 and destructive dilemma (RI_9)
7. $\rceil P$ Idempotent law I_1
8. $S \vee P$ Premise IV
9. S Lines 7,8 and disjunctive syllogism RI_6

The we can conclude S from the given premises.

Ex 18 : Derive S from the following premises using a valid argument :

I. $P \Rightarrow Q$, **II.** $Q \Rightarrow \rceil R$, **III.** $P \vee S$, **IV.** R

Table 3.13 Rules of Inference

Rules of Inference	Implication Form
RI_1 : Addition	
$\dfrac{P}{\therefore P \vee Q}$	$P \Rightarrow (P \vee Q)$
RI_2 : Conjunction	
$\dfrac{Q}{\therefore P \wedge Q}$	$P \wedge Q \Rightarrow P \wedge Q$
RI_3 : Simplification	
$\dfrac{P \wedge Q}{\therefore P}$	$(P \wedge Q) \Rightarrow P$
RI_4 : Modus ponens	
P, $\dfrac{P \Rightarrow Q}{\therefore Q}$	$(P \wedge (P \Rightarrow Q)) \Rightarrow Q$
RI_5 : Modus tollens	
$\rceil Q$, $\dfrac{P \Rightarrow Q}{\therefore \rceil P}$	$(\rceil Q \wedge (P \Rightarrow Q)) \Rightarrow \rceil P$
RI_6 : Disjunctive syllogism	
$\rceil P$, $\dfrac{P \vee Q}{\therefore Q}$	$(\rceil P \wedge (P \vee Q)) \Rightarrow Q$
RI_7 : Hypothetical syllogism	
$P \Rightarrow Q$, $\dfrac{Q \Rightarrow R}{\therefore P \Rightarrow R}$	$((P \Rightarrow Q) \wedge (Q \Rightarrow R)) \Rightarrow (P \Rightarrow R)$
RI_8 : Constructive dilemma	
$(P \Rightarrow Q) \wedge (R \Rightarrow S)$, $\dfrac{P \vee R}{\therefore Q \vee S}$	$(P \Rightarrow Q) \wedge (R \Rightarrow S) \wedge (P \vee R) \Rightarrow (Q \vee S)$
RI_9 : Destructive dilemma	
$(P \Rightarrow Q) \wedge (R \Rightarrow S)$, $\dfrac{\rceil Q \vee \rceil S}{\therefore \rceil P \vee \rceil R}$	$(P \Rightarrow Q) \wedge (R \Rightarrow S) \wedge (\rceil Q \vee \rceil S) \Rightarrow (\rceil P \vee \rceil R)$

Sol :

1. $P \Rightarrow Q$ Premise (I)
2. $Q \Rightarrow \rceil R$ Premise (II)
3. $P \Rightarrow \rceil R$ Lines 1,2 and hypothetical syllogism RI_7
4. R Premise(IV)
5. $\rceil(\rceil R)$ Line 4 and double negation I_7
6. $\rceil P$ Lines 3,5 and modus tollens RI_5
7. $P \vee S$ Premise (III)
8. S Lines 6,7and disjunctive syllogism RI_6

Thus we have derived S from the given premises.

Ex 19 : Check the validity of the following argument :

If Ram has completed B.E. Computer Science or M.B.A., then he is assured of a good job. If Ram is assured of a good job,he is happy. Ram is not happy. So Ram has not completed M.B.A.

Sol : We can name the propositions in the following way:

P denotes " Ram has completed B.E.Computer Science ".

Q denotes" Ram has completed M.B.A "

R denotes " Ram is assured of a good job".

S denotes" Ram is happy".

The given premises are :

I. $(P \vee Q) \Rightarrow R$ ii. $R \Rightarrow S$ iii. $\rceil S$

The conclusion is $\rceil Q$

1. $(P \vee Q) \Rightarrow R$ Premise (I)
2. $R \Rightarrow S$ Premise (II)
3. $(P \vee Q) \Rightarrow S$ Lines1,2 and hypothetical syllogism RI_7
4. $\rceil S$ Premise (III)
5. $\rceil (P \vee Q)$ Lines 3,4 and modus tollens
6. $\rceil P \wedge \rceil Q$ De Morgan's law I_6
7. $\rceil Q$ Line 6 and simplification RI_3

Thus the argument is valid

Ex 20 : Test the validity of the following argument:

If milk is black then every crow is white. If every crow is white then it has four legs. If every crow has four legs then every buffalo is white and birsk. The milk is black. Therefore , the buffalo is white.

Sol : We name the propositions in the following way:

P denotes " The milk is black "
Q denotes " Every crow is white"
R denotes " Every crow has four legs".
S denotes " Every buffalo is white"
T denotes " Every buffalo is brisk ".

The given premises are:

I. $P \Rightarrow Q$, **II.** $Q \Rightarrow R$, **III.** $R \Rightarrow S \wedge T$. **IV.** P

The conclusion is S.

1. P Premise (IV)
2. $P \Rightarrow Q$ Premise (I)
3. Q Modus ponens
4. $Q \Rightarrow R$ Premise (II) RI_4
5. R Modus ponens RI_4
6. $R \Rightarrow S \wedge T$ Premise (III)
7. $S \wedge T$ Modus ponens RI_4
8. S Simplification RI_3

Thus the argument is valid.

11. Predicate Calculus

Consider two propositions "Ram is a student", and " Sam is a student ".As propositions, there is no relation between them, but we know they have something in common.Both Ram and Sam share the property of being a student. We can replace the two propositions by a single statement "x is a student". By replacing x by Ram or Sam(or anyother name), we get many propositions.The common feature expressed by "is a student" is called a predicate. In predicate calculus we deal with sentences involving predicates. Statements involving predicates occur in Mathematics and programming languages. For example , " $2x + 3y = 4z$ ", " IF (D. GE. 0.0) GO TO 20" are statements in Mathematics and FORTRAN,respectively involving predicates.Some logical deductions are possible only by separating predicates'.

12. Predicates

A part of a declarative sentence describing the properties of an object or relation among objects is called a predicate. For example, "is a student " is a predicate. Sentences involving predicates describing the property of objects are denoted by P (x), where P denotes the predicate and x is a variable denoting any object. For example, P (x)can denote " x is a student". In this sentence, x is a variable and P denotes the predicate " is a student". The sentence " x is the father of y" also involves a predicate " is the father of ". Here the predicate describes the relation between two persons. We can write this sentence as F (x,y). Similarly, $2x + 3y = 4z$ can be described by S(x,y,z).

Note : Although, P (x), involving a predicate looks like a proposition, it is not a proposition.As P (x) involves a variable x,we cannot assign a truth value to P (x). However, if we replace x by an individual object, we get a proposition. For example, if we replace x by Ram in P (x), we get the proposition " Ram is a student". (We can denote this proposition by P (Ram).)If we replace x by 'A cat', then also we get a proposition (whose truth value is F).S (2,0,1) is the proposition $2.2 + 3.0 = 4.1$ (whose truth value is T). S (1,1,1) is the proposition $2.1 + 3.1 = 4.1$ (whose truth value is F). The following definition is regarding possible ' values' which can be assigned to variables.

Definition 13: For a declarative sentence involving a predicate, the universe discourse,or simply the universe, is the set of all possible values which can be assigned to variables. For example, the universe of discourse for P (x) : "x is a student", can be taken as the set of all human names ; the universe of discourse for E (n): " n is an even integer", can be taken as the set of all integers (or the set of all real numbers).

Note : In most examples, the universe of discourse is not specified but can be easily given.

13.Universal and Existential Quantifier

The phrase ' for all' (denoted by $\forall$) is called the universal quantifier.Using this symbol, we can write " For all $x, x^2 = (-x)^2$ " as $\forall x Q(x)$, where Q (x) is" $x^2 = (-x)^2$ ".

The phrase ' there exists' (denoted by $\exists$)is called the existential quantifier.The sentence " There exists x such that $x^2 = 5$" can be written as $\exists x R(x)$, where R(x) is ' $x^2 = 5$ '

$P(x)$ in $\forall x P(x)$ or in $\exists x P(x)$ is called the scope of the quantifier $\forall$ or $\exists$.

Note : The symbol $\forall$ can be read as'for every', 'for any','for each',' for arbitrary'.The symbol $\exists$ can be read as' for some'for ' at least one'. When we use quantifiers, we should specify the universe of discourse. If we change the universe of discourse, the truth value may change. For example, consider $\exists x R(x)$ where R (x) is $x^2 = 5$. If the universe of discourse is the set of all integers, then $\exists x R(x)$ is false. If the universe of discourse is the set of all real numbers, then $\exists x R(x)$ is true (when $x = \pm\sqrt{5}, x^2 = 5$).

The logical connectives involving predicates can,be used for declarative sentences involving predicates.The following example illustrates the use of connectives.

Ex 21 : Express the following sentences involving predicates in symbolic form :

1. All students are clever.
2. Some students are not successful.
3. Every clever student is sucessful.
4. There are some successful students who are not clever.
5. Some students are clever and successful.

Sol : As quantifiers are involved, we have to specify the universe of discourse.We can take the universe of discourse as the set of all students.

Let C (x) denote "x is clever"

Let S (x) denote " x is successful".

Then sentence 1 can be written as $\forall x\, C(x)$ .Sentences 2-5 can be written as

$\exists x(\,\rceil S(X))$ $\qquad \forall x(C(x) \Rightarrow S(x))$

$\exists x(\,\rceil S(x) \wedge C(x))$ $\qquad \exists x(C(x) \wedge S(x))$

14. Well - Formed Formulas Of Predicate Calculus

A well-formed formula (wff) of predicate calculus is a string of variables such as $x_1, x_2, \ldots, x_n$, connectives, parentheses and quantifiers defined recursively by the following rules :

I. $P(x_1, \ldots, x_n)$ is a wff, where P is a predicate involving n variables $x_1, x_2, \ldots, x_n$

II. If α is a wff, then $\rceil\alpha$ is a wff.

III. If α and β are wffs, then $\alpha \vee \beta$, $\alpha \wedge \beta$, $\alpha \Rightarrow \beta$, $\alpha \Leftrightarrow \beta$ are also wffs.

IV. If α is a wff and x is any variable, then $\forall x(\alpha)$, $\exists x(\alpha)$ are wffs.

V. A string is a wff if and only if it is obtained by finitely many applications of rules (I) - (IV).

Note : A proposition can be viewed as a sentence involving a predicate with 0 variables. So propositions are wffs of predicate calculus by rule (I).

Definition 14 : Let α and β be two predicate formulas in variables $x_1, \ldots x_n$, and let U be a universe of discourse for α and β .Then α and β are equivalent to each other over U if for every possible assignment of values to each variable in a and b the resulting statements have the same truth values. We can write $\alpha \equiv \beta$ over U.

We say that α and β are equivalent to each other $(\alpha \equiv \beta)$ if $\alpha \equiv \beta$ over U for every universe of discourse U.

Remark : In predicate formulas the predicate variables may or may not be quantified. We can classify the predicate variables in a predicate formula, depending on whether they are quantified or not. This leads to the following definitions.

Definition 15 : If a formula of the form $\exists x(Px)$ or $\forall x(Px)$ occurs as part of a predicate formula α , then such part is called an x-bound part of α , and the occurrence of x is called a bound occurrence of x. An occurrence of x is free if it is not a bound occurrence. A predicate variable in α is free if its occurrence is free in any part of α .

In $\alpha = (\exists x_1 P(x_1, x_2)) \wedge (\forall x_2 Q(x_2, x_3))$, for example, the occurrence of x_1 in $\exists x_1 P(x_1, x_2)$ is a bound occurrence and that of x_2 is free. In $\forall x_2 Q(x_2, x_3)$, the occurrence of x_2 is a bound occurrence the occurrence of x_3 in α is free.

Note : Quantified parts of a predicate formula such as $\forall x P(x)$ or $\exists x P(x)$ are propositions. We can assign values from the universe of discourse only to free variables in a predicate formula α .

Definition 16 : A predicate formula is valid if for all possible assignments of values from any universe of discourse to free variables, the resulting propositions have truth value T.

Definition 17 : A predicate formula is satisfiable if for some assignment of values to predicate variables the resulting proposition has truth value T.

Definition 18 : A predicate formula is unsatisfiable if for all possible assignments of values from any universe of discourse to predicate variables the resulting propositions have truth value F.

We note that valid predicate formulas correspond to tautologies among proposition formulas and unsatisfiable predicate formulas correspond to contradictions.

15. Rules Of Inference For Predicate Calculus

Before discussing the rules of inference, we note that : (a) proposition formulas are also predicate formulas ; (b) predicate formulas where all the variables are quantified are proposition formulas.Therefore, all the rules of inference for proposition formulas are also applicable for predicate calculus wherever necessary. For predicate formulas not involving connectives such as A (x), P (x,y), we can get equivalences and rules of inference similar to those given in Tables 3.11 and 3.13 . For Example, corresponding to I_6 in Table 3.11 we get $\rceil(p(x) \vee Q(x)) \equiv\; \rceil P(x) \wedge \rceil Q(x))$. Corresponding to RI_3 in Table 3.13 $P \wedge Q \Rightarrow P$, we get $P(x) \wedge Q(x) \Rightarrow P(x)$. Thus we can replace propositional variables by predicate variables in Tables 3.11 and 3.13.

Some necessary equivalences involving the two quantifiers and valid implications are given in Table 3.14. Sometimes when we wish to derive some conclusion from a given set of premises involving quantifiers, we may have to eliminate the quantifiers before applying the rules of inference for proposition formulas. Also, when the conclusion involves quantifiers, we may have to introduce quantifiers. The necessary rules of inference for addition and deletion of quantifiers are given in Table 3.15.

Ex 22 : Discuss the validity of the following argument:

I. All graduates are educated.

II. Ram is a graduate. Therefore,

III. Ram is educated.

Table 3.14 Equivalences Involving Quantifiers

I_{13} Distibutivity of $\exists$ over $\vee$:

$\exists x(P(x) \vee Q(x)) \equiv \exists x P(x) \vee \exists x Q(x)$

$\exists x(P \vee Q(x)) \equiv P \vee (\exists x Q(x))$

I_{14} Distributivity of $\forall$ over $\wedge$:

$\forall x(P(x) \wedge Q(x)) \equiv \forall x P(x) \wedge \forall x Q(x)$

$\forall x(P \wedge Q(x)) \equiv P \wedge (\forall x Q(x))$

I_{15} $(\exists x P(x)) \equiv \forall x \ (P(x))$

I_{16} $(\forall x P(x)) \equiv \exists x \ (P(x))$

I_{17} $\exists x(P \wedge Q(x)) \equiv P \wedge (\exists x Q(x))$

I_{18} $\forall x(P \vee Q(x)) \equiv P \vee (\forall x Q(x))$

RI_{10} $\forall x(P(x) \Rightarrow \exists x P(x)$

RI_{11} $\forall x(P(x) \vee \forall x Q(x) \Rightarrow \forall x(P(x) \vee Q(x))$

RI_{12} $\exists x(P(x) \wedge Q(x) \Rightarrow \exists x(P(x) \wedge \exists x Q(x)$

Table 3.15 Rules of Inference for Addition and Deletion of Quantifiers

RI_{13} : Universal instantiation

$$\frac{\forall x P(x)}{\therefore P(c)}$$

c is some element of the universe.

RI_{14} : Existential instantiation

$$\frac{\exists x P(x)}{\therefore P(c)}$$

c is some element for which P (c) is true.

RI_{15} : Universal generalisation

$$\frac{P(x)}{\forall x P(x)}$$

x should not be free in any of the given premises.

RI_{16} : Existential generalisation

$$\frac{P(c)}{\therefore \exists x P(x)}$$

c is some element of the universe.

Sol : Let G (x) denote " x is a graduate ".

Let E (x) denote " x is educated ". Let R denote " Ram ". So the premises are $\forall x(G(x) \Rightarrow E(x))$ and G(R). The conclusion is E(R).

$\forall x(G(x) \Rightarrow E(x))$	Premise (I)
$G(R) \Rightarrow E(R)$	Universal instantiation RI_{13}
G(R)	Premise (II)
$\therefore$ E(R)	Modus ponens RI_{14}

Thus the conclusion, viz. (III), is valid.

Ex 23 : Discuss the validity of the following argument : All graduates can read and write. Ram can read and write. Therefore, Ram is a graduate.

Sol : Let G (x) denote " x is a graduate". Let L (x) denote " x can read and write ". Let R denote " Ram ".

The premises are : $\forall x(G(x) \Rightarrow L(x))$ and L(R).

The conclusion is : G(R)

$((G(R) \Rightarrow L(R)) \wedge L(R)) \Rightarrow G(R)$ is not a tautology.

So we cannot derive G(R). For example, a school boy can read and write and he is not a graduate.

Ex 24 : Discuss the validity of the following argument : All educated persons are well behaved. Ram is educated. No well-behaved person is quarrelsome. Therefore, Ram is not quarrelsome.

Sol : Let the universe of discourse be the set of all educated persons.

Let P (x) denote " x is well- behvaed".
Let y denote " Ram ".
Let Q (x) denote " x is quarrelsome ".

So the premises are :

I. $\forall x P(x)$

II. y is a particular element of the universe of discourse.

III. $\forall x P(x) \Rightarrow Q(x))$.

To obtain the conclusion, we have the following arguments

1. $\forall x P(x)$	Premise (I)
2. P(y)	Universal instantiation RI_{13}
3. $\forall x P(x) \Rightarrow Q(x))$	Premise (III)
4. $P(y) \Rightarrow Q(y)$	Universal instantiation RI_{13}
5. P(y)	Line 2
6. $\rceil Q(y)$	Modus ponens RI_4

$\rceil Q(y)$ means " Ram is not quarrelsome". Thus the argument is valid.

OBJECTIVE QUESTIONS - I

1. Which of the following propositions is a tautology? **[GATE-1997]**
   a) (p v q) → p b) p v (q → p)
   c) p v (p → q) d) p → (p → q)
2. The proposition is p ∧ (~p V q) is **[GATE-1993]**
   a) A tautology b) A contradiction
   c) Logically equivalent to p ∧ q d) None
3. Which of the following is / are tautology: **[GATE-1992]**
   a) a V b → b ∧ c b) a ∧ b → b V c
   c) a V b → (b → c) d) a → b → (b → c)
4. Let p and q be propositions. Using only the truth table decide whether p ⇔ q does not imply p → ⌉ q is **[GATE-1994]**
   a) True b) False
5. (P V Q) ∧ (P → R) ∧ (Q → S) is equivalent to
   a) S ∧ R b) S → R c) S V R d) None
6. Choose the functionally complete set
   a) { ⌉, ∧ , V} b) { ↓, ∧ } c) {↑} d) {→ , ∧ }
7. (P V Q) ∧ (P → R) ∧ (Q → R) is equivalent to
   a) P b) Q c) R d) True ≡ T
8. ⌉ (P→ Q) is equivalent to
   a) P ∧ ⌉ Q b) ⌉P ∧ Q c) ⌉P V Q d) P V Q

9. The logical expression $(A \wedge B)' \Rightarrow (C' \wedge A) \Rightarrow (A \equiv 1)$ is

a) A contradiction b) A tautology

c) A wellformed formula d) None.

10. Which one of the following propositional forms is a tautology? **[UGC NET-2000]**

a) p ∧ ⌉q b) p ⇒(p∨q)

c) p ⇒(p ∧ q) d) (p∨q) ⇒(p ∧ q)

11. The translation of the sentence (given in English) : 'Some cats are black but all buffaloes are black' using the following notations :

B(x) : x is black, C(x) : x is a cat

Bf(x) : x is a buffalo, (∀x) : for all x

(∃x) : for some x.

In logical symbolism is given by **[UGC NET-1995]**

a) (∃x) (C(x) ∧ B(x)) ∧ (∀ x)(B*f*(x) → B(x))

b) (∃x) (C(x) ∧B(x)) ∧ (∀ x)(B*f*(x) ∧ B(x))

c) (∃x) (C(x) →B(x)) ∧ (∀x)(B*f*(x) ∧ B(x))

d) (∃x) (C(x) → B(x)) ∧ (∀ x)(B*f*(x) → B(x))

12. In propositional logic which one of the following is equivalent to p →q **[UGC NET-1998]**

a) ~ p →q b) ~p V q c) ~p V ~q d) p→~q

13. In predicate logic ~ ∀ P(x) is equivalent to **[UGC NET-1998]**

a) ∃ xP(x) b) ∃x~P(x) c) ∀x~P(x) d) ~∃x P(x)

ANSWER KEY - I				
1. c	**2.** c	**3.** b	**4.** b	**5.** c
6. c	**7.** c	**8.** a	**9.** b	**10.** b
11. d	**12.** b	**13.** b		

EXPLANATORY NOTES - I

1. (c) Explanation :

$P \vee (P \to Q) = P \vee (\ P \vee Q) = (P \vee \ P) \vee Q = 1 \vee Q = 1$

$\therefore\ P \vee (P \to Q)$ is a tautology.

2. (c) Explanation :

$P \wedge (\sim p \vee q) = (P \wedge \sim P) \vee (\ P \wedge q) = 0 \vee (\ P \wedge Q) = P \wedge Q$

$\therefore\ P \wedge (\sim p \vee q)$ is logically equivalent to $p \wedge q$

3. (b) Explanation :

$a \wedge b \to b \vee c = (a \wedge b)\ \vee b \vee c$

$\Rightarrow\ a \vee\ b \vee b \vee c =\ a \vee 1 \vee c = 1 \vee c = 1$

$\therefore\ a \wedge b \to b \vee c$ is tautology.

4. (a) Explanation : Let P and q be propositions.

P	Q	Q	$P \to Q$	$p \Leftrightarrow q$
T	T	F	F	T
T	F	T	T	F
F	T	F	T	F
F	F	T	T	T

$\therefore\ P \to\ Q \neq\ P \Leftrightarrow Q$

5. (c) Explanation :

$(P \vee Q) \wedge (\ P \vee R) \wedge (\ Q \vee S)$

$(P \vee\ P) \wedge (\ Q \vee R) \wedge (\ Q \vee S)$

$1 \wedge (\ Q \vee\ Q) \wedge (\ R \vee S)$

$1 \wedge (\ R \vee\ A),\ R \vee S$

7. (c) Explanation :

$(P \vee Q)\ \wedge (\ P \to R),\ (Q \to R)$

$(P \vee Q)\ \wedge (\ P \vee R) \wedge (\ Q \vee R)$

$(P \vee\ P)\ \wedge (\ Q \vee R) \wedge (\ Q \vee R)$

$1\ \wedge (\ Q \vee Q) \wedge (\ R \vee R)$

$(\ Q \vee\ \ Q) \wedge (\ R),\ 1 \wedge R = R$

8. (a) Explanation :

$(P \to Q) =\ \ (\ P \vee Q) = P \wedge\ Q$

9. (b) Explanation :

$(A \wedge B) \Rightarrow (C' \wedge A) \Rightarrow (A \equiv 1)$ is tautology.

10. (b) Explanation :

$P \to (P \vee Q) =\ P \vee (P \vee Q) = (\ p \vee p) \vee Q = 1 \vee Q = 1$

$\therefore\ P \to (P \vee Q)$ is tautology.

11. (d) Explanation :

Some cats are black but all buffaloes are black.

$(\exists x)(C(x) \to B(x)) \wedge (\forall x)(Bf(x) \to B(x))$

12. (b) Explanation :

We know that $p \to q$ is $p \vee q$

13. (b) Explanation :

$\sim \forall\ P(x) = \exists x \sim p(x)$

OBJECTIVE QUESTIONS - II

1. Consider two well-formed formulas in propositional logic $F1: P \to \rceil\ P$ $F2: (P \to \rceil\ P) \vee (\ \rceil P \to P)$ Which of the following statements is correct? **[GATE-2001]**

a) F1 is satisfiable, F2 is valid

b) F1 is unsatisfiable, F2 is satisfiable

c) F1 is unsatisfiable, F2 is valid

d) F1 and F2 are both satisfiable

2. Let a, b, c, d be propositions. Assume that the equivalences a ↔ (b V ⌉ b) and b ↔ c hold. Then the truth value of the formula

(a ∧ b) → ((a ∧ c) V d) is always **[GATE-2000]**

a) True b) False

c) Same as the truth value of b

d) Same as the truth value of d

3. Which of the following if FALSE ? Read ∧ as And, ∨ as OR, ~ as NOT, → as one way implication and ↔ as two way implication ? **[GATE-1996]**

a) ((x → y) ∧ x) → y

b) ((~x → y) ∧ (~x ∧ ~y)) → x

c) (x → (x V ψ)) d) ((x V y) ↔ (~x V ~y))

4. Which of the following predicate calculas statements is /are valid: **[GATE-1992]**

a) (∀ x)P(x) V (∀ x)Q(x)→(∀ x){P(x) V Q(x)}

b) (∃ x)P(x) ∧ (∃ x)Q(x) → (∃ x){P(x) ∧ Q(x)}

c) (∀ x){P(x) V Q(x)} → (∀ x)P(x) V (∀ x)Q (x)

d) (∃ x){P(x) V Q(x)} → ~ (∀ x)P(x) V (∃ x) Q (x)

5. If F1, F2 and F3 are propositional formulae such that F1 $\wedge$ F2 --> F3 and F1 $\wedge$ F2 --> ~ F3 are both tautologies, than which of the following TRUE **[GATE-1991]**

a) Both F1 and F2 are tautologies

b) The conjunction F1 $\wedge$ F2 is not satisfiable

c) Neither is tautologies d) Neither is satisfiable

6. Indicate which of the following well-formed formula(s) are valid **[GATE-1990]**

a) $((P \Rightarrow Q) \wedge (\theta \Rightarrow R)) \Rightarrow (P \Rightarrow R)$

b) $(P \Rightarrow Q) \Rightarrow (P \Rightarrow Q)$ c) $(P \wedge (\rceil P \vee \rceil Q)) \Rightarrow Q$

d) $((P \Rightarrow Q) \vee (\theta \Rightarrow R)) \Rightarrow ((P \vee Q) \Rightarrow R)$

7. Which of the following are tautologies?

a) $((P \vee Q) \wedge Q) \leftrightarrow Q$ b) $(P \vee (P \rightarrow Q)) \rightarrow P$

c) $((P \vee Q) \wedge P) \rightarrow Q$ d) $((P \vee Q) \wedge \rceil P) \rightarrow Q$

8. Identify the valid conclusion from the premises PVQ, Q→R, P→M, ⌉M

a) $P \wedge (Q \vee R)$ b) $P \wedge (Q \wedge R)$

c) $R \wedge (P \vee Q)$ d) $Q \wedge (P \vee R)$

9. Which one of the following statements is true when exactly two of P, Q, or R are TRUE?

a) $(P \wedge \bar{Q} \wedge R) \vee (P \wedge \bar{Q} \wedge \bar{R}) \wedge (\bar{P} \wedge Q \wedge \bar{R})$

b) $(P \vee Q \vee R) \wedge (P \vee \bar{Q} \vee R) \wedge (P \vee Q \vee \bar{R})$

c) $(P \wedge Q \wedge R) \vee (P \wedge \bar{Q} \wedge R) \vee (P \wedge Q \wedge \bar{R})$

d) $(P \wedge \bar{Q} \wedge R) \vee (P \wedge \bar{Q} \wedge \bar{R}) \vee (\bar{P} \wedge Q \wedge \bar{R})$

10. P→(Q →R) is equivalent to

a) $(P \wedge Q) \rightarrow R$ b) $(P \vee Q) \rightarrow R$

c) $(P \vee Q) \rightarrow \rceil R$ d) $(P \wedge Q) \rightarrow \rceil R$

11. What can we correctly say about proposition P1:

P1: $(p \vee \rceil q) \wedge (q \rightarrow r) \vee (r \vee p)$

a) P1 is tautology b) P1 is a contradiction

c) If P is true and q is false and r is false,then P1 is true

d) If P is true and q is true and r is false, then P1 is true

ANSWER KEY - II

1. c	2. a	3. d	4. a,d	5. ?
6. a,c	7. a,d	8. c,d	9. c	10. a
11. c				

EXPLANATORY NOTES - II

1. (c) **Explanation :**

$F_1: P \rightarrow \neg P \quad F_2: (P \rightarrow \neg P) \vee (\neg p \rightarrow P)$

$= \neg P \vee \neg P \quad = (\neg P \vee \neg P) \vee P \vee P$

$= \neg P \quad = \neg P \vee P = T$

$\therefore F_1$ is unsatisfiable and F_2 is valid.

2. (a)

3. (a) **Explanation :** $((x \rightarrow y) \wedge x) \rightarrow y$

$\neg[(\neg x \vee y) \wedge x] \vee y,\ (x \wedge \neg y) \vee \neg x \vee y$

$(\rceil x \wedge \rceil y) \vee y \Rightarrow (\rceil x \vee y) \wedge (\rceil y \vee y),$

$(\neg x \vee y) \wedge T,\ (\neg x \vee y)$

7. (a) **Explanation :**

$[(P \vee Q) \wedge Q \rightarrow Q] \wedge [Q \rightarrow (p \vee Q) \wedge Q]$

$\neg[(P \vee Q) \wedge Q] \vee Q \wedge [\neg Q \vee (p \vee Q) \wedge Q]$

$[(\neg P \wedge \neg Q) \vee \neg Q] \vee Q \wedge \neg Q \vee (P \vee Q) \wedge Q$

$[(\neg P \wedge \neg Q) \vee \neg Q] \vee F \vee (P \vee Q) \wedge Q$

$\neg[(P \vee Q) \wedge Q] \vee (P \vee Q) \wedge Q = T$

10. (a) **Explanation :**

$P \rightarrow (Q \rightarrow R) = \neg P \vee (Q \rightarrow R) = \neg P \vee \neg Q \vee R = \neg(P \wedge Q) \vee R$
$= (P \wedge Q) \rightarrow R$

11. (c) **Explanation :** P1: $(P \vee \neg q) \wedge (q \rightarrow R) \vee (R \vee p)$

$(P \vee \neg Q) \wedge (\neg Q \vee R) \vee (R \vee P),\ (P \vee \neg Q) \wedge (\neg Q \vee R \vee P)$

$(P \vee \neg Q) \wedge (P \vee \neg Q \vee R),\ P \vee \neg Q \vee R$

$\therefore$ if P is true and Q is false and R is false then P 1 is true.

OBJECTIVE QUESTIONS - III

1. Pick the correct prefix formula

a) $\rightarrow P \vee QR \sim S$ b) $\rightarrow P \vee QRS$

c) $\rightarrow\rightarrow PQ \rightarrow\rightarrow QR \rightarrow PR$ d) $\rightarrow P \vee Q \sim QSR$

2. Consider two well formed formulas in propositional logic $F_1: P \rightarrow \sim P; F_2 (P \rightarrow \sim P) \vee (\sim P \rightarrow P)$ Which of the following statements is correct.

a) F_1 is satisfiable, F_2 is valid.

b) F_1 is unsatisfiable, F_2 is satisfiable

c) F_1 is unsatisfiable, F_2 is valid.

d) F_1 and F_2 are both satisfiable.

3. Which of the following sets are functionally complete ?

i) $\{\sim, \rightarrow\}$ ii) $\{\vee, \wedge\}$ iii) $\{\uparrow\}$ iv) $\{\vee, \uparrow\}$

a) i) & iii) b) ii) & iv) c(i) alone d) iii) & iv)

4. The prefix form of $A \wedge (B \vee \sim C)$ is

a) $\wedge AB \vee \sim C$ b) $\wedge A \vee B \sim C$

c) $\sim \wedge \vee AB$ d) $\vee \wedge \sim AB$

5. The prefix notation of $\sim(P \wedge Q) \leftrightarrow (\sim P \vee \sim Q)$

a) $\vee \wedge \sim P \leftrightarrow Q \sim P \sim Q$ b) $\leftrightarrow \sim \wedge PQ \vee \sim P \sim Q$

c) $Pq \wedge \sim P \sim q \sim V \leftrightarrow$ d) none of the above

6. The postfix notation of $(\sim P \wedge (Q \leftrightarrow \sim P)) V \sim Q$ is

a) $\vee \wedge \sim P \leftrightarrow Q \sim P \sim Q$ b) $\leftrightarrow \sim \wedge PQ \vee \sim Q$

c) $Pq \wedge \sim P \sim Q \sim V \leftrightarrow$ d) none of the above

7. Which of the following sentences are propositions.

a) answer this question

b) do not pass go

c) what time is it ?

d) miami is the capital of florida

8. Which of the following are not logically equivalent ?

a) $P \leftrightarrow Q$ and $(P \wedge Q) \vee (\sim P \wedge \sim Q)$

b) $(P \to Q) \to R$ and $P \to (Q \to R)$

c) $\sim P \leftrightarrow Q$ and $P \leftrightarrow \sim Q$ d) $\sim(P \leftrightarrow Q)$ and $\sim P \leftrightarrow Q$

9. $(P \downarrow Q) \downarrow (P \downarrow Q)$ is logically equivalent to

a) P b) $P \vee Q$ c) $P \wedge Q$ d) Q

10. $(P \uparrow Q) \uparrow (P \uparrow Q)$ is logically equivalent to

a) $\sim P$ b) $\sim Q$ c) P d) $\sim(P \uparrow Q)$

11. If X then Y unless Z is represented by which of the following formulas in propositional logic ?

a) $(X \wedge \sim Z) \to Y$ b) $(X \wedge Y) \to \sim Z$

c) $X \to (Y \wedge \sim Z)$ d) $(X \to Y) \wedge \sim Z$

12. "Not all birds can fly". Which of the following formulas represent in predicate calculas.

a) $\sim(\exists x B(x) \wedge F(x))$ b) $\exists x, (B(x) \wedge [\sim F(x)])$

c) $\forall x, B(x) \to \sim F(x)$ d) None

13. The negation of "every complete bipartite graph is not planar" in paredicate calculus is

a) $\forall x[C(x) \wedge B(x) \to P(x)]$

b) $\sim\forall(x)[C(x) \wedge B(x) \to \sim P(x)]$

c) $\exists x[C(x) \wedge B(x) \wedge P(x)]$ d) none of the above

14. Without using universal quantifier, the translated form of "No dogs are intelligent" in predicate calculus is

a) $\exists(x)\{\sim D(x) \wedge I(x)\}$ b) $\sim[\exists(x)\{D(x) \wedge I(x)\}]$

c) $\exists x D(x) \wedge \sim I(x)$ d) none of the above

15. If P has truth values, T, Q and R have truth values F then $(P \wedge \sim Q) \to R, P \to (Q \vee R)$

a) both have same truth Value (F)

b) $(P \wedge \sim Q) \to R$ is true and $P \to (Q \vee R)$ is false

c) both have same truth value T

d) none of the above

16. The number of propositional functions of 2 variables are

a) 4 b) 8 c) 16 d) 2

17. Which of the following propositions are tautologies.

a) $(\sim Q \wedge (P \to Q)) \to \sim P$ b) $((P \vee q) \wedge \sim P) \to q$

c) $(\sim P \vee \sim Q) \to (P \leftrightarrow \sim Q)$ d) $Q \wedge (P \vee \sim Q) \to Q$

a) i) & ii) b) i) & iv) c) iii) & iv) d) ii) & iii)

18. Let $P(x)$ and $Q(x)$ be propositional functions. Then which of the following have same truth values.

i) $\exists x(p(x) \to Q(x))$ and $\forall x P(x) \to \exists x Q(x)$

ii) $\exists x(P(x) \vee Q(x))$ and $\exists x P(x) \vee \exists x Q(x)$

iii) $\forall x(P(x) \wedge Q(x))$ and $\forall x P(x) \wedge \forall x Q(x)$

iv) $\exists x P(x) \wedge \exists x Q(x)$ and $\exists x(P(x) \wedge Q(x))$

a) i), iii), iv) b) i), ii), iii) c) ii), iii), iv) d) i), ii), iii), iv)

19. Which of the following are equivalent

i) $\forall x \forall y P(x,y)$ and $\forall x \forall y p(x,y)$

ii) $\forall y \forall x P(x,y)$ and $\exists x \forall y P(x,y)$

iii) $\exists x \forall y P(x,y)$ and $\forall y \exists x P(x,y)$

iv) $\exists y \forall x P(x,y)$ and $\exists x \exists y P(x,y)$

a) i), iii), iv) b) i), ii), iii)

c) i), ii), iv) d) none of the above

EXPLANATORY NOTES - III

1. (c) **Explanation :** Convert the prefix formula into infix formula

$\to\to PQ \to\to QR \to PR$, $\to\to PQ \to QR \to P \to R$

$\to PQ \to Q \to R \to P \to R$,

$(P \to Q) \to (Q \to R) \to (P \to R)$

So this is true.

2. (c) **Explanation :** $F_1 : P \to \sim P = \sim P \vee \sim P = \sim P$

$\therefore F_1$ is unsatisfiable.

$F_2 : (P \to \sim P) \vee (\sim P \to P) = (\sim P \vee \sim P) \vee (P \vee P) = \sim P \vee P = T$

$\therefore F_2$ is valid.

3. (a) **Explanation :** $\{\sim, \to\}$ we have

$P \to Q \Leftrightarrow \sim P \vee Q$ and $P \vee Q \Leftrightarrow \sim P \to Q$ As $\sim, \to$ is functionaly complete. $\{\uparrow\}$ in order to show that $\uparrow$ functionaly complete, it is sufficient to show that the sets of connectives $\{\wedge, \sim\}$.

Consider $P \uparrow P \equiv \sim(P \wedge P) \equiv \sim P \vee \sim P \equiv \sim P$

Also $(P \uparrow Q) \uparrow (P \uparrow Q) \equiv \sim(P \uparrow Q) = P \wedge Q$

$\therefore \sim, \wedge, \vee$ have been expressed interms of $\uparrow$ alone

$\therefore \uparrow$ is functionaly complete

4. (b) **Explanation :** $A \wedge (B \vee \sim C)$,

$A \wedge (\vee B \sim C)$, $\wedge A \vee B \sim C$

5. (b) **Explanation :** $\sim(\wedge PQ) \leftrightarrow (\vee \sim P \sim Q)$

$\leftrightarrow \sim(\wedge PQ)(\vee \sim P \sim Q)$. $\leftrightarrow \wedge \sim PQ \vee \sim P \sim Q$

6. (a)

7. (d) **Explanation :** $\sim P \wedge (Q \leftrightarrow \sim P) \vee \sim Q$

$\sim P \wedge (Q \sim P \leftrightarrow)) \vee \sim Q$, $\sim P[Q \sim P \leftrightarrow] \vee \sim Q \wedge$

$\sim PQ \sim P \leftrightarrow \sim Q \wedge \vee$

8. (d)

9. (b) **Explanation :** $(P \downarrow Q) \downarrow (P \downarrow Q)$

$= \sim(P \vee Q) \downarrow \sim(P \vee Q) = \sim[\sim(P \vee Q) \vee \sim(P \vee Q)]$

$= (P \vee Q) \wedge (P \vee Q) = P \vee Q$

10. (d) **Explanation :** $\sim(P \wedge Q) \uparrow \sim(P \wedge Q)$,

$\sim[\sim(P \wedge Q) \wedge \sim(P \wedge Q)]$, $(P \wedge Q) \vee (P \wedge Q)$

$(P \wedge Q)$, $\sim[\sim(P \wedge Q)]$, $\sim[P \uparrow Q]$

11. (d) **12.** (b) **13.** (c) **14.** (b)

15. (a) **Explanation :**

$\sim(P \wedge \sim Q) \vee R$	$P \vee (Q \vee R)$
$\sim(T \wedge \sim F) \vee F$	$\sim T \vee (F \vee F)$
$\sim(T \wedge T) \vee F$	$F \vee (F \vee F)$
$\sim(T) \vee F$	$= F$
$F \vee F = F$	

16. (b) **17.** (a) **18.** (b) **19.** (b)

IV. MATHEMATICAL INDUCTION

- **Introduction**
- **Applications of Induction**

1. Introduction

In the Study of Mathematics in order to prove the theorems. We use mathematical Induction. This was first postulated by Piano, an Italian Mathematician. He formulated the properties of the natural numbers in the form of axioms. One of these axioms is the mathematical Induction.

2. Principle Of Finite Mathematical Induction

N = {1 2 3......} is the set of natural numbers.

If S is a subset of the set of natural numbers N such that

I. $1 \in S$ and

II. $k \in S \Rightarrow (k+1) \in S$, then s=N.

It is said to be the principle of Mathematical Induction and also called the principle of Finite Mathematical Induction.

This can also be described in the following way :

Let S (n) be a statement for each $n \in N$. If

I. S (1) is true and

II. S (k) is true $\Rightarrow$ S (k+1) is true, then S(n) is true for all $n \in N$

3. Principle of complete Mathematical Induction

Let S be a subset of N. If

I. $a \in S$ and

II. 1, 2, 3, ... $k \in S \Rightarrow (k+1) \in S$ then S = N. Alternatively,

Let S(n) be a statement for each $n \in N$. If

I. S(1) is true and

II. S(1), S(2), S(3), ... S(k-1), S(k) are true $\Rightarrow$ S(k+1) is true, then S(n) is true for each positive integers n.

4. Theorems of Mathematical Induction

Theorem : Let $\{S(n) \mid n \in N\}$ is a set of statements. If

I. S(1) is true and

II. S(m) is true $\Rightarrow$ S(m+1) is true. then S(n) is true, $\forall n \in N$.

Theorem : Let $\{S(n) \mid n \in N\}$ be a set of statements. If

I. S(1) is true and

II. S(1), S(2), S(3), ... S(m-1) are true $\Rightarrow$ S(m) is true,then S(n) is true $\forall$ n $\in$ N.

Applications Of Mathematical Induction

To find Σn, Σn^2, Σn^3.

Ex 1 : Using the principle of Mathematical Induction show that $1+2+3 \ldots\ldots\ldots\ldots + n = \frac{n(n+1)}{2}$

Sol : Let the statement

S(n) : $1+2+3 \ldots\ldots\ldots\ldots + n = \frac{n(n+1)}{2}$

If n=1 then L.H.S. = 1= $\frac{1.2}{2}$ = R.H.S$\Rightarrow$ S(1) is true.

Assume that S(k) is true.

$\therefore\ 1+2+3 \ldots\ldots\ldots\ldots + k = \frac{k(k+1)}{2}$

Adding (k+1) to both sides

$$1+2+3+\ldots\ldots\ldots\ldots+k+(k+1) = \frac{k(k+1)}{2} + (k+1)$$

$$= \frac{k(k+1)+2(k+1)}{2} = \frac{k^2+3k+2}{2} = \frac{(k+1)(k+2)}{2}$$

$\therefore$ S(k+1) is true.

By the principle of Mathematical Induction, the given statement S(n) is true for all positive integral values of n.

Ex 2 : Using the principle of Mathematical Induction show that $1^2+2^2+3^2+\ldots..n^2 = \frac{n(n+1)(2n+1)}{6}$

Ex : Show that $1^3+2^3+3^3+\ldots..n^3 = \frac{n^2(n+1)^2}{4}$ for all $n \in N$

6. Problems On Divisibility

Ex : 4 Show that $x^n - y^n$ is divisible by (x-y) for all positive integral values of n.

Sol : Let S_n : {$n \in N$, $x^n - y^n$ is divisible by (x-y)}

If n=1, then $x^n - y^n$ = x-y. It is divisible by (x-y)

$\therefore$ S(1) is true. Assume that S(k) is true.

Then $x^k - y^k$ is divisible by x-y. $\Rightarrow x^k - y^k = (x-y)P$

$x^{k+1} - y^{k+1} = x^{k+1} - x^k y + x^k y - y^{k+1}$

$= x^k(x-y) + y\left(\left(x^k - y^k\right)\right) = x^k(x-y) + y(x-y)P$

is divisible by (x-y) $\Rightarrow$ S(k+1) is true.

$\therefore$ By the principle of Mathematical Induction.

S(n) is true for all $n \in N$. Hence $x^n - y^n$ is divisible by (x-y)

Ex 5 : Show that $(49^n + 16n - 1)$ is divisible by 64 for all positive integral values of 1.

Sol : Let S_n "$(49^n + 16n - 1)$ is divisible by 64"

S(1) is true. Assume that S(k) is true.

Thus $49^k + 16k - 1$ is divisible by 64

Hence $49^k + 16k - 1 = 64\,m \rightarrow$ (1) ($m \in Z$ is an integer)

Now $49^{k+1} + 16(k+1) - 1 = 49^k \cdot 49 + 16k + 15$

= (64m - 16k +1)49 + 16k + 15 from (1)

= 64 (49m -12k + 1) $\quad\therefore$ S(k+1) is true.

By the principle of Mathematical Induction S(n) is true for all $n \in N$.

SOLVED EXAMPLES

Ex 1 : I. Find the value of 1 + 3 + 5 + ... + (2n-1)

II. Show that 1.3 + 2.4 + 3.5 + ... n sum to n terms $= \frac{n(n+1)(2n+7)}{6}$

Sol : I. 1, 3, 5, ... (2n-1) are in A.P. first term is 1, least term is (2n-1). There are n terms. Hence

$1+3+ \ldots\ldots +(2n-1) = 1+3+\ldots..+ = \frac{n(1+(2n-1))}{2} = n^2$

II. 3, 4, 5, ... are in A.P.

n^{th} term = 3 + (n-1)1 = n+2

Given problem $S_n = \sum_{m=1}^{n} m(m+2) = \sum_{m=1}^{n} m^2 + 2\sum_{m=1}^{n} m$

$= \frac{n(n+1)(2n+1)}{6} + \frac{2n(n+1)}{2}$

$= \frac{n(n+1)}{6}[2n+1+6] = \frac{n(n+1)(2n+7)}{6}$

Ex 2 : Find the value of $4^3+5^3+6^3+\ldots\ldots\ldots.+10^3$.

Sol : $S_n = (1^3+2^3+3^3+4^3+5^3+\ldots\ldots.+10^3) - (1^3+2^3+3^3)$

$= \frac{10^2(10+1)^2}{4} - \frac{3^2(3+1)^2}{4}$

$= \frac{100(121)}{4} - \frac{9(16)}{4} = 2989$

Ex 3 : Find the nth term of 3+5+9+17+..........

Sol : In the given series, the nth term is denoted by t_n..

$S_n = 3+5+9+17+ \ldots\ldots\ldots.\ t_n \ldots\ldots\ldots.. (1)$

$S_n = 0+3+5+9+17+ \ldots\ldots\ldots\ldots.. t_{n-1} + t_n \ldots\ldots..(2)$

$(1) - (2)\ 0 = 3+2+4+8+ \ldots\ldots\ldots\ldots\ldots. + 2^{n-1} - t_n$

$\therefore n^{th}$ term $t_n = 3+2+4+8+ \ldots. + 2^{n-1}$

$= 3 + \frac{2(2^{n-1}-1)}{(2-1)} = 3 + 2^n - 2 = 2^n + 1.$

Ex 4 : Show that 1.3 + 3.5 + 5.7 + ... n terms

$= \frac{n(4n^2+6n-1)}{3}$

Ex 5 : Show that $\frac{1}{1.4} + \frac{1}{4.7} + \frac{1}{7.10} + \ldots..$n terms $= \frac{n}{3n+1}$

$\forall n \in N$.

Ex 6 : Show that $3.5^{2n+1} + 2^{3n+1}$ is divisible by 17 for all n $\in$N.

Sol : Let "$3.5^{2n+1} + 2^{3n+1}$ is divisible by 17" be S(n).

If n=1 then $3.5^3 + 2^4 = 391$

17 divides 391 $\quad\therefore$ S(1) is true.

Assume that S(k) is true. Then

$3.5^{2k+1} + 2^{3k+1}$ is divisible by 17.

$\therefore\ 3.5^{2k+1} + 2^{3k+1} = 17$ m (for some m$\in$Z) ....(1)

$3.5^{2(k+1)+1} + 2^{3(k+1)+1}$

$= 3[5^{2k+1}.5^2] + 2^{3k+1}.2^3 = 25(3.5^{2k+1}) + 8.2^{3k+1}$

$= 25(17m - 2^{3k+1}) + 8.2^{3k+1}$ (from 1)

$= 425m - 25.2^{3k+1} + 8.2^{3k+1} = 425m - 17.2^{3k+1}$

$= 17(25m - 2^{3k+1}) = 17$ (an integer)

$\therefore$ 17 divides $3.5^{2(k+1)+1} + 2^{3(k+1)+1} \Rightarrow$S(k+1) is true.

$\therefore$ By the principle Mathematical Induction, S(n) is true for all n $\in$N.

Ex 7 : Show that $x^{2n+1} + y^{2n+1}$ is divisible by x+y for all n $\in$N.

Sol : Let S(n) : "$x^{2n+1} + y^{2n+1}$ is divisible by x+y"

If n=1, then $S(1) = x^3 + y^3$ .... (1)

$= (x+y)(x^2-xy+y^2)$

x+y divides (1) $\Rightarrow$ S(1) is true.

$x^5 + y^5 = (x+y)(x^4 - x^3y + x^2y^2 - xy^3 + y^4)$. Hence S(2) is true.

Let S(1), S(2), ...S(k-1), S(k) is true.

Since S(k-1) = $x^{2k-1} + y^{2k-1}$ is divisible by x+y.

$x^{2k-1} + y^{2k-1} = (x+y)m_1$ for some m_1 ...... (2)

Since S(k) is true, $x^{2k+1} + y^{2k+1}$ is divisible by x+y.

$\therefore\ x^{2k+1} + y^{2k+1} = (x+y)m_2$........ (3) Now

$x^{2k+3} + y^{2k+3} = x^{2k+3} + x^2y^{2k+1} + x^{2k+1}y^2 + y^{2k+3} - x^2y^{2k+1} - x^{2k+1}y^2$

$= x^2(x^{2k+1} + y^{2k+1}) + y^2(x^{2k+1} + y^{2k+1}) - x^2y^2(x^{2k-1} + y^{2k-1})$

$= (x^2+y^2)(x^{2k+1}+y^{2k+1}) - x^2y^2(x^{2k-1}+y^{2k-1})$

$= x^2+y^2(x+y)m_2 - x^2y^2(x+y)m_1$ (From (2) & (3))

$= (x+y)[(x^2+y^2)m_2 - x^2y^2m_1]$ .... (4)

(x+y) divides (4) $\Rightarrow$ S(k+1) is true.

$\therefore$ By complete Mathematical Induction. S(n) is true for all n$\in$N.

$\therefore \forall n \in N$, (x+y) divides S(n) = $x^{2n+1} + y^{2n+1}$

OBJECTIVE QUESTIONS - I

1. The nth term of the series 3+7+13+21+...... is

a) 4n-1 b) n^2+2n c) n^2+n+1 d) n^2+2

2. 2+3+5+6+8+9+.....2n terms is

a) $3n^2$+2n b) $4n^2$+2n c) $4n^2$ d) None

3. $\left(\sum n^3\right)\left(\sum n\right) = \left(\sum n^2\right)^2$ Iff

a) n =3 b) n =1 c) n^2=3 d) n =-1

4. 1.2+2.3+3.4+......n terms = **[JNTU- 1986]**

a) $\frac{n(n+1)(n+5)}{3}$ b) $\frac{n(n+1)(n+2)}{3}$

c) $\frac{n(4n^2+6n-1)}{3}$ d) $n(n+1)(n+2)$

5. If n $\in$ N, then n^3+2n is divisible by **[JNTU-1995]**

a) 3 b) 8 c) 9 d) 11

6. $x^n - a^n$ is divisible by x-a for n is any

a) Positive integer b) Integer

c) Odd positive integer d) None

7. If n $\in$ N then x^n-1 is divisible by

a) x-1 b) x-y c) x+y d) y-1

8. x^n-1 is divisible by x-k , then the least positive integral value of k is

a) 1 b) 2 c) 3 d) 4

ANSWER KEY - I

1. c	**2.** a	**3.** b	**4.** b	**5.** a
6. a	**7.** a	**8.** a		

OBJECTIVE QUESTIONS - II

1. If $2^3+4^3+6^3+.....+(2n)^3 = kn^2(n+1)^2$, then K= **[JNTU - 2001]**

a) $\frac{1}{2}$ b) 1 c) $\frac{3}{2}$ d) 2

2. 2.4+4.7+6.10+..... (n-1)terms =

a) $2n^3-2n^2$ b) $\frac{(n^3+3n^2+1)}{6}$

c) $2n^3+2n$ d) None

3. If the natural numbers are divided into groups of {1},{2,3}.{4,5,6}.{7,8,9,10} , ... then the sum of 50th group is **[IIT-1991]**

a) 62525 b) 65225 c) 56255 d) 55625

4. (1)+(2+3+4)+(5+6+7+8+9)+....n brackets = **[IIT-1993]**

a) $\frac{n(n+1)(n+2)}{6}$ b) $\frac{n^2(n^2+1)}{2}$

c) $n(n+1)(n+2)$ d) $\frac{n(n+1)(2n+1)}{6}$

5. If $n \in N$, then 4^n-3n-1 is divisible by

a) 3 b) 8 c) 9 d) 11

6. $7^{2n}+3^{n-1}.2^{2n-3}$ is divisible by **[IIT-1992]**

a) 7 b) 9 c) 25 d) 26

7. The sum of n terms of the series $\frac{1}{2}+\frac{3}{4}+\frac{7}{8}+\frac{15}{16}+$ equals to **[IIT-1998]**

a) 2^n-n-1 b) $n+2^{-n}-1$

c) $1-2^{-n}$ d) 2^n-1

8. If $n \in N$ then 2^n

a) =n b) >n c) <n d) None

9. If n > 1 then

a) $n! = \left(\frac{n+1}{2}\right)^n$ b) $n! < \left(\frac{n+1}{2}\right)^n$

c) $n! > \left(\frac{n+1}{2}\right)^n$ d) None

ANSWER KEY - II

1. d	2. a	3. a	4. b	5. c
6. c	7. -	8. b	9. b	

V. LATTICES AND BOOLEAN ALGEBRA

- **Partial ordering**
- **Lattices**
- **Boolean Algebra**

1. Introduction

Partial ordering : A relation ~ on a set P is said to be partial order if it is

(a) reflexive $x \sim x,\ \forall x \in P$:

(b) antisymmetric : $(x - y) \wedge (y - x) \Rightarrow x = y$

(c) transitive: $(x - y) \wedge (y - z) \Rightarrow x - z$

A partial order is always denoted as $\leq$.

The set P with a partial order $\leq$ is called Partially ordered set, denoted by (P, $\leq$) and it is also called Poset.

Totally ordered Poset : A poset (P, $\leq$) is called totally ordered set if for all x,y $\in$ P , $(x \leq y) \vee (y \leq x)$.

Cover of x : Let x $(P, \leq)$. An element y $\in (P, \leq)$ is called cover of x if $x \leq y$ and there exists no element $z \in P$ such that $x \leq z$ and $z \leq y$.

Hasse Diagram: Hasse diagram is a graph whose points are points of P where the points x and y are connectred by an edge if y covers x. Any finite poset can be represented by a Hasse diagram.

Example: Let P = {1,2,3,4,5,6,25,100}. The Hasse graph is given in figure 5 .1

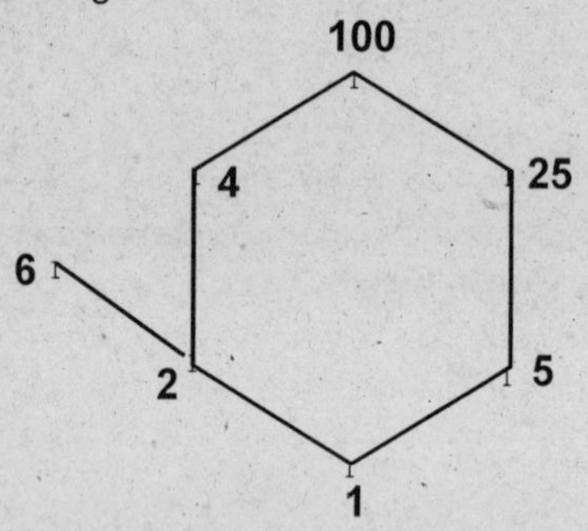

Fig 5.1. Poset

Least element : y is least element of P if $y \leq x$ for all $x \in P$.

Greatest element : y is greatest element of P if $x \leq y$ for all $x \in P$.

Upper bound : Let $A \subseteq P$. An element $x \in P$ is said to be upper bound of A if $a \leq x$ of all $a \in A$

Lower bound : Let $A \subseteq P$. An element $y \in P$ is said to beupper bound of A if $y \leq a$ fpr all $a \in A$.

Least upper bound (LUB) : LUB of A is x is an upper bound and if y is any other upper bound, then $x \leq y$.

Greatest lower bound (GLB) : GLB of A is y if y is a lower bound for A and if b is any other lower bound, then $b \leq y$.

Sometimes GLB is called infimum and LUB is called supremum.

2. Lattice : A lattice is a partially ordered set $(P, \leq)$ in which any two elements $\in$ P has GLB and LUB.

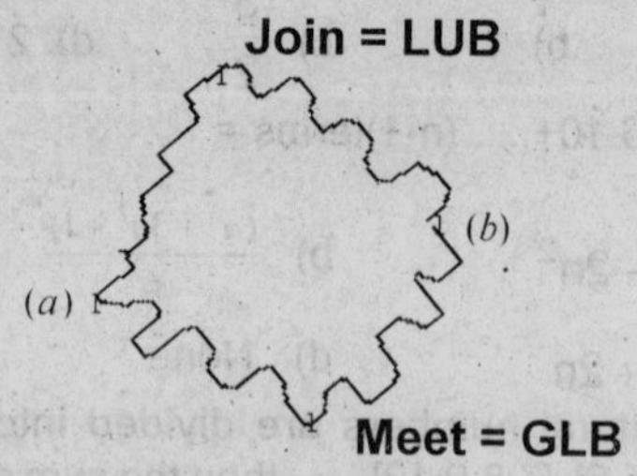

Lattice Fig 5.2

see Figure 5.2 GLB and LUB are shown.They are also called meet and join respectively. Figure 5.3 shows a lattice and non- lattice.

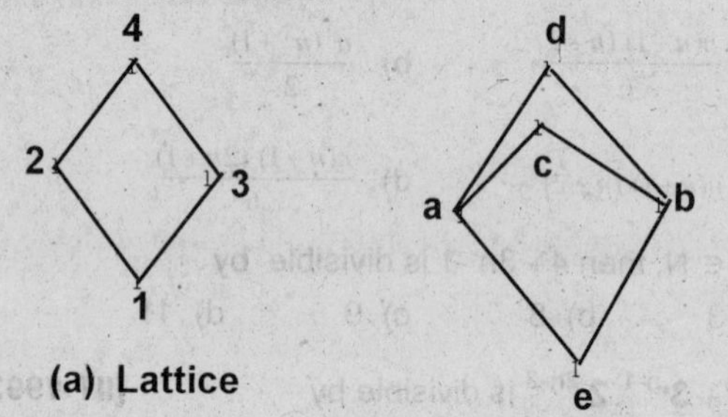

(a) Lattice

(b) Non- lattice

Fig 5.3

The poset shown in Figure 5.3 (b) is not a lattice since a and b have two LUBs namely c and d.

From first lattice, we deduce the following relations : $1 \leq 2, 1 \leq 3, 2 \leq 4$ and $3 \leq 4$. Here 2 and 3 are not compared. If $(P, \leq)$ is a lattice then $(P, \geq)$ is also a lattice. We define $\geq$ as follows: $(x \leq y) \Rightarrow y \geq x$.

In other words, the GLB and LUB are interachanged if we intercahange the relations $\leq$ and $\geq$. In terms of lattices, the operations of meet and join on $(L, \leq)$ become the operations of join and meet on $(L, \leq)$.

Dual lattice : For the lattice $(P, \leq)$ the dual is $(P, \geq)$.

The duals are shown in Figure 5.4 the diagram of $(P, \geq)$ can be obtained from that of $(P, \leq)$ by simply turning it upside down.

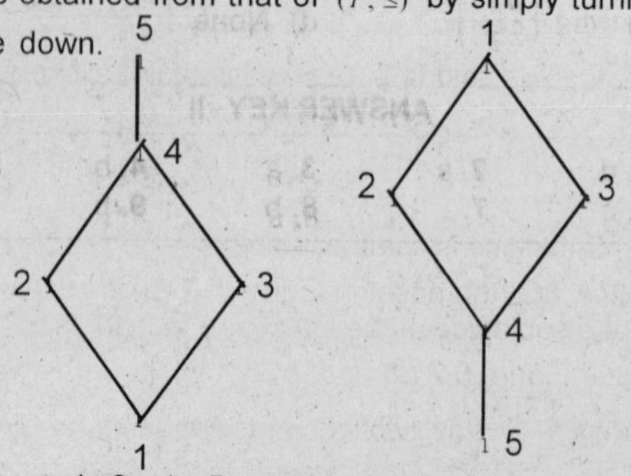

$1 \leq 2 \leq 4 \leq 5$
$1 \leq 3 \leq 4 \leq 5$

$5 \geq 4 \geq 2 \geq 1$
$5 \geq 4 \geq 3 \geq 1$

Dual lattices Fig. 5.4

Let a $*$ b = meet of a and b = GLB
and a$\oplus$b = join of a and b = LUB

Ex : For a, b $\in$a lattice $(P,\leq)$, prove a$\leq$b $\Leftrightarrow$ a$*$b = a $\Leftrightarrow$ a$\oplus$b = b

Sol : Let a$\leq$b. See Figure 5.5.

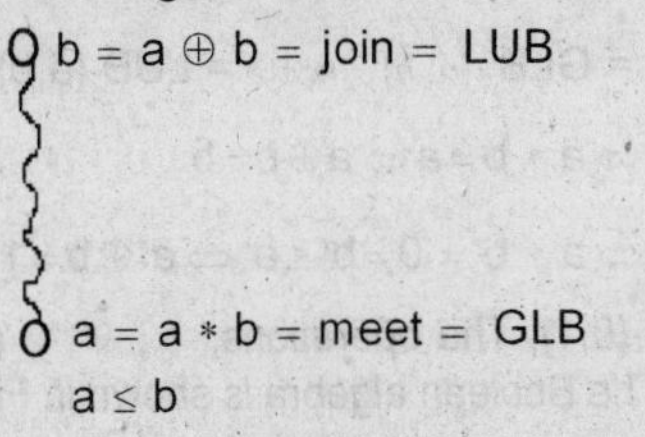

Fig. 5.5

Since a$\leq$a, we have a$\leq$a $*$ b. Since a $*$ b is lower bound for both a and b we have a $*$ b$\leq$a. Hence a = a $*$ b. Proof is similar for remaining part.

Ex : Show that a lattice with less than 4 elements is a chain.

Sol: Let $(L,\leq)$ be a lattice. Obviously if $|L|$ = 1 or 2 it must be a chain.

Figure. 5.6 shows all possible combinations on three elements.

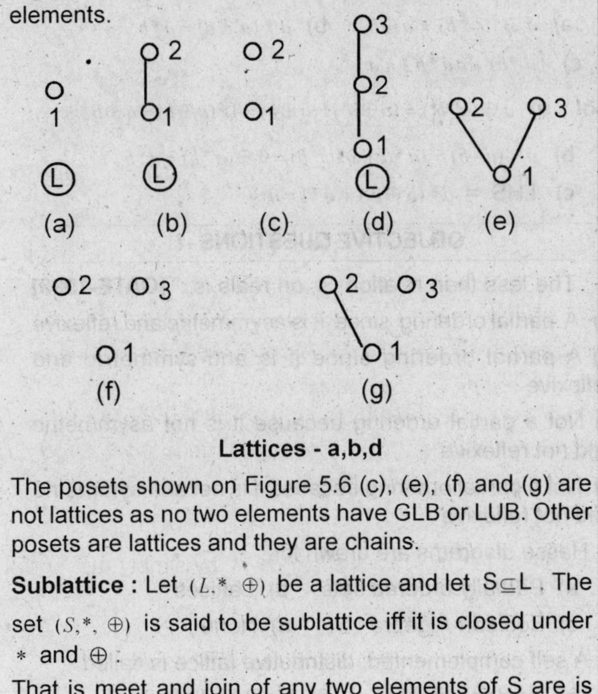

Lattices - a,b,d

The posets shown on Figure 5.6 (c), (e), (f) and (g) are not lattices as no two elements have GLB or LUB. Other posets are lattices and they are chains.

Sublattice : Let $(L,*,\oplus)$ be a lattice and let S$\subseteq$L. The set $(S,*,\oplus)$ is said to be sublattice iff it is closed under $*$ and $\oplus$.

That is meet and join of any two elements of S are is itself a lattice.

Note : Sublattice is itselft a lattice

Ex: Show that the poset in Figure 5.7 (a) is a lattice and it is not a sublattice of the lattice in Figure 5.7 (b).

Sol: See Figure 5.7 (a) and (b).

In poset $(P_1,\leq)$, any two elements have unique GLB and LUB.

Hence $(P_1,\leq)$ is a lattice.

Since GLB (4,5) in P_1 = 1 and LUB (4,5) in P_2 = 8 $\notin P_1$. Hence P_1 is not sublattice of P_2.

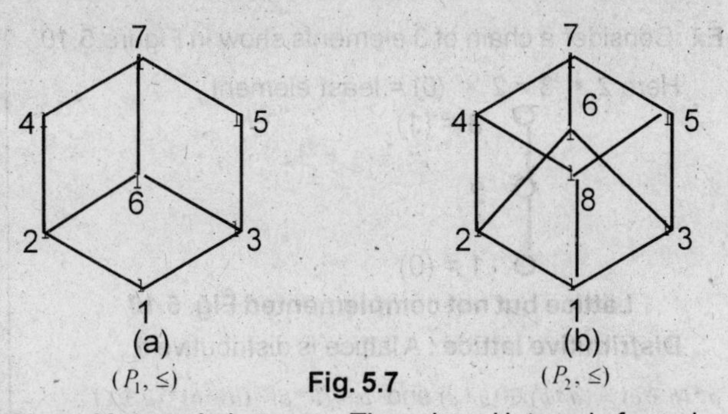

Fig. 5.7

Closed Interval : Let a$\leq$b. Then closed interval of a and b is defined as [a,b] = $\{x \mid a\leq x \text{ and } x\leq b\}$. See Figure 5.8.

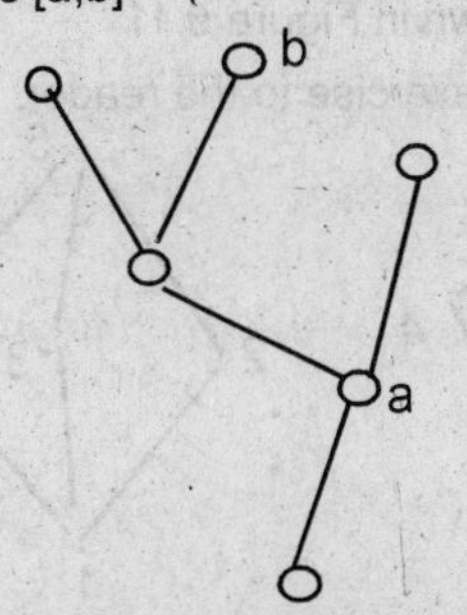

Interval [a,b] Fig 5.8

Clearly any closed interval is a chain.

Complete : A lattice is complete if each of its non-empty subsets has a least upper bound and a greatest lower bound.

Note : Every finite lattice is complete.

Bounds : The least and greatest elements of lattice are called bounds, and denoted by 0 and 1.

Complement : In a bounded lattice $(L,*,\oplus,0,1)$, if a$*$ b = 0 and a$\oplus$b = 1, then b is called complement of a.

Note : If b is complement of a then a is complement of b. See the lattice shown in Figure 5.9

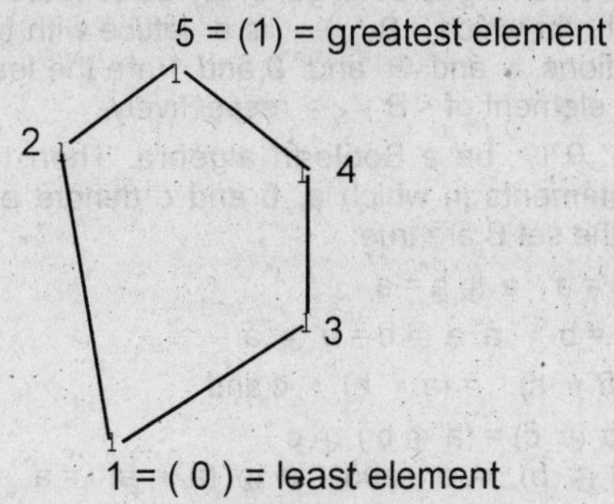

Complements Fig. 5.9

In that lattice, 2 has two complements 4 and 3.

Note : 0 and 1 are complement.

Complemented lattice : A lattice $(L,*,\oplus,0,1)$, is said to be complemented if every element a of L has atleast one complement b (that is a $*$ b = 0 and a$\oplus$b = 1).

Ex : Consider a chain of 3 elements show in Figure 5.10.

Here $2 * 3 = 2 \neq (0)$ = least element

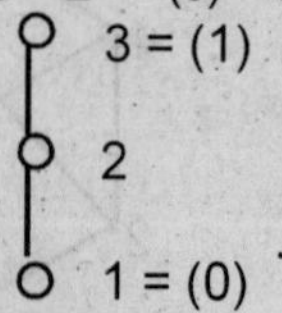

Lattice but not complemented Fig. 5.10

Distributive lattice : A lattice is distributive if

$a*(b\oplus c) = (a*b)\oplus(a*c)$ and $a\oplus(b*c) = (a\oplus b)*(a\oplus c)$

Ex : Give examples of non-distributive lattices.

Sol : See lattices shown in Figure 5.11

Proof is left as an exercise to the reader.

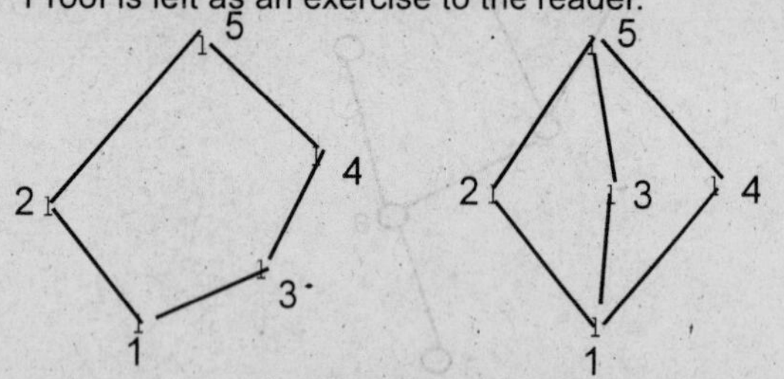

Non-distributive lattices Fig. 5.11

Note : Every chain is a distributive lattice.

Note : Cancellation law can be stated as in a distributive lattice. $(a*b = a*c)\wedge(a\oplus b = a\oplus c)\Rightarrow b = c$

Modular : A lattice is modular if

$a \le c$ then $a\oplus(b*c) = (a\oplus b)*c$

Note : We can show every distributive lattice is modular but the converse is not true.

Note : In a bounded distributive lattice, the elements which have complements form a sublattice.

3. Boolean Algebra

A complemented, distributive lattice is called Boolean Algebra. A Boolean algebra will generally be denoted by $<B,*,\oplus,',0,1>$ in which $<B,*,\oplus>$ is a lattice with two binary operations $*$ and $\oplus$ and 0 and 1 are the least and greatest element of $<B, \le>$ respectively.

Let $<B,*,\oplus,',0,1>$ be a Boolean algebra. Then the following statements in which a, b and c denote any elements of the set B are true:

1. $a * a = a$, $a \oplus a = a$
2. $a * b = b * a$, $a \oplus b = b \oplus a$
3. $a * (b * c) = (a * b) * c$ and $a \oplus (b \oplus c) = (a \oplus b) \oplus c$
4. $a * (a \oplus b) = a$ and $a \oplus (a * b) = a$
5. $a * (b \oplus c) = (a * b) \oplus (a * c)$ and $a \oplus (b * c) = (a \oplus b) * (a \oplus c)$
6. $(a*b)\oplus(b*c)\oplus(c*a) = (a\oplus b)*(b\oplus c)*(c\oplus a)$
7. $(a*b = a*c)\wedge(a\oplus b = a\oplus c)\Rightarrow b = c$

if B is a bounded lattice

8. $0 \le a \le 1$
9. $a*0 = 0$, $a\oplus 1 = 1$
10. $a*1 = a$, $a\oplus 0 = a$

If B is uniquely complemented, then

11. $a*a' = 0$, $a\oplus a' = 1$
12. $0' = 1$, $1' = 0$
13. $(a*b)' = a'\oplus b'$, $(a\oplus b)' = a'*b'$
14. $a*b$ = GLB $\{a, b\}$, $a\oplus b$ = LUB $\{a,b\}$
15. $a \le b \Leftrightarrow a * b = a \Leftrightarrow a \oplus b = b$
16. $a \le b \Leftrightarrow a * b' = 0 = b' \le a' \Leftrightarrow a' \oplus b = 1$

Ex : Let B = {0,1}. The operations, $*$, $\oplus$,$'$, are defined in Figure 5.12. The Boolean algebra is shown in Figure 5.12 (a).

*	0	1
0	0	0
1	0	1

(a)

$\oplus$	0	1
0	0	1
1	1	1

(b)

x	x′
0	1
1	0

(c)

(d)

Fig. 5.12

Ex: Prove the following Boolean identities :

a) $a\oplus(a'*b) = a\oplus b$ b) $a*(a'\oplus b) = a*b$

c) $(a*b)\oplus(a*b') = a$

Sol : a) $a\oplus(a'*b) = (a\oplus a')*(a\oplus b) = 1*(a\oplus b) = a\oplus b$

b) $a*(a'+b) = (a*a')\oplus(a*b) = 0\oplus(a*b) = a*b$

c) LHS $= a*(b\oplus b') = a*1 = a$

OBJECTIVE QUESTIONS - I

1. The less than relation, <, on reals is : **[GATE-1993]**

a) A partial ordering since it is asymmetric and reflexive

b) A partial ordering since it is anti-symmetric and reflexive

c) Not a partial ordering because it is not asymmetric and not reflexive

d) Not a partial ordering because it is not anti-symmetric and not reflexive

2. Hasse diagrams are drawn for

a) Partially ordered sets b) Lattices

c) Boolean Algebra d) None

3. A self complemented, distributive lattice is called

a) Boolean algebra b) Modular lattice

c) Complete lattice d) Self dual lattice

4. Which one of the following lattices are Boolean?

a) 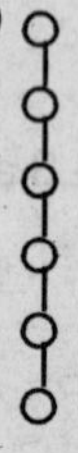 b) 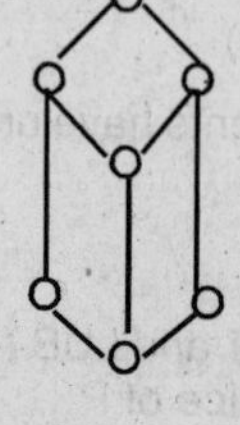 c) 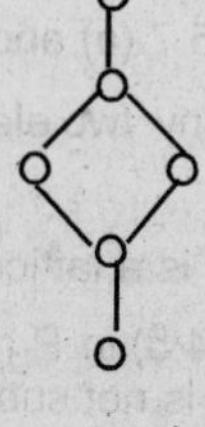 d) None

5. Which one of the following is not a "lattice" ?

a) b) c) d)

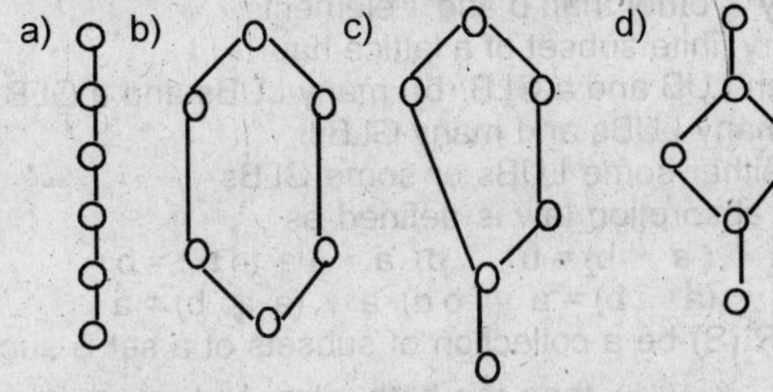

6. Let D_{30}={1, 2, 3, 5, 6, 10, 15, 30} and let the relation I be partial ordering on D_{30} . The all lower bounds of 10 and 15 are

a) 1, 3 b) 1, 5 c) 1, 3, 5 d) None

7. Let D_{30} ={ 1, 2, 3, 5, 6, 10, 15, 30} and let the relation I be a partial ordering on D_{30} The glb of 10 and 15 is

a) 3 b) 5 c) 1 d) None

8. Let D_{30}={1, 2, 3, 5, 6, 10, 15, 30} and let the relation I be a partial ordering on D_{30} The all upper bounds of 10 and 15 is

a) 30 b) 15 c) 10 d) 6

9. Let D_{30}={ 1, 2, 3, 5, 6, 10, 15, 30} and let the relation I be a partial ordering on D_{30} The lub of 10 and 15 is

a) 30 b) 15 c) 10 d) 6

10. Determine the complement of each element of D_6 in $[D_6; \vee, \wedge]$

a) $\overline{1}=2$, $\overline{2}=3$, $\overline{3}=2$, $\overline{6}=1$

b) $\overline{1}=6$, $\overline{2}=3$, $\overline{3}=6$, $\overline{6}=1$

c) $\overline{1}=6$, $\overline{6}=1$, $\overline{3}=2$, $\overline{2}=3$ d) None.

11. The maximal and minimal elements of the Poset are:

a) Maximal 5, 6; Minimal 2

b) Maximal 5, 6; Minimal 1

c) Maximal 3, 5; Minimal 1, 6

d) None

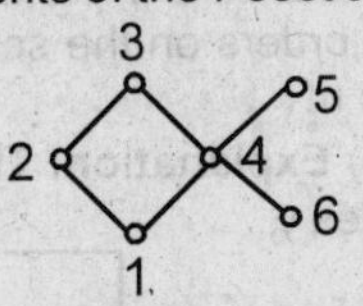

12. The greatest and least elements of the Poset are:

a) Greatest 4, 5; least 1, 2

b) Greatest 5; least 1

c) Greatest None; least None

d) None.

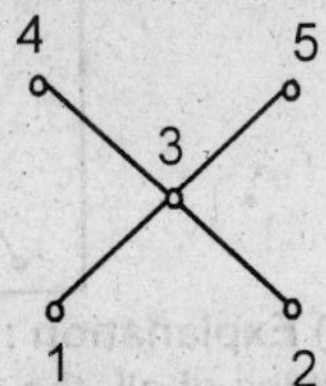

13. Let P (S) be power set of S. The poset $(P(S),\leq)$ is a lattice.

a) True b) False

14. Different partially ordered sets may be represented by the same Hasse diagram if they are

a) same b) lattices with same order

c) isomorphic d) order-isomorphic

15. If $\leq$ is a partial order then $\geq$ is also partial order. If $(x \leq y)$ then $(y \geq x)$) Hasse diagram of $(S,\geq)$ is upside down of Hasse diagram of $(S,\leq)$

a) True b) False

16. The operation meet is associative

a) True b) False

17. In a lattice $(a \geq b) \wedge (a \geq c) \Rightarrow a \geq b \oplus c$

a) True b) False

18. In a lattice, if $a \leq b \leq c$ then $a \oplus b = b * c$.

a) True b) False

19. There exists a lattice with less than four elements which is not a chain.

a) True b) False

20. If $(S,\leq)$ is a sublattice of the lattice $(L,\leq)$ then $(L-S,\leq)$ is not a sublattice of $(L,\leq)$.

a) True b) False

21. If $(S,\leq)$ is a sublattice of $(L,\leq)$ then $(S,\geq)$ is a sublattice of $(L,\geq)$.

a) True b) False

22. Let (a,b) = $\{x \in L / a \leq x,\ x \leq b\}$ The interval [a,b] is always a chain

a) True b) False

23. The inverval [a,b] is a sublattice of $(L,\leq)$

a) True b) False

24. The element 0 has unique complement 1.

a) True b) False

25. Every chain is a distributive lattice.

a) True b) False

26. In Boolean algebra, $a \leq b \Leftrightarrow a * b' = 0$ where b' is complement of b.

a) True b) False

ANSWER KEY - I

1. d	2. a	3. a	4. d	5. b
6. b	7. b	8. a	9. a	10. c
11. c	12. c	13. a	14. d	15. a
16. a	17. b	18. a	19. b	20. b
21. a	22. b	23. a	24. a	25. a
26. a				

EXPLANATORY NOTES - I

6. (b) Explanation: we know that lower bound means let $A \leq p$. An element $y \in p$ is said to be upper bound of A if $y \leq a$ of all $a \in A$. So the all lower bounds of 10 and 15 are 1,5.

7. (b) Explanation:

We know that GLB of A is Y if Y if Y is a lower bound for A and if b is any other lower bound, then $b \leq Y$. So the glb of 10 and 15 is 5.

8. (a) Explanation: We know that let $A \leq P$. An element $x \in p$ is said to be upper bound of A if $a \leq x$ of all $a \in A$. So the all upper bounds of 10 and 15 is 30.

9. (a) Explanation: We know that LUB of A is x if x is an upper bound and if Y is any other upper bound, then $X \leq Y$.

10. (c) Explanation: We know that a is compliment of b then $a = \overline{b}$ or $b = \overline{a}$ So this model of answer is c.

12. (c) Explanation: We know that Y is least element of P if $y \leq x$ for all $x \in p$ and Y is greatest element of P if $x \leq y$ for all $x \in P$. So the given figure greatest is None / least none.

26. (a) Explanation: We know that $a \leq b \Leftrightarrow a * b^1 = 0$ is true because by a known theorem.

OBJECTIVE QUESTIONS - II

1. The simultaneous equations on the Boolean variables x, y, z and w, $x+y+z=1$, $xy=0$, $xz+w=1$, $xy+\bar{z}\bar{w}=0$ Have the following solution for x, y, z and w, respectively. **[GATE-2000]**

a) 0 1 0 0 b) 1 1 0 1 c) 1 0 1 1 d) 1 0 0 0

2. Let L be a set with a relation R which is transitive, antisymmetric and reflexive and for any two elements a, b $\in$ L. Let the least upper bound lub(a, b) and the greatest lower bound glb(a,b) exist. Which of the following is/are TRUE? **[GATE-1999]**

a) L is a Poset b) L is a boolean algebra
c) L is a lattice d) None

3. In the lattice defined by the Hasse diagram given below, how many complements does the element 'e' have **[GATE-1997]**

a) 2 b) 3 c) 0 d) 1

4. A partial order $\leq$ is defined on the set $S=\{x, a_1, a_2, a_3, \ldots, a_n, y\}$ as $x \leq a_i$ for all i and $a_i \leq y$ for all i, where $n \geq 1$. The number of total orders on the set S which contain the partial order $\leq$ is **[GATE-1997]**

a) 1 b) n c) n+2 d) n!

5. Let X={2, 3, 6, 12, 24}, let $\leq$ be the partial order defined by $X \leq Y$ if X divides Y. Number of edges in the Hasse diagram of $(X, \leq)$ is **[GATE-1996]**

a) 3 b) 4 c) 5 d) None

6. What values of A, B, C and D satisfy the following simultaneous boolean equations

$\bar{A} + AB = 0$, $AB=AC$, $AB + A\bar{C} + CD = \bar{C}D$ **[GATE-1995]**

a) A=1, B=0, C=0, D=1 b) A=1, B=1, C=0, D=0
c) A=1, B=0, C=1, D=1 d) A=1, B=0, C=0, D=1

7. Let set A ={1, 2, 3, 4}. Describe the ordered pairs in the relation determined by the Hasse diagram in the figure given below

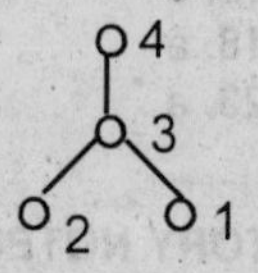

a) {(1,1), (2,2), (3,3), (4,4), (1,3), (1,4), (2,3), (2,4), (3,4)}
b) {(1,1), (2,2), (3,3), (4,4), (1,2),(1,3) (1,4), (2,3), (2,4), (3,4)}
c) Both (a) and (b) d) None

8. Determine the matrix of the partial order whose Hasse diagram is given below.

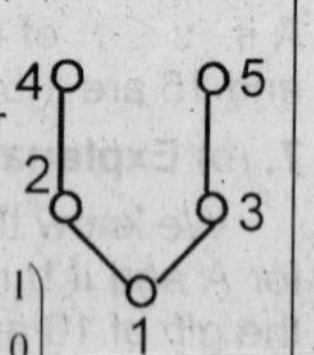

a) $\begin{pmatrix} 0 & 1 & 1 & 1 & 1 \\ 0 & 1 & 0 & 0 & 0 \\ 0 & 0 & 0 & 0 & 0 \\ 0 & 0 & 0 & 0 & 1 \\ 0 & 0 & 0 & 0 & 1 \end{pmatrix}$ b) $\begin{pmatrix} 1 & 1 & 1 & 1 & 1 \\ 0 & 1 & 0 & 0 & 0 \\ 0 & 0 & 1 & 0 & 0 \\ 0 & 0 & 0 & 1 & 0 \\ 0 & 0 & 0 & 0 & 1 \end{pmatrix}$

c) $\begin{pmatrix} 1 & 1 & 1 & 1 & 1 \\ 0 & 1 & 0 & 1 & 0 \\ 0 & 0 & 1 & 0 & 1 \\ 0 & 0 & 0 & 1 & 0 \\ 0 & 0 & 0 & 0 & 1 \end{pmatrix}$ d) None

9. Principle of duality is defined as

a) $\leq$ is replaced by $\geq$ b) LUB becomes GLB
c) all properties not altered when $\leq$ is replaced by $\geq$
d) all properties not altered when $\leq$ is replaced by $\geq$ other than 0 and 1 element

10. Every finite subset of a lattice has

a) an LUB and a GLB b) many LUBs and a GLB
c) many LUBs and many GLBs.
d) either some LUBs or some GLBs

11. The absorption law is defined as

a) $a * (a * b) = b$ b) $a * (a \oplus b) = b$
c) $a * (a\ \ b) = a \oplus b$ d) $a * (a \oplus b) = a$

12. Let R (S) be a collection of subsets of a set S such that if A, $B \in R(S)$ then the both $A \cap B$ and $A \cup B \in R(S)$. Then $(R(S), \leq)$ is a lattice. Here R(S) is called ring of subsets. **a) True** **b) False**

13. If there exists an element x of L such that complement of x is x then

a) $|L| = 2$ b) $|L| \leq 2$ c) $|L| = 1$ d) L doesn't exist

14. If the lattice $(C, \leq)$ is a complemented chain then

a) $|C| \leq 1$ b) $|C| \leq 2$ c) $|C| > 1$ d) C doesn't exist

ANSWER KEY - II

1. c	**2.** a,c	**3.** b	**4.** d	**5.** b
6. a	**7.** a	**8.** c	**9.** c	**10.** a
11. d	**12.** a	**13.** c	**14.** b	

EXPLANATORY NOTES - II

1. (c) **Explanation:** We can substitute c values all the equations are true.

2. (a,c) **Explanation:** By the definition of lattice and poset.

3. (b) **Explanation:** Clearly element e have 3 compliments.

4. (d) **Explanation:** We know that A partial order $\leq$ is defined on the set $S = \{x, a_1, a_2, a_3, \ldots, a_n, y\}$ as $x \leq ai$ for all i, and $ai \leq y$ for all i, when $n \geq 1$. The number of total orders on the set s which contain the partial order $\leq$ is n!

5. (b) **Explanation:** The following is Hasse diagram x divides y.

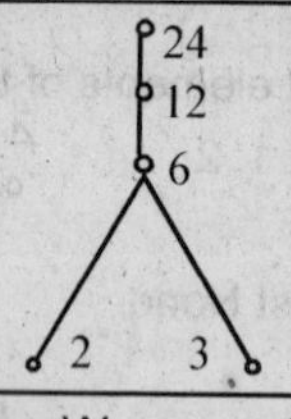

The no of edges is 4

6. (a) **Explanation :** We can substitute a values into the variables of all the equations are true

7. (a) **Explanation :** We know that in a Hasse diagram the path is only moves down to up. If their is a path then we write ordered pairs and every node has its won path.

8. (c) **Explanation :** The following is a Hasse diagram matrix

	1	2	3	4	5
1	1	1	1	1	1
2	0	1	0	1	0
3	0	0	1	0	1
4	0	0	0	1	0
5	0	0	0	0	1

In a Hasse diagram if there is path then we write 1 else we write 0. In the diagram all the nodes are won paths. So the diagonal element are 1. If 1 to 2 there is path then write one and so on.

VI. GROUP THEORY

- Introduction
- Binary Operation
- Properties of Groups

1. Introduction

'Group' is a mathematical system (having an algebraic structure). It has an immense importance in the study of abstract algebra. In the history and development of Mathematics, it has acquired its prime position.

Groups are sets with an algebraic operation. Sets under an algebraic operation satisfy certain rules which we classify and put them under the name of axioms. Most of the sets under discussion here are taken from the number systems.

We use the following notation for some standard sets

N = The set of all natural numbers (positive integers)

Z = The set of all integers

Q = The set of all rational numbers

R = The set of all real numbers

C = The set of all complex numbers

A^* = The set of all nonzero numbers of A where $A \subseteq C$

A^+ = The set of all positive numbers of A where $A \subseteq R$

A^- = The set of all negative numbers of A where $A \subseteq R$.

2. Binary operation

Definition 1 : Let A be a nonempty set. Any mapping from $A \times A$ into A is called a binary operation or binary composition on A.

Note 1 : if f is a binary operation on a nonempty set A and $(x,y) \in A \times A$ then the image f (x,y) of (x,y) under f is denoted by xfy.

Note 2 : Often we use the symobls +, ., $\circ$, * etc., to denoted binary operations on a nonempty set.

Note 3 : If * is a binary operation on a nonempty set A, then for all $x,y \in A$, $x * y$ is a unique element of A.

Definition 2 : Let * be a binary operation on a nonempty set A and $B \subseteq A$. Then B is said to be closed with respect to the binary operation * if $x * y \in B$ for all $x,y \in B$.

Note : If * is a binary operation on a nonempty set A, then A is closed with respect to * If $x,y \in A$ then $x * y$ is called the product of x and y with respect to *.

Note: If A is closed with respect to * then we also say that * is closed in A

We use the following symbols for some usual operations

+ = usual additon

− = usual subtraction

. = usual multiplication

÷ = usual division

Ex : + and . are binary operations on N, Z, Q, R and C.

3. Binary operation tables

Definition 3: If * is a binary operation on a finite set A then the products of elements of A can be represented in a table. These tables are called Cayley binary operation tables or Cayley multiplication tables or Cayley composition tables. If $x,y \in A$ then $x * y$ is shown as follows

*		y	
x		x * y	

Ex 1: Write down the binary operation table for usual multiplication '.' on the set A = {1, −1}.

Sol :

.	1	−1
1	1	−1
−1	−1	1

Ex 2 : Write down the composition table for the set $A = \{1, w, \omega^2\}$ where w is the complex cube roots of unity, under usual multiplication.

Sol:

.	1	ω	ω^2
1	1	ω	ω^2
ω	ω	ω^2	1
ω^2	ω^2	1	ω

4. Types of binary operations

Definition 4 : A binary operation * on a nonempty set A is said to be associative if $(x * y) * z = x * (y * z)$, $\forall$ x, y, z $\in$ A.

Ex: + and . are associative binary operations on N, Z, Q, R and C.

Definition 5 : A binary operation * on a nonempty set A is said to be commutative if $x * y = y * x$, $\forall x, y \in A$.

Ex : + and . are commutative binary operations on N, Z, Q, R and C.

Ex 4: Which of the following are binary operations on N. Also verify which binary operations are associative and commutative

I. $a * b = a + b$ **II.** $a * b = a - b$ **III.** $a * b = a + b + 1$ **IV.** $a * b = ab$ **V.** $a * b = a \div b$ **VI.** $a * b = ab + 1$ **VII.** $a * b = a + b + ab$.

Sol : I. If a, b are natural numbers, then a + b is a natural number.

$\therefore$ * is a binary operation on N in (I).

Addition of natural numbers is both associative and commuative.

II. * is not a binary operation on N, since $3 \in N$, $5 \in N$; but $3 * 5 = 3 - 5 \notin N$.

III. If a, b are natural numbers, then a + b + 1 is also a natural number.

$\therefore$ * is a binary operation on N in (III).

a ∗ b = a + b + 1 = b + a + 1 = b ∗ a. ∴ ∗ is commutative.

(a ∗ b) ∗ c=(a + b + 1) ∗ c=a + b + 1 + c+ 1 = a + (b + c + 1) + 1 = a ∗ (b + c + 1) = a ∗ (b ∗ c). ∴ ∗ is associative.

IV. If a, b are natural numbers, then ab is a natural number.

∴ ∗ is a binary operation on N in (IV).

Multiplication of natural numbers is both associative and commutative.

V. ∗ is not a binary operation in (V), since 3 ∈ N, 5 ∈ N; but 3 ∗ 5 = 3 ÷ 5 ∉ N.

VI. If a, b are natural numbers then ab + 1 is a natural number

∴ ∗ is a binary operation on N in (VI).

a ∗ b = ab + 1 = ba + 1 = b ∗ a. ∴ ∗ is commutative.

(a ∗ b) ∗ c = (ab + 1) ∗ c = (ab + 1) c + 1 = abc + c + 1.

a ∗ (b ∗ c) = a ∗ (bc + 1) = a (bc + 1) + 1 = abc + a + 1.

∴ ∗ is not associative.

VII. If a, b are natural numbers, then a + b + ab is also a natural number.

∴ ∗ is a binary operation on N.

a ∗ b = a + b + ab = b + a + ba = b ∗ a. ∴ ∗ is commutative.

(a ∗ b) ∗ c = (a + b + ab) ∗ c = a + b + ab + c + (a + b+ ab) c

= a + b + c + ab + bc + ac + abc

a ∗ (b ∗ c) = a ∗ (b + c + bc) = a + b + c + bc + a (b + c + bc)

= a + b + c + ab + bc + ac + abc.

∴ ∗ is associative.

5. Algebraic systems or algebraic structures

Definition 6 : If A is a nonempty set and B is a set of binary operations on A then (A, B) is calleld an algebraic system or an algebraic structure.

Note : If ∗ is a binary operation on A then (A, ∗) is an algebraic system.

Ex : (N , +), (N, .), (Z, +), (Z, .), (Q, +), (Q, .), (R, +), (R, .) are algebraic systems

6. Semigroups

Definition 7 : Let S be a nonempty set and ∗ is a binary operation on S. Then the algebraic system (S, ∗) is said to be a semigroup if a ∗ (b ∗ c) = (a ∗ b) ∗ c, ∀ a,b,c ∈ S.

Note: If **S** is a nonempty set and ∗ is an associative binary operation on **S** then (**S**, ∗) is a semigroup.

Note: If (**S**, .) is a semigroup then the element a. b is denoted by ab.

Note: If (**S**, ∗) is a semigroup then we say that **S** is a semigroup under ∗ or a semigroup with respect to ∗ .

Ex: (N, +), (N, .), (Z, +), (Z, .), (Q, +), (Q, .), (R, +), (R, .), (C, +), (C, .) are all semigroups.

Definition 8 : Let (S, ∗) be a semigroup. An element e ∈ S is said to be

I. a left identity of S if e ∗ a = a, ∀ a ∈ S

II. a right identity of S if a ∗ e = a, ∀ a ∈ S

III. a two sided identity or simple identity of S if a ∗ e = e ∗ a = a, ∀ a ∈ S.

Ex: The semigroups (Z, +), (Q, +), (R, +), (C, +) contain '0' as identity. The semigroups (N, .), (Z, .), (Q, .), (R, .), (C, .) contain '1 ' as identity.

Theorem 1 : Let (S, ∗) be a semigroup. If e is a left identity and f is a right identity of S then e = f.

Theorem 2 : If (S, ∗) is a semigroup with identity, then identity in S is unique.

Definition 9 : A semigroup with identity is called a monoid.

Ex: (Z, +) is a monoid and (N, +) is a semigroup but not a monoid.

Definition 10: Let (S, ∗) be a semigroup with identity

I. An element a ∈ S is said to be left invertible if there exists an element b ∈ S such that b ∗ a = e. The element b is called left inverse of a.

II. An element a ∈ S is said to be right invertible if there exists an element c ∈ S such that a ∗ c = e. The element c is called right inverse of a.

III. An element a ∈ S is said to be invertible if a is both left invertible and right invertible.

Ex 1 : - 1 is invertible in the semigroup (**Z**, .).

Theorem 3 : Let (S, ∗) be a semigroup with identity e and a ∈ S. If b is a left inverse and c is a right inverse of a then b = c.

Definition 11 : Let (S, ∗) be a semigroup with identity e and a ∈ S. If a is invertible then the element which is both left inverse and right inverse of a is called inverse of a.

Theorem 4 : Let (S, ∗) be a semigroup with identity e. If a is an invertible element then the inverse of a is unique.

Note : Let (S, ∗) be a semigroup with identity e. If a is an invertible element in S then the inverse of a is denoted by a^{-1}. Now a ∗ a^{-1} = a^{-1} ∗ a = e.

Theorem 5 : Let (S, ∗) be a semigroup with identity e. If a is an invertible element then a^{-1} is also invertible and $(a^{-1})^{-1} = a$.

Theorem 6: Let (S, ∗) be a semigroup with identity e. If a,b are two invertible elements in S then a ∗ b is also ivertible and $(a*b)^{-1} = b^{-1}*a^{-1}$

Ex 9 : Show that $S\left\{\begin{bmatrix} a & b \\ 0 & 0 \end{bmatrix} : a,b \in R\right\}$ is a semigroup under matrix multiplication. Verity that $\begin{bmatrix} 1 & 0 \\ 0 & 0 \end{bmatrix}$ is a left identify in S.

Sol : Let $\begin{bmatrix} a & b \\ 0 & 0 \end{bmatrix}, \begin{bmatrix} c & d \\ 0 & 0 \end{bmatrix} \in S$

$$\begin{bmatrix} a & b \\ 0 & 0 \end{bmatrix}, \begin{bmatrix} c & d \\ 0 & 0 \end{bmatrix} = \begin{bmatrix} ac & ad \\ 0 & 0 \end{bmatrix} \in S$$

∴ Matrix multiplication is a binary operation on S. Since matrix multiplication is associative (S, .) is a semigroup.

Let $\begin{bmatrix} a & b \\ 0 & 0 \end{bmatrix} \in S$

$$\begin{bmatrix}1&0\\0&0\end{bmatrix}\cdot\begin{bmatrix}a&b\\0&0\end{bmatrix}=\begin{bmatrix}a&b\\0&0\end{bmatrix}$$

$\therefore \begin{bmatrix}1&0\\0&0\end{bmatrix}$ is a left identiy of S.

Ex 10 : Let S be a nonempty set and P (S)be the power set of S. Show that $(P(S), \cup)$, $(P(S), \cap)$ are monoids.

Sol : Let A, B $\in$ P(S). Now $A \subseteq S$, $B \subseteq S$ and therefore $A \cup B \subseteq S$, $A \cap B \subseteq S$.

Hence $A \cup B \in P(S)$, $A \cap B \in P(S)$.

We know that $(A \cup B) \cup C = A \cup (B \cup C)$,

$(A \cap B) \cap C = A \cap (B \cap C)$, $\forall$ A,B,C $\in$ p(S).

Further $A \cup \phi = \phi \cup A = A$ and $A \cap S = S \cap A = A$.

$\therefore$ $(P(S), \cup)$ and $(P(S), \cap)$are monoids.

Ex 11 : Let S be a non empty set. Define '.' on S as xy $\in$x, $\forall$ x,y $\in$S. Then (S, .) is a semigroup. This semigroup is called left zero semigroup. In this semigroup S and every element is a right identity of S.

Ex 12 : Let S be a non empty set. Define '.' on S as xy $\in$y, $\forall$ x,y $\in$ S. Then (S, .) is a semigroup. This semigroup is called right zero semigroup. In this semigroup S and every element is a right identity of S.

Ex 13 : Let S = The set of all odd integers. Show that (S, .) is a semigroup with identity (monoid)

Sol : Now S = $\{2n+1 : n \in z\}$. Let 2m+1, 2n+1 $\in$ S

(2m + 1) (2n + 1) = 2(2mn+ m + n) + 1 $\in$ S.

Since multiplication is associative in Z, multiplication is associative in S.

$\therefore$ (S, .) is a semigroup and clearly 1 is the identity of S.

Definition 12 : A semigroup (S, $*$) is said to be

I. left cancellative semigroup if a, b, c $\in$ S, a $*$ b = a $*$ c $\Rightarrow$ b=c.

II. right cancellative semigroup if a, b, c $\in$ S, b $*$ a = c $*$ a $\Rightarrow$ b= c

III. cancellative semigroup if it is both left cancellative and right cancellative semigroup.

Note 1 : a $*$ b = a $*$ c $\Rightarrow$ b= c is called left cancellation law.

Note 2 : b $*$ a = c $*$ a $\Rightarrow$ b= c is called right cancellation law.

Ex : $(Z,+)$, $(Z^*,.)$, $(Q,+)$, $(Q^*,.)$, $(R,+)$, $(R^*,.)$ are cancellative semigroups.

7. Group

A semigroup with identity and in which every element is invertible is called a group. In other words

Definition 1 : Let G be a nonempty set and $*$ be a binary operation on G. Then the algebraic system (G, $*$) is said to be a group if

1. $a*(b*c)=(a*b)*c$, $\forall a,b,c \in G$

2. $\exists e \in G \ni a*e = e*a = a, \forall a \in G$

3. $a \in G \Rightarrow \exists b \in G \ni a*b = b*a = e$

Note 1 : The element e in condition 2 of the definition of a group is called identity element of the group.

Note 2 : The element b corresponditing to a in conditiion 3 of the definition of a group is called inverse of a in the group.

Definition 2 : A group (G, $*$) is said to be an abelian group or commutative group if a $*$ b = b $*$ a, $\forall$ a, b $\in$ G.

Ex : (Z, +), (Q, +), (R, +), (C, +) are all abelian groups.

Ex 1 : Let $G = Q - \{1\}$ If $*$ is defined on G as a $*$ b = a + b - ab then show that (G, $*$) is an abelian group.

Sol : Clearly G is nonempty.

Let a, b $\in$ G.

$a \in G \Rightarrow a \in Q, a \neq 1$

$b \in G \Rightarrow b \in Q, b \neq 1$

$a \in Q, b \in Q \Rightarrow a+b \in Q, ab \in Q \Rightarrow a+b-ab \in Q \Rightarrow a*b \in Q$

If $a * b = 1$ then $a+b-ab = 1 \Rightarrow a-1+b-ab = 0$

$\Rightarrow 1(a-1) - b(a-1) = 0 \Rightarrow (a-1)(1-b) = 0 \Rightarrow a = 1$ or $b = 1$

A contradiction. $\therefore$ a $*$ b $\neq$ 1.

$a * b \in Q, a * b \neq 1 \Rightarrow a * b \in G$.

$\therefore$ $*$ is a binary operation on G.

Let a, b, c $\in$ G.

a $*$ (b $*$ c) = a $*$ (b + C -bc) = a + (b + c - bc) - a(b + c - bc)

= a + b + c - bc - ab - ac + abc.

$$(a*b)*c = (a+b-ab)*c = a+b-ab+c-(a+b-ab)c$$
$$= a+b+c-ab-ac-bc+abc$$

$\therefore a*(b*c) = (a*b)*c$

$\therefore$ $*$ is an associative binary operation on G.

Let e be the identity element of G.Then $a * e = a, \forall a \in G$

$$a*e = a \Rightarrow a+e-ae = a \Rightarrow e(1-a) = 0 \Rightarrow e = 0$$

$\therefore$ e = 0 $\in$ G is the identity element in G.

Let $a \in G$. If b is the inverse of a then

$a * b = 0 \Rightarrow a + b - ab = 0$

$$\Rightarrow b(a-1) = a \Rightarrow b = \frac{a}{a-1} \in G$$

$\therefore$ $b = \dfrac{a}{a-1} \in G$ is the inverse of a in G.Let

$a, b \in G$. Then $a * b = a + b - ab = b + a - ba = b * a$

$\therefore$ $*$ is a cumulative in G. $\therefore (G,*)$ is an abelian group.

Definition 3 : A group (G, $*$) is said to be a nonabelian group if (G, $*$) is not abelian.

8. Some elementary propperties of a group

Theorem 1 : In a group, identity element is unique.

Theorem 2 : In a group, inverse of every element is unique,

Theorem 3 : Let (G,.) be a group and $a \in G$. Then

$(a^{-1})^{-1} = a$

Theorem 4: Let (G,.) be a group and $a, b \in G$. Then

$(ab)^{-1} = b^{-1} a^{-1}$

Theorem 5 : Let (G,.) be a group and

$a_1, a_2, ..., a_n \in G$. Then $(a_1 a_2 .....a_n)^{-1} = a_n^{-1}\, a_{n-1}^{-1}\, ...\, a_1^{-1}$

Theorem 6 : Cancellation laws hold in a group. i.e., If (G,.) is a group then

I. $a,b,c \in G, ab = ac \Rightarrow b = c$

II. $a,b,c \in G, ba = ca \Rightarrow b = c$.

Definition 4 : Let (G,.) be a group and $a \in G, n \in Z$. Then a^n is defined as follows

I. $a^0 = e$ II. If $n > 0$ then $a^1 = a, a^{n+1} = a^n.a$

III. If $n < 0$ then $a^n = (a^{-n})^{-1}$

Theorem 7 : Let (G,.) be a group and $a \in G$ If $m,n \in Z$.

Then I. $a^m.a^n = a^{m+n} = a^n a^m$ II. $(a^m)^n = a^{mn}$

Ex 6: If (G,.) is an abelian group then $ab^n = a^n b^n$, $\forall a,b \in G$ where $n \in Z$

Ex 7 : If (G,.) is a group such that $a^{-1} = a, \forall a \in G$ then (G,.) is an abelian group.

Definition 5 : Let (G,.) be a group. An element $a \in G$ is said to bean idempotent if $a^2 = a$.

Ex 8 : Let (G,.) be a group. Show that an element $a \in G$ is an idempotent iff $a = e$

Definition 6 : A group (G,.) is said to be a finite group if G is a finite set. The number of different elements in G is called the order of the finite group (G,.). It is denoted by o(G).

Definition 7 : A group (G,.) is said to be an infinite group if G is an infinite set. The order of an infinite group is defined to be ∞.

Ex 9 : Let $G = \{1, -1, i, -i\}$ where $i^2 = -1$. Show that G is a finite abelian group under multiplications of complex numbers.

Sol : The composition table for multiplication is as follows.

.	1	-1	i	-i
1	1	-1	i	-i
-1	-1	1	-i	i
i	i	-i	-1	1
- i	-i	i	1	-1

We Know the multiplication of complex numbers is associative and commutative. By observing the table it follows that 1 is the identity element in G and the inverses of 1, -1,i,-i are 1,-1,-i,i respectively. Thus (G,.) is a finite abelian group.

Definition 8 : If S is a finite set containing n elements, then the group of all bijections on S is called a permutation group or symmetric group. It is denoted by P_n or S_n.

Note : $O(S_n) = n!$

Ex 10 : If (G, .) is a group such that $(ab)^n = a^n b^n$ for three consecutive positive integers, $\forall a,b \in G$ then show that (G,.) is an abelian group.

9. Residue classes

Definition 1 : Let n be a positive integer and $a,b \in Z$. Then a is said to be congruent to b modulo n if n divides a - b. It is denoted by $a \equiv b \pmod n$.

Ex 11 : 5 divides $32 - 2 \Rightarrow 32 \equiv 2$

9 divides $-47 - 7 \Rightarrow -47 \equiv 7$

Note : The relation in the above definition is referred to as congruent modulo n.

Theorem 1 : The relation congruent modulo n is an equivalence relation on Z.

Definition 2 : The equivalence classes under the relation congruent modulo a positive integer n on Z are called residue classes modulo n.

Note 1 : The residue class containing an integer a is denoted by [a] or $\overline{a}$.

Note 2 : The set of all residue classes modulo a positive integer n is denoted by Z_n.

Theorem 2 : If n is a positive integer then $Z_n = \{\overline{0}, \overline{1}, \overline{2}, ..., \overline{n-1}\}$ and $O(Z_n) = n$.

Theorem 3 : If $a \equiv b \pmod n$, $c \equiv d \pmod n$ then $a + c \equiv b + d \pmod n$ and $ac \equiv bd \pmod n$.

10. Residue class addition

Definition 3 : The operation $\oplus$ defined on Z_n as $\overline{a} \oplus \overline{b} = \overline{a+b}$, $\forall \overline{a}, \overline{b} \in Z_n$ is called residue class addition or addition modul n. It is also denoted by $\oplus_n$ or $+_n$.

Note : If $\overline{a}, \overline{b} \in Z_n$ then $\overline{a} \oplus \overline{b} = \overline{r}$ where r is the remainder of a + b when divided by n.

Note : The operation, addition modulo n is also defined in {0, 1, 2, ... ,n -1} as $a \oplus b = r$ where r is the ramainder of a + b when divided by n.

Ex : $\overline{5} \oplus_6 \overline{4} = \overline{9} = \overline{3}$, $4 \oplus_7 6 = 3$

Theorem 4 : $(Z_n, \oplus)$ is an abelian group.

11. Residue class multiplication

Definition 4: The operation O defined cn Z_n as $\overline{a} O \overline{b} = \overline{ab}$, $\forall \overline{a}, \overline{b} \in Z_n$ is called residue class multiplication or multiplication modulo n. It is also denoted by O_n or $\cdot_n$ or $\otimes$ or $\otimes_n$ or $\times_n$.

Note : If $\overline{a}, \overline{b} \in Z_n$ then $\overline{a} O \overline{b} = \overline{r}$ where r is the remainder of ab when divided by n.

Note : The operation, multiplication modulo n is also defined in $\{0,1,2,...,n-1\}$ as $a O b = r$ where r is the remainder of ab when divided by n.

Ex : $\overline{5} O_6 \overline{4} = \overline{20} = \overline{2}, 4 O_7 6 = 3$

Theorem 5 : (Z_n, O) is a commutative semigroup with identity.

Theorem 6 : $(Z'_n, \cdot)$ is an abelian group iff n is prime.

OBJECTIVE QUESTIONS - I

1. The matrices $\begin{bmatrix} \cos\theta & -\sin\theta \\ \sin\theta & \cos\theta \end{bmatrix}$ and $\begin{bmatrix} a & 0 \\ 0 & b \end{bmatrix}$ Commute under the multiplication **[GATE 1996]**

a) If a = b (or) $\theta = n\pi$,n is an integer

b) Always c) Never d) If a Cos $\theta \neq$ b Sin θ

2. The set of all Real numbers under the usual multiplication operation is not a group since **[JNTU-1993]**

a) Multiplication is not a binary operation
b) Multiplication is not associative
c) Identity element does not exist
d) Zero has no inverse

3. The set of integers Z with the binary operation '∗' defined as a∗ b= a + b + 1 for a, b $\in$ Z, is a group. The identity element of this group is **[JNTU-1996]**

a) 0 b) 1 c) –1 d) 12

4. If '∗' is defined on R* as a∗ b= $\frac{ab}{2}$ then the identity element in the group (R*, ∗) is

a) 1 b) 2 c) ½ d) 1/3

5. The identity element in the group $G=\left\{\begin{bmatrix} x & x \\ x & c \end{bmatrix} : x\in R,\ x\neq 0\right\}$ With respect to matrix multiplication is **[JNTU-1995]**

a) $\begin{pmatrix} 1 & 0 \\ 0 & 1 \end{pmatrix}$ b) $\begin{pmatrix} -1 & 0 \\ 0 & -1 \end{pmatrix}$

c) $\begin{pmatrix} \frac{1}{2} & \frac{1}{2} \\ \frac{1}{2} & \frac{1}{2} \end{pmatrix}$ d) $\begin{pmatrix} 1 & 1 \\ 1 & 1 \end{pmatrix}$

6. The inverse of –i in the multiplicative group{1,-1, i, -i} is

a) 1 b) –1 c) i d) –i

7. (Z, ∗) is a group with a ∗ b = a + b + 1,va,b $\in$ Z. The inverse of a is **[JNTU-1990]**

a) 0 b) –2 c) a-2 d) –a-2

8. In the group (G, .), the value of $(a^{-1} b)^{-1}$ =

a) a b^{-1} b) b^{-1} c) $a^{-1}b$ d) b a^{-1}

9. If (G, .) is a group such that a^2=e,va $\in$G then G is

a) A semi group b) An ahelian group
c) A non-abelian group d) None

10. If (G, .) is a group, such that $(ab)^2 = a^2 b^2$va,b $\in$ G then G is

a) A commutative semi group
b) An abelian group
c) A non-abelian group d) None

11. If (G, .) is a group such that $(ab)^{-1} = b^{-1} a^{-1}$, va,b $\in$ G then G is

a) A commutative semi group
b) An abelian group
c) A non-abelian group d) None

12. Let G denoted the set of all **n×n** non-singular matrices with matrices with rational numbers as entries. Then under multiplication

a) G is a subgroup b) G is a finite abelian group
c) G is an infinite, non abelian group
d) G is infinite, abelian

ANSWER KEY - I				
1. a	2. d	3. b	4. a	5. c
6. c	7. d	8. b	9. b	10. b
11. b	12. c			

EXPLANATORY NOTES - I

3. (b) **Explanation** : In a groups Z with a multiplication identity is always 1. Identity means Let e be the identity element then $a*e = e*a = a$

$\therefore$ The identity element of group is $(a+b+1)\times 1$

$= 1\times(a+b+1) = a+b+1$

4. (a) **Explanation :**

Let e be the identity element then $a*e = e*a = a$.

The identity element of the group is

$$\frac{ab}{2}\times 1 = 1\times\frac{ab}{2} = \frac{ab}{2}$$

5. (a,c) **Explanation** : The matrix multiplication identity is $\begin{bmatrix} 1 & 0 \\ 0 & 1 \end{bmatrix}$ $\therefore AI = IA = A$

$\therefore \begin{bmatrix} x & x \\ x & x \end{bmatrix}\begin{bmatrix} 1 & 0 \\ 0 & 1 \end{bmatrix} = \begin{bmatrix} 1 & 0 \\ 0 & 1 \end{bmatrix}\begin{bmatrix} x & x \\ x & x \end{bmatrix} = \begin{bmatrix} x & x \\ x & x \end{bmatrix}$ and

$$\begin{bmatrix} x & x \\ x & x \end{bmatrix}\begin{bmatrix} \frac{1}{2} & \frac{1}{2} \\ \frac{1}{2} & \frac{1}{2} \end{bmatrix} = \begin{bmatrix} \frac{1}{2} & \frac{1}{2} \\ \frac{1}{2} & \frac{1}{2} \end{bmatrix}\begin{bmatrix} x & x \\ x & x \end{bmatrix} = \begin{bmatrix} x & x \\ x & x \end{bmatrix}$$

6. (c) **Explanation** : Let e be the identity element then $a*e = e*a = a$

$a*a^{-1} = a^{-1}*a = e$. Therefore a^{-1} is inverse of a. Given problem multiplicative identity is the 1.

i is the inverse of -i because

$i * -i = -i * i = 1$

7. (d) **Explanation** :Let e be the identity element then $a*e = e*a = a$

$a * a^{-1} = a^{-1} * a = e$ Therefore a^{-1} is inverse of a

Given problem multiplicative identity is the 1.

$a+b+1$ is the inverse of $\frac{1}{a+b+1}$ because

$a+b+1\times\frac{1}{a+b+1} = \frac{1}{a+b+1}\times a+b+1 = 1$

8. (b) **Explanation** : We know that a theorem $(G,.)$ is group then $(ab)^{-1} = b^{-1}a^{-1}$

$\therefore$ we can find out a theorem

$\therefore (a^{-1}b)^{-1} = b^{-1}(a^{-1})^{-1} = b^{-1}a$

9. (b) **Explanation** : $(G,.)$ is a abelian group then it satisfy following properties.

1. Associative 2. identity
3. inverse 4. commutative

Clearly given problem is associative. $a^2 = e$

clearly given problem is inverse.

So compulsorily consists identity element.

$a.a = a^2 = e = a^2 = a.a$

$\therefore$ it satisfies the commutative property

So this is an abelian group.

10. (b) **Explanation :** $(G,.)$ is abelian group then it satisfies following properties.

1. Associative 2. Identity
3. Inverse 4. Commutative

$(ab)^2 c^2 = a^2b^2c^2 = a^2(bc)^2$

$\therefore$ given problem is Associative.
Multiplicative identity is always 1.

The inverse of a^2b^2 is $\frac{1}{a^2b^2}$

i.e. $a^2b^2 \times \frac{1}{a^2b^2} = \frac{1}{a^2b^2} \times a^2b^2 = 1$

it satisfies commutative property also.

$(ab)^2 = a^2b^2 \; a^2 = (ba)^2$

11. (b) **Explanation :** $(G,.)$ is abelian group then its satisfies following properties

1. Associative 2. Identity
3. Inverse 4. Commutative

Clearly given problem is associative.
e be the identity element

$(a.a^{-1})^{-1} = a^{-1}(a^{-1}) = a^{-1} \; a = e$

$\therefore$ it satisfies the inverse

$(a.e^{-1})^{-1} = e^{-1}a^{-1} = a^{-1}.e = a^{-1}e^{-1} = (ea)^{-1}$

$\therefore$ it is satisfies commutative properties.

OBJECTIVE QUESTIONS - II

1. Which of the following is TRUE? **[GATE-2002]**
a) The set of all rational negative numbers forms a group undermultiplication
b) The set of all non-singular matrices forms a group under multiplication
c) The set of all matrices forms a group under multiplication
d) Both (b) and (c) are true.

2. Let (Z, ∗) be an algebraic structure, where Z is the set of integers and the operation ∗ is defined by n ∗ m =maximum (n, m). which of the following statements is TRUE for (Z, ∗) ? **[GATE-1997]**
a) (Z, ∗) is a monoid b) (Z, ∗) is an abelian group
c) (Z, ∗) is a group d) None

3. Which of the following statements is FALSE? **[GATE-1996]**
a) The set of rational numbers is an abelian group under addition
b) The set of rational integers is an abelian group under addition
c) The set of rational numbers form an abelian group under multiplication. d) None

4. Let A be the set of all non-singular matrices over real numbers and let ∗ be the matrix multiplication operator. Then **[GATE-1995]**
a) A is closed under ∗ but <A, ∗ > is not a semi group
b) <A, ∗ > is a semi group but not a monoid
c) <A, ∗ > is a monoid but not a group
d) <A, ∗ > is a group but not an abelian group

5. Some group (G, o) is known to be abelian. Then, which one of the following is TRUE for G? **[GATE-1994]**
a) $g = g^{-1}$ for every $g \in G$ b) $g = g^2$ for every $g \in G$
c) $(g \circ h)^2 = g^2 \circ h^2$ for every $g,h \in G$
d) G is of finite order

6. Match the following **[GATE-1990]**

A. Groups	I. Associativity
B. Semi groups	II. Identity
C. Monoids	III. Commutative
D. Abelian Groups	IV. Left inverse

a) A-IV, B-I, C-II, D-III b) A-III, B-I, C-IV, D-I

7. The set of all n^{th} roots of unity, under multiplication of complex numbers form
a) A semi group with identity
b) A commutative semigroup with identity
c) A group d) An abelian group

8. In the group G = { 2, 4, 6, 8 } under multiplication modulo 10, the identity element is **[JNTU-1994]**
a) 6 b) 8 c) 4 d) 2

9. If a,b are positive integers, define $a * b = \alpha$ where $ab \equiv \alpha$ (modulo 7), with this ∗ operation, the inverse of 3 in group G { 1, 2, 3, 4, 5, 6} is **[JNTU-1993]**
a) 3 b) 1 c) 5 d) 4

10. If the binary operation ∗ is defined on a set of ordered pairs of real numbers as $(a,b)*(c,d) = (ad+bc, bd)$ and is associative, then $(1,2)*(3,5)*(3\,4)$ = **[JNTU-1991]**
a) (74, 40) b) (32, 40) c) (23, 11) d) (7, 11)

11. If every element of a group G is its own inverse, then G is
a) finite b) infinite c) cyclic d) abelian

12. A subset H of a group $(G,*)$ is a group iff
a) $a,b \in H \Rightarrow a * b \in H$ b) $a \in H \Rightarrow a^{-1} \in H$
c) $a,b \in H \Rightarrow a * b^{-1} \in H$
d) H contains the identity element

13. G { e, a, b, c} is an abelian group with 'e' as identity element. The order of the other elements are
a) 2, 2, 3 b) 3, 3, 3 c) 2, 2, 4 d) 2, 3, 4

ANSWER KEY - II

1. d	2. d	3. c	4. d	5. c
6. a	7. d	8. a	9. c	10. a
11. d	12. c	13. a		

EXPLANATORY NOTES - II

1. (d) **Explanation:** $(Z,*)$ is not a monoid. So $(Z,*)$ is not a group and $(Z,*)$ is not a abelian group.

5. (c) **Explanation :** (G,o) is known to be abelian group. So $(goh)^2 = g^2 \; oh^2$ for every $g, h \in G$ is true.

8. (a) **Explanation :** In the group $G = \{2, 4, 6, 8\}$ under multiplication modulo 10, the identity element is 6.

9. (c) **Explanation :** if a,b are positive integers define $a*b = \infty \;\; ab = \infty$ (modulo r), with this ∗ operation, the inverse of 3 in group $G\{1,2,3,4,5,6\}$ is 5.

10. (a) **Explanation :** $(1,2)*(3,5)*(3,4) = (3+10, 10)*(3,4)$

$= (39+40, 40) = (79, 40)$

VII. GRAPH THEORY

- Introduction
- Bipartite Graphs
- Operations on Graphs
- Trees and Connectivity
- Covering and Matching

1. Graph

A Graph is a drawing or a diagram consisting of a collection of vertices together with edges joining certain pairs of these vertices. Mathematically , we can write,

A graph G = [V(G), E(G)]

Where V(G) and E(G) are two finite sets defined as

V(G) = vertex set of graph G

E(G) = Edge set of graph G such that each element e of E(G) is assigned an unordered pair of vertices (u,v) called end vertices of e.

Vertices are also some times called points or nodes . If e an edge with end vertices u and v then e is said to join u and v.. It is also possible to have a vertex u joined to itself by an edge, such an edge is called a loop.

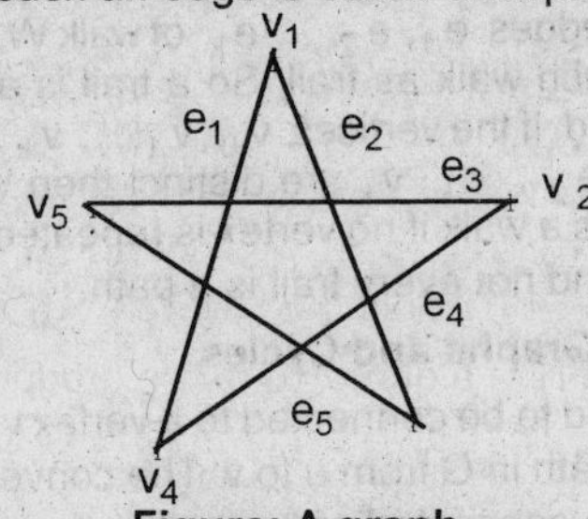

Figure: A graph

2. Some definitions

We here by give some definitions to make you understand some of the basic concepts.

Parallel edges : If two (or more) edges of a graph G have the same end vertices then these edges are called Parallel.

For example

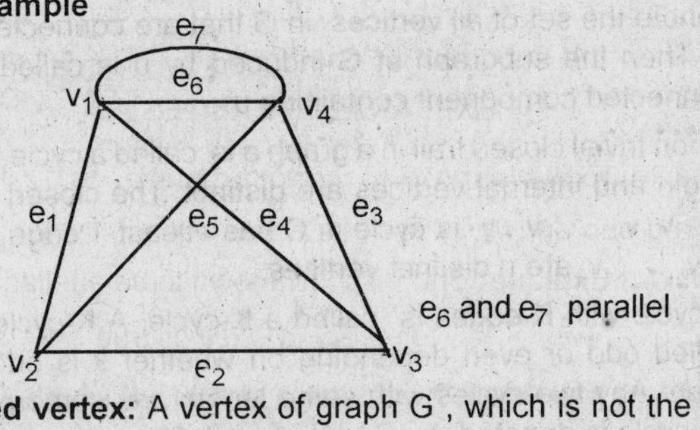

Isolated vertex: A vertex of graph G , which is not the end of any edge is called Isolated.

Neighbours or adjacent Vertex: Two vertices which are joined by an edge are called 'adjacent' or 'neighbours 'The set of all such neighbours of vertex V is called Neighbourhood set of V and is denoted by N(v).

Incidence: An edge e of a graph G is said to be incident with the vertex v if v is an end vertex of e (or v is incident with e). Two edges e and f which are incident with a common vertex v are said to be adjacent. An edge could be incident with either 1 or 2 vertices (1, in case of loop), whereas a vertex could be incident on any finite number of edges.

Degree: The degree d(v) of any vertex v of graph G is the number of edges of G incident with v. Each loop is counted twice or it is the number of times v is an end vertex of an edge.

Theorem 1 : The First Theorem of Graph theory: For any graph G with e edges and n vertices v_1......v_n

$$\sum_{i=1}^{n} d(v_i) = 2e$$

Odd or Even vertex: A vertex of a graph is called odd or even depending on whether its degree is even or odd.

Regular Graph : If for some positive integer k, d(v) = K for every vertex v of the Graph G, then G is called k-regular.

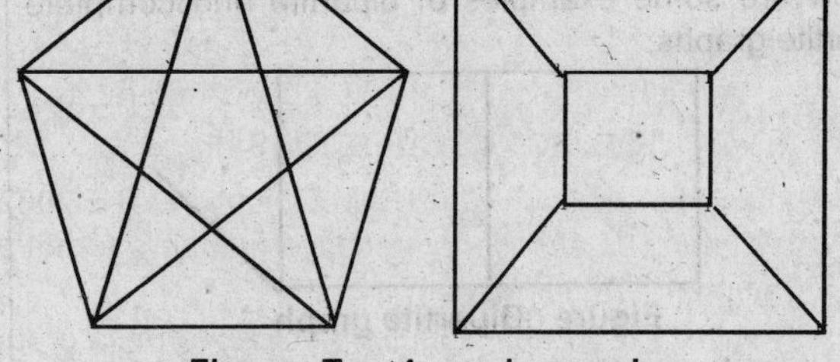

Figure : Two 4-regular graphs

Complete Graph : A complete graph K_n is a simple graph in which each pair of distinct vertices is joined by an edge.

Empty or Trivial Graph : A graph with no edges.

3. Graph Isomorphism

Some graphs have the same structure i.e., same vertices and edges but differ only in the way of representation. This property of commonness is called Isomorphism.

A graph G_1 = (V_1,E_1) is said to be isomorphic with G_2= (V_2,E_2) if there is a one to one correspondence between the edge sets E_1 and E_2 in such a way that if e_1 is an edge with end vertices u_1 and v_1 in G_1 then the corresponding edge e_2 in G_2 has its end points the vertices u_1 and v_2 in G_2 which correspond to u_1 and v_1 respectively. Such a pair of correspondences is called a graph isomorphism. It is basically a possible redrawing of the graphs. Often it is quite difficult to determine whether the two graphs are isomorphic or not. The problem becomes even more harder as the number of vertices and edges of the graphs get larger. A necessary condition for two graphs to be isomorphic is that they should have same number of vertices and same number of edges.

We hereby give some examples of isomorphic graphs.

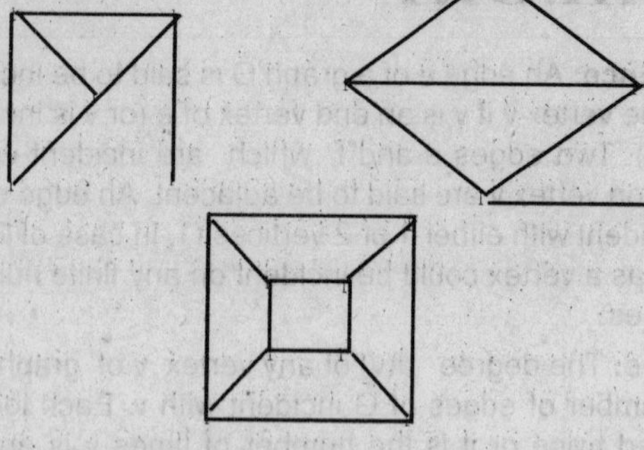

Figure : Some groups of isomorphic graphs

4. Bipartite Graphs

If the vertex set V of a graph G can be partitioned into two non-empty disjoint subsets X and Y in such a way that each edge of G has one end in X and one end in Y then G is called 'bipartite'. The partition $V = X \cup Y$ is called a 'bipartition' of G.

If every vertex in X is joined to every vertex of Y then we call it a 'complete bipartite graph' denoted by $k_{m,n}$, where m and n are vertices in X and Y respectively.

Any complete bipartite graph with a bipartition into two sets of m and n vertices is isomorphic to $k_{m,n}$. Given below are some examples of bipartite and complete bipartite graphs.

Figure :Bipartite graph

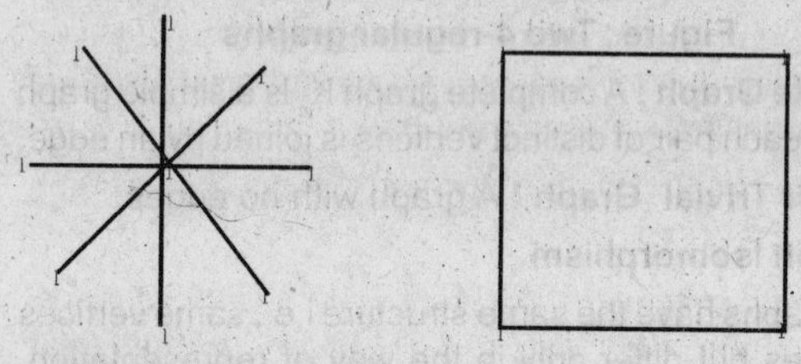

Figure : Some Complete bipartite graphs

5. Subgraphs

If G and H are two graphs with vertex sets V(H), V(G) and edge sets E(H) and E(G) respectively such that $V(H) \leq V(G)$ and $E(H) \leq E(G)$ then we call H as a subgraph of G (or G as a supergraph of H).

If $V(H) \subset V(G)$ and $E(H) \subset E(G)$ then A is a proper subgraph of G and if V(H)=V(G) then we say that H is a spanning subgraph of G.

If from a graph G with vertex set V, we delete a subset U of V and all the edges which have a vertex in U as an end point then (G-U) is called a vertex deleted subgraph and similarly,if a subset F of E is deleted from G then (G-F) denotes the subgraph of G with vertex set V and edge set E-F , then (G-F) is called edge deleted subgraph.

By deleting from a graph G all loops and in each collection of parallel edges all edges but one we obtain a simple spanning subgraph of G, called the underlying simple graph of G.

If G is a graph with vertex set V and U is a subset of V then the subgraph G(U) of G induced by U is defined to be the graph having vertex set U and edge set consisting of those edges of G that have both ends in U. Similarly if F is a non-empty subset of the edge set E of G then the subgraph G(F) of G induced by F is the graph whose vertex set of ends of edges in F and whose edge set is F.

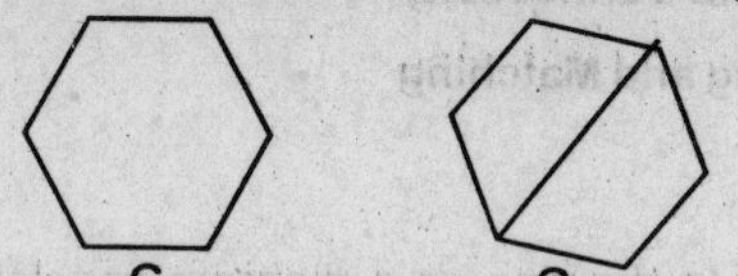

G_1 G_2

G_1 **is a proper spanning subgraph of** G_2

6. Paths and Walks

A walk in a graph G is a finite sequence

$W = v_0 e_1 v_1 e_2 ..... v_{k-1} e_k v_k$

Whose terms are alternately vertices and edges such that, for $1 \leq i \leq k$, the edge has ends v_{i-1} and v_i. The vertex V_0 is called the origin and V_k as the terminus of walk W. They need not be distinct. Vertices $v_1 ...... v_{k-1}$ are called internal vertices. The number of edges in W are called the length of the walk. Vertices and edges can be repeated in a walk. A trivial walk is one containing no edges and has length zero. A walk is closed or open depending on whether origin and terminus are same or different. If the edges $e_1, e_2, ... e_k$ of walk W are distinct then we call such walk as trail. So a trail is a walk if no edge is repeated. If the vertices $v_0, v_1, ... v_k$ of the walk $W = v_0 e_1 v_1 e_2 .... e_k v_k$ are distinct then W is called a path. A path is a walk if no vertex is repeated. So every path is a trail and not every trail is a path.

7. Connected Graphs and Cycles

A vertex u is said to be connected to a vertex v in a graph G if there is a path in G from u to v. The converse is also true as v is also connected with u by the reverse path. If u is connected with v and v is connected with w then u is connected with w.

This process of joining two walks which have a common end point to from a longer walk is called concatenation (stringing together).

A graph G is called connected if every two of its vertices are connected. Given any vertex u of a graph G, let C(u) denote the set of all vertices in G that are connected to u. Then the subgraph of G induced by u is called the connected component containing u.

A non trivial closed trail in a graph a is called a cycle.If its origin and internal vertices are distinct. The closed trail $C = v_1, v_2, ... v_n, v_1$ is cycle if. C has atleast 1 edge and $v_1, v_2, ... v_n$ are n distinct vertices.

A cycle with K edges is called a K-cycle. A K-cycle is called odd or even depending on whether k is odd or even. Any two cycles with same length are isomorphic. An cycle is denoted by C_n. We can define some cycles as

1 cycle → Loop

2 cycle → Pair of parallel edges.

3 cycle → triangle

We will illustrate the concept of trail, path and cycle from the Petersen graph.

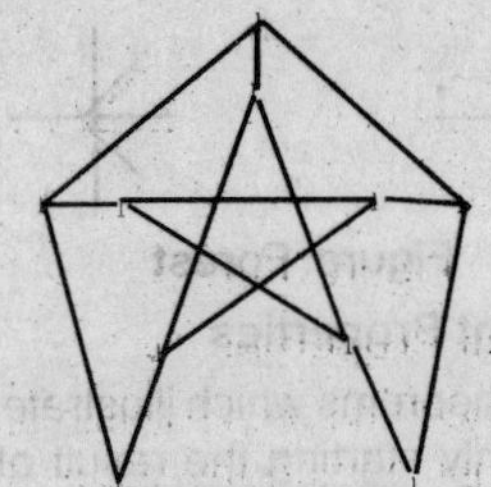

Figure : The Petersen Graph

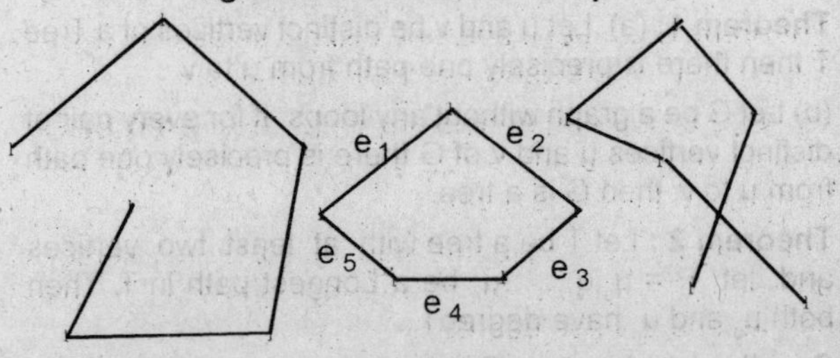

Path of length 5 Cycle of length 5 Trail of length 5

Figure : Showing Path, Cycle and trails

8. Operations On Graphs

In this section we would learn about some operations on the graph and their matrix notation, which is very useful to store a graph inside a computer.

Union: Given two graphs G_1 and G_2 their union will be a graph such that

$V(G_1 \cup G_2) = V(G_1) \cup V(G_2)$

And $E(G_1 \cup E_2) = E(G_1) \cup E(G_2)$

Intersection: Given two graphs G_1 and G_2 with at least one vertex in common then their intersection will be a graph such that

$V(G_1 \cap G_2) = V(G_1) \cap V(G_2)$

$E(G_1 \cap G_2) = E(G_1) \cap E(G_2)$

Complement : The complement $\overline{G}$ of G is defined as a simple graph with the same vertex set as G and where two vertices u and v are adjacent only when they are not adjacent in G.

For example,

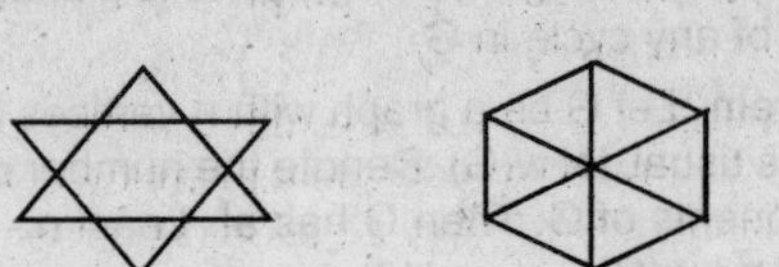

G Complement of G

Product of Graphs: To define the product $G_1 \times G_2$ of two graphs consider any two points $u = (u_1, u_2)$ and $v = (v_1, v_2)$ in $V = V_1 \times V_2$. Then u and v are adjacent in $G_1 \times G_2$ whenever [$u_1 = v_1$ and u_2 adj v_2] or [$u_2 = v_2$ and u_1 adj v_1].

For example,

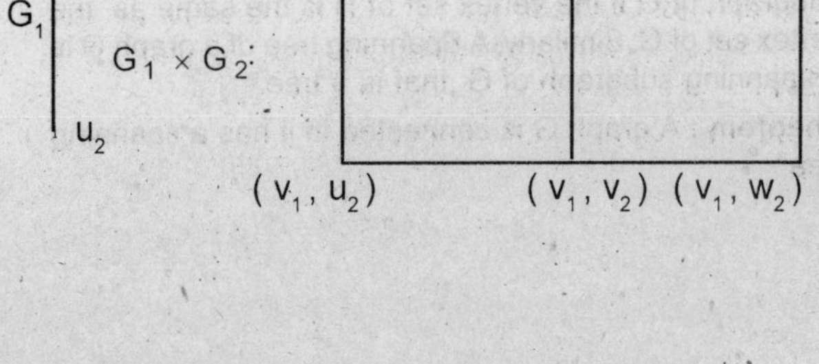

Composition : The composition $G = G_1[G_2]$ also has $V = V_1 V_2$ as its point set and $u = (u_1, u_2)$ is adjacent with $v = (v_1, v_2)$ when ever [u_1 adj v_1] or [$u_1 = v_1$ and u_2 adj v_2].

For example,

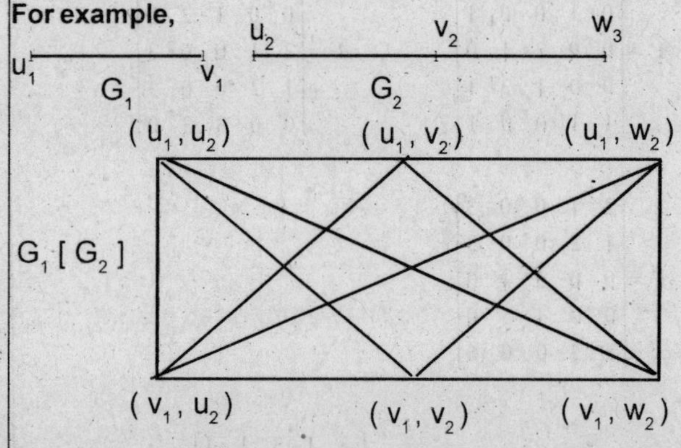

9. Matrix Representation of Graphs

A graph can be represented inside a computer by using the adjacency matrix or the inciden e matrix of a graph.

Let G be a graph with n vertices v_1...........v_n. The adjacency matrix of G with respect to this particular listing of n vertices is the $n \times n$ matrix $A(G) = (a_{ij})$, where a_{ij} is the number of edges joining the vertex v_i to v_j. If G has no loops then all the entries of the main diagonal will be 0 and if G has no parallel edges then the entries of A(G) are either 0 or 1. Given a $n \times n$ symmetric matrix $A = (a_{ij})$, in which all the entries are non-negative integers, one can associate with it a graph G whose adjacency matrix is A by letting G have n vertices and joining vertex i to vertex j by a_{ij} edges. We here by quote two results.

Theorem 1: Let G be a graph with n vertices v_i .....v_n and let A denote the adjacency matrix of G with respect to this listing of the vertices. Let K be any positive integer and let A^k denote the matrix multiplication of A, k times. Then the $(ij)^{th}$ element of A^k is the number of different $v_i - v_j$ walks in G of length K.

We can use the above result to check whether a graph is connected or not.

Theorem 2 : Let G be a graph with n vertices v_1.......v_n and let A denote the adjacency matrix of G with respect to this listing of the vertices. Let $B = (b_{ij})$ be the matrix.

$$B = A + A^2 + \dots\dots\dots + A^{n-1}$$

Then G is a connected graph iff for every pair of distinct indices i and j, we have $b_{ij} \neq 0$ i.e., iff B has no zero entries off the main diagonal.

This result can also be used to check the connectedness of a graph by using its adjacency matrix.

Ex 1 : A graph G has the following adjacency matrix. Check whether it is connected.

$$A(G) = \begin{bmatrix} 0 & 0 & 1 & 0 & 0 \\ 0 & 0 & 0 & 1 & 0 \\ 0 & 0 & 0 & 0 & 1 \\ 0 & 1 & 0 & 0 & 1 \\ 0 & 0 & 1 & 1 & 0 \end{bmatrix}$$

Sol : Since here n = 5, we look at $B = A + A^2 + A^3 + A^4$

We find, A^2, A^3 and A^4 by matrix multiplication

$$A^2 = \begin{bmatrix} 1 & 0 & 0 & 0 & 1 \\ 0 & 1 & 0 & 0 & 1 \\ 0 & 0 & 2 & 1 & 0 \\ 0 & 0 & 1 & 2 & 1 \\ 1 & 1 & 0 & 0 & 1 \end{bmatrix} \quad A^3 = \begin{bmatrix} 0 & 0 & 2 & 1 & 0 \\ 0 & 0 & 1 & 2 & 0 \\ 2 & 1 & 0 & 0 & 3 \\ 1 & 2 & 0 & 0 & 3 \\ 0 & 0 & 3 & 3 & 0 \end{bmatrix}$$

$$A^4 = \begin{bmatrix} 2 & 1 & 0 & 0 & 3 \\ 1 & 2 & 0 & 0 & 3 \\ 0 & 0 & 5 & 4 & 0 \\ 0 & 0 & 4 & 5 & 0 \\ 3 & 3 & 0 & 0 & 6 \end{bmatrix}$$

$$\text{now } = B = A + A^2 + A^3 + A^4 = \begin{bmatrix} 3 & 1 & 3 & 1 & 4 \\ 1 & 3 & 1 & 3 & 4 \\ 3 & 1 & 7 & 5 & 4 \\ 1 & 3 & 5 & 7 & 4 \\ 4 & 4 & 4 & 4 & 8 \end{bmatrix}$$

Since B has no non zero entry off the main diagonal. The graph is connected.

Another matrix associated with graph G is the incidence matrix. Suppose that G has n vertices listed as $v_1,....v_n$ and t edges listed as $e_1,....e_t$. The incidence matrix of a is the n × t matrix $M(G) = (m_{ij})$ where m_{ij} is the number of times that the vertex V_i is incident with the edge e_i, i.e.,

$$m_{ij} \begin{cases} 0 \text{ if } v_i \text{ is not an end of } e_j, \\ 1 \text{ if } v_i \text{ is an end of non loop } e_j \\ 2 \text{ if } v_i \text{ is an end of the loop } e_j \end{cases}$$

Sum of the elements in the ith row of MG gives us the degree of the vertex v_i.

10. Trees

A graph is called acyclic if it contains no cycles. Since loops are cycles of length one while a pair of parallel edges produces a cycle of length two, any acyclic graph must be simple. A graph G is called a tree if it is a connected acyclic graph . As the number of vertices increases the drawing to tree become complicated.

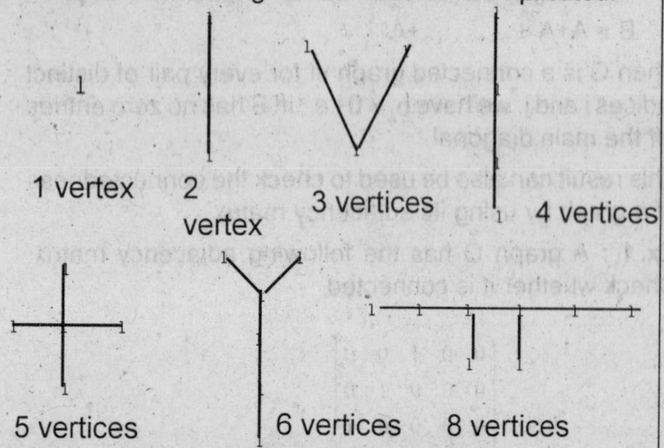

Figure : Trees with different vertices

Let G be an acyclic graph.Then any sub graph of G must also contains no cycles. The connected components of G are also acyclic and so they are trees. This is why an acyclic graph is also called a 'Forest'.

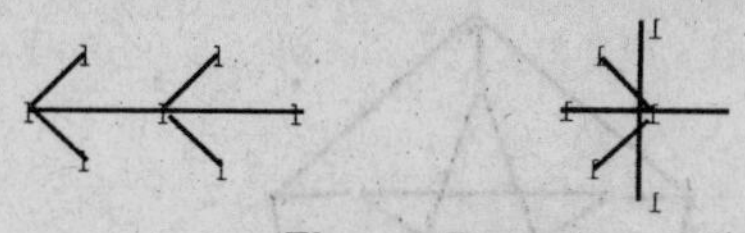

Figure: Forest

11. Some Important Properties

We will give a few theorems which illustrates the property of trees. We are only starting the result of the theorem and leaving the proof.

Theorem 1: (a) Let u and v be distinct vertices of a Tree T then there is precisely one path from u to v.

(b) Let G be a graph without any loops. If for every pair of distinct vertices u and v of G there is precisely one path from u to v, then G is a tree.

Theorem 2 : Let T be a tree with at least two vertices and let $P = u_0, u_1......u_n$ be a Longest path in T. Then both u_0 and u_n have degree1.

Corollary 1: Any tree T with at least two vertices has more than one vertex of degree 1.

Theorem 3: If T is a tree with n vertices then it has precisely (n-1) edges.

Theorem 4 : Let G be an acyclic graph with n vertices and k connected components i.e.,w(G) = k. Then G has n-k edges.

12. Bridge Or Cut Edge

If we remove an edge from a graph then the number of connected components of the graph either remains unchanged or it increases by exactly 1. So, an edge e of a graph G is called a bridge or cut edge. If the subgraph G-e has more connected components than G has.

Theorem : Let e be an edge of the graph G and, as usual,Let G – e be the subgraph obtained by deleting e. Then,

$$\omega(G) \le \omega(G - e) \le \omega(G) + 1$$

where ω(a) is the number of connected components of G

Theorem : An edge e of a graph G is a bridge iff e is not a part of any cycle in G

Theorem : Let G be a graph with n vertices and q edges and, as usual, let w(G). Denote the number of connected components of G. Then G has at Least n - w(G) edges i.e. q ³ n - w(G).

Corollary: A graph with n vertices less than (n – 1) edges can not be connected.

Theorem : Let G be a graph with n vertices then the following three statements are equivalent.

I. G is tree

II. G is an acyclic graph with (n – 1) edges.

III.G is a connected graph with (n – 1) edges.

13. Spanning Trees and Connector Problems

Let G be a graph . A subgraph H of G is called a spanning subgraph of G if the vertex set of H is the same as the vertex set of G. Similarly, A Spanning tree of a graph G is a spanning subgraph of G ,that is a tree.

Theorem : A graph G is connected iff it has a spanning tree.

Cayley's Theorem: The complete graph K_n has n^{n-2} different spanning trees.

Ex: Give all the spanning trees of K_4.

Sol: Here, n=4

So there will be 4^{4-2}=16 spanning trees

A weighted graph is a graph G in which each edge e has been assigned a real number $\omega(e)$, called the weight of e. If H is a subgraph of a weighted graph, the weight $\omega(H)$ of H is the sum of the weights $\omega(e_1) + \ldots\ldots + \omega(e_k)$, where $[e_1, \ldots\ldots\ldots, e_k]$ is the set of edges of H.

Many optimisation problems could be solved by using weighted graphs. We try to find a subgraph of the weighted graph, with minimum (or maximum) weight.

We look at a connected spanning subgraph of G which should not have a cycle i.e., we look at a spanning tree of G and finally we look for a spanning tree with minimum weights such a tree is called a minimal spanning tree or an optimal tree for G. We have two algorithms for finding such a tree. A necessary condition for these algorithms is that no weights should be negative.

14. Algorithm: Kruskal's Algorithm

In this algorithm we choose an edge of G which has smallest weight among the edges of G

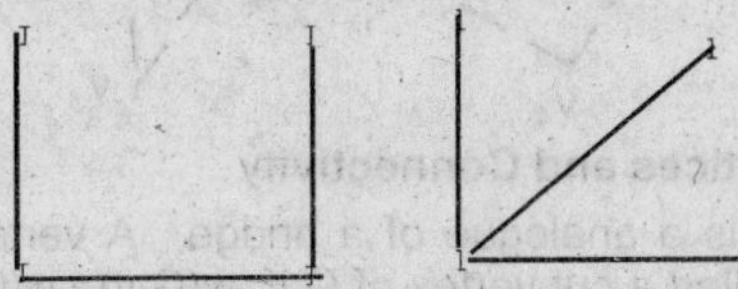

Figure : Some Spanning Trees for K_4

which are not loops. This algorithm gives an acyclic subgraph T of G and the theorem given below proves that T is a minimal spanning tree of G. Following steps are required.

Step 1: Choose e_1, an edge of G, such that weight of e_1, $w(e_1)$ is as small as possible and e_1 is not a loop.

Step 2: If edges $e_1, e_2, \ldots\ldots\ldots e_i$ have been chosen then choose an edge e_{i+1}, not already chosen, such that

I. the induced subgraph $G[(e_1, \ldots\ldots\ldots e_{i+1})]$ is acyclic and

II. $\omega(e_{i+1})$ is as small as possible.

Step 3: If G has n vertices, stop after n-1 edges have been chosen. Otherwise repeat step2.

Theorem: Let G be a weighted connected graph in which the weights of the edges are all non-negative numbers. Let T be a subgraph of G obtained by Kruskal's algorithm. Then T is a minimal Spanning tree of G.

Ex: Given below is a weighted graph G with 5 vertices. Apply Kruskal's Algorithm to determine a minimal spanning tree.

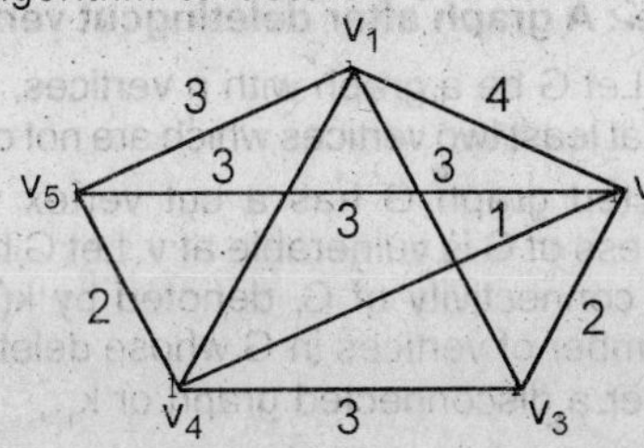

Sol: According to step 1, we will choose, $e=v_2v_4$ as it has the minimum weight.

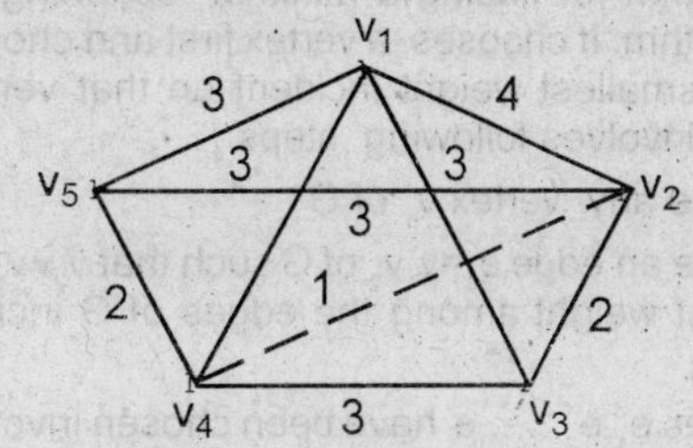

Choose e = v_2 v_3

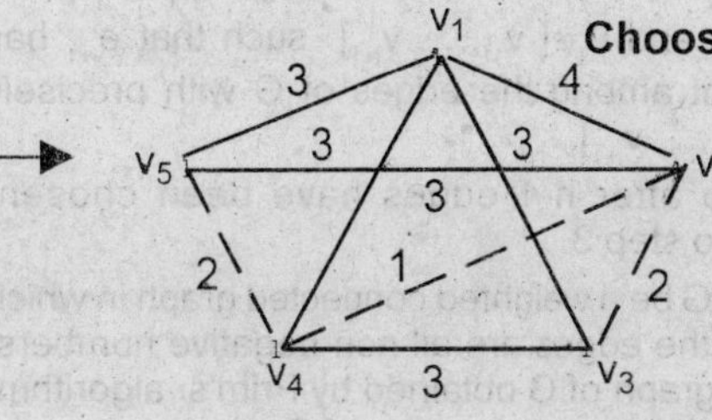

Choose e = v_4 v_5

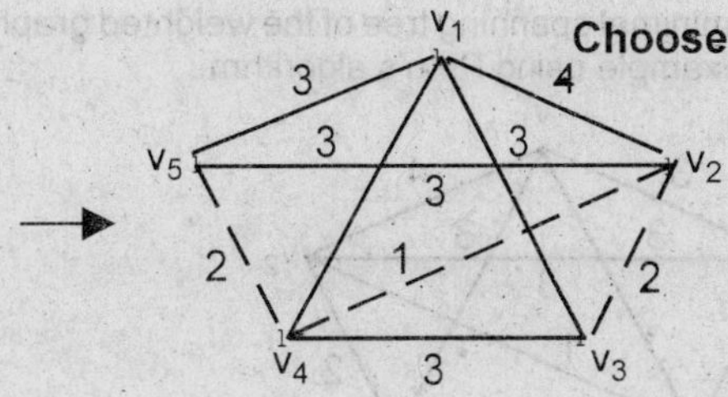

Choose e = v_1 v_4

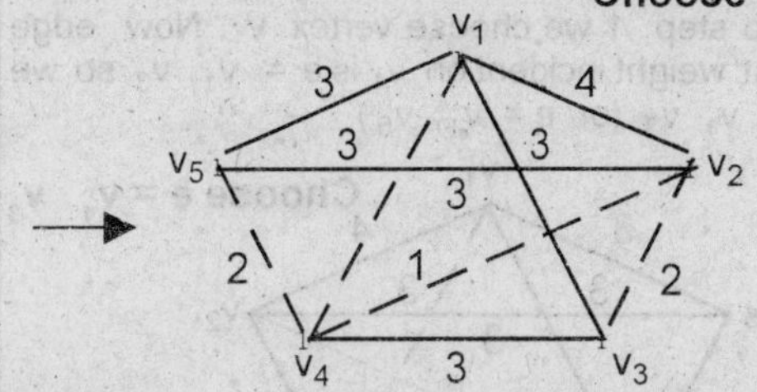

Since vertices are 5 and we have choosen 4 edges stop. The minimal spanning tree of graph is

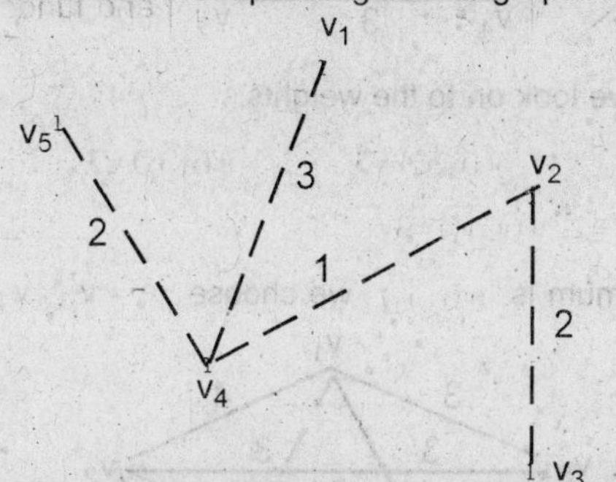

Note : Kruskal's algorithm is an example of a type of algorithm known as greedy. Greedy algorithms are essentially algorithms that proceed by selecting the choice that looks best at the moment. All processes cannot be handled so simply.

15. Algorithm : Prim's Algorithm

Another Algorithm for finding a minimal spanning tree is Prim's Algorithm. It chooses a vertex first and chooses an edge with smallest weight incident on that vertex . The algorithm involves following steps.

Step 1: Choose any vertex v_1 of G

Step 2: Choose an edge $e_1=v_1v_2$ of G such that $v_2\neq v_1$ and e_1 has smallest weight among the edges of G incident with v_1.

Step 3 : If edges $e_1,e_2,\ldots,e_i$ have been chosen involving end points $v_1,v_2,\ldots,v_{i+1}$. Choose an edge $e_{i+1}=v_jv_k$ with $v_j\in\{V_1,\ldots,V_{i+1}\}$ and $v_k\notin[v_1,\ldots,v_{i+1}]$ such that e_{i+1} has smallest weight among the edges of G with precisely one end in $[v_1,\ldots,v_{i+1}]$.

Step 4 : Stop after n-1 edges have been chosen. Otherwise go to step 3.

Theorem : Let G be a weighted connected graph in which the weights of the edges are all non-negative numbers. Let T be a subgraph of G obtained by Prim's algorithm. Then T is a minimal spanning tree of G.

Ex: find the minimal spanning tree of the weighted graph in previous example using Prim's algorithm.

Sol :

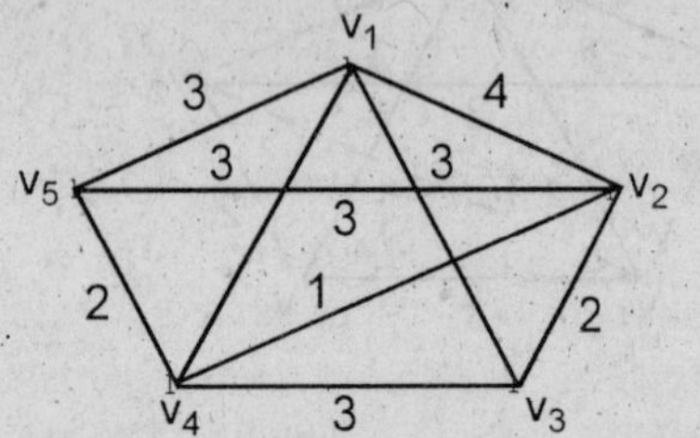

According to step 1 we choose vertex v_1. Now edge with smallest weight incident on v_1 is e = v_1, v_3 so we choose e = v_1 v_3 (or e = v_1, v_5)

Choose e = v_1 v_3

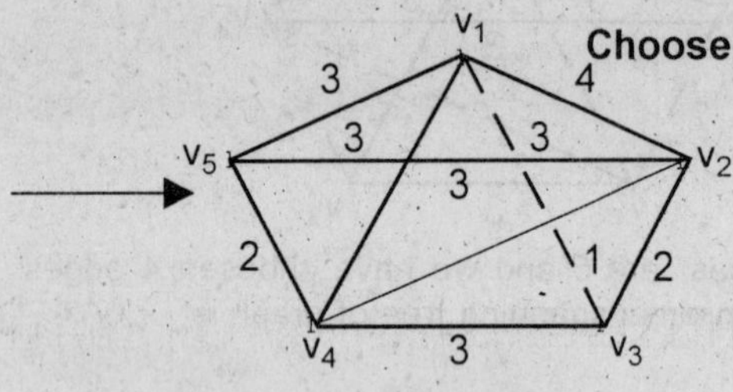

Now we look on to the weights,

$w(v_1,v_2)=4$ $w(v_1,v_4)=3$ $w(v_1,v_5)=3$

$w(v_3,v_2)=2,\ w(v_3,v_4)=3$

Since minimum is $w(v_3,v_2)$, we choose e = v_3 ,v_2

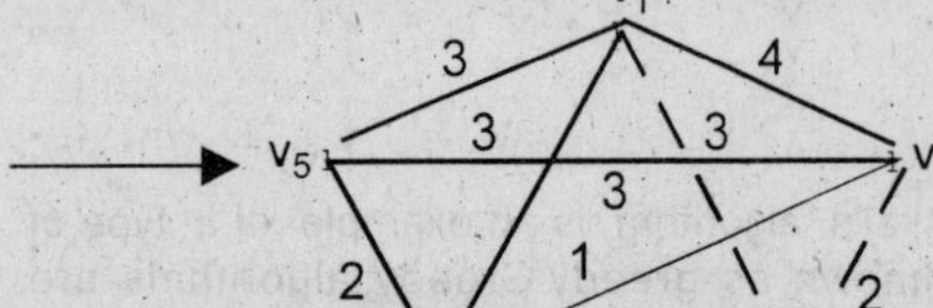

Choose e = v_3 v_2

Again $w(v_2,v_5)=3,\ w(v_2,v_4)=1,$

and $w(v_3,v_4)=3$

Choose e = v_3 . v_2

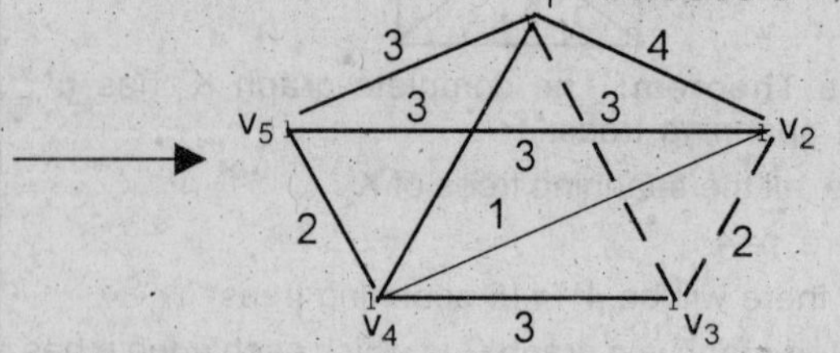

Now , $w(v_4,v_5)=2$

Choose e = v_2 v_4

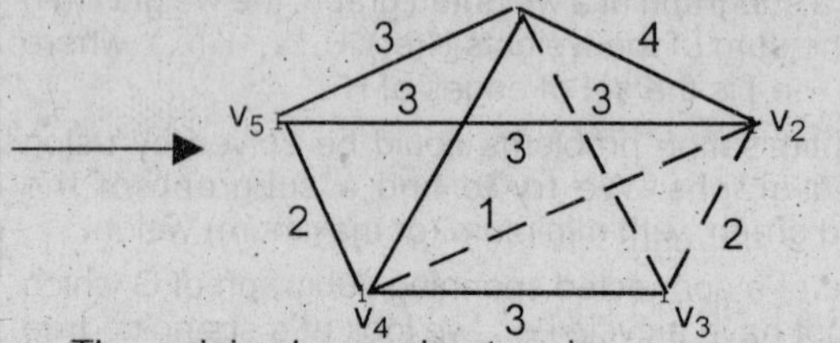

The minimal spanning tree is

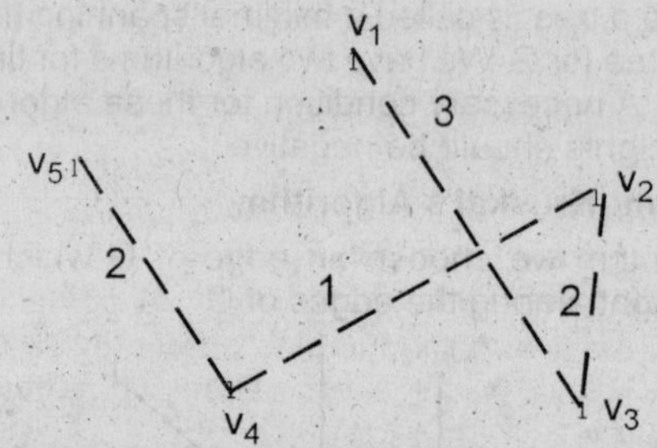

16. Cut Vertices and Connectivity

Cut vertex is a analogue of a bridge. A vertex v of a graph is called a cut vertex of G if w(G-v)> w(G) where w(G) are number of components in graph G. i.e. A cut vertex breaks a graph into a subgraph having more connected components then G has,

For example,

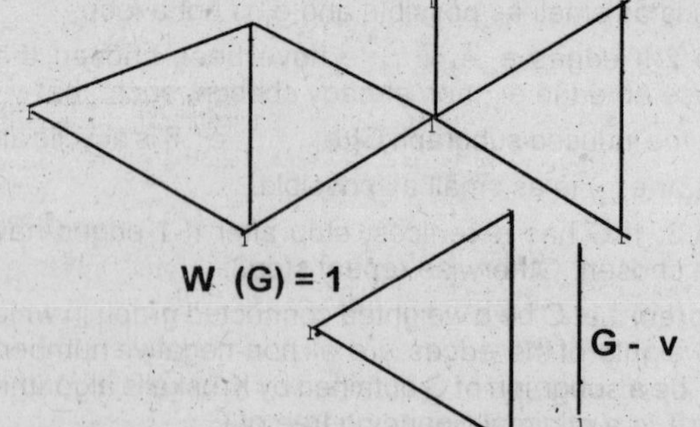

Figure : A graph after deletingcut vertex v

Theorem: Let G be a graph with n vertices, where n³2. Then G has at least two vertices which are not cut vertices.

If a connected graph G has a cut vertex v then the connectedness of G is vulnerable at v. Let G be a simple graph. The connectivity of G, denoted by k(G) , is the smallest number of vertices in G whose deletion from G leaves either a disconnected graph or k_1.

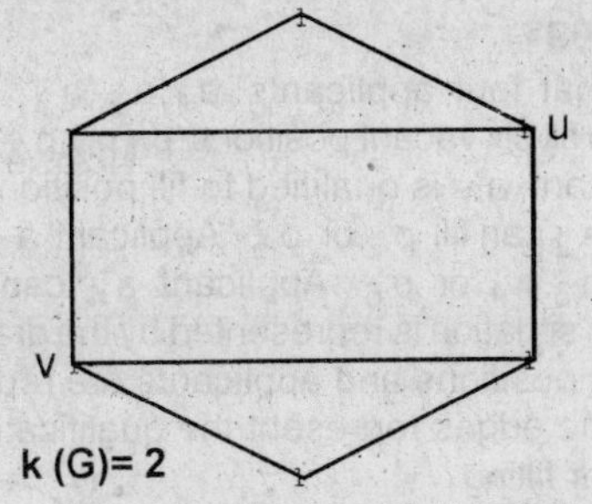

Figure 19 : Showing Connectivity of the Graph (by deleting u and v)

Note : k (K_n) = n-1. It is easy to see that a connected graph has k (G) = 1, iff either G = K_2 o G has a cut vertex. Also k (G) = 0, iff either G = K_1 or G is disconnected.

A simple graph G is called n - connected (n $\geq$ 1) if k (G) $\geq$ n. Let u and v be two vartices of a graph G. A collection $\{P_{(1)}, \ldots\ldots, P_{(m)}\}$ of u- v paths is said to be internaly disjoint if, given any distinct pair $P_{(i)}$ and $P_{(j)}$ in the collection, u and v are the only vertices $P_{(i)}$ and $P_{(j)}$ have in common.

Binary Trees : One of the principle uses of trees in Computer Science is in data representation we use trees to depict the relationship between pieces of information, especially when the usual linear representation seems inadequate. Tree representations are useful when slightly more structure is given to the tree. In this section, all trees will be rooted i.e, they will have a distinguished vertex r (called root) with the property that for every vertex v $\in$ V there is a directed r -v path.

The trees will also be levelled i.e. the root r will constitute level 0, the neighbours of r will constitute level 1 and the neighbours of vertices in level 1 will have level 2 and so on.The neighbours on next level are called children (descendent) and previous level are called.Father (predecessor). A binary tree is a rooted, levelled tree in which any vertex has at most two children. If every vertex of binary tree has either two childern or no children,then we say the tree is a full binary tree.

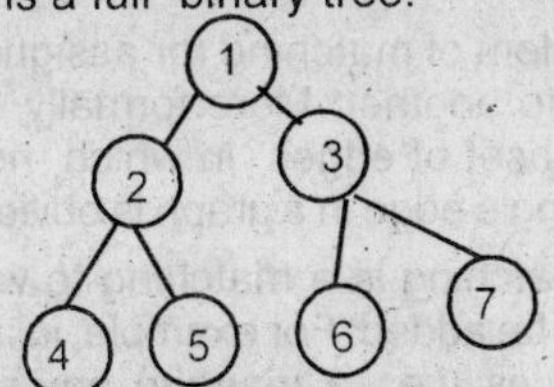

Figure : A full binary tree

We will illustrate the use of binary tree in data representation by an example, say you have to express a + 3b = a+ 3 $\times$ b, then you will draw the following tree:

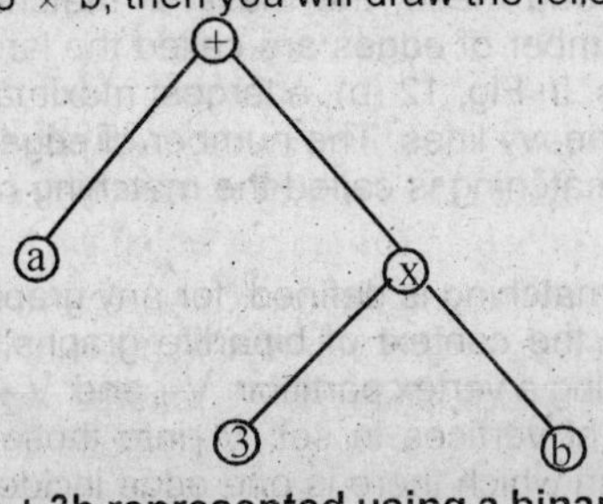

Figure :a + 3b represented using a binary tree

17. EulerTours

We define a trail as Euler trail if it includes every edge of G.So, A trail is Euler if each edge of G is in the trail exactly once.

A tour of G is a closed walk of G which includes every edge of G at least once. An Euler tour of G is a tour which includes each edge of G exactly once. So, An Euler tour is just a closed Euler trail. A graph G is called Eulerian or Euler if it has as Euler tour.

18. Konisberg Bridge Problem

Leonhard Euler considered the problem of the seven Konisberg Bridges on the banks of the river Pregel in Prussian city of Konisberg. The river was crossed by seven bridges, which connected two islands A and B in the river, with each other and with the opposite banks C and D. Townfolks of Konisberg believed that it is not possible to find a route that crossed each bridge just once. Euler showed that there is no prossibility of such a route as the graph has no Euler trail.

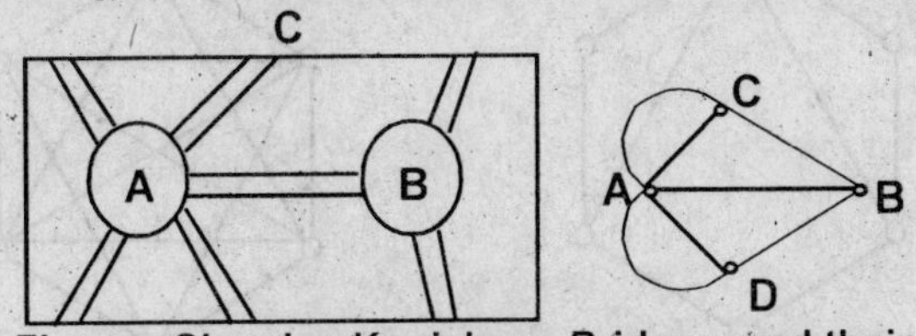

Figure : Showing Konisberg Bridges and their multigraph

Some useful theorems on Euler graph are given below:

Theorem : Let G be a graph in which the degree of every vertex is at least two. Then G contains a cycle.

Theorem: A connected graph G is Euler if and only if the degree of every vertex is even.

Theorem: A connected graph G is Euler iff G has cycles $C_{(1)}, \ldots C_{(m)}$ such that every edge of G belongs to exactly on cycle $C_{(i)}$ i.e. G is the union of edge disjoint cycles.

Theorem : A connected graph G has an Euler Trail iff it has at most two odd vertices i.e. it has either no vertices of odd degree or exactly two vertices of odd degree.

19. Hamiltonian Graphs

Hamiltonian graphs are named after Sir William Hamilton, and Irish mathematician, who invented a puzzle, called the Icosian game. The puzzle involved a dodecahedron on which each of the 20 vertices was labelled by the name of some city. The object of the game was to construct, using the edges of dodecahedron, a tour of all cities which visited each city exactly once, beginning and ending at the same city. The graph of this puzzle is Hamiltonian graph.

A Hamiltonian Path in a Graph G is a path which contains every vertex of G i.e. it contains every vertex of G once and only once. A hamiltonian Cycle in a graph G is a cycle which contains every vertex of G. Now, A graph G is called Hamiltonian if it has a Hamiltonian cycle.

A simple graph G is called maximal non-Hamiltonian if it is not Hamiltonian but the addition of anyedge connecting two non-adjacent vertices forms a Hamiltonian graph.

We give below some theorems on Hamiltonian Graphs.

Theorem 1: (Dirac) If G is a simple graph with n vertices where $n \geq 3$, and the degree $d(v) \geq n/2$ for every vertex v of G, then G is Hamiltonian.

Theorem 2: Let G be a simple graph with n vertices and let u and v be non-adjacent vertices in G such that $d(u) + d(v) \geq n$. If G + uv denote the subgraph of G obtained by joining u and v by an dege. Then G is Hamiltonian iff G + uv is Hamiltonian.

20. Closure Of a Graph

Let G be a simple graph. If there are two non-adjacent vertices u_1 and v_1 in G such that $d(u_1) + d(v_1) \geq n$ in G. Join u_1 and v_1 by an edge to from the supergraph G_1. Then, if two vertices u_2 and v_2 are there such that $d(u_2) + d(v_2) \geq n$ in G_1, join u_2 and v_2 by an edge to form the supergraph G_2. Continue in this way to join pairs of non adjacent vertices whose degree sum is at least n. The final supergraph thus obtained is called the closure of G and is denoted by C (G).

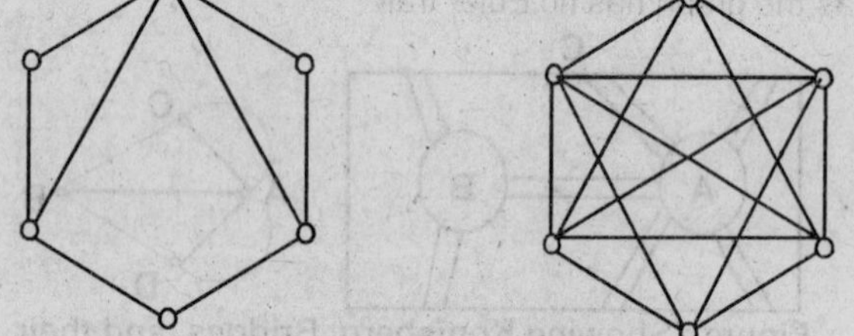

Figure : A graph and its closure C (G)

There is a Theorem due to Bondy and Chvatal which is as follows.

Theorem : A simple graph is Hamiltonian iff its closure C (G) is Hamiltonian.

21. Planar Graphs

A Plane Graph is a graph drawn in the plane in such a way that any pair of edges meet only at their end vertices. A Planar graph is a graph which is isomorphicto a plane graph i.e., it can be redrawn as a plane graph.

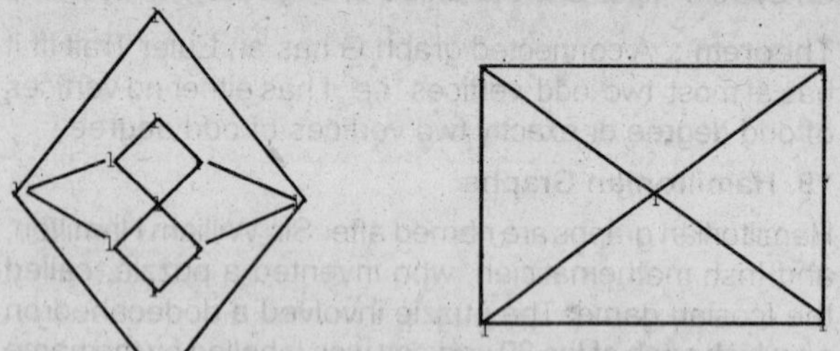

Figure : Some Plane Graphs

A Jordan Curve in the plane is a continuous non-interesecting curve whose origin and terminus coincide.

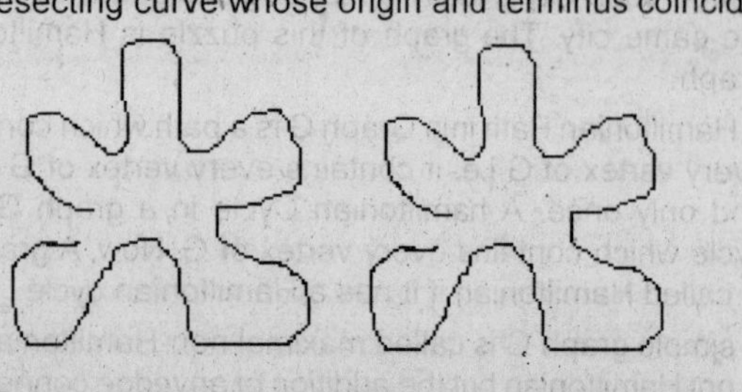

Figure : G_1 is Jordan Curve but G_2 is not

22. Matchings

Suppose that four applicants a_1, a_2, a_3, and a_4 are available to fill six vacant positions p_1, p_2, p_3, p_4, p_5, and p_6. Applicant a_1 is qualified to fill position p_2 or p_5. Applicant a_2 can fill p_2 or p_5. Applicant a_3 is qualified for p_1, p_2, p_3, p_4, or p_6. Applicant a_4 can fill jobs p_2 or p_5. This situation is represented by the graph in Fig.11. The vacant positions and applicants are represented by vertices. The edges represent the qualifications of each applicant for filling

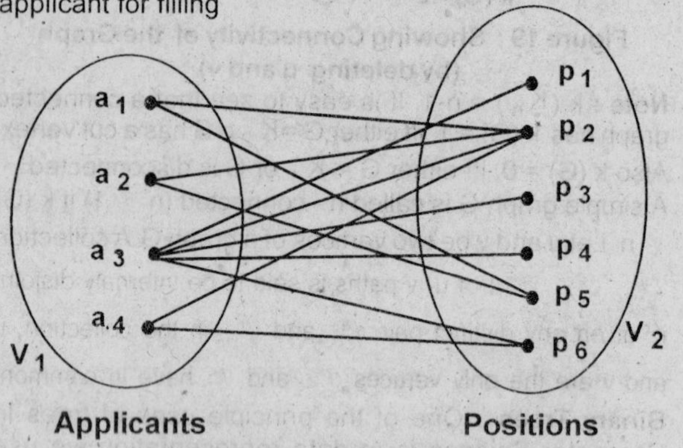

Fig.11 Bipartite graph.

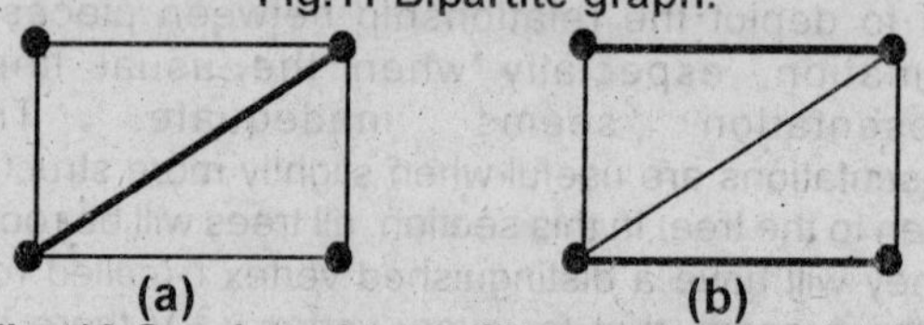

Fig. 12 Graph and two of its maximal matchings.

different positions. The graph clearly is biparite, the vertices falling into two sets $V_1 = \{a_1, a_2, a_3, a_4\}$ and $V_2 = \{p_1, p_2, p_3, p_4, p_5, p_6\}$.

The questions one is most likely to ask in this situation are : Is it possible to hire all the applicants and assign each a position for which he is suitable?

If the answer is no, what is the maximum number of positions that can be filled from the given set of applicants ?

This is a problem of matching (or assignment) of one set of vertices into another. More formally, a matching in a graph is a subset of edges in which no two edges are adjacent. A single edge in a graph is obviously a matching.

A maximal matching is a matching to which no edge in the graph can be added. For example, in a complete graph of three vertices (i.e., a triangle) any single edge is a maximal matching. The edges shown by heavy lines in Fig.12 are two maximal matchings. Clearly, a graph may have many different maximal matchings, and of different sizes. Among these, the maximal matchings with the largest number of edges are called the largest maximal matchings. In Fig. 12 (b), a largest maximal matching is shown in heavy lines. The number of edges in a largest maximal matching is called the matching number of the graph.

Although matching is defined for any graph, it is mostly studied in the context of bipartite graphs.In a bipartite graph having a vertex partition V_1 and V_2, a complete matching of vertices in set V_1 into those in V_2 is a matching in which there is one edge incident with every vertex in V_1. In other words, every vertex in V_1 is matched against some vertex in V_2. Clearly a complete matching (if it exists) is a largest maximal matching, where as the converse is not necessarily true.

For the existence of a complete matching of set V_1 into set V_2, first we must have at least as many vertices in V_2 as there are in V_1. In other words, there must be at least as many vacant positions as the number of applicants if all the applicants are to be hired. This condition, however, is not sufficient. For example, in Fig.11, although there are six positions and four applicants a complete matching does not exist. Of the three applicants a_1, a_2, and a_4, each qualifies for the same two positions p_2, and p_5 and therefore one of the three applicants cannot be matched.

This leads us to another necessary condition for a complete matching: Every subset of r vertices in V_1 must collectively be adjacent to at least r vertices in V_2, for all values of r = 1,2...., $|V_1|$. This condition is not satisfied in Fig.11. The subset $\{a_1, a_2, a_4\}$ of three vertices has only two vertices p_2 and p_5 adjacent to them.

That this condition is also sufficient for existence of a complete matching is indeed surprising. Theorem 1 is a formal statement and proof of this result.

Theorem 1 : A complete matching of V_1 into V_2 in a bipartite graph exists if and only if every subset of r vertices in V_1 is collectively adjacent to or more vertices in V_2 for all values of r.

Let us illustrate this important theorem with an example.

Problem of Distinct Representatives: Five senators s_1, s_2, s_3, s_4, and s_5 are members of three committees, $c_1 c_2$, and c_3. The membership is shown in Fig.13. One member from each committee is to be represented in a super committee. Is it possible to send one distinct representative from each of the committees ?

This problem is one of finding a complete matching of a set V_1 into set V_2 in a bipartite graph. Let us use Theorem 1 and check if r vertices from V_1 are collectively adjacent to at least r vertices from V_2, for all values of r. The result is shown in Table 1 (ignore the last column for the time being).

Thus for this example the condition for the existence of a complete matching is satisfied as stated in Theorem 1. Hence it is possible to from the super committee with one distinct representative from each committee.

The problem of distinct representatives just solved was a small one. A

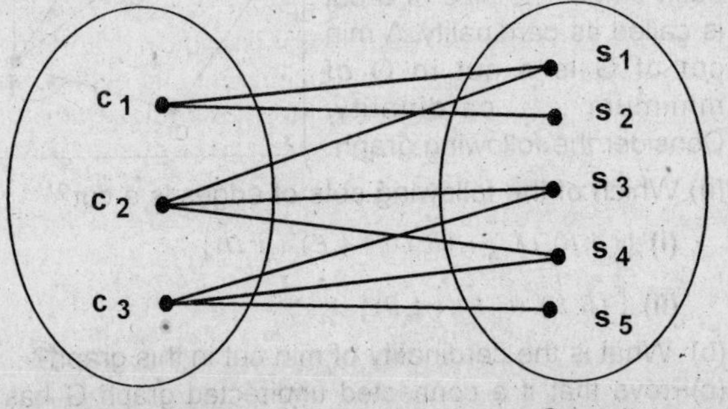

Fig.13 Membership of committees.

	V_1	V_2	r – q
r = 1	$\{C_1\}$	$\{S_1, S_2\}$	–1
	$\{C_2\}$	$\{S_1, S_3, S_4\}$	–2
	$\{C_3\}$	$\{S_3, S_4, S_5\}$	–2
r = 2	$\{C_1, C_2\}$	$\{S_1, S_2, S_3, S_4\}$	–2
	$\{C_2, C_3\}$	$\{S_1, S_3, S_4, S_5\}$	–2
	$\{C_3, C_1\}$	$\{S_1, S_2, S_3, S_4, S_5\}$	–3
r = 3	$\{C_1, C_2, C_3\}$	$\{S_1, S_2, S_3, S_4, S_5\}$	–2

Table 1

larger problem would have become unwieldy. If there are M vertices in V_1, Theorem 1 requires that we take all $2^M - 1$ nonempty subsets of V_1 and find the number of vertices of V_2 adjacent collectively to each of these. In most cases, however, the following simplified version of Theorem 1 will suffice for detection of a complete matching in any large graph.

Theorem 2: In a bipartite graph a complete matching V_1 into V_2 exists if (but not only if) there is a positive integer m for which the following condition is satisfied: degree of every vertex in $V_1 \geq m \geq$ degree of every vertex in V_2 in the bipartite graph of Fig. 13, degree of every vertex in $V_1 \geq 2 \geq$ degree of every vertex in V_2.

Therefore, there exists a complete matching.

In the bipartite graph of Fig.11 no such number is found, because the degree of $p_2 = 4 >$ degree of a_1.

23. Coverings

In a graph G, a set g of edges is said to cover G if every vertex in G is incident on at least one edge in g. A set of edges that covers a graph G is said to be an edge covering, a covering subgraph, or simply a covering of G. For example, a graph G is trivially its own covering. A spanning tree in a connected graph (or a spanning forest in an unconnected graph) is another covering. A hamiltonian circuit (if it exists) in a graph is also a covering.

Just any covering is too general to be of much interest. We have already dealt with some coverings with specific properties, such as spanning trees and

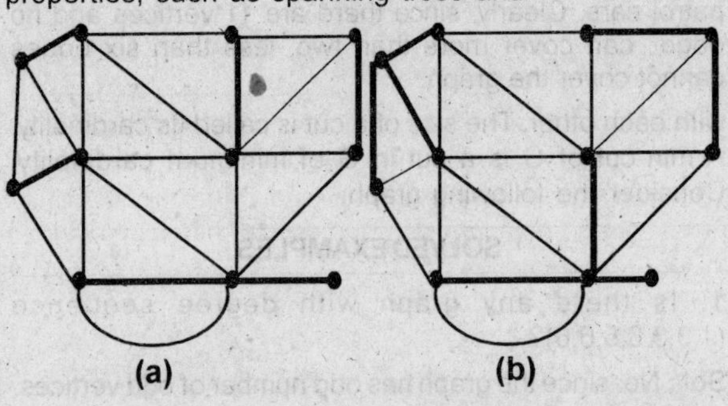

Fig. 14 Graph and two of its minimal coverings.

Hamiltonian circuits. In this section we shall investigate the minimal covering-a covering from which no edge can be removed without destroying its ability to cover the graph. In Fig. 14 a graph and two of its minimal coverings are shown in heavy lines.

The following observations should be made :

1. A covering exists for a graph if and only if the graph has no isolated vertex.

2. A covering of an n - vertex graph will have at least $\lceil n/2 \rceil$ edges.($\lceil x \rceil$ denotes the smallest integer not less than x.)

3. Every pendant edge in a graph is included in every covering of the graph.

4. Every covering contains a minimal covering.

5. If we denote the remaining edges of a graph by (G - g), the set of edges g is a covering if any only if, for every vertex v, the degree of vertex in (G - g) $\leq$ (degree of vertex v in G) - 1.

6. No minimal covering can contain a circuit, for we can always remove an edge from a circuit without leaving any of the vertices in the circuit uncovered. Therefore a minimal covering of an n-vertex graph can contain no more than n - edges.

7. A graph in general has many minimal coverings, and they may be of different sizes (i.e., consisting of different numbers of edges). The number of edges in a minimal covering of the smallest size is called the covering number of the graph.

Theorem 3 : A covering g of a graph is minimal if and only if g contains no paths of length three or more.

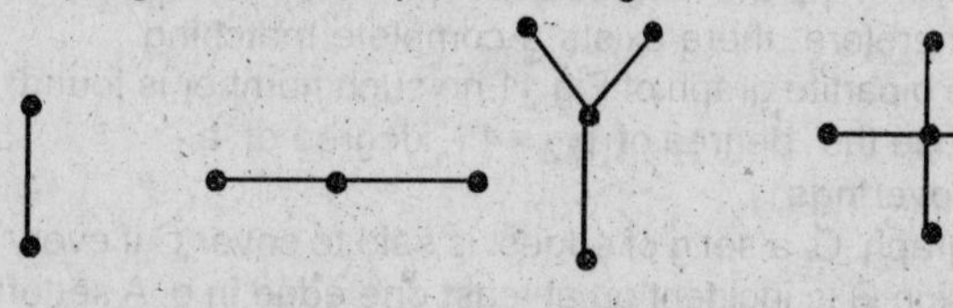

(a) (b) (c) (d)

Fig. 15 Star graphs of one, two, three, and four edges.

Suppose that the graph in Fig.14 represents the street map of a part of a citry. each of the vertices is a potential trouble spot and must be kept under the surveillance of a patrol car. How will you assign a minimum number of patrol cars to keep every vertex covered?

The answer is a smallest minimal covering. The covering shown in Fig. 14 (a) is an answer, and it requires six patrol cars. Clearly, since there are 11 vertices and no edge can cover more than two, less than six edges cannot cover the graph.

with each other. The size of a cut is called its cardinality. A min-cut of G is a cut in G of minimum cardinality. Consider the following graph.

SOLVED EXAMPLES

1. Is there any graph with degree sequence (1,3,3,3,5,6,6)?

Sol : No. since the graph has odd number of odd vertices.

2. Suppose that G is graph with 12 edges. Suppose that G has 6 vertices of degree 3 and the rest have degree less than 3. Determine the mininum number of vertices G can have.

Sol : Suppose G has n vertices. Given that 6 vertices are of degree 3 and the remaining $n-6$ are of degree less than 3. By first theorem of graph theory we have,

$$3.6+3(n-6) \leq 2|E|$$

$$18+3n-18 \leq 2.12 = 24 \text{ , } 3n < 24 \Rightarrow n < 8$$

So minimum number of vertices is 8.

3. Construct a graph on 12 vertices with 2 of them having degree1, three having degree 3, and the remaining seven having degree 10.

Sol : The sum of degrees of all vertices is

1+1+3+3+3+10+10+10+10+10+10+10=81

which is odd. So we cannot have a graph with the given vertices

4. Find the number of vertices in a graph G if it has 21 edges, 3 vertices of degree 4, and the other vertices of degree 3.

Sol : Let n be the number of vertices in G. Given that 3 vertices are of degree 4 and $n-3$ vertices are of degree 3. So by first theorem of graph theory we have

$$3.4+3(n-3)=2.21 \Rightarrow 12+3n-9=42$$

$$\Rightarrow 3n = 39 \Rightarrow n = 13$$

Therefore G has 13 vertices.

5. How many vertices does a regular graph of degree 4 with 10 edges have?

Sol : If G is a graph on n-vertices then $4n = 20$

$$\Rightarrow n = 5$$

6.What is the largest possible number of vertices in a graph with 35edges and all vertices of degree at least3?

Sol : Let n be the number of vertices. From the first theorem of graph theory we know that the degree sum is equal to twice the number of edges. Since all vertices are of degree greater than or equal to 3, the sum of the degrees of vertices is greater than or equal to 3n.

So $2 \times 35 \geq 3n$, or $n \leq \frac{70}{3} \leq 24$

Thus the given graph can have at most 23 vertices.

7. Let G be a connected, undirected graph. A cut in G is a set of edges whose removal results in G being broken into two or more components which are not connected with each other. The size of a cut is called its cardinality. A min cut of G is a cut in G of minimum cardinality. Consider the following graph.

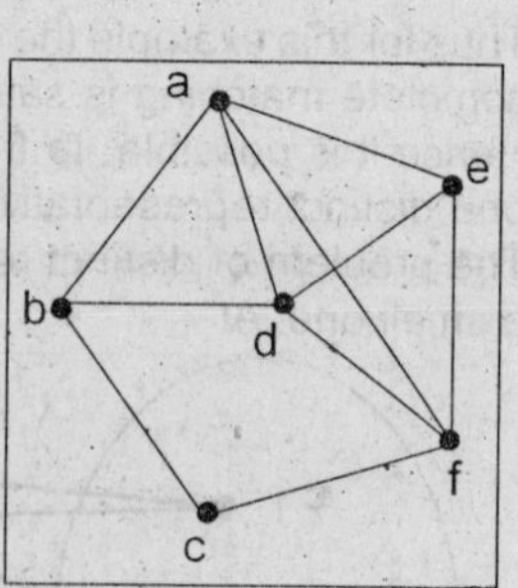

(a) Which of the following sets of edges is a cut?

(i) $\{(A,B),(E,F),(B,D),(A,E),(A,D)\}$

(ii) $\{(B,D),(C,F),(A,B)\}$

(b) What is the cardinality of min cut in this graph?

(c)Prove that if a connected undirected graph G has n vertices has a min cut of cardinality k,thenG has at least $\frac{nk}{2}$ edges.

Sol : (a) On deleting edges BD, CF and AB, the graph gets disconnected

(b) The cardinality of the minimum edge cut in this graph is 2. The minimum edge cut set is the set $\{(B,C),(C,F)\}$

(c)Suppose that the graph G has a min-cut of cardinalityk.

$\Rightarrow$ The degree of each vertex on this graph is at least k. This is because, if the degree was less than k, then by deleting all the edges incident on this vertex, we would be able to disconnect the graph. Which is a contradiction since the cardinality of min cut is k.

Since there are n vertices and the minimum degree of any vertex is k and each edge is incident on 2 vertices, G has at least $\frac{nk}{2}$ edges.

8. Find the length of shortest path between a and z in the given weighted graph.

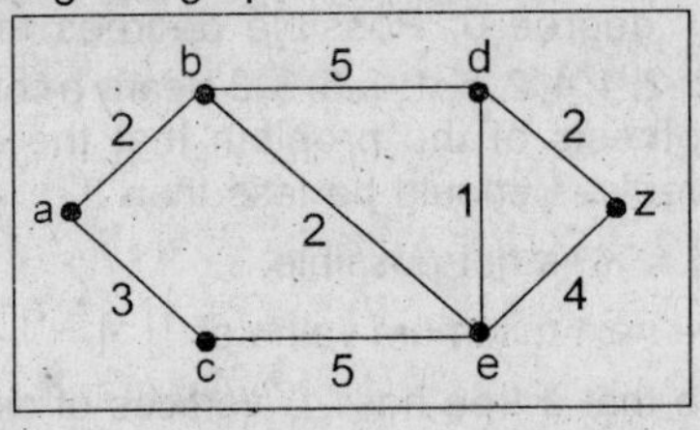

Sol :

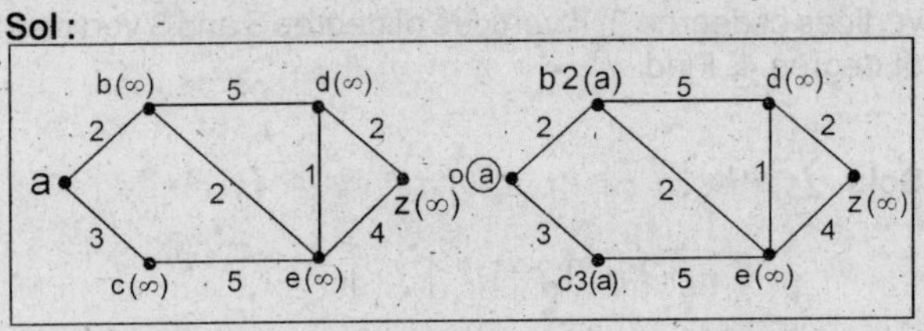

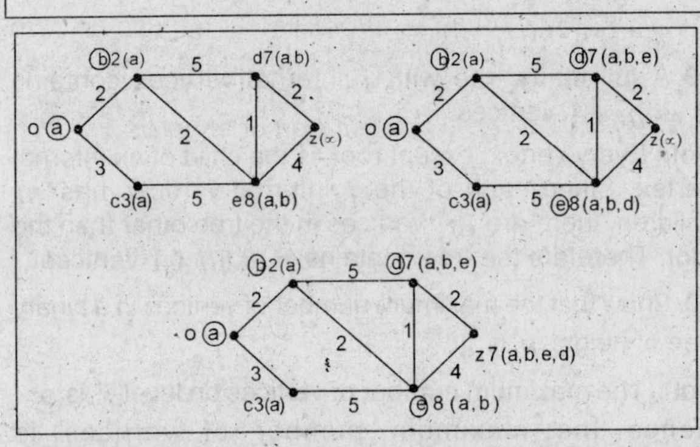

9. Determine whether the graphs shown in the figure are isomorphic or not,

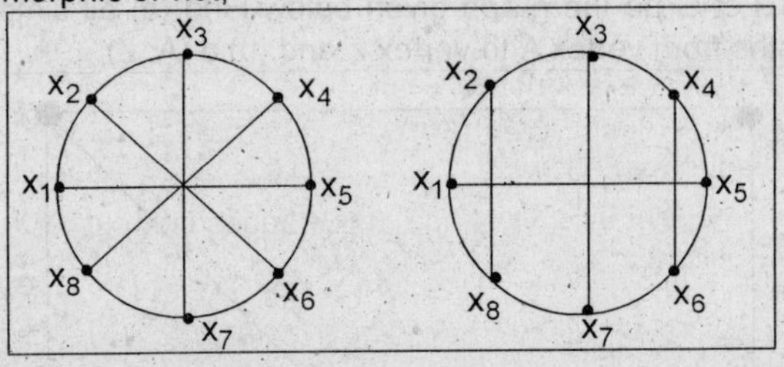

Sol : Both graphs have 8 vertices and 12 edges and also have same degree sequences $\{3,3,3,3,3,3,3,3\}$ .But the graphs are not isomorphic becauseHhas cycles of length 3 (triangles), where as in G there are no cycles of length3.

10. Determine whether the graphs shown in the figure are isomorphic or not.

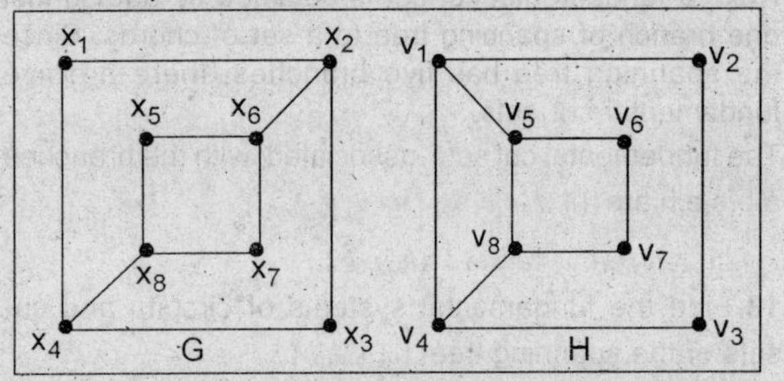

Sol : Both the graphs are of 8 vertices and 10 edges. Also they have the same degree sequence i.e., $\{2,2,2,2,3,3,3,3\}$. But the graphs are not isomorphic since in H,4 vertices of degree 2 are adjancent, where as in G no two vertices of degree 2 are adjacent.

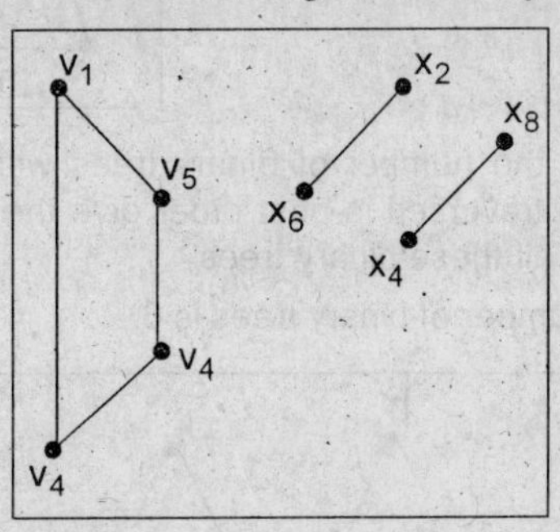

11. Find all the fundamental circuits relative to the spanning tree $\{e_1, e_3, e_5, e_8, e_9\}$ in the following graph.

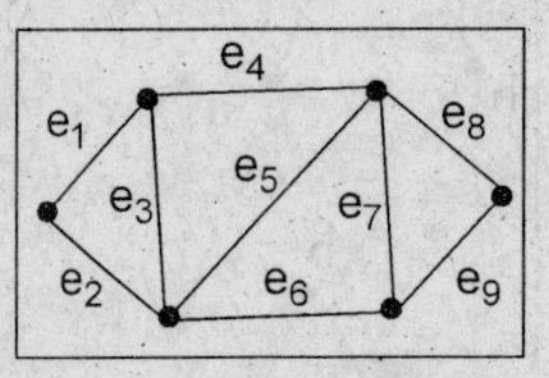

Sol : Fundamental circuits w.r.t to $\{e_2, e_3, e_5, e_8, e_9\}$ are

$\{e_1, e_2, e_3\}$, $\{e_3, e_4, e_5\}$

$\{e_5, e_6, e_8, e_9\}$, $\{e_7, e_8, e_9\}$

12. For the graph shown below, find a spanning tree and obtain all the corresponding fundamental circuits and fundamental cut sets.

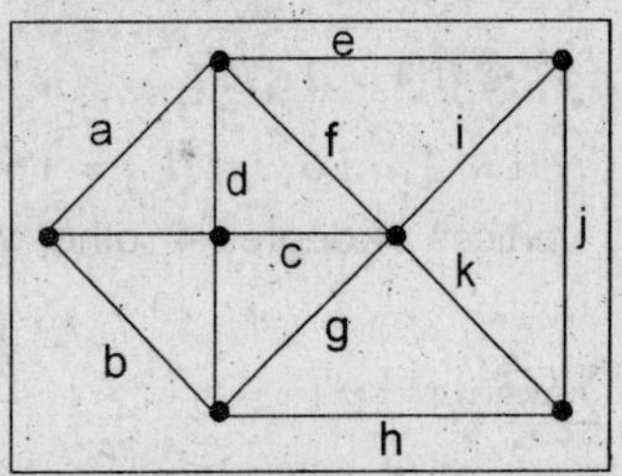

Sol : The given graph has six vertices. Therefore, a spanning tree of graph must have five edges. Consider the following spanning tree.

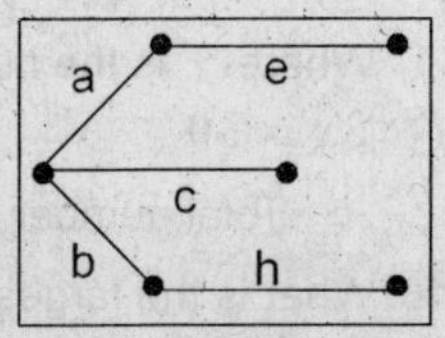

Now relative to this spanning tree, a fundamental circuit by adding only one chord to the spanning tree. Since there are 6 chords, we get six fundamental circuits given by

$\{a,b,d\}, \{a,c,f\}, \{b,c,g\}, \{a,c,e,i\}, \{a,b,e,h,i\}, \{b,c,h,k\}$

Next, a fundamental cut set is obtained by adding just one branch of spanning tree to a set of chords. Since the spanning tree has five branches, there are five fundamental cut sets.

The fundamental cut sets associated with the branches a,b,c,e,h are $\{a,d,f,i,j\}$ $\{b,d,g,j,k\}$

$\{c,f,g,i,k\}$ $\{e,i,j\}$ $\{h,j,k\}$,

13. Find the fundamental systems of circuits and cut sets of the spanning tree $\{a,b,c,d\}$

Sol : Fundamental circuits are

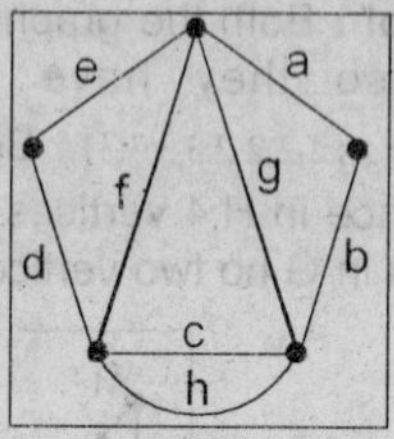

$\{a,b,c,d,e\}$ $\{a,b,g\}$,

$\{a,b,c,f\}$ $\{c,g\}$

Fundamental cut sets

$\{a,g,f,e\}$ $\{b,g,f,e\}$,

$\{c,e,f,h\}$, $\{e,d\}$

14. What is the number of binary trees with 3 nodes which when traversed in post order give the sequence a,b,c. Draw all these binary trees?

Sol : The number of binary trees is 6.

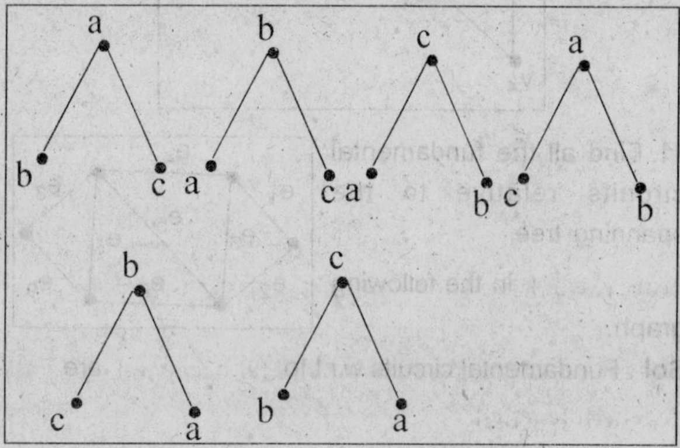

15. How many vertices will the following graphs have ?

(i) If it contain 16 edges and all vertices of degree 2

Sol : $\therefore \delta(G)=\Delta(G)=2$ From the known result

$\delta = \dfrac{2.|E|}{|V|} \Rightarrow \delta\,|V| = 2.|E|$

$2\,|V| = 2\,.\,|E| = 2 \times 16 \;\therefore\; |V| = 16$

(ii) 21 edges, 3 vertices of degree 4, other vertices of degree 3.

Sol : By using $\sum \deg(v_i) = 2.|E|$

$\Rightarrow \quad 3 \times 4 + x \times 3 = 2 \times 21$

Where x is the number of remaining vertices

$3x = 30, \quad x = 10$

$\therefore$ Total number of vertices $(10+3)=13$

16. What is the largest possible number of vertices in a graph with 35 edges and all vertices of degree atleast3?

Sol : We have $\delta(G) \le \dfrac{2|E|}{|V|} \Rightarrow |V| \le \dfrac{2|E|}{\delta}$

Given $\delta = 3, |E| = 35$ then $|V| \le \dfrac{2 \times 35}{3} \le 23.3$

$\therefore$ largest possible number of vertices is 23.

17. Suppose G is an undirected graph with 12 edges and it has 6 vertices of degree 3 and the rest have degree less than 3 determine minimum number of vertices.

Sol : We know $\delta(G) \le \dfrac{2|E|}{|V|} \le \le \Delta(G)$

$\Rightarrow \dfrac{2|E|}{\Delta} \le |V| \le \dfrac{2|E|}{\delta} \Rightarrow 1 \le \dfrac{24}{|V|} \le 3$

(i) consider $\delta(G) = 1, \; \Delta(G) = 3$

$\Rightarrow \dfrac{24}{3} \le |V| \le \dfrac{24}{1} \Rightarrow \quad 8 \le |V| \le 24$

(ii)From (i) and (ii) : possible values of $|V|$ are 8, 12, 24 if $|V| = 8$, we have vertices of degree 3, then remaining vertices are two, which must contribute $(2.12-18=6)$ degree 6. Possible degrees for the two vertices are : 2, 4; 4, 2; 5, 1; 1, 5; 3, 3 clearly a contradiction to the hypothesis of the problem that the degree of remaining vertices should be less than 3.

So $|V| = 8$ is not possible.

Then the next minimum value of $|V| = 12$.

18. Suppose that a tree has d_1 vertices of degree 1, 2 vertices of degree 2, 4 vertices of degree 3 and 3 vertices of degree 4. Find d_1.

Sol: $\sum_{i=1}^{|V|} d(V_i) = 2|E|$

$d_1 + 4 + 12 + 12 = 2\,|E|$

$d_1 + 28 = 2(d_1 + 9 - 1) \Rightarrow d_1 = 12$

19. A full many tree with i internal vertices contains $n = mi + 1$ vertices.

Sol: Every vertex, except root, is the child of an internal vertex. Since each of the i internal vertices has m children, there are mi vertices in the tree other than the root. Therefore the tree contains $n = mi + 1$ vertices.

20. Show that the maximum number of vertices in a binary tree of height n is $2^{n+1} - 1$.

Sol : The maximum number of vertices on level k is 2^k. Hence the maximum number of vertices is $1 + 2 + 2^2 + \ldots + 2^n$ (or) $2^{n+1} - 1$

21. Let G be the graph given below. Find (a) all simple paths from vertex A to vertex Z and (b) d (A, Z).

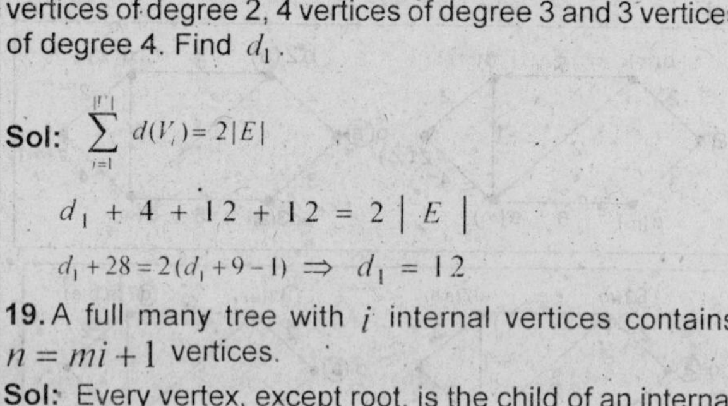

Solution : (a) A path from A to Z is simple if no vertex is repeated.There are four such simple paths as follows.

(A, B, Z), (A, B, Y, Z), (A, X, Y, Z), (A, X, Y, B, Z).

(b) d (A, Z) = 2 ($\because$ the path α = (A, B, Z), of length 2, is the shortest path from A to Z.

22. Suppose you are married and you and your spouse attended a party with three other married couples. Several hand shakes took place. No one shake hands with himself nor with one's own spouse, and no one shake hands with the same person more than once. After all the handshaking was completed. Suppose that you asked each person, including your spouse, how many hands that person had shaken. Each person give a different answer.

(a) How many hands did you shake ?

(b) How many hands did your spouse shake ?

(c) What is the answer to (a) and (b) if there were a total of 5 couples ?

Sol : Here vertices represent the people and dege represents hand shake.

(a) 3 couples (i.e) 6 hand shakes.

(b) 3 couples (i.e) 6 hand shakes.

(c) In general if there are n married and then you and your wife will be shaking hands (n-1) for 5 couples.

(a) = 4 (i.e) 8 hand shakes

(b) = 4 (i.e) 8 hand shakes.

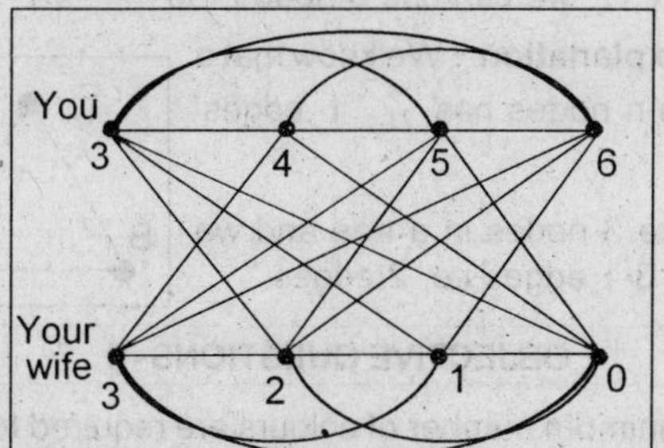

23. What is the largest possible number of vertices in a graph with 35 edges and all vertices of degree at least 3 ?

Sol : We know that the total degree of a graph is

$$\sum_{i=1}^{|V|} d(V_i) = 2|E|$$

Given $\sum_{i=1}^{|V|} 3 \le 2 \times 35$

$3(|V| - 1 + 1) \le 70$

$3|V| \le 70 \Rightarrow |V| \le 23$

$\therefore$ Largest possible number of vertices are 23.

24. Find the chromatic number of the wheel graph ?

Sol :

Case i : Consider a wheel graph with even number if vertices $X(G)=4$. In general $X(G)=4$ if n is even.

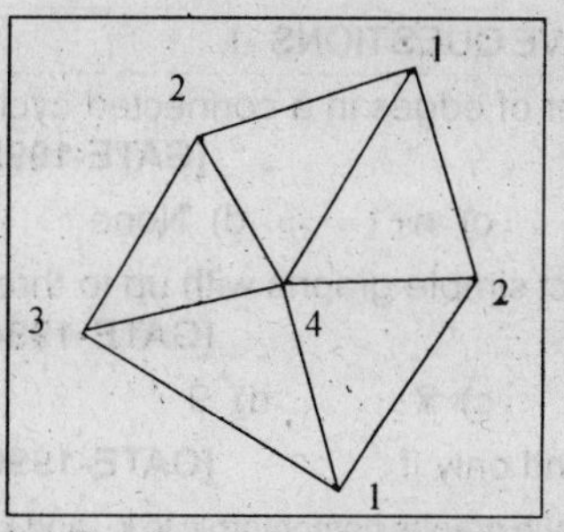

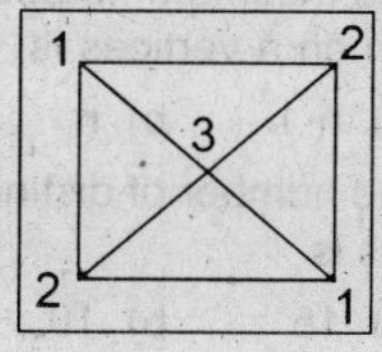

Case ii : Consider a wheel graph of odd number of vertices $X(G)=3$. If n is odd. Hence $X(G)=3$.

25. Find the chromatic number of the complete graph?

Sol : In a complete graph each vertex is adjacent to all the remaining vertices.

Thus if a vertex is give color 1 remaining $n-1$ vertices cannot be given same color. If vertex 2 is given color then remaining $n-2$ vertices cannot be given the same color and so on. If vertex $n-1$ is given color $n-1$, vertex n cannot be given the same color vertex n should be given the same color n. Hence the number of colors required is n. Consider the chromatic number of a complete bipatiate graph $(K_{m,n})$.

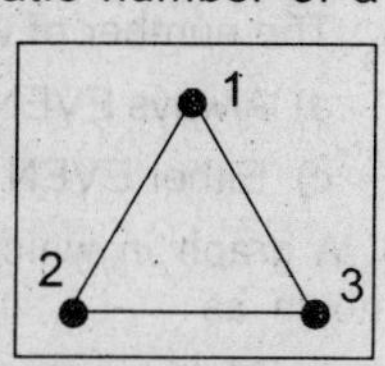

Conider $(K_{m,n})$ in the complete bipartite graph vertices in either partition are not adjancent.ie. vertices in the same set can not be connected among them selves.Hence if we give color 1 to all m vertices in the 1st partition then we give color 2 to all n vertices present in the second partition. Note that no vertex in 2nd partition is adjancent to all vertices in 1st Partition.

$$\therefore \quad X(K_{m,n}) = 2$$

26. What is the max mumber of verices in the entire tree of height ' h' and degree 'K'.

Sol : (i) At level zero there can be only one vertex which is the root therefore $L(0, K) = 1$

(ii) At level on there can be a max of K children for the root since the degree = K. $L(1, K) = K^1 = K$

(iii) At level 2 there can be a max of K children for each vertex at level one. $\therefore L(2, K) = K^2$ and So on.

(iv) At level 'h' each vertex is child of a vertex at level $(h-1)$

$\therefore L(h, K) = L(h-1, K) \,.\; K = K^{h-1} . K = K^{h-1+1} \Rightarrow K^h$

The max number of vertices in the whole tree

$L(0, K) + L(1, K) + L(2, K) + ... L(h, K)$

$$= K^0 + K^1 + K^2 + K^3 + K^h ..... = 1 + K + K^2 + ... + K^h$$

$$\left\{ a + ar + ar^2 + ...ar^h = a.\frac{r^{h+1} - 1}{r - 1} \right\} = 1.\frac{K^{h+1} - 1}{K - 1} = \frac{K^{h+1} - 1}{K - 1}$$

OBJECTIVE QUESTIONS - I

1. The minimum number of edges in a connected cyclic graph on n vertices is **[GATE-1995]**

a) n-1 b) n c) n+1 d) None

2. The number of distinct simple graphs with up to three nodes is **[GATE-1994]**

a) 15 b) 10 c) 7 d) 9

3. A graph is planar if and only if , **[GATE-1990]**

a) It does not contain sub graphs homomorpic to k_3 and $k_{3,3}$

b) It does not contain sub graphs isomorpic to k_5 or $k_{3,3}$

c) It does not contain sub graphs isomorpic to k_5 and $k_{3,3}$

d) It does not contain a sub graphs homomorpic to k_5 or $k_{3,3}$

4. In any undirected graph,

a) The sum of degrees of all the nodes must be EVEN

b) The sum of degrees of all the nodes is twice the number of edges

c) The sum of degrees of all the nodes must be ODD.

d) The sum of degrees of all the nodes need not be EVEN.

5. The number of vertices of ODD degree in a graph is

a) Always EVEN b) Always ODD

c) Either EVEN or ODD d) Always ZERO

6. A graph in which all nodes are of equal degree is known as

a) Multi graph b) Non regular graph

c) Regular graph d) Complete graph

7. The maximum degree of any node in a simple graph with "n" vertices is

a) n-1 b) n c) n/2 d) n-2

8. Two isomorphic graphs must have

a) The same number of vertices

b) The same number of edges

c) An equal number of vertices

d) All of the above

9. A given connected graph G is a Eular graph if and only if all vertices of G are of

a) Same degree b) Even degree

c) Odd degree d) Different degree

10. The total number of edges in a complete graph of n vertices is

a) n b) n/2 c) n(n-1)/2 d) n(n+1)/2

11. A tree with "n" nodes has

a) n/2 edges b) n edges

c) n-1 edges d) n+1 edges

12. The number of circuits in a tree with 'n' nodes is

a) Zero b) One c) n-1 d) n/2

13. A graph is a tree if and only if it

a) Is completely connected b) Is minimally connected

c) Contains a circuit d) Is planner

14. A graph with "n" vertices and n-1 edges that is not a tree, is

a) Connected b) Disconnected

c) Eular d) A circuit

15. A complete graph with "n" vertices is

a) 2-chromatic b) n/2-chromatic

c) (n-1)-chromatic d) n-chromatic

16. The number of different rooted labeled trees with "n" vertices is

a) 2^{n-1} b) 2^n c) n^{n-1} d) n^n

ANSWER KEY - I

1. b	2. c	3. d	4. ?	5. a
6. c	7. a	8. d	9. b	10. c
11. c	12. a	13. b	14. b	15. d
16. c				

EXPLANATORY NOTES - I

1. (b) Explanation : The minimum number of edges in a connected cyclic graph on n vertices is n i.e

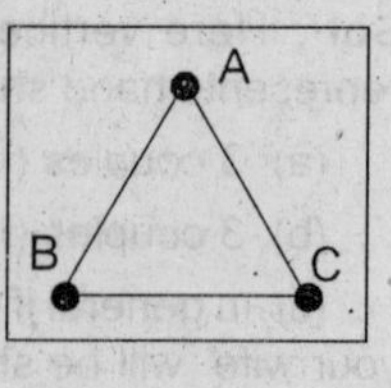

A,B,C are nodes connected cyclically and we can use 3 edges

2. (c) Explanation : We Know that

The number of distinct simple graphs with up to n nodes is $2^n - 1$ $\therefore$ we can use 3 nodes; So $2^3 - 1 = 7$

11. (c) Explanation : We know that a tree with n nodes has $n-1$ edges

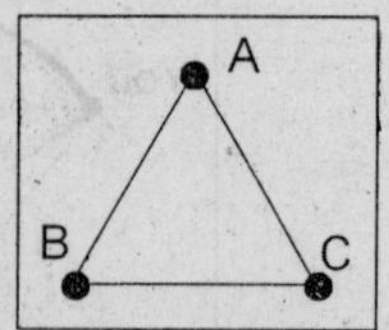

A,B,C are 3 nodes in a tree and we can use 3-1 edges i.e. 2 edges.

OBJECTIVE QUESTIONS - II

1. The minimum number of colours are required to colour the vertices of a cycle with n nodes in such way that no twoadjacent nodes have the same colour is**[GATE-2002]**

a) 2 b) 3 c) 4 d) $n-2\left[\frac{n}{2}\right]+2$

2. Maximum number of edges in a n-node undirected graph without self loops is **[GATE-2002]**

a) n^2 b) n(n-1)/2 c) n-1 d) n(n+1)/2

3. Let G be a graph with 100 vertices numbered 1 to 100. Two vertices i and j are adjacent if | i - j| = 8 or | i – j| = 12. The number of connected components in G is **[GATE-1997]**

a) 8 b) 4 c) 12 d) 25

4. Consider a simple connected graph G with n vertices and n edges (n > 2). Then which of the following statements are TRUE? **[GATE-1993]**

a) G has no cycles. b) G has atleast one cycle

c) The graph obtained by removing any edge from G is not connected d) None

5. Choose the correct statements :

a) If two graphs G1 and G2 are isomorphic, then they should have the same number of vertices and edges.

b) If two graphs have the same number of nodes and edges, they have to be isomorpic

c) Loops can't be present in an isomorphic graph.

d) None

6. T is a graph with "n" vertices. T is connected and has exactly n-1 edges , then

a) T is a tree b) T contains no cycles

c) Every pair of vertices in T is connected by exactly one path

d) The addition of a new edge will create a cycle.

7. The length of a Hamiltanian path (if exists) in a connected graph of n vertices is

a) n-1 b) n c) n+1 d) n/2

8. A simple graph in which there exists an edge between every pair of vertices is called

a) Complete graph b) Eular graph

c) Planner graph d) Regular graph

9. The minimum number of spanning trees in a connected graph with "n" nodes is

a) 1 b) 2 c) n-1 d) n/2

10. If a graph requires k different colours for its proper colouring, then the chromatic number of the graph is

a) 1 b) k c) k-1 d) k/2

11. The number of colours required to properly colour the vertices of every planar graph is

a) 2 b) 3 c) 4 d) 5

ANSWER KEY - II

1. b	**2.** b	**3.** b	**4.** b,c	**5.** ?
6. a,d	**7.** a	**8.** a	**9.** a	**10.** b
11. d	**12.** a	**13.** b	**14.** b	**15.** d
16. c				

EXPLANATORY NOTES - II

1. (b) **Explanation** : The minimum number of colors are required to colour the vertices of a cycle with n nodes in such way that no two adjacent nodes have the same colour is 3

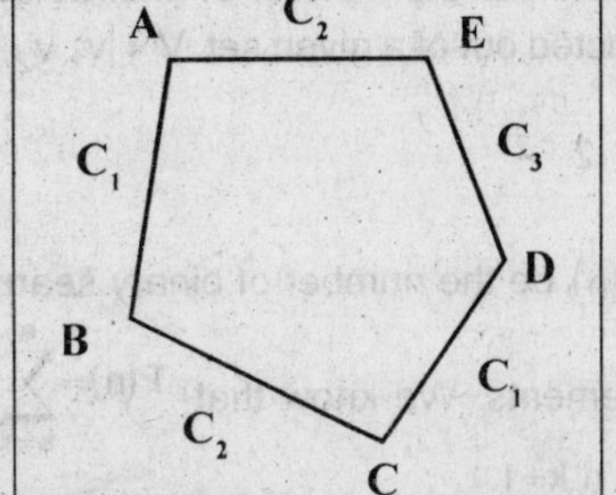

$\therefore$ C_1, C_2, C_3 are colour in alternative positions.

3. (b) **Explanation** : $12 - 8 = 4$

4. (b, c) **Explanation** : A simple connected graph G with n vertices and n edges $(n > 2)$ then G has at least one cycle or the graph obtained by removing any edge from G is not connected.

10. (b) **Explanation** : If a graph requires k different colours for its proper colouring then clearly the chromatic number of the graph is k.

OBJECTIVE QUESTIONS - III

1. The largest possible number of vertices in a graph with 35 edges and all vertices of degree at least 3?

a) 23 b) at most 23 c) at least 23 d)none

2. The number of edges in k-regular graph with n-vertices is

a) Kn b) $\frac{Kn}{2}$ c) $\frac{n}{2}$ d) none

3. Which of the following are true

a) two graphs are isomorphic if and only if their complements are isomorphic.

b) If degree sequences are same then the graphs are isomorphic.

c) there are 4 graphs on 4 vertices which is non isomorphic

d) none

4. The number of vertices in a graph containing 21 edges, 3 vertices of degree 4 and the other vertices of degree 3 are.

a) 10 b) 13 c) 11 d) none

5. Which of the following are true for every graph G

a) there are an odd number of vertices of even degree

b) there are even number of vertices of odd degree.

c) there are an odd number of vertices of odd degree.

d) there are an even number of vertices of even degree.

6. Let G be a simple graph all of whose vertices have degree 3, and $|E| = 2|V| = 3$, What can be said about G.

a) G has 6 vertices and 9 edges.

b) G has 9 vertices and 6 edges.

c) G has 6 vertices and 12 edges

d) none of the above.

7. If G is a simple graph with 15 edges and $\overline{G}$ has 13 edges. How many vertices does G have

a) 6 b) 8 c) 7 d) 5

8. The number of edges in the wheel graph W_n is

a) 2n b) 2n-1 c) 2(n-1) d) n^2

9. The number of non isomorphic simple graphs with five vertices and three edges are

a) 4 b) 3 c) 2 d) 1

10. Suppose that a connected planar simple graph has 20 vertices each of degree 3. Then the number of regions is equal to

a) 10 b) 14 c) 16 d) 12

11. The number of edges in a 50-regular graph with 10 vertices have ?

a) 5000 b) 2500 c) 50 d) none

12. The number of non isomorhpic connected bipartite simple graphs with four vertices are

a) 2 b) 3 c) 4 d) 1

13. The minimum number of vertices necessary for a simple connected graph with 11 edges to be planar are

a) 5 b) 7 c) 6 d) 4

14. Suppose G is a polyhedral graph (degree of each region $\geq$ 3) with 12 vertices and 30 edges, then the degree of each region is

a) 2 b) 3 c) 4 d) 5

15. Suppose that G is a connected planar graph with 35 regions each of degree 6. Then the number of vertices is

a) 70 b) 80 c) 72 d) 62

16. Maximum number of edges in an undirected simple graph with n-vertices are

a) n^2 b) $\frac{n(n-1)}{2}$ c) n-1 d) $\frac{n(n+1)}{2}$

17. The minimum number of colours required to colour the vertices of a cycle with n vertices in such a way that no two adjacent vertices have the same colour is

a) 2 b) 3 c) 4 d) $n-2\left[\frac{n}{2}\right]+2$

18. How many undirected graphs (not necessarily connected) can be constructed out of a given Set $V=\{v_1, v_2.....v_n\}$ of n vertices ?

a) $\frac{n(n-1)}{2}$ b) 2^n c) n! d) $2^{\frac{n(n-1)}{2}}$

19. The chromatic number of tree with n vertices is

a) n-1 b) n+1 c) n d) 2

20. Let $T(n)$ be the number of binary search trees on n distinct elements. Then $T(n)=\sum_{K=1}^{n} T(K-1)T(x)$ where x is

a) n-K+1 b) n-k c) n d) n+1

21. A non planar graph with minimum number of vertices has

a) 9 edges, 6 vertices b) 6 edges, 4 vertices

c) 10 edges, 5 vertices d) 9 edges, 5 vertices

22. The minimum number of edges in a connected cyclic graph on n-vertices

a) n-1 b) n c) n+1 d) none

23. Consider a simple connected graph G with n vertices and n edges $(n>2)$. Then which of the following are true.

a) G has no cycles

b) the graph obtained by removing any edge from G is not connected.

c) G has atleast one cycle.

d) none of the above.

EXPLANATORY NOTES - III

1. Let n be the number of vertices. From the first theorem of graph theory. We know that the degree sum is equal to twice the number of edges. Since all vertices are of degree greater than or equal to 3, the sum of degrees of vertices is greater than or equal to 3n. So $2\times 35\geq 3n$, or $n\leq\frac{70}{3}<24$. Thus the given graph can have atmost 23 vertices.

2. (b) **3.** (a)

4. Let n be the number of vertices in G. Given that 3 vertices are of degree 4 and n-3 vertices are of degree 3. So by first theorem of graph theory we have

$3.4+3(n-3) = 2.21$, $12+3n-9=42$, $3n= 39 \Rightarrow n=13$

5. (b&d) **6.** (a)

7. Suppose G has n vertices. Then G contain ${}^nC_2$ edges. The number of edges in $\overline{G}=\frac{n(n-1)}{2}-m$ where m is the number of edges in G.

$13=\frac{n(n-1)}{2}-m$ where m is the number of edges in G.

$13=\frac{n(n-1)}{2}-15$, $\frac{n(n-1)}{2}=28 \Rightarrow n(n-1)=56, n=8,7$

So, G has 8 vertices.

8. (c) **9.** (a)

10. from the Euler's formula

We have $|V|-|E|+|R|=2\Rightarrow|R|=|E|-|V|+2$, $|V|=20$

We know that sum of degrees of vertices is equal to 2E. (i.e.)

$3+3+3+3+3+3+3+3+3+3+3+3+3+3+3+3+3+3+3+3=2|E|$

$60=2|E|, |E|=30, |R|=30-20+2=12$

11. (b) **12.** (b) **13.** (c)

14. By Euler's formula we have $|V|-|E|+R=2$ the given graph has 12 vertices and 30 edges from Euler's formula we have the number of regions $|R|=2+|E|-|V|$, $2+30-12=20$.

Since G is polyhedral graph, we have degree of each region is greater than or equal to 3. Also in a plane graph G, if the degree of each region is $\geq k$, then $|R|\leq 2|E|$. In particular we have $3|R|\leq 2|E|$.

$\therefore$ degree $(r)=2|E|\geq k|R|$ $r\in R(G)$

$2|E|\geq k|R|\Rightarrow 60\geq K.20\Rightarrow 3\geq k$ or $k\leq 3$ ..... (1)

Since G is polyhedral graph we have $k\geq 3$ ... (2) from (1) and (2) We have k=3. Hence.degree of each region is 3.

15. The gain graph G has 35 regions each of degree 6. Also the degree $(r)=2|E|$, $r\in R(G)$

$\Rightarrow 6(35)=2|E|\Rightarrow|E|=\frac{6\times 35}{2}=105$

The Euler formula is

$|R|=|E|-|V|+2, 35=105-|V|+2, |V|=72$.

16. (b)

17. We know that the minimum number of colours required to colour the vertices of cycle with n vertices in such a way that no two adjacent vertices have the same colour is 3.

18. We know that the number of undirected graphs can be constructed out of a given set $V=\{v_1, v_2, ...v_n\}$ of n vertices is $2^{\frac{n(n-1)}{2}}$.

19. (d)

20. Let $T(n)$ be the number of binary search trees on n distinct elements. We know that $T(n)=\sum_{k=1}^{n} T(k-1)T(x)$ where x is n-k+1.

21. (c) **22.** (b) **23.** (d)

VIII. PRACTICE QUESTIONS

MATHEMATICAL LOGIC

1. Which one of the following statements is true when exactly two of P, Q or R are true ?

(a) $(P \wedge \overline{Q} \wedge R) \vee (P \wedge \overline{Q} \wedge \overline{R}) \wedge (\overline{P} \wedge Q \wedge \overline{R})$

(b) $(P \vee Q \vee R) \wedge (P \vee \overline{Q} \vee R) \wedge (P \vee Q \vee \overline{R})$

(c) $(P \wedge Q \wedge R) \vee (P \wedge \overline{Q} \wedge R) \vee (P \wedge Q \wedge \overline{R})$

(d) $(P \wedge \overline{Q} \wedge R) \vee (P \wedge \overline{Q} \wedge \overline{R}) \wedge (\overline{P} \wedge Q \wedge \overline{R})$

2. $(P \vee Q) \wedge (P \rightarrow R) \wedge (Q \rightarrow R)$ is equivalent to

(a) P (b) Q

(c) R (d) T i.e., true

3. $P \rightarrow (Q \rightarrow R)$ is equivalent to

(a) $(P \wedge Q) \rightarrow R$ (b) $(P \vee Q) \rightarrow R$

(c) $(P \vee Q) \rightarrow \sim R$ (d) $(P \wedge Q) \rightarrow \sim R$

4. $\sim (P \rightarrow Q)$ is equivalent to

(a) $P \wedge \sim Q$ (b) $\sim P \wedge Q$

(c) $\sim P \vee Q$ (d) $P \vee Q$

5. Choose the functionally complete set

(a) $\{\sim \vee V\}$ (b) $\{\downarrow \wedge\}$

(c) $\{\uparrow\}$ (d) $\{\rightarrow, \wedge\}$

6. Pick the correct prefix formula

(a) $\rightarrow P \vee Q R \sim S$

(b) $\rightarrow P \vee Q R S$

(c) $\rightarrow \rightarrow P Q \rightarrow \rightarrow Q R \rightarrow PR$

(d) $\rightarrow P \vee Q \sim Q S P$

7. Consider two well formed formulas in propositional logic. $F_1 : P \rightarrow \sim P$ $F_2 (P \rightarrow \sim P) \vee (\sim P \rightarrow P)$ Which of the following statements is correct..

(a) F_1 is satisfiable, F_2 is valid.

(b) F_1 is unsatisfiable, F_2 is satisfiable

(c) F_1 is unsatisfiable, F_2 is valid

(d) F_1 and , F_2 are both satisfiable

8. Indicate which of the following well-formed formula are valid.

(a) $((P \Rightarrow Q) \wedge (Q \wedge R)) \Rightarrow (P \wedge R)$

(b) $(P \Rightarrow Q) \Rightarrow (\sim P \Rightarrow \sim Q)$

(c) $(P \wedge (\sim P \vee \sim Q)) \Rightarrow Q$

(d) $(P \Rightarrow R) \vee (Q \Rightarrow R) \Rightarrow ((P \vee Q) \Rightarrow R)$

9. Which of the following is/are tautology.

(a) $a \vee b \rightarrow b \wedge c$ (b) $a \wedge b \rightarrow b \vee c$

(c) $(a \vee b) \rightarrow (b \rightarrow c)$ (d) $a \rightarrow b \rightarrow (b \rightarrow c)$

10. Which of the following predicate calculus statements is/are valid.

(a) $(\forall x) P(x) \vee (\forall x) Q(x) \rightarrow (\forall x)\{P(x) \vee Q(x)\}$

(b) $(\exists x) P(x) \wedge (\exists x) Q(x) \rightarrow (\wedge x)\{P(x) \wedge Q(x)\}$

(c) $\forall x \{P(x) \vee Q(x)\} \rightarrow (\forall x) P(x) \vee (\forall x) Q(x)$

(d) $(\exists x)\{P(x) \vee Q(x)\} \rightarrow \sim (\forall x)$

$P(x) \vee (\exists (x) Q(x))$

11. The proposition $P \wedge (\sim P \vee Q)$

(a) a tautology

(b) logically equivalent to $P \wedge Q$

(c) logically equivalent to $P \vee Q$

(d) a contradiction

(e) none of the above.

12. Which of the following is false ? Read $\wedge$ ad AND, $\vee$ as OR, ~ as NOT, $\rightarrow$ as one way implication and $\leftrightarrow$ as two way implication.

(a) $((x \rightarrow y) \wedge x) \rightarrow y$

(b) $((\sim x \rightarrow y) \wedge (\sim x \wedge \sim y)) \rightarrow x$

(c) $(x \rightarrow (x \vee \phi))$

(d) $(x \vee y) \leftrightarrow (\sim x \rightarrow \sim y)$

13. Which of the following propositions as a tautology

(a) $(P \vee q) \rightarrow P$ (b) $P \vee (q \rightarrow P)$

(c) $P \vee (P \rightarrow q)$ (d) $P \rightarrow (P \rightarrow q)$

14. Let a, b, c, d be propositions. Assume that the equivalences $a \not\subset (b \vee 7b)$ and $a \vee b$ hold. Then the truth $(a \wedge b) \rightarrow ((a \wedge c) \vee d)$ is always.

(a) true

(b) false

(c) same as the truth value of b

(d) same as the truth value of d

15. The expression

$(\sim P \wedge (\sim Q \wedge R)) \vee (Q \wedge R) \vee (P \wedge R)$ is equivalent to

(a) P

(b) Q

(c) R

(d) none of the above

16. $P \to (Q \to R)$ is equivalent to

(a) $(P \wedge Q) \to R$ (b) $(P \vee Q) \to R$

(c) $(P \wedge Q \wedge R)$ (d) $7 P \wedge Q \wedge R$

17. Which of the following sets are functionally complete ?

(i) $\{\sim, \to\}$ (ii) $\{\vee, \wedge\}$ (iii) $\{\uparrow\}$ (iv) $\{\vee, \uparrow\}$

(a) (i) & (iii) (b) (ii) & (iv)

(c) (i) alone (d) (iii) & (iv)

18. The principal disjunctive normal form of $P \to Q$ is

(a) $\sim P \wedge Q$

(b) $\sim P \vee Q$

(c) $(P \wedge Q) \vee (\sim P \wedge Q) \vee (\sim P \wedge \sim Q)$

(d) $(P \wedge Q) \vee (\sim P \wedge Q) \vee (P \wedge \sim Q)$

19. The prefix form of $A \wedge (B \vee \sim C)$ is

(a) $\wedge A B \vee \sim C$ (b) $\wedge A \vee B \sim C$

(c) $\sim \wedge \vee A B$ (d) $\vee \wedge \sim A B$

20. The prefix notation of $\sim (P \wedge Q) \leftrightarrow (\sim P \vee \sim Q)$

(a) $\vee \wedge \sim P \leftrightarrow Q \sim P \sim Q$

(b) $\leftrightarrow \sim \wedge P \vee \sim P \sim Q$

(c) $P q \wedge \sim P \sim q \sim \vee \leftrightarrow$

(d) none of the above

21. The postfix notation of $(\sim P \wedge (Q \leftrightarrow \sim P)) \vee \sim Q$ is

(a) $P \sim Q P \sim \leftrightarrow \wedge Q \sim \vee$

(b) $\vee \wedge \sim P \leftrightarrow Q \sim P \sim Q$

(c) $P Q \wedge \sim P \sim Q \sim \vee \leftrightarrow$

(d) none of the above

22. Which of the following sentences are propositions.

(a) answer this question

(b) do not pass go

(c) what time is it ?

(d) miami is the capital of florida

23. Which of the following are not logically equivalent?

(a) $P \leftrightarrow Q\, and\, (P \wedge Q) \vee (\sim P \wedge \sim Q)$

(b) $(P \to Q) \to R\, and\, P \to (Q \to R)$

(c) $\sim P \leftrightarrow Q\, and\, P \leftrightarrow \sim Q$

(d) $\sim (P \leftrightarrow Q)\, and \sim P \leftrightarrow Q$

24. $(P \downarrow Q) \downarrow (P \downarrow Q)$ is logically equivalent to

(a) P (b) $P \vee Q$ (c) $P \wedge Q$ (d) Q

25. $(P \uparrow Q) \uparrow (P \uparrow Q)$ is logically equivalent to

(a) $\sim P$ (b) $\sim Q$

(c) P (d) $\sim (P \uparrow Q)$

26. $(P \downarrow P) \downarrow (Q \downarrow Q)$ is equivalent to

(a) $P \vee Q$ (b) $P \wedge Q$

(c) $\sim Q$ (d) $\sim P \vee Q$

27. $(P \uparrow P) \uparrow (Q \uparrow Q)$ is equivalent to

(a) $P \vee Q$ (b) $P \wedge Q$

(c) $\sim P(d)$ (d) $\sim Q$

28. Which of the following arguments are valid.

(a) $P \to Q, P \to R, Q \to \sim R, P$

(b) $A \to (B \to C), D \to (B \wedge \sim C), A \wedge D$

(c) $\sim P \vee Q, \sim Q \vee R, R \to S \Rightarrow P \to S$

(d) none of the above

29. Which of the following arguments are valid.

(a) $E \to S, S \to H, A \to, \sim H, E \wedge A$

(b) $R \vee M, \sim R \vee S, \sim M, \sim S$

(c) $P \to Q, \sim Q, P$

(d) $P \to Q, Q \to \sim R, R, P \vee (J \wedge S), J \wedge S$

30. If X then Y unless Z is represented by which of the following formulas in propositional logic ?

(a) $(X \wedge \sim Z) \to Y$ (b) $(X \wedge Y) \to \sim Z$

(c) $X \to (Y \wedge \sim Z)$ (d) $(X \to Y) \wedge \sim Z$

31. Which of the following are statements

(i) $(x)(P(x) \vee Q(x)) \wedge R$

(ii) $(x)(P(x) \wedge Q(x)) \wedge (\exists x) S(x)$

(iii) $(x)(P(x) \wedge Q(x)) \wedge S(x)$

(a) (i) (b) (ii) & (iii)

(c) (ii) (d) none of the above

32. "Not all birds can fly". Which of the following formulas represent in predicate calculus.

(a) $\sim (\exists x B(x) \wedge F(x))$

(b) $\exists x, (B(x) \wedge [\sim F(x)])$

(c) $\forall x, B(x) \to \sim F(x)$

(d) none of the above

33. The negation of "every complete bipartite graph is not planar" in predicate calculus is

(a) $\forall x [C(x) \wedge B(x) \to P(x)]$

(b) $\sim \forall (x) [C(x) \wedge B(x) \to \sim P(x)]$

(c) $\exists x [C(x) \wedge B(x) \wedge P(x)]$

(d) none of the above

34. Without using universal quantifier, the translated form of "No dogs are intelligent" in predicate calculus is

(a) $\exists x \{\sim D(x) \wedge I(x)\}$ (b) $\sim [\exists (x) \{D(x) \wedge I(x)\}]$

(c) $\exists x\, D(x) \wedge \sim I(x)$ (d) none of the above

35. If P has truth value T, Q and R have truth value F then $(P\wedge\sim Q)\rightarrow R, P\rightarrow(Q\vee R)$

(a) both have same truth value (F)

(b) $(P\wedge\sim Q)\rightarrow R$ is true and $P\rightarrow(Q\vee R)$ is false

(c) both have same truth value T

(d) none of the above

36. The number of propositional functions of 2 variables are

(a) 4 (b) 8

(c) 16 (d) 2

37. Which of the following propositions are tautologies.

(i) $(\sim Q\wedge(P\rightarrow Q))\rightarrow\sim P$

(ii) $((P\vee q)\wedge\sim P)\rightarrow q$

(iii) $(\sim P\vee\sim Q)\rightarrow(P\overrightarrow{\leftarrow}\sim Q)$

(iv) $Q\wedge(P\vee\sim Q)\rightarrow Q$

(a) (i) & (ii) (b) (i) & (iv)

(c) (iii) & (iv) (d) (ii) & (iii)

38. Let $P(x)$ and $Q(x)$ be propositional functions. Then which of the following have same truth values.

(i) $\exists x(P(x)\rightarrow Q(x))\, and\, "x\, P(x)\rightarrow\exists x\, Q(x)$

(ii) $\exists x(P(x)\vee Q(x))\, and\, \exists x\, P(x)\vee\exists x\, Q(x)$

(iii) $\forall x(P(x)\wedge Q(x))\, and\, \forall x\, P(x)\wedge\forall x\, Q(x)$

(iv) $\exists x\, P(x)\wedge\exists x\, Q(x)\, and\, \exists x(P(x)\wedge Q(x))$

(a) (i), (iii), (iv) (b) (i), (ii), (iii)

(c) (ii), (iii), (iv) (d) (i), (ii), (iii), (iv)

39. Which of the following are not logically equivalent.

(i) $\forall x\, P(x)\vee\forall x\, Q(x)\, and\, \forall x(P(x)\vee Q(x))$

(ii) $\exists x\, P(x)\wedge x\, Q(x)\, and \wedge x(P(x)\wedge Q(x))$

(iii) $\forall x\, P(x)\wedge\exists x\, Q(x)\, and\, \forall x\exists y\, P(x)\wedge Q(y)$

(iv) $\forall x\, P(x)\vee\forall x\, Q(x)\, and\, \forall x\forall y\, P(x)\vee Q(y))$

(a) (i) & (ii) (b) (ii) & (iii)

(c) (iii) & (iv) (d) (i) & (iv)

40. Which of the following are equivalent.

(i) $\forall x\forall y\, P(x,y)\, and\, \forall x\forall y\, P(x,y)$

(ii) $\forall y\forall x\, P(x,y)\, and\, \exists x\forall y\, P(x,y)$

(iii) $\exists x\forall y\, P(x,y)\, and\, \forall y\exists x\, P(x,y)$

(iv) $\exists y\forall x\, P(x,y)\, and\, \exists x\exists y\, P(x,y)$

(a) (i), (iii), (iv)

(b) (i), (ii), (iii)

(c) (i), (ii), (iv)

(d) none of the above

ANSWER KEY

1. (c)	2. (c)	3. (a)	4. (a)	5. (c)
6. (c)	7. (c)	8. (c)	9. (b)	10. (a)&(b)
11. (b)	12. (d)	13. (c)	14. (a)	15. (c)
16. (a)	17. (a)	18. (c)	19. (b)	20. (b)
21.(a)	22. (d)	23. (b)	24. (b)	25. (d)
26. (b)	27. (a)	28. (c)	29. (d)	30. (d)
31. (b)	32. (b)	33. (c)	34. (b)	35.(a)
36. (c)	37. (a)	38. (b)	39. (a)	40. (b)

SETS,RELATIONS,BOOLEAN ALGEBRA & GROUP THEORY

1. If $A=\{a,b,\{a,c\},\phi\}$ then $\{a\}-\{A\}$ is
(a) ϕ (b) $\{a\}$
(c) $\{b,\{a,c\},\phi\}$ (d) $\{\phi\}$

2. Let $A=\{1,2,3,4\}$. Let $\sim=\{(1,2),(1,3),(4,2)\}$ choose the true statement
(a) ~ is not antisymmetric
(b) ~ is not irreflexive
(c) ~ is transitive
(d) ~ is symmetric

3. Let $A=\{1,2,3...\}$ be the set of positive integers. Let $x\sim y$ if x divides y the relation ~ is
(a) symmetric, asymmetric and antisymmetric
(b) not symmetric, not asymmetric and not antisymmetric
(c) symmetric, not asymmetric and antisymmetric
(d) not symmetric, not asymmetric and antisymmetric

4. Let $A=Z$ be the set of integers $=\{...-3,-2,0,1,2....\}$ Let $x\sim y$ if $x<y$ then ~ is
(a) not symmetric, not asymmetric and antisymmetric
(b) not symmetric, asymmetric and not antisymmetric
(c) not symmetric, asymmetric and antisymmetric
(d) symmetric, not asymmetric and antisymmetric

5. Transitive and irreflexive imply
(a) symmetric (b) reflexive
(c) irreflexive (d) asymmetric

6. Necessary condition for existence of 1 - 1 mapping from X to Y is
(a) $|X|=|Y|$ (b) $|X|\le|Y|$
(c) $|X|\ge|Y|$ (d) $f(x)\le Y$

7. Every finite subset of lattice has
(a) and LUB and a GLB
(b) many LUBs
(c) many LUBs and many GLBs
(d) Either some LUBs or some GLBs

8. If the Lattice $(L,\le)$ is complemented chain then
(a) $|L|\le 1$ (b) $|L|\le 2$
(c) $|L|>1$ (d) L does not exist

9. Let A and B be sets with cardinalities m and n respectively. The number of one-one mappings from A to B when $m<n$ is
(a) m^n (b) $n P_m$
(c) $m C_n$ (d) $n C_m$ (e) $m P_n$

10. The number of elements in the power set $P(S)$ of the set $S=\{\phi,1,\{2,3\}\}$ is
(a) 2 (b) 4
(c) 8 (d) none of the above

11. Let $X=\{2,3,6,12,24\}$, Let $\le$ be the partial order defined by $x\le y$. If x divides y. Number of edges in the Hasse diagram of $(X,\le)$ is
(a) 3 (b) 4
(c) 9 (d) none of the above

12. Suppose X and Y are sets and $|X|$ and $|Y|$ are their respective cardinalities. It is given that there are exactly 97 functions from X to Y. From this one can conclude that
(a) $|X|=1,|Y|=97$ (b) $|X|=97,|Y|=1$
(c) $|X|=97,|Y|=97$ (d) none of the above

13. Let R be a non empty relation on a collection of sets defined by ARB if and only if $A\cap B=\phi$ then pick the true statement
(a) R is reflexive and transitive
(b) R is symmetric but not transitive
(c) R is an equivalence relation
(d) R is not reflexive and not symmetric

14. Let A and B be sets and A^C and B^C denote the complements of the sets A and B. The set $(A-B)\cup(B-A)\cup(A\cap B)$ is equal to
(a) $A\cup B$ (b) $A^C\cup B^C$
(c) $A\cap B$ (d) $A^C\cap B^C$

15. In the lattice given by the Hasse diagram in the following below figure how many complements does the elements e have
(a) 2 (b) 3 (c) 0 (d) 1

16. A partial order $\le$ is defined on the set $S=\{x,a_1,a_2...a_n,y\}$ as $x\le a_i$ for all i and $a_i\le y$ for all i when $n\ge 1$. The number of total orders on the set S which contain partial order $\le$ is
(a) $n!$ (b) $n+2$ (c) n (d) 1

17. The number of equivalence relations of the set $\{1,2,3,4\}$ is
(a) 15 (b) 16 (c) 24 (d) 4

18. Suppose A is a finite set with n elements. The number of elements in the largest equivalence relation of A is
(a) n (b) n^2 (c) 1 (d) $n+1$

19.A relation R is defined on the set of integers as xRy if $x+y$ is even. Which of the following is true.
(a) R is not an equivalence relation
(b) R is an equivalence relation having 1 equivalence class
(c) R is an equivalence relation having 2 equivalence classes
(d) R is an equivalence relation having 3 equivalence classes

20. Let $P(S)$ denote the power set of S. Which of the following is always true
(a) $P(P(S))=P(S)$
(b) $P(S)\cap P(P(S))=\phi$
(c) $P(S)\cap S=P(S)$
(d) $S\notin P(S)$

21. Consider the following relations.
aR_1b iff $a+b$ is even over the set of integers.
aR_2b iff $a+b$ is odd over the set of integers.
aR_3b iff $ab>0$ over the set of non-zero relational numbers.
aR_4b iff $|a-b|\leq 2$ over the set of natural numbers.
Which of the following statements are correct
(a) R_1 and R_2 are equivalence relations, R_3 and R_4 are not
(b) R_1 and R_3 are equivalence relations, R_2 and R_4 are not
(c) R_1 and R_4 are equivalence relations, R_2 and R_3 are not
(d) R_1, R_2, R_3, R_4 are all equivalence relations.

22. Let S be an infinite set and $S_1, S_2 ... S_n$ be sets such that $S=S_1\cup S_2\cup ...\cup S_n=S$. Then
(a) at least one of the sets S_i is a finite set
(b) not more than one of the sets S_i can be finite
(c) at least one of the sets S_i is an infinite sets
(d) not more than one of the sets S_i can be infinite

23. The binary relation $R=\{(1,1),(2,1),(2,2),(2,3),(2,4),(3,1),(3,2),(3,3),(3,4)\}$ on the set $A=\{1,2,3,4\}$ is
(a) reflexive, symmetric and transitive
(b) neither reflexive, nor irreflexive but transitive
(c) irreflexive, symmetric and transitive
(d) irreflexive and antisymmetric

24. Let $A=\{1,3,5,7,9\}$, $B=\{2,4,6,8\}$ aRb if and only if $b<a$. Find matrix of R

(a) $\begin{bmatrix}1&0&1&0\\0&1&0&0\\1&0&1&1\\0&1&0&1\\1&0&1&1\end{bmatrix}$ (b) $\begin{bmatrix}1&1&1&1\\0&1&1&1\\0&0&1&1\\0&0&0&1\\0&0&0&0\end{bmatrix}$

(c) $\begin{bmatrix}0&0&0&0\\1&c&0&0\\1&1&0&0\\1&1&1&0\\1&1&1&1\end{bmatrix}$ (d) none of the above

25. If $A=\{1,2,3,4,8\}=B.aRb$ if and only if $a=b$. Then the matrix of R is

(a) $\begin{bmatrix}1&0&0&0&0\\0&1&0&0&0\\0&0&1&0&0\\0&0&0&1&0\\0&0&0&0&1\end{bmatrix}$ (b) $\begin{bmatrix}1&0&1&0&0\\1&0&0&0&0\\1&0&0&0&0\\1&0&1&1&0\\1&0&0&0&1\end{bmatrix}$

(c) $\begin{bmatrix}1&0&0&0&1\\0&1&0&0&1\\1&0&1&1&0\\0&1&1&1&0\\1&0&0&0&0\end{bmatrix}$ (d) none of the above

26. The binary relation $S=\phi$ on set $A=\{1,2,3\}$ is
(a) Neither reflexive on or symmetric
(b) symmetric and reflexive
(c) transitive and reflexive
(d) transitive and symmetric

27. If A and B are any two sets then $A-(A-B)$ is equal to
(a) B (b) ϕ (c) $A\cap B$ (d) $A\cup B$

28. The complement of $(\overline{A}\cap B)\cap(A\cup\overline{B})\cap(A\cup C)$ is
(a) $(A\cup\overline{B})\cup(\overline{A}\cap(B\cup\overline{C}))$
(b) $(\overline{A}\cap B)\cap(\overline{A}\cup(B\cap\overline{C}))$
(c) $(\overline{A}\cup\overline{B})\cup(A\cap(B\cup C))$
(d) none of the above

29. Let A be a set. If $P(A)$ as 256 elements. Then the number of elements in A are
(a) 9 (b) 8 (c) 6 (d) 6

30. Let A be the set of relation numbers. For $a,b \in A$, define $(a,b) \in R$ if $a-b$ is an integer. Then R is
(a) an equivalence relation
(b) reflexive and symmetric but not transitive
(c) reflexive but not symmetric and transitive
(d) none of the above

31. The number of different equivalence relations on a set with n elements for $n=1,2,3,4$ *and* 5 are
(a) 1, 1, 7, 15, 32 (b) 1, 2, 7, 15, 30
(c) 1, 2, 3, 4, 5 (d) 1, 1, 7, 15, 52

32. The number of equivalence classes for the equivalence relation "Two people one equivalent if they are born in the same week" are
(a) 1 (b) 2 (c) 53 (d) none

33. If A, B, C, D are any four sets then $(A \cap B) \times (C \cap D)$ is equal to
(a) $(A \times B) \cup (C \times D)$ (b) $(A \times C) \cap (B \times D)$
(c) $(A \cup B) \times (C \cup D)$ (d) none of the above

34. $(A = \phi) \vee (B = \phi) \vee (A = B)$ is equivalent to
(a) $A \cup B = B \cup A$ (b) $A \cap B = B \cap A$
(c) $A \times B = B \times A$ (d) none of the above

35. If A, B, C are any three sets then $(A-B)-C$ is equal to
(a) $A-(B-C)$ (b) $(A-C)-(B-C)$
(c) $(A-C) \nabla (B-C)$ (d) none of the above

36. Let R and S be two relations on a set of positive integers I:
$R = \{(x, 2x) : x \in I\}$, $S = \{(x, 7x) : x \in I\}$
Then $ROSOR$ is
(a) $\{(x; 14x) : x \in I\}$ (b) $\{(x, 8x) : x \in I\}$
(c) $\{(x, 28x) : x \in I\}$ (d) none of the above

37. The number of partitions of a set of four elements are
(a) 16 (b) 8 (c) 15 (d) 14

38. The number of partitions of a set containing 5 elements are
(a) 32 (b) 5 (c) 52 (d) 64

39. Every interval of a lattice is a
(a) boolean algebra
(b) sublattice
(c) distributive
(d) none of the above

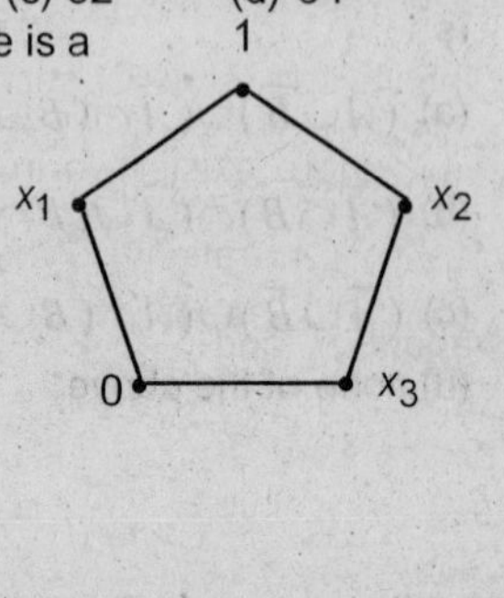

40. The number of complements of x_1 in the above lattice.
(a) zero (b) one
(c) two (d) three

41. Which of the following pairs of elements are in comparable in the poet $(Z^+, 1)$
(a) 5, 15 (b) 6 , 9
(c) 8, 16 (d) 7, 7

42. The maximal elements of the poset
$(\{\{1\},\{2\},\{4\},\{1,2\},\{1,4\},\{2,4\},\{3,4\},\{1,3,4\},\{2,3,4\}\} \subseteq)$
(a) $\{1,2\},\{1,3,4\},\{2,3,4\}$
(b) $\{1,2\},\{1,4\},\{2,4\}$
(c) $\{1,3,4\},\{1,2\},\{3,4\}$
(d) $\{1\},\{2\},\{2,3,4\}$

43. Which of the following posets are lattice.

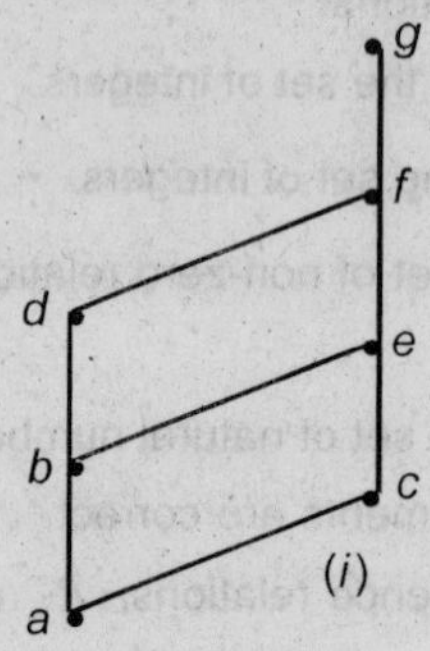

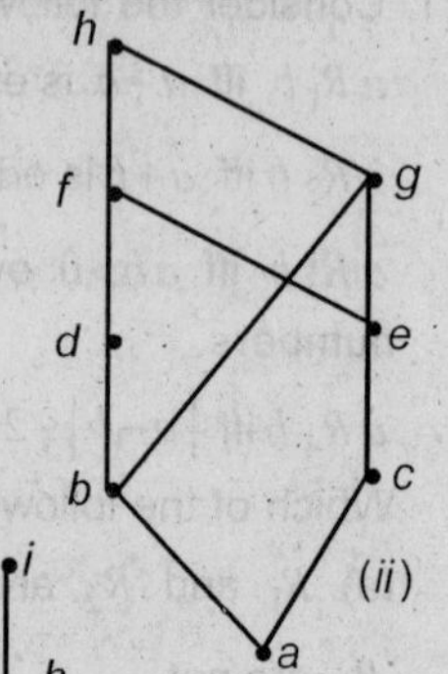

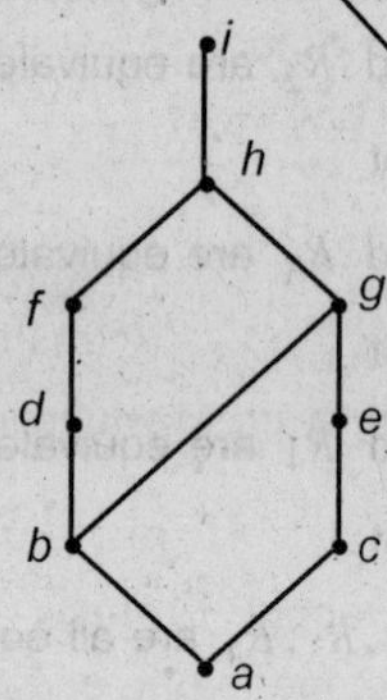

(a) (i) and (ii)
(b) (ii) & (iii)
(c) (i) and (iii)
(d) (ii) and (iii)

44. The number of different equivalence relations with exactly three different equivalence classes are there on a set with five elements
(a) 25 (b) 32 (c) 8 (d) none

45. Let R be the relation represented by the matrix
$$M_R = \begin{bmatrix} 0 & 1 & 0 \\ 0 & 0 & 1 \\ 1 & 1 & 0 \end{bmatrix}$$
Then the matrix represented by R^3 is

(a) $\begin{bmatrix} 1 & 1 & 1 \\ 1 & 1 & 1 \\ 1 & 1 & 1 \end{bmatrix}$ (b) $\begin{bmatrix} 0 & 0 & 1 \\ 1 & 1 & 0 \\ 0 & 1 & 1 \end{bmatrix}$

(c) $\begin{bmatrix} 1 & 1 & 0 \\ 0 & 1 & 1 \\ 1 & 1 & 1 \end{bmatrix}$ (d) $\begin{bmatrix} 0 & 1 & 1 \\ 1 & 1 & 1 \\ 1 & 1 & 1 \end{bmatrix}$

46. How many elements are in $A_1 \cup A_2$ if there are 12 elements in A_1, 18 elements in A_2 and $A_1 \subseteq A_2$

(a) 12 (b) 18
(c) 30 (d) none

47. The number of reflexive relations on a set with 4 elements

(a) 2^{16} (b) 2^{12} (c) 4^2 (d) 4^4

48. The number of relations on a set with n elements that are antisymmetric are

(a) $2^{n(n-1)}$ (b) 2^{n^2}

(c) $2^n 3^{n(n-1)/2}$ (d) none

49. Which of the following collections of subsets are partitions of $\{1,2,3,4,5,6\}$?

(i) $\{1,2\},\{2,3,4\},\{4,5,6\}$

(ii) $\{1\},\{2,3,4\},\{4,5,6\}$

(iii) $\{2,4,6\},\{1,3,5\}$

(iv) $\{1,4,5\},\{2,6\}$

(a) (ii) & (iii) (b) (i), (iii), (iv)
(c) (i), (ii) (d) (iii), (iv)

50. Let f and g are two functions from $N \times N$ to N given by $f(x,y)=x+y, g(x,y)=xy$. Then what we can say about f and g.

(a) both are one-to-one and onto
(b) both are onto but not one to one
(c) both are one to one but not onto
(d) both are not one-one and onto

51. Let $g: I \times I \to I$ where I is the set of integer and $g(x,y)=x*y=x+y-xy$ Then the binary operation * is

(a) commutative and associative
(b) associative not commutative
(c) neither associative nor commutative (d) none of the above

52. Which of the following relations are partial ordered relations.

(i) $A=Z$ and aRb if and only if $a=2b$

(ii) $A=Z$ and aRb iff $b^2 | a$

(iii) $A=Z$ and aRb iff $a=b^k$ for sum k

(a) (i) & (ii) (b) only (iii)
(c) (i), (iii) (d) none

53. Which of the following are true about a poset

(i) A non-empty finite poset has a maximal element
(ii) A non-empty finite poset has a greatest element
(iii) A non-empty finite poset has a minimal element
(iv) A non-empty finite poset has a least element

(a) (i) & (ii) (b) all are true
(c) (i) & (iii) (d) (i) & (iv)

54. Which of the following Hasse diagrams represent lattices ?

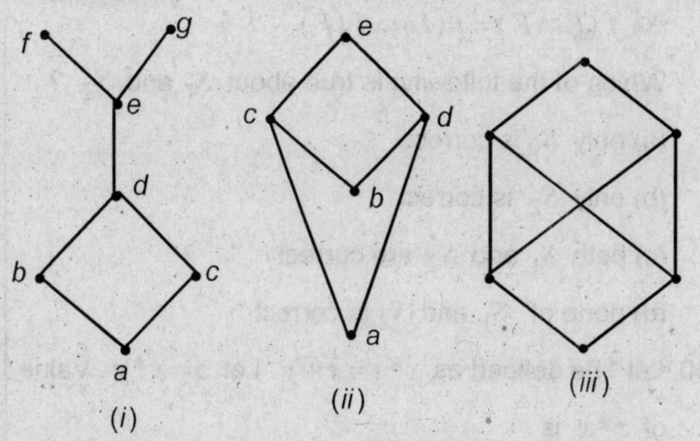

(a) (i) & (ii) (b) (ii) only
(c) (i), (ii), (iii) (d) none of these

55. The following lattice

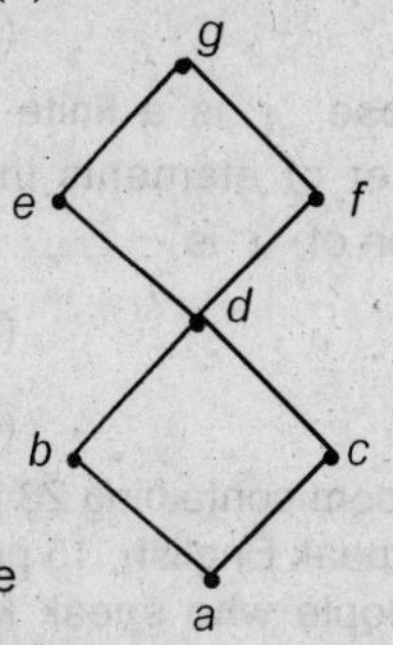

(a) modular
(b) distributive but not complemented
(c) modular but not complemented
(d) complemented lattice

56. Let $A=Z$, aRb if and only if $|a-b|=2$ then R is

(a) irreflexive and symmetric
(b) antisymmetric
(c) transitive
(d) antisymmetric and transitive

57. The relation R on the set A if $A=Z^+$, xRy if and only if $x \le 3y$

(a) reflexive
(b) reflexive, symmetric, transitive
(c) transitive
(d) none of the above

58. Consider the following statements S_1: There exist infinite sets A,B,C such that $A \cap (B \cap C)$ is finite.

S_2: there exists two irrational numbers such that $(x+y)$ is rational. Which of the following is true about S_1 and S_2.

(a) only S_1 is correct

(b) only S_2 is correct

(c) both S_1 and S_2 are correct

(d) none of S_1 and S_2 is correct

59. Let $f:A \to B$ be a function and let E and F be subsets of A. Consider the following statements about images

$S_1: f(E \cup F) = f(E) \cup f(F)$,

$S_2\ f(E \cap F) = f(E) \cap f(F)$

Which of the following is true about S_1 and S_2 ?

(a) only S_1 is correct

(b) only S_2 is correct

(c) both S_1 and S_2 are correct

(d) none of S_1 and S_2 is correct

60. Let * be defined as $x * y = \overline{x} + y$. Let $z = x * y$. Value of $z * x$ is

(a) $\overline{x} + y$ (b) x

(c) 0 (d) $=1$

61. Suppose A is a finite set with n elements the number of elements in the largest equivalence relation of A is

(a) n (b) n^2

(c) 1 (d) $n+1$

62. In a room containing 28 people there are 18 people who speak English, 15 people who speak Hindi and 22 people who speak kannada a persons speak both English and Hindi, 11 persons speak both persons speak both Kannada and English. How many people speak all there languages ?

(a) 9 (b) 8
(c) 7 (d) 6

63. Let A be a finite set of size n. The 21 number of elements in the power set of $(A \times A)$ is

(a) 2^{2^n} (b) 2^{n^2}

(c) $\left(2^n\right)^2$ (d) $\left(2^2\right)^n$

64. Let R be a symmetric and transitive relation on a set A. Then

(a) R is reflexive and hence an equivalence relation

(b) R is reflexive and hence a partial order

(c) R is reflexive and hence not an equivalence relation

(d) none of the above

65. Consider the function $j: N \times N \to N$ such that $h(a,b) = (2a+1)2^b - 1$ where $N = \{0,1,2,3......\}$ is the set of natural numbers. Then

(a) h is one one but not onto

(b) h is neither one one nor onto

(c) h is one one and onto

(d) none of the above

66. Let f and g be defined by $f(x) = 2x+1$ and $g(x) = x^2 - 2$ respectively then $gof(a+2) =$

(a) $2a^2 + 8a + 5$ (b) $4a^2 + 20a + 23$

(c) $(2a+5)^2$ (d) none of the above

67. The number of partitions of the set $S = \{1,2,3\}$ are

(a) 3 (b) 2
(c) 5 (d) none

68. Which of the following are the partitions of the set R of the real numbers

(i) $[\{x : x > 4\}, \{x : x < 5\}]$

(ii) $[\{x : x > 0\}, \{0\}, \{x : x < 0\}]$

(iii) $[\{x : x^2 > 11\}, \{x : x^2 < 11\}]$

(a) (i), (ii), (iii) (b) (ii) only
(c) (i) and (iii) (d) one

ANSWER KEY

1. (b) 2. (c) 3. (d) 4. (b) 5. (a) 6. (c) 7. (a) 8. (b) 9. (b) 10. (c) 11. (b) 12. (a) 13. (b) 14. (a) 15. (b) 16. (a) 17. (a) 18. (b) 19. (c) 20. (b) 21. (d) 22. (a) 23. (b) 24. (c) 25. (a) 26. (d) 27. (c) 28. (a) 29. (b) 30. (a) 31. (d) 32. (c) 33. (b) 34. (c) 35. (b) 36. (c) 37. (c) 38. (c) 39. (b) 40. (c) 41. (b) 42. (a) 43. (c) 44. (a) 45. (c) 46. (b) 47. (b) 48. (c) 49. (a) 50. (b) 51. (a) 52. (b) 53. (b) 54. (b) 55. (b) 56. (a) 57. (a) 58. (c) 59. (a) 60. (b) 61. (b) 62. (d) 63. (b) 64. (d) 65. (c) 66. (b) 67. (c) 68. (b)

GRAPH THEORY(SET - I)

1. The largest possible number of vertices in a graph with 35 edges and all vertices of degree at least 3?
(a) 23 (b) at most 23
(c) at least 23 (d) none

2. The number of edges in $K-$regular graph with $n-$vertices is
(a) Kn (b) $\frac{Kn}{2}$ (c) $\frac{n}{k}$ (d) none

3. Which of the following are true
(a) two graphs are isomorphic if and only if their complements are isomorphic.
(b) If degree sequence are same then the graphs are isomorphic
(c) there are 4 graphs on 4 vertices which is non isomorphic
(d) none

4. The number of vertices in graph containing 21 edges, 3 vertices of degree 4 and the other vertices of degree 3 are.
(a) 10 (b) 13 (c) 11 (d) none

5. Which of the following are true for every graph G
(a) there are an odd number of vertices of even degree
(b) there are even number of vertices of odd degree
(c) there are an odd number of vertices of odd degree
(d) there are an even number of vertices of even degree

6. Let G be a simple graph all of whose vertices have degree 3. and $|E|=2|V|-3$. What can be said about G.
(a) G has 6 vertices and 9 edges
(b) G has 9 vertices and 6 edges
(c) G has 6 vertices and 12 edges
(d) none of the above

7. If G is a simple graph with 15 edges and $\overline{G}$ has 13 edges. How many vertices does G have
(a) 6 (b) 8 (c) 7 (d) 5

8. The number of edges in the wheel graph W_n is
(a) $2n$ (b) $2n-1$
(c) $2(n-1)$ (d) n^2

9. The number of non isomorphic simple graphs with five vertices and three edges are
(a) 4 (b) 3
(c) 2 (d) 1

10. Suppose that a connected planar simple graph has 20 vertices each of degree 3. Then the number of regions is equal to
(a) 10 (b) 14
(c) 16 (d) 12

11. The number of edges in a 50 - regular graph with 10 vertices have ?
(a) 5000 (b) 2500
(c) 50 (d) none

12. The number of non isomorphic connected bipartite simple graphs with four vertices are
(a) 2 (b) 3 (c) 4 (d) 1

13. The minimum number of vertices necessary for a simple connected graph with 11 edges to be planar are
(a) 5 (b) 7 (c) 6 (d) 4

14. Suppose G is a polyhedral graph (degree of each region ≥ 3) with 12 vertices and 30 edges then the degree of each region is
(a) 2 (b) 3 (c) 4 (d) 5

15. Suppose that G is a connected planar graph with 35 regions each of degree 6. Then the number of vertices is
(a) 70 (b) 80 (c) 72 (d) 62

16. What is the minimal number of edges which must be removed from a complete bipartite graph of six vertices $1<(6)$ .So that the remaining graph is planar?
(a) 2 (b) 3 (c) 5 (d) 6

17. Maximum number of edges in a an undirected simple graph with $n-$vertices are
(a) n^2 (b) $\frac{n(n-1)}{2}$
(c) $n-1$ (d) $\frac{n(n+1)}{2}$

18. The minimum number of colours required to colour the vertices of a cycle with n vertices in such a way that no two adjacent vertices have the same colour is
(a) 2 (b) 3
(c) 4 (d) $n-2\left\lfloor\frac{n}{2}\right\rfloor+2$

19. How many undirected graphs (not necessarily connected) can be constructed out of a given set $V=\{v_1,v_2..v_n\}$ of n vertices ?
(a) $\frac{n(n-1)}{2}$ (b) 2^n
(c) $n!$ (d) $2.\frac{n(n-1)}{2}$

20. The chromatic number of tree with n vertices is
(a) $n-1$ (b) $n+1$ (c) n (d) 2

21. A graph $G=(V,E)$ satisfies $|E|\leq 2|V|-6$. The min-degree of G is defined as min $\{\text{degree}(V)\}$. Therefore min-degree G cannot be $v\in V$
(a) 3 (b) 4 (c) 5 (d) 6

22. Let $T(n)$ be the number of binary search trees on n distinct elements. Then $T(n)=\sum_{K=1}^{n} T(K-1)T(x)$

where x is

(a) $n-K+1$ (b) $n-k$
(c) n (d) $n+1$

23. A non planar graph with minimum number of vertices has
(a) 9 edges, 6 vertices
(b) 6 edges, 4 vertices
(c) 10 edges 5 vertices
(d) 9 edges, 5 vertices

24. The minimum number of edges in a connected cyclic graph on n vertices.
(a) $n-1$ (b) n
(c) $n+1$ (d) none

25. Consider a simple connected graph G with n vertices and n edges $(n>2)$. Then which of the following are true.
(a) G has no cycles
(b) the graph obtained by removing any edge from G is not connected.
(c) G has at least one cycle.
(d) The graph obtained by removing any two edges from G is not connected
(e) none of the above

ANSWER KEY

1. (b)	2. (b)	3. (a)	4. (b)	
5.(b)&(d)	6. (a)	7. (b)	8. (c)	9. (a)
10. (d)	11. (b)	12. (b)	13. (c)	14. (b)
15. (c)	16. (d)	17. (b)	18. (b)	19. (d)
20. (d)	21.(d)	22. (a)	23. (c)	24. (b)
25. (d)				

GRAPH THEORY(SET - II)

1. The minimum number of vertices in any height balanced binary tree with height h are
(a) $2^{h+1}-1$
(b) $\frac{1}{\sqrt{5}}\left(\frac{(1+\sqrt{5})}{2}\right)^{h+3}-2$
(c) h
(d) none of above

2. Which of the following is false.
(a) a full binary tree has an odd number of vertices
(b) a complete binary tree is a B-tree of order 2
(c) a complete directed tree of degree 2 is a complete binary tree
(d) the number of vertices in a binary tree is always even.

3. The number of spanning trees of K_n are
(a) $\frac{n(n-1)}{2}$ (b) $n!$
(c) n^{n-2} (d) n^2

4. The value of the expression
$X-+34-72\div12\times3-64$
(a) 5 (b) 4 (c) 3 (d) 1

5. The value of the expression $432\div-5\times42\times5\times3\div\div$ which is post fix notation is
(a) 4 (b) 3
(c) $\frac{15}{16}$ (d) $\frac{16}{15}$

6. The number of pendant vertices in a binary tree with n vertices are
(a) n (b) $\frac{n-1}{2}$
(c) $\frac{n+1}{2}$ (d) $n!$

7. Let G be a graph with K components, where each component is a tree. If n is the number of vertices and m is the number of edges in G then n is equal to
(a) K (b) m
(c) $m-k$ (d)

8. If in a tree, the degree of every non-pendant vertex is 3 then the number of vertices in the tree is
(a) even
(b) odd
(c) $3n$ (n is the no.of vertices)
(d) none of the above

9. The number of leaves in an $m-ary$ tree of height h are
(a) at least m^n (b) at most m^h
(c) m^h (d) none of the above

10. The number of non isomorphic rooted trees with 4 vertices are
(a) 4 (b) 3 (c) 2 (d) 1

11. The number of edges in a full binary tree with 1000 internal vertices are
(a) $1000+1$ (b) 2000
(c) 1011 (d) none

12. The number of edges in a forest of t tree containing n vertices are
(a) n (b) t
(c) $n-t$ (d) $n+t$

13. The number of edges must be removed from a connected graph with n vertices and m edges to produce a spanning tree are
(a) $n-m$ (b) $m-n$
(c) $m-n+1$ (d) m

14. The number of vertices in a full $m-ary$ tree with i internal vertices
(a) $m+1$ (b) $m+i$
(c) $mi+1$ (d) none

15. The number of internal vertices in a full $m-ary$ tree with n vertices
(a) m (b) $m+n$
(c) $\frac{n-1}{m}$ (d) none

16. Which of the following is false
(a) a simple graph G is connected if and only if it has a spanning tree.
(b) the number of different spanning trees of k_4 is 16
(c) every Hamiltonian path is a spanning tree
(d) the number of vertices in the tree is odd

17. Let G be an undirected connected graph with distinct edge weight. Let e_{max} be the edge with maximum weight and e_{max} be the edge with minimum weight. Which of the following statements is false.
(a) every minimum spanning tree of G must contain e_{min}
(b) if e_{max} is in a minimum spanning tree, then its removal must disconnect G
(c) no minimum spanning tree contains e_{max}
(d) G has a unique minimum spanning tree

18. Let G be an undirected graph. Consider a depth first traversal of G, and let T be the resulting depth-first search tree. Let u be a vertex in G and let v be the first new (unvisited) vertex visited after visiting u in the traversal which of the following statements is always true.
(a) $\{u,v\}$ must be an edge in G and u is a descendant of V in T.
(b) $\{u,v\}$ must be an edge in G and v is a descendant of u in T
(c) If $\{u,v\}$ is not an edge in G then u is a leaf in T
(d) If $\{u,v\}$ is not an edge in G then u and v must have the same parent in T.

19. Which of the following statements is false ?
(a) a tree with n nodes has $(n-1)$ edges.
(b) a labelled rotted binary tree can be uniquely constructed given its post order and pre order traversal results.
(c) a complete binary tree with n internal vertices has $(n+1)$ leaves
(d) the maximum number of nodes in a binary tree of height h is $2^{n+1}-1$.

20. The number of rooted binary trees with n vertices is
(a) equal to the number of ways of multiplying $(n+1)$ matrices.
(b) equal to the number of ways of arranging n out of $2n$ distinct elements
(c) equal to $\frac{1}{n+1}(2n_n)$
(d) equal to $n!$

21. Which one of the following statements is false ?
(a) optimal binary search tree construction can be performed efficiently using dynamic programming.
(b) breadth first search cannot be used to find converted components of a graph.
(c) given the prefix and postfix walks over a binary tree, the binary tree cannot be uniquely constructed.
(d) depth first search can be used to find connected components of a graph.

22. A binary tree T has n leaf nodes. The number of nodes of degree 2 in T is
(a) $\log_2 n$ (b) $n-1$ (c) n (d) 2^n

23. A binary search tree is generated by inserting in order the following inters 50, 15, 62, 5, 20, 58, 91, 3, 8, 37, 60, 24. The number of nodes in the left subtree and right subtree of the root respectively is
(a) $(4,7)$ (b) $(7,4)$
(c) $(8,3)$ (d) $(3,8)$

24. In the balanced binary tree in the following figure, how many nodes will become unbalanced when a node is inserted as a child of the node "g".

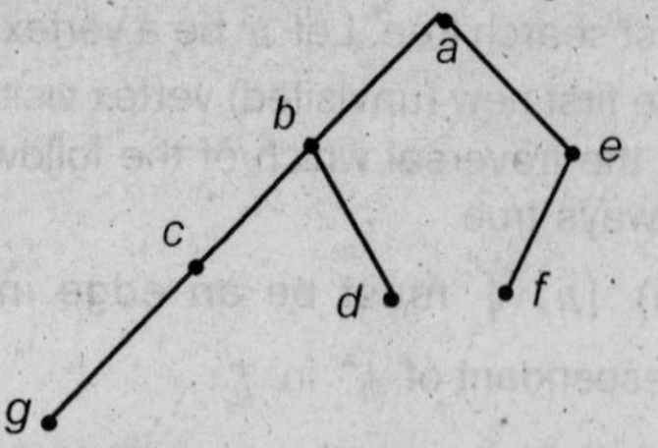

(a) 1

(b) 3

(c) 7

(d) 8

25. Which of the following sequence denotes the post order traversal sequence of the tree of above question.

(a) $f e g c d b a$ (b) $g c b d a f e$

(c) $g c d b f e a$ (d) $f e d g c b a$

26. A binary search tree contains the value 1, 2, 3, 4, 5, 6, 7, 8. The tree is traversed in pre order and the values are printed out. Which of the following sequence is a valid out put.

(a) 5 3 1 2 4 7 8 6 (b) 5 3 1 2 6 4 8 7

(c) 5 3 2 4 1 6 7 8 (d) 5 3 1 2 4 7 6 8

27. A complete $n-ary$ is one in which every vertex has zero or n sons. If x is the number of internal vertices of a complete $n-ary$ tree, the number of leaves in it is given by

(a) $x(n-1)+1$ (b) $xn-1$

(c) $xn+1$ (d) $x(n+1)$

28. Consider an undirected un weighted graph G. Let a breadth - first traversal of G be done starting from a vertex r. Let $d(r,u)$ and $d(r,v)$ be the length of the shortest path from r to u and v respectively in G. If u is visited before v during the breadth first traversal, which of the following statements is correct.

(a) $d(r,u)<d(r,v)$ (b) $d(r,u)>d(r,v)$

(c) $d(r,u)\leq d(r,v)$ (d) none of the above

29. The number of leaf nodes is a rotted tree of a n vertices, with each vertex having 0 or 3 children is

(a) $\frac{n}{2}$ (b) $\frac{n-1}{3}$ (c) $\frac{n-1}{2}$ (d) $\frac{2n+1}{3}$

30. A weight - balanced tree is a binary tree in which for each node, the number of nodes in the left subtree is atleast half and at most twice the number of t notes in the right subtree. The maximum possible height of such tree on n nodes is best described by which of the following.

(a) $\log_2 n$ (b) $\log_{4/3}^{n}$

(c) $\log_3 n$ (d) $\log_{3/2}^{n}$

ANSWER KEY

1. (b)	2. (d)	3. (c)	4. (b)	5. (c)
6. (c)	7. (d)	8. (a)	9. (b)	10. (a)
11. (b)	12. (c)	13. (c)	14. (c)	15. (c)
16. (d)	17. (c)	18. (c)	19. (c)	20. (a)
21. (a)	22. (b)	23. (b)	24. (b)	25. (c)
26. (d)	27. (a)	28. (d)	29. (d)	30. (a)

V. COMBINATORICS AND RECURRENCE RELATIONS

1. Suppose that the number of bacteria in a colony triples every hour, Then the recurrence relation for the number of bacterial after n hours have elapsed is

(a) $a_n = a_{3n}$ (b) $a_n = 3a_{n-1}$

(c) $a_n = 3a_{n-1} - 1$ (d) none

2. The recurrence relation for the number of bit strings of length n that contain a pair of consecutive zeros is

(a) $a_n = a_{n-1} + a_{n-2}$ $for\ n \geq 2$

(b) $a_n = a_{n-2} + 2a_{n-1}$ $for\ n \geq 2$

(c) $a_n = a_{n-1} + a_{n-2} + 2^{n-2}$ $for\ n \geq 2$

(d) none

3. The number of bit strings of length 7 that contain two consecutive 0's are

(a) 94 (b) 72 (c) 80 (d) 128

4. The recurrence relation for the number of bit strings of length n that do not contain three consecutive zeros is

(a) $a_n = a_{n-1} + a_{n-2} + a_{n-3}$ $for\ n \geq 3$

(b) $a_n = a_{n-1} + 3a_{n-2}$ $n \geq 2$

(c) $a_n = a_{n-1} + a_{n-2} + 3^{n-2}$ $n \geq 2$

(d) none

5. A string that contains only 0s, 1s and 2s is called a ternary string. Find a recurrence relation for the number of ternary strings that do no contain two consecutive 0's or two consecutive 1s.

(a) $a_n = a_{n-1} + a_{n-2}$ $n \geq 2$

(b) $a_n = 2a_{n-1} + a_{n-2}$ $n \geq 2$

(c) $a_n = a_{n-1} + 2a_{n-2}$ $n \geq 2$

(d) $a_n = 2a_{n-1}$

6. The recurrence relation for R_n where R_n is the number of regions that a plane is divided into by n lines, if no two of the lines are parallel and no three of the lines go through the same point is

(a) $R_n = nR_{n-1}, R_0 = 1$

(b) $R_n = n + R_{n-1}, R_0 = 0$

(c) $R_n = n + R_{n-1}, R_0 = 1$

(d) none

7. How many bit sequence of length seven contain an even number of 0's ?

(a) 54 (b) 64

(c) 12 (d) 8

8. The solution of the recurrence relation $a_n = a_{n-1} + 2a_{n-2}$ with $a_0 = 2$ and $a_1 = 7$ is

(a) $3.2^n - (-1)^n$ (b) $2^n - (-1)^n$

(c) $3.2^n + 2^n$ (d) none

9. The values of A and B so that $a_n = A_n + B$ is a solution of the recurrence relation $a_n = 2a_{n-1} + n + 5$ are

(a) $A=1, B=7$ (b) $A=7, B=1$

(c) $A=-1, B=-7$ (d) $A=-7, B=-1$

10. The general form of the particular solution of the linear non homogenous recurrence relation $a_n = 8a_{n-2} - 16a_{n-4} + F(n)$ if $F(n) = n^2 4^n$ is

(a) $(an^2 + bn + c)4^n$ (b) $dn^2 4^n$

(c) $dn4^n$ (d) none

11. Suppose that $f(n) = f(n/5) + 3n^2$ where n is divisible by 5 and $f(1) = 4$. Then $f(125) =$

(a) 79 (b) 48, 829

(c) 89 (d) 40

12. A group of 10 people being a chainletter with each person sending the letter to 4 other people. Each of these people sends the letter to 4 additional people. The number of letters sent at the nth stage is

(a) $A_n = 4A_{n-1}$ (b) $A_n = A4_{n-1}$

(c) $A_n = 10.4^n$ (d) none

13. How many positive integers less than 200 are not divisible by three or more primes.

(a) 19 (b) 122 (c) 167 (d) 168

14. If 6 men intend to speak at a convention. How many orders are there with B speaking after A ?

(a) 6! (b) $\frac{6!}{2}$ (c) 2.6! (d) 5!

15. In how many ways can 3 boys shoure 15 different sized apples if each takes 5 ?

(a) $15!\,5!$ (b) $\frac{15!}{5!}$

(c) $\frac{15!}{(5!)^3}$ (d) none

16. Two girls have picked 10 roses, 15 sunflowers and 14 daffodils. What is the number of ways they can divide the followers amongst themselves ?

(a) 1638 (b) 2100

(c) 2640 (d) none

17. The recurrence relation that arises in relation with the complexity of binary search is

(a) $T(n)=T\left(\frac{n}{2}\right)+K$, K constant

(b) $T(n)=2T\left(\frac{n}{2}\right)+K$, K constant

(c) $T(n)=T\left(\frac{n}{2}\right)+\log n$

(d) $T(n)=T\left(\frac{n}{2}\right)+n$

18. Which of the following algorithm design techniques is used in the quick sort algorithm ?

(a) dynamic programming
(b) back tracking
(c) greedy method
(d) divide and conquer

19. How many bit strings are there of length seven ?

(a) 7! (b) 14
(c) 128 (d) none

20. A palindrome is a string whose reversal is identical to the string. How many bit strings of length n are palindromes.

(a) 2^n (b) $2^{\frac{n+1}{2}}$

(c) $2^{\frac{n}{2}}$ (d) 2^{n^2}

21. A group contains n men and n women. How many ways are there to arrange these people in a row if the men and women alternate ?

(a) $n!(n-1)!$ (b) $(n!)^2$

(c) $2(n!)^2$ (d) none

22. How many bit strings contain exactly eight 0s and ten 1's if 0 must be immediately followed by a 1 ?

(a) 15 (b) 40
(c) 45 (d) 10

23. How many ways can one right and one left shoe be selected from 10 pairs of shoes without obtaining a pair.

(a) 70 (b) 80
(c) 900 (d) 90

24. In how many ways can 12 of the 14 people be distributed into three teams of 4 each ?

(a) $\frac{12!}{(4!)^3}$

(b) $C(14,12)\frac{12!}{4!^3}$

(c) $C(14,12)\frac{12!}{(4!)^3\,(3!)}$

(d) none

25. The characteristic equation of the recurrence relation $a_n - a_{n-1} - 6a_{n-2} = 0$ for $n \geq 2$ is

(a) $t^2-t-6=0$ (b) $t^2+t+6=0$

(c) $t-6=0$ (d) none

ANSWER KEY

1. (b)	2. (c)	3. (a)	4. (a)	5. (b)
6. (c)	7. (b)	8. (a)	9. (c)	10. (a)
11. (b)	12. (c)	13. (d)	14. (b)	15. (c)
16. (c)	17. (a)	18. (d)	19. (c)	20. (c)
21. (c)	22. (c)	23. (d)	24. (c)	25. (a)

TEST YOURSELF - I

1. Let G = (V, E) be a graph. Define $\xi(G)=\sum_d i_d \times d$, where i_d is the number of vertices of degree d in G. If S and T are two different trees with $\xi(S)=\xi(T)$, then **[GATE - 10]**

a) $|S|=2|T|$ b) $|S|=|T|-1$

c) $|S|=|T|$ d) $|S|=|T|+1$

2. What is the possible number of reflexive relations on a set of 5 elements ? **[GATE - 10]**

a) 2^{10} b) 2^{15} c) 2^{20} d) 2^{25}

3. Consider the set $S=\{1,\omega,\omega^2\}$, where ω and ω^2 are cube roots of unity. If * denotes the multiplication operation, the structure (S, *) forms **[GATE - 10]**

a) A group (b) A ring

c) An integral domain d) A field

4. The degree sequence of a simple graph is the sequence of the degrees of the nodes in the graph in decreasing order. Which of the following sequences can not be the degree sequence of any graph ?**[GATE - 10]**

I. 7,6, 5, 4, 4, 3, 2, 1 II. 6, 6, 6, 6, 3, 3, 2, 2

III. 7, 6, 6, 4, 4, 3, 2, 2 IV. 8, 7, 7, 6,4, 2, 1, 1

a) I and II b) III and IV

c) IV only d) II and IV

5. Suppose the predicate F(x, y, t) is used to represent the statement that person x can fool person y at time t. Which one of the statements below expresses best the meaning of the formula $\forall x\exists y\exists t(\neg F(x,y,t))$? **[GATE - 10]**

a) Everyone can fool some person at some time

b) No one can fool everyone all the time

c) Everyone cannot fool some person all the time

d) No one can fool some person at some time

6. Which one of the following in NOT necessarily a property of a Group ? **[GATE - 09]**

a) Commutativity

b) Associativity

c) Existence of inverse for every element

d) Existence of identity

7. What is the chromatic number of an n - vertex simple connected graph which does not contain any odd length cycle ? Assume $n \geq 2$. **[GATE - 09]**

a) 2 b) 3 c) n -1 d) n

8. Which one of the following is TRUE for any simple connected undirected graph with more than 2 vertices? **[GATE - 09]**

a) No two vertices have the same degree

b) At least two vertices have the same degree

c) At least three vertices have the same degree

d) All vertices have the same degree.

9. Consider the binary relation R = { (x, y), (x,z), (z,x), (z,y)} on the set { x, y, z}. Which one of the following is TRUE? **[GATE - 09]**

a) R is symmetric but NOT antisymmetric

b) R is NOT symmetric but antisymmetric

c) R is both symmetric and antisymmetric

d) R is neither symmetric nor antisymmetric

10. For the composition table of a cyclic group shown below

*	a	b	c	d
a	a	b	c	d
b	b	a	d	c
c	c	d	b	a
d	d	c	a	b

Which one of the following choices is correct? **[GATE-09]**

a) a, b are generators b) b, c are generators

c) c,d are generators d) d,a are generators

11. Which one of the following is the most appropriate logical formula to represent the statement ? "Gold and silver ornaments are precious ".

The following notations are used :

G(x) : x is a gold ornament

S(x) : x is a silver ornament

P(x) : x is precious **[GATE - 09]**

a) $\forall X(P(X)\rightarrow(G(X)\wedge S(X)))$

b) $\forall X((G(X)\wedge S(X))\rightarrow P(X))$

c) $\exists X((G(X)\wedge S(X))\rightarrow P(X))$

d) $\forall X((G(X)\vee S(X))\rightarrow P(X))$

12. The binary operation □ is defined as follows

P	Q	P □ Q
T	T	T
T	F	T
F	T	F
F	F	T

Which one of the following is equivalent to $P\vee Q$? **[GATE - 09]**

a) $\neg Q \square \neg P$ b) $P \square \neg Q$

c) $\neg P \square Q$ d) $\neg P \square \neg Q$

13. Consider the following well-formed formulae:

I. $\neg\forall x(P(x))$ II. $\neg\exists x(P(x))$

III. $\neg\exists x(\neg P(x))$ IV. $\neg\exists x(\neg P(x))$ **[GATE - 09]**

Which of the above are equivalent ?

a) I and III b) I and IV

c) II and III d) II and IV

14. A set of Boolean connectives is functionally complete if all Boolean functions can be synthesized using those. Which of the following sets of connectives is NOT functionally complete ? **[GATE - 08]**

(a) EX-NOR (b) implication, negation

(c) OR,negation (d)NAND

15. What is the chromatic number of the following graph ? **[GATE - 08]**

(a) 2

(b) 3

(c) 4

(d) 5

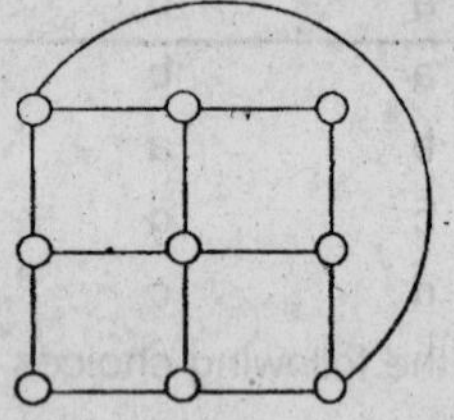

16. What is the size of the smallest MIS(Maximal Independent Set) of a chain of nine nodes? **[GATE - 08]**

(a) 5 (b) 4 (c)3 (d) 2

17. Which of the following first order formulae is logically valid ? Here $\alpha(x)$ is a first order formula with x as a free variable, and β is a first order formula with no free variable. **[GATE - 08]**

(a) $[\beta \rightarrow (\exists x, \alpha(x))] \rightarrow [\forall x, \beta \rightarrow \alpha(x)]$

(b) $[\exists x, \beta \rightarrow \alpha(x)] \rightarrow [\beta \rightarrow (\forall x, \alpha(x))]$

(c) $[(\exists x, \alpha(x)) \rightarrow \beta] \rightarrow [\forall x, \alpha(x) \rightarrow \beta]$

(d) $[(\forall x, \alpha(x)) \rightarrow \beta] \rightarrow [\forall x, \alpha(x) \rightarrow \beta]$

18. Which of the following is the negation of $[\forall x, \alpha \rightarrow (\exists y, \beta \rightarrow \forall u, \exists v, \gamma))]$ **[GATE - 08]**

(a) $[\exists x, \alpha \rightarrow (\forall y, \beta \rightarrow (\exists u, \forall \vee, \gamma))]$

(b) $[\exists x, \alpha \rightarrow (\forall y, \beta \rightarrow (\exists u, \forall \vee, \neg\gamma))]$

(c) $[\forall x, \neg\alpha \rightarrow (\exists y, \neg\beta \rightarrow (\forall u, \exists v \neg\gamma))]$

(d) $[\exists x, \alpha \wedge (\forall y, \beta \wedge (\exists u, \forall \vee, \neg\gamma))]$

19. G is a simple undirected graph. Some vertices of G are of odd degree. Add a node v to G and make it adjacent to each odd degree vertex of G. The resultant graph is sure to be **[GATE - 08]**

(a) regular (b) complete

(c) Hamiltonian (d) Euler

20. Consider the following Hasse diagrams

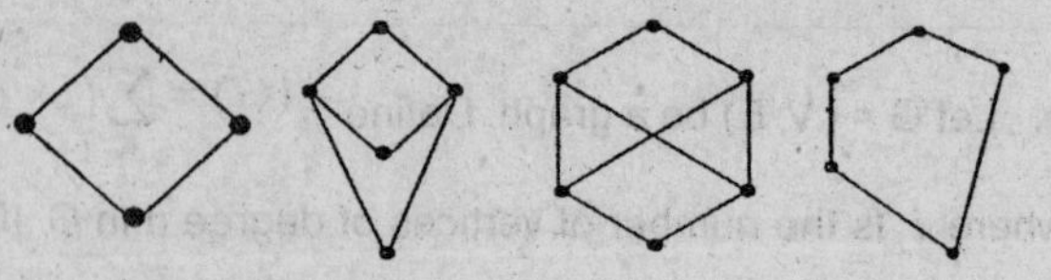

(i) (ii) (iii) (iv)

Which all of the above represent a lattice ?

[GATE - 08]

(a) (i) and (iv) only

(b) (ii) and (iii) only

(c) (iii) only

(d) (i),(ii) and (iv) only

21. If P, Q, R are subsets of the universal set U, then $(P \cap Q \cap R) \cup (P^c \cap Q \cap R) \cup Q^c \cup R^c$ is **[GATE - 08]**

(a) $Q^c \cup R^c$ (b) $P \cup Q^c \cup R^c$

(c) $P^c \cup Q^c \cup R^c$ (d)U

22. Which combination of the integer variables x, y and z makes the variable a get the value 4 in the following expression ?

$$a = (x > y)?((x > z)?x : z) : ((y > z)?y : z)$$

[GATE - 08]

(a) x=3, y=4, z=2 (b) x=6, y=5, z=3

(c) x=6, y=3, z=5 (d) x=5, y=4, z=5

23. Which of the following statements is true for every planer graph on n vertices ? **[GATE - 08]**

(a) The graph is connected

(b) The graph is Eulerian

(c) The graph has a vertex-cover of size at most $\frac{3n}{4}$

(d) The graph has an independent set of size at least $\frac{n}{3}$

24. P and Q are two propositions. Which of the following logical expressions are equivalent ?

I $P \vee \sim Q$

II. $\sim(\sim P \wedge Q)$

III. $(P \wedge Q) \vee (P \wedge \sim Q) \vee (\sim P \wedge \sim Q)$

IV. $(P \wedge Q) \vee (P \wedge \sim Q) \vee (\sim P \wedge Q)$

[GATE - 08]

(a) Only I and II
(b) Only I,II and III
(c) Only I,II and IV
(d) All of I, II, III and IV

Statement for Linked Answer Questions 25 and 26

Let x_n denote the number of binary strings of length n that contain no consecutive 0s.

25. Which of the following recurrences does x_n satisfy ? [GATE - 08]

(a) $x_n = 2x_{n-1}$

(b) $x_n = x_{\left\lfloor \frac{n}{2} \right\rfloor} + 1$

(c) $x_n = x_{\left\lfloor \frac{n}{2} \right\rfloor} + n$

(d) $x_n = x_{n-1} + x_{n-2}$

26. The value of x_5 is [GATE - 08]

(a) 5 (b) 7
(c) 8 (d) 16

27. Which one of these first - order logic formulae is valid ? [GATE - 07]

(a) $\forall x(P(x) \Rightarrow Q(x)) \Rightarrow ((\forall x P(x)) \Rightarrow (\forall x Q(x)))$

(b) $\exists x(P(x) \vee Q(x)) \Rightarrow ((\exists x P(x)) \Rightarrow (\exists x Q(x)))$

(c) $\exists x(P(x) \wedge Q(x)) \Leftrightarrow ((\exists x P(x)) \wedge (\exists x Q(x)))$

(d) $\forall x \exists y p(x, y) \Rightarrow \exists y \forall x\, p(x, y)$

28. A partial order P is defined on the set of natural numbers as follows. Here x/y denotes integer division.

(i) $(0,0) \in P$.

(ii) $(a,b) \in P$ **if and only if** $a\%10 \le b\%10$ **and** $(a/10, b/10) \in P$.

Consider the following ordered pairs :

(i) $(101,22)$ **(ii)** $(22,101)$

(iii) $(145,265)$ **(iv)** $(0,153)$

Which of these ordered pairs of natural numbers are contained in P ? [GATE - 07]

(a) (i) and (iii) (b) (ii) and (iv)
(c) (i) and (iv) (d) (iii) and (iv)

Statement for Linked Answer Questions 29 and 30 :

Consider the sequence $\{x_n\}_{n\geq 0}$ defined by the recurrence relation $x_{n+1} = cx_n^2 - 2$, where c >0

29. Suppose there exists a non - empty open interval (a, b) such that for all x_0 satisfying $a < x_0 < b$, the sequence converges to a limit. The sequence converges to the value [GATE - 07]

(a) $\frac{1+\sqrt{1+8c}}{2c}$ (b) $\frac{1-\sqrt{1+8c}}{2c}$

(c) 2 (d) $\frac{2}{2c-1}$

30. For which of the following values of c, does there exist a non-empty open interval (a, b) such that the sequence x_n converges for all x_0 satisfying $a < x_0 < b$?

(i) 0.25 (ii) 0.35 (iii) 0.45 (iv) 0.5

(a) (i) only [GATE - 07]
(b) (i) and (ii) only
(c) (i), (ii) and (iii) only
(d) (i), (ii), (iii) and (iv)

31. Let S be a set of n elements. The number of ordered pairs in the largest and the smallest equivalence relations on S are [GATE - 07]

(a) n and n (b) n^2 and n
(c) n^2 and 0 (d) n and 1

32. What is the maximum number of different Boolean functions involving n Boolean variables ? [GATE - 07]

(a) n^2 (b) 2^n
(c) 2^{2^n} (d) 2^{n^2}

33. Let G be the non-planar graph with the minimum possible number of edges. Then G has [GATE - 07]

(a) 9 edges and 5 vertices
(b) 9 edges and 6 vertices
(c) 10 edges and 5 vertices
(d) 10 edges and 6 vertices

34. Consider the DAG with $V = \{1,2,3,4,5,6\}$, shown below

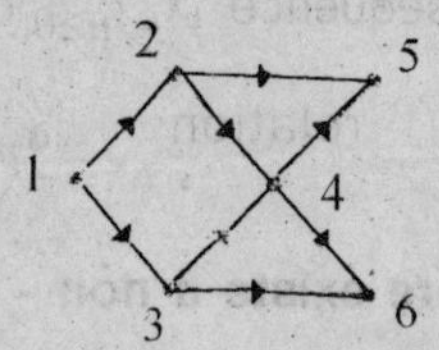

Which of the following is NOT a topological ordering ? **[GATE - 07]**

(a) 123456

(b) 132456

(c) 132465

(d) 324165

35. How many different non-isomorphic Abelian groups of order 4 are there? **[GATE - 07]**

(a) 2 (b) 3 (c) 4 (d) 5

36. Let Graph(x) be a predicate which denotes that x is a graph. Let Connected(x) be a predicate which denotes that x is connected. Which of the following first order logic sentences DOES NOT represent the statement; " Not every graph is connected"? **[GATE - 07]**

(a) $\neg\forall x(Graph(x) \Rightarrow Connected(x))$

(b) $\exists x(Graph(x) \wedge \neg Connected(x))$

(c) $\neg\forall x(\neg Graph(x) \vee Connected(x))$

(d) $\forall x(Graph(x) \Rightarrow \neg Connected(x))$

37. Which of the following graphs has an Eulerian circuit? **[GATE - 07]**

(a) Any k-regular graph where k is an even number

(b) A complete graph on 90 vertices

(c) The complement of a cycle on 25 vertices

(d) None of the above

38. Suppose we uniformly and randomly select a permutation from the 20! permulations of $1,2,3......,20$. What is the probability that 2 appears at an earlier position than any other even number in the selected permutation? **[GATE - 07]**

(a) $\frac{1}{2}$ (b) $\frac{1}{10}$ (c) $\frac{9!}{20!}$ (d) None of these

39. Consider the set $S = \{a,b,c,d\}$. Consider the following 4 partitions $\pi_1, \pi_2, \pi_3, \pi_4$ on $S : \pi_1 = \{\overline{abcd}\}, \pi_2 \{\overline{ab}, \overline{cd}\}, \pi_3 = \{\overline{abc}, \overline{d}\}, \pi_4 = \{\overline{a}, \overline{b}, \overline{c}, \overline{d}\}$. Let $\prec$ be the partial order on the set of partitions $S' = \{\pi_1, \pi_2, \pi_3, \pi_4\}$ defined as follows : $\pi_\eta \prec \pi_j$ if and only if π_i refines π_j. The poset digram for $(S', \prec)$ is **[GATE - 07]**

(a)

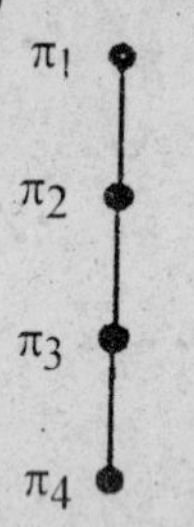

(b)

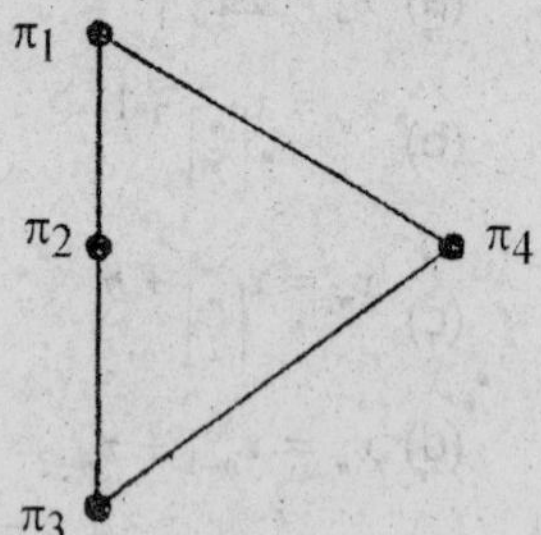

(c)

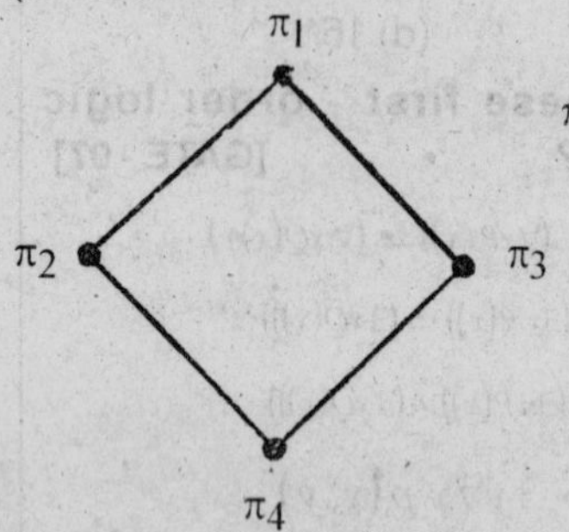

(d)

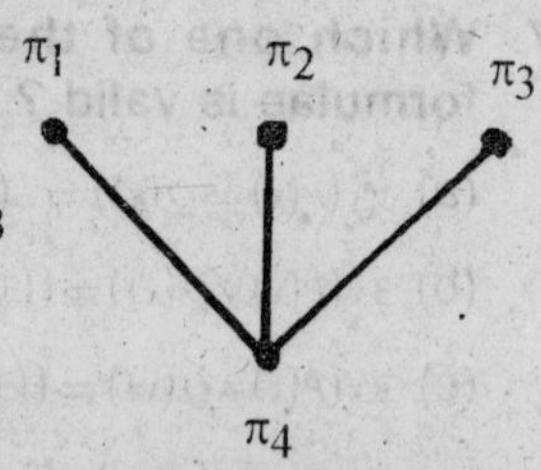

40. Consider the polynomial $p(x) = a_0 + a_1x + a_2x^2 + a_3x^3$, where $a_i \neq 0, \forall i$. The minimum number of multiplications needed to evaluate p on an input x is **[GATE - 06]**

(a) 3 (b) 4 (c) 6 (d) 9

41. Let, X,Y,Z be sets of sizes x,y and z respectively. Let $W = X \times Y$ and E be the set of all subsets of W. The number of functions from Z to E is **[GATE - 06]**

(A) z (B) $z \times 2^{xy}$ (C) z (D) 2^{xyz}

42. The set { 1, 2, 3, 5, 7, 8, 9 } under multiplication modulo 10 is not a group. Given below are four plausible reasons. Which one of them is false ? **[GATE - 06]**

(A) It is not closed

(B) 2 does not have an inverse

(C) 3 does not have an inverse

(D) 8 does not have an inverse

43. A relation R is defined on ordered pairs of integers as follows **[GATE - 06]**

$(x,y)R(u,v)$ *if* $x<u$ **and** $y>v$. **Then** R **is**

(A) Neither a Partial Order nor an Equivalence Relation
(B) A partial Order but not a Total Order
(C) A Total Order
(D) An Equivalence Relation

44. Let $S=\{1,2,3,....,m\}\ m>3$. **Let** $X_1,.....,X_n$ **be subsets of S each of size 3. Define a function f from S to the set of natural numbers as,**

$f(i)$ **is the number of sets** X_j **that contain the element i. That is** $f(i)=|\{j \mid i\in X_j\}|$

Then $\sum_{i=1}^{m} f(i)$ *is* **[GATE - 06]**

(A) 3m (B) 3n
(C) 2m+1 (D) 2n+1

45. Which one of the first order predicate calculus statements given below correctly expresses the following English statement ?

Tigers and lions attack if they are hungry or threatened. **[GATE - 06]**

(A) $\forall x$ [(tiger (x) $\wedge$ lion (x)) $\rightarrow$ {(hungery (x) $\vee$ threatened (x)) $\rightarrow$ attacks (x) }]

(B) $\forall x$ [(tiger (x) $\vee$ lion (x)) $\rightarrow$ {(hungry (x) $\vee$ threatened (x)) $\wedge$ attacks (x) }]

(C) $\forall x$ [(tiger (x) $\vee$ lion (x)) $\rightarrow$ {(attacks (x) $\rightarrow$ (hungry (x) $\vee$ threatened (x))}]

(D) $\forall x$ [tiger (x) $\vee$ lion (x)) $\rightarrow$ {(hungry (x) $\vee$ threatened (x)) $\rightarrow$ attacks (x) }].

46. Consider the following propositional statements

$P1:((A\wedge B)\rightarrow C))\equiv((A\rightarrow C)\wedge(B\rightarrow C))$

$P2:((A\vee B)\rightarrow C))\equiv((A\rightarrow C)\vee(B\rightarrow C))$

Which one of the following is true? **[GATE - 06]**

(A) P_1 is a tautology, but not $P2$

(B) $P2$ is a tautology but not P_1

(C) P_1 and $P2$ are both tautologies

(D) Both P_1 and $P2$ are not tautologies.

47. A logical binary relation $\odot$, is defined as follows

A	B	A ⊙ B
True	True	True
True	False	True
False	True	False
False	False	True

Let ~ be the unary negation (NOT) operator, with higher precedence, then $\odot$. Which one of the following is equivalent to $A\wedge B$? **[GATE - 06]**

(A) $(\sim A\odot B)$

(B) $\sim(A\odot\sim B)$

(C) $\sim(\sim A\odot\sim B)$

(D) $\sim(\sim A\odot B)$

48. For the set N of natural numbers and a binary operation $f: N\times N\rightarrow N$, an element $z\in N$ is called an identity for f, if $f(a,z)=a=f(z,a)$, for all $a\in N$. Which of the following binary operations have a identity?

(I) $f(x,y)=x+y-3$

(II) $f(x,y)=max(x,y)$

(III) $f(x,y)=x^y$ **[GATE - 06]**

a) I and II only b) II and III only

c) I and III only d) None of these

49. Given a boolean function $f(x_1,x_2,....,x_n)$, which of the following equations is NOT true **[GATE - 06]**

a) $f(x_1,x_2,....,x_n)=x_1'f(x_1,x_2....x_n)+x_1 f(x_1,x_2,.....,x_n)$

b) $f(x_1,x_2,....,x_n)=x_2 f(x_1,x_2....x_n)+x_2' f(x_1,x_2,....,x_n)$

c) $f(x_1,x_2,....,x_n)=x_n' f(x_1,x_2,...,0)+x_n f(x_1,x_2,....,1)$

d) $f(x_1,x_2,....,x_n)=f(0,x_2,......x_n)+f(1,x_2,....x_n)$

50. Consider the following first order logic formula in which R is a binary relation symbol.

$\forall x\forall y(Rx,y)\ \Rightarrow R(y,x)$

The formula is **[GATE - 06]**

a) satisfiable and valid
b) satisfiable and so is its negation
c) unsatisfiable but its negation is valid
d) satisfiable but its negation is unsatisfiable

51. Let P, Q and R be sets let Δ denote the symmetric difference operator defined as $P\Delta Q = (P\cup Q)-(P\cap Q)$. Using Venn diagrams, determine which of the following is/are TRUE?

(I) $P\Delta(Q\cap R)=(P\Delta Q)\cap(P\Delta R)$

(II) $P\cap(Q\cap R)=(P\cap Q)\,\Delta\,(P\cap R)$ **[GATE - 06]**

a) I only b) II only
c) Neither I nor II d) Both I and II

52. What is the cardinality of the set of integers x defined below ?

$x = \{n|1 \le n \le 123, n$ **is not divisible by either 2,3 or 5 }** **[GATE - 06]**

a) 28 b) 33 c) 37 d) 44

ANSWER KEY

1. c	2. c	3. a	4. d	5. d
6. a	7. a	8. b	9. d	10. c
11. d	12. b	13. b	14. b	15. b
16. c	17.	18.	19. d	20. a
21. d	22. a	23. a	24. b	25. d
26. c	27. a	28. d	29. d	30. c
31. b	32. c	33. b	34. d	35. b
36. d	37. a	38. d	39. c	40. a
41. d	42. c	43. a	44. b	45. d
46. d	47. d	48. a	49. c	50. a
51. b	52. b			

EXPLANATIONS

2. Total number of reflexive relations are

$$= 2^{n^2-n} = 2^{25-5} = 2^{20}$$

3. A group satisfies four properties

1. Closure 2. Associative 3. Identity 4. Inverse

So let $S = \{(1,\omega,\omega^2), *\}$

1. Closure

$$1*\omega = \omega \in S$$

$$\omega*\omega^2 = 1 \in S$$

$$\omega^2 * 1 = \omega^2 \in S$$

closed under set s

2. Associative

$$a*(b*c) = (a*b)*c$$

$$1*(\omega*\omega^2) \quad (1*\omega)*\omega^2$$

$$= 1*\omega^3 \qquad = \omega*\omega^2$$

$$= \omega^3 \qquad = \omega^3$$

i. e., associative property hold

3. Identity

$$a*e = a \in G,\ \exists e \in G$$

$$\left.\begin{aligned}1*\omega &= \omega \in G\\ 1*\omega^2 &= \omega^2 \in G\\ 1*1 &= 1 \in G\end{aligned}\right\} \textit{Identity hold}$$

4. Inverse

$$x*a = e, x, a, e \in G$$

$$\left.\begin{aligned}1*1 &= 1\\ \omega*\omega^2 &= 1\\ \omega^2*\omega &= 1\end{aligned}\right]$$ hold so given set is a Group

6. Group is an algebraic system having property of closed under operation, associative, contain identity element and exist inverse of every element.

Alternately

A group(G, *) is a monoid with identity e that the additional property that for every element $a \varepsilon$ G there exist an element a' ε G such that a * a' $= a' \times a = e$. Thus group is set together with binary operation *. It follows property

(i) Associativity : $(a*b)*c = a*(b*c)$

(ii) Inverse for every element a*e = e *a for every a ε G

(iii) Existence of identity : a*a' =a' *a

7. Chromatic Number of n - vertex simple connected undirected graph which does not contain any odd length cycle will be 2(n > = 2).

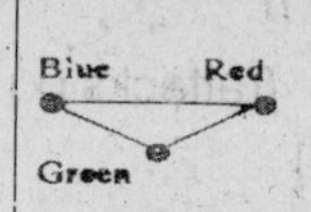

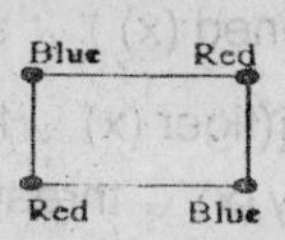

(For n =3 Odd Length Cycle) (For n =4 Even length Cycle)

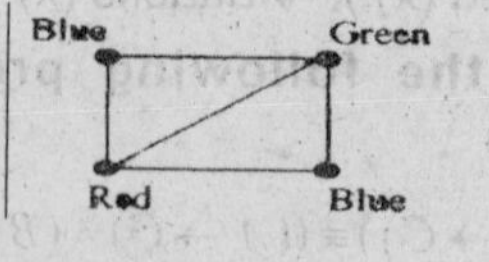

(For n=4 Odd Length Cycle)

8. At least 2 vertices have the same degree.

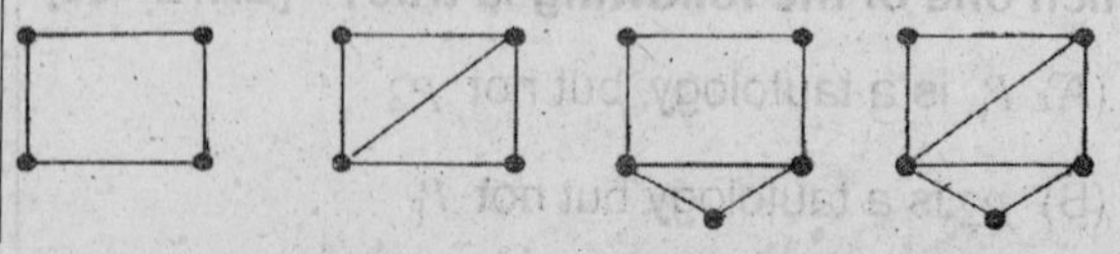

Alternately

For simple corrected graph with more than two vertices at least 2 vertices have the same degree

$n = 3$ 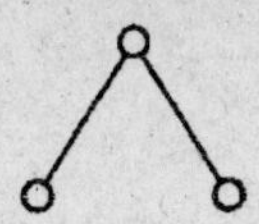 $n = 4$ 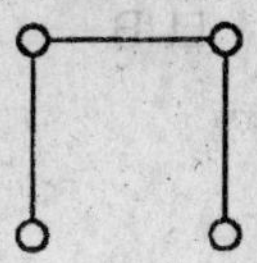

9. $R = \{(x,y),(x,z),(z,x),(z,y)\}$

If (a, b) $\in$ R then $(b,a) \notin R$ to be symmetric but (x,y) $\in R$ and $(y,x) \notin R$

So R is NOT Symmetric

If $(a,b) \in R$ then $(b,a) \in R$ to be Antisymmetric but $(x,z) \in R$ and $(z,x) \in R$

So R is NOT Antisymmetric

10.

*	a	b	c	d
a	a	b	c	d
b	b	a	d	c
c	c	d	b	a
d	d	c	a	b

Hence c, d is generator

Alternately

Every element in Cyclic group can be represented in the power of any generator.

From given table

a*a=a b*b=a

$c*c = c^2 = b \quad c^3 = c*c^2 = c*b = d \quad c^4 = c*c^3 = c*d = a$

$d*d = d^2 = b \quad d^3 = d*d^2 = d*b = c \quad d^4 = d*d^3 = d*c = a$

11. Here given that gold and silver ornament are precious.

We use

gold and silver gold $\vee$ silver

$\forall((G(x) \vee S(x))$

gold and silver $\rightarrow$ precious

$\forall x((G(x) \vee S(x)) \rightarrow P(x)$

Alternately

"Gold and silver ornaments are precious"

Gold ornaments are alone precious.

Silver ornaments are also alone precious.

Gold and silver ornaments are also precious

$\forall x((G(x) \vee S(x)) \rightarrow P(x)$

12.

P	P	P Q	$P \vee Q$	$P \neg Q$
T	T	T	T	T
T	F	T	T	T
F	T	F	T	T
F	F	T	F	F

13. The universal qualification of P(x) is denoted $\forall x P(x)$

The symbol $\forall$ is called the universal quantifier. The existential qualification of predicate P(x) is statement, there exist a value of x for which P(x) is true. The existential qualification of P(x) is denoted by $\exists(x)\ P(x)$. The symbol $\exists$ is called the existential quantier.

so $\neg \forall x P(x) = \exists x(\neg P(x))$

$\neg$ sign is for not

14. As we know, NAND, NOR, XOR, EX-NOR all these are universal gates and all Boolean functions can be synthesized using these gates.

Thus these are functionally complete.

Implication and negation

$\rightarrow$ $\sim$

$P \rightarrow Q$ If P then Q

All boolean functions can't be synthesized using this.

15. B $\rightarrow$ Blue

R $\rightarrow$ Red

G $\rightarrow$ Green

Chromatic Number :

Painting all the vertices of graph with colours such that no two adjacent vertices have the same colour is called proper colouring of a graph.

A graph G that requires K different colors for its proper coloring and no less, is called K-chromatic graph and the number K is called chromatic number of graph G.

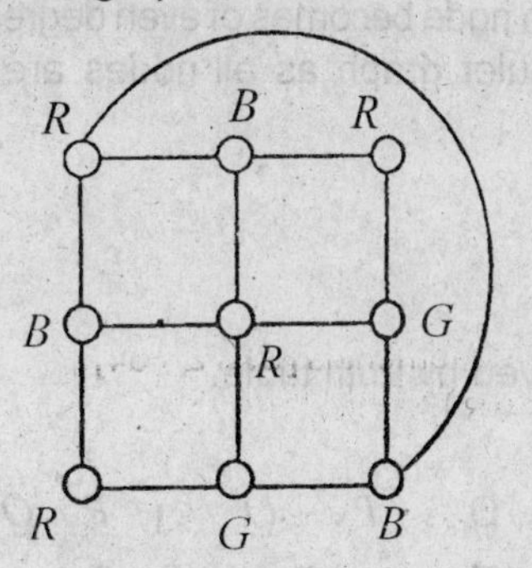

16. **Chain of 9 nodes**

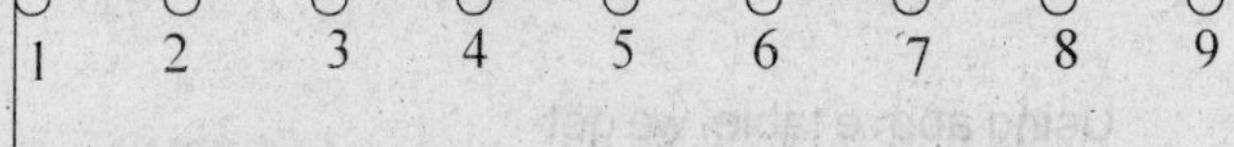

Maximal independent set .

A proper coloring of a graph naturally induces a partitioning of the vertices into different subsets.

No two vertices in any of these are adjacent. Such a subset of vertices is called an independent set.

Thus, a maximal independent set (or maximal internally stable set) is an independent set to which no other vertex can be added without destroying its independence property. e.g(2,5,8): this is the maximal independent set for a chain of 9 nodes.

19. If same closed mark in a graph contains all the edges of the graph, then the walk is called Euler line and the graph that consists of an Euler line is called an Euler graph..

A given connected graph G is an Euler graph if and only if all the vertices of G are of even degree.

e.g consider a graph of 4 vertices.

B and D are of odd degree, i.e 3

Now, add a vertex E, such that it is adjacent to B and D

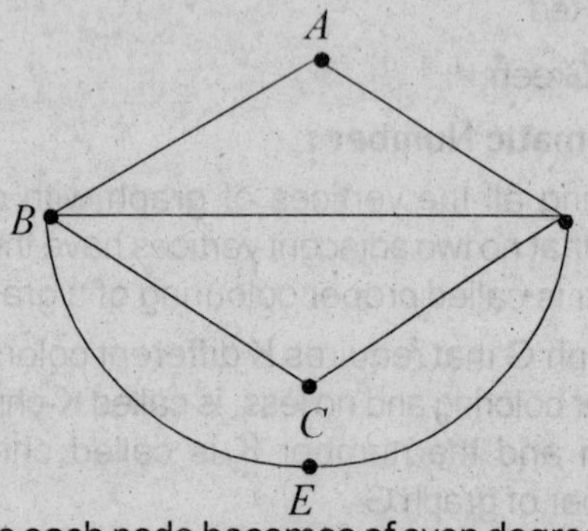

Thus each node becomes of even degree. Hence it is an Euler graph as all nodes are of even degree.

20. Lattice

A lattice is a partially ordered set $(P, \leq)$ in which any two elements EP has single GLB and single LUB

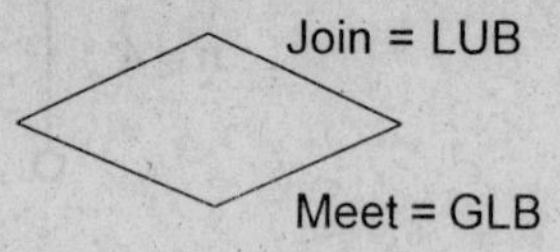

LUB (least upper bound)

LUB of A is x is an upper bound and if y is any other upper bound, then $x \leq y$

GLB (greatest lower bound)

GLB of A is y if y is a lower bond for A and if b is any other lower bound, then $b \leq y$.

From figure we can see that figures (i) and (iv) contain LUB and GLB, so they are lattice.

21.

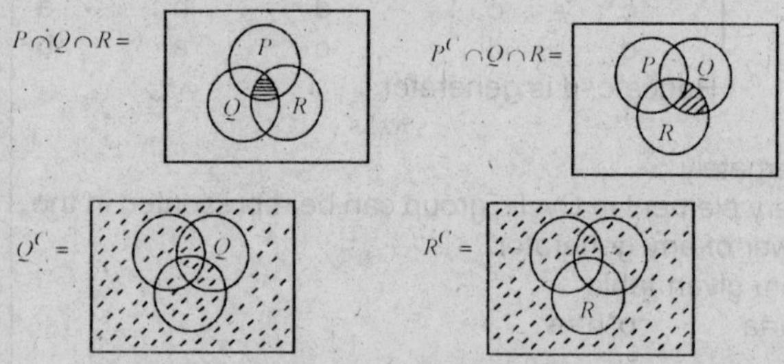

Union of all universal set is U

23. For every planar graph it is considered. It is connected as the isolated nodes does not affect planarity of a given graph.

24. It can solved by truth table.

P	Q	$P \vee \sim Q$	$(\sim P \wedge Q)$	$P \wedge Q$	$P \wedge \sim Q$	$\sim P \wedge \sim Q$
T	T	T	F	T	F	F
T	F	T	F	F	T	F
F	T	F	T	F	F	F
F	F	T	F	T	F	T

Using above table, we get

P	Q	I	II	III	IV
T	T	T	T	T	T
T	F	T	T	T	T
F	T	F	F	F	T
F	F	T	T	T	T

25. $x_1 = 2$ [0, 1]

$x_2 = 3$ [01, 10, 11]

$x_3 = 5$ [010, 011, 101, 110, 111]

$x_4 = 8$ [0101, 0110, 0111, 1010, 1011, 1101, 1110, 1111]

So, $x_n = x_{n-1} + x_{n-2}$

26. $x_5 = 13$ $x_4 + x_3 = x_5$

01010, 01011, 01101, 01110, 01111, 10101, 10110, 10111, 11010, 11011, 11101, 11110, 11111

28. $(0,0) \in P$

$(a,b) \in P$ if and only if $a\%10 \le b\%10$ and $(a/10\ ,\ b/10) \in P$

Option (iv), $(0,153) 0\%10 \le 153\%10$

$0 \le 3$

$(0/10,\ 153/10) = (0,15) \notin P$

$(0,15) \Rightarrow 0 \le 15$ *and* $(0,15/10) = (0,1) \notin P$

$(0,1) \Rightarrow 0 \le 1$ *and* $(0,1/10) = (0,0) \in P$

(0, 153) will be right option

Similarly we check for remaining three and get they are not ordered pass of natural numbers.

31. Smallest equivalance relation for set of n elements contains n ordered pair and largest equivalance relation contains n^2 ordered pairs.

32. For n binary variables, we can obtain 2^n distinct minterm or maxterms.
Since function can be either 1 or 0 for each minterm or maxterm, therefore we can form 2^{2^n} function from n variables.

33. Kurtowskis $3,3(K_{33})$ is non-planer graph with minimum number of edges.
It has 9 edges and 6 vertices.

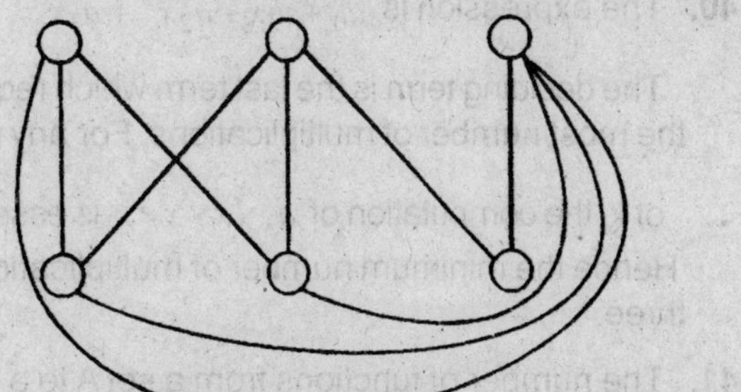

34. here, INDEG (v) = In degree of verter v.

Since, INDEG (1) = 0, INDEG (2) = 1, INDEG (3) = 1 INDEG (4) = 2, INDEG (5) = 2m, INDEG (6) = 2

Node 1 is with zero indegree, so first node in ordering will always 1, so answer (d) is not a topological ordering.

36.

$\forall_x [Graph(x) \Rightarrow 7\, connected\ (x)]$

represents "for every x if x is a graph then it is not connected"

37. A k - regular graph is one in which vertex is of degree k. If k is even then, such a graph will have an Eulerian circuit.

39. Since, $\pi_i\ \alpha\ \pi_j$ only if π_i refines π_j

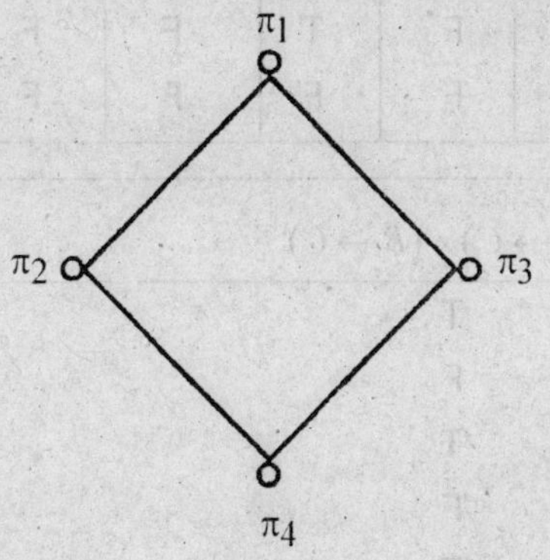

$\pi_1 = \{\overline{abcd}\}$ refines

$\pi_2 = \{\overline{ab}, \overline{cd}\}$

$$\pi_3 = \left\{\overline{abc},\ \overline{d}\right\}$$

and these both refines $\pi_4 = \left\{\overline{a},\overline{b},\overline{c},\overline{d}\right\}$

40. The expression is $a_0 + a_1x + a_2x^2 + a_3x^3$

The deciding term is the last term which requires the most number of multiplications. For any value of x, the computation of $a_3 \times x \times x \times x$ is essential. Hence the minimum number of multiplications is three.

41. The number of functions from a set A to a set B

where, $|A| = m$ and $|B| = n$ is n^m

$$W = X \times Y$$

$$\therefore \quad |W| = xy$$

$$|E| = 2^{xy}$$

$$|Z| = z$$

Hence the number of functions from Z to E is

$$(2^{xy})^z = 2^{xyz}$$

42. The binary operation multiplication modulo 10 is defined as follows

$$a0b = (a \times b)\%10$$

The identity element for the given set is 1.

For closure property, $2\ o\ 7 = 4 \notin set$

$\therefore$ Set is not closed.

2 and 8 do not have an inverse, $3o7 = 1$

$\therefore$ 7 is the inverse of 3.

Hence (c) is false

43. Reflexive Property is an essential condition for Partial order, Total order as well as Equivalence Relation. The given set is not reflexive as $\forall x, y \in I$,

(x,y) is not related to (x,y)

46.

A	B	C	$A \vee B$	$A \wedge B$	$(A \vee B) \to C$	$(A \wedge B) \to C$	$A \to C$	$B \to C$
T	T	T	T	T	T	T	T	T
T	T	F	T	T	F	F	F	F
T	F	T	T	F	T	T	T	T
T	F	F	T	F	F	T	F	T
F	T	T	T	F	T	T	T	T
F	T	F	T	F	F	T	T	F
F	F	T	F	F	T	T	T	T
F	F	F	F	F	T	T	T	T

$(A \to C) \wedge (B \to C)$	$(A \to C) \vee (B \to C)$
T	T
F	F
T	T
F	T
T	T
F	T
T	T
T	T

It is apparent from the truth table that P_1 and P_2 are not tautologies

47.

A	B	A ⊙ B	$\sim(\sim A \odot B)$	$A \wedge B$
T	T	T	$\sim F = T$	T
T	F	T	$\sim T = F$	F
F	T	F	$\sim T = F$	F
F	F	T	$\sim T = F$	F

48. I and II have an identify.

49. For (a)

$$f(x_1, x_2, \ldots, x_n) = x_1' \, f(x_1, x_2, \ldots, x_n) + x_1 \, f(x_1, x_2, \ldots, x_n)$$

$$= f(x_1, x_2, \ldots, x_n)(x_1 + x_1')$$

$$= f(x_1, x_2, \ldots, x_n) \qquad (\because x_1 + x_1' = 1)$$

$\therefore$ L.H.S. = R.H.S.

50. $\forall x \square \forall y \left(R(x, y) \Rightarrow R(y, x) \right)$

is symmetric function.

51. Solving I,

L.H.S. $P \Delta (Q \cap R)$ *R.H.S.* $(P \Delta Q) \cap (P \Delta R)$

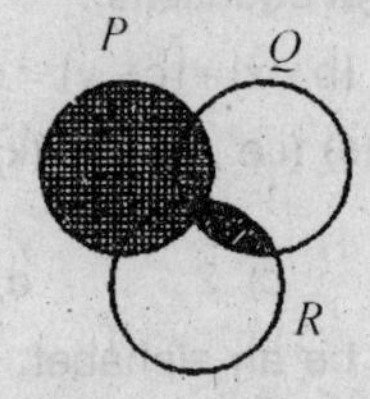

≠

Solving II,

L.H.S. $P \cap (Q \Delta R)$ *R.H.S.* $(P \cap Q) \Delta (P \cap R)$

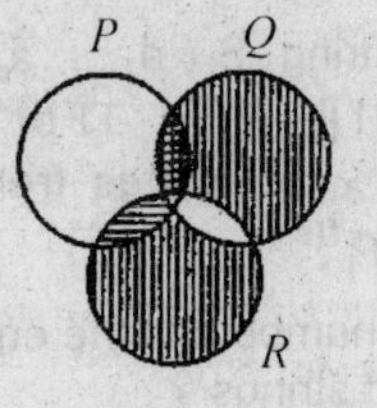

=

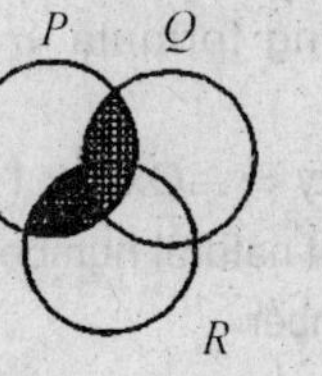

52. Numbers included in set X are

{1, 7, 11, 13, 17, 19, 23, 29, 31, 37, 41, 43, 47, 49, 53, 59, 61, 67, 71, 73, 77, 79, 83, 89, 91, 97, 101, 103, 107, 109, 113, 119}

TEST YOURSELF - II

1. Let $T(n)$ be the number of different binary search trees on n distinct elements. Then

$$T(n) = \sum_{k=1}^{n} T(k-1)T(x)$$, where x is **[GATE - 03]**

a) n - k + 1 b) n - k c) n - k - 1 d) n - k -2

2. Consider the set Σ^* of all strings over the alphabet $\Sigma = \{0,1\}$. Σ^* with the concatenation operator for strings

a) does not form a group **[GATE - 03]**
b) forms a non-commutative group
c) does not have a right identity element
d) forms a group if the empty string is removed from Σ^*

3. Let G be an arbitrary graph with n nodes and k components. If a vertex is removed from G, the number of components in the resultant graph must necessarily lie between **[GATE - 03]**

a) k and n b) k - 1 and k + 1
c) k -1 and n - 1 d) k + 1 and n - k

4. Let $(S, \leq)$ be a partial order with two minimal elements a and b, and a maximum element c. Let $P: S \to$ {True, False} be a predicate defined on S. Suppose that P(a) = True, P(b) = False, and $P(x) \Rightarrow P(y)$ for all $x, y \in S$ satisfying $x \leq y$, where $\Rightarrow$ stands for logical implication. Which of the following statements CANNOT be TRUE ? **[GATE - 03]**

a) P (x) = True for all $x \in S$ such that $x \neq b$
b) P (x) = False for all $x \in S$ such that $x \neq a$ and $x \neq c$
c) P (x) = False for all $x \in S$ such that $b \leq x$ and $x \neq c$
d) p(x) = False for all $x \in S$ such that $a \leq x$ and $b \leq x$

5. Which of the following is a valid first order formula ? (Here α and β are first order formulae with x as their only free variable) **[GATE - 03]**

a) $((\forall x)[\alpha] \Rightarrow (\forall x)[\beta]) \Rightarrow (\forall x)[\alpha \Rightarrow \beta]$
b) $(\forall x)[\alpha] \Rightarrow (\exists x)[\alpha \wedge \beta]$
c) $((\forall x[\alpha \wedge \beta] \Rightarrow (\exists x)[\alpha]) \Rightarrow (\forall x)[\alpha]$
d) $(\forall x)[\alpha \Rightarrow \beta] \Rightarrow ((\forall x)[\alpha] \Rightarrow (\forall x)[\beta])$

6. Consider the followng formula α and its two interpretations I_1 and I_2 **[GATE - 03]**

$\alpha: (\forall x)[Px \Leftrightarrow (\forall y)[Qxy \Leftrightarrow \neg Qxy]] \Rightarrow (\forall x)[\neg Px]$

I_1 : Domain : the set of natural numbers

$P_x \equiv$ ' x is a prime number '

$Qxy \equiv$ ' y divides x '

I_2: same as I_1 except that $P_x \equiv$ 'x is a composite number'.

Which of the following statements is TRUE ?

a) I_1 satisfies α, I_2 does not
b) I_2 satisfies α, I_1 does not
c) Neither I_1 nor I_2 satisfies α
d) Both I_1 and I_2 sastisfy α

7. Consider the following recurrence relation

$T(1) = 1$, $T(n+1) = T(n) + \lfloor \sqrt{(n+1)} \rfloor$ for all $n \geq 1$

The value of $T(m^2)$ for $m \geq 1$ is **[GATE - 03]**

a) $\frac{m}{6}(21m - 39) + 4$ b) $\frac{m}{6}(4m^2 - 3m + 5)$
c) $\frac{m}{2}(3m^{2.5} - 11m + 20) - 5$
d) $\frac{m}{6}(5m^3 - 34m^2 + 137m - 104) + \frac{5}{6}$

8. How many perfect matchings are there in a complete graph of 6 vertices ? **[GATE - 03]**

a) 15 b) 24 c) 30 d) 60

9. Let $f: A \to B$ be an injective (one-to-one) function. Define $g: 2^A \to 2^B$ as : $g(C) = \{f(x); x \in C\}$, for all subsets C of A. Define $h: 2^B \to 2^A$ as: $h(D) = \{x / x \in A, f(x) \in D\}$, for all subsets D of B. Which of the following statements is always TRUE ? **[GATE - 03]**

a) $g(h(D)) \subseteq D$ b) $g(h(D)) \supseteq D$
c) $g(h(D)) \cap D = \phi$ d) $g(h(D)) \cap (B - D) = \phi$

10. Consider the set {a, b, c} with binary operators + and x defined as follows : **[GATE - 03]**

+	a	b	c
a	b	a	c
b	a	b	c
c	a	c	b

×	a	b	c
a	a	b	c
b	b	c	a
c	c	c	b

For example, $a + c = c$, $c + a = a$, $c \times b = c$ and $b \times c = a$

Given the following set of equations :

$$(a \times x) + (a \times y) = c,\ (b \times x) + (c \times y) = c$$

the number of solution(s) (i.e., pair(s) (x, y) that satisfy the equation) is

a) 0 b) 1 c) 2 d) 3

11. Let $\Sigma = \{a, b, c, d, e\}$ be an alphabet. We define an encoding scheme as follows : **[GATE - 03]**

$g(a) = 3, g(b) = 5, g(c) = 7, g(d) = 9, g(e) = 11$

Let p_i denote the i-th prime number $(p_1 = 2)$

For a non-empty string $S = a_1......a_n$, where each $a_i \in \Sigma$, define $f(s) = \Pi_{i=1}^{n} P_i^{g(a_i)}$. For a non-empty sequence $< S_1,......S_n >$ of strings from Σ^+, define $h(< S_1......S_n >) = \Pi_{i=1}^{n} P_i^{f(s_i)}$

Which of the following numbers is the encoding, h, of a non-empty sequence of strings ?

a) $2^7 3^7 5^7$ b) $2^8 3^8 5^8$ c) $2^9 3^9 5^9$ d) $2^{10} 5^{10} 7^{10}$

12. A graph $G = (V, E)$ satisfies $|E| \leq 3|V| - 6$. The min-degree of G is defined as $\min_{v \in V} \{degree(V)\}$. Therefore, min-degree of G cannot be **[GATE - 03]**

a) 3 b) 4 c) 5 d) 6

13. Consider the following logic program P

$A(x) \leftarrow B(x, y), C(y)$ **[GATE - 03]**

$\leftarrow B(x, x)$

which of the following first order sentences is equivalent to P ?

a) $(\forall x)\left[(\exists y)\left[B(x,y)\wedge C(y)\right]\Rightarrow A(x)\right]\wedge\neg(\exists x)\left[B(x,x)\right]$

b) $(\forall x)\left[(\forall y)\left[B(x,y)\wedge C(y)\right]\Rightarrow A(x)\right]\wedge\neg(\exists x)\left[B(x,x)\right]$

c) $(\forall x)\left[(\exists y)\left[B(x,y)\wedge C(y)\right]\Rightarrow A(x)\right]\vee\neg(\exists x)\left[B(x,x)\right]$

d) $(\forall x)\left[(\forall y)\left[B(x,y)\wedge C(y)\right]\Rightarrow A(x)\right]\vee\neg(\exists x)\left[B(x,x)\right]$

14. The following resolutiion rule is used in logic programming. Derive clause $(P\vee Q)$ from clauses $(P\vee R)$, $(Q\vee\neg R)$ which of the following statements related to this rule is FALSE ? **[GATE - 03]**

a) $((P\vee R)\wedge(Q\vee\neg R))\Rightarrow(P\vee Q)$ is logically valid

b) $(P\vee Q)\Rightarrow((P\vee R)\wedge(Q\vee\neg R))$ is logically valid

c) $(P\vee Q)$ is satisfiable if and only if $(P\vee R)\wedge(Q\vee\neg R)$ is satisfiable

d) $(P\vee Q)\Rightarrow \text{FALSE}$ if and only if both P and Q are unsatisfiable.

15. Identify the correct translation into logical notation of the following assertion. **[GATE - 04]**

Some boys in the class are taller than all the girls

Note : taller (x,y) is true if x is taller than y.

a) $(\exists x)(boy(x)\rightarrow(\forall y)(girl(y)\wedge taller(x,y)))$

b) $(\exists x)(boy(x)\wedge(\forall y)(girl(y)\wedge taller(x,y)))$

c) $(\exists x)(boy(x)\rightarrow(\forall y)(girl(y)\rightarrow taller(x,y)))$

d) $(\exists x)(boy(x)\wedge(\forall y)(girl(y)\wedge taller(x,y)))$

16. Consider the binary relation : **[GATE - 04]**

$S=\{(x,y)\mid y=x+1 \text{ and } x,y\in\{0,1,2,.......\}\}$

The reflexive transitive closure of S is

a) $\{(x,y)\mid y>x \text{ and } x,y\in\{0,1,2,.......\}\}$

b) $\{(x,y)\mid y\geq x \text{ and } x,y\in\{0,1,2,......\}\}$

c) $\{(x,y)\mid y<x \text{ and } x,y\in\{0,1,2,.......\}\}$

d) $\{(x,y)\mid y\leq x \text{ and } x,y\in\{0,1,2,.......\}\}$

17. The following propositional statement is **[GATE - 04]**

$(P\rightarrow(Q\vee R))\rightarrow((P\wedge Q)\rightarrow R)$

a) satisfiable but not valid b) valid

c) a contradiction d) none of the above

18. The following is the incomplete operation table of a 4-element group.

The last row of the table is **[GATE - 04]**

*	e	a	b	c
e	e	a	b	c
a	a	b	c	e
b				
c				

a) c a e b b) c b a e c) c b e a d) c e a b

19. The inclusion of which of the following sets into $S=\{\{1,2\},\{1,2,3\},\{1,3,5\},\{1,2,4\},\{1,2,3,4,5\}\}$ **[GATE - 04]** is necessary and sufficient to make S a complete lattice under the partial order defined by set containment ?

a) $\{1\}$ b) $\{1\},\{2,3\}$

c) $\{1\},\{1,3\}$ d) $\{1\},\{1,3\},\{1,2,3,4\},\{1,2,3,5\}$

20. The minium number of colours required to colour the following graph, such that no two adjacent vertices are assigned the same colour, is **[GATE - 04]**

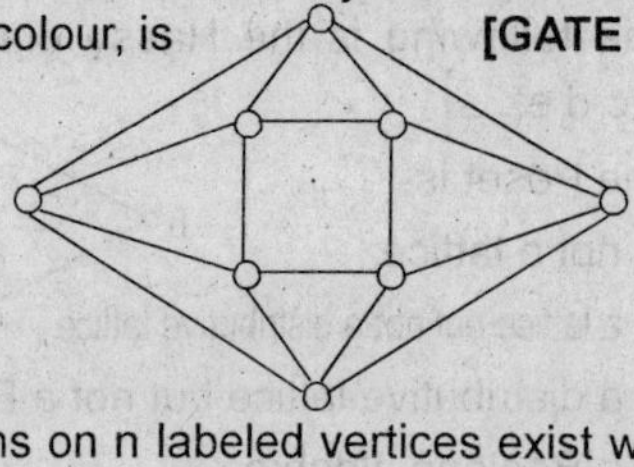

a) 2

b) 3

c) 4

d) 5

21. How many graphs on n labeled vertices exist which have at least $(n^2-3n)/2$ edges ? **[GATE - 04]**

a) ${}^{(n^2-n)/2}C_{(n^2-3n)/2}$ b) $\sum_{k=0}^{(n^2-3n)/2}{}^{(n^2-n)}C_k$

c) ${}^{(n^2-n)/2}C_n$ d) $\sum_{k=0}^{n}{}^{(n^2-n)/2}C_k$

22. Let $a(x,y)$, $b(x,y)$, $c(x,y)$ be three statements with variables x and y chosen from some universe. Consider the following statement : **[GATE - 04]**

$(\exists x)(\forall y)[(a(x,y)\wedge b(x,y))\vee\neg c(x,y)]$

Which one of the following is its equivalent ?

a) $(\forall x)(\exists y)[(a(x,y)\vee b(x,y))\wedge\rightarrow c(x,y)]$

b) $(\exists x)(\forall y)[(a(x,y)\vee b(x,y))\wedge\neg c(x,y)]$

c) $\neg(\forall x)(\exists y)[(a(x,y)\wedge b(x,y))\rightarrow c(x,y)]$

23. Let R_1 be a relation from $A=\{1,3,5,7\}$ to $B=\{2,4,6,8\}$ and R_2 be another relation from B to $C=\{1,2,3,4\}$ as defined below : **[GATE - 04]**

(i) An element x in A is related to an element y in B (under R_1) if $x+y$ is divisible by 3.

(ii) An element x in B is related to an element y in C (under R_2) if $x+y$ is even but not divisible by 3.

Which is the composite relation R_1R_2 from A to C ?

a) $R_1R_2=\{(1,2),(1,4),(3,3),(5,4),(7,3)\}$

b) $R_1R_2=\{(1,2),(1,3),(3,2),(5,2),(7,3)\}$

c) $R_1R_2=\{(1,2),(3,2),(3,4),(5,4),(7,2)\}$

d) $R_1R_2=\{(3,2),(3,4),(5,1),(5,3),(7,1)\}$

24. What is the maximum number of edges in an acyclic undirected graph with n vertices ? **[GATE - 04]**

a) $n-1$ b) n c) $n+1$ d) $2n-1$

25. Let p, q, r and s be four primitive statements. Consider the following arguments : **[GATE - 04]**

$P:[(\neg p\vee q)\wedge(r\rightarrow s)\wedge(p\vee r)\rightarrow(\neg s\rightarrow q)$

$Q:[(\neg p\wedge q)\wedge[q\rightarrow(p\rightarrow r)]\rightarrow\neg r$

$R:[[(q\wedge r)\rightarrow P]\wedge(\neg q\vee p)]\rightarrow r$

$S : [p \wedge (p \to r) \wedge (q \vee \neg r)] \to q$

Which of the above arguments are valid ?

a) P and Q only b) P and R only

c) P and S only d) P, Q, R and S

26. What is the number of vertices in an undirected connected graph with 27 edges, 6 vertices of degree 2, 3 vertices of degree 4 and remaining of degree 3 ? **[GATE - 04]**

a) 10 b) 11 c) 18 d) 19

27. The following is the Hasse diagram of the poset $[\{a,b,c,d,e\}, \leq]$ **[GATE - 05]**

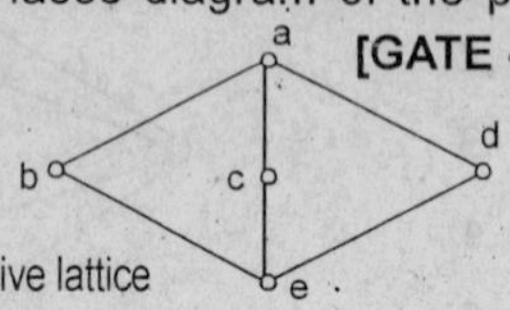

The Poset is

a) not a lattice

b) a lattice but not a distributive lattice

c) a distributive lattice but not a Boolean algebra

d) a Boolean algebra

28. Let G be a simple connected planar graph with 13 vertices and 19 edges. Then, the number of faces in the planar embedding of the graph is **[GATE - 05]**

a) 6 b) 8 c) 9 d) 13

29. Let G be a simple graph will 20 vertices and 100 edges. The size of the minimum vertex cover of G is 8. Then, the size of the maximum indepedent set of G is

a) 12 b) 8 **[GATE - 05]**

c) Less than 8 d) More than 12

30. The set $\{1,2,4,7,8,11,13,14\}$ is a group under multiplication modulo 15. The inverses of 4 and 7 are respectively **[GATE - 05]**

a) 3 and 13 b) 2 and 11 c) 4 and 13 d) 8 and 14

31. Let P, Q and R be three atomic propositional assertions. Let X denote $(P \vee Q) \to R$ and Y denote $(P \to R) \vee (Q \to R)$. Which one of the following is a tautology ? **[GATE - 05]**

a) $X \equiv Y$ b) $X \to Y$ c) $Y \to X$ d) $\neg Y \to X$

32. What is the first order predicate calculus statement equivalent to the following ? **[GATE - 05]**

Every teacher is liked by some student

a) $\forall(x)[\text{teacher}(x) \to \exists(y)[\text{student}(y) \to \text{likes}(y,x)]]$

b) $\forall(x)[\text{teacher}(x) \to \exists(y)[\text{student}(y) \wedge \text{likes}(y,x)]]$

c) $\exists(y)\forall(x)[\text{teacher}(x) \to [\text{student}(y) \wedge \text{likes}(y,x)]]$

d) $\forall(x)[\text{teacher}(x) \wedge \exists(y)[\text{student}(y) \to \text{likes}(y,x)]]$

33. Let R and S be any two equivalence relations on a non-empty set A. Which one of the following statements is TRUE ? **[GATE - 05]**

a) $R \cup S$, $R \cap S$ are both equivalence relations

b) $R \cup S$ is an equivalence relation

c) $R \cap S$ is an equivalence relation

d) Neither $R \cup S$ nor $R \cap S$ is an equivalence relation

34. Let $f : B \to C$ and $g : A \to B$ be two functions and let $h = f \circ g$. Given that h is an onto function. Which one of the following is TRUE ? **[GATE - 05]**

a) f and g should both be onto functions

b) f should be onto but g need not be onto

c) g should be onto but f need not be onto

d) both f and g need not be onto

35. What is the minimum number of ordered pairs of non-negative numbers that should be chosen to ensure that there are two pairs (a,b) and (c,d) in the chosen set such that $a \equiv c \bmod 3$ and $b \equiv d \bmod 5$ **[GATE - 05]**

a) 4 b) 6 c) 16 d) 24

36. Consider the set H of all 3×3 matrices of the type

$$\begin{bmatrix} a & f & e \\ 0 & b & d \\ 0 & 0 & c \end{bmatrix}$$ **[GATE - 05]**

Where a,b,c,d,e and f are real numbers and $abc \neq 0$. Under the matrix multiplication operation, the set H is

a) a group b) a monoid but not a group

c) a semigroup but not a monoid

d) neither a group nor a semigroup

37. Which one of the following graphs is NOT planar ? **[GATE - 05]**

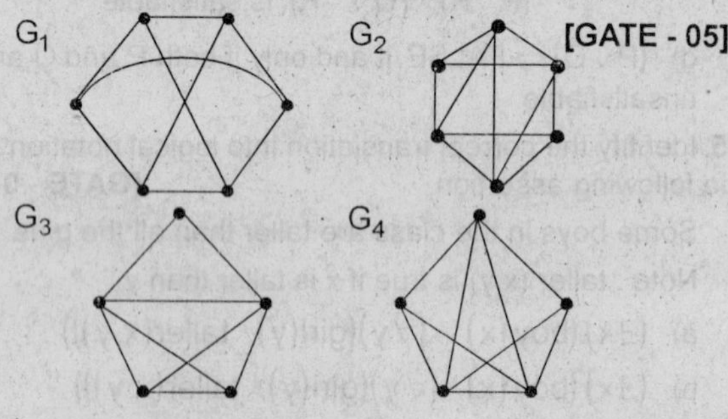

a) G_1 b) G_2 c) G_3 d) G_4

38. Let f be a function from a set A to a set B, g a function from B to C, and h a function from A to C, such that $h(a) = g(f(a))$ for all $a \in A$. Which of the following statements is always true for all such functions f and g ?

a) g is onto $\Rightarrow$ h is onto **[GATE - 05]**

b) h is onto $\Rightarrow$ f in onto

c) h is onto $\Rightarrow$ g is onto

d) h is onto $\Rightarrow$ f and g are onto

39. Let A be a set with n elements. Let C be a collection of distinct subsets of A such that for any two subsets S_1 and S_2 in C, either $S_1 \subset S_2$ or $S_2 \subset S_1$. What is the maximum cardinality of C ? **[GATE - 05]**

a) n b) n+1 c) $2^{n-1}+1$ d) n!

40. Let P(x) and Q(x) be arbitrary predicates. Which of the following statements is always TRUE ? **[GATE - 05]**

a) $((\forall x P(x) \vee Q(x))) \Rightarrow ((\forall x (P(x) \vee (\forall x Q(x)))$

b) $((\forall x P(x) \Rightarrow Q(x))) \Rightarrow ((\forall x (P(x) \Rightarrow (\forall x Q(x)))$

c) $((\forall x P(x)) \Rightarrow (\forall x Q(x))) \Rightarrow ((\forall x (P(x) \Rightarrow Q(x)))$

d) $((\forall x P(x)) \Leftrightarrow (\forall x Q(x))) \Rightarrow ((\forall x (P(x) \Leftrightarrow Q(x)))$

ANSWER KEY

1. a	2. d	3. d	4. c	5. d
6. a	7. b	8. a	9. a	10. d
11. b	12. c	13. c	14. b	15. d
16. b	17. a	18. d	19. a	20. c
21. b	22. c	23. c	24. a	25. c
26. a	27. b	28. b	29. a	30. c
31. b	32. b	33. c	34. b	35.
36. b	37. a	38. c	39. c	40. a

EXPLANATORY NOTES

1. $T(n) = \sum_{k=1}^{n} T(k-1)T(x) = T(x)[T(0) + T(1) + ... + T(n-1)]$

So, $x = n - k + 1$

2. Let A denote the set of all non-empty strings of letters from S. Where, $S = \{\alpha, \beta, \gamma\}$

we have $A = \{\alpha, \beta, \gamma, \alpha\alpha, \alpha\beta, \alpha\gamma .... \alpha\alpha\alpha, \alpha\alpha\beta, ...\}$

3. Vertex is removed, then component between is k + 1 and n - k

4. $(S; \leq)$ is partial order in which S is in reflexive, symmetric and transitive

Given, P(a) = True, P(b) = False

$P(x) \Rightarrow P(y)$

6. $\alpha = (\forall x)[Px \Leftrightarrow (\forall y)[Qxy \Leftrightarrow \neg QPxy]] \Rightarrow (\forall(x)[\neg Px]$

I_1: domain the set of natural numbers

$P_x = x$ is prime numbers = 2, 3, 4, 7, 11, 13, ...

Qxy = y divides x = means y is also prime number

= 2, 3, 5, 7, 11, 13, ...

$I_2 : Px = x$ is composite number means number is not prime number

= 4, 6, 8, 9, 10, 12, 14, ...

Qxy = 2, 3, 4, 5, 7, ... = mean all the numbers

From expression, we know,

I_1 satisfies α, I_2 does not

7. T(1) = 1

$T(n+1) = T(n) + \lfloor \sqrt{(n+1)} \rfloor$ for all $n \geq 1$

If $n = m^2 - 1$

$T(m^2) = T(m^2 - 1) + \lfloor m \rfloor$

$T(2) = T(1) + \lfloor \sqrt{2} \rfloor = 1 + 1 = 2$

$T(3) = T(1) + \lfloor \sqrt{3} \rfloor = 2 + 1 = 3$

T (4) = T(3) + 2 = 5

T (5) = 5 + 2 = 7

T (6) = 7 + 2 = 9

T (7) = 9 + 2 = 11

T (8) = 11 + 2 = 13

T (9) = 13 + 3 = 16

T (10) = 16 + 3 = 19

T (11) = 19 + 3 = 22

T (12) = 22 + 3 = 25

T (13) = 25 + 3 = 28

T (14) = 28 + 3 = 31

T (15) = 31 + 3 = 34

T (16) = 34 + 4 = 38

$T(m^2) = T(m^2 - 1) + \lfloor m \rfloor$

Put m = 2 T(4) = T(3) + 1 = 4

Put m = 3 T(9) = T(8) + 1 = 13 + 1 = 14

(a) $\frac{m}{6}(21m - 39) + 4$

Put m = 2, $\frac{2}{6}(42 - 39) + 4 = 5$

Put m = 3, $\frac{3}{6}(63 - 39) + 4 = 16$

Put m = 4, $\frac{4}{6}(84 - 39) + 4 = \frac{2}{3} \times 45 + 4 = 34$

(b) Put m = 2, $\frac{m}{6}(4m^2 - 3m + 5) = \frac{2}{6}[16 - 6 + 5] = 5$

Put m = 3, $\frac{3}{6}[4 \times 9 - 3 \times 3 + 5] = \frac{1}{2}[32] = 16$

Put m = 4, $\frac{4}{6}[64 - 12 + 5] = \frac{2}{3}[57] = 38$

Hence choice (b) satisfies the solution of $T(m^2)$

8. For complete graph of 6 vertices

number of perfect matching $= \frac{n(n-1)}{2} = \frac{6 \times 5}{2} = 15$

9. $f : A \to B$ be an injective (one-to-one) function.

$A \to B$ means

$A \subseteq B$

Define $g : 2^A \to 2^B$ as :

$g(C) = \{|f(x)| x \in C\}$, for all subset C of A.

$g(C) \subseteq B$

Define $h : 2^B \to 2^A$ as: $h(D) = \{x / x \in A, f(x) \in D\}$, for all subsets D of B.

$2^B \subseteq 2^A$

$h(D) \subseteq 2^A$

$2(h(D)) \subseteq C$

10.

+	a	b	c
a	b	a	c
b	a	b	c
c	a	c	b

×	a	b	c
a	a	b	c
b	b	c	a
c	c	c	b

From given expressions, we write

$a + a = b$, $a + b = a$, $a + c = c$

$b + a = a$, $b + b = b$, $b + c = c$

$c + a = a$, $c + b = c$, $c + c = b$

$a \times a = a$, $a \times b = b$, $a \times c = c$

$b \times a = b$, $b \times b = c$, $b \times c = a$

$c \times a = c$, $c \times b = c$, $c \times c = b$

$a \times x + a \times y = c$ ....(i)

$b \times x + c \times y = c$ ....(ii)

First we see, which equation give result C. These are

$a + c = c$

$b + c = c$

$c + b = c$

For equation (i), we find :

$a \times x = b$ $x = b$ We find

$a \times x = a$ then $x = a$ $a \times y = c$ $y = c$ $x = a,b,c$
$a \times y = c$ then $y = c$ $a \times x = c$ $x = c$ $y = a,b$
$a \times y = b$ $y = b$

For equation (ii), we find

$b \times x = b, c \times y = c$ $b \times x = a$ then $x = c$
$x = a, y = a,b$ $c \times y = c$ then $y = a,b$
$b \times x = c$ then $x = b$ we find
$c \times y = b$ then $y = c$ $x = a,b,c$
$y = a,b,c$

So, total number of solution pairs (x, y) is 3.

11. Let $\Sigma = \{a,b,c,d,e\}$

$g(a) = 3,\ g(b) = 5,\ g(c) = 7,\ g(d) = 9,\ g(e) = 11$

$P_1 = 2$ (1st prime number)

$f(s) = \pi_i^n = iP_i^{g(a_i)} = 2^3 = 8$

So, $h < s_1, \ldots s_n >= \underset{i=1}{\overset{n}{\pi}} P_i^{f(s_i)}$

$= 2^8\ 3^8\ 5^8$

12. Given $|E| \le 3|V| - 6$

we know from graph,

Number of edges $= \dfrac{n(n-1)}{2}$

(a) $V = 3,\quad E = 3$

$|3| \le 9 - 6 \rightarrow$ True

(b) $V = 4,\quad E = 6$

$|6| \le 12 - 6 \rightarrow$ True

(c) $V = 5,\quad E = 10$

$|10| \le 15 - 6 \rightarrow$ Not true

(d) $V = 6,\quad E = 15$

$|15| \le |18| - 6 \rightarrow$ Not true

So, minimum degree of G cannot be 5

13. $A(x) \leftarrow B(x,y), C(y)$

$\leftarrow B(x,x)$

We know, write

$B(x,x) \rightarrow B(x,y), C(y) \rightarrow A(x)$

$B(x,x) \rightarrow \neg(B(x,y), c(y)) \cup A(x)$

$\neg B(x,x) \cup \neg(B(x,y)), C(y)) \cup A(x)$

Finally, we write

$\forall (B(x,x) \cup (B(x,y)), c(y)) \cup A(x)$

Now, we interprete the option (C)

$(\forall x) \neg [(\exists y)[B(x,y) \wedge C(y)] \cup A(x) \cup \neg\exists(x)[B(x,x)]$

$(\forall x)\ [\forall (y) \neg [B(x,y) \wedge C(y)] \cup A(x) \cup (\forall x)[\neg B(x,x)]$

$(\forall\ xy)[\neg B(x,y) \cup \neg C(y) \cup A(x) \neg B(x,x)]$

Now, we write

$\forall [B(x,y), C(y)) \cup A(x)\ B(x,x)]$

14. If P = True, Q = false, R = True

then we find,

$(T \cup F) = ((T \cup T) \wedge (F \cup F))$

$T \Rightarrow (T \wedge F)$

$T \Rightarrow F$

$F \cup F$ = False (is not valid)

18. This table will form the symmetry of * operator with respect of a b c and e. On alyzing the symmetry the table will become.

*	e	a	b	c
e	e	a	b	c
a	a	b	c	e
b	b	c	e	a
c	c	e	a	b

The last row is c e a b.

19. $S = \{\{1,2\}, \{1,2,3\}, \{1,3,5\}, \{1,2,4\}, \{1,2,3,4,5\}\}$

To make S a complete lattice under the partial order defined by set containment of {1} only. It is necessary as well as sufficient.

20. Surrounded by edge and 03 + 1 = 4

3 - colour is required to colour the plane.

Colour is required to colour the all node i.e. 4-colourable.

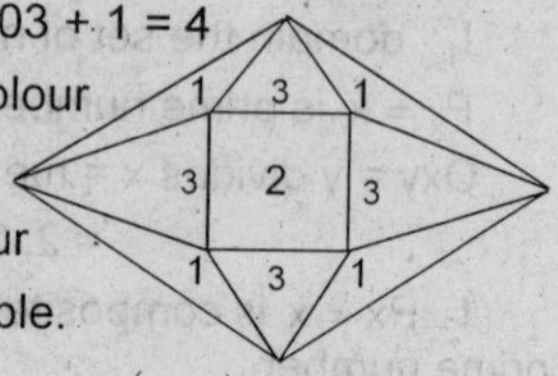

21. Number of graph has k edge is ${}^{(n^2-n)}C_k$ and at least $\dfrac{(n^2-3n)}{2}$ edge

Summation will take from 0 to $(n^2 - 3n)/2$,

Total number $= \displaystyle\sum_{k=0}^{(n^2-3n)/2} {}^{n^2-n}C_k$

22. $(\exists x)(\forall y)[a(x,y) \wedge b(x,y) \wedge \neg c\ (x,y)]$ is same as $\neg[(\forall x)(\exists y)[a(x,y) \wedge b(x,y)\ ) \rightarrow c(x,y)\ ]]$

23. $A = (1,3,5,7)\ B = (2,4,6,8)\ C = (1,2,3,4)$

$R_1 = A \rightarrow B = \{(1,2)(1,8)(3,6)(5,4)(7,2)(7,8)\}$

$R_2 = B \rightarrow C = \{(2,2)(4,4)(6,2)(6,4)(8,2)\}$

The composite relation between

$\Rightarrow R_1$ and R_2 $\quad R_1R_2 = \{(1,2)(3,2)(3,4)(5,4)(7,2)\}$

26. Number of edges = 24

6 vertices with degree 2 ; 3 vertices with degree 4 remaining with degree 3

Number of possible vertices be 10

27. Always found the concept

If a lattice is distributive and complement, it has unique complement

Here, c have two complements b and d

Hence, it is not distributive lattice.

28. Applying the formula

$e - n + 2 = 19 - 13 + 2 = 8$

30. Apply concept of inverse with respect to element and operation.

Let a is an element,

a' is its inverse,

and I an identity element

a * a' = I, where, * is operation.

(4 * 4) mod 15 = 1

(7 * 13) mod 15 = 1

where, 1 is identity with respect to multiplication

31.

P	Q	R
T	T	T
T	T	F
T	F	T
T	F	F
F	T	T
F	T	F
F	F	T
F	F	F

Y

$P \to R$	V	$Q \to R$
T	T	T
F	F	F
T	T	T
F	T	T
T	T	T
T	T	F
T	T	T
T	T	T

(X)

$PVQ \to R$		
T	T	T
T	F	F
T	T	T
T	F	F
T	T	T
F	T	F
T	T	T
F	T	F

X	$\to$	Y
T	T	T
F	T	F
T	T	T
F	T	T
T	T	T
T	T	T
T	T	T
T	T	T
	↑	

Because here, all combinations are true hence, it is a tautology.

33. If R and S both be any two equivalent relation, then $R \cap S$ is reflexive relation (because they have n elements which are related to itself)

Consider an example on set $\{1, 2, 3\}$

$R = \{<1,1>, <2,2>, <3,3>, <1,2>, <2,1>\}$

It is an equivalence relation

$S = \{<1, 1>, <2, 2>, <3,3>, <2, 3>, <3, 2>\}$

It is also equivalence.

$R \cap S = \{<1, 1>, <2, 2>, <3, 3>\}$

It is an equivalence relation.

$R \cup S = \{<1,1>, <2,2>, <3,3>, <1,2>, <2,1> <3,2>, <2,3>\}$

It is not an equivalence relation.

$<1, 3>$ is not true because transitive is not closed under union. Same with symmetric

34. h is $A \to C$

If h have to be onto, then

$f: B \to C$

But then onto only can cover all elements of C.

Now for g:

If g is not onto, then it can't cover all elements of B. Element which are not covered in g are map with f to produce element of C.

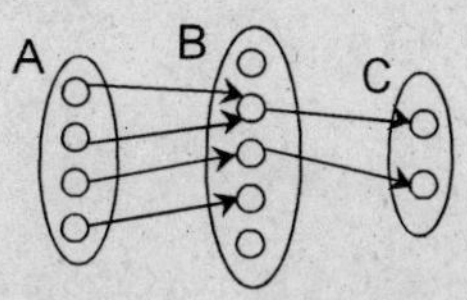

Unit - III
Theory of Computation

UNIT - III THEORY OF COMPUTATION

I. FINITE AUTOMATA AND REGULAR LANGUAGES

- Introduction
- Deterministic Machines
- Transition Systems
- NonDeterministic Machines
- Moore and Mealy Machines
- Regular Expressions
- Closure Properties of Regular languages

1.1 ALPHABET

An alphabet is a finite, non empty set of symbols. It is denoted by Σ . The symbols are called the letters of the alphabet.

Note : The symbols of an alphabet are indivisible.

Ex 1 : $\Sigma = \{0,1\}$, the binary alphabet

Ex 2 : The ASCII alphabet, the set of ASCII characters

Ex 3 : $\Sigma = \{a,b,c,......z\}$ the set of all lower case letters.

Power of Alphabet

Power of alphabet is the set of all strings of length k over alphabet Σ. We use the notation Σ^k.

Ex : Let $\Sigma = \{0,1\}$

$\Sigma^1 = \{0,1\}$ Set of all strings of length 1.

$\Sigma^2 = \{00,01,10,11\}$ Set of all strings of length 2.

$\Sigma^3 = \{000,001,011,100,101,110,111\}$ Set of all strings of length 3.

Note : $\Sigma^0 = \{\in\}$ regardless of what alphabet Σ is, $\in$ is the string having length zero.

1.2 STRINGS

A String over an alphabet Σ is a finite sequence of symbols from alphabet Σ .

Ex 1 : Binary alphabet $(\Sigma) = \{0,1\}$

Strings are 0111100, 111, 000, ...etc.

Ex 2 : $\Sigma = \{a,b,c,......z\}$

aa, bb, abracadabra are strings of this alphabet

Length of a string

The length of the string w is the number of symbols composing the string. It is denoted by $|W|$.

Ex 1 : w = aabb, then $|W|$ = 4

Ex 2 : w = 010110, then $|W|$ = 6

Empty String

The empty string is the string with zero symbols and zero length. It is denoted by '$\in$'.

$$|\in| = 0.$$

Note 1 : The set of all strings over an alphabet Σ is conventionally denoted by Σ^*.

Ex: $\Sigma = \{0,1\}$

$$\Sigma^* = \{\in, 0, 1, 00, 10, 01, 11, 000, ...\}$$

$$\Sigma^* = \Sigma^0 \cup \Sigma^1 \cup \Sigma^2 \cup ....$$

Note 2 : The set of all non-empty strings over an alphabet Σ is denoted by Σ^+.

$$\Sigma^+ = \Sigma^1 \cup \Sigma^2 \cup \Sigma^3 \cup ....$$

$$\Sigma^+ = \Sigma^* - \{\in\}$$

Note 3 : $\Sigma^* = \Sigma^+ \cup \{\in\}$

Prefix and Suffix of a String

A Prefix of a string is any number of leading symbols of that string.

A Suffix is any number of trailing symbols.

Ex1: String : abc

Prefix : $\in, a, ab, abc$

Suffix : $\in, c, bc, abc$

Note : A Prefix or suffix of a string, other than the string itself, is called a proper prefix or suffix.

Operation on strings

Concatenation

Concatenation of two strings is writing the first followed by the second string with no intervening space.

Ex 1: Let x,y be two strings over Σ^*

concatenation of x,y results new string z = xy.

Ex 2: Let x =ab, y = pq ; find xy, yx

xy = abpq, yx = pqab.

Note : The empty string $\in$ is the identity for the concatenation operator. That is, $\in W = W \in = W$, for each string w.

1.3 LANGUAGES

A language L over the alphabet Σ is a subset of Σ^*. That is, a language is a set of strings over the given alphabet.

Note 1 : $L \subseteq \Sigma^*$ L is language over Σ

Note 2 : A language can be empty $L = \phi$

Note 3 : Empty language is not equal to $L = \{\in\}$

Ex 1: Let $\Sigma = \{a,b\}$ then $\Sigma^* = \{\in, a, b, aa....\}$ the set $L = \{a, aa, ab\}$ is a language on Σ.

Ex 2: The set $L = \{a^n, b^n : n \geq 0\}$ is also a language on Σ.

The strings aabb, aaabbb belongs to L.

But string abb is not in L, because the no of a's and b's should be equal.

Operations on Languages

1) Concatenation

The concatenation AB of languages A and B is defined by $AB = \{uv / u \in A, v \in B\}$

Ex : Let $\Sigma = \{0, 1\}$, A, B are languages over S

$A = \{00, 11\}, \; B = \{0, 1\}$

$AB = \{000, 001, 110, 111\}$

$BA = \{000, 011, 100, 111\}$

Note 1 : It should be noted that $AB \neq BA$

Note 2 : The concatenation operation is, however, associative i.e. $A(BC) = (AB)C$

2) Closure or Star Operation

Star Closure : This operation defines on a set S a derived set S^*, having, as members, the empty string and all strings formed by concatenating a finite number of strings in S or, alternatively.

$S^* = S^0 \cup S^1 \cup S^2 \ldots. Where\ S^0 = \in, S^i = S^{i-1}\ S\ for\ i > o$.

Ex : If $S = \{01, 11\}$ then

$S^* = \{\in, 01, 11, 0101, 0111, 1101, 1111, 010101 \ldots\ldots\}$

Positive Closure

The positive closure of a language is defined as

$$S^+ = S^1 \cup S^2 \cup S^3 \ldots..$$

$S^+ = \{01, 11, 0101, 0111, 1101, 1111, 010101 \ldots\ldots\}$

1.4 FINITE STATE MACHINE

The finite automation is a mathematical model of a system with discrete inputs and outputs. Theory of finite automata is a useful design tool for these systems.

A primary example is a switching circuit, such as the control unit of a computer. A switching circuit is composed of a finite number of gates, each of which can be in one of two conditions usually denoted by 0 and 1.

Ex: Consider the shift register as a finite state machine

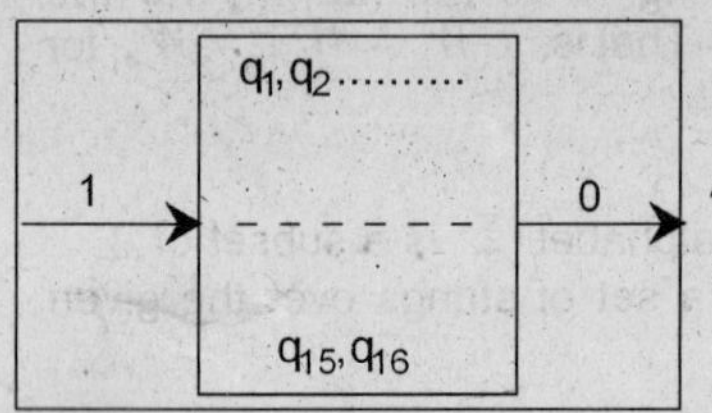

4-bit serial shift register can have $2^4 = 16$ states

1.5 DETERMINISTIC FINITE AUTOMATA

Definition

Analytically, a finite automation can be represented by a 5-tuple $(Q, \Sigma, \delta, q_0, F)$ where

(i) Q is finite non empty set of states;

(ii) Σ is a finite non empty set of inputs called input alphabet;

(iii) δ is a direct transition function which maps

$Q \times \Sigma$ into Q. $\delta : Q \times \Sigma \rightarrow Q$.

(iv) $q_0 \in Q$ is the initial state; and

(v) $F \subseteq Q$ is the set of final states.

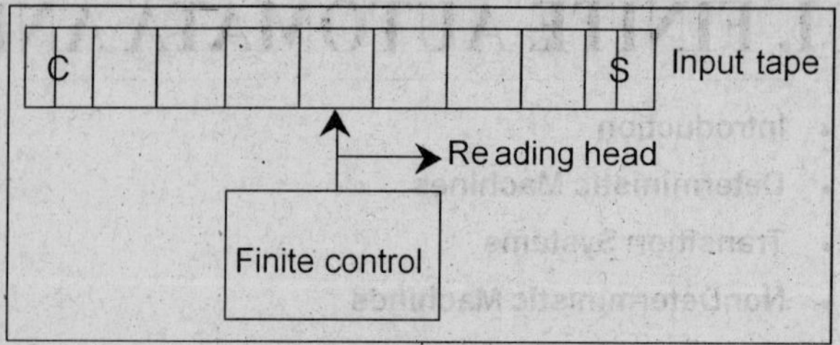

We picture FA as finite control which is in some state from Q. It reads sequence of symbols from Σ written on input tape. In one move the FA in state q and scanning symbol 'a' enters state δ(q,a) and moves its head one symbol to be right. If δ (q,a) is an accepting state, then be FA is deemed to have accepted the string.

1.6 TRANSITION DIAGRAMS

Transition diagram is a directed graph associated with an FA. The vertices of the graph correspond to the states of the FA. If there is a transition from state q to state p on input a, then there is an arc labeled a from state q to state p in the transition diagram. The FA accepts a string x if the sequence of transitions corresponding to the symbols of x leads from the start state to an accepting state.

Features

(a) For each state in Q there is a node in transition diagram

(b) For each state q in Q and each input symbol 'a' in Σ *if* $\delta(q, a) = p$ then there is an arc from node q to node p.

(c) An arrow in to the starting state q_0, labeled start.

(d) Nodes corresponding to accepting states are marked by a double circle.

Ex 1: The transition diagram for DFA accepting the string with sub string 01.

L = { w/w is of the form x01y where x and y all strings consisting of 0's and 1's}

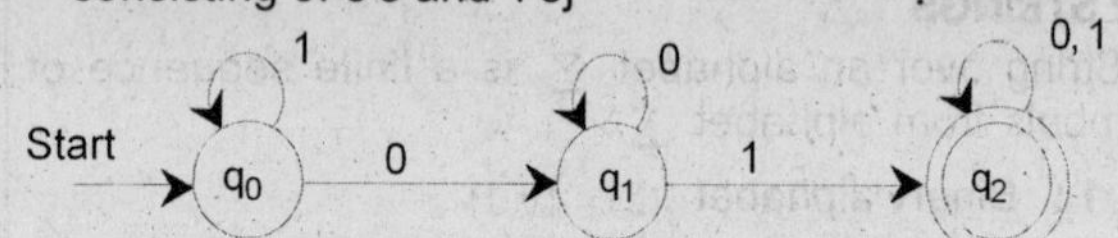

Figure : 1.1

The diagram consists of initial state q_0 labeled start and accepting state q_2 represented by a double circle. Out of each state one arc is labeled 0 and one arc labeled 1.

Ex 2 : Transition diagram of an DFA accepting strings having both even number of 0's and 1's.

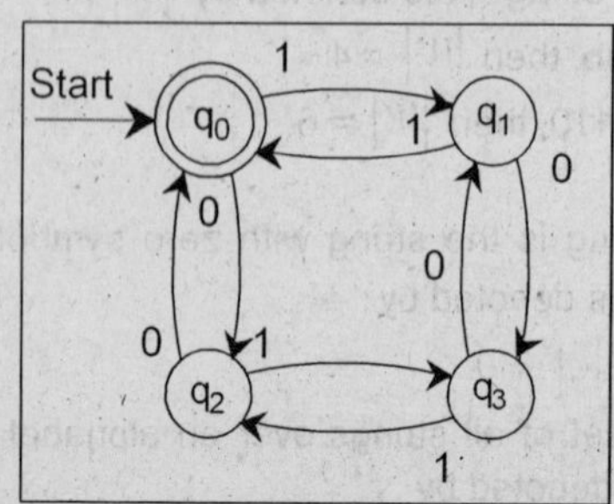

Figure: 1.2

Here q_0 , itself , is the initial state and accepting state. Between the states, there are arcs representing the transitions.

1.7 TRANSITION TABLES

A transition table is a conventional tabular representation of transition function δ that takes two arguments and returns a value.

The rows of the table correspond to the states and columns correspond to the inputs. The entry for the row corresponds to state 'q' and column correspond to input 'a' is the state δ (q,a).

Ex 1: The transition table for fig 1.1

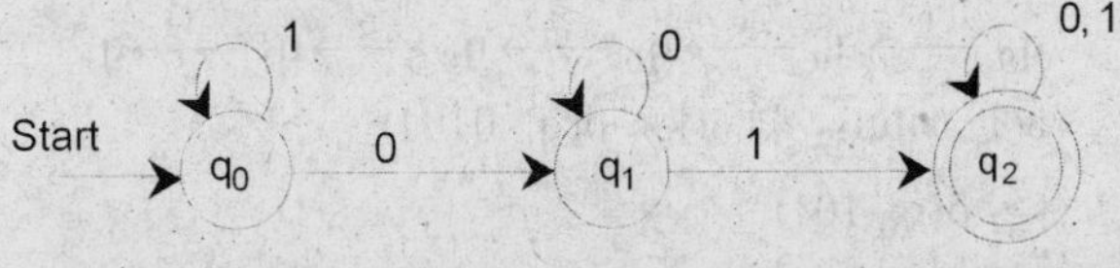

	0	1
$\to q_0$	q_1	q_0
q_1	q_1	q_2
(q_2)	q_2	q_2

The state q_0 is marked with an arrow the final state q_2 is encircled.

Table 1.1

Ex 2: The transition table for Fig. 1.2 is

	0	1
$\to (q_0)$	q_2	q_1
q_1	q_3	q_0
q_2	q_0	q_3
q_3	q_1	q_2

Table 1.2

The initial state q_0 is marked with arrow and also encircled because it is also final state.

Extending the transition function to strings :

The extended transition function is a function that takes a state q and a string w and returns a state p. It is denoted $\hat{\delta}$ is a function mapping from $Q\times\Sigma^*\ to\ Q$

$$\hat{\delta}: Q\times\Sigma^* \to Q.$$

Let W = Xa

$$\hat{\delta}(q,w) = \delta(\hat{\delta}(q,x),a) = P$$

Ex : Construction of $\hat{\delta}$ for table 1.2

Suppose the input w = 101101. Since the string has even number of 0's and 1's both, we expect it is in the language.

$\hat{\delta}(q_0,\in) = q_0$ initial state.

$\hat{\delta}(q_0,101101) = q_0$ final state.

$\hat{\delta}(q_0,101101) = \hat{\delta}(q_1,01101)$

$= \hat{\delta}(q_3,1101)$

$= \hat{\delta}(q_2,101)$

$= \hat{\delta}(q_3,01)$

$= \hat{\delta}(q_1,1)$

$\hat{\delta}(q_0,101101) = q_0$ final state.

1.8 PROPERTIES OF TRANSITION FUNCTION

property 1 : $\delta(q,\in)=q$ in a finite automation. This means the state of the system can be changed only by an input symbol.

Property 2 : For all string w and input symbols a,

$$\delta(q,aw) = \delta(\delta(q,a),w)$$

$$\delta(q,wa) = \delta(\delta(q,a),a)$$

This property gives the state after the automation consumes or reads the first symbol of a string aw and the state after the automation consumes a prefix of the string wa.

1.9 ACCEPTABILITY OF A STRING BY A FA

A string x is accepted by a finite automation M = $(Q,\Sigma,\delta,q_0,F)\ if\ \delta(q_0,x)=q$ for some $q\in F$, This is basically the acceptability of a string by the final state.

HOW A DFA PROCESSES STRINGS

DFA starts with initial state q_0. Let $a_1 a_2 \ldots a_n$ is a sequence of input symbols, δ is transition function.

Step 1: $\delta(q_0,q_1)=q_1$, DFA in state q_0 on input a_1 enters state q_1.

Step 2 : $\delta(q_1,a_2)=q_2\quad DFA\ in\ state\ q_1$ on input a_2 , enters state q_2 similarly $\delta(q_{i-1},a_i)=q_i$ for each i if q_n is a member of F, then input $a_1,a_2,\ldots a_n$ is accepted, otherwise it is rejected.

Ex 1 : Consider the finite state machine whose transition function δ in given in **table 1.2.**

Here Q = $\{q_0, q_1, q_2, q_3\}$

$\Sigma = \{0,1\},\ F = \{q_0\}$

Input string 110101

$$\delta\left(q_0, \overset{\downarrow}{1}10101\right) \vdash \left(q_1, \overset{\downarrow}{1}0101\right)$$

$\vdash \delta(q_0, \overset{\downarrow}{0}101)$

$\vdash \delta(q_2, \overset{\downarrow}{1}01)$

$\vdash \delta(q_3, \overset{\downarrow}{0}1)$

$\vdash \delta(q_1, \overset{\downarrow}{1})$

$\vdash \delta(q_0\ \in)$

$\vdash q_0$ accepting State

Hence,

$$q_0 \xrightarrow{1} q_1 \xrightarrow{1} q_0 \xrightarrow{0} q_2 \xrightarrow{1} q_3 \xrightarrow{0} q_1 \xrightarrow{1} q_0$$

$\therefore$ The given string 110101 is accepted.

Language of DFA

The language of a DFA A = (Q,Σ,δ,q_0,F) is denoted by L(A) is defined as L(A) = $\{w/\delta(q_0,w)\ is\ in\ F\}$

That is, the language of A is the set of strings w that take initial state q_0 to one of the accepting states.

If L is L(A) for some DFA A, then L is called a regular language.

1.10 NON DETERMINISTIC FINITE AUTOMATION (NDFA)

Definition

A non deterministic finite automation is a 5 - tuple, $(Q, \Sigma, \delta, q_0, F)$ where

(i) Q is a finite non empty set of states ;

(ii) Σ is a finite non empty set of inputs ;

(iii) δ is the transition function mapping from $Q \times \Sigma$ into 2^Q which is power set of Q, the set of all subsets of Q;

(iv) $q_0 \in Q$ is the initial state; and

(v) $F \subseteq Q$ is the set of final states.

Note : The difference between the deterministic and non- deterministic automata is only in δ.

For DFA, the outcome is a state, i.e. an element of Q.

For NFA, the outcome is a subset of Q.

Ex 1 : Consider the transition graph

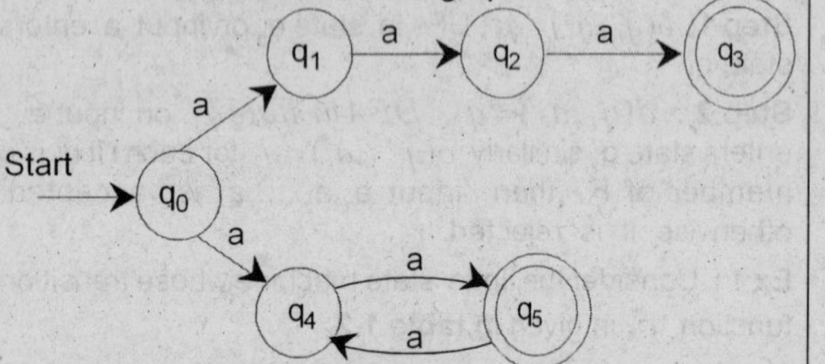

Figure 1.3

It describes a non-deterministic finite automation since there are two transitions labeled a out of q_0.

$$\delta(q_0, a) = \{q_1, q_4\}, \text{a subset of Q.}$$

$$NFA, A = (\{q_0, q_1, q_2, q_3, q_4, q_5\}, \{a\}, \delta, q_0, \{q_3, q_5\})$$

Ex 2 : Consider the NFA

$A = (\{q_0, q_1, q_2\}, \{0,1\}, \delta, q_0, \{q_2\})$

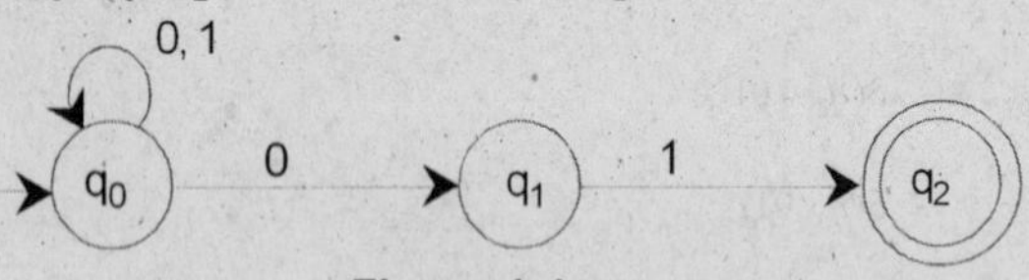

Figure 1.4

An NFA accepting all strings that end in 01. There are two transitions labeled 0 out of q_0.

$\therefore$ NFA has option of going either to q_0 or q_1 from q_0 on input 0.

	0	1
$\rightarrow q_0$	$\{q_0, q_1\}$	$\{q_0\}$
q_1	ϕ	$\{q_2\}$
ⓠq_2	ϕ	ϕ

Table 1.3 Transition table for NFA in Figure 1.4

Processing inputs by NFA

Consider the input sequence 00101 given to NFA in Figure 1.4 Processing of input sequence 00101

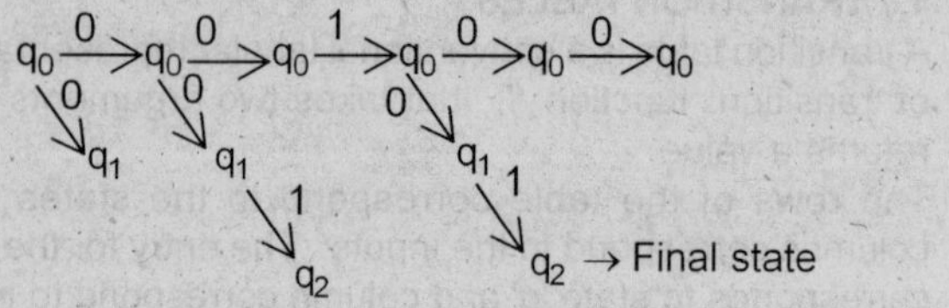

The states of NFA during the processing of input sequence 00101. q_2 is accepted state.

$$q_0 \xrightarrow{0} q_0 \xrightarrow{0} q_0 \xrightarrow{1} q_0 \xrightarrow{0} q_1 \xrightarrow{1} q_2$$

$\delta(q_0, 00101) \quad \vdash \delta(q_0, 0101)$

$\vdash \delta(q_0, 101)$

$\vdash \delta(q_0, 01)$

$\vdash \delta(q_1, 1)$

$\vdash q_2$

$F = \{q_2\}$

$\therefore$ The string is accepted by NFA.

Extended transition function $\hat{\delta}$ is shown below

$\hat{\delta}(q_0, \in) = \{q_0\}$

$\hat{\delta}(q_0, 0) = \delta(q_0, 0) = \{q_0, q_1\}$

$\hat{\delta}(q_0, 00) = \delta(q_0, 0) \cup \delta\{q_1, 0\} = \{q_0, q_1\} \cup \phi = \{q_0, q_1\}$

$\hat{\delta}(q_0, 001) = \delta(q_0, 1) \cup \delta\{q_1, 1\} = \{q_0\} \cup \{q_2\} = \{q_0, q_2\}$

$\hat{\delta}(q_0, 0010) = \delta(q_0, 0) \cup \delta\{q_2, 0\} = \{q_0, q_1\} \cup \phi = \{q_0, q_1\}$

$\hat{\delta}(q_0, 00101) = \delta(q_0, 1) \cup \delta\{q_1, 1\} = \{q_0\} \cup \{q_2\} = \{q_0, q_2\}$

Since $L(A) = \left\{ w / \hat{\delta}(q_0, w) \cap F \neq \phi \right\}$

i.e $\{q_0, q_2\} \cap \{q_2\} \neq \phi$

$\therefore$ The string 00101 is accepted.

Language of NFA

If $A = (Q, \Sigma, \delta, q_0, F)$ is an NFA then

$L(A) = \{ w / \hat{\delta}(q_0, w) \cap F \neq \phi \}$

i.e., L(A) is set of strings w in Σ^* such that $\hat{\delta}(q_0, w)$ contains at least one final state.

1.11 FINITE AUTOMATA WITH $\in$ - MOVES

Definition : NFA with $\in$ - moves is a 5 - tuple $(Q, \Sigma, \delta, q_0, F)$ with all components as before but δ , the transition function maps $Q \times (\Sigma \cup \{\in\})$ to 2^Q.

Ex 1 : NFA with $\in$ - Move

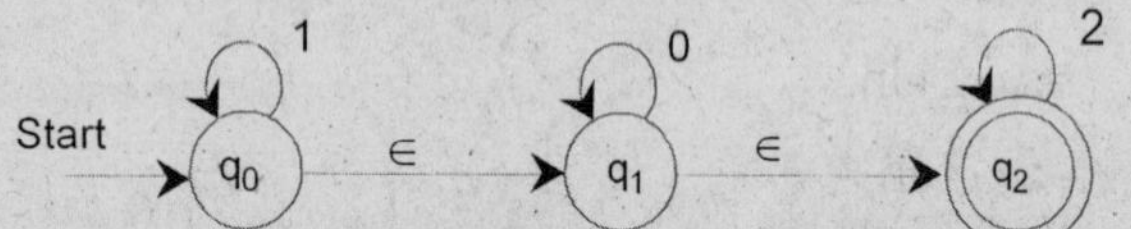

Figure 1.5

This NFA accepts a string having any number of zeroes, followed by 1's and followed by 2's.

$\in-$ closure (q)

It is the set of all vertices P such that there is a Path from q to p labeled $\in$.

EX : $\in-$ closure (q_0) in figure 1.5

$\in-$ closure (q_0) = $\{q_0, q_1, q_2\}$

$\in-$ closure $(q_1) = \{q_1, q_2\}$

$\in-$ closure $(q_2) = \{q_2\}$

1.12 EQUIVALENCE OF NFA WITH AND WITHOUT $\in-$ TRANSITIONS

Theorem : If Lis accepted by an NFA $\in-$ transitions, then L is accepted by an NFA without $\in-$transitions.

1.13 EQUIVALENCE OF TWO FINITE STATE MACHINES

Two Finite Automata over Σ are equivalent if they accept the same set of strings over Σ.

Two Finite Automata over Σ are not equivalent if for some string W over Σ one automaton reaches a final state and the other automaton reaches a non final state.

1.14 MOORE AND MEALY MACHINES

1.14.1 Moore Machine

An automaton in which the output depends only on states of the machine is called a Moore machine.

Definition

The Moore machine is a six - tuple

$(Q, \Sigma, \Delta, \delta, \lambda, q_0)$ where

i) Q is a finite set of state S ;

ii) Σ is the input alphabet ;

iii) Δ is the output alphabet

iv) δ is the transition function $\Sigma \times$ Q into Q;

V) λ is the output function mapping Q into Δ ;

Vi) q_0 is the initial state.

Example : Table 1.4 gives a Moore Machine

Present state	(δ) Next state a = 0	a = 1	Out put (λ)
$\to q_0$	q_0	q_1	0
q_1	q_2	q_0	1
q_2	q_1	q_2	2

Table 1.4 : Moore Machine

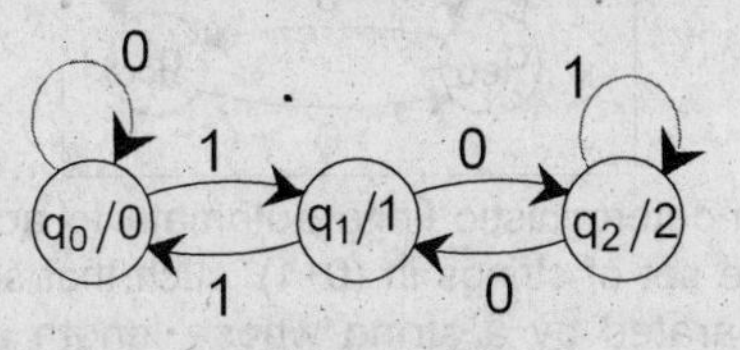

Figure : 1.6

The initial state q_0 is marked with an arrow. The table defines δ and λ.

In Moore machine, the output depends upon present state and not on present input.

For the input string 1000, the transition states is given by $q_0 \to q_1 \to q_2 \to q_1 \to q_2$. The output string is 01212.

1.14.2 MEALY MACHINE

An Automaton in which the output depends on the state and the input at any instant of time is called Mealy Machine.

Definition

A Mealy Machine is a six - tuple $(Q, \Sigma, \Delta, \delta, \lambda, q_0)$ where

i) Q is the finite set of states ;

ii) Σ is the input Alphabet

iii) Σ is the output Alphabet

iv) δ is the transition function $\Sigma \times$ Q into Q

V) λ is the output function mapping $\Sigma \times$ Q into Δ

Vi) q_0 is the initial state.

Ex :

Present state	Next State			
	Input $a = 0$		Input $a = 1$	
	State	O/P	State	Output
$\to q_0$	q_0	0	q_1	1
q_1	q_2	2	q_0	0
q_2	q_1	1	q_2	2

Table : 1.5 Mealy Machine

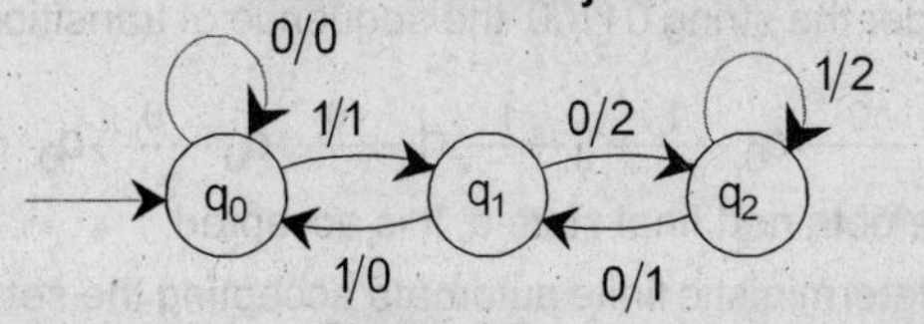

Figure 1.7

In Mealy Machine, the output depends upon the present state and present input. For the input string 1011 the transitions are $q_0 \to q_1 \to q_2 \to q_2 \to q_2$

the output string is 1222.

1.14.3 Differences between mealy and more machine

Mealy machine

1. The output of mealy machine always depends on present state and present input.

2. For the input of length n, the output will be of length n, For example

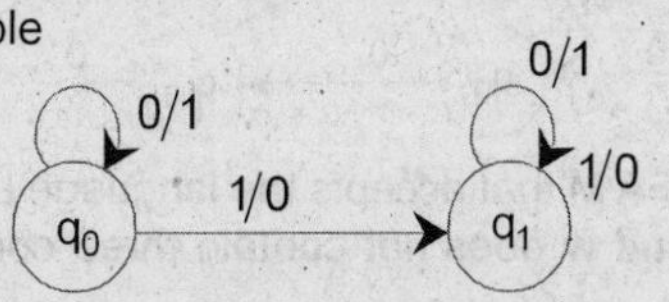

For the input 0011 the output will be 1100. Hence output length is same as that of input

3. For input $\in$, The output is also $\in$ because output is produced only on application of input symbol.

Moore machine

1 The output of Moore machine depends on present state only.

2. For the input of length n, the output will be of length n+1

Example

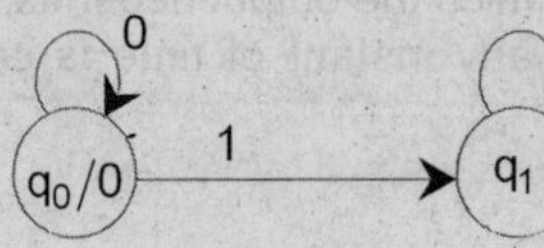

3. For the input $\in$, The output associated with the starting state is produced.

SOLVED EXAMPLES

1. Give deterministic finite automata accepting the set of all strings ending with 00.

Sol : Let q_1 is the initial state, q_F is the final state

Transition table :

	0	1
→ q_1	q_o	q_1
q_0	q_F	q_1
(q_F)	q_F	q_1

Transition diagram :

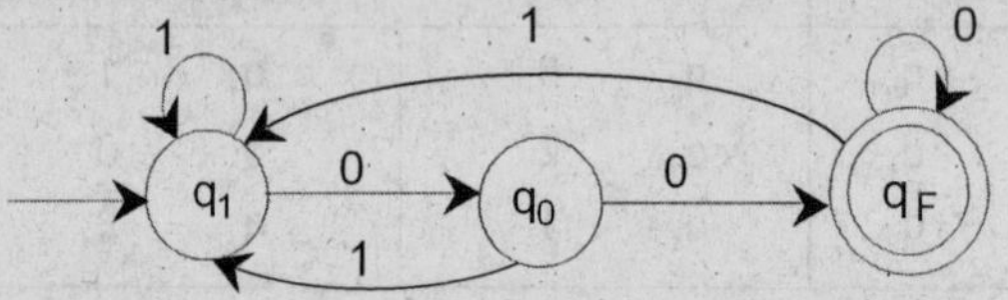

Verification :

1. Consider the string 01100 the sequence of transitions

are : $q_1 \xrightarrow{0} q_0 \xrightarrow{1} q_1 \xrightarrow{1} q_1 \xrightarrow{0} q_0 \xrightarrow{0} q_0 \; q_0$

∴ we obtained final state q_f it is accepted.

2. Give deterministic finite automata accepting the set of all strings with three consecutive 0's.

Sol : Let q_1 is the initial state, q_F is the final state.

Transition table :

	0	1
→ q_1	q_o	q_1
q_0	q_{00}	q_1
q_{00}	q_F	q_1
(q_F)	q_F	q_F

Transition diagram :

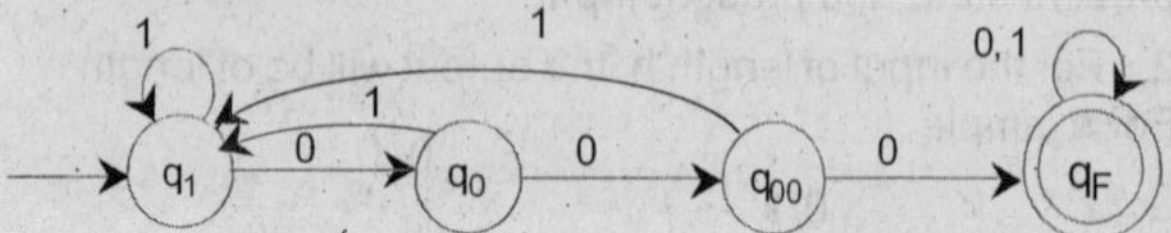

3. Design a DFA M that accepts the language L(M) = {w/ w $\in$ {0, 1}*} and w does not contain three consecutive 0's.

Sol : Previous problem i.e., DFA that accepts the set of all strings with three consecutive 0's

In our present problem we have to design DFA that accepts the set of all strings without three consecutive 0's.

This language is compliment to previous one.

∴ We get the DFA that accept strings without three consecutive 0's by making all non-final states as final states as non final states.

Required DFA is

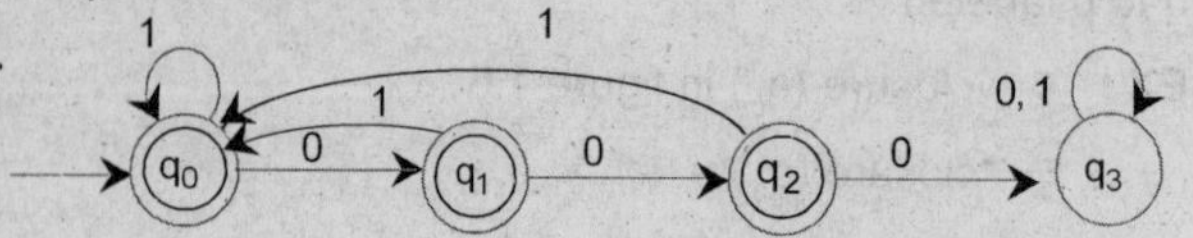

4. Describe in English the sets accepted by the finite automata whose transition diagrams are given below.

(i)

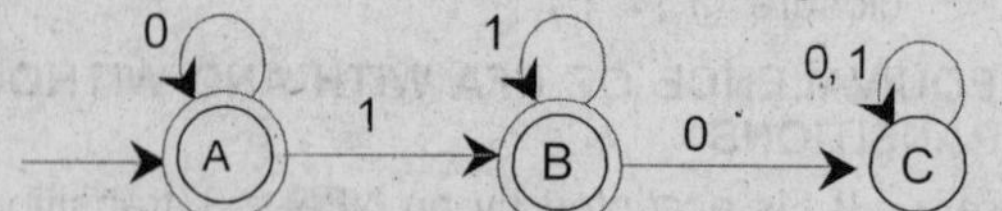

Sol : The set of all strings not containing 10 as a sub string

(ii)

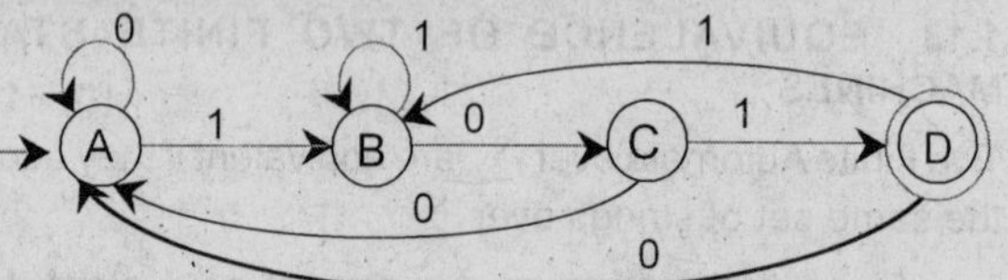

Sol :

The above DFA accepts the set of all strings ending with 101.

5. Construct a DFA accepting the set of strings with an even number of 0's and an even number of 1's over the alphabet {0, 1}

Sol : Let q_{ee} is the state with even number of 0's and 1's.

q_{oe} is the state with odd number of 0's and even number of 1's.

q_{oo} is the state with odd number of 0's and 1's.

Transition table for the above machine is

	0	1
q_{ee}	q_{oe}	q_{eo}
q_{oe}	q_{ee}	q_{oo}
q_{eo}	q_{oo}	q_{ee}
q_{oo}	q_{eo}	q_{oe}

DFA :

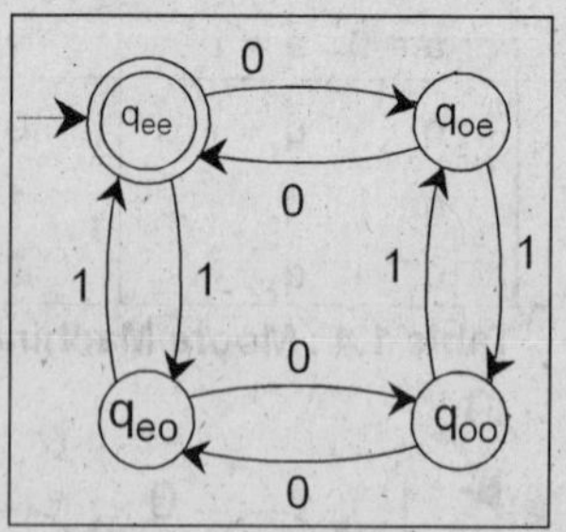

6. Give non-deterministic finite automata to accept the language the set of strings in $(0+1)^*$ such that some two 0's are separated by a string whose length is 4i, for some $i \geq 0$.

Sol : The strings of above language are like this

1001 When i = 0

10|100|01 $|y| = 4$ when i =1

y

$10 \underbrace{01100100}_{y} 01 \quad |y| = 8 \quad$ when i = 2

NDFA :

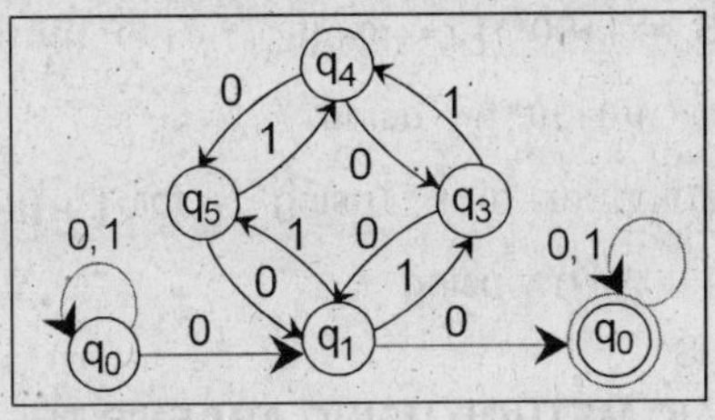

1.15 REGULAR SET

A special class of sets of words over S, called regular sets, is defined recursively as follows (Kleene proves that any set recognized by an FSM is regular conversely, every regular set can be recognized by some FSM).

(i) Every finite set of words over S (including f, the empty set) is a regular set.

(ii) If U and V are regular sets over S, then $U \cup V$ and UV are also regular

(iii) If S is regular set over S, then so is its closure S*.

(iv) No set is regular unless it is obtained by a finite number of applications of definition i) to iii)

i.e the class of regular sets over S is the smallest class containing all finite sets of words over S and closed under union, concatenation, and star operation.

Ex 1: Let $\Sigma = \{a, b\}$, then the set of strings that contain both even number of a's and b's is a regular set.

Ex 2: Let $\Sigma = \{1\}$, then the set of strings {1,11,111, ...} is a regular set.

Ex 3 : Let $\Sigma = \{0, 1\}$, then the set of strings {10,01} is a regular set.

1.16 REGULAR EXPRESSION

Definition

Let Σ be an alphabet. The regular expressions over Σ and the sets that they denote are defined recursively as follows.

(1) ϕ is a regular expression and denotes empty set.

$\in$ is a regular expression and denotes the set $\{\in\}$

For each a in Σ, a is a regular expression and denotes the set {a}. These are called primitive regular expressions.

(2) If r and s are regular expressions denoting the languages R and S, respectively, then (r+s), (rs) and (r*) are regular expressions that denote the sets $R \cup S, RS$, and R*, respectively.

(3) A string is a regular expression if and only if it can be derived from the primitive regular expressions by a finite number of applications of the rules in(2).

Examples

(1) For Σ = {a,b,c}, the string (a+bc)* (C+ ϕ) is a regular expression.

(2) 00 is a regular expression representing {00}.

(3) The regular expression (0+1)* denotes all strings of 0' s and 1's .

(4) The regular expression (1+10)* denotes all strings of 0's and 1's beginning with 1 and not having two consecutive 0's.

1.17 LANGUAGES ASSOCIATED WITH REGULAR EXPRESSIONS

Regular expressions can be used to describe some simple languages. If r is a regular expression, we will let L(r) denote the language associated with ' r '.

Definition

The language L(r) denoted by any regular expression r is defined by the following rules.

(i) ϕ is a regular expression denoting the empty set,

(ii) $\in$ is a regular expression denoting $\{\in\}$,

(iii) For every $a \in \Sigma$, a is a regular expression denoting {a}. If r and s are regular expressions, then

(iv) $L(r+s) = L(r) \cup L(s)$

(v) $L(rs) = L(r)\ L(s)$

(vi) $L((r)) = L(r)$

(vii) $L(r^*) = (L(r))^*$

Ex : For $\Sigma = \{a,b\}$, the expression

$r = (a+b)^*(a+bb)$ is regular , It denotes the language

$L(r) = \{a, bb, aa, ab, ba, bbb, ...\}$

The first part (a+b)* stands for any string of a's and b's. The second part (a+bb) represents either an 'a' or a double 'b' consequently L(r) is the set of all strings on {a,b} terminated by either an 'a' or 'bb'

Ex : Regular expression b* ab* ab* denotes the language containg all words with exactly two a's

1.18 Operations on regular expressions

(i) UNION: Let L = {001,10,111} and $M = \{\in, 011\}$ then

$L \cup M = \{\in, 10, 001, 011, 111\}$

(ii) CONCATENATION :

Let $L = \{001, 10, 111\}$, $M = \{\in, 001\}$

then LM = {001,10,111,001001, 10001,111001}

(iii) Kleen (star) closure :

Positive closure of L is denoted by L^+ is the set $L^+ = \bigcup_{i=1}^{\infty} L^i$

Let L = {0,11} then L* = {011,11110, $\in$, ...}

(iv) COMPLEMENTATION :

Let L_1 = All strings with a double 'a'.

Then $\overline{L_1}$ = All strings that do not contain the substring aa.

(v) INTERSECTION :

Let L_1 = All words that begin with an 'a'

$\therefore r_1 = a(a+b)^*$

$L_2 =$ All words that end with an 'a'

$\therefore r_2 = a(a+b)*a$

$L_1 \cap L_2 =$ All words that begin and end with the letter 'a' $\quad r_1 \cap r_2 = a(a+b)*a + a$

1.19 IDENTITY RULES FOR REGULAR EXPRESSIONS

Two regular expressions p and q are equivalent (we Write p=q) if p and q represent the same set of strings. We now give the identities for regular expressions; these are useful for simplifying regular expressions.

$I_1 : \phi + r = r$

$I_2 : \phi r = r\phi = \phi$

$I_3 : \in r = r \in = r$

$I_4 \ \in * = \in$ and $\phi * = \in$

$I_5 : r + r = r$

$I_6 : r * r* = r*$

$I_7 : rr* = r*r$

$I_8 : (r*)* = r*$

$I_9 : \in + rr* = r* = \in + r*r$

$I_{10} : (pq)*p = p(qp)*$

$I_{12} : (p+q)* = (p*q*)* = (P*+q*)*$

$I_{12} : (p+q)r = pr + qr$ and

$r(p+q) = rp + rq$

1.19.1 Manipulation of regular expressions

Arden's theorem

Let p and q be two regular expressions over Σ. If p does not contain $\in$, then the following equation in r,r=q+rp has a unique solution given by r=qp*

Ex 1: Give an r.e for representing the set L of strings in which every 0 is immediately followed by at least two 1's.

Sol : (a) If is in L, then either (i) W does not contain any 0, or (ii) it contains a 0 preceded by 1 and followed by 11.

So W can be written as $W_1 W_2$ .... Wn, where each W_i is either 1 or 011.

So L is represented by the r.e (1+011)*.

Ex 2: Prove that the r.e $r = \in + 1*(001)*(1*(011)*)*$ describes the same set of strings as that of (1+011)*.

Sol : $r = \in + p_1 p_1*$ where $p_1 = 1*(001)*$

$= p_1^*$ using I_9

$= (1*(011)*)*$

$= (p_2 * p_3*)*$ letting $p_2 = 1, p_3 = 011$

$= (p_2 + p_3)*$ using I_{11}

$= (1+011)*$

Hence proved.

Ex 3: Prove (1+00*1) + (1+00*1) (0+10*1)* (1+10*1) = 0*1(0+10*1)*.

Sol : L.H.S. = (1+00*1) $(\in + (0+10*1*(0+10*1)))$ using I_{12}

$= (1+00*1)(0+10*1)*$ using I_9

$= (\in + 00*1)\ 1\ (0+10*1)*$ using I_{12} for $1+00*1$

$= 0*1\ (0+10*1)*$ using I_9

= R.H.S

ALGEBRAIC METHOD USING ARDEN'S THEOREM

This method is used to find the r.e recognised by a transition system.

The following assumptions are made regarding the transition system:

(i) The transition graph does not have $\in$- moves.

(ii) It has only one initial state, say V_1

(iii) Its vertices are V_1 .... V_n.

(iv) V_i is the r.e representing the set of strings accepted by the system even though V_i is a final state.

(v) α_{ij} denotes the r.e represents the set of lables of edges from V_i to V_j

When there is no such edge, $\alpha_{ij} = \phi$.

Consequently, we can get the following set of equations.

Here '$\in$' Should be added in equation of V_1.

$$V_1 = V_1\alpha_{11} + V_2\alpha_{21} + \ldots\ldots + V_n\alpha_{n1} + \in$$

$$V_2 = V_1\alpha_{12} + V_2\alpha_{22} + \ldots\ldots + V_n\alpha_{n2}$$

$$\vdots$$

$$V_n = V_1\alpha_{1n} + V_2\alpha_{2n} + \ldots\ldots + V_n\alpha_{nm}$$

By repeatedly applying substitutions and Arden's Theorem, We can get V_i in terms of α_{ij}'s.

If there are more final states, then we have to take the 'union' of all V_i's corresponding to final states.

Ex 1: Find r.e that denotes the set accepted by FA whose transition diagram is given below.

Sol : We can apply the above method directly as the transition diagram does not contain more than one initial state and there are no $\in$- moves.

We get the following equations for q_1, q_2, q_3

$q_1 = q_1{}^0 + \in, q_1{}^1 + q_2{}^1, \ q_3 = q_2{}^0 + q_3(0+1)$

By applying arden's theorm to the q_1- equation, we get

$q_1 = \in 0* = 0*$

so, $\quad q_2 = q_1 1 + q_2 1 \ = 0*1 + q_2 1.$

therefore, $q_2 = (0^*1)1^*$

As the final states are q_1 and q_2, we need not solve for q_3:

$q_1 + q_2 = 0^* + 0^*(11^*) = 0^*(\in + 11^*) = 0^*(1^*)$

$\therefore q_1 + q_2 = 0^*(1^*)$

Ex 2: Find r.e., that denotes the set of strings accepted by the given transition system.

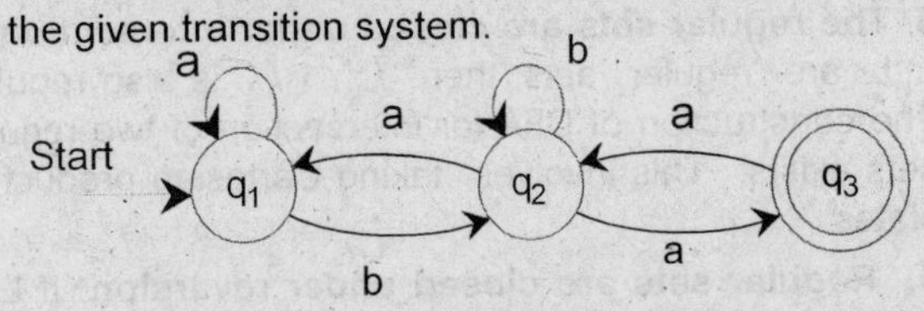

Sol : We can directly apply the above method since the graph does not contain any $\in$- move and there is only one initial state.

The three equations for q_1, q_2 and q_3 can be written as

$q_1 = q_1 a + q_2 b + \in,$

$q_2 = q_1 a + q_2 b + q_3 a$

$q_3 = q_2 a$

By substituting q_3 in q_2 - equation, we get

$q_2 = q_1 a + q_2 b + q_2 aa = q_1 a + q_2(b + aa)$

$q_2 = q_1 a(b + aa)^*$ by Arden's theorem

Substituting q_2 in q_1 we get

$q_1 = q_1 a + q_1 a(b + aa)^* b + \in$

$= q_1(a + a(b + aa)^* b) + \in$

Hence $q_1 = \in(a + a(b + aa)^* b)^*$

$q_2 = (a + a(b + aa)^* b)^* a(b + aa)^*$

$q_3 = (a + a(b + aa)^* b)^* a(b + aa)^* a$

Since q_3 a final state, the set of strings recognised by the graph is given by $(a + a(b + aa)^* b) a(b + aa)^* a$

1.20 Non Regular Languages

The languages which are accepted by some finite automata are called **regular languages.** Here it means that the FA accepts only the words of this language and does not accept any word outside it.

The Languages Which cannot be accepted by any FA are called non **regular languages.** "Not accepted" means that either.

(1) Some of the words of the language are not accepted by FA or

(2) All the words of the language are accepted in addition to that some extra strings are also accepted.

All languages are either regular or non regular, none of the languages are both.

By looking at some of the languages we can say whether they are regular or not:

(i) The languages whose words need some sort of comparision can never be regular.

Ex 1: $L = \{a^n b^n ; n \geq 0\}$

here the number of a's must be equal to number of b's. For each 'a' we check the existence of b which cannot be done using FA.

Ex 2: $L = \{a^m b^n c^m m \geq 0, n \geq 1\}$ this language is also non Regular.

(ii) The languages whose words are in arithmetic progression and need no comparisions will be regular.

Ex 3 : $L = \{a^{2n}, n \geq 1\}$

The words of this language are : aa,aaaa,aaaaaa ......, a^{2n} which are in A.P with period 2. Hence it is a regular language.

Ex 4 : $L = \{a^P : P \text{ is prime}\}$ the words of this language are {a,aa,aaa,aaaa,aaaaaaa,......a^P}

We can see these words are not in A.P. Hence it is not regular.

1.21 PUMPING LEMMA FOR REGULAR SETS

Statement: Let L be a regular set. It is accepted by FA, $M = (Q, \Sigma, \delta, q_0, F)$ with some particular number of states, say n. Let W $\in$ L with $|W| \geq n$ can be decomposed as W=xyz, with $|xy| \leq n$ and $|y| \geq 1$ such that $W_i = xy^iz$, is also in L for all i = 0,1,2,....

1.21.1 Application of Pumping Lemma

This statement can be used to prove that certain sets are not regular. We now give the steps needed for proving that a given set is not regular.

Step 1: Assume L is regular. Let n be the number of states in the corresponding FA.

Step 2: Choose a string w such that $|w| \geq n$. Use pumping lemma to write W = xyz. with $|xy| \leq n$ and $|y| > 0$.

Step 3 : Find a suitable integer i such that $xy^iz \notin$ L. This contradicts our assumption. Hence L is not regular.

NOTE : The crucial part of the procedure is to find i such that $xy^iz \notin$ L. In some cases we may have to use the 'structure' of strings in L.

Example 1 :

Show that L = {0^i1^i/i $\geq$ 1} is not regular.

Sol : Step 1 : Suppose L is regular and we get a contradiction. Let n be the number of states in FA Accepting L.

Step 2: Let w = 0^n1^n. Then $|w| = 2n > n$. By pumping lemma we write w= xyz with $|xy| \leq n$ and $|y| \neq 0$.

Step 3: We want to find i so that $xy^iz \notin L$ for getting a contradiction. The string y can be in any of the following forms.

Case 1 : y has 0's i.e. y = 0^k for some $k \geq 1$.

Case 2: y has only 1's i.e. y = 1^l for some $l \geq 1$.

Case 3: y has both 0's and 1's i.e y=0^k1^j for some k, $j \geq 1$. In case 1, we can take i=0. As xyz = 0^n1^n, xz = $0^{n-k}1^n$ As $K \geq 1,\ n-k \neq n$. So xz $\notin$ L,

In case 2, take i = 0. As before, xz is $0^n 1^{n-l}$ and n$\neq n-l$ So xz $\notin l$

In case 3, take i=2. As xyz = $0^{n-K}0^k1^j1^{n-j}$, xy^{2}z = 0^{n-K} $0^K 1^j 0^K$, $1^J 1^{n-j}$. As xy^{2}z is not of the form 0^i1^i, xy^{2}z $\notin$L.

Thus in all the cases we get a contradiction. There fore, L is not regular.

Ex2: Show that L = {a^p/ p is a prime } is not regular.

Sol: Step 1 : We suppose L is regular and get a contradiction. Let n be the number of states in the FA accepting L.

Step 2 : Let p be a prime number greater than n. Let w=a^p. By pumping lemma, w can be written as w = xyz, with $|xy| \leq n$ and $|y| > 0$.. x,y,z are simply strings of a's. So y = a^m for some m $\geq$ 1(and $\leq$n)

Step 3: Le i = p +1. Then

$$|xy^iz| = |xyz| + |y^{i-1}| = p + (i-1)m = p + pm$$

By pumping Lemma, xy^{i}z $\in$L.

But $|xy^iz| = p + pm = p(1+m)$ and $p(1+m)$ is not a prime. So xy^{i}z $\notin$L. This is a contradiction. Thus L is not regular.

1.22 CLOSURE PROPERTIES OF REGULAR SETS

If a class of languages is closed under a particular operation, we call it closure property of class of languages

1. The regular sets are closed under union.

If L_1 and L_2 are regular sets then L_1 U L_2 is also regular.

Ex : If L_1 = all words with zero or more number of a's

$\therefore$ $r_1 = a^*$

L_2 = all words with zero or more number of b's $r_2 = b^*$

L_1UL_2 = L (r_1+r_2) i.e., L(a^* + b^*)

= all words with 0 or more number of a's or with 0 or more number of b's.

here L_1 and L_2 are regular sets so is L_1UL_2

2. The regular sets are closed under concatenation. If L_1 and L_2 are regular sets then L_1 L_2 is also regular

Ex : Consider languages L_1 and L_2 of previous example.

$$L_1 L_2 = L(r_1r_2)$$
$$= L(a^*b^*)$$

So L_1L_2 = all words with zero or more number of a's followed by 0 or more number of b's

Here L_1 and L_2 are regular so is L_1 L_2.

3. The regular sets are closed under kleen closure. If L1 is a regular set then L_1^* is also regular.

Example, consider $L_1 = a + b$, $L_1^* = (a+b)^*$

L_1 is regular so is L_1^* i.e. $(a+b)^*$

4. The class of regular sets are closed under complementation. If L is a regular set and $L \subseteq \Sigma^*$ then $\Sigma^* - L$ is also regular.

5. The regular sets are closed under intersection. If L_1,L_2 are regular sets, then $L_1 \cap L_2$ is also regular. The construction of DFA for intersection of two regular sets exists. This involves taking Cartesian product of states.

6. Regular sets are closed under reversion. If L is regular the L^T is also regular.

Ex : Let L=a*b*, L^T=b*a* is also regular

7. Regular sets are closed under substitution, homomorphism and inverse homomorphism.

1.23 REGULAR GRAMMARS

Grammar : A grammar is G = (V, T, P, S) where

(i) V $\rightarrow$A finite non-empty set whose elements are called variables.

(ii) T$\rightarrow$A finite non-empty set whose elements are called terminals.

(iii) S$\rightarrow$A special variable (i.e, an element of v) called the start symbol and

(iv) P$\rightarrow$A finite set whose elements are $\alpha \rightarrow \beta$ productions where

α and β are strings on $V \cup T$.

α has at least one symbol from V.

Elements of p are called productions or production rules or rewriting rules.

Ex : Define the four parts of the grammar as following :

$V = \{A, B, S\}$, $T = \{a, b\}$

$P = \{S \rightarrow AB, A \rightarrow Aa, B \rightarrow Bb, B \rightarrow b\}$

with 'S' as the start symbol.

The productions may be abbreviated as

$P = \{S \rightarrow AB, A \rightarrow (Aa, a)\ B \rightarrow Bb, B \rightarrow b\}$

RIGHT - LINEAR GRAMMAR

A grammar G = (V,T,P,S) is said to be right linear if all productions are of the form

$A \rightarrow xB$ (or) $A \rightarrow x$ Where A, B $\in$ V and $x \in T^*$

Ex: The grammar $G_1 = (\{s\}, \{a, b\}, P_1, S)$ where P_1 given as $S \rightarrow abS/a$ is right linear.

The sequences generated by it are

$S \Rightarrow abs \Rightarrow ababs \Rightarrow ababa$.

LEFT - LINEAR GRAMMAR

A grammar is said to be left - linear if all productions are of the form

$A \rightarrow Bx$, (or) $A \rightarrow x$ Where A, B $\in$ V and $x \in T^*$

Ex : The grammar $G_2 = (\{S, S_1, S_2\}, \{a, b\}, p_2, S)$ with

production $S \to S_1 ab,$

$S_1 \to S_1 ab / S_2$

$S_2 \to a$ is left linear.

The sequence is $S \Rightarrow S_1 ab$

$\Rightarrow S_1 abab \Rightarrow S_2 ab\ ab \Rightarrow aab\,ab.$

REGULAR GRAMMAR

A regular grammar is one that is either right - linear or left - linear.

In a regular grammar at most one variable appears on the right side of any production. Further more, that variables must be consistently be either the right most or left most symbol of the right side of any production.

1.24 EQUIVALENCE BETWEEN REGULAR GRAMMAR AND FA

Theorem 1: If L has a regular grammar, then L is a regular set.

Theorem 2 : If L is a regular set, then L is generated by some Left - Linear grammar and by some right - linear grammar.

SOLVED EXAMPLES

1. Write the language for the given regular expression r = (aa)* (bb)* b.

Sol : The given R.E. r =(aa)*(bb)*b denotes the set of all strings with an even number of a's followed by an odd number of b's i.e. L (r) = $\left\{ a^{2n}\, b^{2m+1} : n \geq 0,\, m \geq 0 \right\}$

2. For S = {0,1} , given regular expression r such that L(r) ={$\omega \in$ S*: w has at least one pair of consecutive zeroes}.

Sol : Every string in L(r) must contain 00 some where, but what comes before and what goes after is completely arbitrary.

An arbitrary string on {0,1} can be denoted by (0+1)*

∴ The solution is r = (0+1)*00 (0+1)*.

3. Find a regular expression for the language

L = {w $\in$ {0,1}*: w has no pair of consecutive zeroes}.

Sol : Useful observation is whenever a 0 occurs it must be followed immediately by a1. such a substring may be preceded and followed by an arbitrary no. of 1's i.e., answer involves the repetition of strings of the form.

1 ..... 101 .... 1.

i.e., the language may be denoted by the R.E.

(1*01 1*)*.

Since the strings ending in 0 or consisting of all 1's are unaccounted for. Taking this also into account, the required R.E. is.

r=(1*011*)* (0+ $\in$)+ 1* (0+ $\in$) (or)

If L is taken as repetition of the strings 1 and 01 then the shorter expression is r=(1+01)*(0+ $\in$)

Both are correct.

4. Give regular expressions for the following.

a. The language of all strings of 0's and 1's that have odd length of 1's.

b. The language of all strings of 0's and 1's containing at least one 1.

c. The language of all strings of 0, 1 having length less than or equal to 6.

Sol : a. In the given language, the strings should have odd number of 1's ; so one possible R.E.

(0+11)* 1 and another possibility is 1(0+11)*

∴ required R.E. is r = (0+11)*1+1(0+11)*

b. In the given language, the strings should have at least one 1.

So one possible R.E is (0+1)* 1 and another possible strings are of 1(0+1)*.

∴ required R.E is r = (0+1)* 1+1(0+1)*.

c. R.E for string of length one = (0+1)

R.E for string of length two = (0+1)(0+1)

Similarly for all lengths up to 6.

∴ The required R.E.

r = (0+1) + (0+1) (0+1)+ (0+1) (0+1) (0+1)
+ (0+1) (0+1) (0+1) (0+1)
+ (0+1) (0+1) (0+1) (0+1) (0+1)
+ (0+1) (0+1) (0+1) (0+1) (0+1) (0+1)
+ $\in$.

5. Find regular expression over alphabet {0,1}* for the language of strings containing odd length.

Sol : The RE for strings of length 1 is (0+1)

RE for strings of length 3 is (0+1) (0+1) (0+1)

By similar fashion,

RE for strings containing odd length is

(0+1) [(0+1) (0+1)]*

6. Write regular expressions for each of the following languages over the alphabet {0,1}.

(i) The set of all strings not containing 111.

(ii) The set of all strings in which every pair of adjacent 0's appears before any pair of adjacent 1's

Sol : (i) Strings should not contain 111

Strings can have 0, 10, 110

So r.e is (0+10+110)*

and string can have (0 + 10+ 110)* 1 and (0 + 10+ 110)* 11.

∴ The regular expression for the given set of Strings = (0+10+110)* + (0+10+110)* 1+ (0+10+110)* 11

= (0+10+110)* ($\in$+1+11).

(ii) String have 0011 for every pair of 00

∴ All the strings containing 0011are

(0011+1) 1*

∴ The required R.E. = (01+ (0011+1)1*) (0011+1)1*.

7. Prove that $L = \{ ww / w\ in\ (a+b)^* \}$ **is not regular.**

Sol : Prove the result by the method of contradiction

Step 1 : Suppose L is regular. Let 'n' be the number of states in the automation M accepting 'L'.

Step 2 : Let us consider

ww = $a^n b\ a^n b$ in L $|ww|$ = 2(n + 1) >n apply pumping lemma to write ww = xyz with $|y| \neq 0, |xy| \leq n$.

Step 3 : To find i such that $xy^iz \notin$ L for getting a contradiction. The string 'y' can be in only one of the following forms.

Case 1 : y has no b's i.e., y = a^k for some k ≥ 1

Case 2 : y has only no b

we may note that y cannot have two b's. If so $|y| \geq n+2$

But $|y| \leq |xy| \leq n$, In case 1, we can take i =0

Then xy^0z = xz is of the form $a^m b a^n b$. Where m=n-k<n (or $a^n b a^m b$)

x z cannot be written in the form uu with $u \in \{a, b\}$ and so xz $\notin$ L.

In case 2 also. We can take i=0

Then xy^0z = xz has only one b

So xz $\notin$ L as any element in L should have even number of a's and even number of b's.

Thus in both cases we get Contradiction.

∴ L is not regular

8. Show that the set $L = \{ ww^R / W \in \{a,b\}^* \}$ is not regular.

Sol : We consisting of strings of a's and b's

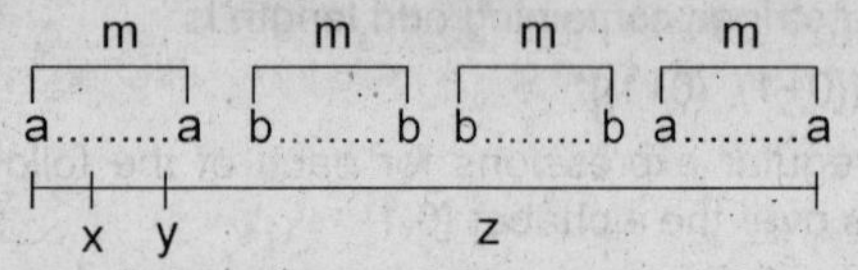

Whatever 'm' we can pick on step 1, we can always choose a w as shown in above figure.

$|xy| \leq m.$

∴ y consisting of entirely of a's.

for i = 0

xy^iz contains fewer number of a's on the left than on the right and so cannot be of the form ww^R. Therefore L is not regular.

OBJECTIVE QUESTIONS - I

1. Can a DFA simulate NFA?

a) No b) Yes

c) Some times d) Depends on NFA

2. Let L be set of strings from alphabet. The kleen closure of L is given as

a) $L^+ = \bigcup_{i=0} L^i$ b) $L^* = \bigcup_{i=0} L^i$

c) $L^* = \bigcup_{i=0} L^i$ d) $L^+ = \bigcup_{i=1} L^i$

3. Which of the following statements is wrong?

a) The languages accepted by finite automata are the languages denoted by regular expresions

b) For every DFA there is a regular expression denoting its language

c) For a regular expression r, there does not exist any NFA with transit that accepts L(r)

d) None of the above

4. Regular expression a/b denotes the set

a) {a} b) {ε,a,b} c) {a,b} d) {ab}

5. Regular expression (a/b)(a/b) denotes the set

a) {a,b,ab,aa} b) {a,b,ba,bb}

c) {a,b} d) {aa,ab,ba,bb}

6. Which of the following regular expressions denote zero or more instances of a or b?

a) a|b b) (a b)* c) (a|b)* d) a*|b

7. Let a and b be two regular expressions then $(a^* \cup b^*)^*$ is equivalent to

a) $(a \cup b)^*$ b) $(b^* \cup a^*)^*$

c) $(b \cup a)^*$ d) $a \cup b$

8. If A = {0, 1} , then the number of possible strings of length 'n' is

a) n! b) n × n c) n^n d) 2^n

9. Choose the correct statements.

a) Moore and mealy machines are FSM's with output capability

b) Any given Moore machine has an equivalent Mealy machine

c) Any given Mealy machine has an equivalent Moore machine

d) Moore machine is not an FSM

10. The major difference between a Moore and a Mealy machine is that

a) The output of the former depends on the present state and present input

b) The output of the former depends only on the present state

c) The output of the former depends only on the present input

d) None of the above

11. Which of the following are not regular?

a) String of 0's whose length is a perfect square

b) Set of all palindromes made up of 0's and 1's

c) Strings of 0's , whose length is a prime number

d) String of odd number of zeroes

12. Set of regular languages over a given alphabet set, is not closed under

a) Union b) Complementation

c) Intersection d) None of the above

13. Which one of the following regular expressions over {0, 1} denotes the set of all strings not containing 100 as a substring?

a) $0^*(1^*0)^*$ b) 0^*1010^*

c) $0^*1^*01^*$ d) $0^*(10+1)^*$

14. The string 1101 does not below the set represented by

a) $110^*(0 + 1)$ b) $1(0 + 1)^*101$

c) $(10)^*(01)^*(00 + 11)^*$ d) $(00 + (11^*)0)^*$

15. Which of the following statements is FALSE?

a) Every finite subset of a non-regular set is regular

b) Every subset of a regular set is regular

c) Every finite subset of a regular set is regular

d) The intersection of two regular sets is regular

16. Which of the following regular expression identities are TRUE? **[UGC NET-1995]**

a) $(r^*)^* = r^*$ b) $(r^* s^*)^* = (r + s)^*$

c) $(r + s)^* = r^* + s^*$ d) $r^* s^* = r^* + s^*$

17. Consider the following regular expression:

R =$(ab/abb)^*$bbab. Which of the following is NOT in the set R?

a) ab ab ab b) ab bb ab

c) abb abb bab d) abba bb abb bab

18. The intersection of two regular languages below:

$L_1=(a + b)^*a$ and $L_2=b(a + b)^*$

a) $b(a + b)^*a$ b) $(a + b)^*ab(a + b)^*$

c) $a(a + b)^*b$ d) $(a + b)^*ab$

19. $(ab)^*a$ is a regular expression defining a language over Σ ={a, b}. another regular expression that defines same language over the same alphabet is

a) $a(ba)^*$ b) $a^*(ba)^*$

c) $a^*(ba)$ d) both(a) and (b)

20. State whether True or False for regular expressions r and s: r(s + t)= rs + rt

a) True b) False

21. Let R_1 and R_2 be regular sets defined over the alphabet Σ then

a) $R_1 \cap R_2$ is not regular b) $R_1 \cup R_2$ is regular

c) $\Sigma^* - R_1$ is regular d) R_1^* is not regular

22. The regular expression have all strings of 0's and 1's, with no two consecutive 0's is

a) (0 | 1) b) (0 | 1)*

c) (0 + ε)(1 + 10)* d) (0 + 1)*011

23. The regular expression with all strings 0's and 1's with atleast two consecutive 0's is

a) (1 + 10)* b) (0 + 1)*00 (0 + 1)*

c) (0 + 1)*011 d) 0*1*2*

24. Which of the following is NOT the set of regular expression: R=(ab/abb)*bbab

a) ab abb bbab b) ab bbab

c) ab abb ab bbab d) aba ba bb

25. Which of the following string can be obtained by the language $L=\{a^ib^{2i} \mid i \geq 1\}$

a) aaabbbbbb b) aabbb

c) abbabbba d) aaaabbbabb

26. An automaton in which the output depends only on the input is called an

a) Automaton with a Finite memory

b) Moore machine

c) Mealy machine

d) Automaton without a Memory

27. An automaton in which the output depends only on the states of the machine is called an

a) Automaton without a Memory

b) Finite memory

c) Moore machine

d) Mealy machine

28. The minimum number of states of a Mealy machine which can output EVEN, ODD according as the total no. of 1's encountered is even or odd. The input symbols are 0 and 1.

a) 4 b) 2 c) 3 d) None of the above

29. The regular expression for the set of all strings of 0's and 1's ending with 00 is

a) 0*1*00 b) (0 + 1)*00

c) 0*1*0* d) None of these

30. There regular cxpression for the set

L ={ε, 11, 1111, 111111,.......} is

a) $(11)^+$ b) (11)* c) 1* d) None of these

31. If P, Q, R are regular expressions such that P does not contain null string, then .... has a unique solution

a) Q = P + QR b) P = Q + PR

c) R = Q + RP d) None of these

32. The set $L =\{a^{i^2} : i \geq 1\}$ is ......

a) Regular b) Not regular

c) Cannot be determined d) None of these

33. Is the set $\{a^{2n} \mid n \geq 1\}$ regular?

a) Yes b) No

c) Cannot be determined d) None of these

34. The smallest finite automaton which accepts the language { x | length of x is divisible by 3} has **[GATE-2002]**

a) 2 states b) 3 states c) 4 states d) 5 states

35. Given an arbitrary non-deterministic finite automation (NFA) with N states, the maximum number of states in an equivalent minimized DFA is at least **[GATE-2001]**

a) N^2 b) 2^N c) 2N d) N!

36. What can be said about a regular language L over {a} whose minimal finite state automaton has two states? **[GATE-2000]**

a) L must be $\{a^n \mid n \text{ is odd}\}$

b) L must be $\{a^n \mid n \text{ is even}\}$

c) L must be $\{a^n \mid n \geq 0\}$

d) Either L must be $\{a^n \mid n \text{ is odd}\}$, or L must be $\{a^n \mid n \text{ is even}\}$

37. Consider the regular expression (0+1) (0+1) ..... n times. The minimum state finite automaton that recognizes the languages represented by this regular expression contains. **[GATE-1999]**

a) n states b) n + 1 states

c) n + 2 states d) None

38. If the regular set A is represented by A =(01 + 1)* and the regular set 'B' is represented by B =((01)* 1*)*, which of the following is TRUE? **[GATE-1998]**

a) $A \subset B$ b) $B \subset A$

c) A and B are incomparable d) A = B

39. The string 1101 does not belong to the set represented by **[GATE-1998]**

a) 110*(0 + 1) b) 1(0 + 1)*101

c) (10)*(01)*(00 + 11)* d) (00 + (11)* 0)*

40. Let L be the set of all binary strings whose last two symbols are the same. The number of states in the minimum state deterministic finite state automation accepting L is **[GATE-1998]**

a) 2 b) 5 c) 8 d) 3

41. Which of the following statements is FALSE? **[GATE-1998]**

a) Every finite subset of a non-regular sets is regular

b) Every subset of a regular set is regular

c) Every finite subset of a regular sets is regular

d) The intersection of two regular sets is regular

42. Let Σ ={0,1}, L = Σ * and R ={$0^n 1^n$ such that n > 0} then the languages $L \cup R$ and R are respectively. **[GATE-1995]**

a) Regular, regular b) Not Regular, Regular

c) Regular, Not Regular d) Not Regular, Not Regular

43. Which of the following languages is not regular?

a) L ={$a^n b^n$ |n ≥ 1} **[UGC NET-1996]**

b) L ={ w ∈ {a, b}* | w has atleast 3 a's}

c) L ={ w ∈ {a, b}* | the second last alphabet of w isb}

d) L ={ w ∈ {a, b}* | w begins with a and ends with b}

44. The following machine will not accept the string **[UGC NET-1999]**

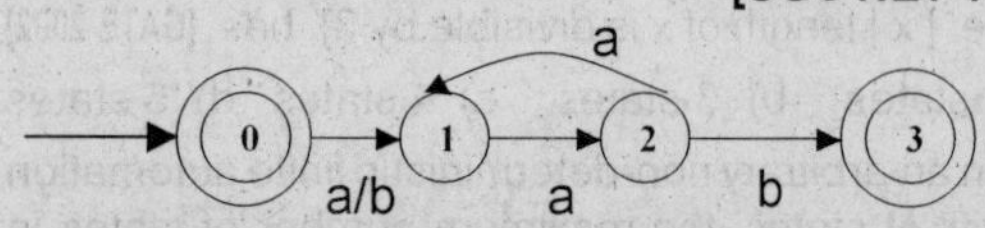

a) aaab b) aabab c) baaaab d) bab

ANSWER KEY - I

1. b	2. b	3. c	4. c	5. d
6. c	7. b	8. d	9. a,b,c	10. b
11. a,b,c	12. d	13. c	14. c	15. a
16. a,b	17. a	18. a	19. a	20. a
21. b,c	22. c	23. b	24. a	25. a
26. d	27. c	28. b	29. b	30. b
31. c	32. b	33. a	34. c	35. c
36. d	37. c	38. d	39. c	40. b
41. a	42. ?	43. a	44. a,b,c	

EXPLANATORY NOTES - I

5. (d) **Explanation** : Given RE= (a/b) (a/b) which denotes the concatenation of two sets

i.e., $\{a,b\}\ \{a,b\} = \{aa, ab, ba, bb\}$

Hence answer 'd' is correct.

6. (c) **Explanation** :-

From the Given, the language is denoted as

$L = \{\in, a, b, aa, ab, ba, ...\}$,indicates the kleen closure operation on a set over the alphabet $\Sigma = \{a,b\}$ i. e., $\{a,b\}$

Hence the required regular expression is

$(a/b)^*$ or $(a+b)^*$

8. (d) **Explanation** : Let $A = \{0, 1\}$

$A^1 = \{0, 1\}$, set of all strings at length 1 (2^1 strings)

$A^2 = \{00, 01, 10, 11\}$, set of all strings at length 2 (2^2 strings)

Hence, the Possible strings of length n is 2^n.

13. (c) **Explanation** : Strings should not contain 100 Strings can have number of 0's followed by number of 1's (or) Number of 2's followed by single 0 (and /or) followed by number of 1's. Hence answer 'c' is correct.

14. (c) **Explanation** : Strings should not Contain 1101 (i.e) two consecutive 1's can not follow the 01. Hence the strings are 01 (or) 10 followed by two consecutive 1's or 0's. So answer 'c' is Correct.

18. (a) **Explanation** : Given $L_1 = (a+b)^* a$

$L_2 = b(a+b)^*$

Then $L_1 \cap L_2 = (a+b)^* \ a \cap b \ (a+b)^*$

$= b\ (a+b)^*\ a$

22. (c) **Explanation** : Let $RE = (0+\in)\ (1+10)^*$

The strings, represented by RE are start with 0 followed by any number of 1's or combination of 10's. Hence no two consecutive 0's occur in the strings.

24. (d) **Explanation** : The sets for the given regular Expression must ends with bbab, and the sets are combination of ab's and abb's. Set 'd' does not follow this rules. Hence choice 'd' is NOT the set of given regular expression.

25. (a) **Explanation** : Given $L = \left\{ a^i\ b^{2i}\ /\ i \geq 1 \right\}$

determines the set of strings i a's followed by 2i b's where $i \geq 1$.

Suppose if i=3 then the string is three a's followed by six b's (aaabbbbbb). Hence answer 'a' is correct.

28. (b) **Explanation** : The following Melay machine generates output in E's and 0's. In this every state input symbol path contains an output symbol.

$Q = \{q_0, q_1\}$

$\Sigma = \{0, 1\}$

$\Delta = \{E, 0\}$

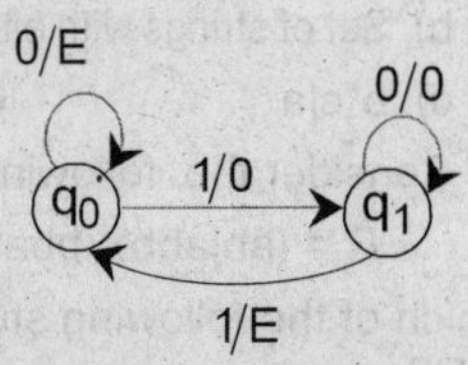

δ and λ is given by

	Present State			
	Input - 0		Input -1	
	input state	output state	input state	output state
$\rightarrow q_0$	q_0	E	q_1	0
q_1	q_1	0	q_0	E

Hence Q has two states. answer 'b' is correct.

29. (b) **Explanation** : The set of all strings of 0's and 1's is represented by RE is $(0+1)$.

Given the RE is ended with 00. Hence the required regular expression is $(0+1)00$.

30. (b) **Explanation** : The set contains an even number of 1's . So the regular expression is either (11) (or) (11). But given set contain $\in$ Hence the regular expression is kleen closure not the Positive closure therefore (11) is correct.

31. (c) **Explanation** : By the Arden's theorem, If P and Q to be regular expressions over $\sum$. If P does not contain $\in$ then the following equation in R

$R = Q + RP$ has unique solution given by

$R = QP^* \Rightarrow$ Hence equation 'c' is correct.

32. (b) **Explanation** : Assume L is regular and n is an integer in Pumping Lemma.

Let $Z = a^{n^2}$

$L = \{a, aaaa, aaaaaaaa, ...\}$

Take any string from the defined languages and write it in uvw form such that $|uv| \leq n$ and $|v| \geq 1$ and apply Pumping Lemma theorem.

Let $Z = \underset{u}{a}\ \underset{v}{aa}\ \underset{w}{a}$

$Z = uv^2w = a(aa)^2a$ $= aaaaaa \notin L$

Which is not belongs to L contradiction,

Hence L is not regular.

34. (b) **Explanation** : Suppose that $\Sigma = \{a, b\}$

L={x/ length of x is divisible by 3, $x \in \{a, b\}$ }.

The finite state automation diagram is show below.

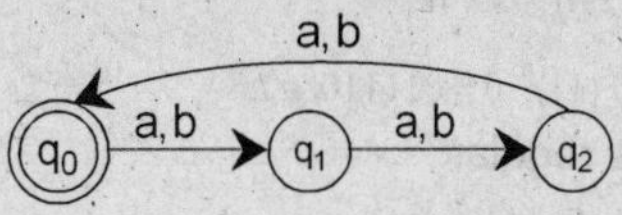

Where $Q = \{q_0, q_1, q_2\}$

Hence the given language required 3 states.

Suppose take string aaa bba (length is 6 is divisible by 3) is accepted by the machine.

36. (d) **Explanation** : Given FSA has two states and input alphabet $\Sigma = \{a\}$. Hence the state diagram is

Where $Q = \{q_1, q_2\}$

$\Sigma = \{a\}$

$F = \{q_1\}$ or $\{q_2\}$

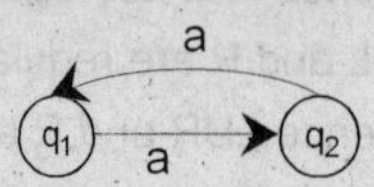

if $F = \{q_1\}$ then L must be $\{a^n / n \text{ is even}\}$ other wise if $F = \{q_2\}$ then $L = \{a^n / n \text{ is odd}\}$ so, choice 'd' is correct.

37. (c) **Explanation** : The finite state automation must basically accept any n bit string. So, we need n+1 states for counting through 0 to n and finally a termination state. Totally, we need n+2 states.

38. (d) **Explanation** : Given $A = (01+1)^*$

$B = ((01)^*\ 1^*)^*$

Let P=01, q=1

Plugging these values in A & B we have

$A = (p+q)^*$

$B = (p^*\ q^*)^*$

Hence from identities for regular expressions A=B

40. (b) **Explanation** : Given L be the set of all binary strings .

Therefore $\Sigma = \{0, 1\}$ the regular expression for the given set is $(0+1)^*\ (00+11)$

The minimum state deteministic finite state automation is

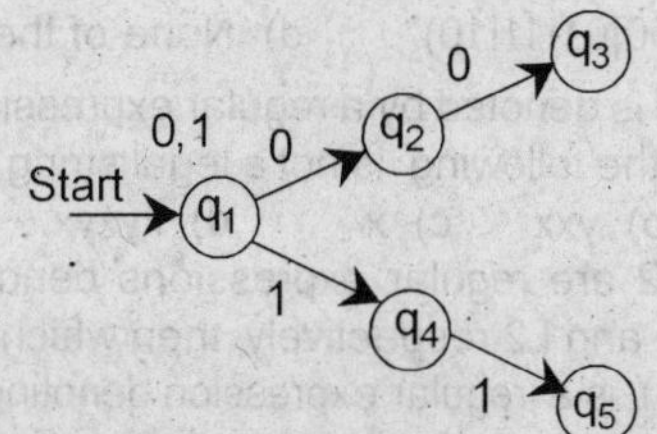

where $Q = \{q_1, q_2, q_3, q_4, q_5\}$

Hence require 5 states.

42. (d) **Explanation :**

The set L is defined as $L=\{\in, 0,1,01,10.....\}$ choose a string from defined language and write it in uvw form and applying pumping Lemma Principle.

Let $Z=\dfrac{0}{u}\ \dfrac{11}{v}\ \dfrac{0}{w}$

$z=uv^2w=0\,(11)^2\,0=011110\in L$

Hence L is regular

The set R is defined as

$R=\{\,01,\ 0011,\ 000111,......\}$

Let $Z=\dfrac{0}{u}\ \dfrac{01}{v}\ \dfrac{1}{w}$

$z=uv^2w=0\,(01)^2 1=001011\notin L$

Hence R is not regular.

If L and R are regular then LUR is regular.

Hence LUR and R are not regulars.

44. (a,b,c) **Explanation** : Let the string aaab

$\xrightarrow[start]{} 0 \xrightarrow[a]{} 1 \xrightarrow[a]{} 2 \xrightarrow[a]{} 1 \xrightarrow[b]{} Halt$

Let the string aabab

Start

$\rightarrow 0 \xrightarrow[a]{} 1 \xrightarrow[a]{} 2 \xrightarrow[b]{} 3 \xrightarrow[a]{} Halt$

Let the string baaaab

start $\rightarrow$ $0 \xrightarrow[b]{} 1 \xrightarrow[a]{} 2 \xrightarrow[a]{} 1 \xrightarrow[a]{} 2 \xrightarrow[a]{} 1 \xrightarrow[b]{}$ Halt

Let the string bab

start $\rightarrow 0 \xrightarrow[b]{} 1 \xrightarrow[a]{} 2 \xrightarrow[b]{} 3 \leftarrow$ accepted.

OBJECTIVE QUESTIONS - II

1. Which of the following regular expressions denote a language comprising all possible strings of even length over the alphabet (0,1)?

a) (0|1)* b) (0|1) (0|1)*

c) (00|01 11|10)* d) (0|1) (0|1)(0|1)*

2. Which of the following regular expression denotes a language comprising all possible strings over the alphabet {a,b} excluding those of length 3?

a) (ε |0|1|00|01|11|10)(0|1)(0|1)*)

b) (ε |0|1| 1 (00|01|11|10)*(0|1)*

c) ε |0|1|(00|01|11|10)* d) None of these

3. A language is denoted by a regular expression L=(x)*(x|yx) which of the following is not a legal string within L?

a) yx b) yxx c) x d) xyxyx

4. If e1 and e2 are regular expressions denoting the languages L1 and L2 respectively, then which is false?

a) (e1) | (e2) is a regular expression denoting L1 U L2

b) (e1) (e2) is a regular expression denoting L1 L2

c) ϕ is not a regular expression

d) {e1} is a regular expression denoting L1*

5. The regular expression (a|b)* denotes the set of all strings

a) With zero or more instances of a or b

b) With one or more instances of a or b

c) Equal to regular expression (a*b*)*

d) Both a and c

6. The string (a)|((b)* (c)) is equivalent to

a) Set of strings with either a or zero or more b's and one c

b) Set of strings with either a or one or more b's and one c

c) b*c|a d) Both (a) and (c)

7. Consider the following regular expression:

R = (ab|abb)* bbab

Which of the following string is NOT in the set denoted by R?

a) ababab b) ababbabbbab

c) abbbab d) abbabbbab

8. Let Δ = (X1,X2,X3) and let

X1=aX2 +bX1 +c

X2=aX3 + bX3

X3=aX1 + bX3

Define the language L, where Σ ={a,b} and c is the empty string. Which of the followng set of regular expressions for X1,X2,X3, provide a solution for the set of expression equations?

I. a*(b*(ab*(a+b)*b)+a*)

II. b*(ab*a(ab*ab*a+b)*ab*+c

III. (ab*ab*a+b)*ab*

IV. (b*(ab*+a)*b)

V. b*a(ab*ab*a+b)*ab*

a) II,I,IV b) II,V,III c) V,III,I d) IV,III,II

9. Suppose (ab)* a is a regular expression that defines a language over the alphabet Σ = {a,b} . The notation for another regular expression that defines the same language over the same alphabet may be:

a) a(ba)* b) a* (ba)*

c) a* ba d) Both a and b

10. Which of the following instances of the post correspondance problem have a viable sequence?

I. {(ab,abb),(ba,aaa),(aa, a)}

II. {(ab,aba), (baa, aa), (aba, baa)}

III. {(b,bb), (bb, bab), (bab, abb),(abb, babb)}

a) I only b) II only

c) I and II only d) I,II and III

11. Choose the correct statements.

a) A Mealy machine generates no language as such

b) A Moore machine generates no language as such

c) A Mealy machine has no terminal state

d) None

12. Which of the following pairs of regular expressions are equivalent?

a) 1(01)* and (10)*1 b) x(xx)* and (xx)*x

c) (ab)* and a*b* d) x^+ and x^*x^+

13. The language of all words (made up of a' s and b's) with atleast 2 a's can be described by the regular expression

a) (a+b)* a(a+b)*a(a+b)* b) (a+b)* ab*a(a+b)*

c) b* ab*a(a+b)* d) a(a+b)* a(a+b)*(a+b)*

14. Which of the following pairs of regular expression are not equivalent

a) (ab)* and a(ba)* b) (a+b)* and (a*+b)*

c) (a*+b)* and (a+b)* d) None of the above

15. Choose the correct statements.

A class of languages that is closed under

a) Union and complementation has to be closed under intersection

b) Intersection and complementation has to be closed under union

c) Union and intersection has to be closed under complementation

d) All of the above

16. Which of the following sets can be recognised by a Deterministic Finite-state automaton?

a) The numbers 1, 2, 4, 8,...., 2^n, ... written in binary

b) The numbers 1, 2, 4, ...2^n, ... Written in unary

c) The set of binary strings in which number of zeros is th same as the number of ones

d) The set {1, 101, 11011, 1110111,.....}

17. Consider the languages

r_1 =(a + b)*a

r_2 = (a + b)* ab (a + b)*

r_3 = (a + b)*(aa + bb)(a + b)*

According the above:

a) The machine that accepts r_1 r_2 is same as the machine which accepts r_2 r_1.

b) The machine which accepts r_1 r_3 as the machine that accepts r_3 r_2.

c) The machine that accepts r_2 is the same as the machine that accepts $b^*aa^*b(a + b)^*$

d) The machine which accepts r_1 r_1 is the same as the one which accepts (r_1^*)

18. Consider the following transition diagram of a finite automaton below:

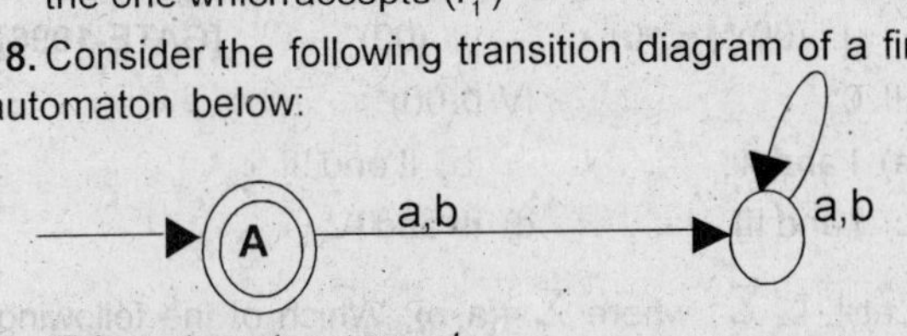

This automaton accepts

a) All words over sigma(a, b) such that symbol a and b alternate

b) All words over sigma(a, b) including ε which corresponds to no output

c) Only the ε, meaning this automation accepts no string of length greater than zero

d) All words over sigma(a, b) except ε

19. Consider the language s *(s followed by Kleene's star)

over the set

s ={aaa, aab, aba, abb, baa, bab, bba, bbb}

Another description of s* may be

a) All words made up of symbol (a, b) of length 3 or more including the word of length 0

b) All words over sigma (a, b) that begin and end with the same symbol

c) All words consisting of symbols{a, b} of even length

d) All words over sigma {a,b} of length divisible by 3 including the word of length zero.

20. Let r =1(1+ 0)*, s =11*0 and t =1*0 be three regular expressions. Which one of the following is true?

a) $L(s) \subseteq L(r)$ and $L(s) \subseteq L(t)$

b) $L(r) \subseteq L(s)$ and $L(s) \subseteq L(t)$

c) $L(s) \subseteq L(t)$ and $L(s) \subseteq L(r)$

d) $L(t) \subseteq L(s)$ and $L(s) \subseteq L(r)$

21. Let $L_1 = \{a^ib^j \mid i > j\}$ and $L_2 = \{a^ib^j \mid i < j\}$. The union of L_1 and L_2 is given by

a) $\{a^ib^j \mid i > j \geq 1\}$ b) $\{a^ib^j \mid i, j \geq 1\}$

c) $\{a^jb^i \mid i > j \geq 1\}$ d) $\{a^ib^i \mid i, j \geq 1, i \neq j\}$

22. Consider two languages

$L_1 = \{a^ib^ic^j \mid i, j \geq 1\}$ and $L_2 = \{a^ib^jc^j \mid i, j \geq 1\}$

The intersection of L_1 and L_2 is given by

a) $\{a^jb^jc^j \mid j \geq 1\}$ b) $\{a^ib^jc^i \mid i \neq j, i, j \geq 1\}$

c) $\{a^ib^ic^i \mid i \geq 1\}$ d) None of the above

23. The regular expression have all strings in which any number of 0's is followed by any number of 1's followed by any number of 2's, is

a) (0 | 1 | 2)* b) 0*1*2*

c) 0*+1 + 2 d) (0 + 1)*2*

24. The length of input string is 'n' in Moore machine. The length of the output string is

a) n + 1 b) n c) n – 1 d) None of the above

25. In Mealy machine , if the input string is of length n, the output string is of length

a) n b) n – 1 c) n + 1 d) None of the above

26. Any two states q_1 and q_2 of an automata are equivalent if

a) $\forall x \in \Sigma^*, \delta(q_1, x) \in F$ and $\delta(q_2, x) \notin F$

b) $\forall x \in \Sigma^*$, either both $\delta(q_1, x)$ and $\delta(q_2, x) \in F$ or both $\delta(q_1, x)$ and $\delta(q_2, x) \notin F$

c) $\forall x \in \Sigma^*, \delta(q_2, x) \in F$ and $\delta(q_1, x) \notin F$

d) None of the above.

27. The automata, whose transition table is shown below is

PS \ NS	Input a	Input b
A	B	C
B	-	D
C	D	-
D	-	-

a) Deterministic automata
b) non-deterministic automata
c) Finite automata d) (b) and (c)

28. If P, Q, R are regular expressions, then

I. $(PQ)^*P = P(QP)^*$ II. $(R^*)^* = R^*$ III. $\varepsilon^* = \phi$

a) Only I and II are correct b) I is correct
c) III is correct d) None of these

29. ......is a regular expression for the set of strings in which every 0 is immediately followed by the atleast two 1's

a) (011)* b) (1 + 011)*
c) (01 + 011)* d) None of these.

30. Is the set L ={ ww / w {a, b}*} regular?

a) Yes b) No
c) Cannot be determined d) None of these

31. The regular expression for the set $\{1^{2n+1} \mid n > 0\}$ is

a) ((11)*)*1 b) 111*
c) 1(11)* d) None of these

32. The regular expression for a regular grammar

· $S \rightarrow aS / a$ is

a) a^+ b) a^*a c) both (a) and (b) d) None of these

33. The Finite state machine described by the following state diagram with A as starting state, where an arc label is x/y and x stands for 1-bit input and y stands for 2-bit output **[GATE-2002]**

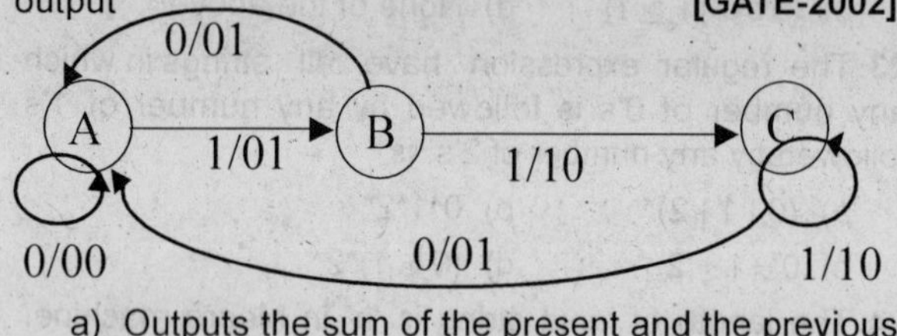

a) Outputs the sum of the present and the previous bits of the input.
b) Outputs 01 whenever the input sequence contains 11
c) Outputs 00 whenever the input sequence contains10
d) None of the above

34. Consider the following two statements:

S1: $\{0^{2n} \mid n \geq 1\}$ is a regular language

S2: $\{0^m 1^n 0^{m+n} \mid m \geq 1 \text{ and } n \geq 1\}$ is a regular language

Whichof the following statements is correct?**[GATE-2001]**

a) Only S1 is correct b) Only S2 is correct
c) Both S1 and S2 are correct
d) None of S1 and S2 is correct

35. Consider a DFA over Σ ={a, b} accepting all strings which have number of a's divisible by 6 and number of b's divisible by 8. What is the minimum number of states that the DFA will have? **[GATE-2001]**

a) 8 b) 14 c) 15 d) 48

36. Consider the following languages:

L1 = { ww | w ∈{a, b}*}
L2 = { ww^R | w ∈{a, b}*, w^R is the reverse of w}
L3 = { 0^{2i} | i is an integer}
L4 = { 0^{i^2} | i is an integer}

Which of the languages are regular? **[GATE-2001]**

a) Only L1 and L2 b) Only L2, L3 and L4
c) Only L3 and L4 d) Only L 3

37. Let S and T be languages over Σ ={a, b} represented by the regular expressions (a + b*)* and (a + b)*, respectively. Which of the following is TRUE? **[GATE-2000]**

a) $S \subset T$ b) $T \subset S$ c) S = T d) $S \cap T = \phi$

38. Which of the following set can be recognized by a Deteriministic Finite-state Automaton? **[GATE-1998]**

a) The numbers 1, 2, 4, 8, ... , 2^n, ... Written in binary
b) The numbers 1, 2, 4, 8, ... , 2^n, ... Written in unary
c) The set of binary strings in which the number of zeros is the same as the number of ones
d) The set {1, 101, 11011, 1110111, ...}

39. Given Σ ={a, b}, which one of the following sets is not countable? **[GATE-1997]**

a) Set of all strings over Σ
b) Set of all languages over Σ
c) Set of all regular languages over Σ
d) Set of all languages over Σ accepted by Turing machines

40. Which one of the following regular expressions over {0, 1} denotes the set of all strings not containing 100 as a sub string? **[GATE-1997]**

a) $0^*(1^+0)^*$ b) 0*1010* c) 0*1*01 d) 0*(10 + 1)*

41. Two of the following four regular expressions are equivalent which two?(ε is the empty string).

I. (00)*(ε +0) II. (00)* **[GATE-1996]**
III. 0* IV. 0(00)*

a) I and II b) II and III
c) I and III d) III and IV

42. Let $L \subseteq \Sigma^*$ where Σ ={a, b}; Which of the following is TRUE? **[GATE-1996]**

a) L ={x | x has an equal number of a's and b's} is regular
b) L ={$a^n b^n$ | n ≥ 1} is regular
c) L ={x | x has more a's than b's} is regular
d) L ={ $a^m b^n$ | m ≥ 1, n ≥ 1} is regular

43. A language with string manipulation facilities uses the following operations

head (s) : first character of a string

tail (s) : all but the first character of a string

Concat (s1, s2) : s1 s2

For the string acbc what will be the output of

Concat (head(s), head(tail (tail(s)))) **[GATE-1995]**

a) ac b) bc c) ab d) cc

44. A finite state machine with the following state table has a single input x and a single output z.

Present state	Next state, z	
	X = 1	X = 0
A	D, 0	B, 0
B	B, 1	C, 1
C	B, 0	D, 1
D	B, 1	C, 0

If the initial state is unknown, then the shortest input sequence to reach the final state C is **[GATE-1995]**

a) 01 b) 10 c) 101 d) 110

ANSWER KEY - II

1. c	**2.** d	**3.** d	**4.** c	**5.** c
6. d	**7.** a	**8.** b	**9.** a	**10.** a
11. a,b,c	**12.** a,b,d	**13.** a,b,c	**14.** d	**15.** a,b
16. a	**17.** c	**18.** c	**19.** d	**20.** ?
21. d	**22.** c	**23.** b	**24.** a	**25.** a
26. b	**27.** d	**28.** a	**29.** b	**30.** b
31. c	**32.** c	**33.** a	**34.** a	**35.** d
36. d	**37.** c	**38.** a	**39.** b	**40.** b
41. c	**42.** d	**43.** c	**44.** b	

EXPLANATORY NOTES - II

5. (c) **Explanation** : The RE (a/b) denotes the strings are $\{a, b, ab, ba\}$, Hence $(a/b)^*$ is equal to the regular expression (a^*b^*)

16. (a) **Explanation** : The set contain the numbers 1,2,4,8,....... 2^n ...... written in binary as.

i.e., $L = \{1, 10, 100, 1000 ....\}$

The DFS to L is

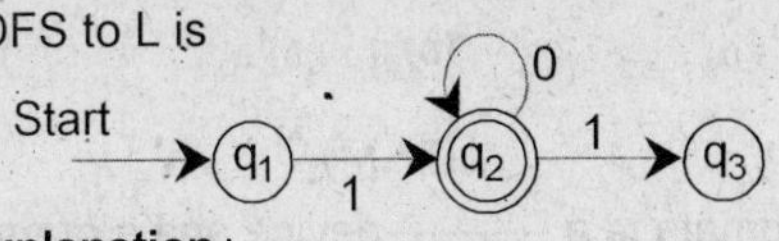

17. (c) **Explanation** :

$r_2 = (a+b)\, ab(a+b)^*$

We know that $(a+b)^*$ is equal to the b^*aa^*b, that means these two sets contains similar strings.

Hence the machine that accepts r_2 is the same as the machine that accepts $b^*\, aa^*\,(a+b)^*$

19. (d) **Explanation** : The set S contains all the strings with the length 3. Hence the set s^* contains all words over $\Sigma = \{a, b\}$ of length divisible by 3 including the word of length zero.

20. (a) **Explanation** : The Languages corresponds to the given regular expressions are

$L(r) = \{1, 11, 10, 110, 101, 111, 1110, ...\}$

$L(s) = \{10, 110, 1110, ...\}$

$L(t) = \{0, 10, 110, 1110, ...\}$

clearly $L(s) \subseteq L(t)$ and

$L(s) \subseteq L(t)$

Hence answer 'a' is correct.

23. (b) **Explanation** : Number of 0's (0^*) followed by number of 1's (1^*) and followed by number of 2's (2^*) is the concatenation of $0^*1^*2^*$.

29. (b) **Explanation** : The strings should have 1's (or) 0's in which every 0 is immediately followed by the at least two 1's (i.e.,) 011. Hence the regular expression corresponding to the above description is $(1+011)^*$

30. (b) **Explanation** :

Let n be the number of states of FA accepting L.

$z = ww(w \; in (a+b)^*)$

Assume set of strings accepted by language L is

$L = \{abab, baba, abaaba, ...\}$

choose a string from accepted language and apply pumping lemma principle.

$$Z = \frac{b}{u}\frac{ab}{v}\frac{a}{w}$$

$uv^2 w = b\,(ab)^2\, a = bababa \notin L$

So L is not regular.

31. (c) **Explanation** : Given Set $\{1^{2n+1} / n > 0\}$

$= \{1^{2n}\, 1 / n > 0\} = \{(1^2)^n\, 1 / n > 0\}$

$= \{(11)^n\, 1 / n > 0\} = 1(11)^*$

32. (c) **Explanation** : The language corresponding to the given regular grammar is $\{a, aa, aaa; ....\} = a^+$

from the definition of positive closure

$$a^+ = a^*a$$

33. (a) **Explanation** : Consider the string as input is 1110. The machine works is as follows

$$A \xrightarrow[01]{1} B \xrightarrow[10]{1} C \xrightarrow[10]{1} C \xrightarrow[01]{0} A$$

Hence the output sequence indicates sum of the present and the previous bits of the input.

34. (a) **Explanation** : Let b be the number of states in the FA accepting S1.

$Z = a^{2n}$

accepted set of strings is

$S1 = \{a, aa, aaaa, aaaaaa, ...\}$

choose a strings.

$\therefore \; |uv| \leq b, \; |v| \geq 1$

$\underline{a}\;\underline{aa}\;\underline{a}$

$u \quad v \quad w$

applying pumping Lemma

$uv^2 w = a(aa)^2 a = aaaaaa \in S1$

Hence S1 is regular.

Similarly we can prove $S2$ is not regular.

There fore only S1 is correct.

37. (c) **Explanation**: Given $S = (a + b^*)^*$

$T = (a+b)^*$

From the identites for regular expressions

$S = \left(a^*(b^*)^*\right)^* \quad \left[\because (p+q)^* = (p^* q^*)^*\right]$

$= (a\,b)^* \quad \left[\because (r^*)^* = r^*\right]$

Similarly $T = (a^* b^*)^*$ Hence $S = T$

41. (c) **Explanation** : The language Corresponding to the regular expression I is

$\{\in, 0, 00, 000, ...\} = 0^*$

43. (c) **Explanation** : Given String is S=acbc

Consider concat(head(s),head(tail(tail(s)))

From the given operations

= Concat (a,head(tail(cbc)))

= concat (a,head(bc))

= concat (a,b)

= ab

OBJECTIVE QUESTIONS - III

1. Which of the following regular expressions denote language Comprising all Possible strings of even length over the alphabet (0,1)?

(a) $(0+1)^*$ (b) $(0+1)(0+1)^*$

(c) $(00+0111+10)^*$ (d) $(0+1)(0+1)(0+1)^*$

2. Palindromes connot be recognized by any FSM because

a) An FSM cannot remember arbitrary large amount of information.

b) An FSM cannot deterministically fix the mid point

c) even if the mid point is known, an FSM cannot find wether the second half of the string matches the first half.

d) all the above.

3. Consider the Production grammer

$S \rightarrow AB \,/\, AS$

$A \rightarrow a \,/\, aA$

$B \rightarrow b$

which of the following regular expressions corresponds to the production grammer?

(a) $(ab)^*$ (b) $a(ab)^* b$

(c) aa^*b^* (d) aa^*b

4. Consider the following state table for a segmented machine. The number of states in the minimized machine will be

Present State \ input	0		1	
A	D	0	B	1
B	A	0	C	1
C	A	0	B	1
D	A	1	C	1

(a) 4 (b) 3 (c) 2 (d) 1

5. The out put generated by equivalent moore and mealy machine are same?

(a) True (b) False

6. How many number of states are required to design a DFSA (using minimum number of states) that recognises the following language $L = \{\omega \in \{0,1\}^* / \omega$ interpreted as binary number (ignoring the leading zero's) is divisible by five}.

(a) 2 (b) 3 (c) 6 (d) 7

7. For which of the following applications, regular expressions cannot be used.

(a) desiging computers (b) designining compilers

(c) both (a) and (b) (d) developing text editors

8. Consider the FA shown in the figure given below, where, '–' is the start state and '+' is the ending state. The language acepted by the FA is

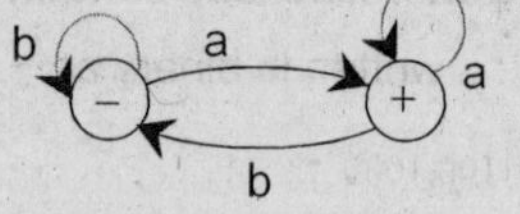

(a) $(a+b)^* b$ (b) $(a+b)^* a$

(c) $a^* b$ (d) $a^* b^*$

9. An automata is a --------- device and a grammer is a ------- device.

(a) generative, cognitive (b) generative, acceptor

(c) acceptor, cognitive (d) cognitive, generative

10. For which of the following applications regular expressions can be used?

(a) Designing compilers (b) Developing text editors

(c) simulating sequential circuits.

(d) both (b) and (c)

11. Any given transition graph has an equivalent

(a) regular expression (b) DFSM

(c) NDFSM (d) all of the above

12. The regular grammer G generating the regular set represented by $a^*b\,(a+b)^*$ is

(a) $A \to aA \,/\, bB \,/\, b$

$B \to bB \,/\, aB \,/\, a \,/\, b$

(b) $A \to aA \,/\, bB \,/\, a$

$B \to bB \,/\, aB \,/\, a \,/\, b$

(c) $A \to Aa \,/\, Bb \,/\, b$

$B \to Bb \,/\, Ba \,/\, a$

(d) none of the above.

EXPLANATORY NOTES - III

1. (c) **Explanation :**

The r.e. for strings of length 2 is (00+11+10+01)

The r.e. for strings of length 4 is

(0+1) (0+1) (0+1) (0+1) = (0000+0101+---)

By similar fashion, From the given answers r.e. for strings containing even length is $(00+0111+10)^*$

2. (d)

3. (d)

4. (b) **Explanation :** FSM given here with out put.

Considering A is strating state and D as final state.

B →	×		
C →	×		
D →	×	×	×
	A	B	C

B and C block is not marked, we can combine

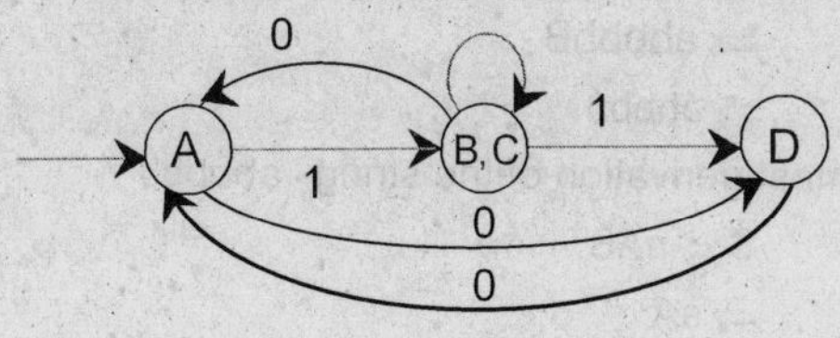

5. (b)

6. (c)

7. (c)

8. (b) **Explanation :** Clearly figure shows that it accepts any number of a's or b's because the state nodes contains loops. To reach final state the machine must take input a. Hence each string in language ends with a.

9. (d)

10. (d)

11. (d)

12. (a) **Explanation :** The NFA Corresponding to given r.e is

$A \to a\,A$

$A \to b\,B$

$A \to b$

$B \to b\,B$

$B \to aB$

$B \to a$

$B \to b$

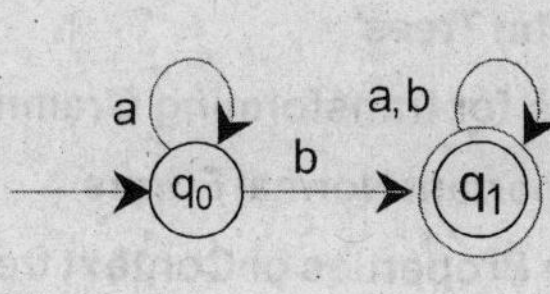

II. CONTEXT FREE GRAMMERS AND PUSHDOWN AUTOMATA

- Context Free Grammars
- Derivation Trees
- Methods for transforming Grammars
- Two important Normal Forms
- Closure Properties of Context free Languages
- Push Down Automata

2.1 Context Free Grammar

Def : A context free grammar is denoted $G=(V,T,P,S)$ Where V is finite set of variables i.e., set of non terminals, T is finite set of terminals

V and T are disjoint

P is a finite set of productions

each production is of the form $A \to \alpha$

where A is a variable

α is a string of symbols from $(V \cup T)^*$

S is a special variable called the start symbol.

A language L is said to be context free if and only if there is a context - free grammar such that $L=L(G)$.

Ex 1 : The grammar $G=(\{S\},\{a,b\},P,S)$, with productions

$S \to aSa$

$S \to bSb$

$S \to \in$

is a context - free grammar

A typical description in this grammar is

$S \Rightarrow aSa$

$\Rightarrow aaSaa$

$\Rightarrow aabSbaa$

$\Rightarrow aabbaa$

$\therefore L(G)=\{ww^R ; w \in \{a,b\}^*\}$ is a context - free language.

2.2 Derivation Trees

Def : Let $G=(V,T,P,S)$ be a CFG. A tree is a derivation (or parse) tree for G if :

i) Every vertex has a label, which is a symbol of $V \cup T \cup \{\in\}$.

ii) The lat el of the root is S.

iii) If a vertex is interior and has label A, then A must be in V.

iv) If n has label A and vertices $n_1, n_2 \ldots\ldots n_k$ are the sons of vertex n, in order from the left, with labels $X_1, X_2 \ldots\ldots X_k$ respectively, then $A \to X_1, X_2 \ldots\ldots X_k$, A must be a production in P.

v) If vertex n has label $\in$, then n is a leaf and is the only son of its father.

Ex : Consider the grammar $G=(\{S,A\},\{a,b\},P,S)$

where P consists of

$S \to aAs / a$

$A \to SbA / SS / ba$

Following figure is an example of derivation tree.

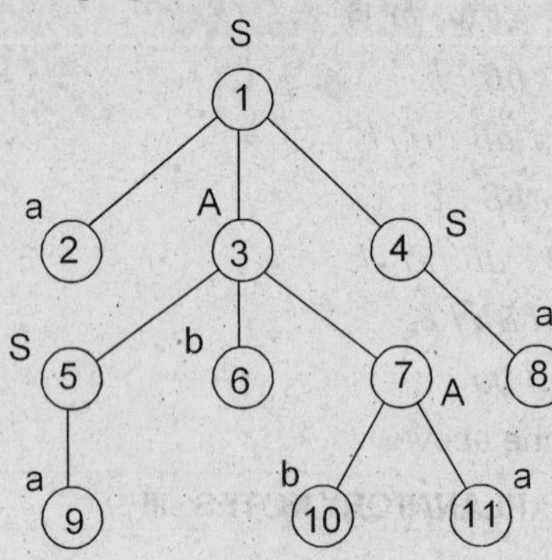

2.3 Leftmost and Rightmost Derivations : Ambiguity

Left most derivation : A derivation is said to be left most if in each step the left most variable in the sentential form is replaced.

Right - most derivation : If in each step the right most variable is replaced, then the derivation is called right most derivation.

Ex : Consider the grammar with productions

$S \to aAB$

$A \to bBb$

$B \to A / \in$

Let - most derivation of the string 'abbbb'

$S \Rightarrow aAB$

$\Rightarrow abBbB$

$\Rightarrow abAbB$

$\Rightarrow abbBbbB$

$\Rightarrow abbbbB$

$\Rightarrow abbbb$

Right - most derivation of the string ' abbbb'.

$S \Rightarrow aAB$

$\Rightarrow aA$

$\Rightarrow abBb$

$\Rightarrow abAb$

$\Rightarrow abbBbb$

$\Rightarrow abbbb$

2.3.1 Ambiguous grammar

Def : A context - free grammar G is said to be ambiguous if there exists some $w \in L(G)$ which has at least two distinct derivation trees is said to be ambiguous.

Ambiguity implies that the word w has more than one left derivation or more than one right - most derivation.

Ex : The grammar with productions

$S \to aSb / SS / \in$ is ambiguous

The sentence aabb has two derivation trees

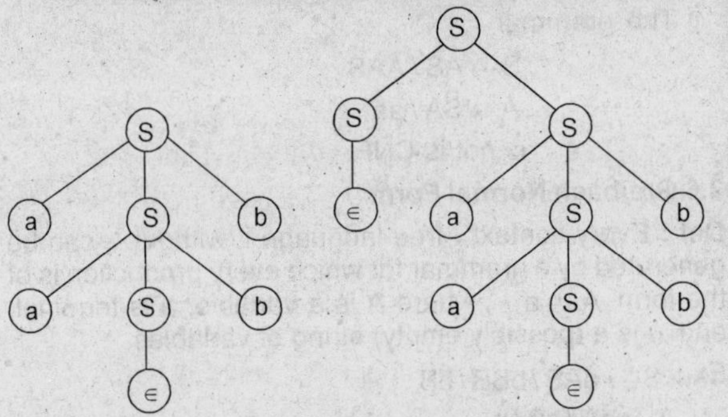

2.4 Simplification of Context - Free Grammars

If L is a non-empty context - free language then it can be generated by a context - free grammar G with the following properties.

i) Each variable and each terminal of G appears in the derivation of some word in L.

ii) There are no productions of the form $A \to B$ where A and B are variables.

2.4.1 Elimination of useless symbols

Useful and useless variables : Let $G=(V,T,P,S)$ be a context - free grammar.

A variable $A \in V$ is said to be useful if and only if there is at least one $w \in L(G)$ such that

$$S \overset{*}{\Rightarrow} xAY \overset{*}{\Rightarrow} w \text{ with x, y in } (V \cup T)^*$$

In other words, a variable is useful if and only if it occurs in at least one derivation.

Note : $s \overset{*}{\Rightarrow} w$ implies that w can be derived from S in zero or more steps.

A variable that is not useful is called useless.

A production is useless if it involves any useless variable.

Ex : In the grammar, with start symbol S and productions

$S \to A$

$A \to aA / \in$

$B \to bA$

The variable B is useless because 'B' doesn't involve in any derivation. And useless production is $B \to bA$. There is no way to achieve $S \overset{*}{\Rightarrow} xBY$

Procedure for removing useless variable : A procedure for removing useless variable and productions is based on recognizing the useless variable on two situations. i.e., a variable is useless :

"Either because it cannot be reached from the start symbol or because it cannot derive a terminal string.

Ex : Eliminate useless symbols and productions from

$G=(V,T,P,S)$ where $V=\{S,A,B,C\}$ and $T=\{a,b\}$ with P consisting of

$S \to aS / A / C$

$A \to a,$

$B \to aa,$

$C \to aCb$

First identify, the set of variables that can lead to a terminal string, Because, $A \to a$ and $B \to aa$, $S \Rightarrow A \Rightarrow a$

Therefore A, B and S belongs to this set,

Where as C is identified as useless. Removing C and its corresponding productions then the grammar becomes G_1, with variables. $V_1=\{S,A,B\}$ and $T=\{a\}$ and production are

$S \to aS / A$

$A \to a,$

$B \to aa$

Next eliminate the variables that cannot be reached from the start variable. For this draw a dependency graph for the variables.

Dependency graphs are a way of visualizing complex relationships and are found in many applications.

For context - free grammars, a dependency graph has its vertices labeled with variable, with an edge between vertices C and D ; if there is a production of the form.

$C \to xDy$

A dependency graph for V_1 is shown below

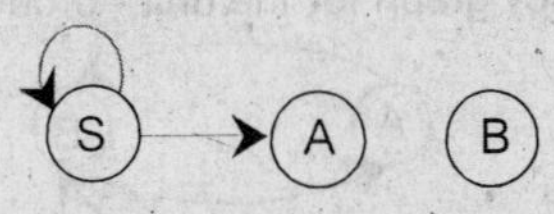

A variable is useful only if there is a path from the vertex labeled S to the vertex labeled with that variable. Above figure, shows B is useless. Removing it and the final grammar with productions listed below.

$\hat{G}=(\hat{V},\hat{T},\hat{P},S)$ with $\hat{V}=\{S,A\}, T=\{a\}$

and productions $S \to aS / A$

$A \to a$

2.4.2 Elimination of $\in$ - Production

$\in$ -Production : Any production of a context free grammar of the form $A \to \in$ is called a $\in$ -Production. Any variable A for which the derivation $A \overset{*}{\Rightarrow} \in$ is possible is called nullable.

A grammar may generate a language not containing $\in$, yet have some $\in$ -productions or nullable variables. In such cases, the $\in$ - productions can be removed.

Ex 1 : Consider the grammar

$S \to aS_1b$

$S_1 \to aS_1b / \in$

The $\in$ - Production $S_1 \to \in$ can be removed after adding new productions obtained by substituting $\in$ for S_1 where it occurs on the right. Doing this we get the grammar.

$S \to aS_1b / ab$

$S_1 \to aS_1b / ab$

Ex2 : Find a context - free grammar without $\in$ - productions equivalent to the grammar defined by

$S \to ABaC$

$A \to BC$

$B \to b / \in$

$C \to D / \in$

$D \to d$

Sol : The $\in$-productions $B \to \in, C \to \in$ can be removed after adding new productions obtained by substituting $\in$ for B, C, A where they occur on right. Doing this we get the grammar. Without $\in$ - productions.

Here B, C are nullable variables.

$A \to BC$, so A is also nullable.

The grammar is

$S \to ABaC / BaC / AaC / ABa / aC / Aa / Ba / a$

$A \to B / C / BC$

$B \to b$

$C \to D$

$D \to d$

2.4.3 Removing Unit Productions

Unit Production : Any production of a CFG of the form $A \to B$ where $A, B \in V$ is called a unit production.

Ex : Remove all unit - productions from

$S \to Aa / B$

$B \to A / bb$

$A \to a / bc / B$

The dependency graph for the unit - production is

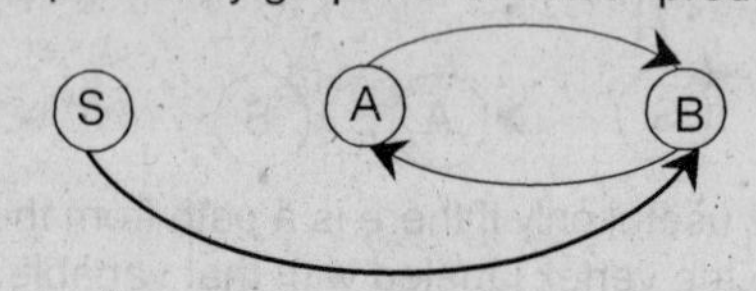

$S \overset{*}{\Rightarrow} A . S \overset{*}{\Rightarrow} B, B \overset{*}{\Rightarrow} A$, and $A \Rightarrow B$. Hence, we add to the original non - units productions.

$S \to Aa$

$A \to a / bc$

$B \to bb$

Now we have unit production from (1)

i) $A \to B$ so add the B productions i.e., bb to A production.

$A \to a / bb / bc$.

ii) $B \to A$ is a unit production so add productions of A to B ; $B \to a / bb / bc$.

iii) similarly $S \to a / bc / bb / Aa$

to obtain the equivalent grammar

$S \to a / bc / bb / Aa$,

$A \to a / bb / bc$,

$A \to a / bb / bc$

Note that the removal of the unit - productions has made B and the associated productions useless.

We can put all these results together to show grammars for context free languages can be made free of useless productions, $\in$-Production and unit - Productions.

2.5 Chomsky Normal Form

Def : Any context - free language without $\in$ is in Chomsky Normal Form, if all productions are of the form

$A \to BC$ (or) $A \to a$; where

A, B and C are variables

a is a terminal.

Ex : The grammar $S \to AS / a$

$A \to SA / b$

is in chomsky normal form

The grammar

$S \to AS / AAS$

$A \to SA / aa$

is not is CNF.

2.6 Greibach Normal Form

Def : Every context - free language L without $\in$ can be generated by a grammar for which every production is of the form $A \to a\alpha$, where A is a variable, a is terminal, and α is a (possibly empty) string of variables.

Ex : $S \to aAB / bBB / bB$

$A \to aA / bB / b$

$B \to b$

is in Greibach normal form.

Ex : The grammar

$S \to AB$

$A \to aA / bB / b$

$B \to b$

is not in Greibach normal form.

2.7 Push Down Automata (PDA)

2.7.1 Def : A push down automation M is a system $(Q, \Sigma, \Gamma, \delta, q_0, Z_0, F)$ where

i) Q is a finite set of states ;

ii) Σ is an alphabet called the input alphabet ;

iii) Γ is an alphabet, called the stack alphabet ;

iv) q_0 in Q is the initial state ;

v) Z_0 in Γ is a particular stack symbol called the start symbol.

vi) $F \subseteq Q$ is the set of final states ;

vii) δ is a mapping from $Q \times (\Sigma \cup \{\in\}) \times \Gamma$ to finite subsets of $Q \times \Gamma^*$

Model : The pushdown automation is in essence a NDFA with $\in$ transitions permitted and one additional capability. A stack on which it can store a string of "stack symbols". The presence of stack, makes PDA can remember "an infinite" amount of information. The PDA can only access the information on its stack in a first - in first - out way. PDA recognize all and only the context - free languages.

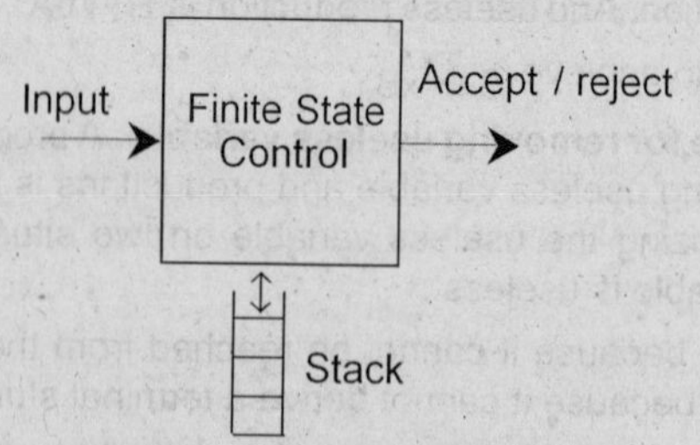

Figure : A PDA is essentially a finite automation with a stack data structure

Finite - state control reads input, one symbol at a time. The PDA is allowed to observe the symbol at the top of the stack and to base its transition on its current state the input symbol, and the symbol at the top of the stack.

In one transition, the PDA

1. Consumes from the input the symbol that it uses in the transition. If $\in$ is used for the input, then no input symbol is consumed.

2. Goes to a new state, which may or may not be the same as previous state.

3. Replace the symbol at the top of the stack by any string. The PDA's Configuration involves both the state and the contents of the stack. Being arbitrarily large, the stack is often more important part of the total configuration of the PDA at any time

Instantaneous descriptions :

The configuration of PDA is represented by a triple (q,w,r) where

1. q is the state
2. w is the remaining input
3. r is the stack contents.

such a triple is called Instantaneous Description ID.

If $M=(Q,\Sigma,\Gamma,\delta,q_0,Z_0,F)$ is a PDA, we say $(q,aw,Z\alpha)$ $\vdash M(p,w,\beta\alpha)$ if $\delta(q,a,Z)$ contains (p,β).

Ex : Let $M=(Q,\Sigma,\Gamma,\delta,q_0,Z_0,F)$ where

$Q=\{q_0,q_1,q_f\},\Sigma=\{a,b\}\ \Gamma=\{a,Z_0\}\ F=\{q_f\}$ δ is given by

$\delta(q_0,a,Z_0)=\{(q_0,a,Z_0)\},\delta(q_1,b,a)=\{(q_1,\in)\}$

$\delta(q_0,a,a)=\{(q_0,a,a)\},\delta(q_1,\in,Z_0)=\{(q_1,\in)\}$

$\delta(q_0,b,a)=\{(q_1,\in)\}$ is a PDA.

2.7.2 Acceptance of CFL

For PDA $(Q,\Sigma,L,\delta,q_0,f)$ we define

i) L (M), the language accepted by final state to be $(w/(q_0,w,z_0))^*\ \vdash (p,\in,\alpha)$ for some P in F. It means that for each word w in α if PDA on string w reaches a final state p then α is said to be accepted. When PDA reaches final state, stack may or may not be empty.

ii) N(M), the language accepted by empty stack (or null stack) to be $(w/(q_0,w,z_0))^*\ \vdash\ (p,\in,\in)$ for some P in Q. When PDA on input w empties its stack (i.e., stack has only stack marker) then it is said to be accepted. Here the set of final states is irrelevant so we usually let final states to be empty set.

Ex 1 : Acceptance of CFL by final state.

$M=(\{q_0,q_1,q_f\},\{a,b,c\},\{a,b,Z_0\},\delta,q_0,Z_0\ \{q_f\})$

is a PDA, where δ is defined as

$\delta(q_0,a,Z_0)=\{(q_0,aZ_0)\},\delta(q_0,b,Z_0)=\{(q_0,bZ_0)\}$.....(1)

$\delta(q_0,a,a)=\{(q_0,aa)\},\delta(q_0,b,a)=\{(q_0,ba)\}$......(2)

$\delta(q_0,a,b)=\{(q_0,ab)\},\delta(q_0,b,b)=\{(q_0,bb)\}$.....(3)

$\delta(q_0,c,a)=\{(q_1,a)\},\delta(q_0,c,b)=\{(q_1,b)\},\delta(q_0,c,Z_0)=\{(q_1,Z_0)\}$ .... (4)

$\delta(q_1,a,a)=\delta(q_1,b,b)=\{(q_1,\in)\}$..........(5)

$\delta(q_1,\in,Z_0)=\{(q_f,Z_0)\}$.......(6)

Sol : Suppose the input string is bacab. An initial configuration is $(q_0,bacab,Z_0)$ we get the following moves.

$(q_0,bacab,Z_0)\vdash(q_0,acab,bZ_0)$ by rule (1)

$\vdash(q_0,cab,abZ_0)$ by rule (3)

$\vdash(q_1,ab,abZ_0)$ by rule (4)

$\vdash(q_1,b,bZ_0)$ by rule (5)

$\vdash(q_1,\in,Z_0)$ by rule (6)

$\vdash(q_f,\in,Z_0)$ by rule (6)

i.e., $(q_0,bacab,Z_0)^*\vdash(q_f,Z_0)$

because after scanning entire string the pda's a has reached final state. String is accepted.

Here we are not concerned about what the stack is containing. Proceeding in a similar way, we can show that $(q_0,wcw^T,Z_0)\vdash^*(q_f,\in,Z_0)$ for $w\in\{a,b\}^*$

$\therefore$ The language wcw^T for all $w\in\{a,b\}^*$ is accepted by the given PDA.

Ex 2 : Acceptance of CFL by empty stack

The language $\{ww^T/w\ in\ (0+1)^*\}$

PDA $M=(\{q_1,q_2\},\{0,1\},\{R,B,G\},\delta,q,R,\phi)$; δ is

1. $\delta(q_1,0,R)=\{(q_1,BR)\}$
2. $\delta(q_1,1,R)=\{(q_1,GR)\}$
3. $\delta(q_1,0,B)=\{(q_1,BB),(q_2,\in)\}$
4. $\delta(q_1,0,G)=\{(q_1,BG)\}$
5. $\delta(q_1,1,B)=\{(q_1,GB)\}$
6. $\delta(q_1,1,G)=\{(q_1,GG),(q_2,\in)\}$
7. $\delta(q_2,0,B)=\{(q_2,\in)\}$
8. $\delta(q_2,1,G)=\{(q_2,\in)\}$
9. $\delta(q_1,\in,R)=\{(q_2,\in)\}$
10. $\delta(q_2,\in,R)=\{(q_2,\in)\}$

Sol : Now consider the string 001100 which is in the form ww^T where w = 001.

Initial configuration $(q_1,001100,R)$

$(q_1,001100,R)\vdash(q_1,01100,BR)$

$\vdash(q_1,1100,BBR)$

$\vdash(q_1,100,GBBR)$

$\vdash(q_2,00,BBR)$

$\vdash(q_2,0,BR)$

$\vdash(q_2,\in,R)$

$\vdash(q_2,\in,\in)$

$\in$ Indicates that stack is empty. Here because after scanning 01100 completely, the stack has become empty it is accepted by Pda.

∴ The language ww^T is accepted by given PDA by empty stack.

2.7.3 Deterministic PDA

A PDA M = $(Q, \Sigma, \Gamma, \delta, q_0, Z_0, F)$ is determinstic if

1) For each q in Q and Z in Γ, whenever $\delta(q, \in, Z)$ is non empty, then $\delta(q, a, Z)$ is empty for all a in Σ.

2) For no q in Q, Z in Γ, and a in $\Sigma \cup \{\in\}$ does $\delta(q, a, Z)$ contain more than one element.

Here deterministic in the sense that at most one move is possible from any instantaneous description. Condition 1 prevents the possibility of a choice between a move independent of the input symbol ($\in$ -move) and a move involving as input symbol condition 2 prevents a choice of move for any (q, a, Z) or $(q, \in, Z)$.

2.7.4. Equivalence of CFL and PDA

Equivalence of acceptance by final state and empty stack :

Theorem 2.1 : If L is $L(M_2)$ for some PDA M_2, then L is $N(M_1)$ for some PDA M_1.

Theorem 2.2 : If L is a context - free language, then there exists a PDA M such that $L = N(M)$

2.8 Enumeration properties of CFL

a) Closure Properties of CFL's :

1) Context - free languages are closed under union, concatenation and kleene closure.

2) The context - free languages are closed under substitution.

3) The CFL's are closed under homomorphism.

4) The context free languages are closed under inverse homomorphism

5) The CFL's are not closed under intersection

6) The CFL's are not closed under complementation.

7) If L is a CFL and R is a regular set, then $L \cap R$ is a CFL.

b) Decision properties of CFL's :

A problem is said to be decidable if there is an algorithm to solve that problem.

1) Emptiness of a CFL is decidable

2) Finiteness of CFL is decidable

3) Infiniteness of a CFL is decidable

4) Membership of a CFL is decidable.

2.9 Pumping Lemma for CFL

2.9.1. Def : Let L be any CFL. Then there is a constant n, depending only on L, such that if z is in L and $|z| \geq n$, then we may write z = uvwxy such that

1) $|vx| \geq 1$

2) $|vwx| \leq n$ and

3) for all $i \geq 0$ $uv^i wx^i y$ is in L.

2.9.2 Application of pumping lemma

Pumping lemma is useful to prove certain languages are not context - free languages.

Ex 1 : Show that $L = \{a^P / P \text{ is a prime}\}$ is not a context free language.

Sol : We use the following property of L : If $w \in L$, then $|w|$ is a prime.

Step 1 : Suppose $L = L(G)$ is context - free. Let n be the natural number obtained by using the pumping lemma.

Step 2 : Let p be a prime number greater than n. Then $z = a^p \in L$. We write z = uvwxy.

Step 3 : By pumping lemma, $uv^0wx^0y = uwy \in L$. So $|uwy|$ is a prime number, say q. Let $|vx| = r$. Then,

$|uV^qWX^qY| = q + qr$ As q + qr is not a prime.

$uv^qwx^qy \notin L$. This is a contradiction. Therefore, L is not context - free.

Ex 2 : Show that $L = \{a^ib^ic^i / i \geq 1\}$ is not context - free.

Sol : Suppose L were context free and let n be the constant of lemma.

Consider = $a^nb^nc^n$ we write z = uvwxy

Since $|vwx| \leq n$, it is not possible for vx to have instances of a's and c's, because that right most a is n+1 positions away from the left most.

If v and x consists of a's only, then uwy (the string uv^iwx^iy units i = 0) has n b's and n c's but fewer than n a's since $|vx| \geq 1$ thus, uwy is not of the form $a^ib^ic^i$. A contradiction the case where v and x consist only of b's (or) only of c's are disposed of similarly, a similar contradiction results.

We conclude that L is not a context - free language.

SOLVED EXAMPLES

1. What is meant by ambiguous grammar ? Show that the grammar. **[JNTU - 2001]**

$E \rightarrow E + E \mid E * E \mid (E) \mid id$ is ambiguous.

Sol : A context - free grammar G is said to be ambiguous if there exists some $w \in L(G)$ which has atleast two distinct derivation trees. Given grammar is

$E \rightarrow E + E \mid E * E \mid (E) \mid id$

The string a+b*c as two different derivation trees.

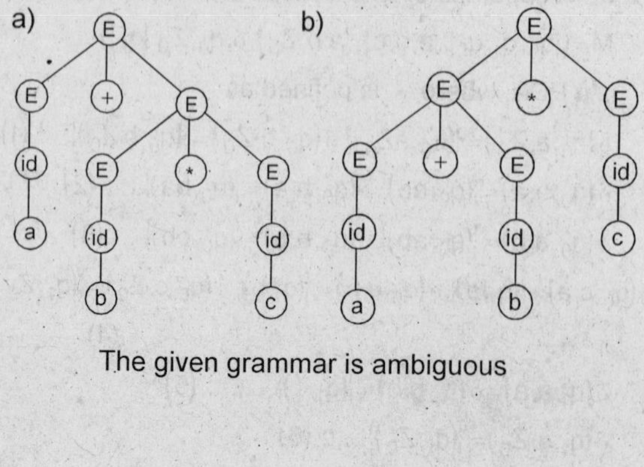

The given grammar is ambiguous

2. Construct an equivalent unambiguous grammar for the grammar. **[JNTU - 2001]**

$$E \to E+E \mid E*E \mid (E) \mid id$$

Sol : In an unambiguous grammar, every string will contain only one derivation tree. To remove ambiguity from the given grammar we use precedence and associativity of operators. The given grammar has only operators.

( ), *, + Precedence among them is

( ) : any thing within parenthesis is evaluated first

*

+

The * and + are left associated which means expressions containing * and + are evaluated from left to right.

i) First we write production for higher precedence operators and terminals. $F \to (E)/id$

ii) Then add production for next higher precedence operator. $T \to T*F/F \to (E)/id$

Here we have used T *F because * is left associative (if it was right associative we should have used F*T)

iii) Now add production for +

$$E \to E+T/T$$
$$T \to T*F/F$$
$$F \to (E)/id$$

This grammar is equivalent to $E \to E+E/E*E, (E)/id$ and it is unambiguous.

3. Consider the CFG, G = {V, T, P, S} where V = {S, A}

$T = \{a, b\}, P = \{S \to AA, A \to AAA, A \to a, A \to bA, A \to Ab\}$

Give leftmost and rightmost derivations for the string babbab. Specify whether the CFG is ambiguous or not.

Sol : Given string is babbab.

The left most derivation for babbab.

$S \Rightarrow AA$
$\Rightarrow bAA$
$\Rightarrow bAbA$
$\Rightarrow babA$
$\Rightarrow babbA$
$\Rightarrow babbAb$
$\Rightarrow babbab$

The right - most derivation for babbab

$S \Rightarrow AA$
$\Rightarrow AAb$
$\Rightarrow AbAb$
$\Rightarrow Abab$
$\Rightarrow bAbab$
$\Rightarrow bAbbab$
$\Rightarrow babbab$

Constructing derivation trees for leftmost and right most derivations.

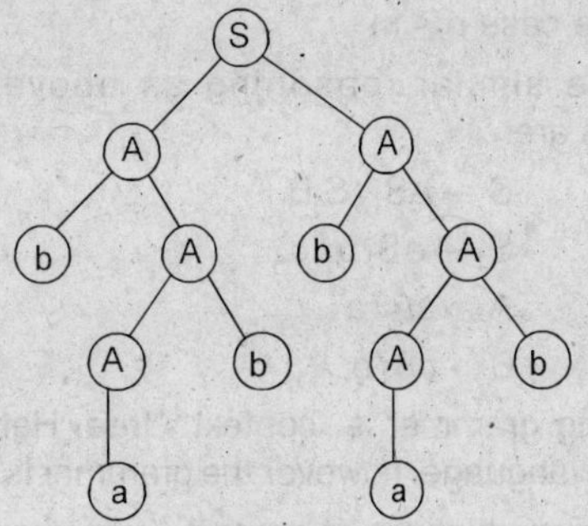

(a) Derivation tree for left most derivation

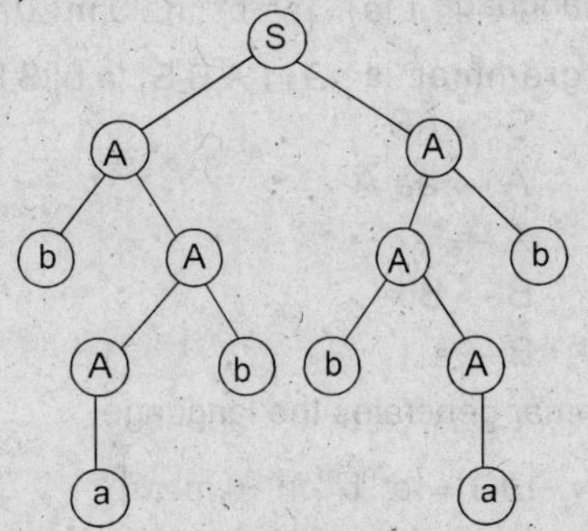

(b) Derivation tree for right - most derivation

Since, two parse trees are different for the string babbab, the given grammar is ambiguous.

4. Construct the grammar generating

$$L = \{a^n b^n c^i \mid n \geq 0, i \geq 0\}$$

Sol : In the given language L, the string will have equal number of a's and b's And any number of c's

Constructing grammar for L

$S \to S_1 S_2$
$S_1 \to a S_1 b / \in$
$S_2 \to C S_2 / \in$

5. Write the productions of a context free grammar G that accepts the language

$$L(G) = \{ab(bbaa)^n bba(ba)^n : n \geq 0\}$$

Sol : The grammar G, $G = (\{S, A, B\}\{a, b\}, P, S)$ where P consists of $S \to abB$
$A \to aaBb$
$B \to bbAa$
$A \to \in$

This grammar accepts the given language L.

6. Show that the language $L = \{a^n b^m : n \neq m\}$ is context free.

Sol : To show this, it is require to produce a context -free grammar for the language.

Take the case n > m.

First generate a string with an equal number of a's and b's then add extra a's on the left. This is done with

$S \to A S_1$
$S_1 \to a S_1 b / \in$
$A \to a A / a$

Take the case n < m

Use the similar reasoning as above and the productions are

$S \to AS_1 / S_1B$

$S_1 \to aS_1b / \in$

$A \to aA / a$

$B \to bB / b$

The resulting grammar is context - free. Hence L is a context free language, however the grammar is not linear.

7. Write the productions of context - free grammar that accepts the language $L(a) = \{a^{2n}b^m : n \geq 0, m \geq 0\}$

Sol : The grammar is $G = (\{A, B, S\}, \{a, b\}, S, P)$ with productions $S \to AB$

$A \to aa\,A$

$A \to \in$

$B \to Bb$

$B \to \in$

The grammar generates the language

$$L(G) = \{a^{2n}b^m : n \geq 0, m \geq 0\}$$

8. Show that the set of all strings over {a,b} consisting of equal number of a's and b's accepted by a PDA

Sol : The PDA $M = (\{q\}, \{a, b\}, \{Z_0, a, b\}, \delta, q, Z_0)$ where δ is defined by the following rules.

$\delta(q, a, Z_0) = \{(q, a, Z_0)\}$ (1)

$\delta(q, b, Z_0) = \{(q, bZ_0)\}$ (2)

$\delta(q, a, a) = \{(q, aa)\}$ (3)

$\delta(q, b, b) = \{(q, bb)\}$ (4)

$\delta(q, a, b) = \{(q, \in)\}$ (5)

$\delta(q, b, a) = \{(q, \in)\}$ (6)

Here the number of occurrences of a and b required i.e., storing a symbol of input string and continue storing until the other symbols occur. If w has equal number of a's and b's then

$(q, w, Z_0) \vdash^* (q, \in, Z_0)$

(or) $(q, \in, \in)$

Consider the string having equal number of a's and b's aabb. Initial configuration.

$(q, aabb, Z_0) \vdash (q, abb, aZ_0)$ by rule (1)

$\vdash (q, bb, aaZ_0)$ by (3)

$\vdash (q, b, aZ_0)$ by (6)

$\vdash (q, \in, Z_0)$ by (6)

↓

Accept.

∴ The set of all strings over {a,b} consisting of equal number of a's and b's is accepted by the above PDA.

9. Find a CFG with no useless symbols equivalent to

$S \to AB \mid CA$ $\quad A \to a$

$B \to BC \mid AB$ $\quad C \to aB \mid b$

Sol : $S \to AB / CA$

$B \to BC / AB$

$A \to a$

$C \to aB / b$

∴ B is useless productions since it does not lead to terminals.

∴ $B \to BC / AB$ are useless productions.

$S \to AB$

$C \to aB$ are also useless.

∴ grammar with no useless symbols is

$S \to CA$

$A \to a$

$C \to b$

10. Give unrestricted grammar for :

$L = \{a^nb^nc^n \mid n \geq 1\}$

Sol : Unrestricted grammar $G = (V, \Sigma, P, S)$ for

$L = \{a^nb^nc^n \mid n \geq 1\}$ is

$V = \{S, A, B\}$, $\Sigma = \{a, b, c\}$

$P: S \to abC / aAbC$

$Ab \to bA$

$Ac \to BbCC$

$bB \to Bb$

$aB \to aa / aaA$

Ex :

$S \Rightarrow aAbc \quad \Rightarrow abAc \quad \Rightarrow abBbcc$

$\Rightarrow aBbbcc \quad \Rightarrow aaAbbcc \quad \Rightarrow aabAbcc$

$\Rightarrow aabbAcc \quad \Rightarrow aabbBbccc$

$\Rightarrow aabBbbccc \quad \Rightarrow aaBbbbccc \quad \Rightarrow aaabbbccc$

11. Give CFG for the set of strings over alphabet {a,b} with exactly twice as many a's and b's

Sol : $G = (V, \Sigma, P, S)$

$V = \{S, A, B\}$ $\Sigma = \{a, b\}$

$S = \{A\}$

$P: S \to aA / Aa$

$A \to bbB / Bbb / bBb$

$B \to S / \in$

12. Give CFG for (a/b)* a b b.

Sol : G = (V, T, P, S)

$V = \{S, A\}$ $T = \{a, b\}$

$P: S \to aA / bA / A$

$A \to aA / bA / abb$

OBJECTIVE QUESTIONS - 1

1. The number of auxiliary memory required for a push down machine (PDM) to behave like a Finite State Machine is

a) 2 b) 1 c) 0 d) 4

2. The intersection of a CFL and a regular language

a) Is always regular b) Is always context free

c) Both (a) and (b) d) Need not be regular

3. Which of the following sentences is generated by the production grammar

S→ aS I bA

A→ d I ccA

a) abbbd b) aabccd c) aadb d) ababccd

4. Which of the following is not accepted by Deterministic Push Down Machine (DPDM) but accepted by Non Deterministic Push Down Machine (NDPDM)?

a) Strings ending with a particular alphabet

b) All strings in which a given symbol is present atleast twice

c) Even palindromes d) None of the above

5. Which of the following statement is wrong?

a) Any regular language has an equivalent context-free grammar

b) Some non-regular languages can't be generated by any context-free grammar

c) The intersection of a context-free language and a regular language is always context-free

d) All languages can be generated by context-free grammar

6. The context-free grammar defined by the regular expression ab* is:

a) S→Sb |a

b) S→XY
X→ aX
Y→bY

c) S→SS|baa|abb| ε

d) S→ aS|bS

7. The class of context-free languages is not closed under:

a) Union b) Concatenation

c) Repeated concatenation

d) Intersection

8. Pumping lemma is generally used for proving

a) A given grammar is regular

b) A given grammar is not regular

c) Whether two given regular expressions are equivalent or not

d) None of the above.

9. The set $\{a^nb^n / n= 1,2,3...\}$ can be generated by the CFG

a) S→ ab | aSb b) S→aaSbb | ab

c) S→ ab | aSb | ε d) S→aaSbb | ab| aabb

10. Choose the correct statements.

a) All languages can be generated by CFG

b) Any regular language has an equivalent CFG

c) Some non-regular languages can't be generated by any CFG

d) Some regular languages can't be generated by any CFG

11. Which of the following CFG's can't be simulated by an FSM?

a) S→Sa | a

b) S→abX
X→ cY
Y→d | aX

c) S→ aSb | ab d) None

12. $L=\{a^nb^na^n \mid n=1,2,3.....\}$ is an example of a language that is

a) Context free b) Not context free

c) Not context free but whose complement is CF

d) Context free but whose complement is not CF

13. P,Q,R are three languages. If P and R are regular and if PQ = R , then

a) Q has to be regular b) Q cannot be regular

c) Q need not be regular d) Q has to be a CFL

14. Which of the following is not possible algorithmically?

a) Regular grammar to context free grammar

b) Non-deterministic FSA to deterministic FSA

c) Non-deterministic PDA to deterministic PDA

d) Non-deterministic Turing machine to deterministic Turing machine

15. Which of the following languages over {a, b, c} is accepted by Deterministic Push Down automata?

a) $\{W \subset W^R/W \in \{a, b\}^*\}$ b) $\{W W^R \mid W \in \{a, b, c\}^*\}$

c) $\{a^n b^n c^n \mid n \geq 0\}$

d) {W/W is a palindrome over {a, b, c}}.

16. Context free languages are closed under

a) Union, intersection b) Union, Kleene closure

c) Intersection, Complement

d) Complement, Kleene closure

17. The grammar G=<{S},{0, 1}, P, S> where P={S → 0S1, S → 0S, S → S1, S → 0} will generate

a) Context free language

b) Regular language c) None of the above

18. Let L be a language recognizable by a finite automaton.

The language REVERSE (L)={ W such that W is the reverse of V where V ∈L} is a

a) Context free language

b) Context sensitive language

c) Regular language d) None of the above

19. Which of the following is a context free language:

a) $L=\{0^n1^n2^n; n>0\}$ b) $L=\{0^n1^n2^k3^k; n,k>0\}$

20. Which of the following statement(s) is (are) wrong?

a) The power of Deterministic Finite State machine (DFSM) and non-deterministic finite state machine are same

b) The class of context free languages is not closed under intersection

c) The power of Deterministic Push Down Machine(DPDM) and Non- Deterministic Push Down Machine are the same

d) The intersection of Context Free Languages(CFL) and a regular languages is always regular.

21. Give a production grammar that specifies language

$L = \{a^i b^{2i} \mid i \geq 1\}$

a) {S → aSbb, S → abb} b) {S → aSb, S → b}

c) {S → aA, S → b, A → b} d) None of the above.

22. Which of the following statement is wrong?

a) The intersection of a CFL and a regular language is always a regular and context free.

b) A Push Down Machine behaves like a Turning Machine when the auxiliary memory it has is 2.

c) A Push Down Machine behaves like a Finite State Machine when the auxiliary memory it has is zero.

d) None of the above.

23. The language generated by G ={{s}, {s}, {s → ss}, s} is

a) L(G) ={ s, ss, sss, ...} b) L(G) = ϕ

c) L(G) can not be determined d) None of these

24. The language generated by

G ={{S}, {a, b}, { S → aS | bS | a | b }, S} is

a) $\{a, b\}^*$ b) $\{a, b\}^+ + \{\varepsilon\}$

c) $\{a, b\}^+$ d) None of these

25. Two grammars of different types....... generate the same language

a) may b) may not c) never d) none of these

26. The language generated by the grammar

S → 0S1 | 0C1, C → 1C | 1 is

a) $\{0^m 1^n \mid n < m \geq 1\}$ b) $\{0^m 1^n \mid n > m \geq 1\}$

c) $\{0^m 1^n \mid n, m \geq 1\}$ d) none of these

27. If G_1 and G_2 are equivalent, then they are of .... types

a) Same b) Different

c) Ambiguous d) None of these

28. The grammar with no useless symbols equivalent to grammar S → AB | a, A → a is

a) S → a b) S → AB, B → a

c) S → B, B → a d) None of these

29. The language generated by S → aS / aSbS / ε is

a) L(G) ={g : each prefix of g has same no. of a's and b's}

b) L(G) ={g : each prefix of g has atleast as many a's and b's}

c) L(G) ={g : each prefix of g has unequal no. of a's and b's}

d) None of these

30. If all productions of a CFG are of the form A → wB or A → w, then L(G) is ...... set

a) Empty set b) Regular

c) Irregular d) None of these

31. An ...... is a CFG with no ε-productions such that no consecutive symbols on the right side of productions are variables

a) Regular grammar b) Ambiguous grammar

c) Operator grammar d) None of these

32. The language accepted by a Pushdown Automaton in which the stack is limited to 10 items is best described as **[GATE-2002]**

a) Context Free b) Regular

c) Deterministic Context Free d) Recursive

33. Which of the following statements is TRUE? **[GATE-2001]**

a) If a language is context free it can always be accepted by a deterministic push-down automata.

b) The union of two context free languages is context free

c) The intersection of two context free languages is context free

d) The complement of a context free language is context free

34. Let L_D is the set of all languages accepted by a PDA by final state and L_E the set of all languages accepted by empty stack. Which of the following is TRUE? **[GATE-1999]**

a) $L_D = L_E$ b) $L_D \supset L_E$

c) $L_D \subset L_E$ d) None of these

35. Regarding the power of recognition of languages, which of the following statements is FALSE? **[GATE-1998]**

a) The non-deterministic finite state automata are equivalent to deterministic finite-state automata

b) Non-deterministic Push-down automata are equivalent to deterministic Push-down automata

c) Non-deterministic Turing machines are equivalent to deterministic Push-down automata

d) Multi-tape Turing machines are equivalent to Single-tape Turing machines

36. If L_1 and L_2 are context free languages and R a regular set, one of the languages below is not necessarily a context free language. Which one? **[GATE-1996]**

a) $L_1 L_2$ b) $L_1 \cap L_2$ c) $L_1 \cap R$ d) $L_1 \cup L_2$

37. Consider the following grammar for defining the "symbol":

< symbol > → {<parent>,<letter>} | <letter>

< parent > → {< letter >|<letter>} | < parent > <letter>

< letter > → A | B | C

Which of the following is not a symbol? **[UGC NET-2000]**

a) C b) {(C), C} c) { B } d) A

ANSWER KEY - I

1. c	2. c	3. b	4. c	5. d
6. a	7. d	8. b	9. d	10. b,c
11. c	12. b,c	13. c	14. c	15. d
16. b	17. b	18. c	19. b	20. c
21. a	22. d	23. b	24. c	25. a
26. b	27. b	28. a	29. b	30. b
31. c	32. b	33. b	34. a	35. b
36. b	37. c			

EXPLANATORY NOTES - I

3. (b) **Explanation :**

$$S \Rightarrow aS$$
$$\Rightarrow aaS$$
$$\Rightarrow aabA$$
$$\Rightarrow aabccA$$
$$\Rightarrow aabccd$$

4. (c) **Explanation :** Even palindromes is accepted only by the NPDA. Consider the language for even palindromes is $L = \{ww^R / w \text{ in } (0+1)^*\}$

The design of NPDA to L is shown below. Here we consider two states in the design. The machine is in q_0 state until middle of the string is reached. If middle of the string is reached M enters into state q_1, and tries to match remaining input symbols with contents of the stack. If inputs is of the form WW^R then M will empty its stack and thus accepts the input string. Here input symbols are (0,1) and stack symbols are (R, B, G), R is initial top of stack.

$$M = (\{q_0, q_1\}, \{0,1\}, \{R,B,G\}, \delta, q_0, R, \phi)$$

δ is defined as

$\delta(q_0, 0, R) = (q_0, BR)$ $\delta(q_0, 1, G) = (q_0, GG), (q_1, \in)$

$\delta(q_0, 1, R) = (q_0, G, R)$ $\delta(q_1, 0, B) = (q_1, \in)$

$\delta(q_0, 0, B) = (q_0, BB), (q_1, \in)$ $\delta(q_1, 1, G) = (q_1, \in)$

$\delta(q_0, 0, G) = (q_0, BG)$ $\delta(q_0, \in, R) = (q_1, \in)$

$\delta(q_0, 1, B) = (q_0, GB)$ $\delta(q_1, \in, R) = (q_1, \in)$

6. (a) **Explanation :** Given RE is ab* means a followed by infinite number of b's. The CFG corresponding to the given RE is $S \rightarrow Sb / a$

consider the string abbb

$$S \Rightarrow Sb$$
$$\Rightarrow Sbb$$
$$\Rightarrow Sbbb$$
$$\Rightarrow abbb$$

The string is accepted by the CFG.

7. (d) **Explanation :** Consider the two languages

$$L_1 = \{a^n b^m c^m ; n \geq 0, m \geq 0\}$$

and $$L_2 = \{a^n b^m c^m ; n \geq 0, m \geq 0\}$$

There are several ways one can show that L_1 and L_2 are context - free for instance, a grammer for L_1 is

$$S \rightarrow S_1 S_2$$
$$S_1 \rightarrow aS_1 b / \lambda$$
$$S_2 \rightarrow c\, S_2 / \lambda$$

Alternatively, we note that L_1 is the concatenation of two context - free languages, it is context - free

$$L_1 \cap L_2 = \{a^n b^n c^n, n \geq 0\}$$

Which is not context - free. Thus, the family of context - free languages is not closed under intersection.

9. (a,d) **Explanation :** Consider a string aaabbb

$$S \Rightarrow aSb$$
$$\Rightarrow aaSbb$$
$$\Rightarrow aaabbb$$

Hence $S \overset{*}{\Rightarrow} aaabbb$

11. (a,b,c) **Explanation :** The steps that are used to text whether CFG is finite (or) infinite is as follows.

1. Convert given CFG in to CNF with no useless symbols.
2. Draw the directed graph for CNF.

The grammer is finite if and only if this graph has no cycles. If the graph contains any cycle then that grammer is infinite.

Suppose the directed graph for choice 'c' is

The CNF of given productions is

$$S \rightarrow c_a D$$
$$D \rightarrow Sc_b$$
$$S \rightarrow c_a c_b$$
$$C_a \rightarrow a$$
$$C_b \rightarrow b$$

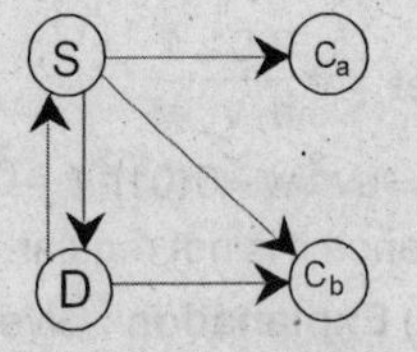

The above graph shows a cycle. So it is infinite.

Similarly we can show remaining grammer (a) and (b) are infinite. Hence no given grammers can't be simulated by an FSM.

12. (b) **Explanation :** Assume L be a context-free. Let i be the constant of pumping lemma.

Let $Z = a^i b^i a^i$ then $|Z| = 3n > n$

Write Z = uvwxy where $|v\,x| \geq 1$ and $|v\,w\,x| \leq n$

If v and x consists of only a's then uwy has i b's and i a's but fewer than i a's. Since $|vx| \geq 1$ thus uwy is not of the form $a^i b^i a^i$.

A contradiction. So L is not context - free.

Similarly if we choose v and x are only b's or only a's then also contrdiction.

$\therefore$ L is not context - free.

15. (a) **Explanation :**

$$M = (\{q_1, q_2\}, \{a,b,c\}, \{R,B,G\}, \delta, q, R, \phi)$$

$\delta(q_1, a, R) = (q_1, BR)$ $\delta(q_1, c, R) = (q_2, R)$

$\delta(q_1, a, B) = (q_1, BB)$ $\delta(q_1, c, B) = (q_2, B)$

$\delta(q_1, a, B) = (q_1, BG)$ $\delta(q_1, c, G) = (q_2, G)$

$\delta(q_1, b, R) = (q_1, GR)$ $\delta(q_2, a, B) = (q_2, \in)$

$\delta(q_1, b, B) = (q_1, GB)$ $\delta(q_2, b, G) = (q_2, \in)$

$\delta(q_1, b, G) = (q_1, GG)$ $\delta(q_2, \in, R) = (q_2, \in)$

Consider a string abcba

$$(q_1, abcba, R) \Rightarrow (q_1, bcba, BR)$$
$$\Downarrow$$
$$(q_1, cba, GBR)$$
$$\Downarrow$$
$$(q_1, ba, GBR)$$
$$\Downarrow$$
$$(q_2, a, BR)$$

$\Downarrow$

$(q_2, \in, BR)$

$\Downarrow$

$(q_2, \in, \in)$

17. (c) **Explanation :** Given $P = \{s \to 0s1/0s/s1/0\}$

The language generated by the given grammar (G) is

$L(G) = \{001, 0001, 011, 0011, ...\}$

means 0's followed by 1's

$\therefore L(G) = \{0^n 1^m / n, m \geq 1, n \neq m\}$

Choose a string from defined language and write it in u v w form and apply Pumping Lemma Principle.

Let $Z = \underset{u}{0}\ \underset{v}{01}\ \underset{w}{1}$

$Z = uv^2 w = 0(01)^2 1 = 001011 \notin L$

Hence it is not regular. Similarly it is not Context free.

18. (c) **Explanation :** Given REVERSE (L) = {w such that w is the reverse of v where $v \in L$ }

Let us assume the language is defined on alphabet $\Sigma = \{0,1\}$. $L = \{01, 110, 1010.....\}$ then

$R(L) = \{10, 011, 0101, ...\}$

Consider the string from defined languages R(L) and write it in uvw form and apply pumping lemma Principles.

Let $Z = \underset{u}{0}\ \underset{v}{11}\ \underset{w}{0}$

$Z = uv^2 w = 0(11)^2 0 = 011110 \in R(L)$

Hence answer 'c' is correct.

19. (b) **Explanation :** Let $L = \{0^n 1^n 2^k 3^k : n, k > 0\}$

L is defined as $L = \{0123, 012233, 00112233, ...\}$.

Choose a string from the defined language and write it is uvwxyz form where $|vx| > 1$ and $|vwx| \leq k$ ($\because$ k is constant)

$\therefore Z = \underset{u}{0}\ \underset{v}{0}\ \underset{w}{1}\ \underset{x}{1}\ \underset{y}{23}$

$uv^2 wx^2 y = 0(0)^2 1(1)^2 23 = 00011123 \in L$

Hence L is context free.

21. (a) **Explanation :**

The productions are $S \to aSbb/abb$, consider the string generated by the given language is aabbbb.

$S \Rightarrow aSbbb$

$\Rightarrow aabbbb$

Hence the string accepted by the grammar.

25. (a) **Explanation :** Consider the language contain strings with number of 1's is 1^+. To generates strings with 1's we can generate productions as

$S \to 1S$

$S \to \in$

Consider string 1111

$S \Rightarrow 1S \Rightarrow 11S \Rightarrow 111S \Rightarrow 1111S \Rightarrow 1111$

clearly this is CFG.

Consider another type grammar for the same language.

$S \to 1S$

$1S \to 1$

Consider the string 1111

$S \Rightarrow 1S \Rightarrow 11S \Rightarrow 111S \Rightarrow 1111S \Rightarrow 1111$

This is context - sensitive grammar.

Hence both grammar's generates same language.

Every context - free language without $\in$ can be generated by a special case of context - sensitive grammar.

26. (b) **Explanation :** Consider two derivations using given productions

1) $S \Rightarrow 0S1$	2) $S \Rightarrow 0S1$
$\Rightarrow 0C11$	$\Rightarrow 00C11$
$\Rightarrow 0111$	$\Rightarrow 0011$

From the above two derivations we can conclude that the language generated by Grammar is $\{0^m 1^n / n > m \geq 1\}$

27. (b) **Explanation :** Two different grammars are equivalent if thy both grammars generates same strings. Hence if G_1 and G_2 are equivalent, then they are of different types.

28. (a) **Explanation :** Given $S \to AB/a$, $A \to a$

A varaible is useless if it cannot derive terminal string and it cannot be reached from start symbol.

In the given grammar S, A are variables are deriving terminal string. But B is not deriving terminal string. So B is identified as useless. Removing B and its corresponding productions the grammar becomes $S \to a$

30. (b) **Explanation :** In given productions B not have terminal string. Hence the reduced productions are $A \to \omega$. $\therefore L(G) = \{\omega\}$

34. (a) **Explanation :** There are two ways to accept a given language by a PDA.

i) Acceptance by empty stack.

ii) Acceptance by final state

However, the language accepted by PDA by empty stack is also have been accepted by PDA by final state. Acceptance of language is proved easily by using empty stack comparing acceptance with final state.

36. (b) **Explanation :** The inter section of two context free languages is not a context - free. This is shown by below example.

Let L_1 is CFG to generate equal number of a's followed by equal number of b's. $L_1 = \{a^n b^n / n \geq 1\}$

Let L_2 is a CFG generate strings in the following format.

$L_2 = \{a^n b^m c^n / n, m \geq 1 \text{ and } n = m \text{ or } n \neq m\}$

$L_1 \cap L_2 = \{a^n b^{2n} a^n / n \geq 1\}$

In L_1 a's and b's are equal, so there are 2n b's in the middle if there are n a's in the front and back.

Hence $L_1 \cap L_2$ is not context - free.

37. (c) **Explanation :** One way to form a symbol is to take parent, letter with in braces another way is only take letter without braces. But choice (c) indicates letter with in braces. Hence it is not a symbol.

OBJECTIVE QUESTIONS - II

1. Consider the production grammar.

   S→AB|AS

   A→ a|aA

   B→ b

   Which of the following regular expressions corresponds to the production grammar?

   a) (ab)* b) a(ab)* b c) aa*b* d) aa*b

2. Consider the following grammar?

   S→ Ax|By

   A→ By|Cw

   B→x|Bw

   C→ y

   Which of these regular expressions describes the same set of strings as the grammar?

   a) xw*y + xw*yx + ywx b) xwy + xw*xyx + ywx

   c) xw*y + xwxyx + ywx d) xwxy + xww* yx + ywx

3. The productions

   E→E+E ,E→E-E, E→E*E, E→E / E, E→id

   a) Generate an inherently ambiguous language

   b) Generate an ambiguous language but not inherently so.

   c) Are unambiguous.

   d) Can generate all possible fixed length valid computations for carrying out addition, subtraction, multiplication, and division which can be expressed in one expression.

4. Any string of terminals that can be generated by the following CFG

   S→XY

   X→aX | bX | a

   Y→Ya|Yb|a

   a) Has atleast one b b) Should end in a 'a'

   c) Has no consecutive a's or b's

   d) Has atleast two a's

5. The following CFG

   S→aB|bA

   A→b|aS|bAA

   B→b|bs|aBB

   Generates strings of terminals that have

   a) Equal number of a's and b's

   b) Odd number of a's and odd number b's

   c) Even number of a's and even number of b's

   d) Odd number a's and even number of a's

6. Let L1 = $\{ a^n b^m a^m \mid n,m=1,2,3...\}$

   L2 = $\{ a^n b^m a^m \mid n,m=1,2,3...\}$

   L3 = $\{ a^n b^n a^n \mid n=1,2,3...\}$

   Choose the correct answers.

   a) L3 = L1 ∩ L2

   b) L1 and L2 are CFL but L3 is not CFL

   c) L1 and L2 are not CFL but L3 is CFL

   d) L1 is a subset of L3

7. Consider the grammar

   S→ PQ | SQ | PS

   P→X

   Q→Y

   To get a string of n terminals, the number of productions to be used is

   a) n^2 b) n +1 c) 2n d) 2n – 1

8. A CFG is said to be in Chomsky Normal Form (CNF), if all the productions are of the form A→BC or A→a. Let G be a CFG in CNF. To derive a string of terminals of length x, the number of productions to be used is

   a) 2x-1 b) 2x c) 2x + 1 d) 2^x

9. If L_1 is a context free language and L_2 is a regular language which of the following is/are false?

   a) $L_1 - L_2$ is not context free

   b) $L_1 \cap L_2$ is context free

   c) $\sim L_1$ is context free

   d) $\sim L_2$ is regular

10. Give a production grammar for the language

    $L=\{a^i b^j \mid i, j, \geq 1, i \neq j\}$

    a) {S → aS, S → aB, B → ab, A → aaB, B → b}

    b) {S → A, S → C, A → aA, A → aB, B → aBb, B → ab, C → Cb, C → Bb}

    c) {S → A, A → aA, A → aB, B → ab}

    d) none of the above

11. Consider the production grammar

    S → AA

    A → aa

    A → bb

    Describe the language specified by the production grammar

    a) L ={aaaa, aabb, bbaa, bbbb}

    b) L ={abab, abaa, aaab, baaa}

    c) L ={aaab, baba, bbaa, bbbb}

    d) L ={aaaa, abab, bbaa, aaab}

12. Give a production grammar for the language

    L={x/x ∈ {a, b}*, the number of a's in x is a multiple of 3}

    a) { S → bS, S → b, S → aA, S → bA, A → aB, B → bB, B → aS, B → a}

    b) {S → aS, S → bA, A → bB, B → bBa, B → bB}

    c) {S → aaS, S → bbA, A → bB, B → ba}

    d) None of the above.

13. The grammar for generating palindromes over {a, b} is

    a) {{S}, {a, b}, S → ε | a | b | aSa | bSb, S}

    b) {{S}, {a, b}, S → ε | a | b | aSa, S}

    c) {{S}, {a, b}, S → ε a | b | aSa | bSb, S}

    d) None of these

14. The strings 001100, 001010, 01010 are generated by

a) $S \to 0S1 \mid 0A1, A \to 0A1 \mid 01$

b) $S \to S1 \mid 0A1, A \to 1A$

c) $S \to 0S1 \mid 0A \mid 0 \mid 1B \mid 1, A \to 0A \mid 0, B \to 1B \mid 1$

d) None of these

15. $S \to aA, A \to aA \mid bA, A \to a$ generates

a) One set of all strings over {a, S} ending in a

b) One set of all strings over {a, S} beginning with a

c) One set of all strings over {a, S} beginning and ending with a

d) None of these

16. If L is a language accepted by some automaton M, which of the following is (are) true?

I. If M is a non-deterministic finite automaton, then L is accepted by some deterministic finite automaton

II. If M is a deterministic push down automation, then L is accepted by some non-deterministic push down automaton.

III. If M is a non-deterministic push down automaton, then L is accepted by some deterministic Turning machine

a) I only b) I and III c) II and III d) I, II and III

17. Let L denote the language generated by the grammar $S \to 0S0 \mid 00$. Which of the following is true? **[GATE-2000]**

a) $L = 0^+$ b) L is regular but not 0^+

c) L is context free but not regular

d) L is not context free

18. Consider the following decision problems:

P(1) Does a given finite state machine accept a given string

P(2) Does a given context free grammar generate an infinite number of strings. Which of the following statements is TRUE? **[GATE-2000]**

a) Both (P1) and (P2) are decidable

b) Neither (P1) and (P2) are decidable

c) Only (P1) is decidable

d) Only (P2) is decidable

19. Which of the following languages over {a, b, c} is accepted by a deterministic pushdown automata? **[GATE-1997]**

a) $\{w \subset w^R \mid w \in \{a, b\}^*\}$ b) $\{ww^R \mid w \in \{a, b, c\}^*\}$

c) $\{a^n b^n c^n \mid n \geq 0\}$ d) {w | w is a palindrome over {a, b, c}}

20. Define for a context free language $L \leq \{0; 1\}$ init (L) ={u / uv ∈ L for some v in {0, 1}}

(in other words, init (L) is the set of prefixes of L)

Let L= {w/w is nonempty and has an equal number of 0's and 1's} Then init(L) is **[GATE-1996]**

a) the set of all binary strings with unequal number of 0's and 1's

b) the set of all binary strings including the null string

c) the set of all binary strings with exactly one more 0's than the number of 1's or one or more 1 than the number of 0's.

d) none of the above

21. The grammar whose productions are

→ if id then <stmt>

→ if id then <stmt> else <stmt>

→ id := id

is ambiguous because **[GATE-1996]**

a) the sentence

if a then if b then c : = d

b) the left most and right most derivations of the sentence

if a then if b then c : = d

give rise top different parse trees

c) the sentence

if a then if b then c : = d else c : = f

has more than two parse trees

d) the sentence

if a then if b then c : = d else c : = f

has two parse trees.

22. Which of the following definitions below generates the same language as L, where $L=\{x^n y^n$ such that $n >= 1\}$?

I. $E \to xEy \mid xy$ II. $xy \mid (x^+ x yy^+)$ III. x^+y^+ **[GATE-1995]**

a) I only b) I and II c) II and III d) II only

23. The set of languages accepted by pushdown automata is the set of **[UGC NET-1996]**

a) Regular languages b) Context-free languages

c) Recursive languages

d) Context-sensitive languages

24. Consider a grammar with the following productions

$S \to xY$

$Y \to yX \mid x \mid y$

$X \to xY \mid x$ **[UGC NET-1998]**

The above grammar is most appropriately classified as

a) context-free b) regular

c) context-sensitive d) LR(k)

ANSWER KEY - II

1. d	2. a	3. b	4. d	5. a
6. a,b	7. d	8. a	9. b,c	10. b
11. a	12. a	13. a	14. c	15. c
16. d	17. b	18. a	19. a	20. b
21. d	22. a	23. b	24. a	

EXPLANATORY NOTES - II

1. (d) **Explanation :** Consider two derivations of strings from the given productions.

$S \Rightarrow AB$ $\quad$ $S \Rightarrow AS$

$\Rightarrow ab$ $\quad$ $\Rightarrow a\ AS$

$\Rightarrow aa\ AB$

$\Rightarrow aaab$

Hence the regular expressions corresponds to the production grammar is aa*b.

2. (a) **Explanation :**

$S \Rightarrow By$ (or) $S \Rightarrow Ax$ (or)

$S \Rightarrow Ax$

$\Rightarrow Bwy$ $\Rightarrow Byx$ $\Rightarrow cwx$

$\Rightarrow Bwwy$ $\Rightarrow Bwyx$ $\Rightarrow ywx$

$\Rightarrow Bwwwy$ $\Rightarrow Bwwyx$

$S \Rightarrow xw * y$ $S \Rightarrow xw * yx$

$\therefore S \Rightarrow ywx$

Finally, the RE is xw*y+xw*yx+ywx

4. (d) **Explanation :** To make S is to take variable X and follow it by Y. But variables X and Y deriving a same terminal string is a. Hence the given grammar produces the string with atleast two a's.

5. (a) **Explanation :** Derivation of two strings using given grammar are

$S \Rightarrow aB$ $S \Rightarrow bA$

$\Rightarrow aaBB$ $\Rightarrow baS$

$\Rightarrow aabb$ $\Rightarrow babA$

$\Rightarrow baba$

Hence the two strings have equal number of a's and b's.

6. (a,b) **Explanation :** Given $L_1 = \{a^n b^m a^m / n, m \geq 1\}$

$L_2 = \{a^n b^m a^m / n, m \geq 1\}$

We have already seen in previous Questions, L_1 and L_2 are context - free grammars and

$L_1 \cap L_2 = \{a^n b^n a^n / n \geq \phi\}$, $\therefore L_3 = L_1 \cap L_2$

Consider the below 2 examples

$S \Rightarrow PQ$ $S \Rightarrow PS$

$\Rightarrow xQ$ $\Rightarrow xS$

$\Rightarrow xy$ $\Rightarrow x PQ$

$\Rightarrow xxQ$

$\Rightarrow xxy$

To get length 2 string used $3(2 \times 2 - 1)$ productions and for length 3 strings used $5(2 \times 3 - 1)$ Productions. Hence answer d is correct.

10. (b) **Explanation :** The language contain strings with number of a's followed by number of b's. Hence the required grammar is

$S \to A / C$

$A \to aA \mid aB$

$B \to aBb \mid ab$

$C \to Cb \mid Bb$

11. (a) **Explanation :**

$S \Rightarrow AA$ $S \Rightarrow Aa$ $S \Rightarrow AA$ $S \Rightarrow AA$

$\Rightarrow aaA$ $\Rightarrow aaA$ $S \Rightarrow bbA$ $\Rightarrow bbA$

$\Rightarrow aaaa$ $\Rightarrow aaBB$ $\Rightarrow bbaa$ $\Rightarrow bbbb$

Hence answer 'a' is correct.

12. (a) **Explanation :** Given

$S \to bS \mid aA \mid bA \mid b$

$A \to aB$

$B \to bB \mid aS \mid a$

Consider a string abaab, Which contain multiple of 3a's. $S \Rightarrow aA$

$\Rightarrow abA$

$\Rightarrow abaB$

$\Rightarrow abaaS$

$\Rightarrow abaab$

Hence the string is accepted by grammer.

13. (a) **Explanation :** Given $P = \{S \to \in \mid a \mid b \mid aSa \mid bSb\}$ consider the string ababa

$S \Rightarrow aSa$

$\Rightarrow abSba$

$\Rightarrow ababa$

The string is accepted by the grammar.

14. (d) **Explanation :**

The grammar is $S \to 0S1 \mid 0A \mid 0 \mid 1B \mid 1$

$A \to 0A \mid 0$

$B \to 1B \mid 1$

Consider the string 001100

$S \Rightarrow 0S1$

$\Rightarrow 00\,S11$

$\Rightarrow 000\,A111$

$\Rightarrow 0000\,1111$

The string does not generated by grammer. In the same way there is no possibility to obtain given string using (a), (b) and (c) grammers.

15. (d) **Explanation :** Given Productions

$S \to aA$

$A \to aA \mid bA \mid a$

To make string must starting with S, is to take a and followed by A, hence every string is starting with a variable A has terminated string a. So the strings ending with a. However all the strings produced by the given grammer beginning and ending with a.

17. (b) **Explanation :** Given grammer produces the language is

$L = \{00, 0000, 000000, \ldots\ldots\}$

i.e., $L = \{0^{2i} / i \geq 1\}$

clearly, L is not 0^+.

Now, we can show L is not 0^+. Now we can show L is regular. Consider a string from the defined language and write it in uvw form where $|uv| \leq n$ and $|v| > 1$ (n is Constant)

$z = \underset{u}{0}\ \underset{v}{00}\ \underset{w}{0}$

$uv^2w = 0(00)^2 0$

$000000 \in L$

Hence L is regular but not 0^+

19. (a) **Explanation :** The design of DPDA corresponding to the language is shown below.

$M=(\{q_1,q_2\},\{a,b,c\},\{R,G,B\},\delta,q,R,\phi)$

The string before 'c' is exactly reversed and is followed by c. In this we assume two states $\{q_1,q_2\}$. State q_1 is continued before reading c. Once we read c we enter into state q_2. On state q_1 reading symbol a with only top stack element symbol B is added to stack. On state q_1 reading symbol b. With any top stack element symbol G is added to stack. Similarly on q_2 state reading Symbol top most element from stack is removed.

$\delta(q_1,a,R)=(q_1,BR)$ $\quad\delta(q_1,c,R)=(q_2,R)$

$\delta(q_1,a,B)=(q_1,BB)$ $\quad\delta(q_1,c,B)=(q_2,B)$

$\delta(q_1,a,G)=(q_1,BG)$ $\quad\delta(q_1,c,G)=(q_2,G)$

$\delta(q_1,b,R)=(q_1,GR)$ $\quad\delta(q_2,a,B)=(q_2,\in)$

$\delta(q_1,b,B)=(q_1,GB)$ $\quad\delta(q_2,b,G)=(q_2,\in)$

$\delta(q_1,b,G)=(q_1,GG)$ $\quad\delta(q_2,\in,R)=(q_2,\in)$

Consider a string abcba

$(q_1,abcba,R)\Rightarrow(q_1,bcba,BR)$

$\Downarrow$

(q_1,cba,GBR)

$\Downarrow$

(q_2,ba,GBR)

$\Downarrow$

(q_2,a,BR)

$(q_2,\in,BR)$

$\Downarrow$

$(q_2,\in,\in)$

The given string is accepted by empty stack.

21. (d) **Explanation :** Given productions are

→ if id then < stmt >

→ If id then < stmt > else < stmt >

→ i d : = id

The sentence

if a then if b then c : = d else c : = f has two parse trees. They are, first one is

if a then (if b then c : = d) else c : = f

secor d one is

if a then (if b then c : = d else c : = f)

Hence given grammer is ambiguous.

22. (a) **Explanation :** Consider the string x x x y y y

$E\Rightarrow xEy$

$\Rightarrow xxEyy$

$\Rightarrow xxxyyy$

The string is accepted by the grammer.

OBJECTIVE QUESTIONS - III

1. Consider the following grammer

$S\to aSbS\,/\,bSaS\,/\in$ is ambiguous ?

a) True b) False

2. Consider the following grammer

$S\to a\,/\,aA\,/\,B\,/\,C$

$A\to aB\,/\in$

$B\to Aa$

$C\to cCd$

$D\to ddd$

What is equivalent grammer to above grammer with no useless productions ?

a) $S\to a\,/\,aA\,/\,B\,/\,C$; $A\to aB\,/\in$; $B\to Aa$; $C\to cCd$

b) $S\to a\,/\,aA\,/\,B\,/\,C$; $A\to aB\,/\in$; $B\to Aa$

c) $S\to a\,/\,aA\,/\,B\,/\,C$; $A\to aB\,/\in$

d) None of the above.

3. Consider the following grammer

$S\to A\,/\,B\,/\,C$

$A\to aAa\,/\,B$

$B\to bB\,/\,bb$

$C\to aCaa\,/\,D$

$D\to baD\,/\,abD\,/\,aa$

After eliminate unit productions the grammer is

a) $S\to aAa\,/\,bB\,/\,bb\,/\,aCaa\,/\,baD\,/\,abD\,/\,aa$

$A\to aAa\,/\,bB\,/\,bb$

$B\to bB\,/\,bb$

$C\to aCaa\,/\,baD\,/\,abD\,/\,aa$

$D\to baD\,/\,abD\,/\,aa$

b) $S\to aAa\,/\,bB\,/\,bb\,/\,aCaa\,/\,baD\,/\,abD\,/\,aa$

$A\to aAa\,/\,bB\,/\,bbb$

$B\to bB\,/\,bb$

$C\to acaa$

$D\to baD\,/\,abD\,/\,aa$

c) $A\to aAa$

$B\to bB\,/\,bb$

$C\to acaa$

$D\to baD\,/\,abD\,/\,aa$

d) None of the above

4. After eliminate null productions from the following grammer

$S\to AB\,/\in$

$A\to aASb\,/\,a$

$B\to bS$ the obtained grammer is

a) $S\to AB$; $A\to aASb\,/\,aAb\,/\,a$; $B\to bS\,/\,b$

b) $S\to AB$; $A\to aAb\,/\,a$; $B\to b$

c) $S\to AB$; $A\to a$; $B\to a$

d) none of the above

5. Let $M=(\{q_0,q_1\},\{0,1\},\{z_0,x\},\delta,q_0,z_0,\phi)$ be a PDA where δ is given by $\delta(q_0,1,z_0)=\{(q_0,xz_0)\}$

$\delta(q_0,\in,z_0)=(\{q_0,\in\})$

$\delta(q_0,1,x)=\{(q_0,xx)\}$

$\delta(q_1,1,x)=\{(q_0,xx)\}$

$\delta(q_1,1,x)=\{(q_1,\in)\}$

$\delta(q_0,0,x)=\{(q_1,x)\}$

$\delta(q_1,0,z_0)=\{(q_0,z_0)\}$

What is the language accepted by this PDA by empty store ?

a) $L(m)=\{0^n 10^n 1 : n>1\}$

b) $L(m)=\{1^n 01^n 0 : n>1\}$

c) $L(m)=\{01^n 01^n : n>0\}$

d) $L(m)=\{10^n 11^n : n>0\}$

6. In a PDA following move indicates

$\delta(q_0,0,z_0)=(q_1,RBG)$

a) RBG is added above z_0 (top of stack)

b) z_0 is replaced by RBG.

c) RBG takes successive locations from top of stack

d) All the above.

7. If a language is accepted by PDA then there exists a CFG.

a) True b) False

8. In the constructions of PDA we can use more than one state as final states.

a) True b) False

9. How many minimum states are required to design a PDA for the following grammer

$S \to 0A$

$A \to 0AB / 1$

$B \to 1$

a) 1 b) 2 c) 3 d) 4

10. Decide whether the following grammer is

$S \to AS / b$

$A \to BC$

$C \to AB$

$A \to ab$

a) Finite b) infinite

11. Which of the following features cannot be captured by context free grammers ?

a) Syntax of if - then - else statements

b) Syntax of recursive procedures

c) whether a variable has been declared before its use

d) Variable names of arbitrary length.

12. The following grammer is

$A \to AB$

$A \to BC / a$

$B \to CC / b$

$C \to a$

a) Finite b) in finite

EXPLANATORY NOTES - III

1. (a) **Explanation :** The string abab generated by given grammer, has two parse trees

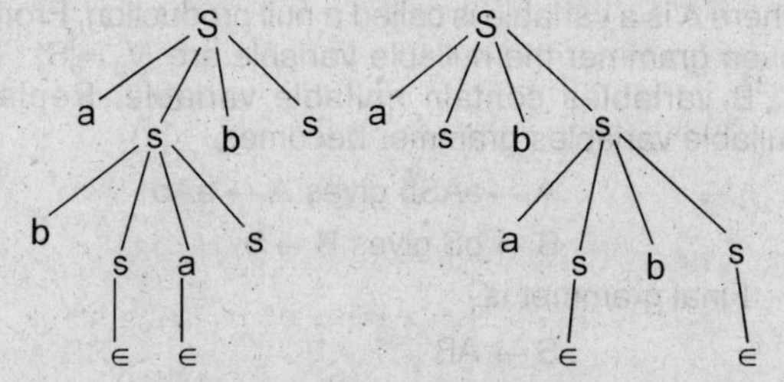

2. (b) **Explanation :** Except C, all variables are deriving terminal strings. In this step the reduced grammer is

$S \to a / aA / B / C$

$A \to aB / \in$

$B \to Aa$

$D \to ddd$

Next determine variables that cannot be reached from the start variable. The dependency graph is drawn as follows

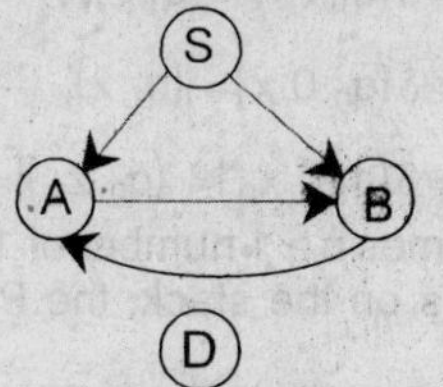

The variable D is not reachable from start symbol. So it is useless symbol. Removing C productions the reduced grammer is

$S \to a / aA / B / C$

$A \to aB / \in$

$B \to Aa$

3. (a) **Explanation :** A production of the form $A \to B, A, B \in V$ are called unit productions. These unit productions in the context - free grammer are undesirable. In given grammer there are no $\in$ - productions. The unit productions are

$S \to A$ $A \to B$

$S \to B$ $C \to D$

$S \to C$

The unit Productions are replaced by

$S \to aAa / bB / bb / aCaa / baD / abD / abD / aa$

$A \to bB / bb$

$C \to baD / abD / aa$

Final productions are

$S \to aAa / bB / bb / aCaa / baD / abD / aa$

$A \to aAa / bB / bb$

$B \to bB / bb$

$C \to aCaa / baD / abD / aa$

$D \to baD / abD / aa$

Hence answer 'a' is correct.

4. (a) **Explanation :** A context free grammer may have productions of the form $A \to \in$. The production $A \to \in$ is just used to erase A. So a production of the fom $A \to \in$, where A is a variable is called a null production. From the given grammer the nullable variable are $V_n = \{S\}$. Here A, B variables contain nullable variable. Replacing nullable variables grammer becomes

$A \to aASb$ gives $A \to aAb$

$B \to bS$ gives $B \to b$

Final grammer is

$S \to AB$

$A \to aASb / aAb / a$

$B \to bS / b$

Hence answer a is correct.

5. (b) **Explanation :**

Let $R_1 = \delta(q_0, 1, z_0) = \{(q_0, x z_0)\}$

$R_2 = \delta(q_0, \in, z_0) = \{(q_0, \in)\}$

$R_3 = \delta(q_0, 1, x) = \{(q_0, xx)\}$

$R_4 = \delta(q_1, 1, x) = \{(q_1, \in)\}$

$R_5 = \delta(q_0, 0, x) = \{(q_1, x)\}$

$R_6 = \delta(q_1, 0, x_0) = \{(q_0, z_0)\}$

i) R_1 and R_2 consumes n > 1 number of 1's and pushing equal number of x's on the stack, the PDA remains in state q_0.

ii) Next R_5 rejectes single zero and PDA moves to state q_1

iii) Next R_4, consumes n number of 1's and popping equal number of x's from the stack and leaving PDA in state q_1.

iv) Now R_6 rejects one more zero, moving PDA to state q_0

v) Using R_2, we can empty the stack

$\therefore L(m) = \{1^n 0 1^n 0 : n > 1\}$

6. (b)

7. (a)

8. (a)

9. (a) **Explanation :**

Defining PDA $M = (\{q\}, \{0,1\}, \{S, A, B\}, \delta, q, S, \phi)$

$S \to 0A \Rightarrow (q, 0, s) \Rightarrow (q, A)$

$A \to 0AB \Rightarrow (q, 0, A) \Rightarrow (q, AB)$

$A \to 1 \Rightarrow (q, 1, A) \Rightarrow (q, \in)$

$B \to 1 \Rightarrow (q, 1, B) \Rightarrow (q, \in)$

10. (b) **Explanation :** Drawing a directed graph from the given productions

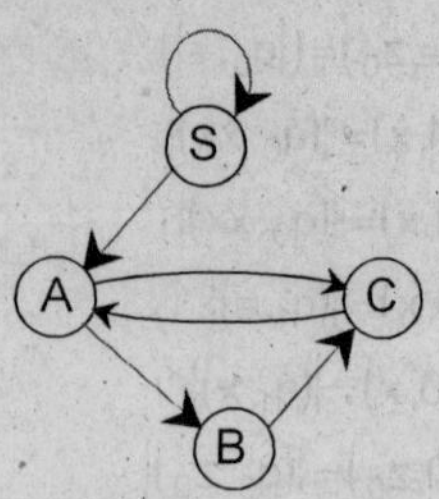

The graph has 2 cycles.

$\therefore$ The given grammer is infinite.

11. (d)

12. (a) **Explanation :** Drawing a directed graph from the given productions

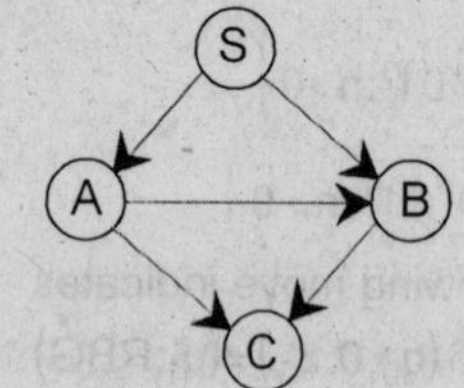

In this graph there are no cycles. Given grammer is finite.

III. TURING MACHINES AND UNDECIDABILITY

- **Turing Machine Model**
- **Techniques for Turing Machine Construction**
- **Non Deterministic Turing Machine**
- **Decidable and undecidable Problems**
- **Chomsky Hierarchy of Languages**

3.1 Turing Machine

Def : A Turing Machine (TM) is denoted

$$M=(Q,\Sigma,\Gamma,\delta,q_0,B,F)$$

Where

1) Q in the finite set of states,
2) Γ is the finite set of allowable tape symbols
3) B, a symbol of Γ, is the blank.
4) Σ, a subset of Γ not including B, is the set of input symbols.
5) δ, is the next move function, a mapping from

$$Q\times\Gamma \text{ to } Q\times\Gamma\times\{L,R\}$$

6) q_0 in Q is the start state.
7) $F\subseteq Q$ is the set of final states.

Instantaneous Description (ID) :

We denote an instantaneous description of the TM M by $\alpha_1 q \alpha_2$ where

$\alpha_1\alpha_2 \rightarrow$ the string in Γ^*

The first symbol of α_2 is the current symbol a under R/W head.

α_2 has all the subsequent symbols of the input string.

α_1 is the substring of the input string formed by all the symbols to the left of a.

Ex :

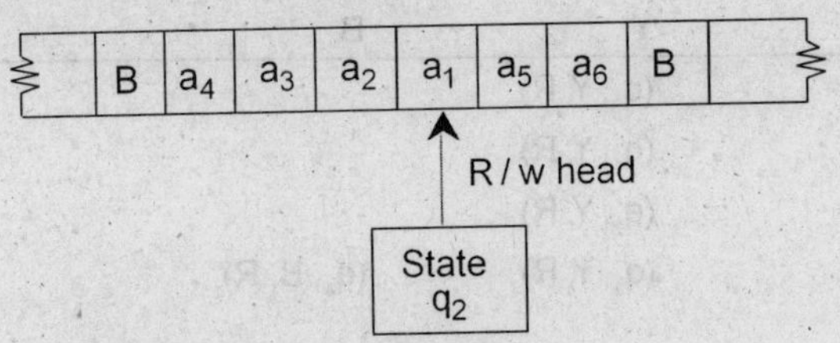

A snapshort of Turing Machine

a_1 is the symbol under R/W head.

$a_4a_3a_2$ is the substring to the left of a.

a_5a_6 is the substring to the right of a.

$\therefore \alpha_1=a_4a_3a_2 \qquad \alpha_2=a_1a_5a_6$

The ID is given as

$$\underbrace{a_4a_3a_2}_{\alpha_1}\ \underbrace{q_2}_{q}\ \underbrace{a_1a_6a_6}_{\alpha_2}$$

Moves in a TM :

$\delta(q,x)$ induces a change in ID of the TM. Suppose $\delta(q,x_i)=(p,y,L)$, the input string to be processed is $x_1x_2.....x_n$ and the present symbol under R/W head is x_i. The change of ID is represented by

$$x_1x_2......x_{i-1}qx_i.....x_n \vdash x_1.....x_{i-2}\,p\,x_{i-1}\,y\,x_{i+1}.....x_n$$

If $\delta(q,x_i)=(p,y,R)$, then change of ID is

$$x_1x_2....x_{i-1}qx_i....x_n \vdash x_1x_2....x_{i-1}\,y\,p\,x_{i+1}....x_n$$

The symbol $\overset{*}{\vdash}$ denotes the reflexive - transitive closure of the relation $\vdash$.

We can denote an ID by I_j for some J

Now, $I_1 \overset{*}{\vdash} I_n$ means $I_1 \vdash I_2 \vdash .... \vdash I_n$

Model of TM : A turing machine is a transducer. This device is essentially a finite automation that has a single tape of infinite length on which it may read and write data. TM is sufficiently simple its configuration can be represented precisely, using a simple notation much like, the ID'S of a PDA. The TM is as shown below.

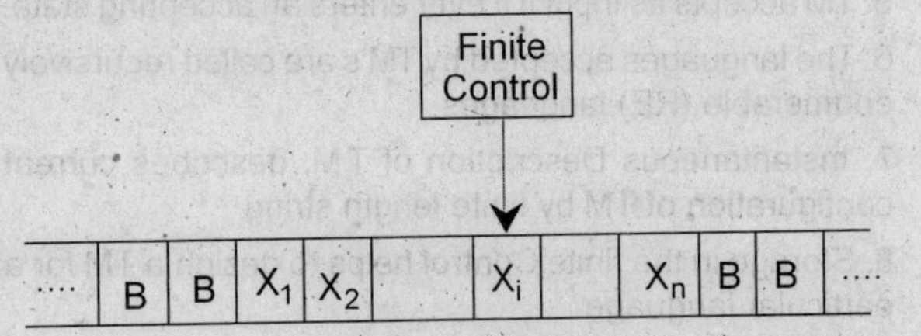

TURING MACHINE

The machine consists of a finite control, which can be in any of a finite set of states. There is a tape divided in to squares or cells; each cell can hold anyone of a finite number of symbols.

Initially, the input, which is a finite-length of string of symbols chosen from the input alphabet, is placed on the tape. All other tape cells, extending infinitely to the left and right initially hold a symbol called blank which is a tape symbol but not input symbol.

There is a tape head that is always positioned at one of the cells, Initially it is at the left most cell that holds the input.

A 'move' of the TM is a function of the state of the finite control and the tape symbol scanned. In one move the TM will

1. Change state
2. Write a tape symbol in the cell scanned.
3. Move the tape head left or right.

3. 2 DIFFERENCES BETWEEN A TM & PDA

PUSHDOWN AUTOMATA:

1. A PDA is a non deterministic finite Automaton coupled with a stack that can be used to store a string of arbitrary length,
2. The stack can be read and modified only at its top.
3. A PDA chooses its next move based on its current state, the next input symbol, and the symbol at the top of the stack.

4. There are two ways in which the PDA may allowed to signal acceptance. One is by entering an accepting state; the other by emptying its stack.

5. ID consisting of the state, remaining input, and stack contents to describe the "Current condition" of a POA.

6. The languages accepted by PDA's either by final state or by empty stack, are exactly the context-free languages.

7. A PDA languages lie strictly between regular languages and CFL's

Turing Machines :

1. The TM is an abstract computing machine with the power of both real computers and of other mathematical definitions of what can be computed.

2. TM consists of a finite-state control and an infinite tape divided in to cells.

3. TM makes moves based on its current state and the tape symbol at the cell scanned by the tape head.

4. The blank is one of tape symbols but not input symbols.

5. TM accepts its input if it ever enters an accepting state.

6. The languages accepted by TM's are called recursively enumerable (RE) languages.

7. Instantaneous Description of T.M. describes current configuration of TM by finite length string.

8. Storage in the finite Control helps to design a TM for a particular language

9. A TM can simulate the storage and control of a real computer by using one tape to store all the locations and their contents.

3.3 Design of Turing Machines

Basic guidelines for designing a Turing machine

(i) Scanning a symbol by R/W head is to 'know' what to do in the future. The machine must remember the past symbols scanned.

(ii) Change the states only when there is a change in the movement of R/W head.

Ex : The design of a TM M to accept the language $L=\{0^n 1^n / n\geq 1\}$

Initially the tape of M contains 0^n1^n followed by an infinity of blanks. Repeatedly, M replaces the leftmost 0 by X, moves right to the leftmost 1, replacing it by Y, moves left to find the right most X, then moves one cell right to the leftmost 0 and repeats the cycle. If however, when searching for a 1, M finds a blank instead, then M halts without accepting. If, after changing a 1 to a Y. M finds a blank instead, then M halts without accepting. If, after changing a1 to a Y, M finds no more 0's then M checks than no more 1's remain, accepting if there are none.

Let $Q=\{q_0,q_1,q_2,q_3,q_4\}, \Sigma=\{0,1\}$,

$\Gamma=\{0,1,X,Y,B\}$ and

$F=\{q_4\}$

Transition graph :

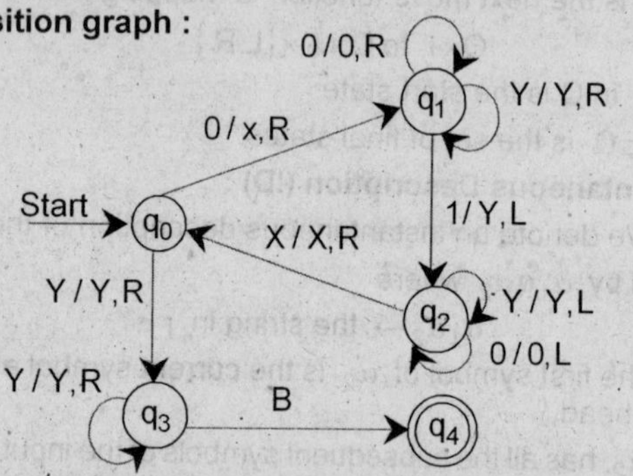

The function of δ

State	0	1	Symbol X	Y	B
q_0	(q_1, X, R)	-	-	(q_3, Y, R)	-
q_1	$(q_1, 0, R)$	(q_2, Y, L)	-	(q_1, Y, R)	-
q_2	$(q_2, 0, L)$	-	(q_0, X, R)	(q_2, Y, R)	-
q_3	-	-	-	(q_3, Y, R)	(q_4, B, R)
q_4	-	-	-	-	-

Computation M on input 0011.

q_0 0011 |- Xq_1011
|- $X0q_111$
|- Xq_20Y1
|- q_2X0Y1
|- Xq_00Y1
|- XXq_1Y1
|- $XXYq_11$
|- XXq_2YY
|- Xq_2XYY
|- XXq_0YY
|- $XXYq_3Y$
|- $XXYYq_3$
|- $XXYYBq_4$

Here because turing machine has reached final state q_4 this input string 0011 is accepted.

3.4 Computable Fuctions

A turing machine is a language acceptor which checks whether a string X is accepted by a languages L. In addition to that it may be viewed as computer which performs computations of functions from integers to integers. In transitional approach an integer is represented in unary, an integer $i\geq 0$ is represented by a string 0^1, For example, 2 is represented as 0^2. If a function has K arguments $i_1,i_2,...i_k$ then these, integers are initially placed on the tape separated by 1's as $0^{i_1}10^{i_2}10^{i_3}....10^{i_k}$

If the TM halts (whether or not in an accepting state) with a tape consisting of 0's for some m, then we say that $f(i_1, i_2, ... i_k) = m$ where f is the function of K arguments computed by this turing machine.

For example, consider a function in C

```
int sum (int i, int j, int k)
{
S = i+j+k ;
return s ;
}
```

Suppose this function is involved using statement

c = sum (2,3,4)

After invoking sum ( ), c will have the value 9. The same computation can be performed by turing machine also. Initially, the turing machine will have the arguments of sum. i.e., 2, 3, 4 on its tape as shown below.

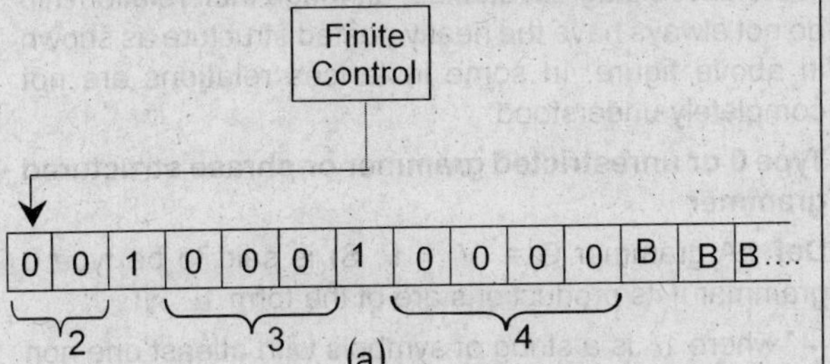

(a)

Before computation

This turing machine performs the sum of these arguments. After some moves it halts with the tape contains value 9 as shown below

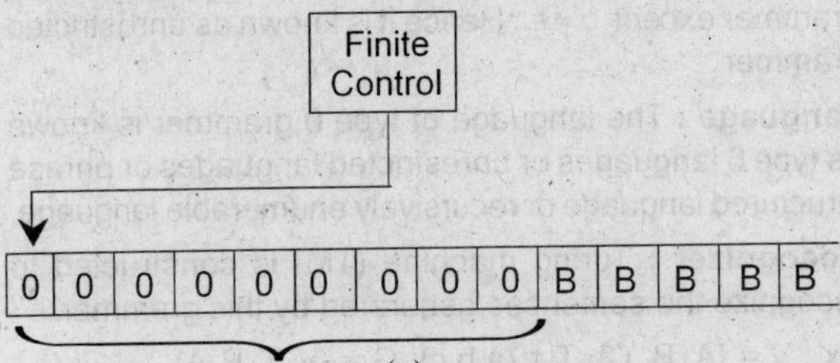

(b)

After Computation

Note that a turing machine may compute a function of one argument, a function of two arguments and so on. The turing machine given in figure (a) can perform sum of two arguments or three arguments or in general sum of any finite number of arguments.

If a TM M computes function f of K arguments i then f need not have a value for all different k-tuples of integers $i_1, i_2, ...i_k$. If $f(i_1, i_2, ...i_k)$ is defined for all, $i_1 ....i_k$ then we say f is total recursive function, otherwise we say f is partial recursive function. Total recursive functions are analogus to recursive language because they are computed by TM that always halts. Partial recursive functions are analogus to recursively enumerable languages. Because they are computed by TM that may or may not halt. Examples of total recursive functions. All common arithmetic functions: on integers, such as n1, multiplication etc are total recursive functions.

Ex : proper subtraction m $\dot{-}$ n is defined to be m -n for $m \geq n$ and

zero for m< n.

3.5 Recursively enumerable Language

A languages L over the alphabet Σ is called recursively enumerable if there is a TM M that accept every word in L and either rejects (crashes) or loops for ever for every word in language L'_e the complement of L.

Accept(M) =L

reject (M)+Loop (M) = L′

when TM M is still running on some input (of recursively enumerable languages) We can never tell whether M will eventually accept if we let it run for long time or M will run for ever (in loop).

Example of recursively enumerable language

Consider a language (a + b) * bb(a + b) *

TM for this language is

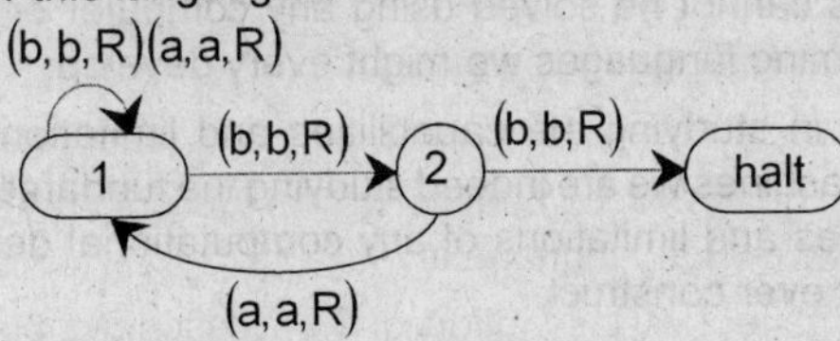

here the inputs are of three types.

i) All words with bb = accept(M)

as soon as TM sees two consecutive b's it halts

ii) All strings without bb but ending in b=reject (M) when TM sees a single b, it enters state 2. If the string is ending with b, TM will halt at state 2 which is not accepting state. Hence it is rejected.

iii) All strings without bb ending in a or blank B= loop (M) here when the TM sees last a it enters state 1. In This state on Blank symbol it loops for ever.

Recursive Language

A language L over the alphabet Σ is called recursive if there is a TM M that accepts every word in L and rejects every word in L′

accept(M) =L

reject (M) = L′

loop (M) = ϕ

Ex : consider a language b(a+b)*. It is represented by TM as

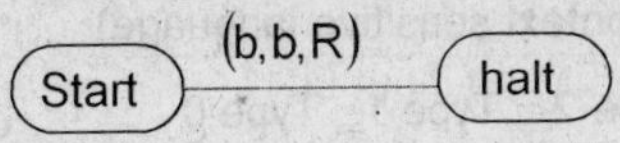

This TM accepts all words beginning with 'b' because it enters halt state and it rejects all words beginning with a because it remains in start state which is not accepting state.

A language accepted by a TM is said to be recursively enumerable languages. The subclass of recursively enumberable sets (r.e) are those languages accepted by at least one TM that halts on all inputs. The languages of this subclass are said to be recursive sets or recursive language.

3.6 Church's hypothesis

According to church's hypothesis all the functions which can be defined by human beings can be computed by turing machine. The turing machine is believed to be ultimate computing machine.

The churches original statement was slightly different because he gave his thesis before machines were actually developed. He said that any machine that can do certain list of operations will be able to perform all algorithms. TM can perform what churches asked so they are possibly the machines which churches described. Churches tied together recursive functions and computable functions together. Every partial recursive function is computable on TM. Computer models such as RAM also give rise to partial recursive functions. So they can be simulated on TM which confirms the validity of churches hypothesis.

Importance of churches hypothesis is as follows:

(1) First we will prove certain problems cannot be solved using TM.

(2) If churches thesis is true this implies these some problems cannot be solved using any computer or any programming languages we might every develop.

(3) Thus in studying the capabilities and limitations of Turing machines we are indeed studying the fundamental capabilities and limitations of any computational device we might ever construct.

It provides a general principle for algorithmic computation and while not provable, gives strong evidence that no more powerful models can be found.

3.7 Chomsky Hierarchy of Languages

We have studied about four classes of languages: regular, context free, context sensitive and recursive enumerable languages. One way of exhibiting relationships among these languages is by chomsky hierarchy. Chomsky provided an initial classification into four languages types: type 0 to type 3.

The type 0 languages are those generated by unrestricted grammars ie., recursive enumerable languages. Type 1 consists of the context sensitive languages, type 2 consists of the context free language and type 3 consists of regular languages.

For $1 \le i \le 3$, every language of type i is a proper subset of type i -1, except that a context free language is a context sensitive only if it does not contain the null string (for any context free language, the set of non null strings in the language is a context sensitive language)

Type 3 $\subseteq$ Type 2 $\subseteq$ Type 1 $\subseteq$ Type 0

The following diagram shows chomsky hierarchy of grammars and their corresponding automata.

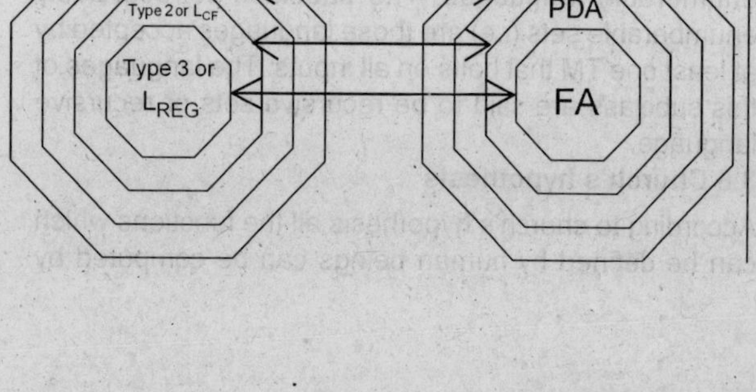

Several other language families can be added to this hierarchy including the families of deterministic context free languages (LDCF) and recursive languages (LREC) the extended hierarchy is shown below.

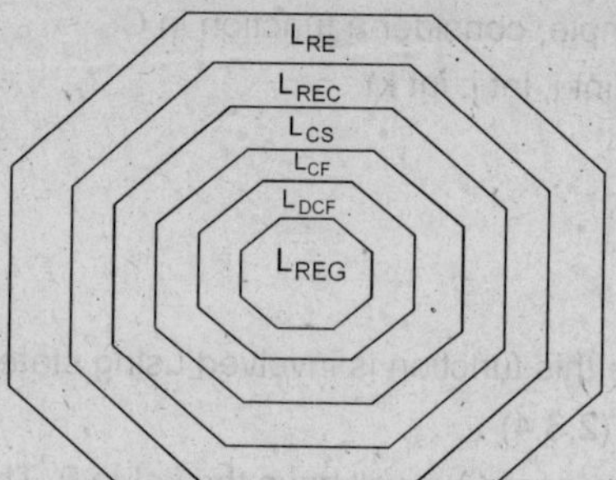

Other language families can be defined and their place in the above diagram studied, although their relationship do not always have the neatly nested structure as shown in above figure. In some instances relations are not completely understood.

Type 0 or unrestricted grammer or phrase structured grammer

Def : A grammar G = (V, T, P, S) is said to be type 0 grammar if its productions are of the form $\alpha \to \beta$.

where α is a string of symbols with atleast one non terminal and $\alpha \ne \in$

β is a string of terminals and non terminals including $\in$ i.e., $\beta \in (V \cup T)^*$

There are no restrictions on production of type 0 grammer except $\alpha \ne \in$. Hence it is known as unrestricted grammer.

Language : The language of type 0 grammer is known as type 0 languages or unrestricted languages or phrase structured language or recursively enumerable language.

Recognizer : Turing machine (TM) is constructed to recognize the sentences generated by this grammer.

Ex : V = {A, B, C}, T = {a,b,c}, G = (V, T, P, A)

A is the start symbol

$A \to AB$, $AB \to BC$, $B \to a$

Type 1 or Context sensitive grammer (CSG)

Def : A grammer G = (V, T, P, S) is said to be context sensitive grammer if the productions are of the form $\alpha \to \beta$

Where α is any string of terminals and nonterminals with atleast one non terminal.

β is any string of terminals and nonterminals with $\beta \ne \in$ with exception for starting symbol where $S \to \in$ is possible. The restriction on the productions is that the length of the consequent β is atleast as much as antecedent α i.e. $|\beta| \ge |\alpha|$

when $S \to \in$ is a production in the grammar then the starting symbol i.e., S should not appear on the right hand side of any production.

The term context sensitive originates from the productions of the from $\alpha_1 A \alpha_2 \to \alpha_1 \beta \alpha_2 (\beta \neq \in)$

Where α_1 is known as left context of A and α_2 is known as right context of A. Replacement of a non terminal A in any production of β depends on the context of α_1 and α_2

Language:

Languages generated by Type 1 grammar is known as type 1 language or context sensitive language (CSL)

Recognizer:

A Linear bounded automata (LBA) which is a TM with its tape restricted as a linear function of input recognizes CSL.

Ex : v = {S,B,C}, T = {a,b,c}, G = (V,T,P,S) ; S is a start symbol.

P consists of

$S \to SB,\ SB \to SC,\ C \to abs$

This grammar is context sensitive grammar

Ex 2 : V = {S, B, C}, T = {a, b, c} ; G = (V,T,P,S) ; S is a start symbol. P consists of $S \to aAB / \in$

$AB \to ab$ Here $S \to \in$ is in P so S does not appear in RHS of any production. Hence it is context sensitive.

Type 2 or Context free grammar

Def: A grammar G = (V, T, P, S) is said to be context free grammar if the productions are of the from $A \to \alpha$

Where A is a nonterminal $A \in V$

$\alpha \in (V \cup T)^*$

The production of the form $A \to \in$ are also permitted and for $S \to \in$. S can appear in RHS of the production.

Language:

The language generated by type 2 grammer is known as type 2 languages or context Free language

Recognizer:

Pushdown automata (PDA) is constructed to recognize the sentences generated by this grammar.

The grammars of most programming languages approximate to this grammar.

Ex : V = {S,A}, T = {a,b,c}, G = (V,T,P,S)

S is the start symbol

Productions P consists of :

$S \to AB,\ A \to abSB | as | b, B \to ab$

The grammar G is type 2 grammar or CFG.

Type 3 (or) Regular grammar

Def : A grammar G (V,T,P,S) is said type 3 grammar if its production are of the form:

1. $A \to xB$ or $A \to x$

Where A is a nonterminal $A, B \in V$

$x \in T^*$

2. $A \to Bx$ or $A \to x$

Where $A, B \in V$ and $x \in T^*$

The grammar consisting of productions only of 1st type is known as right linear grammar and the grammar consisting of productions only of 2nd type is known as left linear grammar.

A grammar which is either left or right linear is known as regular grammar.

Language:

The language of type 3 grammar is known as type 3 language or regular language.

Recognizer:

Regular language is recognized by Finite automation is finite state machine

Example: V = {S,A,B} T = {a,b,c} G =(V,T,P,S)

S is the start symbol and productions P consist of

$S \to aA$

$S \to cB$

$A \to aS$

$B \to a$

This grammar is type 3 or regular grammar.

3.8 Context - Sensitive Languages (CSL)

Context - Sensitive grammars are associated with a language family with the same name is defined as

"A language L is said to be context - sensitive if there exists a context - Sensitive grammar G, such that

$L = L(G) \text{ or } L = L(G) \cup \{\in\}^n$

where as the context sensitive grammar is defined as" A grammar G = (V,T,P,S) is said to be context sensitive if all productions are of the forms.

$x \to y$ Where $x, y \in (V \cup T)$ and $|x| \le |y|$

A context sensitive grammar can never generate a language containing the empty string i.e. $x \to \in$ is not allowed.

Every context free language without $\in$ can be generated by a special case of context- sensitive grammar, say by one in chomsky or Greibach normal form, both of which satisfy the conditions of context - Sensitive grammar. The context free languages is a subset of the family of context sensitive languages if empty string is included in the definition of C.S.L. (but not in grammar)

Ex : The language $L = \{a^n b^n c^n : n \ge 1\}$ is a C.S.L the grammar is

$S \to abc / aAbC$

$Ab \to bA$

$AC \to Bbcc,$

$bB \to Bb,$

$aB \to aa / aaA$

The derivation tree of $a^3b^3c^3$ is looking to be as following

$S \Rightarrow aAbc$

$\Rightarrow abAc$

$\Rightarrow abBbcc \Rightarrow aaAbbcc$

$\Rightarrow aBbbcc$

$\Rightarrow aabAbcc$

$\Rightarrow aabbAcc \Rightarrow aabb\ Bbcc$

$\Rightarrow aabBbbccc$

$\Rightarrow aaBbbbccc$

$\Rightarrow aaabbbccc$

3. 9 Decidability of Problems

A problem whose language is recursive is said to be decidable. "Decidable" is a synonym for "Recursive".

A problem whose language is not recursive is said to be an undecidable problem.

" A class of problems with two output (yes/no) is said to be decidable if there exists some definite algorithm which always terminates (halts) with two outputs (yes/no). Otherwise the class of problems is said to be undecidable."

Ex : It is easy to write a program that will search through all positive integers x,y,z and number n>2 for a solution to the equation $x^n + y^n = z^n$

The problem is to determine whether or not the problem will halt if and when a solution is found.

Some undecidable problems are:

1. Does a TM M accept word w?
2. Is a mathematical statement true?
3. Does a turing machine accept a regular language?
4. Is there a smaller TM accepting the same function?'
5. Are two programs computing the same?

Practicle consequence of undecidability of problems:

For us computer engineers we must recognize that are many arbitary things we cannot do.

1. We cannot write a C program that will determine whether an arbitary C program will halt on a particular input.

2. We cannot write a C program that will determine whether two other C programs solve the same problem.

3.9.1 Properties of Recursive and Recursively enumerable languages:

1. If a language L and its complement $\bar{L}$ are both recursively enumerable, then L (and hence $\bar{L}$) is recursive.

2. The union of two recursive languages is recursive.

3. The union of two recursively enumerable languages is recursively enumerable.

4. The complement of a recursive language is recursive

3.9.2 Rice's Theorem and Undecidable Problems

For a r.e. language that is not recursive, the associated problem is "does M accept w?" is undecidable.

Rice's Theorem : Any non-trivial property ψ of the r.e. languages is undecidable.

The adjective "nontrivial" refers to a property possessed by some but not all recursively enumerable languages.

Some Undecidable Problems:

1. The following properties of r.e. sets are not decidable.

a) emptiness b) finite ness

c) regularity d) context - freedom

3.9.3 Decision Properties of DCFL's

A number of problems that are undecidable for CFL's are decidable for DCFL's

1) Let L be a DCFL and R is a regular set.

The following problems are decidable.

a) Is L = R

b) Is $R \subseteq L$

c) Is $\bar{L} = \phi$

d) Is $\bar{L}$ a CFL

e) Is L regular

2) Let Land L' be arbitary DCFL's, then the following problems are undecidale.

a) Is $L \cap L' = \phi$

b) Is $L \subseteq L'$

c) Is $L \cap L'$ a DCFL

d) Is $L \cap L'$ a CFL

e) Is $L \cup L'$ a DCFL

3.9.4 Turing machine halting problem

There is another notion of "Acceptance" that is commonly used for turing machines: acceptance by halting. A TM halts if it enters a state q, scanning a tape symbol x, and there is no move in this situation; i.e., $\delta(q, x)$ is undefined.

The assumption is that a TM always halts when it is in an accepting state.

Unfortunately, it is not always possible to require that a TM halts even if it does not accept. Those languages with Turing machines that do halt eventually, regardless of whether or not they accept, are called recursive. Turing machines that always halt, regardless of whether or not they accept, are a good model of an "algorithm". If an algorithm to solve a given problem exists, then the problem is "decidable". So TM's that always halt figure importantly into decidability theory.

The halting problem for TM's is a problem similar to one that is RE but not recursive. In fact, the original T.M. of A.M. Turing accepted by halting, not by final state.

Define H(M) for TM M to be the set of inputs w such that M halts for given input w. regardless of whether or not M accepts w. Then, the halting problem is the set of pairs (M,w) such that w is in H(M). This problem or language is another example of one that is RE best not recursive.

3.9.5 Turing Theorem : The halting problem (HP) of Turing machine over $\Sigma = \{0,1\}$ is unsolvable (i.e the problem of determining whether or not an arbitary turing machine M over {0,1} halts for an arbitrary input X in Σ^* is Unsolvable)

3.10 Post's Correspondence Problem:

It was given by Emil post in 1946. An instance if post's correspondence problem (PCP) consists of two lists of strings over some alphabet Σ the two lists must be of equal length; generally refer them as A and B lists, and write $A = w_1, w_2 .... w_k$ and next line $B = x_1, x_2 .... x_k$ for some integer k. For each i, the pair (w_i, x_i) is said to be a corresponding pair.

This instance of PCP has a solution, if there is a sequence of one or more integers $i_1, i_2 ...... i_m$ that, when interpreted as indexes for strings in the A and B lists, yield the same string i.e.

$w_{i1}, w_{i2}, ... w_{im} = x_{i1}, x_{i2}, ... x_{im}$

The sequence $i_1, i_2 .... i_m$ is a solution to this instance of PCP, if so PCP is undecidable.

Ex : Let $\Sigma = \{0,1\}$ let A and B be lists of three strings each.

	list A	list B
i	w_i	x_i
1.	00	0
2.	001	11
3.	1000	011

An instance of PCP

In this case, there can not be any PC - solution simply because any string composed of elements of A will be longer than the corresponding string from B.

∴ In this case, PCP has no solution.

Post correspondence problem can be treated as game of dominoes. Each domino from i^{th} group has string x_i on the top half and string y_i on the bottom half.

X_1	X_2	......	X_i	....	X_m
Y_1	Y_2		Y_i		Y_m

Typical Dominoes

Finding a solution for this instance means lining up dominoes in a horrizontal row, each one positional vertically so that string formed by top half matches with string formed by bottom half as shown below.

xi_1	xi_2	xi_3	....	....	xi_m
yi_1	yi_2	yi_3	.....		yi_m

Example : Let instance of PCP be

1	10111	10
111	10	0

Sequence for it.

The solution is

10111	1	1	10
10	111	111	0

i.e., 101111110

Theorem : The post correspondence problem is undecidable.

SOLVED EXAMPLES

1. Write a type - 2 grammar with productions that generate the language.

$$L(a) = \{0^n 1^n / n \geq 0\}$$

Sol : The grammar is G = (V, T, P, S)

Where G = {S}, T = {0,1}

Productions are

P : $S \to 0\ S1 \quad \to (1)$

$S \to \in$

generates the language

$L(G) = \{0^n 1^n / n \geq 0\}$ Which is a CFG

Consider string $0^2 1^2 \in L(G)$

$S \to 0S1 \quad (1)$

$\to 00\ S11 \quad (1)$

$\to 00 \in 11 \quad (2)$

$\to 0011 = 0^2 1^2$

Accept

2. How the problem "Does turing machine M accepts input w ?" is proved to be undecidable. **[JNTU -1998]**

Sol : Here, both M and w are parameters of the problem. In formalizing the problem as a language we shall restrict w to be over {0,1} and M to have tape alphabet {0,1,B}. As the restricted problem is undecidable, the more general problem is surely undecidable.

Let M = (Q, {0,1}, {0,1,B}, δ, q, B, {q_2}) be a TM with i/p alphabet {0,1} and the blank as the only addition tape symbol.

Assume $Q = \{q_1, q_2 ...... q_n\}$

$L \subseteq (0+1)^*$ is accepted by TM.

It is convenient to call symbols 0,1, and B by synonyms x_1, x_2, x_3 respectively. We also give directions L and R the synonyms D_1 and D_2 respectively.

Then a generic move

$\delta(q_i, x_j) = (q_k, x_l, D_m)$ is encoded by the binary string

$0^i 1 0^j 1 0^k 1 0^l 1 0^m \quad \to (1)$

A binary code i for TM M is

$111 code_1, 11\ code_2\ 11 .. .11\ code_r, 111, \to (2)$

Where code i is a string of form (1)

Every binary string can be interpreted as the code for at most one TM; many binary strings are not the code of any TM. If a string fails the conditions of (2), then TM will be represented by that string.

∴ In general, the problem "Does Turing machine M accept input w ?" is undecidable.

3. Is it possible to have a TM with one internal state. **[JNTU - 2003]**

Sol : Yes it is possible to have a TM with one internal state. For example, a TM is shown which is an inverter - it travels along the tape turning 0's to 1's and is to 0's. This is achieved with only one state and a 2 symbols and blank.

δ function for inverter is

	0	1	B
q	(q, 1, R)	(q, 0, R)	

Note that the function is undefined when blank symbol is read under this circumstance the TM stops.

4. Give a turing machine for the following :

That computes ones complement of a binary number

Sol : Let $Q=\{q_1,q_2\}$, $\Sigma=\{0,1\}$ $=\{0,1,B\}F=\{q_2\}$

	0	1	B
q_1	$(q_1,1,R)$	$(q_1,0,R)$	(q_2,B,R)
q_2	-	-	-

5. Construct a Turing machine that recognize even palindrome over (a,b). Give the type zero grammar for the even pallindrome.

Sol : The language $L\{ww^R \mid w \in (a,b)^*\}$ can be recognized on a single tape turing machine by moving the tape head back and forth on the input, checking symbols from both ends and comparing them.

	a	b	X	Y	B
q_0	(q_1, X, R)	(q_4, Y, R)	(q_f, X, R)	(q_f, Y, R)	-
q_1	(q_1, a, R)	(q_1, b, R)	(q_2, x, L)	(q_2, c, L)	(q_2, B, L)
q_2	(q_2, X, L)	-	-	-	-
q_3	(q_3, a, L)	(q_2, b, L)	(q_0, X, L)	(q_5, Y, L)	(q_5, B, L)
q_4	(q_4, a, R)	(q_4, b, R)	(q_5, X, L)	(q_5, y, L)	(q_5, B, L)
q_5	-	(q_6, Y, R)	-	-	-
q_6	(q_6, a, L)	(q_6, b, L)	-	(q_6, Y, R)	-
q_f	-	-	-	-	-

Type 0 grammer for even pallindrome :

$$S \to aSA \mid bSb \mid \in$$

6. Let a decision problem X be defined as follows :

X : Given a Turing machine M over Σ and any word $w \in \Sigma^*$ does M loop forever on w ? **[GATE- 2001]**

You may assume that the halting problem of Turing machine is undecidable but partially decidable.

(a) Show that X is undecidable.

(b) Show that X is not even partially decidable.

Sol : If Turing machine recognize the word it will reach at the halt state. But it will be in loop forever. Normally a Turing machine should have one or more half states and getting its input, it should reach any of the half state writing the output of the tape. However, if some conditions are not specified then it is not possible to tell the action which would be performed. This situation is known as that a TM is in loop forever.

(a) X is undecidable because if TM reaches in loop there would be no YES/NO answer. In discussion problems if we get an answer in the form of YES/NO, then it is decidable problem, but if we are not getting answer, in the form of YES/NO, then it is undecidable problem. For few situations TM will come in loop, so X is UNDECIDABLE.

(b) X is not even partially decidable because TM is also recognizing some words for $w \in \Sigma^*$.

Therefore, for those words for which it is giving YES/NO response it will be in decidable form.

Hence X is not partially decidable.

OBJECTIVE QUESTIONS - I

1. If every string of a language can be determined, whether it is legal or illegal in finite time,the language is called

a) Decidable b) Undecidable
c) Interpretive d) Non-deterministic

2. Turning machine (TM) is more powerful than FMS because

a) Tape movement is confined to one direction
b) It has no finite state
c) It has the capability to remember arbitrarily long sequences of input symbols
d) None of these

3. A Push Down Machine behaves like a Turning Machine when the number of auxiliary memory it has, is

a) 2 b) 1 c) 0 d) 4

4. Context Sensitive Grammar can be recognized by a

a) Deterministic Push Down Machine (DPDM)
b) Non Deterministic Push Down Machine(NDPDM)
c) Finite State Machine(FSM)
d) Linearly bounded memory machine

5. The number of symbols necessary to simulate a Turning Machine with m symbols and n states is

a) mn b) 2 m(n+m)
c) 4 mn +m d) 8 mn + 4m

6. Recursively enumerable languages are not closed under

a) Union b) Intersection
c) Complementation d) Concatenation

7. The language constructs which are most useful in describing nested structures such as balanced parantheses, matching begin end's, etc., are

a) Regular expressions b) Context-free Grammars
c) Non-Context-free grammars
d) None of the above.

8. The correct hierarchical relationship among context-free, right-linear, and context-sensitive language is given by

a) Context-free / right-linear / context-sensitive
b) Context – free / context – sensitive / right – linear
c) Context-sensitive / right-linear /context-free
d) Right-linear /context – free /context-sensitive

9. Choose the correct statements.

a) $L=\{a^n b^n a^n \mid n = 1,2,3\ldots\}$ is recursively enumerable
b) Recursive languages are closed under union
c) Every recursive language is recursively enumerable
d) Recursive languages are closed under intersection

10. Choose the correct statements.

a) Set of recursively enumerable languages is closed under union

b) If a language and its complement are both regular, then the language must be recursive

c) Recursive languages are closed under complementation

d) None of the above

11. The statement ..."A TM can't solve halting problem " is

a) True b) False

c) Still an open question d) None of the above

12. If there exists a TM which when applied to any problem in the class, terminates if the correct answer is yes and may or may not terminate otherwise, is said to be

a) Stable b) Unsolvable

c) Partially solvable d) Unstable

13. Regarding the power of recognition of languages, which of the following statements is FALSE?

a) The non-deterministic finite state automata is equivalent to deterministic finite-state automata

b) Non-deterministic push-down automata is equivalent to deterministic push-down automata.

c) Non-deterministic turing machine is equivalent to deterministic turing machine.

d) Multi-tape turning machine is equivalent to single tape turning machine.

14. It is undecidable whether

a) An arbitrary turning machine halts on all inputs

b) An arbitrary turning machine halts after 10 steps

c) A turning machine prints a specific letter

d) A turning machine computes the products of two numbers

15. Which one of the following is the strongest correct statement about a finite language over some finite alphabet Σ ?

a) It could be undecidable

b) It is Turning machine recognizable

c) It is a context sensitive language

d) It is a regular language

16. Recursive languages are:

a) A proper superset of context free languages

b) Always recognizable by pushdown automata

c) Also called type and languages

d) Recognizable by Turning machine

17. Which of the following statement is wrong?

a) A turning machine can't solve halting problem

b) Set of recursively enumerable languages is closedunder union

c) An finite state machine with 3 stacks is more powerful than an finite state machine with stacks

d) Context sensitive grammar can be recognized by a linearly bounded memory machine

18. Which of the following statement is wrong?

a) Every recursive language is recursively enumerable

b) A language is accepted by a finite automaton if and only if it is context free

c) Recursive languages are closed under intersection

d) A language is accepted by a finite automaton if and only if it is right linerar

19. Which of the following statement is TRUE?

a) All languages can be generated by context free grammar

b) The number of symbols necessary to simulate a Turning Machine(TM) with m symbols and n states is mn.

c) Any regular language has an equivalent context free grammar

d) The class of context free grammar is not closed under union.

20. A context sensitive language is

a) Recursive and recursively enumerable

b) Recursive but not recursively enumerable

c) Non-recursive but recursively enumerable

d) None of these

21. Each of the classes L_0, L_{cs}, L_{cfl}, L_{rl} is closed under

a) Union b) Transpose

c) Concatenation d) All of these

22. L={abc, bca, cab} can be generated by...... grammar whose terminal set is {a, b, c}

a) Irregular b) Regular

c) Recursive d) None of these

23. If L is a finite subset of Σ*, then L is a ...... language

a) Regular b) Context sensitive

c) Irregular d) Context free

24. Which of the following is TRUE? **[GATE-2002]**

a) The complement of a recursive language is recursive.

b) The complement of a recursively enumerable language is recursively enumerable

c) The complement of a recursive language is either recursive or recursively enumerable

d) The complement of a context-free language is context-free.

25. The C language is **[GATE-2002]**

a) A context free language

b) A context sensitive language

c) A regular language

d) Parsable fully only by a Turing machine

ANSWER KEY - I

1. a	**2.** c	**3.** a	**4.** d	**5.** c
6. c	**7.** b	**8.** d	**9.** a,b,c,d	
10. a,b,c	**11.** a	**12.** c	**13.** b	**14.** a
15. b	**16.** a,d	**17.** c	**18.** b	**19.** c
20. a	**21.** d	**22.** b	**23.** d	**24.** a
25. a				

OBJECTIVE QUESTIONS - II

1. Let L be a language recognizable by finite automaton. The language

REVERSE (L) = {W such that W is the reverse of V where V ∈ L} is a:

a) Regular language b) Context-free language
c) Context-sensitive language
d) Recursively enumerable language

2. Let L be a language recognizable by a finite automaton. The language Front (L)={W such that W is the prefix of V where V ∈ L} is a

a) Context-sensitive language
b) Context-free language
c) Regular language
d) Recursively enumerable language

3. The language L = { a^n b^n a^n where n=1,2,3......} is a

a) Regular language b) Context-free language
c) Non context-free d) None of the above.

4. The language L= { 0^n 1^n 2^R 3^R where n,R >0 } is a :

a) Context-sensitive language
b) Context-free language
c) Regular language
d) Recursively enumerable language

5. Consider a grammar G={{S}, {0,1}, P,S} where elements of P are

S→SS
S→0S1
S→1S0
S→∈

The grammar will generate
a) Regular language
b) Context-free language
c) Context-sensitive language
d) Recursively enumerable language

6. If L(T_i) represents the class of languages that can be generated by the class of T_i grammars where

T_0 is unrestricted grammar
T_1 is context- sensitive grammar
T_2 is context- free grammar
T_3 is regular grammar, then which is TRUE?

a) $L(T_0) \subset L(T_1) \subset L(T_2) \subset L(T_3)$
b) $L(T_0) \supset L(T_1) \supset L(T_2) \supset L(T_3)$
c) All context-free are regular
d) None of the above

7. The language L= { $a^n b^m c^n b^m$ / n>=1,m>=1}

a) Is context-free
b) Is not context free
c) Abstracts the problem of checking number of formal and actual parameters.
d) Both b and c

8. The grammar G = < { S}, {0,1} , P,S>
Where P = {S→0S1, S→0S, S→S1,S→0} Will generate a :

a) Context-free language
b) Context-sensitive language
c) Regular language
d) Recursively enumerable language

9. Which of the following problems is solvable?

a) Determining if an arbitrary Turing machine is a universal Turing machine.
b) Writing a universal Turing machine
c) Determining if a universal Turing machine can be written in fewer than k instructions for some K.
d) Determining if a universal Turing machine and some input will halt.

10. The language L=(0^n 1^n 2^n where n >0) is a:

a) Context-free language
b) Context-sensitive language
c) Regular language
d) Recursively enumerable language

11. The grammar G={{S},{0,1},P,S} where
P={S→SS,S →0S1, S→1S0, S→empty}
Will generate a:

a) Context-free language
b) Context-sensitive language
c) Regular language
d) Recursively enumerable language

12. A Finite state machine can be considered to be a ...... of finite tape length without rewinding capability and unidirection tape movement

a) Turing machine b) Push down automata
c) Context free language d) Regular language

13. Let L_0, L_{csl}, L_{cfl} and L_{rl} denote the family of type 0 languages, context sensitive languages, context free languages and regular languages, then

a) $L_{rl} \subseteq L_{cfl}$, $L_{csl} \subseteq L_0$, $L_{cfl} \subseteq L_0$
b) $L_{cfl} \subseteq L_0$, $L_{cfl} \subseteq L_{rl}$
c) $L_{cfl} \subseteq L_{rl}$, $L_{rl} \subseteq L_0$, $L_0 \subseteq L_{csl}$
d) None of these.

14. To evaluate an expression without any embedded function calls **[GATE-2002]**

a) One stack is enough
b) Two stacks are needed
c) As many stacks as the height of the expression tree are needed
d) A Turing machine is needed in the general case

15. Consider the following problem X.

Given a Turing machine M over the input alphabet Σ, any state q of M and a word $w \in \Sigma^*$, does the computation of M on w visit the state q? Which of the following statements about X is correct? **[GATE-2001]**

a) X is decidable

b) X is undecidable but partially decidable

c) X is undecidable and not even partially decidable

d) X is not a decision problem

16. Which of the following is not decidable? **[GATE-1997]**

a) Given a Turing machine M, a string s and an integer k, M accepts s with in k steps

b) Equivalence of two given Turing machines

c) Language accepted by a given finite state machine is non empty

d) Language generated by a context free grammar is non empty

17. Which of the following statements is FALSE? **[GATE-1996]**

a) The Halting problem of Turing machines is undecidable.

b) Determining whether a context free grammar is ambiguous is undecidable.

c) Given two arbitrary context free grammars G_1 and G_2 it is undecidable whether $L(G_1) = L(G_2)$.

d) Given two regular grammars G_1 and G_2 it is undecidable whether $L(G_1) = L(G_2)$

18. Consider a grammar with the following productions

$S \to a\ \alpha\ b \mid b\ \alpha\ c \mid aB$

$S \to \alpha\ S \mid b$

$S \to \alpha\ bb \mid ab$

$b\ \alpha \to bdb \mid b$

The above grammar is **[GATE-1995]**

a) Context free b) Regular

c) Context sensitive d) LR(k)

19. A language is recursively ennumerable if **[UGC NET-1998]**

a) there is a Turing machine which accepts it

b) there is a finite automata which accepts it

c) there is a push-down automata which accepts it

d) there is a non-deterministic automata which accepts it

ANSWER KEY - II

1. a	**2.** c	**3.** c	**4.** b	**5.** b
6. d	**7.** b	**8.** c	**9.** b	**10.** b
11. a	**12.** a	**13.** a	**14.** a	**15.** b
16. a	**17.** d	**18.** c	**19.** a	

EXPLANATORY NOTES - II

3. (c) **Explanation :** L is not context - free because it is context - sensitive. The CFL grammer corresponding to the L is

$S \to aba / aAba$

$Ab \to bA$

$Aa \to Bbaa$

$bB \to Bb$

$aB \to aa / aaA$

The derivation tree of $a^3b^3a^3$ is looking to be as following.

$S \to aAba$

$\to abAa$

$\to abBbaa$

$\to aBbbaa$

$\to aaAbbaa$

$\to aabAbaa$

$\to aabbAaa$

$\to aabbBbaaa$

$\to aabBbbaaa$

$\to aaBbbbaaa$

$\to aaabbbaaa$

4. (b) **Explanation :** The grammer G corresponding to the given language L is

$S \to S_1S_2$

$S_1 \to 0S_1 1/\epsilon$

$S_2 \to 2S_2 3/\epsilon$

This grammer is context free grammer. Because the rules for CFG is, the left hand side of each production rule is non terminal symbol ie., productions are of the form

$A \to \alpha$ Where A is non - terminal

$\alpha \in (V \cup T)^*$

The productions of the form $A \to \epsilon$ are also permitted, S is allowed to appear on right - hand side of production.

5. (b) **Explanation :** CFG permitting productions of the form $A \to \alpha$

Where A is a non - terminal

$\alpha \in (V \cup T)^*$

The production of the form $A \to \epsilon$ (besides A = S) are also permitted. In addition, S is allowed to appear on right hand side of a production.

Hence given grammer generates the context - free language.

8. (c) **Explanation :** Regular grammer, in each production rule, the left hand side is non - terminal symbol and the right hand side contains atmost one non - terminal symbol which is the right - most (or left most) symbols.

The given grammer is $S \to 0S1$

$S \to 0S$

$S \to S1$

$S \to 0$

Hence this grammer is type 3 or Regular grammer.

IV. PRACTICE QUESTIONS

FINITE AUTOMATA

1. The empty string is the string with:
   (a) no occurrence of symbols
   (b) zero occurrence of symbol
   (c) non zero occurrence of symbol
   (d) none

2. Find the regular expression for the set of all strings having atmost one pair of 0's or atmost one pair of 1's
   (a) $(1+01)^*+(1+01)^*00(1+01)^*+(0+10)^*+(0+10)^*11(0+10)^*$
   (b) $(1+00)^*+(1+01)^*(1+10)^*+(1+11)^*+(0+10)^*11(0+10)^*$
   (c) $(1+01)^*+(1+00)^*(1+10)^*+(1+10)^*+(1+10)^*11(0+10)^*$
   (d) None of the above.

3. Find the regular expression for the set of all strings over $\{a,b\}$ in which there are atleast two occurrences of b between any two occurrences of a.
   (a) $b^*+(b+abb)^*ab^*$ (b) $b^*(aa+bb)^*a^*$
   (c) $(aa)^*ba(bb)^*$ (d) None of the above.

4. Find the regular expression for the set of all strings of the form *vw* where a's occur in pairs in *v* and b's occur in pairs in *w*.
   (a) $(aa+b)^*(bb+a)^*$
   (b) $((aa)^*b)((bb)^*a)$
   (c) $(aabaa)^*(bb+a)^*$
   (d) None of the above.

5. $\in +1^*(011)^*(1^*(011)^*)^* = ?$
   (a) $(1+011)^*$ (b) $1^*(011)^*$
   (c) $1^*01^*(1+011)^*$ (d) None of the above.

6. $(1+00^*1)+(1+00^*1)(0+10^*1)^*(0+10^*1)=?$
   (a) $0^*1(0+10^*1)^*$ (b) $(1+00^*1)(0+10^*1)^*$
   (c) $(0+10^*1)^*0^*1$ d) None of the above.

7. Which of the following regular expressions over $\{0,1\}$ denotes the set of all string not containing 100 as a substring?
   (a) $0^*(1^*0)^*$ (b) 0^*1010^*
   (c) $0^*1^*01^*$ (d) $0^*(10+1)^*$

8. Find the false statement for finite automata,
   (a) $\delta(q,xw)=\delta((q,x),w)$
   (b) $\delta(q,xw)=\delta(q,wx)$
   (c) $\delta(q,\in)=q$
   (d) $if\ \delta(q,y)=\delta(q,x) then\ \delta(q,xz)=\delta(q,yz)$.

9. Given a dfa $A=\langle S,\Sigma,s_0,\delta,F\rangle$, A accepts a word $w\in\Sigma^*$ iff
   (a) $\delta(s,w)\in F,\ Where\ s=s_0$
   (b) $\delta(s,w) doesn't \in F,\ Where\ s=s_0$
   (c) $\delta(s,w)\in F,\ Where\ s\neq s_0$
   (d) $\delta(s,w) doesn't \in F,\ Where\ s\neq s_0$

10. It is given that $\delta(q,x)=\delta(q,y)$, then $\delta(q,xz)=\delta(q,yz)$ for All strings z in:
   (a) Σ^* (b) Σ (c) Σ^+ (d) $\sim\Sigma$

11. dfa has:
   (a) Single final state
   (b) More than one initial states
   (c) Unique path(for a set of inputs) to the final state
   (d) All of the above.

12. Which of the following is there is an FA?
   (a) Input b) State
   (c) State Transition d) All of the above.

13. The language generated by a deterministic finite automata is,
   (a) Regular Language
   (b) Context free language
   (c) Context sensitive language
   (d) Informal language.

14. The basic limitations of Finite state machine is that
   (a) it can't remember arbitrary large information
   (b) it sometimes recognizes non regular language
   (c) it sometimes does not recognizes regular language. (d) all of the above.

15. dfa can recognize
   (a) Any grammar (b) Only CFG
   (c) Only unambiguous grammar
   (d) Only regular language

16. The basic limitation of Finite state machine is that
   (a) it can't remember arbitrary large information
   (b) it sometimes recognizes non regular language
   (c) it sometimes does not recognizes regular language
   (d) all the above.

17. If, in a dfa, $\delta(q_1,x)=q_2\ and\ \delta(q_2,y)=q_1,\ then\ \delta(q_1,xy)$ is
   (a) q_1 (b) q_2
   (c) some state q_3 (d) None of the above.

18. For a deterministic finite automata, $M=(S,\Sigma,\delta,q_0,F);\delta$, the transition function is defined as:
   (a) $\delta: s\times\Sigma u\{\in\}\rightarrow Q$ (b) $\delta: s\times\Sigma\rightarrow\Sigma$
   (c) $\delta: S\times S\rightarrow\Sigma$ (d) $\delta: S\times\Sigma\rightarrow S^+$

19.L is a nonempty language such that any *w* in L has length n, then any dfa accepting L must have

(a) exactly *n* states

(b) atleast $(n+1)$ states.

(c) atmost $(n+1)$ states

(d) exactly $(n+1)$ states.

20. Find the false statement for finite automata.

(a) $\delta(q,xw)=\delta(delta(q,x),w)$

(b) $\delta(q,xw)=\delta(q,wx)$

(c) $\delta(q,\in)=q$

(d) $if\ \delta(q,y)=\delta(q,x)\ then\ \delta(q,xz)=\delta(q,yz)$

21.Find the false statement : An instantaneous description, of the finite-state automation is a singleton *uqv*, where:

(a) q is a state in *S*

(b) *uv* is a string in Σ^*

(c) the configuration is said to be an initial configuration if $u=\varepsilon\ and\ q$ is the initial state

(d)the configuration is said to be a final configuration if $v=\varepsilon\ and\ q$ is the initial state.

22.Consider the FA for a switch with ON/OFF facilities. The automata can be designed with minimum no of states ______?

(a) 1 (b) 2 (c) 3 (d) 4

23.Application of finite automata cannot be found in:

(a) Storage purpose (b) Spelling checkers

(c) Lexical analyzers (d) String matching

24. The regular expression for the machine M below is

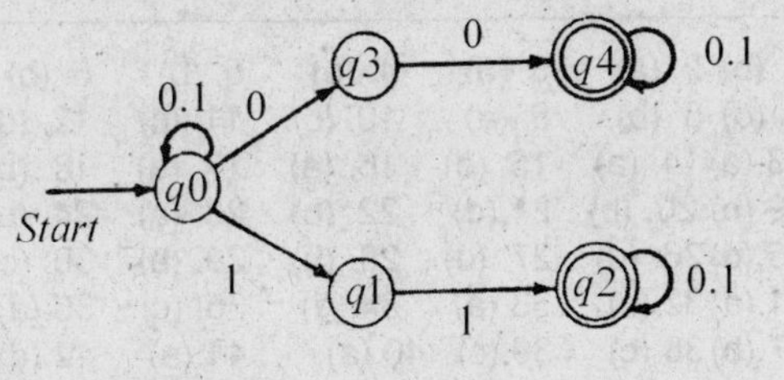

(a) $(1(0^*)1(00+11)^*)+(0(1^*)0(0+1)^*)$

(b) $(0+1)^*(00+11)(0+1)^*$

(c) $(0+1)^*+(00(1^*)(0+1)^*)$

(d) None of the above

25. A nfa computes by reading in an input symbol from a string, and splits into multiple copies of itself, one for each possible transition. If the next input symbol doesn't appear on any of the arrows existing for the current state of a copy of the machine, that copy dics. A nfa accepts an input string when all the input symbols have been read and ....

(a) all of the alive copies of the machine are in an accept state

(b) any one copy of the machine that died was in a reject state.

(c) all copies of the machine that died were in a reject state

(d) any one of the alive copies of the machine are in an accept state.

26.The rules for dfa state that ....

(a) every state of a dfa may have zero, one, or many exiting transition arrow for each symbol in the alphabet.

(b) every state of a dfa may have zero, one, or many exiting transition arrow for each symbol in the alphabet,plus".

(c) every state of dfa must always have at most one exiting transition arrow for each symbol in the alphabet.

(d) every state of a dfa must always have exactly one exiting transition arrow for each symbol in the alphabet.

27. The rules for nfa state that ....

(a) every state of a nfa must always have at most one exiting transition arrow for each symbol in alphabet.

(b) every state of a nfa must always have exactly one exiting transition arrow for each symbol in the alphabet.

(c) every state of a nfa may have zero, one, many exiting transition arrow for each symbol in the alphabet,

(d) every state of a nfa may have zero, one, or many exiting transition arrow for each symbol in the alphabet,plus".

28. Find the true statement

(a) Non-determinism adds to the recognition power of finite-state automata.

(b) Non-determinism does not add to the recognition power of finite-state automata.

(c) it depends from case to case.

(d) There is nothing like non-determinism in finite-state automata.

29. Find the true statement,

(a) Non-determinism adds to the recognition power of finite-state automata.

(b) Non-determinism does not add to the recognition power of finite-state automata.

(c) It depends from case to case.

(d) There is nothing like non-determinism in finite-state automata.

30. For a non-deterministic finite accepter, $M=(S,\Sigma,\delta,q_0,F);\delta$, the transition function is defined as:

(a) $\delta : S \times \Sigma \rightarrow S$

(b) $\delta:S\times\Sigma\rightarrow 2^S$

(c) $\delta:S\times(\Sigma\times\{\in\})\rightarrow 2^S$

(d) None of these.

31.How many of 00, 01001, 10010, 000, 0000 are accepted by the following nfa :

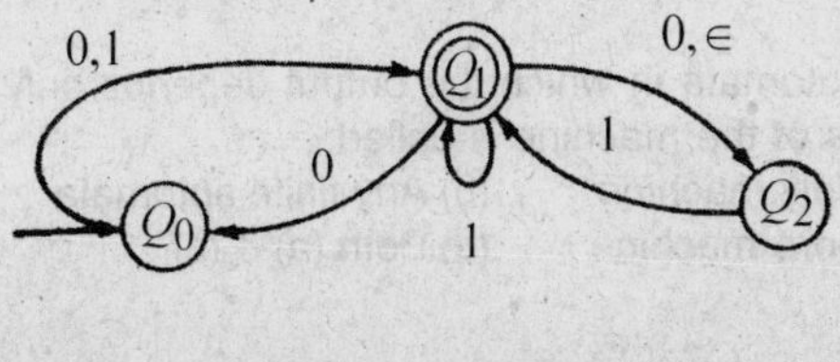

(a) 2 (b) 1 (c) 4 (d) 5

32. If S is the number of states in ndfa then equivalent dfa can have maximum of
(a) S states (b) S-1 states
(c) 2^s states (d) $2^s - 1$ states

33. For text searching applications which of them is used:
(a) nfa (b) dfa
(c) pda (d) npda

34. Consider the following two finite state machine in below figure is.
(a) Both are equivalent
(b) The second finite state machine accepts e-only
(c) The first finite state machine accepts nothing
(d) none of the above.

35. The recognizable property of dfa and ndfa
(a) May be different (b) Must be different
(c) Must be same (d) None of the above.

36. $M = \langle\{q_1, q_2, q_3\}, \{0,1\}, \delta, q_1 \{q_1\}\rangle$ is a nondeterministic finite automation, where delta is given by

$\delta(q_1, 0) = \{q_2, q_3\}$ $\delta(q_1, 1) = \{q_1\}$

$\delta(q_2, 0) = \{q_1, q_2\}$ $\delta(q_2, 1) = \{\phi\}$

$\delta(q_3, 0) = \{q_2\}$, $\delta(q_3, 1) = \{q_1, q_2\}$

An equivalent dfa is given by which one of the following:

(a)

	0	1
q_0	q_1	q_0
q_1	q_2	q_2
q_2	q_3	q_0
q_3	q_3	q_2

(b)

	0	1
q_0	q_1	q_0
q_1	q_2	q_2
q_2	q_0	q_3
q_3	q_3	q_2

(c)

	0	1
q_0	q_1	q_0
q_1	q_2	q_2
q_2	q_3	q_0
q_3	q_3	q_2

(d)

	0	1
q_0	q_1	q_0
q_1	q_2	q_2
q_2	q_0	q_3
q_3	q_2	q_3

37. Given an arbitrary non-deterministic finite automation(nfa) with N states, the maximum number of states is an equivalent minimized dfa is at least
(a) N^2 (b) 2^N
(c) 2^V (d) $N!$

38. Choose the correct statement
(a) A Mealy machine generates no language as scuh
(b) A Moore machine generates no language as such
(c) A Mealy machine has no terminal state
(d) All.

39. The automata in which the output depends only on the status of the machine is called:
(a) Mealy machine (b) Any finite automata
(c) Moore machine (d) Both (a) & (c)

40. Moore machine is
(a) Automaton in which the output depends only on the states
(b) Automaton in which the output depends only on the input
(c) Automaton in which the output depends only on the state and the input.
(d) None of the above.

41. The finite state machine in below figure is a
(a) Mealy machine
(b) Moore machine
(c) Kleene machine
(d) none of the above

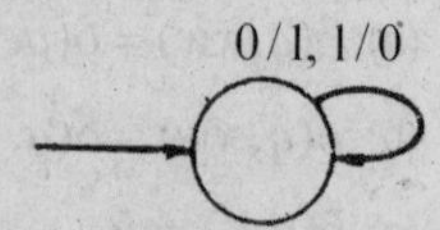

42. Choose the correct statements
(a) Moore and Mealy machines are finite state machines with output capability
(b) Any given Moore machine has an equivalent Mealy machine
(c) Any given Mealy machine has an equivalent Moore machine
(d) All.

43. The major difference between a Moore and Mealy machine is
(a) Output of Moore depends on present state and input
(b) Output of Moore depends on state only
(c) Output of Moore depends on output only
(d) None.

44. Choose the correct statement
(a) A Mealy machine generates no language as such
(b) A Moore machine generates no language as such
(c) A Mealy machine has no terminal state
(d) All

ANSWER KEY

1. (b)	2. (a)	3. (a)	4. (a)	5. (b)	6. (b)
7. (d)	8. (b)	9. (a)	10. (c)	11. (c)	12. (d)
13. (a)	14. (a)	15. (d)	16. (a)	17. (a)	18. (b)
19. (b)	20. (b)	21. (d)	22. (b)	23. (a)	24. (b)
25. (d)	26. (d)	27. (d)	28. (b)	29. (b)	30. (c)
31. (a)	32. (c)	33. (a)	34. (d)	35. (c)	36. (a)
37. (b)	38. (c)	39. (c)	40. (a)	41. (a)	42. (d)
43. (b)	44. (c)				

REGULAR LANGUAGES

1. Let *L* may be language. Define even(w) as the string obtained by extracting from *w* the letters in even numbered positions i.e., if $w = a_1a_2a_3a_4$........, then even(w) $= a_2a_4$...... Corresponding to this, we can define a language : even(*L*) = { even(w) : $w \in L\}$ then given *L* is regular, even(*L*) is
(a) may not be regular (b) must be regular
(c) context free (d) is not context free

2. Which of the following pairs are not equivalent
(a) $1(01)^*$ and $(10)^*1$ (b) $x(xx)^*$ and $(xx)^*x$
(c) $(ab)^*$ and a^*b^* (d) x^+ and x^*x^+

3. The language of all words with at least 2 a's can be described as
(a) $(a+b)^*a(a+b)^*(a+b)^*$
(b) $(a+b)^*.ab^*(a+b)^*$
(c) $b^*ab^*a(a+b)^*$
(d) all

4. Which of the following pairs are equivalent
(a) $(ab)^*a$ and $a(ba)^*$
(b) $(a+b)^*$ and (a^*+b^*)
(c) (a^*+b) and $(a+b)^*$
(d) None

5. The set of regular languages over a given alphabet set is not closed under
(a) union (b) Complement
(c) Intersection (d) none

6. Which one is false
(a) $(r^*)^* = r^*$
(b) $r_1^*(r_1+r_2)^* = (r_1+r_2)^*$
(c) $(r_1+r_2)^* = (r_1^*r_2^*)^*$
(d) none.

7. The intersection of $(a+b)^*a$ and $b(a+b)^*$ is given by
(a) $b(a+b)^*a$ (b) $(a+b)^*ab(a+b)^*$
(c) $a(a+b)^*b$ d) $ab(a+b)^*$

8. A language $L = \{awa : w \in \{a,b\}^+\}$ is
(a) regular (b) context free
(c) context sensitive (d) none of these

9. For $L(r) = \{a, bb, aa, abb, ba, bbb, ...\}, r$ is given by
(a) $r = a(a+bb)^*$
(b) $r = (a+bb)^*$
(c) $r = (aa+b)(a+b)^*$
(d) $r = (a+b)^*(a+bb)$

10. Regular expression for the language, $L = \{w \in \{0,1\}^*$: w has no pair of consecutive zeros} is
(a) $r = (1^*011^*)^*(0+A)+1^*(0+A)$
(b) $r = (1+01)^*(0+A)$
(c) $r = (1+01)^*(0+1^*)$
(d) all of these

11. Let r_1 and r_2 are regular expression and let $\cup$ stands for equivalence in the sense of the language generated, then
(a) $(r_1^2+r_2)^* \equiv (r_1^*r_2^*)^*$
(b) $(r_1^*)^* \equiv r_1$
(c) $r_1(r_1+r_2)^* \equiv (r_1+r_2)^*$
(d) None of these

12. Which of the following is accepted by $L(aa^*+aba^*b^*)$
(a) aaab (b) abba
(c) abab (d) None.

13. Which is the correct order of precedence of regular expression operators in increasing order?
(a) ( ), *, .,...,+ (b) *, ( ),....,+
(c) ( ),., *,+ (d) *, ( ),+,...

14. Which is the correct regular expression for the language : "Set of strings over alphabet $\{a,b,c\}$ containing at least 1 'a' and at least 1 'b'
(a) $(a+b+c)^* - (a^*+b^*+c^*)$
(b) $(a+b+c)^*[a(a+b+c)^*b+b(a+b+c)^*a](a+b+c)^*$
(c) $c^*a(a+c)^*b(a+b+c)^*+c^*b(b+c)^*a(a+b+c)^*$
(d) none of these.

15. Give English description of the language of the regular expression : $(1+\in)(00^*1)^*0^*$
(a) set of all strings of 0's and 1's containing
(b) no pair of consecutive 1's
(c) 0's only in pairs (d) alternating 1's and 0's

16. ϕ^* (Kleene's closure of ϕ)(ϕ is the empty language over Σ) is equivalent to

(a) ϕ (b) $\in$
(c) Σ (d) none of these.

17. Which of the following identities for regular expression does not hold good?

(a) $(R^*)^* = R^*$ (b) $(\in +R^*) = R^*$
(c) $(R^*S^*)^* = (R+S)^*$ (d) $(R+S)^* = R^* + S^*$

18. The value of the relation $\phi + R$ is

(a) R (b) $\in$
(c) ϕ (d) None of the above

19. The value of the relation ϕR is

(a) R (b) $\in$
(c) ϕ (d) None of the above

20. The value of the relation $R\phi$ is

(a) R (b) $\in$
(c) ϕ (d) None of the above

21. The value of the relation $\in R$ is

(a) R (b) $\in$
(c) ϕ (d) None of the above

22. The value of the relation $R \in$ is

(a) R (b) $\in$
(c) ϕ (d) None of the above

23. The value of the relation A^* is

(a) Σ (b) $\in$
(c) ϕ (d) None of the above

24. The value of the relation ϕ^* is

(a) Σ (b) $\in$
(c) ϕ (d) None of the above

25. The value of the relation $(R^*)^*$ is

(a) R (b) R^*
(c) $\in$ (d) None of the above

26. The value of the relation $A + RR^*$ is

(a) R (b) $\in$ (c) ϕ (d) R^*

27. The value of the relation $RR^* + \in$ is

(a) R (b) $\in$ (c) ϕ (d) R^*

28. The value of the relation $R + R$ is

(a) R (b) $\in$ (c) ϕ (d) R^*

29. The value of the relation $(P+Q)^*$ is

(a) Σ^* (b) P^*Q^*
(c) $(P^*Q^*)^*$ (d) $P^* + Q^*$

30. The value of the relation $(P^* + Q^*)^*$ is

(a) Σ^* (b) P^*Q^*
(c) $(P^*Q^*)^*$ (d) None of the above.

31. A solution to the equation $R = Q + RP$ is

(a) $R = QP^*$ (b) $Q = RP^*$
(c) $P = RQ^*$ (d) $R = PQ^*$

32. Which of the following doesn't hold?

(a) $(1+00^*1)+(1+00^*1)(0+10^*1)^*(0+10^*1) = 0^*1(0+10^*1)^*$

(b) $(1+0)^* = 1^*(01^*)^*$

(c) $(111^*)^* = (11+111)^*$

(d) $\in +1^*(011)^*(1^*(011)^*)^* = \in +1^*(0111^*)$

33. How many strings of length t are in Σ^* if Σ is an alphabet of cardinality r.

(a) rt (b) rt (c) tr (d) r + t

34. The regular expression representing the set of all strings over $\{x, y\}$ ending with xx beginning with y

(a) $xx(x+y)^*y$ (b) $yy(x+y)^*x$
(c) $y(x+y)^*xx$ (d) $x(x+y)^*yy$

35. Let L be the language $\{\in,0,10\}$ over $\{0,1\}$. Determine the set $L \cup \overline{L}$.

(a) ϕ (b) same as the given set
(c) $\{0,1\}^*$ (d) None of the above

36. Regular expression for the set $\{a^2, a^5, a^8, ...\}$ is :

(a) $aa(aaa)^*$ (b) $aa(aaaa)^*$
(c) $a(aaa)^*$ (d) $aa(aa)^*$

37. Find the true statement

(a) $\in, \phi$ represents empty language
(b) $\in, \phi$ represents empty word
(c) $\in$ represents empty word, ϕ represents empty language
(d) ϕ represents empty word, $\in$ represents empty language.

38.Find the true statement
(a) If R and S are regular expression then so is $R \cup S$
(b) If R and S are regular expression then so is R.S
(c) If R is regular expression then so is R^*
(d) All of the above.

39.For the following conditions find all the strings over the alphabet {a, } that satisfy the condition, (i) no symbol is repeated in the string and (ii) the length of string is 3.
(a) All possible strings of length 3
(b) Only single string ab
(c) No such string is possible
(d) None of the above.

40.The regular expression representing the set of all strings over $\{a,b\}$ with three consecutive b's
(a) $(a+b)^* bbb(a+b)^*$ (b) $(a+b)bbb(a+b)^*$
(c) $(a+b)^* bb(a+b)^*$ (d) $(a+b)^* bbb(a+b)$

41.In English language the set represented by a^*b+b^*a is:
(a) String containing single a or single b
(b) Strings of a's followed by one b or strings of b's followed by one a.
(c) String containing angle a and single b
(d) Strings of a's followed by one b and strings of b's followed by one a.

42.Which of the following identities doesn't hold?
(a) $(R \cup S)^* = (R^* \cup S^*)^*$
b) $(R^* \cup S^*)^* = (R.S)^*$
(c) $(R \cup S)^* = (R^*.S^*)^*$
d) $R^*.R^* = R^*$

43. $01^* + 0$ is represented by
(a) $((0(1^*)) + 0)$ (b) $((01)^* + 0)$
(c) $(0(1^* + 0))$ (d) $(1^*(0+0))$

44.Let r_1 and r_2 are regular expression which of the following represent $r_1 + r_2$

(a) 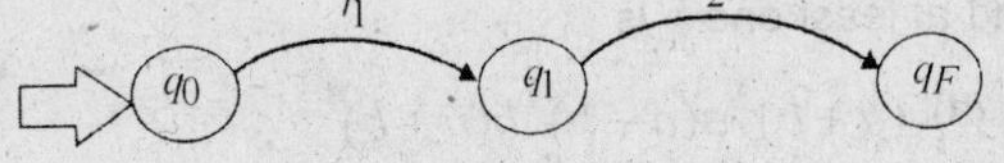

(b) 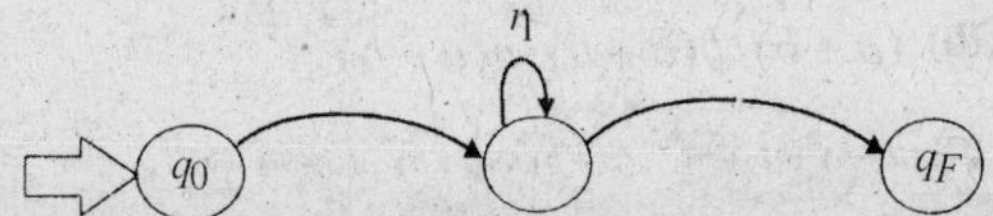

(c) 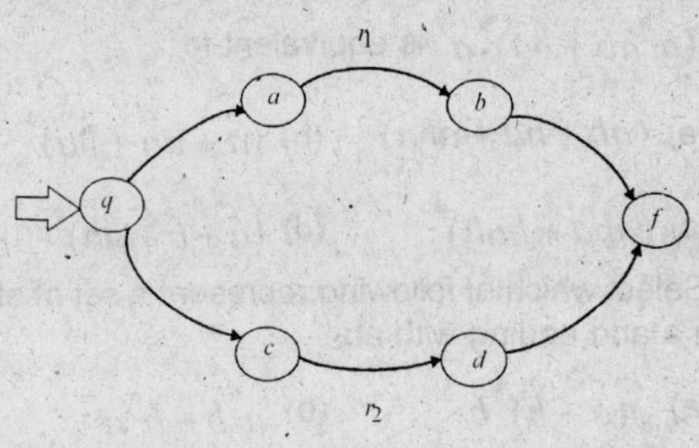

(d) none.

45. Let r_1 and r_2 are regular expression which of the following represent $r_1.r_2$.

(a) 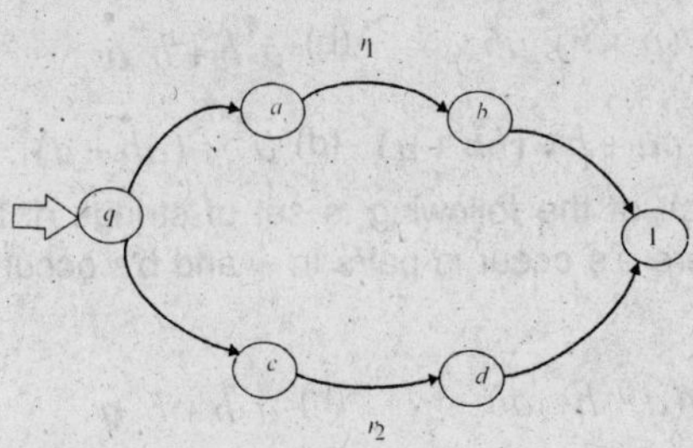

(b) 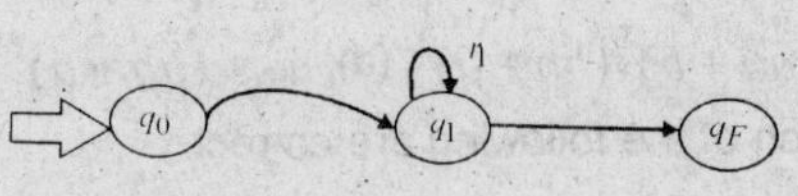

(c) 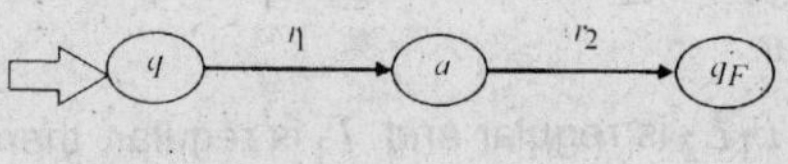

(d) none

46. Which of the following is false
(a) $R+R=R$ (b) $(R^*)^* = R^*$
(c) $RR^* = R^*R$ (d) none.

47. Let r_1 and r_2 are regular expression which of the following represent r_1

(a) 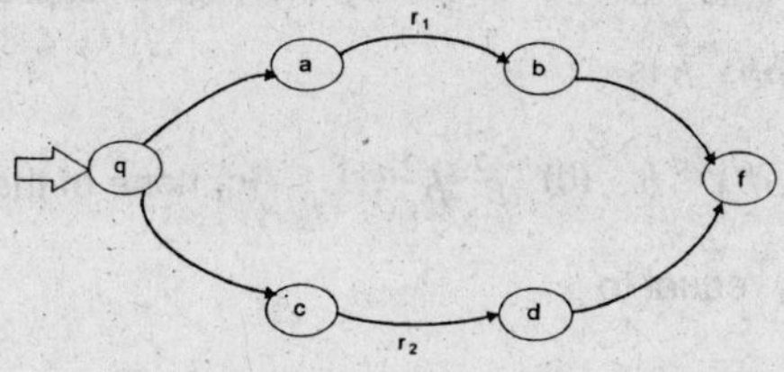

(b) 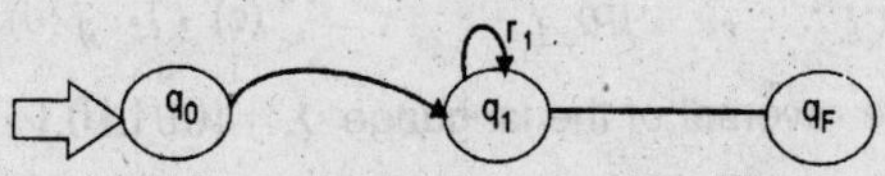

(c) 

(d) none

48. $(a^* ab + ba)^* a^*$ is equivalent to

(a) $(ab + ba + aba)$ (b) $(a + ab + ba)^*$

(c) $(aba + bab)^*$ (d) $(a + b + ab)^*$

49. Select which of following represent a set of all strings with a and ending with ab.

(a) $a(a+b)^* b$ (b) $a^* b + b^* a$

(c) $(aa+b)^* (bb+a)^*$ (d) $a^* + (ab+a)^*$

50. The set of all strings which are either strings of a's followed by one b or strings of b's followed by one a.

(a) $a(a+b)^* ab$ (b) $a^* b + b^* a$

(c) $(aa+b)^* (bb+a)^*$ (d) $a^* + (ab+a)^*$

51. Which of the following is set of strings of the form vw where a's occur in pairs in v and b's occur in pairs in w.

(a) $a(a+b)^* ab$ (b) $a^* b + b^* a$

(c) $(aa+b)^* (bb+a)^*$ (d) $a^* + (ab+a)^*$

52. Which of the following are correct

(a) If $L_1 \cup L_2$ is regular and L_1 is regular, then L_2 is regular,

(b) If $L_1 L_2$ is regular and L_1 is regular, then L_2 is regular.

(c) If L^* is regular then L is regular

(d) all

53. $(1^* 011^*)^* (0 + \in)$ is equivalent to

(a) $(1+01)^* (0+\in)$ (b) $(0+1)(1+10)^*$

(c) none of these

54. The language generated by the regular expression $(aa)^* (bb)^* b$ is

(a) $(ab)^{2n} b$ (b) $a^{2n} b^{2n+1}$ (c) none of these.

55. $(L^*)^*$ equal to :

(a) L^* (b) L^{**} (c) (L^{**}) (d) none

56. The reversal of the language $L = \{001, 10, 111\}$ *is* :

(a) $\{100, 01, 111\}$ (b) $\{111, 10, 001\}$

(c) $\{111, 01, 110\}$ (d) none

57. Write the regular expression for the following: "The set of the strings over alphabet $\{a,b,c\}$ containing at least one a and at least one b"

(a) $c^* a(a+c)^* b(a+b+c)^* + c^* b(b+c)^* a(a+b+c)^*$

(b) $ca^* (a+c)^* b(a+b+c)^* c^* b(b+c)^* a(a+b+c)^*$

(c) $c^* a^* (a+c)^* b(a+b+c) + c^* b(b+c)^* a(a+b+c)$

(d) $ca^* (a+c)^* b(a+b+c)^* c^* b(b+c)^* a(a+b+c)^*$

58. The set all strings of 0's and 1's such that every pair of adjacent 0's appears before any pair of adjacent 1's

(a) $(100)^* (epsilon + 1)(01+1)^* (epsilon + 0)$

(b) $(10+0)^* (epsilon + 1)^* (epsilon + 0)$

(c) $(10+0)^* (epsilon + 1)(01+1)^* (epsilon + 0)$

(d) $(10+0)^* (epsilon + 1)(01+01)^* (epsilon + 0)$

59. The two regular expressions are equivalent i.e., $\in + (a+b)^* b(a+b)^* = [a^* b(a^* ba^* b)^* a^*]^*$

(a) True (b) False.

60. What is the regular expression defining the language of all words with an odd number of b's is

(a) $a^* b + (a^* ba + b)^* + a^*$

(b) $a^* (a^* b)^* a^*$

(c) $a^* b(a^* ba^* b)^* a^*$

(d) None of these.

61. If e_1 *and* e_2 are regular expressions denoting the languages L_1 and L_2 respectively, then which is false?

(a) $(e_1)(e_2)$ is a regular expression denoting $L_1 \cup L_2$

(b) $(e_1)(e_2)$ is a regular expression denoting $L_1 L_2$

(c) ϕ is not a regular expression

(d) $(e_1)^*$ is a regular expression denoting L_1^*

62. Let a and b be two regular expressions then $(a^* \cup b^*)^*$ is equivalent to

(a) $(a \cup b)^*$ (b) $(b^* \cup a^*)^*$

(c) $(b \cup a)^*$ (d) $a \cup b$

63. The language of all words that have at least one a and at least one b is

(a) $(a+b)^* a(a+b)^* b(a+b)^*$

(b) $(a+b)^* b(a+b)^* a(a+b)^*$

(c) $(a+b)^* a(a+b)^* + (a+b)^* b(a+b)^* a(a+b)^*$

(d) $(a+b)^* a(a+b)^* b(a+b)^* + bb^* aa^*$

64.The regular expression representing the set of all strings over $\{a,b\}$ with three consecutive b's:

(a) $(a+b)^* bbb(a+b)^*$ (b) $(a+b)bbb(a+b)^*$

(c) $(a+b)^* bb(a+b)^*$ (d) $(a+b)^* bbb(a+b)$

65.Which of the following identity doesn't hold?

(a) $\in R = R\in = R$ (b) $\in + R = R + \in = R$

(c) $\phi R = R\phi = \phi$ (d) $\phi + R = R + \phi = R$

66.Regular expression generated by the following automaton is given as :

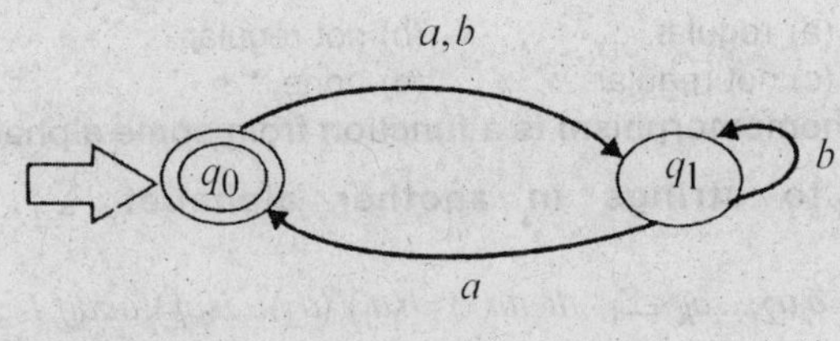

(a) $\in +(a+b)(b+ab+aa)^* a$

(b) $(a+b)(b+ab+aa)^*$

(c) $\in +(a+b)(b+ab+aa)a$

(d) $(a+b)(b+ab+aa)^* a$

67. Regular expression generated by the following automaton is given as

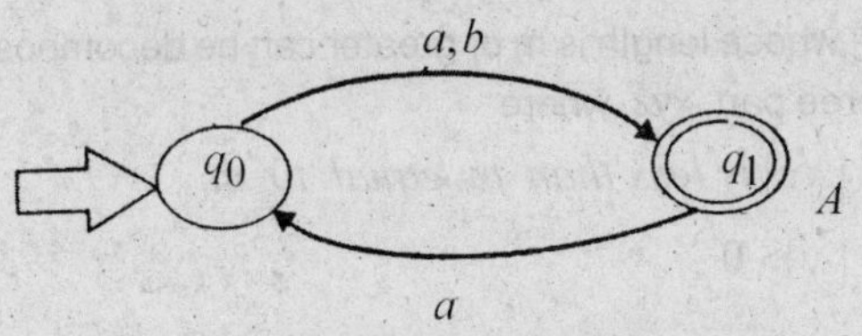

(a) $\in +(a+b)(ab+aa)^*$

(b) $\in +(a+b)(ab+aa)^* a$

(c) $(a+b)(ab+aa)^* a$

(d) $(a+b)(ab+aa)^*$

68.The $R.E. = (10+01+11+00)^*$ represents

(a) All strings over 0 and 1
(b) set of strings with equal 0 and 1's
(c) set of strings with even length
(d)set of strings with at least one 0 and at least one 1

69.Which one is TRUE

(a) $(10)^* = 1^* 0^*$

(b) $(10)^* = (1^* 0^*)^*$

(c) $(10)^* = 1^* + 0^*$

(d) $\{1010\}$ $belongs\ to\ (10)^*$

70. $(1+0)^*$ represents

(a) Set of strings with even number of 1's and 0's
(b) Set of strings with equal number of 1's and 0's
(c) Set of strings starting with 1 and ending with 0
(d) Set of strings over 1 and 0.

71.Which of the following regular expression over $\{0,1\}$ denotes the set of all string not containing 100 as a substring?

(a) $0^*(1^*0)^*$ (b) 0^*1010^*

(c) 0^*101^* (d) $0^*(10+1)^*$

72.The intersection of the two regular languages below:

$L_1 \equiv (a+b)^* a$ and $L_2 = b(a+b)^*$ is given by

(a) $b(a+b)^* a$ (b) $(a+b)^* ab(a+b)^*$

(c) $a(a+b)^* b$ (d) $ab(a+b)^*$

73.The recognizing capability of NDFA and DFA
(a) may be different (b) must be different
(c) must be same (d) none of the above.

74.Which of the following is regular?
(a) String of 0's whose length is a perfect square
(b) String of all palindromes made up of 0's and 1's
(c) Strings of 0's, whose length is a prime number
(d) String of odd number of zeroes.

75.The regular expression $(1+00^*1)+(1+00^*1)\ (0+0+10^*1)^*\ (0+10^*1)$ is equivalent to

(a) $0^*1(0+10^*1)^*$

(b) $(1+00^*1)(0+10^*1)^*$

(c) $(1+00^*1)(0^*(10^*1)^*)^*$

(d) All of the above.

76.Regular expression corresponding to the finite automaton drawn below is given by

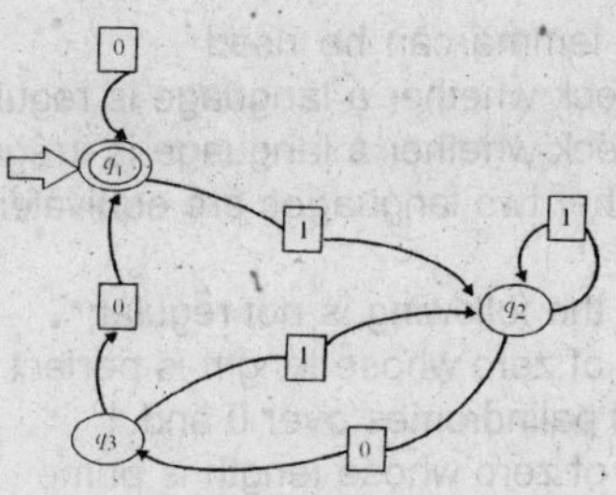

(a) $(0+1(1+01)^*00)^*$

(b) $(0+1(0+10)^*00)^*$

(c) $(0+0(1+01)^*00)^*$

(d) $(0+0(1+10)^*00)^*$

77.Which of the following closure properties hold for regular sets?

(i) If L is regular, then L^T is also regular.

(ii) If L is regular, set over Σ, then $\Sigma^* - L$ is also regular over $\cap$.

(iii) If X and Y are regular, sets over Σ, then X intersection Y is also regular over Σ.

(a) Only (i) (b) Only (ii)
(c) Only (i) and (iii) (d) Only (i), (ii) and (iii).

78. Regular expression corresponding to the FA given below is

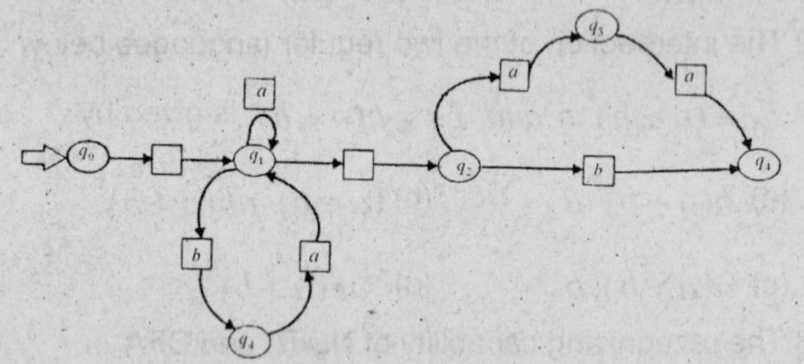

(a) $(ab+a)^*(aa+b)$ (b) $(a^*b+b^*a)^*$

(c) $a^*+(ab+a)^*$ (d) None of the above.

79.Set of all strings over $\{0,1\}$ having atmost one pair of 0's or atmost one pair of 1's.

(a) $(1+01)^*+(1+01)^*00(1+01)^*+(0+10)^*+(0+10)^*11(0+10)^*$

(b) $(1+01)^*+(0+10)^*00(0+10)^*+(0+10)^*+(1+01)^*11(1+01)^*$

(c) $(1^*+(01)^*)^*+(1^*)(01)^*00(1^*(01)^*)^*+(0^*)(10)^*+((0^*)+(10)^*)^*11(0+10)^*$

(d) None of the above.

80.The set all strings over $\{a,b\}$ in which there are atleast two occurrences of b between the two occurrences of a.

(a) $b^*(b+ab)^*b^*$ (b) $b^*+(b+abb)^*ab^*$

(c) $a(bbabb)^*a$ (d) None of the above.

81.Pumping lemma can be used
(a) To check whether a language is regular
(b) To check whether a language is irregular
(c) Whether two languages are equivalent
(d) None.

82.Which of the following is not regular
(a) String of zero whose length is perfect square
(b) Set of palindromes over 0 and 1
(c) String of zero whose length is prime
(d) All

83.Pick the correct statement. The logic of Pumping lemma is a good example of
(a) The Pigeon hole principle
(b) The divide and conquer technique
(c) Recursion
(d) Iteration

84.Finite state machines....... can recognize palindromes
(a) can (b) can't
(c) may (d) may not

85.If L is regular, then {x: reverse (x) in L} is also regular
(a) Yes (b) No
(c) May or may not be (d) None of the above.

86.Suppose h is the homomorphism from the alphabet {0,1,2} to the alphabet {a,b} defined by: $h(0)=a;\ h(1)=ab,\ and\ h(2)=ba$ .What is h(0120)?
(a) aabba (b) aaabb
(c) abbbb (d) ababa

87. The inverse homomorphism of a regular language is :
(a) regular (b) not regular
(c) not regular (d) none

A homomorphism is a function from some alphabet Σ_1 to strings in another alphabet Σ_2. If

$x=a_1a_2,...a_k \in \Sigma_1^*$, $then\ h(x)=h(a_1)h(a_2)...h(a_k), and if\ L \subseteq \Sigma^*$,

then $h(L)=\{h(x)/x \in L\}$.

88. If L is the language $L(01^*2)$, what is h(L):

(a) $a(ab)^*ba$ (b) aab^*ba

(c) $aab(ba)^*$ (d) aba^*ab

89. If L is an infinite regular language, then there exist some positive integer m such that any string $w \in L$ whose length is m or greater can be decomposed into three part, xyz, where

(a) $|xy|$ *is less than or equal to m.*

(b) $|y|>0$

(c) $w=xy^iz$ is also in L for all i = 0,1,2,3,...

(d) All of the above.

90. Which of the following languages cannot be produced by a regular grammar?

(i) $\{a^nb^n : n \geq 0\}$ (ii) $\{a^mb^k : k > n \geq 0\}$

(iii) $\{ww^R : w \in \{a,b\}^*\}$

(a) (i) (b) (i) and (ii)
(c) (ii) and (iii) (d) All of the above.

91.Which of the following sets is regular?

(a) $\{a^{i^2} \mid i \geq 1\}$ (b) $\{a^p \mid p\ is\ a\ prime\}$

(c) $\{ww \mid w \in \{a,b\}^*\}$ (d) $\{a^{2n} \mid n \geq 1\}$

93. Is there a finite automation which accepts all palindromes over {a,b}?
(a) Yes,it can be proved
(b) Yes, but it cannot be proved
(c) No, it can be proved.
(d) No,but it cannot be proved.

94.Let $\Sigma = \{a\}$ and let I be the identify relation on Σ^*.

Let $L = \{\in\} \cup \{a\} \cup \{aa\}$. Then index (I) is

(a) finite (b) infinite
(c) 3 (d) None.

95. Let $\Sigma = \{0,1\}$ and R be the relation defined on Σ^* by $(x,y) \in R$ iff $|x|-|y| = odd$ Then R is

(a) an equivalence relation
(b) a right congruences
(c) not a right congruence
(d) none.

96. Let L be a regular language defined are Σ^*. Then

(a) index (R_L) is finite
(b) index (R_L) may be infinite
(c) index (R_L) may be zero
(d) None.

ANSWER KEY

1. (b)	2. (c)	3. (a)	4. (b)	5. (d)	6. (c)
7. (a)	8. (c)	9.(d)	10.(d)	11.(a)	12.(c)
13.(c)	14.(b)	15.(d)	16.(b)	17.(d)	18.(a)
19.(c)	20.(c)	21.(a)	22.(a)	23.(b)	24.(b)
25(b)	26.(d)	27.(d)	28.(d)	29.(c)	30.(c)
31.(a)	32.(d)	33.(b)	34.(c)	35.(a)	36.(a)
37.(c)	38.(d)	39.(c)	40.(a)	41.(b)	42.(b)
43.(a)	44.(c)	45.(c)	46.(d)	47.(d)	48.(b)
49.(a)	50.(b)	51.(c)	52.(b)	53.(a)	54.(b)
55.(a)	56.(a)	57.(a)	58.(c)	59.(a)	60.(c)
61.(c)	62.(a)	63.(c)&(d)		64.(a)	65.(b)
66.(a)	67.(d)	68.(c)	69.(d)	70.(d)	71.(d)
72.(a)	73.(c)	74.(d)	75.(a)	76.(a)	77.(d)
78.(a)	79.(a)	80.(b)	81.(b)	82.(d)	83.(a)
84.(b)	85.(a)	86.(a)	87.(a)	88.(a)	89.(d)
90.(d)	91.(d)	93.(c)	94.(b)	95.(c)	96.(a)

REGULAR GRAMMARS

1. Which of the following is true?

(a) L_{RL} is subset of L_{CFL}

(b) L_0 is subset of L_{CSL}

(c) L_0 is subset of L_{CFL}.

(d) None of the above.

2. Which of the following is true?

(a) $L_0 \subseteq L_{CSL} \subseteq L_{CFL} \subseteq L_{RL}$

(b) $L_{RL} \subseteq L_{CFL} \subseteq L_{CSL} \subseteq L_0$

(c) $L_{RL} \subseteq L_{CSL} \subseteq L_{CFL} \subseteq L_0$

(d) $L_0 \subseteq L_{CFF} \subseteq L_{CSL} \subseteq L_{RL}$

3. Which of the following statements is true?

(a) $L_{RE} \subset L_{RC} \subset L_{CSL}$

(b) $L_{RE} \supset L_{RC} \supset L_{CSL}$

(c) $L_{RE} \supset L_{RC} \subset L_{CSL}$

(d) $L_{RE} \subset L_{RC} \supset L_{CSL}$

4, Which of the following relationship holds?

(a) $L_{CFL} \subseteq L_{DCS} \subseteq L_{RL}$

(b) $L_{CF} \supseteq L_{DC} \supseteq L_{RL}$

(c) $L_{CF} \supset L_{DC} \subset L_{RL}$

(d) $L_{CF} \subset L_{DC} \supset L_{RL}$

5. Chomsky hierarchy from type 0 to type 3 is:

(a) $L_{RL}, L_{CFL}, L_{CSL}, L_{RE}$

(b) $L_{RE}, L_{CSL}, L_{CFL}, L_{RL}$

(c) $L_{RE}, L_{CFL}, L_{CSL}, L_{RL}$

(d) $L_{RL}, L_{CSL}, L_{CF}, L_{RE}$

6. The set $\{a^n b^n\}$ can be generated by the CFG

(a) $S \to ab \mid aSb$ (b) $S \to aaSbb$

(c) $S \to ab \mid asB \mid E$ (d) $S \to aaSbb$

7. The language $L\{0^n 1^n 2^k 3^k\}$ is a

(a) CFL
(b) CSL
(c) Regular language
(d) Recursively enumerable language.

8. The grammar generated by production rule $S \to aCa, C \to aCab$ is,

(a) $a^n b a^n, n > 0$ (b) $a^n a^n, n \geq 0$

(c) $a^n b a^n, n > 0$ (d) $a^n a^n, n > 0$

9. The grammar generated by production rules $S \to aSBC|abc.\ cB \to Bc.aB \to aB \to aa$ is

(a) $a^n b^n c^n, n > 1$ (b) $a^n b^n c^n, n \geq 0$

(c) $a^n b^n c^n, n > 0$ (d) $a^n b^n c^n, n \leq 0$

10. The grammar having production as $A \to B$, where $A \in B, B \in (V \cup \Sigma)$ is

(a) Type 0 (b) Type 1
(c) Type 2 (d) Type 3

11. Which of the following grammars can generate $w = aabbb$

(a) $S \to AB, A \to aA \mid u, B \to bB \mid b$

(b) $S \to AB, A \to BB \mid a, B \to AB \mid b$

(c) Both
(d) None.

12. Let G be the grammar $S \to aA, A \to A\ bb \mid b$, sentential forms of G are,

(a) $aAb^{2n}, ab^{2n+1}\ where\ , n \geq 0$

(b) $aAb^{2n}, ab^{2n+1}\ where, n \geq 1$

(c) $aAb^{2n}, ab^{2n} where, n \geq 0$

(d) None of above.

13. The grammar that generates $L = \{wew^{t} \mid w \in \{a,b\}^{*}\}\ is,$

(a) $S \to aSa \mid bSb \mid aca \mid bcb$

(b) $S \to aSa \mid bSb \mid c$

(c) Any one of the two
(d) None of the two.

14. The grammar that generates $L = \{a^n b^n c^i \mid n \geq 1, i \geq 0\}$ is,

(a) $S \to \in \mid Sc, \in ab \mid a \in b$

(b) $S \to a \in b \mid Sc, \in \to ab / \in b$

(c) Any one of the two
(d) None of the two.

15. A grammar $G = \langle V, T, S, P \rangle$ is called unrestricted if all the productions are of the form $u \to v$, where,

(a) $u \in (V \cup T)'\ and\ v \in (V \cup T)$

(b) $u \in (V \cup T)^{+}\ and\ v \in (V \cup T)^{*}$

(c) $u \in (V \cup T)^{*}\ and\ v \in (V \cup T)'$

(d) $u \in (V \cup T)\ and\ v \in (V \cup T)^{+}$

16. A grammar $G = \langle V, T, S, P \rangle$ is said to be context sensitive if all Productions are the form $x \to y$, where $x, y \in (V \cup T)^{+}$ and

(a) $|x| \leq |y|$ (b) $|x| \geq |y|$

(c) $|x| = |y|$ (d) $|x| \geq |y|$

17. The grammar having production as $A \to xB$, $Where\ A \in V$, $\mathbf{B} \in (V \cup \in), x \in \Sigma^{*}$ is

(a) Type 0 (b) Type 1
(c) Type 2 (d) Type 3

18. The grammar having production as $A \to B,\ Where\ A \in (V \cup \Sigma)^{+}, \mathbf{B} \in (V \cup \Sigma)^{*}$ is

(a) Type 0 (b) Type 1
(c) Type 2 (d) Type 3

19. Choose the correct statements
(a) all languages can be generated by CFG
(b) Any regular language has not an equivalent CFG
(c) Some non regular languages can't be generated by an CFG
(d) some regular languages can't be generated by an CFG.

20. The set $L = \{a^n b^n c^n\}$ is an example of grammar that is

(a) regular (b) context free
(c) not context free (d) none.

21. The productions: $E \to E + E, E \to E - E, E \to E * E, E \to id$
(a) Generated an inherently ambiguous language
(b) Generate an ambiguous language
(c) Are unambiguous
(d) None.

22. What is the highest type number to grammar given by these production rules $S \to ASB \mid d, A \to aA.$

(a) Zero (b) One
(c) Two (d) Three

23. A production of the form $A \to \alpha$ where $A \in V\ and\ \alpha \in (V \cup \Sigma)^{*}$ represents which type of grammar?

(a) Zero (b) One
(c) Two (d) Three

24. If we have an algorithm to determine whether a given element belongs to a set X or not, then this set is called
(a) Recursive (b) Recursively Enumerable
(c) Complete (d) Context-sensitive

25. Which of the following is true for the language generated by $S \to AB, A \to BB \mid a, B \to AB \mid b$

(a) aabbb doesn't belongs to this language
(b) aabb belongs to this language
(c) ab doesn't belongs to this language
(d) aab belongs to this language.

26. Which of the following grammars can generate $w = aabbb$

(a) $S \to AB, A \to aA \mid a, B \to bB \mid b$

(b) $S \to AB, A \to BB \mid a, B \to AB \mid b$

(c) Both

(d) None.

27. A recursive lenguage is also:

(a) regular (b) recursive enumerable

(c) CFL (d) deterministic

28. A grammar is context-sensitive if all production of the form $x > y$ are such that

(a) $|x| = |y|$ (b) $|x| \leq |y|$

(c) $|x| < |y|$ (d) $|x| > |y|$

29. Which of the following language is regular?

(a) $a^n b^n$ (b) $a^n b^n c^n$

(c) $a^n b^n$ (d) phi

30. Which of the following language is context free?

(a) $a^n b^n c^n$ (b) $a^n b^n$

(c) a (d) $a^n b^i : n = j^2$

31. Type 2 grammar is also called

(a) Context-sensitive (b) regular

(c) context-free (d) none of these.

32. Find language generated by the following grammar $S \to 0S1 \mid 0A1 A \to 1A0 \mid 10$

(a) $\{0^n 1^n 0^n \mid n > 1\}$

(b) $\{1^m 0^n \mid m, n \geq 1\}$

(c) $\{0^n 1^m 0^m 1^n \mid m, n \geq 1\}$

(d) None of the above.

33. Chomsky Hierarchy is representation of

(a) Grammars (b) Machines

(c) Parsers (d) None of these

34. Consider the grammar: $S \to aAB \mid abc,\ A \to aa \mid eB \to b$ then equivalent representation is

(a) $S \to aAB \mid abc,\ A \to aa, B \to b$

(b) $S \to aaaB \mid abc,\ A \to aa, B \to b$

(c) $S \to aaaB \mid aB \mid abc,\ A \to aa, B \to b$

(d) $S \to aB \mid abc,\ A \to aa, B \to b$

35. Language generated by the grammar: $S \to 0A \mid 1s \mid 0 \mid 1, a \to 1A \mid 1S \mid 1\ is$

(a) all strings of 0's and 1's

(b) all strings of 0's and 1's that does no contain 2 consecutive 0's

(c) all strings of 0's and 1's that does no contain 2 consecutive 1's

(d) all strings of 0's and 1's that neither contain 2 consecutive 0's nor contains two consecutive 0's.

36. Which of the following can be generated by: $S \to aS \mid bA, a \to d \mid ccA$

(a) bccddd (b) abbbd

(c) aabccd (d) ababccd

37. The correct relationship is given by

(a) CF < right linear < context sensitive

(b) CF < context sensitive < right linear

(c) context sensitive < CF < right linear

(d) right linear < CF < context sensitive.

38. A language *L* is accepted by a finite automaton if and only if

(a) right linear (b) primitive recursive

(c) context sensitive (d) recursive.

39. A grammar $G = (V, T, S, P)$ is right linear if

(a) $A \to x\mathrm{B} \mid x; A, B \in V\ and\ x \in T^*$

(b) CF < context sensitive < right linear

(c) $A \to x\mathrm{B} \mid x; A, B \in V\ and\ x \in T^*$

(d) None of these

40. A regular grammar is

(a) Both left linear and ritht linear

(b) Right linear and not left linear

(c) Neither left linear nor right linear

(d) Either left linear or right linear

41. Let $G_1 = (V_1, \Sigma, S_1, P_1)$ be right linear and $G_2 = (V_2, \Sigma, S_2, P_2)$ be left linear grammar and assume that V_1 and V_2 are disjoint. Consider, the linear grammar

$G = (\{s\} \cup V_1 \cup V_2, \Sigma, S, P)$, where S is not in $V_1 \cup V_2$

and $P = \{s \to s_1 \mid s_2\} \cup P_1 \cup P_2,\ then\ L(G)\ is$

(a) right linear (b) left linear

(c) regular (d) not regular.

42. $L(G) = \{ww^R : w \in \{a, b\}^*\}\ is$

(a) regular (b) context free

(c) not context free (d) none of these.

43. What is the highest type number to the grammar given by these production rules: $S \to Aa;\ A \to c \mid Ba,\ B \to abc$.

(a) Zero (b) One (c) Two (d) Three.

44. What is the highest type number to grammar given by these production rules: $S \to aS \mid ab$.

(a) Zero (b) One (c) Two (d) Three.

45. A production of the form $A \to a$ or $A \to aB$ represents which type of grammar?

(a) Zero (b) One (c) Two (d) Three.

46.Find language generated by the following grammar: $S \to 0A|1S|1|0 \quad A \to 1A|1S|1$

(a) $\{x \in \{0,1\}^* \mid x$ does not contain any two consecutive zeroes.$\}$

(b) $\{x \in \{0,1\}^*\}$

(c) $\{x \in \{0,1\}^* \mid x$ does not contain any two consecutive zeroes$\}$

(d) None of the above.

47.Give the regular grammar that generates the following set: $\{a^{2n} \mid n \geq 1\}$

(a) $S \to aS_1, S_1 \to aS, S \to aS_2, S_2 \to a$

(b) $S \to aS, S \to a$

(c) $S \to aS_1S_2, S_1 \to aS, S_2 \to a$

(d) None of the above.

48.Give the regular grammar that generates the following set: $\{a^l b^m c^n \mid l, m, n\ n \geq 1\}$

(a) $S \to aS_1, S_1 \to aS_1, S_1 \to bS_2, S_2 \to bS_2, S_2 \to cS_3, S_3 \to c$

(b) $S \cup aSS_1, S_1 \cup aS_1S_2, S_1 \cup bS_2, S_2 \cup bS_2, S_2 \cup cS_3, S_3 \cup c$

(c) $S \to aS_1S_2, S_1 \to aS_1, S_1 \to bS_2, S_2 \to bS_2, S_2 \to cS_2S_3, S_3 \to c$

(d) None of the above.

49.Give the regular grammar that generates the following set : $\{(ab)^n \mid n \geq 1\}$

(a) $S \to aS_1, S_1 \to bS, S \to aS_2, S_2 \to b$

(b) $S \to aS_1S_2, S_1 \to bS, S_2 \to b$

(c) $S \to S_1S_2, S_1 \to bS, S \to aS_2, S_2 \to b$

(d) None of the above.

50.Match the language with the corresponding machine:

Language	Machine
(i) Regular language	(A) Non-deterministic pushdown automaton
(ii) DCFL	(B) Turing machine
(iii) CFL	(C) Deterministic pushdown automaton
(iv) Context-sensitive language	(D) (Non)Deterministic finite-state acceptor
(v) Recursive language	(E) Turing machine that halts
(vi) Recursively enumerable language	(F) Linear-bounded automaton

(a) DCFFAB (b) DCABEF
(c) DCAFEB (d) DACFEB

51.The language generated by the grammar $S \to 0s1 | 0A1 |, A \to 1A | 1$ is

(a) $\{0^m 1^n : m > n > 1\}$ (b) $\{0^m 1^n : m > n > 1\}$

(c) $\{0^m 1^n : n > m \geq 1\}$ (d) $\{0^m 1^n : n > m > 1\}$

52.The language generated by the grammar $S \to 0S1 | 0A1, A \to 1A0 | 10$ is

(a) $\{0^n 1^m 0^m 1^n : m, n > 1\}$

(b) $\{0^n 1^m 0^m 1^n : m, n \geq 1\}$

(c) $\{0^m 1^n : n, m \geq 1\}$

(d) $\{1^m 0^n : n, m \geq 1\}$

53.Find the false statement for k-equivalent states

(a) The k-equivalence is an equivalence relation.

(b) Two states q_1 and q_2 are k-equivalent if both $\delta(q_1, x)$ and $\delta(q_2, x)$ are final states or both non-final states for all strings x of length k or more.

(c) If q_1 and q_2 are $(k+1)$-equivalent then they are k-equivalent.

(d) If q_1 and q_2 are k-equivalent for all $k \geq 0$, then they are equivalent.

54.What language is generated by the following grammar $G = \langle S, \in, B\}, \{a, b\}, P, S \rangle$ where P is :

$S \to aB | bA$

$A \to a | as | bAA$

$B \to b | bS | aBB$

(a) All words consisting of equal numbers of a's and b's.

(b) All words consisting of one a more than the number of b's

(c)All words consisting of one b more than the number of a's.

(d) None of the above

55.When Minimizing a $DFA\ M = \langle Q, \Sigma, \delta, q_0, F \rangle$ we try to minimize

(a) The length of the strings accepted by M.

(b) The out-degree of each vertex.

(c) The cardinality of Q

(d) The cardinality of the language L such that $L = L(M)$

(e) None of the above.

ANSWER KEY

1.(a)	2.(b)	3.(a)	4.(b)	5.(b)	6.(a)
7.(a)	8.(c)	9.(c)	10.(c)	11.(a)	12.(a)
13.(b)	14.(a)	15.(b)	16.(a)	17.(d)	18.(a)
19.(c)	20.(b)	21.(c)	22.(c)	23.(c)	24.(a)
25.(d)	26.(c)	27.(b)	28.(b)	29.(d)	30.(b)
31.(c)	32.(c)	33.(a)	34.(c)	35.(b)	36.(c)
37.(d)	38.(a)	39.(a)	40.(d)	41.(c)	42.(b)
43.(c)	44.(c)	45.(d)	46.(a)	47.(a)	48.(c)
49.(a)	50.(c)	51.(c)	52.(b)	53.(b)	54.(a)
55.(c)					

CONTEXT FREE LANGUAGES

1. A Grammar G has productions of type $A \to aB$ $A \to Ba$ $A \to a$ then G is

(a) regular (b) CF
(c) both (d) none

2. $L=\{ww^R$ where w belongs to $\{0,1\}^*\}$ is

(a) Context sensitive (b) Context Free
(c) both (d) None

3. A grammar G is known to have GNf representation then

(a) we can write G as left linear or right linear
(b) G may be written as left linear or right linear
(c) G can't be written left or right linear
(d) None of these.

4. Consider the language $L=\{a^p : p \text{ is prime}\}$

(a) RL (b) CFL (c) both (d) None

5. Every CFG is ambiguous

(a) true (b) false
(c) some times true (d) some times false

6. Which of the following languages cannot be produced by a Context Free Grammar?

(a) $L=\{a^n b^n : n \geq 0\}$

(b) $L=\{a^n b^k : k > n \geq 0\}$

(c) $L=\{ww^R : w \in \{a, b\}^*\}$

d) $L=\{a^n b^n c^n : n \geq 0\}$

7. The intersection of a CFL and regular language

(a) need not be regular
(b) need not be context free
(c) Is always regular
d) Is always context free

8. CFG is not closed under

(a) Union (b) Kleene star
(c) complementation (d) product

9. The language $L=\{a^n b^m c^n d^m / n \geq 1, m \geq 1\}$

(a) is context free
(b) is not context free
(c) abstracts the problem of checking number of formal and external parameters
(d) (b) and (c)

10. Let two kinds of grammars be defined as follows.

(I) $A \to a$, $A \to BC$

(II) $A \to a\alpha$, where $A, B, C \in V$, $A \in V$ and $\alpha \in V^*$

(a) II denotes Chomsky Normal form and I denotes denote Greibach Normal Form
(b) II denotes Chomsky Normal form but I deos not denote Greibach Normal Form.
(c) I denotes Chomsky Normal form and II denotes Greibach Normal Form.
(d) I denotes Chomsky Normal form but II does not denote Greibach Normal Form.

11. A context free grammar which generates $\{a^l b^m c^n \mid l+m=n\}$ has its production rules given by

(a) $S \to bSc \mid aSc \mid ac \mid a \mid bS_1, S_1 \to S_1 c / c$

(b) $S \to aSc \mid ac \mid bc \mid bS_1 c, S_1 \to bS_1 c / bc$

(c) $S \to cSb, cS \to Sc, bS \to Sb, S \to aS_1 c, S_1 \to aS_1 c, S_1 \to ac$

(d) $S \to aS, S \to aS_1, S_1 \to bS_1, S_1 \to bS_2, S_2 \to bS_3, S_3 \to cS_3 \mid c$

12. Which of the following grammars is ambiguous?

(a) $S \to a \mid abSb \mid aAb, A \to bS \mid aAAb$

(b) $S \to aB \mid ab, A \to aAB \mid a, B \to ABb \mid b$

(c) $S \to S+S \mid S*S \mid a \mid b$

(d) all of the above

13. What is the correct sequence of steps to construct an equivalent reduced grammar?

(a) Construct equivalent Grammar G_1 such that each variable derives some terminal string.
(b) Construct equivalent Grammar G' such that each symbol in G' appears in some sentential form.
(c) Eliminate null productions.
(d) Eliminate unit productions.

(i) C D B A (ii) C D B A
(iii) A B C D (iv) B A D C

14. Let L be a context free language. Then for every $z \in L$, and some strings u, v, w, x and $y \in L$ Then we can find a natural number n such that

(a) with $|z| \geq n$ can be written as uvwxy.

(b) $|vx| \geq 1.\ n \geq |vwx|$

(c) $uv^k wx^k y \in L$ for all $k = 0$

(d) all of the above

15. Which of the following languages is context free?

(a) $L=\{a^m b^m c^n : m \leq n \leq 2^m\}$

(b) $L=\{a^m b^n : n=m^2\}$

(c) $L=\{a^{n^2} : 1 \leq n\}$

(d) None of the above

16. The grammar generated by production rules $S \to aSBc \mid abc, cB \to Bc, aB \to aa$ is

(a) $a^n b^n c^n\ n > 1$ (b) $a^n b^n c^n, n \geq 0$

(c) $a^n b^n c^n, n > 0$ (d) $a^n b^n c^n, n \leq 0$

17. The grammar having production as $A \to B$, where $A \in V$, $B \in (V \cup \Sigma)^+$ is

(a) Type 0 (b) Type 1
(c) Type 2 (d) Type 3

18. A CFL is in CNF if every production is of form $S \to A$ or $S \to C$ where S is in V

(a) A is in Σ^* & B, C are in V

(b) A is in Σ & B, C are in V^*

(c) A is in Σ & B, C are in V

(d) A is in Σ^* & B, C are in V^*

19. Which of the following grammars can generate $w = aabbb$

(a) $S \to AB, A \to aA \mid a, B \to bB \mid b$

(b) $S \to AB, A \to BB \mid a, B \to AB \mid b$

(c) Both

(d) None

20. Let G be the grammar $S \to aA, A \to Abb \mid b$, sentential forms of G are,

(a) aAb^{2n}, ab^{2n+1} where, $n \geq 0$

(b) aAb^{2n}, ab^{2n+1} where, $n \geq 1$

(c) aAb^{2n}, ab^{2n} where, $n \geq 0$

(d) None of above

21. The grammar that generates $L = \{wcw^t \mid w \in \{a, b\}*\}$ is,

(a) $S \to aSa \mid bSb \mid aca \mid bcb$

(b) $S \to aSa \mid bSb \mid c$

(c) Any one of the two

(d) None of the two.

22. The grammar that generates $L = \{a^n b^n c^i \mid n = 1, i = 0\}$ is,

(a) $S \to A \mid Sc, A \to ab \mid aAb$

(b) $S \to aAb \mid Sc, A \to ab \mid aAb$

(c) Any one of the two

(d) None of the two.

23. Match the language with the corresponding machine:

Language	Machine
(i) Regular language	(A) Non deterministic pushdown automaton
(ii) DCFL	(B) Turing machine
(iii) CFL	(C) Deterministic pushdown automaton
(iv) context-sensitive language	(D) (Non) Deterministic finite-state acceptor
(v) Recursive language	(E) Turing machine that halts
(vi) Recursively enumerable language	(F) Linear-bounded automaton

(a) D C F E A B (b) D C A B E F

(c) D C A F E B (d) D A C F E B

24. Which of the following statements is true?

(i) A regular set accepted by a deterministic finite automaton with n states is accepted to final state by a deterministic pda with n states and one pushdown symbol.

(ii) Every regular set accepted by a finite automaton with n states is accepted by a deterministic pda with one state and n pushdown symbols

(iii) If L is accepted by a deterministic pda A, then it can be shown that L is accepted by a deterministic pda A which never adds more than one symbol at a time.

(a) (i) and (ii) only (b) (ii) and (iii) only

(c) (i) and (iii) only (d) (i), (ii) and (iii)

25. The grammar $\{S \to as \mid bS \mid a/b\}$ is not equivalent to

(a) $(a^* + b)^*$ (b) $(a^* + b)^*$

(c) $(a+b)(a+b)^*$ (d) $(a+b)^*(a+b)$

26. Any string of terminals that can be generated by the following CFG

$$S \to XY$$
$$X \to aX \mid bX \mid a$$
$$Y \to Ya \mid Yb \mid a$$

(a) Has at least one b

(b) Should end in a

(c) Has no consecutive a or b

(d) Has atleast two a

28. Choose the correct statements :

(a) all languages can be generated by CFG

(b) any regular language has not an equivalent CFG

(c) some non regular languages can't be generated by an CFG

(d) some regular languages can't be generated by an CFG

29. CFG is not closed under

(a) union (b) Kleene star

(c) Complement (d) Product

30. Which of the following is true

(a) $S \to aSbS \mid bSaS \mid \in$ is unambiguous

(b) $S \to aB \mid ab, A \to aAB \mid a, B \to ABb \mid b$ is ambiguous

(c) $S \to AB \mid aaB, A \to a \mid Aa, B \to b$ is unambiguous

(d) $S \to aB, B \to Bb \mid b$ is ambiguous

31. The CFG for the following language :

The set $\{0^n 1^n \mid n > 1\}$, that is the set of all strings of one or more 0's followed by an equal number of 1's is :

(a) $S \to 0S1 \mid 01$

(b) $S \to 0S1 \mid 01$

(c) $S \to 0S1 \mid 01 \mid$ epsilon

(d) $S \to 0S1 \mid 0 \mid$ epsilon.

32.Let $T=\{0,1,(,),+*,$ phi, e}. We may think of T as set of symbols used regular expressions over alphabet $\{0,1\}$; the only different is that we use e for symbol epsilon, to avoid potential confusion in what follows. The CFG with set of terminals that generates exactly the regular expressions with a alphabet $\{0,1\}$ is :

(a) $S \to S+S \mid SS^* \mid S^* \mid (S) \mid 0 \mid 1 \mid \varphi \mid e$

(b) $S - S+S \mid SS \mid S \mid (S) \mid 0 \mid 1 \mid phi \mid e$

(c) $S \to S+S \mid SS \mid S^* \mid (S) \mid 0 \mid 1 \mid phi \mid e$

(d) none is possible

ANSWER KEY

1.(c)	2.(c)	3.(b)	4.(d)	5.(b)	6.(c)
7.(d)	8.(c)	9.(d)	10.(c)	11.(b)	12.(d)
13.(a)	14.(b)	15.(d)	16.(c)	17.(c)	18.(c)
19.(c)	20.(a)	21.(b)	22.(a)	23.(c)	24.(d)
25.(b)	26.(d)	28.(c)	29.(c)	30.(b)	31.(a)
32.(c)					

PUSH DOWN AUTOMATA

1. Let Ld be the set of all languages accepted by a PDA by final states and Le the set of all languages accepted by empty stack. Which of the following is true?

(a) $L_d = L_e$ (b) $L_d \supset L_e$

(c) $L_d \supset L_e$ (d) None of the above.

2. A pda is said to be deterministic if -

(a) $\delta(q,a,Z)$ is either empty or singleton or $\delta(q,\in,Z)=\phi$ implies $\delta(q,a,Z)\neq\phi$

(b) $\delta(q,a,Z)$ is either empty or singleton & $\delta(q,\in,Z)\neq\phi$ implies $\delta(q,a,Z)=\phi$

(c) $\delta(q,a,Z)$ is either empty or singleton & $\delta(q,\in,Z)=\phi$ implies $\delta(q,a,Z)=\phi$

(d) $\delta(q,a,Z)$ is either empty or singleton or $\delta(q,\in,Z)\neq\phi$ implies $\delta(q,a,Z)=\phi$

3. $A=(\{q_0,q_1,q_f)\},\{a,b,c\},\{a,b,Z_0\},\delta,q_0\ Z_0,\{q_f\})$is a PDA, where ϕ is defined as

$$\delta(q_0,a,Z_0)=\{(q_0,aZ_0)\},\ \delta(q_0,b,Z_0)=\{(q_0,bZ_0)\}$$
$$\delta(q_0,a,a)=\{(q_0,aa)\},\ \delta(q_0,b,a)=\{(q_0,ba)\}$$
$$\delta(q_0,a,b)=\{(q_0,ab)\},\ \delta(q_0,b,b)=\{(q_0,bb)\}$$
$$\delta(q_0,c,a)=\{(q_1,a)\},\ \delta(q_0,c,b)=\{(q_1,b)\},\ \delta(q_0,c,Z_0)=\{(q_1,Z_0)\}$$
$$\delta(q_1,a,a)=\delta(q_1,b,b)=\{(q_1,\in)\}$$
$$\delta(q_1,\in,Z_0)=\{(q_f,Z_0)\}$$

Match the ID that the PDA is in after the strings listed on the LHS are processed by the PDA.

(i) abcba	(A) $(q_0,\in,babaZ_0)$
(ii) abcb	(B) $(q_1,\in,aZ_0)$
(iii) acba	(C) (q_1,ba,aZ_0)
(iv) abab	(D) $(q_f,\in,Z_0)$

(a) A B C D (b) D B C A

(c) D C B A (d) D C A B

4. A PDA behaves like an FA when the number of auxiliary memory it has is

(a) 0 (b) 1 or more

(c) 2 or more (d) none

5. Choose the correct stagements :

(a) the power of DFA and NDFA are almost same

(b) the power of DFA and NDFA are different

(c) the power of NPDA and DPDA are same

(d) the power of NPDA and DPDA are same

6. A regular set accepted by DFA with n states is accepted to final state by a DPDA with n states and at least_____ pushdown symbols

(a) 1 (b) 2 (c) 3 (d) none

7. Every regular set accepted by FA with n states is accepted by a DPDA with______ states and n pushdown symbols
   (a) 1 (b) 2 (c) n (d) none
8. If L is N(M) for some PDA M, then L is a :
   (a) DCFL (b) CFL
   (c) RE (d) none
9. If L is N(M) for some PDA M, the L is a
   (a) Context sensitive language
   (b) Context free language
   (c) Regular grammar
   (d) none.
10. Let L_1 be the set of the languages accepted by a NPDA and L_2 be the set of context free languages. Then :
   (a) $L_1 = L_2$ (b) $L_1 \subseteq L_2$
   (c) $L_2 \subseteq L_1$ (d) none
11. Which of the following languages not is accepted by a NPDA
   (a) ww^r (b) ww
   (c) wcw^r (d) $a^n b^{2n}$
12. Which of the following is false :
   (a) Ever NPDA corresponds to a CFL
   (b) Every CFL corresponds to a NPDA
   (c) None
   (d) Both (a) and (b)
13. Running time of a finite automata (like NPDA) for an input string of length n is
   (a) Linear in n
   (b) Polynomial in n
   (c) Exponential in n
   (d) Can be anything depending on the automata
14. Given two PDAs M & M' :
   where $M = \langle S, \Sigma, \Gamma, \delta, q_0, Z_0, F \rangle$
   and $M' = \langle S', \Sigma, \Gamma, \delta', q'_0, Z_0', F \rangle$
   Which of the following conditions hold if $L(M) = L(M')$.
   (i) There is a one-to-one correspondence between accepting computations of M and M'
   (ii) If M has no $\in$ – moves, then M' has no $\in$ – moves ; If M is unambiguous, then M' is unambiguous.
   (iii) For all $p \in S'$, all $a \in \Sigma \cup \{\in\}$, all $Z \in \Gamma'$, if $(q, y) \in d\delta'(p, a, Z)$ then $q <> q_0'$ and $|y| =< 2$
   (a) Only (i) & (ii) (b) Only (i)
   (c) Only (ii). (d) All of the above.
15. Pushdown automata can recognize -----
   (a) all regular languages, some nonregular languages, all context-free languages.
   (b) all regular languages, all nonregular languages, all context-free languages.
   (c) all regular languages, some nonregular languages, all context-free languages, and some non-context- free languages.
   (d) all regular languges, some nonregular languages, all context-free languages, and all non-context -free languages.
16. If S is the number of states in NDFA then equivalent DFA can have maximum of
   (a) S states (b) S-1 states
   (c) 2^S states (d) $2^S - 1$ *states*
17. For Pushdown automatas, which of the following is true.
   (a) If $(q_1, x, \alpha) \overset{*}{|-} (q_2, \in, \beta)$, then for every $y \in \Sigma^*, (q_1, xy, a) \overset{*}{|-} (q_2, y, \beta)$
   (b) If $(q, x, a) \overset{*}{|-} (q', \in, \gamma)$, the for every $\beta \in \Gamma^*, (q, x, \alpha\beta) \overset{*}{|-} (q', \in, \gamma\beta)$
   (c) Only one out of a or b
   (d) Both a and b.
18. For Pushdown Automatas, which of the following is true.
   (a) If $(q_1, x, a)^* - (q_2, \in, \gamma\beta)$, then for every $y \in \Sigma^*, (q_1, xy, a) \overset{*}{|-} (q_2, y, \beta)$
   (b) If $(q, x, a) \overset{*}{|-} (q', \in, \gamma)$, the for every $\beta \in \Gamma^*, (q, x, \alpha\beta) \overset{*}{|-} (q', \in, \gamma\beta)$
   (c) Only one out of a or b
   (d) Both a and b.
19. Which of the following doesn't hold (in ID of pushdown automata) ?
   (a) If $(q, x, W) \to (t, \in, Z)$ then $(q, xy, W) \to (t, y, Z)$
   (b) If $(q, xy, W) \to (t, y, Z)$ then $(q, x, W) \to (t, \in, Z)$
   (c) If $(q, x, W) \to (t, \in, Z)$ then $(q, x, WU) \to (t, y, U)$
   (d) If $(q, x, WU) \to (t, \in, U)$ then $(q, x, W) \to (t, y, Z)$
20. If for a language L we can find a RE then definitely (choose most appropriate) :
   (a) we can draw NPDA for L
   (b) we can't draw NPDA for L
   (c) we can draw DPDA for L
   (d) none.
21. Which one of the following is true about PDA's
   (a) PDA has an auxillary memory in the form of a stack.
   (b) PDA is represented by 7-tuple $(S, \Sigma, \Gamma, \delta, q_0, Z_0, F)$
   (c) $\delta : S \times (\Sigma \cup \{\in\}) \times \Gamma \to S \times \Gamma^*$ (d) all of the above

22. Match the following :

(a) Type 0 grammar

(i) A grammar $G=(V,T,S,P)$ in which all productions in P are of the form $A \to V$ where $A \in V$ and $X \in (V \cup T)^*$

(b) Type 1 grammar

(ii) A grammar $G=(V,T,S,P)$ in which all productions in P are of the form $x \in y$ where $x, y \in (V \cup T)^+$ and $y \in (V \cup T)^*$ and $|x| \leq |y|$

(c) Type 2 grammar

(iii) A grammar $G=(V,T,S,P)$ in which all productions in P are of the form $x \in y$ where $x \in (V \cup T)^+$ and $y((V(T)^*$

(d) Type 3 grammar

(iv) A grammar $G=(V,T,S,P)$ in which all productions in P are of the form $A \in a$ and $A \in aB$ where $a \in T^*$ and $A, B \in V$

23. Which of the following is accepted by an NPDA and not DPDA

(a) All strings in which a given symbol is present atleast twice

(b) Even palindromes (i.e. palindromes made up if even no of symbols)

(c) String ending with a particular alphabet

(d) None

24. A pda is said to be deterministic if :

(a) $\delta(q,a,Z)$ is either empty or singleton or $(q,\in,Z)=\phi$ implies $\delta(q,a,Z)\neq\phi$

(b) $\delta(q,a,Z)$ is either empty or singleton & $\delta(q,\in,Z)\neq\phi$ implies $\delta(q,a,Z)=\phi$

(c) $\delta(q,a,Z)$ is either empty or singleton & $(q,\in,Z)=\phi$ implies $\delta(q,a,Z)\neq\phi$

(d) $\delta(q,a,Z)$ is either empty or singleton or & $(q,\in,Z)\neq\phi$ implies $\delta(q,a,Z)\neq\phi$

25. Which of the following statements is false

(a) A language is LR iff it can be accepted by a Deterministic PDA.

(b) A language is CFL iff it can be accepted by a Non deterministic PDA.

(c) Inherently ambiguous languages can be modeled LR grammars.

(d) The class of sets accepted by pushdowm automata properly includes the regular sets.

26. A PDA A is deterministic if :

(i) $\delta(q,\in,z)\neq\phi$ implies $\delta(q,a,z)=\phi \forall a \in Z$

(ii) $\delta(q,a,z)$ is

(a) empty (b) singleton

(c) either(a) or (b) (d) none

27. Which of the following can not be accepted by Deterministic PDA

(a) wcw^t

(b) $a^n b^n n \geq 1$

(c) $a^n b^n n \geq 1 \cup a^m b^{2m} m \geq 1$

(d) none

28. The language $\{ww^r\}$ is

(a) Accepted by a DPDA not by a NPDA

(b) Accepted by a NPDA not by a DPDA

(c) None

(d) Cannot say.

29. Let $A=(\{q_0,q_1,q_f\},\{a,b,c\},\{a,b,Z_0\},\delta,q_0,Z_0,\{q_f\})$ is a PDA, where ϕ is defined as

$$\delta(q_0,a,Z_0)=\{(q_0,aZ_0)\}, \delta(q_0,b,Z_0)=\{(q_0,bZ_0)\}$$

$$\delta(q_0,a,a)=\{(q_0,aa)\}, \delta(q_0,b,a)=\{(q_0,ba)\}$$

$$\delta(q_0,a,b)=\{(q_0,ab)\}, \delta(q_0,b,b)=\{(q_0,bb)\}$$

$$\delta(q_0,c,a)=\{(q_1,a)\}, \delta(q_0,c,b)=\{(q_1,b)\}, \delta(q_0 c,Z_0)=\{(q_1,Z_0)\}$$

$$\delta(q_1,a,a)=\delta(q_1,b,b)=\{(q_1,\in)\}$$

$$\delta(q_1,\in,Z_0)=\{(q_f,Z_0)\}$$

Then

(a) A is a NPDA (b) A is a DPDA

(c) A is NPDA but not DPDA (d) none

30. If L_1 & L_2 are two deterministic languages $L_1 \cup L_2$ is

(a) deterministic (b) CFL

(c) RL (d) None

31. $L_1.L_2$ is

(a) deterministic (b) CFL

(c) RL (d) None

32. Complement L_1 is :

(a) deterministic (b) CFL

(c) RL (d) None of these

33. For Deterministic Context Free languages, which of the following hold?

(i) The complement of a DCFL is a DCFL.

(ii) Let L be a DCFL and R is a regular set. Then L/R is a DCFL.

(iii) DCFL's are closed under intersection with a regular set.

(a) (i) and (ii)

(b) (i) and (iii)

(c) (i), (ii) and (iii)

(d) None of the above.

34. Which of the following is accepted by an NPDA and not DPDA :
(a) All strings in which a given symbol is present atleast twice
(b) Even palindromes (i.e. palindromes made up if even no of symbols)
(c) String ending with a particular alphabet
(d) None

35. Given two PDAs M & M' :
where $M = \langle Q, \Sigma, \Gamma, \delta, q_0, Z_0, F \rangle$ and
$M' = \langle Q', \Sigma, \Gamma', d', q_0', Z_0', F' \rangle$
Which of the following conditions hold if $L(M) = L(M')$.
(i) There is a one-to-one correspondance between accepting computations of M and M'
(ii) If M has no $\in-$moves, then M' has no $\in-$moves; If M is unambiguous, then M' is unambiguous.
(iii) For all $p \in Q'$, all $a \Sigma \cup \{\in\}$, all $Z \in \Gamma'$, if $(q, y) \in \delta'(p, a, Z)$, then $q <> q'0$ and $|y| =< 2$
(a) Only (i) & (ii)
(b) Only (i)
(c) Only (ii)
(d) All of the above.

ANSWER KEY

1.(c) 2.(b) 3.(b) 4.(a) 5.(a) 6.(a)
7.(a) 8.(b) 9.(b) 10.(a) 11.(b) 12.(d)
13.(a) 14.(a) 15.(a) 16.(c) 17.(d) 18.(d)
19.(d) 20.(c) 21.(d)
22.(a)-(c) (b) - (b), (c) -(a), (d) - (d)
23.(b) 24.(b) 25.(b) 26.(c) 27.(c) 28.(b)
29.(b) 30.(b) 31.(a) 32.(d) 33.(b)
34.(b) 35.(b)

TURING MACHINES

1. An Finite state machine can be considered to be a TM
(a) a finite tape length, with rewinding and unidirectional movement
(b) a finite tape length, without rewinding capability and unidirectional tape movement
(c) a finite tape length, without rewinding and bi-directional movement
(d) a finite tape length, rewinding capability and bi-directional tape movement.

2. A TM is more powerful than Finite state machine because
(a) tape movement is confined to one direction
(b) it has no finite state control
(c) it has the capability to remember arbitrary long input symbols
(d) none

3. Choose the correct statements
(a) $L\{a^n b^n c^n\}$ is recursively enumerable
(b) Recursive languages are closed under union
(c) Every recursive language is recursively enumerable
(d) All

4. Which of the following classes of Turing machine is not equivalent to the class of standard Turing machine?
(a) Turing machines with stay option
(b) Turing machines with semi-infinite tapes
(c) Non-deterministic Turing machines
(d) All of these

5. Find the false statement.
(a) We construct Turing machine to carry out some algorithm
(b) We construct Turing machine to accept a given language
(c) Turing machine is not capable to performing any calculation which can be performed by computer
(d) Turing machine is simple mathematical model of general purpose computer.

6. The no.of symbols necessary to simulate any TM with m symbols & n states is
(a) $m+n$ (b) $8mn+4m$
(c) mn (d) $4mn+m$

7. Any TM with m symbols & n states can be simulated by another TM with just 2 symbols & less than
(a) $8mn$ states (b) $4mn+8$ states
(c) $8mn+4$ states (d) mn states

8. The class of unrestricted languages corresponds to
(a) TM (b) LBA (c) PDA (d) FA

9. The class of TMs is equivalent to the class of
(a) Type 0 Grammars
(b) Type 1 Grammars
(c) Type 2 Grammars
(d) Type 3 Grammars

10. Comparing TM and computers we find
   (a) Computers have more computational power
   (b) TM have more computational power
   (c) Both are Equivalent
   (d) They cannot be compared

11. Turing machines are similar to finite automaton but have ------
   (a) finite and read-only memory.
   (b) unlimited and read-only memory
   (c) finite and read-write memory
   (d) unlimited and read-write memory

12. Turing machines can do---------
   (a) everything that a real computer can do
   (b) more than a real computer can do.
   (c) less than a real computer can do
   (d) Nothing

13. Turing machines use what as their memory?
   (a) ROM (b) RAM
   (c) finite tape (d) infinite tape

14. Turing machines can move how in memory?
   (a) forward
   (b) backward
   (c) forward and backward
   (d) It cannot move.

15. Turing machines output accept if they enter an accept state. When do Turing machine output reject?
   (a) When they are not in an accept state and halts
   (b) When they never end
   (c) When they enter a reject state
   (d) Never

16. In a Turing machine for a state q and two strings u and υ over the tape alphabet writing '$uq\upsilon$', specifies that the current state is q------
   (a) The tape contents are uqv, and the current head location is the first symbol of u.
   (b) the tape contents are uqv, and the current head location is the first symbol of v.
   (c) the tape contents are uv, and the current head location is the first symbol of u.
   (d) the tape contents are uv, and the current head location is the first symbol of v.

17. A Turing machine computes by going from one configuration to another. We say that configuration C_1 yields configuration C_2 if the Turing Machine can legally move from C_1 to C_2 in ------
   (a) a single step
   (b) a finite number of steps
   (c) an infinite number of steps
   (d) none of the above

18. The definition of Turing machines is robust because.....
   (a) functional testing of turing machines finds no errors.
   (b) turing machines will not crash for any input string.
   (c) certain changes (such as many tapes) result in machines of equivalent power.
   (d) Turing machine has nothing to do with robustness.

19. By giving Turing machine more complex power we can increase the power of the Turing Machine
   (a) Absolutely True (b) May be True
   (c) May not be True (d) Absolutely False

20. Which statement is false?
   (a) we construct Turing machine to carry out some algorithm
   (b) We construct Turing machine to accept a given language
   (c) Turing machine is not capable of performing any calculation which can be performed by computer
   (d) Turing machine is simple mathematical model of general purpose computer.

21. The grammar generated by production rules $S \to A \mid Sc, A \to ab \mid aAb$ is
   (a) $a^n b^n c^i \; n > 0 \; and \; c \geq 0$
   (b) $a^n b^n c^i \; n \geq 0 \; and \; c \geq 0$
   (c) $a^n b^n c^i \; n > 0 \; and \; c > 0$
   (d) $a^n b^n c^i \; n \geq 0 \; and \; c > 0$

22. The grammar generated by production rules $S \to aSBc \mid abc, cB \to Bc, aB \to aa$ is
   (a) $a^n b^n c^n, n > 1$ (b) $a^n b^n c^n, n \geq 0$
   (c) $a^n b^n c^n, n > 0$ (d) $a^n b^n c^n \; n \leq 0$

23. Consider the transition table of a Turing machine :

Present State	Tape symbols		
	b	o	1
q_1	$1L_{q2}$	$0R_{q1}$	
q_2	bR_{q3}	$0L_{q2}$	$1L_{q2}$
q_3	bR_{q4}	bR_{q5}	
q_4	$0R_{q5}$	$0R_{q4}$	$1R_{q4}$
q_5	$0L_{q2}$		

q_5 is the final state.

Computation sequence of string 00 leads to?
   (a) $bb_{q5}00$ (b) $bbb_{q5}000$
   (c) $bbbb_{q5}0000$ (d) Error

24. Consider the Turing Machine M described the transition table.

Present State	Tape Symbols				
	0	1	x	y	b
q_1		xR_{q2}		-	bR_{q5}
q_2	$0bR_{q2}$	yL_{q3}	-	xR_{q2}	

q_3	$0L_{q4}$		xR_{q5}	xR_{q3}	
q_4	$0L_{q4}$		xR_{q1}		
q_5				xR_{q5}	bR_{q6}
q_6					

q_6 is the final state.

Refer to the Turing Machine whose transition diagram is given above. What is the final ID when string 011 is processed?

(a) $xyq_6 1$ (b) $xxyybq_6$

(c) $xyq_6 yx$ (d) $xyq_5 1$

25.Refer to the Turing Machine whose transition diagram as given above in question 24, What is the final ID when string 0011 is processed?

(a) $xyq_6 1$ (b) $xxyybq_6$

(c) $xyq_6 yx$ (d) $xyq_5 1$

26.Which automata is associated with Context Sensitive Language? (Give the best answer)
(a) Turing Machine
(b) Finite Automata
(c) Pushdown Automata
(d) Linear Bounded Automata

27.LBA is
(a) unrestricted T.M.
(b) restricted T.M. from one side
(c) restricted T.M. from both sides (d) none

28.Which of the following is true: Read Wirte head can move
(a)to the left of left endmarker in LBA
(b) to the right of left endmarker in LBA
(c) to the right of right endmarker in LBA
(d) to the left of right endmarker in LBA

29.Find the false statement
(a) Standard Turing machine(TM) is equivalent to multi tape TM
(b) Standard TM is equivalent to non deterministic TM
(c)Standard TM is equivalent to linear bounded automata
(d) None

30.The staement "Standard TM accepts the same languages as are accepted by a stay TM" is
(a) True only if languages is a CFL
(b) True only if languages is regular
(c) True for all languages
(d) Always false.

31.Consider a new type of turing machine where the head can move left and move right but cannot stay put. This new type of turing machine is.........
(a) less powerful than the original Turing mahcine
(b) equivalent in power to the original Turing machine
(c) More powerful than the original Turing mahcine
(d) not comparable.

32.We can show that about this new Turning machine of above question 6 that staying put is the same as moving left and then moving right. This illustrates the general method for comparing new Turing machines to original Turing machines by showing we can.............
(a) simulate one by the other.
(b) make up new Turning machines
(c) give all input strings to one and the result is the same for the other.
(d) Do nothing.

33.Many models of general purpose computation exist. Some are very similar to the original Turing machine, others can be very different than the original.
All of these models are equivalent in power if................
(a) they have unrestricted access to unlimited memory.
(b) satisfy certain reasonable requirements like performing only a finite amount of work in a single step.
(c) they have unrestricted access to unlimited memory, and satisfy certain reasonable requirements like performing only a finite amount of work in a single step.
(d) there is no model if everything is equivalent to everything else!

34.Non-Deterministic Turing Machines are more powerful than deterministic Turing Machine
(a) Absolutely True
(b) May be True
(c) May not be True
(d) Absolutely False

35.The Linear bounded automata is a variant of
(a) Turing Machine
(b) Pushdown Automata
(c) Finite Automata
(d) None of these

36.Let B be a Linear bounded automata. Then grammar corresponding to L(B) is
(a) Context sensitive language
(b) Context free language
(c) Unrestricted grammar
(d) Regular grammar

37.Which of the following is not a variant of the standard Turing Mahine
(a) Linear Bounded Automata
(b) Pushdown Automata
(c) Universal Turing Machine
(d) None of the above.

ANSWER KEY

1.(d)	2.(c)	3.(a)	4.(d)	5.(b)	6.(c)
7.(d)	8.(a)	9.(a)	10.(a)	11.(c)	12.(d)
13.(a)	14.(d)	15.(c)	16.(a)	17.(c)	18.(b)
19.(d)	20.(a)	21.(c)	22.(c)	23.(c)	24.(c)
25.(c)	26.(d)	27.(c)	28.(b,d)	29.(c)	30.(c)
31.(b),(d)		32.(a)	33.(c)	34.(d)	35.(a)
36.(a)	37.(b)				

DECIDABILITY AND UNDECIDABILITY

1. Set of all Turing machines is
   (a) Uncountable (b) Countable but infinite
   (c) Finite (d) Void
2. If there exists a language L, for which there exists a TM, *T* that accepts every word in *L* and either rejects or loops for every word that is not in *L*, is said to be
   (a) recursive (b) recursively enumerable
   (c) NP-HARD (d) none of the above.
3. If we have an algorithm to determine whether a given element belongs to a set X or not, then this set is called.....
   (a) Recursive (b) Recursively Enumerable
   (c) Complete (d) Context-sensitive.
4. If we have a procedure to determine whether a given element belongs to a set X or not, then this set is called...........
   (a) Recursive (b) Recursively Enumerable
   (c) Complete (d) Context-sensitive
5. Find the Correct statement.
   (a) Recurisively enumerable implies recursive
   (b) Recursive implies recursively enumerable
   (c) Recursive is same as recursively enumerable
   (d) None.
6. Find the false statement.
   (a) L is recursively enumerable then so is complement of L
   (b) L is recursive then so is complement of L
   (c) L is recursive then complement of L is recursively enumerable
   (d) If L and complement of L are recursively enumerable then both are recursive.
7. Which of the following is not primitive recursive:
   (a) Addition (b) Division
   (c)LCM (d) Ackerman's function
8. Find the incorrect statement:
   (a) Set of real numbers is not enumerable
   (b) if a set S is infinite countable then its power set is not countable
   (c) if L is not recursively enumerable then its complement cannot be recursive
   (d) family of recursive is a equivalent to family of recursively enumerable.
9. Which of the following statements is true?
   (a) A set X is recursive if we have a algorithm to decide whether a given element belongs to X or not.
   (b) A recursive set is recursively enumerable.
   (c) A context sensitive language is recursive.
   (d) All of the above.
10. Let A= set of recursive languages B=set of recursively enumerable languages. Then
   (a) *A* is a subset of *B*
   (b) *B* is a subset of *A*
   (c) *A* and *B* are the same set
   (d) *A* and *B* are disjoint sets.
11. Any language generated by an unrestricted grammar is
   (a) Recursive (b) Recursively enumerable
   (c) Not recursively enumerable
   (d) None of these.
12. The family of recursive languages is not closed under which of the following operations:
   (a) Union (b) Intersection
   (c) Complementation (d) None of these.
13. Which one is false:
   (a) Every recursive enumerable language is recursive.
   (b) Complement of a recursive language is again recursive
   (c) If *L* & Complement *L* both are recursive Enumerable then L is recursive:
   (d)*L*is recursive then TM halts for every x belongs to *L*.
14. Given the following statements:
   (i) Every monotonic grammar G is equivalent to a type 1 grammar.
   (ii) A context sensitive language is recursive.
   (iii) There exists a recursive set which is not a context sensitive language over {0, 1}.
   Which of the following statements are true?
   (a) Only (i) and (ii) (b) Only (ii) and (iii)
   (c) Only (i) and (iii) (d) All (i), (ii) and (iii)
15. Choose which of the following is not correct:
   (a) Set of recursively enumerable languages is closed under union
   (b) If a language and its complement are both regular, then the language must be recursive
   (c) Recursive languages are closed under complementation
   (d) None of the above.
16. A problem whose language is recursive is said to be:
   (a) decidable (b) undecidable
   (c) can't say (d) none.
17. The complement of recursive is :
   (a) recursive (b) non-recursive
   (c) can't say (d) none.
18. Which of the following statements are false:
   (a) The complement of a recursive language is recursive.
   (b) The union of two recursive langauge is recursive. The union of two recursively enumerable languages is recursively enumerable.
   (c) If a language L and its complement *L* are both recursively enumerable, then L $(and\, hence\, L')$ is recursive.
   (d) None.
19. Let L and L' be a pair of complementary languages then,
   (a) Both L and L' are recursive
   (b) Neither L nor L' is recursively enumerable
   (c) One of L and is recursively enumerable but not recursive & the other is not récursive enumerable
   (d) Any one of the above.
20. Universal Language is:
   (a) recursively enumerable
   (b) non-recursively enumerable
   (c) decidable
   (d) recursive

21.Which of the following properties of recursively enumerable set is recursively enumerable.

(a) $L = \phi$

(b) $L - L_u \neq \phi, (L_u \text{ is the universal language})$

(c) L is recursive

(d) $L = \Sigma^*$

22.Which of the following properties of recursively enumerable set is not recursively enumerable,

(a) $L = \Sigma^*$

(b) $L \neq \phi$

(c) $L - L_u \neq \phi$

(d) L contains atleast 10 members

23.The union of two recursively enumerable languages is:

(a) recursive (b) recursive enumerable
(c) not (b) (d) none

24. Let L be a language which is not recursively enumerbale. Then complement of L must be

(a) Recursive (b) Recursively enumerable
(c) Not recursive (d) None of these.

25. A Language is said to be recursive if:

(a) it is recursively enumerable and its complement is not recursively enumerable.

(b) it is not recursively enumerable and its complement is recursively enumerable

(c) it is recursively enumerable and its complement is recursively enumerable.

(d) it is not recursively enumerable and its complement is not recursively enumerable.

26.What can you say about the membership problem for Type 0 grammars?

(a) Decidable (b) Undecidable
(c) Partially decidable (d) None of the above.

27.Let G be any unrestricted grammar. Then the problem of determining whether or not $L(G) = \phi$ is

(a) Decidable (b) Undecidable
(c) Cannot say (d) None of these.

28.Which of the following subset relation doesn't hold?

(a) $L_r \subseteq L_{cf}$ (b) $L_{cf} \subseteq L_{cs}$

(c) $L_{cs} \subseteq L_0$ (d) $L_0 \subseteq L_r$

29.The problem of determining that a Turing Machine would halt after giving a Yes/No output is

(a) Unsolvable (b) Solvable
(c) Decidable (d) None of the above.

30.Which of the following statement is correct?

(a) If the emptiness problem is decidable for Type 3 grammars, then it is also decidable for Type 0 grammars

(b) If the emptiness problem is decidable for Type 0 grammars, then it is also decidable for Type 3 grammars.

(c) If the emptiness problem is undecidable for Type 0 grammars, then it is also undecidable for Type 3 grammars.

(d) None of the above.

31.Select the false statement:

(a) The Turing Machine Halting problem is Undecidable

(b) The Turing Machine Halting problem is decidable

(c) The blank-tape halting problem is undecidable

(d) None of the above.

32.Decidability in Decidable and Undecidable Problems refers to

(a) Unsolvability

(b) Existence of algorithm which generates the set of solutions.

(c) Existence of algorithm which takes an instance of the problem and determines whether the answer to that instance is "yes" or "no"

(d) Those problems whose answer are only either in "yes" or "no".

33. Find the odd man out

(a) The halting problem of turing machines

(b) State-entry problem.

(c) Blank Tape halting problem

(d) Post Correspondence Problem.

34.It is decidable "whether a given right-linear grammar $G_1 = \{N_1, T_1, S_1, P_1\}$ contains any useless non-terminals?"

(a) True (b) False

35.It is decidable "whether $L(G) = \phi$, where G is any context free grammar?"

(a) True (b) False

36.It is decidable "whether $L(G)$ is infinite, where G is any context-free grammar?"

(a) True (b) False

37.Let G be a context-sensitive grammar. It is decidable "whether any word $x \in L(G)$?"

(a) True (b) False

38.Given a Turing Mahine *T* and word *w*, it is undecidable "whether *T* halts when the string *w* is placed on the input tape?"

(a) True (b) False

39.Given a type 0 grammar G and a word *w*, it is undecidable "whether G generates *w*?"

(a) True (b) False

40.Given an arbitrary Turing machine T, it is undecidable "whether *T* accepts the empty word $\in$?"

(a) True (b) False

41.The problem of determining whether or not a TM over {0,1} will ever print the symbol 1, with a given tape configuration is

(a) Solvable (b) Unsolvable (c) None of these

42.What can you say about the membership problem for Type 0 grammars?

(a) Decidable (b) Undecidable
(c) Partially decidable (d) None of the above.

43.If 2 lists are $x = \{b, bab^3, ba\}$ and $y = \{b^3, ba, a(a)\}$ then solution of PCP problem is:

(a) 2,1,1,1,3 (b) 2,1,1,3

(c) 1,2,3,3 (d) none

44. If we have an algorithm to determine whether a given element belongs to a set X or not, then this set is called..........
(a) Recursive (b) Recursively Enumerable
(c) Complete. (d) Context-sensitive.

45. If we have a procedure to determine whether a given element belongs to a set X or not, then this set is called..............
(a) Recursive (b) Recursively Enumerable
(c) Complete. (d) Context-sensitive.

46. For post Correspondence problem(PCP) and Modified PCP(MPCP)
(a) If there exists PCP solution then there is a MPCP solution
(b) If there exists MPCP solution then there is a PCP solution
(c) Both are unrelated
(d) Both are equivalent

47. Which of the following is decidable?
(a) The halting problem
(b) The problem of determining whether or not $L(G) = \phi$ where G is unrestricted grammar.
(c) The question whether or not $L(M)$ is finite, where M is a Turing Machine.
(d) None of above.

48. Which of the following is decidable?
(a) The modified Post Correspondence Problem
(b) The problem to determine whether a string is in G or not, where G is unrestricted grammar.
(c) The problem to decide whether CFG is ambiguous or not.
(d) None of the above.

49. Which of the following is decidable?
(a) The modified Post Correspondence Problem
(b) The problem to determine whether a string is in G or not, where G is unrestricted grammar.
(c) The problem to decide whether CFG is ambiguous or not.
(d) None of the above.

50. Which of the following instances of the Post Correspondence Problem have a viable sequence?
(i) $\{(ab, abb), (ba, aaa), (aa, a)\}$
(ii) $\{(ab, aba), (baa, aa), (aba, baa)\}$
(iii) $\{(b, bb), (bb, bab), (bab, abb), (abb, babb)\}$
(a) 1 only (b) 2 only
(c) 1 and 2 only (d) 1,2 and 3.

51. PCP is
(a) Decidable
(b) Undecidable
(c) Sometimes undecidable?
(d) None.

52. Which of the following is decidable
(a) The halting problem
(b) The problem of determining whether or not $L(G) = \phi$ where G is unrestricted grammar.
(c) The question whether or not $L(M)$ is finite, where M is a Turing Machine.
(d) None of above.

53. What can you say about the membership problem for Type1 grammars?
(a) Decidable (b) Undecidable
(c) Partially decidable (d) None of the above.

54. Which of the following statements is false?
(a) PCP with two lists $x = (b, bab^3, ba)$ and $y = (b^3, ba, a)$ has no solution.
(b) PCP with two lists $x = (01, 1, 1)$ and $y = (01^2, 10, 1^2)$ has no solution.
(c) PCP over Σ for $|\Sigma| \geq 2$ is unsolvable
(d) None of the above.

55. Which of the following statements is true?
(a) PCP with $\{(01, 011), (1, 10), (1, 11)\}$ has a solution.
(b) PCP with $\{(0, 10), (120, 03), (021, 10)\}$ has a solution
(c) pcp with $x = (b^3, ab^2)$ and $y = (b^3, bab^3)$ has a solution
(d) None of the above.

56. If PCP is decidable then MPCP is
(a) decidable (b) Undecidable
(c) Can't say (d) none

57. Inherent ambiguity for CFLs is:
(a) Decidable (b) Undecidable
(c) Can't say (d) none

58. The statement "halting problem is unsolvable"is
(a) true
(b) false
(c) still an open question
(d) All

59. Which of the following problems is decidable?
(a) Halting problem of TM
(b) Whether L(G) is empty for unrestricted grammar G
(c) Whether L(M) is finite for a TM'M'
(d) None

60. PCP is decidable for two set of strings w and v, if
(a) Corresponding strings in w and v are of equal length
(b) The strings consist of just one character repeated any no of times i.e., input symbol set is a singleton
(c) Both the above.
(d) It is never decidable.

61. The membership problem for Type 0 grammars is ?
(a) Decidable (b) Undecidable
(c) Partially decidable (d) None of the above

62. Which of the following statements is correct?
(a) If the emptiness problem is decidable for Type 3 grammars, then it is also decidable for Type 0 grammars.
(b) If the emptiness problem is decidable for Type 0 grammars, then it is also decidable for Type 3 grammars.
(c) If the emptiness problem is undecidable for Type 0 grammars, then it is also undecidable for Type 3 grammars. (d) None of the above.

63. If L_1 and L_2 are any two context free languages (type 2) over an alphabet Σ and $|\Sigma| \geq 2$, there is no algorithm to determine whether or not

(a) $L_1 \cap L_2 = \phi$

(b) $L_1 \cap L_2$, is a context free language

(c) $L_1 \subseteq L_2$

(d) $L_1 = L_2$

(e) All of the above

(f) None of the above

64. Consider, the following modifications of the Post Correspondence Problem :

(i) There is an MPC-Solution if there is a sequence of integers such that

$w_i w_j....w_k w_1 = \upsilon_i \upsilon_j...\upsilon_k \upsilon_1$

(ii) There is an MPC-Solution if there is a sequence of integers such that $w_1 w_2 w_i w_j....w_k = \upsilon_1 \upsilon_2 \upsilon_i \upsilon_j...\upsilon_k$.

Then,

(a) 1 and 2 are decidable

(b) 1 is decidable but 2 is un-decidable

(c) 1 is un-decidable but 2 is decidable

(d) 1 and 2 are un-decidable

65. Which of the following problems is decidable :

(a) Is a given grammar ambiguous?

(b) Is the intersection of 2 CFls empty?

(c) Are 2 CFls the same?

(d) Is a given CFL equal to Σ^*?

(e) None of the above

66. If a problem P_1 is reducible to P_2 then which of the following does not hold?

(a) If P_1 is unsolvable then so is P_2

(b) If P_1 is non-recursively enumerable then so is P_2.

(c) P_1 is at least as hard as P_2.

(d) If P_2 is solvable then so is P_1.

ANSWER KEY

1.(a)	2.(b)	3.(b)	4.(c)	5.(d)	6.(d)
7.(d)	8.(a)	9.(b)	10.(d)	11.(a)	12.(d)
13.(b)	14.(a)	15.(a)	16.(b)	17.(d)	18.(a)
19.(d)	20.(d)	21.(b)	22.(c)	23.(c)	24.(c)
25.(b)	26.(d)	27.(b)	28.(a)	29.(a)	30.(b)
31.(b)	32.(c)	33.d	34.(a)	35.(a)	36.(a)
37.(a)	38.(a)	39.(a)	40.(a)	41.(b)	42.(c)
43.(b)	44.(a)	45.(b)	46.(d)	47.(d)	48.(b)
49.(d)	50.(a)	51.(b)	52.(d)	53.(a)	54.(d)
55.(d)	56.(c)	57.(a)	58.(b)	59.(c)	60.(d)
61.(a)	62.(b)	63.(e)	64.(d)	65.(e)	66.(c)

TESTYOURSELF - I

1. Let L1 be a recursive language. Let L2 and L3 be languages that are recursively enumerable but not recursive. Which of the following statements is not necessarily true ? **[GATE - 10]**

   a) L2 - L1 is recursively enumerable

   b) L1 - L3 is recursively enumerable

   c) $L2 \cap L1$ is recursively enumerable

   d) $L2 \cup L1$ is recursively enumerable

2. Let $L = \{ w \in (0+1)^* | w$ has even number of 1S $\}$, i.e. L is the set of all bit strings with even number of 1s. Which one of the regular expressions below represents L? **[GATE - 10]**

   a) $(0^*10^*1)^*$ b) $0^*(10^*10^*)^*$

   c) $0^*(10^*1^*)^*0^*$ d) $0^*1(10^*1)^*10^*$

3. Consider the languages $L1 = \{0^i1^j | i \neq j\}$.

   $L2 = \{0^i1^j | i = j\}$. $L3 = \{0^i1^j | i = 2j+1\}$

   $L4 = \{0^i1^j | i \neq 2j\}$. Which one of the following statements is true ? **[GATE - 10]**

   a) Only L2 is context free

   b) Only L2 and L3 are context free

   c) Only L1 and L2 are context free

   d) All are context free

4. Let w be any string of length n in { 0, 1 }*. Let L be the set of all substrings of w. What is the minimum number of states in a non - deterministic finite automaton that accepts L? **[GATE - 10]**

   a) n - 1 b) n c) n+1 d) 2^{n-1}

5. S → aSa | bSb | a | b ; The language generated by the above grammar over the alphabet { a, b } is the set of **[GATE - 09]**

   a) All palindromes

   b) All odd length palindromes

   c) Strings that begin and end with the same symbol

   d) All even length palindromes

6. Which one of the following languages over the alphabet { 0, 1 } is described by the regular expression : (0 + 1)* 0(0+1)*0(0+1)* ? **[GATE - 09]**

   a) The set of all strings containing the substring 00

   b) The set of all strings containing at most two 0's

   c) The set of all strings containing at least two 0's

   d) The set of all strings that begin and end with either 0 or 1

7. Which one of the following is FALSE ? **[GATE - 09]**

   a) There is unique minimal DFA for every regular language

   b) Every NFA can be converted to an equivalent PDA

   c) Complement of every context - free language is recursive

   d) Every nondeterministic PDA can be converted to an equivalent deterministic PDA

8. Given the following state table of an FSM with two states A and B, one input and one output : **[GATE - 09]**

Present State A	Present State B	Input	Next State A	Next State B	output
0	0	0	0	0	1
0	1	0	1	0	0
1	0	0	0	1	0
1	1	0	1	0	0
0	0	1	0	1	0
0	1	1	0	0	1
1	0	1	0	1	1
1	1	1	0	0	1

If the initial state is A = 0, B=0, what is the minimum length of an input string which will take the machine to the state A =0, B=1 with Output =1? **[GATE - 09]**

   a) 3 b) 4 c) 5 d) 6

9. Let $L = L_1 \cap L_2$, where L_1 and L_2 are languages as defined below : $L_1 = \{a^m b^m c a^n b^m | m, n \geq 0\}$

   $L_2 = \{a^i b^j c^k | i, j, k \geq 0\}$

   Then L is **[GATE - 09]**

   a) Not recursive

   b) Regular

   c) Context free but not regular

   d) Recursively enumerable but not context free.

10.

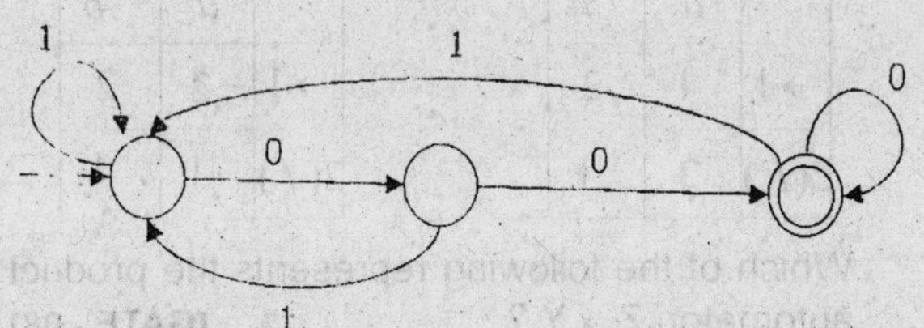

The above DFA accepts the set of all strings over {0, 1} that **[GATE - 09]**

   a) begin either with 0 or 1

   b) end with 0

   c) end with 00

   d) contain the substring 00

11. Which of the following is true for the language $\{a^p$ |p is a prime} ? **[GATE - 08]**

   (a) It is not accepted by a Turning Machine

   (b) Itis regular but not context -free

   (c) It is context-free but not regular

   (d) It is neither regular nor context-free, but accepted by a Turning machine.

12. Which of the following are decidable?

I. Whether the intersection of two regular languages is infinite

II. Whether a given context-free language is regular

III. Whether two push-down automata accept the same language

IV. Whether a given grammar is context-free

[GATE - 08]

(a) I and II (b) I and IV
(c) II and III (d) II and IV

13. If L and $\bar{L}$ are recursively enumerable then L is **[GATE - 08]**

(a) regular (b) context-free
(c) context-sensitive (d) recursive

14. Which of the following statements is false? **[GATE - 08]**

(a) Every NFA can be converted to an equivalent DFA

(b) Every non-deterministic Turing machine can be converted to an equivalent deterministic Turing machine.

(c) Every regular language is also a context-free language

(d) Every subset of a recusively enumerable set is recursive.

15. Given below are two finite state automata ($\rightarrow$ indicates the start state and F indicates a final state)

Y:

	a	b
$\rightarrow 1$	1	2
$2(F)$	2	1

Z:

	a	b
$\rightarrow 1$	2	2
$2(f)$	1	1

Which of the following represents the product automaton Z x Y? **[GATE - 08]**

(a)

	a	b
$\rightarrow P$	S	R
Q	R	S
$R(F)$	Q	P
S	Q	P

(b)

	a	b
$\rightarrow P$	S	Q
Q	R	S
$R(F)$	Q	P
S	P	Q

(c)

	a	b
$\rightarrow P$	Q	S
Q	R	S
$R(F)$	Q	P
S	Q	P

(d)

	a	b
$\rightarrow P$	S	Q
Q	S	R
$R(F)$	Q	P
A	Q	P

16. Which of the following statements are true?

I. Every left-recursive grammar can be converted to a right-recursive grammar and vice-versa

II All ε-productions can be removed from any context-free grammar by suitable transformations.

III The language generated by a context-free grammar all of whose productions are of the form $X \rightarrow w$ or $X \rightarrow wY$ (where, w is a string of terminals and Y is a non-terminal), is always regular

IV. The derivation trees of strings generated by a context-free grammar in Chomsky Normal from are always binary trees. **[GATE - 08]**

(a) I, II, III and IV (b) II, III and IV only
(c) I, III and IV only (d) I,II and IV only

17. Match the following **[GATE - 08]**

(E) Checking that identifiers are declared before their use	(P) L= $\{a^n b^m c^n d^m \mid n \geq 1, m \geq 1\}$
(F) Number of formal parameters in the declaration of a function agrees with the number of actual parameters in a use of that function	(Q) $X \rightarrow XbX \mid XcX \mid dXf \mid g$
(G) Arithmetic expressions with matched pairs of parentheses	(R) $L = \{wcw \mid w \in (a \mid b)^*\}$
(H) Palindromes	(S) $X \rightarrow bXb \mid cXc \mid \varepsilon$

(a) E-P, F-R, G-Q, H-S (b) E-R,F-P,G-S, H-Q
(c) E-R, F-P, G-Q, H-S (d) E-P, F-R, G-S, H-Q

18. Match the following NFAs with the regular expressions they correspond to **[GATE - 08]**

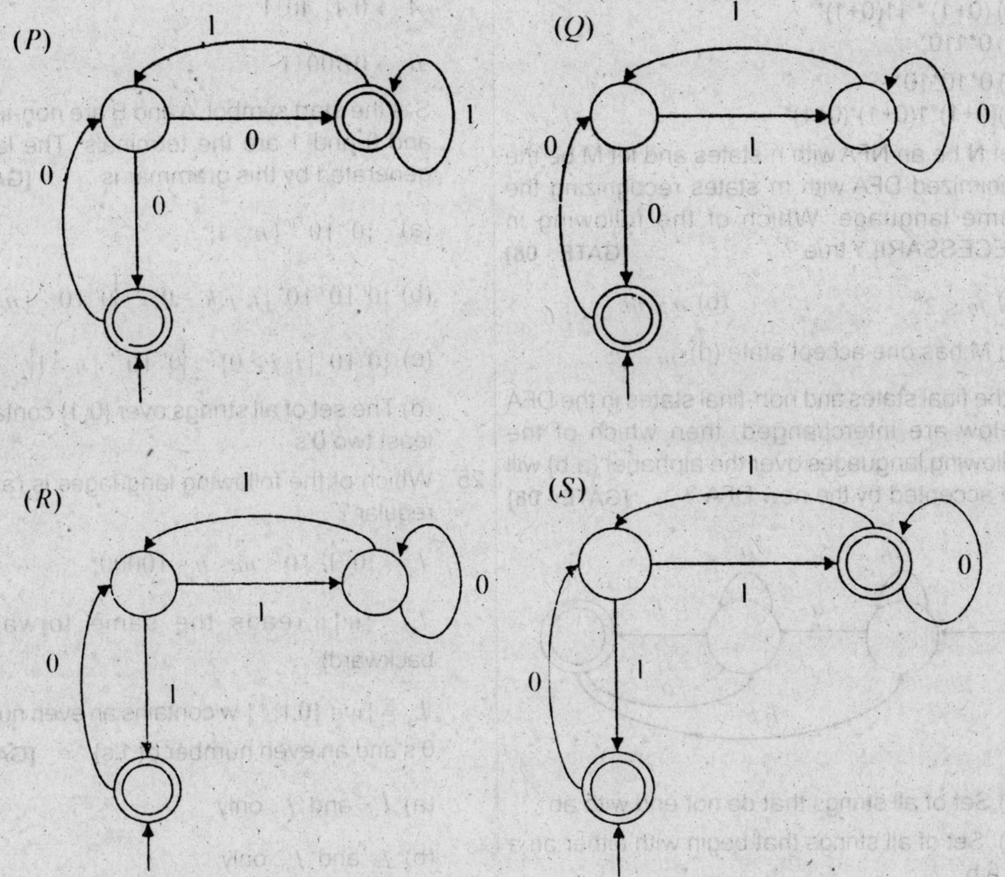

1. ε + 0 (01*1+00)*01*
2. ε + 0 (10*1+00)*0
3. ε + 0 (10*1+10) *1
4. ε + 0 (10*1+10) *10*

(a) P-2, Q-1, R-3, S-4 (b) P-1, Q-3, R-2, S-4

(c) P-1, Q-2, R-3, S-4 (d) P-3, Q-2, R-1, S-4

19. Which of the following are regular sets ?

I. $\{a^n b^{2m} \mid n \geq 0, m \geq 0\}$

II. $\{a^n b^m \mid n = 2m\}$

III. $\{a^n b^m \mid n \neq m\}$

IV. $\{xcy \mid x, y \in \{a,b\}^*\}$ **[GATE - 08]**

(a) I and IV only (b) I and III only (c) I only (d) IV only

20. Which of the following regular expressions describes the language over {0,1} consisting of strings that contain exactly two 1's ? **[GATE - 08]**
(a) (0+1)* 11(0+1)*
(b) 0*110*
(c) 0*10*10*
(d)(0+1)*1(0+1)*(0+1)*

21. Let N be an NFA with n states and let M be the minimized DFA with m states recognizing the same language. Which of the following in NECESSARILY true ? **[GATE - 08]**
(a) $m \le 2^n$ (b) $n \le m$
(c) M has one accept state (d) $m = 2^n$

22. If the final states and non-final states in the DFA below are interchanged, then which of the following languages over the alphabet {a,b} will be accepted by the new DFA ? **[GATE - 08]**

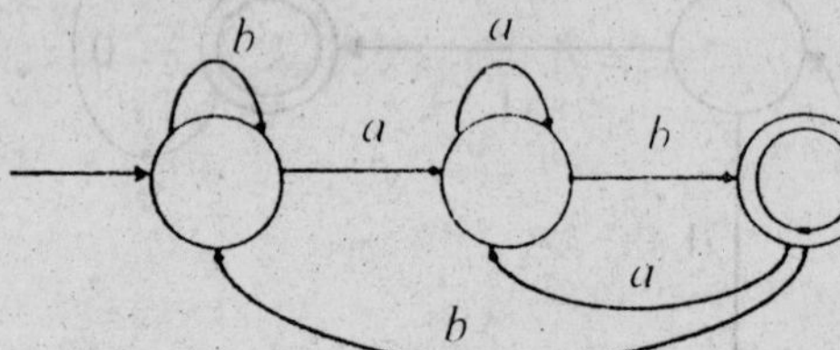

(a) Set of all strings that do not end with ab
(b) Set of all strings that begin with either an a or a b
(c) Set of all strings that do not contain the substring ab
(d) The set described by the regular expression b*aa*(ba)*b*

23. Consider the following languages

$L_1 = \{a^i b^j c^k \mid i = j, k \ge 1\}$

$L_2 = \{a^i b^j \mid j = 2i, i \ge 0\}$

Which of the following is true ? **[GATE - 08]**
(a) L_1 is not a CFL but L_2 is
(b) $L_1 \cap L_2 = \Phi$ and L_1 is non-regular
(c) $L_1 \cup L_2$ is not a CFL but L_2 is
(d) There is a 4-state PDA that accepts L_1, but there is no DPDA that accepts L_2

24. Consider a CFG with the following productions

$S \to AA \mid B$

$A \to 0A \mid A0 \mid 1$

$B \to 0B00 \mid 1$

S is the start symbol, A and B are non-terminals and 0 and 1 are the terminals. The language generated by this grammar is **[GATE - 08]**
(a) $\{0^n 1 0^{2n} \mid n \ge 1\}$
(b) $\{0^i 1 0^j 1 0^k \mid i, j, k \ge 0\} \cup \{0^n 1 0^{2n} \mid n \ge 1\}$
(c) $\{0^i 1 0^j \mid i, j \ge 0\} \cup \{0^n 1 0^{2n} \mid n \ge 1\}$
(d) The set of all strings over {0,1} containing at least two 0's

25 Which of the following languages is (are) non-regular ?

$L_1 = \{0^m 1^n \mid 0 \le m \le n \le 10000\}$

$L_2 = \{w \mid w$ reads the same forward and backward$\}$

$L_3 = \{w \in \{0,1\}^* \mid$ w contains an even number of 0's and an even number of 1's$\}$ **[GATE - 08]**
(a) L_2 and L_3 only
(b) L_1 and L_2 only
(c) L_3 only
(d) L_2 only

26 Consider the following two finite atuomata M_1 accepts L_1 and M_2 accepts L_2
Which one of the following is TRUE ? **[GATE - 08]**

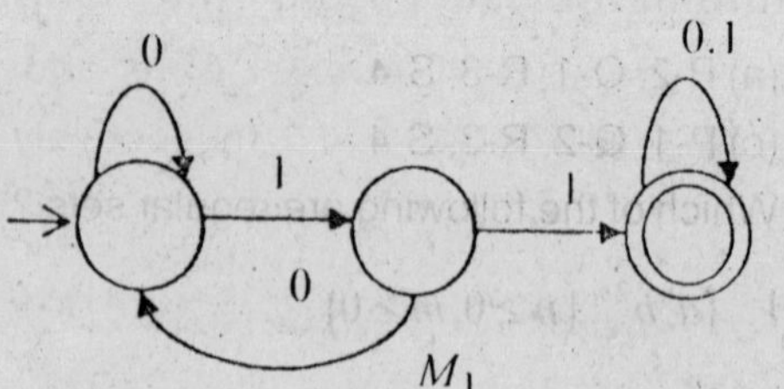

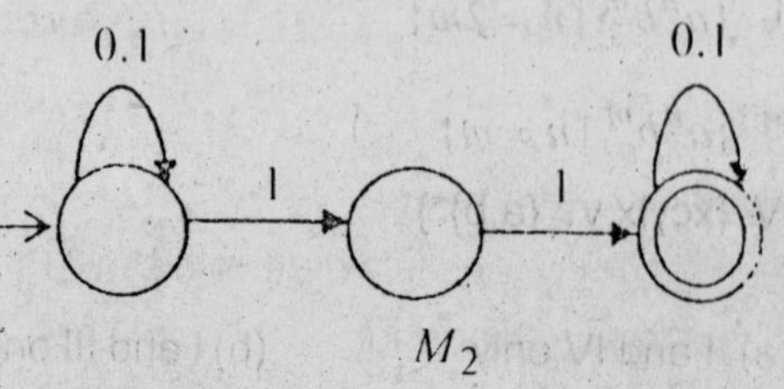

(a) $L_1 = L_2$ (b) $L_1 \subset L_2$

(c) $L_1 \cap \bar{L}_2 = \Phi$ (d) $L_1 \cup L_2 \neq L_1$

Statement for Linked Answer Questions 27 and 28 :

A CFG G is given with the following productions where S is the start symbol, A is a non-terminal and a and b are terminals.

$S \to aS \mid A$

$A \to aAb \mid bAa \mid \in$

27. Which of the following strings is generated by the grammar above ? [GATE - 08]

(a) aabbaba (b) aabaaba
(c) abababb (d) aabbaab

28. For the correct answer in Q.78 how many steps are required to derive the string and how many parse trees are there ? [GATE - 08]

(a) 6 and 1 (b) 6 and 2
(c) 7 and 2 (d) 4 and 2

29. Which of the following problems is undecidable? [GATE - 07]

(a) Membership problem for CFGs
(b) Ambiguity problem for CFGs
(c) Finiteness problem for FSAs
(d) Equivalence problem for FSAs

30. Which of the following is TRUE? [GATE - 07]

(a) Every subset of a regular set is regular
(b) Every finite subset of a non-regular set is regular
(c) The union of two non-regular sets is not regular
(d) Infinite union of finite sets is regular

31. A minimum state deterministic finite automaton accepting the language $L = \{w \mid w \in \{0,1\}^*$, number of 0s and 1s in w are divisible by 3 and 5, respectively$\}$ has [GATE - 07]

(a) 15 states (b) 11 states
(c) 10 states (d) 9 states

32. The language $L = \{0^i 2 1^i \mid i \geq 0\}$ over the alphabet $\{0,1,2\}$ is [GATE - 07]

(a) not recursive
(b) is recursive and is a deterministic CFL
(c) is a regular language
(d) is not a deterministic CFL but a CFL

33. Which of the following languages is regular? [GATE - 07]

(a) $\{ww^R \mid w \in \{0,1\}^+\}$
(b) $\{ww^R x \mid x, w \in \{0,1\}^+\}$
(c) $\{wxw^R \mid x, w \in \{0,1\}^+\}$
(d) $\{xww^R \mid x, w \in \{0,1\}^+\}$

34. Consider the grammar with non-terminals $N = \{S, C, S_1\}$, terminals $T = \{a, b, i, t, e\}$ with S as the start symbol, and the following set of rules.

$S \to iCtSS_1 \mid a$

$S_1 \to eS \mid \varepsilon$

$C \to b$

The grammar is NOT LL(1) because : **[GATE - 07]**

(a) It is left recursive
(b) it is right recursive
(c) it is ambiguous
(d) it is not context - free

Common Data for Questions 35, 36 :

Consider the following Finite State Automation :

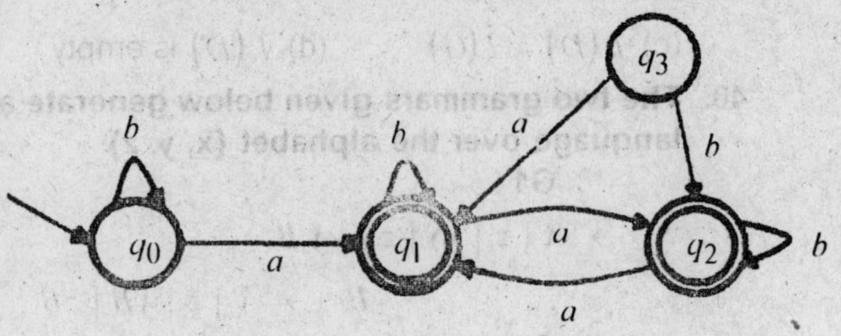

35. The language accepted by this automaton is given by the regular expression [GATE - 07]

(a) $b^*ab^*ab^*ab^*$ (b) $(a+b)^*$
(c) $b^*a(a+b)^*$ (d) $b^*ab^*ab^*$

36. The minimum state automaton equivalent to the above FSA has the following number of states [GATE - 07]

(a) 1 (b) 2 (c) 3
(d) 4

Statement for Linked Answwer Questions 37 to 38 :

Consider the CFG with {S, A, B} as the non-terminal alphabet, {a, b} as the terminal alphabet, S as the start symbol and the following set of production rules :

$S \rightarrow aB$

$S \rightarrow bA$

$B \rightarrow b$

$A \rightarrow a$

$B \rightarrow bS$

$A \rightarrow aS$

$B \rightarrow aBB$

$A \rightarrow bAA$

37. Which of the following strings is generated by the grammar ? [GATE - 07]

(a) aaaabb (b) aabbbb

(c) aabbab (d) abbbba

38. For the correct answer string to Q.37, how many derivation trees are there ?[GATE - 07]

(a) 1 (b) 2 (c) 3 (d) 4

39. Consider an ambiguous grammar G and its disambiguated version D. Let the language recognized by the two grammars be denoted by L(G) and L(D) respectively. Which one of the following is true ? [GATE - 07]

(a) $L(D) \subset L(G)$ (b) $L(D) \supset L(G)$

(c) $L(D) = L(G)$ (d) $L(D)$ is empty

40. The two grammars given below generate a language over the alphabet {x, y, z}

G1 :

$S \rightarrow x \mid z \mid xS \mid zS \mid yB$

$B \rightarrow y \mid z \mid yB \mid zB$

G2 :

$S \rightarrow y \mid z \mid yS \mid zS \mid xB$

$B \rightarrow y \mid yS$

Which one of the following choices describes the properties satisfied by the strings in these languages ? [GATE - 07]

(a) G1 : No y appears before any x
G2 : Every x is followed by at least one y

(b) G1 : No y appears before any x
G2 : No x appears before any y

(c) G1 : No y appears after any x
G2 : Every x is followed by at least one y

(d) G1 : No y appears after any x
G2 : Every y is followed by at least one x

41. Consider the following DFA in which s_0 is the start state and s_1, s_3 are the final states.

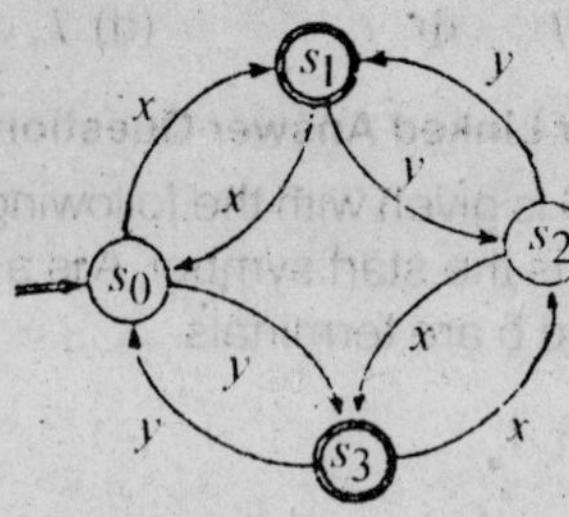

What language does this DFA recognize ? [GATE - 07]

(a) All strings of x and y

(b) All strings of x and y which have either even number of x and even number of y of odd number or x and odd number of y

(c) All strings of x and y which have equal number of x and y

(d) All strings of x and y with either even number of x and odd number of y or odd number of x and even number of y

42. Consider the grammar given below

$S \rightarrow xB \mid yA$

$A \rightarrow x \mid xS \mid yAA$

$B \rightarrow y \mid yS \mid yBB$

Consider the following strings.

(i) xxyyx (ii) xxyyxy (iii) xyxy

(iv) yxxy (v) yxx (vi) xyx

Which of the above strings are generated by the grammar ? [GATE - 07]

(a) (i) , (ii) and (iii) (b) (ii) , (v) and (vi)

(c) (ii), (iii) and (iv) (d) (i), (iii), and (iv)

43. Consider the following grammars. Names representing terminals have been specified in capital letters.

G_1 stmnt → WHILE (expr) stmnt

stmnt → OTHER

expr → ID

G_2 : stmnt → WHILE (expr) stmnt

stmnt → OTHER

expr $\rightarrow$ **expr + expr**

expr $\rightarrow$ **expr * expr**

expr $\rightarrow$ **ID**

Which one of the following statements is true ? **[GATE - 07]**

(a) G_1 is context - free but not regular and G_2 is regular

(b) G_2 is context-free but not regular and G_1 is regular

(c) Both G_1 *and* G_2 are regular

(d) Both G_1 *and* G_2 are context-free but neither of them is regular

44. Consider the following finite automata P and Q over the alphabet { a, b, c}. The start states are indicated by a double arrow and final states are indicated by a double circle. Let the languages recognized by them be denoted by L(P) and L(Q) respectively.

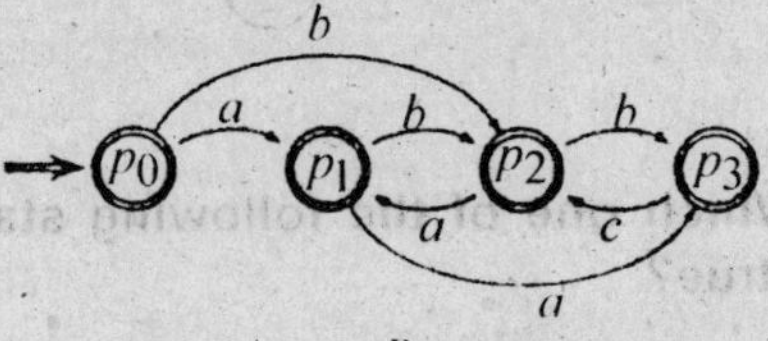

P

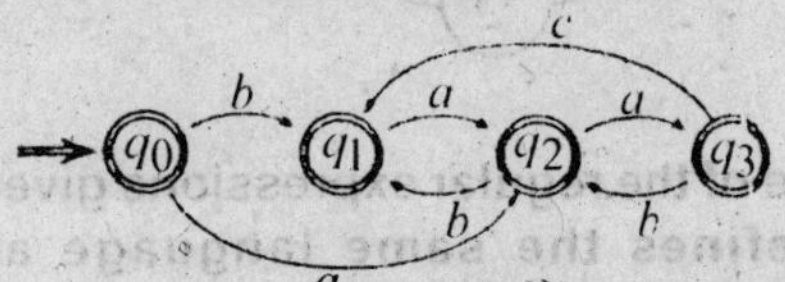

Q

The automation which recognizes the language $L(P) \cap L(Q)$ is **[GATE - 07]**

(a)

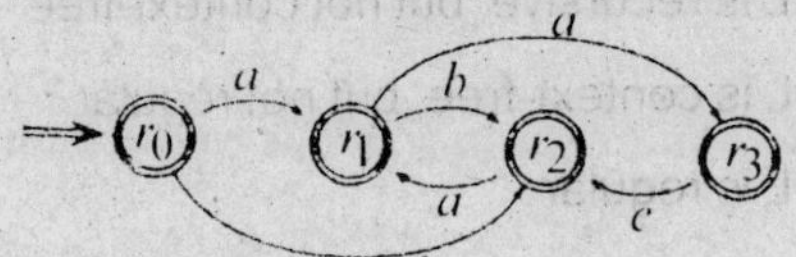

(b)

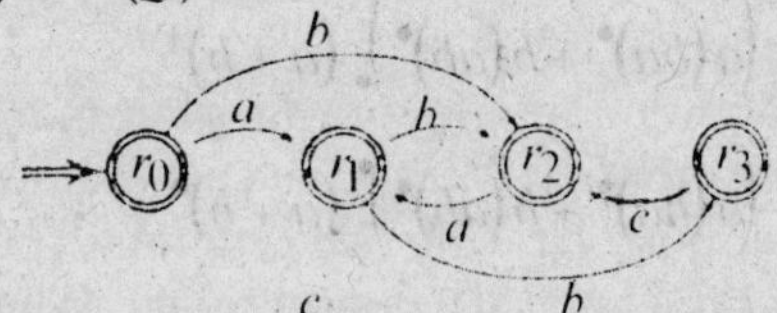

(c)

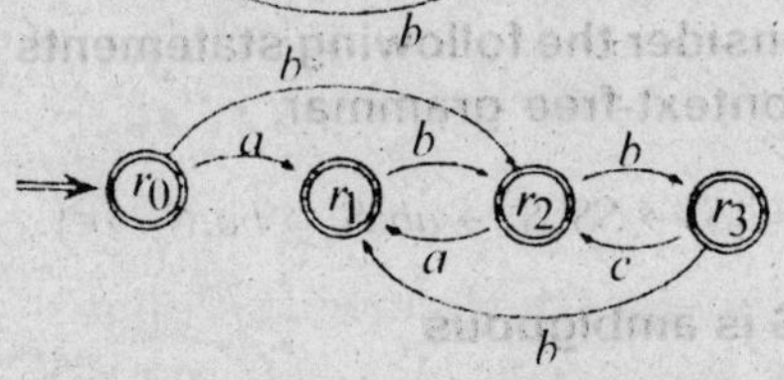

(d)

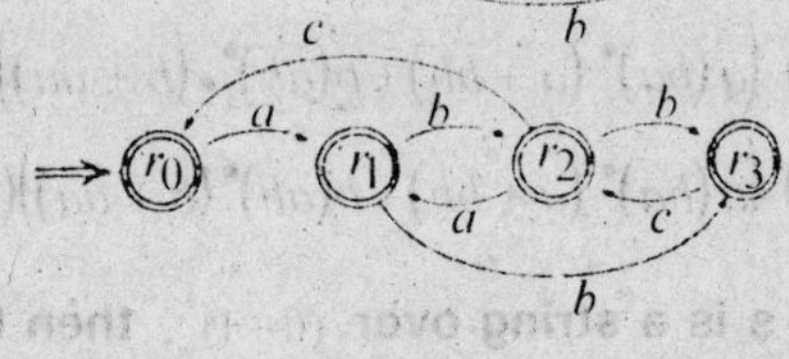

Common data for questions 45, 46, 47 :

Consider the regular expression $R = (a+b)^*(aa+bb)(a+b)^*$

45. Which of the following non-deterministic finite automata recognizes the language defined by the regular expression R ? Edges labeled λ denote transitions on the empty string.

[GATE - 07]

(a)

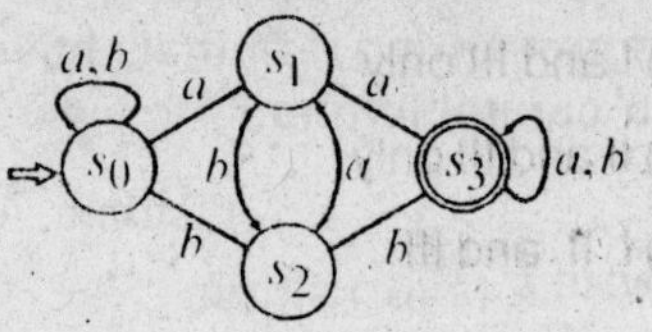

(b)

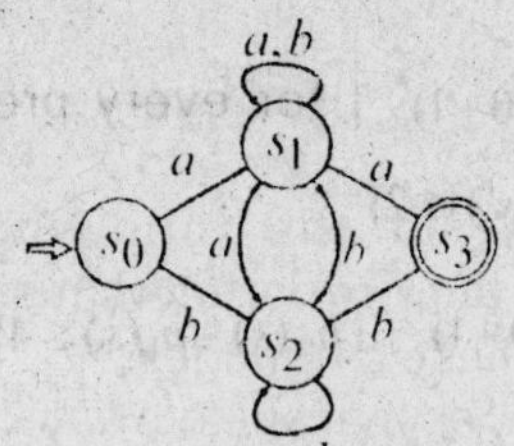

(c)

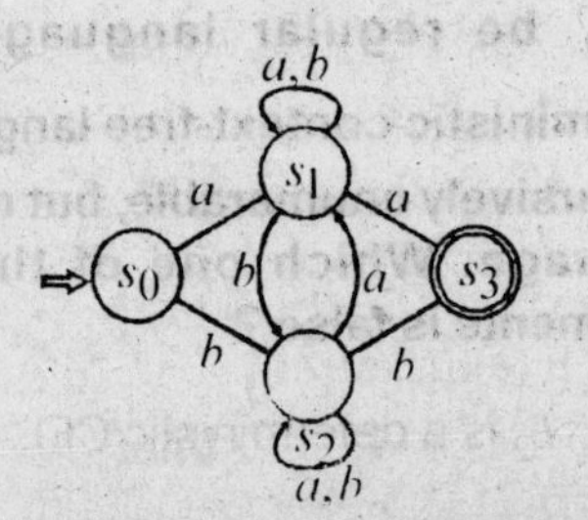

(d)

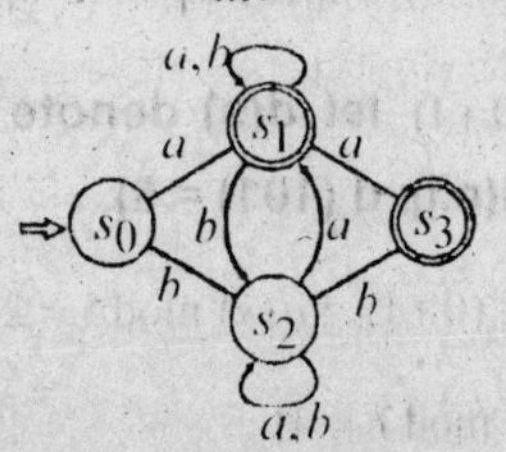

46. Which deterministic finite automaton accepts the language represented by the regular expression R ? **[GATE - 07]**

(a)

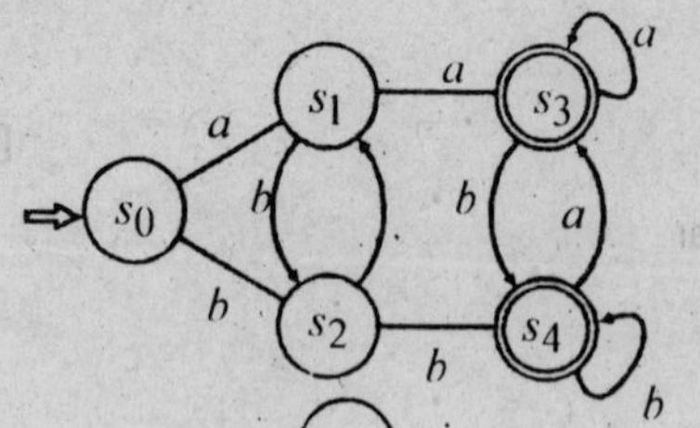

(b)

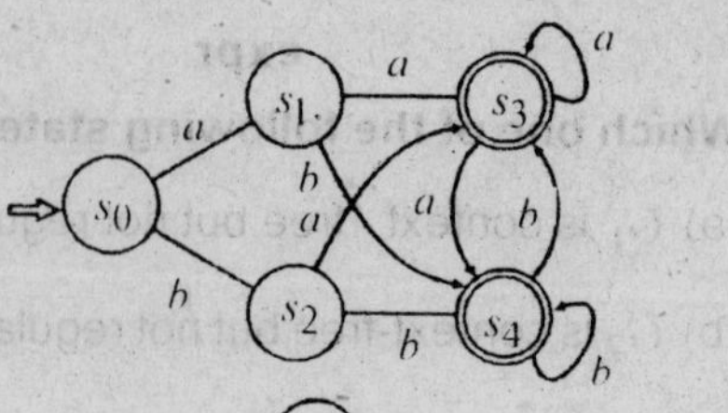

(c)

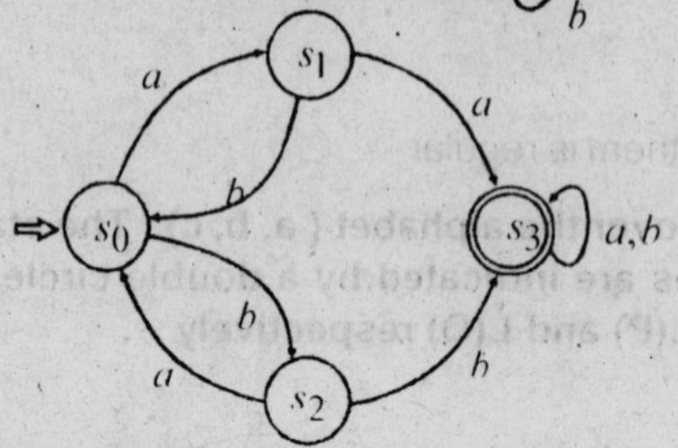

(d)

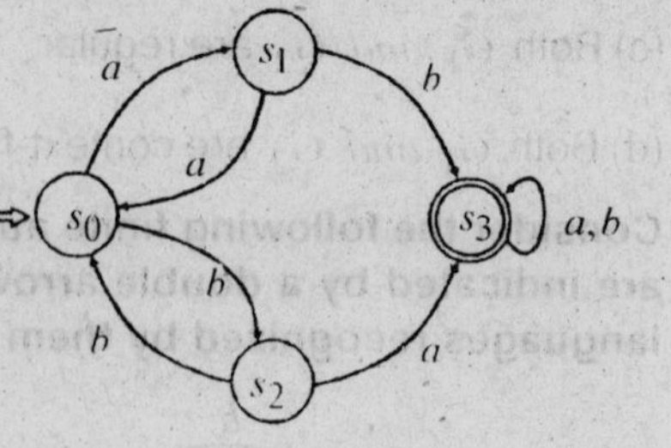

47. Which one of the regular expressions given below defines the same language as defined by the regular expression R ? **[GATE - 07]**

(a) $\left(a(ba)^{\bullet} + b(ab)^{\bullet}\right)_{\bullet}(a+b)^{+}$

(b) $\left(a(ba)^{\bullet} + b(ab)^{\bullet}\right)^{\bullet}(a+b)^{\bullet}$

(c) $\left(a(ba)^{\bullet}(a+bb) + b(ab)^{\bullet}\bullet(b+aa)\right)(a+b)^{\bullet}$

(d) $\left(a(ba)^{\bullet}(a+bb) + b(ab)^{\bullet}(b+aa)\right)(a+b)^{+}$

48. If s is a string over $(0+1)^*$, then let $n_0(s)$ denote the number of $0's$ in s and $n_1(s)$ the number of $1's$ in s. Which one of the following languages is not regular? **[GATE - 06]**

(A) $L = \{s \in (0+1)^* \mid n_0(s)$ is a 3-digit prime}

(B) $L = \{s \in (0+1)^*$ | for every prefixs' of s, $|n_0(s') - n_1(s')| \leq 2\}$

(C) $L = \{s \in (0+1)^* \mid .|\, n_0(s) - n_1(s) \leq 4\}$

D) $L = \{s \in (0+1)^* \mid n_0(s) \bmod 7 = n_1(s) \bmod 5 = 0\}$

49. For $s \in (0+1)^*$ let d(s) denote the decimal value of s(e.g. d (101) = 5).

Let $L = \{s \in (0+1)^* \mid d(s) \bmod 5 = 2$ *and*

d(s) $\bmod 7 \neq 4\}$

Which one of the following statements is true? **[GATE - 06]**

(a) L is recursively enumerable, but not recursive

(b) L is recursive, but not context-free

(c) L is context-free, but not regular

(d) L is regular

50. Consider the following statements about the context-free grammar,

$G = \{S \rightarrow SS, S \rightarrow ab, S \rightarrow ba, S \rightarrow \in\}$

I. G is ambiguous

II. G produces all strings with equal number of a's and b's

III. G can be accepted by a deterministic PDA.

Which combination below expresses all the true statements about G ? **[GATE - 06]**

(A) I only

(B) I and III only

(C) II and III only

(D) I, II, and III

51. Let L_1 be regular language, L_2 be a deterministic context-free language and L_3 a recursively enumerable, but not recursive, language. Which one of the following statements is false ? **[GATE - 06]**

(A) $L_1 \cap L_2$ is a deterministic CFL

(B) $L_3 \cap L_1$ is recursive

(C) $L_1 \cup L_2$ is context free

(D) $L_1 \cap L_2 \cap L_3$ is recursively enumerable

52. Consider the regular language $L = (111 + 11111)^*$. The minimum number of states in any DFA accepting this languages is [GATE - 06]

(A) 3 (B) 5 (C) 8 (D) 9

53. Consider the following grammar

$S \to FR$

$R \to *S \mid \varepsilon$

$F \to id$

In the predictive parser table, M, of the grammar the entries $M[S, id]$ and $M[R, \$]$ respectively [GATE - 06]

(A) $\{S \to FR\}$ and $\{R \to \varepsilon\}$

(B) $\{S \to FR\}$ and $\{\}$

(C) $\{S \to FR\}$ and $\{R \to S*\}$

(D) $\{F \to id\}$ and $\{R \to \varepsilon\}$

Statement for linked answers questions 54 and 55 :

54. Which one of the following grammars generates the language $L = \{a^i b^j : i \neq j\}$? [GATE - 06]

(A) $S \to AC \mid CB$

$C \to aCb \mid a \mid b$

$A \to aA \mid \in$

$B \to Bb \mid \in$

(B) $S \to aS \mid Sb \mid a \mid b$

(C) $S \to AC\ CB$

$C \to aCb \mid \in$

$A \to a\ A \mid \in$

$B \to Bb \mid \in$

(D) $S \to AC \mid CB$

$C \to aCb \mid \in$

$A \to aA \mid a$

$B \to b\ B \mid b$

55. In the correct grammar above, what is the length of the derivation (number of steps starting from S) to generate the string $a^l b^m$ with $l \neq m$? [GATE - 06]

(A) max (l, m)+2 (B) $l + m + 2$

(C) $l + m + 3$ (D) $\max(l, m) + 3$

56. In the automaton below, s is the start state and t is the only final state.

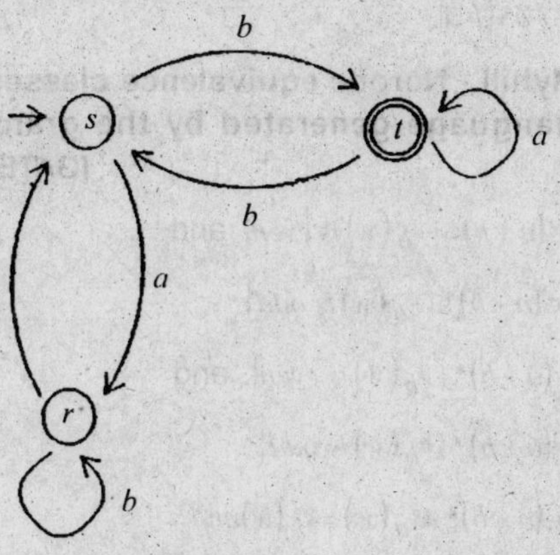

Consider the strings $u = abbaba$, $v = bab$, and $w = aabb$. Which of the following statements is true ? [GATE - 06]

a) The automaton accepts u and v but not w

b) The automaton accepts each of $u, v,$ and w

c) The automation rejects each of $u, v,$ and w

d) The automaton accepts u but rejects v and w

57. In the context-free grammar below, S is the start symbol, a and b are terminals and ε denotes the empty string

$S \to aSa \mid bSb \mid a \mid b \mid \varepsilon$

Which of the following strings is NOT generated by the grammar ? [GATE - 06]

a) aaaa b) baba

c) abba d) babaaabab

58. Which regular expression best describes the language accepted by the non-deterministic automaton below ? [GATE - 06]

a) $(a+b)*a(a+b)b$ b) $(abb)*$

c) $(a+b)*a(a+b)*b(a+b)*$

d) $(a+b)*$

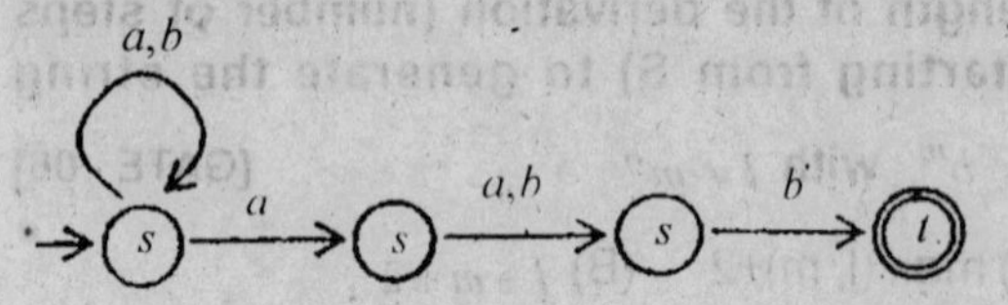

59. Consider the regular grammar below

$S \to bS \mid aA \mid \varepsilon$

$A \to aS \mid bA$

The Myhill - Nerode equivalence classes for the language generated by the grammar are [GATE - 06]

a) $\{w \in (a+b)^* \mid \#_a(w) \text{ is even}\}$ and $\{w \in (a+b)^* \mid \#_a(w) \text{ is odd}\}$

b) $\{w \in (a+b)^* \mid \#_b(w) \text{ is even}\}$ and $\{w \in (a+b)^* \mid \#_b(w) \text{ is odd}\}$

c) $\{w \in (a+b)^* \mid \#_a(w) = \#_b(w) \text{ and }$ $w \in (a+b)^* \mid \#_a(w) \neq \#_b(w)\}$

d) $\{\varepsilon\}, \{wa \mid w \in (a+b)^* \text{ and } \{wb \mid w \in (a+b)^*\}$

60. Which of the following statements about regular languages is NOT true ? [GATE - 06]

a) Every language has a regular superset

b) Every language has a regular subset

c) Every subset of a regular language is regular

d) Every subset of a finite language is regular.

61. Which of the following languages is accepted by a non-deterministic pushdown automaton (PDA) but NOT by a deterministic PDA ? [GATE - 06]

a) $\{a^n b^n c^n \mid n \geq 0\}$

b) $\{a^l b^m c^n \mid l \neq m \text{ or } m \neq n\}$

c) $\{a^n b^n \mid n \geq 0\}$

d) $\{a^n b^n \mid m, n \geq 0\}$

62. Let L be a context-free language and M a regular language. Then the language $L \cap M$ is [GATE - 06]

a) always regular

b) never regular

c) always a deterministic context - free language

d) always a context-free language

63. Consider the pushdown automaton (PDA) below which runs over the input alphabet (a,b,c). It has the stack alphabet $\{Z_0, X\}$ where Z_0 is the bottom-of-stack marker. The set of states of the PDA is $\{s,t,u,f\}$ where s is the start state and f is the final state. The PDA accepts by final state. The transitions of the PDA given below are depicted in a standard manner. For example, the transition $(s,b,X) \to (t,XZ_0)$ means that if the PDA is in state s and the symbol on the top of the stack is X, then it can read b from the input and move to state t after popping the top of stack and pushing the symbols Z_0 and X (in that order) on the stack.

$(s, a, Z_0) \to (s, XXZ_0)$

$(s, \varepsilon, Z_0) \to (f, \varepsilon)$

$(s, a, X) \to (s, XXX)$

$(s, b, X) \to (t, \varepsilon)$

$(t, b, X) \to (t, \varepsilon)$

$(t, c, X) \to (u, \varepsilon)$

$(u, c, X) \to (u, \varepsilon)$

$(u, \varepsilon, Z_0) \to (f, \varepsilon)$

The language accepted by the PDA is [GATE - 06]

a) $\{a^l b^m c^n \mid l = m = n\}$

b) $\{a^l b^m c^n \mid l = m\}$

c) $\{a^l b^m c^n \mid 2l = m + n\}$

d) $\{a^l b^m c^n \mid m = n\}$

64. In the context-free grammar below, S is the start symbol, a and b are terminals, and ε denotes the empty string.

$S \to aSAb \mid \varepsilon$

$A \rightarrow bA \mid \varepsilon$

The grammar generates the language[GATE - 06]

a) $((a+b)^*b)^*$ b) $\{a^m b^n \mid m \le n\}$

c) $\{a^m b^n \mid m = n\}$ d) $a^* b^*$

<u>Statement for linked Answer Questions 65 & 66:</u>

Let L be a regular language : Consider the constructions on L below :

(I) repeat (L) = $\{ww \mid w \in L\}$

(II) prefix (L) = $\{u \mid \exists v : uv \in L\}$

(III) suffix (L) = $\{v \mid \exists u : uv \in L\}$

(IV) half (L) = $\{u \mid \exists v : |v| = |u| \text{ and } uv \in L\}$

65. Which of the constructions could lead to a non-regular language ? [GATE - 06]

a) Both I and IV b) Only I

c) Only IV d) Both II and III

66. Which choice of L is best suited to support your answer above ? [GATE - 06]

a) $(a+b)^*$ b) $\{\varepsilon a, ab, bab\}$

c) $(ab)^*$ d) $\{a^n b^n \mid n \ge 0\}$

ANSWER KEY

1. b	2. b	3. d	4. c	5. b
6. c	7. d	8. a	9. c	10. c
11. d	12. b	13. d	14. d	15. b
16. d	17. c	18. c	19. a	20.c
21. b	22.. a	23. b	24. c	25.
26. a	27. d	28. b	29.b	30.b
31. a	32. b	33. c	34. c	35. c
36. b	37. c	38. b	39. a	40. a
41. d	42. d	43. b	44. b	45. b
46. c	47. c	48. b	49. b	50. d
51. d	52. d	53. a	54. d	55. a,d
56. d	57. b	58. a	59. c	60. a
61. b	62. d	63. b	64. b	65. c
66. d				

EXPLANATORY NOTES

1. Union of two recursive enumerable is not R. E and also the Intersection, $L_2 - L_1$ is r.E is True but $L_1 - L_2$ need not to be recursive enumerable

2. Every string should contain even number of '1' s. min string are ε, 0, 11, 101, 1010 ...... we get those string from the expression 0*(10* 10*)*

In option C and D Min. string is not ε. In option A we don't get, 0,00,000 ....

3. All the given languages can be recognized by the push - Down Automation

4. Ex : Let w = 01

Substring are 0, 1, 01

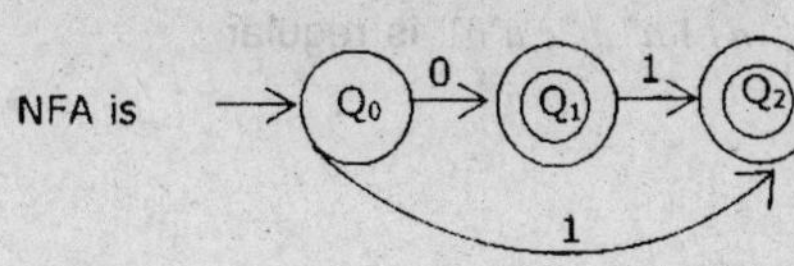

If requires Min 3,

So, If string length is 'n' It required n + 1 states.

5. S -> aSa | bSb | a |b

will generate all odd length palindromes

e.g. S = > aSa => abSba => abbSbba => abbabba

6. R. E = (0+1)*0(0+1)*0(0+1)*

Accepting Language will be

L = { 00, 000, 100, 001, 010, 0000, 0001, 1000,1001,0100, 1100, 0010, 0011, 0110, 0101, 1010, ... }

At least two zeros.

7. (a) There is a unique minimal DFA for every regular language. TRUE

(b) Every NFA can be converted to an equivalent PDA. TRUE

(c) Complement of every context -free language is recursive. TRUE

Because context -free language is not closed under complementation

(d) Every nondeterministic PDA can be converted to an equivalent deterministic PDA. FALSE . Because ww^R can not be accepted by deterministic PDA but nondeterministic PDA can accept it. So there does not exist any equivalent deterministic PDA.

8.

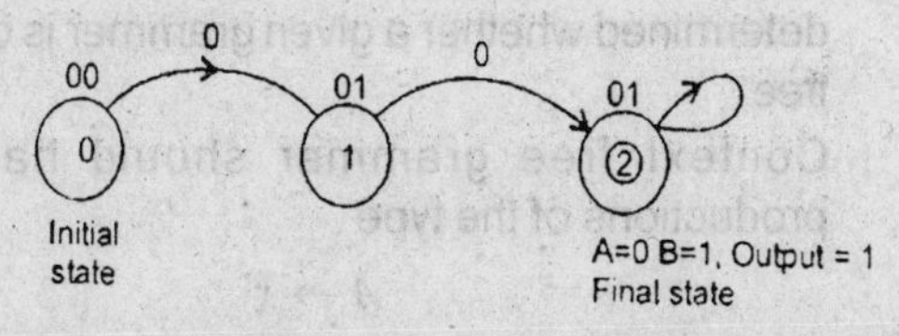

Hence 3 stakes required

Alternately

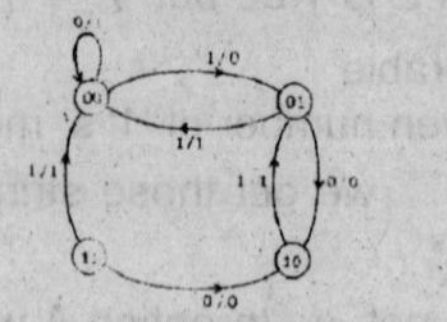

So from state 00 to 01 with output 1 required minimum 3 length of string i.e. 101

9. The inter section of two regular language is regular

Given $L_1 = \{a^m b^m c a^n b^n \mid m, n \geq 0\}$

If n = 1 $\{a,b,c,a,b\}$ not regular
m = 1

If $n / 1\ a^m, b^m c\ a^n b^n$ is regular

Alternately

10. Let A be the starting state

A to A	input 1 0 output 0 0
B to A	input 1 0 output 0 0
C to A	input 1, 0 output 0 0

So end with 0 0

Alternately

R. E. = (0 + 1)*00

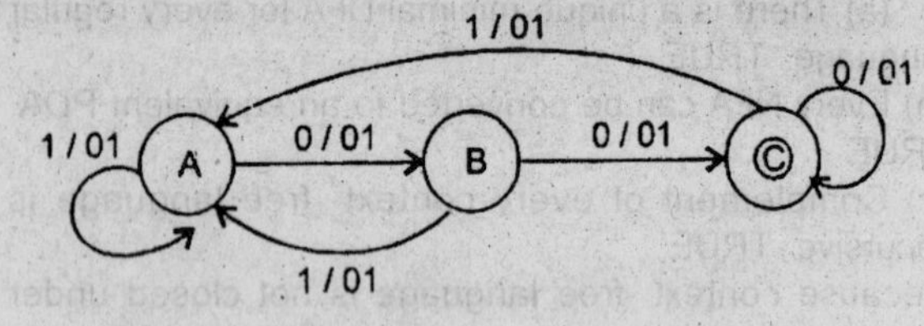

11. L = a^p | p is prime

L is not be regular, can be proved by pumping Lemma, so it cannot be context free language. But it can be accepted by turing machine.

12. Intersection of two regular languages can be determined by a given algorithm, it can be determined whether a given grammer is context free.

Context free grammar should have all productions of the type

$$A \to x$$

where , $A \in V_N$ {set of variable}

$\alpha \in (V_N U \Sigma)^*$ {Σ = set of terminal}

13. If a language is recursively enumerable, and its complement is also, then the language is recursive.

14. I, II and III are true

Instead true statements is, "every recursive set is recusively enumerable"

15

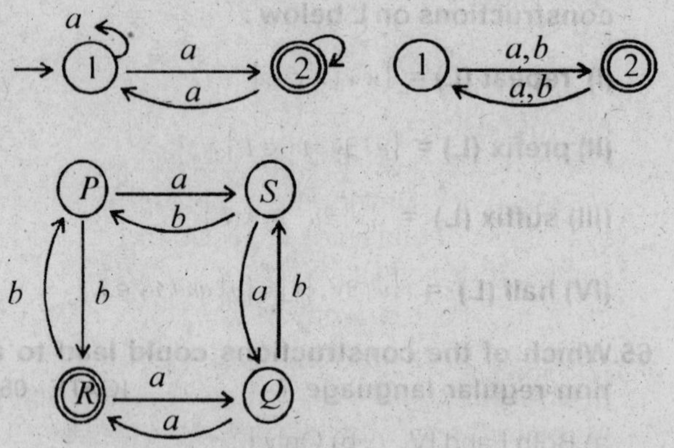

16. Statement III is false. In a context free grammar, right side of the productions should either have single terminal or one non-terminal followed by terminal or λ.

17. Here $X \to bxb | cxc | \varepsilon$ surely represent palindrome and x→x b x | x c x | d x f | g Arithmetic expressions with match pair are simple &L = {wcw/w ε (a/b*)} checking that identifier are declared before their use.

18.

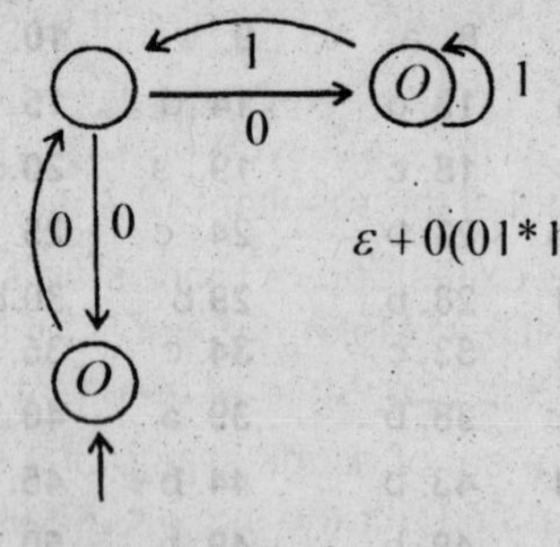

$\varepsilon + 0(01^*1 + 00)^*01^*$

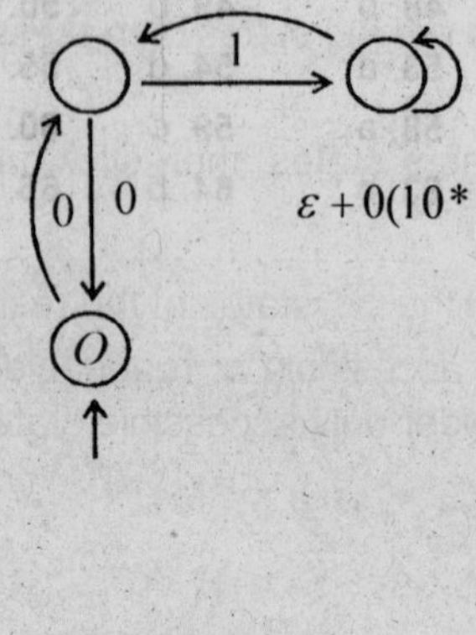

$\varepsilon + 0(10^*1 + 00)^*0$

Alternately

Converting NFA into regular expression. P says ε moves or 0, followed by zero and 1*1 or 00 any number of times followed by zero, again followed by 1 any times.

$$P \rightarrow \varepsilon + 0(01^*1 + 00)^* 01^*$$

Similarly for Q, R & S

19. Only $\{a^n b^{2m} \mid n \geq 0, m \geq o\}$

& $\{x_{cy}, x, y \varepsilon (a,b)^*\}$ are regular sets.

Alternately

In a regular language counting of numbers is not possible but counting that a input is repeated even times or odd times is possible; to get the number (even/odd say 4/7) is not possible.

I can be written as a*(bb)*

IV can be (a+b)* c(a+b)*

while for other regular expressions is not possible as they need the counting as number of times.

20. Regular expression describing the language over (0,1) consisting of strings that contain exactly 2 1's are

(B) 0 * 11 0*

(C) 0* 1 0 * 10 *

and

But 0* 11 0 * is a regular expression containing 2 consecutive 1's, so this is a subset of (C)

21. **Method of conversion of Non-deterministic finite automata to deterministic Finite automata.**

In NDFA, when an input is read, the automata at each step may choose to go to any of the several possible next states.

Step 1. Make all the input symbols of NDFA as the input symbol of the resultant DFA

Step 2. There exists one or more start states for the given NDFA. Make the set of all these start states as the start state of resultant DFA.

Step3. Make a set of all the states of a given NDFA and then find out the subset of this set. All these subsets are the states of the resultant DFA.

Thus if NDFA has n states. then DFA has 2^n states.

Step 4. Out of all the 2^n states of the resultant DFA, all are not accessible or reachable, and we have to consider only accessible states for DFA.

Step 5. Final states of DFA is equal to accessible states of the DFA which contain atleast 1 final state of NDFA.

Thus, if in a NFA, there are n states and in DFA, there are m states, then

$$n \leq m$$

22.

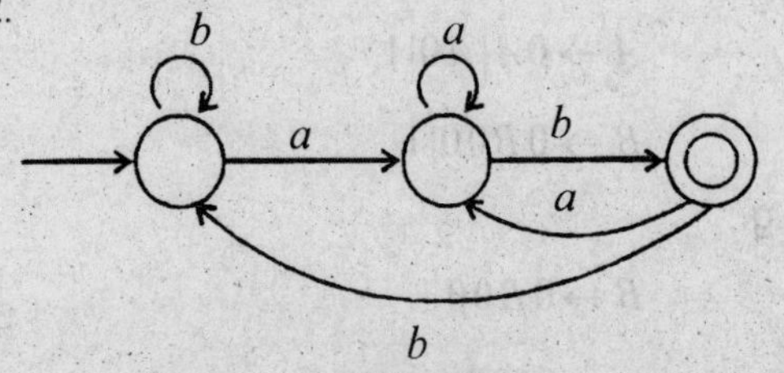

Fig : (i)

Interchanging the final and non final states, we get

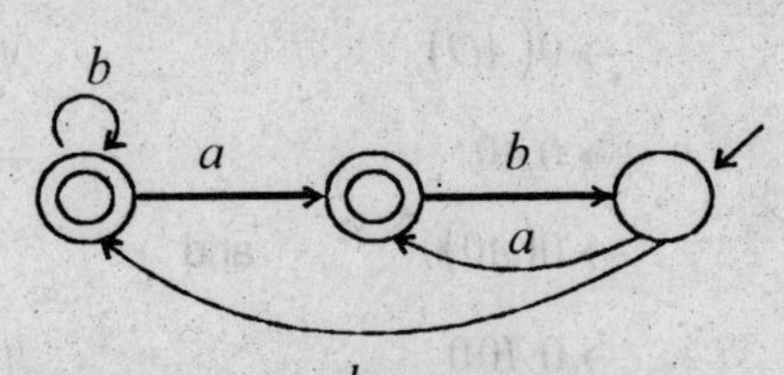

Fig : (ii)

(i) b^*aa^*b

It is the string obtained from fig (1) so (1) will contain all strings ending with ab.

Now, since, final state of (1) has become non final state in (2), so (ii) will satisfy all the strings except ending with ab.

23. $$L_1 = \{a^i b^j c^k \mid i = j, k \geq 1\}$$

eg; a a b b c

$$L_2 = \{a^i b^j \mid j = 2i, i \geq 0\}$$

eg; a b b

Thus, from both the examples, we can see that

$L_1 \cap L_2 = \phi$ as L_1 contains a a b b c

and L_2 contains a b b

e.g. L_1 contains c

if i =0=j, k=1

and L_2 contains ϕ

Thus we see that L_1 will always contains atleast 1c but not L_2

Hence $L_1 \cap L_2 = \Phi$

L_1 is non regular as we can't count the number of i and j.

24. $S \to AA \mid B$

$A \to 0A \mid A0 \mid 1$

$B \to 0B00 \mid 1$

e.g ;

$B \to 0B00$

$\to 0100$

$A \to 0A$ $\quad A \to 1$

$\to 0(AO)$ $\quad A \to 0A$

$\to 0A0$ $\quad \to 01$

$\to 0(A0)0$ and

$\to 0A00$ $\quad A \to A0$

$\to 0100$ $\quad \to 10$

$\therefore$ $A \to 0100 \quad A \to 0A$

and 01 $\quad \to 0A0$

and 10 $\quad \to 010$

and 1

$S \to AA \mid B$

(c) $\{0^i 10^j \mid i, j \geq 0\} \cup \{0^n 10^{2n} \mid n \geq 1\}$

if $i = 0 = j$, then $A \to 1$

$i = 1 = j$, then 010

$i = 1, j = 2$ then 0100

Thus, all these satisfy our grammar

$A \to 0A \mid A0 \mid 1$

26. M_1

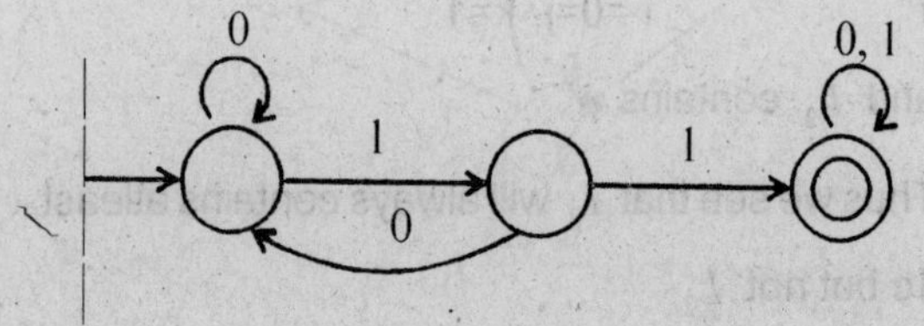

$0*1(00*1)1(0+1)^*$

M_2

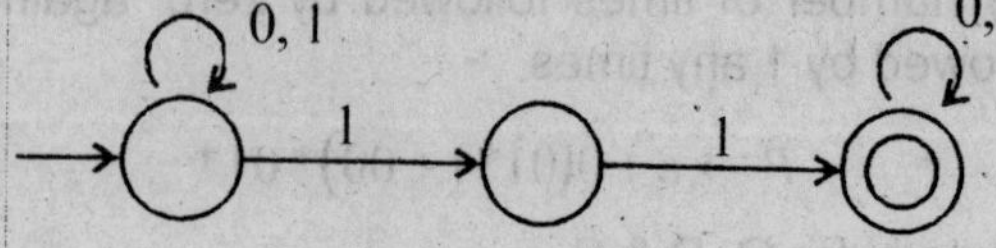

$(0+1)(0+1)*11(0+1)*$

e.g $M_1 \Rightarrow 11, 011,$

01011,...... etc

$M_2 \Rightarrow 11; 011....$ etc.

Thus, both the automata can accept same strings

Hence $L_1 = L_2$

27. S → a(S)

→ a(A) $\quad - I$

→ a(a A b) $\quad - II$

→ a a (A) b

→ a a (b A a) b $\quad - III$

→ a a b (A) a b

→ a a b (bAa) ab $\quad - IV$

→ a a b b ∈ a a b $\quad - V$

→ a a b b a a b $\quad - VI$

28. Total 6 steps are needed and these can be two parse trees which can be made through right production rule and left production rule.

29. As we know that Ambiguity problem for CFGs is undecidable.

30. Every finite subset of a non - regular set is regular.

31. The minimum state deterministic finite automata accepting specified language will have 15 states.

32. Language $L = \{0^i 21^i \mid i \geq 0\}$

Over alphabet $\{0, 1, 2\}$ is recursive and is a deterministic CFL (every CFL is recursive)

33. Language

$$L = \left\{ wxw^R \mid x_1\, w \in \{0,1\}^+ \right\}$$

is a regular language

34. It is ambiguous

35. $b^* a(a+b)^*$ is regular expression which gives language accepted by automata.

36.

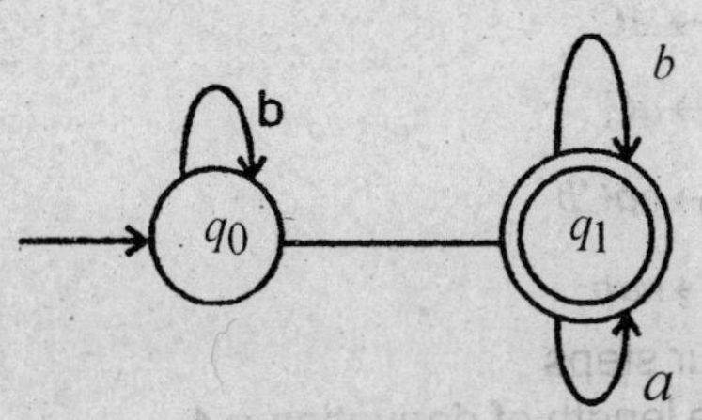

Since state q_3 is isolated and q_1 and q_2 are indistinguishable.

37. String aa bb ab is generated by grammer.

38. Two distinct derivation trees are

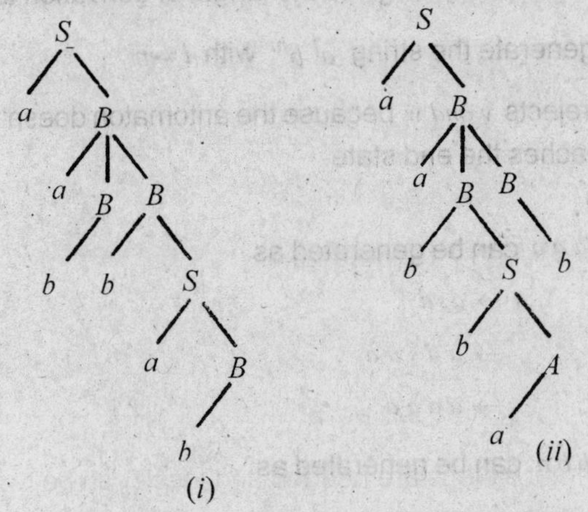

39. Ambiguous grammer is always superset of disambiguous grammer.

So $L(D) \subset L(G)$

40. Because in grammar G1, there is no y which before any x and in grammer G2, every x is followed by at least one y

$$G1: S \Rightarrow yB \Rightarrow yyB \Rightarrow yyzB$$

there is no y appears before any x

$$G2: S \Rightarrow xB \Rightarrow xyS$$

So every x is followed by at least one y in G2.

41. In this DFA, states S_1 *and* S_3 are accepting states and so is start state.

When we move S_0 *to* S_1 it accept strings with odd number of x.

Similarly S_0 *to* S_3 derives string with odd number of y.

So $S_1S_2S_1$ derives string odd number of x and even number of y

So $S_3S_2S_3$ derives string and number of y and even number of x

So DFA recognize all string of x and y with either even number of x and odd number of y or odd number of x and even number of y.

42.

$$S \to xB \mid yA$$

$$A \to x \mid xS \mid yAA$$

$$B \to y \mid yS \mid yBB$$

(i) xxyy

$$A \Rightarrow xS \Rightarrow xxB \Rightarrow xxyS \Rightarrow xxyyA \Rightarrow xxyyx$$

(iii) xyxy

$$S \Rightarrow xB \Rightarrow xyS \Rightarrow xxyB \Rightarrow xyxy$$

(iv) y x x y

$$S \Rightarrow yA \Rightarrow yxS \Rightarrow yxxB \Rightarrow yxxy$$

Hence answer is (d)

43. G1 grammer can be generated using regular grammer

For example a^+

But G2 grammer is context free grammer.

45.

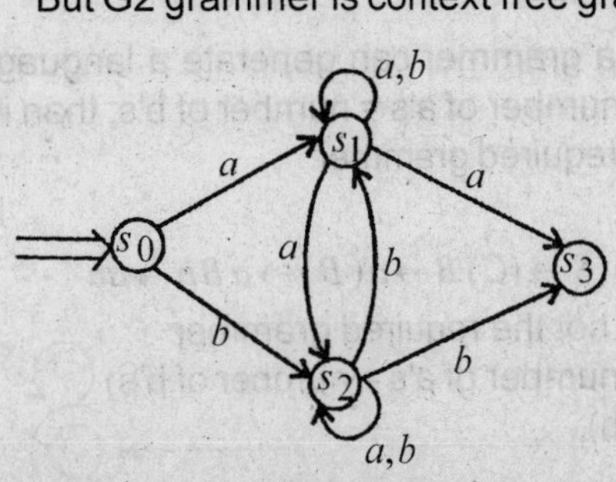

$$R = (a+b)^*(aa+bb)(a+b)^*$$

46.

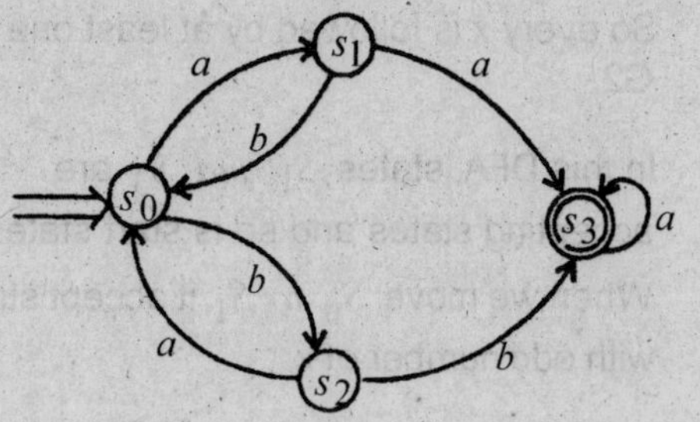

47. Considering only option (c) is right

$$a(ba)^*(a+bb)+b(ab)^*(b+aa)(a+b)^*$$

49. By observation, all the numbers of the form $5x+2, x \in N$ except of the form $7(5n+4)+4n \in N$ belong to the language, since we have an algorithm to check whether a given number belongs be L or not. It is recursive.

50. Let's take a string abba generated by G. The two leftmost derivations are

(i) $S \to SS \to \in S \to \in SS \to SS \to abS \to abba$

(ii) $S \to SS \to abS \to abba$

$\therefore$ G is ambiguou

As any replacement of S with terminals will add equal number of a's and b's to the string

$\therefore$ All strings generated by G will have equal number of a's and b's

G is also accepted by a deterministic PDA.

51. Since L_1 may or may not be content - free, hence options(a) and (c) are ruled out.

Since L_3 is not recursive, hence option (b) is also ruled out.

54. Language $L=\{a^i b^i \mid i \neq j\}$?

i.e. a language in which number of a's $\neq$ number of b's.

And if a grammer can generate a language in which number of a's = number of b's, than it will be the required grammer.

From

$$(a), S \to (C)B \to a(B) \to aBb \to ab$$

So it is not the required grammer

(since number of a's = number of b's)

From (b),

$$S \to aS \to ab\,B$$

not required grammer

From (c),

$$S(C)B \to acbB \to ab\,B \to ab$$

not required grammer.

From (d), In all cases we do not have number of a's = number of b's.

55. Since in above option (d) the derivation is (For string $a^2 b$ suppose)

1. $S \to AC$
2. $S \to aC$
3. $S \to aaCb$
4. $S \to aab$

has four steps

i.e., the length of derivation = 4

Here, $l=2$

$m=1$

so maximum $(l,m)+2=4$

Which satisfies the solution i.e, derivation completes in four steps hence, in general

max. $(l,m)+2$ gives the length of derivation to generate the string $a^l b^m$ with $l \neq m$.

56. It rejects v *and* w because the antomaton doesn't reaches the end state.

57. $a\,a\,a\,a$ can be generated as

$s \to a\,s\,a$

$\to a\,a\,s\,a\,a$

$\to a\,a\,a\,a$

$a\,b\,b\,a$ can be generated as

$s \to a\,s\,a$

$\to a\,b\,s\,b\,a$

$\to a\,b\,b\,a$

$b\,a\,b\,a\,a\,b\,a\,b$ can be generated as

$s \to b\,s\,b$

$\to b\,a\,s\,a\,b$

$\to b\,a\,b\,s\,b\,a\,b$

$\to b\,a\,b\,a\,s\,a\,b\,a\,b$

$\to b\,a\,b\,a\,a\,a\,b\,a\,b$

61. $\left\{a^l\, b^m\, c^n \mid l \neq m \text{ or } m \neq n\right\}$

because when $l \neq m$, no specifications for n are given, same for $m \neq n$.

it cannot be determined by deterministic PDA.

62.

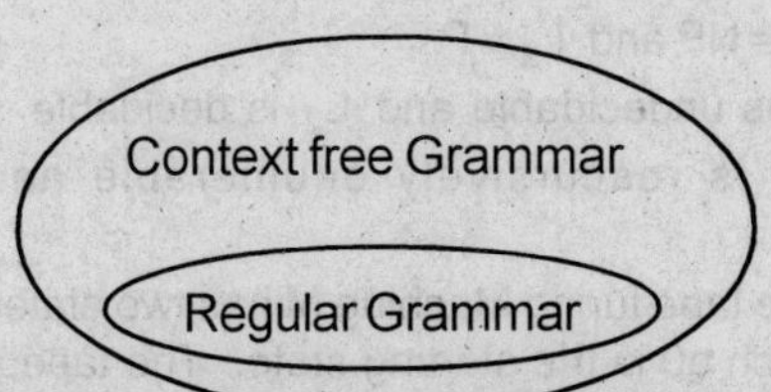

The intersection of a context-free Grammar and regular Grammar is a context free language.

64. Expansion of given productions given

$$
\begin{aligned}
s &\rightarrow a\,S\,A\,b \\
&\rightarrow a\,a\,S\,A\,b\,A\,b \\
&\rightarrow a\,a\,a\,S\,A\,b\,A\,b\,A\,b \\
&\rightarrow a\,a\,a\,A\,b\,A\,b\,A\,b \\
&\rightarrow a\,a\,a\,b\,A\,A\,b\,A\,b \\
&\rightarrow a\,a\,a\,b\,A\,b\,A\,b \\
&\rightarrow a\,a\,a\,b\,b\,A\,b\,A\,b \\
&\rightarrow a\,a\,a\,b\,b\,b\,A\,b\,A\,b \\
&\rightarrow a\,a\,a\,b\,b\,b\,b\,A\,b \\
&\rightarrow a\,a\,a\,b\,b\,b\,b\,b
\end{aligned}
$$

$$a^3\, b^5 \text{ and } 3 \leq 5$$

TEST YOURSELF - II

1. Ram and Shyam have been asked to show that a certain problem Π is NP-complete. Ram shows a polynomial time reduction from the 3-SAT problem to Π, and Shyam shows a polynomial time reduction from Π to 3-SAT. Which of the following can be inferred from these reductions ? **[GATE - 03]**

a) Π is NP-hard but not NP-complete

b) Π is NP, but is not NP-complete

c) Π is NP-complete

d) Π is neither NP-hard, nor in NP

2. No body knows yet if P = NP. Consider the language L defined as follows. $L = \begin{cases} (0+1)^* \text{ if } P = NP \\ \phi \text{ otherwise} \end{cases}$

Which of the following statements is true ? **[GATE - 03]**

a) L is recursive

b) L is recursively enumerable but not recursive

c) L is not recursively enumerable

d) Whether L is recursive or not will be known after we find out if P = NP

3. The regular expression $0^*\left(10^*\right)^*$ denotes the same set as **[GATE - 03]**

a) $(1^*0)^*1^*$ b) $0+(0+10)^*$

c) $(0+1)^*10(0+1)^*$ d) none of the above

4. If the strings of a language L can be effectively enumerated in lexicographic (i.e., alphabetic) order, which of the following statements is true ? **[GATE - 03]**

a) L is necessarily finite

b) L is regular but not necessarily finite

c) L is context free but not necessarily regular

d) L is recursive but necessarily context free

5. Consider the following deterministic finite state automation M. **[GATE - 03]**

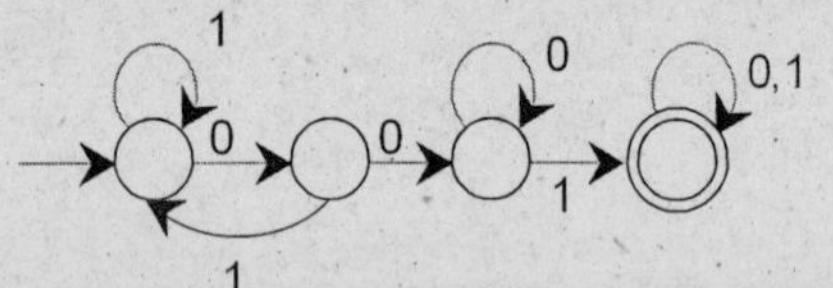

Let S denote the set of seven bit binary strings in which the first, the fourth, and the last bits are 1. The number of strings in S that are accepted by M is

a) 1 b) 5 c) 7 d) 8

6. Let $G=(\{S\},\{a,b\},R,S)$ be a context free grammar where the rule set R is $S \rightarrow aSb \mid SS \mid \epsilon$ **[GATE - 03]**

a) G is not ambiguous

b) There exist $x, y \in L(G)$ such that $xy \notin L(G)$

c) There is a deterministic pushdown automation that accepts L(G)

d) We can find a deterministic finite state automation that accepts L(G)

7. Consider two languages L_1 and L_2, each on the alphabet Σ. Let $f: \Sigma^* \rightarrow \Sigma^*$ be a polynomial time computable bijection such that $(\forall x)[x \in L_1 \text{ iff } f(x) \in L_2]$. Further, let f^{-1} be also polynomial time computable.

Which of the following CANNOT be true ? **[GATE - 03]**

a) $L_1 \in P$ and L_2 is finite

b) $L_1 \in NP$ and $L_2 \in P$

c) L_1 is undecidable and L_2 is decidable

d) L_1 is rescursively enumerable and L_2 is recursive.

8. A single tape turing Machine M has two states q0 and q1, of which q0 is the starting state. The tape alphabet of M is {0, 1, B} and its input alphabet is {0, 1}. The symbol B is the blank symbol used to indicate end of an input string. the transition function of M is described in the following table.

	0	1	B
q0	q1, 1, R	q1, 1, R	Halt
q1	q1, 1, R	q0, 1, L	q0, B, L

The table is interpreted as illustrated below.

The entry (q1, 1, R) in row q0 and column 1 signifies that if M is in state q0 and reads 1 on the current tape square, then it writes 1 on the same tape square, moves its tape head one position to the right and transitions to state q1. Which of the following statements is TRUE about M ? **[GATE - 03]**

a) M does not halt on any string in $(0+1)^+$

b) M does not halt on any string in $(00+1)^*$

c) M halts on all strings ending in a 0

d) M halts on all strings ending in a 1

9. Define languages L_0 and L_1 as follows :

$L_0 = \{<M, w, 0> \backslash M \text{ halts on } w\}$ **[GATE - 03]**

$L_1 = \{<M, w, 1> \backslash M \text{ does not halt on } w\}$

Here <M, w, i> is a triple t, whose first component, M, is an encoding of a Turing Machine, second component, w, is a string, and third component, i, is a bit.

Let $L = L_0 \cup L_1$. Which of the following is TRUE ?

a) L is recursively enumerable, but $\overline{L}$ is not.

b) $\overline{L}$ is recursively enumerable, but L is not

c) Both L and $\overline{L}$ are recursive

d) Neither L nor $\overline{L}$ is recursively enumerable

10. Consider the NFA M shown below. **[GATE - 03]**

Let the language accepted by M be L. Let L_1 be the languae accepted by the NFA M_1, obtained by changing the accepting state of M to a non-accepting state and by changing the non-accepting states of M to accepting states. Which of the following statments is TRUE ?

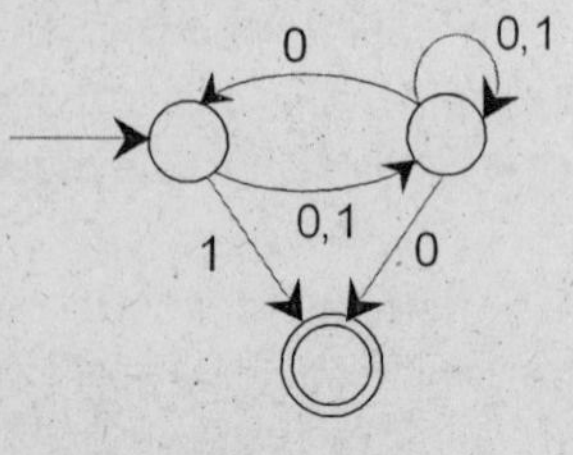

a) $L_1 = \{0, 1\}^* - L$ b) $L_1 = \{0, 1\}^*$

c) $L_1 \subseteq L$ d) $L_1 = L$

11. The following finite state machine accepts all those binary strings in which the number of 1's and 0's are respectively **[GATE - 04]**

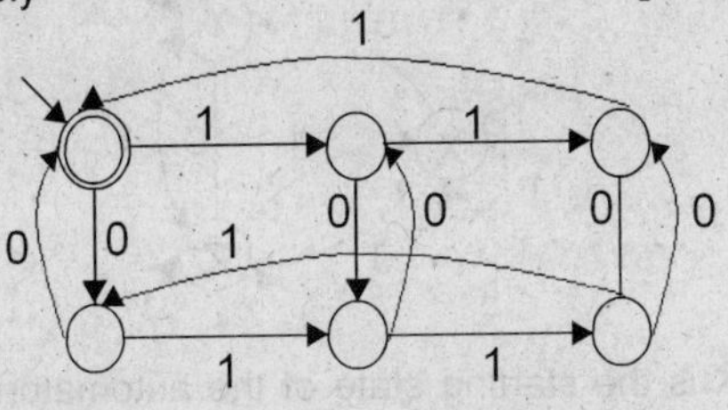

a) divisible by 3 and 2 b) odd and even

c) even and odd d) divisible by 2 and 3

12. The language $\{a^m b^n c^{m+n} \mid m, n \geq 1\}$ is **[GATE - 04]**

a) regular b) context free but not regular

c) context sensitive but not context free

d) type-0 but not context sensitive

13. Consider the following grammar G : **[GATE - 04]**

$S \to bS \mid aA \mid b$

$A \to bA \mid aB$

$B \to bB \mid aS \mid a$

Let $N_a(w)$ and $N_b(w)$ denote the number of a's and b's in a string w respectively. The language $L(G) \subseteq \{a,b\}^+$ generated by G is **[GATE - 04]**

a) $\{w \mid N_a(w) > 3N_b(w)\}$

b) $\{w \mid N_b(w) > 3N_a(w)\}$

c) $\{w \mid N_a(w) = 3k, k \in \{0,1,2,........\}\}$

d) $\{w \mid N_b(w) = 3k, k \in \{0,1,2,........\}\}$

14. L_1 is a recursively enumerable language over Σ. An algorithm A effectively enumerates its words as w_1, w_2, w_3......... Define another language L_2 over $\Sigma \cup \{\#\}$ as $\{w_i \# w_j : w_i, w_j \in L_1,\ i < j\}$. Here # is a new symbol. Consider the following assertions.

S_1 : L_1 is recursive implies L_2 is recursive

S_2 : L_2 is recursive implies L_1 is recursive

Which of the following statements is true ? **[GATE - 04]**

a) Both S_1 and S_2 are true

b) S_1 is true but S_2 is not necessarily true

c) S_2 is true but S_1 is not necessarily true

d) Neither is necessarily true

15. Which one of the following regular expressions is NOT equivalent to the regular expression $(a+b+c)^*$?

a) $(a^*+b^*+c^*)^*$ b) $(a^*b^*c^*)^*$ **[GATE - 04]**

c) $((ab)^*+c^*)^*$ d) $(a^*b^*+c^*)^*$

16. Which one of the following statements is FALSE ?

a) There exist context-free languages such that all the context-free grammars generatingthem as ambiguous

[GATE - 04]

b) An unambiguous context-free grammar always has a unique parse tree for each string of the language generated by it.

c) Both deterministic and non-deterministic pushdown automata always accept the same set of languages

d) A finite set of string from some alphabet is always a regular language

17. Let $M = (K, \Sigma, \Gamma, \Delta, s, F)$ be a pushdown automaton, where $K = \{s, f\}, F = \{f\}, \Sigma = \{a, b\}, \Gamma = \{a\}$ and **[GATE - 04]**

$\Delta = \{((s,a,\varepsilon),(s,a)),((s,b,\varepsilon),(s,a)),((s,a,\varepsilon),(f,\varepsilon),(f,a,a),(f,\varepsilon),(f,b,a),(f,\varepsilon))\}$

Which one of the following strings is not a member of L(M) ?

a) aaa b) aabab c) baaba d) bab

18. Let $M = (K, \Sigma, \delta, s, F)$ be a finite state automation, where $K = \{A, B\}, \Sigma = \{a, b\}, s = A, F = \{B\}$,

$\delta(A, a) = A, \delta(A, b) = B, \delta(B, a) = B$ and $\delta(B, b) = A$

A grammar to generate the language accepted by M can be specified as $G = (V, \Sigma, R, S)$, where $V = K \cup \Sigma$, and $S = A$

Which one of the following set of rules will make $L(G) = L(M)$? **[GATE - 04]**

a) $\{A \to aB, A \to bA, B \to bA, B \to aA, B \to \varepsilon\}$

b) $\{A \to aA, A \to bB, B \to aB, B \to bA, B \to \varepsilon\}$

c) $\{A \to bB, A \to aB, B \to aA, B \to bA, B \to \varepsilon\}$

d) $\{A \to aA, A \to bA, B \to aB, B \to bA, A \to \varepsilon\}$

19. Consider three decision problems P_1, P_2 and P_3. It is known that P_1 is decidable and P_2 is undecidable. Which one of the following is TRUE ? **[GATE - 05]**

a) P_3 is decidable if P_1 reducible to P_3

b) P_3 is undecidable if P_3 is reducible to P_2

c) P_3 is undecidable if P_2 is ruducible to P_3

d) P_3 is decidable if P_3 reducible to P_2 S complement

20. Consider the machine M :

The language recognized by M is :

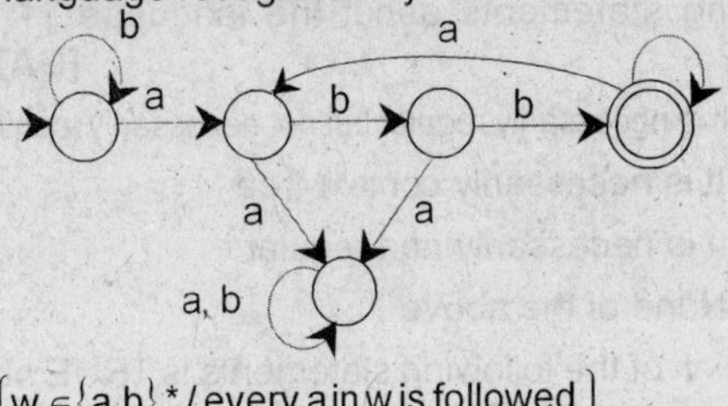

a) $\left\{\begin{matrix} w \in \{a,b\}^* / \text{every a in w is followed} \\ \text{by exactly two b's} \end{matrix}\right\}$

b) $\left\{\begin{matrix} w \in \{a,b\}^* \text{every a in w is followed by} \\ \text{atleast two b's} \end{matrix}\right\}$

c) $\{w \in \{a,b\}^* \text{w contains the substing 'abb'}\}$

d) $\{w \in \{a,b\}^* \text{w does not contain 'aa' as a substring}\}$

21. Let N_f and N_p denote the classes of languages accepted by non-deterministic finite automata and non-deterministic push-down automata, respectively. Let D_f and D_p denote the classes of languages accepted by deterministic finite automata and deterministic push-down automata, respecively. Which one of the following is TRUE ? [GATE - 05]

a) $D_f \subset N_f$ and $D_p \subset N_p$ b) $D_f \subset N_f$ and $D_p = N_p$

c) $D_f = N_f$ and $D_p = N_p$ d) $D_f = N_f$ and $D_p \subset N_p$

22. Consider the languages [GATE - 05]

$L_1 = \{a^n b^n c^m | n, m > 0\}$ and $L_2 = \{a^n b^m c^m | n, m > 0\}$

Which one of the following statements is FALSE ?

a) $L_1 \cap L_2$ is a context-free language

b) $L_1 \cup L_2$ is a context-free language

c) L_1 and L_2 are context-free language

d) $L_1 \cap L_2$ is a context sensitive language

23. Let L_1 be a recursive language, and let L_2 be a recursively enumerable but not a recursive language. Which one of the following is TRUE ? [GATE - 05]

a) $\bar{L_1}$ is recursive and $\bar{L_2}$ is recursively enumerable

b) $\bar{L_1}$ is recursive and $\bar{L_2}$ is not recursively enumerable

c) $\bar{L_1}$ and $\bar{L_2}$ are recursively enumerable

d) $\bar{L_1}$ is recursively enumerable and $\bar{L_2}$ is recursive.

24. Consider the languages [GATE - 05]

$L_1 = \{ww^R \setminus w \in \{0, 1\}^*\}$

$L_2 = \{w\#w^R \setminus w \in \{0, 1\}^*\}$, where # is a special symbol

$L_3 = \{ww \setminus w \in \{0, 1\}^*\}$

Which of the following is TRUE ?

a) L_1 is a deterministic CFL

b) L_2 is a deterministic CFL

c) L_3 is a CFL, but not a deterministic CFL

d) L_3 is a deterministic CFL

25. Let L be a regular language and M be a context-free language, both over the alphabet Σ. Let L^c and M^c denote the complements of L and M respectively. Which of the following statements about the language $L^c \cup M^c$ is TRUE? [GATE - 05]

a) It is necessarily regular but not necessarily context-free

b) It is necessarily context-free

c) It is necessarily non-regular

d) None of the above

26. Which of the following statements is TRUE about the regular expression 01*0 ? [GATE - 05]

a) It represents a finite set of finite strings.

b) It represents an infinite set of finite strings.

c) It represents a finite set of infinite strings.

d) It represents an infinite set of infinite strings.

27. The langugage $\{0^n 1^n 2^n ; 1 \le n \le 10^6\}$ is [GATE - 05]

a) regular

b) context-free but not regular

c) context-free but its complement is not context-free

d) not context-free

28. Consider the non-deterministic finite automaton (NFA) shown in the figure. [GATE - 05]

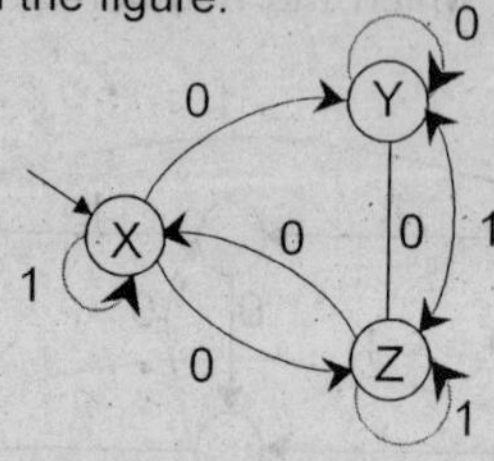

State X is the starting state of the automaton. Let the language accepted by the NFA with Y as the only accepting state be L1. Similarly, let the language accepted by the NFA with Z as the only accepting state be L2. Which of the following statements about L1 and L2 is TRUE ? [GATE - 05]

a) L1 = L2 b) $L1 \subset L2$

c) $L2 \subset L1$ d) None of the above

29. Let P be a non-deterministic push-down automaton (NPDA) with exactly one state, q, and exactly one symbol, Z, in its stack alphabet. State q is both the starting as well as the accepting state of the PDA. The stack is initialized with one Z before the start of the operation of the PDA. Let the input alphabet of the PDA be Σ. Let L(P) be the language accepted by the PDA by reading a string and reaching its accepting state. Let N(P) be the language accepted by the PDA by reading a string and emptying its stack. [GATE - 05]

Which of the following statements is TRUE ?

a) L(P) is necessarily Σ^* but N(P) is not necessarily Σ^*

b) N(P) is necessarily Σ^* but L(P) is not necessarily Σ^*

c) Both L(P) and N(P) is necessarily Σ^*.

d) Neither L(P) nor N(P) are necessarily Σ^*.

30. Consider the regular grammar : [GATE - 05]

$S \to Xa \mid Ya$

$X \to Za$

$Z \to Sa \mid B$

$Y \to Wa$

$W \to Sa$

Where S is the starting symbol, the set of terminals is {a} and the set of non-terminals is {S, W, X, Y, Z}.

We wish to construct a deterministic finite automaton (DFA) to recognize the same language. What is the minimum number of states required for the DFA ?

a) 2 b) 3 c) 4 d) 5

31. A language L satisfies the Pumping Lemma for regular languages, and also the Pumping Lemma for context-free languages. Which of the following statements about L is TRUE ? [GATE - 05]

a) L is necessarily a regular language.

b) L is necessarily a context-free language, but not necessarily a regular language.

c) L is necessarily a non-regular language.

d) None of the above.

Statement for Linked Answer Questions 32 and 33

Consider the context-free grammar **[GATE - 05]**

$E \to E + E$

$E \to (E * E)$

$E \to id$

Where E is the starting symbol, the set of terminals is {id,(,+,),*}, and the set of non-terminals is {E}.

32. Which of the following terminal strings has more than one parse tree when parsed according to the above grammar ?

a) id + id + id + id b) id + (id * (id * id))

c) (id * (id * id)) + id d) ((id * id + id) * id)

33. For the terminal string with more than one parse tree obtained as solution to Questiion 83a, how many parse trees are possible ?

a) 5 b) 4 c) 3 d) 2

ANSWER KEY				
1. c	**2.** d	**3.** d	**4.** a	**5.** a
6. d	**7.** c	**8.** d	**9.** a	**10.** c
11. a	**12.** b	**13.** c	**14.** b	**15.** c
16. c	**17.** c	**18.** b	**19.** a	**20.** b
21. d	**22.** a	**23.** b	**24.** b	**25.** d
26. b	**27.** a	**28.** a	**29.** d	**30.** d
31. a	**32.** a	**33.** a		

EXPLANATORY NOTES

3. $0*(10*) = 0*(1*0*)$

a) $(1*0)*1* = (1*0*)1*$

b) $0+(0+10)* = 0+(0*1*0*)$

c) $(0+1)*10(0+1)* = 0*1*10(0*1*)$

d) none of the above

5.

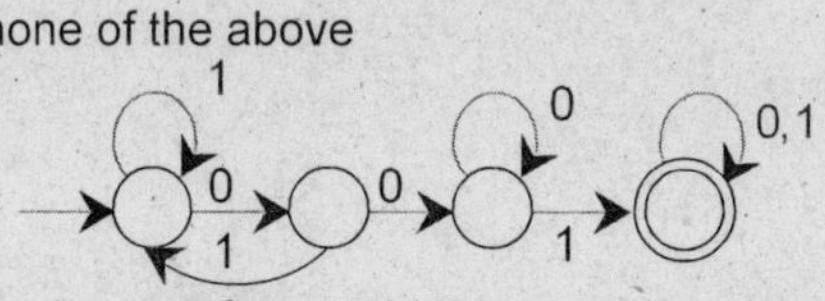

S denote the set of seven bit binary strings in which the first, fourth and last bits are 1.

1 2 3 4 5 6 7

⇒ 1 0 0 1 0 0 1

So number of strings in S that are accepted by M is one.

6. $S \to aSb|SS| \in$

Hence, S is ambiguous

So, we can find a deterministic finite state automation that accepts L(G).

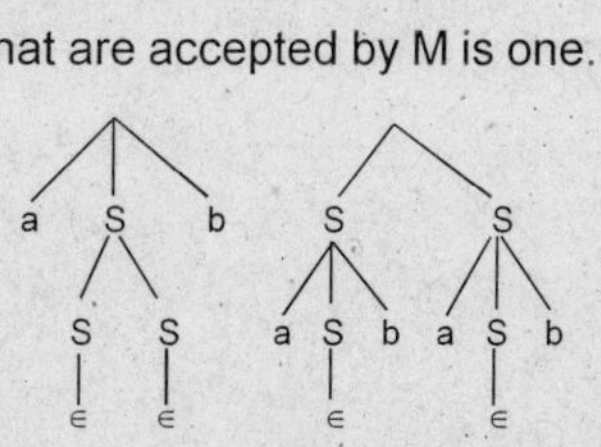

8.

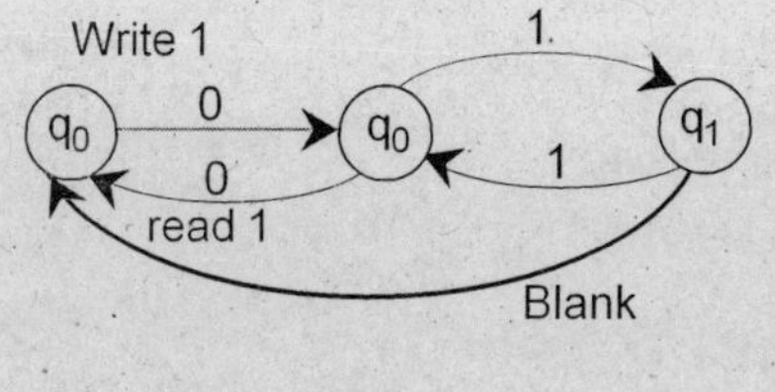

10.

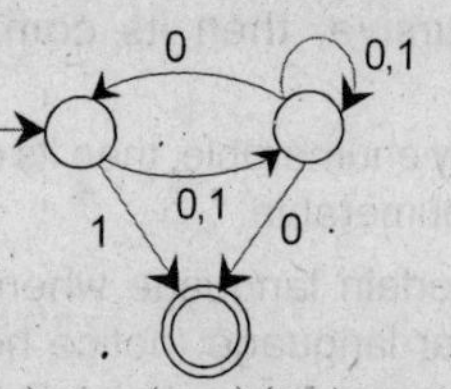

Then $L = (0/1)01 = (0/1)(0,1)*0$

Changing the accepting state to non-accepting state and changing the non-accepting state to accepting state is know as h_1.

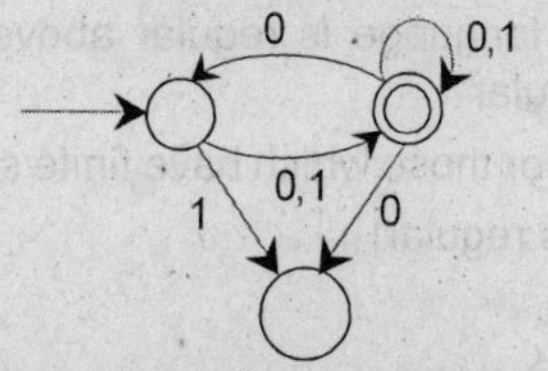

$L_1 = (0/1)(0,1)* = 0(0/1)(0,1)*$

We find from both language L and L_1

So , $L_1 \subseteq L$

11. This automata accept 10 11 11 01

and checking for other input it is clear that it accept number divisible by 3 and 2.

12. $L = \{a^m b^n c^{m+n} \mid m, n \geq 1\}$

The given language is context free but not regular language

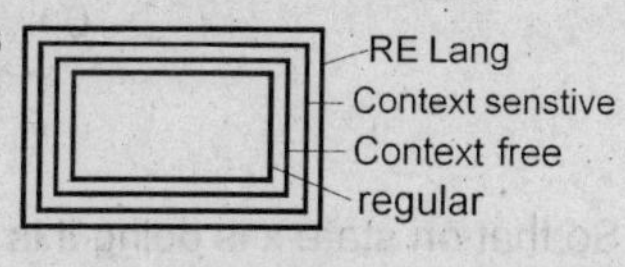

13. $S \to bs \mid aA \mid b$

$A \to bA \mid aB$

$B \to bB \mid aS \mid b$

This grammar will execute

$\{w \mid Na(w) = 3k;\ \ k \in \{0, 1, 2, ...\}\}.$

15. $((ab)* + c*)* = \{e, ab, abab, ab\ ab\ ab + c\}*$

Contains the string which is not present in expansion of $(a+b+c)*$. Hence not equivalent.

16. Acceptability capability of deterministic and non deterministic pushdown automata are different, so both deterministic as well as non-deterministic automata not always accept the same set of language. Hence (c) is not correct

20. (a) and (d) are two possible solutions but best is (b).

Why (a) is wrong : Run this on DFA

a b b b : is stop in final stage while it have three b.

Why not (c) : As **a b b a a** have substring 'a b b'

but, don't stop at final state.

21. There is a language

ww^R which is N.C.F.L. but not D.C.F.L.

There is no language which is accepted by D.F.A. and not by N.D.F.A. or vice versa.

22. Result of these two grammar is only the case where a = b = c.

23. If L_1 is recursive, then its complement is also recursive.

If L_2 is recursively enumerable, then its complement can't be recursively enumerable.

25. If we have certain language where complement is $a^n b^n$ and regular language (notice here that $a^n b^n$ is D.C.F.L. hence its complement also D.C.F.L)

a* b* is complement of certain language, then union is a*b* which is regular but it is not always true.

27. Every finite language is regular above language is finite hence regular

[finite language or those which have finite storing of finite length is always regular]

28. Language accepted with y as an only accepting state is

If we delete state 'Y', then

$x \rightarrow Z$, $Z \rightarrow Z$, $x \rightarrow Z$

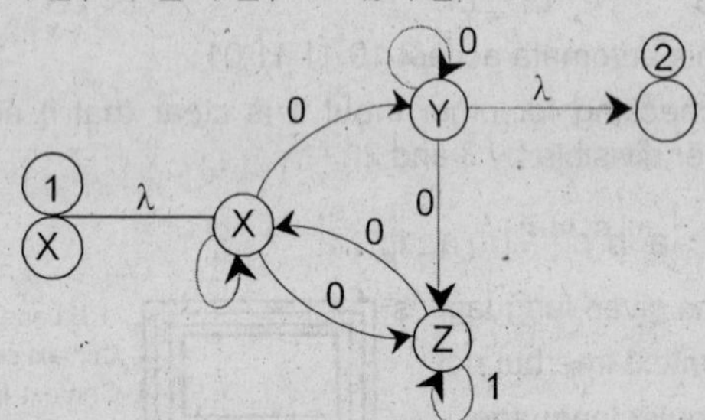

So that on state x is doing it is self loop on Σ/P, when it receive 0. Then we have to go on both Y and Z.

On state Y

If **input** is of only zero, then OK. If looping and goto Z but when **input** is one, you have no way this state is dead.

	0	1
{x}	{y, z}	{y}
{y, z}	{y, z, x}	{y, z}
{y, z, x}	{y, z, x}	{y, z, x}

29. Consider here that language accepted by empty the stack and goes to final state are same.

So, if we not define any transaction on particular L(P) symbol than N(P) is not accept even it is part of Σ^*.

30. Here no non-terminal is useless so all symbol denote to a particular state hence the answer.

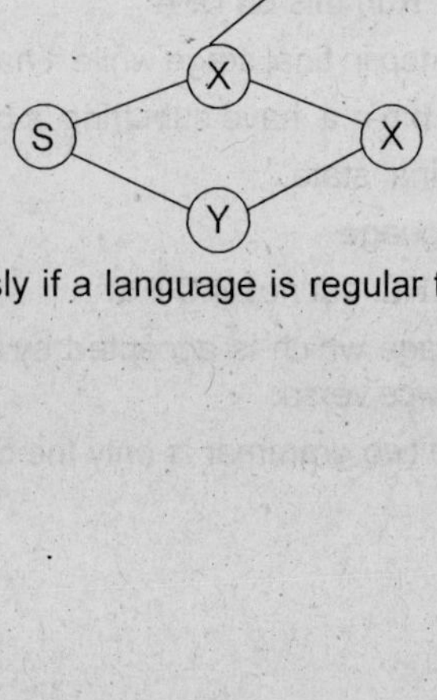

31. Obviously if a language is regular than also context free.

Unit - IV
Digital Logic Circuits

UNIT - IV DIGITAL LOGIC CIRCUITS

I. LOGIC FUNCTIONS AND MINIMISATION TECHNIQUES

- Logic Gates
- Boolean Algebra
- Logic Function Minimisation Techniques
- Digital Integrated Circuits

1. Logic Gates

A logic gate is an electronic circuit which produces a typical output signal depending on its input signal. The output signal of a gate is a simple boolean operation of its input signal. Gates are the basic logic elements. Any boolean function can be represented in the form of gates. In general we can represent gates in three ways. These are :

- Graphic symbols • Algebraic notaion • Truth table.

All these three notations for the six basic gates are shown in the following figure :

Gate Name — Symbol — Notation — Truth Table

AND

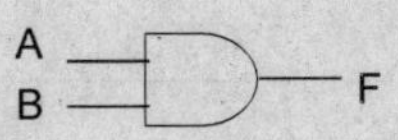

F=A.B

A	B	F
0	0	0
0	1	0
1	0	0
1	1	1

OR

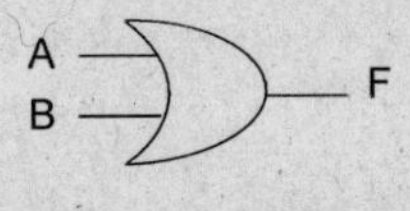

F=A+B

A	B	F
0	0	0
0	1	1
1	0	1
1	1	1

NOT

$F=\overline{A}$

A	F
0	1
1	0

NAND

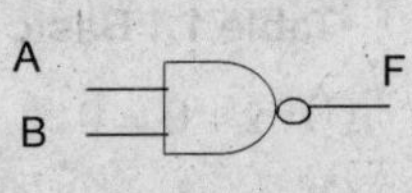

$F = (\overline{AB})$

A	B	F
0	0	1
0	1	1
1	0	1
1	1	0

NOR

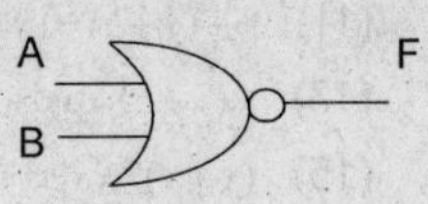

$F = (\overline{A+B})$

A	B	F
0	0	1
0	1	0
1	0	0
1	1	0

XOR

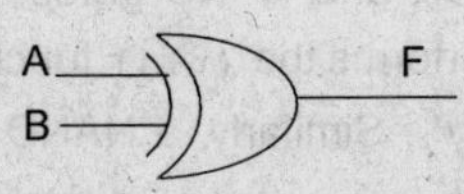

$F=A \oplus B$

A	B	F
0	0	0
0	1	1
1	0	1
1	1	0

2. Boolean Algebra

Boolean algebra is an algebra that deals with binary variables and logic operations. The variables are designated by letters such as A,B,x and y. The three basic logic operations are AND, OR, and complement. A Boolean fuction is an algebraic expression formed with binary variables, the logic operation symbols, parentheses, and equal sign. For a given value of the variables, the Boolean fuction can be either 1 or 0. Consider, for example, the Boolean function

$$F = x + y'z$$

The function F is equal to 1 if x is 1 or if both y' and z are equal to 1; F is equal to 0 otherwise. But saying that $y' = 1$ is equivalent to saying that $y = 0$ since y' is the complement of y. Equivalently, we may say that F is equal to 1 if $x = 1$ or $yz = 01$. The relationship between a function and its binary variables can be represented in a truth table. To represent a function in a truth table we need a list of the 2^n combinations with 1's and 0's of the n binary variables.

As shown in Fig 1.1 (a), there are eight possible distinct combinations for assigning bits to the three variables. The function F is equal to 1 for those combinations where x = 1 or yz = 01; it is equal to 0 for all other combinations.

A Boolean fuction can be transformed from an algebraic expression into

x	y	z	F
0	0	0	0
0	0	1	1
0	1	0	0
0	1	1	0
1	0	0	1
1	0	1	1
1	1	0	1
1	1	1	1

(a) Truth table

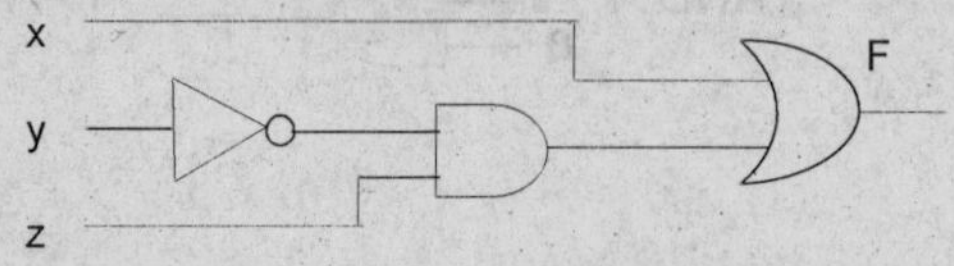

(b) logic diagram

Fig 1.1 Truth table and logic diagram for $F = x + y'z$

A logic diagram composed of AND, OR, and inverter gates. The logic diagram for F is shown in Fig. 1.1 (b). There is an inverter for input y to generate its complement y'. There is an AND gate for the term $y'z$ and an OR gate is used to combine the two terms. In a logic diagram, the variables of the function are taken to be the inputs of the circuit and the variable symbol of the function is taken as the output of the circuit.

The purpose of Boolean algebra is to facilitate the analysis and design of digital circuits. It provides a convenient tool to:

1. Express in algebraic form a truth table relationship between variables.

2. Express in algebraic form the input-output relationship of logic diagrams.

3. Find simpler circuits for the same function.

A Boolean function specified by a truth table can be expressed algebraically in many different ways. By manipulating a Boolean expression according to Boolean algebra rules, one may obtain a simpler expression that will require fewer gates. To see how this is done, we must first study the manipulative capabilities of Boolean algebra.

Table 1.1 lists the most basic relations of Boolean algebra. All the relations can be proven by means of truth tables. The first eight in Table 1.1 show the basic relationship between a single variable and itself, or in conjunction with the binary constants of 1 and 0. The next five relations (9 to 13) are similar to ordinary algebra. Relation 14 does not apply in ordinary algerba but is very useful in manipulating Boolean expressions.Relations 15 and 16 are called De Morgan's theorems and are discussed below. The last relation states that if a variable is complemented twice, one obtains the original value of the variable. The relations listed in the table apply to single variables x,y, and z or to Boolean functions expressed by binary variables x,y,or z.

Table 1.1 Basic relations of Boolean Algebra

(1) $x + 0 = 0$	(2) $x.0 = 0$
(3) $x + 1 = 1$	(4) $x.1 = x$
(5) $x + x = x$	(6) $x.x = x$
(7) $x + x' = 1$	(8) $x.\ x' = 0$
(9) $x + y = y + x$	(10) $xy = yx$
(11) $x+(y+z)=(x+y)+z$	(12) $x(yz)=(xy)z$
(13) $x(y+z)=xy+xz$	(14) $x+yz=(x+y)(x+z)$
(15) $(x+y)'=x'y'$	(16) $(xy)'=x'+y'$
(17) $(x)'=x$	

DeMorgan's theorem is very important in dealing with NOR and NAND gates. It states that a NOR gate that performs the $(x+y)$ function is equivalent to the function xy'. Similarly, a NAND function can be expressed by either (xy) or by $x' + y'$. For this reason, the NOR and NAND gates have two distinct graphic symbols as shown in Figs. 1.2 and 1.3. Instead of representing a NOR gate with an OR graphic symbol followed by a circle, we can represent it by an AND graphic symbol preceeded by circles in all inputs. The invert-AND symbol for the NOR gate follows from DeMorgan's theorem and from the convention that small circles denote complementation. Similarly, the NAND gate has two distinct symbols as shown in fig. 1.3

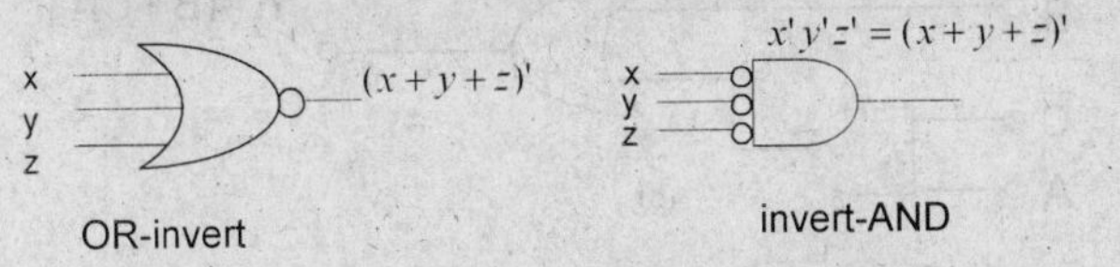

Fig. 1.2 Two graphic symbols for NOR gate

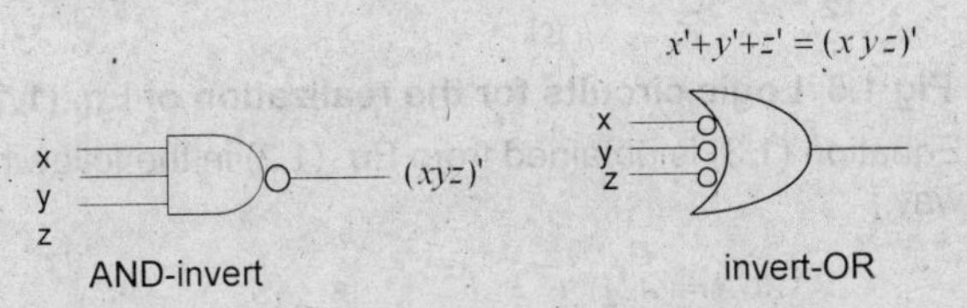

Fig 1.3 Two graphic symbols for NAND gate.

To see how Boolean algebra manipulation is used to simplify digital circuits. Consider the logic diagram of Fig 1.4(a). The output of the first NAND gate is, by DeMorgan's theorem, $(AB')= A'+B$. The output of the circuit is the NAND operation of this term and B' :

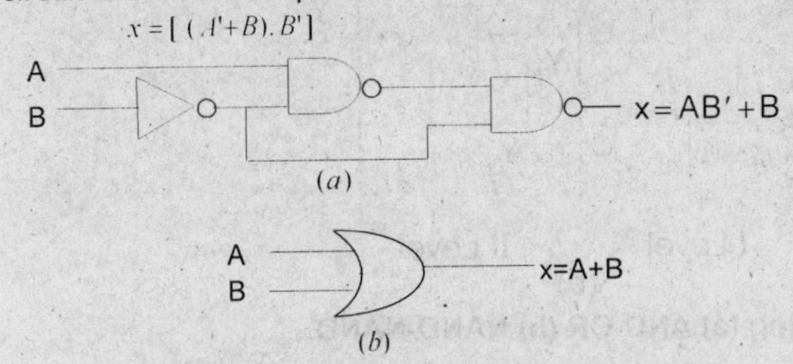

Fig 1.4 Two logic diagrams for the same Boolean function.

Using DeMorgan's theorem twice more, we obtain

$$x = (A'+B)'+B = AB'+B$$

Note that DeMorgan's theorem has been applied three times (to demonstrate its use) but could have been applied only once as follows :

$$x = \left[(AB')'.B'\right]' = AB' + B$$

The expression for x can be simplified by application of the relations listed in Table 1.1.

	by relation
$x = AB' + B$	
$= B + AB'$	(9)
$= (B+A)(B+B')$	(14)
$= (B+A).1$	(7)
$= B+A = A+B$	(4), (9)

The final result produces an OR function and can be implemented with a single OR gate as shown in Fig 1.4(b). One can show that the two circuits produce identical input-output binary relationships by obtaining the truth table of each.

The following methods can be used to simplify the Boolean function:

1. Algebraic method,
2. Karnaugh map technique,
3. Variable entered mapping (VEM) technique, and
4. Quine – McCluskey method.

The algebraic method and the Karnaugh map (K-map) technique have been given here. The K-map is the simplest and most commonly used method. It can be used up to six variables.

The Quine-McCluskey method is a formal method which, in no way depends upon human intuition. It is, suitable for computer mechanization and is seldom used by logic designers manually.

3. Standard Representations For Logical Functions

Logical functions are expressed in terms of logical variables.The values assumed by the logical functions as well as the logical variables are in the binary form. Any arbitrary logic function can be expressed in the following forms :

1. Sum-of-products form (SOP)
2. product-of-sums form (POS)

This does not mean that the logic function cannot be written in any other form. It can be written in various forms but the above two forms are conveniently suited in arriving at the standard methods for designing the circuits which will become clear from the following discussions.

Ex 1.1 : Given the logical equation

$$Y = (A + BC)(B + \overline{C}A) \qquad \textbf{(1.1)}$$

(a) Design a circuit using gates to realize this function.

(b) Find out whether it is possible to design the circuit with only one type of gates (NAND or NOR). If yes, design the circuits.

(c) Find out whether it is possible to simplify this equation. If yes, simplify it.

(d) Now design the circuit using the simplified expression obtained in part (c).

(e) Compare the circuits obtained in parts (a), (b) and (d) from the point of view of number of gates, number of inputs for the gates, types of gates, and propagation delay.

Sol: In Eq. (1.1), there are three input logical variables A,B and C, and Y is the output. The variable C appears as C in one term and as $\overline{C}$ in the other term. A variable in uncomplemented or complemented form is known as a literal.

(a) A circuit using gates can simply be designed by looking at the expression and finding out the basic gates which can be used to realize the various terms and then connect these gates appropriately. We assume that the signals corresponding to each literal are available, that is the variables are available in uncomplemented as well as the complemented form.

I. The first term (A) has only literal A and the second term (BC) has two literals B and C. The second term is recognized as an AND operation and can be realized by using a 2-input AND gate. The combination of these two terms is realized by using a 2-input OR gate. The complete realization of first two terms is shown in Fig 1.5a.

II. The third term (B) is again a single literal term and the fourth term $(\overline{C}A)$ has two literals. The combination of these two terms is similar to the combination of the first two terms and is realized in a similar way. This realization is shown in Fig 1.5b.

III. Now, the complete realization is obtained by using a 2-input AND gate with Y_1 and Y_2 as the inputs and the output of this gate will be the required output Y. This realization is given in Fig 1.5 c. The above design requires three 2-input AND gates and two 2-input OR gates.

(b) (I) Sum-of-Products Form

Equation (1.1) can be written as

$$Y = A(B+\overline{C}A) + (BC)(B+\overline{C}A)$$

$$= AB + A\overline{C}A + BCB + BC\overline{C}A \quad \textbf{(1.2)}$$

$$= AB + A\overline{C} + BC \quad \textbf{(1.3)}$$

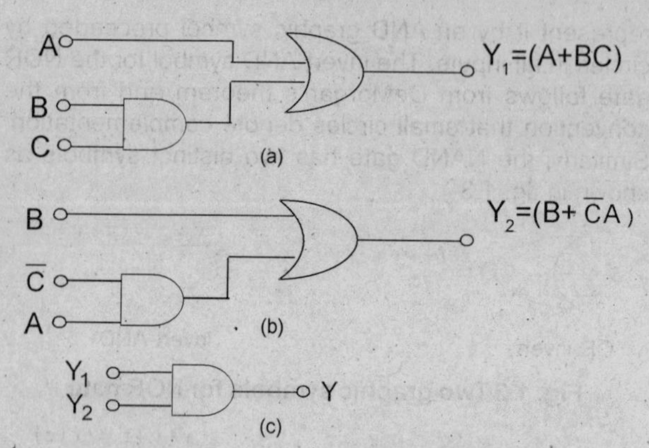

Fig 1.5 Logic circuits for the realization of Eq. (1.1)

Equation (1.3) is obtained from Eq. (1.2) in the following way :

$$A\overline{C}A = (A.A).\overline{C} = A\overline{C}$$

$$BCB = (B.B).C = BC$$

$$BC\overline{C}A = B.(C.\overline{C}).A = B.0.A = 0$$

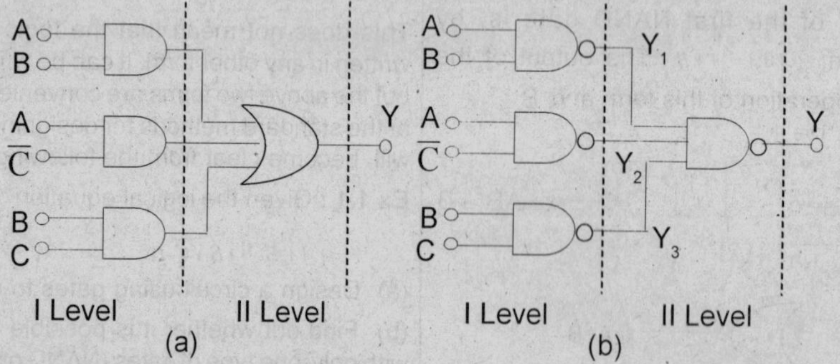

Fig 1.6 Realization of Eq. (1.3) using (a) AND-OR (b) NAND-NAND.

The representation of Eq. (1.3) is known as sum-of-products (SOP) form. This can be realized using AND-OR configuration as shown in Fig 1.6 a. This realization is known as two-level realization.

The first level consists of AND gates and the second level consists of the OR gate. By making use of the De-Morgans theorem we can write Eq. (1.3) as

$$\overline{Y} = \overline{AB + A\overline{C} + BC}$$

$$= \overline{\overline{AB}.\overline{A\overline{C}}.\overline{BC}} \text{ or } Y = \overline{Y_1.Y_2.Y_3} \quad \textbf{(1.4)}$$

whe e $Y_1 = \overline{AB}$, $Y_2 = \overline{A\overline{C}}$, $Y_3 = \overline{BC}$

Equation (1.4) can be realized using NAND gates only. The realization is given in Fig 1.6 b. This is also a two-level realization and in this only NAND gates are used. Therefore, if we express the equation in the SOP form we can always design the circuit using only one type of gates (NAND).

(II) Product-of-Sums Form

It is possible to represent Eq. (1.1) in another form.

$$Y = (A+B)(A+C)(B+\overline{C})(B+A)$$

$$= (A+B)(A+C)(B+\overline{C}) \quad \textbf{(1.5)}$$

The representation of Eq. (1.5) is known as Product-of-sums (POS) form. This can be realized using OR-AND gates as shown in Fig 1.7a. This is also a two-level realization.

Using De-Morgans theorem, we can write Eq.(1.5) as

$$\overline{Y} = \overline{(A+B)(A+C)(B+\overline{C})}$$

$$\overline{(A+B)} + \overline{(A+C)} + \overline{(B+\overline{C})} \text{ or}$$

$$Y = \overline{Y_A + Y_B + Y_C} \quad \textbf{(1.6)}$$

where $Y_A = \overline{A+B}$, $Y_B = \overline{A+C}$, $Y_C = \overline{B+\overline{C}}$

Equation (1.6) can be realized using NOR gates only. This realization is given in Fig 1.7 b which is a two-level realization. Hence, if we express the equation in POS form we can always design the circuit using only one type of gates (NOR).

(c) I. Simplification of Eq. (1.3)

$$Y = AB + A\overline{C} + BC = BC + A\overline{C} \quad \textbf{(1.7)}$$

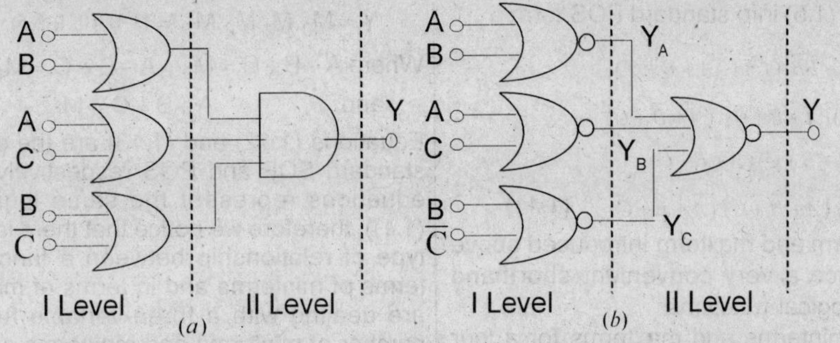

Fig 1.7 Realization of Eq. (1.5) using (a) OR-AND (b) NOR-NOR

II. Simplification of Eq. (1.5),

$$Y = (A+B)(A+C)(B+\overline{C}) = (A+C)(B+\overline{C}) \quad \textbf{(1.8)}$$

(d) Realization of Eqs (1.7) and (1.8) are given in Figs 1.8 a and 1.8 b, respectively.

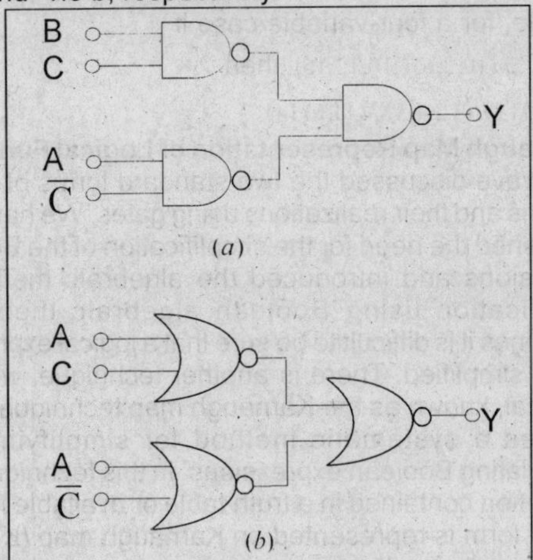

Fig 1.8 Realization of (a) Equation (1.7) (b) Equation (1.8)

(e) The gate requirements corresponding to parts (a), (b), and (d) are given in Table 1.2. The realization of part (a) needs maximum number of gates.

Also, it is a three-level realization which increases the propagation delay time which is same as decrease in speed of operation. Realizations corresponding to parts (b) and (d) are very useful since only one type of gates (NAND/NOR) are required which is very convenient to when we use ICs because a number of similar gates are available in the same package.

Realization corresponding to part (d) requires minimum number of gates. Therefore, the simplification of logical expressions is very useful.

We notice that in the SOP form of Eq. (1.3) and the POS form of Eq. (1.5), all the individual terms do not involve all the three literals.If each term in SOP and POS forms contains all the literals then these are known as standard (or canonical)

Table 1.2 GATE requirements

Part (a)	Part (b)		Part(d)
	AND-OR/NAND-NAND	OR-AND/NOR-NOR	
3,2-input AND or 2,2-input OR	3,2-input AND 1,3-input OR or 3,2-input NAND 1,3-input NAND	3,2-input OR 1,3-inout AND or 3,2-input NOR 1,3-input NOR	3,2-input NAND 3,2-input NAND

SOP and POS, respectively. Each individual term in standard SOP form is called as minterm and in standard POS form as maxterm. The usefulness of the minterm and maxterm representations will become clear from the discussion which follows. SOP form can be converted to standard SOP by ANDing the terms in the expression with terms formed by ORing the variable and its complement which are not present in that term. For example for a three-variable expression with variables A,B and C,if there is a term A, where B and C variables are missing, then we form two terms $(B+\overline{B})$ and $(C+\overline{C})$ and AND them with A. Therfore, we get

$$A.(B+\overline{B}).(C+\overline{C}) = ABC + A\overline{B}C + AB\overline{C} + A\overline{B}\overline{C}$$

Ex 1.2 : Convert Eq. (1.3) into standard SOP form.

Sol : $Y = AB(C+\overline{C}) + A\overline{C}(B+\overline{B}) + BC(A+\overline{A})$

$$= ABC + AB\overline{C} + AB\overline{C} + A\overline{B}\,\overline{C} + ABC + \overline{A}BC$$

$$= ABC + AB\overline{C} + A\overline{B}\,\overline{C} + \overline{A}BC \quad \textbf{(1.9)}$$

Similarly, POS form can be converted to standard POS by ORing the terms in the expression with terms formed by ANDing the variable and its complement which are not present in that term. For example for a three-variable expression with variables A,B, and C, if there is a term A where B and C variables are missing, then we form two terms B. $\overline{B}$ and C. $\overline{C}$ and OR A with these terms. Therefore, we get $A + B\overline{B} + C\overline{C}$

$$= (A + B\overline{B} + C)(A + B\overline{B} + \overline{C})$$

$$= (A+B+C)(A+\overline{B}+C)(A+B+\overline{C})(A+\overline{B}+\overline{C}) \quad \textbf{(1.10)}$$

Ex 1.3 : Convert Eq. (1.5) into standard POS form.

Sol : $=(A+B+C\overline{C})\ (A+B\overline{B}+C)\ (A\overline{A}+B+\overline{C})$

$=(A+B+C)\ (A+B+\overline{C})(A+B+C)\ (A+\overline{B}+C)$

$(A+B+C)(A+B+C)$

$= (A+B+C)(A+B+\overline{C})\ (A+\overline{B}+C)\ (\overline{A}+B+C)$ **(1.11)**

The concept of minterm and maxterm introduced above allows us to introduce a very convenient shorthand notation to express logical functions.

Table 1.3 gives the minterms and maxterms for a four variable logical function where the number of minterms as well as maxterms is $2^4 = 16$. In general, for an n-variable logical function there are 2^n minterms and an equal number of maxterms. While writing a particular minterm or maxterm, we shall always write it with the variables in an orderly way as can be seen from Table 1.3. Each minterm is represented by m_i where the subscript i is the decimal equivalent of the natural binary number corresponding to the minterm with normal (uncomplemented) variables taken as 1's and the complemented variables taken as 0's.

Table 1.3 Minterms/maxterms for four variables

Variable ABCD	Minterm m_i	maxterm M_i
0000	$\overline{A}\,\overline{B}\,\overline{C}\,\overline{D} = m_0$	$A+B+C+D = M_0$
0001	$\overline{A}\,\overline{B}\,\overline{C}\,D = m_1$	$A+B+C+\overline{D} = M_1$
0010	$\overline{A}\,\overline{B}\,C\,\overline{D} = m_2$	$A+B+\overline{C}+D = M_2$
0011	$\overline{A}\,\overline{B}\,C\,D = m_3$	$A+B+\overline{C}+\overline{D} = M_3$
0100	$\overline{A}\,B\,\overline{C}\,\overline{D} = m_4$	$A+\overline{B}+C+D = M_4$
0101	$\overline{A}\,B\,\overline{C}\,D = m_5$	$A+\overline{B}+C+\overline{D} = M_5$
0110	$\overline{A}\,B\,C\,\overline{D} = m_6$	$A+\overline{B}+\overline{C}+D = M_6$
0111	$\overline{A}\,B\,C\,D = m_7$	$A+\overline{B}+\overline{C}+\overline{D} = M_7$
1000	$A\,\overline{B}\,\overline{C}\,\overline{D} = m_8$	$\overline{A}+B+C+D = M_8$
1001	$A\,\overline{B}\,\overline{C}\,D = m_9$	$\overline{A}+B+C+\overline{D} = M_9$
1010	$A\,\overline{B}\,C\,\overline{D} = m_{10}$	$\overline{A}+B+\overline{C}+D = M_{10}$
1011	$A\,\overline{B}\,C\,D = m_{11}$	$\overline{A}+B+\overline{C}+\overline{D} = M_{11}$
1100	$A\,B\,\overline{C}\,\overline{D} = m_{12}$	$\overline{A}+\overline{B}+C+D = M_{12}$
1101	$A\,B\,\overline{C}\,D = m_{13}$	$\overline{A}+\overline{B}+C+\overline{D} = M_{13}$
1110	$A\,B\,C\,\overline{D} = m_{14}$	$\overline{A}+\overline{B}+\overline{C}+D = M_{14}$
1111	$A\,B\,C\,D = m_{15}$	$\overline{A}+\overline{B}+\overline{C}+\overline{D} = M_{15}$

Similar to minterms, each maxterm is represented by M_i, where the subscript i is the decimal equivalent of the natural binary number corresponding to the maxterm with uncomplemented variables taken as 0's and complemented variables taken as 1's. Using these notations the standard SOP (Eq. 1.9) can be written as :

$Y = m_3 + m_4 + m_6 + m_7 = \Sigma m\ (3,4,6,7)$ **(1.12)**

Where $\overline{A}BC = m_3$, $A\overline{B}\,\overline{C} = m_4$, $AB\overline{C} = m_6$ and $ABC = m_7$

Similarly, the standard POS(Eq.(1.11)) can be written as

$Y = M_0.M_1.M_2.M_5 = \Pi M\ (0,1,2,5)$ **(1.13)**

Where $A+B+C = M_0$, $A+B+\overline{C} = M_1$, $A+\overline{B}+C = M_2$ and $\overline{A}+B+\overline{C} = M_5$

Equations (1.12) and (1.13) are the shorthand forms of standard SOP and POS respectively. Since these two equations represent the same logical function (Eq. (1.1)), therefore we notice that there is a complementary type of relationship between a function expressed in terms of minterms and in terms of maxterms. Here, we are dealing with a three-variable function where the number of minterms and maxterms are $2^3 = 8$ and the corresponding decimal numbers are 0 and through 7. Out of this the terms corresponding to decimal numbers 3,4,6 and 7 are minterms, and the terms corresponding to decimal numbers 0,1,2 and 5 are maxterms. Hence, if a logical function is specified in terms of minterm/maxterm, its maxterm/minterm representation can be determined by using this complementary property. For example, for a four-variable case if

$Y = \Sigma m\ (0,3,6,7,10,12,15)$ then

$Y = \Pi M\ (1,2,4,5,8,9,11,13,14)$

4.Karnaugh Map Representation of Logical Functions

We haveve discussed the two standard forms of logical functions and their realizations using gates. We have also established the need for the simplification of the Boolean expressions and introduced the algebraic method of simplification using Boolean algebraic theorems. Sometimes it is difficult to be sure that a logical expression can be simplified. There is another technique, which is graphical, known as the Karnaugh map technique which provides a systematic method for simplifying and manipulating Boolean expressions. In this technique, the information contained in a truth table or available in POS or SOP form is represented on Karnaugh map (K-map). This is perhaps the most extensively used tool for simplification of Boolean functions. Although the technique may be used for any number of variables, it is generally used up to six variables beyond which it becomes very cumbersome.

B \ A	0	1
0	0	2
1	1	3

(a)

C \ AB	00	01	11	10
0	0	2	6	4
1	1	3	7	5

(b)

CD \ AB	00	01	11	10
00	0	4	12	8
01	1	5	13	9
11	3	7	15	11
10	2	6	14	10

(c)

Fig 1.9 Karnaugh-maps (a) two-variable (b) three-variable (c) four-variable

Figure 1.9 shows the K-maps for two, three and four variables. In an n-variable K-map there are 2^n cells. Each cell corresponds to one of the combinations of n variables, since there are 2^n combinations of n variables.

Therefore, we see that for each row of the truth table, for each minterm and for each maxterm there is one specific cell in the K.-map. The variables have been designated as A,B,C and D, and the binary numbers formed by them are taken as AB, ABC, and ABCD for two, three, and four variables respectively. In each map the variables and all possible values of the variables are indicated (the first bit corresponds to the first variable and the second bit corresponds to the second variable) to identify the cells. Gray code has been used for the identification of cells. The reason for using Gray code will become clear when we discuss the application of K-map. You can verify the decimal number corresponding to each cell which is written in the top left corner of the cell as shown in Fig 1.9.

Figure 1.10 shows the minterm/maxterm corresponding to each cell and the term is written inside the cell for clear understanding.

B \ A	0	1
0	$\overline{A}\,\overline{B}$	$A\overline{B}$
1	$\overline{A}B$	AB

(a)

B \ A	0	1
0	$A+B$	$\overline{A}+B$
1	$A+\overline{B}$	$\overline{A}+\overline{B}$

(b)

C \ AB	00	01	11	10
0	$\overline{A}\,\overline{B}\,\overline{C}$	$\overline{A}B\overline{C}$	$AB\overline{C}$	$A\overline{B}\,\overline{C}$
1	$\overline{A}\,\overline{B}\,C$	$\overline{A}BC$	ABC	$A\overline{B}C$

(c)

CD \ AB	00	01	11	10
0	$A+B+C$	$A+\overline{B}+C$	$A+\overline{B}+C$	$\overline{A}+B+C$
1	$A+B+\overline{C}$	$A+\overline{B}+\overline{C}$	$\overline{A}+\overline{B}+\overline{C}$	$\overline{A}+B+\overline{C}$

(d)

CD \ AB	00	01	11	10
00	$\overline{A}\,\overline{B}\,\overline{C}\,\overline{D}$	$\overline{A}B\overline{C}\,\overline{D}$	$AB\overline{C}\,\overline{D}$	$A\overline{B}\,\overline{C}\,\overline{D}$
01	$\overline{A}\,\overline{B}\,\overline{C}D$	$\overline{A}B\overline{C}D$	$AB\overline{C}D$	$A\overline{B}\,\overline{C}D$
11	$\overline{A}\,\overline{B}CD$	$\overline{A}BCD$	$ABCD$	$A\overline{B}CD$
10	$\overline{A}\,\overline{B}C\overline{D}$	$\overline{A}BC\overline{D}$	$ABC\overline{D}$	$A\overline{B}C\overline{D}$

(e)

CD \ AB	00	01	11	10
00	$A+B+C+D$	$A+\overline{B}+C+D$	$\overline{A}+\overline{B}+C+D$	$\overline{A}+B+C+D$
01	$A+B+C+\overline{D}$	$A+\overline{B}+C+\overline{D}$	$\overline{A}+\overline{B}+C+\overline{D}$	$\overline{A}+B+C+\overline{D}$
11	$A+B+\overline{C}+\overline{D}$	$A+\overline{B}+\overline{C}+\overline{D}$	$\overline{A}+\overline{B}+\overline{C}+\overline{D}$	$\overline{A}+B+\overline{C}+\overline{D}$
10	$A+B+\overline{C}+D$	$\overline{A}+\overline{B}+\overline{C}+D$	$\overline{A}+B+\overline{C}+D$	$A+\overline{B}+\overline{C}+D$

Fig 1.10 Maxterm/minterm corresponding to each cell of K-maps.

5. Representation of Truth Table on K-Map

Consider the truth table of the 3-variable logic function given in Table 1.4. The output Y is logical 1 corresponding to the rows 1,2,4 and 7. Corresponding to this we can write the equation in terms of standard SOP as given below :

$$Y = \overline{A}\,\overline{B}\,C + \overline{A}\,B\,\overline{C} + A\,\overline{B}\,\overline{C} + A\,B\,C \qquad \textbf{(1.14)}$$

Equation (1.14) represents the complete truth table in standard SOP form. Similarly, we note that the output Y is logical 0 corresponding to the rows 0,3,5 and 6 and the output Y can be represented in terms of standard POS form as given below:

$$Y = (A+B+C)(A+\overline{B}+\overline{C})\,(A+B+\overline{C})(\overline{A}+\overline{B}+C) \qquad \textbf{(1.15)}$$

Equation (1.15) also represents the complete truth table, and Eqs (1.14) and (1.15) are equivalent. We shall make use of the 3-variable K-map of Fig 1.9 b and enter the value of the output varibale Y (0 or 1) in each cell corresponding to its decimal or minterm or maxterm identification. Figure 1.11 gives the complete K-map of the truth table given in Table 1.4.

The procedure used above is general and is used to represent a truth table on the K-map. On the other hand, if a K-map is given we can make the truth table corresponding to this by following the reverse process. That is, the output Y is logical 1 corresponding to the decimal numbers/minterms represented by cells with entries 1. In all other rows, the output Y is logical 0.

C \ AB	00	01	11	10
0	(0) 0	(2) 1	(6) 0	(4) 1
1	(1) 1	(3) 0	(7) 1	(5) 0

Fig 1.11 K-map for Table 1.4

Table 1.4 Truth table of a 3-variable function

Row No.	Inputs A	B	C	Output Y
0	0	0	0	0
1	0	0	1	1
2	0	1	0	1
3	0	1	1	0
4	1	0	0	1
5	1	0	1	0
6	1	1	0	0
7	1	1	1	1

Ex 1.4 : Prepare the truth table for K-map of Fig. 1.12

CD \ AB	00	01	11	10
00	(0) 1	(4)	(12) 1	(8)
01	(1)	(5) 1	(13)	(9) 1
11	(3)	(7) 1	(15) 1	(11)
10	(2)	(6)	(14) 1	(10)

Fig 1.12 K-map of Ex. 1.4

Sol : The truth table is given in Table 1.5

Note that the 0's are not written in the K-map of fig 1.12. Actually, we need enter either 0's or 1's only in the K-map. If only 1's are entered the empty cells are 0's and if only 0's are entered then the empty cells are 1's.

Table 1.5 Truth table for K-map of Fig 1.12

Row No.	Inputs A	B	C	D	Output Y
0	0	0	0	0	1
1	0	0	0	1	0
2	0	0	1	0	0
3	0	0	1	1	0
4	0	1	0	0	0
5	0	1	0	1	1
6	0	1	1	0	0
7	0	1	1	1	1
8	1	0	0	0	0
9	1	0	0	1	1
10	1	0	1	0	0
11	1	0	1	1	0
12	1	1	0	0	1
13	1	1	0	1	0
14	1	1	1	0	1
15	1	1	1	1	1

6. Representation of Standard SOP Form and K-Map

A logical equation in standard SOP form can be represented on a K-map by simply entering 1's in the cells of the K-map corresponding to each minterm present in the equation.

Ex 1.5 : Write the logical equation in the standard SOP form for the K-map of Fig 1.12

Sol:

$$Y = \overline{A}\,\overline{B}\,\overline{C}\,\overline{D} + AB\overline{C}\,\overline{D} + \overline{A}B\overline{C}D + A\overline{B}\,\overline{C}D + \overline{A}BCD + ABCD + ABC\overline{D}$$

$$= \Sigma m(0,5,7,9,12,14,15) \qquad \textbf{(1.16)}$$

If the equation is in SOP form it can be converted to standard SOP form by using the method discussed earlier and then it can be represented on K-map.

Representation of Standard POS form and K-map

Logical equation in standard POS form can be represented on K-map by entering 0's in the cells of K-map corresponding to each maxterm present in the equation.

Example : Represent Eq. (1.15) on K-map solution corresponding to each maxterm in the equation, there is a cell in the K-map and a 0 is entered in each one of these cells. The K-map will be as shown in Fig.1.11

From a given K-map we can write the logical equation in the standard POS form by ANDing the maxterms corresponding to each 0 entry in the K-map.

Ex 1.6 : Write the logical equation in the standard POS form for the K-map of Fig 1.12

Sol: $Y = (A+B+C+\overline{D})(A+B+\overline{C}+D)(A+B+\overline{C}+\overline{D})$

$(A+\overline{B}+C+D)(A+\overline{B}+\overline{C}+D)(\overline{A}+B+C+D)$

$$(\overline{A}+B+\overline{C}+D)(\overline{A}+B+\overline{C}+\overline{D})(\overline{A}+\overline{B}+C+\overline{D})$$

$$=\Pi M(1, 2, 3, 4, 6, 8, 10, 11, 12) \quad (1.17)$$

If the equation is in POS form, it can be converted into standard POS form by the method discussed earlier and then it can be represented on K-map.

Equation (1.16) represents the K-map of Fig 1.12 in standard SOP form and Eq. (1.17) represents the same K-map in standard POS form. Therefore, these two equations are equivalent. Alternatively stated, an equation in SOP form can be converted into an equivalent POS form and vice-versa.

7. Simplification of Logical Functions Using K-Map

Simplification of logical functions with K-map is based on the principle of combining terms in adjacent cells. Two cells are said to be adjacent if they differ in only one variable. For example, in the two-variable K-maps of Figs 1.10 a and b, the top two cells are adjacent and the bottom two cells are adjacent. Also, the left two cells and the right two cells are adjacent. It can be verified that in adjacent cells one of the literals is same, whereas the other literal appears in uncomplemented form in one and in the complemented form in the other cell.

Similarly, we observe adjacent cells in the 3-variable and 4-variable K-maps. Table 1.6 gives the adjacent cells of each cell in 2-, 3-, and 4- variable K-maps. From this it becomes clear that if the Gray code is used for the identification of cells in K-map, physically adjacent (horizontal and vertical but not diagonal) cells differ in only one variable. Also, the left-most cells are adjacent to their corresponding right-most cells and similarly the top cells are adjacent to their corresponding bottom cells. The simplification of logical function is achieved by grouping adjacent 1's or 0's in groups of 2^i, where i = 1,2,........n and n is the number of variables.

8. Grouping Two Adjacent Ones

If there are two adjacent ones on the map, these can be grouped together and the resulting term will have one less literal than the original two terms.It can be verifiedfor each of the groupings of two ones as given in Table 1.6

Table 1.6 Adjacent cells in K-maps

Cell with decimal number	Decimal numbers of adjacent cells		
	2-variable	3-variable	4-variable
0	1,2	1,2,4	1,2,4,8
1	0,3	0,3,5	0,3,5,9
2	0,3	0,3,6	0,3,6,10
3	1,2	1,2,7	1,2,7,11
4		0,5,6	0,5,6,12
5.		1,4,7	1,4,7,13
6.		2,4,7	2,4,7,14
7		3,5,6	3,5,6,15
8			0,9,10,12
9			1,8,11,13
10			2,8,11,14
11			3,9,10,15
12			4,8,13,14
13			5,9,12,15
14			6,10,12,15
15			7,11,13,14

Ex 1.7 : Simplify the K-map of Fig 1.13

Sol : The standard SOP form of equation can be written by inspection as

$$Y=\overline{A}\,\overline{B}\,\overline{C}+\overline{A}BC+ABC+A\overline{B}\,\overline{C} \quad (1.18)$$

If we combine the ones in adjacent cells (0,4) and (3,7), Eq. (1.18) can be written

$$Y=(\overline{A}+A)\overline{B}\,\overline{C}+(\overline{A}+A)BC \quad (1.19)$$

$$=\overline{B}\,\overline{C}+BC \quad (1.20)$$

Equation (1.20) can be directly obtained from the K-map by using the following procedure:

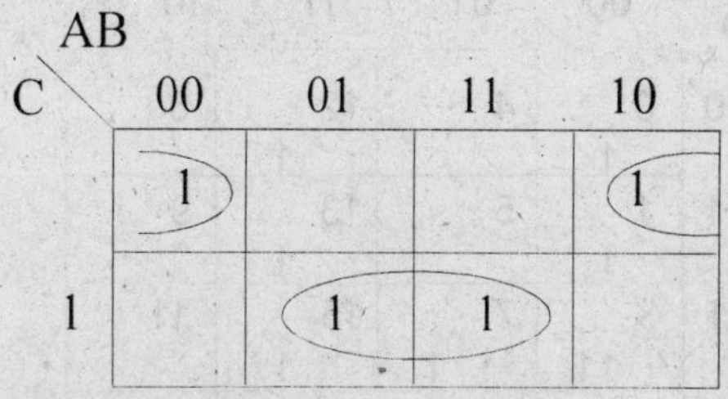

Fig 1.13 K-map of Ex. 1.7

1. Identify adjacent ones, then see the values of the variables associated with these cells. Only one variable will be different and gets eliminated. Other variables will appear in ANDed form in the term, it will be in the uncomplemented form if it is 1 and in the complemented form if it is 0.

2. Determine the term corresponding to each group of adjacent ones. These terms are ORed to get the simplified equation in SOP form.

9. Grouping Four Adjacent Ones

Four cells form a group of four adjacent ones if two of the literals associated with the minterms/maxterms are not same and the other literals are same. Table 1.7 gives all possible groups of four adjacent ones for each cell in a 3-variable map. In case of 2-variable map, there is only one possibility corresponding to entry 1 in all the four cells, and the simplified expression will be Y=1. That is Y always equals 1 (independent of the variables).

On the basis of groupings of 4 adjacent ones given in Table 1.7, we can find the groupings in K-maps of four or more variables. In the case of a four - variable K-map, there are six possible groupings of 4-variables involving any cell. It is left to the reader to verify this fact.

Table 1.7 Groups of four adjacent ones in a 3-variable K-map

Cell with decimal number	Decimal numbers of cells forming groups of adjacent fours
0	(0,2,6,4), (0,1,2,3), (0,1,4,5)
1	(1,0,2,3), (1,3,7,5), (1,0,4,5)
2	(2,0,6,4), (2,3,1,0), (2,3,6,7)
3	(3,1,7,5), (3,2,1,0), (3,2,6,7)
4	(4,6,2,0), (4,5,6,7), (4,5,0,1)
5	(5,1,3,7), (5,4,6,7), (5,4,0,1)
6	(6,0,2,4), (6,7,4,5), (6,7,2,3)
7	(7,1,3,5), (7,6,4,5), (7,6,2,3)

Ex 1.8 : Simplify the K-map of Fig 1.14

Sol: The standard SOP form of equation can be written by inspection as

$$Y = m_0 + m_1 + m_3 + m_7 + m_8 + m_9 + m_{11} + m_{15}$$

$$= (m_0 + m_1 + m_8 + m_9) + (m_3 + m_7 + m_{15} + m_{11}) \quad (1.21)$$

In the K-map of Fig 1.14, there are two groups of four adjacent ones. One corresponding to cells 0,1,8 and 9, and the other one corresponding to 3,7,15 and 11.

CD \ AB	00	01	11	10
00	0 1	4	12 1	8
01	1 1	5	13 1	9
11	3 11	7 1	15 1	11
10	2	6	14	10

Fig 1.14 K-map of Ex. 1.8

In Eq. (1.21), the minterms corresponding to each group are combined. The first term can be written as

$$m_0 + m_1 + m_8 + m_9 = \overline{A}\overline{B}\overline{C}\overline{D} + \overline{A}\overline{B}\overline{C}D + A\overline{B}\overline{C}\overline{D} + A\overline{B}\overline{C}D$$

$$= \overline{B}\overline{C}(\overline{A}\overline{D} + \overline{A}D + A\overline{D} + AD)$$

$$= \overline{B}\overline{C}\,[\,\overline{A}\,(\overline{D} + D) + A(\overline{D} + D)\,] \qquad = \overline{B}\overline{C}\,[\overline{A}.1 + A.1]$$

$$= \overline{B}\overline{C}\,(\,\overline{A} + A) = \overline{B}\overline{C}.1 = \overline{B}\,\overline{C}$$

In the first term of Eq.(1.21) we observe the following:

1. In this group of four minterms, two of the variables appear as $\overline{B}$ and $\overline{C}$ in all the four terms.

2. The variable A appears as A in two and as $\overline{A}$ in the other two minterms.

3. The variable D appears as D in two and as $\overline{D}$ in the other two minterms.

4. The combination of these four minterms results in one term with two literals which are present in all the four terms. Similarly, the second term of Eq. (1.21) is simplified to CD. Therefore, the K-map is simplified to

$$Y = \overline{B}\,\overline{C} + CD \quad (1.22)$$

10. Grouping Eight Adjacent Ones

Eight cells form a group of eight adjacent ones if three of the literals associated with the minterms/maxterms are not same and the other literals are same. In case of 3-variable K-map, there is only one possibility of eight ones appering in the K-map and this corresponds to output equal to 1, irrespective of the values of the input varibales. Table 1.8 gives all possible groups of eight adjacent ones in a 4-variable K-map. From an understanding of this, we can easily find out such combinations for 5- and 6- variable K-maps. When eight adjacent ones are combined, the resulting equation will have only one term with the number of literals three less than the number of literals in the original minterms. Similar to the groupings of adjacent two and four ones, the literals which are common in all the eight minterms will be present and the literals which are nolt same get eliminated in the resulting term.

Table 1.8 Groups of eight adjacent ones in 4-variable K-map

Decimal numbers of cells forming groups of adjacent eights in a 4 variable K map
0,4,12,8,1,5,13,9
0,4,12,8,2,6,14,10
0,1,3,2,4,5,7,6
0,1,3,2,8,9,11,10
1,5,13,9,3,7,15,11
4,5,7,6,12,13,15,14
12,13,15,14,8,9,11,10
3,7,15,11,2,6,14,10

The reader is advised to verify the simplification of eight adjacent ones into a single term with three variables eliminated. For exmaple, let us take the first group of eight adjacent ones in Table 1.8. For all these eight cells, the variable C appears as $\overline{C}$ in the minterms and the other three variables are not same. Therefore, the grouping of these eight cells results in a term $\overline{C}$. Figure 1.15 shows the simplified expression for each of the groupings of eight ones for a 4-variable K-map.

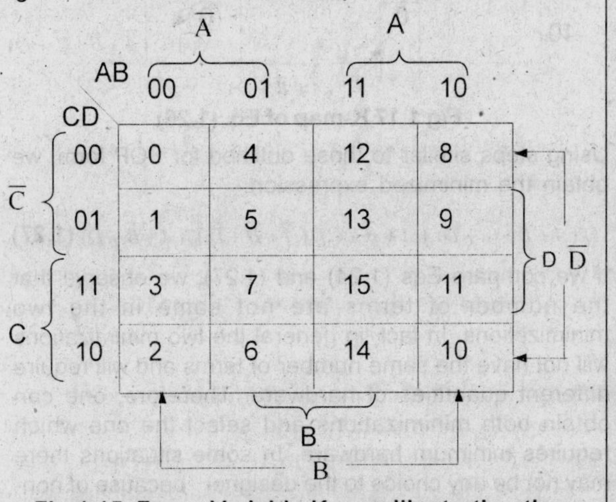

Fig 1.15 Four - Variable K-map illustrating the groupings of eight adjacent ones.

Grouping 2,4 and 8 Adjacent Zeros

In the above discussion, we have considered groups of 2,4 and 8 adjacent ones. Instead of making the groups of ones, we can also make groups of zeros. The procedure is similar to the one used above and is as follows :

1. Group of two adjacent zeros result in a term with one literal less than the number of variables. The literal which is not same in the two maxterms gets eliminated.

2. Group of four adjacent zeros result in a term with two literals less than the number of variables. The two literals which are not same in all the four maxterms get eliminated.

3. Group of eight adjacent zeros result in a term with three literals less than the number of varibales.The three literals which are not same in all the eight maxterms get eliminated.

We have considered groups of 2,4 and 8 adjacent ones and zeros. The same logic can be extended to 16, 32, and 64 adjacent ones and zeros which occur in K-maps with more than 4 variables.

11. Minization of Logical Functions Specified in Minterms/Maxterms or Truth Table

Minimization of SOP Form

We have seen the advantages of simplifying a logical expression. If the expression is simplified to a stage beyond which it can not be further simplified, it will require minimum number of gates with minimum number of inputs to the gates. Such an expression is referred to as the minimized expression:

For minimizing a given expression in SOP form or for a given truth table, we have to prepare the K-map first and then look for combinations of ones on the K-map. We have to combine the ones in such a way that the resulting expression is minimum. To achieve this, the following algorithm can be used which wil definitely lead to minimized expression:

1. Identify the ones which can not be combined with any other ones and encircle them. These are essential prime implicants.

2. Identify the ones that can be combined in groups of two in only one way. Encircle such groups of ones.

3. Identify the ones that can be combined with three other ones, to make a group of four adjacent ones, in only one way. Encircle such groups of ones.

4. Identify the ones that can be combined with seven other ones, to make a group of eight adjacent ones, in only one way. Encircle such groups of ones.

5. After identifying the essential groups of 2,4, and 8 ones, if there still remains some ones which have not been encircled then these are to be combined with each other or with other already encircled ones. Of course, however, we should combine the left-over ones in largest possible groups and in as few groupings as possible. In this, the groupings may not be unique and we should make the groupings in an optimum manner. You can verify that any one can be included any number of times without affecting the expression.

The above algorithm will be used to minimize the logical functions in the examples given.

Ex 1.9 : Minimize the four - variable logic function using K-map.

$$f(A,B,C,D) = \Sigma m(0,1,2,3,5,7,8,9,11,14) \qquad \textbf{(1.23)}$$

Sol : The K-map of Eq. (1.23) is shown in Fig 1.15. The equation is minimized in the following steps :

1. Encircle 1 in cell 14 which can not be combined with any other 1. The term corresponding to this is $ABC\overline{D}$.

2. There are at least two possible ways for every 1 forming groups of two adjacent ones. Therefore, we ignore it for the time being and go to the next step.

3. There is only one possible group of four adjacent ones involving each of the cells 8,11, 5 or 7 and 2, and these are (8,9,0,1), (11,9,13), (5,7, 3,1) and (2,3,1,0), respectively. Encircle these groups. The terms corresponding to these groups are $\overline{B}\,\overline{C}, \overline{B}D, \overline{A}D$, and $\overline{A}\overline{B}$, respectively. Since all the ones have been encircled, therefore, the minimized equation is

$$f(A,B,C,D) = ABC\overline{D} + \overline{B}\overline{C} + \overline{B}D + \overline{A}D + \overline{A}\overline{B} \qquad \textbf{(1.24)}$$

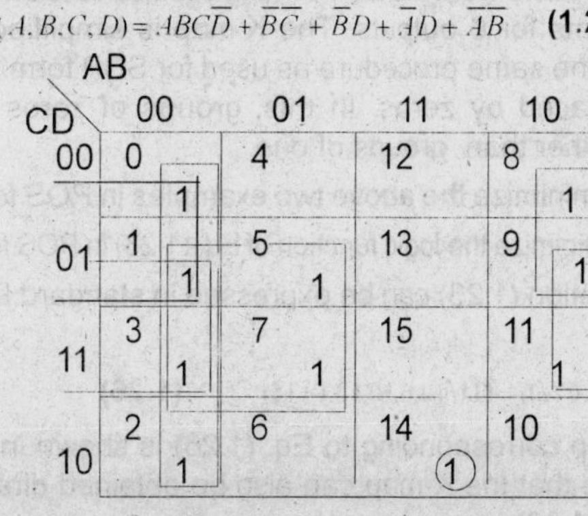

Fig 1.15 K-map for Eq. (1.23)

Ex 1.10 : Determine the minimized expression in SOP form for the truth table given in Table 1.9

Sol : The K-map for the truth table of Table 1.9 is shown in Fig 1.16.

Using the minimization steps, we obtain the minimized expression.

$$Y = \overline{B} + A\overline{C} + \overline{A}CD \quad (1.25)$$

Table 1.9

Inputs				Output
A	B	C	D	Y
0	0	0	0	1
0	0	0	1	1
0	0	1	0	1
0	0	1	1	1
0	1	0	0	0
0	1	0	1	0
0	1	1	0	0
0	1	1	1	1
1	0	0	0	1
1	0	0	1	1
1	0	1	0	1
1	0	1	1	1
1	1	0	0	1
1	1	0	1	1
1	1	1	0	0
1	1	1	1	0

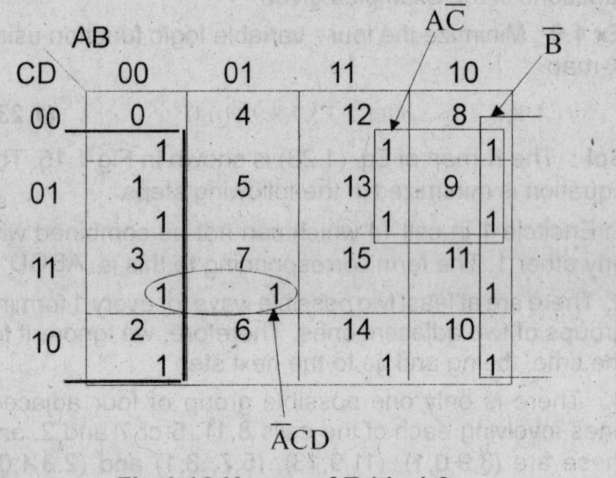

Fig 1.16 K-map of Table 1.9

12. Minimization of POS Form

For minimizing a given expression in POS form or for a given truth table we write zeros in the cells corresponding to maxterms for 0 outputs. The K-map is simplified by following the same procedure as used for SOP form with ones replaced by zeros. In this, groups of zeros are formed rather than groups of one.

We shall minimize the above two examples in POS form.

Ex 1.11 : Minimize the logic function of Eq. (1.23) in POS form.

Sol : Equation (1.23) can be expressed in standard POS form as

$$f(A,B,C,D) = \Pi M(4,6,10,12,13,15) \quad (1.26)$$

The K-map corresponding to Eq. (1.26) is shown in Fig 1.17. Note that the K-map can also be obtained directly from Eq. (1.23)

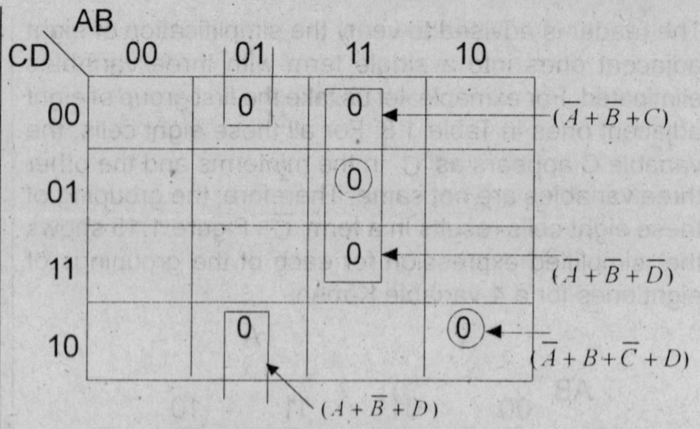

Fig 1.17 K-map of Eq. (1.26)

Using steps similar to those outlined for SOP form, we obtain the minimized expression.

$$f(\overline{A}+B+\overline{C}+D).(\overline{A}+\overline{B}+C).(\overline{A}+\overline{B}+\overline{D}).(A+\overline{B}+D) \quad (1.27)$$

If we compare Eqs (1.24) and (1.27), we observe that the number of terms are not same in the two minimizations. In fact, in general the two minimizations will not have the same number of terms and will require different quantities of hardware. Therefore, one can obtain both minimizations and select the one which requires minimum hardware. In some situations there may not be any choice to the designer because of non-availability of cetain ICs.

Ex 1.12 : Minimize the truth table give in Table 1.9 using maxterms.

Sol : The K-map is given in Fig 1.18.

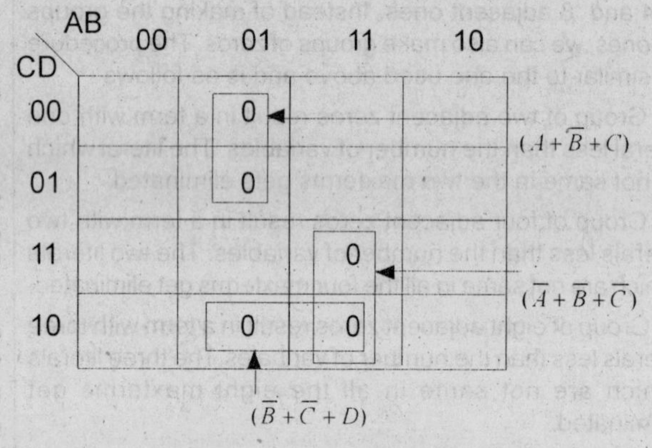

Fig 1.18 K-map of Table 1.9

The simplified expression is

$$Y = (A+\overline{B}+C)\,(\overline{A}+\overline{B}+\overline{C})\,(\overline{B}+\overline{C}+D) \quad (1.28)$$

Comparison of Eqs (1.25) and (1.28) confirms our generalizations made in Ex. 1.11 regarding the hardware requirements in the two methods.

13. Minimization of Logical Functions Not Specified in Minterm/Maxterms

If the function is specified in one of the two standard forms, its K-map can be prepared and the function can be minimized. Now we consider the cases where the functions are not specified in standard forms. In such cases, the equations can be converted into standard forms. Alternately, we can directly prepare K-map using the following algorithm:

1. Enter ones for minterms and zeros for maxterms.
2. Enter a pair of ones/zeros for each of the terms with one variable less than the total number of variables.
3. Enter four adjacent ones/zeros for terms with two variables less than the total number of variables.
4. Repeat for other terms in the similar way.

Once the K-map is prepared the minimization procedure is same as discussed earlier.

The following examples will help in understanding the above procedure:

Ex 1.13 : Minimize the four variable logic function.

$$f(A,B,C,D) = AB\overline{C}D + \overline{A}BCD + \overline{A}\,\overline{B}\,\overline{C} + \overline{A}\,\overline{B}\,\overline{D} + A\overline{C} + A\overline{B}C + \overline{B} \quad \textbf{(1.29)}$$

Sol : The method for obtaining K-map is

1. Enter 1 in the cell with A=1, B=1, C=0, D=1 corresponding to the minterm $AB\overline{C}D$
2. Enter 1 in the cell with A=0, B=1, C=1, D=1 corresponding to the minterm $\overline{A}BCD$
3. Enter 1's in the two cells with A=0, B=0, C=0 corresponding to the term $\overline{A}\,\overline{B}\,\overline{C}$.
4. Enter 1's in the two cells with A=0, B=0, D=0 (one of these is already entered) corresponding to the term $\overline{A}\,\overline{B}\,\overline{D}$.
5. Enter 1's in the two cells with A=1, B=0, C=1 corresponding to the term $A\overline{B}C$.
6. Enter 1's in the four cells with A=1, C=0 (one of them is already entered) corresponding to the term $A\overline{C}$.
7. Enter 1's in the eigth cells with B=0 (all of them except one have already been entered) corresponding to the term $\overline{B}$.

The K-map is given in Fig 1.19 which is same as the K-map of Fig 1.16 and the minimized expression is given by Eq. (1.25).

CD \ AB	00	01	11	10
00	1		1	1
01	1		1	1
11	1	1		1
10	1			1

Fig 1.19 K-map of Eq. (1.29)

Ex 1.14 : Minimize the four variable logic function

$$f(A,B,C,D) = (A+B+\overline{C}+\overline{D}).(\overline{A}+C+\overline{D})$$
$$(\overline{A}+B+\overline{C}+\overline{D}).(\overline{B}+C).(\overline{B}+\overline{C})$$
$$(A+\overline{B}).(\overline{B}+\overline{D}) \quad \textbf{(1.30)}$$

Sol: The K-map cells in which 0's are to be entered corresponding to each term are given in Table 1.10. Even if a cell is involved in more than one terms, a 0 is to be entered only once.

Table 1.10 K-map cells with 0 entries

Term	Cell(s) with 0's
$A+B+\overline{C}+\overline{D}$	A = 0, B = 0, C = 1, D = 1
$\overline{A}+C+\overline{D}$	A = 1, C = 0, D = 1
$\overline{A}+B+\overline{C}+\overline{D}$	A = 1, B = 0, C = 1, D = 1
$\overline{B}+C$	B = 1, C = 0
$\overline{B}+\overline{C}$	B = 1, C = 1
$A+\overline{B}$	A = 0, B = 1
$\overline{B}+\overline{D}$	B = 1, D = 1

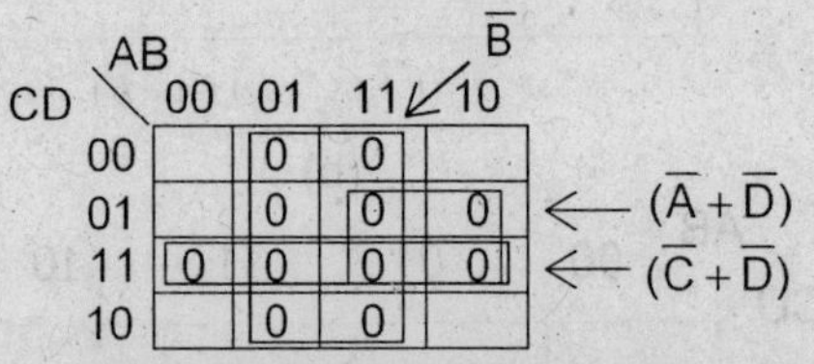

Fig 1.20 K-map of Eq. (1.30)

The K-map is given in Fig1.20.The minimizedexpression is $f(A,B,C,D) = \overline{B}\ .\ (\overline{A}+\overline{D})\,(\overline{C}+\overline{D})$ **(1.31)**

14. Don't-Care Conditions

We enter 1's and 0's in the map corresponding to input varibales that make the function equal to 1 or 0,respectively. The maps are simplified using either 1's or 0's. Therefore, we make the entries in the map for either 1's or 0's. The cells which do not contain 1 are assumed to contain 0 and vice-versa. This is not always true since there are cases in which certain combinations of input variables do not matter. In such situations the designer has a flexibility and it is left to him whether to assume a 0 or a 1 as output for each of these combinations.This condition is known as don't-care condition and can be represented on the K-map as a × mark in the corresponding cell. The × mark in a cell amy be assumed to be a 1 or a 0 depending upon which one leads to a simpler expression. The function can be specified in one of the following ways:

1.In terms of minterms and don't - care conditions. For example,

$$f(A,B,C,D) = \Sigma\, m(1,3,7,11,15) + d(0,2,5) \quad \textbf{(1.32)}$$

Its K-map and the minimized expression are given in Fig 1.21a

2. In terms of maxterms and don't-care conditions.For example,

$$f(A,B,C,D) = \Pi M\ (4,5,6,7,8,12)\ .\ d(1,2,3,9,11,14) \quad \textbf{(1.33)}$$

Its K-map and the minimized expression are given in Fig 1.21b

3. In terms of truth table. For example, consider the truth table of Table 1.11

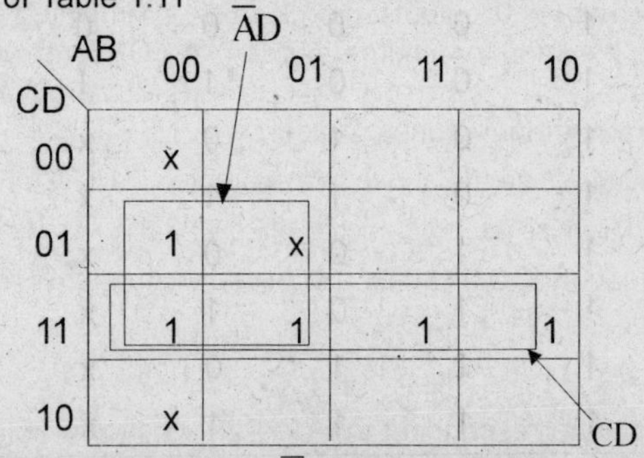

$Y = \overline{A}D + CD$

(a)

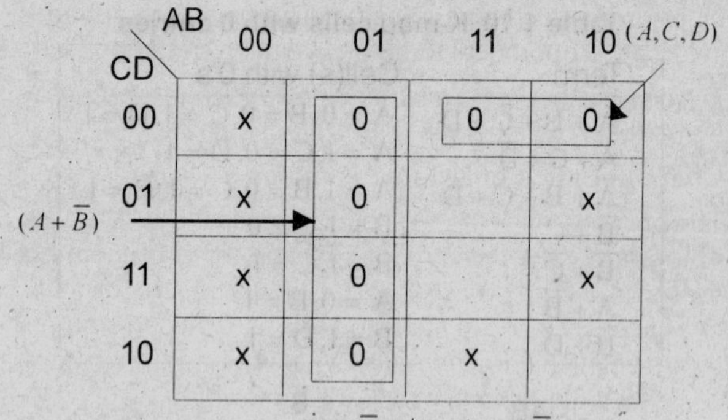

$$Y = (\overline{A} + C + D)\,(A + \overline{B})$$

(b)

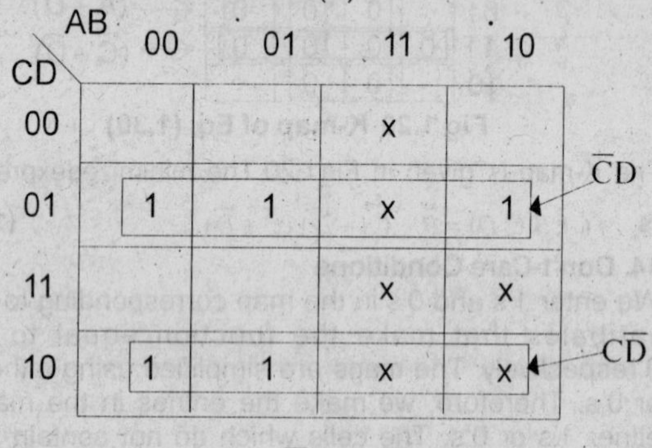

$$Y = \overline{C}D + C\overline{D}$$

(c)

Fig.1.21 K-maps with don't care conditions

Table 1.11

Inputs				Output
A	B	C	D	Y
0	0	0	0	0
0	0	0	1	1
0	0	1	0	1
0	0	1	1	0
0	1	0	0	0
0	1	0	1	1
0	1	1	0	1
0	1	1	1	0
1	0	0	0	0
1	0	0	1	1
1	0	1	0	x
1	0	1	1	x
1	1	0	0	x
1	1	0	1	x
1	1	1	0	x
1	1	1	1	x

In K-map and the minimized expression in SOP form are given in Fig 1.21C.

15. Digital Integrated Circuits

An integrated circuit (IC) is a small silicon semiconductor crystal called a chip containing electrical components such as transistors, diodes, resistors, and capacitors. The various components are interconnected inside the chip to form an electronic circuit. The chip is mounted in a metal or plastic package and connections are welded to external pins to form the IC. Integrated circuits differ from conventional discrete component circuits in that individual components cannot be separated or disconnected and the circuit inside the IC package is accessed only through its external pins. The benefits dervied from IC's as compared to circuits with discrete components are: (1) a substantial reduction in size (2) a substantial reduction in cost, (3) a reduction in power requirements, (4) a higher reliability against failuers, (5) an increase in operating speed, and (6) a reduction of externally wired connections.

Integrated circuits are classified as being in one of two general categories asalog or digital. Analog integrated circuits operate with continuous waveforms to provide electronic functions such as operational amplifiers, voltage comparators, and voltage regulators. Digital integrated circuits operate with binary singals and provide such digital functions as gates, fiip-flops, registers, counters, adders, memories, and arithmetic calculators. Here we will be concerned only with digital integrated circuits.

Because of their many advantages, integrated circuits are used exclusively to provide various digital functions needed in the design of computer systems. In order to understand the architecture of computer systems it is very important that one should be familair with the different digital functions commonly encountered in integrated circuit packages.

Logic Circuit Families

Digital integrated circuits are classified not only by their function but also by their being members of a specific logic circuit family. Each logic family has its own basic electronic circuit upon which more complex circuits and functions are developed. The basic circuit in each family is either a NOR or a NAND gate. The topology of this basic circuit that is the electronic components employed and their interconnection, is usually used to derive the name of the family. Many different logic families of digital integrated circuits have been introduced commercially. The ones that have achieved widespread popularity are :

TTL	Transistor-transistor logic
ECL	Emitter-coupled logic
MOS	Metal-oxide semiconductor
CMOS	Complemetary metal-oxide semiconductor

TTL has an extensive list of digital functions and is the most popular logic family. ECL is used for systems requiring high-speed operation. MOS is used in circuits requiring high component density and CMOS is used in systems requiring low power consumption.

The characteristics of digital logic families are usually compared by analyzing the circuit of the basic gate in each family. The most important parameters that are evaluated and compared are listed below.

1. Fan-out specifies the number of standard loads that the output of a standard gate can drive without impairing its normal operations. A standard load is usually defined to be the load (the amount of current) needed by an input of another similar standard gate.

2. Power-dissipation is the power consumed by the gate which must be available from the power supply.

3. Propagation delay is the average transition delay time for the signal to propagate from input to output when the binary signals change in value. The operating speed is inversely proportional to he propagation delay.

4. Noise margin is the minimum noise voltage that causes an undesirable change in the circuit output.

Characteristics of IC logic Families

The basic circuit of the TTL logic family is the NAND gate. There are many versions of TTL and three of them are listed in Table 1.12 This table gives the general characteristics of the IC logic families. Values listed are representative on a comparison basis. For any one family or version, the values may vary somewhat.

IC logic family	Fan-out	Power dissipation (m W)	Propagation delay (ns)	Noise margin (V)
Standard TTL	10	10	10	0.4
Schottky TTL	10	22	3	0.4
Low-power Schottky TTL	20	2	10	0.4
ECL	25	25	2	0.2
CMOS	50	0.1	25	3

Table 1.12 Typical characteristics of IC logic families

The standard TTL gate was the first version of the TTL family. Additional improvements were added as the technology progressed. The Schottky TTL is a later improvement that reduces that propagation delay but results in an increase in reduced power dissipation. The low-power Schottky TTL version sacrifices some speed for reduced power dissipation. It has the same propagation delay as the standard TTL, but the power dissipation is reduced considerably. The fan-out of the standard TTL is 10 but the low-power Schottky version has a fan-out of 20. Under certain conditions the other versions may also have a fan-out of 20. The noise margin is better than 0.4 V, with a typical value of 1 V.

The basic circuit of the ECL family is the NOR gate. The special advantage of ECL gate is their low propagation delay. Some ECL versions may have a propagation delay as low as 0.5 ns. The power dissipation in ECL gates is comparatively high and the noise margin low. These two parameters impose a disadvantage when choosing ECL over other logic families. However, because of its low propagation delay, ECL offers the highest speed of any family and is the ultimate choice for very fast systems.

The basic circuit of CMOS is the inverter from which both NAND and NOR gates can be constructed. The special advantage of CMOS is its extremely low power dissipation. Under static conditions, the CMOS gate power dissipation is negligible and averages about 10 nW. When the gate signal changes state, there is a dynamic power dissipation which is proportional to the frequency at which the power dissipation in CMOS gates.

The one major disadvantage of CMOS is its high propagation delay. This means that it is not practical for use in systems requiring high-speed operations. The characteristic parameters for the CMOS gate depend on the power supply voltage V_{DD} that is used. The power dissipation increases with increase in voltage supply. The propagation delay decreases with increase in voltage supply, and the noise margin is estimated to be about 40 % of the voltage supply value.

OBJECTIVE QUESTIONS - I

1. Which of the following is the first integrated logic family

a) RTL b) DTL c) TTL d) MOS

2. Which logic family is the fastest ?

a) DTL b) TTL c) ECL d) CMOS

3. The time required for a gate or inverter to change its state is defined as:

a) Rise time b) Delay time

c) Propagation time d) Charging time

4. Generally, the speed of operation of a logic gate is quoted in

a) Seconds b) Milli seconds

c) Micro seconds d) Nano seconds

5. The operations such as AND, OR, NOT, etc., can be performed by using

a) Switches b) Amplifiers

c) Rectifiers d) Oscillators

6. The output of a gate is LOW when atleast one of its inputs is HIGH . It is TRUE for

a) AND b) OR c) NAND d) NOR

7. The output of a gate is HIGH when atleast one of its inputs is LOW. It is TRUE for

a) EX-OR b) NAND c) NOR d) OR

8. The output of a gate is LOW when atleast one of its inputs is LOW. It is TRUE for

a) AND b) OR c) NAND d) NOR

9. The output of a gate is HIGH when atleast one of its input is HIGH. It is TRUE for

a) NAND b) AND c) OR d) EX-OR

10. The output of a gate is HIGH. If and only if all its input are HIGH. It is TRUE for

a) EX-OR b) AND c) OR d) NAND

11. The output of a gate is LOW if and only if all its inputs are HIGH. It is TRUE for

a) AND b) EX-NOR c) NOR d) NAND

12. The output of a gate is HIGH if and only if all its inputs are LOW. It is TRUE for

a) NOR b) EX-OR c) NAND d) EX-NOR

13. The output of a gate is LOW if and only if all its inputs are LOW. It is TRUE for

a) EX-OR b) AND c) OR d) NOR

14. The output of a 2-input gate is HIGH if and only if its inputs are unequal. It is TRUE for

a) OR b) EX-OR c) EX-NOR d) NOR

15. The output of a 2-input gate is LOW if and only if its inputs are equal. It is TRUE for

a) AND b) EX-OR c) OR d) NOR

16. An inverter is also called a **[AUCET-2000]**

a) AND gate b) NAND gate

c) NOT gate d) OR gate

17. The NOR gate is logically equivalent to an OR gate followed by **[AUCET-2000]**

a) Inverter b) AND c) XOR d) XNAND

18. The following logic has the maximum fan out **[AUCET-2000]**

a) CMOS b) NMOS c) RTL d) ECL

19. The most suitable gate for comparing two bits is

a) AND b) OR c) NAND d) EX-OR

20. Which of the following gates can be used as an inverter ?

a) AND b) OR c) EX-OR d) None of these

21. Which of the following gates can not be used as an inverter?

a) NAND b) AND c) NOR d) EX-NOR

22. Which of the following operations is commutative but not associative

a) AND b) OR c) EX-OR d) NAND

23. Which of the following operations is not associative

a) NOR b) OR c) EX-OR d) AND

24. Which of the following operations is an EQUIVALENCE function

a) OR b) AND c) EX-OR d) EX-NOR

25. The output of a logic gate is 0 when all its inputs are at logic 1. The gate is either

a) a NAND or an AND b) a NAND or an EX-OR

c) a NOR or an EX-NOR d) a NAND or an EX-NOR

26. The output of a logic gate is 0 when all its inputs are at logic 0. The gate is either

a) an AND or a NOR b) a NAND or an EX-OR

c) an OR or an EX-OR d) an OR or an EX-NOR

27. Which of the following gates is known as an universal gate **[AUCET-2000]**

a) AND b) NAND c) OR d) XOR

28. Which of the following gates is known as an Universal gate.

a) OR b) NOR c) NOT d) X NAND

29. Any logical expression can be realized by using only **[AUCET-2000]**

a) AND b) OR c) NAND d) EX-OR

30. Which of the following statements is TRUE.

a) AND and NOT gates are necessary and sufficient for the realization of any logical function

b) OR and NOT gates are necessary and sufficient for the realization of any logical function

c) NOR gates are sufficient to realize any logical function

d) NAND gates are not sufficient to realize any logical function

31. Which of the following is not functionally a complete set

a) AND, OR b) AND, OR ,NOT

c) NOR d) NAND

32. The NAND can function as a NOT gate if

a) Inputs are connected together

b) Inputs are left open

c) One input is set to 0 d) One input is set to 1

33. What is the minimum number of two-input NAND gates used to perform the function of two-input OR gate.

a) One b) Two c) Three d) Four

34. How many NAND gates are needed to perform the logic function X.Y ?

a) One b) Two c) Three d) Four

35. The NAND-NAND realization is equivalent to

a) AND-NOT realization b) AND-OR realization

c) OR-AND realization d) NOT-OR realization

36. The NOR-NOR realization is equivalent to

a) AND-OR realization b) NOT-AND realization

c) OR-NOT realization d) OR-AND realization

37. For checking the parity of a digital word, it is preferable to use

a) AND gates b) NAND gates

c) EX-OR gates d) NOR gates

38. The most suitable gates to check whether the number of ones in a digital word is EVEN or ODD is **[AUCET-1999]**

a) EX-OR b) EX-NOR c) NAND d) NOR

39. Which of the folowing expression is not equivalent to $\bar{X}$ **[GATE-1999]**

a) X NAND X b) X NOR X
c) X NAND1 d) X NOR 1

40. Which of the following boolean algebra statements represent distributive law.

a) (A+B)+C=A+(B+C) b) A(B+C)= (AB)+(AC)
c) A(BC)=(AB)C d) None of these

41. Which of the following boolean algebra statements represent commutative law.

a) (A+B)+C=A+(B+C) b) A(B+C)=(AB)+(AC)
c) A+B=B+A d) A+A=A

42. If X is a boolean variable then

a) 0+x=x b) 1+x=x c) x+x=x d) x+ x′ =1

43. If x,y,z are three boolean variables then

a) $x\bar{x}=1$ b) x(y+z)=(x+y)(x+z)
c) x+xz=x d) x+y=y+x

44. The boolean expression x+ **x′** y equals

a) x+y b) x+xy c) y+yx d) y+x y′

45. If A ⊕ B=C,t hen

a) A ⊕ B=B b) B ⊕ C=A
c) A ⊕ B ⊕ C = 0 d) None of these.

46. Let * be a Boolean operation defined by A*B =AB+ **A′B′**. A*A is

a) A b) B c) 0 d) 1

47. The simplified form of the boolean expression (x+y +xy)(x+z)is

a) x+y+z b) xy+yz c) x+yz d) xz+y

48. The boolean expression (A+C) (A$\bar{B}$ +AC) $(\bar{A}\bar{C}+\bar{B})$ simplified to

a) AB+ $\bar{A}$ C b) $\bar{A}$ B+BC c) AB+BC d) A$\bar{B}$

49. $\overline{\bar{A}.\bar{B}.\bar{C}}$ is equivalent to

a) $\bar{A}+\bar{B}+\bar{C}$ b) $\overline{A.B.C}$ c) A+B+C d) A.B.C

50. AB+ $\bar{A}$ C+BC is equivalent to

a) AB+BC b) AB+ $\bar{A}$ C c) $\bar{A}$ C+BC d) AC

51. The dual of a Boolean theorem is obtained by

a) Inter changing all zeros and one's only
b) Changing all zero's to one's only
c) Changing all one's to zero's only
d) Interchanging i) all zero's and one's and ii) + and . signs

52. Any given truth table can be represented by

a) A product-of-sums(POS) of boolean expressions
b) A sum-of-products(SOP) of boolean expression
c) A Karnaugh map d) All of the above

53. How many boolean functions of 3 variables are there

a) 64 b) 256 c) 512 d) 1024

54. Consider the boolean expression $\bar{x}y\bar{z}+\bar{x}\bar{y}z+x(y+z)$. The equivalent product of sums form is:

a) $\bar{x}y\bar{z}+\bar{x}\bar{y}z+xy+xz$
b) $(x+\bar{y}+z)(x+y+\bar{z})(x+y+z)$
c) $(\bar{y}+\bar{z})(\bar{x}+y+\bar{z})$
d) $(y+z)(x+\bar{y}+z)$

55. Which one of the following Boolen expressions is not logically equivalent to all of the rest.

a) $wx\bar{y}+w\bar{z}z+wxyz+w\bar{y}z$
b) $wx+w\bar{y}+w\bar{z}$
c) $w+x+\bar{y}+\bar{z}$ d) $wx+w\bar{y}+w\bar{z}$

56. The boolean expression AB+A$\bar{B}$ + $\bar{A}$ C+AC is independent of the boolean variable

a) A b) B
c) C d) None of these

57. The SOP form of logical expression is most suitable for designing logic circuit using only

a) NOR gates b) NAND gates
c) AND gates d) EX-OR gates

58. The POS form of logical expression is most suitable for designing logic circuits using only

a) NOR gates b) NAND gates
c) AND gates d) EX-OR gates

59. The logic expression Y=AB+A$\bar{C}$ +BC is known as

a) Standard SOP form b) SOP form
c) POS form d) Standard POS form

60. The logic expressionY=(A+B+C)(A+ $\bar{B}$)(B+ $\bar{C}$)(A+C) is known as **[AUCET-1998]**

a) SOP form b) Standard SOP form
c) POS form d) Standard POS form

61. The min term corresponding todecimal number 13 is

a) A+B+ $\bar{C}$ +D b) $\bar{A}+\bar{B}+C+\bar{D}$
c) AB$\bar{C}$D d) $\bar{A}\bar{B}C\bar{D}$

62. The max term corresponding to decimal number 15 is

a) ABCD b) $\bar{A}\bar{B}\bar{C}\bar{D}$
c) A+B+C+D d) $\bar{A}+\bar{B}+\bar{C}+\bar{D}$

63. The minimized expression for the k-map of below fig. is

a) $\bar{A}$.(A+B)
b) $\bar{A}$ +B
c) $\bar{A}$.B
d) $(\bar{A}+\bar{B})(\bar{A}+B)(A+B)$

CD \ AB	00	01	11	10
00	0		0	0
01	0		0	0
11	0		0	0
10	0		0	0

64. The minimized expression for the k-map of below fig. is

CD \ AB	00	01	11	10
00	X	0	0	X
01		0		X
11	0	X	0	X
10		X		X

a) $(C+D)(\bar{C}+\bar{D})(A+\bar{B})$

b) $(C+\bar{D})(\bar{C}+\bar{D})(\bar{A}+\bar{B})$

c) $(\bar{B}+C+D)(A+\bar{B}+C)(A+B+\bar{C}+D)$

d) None of these

65. Given the following k-map, determine the corresponding switching function in its minimal form

yz \ wx	00	01	11	10
00	1		1	1
01				
11		1	1	
10	1		1	1

a) $\bar{x}\bar{z}+\bar{w}\bar{z}+\bar{x}\bar{y}\bar{z}yz$

b) $\bar{x}\bar{z}+w\bar{z}+xyz$

c) $(\bar{x}+\bar{z})(w+\bar{z})(w+x+z)$

d) $xz+\bar{w}z+\bar{x}\bar{y}\bar{z}$

66. The total number of Boolean functions which can be realized with four variables is **[GATE-1987]**

a) 4 b) 16 c) 256 d) 65,536

67. The function represented by the k-map given below is **[GATE-1998]**

A \ BC	00	01	10	11
0	1	0	0	1
1	1	0	0	1

a) AB

b) ABC

c) $\overline{B \oplus C}$

d) AB + BC + CA

68. The output Y of the below circuit is equal to

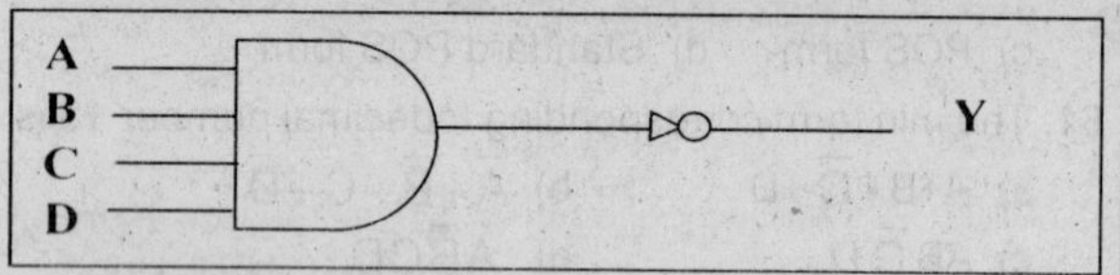

a) $\overline{ABCD}$ b) $\bar{A}+\bar{B}+\bar{C}+\bar{D}$

c) $\overline{A+B+C+D}$ d) $A \oplus B \oplus C \oplus D$

69. The output of Y of the below circuit is equal to

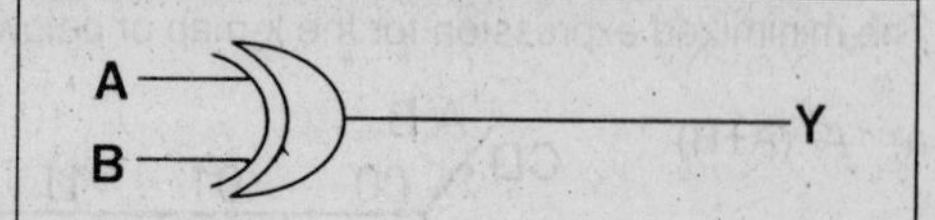

a) $A \oplus B$ b) $\overline{A+B}$

c) $A\bar{B}+\bar{A}B$ d) $\overline{AB}$

70. The Boolean expression, of the following circuit is given by

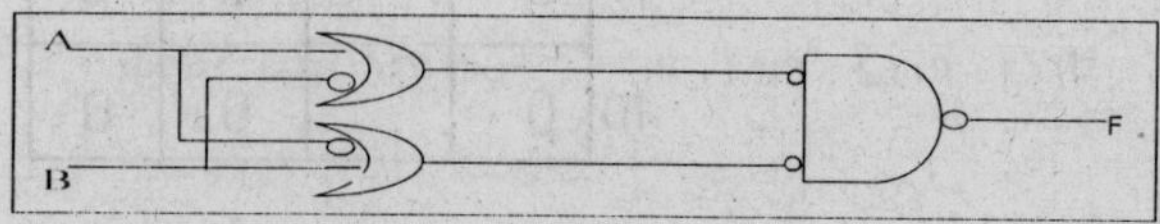

a) Tautology b) Inconsistency

c) Contradiction d) None of these

71. Two NAND gates having open collector outputs are tied together as below. The logic function Y implemented by the circuit is: **[GATE-1990]**

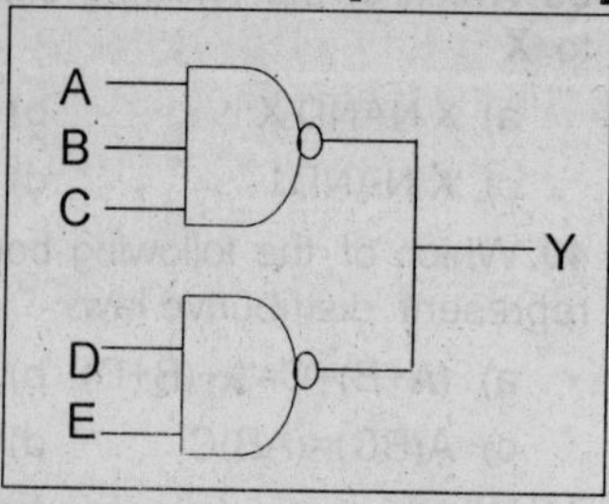

a) Y = ABC + DE

b) $Y = \overline{ABC} + \overline{DE}$

c) $Y = \overline{ABC} . \overline{DE}$

d) None of these

72. The Boolean expression **[GATE-1999]**

NOT A AND NOT B OR C is equivalent to

a) ((NOT A) AND B) OR C

b) NOT A AND (B OR C)

c) (NOT A) AND (B OR C)

d) NOT (A AND B) OR C

73. A POS expression for the function $\mathbf{B\bar{D} + \bar{C}D}$ is **[GATE-1999]**

a) $(B+\bar{C})(\bar{D}+D)$ b) $(B+\bar{D})(C+D)$

c) $(B+D)(\bar{C}+\bar{D})$ d) $(B+\bar{D})(\bar{C}+D)$

74. The reduced form of the Boolean expression A + BC is: **[GATE-1997]**

a) AB + AC + CB b) A + B + C

c) AB + BC d) (A + B) (A + C)

75. Boolean expression for the output of X-NOR logic gate with input's A and B is: **[GATE-1993]**

a) $A\bar{B}+\bar{A}B$ b) $(\bar{A}+B)(A+\bar{B})$

c) $(\bar{A}+\bar{B})(A+B)$ d) None

76. The number of Boolean Functions that can be generated by a variable n, is equal to

a) 2^n b) 2^{2n} c) 2^{n-1} d) None

77. The logic function $f = \overline{(x.\bar{y})\,(\bar{x}.y)}$ is the same as : **[GATE-1999]**

a) $f=(x+y)(\bar{x}+\bar{y})$ b) $f=\overline{(\bar{x}+\bar{y})(x+y)}$

c) $f=(x.y)(\bar{x}.\bar{y})$ d) none of these

78. A Boolean expression for the output of logic circuit below is: **[GATE-1996]**

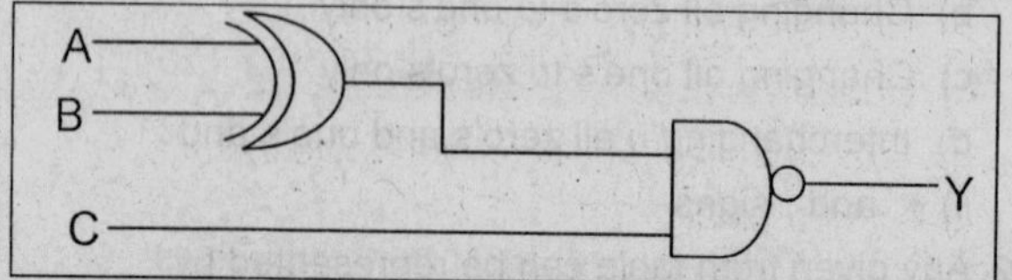

a) Y = AB + AB + C b) Y = $\bar{A}\,\bar{B} + AB + \bar{C}$

79. For the identity AB + $\bar{\mathbf{A}}$ C + BC =AB + $\bar{\mathbf{A}}$ C, the dual form is: **[GATE-1998]**

a) $(A + B)(\bar{A} + C)(B + C) = (A + B)(\bar{A} + C)$
b) $(\bar{A} + \bar{B})(A + \bar{C})(\bar{B} + \bar{C}) = (\bar{A} + \bar{B})(A + \bar{C})$
c) $(A + B)(\bar{A} + C)(B + C) = (\bar{A} + \bar{B})(A + \bar{C})$
d) $\bar{A}\bar{B} + A\bar{C} + \bar{B}\bar{C} = \bar{A}\bar{B} + A\bar{C}$

80. The minimum number of 2-input NAND gates required to implement the Boolean function $Z = A\bar{B}C$, assuming that A, B and C are available is: **[GATE-1998]**

a) three b) four c) five d) six

81. The minimum number of NAND gates required to implement the Boolean Function $A + A\bar{B} + A\bar{B}C$ is equal to **[GATE-1995]**

a) 0 b) 1 c) 2 d) 3

82. The minimal sum-of-products form of $f = A\bar{B}CD + \bar{A}BC + \bar{A}\bar{B}C + BCD$ is **[GATE-1998]**

a) $A\bar{C} + BD$ b) $\bar{A}C + CD$
c) $AC + \bar{B}D$ d) $A\bar{B} + C\bar{D}$

83. CMOS logic gate has : **[GATE-1998]**

a) Highest speed among all logic families
b) Very low quiescent power consumption
c) Very low output impedance

84. The expression $A + \bar{A}B$ is equivalent to: **[GATE-2000]**

a) A + B b) AB + A c) A + B d) AB

85. The output of logic gate in below fig. Is **[GATE-1997]**

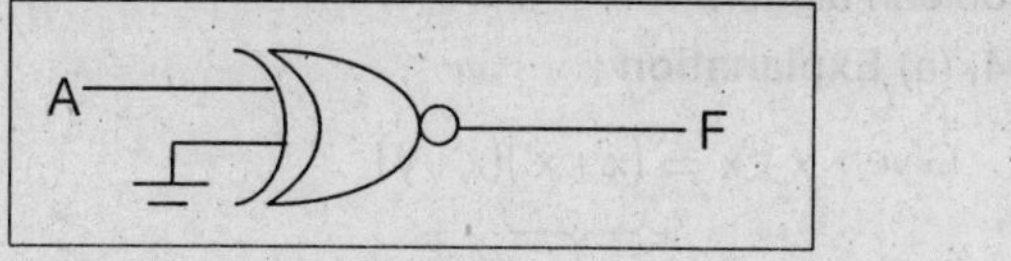

a) 0 b) 1 c) A d) F

86. The output of a logic gate is 1 when all its inputs are at logic 1. The gate is either

a) an OR or an EX-OR b) a NAND or an EX-NOR
c) an AND or a NAND d) an OR or an EX-NOR

87. If C = A*B, the C*A is

a) A b) B c) 0 d) 1

88. An AND gate **[UGC NET - 2000]**

a) implements logic addition
b) is equivalent to a series switching circuit
c) is an any -or-all gate
d) is equivalent to a parallel switching circuit

89. The dual of the statement (A + 1) = 1 is **[UGC NET - 2000]**

a) A . 1= A b) A . 0 = 0 c) A + A = A d) A . A = A

90. The main advantage of Schottly TTL logic family is its **[UGC NET - 2000]**

a) least power dissipation
b) least propagation delay
c) maximum fan - in d) least noice immunity

91. Every boolean function can be expressed in terms of **[UGC NET - 1996]**

a) AND alone b) OR alone
c) NOT alone d) NAND alone

92. In Boolean algebra $\neg(a \wedge b) = \neg a \vee \neg b$ is referred to as **[UGC NET - 1996]**

a) Demorgan's law b) Distributive law
c) Idempotence law d) Absorption law

93. Which of the following digital gates with input 0 and 1 will output 1 ? **[UGC NET - 1996]**

a) Neither OR nor AND b) OnlyAND
c) Only OR d) OR and AND both

94. The expression $X\bar{Y}\bar{Z} + \bar{X}\bar{Y}Z + X\bar{Y}Z + X\bar{Y}Z + XY\bar{Z}$ simplifies to **[UGC NET-1996]**

a) $X + \bar{Y}Z$ b) $Y + XZ$ c) $Z + \bar{X}Y$ d) None

95. MOS logic gates have no current hogging problem because the gate terminal has **[UGC NET-1995]**

a) low input impedance b) zero impedance
c) high impedance d) compensating effect

96. What should be added so that a CMOS gates drives TTL at the same supply voltage? **[UGC NET-1995]**

a) Pull up resistor b) Pull down transistor
c) A high value capacitor d) A buffer

97. Which one of the following expressions is not logically equivalent to all of the rest ? **[UGC NET-1998]**

a) $Wx\bar{y} + W\bar{z} + Wxyz + W\bar{y}z$
b) $Wx + W\bar{y} + Wy\bar{z}$
c) $Wx + W\bar{y} + W\bar{z}$ d) $W + \bar{y} + \bar{z}$

98. The universal building blocks in digital electronic architecture are **[UGC NET-1998]**

a) AND and OR gates b) NAND and OR gates
c) NAND and NOR gates d) OR and NOR gates

99. A NAND gate in positive logic becomes .... Gate in negative logic **[UGC NET-1998]**

a) NAND b) NOR c) OR d) NOT

100. Which of the following logic family has the maximum fan out? **[UGC NET-1998]**

a) RTL b) ECL c) NMOS d) CMOS

101. The simplified form of the boolean expression (X + Y + XY) (X + Z) is **[UGC NET-1999]**

a) X + Y + Z b) XY + YZ
c) X + YZ d) XZ + Y

102. Which of the following boolean algebra statement represent the commutative law **[UGC NET-1999]**

a) (A + B) + CA + (B + C)
b) A . (B + C) (A . B) + (A . C)
c) A + B = B + A d) A + A . A

103. The fastest logic family out of the following is **[UGC NET-1999]**

a) TTL b) RTL c) DTL d) CMOS

104. Which of the following is not equivalent to $\overline{x}$?
o = NAND operator, I = NOR operator **[Hyd.Univ-2001]**

a) x o x b) x I x c) x o 1 d) x I 1

105. The number of 1's in the binary representation of (3*4096+15*256+5*16+3) is **[Hyd.Univ-2001]**

a) 8 b) 9 c) 10 d) 12

106. In a boolean algebra, the expression
z (x + y) + xz′ + yz′ equals **[Hyd.Univ-1998]**

a) y + z b) x + z c) x + y d) x + y + z

107. The minimum number of NAND gates required to implement the boolean function A + AB′ + AB′C, is equal to **[Hyd.Univ-1998]**

a) 0 b) 1 c) 4 d) 7

108. The switching speed of ECL is very high because **[Hyd.Univ-1998]**

a) it uses high speed transistors
b) it uses positive logic
c) it uses negative logic
d) its transitions remain unsaturated

109. Truth table of a logic function **[Hyd.Univ-1995]**

a) summarises its output values
b) displays some of its input conditions
c) displays all its I/O possibilities
d) Tabulates all its input conditions.

110. By simplifying $(AC + AB\overline{C})(A + \overline{A}B + \overline{A}\,\overline{B})$ we get **[Hyd.Univ-1994]**

a) 0 b) 1 c) A(B + C) d) A + B

111. The minimum number of 2-input NAND gates required to implement F = (X + Y) (Z + W) is **[Hyd.Univ-1993]**

a) 4 b) 3 c) 5 d) 6

ANSWER KEY - I

1. a	**2.** c	**3.** c	**4.** d	**5.** a
6. d	**7.** b	**8.** a	**9.** c	**10.** b
11. c	**12.** c	**13.** c	**14.** b	**15.** b
16. b	**17.** a	**18.** a	**19.** d	**20.** c
21. b	**22.** d	**23.** a	**24.** d	**25.** b
26. c	**27.** b	**28.** b	**29.** c	**30.** c
31. a	**32.** a	**33.** c	**34.** b	**35.** b
36. d	**37.** c	**38.** a	**39.** d	**40.** b
41. c	**42.** a	**43.** d	**44.** d	**45.** b
46. d	**47.** c	**48.** d	**49.** c	**50.** b
51. d	**52.** a,b,c	**53.** b	**54.** d	**55.** c
56. b	**57.** b	**58.** a	**59.** b	**60.** c
61. c	**62.** d	**63.** c	**64.** a	**65.** b
66. b	**67.** c	**68.** a,b	**69.** a,c	**70.** a
71. c	**72.** c	**73.** c	**74.** d	**75.** b
76. a	**77.** b	**78.** b	**79.** a	**80.** c
81. a	**82.** b	**83.** c	**84.** c	**85.** c
86. c	**87.** b	**88.** b	**89.** a	**90.** b
91. d	**92.** a	**93.** c	**94.** d	**95.** c
96. a	**97.** d	**98.** c	**99.** c	**100.** d
101. c	**102.** c	**103.** a	**104.** d	**105.** c
106. a	**107.** a	**108.** d	**109.** c	**110.** c
111. a				

EXPLANATORY NOTES - I

33. (c) **Explanation :** 3 two - Input NAND gates are required to perform the function two - input OR gates is shown below.

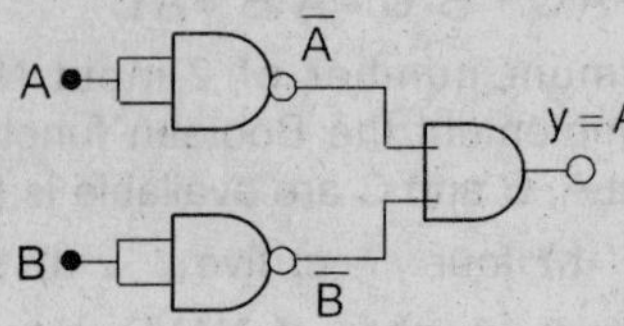

34. (b) **Explanation :** We need 2 two - input (x,y) NAND gates to perform the logical function x,y is shown below.

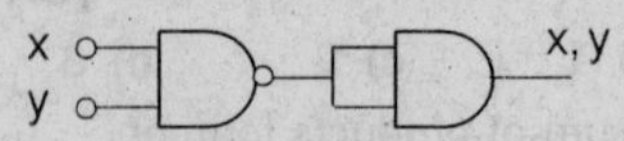

39. (d) **Explanation :**
x NOR 1 $\Rightarrow \overline{X+1} \Rightarrow \overline{X}.\overline{1} \Rightarrow \overline{X}.0 \Rightarrow 0$
Hence answer 'd' is correct.

42. (a,c,d) **Explanation :** From the algebraic rules answers a, c and d are true. (Or)

X	X′	0 + X	X + X′	X + X
0	1	0	1	1
1	0	1	1	1
		= X	= 1	= X

44. (d) **Explanation :** By the commulative law from the boolean algebra X+Y = Y+X

44. (a) **Explanation :**

Given $x + x' \Rightarrow \underbrace{(x + x')}_{1}(x + y)$

$\Rightarrow (x + y)$ [∵ By identity]

$\therefore\; x + x'y = x + y$

45. (b) **Explanation :** Given $A \oplus B = C$. Now, with the help of truth table we can select the correct property among the given properties.

A	B	A⊕B	C	B⊕C
0	0	0	0	0
0	1	1	1	0
1	0	1	1	1
1	1	0	0	1

Hence, If $A \oplus B = C$ then $B \oplus C = A = A$

46. (d) **Explanation :** Given, $A * B = AB + A'B'$
Now, $\therefore A * A = AA + A'A'$
$= A + A'$ [∵ By identity]
$= 1$ [∵ By identity]
Hence answer 'd' is correct.

47. (c) **Explanation :** Given $(x + y + xy)(x + z)$
$\Rightarrow \underbrace{x \quad x}_{x} + xz + xy + yz + xy + xyz$
$\Rightarrow x + xy + xz + yz \Rightarrow x(1 + y) + xz + yz$
$\Rightarrow x + xz + yz \Rightarrow x(1 + z) + yz \Rightarrow x + yz$

48. (d) **Explanation :**
Given $(A + C)(A\overline{B} + AC)(\overline{AC} + \overline{B})$

$$\Rightarrow \left(A\bar{B}+AC+AC\bar{B}+AC\right)\left(\overline{AC}+\bar{B}\right)$$

$$\Rightarrow \underbrace{A\;\;\bar{A}}_{0}\bar{B}\,\bar{C}+A\bar{B}+\underbrace{AC\;\;\bar{A}}_{0}\bar{C}+AC\bar{B}+AC\underbrace{\bar{B}\,\bar{A}}_{0}\bar{C}$$

$$+\,AC\bar{B}\bar{B}+AC\bar{A}\,\bar{C}+AC\bar{B}$$

$$\Rightarrow A\bar{B}+AC\bar{B}\Rightarrow A\bar{B}(1+C)\Rightarrow A\bar{B}$$

49. (c) **Explanation :**

Given $\overline{\bar{A}\,\bar{B}\,\bar{C}}$ By Demorgan's laws

$=\bar{\bar{A}}+\bar{\bar{B}}+\bar{\bar{C}}=A+B+C$

Hence answer 'c' is correct.

50. (b) **Explanation :**

Given $AB+\bar{A}C+BC=AB+\bar{A}C+BC\left(A+\bar{A}\right)$

$=AB+\bar{A}C+ABC+\bar{A}BC$

$=AB(1+C)+\bar{A}C(1+B)\;=AB+\bar{A}C$

51. (b) **Explanation :** The possible functions that can be formed with 'n' variables to be 2^{2^n}, therefore the functions that can be fromed with 3 variables is $2^{2^3}=2^8=256$.

54. (d) **Explanation :** The complement of a function expressed as the sum of minterms equals the sum of maxterms missing from the original function.

To convert from one canonical form to another, interchange the symbols Σ and π and list those numbers missing from the original form.

Given $F=(X,Y,Z)=\Sigma(1,2,5,6,7)$ is expressed in the sum of minterms. It's conversion to product of max terms is $F=(X,Y,Z)=\pi(0,3,4)\;=(\bar{y}+\bar{z})(x+\bar{y}+z)$.

Hence answer 'd' is correct.

55. (d) **Explanation :**

$W+X+\bar{Y}+\bar{Z}$ is the minimized function.

56. (b) **Explanation :** Given $AB+A\bar{B}+\bar{A}C+AC$

$=A\left(B+\bar{B}\right)+C\left(A+\bar{A}\right)\;=A+C$

61. (c) **Explanation :** The 4 - Variable K - Map with minterm numbers are written in each square for reference is shown below.

CD \ AB	00	01	11	10
00	0	4	12	8
01	1	5	13	9
11	3	7	15	11
10	2	6	14	10

The minterm corresponding to decimal number 13 is $AB\bar{C}D$. Hence answer 'c' is correct.

62. (d) **Explanation :** From above K - map representation the max term corresponding to decimal number 15 is $\bar{A}+\bar{B}+\bar{C}+\bar{D}$.

63. (c) **Explanation :** Given

CD \ AB	00	01	11	10
00	0		0	0
01	0		0	0
11	0		0	0
10	0		0	0

The minimised expression for the K-map is $\bar{A}B$

64. (a) **Explanation :** Given

CD \ AB	00	01	11	10
00	X	0	0	X
01		0		X
11	0	X		X
10		X		X

The minimized expression for the given K - map is $(C+D)\left(\bar{C}+\bar{D}\right)\left(A+\bar{B}\right)$

65. (b) **Explanation :** Given

YZ \ WX	00	01	11	10
00	1		1	1
01				
11		1	1	
10	1		1	1

The minimal form is

$(\bar{x}+\bar{z})(w+\bar{z})(w+x+z)$

67. (c) **Explanation :** Given K - map is

A \ BC	00	01	11	10
0	1	0	0	1
1	1	0	0	1

The minimized expression for POS of a given K - map is $\left(B+\bar{C}\right)\left(\bar{B}+C\right)=\overline{B\oplus C}$

68. (a,b) **Explanation :**

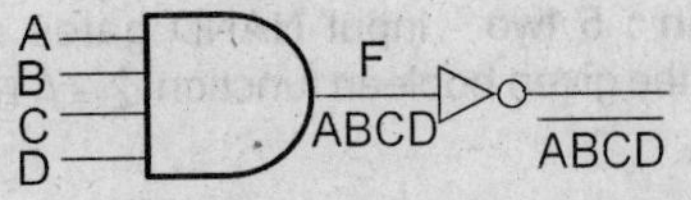

The given circuit is a four - input NAND gate.

69. (a,c) **Explanation :** Given cirucit is the EX - OR gate. The algebraic function of EX - OR is $Y=\bar{A}B+A\bar{B}=A\oplus B$.

70. (a) **Explanation :** From the given circuit the boolean expression for $F=\overline{\left(\overline{A+\bar{B}}\right)\left(\bar{A}+B\right)}=A+\underbrace{\bar{B}+\bar{A}}_{1}+B=1$ (1 1)

$\therefore$ F is true. Hence 'F' is Tautology.

71. (c) **Explanation :** When the two NAND gates tied togehter, perform the wired AND logic. The wired AND gate is not a physical gate but only a symbol to designate the function obtained from the indicated wired connection.

The logical function implemented by the given circuit is $F=\overline{ABC}.\overline{DE}$.

73. (c) **Explanation :** Given sum of products (SOP) expression is $F(B,C,D)=\Sigma(1,4,6,5)$. Now, equalent POS expression is $F(B,C,D)=\pi(0,2,3,7)\;=(B+D)\left(\bar{C}+\bar{D}\right)$.

72. (d) **Explanation :** The operator precedence for evaluating Boolean expression is (1) parentheses (2) Not (3) AND, and (4) OR.

Given, NOT A AND NOT B OR C

$\Rightarrow$ (NOT A AND NOT B) OR C

$\Rightarrow$ NOT (A AND B) OR C

74. (d) **Explanation :** By the distributive law

$A+BC=(A+B).(A+C)$

75. (b) **Explanation :** EX - OR gate followed by an inverter to make an X - NOR gate. The boolean expression for X - NOR gate is $Y=\overline{A\oplus B}=\overline{\bar{A}B+A\bar{B}}=\left(A+\bar{B}\right)\left(\bar{A}+B\right)$

The circuit symbol is

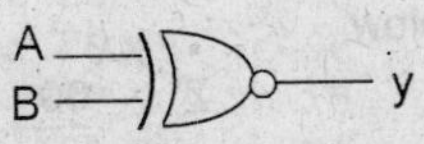

$\therefore y=\overline{A\oplus B}=A\otimes B$.

77. (b) **Explanation :** Given , $f=\overline{(x.\bar{y}+(\bar{x}+y))}$

$$\Rightarrow f'=(x\bar{y})+(\bar{x}y) \;=[x+(\bar{x}y)][\bar{y}+(\bar{x}y)]$$

$$=(x+y)(\bar{x}+\bar{y}) \;\Rightarrow f=\overline{(\bar{x}+\bar{y})(x+y)}$$

78. (b) **Explantion :** From the given circuit obtained boolean expression for y is $Y=(\overline{\bar{A}B+A\bar{B}})C$

The minimised expression is

$$Y=[(A+\bar{B})(\bar{A}+B)]+\bar{C}=AB+\bar{A}\bar{B}+\bar{C}$$

79. (a) **Explanation :** If the dual of an algebraic expression is desired, we simply interchange OR and AND operators and replace 1's by 0's and 0's by 1's. The dual form of a given expression is

$$(A+B)(\bar{A}+C)(B+C)=(A+B)(\bar{A}+C)$$

80. (c) **Explanation :** 5 two - input NAND gates are required to perform the given boolean function $Z=A\bar{B}C$ is shown below.

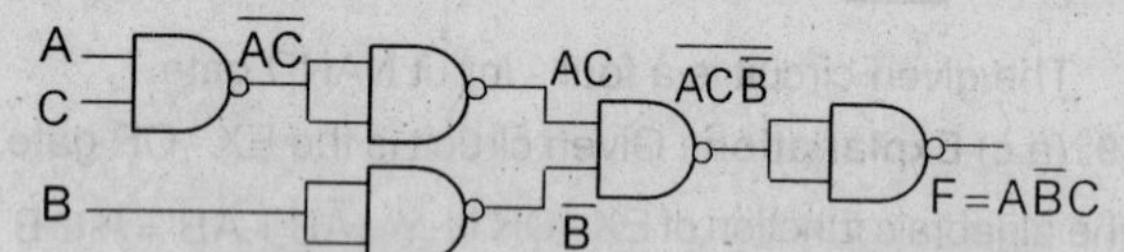

81. (a) **Explanation :** Given Boolean expression

$$A+A\bar{B}+A\bar{B}C \;=A+A\bar{B}(1+C)$$

$$=A+A\bar{B} \;=A(1+\bar{B}) \;=A$$

There is no need of any NAND gates to implement to Given expression. We require only one buffer. Hence answer 'd' is correct.

82. (b) **Explanation :** The minimal SOP forms of given boolean function is obtained by using K - map is shown below.

CD \ AB	00	01	11	10
00				
01				
11	1	1	1	1
10	1	1		

$\therefore f=CD+\bar{A}C$

84. (c) **Explanation :** Given,

$$A+\bar{A}B=(A+\bar{A})(A+B)\ [\therefore \text{by commutative}]$$

$$=A+B\ [\therefore \text{by identity}]$$

85. (c) **Explanation :** In given circuit one input is grounded another input is 'A'. So the output of circuit is F = A.

87. (b) **Explanation :** Given $C=A*B$

$$\Rightarrow C*A=B \text{ or } C*B=A$$

89. (b) **Explanation :** If the dual of boolean expression is derived simply interchange + by × and 0's by 1's and 1's by 0's. So the dual of A+1=1 is A.0=0.

94. (d) **Explanation :** The simplied expression is obtained by using K - map is shown below

Z \ XY	00	01	11	10
0			1	1
1	1		0	1

$x\bar{z}+\bar{y}z$

101. (c) **Explanation :** Given

$$(x+y+xy)(x+z)=x+xz+xy+yz+xy+xyz$$

$$=x+xz+xy+yz+xyz=x(1+z)+xy+yz(1+x)$$

$$=x+xy+yz=x(1+y)+yz=x+yz$$

104. (d) **Explanation :** $x.1=x\,\text{NOR}\,1=\overline{x+1}=0$

105. (c) **Explanation :** Given

$$(3*4096+15*286+5*16+3)$$

$$=\underbrace{3*10^{12}}_{2}+\underbrace{5*2^{8}}_{4}+\underbrace{5*2^{4}}_{2}+\underbrace{3}_{2}$$

No. of 1's required = 2+4+2+2=10

106. (a) **Explanation :** Given $z(x+y)+x'z+yz'$

$$=z[\underbrace{x+y+x'}_{1}]+yz'$$

$$=z(1+y)+yz'=z+yz'=(z+y)(\underbrace{z+z'}_{1})=y+z$$

107. (a) **Explanation :** The minimized expression for given function is A. so there is no need of any NAND gates to implement the given expression.

110. (c) **Explanation :** Given $(AC+ABC\bar{C})(A+\bar{A}B+\bar{A}\bar{B})$

$$=(AC+AB\bar{C})(A+\bar{A}(\underbrace{B+\bar{B}}_{1}))=AC+AB\bar{C}$$

$$=A(C+B\bar{C})=A[(C+B)(\underbrace{C+\bar{C}}_{1})]=A(B+C)$$

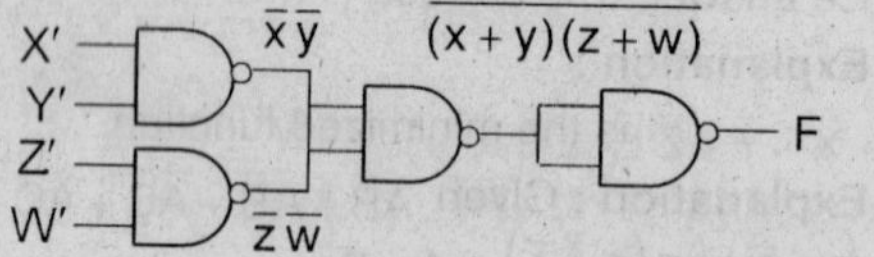

111. (a) **Explanation :** Four 2 - input NAND gates are required to implement $F=(x+y)(z+w)$ is shown above.

OBJECTIVE QUESTIONS - II

1. Standard TTL has a multiple emitter input transistor and a .... Output

a) Totem-pole b) Bipolar
c) Register d) Transistor

2. The main advantage of CMOS as compared to TTL are

a) Lower consumption and better noise margins
b) Lower consumption with low fan-out
c) Higher speed operation with low consumption
d) Higher consumption with high fan-out

3. A gate is logic circuit with one or more input signals and the number of output signals can be **[AUCET-2000]**

a) One b) More than one c) Two d) Zero

4. De Morgan's Theorem – I says that a NOR gate is equivlent to a bubbled **[AUCET-2000]**

a) NOR b) XOR c) XNOR d) AND

5. De Morgan's Theorem – II says that a NAND gate is equivalent to a bubbled **[AUCET-2000]**

a) AND b) OR c) EX-OR d) EX-NOR

6. In the following expression, AND and OR are arithmetic operators: (42 OR 72) AND 55 is The value of the expression is **[AUCET-1999]**

a) 15 b) 34 c) 59 d) 85

7. To set the most significant bit of an 8-bit register without affecting any other bits, perform the following logical operation with an appropriate mark **[AUCET-2001]**

a) OR b) NOT c) XOR d) AND

8. In a typical modern computer system, the outputs of peripheral devices that are connected to the common system bus are **[AUCET-2001]**

a) Totem pole b) Tri-statable
c) Open-collector d) High-inpedance

9. Identify the operation which is commutative but not associative **[GATE-1992]**

a) OR b) NOR c) EX-OR d) NAND

10. The output of a logic gate is LOW when all its inputs are at logic HIGH. The gate is either

a) an AND or a NOR b) an AND or an EX-NOR
c) an OR or a NAND d) a NAND or an EX-OR

11. Consider the expression $Z = A \oplus B \oplus C \oplus \ldots$, where A,B,C,.... are the input variables and Z is the output Variable. Z will be logic 0 if **[AUCET-1999]**

a) An even number of input variables are 1.
b) An odd number of input variables are 1
c) An even number of input variable are 0
d) An odd number of input variables are 0

12. Which of the following represent commutative law, Associate law, and Distributive law

I. A.(B.C) = (A.B).C II. A.(B+C) = A.B+A.C
III. A+B = B+A

a) I,III and II respectively b) II,I and III respectively
c) III ,II and I respectively d) III, I and II respectively

13. The values of A,B,C and D that satisfy the simultaneous boolean equations

$\overline{A}$ +AB=0; AB=AC and AB+A$\overline{C}$ +CD=CD, is

a) A=1,B=0,C=0,D=1 b) A=1,B=1,C=0,D=0
c) A=1,B=0,C=1,D=1 d) A=1,B=0,C=0,D=0

14. The boolean variables A,Band C, that solve the boolean equations AB+$\overline{A}$ C=1 and AC+B=0 simultaneausly is

a) 1,0,0 b) 0,1,1 c) 1,0,1 d) 0,0,1

15. Consider the following sequence of instructions

$a=a \oplus b$, $b=a \oplus b$, $a=b \oplus a$ This sequence

a) Retains the values of a and b.
b) Swaps a and b.
c) Complements the values of a and b
d) Negates values of a and b and then swaps them.

16. The dual of the Boolean theorem A(B+C)= AB+AC is

a) A+(BC)=AB+AC b) A(B+C)=(A+B).(A+C)
c) A+BC=(A+B).(A+C) d) None of these.

17. What values of A,B,C and D satisfy the following simultaneous boolean equations.

$\overline{A}$ +AB=0, AB=AC,AB+A $\overline{C}$ +CD= $\overline{C}\,\overline{D}$ **[GATE-1995]**

a) A=1,B=0,C=0 D = 1 b) A=1,B=1,C=D=0

18. The minimum number of gates require to implement the boolean expression AB+A$\overline{B}$ + $\overline{A}$ C is

a) 1 AND gate and1 OR gate b) 2 NAND gates
c) 3 AND gates and 2 OR gates d) None of these

19. The logic expression Y= $\sum$ m(0,3,6,7,10,12,15) is equivalent to

a) Y= Π M (0,3,6,7,10,12,15)
b) Y= Π M (1,2,4,5,8,9,11,13,14)
c) Y= $\sum$ m (1,2,4,5,8,9,11,13,14)
d) Y= $\sum$ m (0,3,10,12)

20. The logic expression Y= Π M (1,4,6,9,10,11,14,15) is equivalent to

a) Π M (0,2,3,5,7,8,12,13)
b) Π M (0,2,3,4,5,6,12,13)
c) $\sum$ m(1,4,6,9,10,11,14,15)
d) $\sum$ m(0,2,3,5,7,8,12,13)

21. The minimized expression for the k-map of below fig. is

a) $\overline{C}D + \overline{A}\,\overline{B}\,\overline{C} + A\overline{B}\,\overline{C} + \overline{A}\,\overline{B}\,C\overline{D}$

b) $\overline{B}C\overline{D} + \overline{B}\,\overline{C}\,\overline{D} + \overline{C}D$

c) $\overline{B}\,\overline{D} + \overline{C}D$

d) $\overline{A}\,\overline{B}C\overline{D} + A\overline{B}C\overline{D} + \overline{A}\,\overline{B}\,\overline{C}$

CD \ AB	00	01	11	10
00	1			1
01	1	1	1	1
11				
10	1			1

22. The minimezed expression for the k-map of below fig. is

a) $\overline{A}B\overline{C}\,\overline{D} + \overline{A}CD + \overline{A}\,\overline{B}C\overline{D}$

b) $\overline{A}B\overline{C}\,\overline{D} + \overline{A}CD + \overline{A}BC$

c) $B\overline{C}\,\overline{D} + CD + BC$

d) $B\overline{CD} + CD + \overline{AB}C$

CD \ AB	00	01	11	10
00		1	X	
01			X	
11	1	1	X	X
10	1		X	X

23. The k-map for the boolean function F of 4 boolean variables is given below, where A,B,C aredon't care conditions. what values of A,B,C will result in the minimal expression.

		A	
	1	1	
1	B	C	
		1	

a) A=B=C=1

b) B=C=1; A=0

c) A=C=1; B=0

d) A=B=1; C=0

24. What is the equivalent Boolean expression in POS form for the K-map given below **[GATE-1996]**

a) $B\bar{D}+\bar{B}D$

b) $(B+\bar{D})(B+\bar{D})$

c) $(B+\bar{D})(\bar{B}+D)$

d) None of these

CD \ AB	00	01	11	10
00		1	1	
01	1			1
11	1			1
10		1	1	

25. Let $f(x, y, z) = \bar{x} + \bar{y}x + xz$ be a switching function, which one of the following statement is valid. **[GATE-1997]**

a) $\bar{y}x$ is prime implicant of f.
b) xz is a min term of f.
c) xz is an implicant of f.
d) y is a prime implicant of f.

26. Which function does NOT implement the k-map given below **[GATE-2000]**

a) $(w + x)y$

b) $xy+yw$

c) $(w+x)(\bar{w}+y)(\bar{x}+y)$

d) None of these

wz→ xY ↓	00	01	11	10
00	0	x	0	0
01	0	x	1	1
11	1	1	1	1
10	0	x	0	0

27. Given the following k-map, which one of the following represents the minimal SOP of the map? **[GATE-2001]**

a) $xy+y'z$

b) $wx'y'+wy+xz$

c) $w'x+y'z+xy$

d) $xz+y$

YZ \ WX	00	01	11	10
00	0	X	0	X
01	X	1	X	1
11	0	X	1	0
10	0	1	X	0

28. The output F, of the following circuit is given by

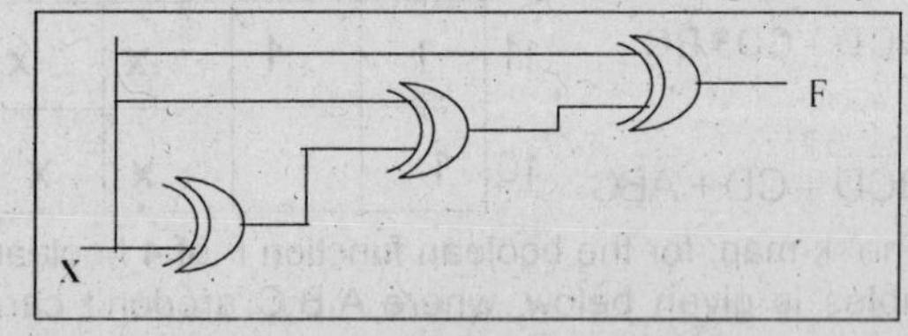

a) 1 b) 0 c) X d) X′

29. Consider the following circuit. In order to make it a tautology the '?' marked box should be replaced by

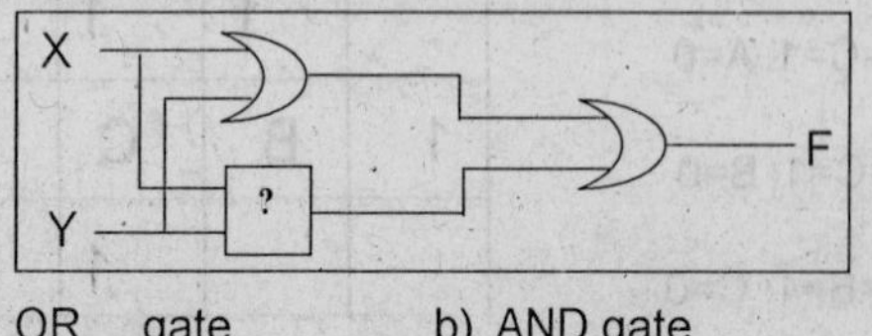

a) OR gate b) AND gate
c) NAND gate d) NOR gate

30. Identify the logic function performed by the circuit shown below. **[GATE-1993]**

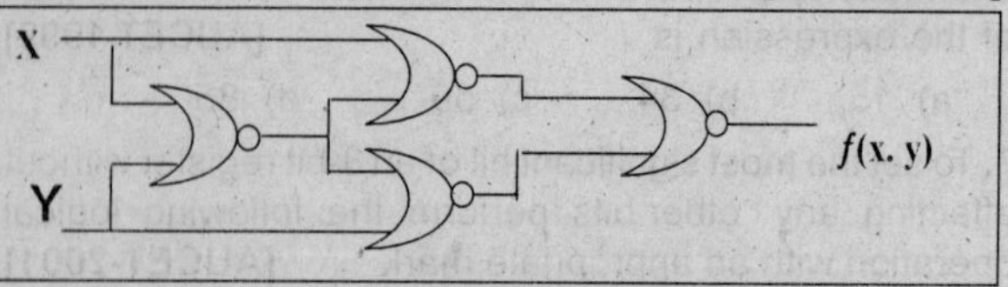

a) Ex-OR b) Ex-NOR c) NAND d) NOR

31. The logic expression for the output of the circuit shown below is **[GATE-1994]**

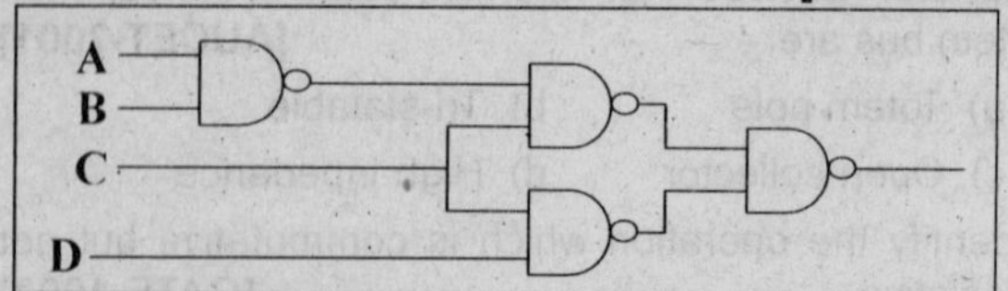

a) $A\bar{C}+B\bar{C}+CD$ b) $\bar{A}\bar{B}+\bar{B}\bar{C}+\bar{C}\bar{D}$
c) $A\bar{C}+B\bar{C}+\bar{C}D$ d) None

32. Consider the logic circuit shown below. The functions f_1, f_2, f (in canonical sum of products form in decimal notation are) **[GATE-1997]**

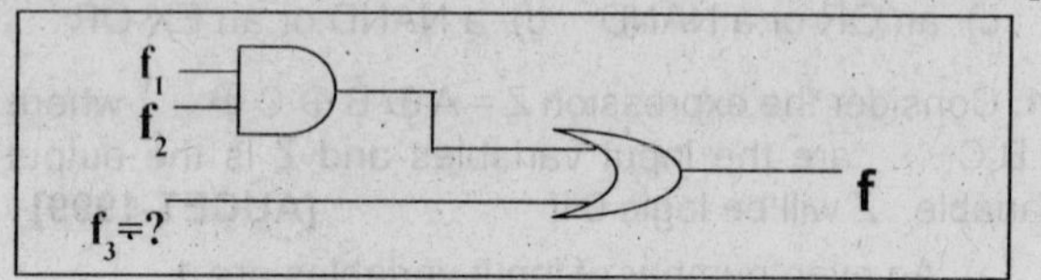

$f_1(w, x, y, z) = \sum (8, 9, 10)$,
$f_2(w, x, y, z) = \sum (7, 8, 12, 13, 14, 15)$,
$f(w, x, y, z) = \sum (8, 9)$. The function f_3 is

a) $\sum (9, 10)$ b) $\sum (9)$
c) $\sum (1, 8, 9)$ d) $\sum (8, 10, 15)$

33. Consider the circuit shown below. In a certain steady state the line Y is at "1". What are the possible values of A, B, C in this state. **[GATE-1999]**

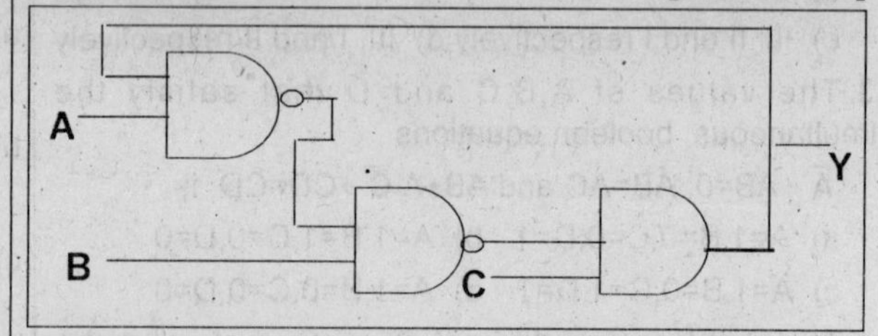

a) A=0, B=0, C=1 b) A=0, B=C=1
c) A=1, B=C=0 d) A=B=1,C=1

34. The k-map for a Boolean function is shown below. The number of essential prime implicants for this function is : **[GATE-1998]**

a) 4

b) 5

c) 6

d) 8

1	1	0	1
0	0	0	1
1	0	0	0
1	0	0	1

35. For the logic circuit shown below, the required input condition (A, B, C) to make the o/p(X) = 1 is : **[GATE-2000]**

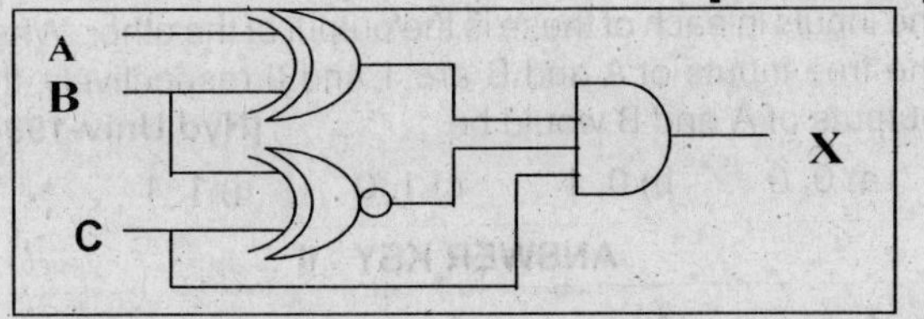

a) 1, 0, 1 b) 0, 0 ,1 c) 1, 1, 1 d) 0, 1, 1

36. From the below truth table, choose the correct expression for Q.

A	0	0	0	0	1	1	1	1
B	0	0	1	1	0	0	1	1
C	0	1	0	1	0	1	0	1
Q	0	0	0	1	0	1	1	1

a) AB + BC + CA b) A + B + C

c) $A\bar{B} + B\bar{C} + C\bar{A}$ d) $AB + \bar{B}\bar{C} + \bar{C}\bar{A}$

37. The output of circuit shown below is equal to: **[GATE-1995]**

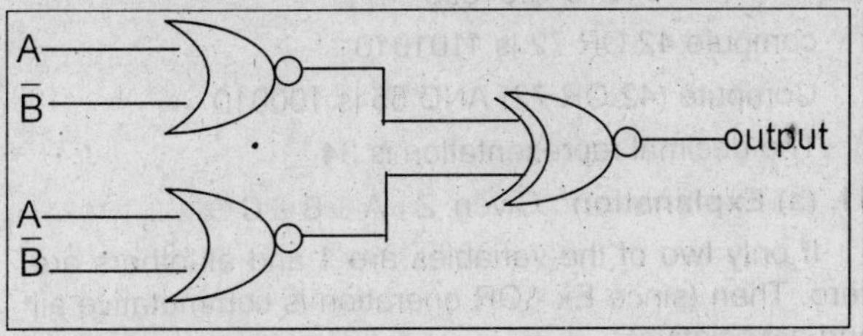

a) 0 b) 1 c) $\bar{A}B + A\bar{B}$ d) None of these

38. The simplified form of the boolean expression

$(X+\bar{Y}+Z).(X+\bar{Y}+Z).(X+Y+Z)$ **[UGC NET - 1997]**

a) $\bar{X}Y + \bar{Z}$ b) $X + \bar{Y}Z$ c) X d) $XY + \bar{Z}$

39. Following Boolean expression **[UGC NET - 1997]**

$Z = \bar{W}\bar{Y}\bar{X} + X\bar{Y}\bar{W} + W\bar{X}\bar{Y} + \bar{X}YW$ can be reduced to

a) $\bar{X}\bar{W} + \bar{Y}W$ b) $\bar{X}W$ c) $\bar{Y}\bar{W}$ d) $\bar{Y}\bar{W} + \bar{X}W$

40. A 3-variable boolean function (or ternary boolen operator) f (A,B,C) is defined by the given table. The operator can be expressed in terms of ¬ (not), ∧ (and), ∨ (or)as **[UGC NET - 1996]**

A	B	C	f (A,B,C)
0	0	0	0
0	0	1	0
0	1	0	0
0	1	1	1
1	0	0	0
1	0	1	1
1	1	0	0
1	1	1	1

a) $A \wedge (\bar{B} \vee C)$ b) $B \wedge (B \vee C)$ c) $C \wedge (B \vee A)$ d) None

41. How do you compare speeds of ECL (Emitter Coupled Logic) and TTL(Transistor - Transistor Logic)? **[UGC NET - 1996]**

a) ECL is faster than TTL b) TTL is faster than ECL

c) Both are comparable d) Cannot be determuned

42. The truth table given adjacent is for the boolean operator ⊕. which of the following equality is correct for a given booleans a and b . **[UGC NET-1995]**

a	b	a ⊕ b
0	0	0
0	1	1
1	0	1
1	1	0

a) b =(a ⊕ b) ⊕ b

b) b =(a ⊕ b) ⊕ ((a ⊕ b) ⊕ b)

c) b =(a ⊕ b) ⊕ (a ⊕ b)

d) b =(a ⊕ b) ⊕((a ⊕ b) ⊕(a ⊕ b))

43. CMOS circuits are characterized by **[UGC NET-1998]**

a) low power dissipation, large fan out, high density

b) high power dissipation, large fan out, high density

c) low power dissipation, small fan out, low density

d) high power dissipation, large fan out, low density

44. If F = (A + B + C) . $(AB+\bar{C})$, then **[UGC NET-1998]**

a) F = 1 b) F = ABC c) $F = A\bar{C} + B$ d) F = AB

45. The circuit shown in figure would work as **[UGC NET-1999]**

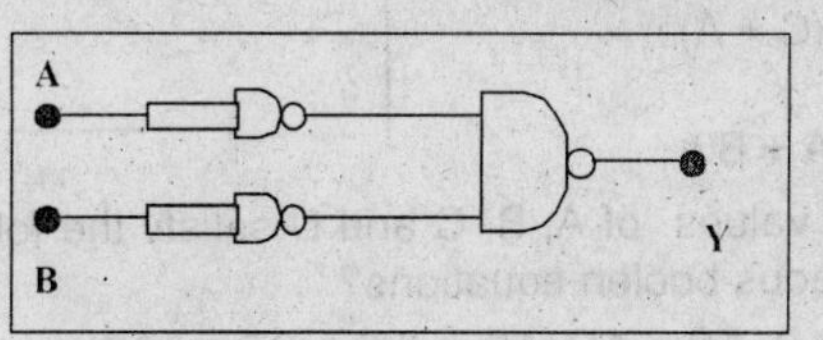

a) NAND gate b) OR gate c) Inverter d) XOR gate

46. The minimum power is being dissipated by the logic family known as

a) DTL b) TTL c) ECL d) CMOS

47. Mark the correct Boolean expression : **[UGC NET - 1998]**

a) $\overline{A+B} = \bar{A}B + A\bar{B}$ b) $A + B = AB + A\bar{B}$

c) $A.B.\bar{A}.\bar{B} = 1$ d) $(A+B)(A+\bar{B}).(\bar{A}+B) = A.B$

48. What values of A, B, C and D satisfy the following simultaneous boolean equations **[Hyd.Univ-2001]**

$\bar{A} + AB = 0; AB = AC; AB + A\bar{C} + CD = \overline{CD}$

a) A = 1, B = 0, C = 0, D = 0

b) A = 1, B = 1, C = 0, D = 0

c) A = 1, B = 0, C = 1, D = 1

d) A = 1, B = 0, C = 0, D = 1

49. A minimum complexity circuit for

$f(x; y, z) = \begin{cases} 1 & Whenever \ x \neq y \\ 0 & otherwise \end{cases}$

If $x \neq y \neq z$, then which of the following equations represents the function f? **[Hyd.Univ-2001]**

a) $x'y + xy'$ b) $x + yz$

c) $x'y'z' + xy'z$ d) $xy + y'z + z'$

50. Given the incomplete Karnaugh map below for the boolean expression, $d'(b' + ac) + bcd$, how many more '1's are needed to complete it? **[Hyd.Univ-2001]**

a) 2

b) 3

c) 4

d) 5

cd \ ab	00	01	11	10
00				
01				
11		1	1	
10			1	

51. Given the following table of values of the boolean function f the concise boolen expression which corresponds to the table is **[Hyd.Univ-2000]**

a) $x_1'x_2 + x_2x_3'$

b) $x_1'x_2' + x_2x_3'$

c) $x_1'x_2 + x_2'x_3$

d) $x_1x_2 + x_2'x_3$

X_1	X_2	X_3	f
1	1	1	1
1	1	0	1
1	0	1	1
1	0	0	0
0	1	1	0
0	1	0	0
0	0	1	1
0	0	0	0

52. The boolean function symbolizing the following circuit is equivalent to **[Hyd Univ.--1998]**

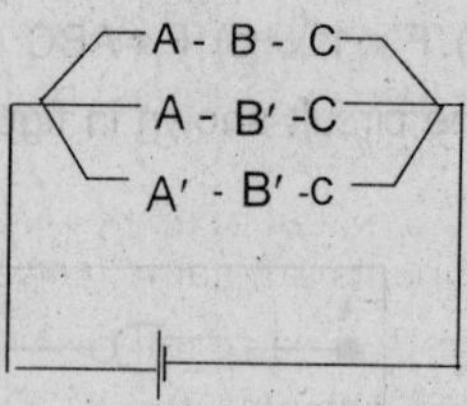

a) $A(B + C')$

b) $B(C + A')$

c) $B'(C + A)$

d) $C(A + B')$

53. What values of A, B, C and D satisfy the following simultaneous boolen equations?

$A' + AB = 0,\ AB = AC,\ AB + A'C + CD = (CD)'$ **[Hyd.Univ-1998]**

a) A = 1, B = 0, C = 0, D = 1

b) A = 1, B = 1, C = 0, D = 0

c) A = 1, B = 0, C = 1, D = 1 d) None

54. Simplify the boolean function:

$F(W, X, Y, Z) = \Sigma m(0, 1, 2, 4, 5, 6, 8, 9, 12, 13, 14)$ **[Hyd.Univ-1997]**

a) $F = \overline{Y} + \overline{W}\,\overline{Z} + X\overline{Z}$ b) $F = Y + \overline{W}Z + \overline{X}Z$

c) $F = \overline{Y} + WZ + X\overline{Z}$ d) $F = \overline{Y} + \overline{W}\,\overline{Z} + \overline{X}Z$

55. $L = \overline{A}\,\overline{B}\,\overline{C}\,\overline{D} + A\,\overline{B}\,\overline{C}\,\overline{D} + \overline{A}BC + \overline{A}C\overline{D} + \overline{A}\,\overline{C}\,\overline{D}$

is simplified to **[Hyd.Univ-1996]**

a) $L = \overline{A}\ \overline{D} + \overline{A}BC + \overline{B}\,\overline{C}\,\overline{D}$

b) $L = \overline{B}CD + BC + \overline{C}\,\overline{D}$

c) $L = A\overline{C} + B\overline{C} + ABD$

d) $L = \overline{A}\,\overline{B}C + \overline{A}C\overline{D} + \overline{A}\,\overline{C}\,\overline{D}$

56. If the expression $X = \overline{A} + BC + (A + B)(A\overline{C} + \overline{B}\,\overline{C})'$ when expressed in a form suitable for realization in two levels gating we get **[Hyd.Univ-1995]**

a) $A + A\overline{B} + \overline{B}\,\overline{C}$ b) $\overline{A} + BC + A\overline{C} + AC + AB\overline{C}$

c) $\overline{A} + BC + A\overline{C} + A\overline{B}\,\overline{C}$ d) $\overline{A} + BC + A\overline{C} + \overline{B}\,\overline{C}$

57. There are two 2-input NAND gates A and B. one of the inputs in each of these is the output of the other. When the free inputs of A and B are 1 and 0 respectively, the outputs of A and B would be **[Hyd.Univ-1994]**

a) 0, 0 b) 0, 1 c) 1, 0 d) 1, 1

ANSWER KEY - II

1. a	**2.** a	**3.** a	**4.** d	**5.** b
6. b	**7.** c	**8.** d	**9.** d	**10.** d
11. a	**12.** d	**13.** a	**14.** d	**15.** b
16. c	**17.** a	**18.** d	**19.** b	**20.** d
21. c	**22.** c	**23.** d	**24.** c	**25.** a
26. d	**27.** a	**28.** b	**29.** c,d	**30.** b
31. c	**32.** b	**33.** a	**34.** a	**35.** d
36. a	**37.** a	**38.** b	**39.** d	**40.** c
41. a	**42.** d	**43.** a	**44.** c	**45.** b
46. d	**47.** d	**48.** a	**49.** a	**50.** c
51. d	**52.** d	**53.** d	**54.** a	**55.** a
56. c	**57.** b			

EXPLANATORY NOTES - II

6. (b) **Explanation :** Binary representation to

42 is 0101010

72 is 1001000

compute 42 OR 72 is 1101010

Compute (42 OR 72) AND 55 is 100010

The decimal representation is 34.

11. (a) **Explanation :** Given $Z = A \oplus B \oplus C \oplus .....$

If only two of the variables are 1 and all others are zero. Then (since Ex - OR operation is commutative all and associative)

$$(1 \oplus 1) \oplus 0 \oplus 0 \oplus ...... = 0 \oplus 0 \oplus 0 \oplus .... = 0$$

In the same way, if we can try higher number of 1's it is obvious that Z = 0, if an even number of variables are 1.

13. (a,d) **Explanation :** Given $\overline{A} + AB = 0 \quad \rightarrow (1)$

$AB = AC \quad \rightarrow (2),\ AB + A\overline{C} + CD = CD \quad \rightarrow (3)$

From equation (2) $B = C \quad \rightarrow (4)$,

From equation (1) $\overline{A} + B = 0 \quad \rightarrow (5)$

From equation (4) & (5) B = C = 0 and A = 1

if C = 0 then D is either 0 (or) 1.

14. (d) **Explanation :** Given $AB + \overline{A}C = 1 \quad \rightarrow (1)$

$AC + B = 0 \quad \rightarrow (2)$

From (2) $(A + B)(B + C) = 0$

$\Rightarrow A + B = 0\ (\text{or})\ B + C = 0$

if B+C=0 then equation (1) is not true,

so $A + B = 0 \Rightarrow A = 0, B = 0$

Applying A & B values in eq (1) Then C = 1

15. (b) **Explanation :** Given $a = a \oplus b = \overline{a}b + a\overline{b}$

$b = a \oplus b = (\overline{a \oplus b})b + (a \oplus b)\overline{b}$

$= (a + \overline{b})(\overline{a} + b)b + a = ab + a = a$

$a = b \oplus a = a \oplus (a \oplus b)$

$= a(a + \overline{b})(\overline{a} + b) + \overline{a}b = ab + \overline{a}b = b$

17. (a) **Explanation :** Given

$\overline{A} + AB = 0 \Rightarrow \overline{A} + B = 0 \quad \rightarrow (1)$

$AB = AC \Rightarrow B = C \quad \rightarrow (2)$

$AB + A\overline{C} + CD = \overline{C}\,\overline{D} \quad \rightarrow (3)$

From (1) & (2), A = 0, B = 0 = C

By substuting these values in eq (3) we can conclude that D = 1.

19. (b) **Explanation :** To convert from one canonical form to another (i.e. POS and SOP) interchange the symbols Σ and π and list those numbers missing from the original form. Hence, the given expression.

$Y = \Sigma m(0,3,6,7,10,12,15)$ is equivalent to

$Y = \pi\, M(1,2,4,5,8,9,11,13,14)$

21. (c) **Explanation :**

From the K - map

the minimized expression is

$\overline{B}\,\overline{D} = \overline{C}\,\overline{D}$.

CD \ AB	00	01	11	10
00	1			1
01	1	1	1	1
11				
10	1			1

23. (a) **Explanation :**

Hence if A=B=C=1 then we can obtained minimal expression.

		A	
	1	1	
1	B	C	
		1	

24. (c) **Explanation :**

The minimized expression in POS form of given K - map is $(B + \overline{D})(\overline{B} + D)$.

CD \ AB	00	01	11	10
00		1	1	
01	1			1
11	1			1
10		1	1	

25. (a) **Explanation :** The terms that are Conditions for inclusion in the simplified function, these terms are called prime implicants.

In the given expression the term $\overline{y}x$ is prime impliment of f. Hence statement 'a' is valid statement.

27. (a) **Explanation :**

From the given K - map the minimized SOP form is $XY + y'Z$.

YZ \ WX	00	01	11	10
00	0	x	0	x
01	x	1	x	1
11	0	x	1	0
10	0	1	x	0

29. (c) **Explanation :** Assume, the '?' marked box is replaced by NAND gate then the output expression of the given circuit is

$$F = \underbrace{(x + y) + (\overline{x} + \overline{y})}_{1} = 1$$

So F is tautology.

31. (d) **Explanation :**

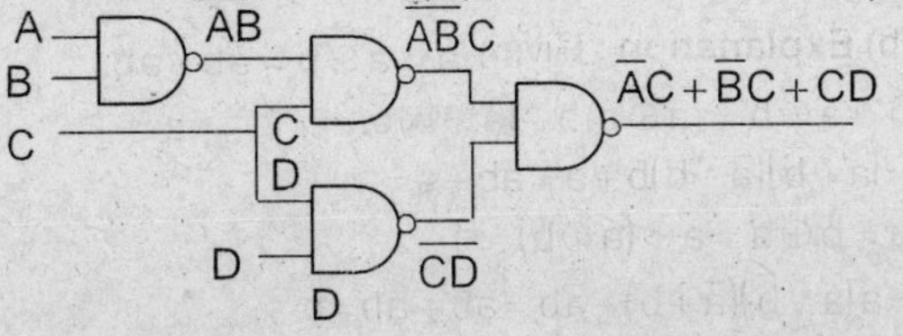

32. (b) **Explanation :** In the given circuit, the functions f_1 and f_2 are inputs to the AND gate, so the output of the AND gate will be

$f_1 \cdot f_2 = \Sigma(8,9,10).\Sigma(7,8,12,13,14,15)$

$= \Sigma(8) \quad [\because m_i.m_j = 0 \text{ if } i \neq j,$

$m_i.n_j = 1 \quad \text{if } i = j]$

Now $f = f_1 f_2 + f_3$

$\Rightarrow \Sigma(8,9) = \Sigma(8) + f_3$

$\Rightarrow f_3 = \Sigma(9) \quad [\because \Sigma m_i + \Sigma m_j = \Sigma(m_i + m_j)]$

33. (a) **Explanation :** the line 'y' is at 1. It must be possible when the inputs of the AND gate at third level must be 1. That means C = 1, and the output of the second level NAND gate is 1. It is also possible when one of the second level NAND gate inputs are zero.

However, the first level NAND gate output is either 0 (or) 1, its depending upon previous output of the circuit. If B = 0 then the desirable output will be produced.

Hence C = 1, B = 0, A = 0.

34. (a) **Explanation :** Prime implicants, that cover minterms with a single cross in their column are called essential prime implicants.

In the given K - map, there are 4 minterms whose column's have a single cross : 0, 2, 8 and 10.

35. (d) **Explanation :** O/P(X) = 1 means output of the AND gate is 1. It must be done when all inputs of AND gate are 1's. $\Rightarrow C = 1$.

The output of the X - NOR gate becomes 1 when the both inputs are same (i.e. either 0's (or) 1's)

$\Rightarrow B = 1 \quad [\because C = 1]$

The output of the X - OR gate becomes 1 when the both inputs are complements $\Rightarrow A = 0 \quad [\because B = 1]$

Hence the required inputs (ABC) are (0,1,1) to make the O/P(X)=1

36. (a) **Explanation :** The K - map corresponding to the given truth table is

The minimum expression is

$Q = AB + BC + CA\overline{B}$

CD \ AB	00	01	11	10
0	0	0	1	0
1	0	1	1	1

37. (b) **Explanation :**

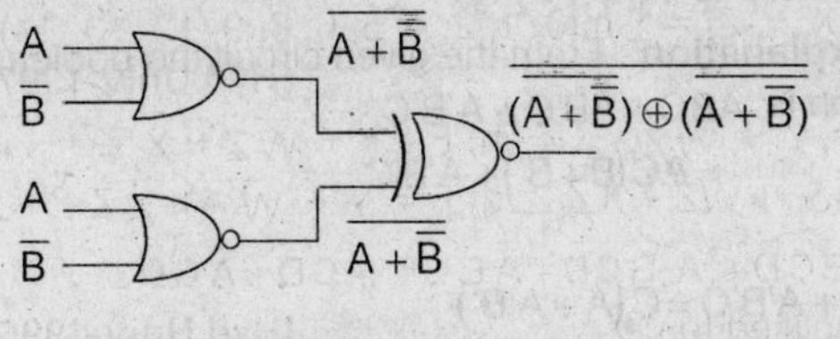

The output is $\overline{\overline{AB} \oplus \overline{AB}}$

$= \overline{0} \quad [\because x \oplus x = 0]$

38. (b) **Explanation :** Given equation is in POS canonical form, the corresponding K - map representation is

The minimal expression is

$(x+z)(x+\overline{y}) = x + x\overline{y} + xz + \overline{y}z$

$= x(1+\overline{y}) + xz + \overline{y}z = x + \overline{y}z$

Z \ XY	00	01	11	10
0	0	0		
1		0		

39. (d) **Explanation :** The simplified expression is obtained by using the K - map representation is shown below.

Y \ WX	00	01	11	10
0	1	1		1
1				1

$\therefore Z = \overline{w}\,\overline{y} + w\overline{x}$

42. (b) **Explanation :** Correct equality is decided by using the truth table is shown below.

a	b	$a \oplus b$	$c = (a \oplus b) \oplus b$	$(a \oplus b) \oplus c$
0	0	0	0	0
0	1	1	0	1
1	0	1	1	0
1	1	0	1	1

$= b$

44. (c) **Explanation :** Given

$F = (A+B+C).(AB+\overline{C})$
$= (A+B+C)(A+\overline{C})(B+\overline{C})$
$= AB + A\overline{C} + AB + B\overline{C} + ABC$
$= AB + A\overline{C} + B\overline{C} + ABC$
$= A\overline{C} + B\overline{C} + AB$

47. (d) **Explanation :**

Let $(A+B)(A+\overline{B})(\overline{A}+B) = (A+B\overline{B})(\overline{A}+B)$
$= A(\overline{A}+B) = AB$

49. (a) **Explanation :** Given f =1 if both x,y are complements. This is the function of EX - OR gate.
$x \oplus y = x'y + xy'$.

50. (c) **Explanation :** Given boolean expression

$d'(b'+ac)+bcd$. To obtain given expression from the given K - map, four more 1's are needed to insert in K - map those minterms are 0,2,8 and 10.

51. (d) **Explanation :** The K - map representation corresponding to the given truth table is

x_3 \ x_1x_2	00	01	11	10
0	0	0	1	0
1	1	0	1	1

Now the boolean expression corresponding to the above K - map is $x_1x_2 + \overline{x}_2x_3$.

52. (d) **Explanation :** From the given circuit the boolean expression is $ABC + AB'C + A'B'C$

$= AC\underbrace{(B+B')}_{1} + A'B'C$
$= AC + A'B'C = C(A + A'B')$
$= f(a,b,c) = C(A+B')$

53. (a) **Explanation :** Given

$A' + AB = 0 \Rightarrow A' + B = 0 \quad \rightarrow (1)$

$AB = AC \Rightarrow B = C \quad \rightarrow (2)$

$AB + A'C + CD = (CD)' \quad \rightarrow (3)$

from equations (2) & (1) A =1, B=0, C=0

By substuting these values in (3) We get D=1.

55. (a) **Explanation :**

CD \ AB	00	01	11	10
00	1	1		1
01				
11		1		
10	1	1		

From the K - map the simplified expression to L is

$L = A\overline{D} + \overline{B}\,\overline{C}\,\overline{D} + \overline{A}BC$

56. (c) **Explanation :**

For two level getting reduce the given expression is shown below.

$X = \overline{A} + BC + (A+B)(A\overline{C} + \overline{B}\,\overline{C})$

$= \overline{A} + BC + A\overline{C} + AB\overline{C} + A\overline{B}\,\overline{C}$

$= \overline{A} + BC + A\overline{C}(1+B) + A\overline{B}\,\overline{C}$

$\therefore X = \overline{A} + BC + A\overline{C} + A\overline{B}\,\overline{C}$

57. (b) **Explanation :** The circuit diagram according to the given problem is shown below.

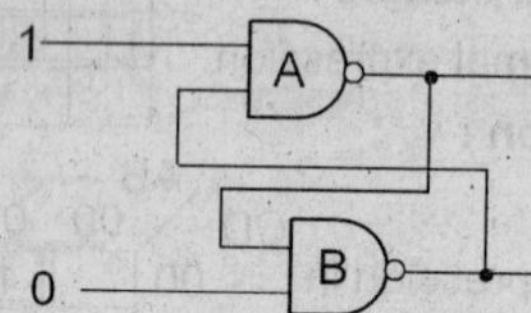

The output of the NAND gate is 1 when at least one input is 0. So the output of the B gate is 1. Now the two inputs of gate A are 1 and 1. therefore the output of the gate A is zero.

II. COMBINATIONAL CIRCUIT DESIGN

- Combinational Circuits
- Arithmetic Circuits
- Code Conversion
- Implementation of Combinational Circuits

1. Combinational Circuits

A combinational circuit is a connected arrangement of logic gates with a set of inputs and outputs. At any given time, the binary values of the outputs are a function of the combination of 1's and 0's at the inputs. A block diagram of a combinational circuit is shown in Fig. 2.1. The n input variables come from an external source; the m output variables go to an external destination; and in between there is an interconnection of logic gates. A combinational circuit transforms binary informationfrom the given input data to the required output data. Combinational circuits are employed in digital systems for generating binary control decisions and for providing digital functions required in data processing. Many digital functions are available commercially in small packages called integrated circuits.

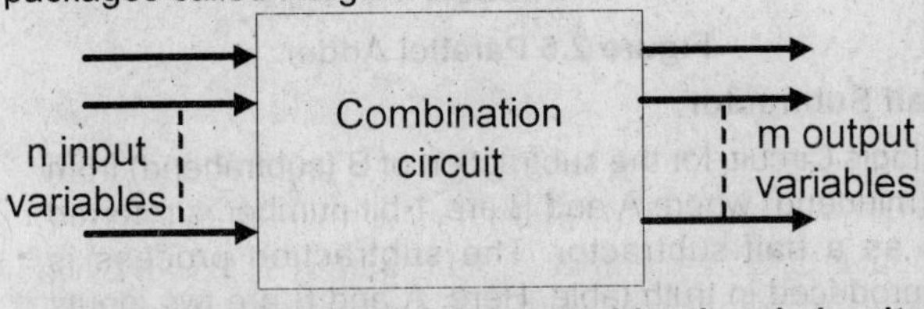

Figure 2.1 Block diagram of a combinational circuit

A combinational circuit can be described by a truth table showing the binary relationships between the n input variables and the m output variables. There are 2^n possible combinations of binary input values. The truth table lists the corresponding output binary values for each of the 2^n input combinations. A combinational circuit also can be specified with m Boolean functions, one for each output variable. Each output function is expressed in terms of the n input variables.

Analysis Procedure

The analysis of a combinational circuit starts with a given logic circuit diagram and culminates with a set of Boolean functions or a truth table.If the digital circuit is accompanied by a verbal explanation of its fuction, then the Boolean functions or the truth table is sufficient for verification.

If the function of the cirucit is under investigation, then it is necessary to interpret the operation of the circuit from the derived. Boolean functions or the truth table. The success of such investigation is enhanced if one has experience and familiarity with digital circuits.The ability to correlate a truth table or a set of Boolean functions with an information processing task is an art that one acquires with experience.

To illustrate with an example, let us investigate the combinational circuit shown in Fig 2.2. The circuit has five input variables and two output variables.

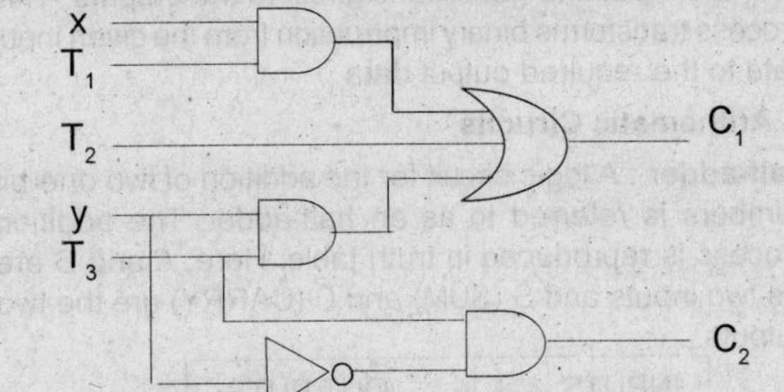

Figure 2.2 Example of Combinational circuit

Let us assume that input variables T_1, T_2 and T_3 are binary signals that occur in a time sequence. Thus, at a given time, variable $T_1 = 1$ while $T_2 = T_3 = 0$. During the next time period $T_1 = 0$, $T_2 = 1$, and $T_3 = 0$. The next time period produces $T_1 = T_2 = 0$ and $T_3 = 1$, and so on. The input binary variables x and y may represent results of previous operations. For example, variable x may be used to indicate that two numbers are equal (x = 1) or unequal (x = 0). Output variable C_1(and C_2) may designate a control variable that enables a certain digital operation when $C_1=1$ and disables the operation when $C_1 = 0$.

The combinational circuit of Fig. 2.2 can be expressed with two Boolean functions as follows:

$$C_1 = xT_1 + T_2 + YT_3, \; C_2 = y'$$

From the diagram or the Boolean functions it is possible to interpret the operation of the circuit. Control variable $C_1 = 1$ if x = 1 when timing variable $T_1=1$; or if timing variable $T_2 = 1$; or if y = 1 when timing variable $T_3 = 1$; otherwise, $C_1 = 0$, Similarly, $C_2 = 1$ if y = 0 and timing variable $T_3 = 1$; otherwise, $C_2 = 0$.

Binary control decisions in digital computers are generated in combinational circuits. The input variables for these circuits are internal timing sequences as well as binary variables from other operations. The digital functions being controlled are circuits that implement various data processing tasks, and these tasks are generated by means of other combinational circuits.

Design Procedure

The design of combinational Circuits starts from the verbal outline of the problem and ends in a logic circuit diagram. The procedure involves the following steps:

1. The problem is stated.
2. The input and output variables are assigned letter symbols.
3. The truth table that defines the relationship between inputs and outputs is derived.
4. The simplified Boolean functions for each output are obtained.
5. The logic diagram is drawn.

Examples : Half adder, full subtractor, code converters Binary adder, decimal adder, magnitude-comparator, Decoder, Demultiplexer, Encoder, Multiplixer, ROM, PLA etc.

A combinational circuit consists of input variables, logic gates and output variables. The logic gates accept signals from the input and generate signals to the outputs. This process transforms binary imprmation from the given input data to the required output data.

2. Arithematic CirucIts

Half-adder : A logic circuit for the addition of two one-bit numbers is referred to as an half-adder. The addition process is reproduced in truth table. Here, A and B are the two inputs and S (SUM) and C (CARRY) are the two outputs.

INPUTS		OUT PUTS	
A	B	Sum	Carry
0	0	0	0
0	1	1	0
1	0	1	0
1	1	0	1

From the truth table, we obtain the logical expression for S and C outputs as

$$S = \bar{A}B + A\bar{B} = A \oplus B,\ C = AB$$

The realization of any half-adder using gates is shown in Fig 2.3

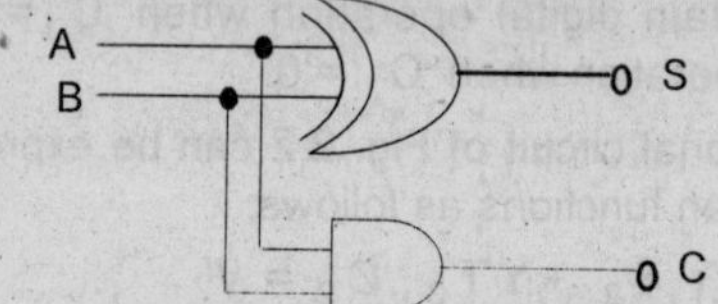

Figure 2.3 realization of Half-Adder

Full Adder : An half adder has only two inputs and there is no provision to add a carry coming from the lower order bits when multibit addition is performed. For this purpose, a third input terminal is added and this circuit is used to add A_n B_n and C_{n-1}, where A_n and B_n are the nth order bit of the numbers A and B respectively and C_{n-1}, is the carry generated from the addition of (n- 1)[th] order bits.

Inputs			outputs	
A_n	B_n	C_{n-1}	S_n	C_n
0	0	0	0	0
0	0	1	1	0
0	1	0	1	0
0	1	1	0	1
1	0	0	1	0
1	0	1	0	1
1	1	0	0	1
1	1	1	1	1

This circuit is referred to as full-adder and its truth table is given above. from the truth table, we obtain the logical expression for the outputs S_n and C_n.

$$\text{Sum} = A \oplus B \oplus C$$

$$\text{Carry} = AB \oplus (A \oplus B)\,C$$

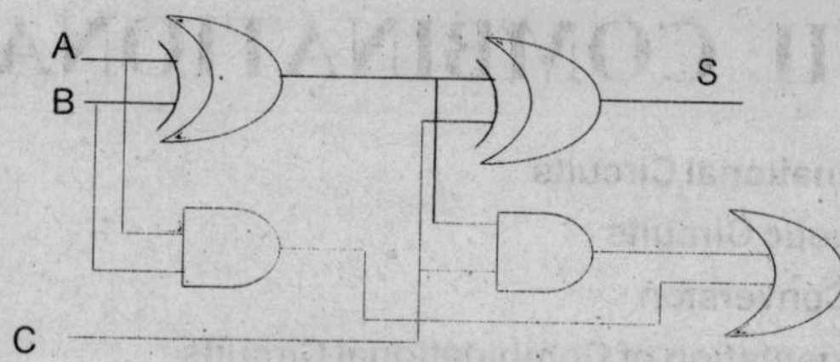

Figure 2.4 Full Adder

Binary Parallel Adder

A binary parallel adder is a digital circuit that produces the arithmetic sum of two binary numbers in parallel.

It consists of full-adders connected in cascade, with the output carry from one full adder connected to the input carry of the next full-adder.

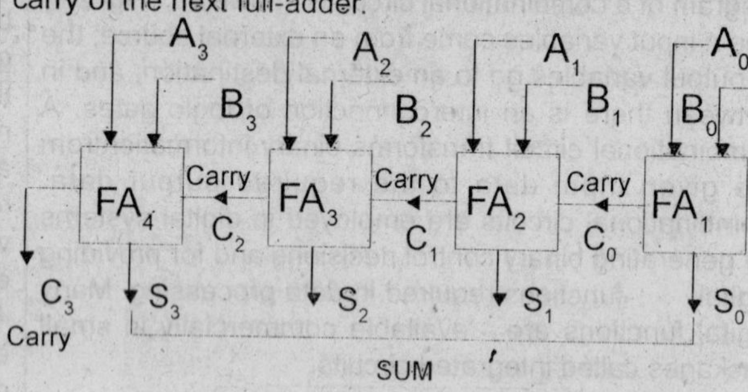

Figure 2.5 Parallel Adder

Half Subtractor

A logic Circuit for the subtraction of B (subtrahend) from A(minuend) where A and B are 1-bit number is referred to as a half-subtractor. The subtraction process is reproduced in truth table. Here, A and B are two inputs and D (difference) and C (borrow) are two outputs.

INPUTS		OUTPUTS	
A	B	D	C
0	0	0	0
0	1	1	1
1	0	1	0
1	1	0	0

From the truth table, the logical expressions for D and C are obtained as $D = \bar{A}B + A\bar{B} = A \oplus B$, $C = \bar{A}B$

The realization of a half-subtractor using gates is shown in Fig 2.6

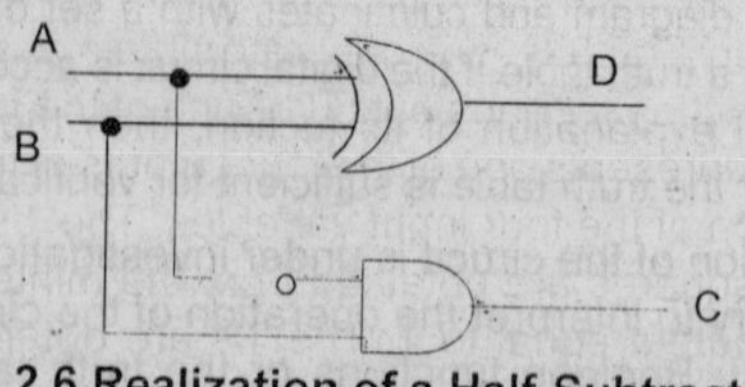

Figure 2.6 Realization of a Half-Subtractor

Full-Subtractor

We require a full-subtractor circuit for performing multibit subtraction wherein a borrow from the previous bit position may also be there. A full-subtractor will have three inputs. A_n (minuend), B_n (subtrahend) and C_{n-1} (borrow from the previous stage) and two outputs, D_n (difference) and C_n (borrow). Its truth table is given below

INPUTS		OUTPUTS		
A_n	B_n	C_{n-1}	D_n	C_n
0	0	0	0	0
0	0	1	1	1
0	1	0	1	1
0	1	1	0	1
1	0	0	1	0
1	0	1	0	0
1	1	0	0	0
1	1	1	1	1

The Simplified expressions for D_n and C_n are

$$D_n = A_n \oplus B_n \oplus C_{n-1}$$

$$C_n = \bar{A}_n B_n + \bar{A}_n C_{n-1} + B_n C_{n-1}$$

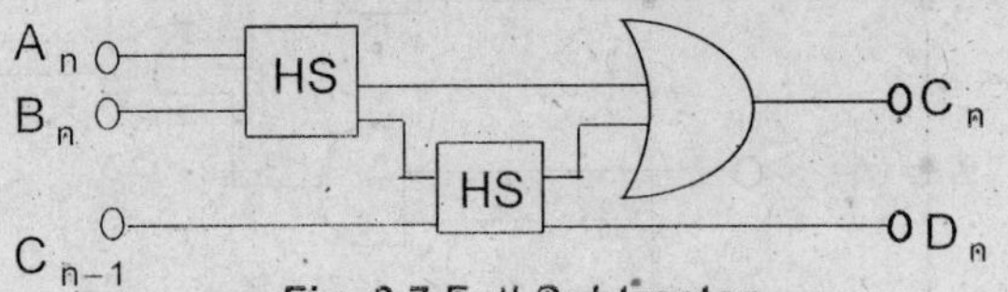

Fig. 2.7 Full Subtractor

3. Code Conversion

The availability of a large variety of codes for the same discrete elements of information results in the use of different codes by different digital systems. It is sometimes necessary to use the output of one system as the input to another. A conversion circuit must be inserted between the two systems if each uses different codes for the same information. Thus, a code converter is a circuit that makes the two systems compatible even though each uses a different binary code.

To convert from binary code A to binary code B, the input lines must supply the bit combination of elements as specified by code A and the output lines must generate the corresponding bit combination of code B. A combinational circuit performs this transformation by means of logic gates. The design procedure of code converters will be illustrated by means of a specific example of conversion from the BCD to the excess-3 code.

The bit combinations for the BCD and excess-3 codes uses four bits to represent a decimal digit, there must be four input variables and four output variables.

Let us designate the four input binary variables by the symbols A, B, C, and D, and the four output variables by w, x, y, and z. The truth table relating the input and output variables is shown in Table 2.1. We note that four binary variables may have 16 bit combinations, only 10 of which are listed in the truth table. The six bit combinations not listed for the input variables are don't-care combinations. Since they will never occur, we are at liberty to assign to the output variables either a 1 or a 0, whichever gives a simpler circuit.

Table 2.1 Truth table for code-conversion example

Input BCD				Output Excess-3 Code			
A	B	C	D	W	X	Y	Z
0	0	0	0	0	0	1	1
0	0	0	1	0	1	0	0
0	0	1	0	0	1	0	1
0	0	1	1	0	1	1	0
0	1	0	0	0	1	1	1
0	1	0	1	1	0	0	0
0	1	1	0	1	0	0	1
0	1	1	1	1	0	1	0
1	0	0	0	1	0	1	1
1	0	0	1	1	1	0	0

The maps in Fig. 2.8 are drawn to obtain a simplified Boolean function for each out put. Each of the four maps of Fig. 2.8 represents one of the four outputs of this circuit as a function of the four input variables. The 1's marked inside the squares are obtained from the minterms that make the output equal to 1. The 1's are obtained from the truth table by going over the output columns one at a time. For example, the column under output z has five 1's; therefore, the map for z must have five 1's. each being in a square corresponding to the minterm that makes z equal to 1. The six don't-care combinations are marked by X's. One possible way to simplify the functions in sum of products is listed under the map of each variable.

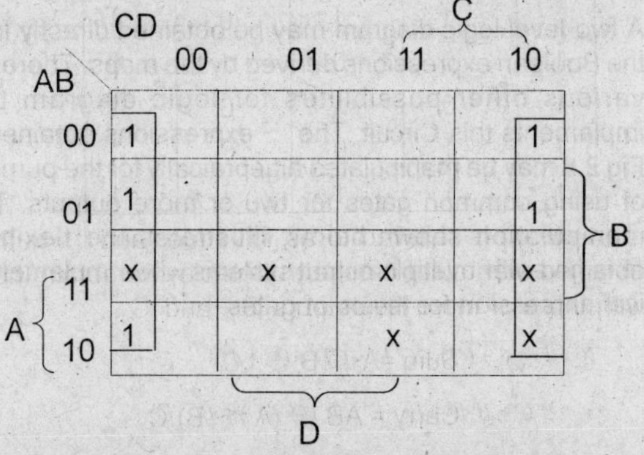

$z = D'$

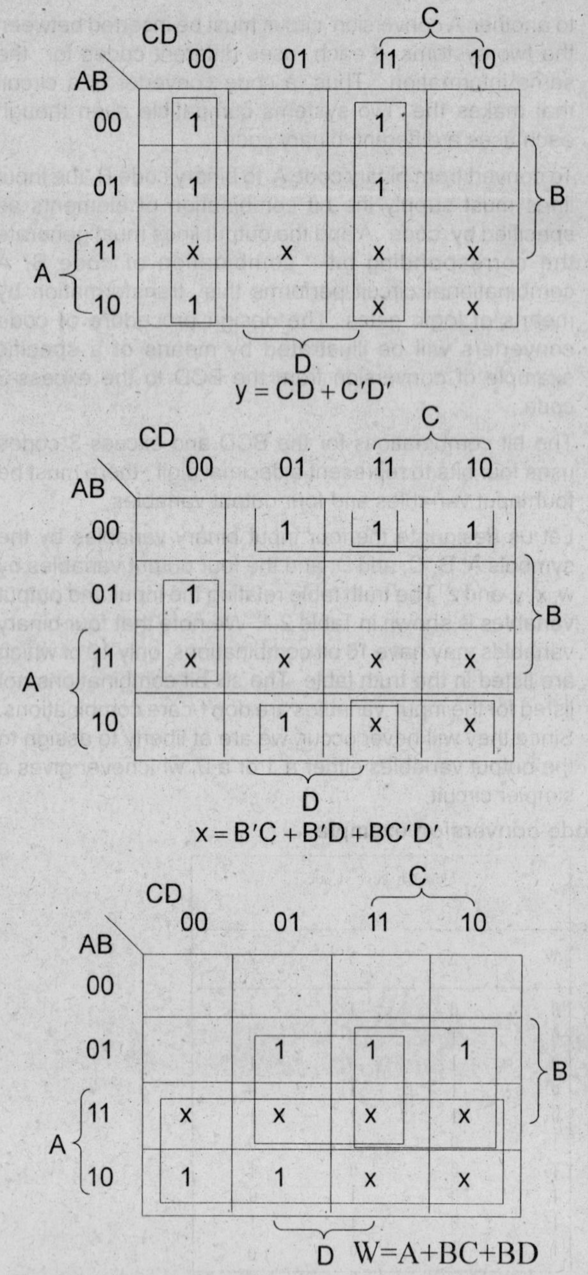

Fig. 2.8. K maps for BCD - to- excess code coverter

A two-level logic diagram may be obtained directly from the Boolean expressions derived by the maps. There are various other possibilites for logic diagram that implements this Circuit. The expressions obtained in Fig 2.8 may be manipulated algebraically for the purpose of using common gates for two or more outputs. This manipulation shown below, illustrates the flexibility obtained with multiple-output systems when implemented with three or more levels of gates.

$Z = D'$, $y = CD + C'D' = CD + (C + D)'$

$x = B'C + B'D + BC'D' = B'(C + D) + BC'D'$

$= B'(C + D) + B(C + D)'$

$W = A + BC + BD = A + B(C + D)$

The logic diagram that implements the above expressions is shown in Fig.2.9. In it we see that the OR gate whose output is C + D has been used to implement partially each of three outputs.

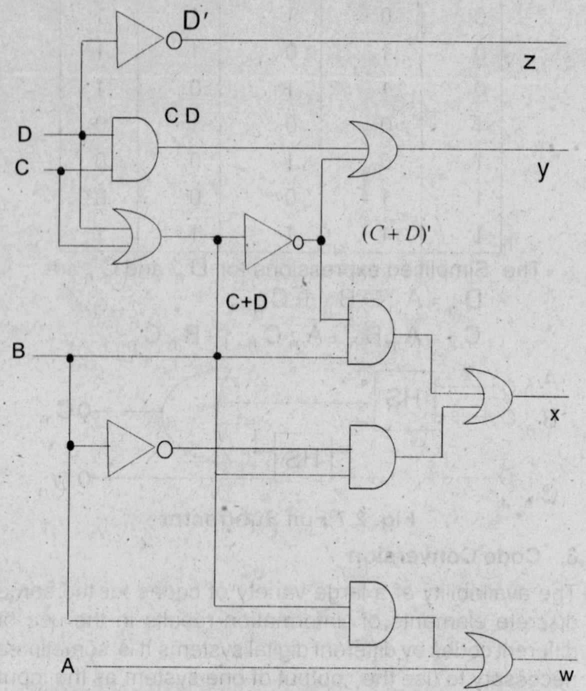

Figure 2.9 Logic diagram for BCD-to-excess -3 code

Not counting input inverters, the implementation in sum of products requires seven AND gates and three OR gates. The implementation of Fig. 2.9 requires four AND gates, four OR gates, and one inverter. If only the normal inputs are available, the first implementation will require inverters for variables B, C, and D, Where as the second implementation requires inverters for variables B and D.

4. Multiplexers And Their Use In Combinational Logic Design

Multiplexer

The multiplexer is a special combinational circuit that is one of the most widely used standard circuits in digital design. The multiplexer (or data selector) is a logic circuit that gates one out of several inputs to a single output. The input selected is controlled by a set of select inputs. Figure 2.10 shows the block diagram of a multiplexer with n input lines and one output line. For selecting one out of n inputs for connection to the output, a set of m select inputs is required, where $2^m = n$. Depending Upon the digital code applied at the select inputs one out of n data sources is selected and transmitted to a single output channel. Normally, a strobe (or enable) input (G) is incorporated which helps in cascading and it is generally active-low, which means it performs its intended operation when it is LOW.

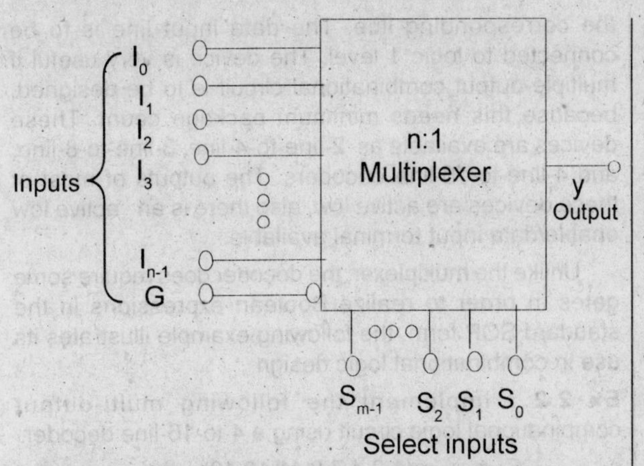

Figure 2.10 Block diagram of a Digital Multiplexer.

Table 2.2 Truth table of a 4:1 multiplexer

Select inputs		Output
S_1	S_0	y
0	0	I_0
0	1	I_1
1	0	I_2
1	1	I_3

Table 2.2 gives the truth table of a 4:1 multiplexer. With LOW input at G the output Y can be expressed as

$$Y=\overline{S}_1\,\overline{S}_0 I_0+\overline{S}_1 S_0 I_1+S_1\overline{S}_0 I_2+S_1 S_0 I_3 \qquad (2.1)$$

Equation (2.1) can be realized using NAND gates and the realization is given in Fig.2.10

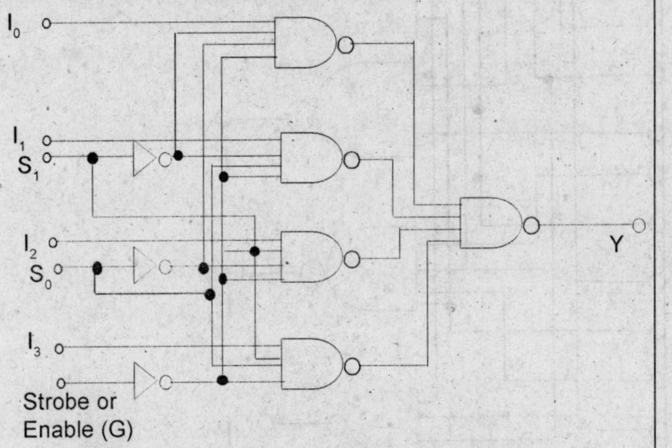

Fig.2.10 A 4 : 1 multiplexer with store input using NAND gates.

5. Combinational Logic Design Using Multiplexers

The multiplexing function discussed above can conveniently be used as a logic element in the design of combinational circuits.

Standard ICs are available for 2:1, 4:1, 8: 1, and 16:1 multiplexers.

Use of multiplexers offers the following advantages:

1. Simplification of logic expression is not required,
2. it minimizes the IC package count, and
3. logic design is simplified.

For using the multiplexer as a logic element, either the truth table or one of the standard forms of Logic expression must be available. The design procedure is given below:

I. Identify the decimal number corresponding to each minterm in the expression. The input lines corresponding to these numbers are to be connected to logic 1 level.

2. All other input lines are to be connected to logic 0 level.

3. The inputs are to be applied to select inputs.

The following examples illustrate the above procedure.

Ex 2.1 : Implement the expression using a multiplexer.

$$f(A, B\,C\,D) = \Sigma\, m\,(0, 2, 3, 6, 8, 9, 12, 14)$$

Sol: Since there are four variables, therefore, a multiplexer with four select inputs is required. The circuit of 16:1 multiplexer connected to implement the above expression is shown in Fig. 2.11. This implementation requires only one IC package. In case the output of the multiplexer is active-low, the logic 0 and logic 1 inputs of Fig. 2.11 are to be interchanged. The reader should verify the validity of this statement.

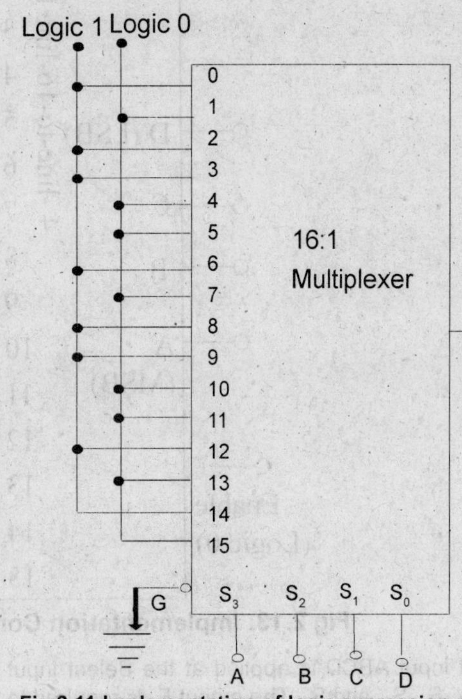

Fig 2.11 Implementation of logic expression of Ex. 2.1

6. Demultiplexers / Decoders And Their Use In Combination Logic Design

The demultiplexer performs the reverse operation of a multiplexer. It accepts a single input and distributes it over several outputs. Figure 2.12 gives the block diagram of a demultiplexer. The select input code determines to which output the data input will be transmitted.

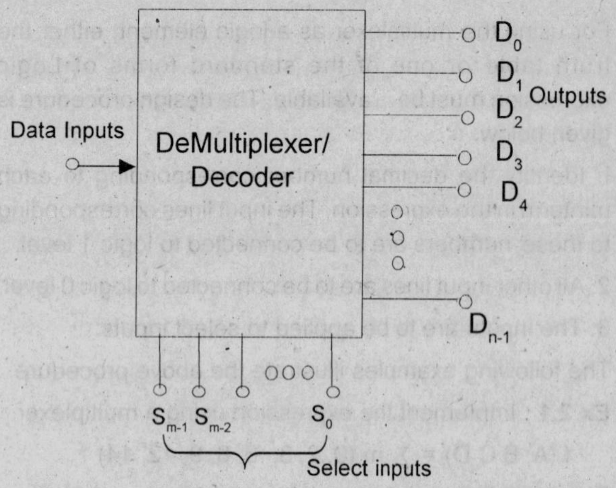

Figure 2.12.Block diagram of a demultiplexer

The number of output lines is n and the number of select lines is m, where $n= 2^m$. This circuit can also be used as binary-to-decimal decoder with binary inputs applied at the select input lines and the output will be obtained on the corresponding line. The data input line is to be connected to logic 1 level. The device is very useful if multiple-output combinational circuit is to be designed, because this needs minimum package count. These devices are available as 2-line-to-4-line, 3-line-to-8-line, and 4-line-to-16-line decoders. The outputs of most of these devices are active low, also there is an active low enable/data input terminal available.

Unlike the multiplexer, the decoder does require some gates in order to realize Boolean expressions in the standard SOP form. the following example illustrates its use in combinational logic design.

Ex 2.2 : Implement the following multi-output combinational logic circuit using a 4-to-16-line decoder.

$$F_1 = \Sigma\ m(1,2,4,7,8,11,12,13)$$
$$F_2 = \Sigma\ m(2,3,9,11)$$
$$F_3 = \Sigma\ m(10,12,13,14)$$
$$F_4 = \Sigma\ m(2,4,8)$$

Sol : The realization is shown in Fig. 2.13

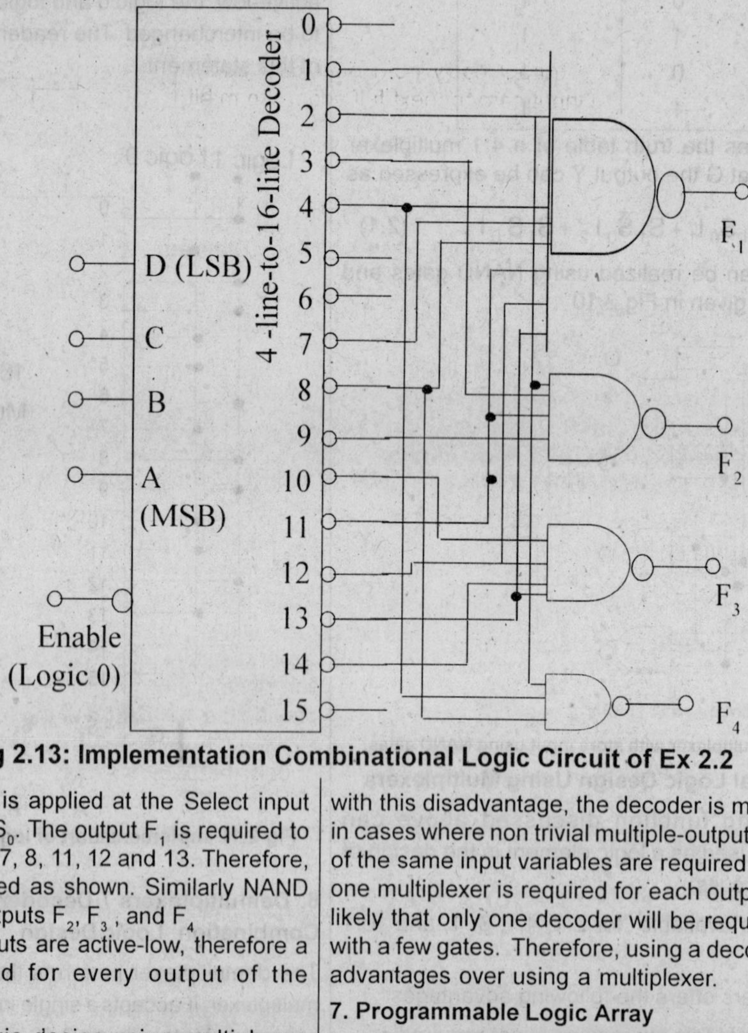

Fig 2.13: Implementation Combinational Logic Circuit of Ex 2.2

The four-bit input ABCD is applied at the Select input terminals S_3, S_2, S_1, and S_0. The output F_1 is required to be 1 for minterms 1, 2, 4, 7, 8, 11, 12 and 13. Therefore, a NAND gate is connected as shown. Similarly NAND gates are used for the outputs F_2, F_3, and F_4. Here, the decoder's outputs are active-low, therefore a NAND gate is required for every output of the combinational circuit.

In the combinational logic design using multiplexer, additional gates are not required, whereas design using demultiplexer requires additional gates. However, even with this disadvantage, the decoder is more economical in cases where non trivial multiple-output expressions of the same input variables are required. In such cases, one multiplexer is required for each output whereas it is likely that only one decoder will be required, supported with a few gates. Therefore, using a decoder could have advantages over using a multiplexer.

7. Programmable Logic Array

A combinational circuit may occasionally have don't-care conditions. When implemented with a ROM, a don't-care condition becomes an address input that will never occur.

The words at the don't-care addresses need not he programmed and may be left in their original state (all 0's or all 1's) The result is that not all the bit patterns available in the ROM are used, which may be considered a waste of available equipment.

Consider, for example, a combinational circuit that converts a 12-bit card code to a 6-bit internal alphanumeric code . The input card code consists of 12 lines designated by 0, 1, 2, ..., 9 11, 12. The size of the ROM for implementing the code converter must be 4096 × 6, since there are 12 inputs and 6 outputs. There are only 47 valid entries for the card code; all other input combinations are don't-care conditions. Thus only 47 words of the 4096 available are used. The remaining 4049 words of ROM are not used and are thus wasted.

For cases where the number of don't-care conditions is excessive it is more economical to use a second type of LSI component called programmable logic array or PLA.

A PLA is similar to a ROM in concept; however, the PLA does not provide full decoding of the variables and does not generate all the minterms as in the ROM. In the PLA, the decoder is replaced by a group of AND gates, each of which can be programmed to generate a product term of the input variables.

The AND and OR gates inside the PLA are initially fabricated with links among them. The specific Boolean functions are implemented in sum of products form by opening appropriate links and leaving the desired connections.

A block diagram of the PLA is shown in Fig .2.14. It consists of n inputs, m outputs, k product terms and m sum terms. The product terms constitute a group of k AND gates and the sum terms constitute a group of m OR gates. Links are inserted between all n inputs and their complement values to each of the AND gates. Links are also provided between the outputs of the AND gates and the inputs of the OR gates.

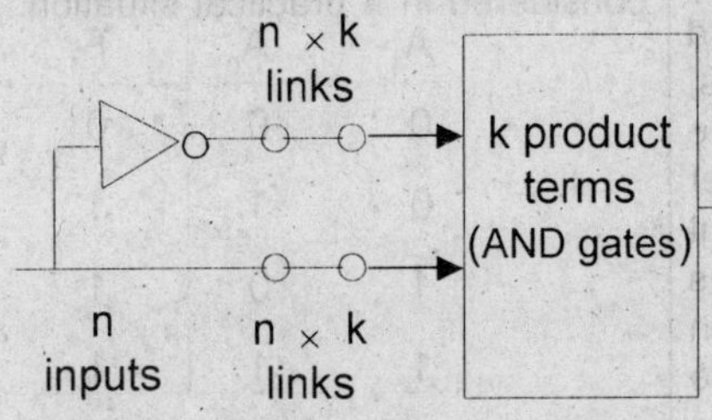

Figure 2.14 PLA Block Diagram

Another set of links in the output inverters allows the output function to be generated either in the AND-OR form or in the AND-OR INVERT form. With the inverter link in place, the inverter is bypassed, giving an AND-OR implementation. With the link broken, the inverter becomes part of the circuit and the function is implemented in the AND-OR-INVERT form.

The size of the PLA is specified by the number of inputs, the number of product terms, and the number of outputs (the number of sum terms is equal to the number of outputs). A typical PLA has 16 inputs, 48 product terms, and 8 outputs The number of programmed links is $2n \times k + k \times m + m$, whereas that of a ROM is $2^n \times m$.

8. Read-Only Memories (ROM)

A read-only memory (ROM), as the name implies, is a memory unit that performs the read operation only; it does not have a write capability. This implies that the binary information stored in a ROM is made permanent during the hardware production of the unit and cannot be altered by writing different words into it. While a RAM is a general-purpose device whose contents can be altered during the computational process, a ROM is restricted to reading words that are permanently stored within the unit.

An m by n ROM is an array of binary cells organized into m words of n bits each.

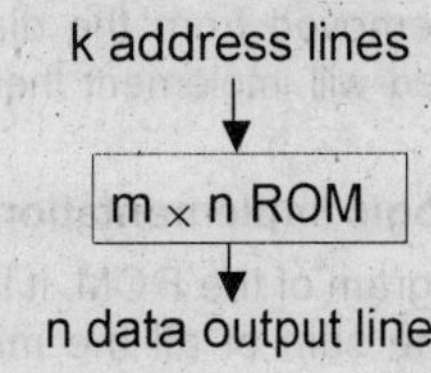

Fig. 2.15 Block diagram of a read-only memory (ROM)

As shown in the block diagram of Fig. 2.15, a ROM has k address lines to select one of 2^k = m words of memory, and n output lines, one for each bit of the word. An IC ROM may also have one or more enable lines for expanding a number of IC packages into a ROM with larger capacity. The ROM does not need a read-control line since at any given time, the output lines automatically provide the 1's and 0's of the n bits of the word selected by the address value. Because the outputs are a function of only the present inputs (the address lines)a ROM is classified as a combinational circuit. In fact a ROM is constructed internally with decoders and a set of OR gates. There is no need for providing storage capabilities as in a RAM, since the values of the bits in the ROM are permanently fixed.

Consider the logic diagram of a 4 by 3 ROM as depicted in Fig. 2.16. The unit contains a 2 by 4 decoder to decode the two address lines. The OR gates provide the three outputs. If each minterm output of the decoder is connected to the input of each OR gate, the circuit outputs will all be 1 no matter what word is selected by the address lines.

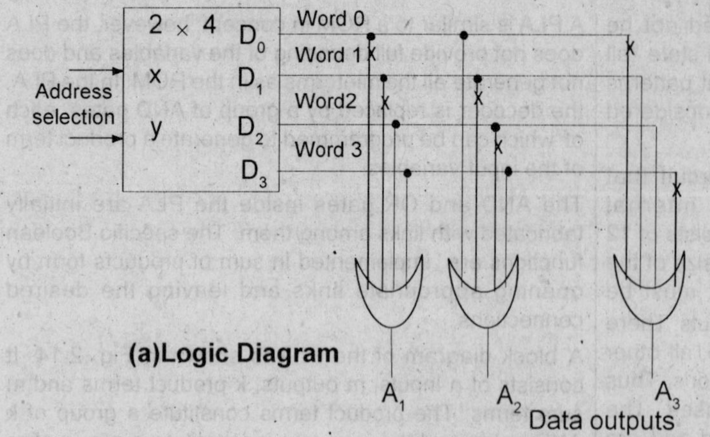

(a)Logic Diagram

Address		Outputs		
x	y	A_1	A_2	A_3
0	0	0	1	0
0	1	0	0	0
1	0	1	0	1
1	1	1	1	0

(b) Truth Table

Fig 2.16 An example of a 4 × 3 ROM

Suppose we want the ROM to contain the bit combinations listed in the truth table accompanying the diagram. The truth table specifies a bit combination of 010 for word 0. This bit combination is obtained by breaking two wires (marked with a cross) between word 0 and the leftmost inputs of the OR gates. In other words, when the input address is 00, the D_0 output of the decoder is equal to 1 and all other outputs of the decoder are equal to 0. Only the OR gate associated with output A_2, receives an input of 1 because the other two wires are broken. Therefore, the output lines will provide an output $A_1A_2A_3 = 010$ (assuming that an open wire to the OR gate behaves as a 0 input). Similarly, all other wires marked with crosses indicate broken wires and when these wires are removed from the diagram, the logic diagram so obtained will implement the truth table listed for the ROM.

Combinational Logic Implementation

From the logic diagram of the ROM, it is clear that each output provides the sum of all the minterms of the n input variables. Remember that any Boolean function can be expressed in sum-of-minterms form. By breaking the links of those minterms not included in the function, each ROM output can be made to represent the Boolean function of one of the output variables in the combinational circuit. For an n-input, m-output combinational circuit, we need a $2^n \times m$ ROM. The opening of the links is referred to as programming the ROM. The designer need only specify a ROM program table that gives the information for the required paths in the ROM. The actual programming is a hardware procedure which follows the specifications listed in the program table.

Let us clarify the process with a specific example. The truth table in Fig. 2.17 (a) specifies a combinational circuit with two inputs and two outputs.

The Boolean functions can be expressed in sum of minterms:

$F_1(A_1, A_0) = \Sigma(1, 2, 3)$, $F_2(A_1, A_0) = \Sigma(0, 2)$

When a combinational circuit is implemented by means of a ROM, the functions must be expressed in sum of minterms or,better yet, by a truth table. If the output functions are simplified,we find that the circuit needs only one OR gate and an inverter.Obviously, this is too simple a combinational circuit to be implemented with a ROM. The advantage of a ROM is in complex combinational circuits. The example merely demonstrates the procedure and should not be considered in a practical situation.

A_1	A_0	F_1	F_2
0	0	0	1
0	1	1	0
1	0	1	1
1	1	1	0

(a) Truth Table

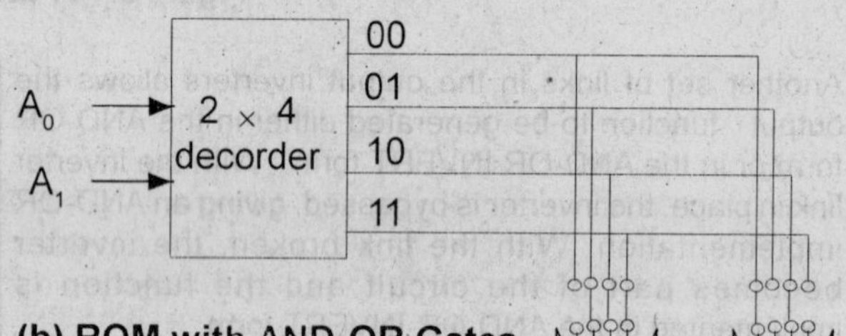

(b) ROM with AND-OR Gates

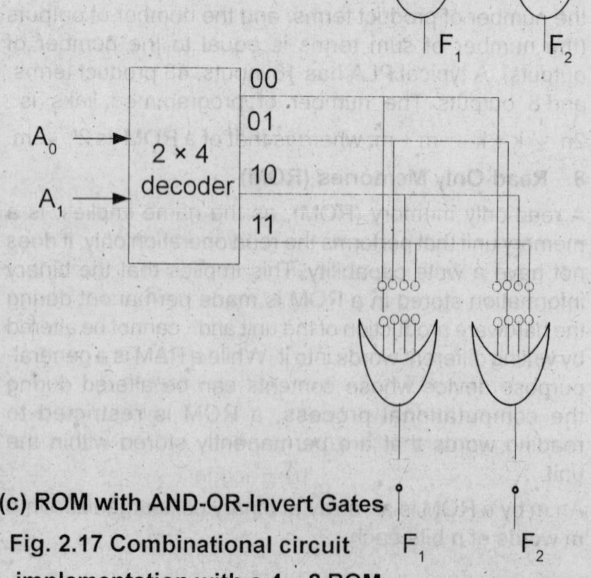

(c) ROM with AND-OR-Invert Gates

Fig. 2.17 Combinational circuit implementation with a 4 × 2 ROM

The ROM that implements the combinational circuit must have two inputs and two outputs; so its size must be 4 × 2. Figure 2.17 (b) shows the internal construction of such a ROM. It is now necessary to determine which of the eight available links must be broken and which should be left in place. This can be easily done from the output functions listed in the truth table. Those minterms that specify an output of 0 should not have a path to the output through the OR gate. Thus, for this particular case the truth table shows three 0's, and their corresponding links to the OR gates must be removed. It is obvious that we must assume here that an open input to an OR gate behaves as a 0 input.

Some ROM units come with an inverter after each of the OR gates and, as a consequence, they are specified as having initially all 0's at their outputs. The programming procedure in such ROMs requires that we open the link paths of the minterms (or addresses) that specify an output of 1 in the truth table. The output of the OR gate will then generate the complement of the function, but the inverter placed after the OR gate complements the function once more to provide the normal output. This is shown in the ROM of Fig.2.17 (c).

OBJECTIVE QUESTIONS - I

1. In a combinational circuit the outputs at any instant of time depend.
   a) Only on the past inputs b) Only on the past outputs.
   c) On past inputs as well as present inputs.
   d) Only on the inputs present at that instant of time.

2. A combinational circuit [AUCET-1998]
   a) Always contains memory elements
   b) Never contains memory elements.
   c) May sometimes contain memory elements.
   d) Contains only memory elements.

3. All digital circuits can be realized using only [GATE-1992]
   a) EX-OR b) Multiplexers
   c) HALF-Adders d) NAND

4. A multiplexer is also known as [AUCET-2000]
   a) Coder b) Decoder
   c) Data Distributor d) Data Selector

5. A Decoder is a
   a) multiplexer b) De-multiplexer
   c) both (a) and (b) d) None of these.

6. In which of the following circuits if there "n" input lines then there will be 2^n output lines.
   a) Multiplexer b) Gates
   c) DeMultiplexer d) None

7. Which of the following is used for converting decimal digits to Binary format.
   a) Decoder b) Encoder
   c) Multiplexer d) DeMultiplexer

8. The device which selects one of the several inputs and transmits it to a single output is.
   a) Decoder b) Multiplexer
   c) DeMultiplexer d) Encoder

9. The circuit used for parallel to serial conversion of data is known as
   a) Decoder b) DeMultiplexer
   c) Multiplexer d) Multivibrator

10. The dynamic hazard problem occurs in
    a) Combinational circuits alone
    b) Sequential Circuits alone
    c) Both Sequential and Combinational Circuits
    d) None of these.

11. A DeMultiplexer is used to
    a) Route the data from single input to one of many outputs.
    b) Select data from several inputs and route it to single output.
    c) Perform serial to parallel conversion
    d) All of the above

12. A multiplexer has
    a) One data input and a number of data outputs.
    b) One data output and a number of data inputs.
    c) One data output , a number of data inputs, and a number of select inputs.
    d) One data output and a number of select outputs.

13. A DeMultiplexer can be used to realize a
    a) Counter b) Shift-register
    c) Combinational circuits d) Display system

14. A Multiplexers with 4-bit data select input is a
    a) 4:1 multiplexers b) 8:1 multiplexers
    c) 16:1 multiplexers d) 32:1 multiplexers

15. In Digital systems subtraction is performed
    a) Using half-adders b) Using half-subtractors
    c) Using adders with ones complement representation of negative numbers d) None of these.

16. Which of the following adders can add three or more numbers at a time.
    a) Parallel adder b) Carry-look-ahead-adder
    c) Carry-save-adder d) Full adder

17. Consider the circuit f implements. [GATE-1996]
    a) $\bar{A}\bar{B}C+\bar{A}B\bar{C}+ABC+A\bar{B}\bar{C}$
    b) A+B+C
    c) AB+BC+CA
    d) None

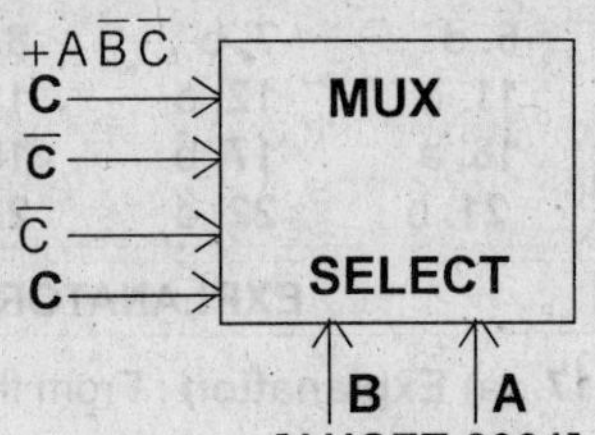

18. A PLA contains [AUCET-2001]
    a) AND and OR gates b) A Micro-Controller
    c) An Array of diodes d) A Phase locked loop

19. The following circuit is.

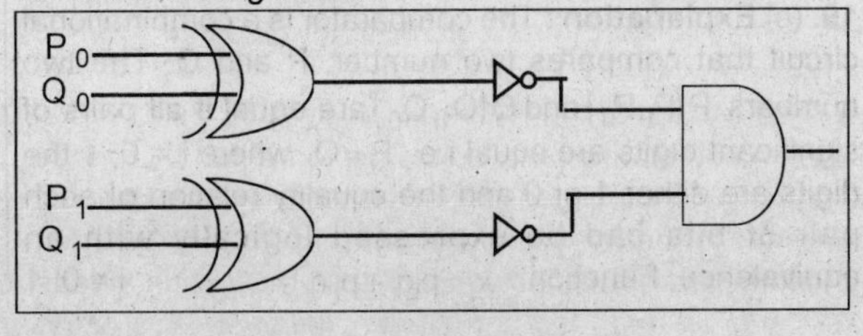

a) Adder b) Full adder c) Comparator

d) Circuit to calculate look-ahead carry

20. A logic circuit which is used to change a BCD number into an equivalent decimal number is:

a) Decoder b) Encoder

c) Multiplexer d) Code converter

21. The logic realized by the circuit shown below **[GATE-1992]**

a) F=A.C

b) F=A+C

c) F=B.C

d) F=B+C

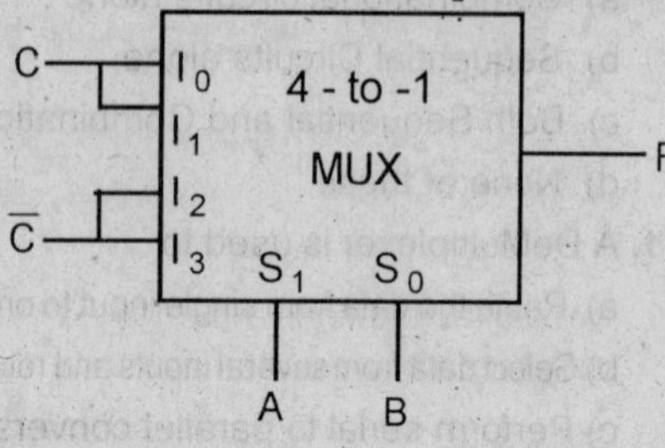

22. A PLA can be used: **[GATE-1994]**

a) as a microprocessor b) as a dynamic memory

c) to realize a sequential logic

d) to realize a combinational logic

23. The function of a multiplexer is **[UGC NET - 1997]**

a) to select 1 out of N input data sources and to transmit it to single channel

b) to transmit data on N lines

c) to perform serial-to-parallel conversion

d) to decode information

24. The minimum number of two-input NAND gates required to realize a full adder is **[UGC NET-1998]**

a) 6 b) 7 c) 9 d) 10

25. The SUM output of an half adder is equivalent to output of a **[UGC NET-1998]**

a) AND circuit b) X-OR circuit

c) NAND circuit d) NOR circuit

ANSWER KEY - I

1. d	**2.** b	**3.** c	**4.** d	**5.** d
6. d	**7.** b	**8.** b	**9.** c	**10.** d
11. c	**12.** c	**13.** c	**14.** c	**15.** c
16. a	**17.** b	**18.** c	**19.** c	**20.** a
21. b	**22.** d	**23.** a	**24.** a	**25.** b

EXPLANATORY NOTES - I

17. (a) **Explanation :** From the circuit

$f = \bar{A}\bar{B}C + A\bar{B}\bar{C} + \bar{A}B\bar{C} + ABC$

A \ BC	00	01	11	10
0			1	1
1	1	1		

19. (c) **Explanation :** The comparator is a combinational circuit that compares two number, P and Q. The two numbers $P(P_1, P_0)$ and $Q(Q_1, Q_0)$ are equal if all pairs of significant digits are equal i.e. $P_i = Q_i$ where i = 0, 1 the digits are either 1 or 0 and the equality relation of each pair of bits can be expressed logically with an equivalence. Function : $x_i = p_i q_i + p_i^{'} q_i^{'}$ i = 0, 1

21. (b) **Explanation :**

From the given circuits boolean function

$f = \bar{A}\bar{B}C + \bar{A}BC + AB\bar{C} + AB\bar{C}$.

The K - map representation

A \ BC	00	01	11	10
0		1		1
1	1		1	

is $\therefore F = A \oplus C$

OBJECTIVE QUESTIONS - II

1. Which of the following statements about circuits is (are)TRUE

I. Combinational circuits may have feedback sequential circuits do not.

II. Combinational circuits may have a "memory less property"sequential circuits donot.

III. Both sequential and combinational circuits must be controlled by an External clock. **[AUCET-1999]**

a) I only b) II only c) III only d) I,II,II

2. Which of the following sets of component(s) is/are sufficient to implement any arbitrary boolean funciton is: **[GATE-1999]**

a) XOR gates, NOR gates b) AND., XOR gates

c) 2 to 1 multiplexers

d) Three input gates that output (A.B)+C for inputs A,B and C.

3. The number of full adders and half adders required to add 16-bit numbers are: **[GATE-1999]**

a) 8 half adders,8 full adders.

b) 1 half adder,15 full adders.

c) 16 half adders, 0 full adders.

d) 4 half adders, 12 full adders.

4. The number of 4-line-to-16-line-decoders required to make an 8-line-to-256-line decoder is

a) 16 b) 17 c) 32 d) 64

5. The logical expression Y= ABC + $\bar{A}$ B + B $\bar{C}$ has to be realized using a 2^n to 1 multiplexer. The minimum value of n is equal to

a) 3 b) 4 c) 2 d) 8

6. A BCD to decimal decoder is designed using a 1 to 2^n demultiplexer. The minimum value of n is equal to.

a) 2 b) 3 c) 4 d) 8

7. How many full adders are required to construct an m-bit parallel adder?

a) m/2 b) m-1 c) m d) m+1

8. Which of the following logic circuits accepts two binary digits on its inputs and produces two binary digits 1 a sum bit and a carry bit, on its outputs?

a) Full adder b) Half adder

c) Decoder d) Multiplexer

9. A logic circuit which is used to change a BCD number into an equivalent decimal number is

a) Decoder b) Encoder

c) Multiplexer d) Code Converter

10. A combinational logic circuit which is used to send data coming from a single source to two or more separate destinations is called as

a) Decoder b) Encoder
c) Multiplexer d) DeMultiplexer

11. Which one of the following Boolean Expression represents the sum output from a HALF ADDER.

a) A.B b) A+B c) $\overline{A}.\overline{B}$ +A.B d) $A.\overline{B} + \overline{A}.B$

12. In below figure, what is the output , If all inputs are 0's If all inputs are 1's.

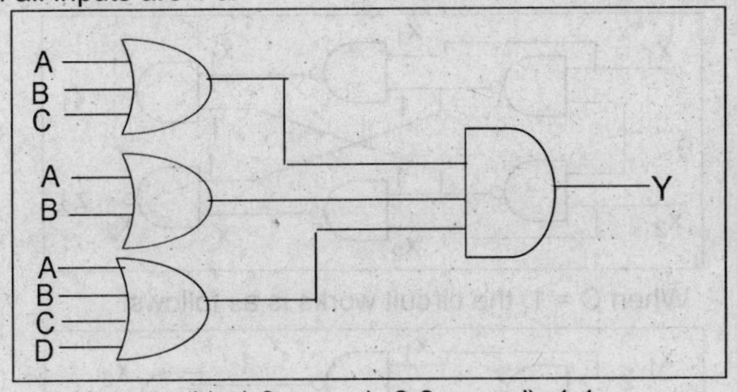

a) 0,1 b) 1,0 c) 0,0 d) 1,1

13. In below figure what is the value of Y for each of the following. I. ABCD = 0011 ; II. ABCD = 0110

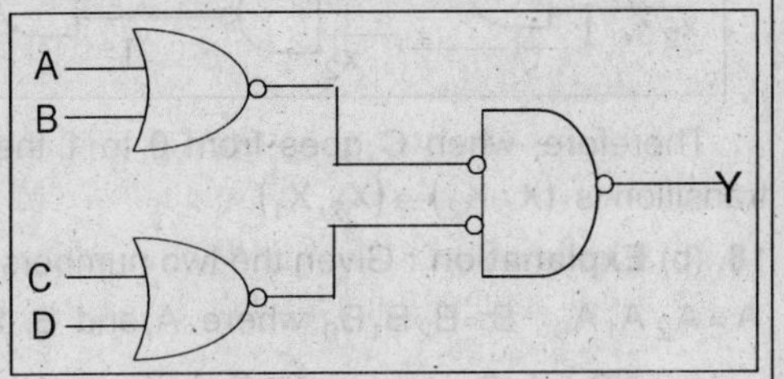

a) 0,1
b) 1,0
c) 0,0
d) 1,1

14. The minimal function that can detect a "divisible by 3" 8421 BCD code digit (representation is D3D4D2D1) is given by **[GATE-1990]**

a) $D_3D_1 + D_4D_2 + D_3D_2D_1$
b) $D_3D_1 + D_4D_2D_1 + \overline{D}_1D_2D_1 + D_3D_4\overline{D}_2\overline{D}_1$

15. A 2 bit binary multiplier can be implemented using **[GATE-1997]**

a) 2 i/p AND gates only
b) 2 i/p XOR gates and 4 i/p AND gates only
c) 2 i/p NOR gates and 1 i/p XNOR gates only

16. For a binary half-subtractor having two i/p's A and B the correct set of logical expressions for the o/p's D(=A minus B) and X(=borrow) are. **[GATE-1999]**

a) $D = AB + \overline{A}B, X = \overline{A}B$ b) $D = \overline{A}B + A\overline{B}, X = A\overline{B}$
c) $D = \overline{A}B + A\overline{B}, X = \overline{A}B$ d) $D = AB + A\overline{B}, X = A\overline{B}$

17. Consider the following combinational circuit with 6 NAND gates :

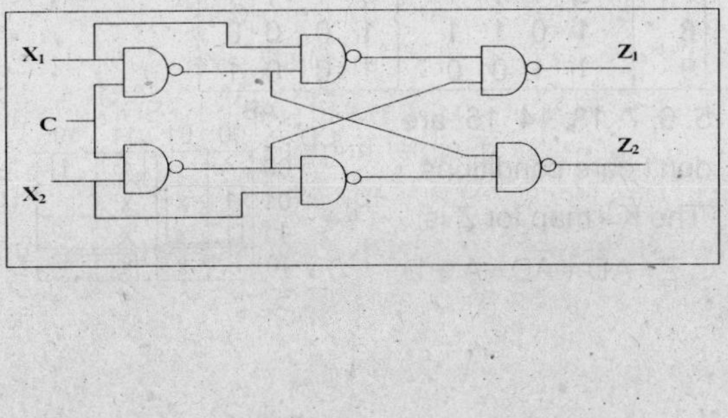

When C goes from 0 to 1, what is a transition in (Z_1, Z_2) **[UGC NET - 2000]**

a) $(\overline{X}_1, \overline{X}_2) \to (\overline{X}_1, \overline{X}_2)$ b) $(X_1, X_2) \to (X_1, X_2)$
c) $(X_1, X_2) \to (\overline{X}_1, X_2)$ d) $(X_1, X_2) \to (X_2, X_1)$

18. The magnitude comparator for 3-bit numbers $A = A_2A_1A_0$ and $B = B_2B_1B_0$, where A_i and B_i being bits, that returns true on A < B, is, using the notations $x_i = A_iB_i + A'_iB'_i$ with i = 0,1,2 and $A'_i = 1 - A_i$, $B'_i = 1 - B_i$, Given by **[UGC NET-1995]**

a) $A_2B'_2 + x_2A_1B'_1 + x_2x_1A_0B'_0$
b) $A'_2B_2 + x_2A'_1B_1 + x_2x_1A'_0B_0$
c) $A'_2B'_2 + x_2A'_1B'_1 + x_2x_1A'_0B'_0$ d) None

19. Two digital functions are enabled by control variables C_1 and C_2. the boolean functions for the control signals are: **[Hyd.Univ-1997]**

$C_1 = ABT_1 + A'B'T_2$; $C_2 = AT_1 + B'T_2$

Under what condition(s) of input variables A, B and timing variables T_1, T_2 will the two digital functions be enabled at the same time? Assume conditions to be in the form of ABT_1T_2.

a) 1110, 01X1 b) 111X, 00X1
c) 11XX, 0001 d) 11X0, 0X10

20. A majority function is generated in a combinational circuit when the output is equal to 1 if the input variables have more 1's than 0's. The output is zero otherwise. For a three input majority function the boolean expression is **[Hyd.Univ-1993]**

a) AB' + BC' + CA' b) AB + BC' + C'A
c) AB + BC + CA d) AB + B'C + CA'

21. A 'code converter' is to be designed to convert from BCD(5421) to the normal BCD(8421). What is the minimized expression for the output Z. **[Hyd.Univ-1998]**

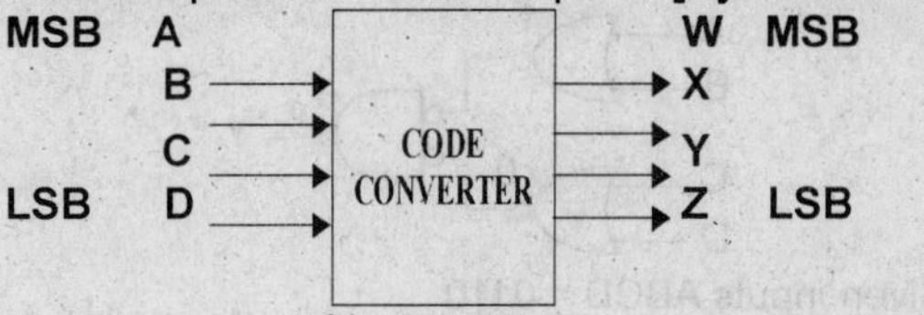

a) $A \oplus D$ b) $C \oplus AD$
c) $A(B + CD)$ d) $A'B + A(B'C' + C'D + B'D')$

22. The maximum number of outputs for a decoder with 6-bit data word would be **[UGC NET-1998]**

a) 64 b) 6 c) 24 d) 256

23. A full adder takes .... inputs and produces .... outputs. **[UGC NET-1999]**

a) one, two b) two, two c) two, three d) three, two

ANSWER KEY - II

1. b	**2.** b,c	**3.** b	**4.** a	**5.** a
6. a	**7.** c	**8.** b	**9.** a	**10.** b
11. d	**12.** a	**13.** b	**14.** b	**15.** b
16. c	**17.** d	**18.** b	**19.** c	**20.** c
21. a	**22.** a	**23.** d		

EXPLANATORY NOTES - II

3. (b) **Explanation :** To add two n - bit numbers we require one - Half adder for adding two LSB bits and (n-1) full adders for adding remaining n-1 bit with carry bits. Therefore, 1 half - adder and 15 full - adders are required to add 16 bit numbers.

5. (c) **Explanation :** If we have a boolean function of n+1 variables, we take 'n' of those variables and connect them to the selection lines of a multiplexer. The remaining single variable of the function is used for the inputs of the multipliexer. If A is the single variable, the inputs of the multiplexer are chosen to be either A' or A or 1 or 0. In this way it is possible to generate any function of n+1 variables with a 2^n - to - 3 multiplexers. Therefore the given boolean function contain 3 variables. From the above discussion we can conclude that the minimum value of n is 2.

6. (a) **Explanation :** To represent BCD number we require 4 bits. So BCD to decimal decoder is designed using a 1 to 4 demultiplexers. Therefore n = 2.

7. Explanation : A binary parallel adder is a digital function that produces the arithametic sum of two binary numbers in parallel. It consists of full - adders connected in cascade with the output carry from one full - adder connected to the input carry of next full adder. An m bit parallel adder requires m full - adders. The below figure shows the inter connection of m full adders.

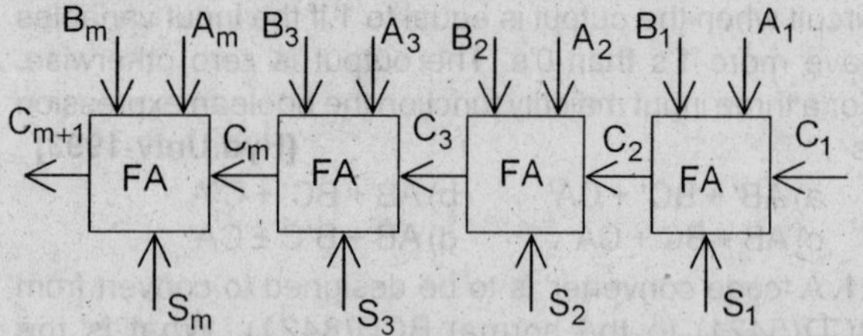

13. (b) **Explanation :** (i) Given input ABCD = 0011

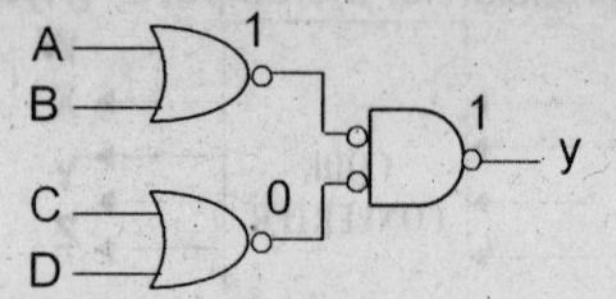

(ii) Given inputs ABCD = 0110

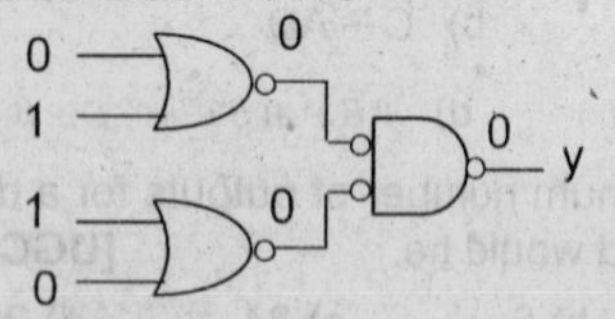

14. (b) **Explanation :**

Let $D_3D_1 + D_4D_2\overline{D}_1 + \underbrace{\overline{D}_1D_2D_1}_{0} + D_3D_4\overline{D}_2\overline{D}_1$

$\Rightarrow D_3D_1 + D_4D_2\overline{D}_1 + D_3D_4\overline{D}_2\overline{D}_1$

$\Rightarrow 1001 + 110 + 1100$

The decimal representation is 9+6+12=27

Hence function b is "divisible by 3".

15. (b) **Explanation :** 2 i/p X - OR gates are used to find sums and the 4 i/p AND gates are used to produce the output carry.

16. (c) **Explanation :** The half-subtractor needs two outputs. One output generates the difference and will be designated by the symbol'D'. The second output designated by the symbol 'X' for borrow, generates the binary signal that informs the next state that a1 has been barrowed. The truth table for the input - output relation ships of a half - subtractor can now be derived as follows. The boolean functions for the two outputs of the half subtractor are derived directly from the truth table

A	B	x
0	0	0
0	1	1
1	0	0
1	1	0

$D = A'B + AB'$, $X = A'B$

17. (d) **Explanation :** When C = 0 the circuit works is as follows.

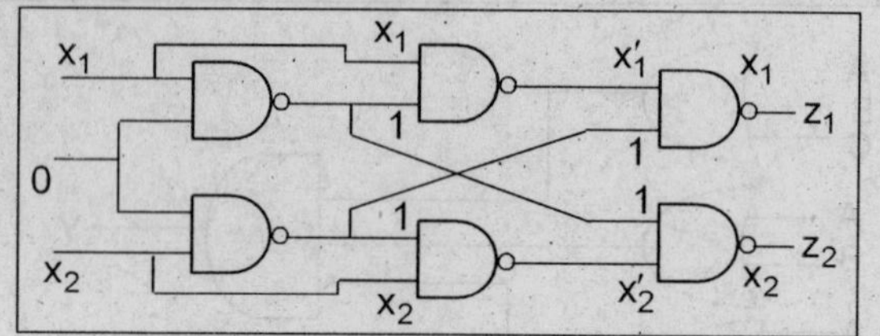

When C = 1, the circuit works is as follows.

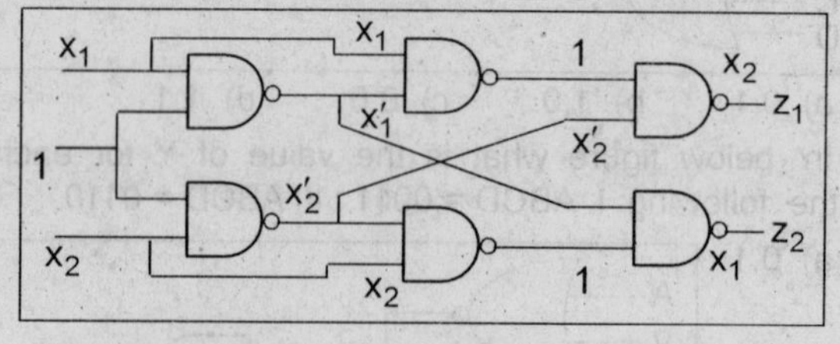

Therefore, when C goes from 0 to 1 the changing transition is $(X_1, X_2) \rightarrow (X_2, X_1)$

18. (b) **Explanation :** Given the two numbers

$A = A_2A_1A_0$, $B = B_2B_1B_0$ where A_i and B_i being bits.

$x_i = A_iB_i + A_i'B_i'$ i = 0, 1, 2

Where $x_i = 1$ only if the pair of bits position i are equal (i.e.,) if both are 2's or both are 0's. If the corresponding digit of A is 0 and that of B is 1, we have that A<B. The sequential comparison can be expressed logically by the following function. $A < B = A_2'B_2 + x_2A_1'B_1 + x_2A_0'B_0$

21. (a) **Explanation :** The truth table for given 'code convertor' is as follows.

Decimal	Input BCD (5421) A	B	C	D	Output BCD (8421) W	X	Y	Z
0	0	0	0	0	0	0	0	0
1	0	0	0	1	0	0	0	1
2	0	0	1	0	0	0	1	0
3	0	0	1	1	0	0	1	1
4	0	1	0	0	0	1	0	0
5	0	1	0	1	0	1	0	1
6	0	1	1	0	0	1	1	0
7	0	1	1	1	0	1	1	1
8	1	0	1	1	1	0	0	0
9	1	1	0	0	1	0	0	1

5, 6, 7, 13, 14, 15, are don't care conditions.

The K - map for Z is

CD \ AB	00	01	11	10
00			1	1
01	1	x	x	
11	1	x	x	
10		x	x	1

$\therefore Z = A\overline{D} + \overline{A}D = A \oplus D$

III. SEQUENTIAL CIRCUIT DESIGN

- Sequential Circuits
- Flip-flops
- Excitation Tables
- Registers
- Counters

1. Sequential Circuits

There are two types of sequential circuits and their classification depends on the timing of their signals. A Synchronous sequential circuit employs storage elements called flip-flops that are allowed to change their binary value only at discrete instants of time. An asynchronous sequential circuit is a system whose outputs depend upon the order in which its input variables change and can be affected at any instant of time. Gate-type asynchronous systems are basically combinational circuits with feedback paths. Because of the feedback among logic gates the system may, at times, become unstable. The instability problems encountered in asynchronous systems impose many difficulties and for this reason, they are very seldom used in the design of digital computer systems.

Synchronous sequential logic systems use gates and flip-flop storage devices. Synchronization is achieved by a timing device called a clock pulse generator. The clock pulses from the generator are distributed throughout the system in such a way that flip-flops are affected only with the arrival of the synchronization pulse. Clocked synchronous sequential circuits do not manifest instability problems and their timing is easily broken down into independent discrete steps, each of which may be considered separately. The sequential circuits discussed in this chapter are exclusively of the clocked synchronous type.

2. Flip-Flops

A flip-flop is a binary cell capable of storing one bit of information. it has two outputs, one for the normal value and one for the complement value of the bit stored in it. A flip-flop maintains a binary state until directed by a clock pulse to switch states. The difference among various types of flip-flops is in the number of inputs they possess and in the manner in which the inputs affect the binary state. The most common types of flip-flops are discussed below.

Basic Flip-Flop or Latch

As asynchronous flip-flop is constructed from two NAND gates or two NOR gates connected back to back. The circuit for the NAND version is shown in Fig. 3.1. The cross-coupled connections from the output of one gate to the input of the other gate constitutes a feedback path. For this reason, the circuit is classified as asynchronous. Each flip-flop has two outputs, Q and Q', and two inputs, set and reset. This type of flip-flop is called a direct-coupled RS flip-flop, the R and S being the first letters of the two input names. Another more popular name for this type of flip-flop is a latch. The latch forms a basic circuit upon which more complicated flip-flop types can be constructed.

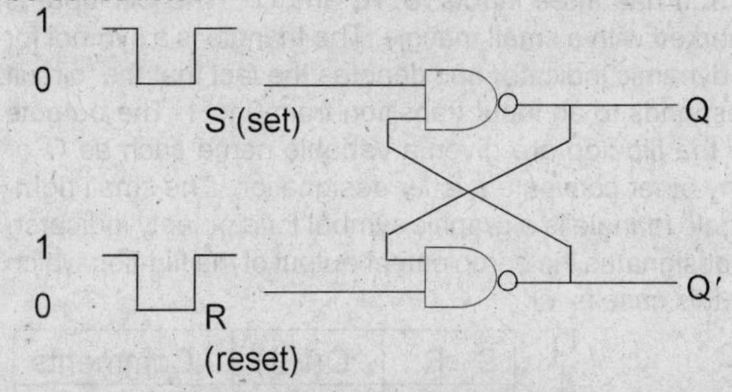

Fig.3.1 Basic flip-flop (latch)

The NAND latch operates with both inputs normally at 1, unless the state of the circuit has to be changed. The application of a momentary 0 to the set input causes output Q to go to 1 and Q' to go to 0. The outputs of the circuit do not change when the set input returns to 1. A momentary 0 applied to the reset input causes an output of Q = 0 and Q' = 1. The state of the flip-flop is always taken from the value of its normal output Q. When Q = 1, we say that the flip-flop stores a 1 and is in the set state. When Q = 0, we say that the flip-flop stores a 0 and is in the clear state.

The latch circuit manifests an undesirable condition if both inputs go to 0 simultaneously. Investigation of the circuit will show that when both inputs are 0, outputs Q and Q' will go to 1, a condition which is normally meaningless in flip-flop operation. If both inputs are then returned to 1, the state of the flip-flop is unpredictable; either state may result, depending on which input remains in the 0 state for a longer period of time before the transition to 1.

RS Flip-Flop

By adding gates to the inputs of the latch, the flip-flop can be made to respond only during the occurrence of a clock pulse.

The clocked RS flip-flop, shown in Fig. 3.2, consists of the basic latch and two other NAND gates. The outputs of gates 3 and 4 remain at 1 as long as the clock pulse (abbreviated CP) is 0, regardless of the S and R input values. When the clock pulse goes to 1, information from the S and R inputs is allowed to reach the latch in gates 1 and 2.

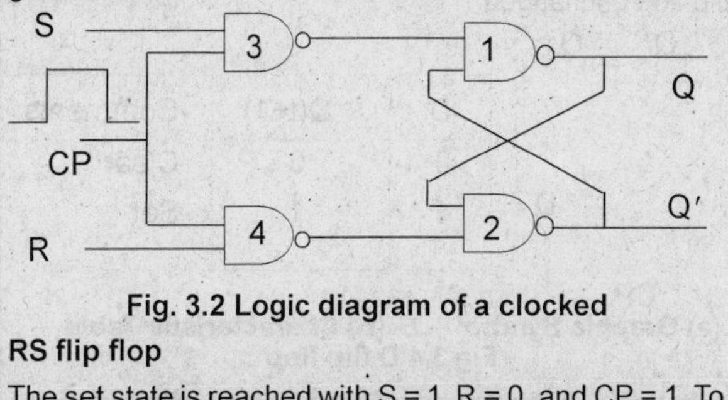

Fig. 3.2 Logic diagram of a clocked RS flip flop

The set state is reached with S = 1, R = 0, and CP = 1. To change to the clear state, the inputs must be S = 0, R= 1, and CP=1. With both S = 0 and R = 0, a CP of 1 does not affect the state of the flip-flop. S and R cannot be 1 during the occurrence of a clock pulse because the next state of the flip-flop is indeterminate.

The graphic symbol of the RS flip-flop is shown in Fig. 3.3. It has three inputs: S, R, and CP. The CP input is marked with a small triangle. The triangle is a symbol for a dynamic indicator and denotes the fact that the circuit responds to an input transition from 0 to 1. The outputs of the flip-flop are given a variable name such as Q or any other convenient letter designation. The small right-angle triangle is a graphic symbol for a polarity indicator. It designates the complement output of the flip-flop which in this case is Q′

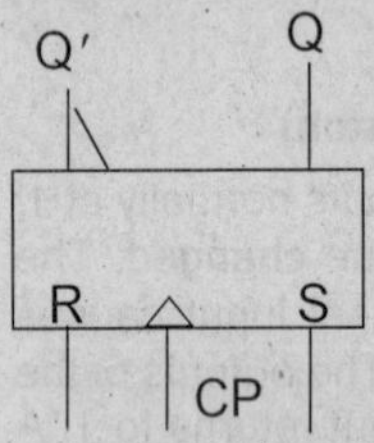

(a) Graphic Symbol

S R	Q(t+1)	Comments
0 0	Q(t)	No change
0 1	0	Clear
1 0	1	Set
1 1	?	Not allowed

(b) Characteristic Table

Fig 3.3 RS flip flop

The characteristic table shown in Fig. 3.3 (b) summarizes the operation of the RS flip-flop in tabular form; Q (t) is the binary state of the flip-flop at a given time (referred to as next state). The S and R columns give the binary values of the inputs. Q (t + 1) is the state of the flip-flop after the occurrence of a clock pulse (referred to as next state). If S = R = 0, a pulse produces no change of state, that is Q (t + 1) = Q (t). If S = 0 and R = 1, the flip-flop goes to the 0 (clear) state. If S=1 and R= 0 the flip-flop goes to the 1 (set) state. An RS flip-flop should not be pulsed when S = R = 1 since it produces an indeterminate next state.

D Flip-Flop

The D (data) flip-flop is a slight modification of the RS flip-flop. An RS flip-flop is converted to a D flip-flop by inserting an inverter between S and R and assigning the symbol D to the S input. The D input is sampled during the occurrence of a clock pulse and if it is 1, the flip-flop goes to the 1 state (because S = 1 and R = 0). If it is 0, a pulse changes the state of the flip-flop to 0 (because S = 0 and R = 1).The graphic symbol and characteristic table of the D flip-flop are shown in Fig. 3.4. Note that no input condition exists that will leave the state of the flip-flop unchanged.

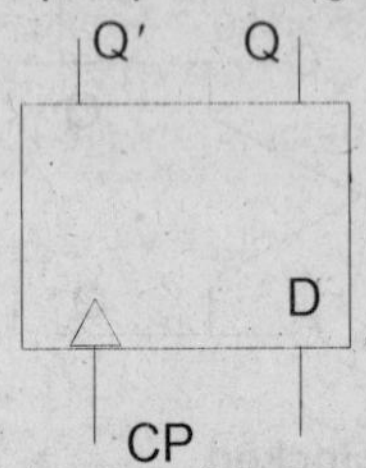

(a) Graphic Symbol

D	Q(t+1)	Comments
0	0	Clear
1	1	Set

(b) Characteristic Table

Fig 3.4 D flip flop

Although a D flip-flop has the advantage of having only one input (excluding the CP), it has the disadvantage that its characteristic table does not have a "no change" condition Q (t +1) = Q (t). The "no change" condition can be accomplished either by disabling the clock pulses with an external AND gate or by feeding the output back into the input so clock pulses keep the state of the flip-flop unchanged.

JK and T Flip-Flops

A JK flip-flop is a refinement of the RS flip-flop in that the indeterminate condition of the RS type is defined in the JK type. Inputs J and K behave like input S and R to set and clear the flip-flop, respectively. When inputs J and K are both equal to 1, a clock pulse switches the outputs of the flip-flop to its complement state, Q(t + 1) = Q′ (t). The graphic symbol and characteristic table of the JK flip-flop are shown in Fig. 3.5. Note that the J input is equivalent to the set condition while the K input produces the clear condition.In addition there are both no change and complement conditions in this type of flip-flop.

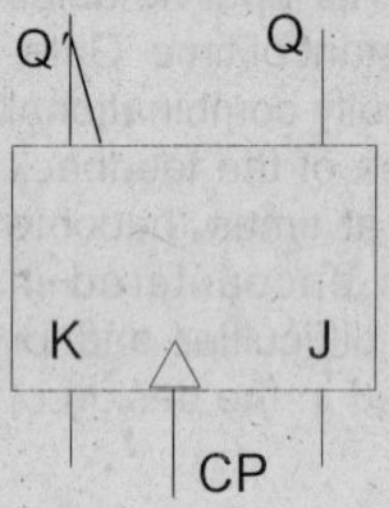

(a) Graphic Symbol

J K	Q(t+1)	Comments
0 0	Q(t)	No change
0 1	0	Clear
1 0	1	Set
1 1	Q′ (t)	Complement

(b) Charachteristic Table

Fig 3.5 JK flip-flop

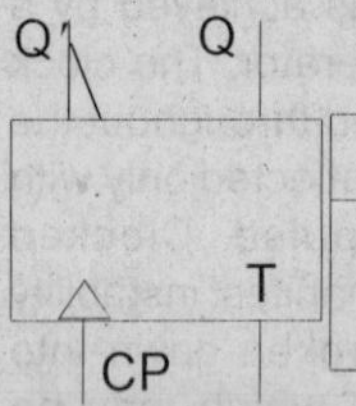

(a) Graphic Symbol

T	Q(t+1)	Comments
0	Q(t)	No change
1	Q′ (t)	Complement

(b) Charachteristic Table

Fig. 3.6 T flip-flop.

Another type of flip-flop found in textbooks is the T (toggle) flip-flop. This flip-flop, shown in Fig.3.6, is obtained from a JK type when inputs J and K are tied together. The T flip-flop therefore, has only two conditions. When T=0 (J= K= 0) a clock pulse does not change the state of the flip-flop. When T=1 (J=K=1) a clock pulse complements the state of the flip-flop.

3. Excitation Tables

The characteristic tables of flip-flops specify the next state when the inputs and the present state are known. During the design process we usually know the required transition from present state to next state and wish to find the flip-flop input conditions that will cause the required transition. For this reason we need a table that lists the required input combinations for a given change of state. Such a table is called a flip-flop excitation table.

Table 3.1 lists the excitation tables for the four types of flip-flops. Each table consists of two columns, Q (t) and Q (t + 1), and a column for each input to show how the required transition is achieved. There are four possible transitions from present state Q (t) to next state Q (t +1). The required input conditions for each of these transitions are derived from the information available in the characteristic tables. The symbol X in the tables represents a don't-care condition; that is, it does not matter whether the input to the flip-flop is 0 or 1.

Table 3.1 Excitation Tables for the Four types of Flip - Flops

Q(t)	Q(t+1)	S	R
0	0	0	X
0	1	1	0
1	0	0	1
1	1	X	0

(a) RS flip-flop

Q(t)	Q(t+1)	J	K
0	0	0	X
0	1	1	X
1	0	X	1
1	1	X	0

(b) JK flip-flop

Q(t)	Q(t+1)	D
0	0	0
0	1	1
1	0	0
1	1	1

(c) D flip-flop

Q(t)	Q(t+1)	T
0	0	0
0	1	1
1	0	1
1	1	0

(d) T flip-flop

The reason for the don't-care conditions in the excitation tables is that there are two ways of achieving the required transition. For example, in an RS flip-flop, a transition from present state of 0 to a next state of 0 can be achieved by having inputs S and R equal to 0 (to obtain no change when the clock pulse is applied), or by letting S = 0 and R = 1 to clear the flip-flop (although it is already cleared). In both casess S must be 0, but R is 0 in the first case and 1 in the second. Since the required transition Will occur in either case, we mark the R input with an X and let the designer choose either 0 or 1 for the R input,whichever is more convenient.

4. The Race-Around Condition

The difficulty of both inputs 1(S= R = 1) being not allowed in an S-R FLIP-FLOP is eliminated in a J-K FLIP-FLOP by using the feedback connection from outputs to the Inputs of the gates.

The inputs do not change during the clock pulse (CK =1) which is not true because of the feedback connections. consider for example, that the inputs are J = K = 1 and Q = 0, and a pulse as shown in Fig. 3.7 is applied at the clock input. After a time interval D t equal to the propagation delay through two NAND gates in series, the output will change to Q = 1. Now we have J = K = 1 and Q = 1 and after another time interval of Δt the output will change back to Q = 0. Hence, we conclude that for the duration t_p of the clock pulse, the output will oscillate back and forth between 0 and 1. At the end of the clock pulse, the value of Q is uncertain. This situation is referred to as the race around condition.

The race-around condition can be avoided if $t_p < \Delta t < T$. However, it may be difficult to satisfy this inequality because of very small propagation delays in ICs. A more practical method for overcoming this difficulty is the use of the master-slave (M-S) configuration discussed below.

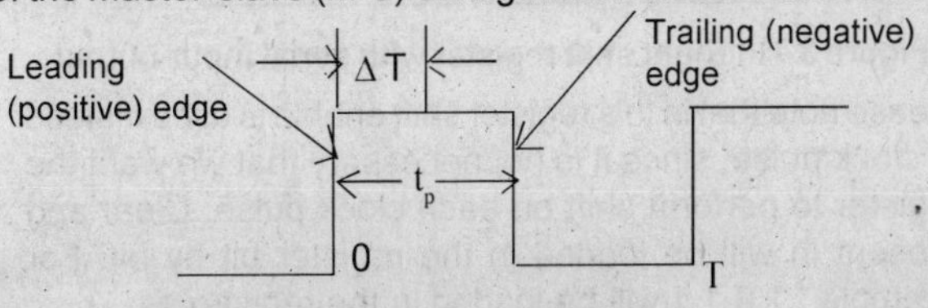

Fig-3.7 A Clock Pulse

5. Sequential Cirucits

A sequential circuit is an interconnection of flip flops and gates. The gates by themselves constitute a combinational circuit but when included with the flip-flops, the overall circuit is classified as a sequential circuit. A block diagram of a clocked sequential circuit is shown in Fig.3.8 it consists of a combinational circuit and two JK clocked flip-flops. In general, any number or type of flip-flops may be encountered. The combinational circuit part receives binary signals from external inputs and from the outputs of flip-flops. It generates binary signals to external outputs and to inputs of flip-flops.

The gates in the combinational circuit determine not only the value of external outputs but also the binary value to be stored in the flip-flops after each clock Pulse. The outputs of flip-flops, in turn, are applied to the combinational circuit inputs and determine the circuit's behavior. This process clearly demonstrates that the external outputs of a sequential circuit are functions of both external inputs and the present state of the flip-flops.

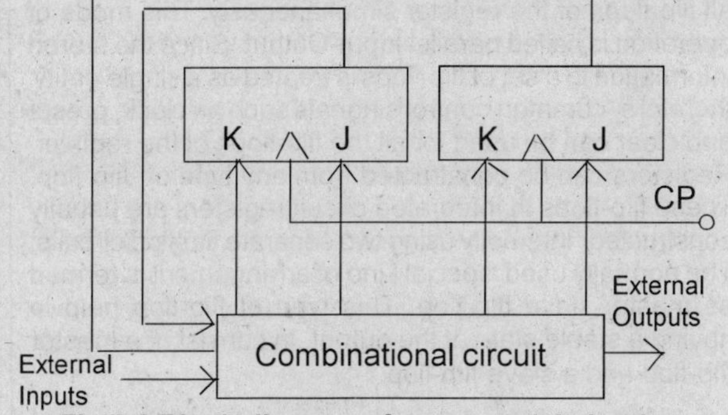

Fig 3.8 Block diagram of a sequential circuit.

Moreover, the next state of flip-flops is also a function of their present state and external inputs. Thus, a sequential circuit is specified by a time sequence of external inputs, external outputs, and internal flip-flop binary states.

Design Procedure

A procedure for the design of sequential circuits. Although intended to serve as a guide for the beginner, this procedure can be shortened with experience. The procedure is first summarized by a list of consecutive recommended steps as follows:

1. The word description of the circuit behavior is stated. This may be accompanied by a state diagram, a timing diagram, or other pertinent information.

2. From the given information about the circuit, obtain the state table.

3. The number of states may be reduced by state reduction methods if the sequential circuit can be characterized by linput - output relationships independent of the number of states.

4. Assign binary values to each state if the state table obtained in step 2 or 3 contains letter symbols.

5. Determine the number of flip-flops needed and assign a letter symbol to each.

6. Choose the type of flip-flop to be used.

7. From the state table, derive the circuit excitation and output tables.

8. Using the map or any other simplification method, derive the circuit output functions and the flip-flop input functions.

9. Draw the logic diagram.

After discussing some of the simple sequential circuits that is flip-flops dicuss some of the complex sequential circuits which can be developed using simple gates and flip-flops.

6. Registers

A register is a binary function which hold the binary information in digital form. Thus, a register consist of a group of binary storage cells. A register consist of one and more flip-flops depending on the number of bits to be stored in a word. A separate flip-flop is used for storing a bit of a word. In addition to storage, register are normally coupled with combinational gates enabling certain data processing tasks. Thus, a register in a broad sense consist of the flip-flops which store binary information and gates which controls when and how information is transferred to the register.

Normally in a register independent data lines are provided for each flip-flop, enabling the transfer of data to and from all flip-flops of the register simultaneously. This mode of operation is called parallel Input-Output. Since the stored information in a set of flip-flops is treated as a single entity, therefore, common control singnals such as clock, preset and clear can be used for all the flip-flops of the register. Registers can be constructed from any type of flip-flop. These flip-flops in integrated circuit registers are usually constructed internally using two separate flip-flop circuits. The normally used special kind of arrangement is termed as master-slave flip-flop. This type of flip-flop help in having a stable state at the output. In consist of a master flip-flop and a slave flip-flop.

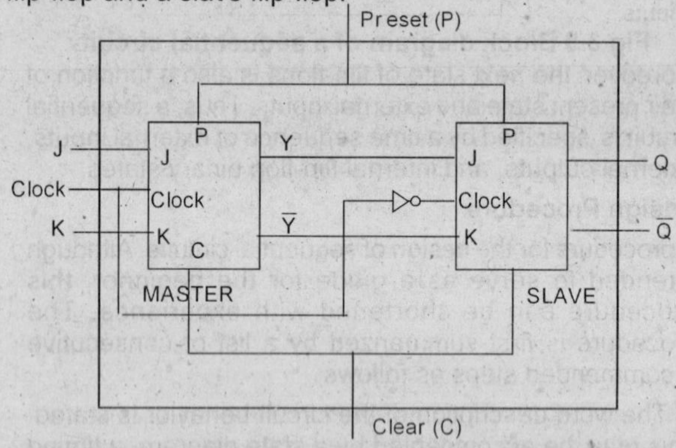

Figure 3.9 : Master-Slave flip-flop using J-K flip-flop

Note : You can construct Master-slave flip-flop with D flip-flop or R-S flip-flop in the same manner.

I. When the clock pulse is 0 the Master is disabled but the slave becomes active and its output Q and $\overline{Q}$ becomes equal to Y and $\overline{Y}$ repectively. Why ? Well the possible combination of the value of Y and $\overline{Y}$ are either Y = 1 which means $\overline{Y}$ = 0, or Y = 0 which implies $\overline{Y}$ = 1. Let us see the characteristic table for these two inputs for the J K flip-flop. The SLAVE flip-flop, thus, can have value either J = 1 and K= 0 which will set the flip-flop that is Q = 1 $\overline{Q}$ = 0 or J = 0, K =1 which will clear the flip-flop. Therefore Q is same as Y.

II. When inputs are applied at J K and clock pulse become 1, only master get activated resulting in intermediate output Y go to state 0 or 1 depending on the input and previous state. Please note during this time slave is is still maintaining it previous state. As the clock pulse become 0, the master become inactive and slave acquires the same state as master.

But why do we require this master-slave combination. There is a major reason for this master-slave form. Consider a situation where output of a flip-flop is going to input of other flip-flop. Here the assumption is that clock pulse inputs of all flip-flops are synchronised and occur at the same time. The change of state of master occurs when clock pulse goes to 1 but during that time the output of slave still has not changed, thus the state of the flip-flops in the system can be changed simultaneously during the same clock pulse even though outputs of flip-flops are connected to the inputs of flip-flops. In other words there is no restrictions on feedback from the registers outputs to its inputs.

Let us come back to the register having parallel input-output. Figure 3.10 shows a 4 bit register with parallel input-output.

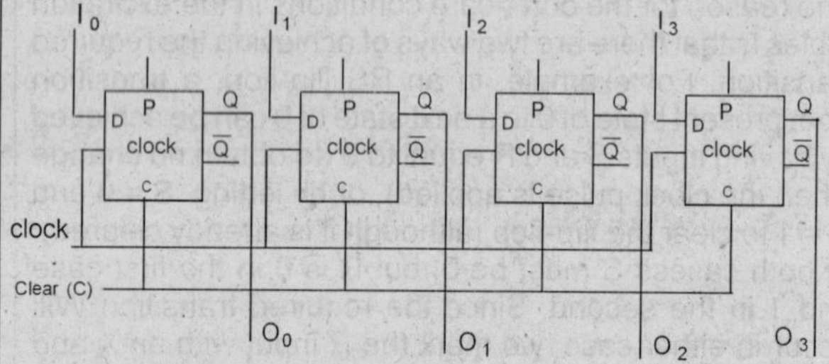

Figure 3.10 : Register with parallel input-output

It is one of the simplest register and contains preset, clear (for clearing the register completely) and clock pulse input in addition to the data value through D input. All the four bits in this register can be input and output simultaneously (as master-slave configuration can be employed in D flip-flop) hence the name parallel input-output.

Another kind of register which is used for shifting the data to the left or right is called a shift register. A shift register may operate in serial input-output mode in which data is entered in the register one bit at a time from one end of the register and can be read from the other end as one bit at a time. Figure 3.11 gives a 4 bit right-shift register with serial input-output. This register is constructed using D flip-flops.

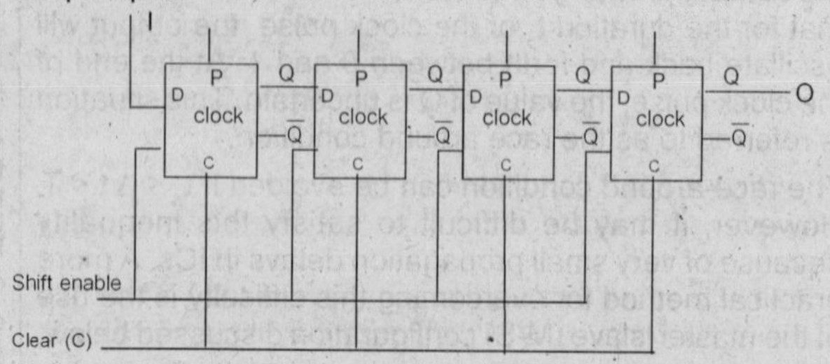

Figure 3.11 : Right shift register with serial input-output

Please note that in this register shift enable is used instead of clock pulse, since it is not necessary that we want the register to perform shift on each clock pulse. Clear and present in will be loaded in the register bit by bit. For example : 1 0 1 1 will be loaded in the register as

Input	Flip Flop 1	2	3	4 (States)	
1	x	x	x	x	On the occurrence of next shift
1	1	1	x	x	enable which may be next
0	0	0	1	1	clock pulse
1	1	0	1	1	

It can be removed in similar fashion from the registers :

Flip Flop 1	2	3	States 4	Output
0	1	0	1	1
0	0	1	0	1
0	0	1	0	0
0	0	0	0	1

shift enable controls the shifting of the data to the right here.

Shift register can be constructed for bidirectional shift with parallel input-output or serial input-ouput. A general structutre may have the facility of parallel data transfer to or from the register. In addition, the facility of left or right shift may also be provided. But this structure will require additional control lines for indicating whether parallel or serial output is desired and left shift or right shift operation is desired. A general symbolic diagram for such a register is given in Figure 3.12 .

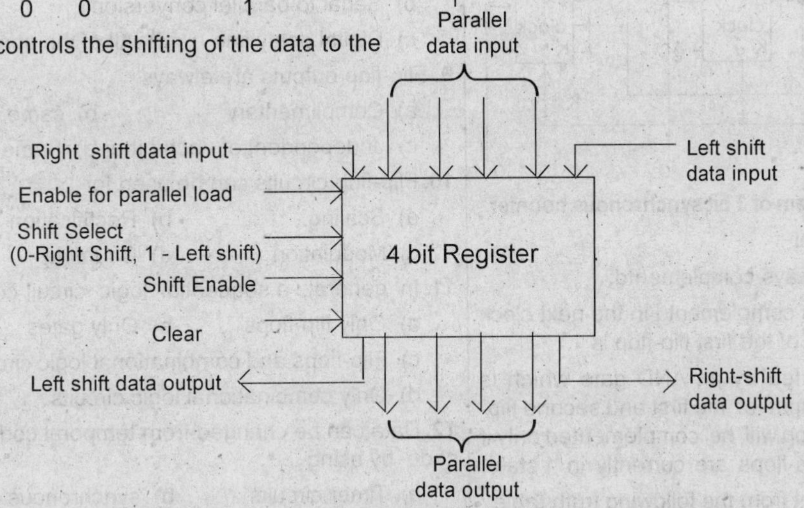

Figure 3.12 : A 4 bit left-right Shift register with parallel load

The basic logic diagram of such a circuit resembles figure 3.10 but require additional combinational circuit before an input is fed. You can refer to further readings for further reference.

These shift registers can be used in number of applications such as storage of serial data, parallel to serial and serial to parallel data conversion and, multiplication and division operations. For example, performing a left-shift for an unsigned binary number is equivalent to multiplication by binary 2.

Similarly a right-shift operation on an unsigned binary number is equivalent to the division by 2. For example, if we multiply

Decimal	Binary	
5	0101	Shift left and place a
× 2	× 0010	zero on the last digit position
10 shifted out →	0 1010 →	shift in

7. Counters

Another useful sequential circuit is the counter. A counter in principle is a register whose value is incremented by 1 on the occurrence of some event. When the value stored in counter reaches the maximum value it can store, the next incremented value becomes zero. The counters are used for counting the number of times of an event occurs and are useful for generating the timing signals for controlling the sequence of operations in digital computers.

There are two types of couters : Asynchronous and Synchronous. This classification is based on the way they operate. In synchronous counter the state of one flip-flop changes at a time while in synchronous counter the state of all flip-flops can be changed at the same time.

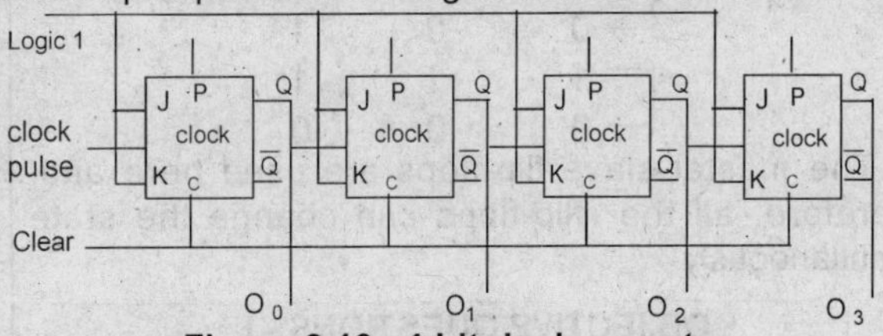

Figure 3.13 : 4 bit ripple counter

Asynchronous Counter : This is also termed as a ripple counter since the change which occurs in order to increment the counter ripples through the counter from one end to other. Figure 3.13 shows an implementation of a 4 bit counter implemented using J-K flip-flops. This counter is imcremented on the occurrence of each clock pulse and it counts from 0 to 15.

The input line to J and K is kept high, i.e. logical 1 in this counter and each time a clock pulse occurs the value of flip-flop is complemented (Refer to characteristic table of J-K flip-flop). Please note that clock inputs to the flip-flops. The first flip-flop is fed with the clock pulse as clock input but second, third and fourth flip-flips are provided with the output of its previous flip-flop as clock signal, implying that these flip-flops will be complemented if the previous flip-flop have a value 1. Thus the effect complement will ripple through these flip-flops.

Synchronous Counter : Take the instance when the state or ripple counter is 0111 now the next state will be 1000 which means change in the state of all the flip-flops, but will it occur simultaneously in ripple counter ? No first O_0 will change then O_1, O_2 and in the last O_3. Thus, a delay is there in changing the value which is proportional to the length of the counter. Therefore, to avoid this delay normally synchronous counters are used, in which all the flip-flops change state at the same time (see figure 3.14).

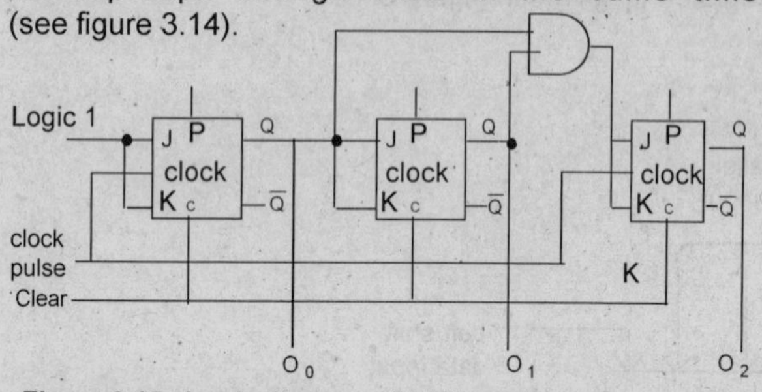

Figure 3.15 : Logic diagram of 3 bit synchronous counter

In a synchronous counter :

1. The first flip-flop is always complementd.

2. The second flip-flop is complementd in the next clock pulse if the current state of the first flip-flop is 1.

3. The third flip-flop is fed by an AND gate which is connected with the output of the first and second flip-flops thus, the third flip-flop will be complemented only if the first AND second flip-flops are currently in 1 state.

This will be more evident from the following truth table :

O_0	O_1	O_2
0	0	0
1	0	0
0	1	0
1	1	0
→ 0	0	1
1	1	1
→ 0	0	0

4. The master-slave flip-flops are used here and, therefore, all the flip-flops can change the state simultaneously.

OBJECTIVE QUESTIONS - I

1. A flip-flop can store **[AUCET-1998]**
   a) One bit of data b) Two bits of data
   c) Three bits of data d) Any number of bits of data.

2. A sequential circuit usually consist of
   a) Only flip-flops b) Only gates
   c) Flip-flops and gates d) None of these.

3. A flip-flop can be made using
   a) Basic gates such as AND,OR, and NOT
   b) NAND gates
   c) NOR gates d) Any of the above.

4. Flip-flops can be used to make.
   a) Latches b) Bounce-elimination switches
   c) Registers d) All of the above

5. In an S-R flip-flop the S-R inputs must not be
   a) S=R=1 b) S=R=0 c) S=0,R=1 d) S=1,R=0

6. When an inverter is placed between both inputs of an SR flip-flop the resulting flip-flop is
   a) JK flip-flop b) D flip-flop
   c) T flip-flop d) Master-Slave JK flip-flop

7. The number of flip-flops required in a decade counter is
   a) 3 b) 4 c) 5 d) 10

8. A shift register can be used for
   a) Parallel to serial conversion
   b) Serial to parallel conversion
   c) Digital delay line d) All of the above

9. Flip-flop outputs are always
   a) Complimentary b) same
   c) Independent of each other d) Same as inputs

10. Flip-flop circuits can be used for
   a) Scaling b) Rectification
   c) Modulation d) Counting

11. In general, a sequential logic circuit consists of
   a) Only flip-flops b) Only gates
   c) Flip-flops and combinational logic circuits
   d) Only combinational logic circuits.

12. Data can be changed from temporal code to special code by using
   a) Timer circuits b) synchronous counters
   c) shift registers d) combinational circuits

13. A ring counter consisting of five flip-flops will have
   a) 5 States b) 10 States c) 32 States d) Infinite States

14. Which of the following flip-flop requires only one input signal
   a) R-S flip-flop b) J-k flip-flop
   c) D flip-flop d) None

15. Asynchronous counters are slower than synchronous counters.
   a) True b) False.

16. Which one of the following is called transparent latch.
   a) J-K flip-flop b) R-S flip-flops
   c) D flip-flop d) Clocked R-S flip-flop

17. A latch is commonly used to
   a) Interface input devices b) Interface memory
   c) Interface output devices d) Interface DMA

18. A flip-flop is a
   a) Synchronous circuit b) Asynchronous circuit
   c) Synchronous Sequential circuit
   d) Asynchronous Sequential circuit

19. In general in master-slave flip-flop the master is negative edge triggered and slave is positive edge triggered.
   a) True b) False

20. With positive clocking the clock signal must be...for the flip-flop to respond.

a) high b) low c) Set d) race

21. Synchronous counter are ....than the ripple counters **[GATE-1994]**

a) Slower b) Faster

22. An SR flip-flop can be converted in a T flip-flop by connecting .... to Q and ... to $\overline{Q}$. **[GATE-1991]**

a) R, S b) S, R

23. The functional difference between SR flip-flop and JK flip-flop is **[UGC NET - 1997]**

a) JK flip-flop is faster than SR flip-flop

b) JK flip-flop is having a feedback path

c) JK flip - flop accepts both input 1

d) JK flip - flop does not require external clock

24. Which one of the following is not a type of flip-flop ? **[UGC NET - 1996]**

a) D-flip b) Jk- flip-flop

c) RS - flip - flop d) P- flip-flop

25. To store an n-bit binary number in a register, the number of flip-flops we require are **[UGC NET-1998]**

a) n b) (n + 1) c) 2^n d) 2^{n+1}

26. The functional difference between SR and JK flip-flop is that **[UGC NET-1999]**

a) JK flip-flop has a feedback path

b) JK flip-flop does not require an external clock pulse

c) JK flip-flop is faster than SR flip-flop

d) JK flip-flop gives valid output when both inputs are logic 1

27. Which of the following flip-flop do not have race problem ? **[UGC NET - 1998]**

a) Master slave flip-flop b) T flip-flop

c) D flip-flop d) Jk flip - flop

28. An R-S latch is a **[Hyd.Univ-1998]**

a) Combinational circuit b) Synchronous circuit

c) One bit memory element d) One clock delay element

29. When a J-K flip-flop is constructed from an S-R flip-flop which of the following is correct? **[Hyd.Univ-1996]**

a) S = JQ,R = KR

b) S = JQ, R = K + J Q′

c) J = S + R′ Q, K = S′ (R + Q′)

d) S = J Q′ , R = KQ

30. If (1, 0), (0, 0), (1, 0) and (1, 1) are given as inputs to a JK flip-flop in the given order the output sequence would be **[Hyd.Univ-1994]**

a) 1010 b) 0101 c) 0001 d) 1110

31. The minimum number of Flip-flops needed in a decade counter to drive a BCD to seven-segment display decoder is? **[Hyd.Univ-1993]**

a) 3 b) 5 c) 4 d) 8

ANSWER KEY - I

1. a	**2.** c	**3.** d	**4.** d	**5.** a
6. b	**7.** b	**8.** d	**9.** a	**10.** d
11. c	**12.** c	**13.** a	**14.** c	**15.** b
16. c	**17.** b	**18.** c	**19.** b	**20.** a
21. b	**22.** a	**23.** c	**24.** d	**25.** a
26. d	**27.** a	**28.** c	**29.** d	**30.** d
31. c				

OBJECTIVE QUESTIONS - II

1. In a Sequential circuit the outputs at any instant of time depend.

a) Only on the inputs present at that instant of time.

b) Only on the past outputs.

c) Only the past outputs and the inputs present at that instant of time. d) Only on the past inputs.

2. In a JK flip-flop when J=K=1 and clock is applied the output Q will

a) No change b) Become 0 c) Become 1

d) Be complement of the output before the clock was applied

3. Race around condition occurs in a JK flip-flop when

a) Both the inputs are 0 b) Both the inputs are 1

c) The inputs are complementary

d) Any one of the above input combinations is present.

4. Master-Slave configuration is used in flip-flops to **[AUCET-1999]**

a) Increase its clocking rate

b) Reduce power dissipation

c) Eliminate race-around condition

d) Improve its reliability.

5. An 'n' stage ripple counter can count up to

a) 2^n b) $2^n - 1$ c) n d) 2^{n-1}

6. The number of flip-flops required in a modulo-N counter is

a) $\log_2 N + 1$ b) $\lceil \log_2 N \rceil$

c) $\log_2(N+1)$ d) $\log_2(N-1)$

7. How many flip-flops are required to divide by 16

a) 2 b) 4 c) 8 d) 16

8. The maximum possible number of states in a ripple counter consisting of four FLIP-FLOPS is

a) 4 b) 8 c) 10 d) 16

9. A full counter with 'n' flip-flops will have ...states.

a) 2^n b) 2^{n-} c) $2^n - 1$ d) $2^n + 1$

10. The divide – by – 10 Johnson counter requires ... flipflops.

a) 4 b) 15 c) 10 d) 11

11. The D flip-flop is called delay flip-flop because

a) $Q_{n+1} = Q_n$ after a clock pulse

b) $Q_{n+k} = Q_n$ for all k >=1.

c) Output is made equivalent to input after a clock pulse.

d) Output is made equivalent to input after a set of clock pulses.

12. If you want to design a boundary counter, you should prefer a flip-flop of

a) D-type b) SR-type c) latch d) JK-type

13. In a ripple counter using edge-triggered JK flip-flops, the pulse input is applied to the

a) Clock input of all flip-flops.

b) Clock input of one flip-flop

c) J and K input of one flip-flop

d) J and K input of all flip-flops

14. How many minimum gates are needed for a 3 bit up-counter using standard binary and using T-flip-flops? Assume unlimited fan-in **[AUCET-1999]**

a) 0 b) 1 c) 2 d) 3

15. The number of flip-flops needed to construct a Johnson counter to produce n timing sequences is **[AUCET-2001]**

a) 2n b) n/2 c) 2^n d) log n

16. A 4-bit modulo-16 ripple counter uses JK flip-flop. If the progresion delay of each flip-flop is 50ms, the maximum clock frequency that can be used is equal to: **[GATE-1990]**

a) 5MHZ b) 10 MHZ c) 15 MHZ d) 20 MHZ.

17. For a flip flop formed from two NAND gates as shown below fig, the unusable state corresponds to: **[GATE-1999]**

a) x = 0, y = 0

b) x = 0, y = 1

c) x = 1, y = 1

d) x = 1, y = 0

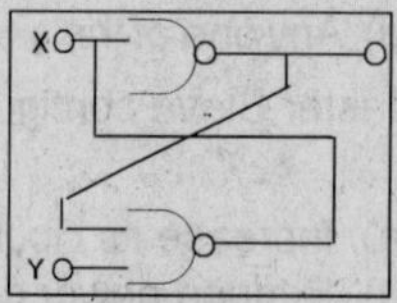

18. The three state Jhonson ring counter shown below is clock at a constant frequency of f_0, from the starting state of $Q_0Q_1Q_2=10$. The frequency of output's will be **[GATE-1991]**

a) f_c 12

b) f_c 13

c) f_c 16

d) None

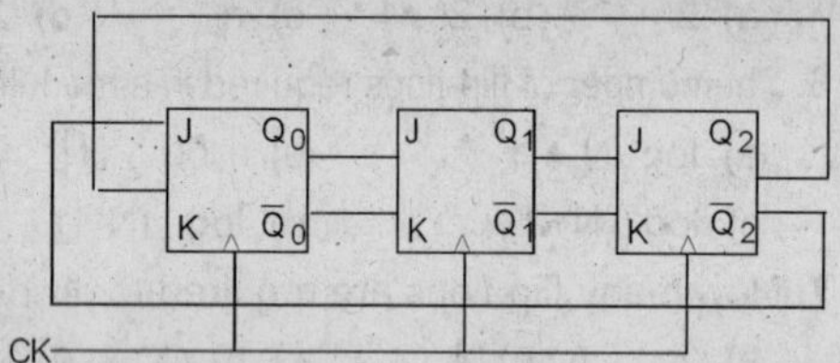

19. In a Jk flip flop we have and k = 1. Assuming the flip flop was initially cleared and then clocked for 6 pulses, the sequence at Q output will be : **[GATE-1997]**

a) 010000

b) 011001

c) 010010

d) 010101

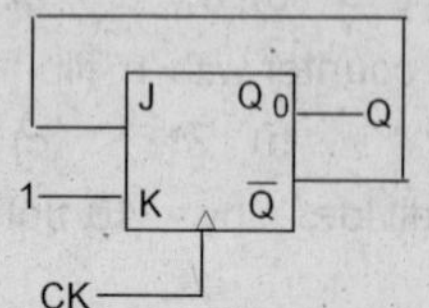

20. The below fig. shows a mod-k counter, here K is equal to : **[GATE-1998]**

a) one

b) two

c) three

d) four

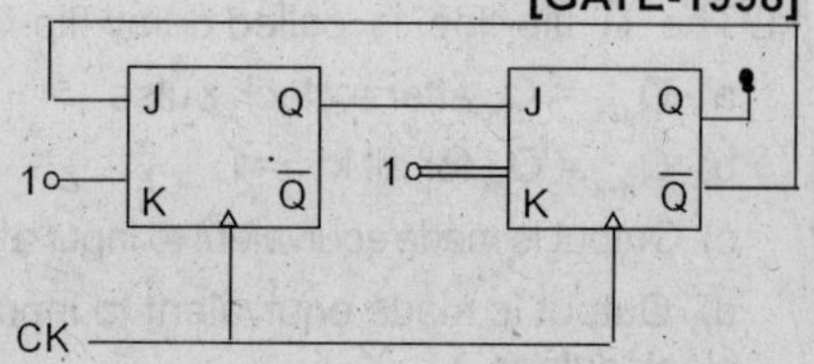

21. A sequential circuit using D flip flop and logic circuit is shown below fig. where x and y are the input's and Z is the output. The circuit is : **[GATE-2000]**

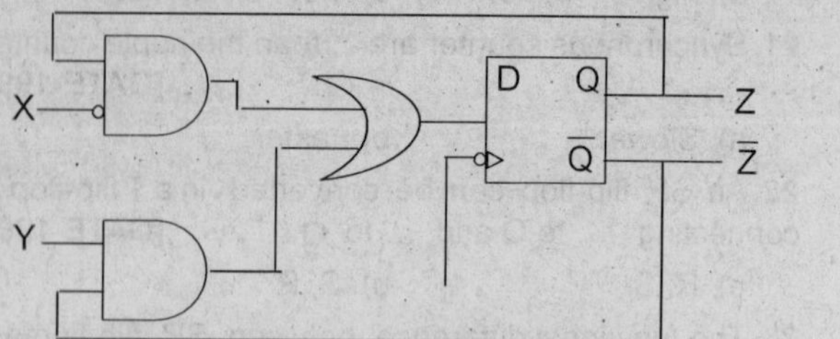

a) SR flip flop with input's X = R and Y = S

b) SR flip flop with input's X = s and Y = R

c) JK flip flop with input's X = J and Y = K

d) JK flip flop with input's X = K and Y = J

22. The Ripple counter shown below works as : **[GATE-1998]**

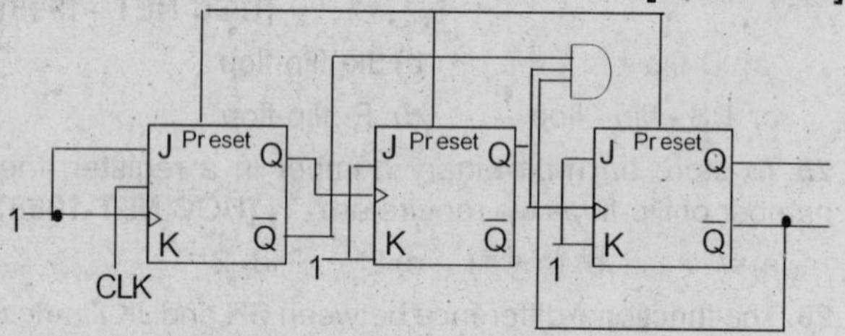

a) mod-3 up counter b) mod-5 up counter

c) mod - 3 down counter d) mod-5 down counter

23. In below figure A = 1 and B = 1. The i/P B is now replaced by a sequence 101010...., the output's x and y will be : **[GATE-1998]**

a) fixed at 0 and 1 respectively

b) x = 1010... while y = 0101...

c) x = 1010.. while y = 1010...

d) fixed at 1 and 0 respectively

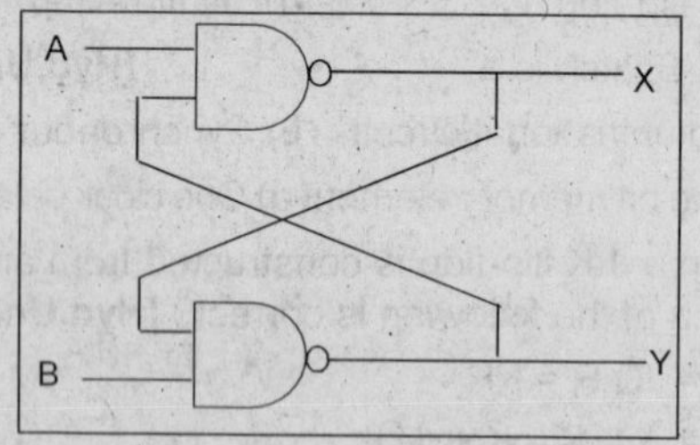

24. A switch-tail Ring counter is made by using a single D flip flop. The resulting circuit is a .... **[GATE-1999]**

a) SR flip flop b) JK flip flop

c) D flip flop d) T flip flop

25. The minimum number of flip flops needed to make mod-2 counter is : **[GATE-1998]**

a) 1 b) 2 c) 3 d) 4

26. The below circuit producer the output sequence is : **[GATE-1987]**

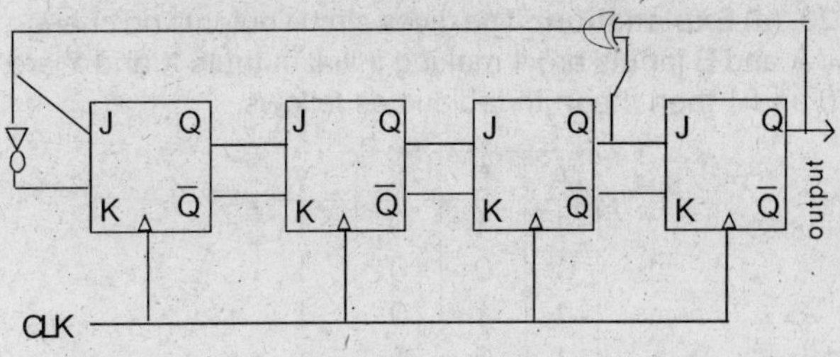

a) 1111 1111 0000 0000 b) 1111 0000 1111 0000
c) 1111 0001 0011 010 d) None

27. For the initial state of 000, the function performed by the arrangement of the JK flip flops is below : **[GATE-1993]**

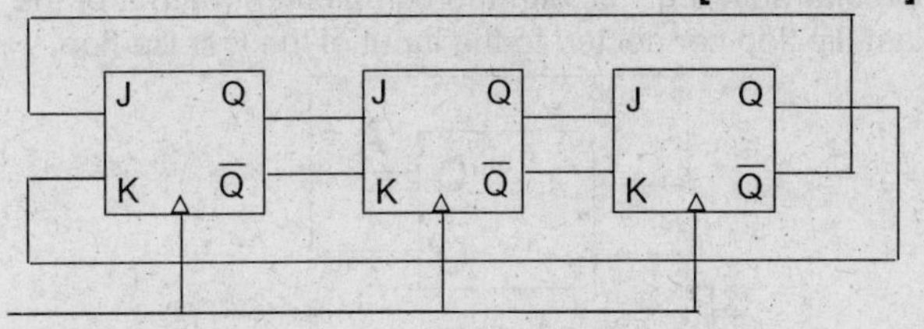

a) shift Reg b) mod-3 counter
c) mod-6 counter d) None

28. The following arrangement of master-slave flip flops : **[GATE-2000]**

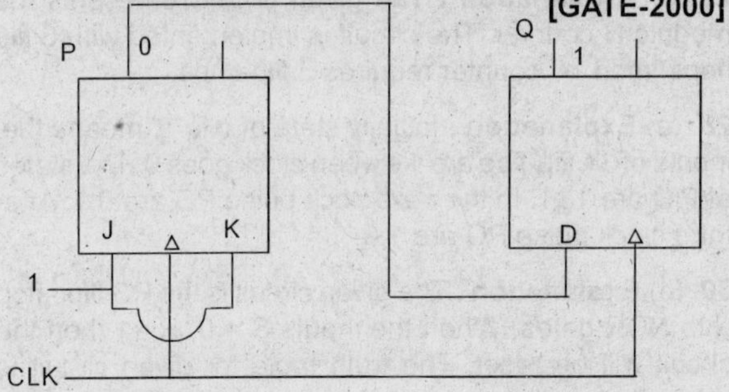

has the initial state of P,Q as 0,1 (respectively), after three clock cycles the output state P,Q is (respectively),

a) 1,0 b) 1,1 c) 0,0 d) 0,1

29. The number of bits required to construct a mod-N counter should be greater than or equal to **[UGC NET-1998]**

a) $2 \log_2 N$ b) $\log_2 N$ c) $N \log_n 2$ d) $\log_n 2$

30. The state of the output Q of the following latch circuit when S=0, R=1 would be **[UGC NET-1998]**

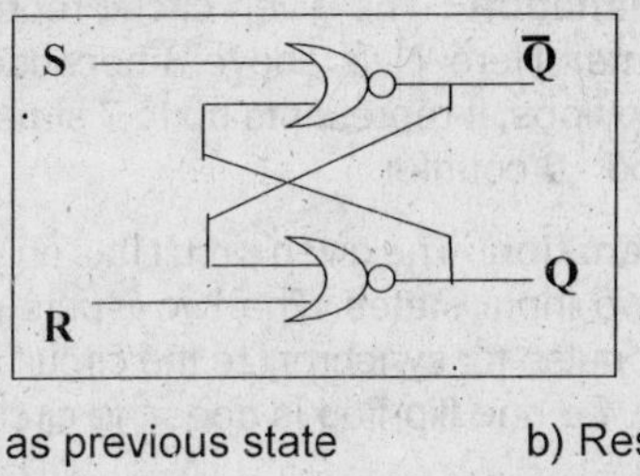

a) Same as previous state b) Reset
c) Set d) Invalid

31. The following sequence of inputs are presented to a synchronous R-S Flip-flop. Indicate the state of the Flip-flop at time T = 8 and 9. **[Hyd.Univ-1998]**

TIME(T)	1	2	3	4	5	6	7	8	9	10
SET	0	0	0	1	1	0	0	1	1	0
RESET	0	0	1	0	0	0	1	1	0	0

a) ?, 0 b) 0,1 c) 1, 0 d) 0, ?

32. For the initial state of 000, the function performed by the arrangement of the J-K flip-flop in the figure below is **[Uni.Hyd-1998]**

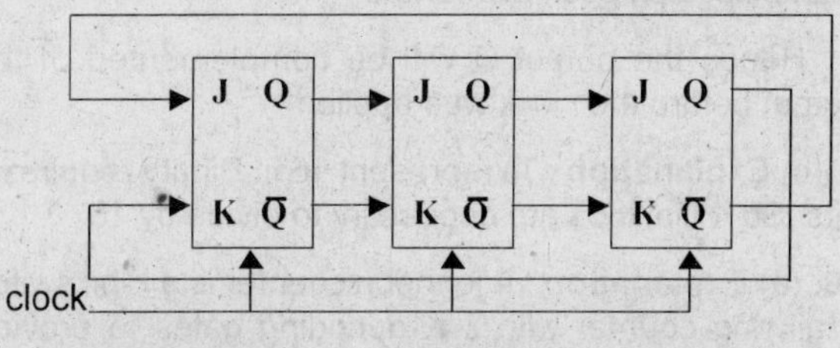

a) Shift Register b) Mod-3 Counter
c) Mod-6 counter d) Mod-2 Counter

Below two questions are based on the following

Given the following state diagram for a circuit, where there are two inputs X, Y and state is denoted by A

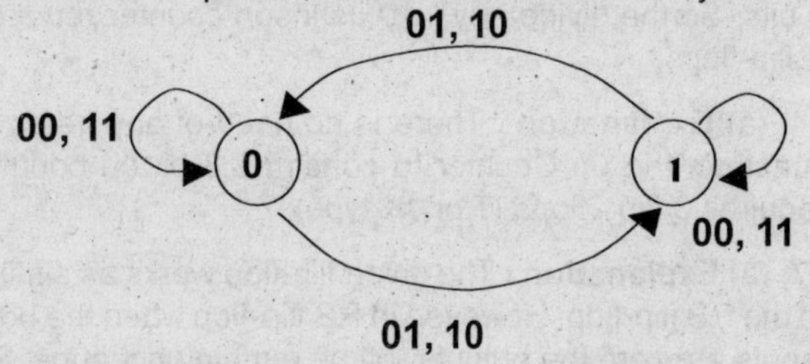

33. How many flip-flops does the circuit use? **[Hyd.Univ-1997]**

a) 0 b) 1 c) 2 d) 3

34. What is the equation that governs the input to the flip-flop? Assume D flip-flop **[Hyd.Univ-1997]**

a) $D = A + X + Y$ b) $D = A \oplus X \oplus Y$
c) $D = A + X \oplus Y$ d) $D = A \oplus X + Y$

ANSWER KEY - II

1. c	**2.** d	**3.** b	**4.** c	**5.** b
6. b	**7.** b	**8.** d	**9.** a	**10.** a
11. a	**12.** a	**13.** d	**14.** a	**15.** a
16. a	**17.** a	**18.** a	**19.** d	**20.** d
21. a	**22.** ?	**23.** a	**24.** d	**25.** b
26. c	**27.** c	**28.** a	**29.** b	**30.** b
31. a	**32.** c	**33.** b	**34.** b	

EXPLANATORY NOTES - II

2. (d) **Explanation :** When both J & K are 1, the clock pulse is transmitted through one AND gate only the one whose input is connected to the flip - flop output which is presently equal to 1. thus if Q =1 the flip - flop is cleared. If Q' =1 the flip-flop is set. In either case the output state of the flip flop is complemented.

Hence the output Q will be complemented of the output before the clock was applied.

7. (b) **Explanation :** To represent 16 in Binary requires 4 bits. So 4 flip flops are necessary to divide by 16.

10. (a) **Explanation :** A johnson counter is a k-bit switch - tail ring counter with 2 K decoding gates to provide outputs for 2K timing signals. The decoding of a K - bit switch - tail ring to obtain 2 K timing sequences follows a regular pattern. The number of flip - flops needed is one - half the number of timing signal.

Hence Now, to represent the decimal 10 we require 4 bits. So the divide - by - 10 Johnson counter requires 4 flip-flops.

14. (a) **Explanation :** There is no need of any gates to construct the up Counter to construct 3 bit up counter requires 3 flip - flops (T or JK type).

17. (a) **Explanation :** The given flip-flop works as similar to the RS flip-flop. However, in RS flip-flop when the both inputs are zero the outputs will be remain unchange. So this stage is unused state.

19. (d) **Explanation :** JK flip-flop is cleared during a clock pulse only if Q was previously 1, if the Q value is zero the input sequences are

	Output Q
001	- 1
101	- 0
011	- 1
101	- 0
111	- 1

20. (d) **Explanation :** A modulo - K (abbrivated mod - K) counter is a counter that goes through a repeated sequence of k - counts. An n - bit binary counter consists of 'n' - flip - flops and can count in binary from 0 to $2^n - 1$. For example a 4-bit binary counter is a mod-16 counter.

From the above discussion, for the given circuit K=2.

21. (a) **Explanation :** The truth table for given circuit is

x	y	z	$\bar{z}$
0	0	No change	
0	1	1	0
1	0	0	1
1	1	Forbidden	

This is similar to SR flip-flop with inputs X = R and Y = S.

23. (a) **Explanation :** The given circuit outputs no change if A and B inputs are 1 making initial outputs X and Y are 0 and 1 then the truth table is as follows.

A	B	X	Y
1	1	0	1
1	0	0	1
1	1	0	1
1	0	0	1
1	1	0	1
1	0	0	1

24. (d) **Explanation :** A switch - tail ring counter is a circular shift register with the complement output of the last flip flop connected to the input of the first flip-flop.

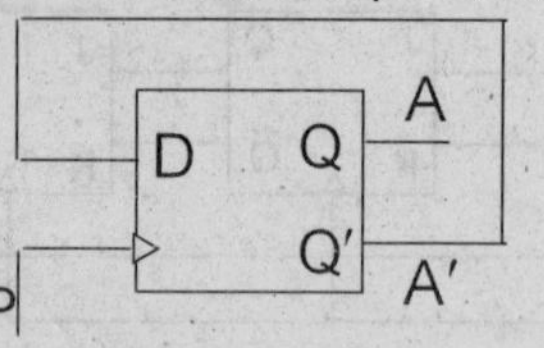

Given, a switch tail ring counter is made by single D flip-flop is shown above.

27. (c) **Explanation :** The given circuit represents the modulo N counter. The circuit is implemented with 3 flip flops, mod - 6 counter requires 3 flip-flops.

28. (a) **Explanation :** Initially state of p is '0' means the, inputs of JK flip-flop are 1's when clock goes 0. The states of P,Q are 1&1. In the next clock pulse PQ are 0,1. After third clock pulse PQ are 1,0.

30. (b) **Explanation :** The given circuit is the RS flip - flop with NOR gates. When the inputs S = 0, R =1 then the circuit will be reset. The truth table for given circuit is shown below.

S	R	Q	$\overline{Q}$
0	0	No change	
0	1	Reset	
1	0	Set	
1	1	X	

32. (c) **Explanation :** The given circuit represnents the Mod-N counter. Here 'N' is above 3 because the circuit contain 3 flip flops, it represents up to 7 states. So given circuit is mod - 6 counter.

33. (b) **Explanation :** The given circuit has only one output state and two input states. The two inputs are may be one is clock pulse for synchronize the circuit and second one is input. So one flip-flop is does the circuit.

34. (b) **Explanation :** In given circuit the state will be go to high when the odd number of 1's in present situation other wise go to state zero. This indicate that works as similar to X-OR function. Hence, $D = A \oplus X \oplus Y$ is correct.

TEST YOURSELF - I

1. The minterm expansion of $f(P,Q,R) = PQ + Q\bar{R} + P\bar{R}$ is **[GATE - 10]**

a) $m_2 + m_4 + m_6 + m_7$ b) $m_0 + m_1 + m_3 + m_5$

c) $m_0 + m_1 + m_6 + m_7$ d) $m_2 + m_3 + m_4 + m_5$

2. P is a 16 - bit signed integer. The 2's complement representation of P is $(F87B)_{16}$. The 2's complement representation of 8*P is **[GATE - 10]**

a) $(C3D8)_{16}$ b) $(187B)_{16}$

c) $(F878)_{16}$ d) $(987B)_{16}$

3. The Boolean expression for the output f of the multiplexer shown below is **[GATE - 10]**

a) $\overline{P \oplus Q \oplus R}$

b) $P \oplus Q \oplus R$

c) $P + Q + R$

d) $\overline{P + Q + R}$

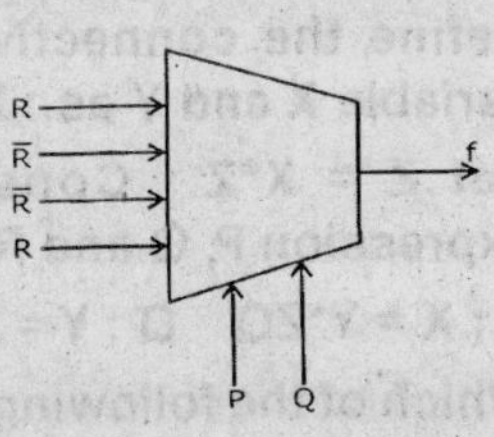

4. What is the Boolean expression for the output f of the combinational logic circuit of NOR gates given below? **[GATE - 10]**

a) $\overline{Q + R}$

b) $P + Q$

c) $\overline{P + R}$

d) $\overline{P + Q + R}$

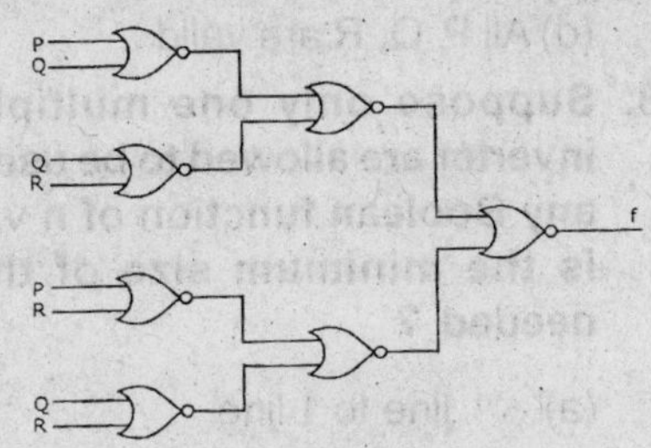

5. In the sequential circuit shown below, if the initial value of the output Q_1Q_0 is 00, what are the next four values of Q_1Q_0? **[GATE - 10]**

a) 11,10,01,00

b) 10,11,01,00

c) 10,00,01,11

d) 11,10,00,01

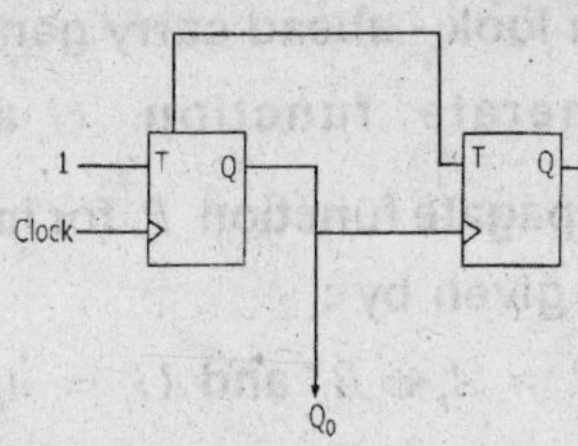

6. $(1217)_8$ is equivalent to **[GATE - 09]**

a) $(1217)_{16}$ b) $(028F)_{16}$

c) $(2297)_{10}$ d) $(0B17)_{16}$

7. What is the minimum number of gates required to implement the Boolean function (AB+C) if we have to use only 2 - input NOR gates ? **[GATE - 09]**

a) 2 b) 3 c) 4 d) 5

8. In the Karnaugh map shown below, X denotes a don't care term. What is the minimal form of the function represented by the Karnaugh map ? **[GATE - 08]**

(a) $\bar{b}.\bar{d} + \bar{a}.\bar{d}$

(b) $\bar{a}.\bar{b} + \bar{b}.\bar{d} + \bar{a}.b.\bar{d}$

(c) $\bar{b}.\bar{d} + \bar{a}.b.\bar{d}$

(d) $\bar{a}.\bar{b} + \bar{b}.\bar{d} + \bar{a}.\bar{d}$

cd \ ab	00	01	11	10
00	1	1		1
01	X			
11	X			
10	1	1		X

9. Let **r** denote number system radix. The only value(s) of **r** that satisfy the equation $\sqrt{121_r} = 11_r$ is/are **[GATE - 08]**

(a) decimal 10 (b)decimal 11

(c) decimal 10 and 11 (d) any value >2

10. Given f_1, f_3 and f in canonical sum of products form (in decimal) for the circuit **[GATE - 08]**

$f_1 = \Sigma m(4,5,6,7,8)$

$f_3 = \Sigma m(1,6,15)$

$f = \Sigma m(1,6,8,15)$

then f_2 is

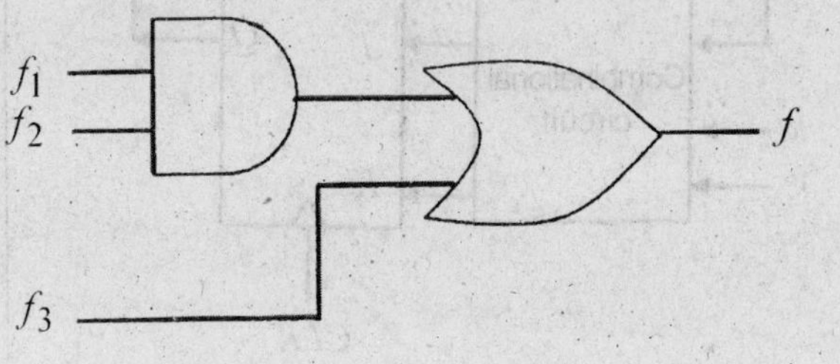

(a) $\Sigma m(4,6)$ (b) $\Sigma m(4,8)$

(c) $\Sigma m(6,8)$ (d) $\Sigma m(4,6,8)$

11. The following bit pattern represents a floating point number in JEEE 754 single precision format

1 10000011 10100000000000000000000

The value of the number in decimal form is **[GATE - 08]**

(a) -10 (b) -13

(c) -26 (d) None of the above

12. Consider the following Boolean function of four variables

$$f(A,B,C,D)=\Sigma(2,3,6,7,8,9,10,11,12,13)$$

The function is **[GATE - 08]**

(a) independent of one variable
(b) independent of two variables
(c) independent of three variable
(d) dependent on all the variables

13. What Boolean function does the circuit below realize ? **[GATE - 08]**

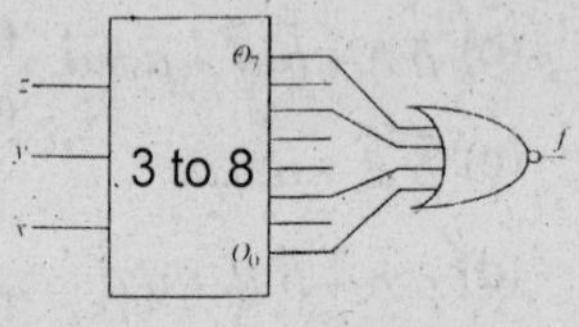

(a) $xz + \overline{xz}$

(b) $x\bar{z} + \bar{x}z$

(c) $\overline{xy} + yz$

(d) $xy + \overline{yz}$

14. Consider the following state diagram and its realization by a JK flip flop **[GATE - 08]**

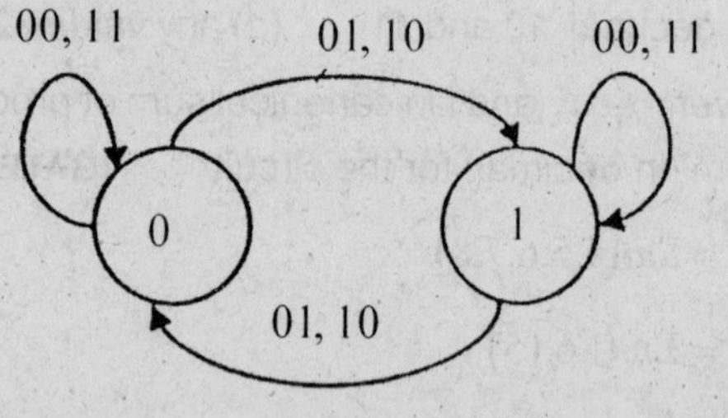

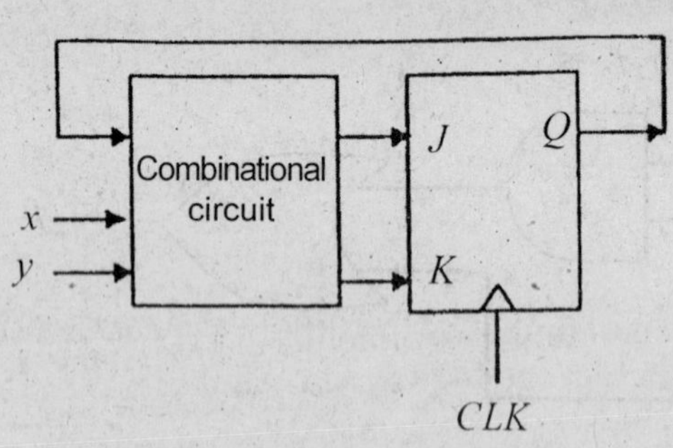

The combinational circuit generates J and K in terms of x,y and Q

The Boolean expressions for J and K are :

(a) $x\oplus y$ and $\overline{x\oplus y}$

(b) $\overline{x\oplus y}$ and $x\oplus y$

(c) $x\oplus y$ and $\overline{x\oplus y}$

(d) $x\oplus y$ and $x\oplus y$

15. How many 3-to-8 line decoders with an enable input are needed to construct a 6-to-64 line decoder without using any other logic gates ? [GATE - 07]

(a) 7 (b) 8
(c) 9 (d) 10

16. Consider the following Boolean function of four variables

$$f(w,x,y,z)=\sum(1,3,4,6,9,11,12,14)$$

The function is [GATE - 07]

(a) independent of one variable
(b) independent of two variables
(c) independent of three variables
(d) dependent on all the variables

17. Define the connective * for the Boolean variable X and Y as : X * Y = XY + X'Y'

Let Z = X*Z. Consider the following expression P, Q and R.

P : X = Y*ZQ Q : Y = X*Z R : X* Y* Z= 1

Which of the following is TRUE ? [GATE - 07]

(a) Only P and Q are valid
(b) Only Q and R are valid
(c) Only P and R are valid
(d) All P, Q, R are valid

18. Suppose only one multiplexer and one inverter are allowed to be used to implement any Boolean function of n variables. What is the minimum size of the multiplexer needed ? [GATE - 07]

(a) 2^n line to 1 line

(b) 2^{n+1} line to 1 line

(c) 2^{n-1} line to 1 line

(d) 2^{n-2} line to 1 line

19. In a look - ahead carry generator, the carry generate function G_i and the carry propagate function P_i for inputs, A_i and B_i are given by :

$P_i = A_i \oplus B_i$ **and** $G_i = A_i B_i$

The expressions for the sum bit S_i and the carry bit C_{i+1} of the look - ahead carry adder are given by :

$S_i = P_i \oplus C_i$ **and** $C_{i+1} = G_i + P_i C_i$**, where C_0 is the input carry.**

Consider a two-level logic implementation of the look-ahead carry generator. Assume that all P_i *and* G_i are available for the carry generator circuit and that the AND and OR gates can have any number of inputs. The number of AND gates and OR gates needed to implement the look-ahead carry generator for a 4-bit adder with S_3, S_2, S_1, S_0 *and* C_4 as its outputs are respectively [GATE - 07]

(a) 6, 3 (b) 10, 4
(c) 6, 4 (d) 10, 5

20. Which of the following input sequences for a cross-coupled R-S flip-flop realized with two NAND gates may lead to an oscillation ? [GATE - 07]

(a) 11, 00
(b) 01, 10
(c) 10, 01
(d) 00, 11

21. The following circuit implements a two-input AND gate using two 2-1 multiplexers.

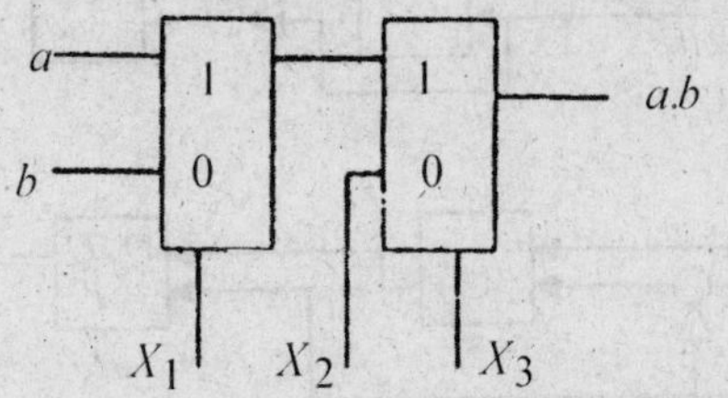

What are the values of X1, X2, X3 ? [GATE - 07]

(a) $X1 = b,\ X2 = 0, X3 = a$

(b) $X1 = b,\ X2 = 1, X3 = b$

(c) $X1 = a,\ X2 = b, X3 = 1$

(d) $X1 = a,\ X2 = 0, X3 = b$

22. The following expression was to be realized using 2 - input AND and OR gates. However, during the fabrication all 2 - input AND gates were mistakenly substituted by 2 - input NAND gates.

$$(a.b).c + (\overline{a}.c).d + (b.c).d + a.d$$

What is the function finally realized ? [GATE - 07]

(a) 1 (b) $\overline{a} + \overline{b} + \overline{c} + \overline{d}$

(c) $\overline{a} + b + \overline{c} + \overline{d}$ (d) $\overline{a} + \overline{b} + c + \overline{d}$

23. What is the final value stored in the linear feedback shift register if the input is 101101 ? [GATE - 07]

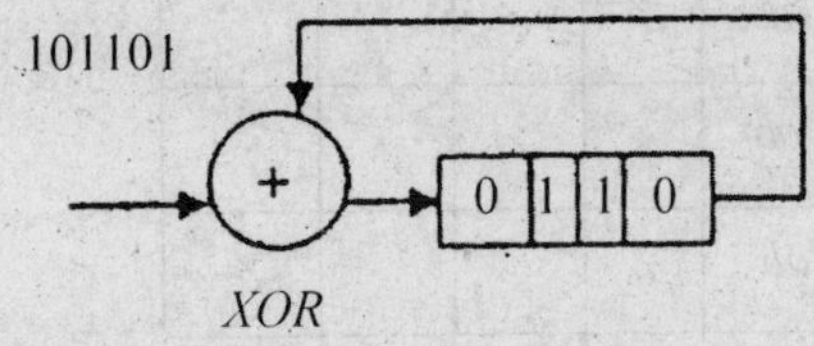

(a) 0110 (b) 1011
(c) 1101 (d) 1111

24. $(C012.25)_H - (10.111001110.101)_B =$ [GATE - 07]

(a) $(135103.412)_O$ (b) $(564411.412)_O$

(c) $(564411.205)_O$ (d) $(135103.205)_O$

25. An error correcting code has the following code words : 00000000, 00001111, 01010101, 10101010, 11110000. What is the maximum number of bit errors that can be corrected ? [GATE - 07]

(a) 0 (b) 1 (c) 2 (d) 3

26. The line T in the following figure is permanently connected to the ground.

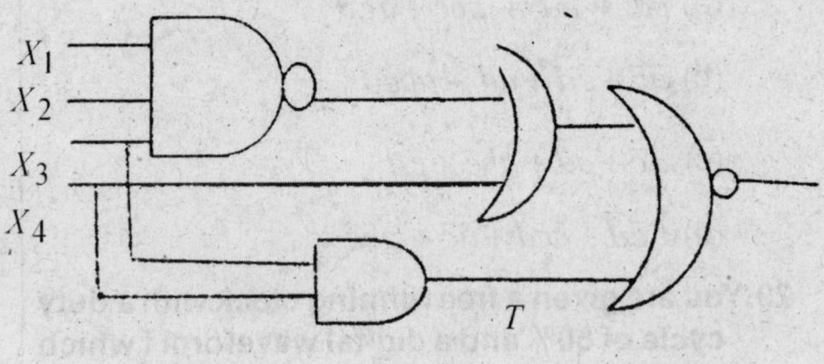

Which of the following inputs (X1 X2 X3 X4) will detect the fault ? [GATE - 07]

(a) 0000 (b) 0111
(c) 1111 (d) None of these

<u>**Statement for Linked Answer Questions 27 and 28 :**</u>

Consider the following expression

$$a\overline{d} + \overline{ac} + b\overline{c}d$$

27. Which of the following Karnaugh Maps correctly represents the expression ? [GATE - 07]

(a)

	$\overline{c}\overline{d}$	$\overline{c}d$	cd	$c\overline{d}$
$\overline{a}\overline{b}$	×	×		
$\overline{a}b$	×	×		
ab	×	×		×
$a\overline{b}$	×			×

(b)

	$\overline{c}\overline{d}$	$\overline{c}d$	cd	$c\overline{d}$
$\overline{a}\overline{b}$	×	×		
$\overline{a}b$	×			
ab	×	×		×
$a\overline{b}$	×	×		×

(c)

	$\overline{c}\overline{d}$	$\overline{c}d$	cd	$c\overline{d}$
$\overline{a}\overline{b}$	×	×		
$\overline{a}b$	×	×		×
ab	×			×
$a\overline{b}$	×			×

(d)

	$\overline{c}\overline{d}$	$\overline{c}d$	cd	$c\overline{d}$
$\overline{a}\overline{b}$	×	×		
$\overline{a}b$	×	×		
ab	×	×		×
$a\overline{b}$	×		×	×

28. Which of the following expressions does not correspond to the Karnaugh Map obtained in Q 27 ? [GATE - 07]

(a) $c\overline{d} + a\overline{d} + ab\overline{c} + \overline{a}\overline{c}d$

(b) $\overline{a}\overline{c} + \overline{c}\overline{d} + a\overline{d} + ab\overline{c}d$

(c) $\overline{a}\overline{c} + a\overline{d} + ab\overline{c} + \overline{c}d$

(d) $\overline{b}\overline{c}\overline{d} + ac\overline{d} + \overline{a}\overline{c} + ab\overline{c}$

29. You are given a free running clock with a duty cycle of 50% and a digital waveform f which changes only at the negative edge of the clock. Which one of the following circuits (using clocked D flip - flops) will delay the phase of f by $180°$? [GATE - 06]

(A)

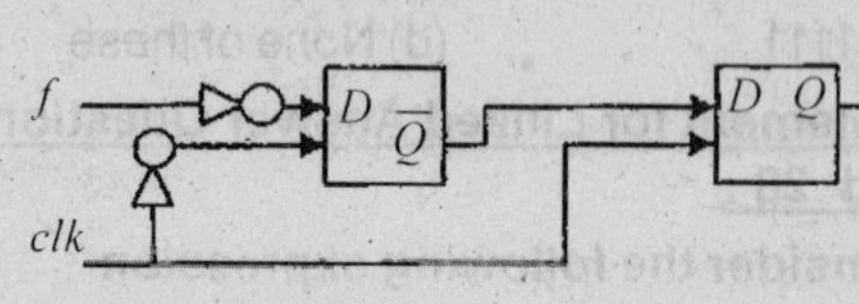

(B)

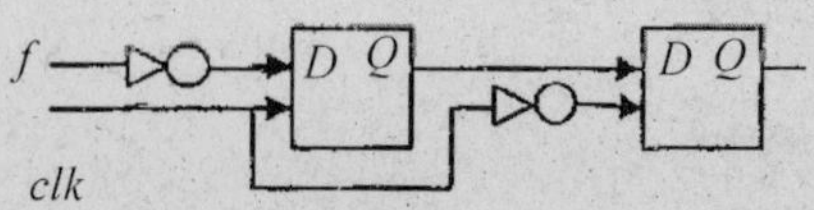

(C)

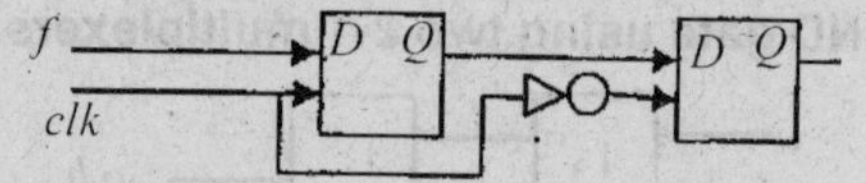

(D)

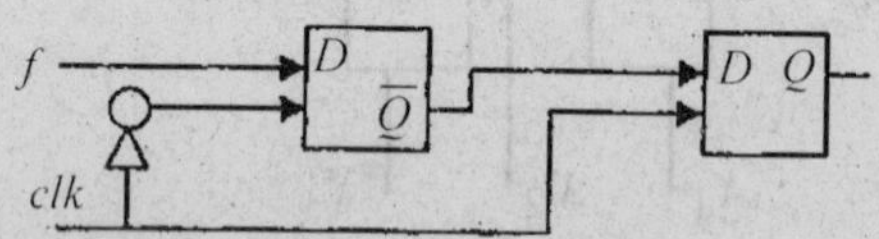

30.

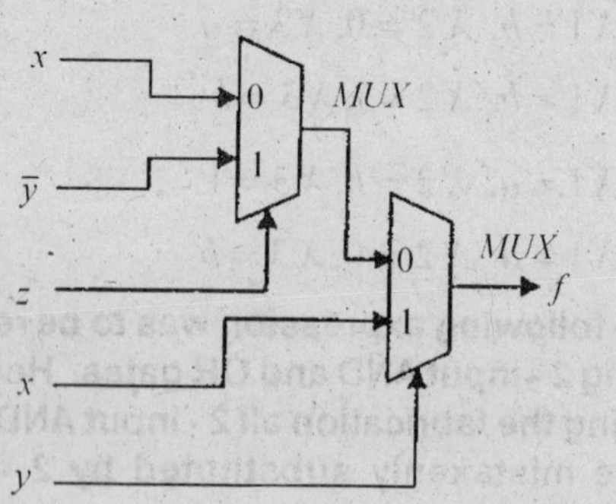

Consider the circuit above. Which one of the following options correctly represents f(x,y,z)? [GATE - 06]

(A) $x\overline{z} + xy + \overline{y}z$ (B) $x\overline{z} + xy + \overline{yz}$

(C) $xz + xy + \overline{yz}$ (D) $xz + x\overline{y} + \overline{y}z$

31. Given two three bit numbers $a_2 a_1 a_0$ and $b_2 b_1 b_0$ and c, the carry in, the function that represents the carry generate function when these two numbers are added is [GATE - 06]

(A) $a_2b_2 + a_2a_1b_1 + a_2a_1a_0b_0 + a_2a_0b_1b_0 + a_1b_2b_1 + a_1a_0b_2b_0 + a_0b_2b_1b_0$

(B) $a_2b_2 + a_2b_1b_0 + a_2a_1b_1b_0 + a_1a_0b_2b_1 + a_1a_0b_2 + a_1a_0b_2b_0 + a_2a_0b_1b_0$

(C) $a_2 + b_2 + (a_2 \oplus b_2)(a_1 + b_1 + (a_1 \oplus b_1)(a_0 + b_0))$

(D) $a_2b_2 + \overline{a_2}a_1b_1 + \overline{a_2 a_1}a_0b_0 + \overline{a_2}a_0\overline{b_1}b_0$ $a_1\overline{b_2}b_1 + \overline{a_1}a_0\overline{b_2}b_0 + a_0\overline{b_2 b_1}b_0$

32. Consider the circuit in the diagram. The $\oplus$ operator represents Ex-OR. The D flip-flops are initialized to zeroes (cleared).

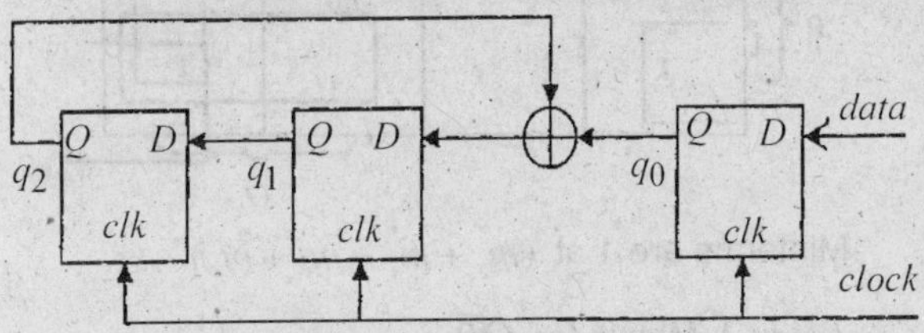

The following data : 100110000 is supplied to the "data" terminal in nine clock cycles. After that the values of $q_2q_1q_0$ are [GATE - 06]

(A) 000 (B) 001

(C) 010 (D) 101

33. Consider a boolean function $f(w,x,y,z)$. Suppose that exactly one of its inputs is allowed to change at a time. If the function happens to be true for two input vectors $i_1 = \langle w_1, x_1, y_1, z_1 \rangle$ and $i_2 = \langle w_2, x_2, y_2, z_2 \rangle$, we would like the function to remain true as the input changes from i_1 to i_2 (i_1 and i_2 differ in exactly one bit position), without becoming false momentarily. Let $f(w,x,y,z) = \sum(5,7,11,12,13,15)$. Which of the following cube covers of f will ensure that the required property is satisfied ? [GATE - 06]

(A) $\overline{w}xz, wx\overline{y}, x\overline{y}z, xyz, wyz$

(B) $wxy, \overline{w}xz, wyz$

(C) $wx\overline{yz}, xz, w\overline{x}yz$

(D) $wzy, wyz, wxz, \overline{w}xz, x\overline{y}z, xyz$

34. We consider the addition of two 2's complement numbers $b_{n-1}b_{n-2}.....b_0$ and $a_{n-1}a_{n-2}.....a_0$. A binary adder for adding unsigned binary numbers is used to add the two numbers. The sum is denoted by $c_{n-1}c_{n-2}.....c_0$ and the carry-out by c_{out}. Which one of the following options correctly identifies the overflow condition ?

(A) $c_{out}\overline{(a_{n-1} \oplus b_{n-1})}$ **[GATE - 06]**

(B) $a_{n-1}b_{n-1}c_{n-1} + a_{n-1}b_{n-1}c_{n-1}$

(C) $c_{out} \oplus c_{n-1}$

(D) $a_{n-1} \oplus b_{n-1} \oplus c_{n-1}$

35. Consider numbers represented in 4-bit gray code. Let $h_3h_2h_1h_0$ be the gray code representation of a number n and let $g_3g_2g_1g_0$ be the gray code of (n+1) (modulo 16) value of the number. Which one of the following functions is correct?

[GATE - 06]

(A) $g_0(h_3h_2h_1h_0) = \sum(1,2,3,6,10,13,14,15)$

(B) $g_1(h_3h_2h_1h_0) = \sum(4,9,10,11,12,13,14,15)$

(C) $g_2(h_3h_2h_1h_0) = \sum(2,4,5,6,7,12,13,15)$

(D) $g_3(h_3h_2h_1h_0) = \sum(0,1,6,7,10,11,12,13)$

36. The addition of 4-bit, two's complement, binary numbers 1101 and 0100 results in

[GATE - 06]

a) 0001 and an overflow

b) 1001 and no overflow

c) 0001 and no overflow

d) 1001 and an overflow

37. The boolean function for a combinational circuit with four inputs is represented by the following Karnaugh map.

RS \ PQ	00	01	11	10
00	1	0	0	1
01	0	0	1	1
11	1	1	1	0
10	1	0	0	1

Which of the product terms given below is an essential prime implicant of the function ?

a) QRS b) PQS **[GATE - 06]**

c) PQ'S' d) Q'S'

38. The majority function is a Boolean function $f(x,y,z)$ that takes the value 1 whenever a majority of the variables x,y,z and 1. In the circuit diagram for the majority function shown below, the logic gates for the boxes labeled P and Q are, respectively,

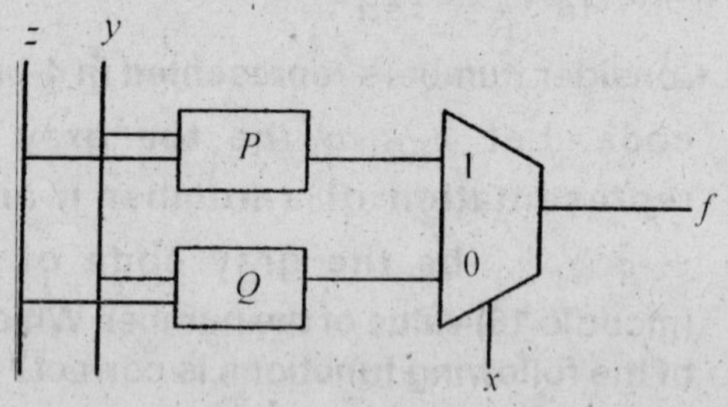

a) $XOR,\ AND$ b) $XOR,\ XOR$ **[GATE - 06]**

c) $OR,\ OR$ d) $OR,\ AND$

39. For a state machine with the following state diagram the expression for the next state S^+ in terms of the current state S and the input variables $x\ and\ y$ is **[GATE - 06]**

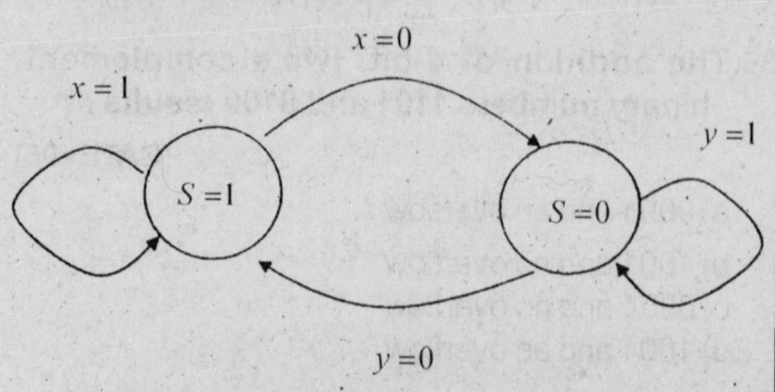

a) $S^+ = S'.y' + S.x$ b) $S^+ = S.x.y' + S'.y.x'$

c) $S^+ = x.y'$ d) $S^+ = S'.y + S.x'$

ANSWER KEY

1. a	2. a	3. b	4. a	5. a
6. b	7. d	8. a	9. d	10. c
11. d	12. a	13. b	14.	15. b
16. b	17. d	18. c	19. c	20. b
21. a	22. b	23. c	24. a	25. d
26. d	27. a	28. a	29. c	30. a
31. a	32. c	33. a	34.	35. b
36. c	37. d	38. c	39. a	

EXPLANATORY NOTES

1.

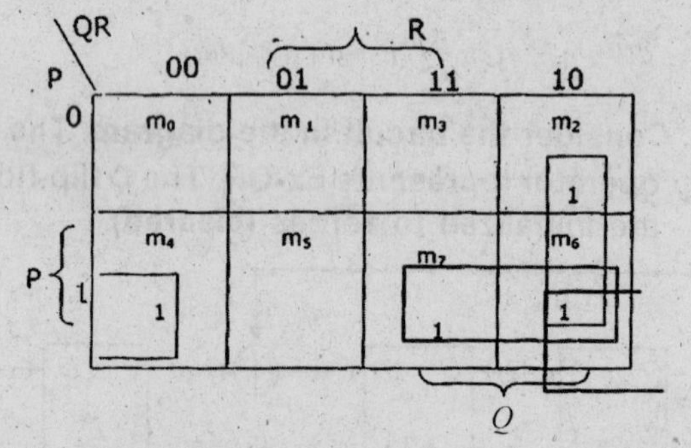

Minterms are 1 at $(m_2 + m_4 + m_6 + m_7)$

$(m_2 m_6)$ Map is for $Q\overline{R}$

$(m_7 m_6)$ Map is for PQ

$(m_4 m_6)$ Map is for $p\overline{R}$

2. $P = (F87B)_{16}$ is -1111 1000 0111 1011 = - 1925

P * 8 = 1925 $\times$ 8 = -15400.

2's complement is 11000011 1101 1000

It is equivalent to $(C3D8)_{16}$

3.

P	Q	F
0	0	R
0	1	$\overline{R}$
1	0	$\overline{R}$
1	1	R

So is

$p'Q'R + p'QR' + pQ'R' + PQR$

$p'(Q'R + QR') + p(Q'R' + QR)$

$p'(Q \oplus R) + p(Q \odot R)$

$= P \oplus Q \oplus R$

4. $\overline{\left(\left(\overline{P+Q}\right)+\left(\overline{Q+R}\right)\right)}+\overline{\left(\left(\overline{P+R}\right)+\left(\overline{Q+R}\right)\right)}$

So it can be written as

$$\left(\left(\overline{P+Q}\right)+\left(\overline{Q+R}\right)\right)\,.\,\left(\left(\overline{P+R}\right)+\left(\overline{Q+R}\right)\right)$$

$$\left[\because (A+B)^{1}=A^{1}B^{1}\right]$$

$$(P^{1}Q^{1}+Q^{1}R^{1}).(P^{1}R^{1}+Q^{1}R^{1})$$

$$P^{1}Q^{1}P^{1}R^{1}+P^{1}Q^{1}Q^{1}R^{1}+Q^{1}R^{1}P^{1}R^{1}+Q^{1}R^{1}Q^{1}R^{1}$$

$$P^{1}Q^{1}R^{1}+P^{1}Q^{1}R^{1}+Q^{1}P^{1}R^{1}+Q^{1}R^{1}[\because xx=x]$$

$$P^{1}Q^{1}R^{1}+Q^{1}R^{1}$$

$$Q^{1}R^{1}[P^{1}+1]\ \Rightarrow Q^{1}R^{1}\ \Rightarrow\left(\overline{Q+R}\right)$$

5. Initially $\begin{matrix} Q_1 & Q_0 \\ 0 & 0 \end{matrix}$

Here T is T - flip flop
If T = 0 the next state is equal to present state
When T = 1 the next state is compliment of present state.
For both flip flops the T input is '1' (constant) the O|P 'Q_0' is given as a clock to the next T - flip flop.
Initially both will be complimented it will become 11,

Next second Q_1 remains 1 since Q_0 became '0' when T i/p is 1, so when clock is not present it remains in same state, ...
So it will go for sequence 11, 10, 01, 00

6. $(1217)_8 = 001\ 010\ 001\ 111$

$0010\,10000\ 1111 = (028F)_{16}$

Alternately

Given digit in octal 1217

Convert them in binary $\overline{001}\,\overline{010}\,\overline{001}\,\overline{111}$

Now convert binary to hexa $\underset{2\quad 8\quad F}{\overline{001010001111}} = (028F)_{16}$

7. For implementing (AB+C) very first we implement AB. We know that for AND operation 3 NOR GATE required

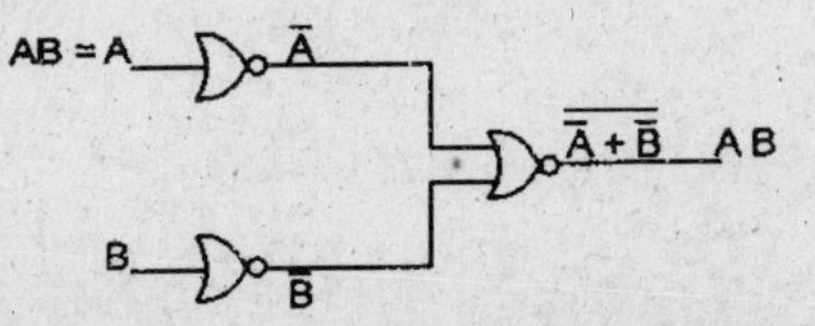

Now for (AB+C)

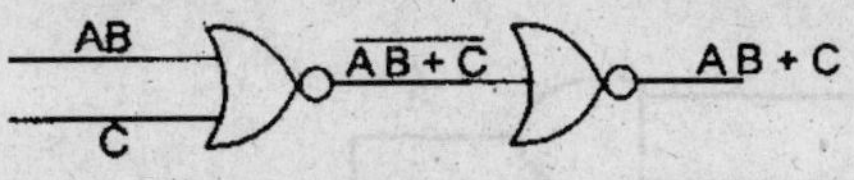

Hence five NOR gates required.

Alternately

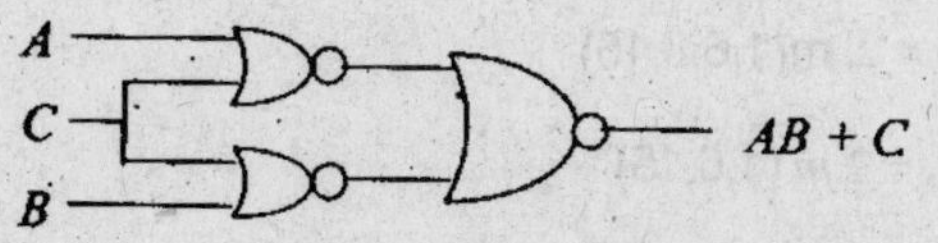

8.

cd \ ab	00	01	11	10
00	1	1		1
01	X			
11	X			
10	1	1		X

$$\bar{a}\bar{b}\bar{d}+\bar{a}b\bar{d}+a\bar{b}\bar{d}+\bar{a}\bar{b}\bar{d}=\bar{b}\bar{d}+\bar{a}\bar{d}$$

9. $\sqrt{121_r}=11_r$

r cannot be 2, so let $r=3$

$$\therefore\ \sqrt{1\times 9+2\times 3+1}=\sqrt{9+6+1}=\sqrt{16}=4$$

$$=3+1=(11)_r$$

So, r can be any value greater than 2,

10.	f_1	f_2	f	f_1f_2	f_2
1	0	1	1	x	
4	1	0	0		
5	1	0	0		
6	1	1	1	x	x
7	1	0	0		
8	1	0	1	1	1
15	0	1	1	x	

$$f_1f_2+f_3=f$$

For $f_o = 1$ any of term in L.H.S must be 4

so $f_2=\Sigma(6,8)$

Alternately

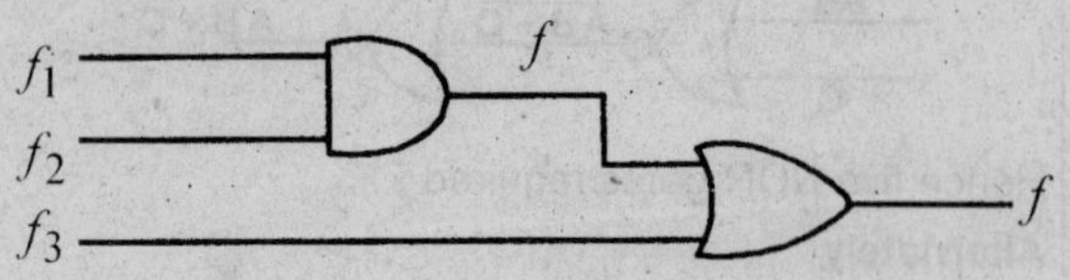

$f = \Sigma$ m(1,6,8,15)

$f_3 = \Sigma m$ (1,6,15)

f' should atleast contain Σm (8) because f' or f_3 gives f. In the given choices, $\Sigma m4$ cannot be in f_2 because if contains $\Sigma m(4)$ then, $f' = \Sigma m(4,8)$ and f will also contain 4

So, $\Sigma m(6,8)$ is correct.

11.

sign	Exponent	Fraction
1	1 0 0 0 0 1 1	1 0 1 0 0 0 0 0 0 0 0 0 0 0 0 0 0 0 0 0 0 0 0

Checking options

(a) -10

sign = -1

$(10)_2 = 1010 = 1.010 \times 2^3$

The exponent is biased by $2^{e-1} - 1$, where e is the number of bits used for the exponent field which is 8.

$\therefore \quad 2^{8-1} - 1 = 127$

For the 32 bit IEEE 754 format, the bias is +127 and so 3+127=130. In binary (130) is encoded as
1 0 0 00010

Thus, it is not equal to exponent field.

(b) -13

sign = -1

$(13)_2 = 1101 = 1.101 \times 2^3$

$\therefore \quad 3 + 127 = 130$

Hence $(130)_2 = 10000010 \neq$ exponent

(c) -26

$(26)_2 = 11010$

$= 1.1010 \times 2^4$

$127 + 4 = 131$ and $(131)_2$

$\therefore$ = 10000011

But fraction part does not match

12. $f(A,B,C,D) = \Sigma(2, 3, 6, 7, 8, 9, 10, 11, 12, 13)$

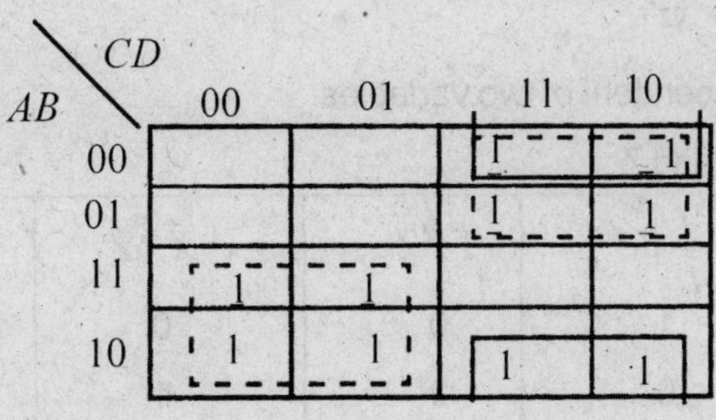

$= A'C + A'C + B'C$

It depends on 3 variables A,B and C

So it is independent of one variable.

13. $f' = x'y'z' + x'yz' + xy'z + xyz$

$= x'z'(y'+y) + xz(y+y')$

$= x'z'(1) + xz(1)$

$= x'z' + xz$

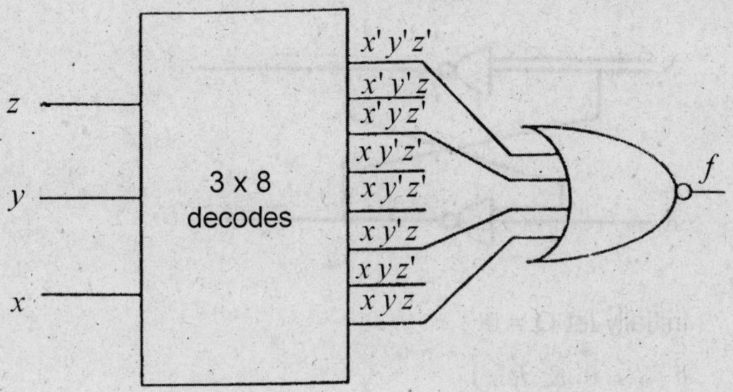

$f = \overline{f'}$

$= \left(\overline{x'z' + xz}\right)$

$= \left(\overline{x'z'}\right)\left(\overline{xz}\right)$

$= (x+z).(x'+z')$

$= xx' + xz' + x'z + zz' \qquad = xz' + x'z$

By Demorgan's theorem

$\left(\overline{A+B}\right) = \overline{A}.\overline{B}$

$\left(\overline{A.B}\right) = \overline{A} + \overline{B}$

and $\quad A.\overline{A} = 0$

$A + \overline{A} = 1$

15. Eight 3 - to - 8 line decoders will be needed to construct a 6 - to 64 line decoder without any other logic gates.

16. Karnaugh map of given function will be

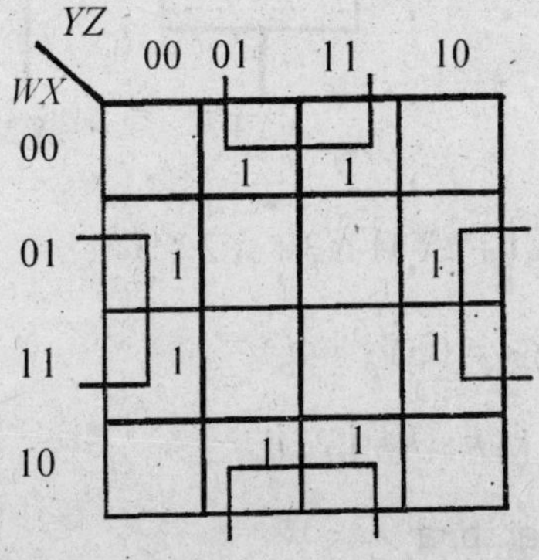

Simplified form of f will be

$$f = x'z + xz'$$

Therefore, it is independent of two variables.

17. Truth table for X, Y and Z

X	Y	$Z = X*Y$	$Y*Z$	$X*Z$	$X*Y*Z$
0	0	1	0	0	1
0	1	0	0	1	1
1	0	0	1	0	1
1	1	1	1	1	1

From above truth table we can say that all P, Q and R are valid.

18. Multiplexes of size 2^{n-1} to 1 will be needed.

20.

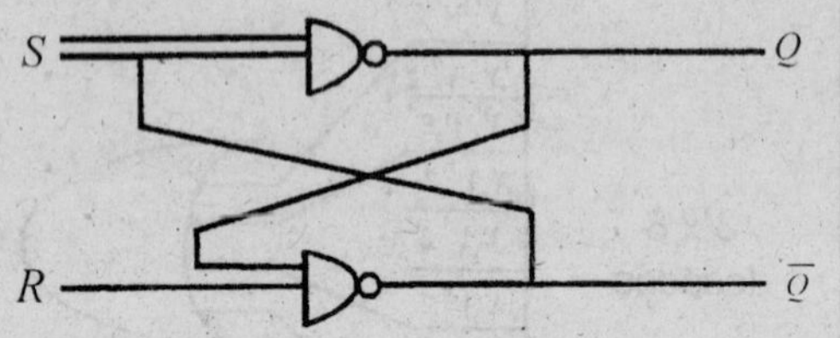

S R	Q_{n+1}
1 0	0
1 1	Q_n
0 0	NA
6 1	1

initially let Q = 0

if $S = 0$ & $R = 1$

then $Q = 1$

after it if $S = 1, \& R = 0,$ then $Q = 0$

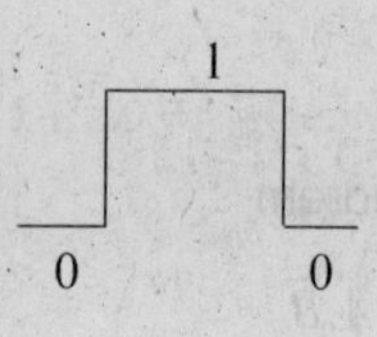

so cycle is formed

for the input 01, 10

21.

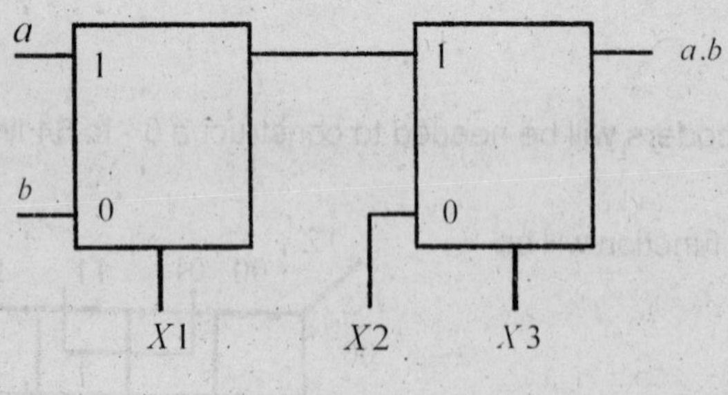

So $a.b = (aX1 + bX'1)\ X3 + X2X'3$

if $X1 = b_1, X_2 = 0, X_3 = a$

then $RHS = (a.b + b.b')a + 0.a'$

= a . b. a

= a. b

23.

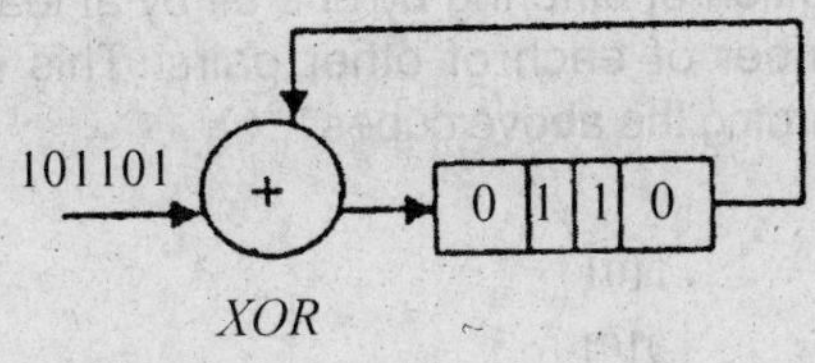

24. $(CO\ 12.25)_H - (10111001110.101)_B$

$(CO12.25)_H = (?)_B$

$(\underline{1100}\ \underline{0000}\ \underline{0001}\ \underline{0010}\ .\ \underline{0010}\ \underline{0101})_B$

Now converting into base 8

$\underline{001}\ \underline{100}\ \underline{000}\ \underline{000}\ \underline{010}\ \underline{010}\ \underline{001}\ \underline{001}\ \underline{010}$

$(140022.112)8$

$(\underline{010}\ \underline{111}\ \underline{001}\ \underline{110}\ \ \underline{101}\ )_B = (?)_8$

$(2\ \ 7\ \ 1\ \ 6\ \ \ 5\ )_8$

$$\begin{array}{r} (140022.112)_8 \\ -(2716.500)_8 \\ \hline 135103.412 \end{array}$$

26.

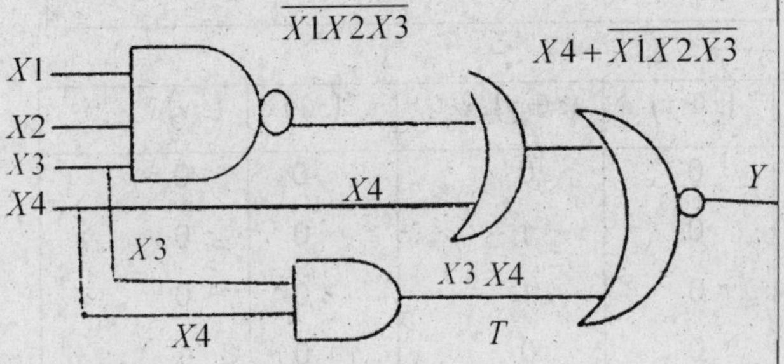

$X3\,X4$	$X4+\overline{X1X2X3}$	Y
0	0	1
0	1	0
1	0	0
1	1	0

Hence the option (d) satisfy this condition.

27. $a\overline{d} + \overline{ac} + b\overline{c}d$

ab \ cd	00	01	11	10
00	1	1		
01	1	1		
11	1	1		1
10	1			1

28. Option (a) does not correspond to K - map option in que 27.

$\overline{c}\overline{d} + \overline{a}\overline{d} + ab\overline{c} + \overline{a}\overline{c}d$

30. Output of the top $MUX = M_1 = x\overline{z} + \overline{y}z$

Output of the second $MUX = M_2 = M_1\overline{y} + xy$

$$= (x\overline{z} + \overline{y}z)\overline{y} + xy$$

$$= x\overline{y}\overline{z} + \overline{y}z + xy$$

$$= \overline{y}z + x\overline{y}\overline{z} + xy(x + \overline{z})$$

$$= \overline{y}z + x\overline{y}\overline{z} + xy\overline{z} + xy\overline{z} + xyz$$

$$= \overline{y}z + x\overline{z}(\overline{y} + y) + xy(z + \overline{z}) = \overline{y}z + x\overline{z} + xy$$

31. $a_2 \quad a_1 \quad a_0$

$b_2 \quad b_1 \quad b_0$

The carry will be generated in the following cases

$a_2 = 1, b_2 = 1$

$a_2 \oplus b_2 = 1$ and $a_1 = b_1 = 1$

$a_2 \oplus b_2 = a_1 \oplus b_1 = 1$ *and* $a_0 = b_0 = 1$

Considering the various options,
in (b), the second term is invalid
as if $a_2 = 1, b_2 = 0, b_1 = 1, b_0 = 1, c \neq 1$
in (c), a_2 is invalid term
In $(d), a_2 a_1, b_1,$ is invalid as, if
$a_2 = 0, a_1 = b_1 = 1, b_2 = 0, c_1 = 1$
so, only (a) encompasses possibilities.

32.

	q_2	q_1	q_0	I/P
Initially	0	0	0	-
	0	0	1	1
	0	1	0	0
	1	0	0	0
	0	0	0	1
	0	0	1	1
	0	1	0	0
	1	0	0	0
	0	0	1	0
	0	1	0	0

33.

wx \ yz	00	01	11	10
00	0	0	0	0
01	0	1	1	0
11	1	1	1	0
10	0	0	1	0

The cube covers of (a) contain the following cells

$\overline{w}\,xz$	0101, 0111
$wx\,\overline{y}$	1101, 1100
$x\,\overline{y}\,z$	0101, 1101
$x\,y\,z$	0111, 1111
$w\,y\,z$	1011, 1111

At least one member of a pair should satisfy the condition of differing by one bit by at least one member of each of other pairs. This is true regarding the above cubes.

36.

$$\begin{array}{lr} & 1101 \\ \textit{Carries} & +0101 \\ \hline & 10001 \end{array}$$

If the two carries are applied to an X-OR gate, an overflow will be detected when the output of the gate is equal to 1.

37. The three sum of minterms for the different solutions of the given Karnangh map are

(i) $S'Q' + R'SP + RSQ$ (ii) $S'Q' + RSP' + SPQ$

(iii) $S'Q' + R'SP + RSP'$

Hence the common term is $S'Q'$

38.

For (a) $\rightarrow$ $P = (z \oplus y).\,x$ $Q = (z\,.\,y)\,\overline{x}$	For (b) $\rightarrow$ $P = (z \oplus y).\,x$ $Q = (z \oplus y).\,\overline{x}$
For (c) $\rightarrow$ $P = (z + y).\,x$ $Q = (z + y).\,\overline{x}$	For (d) $\rightarrow$ $P = (z + y).\,x$ $Q = (z\,.\,y).\,\overline{x}$

x	y	z	$\overline{x}$	$(z+y).\,x$	$(z+y).\,\overline{x}$	$(z\oplus y).\,x$	$(z\oplus y).\,\overline{x}$	$(z.y)\,x$	$(z.y)\,\overline{x}$
0	0	0	1	0	0	0	0	0	0
0	0	1	1	0	1	0	1	0	0
0	1	0	1	0	1	0	1	0	0
0	1	1	1	0	1	0	0	0	1
1	0	0	0	0	0	0	0	0	0
1	0	1	0	1	0	1	0	0	0
1	1	0	0	1	0	1	0	1	0
1	1	1	0	1	0	0	0	1	0

Only option (c) gives thrice the output 1.

39. Consider $S = 0\ as\ S'$ $\quad x = 0\ as\ x'$ $\quad y = 0\ as\ y'$

$S = 1\ as\ S$ $\quad x = 1\ as\ x$ $\quad y = 1\ as\ y$

Only (a) satisfies the above state diagram

TEST YOURSELF - II

1. A 1-input, 2-output synchronous sequential circuit behaves as follows : **[GATE - 03]**

Let z_k, n_k denote the number of 0's and 1's respectively in initial k bits of the input $(z_k + n_k = k)$. The circuit outputs 00 until one of the following conditions holds.

* $Z_k - n_k = 2$. In this case, the output at the k-th and all subsequent clock ticks is 10.

* $n_k - Z_k = 2$. In this case, the output at the k-th and all subsequent clock ticks is 01.

what is the minimum number of states required in the state transition graph of the above circuit ?

a) 5 b) 6 c) 7 d) 8

2. The literal count of a boolean expression is the sum of the number of times each literal appears in the expression. For example, the literal count of $(xy + xz')$ is 4. What are the minimum possible literal counts of the product-of-sum and sum-of-product representations respectively of the function given by the following Karnaugh map ? Here, X denotes "don't care" **[GATE - 03]**

a) (11, 9)
b) (9, 13)
c) (9, 10)
d) (11, 11)

xy \ zw	00	01	11	10
00	X	1	0	1
01	0	1	X	0
11	1	X	X	0
10	X	0	0	X

3. The Boolean function $x'y' + xy + x'y$ is equivalent to **[GATE - 04]**

a) $x' + y'$ b) x+y c) $x + y'$ d) $x' + y$

4. In an SR latch made by cross-coupling two NAND gates, if both S and R inputs are set to 0, then it will result in

a) Q = 0, Q′ = 1 b) Q = 1, Q′ = 0 **[GATE - 04]**
c) Q =1, Q′ = 1 d) Indeterminate states

5. A circuit outputs a digit in the form of 4 bits. 0 is represented by 0000, 1 by 0001, ..., 9 by 1001. A combinational circuit is to be designed which takes these 4 bits as input and outputs 1 if the digit ≥ 5, and 0 otherwise. If only AND, OR and NOT gates may be used, what is the minimum number of gates required ? **[GATE - 04]**

a) 2 b) 3 c) 4 d) 5

6. Which are the essential prime implicants of the following Boolean function ? **[GATE - 04]**

$$f(a,b,c) = a'c + ac' + b'c$$

a) a′c and ac′ b) a′c and b′c
c) a′c only d) ac′ and bc′

7. Consider a multiplexer with X and Y as data inputs and Z as control input. Z=0 selects input X, and Z=1 selects input Y. What are the connections required to realize the 2-variable Boolean function f=T+R, without using any additional hardware ? **[GATE - 04]**

a) R to X, 1 to Y, T to Z b) T to X, R to Y, T to Z
c) T to X, R to Y, 0 to Z d) R to X, 0 to Y, T to Z

8. Consider the partial implementation of a 2-bit counter using T flip-flops following the sequence 0-2-3-1-0, as shown below. **[GATE - 04]**

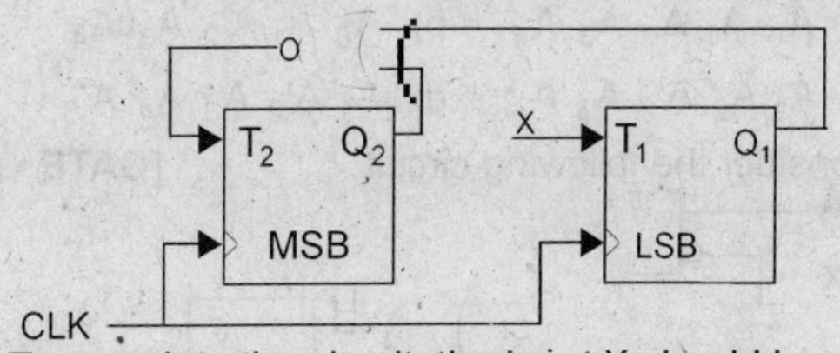

To complete the circuit, the input X should be

a) Q_2' b) $Q_2 + Q_1$
c) $(Q_1 \oplus Q_2)'$ d) $Q_1 \oplus Q_2$

9. A 4-bit carry lookahead adder, which adds two 4-bit numbers, is designed using AND, OR, NOT, NAND, NOR gates only. Assuming that all the inputs are available in both complemented and uncomplemented forms and the delay of each gate is one time unit, what is the overall propagation delay of the adder ? Assume that the carry network has been implemented using two-level AND-OR logic.

a) 4 time units b) 6 time units **[GATE - 04]**
c) 10 time units d) 12 time units

10. What is the minimum number of NAND gates required to implement a 2-input EXCLUSIVE-OR function without using any other logic gate? **[GATE - 04]**

a) 3 b) 4 c) 5 d) 6

11. The function $AB'C + A'BC + ABC' + A'B'C + AB'C'$ is equivalent to **[GATE - 04]**

a) $AC' + AB + A'C$ b) $AB' + AC' + A'C$
c) $A'B + AC' + AB'$ d) $A'B + AC + AB'$

12. Consider the following circuit.

Which one of the following is TRUE ? **[GATE - 05]**

a) f is independent of X
b) f is independent of Y
c) f is independent of Z
d) None of X,Y,Z is redundant

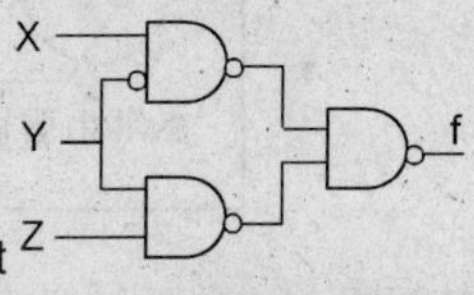

13. The switching expression corresponding to $f(A,B,C,D) = \Sigma\,(1,4,5,9,11,12)$ **[GATE - 05]**

a) $BC'D' + A'C'D + AB'D$ b) $ABC' + ACD + B'C'D$
c) $ACD' + A'BC' + AC'D'$ d) $A'BD + ACD' + BCD'$

14. Consider the following circuit involving a positive edge triggered DFF. **[GATE - 05]**

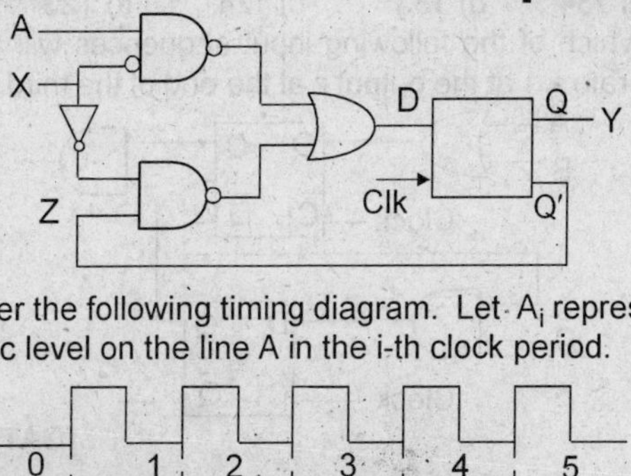

Consider the following timing diagram. Let A_i represent the logic level on the line A in the i-th clock period.

Let A' represent the complement of A. The correct output sequence on Y over the clock periods 1 through 5 is

a) $A_0\ A_1\ A'_1\ A_3\ A_4$ b) $A_0\ A_1\ A'_2\ A_3\ A_4$

c) $A_1\ A_2\ A'_2\ A_3\ A_4$ d) $A_1\ A'_2\ A_3\ A_4\ A'_5$

15. Consider the following circuit **[GATE - 05]**

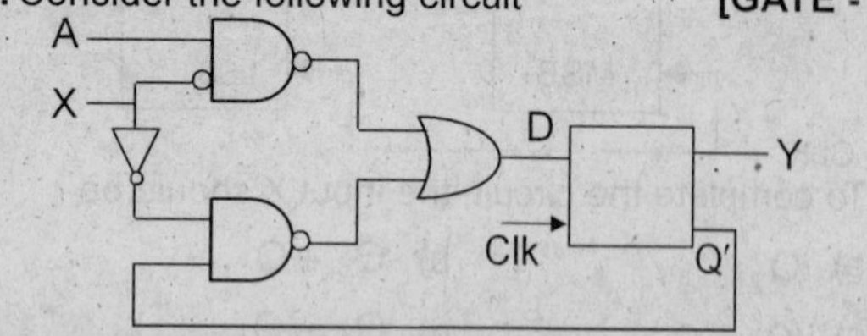

The flip-flops are positive edge triggered D FFs. Each state is designated as a two bit string Q_0Q_1. Let the initial state be 00. The state transition sequence is :

a) $00 \rightarrow 11 \rightarrow 01$ b) $00 \rightarrow 11$

c) $00 \rightarrow 10 \rightarrow 01 \rightarrow 11$ d) $00 \rightarrow 11 \rightarrow 01 \rightarrow 10$

16. Which of the following expressions is equivalent to $(A \oplus B) \oplus C$ **[GATE - 05]**

a) $(A+B+C)\left(\overline{A}+\overline{B}+\overline{C}\right)$ b) $(A+B+C)\left(\overline{A}+\overline{B}+C\right)$

c) $ABC+\overline{A}(B\oplus C)+\overline{B}(A\oplus C)$

d) None of the above

17. A two-way switch has three terminals a,b and c. In ON position (logic value 1), a is connected to b, and in OFF position, a is connected to c. Two of these two way switches S1 and S2 are connected to a bulb as shown below.

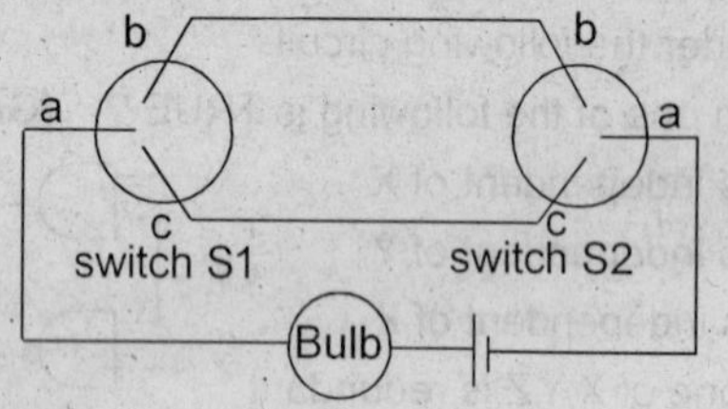

Which of the following expressions, if true, will always result in the lighting of the bulb ? **[GATE - 05]**

a) $S1.\overline{S2}$ b) $S1+S2$ c) $\overline{S1\oplus S2}$ d) $S1\oplus S2$

18. How many pulses are needed to change the contents of a 8-bit upcounter from 10101100 to 00100111 (rightmost bit is the LSB)? **[GATE - 05]**

a) 134 b) 133 c) 124 d) 123

19. Which of the following input sequences will always generate a 1 at the output z at the end of the third cycle?

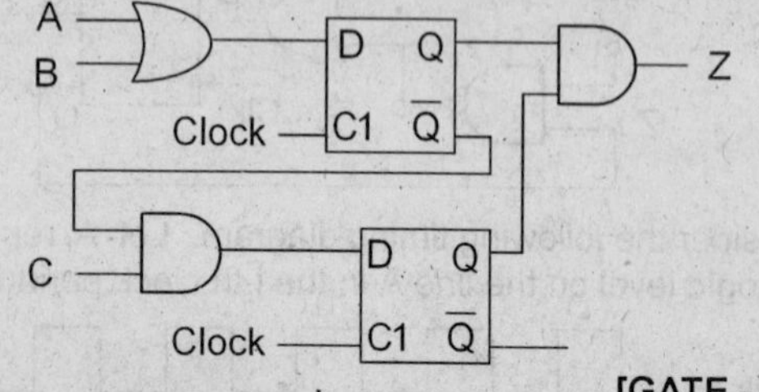

[GATE - 05]

(a)	A	B	C
	0	0	0
	1	0	1
	1	1	1

(b)	A	B	C
	1	0	1
	1	1	0
	1	1	1

(c)	A	B	C
	0	1	1
	1	0	1
	1	1	1

(d)	A	B	C
	0	0	1
	1	1	0
	1	1	1

20. The circuit shown below implements a 2-input NOR gate using two 2-1 MUX (control signal 1 selects the upper input). What ar the values of signals x, y and z ?

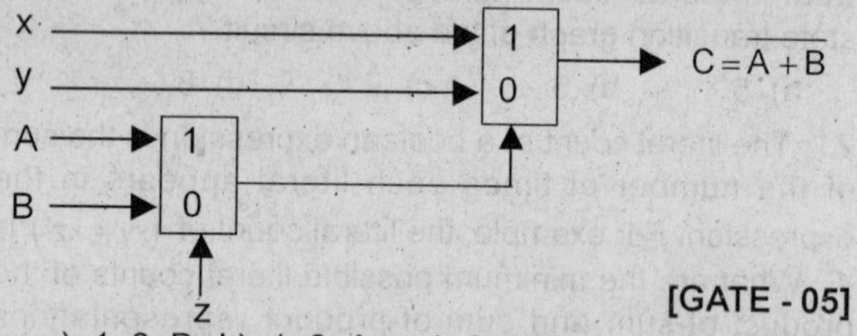

[GATE - 05]

a) 1, 0, B b) 1, 0, A c) 0, 1, B d) 0, 1, A

ANSWER KEY

1. c	**2.** a	**3.** d	**4.** d	**5.** c
6. a	**7.** a	**8.** a	**9.** a	**10.** c
11. b	**12.** a	**13.** a	**14.** a	**15.** d
16. c	**17.**	**18.** b	**19.** b,c	**20.** a

EXPLANATORY NOTES

1. $z_k = 00.........k$

$n_k = 1+......k$

$z_k + n_k = k$

Number of transition states $= 2^3 - 1 = 7$

2. When we take sum of Product

literal = $wy + x\overline{z}\,\overline{w} + x\,\overline{z}\,\overline{y} + \overline{w}\,\overline{x}\,\overline{y}$

Total number of literal count = 9

xy \ zw	00	01	11	10
00	x	1	0	1
01	0	1	x	0
11	1	x	x	0
10	x	0	0	x

When we take product of Sum

literal $= (z+w+x)\,(\overline{z}+\overline{w})(\overline{z}+\overline{y})\,(\overline{x}+y)$

Total number of literals = 11

Minimum possible literal count is (11, 9)

xy \ zw	00	01	11	10
00	x	1	0	1
01	0	1	x	0
11	1	x	x	0
10	x	0	0	x

3. $x'y' + xy + x'y$

$= x'(y'+y) + xy = x' + xy$

$= (x'+x)(x'+y) = x' + y$

4.

Input		Output
S	R	
0	0	indeterminate
1	0	1
0	1	0
1	1	? ⟶ Indeterminate

Correct answer is (d).

5. Output in the form of 4 bit and Output is 1 when input is greater than 5, i.e. Output will be 1 in case of

0101 - 5 0110 - 6 0111 - 7 1000 - 8 1001 - 9

This can be implemented with the help of 4 logic gates of type AND, OR and NOT.

6. The K-map of given function is

c \ ab	00	01	11	10
0			1	1
1	1	1		1

The essential prime implicant of this boolean function is

001, 011 and 110, 100 which is ac' and $a'c$

8. The flip - flop shown are T-type. To implement this as a 2-bit counter, the input X should be input of X - OR of two input Q_1 and Q_2. Hence $Q_1 \oplus Q_2$

9. We are adding two 4-bit numbers and delay of each gate is one time unit. The delay of adder is {4 × delay in one gate} Unit = 4 × 1 = 4 time unit

10. $A \oplus B = \overline{A}B + \overline{B}A$

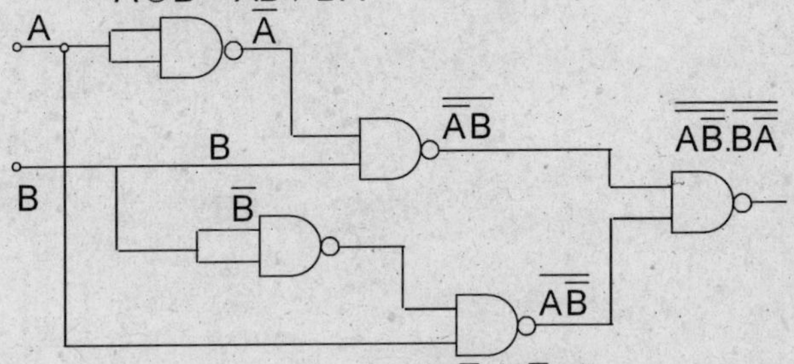

$= A\overline{B} + B\overline{A}$ (De Morgan's law)

Hence five NAND gates are used.

11. $AB'C + A'BC + ABC' + A'B'C + AB'C'$

$= AB'(C + C') + A'C(B + B') + ABC'$

$= AB' + A'C + ABC' = A(B' + BC') + A'C$

$= A(B' + B)(B' + C') + A'C = A(B' + C') + A'C = AB' + AC' + A'C$

12. $\overline{\overline{X\overline{Y}}.\ YZ} = F$

$X\overline{Y} + \overline{YZ}$

$X\overline{Y} + \overline{Y} + \overline{Z}$

$\overline{Y} + \overline{Z}$

13. Given : $f(A, B, C, D) = \Sigma(1, 4, 5, 9, 11, 12)$

AB \ CD	00	01	11	10
00		1		
01	1	1		
11	1			
10		1	1	

$= \overline{A}\ \overline{C}D + B\overline{C}\ \overline{D} + A\overline{B}D$

14. $D = A.X + \overline{X}\ \overline{Q}$

Consider for first two cases,

$\overline{X}\ \overline{Q} = 0$

because X = 1, So $0.\overline{Q} = 0$

So, answer depend only on A. X

For first clock pulse '0'

$\Rightarrow D = A_0$

For clock pulse '1'

$\Rightarrow$ output $= A_0$

For clock pulse '1'

$\Rightarrow D =$ again A_1

For clock pulsel '2'

$\Rightarrow$ ouput $= A_1$

For clock pulse '2'

$D = \overline{A}_1$

because $A.X = 0$ and $X = 0$

So, $\overline{X} = 1$

$\overline{X}\ \overline{Q} = \overline{Q}$

15. $D_1 = Q_0 \oplus \overline{Q_1}$

$D_1 = \overline{Q}_0$

Q_0	Q_1	Next state	
0	0	1	1
1	1	0	1
0	1	1	0
1	0	0	0

16. $(A \oplus B) \oplus C$

$(A\overline{B} + B\overline{A}) \oplus C$

$\overline{(A\overline{B} + B\overline{A})}C + \overline{C}(A\overline{B} + B\overline{A})$

$(\overline{A\overline{B}})(\overline{B\overline{A}})C + \overline{C}(A\overline{B} + B\overline{A})$

$(\overline{A} + B)(\overline{B} + A)C + \overline{C}(A\overline{B} + B\overline{A})$

$(\overline{A}\overline{B} + BA)C + \overline{C}A\overline{B} + \overline{C}B\overline{A}$

$\overline{A}\ \overline{B}C + ABC + A\overline{B}\ \overline{C} + \overline{A}\ \overline{C}B$

$ABC + \overline{A}(B \oplus C) + \overline{B}(A \oplus C)$

$ABC + \overline{A}(\overline{B}C + \overline{C}B) + \overline{B}(\overline{A}C + \overline{C}A)$

$ABC + \overline{A}\ \overline{B}C + \overline{A}B\overline{C} + \overline{B}\ \overline{A}C + A\overline{B}\ \overline{C}$

$ABC + \overline{A}\ \overline{B}C + \overline{A}B\overline{C} + A\overline{B}\ \overline{C}$

18. Take the binary value of both numbers and subtract, required answer is obtained.

19.
$$Q_1 = A + B$$
$$Q_2 = \overline{Q}_1 . C$$
$$Z = Q_1 . Q_2$$

So for producing one at end of third clock cycle Q_1 and Q_2 must be one.

		1	2	3
(a)	DA	0	1	1
	DB	0	1	0
(b)	DA	1	1	1
	DB	1	0	1
(c)	DA	1	1	1
	DB	1	0	0
(d)	DA	0	1	1
	DB	1	0	1

So, here are two possible solutions.

20. $\overline{(A+B)}$

A	B	A + B	$\overline{A+B}$
0	0	0	1
0	1	1	0
1	0	1	0
1	1	1	0

x = 1, y = 0, z = B

Solving equation, we get

$$Z = A.\overline{B} + B$$
$$C = 1\left(\overline{A.\overline{B} + B}\right) + 0\left(A.\overline{B} + B\right)$$
$$= \overline{A.\overline{B}}.\overline{B} = \left(\overline{A} + B\right).\overline{B} = \overline{A}.\overline{B}$$

and $\left(\overline{A+B}\right) = \overline{A}.\overline{B}$ both are equal

Unit - V
Computer Organisation

UNIT - V COMPUTER ORGANISATION

I. NUMBER REPRESENTATION AND ARITHMETIC

- Binary Number System
- Octal Number System
- Hexadecimal System
- Codes
- Representation of Negative Numbers
- Fixed Point Representation
- Floating Point Representation

1. Number Systems

In general, in any number system there is an ordered set of symbols known as digits with rules defined for performing arithmetic operations like addition, multiplication, etc. A collection of these digits makes a number which in general has two parts—integer and fractional, set apart by a radix point (.), that is

$$(N)_b = \underbrace{d_{n-1}d_{n-2}.....d_i.....d_1\, d_0}_{\text{Integer portion}} \;.\; \underbrace{d_{-1}d_{-2.....}\,d_{-f}......d_{-m}}_{\text{Fractional portion}} \quad (1.1)$$

(the "." is the Radix point)

where N = a number

b = radix or base of the number system

n = number of digits in integer portion

m = number of digits in fractional portion

d_{n-1} = most significant digit (msd)

d_{-m} = least significant digit (lsd)

and $0 \leq (d_i \text{ or } d_{-j}) \leq b-1$

The digits in a number are placed side by side and each position in the number is assigned a weight or index of importance by some predesigned rule. Table 1.1 gives the details of commonly used number systems

Number system	Base or Radix (b)	Symbols used (d_i or d_{-f})	weight assigned to position (i)	(-f)	Example
Binary	2	0,1	2^i	2^{-f}	1011.11
Octal	8	0,1,2,3,4,5,6,7	8^i	8^{-f}	3567.25
Decimal	10	0,1,2,3,4,5,6,7 8,9	10^i	10^{-f}	3974.57
Hexadecimal	16	0,1,2,3,4,5,6,7 8,9,A,B,C,D,E,F	16^i	16^{-f}	3FA9.56

Table 1.1 Characteristics of commonly used number systems

2. Binary Number System

The number system with base (or radix) two is known as the binary number system. Only two symbols are used to represent numbers in this system and these are 0 and 1. These are known as bits. This system has the minimum base (0 is not possible and 1 is not useful). It is a positional system, that is every position is assigned a specific weight.

Table 1.2 illustrates counting in binary number system. The corresponding decimal numbers are given in the right-hand column. Similar to decimal number system, the left-most bit is known as the most significant bit (MSB) and the right-most bit is known as the least significant bit (LSB). Any number of 0s can be added to the left of the number without changing the value of the number . In the binary number system, a group of four bits is known as a nibble, and a group of eight bits is known as a byte.

Table1. 24 - bit binary number sand their corresponding decimal numbers

Binary number				Decimal number	
B3	B2	B1	B0	D1	D0
0	0	0	0	0	0
0	0	0	1	0	1
0	0	1	0	0	2
0	0	1	1	0	3
0	1	0	0	0	4
0	1	0	1	0	5
0	1	1	0	0	6
0	1	1	1	0	7
1	0	0	0	0	8
1	0	0	1	0	9
1	0	1	0	1	0
1	0	1	1	1	1
1	1	0	0	1	2
1	1	0	1	1	3
1	1	1	0	1	4
1	1	1	1	1	5

3. Binary-to-Decimal Conversion

Any binary number can be converted into its equivalent decimal number using the weights assigned to each bit position as given in Table 1.1.

Ex : Determine the decimal numbers represented by the following binary numbers :

110101

Sol:

$(110101)_2 = 1 \times 2^5 + 1 \times 2^4 + 0 \times 2^3 + 1 \times 2^2 + 0 \times 2^1 + 1 \times 2^0$

$= 32 + 16 + 0 + 4 + 0 + 1 = (53)_{10}$

Ex : Determine the decimal numbers represented by the following binary number :

101101.10101

Sol:

$(101101.10101)_2 = 1 \times 2^5 + 0 \times 2^4 + 1 \times 2^3 + 1 \times 2^2 + 0 \times 2^1 + 1 \times 2^0 + 1 \times 2^{-1} + 0 \times 2^{-2} + 1 \times 2^{-3} + 0 \times 2^{-4} + 1 \times 2^{-5}$

$= 32 + 0 + 8 + 4 + 0 + 1 + \frac{1}{2} + 0 + \frac{1}{8} + 0 + \frac{1}{32}$

$= (45.65625)_{10}$

4. Decimal-to-Binary Conversion

Any decimal number can be converted into its equivalent binary number. For integers, the conversion is obtained by continuous division by 2 and keeping track of the remainders, while for fractional parts, the conversion is affected by continuous multiplication by 2 and keeping track of the integers generated. The conversion process is illustrated by the following examples.

Ex : Convert $(13)_{10}$ to an equivalent base-2 number.

Sol :

	Quotient	Remainder
$\frac{13}{2}$	6	1
$\frac{6}{2}$	3	0
$\frac{3}{2}$	1	1
$\frac{1}{2}$	0	1

1 1 0 1

Thus $(13)_{10} = (1101)_2$

Ex : Convert $(0.65625)_{10}$ to an equivalent base-2 number.

Sol :

0.65625	0.31250	0.62500	0.25000	0.50000
× 2	× 2	× 2	× 2	× 2
1.31250	0.62500	1.25000	0.50000	1.00000
1	0	1	0	1

Thus $(0.65625)_{10} = (0.10101)_2$

Ex : Express the following decimal number in the binary form

25.5

Sol : Integer part

	Quotient	Remainder
$\frac{25}{2}$	12	1
$\frac{12}{2}$	6	0
$\frac{6}{2}$	3	0
$\frac{3}{2}$	1	1
$\frac{1}{2}$	0	1

1 1 0 0 1

Therefore, $(25)_{10} = (11001)_2$

Fractional part 0.5
× 2
1.0
↓
1

i.e, $(0.5)_{10} = (0.1)_2$

Therefore $(25.5)_{10} = (11001.1)_2$

5. Binary Arithematic

We all are familiar with the arithmetic operations such as addition, subtraction, multiplication, and division of decimal numbers. Similar operations can be performed on binary numbers; infact, binary arithmetic is much simpler than decimal arithmetic because here only two digits, 0 and 1 are involved.

Binary Addition

The rules of binary addition are given in Table 1.3

Table 1.3 Rules of binary addition

Augend	Addend	Sum	Carry	Result
0	0	0	0	0
0	1	1	0	1
1	0	1	0	1
1	1	0	1	10

In the first three rows above, there is no carry, that is, carry = 0, whereas in the fourth row a carry is produced (since the largest digit possible is 1), that is carry = 1, and similar to decimal addition it is added to the next higher binary position.

Ex : Add the binary number :

0101 and 1111

Sol :

		(1)	(1)	(1)	← carry
		0	1	0	1
+	1	1	1	1	
1	0	1	0	0	

↑
carry

From the above example, we observe the following :

I. if the number of 1's to be added in a column is even then the sum bit is 0, and if the number of 1's to be added in a column is odd then the sum bit is 1.

II. Every part of 1's in a column produces a carry (1) to be added to the next higher bit column.

Binary Subtraction

The rules of binary subtraction are given in Table 1.4

Table1.4 Rules of binary subtraction

Minuend	Subtrahend	Difference	Borrow
0	0	0	0
0	1	1	1
1	0	1	0
1	1	0	0

Except in the second row above, the borrow = 0. When the borrow = 1, as in the second row, this is to be subtracted from the next higher binary bit as it is done in decimal subtraction.

Ex : Perform the following subtraction :

```
   1 0   1 1
-  0 1   1 0
```

Sol :

```
                    column 4
                       3
                       2
                       1
    1    0    1    1      Minuend
(-) 0    1    1    0      Subtrahend
    0    1    0    1      Difference
```

Here, in columns 1 and 2, borrow = 0 and in column 3 it is 1. Therefore, in column 4 first subtract 0 from 1 and from this result obtained subtract the borrow bit.

Binary Multiplication

Binary multiplication is similar to decimal multiplication. In binary, each partial product is either zero (multiplication by 0) or exactly same as the multiplicand (multiplication by 1). An example of binary multiplication is given below:

Ex : Multiply 1001 by 1101.

Sol :

```
      1001         Multiplicand
    × 1101         Multiplier
    -------
      1001   } I
      0000   } II  Partial Products
      1001   } III
      1001   } IV
    -------
   1110101         Final Product
```

In a digital circuit, the multiplication operation is performed by repeated additions of all partial products to obtain the full product.

Binary Division

Binary division is obtained using the same procedure as decimal division. An example of binary division is given below :

Ex 1.10 : Divide 1 1 1 0 1 0 1 by 1 0 0 1

Sol :

```
                      1101      ← Quotient
divisor →   1001)   1110101     ← Dividend
                    1001
                    1011
                    1001
                    001001
                      1001
                      0000
```

Ans: 1101

6. Octal Number System : The number system with base (or radix) eight is known as the octal number system. In this system, eight symbols, 0,1,2,3,4,5,6 and 7 are used to represent numbers. Similar to decimal and binary number systems, it is also a positional system and has, in general, two parts: integer and fractional, set apart by a radix (octal) point (.) . Any number can be expressed in the form of Eq.(1.1) with b = 8 and $0 \le (d_i or\, d_{-1}) \le 7$. The weights assigned to the various positions are given in Table 1.1,For example, $(6327.4051)_8$ is an octal number. Using the weights given in Table 1.1, it can be written as

$$(6327.4051)_8 = 6 \times 8^3 + 3 \times 8^2 + 2 \times 8^1 + 7 \times 8^0 + 4 \times 8^{-1} + 0 \times 8^{-2} + 5 \times 8^{-3} + 1 \times 8^{-4}$$

$$= 3072 + 192 + 16 + 7 + \frac{4}{8} + 0 + \frac{5}{512} + \frac{1}{4096}$$

$$= (3287.5100098)_{10}$$

Using the above procedure, an octal number can be converted into an equivalent decimal number or a base-8 number can be converted into an equivalent base-10 number.

The conversion from decimal to octal (base-10 to base-8) is similar to the conversion procedure for base-10 to base-2 conversion. The only difference is that number 8 is used in place of 2 for division in the case of integers and for multiplication in the case of fractional numbers.

Ex : Convert $(3287.5100098)_{10}$ into octal

Sol : Integer part

	Quotient	Remainder
3287/8	410	7
410/8	51	2
51/8	6	3
6/8	0	6

6 3 2 7

Thus $(3287)_{10} = (6327)_8$

Fractional part

0.5100098	0.0800784	0.6406272	0.1250176
× 8	× 8	× 8	× 8
4.0800784	0.6406272	5.1250176	1.0001408
↓	↓	↓	↓
4	0	5	1

Thus $(0.5100098)_{10} \approx (0.4051)_8$

Therefore, $(3287.5100098)_{10} = (6327.4051)_8$

7. Octal-to-Binary Conversion

Octal numbers can be converted into equivalent binary numbers by replacing each octal digit by its 3-bit equivalent binary. Table 1.5 gives octal numbers and their binary equivalents for decimal numbers 0 to 15.

Ex : Convert $(736)_8$ into an equivalent binary number.

Sol : From Table 1.5, we observe the binary equivalents of 7,3 and 6 as 111, 011 and 110, respectively.

Therefore, $(736)_8 = (111\ 011\ 110)_2$

8. Binary-to-Octal Conversion

Binary numbers can be converted into equivalent octal numbers by making groups of three bits starting from LSB and moving towards MSB for integer part of the number and then replacing each group of three bits by its octal representation. For fractional part, the groupings of three bits are made starting from the binary point.

Ex : Convert $(1001110)_2$ to its octal equivalent.

Sol : $(1001110)_2 = (\underline{001}\ 001\ 110)_2$
$= (1\ 1\ 6)_8$

Ex : Convert $(0.10100110)_2$ to its equivalent octal number.

Sol: $(0.10100110)_2 = (0.101\ 001\ 100)_2$
$= (0.514)_2$

From the above examples we observe that in forming the 3-bit groupings, 0's may be required to complete the first (most significant digit) group in the integer part and the last (least significant digit) group in the fractional part.

9. Octal Arithmetic

Octal arithmetic rules are similar to the decimal or binary arithmetic. Normally, we are not interested in performing octal arithmetic operations using octal representation of numbers. This number system is normally used to enter long strings of binary data into a digital system like a microcomputer. This makes the task of entering binary data in a microcomputer easier. Arithmetic operations can be performed by converting the octal numbers and then using the rules of binary arithmetic.

Ex : Add $(23)_8$ and $(67)_8$.

Sol :
23 = 010011
(+) 67 = 110111
$(112)_8$ = 1001 010

Ex : Subtract (a) $(37)_8$ from $(53)_8$
(b) $(75)_8$ from $(26)_8$

Sol : Using 8-bit representation

(a) $(53)_8$ = 00101011
$-(37)_8$ = (+) 11100001 Two's complement of $(37)_8$
$(14)_8$ = 100 001 100
Discard carry

(b) $(26)_8$ = 00010110
$-(75)_8$ = (+)11000011 Two's complement of $(75)_8$
$-(47)_8$ = 110110 01 Two's complement of result

Two's Complement of 110110 01= 00 100 111 = $(47)_8$

Multiplication and division can also be performed using the binary representation of octal numbers and then making use of multiplication and division rules of binary numbers.

10. Hexadecimal Number System

Hexadecimal number system is very popular in computer uses. The base for hexadecimal number system is 16 which requires 16 distinct symbols to represent the numbers. These are numerals 0 through 9 and alphabets A through F.

Since numeric digits and alphabets both are used to represent the digits in the hexadecimal number system, therefore, this is an alphanumeric number system. Table 1.5 gives hexadecimal numbers with their binary equivalents for decimal numbers 0 through 15. From the table, it is observed that there are 16 combinations of 4-bit binary numbers and sets of 4-bit binary numbers can be entered in the computer in the form of hexadecimal (hex.) digits. These numbers are required to be converted into binary representation, using hexadecimal-to-binary converter circuits before these can be processed by the digital circuits.

Table 1.5 Binary and decimal equivalents of hexadecimal numbers

Hexadecimal	Decimal	Binary
0	0	0000
1	1	0001
2	2	0010
3	3	0011
4	4	0100
5	5	0101
6	6	0110
7	7	0111
8	8	1000
9	9	1001
A	10	1010
B	11	1011
C	12	1100
D	13	1101
E	14	1110
F	15	1111

11. Hexadecimal-to-Decimal Conversion

Using Eq. (1.1), hexadecimal numbers can be converted to their equivalent decimal numbers.

Ex : obtain decimal equivalent of hexadecimal number $(3A.2F)_{16}$

Sol : Using Eq.(1.1),

$$(3A.2F)_{16} = 3 \times 16^1 + 10 \times 16^0 + 2 \times 16^{-1} + 15 \times 16^{-2}$$

$$= 48 + 10 + \frac{2}{16} + \frac{15}{16^2} = (58.1836)_{10}$$

The factional part may not be an exact equivalent and therefore, may give a small error.

12. Decimal-to-Hexadecimal Conversion

For conversion from decimal to hexadecimal, the procedure used in binary as well as octal systems is applicable, using 16 as the dividing (for integer part) and multiplying (for fractional part) factor.

Ex : Convert the following decimal number into hexadecimal number. 675.625

Sol : Integer part

	Quotient	Remainder
675/16	42	3
42/16	2	10
2/16	0	2

2 A 3

Thus $(675)_{10} = (2A3)_{16}$

Fractional part 0.625

× 16

10.000

A Thus, $(0.625)_{10} = (0.A)_{16}$

Therefore $(675.625)_{10} = (2A3.A)_{16}$

13. Hexadecimal-to-Binary Conversion

Hexadecimal numbers can be converted into equivalent binary numbers by replacing each hex digit by its equivalent 4-bit binary number.

Ex : Convert $(2F9A)_{16}$ to equivalent binary number.

Sol : Using Table 1.5 find the binary equivalent of each hex digit.

$$(2F9A)_{16} = (0010\ 1111\ 1001\ 1010)_2$$

$$= (0010111110011010)_2$$

14. Binary-to-Hexadecimal Conversion

Binary numbers can be converted into the equivalent hexadecimal numbers by making groups of four bits starting from LSB and moving towards MSB for integer part and then replacing each group of four bits by its hexadecimal representation.

For the fractional part, the above procedure is repeated starting from the bit next to the binary point and moving towards the right.

Ex : Convert the following binary numbers to their equivalent hex numbers.

(a) 10100110101111 (b) 0.00011110101101

Sol :

(a) $(10100110101111)_2$ = 0010 1001 1010 1111

2 9 A F

$$\therefore (10100110101111)_2 = (29.AF)_{16}$$

(b) 0.00011110101101= 0.0001 1110 1011 0100

1 E B 4

$$\therefore (0.00011110101101)_2 = (0.1EB4)_{16}$$

15. Conversion from Hex-to-Octal and Vice-Versa

Hexadecimal numbers can be converted to equivalent octal numbers and octal numbers can be converted to equivalent hex numbers by converting the hex/octal number to equivalent binary and then to octal/hex, respectively.

Ex : Convert the following hex numbers to octal numbers.

(a) A72E (b) 0.BF85

Sol :

(a) $(A72E)_{16}$ = $(1010\ 0111\ 0010\ 1110)_2$

= 001 010 011 100 101 110

$=(123456)_8$

(b) $(0.BF85)_{16}$ = $(0.1011111110000101)_2$

= 0.101 111 111 000 010 100

$=(0.577024)_8$

Ex : Convert $(247.36)_8$ to equivalent hex number.

Sol :

$(247.36)_8 = (010\ 100111.011110)_2$

$= (0\ 1010\ 0111.0111\ 1000)_2$

$= (A7.78)_{16}$

16. Hexadecimal Arithmetic

The rules for arithmetic operations with hexadecimal numbers are similar to the rules for decimal, octal and binary systems.The information can be handled only in binary form in a digital circuit and it is easier to enter the information using hexadecimal number system. Since arithmetic operations are performed by the digital circuits on binary numbers, therefore hexadecimal numbers are to be first converted into binary numbers. Arithmetic operations will become clear from the following examples.

Ex : Add $(7F)_{16}$ and $(BA)_{16}$

Sol :

	7F	= 01111111
	(+)BA	= 10111010
	$(139)_{16}$	= 10011 1001

Ex : Subtract (a) $(5C)_{16}$ from $(3F)_{16}$

(b) $(7A)_{16}$ from $(C0)_{16}$

Sol : (a) 3F = 00111111

5C =(+)10100100 Two's complement of $(5C)_{16}$

-1D =11100011 Two's complement of result

Two's complement of 1100011 = 0001 1101 = $(1D)_{16}$

(b) C0 = 11000000

-7A = (+)10000110 Two's complement of $(7A)_{16}$

46 = 10100 0110

Discard carry ↑

Multiplication and division can also be performed using the binary representation of hexadecimal numbers and then making use of multiplication and division rules of binary numbers.

17. Codes

Computers and other digital circuits process data in the binary format. Various binary codes are used to represent data which may be numeric, alphabets or special characters. Although, in every code used the information is represented in binary form, ten interpretation of this binary information is possible only if the code in which this information is available is known. For example, the binary number 1000001 represents 65 (decimal) in straight binary, 41 (decimal) in BCD and alphabet A in ASCII code. A user must be very careful about the code being used while interpreting information available in the binary format. Codes are also used for error detection and error correction in digital systems.

Some of the commonly used codes are given below.

Straight Binary Code

This is used to represent numbers using natural (or straight) binary form . Various arithmetic operations can be performed in this form. Binary codes for decimal numbers 0 to 15 are given in Table 1.6.

Natural BCD Code

In this code, decimal digits 0 through 9 are represented (coded) by their natural binary equivalents using four bits and each decimal digit of a decimal number is represented by this four bit code individually. For example, $(23)_{10}$ is represented by 0010 0011 using BCD code, rather than $(10111)_2$ .From this it is observed that it requires more number of bits to code a decimal number using BCD code than using the straight binary code. However, in spite of this disadvantage it is very convenient and useful code for input and output operations in digital systems.

This code is also known as 8-4-2-1 code or simply BCD code, 8,4,2 and 1 are the weights of the four bits of the binary code of each decimal digit similar to straight binary number system.Therefore, this is a weighted code and arithmetic operations can be performed using this code. BCD codes for decimal digits 0 through 9 are given in Table 1.6.

Table 1.6 Various binary codes

Decimal Number	Binary $B_3\ B_2\ B_1\ B_0$	BCD D C B A	Exess-3 $E_3\ E_2\ E_1\ E_0$	Gray $G_3\ G_2\ G_1\ G_0$
0	0 0 0 0	0 0 0 0	0 0 1 1	0 0 0 0
1	0 0 0 1	0 0 0 1	0 1 0 0	0 0 0 1
2	0 0 1 0	0 0 1 0	0 1 0 1	0 0 1 1
3	0 0 1 1	0 0 1 1	0 1 1 0	0 0 1 0
4	0 1 0 0	0 1 0 0	0 1 1 1	0 1 1 0
5	0 1 0 1	0 1 0 1	1 0 0 0	0 1 1 1
6	0 1 1 0	0 1 1 0	1 0 0 1	0 1 0 1
7	0 1 1 1	0 1 1 1	1 0 1 0	0 1 0 0
8	1 0 0 0	1 0 0 0	1 0 1 1	1 1 0 0
9	1 0 0 1	1 0 0 1	1 1 0 0	1 1 0 1
10	1 0 1 0			1 1 1 1
11	1 0 1 1			1 1 1 0
12	1 1 0 0			1 0 1 0
13	1 1 0 1			1 0 1 1
14	1 1 1 0			1 0 0 1
15	1 1 1 1			1 0 0 0

Excess-3 Code

This is another form of BCD code, in which each decimal digit is coded into a 4-bit binary code. The code for each decimal digit is obtained by adding decimal 3 to the natural BCD code of the digit. For example, decimal 2 is coded as 0010+0011 = 0101 in Excess-3 code. It is not a weighted code. This code is a self-complementing code, which means 1's complement of the coded number yields 9's complement of the number itself. For example, Excess-3 code of decimal 2 is 0101, its 1's complement is 1010 which is Excess-3 code for decimal 7, which is 9's complement of 2. the self complementing property of this code helps considerably in performing subtraction operation in digital systems. Excess-3 codes for decimal digits 0 through 9 are given in Table 1.6

Gray Code

It is a very useful code in which a decimal number is represented in binary form in such a way so that each Gray-code number differs from the preceding and the succeeding number by a single bit. For example, the Gray code for decimal number 5 is 0111 and for 6 is0101.

These two codes differ by only one bit position (third from the left). This code is used extensively for shaft encoders because of this property. It is not a weighted code. The gray code is a reflected code and can be constructed using this property as given below.

I. A 1-bit Gray code has two code words 0 and 1 representing decimal numbers 0 and 1 respectively

II. An n-bit ($n \geq 2$) Gray code will have first 2^{n-1} Gray codes of (n-1) bits written in order with a leading 0 appended.

III. The last 2^{n-1} Gray codes will be equal to the Gray code words of an (n-1)-bit Gray code, written in reverse order (assuming a mirror placed between first 2^{n-1} and last 2^{n-1} Gray codes) with a leading 1 appended.

Octal Code

It is a 3-bit binary code, in which each of the octal digits 0 through 7 is coded into 3-bit straight binary number. For example, code for octal digit 4 is 100. Using this code, octal numbers can be coded into straight binary form or the binary numbers can be represented by octal numbers.

This code is used for binary inputs in digital computers, microprocessors, etc.

Hexa Decimal Code: It is a 4-bit binary code used for input/output in digital computers. microprocessors, etc.

Ex : Represent the decimal number 27 in binary form using

I. Binary Code **II.** BCD code **III.** Excess-3 code
IV. Gray code **V.** Octal code **VI.** Hexadecimal code

Sol :

I. The decimal number is converted into straight binary form. Its value is 11011

II. Each digit of the decimal number is coded using 4-bit BCD code as given 0010 0111

III. Each digit of the decimal number is coded using 4-bit Excess-3 code as given 0101 1010

IV. 5 bits are required to represent 27, therefore, 5-bit Gray code is constructed and 27 is represented as 10110.

V. $(27)_{10} = (33)_8 = 011\ 011$

VI. $(27)_{10} = (1B)_{16} = 0001\ 1011$

18. Representation of Negative Numbers

The 2's complement is used to represent the negative of a binary number. Before understanding 2's complement in the binary number system. Let us understand 9's and 10's complement in the decimal number system.

9's Complement.

To obtain the 9's complement of a decimal number each digit of the number is subtracted from 9.

For example, the 9's complement of 45 is (99 – 45) = 54,

The 9's complement of 523 is (999 – 523) = 476.

10's Complement

The 10's complement of a decimal number = its 9's complement +1.

The 10's complement of 45 = 54 +1 = 55

The 10's complement of 523 = 476 + 1 = 477.

Now let us add the number and its 10's complement and examine the sum.

Ex :

```
    4   5     (decimal number)
  + 5   5     (its 10's complement)
  ---------
    0   0
  carry = 1
```

In this case the decimal number is of two digits. If we consider the sum of the number and its 10's complement only upto two digits, it is zero. In other words we are neglecting the carry of the last stage.

Ex :

```
   5   2   3     (decimal number)
 + 4   7   7     (its 10's complement)
 -------------
       0   0
 carry = 1
```

The decimal number 523 is of three digits. If we consider the sum of the number and its 10's complement only upto three digits, it is zero. Thus we see that the 10's complement gives the negative value of the number.

The 10's complement of a decimal number
= - decimal number.

1's Complement.

The 1's complement in the binary number system is similar to the 9's complement in the decimal system. The 1's complement of a binary number is obtained by subtracting each bit of the number from 1. The 1's complement of 01 is 10. The 1's complement of 111 is 000. Thus we see that the 1's complement of a binary number can be obtained by simply changing each bit 1 to 0 and 0 to 1

Ex : Find 1's complement of 100110.

The 1's complement of 100110 = 011001.

2's Complement.

The 2's complement in the binary number system is similar to the 10's complement in the decimal number system.

The 2's complement of binary number
= its 1's complement +1.

Ex : Find 2's complement of 10011.

The 2's complement of 10011 = 01100+1= 01101.

Now let us examine the sum when a binary number is added to its 2's complement.

```
Ex : Binary number              = 1001
     Its 1's complement         = 0110
     Its 2's complement         = 0110 + 1
                                = 0111.
     Number + its 2's complement  = 1001
                                 + 0111
                                 ------
                                   0000
                                 carry
```

The last carry is lost if the processor is of 4 bits or sum is considered only up to 4 bits. For 8-bit processor the number and 2's complement will be written in 8 bits as given below :

Number	=	00001001
1's complement	=	11110110
2's complement	=	11110110 + 1
	=	11110111
Number	=	00001001
+2's complement	=	11110111
		0000000
		carry

The last carry will be neglected , if the 8 bit sum is considered.

Thus we see that the 2's complement of a binary number represents its negative.

19. Binary Subtraction Using 2's Complement.

The addition of 2's complement of a number is equivalent to the subtraction of the number. Suppose, we want to subtract 0010. (2 decimal) from 0101 (5 decimal). If the 2's complement of 0010. (2 decimal) is added to 0101 (5 decimal) the sum will be 0011 (3 decimal). It is equal to 0101 (5 decimal) –0010 (2 decimal) = 0011 (3 decimal).

Ordinary binary subtraction:

0101	(5 decimal)
-0010	(-2 decimal)
0011	(3 decimal)

Subtraction using 2's complement :

1's complement of 0010 (2 decimal) = 1101

2's complement of 0010 = 1101 + 1

= 1110

0101	(5 decimal)
+1110	(+ 2's complement of 2)
0011	(3 decimal)

The carry of the last stage is neglected.

20. Fixed Point Representation

Numbers used in scientific calculations are designated by a sign, the magnitude of the number and sometimes a decimal point. The sign is needed for arithmetic operations as it shows whether the number is positive or negative. The position of the decimal (or binary) point is needed to represent fractions or mixed integer-fraction numbers.

The sign of a number may be considered as a set of two elements, plus and minus. this two-element set can be assigned a binary code of one bit. the convention is to represent a plus with a 0 and a minus with a 1. To represent a signed binary number in a register we need n+1 bits; n bits for the number and one bit for the sign. The sign bit is customarily placed in the leftmost position of the register.

The representation of the decimal (or binary) point in a register is complicated by the fact that it is characterized by a position between two flip-flops in the register. There are two ways of specifying the position of the decimal point in a register; by giving it a fixed position or by employing a floating-point representation. The fixed-point method assumes that the decimal point is always fixed in one position. The two positions most widely used are: (1) a decimal point in the extreme left of the register to make the stored number a fraction, and (2) a decimal point in the extreme right of the register to make the stored number an integer.

In either case the decimal point is not actually present but its presence is assumed from the fact that the number stored in the register is treated as a fraction or as an integer. The floating-point representation uses a second register to store a number that designates the position of the decimal point in the first register.

21. Binary Fixed-Point Representation.

When a fixed-point binary number is positive, the sign is represented by 0 and the magnitude by a positive binary number. When the number is negative, the sign is represented by 1 but the rest of the number may be represented in one of three possible ways. These are:

1. Signed-magnitude representation
2. Signed-1's complement representation
3. Signed-2's complement representation.

In the signed-magnitude representation of a negative number the maginitude of the number is inserted next to its negative sign. In the other two representations, the negative number is represented as either the 1's or 2's complement of its position value designation. As an example, consider the number 9 stored in a 7-bit register. +9 is represented by a sign bit of 0 in the leftmost position followed by the binary equivalent of 9:0 001001. Note that each of the seven bits of the register must have a value and therefore, 0's must be inserted in the two most significant positions following the sign bit. Although there is only one way to represent +9, there are three different ways to represent -9. These representations are shown below.

In signed-magnitude representation 1 001001

In signed-1's complement representation 1 110110

In signed-2's complement representation 1 110111

The signed-magnitude represetnation of -9 is obtained from +9 (0 001001) by complementing only the sign bit. the signed 1's complement representation of -9 is obtained by complementing all the bits of 0 001001 (+9), including the sign bit. The signed-2's complement designation is obtained by taking the 2's complement of the positive number, including its sign bit.

22. Real or Floating-Point Representation of Numbers

A real number consists of two parts, an integer part and a fraction. The examples of a real decimal numbers are 125.46, 0.586 and -185.98. Positive or negative sign is also associated with the number. Examples of real numbers in binary system are 1101.1001, 0.11011 and 1001.0011.

The real decimal number can also be represented in the form 1.254×10^{8}, 0.146×10^{4} and 2.863×10^{-7}. This form of representation is a scientific form of representation.

A real or floating-point number is represented in a computer in two parts, the first part represents a signed fixed-point number called the mantissa. The second part indicates the position of the decimal (or binary) point, and it is called exponent. For example a decimal number 5834.876 is represented in floating-point representation as given below.

Sign		Sign	
0	.5834876	0	04
	Mantissa	Exponent	

Zero in the leftmost position of the mantissa denotes its plus sign. The mantissa has been taken as a fraction. The decimal point shown above is an assumed decimal point. Exponent is +4 which shows that the actual position of the decimal point is four decimal positions to the right of the assumed decimal point.

To represent the above mentioned decimal number in BCD form 29 flip-flops are required to store the mantissa: four flip-flops for each decimal digit and one flip-flop for the sign. 9 flip-flops will be required to store the exponent. The decimal point is not physically present in the register. It is only assumed to be there . In this representation the mantissa has been taken as a fraction. Some computers are designed to have mantissa as an integer.

The decimal number 0.0058346=0.58346 $\times 10^{-2}$ can be represented as shown below:

Sign		Sign	
0	.5834600	1	02
	Mantissa	Exponent	

Again, a negative number - 893.4843 can be represented as

Sign		Sign	
1	.8934843	0	03
	Mantissa	Exponent	

A negative fraction -0.000386984 = -0.3869840 $\times 10^{-3}$ can be represented as

Sign		Sign	
0	.3869840	1	03
	Mantissa	Exponent	

The floating point binary number is also represented in a similar manner. For example, the binary number 1101.0110 can be represented in a 16- bit register as shown below:

Sign		Sign	
0	.110101100	1	00100
	Mantissa	Exponent	

The mantissa occupies 10 bits (including sign bit). The binary point is not physically indicated but it is assumed to be there. The exponent occupies 6 bits including sign bit. In a general form a real number can be represented as $M \times R^e$

Where M represents mantissa which may be either a fraction or an integer. R is the radix of the number system used, e represents exponent.

The mantissa M and the exponent e are physically present in the register. The radix R and the radix-point (such as decimal point or binary point) are not present in the register. These are assumed things which are taken into account by the electronic circuitry which is doing manipulation.

23. Normalized Floating-Point Number

A floating-point number is called normalized if the most significant position has a non-zero digit. For example 0.005384×10^{8} and 0.086934×10^{-10} are not normalised numbers. The corresponding normalized numbers will be 0.5384×10^{6} and 0.86934×10^{-11} respectively.

Similarly, in binary number system 0.11011×2^{10} and 0.1000×2^{8} are normalized binary numbers.

The binary numbers 0.0011011×2^{12} and 0.01000×2^{-7} are not normalized binary numbers. While storing numbers the mantissa is shifted left till its most significant bit becomes non-zero, and the exponent adjusted accordingly. For example, the binary number 0.000011001 will be stored as 0.11001×2^{-4}. The shifting of mantissa left to make its most significant bit non-zero is called normalization. A zero cannot be normalized because it does not contain a nonzero digit. A zero is represented in floating-point by all 0's in the mantissa and exponent, including their signs. It is then necessary to check for an all 0 quantity before deciding whether the number can be normalized.

24. Range of Numbers to be stored.

The range of numbers to be stored depends on the word length used to store the number and the number of bits allotted for mantissa and exponent. Suppose, a 16-bit word is used to store numbers. Out of this 9 bits are for mantissa including sign bit and 7 bits for exponent including, sign bit. The maximum value of the number which can be stored is shown in Fig.1.1.

The maximum value = 0.11111111 E0111111

where E represents exponent; the 0 after E represents that the exponent is positive.

The maximum value $= (1-2^{-8}) \times 2(2^{6}-1)$

$$= (1-2^{-8}) \times 2^{63} \cong 1 \times 2^{63} = 2^{63} \cong 10^{19}$$

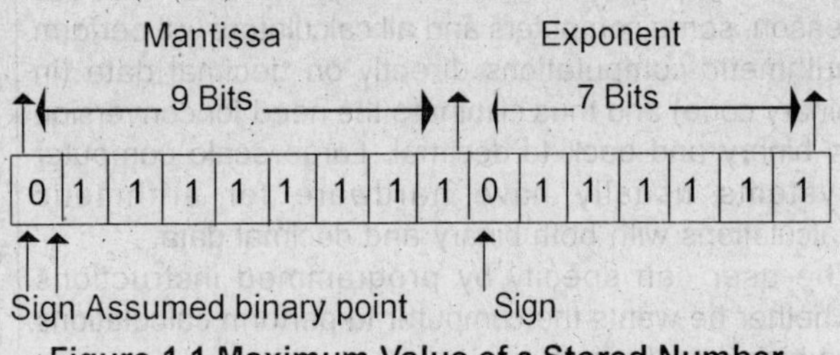

Figure 1.1 Maximum Value of a Stored Number.

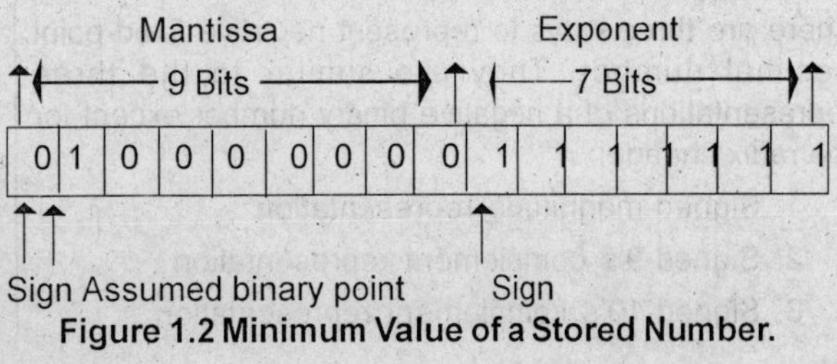

Figure 1.2 Minimum Value of a Stored Number.

The minimum value is shown in Fig 1.2:

Minimum Value = 0.10000000 E 1111111

$= (2^{-1}) \times 2^{-(2^8-1)}$

$= \frac{1}{2} \times 2^{-63} = 2^{-64} \cong 10^{-19}$

The range $= 2^{-64}$ to 2^{63}

$= 10^{-19}$ to 10^{19}

Similarly, the range can be obtained for 32 bit word length. Suppose 24 bits are for the mantissa including the sign bit and 8 bits for exponent including the sign bit.

Mazimum value $= (1-2^{-23}) \times 2^{(2^7-1)} = 2^{127} \cong 10^{38}$

Minimum Value $= (2^{-1}) \times 2^{(2^7-1)}$

$= \frac{1}{2} \times 2^{-127} = 2^{-128} \cong 10^{-38}$

25. Decimal Fixed-Point Representation

The representation of decimal numbers in registers is a funciton of the binary code used to represent a decimal digit. A 4-bit decimal code requires four flip-flops for each decimal digit. The representation of +4385 in BCD requires at least 17 flip-flops, one flip-flop for the sign and four for each digit. This number will be represented in a register with 25 flip-flops as follows:

+	0	0	4	3	8	5
0	0000	0000	0100	0011	1000	0101

By representing numbers in decimal we are wasting a considerable amount of storage space since the number of flip-flops needed to store a decimal number in a binary code is greater than the number of flip-flops needed for its equvalent binary representation. Also the circuits required to perform decimal arithmetic are more complex. However, there are some advantages in the use of decimal representation because computer input and output data are generated by people who use the decimal system. some applications such as electronic calculators or business data processing require small amounts of arithmetic computations compared to the amount requred for input and output of data. For this reason, some computers and all calculators will perform arithmetic computations directly on decimal data (in binary code) and thus eliminate the need for conversion to binary and back to decimal. Large scale computer systems usually have hardware for arithmetic calculations with both binary and decimal data.

The user can specify by programmed instructions whether he wants the computer to perform calculations on binary or decimal data.

There are three ways to represent negative fixed-point decimal number. They are similar to the three representations of a negative binary number except for the radix change:

1. Signed-magnitude representation
2. Signed-9's complement representation
3. Signed-10's complement representation.

For all three representations,a positive decimal number is represented by 0 for plus followed by the magnitude of the number. It is in regard to negative numbers that the representation differ. The sign of a negative number is represented by 1 and the magnitude of the number is positive in signed-magnitude representation. In the other two representations the number is represented by the 9's or 10's complement form. The sign of the decimal number is sometimes represented by a 4-bit code to conform with the 4-bit representation of digits. For example, the code 1100(not assigned to a BCD digit) may be used to represented a plus and 1101 to represent a minus.

26. Alphanumeric Codes

The ASCII code is the standard code commonly used for the transmission of binary information.

Each echaracter is represented by a 7-bit code and usually an eighth bit is inserted for parity. The code consists of 128 cahracters. 95 characters represent graphic symbols that include upper and lower case letters, numerals zero to nine, punctuation marks and special symbols. Twenty-three characters represent format effectors which are functional characters for controling the layout of printing or display devices such as carriage rcturn, line feed, horizontal tabulation ,and back space. The other 10 characters are used to direct the data communication flow and report its status.

Another alphanumeric (sometimes called alphanumeric) code used in IBM equipment is the EBCDIC(Extended BCD Interchange Code). It uses eight bits for each character(and a ninth bit for parity). EBCDIC has the same character symbols as ASCII but the bit assignment to characters is different.

When alphanumeric characters are used internally in a computer for data processing (not for transmission purposes) it is more convenient to use a 6-bit code to represent 64 characters. A 6-bit code can specify the 26 upper-case letters of the alphabet, numerals zero to nine, and up to 28 special characters. This set of cahracters is usually sufficient for data processing purpose. Using fewer bits to code characters has the advantage of reducing the memory space needed to store large quantities of alphanumeric data.

27. Error Detection Codes

Binary information transmitted through some form of communication medium is subject to external noise that could change bits from 1 to 0, and vice versa. An error detection code is a binary code that detects digital errors during transmission. The detected errors cannot be corrected but their presence is indicated. The usual procedure is to observe the frequency of errors. If errors occur infrequently at random, the particular erroneous information is transmitted again. If the error occurs too often the system for malfunction.

The most common error detection code used is the parity bit. A parity bit is an extra bit included with a binary message to make the total number of 1's either odd or even. A message of three bits and two possible parity bits is shown in Table1.7.

Table 1.7 Parity Bit Generation

Message xyz	P(odd)	P(even)
000	1	0
001	0	1
010	0	1
011	1	0
100	0	1
101	1	0
110	1	0
111	0	1

The P(odd) bit is chosen in such a way as to make the sum of 1's (in all four bits) odd. The P(even) bit is chosen to make the sum of all 1's even. In either case, the sum is taken over the message and the P bit. In any particular application, one or the other type of parity will be adopted. The even parity scheme has the disadvantage of having a bit combination of all 0's, while in the odd parity there is always one bit (of the four bits that constitute the message and P) that is 1. Note that the P(odd is the complement of the P(even).

Duringtransfer of information from one locationtoanother, the parity bit is handled as follows. At the sending end, the message(in this case three bits) is applied to a parity generator, where the required parity bit is generated. The message, including the parity bit, is transmitted to its destination. At the receiving end, all the incoming bits (in this case, four) are applied to a parity checker that checks the proper parity adopted (odd or even). An error is detected if the checked parity does not conform to the adopted parity.The parity method detects the presence of one,three,or any odd number of errors.An even number of errors is not detected.

Parity generator and checker networks are logic circuits constructed with exclusive-OR functions. The exclusive-OR function of three or more variables is by definition an odd function. An odd function is a logic function whose value is binary 1 if, and only if, an odd number of variables are equal to 1.

According to this definition, the P(even) function is the exclusive-OR of x,y,and z because it is equal to 1 when either one or all three of the variables are equal to 1(Table 1.7).The P(odd) funciton is the complement of the P(even) function.

As an example, consider a 3-bit message to be transmitted with an odd parity bit. At the sending end, the odd parity bit is generated by a parity generator circuit. As shown in Fig 1.3, this circuit consists of one exclusive-OR and one exclusive-NOR gate.

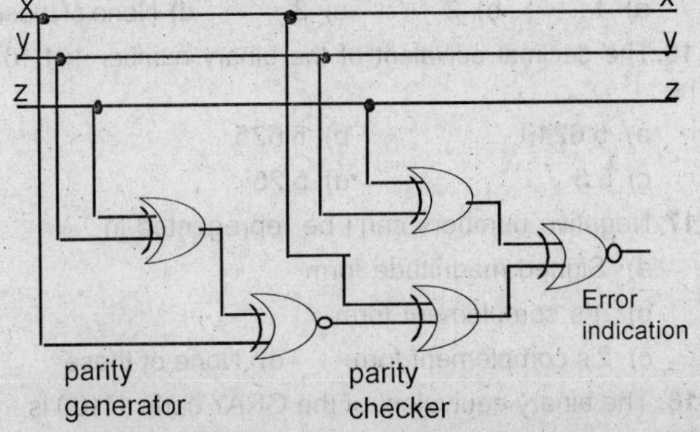

Fig.1.3 Error detection with odd parity bit.

Since P(even) is the exclusive-OR of x,y,z,and P(odd) is the complement of P(even), it is necessary to employ an exclusive-NOR gate for the needed complementation. The message and the odd parity bit are transmitted to their destination where they are applied to a parity checker. An error has occurred during transmision if the parity of the four bits received is even, since the binary information transmitted was originally odd. The output of the parity checker would be 1 when an error occurs, that is, when the number of 1's in the four inputs is even. Since the exclusive-OR function of the four inputs is an odd function, we again need to complement the output by using an exclusive-NOR gate.

SOLVED EXAMPLES

1. Convert the following numbers with the indicated bases to decimal : $(12121)_3$; $(4310)_5$; $(50)_7$; and $(198)_{12}$.

Sol : $(12121)_3 = 3^4+2\times 3^3+3^2+2\times 3+1$

$= 81+54+9+6+1 = 151$

$(4310)_5 = 4\times 5^3+3\times 5^2+5$

$= 500+75+5=580$

$(50)_7 = 5\times 7 = 35$

$(198)_{12} = 12^2+9\times 12+8$

$=144+108+8=260$

2. What is the radix of the numbers if the solution to the quadratic equation $x^2-10x+31=0$ is x = 5 and x = 8 ?

Sol : $(x^2-10x+31)_r = [(x-5)(x-8)]_{10}$

$= x^2-(5+8)_{10}\,x+(40)_{10}x$

Therefore $(10)_r = (13)_{10}$

Radix = 13

Also, $(31)_r$ = 3 x 13 + 1 = $(40)_{10}$

Radix = 13

3. Show the value of all bits of a 12 - bit register that hold the number equivalent to decimal 215 in (a) binary ; (b) binary-coded octal ; (c) binary-coded hexadecimal ; (d) binary coded decimal (BCD).

Sol : $(215)_{10}$ = 128 + 64 + 16 + 7 = $(110\ 10\ 111)_2$

a. 000011010111 Binary

b. 000 011 010 111 Binary coded octal
0 3 2 7

c. 0000 1101 0111 Binary coded hexadecimal
0 2 7

d. 0010 0001 0101 Binary coded decimal
2 1 5

4. Show the bit configuration of a 24-bit register when its content represents the decimal equivalent of 295 : (a) in binary ; (b) in BCD; (c) in ASCII using eight bits with even parity.

Sol : $(295)_{10} = 256+32+7= (100100111)_2$

a. 0000 0000 0000 0001 0010 0111

b. 0000 0000 0000 0010 1001 0101

c. 10110010 00111001 00110101

5. A 36-bit floating point binary number has eight bits plus sign for the exponent and 26 bits plus sign for the mantissa. The mantissa is a normalized fraction. Numbers in the mantissa and exponent are in signed magnitude representation. What are the largest and smallest positive quantities that can be represented, excluding zero ?

Sol : Mantissa **Exponent**

+	26 bits		S	8 bits	36 bits

Largest :+ 0.1111....1   + 11111111

$1 - 2^{-26}$ + 255 $(1-2^{-26}) \times 2^{+255}$

Smallest: +0.1000...0   - 11111111

(normalized) 2^{-1} - 255 2^{-256}

6. Represent the number $(+46.5)_{10}$ as a floating-point number with 24 bits. The normalized fraction mantissa has 16 bits and the exponent has 8 bits.

$46.5 = 32 + 8 + 4 + 2 + 0.5 = (101110.1)_2$

Sign

0 101110100000000   000001   (+6)

24 - bit mantissa   8 - bit exponent

7. X and Y are successive digits in a positional number system. Also, $XY = 25_{(10)}$ and $YX = 31_{(10)}$. Determine the radix value of the system and values of X and Y.

[IES-1994]

Sol : $XY = 25_{(10)}$, $YX = 31_{(10)}$

$$7\,|\,25 \qquad 3\,|\,4$$ Hence, $(25)_{10} = (34)_7$

Similarly, $$7\,|\,31 \qquad 4\,|\,3$$ Hence, $(31)_{10} = (43)_7$

Therefore, the radix value of the system is 7 and the values of 'X' and 'Y' are 3 and 4 respectively.

OBJECTIVE QUESTIONS

1. Binary-coded-Decimal(BCD) numbers express each digit as: **[AUCET -2000]**

a) 1 bit b) Nibble c) Byte d) None of these

2. The one's complement of binary numbers :

0000 1111 0010 1101 is **[AUCET -2000]**

a) 1111 0000 1101 0010

b) 0000 1111 1100 0011

3. The decimal equivalent of 2^{12} is : **[AUCET -2000]**

a) 16 b) 1000 c) 1024 d) 4096

4. The hexadecimal equivalent of decimal number 511 is : **[AUCET -2000]**

a) FF1 b) FF3 c) 3FF d) 1FF

5. The exponent of a floating point number is repragented in Excess-N code so that : **[GATE-1987]**

a) Dynamic range is large

b) Overflow is avoided

c) Precision is high

d) The samllest number is repragented by all zeroes.

6. What is the distance of the following code :

000000 , 010101 , 000111 , 011001, 111111 **[GATE -1995]**

a) 2 b) 3 c) 4 d) 5

7. Given $\sqrt{(224)_r} = (13)_r$ the value of radix "r" is :

[GATE -1997]

a) 5 b) 6 c) 8 d) 10

8. ZERO has two repragentations in : **[GATE - 1999]**

a) Sign magnitude b) 1's complement

c) 2's complement d) None

9. The number 43 in 2's complement repragentation is **[GATE -2000]**

a) 01010101 b) 11010101

c) 00101011 d) 10101011

10. The 2's complement repragentation of $(-539)_{10}$ in hexadecimal is : **[GATE -2001]**

a) ABE b) DBC c) DE5 d) 9E7

11. The decimal value 0.25 **[GATE -2002]**

a) Is equivalent to the binary value 0.1

b) Is equivalent to the binary value 0.01

c) Is equivalent to the binary value 0.00111.........

d) Cannot be repragented precisely in binary

12. The 2's complement repragentation of the decimal value - 15 is : **[GATE -2002]**

a) 1111 b) 11111 c) 111111 d) 10001

13. Sign extension is a step in **[GATE-2002]**

a) Floating point multiplication

b) Signed 16-bit integer addition.

c) Arithmetic left shift

d) Converting a signed integer from one size to another.

14. When two n-bit binary numbers are added,the sum will contain at the most

a) n bits b) n+1 bits

c) n+2 bits d) n+3 bits

15. If $(123)_5=(x3)_y$, then the numberof possible values of x is

a) 1 b) 2 c) 3 d) None of these

16. The decimal euivalent of the binary number 101.101 is

a) 5.6249 b) 5.625

c) 5.5 d) 5.25

17. Negative numbers can't be repregented in

a) Signed magnitude form

b) 1's complement form

c) 2's complement form d) None of these

18. The binary equivalent of the GRAY code 11100 is

a) 10111 b) 00111 c) 01011 d) 10101

19. Which of the following is the minimum ERROR code?

a) Octal code b) Binary code
c) Gray code d) Excess -3 code

20. Which of the following 4-bit numbers equal its 2's complement:

a) 1010 b) 0101
c) 1000 d) No such number exists

21. Which of the following 4- bit number equals it's 1's complement?

a) 1010 b) 1000
c) No such number exists d) None of these

22. The number of 1's in the binary reperagentation of the value of decimal expression 16 × 9+16 × 7+16 × 5+3

a) 6 b) 9 c) 12 d) 15

23. One's complement of the binary number 10001011 is:

a) 01110101 b) 01110111
c) 01110100 d) 11110100

24. Two's complement of the binary number 10010100 is:

a) 01101011 b) 01101100
c) 11101100 d) 10001011

25. The Parity of the binary number 11001110

a) Is even b) Is not known
c) Is odd d) Is same as the number of zero's

26. In which code the succesive code characters differ in only one bit position?

a) 8421 code b) Gray code
c) Excess-3 code d) None of these

27. CYCLIC codes are used in

a) Data transfer
b) Arithmetic and logic computation
c) Continously varying signal repragentation
d) None of these

28. When an ODD number is converted into the binary number, the least signifi cant digit(LSD) is

a) 0 b) 1 c) 0 or 1 d) None of these

29. When the decimal number 16 is converted to the binary number, how many digits are needed:

a) 3 b) 4 c) 5 d) 7

30. The GRAY code for decimal 7 is

a) 0111 b) 1011 c) 0100 d) 0101

31. The EXCESS-3 code is also kown as

a) Cyclic redundancy code
b) Weighted code
c) Self complimenting code
d) Algebraic code

32. Which of the following is an unweighted code:

a) 8421 code b) Excess-3 code
c) 2421 code d) 5211 code

33. Which of the following codes is non-Weighted:

a) Decimal number system
b) Excess-3 code
c) Binary number system
d) Roman number system

34. Which of the following is the minimum error code

a) Binary code b) Gray code
c) Excess-3 code d) Octal code

35. Floating-point numbers are normally used to repragent

a) Very large or small numbers
b) Only large positive numbers
c) Only large negative numbers
d) Negative numbers within the computers

36. The INTERNET code is

a) ASCII b) EBCDI c) UNICODE d) None

37. The number 1.5 in decimal number system is represented in octal as **[UGC NET - 2000]**

a) 1.5 b) 1.4 c) 1.1 d) 1.6

38. A device which converts from decimal to binary number is called **[UGC NET - 1999]**

a) Decoder b) Encoder
c) CPU d) Converter

39. The octal equivalent of decimal 324.987 is **[UGC NET - 1999]**

a) 640.781 b) 815.234
c) 70.771 d) 504.771

40. The binary equivalent of $(2568)_{10}$ is **[UGC NET - 1998]**

a) 1010 0001 1000 b) 1010 0000 1000
c) 1010 0000 1010 d) 1000 0000 1010

41. The BCD addition of two BCD digits A (0101) and B (0110) gives **[UGC NET - 1998]**

a) 0001 b) 1011 c) 1000 d) 1010

42. CPU register that holds the address of the next instruction for execution is called **[UGC NET - 1996]**

a) Stack Pointer b) Program Counter
c) Accumulator d) None

43. The number of ternary boolean functions is **[UGC NET - 1996]**

a) 8 b) 81 c) 256 d) 729

44. A parity bit **[UGC NET - 1996]**

a) is a check bit added to each 7-bit character represented in storage
b) always produces an odd numbers of 1 bits in a stored character
c) is used to detect human being error
d) is always1

45. The decimal number 1.3125 is represented as 1.5 in the number system **[UGC NET - 1995]**

a) 16 b) 8 c) 18 d) 12

46. While adding two numbers in signed 2's complement, the overflow occurrence happens when **[UGC NET-1995]**

a) Carry-in and carry-out of the sign -bit equal 1
b) Carry-in and carry-out of the sign -bit equal 0
c) Carry - in and carry-out bits are unequal
d) None

47. The result of $\mathbf{60}_{10} + \mathbf{F1}_{16} - \mathbf{1001001}_2$ is **[UGC NET - 1995]**

a) $(108)_{10}$ b) $(16\,D)_{16}$ c) $(228)_{10}$ d) $(161)_8$

48. Convert the haxadecimal number 'FAFAFA' into octal, **[Hyd.Univ-2001]**

a) 76767676 b) 76737672
c) 76575372 d) 76727672

49. The 10's complement of $(715)_8$ is **[Hyd.Univ-1999]**

a) 538 b) 539 c) 284 d) 285

50. $(41.6875)_{10}$ converted to base 2 is: **[Hyd.Univ-1997]**

a) 111001.1011 b) 101001.1011
c) 101001.1010 d) 101011.1011

51. Which of the following is not an octal number **[Hyd.Univ-1995]**

a) 28 b) 75 c) 16 d) 102

52. Which of the following is not a hexadecimal number? **[Hyd.Univ-1994]**

a) BED b) CAB c) AGE d) DAD

53. The number of 1's in the binary representation of (3 * 4096 + 15 * 256 + 5 * 16 + 3) is : **[Hyd.Univ-1997]**

a) 9 b) 10 c) 11 d) 12

ANSWER KEY - I

1. b	2. a	3. d	4. a	5. d
6. a	7. a	8. a	9. c	10. c
11. b	12. d	13. a	14. b	15. b
16. b	17. d	18. a	19. c	20. c
21. c	22. b	23. c	24. b	25. c
26. b	27. c	28. b	29. c	30. c
31. c	32. b	33. d	34. b	35. a
36. c	37. b	38. d	39. d	40. b
41. a	42. b	43. c	44. a	45. a
46. c	47. c	48. c	49. b	50. b
51. a	52. c	53. a		

EXPLANATORY NOTES - I

2. (a) Explanation : The 1's compliment of binary number is obtain by subtracting each bit of the number from 1. The given no's are

0000	1111	0010	1101
1111	0000	1101	0010

3. (d) Explanation : We know that 2^{10} is 1024.

$2^{12} = 2^{10+2} = 2^{10} \times 2^2 = 1024 \times 4 = 4096$

4. (d) Explanation :

16	511
16	31 - F
16	1 - F
16	0 - 1

$(511)_{10} = (1FF)_{16}$

6. (a) Explanation : The given code is in binary code. So the distance is 2.

7. (a) Explanation : $(224)_r = \left[(13)_r\right]^2$

$4 \times r^0 + 2 \times r^1 + 2 \times r^2 = \left[3 \times r^0 + 1 \times r^1\right] \times \left[3 \times r^{-0} + 1 \times r^1\right]$

$4 + 2r + 2r^2 = (3 + r)(3 + r)$

$r^2 - 4r - 5 = 0,\ r^2 - 5r + r - 5 = 0$

$r(r-5) - 1(r-5) = 0 \Rightarrow r = 5, 1$

10. (c) Explanation :

2	539
2	269 - 1
2	134 - 1
2	67 - 0
2	33 - 1
2	16 - 1
2	8 - 0
2	4 - 0
2	2 - 0
	1 - 1

$(539)_{10} = (1000011011)_2$

2's complement signed number is

0 1 1 1 1 0 0 1 0 0
1

1 1 0 1 | 1 1 1 0 | 0 1 0 1
D E 5

11. (b) Explanation : $0.25 \times 2 = 0.50 - - - - - 0$

$0.50 \times 2 = 1.00 - - - 1,\ (0.25)_{10} = (0.01)_2$

15. (d) Explanation :

$3 \times 5^0 + 2 \times 5^1 + 1 \times 5^2 = 3 \times y^0 + x \times y$

$3 + 10 + 25 = 3 + xy$, $xy = 35$

The possible values of x are 7, 5, 1, 35

So the possible values are 4.

16. (b) Explanation :

$(101.101)_{10} = \left[1 \times 2^2 + 0 \times 2^1 + 1 \times 2^0\right] +$

$\left[1 \times 2^{-1} + 0 \times 2^{-2} + 1 \times 2^{-3}\right]$

$= [4 + 0 + 1] + [0.5 + 0 + 0.125] = 5.625$

22. (c) Explanation :

$10000 \times 1001 + 10000 \times 0111 + 10000 \times 101 + 011$

The number of 1's are 12.

25. (c) Explanation :

The number of 1's in the number is 5

5 is odd number. So this is odd parity.

29. (c) **Explanation :**

$$\begin{array}{r|l} 2 & 16 \\ \hline 2 & 8 - 0 \\ \hline 2 & 4 - 0 \\ \hline 2 & 2 - 0 \\ \hline & 1 - 0 \end{array} \quad (16)_{10} = (10000)_2$$

The number of digits are needed is 5.

37. (b) **Explanation :** The number 1 is represented octal as 1 $\quad 0.5 \times 8 = 4.0 - - - - 4 \Rightarrow 1.4$

39. (d) **Explanation :**

$$\begin{array}{r|l} 8 & 324 \\ \hline 8 & 40 - 4 \\ \hline & 5 - 0 \end{array} \quad (324)_8 = (504)_{10}$$

$.987 \times 8 = 7.896$

$.896 \times 8 = 7.168 \quad (0.987)_8 = (0.771)_{10}$

$.168 \times 8 = 1.344 \quad \therefore (324.987)_8 = (504.771)_{10}$

47. (c) **Explanation :**

$$60 + 1 \times 16^0 + 15 \times 16^1 - \left[1 \times 2^0 + 1 \times 2^1 + 1 \times 2^2 + 1 \times 2^3 + 0 \times 2^4\right]$$

$60+241-1+8+64, \; 301-73 = (228)_{10}$

48. (c) **Explanation :**

111|110|101|111|101|011|111|010

7 6 5 7 5 3 7 2

OBJECTIVE QUESTIONS - II

1. Which of the following statements about floating-point arithmetic is not TRUE ? **[AUCET -1999]**

a) It is inherently non-associative because some numbers have no exact repragentation.

b) It is inherently non-associative because there have to be upper and lower bounds on the size of numbers.

c) Associativity can be achieved with appropriate round of conventions.

d) Some rational numbers have no exact repragentation.

e) No irrational number has an exact repragentation.

2. Which of the following pairs of 8-bit, two's complement numbers will result in overflow when the members of the pairs are added : **[AUCET -1999]**

a) 11111111 , 00000001 b) 00000001 , 10000000

c) 11111111 , 10000001 d) 10000001 , 10101010

3. Multiplying a number with -31, using BOOTH's algorithm requires : **[AUCET - 2001]**

a) One subtraction and one addition

b) One subtraction and two additions

c) Two subtractions and one addition

d) Two subtractions and two additions

4. In floating-point repragentation of numbers, excess repragentation is used for exponent to facilitate **[AUCET -2001]**

a) Easy comparison of floating point numbers.

b) Reduction of the number of bits used for exponent.

c) Repragentation of zero.

d) Repragentation of Infinity

5. With 2's complement repragentation, the range of values that can be repragented on data bus of an 8-bit microprocessor is given by : **[GATE -1990]**

a) -128 to +127 b) -128 to + 128

c) -127 to +128 d) -256 to + 256

6. 2's complement repragentation of a 16-bit number (one sign bit and 15 magnitude bits) is FFFF. It's magnitude in DECIMAL repragentation is : **[GATE -1993]**

a) 0 b) 1 c) 32767 d) 65535

7. A signed integer has been stored in a byte using the 2's complement format. We wish to store the same integer in a 16-bit word. we should **[GATE -1997]**

a) Copy the original byte to the less significant byte of word and fill the more significant with zeroes.

b) Copy the original byte to the more significant byte of word and fill the less significant with zeroes.

c) Copy the original byte to the less significant byte of word and make each bit of the more significant byte equal to the most significant bit of the original byte.

d) None.

8. Booth's algorithm for integer multiplication gives wrost performance when the multiplier pattern is : **[GATE -1996]**

a) 101010......1010 b) 100000.......0001

c) 011111........1110 d) None

9. Consider the floating-point number repragentation :

31 24	23 0
EXPONENT	MANTISSA

The exponent is in 2's complement repragentation and mantissa is in the sign magnitude repragentation. The range of the magnitude of the normalized number in this repragentation is **[GATE -1996]**

a) 0 to 1 b) 0.5 to 1 c) 0.5 to $(1-2^{-23})$ d) None

10. Consider the values A= 2.0 × 10^{30} , B=-2.0 × 10^{30} , c=1.0 and the sequence **[GATE - 2000]**

x := A+B y := A + C x := x+C y:=y+B

Executed on a computer where floating point numbers are repragented with 32 bits. The values for x any y will be :

a) x=1.0 , y=1.0 b) x=1.0 , y=0.0

c) x=0.0 , y=1.0 d) x=0.0 , y=0.0

11. In 2's complement addition, overflow **[GATE - 2002]**

a) Is flagged whenever there is carry from sign bit addition.

b) Cannot occur when a positive value is added to a negative value.

c) Is flagged when the carries from sign bit and previous bit match

d) None of these.

12. Even parity is being used for error checking computer data,if the data word to be transmitted is 0101 0111,the value of the parity will be: **[AUCET- 2001]**

a) 0 b) 01 c) 1 d) 10

13. The greatest negative number which can be stored in a computer that has 8-bit word length and users two's comple-ment arithmetic is

a) -256 b) -255 c) -128 d) -127

14. A number system uses 20 as the radix.The excess code that is neccessary for its equivalent binary code repregentation is

a) 4 b) 5 c) 6 d) 7

15. A decimal number has 25 digits.The nuber of bits needed for its equvalent binary repragentation is,approximately.

a) 50 b) 60 c) 70 d) 75

16. Most of the digital computers do not have floating-point hardware because

a) It is costly
b) It is slower than Software
c) Floating point addition can't be performed by hardware d) None of these

17. Let $a_n a_{n-1}....a_1 a_0$ be the binary repragentation of an integer b.the integer b is divisible by 3 if

a) The number of one's is divisible by 3
b) The number of one's is divisible by 3,but not by 9
c) The number of zeros is divisible by 3
d) The diffrence of alternate sum,
i.e $(a_0+a_2+.......)-(a_1+a_2+........)$ is divisible by 3.

18. Suppose the largest n-bit binary number requires "d" digits in decimal repregentation.Which of the following relations between 'n' and 'd' is apporoximately correct?

a) $d=2^n$ b) $n=2^d$ c) $d<n\log_{10}2$ d) $d>n\log_{10}2$

19. Hexa decimal number system is used in digital computers and digital systems to

a) Perform arithmetic operations
b) Input binary data into the system
c) Perform logical operations
d) Perform arithmetic and logical operations

20. A computer with a 32-bit word size uses 2's cmplement to repragent numbers.The range of integers that can be repragented by this computer is:

a) -2^{32} to 2^{32} b) -2^{32} to $+2^{31}$
c) -2^{31} to $2^{31}-1$ d) -2^{32} to $2^{31}-2$

21. ASCII code is used as

a) an alphabetic code b) an alphanumeric code
c) a weighted code d) a cyclic code

22. ASCII code is

a) a 4-bit code b) a 6-bit code
c) a 7-bit code d) a 8-bit code

23. A number $N = b_n b_{n-1} ......b_0$ is said to be negabinary if and only if $N=\sum_{i=0}^{n}(-2)^i b_i$ (where b'_i s are negabinary digits 0,1 as shown above). Thus,negabinary number 111 is equivalent to 3 in decimal, and negabinary number 1011 is equivalent to -9 in decimal. The negabinary number 101010 is equivalent to **[UGC NET - 2000]**

a) 42 b) - 37 c) -42 d) 37

24. Which one of the following statements is true about the complement forms? **[UGC NET - 2000]**

a) In 2's complement form, there are two representations for zero
b) In1's complement form, there are two representations for zero
c) In1's complement form, magnitude of largest positive number is one less than the magnitude of largest negative number .
d) In 2's complement form,magnitude of largest positive and largest negative numbers are equal

25. The IEEE - 754 32 - bit representation of 2.75 is **[UGC NET - 1998]**

a) 0 0 0 0 0 0 0 1 1 1011 ------- 0 0 0
b) 0 0 0 0 0 0 0 10 01100 ------- 0 0 0
c) 0 0 0 0 0 0 0 01 01100 ------- 0 0 0
d) 0 1 0 0 0 0 0 0 0 01100 ------- 0 0 0

26. A normalized floating - point number is one which has **[UGC NET - 1997]**

a) number of digits in mantissa is equal to exponent
b) both mantissa and exponent as positive numbers
c) positive mantissa and negative exponent.
d) most significant digit of mantissa is non-zero

27. Consider a hypothetical 4-bit machine. Using 2's complement representation, the range of values that can be stored in the system **[UGC NET - 1996]**

a) - 8 to 8 b) - 8 to 7 c) -7 to 8 d) -7to 7

28. Which of the following is not a theorem of Boolean algebra ? **[UGC NET - 1996]**

a) $X(1+Y) = X.Y$ b) $\overline{X.Y} = \overline{X}+\overline{Y}$
c) $X.(Y+Z)= X.Y+X.Z$ d) $X+X= X$

29. The number system with negative base (-2) is called Negabinary system.The negasbinary representation of the decimal number 41 given by **[UGC NET - 1995]**

a) $(101001)_{(-2)}$ b) $(010110)_{(-2)}$
c) $(0000110)_{(-2)}$ d) $(1111001)_{(-2)}$

30. Consider a hypothetical 5 - bit machine, using 2's complement representation the range of values that can be stored in the system is **[UGC NET - 1995]**

a) - 16 to 16 b) - 16 to 15
c) - 15 to 15 d) None

31. The decimal number 1.2 is represented as 1.3333.......3.... in the number system with base (or radix) **[UGC NET - 1995]**

a) 8 b) 16 c) 18 d) 12

32. The number system with -ve base (-2) is called Negabinary system. The decimal value of the negabinary number $(1100101)_{-2}$ is **[UGC NET - 1995]**

a) - 37 b) 37 c) - 101 d) 101

33. If a computer with a 32 bit word size used 2s complement representation for numbers, that is the range of integers that may be represented? **[Hyd.Univ-2001]**

a) -2^{32} to 2^{32} b) -2^{31} to $2^{31}-1$
c) -2^{31} to 2^{32} d) -2^{32} to 2^{31}

34. The exponent in floating point number representations is biased to **[Hyd.Univ-2000]**

a) Enhance the range of representation

b) Avoid comparing sign bits of exponent in floating point arithmetic operations

c) To facilitate representation of zeros

d) All of the above

35. The excess-3 code is a self-computing code, which means **[Hyd.Univ-2000]**

a) The complement can be generated by inverting each bit pattern

b) It is a weighted code

c) The 1's complement of an excess-3 number is equal to its 9's complement of the decimal number

d) Only and option A and B are true

ANSWER KEY - II

1. d	2. c	3. c	4. c	5. a
6. c	7. c	8. c	9. c	10. b
11. b	12. a	13. c	14. c	15. d
16. a	17. d	18. d	19. b	20. b
21. b	22. c	23. c	24. b	25. ?
26. d	27. a	28. a	29. d	30. b
31. ?	32. b	33. a	34. b	35. a

EXPLANATORY NOTES - II

2. (a) **Explanation :**

```
  1 1 1 1 1 1 1 1
  0 0 0 0 0 0 0 1
  1 1 1 1 1 1 1
1 0 0 0 0 0 0 0 0
```

All the 1's are added to the 0 0 0 0 0 0 0 1 then all the 0's will appear. So they are overlaped.

4. (a) **Explanation :** We know that with 2's complement repregentation, the range of values that can be repregented on data bus of n-bit micro processor is given by -2^{n-1} to $2^{n-1}-1$.

We can use 8 bits so -2^{8-1} to $2^{8-1}-1$

- 128 to 127

6. (b) **Explanation :** F F F F is hexadecimal number. It is converted to the binary no is 1111 1111 1111 but removing sign bit and the magnitude of the number is 111 1111 1111.

```
2's complement      1 1 1 1 1 1 1 1 1 1 1
                                        1
1's complement      1 1 1 1 1 1 1 1 1 1 0
ordinary binary     0 0 0 0 0 0 0 0 0 0 1
```

but it canbe converted to ordinary binary number is also 1.

9. (c) **Explanation :** In a floating point number the exponent is in 2's complement repragentation and mantissa is in a sign magnitude repragentation. The range of the magnitude of the normalized number in this repragentation is 0.5 to $1-2^{-23}$.

10. (a) **Explanation :**

$X = A + B = 2.0\times10^{30} - 2.0\times10^{30} = 0$

$Y = A + C = 2.0\times10^{30} + 1.0$

$X = X + C = 0 + 1.0 = 1.0$

$Y = Y + B = 2.0\times10^{30} + 1.0 - 2.0\times10^{30} = 1.0$

12. (c) **Explanation :** The transmitted data consists number of 1's is 5. But we can use even parity so we can add extra 1.

17. (d) **Explanation :** For example 90 is divisible by 3. In a binary representation 90 is $(1011110)_2$.

The number of 1's is 5 is not divisible by 3. So answers 'a', 'b' are not correct.

The number of zero's is 2. So 2 is not divisible by 3. So 'c' is not correct. So we can prefer answer 'd'.

23. (c) **Explanation:**

Given $N=\sum_{i=0}^{n}(-2)^i b_i$

1 1 1 = 3 i.e., $1\times-2^0+1\times-2^1+1\times-2^2=3$

$1011=-9$ i.e., $1\times-2^0+1\times-2^1+0\times-2^2+1\times-2^3=-9$

$101010=0\times-2^0+1\times-2^1+0\times-2^2+1\times-2^3+0\times-2^4+1\times-2^5$

= 0 - 2 + 0 - 8 + 0 - 32 = - 42.

29. (d) **Explanation :** The formula of the negabinary system is $N=\sum_{i=0}^{n}(-2)^i b_i$ where b_i's are negabinary numbers 0,1.

$1\times2^0+0\times-2^1+0\times-2^2+1\times-2^3+1\times-2^4+1\times-2^5+1\times-2^6$

= 1 + 0 + 0 - 8 + 16 - 32 + 64 = 81 - 40 = 41.

II. CPU ORGANISATION

- **Processor Bus Organisation**
- **Arithmetic Logic Unit**
- **Instructions and Addressing Modes**
- **Parallel Processing**
- **Hardwired and Micro-Programmed Control**

1. Processor Bus Organization

The part of a computer that performs the bulk of data processing operations is called the central processor unit and is referred to as the CPU contains the hardware components for processing instructions and data. It is comprised of a control unit and a processor unit which together supervise and implement the various data processing tasks in the central part of a computer system. Computers with a limited number of registers in the CPU employ a single accumulator for implementing micro-operations. With the availability of integrated circuits, registers and other digital circuits are not as expensive as when constructed with discrete components. Consequently, most computers employ a large number of processor registers and route information among them through common buses.

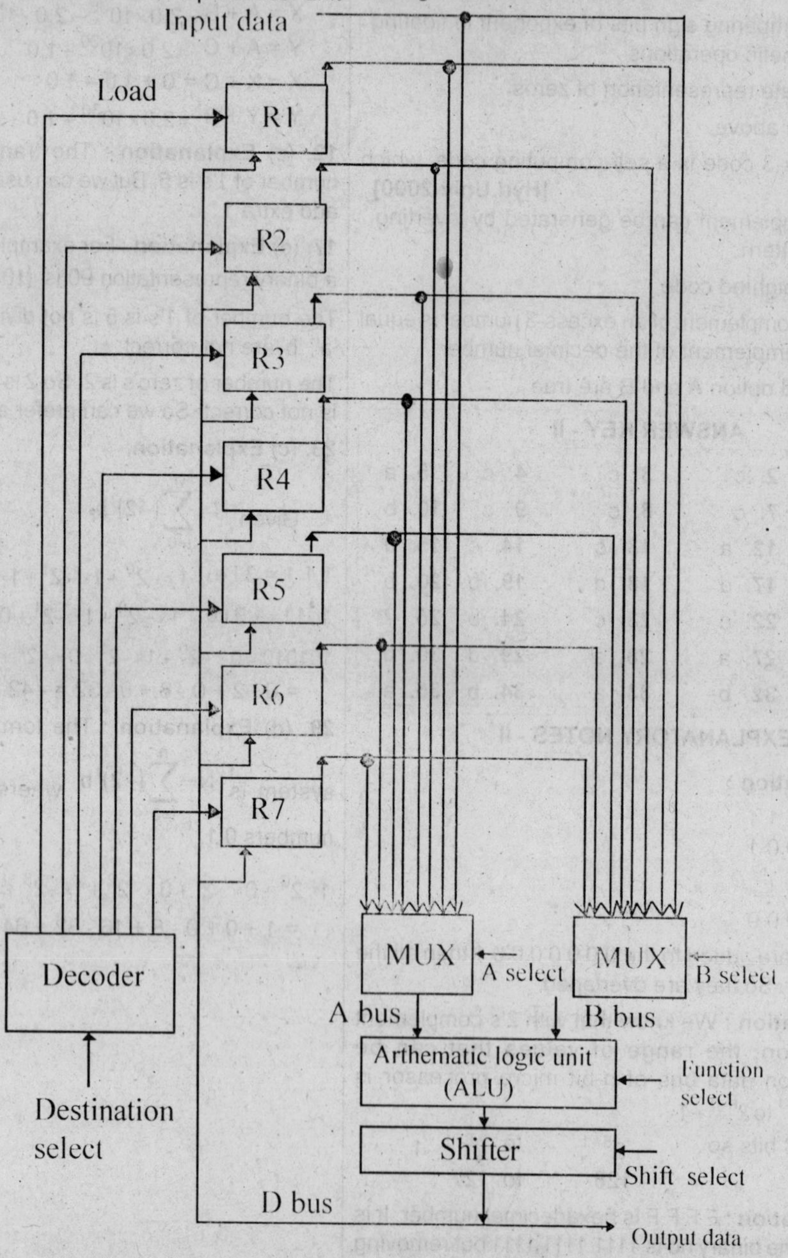

Fig 2.1 CPU registers connected with common buses

A bus organization for seven CJPU registers is shown in Fig.2.1. The output of each register is connected to two multiplexers (MUX) to form input buses A and B. The selection lines in each multiplexer select one register or the input data for the particular bus. The A and B buses form the inputs to a common arithmetic logic unit (ALU). The function selected in the ALU determines the arithmetic or logic micro-operations that is to be performed. The shift micro-operations are implemented in the shifter. The shifter unit is attached to the output of the ALU to provide a postshift capability. The shifter unit could be placed in the input of the ALU to provide a preshift capability,or it may be independent of the ALU to facilitate parallel operations. The result of the micro-operation is available for output data and also goes through the destination bus and into the inputs of all registers. The destination register that receives the information from the D bus is selected by a decoder. The decoder activates one of the register load inputs, thus providing a transfer path between the data in the D bus and the inputs of the selected destination register.

The control unit that operates the CPU bus system directs the information flow through the ALU by selecting the various components in the system. For example, to perform the micro-operation R1 $\leftarrow$ R2+R3 the control must provide binary selection variables to the following selector inputs:

1. MUX A selector: to place the contents of R2 into bus A.
2. MUX B selector: to place the contents of R3 into bus B.
3. ALU function selector: to provide the arithmetic operation A+B.
4. Shift selector: for direct transfer from the output of the ALU into output bus D (no shift).
5. Decoder destination selector: to transfer the contents of bus D into R1.

The five control selection variables must be generated simultaneously and must be available during one clock pulse interval. The binary information from the two source registers propagates through the combinational gates in the multiplexers, the ALU, and the shifter, to the output bus, and into the inputs of the destination register, all during one clock pulse interval. Then, when the next clock pulse arrives, the binary information on the output bus is transferred into R1. To achieve a fast response time, the ALU is constructed with high-speed circuits and the shifter is implemented with combinational gates.

2. Arithematic Logic Unit (ALU)

An arithmetic logic unit is a digital circuit that performs a set of arithmetic micro -operations and a set of logic micro-operations. The ALU has a number of selection lines to select a particular micro-operation in the unit. The selection variables are decoded within the ALU so that k selection variables can specify up to 2^k distinct micro-operations.

The internal operation of an ALU is better understood if it is separated into its arithmetic and logic parts. We will proceed by first explaining the operation of a typical arithmetic circuit that performs eight arithmetic micro-operations. A second circuit will then be shown for implementing four logic micro-operations. Finally, the two circuits will be combined to form an arithmetic logic unit.

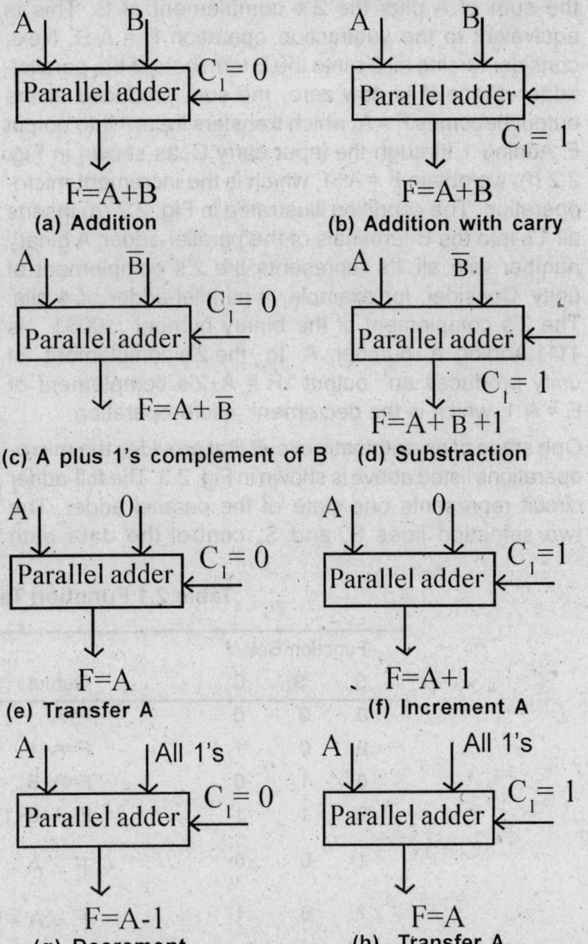

(a) Addition (b) Addition with carry (c) A plus 1's complement of B (d) Substraction (e) Transfer A (f) Increment A (g) Decrement (h) Transfer A

Fig.2.2 Arithmetic micro-operations obtained by controlling one set of inputs to a parallel adder.

Arithmetic Circuit

The binary parallel-adder, is an arithmetic circuit that adds two binary numbers. Additional arithmetic micro-operations can be obtained by augmenting the parallel-adder with certain selection logic circuits.

Figure 2.2 demonstrates the arithmetic micro-operations obtained when one set of inputs to a parallel-adder is controlled externally. The straight arithmetic addition is achieved when one set of inputs receives a binary number A, the other set of inputs receives a binary number B, and the input carry C_1 is maintained at 0. This case is shown in Fig. 2.2 (a), where the output of the circuit designated by F produces the add micro-operation F = A+ B. By making the input carry C_1 = 1 as in Fig. 2.2 (b), it is possible to add 1 to the sum to produce an output F = A+B+1.

Different arithmetic micro-operations can be obtained by controlling the data entering the inputs at B as indicated in the other six diagrams of Fig. 2.2. Complementing all the bits of input B results in a micro-operation $F = A + \overline{B}$, which is the sum of A plus the 1's complement of B. Adding 1 to this sum by making the input carry $C_1 = 1$, we obtain $F = A + \overline{B} + 1$, which produces the sum of A plus the 2's complement of B. This is equivalent to the subtraction operation F = A-B. Next, consider forcing all 0's into the B terminals of the parallel-adder. Since B is now zero, the sum produced at the output becomes F = A, which transfers input A into output F. Adding 1 through the input carry C_1 as shown in Fig. 2.2 (f), we obtain F = A+1, which is the increment micro-operation. The condition illustrated in Fig. 2.2 (g) inserts all 1's into the B terminals of the parallel-adder. A binary number with all 1's represents the 2's complement of unity. Consider, for example, a parallel-adder of 4 bits. The 2's complement of the binary number 0001 is 1111. Adding a number A to the 2's complement of unity produces an output F = A+2's complement of F = A-1, which is the decrement micro-operation.

One stage of an arithmetic circuit that provides the micro-operations listed above is shown in Fig. 2.3. The full-adder circuit represents one state of the parallel adder. The two selection lines S_1 and S_0 control the data path between the B terminal and one input of the full-adder circuit. When $S_1S_0 = 00$, the controlled input of the full-adder is always 0. When $S_1S_0 = 01$, the input receives the value of B_i. When $S_1S_0 = 10$, the input receives the complement value of B_i. With $S_1S_0 = 11$, the input is always equal to 1. These conditions can be verified by deriving the truth table of the selection logic. Note that the two selection variables S_1 and S_0 control the input path in the B terminal according to the four conditions illustrated in Fig.2.2.

The function selection lines S_1, S_0, and C_1 select the arithmetic micro-operation and the F terminals generate the output function. Note that C_1 is the input carry to the first stage and its effect is to add 1 to the low-order pair of bits. S_1 and S_0 control the inputs to the B terminals in each stage.

The micro-operations implemented in the arithmetic circuit are listed in Table 2.1:

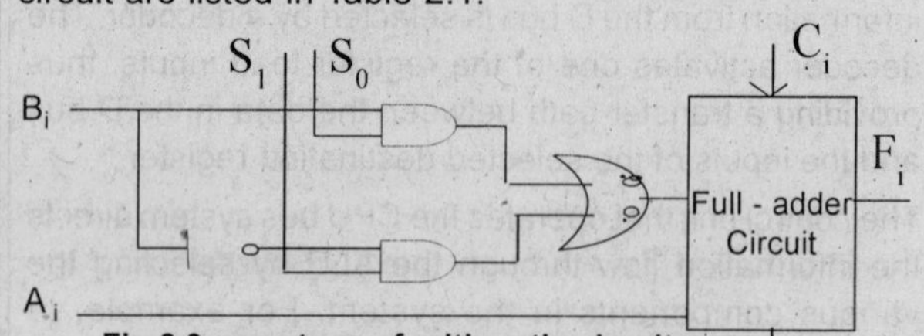

Fig.2.3 one stage of arithmetic circuit.

Table 2.1 Function Table for Arithmetic Circuit

Function Select S_1	S_0	C_1	Output	correspond to Fig 2.2 Function Diagram
0	0	0	F=A	(e) Transfer A
0	0	1	F=A+1	(f) Increment A
0	1	0	F=A+B	(a) Addition
0	1	1	F=A+B+1	(b) Addition with carry
1	0	0	$F = A + \overline{B}$	(c) A plus1's complement of B
1	0	1	$F = A + \overline{B} + 1$	(d) Subtraction
1	1	0	F=A - 1	(g) Decrement A
1	1	1	F=A	(h) Transfer A

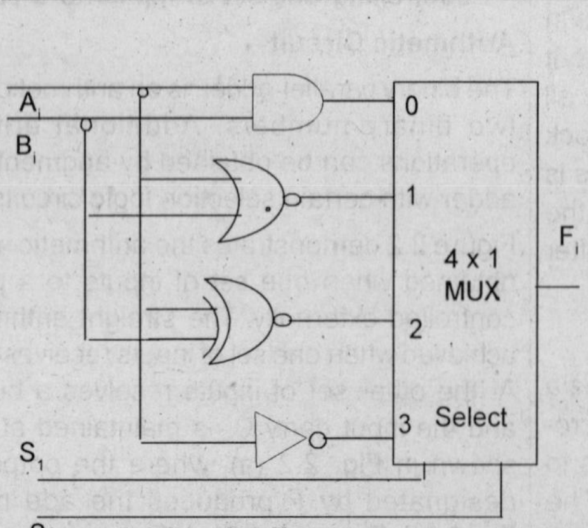

(a) Logic diagram

S_1 S_0	Output	Micro-operation
0 0	$F = A \wedge B$	AND
0 1	$F = A \vee B$	OR
1 0	$F = A \oplus B$	XOR
1 1	$F = \overline{A}$	Complement

(b) Function table

Fig. 2.4 One stage of logic circuit.

Logic Circuit

The logic micro-operations manipulate the bits of the operands by treating each bit as a binary variable. There is a list of 16 logic micro-operations that can be performed with two binary variables. The 16 logic operations can be generated in a circuit and each operation can be selected by means of four selection variables. However, there are essentially only four basic logic operations from which all others can be derived by means of Boolean algebra. They are the AND, OR, XOR (exclusive OR), and complement micro-operations.

Figure 2.4(a) shows one stage of a logic circuit. It consists of four gates and a multiplexer. Each of the four logic operations AND, OR, XOR, and complement are generated with a gate that performs the required logic,. The outputs of the gates are applied to a multiplexer with two selection variables S_1 and S_0.

These selection variables choose one of the multiplexer inputs and direct its value to the output. The diagram shows one typical stage with subscript i. For a logic circuit with n bits, the diagram of Fig. 2.4 (a) must be repeated n times for i = 1,2,3,....,n. The selection variables must be applied to all stages. The function table in

Fig. 2.4 (b) lists the logic micro-operation obtained for each combination of the selection variables.

Arithmetic Logic Unit

The logic circuit can be combined with the arithmetic circuit to produce an arithmetic logic unit commonly referred to as an ALU. Selection variables S_1 and S_0 can be common to both circuits provided that we use a third selection variable to differentiate between the two. The configuration of a one-stage ALU is illustrated in

Fig. 2.5. The outputs of the arithmetic and logic circuits in each stage are applied to a multiplexer with selection variable S_2. When $S_2 = 0$, the arithmetic output is selected, but when $S_2 = 1$, the logic output is selected. Note that the diagram shows just one typical stage of the ALU. The circuit of Fig. 2.5 must be repeated n times for an n-bit ALU.

The output carry C_{i+1} of a given stage must be connected to the input carry C_i of the next-higher-order stage. The input carry to the first stage is designated by C_1 and provides a selection variable for the arithmetic micro-operations. Input C_1 has no effect on the logic micro-operations.

An ALU with individual stages as defined in Fig.2.5 will provide 8 arithmetic and 4 logic micro-operations, for a total of 12 micro-operations. Each micro-operation is selected through the variables S_2, S_1, S_0 and C_1. The input carry C_1 is used for the selection of arithmetic micro-operations only.

Table 2.2 lists the 12 micro-operations of the ALU. The first eight listed (with $S_2 = 0$) are identical to the arithmetic micro-operations of Table 2.1. the input carry during the four logic operations (when $S_2 = 1$) has no effect on the operation of the unit, so the entries under the C_1 column for the logic micro-operations are marked with don't-care X's.

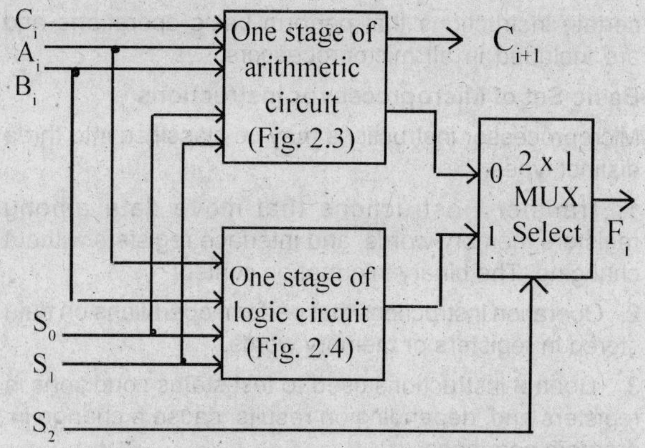

Fig. 2.5 One stage of arithmetic logic unit (ALU)

Table 2.2 Function Table for ALU

Function Select					
S_2	S_1	S_0	C_1	Output	Function
0	0	0	0	$F = A$	Transfer A
0	0	0	1	$F = A + 1$	Increment A
0	0	1	0	$F = A + B$	Addition
0	0	1	1	$F = A + B + 1$	Add with carry
0	1	0	0	$F = A + \overline{B}$	A plus 1's complement of B
0	1	0	1	$F = A + \overline{B} + 1$	Subtraction
0	1	1	0	$F = A - 1$	Decrement A
0	1	1	1	$F = A$	Transfer A
1	0	0	X	$F = A \wedge B$	AND
1	0	1	X	$F = A \vee B$	OR
1	1	0	X	$F = A \oplus B$	XOR
1	1	1	X	$F = \overline{A}$	Complement A

3. Instructions And Addressing Modes

The logical structure of microprocessors is described in reference manuals provided by the manufacturer. A reference manual for a particular microprocessor describes the internal organization of the CPU, the function of all input and output terminals, and the processor registers available from the user's point of view.

The manual describes all the instructions available in the computer and explains their functions. It also shows how status bits are affected by each instruction. The internal code for each instruction is listed in binary, octal, or hexadecimal. In most cases, either the octal or hexadecimal equivalent code is adopted because these codes need fewer digits than the binary representation. When a program is written for the computer, each instruction is assigned a symbolic name to identify it.

Symbolic names and codes assigned to instructions used in one microprocessor are different from the names and codes used in a different microprocessor, even for similar instructions. For this reason, the user must study and remember the set of instructions and their symbolic names every time a different microprocessor is employed. Although the instruction sets of different microprocessors differ from one another, there are certain instructions that perform basic operations and are included in all microprocessors.

Basic Set of Microprocessor Instructions

Microprocessor instructions may be classified into three distinct types.

1. Transfer instructions that move data among registers,memory words, and interface registers without changing. The binary information content.

2. Operation instructions that perform operations on data stored in registers or memory words.

3 Control instructions used to test status conditions in registers and, depending on results, cause a change in program sequence.

The instruction set of a particular microprocessor specifies the register-transfer operations and control decisions that are available in the microcomputer system. A specific program for a microcomputer is equivalent to specifying the sequence of operations for the particular digital system that the microcomputer implements.

Transfer-type instructions in microprocessors are indicated by various names. A move instruction causes a transfer of data from source to destination. Either the source or the destination may be a processor register or a memory location. The load and store instructions are similar to the move instruction, except that they usually refer to transfers from and to memory and accumulator.

The exchange instruction swaps information between two registers or between a register and a memory word. The push and pop instructions transfer data among processor registers and a memory stack. The input and output instructions transfer data among processor registers and interface registers.

Operation-type instructions perform arithmetic, logic, and shift operations among processor registers or memory words.They also set, clear, or complement status or flag bits. Typical operation instructions are add, subtract, AND, OR, complement, and set carry. Most of the operation type instructions also change the status bits in the status register of the processor.

Control-type instructions provide decision-making capabilities and change the path taken by the program when executed in the computer. Instructions are stored in consecutive memory locations and are executed one after the other in sequence.

The programmer inserts a control instruction every time that control must be transferred to an instruction out of normal sequence. control instructions may be conditional or unconditional. A conditional control instruction causes a branch out of normal program sequence only when a specified status condition is detected. An unconditional control instruction causes a branch unconditionally. The branch out of normal program sequence is accomplished by changing the program counter so it contains the address of the instruction that is to be executed next.

There are three types of control instructions and each type may be conditional or unconditional:

1. Jump or branch instructions.
2. Call to and return from subroutine instructions.
3. Skip instructions.

The words jump and branch are used interchangeably to mean the same thing, but sometimes they are used to denote different addressing modes. These instructions are associated with an address that specifies where the jump or branch is to be made. A skip instruction skips the next instruction in sequence. By placing an unconditional branch instruction after the skip instruction, it is possible to branch to one of two possible locations, depending on the value of a specified status-bit condition.

Instructions for Microprocessor

The number of distinct instructions in a particular microprocessor may range from 50 to 250.These instructions must be studied and memorized by the user who writes programs for the microcomputer. A partial list of instructions formulated for the microprocessor is given in below table 2.3 The instructions are divided into three sections to give examples of transfer-, operation-, and control type instructions.

The hexadecimal code listed in the table is a 2-digit number equivalent to the 8-bit operation code assigned to the instruction. The symbolic name for each instruction is a 2- to 4-letter designation followed by one or two symbols for a register, an operand, or a memory address. The description column explains the instruction in words, and the function Column defines the instruction precisely with a register-transfer statement. Note that computer instructions specify macrooperations for the computer and can be symbolized with appropriate statements in the register transfer method.

However ,for various practical reasons, computer instructions are written with specialzed symbols as in the second column of the table. These special symbols are assigned by the computer manufacturer and tend to be different in different computers.

The first four instructions in Table 2.3 are move instructions that transfer information from a given source to a given destination. The next three are load and store instructions that accomplish a similar function. A representative number of operation-type instructions are listed in the second part of the table.The last section lists a few control instructions.

The move with register-indirect instruction:

MOV A, FG

symbolizes the register - transfer operation $A \leftarrow M[FG]$. It transfers the memory byte whose address is in register pair FG into register A. This is called a register-indirect instruction because register pair FG specifies the address of the operand rather than the operand itself.

The load Immediate instruction:

LXI FG, D16

symbolizes the register-transfer operation $FG \leftarrow D16$, where D16 is a 2-byte number that may represent an address. This instruction can be used to transfer an address into register pair FG. When used in this manner,

Hexa decimal code	instruction	Symbol	Description	Function*
78	MOV	A, B	Move B to A	A ← B
3E	MVI	A. D8	Move immediate operand to A	A ← D8
7E	MOV	A, FG	Move to A with register indirect	A ← M[FG]
77	MOV	FG, A	Move A with register indirect	M[FG] ← A
3A	LDA	AD,16	Load A direct	A ← M [A D16]
32	STA	AD16	Store A direct	M [AD 16] ← A
01	LX1	FG, D16	Load register pair immediate	FG ← D 16
8o	ADD	B	Add B to A	A ← A + B
C6	ADI	D8	Add immediate operand to A	A ← A + D8
86	ADD	FG	Add to A with register indirect	A ← A + M [FG]
90	SUB	B	Subtract B from A	A ← A — B
A0	ANA	B	AND B to A	A ← A ∧ B
B0	ORA	B	OR B to A	A ← A VB
04	INR	B	Increment B	B ← B + 1
05	DCR	B	Decrement B	B ← B - 1
03	INX	BC	Increment register pair BC	BC ← BC + 1
0B	DCX	BC	Decrement register pair BC	BC ← BC - 1
2F	CMA		Complement A	A ← $\overline{A}$
07	RLC		Rotate A left with carry	A ← clc A
OF	RRC		Rotate A right with carry	A ← crc A
37	STC		Set carry bit to 1	C ← 1
C3	JMP	AD16	Jump unconditionally	PC ← AD 16
DA	JC	AD16	Jump on carry	If (C = 1) then (PC ← AD 16)
C2	JNZ	AD16	Jump on nonzero	If (Z = 0) then (PC ← AD 16)
CD	CALL	AD16	Call subroutine	Stack ← PC, PC ← AD16
C9	RET		Return from subroutine	PC ← stack
76	HLT		Halt processor	

Table 2.3. A Partial list of instructions for microprocessor

* A = accumulator register; B=B register : FG = register pair F and G; BC = register pair B and C; D8= 8-bit data operand (1byte); D16= 16-bit data operand (2 bytes); AD16=16-bit address (2bytes).

register pair FG constitutes a data counter or a pointer that points to an address in memory where an operand is stored. FG can be incremented with the increment-register-pair instruction:

INX FG

which symbolizes the register-transfer operation FG ← FG + 1.

In this way, the data counter or pointer can be incremented to point at consecutive addresses in memory where the programmer stores a table of consecutive data bytes.

The operation instructions provide the common arithmetic, logic, and shift operations. Note that more instructions of the same type can be formulated if we specify one of the other five processor registers C, D, E, F, or G instead of specifying register B. Similarly, an instruction that specifies a register pair can be duplicated by using any one of the three possible register pairs BE, DE, or FG.

The last six instructions in the table are control instructions. The jump and call instructions need a 16-bit address symbolized by AD 16. The return and halt instructions are one-byte instructions. Those that include the symbol AD 16 or D 16 are three-byte instructions and those that use a D8 symbol are two-byte instructions.

All others are one-byte instructions whether they do or do not specify a register.

Addressing Modes

The operation field of an instruction specifies the operation to be performed. This operation must be executed on some data stored in computer registers or memory words. The way the operands are chosen during program execution is dependent on the addressing mode of the instruction.The addressing mode specifies a rule for interpreting or modifying the address field of the instruction before the operands actually referenced.

Computers use addressing mode techniques for the purpose of accommodating one or both of the following provisions:

1. To give programming versatility to the user by providing such facilities as pointers, counters, indexing, and program relocation.

2. To reduce the number of bits in the address field of the instruction.

To the inexperienced user, the variety of addressing modes in some computers may seem excessively complicated. However, the availability of different addressing schemes gives the experienced programmer flexibility for writing programs that are more efficient with respect to the number of instructions and execution time.

Although most addressing modes modify the address field of the instruction, there are two modes that, need no address field at all. These are the implied and immediate modes.

Implied Mode

In this mode the operands are specified implicitly in the definition of the instruction. For example, the instruction "complement accumalator" is an implied-mode instruction because the operand in the accumulator register is implied in the definition of the instruction. In fact, all register reference instructions that use an accumulator are implied-mode instructions. Zero-address instructions in a stack-organized computer are implied-mode instructions since the operands are implied to be on top of the stack.

Immediate Mode

In this mode the operand is specified in the instruction itself. In other words, an immediate-mode instruction has an operand feild rather than an address field.

The operand field contains the actual operand to be used in conjunction with the operation specified in the instruction. Immediate-mode instructions are useful for initializing registers to a constant value.

Register Mode

In this, mode the operands are in registers which reside within the CPU. The particular register is selected from a register field in the instruction. A k-bit field can specify any one of 2^k register .

Register-Indirect Mode

In this mode the instruction specifies a register in the CPU whose contents give the address of the operand in memory. In other words, the selected register contains the address of the operand rather than the operand itself. Before using a register-indirect mode instruction, the programmer must ensure that the memory address of the operand is placed in the processor register with a previous instruction. A reference to the register is then equivalent to specifying a memory address. The advantage of a register- indirect mode instruction is that the address field of the instruction uses fewer bits to select a register than would have been required to specify a memory address directly.

Auto increment or Auto decrement Mode

This is similar to the register-indirect mode except that the register is incremented or decremented after (or before) its value is used to access memory. When the address stored in the register refers to a table of data in memory, it is necessary to increment or decrement the register after every access to the table. This can be achieved by using the increment or decrement instruction. However, because it is such a common requirement, some computers incorporate a special mode which automatically increments or decrements the content of the register after data accuse.

Direct-Address Mode

In this mode the effective address is equal to the address part of the instruction. The operand resides in memory and its address is given directly by the address field of the instruction. In a branch-type instruction the address field specifies the actual branch address

Indirect-Address Mode

In this mode the address field of the instruction gives the address where the effective address is stored in memory. Control fetches the instruction from memory and uses its address part to access memory again to read the effective address.

A few addressing modes require that the address field of the instruction be added to the content of a specific register in the CPU. The effective address in these modes is obtained from the following computation:

Effective address = Address part of instruction
+ Content of CPU register

The CPU register used in the computation may be the program counter, an index register, or a base register. In either case we have a different addressing mode which is used for a different application.

Relative Address Mode

In this mode the content of the program counter is added to the address part of the instruction in order to obtain the effective address.The address part of the instruction is usually a signed number (in 2's representation) which can be either positive or negative. When this number is added to the content of the program counter, the result produces an effective address whose position in memory is relative to the address of the next instruction.

To clarify with an example, assume that the program counter contains the number 825 and the address part of the instruction contains the number 24. The instruction at location 825 is read from memory during the fetch cycle and the program counter is then incremented by one to 826. The effective address computation for the relative address mode is 826 + 24 = 850. This is 24 memory locations forward from the address of the next instruction.

Relative addressing is often used with branch-type the instructions when the branch address is in the area surrounding the instruction word itself.

Indexed Addressing Mode

In this mode the content of an index register is added to the address part of the instruction to obtain the effective address. The index register is a special CPU register that contains an index value. The adress field of the instruction defines the beginning address of a data array in memory. Each operand in the array is stored in memory relative to the beginning address.The distance between the beginning address and the address of the operand is the index value stored in the index register. Any operand in the array can be accessed with the same instruction provided the index register contains the correct value.

Base Regisster Addressing Mode

In this mode the content of a base register is aded to the address part of the instruction to obtain the effective address. This is similar to the indexed addressing mode except that the register is now called a base register instead of an index register.The difference between the two modes is in the way they are used, rather than in the way that they are computed. An index register is assumed to hold an index number which is relative to the address part of the instruction. A base register is assumed to hold a base address and the address field of the instruction gives a displacement relative to this base address.

The base register addressing mode is used in computers to facilitate the relocation of programs in memory. When programs and data are moved from one segment of memory to another, as required in multiprogramming systems, the address values of instructions must reflect this change of position. With a base register, the displacement values of instructions do not have to change. Only the value of the base register requires updating to reflect the beginning of a new memory segment

4. Parallel Processing

Parallel processing is a term used to denote simultaneous computations in the CPU for the purpose of increasing its computational speed.Instead of processing each instruction sequentially as in conventional computers, a parallel processor performs concurrent data processing tasks to achieve faster execution time. For example, while an instruction is being executed in the ALU, the next instruction can be read from memory. If two or more ALUs are included within the processor unit, it may be possible to execute two or more instructions at the same time.The purpose of parallel processing is to speed up the computer processing capability and increase its throughput, that is, the amount of processing that can be accomplished during a given interval of time. The amount of Hardware increases with parallel processing and with it the cost of the system increases. However, tecnhological developments have reduced hardware costs to the point where certain parallel processing techniques are economically feasible .

The normal operation of a CPU is to fetch instructions from memory and execute them in the processor. The steps associated with reading instructions from memory can be viewed as forming an instruction stream. The various operations performed on the data in the processor form a data stream. Parallel processing may occur in the instruction stream, in the data stream, or both. If a single instruction stream causes parallel execution of the incoming data, the system is said to operate with a single instruction stream and multiple data stream or SIMD. Another possibility is MISD, which is a system with a multiple instruction stream and a single data stream. In most cases, parallel processing exhibits parallelism in both the instruction and data streams, forming a multiple-instruction multiple data stream or MIMD.

Pipeline Processing

Pipeline is a technique of decomposing a sequential process into sub processes with each sub process being executed in a special dedicated segment that operates concurrently with all other segments. A pipeline can be visualized as a collection of processing segments through which binary information flows. Each segment performs partial processing dictated by the way the task is partitioned. The result obtained from the computation in each segment is transferred to the next segment in the pipeline. The final result is obtained after the data have passed through all segments.

The name "pipeline" implies a flow of information analogous to an industrial assembly line. It is characteristic of pipelines that several computations can be in progress in distinct segments at the same time. The overlapping of computation is made possible by associating a register with each segment in the pipeline. The registers provide isolation between each segment so that each can operate on distinct data simultaneously The Pipeline concept will be demonstrated by means of a very simple example This example-is of no practical value and is used merely for illustrative purposes.

Suppose that we want to subtract a series of numbers:

$$C_i = A_i - B_i \quad \text{for } i = 1, 2, 3, \ldots, n$$

Each subtraction will be performed with a sequence of three suboperations: (1) complement B_i ; (2) inerernent to get the 2's complement ; and (3) add the 2's complement to A_i. Each suboperation is to be implemented in a segment within a pipeline.

Each segment has one or two registers and a combinational circuit as shown in Fig 2.6 (a). R1 through R5 are registers that receive new input data with every clock pulse. The complementer, incrementer and adder are combinational circuits. The suboperations performed in each segment of the pipeline are as follows:

$R1 \leftarrow B_i$,	Input B_i
$R2 \leftarrow \overline{R1}$	Complement
$R3 \leftarrow R2 + 1$, $R4 \leftarrow A_i$	Increment and, input A_i
$R5 \leftarrow R3 + R4$	Add to get $A_i + \overline{B_i} + 1$

The five registers are loaded with new data every clock pulse. The effect of each clock pulse is shown in Fig. 2.6 (b). It takes four clock pulses to fill up the pipe and retrieve the first output from register R5. From there on, each clock pulse produces a new output and moves the data one step down the pipeline. This happens as long as new input data flow into the system.When no more input data are available, the clock pulses must continue until the last output emerges out of the pipeline.

Any operation that can be decomposed into a sequence of suboperations of about the same complexity can be implemented by a pipeline processor .

The procedure is efficient only in those applications where the same computation must be repeated on a stream of input data. Pipeline processing has been applied mostly to floating-point arithmetic operations. These operations are easily decomposed into consecutive suboperations

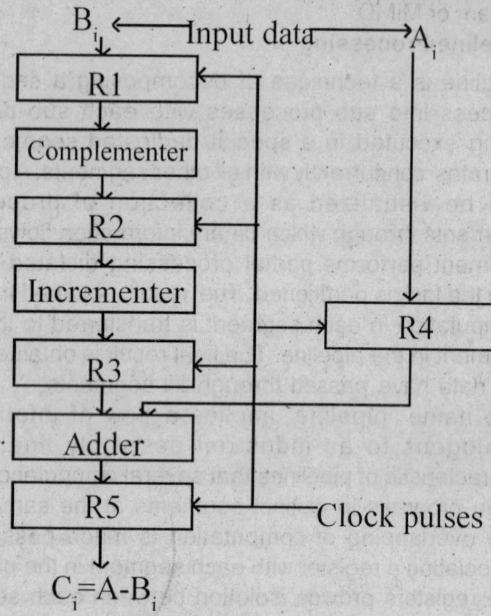

(a)Segmented pipeline hardware

Pulse No	R1	R2	R3	R4	R5
1	B_1	–	–	–	–
2	B_2	$\overline{B_1}$	–	–	–
3	B_3	$\overline{B_2}$	$\overline{B}_1+1$	A_1	–
4	B_4	$\overline{B_3}$	$\overline{B}_2+1$	A_2	C_1
5	B_5	$\overline{B_4}$	$\overline{B}_3+1$	A_3	C_2
6	B_6	$\overline{B_5}$	$\overline{B}_4+1$	A_4	C_3
7	B_7	$\overline{B_6}$	$\overline{B}_5+1$	A_5	C_4
8	–	$\overline{B_7}$	$\overline{B}_6+1$	A_6	C_5
9	–	–	$\overline{B}_7+1$	A_7	C_6
10	–	–	–	–	C_7

(b) Contents of registers after each clock pulse

Figure 2.6 A simple example demonstrating pipeline processing.

For example , the floating- point addition can be divided into three consecutive suboperations.

1. Align the mantissas.
2. Add the mantissas.
3. Normalize the result.

The floating-point add hardware can be structured as a pipeline with three segments . This is demonstrated in Fig.2.7. The internal operation of each segment is demonstrated by means of a numerical example. Two floating-point numbers X and Y are

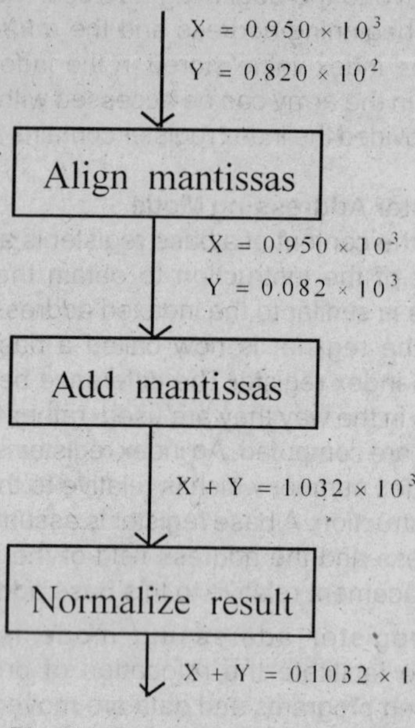

Figure 2.7. Pipeline Processing of floating point addition

aligned by shifting one of their mantissas until their two exponents are equal.Adding the mantissas may cause an overflow. This is adjusted by normalizing the result so that the sum is always a fraction with a nonzero first digit. New sets of numbers can enter the pipeline as the information moves along the three segments to form the sum.

Instruction Pipeline

Pipeline Processing can occur not only in the data stream but in the instruction stream as well. An instruction pipeline reads consecutive instructions from memory while previous instructions are being executed in the processor. This causes the instruction fetch and execute cycles to overlap and perform simultaneous operations.An instruction pipeline is also known by the name of instruction look-ahead buffer, as it reads instructions ahead of their use by the processor

An instruction Pipeline can be implementd by means of a first-in first- out (FIFO) buffer.This is a type of unit that forms a queue rather than a stack.When ever the execute cycle is not using the memory, the control increments the program counter and uses its address value to read consecutive instructions from memory. The instructions are inserted in the FIFO buffer so that they can be executed on a first-in first-out basis. Thus, an instruction stream can be placed in a queue, waiting for decoding and execution by the processor.

The instruction stream queuing mechanism provides an efficient way for reducing the average access time to memory for reading instructions. Whenever there is space in the buffer, the processor executes an instruction fetch cycle. The buffer acts as a queue from which control extracts instructions for execution. If the CPU executes an instruction that transfers control to a location out of normal sequence, the buffer is reset and the pipeline is declared empty. Control then fetches the instruction from the branch address and begins to refil the buffer from the new location. Often branch instructions form a small fraction of the total number of instructions executed, so throughput can be increased by this kind of instruction pipeline.

Memory Interleaving

In order to carry out two or more simultaneous accesses to memory, the memory must be partitioned into separate modules vvhich are independent of each other. A memory module is a memory array together with its own address and buffer registers. Figure 2.8 shows a memory unit organized in four modules. Each memory array has its own address register MAR and buffer or data register MBR. The address registers receive information from a common address bus and the buffer registers communicate with a bidirectional data bus. The two least significant bits of the address can be used to distinguish between the four modules. The modular system permits one module to be accessed while other modules are in a process of reading or writing a word since each module can honor a memory request independent of the state of other modules.

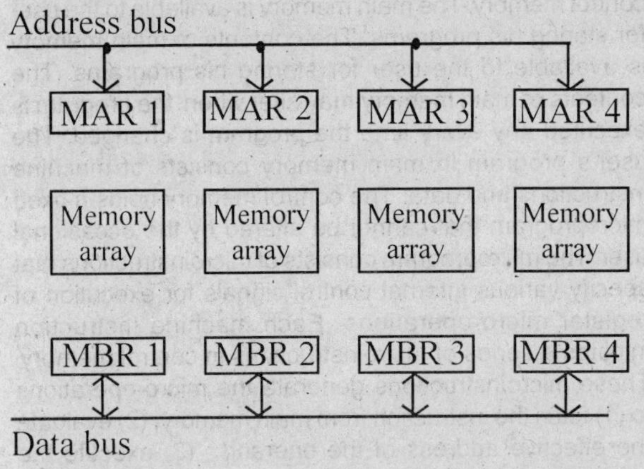

Fig. 2.8 Multiple module memory organization

The advantage of a modular memory is that it allows the use of a technique called interleaving . In an interleaved memory, consecutive addresses are assigned to different memory modules.

For example, in a two-module memory system, the even addresses are in one module and the odd addresses in the other. The control can fetch two consecutive instructions from two modules and place them in an instruction buffer.with m- way interleaving m consecutive instructions can be fetched from m different modules. By staggering the memory access, the effective memory cycle time can be reduced by a factor somewhat less than the number of modules.

A modular memory is also useful in systems with pipeline processing. If instructions and operands are placed in two different-modules, the instruction fetch cycle and reading of the operand may be overlapped.A pipelined organized CPU can take advantage of multiple memory modules so that each segment in the pipeline can access memory independent of memory requests from other segments.

Multiple Functional Units

Parallel processing can be achieved through separate functional units that perform identical or different operations simultaneously. Parallel processing is established by distributing the data among the multiple functional units.

For example, floating-point, arithmetic operations can be implemented using two functional units, one for the exponents and the other for the mantissas. The multiplication of two floating-point numbers requires that we multiply the mantissas and add the exponents. These two operations can be done concurrently in the two separate functional units.

Another possibility is to separate the arithmetic, logic, and shift operations into separate units and divert operands to each unit under the supervision of a special control unit. Any possible parallelism associated with programs can be exploited by using a multifunction processor unit.

Vector Processor

A vector processor includes vector operations in its instruction set in addition to the conventional set of instructions normally found in a CPU. A typical vector instruction has the form $C(I) = A(I) * B(I)$ for I = 1 through n, where the symbol * represents the operation of the vector instruction. Typical vector instructions include arithmetic operations with floating-point data, logical operations, compare and test, matrix operations such as dot product, and search instructions for maximum or minimum values. Vector processors are especially designed to execute complex instructions quickly and efficiently.This is done by an internal organization that combines multiple functional units operating in parallel together with pipeline processing within the functional units.

As an example of a vector instruction execution consider the following DO loop:

```
DO I = 1, 50
C(I) = A(I) + B(I)
```

This defines a program loop that adds 50 sets of numbers. For most computers, the machine-language equivalent of the above DO loop is a sequence of instructions that read a pair of operands from the A and B arrays and perform a floating-point addition.

The loop control variables are then updated and the steps repeated 50 times. A vector processor eliminates the overhead associated with the time it takes to fetch and execute the instructions in the program loop. This is because a vector instruction includes the initial addresses of the operands, the length of the vectors, and the operation to be performed, all in one composite instruction.

Two operands A(I) and B(I) are read simultaneously from an interleaved memory and directed to a pipeline processor that performs the floating-point addition. After the first sum C(1) emerges from the pipeline, it is stored in a third interleaved memory module while succeeding operations are being performed within the pipeline.

A processor suitable for vector processing sometimes called an array processor. This is because the processor is capable of performing parallel computations on large arrays of data. The name "array" is also used to signify a special type of parallel organization. Here the array designates a hardware connection of a number of identical processing units connected together with direct paths between each unit and its neighbours.

Vector processors are not suitable for normal everyday processing in a typical computer installation. They are limited to a small number of scientific applications, such as nuclear research, seismic analysis, and weather prediction and modeling.The number of vector processors available commercially is very small. The three most commonly cited examples are Gray-1, Cyber 205, and Burroughs Scientific processor

5. Hardwired and Micro-Programmed Control

The function of the control unit in a digital system is to initiate sequences of micro operations. The number of different micro-operations available in a given system is finite. The complexity of the digital system is derived from the fact that a large number of micro-operations are performed in a given time sequence. When the control functions are generated by hardware using conventional logic design techniques, the control unit is said to be hard-wired. Microprogramming is a second alternative for designing the control unit of a digital computer.

The principle of microprogramming is an elegant and systematic method for generating the micro-operation sequences in a digital system.

The control function that specifies a micro-operation is a binary variable. When it is in the 1 state, the corresponding micro-operation is executed. A control variable in its 0 state does not affect the state of the registers in the system. There are occasions when the state of control variables are defined in reverse with the 0 state initiating the micro-operation and the 1 state being the idle state. In a bus-organized system, the control signals that specify micro-operations are groups of bits that select the paths in multiplexers, decoders, and arithmetic logic units.

The control unit initiates a series of sequential steps of micro-operations. During any given time interval, certain micro-operations are to be initiated while all others remain idle. Thus the micro-operation steps in each time interval can be represented by a string of 1's and 0's called a control word. As such, control words can be programmed to initiate the various components of the system in a prescribed fashion. A control unit whose micro-operation steps are stored in a memory is called a micro-programmed control unit. Each control word of memory is called a microinstruction and a sequence of words is called a microprogram. Since alterations of the microprogram are seldom needed, the memory can be a read-only memory (ROM). The concept of micro programming was not practical to implement until fast read-only memories became available commercially. The utilization of a microprogram involved placing all micro-operation steps in words of ROM for use by the control unit through successive read operations. The content of the words in ROM are fixed and cannot be altered by simple programming since no writing capability is a available in the ROM. ROM words are made permanent during the hardware production of the unit. Although sometimes programmable read-only memories (PROM) are used, the programming procedure is a hardware process requiring the opening of used links. Once a PROM is programmed it cannot be altered by simple programming.

A more advanced development known as dynamic microprogramming permits a prescribed microprogram to be loaded initially from either the computer console or from an auxiliary memory. Control units that use dynamic microprogramming employ a writable control memory (WCM). This memory has the capability of writing (to change the microprogram) but is used mostly for reading. A read-only memory or a writable control memory when used in the control unit of a computer is referred to as a control memory.

A computer that employs a microprogrammed control unit will have two separate memories : a main memory and a control memory. The main memory is available to the user for storing his programs. The contents of main memory is available to the user for storing his programs. The contents of main memory may alter when the program is executed any every time the program is changed. The user's program in main memory consists of machine instructions and data. The control memory holds a fixed microprogram that cannot be altered by the occasional user. The microprogram consists of microinstructions that specify various internal control signals for execution of register micro-operations. Each machine instruction initiates a series of microinstructions in control memory. These microinstructions generate the micro-operations to (1) fetch the instruction from main memory; (2) evaluate the effective address of the operand ; (3) execute the operation specified by the instruction; and (4) return control to the beginning of the fetch cycle to repeat the sequence again for the next instruction.

Figure 2.9 illustrates the general configuration of the microprogrammed control unit. The control memory is assumed to be a ROM, within which all control information is permanently stored. The control address register specifies the word read from control memory. It must be realized that a ROM operates as a combinational circuit

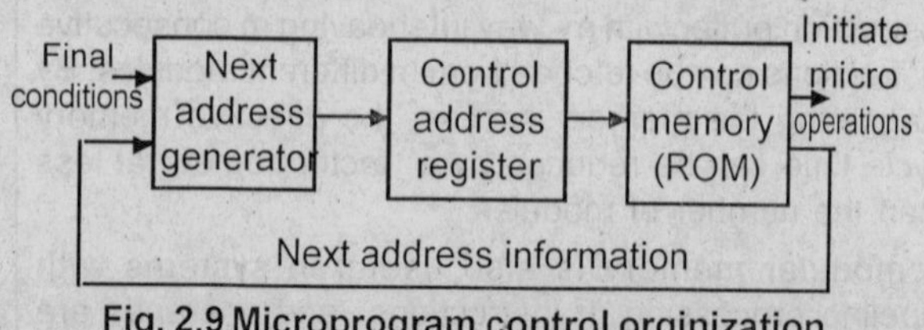

Fig. 2.9 Microprogram control orginization

with theaddress value as the input and the corresponding word as the output.The content of the specified word remains in the output wires as long as the address value remains in the address register. No read signal is needed as in a random-access memory. The word content of ROM should be transferred to a buffer register if the address register changes while the ROM word is still in use. If the change in address and ROM word can occur simultaneously, no buffer register is needed.

The word read from control memory represents a microinstruction. The microinstruction specifies one or more micro-operations for the components in the computer. Once these micro-operations are executed, the control unit must determine its next, address. The location of the next microinstruction may be the one next in sequence or, it may be located somewhere else in the control memory. For this reason it is necessary to use some bits of the microinstrucion to control the generation of the address for the next microinstruction. A given microinstruction may also have special bits that specify explicitly the address of the next microinstruction. While the micro-operations are being executed in the processor the next address is computed in the next address generator circuit and then transferred into the control address register to read the next microinstruction. Thus, a microinstruction contains bits for controlling the micro-operations in the digital system and bits that determine the address sequence of the control memory itself.

SOLVED EXAMPLES

1. A digital computer has a common bus system for 16 registers of 32 bits each. The bus is constructed with multiplexers.

a. How many selection inputs are there in each multiplexer?

b. What size of multiplexers are needed ?

c. How many multiplexers are there in the bus ?

Sol : a. 4 selection lines to select one of 16 registers.

b. 16 × 1 multiplexers are needed.

c. 32 multiplexers, one for each bit of the registers.

2. A computer uses a memory unit with 256K words of 32 bits each. A binary instruction code is stored in one word of memory. The instruction has four parts : an indirect bit, an operation code, a register code part to specify one of 64 registers, and an address part.

a. How many bits are there in the operation code, the register code part, and the address part ?

b. Draw the instruction word format and indicate the number of bits in each part.

c. How many bits are there in the data and address inputs of the memory ?

Sol : $256\text{ K} = 2^8 \times 2^{10} = 2^{18}$

$64 = 2^6$

a. Address : 18 bits

Register code : 6 bits

Indirect bit : 1 bit

25

32 - 25 = 7 bits for opcode,

b.

1	7	6	18
1	Opcode	Register	Address

= 32 bits

c. Data : 32 bits;

address : 18 bits.

3. The control memory has 4096 words of 24 bits each.

a. How many bits are there in the control address register ?

b. How many bits are there in each of the inputs shown going into the mulitplexers ?

c. What are the number of inputs in each multiplexer and how many mulitplexers are needed ?

Sol : Control memory = $2^{12} \times 24$

a. 12 bits b. 12 bits

c. 12 multiplexers, each of size 4-to-1 line.

4. A bus-organized CPU similar to figure given below has 16 registers with 32 bits in each, an ALU, and a destination decoder.

a. How many multiplexers are there in the A bus, and what is the size of each multiplexer ?

b. How many selection inputs are needed for MUX A and MUX B ?

c. How many inputs and outputs are there in the decorder ?

d. How many inputs and outputs are there in the ALU for data, including input and output carries ?

e. Formulate a control word for the system assuming that the ALU has 35 operations.

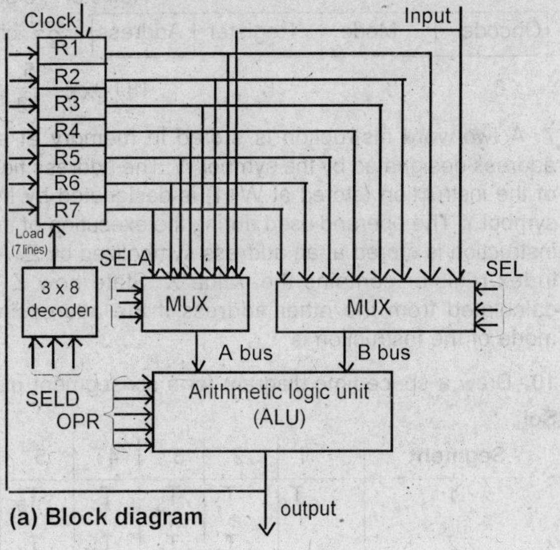

(a) Block diagram

3	3	3	3
SELA	SELB	SELD	OPR

(b) Control word

Figure : Register set with common ALU.

Sol : a. 32 multiplexers, each of size 16 x 1.

b. 4 inputs each, to select one of 16 registers.

c. 4-to-16 line decoder.

d. 32 + 32 +1 = 65 data input lines.

e. 32 + 1 = 33 data output lines.

4	4	4	6 = 18 bits
SELA	SELB	SELD	OPR

5. A computer has 32 - bit instructions and 12-bit address. If there are 250 two-address instructions, how many one-address instructions can be formulated ?

Sol :

8	12	12	= 32 bits
Opcode	Address	Address	

(Two address instructions)

2^8 = 256 combinations

256 - 250 = 6 combinations can be used for one address.

Opcode	Address

6×2^{12} (one address instructions)

Maximum number of one address instructions :

$= 6 \times 2^{12} = 24{,}576.$

6. The memory unit of a computer has 256 K words of 32 bits each. The computer has an instruction format with four fields : an operation code field, a mode field to specify one of seven addressing modes, a register address field to specify one of 60 processor registers, and a memory address. Specify the instruction format and the numbe of bits in each field if the instruction is in one memory word.

Sol : 256 K = $2^8 \times 2^{10} = 2^{18}$ Address = 18 bits

Mode = 3 bits

Register= 6 bits

27 bits

5

Opcode $\overline{32}$ bits

Opcode	Mode	Register	Address
5	3	6	18

7. A two-word instruction is stored in memory at an address designated by the symbol W. The address field of the instruction (stored at W+1) is designated by the symbol Y. The operand used during the execution of the instruction is stored at an address symbolized by Z. An index register contains the value X. State how Z is calculated from the other address if the addressing mode of the instruction is

a. direct PC

b. indirect

c. relative XR =X

d. Indexed

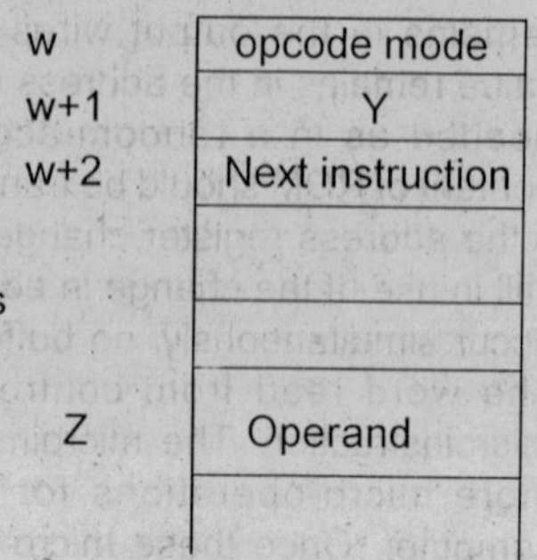

Sol : Z= Effective address

a. Direct : Z = Y

b. Indirect : Z = M[Y]

c. Relative : Z = Y+W+2

d. indexed : Z= Y+X

8. An instruction is stored at location 300 with its address field at location 301. The address field has the value 400. A processor register R1 contains the numbe 200. Evaluate the effective address if the addressing mode of the instruction is (a) direct ; (b) immediate; (c) relative; (d) register indirect; (e) index with R1 as the index register.

Sol : Effective address

a. Direct : 400

b. Immediate : 301

c. Relative : 302 + 400 = 702 R1=200

d. Register indirect : 200

e. Indexed : 200+400=600

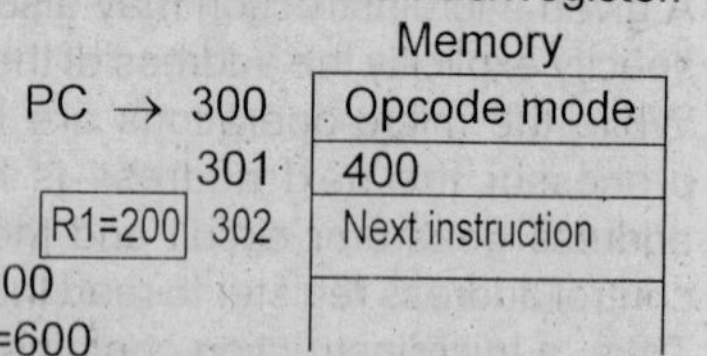

9. In certain scientific computations it is necessary to perform the arithmetic operation $(A_i + B_i)(C_i + D_i)$ with a stream of numbers. Specify a pipeline configuration to carry out this task. List the contents of all registers in the pipeline for i =1 through 6.

Sol :

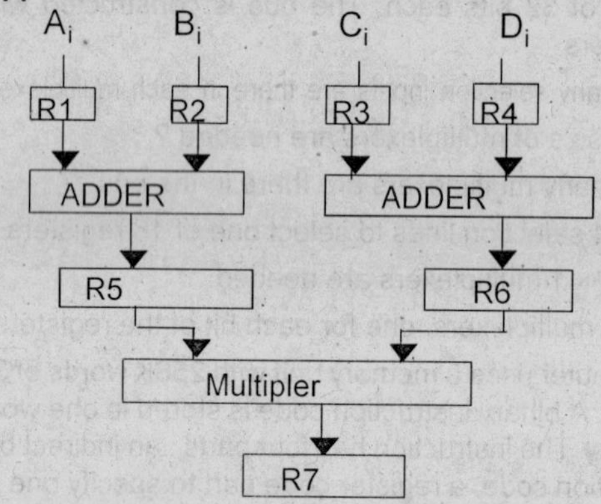

10. Draw a space-time diagram for a six-segment pipeline showing the time it takes to process eight tasks.

Sol :

Segment	1	2	3	4	5	6	7	8	9	10	11	12	13
1	T_1	T_2	T_3	T_4	T_5	T_6	T_7	T_8					
2		T_1	T_2	T_3	T_4	T_5	T_6	T_7	T_8				
3			T_1	T_2	T_3	T_4	T_5	T_6	T_7	T_8			
4				T_1	T_2	T_3	T_4	T_5	T_6	T_7	T_8		
5					T_1	T_2	T_3	T_4	T_5	T_6	T_7	T_8	
6						T_1	T_2	T_3	T_4	T_5	T_6	T_7	T_8

↑

$(k+n-1)t_p = 6+8-1 = 13$ cycles.

11. Determine the number of clock cycles that it takes to process 200 tasks in a six-segment pipeline.

Sol : k = 6 segment

n = 200 tasks $\Rightarrow$ (k + n - 1) = 6 + 200 - 1 = 205 cycles.

12. A nonpipeline system takes 50 ns to process a task. The same task can be processed in a six-segment pipeline with a clock cycle of 10 ns. Determine the speedup ratio of the pipeline for 100 tasks. What is the maximum speedup that can be achieved ?

Sol : $t_n = 50$ ns ; $S = \frac{nt_n}{(k+n-1)} = \frac{100\times 50}{(6+99)\times 10} = 4.76$

$k=6,\ t_p=10ns,\ n=100$

$$S_{max} = \frac{t_n}{t_p} = \frac{50}{10} = 5$$

13. The pipeline of figure shown below has the following propagation times: 40 ns for the operands to be read from memory into registers R1 and R2, 45 ns for the signal to propagate through the multiplier, 5 ns for the transfer into R3, and 15 ns to add the two numbers into R5.

a. What is the minimum clock cycle time that can be used ?

b. A nonpipeline system can perform the same operation by removing R3 and R4. How long will it take to multiply and add the operands without using the pipeline ?

c. Calculate the speedup of the pipeline for 10 tasks and again for 100 tasks.

d. What is the maximum speedup that can be achieved ?

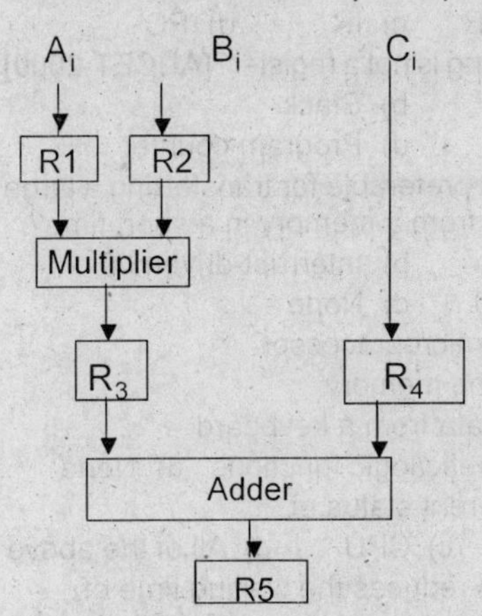

Figure - Example of pipeline processing

Sol : a. $t_p = 45 + 5 = 50ns$; k=3

b. $t_n = 40 + 45 + 15 = 100ns$

c. $S = \frac{nt_n}{(k+n-1)t_p} = \frac{10\times 100}{(3+9)\ 50} = 1.67$ for n= 10

$= \frac{100\times 100}{(3+99)\ 50} = 1.96$ for n=100

d. $S_{max} = \frac{t_n}{t_p} = \frac{100}{50} = 2$

14. Consider a computer with four floating point pipeline processors. Suppose that each processor uses a cycle time of 40 ns. How long will it take to perform 400 floating-point operations ? Is there a difference if the same 400 operations are carried out using a single pipeline processor with a cycle time of 10 ns ?

Sol : Divide the 400 operations into each of the four processors. Processing time is :

$$\frac{400}{4}\times 40 = 4000\,n\,sec.$$

Using a single pipeline, processing time is

$$= 400\times 10 = 4000n\,sec.$$

OBJECTIVE QUESTIONS - 1

1. The sequence of events that happen during a typical fetch operation is
a) PC $\rightarrow$ MAR $\rightarrow$ Memory $\rightarrow$ MDR $\rightarrow$ IR
b) PC $\rightarrow$ Memory $\rightarrow$ MDR $\rightarrow$ IR
c) PC $\rightarrow$ Memory $\rightarrow$ IR
d) PC $\rightarrow$ MAR $\rightarrow$ Memory $\rightarrow$ IR

2. The advantage of a single bus over a multi-bus is the
a) Low cost
b) Hign operating speed
c) Flexibility in attaching peripheral devices
d) All of the above

3. A micro-programmed control unit
a) Is faster than a hard-wired control unit.
b) Facilitates easy implementation of new instructions
c) Is useful when very small programs are to be run
d) Usually refers to the control unit of a microprocessor

4. Programming in a language that actually controls the path of signals or data within the computer is called
a) Micro programming
d) Systems programming
c) Assembly language programming
d) Machine language programming

5. Most of the digital computers do not have floating-point hardware because
a) It is costly b) It is slower than software
c) floating point addition cannot be performed by hardware d) None

6. Micro program is
a) Name of source program in micro computers
b) The set of instructions indicating the primitive operations in a system
c) Primitive form of macros used in assembly language programming
d) Program of very small size

7. The three main components of a digital computer system are
a) Memory , I/O , DMA b) ALU , CPU , Memory
c) Memory , CPU , I/O d) None

8. A subtractor is not usually present in a computers because
a) It is expensive
b) It is not possible to design it
c) The adder will take care of subtraction
d) None

9. The stack is nothing but a set of
a) Reserved ROM address space
b) Reserved RAM address space
c) Reserved I/O address space
d) None

10. In a vectored Interrupt
a) The branch address is assigned to a fixed location in memory
b) The branch address is obtained from a register in the processor
c) The interrupting source supplies the branch information to the processor througn an interrupt vector. d) None

11. Bit - slice processors
a) Can be cascaded to get any desired word length processor
b) Speed of operation is independent of the word length configured.
c) Contain only the data path of a normal CPU
d) None

12. If a particular idea can be implemented in hardware/ software, the factor(s) that favour hardware implementation is/are :
a) Cost effectiveness b) speed of operation
c) reliability d) frequent changes expected

13. To achieve parallelism, one needs a minimum of
a) 2 processors b) 3 processors
c) 4 processors d) None

14. A processor needs s/w interrupt to **[GATE-2001]**
a) Test the interrupt system of the processor
b) Implement co-routines
c) Obtain system services which need execution of privileged instructions
d) Return from sub routines.

15. System calls are usually invoked by using **[GATE-1999]**
a) A s/w interrupt b) An indirect jump
c) Polling d) A privileged instruction

16. Which of the following is TRUE? **[GATE-1998]**
a) Unless enabled,a CPU will not be able to process interrupts
b) Loop instructions can't be interrupted till they complete
c) A processor checks for intrrupts before executing a new instruction
d) Only level triggered intrrupts are possible on microprocessors

17. A fech cycle
a) Reads data b) Reads instruction
c) Reads address of operand
d) Reads operand

18. Daisy chain is a device for
a) Connecting a no.of devices to a controller
b) Connecting a number of controllers to a device
c) Inter-connecting number off devices to number of controllers d) None

19. Which is not characterstic of a daisy chain priorty control scheme? **[GRE-CS]**
a)It is relatively easy to add more devices to the chian
b) The failure of one device may affect other devices on the chain
c) Priority is programmable d) None

20. When will DMA most probably be used? **[GRE-CS]**
a) The performance of a system is to be increased
b) Sevral CPU's in a multi-processing system share the same memory
c) The CPU most control a variety of the devices
d) All of the above

21. Pipelining is used to increase the throughput of **[AUCET-2000]**
a) CPU b) ALU c) Both a and b d) None

22. An instruction pipeline is called when the following instructions occurs **[AUCET-2000]**
a) Register Transfer b) Branch
c) Stack Operations d) Arithmetic operations

23. Micro programmed control provides, execution of machine instructions. **[AUCET-2000]**
a) Faster b) Inexpensively
c) In parallel d) Concurrently

24. Which of the following is responsible for coordinating various operations using timing signals?
a) ALU b) Control unit
c) Memory unit d) I/O unit

25. Arithmetic logic unit
I. Perform arithmetic operations
II. Store data
III. Perform comparisons
IV. Communicate with input devices
a) I only b) III only c) I and II only d) I and III only

26. Which of the following registers is used to keep track of address of the memory location where the next instruction is located?
a) MAR b) MDR c) IR d) PC

27. Which of the following is not a register?**[AUCET-2000]**
a) Buffer b) Stack
c) Accmulator d) Program counter

28. Which technique is preferable for transferring a large amount of data to and from a memory in a short time?
a) DMA b) Interrupt-driven I/0
c) Programmed I/0 d) None

29. The control unit of microprocessor
a) Stores data in the memory
b) Accepts input data from a keyboard
c) Performs arithmetic/logic functions d) None

30. PSW contains different status of
a) ALU b) CU c) CPU d) All of the above

31. Instruction perfetch reduces the waiting time of
a) Memory b) CPU c) ALU d) CU

32. Speed of the computer depends on speed of
a) The clock (frequency)
b) The CPU(μp) c) The CPU and memory
d) The CPU, memory and peripherals

33. When an instruction operates an only one particular operand, the addressing mode used is
a) Register b) Direct
c) Immediate d) Implicit

34. Which of the following addressing modes permits relocation without any change what ever in the code **[GATE-1998]**
a) Indirect addressing
b) Indexed addressing
c) Base Register addressing
d) PC relative addressing

35. An instruction which contains the data in it uses
a) Register Addressing b) Immediate Addressing
c) Relative Addressing d) Implicit Addressing

36. Efficient use of addressing modes
a) Speeds up execution
b) reduces the number of instructions
c) reduces the size of instructions
d) All of the above

37. In which addressing mode, the effective address of the operand is generated by adding a constant value to the contents of a register?
a) Absolute mode b) Indirect mode
c) Immediate mode d) Index mode.

38. Index register in a microprocessor is used for
a) Indirect Addressing
b) Pointing to the Stack address
c) Address modification
d) keeping track of the number of times a loop is executed

39. In which addressing mode the address of the location of the operand is given explicitly as a part of the instruction?
a) Absolute mode b) Immediate mode
c) Indirect mode d) Index mode

40. In which addressing mode the operand is given explicitly in the instruction?
a) absolute mode b) Immediate mode
c) Indirect mode d) Index mode

41. In which addressing mode the effective address of the operand is generated by adding a constant value to the contents of a register?
a) Absolute mode b) Indirect mode
c) Immediate mode d) Index mode

42. The most relevant addressing mode to write position independent code is
a) Direct mode b) Indirect mode
c) Relative mode d) Indexed mode

43. Which one of the following is an incorrect register transfer statement ? **[UGC NET - 2000]**
a) xT : PC ← PC+ 1, PC AR
b) xT : AR ← M [AR]
c) xT : AR ← AR + 1
d) xT : DR ← AC, → AC ← DR

44. The CPU register (for 8 bit microprocessors)that holds the address of next for execution instruction is called **[UGC NET - 1999]**
a) Index register b) Program counter
c) Stack pointer d) Accumulator

45. For each instruction in program memory , the CPU goes through a **[UGC NET - 1999]**
a) fetch-decode-execute sequence
b) decode-fetch-execute sequence
c) execute -store- decode sequence
d) fetch-execute-decode sequence

46. The operand of an instruction with an Indirect Addressing mode is **[UGC NET - 1998]**
a) the contents of a CPU register
b) within the instructions
c) in a memory location whose address is in a CPU register
d) in a memory location whose address is given in the instruction

47. A RISC processor has **[UGC NET - 1998]**
a) variable size instructions
b) complex memory addressing modes
c) fixed size instruction with onlyload/store memory access
d) micro-programmed control unit

48. TheVon-Neumann computer architecture is basedon
a) one CPU and a memory **[UGC NET - 1998]**
b) many CPUs, a memory and many I/Os
c) one CPU,a memory hierarchy and many I/Os
d) one CPU, many memories and one I/O

49. The horizontal microprogramming **[UGC NET - 1998]**
a) reduces the size of micro instruction
b) increases size of control memory
c) needs decoding circuits
d) reduces size of control memory

50. The CPU consists of **[UGC NET - 1997]**
a) input, output and processing
b)control unit,primary storage and secondary storage
c) control unit, arithmetic-logic unit and primary storage d) None

51. The function of control unit in the CPU is to
a) decode program instructions **[UGC NET - 1997]**
b) transfer data to primary storage
c) perform logical operations
d) store program instructions

52. A CPU's processing power is measured in **[UGC NET - 1997]**
a) IPS b) CIPS
c) MIPS d) Nano-seconds

53. The central processing unit of a computer consists of
a) Arithmetic and Logic Unit **[UGC NET - 1996]**
b) Primary Storage
c) Control Unit d) All of these

54. Program counter contents indicate **[UGC NET - 1996]**
a) the time elapsed since execution began
b) the address where next instruction is stored
c) the count of program being executed after switching the power ON d) None

55. Which one of the following is not an addressing technique ? **[UGC NET - 1996]**
a) Direct addressing b) Indirect addressing
c) Semi-direct addressing d) Immediate addressing

56. The storage place that holds data and instruction temporarily within the CPU is called **[UGC NET - 1996]**
a) Accumulator b) Bus
c) Register d) Addresses

57. CPU register that holds the address of the next instruction for execution is called **[UGC NET - 1996]**
a) Stack Pointer b) Program Counter
c) Accumulator d) None

58. During a fetch cycle the CPU **[Hyd.Univ-2000]**
a) Performs the operation specified in the instruction's opcode
b) Retrieves from memory the address of the last instruction before a jump
c) Looks for a new jump location as pointed to by the program counter
d) Retrieves from memory the instruction to be executed

59. How many address lines are there in the address bus of the CPU **[Hyd.Univ-1998]**
a) 12 b) 14 c) 16 d) 18

60. The instruction register contains **[Hyd.Univ-1994]**
a) the address of the instruction to be fetched
b) the data item to be written to memory
c) the data item most recently retrieved from memory
d) the instruction most recently fetched

ANSWER KEY - 1

1. a	2. a,c	3. b	4. a	5. a
6. c	7. c	8. c	9. b	10. a
11. a	12. c	13. d	14. c	15. a
16. a	17. b	18. a	19. c	20. d
21. a	22. d	23. a	24. b	25. d
26. d	27. a	28. a	29. d	30. c
31. b	32. b	33. d	34. c	35. b
36. d	37. d	38. c	39. a	40. b
41. d	42. c	43. a	44. b	45. a
46. c	47. c	48. a	49. a	50. c
51. a	52. c	53. d	54. b	55. c
56. c	57. b	58. d	59. c	60. c

EXPLANATORY NOTES - I

6. (c) **Explanation :** The micro operation $F=A+\overline{B}+1$, which produces the sum of A plus the 2's complement of B. This is equivalent to the subtraction operation F = A-B. In such way there is no need of a subtractor in computer's, the adder will take care of subtraction.

19. (c) **Explanation :** Priority in daisy chaining is determined by the physical position of the device in the chain, and it is not programmable.

20. (d) **Explanation :** Direct memory Access (DMA) is a mean of transferring data between memory and some external device without assistance of the CPU. Therefore, DMA can be used in all the given situations.

43. (a) **Explanation :** Can not transfer a new value in to a register (P.C) and increment the original value by one at the same time. Hence answer 'a' is the incorrect register transfer statement.

OBJECTIVE QUESTIONS - II

1. Which of the following are typical characteristics of a RISC machine ?
a) Instruction taking multiple cycles
b) Highly pipelined
c) Instructions interpreted by micro-programs
d) Multiple register sets

2. On receiving an interrupt from an I/O device, the CPU
a) Halts for a predetermined time
b) Hands over control of address bus and data bus to the interrupting device
c) Branches off to the interrupt service routine immediately.
d) Branches off to the interrupt service routine after completion of the current instruction

3. A certain machine uses expanding OPCODE. It has 16 bit instructions and 6 bit addresses. It supports one address and two address instructions only. If there are 'n' two address instructions the maximum number of one address instructions are :
a) 2^{10} b) 2^{10} - n c) 2^{16} - n d) $(2^4 - n) \times 2^6$

4. Assume a system having two CPU's, each executing different programs. At the same time one of the CPU's tries to increment a variable x, while the others CPU tries to decrement the same x. The final value of x can never be
a) It's original value b) It's original value + 1
c) It's original value - 1 d) None

5. The performance of a pipelined processor suffers if **[GATE-2002]**
a) The pipeline stages have different delays
b) Consecutive instructions are dependent on each other
c) The pipeline stages share h/w resources
d) All of the above

6. Horizantal microprogramming **[GATE-2001]**
a) Does not require use of signal decoders
b) Results in larger sized microinstructions than vertical micro-programming
c) Uses one bit for each control signal
d) All of the above

7. Suppose a processor does not have any stack pointer registers. Which of the following statements is true ? **[GATE-2000]**
a) It cannot have subroutine call instruction
b) It can have subroutine call instruction, but no nested subroutine calls.
c) Nested subroutine calls are possible, but interrupts are not
d) All sequence of subroutine calls and also interrupts are possible.

8. A CPU has two modes - privileged and non-privileged. In order to change the mode from privileged to non-privileged. **[GATE-2001]**
a) A hardware interrupt is needed
b) A software interrupt is needed
c) A privileged Instruction (Which does not generate on interrupt) is needed
d) A non - privileged instruction(Which does not generate on interrupt) is needed

9. Arrange the following configurations for CPU in decreasing order of operating speeds : Hard wired control, Vertical micro-programming,horizontal micro-proramming. **[GATE-1999]**
a) Hardwired control,vertical micro-programming Horizontal micro-programming
b) Hardwired control,Horizontal micro-programming Vertical micro-programming
c) Horizontal micro-programming,Vertical microProgramming, Hardwired control
d) Vertical microprogramming,horizontalmicro-programming Hardwired cotrol

10. The main difference(s) between a CISC and a RISC processor is/are that a RISC processor Typically: **[GATE-1999]**
a) Has fewer instructions
b) Has fewer addressing modes
c) Has more registers
d) Is easier to implement using hard-wired control logic

11. The correct matching for the following Pairs is: **[GATE-1997]**

a. DMA I/O	I. High speed RAM
b. Cache	II. Disk
c. Intrrupt I/O	III. Printer
d. Condition code Register	IV. ALU

a) a-IV,b-III,c-I,d-II b) a-II,b-I,c-III,d-IV
c) a-IV,b-III,c-II,d-I d) a-II,b-III,c-IV,d-I

12. A micro instruction is to be designed to specify **[GATE-1997]**
I. None or one of the three micro operations of one kind
II. None or upto six microperations of another kind
The minimum no.of bits in the micro-instruction is:
a) 5 b) 8 c) 9 d) None

13. For the daisy chain scheme of connecting I/O devices,Which of the following statments is TRUE? **[GATE-1996]**
a) It gives non-uniform Priority to various devices
b) It gives uniform priority to all divices
c) It is only useful for connecting slow devices to a processor device
d) It requires a separate interrupt Pin on the processor device

14. A micro program control unit is required to genrate a total of 25 control signals. Assume that during any microinstruction, at most two control signals are active.Minimum no.of bits required in the control word to generate the required control signal will be **[GATE-1996]**
a) 2 b) 2.5 c) 10 d) 12

15. Which case/cases following leads to a pipelined computer architechture?
I. The evaluation of each basic related function is relatively independent of the previous one
II. The sub functions are closely related to each other
III.The evaluation of each sub function requires approxmately the same sequence
a) I only b) II only c) III only d) I,II and III

16. A certain 8-bit microprocessor executes "SUB A,B" by loading B to the ALU,then loading A to the ALU,subtracting A from B,and storing the result in B.Which of the following instructions would cause the zero FLAG to be set? **[GRE-CS]**
a) SUB -(X), (X)+ b) SUB (X) , (X)+
c) SUB -(X), (X) d) All of the above

17. Which are the characterstics of horizontal microstructions? **[GRE-CS]**
I. Longer control word than vertical microinstructions
II. High degree of parallelism
III. Their execution slower than vertical instructions
a) I only b) II only c) III only d) I and II only

18. The ALU of a microprocessor performs operations on 8-bit two's complement operands .What happens when the operation $7A_{16}$-$A2_{16}$ is performed? **[GRE-CS]**
a) Result=$D8_{16}$,overflow and negative flags set
b) Result=$D8_{16}$,negative flag is set
c) Result=$D8_{16}$,no flags set
d) Result=28_{16},Overflow flag set

19. The scheduling procedures for a two-stage pipelined computer are repragented by the following reservation table:What is the shortest latency time causing collision? **[GRE-CS]**

	0	1	2	3	4	5	6
STAGE I	X						X
STAGE II			X		X		
STAGE III		X		X		X	

a) 1 b) 2 c) 3 d) 4

20. Which case(s) following leads to a pipelined computer architecture?
I. The evaluation of each basic function is relatively inde pendent of the previous one.
II. The subfunctions are closely related to each other.
III. The evaluation of each subfunction requires approximately the same sequence. **[GRE-CS]**
a) I only b) II only c) I and III only d) I,II and III

21. A primitive computer uses a single register. Suppose that the fragment of assembler below were written for such a machine: **[GRE-CS]**

```
LOAD X
MULT Y
STORE T1
MULT T1
STORE T1
LOAD Z
ADD Y
ADD T1
STORE R
```

What expression is evaluated ?
a) R:= (XY) + Y + Z b) R:= $(XY)^2$ + Y + Z
c) R:= XY^2 + Y + Z d) R:= XY^2 (Y + Z)

22. Which of the following statements is incorrect: **[GRE-CS]**
a) A collision in a pipelined architecture is an attempt by two different initiations to use the same stage at the same time.
b) Latency is the number of time units between two initiations in a pipelined architecture.
c) If initiations are of different but fixed reservation tables, the architecture is known as static pipelined configuration

23. Less hardware is needed to handle the execution of micro-instructions in the case of **[AUCET-2000]**
a) Horizontal organization b) Vertical organization
c) Hardwired organization d) None of these

24. Super Scalar processors are capable of executing **[AUCET-2000]**
a) More than one instruction per clock cycle
b) At a rate of morethan 100 MIPS
c) Morethan one micro-instruction per clock cycle
d) None of these

25. Which of the following statements about horizontal versus vertical micro architecture is (are) TRUE? **[AUCET-2001]**
I. Program for horizontal architecture require more time steps than those for vertical architectures.
II. Horizontal micro instructions are unceded.
III.Horizontal micro instructions usually have a single opcode and multiple operand specifiers.
a) I only b) II only c) II and III only d) III only

26. A small microprocessors micro instruction has 6 control fields, each one triggering the following number of control lines: **[AUCET-1999]**

Control field:	1	2	3	4	5	6
No.of control Lines	7	4	5	2	1	3

What is the fewest number of control bits required to describe all the control fields?

a) 6 b) 7 c) 12 d) 14

27. A computer can execute 1,000,000 instructions per second. Program running on this computer performs on average a one sector read and a one sector write for every 200 instructions that it executes. The disk drive handling the I/0 transfers requires 0.00010 seconds each to perform the read and write operations. Assuming no overlap of these operations, the percent of CPU time spent in the wait state, is **[AUCET- 2001]**

a) 12% b) 39% c) 57% d) 91%

28. The most appropriate matching for the following pairs

X : Indirect Addressing I : loops
Y : Immediate addressing II : pointers
Z : Autodecrement addressing III : constants

Is: **[GATE-2000]**

a) X-III,Y-II,Z-I b) X-I,Y-III,Z-II
c) X-II,Y-III,Z-I d) X-III,Y-I,Z-II

29. Which is the most appropriate match for the items in the first column with the items in second column?

X : Indirect Addressing I : Array Imple mentation
Y : Indexed Addressing II : Writing relocatable code.
Z : Base Register addressing III: Passing array as parameter. **[GATE-2001]**

a) X-III,Y-I, b) X-II,Y-III,Z-I
c) X-III,Y-II,Z-I d) X-I,Y-III,Z-II

30. A certain processor supports only the immediate and the direct addressing modes. Which of the following programming language features cannot be implemented on this processor ? **[GATE-1999]**

a) Pointers b) arrays c) Records
d) Recursive procedures with local variable

31. Which of the following addressing modes facilitates access to an operand whose location is defined relative to the beginning of the data structure in which it appears?

a) Absolute b) Immediate c) Index d) Indirect

32. For a computer system,we have memory size of 2^{16} bytes, the number of distinct operations (machine instructions) is 107. The computer memory word may store instruction, or data, as usual. Which of the following is the most suitable word size,where the instructions may be one-address instructions, or two-address instructions, or zero-address instructions? **[UGC NET - 2000]**

a) 32 b) 40 c) 48 d) 24

33. Select the statement which is not correct ? **[UGC NET - 1996]**

a) RF modulator is a device which can be used with a microcomputer to make an ordinary TV set on output device
b) The three main buses are control bus, data bus and address bus
c) PROM means Programmable Read Only Memory
d) Program Counter is a type of check fitted in the CPU

34. The number of instructions which may possibly be executed by a computer depends upon the size of **[UGC NET - 1996]**

a) Opcode b) Program counter
c) Memory Address Register d) A word

35. What is the address space allocated to the 8K RAM **[Hyd.Univ-1998]**

a) 1000-2FFF b) 5000-6FFF
c) 5000-5FFF d) 6000-7FFF

36. A 64 register memory contains registers R0 to R63 (all register numbers in decimal) arranged in banks of 8 registers each. What is the required address value to address the third register in any bank of eight registers. (X indicates don't care) **[Hyd.Univ-1993]**

a) XXX011 b) XXX010
c) 010XXX d) 011XXX

ANSWER KEY - II

1. b,d	2. d	3. d	4. d	5. d
6. c	7. d	8. d	9. b	
10. a,b,c,d	11. b	12. b	13. a	14. c
15. d	16. a	17. d	18. a	19. b
20. d	21. b	22. c	23. c	24. b
25. ?	26. c	27. d	28. c	29. a
30. a,b,c,d	31. c	32. ?	33. d	34. a
35. d	36. a			

EXPLANATORY NOTES - II

3. (d) **Explanation**

4 6 6 = 16 bits

OP	Address	Address

(two address instructions)

$2^4 = 16$ combinations

Given, there are n - two address instructions.

$\therefore 2^4 - n$ combinations can be used for one address.

OP	address

(One address instruction)

$\therefore$ Maximum number of one address instructions are $\left(2^4 - n\right) \times 2^6$

4. (d) **Explanation :** Given condition indicates the race condition situation. Hence answer 'd' is correct.

9. (b) **Explanation :** The complexity of the digital system is derived from the fact that a large number of micro - operations are performed in a given time sequence.

When the control functions are generated by hardware using conventional logic design techniques, the control unit is said to be hard - wired micro programing is a second alternative for designing the control until of digital computer.

12. (a) **Explanation :** Given the micro - op filed is subdivided into 2 subfields F1 (To specify I), F2 (to specify II) of 2 bits and 3 bits each respectively.

A field of 2 bits can specify $2^2 - 1 = 3$ micro - ops

A field of 3 bits can specify $2^3 - 1 = 7$ micro - ops

5 bits

$\therefore$ Minimum no. of bits in the micro-instruction is 5.

15. (d) **Explanation :** All three cases are characterstics of a pipelined computer architecture.

16. (a) **Explanation :** Note that

(X)+ Signifies "locate the operand at location X, and then increment X by 1".

-(X) Signifies "decrement X by 1, and then locate the operand at location old - X minus one".

Following is a diagram of a section of linear memory containing the location X.

Address	Memory Contents
x-1 :	0
x :	
x+1 :	

Consider the instruction

SUB - (X), (X) +

Observe that at both times the same operand, namely a, is being accessed. Since a-a=0, the zero flag will be set.

17. (d) **Explanation :** A horizontal micro instruction control word employs no, or very little, encoding of bit fields, where as a vertical micro instruction control word is highly encoded. Therefore, a horizontal micro instruction requires a longer control word and permits a high degree of parallelism. It is also faster than vertical micro instruction because very little, if any, decoding is needed.

18. (a) **Explanation :**

7A	0111	1010	0111	1010
-A2	-1010	0010	+0101	1110

Result = $D8_{16}$

Sign bit = 1, so negative flag set. The sign of the result is the same as the sign of the number subtracted, thus overflow flag set.

19. (b) **Explanation :** From the table provided it can be seen that at a latency of two time units, there will be a collision.

21. (b) **Explanation :**

Instruction	Reg. Contents	Comments
LOAD X	X	
MULT Y	X * Y	
STORE T1	X * Y	T1 : = X * Y
MULT T1	$(X+Y)^2$	
STORE T1	$(X+Y)^2$	T1 : = $(X+Y)^2$
LOAD Z	Z	
ADD Y	Z+Y	
ADD T1	$(X+Y)^2+Y+Z$	
STORE R		

22. (c) **Explanation :** Static Pipeline configuration is one in which all intiations are the same reservation table.

26. (c) **Explanation :**

To describe control field 1 requires 3 bits
To describe control field 2 requires 2 bits
To describe control field 3 requires 3 bits
To describe control field 4 requires 1 bits
To describe control field 5 requires 1 bits
To describe control field 6 requires 2 bits
12 bits

Hence 12 bits are required to describe all the control fields.

32. (a) **Explanation :**

A field of 6 bits can specify $2^6 - 1 = 63$ micro - ops
A field of 5 bits can specify $2^5 - 1 = 31$
A field of 3 bits can specify $2^3 - 1 = 7$
A field of 2 bits can specify $2^2 - 1 = 3$
A field of 2 bits can specify $2^2 - 1 = 3$
18 bits 107 micro - ops

Given memory size is 2^{16} bytes

$\therefore$ address field = 16 bits
control field = 18 bits
34 bits

Hence the word size is 34 bits.

35. (a) **Explanation :** Given 8K RAM = 2^{13} RAM

Hence in 16 line address bus 13 bits are used to address given RAM i.e. 0001 1111 1111 1111 = 1 FFF (Hexa decimal). The address space is start from 0000. Hence answer a is correct.

36. (a) **Explanation :** To select the one of the 64 registers required 6 bits. Given that the 64 register memory arranged in banks of 8 registers each. However, there are 8 register banks each with 8 registers. Hence in 6 bits the 3 LSB a represent the one of the 8 registers in each bank similarly remaining 3 MSBs represent the one of 8 register banks. Hence answer 'a' is correct.

III. MEMORY ORGANISATION

- **Memory System**
- **Random Access Memory**
- **External/Auxiliary Memory**
- **High Speed Memories**

1. Memory System

Memory in a computer system is required for storage and subsequent retrieval of the instructions and data. A computer system uses variety of devices for storing these instructions and data which are required for its operations. Normally we classify the information to be stored on computer in two basic categories : Data and the Instructions.

The storage devices along with the algorithm or information on how to control and manage these storage devices constitute the memory system of a computer.

A memory system is a very simple system yet it exhibits a wide range of technology and types. The basic objective of a computer system is to increase the speed of computation. Like wise the basic objective of a memory system is to provide fast, uninterrupted access by the processor to the memory such that the processor can operate at the speed it is expected to work.

But does this kind of technology where there is no speed gap between processor and memory speed exist?. The answer is yes they do, but unfortunately as the access time (time taken by CPU to access a location in memory) becomes less and less the cost per bit of memory becomes increasingly higher. In addition, normally these memories require power supply till the information need to be stored. Both these things are not very convenient, but on the other hand the memories with smaller cost have very high access time which will result in slower operation of the CPU. Thus, the cost vs access time anomaly has lead to a hierarchy of memory where we supplement fast memories with larger, cheaper, slower memories. These memory units may have very different physical and operational characteristics, therefore, making the memory system very diverse in type, cost, organisation, technology and performance. This memory hierarchy will work only if the frequency of access to the slower memories are significantly less than the faster memories.

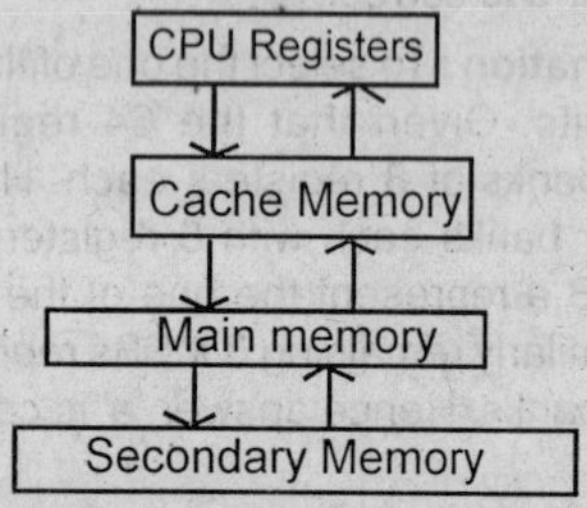

Figure 3.1 : The Memory Hierarchy

Thus, a memory system can be considered to consist of three group of memories . These are:

a) Internal Processor Memories: These consist of the small set of high speed registers which are internal to a processor and are used as temporary locations where actual processing is done.

b) Primary Memory or Main Memory: It is a large memory which is fast but not as fast as internal processor memory. This memory is accessed directly by the processor. It is mainly based on integrated circuits.

c) Secondary Memory/Auxiliary Memory/Backing Store: Auxiliary memory in fact is much larger in size than main memory but is slower than main memory. It normally stores system programs (Programs which are used by system to perform various operational functions), other instructions, programs and data files. Secondary memory can also be used as an overflow memory in case the main memory capacity has been exceeded. Secondary memories cannot be accessed directly by a processor. First the information of these memories is transferred to the main memory and then the information can be accessed as the information of main memory.

There is another kind of memory which is increasingly being used in modern computers, this is called Cache memory. It is logically positioned between the internal memory (registers) and main memory. It stores or catches some of the content of the main memory which is currently in use of the processor.

2. Characteristics Terms for Various Memory Devices

The following terms are most commonly used for identifying comparative behavior of various memory devices and technologies.

Storage Capacity : It is a representative of the size of the memory.

The capacity of internal memory and main memory can be expressed in terms of number of words or bytes. The storage capacity of external memory is normally measured in terms of bytes

Unit of transfer : Unit of transfer is defined as the number of bits read in or out of the memory in a single read or write operation .For main memory and internal memory ,the normal unit of transfer of information is equal to the word length of a processor in fact it depends on number of data lines in and out of the memory module. In general ,these lines are kept equal to the word size of the processor. The unit of transfer of external memory is normally quite large and is referred to as block of data

Access modes : once we have defined the unit of transfer next important characteristics is the access mode in which the information is accessed from the memory. A memory is considered to consist of various memory locations.

The information from memory devices can be accessed in the following ways:

Random Access Memory (RAM): It is the mode in which any memory location can be accessed in any order in the same amount of time. Ferrite and Semiconductor memories which generally constitute main memory are of this nature. The storage locations can be accessed independently and there exist separate access mechanism for each location.

Sequential Access : On the other hand we have memories which can be accessed in a pre-defined sequences for example, the songs stored on a cassette can be accessed only one by one. The example of sequential access memory is Magnetic Tape. Here the access mechanism need to be shared among different locations. Thus, either the location or the read/write head or both should be moved to access the desired location.

Direct Access: In certain cases the information is neither accessed randomly nor in sequence but something in between.

In these kind of access, a separate read/write head exist for a track and on a track the information can be accessed serially. This semi-random mode operation exist in magnetic disks.

Access Time: The access time is the time required between the request made for a read or write operation till the time the data is made available or written at the requested location.Normally it is measured for read operation. The access time depends on the physical characteristis and access mode used for that device.

Permanence of Storage: Is it possible to lose information by the memories over a period of time? If yes then what can be the reasons of the loss of information and what should be done to avoid it?

There are several reasons for information destruction, these are destructive readout, dynamic storage, volatility and hardware failure.

If for a particular memory the reading process destroys the stored information, we call it Destructive readout. In such memories the information has to be written back on the same location from which it had been read after each read operation. The reading process where the data is not destroyed on reading are referred to as Non-destructive readout.

There can be some memories where the stored 1 looses its strength to become 0 over a period of time. These kind of memories require refreshing. The memories which require refreshing are termed as dynamic memories.In contrast, the memories which do not require refreshing are called static memories.

Another factor which can destroy the contents is the presence and absence of electricity. The memories which loses their content on failure of power are termed as volatile memories, those which do not are called non-volatile. Magnetic memories are non-volatile and semiconductor main memories are volatile in nature.

Cycle Time: It is defined as the minimum time elapsed between two consecutive read requests. Is it equal to access time? Yes, for most of the memories except the ones in which destructive readout is encountered. Cycle time for such memories is the access time (time elapsed when a read request is made available) plus writing time as after the data has been made available the information has to be written back in the same memory location as the previous value has been destroyed by reading. But for most of the commonly used semi-conductor memories cycle time is equal to the access time.

Data Transfer Rate: The amount of information that can be transferred in or out of the memory in a second is termed as data transfer rate or bandwidth. It is measured in bits per second. Maximum number of bits that can be transferred in a second depends on how many bits can be transferred in or out of the memory simultaneously and thus the data bus width become one of the controlling factors.

3. **Physical Characteristics**: In this respect the memory devices can be categorised into four main categories viz. electronic, magnetic, mechanical and optical. One of the requirement for a storage device is that it should exhibit two well-defined physical states, such that 0 and 1 can be represented in those two states. The data transfer rate of the memory depends on the how quickly the state can be recognised and altered. The following table lists some of the memory technologies along with their physical and other important characteristics.

Some of the main attributes which are considered for the storage devices are physical size,energy consumption and reliability.

The physical size also depends on the storage density of the memories.

Technology	Access Time (in sec)	Access Mode	Permanence of Storage	Physical nature of Storage medium	Average (Rs / bit) (Appox)
Semiconductor Memories	10^{-8}	Random	Volatile	Electronic	10^{-2}
Magnetic disk	10^{-2}	Direct	Non-Volatile	Magnetic	10^{-5}
Magnetic tape	10^{-1}	Sequential	Non-volatile	Magnetic	10^{-6}
Compact disk ROM	1	Direct	Non-volatile	Optical	10^{-7}

Figure 3.2: Characteristics of some memory technologies

This is also coupled with the question of portability of memory. The energy consumption is another key factor which determines the power consumption of the computer and cooling system requirements.

The higher is the power consumption, more costly equipment is required for internal cooling of computer.Reliability is measured as Mean Time To Failure. The storage devices which require mechanical motion e.g. hard disk are more prone to failure rather than the semiconductor memories which are totally electronic in nature. Very high speed semiconductor memories are also prone to failure as technology is moving towards its limits.

Cost : Another key factor which is of prime concern in a memory system is cost. It is normally expressed per bit. The cost of a memory system includes not only the device but the cost of access circuitry and peripherals essential for the operation of the memory. Figure 3.2 also shows per bit cost of these memories. Please note as the access time for memories is increasing, the cost is decreasing.

4. Random Access Memory : Here we will confine out discussions in general to the Random Access Memory only which also contains the discussion on Main Memory. Main memory is a random access memory. It is normally organised (logically) as words of fixed length. The length of a word is called word length.Each of these memory words have an independent address and each have same number of bits. Normally the total number of words in memory are some power 2.

The main memory can be both read and written into, therefore, is called read-write memory.

The access time and cycle time in RAMs are constant and independent of the location accessed.

5. Semiconductor Memories

The semiconductors were used for making high speed CPU registers since 1950, but it was made economic to construct RAM chip only in 1970 with the development of VLSIs. At present RAMs are manufactured in a wide range of sizes i.e. from a few hundred bits to a megabit or more. The present limit on technology do not allow to construct one giga bit on a single chip. To construct large memories the small IC RAMs can be combined. The semiconductor memories falls under two main technologies bipolar. semiconductor memories and Metal Oxide Semiconductor (MOS) transistor semiconductor memories. For larger RAM chips normally MOS is used at present. There are two main categories of semiconductor RAMs static and dynamic. The larger chips are normally constructed as dynamic RAMs because the dynamic RAM require less number of transistors than static RAM and hence can be packed more densely on a single chip. Thus, higher storage density can be achieved by dynamic RAMs.

On the other hand, dynamic RAMs tend to loose their charge with time and needs periodic refreshing. Thus, requiring extra control circuitry and interleaving of normal memory access with refreshing operation.

Thus, the dynamic RAMs although can be packed more densely yet are more difficult to use than that of static RAMs. In contrast to ferrite core memory the semiconductor RAMs are volatile in nature.

These memories which we have discussed are both read/write type but what about a memory where we have only one of the operation possible, e.g. if we only allow writing in the memory and no reading then how can we use that memory probably it is a wastage of memory, but what about having a memory where we cannot change the information in normal case and only read the information from the memory. This memory might have plenty of uses, for example an important bit of computer's operating system which normally do not change can be stored in these kinds of memory so that one cannot change the operating system accidentally. These memories are called ROMs (Read Only Memories).

6. Read Only Memories

A ROM is basically a combinational circuit and can be constructed as:

Input		Output	
I_0	I_1	O_0	O_1
0	0	0	1
0	1	1	0
1	0	1	1
1	1	0	0

(a) Truth table

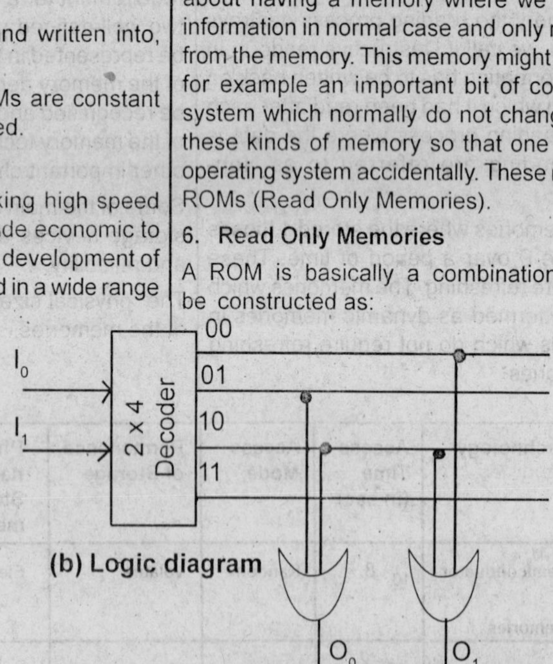

(b) Logic diagram

I_0 and I_1 perform the task of address selection

Figure 3.3: A Sample ROM

Thus, using this hardwired combinational circuit we can create ROM. Please note that on applying an input I_0 = 0, I_1 = 0 we will get O_0=0 and O_1=1; on applying I_0 = 0, and I_1=1 we will get O_0=1 an O_1=0 as 01 line of decoder will be selected. This same logic can be used for constructing larger ROMs.

ROMs (Read Only Memories) are the memories on which it is not possible to write the data when they are on-line to the computer. They can only be read. The ROMs can be used in storing Micro programs, Systems programs, subroutines. ROMs are non-volatile in nature and need not be loaded in a secondary storage device. ROMs are fabricated in large number in a way where there is no room for even a single error. But this is an inflexible process and require mass production, therefore, a new kind of ROM called PROM was designed which is also non-volatile and can be written only once and hence the name programmable ROM (PROM). The writing process in PROM can be performed electrically by the supplier or the customer. A special equipment is needed to perform this writing operation. Therefore, PROMs are more flexible and convenient than ROMs.

The ROMs/PROMs can be written just once (in ROMs at the time of manufacture and PROMs at any time later also), but in both the cases once whatever is written on, cannot be changed. But what about a case where you read mostly but write only very few times. This lead to the concept of Read mostly memories and the best examples of these are EPROMs (Erasable PROMs) and EEPROMs (Electrically Erasable ROMs). The EPROMs can be read and written electrically. But, the write operation is not simple. It requires erasure of whole storage cells by exposing the chip to ultra violet light, thus bring them to same initial state. This erasure is a time consuming process. Once all the cells have been brought to same initial state, then the EPROM can be written electrically. EEPROMs are becoming increasingly popular as they do not require prior erasure of previous contents. However, in EEPROMS the writing time is considerably higher than reading time. The biggest advantage of EEPROM is that it is non-volatile memory and can be updated easily, while the disadvantages are the high cost and at present they are not completely non-volatile and the write operation takes considerable time. But all these advantages are disappearing with the growth in technology. In general, ROMs are considered slower than RAMs. Figure 3.4 summarises the features of these read only and read mostly memories.

Memory Type	Write Time	Order of Read Time	Number of Write Cycles allowed
ROM	Once	Nano seconds	ONE
PROM	Hours	Nano seconds	TENS
EPROM	Minutes (Including Time of erasure)	Nano seconds	HUNDREDS
EEPROM	Milliseconds	Nano seconds	THOUSANDS

Common Features

- Non-destructive read out
- Long data life
- Non-volatile

Figure 3.4: Features of Read Only and Read Mostly memories

One of the new memory technology is flash memory. These memories can be reprogrammed at high speed and hence the name flash. The flash memory characteristics such as cost and write time, etc. fall in between of EPROM and EEPROM. In flash memoies the entire memory can be erased in few seconds (compare it to EPROM) by using electrical erasing technology. There is another possibility in flash memory which is erasure of a block is possible in it.

7. External/Auxiliary Memory

As discussed earlier the cost of RAM is very high and the semi-conductor RAMs are mostly volatile, therefore, it is highly likely that a secondary cheap media should be used which should show some sort of permanence of storage and should be relatively inexpensive. The magnetic material was found to be inexpensive and quite long lasting material, therefore, became an ideal choice to do so. Magnetic tape and magnetic disks are commonly used as storage media. With the advancements in the optical technology now the optical disks are trying to make inroads as one of the major external memory.

8. Magnetic Disk

A magnetic disk is a circular platter of plastic which is coated with magnetizable material. One of the key component of a magnetic disk is a conducting coil named as Head which performs the job of reading and writing on the magnetic surface.

The head remains stationary while the disk rotates below it for reading or writing operation.

For writing data on the magnetic disk current is passed through the head which produce the magnetic field. This magnetic field causes magnetic patterns to be recorded on the magnetic surface. These recorded magnetic patterns depend on the direction of the current in the head. While for reading the magnetic field recorded on the disk surface moves relative to the head resulting in generation of electric current of the same polarity in the head. The capacities of magnetic disk range from 1 MB (for floppy disks) to hundreds of MB.

Data Organisation and Format

The head of disk is small coil and reads or writes on the position of the disk rotating below it, therefore, the data is stored in concentric set of rings (refer figure 3.5). These are called tracks. The width of a track is equal to the width of the head. To minimise the interference of magnetic fields and to minimise the errors of misalignment of head, the adjacent tracks are separated by inter track gaps.

As we go towards the outer tracks the size of a track increases but to simplify electronics same number of bits are stored on each track. Therefore, the linear storage density from outer to inner track increases per linear inch. Figure 3.5 gives the details of a disk.

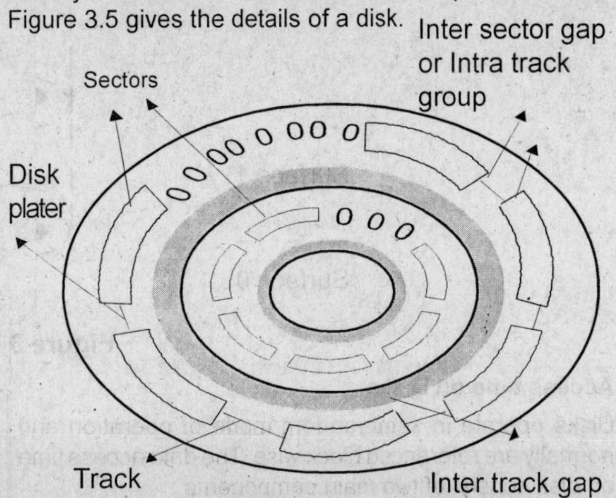

Figure 3.5: Logical layout of Magnetic Disk

The data is transferred from and to the disks in blocks. Block is a section of disk data and is normally equal to a sector. A track is divided into 10-100 sectors and these sectors should be either fixed or variable length sectors. Two adjacent sectors are separated by intra-track gaps. This help in reducing the precision requirements of sectors. To identify the sector position normally there may be a starting point of a track and a starting and end point of each sector. But how does a sector of a track is recognised ? A disk is formatted to record control data on it such that some extra data is stored on it for identification purpose. This control data is accessible only to the disk drive and not to the user.

Various characteristics of disks can be classified as:

Fixed head/ Movable head disks: The disks in which the head do not move are called fixed head. For such disks we require one read/write head per track. Heads can be mounted on a rigid arm which extends to the centre of the disk. On the movable head disk normally we have one read/write head which is fitted on an arm. This arm allows head to position on any track of the platter. Figure 3.6 shows these fixed and movable head schemes.

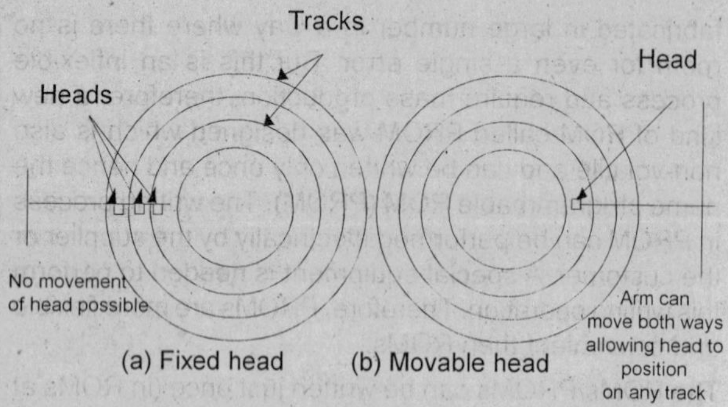

Figure 3.6: Fixed and Movable head disks

Removable and Non Removable Disks: The disks are normally mounted in a disk drive which consist of an arm and a shaft along with the electronic circuitry for input-output of data. The disk rotates along with the shaft. A non-removable disk is permanently mounted on the disk drive. Example of a non-removable disk is the hard disk of a PC-XT. The removable disks can be replaced by another similar disk on to same or different computer, thus, providing enormous data storage which is not limited by the size of the disk. Example of such disks are floppy diskettes.

Sides: The magnetic coating if applied to both the sides of the platter is called as double sided disks. The data can be recorded on either side of these disks. Some inexpensive disks were initially single sided.

Platters: Some disks have single platter e.g. floppy disks while some disks have multiple platters which are stacked vertically normally at a distance of an inch. This is known as disk pack. In disk pack one additional term cylinder is defined which is the ring of all co-centric tracks (Figure 3.7). A disk pack can contain multiple heads mounted with the same arm.

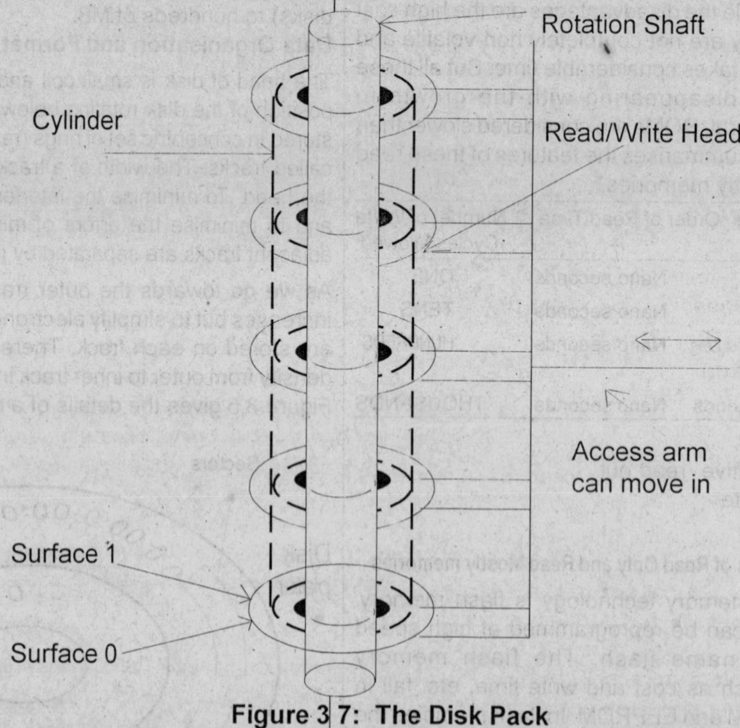

Figure 3 7: The Disk Pack

Access time on Disks

Disks operate in semi-random mode of operation and normally are referenced block wise. The data access time on disk consist of two main components :

Seek time: Time to position the head on a specific track. On a fixed head disks it is the time taken by electronic circuit to select the required head while in movable head disks it is the time required to move the head to a particular track.

Latency time: The time required by a sector to reach below the read/write head. On an average it is half of the time taken for a rotation by the disk.

In addition to these two times the taken to read block of word can be considered but normally it is too small in comparison to latency and seek time and in general the disk access time is considered to be sum of seek time and latency time. Since, access time to disks are large, therefore, it is advisable to read a sizable portion of data in a single go and that is why the disks are referenced block wise.

9. Magnetic Tapes

Magnetic tape consist of tape constructed from a plastic and are covered with magnetic oxide layers. These tapes are mounted on reels.

Storage Format: The data is stored as one byte at a time. Normally data representation on tapes allow 9 bits inclusive of one parity bit with each byte, therefore, a tape has 9 tracks, each track representing a bit position. Figure 3.8 shows the format of a magnetic tape. In addition to tracks, data on tapes are written in contiguous blocks. This is also termed as a physical record on a tape. These records are separated by a gap known as inter record gaps.

Characteristics

The tape drive is termed as sequential access device, that is, to read a record N, if tape is at present at record 1 then it has to pass through all the records till N-1 sequentially. In case the tape head is positioned beyond N[th] record then the tape need to be re-wound to a particular distance and then forward reading is done to read N[th] record. Tape moves only if a read or write operation is requested.

The tapes are one of the earliest storage devices. They are low cost, low speed, portable and are still widely used because of their low cost.

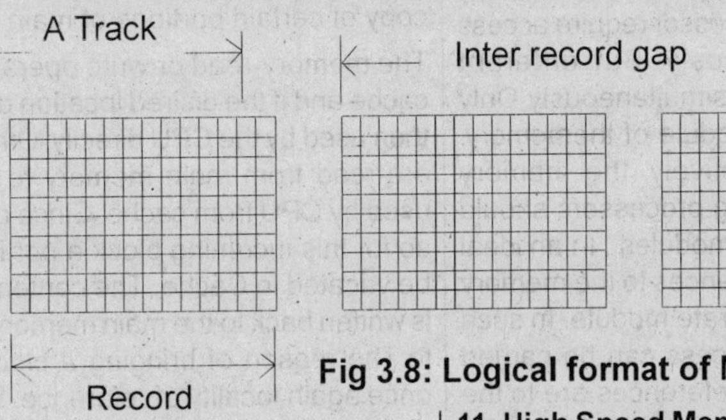

Fig 3.8: Logical format of Magnetic Tape

10. Optical Memories

Optical memories are alternate mass storage devices with huge capacity. The advent of compact disk digital audio system, a non-erasable optical disk, paved the way for the development of a new low cost storage technology. In optical storage devices the information is written using laser beam.These memories can store large amount of data

CD-ROM

The CD-ROM (Compact disk read - only memory) is a direct extension of audio CD. CD- ROM players are more rugged and have error correction facility. This ensures proper data transfer from CD-ROM to the maim memory of the computer. The CD-ROM disk is normally formed from a resin named polycarbonate which is coated with aluminum to form a highly reflective surface. The information on CD-ROM is stored as a series of microscopic pits on this reflective surface. A high-intensity laser beam is focused to create pits on the master disk. This master disk is then used to make a die which is used to make copies. A top coat of clear lacquer is applied on the CD-ROMs surface to protect it from dust and scratches.

For information retrival from a CD-ROM, a low-powered laser, which is generated in an optical disk drive unit, is used. The disk is rotated and the laser beam is aimed at the disk on a particular track. The intensity of the reflected laser beam changes as it encounters a pit. A photosensor detects the charge in intensity, thus recognises the digital signals recorded on the CD-ROM surface.

11. High Speed Memories

The Need: Why the high speed memories? Is the main memory not a high speed memory? The answer to second question is definitely "No", but why so, well for this we have to go to the fundamentals of semiconductor technology which is beyond the scope of the unit. Then, if the memories are slower then how slow are they? On an average it has been found that the operating speed of main memories lack by a factor of 5 than that of the speed of processors (such as CPU or Input Output Processors). In addition each instruction require several memory accesses (It may range, from 2 to 7 or even more sometimes).

If an instruction require even 2 memory accesses even then almost 80% time of executing an instruction, processor waits for memory access.

What can be done to increase this processor- memory interface bandwidth?

There are four possible answers to the questions. These are-

a) Decrease the memory access time, use a faster but expensive technology for main memory probably it will be feasible after few years.

b) Access more words in a single memory access cycle. That is instead of accessing one word from the memory in a memory access cycle, access more words.

c) Insert a high speed memory termed as Cache between the main memory and processor.

d) Use associative addressing in place of random access.

Hardware researchers are taking care of the first point. Let us discuss the schemes for rest three points

12. Interleaved Memories

The interleaved memories are the implementation of the concept of "accessing more words in a single memory access cycle". But the question is how can it be achieved? By partitioning the memory into say N separate memory modules. Thus, N accesses can be carried out to the memory simultaneously.

Figure 3.9 gives a block diagram for interleaved memory system.But what are the requirements for such a system? For such a system we will be requiring.

F Independent addressing ciruitry for each of the N modules.

F An appropriate bus control mechanism, in case, the physical buses between the processor and memory are shared by the modules.

F sufficient buffer storage in the 'processor' to control the increased flow of information.

Where this memory can be useful? Well, this kind of interleaved memory has found application in multi-processor systems where many processor require access to common memory simultaneously. But different processor can access the memory simultaneously Only if they desire access to different module of thememory. For utilising such a system effectively, the memory references generated by referencing processors should be distributed evenly among the N modules . In an ideal case for a set of N consecutive references to the memory every reference should be to a separate module. In such a case simultaneously memory access can be carried out. In case, two or more than two references are to the same module then these memory accesses cannot be carried out simultaneously. The maximum processor memory bandwidth in interleaved memory can be equal to the number of modules i.e. N words per memory cycle.

To achieve the address interleaving consecutive addresses are distributed among the N inter leaved modules. For example, if we have consecutive addresses and 4 interleaved memory modules then the 1st, 5th, 9th, 13th, ...address will be assigned to the first memory modules and so on.

1, 5, 9, 13, 17.... Addresses to Memory Module 1
2, 6, 10,14, 18.... Addresses to Memory Module 2
3, 7, 11, 15, 19.... Addresses to Memory Module 3
4, 8, 12, 16,20.... Addresses to Memory Module 4

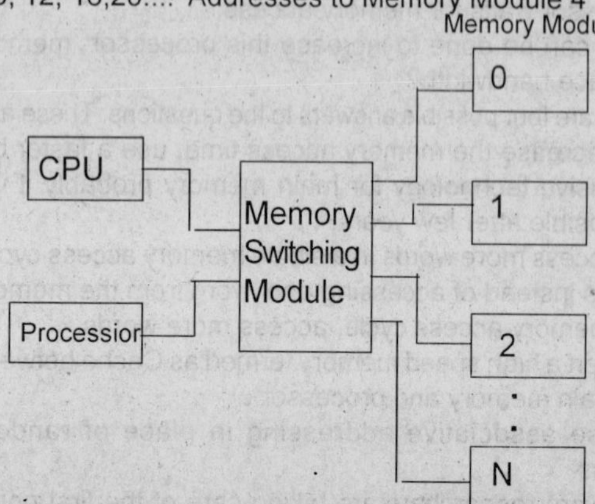

Figure 3.9 : Interleaved Memories

13. Cache Memory

These are small fast memories placed between the processor and the main memory . Caches are faster than main memory.

Then why do we need main memory at all? Well once again it is the cost which determines this. The caches although are fast yet are very expensive memories and are used in only small sizes. For example, caches of sizes 64K, 1218K etc. are normally used in typical PC-386 and PC-486 while they can have 1 to 8 MB RAMS or even more. Thus, small cache memories are intended to provide fast speed of memory retrieval without sacrificing the size of memory (because of main memory size). If we have such a small size of fast memory how could it be advantageous in increasing the overall speed of memory references? The answer lies in the principle of locality, which says that if a particular memory location is accessed at a time then it is highly likely that its near-by locations will be accessed in the near future. Cache contains a copy of certain portions of main memory.

The memory-read or write operation is first checked with cache and if the disired location data is available in cache then used by the CPU directly. Otherwise, a block of words are read from main memory to cache and the word is used by CPU from cache. Since cache has limited space, so for this incoming block a portion called a slot need to be vacated in Cache. The contents of this vacating block is written back to the main memory at the position it belong to.The reason of bringing a block of words to cache is once again locality of reference. We expect that next few addresses will be close to this address and, therefore, the block of word is transferred from main memory to cache. Thus, for the word which is not in cache access time is slightly more than the access time for main memory without cache. But, because of locality of references, the next few words may be in the cache, thus, enhancing the overall speed of memory references. For example, if memory read cycle takes 100 ns and a cache read cycle takes 20 ns, then for four continuous references (first one brings the main memory content to cache and next three from cache).

The time taken with cache

= (100+20) + 20 × 3

for the first for the last three

read operation read operation

= 120 + 60 = 180

Time taken without cache = 100 × 4 = 400 ns.

Thus, the closer are the reference better is the performance of cache and that is why structured code is considered to be a good programming practice, since it provides maximum possible locality.

The performance of cache is closely related to the nature of the programs being executed, therefore, it is very difficult to say what should be the optimum size of cache, but in general a cache size between 1k to 128k is considered to be optimum for most of the cases. Another important aspect in cache is the size of the block (refer figure 3.10) which is normally equivalent to 4 to 8 addressable units (which may be words,bytes etc.). Figure 3.10 gives, the cache memory organisation.

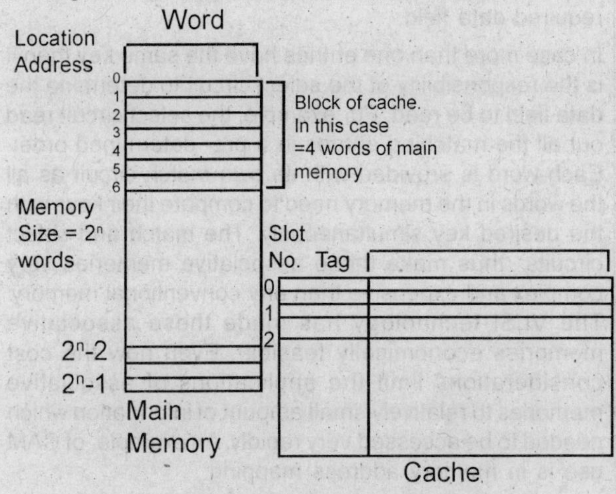

Figure 3.10: Cache Memory Organization

A Cache consist of a number of slots. As mentioned earlier, the cache size is smaller than that of main memory, therefore, there is no possibility of one is to one mapping of the contents of main memory to cache. So how the computer will identify which block is residing in cache? well this is achieved by storing the block address to the tag number. In the above figure four consecutive words are put in a single block. Thus if we have 2^n words in main memory then the total number of blocks which exist in main memory are = $2^{n/4}$ = 2^{n-2} (size of one block is 4 byte). Therefore, we need at least (n-2) bits as size of the tag field. The important feature in cache memory is how the main memory block can be mapped in cache. There are three ways of doing so: Direct, Associative and Set Associative.

Direct Mapping: In this mapping each block of memory is mapped in a fixed slot of cache only. For example, if a cache has four slots then the main memory blocks 0 or 4 or 8 or 12or 16... can be found in slot 0, while 1 or 5 or 9 or 13 or 17... can be found in slot 1; 2 or 6 or 10 or 14 or 18.... in slot 2: and 3 or 7 or 11 or 15 or 19... in slot 3. This can be mathematically defined as

Cache slot number = Block number of main memory MOD Total number of slots in cache.

In this technique, it can be easily determined whether a block is in cache or not. (How?)

This technique is simple; but there is a disadvantage of this scheme suppose two words which are referenced alternately repeatedly are falling in the same slot then the swapping of these two blocks will take place in the cache, thus, resulting in reduced efficiency of the cache.

14. Associative Mapping: In associate mapping, any block of the memory can be mapped on to any location of the cache. But here the main difficulty is to determine "Whether a block is in cache or not?", This process of determination is normally carried out simultaneously. The main disadvantage of this mapping is the complex circuitry required to examine all the cache slots in parallel to determine the presence or absence of a block in cache.

Set Associative Mapping: This is a compromise between the above two types of mapping.Here the advantages of both direct and associative cache can be obtained.

The cache here is divided in some sets, lets say A. The scheme is that a direct mapping is used to map the main memory blocks in one of the. A sets and with in this set any slot can be assigned to this block. Thus, we have associative mapping inside each set of 'A' sets.

Another important feature for cache is the replacement algorithm. For direct mapping no algorithm is needed since only one slot can be occupied by a block in cache but in associative and set associative mapping many slots may be used by a block. So which slot should be vacated for this new block. One of the common strategies used is to vacate the least recently used slot for this new block. The reason is that the probability of accessing a block which was used quite long back is less in compared to blocks which are used afterwards. This scheme is also derived from the principle of locality.

Other schemes which may not as effective as least recently used can be.

First in First out: Replace the block which has entered first in cache and free its slot for the incoming block.

Random:Replace any block at random.

Least frequently used: Replace block which is referenced least number of times.

Write Policy: The data in cache and main memory can be written by processors or Input/Output devices. The main problems faced in writing with cache memories are:

- The content of cache and main memory can be altered by more than one devices e.g. CPU can write to caches and Input/Out put module can directly write to main memory. This can result in inconsistencies in the values of cache and main memory,
- In the case of multiple CPU's with different caches a word altered in one cache automatically invalidate the word in other cache.

The suggested techniques for writing in system with caches are:

Write through: Write the data in cache as well as main memory. The other CPUs - Cache combination (in multiprocessor system) has to watch traffic to the main memory and make suitable amendments in the contents of cache.The disadvantage of this technique is that a bottleneck is created due to large number of accesses to main memory by various CPUs.

Write block: In this method updates are made only in the cache setting a bit called Update bit. Only those block whose update bit is set is replaced in the main memory. But here all the accesses to main memory whether from other CPUs or Input /Output modules need to be from the cache resulting in complex circuitry.

Research shows that only 15% of memory references are write references and, therefore, write through policies being simple can be utilised.

Instruction Cache: An instruction cache is the one which is emp loyed for accessing only the instructions and nothing else. The advantage of such a cache is that as the instructions do not change we need not write the instruction cache back to memory this is unlike data storage cache.

Associative Memories

In associative memories any stored item can be accessed directly by assigning the contents of the item in question, such as name of a person, account number, number etc., as an address. Associative memories are also known as content addressable memories (CAMs). The entity chosen to address the memory is known as the key.

Figure 3.11 shows the structure of a simple associative memory.The information is-stored in CAMs as fixed-length words. Any entity of the word may be chosen as the key field. The desired key is indicated by the mask register. Then, the key is compared simultaneously with all stored words. The words that match the key issue a match signal. This match signal then entersa select circuit. The select circuit in turn helps for the access of required data field.

In case more than one entries have the same key then it is the responsibility of the select circuit to determine the data field to be read. For example, the select circuit read out all the matching entries in a pre- determined order. Each word is provided with its own match circuit as all the words in the memory need to compare their keys with the desired key simultaneously. The match and select circuits thus make these associative memories very complex and expensive than any conventional memory. The VLSI technology has made these associative memories economically feasible. Even now the cost considerations limit the applications of associative memories to relatively small amount of information which needed to be accessed very rapidly. An example, of CAM use is in memory address mapping.

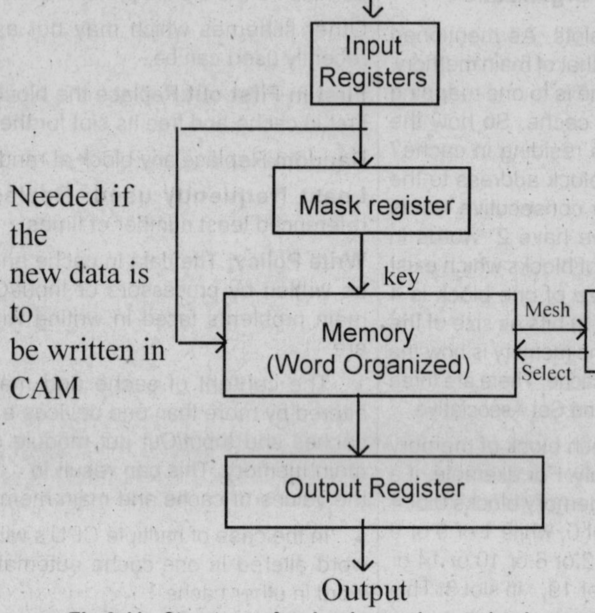

Fig 3.11 : Structure of a simple word-organised associative memory

SOLVED EXAMPLES

1. a. How many 128×8 RAM chips are needed to provide a memory capacity of 2048 bytes ?

b. How many lines of the address bus must be used to access 2048 bytes of memory ? How many of these lines will be common to all chips ?

c. How many lines must be decoded for chip and select? Specify the size of the decorders.

Sol : a. $\frac{2048}{128} = 16$ chips

b. $2048 = 2^{11}$; 11 lines to address 2048 bytes.

$128 = 2^7$; 7 lines to address each chip and 4 lines to decoder for selecting 16 chips

c. 4×16 decoder.

2. A computer uses RAM chips of 1024×1 capacity.

a. How many chips are needed, and how should their address lines be connected to provide a memory capacity of 1024 bytes ?

b. How many chips are needed to provide a memory capacity of 16K bytes ? Explain in words how the chips are to be connected to the address bus.

Sol: a. 8 chips are needed with address lines connected in parallel.

b. $16 \times 8 = 128$ chips .Use 14 address lines $(16K = 2^{14})$

10 lines specify the chip address.

4 lines are decoded into 16 chip-select inputs.

3. An 8 bit computer has a 16-bit address bus. The first 15 lines of the address are used to select a bank of 32K bytes of memory. The high order bit of the address is used to select a register which receives the contents of the data bus. Explain how this configuration can be used to extend the memory capacity of the system to eight banks of 32K bytes each, for a total of 256K bytes of memory.

Sol : The processor selects the external register with an address 8000 hexadecimal. Each bank of 32 K bytes are selected by address 0000-7FFF. The processor loads an 8-bit number into the register with a single 1 and 7 0's. Each output of the register selects one of the 8 banks of 32 K bytes through 2 chip-select input. A memory bank can be changed by changing the number in the register.

4. A magnetic disk system has the following parameters:

T_s = average time to position the magnetic head over a track.

R = rotation speed of disk in revolutions per second.

N_t = number of bits per track

N_s = number of bits per sector

Calculate the average time T_a that it will take to read one sector.

Sol : Average time = T_s + time for half revolution + time to read a sector.

$$T_a = T_s + \frac{1}{2R} + \frac{N_s}{N_t} \times \frac{1}{R}$$

5. What is the transfer rate of an eight track magnetic tape whose speed is 120 inches per second and whose density is 1600 bits per inch ?

Sol : An eight-track tape reads 8 bits (one character) at the same time. Transfer rate = $1600 \times 120 = 192{,}000$ character/sec.

6. A two way set associative cache memory uses blocks of four words. The cache can accommodate a total of 2048 words from main memeory. The main memory size is 128K × 32.

a. Formulate all partinent information required to construct the cache memory.

b. What is the size of the cache memory ?

Sol : 128K = 2^{17} ; For a set size of 2, the index address has 10 bits of accomodate $\frac{2048}{2} = 1024$ words of cache.

a. 7 bits 10 bits

TAG	INDEX

← Block → → word ←

8 bits 2 bits

b. Tag1 Data 1 Tag 2 Data 2

← 7 → ← 32bits → ← 7 → ← 32bits →

1024				

Size of cache memory is 1024 × 2(7+32) = 1024 × 78

7. The access time of a cache memory is 100 ns and that of main memory 1000 ns. It is estmated that 80 percent of the memory requests are for read and the remaining 20 percent for write. The hit ratio for read accesses only is 0.9. A write-through procedure is used.

a. What is the average access time of the system considering only memory read cycles ?

b. What is the average access time of the system for both read and write requests ?

c. What is the hit ratio taking into consideration the write cycles?

Sol : a.

$$\underbrace{0.9 \times 100}_{\text{cache access}} + \underbrace{0.1 \times 11000}_{\text{cache + memory access}} = 90 + 110 + 200\,\text{nsec.}$$

b. $\underbrace{0.2 \times 1000}_{\text{write access}} + \underbrace{0.8 \times 200}_{\text{read access from (a)}} = 200 + 160 = 360\,\text{nsec}$

c. Hit ratio = $0.8 \times 0.9 = 0.72$

8. A digital computer has a memeory unit of 64K × 16 and a cache memory of 1K words. The cache uses direct mapping with a block size of four words.

a. How many bits are there in the tag, index, block, and word fields of the address format ?

b. How many bits are there in each word of cache, and how are they divided into functions ? Include a valid bit.

c. How many blocks can the cache accommodate ?

Sol : 64K × 16 : 16 bit address; 16 bit data.

a. 6 8 2 = 16 bit address

Tag	BLOCK	WRD

INDEX = 10 bit cache address.

b. 1 6 16 = 23 bits in each word of cache

V	TAG	DATA

c. $2^8 = 256$ blocks of 4 words each.

9. An address space is specified by 24 bits and the corresponding memory space by 16 bits.

a. How many words are there in the address space ?

b. How many words are there in the memory space ?

c. If a page consists of 2K words, how many pages and blocks are there in the system ?

Sol : a. Address space = 24 bits ; 2^{24} = 16M words

b. Memory space = 16 bits ; 2^{16} = 64 K words

c. $\frac{16M}{2K} = 8k$ pages $\frac{16M}{2K} = 32$ blocks.

10. If a cache has a capacity of 16 KB and a line length of 128 bytes, how many sets does the cache have if it is 2-way, 4-way, or 8-way set-associative?

Sol : With 128-byte lines, the cache contains a total of 128 lines. The number of sets in the cache is the number of lines divided by the associativity. So the cache has 64 sets if it is 2-way set-associative, 32 sets if 4-way set-associative, and 16 sets if 8-way set-associative.

11. A cache has 64-KB capacity, 128-byte lines, and is 4-way set-associative. The system containing the cache uses 32-bit addresses.

a. How many lines and sets does the cache have?

b. How many entries are required in the tag array?

c. How many bits of tag are required in each entry in the tag array?

d. If the cache is write-through, how many bits are required for each entry in the tag array, and how much total storage is required for the tag array if an LRU replacement policy is used? What if the cache is write-back?

Sol: a. 64 KB/ 128 B = 512, so the cache has 512 lines. Since it is 4-way set-associative, the cache has 512/4 = 128 sets.

b. One tag array entry is required for each line, so the tag array needs 512 entries.

c. Since there are 128 sets,.7 bits of the address will be used to select a set out of the array. An additional 7 bits will be used to select a byte within each line, because lines are 128 bytes long. Therefore, 18 (32-14) bits of tag are required in each entry in the tag array.

d. Because the cache uses LRU replacement and is 4-way set-associative, 2 bits are required in each tag array entry to hold the age of the line. Write-through caches require a valid bit in each tag array entry but no dirty bit, bringing the size of each tag array entry to 21 bits (18 + 2 + 1). Since there are 512 lines in the cache, the size of the tag array is 10,752 bits. Write-back caches require a dirty bit in each tag array entry in addition to the bits required in a similar write-through cache, so each entry in the tag array would require 22 bits, for a total array size of 11,264 bits.

12. If a processor has 32-bit virtual addresses, 28-bit physical addresses, and 2-KB pages, how many bits are required for the virtual and physical page numbers?

Sol: The base-2 logarithm of 2048 is 11, so 11 bits of each address are required to specify the offset within the page. Therefore, 21 bits (32 - 11) are required to specify the virtual page number, and 17 bits are required to specify the physical page number.

13. A system has 48-bit virtual addresses, 36-bit physical addresses, and 128 MB of main memory. If the system uses 4096-byte pages, how many virtual and physical pages can the address spaces support? How many page frames of main memory are there?

Sol: 4096 = 2^{12}, so 12 bits of the virtual and physical addresses are used for the offset within a page. Therefore, the VPN is (48 - 12) = 36 bits long, so the virtual address space can support 2^{36} virtual pages. The PPN is (36 - 12) = 24 bits long, so the physical address space can support 2^{24} physical pages.

To get the number of page frames in the system, we divide the amount of main memory by the size of each page to get the number. of pages that can fit in the main memory at the same time. 128 MB/4 KB = 32,768 page frames. Thus, the amount of data that the operating system on this particular computer can map into the physical address space simultaneously is 32,768 pages, despite the fact that the physical address space would allow many more pages to be mapped, because there must be space in the main memory for every virtual page that is mapped onto a physical page.

14. A system has 32-bit virtual addresses, 24-bit physical addresses, and 2 KB pages.

a. How big is each page table entry if a single-level page table is used? (Round up to the nearest byte.)

b. How many page table entries are required for this system?

c. How much storage is required for the page table (single-level page table)?

Sol: a. A page size of 2 KB means that 11 bits are required for the offset field in the virtual and physical addresses. Therefore, 13 bits are required for the PPN. Each page table entry needs to hold the PPN for its page and the valid and dirty bits, for a total of 15 bits, which is rounded up to 16 bits, or 2 bytes.

b. With a page size of 2 KB, the virtual address space can hold 2^{21} pages, requiring 2^{21} page table entries.

c. Each page table entry requires 2 bytes of storage, so the page table takes up 2^{22} bytes, or 4MB.

OBJECTIVE QUESTIONS - I

1. The memory which is programmed at the time of it is manufactured is: **[AUCET-2000]**

a) RAM b) ROM c) PROM d) EPROM

2. The following memory has the shortest access time:

a) Cache memory b) RAM **[AUCET-2000]**
c) ROM d) EPROM

3. The number of bits that are typically stored on each track of a magnetic disk is usually: **[AUCET-2000]**

a) Fifty b) Same c) Different d) None

4. How many RAM chips of size(256k × 1 bit)are required to build 1MB memory **[AUCET-2000]**

a) 13 b) 15 c) 16 d) 32

5. The property of locality of reference may fail If a program has: **[AUCET-2000]**

a) Many operands b) Many unconditional jumps
c) Many conditional jumps d) All of these

6. Which of the following is not a Register : **[AUCET-2000]**

a) Buffer b) Stack
c) PC d) Accumulator

7. The Principal advantage of Dynamic shift registers over static shift register's is that : **[AUCET-2000]**

a) Less expensive
b) They can be shifted faster.
c) They lose their data If shifted too slowly
d) They provide more bits in smaller area.

8. The shifting of a program from one location to another in the memory is called: **[AUCET-2000]**

a) Data collection b) Data transmission c) Relocation

9. Blocking factor refers to: **[AUCET-2001]**

a) The no.of blocks that can be accommodated in a sector.
b) The no.of records thatcan be accommodated in a block
c) The no.of blocks into which the file is divided.

10. For an Ideal DRAM, the refresh rate must be: **[AUCET-2001]**

a) Very large b) Very small

c) Medium d) Proportional to the access time.

11. Memory Interleaving is done to: **[AUCET-2001]**

a) Increase the amount of logical memory

b) Reduce memory access time

c) Simplify memory interfacing

d) Reduce Page faults.

12. A 16K x 4 SRAM is internally organized as four 64×256 cell arrays. Such an arrangement is referred to as: **[AUCET-2001]**

a) 3-D memory b) 1-D memory

c) 2-D memory d) 2 1/2 -D memory

13. The average seek time can be reduced in a hard disk drive If **[AUCET-2001]**

a) The speed of rotation is increased

b) The recording density is increased

c) The no.of surfaces of the hard disk are increased

d) The no.of cylinders of the hard disk are reduced.

14. A file is to be printed on a serial printer from the hard disk. This requires: **[AUCET-2001]**

a) A DMA controller b) Bus arbitor

c) An Interrupt controller

d) Both an interrupt controller and DMA controller

15. The following device is called Bus master: **[AUCET-2001]**

a) Bus Arbitor b) Microprocessor

c) DMA controller d) None

16. While DMA transfer is taking place, the Processor

a) Has to remain idle in a tight loop **[AUCET-2001]**

b) Has to be in HOLD state

c) Can work on other programs

d) Must be halted

17. The transfer of data between I/O device and memory without the use of microprocessor is called as:

a) Polling b) DMA **[AUCET-2000]**

c) Direct I/O d) None of these

18. A memory has 16-bit address bus. The number of locations in this memory are: **[AUCET-1998]**

a) 16 b) 32 c) 1024 d) 65536

19. It is desired to have a 64 × 8 memory. The memories available are of 16 × 4 size. The number of memories will be: **[AUCET-1998]**

a) 2 b) 4 c) 6 d) 8

20. One of the following memories must be refreshed many times per second: **[AUCET-1997]**

a) SRAM b) DRAM c) EPROM d) EEPROM

21. A 12-address line microprocessor is capable of addressing .... locations **[AUCET-1997]**

a) 64K b) 1024 c) 2048 d) 4096

22. Advantage of the DRAM over SRAM is: **[AUCET-1997]**

a) Lower bit density and higher power consumption

b) Higher bit density and lower power consumption

c) Higher bit density and higher power consumption

d) Lower bit density and lower power consumption

23. Dynamic memory stores its information on the following element: **[AUCET-1997]**

a) Capacitors b) Transistors

c) Resistors d) None

24. The desirable charaterstic(s) of a memory system is (are):

a) Speed and reliability

b) Low power consumption

c) Durability and compactness

d) All of these

25. Which memory is volatile:

a) RAM b) ROM c) EPROM d) PROM

26. The minimum time delay required between the initiation of two successive memory operations is known as:

a) Memory cycle time b) Memory access time

c) Transmission time d) Waiting time

27. For a memory system, the cycle time is

a) Same as the access time

b) Longer than the access time

c) Shorter than the access time

d) Sub multiple of the access time

28. Which of the following memory medium is not used as main memory system:

a) Magnetic core b) Semiconductor

c) magnetic tape

d) Both semi conductor and magnetic tape

29. The refreshing rate of DRAM is approximately once in **[GATE-1987]**

a) Two micro seconds b) Two milli seconds

c) Fifty milliseconds d) Two seconds

30. The disadvantage of DRAM over SRAM is:

a) Higher power consumption

b) Variable speed c) Higher bit density

d) Need to refresh the capacitor charge every once in two milliseconds

31. Which of the following storage is volatile:

a) Floppy Disk b) Bubble memory

c) Core memory d) Semiconductor memory

32. In Comparison to the Internal (main) memory, tape or disk memory is:

a) Slower and more expensive

b) Slower and less expensive

c) Faster and more expensive

d) Faster and less expensive

33. The fastest type of memory from the following list is:

a) Tape b) Semi conductor memory

c) Disk d) Bubble memory

34. Which of the following memories allows simultaneous read and write operations?

a) ROM b) RAM

c) EPROM d) None of these

35. The Idea of cache memory is based on

a) The property of locality of reference

b) The heuristic 90-10 rule

c) The fact that only a small portion of a program is referenced relatively frequently.

d) All of these.

36. The search concept used in associate memory is

a) Parallel search b) Sequential search

c) Binary search d) Selection search

37. Memory refreshing may be done

a) By the CPU that contains a special regress counter, only

b) By an external refresh controller, only

c) Either by the CPU or by an external refresh controller

d) None of these.

38. Semi conductor memories are widely used because of

a) Their small size b) Their low cost

c) Their Compatibility with microprocessors

d) All of these.

39. A PLA is

a) Mask progr^^mmable

b) Field programmable

c) Can be programmed by a user

d) Can be erased and programmed.

40. The instructions for starting the computer are housed on

a) RAM b) ROM c) CD-ROM d) All of these

41. A byte consists of one group of 8 bits put together. What is the largest decimal number which can be represented by a BYTE:

a) 128 b) 255 c) 256 d) 520

42. What is the number of bits needed for an address in a 4k memory ?

a) 4 b) 8 c) 12 d) 16

43. The Data transfer rate of a double-density floppy disk system is about: [GATE-1987]

a) 5k bits/sec. b) 50k bits/sec

c) 500 bits/sec. d) 5000k bits/sec.

44. The total size of address space in a virtual memory system is limited by [GATE-1991]

a) The length of MAR

b) The available main memory

c) The available secondary storage

d) All of the above e) None of the above

45. The principle of locality justifies the use of [GATE-1995]

a) Interrupts b) DMA

c) Polling d) Cache memory.

46. A ROM is used to store the table for multiplication of two 8-bit unsigned integers. The size of ROM required is [GATE-1996]

a) 256×16 b) $64k \times 8$

c) $4k \times 16$ d) $64k \times 16$

47. Formatting of a floppy disk refers to [GATE-1998]

a) Arranging the data on the disk in contiguous fashion

b) Writing the directory

c) Erasing the system area.

d) Writing identification information on all tracks and sectors.

48. More than one word are put in one cache block to [GATE-2001]

a) Exploit the temporal locality of reference in a program

b) Exploit the spatial locality of reference in a program

c) Reduce the miss penality

d) None of the these.

49. Which of the following requires a device driver:

a) Register b) Cache [GATE-2001]

c) Main Memory d) Disk

50. Which of the follwing is not a form of memory? [GATE-2002]

a) Instruction cache b) Instruction Register

c) Instruction Opcode d) Translation Lookaside buffer

51. When we move from the outermost track to the innermost track in a magnitic disk, the density (bits per linear inch) [UGC NET - 1999]

a) Increases b) decreases

c) remains the same

d) either remains constant or decreases

52. Identify the correct sequence of memories according to access time: [UGC NET - 1998]

a) Disk memory, Main memory, Cache, Register

b) Cache, Register, Main memory, Disk memory

c) Disk memory, Main memory, Register, Cache

d) Main memory, Disk memory, Cache, Register

53. Memory access by its contents is known as [UGC NET- 1998]

a) Associative memory b) Cache memory

c) RAM d) Virtual memory

54. In computer system, the information is stored on the memory in the format of [UGC NET - 1998]

a) Octal b) Binary c) Hexadecimal d) Decimal

55. A typical modern computer uses [UGC NET - 1997]

a) magnetic core for secondary storage

b) VLSI chips

c) magnetic tapes for primary memory d) None

56. Which of the following is NOT a way to improve cache bandwidth ? **[UGC NET - 1997]**

a) Increasing the data - path width

b) Reducing the word-size

c) Decreasing the access time

d) Interleaving the cache for concurrency

57. Compared with secondary storage, primary storage is **[UGC NET - 1997]**

a) slow and inexpensive b) fast and inexpensive

c) fast and expensive d) slow and expensive

58. Which one of the following is optical memory ? **[UGC NET - 1997]**

a) CD-ROM b) Floppy disk

c) Bubble memory

d) Charge coupled memory

59. The larger the RAM of a computer, the faster its processing speed is, since it eliminates **[UGC NET - 1997]**

a) need of ROM b) need for external memory

c) frequent disk I / Os d) need for wider data path

60. 'Volatile memory' means the memory which **[UGC NET - 1996]**

a) can be expanded

b) cannot be expanded

c) retains information even after the power is switched off

d) cannot retain the information if power is cut-off

61. Buffer caches are used **[UGC NET - 1996]**

a) to handle interrupts

b) to speed up main memory read operation

c) to increase the capacity of the main memory

d) to improve disk performance.

62. The sizes of srandard floppy disks are **[UGC NET - 1996]**

a) $5\frac{1''}{4}$ b) $3\frac{1''}{2}$

c) $5\frac{1''}{4}$ and $3\frac{1''}{2}$ d) None

63. Which of the following are not likely to be found on a PC mother board? **[UGC NET - 1995]**

a) RAM chips

b) Microprocessor chips

c) ROM chips

d) Liquid-crystal display clocks

64. A buffer is **[Hyd.Univ-2000]**

a) Storage used to compensate for a difference in rate of flow of data, or time of occurrence of events, when transferring data from one device to another

b) A shared communication path consisting of a one or more collection of parallel lines

c) Always external to a CPU, but lock-step synchronous to a data interface

d) A shared communication path where conflict of resource allocation is decided by a priority scheme

ANSWER KEY - I

1. b	2. a	3. b	4. d	5. d
6. a	7. d	8. c	9. a	10. d
11. b	12. ?	13. a	14. a	15. b
16. c	17. b	18. d	19. d	20. b
21. d	22. b	23. a	24. a	25. a
26. a	27. b	28. c	29. b	30. d
31. d	32. b	33. b	34. d	35. d
36. d	37. c	38. d	39. a	40. b
41. b	42. c	43. ?	44. d	45. d
46. d	47. d	48. b	49. d	50. c
51. a	52. a	53. a	54. b	55. a
56. ?	57. c	58. a	59. c	60. d
61. b	62. c	63. d	64. a	

EXPLANATORY NOTES - I

4. Required chips $=\frac{1MB}{256K}\times 8$

$$=\frac{1024\times 1024}{256\times 1024}\times 8=32$$

18. With n bit address bus can address the 2^n locations (bytes). Hence the number of locations in memory is determined by the address bus. However the size of memory with 16 bit address bus is $2^{16}(=65536)$ locations.

19. Given, the size of available memory chips are 16×4, Desired memory is 64×8 size required 16×4 size chips to desired memory are $\frac{64}{16}\times 2=8$

21. n - address line μp is capable of addressing 2^n locations. Hence with 12 address lines capable of addressing $2^{12}(=4096)$ locations.

41. With n bits we can represent maximum defined number is 2^n-1.

Hence largest number, which can be described by the BYTE is $2^8-1=255$.

42. Given memory is $4K=4\times 1024=2^{12}$

Hence 12 bits are needed to address 4K memory.

46. The maximum decimal numbers with two 8 - bit multiplication table is $(2^8\times 2^8)=2^{16}=64K$

Hence the capacity of the ROM is 64K words each with 16 bits. The size of ROM is $64K\times 16$.

OBJECTIVE QUESTIONS - II

1. A computer can execute 1,000,000 instructions per second. A program running on this computer performs on average a one sector read and a one sector write for every two hundred instructions that it executes. The Disk drive handling the I/O transfers requires 0.00010 seconds, each to perform the read and write operations. Assuming no overlapping of these operations, the % of CPU time spent in the wait state is: **[AUCET-1999]**

a) 12 b) 39 c) 57 d) 91

2. In a two-level memory hierarchy,the cache memory access time is 10 ns and the main memory access time is 60 ns. To maintain an average access time of less than 12 ns, the cache hit ratio must be: **[AUCET-2001]**

a) Greater than 0.96 b) Less than 0.96
c) Greater than 0.04 d) Less than 0.04

3. A hard disk has 2048 cylinders and 16 surfaces. Each track of the disk has 128 sectors.The disk rotates at the rate of 3600 rpm. The average latency time for the disk is: **[AUCET-2001]**

a) 1.28 ms b) 160ns c) 10.24ms d) 8.33ms

4. Serial access memories are useful in applications where

a) Data consists of numbers
b) Short access time is required
c) Each stored word is processed differently
d) Data naturally needs to flow in and out in serial form.

5. What is the correct sequence of time delays that happen during a data transfer from a disk to memory:

a) Seek time, access time, transfer time
b) Seek time, latency time, transfer time
c) Latency time, seek time, transfer time
d) Latency time, access time, transfer time

6. The seek time of a disk is 30ms. It rotates at the rate of 30 rotations/second. The capacity of each track is 300 words. The access time (approximately)is:

a) 62ms b) 60ms c) 50ms d) 47ms

7. If each address space represents one byte of storage space, how many address lines are needed to access RAM chips arranged in a 4×6 array, where each chip is 8k×4 bits:

a) 13 b) 15 c) 16 d) 17

8. In a memory system, four 256 × 8 PROM chips are used to make total memory of size 1024 × 4. What is the number of address bus lines:

a) 4 b) 8 c) 10 d) 16

9. Consider a high speed 40ns memory cache with a successful hit ratio of 80% . The regular memory has an access time of 100ns. What is the average effective time for CPU to access memory:

a) 52ns b) 60ns c) 70ns d) 80ns

10. What is the hit ratio of a cache If a system performs memory access at 30 nano seconds with the cache and 150 nano seconds without it. Assume that the each uses 20 nano seconds memory, choose the closest approximate.

a) 81% b) 75% c) 92% d) 87%

11. Consider a disk with the following specifications:

Track Size= 10,000 bytes

Rotational latency=10ms/revolution

Block Size= 1,000 bytes

What is the maximum transfer rate per track measured in bits per second as is conventional for this disk unit:

a) 400 MBPS b) 8MBPS
c) 6400 MBPS d) 4250 MBPS

12. If the cache needs an access time of 20ns and the main memory 120ns, then the average access time of a CPU is (assume hit ratio is 80%)

a) 30ns b) 40ns c) 35ns d) 45ns

13. The capacity of a memory unit is defined in terms of the number of words multiplied by the number of bits / word. The number of separate address and data lines needed for a memory of 4k × 16 is

a) 10,16 b) 11,8 c) 12,16 d) 12,12

14. Transferring data in blocks from the main memory to the cache memory enables an interleaved main memory unittooperate unit at its maximumspeed. **[GATE-1999]**

a) True b) False

15. A ROM is used to store the truth table for a binary multiplier unit that will multiply two 4-bit numbers. The size of the ROM (number of words × number of bits) that is required to accommodate the truth table is M words × N bits. What are the values M and N. **[GATE-1993]**

a) 256,8 b) 128,8 c) 256,16 d) 128,16

16. A certain moving arm disk storage, with one head, has the following specifications.

Number of tracks/recording surface=200.

Disk rotation speed = 2400 RPM.

Track Storage Capacity = 62,500 bits.

The average latency of this device is P msec and the data transfer rate is Q bits/sec.What are the values of P and Q: **[GATE-1993]**

a) 12.0,2.5 $\times 10^6$ b) 12.5, 2.5 $\times$ 10^6
c) 12.0, 2.5 $\times$ 10^4 d) 11.5,2.6 $\times$ 10^5

17. The capacity of a memory unit is defined by the number of words multiplied by the number of bits/word. How many separate address and data lines are needed for a memory of 4k × 16 : **[GATE-1995]**

a) 10 address,16 data lines
b) 11 address,8 data lines
c) 12 address, 16 data lines
d) 12 address,12 data lines

18. A computer system has a 4k word Cache organized in block associative manner with 4 blocks per set, 64 words per block. The number of bits in the SET and WORD fields of the main memory address format is **[GATE-1995]**

a) 15,40 b) 6,4 c) 7,2 d) 4,6

19. The correct matching for the following Pairs is

A. DMA I/O	I. High speed RAM
B. Cache	II. Disk
C. Interrupt I/O	III. Printer
D. Condition code Register	IV. ALU

[GATE-1999]

a) A-IV,B-III,C-II,D-I b) A-II,B-I,C-III,D-IV
c) A-IV,B-III,C-II,D-I d) A-II,B-III,C-IV,D-I

20. Suppose the domain set of an attribute consists of signed four digit numbers. What is the percentage of reduction in storage space of this attribute if it is stored as an integer rather than in character form: **[GATE-1998]**

a) 80% b) 20% c) 60% d) 40%

21. The main memory of a computer has 2 C.M.blocks while the cache has 2c blocks.If the cache uses the set associative mapping scheme with 2 blocks per set,then block k of the main memory maps to the set: **[GATE-1999]**

a) (k mod m) of the cache b) (k mod c) of the cache
c) (k mod 2c)of the cache d)(k mod 2cm) of the cache

22. Select the incorrect statement. **[UGC NET - 1996]**
a) Ferrite core memory is a non-volatile memory
b) Information from auxiliary memory is transferred to ALU via main memroy
c) Main memory unit of a computer supervises the working of all other units
d) The computer hardware units are of no use without a program

23. Which one of the following is TRUE?**[UGC NET - 1995]**
a) Magnetic core memory has destructive - read property but provides a non-volatile storage medium
b) Semi-conductor storage has destructive-read property but provides non-volatilestorage medium
c) Magnetic core memory has non-destructive-read property and provides a non-volatile storage medium
d) Semi-conductor storage has non-destructive-read property and magnetic core provides volatile storage

24. The time elapsed between the initiation of a memory operation and the completion of that operation is known as **[Hyd.Univ-1998]**
a) memory cycle time b) memory access time
c) transfer time d) skip time

25. The number of records contained within a block of data on magnetic tape is defined by the **[Hyd.Univ-1998]**
a) block definition b) record contain clause
c) blocking factor d) record per block factor

ANSWER KEY - II

1. d	2. d	3. d	4. d	5. b
6. d	7. d	8. c	9. a	10. c
11. b	12. b	13. c	14. a	15. a
16. b	17. c	18. d	19. b	20. c
21. b	22. c	23. ?	24. a	25. c

EXPLANATORY NOTES - II

2. To solve this problem, We use the formula for average access time, and solve for the hit rate. For each level, average access time = (hit rate × access time for that level) + ((1-Hit rate) × access time for the next level)
Plugging the given values into the formula gives
10 ns = (hit rate × 60 ns) + (mis rate × 12 ns)
= (hit rate × 60 ns) + ((1-hit rate) × 12 ns)

$\therefore$ hit rate $= -\frac{2}{48} ns = -0.04$

Hence the hit ratio must be less than 0.04.

3. Given disk rotates at the rate of 3,600 RPM.
At 3,600 rp_m each rotation takes

$\frac{1}{3{,}600} \times 60 = 0.01666 = 16.66\,ms$

The average latency is half the rotation time or 8.33 ms.

6. The seek time of a disk is 30 ms. Disk rotates at the rate of 30 rps. At 30 rps, each rotation takes 1/30 = 33.33ms. The rotational latency is half the rotation rate, or 33.33/2 = 16.6ms

The trasfer time is $\frac{33.33}{300} = 0.1ms$

Hence the access time is sum of seek time, latency time and transfer time = 30 + 16.6+0.1=46.7=47 ms.

8. Required bus lines are 10. Because 8 lines to select the chip address, 2 lines are decoded into 4 chip select inputs.

9. The average effective time for C.P.U to access memory is $(0.8 \times 40) + ((1-0.8) \times 100) \Rightarrow 32 + 20 = 52$ ns

10. Average access time
= (hit rate × cache access time) + ((1-hit rate) × memory access time)
$\Rightarrow$ 20 ns = (hit rate × 30 ns) + ((1-hit rate) × 150 ns)
Hence the hit ratio is approximate to 92%

11. transfer rate $= \frac{\text{Rotational latency}}{\text{no. of sec tors per track}}$

$= \frac{10}{(10{,}000/1000)} MB/s = 1 \times 8 MB/S = 8$ MBPS

12. Average access time $= (0.8 \times 20) + ((1-0.8) \times 120)$
= 16 + 24 = 40 ns

13. Given memory size is 4 K $\times 16 = 10^{12} \times 16$
Hence 12 address lines and 16 data lines are needed to reference given memory.

15. The result of multiplication of two n - bit numbers consist of maximum of 2n bits i.e., it represents the m $(m = 2^{2n})$ words.
However the multiplication of two 4 bit numbers consist of 258 words each with 8 bits. Hence the ROM size is 258×8.

16. Rotation rate $= \frac{1}{2{,}400} \times 60 = 0.025 = 25$ ms

$\therefore$ Rotational latency (P) = Half the rotation rate $= \frac{25}{2} = 12.5$

tranfer rate $(Q) = \frac{\text{Rotation rate}}{\text{no. of sec tors per track}} = 2.5 \times 10^6$

17. Given memory size is $4K \times 16 = 2^{12} \times 16$
means 4096 words with 16 bit each. Hence 12 address lines and 16 data lines are need for reference given memory.

18. A block contains 64 words. i.e., WORD field contain 6 bits.
Total cache is $4K (= 2^{12})$ words.
Hence $2^{12} = 4 \times \text{no of sets} \times 64$
$\therefore$ no. of sets $= 2^4 = 16$.
$\therefore$ SET field contain 4 bits.

IV. I/O ORGANISATION

- **Input/Output Module**
- **Input/Output Techniques**
- **Asynchronous Data Transfer**

1. Input/Output Module

The input/output module (I/O module) is normally connected to the system Bus of the system on one end and one or more Input/Output devices on the other. Normally, an I/O module can control one or more peripheral devices.

Is the I/O module mere a connector of Input/Output(I/O) devices to system bus?

No it performs additional functions such as communicating with the CPU as well as with the peripherals. But why to use I/O module at all why not connect a peripheral directly to the system bus? Well there are three main reasons for this:

(a) Diversity and variety of I/O devices makes it difficult to incorporate all the peripheral device logic (i.e. its control commands, data format etc.) into CPU. This in turn will also reduce flexibility of using any new development.

(b) The I/O devices are normally slower than that of memory and CPU, therefore, it is suggested not to use them on high speed bus directly for communication purposes.

(c) The data format and word length used by the peripheral may be quite different than that of a CPU.

2. Functions of I/O Module

An I/O module is a mediator between the processor and an I/ O device/devices. It controls the data exchange between the external devices and main memory; or external devices and CPU registers. Therefore, an I/O module provide an interface internal to the computer which connect it to CPU and main memory and an interface external to the computer connecting it to external device or peripheral . The I/O module should not only communicate the information from CPU to I/O device, but it should also coordinate these two. In addition since there are. Speed differences between CPU and I/O devices, the I/O module should have facilities like buffer (storage area) and error detection mechanism. Therefore, the functional requirements of an I/O module are:

(a) It should be able to provide control and timing signals.

The need of I/O from various I/O devices by the CPU is quite unpredictable.

In fact it depends on I/O needs of particular programs and normally do not follow any pattern.

Since, the I/O module also, share system Bus and memory for data Input/Output., Therefore, control and timing are needed to coordinate the flow of data from, to external devices to/from CPU or memory.

A typical control cycle for transfer of data from I/O device to CPU is:

- Enquire status of an attached device from I/O module. The status can be busy ,ready or out of order..
- I/O module responds back with the status of the device.
- If device is ready, CPU commands I/O module to transfer data from the I/O device. .
- I/O module accept data from the I/O device.
- The data is then transferred from I/O module to theCPU.

(b) It should communicate with the CPU.

The example given above clearly specifies the need of communication between the CPU and I/O module. This communication can be: commands such as READ SECTOR, WRITE SETOR, SEEK track number (which are issued by CPU to I/O module); or Data (which may be required by the CPU or transferred out); or Status information such as BUSY or READY or any error condition from I/O modules. The status recognition is neces-sary because of the speed gap between the CPU arid I/O device. An I/O device might be busy in doing the I/O of previous instruction when it is needed again.

Another important communication from the CPU is the unique address of the peripheral from which I/O is expected or is to be controlled.

(c) It should communicate with the I/O device.

Communication between I/O module and I/O device is needed to complete the I/O operation.This communication involves commands, status or data.

(d) It should have a provision for data buffering.

Data buffering is quite useful for the purpose of smoothing out the gaps in speed of CPU and I/O devices. The data buffers are registers which hold the I/O information temporarily.

The I/O is performed in short bursts in which data is stored in buffer area while the device can take its own time to accept it.On the I/O device to CPU transfer data is first transferred to the buffer and then passed on to CPU from these buffer registers. Thus, the I/O operation does not tie up bus for the slower I/O devices.

(e) Error detection mechanism should be in built. The error detection mechanism may involve checking the mechanical as well as data communication errors. These errors should be reported to the CPU. The examples of the kind of mechanical errors that can occur in devices are paper jam in printer, mechanical failure, electrical failure etc. The data communication errors may be checked by using parity bit.

3. Input/Output Techniques

The Input/Output operations can be performed by three basic techniques. These are:

- Programmed Input/Output
- Interrupt driven Input/Output
- Direct Memory Access .

Figure 4.1 give an overview of these three techniques.

	Interrupt Required	I/O module to/ from memory transfer
Programmed I/O	No	Through CPU
Interrupt driven I/O	Yes	Through CPU
Direct Memory Access	Yes	Direct transfer

Figure 4.1: Overview of the three Input/Output Techniques

In programmed I/O the I/O operations are completely controlled-by the CPU. The CPU executes programs that initiate, directs and terminate an I/O operation.

It requires a little special I/O hardware, but is quite time consuming for the CPU , since CPU has to wait for slower I/O operations to complete.

Another technique suggest to reduce the waiting by CPU is interrupt driven I/O.The CPU issues the I/O command to I/O module and starts doing other work which may be execution of a separate program, When the I/O operation is complete, I/O module interrupts CPU by informing the CPU that I/O has finished. CPU, then, may proceed execution of this program.

In both programmed I/O and interrupt driven I/O CPU is responsible for extracting data from the memory for Output and storing data in memory for input. Such a requirement does not exist in DMA where the memory can be accessed directly by I/O module. Thus, the I/O module can store or extract data in/from the memory.

In case the I/O is solely looked after by a separate dedicated processor, then this is referred to as I/Ochannel or I/O processor. The basic advantage of these devices are that they free CPU of the burden of Input/Output. Thus, during this time CPU can do other work, therefore, effectively increasing the CPU utilisation.

4. Programmed Input/Output

Programmed Input/Output are useful I/O method for computers where hardware costs need to be minimised. The Input or Output operation in such cases may involve:

- Transfer of data from I/O device to the CPU registers.
- Transfer of data from CPU registers to memory.

In addition, in a programmed I/O method the responsibiliy of CPU is to constantly check the status of the I/O device to check whether it has become free (in case output is desired) or it has finished inputting the current series of data (in case input is going on). Thus, Programmed I/O is a very time consuming method where CPU wastes lot of lime for checking and verifying the status of an I/O device.

Let us now try to focus how this Input Output is performed. Figure 4.2 (a) gives the block diagram of transferring a Block of data word by word using programmed I/O technique

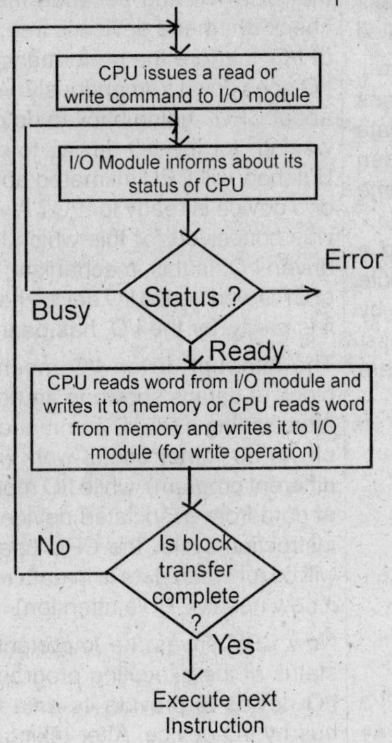

(a) Programmed I/O

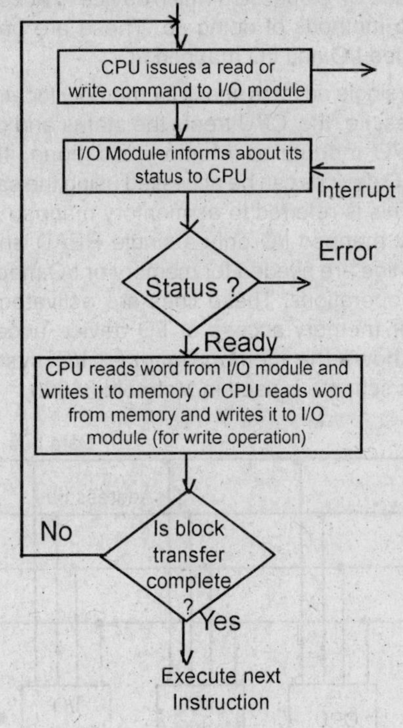

(b) Interrupt drive I/O

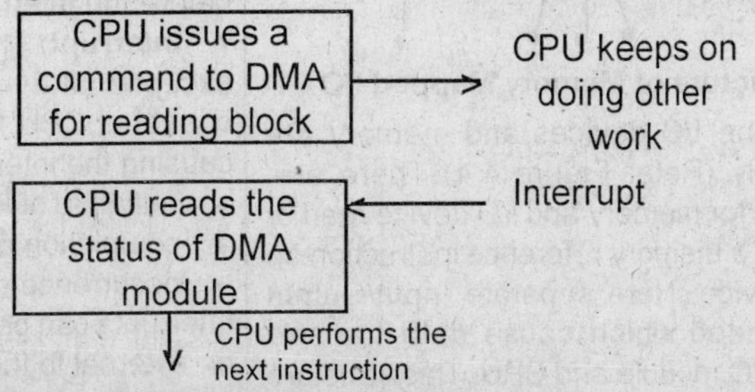

(c) Direct memory access

Fig 4.2 : Threee techniques of Input/ Output

5. I/O Instructions: To carry out Input/Output CPU issues I/O related instructions. These instructions consist of two components:

- The address of the Input/Output device specifying the I/O device and I/O module; and
- An Input/Output command.

There are four types of I/O commands which can be classified as: control,test,read and write.

Control commands are device specific and are used to control the specific instructions to the device e.g., a magnetic tape requires rewinding or moving forward by a block. Test commands checks the status such as, if a device is ready or not or is in error condition. The read command is used for input of data from input device and write command is used for output of data to output device.

The other part of I/O instruction is the address of the I/O device. In systems with programmed I/O the I/O module, the main memory and the CPU normally share the system bus.Thus, each I/O module should interpret the address lines to determine if the command is for itself. Or in other words How does CPU specifies which device to access? There are two methods of doing so. These are called memory mapped I/O and I/O mapped I/O.

If we use the single address space for memory locations and I/O devices, i.e. the CPU treats the status and data registers of I/O module as memory locations, then memory and I/O devices can be accessed using the same instructions. This is referred to as memory mapped I/O. For a memory mapped I/O only a single READ and a single WRITE line are needed for memory or I/O module read or write operations. These lines are activated by CPU for either memory access or I/O device access. Figure 4.3 shows the memory mapped I/O system structure. This scheme is used in Motorola 68000.

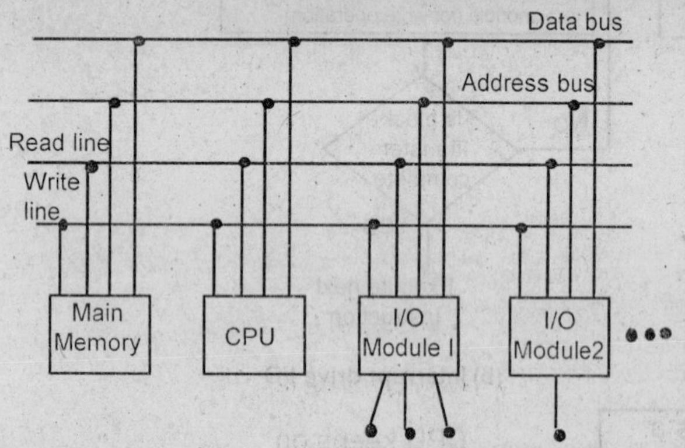

Figure 4.3 : Structure of Memory Mapped I/O

In I/O mapped I/O the I/O devices and memory are addressed separately (Refer Figure 4.4) There are, separate control lines for memory and I/O device read or write operations, thus a memory reference instruction do not affect an I/O device, Here separate Input/Output instructions are needed which cause data transfer between addressed I/O module and CPU. This structure is used in Intel 8085 and 8086 series,

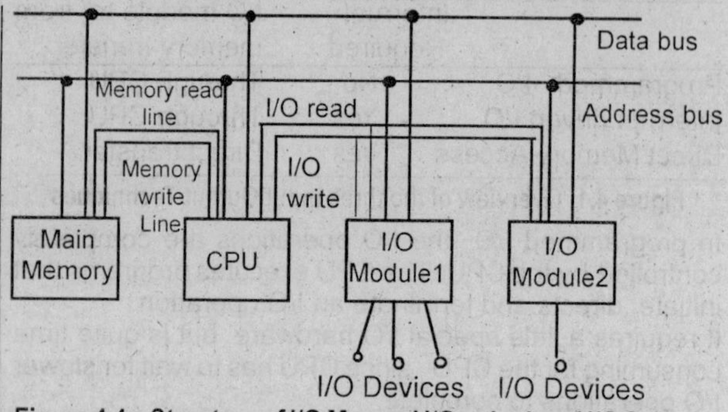

Figure 4.4 : Structure of I/O Mapped I/O or Isolated I/O Scheme

Please note the difference of requirements as in the case of memory mapped I/O the READ Instruction may bring data to or from memory or I/O module, while in I/O mapped I/O we need to have separate instruction for Input/Output.

6. Interrupt Driven Input/Output

What are the basic drawback of programmed I/O? The speed of I/O devices is much slower in comparison to that of CPU, and because the CPU has to repeatedly check whether a device is free, or wait till the completion of I/O, therfore the performance of CPU in programmed I/O goes down tremendously. What is the solution? What about CPU going back to do other useful work without waiting for the I/O device to complete or get freed up. But how will CPU intimated about the completion of I/O or a device is ready for I/O? A well designed mechanism was conceived for this which is referred to as Interrupt driven I/O. In this mechanism, provisions of interruption of CPU work, once I/O device has finished the I/O or when it is ready, for the I/O, has been provided.

The interrupt driven I/O mechanism for transferring-a block of data is shown in figure 4.2(b). Please note that after issuing a READ command (for input) the CPU goes off to do other useful work (it may be execution of a different program). while I/O module proceeds for reading of data from associated device. At the completion of an instruction cycle the CPU checks for interrupts (which will occur when data is in data register of I/O module and it now needs CPU's attention).

Now CPU saves the important register and processor status of the executing program in a stack and request I/O device to provide its data which is placed on data bus by I/O device. After taking required action with the data, the CPU can go back to program it was executing before the interrupt

7. **Interrupt:** The term interrupt loosely is used for any exceptional event that causes temporary transfer of control of CPU from one program to the other which is causing the interrupt. Interrupts are primarily issued on:

- initiation of Input/Output operation
- completion of an Input/Output operation
- occurrence of hardware or software errors.

Interrupts can be generated by various sources internal or external to the CPU.An interrupt generated internally by CPUis sometimes termed as Traps.The traps are normally result of programming errors such as division by zero while execution of a program.

The two key issues in Interrupt driven Input/Output are

- to determine the device which has issued an interrupt
- in case of occurrence of multiple interrupts which one to be processed first.

There are several solutions to these problems. The simplest of them is to provide multiple interrupt lines which will result in immediate recognition of the interrupting device. The priorities can be assigned to various interrupts and the interrupt with highest priority should be selected for service in case multiple interrupt occurs. But providing multiple interrupt lines is an impractical approach because only a few lines of the system bus can be devoted for the interrupt. Other methods for this are software poll, daisy chaining and bus arbitration.

8. Software Poll: In this scheme on occurrence of an interrupt, CPU starts executing an software routine termed as interrupt service program or routine which poll to each I/O module to determine which I/O module has caused the interrupt. This may be achieved by reading the status register of the I/O modules. The priority here can be implemented easily by defining the polling sequence, since the device polled first will have higher priority. Please note that after identifying the device the next set of instructions to be executed will be the device service routines of that device, resulting in the desired input or output.

As far as daisy chaining is concerned, we have one Interrupt Acknowledge line which is chained through various interrupt devices. There is just one Interrupt Request line. On receiving an Interrupt Request the Interrupt Acknowledge line is activated which in turn passes this signal device by device. The first device which has made the interrupt request thus graps the signal and responds by putting a word which is normally an address of interrupt servicing program or a unique identifier on the data lines. This word is also referred to as interrupt vector. This address or identifier in turns is used for selecting an appropriate interrupt servicing program. The daisy chaining has an in-built priority scheme which is determined by the sequence of devices on interrupt acknowledge lines. In bus arbitation technique the I/O module first need to control the bus and only after that it can request for an interrupt. In this scheme since only one of the module can control the bus therefore, only one request can be made at a time. The interrupt request is acknowledged by the CPU on response of which I/O module places the interrupt vector on the data lines.

9. Direct Memory Access (DMA)

When large amount of data is to be transferred from CPU, a DMA module can be used. But why? In both Interrupt driven and programs I/O the CPU is tied up in executing input/output instructions while DMA acts as if it has taken over control from the CPU. The DMA operates in the following way :

- When an I/O is requested the CPU instructs the DMA module about the operation by providing the information:
  - which operation (Read or Write) to be performed.
  - the address of I/O device which is to be used.
  - the starting location on the memory where the information will be read or written to.
  - the number of words to be wirtten or to be read.
- The DMA module transfer the requested block byte by byte directly to the memory without intervening the CPU.
- On completion of the request DMA module sends an interrupt signal to the CPU.

Thus in DMA the CPU involvement can be restricted at the beginning and end of the transfer. But what do CPU do while DMA is doing Input/Output, it may execute another program or may be another part of the same program. Figure 4.5 shows register of general DMA module. Please note that it contains additional registers for counting the data bytes. Please note that address register and data count register are fed with the data lines.

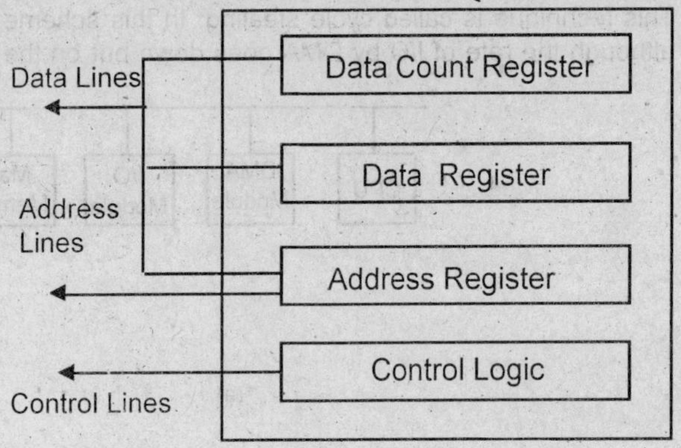

Figure 4.5 : The Registers on DMA Module

Let us now see how this DMA mechanism works. Since DMA module share the system Bus, therefore it need to have some way to take control of the bus such that the data is transferred to/from memory from/to the DMA

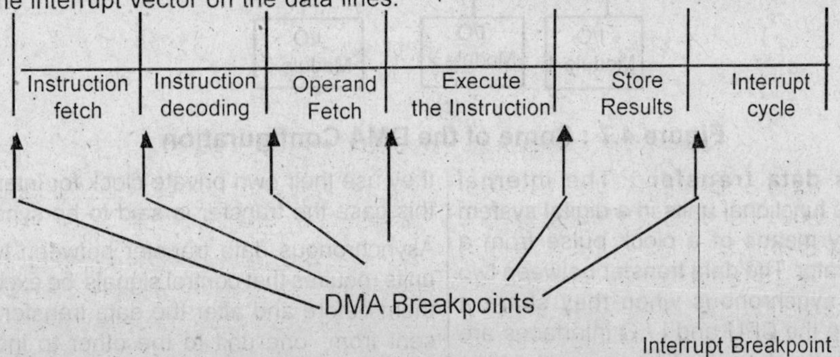

Figure 4.6 : DMA and Interrupt Breakpoints

module. A DMA transfer an entire block of data or one word at a time directly to/from memory. But when should the DMA take control of the bus ?

For this we will recall the phenomena of execution of an instruction by the CPU. Figure 4.6 shows the five cycles for an instruction execution. The figure also shows the five points where a DMA request can be responded and a point where the interrupt request can be responded. Please note that an interrupt request is acknowledged only at one point of an instruction cycle.

Let us now discuss the data transfer modes of DMA. The first mode is the block transfer of data after taking control of the bus. CPU may not use the bus during this time. In such mode a complete block of data for example a complete sector in a disk is transferred, in a single continuous burst. During this transfer the DMA controller/ module remain master of the bus.Such a transfer is quite useful in cases where we have fast secondary memory such as magnetic dist where a delay of one or two pulse may result in a delay by a rotation, thus, the burst of data is preferred through DMA. The drawback in such scheme is that the CPU has to wait for the bus for quite some time.

Alternatively, instead of transferring a complete block, a few words or a single word is transferred through system bus to the memory through DMA. In this mode the DMA forces the the CPU to suspend its operation temporarily. After this transfer the control is returned back to the CPU. This technique is called cycle stealing. In this scheme although the rate of I/O by DMA goes down but on the other hand it reduces the interference caused by the DMA to CPU operation.

It is possible to minimize this interference of CPU by DMA controller such that the cycles are stolen only when CPU is not using system bus. This is termed as transparent DMA.

Finally let us discuss how DMA can be configured ? Well there are many ways we will discuss some of them.

The simplest possibility is to allow DMA, I/O and all the modules to share the sytem bus. This structure is shown in figure 4.7 (a). In this kind of configuration DMA may act as a supportive processor and can use programmed I/O for exchanging data between memory and I/O module through DMA module. But once again this spoils the basic advantage of DMA of not using extra cycles for transferring information from memory to/from DMA and DMA from /to I/O module.

The Figure 4.7 (b) configuration suggest a clear cut advantages over the one shown in figure 4.7 (a). In these systems a path is provided between I/O module and DMA module which does not include system bus. The DMA logic may become part of an I/O module and can control one or more I/O module. In an extended concept an I/O bus can be connected to this DMA module. Such a configuration (shown in Figure 4.7 (c)) is quite flexible and can be extended very easily. In both these configurations the added advantage is that the data between I/O module and DMA module is transferred off the system bus. Thus, eleminately the disadvantage we have witnessed for the first configuration.

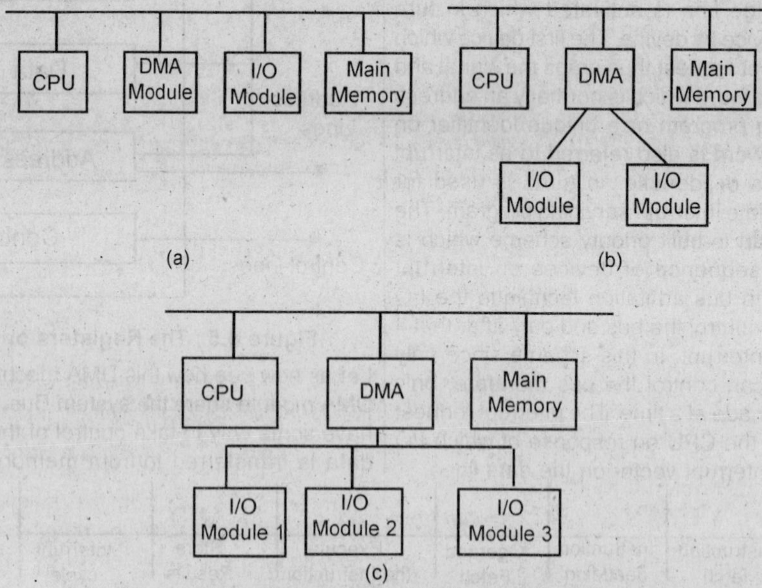

Figure 4.7 : Some of the DMA Configuration

10. Asynchronous data transfer : The internal operations of various functional units in a digital system are synchronized by means of a clock pulse from a common clock generator. The data transfer between two units is said to be synchronous when they share a common clock. Since the CPU and I / O interfaces are designed independently of each other, in most cases they use their own private clock for internal registers. In this case the transfer is said to be synchronous.

Asynchronous data transfer between two independent units requires that control signals be exchanged between them before and after the data transfer. A signal that is sent from one unit to the other to indicate when the transfer should occur is called strobe. The strobe signal can be source initiated or destination initiated as shown in Figure 4.8. The receiver after receiving the data may send an acknowledgement to the sender. This kind of exchange of signals before and after the data transfer is called handshaking. Data transfer between an interface and the CPU is a strobe controlled transfer and the data transfer between the interface and an I / O device is commonly controlled by a set of handshaking signals.

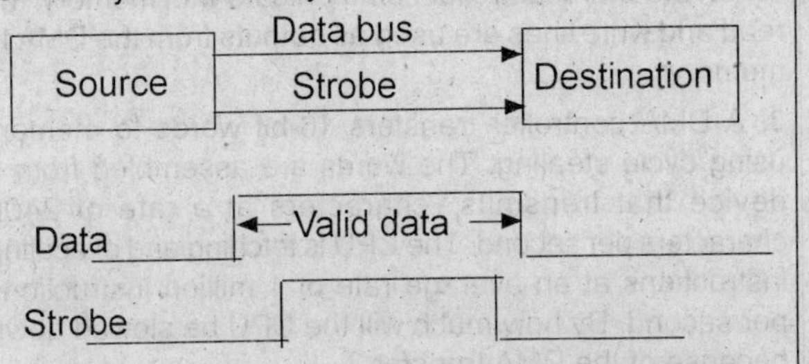

Figure 4.8 a: Source initiated strobe for data transfer

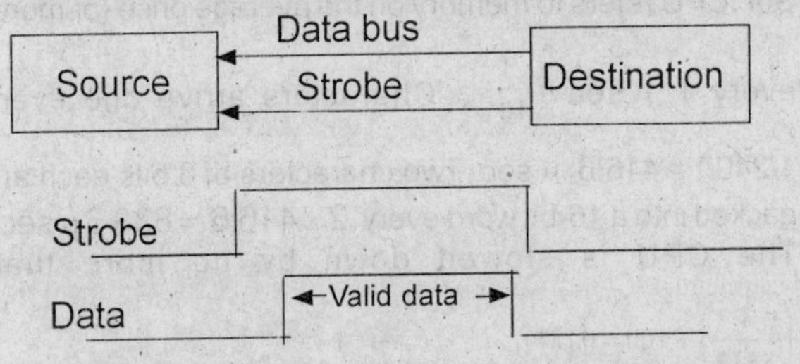

Figure 4.8 b: Destination initiated strobe for data transfer

The disadvantage of the strobe control is that, if it is source initiated, there is no way for the source of knowing whether the destination has received the data or not.

Similarly if it is destination initiated, there is no way for the destination to know whether the source has placed the data on the data bus or not. This problem is removed by using what is called two-wire handshaking. In two wire handshaking, as shown if Figure 4.9, the source sends a data valid signal after placing the data in the data buffer. The destination after accepting the data sends a "data accepted" signal. Figure 4.9 and 4.10 show two kinds of handshake data transfer along with the sequence of events that take place.

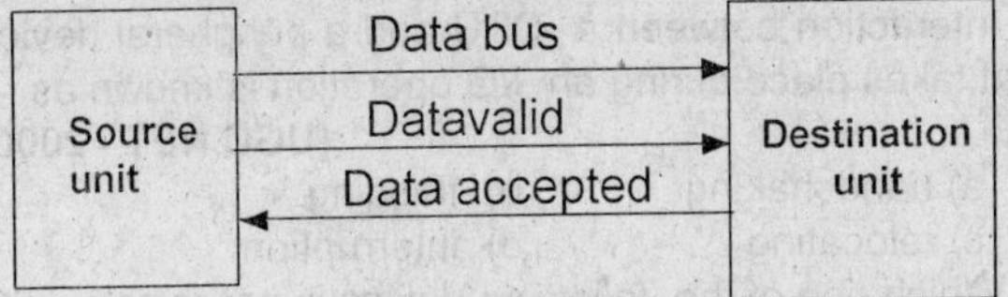

4.9 (a) Block diagram

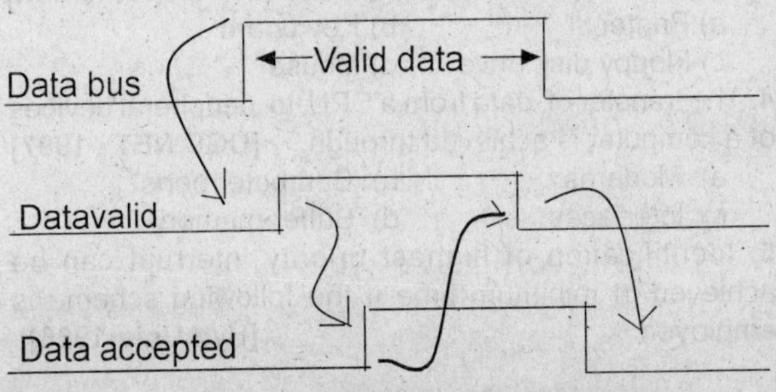

4.9 (b) Timing diagram

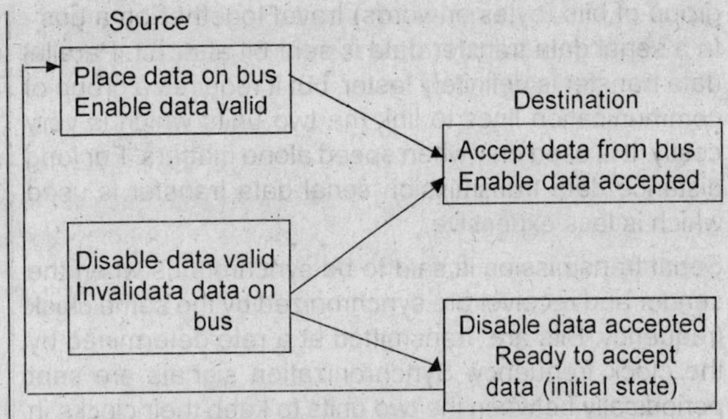

4.9 (c) Sequence of Events

Figure 4.9 Source initiated transfer using handshaking

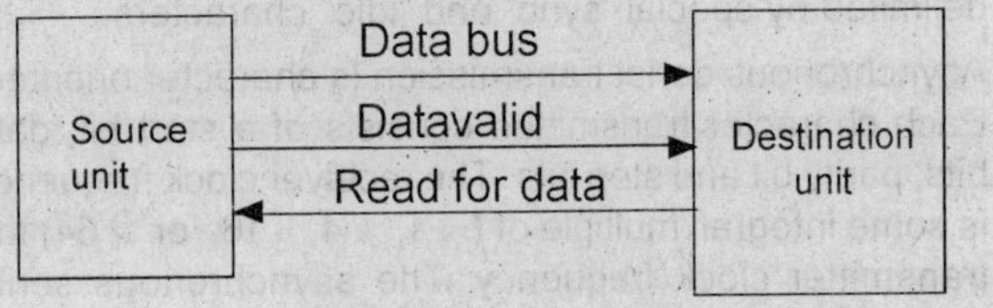

4.10 (a) Block diagram

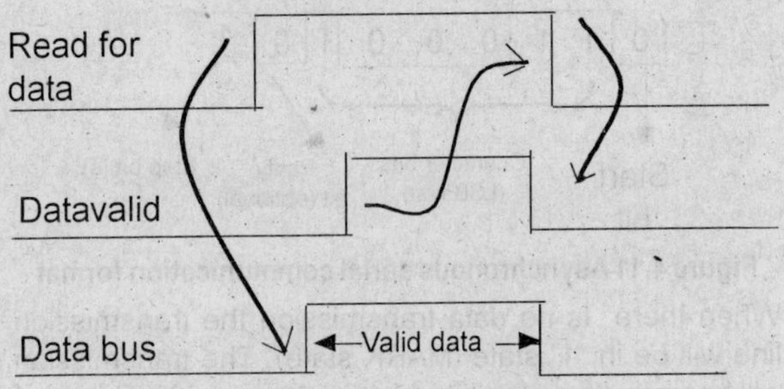

4.10 (b) Timing diagram

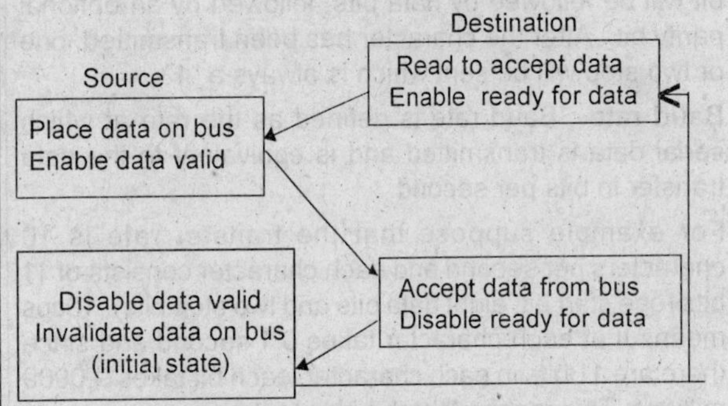

4.10 (c) Sequence of Events

Fig 4.10 : Destination initiated transfer using hand shaking

The handshaking scheme provides a high degree of flexibility and reliability because of the participation of both units. This will ensure successful data transfer. Even if one unit is faulty the transfer will not be completed. A faulty unit can be detected by means of a "time out" alarm which can be activated if the transfer is not complete within a prespecified time.

The timeout signal can be used to interrupt the processor which can execute a service routine that takes appropriate action to recover from the error.

11. Asynchronous Serial Transfer : Data transfer between a source and a destination can be parallel data transfer or serial data transfer. In parallel data transfer group of bits (bytes or words) travel together on a bus . In a serial data transfer data is sent bit after bit. Parallel data transfer is definitely faster, but it requires a group of communication lines to link the two units which is very costly. It is used only when speed alone matters. For long distance data transmission serial data transfer is used which is less expensive.

Serial transmission is said to be synchronous when the sender and receiver are synchronized by the same clock frequency. Bits are transmitted at a rate determined by the clock frequency. Synchronization signals are sent periodically between the two units to keep their clocks in step. Blocks of characters are transmitted continuously in serial synchronous tranmission. The blocks are delimited by special "sync" and "idle" characters.

Asynchronous serial transmission is character oriented. Each character transmitted consists of a start bit, data bits, parity bit and stop bits. The receiver clock frequency is some integral multiple of ($\times$ 1, $\times$ 4, $\times$ 16, or $\times$ 64) the transmitter clock frequency. The asynchronous serial transmission format is shown below figure

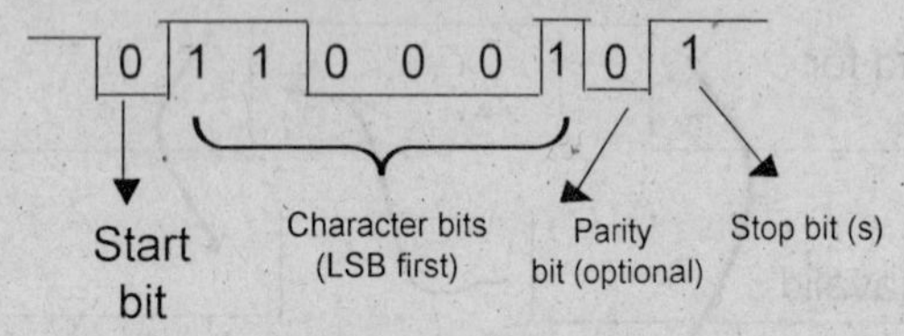

Figure 4.11 Asynchronous serial communication format

When there is no data transmission the transmission line will be in "1" state (MARK state). The transmission will begins with a start bit which is always a "0". This start bit will be followed by data bits, followed by an optional parity bit. After the character has been transmitted, one or two stop will be sent which is always a "1".

Baud rate : Baud rate is defined as the rate at which serial data is transmitted and is equivalent to the data transfer in bits per second.

For example suppose that the transfer rate is 10 characters per second and each character consists of 11 bits (one start bit, eight data bits and two stop bits). 10cps means that each character takes 0.1 second and since there are 11 bits in each character, each bit takes 9.0909 millisec. This means that the baud rate is

1 / .0090909 = 110 baud.

SOLVED EXAMPLES

1. How many characters per second can be transmitted over a 1200-baud line in each of the following modes ? (Assume a character code of eight bits).

a. Synchronous serial transmission
b. Asyncronous serial transmission with two stop bits.
c. ASynchronous serial transmission with one stop bit.

Sol : a. $\frac{1200}{8} = 150$ characters per second (cps)

b. $\frac{1200}{11} = 109\,\text{cps}$ **c.** $\frac{1200}{10} = 120\,\text{cps}$

2. Why are read and write control lines in a DMA controller bidirectional ? Under what condition and for what purpose are they used as inputs ? Under what condition and for what purpose are they used as outputs ?

Sol : When the CPU communicates with the DMA controller the read and write lines are used as inputs from the CPU to the DMA controller.

When the DMA controller communicate with memory, the read and write lines are used as outputs from the DMA to memory.

3. A DMA controller transfers 16-bit words to memory using cycle stealing. The words are assembled from a device that transmits characters at a rate of 2400 characters per second. The CPU is fetching and executing instructions at an average rate of 1 million instructions per second. By how much will the CPU be slowed down because of the DMA transfer ?

Sol :CPU refers to memory on the average once (or more) every 1 μ sec $\left(\frac{1}{10^6}\right)$. Characters arrive one every 1/2400 = 416.6 μ sec. Two characters of 8 bits each are packed into a 16-bit word every $2 \times 416.6 = 833.3\,\mu$ sec. The CPU is slowed down by no more than $\left(\frac{1}{833.3}\right) \times 100 = 0.12\%$.

4. The address of a terminal connected to a data communications processor consists of two letters of the alphabet or a letter followed by one of the 10 numerals. How many different addresses can be formulated.

Sol : There are 26 letters and 10 numerals.

$26 \times 26 + 26 \times 10 = 936$ possible addresses.

OBJECTIVE QUESTIONS - I

1. Interrupt driven I/O is a type of I/O transfer that
a) relies on hardware only **[UGC NET - 2000]**
b) relies on software only
c) relies on memory only
d) relies on both hardware and software

2. Interaction between a CPU and a peripheral device that takes place during an I/O operation is known as **[UGC NET - 2000]**
a) handshaking b) flagging
c) relocating d) interruption

3. Which one of the following device is accessible with the help of Direct Memory Access (DMA) ?**[UGC NET - 1999]**
a) Printer b) Key board
c) Floppy disk drive d) Mouse

4. The transfer of data from a CPU to peripheral devices of a computer is achieved through **[UGC NET - 1997]**
a) Modems b) Computer ports
c) Interfaces d) Buffer memory

5. Identification of highest priority interrupt can be achieved in minimum time if the following scheme is employed **[Hyd.Univ-1996]**

a) Hardware polling
b) Priority encoder circuit
c) Software polling
d) Subdivision of priority register into a number of sub registers checked parallely

6. Find the odd one among the following.Monitor, Printer, Key board, hard-disk, teletypewriter **[Hyd.Univ-1993]**
a) Monitor b) hard-disk
c) Printer d) keyboard

7. The Data transfer between memory and I/O devices using programmed I/O is faster than interrupt driven I/O **[GATE-1990]**
a) True b) False

8. Start and Stop bits do not contain an information but are used in serial communication for **[GATE-1992]**
a) Error Detection b) Error correction
c) Synchronization d) None

9. Purpose of a start bit in RS-232 serial communication protocol is : **[GATE-1997]**
a) To Synchronize receiver for receiving every byte
b) To Synchronize receiver for receiving a sequence of bytes
c) To Synchronize receiver for receiving the last byte
d) A parity bit.

10. In a microprocessor based system. If a bus (DMA) request and an interrupt requests arrive simultaneously the microprocessor attends first to the bus request **[GATE-1987]**
a) True b) False

11. Which of the following devices should get higher priority in assigning interrupts **[GATE-1998]**
a) Hard disk b) Floppy disk
c) Printer d) Keyboard

12. Parallel printer uses
a) RS- 232 interface b) Centronics interface
c) hand-shake mode
d) Synchronous data transfer mode

13. In a vectored interrupt
a) The branch address is assigned to a fixed location in memory
b) The interrupting source supplies the branch information to the processor through an interrupt vector.
c) The branch address is obtained from a register in the processor d) None

ANSWER KEY - I

1. b	2. a	3. a	4. c	5. b
6. b	7. b	8. c	9. a	10. ?
11. d	12. b,c	13. a		

OBJECTIVE QUESTIONS - II

1. The CPU,after receiving an interrupt from an I/O device **[UGC NET - 1998]**
a) hands over control of address and data buses to the interrupting device
b) halts for a predetermined time
c) branches off to the interrupt service routine immediately
d) branches off to the interrupt service routine after completion of the current instruction

2. For various type of Commands e.g., a test, control or data transfer which one of the following is incorrect ? **[UGC NET - 1995]**
a) "Checking status of flag" is a test command.
b) "Seeking a given record on a magnetic disk" is control command.
c) "Moving paper tape to next character position" is data transfer command.
d) Read status register is a combination of data transfer and test command.

3. The method of performing input/output operations where a program explicitly controls all the phases of input/ output operations performed is called **[Hyd.Univ-2000]**
a) Memory mapped input/output
b) Phased input/output
c) Programmed input/output
d) Direct Memory Access

4. The HOLD signal **[Hyd.Univ-1994]**
a) requests the control unit to indicate whether read or write operation is taking place
b) requests the CPU to relinquish control and use the external system bus
c) requests the I/O module to relinquish and use ALE
d) acknowledges the signal and indicates that the bus is now available

5. Data Transfer between a microprocessor and an I/O device is usally faster in memory-mapped I/O scheme than in I/O mapped I/O scheme **[GATE-1987]**
a) True b) False

6. Assume that each character code consists of 8 bits. The number of characters that can be transmitted per second through an a synchronous serial line at 2400 baud rate, and with two stop bits is : **[GATE-1993]**
a) 109 b) 216 c) 218 d) 240

7. On receiving an interrupt from an I/O device the CPU
a) halts for a predetermined time.
b) hands over control of address bus and data bus to the interrupting device.
c) branches off to the interrupt service routine immediately
d) branches off to the interrupt service routine after completion of the current instruction.

8. How many characters per sec (7 bits + parity) can be transmitted over a 2400 - bps line in asynchronous mode?
a) 300 b) 240 c) 320 d) None

9. Choose the incorrect statement. For a microprocessor system use I/O mapped I/O
a) Memory space available is greater
b) Not all data transfer instructions are available.
c) I/O and memory address spaces are distinct
d) I/O address space is greater.

ANSWER KEY - II

1. c	2. a	3. c	4. b	5. ?
6. c	7. d	8. b	9. d	

OBJECTIVE QUESTIONS - III

1. If a disk spins at 10,000 r/min, what is the average rotational latecy time of a request ? If a given track on the disk has 1024 sectors, what is the transfer time for a sector?
a) 6,6 b) 6,9 c) 9,6 d) 3,6

2. Suppose a given bus protocal requires 10ns for devices to make requests, 15ns for arbitration, and 25ns to complete each operation, How many operations can be completed per second ?

a) 20 million b) 20 billion
c) 25 million d) 25 billion

3. A given processor has eight interrupt lines (numbered 0-7) and a policy that low numbered interrrupts have priority over higher - numbered ones, the processor starts with no interrupts pending, and the following sequence of interrupts occurs. 4,7,1,3,0,5,6,4,2,1. Assume that handling any interrupt key enough time that two more interrupts arrive while the first interrupt is being handled, until all the interrupts have arrived, and that interrupts cannot interrupt each other. What order are the interrupts handled in ?

a) 0,1,1,2,3,4,4,5,6,7 b) 4,1,3,0,5,6,4,2,1,7
c) 4,1,0,3,2,1,4,5,6,7 d) 4,1,0,3,4,2,1,5,6,7

4. A hard disk with 5 platters has 2048 tracks/platter, 1024 secotrs/track (fixed number of sectors per track), and 512 byte sector. What is its total capacity ?

a) 3 GB b) 5 GB c) 6 GB d) 9 GB

5. A manufacturer wishes to design a hard disk with a capacity of 30 GB or more (using the standard definition of 1 GB = 2^{30} bytes). If the technology used to manufacture the disks allows 1024 byte sectors, 2048 sectors/ track, 4096 tracks/platter, how many platters are required.

a) 2 b) 3 c) 4 d) 5

Question number from 6 to 9 are based on the following information.

A hard disk with one platter rotates at 15,000 r/min and has 1024 tracks, each with 2048 sectors. The disk head starts at track 0 (tracks are numbered) from 0 to 1023. The disk then receives a request to acces a random sector on a random track. If the seek time of the disk head is 1ms for every 100 tracks it must cross :

6. What is the average seek time ?

a) 5ms b) 7ms c) 9ms d) 10ms

7. What is the average rotational latency ?

a) 4ms b) 2ms c) 5ms d) 3ms

8. What is the transfer time for a sector ?

a) 2 μ s b) 1 μ s c) 3 μ s d) 4 μ s

9. What is the total average time to resolve a request ?

a) 5ms b) 6ms c) 7ms d) 8ms

Question number from 10 to 12 are based on the following information.

A Program repeatedly executes a loop that has 120 iterations. Each iteration takes 10,000 cycles. on multiprocessor systems, 50,000 cycles are required to synchronize the processors once all iterations of the loop have completed.

10. What is the execution time of each loop on a uniprocessor system ?

a) 12×10^5 cycles b) 12×10^4 cycles

c) 12×10^3 cycles d) 12×10^2 cycles

11. What is the execution time of each loop on a 2-processor system, and what is the spped up over the uniprocessor system.

a) 65×10^4 cycles 1.85 b) 60×10^4 cycles 2.85

c) 60×10^4 cycles 1.85 d) 65×10^4 cycles 2.85

12. What is the execution time of each loop on a 4-processor system, and what is the speedup over the uniprocessor system ?

a) 3.43 b) 2.43 c) 5.40 d) 4.43

EXPLANATORY NOTES - III

1. At 10,000 r/min, it takes 6ms for a complete rotation of the disk. On average, the read/write head will have to wait for half a rotation before the needed sector reaches it, so the average rotational latency will be 3ms. Since there are 1024, or approximately 6ms.

2. The total time for an operation is the sum of the request, arbitration, and completion times, which is 20ns. Therefore, the bus can complete 20 million operations/s.

3. A good way to approach this sort of problem is to create a list of pending events (such as interrupts) whenever the processor is free, it handles the highest priority pending event, removing it from the list, thus interrupt 4 is handled first, because it arives first. When the processor is done with interrupt 4, interrupts 7 and 1 are pending, so interrupt 1 gets handled. Repeating this process gives the following order of interrupt handling 4,1,0,3,2,1,4,5,6,7. Hence answer 'c' is correct.

4. 512 bytes $\times$ 1024 sectors = 0.5 MB/track

Multiplying by 2040 tracks/platter gives 1 GB/platter, or 5GB capacity in the drive. Hence answer 'b' is correct.

5. Multiplying bytes per sector times sectors per track times tracks per platter gives a capacity of 8 GB $(8\times2^{30}\text{bytes})$ per platter, therefore, 4 platters will be required to give a total capacity of $\geq$ 30 GB.

6. Since the disk head starts at track 0, it will have to travel 0 tracks to handle a request to track 0, 1 track to handle a request to track 1, and so on, up to 1023 tracks to request to tracks 1023. on average, the head will have to travel half of the way to the outermost track, or 511.1 tracks. At 100 tracks/ ms, this gives an average seek time of 5.115 ms.

7. At 15,000 rpm, each rotation takes 4 ms, the average rotational latency is half the rotation time, or 2ms.

8. Each rotation takes 4 ms. there are 2048 sectors per track, so each sector takes 4ms / 2048 = 1.95 μ s to pass under the read/write head. Therefore, the transfer time is 1.95 micro seconds.

9. The average access time is just the sum of the three (seek time, transfer time, rotational latency) components, or 7.117 ms. Hence answer 'c' is correct.

10. 120 iterations $\times$ 10,000 cycles = 12,00,000 cycles.
= 12 $\times$ 10^5 cycles.

11. On a two-processor system, each processor would handle 60 iterations, for a base execution time of 600,000 cycles. Adding the communication time gives an execution time of 650,000 cycles and speed up of 1.85. Hence answer 'a' is correct.

12. Each processor executes 30 iterations. for an execution time of 350,000 cycles and a speedc up of 3.43.

TESTYOURSELF - I

1. A main memory unit with a capacity of 4 megabytes is built using $1M \times 1-bit$ DRAM chips. Each DRAM chip has 1K rows of cells with 1K cells in each row. The time taken for a single refresh operation is 100 nanoseconds. The time required to perform one refresh operation on all the cells in the memory unit is **[GATE - 10]**

a) 100 nanoseconds

b) $100 * 2^{10}$ nanoseconds

c) $100 * 2^{20}$ nanoseconds

d) $3200 * 2^{20}$ nanoseconds

2. A 5 - stage pipelined processor has Instruction Fetch(IF), Instruction Decode(ID), Operand Fetch(OF), Perform Operation (PO) and Write Operand(WO) stages. The IF, ID, OF and WO stages take 1 clock cycle each for any instruction. The PO stage takes 1 clock cycle for ADD and SUB instructions, 3 clock cycles for MUL instruction, and 6 clock cycles for DIV instruction respectively. Operand forwarding is used in the pipeline. What is the number of clock cycles needed to execute the following sequence of instructions ? **[GATE - 10]**

Instruction	**Meaning of Instruction**
$I_0: MUL R_2, R_0, R_1$	$R_2 \leftarrow R_0 * R_1$
$I_1: DIV\ R_5, R_3, R_4$	$R_5 \leftarrow R_3 / R_4$
$I_2: ADD\ R_2, R_5, R_2$	$R_2 \leftarrow R_5 + R_2$
$I_3: SUB\ R_5, R_2, R_6$	$R_5 \leftarrow R_2 - R_6$

a) 13 b) 15 c) 17 d) 19

3. The program below uses six temporary variables a,b,c,d,e,f.

```
a = 1
b = 10
c = 20
d = a + b
e = c + d
f = c + e
b = c + e
e = b + f
d = 5 + e
return d + f
```

[GATE - 10]

Assuming that all operations take their operands from registers, what is the minimum number of registers needed to execute this program without spilling ?

a) 2 b) 3 c) 4 d) 6

Common Data Questions Q.4 & 5

A computer system has an L1 cache, an L2 cache, and a main memory unit connected as shown below. The block size in L1 cache is 4 words. The block size in L2 cache is 16 words. The memory access times are 2 nanoseconds. 20 nanoseconds and 200 nanoseconds for L1 cache, L2 cache and main memory unit respectively.

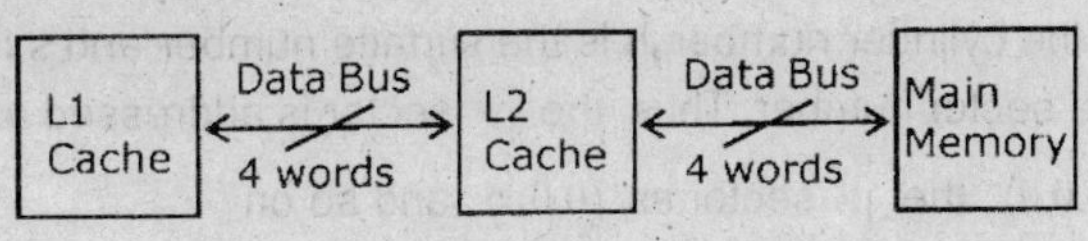

4. When there is a miss in L1 cache and a hit in L2 cache, a block is transferred from L2 cache to L1 cache. What is the time taken for this transfer ? **[GATE - 10]**

a) 2 nanoseconds b) 20 nanoseconds

c) 22 nanoseconds d) 88 nanoseconds

5. When there is a miss in both L1 cache and L2 cache, first a block is transferred from main memory to L2 cache, and then a block is transferred from L2 cache to L1 cache. What is the total time taken for these transfers ? **[GATE - 10]**

a) 222 nanoseconds b) 888 nanoseconds

c) 902 nanoseconds d) 968 nanoseconds

6. How many 32 K x 1 RAM chips are needed to provide a memory capacity of 256K - bytes ? **[GATE - 09]**

a) 8 b) 32 c) 64 d) 128

7. A CPU generally handles an interrupt by executing an interrupt service routine **[GATE - 09]**

a) As soon as an interrupt is raised

b) By checking the interrupt register at the end of fetch cycle

c) By checking the interrupt register after finishing the execution of the current instruction.

d) By checking the interrupt register at fixed time intervals.

8. Consider a 4 stage pipeline processor. The number of cycles needed by the four instructions I1, I2, I3, I4 in stages S1, S2, S3, S4 is shown below : **[GATE - 09]**

	S1	S2	S3	S4
I1	2	1	1	1
I2	1	3	2	2
I3	2	1	1	3
I4	1	2	2	2

What is the number of cycles needed to execute the following loop ? **[GATE - 09]**

For (i = 1 to 2) { I1; I2; I3; I4; }

a) 16 b) 23 c) 28 d) 30

9. Consider a 4-way set associative cache (initially empty) with total 16 cache blocks. The main memory consists of 256 blocks and the request for memroy blocks is in the following order :

0, 255, 1, 4, 3, 8, 133, 159, 216, 129, 63, 8, 48, 32, 73, 92, 155.

Which one of the following memory block will NOT be in cache if LRU replacement policy is used ? **[GATE - 09]**

a) 3 b) 8 c)129 d) 216

Common Data Questions : 10 & 11

A hard disk has 63 sectors per track, 10 platters each with 2 recording surfaces and 1000 cylinders. The address of a sector is given as a triple $\langle c,h,s \rangle$, where c is the cylinder number, h is the surface number and s is the sector number. Thus, the 0th sector is addressed as $\langle 0,0,0 \rangle$, the 1st sector as $\langle 0,0,1 \rangle$, and so on

10. The address < 400, 16, 29> corre4sponds tp sector number : **[GATE - 09]**

a) 505035 b) 505036
c) 505037 d) 505038

11. The address of the 1039th sector is **[GATE - 09]**

a) $\langle 0,15,31 \rangle$ b) $\langle 0,16,30 \rangle$
c) $\langle 0,16,31 \rangle$ d) $\langle 0,17,31 \rangle$

12. In the IEEE floating point representation the hexadecimal value 0x00000000 corresponds to **[GATE - 08]**

(a) the normalized value 2^{-127}
(b) the normalized value 2^{-126}
(c) the normalized value +0
(d) the special value +0

13. For a magnetic disk with concentric circular tracks, the seek latency is not linearly proportional to the seek distance due to **[GATE - 08]**

(a) non-uniform distribution of requests
(b) arm starting and stopping inertia
(c) higher capacity of tracks on the periphery of the platter
(d) use of unfair arm scheduling policies

14. Which of the following is/are true of the auto-increment addressing mode?

I. It is useful in creating self-relocating code

II. If it is included in an Instruction Set Architecture, then an additional ALU is required for effective address calculation

III. The amount of increment depends on the size of the data item accessed **[GATE - 08]**

(a) I only (b) II only
(c) III only (d) II and III only

15. Which of the following must be true for the RFE(Return from Exception) instruction on a general purpose processor?

I. It must be a trap instruction

II. It must be a privileged instruction

III. An exception cannot be allowed to occur during execution of an RFE instruction **[GATE - 08]**

(a) I only (b) II only
(c) I and II only (d) I, II and III only

16. For inclusion to hold between two cache levels L1 and L2 in a multi-level cache hierarchy, which of the following are necessarily ? **[GATE - 08]**

I. L1 must be a write-through cache

II. L2 must be a write-through cache

III. The associativity of L2 must be greater than that of L1

IV. The L2 cache must be at least as large as the L1 cache

(a) IV only (b) I and IV only
(c) I,III and IV only (d) I,II,III and IV

17. Which of the following are NOT true in a pipelined processor ?

I. Bypassing can handle all RAW hazards.

II. Register renaming can eliminate all register carried WAR hazards

III. Control hazard penalties can be eliminated by dynamic branch prediction. **[GATE - 08]**

(a) I and II only (b) I and III only
(c) II and III only (d) I,II and III

18. The use of multiple register windows with overlap causes a reduction in the number of memory accesses for

I. function locals and parameters

II. register saves and restores

III. instruction fetches. **[GATE - 08]**

(a) I only (b) II only
(c) III only (d) I,II and III

19. In an instruction execution pipeline, the earliest that the data TLB (Translation Look aside Buffer) can be accessed is **[GATE - 08]**

(a) before effective address calculation has started
(b) before effective address calculation
(c) after effective address calculation has completed
(d) after data cache lookup has completed.

Common Data Questions :

Common Data for Questions 20,21 and 22 :

Consider a machine with a 2-way set associative data cache of size 64 K bytes and block size 16 bytes. The cache is managed using 32 bit virtual addresses and the page size is 4 Kbytes. A program to be run on this machine begins as follows :

```
double Arr [1024] [1024];
 int i, j ;
 /* Initialize array ARR to 0.0*/
for ( i = 0;  i < 1024;  i++)
for ( j = 0;  j < 1024;  j++)
ARR [i ] [j]  = 0.0 ;
```

The size of double is 8 bytes. Array ARR is located in memory starting at the beginning of virtual page 0xFF000 and stored in row major order. The cache is initially empty and no pre-fetching is done. The only data memory references made by the program are those to array ARR.

20. The total size of the tags i the cache directory is **[GATE - 08]**

(a) 32 Kbits (b) 34 Kbits
(c) 64 Kbits (d) 68 Kbits

21. Which of the following array elements has the same cache index as ARR [0] [0]? **[GATE - 08]**

(a) ARR [0] [4] (b) ARR [4] [0]
(c) ARR [0] [5] (d) ARR [5] [0]

22. The cache hit ratio for this initialization loop is **[GATE - 08]**

(a) 0% (b) 25%
(c) 50% (d) 75%

Statement for Linked Answer Questions 23 and 24

Delayed branching can help in the handling of control hazards

23. For all delayed conditional branch instructions, irrespective of whether the condition evaluates to true or false. **[GATE - 08]**

(a) the instruction following the conditional branch instruction in memory is executed

(b) the first instruction in the fall through path is executed

(c) the first instruction in the taken path is executed

(d) the branch takes longer to execute than any other instruction.

24. The following code is to run on a pipelined processor with one branch delay slot

11: ADD R2 $\leftarrow$ R7 + R8

12: SUB R4 $\leftarrow$ R5 - R6

13: ADD R1 $\leftarrow$ R2 + R3

14: STORE Memory [R4] $\leftarrow$ R1

BRANCH to Label if R1==0

Which of the instructions I1, I2, I3 or I4 can legitimately occupy the delay slot without any other program modification? **[GATE - 08]**

(a) I1 (b) I2
(c) I3 (d) I4

25. A processor that has carry, overflow and sign flag bits as part of its program status word (PSW) performs addition of the following two 2's complement numbers 01001101 and 11101001. After the execution of this addition operation, the status of the carry, over-flow and sign flags, respectively will be: **[GATE - 08]**

(a) 1,1,0 (b) 1,0,0
(c) 0, 1, 0 (d) 1, 0, 1

26. A paging scheme uses a Translation Look-aside Buffer (TLB). A TLB-access takes 10 ns and a main memory access takes 50 ns. What is the effective access time (in ns) if the TLB hit ratio is 90% and there is no page-fault ? **[GATE - 08]**

(a) 54 (b) 60 (c) 65 (d) 75

27. Assume that EA=(X)+ is the effective address equal to the contents of location X. with X incremented by one word length after the effective address is calculated; EA=-(X) is the effective address equal to the contents of location X, with X decremented by one word length before the effective address is calculated; EA=(X)- is the effective address equal to the contents of location X with X decremented by one word length after the effective address is calculated. The format of the instruction is (opcode, source, destination), which means (destination $\leftarrow$ source op destination) Using X as a stack pointer, which of the following instructions can pop the top two elements from the stack, perform the addition operation and push the result back to the stack. **[GATE - 08]**

(a) ADD (X)-, (X) (b) ADD (X), (X)-
(c) ADD-(X), (X)+ (d) ADD - (X), (X)

28. Consider a CPU where all the instructions require 7 clock cycles to complete execution. There are 140 instructions in the instruction set. It is found that 125 control signals are needed to be generated by the control unit. While designing the horizontal microprogrammed control unit, single address field format is used for branch control logic. What is the minimum size of the control word and control address register? **[GATE - 08]**

(a) 125, 7 (b) 125,10 (c) 135, 9 (d) 135, 10

29. A nonpipelined single cycle processor operating at 100 MHz is converted into a synchronous pipelined processor with five stages requiring 2.5 nsec, 1.5 nsec, 2nsec, 1.5 nsec and 2.5 nesec, respectively. The delay of the latches is 0.5 nsec. The speedup of the pipeline processor for a large number of instructions is **[GATE - 08]**
(a) 4.5 (b) 4.0 (c) 3.33 (d) 3.0

30. Assume that a main memory with only 4 pages, each of 16 bytes, is initially empty. The CPU generates the following sequence of virtual addresses and uses the Least Recently Used (LRU) page replacement policy.
0,4,8,20,24,36,44,12,68,72,80,84,28,32,88,92
How many page faults does this sequence cause ? What are the page numbers of the pages present in the main memory at the end of the sequence ? **[GATE - 08]**
(a) 6 and 1,2,3,4 (b) 7 and 1,2,4,5
(c) 8 and 1,2,4,5 (d) 9 and 1,2,3,5

31. The two numbers given below are multiplied using the Booth's algorithm.
Multiplicand :0101 1010 1110 1110
Multiplier : 0111 0111 1011 1101
How many additions/Subtractions are required for the multiplication of the above two numbers? **[GATE - 08]**
(a) 6 (b) 8 (c) 10 (d) 12

Statement for Linked Answer Questions 32 and 33 :
Consider a computer with a 4-ways set-associative mapped cache of the following characteristics: a total of 1 MB of main memory, a word size of 1 byte, a block size of 128 words and a cache size of 8 KB.

32. The number of bits in the TAG, SET and WORD fields, respectively are : **[GATE - 08]**
(a) 7, 6, 7 (b) 8, 5, 7
(c) 8,.6,6 (d) 9,4,7

33. While accessing the memory location 0C795H by the CPU, the contents of the TAG field of the corresponding cache line is **[GATE - 08]**
(a) 000011000 (b) 110001111
(c) 00011000 (d) 110010101

34. Consider a 4-way set associative cache consisting of 128 lines with a line size of 64 words. The CPU generates a 20-bit address of a word in main memory. The number of bits in the TAG, LINE and WORD fields are respectively **[GATE - 07]**
(a) 9, 6, 5 (b) 7, 7, 6
(c) 7, 5, 8 (d) 9, 5, 6

35. Consider a disk pack with 16 surfaces, 128 tracks per surface and 256 sectors per track 512 bytes of data are stored in a bit serial manner in a sector. The capacity of the disk pack and the number of bits required to specify a particular sector in the disk are respectively **[GATE - 07]**
(a) 256 Mbyte, 19 bits (b) 256 Mbyte, 28 bits
(c) 512 Mbyte, 20 bits (d) 64 Gbyte, 28 bits

36. The control signal functions of a 4-bit binary counter are given below (where X is "don't care") :

Clear	**Clock**	**Load**	**Count**	**Function**
1	**X**	**X**	**X**	**Clear to 0**
0	**X**	**0**	**0**	**No change**
0	↑	**1**	**X**	**Load input**
0	↑	**0**	**1**	**Count next**

The counter is connected as follows :

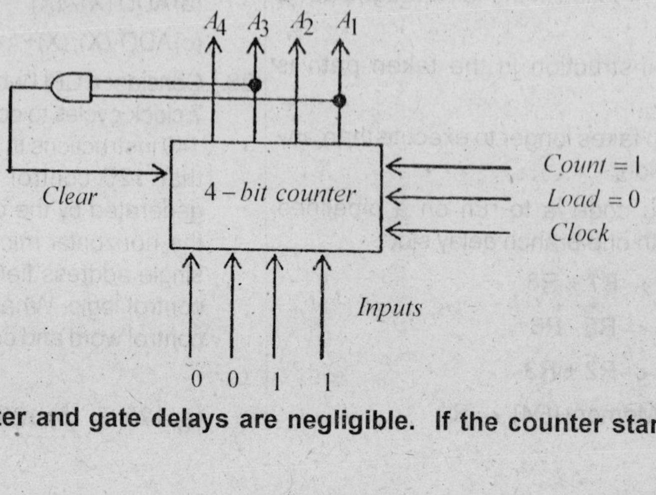

Assume that the counter and gate delays are negligible. If the counter starts at 0, then it

cycles through the following sequence [GATE - 07]

(a) 0, 3, 4 (b) 0, 3, 4, 5
(c) 0, 1, 2, 3, 4 (d) 0, 1, 2, 3, 4, 5

37. Consider a pipelined processor with the following four stages :

IF : Instruction Fetch
ID : Instruction Decode and Operand Fetch
EX : Execute
WB : Write Back

The IF, ID and WB stages take one clock cycle each to complete the operation. The number of clock cycles for the EX stage depends on the instruction. The ADD and SUB instructions need 1 clock cycle and the MUL instruction need 3 clock cycles in the EX stage. Operand forwarding is used in the pipelined processor. What is the number of clock cycles taken to complete the following sequence of instructions?

ADD	R2,	R1, R0	$R2 \leftarrow R1 + R0$
MUL	R4,	R3, R2	$R4 \leftarrow R3 * R2$
SUB	R6,	R5, R4	$R6 \leftarrow R5 - R4$

[GATE - 07]

(a) 7 (b) 8 (c) 10 (d) 14

38. In a simplified computer the instructions are

$OP\ R_j, R_i$ **- Performs $R_j\ OP\ P_i$ and stores the result in register R_i**

$OP\ m, R_i$ **- Performs val $OP\ R_i$ and stores the result in R_i. val denotes the content of memory location m.**

$MOV\ m, R_i$ **- Moves the content of memory location m to register R_i**

$MOV\ R_i, m$ **- Moves the content of register R_i to memory location m.**

The computer has only two registers, and OP is either ADD or SUB. Consider the following basic block :

$$t_1 = a + b$$
$$t_2 = c + d$$
$$t_3 = e - t_2$$
$$t_4 = t_1 - t_3$$

Assume that all operands are initially in memory. The final value of the computation should be in memory. What is the minimum number of MOV instructions in the code generated for this basic block ?

[GATE - 07]

(a) 2 (b) 3 (c) 5 (d) 6

Common data for questions 39, 40, 41 :

Consider the following program segment. Here R1, R2, and R3 are the general purpose registers.

	Instruction	Operation	Instruction size (no.of words)
	MOV R1, (3000)	$R1 \leftarrow M[3000]$	2
LOOP :	MOV R2, (R3)	$R2 \leftarrow M[R3]$	1
	ADD R2, R1	$R2 \leftarrow R1 + R2$	1
	MOV (R3), R2	$M[R3] \leftarrow R2$	1
	INC R3	$R3 \leftarrow R3 + 1$	1
	DEC R1	$R1 \leftarrow R1 - 1$	1
	BNZ LOOP	Branch on not zero	2
	HALT	Stop	1

Assume that the content of memory location 3000 is 10 and the content of the register R3 is 2000. The content of each of the memory locations from 2000 to 2010 is 100. The program is loaded from the memory location 100. All the numbers are in decimal.

39. Assume that the memory is word addressable. The number of memory references for accessing the data in executing the program completely is [GATE - 07]

(a) 10 (b) 11 (c) 20 (d) 21

40. Assume that the memory is word addressable. After the execution of this program, the content of memory location 2010 is [GATE - 07]

(a) 100 (b) 101 (c) 102 (d) 110

41. Assume that the memory is byte addressable and the word size is 32 bits. If an interrupt occurs during the execution of the instruction "INC R3", what return address will be pushed on to the stack? [GATE - 07]

(a) 1005 (b) 1020
(c) 1024 (d) 1040

Statement for Linked Answer Questions 80 and 81: Consider a machine with a byte addressable main memory of 2^{16} bytes. Assume that a direct mapped data chache consisting of 32 lines of 64 bytes each is used in the system. A 50×50 two - dimesional array of bytes is stored in the main memory starting from memory location 1100H. Assume that the data cache is initially empty. The complete arrary is accussed twice. Assume that the contents of the data cache do not change in between the two accesses.

42. How many data cache misses will occur in total ? [GATE - 07]

(a) 48 (b) 50 (c) 56 (d) 59

43. Which of the following lines of the data cache will be replaced by new blocks in accessing the array for the second time? [GATE - 07]

(a) line 4 to line 11 (b) line 4 to line 12
(c) line 0 to line 7 (d) line 0 to line 8

44. Which of the following systems is a most likely candidate example of a pipe and filter architecture ? [GATE - 07]

(a) Expert system
(b) DB repository
(c) Aircraft flight controller
(d) Signal processing

45. A processor takes 12 cycles to complete an instruction 1. The corresponding pipelined processor uses 6 stages with the execution times of 3, 2, 5, 4, 6 and 2 cycles respectively. What is the asymptotic speedup assuming that a very large number of instructions are to be executed ? [GATE - 07]

(a) 1.83 (b) 2
(c) 3 (d) 6

46. The floating point unit of a processor using a design D takes 2t cycles compared to t cycles taken by the fixed point unit. There are two more design suggestions D1 and D2. D1 uses 30% more cycles for fixed point unit but 30% less cycles for floating point unit as compared to design D. D2 uses 40% less cycles for fixed point unit but 10% more cycles for floating point unit as compared to design D. For a given program which has 80% fixed point operations and 20% floating point operations, which of the following ordering reflects the relative performances of three designs ?

($Di > Dj$ denotes that Di is faster than Dj) [GATE - 07]

(a) $D1 > D > D2$

(b) $D2 > D > D1$

(c) $D > D2 > D1$

(d) $D > D1 > D2$

47. Consider a Direct Mapped Cache with 8 cache blocks (numbered 0 - 7). If the memory block requests are in the following order

3, 5, 2, 8, 0, 63, 9, 16, 20, 17, 25, 18, 30, 24, 2, 63, 5, 82, 17, 24.

Which of the following memory blocks will not be in the cache at the end of the sequence? [GATE - 07]

(a) 3 (b) 18 (c) 20 (d) 30

48. Data forwarding techniques can be used to speed up the operation in presence of data dependencies. Consider the following replacements of LHS with RHS.

(i) $R1 \rightarrow Loc$, $Loc \rightarrow R2 = R1 \rightarrow R2$, $R1 \rightarrow Loc$

(ii) $R1 \rightarrow Loc$, $Loc \rightarrow R2 = R1 \rightarrow R2$

(iii) $R1 \rightarrow Loc$, $R2 \rightarrow Loc = R1 \rightarrow Loc$

(iv) $R1 \rightarrow Loc$, $R2 \rightarrow Loc = R1 \rightarrow Loc$

In which of the following options, will the result of executing the RHS be the same as executing the LHS irrespective of the instructions that follow ? [GATE - 07]

(a) (i) and (iii) (b) (i) and (iv)
(c) (ii) and (iii) (d) (ii) and (iv)

49. Following table indicates the latencies of operations between the instruction producing the result and instruction using the result.

Instruction production the result	Instruction using the result	Latency
A L U Operation	ALU Operation	2
ALU Operation	Store	2
Load	ALU Operation	1
Load	Store	0

Consider the following code segment.

Load R1, Loc 1 ;	Load R1 from memory location Loc 1
Load R2, Loc 2 ;	Load R2 from memory location Loc 2
Add R1, R2, R1 ;	Add R1 and R2 and save result in R1
Dec R2 ;	Decrement R2
Dec R1 ;	Decrement R1
Mpy R1, R2, R3 ;	Multiply R1 and R2 and save result in R3
Store R3, Loc 3 ;	Store R3 in memory location Loc 3

What is the number of cycles needed to execute the above code segment assuming each instruction takes one cycle to execute ? **[GATE - 07]**

(a) 7 (b) 10
(c) 13 (d) 14

50. A hard disk system has the following parameters :

Number of tracks = 500

Number of sectors / track = 100

Number of bytes / sector = 500

Time taken by the head to move from one track to adjacent track = 1 ms

Rotation speed = 600 rpm.

What is the average time taken for tranferring 250 bytes from the disk ? [GATE - 07]

(a) 300.5 ms (b) 255.5 ms (c) 255 ms (d) 300 ms

51.A CPU has 24 - bit instructions. A program starts at address 300 (in decimal). Which one of the following is a legal program counter (all values in decimal) ? [GATE - 06]

(A) 400 (B) 500 (C) 600 (D) 700

52.A CPU has a cache with block size 64 bytes. The main memory has k banks, each bank being c bytes wide. Consecutive c-byte chunks are mapped on consecutive banks with warp-around. All the k banks can be accessed in parallel, but two accesses to the same bank must be serialized. A cache block access may involve multiple iterations of parallel bank accesses

depending on the amount of data obtained by accessing all the k banks in parallel. Each iteration requires decoding the bank numbers to be accessed in parallel and this takes k/2 ns. The latency of one bank access is 80 ns. If c = 2 and k = 24, then latency of retrieving a cache block starting at address zero from main memory is [GATE - 06]

(A) 92 ns (B) 104 ns

(C) 172 ns (D) 184 ns

53. A CPU has five-stage pipeline and runs at 1 GHz frequency. Instruction fetch happens in the first stage of the pipeline. A conditional branch instruction computes the target address and evaluates the condition in the third stage of the pipeline. The processor stops fetching new instructions following a conditional branch until the branch outcome is known. A program executes 10^9 instructions out of which 20% are conditional branches. If each instruction takes one cycle to complete on average, then total execution time of the program is [GATE - 06]

(A) 1.0 second (B) 1.2 seconds

(C) 1.4 seconds (D) 1.6 seconds

54. Consider a new instruction named branch-on-bit-set (mnemonic bbs). The instruction "bbs reg, pos, label" jumps to label if bit in position pos of register operand reg is one. A register is 32 bits wide and the bits are numbered 0 to 31, bit in position 0 being the least significant. Consider the following emulation of this instruction on a processor that does not have bbs implemented.

$temp \leftarrow reg \text{ and } mask$

Branch to label if temp is non-zero

The variable temp is a temporary register. For correct emulation, the variable mask must be generated by [GATE - 06]

(A) $mask \leftarrow 0x1 << pos$

(B) $mask \leftarrow 0x\,ffffffff >> pos$

(C) $mask \leftarrow pos$

(D) $mask \leftarrow 0xf$

Common data for questions 55, 56 :

Consider two cache organizations : The first one is 32 KB 2-way set associative with 32-byte block size. The second one is of the same size but direct mapped. The size of an address 32 bits in both cases. A 2- to-1 multiplexer has latency of 0.6 ns while a k-bit comparator has a latency to k/10ns. The hit latency of the set associative organization is h_1 while that of the direct mapped one is h_2.

55. The value of h_1 is [GATE - 06]

(A) 2.4 ns (B) 2.3 ns

(C) 1.8 ns (D) 1.7 ns

56. The value of h_2 is [GATE - 06]

(A) 2.4 ns (B) 2.3 ns

(C) 1.8 ns (D) 1.7 ns

Statement for Linked Answer Questions 57 and 58 :

A CPU has a 32 KB direct mapped cache with 128-byte block size. Suppose A is a two dimensional array of size 512×512 with elements that occupy 8-bytes each. Consider the following two C code segments, P1 and P2

```
P1 : for (i = 0; i < 512; i++) {
        for (j = 0; j < 512; j++) {
            x += A[i][j];
        }
}

P2 : for (i = 0; i < 512; i++) {
        for (j = 0; j < 512; j++) {
            {x += A[j][i];}
        }
}
```

P1 and P2 are executed independently with the same initial state, namely, the array A is not in the cache and i,j, x are in registers. Let the number of cache misses experienced by P1 be M_1 and that for P2 be M_2.

57. The value of M_1 is [GATE - 06]

(A) 0 (B) 2048 (C) 16384 (D) 262144

58. The value of the ratio M_1 / M_2 is [GATE - 06]

(A) 0 (B) $\frac{1}{16}$ (C) $\frac{1}{8}$ (D) 16

59. Which of the following DMA transfer modes and interrupt handling mechanisms will enable the highest I/O band-width? [GATE - 06]

a) Transparent DMA and Polling interrupts
b) Cycle-stealing and Vectrored interrupts
c) Block transfer and Vectored interrupts
d) Block transfer and Polling interrupts

60. When multiplicand Y is multiplied by multiplier $X = x_{n-1}\ x_{n-2} \ldots x_0$ using bit-pair recoding in Booth's algorithm, partial products are generated according to the following table.

Row	x_{i+1}	x_i	x_{i-1}	Partial Product
1	0	0	0	0
2	0	0	1	Y
3	0	1	0	Y
4	0	1	1	2Y
5	1	0	0	?
6	1	0	1	-Y
7	1	1	0	-Y
8	1	1	1	?

The partial products for rows 5 and 8 are [GATE - 06]

a) $2Y$ and Y

b) $-2Y$ and $2Y$

c) $-2Y$ and 0

d) 0 and Y

61. Which of the following statements about relative addressing mode is FALSE ? [GATE - 06]

a) It enables reduced instruction size
b) It allows indexing of array elements with same instruction
c) It enables easy relocation of data
d) It enables faster address calculations than absolute addressing

62. The memory locations 1000, 1001 and 1020 have data values 18, 1 and 16 respectively before the following program is executed.

```
MOV1    Rs, 1          ; Move immediate
LOAD    Rd,1000(Rs)    ; Load from memory
ADD1    Rd, 1000       ; Add immediate
STOREI  0(Rd), 20      ; Store immdiate
```

Which of the statements below is TRUE after the program is executed ? [GATE - 06]

a) Memory location 1000 has value 20
b) Memory location 1020 has value 20
c) Memory location 1021 has value 20
d) Memory location 1001 has value 20

63. The data path shown in the figure computes the number of 1s in the 32 - bit input word corresponding to an unsigned even integer stored in the shift register.
The unsigned counter, initially zero, is incremented if the most significant bit of the shift register is 1.

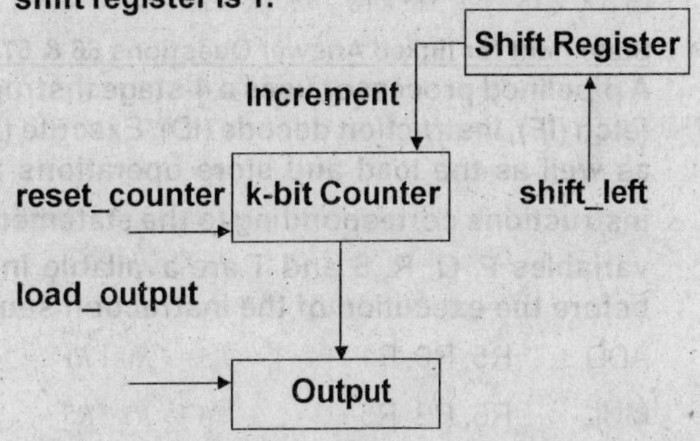

The microprogram for the control is shown in the table below with missing control words for microinstructions $I_1, I_2, \ldots I_n$.

Microinstruction	reset_counter	shift_left	load_output
BEGIN	1	0	0
I_1	?	?	?
:	:	:	:
I_n	?	?	?
END	0	0	1

The counter width (k), the number of missing microinstructions (n), and the control word for microinstructions $I_1, I_2, \ldots I_n$ are, respectively. [GATE - 06]

a) 32, 5, 010 b) 5, 32, 010

c) 5, 31, 011 d) 5, 31, 010

64.A cache line is 64 bytes. The main memory has latency 32ns and bandwidth 1G. Bytes/s. The time required to fetch the entire cache line from the main memory is

[GATE - 06]

a) 32 ns b) 64 ns c) 96 ns d) 128 ns

65.A computer system has a level - 1 instruction cache (1-cache), a level-1 data cache (D-cache) and a level-2 cache (L2-cache) with the following specifications :

	Capacity	Mapping Method	Block size
1 - cache	4K words	Direct mapping	4 words
D-cache	4K words	2-way set-associative mapping	4 words
L2-cache	64K words	4-way set-associative mapping	16 words

The length of the physical address of a word in the main memory is 30 bits. The capacity of the tag memory in the I-cache, D-cache and L2-cache is, respectively, **[GATE - 06]**

a) $1K \times 18-bit, 1K \times 19-bit, 4K \times 16-bit$

b) $1K \times 16-bit, 1K \times 19-bit, 4K \times 18-bit$

c) $1K \times 16-bit, 512 \times 18-bit, 1K \times 16-bit$

d) $1K \times 18-bit, 512 \times 18-bit, 1K \times 18-bit$

Statement for linked Answer Questions 66 & 67:

A pipelined processor uses a 4-stage instruction pipeline with the following stages : Instruction fetch (IF), Instruction decode (ID), Execute (EX) and Writeback (WB). The arithmetic operations as well as the load and store operations are carried out in the EX stage. The sequence of instructions corresponding to the statement $X = (S - R*(P+Q))/T$ is given below. The values of variables P, Q, R, S and T are available in the registers R0, R1, R2, R3 and R4 respectively, before the execution of the instruction sequence.

```
ADD    R5, R0, R1   ;  R5 → R0 + R1
MUL    R6, R2, R5   ;  R6 → R2 * R5
SUB    R5, R3, R6   ;  R5 → R3 − R6
DIV    R6, R5, R4   ;  R6 → R5 / R4
STORE  R6, X        ;  X ← R6
```

66.The number of Read-After-Write (RAW) dependencies, Write-After-Read (WAR) dependencies, and Write-After-Write (WAW) dependencies in the sequence of instructions are, respectively,

[GATE - 06]

a) 2, 2, 4 b) 3, 2, 3 c) 4, 2, 2 d) 3, 3, 2

67.The IF, ID and WB stages take 1 clock cycle each. The EX stage takes 1 clock cycle each for the ADD, SUB and STORE operations, and 3 clock cycles each for MUL and DIV operations. Operand forwarding from the EX stage to the ID stage is used. The number of clock cycles required to complete the sequence of instructions is **[GATE - 06]**

a) 10 b) 12 c) 14 d) 16

ANSWER KEY

1. d	2. b	3.	4. c	5. a	6. c	7. c	8. b	9. d	10. c
11. c	12. d	13. b	14. c	15. d	16. b	17. a	18. b	19. b	20. c
21. b	22. b	23. d	24. d	25. b	26. c	27.	28.	29.	30. b
31. b	32.	33.	34. b	35. a	36. d	37.b	38. b	39. d	40.a
41. a	42. c	43. c	44. d	45. a	46. b	47. b	48. b	49. b	50. b
51. c	52. d	53. b	54. a	55. a	56. d	57. c	58. b	59. d	60.
61. b	62. d	63.	64. c	65. d	66. c	67. b			

EXPLANATORY NOTES

2. Clocks.

	1	2	3	4	5	6	7	8	9	10	11	12	13	14	15	16
I_0	IF	ID	OF	PO	PO	PO	WO									
I_1		IF	ID	OF	Stall	Stall	PO	PO	PO	PO	PO	PO	WO			
I_2			IF	ID	OF	-	-	-	-	-	-	-	PO	WO		
I_3				IF	ID	OF	-	-	-	-	-	-	-	PO	WO	

So total 15 cycles required

4. A block to access in L_2 cache requires 20 nanoseconds, and 2 sec to place in L_1 - cache, so is total 22 nanoseconds

5. When there is a miss in both L_1, L_2, It has to access in Main which takes 200 nanoseans, and 20 nanoseconds to transfer to L_2 cache and 2 for L_1 cache, Total = 200 + 20 + 2 = 222

6. Number of chips required $= \frac{256*1024*8}{32*1024*1} = 64$ chips

7. A CPU generally handles an interrupt by executing ISR by checking the interrupt register after finishing the execution of the current instruction.

8.

	1	2	3	4	5	6	7	8	9	10	11	12	13	14	15	16	17	18	19	20	21	22	23
S_1	I_1		I_2	I_3		I_4	I_1		I_2	I_3		I_4											
S_2			I_1	I_2			I_3	I_4		I_1	I_2			I_3	I_4								
S_3				I_1			I_2		I_3	I_4		I_1		I_2		I_3	I_4						
S_4					I_1				I_2		I_3			I_4		I_1	I_2		I_3			I_4	

9.

Set	Blocks
Set 0	~~0~~48
	4 32
	8
	~~2~~16 92
Set 1	1
	133
	129
	73
Set 2	
Set 3	~~255~~ 155
	3
	159
	63

10. $\langle 400,16,29\rangle = 400*20*63+16*63+29 = 505037^{th}$ Sector

11. $\langle 0,16,31\rangle = 0*20*63+16*63+31 = 1039^{th}$ Sector

12. 0 x 00000000 represents special value of + 0

13. 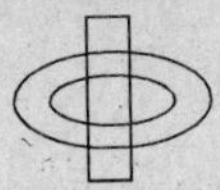 co - centric circular tracks

The seek latency is not linearly proportional to the seak distance due to arm starting and stopping inertia.

14. Auto increment add mode is basically used for incrementing a data. It depends on size of data.

15. RFE. For any exception, while executing RFE, no exception is allowed (no interrupts). It is a privileged instruction, as soon as exception occurs RFE is called (executed)

16. For inclusion to hole between two cache levels cache hierarchy

L1 must be write through cache

and L2 cache must be at least as large as L1 cache

17. In Pipelined processor bypassing can handle all RAW hazards and register remaining can eliminate all register carried WAR hazard.

18. The use of multiple windows with overlap causes a reduction in the number of memory accesses for functions local and parameters, register saves and restores and instruction fetches.

19. The Translations Lookaside Buffer can be accessed during effective address calculations.

20. Total size of cache $2^{16} = 64$ Kbits

21. Because of page size is 4 KB. 1 row contain 1024 elements means 2^{10} location.

So ARR [0] [0] same as ARR [4] [0] which is some location in next page.

22. Hit ratio = $\frac{number\ of\ hits}{number\ of\ hits + number\ of\ rrus}$

$= \frac{4}{16} = 0.25 = 25\%$

23. Using delay instruction causes the branch to take more time to execute

24. Clearly I 4 occupy delay slot

25.

L 0 1 0 01101

1 1 1 01001

0 0 1 10110

In case of signed numbers, the left most bit always represents the sign. When two signed numbers are added, the sign bit is treated as part of the numbers and the end carry does not indicate an overflow.

Thus here,

carry = 1

overflow = 0

sign bit = 0

26. The percentage of times that a particular page number is found in the TLB is called the hit ratio.

An 90% bit ratio means that we find the desired page number in the TLB 90% of the time. If it takes 10 ns to search the TLB and 50 ns to access memory, then a mapped memory access takes 60 ns when the page number is in the TLB. If we fail to find the page number in the TLB (10 ns), then we must find access memory for the page table and frame number (50 ns) and then access the desired byte in memory (50ns), for a total of 110 ns.

To find effective memory access time,

effective memory access = 90 x 60 + 0.10 x 110

= 54 + 11

= 65

30. LRU (Least Recently Used) page replacement policy replaces that page from the memory that has not been used for the longest period of line.

(i) 0 - 15

(ii) 16 - 31

(iii) 32 - 47

(iv) 48 - 63

(v) 64 - 79

(vi) 80 - 95

(vii) 96 - .....

Pages

0
4
8
20

1	2	3	5

6	2	3	5

4 page faults

31. Booth's algorithm requires examination of the multiplier bits and shifting of the partial product. Prior to the shifting, the multiplicand may be added to the partial product, subtracted from the partial product or left unchanged according to the following rules.

1. The multiplicand is subtracted from the partial product upon encountering the first least significant 1 in a string of 1's in the multiplier.
2. The multiplicand is added to the partial product upon encountering the first 0 in a string of 0's in the multiplier.
3. The partial product does not change when the multiplier bit is identical to the previous multiplier bit.

As in multiplier we see that a change occurs in bit as 0 changes to 1 and 1 changes to 1

Multiplier:

0 1 1 1 0 1 1 1 1 0 1 1 1 1 0 1

1 1 1 1 1 1 1 ↑

1 count

1+1+1+1+1+1+1+1=8

34. For identifying each of 128 lines uniquely, number of bits required = 7 bits.

Since each line is of size 64 words, so number of bits needed to identify each word uniquely = 6 bits.

Now, since there are 20 bits in address generated by CPU

$\therefore$ bits needed for tag $= 20 - (7+6) = 7\ bits$

35. Size of disk = number of surfaces × number of tracks per surface

× number of sectors per track × capacity of each sector

$= 16 \times 128 \times 256 \times 512\ bytes$

$= 256$ Mega bytes (MB)

Since, total number of sector in disk

$= 16 \times 128 \times 256 = 512\ K$

$= 2^{19}$ Sector

therefore, it requires 19 bits to specify a particular sector.

36. Counter will start with 0

for Load = 0, Count = 1 and clock it will go to 1, then 2, then 3, then 4 and at last at 5. As soon as it will reach to 5, the clear will becone one (1) which will clear the counter to zero.

37. For each of the three instruction four stages will be there. Since, it is a pipelined processor, so one instruction may be fetched while other is being decoded or executed or written back action performed in each clock cycle may be represented as

Clock Cycle	ADD	MUL	SUB
1	IF		
2	ID	IF	
3	EX	ID	IF
4	WB	EX	ID
5		EX	
6		EX	
7		WB	EX
8			WB

Only, Two instructions cannot be executed simultaneously.

38. Instructions will be following :

MOV	b,	R_1
ADD	a,	R_1
MOV	d,	R_2
ADD	c,	R_2
SUB	e,	R_2
SUB	R_1,	R_2
MOV	R_2,	t_4

It takes 3 MOV instructions.

39. Operation $R_1 \leftarrow M\left[3000\right]$ executed once whereas

$R_2 \leftarrow M\left[R_3\right]$ *and* $M[R_3] \leftarrow R_2$
executed ten times each

Therefore total memory reference

$$= 2 \times 10 + 1 = 21$$

40. Content at memory location 2010 will be still 100.

It doesn't get altered because after address in R_3 became 2010, value in R_1 became 0, so loop exits.

41. When interrupt will occur instruction "INC R3" will be on address 1005, so 1005 will be placed on stack.

44. Signal processing is the most likely condidate example of pipe and filter architecture.

45. 12 cycles to complete instruction.

Pipelined processor uses 6 stages with execution times 3, 2, 5, 4, 6 and 2 cycles.

So, total time = 3 + 2 + 5 + 4 + 6 + 2 = 22

So, asymptotic speed up = $\frac{12}{22} = 1.83$

46.

	Fixed	Float		
For D	t	2t		
For D1	1.3 t	1.4 t	t + 30% of t	2t - 30% of t
For D2	0.6 t	2.2 t	t - 40% of t	2t + 10% of t

$\because$ 80% fixed point operation and 20% floating point operations.

For D performance = $1.3 \times 0.8t + 2t \times .2 = 1.2t$

Similarly $PD1 = 1.28t$ *and* $PD2 = 0.92t$

Hence $D2 > D > D1$

47.

0	8, 0, 16, 24
1	9, 17, 25, 17
2	2, 18, 2, 82
3	3
4	20
5	5
6	30
7	63

So 18 is not present in final map show of cache memory.

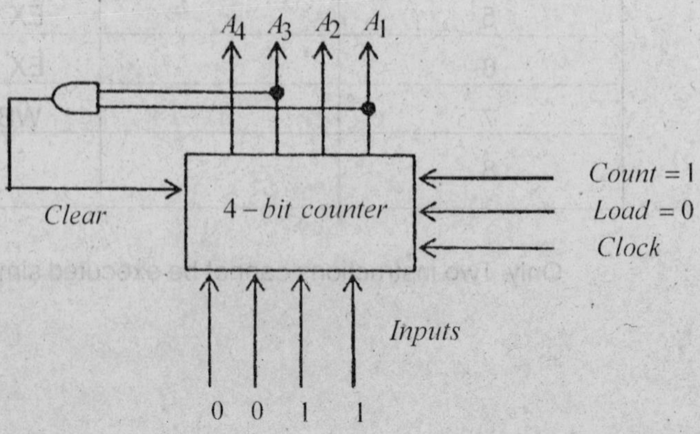

48. (i) $R1 \to Loc,\ Loc \to R2$ is same as

$R1 \to R2,\ R \to Loc$, so true

(ii) $R1 \to Loc,\ Loc \to R2$ in it Loc is not update, so false

(iii) $R1 \to Loc,\ R2 = Loc \equiv R1 \to Loc$. If h incorrect because last value in Loc in R2 but RHS show that Loc has R1 value.

(iv) $R1 \to Loc,\ R2 \to Loc \equiv R2 \to Loc$. it correct

Hence answer is (i) and (iv)

49. All seven instruction takes 1 clock cycles. But 1 clock cycle latency needed between load R2, Loc2, and

add R1, R2, R1.

Also 2 clock cycle Latency needed between

Mpy R1, R2, R3 ; and

Store R3, Loc 3

So $7+1+2=10$

51. Each address corresponds to one byte in the memory. Each instruction consists of 24 bits i.e., 3 bytes. Therefore, from 300 onwards only a multiple of 3 will designate a valid program counter.

$(400-300), (500-300)\ and\ (700-300)$ are not multiples of 3.

But 600-300=300 is a multiple of 3.

52. The main memory consists of 24 banks, each of 2 bytes. Since parallel accesses to all banks are possible, only two parallel accesses of all banks are needed to traverse the whole data.

For one parallel access,

Total time = Decoding Time + Latency Time

$$=\frac{24}{2}+80=92\,ns$$

Hence, for 2 such accesses, time

$$=2\times 92=184\,ns$$

53. For 80% of 10^9 instruction a single cycle is sufficient as there is no conditional branch. For the rest 20%, an extra cycle will be needed making the number of cycles required to 2.

Total time will be computed as below

One cycle takes $=\frac{1}{1G}=10^{-9}\,s$

$\therefore$ Total time

$$=10^{-9}\times\left(\frac{80}{100}\times 10^9\times 1+\frac{20}{100}\times 10^9\times 2\right)$$

$$=\frac{4}{5}+\frac{2}{5}=\frac{6}{5}=1.2 \text{ seconds}$$

54. Since only the bit with position pos is the deciding factor in jumping to the label, we have to set all other bits to 0 in temp. The mask register required for this purpose just has to have 1 in position pos. This can be achieved through pos number of left shifts over 1.

57. Cache is 32 kB with 128 bytes block size
P1 : When the first element A [0] [0] is accessed,

128 byte block consisting of $\frac{128}{8}=16$.

Array elements is brought into the cache.
For the next 15 accesses, there will be hits as A [0] [0] to A [0] [15] are there in the cache.
Then again there will be a miss at A [0] [16] and bits at next 15 accesses.

Thus for every $\frac{512\times 512}{16}=16384$ block transfer during P1, there will be one miss and 15 hits

58. When A [0] [0] is accessed; block from A [0] [0] to A [0] [15] is brought into the cache but it is of no use as the next element required to access will be A [1] [0].
Thus there will not be a single hit and all $512\times 512=262144$ accesses will be misses.

$$\therefore\quad \frac{M_1}{M_2}=\frac{16384}{262144}=\frac{1}{16}$$

62. Contents of registers as program will be executed are

$Rs=1$

$Rd=$ content of 1000(1) = 1000 + 1 = 1001 memory address.

$\therefore\quad Rd \to 1$

Add I. Rd, 1000

$\therefore Rd = 1001$

Store 0 (1001), 20 means store 20 in the address value 1001.

64. Latency time = 32 ns.,

Bandwidth = 1 GB / sec.

$\therefore \quad 1\,sec \rightarrow 10^9$ bytes

or, $10^9\ bytes \rightarrow 1\ sec$

$64\ bytes \rightarrow 64\ ns$

$\therefore$ Total time = $64 + 32 = 96\ ns$.

65. For 1 cache : Number of blocks = 1K.

10 bits can be used to distinguish between blocks

$\Rightarrow$ 1 block = 4 words

2 bits to recognize between words.

30 (over) | 18 | 10 | 2 |

tag memory

$\therefore$ Capacity of I-cache = $18 \times 1K$.

D - cache : Number of blocks = $\frac{4k}{4} = 1 blocks$

$\therefore$ 10 bits are enough to recognize between blocks.

2 bits are required to recognize among 4 words within a block. and it uses 2-way set mapping.

| 18 | 10 | 2 |

tag memory

$\therefore$ Capacity of cache = $18 \times 512 (\because 1/2\ K = 512)$

66. $R_5 \leftarrow R_0 + R_1$ .....(i)

$R_6 \leftarrow R_2 + R_5$ .....(ii)

$R_5 \leftarrow R_3 = R_6$ .....(iii)

Write $R_6 \leftarrow R_5 / R_4$ .....(iv)

after write $X \leftarrow R_6$ .....(v)

From (ii) , (iii) and (iv), 2 write after read.

from (i), (ii), (iii), (iv) and (v), 4 read after write.

67. IF ID EX WB

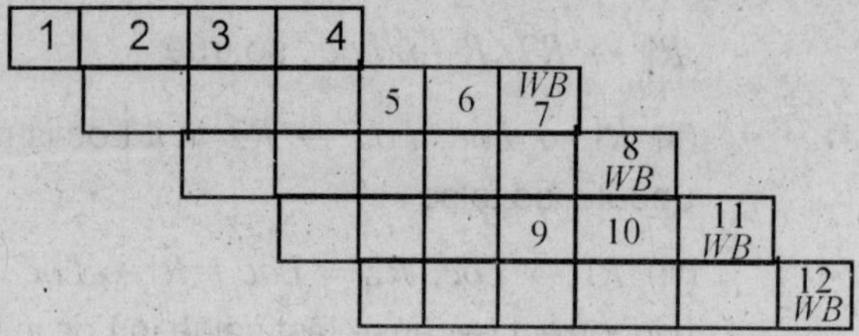

$\therefore$ 12 clock cycles required.

TEST YOURSELF - II

1. Assuming all numbers are in 2's complement representation, which of the following numbers is divisible by 11111011 ? **[GATE - 03]**

a) 11100111 b) 11100100
c) 11010111 d) 11011011

2. For a pipelined CPU with a single ALU, consider the following situations. **[GATE - 03]**

I. The (J+1)'st instructiion uses the result of the (J)'th instruction as an operand
II. The executiion of a conditional jump instruction
III. The (J)'th and (J + 1)'st instructions require the ALU at the same time.

Which of the above can cause a hazard ? **[GATE - 03]**

a) I and II only b) II and III only
c) III only d) All the three

3. Consider an array multiplier for multiplying two n bit numbers. If each gate in the circuit has a unit delay, the total delay of the multiplier is : **[GATE - 03]**

a) $\Theta(1)$ b) $\Theta(\log n)$
c) $\Theta(n)$ d) $\Theta(n^2)$

4. The following is a scheme for floating point number representation using 16 bits **[GATE - 03]**

Bit Position	15	14 ... 9	8 ... 0
	s	e	m
	Sign	Exponent	Mantissa

Let s, e and m be the numbers represented in binary in the sign, exponent, and mantissa fields respectively. then the floating point number represented is :

$$\begin{cases}(-1)^2\left(1+m\times 2^{-9}\right)2^{e-31}, \text{ if the exponent} \neq 111111 \\ 0, \text{ otherwise}\end{cases}$$

What is the maximum difference between two successive real numbers representable in this system ?

a) 2^{-40} b) 2^{-9} c) 2^{22} d) 2^{31}

5. Consider the ALU shown below. **[GATE - 03]**

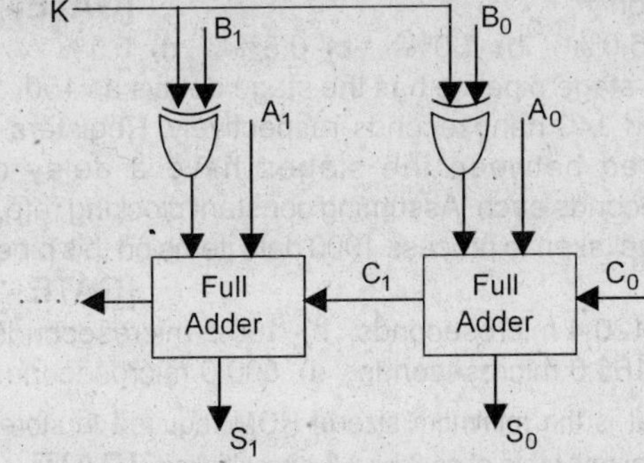

If the operands are in 2's complement representation, which of the following operations can be performed by suitable setting the control lines K and C_0 only (+ and - denote addition and subtraction respectively) ?

a) A+B, and A-B, but not A+1
b) A+B, and A+1, but not A-B
c) A+B, but not A-B or A+1
d) A+B, and A-B, and A+1

6. Consider the following circuit composed of XOR gates and non-inverting buffers. **[GATE - 03]**

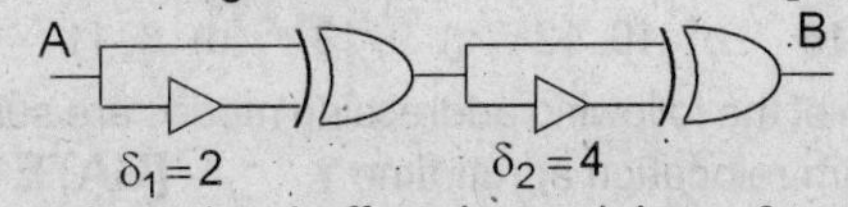

The non-inverting buffers have delays $\delta_1 = 2$ns and $\delta_2 = 4$ns as shown in the figure. Both XOR gates and all wires have zero delay. Assume that all gate inputs, outputs and wires are stable at logic level 0 at time 0. If the following waveform is applied at input A, how many transaction(s) (change of logic levels) occur(s) at B during the interval from 0 to 10 ns ?

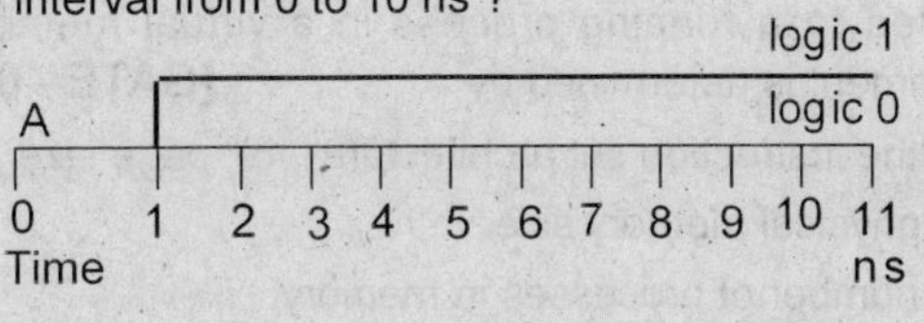

a) 1 b) 2 c) 3 d) 4

The following information pertains to Q.7-8

Consider the following assembly language program for a hypothetical processor. A, B and C are 8 bit registers. The meaning of various instructions are shown as comments.

```
     MOV B, # 0;     B ← 0
     MOV C, # 8  ;   C ← 8
Z    CMP, C # 0;     Compare C with 0
     J Z  X     ;    jump to X if zero flag is set
     SUB C, #1  ;    C ← C-1
     RRC A, #1  ;    right rotate A through carrry
                     by one bit. Thus
                ;    if the initial values of A and
                     the carry flag are a7....a0 and
                ;    c0 respectively, their values after
                     the execution of this
                     instruction will be c0a7....a1 and
                     a0 respectively.
     JC Y       ;    jump to Y if carry flag is set
     JMP Z      ;    jump to Z
y :  ADD B, #1  ;    B ← B + 1
     JMP Z      ;    jump to Z
X :
```

[GATE - 03]

7. If the initial value of register A is A_0 the value of register B after the program execution will be **[GATE - 03]**

a) the number of 0 bits in A_0
b) the number of 1 bits in A_0
c) A_0 d) 8

8. Which of the following instructions when inserted at location X will ensure that the value of register A after program execution is the same as its initial value ? **[GATE - 03]**

a) RRC A, #1
b) NOP ; no operation
c) LRC A, #1 ; left rotate A through carry flag by one bit
d) ADD A, #1

9. If 73_x (in base-x number system) is equal to 54_y (in base-y number system), the possible values of x and y are **[GATE - 04]**

a) 8, 16 b) 10, 12 c) 9, 13 d) 8, 11

10. Which of the following addressing modes are suitable for program relocation at run time ? **[GATE - 04]**

i) Absolute addressing ii) Based addressing

iii) Relative addressing iv) Indirect addressing

a) i) and iv) b) i) and ii)

c) ii) and iii) d) i), ii) and iv)

11. The minimum number of page frames that must be allocated to a running process in a virtual memory environment is determined by **[GATE - 04]**

a) the instruction set architecture b) page size

c) physical memory size

d) number of processes in memory

12. How many 8-bit characters can be transmitted per second over a 9600 baud serial communication link using asynchronous mode of transmission with one start bit, eight data bits, two stop bits, and one parity bit ? **[GATE - 04]**

a) 600 b) 800 c) 876 d) 1200

13. What is the result of evaluating the following two expressions using three-digit floating point arithmetic with rounding ? **[GATE - 04]**

(113. + -111.) + 7.51

113. + (-111. + 7.51)

a) 9.51 and 10.0 respectively

b) 10.0 and 9.51 respectively

c) 9.51 and 9.51 respectively

d) 10.0 and 10.0 respectively

The Following Information Pertains to Q. 14-15

Consider the following program segment for a hypothetical CPU having three user registers R1, R2 and R3. **[GATE - 04]**

Instruction	Operation	Instruction Size (in words)
MOV R1, 5000	; R1 ← Memory [5000]	2
MOV R2, (R1)	; R2 ← Memory [(R1)]	1
ADD R2, R3	; R2 ← R2+R3	1
MOV 6000, R2	; Memeory [6000] ← R2	2
HALT	; Machine halts	1

14. Consider that the memory is byte addressable with size 32 bits, and the program has been loaded starting from memory location 1000 (decimal). If an interrupt occurs while the CPU has been halted after executing the HALT instruction, the return address (in decimal) saved in the stack will be

a) 1007 b) 1020 c) 1024 d) 1028

15. Let the clock cycles required for various operations be as follows :

Register to/ from memory transfer : 3 clock cycles

ADD with both operands in register : 1 clock cycle

Instruction fetch and decode : 2 clock cycles per word

The total number of clock cycles required to execute the program is

a) 29 b) 24 c) 23 d) 20

16. Consider a small two-way-set-associative cache memory, consisting of four blocks. For choosing the block to be replaced, use the least recently used (LRU) scheme. The number of cache misses for the following sequence of block addresses is 8, 12, 0, 12, 8 **[GATE - 04]**

a) 2 b) 3 c) 4 d) 5

17. Let A = 1111 1010 and B = 0000 1010 be two 8-bit 2's complement numbers. Their product in 2's complement is **[GATE - 04]**

a) 1100 0100 b) 1001 1100

c) 1010 0101 d) 1101 0101

18. The microinstructions stored in the control memory of a processor have a width of 26 bits. Each microinstruction is divided into three fields: a micro-operation field of 13 bits, a next address field (X), and a MUX select field (Y). There are 8 status bits in the inputs of the MUX. **[GATE - 04]**

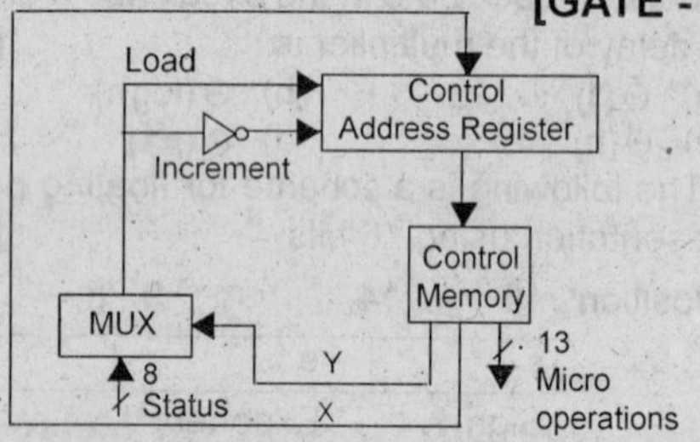

How many bits are there in the X and Y fields, and what is the size of the control memory in number of words ?

a) 10, 3, 1024 b) 8, 5, 256

c) 5, 8, 2048 d) 10, 3, 512

19. A hard disk with a transfer rate of 10 Mbytes/second is constantly transferring data to memory using DMA. The processor runs at 600 MHz, and takes 300 and 900 clock cycles to initiate and complete DMA transfer respectively. If the size of the transfer is 20 Kbytes, what is the precentage of processor time consumed for the transfer operation ? **[GATE - 04]**

a) 5.0% b) 1.0% c) 0.5% d) 0.1%

20. A 4-stage pipeline has the stage delays as 150, 120, 160 and 140 nanoseconds respectively. Registers that are used between the stages have a delay of 5 nanoseconds each. Assuming constant clocking rate, the total time taken to process 1000 data items on this pipeline will be **[GATE - 04]**

a) 120.4 microseconds b) 160.5 microseconds

c) 165.5 microseconds d) 590.0 microseconds

21. What is the minimum size of ROM required to store the complete truth table of an 8-bit × 8-bit multiplier ? **[GATE - 04]**

a) 32K × 16 bits b) 64K × 16 bits

c) 16K × 32 bits d) 64K × 32 bits

22. What is the bit rate of a video terminal unit with 80 characters/line, 8 bits/character and horizontal sweep time of 100 μs (including 20 μs of retrace time)? **[GATE - 04]**

a) 8 Mbps b) 6.4 Mbps

c) 0.8 Mbps d) 0.64 Mbps

23. Consider a system with 2 level cache. Access times of Level 1 cache, Level 2 cache and main memory are 1 ns, 10ns, and 500 ns, respectively. The hit rates of Level 1 and Level 2 caches are 0.8 and 0.9 respectively. What is the average access time of the system ignoring the search time within the cache ? **[GATE - 04]**

a) 13.0 ns b) 12.8 ns
c) 12.6 ns d) 12.4 ns

24. Let x be an integer which can take a value of 0 or 1. The statement If $(x == 0)\, x = 1;$ else $x = 0;$ is equivalent to which one of the following ? **[GATE - 04]**

a) $x = 1 + x;$ b) $x = 1 - x;$
c) $x = x - 1;$ d) $x = 1\%x;$

25. Using a 4-bit 2's complement arithmetic, which of the following additions will result in an overflow ? **[GATE - 04]**

i) 1100 + 1100 ii) 0011 + 0111 iii) 1111 + 0111

a) i) only b) ii) only c) iii) only d) i) and iii) only

26. The number $(123456)_8$ is equivalent to **[GATE - 04]**

a) $(A72E)_{16}$ and $(22130232)_4$
b) $(A72E)_{16}$ and $(22131122)_4$
c) $(A73E)_{16}$ and $(22130232)_4$
d) $(A62E)_{16}$ and $(22120232)_4$

27. A serial transmission T1 uses 8 information bits, 2 start bits, 1 stop bit and 1 parity bit for each character. A synchronous transmission T2 uses 3 eight-bit sync characters followed by 30 eight-bit information characters. If the bit rate is 1200 bits/second in both cases, what are the transfer rates of T1 and T2 ? **[GATE - 04]**

a) 100 characters/sec, 153 characters/sec
b) 80 characters/sec, 136 characters/sec
c) 100 characters/sec, 136 character/sec
d) 80 characters/sec, 153 characters/sec

28. If we use internal data forwarding to speed up the performance of a CPU (R1, R2 and R3 are registers and M[100] is a memory reference), then the sequence of operations **[GATE - 04]**

$R1 \rightarrow M[100]$
$M[100] \rightarrow R2$
$M[100] \rightarrow R3$ can be replaced by

a) $R1 \rightarrow R3$, $R2 \rightarrow M[100]$
b) $M[100] \rightarrow R2$, $R1 \rightarrow R2$, $R1 \rightarrow R3$
c) $R1 \rightarrow M[100]$, $R2 \rightarrow R3$
d) $R1 \rightarrow R2$, $R1 \rightarrow R3$, $R1 \rightarrow M[100]$

29. Consider a pipeline processor with 4 stages S1 to S4. We want to execute the following loop for $(i = 1; i <= 1000; i++)\{I1, I2, I3, I4\}$ where the time taken (in ns) by instructions I1 to I4 for stages S1 to S4 are given below :

	S1	S2	S3	S4
I1 :	1	2	1	2
I2 :	2	1	2	1
I3 :	1	1	2	1
I4 :	2	1	2	1

[GATE - 04]

The output of I1 for i=2 will be available after

a) 11ns b) 12 ns c) 13 ns d) 28 ns

30. Consider a fully associative cache with 8 cache blocks (numbered 0-7) and the following sequence of memory block requests :

4, 3, 25, 8, 19, 6, 25, 8, 16, 35, 45, 22, 8, 3, 16, 25, 7

If LRU replacement policy is used, which cache block will have memory block 7 ? **[GATE - 04]**

a) 4 b) 5 c) 6 d) 7

31. A CPU has only three instructions I1, I2 and I3, which use the following signals in time steps T1-T5 :

I1 : T1 : Ain, Bout, Cin
T2 : PCout, Bin
T3 : Zout, Ain
T4 : PCin, Bout
T5 : End

I2 : T1 : Cin, Bout, Din
T2 : Aout, Bin
T3 : Zout, Ain
T4 : Bin, Cout
T5 : End

I3 : T1 : Din, Aout
T2 : Din, Bout
T3 : Zout, Ain
T4 : Dout, Ain
T5 : End

Which of the following logic functions will generate the hardwired control for the signal Ain? **[GATE - 04]**

a) T1.I1 + T2.I3 + T4.I3 + T3
b) (T1 + T2 + T3). I3 + T1. I1
c) (T1 + T2). I1 + (T2 + T4). I3 + T3
d) (T1 + T2). I2 + (T1 + T3) . I1 + T3

32. In an enhancement of a design of a CPU, the speed of a floating point unit has been increased by 20% and the speed of a fixed point unit has been increased by 10%. What is the overall speedup achieved if the ratio of the number of floating point operations to the number of fixed point operations is 2:3 and the floating point operation used to take twice the time taken by the fixed point operation in the original design ? **[GATE - 04]**

a) 1.155 b) 1.185 c) 1.255 d) 1.285

33. The storage area of a disk has innermost diameter of 10 cm and outermost diameter of 20 cm. The maximum storage density of the disk is 1400 bits/cm. The disk rotates at a speed of 4200 RPM. The main memory of a computer has 64-bit word length and 1 μs cycle time. If cycle stealing is used for data transfer from the disk, the percentage of memory cycles stolen for transferring one word is **[GATE - 04]**

a) 0.5% b) 1% c) 5% d) 10%

34. The range of integers that can be represented by an n bit 2's complement number system is **[GATE - 05]**

a) -2^{n-1} to $(2^{n-1}-1)$ b) $-(2^{n-1}-1)$ to $(2^{n-1}-1)$
c) -2^{n-1} to 2^{n-1} d) $-(2^{n-1}+1)$ to $(2^{n-1}-1)$

35. The hexadecimal representation of 657_8 is **[GATE - 05]**

a) 1AF b) D78 c) D71 d) 32F

36. Which one of the following is TRUE for a CPU having a single interrupt request line and a single interrupt grant line ? **[GATE - 05]**

a) Neither vectored interrupt nor multiple interrupt devices are possible

b) Vectored interrupts are not possible but multiple interrupting devices are possible.

c) Vectored interrupts and multiple interrupting devices are both possible

d) Vectored interrupt is possible but multiple interrupting devices are not possible

37. Normally user programs are prevented from handling I/O directly by I/O instructions in them. For CPUs having explicit I/O instructions, such I/O protection is ensured by having the I/O instructions priviliged. In a CPU with memory mapped I/O, there is no explicit I/O instruction. Which one of the following is TRUE for a CPU with memory mapped I/O ? **[GATE - 05]**

a) I/O protection is ensured by operationg system routine (s)

b) I/O protection is ensured by a hardware trap

c) I/O protection is ensured during system configuration

d) I/O protection is not possible

38. The following diagram represents a finite state machine which takes as input a binary number from the least significant bit.

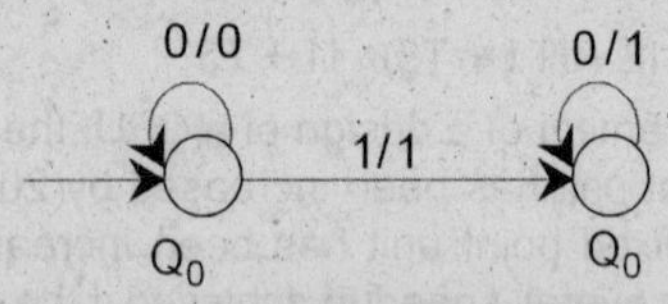

Which one of the following is TRUE ? **[GATE - 05]**

a) It computes 1's complement of the input number

b) It computes 2's complement of the input number

c) It increments the input number

d) It decrements the input number

39. Consider a three word machine instruction

ADD A[R0], @ B

The first operand (destination) "A[R0]" uses indexed addressing mode with R0 as the index register. The second operand (source) "@ B" uses indirect addressing mode. A and B are memory addrsses residing at the second and the third words, respectively. The first word of the instruction specifies the opcode, the index register designation and the source and destination addressing modes. During execution of ADD instruction, the two operands are added and stored in the destination (first operand). **[GATE - 05]**

The number of memory cycles needed during the execution cycle of the instruction is

a) 3 b) 4 c) 5 d) 6

40. Match each of the high level language statements given on the left hand side with the most natural addressing mode from those listed on the right hand side. **[GATE - 05]**

(1) A[I] = B[J] ;	(a) Indirect addressing
(2) While [*A++];	(b) Indexed addressing
(3) int temp = *x;	(c) Autoincrement

a) (1, c), (2, b), (3, a) b) (1, a), (2, c), (3, b)

c) (1, b), (2, c), (3, a) d) (1, a), (2, b), (3, c)

41. Consider a direct mapped cache of size 32 KB with block size 32 bytes. The CPU generates 32 bit addresses. The number of bits needed for cache indexing and the number of tag bits are respectively. **[GATE - 05]**

a) 10, 17 b) 10, 22 c) 15, 17 d) 5, 17

42. A 5 stage pipelined CPU has the following sequence of stages :

IF - Instruction fetch from instruction memory,

RD - Instruction decode and register read,

EX - Execute:ALU operation for data and address computation,

MA - Data memory access - for write access, the register read at RD stage is used,

WB - Register write back.

Consider the following sequence of instructions :

I_1: L R0, 1oc1; R0 <= M[1oc1]

I_2: A R0, R0 ; R0 <= R0 + R0

I_3: A R2, R0; R2 <= R2 - R0

Let each stage take one clock cycle.

What is the number of clock cycles taken to complete the above sequence of instructions starting from the fetch of I_1 ? **[GATE - 05]**

a) 8 b) 10 c) 12 d) 15

43. A device with data transfer rate 10 KB/sec is connected to a CPU. Data is transferred byte-wise. Let the interrupt overhead be 4 μ sec. The byte transfer time between the device interface register and CPU or memory is negligible. What is the minimum performance gain of operating the device under interrupt mode over operating it under program controlled mode? **[GATE - 05]**

a) 15 b) 25 c) 35 d) 45

44. Consider a disk drive with the following specifications: 16 surfaces, 512 tracks/surface, 512 sectors/track, 1KB/ sector, rotation speed 3000 rpm. The disk is operated in cycle stealing mode whereby whenever one byte word is ready it is sent to memory; similarly, for writing, the disk interface reads a 4 byte word from the memory in each DMA cycle. Memory cycle time is 40 nsec. The maximum percentage of time that the CPU gets blocked during DMA operation is **[GATE - 05]**

a) 10 b) 25 c) 40 d) 50

Common Data for Q.45 and Q.46

Consider the following data path of a CPU

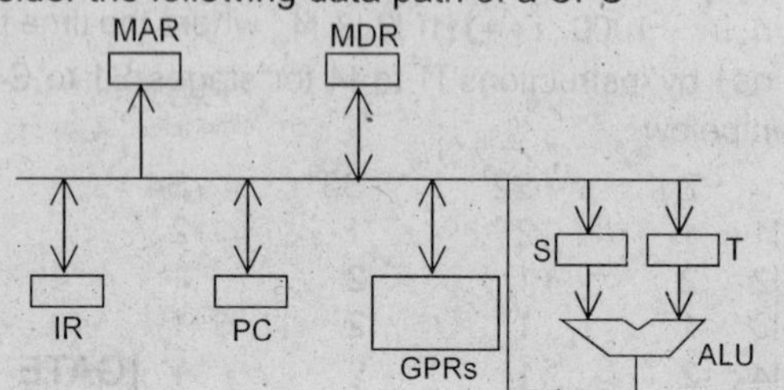

The ALU, the bus and all the registers in the data path are of identical size. All operations including incrementation of the PC and the GPRs are to be carried out in the ALU. Two clock cycles are needed for memory read operation - the first one for loading address in the MAR and the next one for loading data from the memory bus into the MDR. **[GATE - 05]**

45. The instruction "add R0, R1" has the register transfer interpretation R0 < = R0 + R1. The minimum number of clock cycles needed for execution cycle of this instruction is **[GATE - 05]**

a) 2 b) 3 c) 4 d) 5

46. The instruction "call Rn, sub" is a two word instruction. Assuming that PC is incremented during the fetch cycle of the first word of the instruction, its register transfer interpretation is : **[GATE - 05]**

Rn < = PC + 1 ;

Pc < = M[PC];

The minimum number of CPU clock cycles needed during the execution cycle of this instruction is :

a) 2 b) 3 c) 4 d) 5

Statement for Linked Answer Questions 47 and 48

Consider the following floating point format.

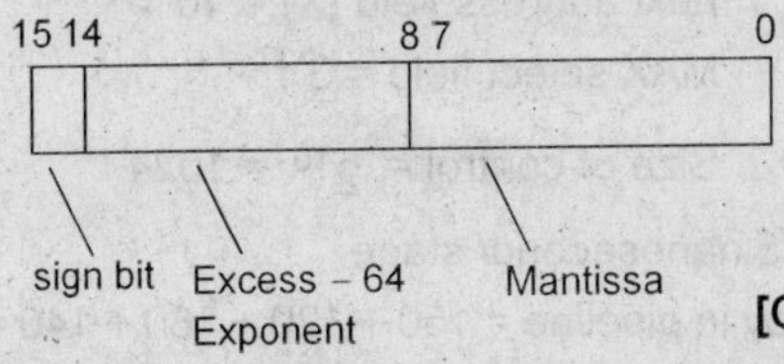

[GATE - 05]

Mantissa is a pure fraction in sign-magnitude form.

47. The decimal number 0.239×2^{13} has the following hexadecimal representation (without normalizatiion and rounding off

a) 0D 24 b) 0D 4D c) 4D 0D d) 4D 3D

48. The normalized representation for the above format is specified as follows.

The mantissa has an implicit 1 preceding the binary (radix) point. Assume that only 0's are padded in while shifting a field.

The normalized representation of the above number $\left(0.239 \times 2^{13}\right)$ is :

a) 0A 20 b) 11 34 c) 4D D0 d) 4A E8

49. Using Booth's Algorithm for multiplication, the multiplier -57 will be recoded as **[GATE - 05]**

a) 0 -1 0 0 1 0 0 -1 b) 1 1 0 0 0 1 1 1

c) 0 -1 0 0 1 0 0 0 d) 0 1 0 0 -1 0 0 1

50. A dynamic RAM has a memory cycle time of 64 nsec. It has to be refreshed 100 times per msec and each refresh takes 100 nsec. What percentage of the memory cycle time is used for refreshing ? **[GATE - 05]**

a) 10 b) 6.4 c) 1 d) 0.64

51. We have two designs D1 and D2 for a synchronous pipeline processor. D1 has 5 pipeline stages with execution times of 3 nsec, 2 nsec, 4nsec, 2nsec and 3 nsec while the design D2 has 8 pipeline stages each with 2 nsec execution time. How much time can be saved using design D2 over design D1 for executing 100 instructions ? **[GATE - 05]**

a) 214 nsec b) 202 nsec

c) 86 nsec d) 200 nsec

52. A hardwired CPU uses 10 control signals S1 to S10 in various time steps T1 to T5 to implement 4 instructions I1 to I4 as shown below.

	T1	T2	T3	T4	T5
I1	S1, S3, S5	S2, S4, S6	S1, S7	S10	S3, S8
I2	S1, S3, S5	S8, S9, S10	S5, S6, S7	S6	S10
I3	S1, S3, S5	S7, S8, S10	S2, S6, S9	S10	S1, S3
I4	S1, S3, S5	S2, S6, S7	S5, S10	S6, S9	S10

Which of the following pairs of expressions represent the circuit for generating control signals S5 and S10 respectively [(Ij+Ik)Tn indicates that the control signal should be generated in time step Tn if the instruction being executed is Ij or Ik] ? **[GATE - 05]**

a) S5 = T1 + I2 . T3 & S10 = (I1+I3) . T4 + (I2+I4) . T5

b) S5 = T1 + (I2 + I4) . T3 & S10 = (I1 + I3) . T4 + (I2 + I4). T5

c) S5 = T1 + (I2 + I4).T3 & S10 = (I2 + I3 + I4) . T2 + (I1+I3) . T4 + (I2 + I4) . T5

d) S5 = T1 + (I2 + I4).T3 & S10 = (I2 + I3).T2 + I4.T3 + (I1+I3).T4 + (I2+I4).T5

53. A line L in a circuit is said to have a stuck-at-0 fault if the line permanently has a logic value 0. Similarly a line L in a circuit is said to have a stuck-at-1 fault if the line permanently has a logic value 1. A circuit is said to have a multiple stuck-at fault if one or more lines have stuck at faults. The total number of distinct multiple stuck-at faults possible in a circuit with N lines is **[GATE - 05]**

a) 3^N b) $3^N - 1$ c) $2^N - 1$ d) 2^N

54. $(34.4)_8 \times (23.4)_8$ evaluates to **[GATE - 05]**

a) $(1053.6)_8$ b) $(1053.2)_8$

c) $(1024.2)_8$ d) None of these

55. An instruction set of a processor has 125 signals which can be divided into 5 groups of mutually exclusive signals as follows : **[GATE - 05]**

Group 1 : 20 signals, Group 2 :70 signals, Group 3:2 signals, Group 4:10 signals, Group 5 : 23 signals.

How many bits of the control words can be saved by using vertical micro-programming over horizontal microprogramming ?

a) 0 b) 103 c) 22 d) 55

ANSWER KEY

1. a	**2.** d	**3.** d	**4.** c	**5.**
6. c	**7.** a	**8.** a	**9.** d	**10.** c
11. d	**12.** b	**13.** a	**14.** a	**15.** b
16. b	**17.** a	**18.** a	**19.** d	**20.** b
21. b	**22.** d	**23.** c	**24.** b	**25.** b
26. d	**27.** c	**28.** d	**29.** c	**30.**
31. a	**32.** d	**33.** c	**34.** a	**35.** a
36. b	**37.** a	**38.** b	**39.** c	**40.** c
41. a	**42.** a	**43.** b	**44.** b	**45.** a
46. d	**47.** d	**48.** d	**49.** a	**50.**
51. a	**52.** d	**53.** c	**54.** d	**55.** b

EXPLANATORY NOTES

1. 2's complement → 00000101 = 5

2's complement of 11100111 is 00011001 = 25

3. $1+2+....+n = \frac{n(n+1)}{2} = O(n^2)$

4. 15 14....................9 8...............0

s	e	m
Sign	Exponent	Mantissa

s = sign bit

e = exponent = $2^6 = 0$ to 63

$m = 2^8 = 0$ to 255.

Floating point representation is

$(-1)^s(1+m\times 2^{-9})2^{e-31}$ if the exponent ≠ 111111

$(-1)^s(1+2^9\ 2^{-9})\ 2^{e-31}$

$= (-1)^s(1+2^0)(2^{e-31})$

If exponent is 111110 or 111101

then $e_1 = 62$

$e_2 = 61$

Then successive difference $= 2^{31}$

5.

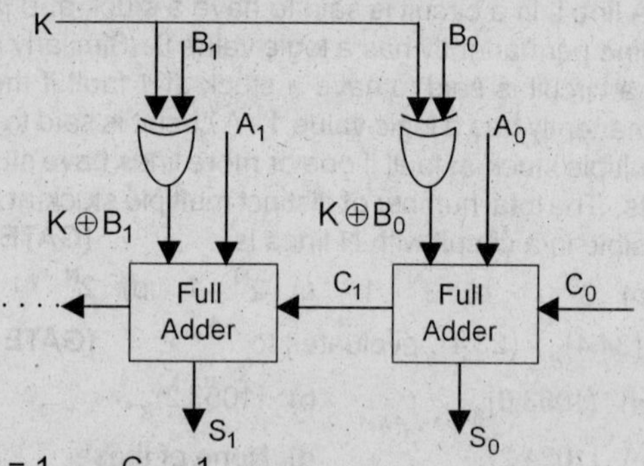

If k = 1, $C_0 = 1$

$S_0 = \overline{B}_0 \oplus A_0 \oplus C_0 = \overline{B}_0 \oplus \overline{A}_0 = A_0\overline{B}_0 + B_0\overline{A}_0$

$C_1 = \overline{B}_0 A_0 + (\overline{B}_0 \oplus A_0)C_0$

$= \overline{B}_0\ A_0 + B_0A_0 + \overline{A}_0\overline{B}_0 = A_0 + \overline{A}_0\ \overline{B}_0$

$S_1 = A_1 \oplus \overline{B}_1 \oplus C_1$

If k = 0, $C_0 = 1$

$S_0 = A_0 \oplus \overline{B}_0 \oplus 1 = A_0 \oplus \overline{B}_0 = A_0B_0 + \overline{A}_0\overline{B}_0$

$C_1 = A_0B_0 + (A_0 \oplus B_0) = A_0 + \overline{A}_0B_0$

⇒ A + B and A - B and A + 1

6. Delay $\delta_1 = 2$ ns

$\delta_2 = 4$ ns

at 1 ns transition logic 0 to logic 1

at 2 ns some transition occur

at 4 ns some transition occur

9. $(73)_X = (54)_Y$

The possible value of x and y is 8 and 11 evaluated by putting the value alternative option.

12. $= \frac{9600}{(\text{start bit} + \text{data bit} + \text{end bit} + \text{parity bit})}$

$= \frac{9600}{12} = 800$

16.

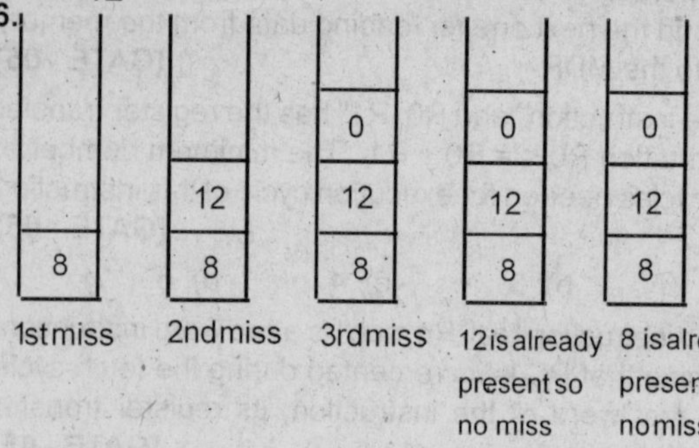

So, total number of miss is (3).

17. A = 11111010

$= -2^7 + 2^6 + 2^5 + 2^4 + 2^3 + 2^1 = -6$

$B = 2^3 + 2^1 = 8 + 2 = 10$

AB = - 60 which is 11000100

18. Control memory processor has 26 bit microinstruction divided into three field microperation field is 13.

So next address field (X) = 10

MAX select field = 3.

Size of control = $2^{10} = 1024$

20. Delay is 5 nanosecond/ stage.

Total delay in pipeline = 150 + 120 + 160 + 140 = 570

Delay due to stage delay = 15

Total delay = 585

Total time taken = $\frac{1000\ (\text{data item size})}{585}$ (ns)

= 165.5 microsecond

21. Minimum size of ROM required to store the complete truth table of 8 bit×8bit multiplexer is 64 K × 16 bit, because it has 18 address lines and 16 data lines.

22. 8 character/line and 8 bit/character.

So bit per line = 8 × 8 = 64 bit/ μ sec

Bit rate = $\frac{64}{100\ \mu}$ = 0.64 Mbps

23.

Level	Access time
1	1 ns
2	10 ns
main memory	500 ns

So average access time =

$0.8 \times 1 + 0.2 \times 0.9 \times 10 + 0.2 \times 0.1 \times 500 = 12.6$

Hence correct answer is (c). 0.2 and 0.1 is miss rate of level 2 and level 1 respectively.

25. i) 1100 ii) 0011 iii) 1111

i)	ii)	iii)
1100	0011	1111
<u>1100</u>	<u>0111</u>	<u>0111</u>
1000	1010	0110

In case of i), no sign change takes place. So no overflow.

In case of ii) sign changes from (+ve) to (-ve). Hence overflow.

In case of iii), one member is positive and other is negative, hence there is no chance of overflow.

26. Binary equivalent of

$$(123456)_8 = (00\,001010\,011100\,101110)_2$$

So for converting it into hexadecimal group, each four bit together $= (A72E)_{16}$

Similarly for converting it into base 4 group to bit $= (22130232)$

27. T_1 : Character size $= (8+2+1+1) = 12$

Transfer rate = 100 characters/sec.

T_2 : 240 ——— 24

$\therefore$ 1 ——— $\frac{24}{240} \times 1200 = 120$

$\therefore$ Character transmitted $= \frac{1200 - 120}{8} = 135$

28. R_1, R_2, and R_3 are registers

Given sequence of instruction are
$R_1 \rightarrow M[100]$
$M[100] \rightarrow R_2$
$M[100] \rightarrow R_3$

Content of R1 is loaded to M [100] as well as R_2 and R_3. So, equivalent instruction can be written as $R_1 \rightarrow R_2$
$R_1 \rightarrow R_3$
$R_1 \rightarrow M[100]$ in any sequence.

30.

7 →	35
6 →	16
5 →	6
4 →	19
3 →	8
2 →	25
1 →	3
0 →	4

After that when 45 encounter it will replace 4. The new sequence will be

	when 22 encounters, it will replace 3 →		eight already memory Now when 3 encounters →	
35		35		35
16		16		16
6		6		6
19		19		3
8		8		8
25		25		25
3		22		22
45		45		45

Now when 7 encounter it will replace 6 →

35
16
7
3
8
25
22
45

So 7 will be in memory block 5.

32. Required speed up

$$= \frac{2 \times \frac{2}{5} \times 120 + 1 \times \frac{3}{5} \times 110}{100} = \frac{2 \times 2 \times 120 + 330}{500} = 1.62 \approx 1.55$$

34. Negative numbers are represented by the forms

1. sign magnitude
2. one's complement
3. two's complement

For sign one's complement representation, we have two zero's -ve and +ve, while in case of two's complement +ve is only one zero.

This additional code which is waste in top two representation is used to assign one more negative code.

35. Given, $(657)_8$

First we find its decimal.

$(6 \times 8^2 + 5 \times 8^1 + 7 \times 8^0)$

$= 6 \times 64 + 5 \times 8 + 7 \times 1$

$= 384 + 40 + 7 = 431$

256	128	64	32	16	8	4	2	1
0001	1	0	1	0	1	1	1	1
1	A				F			

Hence $(657)_8 = (1AF)_{16}$

37. We have two regestion limit and base.

So, when we refer memory location which is not in range provided by these two regestor is always cause OS to interface.

38. Given its input 1 0 1 1 0 →

Output 0 1 0 1 0

It follows the concept that whenever you have to find two's complement, take number from the right end (L.S.B) go to left side and print same number as input till first '1' encounter after that complement each number

↓
0 0 1 1 1 0 <u>0</u>

Number which, we take are underlined at each step and marks is used to give mark (arrow) of victim one.

↓
0 0 1 1 1 <u>0</u> 0

↓
0 0 1 1 <u>1</u> 0 0

↓
0 0 1 <u>1</u> 1 0 0

↓
0 0 1 0 1 0 0
↓
0 0 0 0 1 0 0
↓
0 1 0 0 1 0 0
↓
1 1 0 0 1 0 0

39. Notice the point here that we have to calculate only execution cycle not fetching cycle.

So during fetching cycle we have addresses of 'A' and 'B' (read question carefully).

Now, we have to calculate index addresses by reading value of $A \to 1$

Then, we have to add R0, we get the address of first source of and we have to read its value $\to (2)$

* Notice here we have address of B for getting actual value

First, we have to read contents of $B \to (3)$

then value at contents of $B \to (4)$

then we have to write actual answer $\to (5)$

41. Here, we have 32 bit virtual address space.

32 KB cache memory hence, we need 15 bit for occuring them (not all for indexing explained next).

So 17 used for tag.

But notice here that cache are in block of 32 bytes. So, there actually 1 K (blocks) for a indexing 1 K location, we need 10 bits.

Remainder five are used for selecting byte

17 bit	10 bit	5 bit
⇓	⇓	⇓
for tag	for cache index	for selecting byte within word (32 bytes)

43. If overhead of a single byte transfer with interface mode $= 4\,\mu$ sec, then actual data transfer can be = 250000

Actually transfer take place at the rate $\to$ 10000

Now 25 is the minimum performance gain.

44. Round per minute (R.P.M.) = 3000

Round per sec $= \dfrac{3000}{60} = 50$

In one round data read = 512 * 1 KB = 512 KB

(Notice here that in round)

We can read one track $= \dfrac{2^{19}}{4} = 2^{17}$

In one second $= 2^{17} * 50$

Hard disk interrupt = 6553600 C.P.U.

(DMA at most)

Each intruption take = 4nsec

So, 6553600 Intrupts takes = 0.262144 s

So, percentage $= \dfrac{0.262144}{1} * 100 = 26.2144$ (approx)

In case, we use 1 KB = 1000, then answer is 25.

49. -57 in two complement

1 1 0 0 0 1 1 1

in booth's form

0 -1 0 0 + 0 0 -1

52. Notice here that S5 have to be activated on T1 for all instructions and for I2 and I3 and T3.

By S10 have to be activated.

for I2 and I4 on T5

for I1 and I3 on T4

for I4 and T3

for I2 and I3 on T2

53. Apply the concept of total number of substuck of fault, there are 2^N. Where '1' subset ϕ required Q no one is line stuck

55. For horizontal, we need 125 signals. For vertical, we need $5(20)+7(70)+1(2)$,

$$\left(2^5\right)+\left(2^7\right)+\left(2^1\right)+4(10)+5(23)\left(2^4\right)$$

$= 22$; $2^5 = 125 - 22 = 103$

Unit - VI
Programming Methodology

UNIT - VI PROGRAMMING METHODOLOGY

I. FUNDAMENTALS OF 'C'

- **Introduction**
- **Data Types In C**
- **Operators In C**
- **Operator Precedence**
- **Input-Output Operators**

1. Introduction

a. C is a general purpose, structured programming language.

b. C was designed by Dennis.M.Ritchie in 1972 AT & T, Bell Labs, USA.

c. C is originated from a language called 'B' which was an improved version of the language called BCPL (Basic Combined Programming Language) developed by Martin Richards in 1960's.

Applications of C

C is used for writing

- Operating systems
- Compilers
- Interpreters
- Assemblers
- Editors
- Spreadsheets
- Database Management Systems
- CAD, COM and Animators

Advantages of C

- Structured language. It has few fundamental data types and keywords. This makes the language simple and small.
- Middle level language, hence suitable for developing application software as well as system software.
- Portable language, i.e., programs written in C in one machine can be easily run on some other machine without many modifications.
- It has wide variety of derived data structures like arrays, pointers, structures and unions apart from fundamental data types like integers, floating point numbers and characters.
- Provides a rich set of built-in functions.
- Its compiler is compact and can translate a program into efficient machine instructions.
- Fast execution of programs when compared to other languages.
- It has the ability to deal efficiently with bits, bytes, words, addresses etc.,

Structure of a 'C' Program

```
    Part-1;
Comment lines (name of the program, author...)
Link section (to link from system library)
Symbolic constant definitions
Global declarations
    Part-2;
main( ) function
{
    variable declaration
    statements
}
    Part-3;
Sub programs (User-defined functions)
```

Note : All parts are optional except main( ) function

Punctuations in C

- Usually all C statements are entered in small case letters.
- Every statement in C ends with a semicolon(;)
- Single quotes are used to enclose a characterconstant.
- Double quotes are used to enclose a string constant.
- A function name is always followed by a pair of parenthesis, i.e( )
- Compound statements are enclosed with in curly braces, i.e., { }
- When more than one variable is declared in single statement, they must be separated by commas

The C Character Set

Alphabets :	A-Z and a-z
Digits :	0-9
Special Symbols :	~!@#$%^&*()_-+/|\{}[]:;"'<>,.?/

2. Data Types

A Data Type defines a set of values and the operations that can be performed on them. Every data type item (constant, variable etc..) in a C-program has a data type associated with it. C supports several different types of data, each of which may be represented differently within the computer's memory.

The following diagram summarizes the various basic data types that are available in C.

All C compiles support four fundamental data types, namely integer (int), character (char), floating point (float), and double-precision floating point (double). Many of them also offer extended data types such as long int and long double.

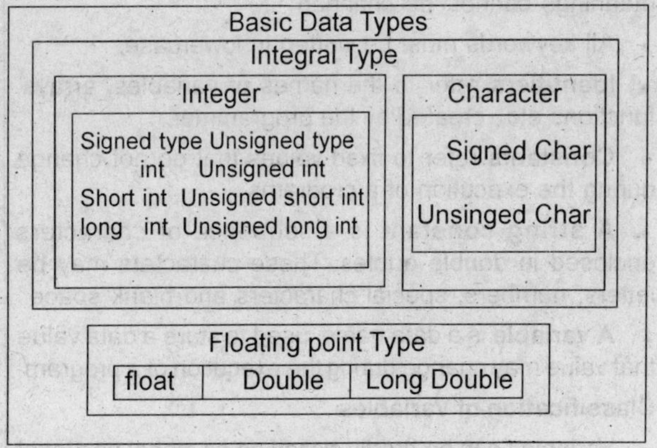

Fig : Basic data types in C

The RANGE of the basic four types are shown below table.

Data type	Range of values
Char	- 128 to 127
int	- 32,768 to 32,767
float	3.4 e - 38 to 3.4 e + 38
double	1.7 e - 308 to 1.7e + 308

Fig : Size and Range of Basic Data types

Integer Types : Integers are whole numbers with a range of values supported by a particular machine. Generally, integers occupy one word of storage, and since the word sizes of machines vary, the size of an integer that can be stored depends on the computer. If we use a 16-bit word length, the size of the integer value is limited to the range - 32768 to + 32767 (that is - 2^{15} to + 2^{15} -1). A signed integer uses one bit for sign and 15 bits for the magnitude of the number.

In order to provide some control over the range of numbers and storage space, C has three classes of integer storage, namely short int, int and long int, in both signed and unsigned forms.

For example, short int repragents fairly small integer values and requires the half amount of storage as a regular int number uses.

Unlike signed integers, unsigned integers use all the bits for the magnitude of the number and are always positive. Therefore, for a 16 bit machine, the range of unsigned integer numbers will be from 0 to 65,535.

We declare long and unsigned integers to increase the range of values. The use of qualifier signed on integers is optional because the default declaration assumes a signed number.

Floating Point Types : Floating point (or Real) numbers are stored in 32 bits, with 6 digits of precision. Floating point numbers are defined in C by the keyword float. When the accuracy provided by a float number is not sufficient, the type double can be used to define the number. A double data type number used 64 bits giving a precision of 14 digits. These are known as double precision numbers. To extent the precision further, we may use long double which uses 80 bits.

Character Types : A single character can be defined as a character (char) type data. Characters are usually stored in 8 bits of internal storage. The qualifier signed (or) unsigned may be explicitly applied to char. While unsigned characters have values between 0 to 255, signed characters have values from -128 to 127.

C Tokens : The smallest individual units of a C program are called C tokens.C programs are written using these tokens and the syntax of the language.

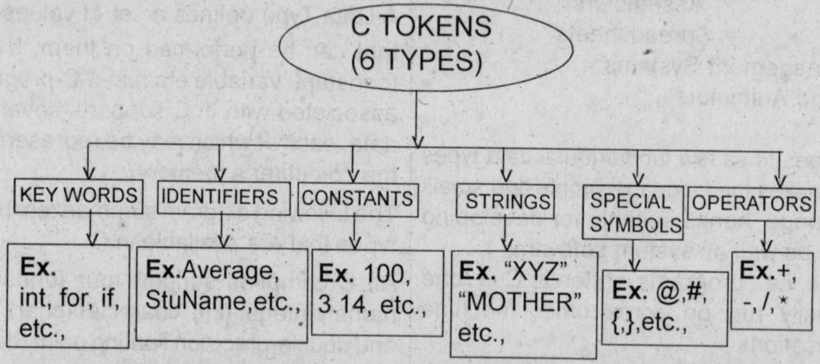

- All **keywords** have pre-defined meaning and these meanings cannot be changed
- All keywords must be written in lowercase.
- **Identifiers** refer to the names of variables, arrays, functions etc., created by the programmer.
- **Constants** refer to fixed values that do not change during the execution of a programs.
- **A string constant** is a sequence of characters enclosed in double quotes. These characters may be letters, numbers, special characters and blank space.
- **A variable** is a data name, used to store a data value that value may change during the execution of a program.

Classification of Variables

- Variables can be brodly classified as global and local variables.
- **Global variables** are the variables accessible anywhere in a program.
- The variables used within a function and which cannot be accessed outside the function are called **local variables.**

3. Operators in C

1. Arithmetic [+, -, ∗ , /, %(modulo division)]
2. Assignment [=, +=, -=, ∗ =, /=, %=]
3. Conditional [<expression1>?<expression2>:<expression 3>]
4. Relational [>, <, <=, >=, = =, !=]
5. Logical [&&, ||, !]
6. Increment and Decrement [++, --]
7. Bitwise [~, >>, <<, &, |, ^]

Note ~ : is complement

>> : is right shift

<< : is left shift

& : bitwise AND

| : bitwise OR

^ : bitwise Exclusive OR (XOR)

Miscellaneous

a) The comma (,) operator

The comma operator separates the expressions.

Syntax: **exp1, exp2, ... expn** Where exp1, exp2, ... expn are expressions.
Ex. k=(i=5,j=10,i+j);

Here, first the value 5 is assigned to i, then 10 is assigned to j and the result i + j is assigned to i and k.

Generally, the comma operator is used in for loops.

Ex. for(i=0,j=1;i<=10;i++,j++)

b) size of operator is used to obtain the size of any object in a C program.

Syntax: sizeof(object); Where object may be type, variable or expression
Ex.1. n1=**sizeof(int);** **2.** n2=**sizeof**(name);

In the first example:

Assingns the size of the basic data types int in terms of bytes to the variable n1.

In the second example:

Assigns the size of the variable name in terms of bytes to the variable n2.

Unary operators : The unary operators require only a single expression to produce a line. Unary operators usually precede their single operands. Sometimes, some unary operators may be followed by the operands such as increment and decrement.

The most common unary operation is unary minus, where a minus sign precedes a numerical constant, a variable, or an expression.

The following operators are the unary operators in C.

Operator	Meaning
-	Negative value (minus)
!	Negation
~	Bitwise complement
- -	Decrementer
+ +	Incrementer
*	Contents of the storage field to which a pinter is pointing
&	Address of a variable

Increment and decrement operators : Two special operators are used in C, namely, incrementer and decrementer. These operators are used to control the loops in an effective manner.

The syntax of the operators is given below :

```
++ <variable name>          - - <variable name>
<variable name>++           <variable naem>- -
```

The operator ++ adds 1 to the operand and - - subtracts from the operand. These operators manifest in two forms: prefix and postfix.

For example, the ++ operator can be used in two ways:

++m and m++

These are two different expressions. The first expression immediately increments the value of m by 1.

For example, the statements,

```
. . . .
int t,m=1 ;
t = ++m ;
printf('prefix operator : t=%d,m=%d\n",t,m);
```

Will print

prefix operator : t = 2, m = 2

The expression ++m will cause the value of m to be incremented first, and then this value is asigned to t, resulting both having the same value.

On the other hand, the statements,

```
. . .
int t,m = 1
t = m++;
printf('postfix operator:t = %d,m = %d\n",t,m);
```

will output

postfix operator:t = 1, m = 2

The expression m++ will first evaluate the value of m (which is 1),

Conditional operator (? :) C includes a very special operator called the ternary (or) conditional operator. It is called ternary because it usesx three expressions. The ternary operator acts like a shorthand version of the if-else construction.

The general format of the ternary operator is :

expression 1 ? expression 2 : expression 3 ;

Which results in either expression 2 (or) expression 3 being evaluated. If expression 1 is true, then expression 2 evaluated; otherwise, expression 3 is evaluated. The conditional operator is used in place of a single if-else to make an assignment.

For example,

```
If (first > second)
        max = first;
else
    max = second;
```

This can be written as,

max = (first > second) ? first : second;

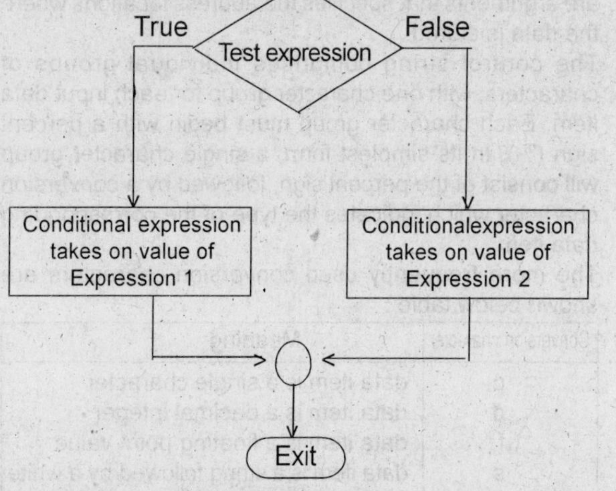

Fig : Flowchart of the ternary operator

The expression (first > second) is evaluated first. If it is true, the value of first is assigned to max. Otherwise, the value of second is assigned to max.

4. Operator precedence

Operators have rules of precedence and associativity that are used to determine how expressions are evaluated. Precedence of operators is a vital aspect in executing arithmetic statements, having two (or) more operators.An arithmetic expression with having paranthesis(or)brackets will be evaluated from left to right,using the rules of precedence of operators ODMAS.

These are two distinct priority levels :

Highest priority : * , / , %

Low priority : + , -

An expression will be evaluated from left to right within a maximum of two passes. During first pass it will scan from left to right, so that the highest priority operators will be executed as they come. During second pass the low priority operators are executed as they come.

Consider the following example :

x = 2 * 3 + 5 - 3 * 6

I scan : x = 6 + 5 - 18

II scan : x = 11 - 18 = - 7

The order of evaluation is going to be changed if we are using a paranthesis. The expression with in the paranthesis will have highest priority above all.

5. Input-Output Operators

Formatted Input : Input data can be entered into the computer from a standard input device by means of the C library function scanf. This function can be used to enter any combination of numerical values, single characters and strings. The function returns the number of data items that have been entered successfully.

In general terms, the scanf function is written as

scanf (control string, arg1, arg2, ..., argn);

Where control string refers to a string containing certain required formatting information, and arg1, arg2, . . . , argn are arguments that specifies the address locations where the data is stored.

The control string comprises individual groups of characters, with one character group for each input data item. Each character group must begin with a percent sign (%). In its simplest form, a single character group will consist of the percent sign, followed by a conversion character which indicates the type of the corresponding data item.

The more frequently used conversion characters are shown below table :

Conversion character	Meaning
c	data item is a single character
d	data item is a decimal integer
f	data item is a floating-point value
s	data item is a string followed by a white space character
[....]	data item is a string which may include white space characters

The arguments are written as variables (or) arrays, whose types match the corresponding character groups in the control string. Each variable name must be preceded by an ampersand (&). However, array names should not begin with as ampersand.

Example : Here is a typical application of a scanf function.

```
# include <stdio.h>
main ( )
{
    char item [20];
int partno;
float cost;
. . .
scanf("%s %d %f", &item, &partno, &cost);
. . .
}
```

Within the scanf function, the control string is "%s %d %f". It contains three character groups. The first character group %s, indicates that the first argument (item) repragents a string. The second character group, %d, indicates that the second argument (&partno) repragents a decimal integer value, and the third character group, %f, indicates that the third argument (&cost) repragents a floating point value.

Formatted output : Output data can be written from the computer onto a standard output device using the library funtion printf. This function can be used to output any combination of numerical values, single characters and strings. Simply, the printf function moves data from the computers memory to the standard output device (monitor).

In general terms, the printf function is written as

printf (control string, arg1, arg2, ..., argn);

Where control string refers to a string that contains formatting information, and arg1, arg2, . . . , argn are arguments that repragent the individual output data items.

Control string consists of 3 types of items :

1) Characters that will be printed on the screen as they appear.

2) Format specifications that define the output format display of each item.

3) Escape sequence characters such as \n \t \b.

Formatting : The printf function allows you to format your output. The printf function provides field width specifier, so that, one can control how printed values will appear on the output media.

The syntax is : % W.DF

where % = Indicates the format specification

W = Width of the field

D = Number of digits to the sight of the decimal

F = Format specifier

Example : printf("The number %5.2f uses them", 6.0);

would display

The number 6.00 uses them.

Note that, the width first indicates how the number is justified (five spaces), and then the number of digits following the decimal point are specified.

In case of integers and characters, the number of decimal places Din the field width specifier is not needed.

OBJECTIVE QUESTIONS - I

1. C-language is a: **[AUCET-2000]**

a) high level language b) low level language

c) High level language with low level features.

2. What is the value of d in the following C-language assignment statement when a is 8, d is 10 and b is 16.

{d+=(a>0&&a<=10)?++a:a/b;} **[AUCET-2000]**

a) 19 b) 18 c) 9 d) 0.5

3. The value of the arithmetic expression in C

2 * {i%5) * (4+ (j-3)/(k+2))} **[AUCET-2000]**

When i is 8 , j = 15 and k is 4 is

a) 12 b) 19.2 c) 36 d) 57.6

4. Equivalent hexadecimal number for 69 decimal number in C-program is : **[AUCET-2000]**

a) 0x45 b) 0x69 c) 0x54 d). 0x96

5. In C, which of the following type of operators have the highest precedence: **[AUCET-1997]**

a) Relational operator b) logical operator

c) Equality operator d) Arithmetic operator

6. The operator precedence determines which operator **[AUCET-2000]**

a) is slowest b) is fastest

c) is executed first d) has highest priority

7. Identify which is not data type Qualifier **[AUCET-2000]**

a) auto b) long int c) double d) short int

8. The range of numbers that can be stored in an integer type variable, where n is number of bits **[AUCET-2000]**

a) -2^n to $+2^n$ b) -2^{n-1} to $+2^n$

c) -2^n to $+2^n-1$ d) -2^{n-1} to $+2^{n-1}$

9. An Identifier is a name given to ---- in C. **[AUCET-2000]**

a) a constant b) a variable

c) procedure d) All of the above

10. The range of values for the long type data on a 16-bit machine is **[TCS]**

a) -2^{31} to $+2^{31}-1$ b) -2^{64} to $+2^{64}$

c) -2^{63} to $+2^{63}-1$ d) -2^{32} to $+2^{32}-1$

11. Which of the following repragent a hexadecimal number **[TCS]**

a) 570 b) (hex) 5 c) 0x9F d) 0x5

12. Which of the following assignments are valid **[TCS]**

a) float x=123.4 b) long m = 023

c) int n=(int) false d) double y=0x756

13. If m and n are int type variables , what will be the result of the expression m % n when m=5 and n=2? **[TCS]**

a) 0 b) 1 c) 2 d) None of the above

14. Which of the following are correct: **[TCS]**

a) A relational expression yields an int type result.

b) A relational expression yields a bool type result.

c) A logical operator compare two bool type values.

d) A logical operator combines two bool type values.

15. Which of the following will produce a value of 22 if x=22.9 **[TCS]**

a) ceil(x) b) log(x) c) abs(x) d) floor(x)

16. Which of the following expressions are Illegal: **[TCS]**

a) (10/5) b) (false && true)

c) float y=12.34 d) None of these

17. In C-language "size of" is: **[AUCET-2001]**

a) predefined constant b) Library function

c) an operator d) predefined function

18. By default any real number in "C" is treated as :

a) a float b) a double c) a long double

d) depends upon the memory model that you are using.

19. The rule for implicit type conversion in "C" is

a) int < unsigned < float < double.

b) unsigned< int < float < double.

c) int < unsigned < double < float.

d) unsigned < int < double < float..

20. Integer Division in a "C" program results in

a) truncation b) rounding

c) overflow d) None of the above

21. If y is of integer type ,then the expressions 3 * (y-8)/9 and (y-8)/9 * 3 yield the same value if

a) y is an even numberb) y is an odd number.

c) (y-8) is an integral multiple of 9

d) (y-8) is an integral multiple of 3.

22. If y is of integer type, then the expressions 3 * (y-8)/9 and (y-8)/9 * 3

a) must yield the same value.

b) Must yield different values.

c) May or may not yield the same value.

d) None of the above.

23. Which of the following comments about the ++ operator is (are) correct:

a) It is a unary operator.

b) The operand can come before or after the operator

c) It cannot be applied to an expression.

d) It associates from the right

24. When a variable of data type double is converted into float then

a) rounding takes place b) truncation takes place

c) the lower order bits are dropped d) None of these

25. The maximum length of a string in C is:

a) 256 characters b) 1024 characters

c) 64 characters d) Infinite.

26. The typical value of EOF is

a) 0 b) -1 c) 1 d) None

27. Which of the following does not have a unary operator:

a) –7 b) ++i c) j d) All of the above

28. If t both a and b have the value 2, then the statement a * =b+4;

a) Gives an error message

b) a is assigned the value 12

c) a is assigned the value 8

d) a is assigned the value 13.

29. If a is an integer variable, a=5/2; will return a value

a) 2.5 b) 3 c) 2 d) 0

30. Hierarchy decides which operator

a) is most important b) is used first

c) is fastest d) operates on larget numbers

31. "ungetc" function in C-language: **[AUCET- 2001]**

a) Pushes back the given character on to the i/p stream.

b) Skips one character from the i/p stream

c) Reads the next character from the file d) is same as "putc"

32. getchar( ) function can be alternatively expressed as: **[AUCET- 1999]**

a) getc(stream) b) scanf(%c)

c) getc(stdin) d) None

33. Where does 'OS' search for

include<conio.h>

a) Current directory b) All sub directories of C

c) All Sub directories under DOS d) None

34. Printf("%c",100);

a) prints 100b) Prints the ASCII equivalent of 100

c) prints garbage d) None

35. A header file is:

a) a file that contains standard library functions

b) a file that contains definitions and macros.

c) a file that contains user defined functions

d) a file that is present in current working directory.

36. Which of the following ' C ' type is not a primitive data structure ?

a) Int b) Float c) Char d) None

37. A typecast is used to : **[DOEACC- 2001]**

a) Define a new data type

b) Force a value to be of a particular variable type

c) Rename an old type d) None

38. The Ternary operator(? :) and sizeof operator have: **[DOEACC- 2001]**

a) Same associativety b) Different associativety

c) No associativety d) None

39. Which of the following is returned by the function "strcmp" if the strings that are compared are identical? **[DOEACC- 2001]**

a) 0 b) 1 c) 2 d) True

ANSWER KEY - I

1. c	**2.** a	**3.** c	**4.** a	**5.** d
6. d	**7.** a	**8.** c	**9.** b	**10.** a
11. c	**12.** a	**13.** b	**14.** a	**15.** c
16. b	**17.** c	**18.** b	**19.** a	**20.** a
21. c	**22.** b	**23.** c	**24.** b	**25.** d
26. b	**27.** c	**28.** b	**29.** c	**30.** b
31. a	**32.** b	**33.** a	**34.** b	**35.** a
36. d	**37.** b	**38.** c	**39.** a	

OBJECTIVE QUESTIONS - II

1. What will be the values of a and b after execution of the following: n=5; a=n++; b=--n ; **[AUCET-1997]**

a) 5,4 b) 5,5 c) 6,5 d) 6,6

2. The expression is formed using **[AUCET-2000]**

```
float     a;
long int  b;
short int c;
char      d;
          d+c * b/a ;
```

What is the type of the result of evaluation of the above expression

a) float b) long int c) double d) short int

3. What will be the values of x, m, and n after execution of the following statements: **[TCS]**

```
int  x,m,n;
     m = 10;
     n = 15;
     x = ++m+n++;
```

a) x = 25, m = 10, n = 15 b) x = 27, m = 10, n = 15

c) x = 26, m = 11, n = 16 d) x = 27, m = 11, n =16

4. Consider the following statements:

```
int  x = 10, y = 15;
     x = ((x<y) ? ( y+x) : (y-x));
```

What will be the value of x after execuing these statements

a) 10 b) 25 c) 15 d) 5

5. Given the declarations **[TCS]**

```
bool b;
int  x1=100,x2=200,x3=300;
```

Which of the following statements are evaluated to TRUE:

a) b=x1 * 2 == x2; b) b=x1+x2 ! = 3 * x 1 ;

c) b=(x3-2 * x2<0)||((x3=400)<2 * x2);

d) b=(x3-2 * x2>0)||((x3=400)<2 * x2);

e) All are correct

6. When the following code is executed, what will be the values of x and y: **[TCS]**

```
int x = 1,y = 0;
    y = x++;
```

a) 2,1 b) 2,2 c) 1,1 d) None

7. In C, program

```
int   x = 5,y = 10;
char  a = 'a',b = 'b';
```

Then, what will be the value of the following expression (5 * x-y) % (2 * a-b)

a) 10 b) 12 c) 15 d) 20

8. What is the output of the below program, **[AUCET- 1999]**

```
#include<stdio.h>
main( )
{
int passing, failing, grade_average;
passing=failing=0;
grade_average=75;
(grade_average>=70)? passing ++:failing ++;
printf("%d%d\n",passing,failing);
}
```

a) 0,0 b) 1,0 c) 0,1 d) none

9. The output of the following program is: **[AUCET- 1999]**

```
main( )
{
 int i, j;
 char string[9];
     strcpy("string,"14A05");
     i=atoi(string);
     printf("%d\n",i);
}
```

a) 14 b) 15005 c) "14A05" d) none

10. What is the output of the following code:

```
int i = 7;
printf("% d",i ++ * i --) ;
```

a) 49 b) 56 c) 7 d) 48

11. Which of the following I/O functions in C is used to read formatted data from Keyboard: **[AUCET- 2000]**

a) scanf b) sscanf c) getc d) fscanf

12. The o/p of the program **[AUCET- 1999]**

```
int x=4,y=3,z=6;
return[(x>y<<3!(x<z)<<1];
```

a) 1001 b) 1010 c) 1100 d) 0011

13. What is the output : **[AUCET- 1998]**

```
printf("%d",printf( "tim" ));
```

a) results in a syntax error b) outputs tim3
c) outputs garbage
d) prints tim and terminates abruptly

14. What is the output of the following c-code segment **[AUCET- 2000]**

```
int t=467;
float p;
   p=t/500.0 * 100;
   printf("%d",p);
```

a) 93.0 b) 93.4 c) 0.930 d) 0

15. What is the output of the following C-code segment **[AUCET- 2000]**

```
int x,y,z,a,b,i;
   i=5; x=++i; y=i++; z= --i;
   a=i++;  b=i;
   printf("%d",b);
```

a) 5 b) 6 c) 7 d) 8

16. when the below program segment is run, it prints **[DOEACC- 1993]**

```
main( )
{ int i,j,r;
     i=100; j=5; r=20;
     printf("value of i+j-r is % o \n",i+j-r);
}
```

a) 55 b) 85 c) 125 d) None

17. Consider the program segment: **[DOEACC- 1994]**

```
main( )
{
  printf("hello\n");
  main( );
        }
```

Which of the following is TRUE for the above program segment:

a) Prints only once b) Infinite loop
c) Compilation error because of main with in main module

18. What would be the output of the following

```
main( )
{
 int x=1,y=2,z=3;
    printf("%d",x+=(x+=,5,x));
}
```

a) 6 b) 8 c) 9 d) 12

19. What is the output of the following code:

```
int x, y=2, z, a;
    x=(y * =2)+(z=a=y);
    printf("%d",x);
```

a) 7 b) 6 c) 8 d) Syntax error

20. The following program fragment

```
int k=-7;
printf("%d",0<!k);
```

a) prints 0 b) prints a non-zero value
c) is illegal d) prints an unpredictable value.

21. If the following variables are set to the values as shown below, then what is the value of the expression following it? **[DOEACC- 2000]**

```
answer=2;
marks=10;
!((answer<5)&&(marks>2))
```

a) 0 b) 1 c) –1 d) 2

22. What would be the value of i and k at the end of the following program segment **[DOEACC- 2000]**

```
main( )
{
 int i,j,k;
  j=5;
  i=2 * j/2;
  k=2 * ( j/2);
}
```

a) i=5,k=5 b) i=4,k=4 c) i=5,k=4 d) i=4,k=5

23. Consider the following program fragment [DOEACC- 2000]

```
main( )

{
 int a,b,c;
  b=2, a=2;
  a=2 * (b++);
  c=2 * (++b);
}
```

The correct values are

a) a=4,c=6 b) a=3,c=8 c) a=4,c=8 d) a=3,c=6

24. Consider the following statements **[DOEACC- 2002]**

```
int a=2,b=3, c=4;
  a=(b++)+c;
  b=a+(++c);
```

what are the values of a,b and c respectively

a) 8,12,5 b) 8,13,5 c) 7,12,5 d) 7,11,5

25. What is the value of j at the end of the following code segment if the user types in '4' for i ? **[Hyd.Univ-2000]**

```
int i, j;
  printf("Enter number of runs:");
  j=scanf("%d",&i);
```

a) 1 b) 2 c) 0 d) 4

26. a = b = c = 0 is possible in 'C' because **[Hyd.Univ-1999]**

a) Assignment is a side-effect

b) Richie said it must be so!

c) Assignment operation returns a value

d) Integer operations always return a value

27. Suppose that i = 0 before the following printf statement. What would be the output? **[Hyd.Univ-1999]**

printf("%d %d %d \n";i++,++i,++i * 2) ;

a) 1 2 6 b) 2 2 2 c) 3 1 4

d) Can not be determined; output depends on the complier

28. In the C programming language which of the following operators has the highest precedence? **[Hyd.Univ-1998]**

a) unary + b) * c) >= d) ==

29. A C-program contains the following statements: **[Hyd.Univ-1996]**

```
#include <stdio.h>
int    i, j ;
long   lx;
short  s;
unsigned u;
float  x;
double dx;
char   c;
```

A scanf function to be written so as to read the group of data items (c, x, dx, s) assuming that all the integers to be read as decimal quantities. Which of the following is correct?

a) scanf("%c %c %f %f", &c, &dx, &x, &s)

b) scanf("%f %c %f %c", &c, &x, &dx, &s)

c) scanf("%c %f %lf %hd", &c, &x, &dx, &s)

d) scanf("%4c", &c, &dx, &x, &s)

30. The value of the C expression x = 1,024 is **[Hyd.Univ-1994]**

a) 1024 b) 24 c) 20 d) Syntax Error

31. The value of the C expression

10 > 4 ? 1 < - 4 ? 15 : 7 * 3 : 5 / 4 Is **[Hyd.Univ-1994]**

a) 1 b) 21 c) 15 d) 0

32. What does the following program segment do? **[Hyd.Univ-1994]**

```
int a, b;
a = a + b;
b = a – b;
a = a – b;
```

a) finds a + b b) finds a – b c) finds b – a

d) exchanges the values of a and b

33. What is the final value of x in the following C program segment? **[Hyd.Univ-1993]**

```
x = y = z = 2
z = y - - * x;
x = (x != y)? z : y;
```

a) 4 b) 2 c) 1 d) 0

ANSWER KEY - II

1. b	**2.** a	**3.** c	**4.** b	**5.** e
6. a	**7.** c	**8.** b	**9.** a	**10.** b
11. a	**12.** ?	**13.** b	**14.** d	**15.** c
16. c	**17.** b	**18.** d	**19.** c	**20.** a
21. a	**22.** c	**23.** c	**24.** b	**25.** a
26. b	**27.** b	**28.** a	**29.** c	**30.** d
31. b	**32.** d	**33.** b		

II. PROGRAM CONTROL

- **Control Structures**
- **Functions**
- **Parameter Passing Techniques**
- **Recursion**
- **Arrays**
- **Strings**

1. Control Structures

The control structures can be classified as

1. Sequential. 2. Decision or selection

3. Repetition or iteration

Sequential structure

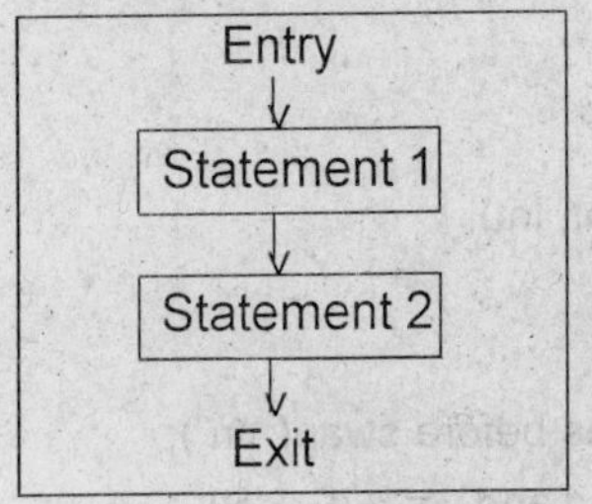

Execution follows the order in which the statements are written.

Decision (or) Selection structure

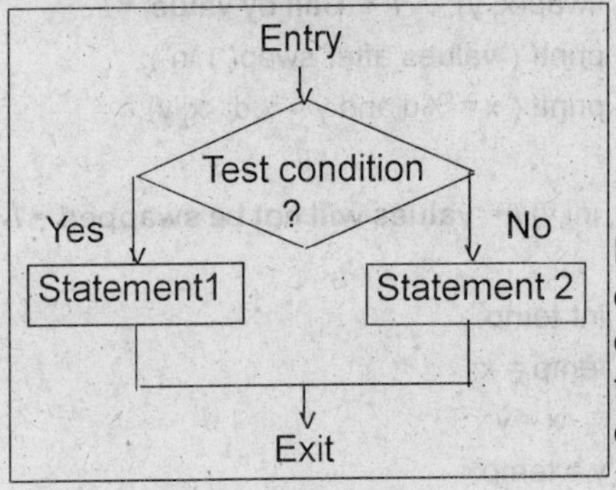

Here the test condition decides the flow of execution

Decision Structures Available in C

1. if
2. if ... else
3. switch

Repetition (or) Iteration Structures

These structures are used to execute a group of instructions repeatedly until some logical condition is satisfied.

The loop structures in C:

1. for loop (or Entry Control loop)
2. while loop (or Entry Control loop)
3. do...while loop (or Exit Control loop)

Use of Loops

- The for loop is used when the number of iterations can be specified earlier explicitly.
- The while, do...while loops are used when it is not known when the loop terminates.
- If it is not required to execute a loop even once under certain conditions while is used
- When the loop is required to be executed atleast once, the do...while is used.

Infinite Loops

- On certain situations, there will be a need to use infinite loops. Most commonly used methods are:

```
while(1)
statement;
```

and

```
for( ; ; )
statement;
```

- To get out an infinite loop, you need to abort the program manually.

Jumping Statements

goto Statement: (Unconditional branching) is used to transfer the control form one point to any other point in a function.

Syntax: goto label;

This statement provides an unconditional jump to the statement is having the label.

- No declaration is required for the label.
- The syntax of label requires a colon (:) after the label

Note: The structured **programming,** as far as possible, goto statements are to be avoided otherwise program debugging becomes difficult.

- The **break** statement causes the immediate terminations of the loop that contains it.
- The **continue** statement terminates only the current iteration of the loop that encloses it, thus causing execution to resume with the next iteration.
- The **exit( )** function

This function causes immediate termination of the program and execution control return to the operation system.

In general, the argument to exit( ) function is 0 (zero) to indicate that termination is normal. Other argument may be used to indicate some sort of an error.

2. Functions

- Functions are building blocks of C.
- The general form of C (user-defined) function is

```
return-type function-name(parameter list)
        Parameter declaration;
{
        body of function;
}
```

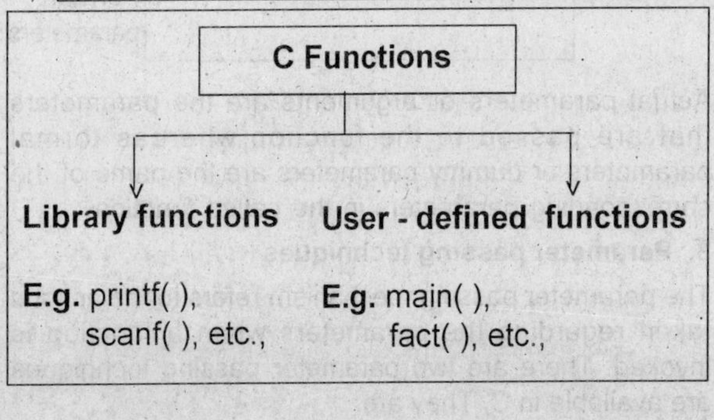

For the library functions, user needs to write the code whereas a user-defined function has to be developed by the user at the time of writing a program.

If a program is divided into functional parts, then each part may be independently coded and later combined into a single unit. These subprograms called 'functions' are much easier to understand, debug and test.

Purpose of Functions

- Using functions avoids rewriting the same code over and over.
- It is simpler to write a function that does a specific job.
- Programs with functions are compact.
- It facilitates top-down modular programming.
- Debugging the code is much simpler because errors can be localized and corrected.
- The length of the source program can be reduced.
- It increases program readability and helps documentation.
- Many other programs may use a function.

Characteristics of a Function

- A function can return only one value.
- Return statement indicates exit from the function and return to the point from where the function was invoked.
- When a function is not returning any value, void type can be used as return type.
- Parameter argument list is optional.
- A call to the function must end with a semicolon.
- A function can call any number of times.
- A C function cannot be defined in other function.
- C allows recursion i.e., a function can call itself.

Actual and Formal Parameters (arguments)

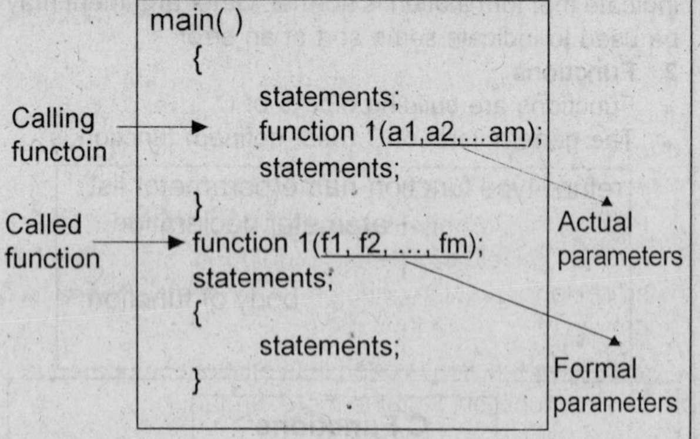

Actual parameters or arguments are the parameters that are passed to the function whereas formal parameters or dummy parameters are the name of the corresponding parameters in the called function.

3. Parameter passing techniques

The parameter passing mechanism refers to the actions taken regarding the parameters when a function is invoked. There are two parameter passing techniques are available in C. They are

a) Call by value

b) Call by Reference

1) Call by value : Whenever a portion of the program invokes a function with formal arguments, control will be transferred from the main to the calling function and the value of the actual argument is copied to the function. Within the function, the actual value copied from the calling portion of the program may be altered (or) changed. When the control is transferred back from the function to the calling portion of the program, the altered values are not transferred back. This type of passing formal arguments to a function is technically known as call by value.

Example : A program to exchange the contents of two variables using a call by value.

```
# inclue <stdio.h>
main ( )
{
        int x, y;
        void swap(int, int);
        x = 10;
        y = 20;
        printf ("values before swap( )\n");
        printf ("x = %d and y = %d", x, y);
        swap(x, y);   / * Call by value * /
        printf ("values after swap( ) \n");
        printf ("x = %d and y = %d", x, y);
}
void swap(int x, int y) / * values will not be swapped * /
{
        int temp;
        temp = x;
          x = y;
        y = temp;
}
```

Output of the above program

values before swap( )

x = 10 and y = 20

values after swap( )

x = 10 and y = 20

2) Call by Reference : When a function is called by a portion of a program, the address of the actual arguments are copied onto the formal arguments, though they may be referred by different variable names. The content of the variables that are altered within the function block are returned to the calling portion of a program in the altered form itself, as the formal and the actual arguments are referencing the same memory location (or) address. This is technically known as call by reference (or) call by address.

Ex : A program to exchange the contents of two variables using a call by reference.

```
# include <stdio.h>
main( )
{
        int x, y;
        void swap(int * x, int * y );
        x = 10;
        y = 20;
        printf ("values before swap( )\n");
        printf ("x = %d and y = %d \n", x, y);
        swap(&x, &y);  /* Call by reference */
        printf ("values after swap( ) \n");
        printf ("x = %d y = %d \n", x, y);
}
void swap(int * x, int * y) /* values will be swapped */
{
        int temp;
        temp = * x;
        * x = * y;
        * y = temp;
}
```

Output of the above program :

```
values before swap( )
x = 10 and y = 20
values after swap( )
x = 20 and y = 10
```

4. Recursion

Recursion is a process by which a function calls itself repeatedly, until some specified condition has been satisfied. The process is used for repetitive computations in which each action is stated in terms of a previous result. Many iteration (i.e.repetitive) problems can be written in this form.

In order to solve a problem recursively two conditions must be satisfied. First, the problem must be written in a recursive form, and second, the problem statement must include a stopping condition.

For example,

```
# include <stdio.h>
main ( )
{
void funct1();        /* function declaration */
        ---------------
funct1( );      /* function calling */
}
void funct1( )        /* function definition */
{
        ---------------
        ---------------
funct1( ); /* function calls recursively */
}
```

Ex : Suppose, for example we wish to calculate the factorial of a positive integer quantity. We would normally express this problem as $n! = 1 \times 2 \times 3 \times . . . \times n$, where n is the specified positive integer.

However, we can also express this problem in another way; by writing $n! = n \times (n - 1)!$. This is a recursive statement of the problem, in which the desired action is expressed in terms of a previous result. Also, we know that $1 ! = 1$ by definition. This last expression provides a stopping condition for the Recursion.

Ex : int factorial (int n)

```
{
        if (n < = 1)
        return (1);
        else
        return (n * factorial (n - 1));
}
```

The use of recursion is not necessary the best way to approach a problem, even though the problem definition may be recursive in nature. A non recursive implementation may be more efficient in terms of memory utilization and execution speed.

Mutual Recursion

Two or more functions, which invoke (call) each other, are called mutually recursive functions.

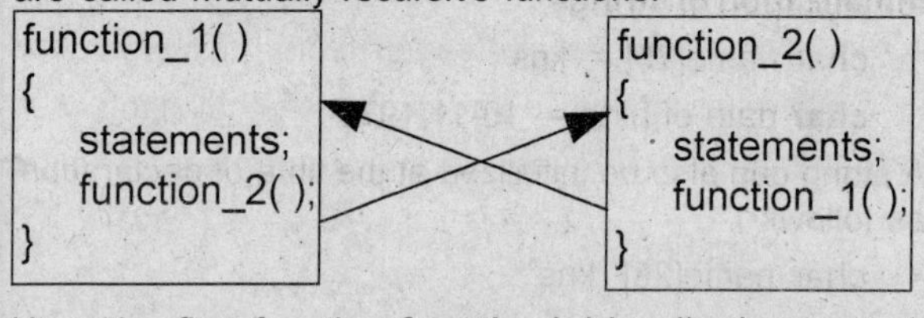

Here the first function **function1 ( )** calls the second function; **function2 ( )** which in turn calls the first function **function1 ( )** again. These two functions are therefore called **mutually recursive functions.**

- For certain problems, recursive solution straight forward and single as in case of factorial of a number, tower of Hanoi problem, finding GCD, finding fibonacci series, etc.,
- A recursive function is often less efficient compared to an iterative function. But it is more elegant.
- An iterative function is preferred when its recursive equivalent is complex.

5. Arrays

- An array is a collection of elements of the same data type.
- An array can be of single dimensional or multi-dimensional type.
- Single dimensinal array (vector) represents the

number of consecutive memory locations in which data are stored and each element can be accessed through an index.

Ex. int marks[100], char name[10]

- A multi-dimensional array can be considered as an array of single dimensional arrays

Ex. int matrix[10][10], char names[20][20]

- There is no restrriction on the number of dimensions in C but it is limited only by the amount of memory available to the program.

6. Strings

- A group of characters can be stored in a character array. These character arrays are called strings.
- To recongnize a character array should end with a null character ('\0')

Ex. the string student could be stored as

'S'	'T'	'U'	D'	'E'	'N'	'T'	'\0'

- The length of a string is the number of characters it contains excluding null character.
- So, to store a string, we need one more location than the length of the string.

The syntax of string declaration :

char string-name[size];

Where string-name is the name of a string and size is the maximum number of characters the string-name can contain.

Ex. char name[25];

char date-of-birth[10];

Initialization of strings

char name[25] = "kris";

char date-of-birth = "10-11-1976";

A string can also be initialized at the time of declaration as follows :

char name[25] "kris";

(or)

char name[25] = {'k', 'r', 'i', 's', 'h', 'n', 'a'};

(or)

char name[ ]="kris";

(or)

char name[ ]={'k', 'r', 'i', 's', 'h', 'n', 'a'};

Note:

1. When initialize a character array be listing its elements, we must explicitly specify the null terminator.

2. When a character using is assigned to a character arrays C adds the null character at the end of the string automatically.

Reading and Writing Strings

The scanf( ) and printf( ) functions are used with %s format specification to read and print a string respectively.

Ex. char name[25];

scanf("%s",name);

printf("%s",name);

Note: To read a string, ampersand (&) is not required before the string variable.

The **gets( )** function reads a string of characters terminates by a newline character ('\n') from the standard input device. The newline character will replace a null character ('\0'), thus creating a string.

Ex. gets(string);

Where string is the name of the string variable.

- The puts( ) function writes the character string is supplied as a parameter to the standard output device.

Ex. puts(string);

Where string contains the address of the string that is to be outputted.

String Handling Functions (string.h)

Function	Meaning
Strcat( )	String concatenate (Append one string to another)
Strncat( )	Appends first n characters of a string at the end of another
Strcmp( )	Compares two strings
Strncmp( )	Compares first n characters of two strings.
Strcpy( )	Copies a string into another.
Strncpy( )	Copies first n characters of one string into another string.
Strcmpi( )	Compare two strings without case sensitivity.
Strchr( )	Finds first occurrence of a given string in another
Strlen( )	Returns the length of the string not counting the null character
Strlwr( )	Converts a string to lowercase.
Strupr( )	Converts a string to uppercase.
Strdup( )	Copies a string into a newly created location.
Strstr( )	Finds first occurrence of a given string in a another string.
Setrev( )	Reverse a string.

Functions for Reading Single Character Variables

getch( ) and **getche( )** are input functions which return the character that has been most recently typed. The 'e' in getche( ) function means it echoes (displays) the character that you typed to the screen.

getchar( ) works similarly and echoes the character that you typed on the screen, but it requires enter key to be typed following the character that you typed.

OBJECTIVE QUESTIONS - I

1. Repeated execution of simple computations may cause compounding of **[AUCET-1999]**

   a) Syntax Errors b) Runtime Errors

   c) Round Off Errors d) Logical Errors.

2. Which of the following language features requires stack based storage allocation be used rather than static allocation: **[AUCET- 1999]**

   a) Reference parametersb) Integer-valued functions

   c) Two-dimensional arrowsd) Recursive procedures

3. The default parameter passing mechanism in C is: **[AUCET- 1999]**

   a) call by reference b) call by value result

   c) call by value d) both (a) and (b)

4. The statements that can be used to change the flow of control are: **[AUCET- 2000]**

a) break and continue b) if and switch

c) if and while d) do...while and switch

5. Consider the following program fragment:

```
if(a>b)
  if(b>c)
    k1;
  else
    k2;
```

k2 will be executed if

a) b>c b) a>b and b<=c

c) a<b and b>c d) b<=c

6. Break statement can be simulated by

a) return b) exit

c) goto d) Any of the above

7. Use of the "goto" statement is harmful because of

a) It makes it more difficult to modify existing program

b) It makes it more difficult to verify programs

c) It results in the compiler generating longer machine code

d) It makes programs more inefficient

8. When you pass an array as an argument to a function, what actually gets passed:

a) address of the array.

b) values of the elements of the array.

c) address of the first element of the array.

d) number of elements of the array.

9. If you don't initialize a static array, what will be the elements set to :

a) 0 b) an undetermined value.

c) a floating point number d) None.

10. The results returned by functions under value-result nd reference parameter passing conventions **[GATE-2002]**

a) Do not differ b) Differ in all cases

c) Differ in the presence of loops

d) may differ in the presence of exceptions.

11. When the break statement is encountered inside a loop, which one of the followng occurs: **[TCS]**

a) Control goes to the end of the program.

b) Control leaves the function that contains the loop.

c) Causes an exit from the innermost loop containing it.

d) Causes an exit from all the nested loop.

12. When the continue statement is executed within a loop the control goes to **[TCS]**

a) the next statement in the loop.

b) the top of the loop.

c) the statement immediately after the loop.

d) the beginning of the program.

13. What is the output of the following code:

```
if(a=0)
  printf("a is zero");
else
  printf("a is not zero");
```

a) a is zero b) a is not zero

c) nothing d) garbage

14. The following program fragment

```
int x [ 5 ] [ 5 ], i, j;
  for (i=0;i<5;++i)
    for ( j=0;j<5;j++)

      x [ i ] [ j ] = x [ j ] [ i ];
```

prints:

a) transposes the given matrix x.

b) makes the given matrix x,symmetric.

c) does'nt alter the matrix x.

d) None.

15. The following program

```
main( )
{
  int a=4;
    change(a);
    printf("%d",a);
}
change(a)
  int a;
{
  printf("%d",++a);
}
```

outputs:

a) 5 5 b) 4 5 c) 5 4 d) 4 4

16. If int s[5] is a one dimensional array of integers, which of the following refers to the third element in the array:

a) * (s+2) b) * (s+3) c) s+3 d) s+2

17. The following expression

```
k= (x>=y) ? x : y ;
```

is equivalent to

```
a) if (x>=y)    k = x;
   else         k = y;
b) if (x>y)     k = x;
if ( x == y)  k = y;
c) if ( x > = y)   k = y;
   else         k = x ;
```

18. All the elements in the array must be: **[DOEACC -1999]**

a) Initialized b) Defined

c) Initialized and defined d) None

19. Arrays can't be initialized if they are **[DOEACC -1999]**

a) Automatic b) External c) Static d) None

20. It is necessary do declare the type of a function in the calling program if **[DOEACC -1999]**

a) The function returns an integer

b) The function returns a non- integer value

c) The function is not defined in the same file

d) None

21. A collection of recursive functions is executed in a **[DOEACC -1999]**

a) Last in first out order b) First in first out order

c) Parallel fashion d) None

22. Consider the following program segment **[DOEACC -1999]**

```
if("India"= ="India")
   printf("Equal");
else
   printf("UnEqual");
```

what will be printed by the above code:

a) Equal b) UnEqual c) India

d) The program will cause run-time error

23. How many times will the following loop be executed: **[DOEACC -1999]**

```
ch='b';
while(ch>='a'&&ch<='z')
ch++;
```

a) 0 b) 1 c) 25 d) 26

24. Arrays can be initialized only if they are:

a) Automatic b) Static

c) External d) Both b and c

25. If you want to exchange two rows in a two-dimensional array, the fastest way is to : **[DOEACC -2001]**

a) Exchange the elements of the two rows

b) Exchange the addresses of each element in the two rows

c) Store the addresses of the rows in an array of pointers and exchange the pointers

d) None

26. If you do not initiate a static array, what will be the elemen s set to? **[DOEACC-2001]**

a) Zero b) An undetermined value

c) Floating point number d) None

27. Study the following statements: **[DOEACC- 2001]**

I. Continue statement may also be used in a for loop

II. Break statement may also be used in a if statement pick out the most correct answer from the following

a) Only statement (I) is correct

b) Only statement (II) is correct

c) Both statements are correct d) None

28. The library function exit( ) causes an exit from

a) The loop in which in which it occurs

b) The block in which it occurs

c) The function in whihc it occurs d) None

ANSWER KEY - I

1. c	**2.** d	**3.** c	**4.** a	**5.** b
6. d	**7.** d	**8.** c	**9.** a	**10.** b
11. c	**12.** b	**13.** b	**14.** a	**15.** c
16. a	**17.** a	**18.** b	**19.** ?	**20.** b
21. a	**22.** b	**23.** c	**24.** a	**25.** c
26. a	**27.** d	**28.** d		

OBJECTIVE QUESTIONS - II

1. Suppose an array x contains the integer values [30,21,-10,5,0,-1,-100,0,2,-6]. The output of the following program segment is: **[AUCET- 2001]**

```
for(i=1;i<10;i++)
{
  if(x[i]<0)
    continue;
  if(x[i]==0)
    break;
  printf("%d",x[i]);
}
```

a) 30 21 –10 5 0 –1 –100 0 2 6

b) –10 -1 -100 c) 21 5 d) 30 21 5

2. The output of the following program segment is : **[AUCET- 2001]**

```
int i,j,k;
i=10;
while(i>0)
{
 switch(i)
 { case8:case5:
      i=0;
      break;
 }
 printf("%d",i--);
}
```

a) 10 9 0 7 6 0 4 3 2 1 b) 0 0

c) 10 9 8 d) 10 9 0

3. Consider the following C-language program segment: **[AUCET- 2000]**

```
for(i=0;i<100;i++)
for( j=0;j<100;++j)
if(i+j<100)
break;
```

How many number of times the "if" statement will get executed:

a) 100 b) 1010 c) 5050 d) 1000

4. The output of the program is: **[AUCET- 1999]**

```
if(ch>='A' and ch<='Z')
ch=ch+'a'-'A';
```

a) upper case to lower case conversion
b) lower case to upper case conversion
c) character to integer conversion d) None

5. What will be the output of the following program **[DOEACC- 1993]**

```
main( )
{
  int i; i=200;
    if(i=5)
      printf("\n you entered 5");
    else
   printf("\n you entered something other than 5");
}
```

a) you entered 5
b) you entered something other than 5
c) 200 d) None

6. What will be the output of the following program : **[AUCET- 2002]**

```
main( )
{
 int i=2;
   switch(i)
   { case 1: printf("In case1 \n");
     case 2: printf("In case2\n");
     case 3: printf("In case3\n");
      default:printf("In default\n");
   }
}
```

a) In case2 In case3 In default
b) In case1 In case2 In case3 In default
c) In case2
d) None

7. What will be the output of the following program segment: **[DOEACC- 1993]**

```
main( )
{ int i=1;
   while(i<=10)
   { pirntf("%d",i);
     i=i+1;
   }
}
```

a) prints 1 to 10 b) prints 1 to 11
c) It's an idefinite loop d) None

8. What will be the output of the following program:

```
main( )
{ float a=12.25, b=13.65;
    if(a=b)
      printf("a and b are equal");
    else
 printf("a and b are notequal");
}
```

a) a and b are equal b) a and b are not equal
c) None

9. What will be the output of the following program:

```
main( )
{
  float a=0.7;
     if(a<0.7)
        printf("RAMESH");
 else
        printf("RAJA");
}
```

a) RAMESH b) RAJA c) None

10. In the below program, the output is: **[DOEACC- 1993]**

```
main( )
{
  int counter = 40;
     for(;counter;)
       counter--;
       printf("%d\n",counter);
}
```

a) 0 b) 40 c) –40 d) None

11. What is the output of the following program: **[DOEACC- 1994]**

```
#define ROWS 3
#define COLUMNS 4
int z[ROWS][COLUMNS]
            ={1,2,3,4,5,6,7,8,9,10,11,12}
main( )
{
 int a,b,c=999;
   for(a=0;a<ROWS;++a)
     for(b=0;b<COLUMNS;++b)
       if(z[a][b]<c)
         c=z[a][b];
         printf("%d",c);
}
```

a) 1 b) 12 c) 999 d) None

12. Consider the following program segment: **[DOEACC- 1994]**

```
main( )
{
  int fa;
    fa=fac(s);
    /* more program lines */
}
fac(num)
 int num;
{
   if(num==1)
    return 1;
  return(num * fac(num-1));
}
```

The above program segment
a) Assigns 5 to fa b) Assigns 120 to fa
c) Assigns1 to fa d) None

13. Consider the program segment: **[DOEACC- 1994]**

```
main( )
{
 int i,j=1;
   for(i=j-2;i<5;i++)
     j+=j;
     printf("The value of j is:%d\n",J);
}
```

The output of the above program segment is

a) 5 b) 12 c) 13 d) None

14. What is the output of the following program segment: **[DOEACC- 1994]**

```
main( )
{
 int val,x;
   val = 5;
   call_fun1(val);
   printf("The output is:%d\n",val);
}
call_fun1(val)
 int val;
{
   return(call_fun2(val * 2));
}
call_fun2(val)
 int val;
{
   val * =2;
   return(val);
}
```

a) 5 b) 10 c) 20 d) None

15. Consider the declaration

```
main( )
{
 float f1=0.4,f2=0.5;
   if(f2<0.5)
    if(f1<0.4)
printf("JAVA");
   else
     printf("ORACLE");
   else
     printf("VB");
}
```

a) JAVA b) ORACLE c) VB d) None

16. What will be the output of the following program:

```
main ( )
{
 int i=5,j=2;
 junk(i,j);
 printf("\n%d%d",i,j);
}
junk(int i, int j)
{
 i=i * i;
 j=j * j;
}
```

a) 5,2 b) 2,5 c) 5,5 d) None

17. Consider the following declaration of a two-dimensional array in C:

```
char a[100][100];
```

Assuming that the main memory is byte-addressable and that the array is stored starting from memory address 0, theaddress of a[40][50] is **[GATE-2002]**

a) 4040 b) 4050 c) 5040 d) 5050

18. Consider the following C function definition

```
int trial(int a, int b, int c)
{
  if((a>=b) && (c<b)) return b:
  else
   if(a>=b)    return trial(a,c,b);
  else
           return trial(b,a,c);
}
```

The function trial: **[GATE-1999]**

a) finds the maximum of a,b and c

b) finds the minimum of a,b and c

c) finds the middle number of a,b and c d) None

19. Consider the following code: **[TCS]**

```
if(number>=0)
    if(number>0)
     printf(" number is Positive \ n")
    else
     printf(" number is negative \ n")
```

What will be the output if number is equal to 0 :

a) Number is negative b) Number is positive

c) Both (a) and (b) d) None.

20. What will be the value of test after the following code is executed:

```
int m=10, n=-1,test=1;
if(m<15)
 if(n>1)
  test=2;
 else
  test=3;
```

a) 3 b) 4 c) 2 d) None

21. What is the O/P of the following code segment: **[TCS]**

```
int n=0;
int i=1;
  do
   {
     printf("%d",i);
     i++;
   }
        while (i<=n)
```

a) 0 b) 10 c) 1 d) 5

22. What will be the output of the following program segment: **[TCS]**

```
int x=1234;
int d,y=0;
  while (x>0)
  {
    d=x%10;
    x/=10;
    y=10 * y+d;
  }
  printf("%d\n",y);
```

a) 4 3 2 1 b) 1 2 3 4 c) 4 3 2 0 d) None

23. What is the output of the following code: **[TCS]**

```
int m=100;
  while(true)
  {
    if(m<10)
      break;
      m=m-10;
  }
  printf("%d",m);
```

a) 0 b) 2 c) 1 d) None

24. What is the output of the following code:

```
int m=100;
  while(true)
  {
    if(m<10)
      continue;
      m=m-10;
  }
  printf("%d",m);
```

a) 4 b) 3 c) 0 d) No output, infinite loop.

25. What is the output of the following code:

```
for(i=1;i<5;++i)
  if(i==3)
continue;
  else
   printf("%d",i);
```

a) 1 2 3 4 b) 1 2 4 c) 2 4 5 d) None

26. The following while loop will **[DOEACC -1998]**

```
int x=0;
  while(1)
  {
    x++;
  }
```

a) Run for only once b) Run for infinite no. of times
c) Not run at all d) Give compile time error

27. The following program **[DOEACC -1998]**

```
main( )

{
  int i,index=0;
   for(i=0 i<10;i++)
   {
     int i=0;
       while(i++<5)
         index++;
   }
}
```

a) will run for infinite number of times
b) will have index=5 at the end of the execution
c) will given compilation error
d) will have index=50 at the end of the execution

28. The following program **[DOEACC -1998]**

```
main( )
{
  int x=0;
   while(x<=50)
   for( ; ; )
    if(++x%50 = = 0)
     break;
    printf("x=%d",x);
}
```

a) will print x=50 b) will give compilation error
c) will give runtime error d) will print x=100

29. The following program **[DOEACC- 1998]**

```
main( )
{
  int a=12,b=154;
   printf("%d",f(a,b));
}
int f(int a,int b)
{
  if(a<b)
   return(f(b,a));
  if(b==0)
```

```
        return(a);
        return(f(b,a%b));
}
```

a) will print 1 b) will print 2

c) will give compilation error d) will give runtime error

30. Consider the following two functions **[DOEACC- 1998]**

```
int f1 (int a)
{
  int x=0;
    do
      x+=a--;
      while(a>0)
        return(x);
}
int f2(int a)
{
  int x=0;
    while(a>0)
      x+=a--;
      return(x);
}
```

state which of the following statement(s) is TRUE.

a) f1( ) and f2( ) are identical functions

b) The value of each of f1(10) and f2(10) is 55

c) The value of of f1(-1) and f2(1) is true

d) All of the above statements are false

31. Consider the following program: **[DOEACC -1998]**

```
main( )
{
  int i,j,x=0;
    for(i=0;i<5;i++)
      for( j=i;j>0;j--)
        x+=i+j-1;
        printf("X=%d",x);
}
```

The output of the above program will be:

a) x=18 b) x=39 c) x=40 d) None

32. Study the following program to reverse any given string of maximum length 80. **[DOEACC -2001]**

```
void main( )
{
  char ch,st[80];
  int j,len;
    gets(st);
    len=strlen(st);
    for( j=0;j<len/2;j++)
    {
      ch=st[ j];
      ......statement s1 ......
      st[ j]=ch;
    }
}
```

To complete the above program , choose the appropriate expression that is required as statement s1:

a) st[j]=st[len-j-1] b) st[j]=st[len-j+1]

c) st[j]=st[len+j+1] d) None

33. Consider the following c program **[DOEACC- 2001]**

```
main( )
{
  int a=7,b=5;
  switch(a/a%b)
  {
    case 1 :    a=a-b;
    case 2 :    a=a+b;
    case 3 :    a=a * b;
    case 4 :    a=a/b;
    default:    a=a;
  }
}
```

on the execution of the above program, what will be the value of the variable a:

a) 2 b) 5 c) 7 d) None

34. Consider the following function **[DOEACC- 2001]**

```
int f(int x)
{
  int i,j=0;
    if(x==0)
      return 1;
    for(i=0;i<x/2;i++)
      j+=f(i);
      return j;
}
```

what will be the result of invoking f(4)?

a) 0 b) 1 c) 2 d) 3

35. How many times is the word "done" printed by the following program segment ? **[Hyd.Univ-2001]**

```
for(i=1; i<=2; i++)
  for( j=1; j<=2 * i; j++)
    for( j=1; j<=2 * i; j++)
      printf("done\n");
```

a) 6 b) 9 c) 10 d) 12

36. What happens if the user types 'a' for marks the following code segment ? Assume that j and marks are both declared as int. **[Hyd.Univ-2000]**

```
marks=0
  while(marks!=-1)
  {
```

```
    printf("Enter Marks :");
    j = scanf("%d",&marks);
}
```

a) j = 0, marks = 0 and program waits for next input from user

b) j = -1, marks = 0 and program waits for next input from user

c) j = 1, marks = 97 and program goes into an infinite loop

d) j = 0 and the program goes into an infinite loop

37. Which of the following is FALSE about the C language? **[Hyd.Univ-2000]**

a) Parameters are passed only by value

b) Printf() is not a part of the language but a part of a library of input/output functions

c) Functions may be defined within other functions

d) Any function including main( ) can call itself recursively

38. Consider the following C program segment: **[Hyd.Univ-2000]**

```
if (a < L)
    a = 3 * a;
if (a < M)
    a = 2 * a;
if (a < N)
    a = 6 * a;
```

L, M and N are integer constants such that L<M<N. what is the output if L<a<M prior to entry into this code segment?

a) 3a b) 6a c) 12a d) 36a

39. Which of the following statements is FALSE about C and Pascal? **[Hyd.Univ-2000]**

a) Pascal has more ways of passing parameters than C

b) Pascal and C both distinguish between functions and sub-programs

c) Pascal allows local, Non -local and global variables while C permits only local and global variables

d) The case statement in Pascal is more flexible than the switch statement in C

40. Which of the following statements is TRUE about 'C'? **[Hyd.Univ-1999]**

a) Parameters are passed both by value and by reference

b) Parameters are passed only by value

c) Parameter are passing by value-result is used for arrays

d) Parameter passing by reference is used for arrays

Below two questions are based on the following

Given A and B are two symmetric N × N matrices, and the C program segment:

```
for ( j=0;j<n;j++)
  for (k=j;k<n;k++)
  { D = 0;
    for (l=0;l<n;l++)
      D=D+A[ j][l] * B[l][k]; ....(* )
  {
      C[ j][k]=D
      C[k][ j]=D;
  }
}
```

41. What does the above program segment do? **[Hyd.Univ-1997]**

a) Finds A^T and B^T

b) Computes the product A ´ B and store it in C

c) Computes C = A^T ´ B d) Computes C = B^T ´ A

42. How many times does the statements marked (*) get executed **[Hyd.Univ-1997]**

a) $O(N^3)$ b) $O(N^2)$ c) $O(N^4)$ d) O(N)

43. If in a language, there is a construct: **[Hyd.Univ-1997]**

```
if B1 then S1
   B2 then S2
   Bn then Sn
else Sn+1
```

This mans "if B1 then S1 else if B2 then S2 ... else Sn + 1. What is the equivalent in Pascal for the following:

```
if B1 then S1
   B2 then S2
else S3
```

a) case <condition> of	b) case <condition> of
B1 : S1;	B1 : S1;
B2 : S2;	B2 : S2;
end ;	end ;
S3;	if (not(B1) and not (B2))
	then S3;

c) both A and B above are correct

d) we cannot write an equivalent statement in Pascal

44. In a C program loop, if break and continue are used which is (are) TRUE?

I. break and continue have the same effect

II. break causes an immediate exit from the loop, resuming at the statement following loop

III.continue causes a halt at the continue and return immediately to the loop **[Hyd.Univ-1996]**

a) I only b) II only c) III only d) II and III only

45. Consider the following C program with input as arg. **[Hyd.Univ-1995]**

```
long xxxx ( arg )
```

```
long arg;
{
long fox =1, j;
for ( j = 1; j<arg; j++)
fox = fox * j;
return fox;
}
```

For a given input arg =n what does it compute?

a) n! b) n^2 c) n (n + 1) d) None of the above

46. Consider the following C program segment **[Hyd.Univ-1995]**

```
if ( j == 1);
( j = 0)
```

Assume that just before the segment, j has value 1. What will be its value after the segment?

a) j = 12 b) j = 1 c) j = 0 d) There is some error in the program

47. Which of the following is syntactically invalid in C ? **[Hyd.Univ-1995]**

a) do this-same-statement while(1);

b) for (i =0; i = -1; i++);

c) if (cond) var1 = else var2 = 5

d) while (and = true) and = false

48. If the value of s is 1, which sequence of statements is executed in the following C segment **[Hyd.Univ-1994]**

```
switch(s)
{
  case 0 :
      statement 1;
      statement 2;
  case 2 :
      statement 3;
      statement 4;
  case 1 :
      statement 5;
      statement 6;
  case 4 :
      statement 7;
}
```

a) 5, 6, 7, 3, 4 b) 3, 4, 5, 6, 7

c) 5, 6 d) 5, 6, 7

49. What does the following program print? **[Hyd.Univ-1993]**

```
#include<stdio.h>
char input[ ]= "SSSWILTECH\1\11W\1WALLMP1";
main ( )
{
int i, c;
for (i=2; (c=input[i]) != '\0'; i + +)
{
switch(c)
{
  case 'a' : putchar('i');
          continue;
  case'1': break;
  case 1 : while((c=input[++ i])!='\1'&& c!= '\0');
  case 9 : putchar('S');
  case'E':
  case 'L' :
          continue;
  default: putchar(c);
          continue;
  }
  putchar(' ');
}
  putchar('\n');
}
```

a) SWITCH SWAMP b) SWILT WALL
c) WILTECH SWAMP d) SWITCH WALL

ANSWER KEY - II

1. c	**2.** d	**3.** a	**4.** a	**5.** a
6. a	**7.** a	**8.** a	**9.** a	**10.** a
11. b	**12.** b	**13.** d	**14.** c	**15.** c
16. a	**17.** b	**18.** c	**19.** a	**20.** a
21. c	**22.** a	**23.** a	**24.** d	**25.** b
26. b	**27.** a	**28.** d	**29.** d	**30.** b
31. d	**32.** d	**33.** c	**34.** b	**35.** a
36. c	**37.** a	**38.** c	**39.** c	**40.** a
41. ?	**42.** ?	**43.** d	**44.** d	**45.** d
46. c	**47.** c	**48.** d	**49.** a	

III. ADVANCED PROGRAMMING IN 'C'

- Scope of Variables
- Storage classes
- Structures and unions
- Pointers

1. Scope of Variables

Local variables : These are also called normal local variables. These are declared in functions. These variables retains its values in between the function execution. (i.e.) Memory is allocated when the function starts executing and released when the function returns. These variables are lost once they are used in a function. These are needed no longer.

For example,

```
void funct (int i, int j)
{
Int k,m; /* local variables */
----
---
/* body of the function */
}
```

The integer variables k and m are defined within a function block of the funct ( ). All the variables to be used within a function block must be either defined at the begining of the block itself (or) before using in a statement. Local variables are referred only the particular part of a block (or) a function.

Global variables : Global variables are variables defined outside the main function block. These variabels are referred by the same data type and by the same name through out the program in both calling portion of a program and in the function block.

Ex : int x, y = 4 ; / * **global declarations** * /

```
main ( )
{
void function ( ) ;
x = 10 ;
----
-----
function ( );
}
function ( )
{
int sum ;
sum = x + y
-----
----
}
```

2. Storage classes in C

The storage class of a variable dictates how, when and where storage will be allocated for the variable. The different storage classes available are :

1) Automatic variables 2) Register variables
3) Extern variables 4) Static variables

1) **Automatic variables :** Automatic variables are declared inside a function in which they are to be utilized. They are created when the function is called and destroyed automatically when the function is exited, hence the name automatic. Automatic variables are therefore private (or local) to the function in which they are declared. Because of this property, automatic variables are also referred to as local (or) internal variables.

A Variable declared inside a function without storage class specification is, by default, an automatic variable. For example, the storage class of the variable number in the example below is automatic.

```
main ( )
{ int number;
  ........
  ........
}
```

We may also use the keyword auto to declare automatic variables explicitly.

```
main ( )
{ auto int number;
  ........
  ........
}
```

One important feature of automatic variables is that their value cannot be changed accidentally by what happens in some other function in the program.

2) **Register variables :** We can tell the compiler that a variable should be kept in one of the machine's registers instead of keeping in the memory (where normal variables are stored). Since a register access is much faster than a memory access. Keeping the frequently accessed variables (ex : loop control variables) in the register will lead to faster execution of programs.

This is done as follows :

```
register int count;
```

Since only a few variables can be placed in the register, it is important to carefully select the variables for this purpose. However, C will automatically convert register variables into nonregister variables once the limit is reached.

3) **Extern variables :** Variables that are both alive and active throughout the entire program are known as external variables. They are also known as global

variables. Unlike local variables, global variables can be accessed by any function in the program.

External variables are declared outside a function.

For example, the external declaration of integer number and float length might appear as :

```
int number;
float length = 7.5;
main ( )
{
  ------
  ------
}
function1 ( )
{
  ------
  
}
function2 ( )
{
  ------
  ------
}
```

The variables number and length are available for use in all the three functions. In case a local variable and global variable have the same name, the local variable will have precedence over the global one in the function where it is declared..

4) Static variables : As the name suggests, the value of static variables will be retained until the end of the program. A static variable may be any one of the two types.

a) Internal b) External

Internal static variables are declared inside a function. Their scope is limited to the function in which they are declared. They are used to retain values between calls. Thus the function is said to retain the information in static variable permanently through the execution of the program.

As external static variable is declared outside of all functions and is available to all the functions in that program. This external static variable is available only with in the file where it is declared, while external can be accessed by other files.

```
#include<studio.h>
main( )
{
        increment ( );
        increment ( );
        increment ( );
}
increment ( )
{
        static int i = 1;
        printf ("%\n",i );
        i = i + 1;
}
```

The output of the above program is:

1

2

3

3. Structures and Unions

- A structure is a collection of variables of different data types that are logically grouped together and referenced under one name i.e., a structure provides a compound data type
- We can build more complex data types using structures.

Syntax:

```
struct structure-name (or tag)
{
        datatype 1        variable 1;
        datatype 2        variable 2;
        -                 -
        -                 -
        -                 -
        datatype n        variable n;
} structure-variables;
```

Here, **struct** is a keyword, which tells the compiler that it is a structure. Tag is an identifier, which represents a structure, will follow variables (or members or elements).

- Structures are used to hold related data belonging to different data types.

```
Estruct book
{
  char bookname[20];
  float price;
  int pages;
};
struct b1, b2, b3, b4;
```

```
struct book
{
  char bookname[20];
  float price;
  int pages;
} b1, b2, b3, b4;
```

Here the variables b1, b2, b3, b4 are of the data type struct book.

- When structure book is declared, it only defines the data type book. At this stage the compiler will not allocate any memory.

- The memory is allocated actually when variables of that type are declared, in this example b1, b2, b3 and b4.
- A member of a structure can be referenced by using dot operator (.) as follows.

Strucrture-Variable-name. Member-name

Ex. To refer the price of b1 structure is b1.price

Unions

Union is similar to the structure i.e., it contain members of different data types.

Difference between structure and Union

- The members of a structure occupy separate areas in memory while members of a union occupy the same area in memory.
- In general, the **Unions** are used to give the memory area occupied by the program.
- Note that Union declaration has the same form as the structure declaration except the keyword **union.**
- In **Unions** the variables will have only one member at any time. Hence, any reference to the wrong type of data will produce meaningless results.
- The amount of storage space occupied for a union variable is the amount of storage required for the largest member of the union.
- All union members are stored in the same memory space, with the same starting address, but not at the same time.
- Unions are used when all members need not exist in the memory at thes same time.

4. Pointers

A pointer is a simple data type which contains the address of a variable it is pointing to.

Pointers are used for

- Accessing variables.
- Accessing arrays.
- Passing parameters (arguments) to a function when the original parameter (argument) needs to be modified.
- Passing arrays to functions.
- Creation of linked lists.
- Dynamic storage allocation and hence reducing the overall memory requirement.
- Reducing the size and complexity of the program.
- Increasing the speed and efficiency of execution.
- An ampersand(&)is called the address operator inC
- &x gives the address of variable x.
- An asterisk (∗) is called indirection operator in C.
- An asterisk (∗) followed by a variable name of a data type represents a pointer to that data type.
- The pointer to void can point to any data type.

OBJECTIVE QUESTIONS - I

1. For "C" programming language, which of the following statements is (are) TRUE:

a) Constant expressions are evaluated at compile time.

b) String constants can be concatenated at compile time.

c) Size of array should be known at compile time.

d) All of the above.

2. Scope of a Register variable in C-language is **[AUCET- 2001]**

a) Within the CPU

b) Within the function in which it is defined

c) Within the block in which it is defined

d) Within the function "main"

3. Choose the correct answer: **[AUCET- 1998]**

a) It is possible for a function to access variable that is local to another function

b) Two local variables may have the same name

c) The scope of a local variable should be a function

d) None of these

4. The parameter passing mechanism for an array is: **[AUCET- 1998]**

a) Call by value b) Call by value result

c) Call by reference d) None

5. In the C language **[GATE-2002]**

a) At most one activation record exists between the current activation record and the activation record for the main.

b) The number of activation records between the current activation record and the activation record for the main depends on the actual function calling sequence.

c) The visibility of global variables depends on the actual function calling sequence.

d) Recursion requires the activation record for the recursive function to be saved on a different stack before the recursive function can be called.

6. Consider int ∗ a [100] ;Here a is: **[AUCET-2001]**

a) Improperly declared.

b) A pointer to an array of integers.

c) A pointer to an array of pointers.

d) An integer variable.

7. Consider the following declaration **[AUCET-2001]**

```
int * a, * b;
a=&10; b=&20;
a=b; b=a;
```

The above program segment results in

a) exchange of values 10 and 20

b) syntax error

c) the pointer to value 10 is overwritten

8. The statement int * f [4]( ) ; declares: **[AUCET-2001]**

a) a pointer to 4th location of the array f.

b) an array of 4 functions that return integer pointers.

c) generates a syntax error

d) a function that returns a pointer to an array.

9. Consider the following C-declarations:

satic int a[10]={10,9,8,7,6,5,4,3,2,1};

What is the value of * (a+3)+4 is : **[AUCET-000]**

a) 10 b) 11 c) 12 d) 14

10. Declare an array of P pointers to functions returning to pointers to functions returning pointers to characters (in C-program): **[AUCET-1999]**

a) char(+(+o[p])) b) char(-(+o[p]))

c) both a) and b) d) None

11. Pointers are of **[AUCET-1999]**

a) integer data type b) character data type

c) integers d) None

12. Invalid operation on a pointer x in C is: **[AUCET-2001]**

a) ++x b) x-- c) !(x) d) * x

13. Pick the correct answer(s) If a is one dimensional array, then

a) &a[i] is same as a+1 b) * (a+i) is same as * (&a[i])

c) * (a+i) is same as a[i] d) * (a+i) is same as * a+i

14. The following program fragment

```
int * a;
* a=7;
```

a) assigns 7 to a b) results in compilation error

c) results in runtime d) is a dangling reference

15. The declaration int(*p)[5]; means

a) p is a one dimensional array of size 5,of pointer to integer

b) p is a pointer to a 5 element integer array.

c) The same as int * p[5]; d) None.

16. Consider the following C program fragment

```
main( )
{
  static a,b;
      while(a>b++);
}
```

what will be the value of a and b on the execution of the above program? **[AUCET-2001]**

a) a=0,b=1 b) a=0,b=0 c) a=1,b=1 d) None

17. Consider the following program fragment **[DOEACC- 1998]**

```
main( )
{
  int x[10][20];
     scanf("%d",( * (x+2)+5));
}
```

The value will be read into

a) x[2][5] b) x[5][2] c) x[7][0] d) None

18. Consider the following variable declaration

```
union un{
     int x;
     float y;
     char z;
     }u;
```

If the size of the integer , floating point and character variables are 2 bytes , 4 bytes and 1 byte respectively then the size of the variable u is? **[DOEACC- 1998]**

a) 2 bytes b) 4 bytes c) 7 bytes d) None

19. Consider the following statements **[DOEACC- 2002]**

```
int a[5]={2,3,4,1,9};
int * pb=a;
int i=4;
```

which of the following is not a valid statement in C is:

a) * (pb+3) =9 b) pb[3]=5

c) pb=pb+2 d) pb=pb * 1

ANSWER KEY - I

1. d	2. c	3. c	4. c	5. a
6. b	7. c	8.b	9. b	10. d
11. c	12. c	13. c	14. c	15. b
16. a	17. a	18. b	19. d	

OBJECTIVE QUESTIONS - II

1. What will be the output of the following program:

```
main ( )
{
   increment( );
   increment( );
   increment( );
}
increment( )
{
   auto int i=1;
       printf("%d",i);
       i=i+1;
}
```

a) 1 1 1 b) 1 2 3 c) 1 2 2 d) None

2. What will be the output of the following program:

```
main ( )
{
   increment( );
   increment( );
   increment( );
}
increment( )
{
   static int i=1;
```

```
        printf("%d",i);
        i=i+1;
    }
```
a) 1 1 1 b) 1 2 3 c) 2 2 2 d) None

3. What will be the output of the following program:
```
int x=10;
main( )
{
  int x=20;
    printf("%d",x);
    display( );
}
display( )
{
    printf("%d",x);
}
```
a) 20 10 b) 10 20 c) 10 10 d) 20 20

4. The value of j at the end of the execution of the following C program **[GATE-2000]**
```
int incr(int i)
{
   static int count=0;
        count=count+i;
        return(count);
}
main( )
{
   int i,j;
     for(i=0;i<4;i++)
        j=incr(i);
}
```
is:

a) 4 b) 6 c) 7 d) 10

5. The following program
```
main( )
{
   int abc( );
     abc( );
     (* abc)( );
}
int abc( )
{
     printf("come");
}
```
a) results in a compilation error.

b) prints comecome.

c) results in a runtime. d) prints come come.

6. Consider the following C-langauge declarations: **[AUCET-2001]**
```
union address_type
{
   char hostel_address[50];
   char home_address[60];
}
struct student_type
{
     long int roll_no;
char name[32];
char hostler;
union address_type address;
}
```
The storage requirement for a structure identifier of student_type is(in bytes) :

a) 98 b) 100 c) 101 d) 151

7. The output of the following program is:
```
main( )
{ int x,y,temp, * ptr1, * ptr2;
    x=23;    y=37;
    ptr1=&x; ptr2=&y;
    temp= * ptr2;
    * ptr1= * ptr2;
    * ptr2=temp;
printf("x is %d while y is %d",x,y);
}
```
a) 23,37 b) 37,23 c) None

8. What is the output of the following program: **[AUCET-1999]**
```
main( )
{
   int i=2,j=5;
     function(&i,j);
     printf("i=%d, j=%d \n",i,j);
}
function(x,y)
  int * x,y;
{
     y+=10;
     x=10+y;
}
```
a) 25,15 b) 25,5 c) 2,5 d) 25,15

9. Consider the following program: **[AUCET-1999]**
```
int one_d[ ]={1,2,3}
main( )
{
   int * ptr;
     ptr=one_d;
     ptr+=3;    /* statement a */
     printf("%d\n", * ptr);
}
```
After the statement A is executed the new address of Ptr will be :

a) 3 bytes more than the old value.

b) 21 bytes more than the old value.

c) 6 bytes more than the old value.

d) (sizeof(integer) × 3) bytes more than old value.

10. Consider the following three C functions:**[GATE-2001]**
```
[p1]   int * g(void)
       {
         int x=10;
```

```
        return(&x);
        }
[p2]    int * g(void)
        {
          int * px;
          * px=10;
            return px;
        }
[p3]    int * g(void)
        {
            int * px;
px=(int * )malloc (sizeof(int));
            * px=10;
            return px;
        }
```

Which of the above three functions are likely to cause problems with pointers:

a) only p3   b) only p1 and p3
c) only p1 and p2   d) p1,p2 and p3

11. The following C declarations **[GATE-2000]**

```
struct node
{  int i;
   float j;
};
struct node * s[10];
```

define s to be

a) An array, each element of which is a pointer to a structure of type node

b) A structure of 2 fields, eachafield being a pointer to an array of 10 elements.

c) A structure of 3 fields: an integer , a float , and an array of 10 elements.

d) An array , each element of which is a structure of type node.

12. Consider the following C-declaration **[GATE-2000]**

```
struct
  { short s[5];
    union
    { float y;
      long z;
    }u;
  }t;
```

Asume that objects of the type short, float and long occupy 2 bytes , 4 bytes and 8bytes respectively The memory requirments for variable t, ignoring alignment considerations, is

a) 22 bytes   b) 14 bytes
c) 18 bytes   d) 10 bytes.

13. Consider the following program: **[DOEACC-1993]**

```
char oadd(char c)
{
  return &c;
}
void ost(char * cad,char val)
{
  * cad=val;
}
main( )
{
  char a='A', b='B';
     ost(oadd(a),b);
  printf("value of a=%c\n",a);
}
```

which of the following is TRUE for the above program:

a) this program contains a syntax error
b) program prints "value of a=A"
c) program prints "value of a =B"   d) None

14. Consider the following program segment: **[DOEACC-1993]**

```
main ( )
{
  int buf[5];
  int * btr=&buf[2];
    btr[-1]=20;
    btr[-2]=10;
    btr[-1]=0;
```

the above program segment, the output is:

a) buf[1]=20   b) buf[1]=10
c) buf[1]=0   d) None

15. Consider the program segment: **[DOEACC-1994]**

```
main( )
{
   char c_arr[]={'A','B','C','D'}, * ptr;
     * ptr=c_arry;
        * ++ptr= * (c_arry+2);
     printf("c=%c", * ptr+1);
}
```

The output of the above program segment is:

a) A   b) D   c) C   d) None

16. Consider the program segment: **[DOEACC-1994]**

```
main ( )
{
  int n=2, * ptr;
    ptr=&n;
    n * =3;
    printf("%d\n", * ptr * * ptr);
}
```

The above program segment prints

a) 4   b) 6   c) 36   d) None

17. Consider the following program segment:

```
func (a,b)
 int a,b;
{
 return(a=(a==b));
}
main( )
{
 int process( ),func( );
printf("Thevalueofprocess:%d\n",process(func,3,4));
}
```

```
process (pf,val1,val2)
  int ( * pf)( );
  int val1,val2;
  {
     return(( * pf)(val1,val2));
  }
```

The above program segment prints:

a) 0 b) 3 c) 4 d) None

18. What will be the value of 'sum' after the following program is executed? **[DOEACC -1998]**

```
main( )
{
  int sum=0,i;
for(i=0;i<10;i++)
      sum=f( );
}
int f( )
{
  static int a=0;
  int i;
for(i=0;i<5;i++)
          a++;
          return(a);
}
```

a) 5 b) 0 c) 50 d) None

19. What is the output of the following program: **[DOEACC -1999]**

```
main( )
{
  int a=0;
    while(a<5)
    {
      sum(a);
      a++;
    }
}
void sum(int a)
{
  static s;
    printf("%d ",s+a);
    s++;
}
```

a) 0 2 4 6 8 b) 0 1 2 3 4
c) 1 3 5 7 9 d) 1 2 3 4 5

20. Consider the following program **[DOEACC- 2002]**

```
int f (int x)
{
  static int y;
      y+=x;
      return(y);
}
main( )
{
  int a,i;
    for(i=0;i<6;i++)
          a=f(i);
          printf("%d",a);
}
```

what would be printed if the above program is run?

a) 6 b) 8 c) 10 d) 15

21. The following program **[DOEACC- 1998]**

```
main( )
  {
    struct s {
          int x;
          float y;
        }
          s1={5,2.5};
  struct t{
        int x;
        float y;
      }
        t1;
  } t1 = s1;
```

a) will produce compilation error
b) will give runtime error
c) will give warning d) will make t1.x=5,t1.y=2.5

22. What is the missing statement in the following function which copies string t into string **[DOEACC- 1998]**

```
void strcpy(char * s,char * t)
    {
      while( * t!='\0')
          / * missing statement * /
        * s='\0';
    }
```

a) s=t b) * s++= * t++
c) (* s)++=(* t)++ d) None

23. Which of the following statements is TRUE after execution of the following program? **[DOEACC- 1999]**

```
int a[10],i, * p;
  a[0]=1;
  a[1]=2;
  p=a;
  ( * p)++;
```

a) The value of a[0] will be 2
b) The value of a[1] will be 2
c) The value of a[1] will be 3 d) None

24. Consider the following program **[DOEACC- 1999]**

```
main( )
  {
    int  * iptr,i;
        i=12;
      * iptr=i * i;
        ++i;
      printf("%d%d\n",i, * iptr);
  }
```

what will be printed by the above code:

a) 12 144 b) 13 144 c) 13 0 d) None

25. What will be the value of u1 and u2 ? **[DOEACC- 2000]**

```
int u1,u2;
int v=3;
```

```
int * pv;
    pv=&v;
    u1=2 * (v+5);
    u2=2 * ( * pv+5);
```

a) u1=16,u2=16 b) u1=8, u2 = 16
c) u1=16, u2=3 d) u1=8, u2=3

26. Consider the following code

```
main( )
{
   char * p=NULL;
      strcpy(p," ");
      printf("%s\n",p);
}
```

what will be printed if the above program is run?

a) NULL b) Undefined
c) Compiler can't recognize " " in the strcpy function
d) Run-time error as "p" has not been allocated

27. In the following segment of c code how much of space is allocated for the union variable if integers need 2 bytes, characters take 1 byte and doubles take 8 bytes on a computer system? **[Hyd.Univ-2001]**

```
union u { struct
      {
         int n;
         char ar[10];
      } s;
   double d;
   char str[100];
      } x;
```

a) 100 b) 120 c) 20 d) 10

28. Complete the statement: "Static variables in C are ...." **[Hyd.Univ-2001]**

a) are automatically intialised to zero at compile time
b) are local to the function in which they are declared
c) retain values from one function call to the next
d) all of the above are true

29. Let variable represent data item and Pv be the pointer to v. Which one of the following indicates the value stored in v? **[Hyd.Univ-1996]**

a) &v b) * Pv c) * v d) &Pv

30. What is the output of the following sample C program: **[Hyd.Univ-1996]**

```
#include<stdio.h>
main( )
{
   int v = 3;
   int * pv;
      pv =&v;
      * pv = 0;
printf("\n\n * pv =%d v = % d ", * pv, v);
}
```

a) * pv = 3, v = 3 b) * pv = 1, v = 1
c) * pv = 0, v = 3 d) * pv = 0, v = 0

31. What will be the output of the following C program: **[Hyd.Univ-1996]**

```
#include<stdio.h>
   main( )
   {
   int    a, b = 0;
static int c[10]={1,2,3,4,5,6,7,8,9,0}
   for (a =0; a<10; ++a)
      if((a%2)==0)b+=c[a];
         printf("%d",b);
   }
```

a) 20 b) 25 c) 45 d) 12

32. The following program prints **[Hyd.Univ-1994]**

```
main ( )
{
   char name[30] = 'RAMA';
   char * P;
      for(P = name ; P; P++)
         printf("%c", * P);
      printf("\n");
}
```

a) RAMA and 26 other characters and stops
b) RAMA and 25 other characters and stops
c) RAMA and stops
d) RAMA and continues to print other characters

33. Out of the two C functions **[Hyd.Univ-1994]**

```
I. void abcd(s1, s2)
      char * s1, s2;
   {
   while ( * s1 + + = * s2 + +);
   }
II.void abcd (s1, s2)
      char * s1, s2;
   {
         if (s1 && s2)
      {
         do
         {
   if ( * s1 + + = * s2 + +);
      } while ( * s2 != '\0');
   }  }
```

a) I is syntactically wrong
b) II is syntactically wrong
c) both are syntactically wrong
d) I and II given the same result

ANSWER KEY - II

1. a	**2.** b	**3.** a	**4.** d	**5.** b
6. a	**7.** a	**8.** c	**9.** d	**10.** d
11. a	**12.** d	**13.** a	**14.** c	**15.** b
16. c	**17.** a	**18.** c	**19.** a	**20.** d
21. d	**22.** b	**23.** a	**24.** b	**25.** a
26. a	**27.** a	**28.** d	**29.** b	**30.** d
31. b	**32.** a	**33.** c		

IV. LOW LEVEL PROGRAMMING IN 'C'

- User - Defined Data Types
- Enumerations
- Bitwise Operators
- Dynamic Memory Allocation
- PreProcessor Directives
- Command Line Arguments

1. User-defined Data Types

The typedef feature allows users to define new data types that are equivalent to existing data types. Once a user-defined data type has been established, then new variables, arrays, Structures and so on, can be declared in terms of this new data type.

The general format of the user defined data types

typedef	datatype	newtype

Where typedef is a keyword for declaring the new data items and data type is an existing data type being converted to the new name.

For example,

```
typedef int integer;
typedef float real;
integer i, j ;
real a , b ;
```

Where the declaration is same

```
int i,j;
float a,b;
```

The typedef is used in a program to make it readable and portable.

2. Enumerations : An enumeration data type is a set of values repragented by identifiers called enumeration constants. The enumeration constants are specified when the type is defined.

The general format of the enumeration data types is

```
enum user- defined-name {
member - 1;
member - 2;
------
------
member-n;
};
```

Where enum is a keyword for defining the enumeration data type and the braces are essential. The memebers of the enumeration data type such as member-1, member-2, and member-n are the individual identifiers. Once the enumeration data type is defined, it can be declared in the following ways:

storage -class enum user-defined-name V1, V2, ...Vn ;

Where the storage class is optional.

The enumerated variables V1, V2, ...Vn can only have one of the values member-1, member-2,....,member-n.

The assignments of the following types are valid:

```
V1 = member - 3;
V5 = member - 1;
```

For example, following are valid enumeration data type declarations :

```
enum day {Monday, Tuesday, ...., Sunday} ;
enum day week-st, week-end;
week -st = Monday
week - end = Sunday;
```

The compiler automatically assigns integer digits begining with 0 to all the enumeration constants. That is, the enumeration constant member-1 is assigned 0, member-2 is assigned 1, and so on. However, the automatic assignments can be overridden by assigning values explicitly to the enumeration constants.

For example ,

```
enum day {Monday = 0, Tuesday = 1, ...., Sunday};
```

Here, the constant monday is assigned the value of 0, Tuesday is assigned the value of 1, remaining constants are assigned values that increase successively by 1.

The definition and declaration of enumerated variables can be combined in one statement.

For example ,

```
enum day {Monday, Tuesday,....., Sunday} week-st, week-end;
```

3. Bitwise operators : One of C's powerful features is a set of bit manipulation operators. These permit the programmer to access and manipulate individual bits with in a piece of data. The various Bitwise operators available in C are shown below table 1.

Operator	Meaning
~	One's complement
>>	Right shift
<<	Left shift
&	Bitwise AND
/	Bitwise OR
^	Bitwise XOR (Exclusive OR)

Table 1 : Bitwise operators

These operators can operate upon ints and chars but not on floats and doubles.

a) Bitwise complement : The operator ~ is called the one's complement operator, or the boitwise complement operator. It inverts the bit string repragentation of its agrument; the 0's becomes 1's and the 1's becomes 0's

For example, the declaration

int a = 65;

The binary repragentation of a is

00000000 01000001

The expression ~ a is the bitwise complement of a, and this expression has the binary repragentation

11111111 10111110

b) Bitwise logical operators : The following are the bitwise logical operators.

I. bitwise AND

II. bitwise OR

III. bitwise exclusive OR

1. Bitwise AND : The bitwise AND operates on two operands. While opeating upon these two operands they are compared on a bit-by-bit basis. Hence both the operands must be of the same type (either char (or) int.) The second operands is often called an AND mask. The bitwise AND operator operates on a pair of bits to yield a resultant bit. The symbol '&' repragents the bit wise AND the rules that decide the value of the resultant bits are shown below :

a	b	a & b
0	0	0
0	1	0
1	0	0
1	1	1

For example,

	7	6	5	4	2	1	0
x = 5	0	0	0	0	1	0	1
y = 2	0	0	0	0	0	1	0
x&y = 0	0	0	0	0	0	0	0

II. Bitwise OR : The bitwise OR operations are similar to the bitwise AND and the results is 1 if any of the bit value is 1. The symbol 'I' repragents the bitwise OR.

The rules that decide the value of the resultant bits are shown below:

a	b	a | b
0	0	0
0	1	1
1	0	1
1	1	1

For example ,

	7	6	5	4	2	1	0
x = 5	0	0	0	0	1	0	1
y = 2	0	0	0	0	1	1	1
x | y=7	0	0	0	0	1	1	1

III. Bitwise exclusive OR : The bitwise exclusive OR (XOR) operators is repragented as ^. The OR operator returns 1, when any one of the two bits (or) both the bits are 1, whereas XOR returns 1 only if the one of the two bits is 1.

The rules that decide the values of the resultant bits are shown below :

a	b	a ^ b
0	0	0
0	1	1
1	0	1
1	1	0

For example,

	7	6	5	4	2	1	0
x = 5	0	0	0	0	1	0	1
y = 2	0	0	0	0	0	1	0
x ^ y= 7	0	0	0	0	1	1	1

c) Shift operations : The shift operations take binary patterns and shift the bits to the left or right, keeping the same number of bits by dropping shifted bits off the end and filling in with zeros from the other end. C provides two types shift operations, left shift and right shift.

I. Right shift operators : It is repregented by >> and it shifts each bit in the operand to the right. The number of places the bits are shifted depends on the number following the operand.

Thus, ch >> 3 would shift all bits in ch three places to the right. Similarly, ch >> 5 would shift all bits 5 places to the right.

For example ,

	7	6	5	4	3	2	1	0	
x = 44	0	0	1	0	1	0	0	1	(8 bits)

x >> 3, the bit pattern of the x value is right shifted by thrice.

0 0 1 0 1 0 0 1

0 0 0 1 0 1 0 0 1

0 0 0 0 1 0 1 0 0

0 0 0 0 0 1 0 1 0

The result bit pattern will be

	7	6	5	4	3	2	1	0
x >> 33	0	0	0	0	0	1	0	1

II. Left shift : It is repragented by << and it shifts each bit in the operand to the left. The number of places the bits are shifted depends on the number following the operand.

Thus, ch << 3 would shift all bits in ch three places to the left. Similarly, ch << 5 would shift all bits 5 places to the left.

For example,

	7	6	5	4	3	2	1	0	
x = 33	0	0	0	0	0	1	0	1	(8 bits)

x << 3, the bit pattern of the x value is left shifted by thrice.

0 0 1 0 0 0 0 1

0 0 1 0 0 0 0 1 0

0 1 0 0 0 0 1 0 0

1 0 0 0 0 1 0 0 0

The resultant bit pattern will be

	7	6	5	4	3	2	1	0	
x >> 33	0	0	0	0	1	0	0	0	(8 bits)

4. Dynamic Memory Allocation

The process of allocating memory at run time is known as dynamic memory allocation.

Dynamic memory management techniques allow us to allocate additional memory space or to release unwanted space at runtime, thus optimizing the use of storage space.

The following **memory allocation functions** are used in dynamic memory allocation

Function	Task
1. calloc( )	Allocates space for an array of elements, initializes them to zero and then returns a pointer to the memory.
2. free ( )	Frees previously allocated space.
3. malloc( )	Allocates requested size of bytes and returns a pointer to the first byte of the allocated space.
4. realloc( )	Modifies the size of previously allocated space.

Note: In a single call to **malloc ( )**, maximum we can allocate 64 KB of memory.

5. Preprocessor Directives

Preprocessing (preprocessor directives) is a unique in C, which tells the compiler to insert another file into the source file. Through this facility the capability of a program can be continuously enhanced.

These directives can be divided into three categories.

1. Macro substitution (#define).
2. File inclusion directives (#include).
3. Compiler control directives (#if def, #if, etc.,).

Using these preprocessor directives, we can make our program more compact and easily readable, easily modifiable, portable, and more efficient.

Syntax Rules for preprocessor directives:

- It starts with the symbol # (hash).
- # should be the first non-white space character.
- There should be no semicolon at the end of the statement.
- If the directive is to be continued in different lines then each line except the last one should end with backslash(\).

Preprocessor directives

Directive	Function
#define	Defines a macro substitution.
#undef	Undefines a macro.
#include	Specifies the file to be included.
#ifdef	Tests for a macro definition.
#endif	Specifies the end of # if.
#ifndef	Tests whether a macro is not defined.
#if	Tests a compile-time condition.
#else	Specifies alternatives when # if test fails.

6. Macros Vs Functions

Macros and functions are similar in that they both enable a set of operations to be represented by a single name. But the following points gives the advantages and disadvantages of macros compared to functions.

Advantages

- Macros are faster than functions since they avoid the function call overhead.
- The number of macro arguments is checked to match the definition (ANSI C does this for function, but some compilers may not check the prototyping syntax).
- No type restriction is placed on parameters (arguments) so that one macro may serve for several data types.

Disadvantages

- Function bodies are compiled once so that multiple calls to the same function can share the same code without repeating each time. Macros, on the other hand, are expanded each time they appear in a program. As a result, a program with many large macros may be longer than a program that uses functions in place of the macros.
- Macro arguments are reevaluated at each mention in the macro body, which can lead to unexpected behavior if an argument contains side effects.
- Macros do not check the argument types. But ANSI function prototype checks both the number of arguments and the argument types.
- It is more difficult to debug programs that contain macros because the source code goes through an additional layer of translation, making the object code even further removed from the source code.

7. Command line Arguments (argc and argv)

The compiler treats the main( )function like any other function, except that at runtime the host environment is responsible for providing two arguments (argc and argv)

- The **agrc** (argument counter) is an integer that represents the number of arguments that are present on the command line when the program is invoked.
- The **argv** (argument vector) is an array of pointers to the command line arguments.

Consider the following command line to understand the concept of command line arguments.

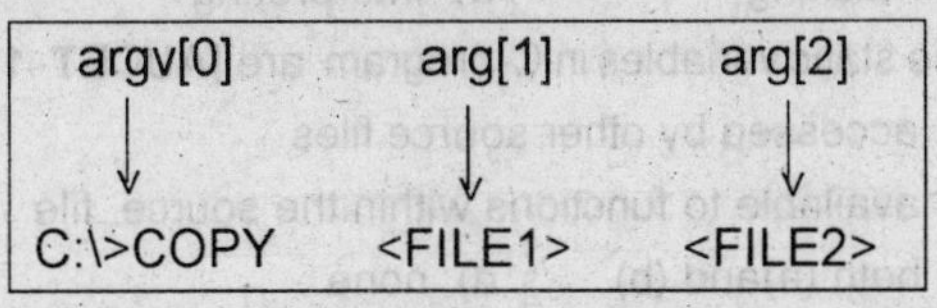

Here the **COPY** will performs the operation in **FILE 1** and **FILE 2** i.e., the contents of **file1** will copy into the **FILE 2.**

- Both the operating system understands that the **copy** is the filename where the executable code of the program is stored, **file1** and **file2** are command line arguments.

For the above command line, **argc** is three and **argv** is an array of three pointers to strings as shown below :

```
1. argv[0] = copy  ─┐
2. argv[1] = file1 ─┼──> argc=3
3. argv[2] = file2 ─┘
```

To access the command line agruments, the main( ) should be as follows:

```
main(argc, argv)
int argc;
char * argv[ ];
{
    statements;
}
```

OBJECTIVE QUESTIONS - I

1. In C-language the Bit-wise operators can be applied to which of the following operands: **[AUCET-1997]**

a) char b) int c) short long d) All

2. What will be the result of the expression 13& 25: **[TCS]**

a) 38 b) 25 c) 9 d) 12

3. What will be the result of the expression 9 | 9: **[TCS]**

a) 1 b) 18 c) 9 d) None of these

4. Which of the following are correct: **[TCS]**

a) int a=16, a>>2 = 4 b) int b=-8, b>>1=-4

c) int a=16, a>>>2=4 d) All the above

5. What is the result of the expression(1& 2) + (3|4) in base ten: **[TCS]**

a) 1 b) 2 c) 3 d) 0

6. In which of the following code fragements, the variable x is evaluated to 8. **[TCS]**

a) int x=32; x=x>>2; b) int x=33; x=x>>2;

c) int x= 35; x=x>>2; d) int x=16 x=x>>1;

7. The process of transforming one bit pattern into another using bit-wise operations is known as

a) Shifting b) Chopping

c) masking d) Interpreting

8. The static variables in C-program are: **[AUCET- 1999]**

a) accessed by other source files

b) available to functions within the source file

c) both (a)and (b) d) none

9. The difference between "malloc" and "calloc" functions is: **[AUCET-2001]**

a) malloc returns a pointer , where as calloc returns an array.

b) malloc returns a pointer to uninitialized storage.

10. Command line arguments passed to the main are stored in: **[AUCET-1999]**

a) argc b) * argv[ ]

c) place holder d) actual parameter.

11. Which of the following functions in C-language allocates a block of memory **[AUCET-1999]**

a) new b) malloc c) calloc d) None

12. The use of macro in the place of function: **[AUCET-1998]**

a) reduces execution time

b) reduces code size

c) increases execution time and code size.

d) None

13. The command used to redirect output of a program at command line of the operating system. **[AUCET-2000]**

a) < or > b) [and] c) /* and */ d) (and)

14. # preprocessor directive must be written: **AUCET-2000]**

a) in the main ( ) function

b) as the last statement of a function

c) any where in the C program

d) as first statement of the C program

15. What is a preprocessor directive

a) a message from compiler to the programmer.

b) a message from compiler to the linker.

c) a message from programmer to the preprocessor.

d) a message from programmer to the microprocessor.

ANSWER KEY - I

1. d	2. c	3. a	4. a,b	5. d
6. a,b,c,d	7. a	8. c	9. a	10. b
11. b	12. c	13. a	14. d	15. c

OBJECTIVE QUESTIONS - II

1. Consider the following program segment **[AUCET- 1999]**

```
main( )
{ int x,y,z;
    x=2; y=3;
    z=x<<y;
     printf("%d\n", z);
    z=x&y;
     printf("%d\n",z);
}
```

The output of the above program

a) 8,2 b) 8,3 c) 16,3 d) none

2. What is the output by the following C-program: **[AUCET-2000]**

```
#define SQR(x) x * x
main( )
{
  printf("%d\n",SQR(2+3));
}
```

a) 10 b) 14 c) 17 d) 25

3. Consider the declaration **[AUCET-1998]**

```
static struct
{
  unsigned a:5;
  unsigned b:5;
  unsigned c:5;
  unsigned d:5;
}v={1,2,3,4};
```

v occupies:

a) 4 words b) 2 words c) 1 word d) None

4. To execute the program to copy the contents of a file named X-FILE to another one named Y-FILE then we may use a command line like: **[AUCET-1998]**

a) C>PROGRAM X-FILE=Y-FILE

b) C>PROGRAM X-FILE Y-FILE

c) C<PROGRAM X-FILE=Y-FILE

d) None

5. Match the following Pairs: **[GATE-2000]**

X:m=malloc (5) ; m=NULL;

Y:free(m);m → value(5);

Z:char ∗ p; ∗ p='a';

1:using dangling pointers

2:using uninitialized pointers

3:lost memory

Is:

a) X-1,Y-3,Z-2 b) X-2,Y-1,Z-3

c) X-3,Y-2,Z-1 d) X-3,Y-1,Z-2

6. What would be the output of the following program:

```
#define MAX(x,y) (x>y ? x:y)
main( )
{ int i;
    i= MAX(6+4,8-2);
    printf("%d",i);
}
```

a) 6 b) 8 c) 10 d) 11

7. Consider the program segment: **[DOEACC-1993]**

```
#define SQUARE(x)=(x * x)
#define csi(m,n)(SQUARE(m)> SQUARE(n) ? n:m)
main( )
{
  int s;
    s=csi(2,-3);
    printf("%d\n",a);
}
```

In the above program segment, the out put is:

a) 2 b) –3 c) 4 d) None

8. What will be the output of the following program:

```
#define PRODUCT(x) (x * x)
main( )
{
  int i=3,j,k;
    j=PRODUCT(i++);
    k=PRODUCT(++i);
    printf("\n%d %d ",j,k); }
```

a) 12,42 b) 49,9 c) 10,49 d) 9,49

9. Consider the following program **[DOEACC- 1998]**

```
#define MAX   100
#define TOTAL MAX+500
main( )
{
  int x=TOTAL/50;
  int y=1000-TOTAL;
}
```

The correct values are

a) x=12,y=1400 b) x=110,y=400

c) x=110,y=1400 d) x=12,y=400

10. Consider the following c program **[DOEACC- 2000]**

```
#include<stdio.h>
#define
F(a,b) (a+b) * (a * a+2 * a * b+b * b)
main( )
{
  int i,j,sum;
    i=j=sum=1;
    for(i=0;i<2;i++)
      sum=sum+F(i,j);
        printf("\n sum =%d",sum);
}
```

what will be the value of the variable sum after the execution of above program:

a) 10 b) 11 c) 12 d) None

11. What will be the output of the following C program **[DOEACC- 2001]**

```
#include<stdio.h>
#define f(x,y) x * y-y * y+1
#define g(x,y) f(f(1,-1),f(1,-1)
main( )
{
  int diff;
    diff=f(1,-1)+g(1,-1);
      printf("\n the difference is= %d",diff);
}
```

a) 12 b) –1 c) 1 d) None

ANSWER KEY - II

1. d	**2.** d	**3.** b	**4.** b	**5.** c
6. c	**7.** d	**8.** d	**9.** c	**10.** a
11. d				

TEST YOURSELF - I

1. What does the following program print ?

```
# include <stdio.h>
void f(int *p, int *g) {
        p = q ;
        *p=2;
        }
int i =0, j = 1;
int  main( ) {
        f ( &i, &j);
        print f ("%d %d\n", i, j );
        return 0;
}
```

[GATE - 10]

a) 2 2   b) 2 1   c) 0 1   d) 0 2

2. What is the value printed by the following C program?

```
#include<stdio.h>
int f(int  *a, int n)
{
        if (n<=0) return 0;
        else if(*a%2 = = 0) return *a+f(a+1,n-1);
        else return *a-f(a+1,n-1);
}
int main( )
{
        int a[ ] = { 12, 7, 13, 4, 11, 6 } ;
        printf("%d", f(a, 6));
        return 0;
}
```

[GATE - 10]

a) -9   b) 5   c) 15   d) 19

3. The following C function takes a simply - linked list as input argument. It modifies the list by moving the last element to the front of the list and returns the modified list. Some part of the code is left blank.

```
typedef struct node {
  int value ;
  struct node *next ;
} Node ;
Node * move_to_front(Node * head)  {
  Node *p ,  *q;
  if (( head = = NULL: || (head->next= = NULL)) return
                                        head ;
  q = NULL: p = head ;
  while (p-> next ! =NULL) {
        q = p;
        p = p -> next ;
}
______________________
return head ;
}
```

[GATE - 10]

Choose the correct alternative to replace the blank line.

a) q =NULL;p->next=head; head=p;

b) q->next=NULL;head=p; p->next=head;

c) head=p; p->next=q; q->next=NULL;

d) q->next =NULL;p->next=head;head=p;

4. Consider the program below :

```
# include <stdio.h>
int fun (int n, int * f_p) {
        int t, f ;
        if ( n < = 1 ) {
                *f_p=1;
                return 1;
        }
        t = fun (n-1, f_p);
        f = t + *f _ p ;
        * f _ p = t ;
        return f ;
}
int main( ) {
        int x = 15 ;
        printf("%d\n", fun (5, &x));
        return 0;
}
```

The value printed is **[GATE - 09]**

a) 6   b) 8   c) 14   d) 15

5. What is printed by the following C program ?

```
int f (int x, int *py, int **ppz)      void main ( )
  {                                      {
  int y, z;                              int c, *b, **a;
  **ppz + =1; z = *ppz;                  c =4 ; b = &c; a = &b
  *py + =2 ; y = -py                     printf( "%d", f(c,b,a);
  x + = 3;                               }
  return x + y + z;
}
```

[GATE - 08]

(a) 18   (b) 19

(c) 21   (d) 22

6. Choose the correct option to fill ?1 and ?2 so that the program below prints an input string in reverse order. Assume that the input string is terminated by a newline character.

```
        void reverse (void) {
        Int c;
        If (?1) reverse () ;
        ?2
  }
  main  ( ) {
```

```
        printf ( "Enter Text") ; printf ("\n");
        reverse ( ) ; printf ("\n");
}
```

[GATE - 08]

(a) ?1 is (getchar( ) != '\n')

?2 is getchar (c);

(b) ?1 is (c = getchar( )) != '\n')

?2 is getchar (c);

(c) ?1 is (c! = '\n')

?2 is putchar (c);

(d) ?1 is ((c = getchar( )) != '\n')

?2 is putchar (c);

7. The following C function takes a singly-linked list of integers as a parameter and rearranges the elements of the list. The function is called with the list containing the integers 1,2,3,4,5,6, 7 in the given order. What will be the contents of the list after the function completes execution ?

```
struct node {
  int value ;
  struct node *next;
};
void rearrange (struct node *list) {
  struct node *p, *q;
  int temp ;
  if (! list || ! list -> next) return ;
  p = list ; q = list -> next;
  while (q) {
    temp = p -> value; p - > value = q -> value;
    q → value = temp; p = q → next ;
    q = p ? p -> next : 0 ;
  }
}
```

[GATE - 08]

(a) 1,2,3,4,5,6,7 (b) 2,1,4,3,6,5,7

(c) 1,3,2,5,4,7,6 (d)2,3,4,5,6,7,1

Common Data for Questions 8 and 9 :

Consider the following C functions.

```
int f1 ( int n)
{
  if (n == 0 | | n == 1)
    return n;
  else
    return (2*f1 (n-1) + 3* f1 (n-2));
}
int f2 (int n)
{
  int i;
  int X[N], Y[N], Z[N];
  X[0] = Y[0] = Z[0] = 0;
  X[1] =1;  Y[1] =2;  Z[1] = 3;
  for ( i = 2;  i <= n; i++)    {
    X [i] = Y [i - 1] + z [i - 2];
    Y [i ] = 2 * X [ i ] ;
    Z [i ] = 3 * X [ i ] ;
  }
  return X[n];
}
```

8. The running time of f1 (n) and f2(n) are **[GATE - 08]**

(a) Θ (n) and Θ (n)

(b) Θ (2^n) and Θ (n)

(c) Θ (n) and Θ (2^n)

(d) Θ (2^n) and Θ (2^n)

9. f1(8) and f2(8) return the values **[GATE - 08]**

(a) 1661 and 1640

(b) 59 and 59

(c) 1640 and 1640

(d) 1640 and 1661

Statement for Linked Answer Questions 10 and 11

Consider the following C program that attempts to locate an element x in an array Y[] using binary search. The program is erroneous.

```
f (int Y[10], int x ) {
  int i, j, k;
  i = 0;  j = 9;
  do {
      k = (i + j) /2
      if ( Y[k] < x) i = k; else j = k;
  } while (( Y [k]  !=x ) && ( i < j)) ;
  if ( Y[k] == x) printf ("x is in the array ");
  else printf ("x is not in the array ") ;
}
```

10. On which of the following contents of Y and X does the program fail ? **[GATE - 08]**

(a) Y is [1 2 3 4 5 6 7 8 9 10] and x < 10

(b) Y is [1 3 5 7 9 11 13 15 17 19] and x < 1

(c) Y is [2 2 2 2 2 2 2 2 2 2] and x >2

(d) Y is [2 4 6 8 10 12 14 16 18 20] and 2 < x < 20 and x is even.

11. The correction needed in the program to make it work properly is **[GATE - 08]**

(a) change line 6 to : if (Y[k] < x) i = k + 1; else j = k - 1;

(b) change line 6 to : if (Y[k] < x) i = k - 1; else j = k + 1;

(c) change line 6 to : if (Y[k] <= x) i = k; else j = k ;

(d) change line 7 to : } while (Y[k] = = x) && (i<j));

12. What is the output printed by the following C code ?

```
# include <stdio.h>
int main ( )
{
  char a[6] = "world";
  int i, j ;
  for (i = 0, j=5; i<j;  a[i++] =a[j--])
  printf ("%s\n", a);
}
```

[GATE - 08]

(a) dlrow (b) Null string
(c) dlrld (d) worow

13. Consider the C program below. What does it print ?

```
# include <stdio.h>
# define swapl (a,b) tmp=a; a=b; b=tmp
void swap2 ( int a , int b)
{
  int tmp;
  tmp=a; a=b; b=tmp;
}
void swap3 (int *a, int *b)
{
  int tmp;
  tmp = *a;  *a=*b;  *b=tmp;
}
int main ( )
{
 int num1 =5, num2 = 4, tmp;
 if (num1<num2) {swap1 (num1, num2); }
 if (num1<num2) {swap2 (num1+1, num2); }
 if (num1>=num2) {swap3 (&num1, &num2); }
 printf("%d, %d", num 1, num 2 ) ;
}
```

[GATE - 08]

(a) 5,5 (b) 5,4 (c) 4,5 (d) 4,4

14. Consider the C program given below. What does it print ?

```
# include <stdio.h>
int main ( )
{
int i, j;
  int a[8] = {1, 2, 3, 4, 5, 6, 7, 8};
  for (i=0;  i<3;  i++) {
   a[i] = a[i]  + 1;
   i++;
  }
  i-- ;
  for (j=7; j>4, j--) {
  int i=j/2;
  a[i] = a[i] -1;
  }
  printf ("%d, %d", i, a[i]);
}
```

[GATE - 08]

(a) 2,3 (b) 2,4
(c) 3,2 (d) 3,3

15. A C program is given below:

```
# include <stdio.h>
int main ( )
{
  int i, j;
  char a[2] [3] = {{'a', 'b',  'c'}, {'d', 'e', 'f'} };
  char b[3] [2];
  char *p = *b;
  for (i=0; i<2;  i++0  {
       for ( j = 0; j < 3 ;  j ++  ) {
       * (p + 2 *j +1) = a [i] [j];
       }
  }
}
```

What should be the contents of the array b at the end of the program ? **[GATE - 08]**

(a)
a b
c d
e f

(b)
a d
b e
c f

(c)
a c
e b
d f

(d)
a e
d c
b f

Statement for Linked Answer Questions 16 and 17 :

Consider the code fragment written in C below :

```
Void f (int n )
{
 if (n<=f) {
 printf("%d", n);
 }
 else {
 f (n/2);
 printf("%d", n%2);
  }
}
```

16. What does f(173) print ? **[GATE - 08]**

(a) 010110101 (b) 010101101
(c) 10110101 (d) 10101101

17. Which of the following implementations will produce the same output for f(173) as the one from Q.16?

P1:

```
Void f (int n)

{
    if (n/2) {
      f (n/2);
}
printf("%d", n%2);
}
```

P2:

```
Void f (int n)
{
    if (n<= 1) {
     printf ("%d", n);
}
else {
  printf("%d", n%2);
f (n/2);
  }
}
```

[GATE - 08]

(a) Both P1 and P2
(b) P2 only
(c) P1 only
(d) Neither P1 nor P2

18. The function f is defined as follows :

```
int f (int n) {
if (n < = 1 ) return 1 ;
else if (n % 2 = = 0 ) return f (n/2) ;
else return f (3n - 1) ;
```

Assuming that arbitrarily large integers can be passed as a parameter to the function, consider the following statements.

(i) The function f terminates for finitely many different values of $n \geq 1$.

(ii) The function f terminates for infinitely many different values of $n \geq 1$

(iii) The function f does not terminate for finitely many different values of $n \geq 1$.

(iv) The function f does not terminate for infinitely many different values of $n \geq 1$.

Which one of the following options is true of the above ? [GATE - 07]

(a) (i) and (iii) (b) (i) and (iv)
(c) (ii) and (iii) (d) (ii) and (iv)

19. Consider the C program given below :

```
# include <stdio.h>
int main ( )     {
  int  sum = 0, maxsum = 0 ; i ,  n  =  6 ;
  int a ( ) = { 2, -2, -1, 3, 4, 2 } ;
  for ( i = 0; i < n ; i + + )         {
    if ( i = = 0 | | a [i] < 0  | | a [i] < a [ i -1] )     {
      if (sum > maxsum) maxsum  =  sum ;
      sum = ( a [i]  >  0 )  ? a [i]  :  0  ;
    }
    else sum + = a [i] ;
    if (sum > maxsum)  maxsum = sum ;
    print f (11 %d\n" , maxsum) ;
}
```

What is the value printed out when this program is executed? **[GATE - 07]**

(a) 9 (b) 8
(c) 7 (d) 6

20. Consider the following C program :

```
# include  <stdio . h >
# define EOF - 1
Void push (int) ;  / * push the argument on
the stack * /
int pop (void)   ;  /* pop the top of the
                        stack */
void flagError ( ) ;
int main ( )
{  int c, m, n, r ;
   while ( ( c = getchar ( ) )   ! =  EOF)
              {if    (isdigit  (c) )
                    push (c) ;
              else if ( ( c = = '+' )  | | ( c =
= ' * ' ) )
              {   m  =  pop  ( ) ;
                  n  =  pop  ( ) ;
                  r  =  ( c == '+'  )  ?  n +
m  :  n * m  ;
```

```
            push (r) ;
        }
        else if (c ! = ' ' )
            flagError ( ) ;
    }
    print f ( " % c " , pop ( ) ) ;
}
```

What is the output of the program for the following input ?5 2 * 3 3 2 + * +

[GATE - 07]

(a) 15 (b) 25
(c) 30 (d) 150

21. Consider the program below in a hypothetical language which allows global variable and a choice of call by reference or call by value methods of parameter passing.

```
int i ;
program main ( )
{  int j = 60 ;
   i = 50 ;
   call f ( i, j ) ;
   print i, j ;
}
procedure f (x , y )
{  i = 100 ;
   x = 10 ;
   y = y + i ;
}
```

Which one of the following options represents the correct output of the program for the two parameter passing mechanisms ? **[GATE - 07]**

(a) Call by value : i = 70, j = 10 ; Call by reference : i = 60 , j = 70

(b) Call by value : i = 50, j = 60 ; Call by reference : i = 50 , j = 70

(c) Call by value : i = 10, j = 70 ; Call by reference : i = 100 , j = 60

(d) Call by value : i = 100, j = 60 ; Call by reference : i = 10 , j = 70

22. Consider the program below in a hypothetical programming language which allows global variables and a choice of static or dynamic scoping.

```
int i ;
Program main ( )
{    i = 10 ;
     call f ( ) ;
}
procedure f ( )
{    int i = 20 ;
     call g ( ) ;
}
procedure g ( )
{    print i ;
}
```

Let x be the value printed under static scoping and y be the value printed under dynamic scoping. Then, x and y are

[GATE - 07]

(a) $x = 10,\ y = 20$

(b) $x = 20,\ y = 10$

(c) $x = 10,\ y = 10$

(d) $x = 20,\ y = 20$

23. Consider the following C function :

```
int f (int n)
{static int r = 0 ;
if (n < = 0)     return 1 ;
if (n > 3)
   (r = n ;
   return f (n-2) + 2 ;
   }
return f (n-1) + r ;
}
```

What is the value of f (5) ? **[GATE - 07]**

(a) 5 (b) 7 (c) 9 (d) 18

24. Consider the following C-function in which a[n] and b[m] are two sorted integer arrays and $c[n+m]$ be another integer array.

```
void xyz (int a [ ], int b [ ], int c[ ] ) {
int i, j, k;
i = j = k = 0;
while ((i < n) && (j < m))
  if (a[i] < b[j])c[k++] = a[i++];
    else c[k++] = b[j++];
}
```

Which of the following condition (s) hold(s) after the termination of the while loop?

(i) $j < m, k = n + j - 1,$ **and** $a[n-1] < b[j]\ if\ i = n$

(ii) $i < n, k = m + i - 1,$ **and** $b[m-1] \le a[i]\ if\ j = m$

[GATE - 06]

(A) only (i)

(B) only (ii)

(C) either (i) or (ii) but not both

(D) neither (i) nor (ii)

25.Consider this C code to swap two integers and these five statements : the code

```
void swap (int* px, int* py){
*px = *px - *py;
*py = *px + *py;
*px = *py - *px;
}
```

S1 : will generate a compilation error

S2 : may generate a segmentation fault at runtime depending on the arguments passed

S3 : Correctly implements the swap procedure for all input pointers referring to integers stored in memory locations accessible to the process

S4 : implements the swap procedure correctly for some but not all valid input pointers

S5 : may add or subtract integers and pointers

[GATE - 06]

(A) S1 (B) S2 and S3

(C) S2 and S4 (D) S2 and S5

26. Which one of the choices given below would be printed when the following program is executed ?

```
# include <stdio.h>
struct test {
    int i ;
    char * c ;
} st [ ] = {5, "become", 4, "better", 6, "jungle", 8,
"ancestor", 7,    "brother"};
main ( )
{ struct test *p=st ;
p+=1 ;
    ++p -> c ;
printf("%s," , p++->c) ;
printf("%c,", *++p->c);
printf("%d,", p[0].i);
    printf("%s\n", p--->c) ;
```

[GATE - 06]

a) jungle, n, 8, ncestor

b) etter, u, 6, ungle

c) cetter, k, 6, jungle

d) etter, u, 8, ncestor

27.Which one of the choices given below would be printed when the following program is executed?

```
# include <stdio.h>
void swap (int *x, int *y)
{
        static int *temp;
        temp = x ;
        x = y ;
        y = temp ;
}
void printab ()
{
        static int i, a = -3, b = -6;
        i = 0 ;
        while (i<=4)
        {
        if ((i++)  %2 ==1) continue ;
        a = a+i ;
        b = b+i ;
}
swap (&a, &b) ;
printf("a =%d, b=%d\n", a,b);
}
main ( )
{
        printab ( ) ;
        printab ( ) ;
}
```

[GATE - 06]

a) $a = 0, b = 3$
$a = 0, b = 3$

b) $a = 3, b = 0$
$a = 12, b = 9$

c) $a = 3, b = 6$
$a = 3, b = 6$

d) $a = 6, b = 3$
$a = 15, b = 12$

28. Which one of the choices given below would be printed when the following program is executed ?

```
# include <stdio.h>
int al[] = {6, 7, 8,18, 34, 67} ;
int a2[] = {23, 56, 28, 29};
int a3[] = {-12, 27, -31};
int *x[] = {a1, a2, a3 };
void print ( int *a[] )
{
    printf("%d,", a[0] [2] ) ;
    printf("%d,", *a[2] ) ;
    printf("%d,", *++a[0] ) ;
    printf("%d,", *(++a) [0] ) ;
    printf("%d\n",  a[-1] [+1] ) ;
}
main ()
{
    print(x) ;
}
```

[GATE - 06]

a) $8,-12,7,23,8$ b) $8,8,7,23,7$

c) $-12,-12,27,-31,23$ d) $-12,-12,27,-31,56$

29. The following function computes the value of $\binom{m}{n}$ correctely for all legal values m and n $(m \geq 1, n \geq 0 \; and \; m > n)$

```
int func (int m, int n)
{
    if (E) return 1 ;
    else return (func (m-1, n) +
                 func(m-1, n-1)) ;
}
```

In the above function, which of the following is the correct expression for E ? **[GATE - 06]**

a) $(n==0) || (m==1)$

b) $(n==0) \&\& (m==1)$

c) $(n==0) || (m==n)$

d) $(n==0) \&\& (m==n)$

<u>Common Data for Questions 30,31</u>

Consider the following program module :

```
void swap (float*A1, float * A2)
{
    float temp ;
    if (*A1 = = *A2) return ;
    temp = *A1 ;
    *A1 = *A2 ;
    *A2 = temp ;
    return ;
}
```

30. The program volume for the above module using Halstead's method is [GATE - 06]

a) 60 b) 63

c) 66 d) 69

31. The program effort for the above module using Halstead's method is [GATE - 06]

a) 315 b) 330

c) 393 d) 403

ANSWER KEY

1. d	2. c	3. d	4. b	5. d
6. d	7. b	8. b	9. c	10. c
11. a	12. b	13. c	14. a	15. b
16. d	17. c	18. c	19. c	20. b
21. d	22. c	23. d	24. c	25. b
26. a	27. d	28. a	29. c	30.
31.				

EXPLANATORY NOTES

2.

12	7	13	4	11	6
A = 100	102	104	106	108	110

$f(a,6) \rightarrow 12 + f(a+1, n+1)$

$\downarrow$

$7 - f(a+1, n-1) \rightarrow 13 - f(a+1, n-1)$

$\downarrow$

$4 + f(a+1)n+1)$

$\downarrow$

$11 - f(a+1, n-1)$

$(12+(7-(13-(4+(11-(6+0)))))) \downarrow$

$= 15$ $\quad 6 + f(a+1, 0)$

$\downarrow$

0

4. Stack Positions :

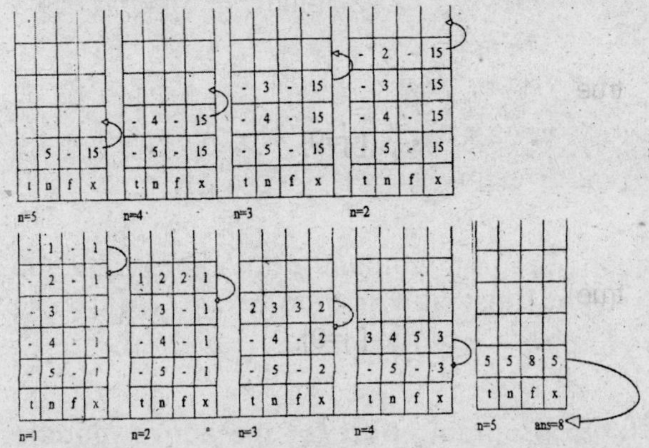

5. Suppose 100 [4] b [100] 4 [1000]

C 1000

**ppz + = 1 c=5

z =5

*py + =2 c=7

y =7

x + = 3

x+y+z =22

6. Whenever the character input is not return key (in)

Its accept character input ? 1 (c=getchar()) =! (in) ? 2 for bringing all character after reverse functions so ? 2 is putchar (c).

Alternately

To print the reverse of the input, this programme basically collects the input in buffer with the help of some f'' i.e f'' is getchar()) and by using stack, then where ↵ is pressed stack is popped off and characters are printed in reverse order as that of the process call i.e last process call to reverse is processed first because of stack.

7. In function rearrange the string of integer 1,2,3,4,5,6,7 converted into 2,1,4,3,6,5,7 we can check this by execute this program.

Alternately

Initially

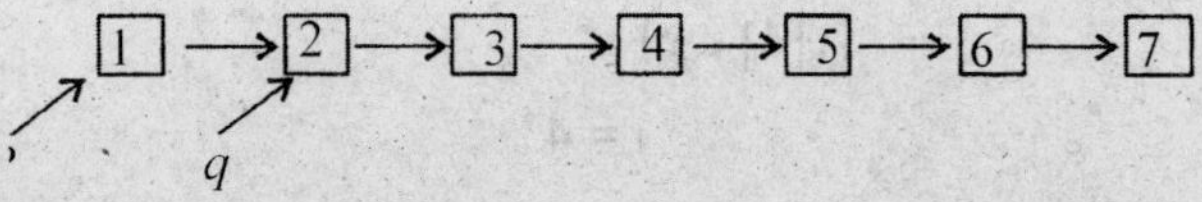

1st iteration of while

$P = (q \rightarrow next)$ $q = (p \rightarrow next)$

2nd iteration

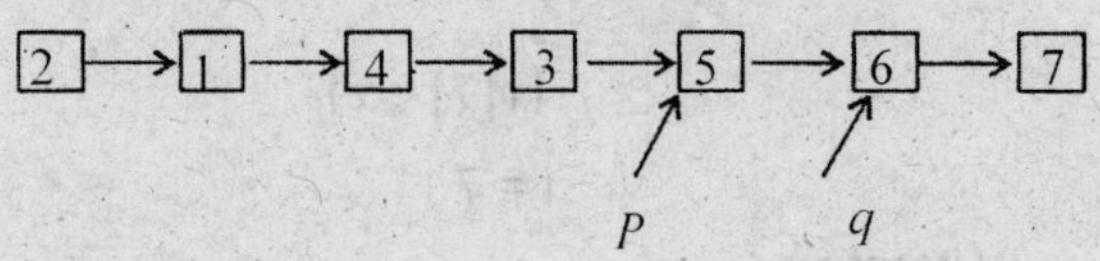

3rd iteration

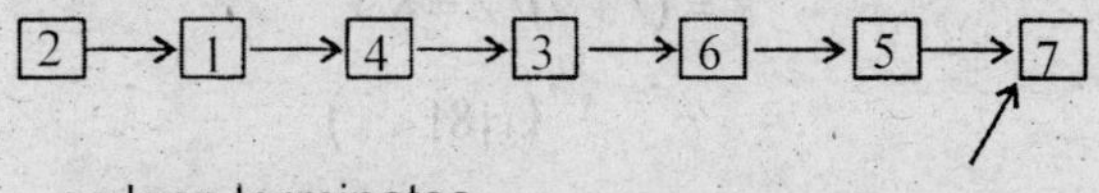

so loop terminates P $q = 0$

8. Function f1 contain recursive function so its running time must be $0(2^n)$ but $f2$ have not.

Alternately

$f1$ has complexity of $0(2^n)$, while $f2$ has a loop from 2 to N, so the complexity of it is O(n).

9. When solved

$$f1(0) = 1$$

$$f_1(1) = 1$$

$$f(2) = 2f_1(1) + 3f_1(0) = 2$$

$$f(3) = 2f_1(2) + 3f_1(1) = 7$$

$$f(4) = 2f_1(3) + 3f_1(2) = 20$$

$\vdots$

$$f_1(8) = 1640$$

Now for f_2

$$x[0] = 0 \quad y[0] = 0 \quad z[0] = 0$$

$$x[1] = 1 \quad y[1] = 2 \quad z[1] = 3$$

In loop

$$x[2] = 2 \quad y[2] = 4 \quad z[2] = 6$$

$$x[3] = 7 \quad y[3] = 14 \quad z[3] = 21$$

$$x[4] = 20 \quad y[4] = 40 \quad z[4] = 60$$

$\vdots$

$$x(8) = 1640$$

10. In C option when we give x=4 (say) the working is as follows

$k = (0+9)/2 = 4$

$(y[4] < x)$ true

i = 4 [j=9]

II iteration

$k = (4+9)/2 = 6$

(y[6]<x) true

i = 6 [j=9]

III Iteration

$k = (6+9)/2 = 7$

$(y[7] < x)$ true

i = 7 [j=9]

IV iteration

$k = (7+9)/2 = 8$

$(y[8] < x)$ true

i =8 [j=9]

V iteration

$k = (8+9)/2 = 8$

$(y[8] < x)$ true

$i = 8$ [j=9]

remains in loop for ever as condition (4[8]! =x & i <j) is always true

11. If we increase i to k +1 when y[k] < x and decrease j to k- 1 otherwise, above scenario does not happen and moreover the element at k position will be checked for equality in the while condition.

12. Initially, a[6] = "World"

w	o	r	l	d	\n

array of a size 6

When the for loop executes first time \n comes in place of w.

a [0] = a[5]

a [0] = '\n'

\n	o	r	l	d	\n

so, printf ("%s",a);

prints the string starting at address a, which starts with '\n'. And '\n' indicates the end of string, so it will print' null string'

13. Initially num 1 =5

num 2 =4

if (num>num2)

{swap 1 (num1, num2);}

will execute and will swap the value of num 1 and num 2 using macro function call.

Now, Num 1 =4

Num 2 = 5

Now

if (num 1 < num2)

{

swap 2 (num 1 + 1 num2);

}

Now, call to swap 2 function is made. It is call by value and not call by reference. So, values of num 1 and num 2 do not change.

$\therefore$ num 1 =4

num 2 = 5

so, 3rd if condition is not satisfied

Hence finally, num 1 =4

num 2 = 5

14. From the first for loop, we get

i = 4

and a[8] = {2, 2, 4, 5, 6, 7, 8}

From second for loop, we get

i = 2

and a[8] = {2, 2, 3, 3, 5, 6, 7, 8}

∴ printf ("%d %d", i, a[i]);

↓

2, a [2]

2, 3

15. Initially, we have

a →

a	b	c
d	e	f

Now, from for loop

*(p+0+0)=a[0] [0] i =0

*(p)=a j=0

i = 0 *(p+2+0) = a[0] [1]

j = 1 *(p+2) = b

i = 0 *(p+4+0) = a[0] [2]

j = 2 *(p+4) = c

i = 1 *(p+0+1) = a[1] [0]

j = 0 *(p+1) = d

i = 1 *(p+2+1) = a[1] [1]

j = 1 *(p+3) = e

i = 1 *(p+4+1) = a[1] [2]

j = 2 *(p+5) = f

so final result is

a d

b e

c f

16. It is the recursive program to print the binary form of a number

2 173

2 86 - 1

2 43 - 0

2 21 - 1

2 10 - 1

2 5 - 0

2 2 - 1

2 1 - 0

∴ 10101101

17. Here, P_1 is also the program to print the binary form of any number but P_2 will print reverse of binary.

18. For n = 5 this function terminates infinitely. So statement (ii) and (iii) are true.

19.

	initial	i = 0	i = 1	i = 2	i = 3	i = 4	i = 5
Sum	0	0	2	0	3	7	2
MaxSum	0	0	2	2	2	2	7

The value of maxsum = 7

20. 52 * 332 + * +

= 10 332 + * +

= 10 35* +

= 10 15 +

= 25

21. **Call by value :** since i is global variable

When procedure is invoked value of i is changed in procedure. It modifies the value of i in main program. So value of i = 100 and value of i does not change because modification in value of j does not reflect the main program.

Call by reference : In this mechanism x and y is passed as a reference of i and j respectively, so modification of x and j reflects in procedure reflects the change in main program.

x = 10

y = y + i = 60 + 10 = 70

23. Calbing sequence will be

Sequence	returns	
$f(5)$	$f(3)+2$	= 18
↓	↑	
$f(3)$	$f(2)+5$	= 16
↓	↑	
$f(2)$	$f(1)+5$	= 11
↓	↑	
$f(1)$	$f(0)+5$	= 6
↓	↑	
$f(0)$	1= 1	

24. (i) will occur, if $i=n$ occurs before $j=n$ so that the while loop terminates till that time $j-1$ elements of b and n elements of a would have been inserted in c. $i=n$ would have occured only if i would have been incremented i.e,

$a[i]<b[j]$ for the last i.

$\Rightarrow \quad a[n-1]<b[j]$

$\therefore \quad k=j-1+n$

25. S_3 is true S_4 and S_5 are false as the program will prints 3 and 8. The option left is (a).

29. The function computes $^{m}C_n=\frac{m!}{n!(m-n)!}$

1 is the returned, if $m=n$. $\because \quad \frac{n!}{n!(m-n)!}=\frac{m!}{m!0!}=1$

or if $n=0 \quad \because \quad \frac{n!}{n!(m-n)!}=\frac{m!}{0!m!}=1$

TEST YOURSELF - II

1. Consider the following C function. **[GATE - 03]**

```
float f (float x, int y)      {
    float p, s; int i ;
    for (s = 1, p = 1, i = 1; i < y ; i++) {
        P * = x / i;
        S + = p ;
    }
    return s ;
    }
```

For large values of y, the return value of the function f best approximates

a) x^y b) e^x c) $\ln(1+x)$ d) X^x

2. Assume the following C variable declaration

int * A [10], B[10] [10]; **[GATE - 03]**

Of the following expressions

I. A [2] II. A [2] [3] III. B [1] IV. B [2] [3]

Which will not give compile-time errors if used as left hand sides of assignments statements in a C program?

a) I, II and IV only b) II, III and IV only

c) II and IV only d) IV only

The following information pertains to Q.3 - 4

The following program fragment is written in a programming language that allows global variables and does not allow nested declarations of functions.

[GATE - 03]

```
global int i = 100, j = 5;
void P(X) {
    int i = 10 ;
    Print (x + 10) ;
    i = 200 ;
    j = 20 ;
    Print (X) ;
}
main ( ) { P(i+j) ; }
```

3. If the programming language uses static scoping and call by need parameter passing mechanism, the values printed by the above program are

a) 115, 220 b) 25, 220

c) 25, 1 d) 115, 105

4. If the programming language uses dynamic scoping and call by name parameter passing mechanism, the values printed by the above program are

a) 115, 220 b) 25, 220

c) 25, 15 d) 115, 105

5. Consider the following class definitions in a hypothetical Object Oriented language that supports inheritance and uses dynamic binding. The language should not be assumed to be either Java or C++, though the syntax is similar. **[GATE - 03]**

```
Class P {                     Class Q subclass of P {
  Void f(int i) {                      Void f(int i) {
    print (i) ;                          print (2*i) ;
  }                                    }
}                                    }
```

Now consider the folowing program fragment :

PX = new Q ( );

Qy = new Q ( );

Pz = new Q ( );

x. f(1); ((P) Y) denotes a typecast of y to P. the output produced by executing the above program fragment will be

a) 121 b) 211 c) 212 d) 222

6. In the following C program fragment, j, k, n and Two log_n are integer variables, and A is an array of integers. The variable n is initialized to an integer ≥ 3, and Two Log_n is initialized to the value of $2 * \lceil \log_2(n) \rceil$

```
for (k = 3; k < =n ; k++)
    A [k] = 0 ;
for ( k = 2; k<= Two Log_n ; k++)
    for ( j = k+1; j <=n ; j++)
    A [ j ] = A [ j ]  | | ( j%k) ;
for ( j = 3; j <=n; j++)
    if (! A [ j ] ) printf ("%d", j);
```

[GATE - 03]

The set of numbers printed by this program fragment is

a) $\{m/m \leq n, (\exists i)\ [m = i!]\}$ b) $\{m/m \leq n, (\exists i)\ [m = i^2]\}$

c) $\{m/m \leq n, \text{m is prime}\}$ d) $\{\ \}$

7. Consider the c program shown below.

```
# include <stdio.h>
# define print(x) printf("%d, x)
int x;
Void Q (int z) {
    z + = x; print (z) ;
}
void P (int*y)     {
    int x = *y+2 ;
    Q(x);  *y = x - 1 ;
    print (x) ;
}
main (void)        {
    x = 5 ;
    p (&x) ;
    print (x) ;
}
```

The output of this program is **[GATE - 03]**

a) 12 7 6 b) 22 12 11 c) 1 4 6 6 d) 7 6 6

8. The goal of structured programming is to **[GATE - 04]**

a) have well indented programs

b) be able to infer the flow of control from the compiled code

c) be able to infer the flow of control from the program text

d) avoid the use of GOTO statements

9. Consider the following C function **[GATE - 04]**

```
void swap (int a, int b)
{ int temp ;
temp = a ;
a = b ;
b = temp ;
}
```

In order to exchange the values of two variables x and y.

a) call swap (x, y) b) call swap (&x, &y)

c) swap (x, y) cannot be used as it does not return any value

d) swap (x, y) cannot be used as the parameters are passed by value

10. Consider the following C function : **[GATE - 04]**

```
int f (int n)
{ static int i = 1 ;
  if (n>=5) return n ;
  n = n+i ;
  i ++;
  return f (n) ; }
```

The value returned by f(1) is

a) 5 b) 6 c) 7 d) 8

11. Consider the following program fragment for reversing the digits in a given integer to obtain a new integer. Let $n=d_1d_2....d_m$ **[GATE - 04]**

```
int n, rev ;
rev = 0 ;
while (n>0) {
    rev = rev * 10 + n % 10 ;
    n = n / 10 ;
}
```

The loop invariant condition at the end of the ith iteration is :

a) $n=d_1d_2.......d_{m-1}$ and $rev=d_md_{m-1}........d_{m-i+1}$

b) $n=d_{m-i+1}....d_{m-1}d_m$ or $rev=d_{m-i}.....d_2d_1$

c) $n \neq rev$

d) $n=d_1d_2.....d_m$ or $rev=d_m......d_2d_1$

12. Consider the following C program segment :

```
char p [20] ;
char * s = "string" ;
int length = strlen (s) ;
for (i = 0 ; i < length ; i ++)
  p [ i ] = s [length - i] ;
printf ("%s", p) ;
```

The output of the program is **[GATE - 04]**

a) gnirts b) string

c) gnirt d) no output is printed

13. It is desired to design an object-oriented employee record system for a company. Each employee has a name, unique id and salary. Employees belong to different categories and their salary is determined by their category. The functions get Name, getId and compute Salary are required. Given the class hierarchy below, possible locations for these functions are : **[GATE - 04]**

i) getId is implemented in the supercalss

ii) getId is implemented in the subclass

iii) getName is an abstract function in the superclass

iv) getName is implemented in the superclass

v) getName is implemented in the subclass

vi) getSalary is an abstract function in the superclass

vii) getSalary is implemented in the superclass

viii) getSalary is implemented in the subclass

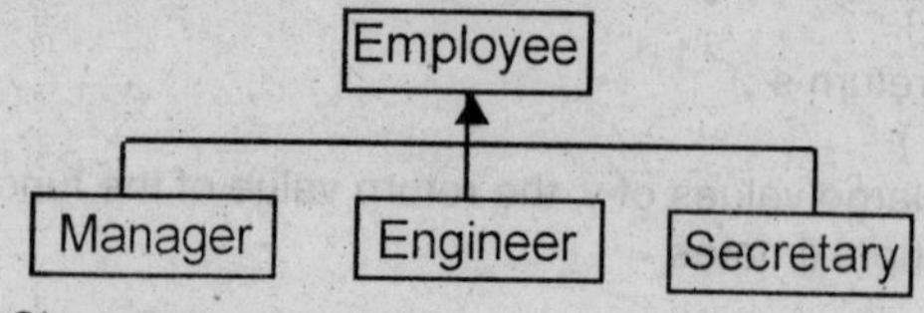

Choose the best design

a) i), iv), vi), viii) b) i), iv), vii)

c) i), iii), v), vi), viii) d) ii), v), viii)

14. Choose the best matching between the programming styles in Group 1 and their characteristics in Group 2. **[GATE - 04]**

Group - 1	Group - 2
P. Functional	1. Command-based, procedural
Q. Logic	2. Imperative, abstract data types
R. Object-oriented	3. Side-effect free, declarative, expression evaluation
S. Imperative	4. Declarative, Clausal representation, theorem proving

a) P-2, Q-3, R-4, S-1 b) P-4, Q-3, R-2, S-1

c) P-3. Q-4, R-1, S-2 d) P-3, Q-4, R-2, S-1

15. Consider the following C program which is supposed to compute the transpose of a given 4 × 4 matrix M. Note that, there is an X in the program which indicates some missing statements. Choose the correct option to replace X in the program. **[GATE - 04]**

```
#include<stdio.h>
#define ROW 4
#define COL 4
int M[ROW][COL] =
    {1,2,3,4,5,6,7,8,9,10,11,12,13,14,15,16} ;
main ( )
{
int i, j, t ;
for (i=0;i<4; ++i)
{
    X
}
for (i=0 ; i<4; ++i)
for ( j=0; j<4; ++j)
printf("%d", M[ i ] [ j ]);
}
```

a) for(j=0; j<4; ++=j){
t = M [i] [j] ;
M [i] [j] = m [j] [i]
M [j] [i] = t ;
}

b) for (j=0; j<4; ++j){
M [i] [j] = t;
t = M [j] [i];
M [j] [i] = M [i] [j]
}

c) for(j=i; j<4; ++j){
t = M [j] [i];
M [i] [j] = M [j] [i] ;
M [j] [i] = t;
}

d) for(j=i; j<4; ++j){
M [i] [j] = t;
t = M [j] [i];
m [j] [i] = M [i] [j];
}

16. What is the output of the following program ?

```
#include<stdio.h>
int funcf (int x);
int funcg (int y);
main( )
{
int x=5, y=10, count ;
for (count = 1; count <=2; ++ count) {
y + =funcf(x) + funcg(x);
printf("%d",y);
}}
funcf (int x) {
int y ;
y = funcg(g);
return (y);
}
funcg (int x) {
static int y = 10;
y +=1;
return (y+x);
}
```

[GATE - 04]

a) 43 80 b) 42 74 c) 33 37 d) 32 32

17. Choose the correct option to fill the ? 1 and ?2 so that the program prints an input string in reverse order. Assume that the input string is terminated by a new line character. **[GATE - 04]**

```
#include<stdio.h>
void wrt_it(void);
int main (void)
{
printf("Enter Text");
printf("\n");
wrt_it( );
printf("\n");
return 0 ;
}
void wrt_it (void)
{
int c;
if (?1)
wrt_it( );
?2
}
```

a) ?1 is getchar( ) !='\n'
?2 is getchar(c);

b) ?1 is (c=getchar( )) !='\n'
?2 is getchar(c);

c) ?1 is c !='\n'
?2 is putchar(c);

d) ?1 is (c = getchar( )) !='\n'
?2 is putchar(c);

18. Consider the following C program : **[GATE - 04]**

```
#include <stdio.h>
typedef struct {
char *a;
char *b;
} t;
void f1 (t s);
void f2 (t *p);
main ( )
{
static t s = {"A", "B"} ;
printf ("%s %s\n", s.a, s.b) ;
f1 (s) ;
printf ("%s %s\n", s.a, s.b);
f2(&s);
}
void f1(t s)
{
s.a = "U";
s.b = "V",
printf (%s %s\n", s.a, s.b);
return;
}
void f2 (t *p)
{
p → a = "V";
p → b = "W";
printf(%s %s\n", p → a, p → b);
return;
}
```

What is the output generated by the program ?

a)	b)	c)	d)
A B	A B	A B	A B
U V	U V	U V	U V
V W	A B	U V	V W
VW	VW	VW	U V

19. What does the following C-statement declare ? int (*f) (int*); **[GATE - 05]**

a) A function that takes an integer pointer as argument and returns an integer

b) A function that takes an integer as argument and returns an integer pointer

c) A pointer to a function that takes an integer pointer as argument and returns an integer

d) A function that takes an integer pointer as argument and returns a function pointer.

20. An abstract Data Type (ADT) is **[GATE - 05]**

a) same as an abstract class

b) a data type that cannot be instantiated

c) a data type for which only the operations defined on it can be used, but none else

d) all of the above

21. A common property of logic programming languages and functional languages is **[GATE - 05]**

a) both are procedural languages

b) both are based on λ - calculus

c) both are declarative

d) both use Horn-clauses

22. Which one of the following are essential features of an object-oriented programming language ? **[GATE - 05]**

(i) Abstraction and encapsulation

(ii) Strictly typedness

(iii) Type safe property coupled with subtype rule

(iv) Polymorphism in the presence of inheritance

a) (i) and (ii) only b) (i) nad (iv) only

c) (i), (ii) and (iv) only d) (i), (iii) and (iv) only

23. Consider the following C-program **[GATE - 05]**

```
Void foo (int n, int sum) {
    int k = 0, j = 0;
    if (n ==0) return;
    k = n% 10 ; j = n/10;
    sum = sum + k;
    foo ( j, sum);
    printf ("%d," k);
}
int main ( ) {
int a = 2048, sum = 0;
foo (a, sum);
printf ("% d\n", sum);
}
```

What does the above program print ?

a) 8,4,0,2,14 b) 8,4,0,2,0

c) 2,0,4,8,14 d) 2,0,4,8,0

24. Consider the following C-program **[GATE - 05]**

```
double foo (double) ; /*Line 1*/
int main ( ) {
double da,db;
// input da
db = foo (da);
}
double foo (double a) {
  return a;
}
```

The above code compiled without any error or warning. If Line 1 is deleted, the above code will show

a) no compile warning or error

b) some compiler-warnings not leading to unintended results

c) some compiler-warnings due to type-mismatch eventually leading to unintended results

d) Compiler errors

25. The following C function takes two ASCII strings and determines whether one is an anagram of the other. An anagram of a string s is a string obtained by permuting the letters in s. **[GATE - 05]**

```
int anagram (char *a, char *b) {
int count [128], j;
for ( j = 0; j<128; j++) count [ j ] = 0;
j = 0;
while (a[ j ] && b[ j ]) {
    A;
    B;
}
for ( j = 0; j <128 ; j++) if (count[ j ]) return 0;
return 1;
}
```

Choose the correct alternative for statements A and B.

a) A : count [a[j]]++ and B : count [b[j]] - -

b) A : count [a[j]]++ and B : count [b[j]]++

c) A : count [a[j++]]++ and B : count [b [j]] - -

d) A : count [a[j]]++ and B : count [b[j++]] - -

26. What is the output printed by the following program? **[GATE - 05]**

```
# include<stdio.h>
int f(int n, ink k) {
    if (n = = 0) return 0;
    else if (n% 2) return f(n/2, 2*k) + k;
    else return f(n/2, 2*k) - k;
    int main ( ) {
    printf("%d", f(20, 1));
    return 0;
```

a) 5 b) 8 c) 9 d) 20

27. Let a be an array containing n integers in increasing order. The following algorithm determines whether there are two distinct numbers in the array whose difference is a specified number S > 0. **[GATE - 05]**

```
i = 0; j = 1;
while ( j < n) {
   if (E) j++;
   else if (a[ j ] - a[ i ] ==S) break;
```

else i++;

}

if (j < n) printf("yes") else printf ("no");

Choose the correct expression for E.

a) a[j] - a[i] > S b) a[j] - a[i] < S

c) a[i] - a[j] < S d) a[i] - a[j] >S

28. Let a and b be two sorted arrays containing n integers each, in non-decreasing order. Let c be a sorted array containing 2n integers obtained by merging the two arrays a and b. Assuming the arrays are indexed starting from 0, consider the following four statements.

I. $a[i] \geq b[i] \Rightarrow c[2i] \geq a[i]$ II. $a[i] \geq b[i] \Rightarrow c[2i] \geq b[i]$

III. $a[i] \geq b[i] \Rightarrow c[2i] \leq a[i]$ IV. $a[i] \geq b[i] \Rightarrow c[2i] \leq b[i]$

Which of the following is TRUE **[GATE - 05]**

a) only I and II b) only I and IV

c) only II and III d) only III and IV

ANSWER KEY

1. c	**2.** d	**3.** d	**4.** a	**5.** d
6. b	**7.** a	**8.** c	**9.** d	**10.** c
11. a	**12.** a	**13.** a	**14.** d	**15.** c
16. a	**17.** d	**18.** b	**19.** c	**20.** c
21. c	**22.** b	**23.** d	**24.** d	**25.** a
26. c	**27.**	**28.** b		

EXPLANATORY NOTES

1. For larger values of y

$$P = P * \frac{x}{i} = \frac{x}{i} \Rightarrow S = S + P = 1 + \frac{x}{i}$$

when $i < y$ and i ++

$$S = 1 + \frac{x}{i} + \frac{x}{i+1} + \ldots\ldots\ldots$$

$$= 1 + x + \frac{x}{2} + \frac{x}{3} + \ldots\ldots.$$

Hence for larger value of y, ln (1 + x)

2. B[2] [3] gives the element of second rows from third column.

3. P(i + j)

x = i + j = 105 (global value of i and j)

print (x + 10) = 115

print (x) = 105

because static value store previous value stored at place of x.

4. By dynamic and coping we take local value

P(i + j), x = i + j = 105 (global value of i and j)

print (x + 10) = 115

print (x) = 200 + 20

= 220 (due to local value)

5. x. f(1)

f(1) copy to f(int i)

So i = 1

print (2 * i) = 2

Same as, ((P)y).f(1); z.f(1) = 2

7. * y = &x = 5

x = *y + 2 = 7

Q(x);

z = x = 7

z = z + x = 12

print (z) = 12

* y = x - 1 = 6

print (x) = 7

print (x) = 6

10. Program is

```
int f (int n)
{
static int i = 1
 if (n>=5) return n      ......... (1)
 n = n  + i ;
 i ++;
 return f(n) ;
}
```

as 1 does not fullfill the condition of if statement so execute next instruction, the iteration will be, as

iteration	n	i
1	1+1=2	2
2	2+2=4	2+1=3
3	4+3=7	3+1

as 7 is greater than 5 so it will return to the main program.

11.

```
int n, rev ;
rev = 0;
while (n>0)
{
rev = rev* 10+n% 10 ;
n=n /10 ;}
```

If $n = d_1 d_2 d_3 - - - d_{m-i}$

then after entering into the loop

loop	n	rev
1	$d_1 d_2 \ldots\ldots d_{m-i-1}$	d_{m-i}
2	$d_1 d_2 \ldots d_{m-i-2}$	$(d_{m-i})(d_{m-i-1}) + (d_{m-i-1})$

After executing m times within the loop rev will be

$rev = d_m \, d_{m-1} - - d_{m-i+1}$

12. After executing within the loop

$p[0]=g \;\; p[1]=n \;\; p[2]=i \;\; p[3]=r \;\; p[4]=t \;\; p[5]=s$

length will be 6 from the statement length = strlen (s) ; so after that it will go out of the loop, then p = gnirts

15. For finding the swap the element in row with column, i.e., M[i][j] with M[j][i] using third variable t (as defined as int type)

So, block at palce of X will be

```
for        (j = i ; j < 4 ; + +j)
               {t = M[i][j]
           M[i][j]=M[j][i]
           M [j][i] = t }
```

23. Consider 'print f' statement in main ( )

Sum is initilize to 0

So final answer is always here

only option's have 0 in last digits are (b) and (d)

n	sum	K	J
2048	0	0	0
		8	204
	8		

Call to 2nd instance

n	sum	K	J
204	8	0	0
		4	20
	12		

Call to 3rd instance

n	sum	K	J
20	12	0	0
		0	2
	12		

Call to 4th instance

n	sum	K	J
2	12	0	0
		2	0
	14		

Call to 5th instance

n	sum	K	J
0	14		

2 0 4 8 0

25. This is a tricky question

First note that ASCII value range to 0-127 than why count is of size 128. In permutation we can place a letter any where but number of letters and number of counts of each letter should be same. here a [j] or b [j] gives us the ASCII value of character, so we increase index of that character a [j] by 1 and then decrease the index of coresponding index of b's by one. At final, we must have all values of count zero.

26.

	n	K
→ 9		
	20	1
→ +10		−1
	10	2
→ +12		−2
	5	2
→ +8		+4
	2	8
→ +16		−8
	1	16
→ 0		+16
	0	32

27. Key point here is that number are stored in increasing order and S > 0. If we use a[i] -a[j] while i < j than result is always negative.

So we have to choose option from first two options.

If a [j] - a [i] < S

This mean that we have to increase j so next time we compare with higher number.

If a [j] = a [i] we have done it else.

If a [j] = a [i] > S, then we have to increase index of [i].

Unit - VII
Data Structures

UNIT - VII DATA STRUCTURES

I. ARRAYS, STRINGS, STRACKS, QUEUES AND LINKED LISTS

- **Implementation of Arrays**
- **Manipulation of Strings**
- **Implementation of Stacks and Queues**
- **Types of Linked Lists**

1. Arrays

Arrays form an important part of almost all-programming languages. It provides a powerful feature and can be used as such or can be used to form complex data structures like stacks and queues. An array can be defined as an infinite collection of homogeneous (similar type) elements. This means that an array can store either all integers, all floating point numbers, all characters (array of character or string), or any other complex data type, but all of same type. There are some important points to be pointed out about arrays. These are as follows:

I. Arrays are always stored in consecutive memory locations.

II. An array can store multiple values which can be referenced by a single name unlike a simple variable which store one value at a time (i.e., arrays are always stored under a common heading or variable name) followed by an index or subscript, specified inside a square bracket.

III. Array name is actually a pointer to the first location of the memory block allocated to the name of the array.

IV. An array either be a integer, character, or floating data type can be initialized only during declaration time, and not afterwards.

V. There is no bound checking concept for arrays in C. It means you can attempt to enter any number of values irrespective of the integer index specified inside square brackets, during declaration of arrays. It is upto programmer to check upper bound of array.

2. One- Dimensional Arrays

A one – dimensional array is one in which only one subscript specification is needed to specify a particular element of the array. One – dimensional array can be declared as follows:

data - type var_name [Expression] ;

Data type is the type of elements to be stored in the array, Var_name specifies the name of array, it may be given any name like other simple variables. The Expression or subscript specifies the number of values to be stored in the array. Arrays are also called subscripted values. The subscript (also known as length or size of array) must be an integer value. The subscript of the array starts from 0 onwards. For example if the array declaration looks like follows:

int num [10] ;

3. Implementation Of One-Dimensional Array In Memory

The Address of a particular element in a one dimensional array is given by the relation:

Address of element a[k] = B + W ∗ k

Where B is the base address of the array, W is the size of each element of the array, and k is the number of required elements in the array (index of element) which should be a integer quantity.

For example:

Let the base address of the first element of the array is 2000 (i.e.,base address B is = 2000), and each element of the array occupies four bytes in the memory, then address of fifth element of a one-dimensional array a[10] will be given as:

Address of element a[5] = 2000+4 ∗ 5 =2000+20 = 2020.

4. Multi-Dimensional Arrays

When we talk of multi-dimensional arrays, we generally talk of two-dimensional arrays, as more than two-dimensions is rarely needed. Two-dimensional array can be declared in the same manner as one-dimensional arrays, except that a separate pair of braces is required for second dimension as shown below :

```
int  a [ 3 ] [ 3 ];
float b [ 3 ] [ 4 ];
char names[3][20];
```

Matrix is a best example of two-dimensional arrays. The length of any dimension in a multidimensional array can be calculated as follows:

Length = Upper bound – Lower bound + 1

For example we have an array a [3] [2], then we can represent this array as follows:

a [0.. 2, 0.. 1] ;

This indicates that the array a has three rows (0, 1 and 2) and two columns (0, and 1). As usual figures inside first square brackets indicate the number of rows and those inside second square brackets indicate number f columns. Hence the length of row will be:

Row = Upper bound –Lower bound +1

Row = 2 – 0 + 1 = 3

And the length of columns will be :

Length = Upper bound – Lower bound + 1

Length = 1 – 0 + 1 = 2

These matrices can be called a 3 × 2 (pronounced as three by two) matrices. Note that for rows here, 2 is called the upper bound (denoted by U_1), and 0 is called the lower bound (denoted by L_1), similarly 1 will be the upper bound (denoted by U_2) and 0 will be the lower bound (denoted by L_2) for columns.

5. Implementation Of Two –Dimensional Arrays

A two-dimensional array can be implemented in a programming language in two ways :

1. Row-major implementation
2. Column-major implementation

Row-Major Implementation

Row-major implementation is a linearization technique in which elements of array are reader from the keyboard row-wise i.e., the complete first row is stored, then the complete second row is stored and so on. For example, an array a [3] [3] is stored in the memory as shown in Fig.1.1 below :

a_{00}	a_{01}	a_{02}	a_{10}	a_{11}	a_{12}	a_{20}	a_{21}	a_{22}

← Row 1 → ← Row 2 → ← Row 3 →

Fig. 1.1

The storage can be clearly understood by arranging array as matrix as shown below :

$$a = \begin{pmatrix} a_{00} & a_{01} & a_{02} \\ a_{10} & a_{11} & a_{12} \\ a_{20} & a_{21} & a_{22} \end{pmatrix} \begin{matrix} \text{Row 1} \\ \text{Row 2} \\ \text{Row 3} \end{matrix}$$

Above figure shows the actual physical storage of elements of the array, whereas the matrix form is logical representation of array (and is not the actual way in which two-dimensional array is stored).

Address of elements in Row major implementation

The computer does not keep the track of all elements of the array, rather, it keeps a base address (i.e., the address of first element in the array), and calculates the address of required element when needed. It calculates this (in row major implementation) by the following relation:

Address of element a [i][j] = $B+W(n(i-L_1)+(j-L_2))$

Where B is the base address of the array, W is size of each array element, n is the number of columns (i.e., $U_2 - L_2$). L_1 the lower bound of row, L_2 is bound of column. Let us study an example to get a clear idea of row major implementation :

Ex : A two-dimensional array defined as a [4.. 7, - 1.. 3] requires 2 bytes of storage space for each element. If the array is stored in row-major form, then calculate the address of element at location a [6,2]. Given that the base address is 100.

Sol: The following information is given :

Base address B = 100

Size of each element in the array W = 2bytes

Lower bound of row L_1 = 4

Lower bound of column L_2 = -1

Upper bound of row U_1 = 7

Upper bound of column U_2 = 3

Row number of the required element i = 6

Column number of required element j = 2.

Now the number of columns n will be :

$U_2 - L_2 + 1 = 3 - (-1) + 1 = 5$

As the address of a particular element in a two-dimensional array (in row major implementation) is given by the relation :

Address of a[i][j] = $B + W(n(i-L_1)+(j-L_2))$

Therefore address of element a [6,2] or a[6] [2] will be:

Address of a [6] [2] = 100 + 2 (5 (6-4) + (2- (-1))

= 100 + 2(5 * 2 + 3) = 100 + 26 = 126.

Column Major Implementation

In column major implementation memory allocation is done column by column i.e., first the elements of the complete first column is stored, then elements of complete second column is stored, and so on. For example an array a[3][3] is stored in the memory as shown in the Fig. 1.2 below :

a_{00}	a_{01}	a_{02}	a_{10}	a_{11}	a_{12}	a_{20}	a_{21}	a_{22}

← Column 1 → ← Column 2 → ← Column 3 →

Fig. 1.2

Address of element in Column major implementation

In column major implementation address of an element a[i][j] is given by the relation :

Address of element a[i] [j] = B + W(m (j – L 2) + (i - L1))

Where m is the number of rows (i.e., $U_1 - L_1$).

Ex : Each element of an array a [-20..20, 10..35] requires one byte of storage. If the array is column major implemented, and the beginning of the array is at location 500 (base address), determine the address of element a [0,30] or a [0] [30].

Sol : The following information is given :

Base address	B = 500
Size of each element	W = 1 byte
Lower bound of row	L_1 = -20
Lower bound of column	L_2 = 10
Upper bound of row	U_1 = 20
Upper bound of column	U_2 = 35

Row number of the required element i = 0

Column number of required element j = 30

Number of rows m for this configuration will be :

$U_1 - L_1 + 1 = 20 - (-20) + 1 = 41$

As address of a particular element in two-dimensional array (column major implemented) is given by the relation:

Address of a[i] [j] = B + W (m (j – L2) + (i – L1))

Therefore address of element a[0][30] will be

Address of a[0] [30] = 500 + 1 (41(30 – 10) + (0 – (-20)))

= 500 + 1 (41 * 20 + 20)

= 500 + 1 (820 + 20)

= 500 + 840 = 1340

6. Stack

Stack is a linear list in which insertions (also called additions) and deletions take place at the same end. This end is called the top. The other end of the list is called the bottom.

Figure 1.3 (a) shows a stack with four elements, suppose we wish to add the element E to the stack of fig 1.3 (a). This element will have to be placed on top of the element D, giving us the configuration of figure 1.3 (b), it will be the element E.Following the deletion, the configurationof figure1.3(a)results.If we perform three successive deletions on the stack of figure1.3(b),the stack figure1.3 (c) results.

(a)		(b)		(c)	
		E ← top			
D ←	top	D			
C		C			
B		B		B← top	
A ←	bottom	A ←bottom		A ←bottom	

Fig 1.3 : Stack configurations

Operations on a stack : The following operations that are usually performed on a stack.

1) Create ( ) : Create an empty stack.

2) IsEmpty( ) : Return true if stack is empty, return false other wise.

3) IsFull ( ) : Return true if stack is full, return false otherwise.

4) Top ( ) : Return top element of stack.

5) Push (x) : Add element x to the stack.

6) Pop (x) : Delete top element from stack and put it in x.

Algorithms For **PUSH** And **POP**

Algorithm

I. Algorithm for Inserting an item into the stack (PUSH)

```
Let stack [ Max size ] is an array for implementing the
stack.
1. [ Check for stack overflow ? ]
If  Top= Max size -1, then : print : overflow and exit.
2. Set TOP = TOP +1 [ increase TOP by 1 ]
3. Set stack [ TOP ] := ITEM [inserts item in new TOP
position]
4. Exit
```

II. Algorithm for Deleting an element from the stack

This algorithm deletes the top element of stack and assign it to variable Item.

```
1. [ Check for the stack underflow ]
   If TOP < 0, then
   Print " Stack underflow " and exit.
   Else [ remove the TOP element ]
   Set item = stack [ TOP ]
2. Decrement the stack TOP
   Set TOP = TOP - 1
3. Return the deleted item from the stack
4. Exit
```

7. Queue

A queue is a linear list in which additions and deletions take place at different ends. The end at which new elements are added is called the **rear**, and that from which old elements are deleted is called the **front**.

A queue with three elements is shown in figure 1.4 (a). the first element we delete from the queue of figure 1.4(a), is A. Following the deletion, the configuration of figure 1.4 (b) results. To add the element D to the queue of figure 1.4 (b) we must place it after element C. The new configuration is shown in figure 1.4 (c).

(a)			(b)		(c)		
front ↓		rear ↓	front ↓	rear ↓	front ↓		rear ↓
A	B	C	B	C	B	D	C

Fig 1.4 : Sample queues.

Operations on a Queue : The following operations that are usually performed on a queue.

1) Create ( ) : Create an empty queue.

2) IsEmpty ( ) : Return true if queue is empty, return false other wise.

3) IsFull ( ) : Return true if queue is full, return false other wise.

4) First ( ) : Return first element of queue

5) Last ( ) : Return last element of queue.

6) Add (x) : Add element x to the queue.

7) Delete (x) : Delete front element from queue and put it in x.

Algorithms For Insertion And Deletion In Queue

Let Queue be the array of some specified size say MAXSIZE, then the insertion algorithm will be as given below. While implementing a queue using arrays we must check underflow and overflow conditions for queue.

Algorithm for Addition in a Queue

```
1. Initialize Front = -1 and Rear = -1
2. If Rear > = MAXSIZE  [ check overflow condition]
   Write Queue Overflow and return
   else :
   Set Rear = Rear + 1
3. Queue [ Rear ] = Item
4. If Front = -1           [ set the front pointer ]
5. Return.
```

Algorithm for Deletion From a Queue

```
1. if front = 0[ Check Underflow condition ]
   Write Queue is Empty and return
2. else : item = Queue [ Front ] [ Remove Element
   From Front ]
3. Find new value of front.
   If ( Front = Rear ) [ Checking For Empty Queue]
   Set front = 0, Rear = -1[ Re-Initialize the pointers]
   else :
   Front = Front + 1
```

8. Variations in a Queue

There are several variations we can made to a simple queue, to suit our requirement. Three of the major variations made in a simple queue are :

1. Circular queue.
2. Double-ended queue (de-queue).
3. Priority queue.

9. Circular Queue

A circular queue is one in which the insertion of a new element is done at the very first location of the queue if the last location of the queue is full. In other words if we have queue Q of say n elements, then after inserting an element last (i.e., in the n-1th) location of the array the next element will be inserted at the very first location (i.e., location with subscript 0) of the array. It is possible to insert new elements, if and only if those locations (slots) are empty. We can say that a circular queue is one in which the first element comes just after the last element.It can be viewed as a mesh or loop of wire, in which the two ends of the wire are connected together. Fig 1.5 shows a empty circular queue Q[5] which can accommodate five elements.

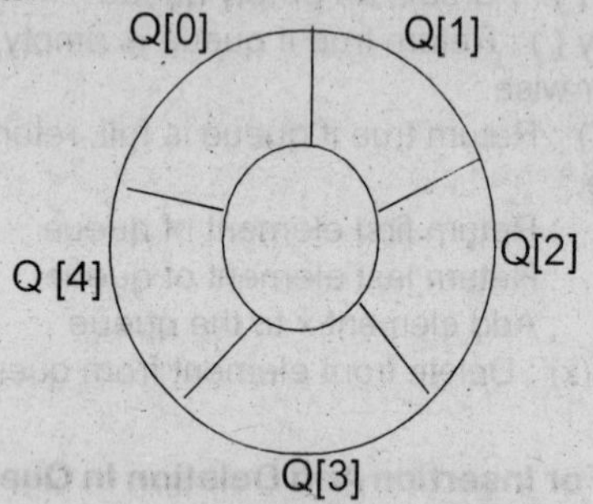

Figure 1.5 Circular Queue

A circular queue overcomes the problem of unutilized space in linear queues implemented as arrays. A circular queue also have a Front and Rear to keep the track of the elements to be deleted and inserted and therefore to maintain the unique characteristic of the queue. The following assumptions are made :

1. Front will always be pointing to the first element(as in the linear queue).
2. If Front = Rear the queue will be empty.
3. Each time a new element is inserted into the queue the Rear is incremented by one.

   Rear = Rear + 1

4. Each time an element is deleted from the queue the value of Front is incremented by one. Front = Front +1

10. Double Ended Queue

It is also a homogeneous list of elements in which insertion and deletion operations are performed from both the ends. That is, we can insert elements from the rear end or from the front ends. Hence it is called double-ended queue. It is commonly referred to as deque.

There are two types of deques. These two types are due to the restrictions put to perform either the insertions or deletions only at one end. They are,

1. Input-restricted deques.
2. Output-restricted deque.

Figure shows a deque of 5 elements.

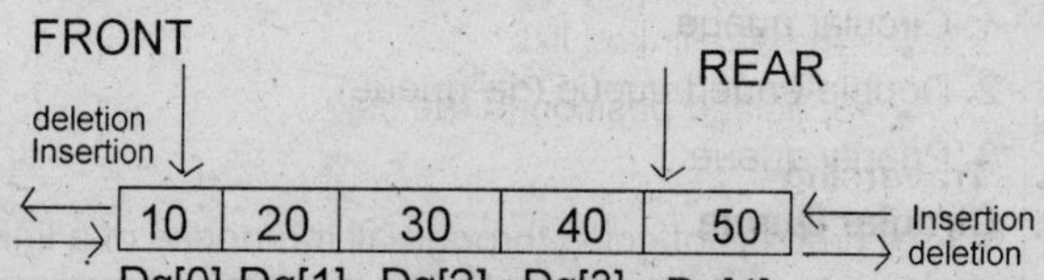

Fig 1.6 : A Dequeue

Since both insertion and deletion are performed from either end, it is necessary design algorithm to perform the following four operations.

11. Priority Queues

Often the items added to a queue have a priority associated with them : this priority determines the order in which they exit the queue-highest priority items are removed first.

This situation arises often in process control systems.Imagine the operator's console in a large automated factory. It receives many routine messages from all parts of the system : they are assigned a low priority because they just report the normal functioning of the system-they update various parts of the operator's console display simply so that there is some confirmation that there are no problems. It will make little difference if they are delayed or lost.

However, occasionally something break fails or alarm messages are sent. These have high priority because some action is required to fix the problem (even if it is mass evacuation because nothing can stop the imminent explosion !).

Typically such a system will be composed of many small units, one of which will be a buffer for messages received by the operator's console. The communications system places messages in the buffer so that communications links can be freed for further messages while the console software is processing the messages. The console software extracts messages from the buffer and updates appropriate parts of the display system. Obviously we want to sort messages on their priority so that we can ensure that the alarms are processed immediately and not delayed behind a few thousand routine messages while the plant is about to explode.

12. Applications of Queues

1. Round robin technique for Processor scheduling is implemented using queues.
2. All types of customer services (like railway ticket reservation) center software's are designed using queues to store customers information.
3. Printer server routines are designed using queues. A number of users share a printer using printer server (a dedicated computer to which a printer is connected), the printer server then spools all the jobs from all the users , to the server's hard disk in a queue. From here jobs are printed one-by-one according to their number in the queue.

13. Linked Lists

Linked lists are special list of some data elements linked to one another. The logical ordering is represented by having each element pointing to the next element. Each element is called a node, which has two parts. INFO part which stores the information and POINTER which points to the next element.

Following Figure 1.7 shows both types of lists(singly linked list and doubly linked list)

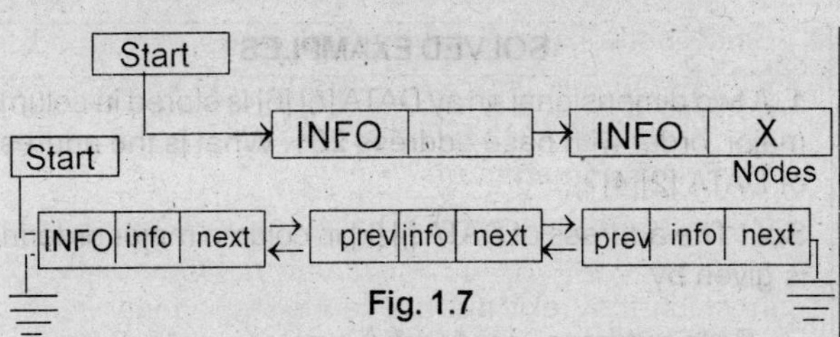

Fig. 1.7

In singly linked list nodes have one pointer (next) pointing to the next node, whereas nodes of doubly linked lists have two pointers (prev and next). Prev points to the previous node and next points to the next node in the list.

Advantages

Linked lists have many advantages. Some of the very important advantages are:

I. Linked lists are dynamic data structures : That is, they can grow or shrink during the execution of a program.

II. Efficient memory utilization : Here,memor is not preallocated.Memory isallocated whenever it is required. And it is deallocated (removed) when it is no longer needed.

III. Insertion and deletions are easier and efficient: Linked lists provide flexibility in inserting a data item at a specified position and deletion of a data item from the given position.

IV. Many complex applications can be easily carried out with linked lists.

Disadvantages

I. More memory : if the number of fields are more, then more memory space is needed.

II. Access to an arbitrary data item is little bit cumbersome and also time-consuming.

Key Terms

As you know, a linked list is a non-sequential collection of data items called nodes. These nodes in principle are structures containing fields. Each node in a linked list basically two fields.

1. Data field
2. Link field

The Data field contains an actual value to be stored and processed. And, the link field contains the address of the next data item in the linked list. The address used to access a particular node is known as a pointer. Therefore, the elements in a linked list are ordered not by their physical placement in memory but their logical links stored as part of the data within the node itself. In other words, the logical and physical ordering of data items in a linked list need not be the same. But in sequential representation these ordering are the same.

Null Pointer

Note that, the link field of the last node contains NULL rather than a valid address. It is a null pointer and indicates the end of the list.

External Pointer

It is a pointer to the very first node in the linked list it enables us to access the, entire linked list.

Empty List

If the nodes are not present in a linked list, then it is called an empty linked list or simply empty list. It is also called the null list.

The value of the external pointer will be zero for an empty list. A linked list can be made an empty list by assigning a NULL value to the external pointer. That is, in our case.

Start = NULL ;

Notations

Let P be s pointer to node in a linked list. Then, we can form the following notations. To do something on linked lists.

- Node (p) : a node pointed to by the pointer p.
- Data (p) : data of the node pointed to by p.
- Link (p) : address of the next node that follows the node pointed to by the pointer p.

suppose, if the value P = 2000, then

Node (p) = Node (2000)

Is the second node, and

Data (p) = Data (2000) = 20

Link (p) = Link (2000) = 3000

Which is the address of the third node.

14. Operations On Linked List

The basic operations to be performed on the linked lists are as follows :

1. Creation 2. Insertion 3. Deletion 4. Traversing
5. Searching 6. Concatenation 7. Display

Let us define these terminology before applying them to any of the four types of linked lists.

Creation

This operation is used to create a linked list. Here, the constituent node is created as and when it is required and linked to the list to preserve the integrity of the list.

Insertion

This operation is used to insert a new node in the linked list at the specified position. A new node may be inserted

- At the beginning of a linked list
- At the end of a linked list
- At the specified position in a linked list
- If the list itself is empty, then the new node is inserted as a first node.

Deletion

This operation is used to delete an item (a node) from the linked list. A node may be deleted from the

- Beginning of a linked list
- End of a linked list
- Specified position in the list.

Traversing

It is a process of going through all the nodes of a linked list from one end to the other end. If we start traversing from the very first node towards the last node, it is called forward traversing. If the desired element is found, we signal operation "SUCCESSFULL". Otherwise, we signal it as "UNSUCCESSFULL".

Concatenation

It is the process of appending (joining) the second list to the end of the first list consisting of m nodes. When we concatenate two lists, the second list has n nodes, then the concatenated list will be having (m+n) nodes. The last node of the list is modified so that it is now pointing to the first node in the second list.

Display

This operation is used to print each and every node's information. We access each node from the beginning (or the specified position) of the list and output the data housed there.

15. Types Of Linked Lists

Basically, we can put linked lists into the following four types:

- Singly-linked list
- Doubly linked list
- Circular linked list
- Circular doubly linked list

1. A singly linked list is one in which all nodes are linked together in some sequential manner. Hence, it is also called linear linked list. Clearly it has the beginning and the end. The problem with this list is that we cannot access the predecessor of node from the current node. This can be overcome in doubly linked lists.

2. A doubly linked list is one in which all nodes are linked together by multiple links which help in accessing both the successor node (next node) and predecessor node (previous node) for any arbitrary node within the list. Therefore each node in a doubly linked list fields (pointers) to the left node (previous) and the right node (next). This helps to traverse the list in the forward direction and backward direction. A doubly linked list is shown in Figure 1.8

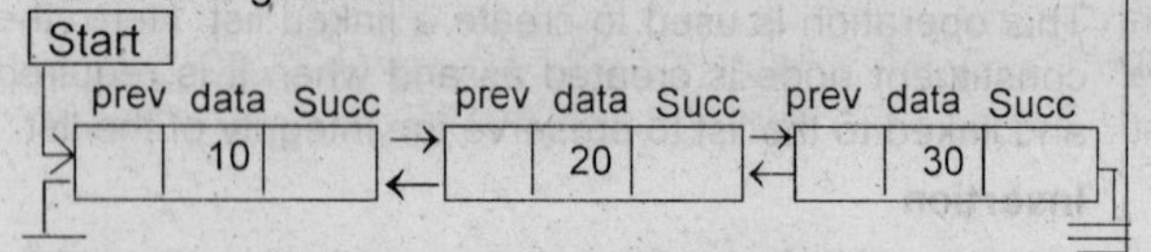

Fig. 1.8 A doubly linked list

3. A circular linked list is one which has no beginning and no end. A singly linked list can be made a circular linked list by simply sorting the address of the very first node in the link field of the last node. A circular linked list is shown in Figure 1.9

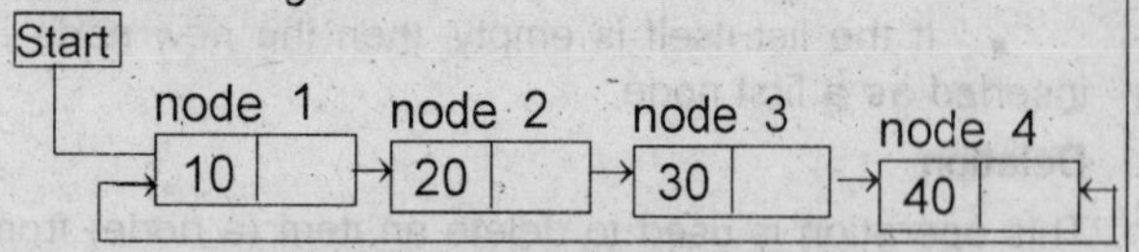

Fig. 1.9. A circular Linked list

4. A circular doubly linked list : is one which has both the succesor pointer and predecessor pointer in circular manner. It is shown in Figure 1.10.

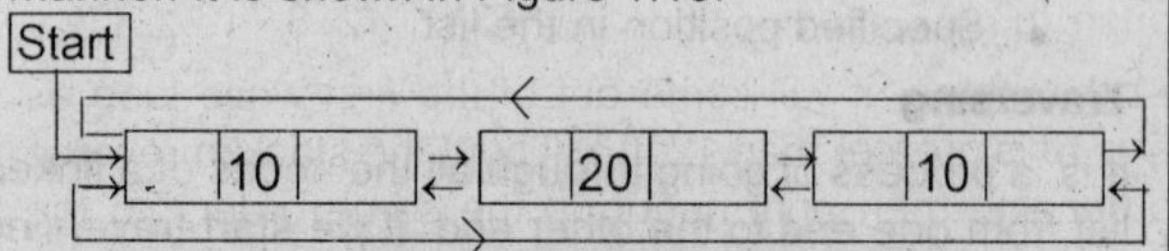

Fig. 1.10. A circular doubly linked list.

SOLVED EXAMPLES

1. A two dimensional array DATA [5] [6] is stored in column major order with base address 201. What is the address of DATA [2][4]?

Sol : The address of DATA [j] [k] in column major ordering is given by

$$\text{Base address } + w*(m*k+j)$$

where w is the size of each element in the array and m is the number of rows in the array.

Assuming w = 1 we can calculate the address of DATA [2] [4] as

$$= 301 + 1\times(5\times4+2) \quad = 301+22 = 323$$

2. Write down the equivalent infix expression for the following post-fix expression

ABC−/ADE+*−

Sol : ABC−/ADE+*−

$$= A(B-C)\ /\ ADE+*-$$

$$= (A/(B-C))ADE+*-$$

$$= (A/(B-C))\ \ A(D+E)+*-$$

$$= (A/(B-C))\ \ (A*(D+E))-$$

$$= ((A/(B-C))-(A*(D+E)))$$

3. A recursive definition of multiplication is as follows:

$$a*b = \begin{cases} a & if\ b=1 \\ a*(b-1)+a & if\ b>1 \end{cases}$$

How many times this recursive multiplication will be called for multiplying **9*4?**

Sol : To multiply 9*4, we call this definition four times.

9*4	(first call)
=9*3+9	(second call)
=9*2+9+9	(third call)
=9*1+9+9+9	(fourth call)
=9+9+9+9	

4. Evaluate the following postfix expression **5, 7, 9, *, +, 4, 9, 3, /, +, −**

Sol : Symbol scanned	Stack
5	5
7	5, 7
9	5, 7, 9
*	5,63
+	68
4	68, 4
9	68, 4, 9
3	68, 4, 9, 3
/	68, 4, 3
+	68, 7
-	61

Therefore, the value of the given postfix expression is 61.

5. Given the following definitions

```
int i = 10 ; j = 5 ;
int * p ;
p = & i ;
j = * p ;
```

and that i is stored at location 500. What are the values of i, p, j ?

Sol : The value of i is 10, p is 500 and j is 10.

6. What is the output of the following slice of code ?

```
char c = 'S' ;
char d = 'T' ;
char * p, * q, * r ;
p = &c ;
q = &d;
r = p ;
p = q;
q = r;
printf("valuesof * p, * qand * rare:");
printf("%c%c%c", * p, * q, * r);
```

Sol : The output of this code is as follows :

Values of * p, * q and * r are T S S

Initially p is pointing to c and q is pointing to d which contains S and T respectively. Statement r=p causes r to point to c. Because of the statement p=q, p starts pointing to d and q = r causes q to point to c. At this point p is pointing to d and q and r are pointing c.

7. Show what is written by the following segments of code. S is declared to be a stack of integers.

```
int x = 3 ;
int y = 4 ;
int z = 5 ;
clear (s) ;
push (s,x) ;
push (s,6) ;
pop (s,z) ;
push (s,y) ;
push (s,2) ;
push (s,z) ;
pop (s,x) ;
push (s,z) ;
while ( ! empty (s) ) {
pop (s, x) ;
print f (" % d", x)
}
```

Sol : The given segment of code will write

6 2 4 3

The stack for various statements will be as follows.

Statements	x	y	z	Stack
clear (s)		3	4	5
push (s,x)	3	4	5	3
push (s, 6)	3	4	5	3, 6
pop (s,z)	3	4	6	3
push (s,y)	3	4	6	3,4
push (s,2)	3	4	6	3,4,2
push (s,z)	3	4	6	3,4,2,6
pop (s,x)	6	4	6	3,4,2
push (s,z)	6	4	6	3,4,2,6

So when the while loop is executed, it pops the top element and prints them in the reverse order as 6, 2, 4, 3.

8. Let s and q are a stack and a queue respectively, of integers and a, b and c are intergers. What is the output of the following segment of code.

```
clear (s) ;
clearq (q) ;
a = 0 ;
b = 1 ;
c = a + b ;
while ( c < 12)
{
        if (c % 2) = 0
        push (s,c) ;
        else
        enq (q,c) ;
        a = b ;
        b = c ;
        c = a+b ;
}
while not empty (s)
{
pop (s,c) ;
printf("%d", c) ;
}
while not emptyq (q)
{
deq (q,c) ;
printf("%d", c) ;
}
```

Sol : The stack and queue for the various statements will be as follows :

Statement	a	b	c	stack	queue
clear (s)					
clear (q)	0	1	1		1
enq (q,c)	1	1	2		1
push (s,c)	1	2	3	2	1
enq (q,c)	2	3	5	2	1, 3
enq (q,c)	3	5	8	2	1, 3, 5
push (q,c)	5	8	13	2, 8	1,3,5

At this stage, it will come out of the first while loop as C = 13 is greater than 12. It will print the stack in reverse order and then print queue in the same order. So the output from this segment is .

8 2 1 3 5

9. Write a single C statement to perform the following task. Each time restart with the original list as given below.

(a) p should point to the first node in the list.

(b) print the value of the next node pointed to by p.

(c) q should point to the second node in the list.

(d) delete the node which is next to the node pointed to by q.

Sol :

(a) p = list

(b) print f ("%d", p -> next)

(c) q = list -> next

(d) q -> next = q -> next -> next

10. What is wrong in the following code.

(a)

```
void successor (NODEPTR p)
{
        while (p!= null)
        {
        print f (p ->info, 'is followed by'
        p -> next -> info ) ;
        }
}
```

(b)

```
int pop (NODEPTR p)
{
        int x ;
        free (p) ;
        x = p -> info ;
        p = p -> next ;
        return (x) ;
}
```

Sol : (a) The loop checks for the condition p ! = null, but the code accesses p $\rightarrow$ next $\rightarrow$ info. To correct this, the loop condition should be changed top $\rightarrow$ next ! = null.

(b) The code tries to access p $\rightarrow$ info and p $\rightarrow$ next after p is freed which is illegal. Therefore, the code should first copy the contents of p and assign it a temporary pointer before freeing it so that the code looks as follows:

```
int pop (NODEPTR p)
{
        int x ;
        NODEPTR q ;
        x = p -> info ;
        q = p ;
        p = p -> next ;
        free (q) ;
        return (x) ;
}
```

11. Show what would be returned by the following recursive function after we call

(a) test (0, 3)

(b) test (10, -7)

(c) text (5,6)

The recursive function is as follows :

```
int test (int a, int b)
{
        if a = = b
        return (1)
        else
        if ( a > b)
        return (0)
        else
        return (a + test ( a+1, b)
}
```

Sol : (a) test (0,3)

a = 0, b = 3

since a < b, it will call

0 + test (1,3)

Again 1 < 3, it will call

1 + test (2,3)

Again 2 < 3, it will call

2 + test (3,3)

Now since a = b, it will return 1 to the previous call. So the output is 4.

(b) test (10, -7)

a = 10, b = -7

Since a > b it will return 0. So the output is 0.

(c) test (6,6)

a = b, b = 6

Since a = b, it will return 1. So the output is 1.

OBJECTIVE QUESTIONS - I

1. A Linear List in which elements can be added or removed either end but not in the middle is known as: **[AUCET-2000]**

a) Stack b) Queue c) Tree d) DeQueue

2. The data structure needed to convert infix notations to postfix form: **[AUCET-2000]**

a) Queue b) Stack c) Tree d) Linked List

3. What type of data access is possible with stack: **[AUCET-2000]**

a) FIFO b) FILO c) LIFO d) None

4. An ordered arrangement of data (such as a table of numbers) identified by a single name is called a/an: **[AUCET-1998]**

a) Stack b) Queue c) Array d) Tree

5. The address of a node in a data structure is called: **[AUCET-1998]**

a) Link b) Pointer
c) Referencer d) any of the above

6. A Queue is served on the principle of: **[AUCET-1997]**
a) FIFO b) LIFO c) FILO d) None

7. An abstract data type defines **[AUCET-2001]**
a) The data objects and their values
b) Data stored in virtual memory
c) Logical data
d) The data objects and associated operations.

8. A stack is a special form of **[AUCET-2001]**
a) a Linear List b) an array
c) a Queue d) a non linear list

9. The number of links to be modified to insert a new node into a circular linked list is: **[AUCET-2001]**
a) 1 b) 2 c) 3 d) 4

10. The Data structure that is used to implement recursion : **[AUCET-2000]**
a) Binary Tree b) Doubly Linked List
c) Stack d) None

11. Elements can be added and removed at the both ends in the following repragentations: **[AUCET-2000]**
a) Doubly Linked List b) Binary Tree
c) Degree d) Bi-connected Graph

12. The minimum number of fields in a node of doubly linked list : **[AUCET-2000]**
a) 1 b) 2 c) 3 d) 4

13. Elements of array are accessed by:
a) accessing function in built-in in data structure
b) mathematical function
c) Index d) None

14. Which of the following permutations can be obtained in the output(in the same order) using a stack assuming that the input is the sequence 1,2,3,4,5 in that order: **[GATE-1994]**
a) 3,4,5,1,2 b) 3,4,5,2,1
c) 1,5,2,3,4 d) 5,4,3,1,2

15. The number of sub strings(of all lengths inclusive) that can be formed from a character string of length n is: **[GATE-1994]**
a) n b) n^2 c) $\frac{n(n-1)}{2}$ d) $\frac{n(n+1)}{2}$

16. In a circular linked list organization, insertion of a record involves modification of : **[GATE-1987]**
a) one pointer b) two pointers
c) multiple pointers d) None

17. Which of the following operations is performed more efficiently by doubly linked list than by linear linked list:
a) Deleting a node whose location is given
b) Searching an unsorted list for a given item
c) Inserting a node after the node with a given location.
d) Traversing the list to process each node.

18. How many values can be held by an array A(-1..m,1..m):
a) m b) m^2 c) m(m+1) d) m(m+2)

19. The initial configuration of Queue is a,b,c,d ('a' is at the front). To get the configuration d,c,b,a. One needs a minimum of :
a) 2 deletions and 3 additions.
b) 3 deletions and 2 additions
c) 3 deletions and 3 additions
d) 3 deletions and 4 additions.

20. Linked Lists are not suitable for which of the following problems:
a) Polynomial manipulation b) Binary search
c) Insertion sort d) Quick sort

21. Which of the following is essential for converting an infix expression to the post fix form:
a) a parse tree b) an operator stack
c) an operand stack d) None

22. Queue can be used to implement:
a) DFS b) BFS c) Radix sort d) Recursion

23. The condition FRONT=REAR indicates "QUEUE FULL" in the following Data structure:
a) Linked List b) Queue
c) Stack d) Circular Queue.

24. The process of accessing data stored on a tape is similar to manipulating data on a
a) stack b) Queue c) List d) Tree

25. Stack is:
a) Static Data structure b) Dynamic Data structure
c) Inbuilt Data structure d) None

26. Direct or random access of element is not possible in:
a) Linked List b) Array c) String d) None

27. Searching a Linked List requires Linked List be created:
a) In sorted order only b) In any order
c) Without underflow condition d) None

28. Inserting a node after a given node in doubly Linked List requires:
a) one pointer is changed.
b) Two pointers are changed.
c) Three pointers are changed. d) None.

29. A technique which collects all deleted space onto free storage List is called:
a) Static memory allocation b) Garbage collection
c) Dynamic memory allocation d) None

30. When a new element is inserted in the middle of a Linked List:
a) only the elements that appear after the new element need to be moved.
b) only the elements that appear before the new element need to be moved.
c) The elements before and after the new element need to be moved. d) None

31. A Data structure in which insertion and deletion can take place at both the end is called:

a) DeQueue b) Stack c) Circular Queue d) None

32. The following Data structure can suffer from overflow problem:

a) Stack implemented using array b) Linked list
c) Circular Linked List d) None

33. Using arrays most efficient implementation of Queue is as :

a) Linear Queue b) Priority Queue
c) Circular Queue d) None

34. We can efficiently reverse a string using a :

a) Linear Queue b) Circular Queue
c) Stack d) Doubly Linked List

35. Which one of the following statements about records and arrays is not correct? **[Hyd.Univ-1993]**

a) Each is a linear and static data structure

b) References to components of array are resolved at run time whereas those for records are resolved at translation time

c) Record is a homogeneous structure whereas array is a heterogeneous structure

d) Each can be multidimensional

36. A famous quote of Niklaus Wirth states: Algorithm + Data Structure = ? **[UGC NET-1997]**

a) Program b) Computer c) Software d) None

37. FORTRAN implementations do not permit recursion because **[UGC NET-1997]**

a) they use static allocation for variables

b) they use dynamic allocation for variables

c) recursion cannot be implemented on all machines

d) FORTRAN is an old language

38. Linked lists are more efficient than arrays for **[UGC NET-1997]**

a) Insertion/deletion b) Retrieval
c) Updation d) None of these

39. In Linked Lists as compared to Arrays **[UGC NET-1996]**

a) Insertions/Deletions are faster, but searching is slower

b) Searching is faster, but insertions/deletions are slower

c) Insertion/Deletions are performed at comparable speed, but searching is slower

d) Searching is performed at comparable speed, but insertion/deletions is slower

40. 'Garbage Collection' is a technique of **[UGC NET-1996]**

a) Memory management b) Processor management
c) I/O management d) File management

41. The data structure used in subroutine calls is **[UGC NET-1996]**

a) Queue b) Linked List c) Stack d) Array

42. Which one of the following is not a data structure? **[UGC NET-1996]**

a) B-tree b) Heap c) Stacks d) Integer

ANSWER KEY - I

1. d	2. b	3. c	4. c	5. d
6. a	7. d	8. a	9. b	10. c
11. a	12. c	13. c	14. b	15. d
16. b	17. a	18. d	19. c	20. b
21. b	22. c	23. b	24. b	25. c
26. a	27. b	28. b	29. b	30. d
31. a	32. a	33. a	34. c	35. c
36. a	37. a	38. a	39. c	40. a
4 41. c	42. d			

EXPLANATORY NOTES - I

5. In any data structure, data is arranged in a specific order and each data item maintains certain communication with other items in terms of address. Such an address can be reffered as Link, pointer and referencer.

7. An abstract datatype is different from primitive datatypes. It means data and its corresponding operations that operate upon the data. It is also termed as ADT.

9. A circular linked list is basically to provide flexibility in tranversing throughout the list for many times. It also avoids the 'null' pointer.

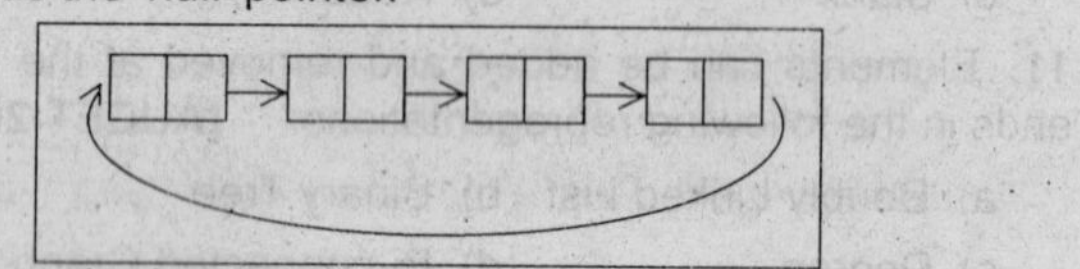

i.e., the last node consists of the address of the first node of the list.

To insert a new node any where in the list, you need 2 pointers to be modified.

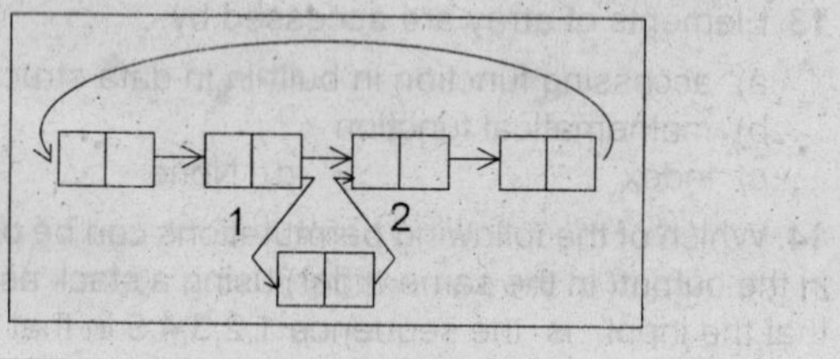

14.

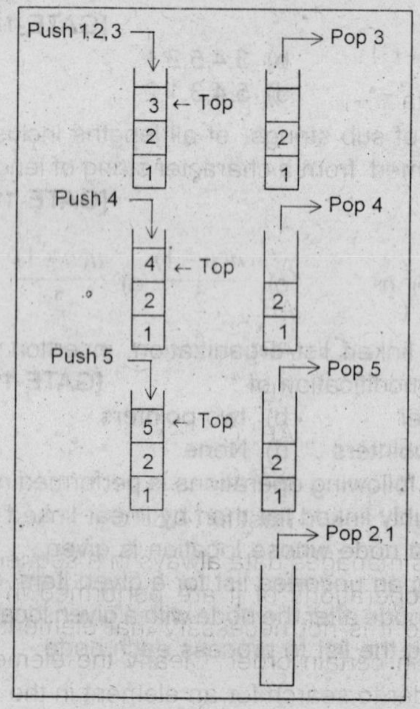

15. Consider a string "Rama". Whose length n = 4 so the substrings are R, a, m, a, Ra, am, ma, Ram, ama, Rama. Total 10 substrings means $\frac{4(4+1)}{2}=10$. Hence the number of substrings that can be formed from a character string of length n is $\frac{n(n+1)}{2}$

17. Deletion and insertion of a node at a given location, traversing the list to process each node are the general operation that are performed by a linked list irrespective of its type. But searching process is more efficiently performed by Doubly linked list compare to single L.L. because S.L.L. uses linear search, doubly L.L. can make use binary search.

Obviously Binary search is better than linear search.

18. Here array size includes two values called row and column. In the given array the row index starts from -1 and ends with m which include '-1' and '0'. Clearly the values are -

-1, 0, 1, 2 .... m

For column it is 1, 2 ..... m

The no. of elements in array with 2 dimension is m ×n. Here m represents no. of rows.

n represents no. of columns.

$\Rightarrow$ m row

$\underbrace{-1,0,}_{2}\underbrace{1,2,\ldots\ldots m}_{m} \Rightarrow m+2$ elements

m column

$\underbrace{1,2,\ldots\ldots m}_{m} \Rightarrow m$

∴ Total no. of elements = $m(m+2)$.

21. An operator stack is required.

for Ex.

Infix : $((A*B)-C)$

Char	Stack	Output
(	(	
(	((	
A	(, (	A
*	(, (*	A
B	(, (, *	AB
)	(	AB*
-	(, -	AB*
C	(, -	AB* C
)		AB* C -

Post fix : AB * C -

27. A linked list is manages data always in a sequential manner. All the operations of it are performed in the same manner. So it is not necessary that elements in the list must be in certain order. Clearly the elements can be in any order to search for an element in the list.

28. Similar to a linked list, in doubly linked list we need 2 pointers to be modified to inset a new node at a given location.

See the following :

List : head

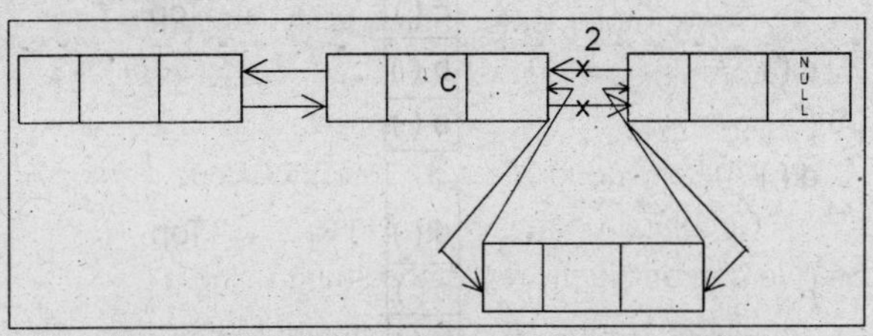

in the diagram the 1,2 pointers are to be modified.

32. A data structure 'stack' can be implemented in two different ways. They are using

i) array method ii) Linked list method

If the implementation of array method, as the array size is fixed, there may be a possibility of OVERFLOW problem.

33. Linear queues can be implemented in a most efficient way using Arrays because an array doesn't require any additional overheads to implement a queue of linear type.

34. Stack follows LIFO manner in data access. i.e. the lastly inserted element comes out first. In reversing a string. The efficient data structure is nothing but a stack.

Consider an example of a string. "GATE". Push Each character into stack

Then pop all the characters and display.

ETAG Answer.

35. A record is a collection of either homogeneous or heterogeneous data items where as an array is a collection of homogeneous data items only. An array is fixed is size where as a record is not.

38. One of the major advantages of linked lists over arrays is nothing but the operations : insertion and deletion.

In arrays inserting a new element at a given location may lead to reshufling entire elements from their original positions. But it is not the case with linked list. They perform just address transformations. So is the case even in deleting an element.

39. In linked list, the operations like insertion/deletion made easy and comparably faster than in arrays. But the process of searching is slower as the linked list follows sequential order in performing so.

40. Garbage collection is a process of gathering the unused memory spaces allocated to certain objects as a seperate list. So that such space can be reutilized further. In other words best utilization of memory is possible through garbage collection. This process is part and parcel of memory management of an operating system.

41. Ex :

```
a ()
{
b ();
}
```

Stack

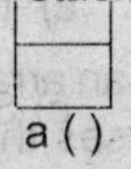

← Top

```
b ()                 | b () |  ← Top
{                    | a () |
c ()
}                    | c () |  ← Top
                     | b () |
c ()                 | a () |
{
d ()                 | d () |  ← Top
                     | c () |
}                    | b () |
                     | a () |
```

42. 'Integer' is basically a data type but not a data structure. A data structure is able to handle data in a more efficient manner. Where as a data type tells about just the type of the data being used in the application.

OBJECTIVE QUESTIONS - II

1. What are the contents of stack (initially stack is empty) after the following operations: **[AUCET-2000]**

PUSH(A) ; PUSH (B) ; PUSH(C) ; POP ;
PUSH (O) ; POP; POP ; PUSH (E) ; PUSH (F); POP?

a) ABE b) AE c) A d) ABCE

2. The five items A,B,C,D and F are pushed in a stack, one after the other starting from A. The stack is popped four times and each element is inserted in a Queue.The elements are deleted from the Queue and pushed back on the stack. Now one item is Popped from the stack The Popped item is.... **[AUCET-1998]**

a) A b) B c) C d) D

3. Stack A has the entries a,b,c (with 'a' an top) Stack B is empty. An entry popped out of stack A can be printed immediately or pushed to stack B, An entry Popped out of Stack B, can only be printed . In this arrangement which of the following permutations of a,b,c are not possible: **[AUCET-1997]**

a) bac b) bca c) cab d) abc

4. Dummy node or head node is an additional node introduced into a Linked list to: **[AUCET-2001]**

a) Duplicate the nodes
b) Ensure a single algorithm for all linked list manipulations
c) Reduce order of complexity
d) Ensure simpler algorithms for linked list manipulations.

5. Pick out invalid statement from the following : Queues can be used for

a) The line printer b) Access to disk storage
c) function call d) None

6. What is the minimum number of stacks of size 'n' required to implement a Queue of size 'n'?**[GATE-2001]**

a) one b) two c) three d) four

7. Suppose you are given an array s[1..n] and a procedure reverse(s,i,j)which reverses the order of elements in s between positions i and j (both inclusive) what does the following sequence do where1 $\leq$ k < n:**[GATE-2000]**

reverse(s,1,k);
reverse(s,k+1,n);
reverse(s,1,n);

a) Rotates s left by k positions.b) Leaves s unchanged
c) Reverses all elements of s. d) None

8. Suppose we want to arrange the n numbers stored in an array such that all negative values occur before all positive ones. Minimum number of exchanges required in the worst case is: **[GATE-1999]**

a) n-1 b) n c) n+1 d) None

9. Let A be a two dimensional array declared as follows:

A: array[1..10][1..15] of integer;

Assuming that each integer takes one memory location, the array is stored in row-major order and the first element of the array is stored at location 100, whatis the address of the element A[i] [j] : **[GATE-1998]**

a) 15i + j + 84 b) 15j + i + 84
c) 10i + j + 89 d) 10j + i + 89

10. Consider the following declaration of a two-dimensional array in C:

Char a[100][100];

Assuming that the main memory is byte addressable and that the array is stored starting from memory address 0, the address of a[40][50] is **[GATE-2002]**

a) 4040 b) 4050 c) 5040 d) 5050

11. A priority Queue Q is used to implement a stack S that stores characters PUSH (C) is implemented as INSERT(Q,C,K) where K is an appropriate integer key chosen by the implementation. POP is implemented as DELETEMIN(Q) For a sequence of operations the keys chosen are in **[GATE-1997]**

a) non-increasing order b) non-decreasing order
c) strictly increasing order d) strictly decreasing order

12. Consider the following statements:

I. FIFO types of computations are effciently supported by STACKS

II.Implementing LISTS on Linked Lists is more efficient than implementing LISTS on an array for almost all the basic LIST operations

III.Implementing QUEUES on a circular array is more efficient than implementing QUEUES on a Linear array with two indices.

IV.LIFO type of computations are efficiently supported by QUEUES. **[GATE-1996]**

a) II and III are true b) I and II are true
c) III and IV are true d) II and IV aretrue

13. In a compact single dimensional array repragentation for lower triangular matrices(i.e., all the elements above the diagonal are zero) of size n $\times$ n, non-zero elements (i.e., elements of the lower triangle of each row are stored one after another, starting from the first row, the index of the (i,j)th element of the lower triangular matrix in this new repragentation is: **[GATE-1994]**

a) i + j b) i + j – 1 c) $j+\frac{i(i-1)}{2}$ d) $i+\frac{j(j-1)}{2}$

14. The following sequence of operations is performed on a stack:

PUSH(10), PUSH(20) ,POP, PUSH(10)
PUSH (20) , POP,POP ,POP, PUSH(20), POP

The sequence of values Popped out is : **[GATE-1991]**

a) 20,10,20,10,20 b) 20,20,10,10,20
c) 10,20,20,10,20 d) 20,20,10,20,10

15. Which of the following arrays is defined in the statements:

```
# define A 66
# define B 132
.........
.........
char memo[A][B];
```

a) Memo is one dimensional 8712-element character array.
b) Memo is two-dimensional 8712-element character array.
c) Memo is two-dimensional 8712-element integer array.
d) None

16. Availnode is a pointer variable thatpoints to the next available node in a singly linked list of available nodes: If P points to a node currently being accessed in a program, then the program fragment.

```
if p=NIL then
error;
else
begin
p^.next:= availnode;
availnode:=p;
end
```

a) Returns to the pascal heap the space that is occupied by thenode to which P is pointed.
b) Swaps the nodes to which P and availnode are pointing
c) Replaces the information stored in the node pointed to byavailnode with the information stored in the node pointed to by P.
d) Disposes of P by inserting it into thefront of the availnode list.

17. A one-dimensional array A has indices 1..75 . Each element is a string and takes up three memory words. The array is stored starting at location 1120 decimal . The starting address of A[49] is:

a) 1164 b) 1264 c) 1386 d) 1451

18. When we concatenate two strings S1and S2 of size m and n, size of the resultant string is always

a) of size m+n b) of size lessthan m+n
c) of size m ∗ n d) of size greaterthan m+n

19. A is an array of size m × n, stored in the row major order. If the address of the first element in the array is m ,the address of the element A(i,j) (A[0,0] is the first element of the array of the array and each element occupies one location in memory) is :

a) m + (i-1) ∗ m + j-1 b) m + i ∗ m + j
c) m + (j – 1) ∗ m + i – 1 d) m + (i – 1) ∗ n + j – 1

20. Select the statement which best completes the following sentence: “Abstract data type is ....” **[Uni.Hyd-2001]**

a) a data type which is abstract in nature
b) a kind of data type c) a data structure
d) a mathematical model together with a set of operations defined on it

21. Suppose you want to delete the name that occurs before ‘Wanjerkhede’ in an alphabetical listing. Which of the following data structure offers the best efficiency for this operation ? **[Hyd. Univ-2000]**

a) Circular Linked List b) Linked List
c) Doubly Linked List d) Dequeue

22. Suppose the data items numbered 1, 2, 3, 4, 5, 6 come in the input stream in this order. By using a queue which of the following rearrangements can be obtained in the output order? **[Hyd. Univ-2000]**

a) 123456 b) 243651 c) 152436 d) 421356

23. If a circular queue is maintained in an array A[0…n-1] and F and R indicate the front and rear indices, the size of the queue is given by **[Hyd.Univ-1994]**

a) R – F b) R – F + 1
c) R – F + 1 (mod n + 1) d) R – F + 1 (mod n)

24. A singly linked list is implemented in two arrays DATA and POINTER in which POINTER(I) points to the successor of DATA(I). If an element not initially in the list is assigned to DATA(J) then the program fragment

POINTER (J) ← POINTER (I)
POINTER (I) ← J

is one that **[Hyd.Univ-1993]**

a) inserts DATA(J) before DATA(I) in the list
b) inserts DATA(J) after DATA(I) in the list
c) replaces DATA(J) by DATA(I) in the list
d) replaces DATA(I) by DATA(J) in the list

25. A stack A of books (numbered 1 to 6) is lying on a table such that 1 is the bottom most book and 6 is the topmost book. Assume that there is place on the table for just two more stacks B and C of books. Let the stack operation consist of removing the topmost book of a stack and placing it on top of another stack.

If our operation is constrained so that it is only possible to move top-most books from stack A to stack B, or from stack B to stack C, or from stack C to stack A then what is the smallest number of operations required to obtain a stack of books in the order 3,4,5,6,1,2? **[UGC NET-2000]**

a) 5 b) 6 c) 10 d) 12

26. Given a 3-dimentional matrix A(10, 20, 30) of integers in which the starting address of the first element A(1, 1, 1) is 101 and each integer occupies one word of a word-addressable memory. Given that the elements A(6, 11, 15) and A(6, 12, 16) are 301 words a part, what is the address of the element A(4,17,27)? **[UGC NET-2000]**

a) 5017 b) 2407 c) 5464 d) None of the above

Note: For the next two questions, read the following:

Given a 3-dimensional matrix A(10, 20, 30) of integers in which starting address A(1, 1, 1) is 101 and each integer occupies one word of word-addressible memory. Given that the elements A(6, 11, 15) and A(6, 12, 16) are 201 words apart. Then

27. The address of the element A(4, 17, 27) is [UGC NET-1995]

a) 4959 b) 5377 c) 1797 d) 5416

28. The address 5001 corresponds to **[UGC NET-1995]**

a) A(1, 17, 6) b) A(17, 1, 6)

c) A(6, 1, 25) d) A(6, 17, 1)

Note : For next two questions, read the following:

A stack A of books (numbered 1 to 6) is lying on a table such that 1 is the bottom-most and 6 is the top-most book.

Assume that there is place on the table for just more stacks say B and C of books. Let a stack operation consist of removing the top-most book of a stack and placing it on the top of another stack.

29. What is the smallest number of operations required to get a stack of books in the order 1, 2, 5, 4, 3, 6? **[UGC NET-1995]**

a) 12 b) 9 c) 18 d) 13

30. What is the smallest number of operations required to have a stack 6, 5, 4, 3, 1, 2? **[UGC NET-1995]**

a) 7 b) 12 c) 13 d) 10

ANSWER KEY - II

1. b	**2.** d	**3.** ?	**4.** d	**5.** a
6. b	**7.** a	**8.** d	**9.** b	**10.** b
11. d	**12.** a	**13.** c	**14.** b	**15.** b
16. d	**17.** b	**18.** b	**19.** ?	**20.** d
21. c	**22.** a	**23.** b	**24.** a	**25.** d
26. b	**27.** c	**28.** b	**29.** b	**30.** a

EXPLANATORY NOTES - II

1. Consider an empty stack STK. TOP =0.

Then PUSH (A);
PUSH (B);
PUSH (C);

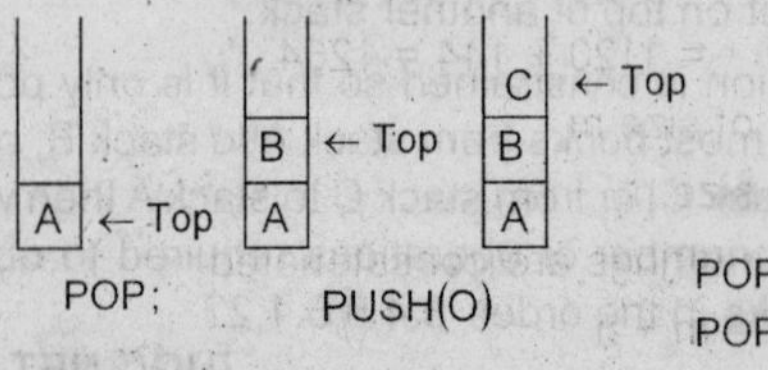

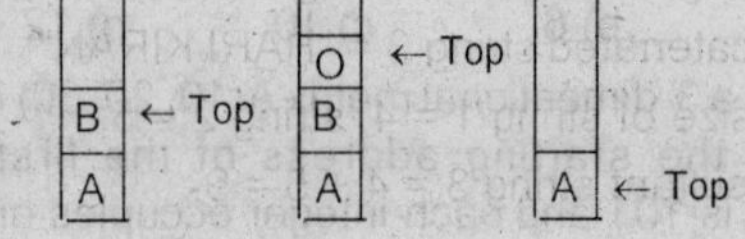

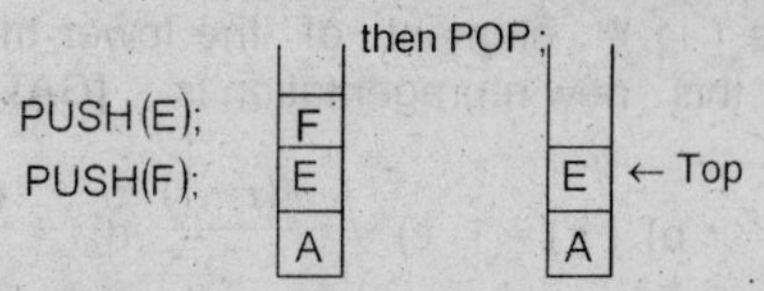

Now contents are - AE.

2. Consider an empty stack 'STK'

Push A, B, C, D, F into STK.

Now, Stk looks like this

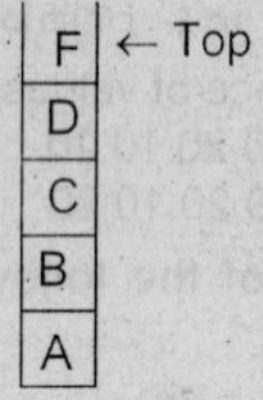

POP 4 items from 'STK' and insert into an empty 'Queue'. Now contents of 'STK' and 'Queue' are -

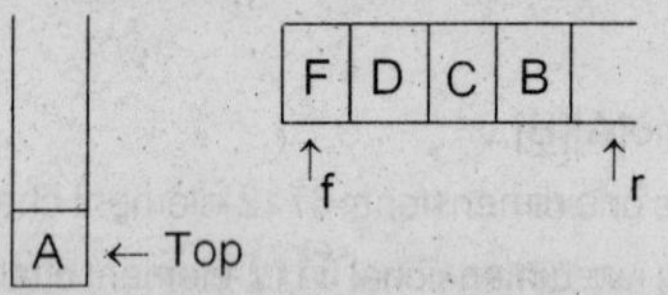

Delete all elements from queue and push back onto the STK. Then the status is as follows.

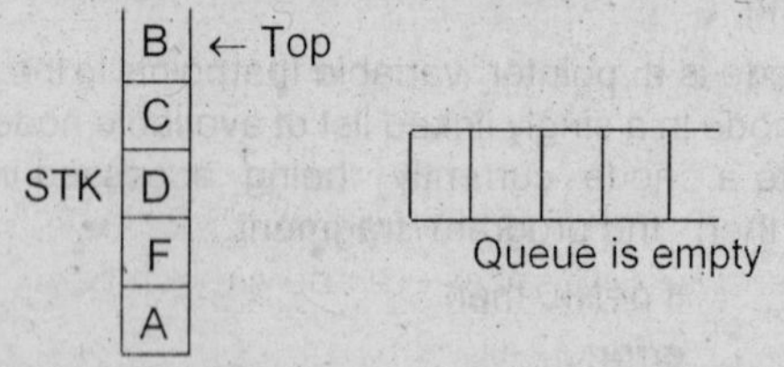

Now POP one item from 'STK' and the item is 'B'.

3. Assume stack A organization is

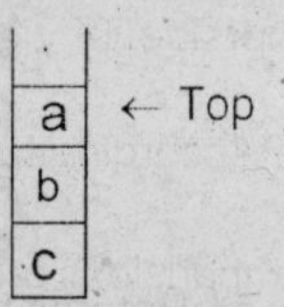

To print c must POP a, b elements and Push them into B. Diagrams shows these operations.

Now, the printed sequence is cba. So only the possible sequence to print start with C is cba. Hence cab is not possible.

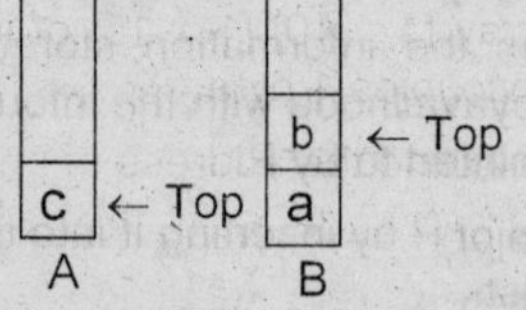

4. Some times it is desirable to keep an extra node at the front of a list. Such a node doesn't hold an item of list and is called header node (or) list header. In general, header node holds global information regarding entire list such as no. of nodes in the list etc.,

i.e. we need not perform 'count' operation on the list to calculate no. of items/ nodes as the list nodes consists of the same.

So, we say that an additional node to the list may ensure simpler algorithms like count ( ), for linked list manipulations.

6. Consider a stack with elements 1, 2,3, 4, 5 push all elements

⇒ | 1 | 2 | 3 | 4 | 5 | ↑ Top

if the same order in which the elements are pushed in required to be the output.

i) Take another stacks.

ii) Pop all elements from first and push into second.

⇒ | | | | | | ↑ Stack A Top | 5 | 4 | 3 | 2 | 1 | Stack B ↑ Top

ii) POP all from stack and display the elements.

⇒ 1, 2, 3, 4, 5

Hence we require 2 stacks.

7. Let us consider an array with a set of elements.

1	2	3	4	5	6

← S →

Apply the routines :

reverse (S, 1, K) ;
reverse (S, K+1, n) ;
reverse (S, 1, n) ;

Assume K = 3

For First call, items of array are	3	2	1	4	5	6
For Second call, items are	3	2	1	6	5	4
For Third call, items are	4	5	6	1	2	3

Here the array, S is rotated left by K positions.

8. Consider a list with set of +ve and -ve numbers.

5	−1	4	−2	−3	6

To set of all +ve no's to one side and all -ve no's to the otherside, any general sorting technique is needed.

According to any sorting technique a min. of $\frac{n(n+1)}{2}$ comparisions are required and almost nearest to $\frac{n(n+1)}{2}$ swapping may be needed.

It maynot be equavalent to n-1, n, n+1.

9. Given A : array [1. . 10] [1. . 15] of integer; Assume each element occupies 1 byte.

∃ a formula to find the address of an element,

(For row major)

$\Rightarrow B+W(n(i-L_1)+(j-L_2))$

Here B = base address = 100

W = Size of an element = 1 byte

n = no. of columns = 15

L_1 = Lower bound of row = 1

L_2 = Lower bound of column =1

Substitute the values.

$\Rightarrow 100+1(15(i-1)+(j-1))$

= 100+15 i - 15 + j - 1 = 15 i + j + 84.

10. Base Address = 0

W = 1 bytes

n = 100

$L_1=0, L_2=0$

Apply value in $B+W(n(i-L_1)+(j-L_2))$

⇒ 0 + 1 (100 (40 - 0) + (50 - 0))

= 100 * 40 + 50 = 4000 + 50 = 4050

11. Priority queue is used to implement stack means highest elements are pushed into stack first. Means stack elements are organised in ascending order from top to bottom. So the POP operation is strictly decreasing order.

14. The symbolic representation for the following operation on a stack is

PUSH (10) Push (20), POP, Push (10), Push (20) POP, POP, POP, Push (20), POP.

| 10 | | 20 / 10 | Poped element is 20

| 10 / 10 | | 20 / 10 / 10 | Then the poped element's list is 20 20 10 10

| 10 | the poped elements are 20 20 10 10 20

15. Char memo [A] [B]; is declared where A = 66, B = 132. So Memo is a two dimensional 8712 element character array because the no. of elements

$=66\times132=8712$

16. Here avail node is the next node address of a linked list. P is the address of the current node in use. If the given code is executed then

Suppose P availnode

| info | | info | → . . . . NULL

P availnode

| info | → | | → . . . . . NULL

then availnode, P

| info | → | | → . . . . NULL

i.e., the P is inserted before the availnode.

17. Loc (Index) = Base + W [i - 1]

= 1120 + 3 [49 - 1]

= 1120 + 144 = 1264.

18. Strip 1 is of size m

Strip 2 is size n

If both the strings are concatenated

total size = m + n

Ex. string 1 = "HARI" ; string 2 = 'KIRAN' ;

If concatenated string 3 = "HARI KIRAN"

Here size of string 1 = 4, string 2 = 5.

the resultant string 3 = 4 + 5 = 9.

19. Actually $\Rightarrow m+1(j(i-0)+(j-0))$

$$=m+j*i+j=m+j(i+1)$$

20. An abstract data type is to define data and corresponding set of operations together as a single unit. An example for abstract datatype is nothing but a class.

21. Assume ∃ a doubly linked list that holds a set of names as data and according to the problem there is a name before 'Wanjerkhede'. It becomes easy to delete the previous name because there ∃ a link from one name to other in both the directions. Such link never be existed in other data structures.

22. As a queue follows FIFO (First in First out elements are inserted in the order 1, 2, 3, 4, 5, 6 and deleted in the same order. So the order is 1,2,3,4,5,6.

23. According to the diagram the size of the circular queue is R - F + 1
i.e., 4 - 0 + 1 = 5.

24. (a) Data (i) is the data list

Pointer (i) is the address list (pointer)

According to the program segment

The pointer (i) is assigned to pointer (j)

i.e., next location.

The value j is assigned to pointer list (i).

⇒ An item 'J' is inserted before 'i' is data list.

25. Explanation :

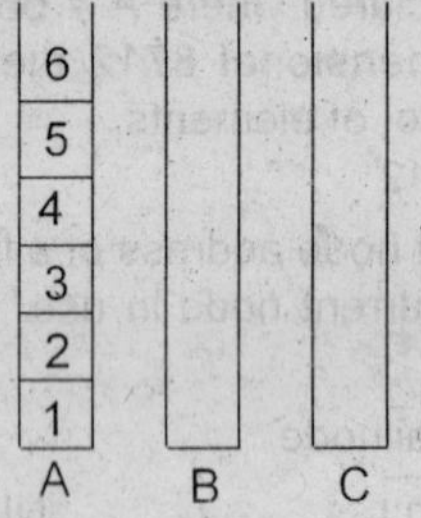

i) Push 4 items of stack A into stack B. (4 operations)

ii) Push are item to B to C (4 operations)

iii) Push remaining items of A to B.

iv) Again push items from B to C (2).

⇒

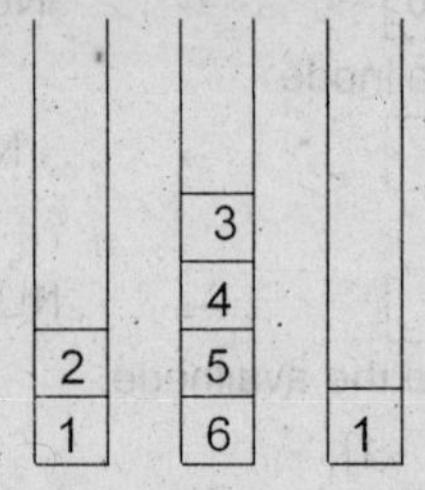

= 4 + 4 + 2+ 2

= 12 operations.

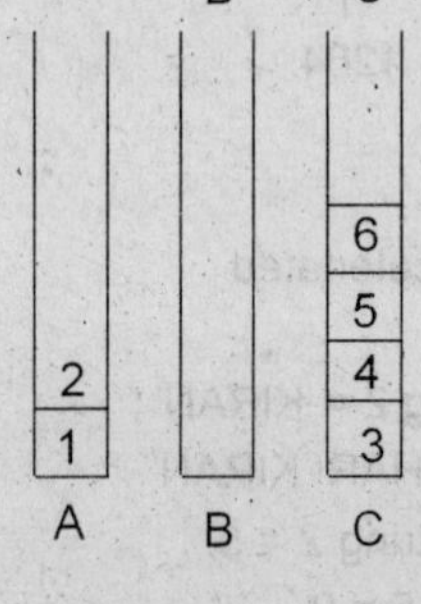

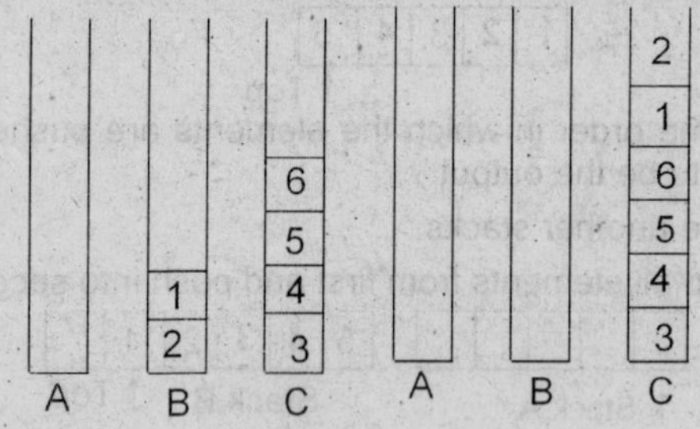

29. Consider the stacks A, B, C.

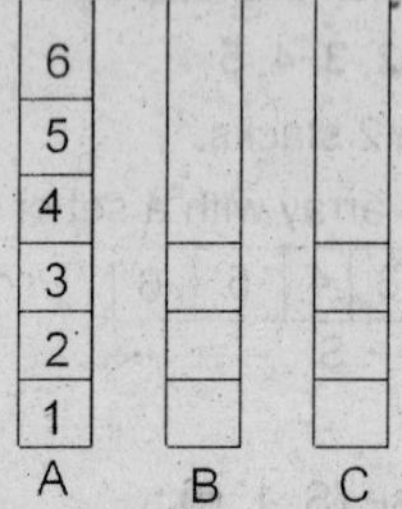

POP 6 and insert to C

POP 5, and push to B, POP 4,3 and push to C.

⇒

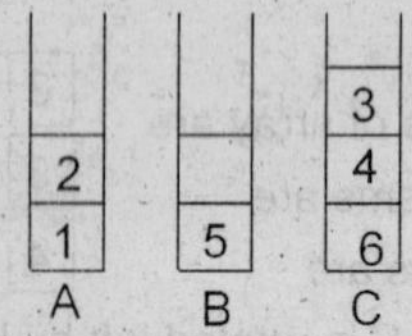

POP 5 from B and push to A

POP 3 and push to B

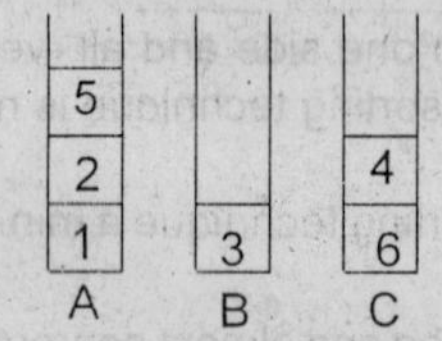

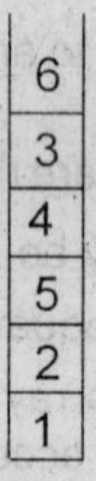

POP 4 from C and push to A

POP 3 from B and push to A

POP 6 from C and push to A

⇒ Total 9 operations.

30. Push 4 items of 'A' to B i.e. 4 operations

Push 1 item of A to C i.e. 1 operations

Push 1 item of A to B i.e. 1 operations

Push one item of C to B i.e. 1 operations

4 + 1 + 1+ 1 = 7

II. TREES

- Tree
- Binary Tree
- Binary Search Tree
- Traversing of a Binary Tree
- Operations on Binary Tree

1. Tree

A tree is a non-linear data structure in which items are arranged in a sorted sequence. It is used to represent hierarchical relationship existing amongst several data items.

The graph theoretic definition of tree is : it is a finite set of one or more data items (nodes) such that

1. There is a special data item called the root of the tree.
2. And its remaining data items are partitioned into number of mutually exclusive (i.e., disjoint) subsets, each of which is itself a tree. And they are called subtrees.

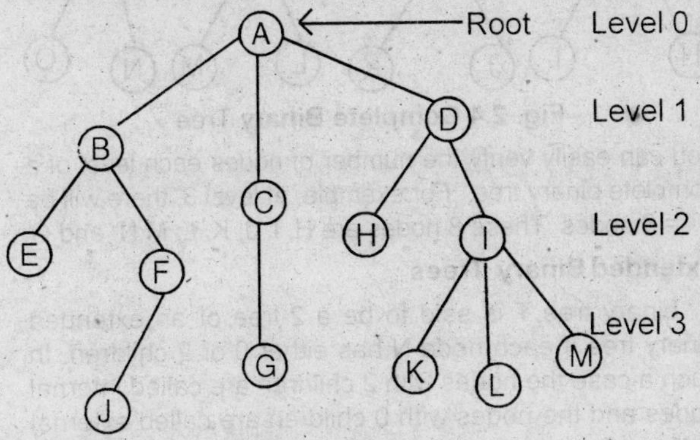

Fig. 2.1 A Tree

Natural tress grow upwards from the ground into the air. But, tree data structure grows downwards from top to bottom. It is an universally practiced convention for trees.

2. Tree Terminology

There are number of terms associated with the trees which are listed below :

1. Root	2. Node	3. Degree of a node	
4. Degree of a tree		5. Terminal node (s)	
6. Non-terminal node (s)		7. Siblings	8. Level
9. Edge	10. Path	11. Depth	12. Forest.

Root

It is specially designed data item in a tree. It is the first in the hierarchical arrangement of data items. In the above tree, A is the root item.

Node

Each data item in a tree is called a node. It is the basic structure in a tree. It specifies the data information and links (branches) to other data items. There are 13 nodes in the above tree.

Degree of a node

It is the number of subtrees of a node in a given tree. In the above tree.

- The degree of node A is 3
- The degree of node C is 1
- The degree of node B is 2
- The degree of node H is 0
- The degree of node I is 3.

Degree of a tree

It is the maximum degree of nodes in a given tree. In the above tree the node A has degree 3 and another node I is also having its degree 3. In all this value is the maximum. So, the degree of the above tree is 3.

Terminal node(s)

A node with degree zero is called a terminal node or a leaf. In the above tree, there are 7 terminal nodes . They are E, J, G, H, K, L, and M.

Non-terminal node(s)

Any node (except the root node) whose degree is not zero is called non-terminal node. Non-terminal nodes are the intermediate nodes in traversing the given tree from its root node to the terminal nodes (leaves). There are 5 non-terminal nodes.

Siblings

The children nodes of a given parent node are called siblings. They are also called brothers . In the above tree,

- E and F are siblings of parent node B.
- K, L and M are siblings of parent node I.

Level

The entire tree structure is levelled in such a way that the root node is always at level 0.Then, its immediate children are at level 1 and their immediate are at leve2 and so on up to the terminal nodes. In general, if a node is at level n, then its children will be at level n+1.

Edge

It is connecting line of two nodes. That is the line drawn from one node to another node is called an edge.

Path

It is a sequence of consecutive edges form the source node to the destination node. In the above tree the path between A and J is given by the node pairs.

(A,B), (B,F) and (F,J)

Depth

It is the maximum level of any node in a given tree. In the above tree the root node A has the maximum level. That is the number of levels one can descend the tree from its root to the terminal nodes (leaves). The term height is also used to denote the depth.

Forest

It is a set of disjoint trees. In a given tree, if you remove its root node then it become a forest. In the above tree, there is forest with three trees.

3. Binary Trees

A binary tree is a finite set of data items which is either empty or consists of a single item called the root and two disjoint binary trees called the left subtree and right subtree.

A binary tree is a very important and the most commonly used non linear data structure. In a binary tree the maximum degree of any node is at most two. That means there may be a zero degree node (usually an empty tree) or a one degree node and two degree nodes. Fig.2.2 shows a binary tree consisting of 10 elements.

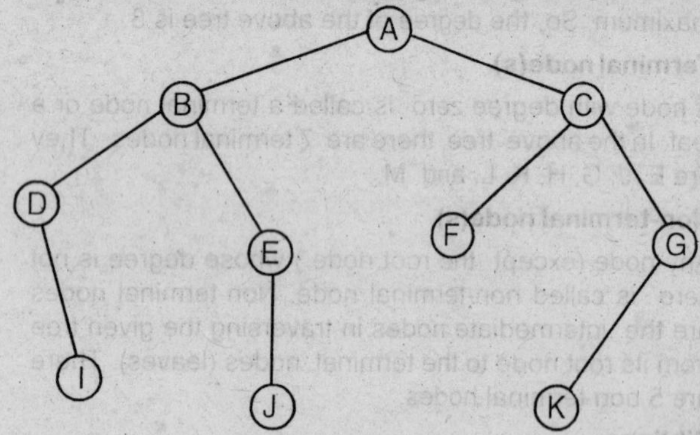

Fig. 2.2 A binary tree

In the above binary tree, A is the root of the tree. The left subtree consists of the tree with root B. And the right sub tree consists of the tree with root C. Further B has its left subtree with root D and right subtree with root E. Similarly, C has its left subtree with root F and its right subtree with root G.In the next level, D has an empty left subtree and its right subtree with root J. F has neither its left subtree nor right subtree K as its root and has an empty subtree.

Strictly Binary Trees

If every non terminal node in a binary tree consist of non empty left subtree and right subtree, then such a tree is called strictly binary tree.

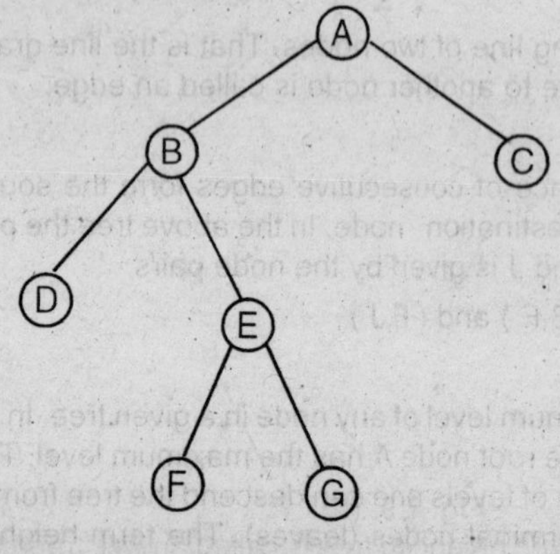

Fig. 2.3. A strictly binary tree.

In the above binary tree all the non-terminal nodes such as B and E having non-empty left and right subtrees.

Complete Binary Tree

A binary tree with n nodes and of depth d is a strictly binary tree all of whose terminal nodes are at level d. In a complex binary tree. There is exactly one node at level 0 two nodes at level 1, and four nodes at level 2 and so on.

If there are m nodes at level i then a binary tree contains at most 2m nodes at level i+1. A binary tree has exactly one node at the root level and has at most 2^i nodes at level i. Taking into consideration of this property, we can show further that a complete binary tree of depth d contains exactly 2^i nodes at each level. The value i ranges from 0 to d.

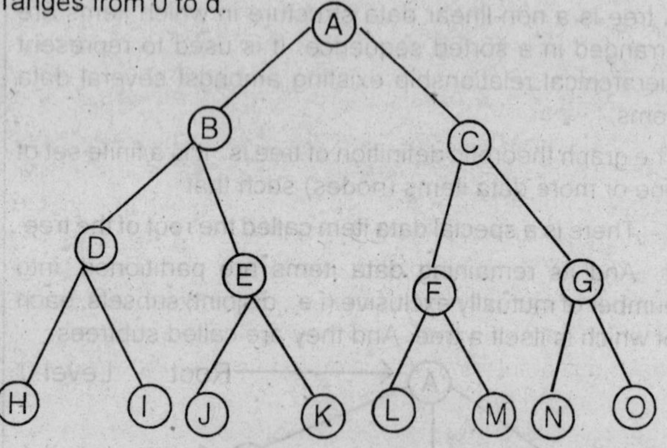

Fig. 2.4 Complete Binary Tree

You can easily verify the number of nodes each level of a complete binary tree. For example, at level 3, there will be $2^3 = 8$ nodes. These 8 nodes are H, I, J, K, L, M,N and O

Extended Binary Trees

A binary tree T is said to be a 2-tree or an extended binary tree if each node N has either 0 or 2 children. In such a case the nodes with 2 children are called internal nodes and the nodes with 0 children are called external nodes. Sometimes the nodes are distinguished in diagrams by using circles for internal nodes and squares for external nodes.

The term extended binary tree comes form the following operation. Consider any binary tree T,such as the tree in fig. 2.5. Then T may be "converted" into a 2-tree by replacing each empty subtree by a new node, as pictured in Fig.2.5. Observe that the new tree is indeed, a 2-tree. New nodes are the external nodes in the external nodes in the extended tree.

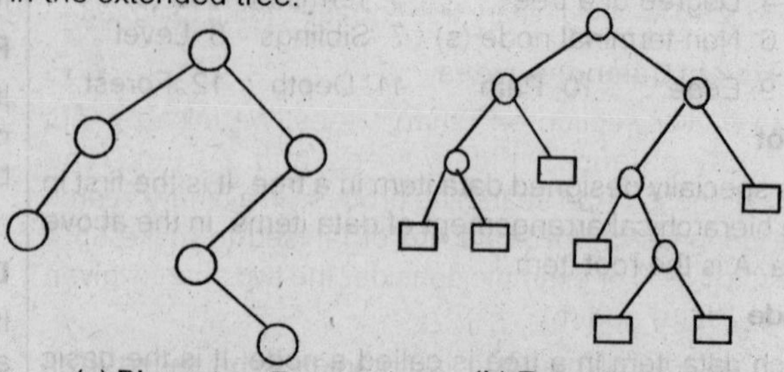

(a) Binary tree T (b) Extended 2 - tree

Fig. 2.5 Converting a binary tree T into a 2-tree

Threaded Binary Trees

In a linked representation of a binary trees, the number of null links (null pointers) are actually more than non-null pointers. Consider the following binary tree.

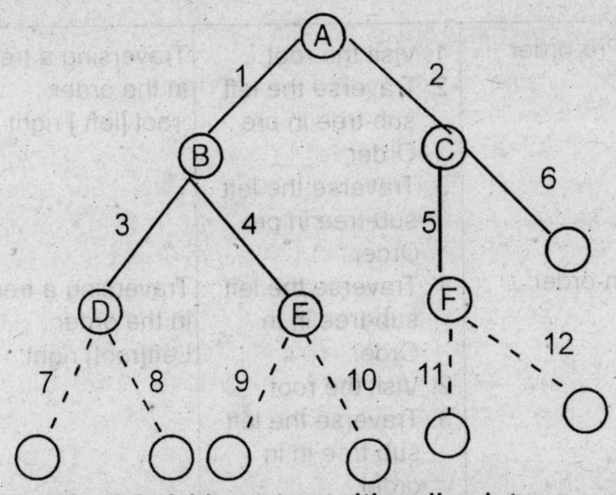

Fig. 2.6. A binary tree with null pointers.

In the above binary tree there are 7 null pointers (number 6 through 12 and shown with dotted lines) and 5 actual pointers. In all there are 12 pointers. We can generalize it that for any binary tree with n nodes there will be (n + 1) null pointers and 2n total pointers. The objective here is to make effective use of these null pointers. A.J.Perlis and C. Thornton jointly proposed an idea to make use of these null pointers. According to this idea we are going to replace all the null pointers by the appropriate pointer values called threads. A binary tree is threaded is threaded according to a particular traversal method. And the null pointer replacement scheme is defined as follows.

If the left link (l child) of a node p is normally equal to NULL then this link is replaced by a pointer to the node which immediately precedes node p in inorder traversal. And similarly if the right link (rchild)of a node p is normally equal to NULL then this link is replaced by a pointer to its inorder successor node. A threaded binary tree is shown in Fig.2.7.

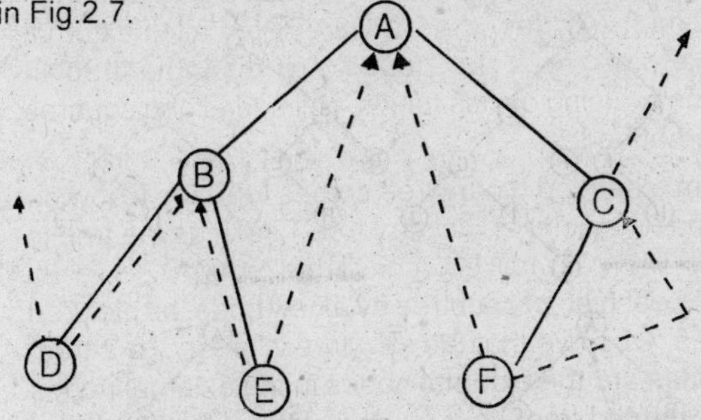

Fig. 2.7.Threaded binary tree corresponding to fig. 2.6

6. Height Balanced Trees

In an earlier section on Binary Trees, we talked about perfectly balanced binary trees and 2-trees. The effectiveness of the searching process in a binary search tree depends on low data are organised to make up a specific tree. For example consider the two shapes given in fig 2.8 (a) and (b)

A binary tree of height h is completely balanced if all leaves occur at nodes of level h or h-1 and if all nodes at levels lower than h-1 have two children. Acccording to this definition, the tree in fig 2.8 (a) is balanced, because all leaves occur at level 3 considering at level 1 and all nodes at levels 1 and 2 have two childern. Intuitively we might consider a tree to be well balanced if , for each node, the longest paths from the left of the node are about the same length as the longest paths on the right.

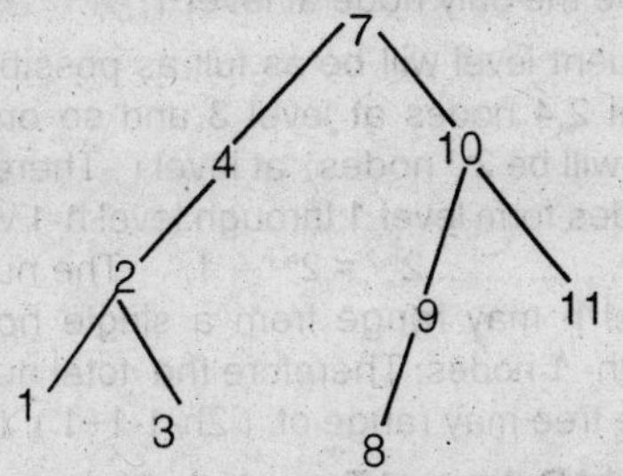

Fig. 2.8. (a) A nearly full binary tree

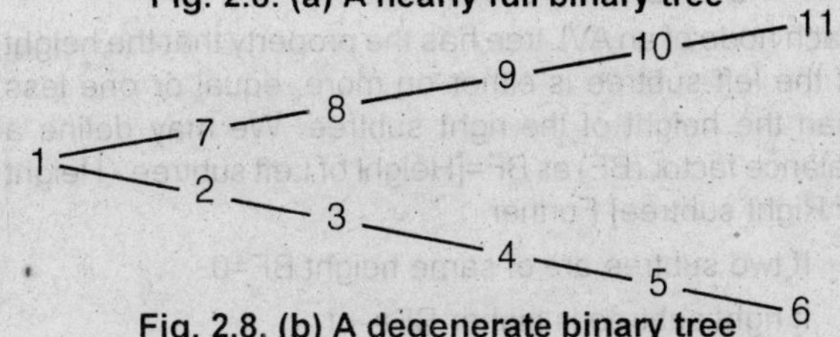

Fig. 2.8. (b) A degenerate binary tree

More precisely, a tree is height balanced if, for each node in the tree, the height of the left subtree differs from the height, of the right subtre by no more than1. The tree in fig 2.9 (a) height balanced, but it is not completely balanced. On the other hand, the tree in fig 2.9 (b) is completely balanced tree.

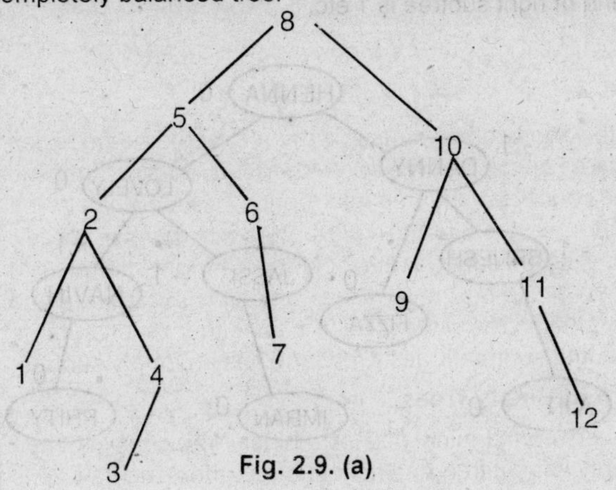

Fig. 2.9. (a)

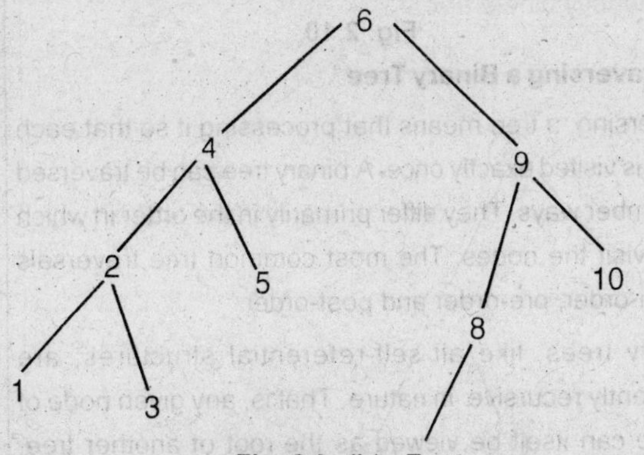

Fig. 2.9. (b) 7

An almost height balanced tree is called an AVL tree after the Russian mathematician G.M. Adelson-Velskii and E.M Lendis, who first defined and studied this form of a tree. AVL Tree may or may not be perfectly balanced.

Let us determine how many nodes might be there in a balanced tree of height h.

The root will be the only node at level 1;

Each subsequent level will be as full as possible i.e., 2 nodes at level 2,4 nodes at level 3 and so on, i.e., in general there will be 2^{i-1} nodes at level i . Therefore the number of nodes form level 1 through level h-1 will be $1 + 2 + 2^2 + 2^3 + ...2^{h-2} = 2^{h-1} - 1$ The number of nodes at level h may range from a single node to a maximum of 2h- 1 nodes. Therefore the total number of nodes n of the tree may range of (2h-1-1+1) to 2h - 1

Building Height Balanced Tree

Each node of an AVL tree has the property that the height of the left subtree is either on more, equal or one less than the height of the right subtree. We may define a balance factor (BF) as BF=[Height of Left subtree - Height of Right subtree] Further

If two subtree are of same height BF=0

If right subtree is higher BF= +1

If left subtree is higher BF= -1

For example balance factor are stated near the nodes is Fig. 2.10. BF of the root node is zero because height of the right subtree and the left subtree is three. The BF at the DIN si -17 because the height of its left subtree is 2 and of right subtree is 1 etc.

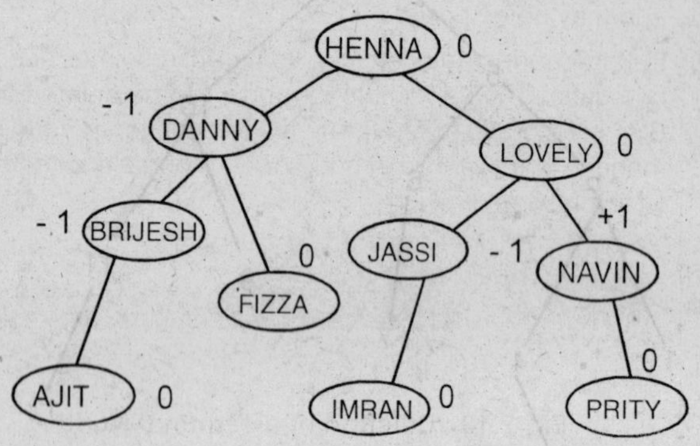

Fig. 2.10

5. Traversing a Binary Tree

Traversing a tree means that processing it so that each node is visited exactly once. A binary tree can be traversed in number ways. They differ primarily in the order in which they visit the nodes. The most common tree traversals are in-order, pre-order and post-order.

Binary trees, like all self-referential structures, are inherently recursive in nature. That is, any given node of a tree can itself be viewed as the root of another tree. That makes the use of recursive calls relatively straightforward for working with trees. The recursive procedures for pre-order, in-order and post-order traversals are given below.

Pre-order	1. Visit the root. 2. Traverse the left sub-tree in pre Order 3. Traverse the left sub-tree in pre Order.	Traversing a tree in the order : root [left] right
In-order	1. Traverse the left sub-tree in in Order. 2. Visit the root 3. Traverse the left sub-tree in in order	Traversing a tree in the order: Left[root] right
post-order	1. traverse the left sub-tree in post order 2. Traverse the right sub-tree in post order 3. Visit the root	Traversing a tree in the order: left [right]root

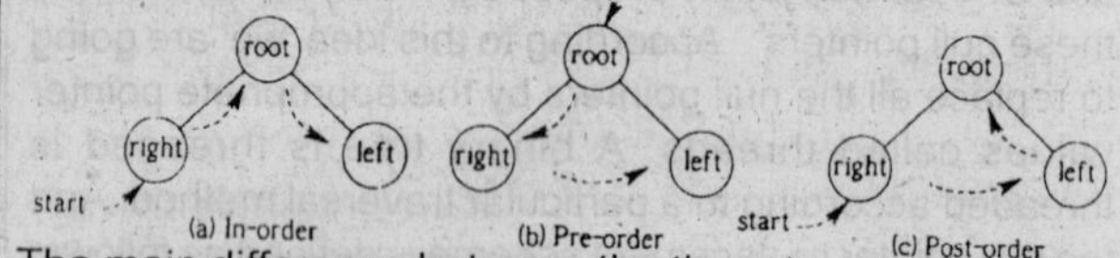

The main difference between the three traversals is that when the root node is processing. Fig presents how the nodes in a tree are visited in pre-, in-, and post-order traversal.

During the traversal, if an empty subtree is encountered meaning that it was traversed that subtree completely.

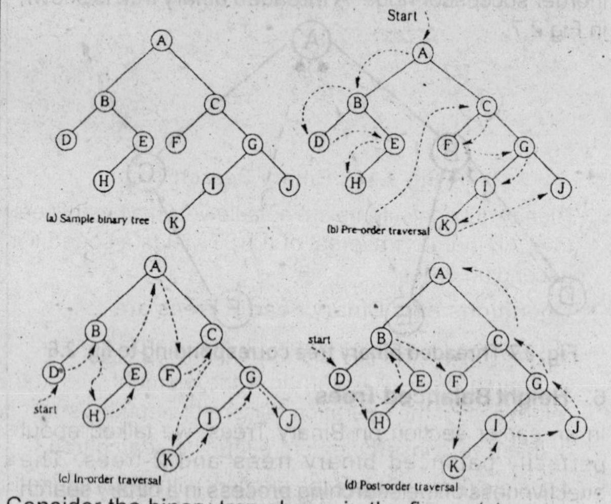

Consider the binary tree shown in the figure (a).

The pre-order traversal is (Fig. (b)) : ABDEHCHGIKJ

The in-order traversal is (Fig. (c)) : DBHEAFCJIGJ

The post-order traversal is (Fig.(d)) : DHEBFKIJGCA

Properties Of Binary Trees

Some of the important properties of the binary trees are:

1. The maximum number of nodes at level 1 of a binary trees 2^1, where $1 \geq 0$.

2. The maximum number of nodes in a binary tree of depth d is 2^{d+1} 1where d $\geq$ 1

3. The total number of edges in a complete binary tree with n terminal nodes is 2(n-1)

4. A binary tree with n nodes has exactly (n+1) null branches.

Disadvantages:

F Not suitable for normal binary trees. It is only ideal for the complete binary trees

F More memory is unnecessarily wasted.

F Insertion and deletion at the middle of the tree is cumbersome.

All these problems can be casily overcome by using linked list representation. Hence, here onwards we use only linked list representation for binary trees.

6. Binary Search Tree

A binary search tree, BST, is an ordered tree T such that either it is an empty tree or

- Each data value in its left subtree less than root value.
- Each data value in its right subtree greater than the root value, and
- Left and right value in its right subtrees are again search trees.

The below fig. 2.11 represents a sample binary search tree.

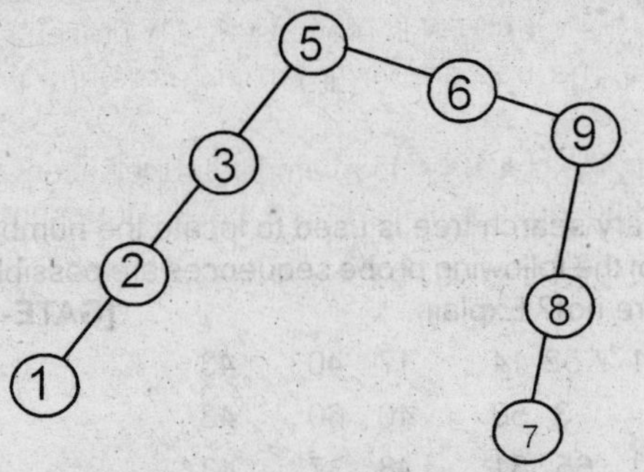

Fig 2.11 : Binary Search Tree

Clearly duplicate items are not allowed in a binary search tree. An inorder traversal of a BST yields a sorted list in ascending order.

Operations on a Binary Search Trees are :

1. Create : Create an empty binary search tree.
2. Search (K,e) : Return in e the element with key K, return flase if the operation fails, return true if it succeds.
3. Insert (e) : Insert element e into the search tree.
4. Delete (K,e) : Delete the element with key K and return it in e
5. Ascend ( ) : Output all elements in ascending order of key.

Insertion in a BST : To insert a node in a BST, we must check whether the tree already contains any nodes. If tree is empty, the nodes is placed in the root node. If the tree is not empty, then the proper location is found and the added node becomes either a left or right child of an existing node. Let us consider a BST given in fig 2.12.

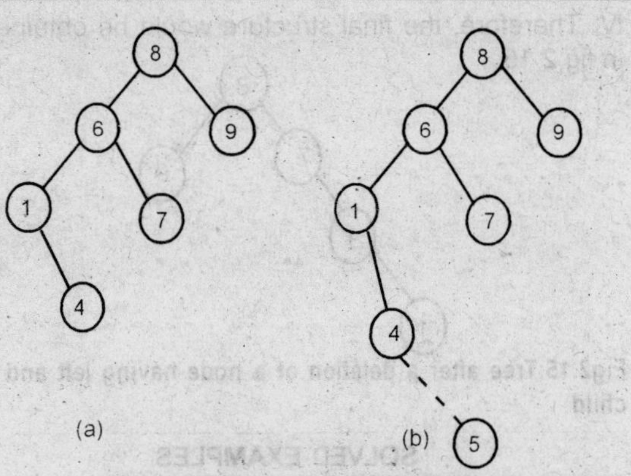

Fig 2.12 : Insertion in a BST

If we want to insert 5 in the above BST, we first search the tree. If the key to be inserted is found in tree, we do nothing (since duplicates are not allowed), otherwise a NIL is returned. In case a NIL is returned, we insert the data at the last point traversed. In the above example a search operation will return NIL on not finding a right subtree of tree rooted at 4. Therefore 5 must be inserted as a right child of 4.

Deletion of a Node : Once again the node to be deleted is searched in BST. If found, we need to consider the following possibilites :

I. If node is a leaf, it can be deleted by making its parent pointing to NIL. The deleted node is now unreferenced and may be disposed off.

II. If the node has one child, its parent's pointer need to be adjusted. For example for node 1 to be deleted from BST given in figure 2.13 (a) ; the left pointer of node 6 is made to point to child of node 1. i.e., node 4 and the new structure would be

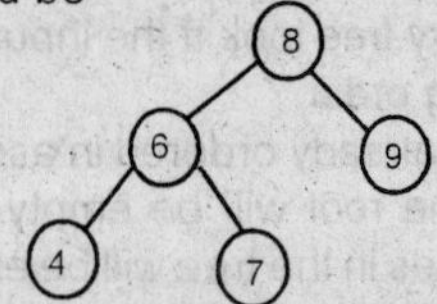

Fig 2.13 : Deletion of a Terminal Node

III. If the node to be deleted has two children; then the value is replaced by the smallest value in the right subtree or the largest key value in the left subtree; subsequently the empty node is recursively deleted.

For example the node 6 is to be deleted from BST given in figure 2.12 (a) then first its value is replaced by smallest value in its subtree, i.e.,by 7 and the new structure would be

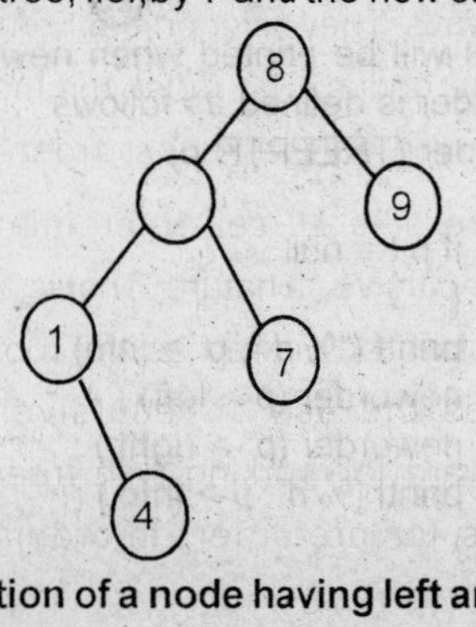

Fig 2.14 : Deletion of a node having left and right child

IV. Therefore, the final structure would be obtained as in fig 2.15.

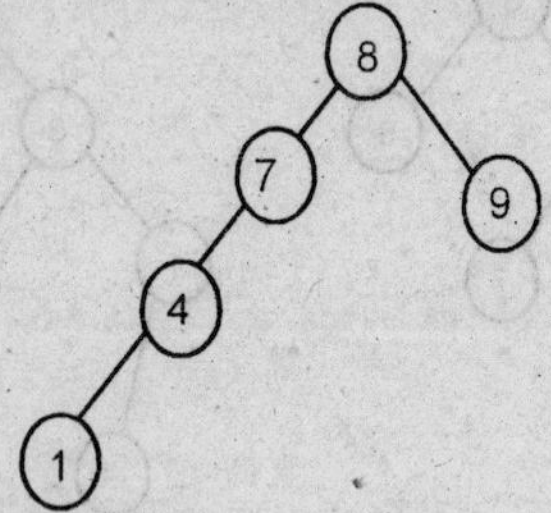

Fig2.15:Tree after a deletion of a node having left and right child

SOLVED EXAMPLES

1. Create a binary search tree for the following sequence

Pinki, Dolly, Gita, Sita, Hari, Alka, Tina

Sol :

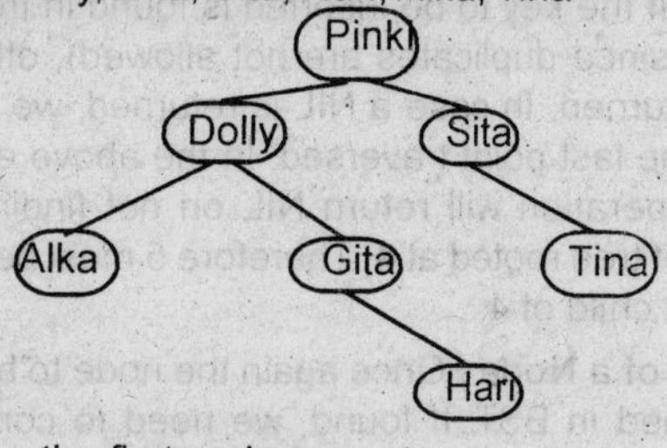

- Place the first node as root.

- If coming node is smaller than root, move towards the left from root, otherwise move towards right. Place the node where you find null pointer along the path.

2. What is the maximum possible number of nodes in the binary tree at level 5 ?

Sol : Maximum number of nodes at any level n in a binary tree is given by 2^n. Therefore, the maximum possible number of nodes in the binary tree at level 5 is 32.

3. How would a binary tree look if the input were already ordered in ascending order ?

Sol: If the input were already ordered in ascending order, the left subtree of the root will be empty. Also, the left subtree of all the nodes in the tree will be empty. The tree will then look like as follows for input 5, 10, 15, 20, 25.

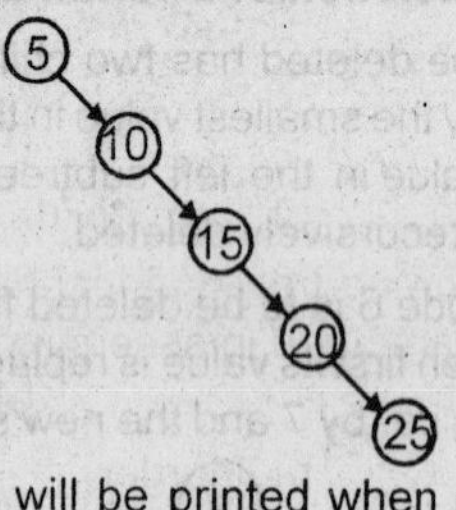

4. What output will be printed when neworder (root) is called ? neworder is defined as follows.

```
void neworder (TREEPTR p) ;
{
        if p ! = null
        {
        printf ("% d ", p -> info) ;
        neworder (p-> left) ;
        neworder (p -> right ) ;
        printf (% d", p-> info ) ;
        }
}
```

root is pointing to the root node of the following tree.

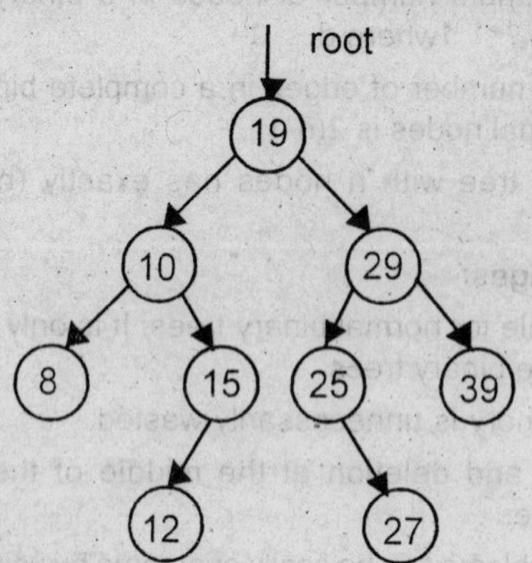

Sol : When neworder (root) is called, it will print the following result.

19, 10, 8, 8, 15, 12, 12, 15, 10, 29, 25, 27, 27, 25, 39, 39, 29, 19

5. What is the number of binary trees with 3 nodes which when traversed in post order give the sequence A, B, C ? Draw all these binary trees. **[GATE-1995]**

Sol : The number of binary trees is 6

In post order sequence → Left Right Root

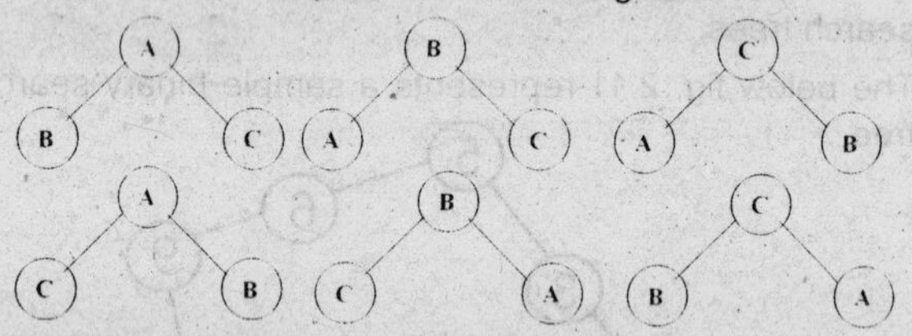

6. A binary search tree is used to locate the number 43. Which of the following probe sequences are possible and which are not? Explain. **[GATE-1996]**

a) 61 52 14 17 40 43

b) 2 3 50 40 60 43

c) 10 65 31 48 37 43

d) 81 61 52 14 41 43

e) 17 77 27 66 18 43

Sol : In binary search tree, the left child of a node is always less than the node and the right child is always larger than the node on applying this rule

a) Possible

b) Not possible, because 60 violates the search tree. This is because 60 can not be in the left subtree of 50

c) Possible

d) Possible

e) Not possible. Element 18 violates the search tree as 18 can not be in the right subtree of 27.

7. Draw the binary tree with node labels a, b, c, d, e, f, and g for which the inorder and postorder traversals result in the following sequences. **[GATE-1998]**

Inorder a f b c d g e

Postorder a f c g e d b

Sol : The required tree is

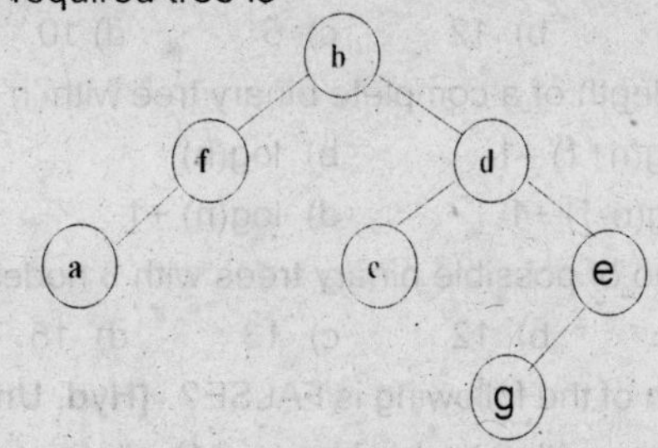

8. (a) Derive a recurrence relation for the size of the smallest AVL tree with height h.

(b) What is the size of the smallest AVL tree with height 8? **[GATE-1998]**

Sol : (a) Let h be the height of the AVL tree. By definition of AVL tree height of right subtree and left subtree differ by almost 1.

They are of the height (h – 1) and (h – 2) and are AVL trees themselves.

Assume f(h) denotes the minimum number of nodes in AVL tree. And this has been constructed by adding a root node with left AVL subtree and right AVL subtree.

$\therefore$ f(h) = f(h – 1) + f(h – 2) + 1

The AVL trees with heights 0, 1, 2 will have minimum 0, 1, 2 nodes

$\therefore$ Required recurrence relation is

f(h) = f(h – 1) + f(h – 2) + 1

f(0) = 0

f(1) = 1

f(2) = 2

(b) Putting h = 8, we have

f(8) = f(7) + f(6) + 1

or solving we get the following table

f(0)	f(1)	f(2)	f(3)	f(4)	f(5)	f(6)	f(7)	f(8)
0	1	2	4	7	12	20	33	54

OBJECTIVE QUESTIONS - I

1. The number of leaf nodes in a rooted tree of n nodes, with each node having 0 or 3 children is: **[GATE-2002]**

a) n/2 b) (n-1)/3 c) (n-1)/2 d) (2n+1)/3

2. Which of the following statements is FALSE: **[GATE-1998]**

a) A tree with n nodes has (n-1) edges

b) A labeled rooted binary tree can be uniquely constructed, given its postorder, preorder traversal results.

c) The complete binary tree with n internal nodes has(n+1)leaves.

d) The maximum number of nodes in a binary tree of height h is $(2^{h+1}-1)$.

3. A Binary tree T has n leaf nodes. The number of nodes of degree 2 in T is: **[GATE-1995]**

a) $\log_2^n$ b) n-1 c) n d) 2^n

4. The post fix notation of A/B * * C+D * E-A * C **[AUCET-2000]**

a) AB ** C/DE * +AC - b) ABC ** D/E * +AC+ -

c) ABC ** /DE * AC+ - d) ABC ** /DE*+AC+ -

5. A full binary tree with n leaves contains **[AUCET-2000]**

a) 2^n nodes b) 2^n-1 node

c) n nodes d) $\log_2^n$ nodes

6. The number of nodes in a complete binary tree of level 5 is: **[AUCET-2000]**

a) 25 b) 15 c) 63 d) 71

7. The maximum number of nodes in a binary tree at level i is: **[AUCET-2000]**

a) 2i b) 2i-1 c) 2^{i-1} d) 2^i-1

8. The maximum number of nodes in a binary tree of depth k is: **[AUCET-2000]**

a) 2k-1 b) 2^{k-1} c) 2^k d) 2^k-1

9. Non-recursive traversal of a binary tree is: **[AUCET-2001]**

a) Impossible b) possible, but requires a stack

c) possible, but requires use of lowlevel programming

d) Impossible, without the use of functions procedures.

10. In the worst case, how many nodes are to be traversed, to insert a new node, into a binary search tree of n nodes is: **[AUCET-2001]**

a) 1 b) n c) logn d) n log n

11. A heap is: **[AUCET-2001]**

a) A complete binary tree.

b) Invariably a full binary tree.

c) a special case of a binary search tree.

d) a special type of memory module

12. The infix expression A-(B+C) * (D/E) is equivalent to which of the following postfix expressions: **[AUCET-1999]**

a) ABC+-DE/ * b) ABC+DE/ * -

c) ABC+ * DE/- d) ABC+D/ * E-

13. If n is the no.of leaf nodes in a full binary tree then its height is: **[AUCET-2000]**

a) $\log_2^n$ b) $\log_2^{(2n)+1}$ c) $\log_2^{(2n+1)}$ d) $\log_2^{(2n+1)+1}$

14. Which of the following traversal techniques lists the nodes of a binary search tree in ascending order: **[AUCET-1999]**

a) postorder b) Inorder

c) preorder d) None

15. The number of null pointers of a binary tree of n nodes is:

a) n+1 b) n(n+1) c) n^2 d) 2n

16. A complete binary tree of level 5 has how many nodes:

a) 20 b) 30 c) 63 d) 73

17. If each node in a tree has value greater than every value in its left subtree and has value less than every value in its right subtree, the tree is known as:

a) complete tree b) full binary tree

c) binary search tree d) threaded binary tree

18. In what tree, for every node the height of its left subtree and right subtree differ atleast by one is:

a) Binary search tree b) AVL tree
c) Complete tree d) Threaded binary tree

19. A complete binary tree with the property that the value at each node is atleast as large as the values at its children is known as:

a) Binary search tree b) AVL tree
c) Complete balanced tree d) Heap

20. A full binary tree with n non-leaf nodes contains

a) $\log_2^n$nodes b) n+1 nodes
c) 2n nodes d) 2n+1 nodes

21. A tree with n nodes:

a) n/2 edges b) n-1 edges
c) n edges d) n+1 edges

22. The number of paths between any pair of nodes in a tree on n nodes is:

a) 0 b) 1 c) n-1 d) n

23. The number of elements (edges) in a cutset of a tree with n vertices is:

a) 1 b) n/2 c) n-1 d) n

24. The number of different rooted labeled trees with n vertices is

a) 2^{n-1} b) 2^n c) n^{n-1} d) n^n

25. The tree shown in figure given below, represents the postorder expression:

a) ABDGCEJHIF
b) DGBAHEICF
c) GDBHIEFCA
d) ABHIEFCDG

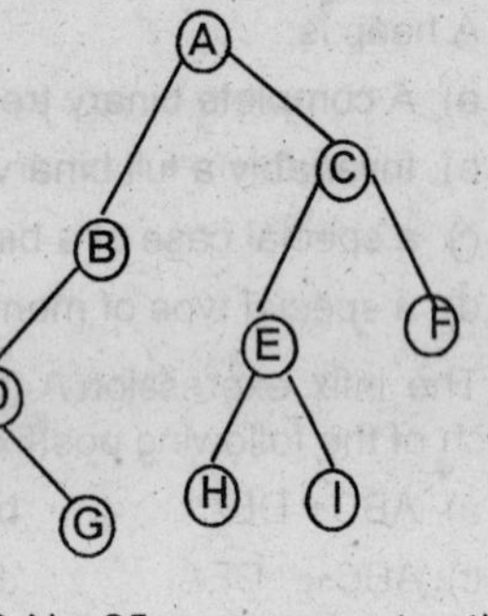

26. The tree shown in Q.No 25 , represents the preorder expression

a) ABDGCEHIF b) DGBAHEICF
c) GDBHIEFCA d) ABHIEFCDG

27. The tree shown in Q.No. 25, represents the Inorder Expression

a) ABDGCEHIF b) DGBAHEICF
c) GDBHIEFCA d) ABHIEFCDG

28. In a complete binary tree of n nodes how far are the most distant two nodes ? Assume each edge in the path counts as 1.

a) about $\log_2^n$ b) about $2\log_2^n$
c) about $3\log_2^n$ d) about $4\log_2^n$

29. Evaluate the postfix expression: **[GRE-CS]**

$$623+382/+*2\uparrow 3+$$

a) 244 b) 52 c) 131 d) 168

30. The number of possible ordered trees with 3 nodes A,B,C is

a) 16 b) 12 c) 6 d) 10

31. The depth of a complete binary tree with 'n' nodes is

a) log(n+1) –1 b) log(n)
c) log(n-1) +1 d) log(n) +1

32. The no.of possible binary trees with 3 nodes is

a) 5 b) 12 c) 13 d) 15

33. Which of the following is FALSE? **[Hyd. Univ-2001]**

a) A tree with n nodes has (n – 1) edges
b) A labelled rooted binary tree can be uniquely constructed given its preorder and postorder numberings
c) A complete binary tree with n internal nodes has n+1 levels d) None

34. The terminal nodes of a binary tree occur in the same relative position in **[Hyd.Univ-1998]**

a) preorder, inorder b) inorder, postorder
c) preorder, postorder d)preorder, inorder, postorder

35. In a binary tree with n nodes, the number of null links are given by: **[Hyd.Univ-1997]**

a) 2n b) n – 1 c) n d) n + 1

36. The prefix expression of (A + B) * C – D / E is : **[Hyd.Univ-1997]**

a) - * + A B C / D E b) - + * A B C/ D E
c) + * A B C - / D E d) * + A B – C / D E

37. Given the following postfix form of an arithmetic expression ABC * * CDE+- the equivalent infix form is **[Hyd.Univ-1996]**

a) ((A * (B * C))-(C/(D+E)) b) (A * (B * C)-C)/(D+E)
c) ((A * (B * C))-(C/D+E)) d) ((A-(B * C)/C * D+E))

38. For the arithmetic expression a + b * c * d – e the equivalent expression in postfix form is **[Hyd.Univ-1993]**

a) - + a * * b c d e b) - + a * b c * d e
c) a b c * d * + e – d) a b + c * d * e -

39. The number of edges in a binary tree having 'n' nodes is **[Hyd.Univ-1993]**

a) n b) n + 1 c) n – 1 d) n – 2

40. The number of distinct binary trees with three elements is **[Hyd.Univ-1994]**

a) 2 b) 3 c) 4 d) 6

41. The infix expression A + ((B –C) * D) is correctly represented in prefix notation as **[UGC NET-1997]**

a) A + B – C * D b) A + BC – D *
c) + A * - B C D d) A B C – D * +

42. The number of nodes in a complete binary tree of depth 'd' is (the root is at level '0') **[UGC NET-1999]**

a) 2^d b) 2^{d+1} c) $2^{d+1}+1$ d) $2^{d+1}-1$

43. What is the minimum number of modes in an AVL tree of height 3 (the root is at level '0') **[UGC NET-1999]**

a) 7 b) 6 c) 5 d) 4

44. What is the infix form of prefix expression $* + ABC$? **[UGC NET-1999]**

a) $A + B * C$ b) $(A + B) * C$

c) $A * (B + C)$ d) None of the above

ANSWER KEY - I				
1. d	**2.** c	**3.** b	**4.** a	**5.** b
6 . c	**7.** c	**8.** d	**9.** b	**10.** b
11. c	**12.** a	**13.** a	**14.** b	**15.** a
16 c	**17.** c	**18.** b	**19.** d	**20.** d
21. b	**22.** b	**23.** a	**24.** c	**25.** c
26. a	**27.** b	**28.** b	**29.** b	**30.** b
31. a	**32.** a	**33.** d	**34.** a	**35.** d
36. a	**37.** a	**38.** c	**39.** c	**40.** d
41. c	**42.** d	**43.** a	**44.** b	

EXPLANATORY NOTES - I

1. Consider a tree

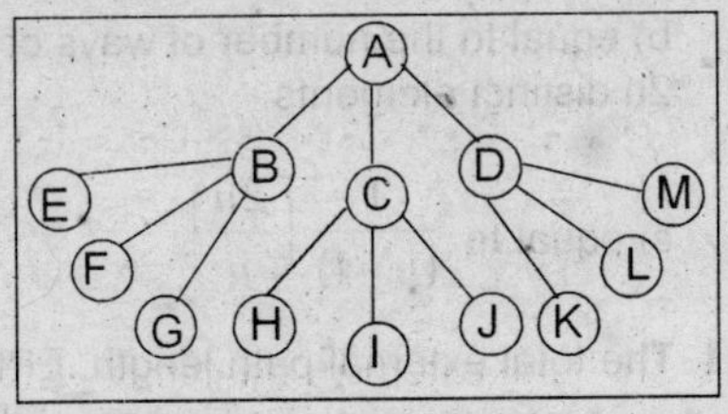

Total no. of nodes = 13

Substitute in $\frac{2n+1}{3} \Rightarrow \frac{2 \times 13+1}{3} \Rightarrow 27/3 = 9$

The no. of leaf nodes = 9.

2. The statement a) that complete binary tree with n internal nodes has $(n-1)$ leaves is wrong.

3. From the binary tree definition it will be true.

4. $(((A/B**C)+(D*E))-(A*C))$

$ABC**/DE*+AC*-$

operator precedence is 1)** 2) *, / 3) +, -

6. The no. of nodes at level r is 2^r

$\Rightarrow r=5=2^5=32 \Rightarrow r=4=2^4=16$

$\Rightarrow r=3=2^3=8 \Rightarrow r=2=2^2=4 \Rightarrow r=1=2^1=2$

$\Rightarrow r=0=2^0=1=1+2+4+8+16+32=63$

7. The max no. of nodes in a binary tree at level i is 2^i.

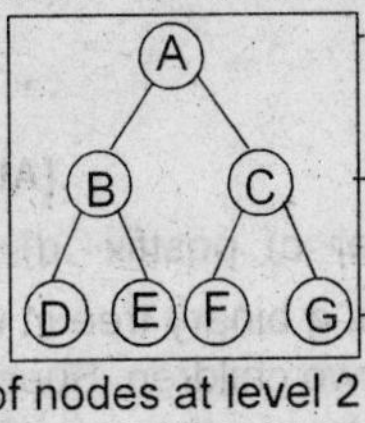

level 0

level 1

level 2

The no. of nodes at level 2 is $2^2=4$.

12. Infix $A-(B+C)*(D/E)$, postfix notation is

$\Rightarrow (A-(BC+))*(DE/) \Rightarrow (ABC+-)*(DE/)$

$\Rightarrow ABC+-DE/*$

14. Consider a binary search tree

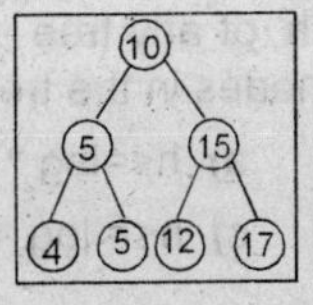

The in order traversal of the tree is

4 5 8 10 12 15 17

15. Consider a binary tree

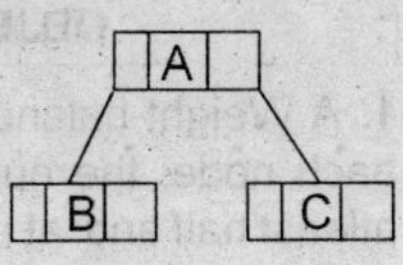

Here the no. of null pointers = 4

Total no. of nodes = 3

$\Rightarrow 3+1=n+1$

16. A complete binary tree of level i will have $2^{i+1}-1$

$\Rightarrow$ According the problem i = 5

$\Rightarrow 2^{5+1}-1=2^6-1=64-1=63$

20. A full binary tree with 'n' non - leaf nodes contains 2n+1 nodes. For Ex .

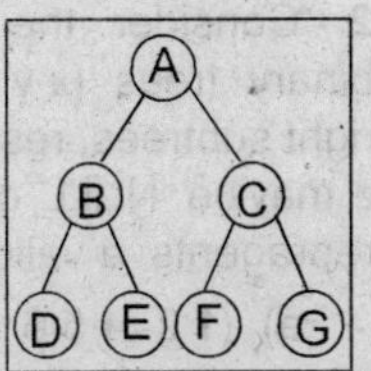

Here no. of non - leaf nodes n = 3.

Substitute 'n' in 2n+1

$\Rightarrow (2*3)+1=7.$= no. of nodes.

32. The no. of possible binary trees with 3 nodes in '5'

Ex : A, B, C are 3 nodes.

The possible binary trees are -

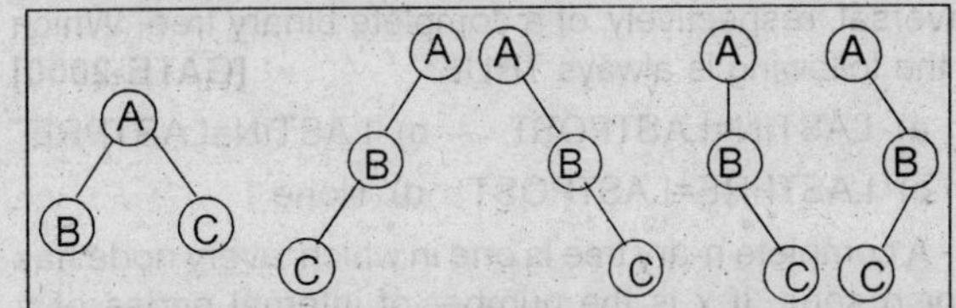

35. A binary tree with n nodes, the no. of null links are n+1. For Ex :

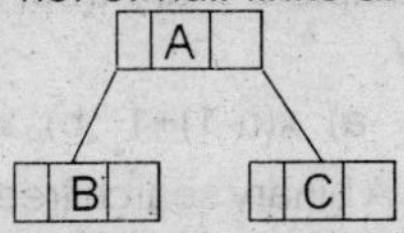

Here no. of nodes = 3.

null pointers = 4 i.e., n+1

36. The prefix for (A+B)*C-D/E is

$(((A+B)*C)-(D/E))$

$(((+AB)*C)-(/DE))$

$((*+ABC)-(/DE))$

$-*+ABC/DE$

38. Post fix notation for a+b*c*d - e

priority of operators are *, +, -

$\Rightarrow a+(b*c)*d-e$

$\Rightarrow a+((b*c)*d)-e$

$\Rightarrow ((a+((b*c)*d))-e)$

$\Rightarrow ((a+((bc*)*d))-e) \Rightarrow ((a+(bc*d*))-e)$

$\Rightarrow (abc*d*+)-e \Rightarrow abc*d*+e-$

41. The prefix notation for the following infix notation is

$=A+((B-C)*d)=A+((-BC)*d)$

$A+(*-BCd)=+A*-BCD$

43. for h = 0, N(h) = 1 no. of nodes

for h =1, N(h) = 2.

$\Rightarrow$ for a height 'h' = no. of left subtree nodes + no. of nodes in right subtree + 1

$N(h)=N(h-1)+N(h-2)+1$

for h = 3 ; $N(3)=N(2)+N(1)+1$

$=(N(1)+N(0)+1)+N(1)+1=4$

OBJECTIVE QUESTIONS - II

1. A Weight-balanced tree is a binary tree in which for each node, the number of nodes in the left subtree is atleast half and at most twice the number of nodes in the right subtree The maximum possible height(number of nodes on the path from the root to the furthest leaf) of such a tree on n nodes is best described by which of the following: **[GATE-2002]**

a) $\log_2{}^n$ b) $\log_{4/3}{}^n$ c) $\log_3{}^n$ d) $\log_{3/2}{}^n$

2. Consider the following nested repragentation of binary trees :(x y z) indicates y and z are the left and right subtrees, respectively, of node x. Note that y and z may be NULL, or furtrher nested.Which of the following repragents a valid binary tree: **[GATE-2000]**

a) (1 2 (4 5 6 7)) b) (1((2 3 4) 5 6) 7)
c) (1(2 3 4)(5 6 7)) d) (1(2 3 NULL)(4 5))

3. Let LASTPOST,LASTIN and LASTPRE denote the last vertex visited in a postorder, inorder, and preorder traversal, respectively, of a complete binary tree. Which of the following is always TRUE: **[GATE-2000]**

a) LASTIN=LASTPOST b) LASTIN=LASTPRE
c) LASTPRE=LASTPOST d) None.

4. A complete n-ary tree is one in which every node has 0 or n sons. If x is the number of internal nodes of a complete n-ary tree, the number of leaves in it is given by: **[GATE-1998]**

a) x(n-1)+1 b) x n-1 c) x n+1 d) x(n+1)

5. A binary search tree contains the values 1,2,3,4,5,6,7,8. The tree is traversed in pre-order and the values are printed out. Which of the following sequences is a valid output: **[GATE-1997]**

a) 5 3 1 2 4 7 8 6 b) 5 3 1 2 6 4 8 7
c) 5 3 2 4 1 6 7 8 d) 5 3 1 2 4 7 6 8

6. In the balanced binary tree given below ,how many nodes will become unbalanced when a node is inserted as a child of the node "g": **[GATE-1996]**

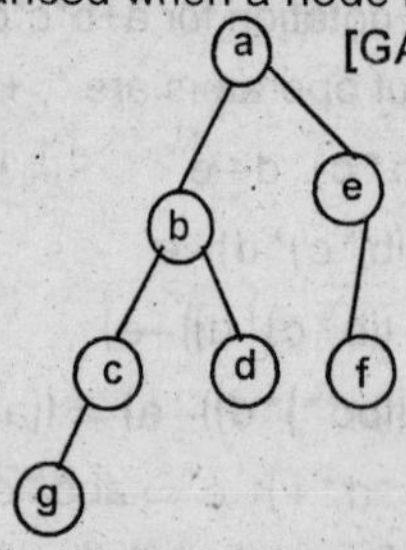

a) 1

b) 3

c) 7

d) 8

7. A Binary search tree is generated by inserting in order the following integers: 50, 15, 62, 5, 20, 58, 91, 3, 8, 37, 60, 24. The number of nodes in the left subtree and right subtree of the root respectively is: **[GATE-1996]**

a) (4,7) b) (7,4) c) (8,3) d) (3,8)

8. A 2-3 tree is a tree such that
I. All internal nodes have either 2 or 3 children.
II. All paths from root to the leaves have the same length.
The number of internal nodes of a 2-3 tree having 9 leaves could be: **[GATE-1992]**

a) 4 b) 5 c) 6 d) 7

9. The weighted external path length of the binary tree given below is: **[GATE-1991]**

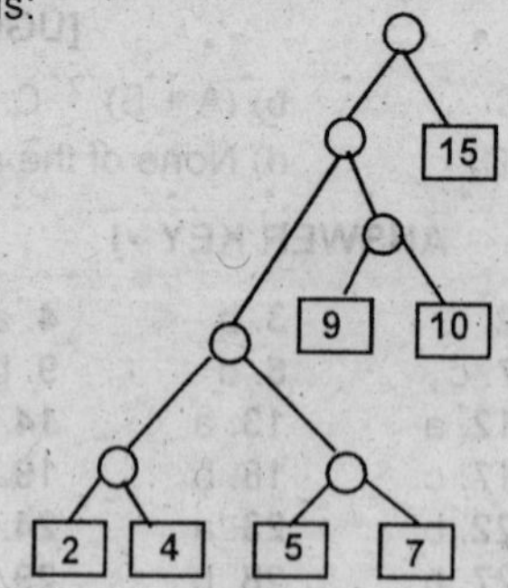

a) 20

b) 50

c) 36

d) None

10. The number of rooted binary trees with n nodes is : **[GATE-1990]**

a) equal to the number of ways of multiplying (n+1) matrices.

b) equal to the number of ways of arranging n out of 2n distinct elements.

c) equal to $\frac{1}{(n-1)}\binom{2n}{n}$ d) equal to n!

11. The total external path length, EPL, of a binary tree with n external nodes is, (EPL= $\sum_w I_w$ where I_w is the path length of external node w). **[GATE-1990]**

a) $\leq n^2$ always b) $\geq n\log_2{}^n$ always
c) equal to n^2 always d) o(n) for some special trees

12. The worst case height of a AVL tree with n nodes is: **[AUCET-2001]**

a) 2logn b) nlog(n+1)
c) 1.44 log(n+2) d) 1.44 nlogn

13. The following recursive procedure implements which traversal scheme procedure

```
LookAtIt(current:Treeptr);
BEGIN
IF current^.left<>NIL
THEN LookAtIt(current^.left);
IF current^.right<>NIL
THEN
lookAtIt(current^.right);
Writeln(current^.word);
END;
```

[AUCET-1999]

a) postorder b) Inorder c) postfix d) infix

14. A "strictly" binary tree is a binary tree in which every node that is not a leaf has two children. Suppose that for a class of strictly binary trees there exists c >0 such that, for any tree in the class , the ratio of the lengths of any two root - to - leaf paths is bounded above b to c. Which of the following best characterizes the height 'h' of any tree in this class, where N is the number of nodes in the tree and n>1: **[AUCET-1999]**

a) $h<=\log_2{}^n$ b) $h=1/c \log_2{}^n$
c) $h<= \log_2{}^{(cn)}$ d) $h>c\log_2{}^n$

15. What is the pre-order traversal of the following binary tree is: **[AUCET-2000]**

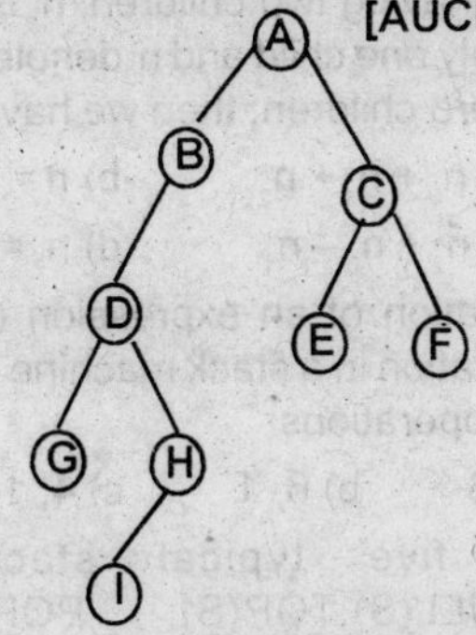

a) IGHDEFBCA

b) ABCDEFGHI

c) GDIHBAECF

d) None

16. Which of the following statements is FALSE: **[AUCET-1997]**

a) Every tree is a Bipartite graph
b) A tree contains n cycles
c) A tree with n nodes contains n-1 edges
d) A tree is a connected graph.

17. A binary tree in which if all its level except possibly the last, have the maximum number of nodes and all the nodes at the last level appear as for left as possible is known as:

a) full binary tree b) 2-tree
c) threaded tree d) complete binary tree

18. A list of integers is read in, one at a time, and a binary search tree is constructed. Next the tree is traversed and the integers are printed. Which traversal would result in a printout which duplicates the original order of the list of integers:

a) preorder b) postorder
c) Inorder d) None

19. Which of the following lists of nodes correspond to a postorder traversal of the binary tree shown in below fig. is:

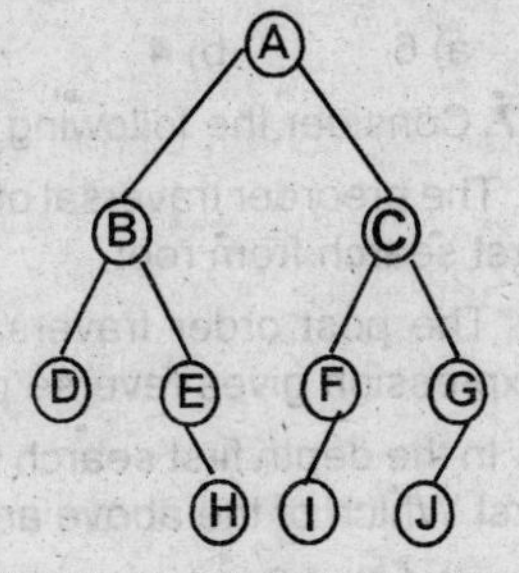

a) A B C D E F G H I J

b) A B D E H C F I J

c) D H E B I F J G C A

d) D B E H A I F C J G

20. Consider Fibonacci Tree. The depth of a node i in a binary tree is the length of the path from the root node to i. The depth of the root node is 0. If d is the depth of a fibonacci tree, and F_d is the dth number in the fibonacci sequence, then the number of nodes in fibonacci tree is given by $|F_d| = |F_{d-1}| + |F_{d-2}| + 1$

$|F_d|$ is also given by the following relationship:

a) $|F_d|=F_d$ b) $|F_d| = F_d + d$
c) $|F_d|=F_{d+3}-1$ d) $|F_d|=F_d*d$

21. A digital search tree is implemented as a tree with n nodes each of which can contain m pointers, corresponding to the m possible symbols in each position of the key. The number of nodes that must be accessed to find a particular key is:

a) m b) n c) n^m d) m^n

22. A binary tree is to be used to sort postitive integers. The partial tree shown below is formed:

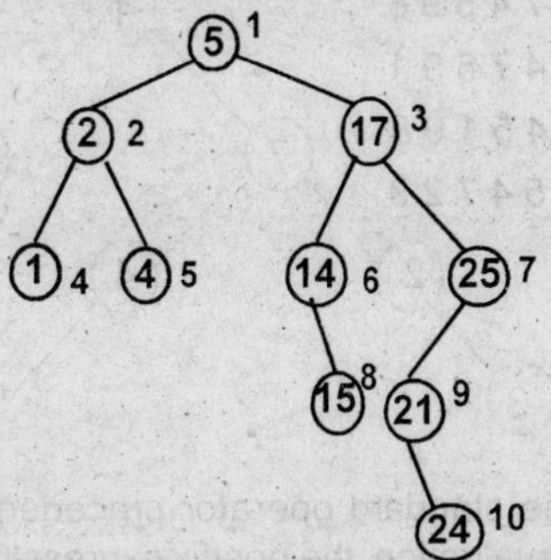

The next number in the file is 13; where should this be placed such that inorder traversal of the tree yields the sorted file? **[GRE-CS]**

a) right son of node 5 b) left son of node 6
c) right son of node 7 d) left son of node 8

23. The binary tree below is to be right in-threaded; to which node would the successor pointer of "Q" point? **[GRE-CS]**

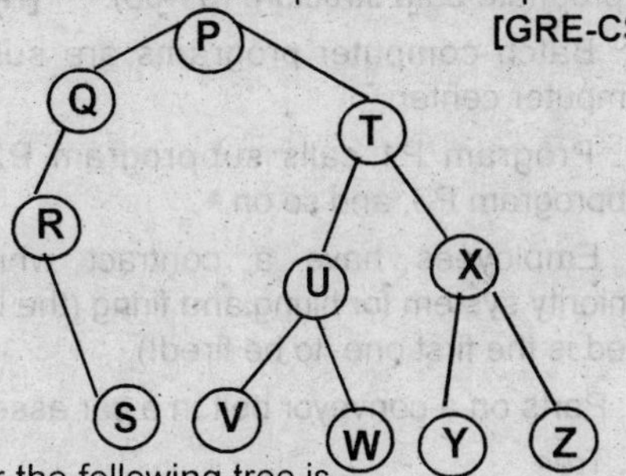

a) R
b) S
c) T
d) P

24. Tell whether the following tree is

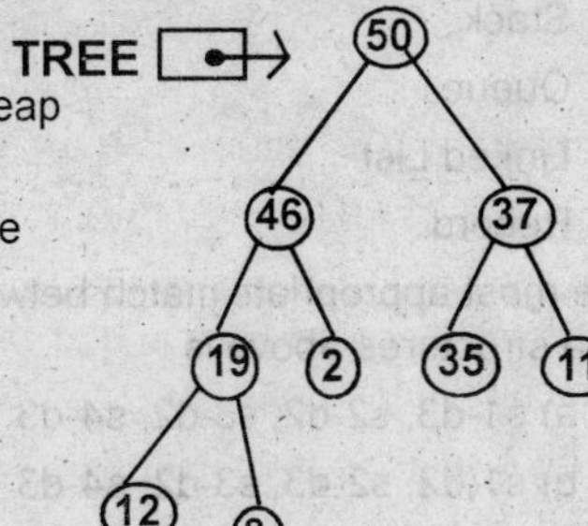

a) complete and heap

b) full and complete

c) full

d) heap

25. The minimum no.of interchanges needed to convert the array 89,19,40,17,12, 10,2,5,7,11,6,9,70 into a heap with the maximum element at the root is

a) 0 b) 1 c) 2 d) 3

26. A binary tree in which every non-leaf node has non-empty left and right subtrees is called a strictly binary tree. Such a tree with 10 leaves.

a) cannot have more than 19 nodes
b) has exactly 19 nodes
c) has exactly 17 nodes
d) cannot have more than 17 nodes.

27. A 3-ary tree is a tree in which every internal node has exactly 3 children. The no.of leaf nodes in such a tree with 6 internal nodes will be

a) 10 b) 13 c) 17 d) 23

28. What is the postorder sequence of the following tree **[Hyd. Univ-2001]**

a) 1 2 3 7 4 5 9 8

b) 3 2 5 4 7 8 9 1

c) 3 2 7 4 5 1 9 8

d) 9 8 1 5 4 7 2 3

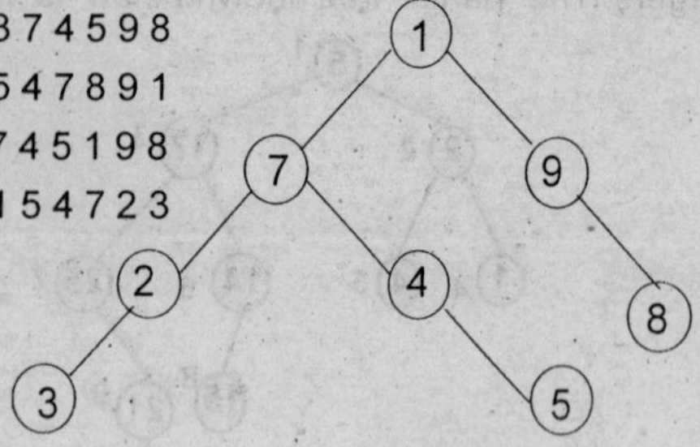

29. Using the standard operator precedence rules and left-to-right evaluation, the postfix expression for the infix expression A + B ∗ (C + D) / F + D ∗ E is **[Hyd. Univ-2001]**

a) AB+CD+ * F/D+E * b) ABCD+ * F/DE * ++

c) A * B+CD/F * DE++ d) A+ * B CD/F * DE++

30. Match each of the following situations (s1-s4) with the appropriate data structure (d1-d5). **[Hyd. Univ-2000]**

s1. Batch computer programs are submitted to the computer center

s2. Program P1 calls subprogram P2, which calls subprogram P3, and so on

s3. Employees have a contract which calls for a seniority system for hiring and firing (the latest one to be hired is the first one to be fired!)

s4. Parts on a conveyor belt in a car assembly line.

d1. Tree

d2. Stack

d3. Queue

d4. Linked List

d5. Record

The most appropriate match between the situations and data structures above is

a) s1-d3, s2-d2, s3-d2, s4-d3

b) s1-d2, s2-d3, s3-d2, s4-d3

c) s1-d4, s2-d1, s3-d5, s4-d2

d) s1-d3, s2-d2, s3-d4, s4-d5

31. Efficient sequential storage for a binary tree uses a single linear array where the root is stored as the first element and if a node occupies kth element, then its left child is stored as 2kth and its right child is stored as (2k+1)st elements. What is the height of the following binary tree stored in an array A?

A=[45, 22, 77, 11, 30, 12, -, 15, -, 21, 5, 6]

Note : Empty cells are denoted by – **[Hyd. Univ-2000]**

a) 12 b) 24 c) 5 d) 4

32. In a binary tree of height h having t terminal nodes, which of the following best describes the relation between h and t : **[Hyd.Univ-1997]**

a) $\log_2 t > h$ b) $\log_2 t \leq h$

c) $\log_2 h \geq t$ d) $\log_2 h^3 t$

33. If m a binary tree with n nodes, n_2 denotes the number of nodes having two children, n_1 is the number of nodes having only one child and n_0 denotes the number of nodes having zero children, then we have **[Hyd.Univ-1997]**

a) $n = n_0 + n_1 + n_2$ b) $n = n_2 - n_1 - n_0$

c) $n = n_2 + n_0 - n_1$ d) $n_2 = n - n_1 + n_0$

34. Evaluation of an expression (a*b)+(x*y) by reverse polish notation in a stack machine needs how many push and pop operations **[Hyd.Univ-1996]**

a) 4, 3 b) 6, 1 c) 4, 1 d) 4, 2

35. The five typical stack operation are MAKENULL(S),TOP(S), POP(S), PUSH(x, S), EMPTY(S), with their usual meaning. A text editor processes a line of text using the stack.

Consider the following pascal-like program segment which reads characters 'c'

```
MAKENULL(S)
if c = '#' then
POP(S)
else if c ='@' then
        MAKENULL(S);
        else PUSH(c, S);
end
print S in reverse order
end
```

[Hyd.Univ-1995]

If the input is abc#d##cfg# what does it return?

a) abef b) aef c) afg d) abc

36. (A, (B, (E, F)), (C, (G)), (D, (H, I, J))) represents a tree. The number of leaf nodes in the binary tree representation of this tree is **[Hyd.Univ-1994]**

a) 6 b) 4 c) 3 d) 5

37. Consider the following statements.

1. The preorder traversal of a tree is equivalent to breadth first search from root

2. The post order traversal of a tree representation of expression gives reverse polish notation

3. In the depth first search the leaves of a tree are visited first Which of the above are correct? **[Hyd.Univ-1993]**

a) 1 and 2 only b) 3 only

c) 1 and 3 only d) All

38. For the binary tree given below, which one of the following gives correct order of vertices visited under various traversal schemes **[UGC NET-2000]**

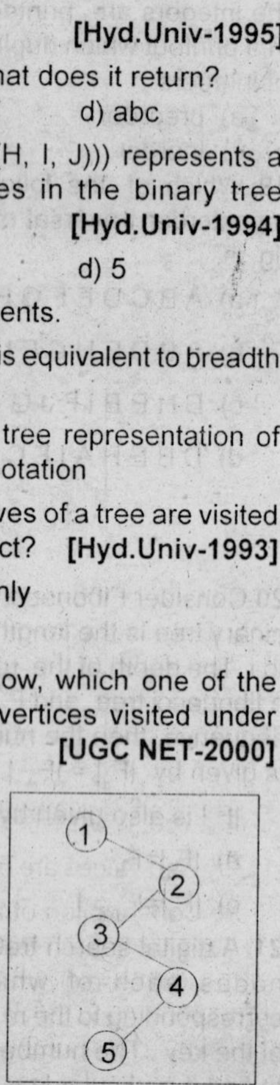

a) Pre-order: 1, 2, 3, 4, 5
Post-order: 4, 5, 2, 3, 1

b) Pre-order : 1, 3, 2, 4, 5
In-order : 1, 3, 5, 4, 2

c) Post-order : 5, 4, 3, 2, 1
In-order : 1, 3, 5, 4, 2

d) None of the above

39. A full binary tree (with root at level 0) of height h has a total number of nodes equal to **[UGC NET-1998]**

a) 2^h b) $2^{h+1}-1$ c) 2^h-1 d) 2^{h-1}

ANSWER KEY - II

1. a	**2.** c	**3.** b	**4.** a	**5.** d
6. b	**7.** b	**8.** a	**9.** ?	**10.** c
11. a,b	**12.** c	**13.** a	**14.** c	**15.** b
16. b	**17.** d	**18.** d	**19.** c	**20.** c
21. a	**22.** b	**23.** d	**24.** a	**25.** c
26. b	**27.** b	**28.** b	**29.** b	**30.** d
31. ?	**32.** a	**33.** a	**34.** a	**35.** b
36. d	**37.** d	**38.** c	**39.** b	

EXPLANATORY NOTES - II

5. The values of binary search tree are 1,2,3,4,5,6,7,8 i.e., the inorder traversal of binary search is nothing but the above sequence.

Hence the tree is as follows :

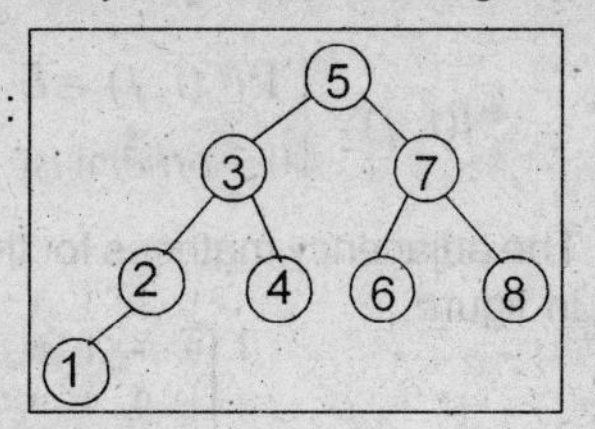

The preorder traveral is

53214768

6. The balanced binary tree is

i.e., 3 nodes c,g,m are unbalanced.

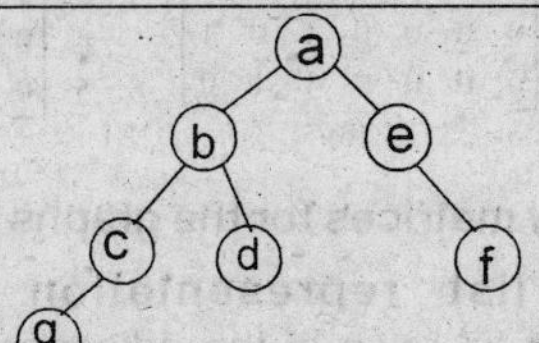

Let us insert 'm' to left of g

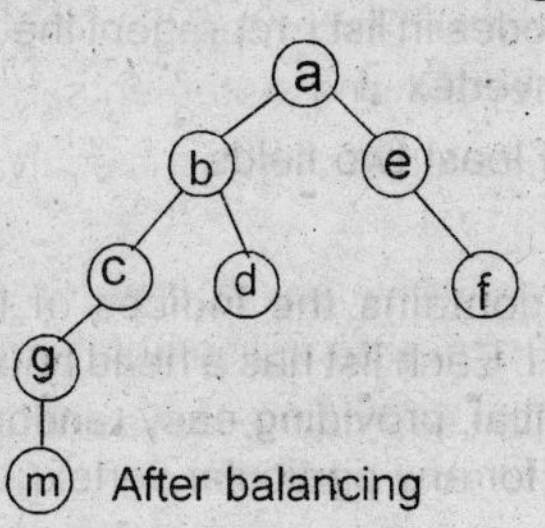

After balancing

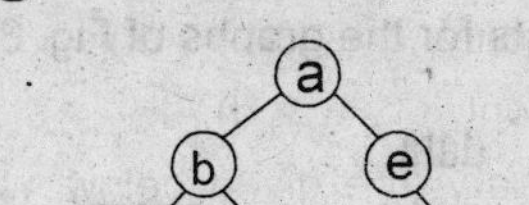

7. Given values are 50, 15, 62, 5, 20, 58, 91, 3, 8, 37, 60, 24

Construction of a binary search tree as follows :

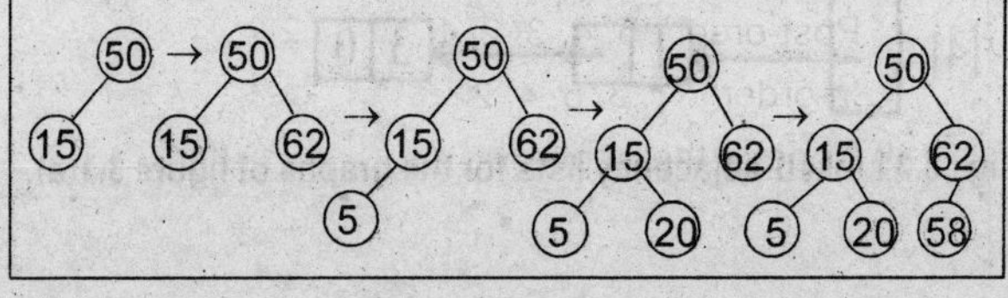

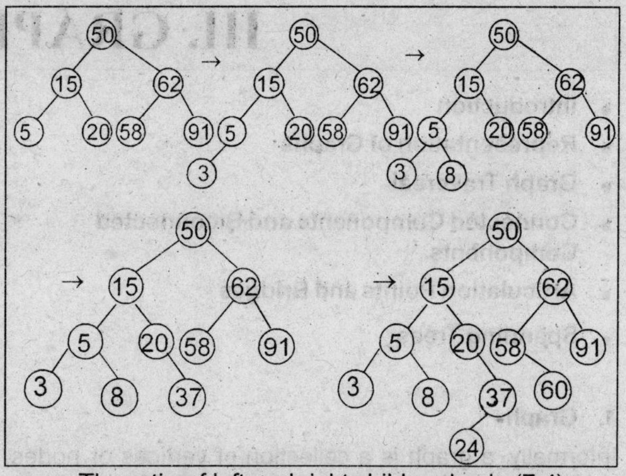

⇒ The ratio of left and right child nodes is (7,4)

8. Consider a 2-3 tree that 3 children.

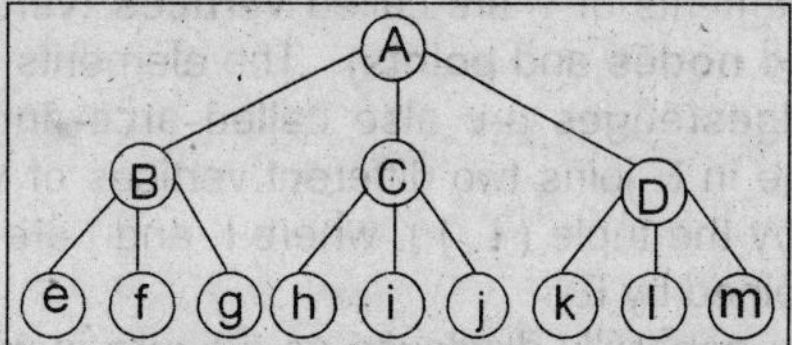

As per the above tree,we have 19 internal nodes and all paths from root to the leaves have the same length.

Here the no. of leave nodes = 9.

⇒ The no. of internal nodes = 4.

9. Weighted path length

$P = w_1p_1 + w_2p_2 + ... + w_np_n$

Here $w_1, w_2, ...w_n$ are the weights of each node.

$p_1, p_2, ... p_n$ are the path lengths.

For the given tree

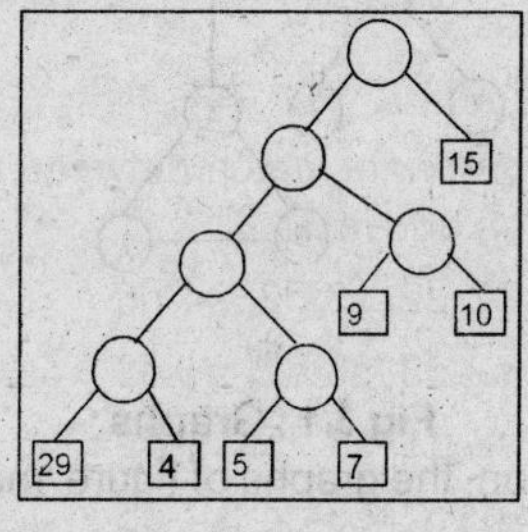

$P = 15 \times 1 + 9 \times 3 + 10 \times 3 + 5 \times 4 + 7 \times 4 + 2 \times 4 + 4 \times 4$

$= 15 + 27 + 30 + 20 + 28 + 8 + 16 = 144$

III. GRAPH THEORY

- **Introduction**
- **Representation of Graphs**
- **Graph Traversal**
- **Connected Components and Biconnected Components**
- **Articulation Points and Bridges**
- **Spanning Trees**

1. Graph

Informally, a graph is a collection of vertices or nodes, pairs of which are joined by lines or edges. More formally, a **graph G = (V,E)** is an ordered pair of finite sets V and E. The elements of V are called **vertices** (vertices are also called **nodes** and **points**). The elements of E are called edges(edges are also called arcs and lines). Each edge in E joins two different vertices of V and is denoted by the tuple (i , j), where i and j are the two vertices joined by E.

A graph is generally displayed as a figure in which the vertices are represented by circles and the edges by lines. Examples of graphs appear in figure 3.1. Some of the edges in the figure are oriented (i.e., they have arrow heads), while others are not. An edge with an orientation is a **directed** edge while an edge with no orientation is an **undirected** edge. The undirected edges (i,j) and (j,i) are the same; the direct edge (i,j) is different from the directed edge (j,i), the former being oriented from i to j and the latter from j to i.

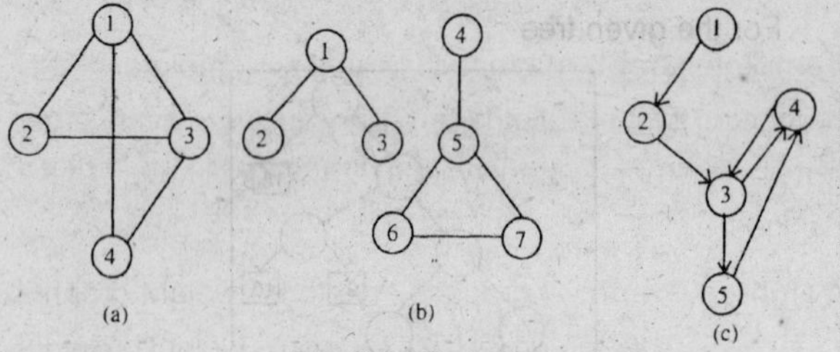

Fig.3.1 : Graphs

Using set notation, the graphs of figure may be specified as $G_1 = (V_1, E_1)$; $G_2 = (V_2, E_2)$ and $G_3 = (V_3, E_3)$ where

$V_1 = \{1,2,3,4\}$; $E_1 = \{(1,2),(1,3)(2,3),(1,4),(3,4)\}$

$V_2 = \{1,2,3,4,5,6,7\}$

$E_2 = \{(1,2),(1,3)(4,5),(5,6),(5,7),(6,7)\}$

$V_3 = \{1,2,3,4,5\}$

$E_3 = \{(1,2),(2,3)(3,4),(4,3),(3,5),(5,4)\}$

If all the edges in a graph are undirected, then the graph is an **undirected** graph. The graphs of figures (a) and (b) are undirected graphs. If all the edges are directed, then the graph is **directed** graph. The graph of figure (c) is a directed graph.

2. Representation of graphs and diagraphs

The most frequently used representation schemes for graphs and diagraphs are adjacency based : Adjacency matrices and Adjacency lists.

1) Adjacency matrix representation : The adjacency matrix of an single line graph G = (V,E) is an n × n matrix A. Each element of A is either zero or one. We shall assume that V = {1,2,...,n}. If G is an undirected graph, then the elements of A are defined as follows.

$$A(i,j) = \begin{cases} 1, if\ (i,j) \in E\ or\ (j,i) \in E \\ 0 \quad otherwise \end{cases} \quad \rightarrow (1)$$

If G is a digraph, then the elements of A are defined as follows :

$$A(i,j) = \begin{cases} 1, if\ (i,j) \in E \\ 0 \quad otherwise \end{cases} \quad \rightarrow (2)$$

The adjacency matrices for the graphs of figure 3.1 appear in figure.

$$\begin{array}{c|cccc} & 1 & 2 & 3 & 4 \\ \hline 1 & 0 & 1 & 1 & 1 \\ 2 & 1 & 0 & 1 & 0 \\ 3 & 1 & 1 & 0 & 1 \\ 4 & 1 & 0 & 1 & 0 \end{array} \qquad \begin{array}{c|ccccccc} & 1 & 2 & 3 & 4 & 5 & 6 & 7 \\ \hline 1 & 0 & 1 & 1 & 0 & 0 & 0 & 0 \\ 2 & 1 & 0 & 0 & 0 & 0 & 0 & 0 \\ 3 & 1 & 0 & 0 & 0 & 0 & 0 & 0 \\ 4 & 0 & 0 & 0 & 0 & 1 & 0 & 0 \\ 5 & 0 & 0 & 0 & 1 & 0 & 1 & 1 \\ 6 & 0 & 0 & 0 & 0 & 1 & 0 & 1 \\ 7 & 0 & 0 & 0 & 0 & 1 & 1 & 0 \end{array} \qquad \begin{array}{c|ccccc} & 1 & 2 & 3 & 4 & 5 \\ \hline 1 & 0 & 1 & 0 & 0 & 0 \\ 2 & 0 & 0 & 1 & 0 & 0 \\ 3 & 0 & 0 & 0 & 1 & 1 \\ 4 & 0 & 0 & 1 & 0 & 0 \\ 5 & 0 & 0 & 0 & 1 & 0 \end{array}$$

(a) (b) (c)

Fig3.2:Adjacency matrices for the graphs of figure 3.1

2) Adjacency list representation : In this representation the 'n' rows of the adjacency matrix are repragented as n linked lists.There is one list for each vertex in G. The nodes in list i repragent the vertices that are adjacent from vertex i .

Each node has at least two fields :

Vertex and Link.

The vertex field contains the indices of the vertices adjacent to vertex i. Each list has a head node. The head nodes are sequential, providing easy random access to the adjacency list for any particular vertex.

The adjacency lists for the graphs of Fig. 3.1 (a) appear in Fig. 3.3.

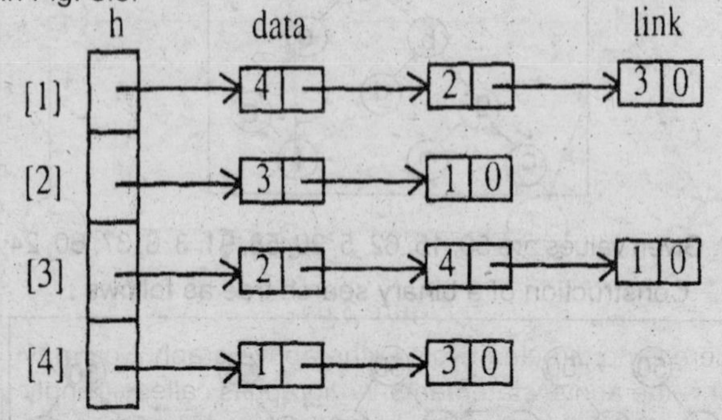

Fig : 3.3 Linked-adjacency lists for the graphs of figure 3.1(a).

3. Basic Terminalogy

Adjacent Verities : Vertex v_1 is said to be adjacent to a vertex v_2 if there is an edge (v_1,v_2) or (v_2,v_1).

Let us consider graph in Fig 3.4

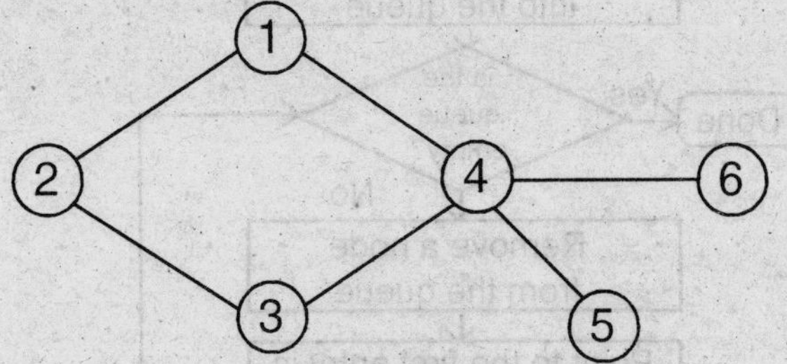

Figure 3.4

Vertices adjacent to node 2 are 1 and 3 and that to Node 4 is 1,3,6 and 5.

Path: A path from vertex w is a sequence of vertices, each adjacent to the next. Consider the above example again 1,2,3 is a path and 1,4 and 3 is also a path.

Cycle: A cycle is a path in which first and last vertices are the same. In the above example(1,2,3,4,1) is a cycle.

Connected Graph: A graph is called connected if there exists a path from any vertex to any other vertex. These are graphs which are unconnected. Consider the graph in fig 3.5

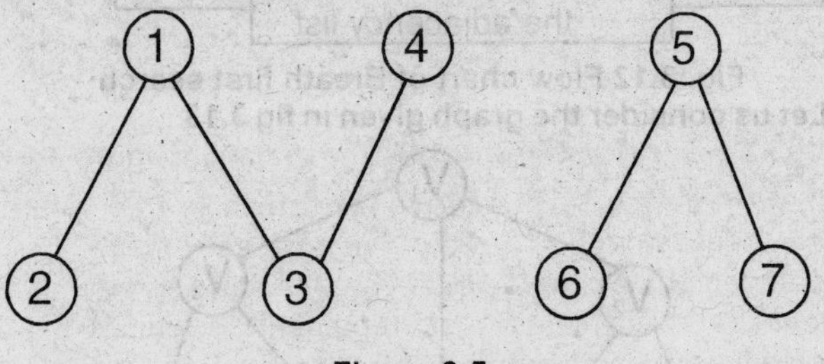

Figure 3.5

It is an unconnected graph. You may say that these are to graphs and not one. But it is one graph having two unconnected components. Since there are unconnected components, it is an unconnected graph.

So far we have talked about paths, cycles and connectivity of undirected graph. In a Digraph the path is called a directed path and cycle is called as directed cycle.

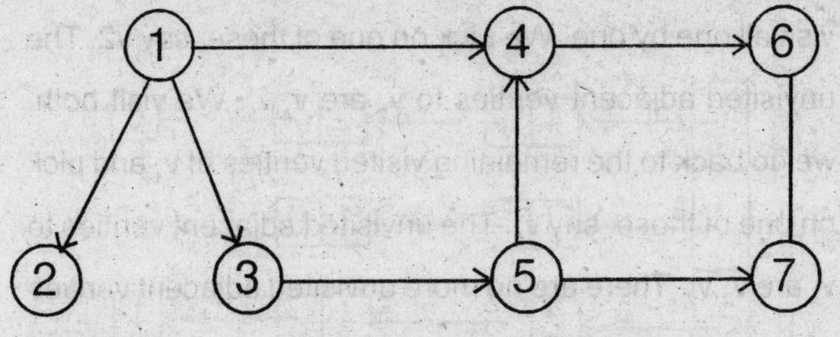

Figure 3.6

There is no directed cycle in the above graph. You may verify the above statements. A digraph is called strongly connected if there is a directed path from any vertex to any other vertex consider the graph given below in fig: 3.7

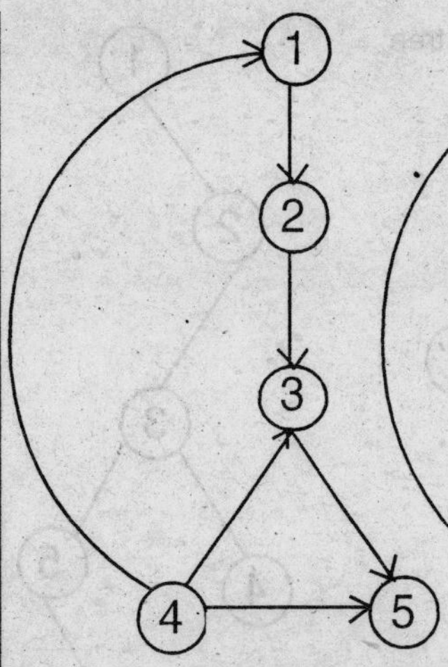

Fig. 3.7

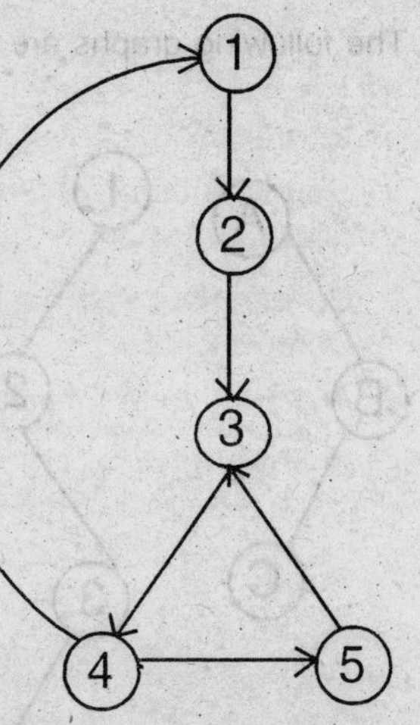

Fig. 3.8

There dos not exist a directed graph from vertex 1 to vertex 4, also from vertex 5 to other vertices and so on. There fore, it is a weekly connected graph. To make it strongly connected there are certain changes that are require given as above in fig: 3.8

Degree: The number of edges incident on a vertex determine its degree. The degree of vertex u, is written as degree (u).If degree (u) = 0 , this means that vertex u does not belong to any edge ,then vertex u is called isolated vertex.

In a Directed graph (Digraph), we attach an in degree and an out degree to each of the vertices.

Complete Graph: A graph G is said to complete or fully connected if there is path from every vertex to every other vertex. A complete graph with n vertices will have n(n-1)/2 edges.

Weighted Graph: A graph is said to be weight graph if every edge in the graph is assigned some weight or value. The weight of the edge is a positive value that may be representing the cost of moving along the edge or distance between the vertices.

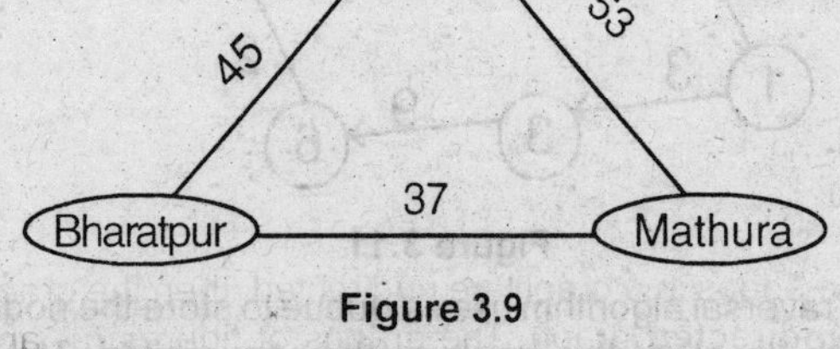

Figure 3.9

Tree. A graph is a tree if it has two properties:

1. It is connected , and
2. There are no cycles is the graph.

The following graphs are tree:

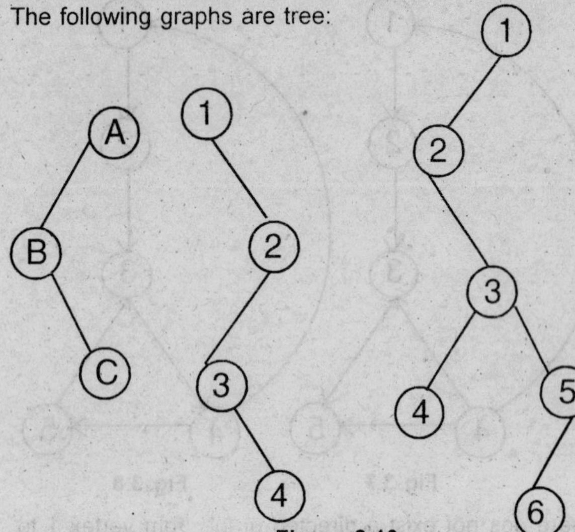

Figure 3.10

4. Graph Traversal

A graph traversal means visiting all the nodes of the graph. Graph traversal may be needed in many of application areas and there may be many methods for visiting the verities of the graph there are two graph traversal methods.

I. Breadth first traversal (BFS)

II. Depth First Traversal (DFS)

Breadth – First Traversal

In breadth first traversal of graph, one node is selected as the start position. It is visited and marked, then all unvisited nodes adjacent to the next node are visited and marked in some sequential order. Finally, the unvisited nodes immediately adjacent to these nodes are visited and marked and so forth, until the entire graph has been traversed.

Breadth first traversal of the graph fig.3.11 results in visiting the nodes in the following order 1,2,3,4,5,6,7,8. The sequence 1,3,2,6,5,4,7,8 is also a valid, breadth first traversal visitation order.

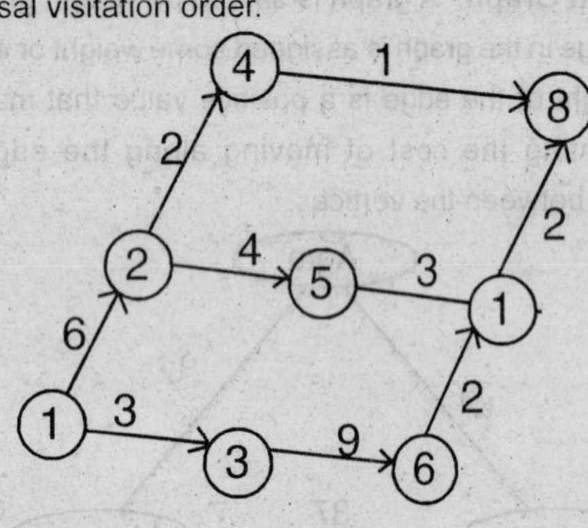

Figure 3.11

The Traversal algorithm uses a queue to store the nodes of each "level" of the graph as they are visited. These stored nodes are then treated one by one by one and their adjacent nodes are visited. These stored nodes have been visited. This terminating condition is reached when the queue is empty.

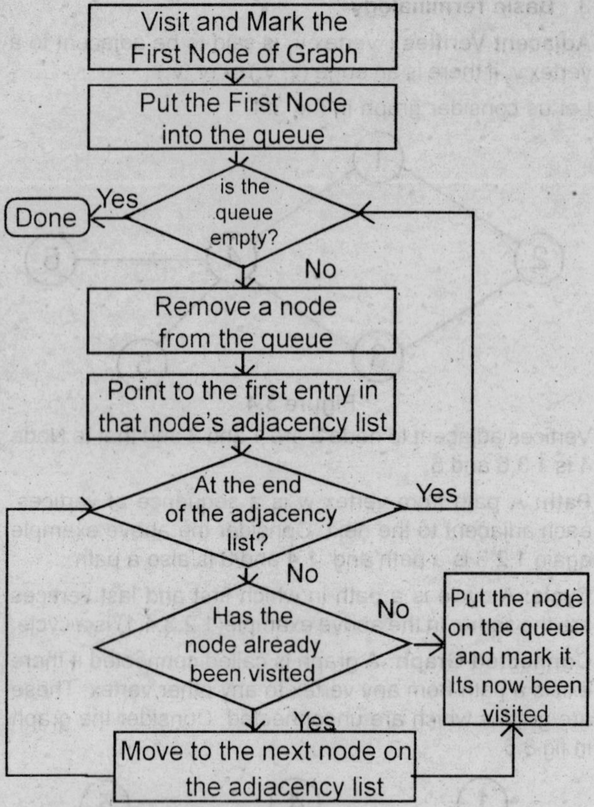

Fig. 3.12 Flow chart of Breath first search

Let us consider the graph given in fig 3.13

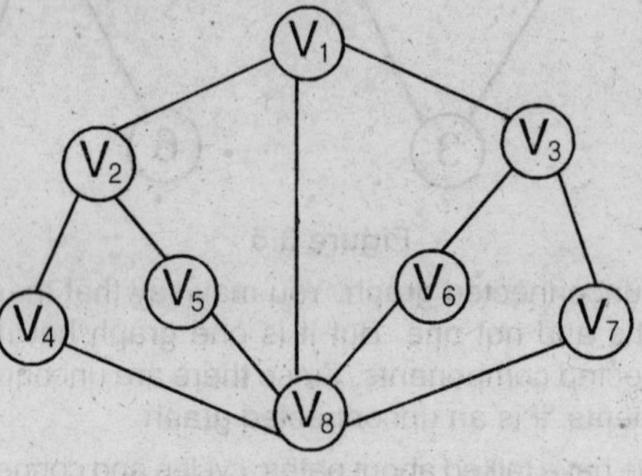

Fig 3.13

We start, say, with v_1 its adjacent verities are v_2,v_8,v_3.we visit all one by one. We pick on one of these, say v2. The unvisited adjacent verities to v_2 are v_4,v_5. We visit both, we go back to the remaining visited verities of v_1 and pick on one of those, say v_3. The unvisited adjacent verities to v_3 are v_6,v_7. There are no more unvisited adjacent verities of v_8,v_4,v_5,v_6 and v_7.

Thus, the sequence so generated is $v_1,v_2,v_8,v_3,v_4,v_5,v_6,v_7$.Here we need a queue and add unvisited verities adjacent to the one just visited at the rear and read at from to find the next vertex to visit.

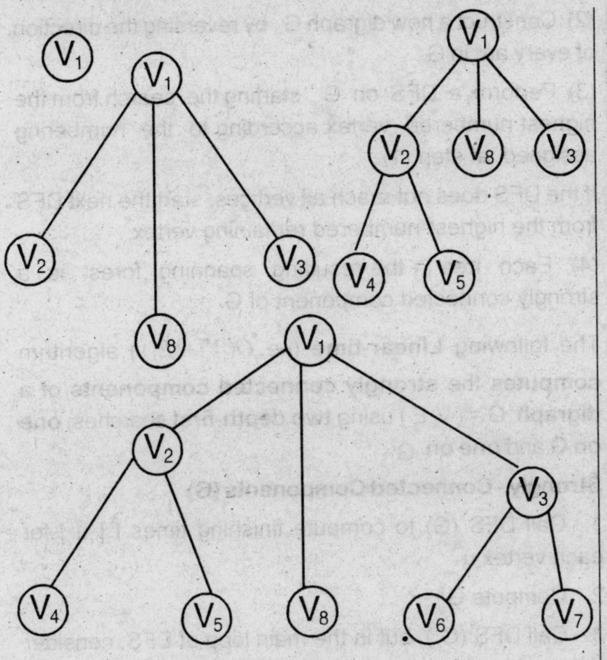

Figure 3.14

Depth First Traversal

Breadth first traversal of the graph proceeds level-by-level, depth first traversal follows first a path from the starting Node to an ending Node, then another path from the start to the end, and so forth until all nodes have been visited.

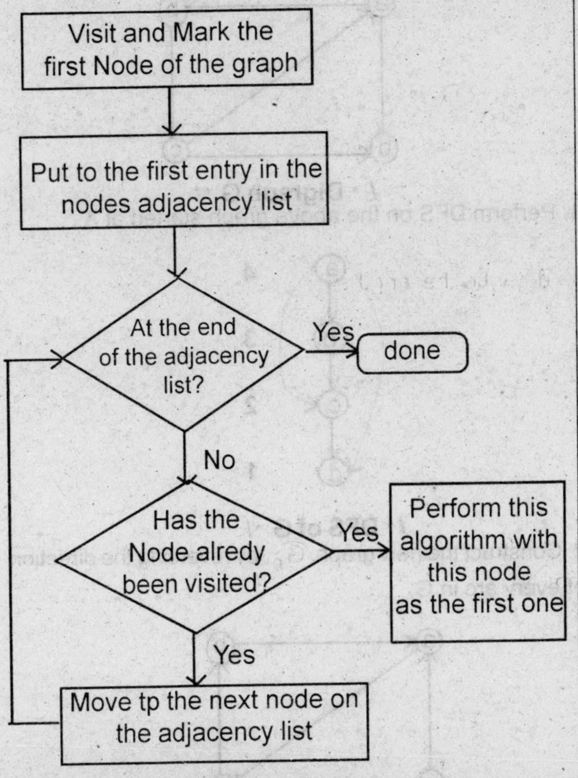

Fig. 3.15 Flow chart of recursive depth first graph traversal algorithm

Depth first traversal of the graph of fig:3.11 results in visiting the nodes in the following order:1,2,4,8,5,7,3,6. A path is pursued until no unvisited nodes remain reachable, then the algorithm backs up to results from depth- first traversal of the example graph is 1,3,6,7,8,2,5,4.

For the graph of fig 3.13 we start with say , vertex V. And adjacent vertex is selected and a Depth first search is initiated from it. i.e, let $V_1, V_2, \ldots, V_k$ are adjacent verities of vertex V. We may select any vertex from this list. Say we select V_1 .Now all the adjacent verities of V_1 are identified and all of those are visited; next V_2 is selected and all its adjacent verities visited and so on. This process continues till all the verities are visited.

It is very much possible that we reach a traversed vertex second time. Therefore we have to set a flag some where to check if the vertex is already visited. Let us see it through an example.

For graph Fig.3.13.

Let us start with V_1

Its adjacent vertices are V_2, V_8 and V_3 . Let us pick on V_2.

Its adjacent vertices are V_1, V_4, V_5, V_1 is already visited. Let us pick on V_4 .

Its adjacent vertices are V_2, V_8 .

V_2 is already visited. Let us visit V_8 .Its adjacent vertices are V_4, V_5, V_1, V_6, V_7 V_4 and V_1 are already visited. Let us traverse V_5 .

Its adjacent vertices are V_2, V_8 .Both are already visited. Therefore, we back track. We had V_6 and V_7 unvisited in the list of V_8 .We may visit any. We visit V_6 .

Its adjacent are V_8 and V_3 Obviously the choice is V_3

Its adjacent vertices are V_1, V_7 .We visit V_7 .

Therefore the sequence of traversal is $V_1, V_2, V_4, V_8, V_5, V_6, V_3, V_7$.

5. Connected Graphs Disconnected Graphs and Components

A graph is connected, If we can reach any vertex from any other vertex by traveling along the edges.

More Formally : A graph G is said to be connected if there is atleast one path between every pair of vertices in G, Otherwise G is disconnected.

For example, the graph below is connected.

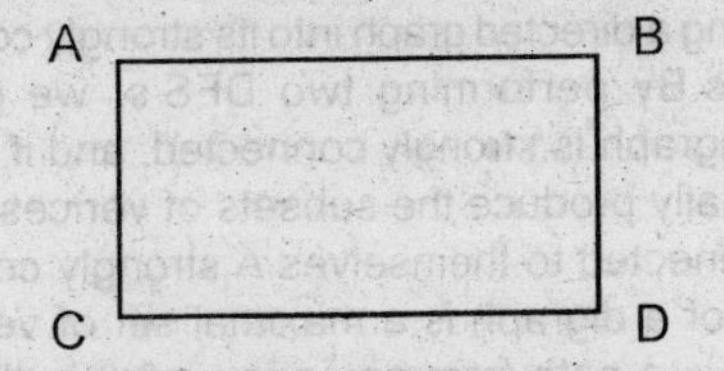

The below graph is disconnected.

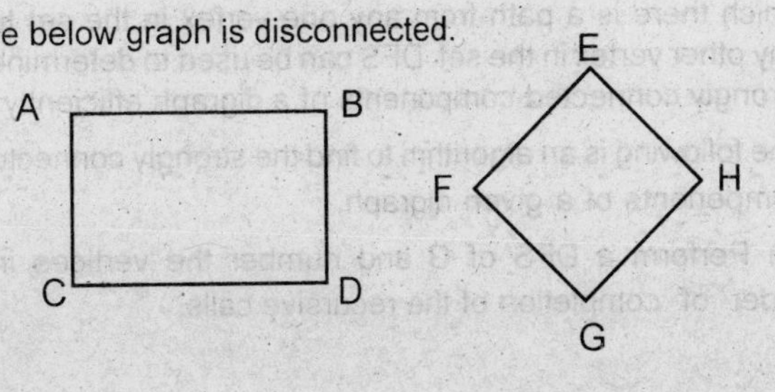

It is easy to see that a disconnected graph consists of two or more connected graphs.

Each of these connected sub graphs is called a component. The graph below has two components.

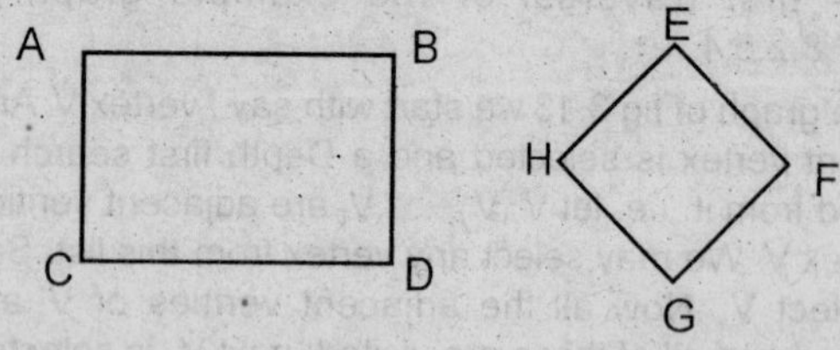

/ * disconnected graph with two components * /

Connected Digraphs : In a diagraph there are two different types of paths.Consequently we have two different types of connectedness in digraphs.A Digraph G is said to be strongly connected if there is atleast one directed path from every vertex to every other vertex. A Digraph G is said to be weakely connected if its corresponding undirected graph is connected but G is not strongly connected.

In below fig. one is strongly connected and other one is not strongly connected.

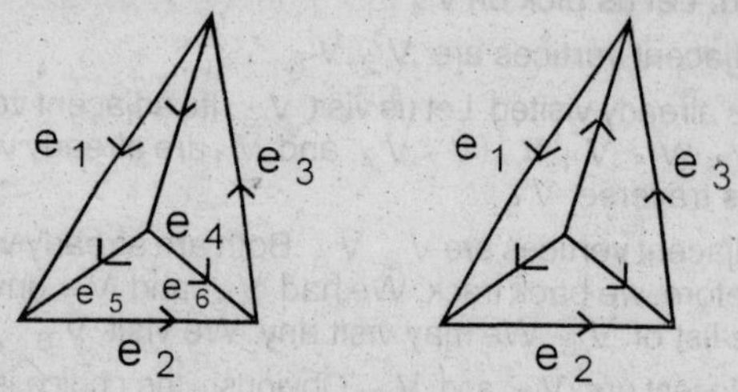

The statement that a digraph G is connected simply means that its corresponding undirected graph is connected; and thus G may be strongly (or) weakly connected.

Since there are two types of connectedness in a digraph, we can define two types of components also.Each maximal connected (weakly or strongly) subgraph of a digraph G will still be called a component of G.But with in each component of G the maximal strongly connected subgraphs will be called the fragments (or strongly connected fragments) of G.

6. Strongly Connected Components

Decomposing a directed graph into its strongly connected components.By performing two DFS's, we can test whether a digraph is strongly connected, and if it is not, we can actually produce the subsets of verices that are strongly connected to themselves.A strongly connected component of a digraph is a maximal set of vertices in which there is a path from any one vertex in the set to any other vertex in the set. DFS can be used to determine strongly connected components of a digraph efficiently.

The following is an algorithm to find the strongly connectd components of a given digraph.

(1) Perform a DFS of G and number the vertices in order of completion of the recursive calls.

(2) Construct a new digraph G_r, by reversing the direction of every arc in G.

(3) Perform a DFS on G_r, starting the search from the highest numbered vertex according to the numbering assigned at step (1).

If the DFS does not reach all vertices, start the next DFS from the highest-numbered remaining vertex.

(4) Each tree in the resulting spanning forest is a strongly connected component of G.

The following **Linear-time** (i.e. $O(|V|+|E|)$) algorithm **computes** the **strongly connected components** of a **digraph** G = (V, E) using **two depth-first** searches, **one on** G and **one on** G'.

Strongly - Connected Components (G)

1. Call DFS (G) to compute finishing times f [u] for each vertex u
2. Compute G^T
3. Call DFS (G^T), but in the main loop of DFS, consider the vertices in order of decreasing f(u) (as computed in line)
4. Output the vertices of each tree in the depth-first forest of step 3 as a separate strongly connected component.

Ex: Find the strong components of below graph using above algorithm.

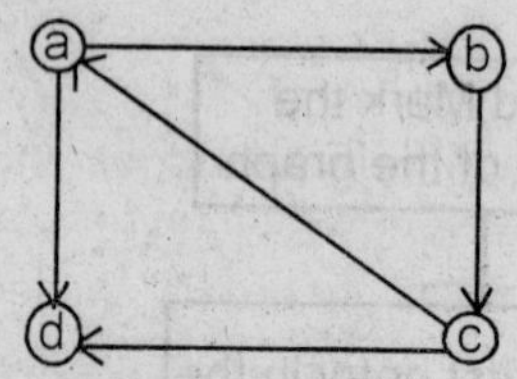

/ * Digraph G */

1. Perform DFS on the above graph started at A.

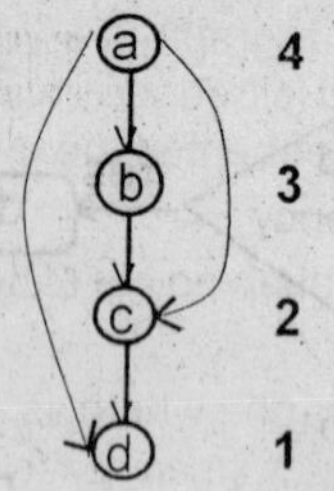

/* DFS of G */

2. Construct the new graph G_r, by reversing the direction of every arc in G.

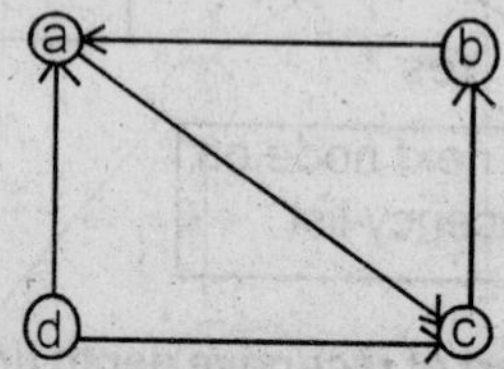

/*G_r */

3. Now perform DFS on G_r starting from high numbered vertex in step 1.

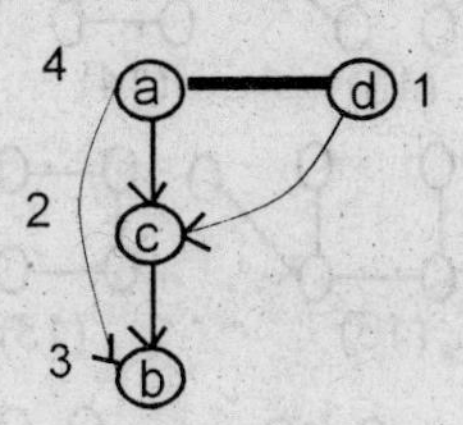

/* **DFS of G_r** */

4. Each of the trees in the above depth first spanning forest forms a strongly connected component.

So, the strongly connected components are :{a, c, b} , {d}

7. Articulation points, Bridges, Biconnected Components

Let G = (V, E) be a connected, undirected graph.

F An articulation point of G is a vertex whose removal disconnects G.

F A brdige of G is an edge whose removal disconnects G.

F A Biconnected component of G is a maximal set of edges such that any two edges in the set lie on a common simple cycle.

We can determine articulation Points, bridges and biconnected components using DFS.

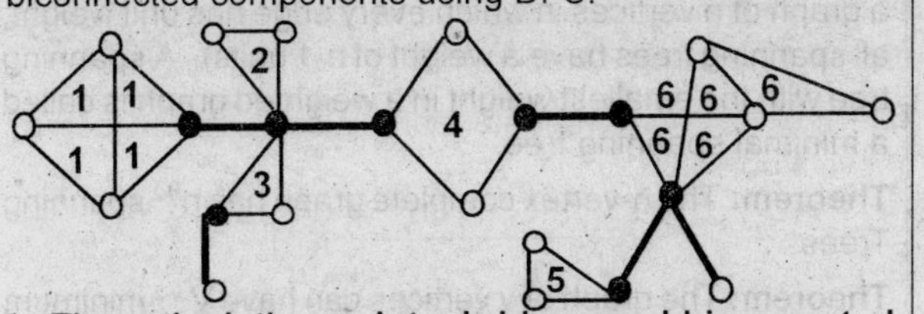

/* **The articulation points, bridges and biconnected components of a connected, undirected graph** */.

The articulation points are heavily shaded vertices.

The brdiges are the heavily shaded edges.

The biconnected components are the edges in numbering regions.

An Articulation point of a graph is a vertex u such that when we remove u and all edges incident upon V, we break a connected component of the graph into two (or) more pieces.

For example, the articulation points of below graph are a and c.

If we delete a, the graph, which is one connected component, is divided into two triangles {b, d, e} and {c, f, g}. If we delete C, we divide the graph into {a, b, d, e} and {f, g}

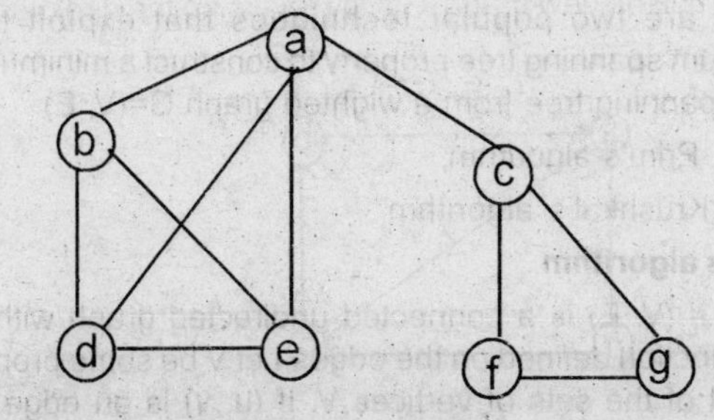

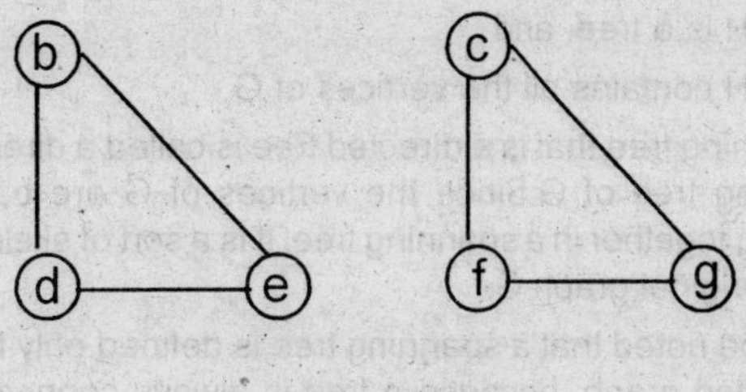

/* **deleting a** */

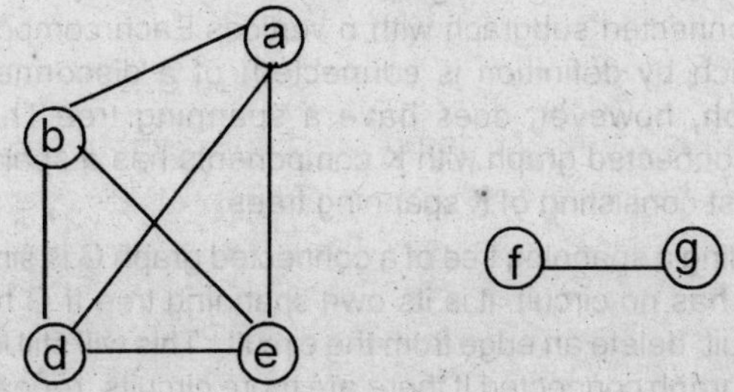

/* **deleting c** */

However, if we delete any one of the other vertices from the above graph, we do not split the connected component.

Biconnected : A connected graph with no articulation points is said to be Biconnected.

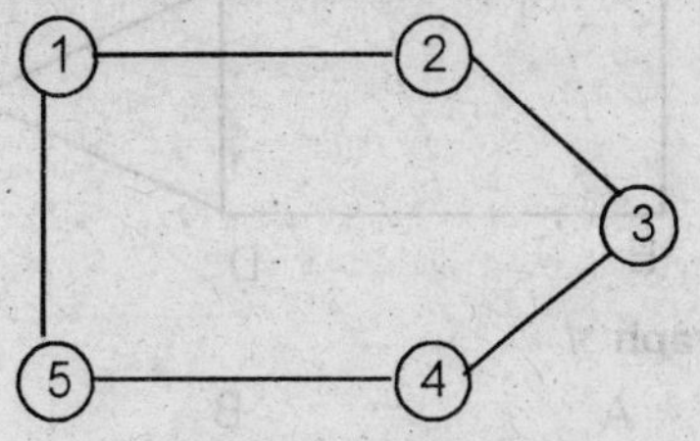

/* **A Biconnected Graph** */

DFS is particularly useful in finding the Biconnected components of a graph. The problem of finding articulatin points is the simplest of many important problems concerning the connectivity of graphs.The presence of articulatin points in a connected graph can be an undesirable feature in many cases.

For example if G reparagents communication network with the vertices repragenting communication stations and the edges communication lines then the failure of a communication station i which is an articulation point would result in loss of communication to points otherthan i too.On the otherhand, if G has no articulation point, then if any station i fails, we can still communicate between every pair of stations not including station i.

8. Spanning Trees

The tree as a subgraph of another graph. A given graph has numerous subgraphs from e edges, 2^e distinct combinations are possible. Obviously, some of these subgraqphs will be trees out of these trees we are particularly interested in certain types of trees, called spanning trees.

Definition : A sub graph H of a graph G is called a spanning tree of G if

(a) H is a tree, and

(b) H contains all the vertices of G.

A spanning tree that is a directed tree is called a directed spanning tree of G.Since the vertices of G are barely hanging together in a spanning tree; it is a sort of skeleton of the original graph G.

It is to be noted that a spanning tree is defined only for a connected graph, because a tree is always connected, and in a disconnected graph of n vertices we cannot find a connected subgraph with n vertices.Each component (which by definition is connected) of a disconnected graph, however, does have a spanning tree.Thus a disconnected graph with K components has a spanning forest consisting of K spanning trees.

Finding a spanning tree of a connected graph G is simple. If G has no circuit, it is its own spanning tree.If G has a circuit, delete an edge from the circuit. This will still leave the graph connected.If there are more circuits, repeat the operation till an edge from the last circuit is deleted - leaving a connected, circuit-free graph that contains all the vertices of G.

Thus we have:

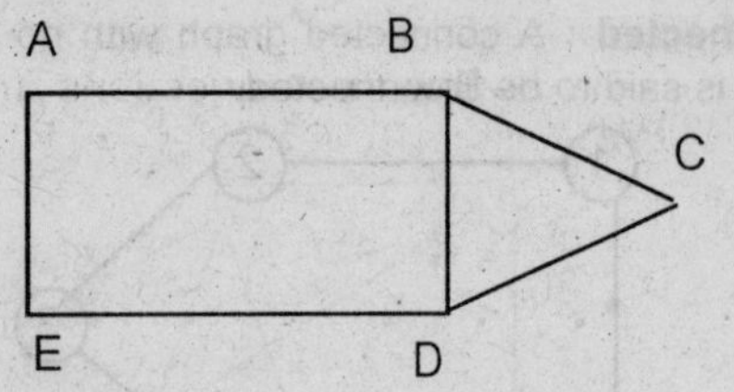

/* given graph */

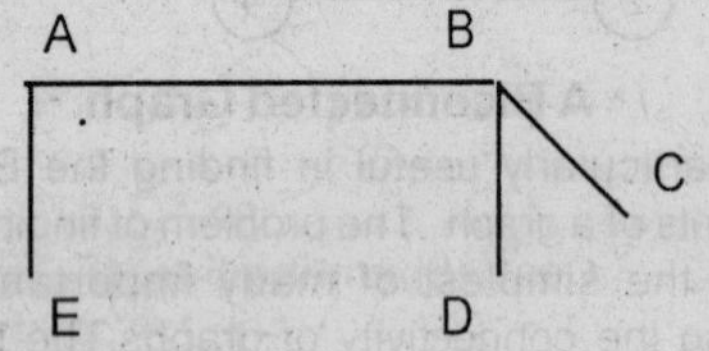

/* spanning tree */

- Every Connected graph has atleast one spanning tree

Finding All Spanning Trees Of a Graph

Usually, in a given connected graph there are a large number of spanning trees.In many applications we require all spanning trees.

Ex: How many minimum spanning trees does the following graph have? Draw them. **[GATE-1995]**

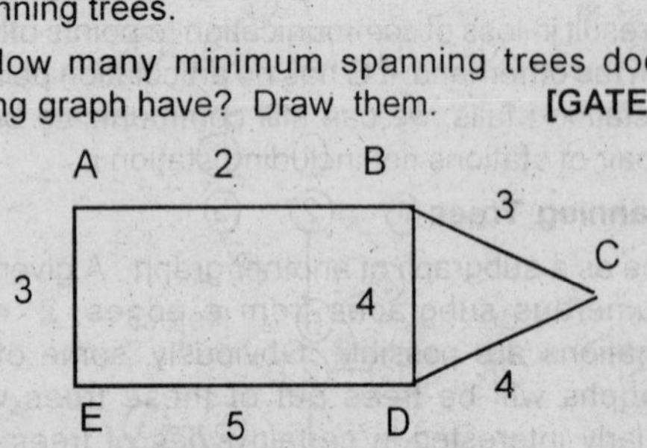

Sol : There are two minimum spanning trees with 12 as its weight. There are 11 spanning trees and their width's are given below:

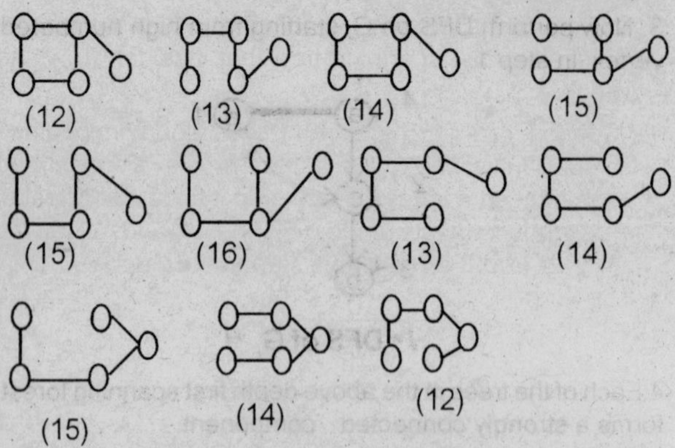

Spanning Trees In a Weighted Graph

A spanning tree in a graph G is a minimal subgraph connecting all the vertices of G.If graph G is a weighted graph (i.e. if there is a real number associated with each edge of G), then the weight of a spanning tree T of G is defined as the sum of the weights of all the branches in T. In general different spanning trees of G will have different weights.

Among all the spanning trees of G, one with the smallest weight is of practical significance. (There may be several spanning trees with the smallest weight; for instance, in a graph of n vertices in which every edge has unit weight, all spanning trees have a weight of n-1 units). A spanning tree with the smallest weight in a weighted graph is called a minimal spanning tree.

Theorem: The n-vertex complete graph has n^{n-2} spanning Trees

Theorem: The graph of v vertices can have V^{v-2} minimum spanning Trees.

9. Minimum spanning Tree

We will consider is that of finding a minimum spanning tree in an undirected graph.The problem make sense for directed graphs but appears to be more difficult.Informally, a minimum spanning tree of an undirected graph G is a tree formed from graph edges that connects all the vertices of G at lowest total cost.

A Minimum spanning tree exists if and only if G is connected. Although a robust algorithm should report the case that G is unconnected, we will assume that G is connected.Notice that the number of edges in the minimum spanning tree is $|V| - 1$.The minimum spanning tree is a tree because it is acyclic, it is spanning because it covers every vertex, and it is minimum for the obvious reason.

There are two popular techniques that exploit the minimum spanning tree property to construct a minimum-cost spanning tree from a wighted graph G=(V, E)

- Prim's algorithm
- Krushkal's algorithm

Prim's algorithm

Let G = (V, E) is a connected undirected graph with a cost function defined on the edges.Let V be some proper subset of the sets of vertices V. If (u, v) is an edge of lowest cost such that $u \in U$ and $v \in V-U$ then there is a minimum-cost spanning tree that includes (u, v) as an edge.

Suppose V = { 1,2,3,...... n}. Prim's algorithm begins with a set U initialized to {1}. It then "grows" a spanning tree, one edge at a time. At each step it finds a shortest edge (u,v) that connects u and u-v and then adds V, the vertex in V -U to U.It repeats this step until u = v.

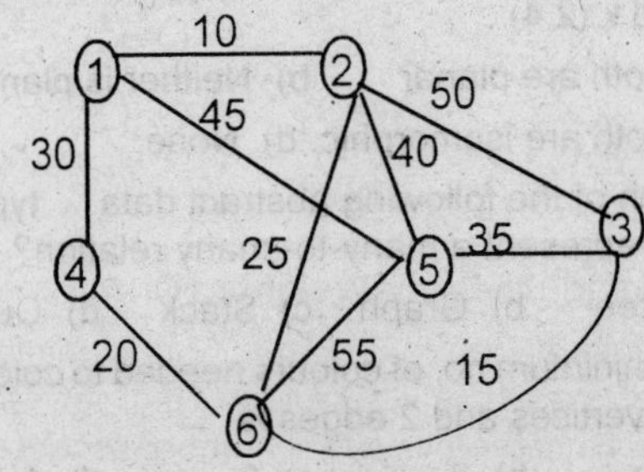

Edge	Cost	Spanning Tree
(1, 2)	10	
(2, 6)	25	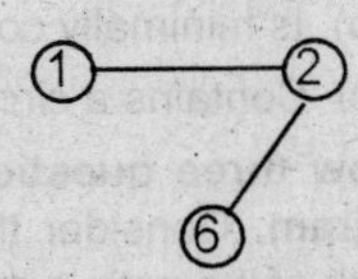
(3, 6)	15	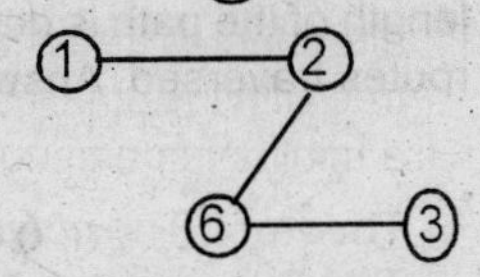
(6, 4)	20	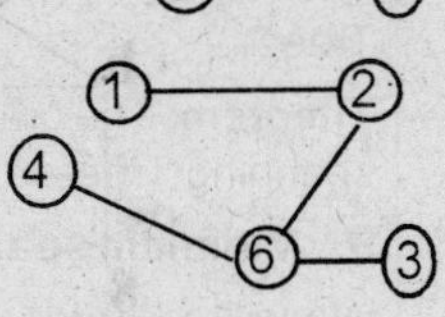
(1, 4)	30	(Reject)
(3,5)		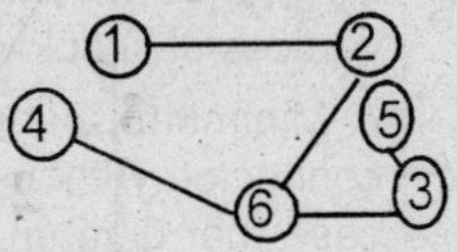

/* stages in prim's algorithm */

Krushkal's Algorithm: The time complexity of prim's algorithm is $O(n^2)$ where n is number of vertices in the given undirected weighted graph.As n gets large the performance of this algorithm may become unsatisfactory.

We now given another algorithm due to Krushkal for finding minimum cost spanning trees whose performance is at most O (e loge) where 'e' is the no.of edges in the given graph.If e is much less than n^2, Krushkal algorithm is superior, although if e is about n^2, we would prefer prim's algorithm.

For clarity, Krushkal's algorithm is written out more formally in below algorithm.

Initially E is the set of all edges in G. The only functions we wish to perform an this set are: I. Determine an edge with minimum cost (line 3) and

II. Delete this edge (line 4).

Both these functions can be performed efficiently if the eges in E are maintained as a sorted sequential list. Actually, it is not essential to sort all the edges so long as the next edge for line 3 can be determined easily. If the edges are maintained as a minimum heap then the next edge to consider can be obtained in O (e loge) time if g has e edges.

The construction of the heap itself takes O (e) time. The computing time is, therefore, determined by the time for lines (3) and (4) which in the worst case is O (e loge).

1. $T \leftarrow \phi$
2. While T contains fewer than n-1 edges and $E \neq \phi$ do
3. Choose an edge (v,w) from E of lowest cost.
4. delete (v,w) from E
5. If (v,w) does not create a cycle in T.
6. then add (v, w) to T
7. else discard (v, w)
8. end if
9. repeat

Example :

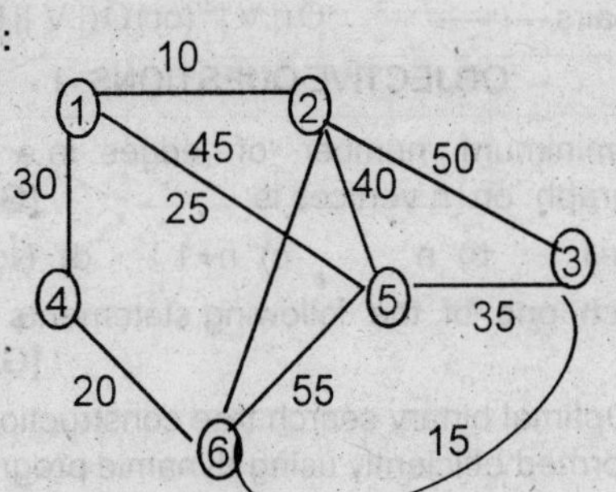

Edge	Cost	Spanning Forest
(1,2)	10	
(3,6)	15	
(4,6)	20	
(2,6)	25	
(1,4)	30	Reject
(3,5)	35	

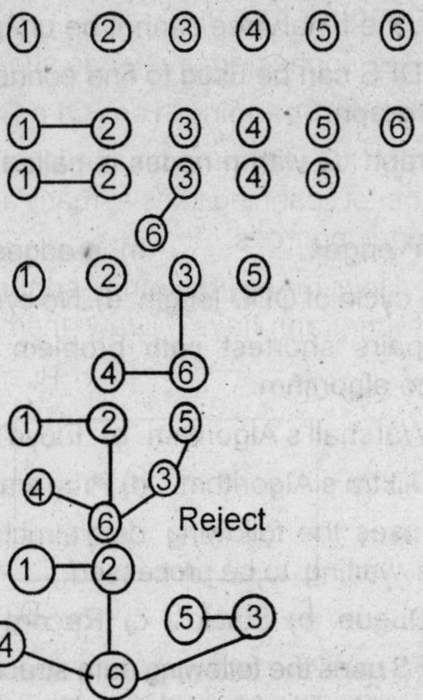

/* Stages in kruskal's algorithm */

11. Performance Of Graph Algorithms

Problem	Run-Time Bounds
DFS-------------------------	$O(\|E\|+\|V\|)$
Connected Components------	$O(\|E\|+\|V\|)$
Strongly Components---------	$O(\|E\|+\|V\|)$
Euler Circuits------------------	$O(\|E\|+\|V\|)$
Biconnectivity------------------	$O(\|E\|+\|V\|)$
Topological Sorting-------------	$O(\|E\|+\|V\|)$
Spanning Tree------------------	$O(\|E\|+\|V\|)$
Test For Cycle------------------	O(| V |)
BFS ------------------------------	$O(\|E\|+\|V\|)$
minimum spanning tree	
Prim's algorithm ----------------	$O(\|V^2\|)$
Krushkal's algorithm ------	O(|E| logE) (or) O(|E |log| V |)
Shortest path between two vertices ------------	$O(\|V^2\|)$ (or) $O(\|E\|\log\|V\|)$
Shortest paths between all vertex-pairs ---------	O(| V |^3 (or) O(| V || E | log | V |)

OBJECTIVE QUESTIONS - I

1. The minimum number of edges in a connected cyclic graph on n vertices is: **[GATE-1995]**

a) n-1 b) n c) n+1 d) None

2. Which one of the following statements is FALSE? **[GATE-1994]**

a) Optimal binary search tree construction can be performed efficiently using dynamic programming.

b) BFS cannot be used to find connected components of a graph.

c) Given the prefix and postfix walks over a binary tree, the binarytree cannot be uniquely constructed.

d) DFS can be used to find connected components of a graph.

3. A graph G with n nodes is called a bipartite then it has : **[AUCET-2000]**

a) n^2 edges. b) n edges.

c) A cycle of ODD length. d) No cycle of ODD length.

4. All pairs shortest path problem can be solved by using the algorithm: **[AUCET-2000]**

a) Warshall's Algorithm. b) Floyd's algorithm.

c) Dijktra's Algorithm. d) Prim's algorithm.

5. BFS uses the following data structure to hold nodes that are waiting to be processed: **[AUCET-2001]**

a) Queue b) Stack c) Record d) File

6. A DFS uses the following data structure to hold nodes that are waiting to be processed:

a) Queue b) Stack c) Record d) File

7. Preorder is nothing but

a) DFS b) BFS

c) Topological order d) None.

8. What is the number of nodes in the largest maximal independent set of the complete bipartite graph k(4,2) is:

a) 2 b) 3 c) 4 d) 6

9. What is true for the complete bipartite graphs k (3,3) and k (2,4) :

a) Both are planar b) Neither is planar.

c) Both are isomorphic. d) None.

10. Which of the following abstract data types can be used to represent a many-to–many relation?

a) Tree b) Graph c) Stack d) Queue.

11. The minimum no. of colours needed to colour a graph of n(>3) vertices and 2 edges is

a) 1 b) 2 c) 3 d) 4

12. A graph is a tree if and only if

a) Is completely connected.

b) Is minimally connected.

c) Contains a circuit. d) Is planar.

Below three questions are based on the following diagram. Consider the following road map where the length of the path is defined to be sum of the costs of the routes traversed. Answer the following questions.

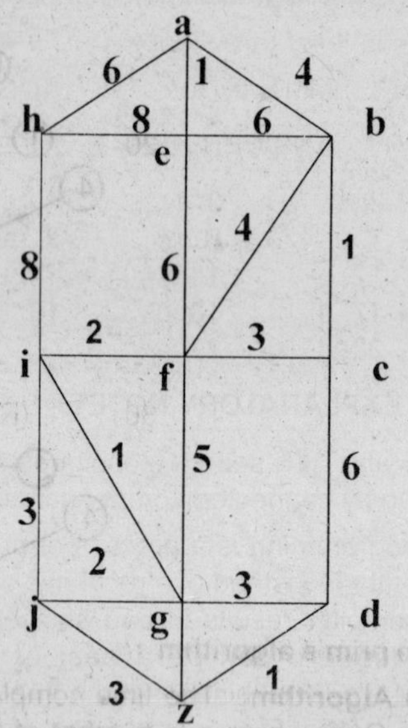

13. Length of the shortest path from 'a' to 'i' is : **[Hyd.Univ-1997]**

a) 14 b) 10 c) 9 d) 8

14. Length of the shortest path from 'a' to 'z' is: **[Hyd.Univ-1997]**

a) 9 b) 12 c) 13 d) 11

15. Length of shortest path from 'a' to 'z' passing through 'c' is : **[Hyd.Univ-1997]**

a) 9 b) 12 c) 13 d) 11

16. Breadth First Search (BFS) algorithm uses ¾¾ data structure **[Hyd.Univ-1994]**

a) Array b) Stack

c) Queue d) Linear linked list

17. Spanning tree of any graph can be obtained using the following algorithm **[UGC NET-1997]**

a) Breadth-first search b) Kruskal's algorithm

c) Prim's algorithm d) All of the above

18. If G is an undirected planar graph on n–vertices with e edges, then **[UGC NET-1996]**

a) $e \le n$ b) $e \le 2n$ c) $e \le 3n$ d) None of these

19. In an undirected graph, the number of nodes with odd degree must be **[UGC NET-1996]**

a) Zero b) even c) odd d) prime

20. A graph is planar, **[UGC NET-1996]**

a) K_5 is not a sub graph b) $K_{3,3}$ is not a sub graph

c) K_5 and $K_{3,3}$ are not a subgraphs

d) K_4 is not a sub graph

21. In a directed graph G(V, E); always **[UGC NET-1998]**

a) $E \subseteq V \times V$ b) $E \not\subseteq V \times V$ c) $E = V \times V$ d) $E \neq V \times V$

22. The number of edges in a complete graph with 7 vertices is **[UGC NET-1998]**

a) 7 b) 14 c) 128 d) 21

23. A connected graph is the one which **[UGC NET-1998]**

a) cannot be partitioned without removing an edge

b) contains at least three loops

c) does not contain a cycle d) is not simple

ANSWER KEY - I

1. b	2. a	3. c	4. b	5. a
6. b	7. a	8. c	9. d	10. b
11. b	12. b	13. c	14. b	15. b
16. b	17. d	18. c	19. c	20. c
21. a	22. d	23. a		

EXPLANATORY NOTES - I

2. i) Both DFS and BFS search techniques can be used to find the strongly connected components of a graph.

ii) Dynamic programming is a process of providing non recursive algorithms instead of a recursive algorithms in case of the compiler results in bad answer. So a non recursive procedure can be a best of thought of construction of a optimal binary search tree.

iii) By Means of the prefix and postfix, we can construct unique binary tree.

9. For complete biapartite graphs K (3,3) and K (2,4), the condition applies is $m \le 3$ and $n \le 3$ $[\because K_{mn}]$.

According to the condition K (3,3) is planar and K (2,4) is non planar.

10. In graphs only we can have many - to - many relationships. Because each vertex can have multiple indegrees and out degrees.

No other data structure (or) ADT follows this except graphs.

12. A Graph can be a tree if and only if it is minimally connected.

→ If it is completely connected, ∃ a cycle.

A tree graph must not contain a cycle.

→ If it is planar, No two edges must not cross over one another. It may lead to a cycle.

→ If it contains a circuit, it can't be a tree graph.

13. Shortest path from 'a' to 'i' is

length (a,e) + lenght (e,f)+length (f,i) = 1+6+2=9.

15. Length of shortest path from 'a' to 'z' pushing through 'c' is 12.

i.e., $L(a,b)+L(b,c)+L(c,d)+L(d,z)$

= 4+1+6+1 = 12

20. A graph K_{mn} is planar if $m \le 3$ and $n \le 3$

The graphs K_5 is not planar. K_{33} is planar.

23. → A Connected graph may or may not have atleast 3 loops.

→ A Connected graph certainly contains atleast one cycle

→ A connected graph may be simple.

But a connected graph cannot be partitioned without removing an edge.

OBJECTIVE QUESTIONS - II

1. The minimum number of colours required to colour the vertices of a cycle with n nodes in such way that no two adjacent nodes have the same colour is the **[GATE-2002]**

a) 2 b) 3 c) 4 d) $n-2\left(\frac{n}{2}\right)+2$

2. Maximum number of edges in a n- node undirected graph without self loops is : **[GATE-2002]**

a) n^2 b) n(n-1)/2 c) n-1 d) n (n+1)/2

3. Consider an undirected unweighted graph G. Let a breadth-first traversal of G be done starting from a node r. Let d(r, u) and d(r, v)be the lengths of the shortest paths from r to u and v respectively in G. If u is visited before v during the Bredth-first traversal. Which of the following statements is correct. **[GATE-2001]**

a) d(r, u) < d(r, v) b) d(r, u) > d(r, v)

c) $d(r, u) \le d(r, v)$ d) None

4. How many undirected graphs(not necessarily connected) can be constructed out of a given set $V = \{v_1, v_2, \ldots, v_n\}$ of n vertices **[GATE-2001]**

a) n (n-1)/2 b) 2^n c) n! d) $2^{n(n-1)/2}$

5. The most appropriate matching for the following pairs **[GATE-2000]**

X. DFS	I. Heap
Y. BFS	II. Queue
Z. Sorting	III. Stack

a) X-I, Y-II, Z-III b) X-III, Y-I, Z-II

c) X-III, Y-II, Z-I d) X-II, Y-III, Z-I

6. Let G be an undirected connected graph with distinct edge weights. Let e_{max} be the edge with maximum weight and e_{min} the edge with minimum weight. Which of the following statements is FALSE: **[GATE-2000]**

a)Every minimum spanning tree of Gmust contain e_{min}

b) If e_{max} is in a minimum spanning tree, then its removal must disconnect G.

c) No minimum spanning tree contains e_{max}.

d) G has a unique minimum spanning tree.

7. Let G be an undirected graph. Consider a depth-first traversal of G, and let T be the resulting depth-first Search tree. Let u be a vertex in G and let v be the first new (unvisited) vertex visited after visiting u in the traversal. Which of the following statements is always TRUE. **[GATE-2000]**

a) {u, v} must be an edge in G, and u is a descendant of v in T.

b) {u, v} must be an edge in G, and v is a descendant of u in T.

c) If {u, v} is not an edge in G then u is a leaf in T.

d) If {u, v} is not an edge in G then u and v must have the same parent in T.

8. The number of articulation points of the following graph is: **[GATE-1999]**

a) 0

b) 1

c) 2

d) 3

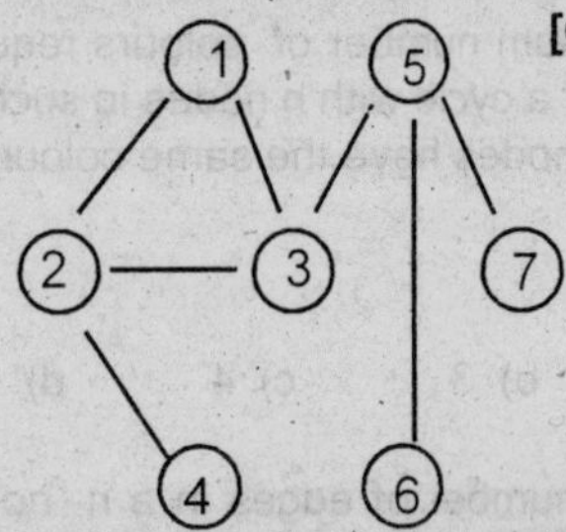

9. Number of edges in the minimum spanning tree of a graph containing n nodes is: **[AUCET-2001]**

a) 2^n b) n+1 c) n-1 d) log n

10. Which of the following statements is FALSE:

a) Every tree is a bipartite graph.

b) A tree contains cycle.

c) A tree with n nodes contain n-1 edges.

d) A tree is a connected graph.

11. What is the minimum number of edges which must be removed from a complete bipartite graph of six nodes k(6) so that the remaining graph is planar:

a) 2 b) 3 c) 4 d) 6

12. How many colours are needed to colour the below graph is:

a) 3

b) 4

c) 5

d) 6

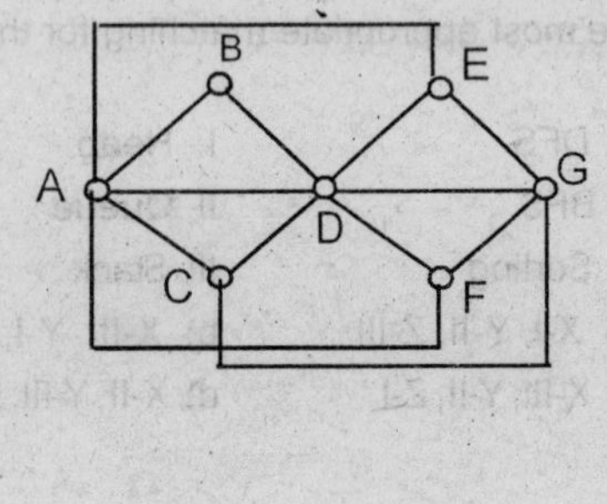

13. Consider the graph shown below.Which of the following is a valid strong component:

a) b, c, d

b) a, c, d

c) a, b, d

d) a, b, c

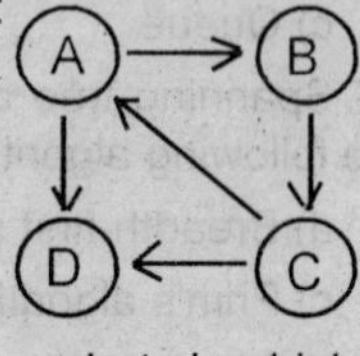

14. Consider the graph shown below, what should be the labels of nodes marked 1 and 2. If the BFS yields the list A, B, C, D, E.

a) D and E

b) E and D

c) Unpredictable

d) None .

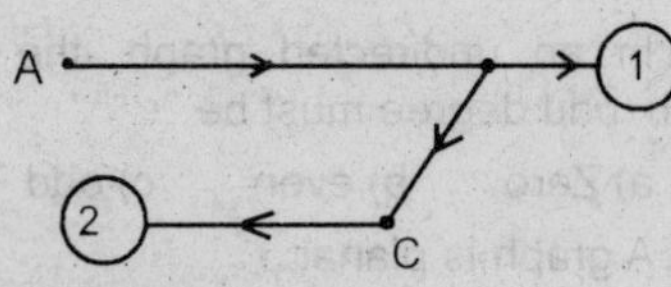

15. Consider the graph shown below:

Which of the following is a valid topological sorting?

a) A B C D

b) B A C D

c) B A D C

d) A B D C.

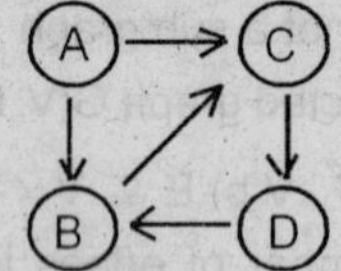

16. Consider the Graph shown below :

The minimum cost spanning tree has the cost

a) 18

b) 20

c) 22

d) 24

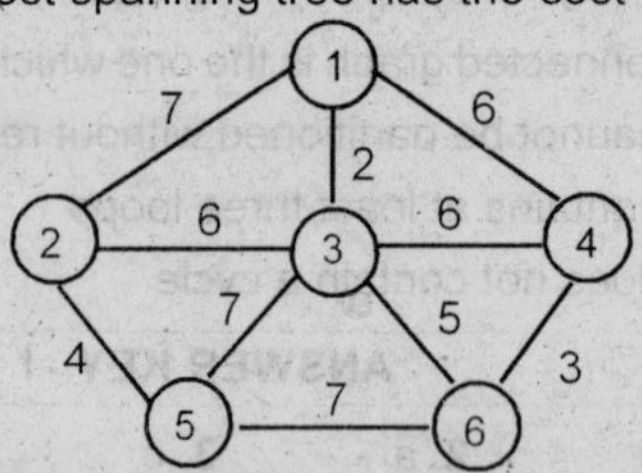

17. Which of the following is a Non linear data structure ? **[Hyd. Univ-2000]**

I. Stack II. Queue

III. Tree IV. Graph

V. Doubly Linked List

a) All of the above b) III, IV and V

c) III and IV d) I, II and V

18. The maximum depth that a tree of N nodes can have is **[Hyd. Univ-1999]**

a) N b) N/2 c) $\lceil \log_2 (N+1) \rceil$ d) 1

19. If a tree has n_1 nodes of degree 1, n_2 nodes of degree 2, ..., n_m nodes of degree m, then how many terminal nodes does it have? **[Hyd.Univ-1998]**

a) $1 + n_2 + 2n_3 + ... + (m-1)n_m$

b) $n_1 + 2n_2 + 3n_3 + ... + mn_m$

c) $n_1 + n_2 + n_3 + ... + n_m$

d) $1 + n_1 + 2n_2 + ... + mn_m$

20. The particular spanning tree for a graph depends on the criteria used to generate it. Consider the graph G given in figure(i) and its sub graph G′ as given in figure(ii) **[Hyd.Univ-1995]**

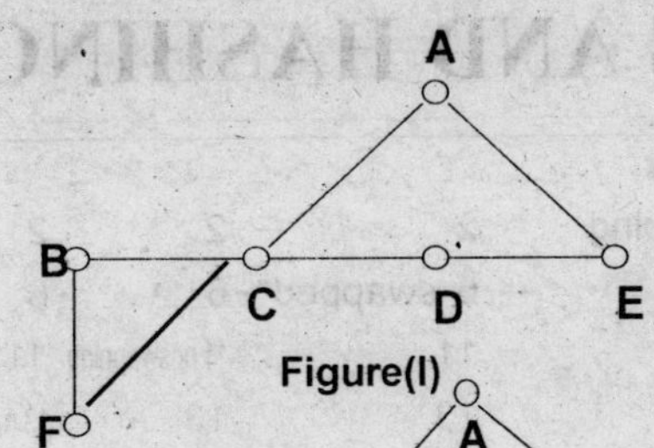

Figure(I)

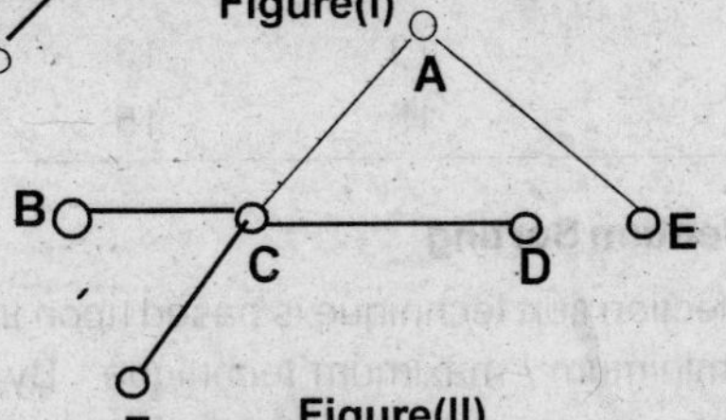

Figure(II)

a) G′ is the minimum spanning tree

b) G′ is the DFS spanning tree

c) G′ is the BFS spanning tree

d) G′ is the spanning tree obtained by greedy method

21. The graph of fig. shown below is **[UGC NET-1996]**

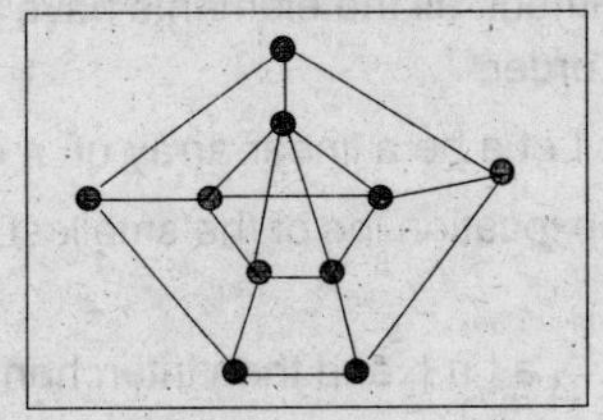

a) planar b) Bipartite c) Eulerian d) Regular

ANSWER KEY - II

1. b	**2.** b	**3.** c	**4.** d	**5.** c
6. c	**7.** c	**8.** d	**9.** c	**10.** b
11. d	**12.** a	**13.** d	**14.** a	**15.** d
16. b	**17.** c	**18.** a	**19.** a	**20.** c
21. c				

EXPLANATORY NOTES - II

6. G is undirected connected graph. e_{min} is the min. Weighted distinct edge e_{max} is the max. Weighted distinct edge. So every min spanning tree must include e_{min}. As it is undirected connected graph, there may be multiple min. spanning tress. Hence any one of the spanning trees may contain e_{max} edge.

7. G be an undirected graph and T is the DFS tree.

According to the situation, as the DFS makes use of stack, the (u,v) may not be an edge in G but it can be a leaf node of DFS tree T.

10. A tree must not have any cycles. And it can be referred as a connected graph because every node is reachable from any other node. Even a tree can be refereed as a bipartite graph. Every tree has (n-1) edges. Hence the statement.

'A tree contains cycles' is false.

12. Order the vertices according to the degrees in decreasing order ⇒ DAGCEFB

Assign 1st color to first vertex and its non - neighbors assign 2nd color to the Subsequence of non colored vertices. Do the above process until all the vertices are colored.

According to this color 1 to D

Color 2 to A, G

Color 3 to C, E, F, B.

⇒ No. of colors = 3.

13. i) Apply DFS over the graph

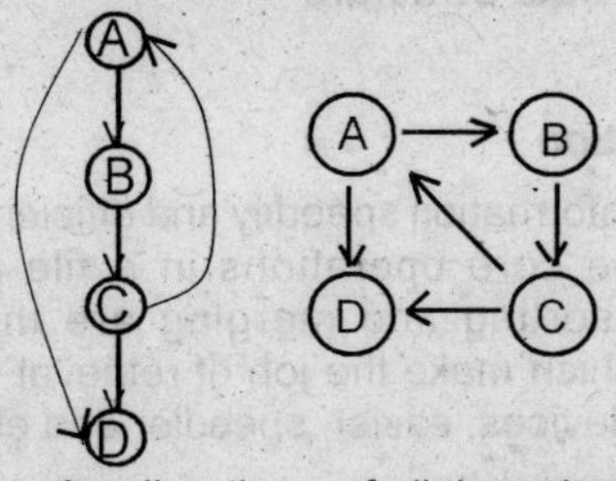

ii) Reverse the directions of all the edes

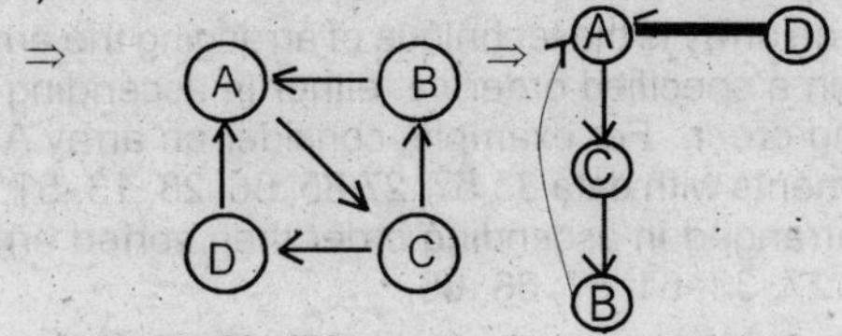

⇒ {A, C, B}, {D} are strongly connected.

14.

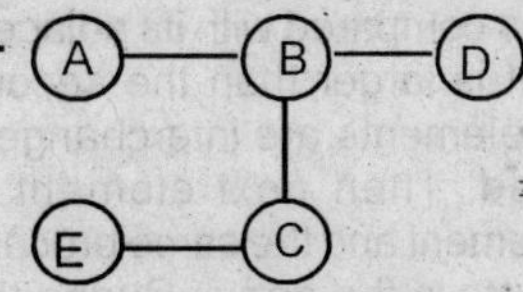

⇒ Ans D & E

15. Consider the graph

i) find the indegree of each node

A = 0, B = 2, C = 2, D =1

Queue = A

ii) Neighbors of A = B, C

Decrease the indegree of each neighbor

⇒ B = 2-1 = 1, C = 2-1 = 1

Queue = A

iii) Do ii) again

⇒ Indegree (B) = 1 - 1 = 0

Indegree (C) = 1 - 1 = 0

Hence insert B, C into queue

Queue : - A, B, C

iv) find neighbors of B ⇒ C

C already existed. So find the neighbor of 'C'

⇒ B is the neighbor

Insert 'D' into Queue

⇒ ABCD is the toplogical order.

16. Explanation : Consider a graph

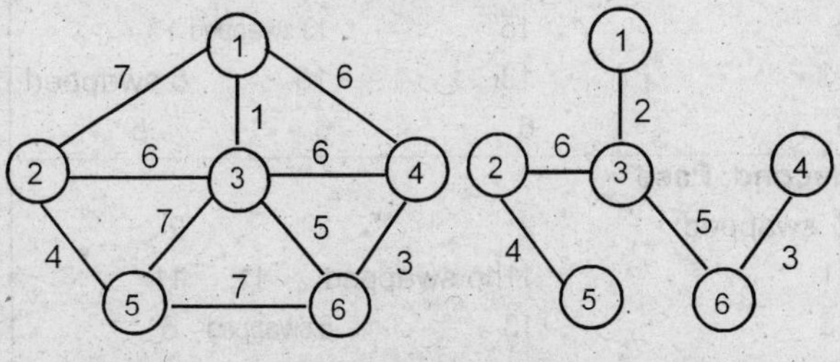

Min. cost spanning tree for the above graph is as follows:

= 2 + 6 + 5 + 4 + 3 = 20

IV. SORTING, SEARCHING AND HASHING

- Sorting Techniques
- Searching Techniques
- Hashing Data Structure

1. Introduction

Retrieval of information speedily and efficiently from files is one of the core operations in a file processing. Searching, sorting and merging are three related operations which make the job of retrieval of data from the storage devices, easier, speedier and efficient.

2. Sorting Array

Sorting of an array is the technique of arranging the array elements in a specified order i.e. either in ascending or descending order. For example, consider an array A of size 8 elements with data 33,57, 27,85, 66, 23, 13, 51. If they are arranged in ascending order then sorted array is :13, 23, 27, 33, 51, 57, 66, 85.

3. Bubble Sorting

In bubble sort, each element is compared with its adjacent element. If the first element is larger than the second one then the position of the elements are interchanged, otherwise it is not changed. Then next element is compared with its adjacent element and the same process is repeated for all the elements in the array. During the next pass the same process is the largest leaving the largest element. During the pass, the second largest element occupies the second last position. During the next pass, the same process is repeated leaving the largest element. During this pass, the largest element occupies the n-1 position. The same process is repeated until no more elements are left for comparison. Finally the array is sorted one.

The various steps involved in sorting of an array of 5 elements is given below:

Initial Elements (without sorting)

11
15
2
13
6

First Pass

11 no swapping	11	11	11
15	2 swapped	2	2
2	15	13 swapped	13
13	13	15	6 swapped
6	6	6	5

Second Pass

2 swapped	2	2	2
11	11no swapped	11	11
13	13	6 swapped	6
6	6	13	13 no swapping
15	15	15	15

Third Pass

2 no swapping	2	2	2
11	6 swapped	6	6
6	11	11 noswapping	11
13	13	13	13 no swapping
15	15	15	15

4. Selection Sorting

The selection sort technique is based upon the extension of the minimum / maximum technique. By means of a nest of for loops, a pass through the array is made to locate the minimum value. Once this found, it is placed in the first position of the array (position 0). Another the remaining elements is made to find the next smallest element, which is placed in the second position (position 1), and so on. Once the next to last element is compared with the last element, all the elements have been sorted into ascending order.

More precisely: Let a be a linear array of n elements.

Pass 1. Find the location loc of the smallest in the list of n elements.

a [1], a [2] ......, a [n], and then interchanged a [loc] and a [1].

Pass 2. Find the location loc of the smallest in the sublist of n-1 elements

a[2],a[3] ...., , a[n] and then interchanged a [loc] and a[2]. Then: a[1], a[2] is stored, since a[1] $\leq$ a[2].

Pass 3. Find the location loc of the smallest in the sublist of n-2 elements

a[3], a[4], ..., a[n], and then interchanged a[loc] and a[3] Then: a[1], a[2],.., a[3] is stored, since a[2] $\leq$ a[3].

....

....

Pass n – 1. Find the location loc of the smaller of the elements

a[n-1], a[n], and then interchanged a[loc] and a[n] –1] Then:

a[1], a[2],..., a[n] is stored, since a[n-1] $\leq$ a[n].

Thus a is sorted after n-1 passes.

For example, suppose we have a list, which contains the elements: 40,30,50,20 and 10. If we want to sort this list of elements using the selection sort technique, the details of this technique are illustrated in Fig,4.1(a) to Fig 4.1(c). In Fig. 4.1(a), Fig,.4.1(b) and figure 4.1 (c) the parts(I), (ii), (iii) and (iv) are generated during the first iteration, similarly (v), (vi) and (vii) in the second iteration and in the third iteration (viii) (ix) and (x) are produced.

I	II	III	IV
40	30	30	20
30	40	40	40
50	50	50	50
20	20	20	30
10	10	10	10

Fig. 4.1 (a) Selection Sort

V	VI	VII	VIII
10	10	10	10
40	40	30	20
50	50	50	50
30	30	40	40
20	20	20	30

Fig. 4.1 (b) Selection Sort

IX	X	XI
10	10	10
20	20	20
40	30	30
50	50	40
30	40	50

Fig. 4.1 (c) Selection Sort

5. Insertion Sort

An insertion sort is one that sorts a set of values by inserting values into an existing sorted file.

Suppose an array a with n elements a[1], a[2], ….., a[n] is in memory. The insertion sort algorithm scans a from a[1] to a[n], inserting each element a[k] into its proper position in the previously sorted subarray a[1], a[2],……, a[k-1]. That is:

Pass 1. a[1] by itself is trivially sorted.

Pass 2. a[2] is inserted either before or after a [1] so that: a[1], a[2] is sorted.

Pass 3. a[3] is inserted into its proper place in a[1], a[2], that is, before a[1], between a [1] and a [2], or after a [2], so that: a[1], a[2], a[3] is sorted.

Pass 4. A[4] is inserted into its proper place in a[1], a[2], a[3] so that: a[1], a[2], a[3], a[4] is sorted.

………………………………………………………………

Pass N. a[n] is inserted into its proper place in a[1], a[2],….a[n-1] so that: A[1], a[2],……a[n] is sorted.

To illustrate the insertion sort method, consider the following array a with 7 elements:

25,15,30,9,99,20,26

Array a of 7 elements

0	1	2	3	4	5	6
25	15	30	9	99	20	26

Pass I : a[1]<a[0], interchanging the position of elements, we get.

0	1	2	3	4	5	6
15	25	30	9	99	20	26

Pass II : a[2]> a[1], position of elements remain same.

0	1	2	3	4	5	6
15	25	30	9	99	20	26

Pass III :a[3] is less than a[0], a[1], so insert a[3] before a[0],we get a[5] before a[2], we get

0	1	2	3	4	5	6
9	15	25	30	99	20	26

Pass IV:a[4]> a[3], no change is performed.

0	1	2	3	4	5	6
9	15	20	25	30	99	26

Pass V: a[5] is less than a[2], a[3] and a[4], therefore insert a[5] before a[2], we get

0	1	2	3	4	5	6
9	15	20	25	26	30	99

Pass VI: Now a[6] is less than a[4], a[5] therefore insert a[6] before a[4], giving the following array:

0	1	2	3	4	5	6
9	15	20	25	26	30	99

After the Pass VI, we get array with sorted elements.

6. Quick Sort

It is one of the most popular sorting technique. Quick sort possesses a very good average case behavior among all the sorting techniques.This is developed by C.A.R. Hoare.

The quicksort algorithm works by partitioning the array to be sorted, and each partition is in turn sorted recursively. In partition, one of the array elements is chosen as a key value. This key value can be the first element of an array. That is, if a is an array then key =a [0]. And rest of the array elements are grouped into two partitions such that

- One partition contain selements smaller than thekey value.
- Another partition contains elements larger than the key value.

Figure 4.2 shows a key and partitions of element in an array.

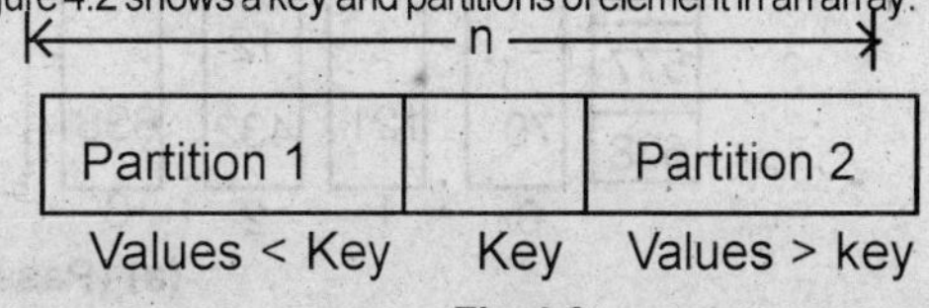

Fig.4.2

As an example consider an array with the following elements.

45 26 77 14 68 61 97 39 99 90

Here 45 is selected as a key value. Then two indices namely, low and high are used to indicate the element (key+1) and the last element. The low index starts on the left and selects an element that is greater than the key value. Then these elements are interchanged.

This process is repeated until all elements to the left of the key are smaller than the key value. And all elements to the right of the key are greater than the key value. The steps involved in placing the key value, 45 in its proper position in the array are given below,. The elements being compared are encircled on each line.

(45)	(26) low	77	14	68	61	97	39	99	90	
(45)	26	(77) low	14	68	61	97	39	99	90	77 > 5
(45)	26	77	14	68	61	97	39	99	(90) high	
(45)	26	77	14	68	61	97	39	(99) high	90	39 < 45 interchange 77 and 39
(45)	26	77	14	68	61	97	(39) high	99	90	
(45)	26	77	(14) low	68	61	97	39	99	90	
(45)	26	77	14	(68) low	61	97	39	99	90	
(45)	26	77	14	68	61	(97) high	39	99	90	
(45)	26	77	14	68	(61) high	97	39	99	90	
(45)	26	77	14	(68) high	61	97	39	99	90	14 < 45 interchange 45 and 14
(45)	26	77	(14) high	68	61	97	39	99	90	
14	26	39	45	68	61	97	77	99	90	low ≥ high

Value < 45 (14 26 39)   Value > 45 (68 61 97 77 99 90)

The given array has been partitioned into two sub arrays, The first sub array is {14,26,39} and the second subarray is {68,61,97,77,99,90}. We can repeatedly apply this procedure on each of these sub arrays until the entire array is sorted. Since the array elements are partitioned. Since the array elements are partitioned and exchanged, this technique is called partition-exchange technique.

In the quick sort technique, each time the given array is partitioned into two smaller subarrays, each of these subarrays can be processed either iteratively or recursively.

7. Bucket or Radix Sort.

Bucket sort or radix sort, is a method can be used to sort a list of names alphabetically. Here the base or radix is 26,(the 26 letters of the alphabetic).

First of all the list of names is sorted according to the first letter of each name thus the names are arranged in 26 buckets. In second pass, names are arranged according to the second letter of each name and so on this process depend on the length of the names with maximum letters. Suppose if no name contains more than 15 letters, the names are alphabetized with at most 15 passes.

To sort decimal numbers, where radix or base is 10, we need ten buckets. These buckets are numbered 0,1,2,3,4,5,6,7,8,9.Un like, sorting names, decimal numbers are sorted from right to left.

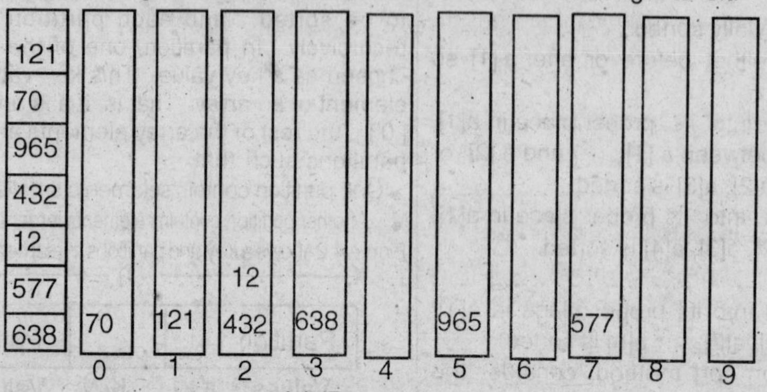

(a) (Pass - I)

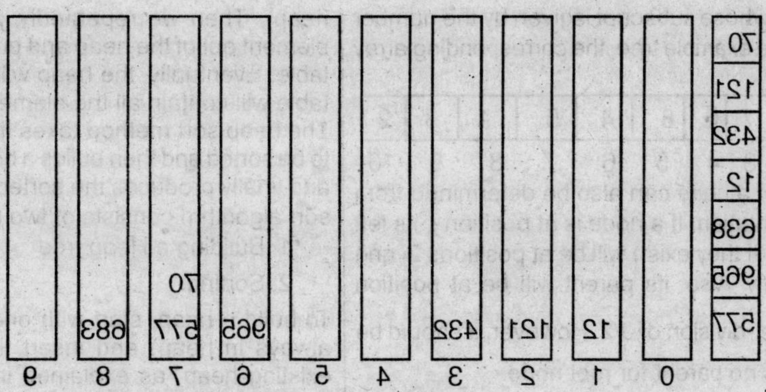

(b) (Pass - II)

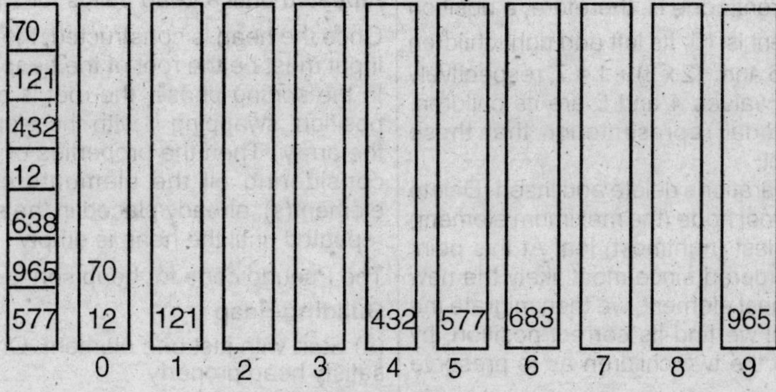

(c) (Pass - III)

After pass III, when the numbers are checked, they are in ascending order.

12 70 121 432 577 683 965

In this example, there are three passes, depending upon the number of digits in largest number 965.

8. Merge Sort

Merge sort is a sorting technique which divides the array into subarray of size 2 and merge adjacent pairs. We then have approximately n/2 arrays of size 2.This process is repeated until there is only one array remaining of size n.

Suppose an array a with n elements a[1],a[2],..............,a[N] is in memory. The merge-sort algorithm which sorts a will first be described by means of a specific example.

Example: suppose the array a contains 12 elements as follows:

85,76,46,92,30,41,12,19,93,3,50,11

Each pass of the merge sort algorithm will start at the beginning of the array A and merge pairs of sorted sub arrays as follows:

Pass 1: Merge each pair of elements to obtain the following list of sorted pairs.

76	85		46	92		30	41		12	19		3	93		11	50

Pass II: Merge each pair of pairs to obtain the lists of sorted elements.

46	76	85	92		12	19	30	41		3	11	50	93

Pass III: Again Merge the two subarrays to get two lists.

12	19	30	41	46	76	85	92		3	11	50	93

Pass IV: Merge the above two lists, we get

3	11	12	19	30	41	46	50	76	85	92	93

In pass IV, we get sorted list of elements of array a with the size 12.

9. Heap Sort

This method sorts by building a heap, and then repeatedly extracting the maximum item.

Heap is a complete binary tree with the property that a parent is always greater than or equal to either of its children (if they exist). Thus, the root node always contains the largest element. Figure 4.3 illustrates an example for a heap.

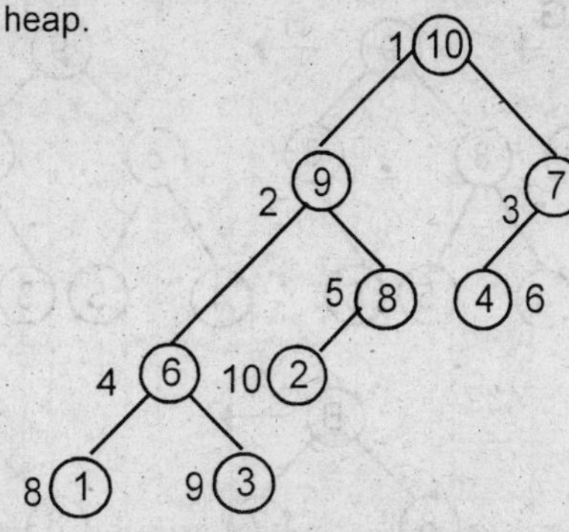

Figure : 4.3 An illustration of Heap

Binary Trees can be represented as arrays, by first numbering the nodes (starting from the in root) from left to right. The key values of the nodes are then assigned to array positions whose subscript isgiven by the number of the node. For the example tree, the corresponding array would be

Array	10	9	7	6	8	4	5	1	3	2
Subscript	1	2	3	4	5	6	7	8	9	10

The relationship of a node can also be determined from this array representation. If a node is at position j, its left and right children (if they exist) will be at positions 2j and 2j + 1, respectively. Also, its parent will be at position $\lfloor J/2 \rfloor$, i.e., integer division of J/2. However, it should be noted that there is no parent for root node.

For instance, consider the node having key Value 7. It is at the position 3. Its parent node is, therefore, a position $\lfloor 3/2 \rfloor$ = 1 i.e. the parent is 10. Its left and right children are at positions 2 x 3 = 6 and (2 x 3) + 1 = 7, respectively i.e. the nodes with key values 4 and 5 are its children. We see from the pictorial representation that these relationships are correct.

The heap supports 2 operations delete and insert. Delete operation removes the root node (the maximum element) and replace it with the last (rightmost) leaf At this point the tree might not be ordered since most likely the new node is not the next largest element; we then migrate the root down the tree until we find its correct position, by swapping it with one of the two children as to preserve the heap property.

Insert operation inserts the new element as the last rightmost node in the binary tree. At this point, the new node will most likely not be in the correct position, so we move it up the tree structure (by swapping it with its parent node if it has a higher priority) until its correct position is attained.

Heap sort simply uses the Heap structure and operations. First it initialized the heap by putting the data from array onto a heap (repeatedly call the insert operation on the heap). Then we repeatedly yank the root (maximum) element out of the heap and put it at the end of the sorted table. Eventually, the heap will be empty and the sorted table will contain all the elements.

The heap sort method takes the input array of elements to be sorted and then builds a heap based on the elements and finally produces the sorted elements. Thus the heap sort algorithm consists of two phases :

1. Building a Heap tree
2. Sorting.

To build a heap, start with one element (single node is always in heap) and insert the next element into the existing heap, as explained in the insert operation, so that again heap is constructed. The insertion process is repeated until a heap builds for all input data elements.

Once the heap is constructed, the largest element in the input must be the root of the heap, located at position 1. In the sorting phase, the root is placed into the correct position, swapping it with the element at the bottom of the array. Then the properties of the heap are restored considering all the elements excluding the largest element(s), already placed in the sorted position. This is repeated until the heap is empty.

The Pseudo code for heap sort is :

Building Heap

(a) Start with just one element. One element will always satisfy heap property

(b) Insert next elements and make this a heap

(c) Repeat step b, until all elements are included in the heap

Sorting

(a) Exchange the root and the last element in the heap

(b) Make this a heap again, but this time do not include the last node.

(c) Repeat steps a and b until there is no element left.

An example of the heap sort for the input elements 4, 8, 3, 9, 5, 6, 7

Building Heap

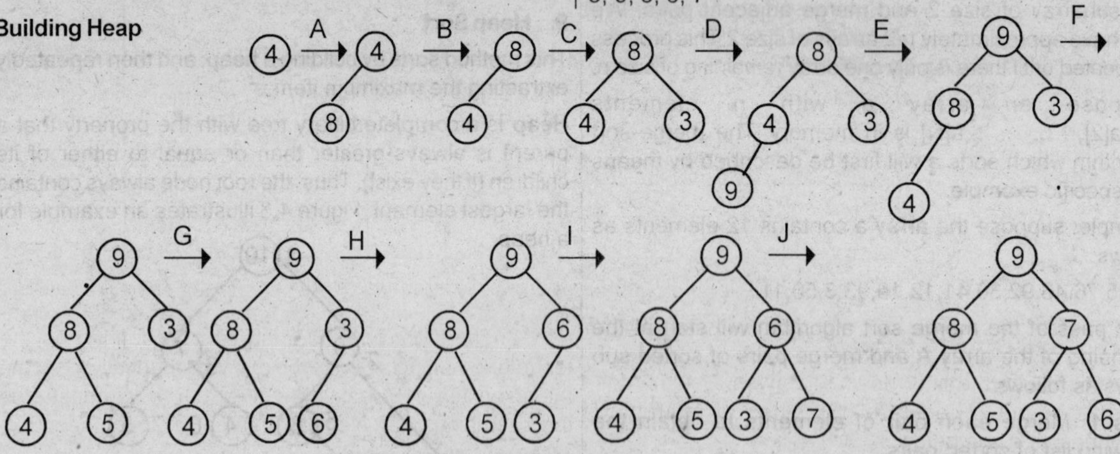

Sorting

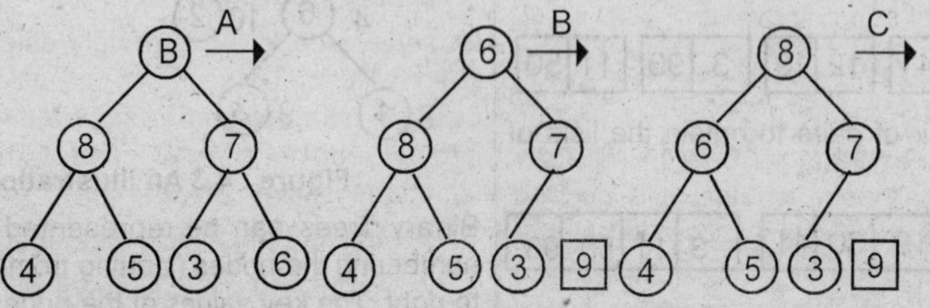

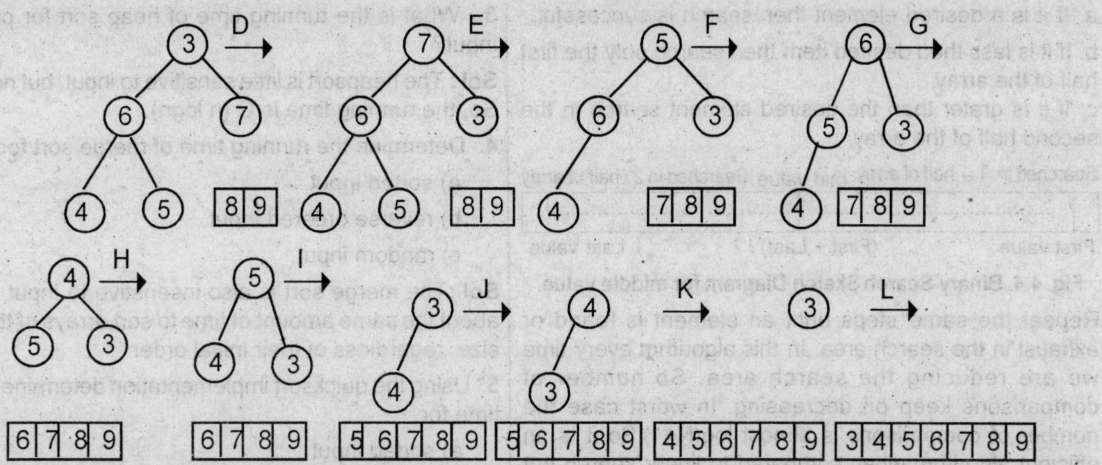

10. Summary of sorting methods:

For most general internal sorting applications, either insertion sort, shell sort (or) quick sort will be the method of choice, and the decision of which to use will depend mostly on the size of the input.

Insertion sort is useful only for small (or) very nearly sorted inputs.

We have not included merge sort, because its performance is not as good as quick sort for main memory sorts and it is not any simpler to code. We have seen however, that merging is the central idea of external sorts.

In summary, the selection (or) bubble sort can be used if the number of records in the table is small. If n is large and the keys are short, the radix sort can perform well. With a large n and long keys, quick sort, heapsort (or) merge sort can be used. If the table is, initially, almost sorted, then quick sort should be avoided.

When the keys, after hashing are uniformly distributed over the interval [1, m] then an address-calculation sort is a very good method to use.

Comparison of sorting methods

Algotithm	Average case	Worst case	space usage
insertion-sort	n^2	n^2	Extra Array
selection-sort	$n^2/4$	$n^2/4$	In place
Bubble-sort	$n^2/4$	$n^2/4$	In place
merge-sort	$O(n\log_2 n)$	$O(n\log_2 n)$	Extra n entries
quick-sort	$O(n\log_2 n)$	$n^2/2$	Extra $\log_2 n$ entries
heap-sort	$O(n\log_2 n)$	$O(n\log_2 n)$	In place
Radix-sort	$O(m+n)$	$O(m+n)$	Extra space for links
Address-sort	$O(n)$	$O(n^2)$	Extra space for links

Where, n is the size of the input file and m denotes the number of digits in a key.

11. Searching Of Array Element

Searching is used to find the location where element is available or not. There are two types of searching techniques as given below:

1. Linear or Sequential searching.
2. Binary searching.

12. Linear or Sequential Searching

In linear search, we access each element of an array one by one sequentially and see whether it is desired element or not. A search will be unsuccessful if all the elements are accessed and the desired element is not found. In the worst case, the number of average case we may have to scan half of the size of the array (N / 2). Therefore linear search can be defined as the technique which traverses the array sequentially to locate the given item. The algorithm given below demonstrates the searching of an element of the array using search.

13. Binary Searching

Binary search is an extremely efficient algorithm. This search technique searches the given item in minimum possible comparisons. To do the binary search, first we had to sort the array elements. The logic behind this technique is given below:

1. First find the middle element of the array.
2. Compare the mid element with an item
3. There are three cases.

a. If it is a desired element then search is successful.

b. If it is less than desired item then search only the first half of the array.

c. If it is grater than the desired element search in the second half of the array.

Searched in 1_{st} half of array	mid value	Searched in 2nd half of array
First value	(First + Last) / 2	Last value

Fig. 4.4. Binary Search Sketch Diagram for middle value.

Repeat the same steps until an element is found or exhaust in the search area. In this algorithm every time we are reducing the search area. So number of comparisons keep on decreasing. In worst case the number of comparisons is almost log(N+1).So it is an efficient algorithm when compared to linear search but the array has to be sorted before doing binary search.

Suppose we have an array a of 7 elements.

9	12	24	30	36	45	70
0	1	2	3	3	5	6

The steps to search 45 using binary search in array a [7] are:

Step I: The given array is in ascending order ; item to be searched for is 45.

beg =0, last = 6

mid =int ((beg + last)/2 = int(0 + 6/2) = int (3) = 3

beg 0	1	2	3	3	5	last 6
9	12	24	30	36	45	70

Step II: a[mid] i.e. a[3] is 30

30 < 45 then

beg = mid+1 = 3+1 = 4

Step III: mid = int((beg+last)/2) = int (4+6)/2 = int (5) = 5

beg 4	5	last 6
9	12	24

mid

a[mid] i.e. a[5] is 45. 45 = 45

Search Successful !! At location numbers 5 (element number 6.)

14.Some important properties of sorting and searching:

1. What is the running time of insertion sort if all keys are equal ?

Sol : $O(n^2)$

2. Determine the running time of shell sort for

a) Sorted input b) reverse-ordered input.

Sol : Shell sort is insensitive to input; it takes about the same amount of time to sort arrays of the same size; regardless of their initial order.

3. What is the running time of heap sort for presorted input?

Sol : The heapsort is little sensitive to input, but not much. So, the running time is O (n logn).

4. Determine the running time of merge sort for

a) sorted input.

b) reverse ordered input

c) random input.

Sol : The merge sort is also insenstive to input; it takes about the same amount of time to sort arrays of the same size, regardless of their initial order.

5. Using the quicksort implementation determine running time for

a) sorted input

b) reverse ordered input

c) random input

Sol : The Quick sort is quite sensitive to input. The Quick sort will degrade into an O (n^2) algorthm in the special cases where the array is initially sorted in either order. That is because the pivot element will always be an extreme value within its sub array, so the partitioning splits the subarray very unevenly, there by requiring n steps instead of log n

6. For the quick sort, what is the running time when all keys are equal?

Sol : O(N^2).

7. In how many ways can two sorted arrays of N elements be merged?

Sol : Two ways.

8. A sorting algorithm is stable if elements with equal keys are left in the same order as they occur in the input. (or) A sorting procedure is said to be stable if when ever two items which have the same value will be in the same order in the sorted list as in the unsorted one.

1. Quick sort is an unstable sorting method.
2. Merge sort is an stable sorting method.
3. Heap sort is an unstable sorting method.
4. Shell sort is an unstable sorting method.
5. Selection sort is stable sorting method.
6. Insertion sort is stable sorting method.
7. Bubble sort is stable sorting method.
8. Radix sort is unstable sorting method.

9. Suppose you are to sort a list L consisting of a sorted list followed by a few "random" elements, which of the sorting methods discussed so far would be especially suitable for such a task?

Sol: Insertion sort is useful only for small (or) very nearly sorted inputs.

10. A sorting procedure is called to sort a list of 100 integers that have been read from a file. If all 100 values are zero, what would the execution requirements (interms of Big-0) be if the sort used was

a) Quick sort, with the first element used as the split value

b) Selection sort?

c) Heap sort?

d) Merge sort?

Sol: (a) O (N^2)

(b) O(N^2)

(c) O ($N\log_2 N$)

(d) O ($N\log_2 N$)

11. A very large array of elements is to be sorted. The program is to be run on a personal computer with limited memory. Which sort would be a better choice: Heap sort (or) merge sort. why?

Sol: Heap sort would be a better choice than merge sort, because, merge sort needs a copy of the array for its processing where as Heap sort sorts with in the original array.

12. Give arguments for and against using procedures (such as swap) to encapsulate frequently used code in a sorting procedure?

Sol: For: Making swap a procedure will make the code easier to read and understand.

Against: It would be more efficient, in terms of execution time, to code the swap algorithm "in line" instead of requiring a procedure call, which takes extra time to execute.

13. Name an advantage and disadvantage to use selection sort that does not make any procedure calls?

Sol. Advantage: Because there are no procedure (or) function calls, the sort should execute more quickly.

Disadvantage : The code is not readable.

14. The element being searched for is not in an array of 100 elements. What is the average number of comparisons needed in a sequential search to determine that the element is not there?

a) If the elements are completely un-ordered?

b) If the elements are ordered from smallest to largest?

c) If the elements are ordered from largest to smallest?

Sol:(a) 100 (b) 50 (c) 50

15. The Bubble sort, the selection sort, and the Insertion sort are all O (N^2) algorithms, which is the fastest, and which is the slowest among them?

Sol: The Bubble sort is slower than the selection sort, and the insertion sort (in most cases) is a little faster.

16. How does the Bubble sort perform on:

a) an array that is already sorted?

b) an array that is sorted in reverse order?

Sol: The bubble sort is insensitive to input.That means that it will execute the same number n(n-1)/2 of comparisons regardless of the original order of the elements in the array. so it doesn't matter whether the array is already sorted (or) whether it is sorted in reverse order; it is still very slow.

15. Hashing

In tree tables, the search for an identifier key is carried out via a sequence of comparisons.Hashing differs from this in that the address (or) location of an identifier x, is obtained by computing some arithmetic function, f, of x f (x) gives the address of x in the table. This address will be referred to as the hash (or) home address of x.

The memory available to maintain the symbol table is assumed to be sequential. This memory is referred to as the hash table, HT.

Binary search algorithms, while fast, can only operate on tables that are ordered and packed.i.e. tables that have adjacent items ordered by key words. It is possible to do considerably better with an unpacked, unordered table, provided it is sparse i.e. the number of storage spaces allocated to it exceeds the number of items to be stored.

16. General Idea

The ideal hash table data structure is merely an array of some fixed size, containing the keys. Typically, a key is a string with an associated value for instance, salary information.

We will refer to the table size as Tablesize, with the understanding that this is part of a hash data structure and not merely some variable floating around globally. The common convention is to have the table run from 0 to Tablesize-1. Each key is mapped into some number in the range 0 to Tablesize-1 and placed in the appropriate cell. The mapping is called a hash function which ideally should be simple to compute and should ensure that any two distinct keys get different cells. Since there are a finite number of cells and a virtually inexhaustible supply of keys, this is clearly impossible, and thus we seek a hash function that distributes the keys evently among the cells.

The following fig. is typical of a perfect situation. In this example, John hashes to 3. Phil hashes to 4, dave hashes to 6, and mary hashses to 7.

/* An ideal hash table */

0	
1	
2	
3	John 25000
4	Phil 31250
5	
6	Dave 27500
7	Mary 28200
8	
9	

This is the basic idea of hashing. The only remaining problems deal with choosing a function, deciding what to do when two keys hash to the same value (this is known as collision), and deciding on the table size.

17. Hashing Functions

The Divison Method : The most common hash functions use the division method (MOD) to compute hash addresses. The general function is :

key MOD Tablesize (or KEY MOD Tablesize + 1 to index the table begining at 1, ratherthan at 0).

It has been found, that better results are produced with the division method when the table size is prime.The operator MOD denotes the Modulo arithmetic System.In this sysem the term x MOD m has a value which is equal to the remainder of dividing x by m.

For example, it x = 35 and m =11, then H(35)=35 MOD 11+1 = 2+1=3.

The division method yields a "hash value" which belongs to the set {1, 2, ...M}.The advantage of the division hash function is simplicity.

Sometimes, however, it is necessary to use a more complicated (or even exotic) hash function to get a good distribution of hash addresses.

In general, it is uncommon for a number of keys to yield the same remainder when m is a large prime number.

In practice it has been found that ODD divisors without factors lessthan 20 are also satisfactory. In particular, divisors which are even numbers are to be avoided since even ODD keys would be mapped to odd and even addresses, respectively (assuming an addresses space of {1, 2, ...m}).

The Midsquare Method : Another hashing function which has been widely used in many applications is the mid square method.In this method a key is multiplied by itself and the address is obtained by selecting an appropriate number of bits (or) digits from the middle of the square. Usually, the number of bits (or) digits chosen depends on the table size and, consequently, can fit into one computer word of memory.The same positions in the square must be used for all products.

As an example, consider a six-digit key, 123456 squaring this key results in the value 15241383936. If a three-digit address is required, positions 5 to 7 could be chosen, giving address 138.

The Folding Method : In the folding method a key is partitioned into a number of parts, each of which has the same length as the required address with the possible exception of the last part. The parts are then added together, ignoring the final carry to form an address.If the keys are in binary form, the "exclusive-or" operation may be substituted for addition.

As an example, assume that the key 356942781 is to be transformed into a three-digit address.In the fold-shifting method, 356, 942, and 781 are added to yield 079.

A variation of the basic method involves the reversel of the digits in the outermost partitions.This variation is called the fold-boundary method.

In the previous example, 653, 942 and 187 are added, together, yielding 782.

Folding is a hashing function which is also useful in converting multiword keys into a single word so that other hashing functions can be used.

How can we use hashing if the key yield is a string instead of an integer?

One approach is to use the ordinal values (ORD) of the string's characters to create a number that can be used as an index.For instance, here is a simple hash function that converts a key of string type to an index.

18. Collision-Resolution Techniques

A hashing function can in general, map several keys into the same address.When this situation arises, the colliding records must be sorted and accessed as determined by a collision-resolution technique.(or) it, when an element is inserted, it hashes to the same value as an already inserted element, then we have a collision and need to resolve it.

There are two broad classes of such techniques:

- Open Addressing.
- separate chaining.(Closed Addressing)

Open Addressing : If the first probe gives a position K and that position is filled, then the next location K+1 is probed, and so on until a free position is found. If the search runs off the bottom of the table, then it is renewed at the top (i.e., the table is considered to be cyclic) consider a table of 17 positions (N=17) in which the following twelve numbers are to be stored : 19, 13, 05, 27, 01, 26, 31,16, 02, 09, 11, 21. These items are to be entered in the table at a position defined by the remainder after division by 17; if that position is filled, then the next position is examined etc.

The following table shows the progress of entry for the 12 items.

Position	Item	Probes to Find	Probes to Find not
0			1
1	01	1	6
2	19,02*	1	5
3	02	2	4
4	21	1	3
5	05	1	2
6			1
7			1
8			1
9	26,09*	1	7
10	27,09*	1	6
11	09,11*	3	5
12	11	2	4
13	13	1	3
14	31	1	2
15			1
16	16	1	1
		16	54

/* Example of open addressing */

notice the resolution of conflicts on items 02, 09 and 11. The column 'probes to find' gives the number of probes necessary to find the corresponding items in the table; thus it takes 3 probes to find items 09 and 1 to find item 26.

The column 'probes to find not' gives the number of probes necessary to determine that an item is not in the table; thus the search for the number 54 would give an initial position of 3 and it would take 4 probes to find that the item is not present.The item is known not to be present when a void position is encountered (positions 6 in this case).

Notice here that the following figures hold:

length of table	N=17
items stored	M=12
density	p=12/7 = 0.705
probes to store	T_s=16
Avg.probes to Find	T_p=16/12=1.33
Avg.probes to Findnot	T_n=54/16=3.37

Thus, it would appear that the open addressing scheme holds considerable advantage in speed, but it pays for this by having a table nearly 50 percent longer than necessary.

Further more, the table can't be compressed after its initial assigment nor can the assigned area be easily shared among several tables.

One final very serious disadvantage is that it is very difficult to delete any item from the table-one cannot simply zero out that location because this might break the addressing chain.

Linear Probing : In linear probing, F is a linear function of i typically F(i) = i. This amounts to trying cells sequentially in search of an empty cell.

The below figure shows the result of insering keys {89, 18, 49, 58, 69} into a hash table using the same hash function as before and the collison resolution strategy, F(i) = i.

	Empty table	After 89	After 18	After 49	After 58	After 69
0				49	49	49
1					58	58
2						69
3						
4						
5						
6						
7						
8			18	18	18	18
9		89	89	89	89	89

/ * Open addressing hash table with linear Probing, after each insertion * /

The first collision occurs when 49 is inserted; it is put in the next available spot, namely spot 0, which is open. The key 58 collides with 18, 89 and then 49 before an empty cell is found three away.The collision for 69 is handled in a similar manner. As long as the table is big enough, a free cell can always be found, but the time to do so can get quite large.

Worse, even if the table is relatively empty, blocks of occupied cells start forming.This effect, known as primary clustering, means that any key that hashe in to the cluster will require several attempts to resolve the collision, and then it will add to the cluster.

19. Clustering : One problem with linear probing is that it results in a situation called Clustering.

Order of insertion :

14001

00104

50003

77003

42504

33099

[00]	EMPTY
[01]	14001
[02]	EMPTY
[03]	50003
[04]	001004
[05]	77003
[06]	425004
[07]	EMPTY
[08]	EMPTY
[09]	EMPTY
	:
	:
[99]	element with key 33099

/ * A Hash table with linear probing * /

Consider the hash table above, only a record whose key produces the hash address 8 would be inserted into array slot [08]. However, any records with keys that produce the hash addresses 3, 4, 5, 6 (or) 7 would be inserted into array slot [07].That is, array slot [07] is five times as likely as array slot [08] to be filled.Clustering results in inconsistent efficiency of insertion and retrieval operations.

Quadratic Probing : It is a collision resolution method that eliminates the primary clustering problem of linear probing, Quadratic probing is what you would expect - the collision function is Quadratic.

The popular choice is $F(i) = i^2$.

The following figure shows the resulting open addressing hash table with this collision function on the input {89, 18, 49, 58, 69}

	Empty table	After 89	After 18	After 49	After 58	After 69
0				49	49	49
1					58	58
2						69
3						
4						
5						
6						
7						
8			18	18	18	18
9		89	89	89	89	89

/ * Open addressing hash table with Quadratic probing, after each insertion * /

When 49 collides with 89, the next position attempted is one cell away. This cell is empty, so 49 is placed there.

Next 58 collides at position 8. Then the cell one away is tried, but another collision occurs.A vacant cell is found at the next cell tried, which is $2^2 = 4$ away.The key 58 is thus placed in cell 2. The same thing happens for 69.

For linear probing it is a bad idea to let the hash table get nearly full, because performance degrades.

For quadratic probing, the situation is even more drastic: There is no guarantee of finding an empty cell once the table gets more than half full, or even before the table gets half full If the table size is not prime.This is because at most half of the table can be used as alternative locations to resolve collisions. Indeed we prove that if quadratic probing is used, and the table size is prime, then a new element can always be inserted if the table is atleast half empty.

If the table is even one morethan half full, the insertion could fail. Therefore, it is important to keep this in mind. It is also crucial that the table size be prime. If the table size is not prime, the number of alternative locations can be severely reduced.

As an example, if the table size were 16, then the only alternative locations would be at distances 1, 4 or 9 away. Standard deletion can't be performed in an open addressing hash table, because the cell might have caused a collision to go past it.

For instance, if we remove 89, then virtually all of the remaining finds will fail. Thus, open addressing hash tables require lazy deletion, although in this case there really is no laziness implied. Although quadratic probing eliminates primary clustring, elements that hash to the same position will probe the same alternative cells. This is known as secondary clustering.

20. Separate Chaining

One of the most popular methods of handling overflow records is called separate chaining. In this method the colliding records are chained into a special overflow area which is distinct from the primary Area. This area contains that part of the table into which records are initially hashed. A separate linked list is maintained for each set of collding records. Therefore, a pointer field is required for each record in the primary and overflow areas.

Collision Resolution By Chaining: Upto now we have implicitly assumed that we are using only contiguous storage while working with hash tables. Contiguous storage for the hash table itself is, in fact, the natural choice, since we wish to be able to refer quickly to random positions in the table, and the linked storage is not suited to random access. There is however, no reason why linked storage should not be used for the records themselves.

We can take the hash table itself as an array of pointers to the records, that is, as an array of linked lists. We assume that the keys are the first 10 perfect squares and that the hashing function is simply Hash (x) = x MOD 10.

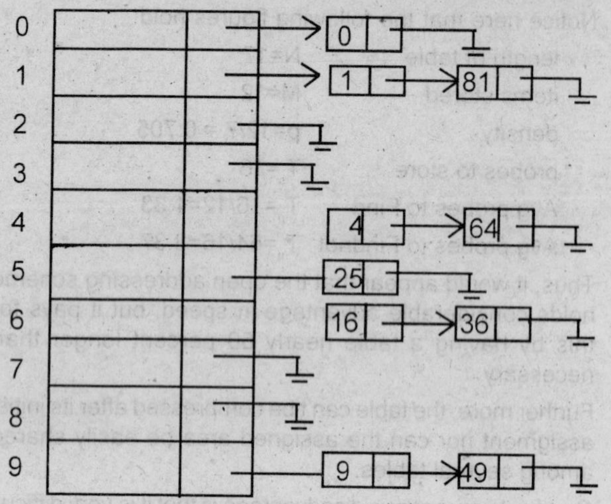

/ * A seperate chainig hash table * /

It is traditional to refer to the linked lists from the hash table as chains and call this method collisiion resolution by chaining.

21. Time Complxity Of Hashing

Searching like sorting is a topic that is closely tied to the goal of efficiency.

We speak of a sequential search as on O(N) search, because it may require upto N comparisons to locate on element. (N refers to the number of elements in the list).

Binary searches are considered to be $O(\log_2 N)$ and are appropriate for arrays only if they are sorted.A binary search tree may be used to allow binary searches on a linked structures.

The goal of hashing is to produce a search that approaches to O(1). Because of collisions of hash addressed, some searching (or) rehashing is usually necessary.

A good hash function minimizes collisions and distributes the elements randomly throughout the table. Hash tables can be used to implement the insert and Find operations in constant average any time.

OBJECTIVE QUESTIONS - I

1. A sorting technique is called stable if : **[GATE-1999]**
   a) It takes O(n log n) time .
   b) It maintains the relative order of occurrence of non-distinct elements.
   c) It uses divide-and-conquer paradigm.
   d) It takes O(n)space.

2. Suppose we want to arrange the n numbers stored in an array such that all negative values occur before all positive ones. Minimum number of exchanges required in the worst case is: **[GATE-1999]**
   a) n b) n-1 c) n+1 d) None.

3. An advantage of Chained Hash table (external hashing) over the open addressing scheme is: **[GATE-1996]**
   a) Worst case complexity of search operations is less.
   b) Space used is less.
   c) Deletion is easier. d) None.

4. The minimum number of interchanges needed to convert the array 89, 19, 40, 17,12, 10, 2, 5, 7, 11, 6, 9, 70 into a heap with maximum element at the root is : **[GATE-1996]**

a) 0 b) 1 c) 2 d) 3.

5. The average number of key comparisons done on a successful sequential search in list of length n is : **[GATE-1996]**

a) log n b) $\frac{n-1}{2}$ c) $\frac{n}{2}$ d) $\frac{n+1}{2}$

6. For merging two sorted lists of sizes m and n into a sorted list of size m + n, we require comparisions of : **[GATE-1995]**

a) O(m) b) O(n)
c) O(m + n) d) O(log m + log n).

7. Pick out the unstable sorting method: **[AUCET-2000]**

a) Insertion sort. b) Quick sort
c) Selection sort. d) Shell sort.

8. The slowest sorting procedure is : **[AUCET-2000]**

a) Bubble sort. b) Heap sort
c) Shell sort. d) Quick sort

9. A hashing function that ensures that there are no collisions is called: **[AUCET-2001]**

a) Golden hashing function.
b) Pure hashing function.
c) Perfect hashing function.
d) Optimal hashing function.

10. In bubble sort, the number of comparisons in the best case for a problem of size n is: **[AUCET-2001]**

a) n-1 b) log n c) n^2 d) n log n.

11. A characterstic of the data that binary search uses but the linear search ignores is the **[AUCET-1999]**

a) Length of the list. b) Order of the list.
c) Maximum value of the list. d) Mean of the data values.

12. The number of comparisons in the worst case of a bubble sort on a file of n elements is : **[AUCET-1999]**

a) n b) n-1 c) n^2 d) n (n-1)

13. A sort which compares adjacent elements in a list and switches when necessary is: **[AUCET-1999]**

a) Selection sort. b) Insertion sort.
c) Heap sort. d) Bubble sort.

14. Pick up the best average behavior of sorting technique.

a) Quick sort. b) Selection sort.
c) Heap sort d) Merge sort.

15. Which sorting technique is stable? Insertion sort or Heap sort

a) Insertion sort b) Heap sort.

16. In binary search the location of key is less than the mid then **[AUCET-1999]**

a) low = mid +1 b) high = mid – 1
c) low = mid – 1 d) high = mid + 1

17. A sort which iteratively passes through a list to exchange the first element with any element less than it and then repeats with a new first element is called :

a) Insertion sort. b) Selection sort.
c) Heap sort. , d) Quick sort.

18. A sort which uses the binary tree concept such that any number is largerthan all the numbers in the subtree below is called:

a) Insertion sort. b) Selection sort.
c) Heap sort. d) Quick sort.

19. The number of swappings needed to sort the numbers 8, 22, 7, 9, 31, 19, 5, 13 in ascending order, using bubble sort is :

a) 11 b) 12 c) 13 d) 14

20. How many comparisons are needed to sort an array of length 5. If a straight selection sort is used and array is already in the opposite order :

a) 1 b) 10 c) 15 d) 20.

21. Which of the following sorting method is stable?

a) Straight insertion sort. b) Binary insertion sort.
c) Shell sort. d) Heap sort.

22. Which of the following sorting algorithm (s) has a worst case running time of $O(n^r)$ where 1< r < n ?**[GRE-CS]**

a) Bubble sort. b) Insertion sort.
c) Quick sort. d) Shell sort.

23. Which of the following sorting algorithms does not have a worst case running time of $O(n^2)$? **[GRE-CS]**

a) Insertion sort. b) Merge sort.
c) Heap sort. d) Quick sort.

24. The average search time of Hashing, with linear probing will be less. If the load factor is

a) is far less than one b) Equals one
c) is far greater than one d) None.

25. A hash table has space for 100 records. What is the probability of collision before the table is 10% Full?

a) 0.45 b) 0.5 c) 0.3 d) 0.34 (Appr)

26. Choose the correct statements

a) Internal sorting is used if the number of items to be sorted is very large.

b) External sorting is used if the number of items to be sorted is very large.

c) External sorting needs auxiliary storage.

d) Internal sorting needs auxiliary storage.

27. A sorting technique that guaranteesthat records with the same primary key occurs in the in the same order in the sorted list as in the original unsorted list is said to be

a) Stable b) Consistent c) External d) Linear

28. Select the statement which best completes the following sentence: "The binary search technique ... " **[Hyd. Univ-2001]**

a) can be applied only to a list of integers

b) requires that the list is sorted prior to searching

c) uses a binary tree data structure for the search

d) is another name for linear search but for binary numbers

29. What is the order of searching a table of size N using the worst possible hash function? **[Hyd. Univ-2001]**

a) 1 b) log 2N c) N d) N!

30. Merge sort uses **[Hyd.Univ-1999]**

a) Divide and conquer b) Backtracking approach

c) Heuristic approach d) Greedy approach

31. The goal of hashing on an n-element list is produce a search that takes **[Hyd.Univ-1999]**

a) $O(n^2)$ b) $O(1)$ c) $O(\log_2 n)$ d) $O(n \log_2 n)$

32. Which of the following sorting algorithm does not have worst case running time of $O(n^2)$ **[Hyd.Univ-1996]**

a) Merge sort b) Bubble sort

c) Insertion sort d) Quick sort

33. Which of the following will be fastest if the input is already sorted? **[Hyd.Univ-1993]**

a) Quick sort b) heap sort

c) bubble sort d) two-way merge sort

34. Every comparison based sorting algorithm sorts n objects i **[UGC NET-1997]**

a) O(n) steps b) O(log n) steps

c) $O(\log^2 n)$ steps d) O(n log n) steps

35. Search tables used by compilers for efficient searching generally use

a) Hash table b) Linear list of records

c) Binary search tables d) Binary search tree

ANSWER KEY - I

1. b	**2.** d	**3.** c	**4.** c	**5.** b
6. c	**7.** c	**8.** a	**9.** d	**10.** a
11. b	**12.** c	**13.** d	**14.** a	**15.** a
16. b	**17.** b	**18.** c	**19.** d	**20.** d
21. a	**22.** d	**23.** b	**24.** a	**25.** a
26. b,c	**27.** a	**28.** b	**29.** b	**30.** a
31. b	**32.** a	**33.** c	**34.** d	**35.** a

EXPLANATORY NOTES - I

3. Chained hash table, infact, uses a linked list concept. So, obviously deletions made easy with external hashings over open addressing scheme.

5. The average no. of comparisions on a sequential search method is

$$f(n)=1.\frac{1}{n}+2.\frac{1}{n}+3.\frac{1}{n}+......+n.\frac{1}{n}$$

$$=\frac{1}{n}(1+2+.....+n)=\frac{n(n+1)}{2}.\frac{1}{n}=\frac{n+1}{2}$$

16. Explanation :

10	20	30	40	50	60	70
0	1	2	3	4	5	6

Mid $= \frac{0+6}{2} = 3$ Low = 0

Assume Loc. key = 1 High = 6

Here Lockey < mid

Hence the searching is done between '0' and '3' locations.

So high = mid - 1

20. Consider a list with size 5 in opposite order.

10 9 6 4 3

Apply straight selection sort

						Comparison
(10)	9	6	4	3	→	4
3	(9)	6	4	10	→	3
3	4	(6)	9	10	→	2
3	4	6	(9)	10	→	1
3	4	6	9	10	→	0
						10

Merge sort doesn't have worst case running because there is no other way inputing the values. The 2 sublists must be in sorted order.

26. Infact internal sorting is performed over the oftens available in main memory where as external sorting is performed in some records are existed in auxilury storage more over records can be kept in auxilary storage device if no. of records are more.

OBJECTIVE QUESTIONS - II

1. Consider the following algorithm for searching for a given number x in an unsorted array A[1....n] having n distinct values:

I. Choose an i uniformly at random from 1.. n;

II. If A[x]=x then stop else Goto I;

Assuming that x is present in A, what is the expected number of comparisons made by the algorithm before it terminates ? **[GATE-2002]**

a) n b) n-1 c) 2n d) n/2.

2. If one uses straight two-way merge sort algorithm to sort the following elements in ascending order :

20, 47, 15, 8 ,9, 4, 40, 30, 12, 17.

Then the order of these elements after the second pass of the algorithm is : **[GATE-1999]**

a) 8, 9, 15, 20, 47, 4, 12, 17, 30, 40.

b) 8, 15, 20, 47, 4, 9, 30, 40, 12, 17.

c) 15, 20, 47, 4, 8, 9, 12, 30, 40, 17.

d) 4, 8, 9, 15, 20, 47,12, 17, 30, 40.

3. The minimum number of record movements required to merge five files A(with 10 records), B(with 20 records),C(with 15 records),D(with 5 records) and E(with 25 records) is: **[GATE -1999]**

a) 65 b) 75 c) 90 d) 165.

4. Quicksort is run on two inputs shown below to sort in ascending order:

I. 1, 2, 3, ...,n II. n, n-1,.........,2,1.

Let c1 and c2 be the number of comparisons made for the inputs (1) and (2) respectively. Then, **[GATE-1996]**

a) c1 < c2 b) c1 > c2 c) c1 = c2 d) None.

5. Which of the following statements is TRUE : **[GATE-1995]**

I. As the number of entries in a hash table increases the number of collisions increases.

II. Recursive programs are efficient.

III. The worst case complexity of Quicksort is $O(n^2)$.

IV. Binary search using a linear linked list is efficient.

a) I and II b) II and III c) I and IV d) I and III.

6. The number of comparisons are needed to sort an array of length 5 if a straight selection sort is used and array is already in the opposite order: **[AUCET-2000]**

a) 1 b) 10 c) 15 d) 20.

7. The preferred hashing technique to be used when the total number of data items keeps changing is: **[AUCET-2001]**

a) Dynamic hashing b) Non-uniform hashing.

c) Double hashing d) Linear rehashing .

8. The decision to choose a particular sorting algorithm should be made based on **[AUCET-1999]**

I. Order of magnitude of the sort (O(n), O(n log n), etc..)

II. Size of the file.

III. Original sequence of data.

a) I only b) II only c) III only d) II and III only

9. An algorithm for searching an item x in an array A of n elements stops if for some i ,

a) x = A[i] b) i >n

c) x = A[i] and i > n. d) x = A[i] or i > n.

10. The initial sequence 1, 2, 3, 4, 5, 6, 7, 8, 9 is to be sorted into ascending order using quick sort. The number of iterations required is proportional to: **[AUCET-2000]**

a) 10 b) 100 c) $\log_2 100$ d) $10 \log_2 10$

11. Given two sorted lists of size 'm' and 'n' respectively. The number of comparisons needed in the worst case by the merge sort algorithm will be : **[AUCET-1998]**

a) m n b) max (m , n)

c) min (m , n) d) m + n –1.

12. In which of the following sorting algorithm, the number of comparisons needed is the minimum, If the items are initially in reverse order and is the maximum, If the items are in order :

a) Straight Insertion sort. b) Binary Insertion sort.

c) Heap sort. d) Bubble sort.

13. A search procedure which associates an address with a key value provides a mechanism for dealing with two or more values assigned to the same address is called :

a) Linear search b) Binary search

c) Radix search d) Hash coded search.

14. Let A be a sorted array of n=10 elements. Assume that only one comparison is required to determine whether the target is equal to, less than, or greaterthan A[i]. Which of the following denotes the average successful time of finding an arbitrary element x in A using Binary search:

a) 1.6 b) 2.9 c) 4.2 d) 5.5

15. A Hash function f is defined as f (key) = key MOD 7, with linear probing is used to insert the keys 37, 38, 72, 48, 98, 11,56 into a table indexed from 0 to 6. In which location the element 11 will be stored :

a) 3 b) 4 c) 5 d) 6

16. The Hash function Hash := key MOD size, and linear probing is used to insert keys 37, 38, 72, 48, 96, 11, 56 into the Hash table with indices 0...6. The order of the keys in the array are given by

a) 98, 11, 37, 38, 72, 56, 48

b) 98, 56, 37, 38, 72, 11, 48

c) 11, 48, 37, 38, 72, 98, 56

d) 72, 11, 37, 38, 56, 98, 48

17. An internal hash table has 5 buckets, numbered 0, 1, 2, 3,4 . Keys are integers and the hash function h [I]=I MOD 5 is used,with linear solution of collisions(i.e, If bucket h [I] is filled, the buckets h [I]+1, h [I] + 2, ... are tried successively with all bucket numbers computed module 5). If elements with 13, 8, 24, 10 and 3 are inserted, in that order , into an initially blank table, then the content of the bucket numbered 2 is :**[AUCET-1999]**

a) 3 b) 8 c) 10 d) 13.

18. A radix sort is to be used to sort the file of non negative integers shown below into ascending order.What would the order of the numbers be after one pass of the algorithm?

12, 37, 42, 9, 5, 7, 50, 40, 45, 92 **[GRE-CS]**

a) 12, 37, 42, 40, 45, 5, 50, 7, 9, 92.

b) 50, 40, 12, 42, 92, 5, 45, 37, 7, 9.

c) 9, 5, 7, 12, 37, 42, 40, 45, 50, 92.

d) 5, 7, 9, 12, 37, 42, 50,40, 45, 92.

19. A natural merge sort is to be used to sort the file of integers shown below into ascending order.What would the order of the numbers be after one pass of the sort?

12, 37, 42, 9, 5, 7, 50, 40, 45, 92 **[GRE-CS]**

a) 12, 37, 9, 42, 5, 7, 40, 50, 45, 92.

b) 9, 12, 37, 42, 5, 7, 40, 50, 45, 92.

c) 9, 12, 37, 42, 5, 7, 40, 45, 50, 92.

d) 5, 7, 9, 12, 37, 42, 40, 45, 50, 92.

20. A hash table with 10 buckets with one slot per bucket is depicted. The symbols, s1 to s7 are initially entered using a hashing function with linear probing. The maximum number of comparisons needed in searching an item that is not present is :

S7	S1		S4	S2		S5		S6	S3
0	1	2	3	4	5	6	7	8	9

a) 4 b) 5 c) 6 d) 3

21. A hash table can store a maximum number of 10 records. Currently there are records in locations 1, 3, 4, 7, 8, 9, 10. The probability of a new record going into location 2, with a hash function resolving collisions by linear probing is

a) 0.6 b) 0.1 c) 0.2 d) 0.5

22. Consider a Hashing function that resolves collision by Quadratic probing. Assume the address space is indexed from 1 to 8. If a collision occurs at position 4, then which of the following locations will never be probed

a) 4 b) 5 c) 8 d) 2

23. A hash function randomly distribute records one by one in a space that can hold x number of records. The probability that the m th record is the first record to result in collision is

a) $(x-1)(x-2)\ldots\ldots(x-(m-2))(m-1) / x^{m-1}$

b) $(x-1)(x-2)\ldots\ldots(x-(m-1))(m-1) / x^{m-1}$

c) $(x-1)(x-2)\ldots\ldots(x-(m-2))(m-1) / x^{m}$

d) $(x-1)(x-2)\ldots\ldots(x-(m-1))(m-1) / x^{m}$

24. If the hashing function is the remainder on division, then clustering is more likely to occur. If the storage space is divided into 40 sectors rather than 41. This conclusion is

a) More likely to be false b) More likely to be true

c) Is always false d) None.

25. Every comparison based sorting algorithm must perform **[UGC NET-1996]**

a) at least n^2 comparisons

b) at least $O(n \log n)$ comparisons

c) at most n comparisons

d) at most $\log_n$ comparisons

ANSWER KEY - II

1. a	**2.** b	**3.** a	**4.** c	**5.** d
6. b	**7.** a	**8.** d	**9.** d	**10.** d
11. d	**12.** b	**13.** d	**14.** b	**15.** c
16. b	**17.** a	**18.** b	**19.** c	**20.** b
21. a	**22.** d	**23.** a	**24.** b	**25.** b

EXPLANATORY NOTES - II

1. The list with n distinct elements and moreover it is unsorted. If an element x is to be found in the list, according to the given algorithm, it may take 'n' comparisions because the elements are in random order. Hence expected no of searchings in 'n'.

2.

Pass I 20 47 15 8 9 4 40 30 12 17

Pass II 20 47 8 15 4 9 30 40 12 17

8 15 20 47 49 30 40 12 17

3. The min no. of record movements required to merge 5 files A (with 10 records), B (with 20 records) C (15 records), D (5 records), E (25 records)

$\Rightarrow$ 10+20+15+5+25 = 65 (min)

5. i) Recursive programs are complex and increase burden over the system and programmer as well

ii) Apply binary search over linked list it self is tedious job. But, time complexity is worst case of quick sort is $o(n^2)$. The no. of collisions will be more if the entries into hash table are more.

10. The no. of elements = 10.

Running time complexity of quick sort is $n \log_2^n$

$\Rightarrow 10 \log_2^{10}$

11. In merge sort, if two lists are manipulated with size m,n then in worst case we need (m+n+1) Comparisions.

13. Linear binary and radix search techniques do not require any address manipulations for dealing 2 or more values assigned to the same address.

15. Hash function f = key MOD 7 using linear probing

Insert the keys 37, 38, 72, 48, 98, 11, 56 into a table indexed 0 to 6 $\Rightarrow$

0	1	2	3	4	5	6

i) f (37) = 37 mod 7 = 2 insert '37' at loc. 2

ii) f (38) = 38 mod 7 = 3 insert '38' at loc. 3

iii) f(72)=72 mod 7 = 2 loc. 2 is filled 30 insert '72' in the next available loc. i.e.4

iv) f(48)=48 mod 7=6 insert 48 at loc. 6

v) f(98)=98 mod 7 = 0 insert 98 at loc. 0

vi) f(11)=11 mod 7 = 4 loc. is 4 is already filled 30. insert 11 at next available loc. i.e., 5

$\Rightarrow$

96	56	37	38	72	11	48
0	1	2	3	4	5	6

16. 37 38 72 48 96 11 56

			32			
0	1	2	3	4	5	6

		37	38			
0	1	2	3	4	5	6

		37	38	72		
0	1	2	3	4	5	6

		37	38	72		48
0	1	2	3	4	5	6

96		37	38	72		48
0	1	2	3	4	5	6

96		37	38	72	11	48
0	1	2	3	4	5	6

96	56	37	38	72	11	48
0	1	2	3	4	5	6

19. Merge sort on the following list

after pass I is ——

12 37 42 9 5 7 50 40 45 92

12 37 9 42 5 7 40 50 45 92

22. If a collision occured at location 4, in a hash table indexed 1 to 8, then we proceed forward to probe the new locations but not is backward direction.

V. ALGORITHM ANALYSIS AND DESIGNING

- Introduction
- Analysis of Algorithms
- Designing of Algorithms

1. Introduction

An algorithm is composed of a finite set of steps each of which may require one(or) more operations.The possiblity of a computer carrying out these operations necessitates that certain constraints be placed on the type of operations on algorithm can include.

For example,each operation must be definite meaning that it must be perfectly clear what should be done. One Important criterion we will assume about algorithms is that terminate after a finite number of operations.

2. How To Analyze Algorithms

This field of study is called analysis of algorithms. As an algorithm is executed,it makes use of the computers central processing unit(cpu) to perform operations and it uses the memory (both imeediate and auxiliary)to hold the program and its data.

Analysis of algorithms refers to the process of determining how much computing time and storage an algorithm wil require.This area is a challenging one which some times requires great mathematical skill.one important result of this study is that it allows one to make quantitative judgments about the value of one algorithm over another. Another result is that it allows us to predict if our software will meet any efficiency constraints which may exist.

Questions such as how well does an algorithm perform in the best case,in the worst case or on the average are typical.

Analyzing Algorithms : Why do we bother to analyze an algorithm?

For some of us analyzing algorithms is an intellectual activity that is fun. Another reason is the challenge of being able to predict the future and even though we are narrowing our predictions to algorithms,it is gratifying when we succeed. A third reason is because computer science attracts many people who enjoy being efficiency experts.Analyzing algorithms gives these people a chance to exhibit their skills by devising new ways of doing the same task leven faster. This tendecy has a large payoff in computing where time means money and efficiency save dollars.

Given an algorithm to be analyzed, the first task is to determine which operations are employed and what their relative costs are. These operations may include the four basic arithmetic operations on integers: addition, subtraction, multiplication and division. Other basic operations might include arithmetic on floating point numbers, comparisions, assigning, values to variables and executing procedure calls. These operations typically take no more than a fixed amount of time and so we say that their time is bounded by a constant.

The second task is to determine a sufficient number of data sets which cause the algorithm to exhibit all possible patterns of behaviour. This is one of the important and creative tasks of algorithm analysis it requires us to understand the workings of the algorithm well enough to concoct the data configurations which produce the best (or) worst (or) typical behaviour

In producing a complete analysis of the computing time of an algorithm, we distinguish between two phases:

- A priori Analysis and
- A posterior testing.

In a priori Analysis we obtain a function (of some relavant parameters) which bounds the algorithms computing time.

In a posteriori testing we collect actual statistics about the algorithm's consumption of time and space, while it is executing.

Supose there is the statement $x \leftarrow x + y$ some where in the middle of a program.We wish to determine the total time that statement will spend executing, given some initial state of input data. This requires essentially two items of information, the staements frequency count (i.e. the number of times the statement will be executed) and the time for one execution. The product of these two numbers is the total time. Since the time per execution depends on both the machine being used and the programming language together with its compiler, an a priori analysis limits itself to determing the frequency count of each statement. This number can be determined directly from the algorithm. independent of the machine it will be executed on and the programming language the algorithm is written in.

For example consider the three program segments a, b, c:

```
              for i ← 1 to n do        for i ← 1 to n do
x ← x+y       x ← x+y                  for j ← 1 to n do
              repeat                       x ← x+1
                                           repeat
  (a)            (b)                   repeat (c)
```

For each segment we assume the statement $x \leftarrow x+y$ is contained with in no other loop than what is already visible. Thus for segment (a) the frequency count of this statement is 1.

For segment (b) the count is n and for segment (c) it is n^2. These frequencies 1, n, n^2 are said to be different, increasing orders of magnitude.

Given three algorithms for solving the same problem whose orders of magnitude are n, n^2 and n^3, naturally we will prefer the first since the second and third are progressively slower.

For example, if n=10 then these algorithms will require 10, 100 and 1000 units of time to execute respectively (assuing all basic operations are of equal duration).

Determining the order of magnitude of an algorithm is very imprtant and producing an alogirthm which is faster by an order of magnitude is a significant accomplishment.

The a priori analysis of algorithms is concerned chiefly with order of magnitude determination.

3. Asymptotic notation

An a priori analaysis of computing time ignores all of the factors which are machine (or) programming language dependent and concentrates on determing the order of magnitude of the frequency of execution of statements.

There are several kinds of mathematical notation which are very useful for this kind of analysis. One of these is the O-notation.

Definition : f (n) = O(g(n)) (read as "f of n equals big oh of g of n"). Iff there exist two positive constants c and n_0 such that $|f(n)| \le c|g(n)|$ for all $n \ge n_0$.

Suppose we are determining the computing time, f(n), of some algorithm. The variable n might be the number of inputs (or) outputs their sum (or) the magnitude of one of them. Since f(n) is machine dependent, an a priori analaysis will not suffice to determine it. However, on a priori analaysis can be used to determine a g(n) such athat f(n)=O (g(n)).

When we say that an algorithm has computing time O (g(n)) we mean that, if the algorithm is run on some computer on the same type of data but for increasing values of n, the resulting times will always be lessthan some constant times $|g(n)|$ when determing the order of magnitude of f(n) we shall always try to obtain the smallest g(n) suchthat f(n) = O (g(n)).

Theorem 1 : If $A(n) = a_m n^m + \ldots + a_1 n + a_0$ is a polynomial of degree m then $A(n) = O(n^m)$.

Theorem : 1 says that if we can describe the frequency of execution of a statement in an algorithm by a polynomial such as A(n), then that statements computing time is $O(n^m)$. However the constant in the above theorem is not the best possible. Actually we can show that any constant greaterthan $|a_m|$ can be used (for sufficiently large n.)

If an algorithm has k statements whose orders of magnitude are $c_1n^{m1}, c_2n^{m2} \ldots\ldots c_kn^{mk}$ then the order of magnitude of the entire algorithm is given by $c_1n^{m1} + \ldots\ldots + c_kn^{mk}$ is by theorem.1 is equal to o(n^m) where m=max{ m_i}, $1 \le i \le k$.

The most common computing times for algorithms we will see here are:

$$O(1) < O(\log) < O(n) < O(n\log n) < O(n^2) < O(n^3) < O(2^n) < O(n!) < O(n^n)$$

O (1) means that the number of executions of basic operations is fixed and hence the total time is bounded by a constant. The first six orders of magnitude have an important property in common, they are bounded by a polynomial.

O (n), O (n^2) and O (n^3) are themselves polynomials referred to by their degrees: linear, quadratic, and cubic. However there is no integer m such that n^m bounds 2^n, or $2^n \ne O(n^m)$ for any integer m. The order of this formula is O (2^n).below fig. show how the computing times for six of the typical functions grow with a constant equal to one.

Notice how the time O (n) and O (n logn) grow much more slowly than the others. For large data sets, algorithms with a complexity greaterthan O (nlogn) are often inpractical.

An algorithm which is exponential will be practical only for very small values of n and even if we descrease the leading constant, say by a factor of 2 (or) 3, we will not improve the amount of data we can handle be very much.

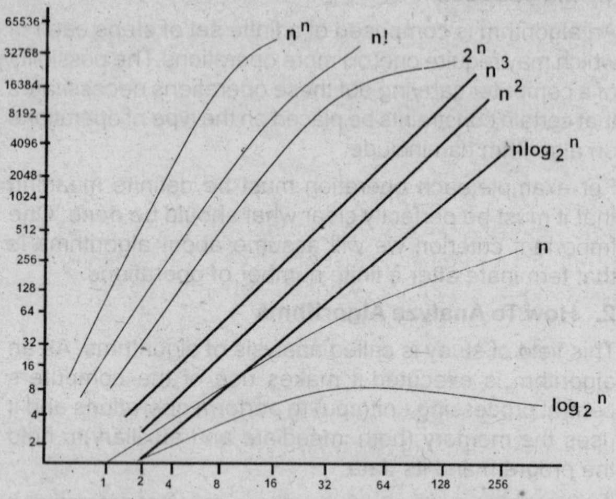

/* Rate of growth of common computing time functions */

logn	n	nlogn	n^2	n^3	2^n
0	1	0	1	1	2
1	2	2	4	8	4
2	4	8	16	64	16
3	8	24	64	512	256
4	16	64	256	4096	65536
5	32	160	1024	32768	4294967296

/* values for computing functions */

scfar we have concentrated on O-notatin as a means for describing an algorithms performance. Where as O-notation is used to express an upper bound, we might also with to determine a function which is a lower bound. What is needed is a mathematical notation for expressing a formula which is a lower bound on the computing time of an algorithm to with in a constant.

Definition : $f(n) = \Omega(g(n))$, (read as f of n equals omega of g(n)) iff there exist positive constants c and n_0 such that for all $n > n_0$, $|f(n)| \ge c|g(n)|$

In some cases the time for an algorithm, f(n), will be such that $f(n) = \Omega(g(n))$ and f(n) = O(g(n)). For this circumstance we will use the following notation.

Definition : f(n) = θ (g (n)) if f there exist positive constants c1, c2 and n_0 such that for all $n > n_0$, $c_1|g(n)| \le c_2|g(n)|$.

If $f(n) = \theta(g(n))$ then g(n) is both an upper and lower bound on f(n). This means that the worst and best cases require the same amount of time to within a constant factor.

As an example consider the algorithm which finds the maximum. of n elements, algorithm

1. The computing time for this algorithm is both O (n) and $\Omega(n)$ since the for loop always makes n-1 iterations. Thus, we say that its time is $\theta(n)$. The procedure of linear search, an array of n elements for a single value. It has a computing time which is O (n) but $\Omega(1)$

In the best case it might find the value on the first comparision, but in the worst case it will look at all elements once.

An even stronger mathematical notation is given by the following.

Definition : f(n) ~ O(g(n)) (read as "f of n is asymptotic to g(n) ") iff limit $\frac{f(n)}{g(n)} \rightarrow 1$ as $n \rightarrow \infty$

since the ratio in the limit is one, the functins f(n) and g(n) must agree even closer than by a constant factor. If there is an algorithm whose exact computing time is f(n) and we can determine a g(n) such that f is asymptotic to g, then we will have a more precise description of the computing time than if we had used the big-O notation.

In practice it implies we will know both the order of the leading term and its constant.

for example if $f(n) = a_k n^k + \ldots\ldots + a_0$, then

$$f(n) = O(n^k)$$

and $$f(n) = O(a_k n^k)$$

4. The Important Rules are

Rule 1 :

If $T_1(N) = O(f(N))$ and $T_2(N) = O(g(N))$, then

(a) $T_1(N) + T_2(N) = \max(O(f(N)), O(g(N)))$

(b) $T_1(N) * T_2(N) = O(f(N)) * g(N))$

Rule 2 : If T (N) is a polynomial of degree k, then

$T(N) = \theta(N^k)$

Rule 3 : $\log^k N = O(N)$ for any constant K. This tells us that logarithms grow very slowly.

5. What To Analyze

The most important resource to analyze is generally the running time. Several factors affect the running time of a program, some such as the compiler and computer used, are obviously beyond the scope of any theoretical model, so, although they are important, we cannot deal with them here. The other main factors are the algorithm used and the input to the algorithm.

Typically, the size of the input is the main consideration.

We define two functions, $T_{avg}(N)$ and $T_{worst}(N)$, as the average and worst case running time, respectively used by an algorithm on input of size N. Clearly $T_{avg}(N) \leq T_{worst}(N) \leq$.

If there is more than one input, these functions may have more than one argument. Generally, the quantity required is the worst-case time, unless otherwise specified. One reason for this is that it provides a bound for all input, including particularly bad input, which on avg-case analysis does not provide. The other reason is that avg-case bounds are usually much more difficult to compute.In some instances, the definition of "average" can affect the result.

6. Running Time Calculations

There are several ways to estimate the running time of a program.

If two programs are expected to take similar times. Probably the best way to decide which is faster is to code them both up an run them.

To simplify the analysis, we will adopt the convention that there are no particular units of time.

Thus, we throw away leading constants. We will also throw away low-order terms, so what we are essentially doing is computing a Big-oh running time.

Since Big-oh is an upperbound, we must careful never to underestimate the running time of the program.

In effect, the answer provided is a guarantee that the program will terminate with in a certain time period.

The program may stop earlier than this, but never later.

A Simple Example : Here is a simple program fragment to calculate $\sum_{i=1}^{N} j^3$;

```
int Sum ( int N )
{       int i, Partialsum ;
/ * 1 * /   Partialsum = 0 ;
/ * 2 * /   for ( i = 1; i < = n ; i + +)
/ * 3 * /   Partialsum + = i * i * i ;
/ * 4 * /   return Partialsum;
}
```

The analysis of this program is simple. The declarations count for no time. Lines 1 and 4 count for one unit each. Line 3 counts for four units per time executed (two multiplications, one addition, and one assignment) and is executed N times, for a total of 4N units. Line 2 has the hidden costs of initializing i, testing i≤N and incrementing i.

The total cost of all these is 1 to initialize. N + 1 for all the tests, and N for all the increments, which is 2N + 2

We ignore the costs of calling the function and returning for a total of 6N + 4. Thus, we say that this function is O (N)

7. General Rules

Rule : 1 ← For loops:

The running time of a for loop is at most the running time of the statements inside the for loop (including tests) times the number of interations.

Rule : 2 ← Nested for loops :

Analyze these inside out. The total running time of a statement inside a group of nested loops is the running time of the statement multiplied by the product of the sizes of all the for loops.

As an example, the following program fragment isO (N^2):

```
for ( i = 0; i < N ; i + +)
  for  ( j = 0 ; j < N; j + +)
    K + +;
```

Rule : 3 ← Consecutiue Statements :

These just add (which means that the max is the one that counts)

As an example, the following program fragment which has O (N) work follwoed by O (N^2) work, is also O (N^2).

```
for ( i = 0; i < n ; i + +)
      A [ i ] = 0;
for  ( i = 0; i < n; i + +)
for ( j = 0; j < n ; j + +)
         A [ i ] + = A [ J ] + i + J;
```

Rule : 4 ← If / Else : For the fragment

```
If (condition)
   S1
else
   S2
```

The running time of an if/else statement is never morethan the running time of the test plus the larger of the running times of S1 and S2

Clearly, this can be an overestimate in some cases, but it is never an understimate

8. Logarithms in the running time

The most confusing aspect of analyzing algorithms probably centers around the logarithm. We have already seen that some divide-and-conquer algorithms. will run in O (NlongN) time.Besides divide-and-conquer algorithms, the most frequent appearance of logarithms centers around the following general rule:

An algorithm is O (logN) if it takes constant (o(1)) time to cut the problem size by fraction (which is usually 1/2).On the otherhand, if constant time is required to merely reduce the problem by a constant amount (such as to make problem smaller by1), then the algorithm is O (N).

It should be obvious that only special kinds of problems can be O (log N). For instance, if the input is a list of N numbers, an algorithm must take Ω (N) merely to read the input in.Thus , when we talk abount O (log N) algorithms for these kinds of problems, we usually presume that the input is pre-read.

9. Algorithm Design Techniques

Sofar, we have been concerned with the efficient implementation of algorithms. we have seen that when an algorithm is given, the actual data structures need not be specified. It is up to the programmer to choose the appropriate data structure in order to make the running time as small as possible.

In this section, we switch our attention from the implementation of algorithms to the design of algorithms. Most of the algorithms that we have see so far are straight forward and simple. In this section, we will focus on four of the common types of algorithms used to solve problems. For many problems, it is quite likely that at least one of these methods will work. Specifically, for each type of algorithm we will

- See the general approach
- Look at several examples
- Discuss, in general terms, the time and space complexity, where appropriate.

10. Divide-and-Conquer Method

Perhaps the most important, and most widely applicable technique for designing efficient algorithms is a strategy called divide-and conquer. It consits of breaking a problem of size n into smaller problems in such a way that from solutions to the smaller problems we can easily construct a solution to the entire problem.

The General method: Given a function to compute on n inputs the divide-and-conquer strategy suggests splitting the inputs into K distinct subsets, $1 < K \leq n$ yielding k subproblems. These subproblems must be solved and then a method must be found to combine subsolutions into a solution of the whole.If the subproblems are still relatively large, then the divide-and-conquer strategy may possibly be reapplied.Often the subproblems resulting from a divide-and-conquer design are of the same type as the original problem. For those cases the reapplication of the divide-and-conquer principle is naturally expressed by a recursive procedure. Now smaller and smaller subproblems of the same kind are generated, eventually producing subproblems that are small enough to be solved without splitting.

To be more precise suppose we consider the divide-and-conquer strategy when it splits the input into two subproblems of the same kind as the original problem. This splitting is typical of many of the problems we will see here.If the sizes of the two subproblems are approximately equal then the computing time of divide - and - conquer is naturally described by the recurrence relation.

$$T(n) = \begin{cases} g(n), & n \text{ small} \\ 2T(n/2) + f(n), & \text{otherwise} \end{cases}$$

where T(n) is the time for divide - and - conquer on n inputs, g(n) is the time to compute the answer directly for small inputs and f(n) is the time for Divide and Combine. Recurrence relations will often arise for divide-and-conquer based algorithms and we will see how to work with them as they arise. For divide-and-conquer based algorithms which produce subproblems of the same type as the original problem it is very natural to first describe such an algorithm using recursion. But to gain efficiency it may be desirable to translate the resulting program into iterative form.

Divide and conquer Algorithms consist of two parts

Divide : Smaller problems are solved recursively.

Conquer: The solution to the original problem is then formed from the solutions to the subproblems.

Traditionally, routines in which the text contains at least two recursive calls are called divide-and-conquer algorithms, while routines whose text contains only one recursive call are not. We generally insist that the subproblems be disjoint (that is, essentially nonoverlapping).

Running Time Of Divided Conquer algorithms: All the efficient divide and conquer algorithms we will see divide the problems into subproblems each of which is some fraction of the original problem, and then perform some additional work to compute the final answer.

As an example, we have seen that merge sort operaters on two problems, each of which is half the size of the original, and then uses O(N) additional work.

This yields the running time equation

$$T(N) = 2T(N/2) + O(N)$$

The following theorem can be used to determine the running time of most divide and conquer algorithms.

Theorem: The solution to the equation $T(N) = aT(N/b) + \theta(N^k)$, where $a \geq 1$ and b>1 is

$$T(N) = \begin{cases} O\left(N^{\log_b^a}\right) & if \quad a > b^k \\ O(N^k \log N) & if \quad a = b^k \\ O(N^k) & if \quad a < b^k \end{cases}$$

As an example, merge sort has a=b=2 and k=1. The second case applies, giving the answer O(N log N).

If we solve three problems, each of which is half the original size, and combine the solutions with O(N) additional work, then a=3, b=2 and k=1.

Case 1 applies here, giving a bound of $O\left(N^{\log_2^3}\right) = O\left(N^{1.59}\right)$. An algorithm that solved three half-sized problems, but required $O(N^2)$ work to merge the solution, would have an $O(N^2)$ running time, since the third case would apply.

There are two important cases that are not covered by above theorem.

Theorem: The solution to the equation

$$T(N) = aT(N/b) + \theta(N^k \log^p N)$$

where $a \geq 1$, $b > 1$ and $p \geq 0$ is

$$T(N) = \begin{cases} O\left(N^{\log_b^a}\right) & if \ a > b^k \\ O(N^k \log^{p+1} N) & if \ a = b^k \\ O(N^k \log^p N) & if \ a < b^k \end{cases}$$

Theorem: $\sum_{i=1}^{k} \alpha_i < 1$, then the solution of the equation.

$$T(N) = \sum_{i=1}^{k} T(\alpha_i N) + O(N) \ is \ T(N) = O(N)$$

The natural divide-and Conquer algorithms are :

- Merge Sort Algorithm
- Quick Sort Algorithm
- Binary Search Algorithm
- Towers of Hanoi Problem
- The problem of multiplying long Integers.

11. Dynamic Programming

Dynamic programming, like the divide-and-conquer method solves problems by combining the solutions to subproblems. Divide-and-conquer algorithms partition the problem into independent subproblems, solve the subproblems recursively, and then combine their solutions to solve the original problem.In constrast dynamic programming is applicable when the subproblems are not independent, that is, when subproblems share sub sub-problems. In this context, a divide-and-conquer algorithm does more work than necessary, repeatedly solving the common subsubproblems.

A dynamic-programming algorithm solves every sub subproblem just once and then saves its answer in a table, there by avoiding the work of recomputing the answer every time the subsubproblem is encountered.

Dynamic programming is typically applied to Optimization problems. In such problems there can be many possible solutions. Each solution has a value, and we wish to find a solution with the optimal (min (or) max) value. We call such a solution an ooptimal solution to the problem, as opposed to the optimal solution, since there may be several solutions that achieve the optimal value.

The development of a dynamic-programming algorithm can be broken into a sequence of four steps:

1. Characterize the structure of an optimal solution.
2. Recursively define the value of an optimal solution.
3. Compute the value of an optimal solution in a bottom-up fashion.
4. Construct an optimal solution from computed information.

Steps 1-3 form the basis of a dynamic-programming solution to a problem. Step 4 can be omitted if only the value of an optimal solution is reuired. When we do perform step 4, we sometimes maintain additional information during the computation in step 3 to case the construction of an optimal solution.

Divide-and-Conquer is a top-down method, when a problem is solved by divide-and-conquer, we immediately attack the complete instance, which we then divide into smaller and smaller subinstances as the algorithm. progresses. Dynamic programming, on the other hand is a Bottom-up technique. We usually start with the smallest, and hence the simplest, sub instances. By combining their solutions, we obtain the answers to subinstances of increasing size, until finally we arrive at the solution of the original instance.

Any recursive mathematical formula could be directly translated to a recursive algorithm, but the underlying reallity is that often the compiler will not do justice to the recursive algorithm, and an inefficiant program results. When we suspect that this is likely to be the case, we must provide a little more help to the compiler, by rewriting the recursive algorithm as a nonrecursive algorithm. that systematically record the answers to the subproblems in a table. One technique that makes use of this approach is known of dynamic programming.

The natural Dynamic Algorithms are :

- All-pairs shortest path problem (Floyd's) is the best example as Dynamic programming.
- Optimal binary search tree construction is performed by Dynamic programming technique.
- Chained matrix multiplication is well performed by using Dynamic programming technique.

12. Greedy Algorithms

Algorithms for optimization problems typically go through a sequence of steps, with a set of choices at each step. For many optimization problems, using Dynamic programming to determine the best choices is overkill;simpler, more efficient algorithms will do.

A Greedy algorithm always makes the choice that looks best at the moment. That is, it makes a locally optimal choice in the hope that this choice will lead to a globally optimal solution. Greedy algorithms do not always yield optimal solutions, but for many problems they do.

The greedy method is quite powerful and works well for a wide range of problems. Greedy algorithms work in phases. In each phase, a decision is made that appears to be good, without regard for future consequences. Generally, it is means that some local optimum is chosen.This "take what you can get now " strategy is the source of the name for this class of algorithms. When the algorithm terminates, we hope that the local optimum is equal to the global optimum. If this is the case, then the algorithm is correct; otherwise, the algorithm has produced a sub optimal solution.

If the absolute best answer is not required, then simple greedy algorithms are some times used to generate approximate answers, rather than using the more complicated algorithms, generally required to generate an exact answer.

The natural greedy algorithms are:

- Huff man algorithm
- A simple sheduling problem
- Dijkstra's algorithm
- prim's algorithm
- Krushkal's algorithm
- Knapsack problem
- Optimal merge patterns.

13. Backtracking Algorithms

The last algorithm design technique we willl examine is backtracking. In many cases, a backtracking algorithm amounts to a clever implementation of exhaustive search, with generally unfavorable performance. This is not always the case, however, and even so, in some cases, the savings over a brute force exhaustive search can be significant. Performance is, of course, relative : An $O(n^2)$ algorithm for sorting is pretty bad, but an $O(n^5)$ algorithm for the traveling salesman (or any NP-complete) problem would be a landmark result.

OBJECTIVE QUESTIONS - I

1. Which one of the following algorithm design techniques is used in finding all pairs of shortest distance in a graph? **[GATE-1998]**

   a) Dynamic programming b) Backtracking
   c) Greedy d) Divide and conquer

2. Which of the following is FALSE? **[GATE-1996]**

   a) 100 nlog n = $O\left(\frac{n\log n}{100}\right)$
   b) $\sqrt{\log n}$ =O(loglogn)
   c) If 0<x<y then n^x =O(n^y) d) 2n $\neq$ O(nk)

3. The recurrence relation **[GATE-1996]**

   T(1) =2 T(n) = 3T(n/4) +n

   Has the solution T(n) equal to:

   a) O(n) b) O(log n) c) $O(n^{3/4})$ d) None

4. Merge sort uses **[GATE-1995]**

   a) Divide and conquer strategy
   b) Greedy approach
   c) Backtracking approach
   d) Heuristic search

5. The recurrence relation that arises in relation with the complexity of binary search is: **[GATE-1994]**

   a) T(n) =T(n/2)+k, k is a constant
   b) T(n) =2T(n/2)+k, k is a constant
   c) T(n)=T(n/2)+log n d) T(n) =T(n/2)+n

6. Krushkal's algorithm for finding a minimum spanning tree of a weighted graph G with n vertices and m edges has the time complexity of: **[GATE-1991]**

   a) O(n^2) b) O(mn) c) O(m+n) d) O(m log n)

7. The time complexity of comparison based sorting algorithms Is: **[GATE-1989]**

   a) O(nlogn) b) O(n) c) O(n^2) d) O(n^3)

8. The time complexity of linear search is **[AUCET-2000]**

   a) O($\log_2{}^n$) b) O(n) c) O(nlogn) d) O(n^2)

9. The time complexity of binary search is **[AUCET-2000]**

   a) O($\log_2{}^n$) b) O(n) c) O(nlogn) d) O(n^2)

10. The worst case complexity of bubble sort is **[AUCET-2000]**

    a) O(n) b) O($\log_2{}^n$) c) O(n) d) O(n^2)

11. The asymptotic complexity for f(n)= $\sum_{i=1}^{n} i$ is: **[AUCET-2000]**

    a) O(n) b) O(i) c) O(n^2) d) O(1)

12. To speed up search for an optimal solution, bounding function can be used in the following strategy: **[AUCET- 2000]**

a) Branch and Bound b) Greedy method
c) Dynamic programming d) Divide and conquer

13. Which of the following sorting algorithm has average-case and worst - case running time of O(nlogn): **[AUCET- 1999]**

a) Bubble sort b) Insertion sort
c) Merge sort d) None

14. Which of the following sorting algorithm uses in divide-and-conquer method: **[AUCET- 2000]**

a) Selection sort b) Bubble sort
c) Quick sort d) None

15. The stack space required for quick sort is :

a) O(logn) b) O (n) c) O(nlogn) d) None

16. There are four different algorithms A1,A2,A3,A4 to solve a given problem with the order logn , log logn , n log n /logn respectively. Which is the best algorithm.

a) A1 b) A2 c) A3 d) A4

17. The time complexity of an algorithm T(n) ,where 'n' is the input size is given by

T(N)=T(n-1)+1/n,if n>1
=1, otherwise

the order of this algorithm is :

a) logn b) n c) n^2 d) n^n

18. The running time of an algorithm is given by

T(n) =T(n-1)+T(n-2)-T(n-3),if n > 3
=n ,otherwise

the order of this algorithm is :

a) n b) logn c) n^n d) n^2

19. In the above QNo.18, what should be the relation between T(1),T(2),T(3) gives an algorithm whose order is constant :

a) T(1)=T(2)=T(3) b) T(1)+T(3) =2T(2)
c) T(1)-T(3)=T(2) d) T(1)+T(2)=T(3)

20. What is the time required to search an element in a linked list of length 'n'?

a) O(n) b) $O(\log_2 n)$ c) O(1) d) $O(n^2)$

21. The time required to search an element in a binary search tree having n elements is:

a) O(1) b) $O(\log_2 n)$ c) O(n) d) $O+(n\log_2 n)$

22. What is the worst case time complexity of shell sort algorithm to sort n elements?

a) O(n) b) $O(n\log_2 n)$ c) $O(n^{1.2})$ d) $O(n^2)$

23. What is the worst case time complexity of straight insertion sort algorithm to sort n elements?

a) O(n) b) $O(n\log_2 n)$ c) $O(n^{1.2})$ d) $O(n^2)$

24. What is the worst case time complexity of binary insertion sort algorithm to sort n elements?

a) O(n) b) $O(n\log_2 n)$ c) $O(n^{1.2})$ d) $O(n^2)$

25. Which of the following sorting algorithms does not have a worst case running time of $o(n^2)$:

a) Insertion sort b) Merge sort
c) Quick sort d) Bubble sort

26. Which of the following sorting algorithms has a worst case running time of $o(n^2)$ where 1<r<2 is:

a) Insertion sort b) Merge sort
c) Quick sort d) Shell sort

27. Which sort will operate in quadratic time relative to the number of elements in the array(on the average):

a) Quick sort b) Merge sort
c) Heap sort d) Bubble sort

28. What is the worst case time required to search a given element in a sorted linked list of length n.

a) O(1) b) $O(\log_2 n)$ c) O(n) d) $O(n\log_2 n)$

29. Consider a linked list of n elements What is time taken to insert an element after an element pointed by some pointer?

a) O(1) b) $O(\log_2 n)$ c) O(n) d) $O(n\log_2 n)$

30. The recurrence relation $T(n)=mT(n/2)+an^2$ is satisfied by

a) $T(n)=O(n^m)$ b) T(n)=O(n log m)
c) T(n)=O(mlogn) d) None

31. The time required to find shortest path in a graph with n vertices and e edges is:

a) O(e) b) O(n) c) $O(e^2)$ d) $O(n^2)$

32. The goal of hashing is to produce a search that takes:

a) O(1) time b) $O(n^2)$time
c) O(logn)time d) O(nlogn)time

33. Example(s) of O(1) algorithms are:

a) Incrementing a variable
b) Printing a character To the screen
c) Adding two numbers together
d) All of the above

34. The average computing time of heap sort is:

a) $O(n^2)$ b) O(nlogn) c) O(log n) d) $O(n^3)$

35. The algorithm which required a fixed amount of storage is:

a) Heap sort b) Quick sort

36. Examples of $O(n^2)$ algorithms is(are):

a) Initializing all the elements in a two dimensional array to zero
b) Printing out all the elements in a two dimensional array
c) Searching for the smallest element in an unsorted two dimensional array
d) All of the above

37. Consider a linked list implementation of a queue with two pointers: front and rear. What is the time needed to insert an element in a queue of length n? **[Hyd.Univ-1998]**

a) O(1) b) $O(\log_2 n)$ c) O(n) d) $O(n\log_2 n)$

38. The complexity of inserting an element in a stack with linked list representation is **[UGC NET-1997]**

a) O(1) b) O($\log_2$ n) c) O(n) d) O(n $\log_2$ n)

39. An undirected graph with n vertices and e edges are represented by adjacency matrix. The time required to determine the degree of any vertex is **[UGC NET-1997]**

a) O(e) b) O(n) c) O(n^2) d) O(e+n)

40. The travelling Sales person problem cannot be solved using **[UGC NET-1998]**

a) Dynamic programming b) Branch and bound

c) Greedy algorithm d) Back tracking

ANSWER KEY - I

1. a	**2.** b	**3.** a	**4.** a	**5.** a
6. d	**7.** c	**8.** b	**9.** a	**10.** d
11. a	**12.** c	**13.** c	**14.** c	**15.** a
16. b	**17.** a	**18.** a	**19.** a	**20.** a
21. b	**22.** c	**23.** d	**24.** d	**25.** b
26. d	**27.** d	**28.** c	**29.** a	**30.** b
31. d	**32.** a	**33.** d	**34.** b	**35.** a
36. d	**37.** a	**38.** a	**39.** c	**40.** a

EXPLANATORY NOTES - I

1. In all pairs shortest paths algorithm we find several path, from source vertex to destination vertex. Among all those paths, the shortest paths is to be found.

Hence each 'path finding' is a subproblem and all the subproblems are to be recursively executed. So the design of such algorithms require dynamic programming.

4. Merge sort uses Divide and conquer strategy. The Divide and conquer means dividing a problem into several sub problems and solving each sub problem seperately.

8. The time complexity of linear search is O(n), Because a list with n elements perform n comparisons and hence 1 + 2 + 3 + ...+ n - 1= O (n)

9. The time complexity of binary search is O (log n) the complexity is measured by the number f(n) of comparisons to locate an item in list where the no of elements = n.

Each comparison reduces the list size to half

Hence atmost f (n) comparisons are required to locate an item where $2^{f(n)} > n \Rightarrow f(n) > \log_2 n$

$\Rightarrow O f(n) = O(\log n)$

10. The worst case complexity of bubble sort is

Assume no. of elements = n. In worst case,

No. of comparisons $f(n) = n + (n-1) + (n-2) + ... + 1$

$$= \frac{n(n+1)}{2} = O(n^2)$$

11. A symptotic complexity means a prior analysis of computing time by ignoring the factors which are machine dependent.

f (n) = function $\quad f(n) = \sum_{i=1}^{n} i$

= 1 + 2 + 3 + 4 + ... + n = O (n)

12. The optimal binary search tree generates an optimal solution to speed up search. To do so, one can adopt Dynamic programming. Because it provides alternative procedures when a process / procedure is not doing well.

13. Merge sort has average case, worst case running time as O (n log n).

⇒ Assume no of elements = n

Each pass requires 'log n' comparisons and each comparison requires n elements. Infact, according to the merge technique, no. of comparisons is < n ⇒ log n.

Running time = n log n ⇒ O (n log n)

15. The stack space required is always less than the no of elements in quick sort.

Hence it may not belongs to O (n).

It may be equal to O (log n). Since log n < n

16. According to the problem, the algorithm $A_1 A_2 A_3$ and A_4 solves a gives problem with order of log n, log log n, n log n n/ log n respectively.

Here, infact, log log n < n/logn < n log n < log n

⇒ O (log (log n)) will be the best choice.

⇒ A_2 algorithm is best.

18. The running time of an algorithm is

$T(n) = T(n-1) + T(n-2) + T(n-3)$ if n > 3

= n other wise

⇒ for n = 1 = o (1), n = 2 = o (2)

⇒ for n = o (n)

19. For T (1), T (2), T(3) the order is constant i.e., O (n). If the running times are equal.

T (1) = T (2) = T (3).

20. A linked list follows sequential / linear search i.e.,

An item is to be compared with all the elements of the entire list.

⇒ For each element in the list, we perform 'one' comparison.

⇒ Total n comparisons are required.

1 + 1 + 1 .... + 1 (n times) ⇒ O (n).

28. In worst case, the time required to search for a given element in a sorted linked list of length n is O (n).

Because, irrespective of the fact that the list is in sorted order or not, the linked list follows sequential search.

Each element must be compared with the search element.

⇒ n comparisons are required ⇒ O (n).

33. Because

1) Incrementing a variable require one unit of time.

2) prinitng a character to seven consumes one unit of time.

3) Even addition of 2 no.s uses one unit of time.

34. Suppose H is a heap. Observe that the no. of comparisons to find the appropriate place of an element 'item' in H cannot exceed the depth of H.

Since H is a complete tree, its depth is bounded by $\log_2^m$ where m is the no. of elements in H.

Accordingly the total number g(n) of comparison of n elements $=O\left(n\log\frac{n}{2}\right)$.

36. Examples of $O\left(n^2\right)$ algorithm is (are) all the given problems.

Because each problem relates to a two dimensional array. Moreover, all the elements of array are involved. So, m × n elements are manipulated.

39. Consider a graph

The corresponding adjacency matrix is

$$\begin{bmatrix} 0 & 1 & 1 & 1 \\ 1 & 0 & 1 & 1 \\ 1 & 1 & 0 & 1 \\ 1 & 1 & 1 & 0 \end{bmatrix}$$

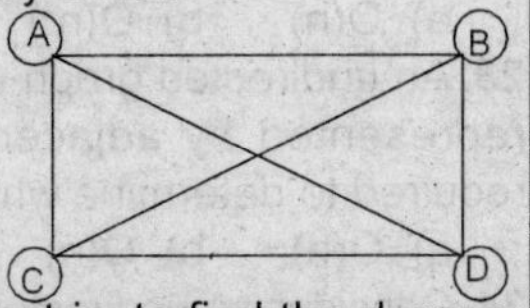

With the above adjacency matrix, to find the degree of the graph, the time required is $O\left(n^2\right)$ [∴ the matrix is a two dimensional Array].

OBJECTIVE QUESTIONS - II

1. The solution to the recurrence equation $T(2^k)=3T(2^{k-1})+1$, T(1)=1 is : **[GATE-2002]**

a) 2^k b) $(3^{k+1}-1)/2$ c) $3^{\log_2 k}$ d) $2^{\log_3 k}$

2. The running time of the following algorthim **[GATE-2002]**

Procedure A(n)
if n<=2
return(1)
else
return $\left(A\left(\left\lceil\sqrt{n}\right\rceil\right)\right)$;

is best described by

a) o(n) b) o(logn) c) o(log logn) d) o(1)

3. Randomized Quick sort is an extension of Quick sort where the pivot is chosen randomly. What is the worst case complexity of sorting n numbers using randomized Quick sort? **[GATE-2001]**

a) o(n) b) o(n log n) c) $O(n)^2$ d) o(n!)

4. let $f(n)=n^2\log n$ and $g(n)=n(\log n)^{10}$ be two positive functions of n . Which of the following statements is correct: **[GATE-2001]**

a) f(n)=o(g(n))and g(n) ≠ o(f(n))
b) g(n)=o(f(n)) and f(n) ≠ o(g(n))
c) g(n) ≠ o(g(n))and g(n) ≠ o(f(n))
d) f(n)=o(g(n)) and g(n)=o(f(n))

5. Let s be a sorted array of n integers . Let T(n) denote the time taken for the most efficient algorithm to determine if there are two elements with sum less than 1000 in s . Which of the following statements is TRUE: **[GATE-2000]**

a) T(n) is O(1) b) $n <= T(n) <= n\log_2 n$

c) $n\log_2 n <= T(n) < \binom{n}{2}$ d) $T(n) = \binom{n}{2}$

6. Consider the following functions

$f(n) = 3\ n^{\sqrt{n}}$
$f(n) = 2^{\sqrt{n}\ \log_2 n}$
h(n) = n!

which of the following is TRUE? **[GATE-2000]**

a) h(n) is O(f(n)) b) h(n) is O(g(n))
c) g(n) is not O(f(n)) d) f(n) is O(g(n))

7. If T_1 =O(1), give the correct matching for the **[GATE-1999]**

M. $T_n=T_{n-1}+n$ U. $T_n=O(n)$
N. $T_n=T_{n/2}+n$ V. $T_n=O(n \log n)$
O. $T_n=T_{n/2}+n\log n$ W. $T_n=O(n^2)$
P. $T_n=T_{n-1}+\log n$ X. $T_n=O(\log^2 n)$

a) M-W,N-V,O-U,P-X b) M-W,N-U,O-V,P-X
c) M-V,N-W,O-X,P-U d) M-W,N-U,O-X,P-V

8. Give the correct matching for the following pairs: **[GATE-1998]**

A. O(log n) I. selection sort
B. O(n) II. insertion sort
C. O(nlogn) III. binary search
D. $O(n^2)$ IV. merge sort

a) A-III,B-I,C-II,D-IV b) A-III,B-I,C-IV,D-II
c) A-I,B-III,C-IV,D-II d) A-I,B-IV,C-III,D-II

9. The correct matching for the following pairs is: **[GATE-1997]**

A. All pairs shortest paths I. Greedy
B. Quick sort II. Depth-First search
C. minimum weighted Spanning tree III. Dynamic programming
D. connected components IV. Divide and conquer

a) A-II,B-IV,C-I,D-III b) A-III,B-IV,C-I,D-II
c) A-III,B-IV,C-II,D-I d) A-IV,B-I,C-II,D-III

10. Let T(n) be the function defined by T(1)=1,

$T(n)=2T(\lfloor n/2 \rfloor)+\sqrt{n}$ for n ≥ 2

which of the following statements is TRUE ? **[GATE-1997]**

a) $T(n)=O(\sqrt{n})$ b) T(n) =O(n)
c) T(n) =O(logn) d) None

11. Consider the following two functions : **[GATE-1994]**

$$g_1(n)=\begin{cases} n^3 & for\ \ 0 \le n < 10000 \\ n^2 & for\ \ n \ge 10000 \end{cases}$$

$$g_2(n)=\begin{cases} n & for\ \ 0 \le n \le 10000 \\ n^3 & for\ \ n > 10000 \end{cases}$$

which of the following is TRUE:

a) $g_1(n)$ is $O(g_2(n))$ b) $g_1(n)$ is $O(n^3)$
c) $g_2(n)$ is $O(g_1(n))$ d) $g_2(n)$ is O(n)

12. Match the following **[GATE-1990]**

A. Strassen's matrix multiplication algorithm I. Greedy method
B. Krushkal's minimum spanningTree algorithm II. Dynamic Programming
C. Biconnected components III. Divide and conquer
D. Floyd's shortest path IV. Depth First Search Algorithm

a) A-III,B-I,C-IV,D-II b) A-I,B-III,C-II,D-IV

13. Match the following **[GATE-1989]**

A. O(logn) I. Heap sort
B. O(n) II. DFS
C. O(n log n) III. Binary search
D. $O(n^2)$ IV. Selection of the kth smallest element In a set of n elements

a) A-III,B-II,C-I;D-IV b) A-IV,B-I,C-III,D-II
c) A-III,B-I,C-II,D-IV d) A-III,D-IV,B-I,C-II

14. Let P be a quick sort program to sort numbers in ascending order. Let T_1 and T_2 be the time taken by the program for the inputs [1 2 3 4 5] and [5 4 3 2 1]. Which of the following holds **[GATE-1987]**

a) $T_1 = T_2$ b) $T_1 > T_2$
c) $T_1 < T_2$ d) None

15. A principle that states " no matter what the first decision, the remaining decisions" much optimal with respect to the state that results from the first decision is followed in the following strategy: **[AUCET-2001]**

a) Greedy method b) Divide and conquer
c) Back tracking d) Dynamic programming

16. What function exactly satisfies : **[AUCET-1999]**

$T(n) = T(n/2)+2^T(n-1)-2^n$

Assume that k's are constants

a) Logarithmic function k_1 logn
b) Linear function $k_1{}^n+k_0$
c) Quadratic function $k_2n^2+k_1n+k_0$
d) Exponential function k_1n^{kc}

17. The Ackermann function

A(0,y)=y+1
A(x+1,0)=A(x,1)
A(x+1,y+1)=A(x,A(x+1,y))

is an example of a function which : **[AUCET-1999]**

a) Operates in cubic time (in x and y)
b) Operates in quadratic time (in x and y,i.e.,$O(x^4 y^4)$)
c) Requires exponential time to be computed (i.e., $O(k^{xy})$ for some constant k) d) Is not computable

18. Assume that you have an algorithm that operates on a set of data with 'n' elements. If the recurrence formula that computes the time requirement for the algorithm is

given by $T(n) = \begin{cases} 8T(n/2)+2n & \text{if } n>1 \\ P & \text{if } n=1 \end{cases}$

Where p and q are constants , which of the following gives the order of complexity , O , of the algorithm: **[AUCET -1999]**

a) $n\log_2 n$ b) n^3 c) n^2 d) 2^n

19. The running time T(n) where 'n' is the input size of a recursive algorithm is given as follows: **[AUCET-1999]**

T(n)=c+T(n-1),if n>1
= d, if n<=1

the order of the algorithm is :

a) n^2 b) n c) n^3 d) n^n

20. For a linear search in an array of n elements the complexity for best, average and worst cases are ..... , ... and ..... respectively.

a) O(n),O(1),O(n/2) b) O(1),O(n),O((n+1)/2)
c) O(1),O((n+1)/2),O(n) d) None

21. Which of the following shows the correct relationship among some of the more common computing times for algorithm:

a) O(logn) < O(n) < O(nlogn) < $O(2^n)$ < $O(n^2)$
b) O(n) < O(logn) < O(nlogn) < $O(2^n)$ < $O(n^2)$
c) O(n) < O(logn) < O(nlogn) < $O(n^2)$ < $O(2^n)$
d) O(logn) < O(n) < O(nlogn) < $O(n^2)$ < $O(2^n)$

22. Consider the following program segment:

```
FOR item:=n down to 2 DO
BEGIN
large:=list[1];
index:=1;
for i:=2 to item do
IF list[i]> large THEN
BEGIN
large:=list[i];
index:=i;
END;
list[index]:=list[item];
list[item]:=large;
END;
```

The number of comparisons made by this:

a) O(n) b) $O(n^2)$ c) $O(n^3)$ d) O(n logn)

23. An undirected graph with n vertices and e edges are represented by adjacency matrix. What is the time required to determine whether the graph is connected:

a) O(e) b) O(n) c) $O(n^2)$ d) O(e+n)

24. Consider an undirected graph G with n vertices and e edges. What is the time taken by DFS if the graph is represented by (i) adjecency matrix and (ii) adjacency list:

a) $O(n^2)$,O(n) b) $O(n^2)$,O(e)
c) O(e),$O(n^2)$ d) O(e+n),0(e)

25. Three algorithms do the same task . Algorithm1 is $O(\sqrt{n})$,algorithm2 is O(n), algorithm3 is $O(\log_2 n)$. Which algorithm should execute fast for large values of n?

a) $O(\sqrt{n})$ b) O(n) c) $O(\log_2 n)$ d) None

26. The big 'O' notation is used in estimating the time complexity of an algorithm. Which of the following statements is true using this notation for n > 1? **[Hyd. Univ-2001]**

a) $n! = O(2^n)$ b) $2^n = O(n)^2$
c) $\log_2 (n!) = O(n \log_2 n)$ d) $n^2 = O(n + \log_2 n)$

27. An undirected graph with N vertices and E edges is represented by an adjacency matrix. What is the time required to determine if the graph is connected? **[Hyd.Univ-1999]**

a)O(E) b) O(N) c) $O(N^2)$ d) O(E + N)

28. Elements in the first column have to be matched with the elements in the second column. **[Hyd.Univ-1999]**

I. O(log n) A) Heap Sort
II. O(n) B) Depth First Search
III. O(n log n) C) Binary Search
IV. $O(n^2)$ D) Selecting k^{th} smallest element in set of n elements

a) I-D, II-A, III-B, IV-C b) I-C, II-D, III-A, IV-B
c) I-C, II-B, III-A, IV-D d) I-C, II-B, III-D, IV-A

29. Let :

F(0) =1, F(1) = 3,
$F(x) = 3F(x-1) + F(x-2)$ for $x \geq 2$

What is a value of F(5)? **[UGC NET-2000]**

a) 150 b) 7690 c) 723 d) 360

Note : For the next two questions, we need the following recursive definition of Josephus function:

J (1) = 1
J(2 n) = 2 J(n) – 1
J(2n + 1) = 2 J(n) + 1

For all natural numbers n.

30. Which of the following is FALSE? **[UGC NET-1995]**

a) J (9) = 3 b) J (11) = 7
c) J (2n + 1) – J (2n) = 2 and hence J is monotonically increasing functions, for all positive integers n. d) None

31. The function H is defined in terms of the function J defined earlier as follows:

H(n) = J(n+1) – J(n), n ∈N.

Then which of the following is incorrect? **[UGC NET-1995]**

a) H(2n) = 2 for all natural numbers n.

b) H(2n + 1) = 2H(n) – 2 for all n ÎN.

c) If (a) and (b) above are true then by principle of mathematicaly induction H(n) = 2 for all n.

d) None of the above

ANSWER KEY - II

1. b	2. a	3. c	4. b	5. a
6. d	7. b	8. b	9. b	10. b
11. c	12. a	13. a	14. a	15. d
16. d	17. d	18. b	19. b	20. c
21. d	22. b	23. c	24. b	25. c
26. b	27. c	28. b	29. b	30. c
31. a				

OBJECTIVE QUESTIONS - III

1. Which of the following graph is planar ? **[UGCNET-95]**

a) $K_{3,3}$ b. $K_{5,5}$ c. K_5 d. $K_{2,3}$

2. Greedy algorithm works for the following graph problems **[UGCNET-98]**

a) Minimum Spanning Tree b) Shortest Path

c) Both (a) and (b) d) Neither (a) and (b)

3. Taking a list of 8 elements $x_1, x_2, ...x_8$ odd elements are pushed on to a stack and even elements are pushed into a queue alternatively. Then with successive alternative deletions from the stack and queue results in a sequence. **[UGCNET - 98]**

a) $x_7x_6x_5x_4x_3x_2x_1$ b) $x_7x_2x_5x_4x_3x_6x_1x_8$

c) $x_2x_1x_4x_2x_6x_5x_8x_7$ d) $x_8x_1x_6x_3x_4x_5x_2x_7$

4. The in-order and pre-order traversal of a binary tree Tield the following orders

in - order : DGBAHEICF

pre-order : ABDGCEHIF

Then the post-order traversal would be **[UGCNET-98]**

a) DBHIGEFCA b) HIGEFDBCA

c) GDBHIEFCA d) GIHDBEFCA

5. The list 3,1,4,5,9,2,6,8,7 is sorted into ascending order using bubble sort. How many times is the number '8' moved during the entire process ? **[Hyd. Univ-2000]**

a) 1 b) never moved c) 2 d) 5

6. How many elements are there in the multidimensional array A (-75 : 124, 42:46, -20 : -1) ?

a) 10,000 b) 20,000 c) 19,997 d) 200,000

7. The height of a complete 4-way tree with N nodes is **[Hyd. Univ - 97]**

a) $\lfloor \log_4 (N+1) \rfloor$ b) $\lfloor \log_4 N \rfloor - 1$ c) $\lfloor \log_4 N \rfloor$ d) $\lfloor \log_4 N \rfloor + 1$

8. The following list is input to a bubble sort algorithm : 28,3,9,6,12,1,45,32,29,15,40,5. The List if output at the end of each complete pass through the list. Which of the following is NOT a valid intermediate list ? **[Hyd. Univ-99]**

a) 3,9,6,12,1,28,32,29,15,40,5,45

b) 3,6,9,12,1,28,32,29,15,5,40,45

c) 3,6,9,1,12,28,29,15,32,5,40,45 d) None

9. If 4*5/2+10-2=2, then what is the value of the expression 2+4*5/2+10-2= ? **[Hyd. Univ - 97]**

a) 4 b) 3 c) 23 d) 20

The below two questions are based on the following information.

Let T be a tree with n nodes $a_1, a_2.......a_n$ and let d_i be the length of the unique directed path from the root of T to node a_i. Then :

10. The total path length of T dentoed by L_T is given by **[Hyd. Univ - 97]**

a) $L_T = \sum_{i=1}^{n} (d_i)^2$ b) $L_T = \sum_{i=1}^{n} d_i$ c) $L_T < \sum_{i=1}^{n} d_i$ d) None

11. If T is a binary tree and if $h \geq 0$ is the height of T, then L_T and h are related as :

a) $L_T = \sum_{i=1}^{n} (d_i)^2$ b) $L_T = \sum_{i=1}^{n} d_i$

c) $L_T < \sum_{i=1}^{n} d_i$ d) None

EXPLANATORY NOTES - III

1. (a) A Graph $K_{m,n}$ is planar if $m \leq 3$ & $n \leq 3$

$\therefore K_{3,3}$ is planar. Hence answer 'a' is correct.

2. (d)

3. (b) After inserting elements into stack and queue, the arrangments are shown below.

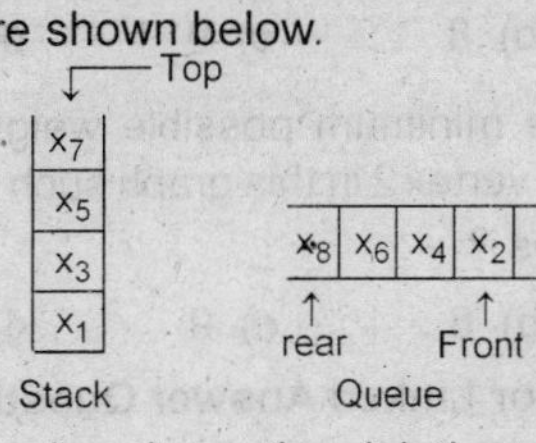

Now, the successive alternative deletions from the stack and queue, the sequence is $x_7x_2x_5x_4x_3x_6x_1x_8$

5. (c)

6. (b) The length L_i of dimension i of n dimentional array can be calculated by using following formula.

L_i = upper bound - lower bound + 1

Hence $L_1 = 124 - (-75) + 1 = 200$

$L_2 = 46 - 42 + 1 = 5$, $L_3 = -1 - (-20) + 1 = 20$

∴ Total number of elements in A is

$L_1 \times L_2 \times L_3 = 200 \times 5 \times 20 = 20{,}000$

7. (a)

8. (b)

9. (b) Given 4*5/(2+10)-2

The expression is evaluated by using operator recedance (lowest to highest) from left to right. That means = 4*5/(2+10) -2

= 4*5/ (12-2) = 4*5/10 = 20/10=2

Now, 2+4*5/2+10-2

= 6*5/2+10-2= 6*5/12-2= 6*5/10= 30/10 = 3

10. (b)

11. (a)

TEST YOURSELF - I

1. In a binary tree with n nodes, every node has an odd number of descendants. Every node is considered to be its own descendant. What is the number of nodes in the tree that have exactly one child ? **[GATE - 10]**

a) 0 b) 1 c) (n-1)/2 d) n - 1

Common Data Questions : 2 & 3

Consider a complete undirected graph with vertex set { 0, 1, 2, 3, 4}. Entry W_{ij} in the matrix w below is the weight of the edge { i, j } .

$$W = \begin{pmatrix} 0 & 1 & 8 & 1 & 4 \\ 1 & 0 & 12 & 4 & 9 \\ 8 & 12 & 0 & 7 & 3 \\ 1 & 4 & 7 & 0 & 2 \\ 4 & 9 & 3 & 2 & 0 \end{pmatrix}$$

2. What is the minimum possible weight of a spanning tree T in this graph such that vertex 0 is a leaf node in the tree T ? **[GATE - 10]**

a) 7 b) 8 c) 9 d) 10

3. What is the minimum possible weight of a path P from vertex 1 to vertex 2 in this graph such that P contains at most 3 edges ? **[GATE - 10]**

a) 7 b) 8 c) 9 d) 10

Statement for Linked Answer Questions : 4 & 5

A hash table of length 10 uses open addressing with hash function h(k)=k mod 10, and linear probing. After inserting 6 values into an empty hash table, the table is as shown below

0	
1	
2	42
3	23
4	34
5	52
6	46
7	33
8	
9	

4. Which one of the following choices gives a possible order in which the key values could have been inserted in the table ? **[GATE - 10]**

a) 46,42, 34, 52, 23, 33

b) 34, 42, 23, 52, 33, 46

c) 46, 34, 42, 23, 52, 33

d) 42, 46, 33, 23,34, 52

5. How many different insertion sequences of the key values using the same hash function and linear probing will result in the hash table shown above ? **[GATE - 10]**

a) 10 b) 20 c) 30 d) 40

6. What is the number of swaps required to sort n elements using selection sort, in the worst case ? **[GATE - 09]**

a) $\theta(n)$ b) $\theta(n \log n)$

c) $\theta(n^2)$ d) $\theta(n^2 \log n)$

7. Which of the following statement(s) is / are correct regarding Bellman - Ford shortest path algorithm ?

P. Always finds a negative weighted cycle, if one exists.

Q. Finds whether any negative weighted cycle is reachable from the source. **[GATE - 09]**

a) P only b) Q only

c) both P and Q d) Neither P nor Q

8. The keys 12, 18, 13, 2, 3, 23, 5 and 15 are inserted into an initially empty hash table of length 10 using open addressing with hash function h(k) = k mod 10 and linear probing. What is the resultant hash table ? **[GATE - 09]**

Index	(A)	(B)	(C)	(D)
0				
1				
2	2	12	12	12,2
3	23	13	13	13,3,23
4			2	
5	15	5	3	5,15
6			23	
7			5	
8	18	18	18	18
9			15	

9. What is the maximum height of any AVL - tree with 7 nodes? Assume that the height of a tree with a single node is 0 . **[GATE - 09]**

a) 2 b) 3 c) 4 d) 5

10. Consider the following graph :

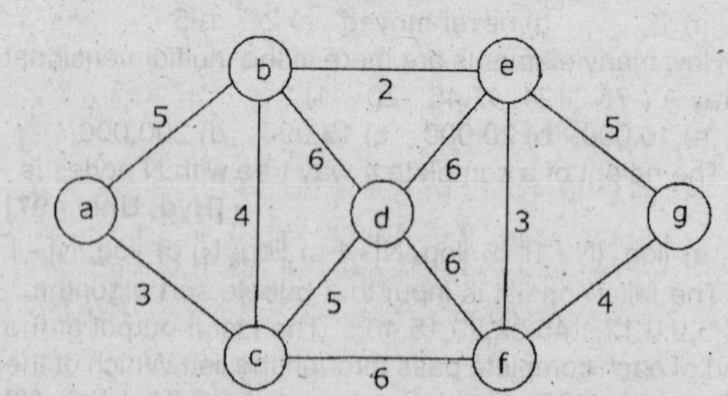

Which one of the following is NOT the sequence of edges added to the minimum spanning tree using Kruskal's algorithm ? **[GATE - 09]**

a) (b,e) (e, f) (a, c), (b, c), (f,g) (c, d)

b) (b, e) (e, f) (a, c) (f,g) (b, c) (c,d)

c) (b, e) (a,c) (e, f) (b, c) (f,g) (c,d)

d) (b,e) (e, f) (b,c) (a,c) (f,g) (c,d)

11. In quick sort, for sorting n elements, the $(n/4)^{th}$ smallest element is selected as pivot using an O(n) time algorithm. What is the worst case time complexity of the quick sort ? **[GATE - 09]**

a) $\theta(n)$ b) $\theta(n \log n)$

c) $\theta(n^2)$ d) $\theta(n^2 \log n)$

STATEMENT FOR Linked Answer Questions : 12 & 13

Consider a binary max - heap implemented using an array.

12. Which one of the following array represents a binary max - heap ? **[GATE - 09]**

a) { 25, 12, 16, 13, 10, 8, 14 }

b) { 25, 14, 13, 16, 10, 8, 12 }

c) { 25, 14, 16, 13, 10, 8, 12 }

d) { 25, 14, 12, 13, 10, 8 16 }

13. What is the content of the array after two delete operations on the correct answer to the previous question? **[GATE - 09]**

a) { 14, 13, 12, 10, 8 } b) { 14, 12, 13, 8, 10 }

c) { 14, 13, 8, 12, 10 } d) { 14, 13, 12, 8, 10 }

14. The Breadth First Search algorithm has been implemented using the queue data structure, One possible order of visiting the nodes of the following graph is **[GATE - 08]**

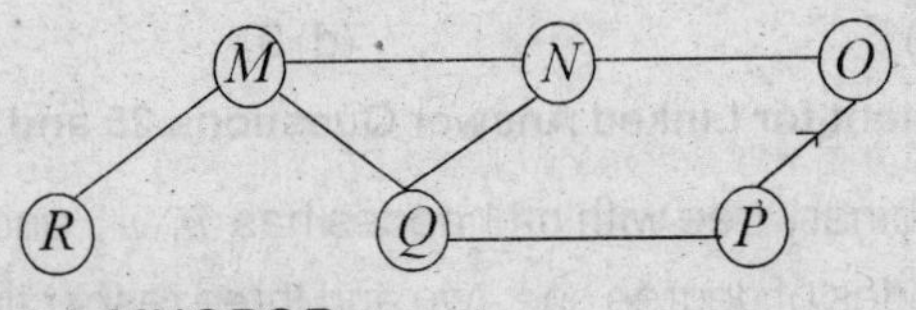

(a) MNOPQR

(b) NQMPOR

(c) QMNPRO

(d) QMNPOR

15. A B -tree of order 4 is built from scratch by 10 successive insertions. What is the maximum number of node sphitting operations that may take place? **[GATE - 08]**

(a) 3 (b) 4

(c) 5 (d) 6

16. G is a graph on n vertices and 2n-2 edges. The edges of G can be partitioned into two edge-disjoint spannint trees. Which of the following is NOT true for G ? **[GATE - 08]**

(a) For every subset of k vertices, the induced subgraph has at most 2k-2 edges

(b) The minimum cut in G has at least two edges

(c) There are two edge-disjoint paths between every pair of vertices

(d) There are two vertex--disjoint paths between every pair of vertices

17. Dijkstra's single source shortest path algorithm when run from vertex a in the above graph, computes the correct shortest path distance to **[GATE - 08]**

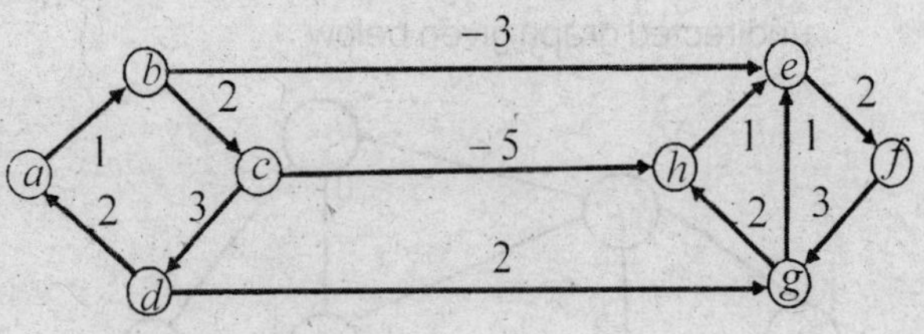

(a) only vertex a

(b) only vertices a,e,f,g,h

(c) only vertices a,b,c,d

(d) all the vertices.

18. For the undirected, weighted graph given below, which of the following sequences of edges represents a correct execution of Prim's algorithm to construct a Minimum Spanning Tree? **[GATE - 08]**

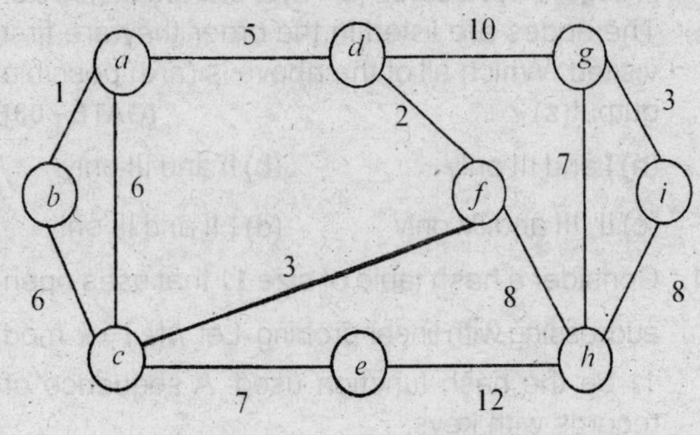

(a) (a,b), (d,f), (f,c), (g,i), (d,a), (g,h), (c,e) , (f,h)

(b) (c,e), (c,f), (f,d), (d,a), (a,b),(g,h), (h,f), (g,i)

(c) (d,f), (f,c), (d,a), (a,b), (c,e), (f,h), (g,h),(g,i)

(d) (h,g), (g,i), (h,f) , (f,c), (f,d), (d,a), (a,b), (c,e)

19. The following three are known to be the preorder, inorder and postorder sequences of a binary tree. But it is not known which is which

I. MBCAFHPYK II. KAMCBYPFH

III. MABCKYFPH

Pick the true statement from the following **[GATE - 08]**

(a) I and II are preorder and inorder sequences, respectively

(b) I and III are preorder and postorder sequences, respectively

(c) II is the inorder sequence, but nothing more can be said about the other two sequences

(d) II and III are the preorder and inorder sequences, respectively.

20. Consider the following sequence of nodes for the undirected graph given below

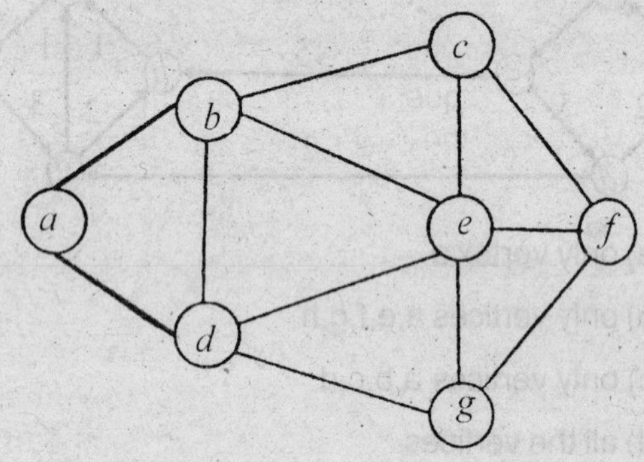

I. a b e f d g c II. a b e f c g d

III. a d g e b c f IV. a d b c g e f

A depth First Search (DFS) is started at node a. The nodes are listed in the order they are first visited. Which all of the above is (are) possible output(s) ? **[GATE - 08]**

(a) I and III only (b) II and III only

(c) II,IIII and IV only (d) I,II and III only

21. Consider a hash table of size 11 that uses open addressing with linear probing. Let $h(k) = k$ mod 11 be the hash function used. A sequence of records with keys.

43 36 92 87 11 4 71 13 14

is inserted into an initially empty hash table, the bins of which are indexed from zero to ten. What is the index of the bin into which the last record is inserted ? **[GATE - 08]**

(a) 3 (b) 4
(c) 6 (d) 7

Common Data for Questions 22, 23 and 24 :

A Binary Search Tree (BST) stores values in the range 37 to 573. Consider the following sequence of keys.

1. 81, 537, 102, 439, 285, 376, 305

II. 52, 97, 121, 195, 242, 381, 472

III. 142, 248, 520, 386, 345, 270, 307

IV. 550, 149, 507, 395, 463, 402, 270

22. Suppose the BST has been unsuccessfully searched for key 273. Which all of the above sequences list nodes in the order in which we could have encountered them in the search ? **[GATE - 08]**

(a) II and III only (b) I and III only
(c) III and IV only (d) III only

23. Which of the following statements is TRUE ? **[GATE - 08]**

(a) I, II and IV are inorder sequences of three different BSTs

(b) I is a preorder sequence of some BST with 439 as the root

(c) II is an inorder sequence of some BST where 121 is the root and 52 is a leaf

(d) IV is a postorder sequence of some BST with 149 as the root

24. How many distinct BSTs can be constructed with 3 distinct keys ? **[GATE - 08]**

(a) 4 (b) 5
(c) 6 (d) 9

Statement for Linked Answer Questions 25 and 26 :

A binary tree with n>1 nodes has n_1, n_2 and n_3 nodes of degree one, two and three respectively. The degree of a node is defined as the number of its neighbours.

25. n_3 can be expressed as : **[GATE - 08]**

(a) $n_1 + n_2 - 1$ (b) $n_1 - 2$

(c) $\left[\frac{n_1 + n_2}{2}\right]$ (d) $n_2 - 1$

26. Starting with the above tree, while there remains a node v of degree two in the tree, add an edge between the two neighbours of v and then remove v from the tree. How many edges will remain at the end of the process ? **[GATE - 08]**

(a) $2 * n_1 - 3$ (b) $n_2 + 2 * n_1 - 2$

(c) $n_3 - n_2$ (d) $n_2 + n_1 - 2$

27. Consider a weighted undirected graph with positive edge weights and let uv be an edge in the graph. It is known that the shortest path from the source vertex s to u has weight 53 and the shortest path from s to v has weight 65. Which one of the following statements is always true? [GATE - 07]

(a) weight $(u, v) < 12$

(b) weight $(u, v) \leq 12$

(c) weight $(u, v) > 12$

(d) weight $(u, v) \geq 12$

28. A depth-first search is performed on a directed acyclic graph. Let $d[u]$ denote the time at which vertex u is visited for the first time and $f[u]$ the time at which the dfs call to the vertex u terminates. Which of the following statements is always true for all edges (u, v) in the graph? [GATE - 07]

(a) $d[u] < d[v]$ (b) $d[u] < f[v]$

(c) $f[u] < f[v]$ (d) $f[u] > f[v]$

29. What is the largest integer m such that every simple connected graph with n vertices and n edges contains at least m different spanning trees ? [GATE - 07]

(a) 1 (b) 2 (c) 3 (d) n

30. Consider a hash function that distributes keys uniformly. The hash table size is 20. After hashing of how many keys will the probability that any new key hashed collides with an existing one exceed 0.5. [GATE - 07]

(a) 5 (b) 6

(c) 7 (d) 10

31. When searching for the key value 60 in a binary search tree, nodes containing the key values 10, 20, 40, 50, 70, 80, 90 are traversed, not necessarily in the order given. How many different orders are possible in which these key values can occur on the search path from the root to the node containing the value 60? [GATE - 07]

(a) 35 (b) 64

(c) 128 (d) 5040

32. Suppose you are given an implementation of a queue of integers. The operations that can be performed on the queue are :

(i) is Empty (Q) - returns true if the queue is empty, false otherwise.

(ii) delete (Q) - deletes the element at the front of the queue and returns its value.

(iii) insert (Q, i) - inserts the integer i at the rear of the queue.

Consider the following function :

```
void f (queue Q) {
int i ;
if (! is Empty (Q) ) {
  i = delete (Q) ;
  f (Q) ;
  insert (Q, i) ;
  }
}
```

What operation is performed by the above function f ? **[GATE - 07]**

(a) Leaves the queue Q unchanged

(b) Reverses the order of the elements in the queue Q

(c) Deletes the element at the front of the queue Q and inserts it at the rear keeping the other elements in the same order

(d) Empties the queue Q

33. The height of a binary tree is the maximum number of edges in any root to leaf path. The maximum number of nodes in a binary tree of height h is [GATE - 07]

(a) 2^h (b) $2^{h-1} - 1$

(c) $2^{h+1} - 1$ (d) 2^{h+1}

34. The maximum number of binary trees that can be formed with three unlabeled nodes is [GATE - 07]

(a) 1 (b) 5

(c) 4 (d) 3

35. Which of the following sorting algorithms has the lowest worst - case complexity? [GATE - 07]

(a) Merge sort

(b) Bubble sort

(c) Quick sort

(d) Selection sort

36. The following postfix expression with single digit operands is evaluated using a stack

$$8\,2\,3\,\wedge\,/\,2\,3\,*\,+\,5\,1\,*\,-$$

Note that ^ is the exponentiation operator. The top two elements of the stack after the first * is evaluated are [GATE - 07]

(a) 6, 1 (b) 5, 7 (c) 3, 2 (d) 1, 5

37. The inorder and preorder traversal of a binary tree are

d b e a f c g and a b d e c f g respectively.

The postorder traversal of the binary tree is **[GATE - 07]**

(a) d e b f g c a (b) e d b g f c a

(c) e d b f g c a (d) d e f g b c a

38. Consider a hash table of size seven, with starting index zero, and a hash function (3x + 4) mod 7. Assuming the hash table is initially empty, which of the following is the contents of the table when the sequence 1, 3, 8, 10 is inserted into the table using closed hashing? Note that - denotes an empty location in the table. **[GATE - 07]**

(a) 8, -, -, -, -, -, 10 (b) 1, 8, 10, -, -, -, 3

(c) 1, -, -, -, -, -, 3 (d) 1, 10, 8, -, -, -, 3

39. In an unweighted, undirected connected graph, the shortest path from a mode S to every other node is computed most efficiently, in terms of time complexity, by **[GATE - 07]**

(a) Dijkstra's algorithm starting from S

(b) Warshall's algorithm

(c) Performing a DFS starting from S

(d) Performing a BFS starting from S

40. A complete n-ary tree is a tree in which each node has n children or no children. Let I be the number of internal nodes and L be the number of leaves in a complete n-ary tree. If L=41, and I = 10, what is the value of n **[GATE - 07]**

(a) 3 (b) 4

(c) 5 (d) 6

41. Consider the following C program segment where Cell Node represents a node in a binary tree

```
struct CellNode {
struct cellNode * leftchild ;
int element ;
struct CellNode * right child ;
} ;
int GetValue (struct CellNode * ptr) {
int value = 0 ;
if (ptr ! = NULL) {
  if ( ( ptr - > leftchild = = NULL) &&
        (ptr -> rightchild = = NULL) )
  value = 1 ;
else
  value = value + GetValue (ptr -> leftchild)
  + GetValue (ptr -> rightchild) ;
  }
  return (value) ;
  }
```

The value returned by Get Value when a pointer to the root of a binary tree is passed as its argument is **[GATE - 07]**

(a) the number of nodes in the tree

(b) the number of internal nodes in the tree

(c) the number of leaf nodes in the tree

(d) the height of the tree

42. Let w be the minimum weight among all edge weights in an undirected connected graph. Let e be a specific edge of weight w. Which of the following is FALSE ? **[GATE - 07]**

(a) There is a minimum spanning tree containing e.

(b) If e is not in a minimum spanning tree T, then in the cycle formed by adding e to T, all edges have the same weight.

(c) Every minimum spanning tree has an edge of weight w

(d) e is present in every minimum spanning tree

43. An array of n numbers is given, where n is an even number. The maximum as well as the minimum of these n numbers needs to be determined. Which of the following is TRUE about the number of comparisons needed ? **[GATE - 07]**

(a) At least 2n-c comparisons, for some constant c, are needed.

(b) At most 1.5 n-2 comparisons are needed.

(c) At least $n \log_2 n$ comparisons are needed

(d) None of the above

44. In a binary max heap containing n numbers, the smallest element can be found in time **[GATE - 06]**

(A) $\theta(n)$ (B) $\theta(\log n)$

(C) $\theta(\log \log n)$ (D) $\theta(1)$

45. Consider a weighted complete graph G on the vertex set $\{V_1, V_2, ..., V_n\}$ such that the weight of the edge (V_i, V_j) is $2|i-j|$. The weight of a minimum spanning tree of G is **[GATE - 06]**

(A) $n-1$ (B) $2n-2$

(C) $\binom{n}{2}$ (D) n^2

46. To implement Dijkstra's shortest path algorithm on unweighted graphs so that it runs in linear time, then data structure to be used is [GATE - 06]

(A) Queue (B) Stack
(C) Heap (D) B-Tree

47. A scheme for storing binary trees in an array X is as follows. Indexing of X starts at 1 instead of 0. The roots is stored at X[1]. For a node stored at X[i], the left child, if any, is stored in x[2i] and the right child, if any, in $X[2i+1]$. To be able to store any binary tree on n vertices, the minimum size of X should be [GATE - 06]

(A) $\log_2 n$ (B) n (C) $2n+1$ (D) 2^n-1

48. Which one of the following in place sorting algorithms needs the minimum number of swaps ? [GATE - 06]

(A) Quick sort (B) Insertion sort
(C) Selection sort (D) Heap sort

49. Consider the following graph :

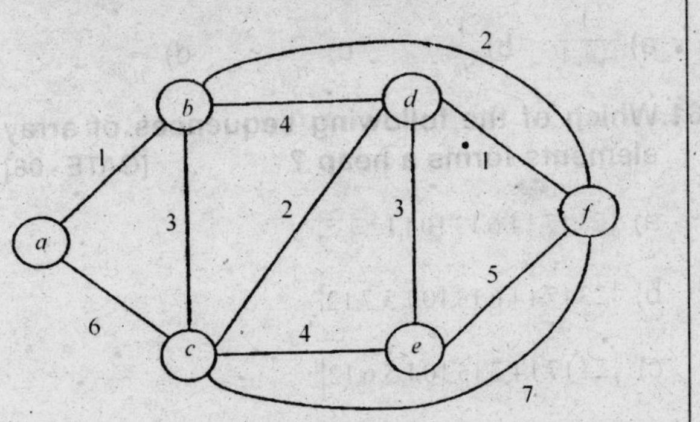

Which one of the following cannot be the sequence of edges added, in that order, to a minimum spanning tree using Kruskal's algorithm ? [GATE - 06]

(A) $(a-b),(d-f),(b-f),(d-c),(d-e)$

(B) $(a-b),(d-f),(d-c),(b-f),(d-e)$

(C) $(d-f),(a-b),(d-c),(b-f),(d-e)$

(D) $(d-f),(a-b),(b-f),(d-e),(d-c)$

50. Let T be a depth first search tree in a undirected graph G. Vertices u and v are leaves of this tree T. The degrees of both u and v in G are at least 2. Which one of the following statements is true ? [GATE - 06]

(A) There must exist a vertex w adjacent to both u and v in G

(B) There must exist a vertex w whose removal disconnects u and v in G

(C) There must exist a cycle in G containing u and v

(D) There must exist a cycle in G containing u and all its neighbours in G

51. An Implementation of a queue Q, using two stacks S1 and S2, is given below

```
Void insert (Q, X) {
    Push (S1, x);
}
void delete (Q) {
    if (stack-empty (S2)) then
        if (stack-empty (S1)) then {
    print (Q is empty ");
    return;
}
else while (! (stack-empty) (S1))) {
    x = pop (S1);
    push (S2, x);
}
    x=pop (S2);
}
```

Let n insert and $m(\le n)$ delete operations be performed in an arbitrary order on an empty queue Q. Let x and y be the number of push and pop operations performed respectively in the process. Which one of the following is true for all m and n ? [GATE - 06]

(A) $n+m \le x < 2n$ and $2m \le n+m$

(B) $n+m \le x < 2n$ and $2m \le y \le 2n$

(C) $2m \le x < 2n$ and $2m \le y \le n+m$

(D) $2m \le x < 2n$ and $2m \le y \le 2n$

52. A set X can be represented by an array x[n] as follows

$$x[i] = \begin{cases} 1 & \text{if } i \in X \\ 0 & \text{otherwise} \end{cases}$$

Consider the following algorithm in which x,y, and z are boolean arrays of size n:

```
algorithm zzz (x[ ], y[ ], z[ ] {
int i;
for (i = 0; i < n; ++i)
    z[i] = (x[i] ∧ ~ y[i]) ∨ (~ x[i] ∧ y[i])
}
```

[GATE - 06]

The set Z computed by the algorithm is

(A) $(X \cup Y)$ (B) $(X \cap Y)$

(C) $(X-Y)\cap(Y-X)$ (D) $(X-Y)\cup(Y-X)$

Common data for Question 53, 54, 55 :

The 2^n vertices of graph G correspond to all subsets of a set of size n, for $n \geq 6$. Two vertices of G are adjacent if and only if the corresponding sets intersect in exactly two elements

53. The number of vertices of degree zero in G is [GATE - 06]

(A) 1 (B) n (C) $n+1$ (D) 2^n

54. The maximum degree of a vertex in G is [GATE - 06]

(A) $\left(2^{n/2}\right)2^{n/2}$ (B) 2^{n-2}

(C) $2^{n-3} \times 3$ (D) 2^{n-1}

55. The number of connected components in G is [GATE - 06]

(A) n (B) $n+2$

(C) $2^{n/2}$ (D) $\frac{2^n}{n}$

Statement for Linked Answer Questions 56 and 57 :

A 3-ary max heap is like a binary max heap, but instead of 2 children, nodes have 3 children. A 3-ary heap can be represented by an array as follows : The root is stored in the first location, $a[0]$, nodes in the next level, from left to right, is stored from $a[1]$ to $a[3]$. The nodes from the second level of the tree from left to right are stored from $a[4]$ location onward. An item x can be inserted into a 3-ary heap containing n items by placing x in the location $a[n]$ and pushing it up the tree to satisfy the heap property.

56. Which one of the following is a valid sequence of elements in an array representing 3-ary max heap ? [GATE - 06]

(A) 1, 3, 5, 6, 8, 9
(B) 9, 6, 3, 1, 8, 5
(C) 9, 3, 6, 8, 5, 1
(D) 9, 5, 6, 8, 3, 1

57. Suppose the elements 7, 2, 10, and 4 are insetrted, in that order, into the valid 3-ary max heap found in the above question, Q.56. Which one of the following is the sequence of items in the array representing the resultant heap ? [GATE - 06]

(A) 10, 7, 9, 8, 3, 1, 5, 2, 6, 4
(B) 10, 9, 8, 7, 6, 5, 4, 3, 2, 1
(C) 10, 9, 4, 5, 7, 6, 8, 2, 1, 3
(D) 10, 8, 6, 9, 7, 2, 3, 4, 1, 5

58. In a binary tree, the number of internal nodes of degree 1 is 5, and the number of internal nodes of degree 2 is 10. The number of leaf nodes in the binary tree is [GATE - 06]

a) 10 b) 11 c) 12 d) 15

59. If all the edge weights of an undirected graph are positive, then any subset of edges that connects all the vertices and has minimum total weight is a [GATE - 06]

a) Hamiltonian cycle b) grid
c) hypercube d) tree

60. Consider the undirected graph G defined as follows. The vertices of G are bit strings of length n. We have an edge between vertex u and vertex v if and only if u and v differ in exactly one bit position (in other words, v can be obtained from u by flipping a single bit). The ratio of the chromatic number of G to the diameter of G is [GATE - 06]

a) $\frac{1}{2^{n-1}}$ b) $\frac{1}{n}$ c) $\frac{2}{n}$ d) $\frac{3}{n}$

61. Which of the following sequences of array elements forms a heap ? [GATE - 06]

a) $\{23,17,14,6,13,10,1,12,7,5\}$

b) $\{23,17,14,6,13,10,1,5,7,12\}$

c) $\{23,17,14,7,13,10,1,5,6,12\}$

d) $\{23,17,14,7,13,10,1,12,5,7\}$

62. Suppose that we have numbers between 1 and 100 in a binary search tree and want to search for the number 55. Which of the following sequencies CANNOT be the sequence of nodes examined ? [GATE - 06]

a) $\{10,75,64,43,60,57,55\}$

b) $\{90,12,68,34,62,45,55\}$

c) $\{9,85,47,68,43,57,55\}$

d) $\{79,14,72,56,16,53,55\}$

63. Which of the following is the correct decomposition of the directed graph given below into its strongly connected components? [GATE - 06]

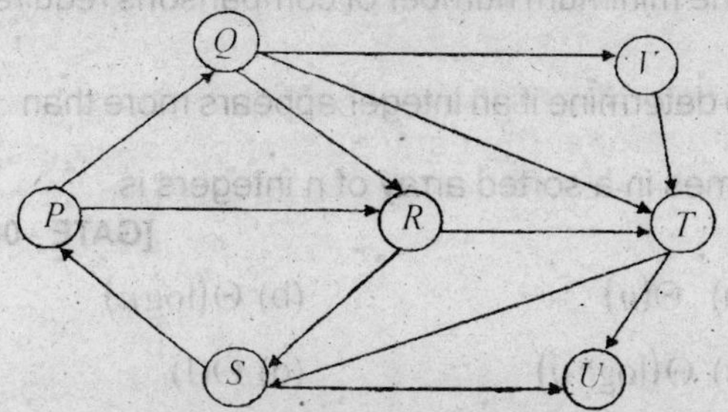

a) $\{P,Q,R,S\},\{T\},\{U\},\{V\}$ b) $\{P,Q,R,S,T,V\},\{U\}$

c) $\{P,Q,S,T,V\},\{R\},\{U\}$ d) $\{P,Q,R,S,T,U,V\}$

64. Consider the depth-first search of an undirected graph with 3 vertices P, Q, and R. Let discovery time $d(u)$ represent the time instant when the vertex u is first visited, and finish time $f(u)$ represent the time instant when the vertex u is last visited. Given that

$d(P)=5$ units $f(P)=12$ units

$d(Q)=6$ units $f(Q)=10$ units

$d(R)=14$ units $f(R)=18$ units

Which one of the following statements is TRUE about the graph [GATE - 06]

a) There is only one connected component

b) There are two connected components, and P and R are connected

c) There are two connected components, and Q and R are connected

d) There are two connected components, and P and Q are connected.

65. The characters a to h have the set of frequencies based on the first 8 Fibonacci numbers as follows

$a:1, b:1, c:2, d:3, e:5, f:8, g:13, h:21$

A Huffman code is used to represent the characters. What is the sequence of characters corresponding to the following code ?

110111100111010 [GATE - 06]

a) fdheg b) ecgdf

c) dchfg d) fehdg

Common Data for Questions 66,67,68

An array X of n distinct integers is interpreted as a complete binary tree. The index of the first element of the array is 0.

66. The index of the parent of element $X[i], i \neq 0$, is [GATE - 06]

a) $\lfloor i/2 \rfloor$ b) $\lceil (i-1)/2 \rceil$

c) $\lceil i/2 \rceil$ d) $\lceil i/2 \rceil - 1$

67. If only the root node does not satisfy the heap property, the algorithm to convert the complete binary tree into a heap has the best asympototic time complexity of [GATE - 06]

a) $O(n)$ b) $O(\log n)$

c) $O(n \log n)$ d) $O(n \log \log n)$

68. If the root node is at level 0, the level of element $X[i], i \neq 0$, is [GATE - 06]

a) $\lfloor \log_2 i \rfloor$ b) $\lceil \log_2 (i+1) \rceil$

c) $\lfloor \log_2 (i+1) \rfloor$ d) $\lceil \log_2 i \rceil$

69. The weight of a sequence $a_0, a_1, \ldots, a_{n-1}$ of real numbers is defined as $a_0 + a_1/2 + \ldots + a_{n-1}/2^{n-1}$. A subsequence of a sequence is obtained by deleting some elements from the sequence, keeping the order of the remaining elements the same. Let X denote the maximum possible weight of a subsequence of $a_0, a_1, \ldots, a_{n-1}$. Then X is equal to [GATE - 10]

a) $\max(Y, a_0 + Y)$ b) $\max(Y, a_0 + Y/2)$

c) $\max(Y, a_0 + 2Y)$ d) $a_0 + Y/2$

70. Let π_1 be a problem that belongs to the class NP. Then which one of the following is TRUE ? [GATE - 09]

a) There is no polynomial time algorithm for π_1

b) If π_1 can be solved deterministically in polynomial time, then P = NP

c) If π_1 is NP - hard, then it is NP - complete

d) π_1 may be undecidable

71. The running time of an algorithm is represented by the following recurrence relation :

$$T(n) = \begin{cases} n & n \leq 3 \\ T\left(\dfrac{n}{3}\right) + cn & otherwise \end{cases}$$

Which one of the following represents the time complexity of the algorithm ? **[GATE - 09]**

a) $\theta(n)$ b) $\theta(n \log n)$

c) $\theta(n^2)$ d) $\theta(n^2 \log n)$

Common Data Questions 72 & 73

A sub-sequence of a given sequence is just the given sequence with some elements (possibly none or all) left out. We are given two sequences X[m] and Y[n] of lengths m and n, respectively, with indexes of X and Y starting from 0.

72. We wish to find the length of the longest common sub - sequence (LCS) of X[m] and Y[n] as l(m,n), where an incomplete recursive definition for the function l(i,j) to compute the length of the LCS of X[m] and Y[n] is given below :

l(i,j)=0, if either i=0 or j=0
= expr1, if i, j>0 and X[i-1]= Y[j-1]
= expr2, if i, j>0 and X[i-1]= Y[j-1]

Which one of the following options is correct? **[GATE-09]**

a) $\exp r1 \equiv l(i-1, j)+1$

b) $\exp r1 \equiv l(i, j-1)$

c) $\exp r2 \equiv \max(l(i-1, j), l(i, j-1))$

d) $\exp r2 \equiv \max(l(i-1, j-1), l(i, j))$

73. The values of l(i, j) could be obtained by dynamic programming based on the correct recursive definition of l(i, j) of the form given above, using an array L[M,N], where M=m+1 and N=n+1, such that L[i,j] =l(i,j).

Which one of the following statements would be TRUE regarding the dynamic programming solution for the recursive definition of 1 (i, j) ? **[GATE - 09]**

a) All elements L should be initialized to 0 for the values of i(i, j) to be properly computed

b) The values of l(i,j) may be computed in a row major order or column major order of L(M,N)

c) The values of l(i,j) cannot be computed in either row major order or column major order of L(M,N)

d) L[p,q] needs to be computed before L[r,s] if either p<r or q<s.

74. The most efficient algorithm for finding the number of connected components in an undirected graph on n vertices and m edges has time complexity **[GATE - 08]**

(a) Θ (n) (b) Θ (m)

(c) Θ (m+n) (d) Θ (mn)

75. Consider the following functions:

$f(n)=2^n$

$g(n)=n!$

$h(n)=n^{\log n}$

Which of the following statements about the asymptotic behaviour of f(n), g(n) and h(n) is true? **[GATE - 08]**

(a) f(n)=O(g(n)); g(n)= O(h(n))

(b) f(n)= Ω (g(n)); g(n)=0(h(n))

(c) g(n)= O(f(n)); h(n)= O(f(n))

(d) h(n)=O(f(n)); g(n)= Ω (f(n))

76. The minimum number of comparisons required to determine if an integer appears more than $\frac{n}{2}$ times in a sorted array of n integers is **[GATE - 08]**

(a) $\Theta(n)$ (b) $\Theta(\log n)$

(c) $\Theta(\log^* n)$ (d) $\Theta(1)$

77. Consider the Quicksort algorithm. Suppose there is a procedure for finding a pivot element which splits the list into two sub-lists each of which contains at least one-fifth of the elements. Let T(n) be the number of comparisons required to sort n elements. Then **[GATE - 08]**

(a) $T(n) \le 2T\left(\frac{n}{5}\right)+n$

(b) $T(n) \le T\left(\frac{n}{5}\right)+T\left(\frac{4n}{5}\right)+n$

(c) $T(n) \le 2T\left(\frac{4n}{5}\right)+n$

(d) $T(n) \le 2T\left(\frac{n}{2}\right)+n$

78. The subset-sum problem is defined as follows: Given a set S of n positive integers and a positive integer W, determine whether there is a subset of S whose elements sum to W. **[GATE - 08]**

An algorithm Q solves this problem in O(nW) time. Which of the following statements is false?

(a) Q solves the subset-sum problem in polynomial time when the input is encoded in unary

(b) Q solves the subset-sum problem in polynomial time when the input is encoded in binary

(c) The subset sum problem belongs to the class NP

(d) The subset sum problem is NP-hard

79. You are given the postorder traversal, P, of a binary search tree on the n elements 1,2...n. You have to determine the unique binary search tree that has P as its postorder traversal. What is the time complexity of the most efficient algorithm for doing this ? **[GATE - 08]**

(a) Θ (log n)

(b) Θ (n)

(c) Θ (nlog n)

(d) none of the above, as the tree cannot be uniquely determined.

80. We have a binary heap on n elements and wish to insert n more elements (not necessarily one after another) into this heap. The total time required for this is **[GATE - 08]**

(a) Θ (log n) (b) Θ (n)

(c) Θ (nlog n) (d) Θ (n^2)

Statement for Linked Answer Questions 81 to 82.

The subset-sum problem is defined as follows. Given a set of n positive integers, $S=\{a_1,a_2,a_3...a_n\}$ and a positive integer W, is there a subset of S whose elements sum to W? A dynamic program for solving this problem uses a 2-dimensional Boolean array, X, with n rows and W+1 columns. X[i, j], $1\le i\le n, 0\le j\le W$, is TRUE if and only if there is a subset of $\{a_1,a_2,....,a_i\}$ whose elements sum to j.

81. Which of the following is valid for $2\le i\le n$ and $a_i\le j\le W$? **[GATE - 08]**

(a) X [i, j] = X [i-1, j] $\vee$ X[i, j-a_i]
(b) X [i, j] = X [i-1, j] $\vee$ X[i-1, j-a_i]
(c) X [i, j] = X [i-1, j] $\wedge$ X[i, j-a_i]
(d) X [i, j] = X [i-1, j] $\wedge$ X[i-1, j-a_i]

82. Which entry of the array X, if TRUE, implies that there is a subset whose elements sum to W? **[GATE - 08]**

(a) X [1, W] (b) X [n,0]
(c) X[n, W] (d) X[n-1, n]

83. Arrange the following functions in increasing asymptotic order.

A. $n^{1/3}$ B. e^n

C. $n^{7/4}$ D. $n\log^9 n$

E. 1.0000001^n **[GATE - 08]**

(a) A,D,C,E,B (b) D,A,,C,E,B
(c) A,C,D, E,B (d) A,C,D,B,E

84. For problems X and Y, Y is NP-complete and X reduces to Y in polynomial time. Which of the following is TRUE ? **[GATE - 08]**

(a) If X can be solved in polynomial time, then so can Y
(b) X is NP-complete
(c) X is NP-hard
(d) X is in NP, but not necessarily NP-Complete

85. Which of the following is TRUE ? **[GATE - 08]**

(a) The cost of searching an AVL tree is $\theta(\log n)$ but that of a binary search tree is O(n)

(b) The cost of searching an AVL tree is $\theta(\log n)$ but that of a complete binary tree is $\theta(n\log n)$

(c) The cost of searching a binary search tree is O(log n) but that of an AVL tree is $\theta(n)$

(d) The cost of searching an AVL tree is $\theta(n\log n)$ but that of a binary search tree is O(n).

86. If we use Radix Sort to sort n integers in the range $\left(n^{k/12}, n^k\right)$, for some k>0 which is independent of n, the time taken would be **[GATE - 08]**

(a) $\theta(n)$ (b) $\theta(kn)$

(c) $\theta(n\log n)$ (d) $\theta(n^2)$

87. When $n=2^{2k}$ for some $k\ge 0$ the recurrence relation

$T(n)=\sqrt{2}\,T(n/2)+\sqrt{n}, T(1)=1$

evaluate to **[GATE - 08]**

(a) $\sqrt{n}(\log n+1)$ (b) $\sqrt{n}\log n$

(c) $\sqrt{n}\log\sqrt{n}$ (d) $n\log\sqrt{n}$

Statement for Linked Answer Questions 88 and 89

Let $P_1, P_2, ..., P_n$ be n points in the xy - plane such that no three of them are collinear. For every pair of points P_i and P_j, let L_{ij} be the line passing through them. Let L_{ab} be the line with the steepest gradient amongst all $n(n-1)/2$ lines.

88. Which one of the following properties should necessarily be satisfied ? [GATE - 07]

(a) P_a and P_b are adjacent to each other with respect to their x - coordinate

(b) Either P_a or P_b has the largest or the smallest y - coordinate among all the points

(c) The difference between x - coordinatef P_a and P_b is minimum

(d) None of the above

89. The time complexity of the best algorithm for finding P_a and P_b is [GATE - 07]

(a) $\theta(n)$ (b) $\theta(n \log n)$

(c) $\theta(n \log^2 n)$ (d) $\theta(n^2)$

90. Consider the following segment of C - code

```
int J , n ;
  j = 1 ;
while ( j < =  n )
    j = j * 2 ;
```

The number of comparisons made in the execution of the loop for any $n > 0$ is [GATE - 07]

(a) $\lceil \log_2 n \rceil + 1$ (b) n

(c) $\lceil \log_2 n \rceil$ (d) $\lfloor \log_2 n \rfloor + 1$

91. Consider the process of inserting an element into a Max Heap, where the Max Heap is represented by an array. Suppose we perform a binary search on the path from the new leaf to the root to find the position for the newly inserted element, the number of comparisons performed is [GATE - 07]

(a) $\Theta(\log_2 n)$ (b) $\Theta(\log_2 \log_2 n)$

(c) Θn (d) $\Theta(n \log_2 n)$

92. Consider the following C code segment :

```
int IsPrime (n)
{
        int i, n ;
        for ( i = 2 ; i < = sqrt (n) ;
        i++ )
          if ( n% i = = 0)
            {printf ('Not Prime \ n"};
          return 0 ;)
          return 1 ;
        }
```

Let T(n) denote the number of times the for loop is executed by the program on input n. Which of the following is TRUE ? [GATE - 07]

(a) $T(n) = O(\sqrt{n})$ and $T(n) = \Omega(\sqrt{n})$

(b) $T(n) = O(\sqrt{n})$ and $T(n) = \Omega(1)$

(c) $T(n) = O(n)$ and $T(n) = \Omega(\sqrt{n})$

(d) None of these

93. Consider the following C - program fragment in which i, j, and n are integer variables.

for $(i = n, j = 0; i > 0; i /= 2, j += i)$; [GATE - 06]

Let val(j) = denote the value stored in the variable j after termination of the for loop. Which one of the following is true?

(A) $val(j) = \theta(\log n)$ (B) $val(j) = \theta(\sqrt{n})$

(C) $val(j) = \theta(n)$ (D) $val(j) = \theta(n \log n)$

94 Let S be an NP-complete problem Q and R be two other problems not known to be in NP. Q is polynomial-time reducible to S and S is polynominal-time reducible to R. Which one of the following statements is true ? [GATE - 06]

(A) R is NP-complete (B) R is NP-hard

(C) Q is NP-complete (D) Q is NP-hard

95. An element in an array X is called a leader if it is greater than all elements to the right of it in X. The best algorithm to find all leaders in an array [GATE - 06]

(A) Solves it in linear time using a left to right pass of the array

(B) Solves in linear time using a right to left pass of the array

(C) Solves it using divide and conquer in time $\theta(n \log n)$

(D) Solves it in time $\theta(n^2)$

96. Let $SHAM_3$ be the problem of finding a Hamiltonian cycle in a graph $G = (V, E)$ with $|V|$ divisible by 3 and $DHAM_3$ be the problem of determining if a Hamiltonian cycle exists in such graphs. Which one of the following is true ? [GATE - 06]

(A) Both $DHAM_3$ and $SHAM_3$ are NP-hard

(B) $SHAM_3$ is NP-hard, but $DHAM_3$ is not

(C) $DHAM_3$ is NP-hard, but $SHAM_3$ is not

(D) Neither $DHAM_3$ nor $SHAM_3$ is NP-hard

97. Consider the following recurrence

$T(n) = 2T(\lceil \sqrt{n} \rceil) + 1, T(1) = 1$

Which one of the following is true? [GATE - 06]

(A) $T(n) = \theta(\log \log n)$ (B) $T(n) = \theta(\log n)$

(C) $T(n) = \theta(\sqrt{n})$ (D) $T(n) = \theta(n)$

98. The median of n elements can be found in $O(n)$ time. Which one of the following is correct about the complexity of quick sort, in which median is selected as pivot? [GATE - 06]

(A) $\theta(n)$ (B) $\theta(n \log n)$

(C) $\theta(n^2)$ (D) $\theta(n^3)$

99. Given two arrays of numbers $a_1,......,a_n$ and $b_1,.....,b_n$ where each number is 0 or 1, the fastest algorithm to find the largest span (i,j) such that $a_i + a_{i+1} + ... + a_j = b_i + b_{i+1} + ... + b_j$, or report that there is no such span, [GATE - 06]

(A) Takes $O(3^n)$ and $\Omega(2^n)$ time if hashing is permitted

(B) Takes $O(n^3)$ and $\Omega(n^{2.5})$ time in the key comparison model

(C) Takes $\Theta(n)$ time and space

(D) Takes $O(\sqrt{n})$ time only if the sum of the 2n elements is an even number

100. Consider these two functions and two statements S1 and S2 about them.

```
int work 1 (int * a, int i, int j )
{
    int x = a[i+2];
    a[j] = x+1;
    return a[i+2]-3;}
}
```

```
int work 2 (int * a, int i, int j )
{
    int t1 = i+2;
    int t2 = a[t1];
    a[j] = t2+1;
    return t2-3;
}
```

S1: The transformation from work 1 to work 2 is valid, i.e., for any program state and input arguments, work 2 will compute the same output and have the same effect on program state as work 1.

S2 : All the transformations applied to work 1 to get work 2 will always improve the performance (i.e. reduce CPU time) of work 2 compared to work 1 [GATE - 06]

(A) S1 is false and S2 is false

(B) S1 is false and S2 is true

(C) S1 is true and S2 is false

(D) S1 is true and S2 is true

101. Consider the following code written in a pass-by-reference language like FORTRAN and these statements about the code.

```
      Subroutine swap (ix, iy)
      it = ix
L1:   ix = iy
L2:   iy = it
   end
   ia = 3
   ib = 8
   call swap (ia, ib +5)
   print * , ia, ib
  end
```

S1: The compiler will generate code to allocate a temporary nameless cell, initialize it to 13, and pass the address of the cell to swap

S2: On execution the code will generate a runtime error on line L1

S3: On execution the code will generate a runtime error on line L2

S4: The program will print 13 and 8

S5 : The program will print 13 and -2

Exactly the following set of statement (s) is correct : [GATE - 06]

(A) S1 and S2 (B) S1 and S4

(C) S3 (D) S1 and S5

102. A problem in NP is NP-complete if [GATE - 06]

a) it can be reduced to the 3 - SAT problem in polynomial time

b) the 3-SAT problem can be reduced to it in polynomial time

c) it can be reduced to any other problem in NP in polynomial time

d) some problem in NP can be reduced to it in polynomial time

ANSWER KEY

1. a	2. d	3. b	4. c	5.
6. a	7. b	8. c	9. b	10. d
11. b	12. c	13. d	14. c	15. a
16. d	17. d	18. c	19. d	20. b
21. d	22.	23. c	24. b	25. b
26. a	27. d	28. a	29. d	30. d
31. c	32.b	33. c	34. b	35. a
36. a	37. a	38. b	39. d	40. c
40. c	42. d	43. d	44. b	45. b
46. c	47. d	48. c	49. d	50. d
51. a	52. d	53. c	54. a	55. b
56. d	57. a	58. b	59. b	60.
61. c	62. c	63. c	64. d	65.
66. a	67. c	68. a	69. a	70.c
71. a	72.c	73. b	74. c	75. d
76. a	77. b	78. b	79. b	80.a
81. c	82. c	83. c	84. b	85. a
86. c	87. a	88. b	89. a	90. d
91. a	92. b	93. c	94. c	95. c
96. a	97. a	98. b	99. c	100.c
101. a	102. c			

EXPLANATORY NOTES

4. 46, 34, 42, 23, 52, 33

hash function h(k) = k mod 10

sol, 46, h(46) =46 mod10=6.(index)

so, 46 will be stored on index 6

Similarly, 34

h(34) = 34 mod 10 = 4

$42 \rightarrow 2$

$23 \rightarrow 3$

$52 \rightarrow 2$ (location is occupied)

so it will be stored next at next rempty location

$33 \rightarrow 3$ (location is occupied)

so it will be stored next at next rempty location

6. If there are n elements then in worst case, total swaps will be (n-1) in selection sort. So the number of swaps is $\theta(n)$.

Alternately

The selection sort is similar to bubble sort except it does not swap elements with every move. The sorting algorithm first finds the smallest element in the list and then puts it in to place in single swap. So n swap required for n elements.

7. In Bellman Ford shortest path algorithm, we choose a node first and then we find any negative weighted cycle is reachable from source.

8. In linear probing, if there is already an element then the new element is fed into the next blank value field in the table.

9.

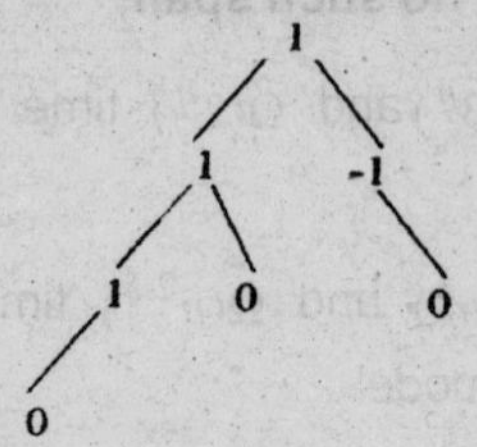

Maximum Height = 3

10. In Kruskal's Algorithm we choose on edge of G which has smallest weight among the edges of G.

So (b,e)=2,(e,f)=3,(b,c)=4, (a,c)=8, (f,g)=4 (c,d)=5

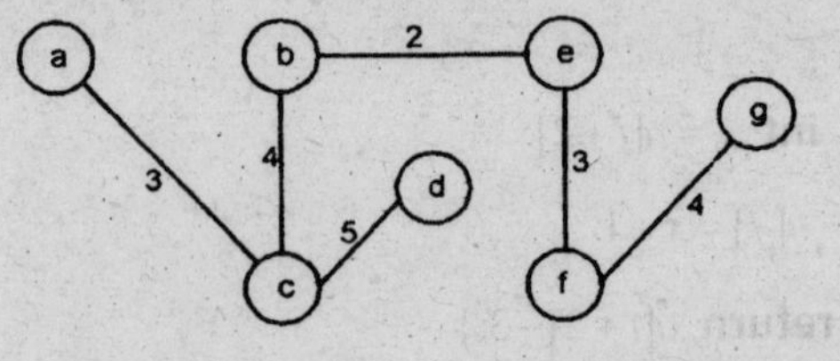

Alternately

Order the edges in non - decreasing order and pick edge one by one until all the nodes are completed with no edge making cycle on addition.

11. Average complexity of quick sort is $\theta(n \log n)$.

12. In binary max heap, the difference between two level mode or child of a single mode is atleast differ by one so the correct array with represent binary max - heap.

13. The deletion from binary max heap always contain greater element. In above question the correct array

{ 25, 14, 16, 30, 10, 12, 18 }

{ 14, 16, 13, 10, 12, 18 }

after one deletion { 14, 13, 10, 12, 8 }.

after two deletions After Heapifying

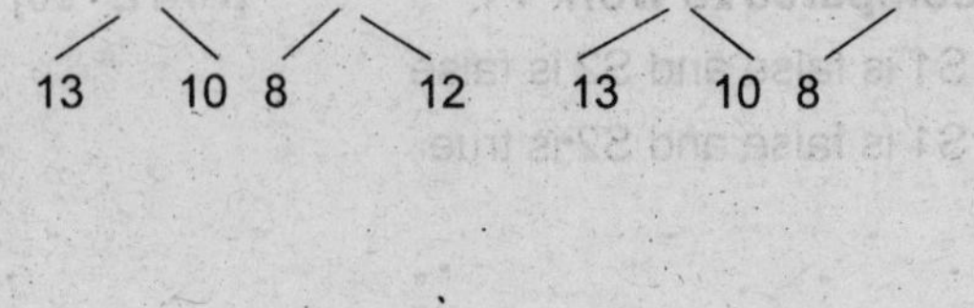

Deleting 25 Deleting 16

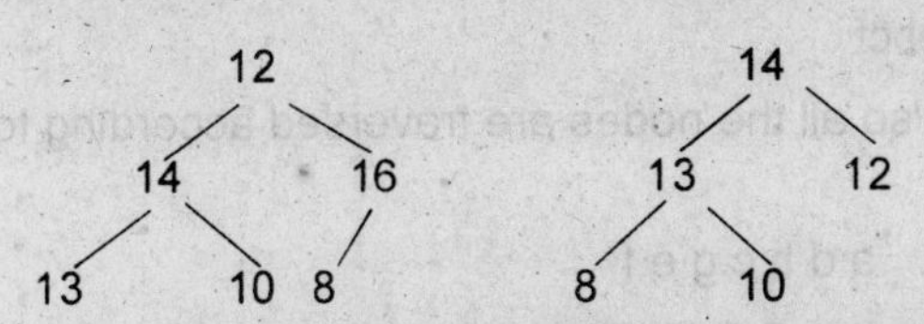

So ans { 14, 13, 12, 8, 10 }

14. Applying BFS Algorithms

Starting from Q

Queue Q

Making Q's child's status 1

Queue

Status

Q	M	N	O	P	R
2	1	1		1	
2	2	1		1	1
2	2	2	1	1	1

Q | M | N | P

Making M's child 1 (which are not 1)

Q | M | N | P | R

Making N's Child 1

Q | M | N | P | R | O

(other are already 1)

15. A B -tree of order 4. All internal nodes except root have at most 4 children and at least 2 children.

For 10 instruction it will stake 3 node splitting operation.

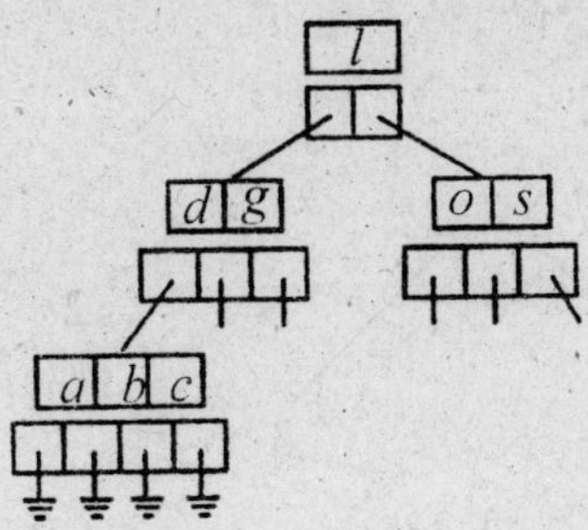

Alternately

Order 4-B tree can have maximum 3 keys in one node and each node have max 4 children

so lets take values 84, 82, 29, 99, 65, 12, 50, 28, 58, 71

Inserting 84, 82, 29

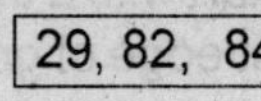

1st Splitting

Inserting 99

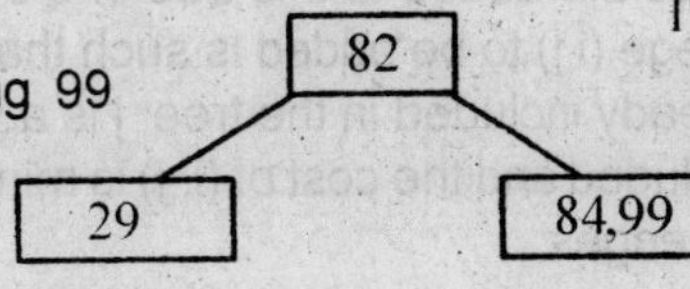

Inserting 65, 12

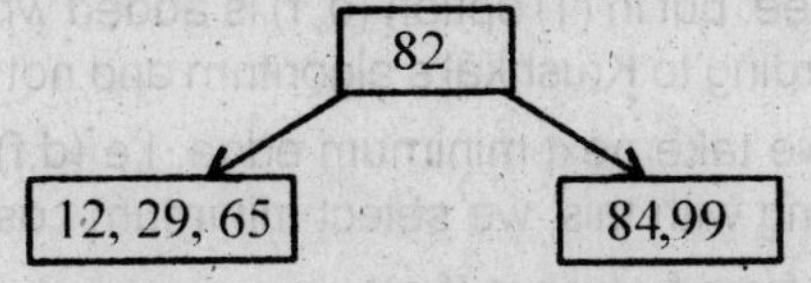

Inserting 50

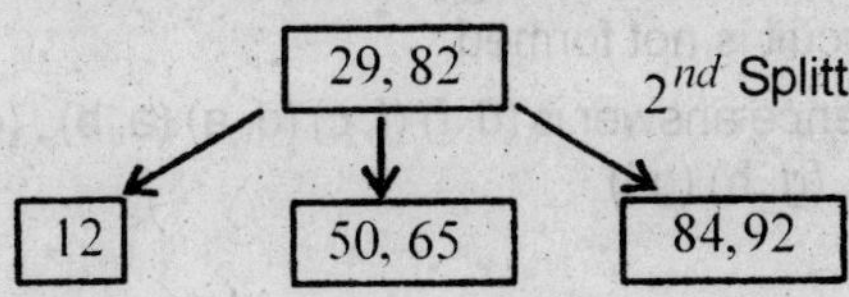

2nd Splitting

Inserting 28, 58

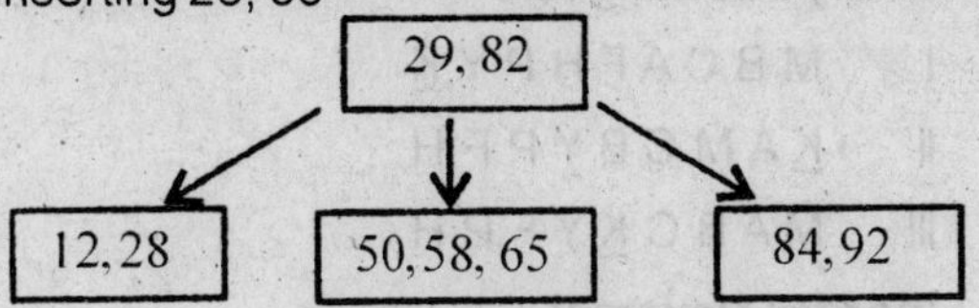

Inserting 71

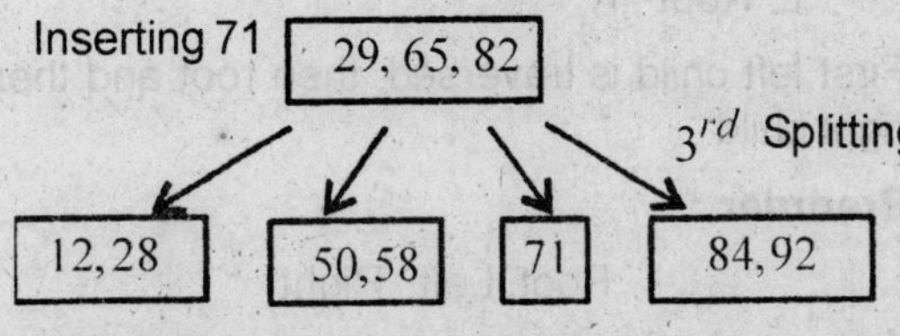

3rd Splitting

For any sequency of insertion, maximum splitting is 3.

16. The minimum number of edges

for a connected graph =n

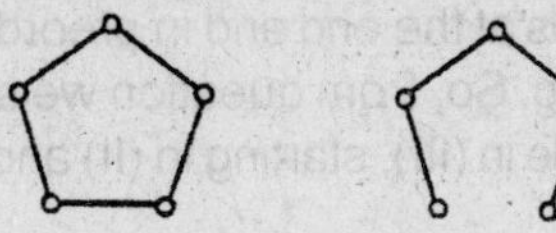

and for a connected tree = n-1

Here given that 2n -2 edges and the edges of graph is partitioned into two edge disjoint trees. So there are two vertex disjoint paths between every pair of vertices.

18. Prim's Algorithm

This algorithm will start with a tree that includes only a minimum cost edge of graph G.

Then edges are added to this tree one by one. The next ege (i,j) to be added is such that i is a vertex already included in the tree, j is a vertex not yet included and the cost of (i, j) is minimum among all edges.

Here, so when we start with the minimum cost edge (a,b) so after (a, b) (b,c) must be added to the tree, but in (1) option (d, f) is added which is according to Krushkal's algorithm and not Prim.

So, we take next minimum edge, i.e (d,f) Now starting with this, we select minimum cost edge from f which is (f,c)

Thus, in this way, we move on selecting next minimum cost edge at each node such that circuit is not formed.

Hence answer is (d, f) (f, c) (d, a) (a, b) , (c,e), (f, h), (g, h) (g,i)

19. Preorder , Postorder and Inorder sequences of a binary tree are given:

I M B C A F H P Y K̲

II K̲ A M C B Y P F H

III M A B C K̲ Y F P H

Inorder traversal.

L, Root , R

First left child is traversed, then root and then right child.

Preorder :

Root, Left, Right

Postorder :

Left, Right, Root

Thus we see that in inorder, root is in the middle. In postorder, it is at the end and in preorder, root is at the starting. So, from question we see that k̲ is in the middle in (III), starting in (II) and at end in (I).

So

I Postorder II. Preorder III Inorder

20. **Checking options**

I a b e f d g e

d can't be traversed after f and before g and c

II a b e f c g d

Here, all the nodes are traversed according to D F S

III. adgebcf

Here, also all the nodes are traversed according to DFS

IV. a d b c g e f

It is B F S (Bredth first search)

21. h(k) = K mod 11

43 mod 1 1 → 10

36 mod 1 1 → 3

92 mod 1 1 → 4

87mod 1 1 → 10

11 mod 1 1 → 0

4 mod 1 1 → 4

71 mod 1 1 → 5

13 mod 1 1 → 2

14 mod 1 1 → 3

Hash table

0	87
1	11
2	13
3	36
4	92
5	4
6	71
7	14
8	
9	
10	43

23. Option II. 52, 97, 121, 195, 242, 381, 472 only option (II) is in increasing order, so it is the inorder traversal. In order traversal, elements, are in increasing order.

24. Let three distinct keys be 1,2,3

Total ways of writing these 3 distinct values = 3! = 6

1, 2, 3
1, 3, 2
2, 1, 3
2, 3, 1
3, 1, 2
3, 2, 1

BST :

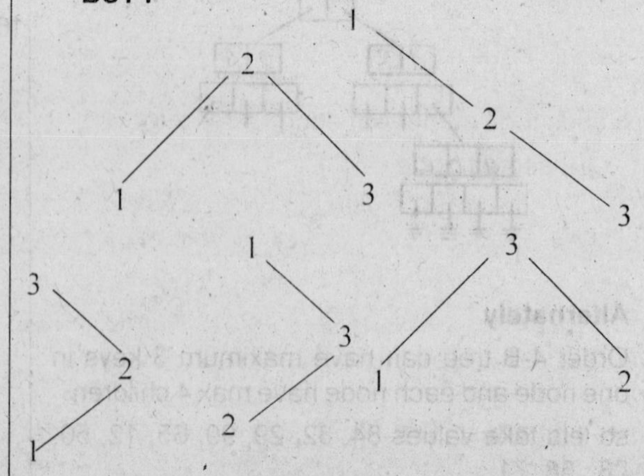

Total 5 BST's can be formed with 3 distinct keys

25 e.g.

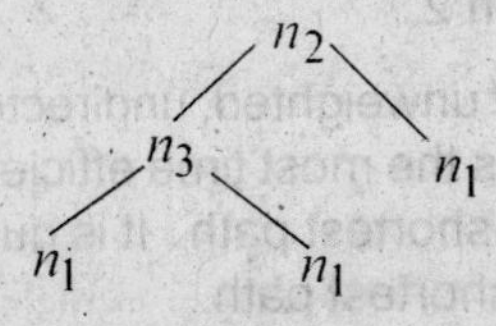

n_2 will be 1 always

$\therefore$ n_3 will never depend on n_2

and here $n_1 = 3,\ n_2 = 1, n_3 = 1$

these satisfy option (b).

26. Two degree mode is n_2 and n_2 =1

So, removing n_2, 2 edges will be removed. And according to question, 1 new edge is added.

Total number of edges in the Binary tree = $\dfrac{2n_2 + 3n_3 + n_1}{2}$

Here $n_2 = 1$

$n_3 = n_1 - 2$

n_1

$\therefore$ Number of edges $= \dfrac{2 + 3(n_1 - 2) + n_1}{2}$

$= \dfrac{2 + 3n_1 - 6 + n_1}{2}$

$= \dfrac{4n_1 - 4}{2}$

$= 2(n_1 - 1)$

Thus total number of edges after removing n_2 i.e 2 degree node

$= (2n_1 - 2) - 1$

$= 2n_1 - 3$

27. Let (u, v) be an edge in graph

Shortest path $(s, u) = 53$

$\because$ Shortest parth $(s, v) = 65$ it means if edge (u, v) is included in the shortest path between s and v, then weight (u, v) should be equal to 12.

If it is not included it means weight $(u, v) \geq 12$

So weight $(u, v) \geq 12$

29. Take an example of three node

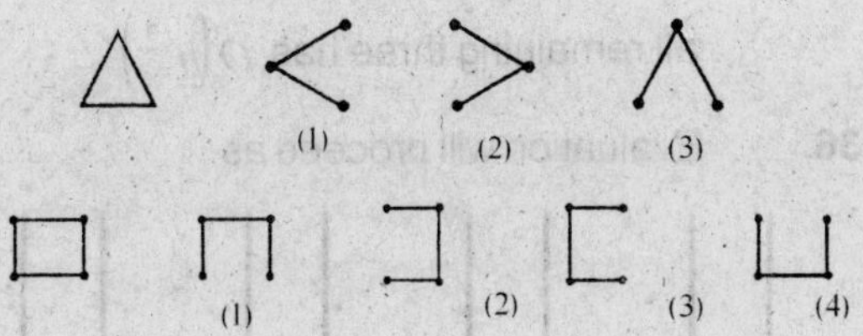

So if simple connected graph with n vertices and n edges, then it contains atleast n different spanning tree.

31. Search key value 60

Tree containing following key values

10, 20, 40, 50, 70, 80, 90 (7 nodes)

So it have 2^7 different orders in which these key value can occur on the search path from the root to the node containing value 60.

32. In this program if statement execute only queue is not empty. This function is called recursively untill queue becomes empty after it. Flow of control is transfered to next statement and insert the element in reverse of the order of the elements in the queue.

33. Maximum nodes in binary tree of height

$0 = 1 = (2^{0+1} - 1)$

Maximum nodes in binary tree of height

$1 = 3 = (2^{1+1} - 1)$

$\therefore$ MAximum nodes in binary tree of height

$h = (2^{h+1} - 1)$

34. 5 binary trees can be formed with three unlabeled nodes as shown below.

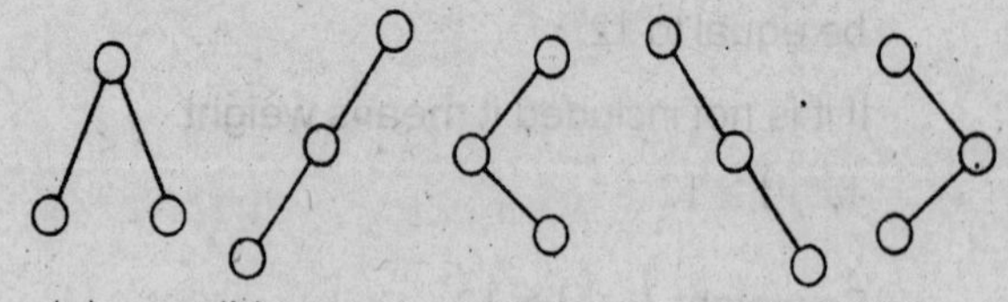

(a) (b) (c) (d) (e)

35. Merge sort has lowest worst - case complexity, that is, $O(n \log n)$, whereas all remaining three has $O(n^2)$

36. Evaluation will proceed as

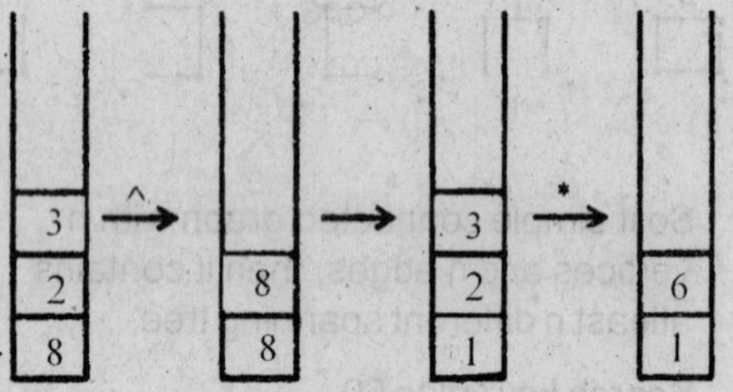

So after evaluation of first

* top elements will be 6, 1.

37. From given inorder and preorder traversal, the binary tree can be formed.

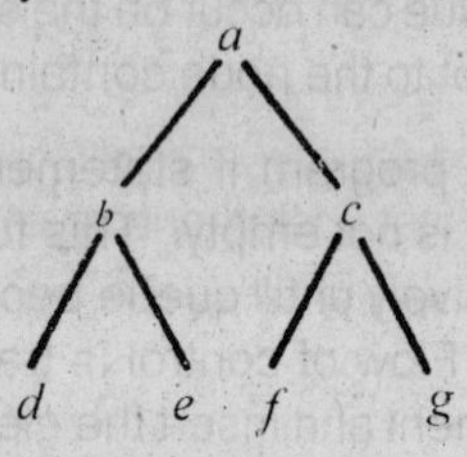

Since a is first node traversed in preorder, so it will be root of tree elements (nodes) left to a and right to a in inorder traversal will be left subtree and right subtree respectively of the tree. In left subtree elements d, b, e since b comes first in preorder, so b will be root of left subtree d is left to b in inorder, so it will be left child of d and e will be right child of d. Tree will be like

38. 1 will occupy location 0.

3 will occupy location 6.

8 hashed to location 0 which is already occupied, so it will be hashed to one location next to it (0) i.e. to location 1.

Since 10 also clashes, so it will be hashed to location 2.

39. In case of unweighted, undirected graphs, BFS gives the most time efficient computation for shortest path. It is guaranteed to find first shortest path.

40.

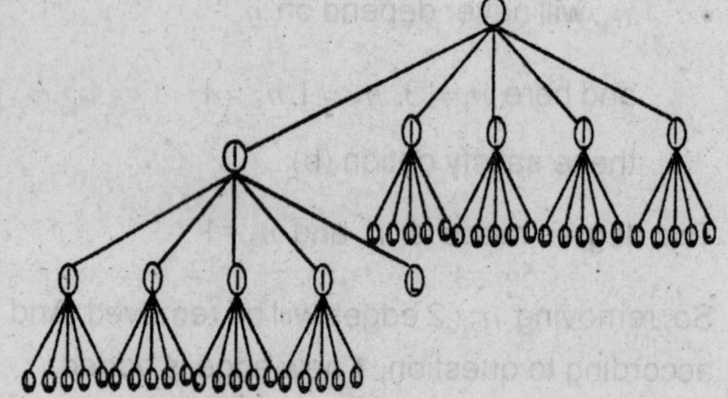

41. Since value = 1 only if nodes has no left as well as right node otherwise value = 0. Therefore it returns number of leg nodes in tree.

42. e is need not to be present in every spanning tree since there may be edges (in cycle formed by adding e), which has same weight as e has.

43. Since, 2n - c = average number of comparison needed.

1.5 n - 2 = number of comparison in case.

$n \log_2 n$ = also doesn't conform with number of comparison needed.

45. The minimum spanning tree will be formed when $\forall\, i. \in .\{1, 2, ...n\}$

V_i is connected to V_{i+1}, as in that case each edge will have the minimum weight of 2. For example, for n = 4, the minimum spanning tree is as follows

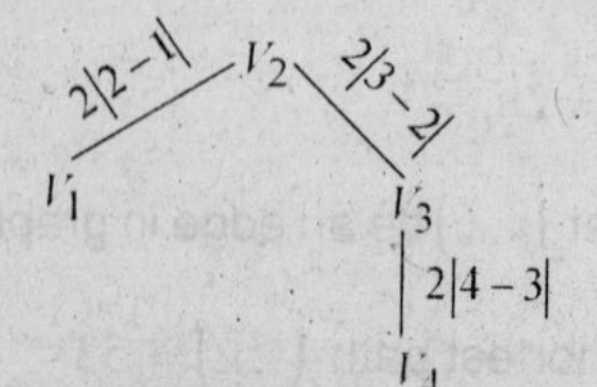

In such a tree the weight of tree consisting of n-1 edges each of weight 2 will be

$2(n-1) = 2n-2$

47. In order to find the minimum size of X capable of storing any kind of binary tree, the worst cost size of the tree has to be taken. In the worst case size of the tree has to be taken. In the worst case only a single node will contain the data at each level of the tree.

For example, for 4 data items, the worst case binary tree will be as follows

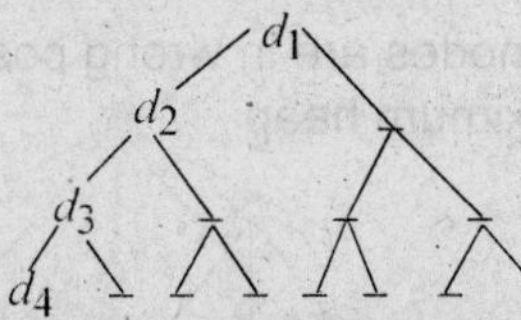

Generalizing the concept, for n vertices, the levels of tree that are required will be n.

The memory required at each level (starting from root) will be : $2^0, 2^1, 2^2, ...2^{n-1}$

$\therefore$ Minimum size $= 2^0 + 2^1 ....2^{n-1}$

$$= \frac{2^0(2^n - 1)}{2-1} = 2^{n-1}$$

49. Ordering the weights, the sequence obtained is 1, 1, 2, 2, 3, 3, 4, 4, 5, 6, 7 now, in order to obtain the minimum spanning tree using Kruskal's Algorithm we have to add edges with weights in increasing order such that the weight of the spanning tree is minimum

In (d) the weights of the edges are 1, 1, 2, 3, 2 which contradicts Kruskal's Algorithm.

50.

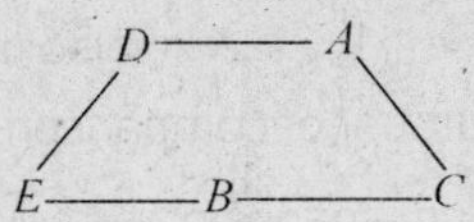

For the above graph, conducting depth First scarch :-

Stack : A

Stack : DBC Tree : A

Stack : EBC Tree : A
/
D

Stack : BC Tree : A
/
D
/
E

Stack : BC

Stack : ϕ Tree :

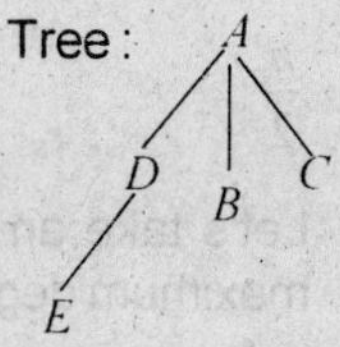

Considering leaf nodes E and C, the given condition is satisfied as their degrees are more than one. But none of $(a), (b)$ or (c) hold

But a cycle consisting of E and all its neighbour hold.

51. The number of push operations wll be minimum when m alternate insert and delete op are performed with $n - m$, incerts in the end.

For each alternate insert and delete op 1 push is needed

$\therefore$ $2m$

Rest of the $n - m$ inserts need $n - m$ push

$\therefore$ Minimum push $= 2m + n^{-n}$

$= m + n$

The number of pop operations will be maximum when n insertions are done consecutively and then m deletions are performed. For the first deletion, first n element have to be popped from S_1, then all the m deletions can be performed each by a single pop.

$\therefore$ Maximum $pop = n + m$

52. The statement inside the for loop is similar to $X - OR$ operation such that the set obtained is $(X \cap Y') \cup (X' \cap Y)$.

It is equivalent to $(X - Y) \cup (Y - X)$.

53. The vertices of degree zero will be those subsets which do not have two elements common with any other subset. This is true only for the subsets of cardinality 0 and 1.

Number. of sets with cardinality $1 = n$

$\therefore$ Number of vertices of degree $0 = n + 1$

54. A little observation of the adjacent vertices of different subsets reveals that the maximum degree will be possessed by a subset of size $\frac{n}{2}$.

Let's take an example for computation of the maximum degree.

Set is $\{1,2,3,4,5,6\}$

Take a subset of size $\frac{n}{2}$ $\{1,2,3\}$

Now, lets compute its degree

The number of adjacent subsets of size $2 = {}^3C_2 = 3$

For each one of $\{1,2\}$ $\{2,3\}$ and $\{1,3\}$ there are 3 subset of size 3 which are adjacent to $\{1,2,3\}$

e.g. $\{1,2,4\},\{1,2,5\},\{1,2,6\}$

$\{2,3,4\},\{2,3,5\},\{2,3,6\}$

Similarly, for each of $\{1,2\},\{2,3\}$ and $\{1,3\}$ there are ${}^3C_2$ subsets of size 4 which are adjacent to $\{1,2,3\}$

There will be one adjacent subset of size 5 for each of the three.

The set itself would not be adjacent to $\{1,2,3\}$

$\therefore$ Degree $=3+3\times3+3\times3+3=24$

$\therefore$ $\left(\frac{n/2}{2}\right)2^{n/2} = {}^3C_2\,2^3 = 3\times8 = 24$

56. In a maximum heap, root in the larger element than all to its children, i.e. in general every level nodes are greater than to the nodes of lower level (s).

In option (a) 1, 3, 5, 6, 8, 9, the heap will be

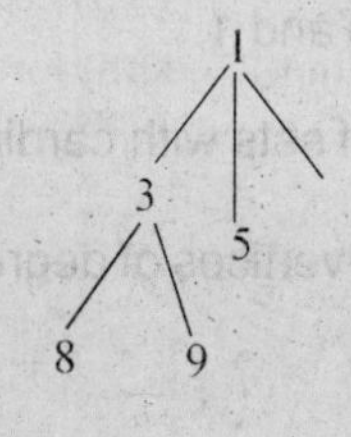

Which is definitely not a maximum heap.

From (b)

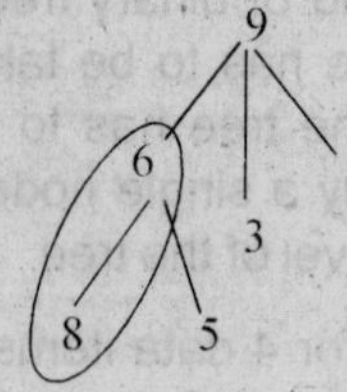

The circled nodes are in wrong position hence its not a maximum heap

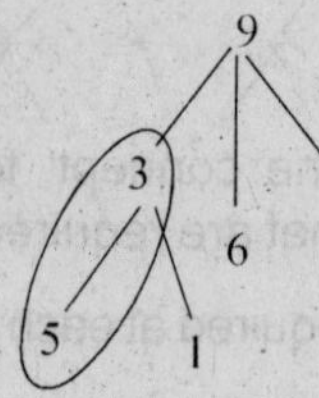

Its again not a maximum heap

From (d)

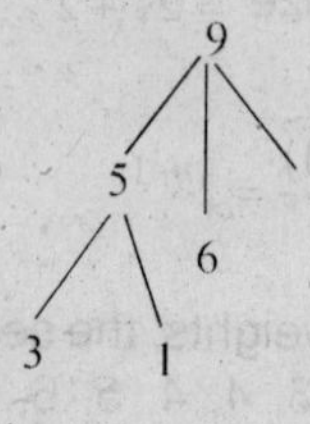

which is definitely or maximum heap.

57. To insert into a maximum heap we add element to last position and then let is raise to correct position, so that the resulting heap is a maximum heap.

Insert 7 in heap 9, 5, 6, 8, 3, 1

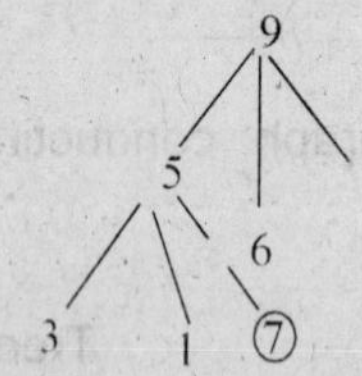

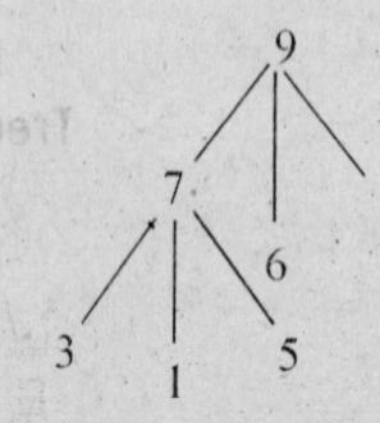

Which is a maximum heap.

Insert - 2

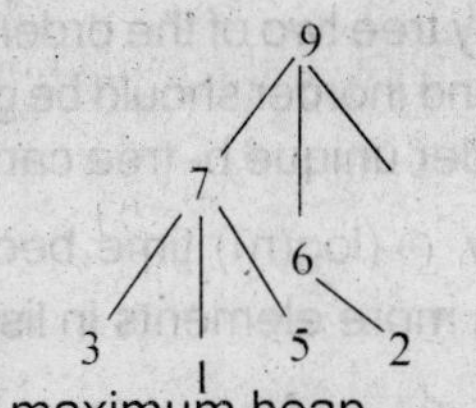

Which is maximum heap.

Insert - 10

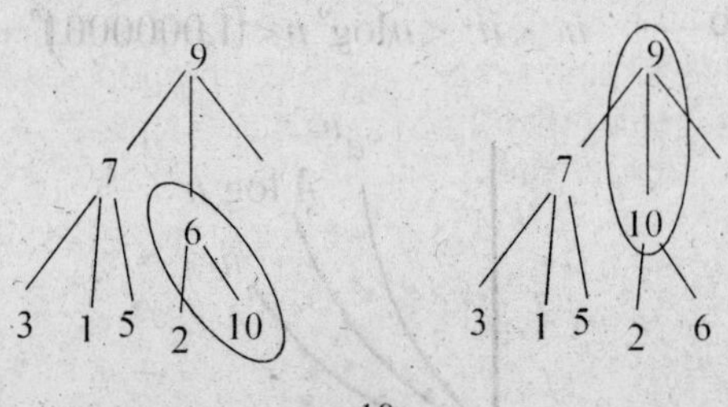

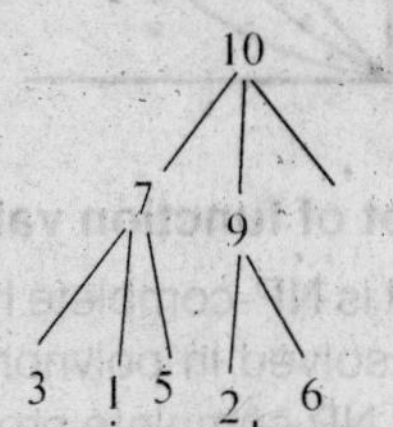

Which is a maximum heap.

Insert - 4

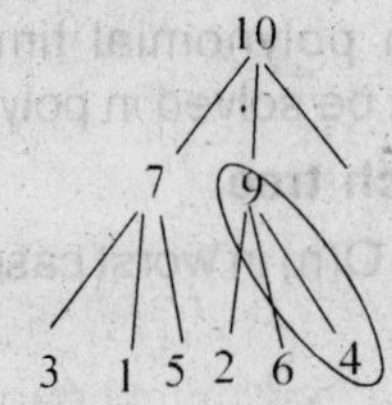

Which is a maximum heap.

Hence the maximum heap after insertion's is
10, 7, 9, 8, 3, 1, 5, 2, 6, 4

58.

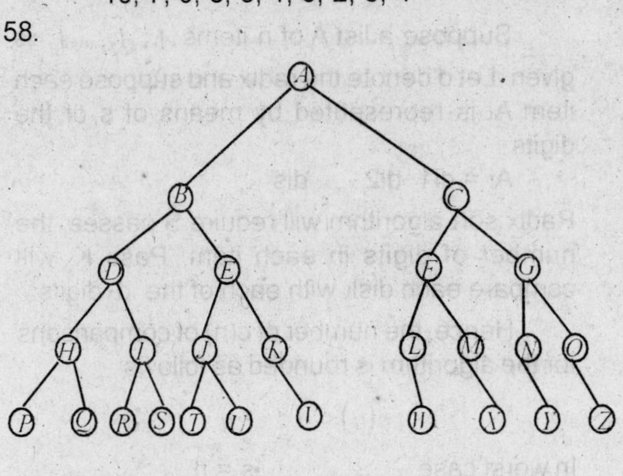

59. Grid connects all the vertices like Hamiltonian cycle but has minimum total weight unlike Hamiltonian cycle.

61. It is called a heap or a maxheap, if each node N has the following property :

The value at N is greater than or equal to the value at each of the children of N.

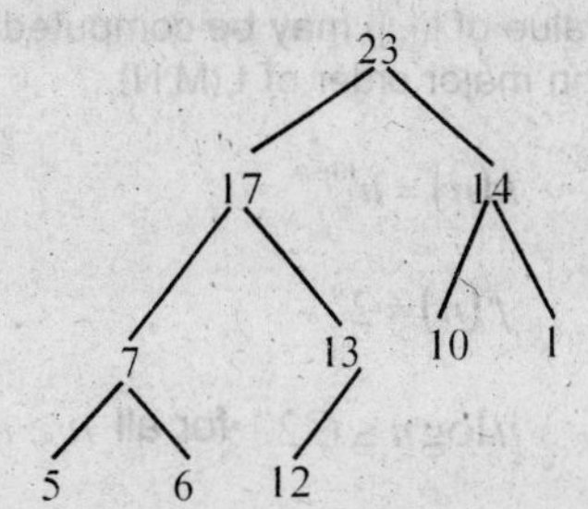

62. The inorder traversal of binary search true is sorted array of elements.

Option (c) doesn't gives the value in sorted order.

63. Directed graph is strongly connected, if there is a path from U to V and V to U.

Removal of R and U will leave the above graph as strongly connected.

64. The discovery time of $Q > P$.

$\therefore$ P's immediate neighbour is Q, after Q is finished visited, P is finished visited and then R is discovered, which clearly shows P and Q are connected.

67. As the root node itself doesnot satisfy the property of heap, all the nodes needs to be visited least once to convert it into heap tree.

$\therefore$ Complexity would by multiple of n.

70. If π_A is NP - hard, then it is NP - compiete

Alternately

π_A be a problem (given) in class NP. we say that π_A is NP complete if the following statements are true about L.

(1) π_A is in NP

(2) For every π_A' in NP there is polynomial time reduction of π_A' to π_A.

71. Applying Mater's Theorem

$cn > n^{\ln 1}$;

$cn > n^0$

Now Checking af(n/b) $\leq$ kf(n) for some k<1

1*cn/3 $\leq$ kcn

This is true for k > 1/3

so the solution is $\theta(n)$.

72. expr 1 ≡ 1 (i - 1 , j - 1) + 1
expr 2 ≡ max (1(i-1,j), 1(i, j-1))

73. Here the value of l(i, j) may be computed row major order or column major order of L(M,N).

75. $h(n) = n^{\log n}$

$f(n) = 2^n$

$n \log n \le C2^n$ for all $n \ge n_o$

C = 1 and $n_o = 2$

$2 \log 2 \le 2^2$

So $h(n) = 0(f(n))$

$g(n) = n!$

$f(n) = 2^n$

$n! \ge C2^n$ for all $n \ge n_o$

C = 1 and n=4

$4! = 24 \ge 2^4$

≥ 16

$g(n) = \Omega(f(n))$

76. To find an integer appears more than n/2 times minimum comparisons are of order O(n).

77. In quicksort a set of number is reduced to sorting two smaller set. We take first element as key value and combine all other with this. A pivot element which splits list into two sublists each of which at least one fifth of element only (B), i.e

$T(n) \le T(n/5) + T(4n/5) + n$ the problem

Alternately

If one sublist contains 1/5 elements other contains 4/5 elements

If T(n) number of comparisons for sorting n elements

So, for 1/5 elements = T(1/5 n)

and for 4/5 elements =T(4n/5)

So, $T(n) \le T(n/5) + T(4n/5) + n$

Here, n = time to spilt

78. Given a set S of n positive integer and a positive integer W. And also a algorithm Q solves in 0(xW) time. So only (b) is false.

79. To find binary tree two of the orders, i.e preorder , postorder and inorder should be given. By using only post order unique b- tree cannot be found.

80. It takes only Θ (log(n)) time because it is not addition of n more elements in list

83. Here, $n < n \log n < e^n$

so $n^{\frac{1}{3}} < n^{\frac{7}{4}} < n \log^9 n < (1.000000)^n <$

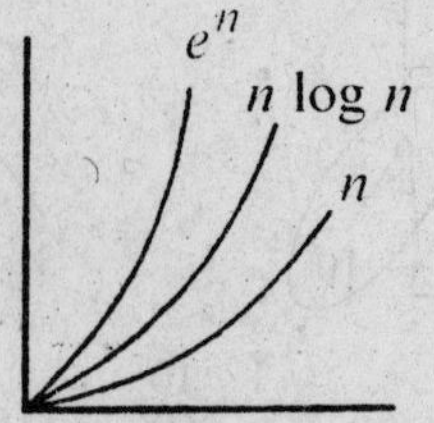

Fig. Plot of function values

84. A problem that is NP-complete has the property that it can be solved in polynomial time if and only if all other NP complete problems can also be solved in polynomial time. If an NP hard can be solved in polynomial time, then all NP complete can be solved in polynomial time.

85. **Binary Search tree.**

Complexity is O(n) in worst case and O(log n) in average case.

AVL Tree

Complexity in worstcase is O(n)

86. Complexity of Radix sort.

Suppose a list A of n items $A_1, A_2 .... A_n$ is given. Let d denote the radix and suppose each item Ai is represented by means of s of the digits :

Ai = di1 di2 .... dis

Radix sort algorithm will require S passes, the number of digits in each item. Pass K. will compare each disk with each of the d digits.

Hence, the number of c(n) of comparisons for the algorithm is rounded as follows:

$$c(n) \le d * s * n$$

In worst case s = n

and d is independent of n, c(n) = n_2

In best case, $s = \log n$,

so $c(n) = 0(n \log n)$

87. $T(n)=\sqrt{2}\,T(n/2)+\sqrt{n}$

$T(1)=1$

Comparing it with $T(n)=aT(n/b)+f(n)$

we get, $a=\sqrt{2}; b=2; f(n)=\sqrt{n}$

$\log_b a=\log_2\sqrt{2}=\frac{1}{2}$

$h(n)=\frac{f(n)}{n^{\log_b a}}$ (given formula)

$=\frac{\sqrt{n}}{\sqrt{n}}=1$

$\therefore\ \mu(n)=\frac{\theta(\log n)^{i+1}}{i+1}$, where i =0

$\therefore\ \mu(n)=\theta(\log n)$

$\therefore\ T(n)=f(n)[T(1)+\mu(n)]$

$=\sqrt{n}[1+\log n]$

90. Since, j increases in power of 2's.

if statement $j = j*2$ executes k times, then

$2^k\le n$

$\Rightarrow k\le\log_2 n$

Since k will be integer, total number of comparison = $\lfloor\log_2 n\rfloor+1$ (when loop exits)

91. Since tree for Max Heap is binary tree and binary search is being performed.

92. Upper bound may be $\sqrt{n}$ times it gets executed.

Lower bound may be that if gets executed only once or twice.

93. If n is a power of 2, then

$j=n+\frac{n}{2}+\frac{n}{2}+........1\log_2 n$ term

If n is not a power of 2, then there will be minor differences of 1 at $\frac{n}{2^i}$ wherever $\frac{n}{2^{i-1}}$ is odd.

Hence the val(j) computed on the basis of $n=2^n$ will give a fair answer

$j=n+\frac{n}{2}+........1\,G.P.$

$$=\frac{n\left(\left(\frac{1}{2}\right)^{\log n}-1\right)}{\frac{1}{2}-1}=2n\left(1-\frac{1}{2^{\log n}}\right)=2n\left(1-\frac{1}{n}\right)$$

96. C_{out} will be generated, if either a_{n-1} *and* b_{n-1} both are one or c_{n-1} comes out to be one in spite of both a_{n-1} and b_{n-1} being zero.

$\therefore$ Required case is designated by,

$a_{n-1}b_{n-1}\overline{C}_{n-1}+\overline{a}_{n-1}+\overline{b}_{n-1}C_{n-1}$

98. After the median is found out in $O(n)$ time, each reduction step divides the list into 2 sublists so that at each of the $\log n$ reduction steps n comparisons will be needed at most.

$\therefore$ Time taken $=O(n)+O(n\log n)$

$=O(n\log n)=\theta(n\log n)$

99. $a_1, a_2 ....a_n$

$b_1, b_2 ....b_n$

In order to find out the largest span.

$\therefore$ Check the sums of $a_i+......a_j$ and $b_i+......b_j$ at each step.

If $a_1+a_2=b_1+b_2$ go on to check $a_1+a_2+a_3$ and $b_1+b_2+b_3$.

If not, then check a_2+a_3 and b_2+b_3

Similarly a check is done at each of the n places during traversal. A separate variable has to be kept that contains the maximum span observed hitherto

$\therefore$ The fastest algorithm computes with $(\sim)(n)$ time and space

100. Only an extra variable $t1$ has been added in work 2 instead of directly computing the subscript as in work 1. The output wll be the same. S1 is true.

The addition of variables $t1$ and $t2$ will not improve the performance in any way. i.e. S2 is false.

101. $S1$ is true S_4 and S5 are false as the program will print 3 and 8. The option left is (a)

TEST YOURSELF - II

1. Suppose the numbers 7, 5, 1, 8, 3, 6, 0, 9, 4, 2 are inserted in that order into an initially empty binary search tree. The binary search tree uses the usual ordering on natural numbers. What is the in-order traversal sequence of the resultant tree ? **[GATE - 03]**

a) 7 5 1 0 3 2 4 6 8 9 b) 0 2 4 3 1 6 5 9 8 7

c) 0 1 2 3 4 5 6 7 8 9 d) 9 8 6 4 2 3 0 1 5 7

2. Consider the following three claims **[GATE - 03]**

I. $(n+k)^m = \Theta(n^m)$, where k and m are constants

II. $2^{n+1} = O(2^n)$ III. $2^{2n} = O(2^n)$

Which of these claims are correct ?

a) I and II b) I and III

c) II and III d) I, II and III

3. Consider the following graph

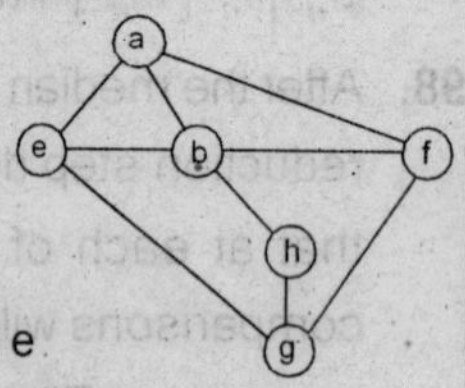

Among the following sequences

I. a b e g h f

II. a b f e h g

III. a b f h g e IV. a f g h b e

Whcih are depth first traversals of the above graph ?

a) I, II and IV only b) I and IV only **[GATE - 03]**

c) II, III and IV only d) I, III and IV only

4. The usual $\Theta(n^2)$ implementation of Insertion Sort to sort an array uses linear search to identify the position where an element is to be inserted into the already sorted part of the array. If, instead, we use binary search to identify the position, the worst case running time will **[GATE - 03]**

a) remain $\Theta(n^2)$ b) become $\Theta(n(\log n)^2)$

c) beocme $\Theta(n \log n)$ d) become $\Theta(n)$

5. In a heap n elements with the smallest element at the root, the 7th smallest element can be found in time **[GATE - 03]**

a) $\Theta(n\log n)$ b) $\Theta(n)$ c) $\Theta(\log n)$ d) $\Theta(1)$

The following information pertains to Q.6-7

In a permutation $a_1.....a_n$ of n distinct integers, an inversion is a pair (a_i, a_j) such that $i < j$ and $a_i > a_j$

6. If all permutations are equally likely, what is the expected number of inversions in a randomly chosen permutation of 1.....n ? **[GATE - 03]**

a) $n(n-1)/2$ b) $n(n-1)/4$

c) $n(n+1)/4$ d) $2n\lfloor \log_2 n \rfloor$

7. What would be the worst case time complexity of the Insertion Sort algorithm, if the inputs are restricted to permutations of 1........n with atmost n inversions ?

a) $\Theta(n^2)$ b) $\Theta(n\log)$ c) $\Theta(n^{1.5})$ d) $\Theta(n)$

8. A data structure is required for storing a set of integers such that each of the following operations can be done in O(log n) time, where n is the number of elements in the set. **[GATE - 03]**

I Deletion of the smallest element

II Insertion of an element if it is not already present in the set.

Which of the following data structures can be used for this purpose ?

a) A heap can be used but not a balanced binary search tree

b) A balanced binary search tree can be used but not a heap

c) Both balanced binary search tree and heap can be used

d) Neither balanced binary search tree nor heap can be used

9. Let S be a stack of size $n \geq 1$. Starting with the empty stack, suppose we push the first n natural numbers in sequence, and then perform n pop operations. Assume that Push and Pop operations take X seconds each, and Y seconds elapse between the end of one such stack operation and the start of the next operation. For $m \geq 1$, definie the stack-life of m as the time elapsed from the end of Push(m) to the start of the pop operation that removes m from S. The average stack-life of an element of this stack is **[GATE - 03]**

a) $n(X+Y)$ b) 3Y+2X

c) n(X+Y) - X

d) Y + 2X

10. Consider the following 2-3-4 tree (i.e. B-tree with a minimum degree of two) in which each data item is a letter. The usual alphabetical ordering of letters is used in constructing the tree. What is the result of inserting g in the above tree ? **[GATE - 03]**

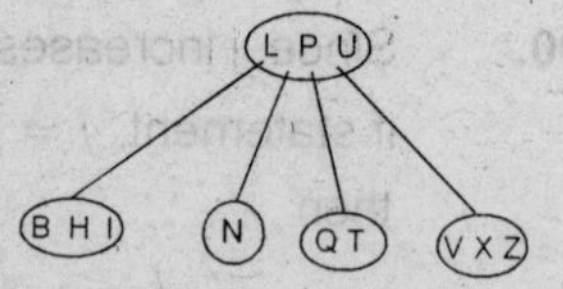

a)

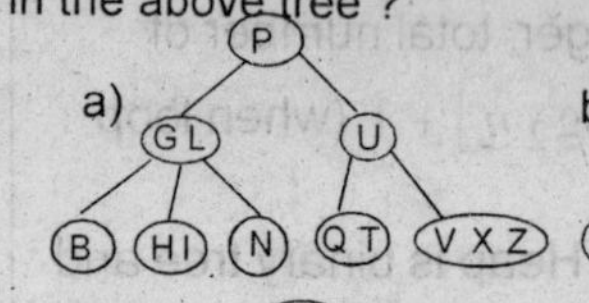

b)

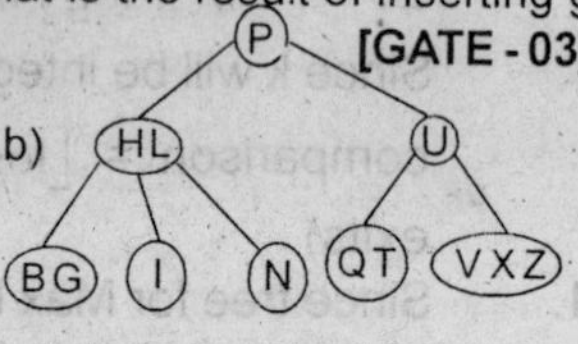

c)

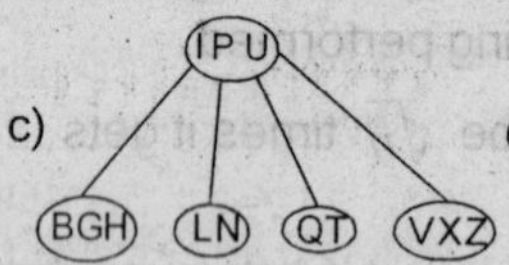

d) None of the above

11. The cube root of a natural number n is defined as the largest natural number m such that $m^3 \leq n$. The complexity of computing the cube root of n (n is represented in binary notation) is **[GATE - 03]**

a) O(n) but not $O(n^{0.5})$

b) $O(n^{0.5})$ but not $O((\log n)^k)$ for any constant $k > 0$

c) $O((\log n)^k)$ for some constant k>0, but not $O((\log \log n)^m)$ for any constant $m > 0$

d) $O((\log \log n)^k)$ for some constant k > 0.5 ; but not $O((\log \log n)^{0.5})$

12. Let G = (V, E) be an undirected graph with a subgraph $G_1 = (V_1, E_1)$. Weights are assigned to edges of G as follows.
[GATE - 03]

$$w(e) = \begin{cases} 0 \text{ if } e \in E_1 \\ 1 \text{ otherwise} \end{cases}$$

A single source shortest path algorithm is executed on the weighted graph (V, E, w) with an arbitrary vertex v_1 of V_1 as the source. Which of the following can always be inferred from the path costs computed ?

a) The number of edges in the shortest paths from v_1 to all vertices of G
b) G_1 is connected.
c) V_1 forms a clique in G d) G_1 is a tree.

13. What is the weight of a minimum spanning tree of the following graph ? **[GATE - 03]**

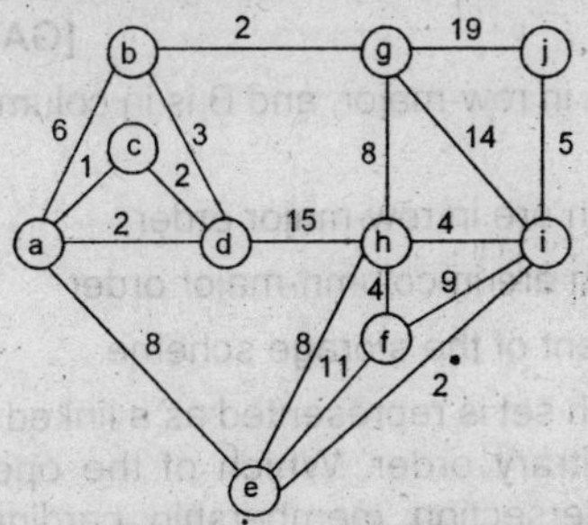

a) 29 b) 31 c) 38 d) 41

14. The following are the starting and ending times of activities A, B, C, D, E, F, G and H respectively in chronological order.
"$a_s b_s c_s d_s c_e e_s f_s b_e d_e g_s e_e f_e h_s g_e h_e$". Here, x_s denotes the starting time and x_e denotes the ending time of activity X. We need to schedule the activities in a set of rooms available to us. An activity can be scheduled in a room only if the room is reserved for the activity for its entire duration. What is the minimum number of rooms required ? **[GATE - 03]**

a) 3 b) 4 c) 5 d) 6

15. Let G=(V, E) be a directed graph with n vertices. A path from v_i to v_j in G is a sequence of vertices $(v_i, v_{i+1}, ..., v_j)$ such that $(v_k, v_{k+1}) \in E$ for all k in i through j - 1. A simple path is a path in which no vertex appears more than once. **[GATE - 03]**

Let A be an $n \times n$ array initialized as follows.

$$A[j, k] = \begin{cases} 1 \text{ if } (j, k) \in E \\ 0 \text{ otherwise} \end{cases}$$

Consider the following algorithm

```
for i = 1 to n
   for j = 1 to n
      for k = 1 to n
```

$$A[j, k] = \max\left(A[j, k], A[j, i] + A[i, k]\right);$$

which of the following statements is necessarily TRUE for all j and k after termination of the above algorithm ?

a) $A[j,k] \le n$
b) If $A[j,j] \ge n-1$, then G has a Hamiltonian cycle
c) If there exists a path from j to k, A [j, k] contains the longest path length from j to k
d) If there exists a path from j to k, every simple path from j to k contains at most A [j, k] edges.

16. Consider the function f defined below.

```
Struct item {
      int data ;
      struct item * next ;
}
int f(struct item*p)          {
   return ((p==NULL || (p - > next ==NULL)    ||
         (( p->data<=p->next->data) &&
            f(p->next) ) ) ;
```

For a given linked list p, the function f returns 1 if and only if **[GATE - 03]**

a) the list is empty or has exactly one element
b) the elements in the list are sorted in non-decreasing order of data value
c) the elements in the list are sorted in non-increasing order of data value
d) not all elements in the list have the same data value.

17. A single array A [1..MAXSIZE] is used to implement two stacks. The two stacks grow from opposite ends of the array. Variables top 1 and top 2 (top 1< top 2) point to the location of the topmost element in each of the stacks. If the space is to be used efficiently, the condition for "stack full" is **[GATE - 04]**

a) (top1 = MAXSIZE/2) and (top2 = MAXSIZE/2+1)
b) top1 + top2 = MAXSIZE
c) (top1 = MAXSIZE/2) or (top2 = MAXSIZE)
d) top1 = top2 -1

18. The following numbers are inserted into an empty binary search tree in the given order : 10, 1, 3, 5, 15, 12, 16. What is the height of the binary search tree (the height is the maximum distance of a leaf node from the root) ? **[GATE - 04]**

a) 2 b) 3 c) 4 d) 6

19. The best data structure to check whether an arithmetic expression has balanced parentheses is a **[GATE - 04]**

a) queue b) stack c) tree d) list

20. Level order traversal of a rooted tree can be done by starting from the root and performing **[GATE - 04]**

a) preorder traversal b) inorder traversal
c) depth first search d) breadth first search

21. Given the following input (4322, 1334, 1471, 9679, 1989, 6171, 6173, 4199) and the hash function x mod 10, which of the following statements are true ? **[GATE - 04]**

i) 9679, 1989, 4199 hash to the same value
ii) 1471, 6171 hash to the same value
iii) All elements hash to the same value
iv) Each element hashes to a different value

a) i) only b) ii) only

c) i) and ii) only d) iii) or iv) only

22. The tightest lower bound on the number of comparisons, in the worst case, for comparison-based sorting is of the order of **[GATE - 04]**

a) n b) n^2 c) nlogn d) $n\log^2 n$

23. The problems 3-SAT and 2-SAT are **[GATE - 04]**

a) both in P b) both NP-complete

c) NP-complete and in P respectively

d) undecidable and NP-complete respectively

24. Consider the label sequences obtained by the following pairs of traversals on a labeled binary tree. Which of these pairs identify a tree uniquely ? **[GATE - 04]**

i) preorder and postorder

ii) inorder and postorder

iii) preorder and inorder

iv) level order and postorder

a) i) only b) ii), iii) c) iii) only d) iv) only

25. A circularly linked list is used to represent a Queue. A single variable p is used to access the Queue. To which node should p point such that both the operations enQueue and deQueue can be performed in constant time ?

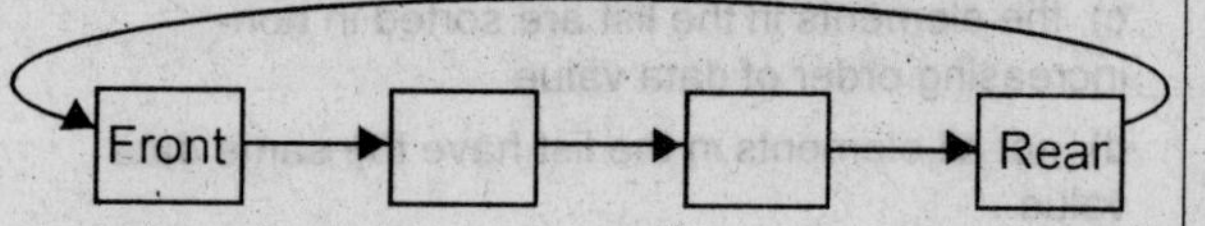

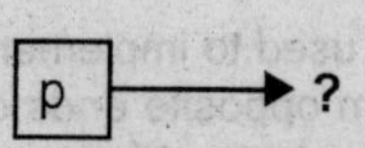

a) rear node b) front node **[GATE - 04]**

c) not possible with a single pointer

d) node next to front

26. The elements 32, 15, 20, 30, 12, 25, 16 are inserted one by one in the given order into a maxHeap. The resultant maxHeap is **[GATE - 04]**

a)

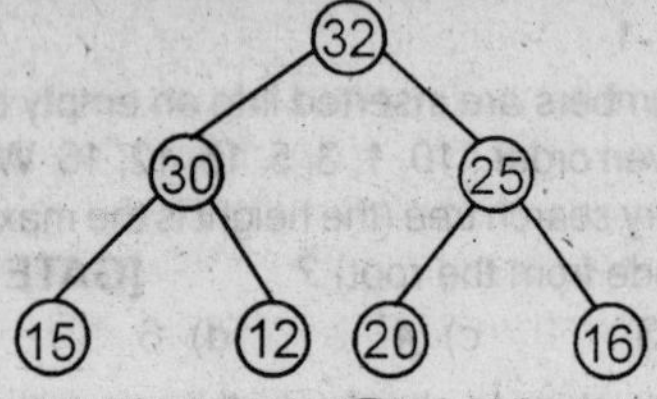

b)

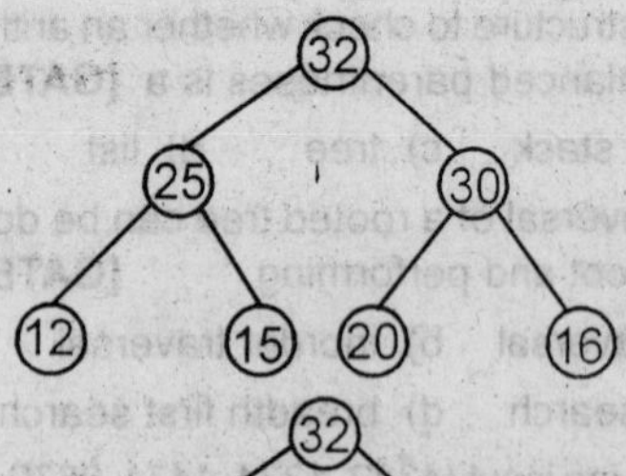

c)

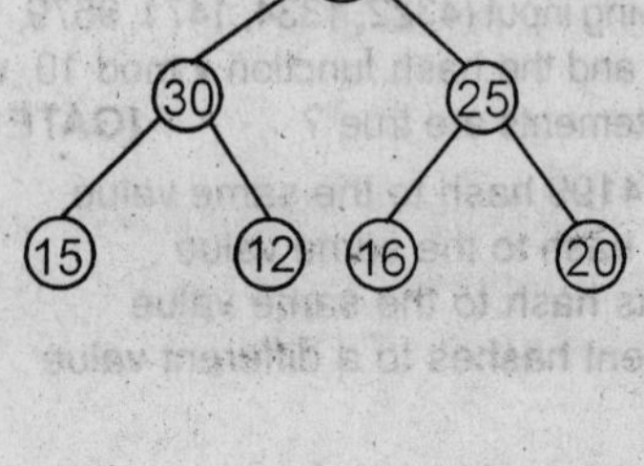

d)

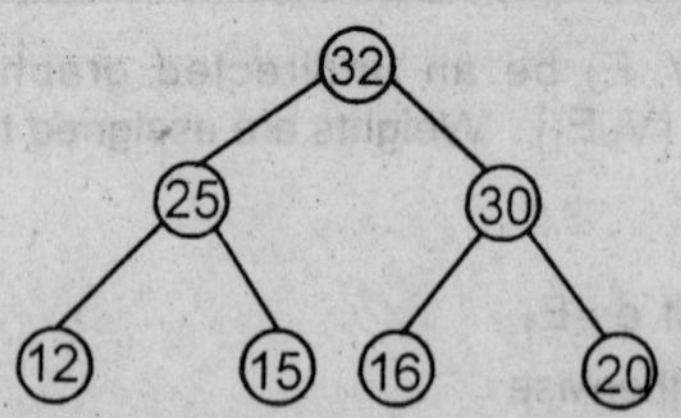

27. Assume that the operators +, -, × are left associative and ∧ is right associative. The order of precedence (from highest to lowest) is ∧, ×, +, -. The postfix expression corresponding to the infix expression $a + b \times c - d$^e^f is **[GATE - 04]**

a) abc×+def^^− b) abc×+de^f^−

c) ab+c×d−e^f^ d) −+a×bc^^def

28. Two matrices M_1 and M_2 are to be stored in arrays A and B respectively. Each array can be stored either in row-major or column-major order in contiguous memory locations. The time complexity of an algorithm to compute $M_1 \times M_2$ will be **[GATE - 04]**

a) best if A is in row-major, and B is in column-major order

b) best if both are in row-major order

c) best if both are in column-major order

d) independent of the storage scheme

29. Suppose each set is represented as a linked list with elements in arbitrary order. Which of the operations among union, intersection, membership, cardinality will be the slowest ? **[GATE - 04]**

a) union only b) intersection, membership

c) membership, cardinality

d) union, intersection

30. Consider the following C program

```
main ( )
    { int x, y, m, n ;
      scanf ("%d %d", &x, &y) ;
      / * Assume x>0 and y>0 * /
      m = x ; n = y ;
      while (m!=n)
       { if (m > n)
           m = m - n ;
         else
           n = n - m ;
       }
      printf("%d", n);
    }
```

The program computes **[GATE - 04]**

a) x ÷ y using repeated subtraction

b) x mod y using repeated subtraction

c) the greatest common divisor of x and y

d) the least common multiple of x and y

31. What does the following algorithm approximate ? (Assume m > 1, ∈ >0). **[GATE - 04]**

```
x = m ;
y = 1 ;
while (x - y > ∈ )
    { x = (x + y) / 2 ;
      y = m / x ;
    }
print (x) ;
```

a) log m b) m^2 c) $m^{1/2}$ d) $m^{1/3}$

32. Consider the following C program segment

```
struct CellNode {
  struct CellNode *leftChild ;
  int element ;
  struct CellNode *rightChild ;
  };
int DoSomething (struct CellNode *ptr)
{
  int value = 0 ;
  if (ptr ! = NULL)
    { if (prt -> leftChild != NULL)
        value = 1 + DoSomething (ptr-> leftChild) ;
      if (ptr-> rightChild ! = NULL)
Value = max (value, 1 + DoSomething (ptr - > rightChild));
    }
    return (value) ;
}
```

The value returned by the function DoSomething when a pointer to the root of a non-empty tree is passed as argument is **[GATE - 04]**

a) The number of leaf nodes in the tree
b) The number of nodes in the tree
c) The number of internal nodes in the tree
d) The height of the tree

33. Suppose we run Dijkstra's single source shortest-path algorithm on the following edge-weighted directed graph with vertex P as the source.

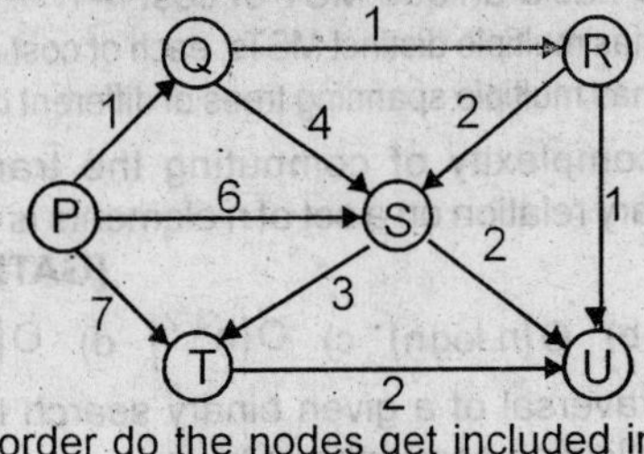

In what order do the nodes get included into the set of vertices for which the shortest path distances are finalized? **[GATE - 04]**

a) P, Q, R, S, T, U b) P, Q, R, U, S, T
c) P, Q, R, U, T, S d) P, Q, T, R, U, S

34. Let $G_1 = (V, E_1)$ and $G_2 = (V, E_2)$ be connected graphs on the same vertex set V with more than two vertices. If $G_1 \cap G_2 = (V, E_1 \cap E_2)$ is not a connected graph, then the graph $G_1 \cup G_2 = (V, E_1 \cup E_2)$ **[GATE - 04]**

a) cannot have a cut vertex
b) must have a cycle
c) must have a cut-edge (bridge)
d) has chromatic number strictly greater than those of G_1 and G_2

35. Let $A[1 \ldots\ldots n]$ be an array storing a bit (1 or 0) at each location, and $f(m)$ is a function whose time complexity is $\theta(m)$. Consider the following program fragment written in a C like language: **[GATE - 04]**

```
counter = 0 ;
for (i=1; i<=n ; i++)
{ if (A [i] ==1) counter ++
   else {f (counter) ; counter =0 ; }
}
```

The complexity of this program fragment is

a) $\Omega(n^2)$ b) $\Omega(n \log n)$ and $O(n^2)$
c) $\theta(n)$ d) $o(n)$

36. The time complexity of the following C function is (assume n>0) **[GATE - 04]**

```
int recursive (int n) {
   if (n == 1)
     return (1) ;
   else
 return (recursive (n-1) + recursive (n-1)) ;
}
```

a) $O(n)$ b) $O(n \log n)$ c) $O(n^2)$ d) $O(2^n)$

37. The recurrence equation

$T(1) = 1$

$T(n) = 2T(n-1) + n, n \geq 2$

evaluates to **[GATE - 04]**

a) $2^{n+1} - n - 2$ b) $2^n - n$
c) $2^{n+1} - 2n - 2$ d) $2^n + n$

38. A program takes as input a balanced binary search tree with n leaf nodes and computes the value of a function g(x) for each node x. If the cost of computing g(x) is min{number of leaf-nodes in left-subtree of x, number of leaf-nodes in right-subtree of x} then the worst-case time complexity of the program is **[GATE - 04]**

a) $\Theta(n)$ b) $\Theta(n \log n)$ c) $\Theta(n^2)$ d) $\Theta(n^2 \log n)$

39. Let P be a singly linked list. Let Q be the pointer to an intermediate node x in the list. What is the worst-case time complexity of the best known algorithm to delete the node x from the list ? **[GATE - 04]**

a) $O(n)$ b) $O(\log^2 n)$
c) $O(\log n)$ d) $O(1)$

40. Which one of the following is NOT shared by the threads of the same process ? **[GATE - 04]**

a) Stack b) Address Space
c) File Descriptor Table d) Message Queue

41. A program attempts to generate as many permutations as possible of the string 'abcd' by pushing the characters a, b, c, d in the same order onto a stack, but it may pop off the top character at any time. Which one of the following strings CANNOT be generated using this program ? **[GATE - 04]**

a) abcd b) dcba c) cbad d) cabd

42. An array of integers of size n can be converted into a heap by adjusting the heaps rooted at each internal node of the complete binary tree starting at the node $\lfloor (n-1)/2 \rfloor$, and doing this adjustment up to the root node (root node is at index 0) in the order $\lfloor (n-1)/2 \rfloor, \lfloor (n-3)/2 \rfloor, ..., 0$. The time required to construct a heap in this manner is

a) $O(\log n)$ b) $O(n)$ **[GATE - 04]**

c) $O(n \log \log n)$ d) $O(n \log n)$

43. Which one of the following binary trees has its inorder and preorder traversals as BCAD and ABCD, respectively? **[GATE - 04]**

a) 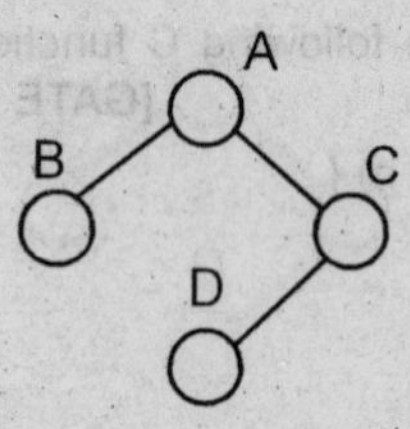

b) 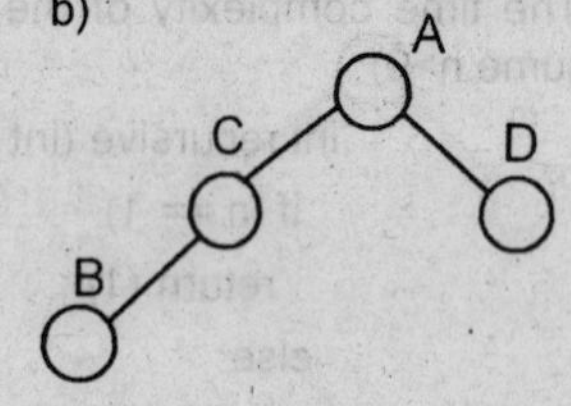

c) 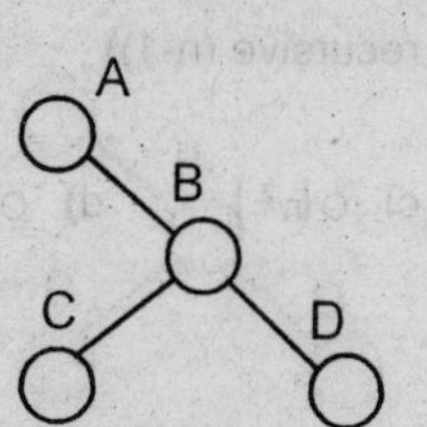

d) 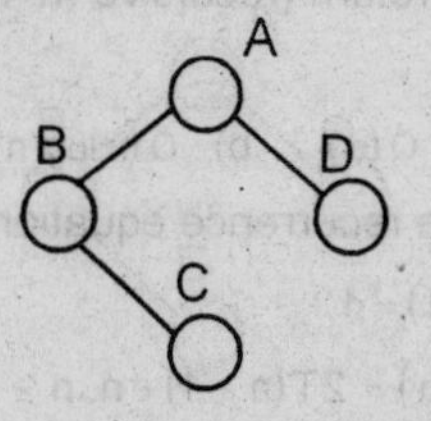

44. Let $f(n), g(n)$ and $h(n)$ be functions defined for positive integers such that **[GATE - 04]**

$f(n) = O(g(n)), g(n) \neq O(f(n)), g(n) = O(h(n))$, and $h(n) = O(g(n))$

Which one of the following statements is FALSE ?

a) $f(n) + g(n) = O(h(n) + h(n))$

b) $f(n) = O(h(n))$

c) $h(n) \neq O(f(n))$

d) $f(n)h(n) \neq O(g(n)h(n))$

45. Consider the undirected graph below :

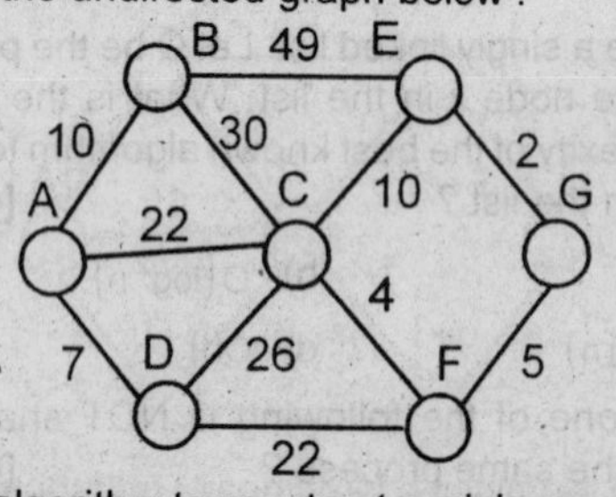

Using Prim's algorithm to construct a minimum spanning tree starting with node A, which one of the following sequences of edges represents a possible order in which the edges would be added to construct the minimum spanning tree ? **[GATE - 04]**

a) (E,G), (C,F), (F,G), (A,D), (A,B), (A,C)

b) (A,D), (A,B), (A,C), (C,F), (G,E), (F,G)

c) (A,B), (A,D), (D,F), (F,G), (G,E), (F,C)

d) (A,D), (A,B), (D,F), (F,C), (F,G), (G,E)

46. Consider a list of recursive algorithms and a list of recurrence relations as shown below. Each recurrence relation corresponds to exactly one algorithm and is used to derive the time complexity of the algorithm.

Recursive Algorithm	Recurrence Relation
P. Binary search	I. $T(n) = T(n-k) + T(k) + cn$
Q. Merge sort	II. $T(n) = 2T(n-1) + 1$
R. Quick sort	III. $T(n) = 2T(n/2) + cn$
S. Tower of Hanoi	IV. $T(n) = T(n/2) + 1$

Which of the following is the correct match between the algorithms and their recurrence relations ? **[GATE - 04]**

a) P-II, Q-III, R-IV, S-I b) P-IV, Q-III, R-I, S-II

c) P-III, Q-II, R-IV, S-I d) P-IV, Q-II, R-I, S-III

47. A program P reads in 500 integers in the range (0, 100) representing the scores of 500 students. It then prints the frequency of each score above 50. What would be the best way for P to store the frequencies ?

a) An array of 50 numbers **[GATE - 05]**

b) An array of 100 numbers

c) An array of 500 numbers

d) A dynamically allocated of 550 numbers

48. An undirected graph G has n nodes. Its adjacency matrix is given by an n x n square matrix whose

(i) diagonal elements are 0's and **[GATE - 05]**

(ii) non-diagonal elements are 1's. Which one of the following is TRUE ?

a) Graph G has no minimum spanning tree (MST)

b) Graph G has a unique MST of cost n-1

c) Graph G has multiple distinct MSTs, each of cost n-1

d) Graph G has multiple spanning trees of different costs.

49. The time complexity of computing the transitive closure of a binary relation on a set of n elements is known to be **[GATE - 05]**

a) $O(n)$ b) $O(n \log n)$ c) $O(n^{3/2})$ d) $O(n^3)$

50. Postorder traversal of a given binary search tree, T produces the following sequence of keys

10, 9, 23, 22, 27, 25, 15, 50, 95, 60, 40, 29

Which one of the following sequences of keys can be the result of an inorder traversal of the tree T ? **[GATE - 05]**

a) 9,10, 15, 22, 23, 25, 27, 29, 40, 50, 60, 95

b) 9, 10, 15, 22, 40, 50, 60, 95, 23, 25, 27, 29

c) 29, 15, 9, 10, 25, 22, 23, 27, 40, 60, 50, 95

d) 95, 50, 60, 40, 27, 23, 22, 25, 10, 9, 15, 29

51. A Priority-Queue is implemented as a Max-Heap. Initially, it has 5 elements. The level-order traversal of the heap is given below : **[GATE - 05]**

10, 8, 5, 3, 2

Two new elements '1' and '7' are inserted in the heap in that order. The level order traversal of the heap after the insertion of the elements is

a) 10, 8, 7, 5, 3, 2, 1 b) 10, 8, 7, 2, 3, 1, 5
c) 10, 8, 7, 1, 2, 3, 5 d) 10, 8, 7, 3, 2, 1, 5

52. How many distinct binary search trees can be created out of 4 distinct keys ? **[GATE - 05]**

a) 5 b) 14 c) 24 d) 42

53. In a complete k-ary tree, every internal node has exactly k children. The number of leaves in such a tree with n internal nodes is **[GATE - 05]**

a) nk b) $(n-1)k+1$ c) $n(k-1)+1$ d) $n(k-1)$

54. Suppose $T(n)=2T(n/2)+n$, $T(0)=T(1)=1$

Which one of the following is FALSE ? **[GATE - 05]**

a) $T(n)=O(n^2)$ b) $T(n)=\theta(n\log n)$
c) $T(n)=\Omega(n^2)$ d) $T(n)=O(n\log n)$

55. Let G (V,E) an undirected graph with positive edge weights. Dijkstra's single source shortest path algorithm can be implemented using the binary heap data structure with time complexity : **[GATE - 05]**

a) $O(|V|^2)$ b) $O(|E|+|V|\log|V|)$
c) $O(|V|\log|V|)$ d) $O((|E|+|V|)\log|V|)$

56. Suppose there are log n sorted lists of n log n elements each. The time complexity of producing a sorted list of all these elements is : **[GATE - 05]**

(Hint : Use a heap data structure)

a) $O(n\log\log n)$ b) $\theta(n\log n)$
c) $\Omega(n\log n)$ d) $\Omega(n^{3/2})$

Statement for Linked Answer Questions 57 and 58

Consider the following C - function : **[GATE - 05]**

```
double foo (int n) {
    int i;
    double sum;
    if (n==0) return 1.0;
    else {
        Sum = 0.0;
        for (i =0; i < n; i++)
            sum + = foo (i) ;
        return sum;
    }}
```

57. The space complexity of the above function is

a) O(1) b) O(n) c) O(n!) d) $O(n^n)$

58. Suppose we modify the above function foo ( ) and store the values of foo (i), 0 < = i < n, as and when they are computed. With this modification, the time complexity for function foo ( ) is significantly reduced. The space complexity of the modified function would be :

a) O(1) b) O(n) c) $O(n^2)$ d) O(n!)

Statement for Linked Answer Questions 59 and 60

Let s and t be two vertices in a undirected graph G + (V, E) having distinct positive edge weights. Let (X, Y) be a partition of V such that $s \in X$ and $t \in Y$. Consider the edge e having the minimum weight amongst all those edges that have one vertex in X and one vertex in Y.

59. The edge e must difinitely belong to : **[GATE - 05]**

a) the minimum weighted spanning tree of G
b) the weighted shortest path from s to t
c) each path from s to t
d) the weighted longest path from s to t

60. Let the weight of an edge e denote the congestion on that edge. The congestion on a path is defined to be the maximum of the congestions on the edges of the path. We wish to find the path from s to t having minimum congestion. Which one of the following paths is always such a path of minimum congestion ?

a) a path from s to t in the minimum weighted spanning tree
b) a weighted shortest path from s to t
c) an Euler walk from s to t
d) a Hamiltonian path from s to t

61. The numbers 1,2, ... n are inserted in a binary search tree in some order. In the resulting tree, the right subtree of the root contains p nodes. The first number to be inserted in the tree must be **[GATE - 05]**

a) p b) p + 1 c) n - p d) n - p + 1

62. A function f defined on stacks of integers satisfies the following properties. $f(\phi)=0$ and f(push (S, i)) = max (f(S),0) + i for all stacks S and integers i. If a stack S contains the integers 2, -3, 2, -1, 2 in order from bottom to top, what is f(S) ? **[GATE - 05]**

a) 6 b) 4 c) 3 d) 2

63. In a depth-first traversal of a graph G with n vertices, k edges are marked as tree edges. The number of connected components in G is **[GATE - 05]**

a) k b) k + 1 c) n - k -1 d) n - k

64. In the following table, the left column contains the names of standard graph algorithms and the right column contains the time complexities of the algorithms. Match each algorithm with its time complexity. **[GATE - 05]**

1. Bellman-Ford algorithm	A : O (mlog n)
2. Kruskal's algorithm	B : $O(n^3)$
3. Floyd-Warshall algorithm	C : O (nm)
4. topological Sorting	D : O (n + m)

a) $1\to C, 2\to A, 3\to B, 4\to D$
b) $1\to B, 2\to D, 3\to C, 4\to A$
c) $1\to C, 2\to D, 3\to A, 4\to B$
d) $1\to B, 2\to A, 3\to C, 4\to D$

65. A hash table contains 10 buckets and uses linear probing to resolve collisions. The key values are integers and the hash function used is key % 10. If the values 43, 165, 62, 123, 142 are inserted in the table, in what location would the key value 142 be inserted ? **[GATE - 05]**

a) 2 b) 3 c) 4 d) 6

66. In a binary tree, for every node the difference between the number of nodes in the left and and right subtrees is at most 2. If the height of the tree is h>0, then the minimum number of nodes in the tree is **[GATE - 05]**

a) 2^{h-1} b) $2^{h-1}+1$ c) 2^h-1 d) 2^h

67. Let T(n) be a function defined by the recurrence

$T(n) = 2T(n/2) + \sqrt{n}$ for $n \geq 2$ and **[GATE - 05]**

$T(1) = 1$

Which of the following statements is TRUE ?

a) $T(n) = O(\log n)$ b) $T(n) = \theta(\sqrt{n})$

c) $T(n) = \theta(n)$ d) $T(n) = \theta(n \log n)$

68. Let G be a weighted undirected graph and e be an edge with maximum weight in G. Suppose there is a minimum weight spanning tree in G containing the edge e. Which of the following statements is always TRUE ? **[GATE - 05]**

a) There exists a cutset in G having all edges of maximum weight.

b) There exists a cycle in G having all edges of maximum weight.

c) Edge e cannot be contained in a cycle.

d) All edges in G have the same weight.

69. The following C function takes a single-linked list of integers as a parameter and rearranges the elements of the list. The list is represented as pointer to a structure. The function is called with the list containing the integers 1,2,3,4,5,6,7 in the given order. What will be the contents of the list after the function completes execution ?

```
struct node {int value; struct node*next;};
void rearrange (struct node*list) {
   struct node *p, *q;
   int temp;
   if ( ! list  II !list → next) return;
   p = list; q =list → next;
   while (q) {
   temp = p → value;
   p → value = q → value;
   q → value = temp;
   p = q → next;
   q = q → next;
      }
  }
```

[GATE - 05]

a) 1, 2, 3, 4, 5, 6, 7 b) 2, 1, 4, 3, 6, 5, 7

c) 1, 3, 2, 5, 4, 7, 6 d) 2, 3, 4, 5, 6, 7, 1

70. A binary search tree contains the numbers 1, 2, 3, 4, 5, 6, 7, 8. When the tree is traversed in pre-order and the values in each node printed out, the sequence of values obtained is 5, 3, 1, 2, 4, 6, 8, 7. If the tree is traversed in postorder, the sequence obtained would be **[GATE - 05]**

a) 8, 7, 6, 5, 4, 3, 2, 1 b) 1, 2, 3, 4, 8, 7, 6, 5

c) 2, 1, 4, 3, 6, 7, 8, 5 d) 2, 1, 4, 3, 7, 8, 6, 5

Statement for Linked Answer Questions 71 and 72

A sink in a directed graph is a vertex i such that there is an edge from every vertex $j \neq i$ to i and there is no edge from i to any other vertex. A directed graph G with n vertices is represented by its adjacency matrix A, where A [i] [j] = 1 if there is an edge directed from vertex i to j and 0 otherwise. The following algorithm determines whether there is a sink in the graph G. **[GATE - 05]**

```
i = 0;
do {
      j = i + 1;
      while (( j < n) && E1) j++;
      if ( j < n) E2;
} while ( j < n)
flag = 1 ;
for ( j = 0; j < n; j++)
if ( j! = i)&& E3) flag = 0;
if (flag) printf("Sink exists") else printf ("Sink does not exist");
```

71. Choose the correct expressions for E_1 and E_2

a) $E_1 : A[i][j]$ and $E_2 : i = j;$

b) $E_1 : !A[i][j]$ and $E_2 : i = j+1;$

c) $E_1 : !A[i][j]$ and $E_2 : i = j;$

d) $E_1 : A[i][j]$ and $E_2 : i = j+1;$

72. Choose the correct expression for E_3

a) (A [i] [j] &&!A [j] [i]) b) (!A [i] [j] && !A [j] [i])

c) (!A [i] [j] | | A [j] [i]) d) (A [i] [j] | | !A [j] [i])

Statement for Linked Answer Questions 73 and 74

Consider a simple graph with unit edge costs. Each node in the graph represents a router. Each node maintains a routing table indicating the next hop router to be used to relay a packet to its destination and the cost of the path to the destination through that router. Initially, the routing table is empty. The routing table is synchronously updated as follows. In each updation interval, three tasks are performed. **[GATE - 05]**

(i) A node determines whether its neighbours in the graph are accessible. If so, it sets the tentative cost to each accessible neighbour as 1. Otherwise, the cost is set to ∞.

(ii) From each accessible neighbour, it gets the costs to relay to other nodes via that neighbour (as the next hop).

(iii) Each node updates its routing table based on the information received in the previous two steps by choosing the minimum cost.

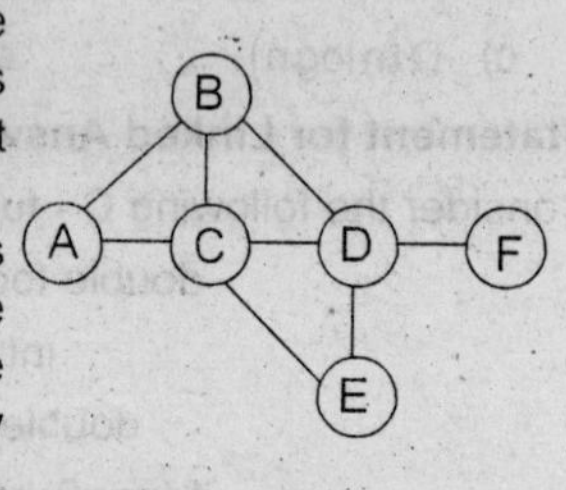

73. For the graph given above, possible routing tables for various nodes after they have stabilized, are shown in the following options. Identify the correct table.

(a) Table for node A

A	-	-
B	B	1
C	C	1
D	B	3
E	C	3
F	C	4

(b) Table for node C

A	A	1
B	B	1
C	-	-
D	D	1
E	E	1
F	E	3

(c) Table from node B

A	A	1
B	-	-
C	C	1
D	D	1
E	C	2
F	D	2

(d) Table for node D

A	B	3
B	B	1
C	C	1
D	-	-
E	E	1
F	F	1

74. Continuing from the earlier problem, suppose at some time t, when the costs have stabilized, node A goes down. The cost from node F to node A at time (t+100) is:

a) >100 but finite b) ∞

c) 3 d) >3 and $\leq$ 100

ANSWER KEY

1. b	**2.** a	**3.** d	**4.** c	**5.** c
6. b	**7.** b	**8.** b	**9.** d	**10.** c
11. c	**12.** b	**13.** b	**14.** b	**15.** d
16. d	**17.** d	**18.** b	**19.** b	**20.** d
21. c	**22.** b	**23.** c	**24.** b	**25.** c
26. a	**27.** a	**28.** a	**29.** d	**30.** c
31. c	**32.** d	**33.** b	**34.** c	**35.** c
36. d	**37.** a	**38.** a	**39.** a	**40.** a
41. d	**42.** b	**43.** d	**44.** d	**45.** a
46. b	**47.** a	**48.** c	**49.** d	**50.** a
51. d	**52.** b	**53.** c	**54.** b	**55.** b
56. a	**57.** c	**58.** b	**59.** a	**60.** b
61. d	**62.** c	**63.** c	**64.** a	**65.** b
66. b	**67.** d	**68.** c	**69.** b	**70.** d
71. a	**72.** c	**73.** c	**74.** b	

EXPLANATORY NOTES

1. In a Binary search tree, left child is less than Right child.

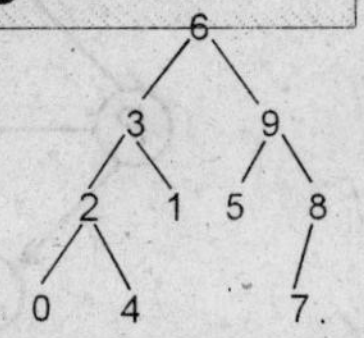

In order $\rightarrow$ 0 2 4 3 1 6 5 9 8 7

2. $(n+1)^m = \Theta(n^m)$, above k and m are constants

$2^{n+1} = O(2^n)$

3. Depth first traversals of the above graph: depth first traversals is also called **pre-order**

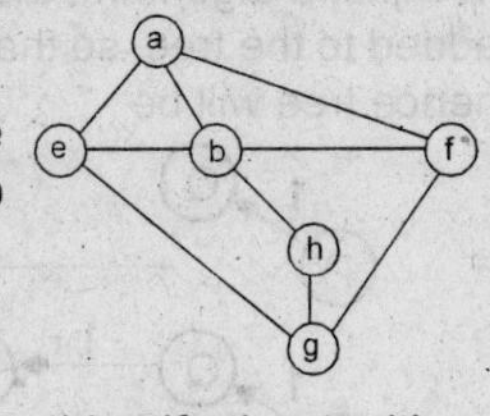

I. a b e g f h II. ab f h ge

IV. af g hbe

4. As array uses linear search to identify the position, where an element is to be inserted into the already started part of the array. Hence worst case running time will become $\Theta(n \log n)$

6. Expected number of inversion $= \frac{n(n-1)}{2}$

For randomly $= \frac{n(n-1)}{2} \times \frac{1}{2} = \frac{n(n-1)}{4}$

8. Deletion of the smallest element or randomly any element is insertion or deletion, complexity is $O(\log_2 n)$.

9. Push & Pop take time = 2 X

Delay between one push and one pop = Y

So, average stack life of an element (m)

$$= \frac{2Xm + mY}{m} = 2X + Y$$

13. Spanning tree is minimum, when all weight assigned to its edges between two vertices where every vertex is connected.

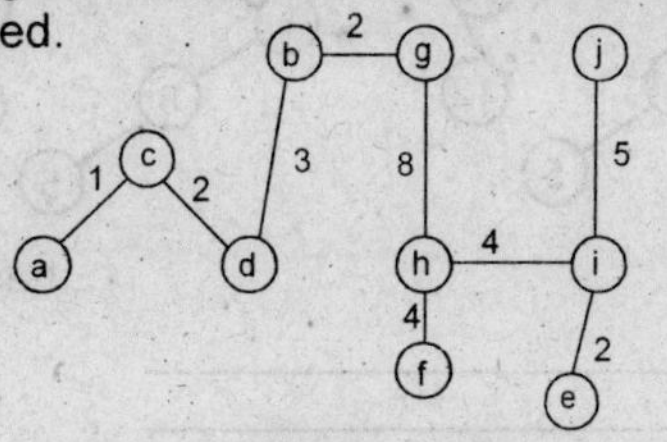

$\therefore$ Height of minimum spanning tree

$= 1 + 2 + 3 + 2 + 8 + 4 + 4 + 2 + 5 = 31$

14. Maximum room required = 4

a_s	b_s	c_s		a_e is finish from room
d_s	b_s	c_s		entry of a_s in place of a_e in room
d_s	d_s	e_s	c_e	c_s is out and e_s in available room
d_s	b_s	e_s	f_s	f_s is in next room
		e_s	f_s	b_e and d_e are vacant due to end
g_s		e_s	f_s	g_s is start in available room
g_s				e_s and f_s vacant the room
g_s	h_s			h_s is start in vacant room
				all vacant room

17.

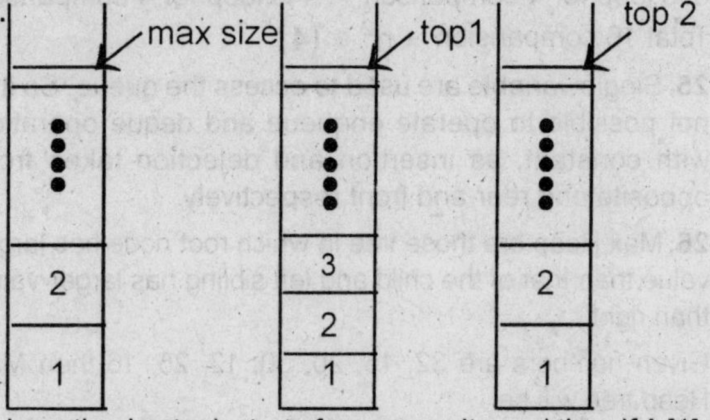

An insertion in stack starts from opposite end then if A [1] is inserted into stack 1 and A [max size] is inserted into the other stack, stack 2 as all element of array are placed into stack

so top + 1 = top 2 $\Rightarrow$ top 1 = top 2 -1

18. Binary search tree formed after inserting into empty binary search tree as

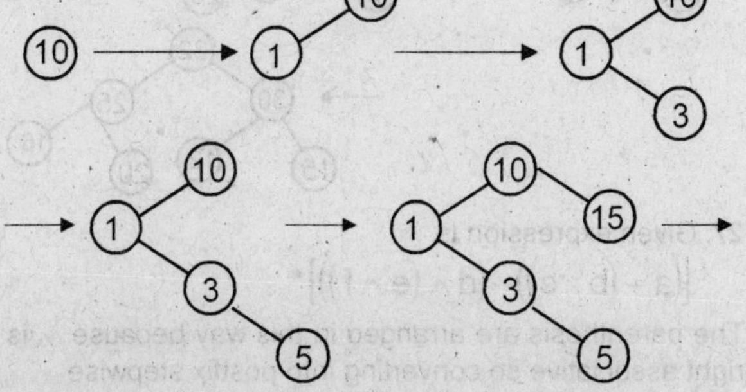

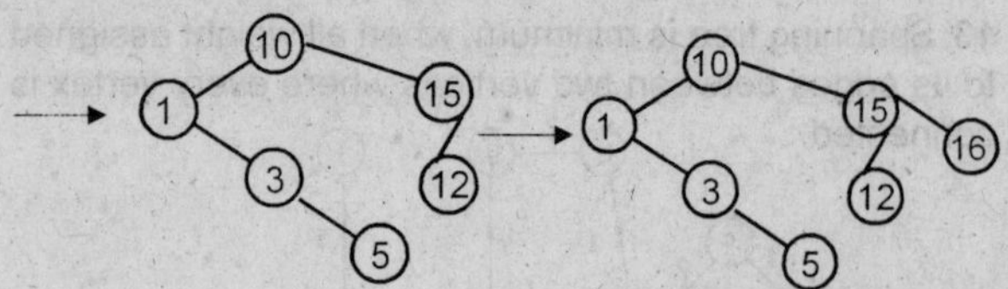

21.

index		hash value
4322		2
1334		4
1471		1
9679		9
1989		9
6171		1
6173		3
4199		9

9679, 1989 and 4199 has same hash value and 1471 and 6171 has the same hash value as 9 and 1 respectively

22. In every pass of the comparison, each element is compared with other elements.

List 1	List 2
1	1
2	2
3	3
4	4

There are two lists every element of list 1 compared with every element of list 2 in sequence.

1 is loop for 4 comparison 2 is loop for 4 comparison
3 is loop for 4 comparison 4 is loop for 4 comparison

Total 16 comparision $= n^2 = (4)^2$

25. Single variable are used to access the queue. So it is not possible to operate enqueue and deque operation with constant, as insertion and delection takes from opposite end rear and front respectively.

26. Max Heap are those tree in which root node has larger value than that of the child and left sibling has larger value than right.

Given numbers are 32, 15, 20, 30, 12, 25, 16 then Max Heap tree will be

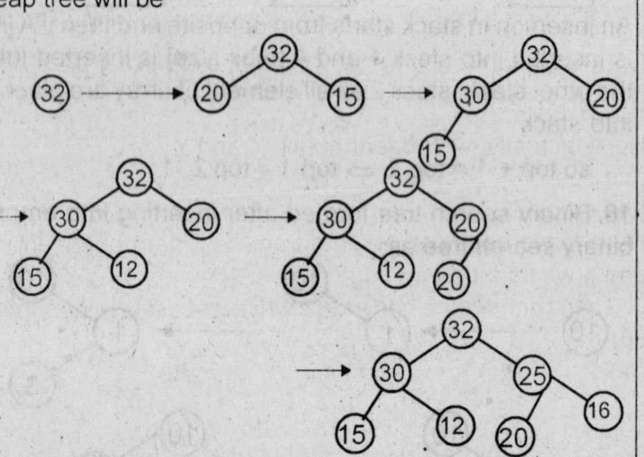

27. Given expression is

$$[(a+(b\times c))-(d\wedge(e\wedge f))]^*$$

The parenthesis are arranged in this way because $\wedge$ is right associative so converting into postfix stepwise

$$[(a+(b\times c))-(d^\wedge(e^\wedge f))] = (a+(bc\times))-(def^{\wedge\wedge})$$

$$=(abc+)-(def^{\wedge\wedge}) = abc\times+def^{\wedge\wedge}-$$

28. If first matrix are stored in row major, and second in column major it will take less time.

29. Set are arranged in arbitrary order and represented by linked list. Traversal of linked list are in sequential fashion so every traversal starts from the start node and end upto required node. So union and intersection will take more time.

31. The given program will find out the square root of m, suppose that m = 2

	X-Y	X	Y
Ist looping	2	3/2=1.5	2/1.5=1.33
2nd looping	0.16	$\frac{1.5+1.3}{2}=1.415$	$\frac{2.0}{1.415}=1.413$
3rd looping	0.02	1.414	1.414
4th looping	0		

as x-y = 0 exit from loop. Hence print 1.414 which is root of 2.

33.

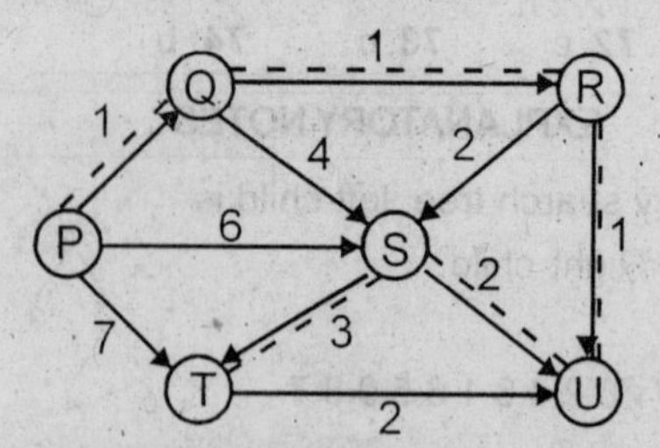

In Dijkstra algorithm, the edge with minimum weight are added to the tree, so that no node remain disconnected, hence tree will be

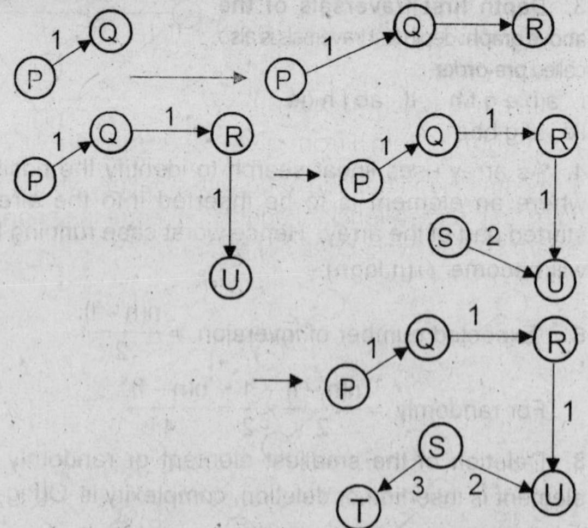

34. $G_1=(\upsilon_1 E_1)$, $G_2=(\upsilon_1 E_2)$

$G_1\cap G_2=(\upsilon, E_1\cap E_2)$ is not connected graph, then union of these two graphs have a cutting edge.

Hence $(G_1\cup G_2)=(\upsilon_1, E_1\cup E_2)$ must have a cut edge (bridge).

35. It contains only one loop so compute with linear complexity, hence complexity will be $\theta(n)$.

37. $T(1)$

$T(n)=2T(n-1)+n$ for evaluating $T(n)$

$T(n)=2T(n-1)+n$

$=2(2(n-2)+(n-1)+n)$

$\vdots$

This will reduce to $2^{n+1}-n-2$

39. Required complexity = (complexity of searching (X)) × {delete (X)}

41. Elements are abcd

In stack (based on LIFO) always top element are poped of f. So

adjacent of a be b or d

adjacent of b be a or c

adjacent of c be b or d

adjacent of d be c or d

So **cabd** order cannot be possible.

43. Checking for each choice

a) In order __	BADC
Preorder __	ABCD
b) Inorder __	BCAD
Postorder__	ABCD
c) Inorder __	ACBD
Preorder __	ABCD
d) Inorder __	BCAD
Preorder __	ABCD

44. (d) is false statement because

$O(f(n))\,O(h(n)) = O(g(n)\,h(n))$;. $f(n) = O(g(n))$;

$h(n) = O(g(n))$

$f(n)\,h(n) = (O(g(n)))(O(g(n)))$

$= O(g(n))\,O(h(n))$ from given assumption

45. In prims algorithm, the edge with smallest weight are added first, so sequence of addition of edge in spanning tree are so that cycle does not exist.

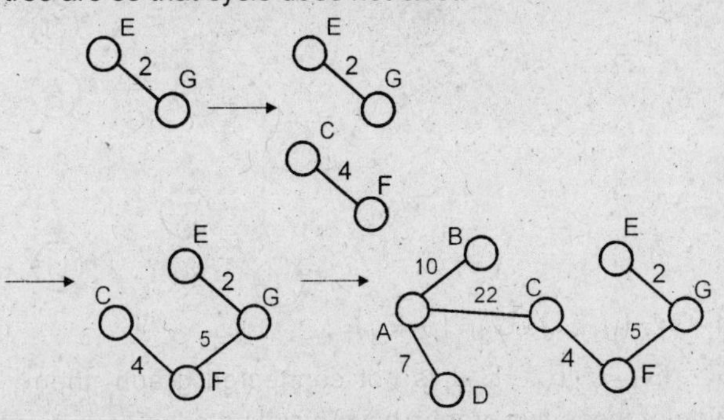

47. Apply concept of hashing where hash function return index where to increment corresponding entry

Index = (number - 51)

(no.is given no.)

So mapping goes through 0 - 49

Hence, an array of 50 numbers

48. Draw correspond matrix :

$$\begin{bmatrix} 0 & 1 & 1 & 1 & 1 \\ 1 & 0 & 1 & 1 & 1 \\ 1 & 1 & 0 & 1 & 1 \end{bmatrix}$$

is the corresponding matrix, and you can find many spanning trees.

51. 10, 8, 5, 3, 2

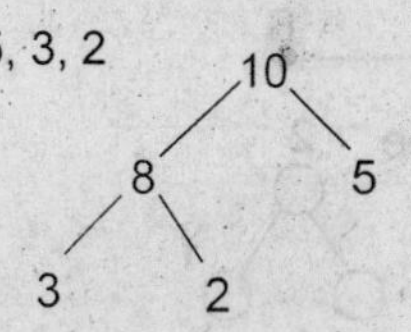

after insertion of 1 and 7

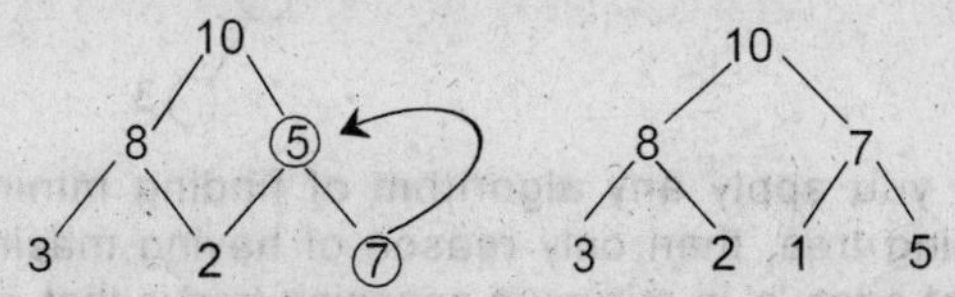

Hence the level order traversal is 10, 8, 7, 3, 2, 1, 5.

57. The foo i will be called for n! times for any n that's why space complexity is 0 (n!).

58. When we declare array of size n, this will increase complexity by O (n).

For all previous value we always find corresponding value in array that we declared.

59. Consider an example

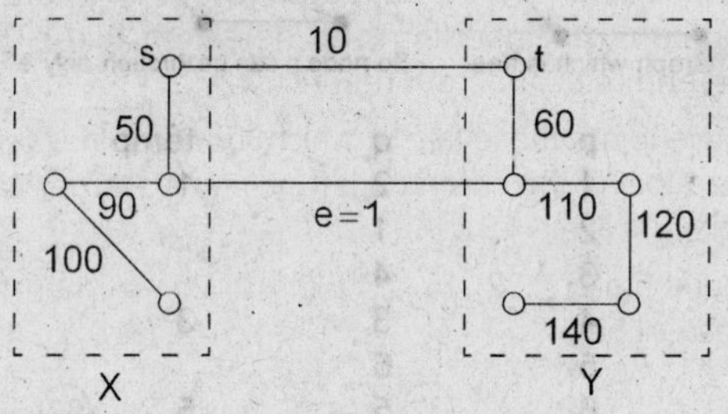

Here, minimum cost of edge joining X and Y is one (1), we called as e.

Here, actually two edge that join X and Y one cost '10' and e cost '1'

Here, if we consider weighted shortest path between s and t, we find it as an edge of cost 10.

We can't consider e edge here while its cost is minimum.

60.

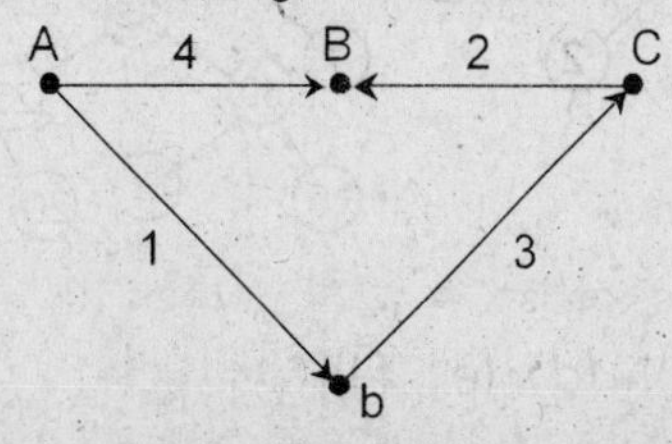

Consider here a path with minimum congestion between A and B

We follow any minimum spanning algorithm
Hence, Resultant tree is

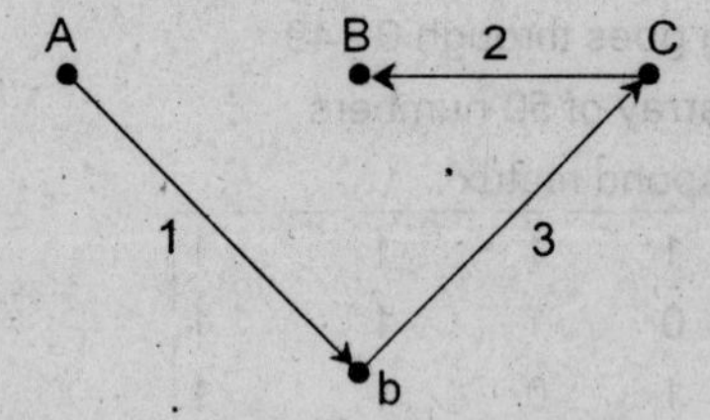

While, a minimum path between A and B is 4

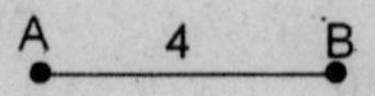

66. Consider the given tree

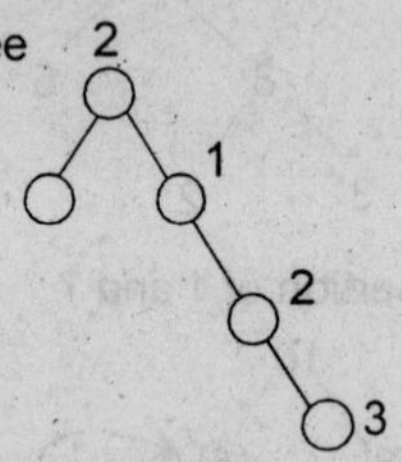

68. If you apply any algorithm of finding minimum spaning tree, then only reason of having maximum weight edge 'e' in minimum spanning tree is that graph itself a tree means only n - 1 edges are there with n nodes.

or there is only one edge for a particular node (vertex)

In the all above cases, there is no cycle.

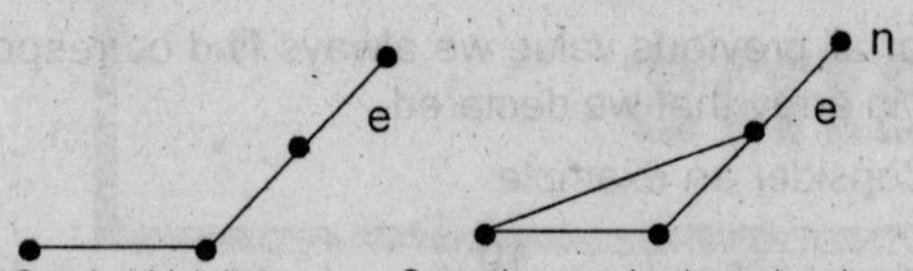

Graph which is tree So node n can be through only e

69.

p	q	temp
1	2	1
2	1	
3	4	
4	3	3
5	6	
6	5	5
7		

So final link list will be 2, 1, 4, 3, 6, 5, 7

70.

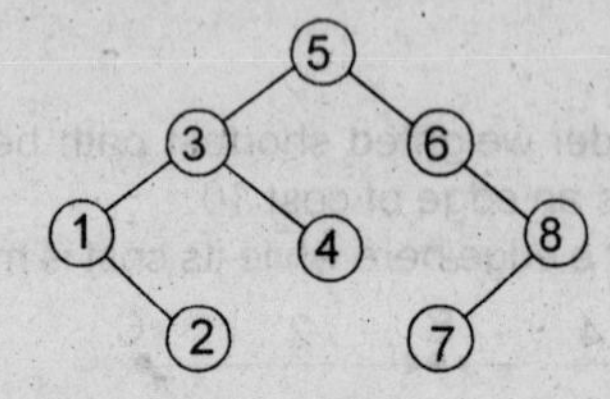

Unit - VIII
Operating Systems

UNIT - VIII OPERATING SYSTEMS

I. INTRODUCTION TO OPERATING SYSTEM

- Introduction
- Functions of an Operating System
- Services of an Operating System
- Types of Operating System

1. Introduction

An Operating System is a program that acts as a intermediary between a user of a computer and the computer hardware. The purpose of an operating system is to provide an environment in which a user can execute programs. The primary goal of an operating system is thus to make the computer system convenient toS use. A secondary goal is to use the computer hardware in an efficient manner.

A computer system can be divided roughly into four components: the hardware, the operating system, the application programs and the users.

The hardware — the central processing unit (CPU), the memory, and the input/output(I/O) devices — provides the basic computing resources. The application programs-such as compilers, database systems, games and business programs-define the ways in which these resources are used to solve the computing problems of the users.

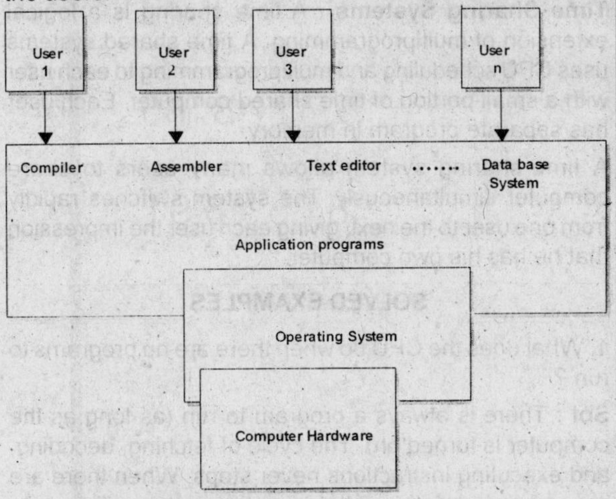

Fig : Abstract view of the components of a computer system.

We can view an operating system as a resource allocator. The primary goal of an operating systems is convenience for the user. Operating systems exist because they are supposed to make it easier to compute with them than without them. This view is particularly clear when you look at operating systems for small personal computers.

A secondary goal is efficient operation of the computer system. This goal is particularly important for large, shared multiuser systems.

Given below are several advantage points from which to view an operating system.

I. By examining the services it provides.

II. By looking at the interface it makes available to users and programmers

III. By disassembling the system into components and their interconnections.

An operating system manages four major resources

I. memory

II. processor

III. I/O devices

IV. Files

Hence an operating system consists of four major subroutines which does:

a) Memory management

b) Processor management

c) Device management

d) File management

2. Process Management

A process needs certain resources, including CPU time, memory, files, and I/O devices, to accomplish its task. These resources are either given to the process when it is created, or allocated to it while it is running. The operating system is responsible for the following activities in connection with process management:

I. creation and deletion of both user and system processes

II. suspension and resumption of processes

III. provision of mechanisms for process synchronization

IV. provision of mechanisms for process communication

V. provision of mechanisms for deadlock handling

3. Main-Memory Management

The Operating system is responsible for the following activities in connection with memory management:

I. Keep track of which parts of memory are currently being used and by whom

II. Decide which processes are to be loaded into memory when memory space becomes available

III. Allocate and deallocate memory space as needed

4. File Management

A file is a collection of related information defined by its creator. Commonly, files represent programs(both source and object forms) and data. Data files may be numeric, alphabetic, or alphanumeric. Files may be freeform, such as text files, or may be formatted rigidly. A file consists of a sequence of bits, bytes, lines, or records whose meanings are defined by their creators.

In file management, the Operating system is responsible for the following activities in connection with file management:

I. creation and deletion of files

II. creation and deletion of directories

III. support of primitives for manipulating files and directories

IV. mapping of files onto secondary storage

V. backup of files on stable (nonvolatile) storage media

5. I/O System Management

The I/O subsystem consists of

I. A memory management component including buffering, caching, and spooling

II. A general device-driver interface

III. Drivers for specific hardware devices

Only the device driver knows the peculiarities of the specific device to which it is assigned.

6. Secondary-Storage Management

The main purpose of a computer system is to execute programs. These programs, with the data they access, must be in main memory (primary storage) during execution. Because main memory is too small to accommodate all data and programs, and its data are lost when power is lost, the computer system must provide secondary storage to back up main memory.

In disk management the Operating system is responsible for the following activities:

I. Free-space management

II. Storage management

III. Disk scheduling

7. Protection System

Protection refers to a mechanism for controlling the access of programs, processes, or users to the resources defined by a computer system. The mechanism must provide a means for specification of the controls to be imposed, together with a means of enforcement.

8. Operating-System Services

An Operating system provides an environment for the execution of programs. Services are provided for the convience of the user.

Program execution: The system should be able to load the program, run it and end it either normally or abnormally

I/O operations : This may require a files as an I/O device. Specific devices require special functions, therefore Operating system should provide some means to do these operations.

File System Manipulation : This operation requires creating and deleting files. The programs need to read and write files

Error Detection : Errors may occur in CPU, MEMORY HARDWARE, I/O DEVICES or in USER PROGRAMS. Therefore, required action should be taken so that correct computing can take place.

For ensuring smooth and efficient operation of the systems, there exists another set of Operating system functions viz RESOURCE ALLOCATION, ACCOUNTING, PROTECTION.

System Calls : System calls provide the interface between a process and the Operating system

9. Batch Processing , Multiprogramming and Time Sharing Systems

Batch Processing : Jobs with similar needs are batched together. Processing all jobs of a batch considering them as a single process is called batch processing. In earlier systems, programmers left their programmes with the operator. The operator sort the programmes into batches of similar requirements. Then each batch was run and output sent to the programmer.

Multiprogramming systems : Multiprogramming is the concurrent execution of more than one processes. The operating system picks and begins to execute one of the jobs from job pool. The job may have to wait for some task such as I/O operation to complete. Then in a multiprogramming system the operating system switches to and executes another job.

In a multiprogramming system

I. To have several jobs ready to run, the system must keep all of them simultaneously in memory. This requires memory management.

II. If several jobs are ready to run at the same time, the system must choose among them. This is CPU scheduling.

III. When multiple jobs are running concurrently their ability to affect each other is to be controlled in all phases of the operating system including process scheduling, disk storage and memory management.

Time Sharing Systems : A time sharing is a logical extension of multiprogramming. A time shared systems uses CPU scheduling and multiprogramming to each user with a small portion of time shared computer. Each user has separate program in memory.

A time-sharing system allows many users to share computer simultaneously. The system switches rapidly from one user to the next, giving each user the impression that he has his own computer.

SOLVED EXAMPLES

1. What does the CPU do when there are no programs to run ?

Sol : There is always a program to run (as long as the computer is turned on). The cycle of fetching, decoding, and executing instructions never stops. When there are no user programs to run, the operating system will execute in a loop that does nothing (called a busy - wait loop or idle loop) until an interrupt occurs.

2. What characteristic is common to traps, interrupts, supervisor calls, and subroutine calls ?

Sol : Traps, interrupts, supervisor calls, and subroutine calls all save the current value of the program counter and branch to a new location in memory.

3. Which of the following instructions should be privileged (can only be executed in kernel mode)?

a) Change memory management registers.

b) Write the program counter

c) Read the time - of - day clock

d) Set the time - of - day clock

e) Change processor priority

Sol : a) Yes. Changing memory management registers would allow a process to access memory locations it was not authorized to access.

(b) No. Writing the program counter is no different than executing on unconditional branch.

(c) No. Although direct access to devices is usually unwise, read access of the clock should not be harmful.

d) Yes. Changing the clock could disrupt scheduled events and is typically not a right granted to a user process.

(e) Yes. Changing the processor's priority could cause interrupts to be lost.

4. Classify the following applications as batch-oriented or interactive.

a) Word processing

b) Generating monthly bank statements

c) Computing pi to a million decimal places,

Sol : a) Interactive b) Batch c) Batch

This job is almost 100% CPU-bound. A batch system without multiprogramming would attain nearly 100% CPU utilization running this program.

OBJECTIVE QUESTIONS - I

1. Where does the swap space reside? **[GATE-2001]**

a) RAM b) Disk c) ROM d) On-Chip Cache

2. Which of the following requires a device driver? **[GATE-2001]**

a) Register b) Cache

c) Main memory d) Disk

3. System calls are usually invoked by using: **[GATE-1999]**

a) A Software interrupt b) Polling

c) An indirect jump d) A privileged instruction

4. Which of the following is an example of spooled device? **[GATE-1996]**

a) A line printer used to print the output of a number of jobs.

b) A terminal used to enter input data to a running program

c) A secondary storage device in a virtual memory systems.

d) A graphic display device.

5. Which of the following are real-time systems:

a) An on-line railway reservation system

b) A process control system

c) Aircraft control system

d) Payroll processing system

6. Kernel is:

a) Considered as the critical part of the OS

b) The Software which monitors the OS

c) The set of primitive functions upon which the rest of the OS functions are built-up. d) None.

7. Fense register is used for:

a) CPU protection b) Memory protection

c) File protection d) None

8. Which of the following service is not supported by the operating system?

a) Protection b) Accounting

c) Compilation d) I/O operation

9. which of the following is a single user operating system

a) MS-DOS b) UNIX

c) XENIX d) OS/2

10. Distributed systems should

a) Meet prescribed time constraints

b) Aim better resource sharing

c) Aim better system utilization

d) Aim low system overhead.

11. The primary purpose of an operating computer is:

a) To make the most efficient use of the computer hardware

b) To allow people to use the computer

c) To keep system programmers employed

d) To make computers easier to use.

12. The operating system manages

a) Memory b) Processor

c) Disk and I/O devices d) All of the above.

13. Find ODD man out:

a) Executive b) Task manager

c) Kernel d) Spooler

14. Operating system is a machine dependent software:

a) True b) False

15. Command interpreter is a part of **[UGC NET-2000]**

a) Operating system b) compiler

c) loader d) linker

16. Identify the false statement in reference to a distributed system **[UGC NET-2000]**

a) processors normally do not share memory

b) processors always share memory

c) processors do not share a clock

d) processors in a distributed system may vary in size and function

17. Which one of the following is not a function of bootstrap program? **[UGC NET-2000]**

a) Initializing the CPU registers

b) Initializing the memory contents

c) Loading operating system

d) Loading the compilers

18. What is the name given to the process of initializing a microcomputer with its operating system? **[UGC NET-1997]**

a) Cold booting b) Booting

c) Warm booting d) Boot recording

19. The primary job of the operating system of a computer is to **[UGC NET-1997]**

a) mange resources b) command resources

c) provide utilities d) be user friendly

20. Booting a computer means **[UGC NET-1996]**

a) loading the operating system into its memory

b) loading any program from disk to memory

c) starting the work of processing

d) illuminating the video screen

21. Time-sharing operating system have **[UGC NET-1996]**

a) high through put b) low execution time
c) faster I/Od) none of these

22. Spooling is most beneficial in a multiprogramming environment when **[Hyd.Univ-2001]**

a) most jobs are I/O bound
b) most jobs are CPU bound
c) jobs are evenly distributed between I/O bound and CPU bound
d) there is limited memory with clear requirement for secondary memory

ANSWER KEY - I

1. b	**2.** d	**3.** a	**4.** a	**5.** b,c
6. c	**7.** b	**8.** c	**9.** a	**10.** b
11. a	**12.** d	**13.** b	**14.** b	**15.** a
16. b	**17.** d	**18.** b	**19.** d	**20.** a
21. b,d	**22.** a			

OBJECTIVE QUESTIONS - II

1. Which combination of the following features will suffice to characterize an OS as a multiprogrammed OS?
I. More than one program may be loaded into main memory at the same time for execution.
II. If a program waits for certain events such as I/O , another program is immediately scheduled for execution.
III.If the execution of a program terminates, another program is immediately scheduled for exectuion. **[GATE-2002]**

a) I only b) I and II only
c) I and III only d) I,II and III

2. Listed below are some operating system abstractions and the h/w components or matching of pairs is correct: **[GATE-1999]**

A. Thread	I. Interrupt
B. Virtual address Space	II. Memory
C. File System	III. CPU
D. Signal	IV. Disk

a) A – II,B – IV,C- III,D – I
b) A – I, B – II,C – III, D – IV
c) A – III, B – II,C – IV,D – I
d) A – IV,B – I,C – II, D – III

3. A multi-user , multi-processing operating system cannot be implemented on h/w that does not support: **[GATE-1999]**

a) Address translation b) DMA for disk transfer
c) At least two modes of CPU execution (privileged and non-privileged) d) Demand paging.

4. When an interrupt occurs, an operating system: **[GATE-1997]**

a) Ignores the interrupt
b) Always changes state of interrupted process after processing the interrupt
c) Always resumes execution of interrupted process after processing the interrupt
d) May change state of interrupted process to 'blocked' and schedule another process.

5. In a multiprogramming environment

a) The processor executes more than one process at a time
b) The programs are developed by morethan one person
c) More than one process is resident in the memory
d) A single user can execute many programs in the same time

6. For implementing a multiprogramming OS,

a) Special support from processor is essential
b) Special support from processor is not essential
c) Cache memory must be available.
d) More than one processor must be available.

7. Memory protection is normally done by

a) The processor and the associated h/w
b) The operating system
c) The compiler d) The user program

8. Which of the following is well suited for batch processing:

a) Process control b) Video game control
c) Preparing pay bills of employees.
d) Processing mailing addresses.

9. Which of the following is not a part of the operating system?

a) Supervisor b) Performance monitor
c) Job – control program
d) Input-Output control program

10. All the time a computer is switched on, it's operating system software has to stay in

a) Main storage b) Primary storage
c) Floppy disk d) Disk drive

11. Which of the following is a multi-user OS

a) Windows 2000 b) MS-DOS
c) UNIX d) Windows NT

12. Which one of the following is not a main function of an Operating System? **[UGC NET-1996]**

a) Memory management b) Processor management
c) Program management d) I/O management

13. Time sharing increases **[UGC NET-1996]**

a) processing time b) through put
c) execution time d) I/O time

14. What type of system software provides information on processor utilisation? **[UGC NET-1996]**

a) Utility program b) Communication software

c) Performance monitoring software

d) Operating system

15. The Indian railway reservation system is **[UGC NET-1998]**

a) Batch system b) On-line system

c) Real-time system d) Knowledge based system

16. Windows 95 Operating System is a **[UGC NET-1998]**

a) multiuser multitasking

b) single user multitasking

c) single user single tasking

d) multiuser single tasking

17. The BIOS **[UGC NET-1998]**

a) is the interface between hardware and software

b) is called firmware because it is integrated with hardware

c) activates the PC which otherwise remains dead

d) all the above

18. Software that measures, monitors, analyzes and controlreal-world events is called: **[UGC NET-1998]**

a) system software b) scientific software

c) real-time software d) on-line software

19. In a distributed operating system, which of the following is not true? **[UGC NET-1999]**

a) There are more than one CPU

b) There exists more than one memory unit

c) There exists more than one hard disk

d) There exists only one CPU

ANSWER KEY - II

1. d	2. c	3. a,b,c	4. d	5. c
6. a	7. b	8. c	9. b	10. d
11. c	12. b	13. b	14. c	15. c
16. b	17. d	18. c	19. d	

II. PROCESS MANAGEMENT

- Process Concept
- Cooperating Processes
- Threads
- CPU Scheduling
- Concurrent Processes
- Process Synchronization
- Deadlocks

1. Process Concept : A process is more than the program code (some times as the text section). It also includes the current activity, as represented by the value of the program counter and the contents of the procesor's registers. A process generally also includes the process stack, containing temporary data (such as subroutine parameters, return addresses, and temporary variables), and a data section containing global variables.

process state : As a process executes, it changes state. The state of a process is defined in part by the current activity of that process. Each process may be in one of the following states:

I. New : The process is being created

II. Running : Instructions are being executed

III. Waiting : The process is waiting for some event to occur (such as an I/O completion or reception of a signal).

IV. Ready : The process is waiting to be assigned to a processor.

V. Terminated : The process has finished execution.

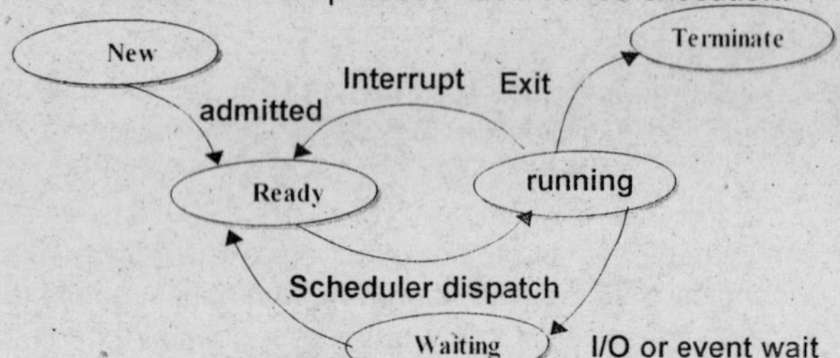

Fig 2.1: Diagram of process state

Process Control Block : Each process is represented in the Operating system by process a control block (PCB) — also called a task control block. A PCB is shown in the figure 2.2 shown below:

Pointer	Process State
Process number	
Program counter	
Register	
Memory limits	
List of open files	
:	

Fig 2.2 : process control block

It contains following pieces of information associated with a specific process, including:

- **Process state :** The state may be new, ready, running, waiting, halting and so on
- **Program counter :** The counter indicates the address of the next instruction to be executed for this process.
- **CPU register :** The registers vary in number and type, depending on the computer architecture. They include accumulators, index registers, stack pointers, and general-purpose registers, plus any condition-code information. Along with the program counter, this state information must be saved when an interrupt occurs, to allow the process to be continued correctly afterward.
- **CPU scheduling information :** This information includes a process priority, pointers to scheduling queues, and any other scheduling parameters.
- **Memory-management information :** This information may include such information as the value of the base and limit registers, the page tables, or the segment tables depending on the memory system used by the Operating system
- **Accounting information :** This information includes the amount of CPU and real time used, time limits, account numbers, job or process numbers, and so on
- **I/O status information :** The information includes the list of I/O devices (such tape drives) allocated to this process, a list of open files, and so on.

2. Process Scheduling : The objective of multiprogramming is to have some process running to all times, to maximize, CPU utilization.

Scheduling Queues : As process enter the system, they are put into a job queue. The queue consists of all processes in the system. The processes that are residing in main memory and are ready and waiting to execute are kept on a list called the ready queue.

A common representation for a discussion of process scheduling is a queueing diagram, such as shown in figure 2.3 below

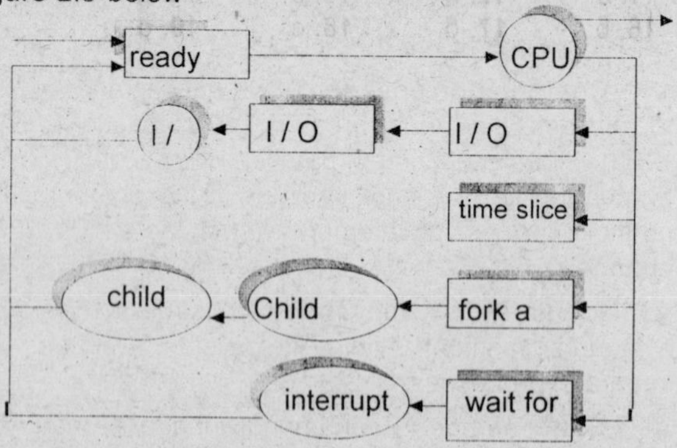

Fig 2.3 : Queueing-diagram representation of process scheduling

Schedulers : A process migrates between the various scheduling queues throughout its lifetime. The Operating system must select, for scheduling purposes, processes from these queues in some fashion. The selection process is carried out by the appropriate scheduler.

The long-term scheduler (or job scheduler)

The short-term scheduler (or CPU scheduler)

The primary distinction between these two schedulers is the frequency of their execution. The short-term scheduler must select a new process for the CPU quite frequently. A process may execute for only a few milliseconds before waiting for an I/O request. Often, the short-term scheduler executes at least once every 100 milliseconds.

The long-term scheduler, on the other hand, executes much less frequently. There may be minutes between the creation of new processes in the system. The long-term scheduler controls the degree of multiprogramming (the number of processes in memory).

Context Switch : Switching the CPU to another process requires saving the state of the old process and loading the saved state for the new process. This task is known as a context switch.

Operation On Processes : The processes in the system can execute concurrently, and must be created and deleted dynamically. Thus, the Operating system must provide a mechanism for process creation and termination.

3. **Cooperating Processes :** The concurrent processes executing in the operating sytem may be either independent processes or cooperating processes. A Process is independent if it cannot affect or be affected by the other processes executing in the system. Clearly, any process that does not share any data (temporary or persistent) with any other process is independent. On the other hand, a process is cooperating if it can affect or be affected by the other processes executing in the system. Clearly, any process that shares data with other processes is a cooperating process.

We may want to provide an environment that allows process cooperation for several reasons.

- **Information sharing :** Since several users may be interested in the same piece of information (for instance, a shared file), we must provide an environment to allow concurrent access to these types of resources.
- **Computation speedup:** If we want a particular task to run faster, we must break it into subtasks, each of which will be executing in parallel with the others. Such a speedup can be achieved only if the computer has multiple processing elements (such as CPUs or I/O channels).
- **Modularity :** We may want to construct the system in a modular fashion, dividing the system functions into separate processes.
- **Convenience :** Even an individual user may have many tasks on which to work on at one time. For instance, a user may be editing, printing, and compiling in parallel.

Concurrent execution of cooperating processes requires mechanisms that allow processes to communicate with one another and to synchronize their actions .

To illustrate the concept of cooperating processes, let us consider the producer-consumer problem, which is a common paradigm for cooperating processes. A producer process produces information that is consumed by a consumer process. For example, a print program produces characters that are consumed by the printer driver. A compiler may produce assembly code, which in consumed by an assembler. The assembler, in turn, may produce object modules, which are consumed by the loader.

To allow producer and consumer processes to run concurrently, we must have available a buffer or items that can be filled by the producer and emptied by the consumer. A producer can produce one item while the consumer is consuming another item. The producer and consumer must be synchronized, so that the consumer does not try to consume an item that has not yet been procuced. In this situation, the consumer must wait until an item is produced.

The unbounded -bufffer producer-consumer problem places no practical limit on the size of the buffer. The consumer may have to wait for new items, but the producer can always produce new items. The bounded-buffer producer-consumer problem assumes a fixed buffer size. In this case, the consumer must wait if the buffer is empty, and the producer must wait if the buffer is empty.

The buffer may either be provided by the operating system through the use of an interprocess-communication (IPC) facility, or by explicitly coded by the application programmer with the use of shared memory. Let us illustrate a shared-memory solution to the bounded-buffer problem. The producer and consumer processes share the following variables.

```
# define  BUFFER SIZE 10
typedef struct {

} item ;
item  buffer [BUFFER SIZE];
int in  = 0 ;
int out = 0 ;
```

The shared buffer is implemented as a circular array with two logical pointers: in and out. The variable in points to the next free position in the buffer; out points to the first full position in the buffer. The buffer is empty when in = = out; the buffer is full when ((in + 1) % BUFFER_ SIZE) == out.

The code for the producer and consumer processes follows. The producer process has a local variable next produced in which the new item to be produced is stored:

```
while        (1)  {
/* produce an item in nextproduced */
while ((( in + 1 ) % BUFFER _ SIZE) = = out)
; /* do nothing */
buffer [in] = nextProduced;
in = (in + 1) % BUFFER _ SIZE;
}
```

The consumer process has a local variable next Consumed in which the item to be consumed isstored:

```
while (1) {
while (in = = out)
  ; // do nothing
next Consumed = buffer [out] ;
out = (out + 1) % BUFFER _ SIZE;
/* consume the item in next Consumed */
}
```

This scheme allows at most BUFFER_SIZE - 1 items in the buffer at the same time. We leave it as an exercise for you to provide a solution where BUFFER_SIZE items can be in the buffer at the same time.

4. Threads : A thread, sometimes called a lightweight process (LWP), is a basic unit of CPU utilization; it comprises a thread ID, a program counter, a register set, and a stack. It shares with other threads belonging to the same process its code section, data section, and other operating-system resources, such as open files and signals. A traditional (or heavyweight) process has a single thread of control. If the process has multiple threads of control, it can do more than one task at a time. Figure 2.4 illustrates the difference between a traditional single-threaded process and a multithreaded process.

Motivation: Many software packages that run on modern desktop PCs are multithreaded. An application typically is implemented as a separate process with several threads of control. A web browser might have one thread display images or text while another thread retrieves data from the network. A word processor may have a thread for displaying graphics, another thread for reading keystrokes from the user, and a third thread for performing spelling and grammar checking in the background.

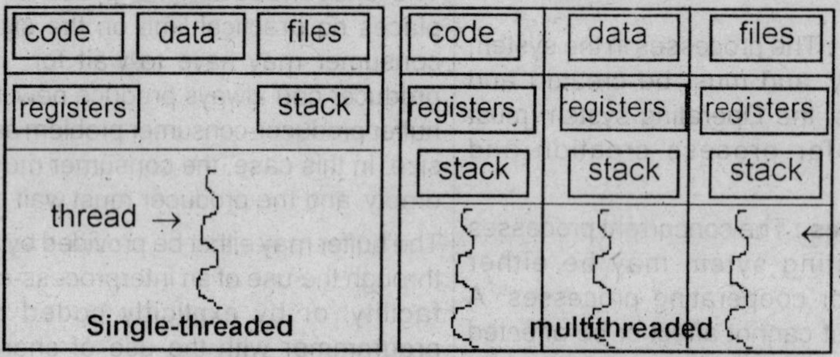

Fig 2.4 Single and Multithreaded Processes

In certain situations a single application may be required to perform several similar tasks. For example, a web server accepts client requests for web pages, images, sound, and so forth. A busy web server may have several (perhaps hundreds) of clients concurrently accessing it. If the web server ran as a traditional single threaded process, it would be able to service only one client at a time. The amount of time that a client might have to wait for its r quest to be serviced could be enormous.

One solution is to have the server run as a single process that accepts requests. When the server receives a request, it creates a separate process to service that request. In fact, this process-creation method was in common use before threads became popular. Process creation is very heavyweight . If the new process will perform the same tasks as the existing process, why incur all that overhead ? It is generally more efficient for one process that contains multiple threads to serve the same purpose. This approach would multithread the web-server process. The server would create a separate thread that would listen for client requests; when a request was made rather than creating another process, it would create another thread to service the request.

Threads also play a vital role in remote procedure call (RPC) systems. RPCs allow interprocess communication by providing a communication mechanism similar to ordinary function or procedure calls.Typically, RPC servers are multithreaded. When a server receives a message, it services the message using a separate thread. This allows the server to service several concurrent requests.

Benefits : The benefits of multithread programming can be broken down into four major categories:

1. Responsiveness : Multithreading an interactive application may allow a program to continue running even if part of it is blocked or is performing a lengthy operation, thereby increasing responsiveness to the user. For instance, a multithreaded web browser could still allow user interaction in one thread while an image is being loaded in another thread.

2. Resource sharing : By default, threads share the memory and the resources of the process to which they belong. The benefit of code sharing is that it allows an application to have several different threads of activity all within the same address space.

3. Economy : Allocating memory and resources for process creation is costly. Alternatively, because threads share resources of the process to which they belong, it is more economical to create and context switch threads. It can be difficult to guage empirically the difference in overhead for creating and maintaining a process rather than a thread, but in general it is much more time consuming to create and manage processes than threads.

4. Utilization of multiprocessor architectures : The benefits of multithreading can be greatly increased in a multiprocessor architeture where each thread may be running in parallel on a different processor. A single threaded process can only run on one CPU, no matter how many are available. Multithreading on a multi-CPU machine increases concurrency. In a single processor architecture, the CPU generally moves between each thread so quickly as to create an illusion of parallelism, but in reality only one thread is running at a time.

5. CPU Scheduling : CPU scheduling is the basis of multiprogrammed operating systems. By switching the CPU among processes, the operating system can make the computer more productive.

The idea of multiprogramming is relatively simple. A process is executed until it must wait, typically for the completion of some

I/O request. In a simple computer system, the CPU would then just sit idle. All this waiting time is wasted; no useful work is accomplished. With multiprogramming, we try to use this time productively. Several processes are kept in memory at one time. When one process has to wait, the operating system takes the CPU away from that process and gives the CPU to another process. This pattern continues. Every time one process has to wait, another process may take over the use of the CPU.

Scheduling is a fundamental operating system function

CPU-I/O Burst Cycle : The success of CPU scheduling depends on the following observed property of process: Process execution consists of a cycle of CPU execution and I/O wait. Processes alternate back and forth between these two states. Process execution begins with a CPU burst. That is followed by an I/O burst, which is followed by another CPU burst, then another I/O burst, and so on Eventually, the last CPU burst will end with a system request to terminate execution, rather than with another I/O burst.

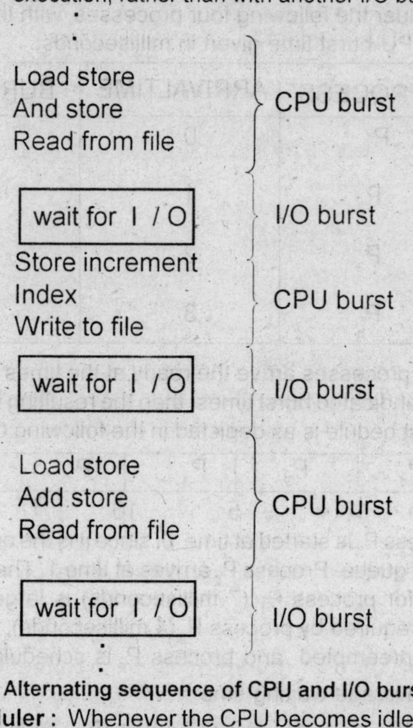

Fig 2.5 : Alternating sequence of CPU and I/O bursts

CPU Scheduler : Whenever the CPU becomes idle, the Operating system must select one of the processes in the ready queue to be executed. The selection process is carried out by the short-term scheduler(or CPU scheduler)

The ready queue is not necessarily a first-in, first-out (FIFO) queue. When we consider the various scheduling algorithms, a ready queue may be implemented as a FIFO queue, a priority queue, a tree, or simply an unordered liked list. Queue are linked up waiting Queues are generally PCBs.

Preemptive Scheduling : CPU scheduling decisions may take place under the following four circumstances:

I. When a process switches from the running state to the waiting state, I/O request, or invocation of wait for the termination of one of the child processes

II. When a process switches from the running state to the ready state, when an interrupt occurs

III. When a process switches from the waiting state to the ready state e.g. completion of I/O

IV . When a process terminates

Under these circumstances, there is no choice in terms of scheduling. A new process (if one exists in the ready queue) must be selected for execution. There is a choice, however, for circumstances II and III.

When a scheduling takes place only under circumstances (I) to (IV), the scheduling scheme is non preemptive; otherwise, the scheduling scheme is preemptive. Under non preemptive scheduling, once the CPU has been allocated to a process, the process keeps the CPU until it releases the CPU either by terminating or by switching to the waiting state.

Unfortunately, preemptive scheduling incurs a cost. Consider the case of two processes sharing data. One may be in midst of updating the data when it is preempted and the second process is run. The second process may try to read the data, which are currently in an inconsistent state.

Dispatcher : Another component involved in the CPU scheduling function is the dispatcher. The dispatches is the module that gives control of the CPU to the process selected by the short-term scheduler. This function involves:

I. Switching context

II. Switching to user mode

III.Jump it to the proper location in the user program to restart that program

The dispatcher should be as fast as possible, given that it is invoked during every process switch. The time it takes for the dispatcher to stop one process and start another running is known as the dispatcher latency.

Scheduling Criteria : The scheduling criteria that are used includes

I. CPU utilization : The CPU should be kept as busy as possible. CPU utilization may range from 0 to 100 percent. In a real system, it should range from 40 percent (for a lightly loaded system) to 90 percent (for a heavily used system).

II. Throughput : If the CPU is busy executing processes, then work is being done. One measure of work is the number of processes that are completed per time unit, called throughput. For long processes, this rate may be done

III.Turnaround time : The interval from the time of submission of a process to the time of completion is the turnaround time.

IV.Waiting time : The amount of time that a process spends waiting in the ready queue

V. Response time : Response time, is the amount of time it takes to start responding, but not the time that it takes to output that response. The Turnaround time is generally limited by the speed of the output device.

6. Scheduling Algorithms

First-Come, First-Served Scheduling (FCFS)

The average waiting time under the FCFS policy, however, is often quite long. Consider the following set of processes that arrive at time 0, with the length of the CPU-burst time given in milliseconds:

PROCESS	BURST TIME
P_1	24
P_2	3
P_3	3

If the processes arrive in the order P_1,P_2,P_3 and are served in FCFS order, we get the result shown in the following Gantt chart:

P_1	P_2	P_3

0 24 27 30

Now, the waiting time is 0 milliseconds for process P_1, 24 milliseconds for process P_2 and 27 milliseconds for process P_3. Thus, The average waiting time

$$= \frac{0+24+27}{3} = 17 \text{ milliseconds.}$$

If the processes arrives in the order P_2,P_3,P_1 however, the results will be as shown in the following Gantt chart:

P_2	P_3	P_1

0 3 6 30

Now, the average waiting time

$$= \frac{6+0+3}{3} = 3 \text{ milliseconds.}$$

This reduction is substantial. Thus, the average waiting time under a FCFS policy is generally not minimal. The FCFS scheduling algorithm is non preemptive. Once the CPU has been allocated to a process, that process keeps the CPU until it releases the CPU, either by terminating or by requesting I/O.

Shortest-Job-First Scheduling (SJF)

Consider the following set of processes, with the length of the CPU burst time given in milliseconds:

PROCESS	BURST TIME
P_1	6
P_2	8
P_3	7
P_4	3

Using SJF scheduling, we would schedule these processes according to the following Gantt Chart :

P_4	P_1	P_3	P_2

0 3 9 16 24

The waiting time is 3 milliseconds for process P_1, 16 milliseconds for process P_2, 9 milliseconds for process P_3, and 0 milliseconds for process P_4.

Thus, the average waiting time

$$= \frac{3+16+9+0}{4} = 7 \text{ milliseconds.}$$

If we were using the FCFS scheduling scheme, then the average waiting time would be 10.25 milliseconds. The SJF scheduling algorithm is provably optimal, in that it gives the minimum average waiting time for a given set of processes.

If two processes have the same length next CPU burst, FCFS scheduling is used to break the tie. The real difficulty with the SJF algorithm is knowing the length of the next CPU request.One approach is to try approximate SJF scheduling.

The SJF algorithm may be either preemptive or non preemptive. The choice arises when a new process arrives at the ready queue while a previous process is of the currently executing process. A preemptive SJF algorithm will preempt the currently executing process, whereas a nonpreemptive SJF algorithm will allow the currently running process to finish its CPU burst. Preemptive SJF scheduling is sometimes called shortest-remaining-time-first scheduling.

Consider the following four processes, with the length of the CPU-burst time given in milliseconds:

PROCESS	ARRIVALTIME	BURST TIME
P_1	0	8
P_2	1	4
P_3	2	9
P_4	3	5

If the processes arrive the ready at the times whose and need indicated burst times, then the resulting preemptive SJF schedule is as depicted in the following Gantt chart:

P_1	P_2	P_4	P_1	P_3

0 1 5 10 17 26

Process P_1 is started at time 0, since it is the only process in the queue. Process P_2 arrives at time 1. The remaining time for process P_1 (7 milliseconds) is larger than the time required by process P_2 (4 milliseconds), so process P_1 is preempted, and process P_2 is scheduled.

The average waiting time =

((10 – 1) + (1 – 1) + (17 – 2) + (5 – 3)) /4

= 26/4 =6.5 milliseconds.

A nonpreemptive SJF scheduling would result in an average waiting time of 7.75 milliseconds.

Priority Scheduling : The SJF algorithm is a special case of the general priority scheduling algorithm. A priority is associated with each process, and the CPU is allocated to the process with the highest priority.3

Equal-priority processes are scheduled in FCFS order. Consider the following set of processes, assumed to have arrived at time 0, in the order $P_1, P_2, \ldots\ldots P_5$, with the length of the CPU-burst time given in milliseconds:

PROCESS	BURST TIME	PRIORITY
P_1	10	2
P_2	1	1
P_3	2	3
P_4	1	4
P_5	5	2

Using priority scheduling, we would schedule these processes according to the following Gantt chart :

P_2	P_5	P_1	P_3	P_4

0 1 6 16 18 19

The average waiting time is 8.2 milliseconds.

A major problem with priority scheduling algorithms is indifinite blocking or starvation. A process that is ready to run but lacking the CPU can be considered blocked, waiting for the CPU. A priority scheduling algorithm can leave some low-priority processes waiting indefinitely for the CPU.

A solution to the problem of indefinite blockage of low-priority processes is aging. Aging is a technique of gradually increasing the priority of processes that wait in the system for a long time.

Round-Robin Scheduling : The round-robin (RR) scheduling algorithm is designed especially for time sharing systems. It is similar to FCFS scheduling, but preemption is added to switch between processes. A small unit of time, called a time quantum, or time slice, is defined. A time quantum is generally from 10 to 100 milliseconds. The ready queue is treated as a circular queue. The CPU scheduler goes around the ready queue, allocating the CPU to each process for a time interval of up to 1 time quantum.

The average waiting time under the RR policy, however, is often quite long. Consider the following set of processes that arrive at time 0, with the length of the CPU-burst time given in milliseconds.

PROCESS	BURST TIME
P_1	24
P_2	3
P_3	3

If we use a time quantum of 4 milliseconds, then process P_1 gets the first 4 milliseconds. Since it requires another 2 milliseconds, it is preempted after the first time quantum, and the CPU is given to the next process in the queue, process P_2. Since process P_2 does not need 4 milliseconds, it quits before its time quantum expires. The CPU is then given to the next process, process P_3. Once each process has received 1 time quantum, the CPU is returned to process P_1 for an additional time quantum. The resulting RR schedule is

P_1	P_2	P_3	P_1	P_1	P_1	P_1	P_1

0 4 7 10 14 18 22 26 30

The average waiting time = $\frac{17}{3} = 5.66$ milliseconds.

If there are n processes in the ready queue and the time quantum is q, then each process gets 1/n of the CPU time in chunks of at most q time units. Each process must wait no longer than $(n-1) \times q$ time units until its next time quantum. For example, if there are five processes, with a time quantum of 20 milliseconds, then each process will get up to 20 milliseconds every 100 milliseconds.

The performance of the RR algorithm depends heavily on the size of the time quantum.

The time quantum to is required to large with respect to the context-switch time. If the context-switch time is approximately 10 percent of the time Quantum, then about to percent of the CPU will be spent in context switch. Turnaround time also depends on the size of the time quantum.

Multilevel Queue Scheduling : Another class of scheduling algorithms has been created for situations in which processes are easily classified into different groups. For example, a common division is made between foreground (interactive) processes and background (batch) processes. These two types of processes have different response-time requirements, and so might have different scheduling needs. In addition, foreground processes may have priority (externally defined) over black ground processes.

A multilevel queue-scheduling algorithm partitions the ready into several separate queues (Figure. 2.6). The processes are permanently assigned to.

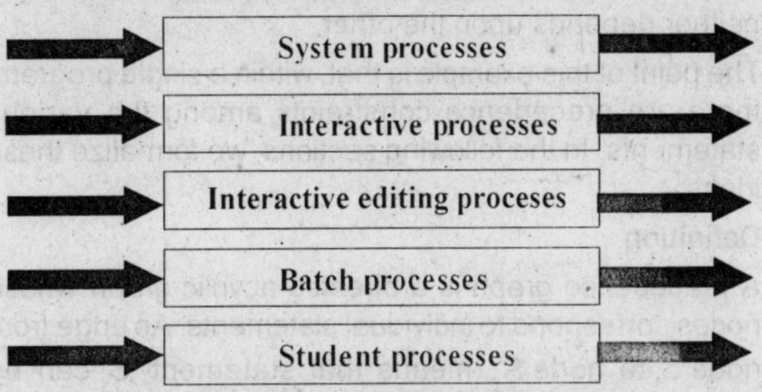

Fig 2.6: Multilevel queue scheduling

Multilevel Feedback Queue Scheduling : Multilevel feedback queue scheduling, however, allows a process to move between queues. The idea is to separate processes with different CPU-burst characteristics. If a process uses too much CPU time, it will be moved to a lower-priority queue. This scheme leaves I/O-bound and interactive processes in the higher priority queues. Similarly, a process that waits too long in a lower-priority queue may be moved to a higher-priority queue. This form of aging prevents starvation.

In general, a multilevel feedback queue scheduler defined by the following parameters:

I. The number of queues

II. The scheduling algorithm for each queue

III.The method used to determine when to upgrade a process to a higher-priority queue

IV.The method used to determine when to demote a process to a lower-priority queue

V. The method used to determine which queue a process will enter when that process needs service

The definition of a multilevel feedback queue scheduler makes it the most general CPU scheduling algorithm.

7. Concurrent Processes

Precedence Graphs

Consider the following program segment, Program 1, which performs some simple arithmetic operations.

```
a := x + y;
b := z + 1;
c := a - b ;
w := c + 1;
```

Progarm 1

Suppose we want to execute some of these statements concurrently. We may have multiple functional units (such as adders) in our processor, or multiple cpus. The "addition" and "subtraction" might be operations on matrices or vectors, or on very large sets (union and intersection), or on files (concatenation and difference). With multiple cpus we may be able to execute some statements concurrently with others, reducing our total execution time.

Clearly, the statement c := a – b cannot be executed before both a and b have been assigned values. Similarly, w := c + 1 cannot be executed before the new value of c has been computed. On the other hand, the statements a := x + y and b := z + 1 could be executed concurrently since neither depends upon the other.

The point of this example is that, within a single program, there are precedence constraints among the various statements. In the following sections, we formalize these ideas.

Definition

A precedence graph is a directed acyclic graph whose nodes correspond to individual statements. An edge from node S_i to node S_j means that statement S_j can be executed only after statement S_i has completed execution.

For example, in the precedence graph depicted in Figure 2.7, the following precedence relations exist:

- S_2 and S_3 can be executed after S_1 completes.
- S_4 Can be executed after S_2 completes.
- S_5 and S_6 can be executed after S_4 completes.
- S_7 Can execute only after S_5, S_6 and S_3 complete.

Note that S_3 can be executed concurrently with S_2, S_4, S_5 and S_6

The precedence graph must be acyclic. The graph in Figure 2.8 has the following precedence constraints: S_3 can be executed only after S_2 has completed and S_2 can be executed only after S_3 has completed. Obviously, these constraints cannot both be satisfied at the same time.

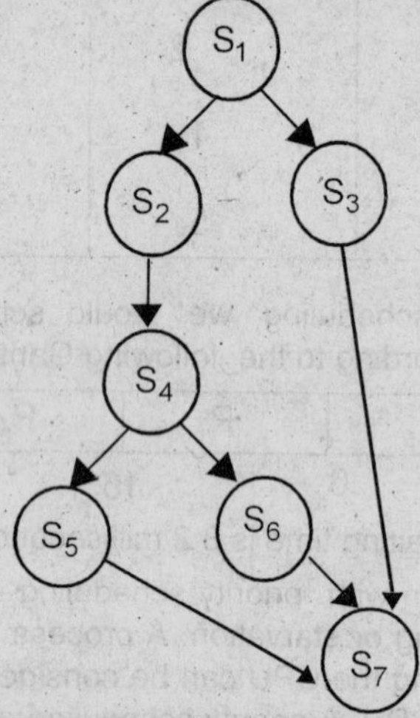

Figure 2.7 Precedence graph

Concurrency Conditions

When can two statements in a program be executed concurrently and still produce the same results ? This is, should there be an edge from S_1 to S_2 in the precedence graph corresponding to this program? Before we answer this question, let us first define some notation.

- $R(S_i) = \{a_1, a_2, ...a_m\}$ the read set for S_i ,is the set of all variables whose values are referenced in statement S_i during its execution.
- $W(S_i) = \{b_1, b_2, ...b_n\}$,the write set for S_i is the set of all variables whose values are changed (written) by the execution of statement S_i .

To illustrate this notation, consider the statement c := a – b. The values of the variables a and b are used to compute the new value of c. Hence a and b are in the read set. The (old) value of c is not used in the statement, but a new value is defined as a result of the execution of the statement. Hence c is in the write set, but not the read set.

R(c := a – b) = {a, b}
W(c := a – b) = {c}

For the statement, w := c + 1 the read and write sets are:

R(w := c + 1) = {c}
W(w := c + 1)] {w}

The intersection of R(S_i) and W(S_i) need not be null. For example, in the statement x := x + 2, R(x := x + 2) = W(x := x + 2) = {x}.

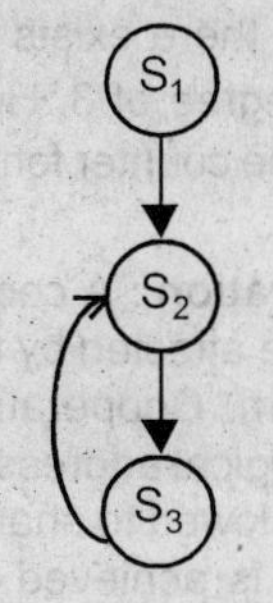

Figure 2.8 Precedence graph with cycles

As another example, consider the statement read (a). Notice that a is being read into, thus its value is changing. The read and write sets are :

R(read(a)) = { }

W(read(a)) = {a}

The following three conditions must be satisfied for two successive statements S_1 and S_2 to be executed concurrently and still produce the same result. These conditions were first stated by Bernstein and are commonly known as Bernstein's conditions.

1. $R(S_1) \cap W(S_2) = \{\ \}$
2. $W(S_1) \cap R(S_2) = \{\ \}$
3. $W(S_1) \cap W(S_2) = \{\ \}$

As an illustration, consider S_1 : a := x + y and : S_2 b := z + 1. These two statements can be executed concurrently because

$R(S_1)$ = {x, y}

$R(S_2)$ = {z}

$W(S_1)$ = {a}

$W(S_2)$ = {b}

However, S_2 cannot be executed concurrently with

S_3 : c := a – b, since $W(S_2) \cap R(S_3) = \{b\}$

Specification

The precedence graph is a useful device for defining the precedence constraints of the parts of a computation. However, a precedence graph would be difficult to use in a programming language, since it is a two-dimensional object. Other means be provided to allow the programmer to specify the precedence relations among the various statements in a program.

The Fork and Join Constructs

The fork and join instructions were introduced by Conway and Dennis and Van Horn. They were one of the first language notations for specifying concurrency.

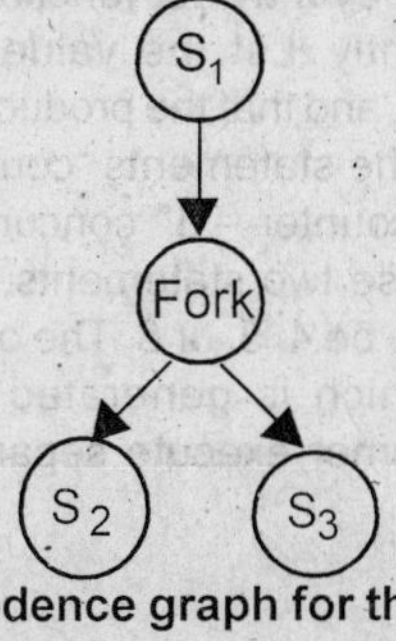

Figure 2.9 Precedence graph for the fork consturct

The fork L instruction produces two concurrent executions in a program. One execution starts at the statement labeled L, while the other is the continuation of the execution at the statement following the fork instruction.

To illustrate this concept, consider the following program segment :

```
S1;
fork L;
S2;
.
.
.
L:S3;
```

Part of the precedence graph corresponding to this program is presented in Figure 2.9 When the **fork** L statement is executed, a new computation is started at S_3; .This new computation executes concurrently with the old computation, which continues at S_2

Note that the execution of a **fork** statement splits one single computation into two independent computations; hence the name **fork**.

The join instruction provides the means to recombine two concurrent computations into one. Each of the two computations must request to be joined with the other. Since computations may execute at different speeds, one may executed the join before the other. In this case, the computation which executes the **join** first is terminated, while the second computation is allowed to continue. If there were three computations to be joined, the first two to execute the join are terminated, while the third is allowed to continue.

Figure 2.10 Precedence graph for the join construct

We need to know the number of computations which are to be joined; so that we can terminated all but the last one. The **join** instruction has a parameter to specify the number of computations to join. Thus the execution of the join instruction with a parameter count has the following effect:

Count := count – 1;

If count $\neq$ 0 **then** quit;

Where count is a non-negative integer variable, and quit is an instruction which results in the termination of the execution. For 2 computations, the variable count would be initialized to 2.

The join instruction must be executed atomically; that is, the concurrent execution of two join statements is equivalent to the serial execution of these two statements, in some undefined order.

To illustrate this concept, consider the following program segment :

```
Count := 2;
Fork L1;
.
.
.
s;
go to L2;
L1 : S2;
L2 : join count;
```

Part of the precedence graph corresponding to this program is presented in Figure 2.10.

Note that the execution of the **join** statement merges several concurrent executions; hence the name join.

Let us further illustrate these concepts through additional examples. Consider again Program 1. To allow the concurrent execution of the first two statements, this program could be rewritten using **fork** and **join** instructions :

```
      count := 2;
      fork L1;
      a := x + y;
      go to L2;
L1 : b := z + 1;
L2 : join count;
      c := a – b;
      w := c + 1;
```

Now return to the precedence graph of Figure 2.7. The corresponding program using the **fork** and **join** instructions is :

```
S1;
count := 3;
fork L1;
S2;
S4;
fork L2;
S5;
go to L3;
L2 : S6;
go to L3;
L1 : S3;
L3 : join count;
S7;
```

Note that in Figure 2.7 there exists only one **join** node S_7, which has an in-degree of 3. Hence, only one **join** statement is needed. The counter for this **join** is initialized to 3.

8.Process Synchronization : A cooperating process is one that can affect or be affected by the other processes executing in the system. Cooperating processes may either directly share a logical address space (that is, both code and data), or be allowed to share data only through files. The former case is achieved through the use of lightweight process or threads. Concurrent access to shared data may result in data inconsistency.

Background : The shared-memory-solution to the bounded-buffer problem:

Our solution allows at most n – 1 items in the buffer at the same time. Suppose that we wanted to modify the algorithm to remedy this deficiency. One possibility is to add an integer variable counter, initialized to 0. Counter is incremented every time we add a new item to the buffer, and is decremented every time we remove one item from the buffer. The code for the producer process can be modified as follows:

```
repeat
......
produce an item in nextp
......
while counter = n do no-op;
buffer[in] := nextp;
in := in + 1 mod n;
counter := counter + 1;
until false;
```

The code for the consumer process can be modified as follows:

```
repeat
While counter = 0 do no-op;
nextc := buffer[out];
out := out + 1 mod n;
counter := counter – 1;
.......
Consume the item in nextc
.......
until false;
```

Although both the producer and consumer routines are correct separately, they may not function correctly when executed concurrently. Let the value of the variable counter is currently 5, and that the producer and consumer processes execute the statements "counter := counter + 1" and "counter := counter – 1" concurrently. Following the execution of these two statements, the value of the variable counter may be 4, 5, or 6. The only correct result is counter = -5, which is generated correctly if the producer and consumer execute separately.

9. The Critical-Section Problem : Consider a system consisting of n processes {P_0, P_1,..., P_{n-1}}. each process has a segment of code, called a critical section, in which the process may be changing common variables, updating a table, writing a file, and so on. The important feature of the system is that, when one process is executing in its critical section, no other process is to be allowed to execute in its critical section. Thus, the execution of critical sections by the process is mutually exclusive in time.

The section of code implementing this request is the entry section. The critical section may be followed by an exit section. The remaining code is the remainder section. A solution to the critical section problem must satisfy the following three requirements:

1. Mutual exclusion: If process P_i is executing in its critical-section, then no other processes can be executing in their critical section.

2. Progress: If no process is executing in its critical section and there exist some processes that wish to enter their critical sections, then only those processes that are not executing in their remainder section can participate in the decision of which will enter its critical section next, and this selection cannot be postponed indefinitely

3.Bounded waiting : There exist a bound on the number of times that other processes are allowed to enter their critical sections after a process has made a request to enter its critical section and before that request is ganted.

Two process solution:

The process share two variables:

```
var flag : array [0...1] of boolean;
    turn : 0......1;
```

Initially flag [0] = flag [1] = false, and the value of turn is immaterial(but is either 0 or 1). The structure of process P_i is shown in Figure 2.11. below.

To enter the critical section, process P_i first sets falg[i] to be true, and then asserts that it is the other process turn to enter if appropriate (turn = j). If both processes try to enter at the same time, turn will be set to both i and j at roughly the same time. Only one of these assignments will last ; the other will occur; but will be overwritten immediately. The eventual value of turn decides which of the two processes is allowed to enter its critical section first.

To prove that this solution is correct, it is needed to show that:

1.Mutual exclusion is preserved

2.The progress requirement is satisfied

3.The bounded-waiting requirement is met

```
Repeat
    flag [ i ] := true ;
    true := j;
    While ( flag [ j ] and turn = j)do no-op
              Critical section
              flag [ j ] : = false ;
              remainder section
Until false;
```

Fig 2.11: Structure of process P_i.

Multiple-Process Solutions : An algorithm is developed for solving the critical-section problem for n processes, known as the bakery algorithm, and it is based on a scheduling algorithm commonly used in bakeries, ice-cream stores, meat markets, motor-vehicle registeries, and other locations where order must be made out of chaos. On entering the store, each customer receives a number. The customer with the lowest number is served next. P_i and P_j receive the same number and if i < j, then P_i is served first. Since process names are unique and totally ordered, our algorithm is completely deterministic.

The common data structures are:

```
var     choosing : array[0....n-1] of boolean;
        number   : array [0...n-1] of integer;
```

Initially, these data structures are initialized to false and 0, respectively.

For convenience, we define the following notation:

(a, b) < (c, d) if a < c or if a = c and b < d.

This structure of process P_i used in the bakery algorithm, shown in figure 2.12.

To prove that the bakery algorithm is correct, we need first to show that, if P_i is in its critical section and P_k trying (k ≠ i)has already chosen its number[k] ≠0, then (number [i] , i)< (number [k], k).

Now it is simple to show that mutual exclusion is observed. Indeed,consider P_i in its critical section and P_k trying to enter the P_k critical section. When process P_k executes the second while statement for j = i, if finds that

- number [i] ≠ 0;
- (number [i], i) < (number [k], k)

Thus, it continues looping in the while statement until P_i leaves the P_i critical section. If we wish to show that the progress and bounded-waiting requirements are preserved, and that the algorithm ensures fairness, it is sufficient to observe that the processes their critical section on a first-come, first-served basis order.

```
repeat
  choosing [ i ] : = true
  number [ i ] : = max ( number [ 0 ], number [1], ...,
  number [ n - 1 ] + 1)
  choosing [ i ] : = false ;
  for j : = 0 to n - 1
              do
              begin
  while choosing [ j ] do no-op ;
  while number [ j ]  ≠ 0
  and ( number [ j ], i ) < ( number [ i ], i ) do no-op;
end
Critical-section
  number [i] := 0;
remainder section
  until false;
```

Fig 2.12:The structure of process P_i in the bakery algorithm

10. Semaphores : To generalise the solutions to the critical-section problem we can use a synchronization tool called a semaphores. A semaphore S is an integer variable that, apart from initialization, is accessed only through two standard atomic operations; wait and signal. These operations were originally termed P (for wait; from the Dutch proberen, to test) and V(for signal; from verhogen, to increment).

The classical definitions of wait and signal are

```
wait (S) : while S <= 0 do no-op;
    S := S – 1
    signal (S) : S := S + 1;
```

Usage : We can use semaphores to deal with the n-process critical-section problem. The n processes share a semaphore, mutex (standing for mutual exclusion), initialized to 1. Each process P_i is organized as shown below figure 2.13. We can also use semaphores to solve various synchronization problems.

Consider two concurrently running processes : P_1 with a statement S_1, and P_2 with a statement S_2. Suppose that we require that S_2 be executed only after S_1 has completed. We can implement this scheme readily by letting P_1 and P_2 share a common semaphore synch, initialized to 0, and by inserting the statements

```
repeat
wait (mutex);
critical section
signal (mutex);
remainder section
until false;
```

Fig 2.13 : Mutual-exclusion implementation with semaphores

S_1;
signal(synch);
in process P_1, and the statements
wait(synch);
S_2;

In process P_2. Because synch is initialized to 0, P_2 will execute S_2 only after P_1 has invoked signal(synch), which is after S_1.

Implementation : The main disadvantage of the mutual-exclusion solutions is that they all require busy waiting. While a process is in its critical section, any other process that tries to enter its critical section must loop continuously in the entry code. This continual looping is clearly a problem in a real multiprogramming system, where a single CPU is shared among many processes. Busy waiting wastes CPU cycles that some other process might be able to use productivety. This type of semaphore is also called a spinlock(because the process "spins" while waiting for the lock). Spinlocks are useful in multiprocessor systems.

To overcome the need for busy waiting, we can modify the definition of the wait and signal semaphore operations. When a process executes the wait operation and finds that the semaphore value is not positive, it must wait. However, rather than busy waiting, the process can block itself. The block operation places a process into a waiting queue associated with the semaphore and the state of the process is switched to the waiting state. Then, control is transferred to the CPU scheduler, which selects another process to execute.

A process that is blocked, waiting on a semaphore S, should be restarted when some other process executes a signal operation. The process is restarted by a wakeup operation, which changes the process from the waiting state to the ready state. The process is then placed in the ready queue.

To implement semaphores under this definition, we define a semaphore as a record:

```
type semaphore = record
value : integer;
L : list of processes;
end;
```

Each semaphore has an integer value and a list of processes. When a process must wait on a semaphore, it is added to the list of processes. A signal operation removes one process from the list of waiting processes, and awakens that process.

The semaphore operations can now be defined as

```
wait (S) : S.value := S.value - 1;
if S.value < 0
then begin
addition this process to S.L;
block;
end;
signal(S) : S.value := S.value + 1;
if S.value ≤ 0
then
begin
remove a process P from S:L;
wakeup(P);
end;
```

This block operation suspends the process that invokes it. The wakeup(P) operation resumes the execution of a blocked process P. these two operations are provided by the Operating system as basic system calls.

Deadlocks and Starvation : The implementation of a semaphore with a waiting queue may result in a situation where two or more processes are waiting indefinitely for an even that can caused by only one of the waiting processes. The event in question is the execution of a signal operation. When such a state is reached, these processes are said to be deadlocked.

Binary Semaphores : The semaphore construct is commonly known as a counting semaphore, since its integer value can range over an unrestricted domain. A binary semaphore is a semaphore with an integer value that can range only between 0 and 1.

11. Classical Problems of Synchronization

The Bounded-Buffer Problem

The bounded-buffer problem is commonly used to illustrate the power of synchronization primitives; We present here a general structure of this scheme, without committing ourselves to any particular implementation. We assume that the pool consists of n buffers, each capable of holding one item. The mutex semaphore provides mutual exclustion for accesses to the buffer pool and is initialized to the value 1. The empty and full semaphores count the number of empty and full buffers, respectively. (Semaphores with initial values other than 1 are sometimes known as counting semaphores.) The semaphore empty is initialized to the value n; the semaphore full is initialized to the value 0.

```
repeat
    ...
    produce an item in nextp
    ...
    wait (empty);
    wait (mutex);
    ...
    add nextp to buffer
    ...
    signal (mutex);
    signal (full);
until false;
```

Figure 2.14 The structure of the producer process.

The code for the producer process is shown in Figure 2.14; the code for the consumer process is shoun in Figure 2.15. Note the symmetry between the producer and the consumer. We can interpret this code as the producer producing full buffers for the consumer, or as the consumer producing empty buffers for the producer.

```
repeat
        wait(full);
        wait(mutex);
        ...
        remove an item from buffer to nextc
        ...
        signal(mutex);
        signal(empty);
        ...
        consume the item in nextc
        ...
        until false;
```

Figure 2.15 The structure of the consumer process.

The Readers and Writers Problem

A data object (such as a file or record) is to be shared among several concurrent processes. Some of these processes may want only to read the content of the shared object, whereas others may want to update (that is, to read and write) the shared object. We distinguish between these two types of processes by referring to those processes that are interested in only reading as readers, and to the rest as writers. Obviously, if two readers access the shared data object simultaneously, no adverse effects will result. However, if a writer and some other process (either a reader or a writer) access the shared object simultaneously, chaos may ensure.

To ensure that these difficulties do not arise, we require that the writers have exclusive access to the shared object. This synchronization problem is referred to as the readers-writers problem. Since it was originally stated, it has been used to test nearly every new synchronization primitive. The readers-writers problem has several variations, all involving priorities. The simplest one, referred to as the first readers-writers problem, requires that no reader will be kept waiting unless a writer has already obtained permission to use the shared object. In other words, no reader should wait for other readers to finish simply

```
wait(wrt);
...
writing is performed
...
signal(wrt);
```

Figure 2.16 The structure of a writer process

because a writer is waiting. The second readers-writers problem requires that, once a writer is ready, that writer performs its write as soon as possible. In other words, if a writer is waiting to access the object, no new readers may start reading.

We note that a solution to either problem may result in starvation. In the first case, writers may starve; in the second case, readers may starve. For this reason, other variants of the problem have been proposed. In this section, we present a solution to the first readers-writers problem.

In the solution to the first readers-writers problem, the reader processes share the following data structures.

```
var mutex, wrt: semaphore;
readcount : integer;
```

The semaphores mutex and wrt are initialized to 1; readcount is initialized to 0. The semaphore wrt is common to both the reader and writer processes. The mutex semaphore is used to ensure mutual exclusion when the variable readcount is updated. Readcount keeps track of how many processes are currently reading the object. The semaphore wrt functions as a mutual exclusion semaphore for the writers. It also is used by the first or last reader that enters or exits the critical section. It is not used by readers who enter or exit while other readers are in their critical sections.

The code for a writer process is shown in Figure 2.16 the code for a reader process is shown in Figure 2.17. Note that, if a writer is in the critical section and n readers are waiting, then one reader is queued on mutex. Also observe that, when a writer executes signal(wrt), we may resume the execution of either the waiting readers or a single waiting writer. The selection is made by the scheduler.

```
wait(mutex);
readcount := readcount + 1;
if readcount = 1 then wait(wrt);
signal(mutex);
...
reading is performed
...
wait(mutex);
readcount := readcount - 1;
if readcount = 0 then signal (wrt);
signal(mutex);
```

Figure 2.17 The structure of a reader process.

The Dining-Philosophers Problem

Consider five philosophers who spend their lives thinking and eating. The philosophers share a common circular table surrounded by five chairs, each belonging to one philosopher. In the center of the table there is a bowl of rice, and the table is laid with five single chopsticks (Figure 2.18). When a philosopher thinks, she does not interact with her colleagues. From time to time, a philosopher gets hungry and tries to pick up the two chopsticks that are closest to her (the chopsticks that are between her and her left and right neighbours). A philosopher may pick up only one chopstick at a time. Obviously, she cannot pick up a chopstick that is already in the hand of a neighbor. When a hungry philosopher has both her chopsticks at the same time, she eats without releasing her chopsticks. When she is finished eating, she puts down both of her chopsticks and starts thinking again.

The dining-philosophers problem is considered a classic synchronization problem, neither because of its practical improtance, nor because computer scientists dislike philosophers, but because it is an example for a large class of concurrency-control problems. It is a simple

Figure 2.18 The situation of the dining philosophers.

representation of the need to allocate several resources among several processes in a deadlock and starvation-free manner.

One simple solution is to represent each chopstick by a semaphore. A philosopher tries to grab the chopstick by executing a wait operation on that semaphore; she releases her chopsticks by executing the signal operation on the appropriate semaphores. Thus, the shared data are

var chopstick: **array** [0..4] **of** semaphore;

where all the elements of chopstick are initialized to 1. The structure of philosopher i is shown in Figure 2.19.

Although this solution guarantees that no two neighbors are eating simultaneously, it nevertheless must be rejected because it has the possibility of creating a deadlock. Suppose that all five philosophers become hungry sinultaneously, and each grabs her left chopstick. All the elements of chopstick will now be equal to 0. When each philosopher tries to grab her right chopstick, she will be delayed forever.

Several possible remedies to the deadlock problem are listed next. we present a solution to the dining-philosophers problem that ensures freedom from deadlocks.

- Allow at most four philosophers to tbe sitting simultaneously at the table.
- Allow a philosopher to pick up her chopsticks only if both chop - sticks are available (note that she must pick them up in a critical section).
- Use an asymmetric solution; that is, an odd philosopher picks up first her left chopstick and then her right chopstick, whereas an even philosopher picks up her right chopstick and then her left chopstick.

```
repeat
    wait(chopstick[i]);
    wait(chopstick[i+1 mod 5]);
    ...
    eat
    ...
    signal(chopstick[i]);
    signal(chopstick[i+1 mod 5]);
    ...
    think
    ...
until false;
```

Figure 2.19 The structure of philosopher i.

Finally, any satisfactory solution to the dining-philosophers problem must guard against the posssibility that one of the philosophers will starve to death. A deadlock-free solution does not necessarily eliminate the possibility of starvation.

12. Deadlocks

In a multiprogramming environment, several processes may compete for a finite number of resources. Deadlocks can occur when processes have been granted exclusive access to devices, files etc.,. We refer all the devices, files, databases are Resources. A computer normally have may different resources that can be acquired.

A resource is anything that can be only used by a single process at any time. The sequence of events required to use a resource are

1. Request the resource.
2. Use the resource
3. Release the resource.

A process must request a resource before using it and release it after using it. The request and release of resources are purely done by system calls. The operating system maintains a table which has a record of whether each resource is free or allocated, and if allocated, to which process. If a request for an allocated resource is made then it will be added to a queue of process waiting for resource.

Definition of Deadlock : A process in a multiprogramming system is said to be in a state of Deadlock (or deadlocked) if it is waiting for a particular event that will not occur.

Example : A process requests resources, and if the resources are not available at that time, the process enters a wait state. It may happen that waiting processes will never again change state, because the resources they have requested are held by other waiting processes. For example, this situation occurs in a system with four tape drives and two processes. If each process holds two tape drives but needs three, then each will wait for the other to release its tape drives. This situation is called deadlock.

The Deadlock problem : To illustrate the deadlock problem let us consider the problem of crossing a river that has number of stepping stones by which we can go to the other side of the river. Let us assume that the stepping stone used by the man are resources. A deadlock occurs when two persons stand crossing the river opposite sides and meet in the middle as shown in the figure 2.20.

Figure 2.20 Deadlock situation in river crossing

To step on a stone is, request and using a resource and taking the foot is releasing the resource.

One possibility to avoid deadlock is to Roll back, i.e., to step back. Either one process has to release its resource for the other process to step up and go to the opposite bank. Moreover when one person stands from one point must also check that no one is standing from the another end. What is needed is that, to follow upon an agreed protocol. If no one comes from another side then he can cross, otherwise he must wait.

Several remarks should be made concernig Agreed protocol.

1. Mechanism to find any person from otherside of the river.
2. Assign high priority to one side of the river.
3. If more than one process are to wait indefinitely then the situation is called Starvation. Instead of higher-priority to one bank alternate priorities from time-to-time can be arranged and designed.

Deadlock characterization : Deadlock at all times is undesirable. Deadlock occurs only if the following four conditions hold simultaneously in a system. The necessary conditions are :

1. Mutual exclusion : Only one process at a time can use a resource. Other requesting processes must wait until the resource is released.

2. Hold and wait : Processes currently holding resources granted earlier can request new resources and keep waiting for the additional resoruce.

3. No preemption : Resources previously granted to a process cannot be forcibly taken away from it. They (resources) must be explicitly released by the process holding them.

4. Circular wait : There must be a circular chain of two or more processes each of which is waiting for a resource held by the next number of chain.

Example :

Set of processes $\{P_0, P_1........P_n\}$

P_0 waiting for resource held by P_1

P_1 waiting for resource held by P_2

...

...

...

P_{n-1} waiting for resource held by P_n

P_n waiting for resource held by P_0

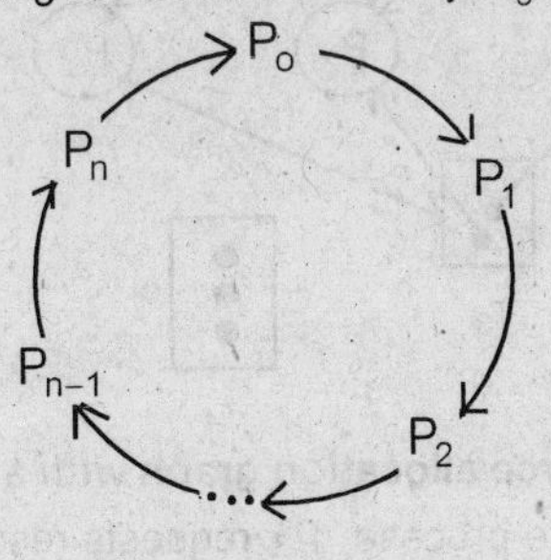

Fig: 2.21 Circular wait

Resource Allocation Graph : Deadlock can be effectively described in terms of a Directed Graph called a system Resource Allocation graph.

The graph has a pair $G = (V, E)$

where V is set of vertices.

E is set of edges.

In our resources graph the set of vertices is separated into two types.

$P = \{p_1, p_2, p_3............p_n\}$ whcih is the set consisting of all the process in the system.

$R = \{ r_1, r_2, r_3 .... r_n \}$ which is the set consisting of all resources types in the system.

Each elements in the set, E of edges is an Ordered pair (p_i, r_j) or (r_j, p_i) where p_i is a process and r_j is a resource type.

1. If $(p_i, r_j) \in E$ then it means that process p_i requested for resource r_j and currently waiting for that resoruce (Request Edge).

2. If $(r_j, p_i) \in E$ then it means that resource type r_j has been allocated to process p_i (Assignment Edge)

In graphical form process p_i is represented as a Circle and r_j is represented as a Square.

In a resource graph if there is no cycle then no process in the system is deadlocked. If each process is involved in a cycle then it is said to be a deadlock situation.

The figure 2.22 shows the resource graph with no deadlock.

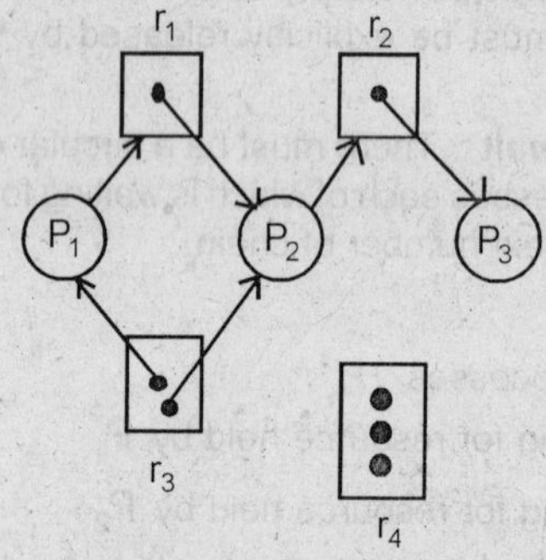

Figure 2.22 Resource allocation graph

The figure 2.23 shows the resource graph with deadlock.

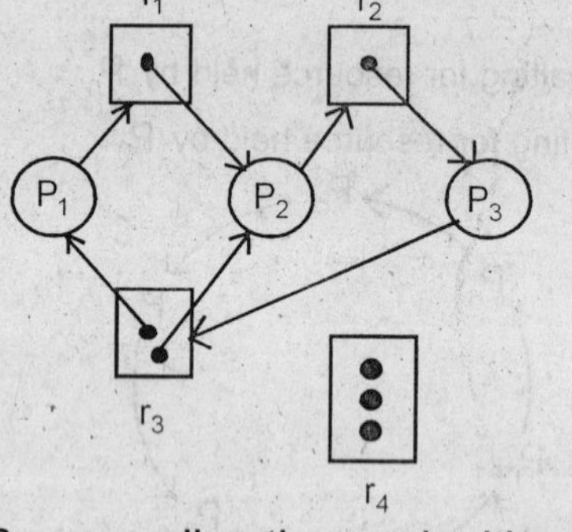

Fig. 2.23. Resource allocation graph with a deadlock

In the above figure process p_3 requests resource type r_3. Therfore, a request edge (p_3, r_3) is added in the figure 2.22

p_2 waits for r_3 which is held by p_3 and p_3 waiting for r_2 and r_3 which are held by p_2 and p_1 respectively.

Thus there is a cycle and this situation is a deadlock which can be identified by the use of resource graphs.

There are also other methods for handling deadlocks. They are

I. Deadlock Prevention

II. Deadlock Avoidance.

13. Deadlock Prevention : Deadlock can be prevented by ensuring that atleast one of the four conditions stated by Coffman cannot hold.

First let us attack the Mutual Exclusion condition. If no resource was ever be asssigned exclusively to a single process, we would never have deadlocks. Mutual exclusion can be guaranteed in printers when the process are spooled and manged by printer daemon. The printer daemon never request any other resources.

The second condition is Hold and Wait. If we can prevent processes that hold resources from waiting for more resources we can eliminate deadlocks. One way to achieve this goal is to require all processes to request all their resources before starting execution. But the problem is that the requirement of resources for some processes may not be predicted beforehand. An alternate approach to break hold-and wait condition is to require a process requesting a resource to first temporarily release all the resources it currently holds. Then after the request is successful it can get the original resources back.

The third condition is No Preemption. No resource should be forcibly taken away when it is used by one process. For example, a printer cannot be taken away from one process in the middle and be given to another process as this may lead to confusion.

The fourth condition is No Circular wait. One way to attack this is to have a rule saying that a process is entitled only to a single resource at any moment. If it needs a second resource then it must release the first one. But in some job it is unacceptable and not possible. Another way is that to provide a global numbering of all the resources follows.

Card reader - 1

Printer - 2

Plotter - 3

Tape drive - 4

Card punch - 5

The rule is that : The processes can request resources whenever they want but all requests must be made in numerical order. A process may request first a printer and then a tape drive, but not a printer and then a tape drive, but not a plotter and then a printer. With this rule the resource allocation graph can never have cycles. For example, assuming i and j are distinct resources they will have different numbers. If i>j then A is not allowed to request j. If i < j then B is not allowed to request i. Eitherway deadlock is impossible.

14. Deadlock Avoidance : To avoid deadlock additional information are required about how resources are to be requested. That is the order in which the request of resources are to be made. With a knowledge of complete sequence of requests and releases for each processes we can decide whether a process should wait or not. There are several algorithms to decide upon whether the current request can be satisfied or it should wait.

The Banker's Algorithm for a single resource : The Banker;s Algorithm is easily extendable to pools of resources of several different types. For example, consider the allocation of a quantity, t, of identical tape drives.

An operating system shares a fixed number of equivalent tape drives, t, among a fixed number of users, u. Each user specifies in advance the maximum number of tape drives needed during execution. The operating system will accept a user's request if that user's maximum need for tape drives does not exceed t. A user may obtain or release tape drives one by one. Sometimes, a user may have to wait to obtain an additional tape drive, but the operating system guarantees a finite wait. The current number of tape drives allocated to a user will never exceed that user's stated maximum need. If the operating system is able to satisfy a user's maximum need for tape drives, then the user guarantees that the tape drives will be used and released to the operating system within finite time.

The current state of the system is called safe if it is possible for the OS to allow all current users to complete their jobs within a finite time. If not, then the current system state is called unsafe.

Now suppose there are n users.

Let l (i) represent user i's current loan of tape drives. If user 5, for example, has been allocated four tape drives, then l (5) = 4. Let m(i) be the maximum need of user i so that if user 3 has a maximum need of two tape drives then m(3) = 2. Let c(i) be the current claim of a user where a user's claim is equal to his maximum need minus his current loan. If, for example, user 7 has a maximum need of six tape drives and a current loan of four tape drives, then we have c(7) = m(7) - l (7) = 6 - 4 = 2

The operating system controls t tape drives. Let a be the number of tape drives still available for allocation. Then a is t minus the sum of the loans to the various users.

Dijkstra's Banker's algorithm says to allocate tape drives to users only when the allocations result in safe states rather than in unsafe states. A safe state is one in which the total resource situation is such that all users would eventually be able to finish. An unsafe state is one that might eventually lead to deadlock.

Example of Safe State

Suppose a system has 12 equivalent tape drives, and 3 users sharing the drives as in State I

State I

	Current loan	Maximum need
User (1)	1	4
User (2)	4	6
User (3)	5	8
Available	2	

This state is "safe" because it is still possible for all three users to finish. Note that User (2) currently has four tape drives allocated to him and will eventually need a maximum of six, or two additional drives.

The system has 12 drives of which 10 are currently in use and two are available. If the 2 available drives are given to User (2)'s maximum need, then User (2) may run to completion. User User (2) fulfilling (2) upon finishing will release all six tape drives that the system may assign to User(1) and User(3). User(1) has one tape drive and will need three more eventually. User(3) has five and will need three more eventually. If User(2) returns six, then three may be given to User(1) who may then finish and return four tape drives to the system. The system may then give three drives to User(3) who may now also finish. Thus the key to a state being safe is that there is a way for all users to finish.

Example of an Unsafe State

Assume a system's 12 tape drives are allocated as in State II.

State II

	Current loan	Maximum need
User (1)	8	10
User(2)	2	5
User (3)	1	3
Available	2	

Here eleven of the system's twelve tape drives are currently in use and only one drive is available for allocation. At this point, no matter which user requests the available drive, we cannot guarantee that all three users will finish. In fact, suppose User(1) requests and is granted the last available drive. A three-way deadlock could occur if indeed each process needs to request at least one more drive before releasing any drives to the pool.

Example of Safe State to Unsafe state Transition

That a state is known to be safe does not imply that all future states will be safe. Our resource allocation policy must carefully consider all resource requests before granting them. For example, suppose the current state of a system is as in State III.

State III

	Current loan	Maximum need
User (1)	1	4
User (2)	4	6
User (3)	5	8
Available	2	

Now suppose that User(3) requests an additional resource. If the system were to grant this request, then the new state would be State IV.

State IV

	Current loan	Maximum need
User (1)	1	4
User (2)	4	6
User (3)	6	8
Available	1	

Certainly, State IV is not necessarily deadlocked. But the state has gone from a safe one to an unsafe one. The Sate IV characterizes a system in which completion of all user processes cannot be guarnteed. One resource is available. A minimum of two resources must be available to ensure that either User (2) or User (3) could complete, return its resources to the system, and ultimately allow the other users to finish.

Weaknesses in the Banker's Algorithm

- Fixed number of resources are allowed to allocate.
- Fixed number of Users are allowed.
- All requests are granted within finite time, but for real systems much more better guaranting time is needed.
- Jobs guarantee that loans will be repaid (ie., resources will be returned) with in finite time, but for real systems much better guarantees than this are needed.
- The user must state that their maximum needs in advance.

The Bankers Algorithm for Multiple Resources : When arbitrary number of processes and arbitrary number of resources classes are used it is very difficult to perform allocation each with multiple instances. The banker's algorithm for single resource can be extended to multiple resources.

Resources Assigned

A	3	0	1	1
B	0	1	0	0
C	1	1	1	0
D	1	1	0	1
E	0	0	0	0

Column 1 : Processes
Column 2 : Tape drives
Column 3 : Plotters
Column 4 : Printers

A	1	1	0	0
B	0	1	1	2
C	3	1	0	0
D	0	0	1	0
E	2	1	1	0

Column 1 : Processes
Column 2 : Tape drives
Column 3 : Plotters
Column 4 : Printers
Column 5 : Punches

P = (5 3 2 2)   E = (6 3 4 2)

Resources still needed A = (1 0 2 0)

In the above diagram we have two martices. The first matrix shows how many resources is currently assigned to each of the five processes (A,B,C,D,E). The Second matrix shows how many processes still needs in order to complete.

The three vectors E,P,A shows the Existing resource (E), the possessed resource(P), and the Available resource(A) respectively.

E = (6 5 4 2) is the system has 6 tape drives, 3 plotters, 4 printers and 2 punches.

P = (5 3 2 2) is resources possessed by processes are 5 tape drives, 3 plotters, 2 printers and 2 punches, that are currently assigned.

A = (1 0 2 0) is the available resource vector which is simply the difference between what the system has and what is currently in use, i.e., [E-P = A]. The available resources after allocation is 1 tape drive, no plotter, 2 printers, no punches.

This is the State at present.

The algorithm for checking to see if a state is safe is :

1. Look for a row, R, whose unmet resource needs are smaller than A. If no such rows exists, the system is deadlocked since no process can run to completion.
2. If row chosen has met all the requests it needs and finishes, mark that process as Terminated and add all its resources to A vector (Available).
3. Repeat steps 1 and 2 until either all processes are marked terminated, in which case the initial state was safe, or until a deadlock occurs, in which case it was unsafe.

In the above example matrix, the current state is safe. Now B process requests a printer. This request can be granted. The state is safe now. Now E wants the last printer. Granting it the available 'A' vector will the (1 0 0 0) which leads to deadlock. E's request must be deferred (to wait) for a while.

For specific application many excellent special-purpose algorithms are known and used.

15. Deadlock Detection and Recovery : Deadlock detection is the process to check whether the deadlock exists or not. If deadlock exists, then identify the processes and resources that are involved in the deadlock. Graph reduction technique is used for detecting deadlocks. This technique detemines the processes that are completed and the processes that will remain deadlocked.

In the resource allocation graphs, squares represent processes and large circles represent classes of identical resources. Small circles drawn inside large circles indicate the number of identical resources of each class. For example, a large circle labeled "R1" containing three small circles indicates that there are three equivalent resources of type R1 available for allocation in this system.

If a process's resource requests may be granted, then we say that a graph may be reduced by that process. This reduction is equivalent to showing the graph if this process is allowed to complete and return its resources to the system. The reduction of a graph by a particular process is shown by removing the arrows to that process from resources (i.e., resources allocated to that process), and by removing arrows from that process to resources (i.e., current resource requests of that process). If a graph can be reduced by all its processes, then there is no deadlock. If a graph cannot be reduced by all its processes, then the "irreducible" processes constitute the set of deadlocked processes constitute the set of deadlocked processes in the graph. Figure 2.24 shows a series of graph reductions that eventually demonstrates that a particular set of processes is not deadlocked.

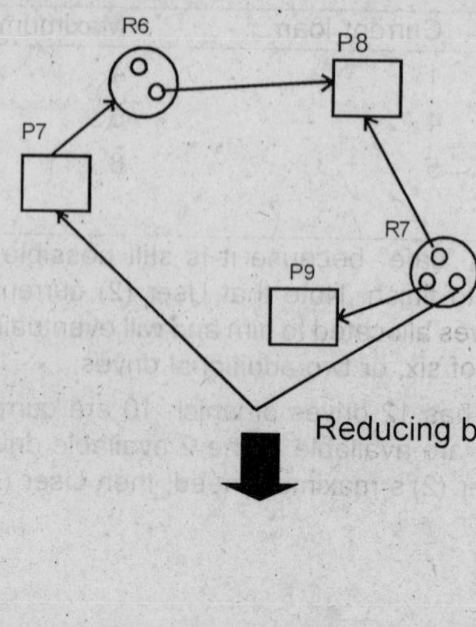

Reducing by P9

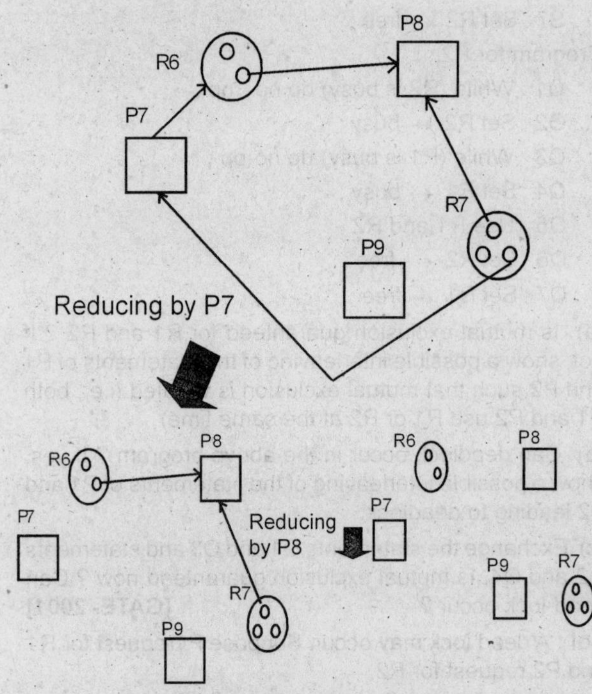

Figure 2.24 Graph Reductions

Figure 2.25 : shows a set of processes that is irreducible and thus constitutes a deadlocked system.

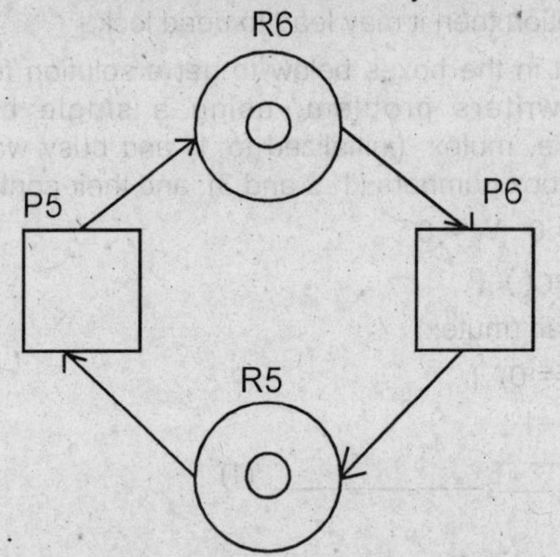

Process P5 has been allocated resource R5 that is being requested by process P6 that has been allocated resource R4 that is being requested by process P5 (the "circular wait")

Figure 2.25 Resource allocation and request graph.

16. Deadlock Recovery : Once a system has become deadlocked, the deadlock must be broken by removing one or more of the necessary conditions. Usually, several processes will lose some or all of the work they have accomplished.

The recovery is ordinarily performed by forcibly removing a process from the system and reclaiming its resources. The removed process is lost, but the remaining processes may now be able to complete. Sometimes it is necessary to remove several processes until sufficient resources have been reclaimed to allow the remaining processes to finish. Processes may be removed according to some priority order.

SOLVED EXAMPLES

1. The concurrent programing constructs fork and join are as below:

fork < label > which creates a new process executing from the specified label

Join < variable > which decrements the specified synchronisation variable (by 1) and terminates the process if the new value is not 0.

Show the precedence graph for S1, S2, S3, S4 and S5 of the concurrent program below.

```
N = 2
M = 2
fork L3
fork L4
S1
L1 : join N
S3
L2 : join M
S5
L3 : join S2
goto  L1
L4 : S4
goto L2
next :
```

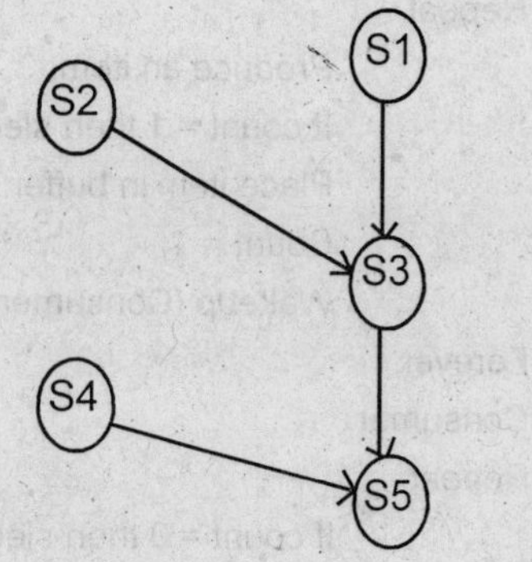

Sol: Precedence graph for the program is

In-degree of S_1, S_2 and S_4 = 0

In-degree of S_3,S_5 = 2

2. Is it possible to have deadlock when there is only one process? Explain.

Sol : Yes. If a process is given the task of waiting for an event to occur, and if the system includes provision for signalling the event, then we have a one-process deadlock.

e.g. let us consider two processes P_1 and P_2 and two resources A and B. Consider the following situation :

(a) A is allocated to P_1

(b) But P_1 requests B

(c) B is allocated to P_2

(d) But P_2 requests A and

(e) B will be released only when P_2 gets A and A will be released only when P_1 gets B.

Figure below shows resource allocation graph.

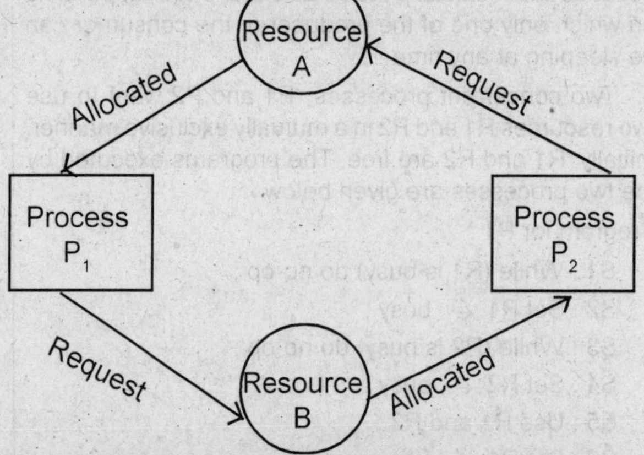

Fig. Deadly embrace deadlock (circular wait)

3. (a) A certain processor provides a 'test and set' instruction that is used as follows.

TEST register, flag

This instruction automatically copies flag to register and sets flag to 1. Give pseudocode for implementing the entry and exit code to a critical region using this instruction.

(b) Consider the following solution to the producer-consumer problem using a buffer of size 1. Assume that the initial value of count is 0. Also assume that the testing of count and assignment to count are atomic operations.

```
Producer :
Repeat
        Produce an item;
        If count = 1 then sleep;
        Place item in buffer:
        Count = 1;
        Wakeup (Consumer):
Forever.
Consumer
Repeat
        If count = 0 then sleep;
        Remove item from buffer;
        Count = 0;
        Wakeup (producer);
        Consume item;
Forever;
```

Show that in this solution it is possible that both the processes are sleeping at the same time. **[GATE-1999]**

Sol :

(a) loop : TEST R1, lock

JNZ loop ; ie. If lock was already true, continue to remain in loop, else exit

loop

Critical Section

MOV lock, 1

(b) In the given program segment, it is not possible for both processors to be sleeping at the same time. This is because count can take the value either 0 or 1, depending on which only one of the producer or the consumer can be sleeping at any time

4. Two concurrent processes P1 and P2 want to use two resources R1 and R2 in a mutually exclusive manner. Initially, R1 and R2 are free. The programs executed by the two processes are given below .

Program for P1 :

```
S1 : While (R1 is busy) do no-op ;
S2 : Set R1 ← busy ;
S3 : While (R2 is busy) do no-op ;
S4 : Set R2 ← busy ;
S5 : Use R1 and R2 ;
S6 : Set R1 ← free ;
S7 : Set R2 ← free ;
```

Program for P2 ;

```
Q1 : While (R2 is busy) do no - op ;
Q2 : Set R2 ← busy ;
Q3 : While (R1 is busy) do no-op ;
Q4 : Set R1 ← busy ;
Q5 : Use R1 and R2 ;
Q6 : Set R2 ← free ;
Q7 : Set R1 ← free ;
```

(a) Is mutual exclusion guaranteed for R1 and R2 ? If not, show a possible interleaving of the statements of P1 and P2 such that mutual exclusion is violated (i.e., both P1 and P2 use R1 or R2 at the same time)

(b) Can deadlock occur in the above program ? If yes, show a possible interleaving of the statements of P1 and P2 leading to deadlock.

(c) Exchange the statements Q1 and Q3 and statements Q2 and Q4. Is mutual exclusion guaranteed now ? Can dead-lock occur ? **[GATE- 2001]**

Sol : A dead lock may occur. Suppose P1request for R1 and P2 request for R2.

For P1 R1→busy

For P2 R2→busy

and if process P1or P2requires other resource in order to completion then it may lead to dead lock.

5. (a) Fill in the boxes below to get a solution for the readers-writers problem, using a single binary semaphore, mutex, (initialized to 1) and busy waiting. Write the box numbers (1, 2 and 3), and their contents

```
int R = 0,  W = 0 ;
Reader ( ) {
L1 : wait (mutex) ;
if  (w == 0) {
R = R +1 ;
[          ] ____________ (1)
}
else {
[          ] ____________ (2)
goto L1 ;
}
... / * do the read * /
wait (mutex) ;
R = R - 1 ;
signal (mutex) ;
}
Writer ( ) {
L2 :        wait (mutex) ;
if ([          ]) {     (3)
signal (mutex) ;
goto L2 ;
}
```

w = 1 ;

isgnal (mutex) ;

... / * do the write * /

wait (mutex) ;

w = 0 ;

signal (mutex) ; **[GATE- 2000]**

(b) Can the above solution lead to starvation of writers ?

Sol:

(a) (1) signal (mutex)

(2) signal (mutex)

(3) R ! = or W ! = 0

or

not (R = 0 and W = 0)

(b) Yes, writers may starve.

6. On a system with n CPU's what is the maximum number of processes that can be in the ready, run, and blocked states ?

Sol : There is no limit to the number of processes that can be in the ready and blocked staes. Atmost, n processes may be in the run state, since a process in the run state must be allocated to a CPU, and there are only n CPUs.

7. On a system with n CPUs, what is the minimum number of processes that can be in the ready, run, and blocked states ?

Sol : There can be zero processes in any of the three states. It is possible for all the processes to be blocked while waiting on I/O operations, leaving no processes in the ready and running states. All processes can be either in the ready or run state, leaving no processes blocked.

8. For each of the following transitions between process states, indicate whether the transition is possible. If it is possible, given an example of one thing that would cause it.

a) Run → ready
b) Run → blocked
c) Run → swapped - blocked
d) Blocked → run
e) Run → terminated

Sol :

a) Possible, when a process's time quantum expires

b) Possible, when a process issues an I/O request

c) Not possible, although process could first go to the blocked state, then the swapped - blocked.

d) Not possible, although process could first go to the ready state, then to the run state.

e) Possible, when a process terminates itself.

9. Given a system with n processes, how many possible ways can those processes be scheduled ?

Sol : This is a combinatorics problem of how many ways n objects can be ordered. When selecting the first object, one has n possible choices. For the second object, n - 1. Each subsequent selection involves one less choice. The total number of possibilities is computed by multiplying all the possibilities at each point, making the answer n !.

10. On a system using round - robin scheduling. What would be the effect of including one process twice in the list of processes ?

Sol : In round robin, all ready processes are given equal access to the CPU. CPU time is divided equally among the processes. If a process was listed twice, it would be granted twice the CPU time of the other processes.

11. On a system using multilevel feedback queues, a totally CPU-bound process requires 40 seconds to execute. If the first queue uses a time quantum of 2 and at each level the time quantum increases by 5 time-units, how many times will the job be interrupted and on what queue will it be when it terminates ?

Sol : The process will be interrupted at time 2 in the first queue, time 2 + 7 in the second queue, time 9 + 12 the third queue, time 21 + 17 the fourth queue, and will be running in the fifth queue when it terminates. There were four interrupts.

12. To what degree do the following algorithms favor short processes ?

a) First - Come First - Served

b) Shortest Job First

c) Shortest Remaining Time

d) Round Robin

e) Multilevel Feedback Queues

Sol : a) First-come first-served is relatively unfriendly to short jobs. Short jobs that follow long jobs must wait until the long job releases the CPU.

b) Clearly favors short jobs.

c) Favors short jobs even more than shortest job first since it will preempt a longer job currently running on the CPU

d) Neutral toward short jobs, treating all jobs equally.

e) Generally neutral toward short jobs, although perhaps they fare slightly better in multilevel feedback queues than in round robin. Large CPU - bound jobs move to lower - priority queues where they can only execute if no other processes are ready.

13. Gridlock is a traffic problem (Fig.1) in which further movement by traffic is impossible. Show that the four conditions of deadlock apply.

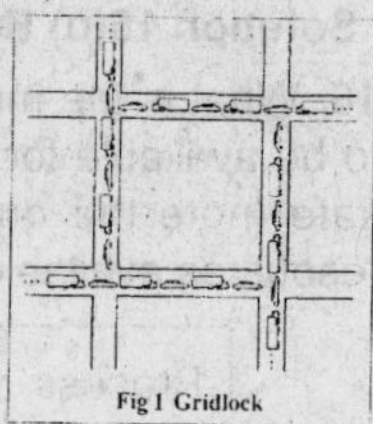

Fig I Gridlock

Sol : Mutual Exclusion : Only one car may be occupying a particular spot on the road at any instant.

Hold and Wait : No car ever backs up.

No Preemption : No car is permitted to push another car out of the way.

Circular Wait : Each corner of the city block contains vehicles whose movement depends on the vehicles blocking the next intersection.

14. Given process resource usage and availability as described in Table 1., draw the resource allocation graph

Table 1. Resource Usage

Process	Current Allocation			Outstanding Requests			Resources Available		
	R_1	R_2	R_3	R_1	R_2	R_3	R_1	R_2	R_3
P_1	2	0	0	1	1	0	0	0	0
P_2	3	1	0	0	0	0			
P_3	1	3	0	0	0	1			
P_4	0	1	1	0	1	0			

Sol :

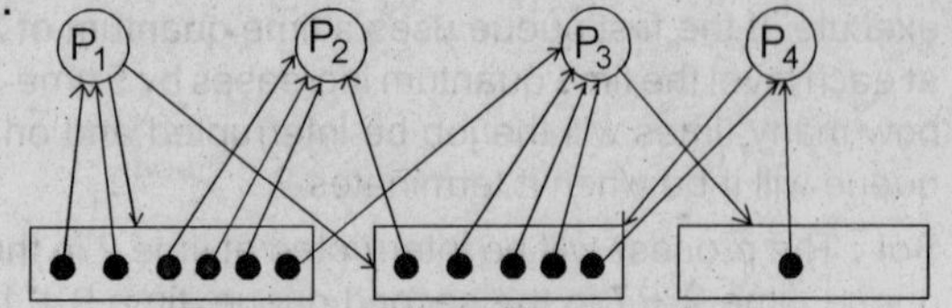

Resource allocation graph

15. Given process resource usage and availability as described in Table 1., draw the reduced resource allocation graph.

Sol : P_2 has only links pointing toward it, so all those links may be erased. The requests for R_1 and R_2 by P_1 may now be granted, leaving P_1 with arcs that only point toward it. Those arcs can be erased leaving the graph looking like this.

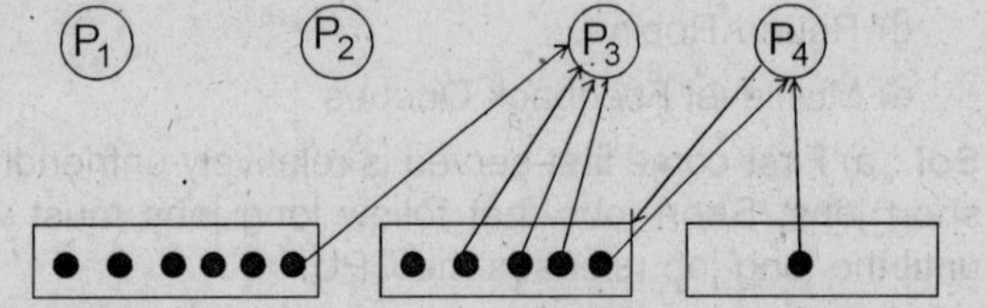

Sol : 15 (a) Partially reduced resource allocation graph

The request for R_3 by P_4 may now be granted and all the arcs pointing toward P_4 may be erased. Similarly, the arcs for P_3 can be erased and the reduced graph has no arcs.

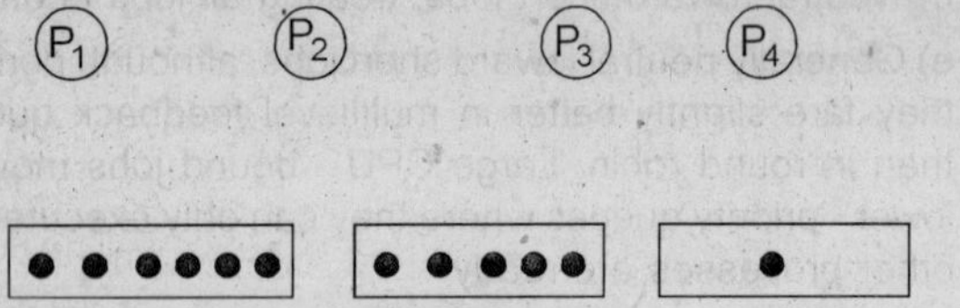

Solution 15(b) Reduced resource allocation graph

16. What is the minimum number of resources needed to be available for the state described in Table 2. to be safe (note the question is asking for the number of resources available not how many resources exist). :

Process	Current Allocation	Maximum Allocation
	R_1	R_1
P_1	1	3
P_2	1	2
P_3	3	9
P_4	2	7

Table : 2

Sol : If one instance of R_1 was available, process P_2 could be guaranteed of running to completion. It could then release the resource it is currently using, making two of R_1 available. This would allow P_1 to run to completion, increase the number of R_1's available to three. With only three R_1's P_3 and P_4 could still deadlock if they both request their maximum potential allocation. If there were three R_1's to start instead of one, P_4 would now have five R_1's available to it and could run to completion. Adding its two R_1's to the pool allows P_3 to run. The minimum number of avalilable resources needed for the state to be safe is three.

17. Is the state described in Table 3. safe or unsafe ?

Table 3

Process	Current Allocation		Maximum Allocation		Resource Available	
	R_1	R_2	R_1	R_2	R_1	R_2
P_1	1	2	4	2	1	1
P_2	0	1	1	2		
P_3	1	0	1	3		
P_4	2	0	3	2		

Sol : The state is safe. Allocating both available resources allows P_2 to run to completion, leaving one R_1 available and two R_2's. This allows P_4 to run, leaving three R_1's and two R_2's. P_1 may now run, leaving four R_1's and four R_2's. P_3 may now run, so the state is safe.

18. Given the resource allocation graph in Fig. 2, fill in the resource usage table (see table 4).

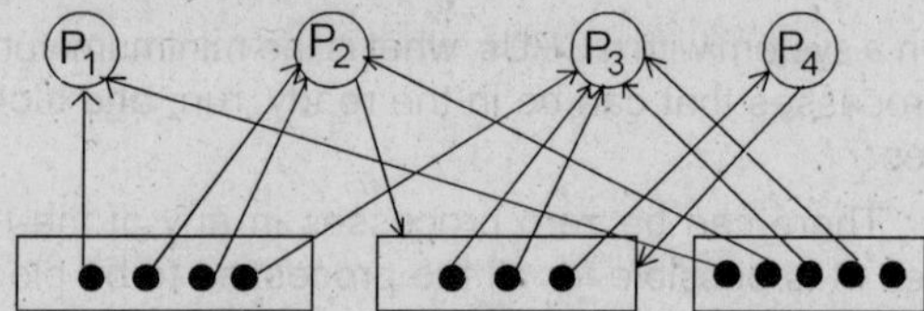

Fig 2. Resource allocation graph.

Sol : **Table 4.**

Process	Current Allocation			Outstanding Requests			Resources Available		
	R_1	R_2	R_3	R_1	R_2	R_3	R_1	R_2	R_3
P_1	1	0	1	0	0	0	0	0	1
P_2	2	0	1	0	1	0			
P_3	1	2	2	0	0	0			
P_4	0	1	0	0	1	0			

19. Is the following state described in Table 5. safe or unsafe ? **Table 5.**

Process	Current Allocation		Maximum Allocation		Resources Available	
	R_1	R_2	R_1	R_2	R_1	R_2
P_1	7	2	9	5	2	1
P_2	1	3	2	6		
P_3	1	1	2	2		
P_4	3	0	5	0		

Sol : No

20. For the system described in Table 6., draw the resource allocation graph.

Table 6., Resource Usage.

Process	Current Allocation			Outstanding Requests			Maximum Allocation			Resources Allocation		
	R_1	R_2	R_3	R_1	R_2	R_3	R_1	R_2	R_3	R_1	R_2	R_3
P_1	2	0	0	1	0	0	2	0	1	0	2	0
P_2	1	2	0	0	0	1	2	5	2	0	2	0
P_3	0	1	1	0	0	0	1	4	2	0	2	0
P_4	0	0	1	0	0	1	2	0	1	0	2	0

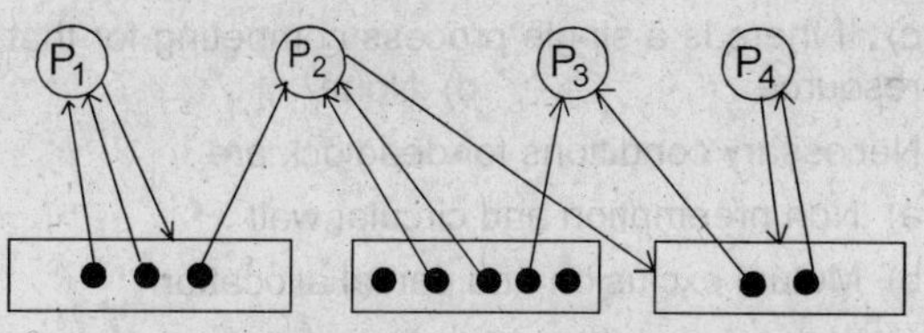

Solution 20. Resource allocation graph.

21. For the system, described Table 6, draw the reduced resource allocation graph.

Sol :

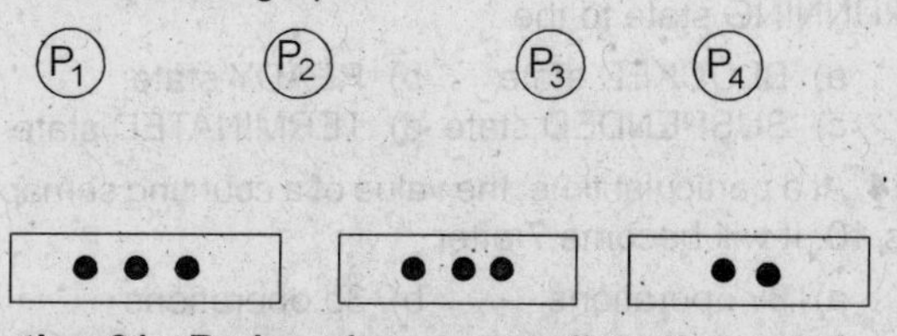

Solution 21 : Reduced resource allocation graph

22. For the system described in Table 6., is the system dead locked ?

Sol : No.

OBJECTIVE QUESTIONS - I

1. Which of the following scheduling algorithms is non-preemptive ? **[GATE-2002]**

a) Round Robin b) First-In First Out

c) Multilevel Queue scheduling

d) Multilevel Queue scheduling with feed back

2. Which of the following does not interrupt a running process? **[GATE-2001]**

a) A device b) Timer

c) Scheduler Process d) Power failure

3. Which of the following need not necessarily be saved on a context switch between processes? **[GATE-2000]**

a) General purpose registers

b) Translation lookaside buffer

c) Program counter d) All of the above

4. Which of the following is NOT a valid deadlock prevention scheme? **[GATE-2000]**

a) Release all resources before requesting a new resource.

b) Never request a resource after releasing any resource.

c) Number the resources uniquely and never request a lower numbered resource than the last one requested.

d) Request and be allocated all required resources before execution.

5. Which of the following actions is/are typically not performed by the OS when switching context from process A to process B: **[GATE-1999]**

a) Saving current register values and restoring saved register values for process B

b) Changing address translation tables.

c) Swapping out the memory image of process A to the disk.

d) Invalidating the translation look-aside buffer.

6. When the result of a computation depends on the speed of the processes involved, there is said to be: **[GATE-1998]**

a) Cycle stealing b) Race condition

c) A time lock d) A deadlock

7. An operating system contains 3 user processes each requiring 2 units of resource R. The minimum number of units of R such that no deadlocks will ever arise is: **[GATE-1997]**

a) 3 b) 4 c) 5 d) 6

8. The process state transition diagram in below figure is representative of: **[GATE-1996]**

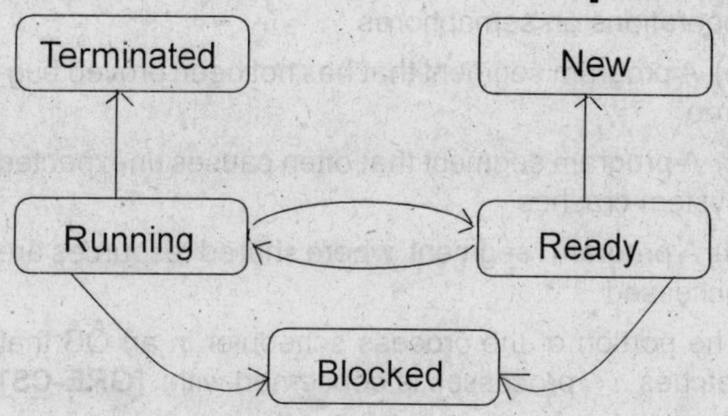

a) A batch operating system

b) An operating system with a preemptive scheduler

c) An operating system with a non-preemptive scheduler

d) A uni -programmed operating system.

9. A critical section is a program segment: **[GATE-1996]**

a) Which should run in a certain specified amount of time

b) Which avoids deadlocks

c) Where shared resources are accessed.

d) Which must be enclosed by a pair of semaphore operations, P and V

10. Which scheduling policy is most suitable for a time-shared operating system: **[GATE-1995]**

a) Shortest job first b) Round robin

c) First come first serve d) None

11. Consider a system having m resources of the same type. These resources are shared by 3 processes A,B and C which have peak demands of 3,4 and 6 respectively. For what value of m deadlock will not occur: **[GATE-1993]**

a) 7 b) 9 c) 10 d) 13

12. Assume that the following job's are to be executed on a single processor system: **[GATE-1993]**

Job Id	CPU Burst Time
P	4
Q	1
R	8
S	1
T	2

The job's are assumed to have arrived at time 0^+ and in the order P,Q,R,S and T.Calculate the departure with time slice (completion time) for Job P It Scheduling is round robin with time slices ?

a) 4 b) 10 c) 11 d) 12

13. The highest-response ratio next scheduling policy favors ....Jobs, but is also limits the waiting time of ....Jobs. **[GATE-1990]**

a) Shorter,longer b) Longer,shorter

14. Semaphore operations are atomic because they are implemented with is the OS .... **[GATE-1990]**

a) Kernel b) Shell

15. A critical region is : **[GATE-1987]**

a) One which is enclosed by a pair of P and V operations on semaphores

b) A program segment that has not been proved bug-free

c) A program segment that often causes unexpected system crashes

d) A program segment where shared resources are accessed

16. The portion of the process scheduler in an OS that dispatches processes is concerned with: **[GRE-CS]**

a) Assigning ready processes to the CPU

b) Activating suspended I/O bound processes

c) Temporarily suspending processes when the CPU load is too great

d) All of the above

17. Dijkstra's banker's algorithm in an OS solves the problem of **[GRE-CS]**

a) Deadlock avoidance b) Deadlock recovery

c) Mutual exclusion d) Indefinite postponement

18. The only state transition initiated by the user process itself in an OS is: **[GRE-CS]**

a) Dispatch b) Block

c) Timer runout d) Wake up

19. Concurrent process are :

a) Processes that don't overlap in time

b) Processes that overlap in time

c) Processes that are executed by a processor at the same time d) None

20. Critical region is:

a) A part of the OS which is not allowed to be accessed by any process

b) A set of instructions that access common shared resource which exclude one another in time

c) The portion of the main memory which can be accessed only by one process at a time

d) None

21. With a single resource, deadlock occurs

a) If there are more than two processes competing for that resource

b) If there are only two processes competing for that resource

c) If there is a single process competing for that resource d) None

22. Necessary conditions for deadlock are

a) Non preemption and circular wait

b) Mutual exclusion and partial allocation

c) both (a) and (b) d) None

23. In a time sharing OS, when the time slot given to a process is completed, the process goes from the RUNNING state to the

a) BLOCKED state b) READY state

c) SUSPENDED state d) TERMINATED state

24. At a particular time, the value of a counting semaphore is 10. It will become 7 after

a) 3v operations b) 3p operations

c) 5v operations and 2p operations

d) 13p operations and 10v operations

25. Semaphores are used to solve the problem of

a) Race condition b) Process synchronization

c) Mutual exclusion d) None

26. Pre-emptive scheduling is the strategy of temporarily suspending a running process

a) Before the CPU time slice expires

b) To allow starving processes to run

c) When it requests I/O d) None

27. Mutual exclusion problem occurs between

a) Two disjoint processes that don't interact

b) Processes that share resources

c) Processes that do not use the same resource

d) None

28. In Round-Robin CPU scheduling, as the time quantum is increased, the average turn around time

a) Increases b) Decreases

c) Remains constant d) Varies irregularly

29. In which of the following scheduling policies does context switching never takes place?

a) Round-Robin b) Shortest-Job-First

c) Pre-emptive d) First-Cum-First-Served

30. A system has 3 processes sharing 4 resources. If each process needs a maximum of 2 units then

a) Deadlock can never occur

b) Deadlock may occur

c) Deadlock has to occur

d) None

31. In real-time operating systems, which of the following is the most suitable scheduling scheme?

a) Round-Robin

b) First-Cum-First-Served

c) Pre-emptive Scheduling

d) Random Scheduling

32. Which of the following schemes gives minimum average waiting time?

a) FCFS b) SJF

c) Round-Robin d) Priority

33. Consider the following resource allocation graph. This system is in a deadlock state. This remark is

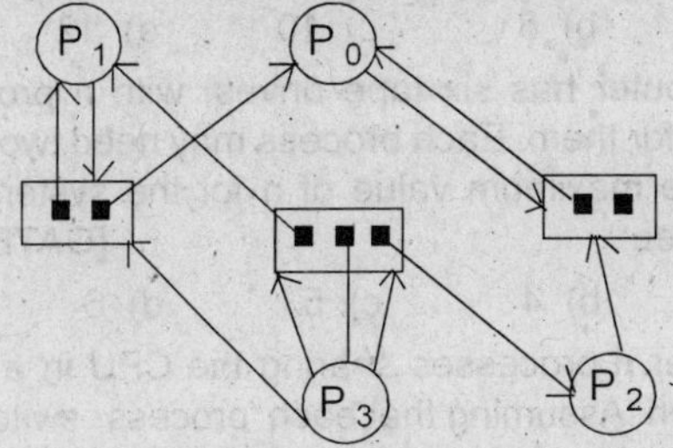

a) True b) False
c) Impossible to determine
d) Unpredictable

34. Which of the following is a safe sequence?

a) P_0,P_1,P_2,P_3 b) P_1,P_0,P_3,P_2
c) P_2,P_0,P_1,P_3 d) None

35. First-In-First-Out scheduling is:

a) Pre-emptive scheduling
b) Non-pre-emptive scheduling
c) Dead line scheduling
d) None

36. Race around condition occurs when

a) Two processes unknowingly wait for resources that are held by each other
b) Two processes wait for the same resource
c) All resources are shared
d) Two processes share the same shared resource

37. Round-Robin scheduling

a) Is Quite complex to implement
b) Allows interactive tasks access to the processor
c) Allows processor-bound tasks more time in the processor
d) Gives each task the same chance at the processor

38. Which of the following resources can cause Deadlocks?

a) Read only files b) Shared Programs
c) Printers e) All of the above

39. To avoid the race condition the number of processes that may be simultaneously inside their critical section is

a) 0 b) 1 c) 2 d) 4

40. Producer-Consumer problem can be solved using

a) Semaphores b) Event counters
c) Monitors d) All of the above

41. Moving process from main memory to disk is called

a) Scheduling b) Caching
c) Swapping d) Spooling

42. The main function of the dispatcher (the portion of the process scheduler) is:

a) Swapping a process to the disk
b) Assigning ready process to the CPU
c) Suspending some of the processes when the CPU load is high
d) Bring processes from the disk to the main memory

43. Round- Robin scheduling is essentially the preemptive version of

a) FIFO b) Shortest Job First
c) Shortest remaining time d) Longest time first

44. Which scheduling algorithm is inherently preemptive

a) FCFS b) SJF
c) RR d) Priority scheduling

45. Total time taken by a process to complete execution is

a) Waiting time b) Turn-around time
c) Response time d) Through put

46. The time spent by a process in the ready queue is

a) Waiting time b) Turn-around time
c) Response time d) Through put

47. Swapping is done by

a) Long term scheduler b) Medium term scheduler
c) Short term scheduler d) None

48. The optimal scheduling algorithm is

a) FCFS b) SJF
c) RR d) Multi-level queue scheduling

49. The performance of round-robin algorithm depends heavily on

a) Size of the process
b) The I/O bursts of the process
c) The CPU bursts of the process
d) The size of the time quantum

50. The technique of gradually increasing the priority of a process that wait in a system for a long time is known as:

a) Blocking b) Starvation
c) Aging d) Convoy effect

51. When a process is rolled back as a result of deadlock the difficulty which arises is

a) Starvation b) System through put
c) Convoy effect d) None

52. Which of the following CPU scheduling algorithms gives the minimum average waiting time? **[UGC NET-2000]**

a) First-Come-First-Served
b) Shortest job First
c) Round Robin
d) Multilevel Queue Scheduling

53. A relation between processes such that each has some part which must not be executed while the critical section of another is being executed is known as **[UGC NET-1996]**

a) Mutual Exclusion b) Semaphore
c) Multitasking d) Multiprogramming

54. What problem is solved by Dijkstra's Banker's Algorithm? **[UGC NET-1995]**

a) Deadlock recovery b) mutual exclusion
c) Cache coherence d) Deadlock avoidence

ANSWER KEY - I

1. b	2. c	3. b	4. b	5. c,d
6. b	7. b	8. b	9. c	10. b
11. d	12. c	13. a	14. a	15. d
16. a	17. a	18. b	19. b	20. b
21. d	22. c	23. b	24. b,d	25. b,c
26. a	27. b	28. d	29. b,d	30. a
31. c	32. b	33. b	34. c	35. b
36. b	37. d	38. c	39. b	40. d
41. c	42. b	43. a	44. c	45. b
46. a	47. b	48. b	49. d	50. c
51. c	52. b	53. a	54. d	

OBJECTIVE QUESTIONS - II

1. Consider a set of n tasks with known run times $r_1, r_2, \ldots\ldots r_n$ to be run on a uniprocessor machine. Which of the following processor scheduling algorithms will result in the maximum throughput? **[GATE-2001]**

a) Round-Robin b) Shortest- Job-First

c) Highest-Response-Ratio-Next

d) First-come-First-served

2. Consider Peterson's algorithm for mutual exclusion between two concurrent processes I and J . The program executed by processor I is shown below. **[GATE-2001]**

```
Repeat
        Flag[I] = true;
        Turn = J;
        While(P) do
        No-op;
        Enter
        Critical section .
        perform actions,
        Then exit critical section
        Flag[I] = false;
    Perform other non-critical section actions.
        Until false;
```

For the program to guarantee mutual exclusion, the predicate P in the while loop should be:

a) flag[J] = true and turn = I;

b) flag[J] = true and turn = J;

c) flag[I] = true and turn = J;

d) flag[I] = true and turn = I;

3. Let m[0],...... m[4] be mutexes (binary semaphores) and P[0],......P[4] be processes .Suppose each process P[I] executes the following: **[GATE-2000]**

wait (m[I]; wait (m[(I+1)mod 4])

.....

release (m[I]); release (m[(I+1) mod 4])

This could cause:

a) Thrashing b) Deadlock

c) Starvation, but not deadlock d) None

4. A counting semaphore was initialized to 10. Then 6P(wait) operation and 4V(signal) operations were completed on this semaphore . The resulting value of the semaphore is: **[GATE-1998]**

a) 0 b) 8 c) 10 d) 12

5. A computer has six tape drives, with n processes competing for them. Each process may need two drives. What is the maximum value of n for the system to be deadlock free: **[GATE-1998]**

a) 3 b) 4 c) 5 d) 6

6. Consider n processes sharing the CPU in a round-robin fashion. Assuming that each process switch take 5 seconds, what must be the quantum size q such that the overhead resulting from process switching is minimized but, at the same time each process is guaranteed to get it's turn at the CPU at least every t seconds ? **[GATE-1998]**

a) $q \le \frac{t-ns}{n-1}$ b) $q \ge \frac{t-ns}{n-1}$

c) $q \le \frac{t-ns}{n+1}$ d) $q \ge \frac{t-ns}{n+1}$

7. Each process P_i ,i =1,2,...9 is coded as follows : **[GATE-1997]**

```
REPEAT
        P(mutex)
        {critical section }
        V(mutex)
        FOREVER
```

The code for P_{10} is identical except that it uses V(mutex) in place of P(mutex). What is the largest number of processes that can be inside the critical section at any moment

a) 1 b) 2 c) 3 d) None

8. A solution to the dining philosophers problem which avoids deadlock is: **[GATE-1996]**

a) Ensure that all philosophers pickup the left fork before the right fork.

b) Ensure that all philosophers pickup the right fork before the left fork.

c) Ensure that one particular philosopher picks up the left fork before the right fork, and that all other philosophers pickup the right fork before the left fork.

d) None.

9. Four jobs to be executed on a single processor system arrive at time 0^+ in the order A,B,C,D. Their burst CPU time requirements are 4,1,8,1 time units respectively . The completion of A under round robin scheduling with time slice of one time unit is: **[GATE-1996]**

a) 4 b) 8 c) 9 d) 10

10. The sequence ....is an optimal non-preemptive scheduling sequence for the following job's which leaves the CPU idle for..... unit(s) of time. **[GATE-1995]**

Job	Arrival Time	Burst Time
1	0.0	9
2	0.6	5
3	1.0	1

a) { 3,2,1} , 1 b) { 2,1,3}, 0

c) { 3,2,1 } , 0 d) { 1,2,3 } , 5

11. At a particular time of computation the value of a counting semaphore is 7. Then 20 p operation and 15v operations were completed on this semaphore . The resulting value of the semaphore is: **[GATE-1992]**

a) 2 b) 7 c) 12 d) 42

12. A computer system has 6 tape drives, with n. processes competing for them. Each process may need 3 tape drives. The maximum value of n for which the system is guaranteed to be deadlock free is: **[GATE-1992]**

a) 1 b) 2 c) 3 d) 4

13. Match the following : **[GATE-1990]**

A. Critical Region	I. Hoare's monitor
B. Wait/signal	II. Mutual exclusion
C. Working set	III. Principle of locality
D. Deadlock	IV. Circular wait

a) A-II,B-I,C-III,D-IV b) A-I,B-II,C-IV,D-III

c) A-II,B-III,C-I,D-IV d) None

Question 14 and 15 refer to the following information.

An operating system makes use of Dijkstra's banker's algorithm to allocate twelve printers. Printers will be allocated to a process requesting them only if there are enough available printers to allow it to run to completion.

14. User1 is using seven printers and will need at most a total of ten printers. User2 is using one printer and will need at most four printers. User3 is using two printers and will need at most four printers. Each user is currently requesting one more printer. Which of the following is True? **[GRE-CS]**

a) The OS will grant a printer to User1 only.

b) The OS will grant a printer to User2 only.

c) The OS will grant a printer to User3 only.

d) The OS will not grant any more printers until some are relinquished

e) Dijkstra's banker's algorithm will prevent this situation from occurring

15. User1 is using x printers and will need a total of y printers. Users2 is using m printers and will need a total of n printers. This state is safe if and only if (y<=12,n<=12,x<=y,m<=n): **[GRE-CS]**

a) x+n<=12 and y+m<=12 and x+m<=12

b) x+n<=12 and y+m<12 and x+m<=12

c) x+n<=12 or y+m<=12 and x+m<=12

d) x+m <=12

16. A state is safe if the system can allocate resources to each process (up to its maximum) in some order and still avoid deadlock.Which of the following is/are True?

a) Deadlocked state is unsafe

b) Unsafe state may lead to a deadlock situation

c) Unsafe state must lead to a deadlock situation

d) Deadlock state is a subset of unsafe state

17. 'm' processes share 'n' resources o f the same type. The maximum need of each process don't exceed 'n' and the sum all the their maximum needs is always less than m+n In this set up

a) Deadlock can never occur

b) Deadlock may occur

c) Deadlock has to occur d) None

18. When the result of a computation depends on the speed of the processes involved there is said to be:

a) Cycle stealing b) Race condition

c) A time lock d) A deadlock

19. Match the following

A. Disk Scheduling	I. Round-Robin
B. Batch processing	II. SCAN
C. Time scheduling	III. LIFO
D. Interrupt Processing	IV. FIFO

a) A-II,B-IV,C-I,D-III b) A-I,B-II,C-III,D-IV

c) A-I,B-III,C-IV,D-II d) None

20. Imposing a linear order on all resource types, and letting processes request resources in increasing order of enumeration is an example of:

a) Deadlock avoidance where the system will never enter an'unsafe' state

b) Deadlock prevention where the maximum needs of each job must also be taken into account

c) Deadlock avoidance where hold and wait condition can't occur

d) Deadlock prevention where circular wait for resources can never takes place

21. If the resource allocation graph for a system (with multiple instances for each resource type) has a cycle then

a) The system is not deadlocked

b) The system may be deadlocked

c) The system is deadlocked

d) None

ANSWER KEY - II

1. d	2. b	3. b	4. b	5. c
6. b	7. d	8. b	9. c	10. a
11. a	12. b	13. a	14. c	15. c
16. c	17. a	18. b	19. a	20. d
21. b	22. c			

III. MEMORY MANAGEMENT

- Introduction
- Swapping
- Paging
- Segmentation
- Virtual Memory
- Demand Paging

1. Introduction :

As a result of CPU scheduling, we can improve both the utilization of the CPU and the speed of the computer's response to its users. To realize this increase in performance, however, we must keep several processes in memory, we must share memory.

Memory is a large array of words of bytes, each with its own address. The CPU fetches instructions from memory according to the value of the program counter. These instrucrtions may cause additional loading from and storing to specific memory addresses.

A typical instruction execution cycle, for example, will first fetch an instruction from memory. The instruction is then decoded and may cause operands to be fetched from memory. After the instruction has been executed on the operands, results may be stored back in memory. The memory unit sees only a stream of memory addresses; it does not know how they are generated (the instruction counter, indexing, indirection, literal addresses, and so on) or what they are for (instructions or data).

2. Address Binding : Most systems allow a user process to reside in any part of the physical memory. Thus, although the address space of the computer starts at 00000, the first address of the user process does not need to be 0000. This arrangement affects the addresses that the user program can use. In most cases, a user program will go through several steps (some of which may be optional) before being executed.

Classically, the binding of instruction and data to memory addresses can be done at any step along the way :

I. Compile Time : If it is known at compile time where the process will reside in memory, then absolute code can be generated. For example, if it is known a priori that a user process resides starting at location R, then the generated compiler code will start at that location and extend up from there. If, at some later time, the starting location changes, then it will be necessary to recompile this code. The MS-DOS. COM- format programs are absolute code bound at compile time.

II. Load Time : If it is not known at compile time where the process will reside in memory, then the compiler must generate relocatable code. In this case, final binding is delayed until load time. If the starting address changes, we need only to reload the user code to incorporate this changed value.

III. Execution Time : If the process can be moved during its execution from one memory segment to another, then binding must be delayed until run time. Special hardware must be available for this scheme to work.

Dynamic Loading : Dynamic loading is used to obtain better memory-space utilization. With dynamic loading, a routine is not loaded until it is called. All routines are kept on disk in a relocatable load format. The main program is loaded into memory and is executed. When a routine needs to call another routine, the calling routine first checks to see whether the other routine has been loaded.

The advantage of dynamic loading is that an unused routine is never loaded. This scheme is particularly useful when large amounts of code are needed to handle infrequently occuring cases, such as error routines. In this case, although the total program size may be large, the portion that is actually used (and hence actually loaded) may be much smaller.

Dynamic Linking : Most operating systems support only static linking, in which system language libraries are treated like any other object module and are combined by the loader into the binary program image.

The concept of dynamic linking is similar to that of dynamic loading. Rather than loading being postponed until execution time, linking is postponed. This feature is usually used with system libraries, such as language subroutine libraries. Without this facility, all programs on a system need to have a copy of their language library (or at least the routines referenced by the program) included in the executable image. This requirement wastes both disk space and main memory. With dynamic linking, a stub is included in the image for each library routine reference. This stub is a small piece of code that indicates how to locate the appropriate memory resident library routine, or how to load the library if the routine is not already present.

Overlays : The entire program and data of a process must be in physical memory for the process to execute. The size of a process is limited to the size of physical memory. So that a process can be larger than the amount of memory allocated to it, a technique called overlays is sometimes used. The idea of overlays is to keep in memory only those instructions and data that are needed at any given time. When other instructions are needed, they are loaded into space that was occupied previously by instructions that are no longer needed.

3. Logical Versus Physical Address Space : An address generated by the CPU is commonly referred to as a logical address, whereas an address seen by the memory unit (that is, the one loaded into the memory address register of the memory) is commonly referred to as a physical address.

The run-time mapping from virtual to physical address is done by the memory-management unit (MMU), which is a hardware device.

As shown in Fig.3.1, this scheme requires hardware support slightly different from the hardware configuration. The base register is now called a relocation register. The value in the relocation register is added to every address generated by a user process at the time it is sent to memory. For example, if the base is at 14,000 then an attempt by the user to address location 0 is dynamically relocated to location 14,000; and access to location 346 is mapped to location 14,346.

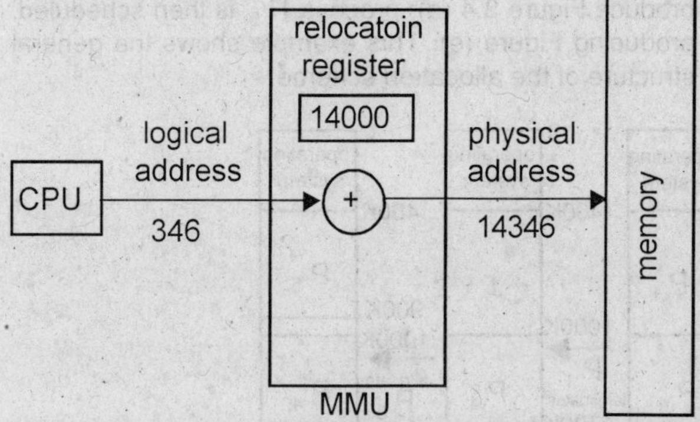

Fig. 3.1 Dynamic relocation using a relocation register

We have two different types of addresses :

Logical addresses (in the range 0 to max)

and physical addresses (in the range R +0 to R+ max for a base value R).

The user generates only logical addresses and thinks that the process runs in locations 0 to max. The user program supplies logical addresses; these logical addresses must be mapped to physical addresses before they are used.

4. Swapping : A process needs to be in the memory to be executed. A process, however, can be swapped temporarily out of memory to a backing store, and then brought back into memory for continued execution. For example, assume a multiprogramming environment with a round-robin CPU-scheduling algorithm. When a quantum expires, the memory manager will start to swap out the process that just finished, and to swap in another process to the memory space that has been freed.

Swapping requires a backing store. The backing store is commonly a fast disk. It must be large enough to accommodate copies of all memory images for all users, and must provide direct access to these memory images.

The context-switch time in such a swapping system is fairly high. To get an ideal of the context-switch time, assume that the user process is of size 100 K and the backing store is a standard hard disk with a transfer rate of 1 megabyte per second. The actual transfer of the 100K process to or from memory takes

100K/1000K per second = 1/10 second = 100 milliseconds

Assuming that no head seeks are necessary and an average latency of 8 milliseconds, the swap time takes 108 milliseconds. Since we must both swap out and swap in, the total swap time is then about 216 milliseconds.

For efficient CPU utilization, we want out execution time for each process to be long relative to the swap time. Thus, in a round-robin CPU-scheduling algorithm, for example, the time quantum should be substantially larger than 0.216 seconds.

There are other constraints on swapping. If we want to swap a process, we must be sure that it is completely idle. Of particular concern is any pending I/O. If a process is waiting for an I/O operation, we may want to swap that process to free up its memory. However, if the I/O is asynchronously accessing the user memory for I/O buffers, then the process cannot be swapped. Assume that the I/O operation was queued because the device was busy. Then, if we were to swap out process P_1 and swap in process P_2, the I/O operation might then attempt to use memory that now belongs to process P_2. The two main solutions to this problem are

I. never to swap process with pending I/O, or

II. to execute I/O operations only into operating-system buffers. Transfers between operating-system and process memory then occur only when the process is swapped in.

5. Contiguous Allocation : The main memory must accommodate both the operating system and the various user processes. The memory is usually divided into two partitions.

One for the resident operating system, and one for the user process. It is

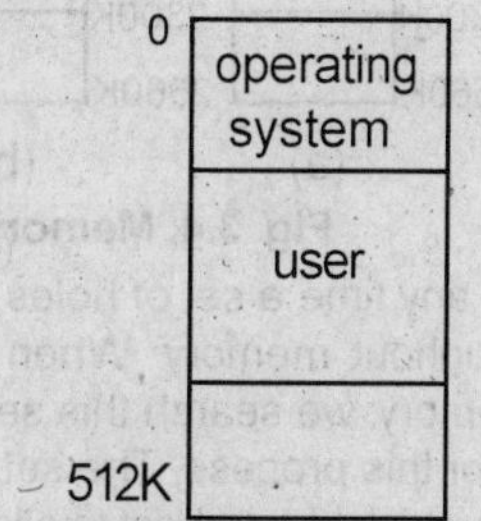

Fig.3.2 Memory partition.

Single-Partition Allocation : If the operating system is residing in low memory and the user processes are executing in high memory, it is required to protect the operating-system code and data from changes (accidental or malicious) by the user processes. It is also required to protect the user processes from one another. This protection can be provided by using a relocation register. The operating system keeps a table indicating which parts of memory are available and which are occupied. Initially, all memory is available for user processes, and is considered as one large block of available memory, a hole. When a process arrives and needs memory, search for a hole large enough for this process. If we find one, we allocate only as much memory as is needed, keeping the rest available to satisfy future requests.

Multiple-Partition Allocation : Since it is desirable, that there be several user processes residing in memory at the same time, we need to consider the problem of how to allocate available memory to the various processes that are in the input queue waiting to be brought into memory. One of the simplest schemes for memory allocation is to divide memory into a number of fixed-sized partitions.

For example assume that we have 2560K of memory available and a resident operating system of 400K. This situation leaves 2160 K for user processes, as shown in Figure 3.3. Given the input queue in the figure,

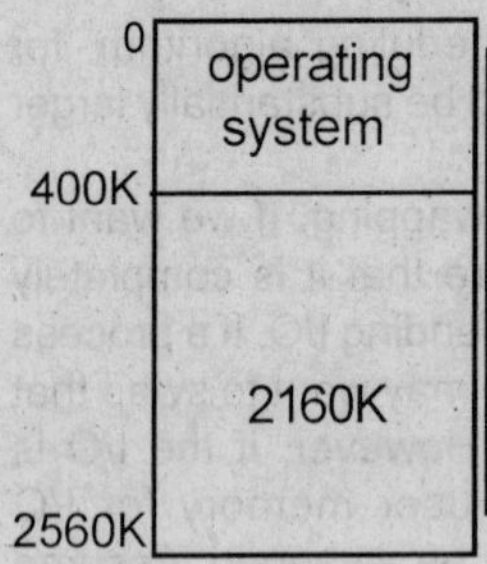

(a)

job queue		
process	memory	time
P_1	600K	10
P_2	1000K	5
P_3	300K	20
P_4	700K	8
P_5	500K	15

(b)

Fig. 3.3 Scheduling example

And FCFS job scheduling, we can immediately allocate memory to processes P_1, P_2 and P_3, creating the memory map of Figure (a). We have a hole of size 260K that cannot be used by any of the remaining processes in the input queue. Using a round-robin CPU-scheduling with a quantum of 1 time unit, process P_2 will terminate 14, releasing its memory. This situation is shown Figure (b). We then return to out job queue and schedule the next process, process P_4, to produce the memory map of Figure (c). Process P_1 will terminate at time 28 to produce Figure 3.4 (d); process P_5 is then scheduled, producing Figure (e). This example shows the general structure of the allocation scheme.

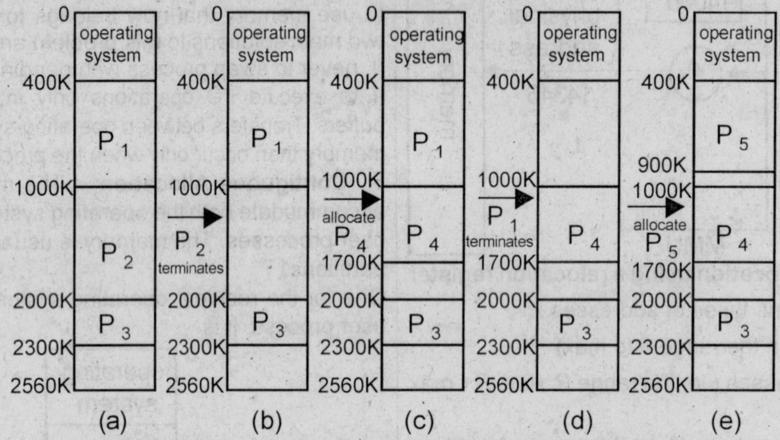

Fig. 3.4. Memory allocation and long-term scheduling.

In general, there is at any time a set of holes, of various sizes, scattered throughout memory. When a process arrives and needs memory, we search this set for a hole that is large enough for this process. The set of holes is searched to determine which hole is best to allocate. First-fit and Worst-fit are the most common strategies used to select a free hole from the set of available holes.

- **First-fit :** Allocate the first hole that is big enough. Searching can start either at the beginning of the set of holes or where the previous first-fit search ended. We can stop searching as soon as we find a free hole that is large enough.
- **Best-fit :** Allocate the smallest hole that is big enough. We must search the entire list, unless the list is kept ordered by size. This strategy produces the smallest leftover hole.
- **Worst-fit :** Allocate the largest hole. Again, we must search the entire list, unless it is sortee by size. This strategy produces the largest leftover hole, which may be more useful than the smaller leftover hole from a best-fit approach.

6. External and Internal Fragmentation : As processes are loaded and removed from memory, the free memory space is broken into little pieces. External fragmentation exists when enough total memory space exists to satisfy a request, but it is not contiguous; storage is fragmented into a large number of small holes.

This fragmentation problem can be servere. In the worst cast we could have a block of free (wasted) memory between every two processes. If all this memory were in one big free block, we might be able to run several more processes.

One solution to the problem of external fragmentation is compaction. The goal is to shuffle the memory contents to place all free memory together in one large block. Resolution is done at assembly or load time, compaction cannot be done; compaction is possible only if relocation is dynamic, and is done at execution time.

Swapping can also be combined with compaction. A process can be rolled out of main memory to a backing store and rolled in again later. When the process is rolled out, its memory is released, and perhaps is reused for another process. If static relocation is used, the process must be rolled into the exact same memory locations that it occupied previously. This restriction may require that other processes be rolled out to free that memory.

If dynamic relocation (such as with base and limit registers) is used, then a process can be rolled into a different location.In this case, we find a free block, compacting if necessary, and roll in the process.

7. Paging : Paging is a memory management technique that permits a program's memory to be non-contiguous into physical memory thus allowing a program to be allocated physical memory wherever it possible.

When a programmer wants to transfer a data from one location to another, he might write for example : move R_1 2000 (i.e. moving the content of memory address 2000 to RegisterR_1). Address of memory location can be generated using indexing, base registers, segment registers and other ways.

These programs generated addresses are called virtual addresses and form the virtual address space. There has to be mapping between virtual address space and physical address space. In this section we will see how this mapping is done through paging.

Address mapping in a paging system : The physical memory is conceptually divided into a number of fixed size blocks called frames (or page frames). The virutal address space or logical memory of a process is also broken into blocks of the same size called pages. When a program is to be run, its pages are loaded into any frame from the disks.

The figure 3.5 illustrates the address mapping in a paging system.

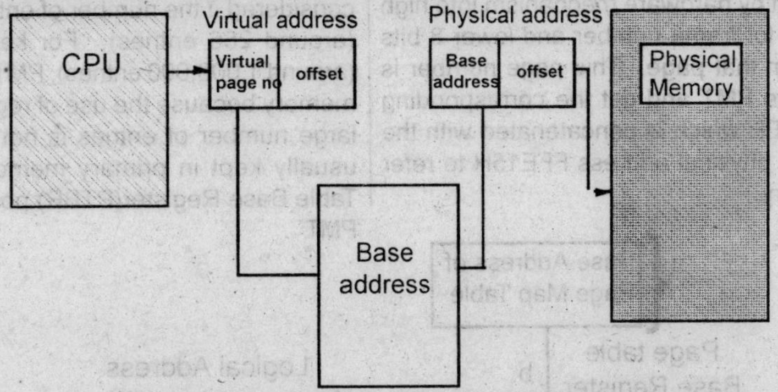

Figure 3.5 : Address mapping in a paging system

An important component of paging operation is a page map table which contains starting address or base address of each page stored in physical memory.

As shown in the figure 3.5, every address generated by CPU contains two components : vitual (or logical) page number and a page offset into that page. The page number works as an index into the page map table. To define a physical memory address, the base address in PMT is added withoffset and sent to physical memory unit.

Page size(and frame size) is defined between 512 bytes to 4K byte depending on computer architecture.

The size of a page is typically a power of 2 (256 is equivalent to 2^8) makes a translation of a virtual address into a page number and page offset particularly easy. For a concrete example, consider the following figure 3.6. It illustrates a sample space of 1-MB physical memory where virtual and physical address is 20 bits long each. The page size is assumed to be 256 bytes. Thus, the physical memory will have maximum (1MB/256 bytes) 12 frames (12pages) of 256 bytes each.

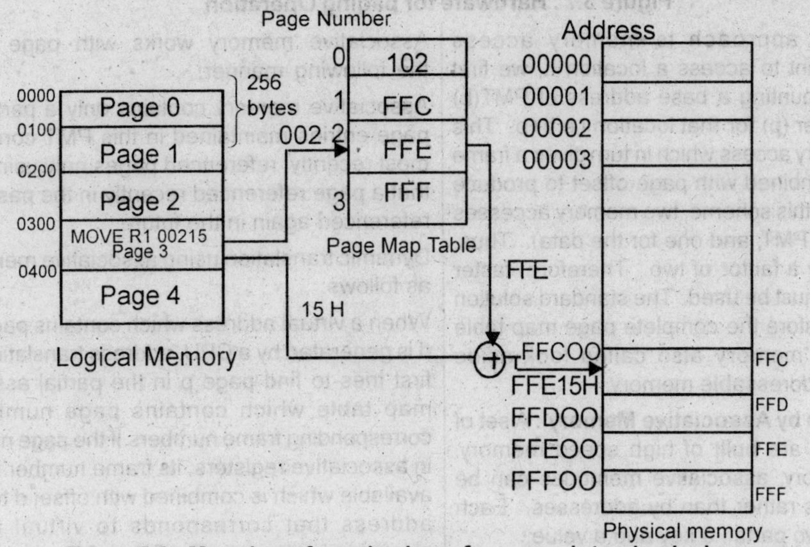

Figure 3.6 : Mapping of user's view of memory into physical memory

Physical memory

Since pages and page frames have identical sizes, offsets within each are identical and need not be mapped. In this example, each 20 bit virtual address may be regarded as a 12 bit page number (high order bits) and 8-bit offset within a page. This is sufficient enough for addressing any frame (page) and any location within page. As said earlier , PMT contains frame number of physical memory where the corresponding virtual page is plced. Since offsets do not require any mapping, PMT contains only the high-order bits of physical base address, that is frame number. Figure 3.6 shows how users view of memory can be mapped into physical memory.

For example virtual page is assumed to be placed in the physical frame whose starting address is FFF00H. This value is stored in the first entry of the PMT. All other PMT entries are filled with frame address where the corresponding pages are loaded.

The logic of address translations is also explained through an assembly language statement MOVE R_1 00215. The virtual address is split by hardware mechanism into high order 12 bits (002H) for frame number and lower 8 bits (15H) for offset within that page. This page number is used as an index into PMT and get the corresponding physical address FFEH which is concatenated with the offset to produce the physical address FFE15H to refer to any data in memory.

Hardware Support for Paging : The main objectives for hardware support for paging (illustrated in figure 3.7) is to store page map table and make virtual to physical address translation more efficient since every access to memory must go through the PMT, efficiency is a major consideration.

To reduce the access time, the use of registers can be considered if the number of entries in PMT is quite small (around 256 entries). For keeping very large entries (around 1,000,000 entries), PMT is usually kept in primary memory because the use of registers for keeping such a large number of entries is not feasible. Thus, PMT is usually kept in primary memory and there is a page Table Base Register(PTBR) pointing to the beginning of PMT.

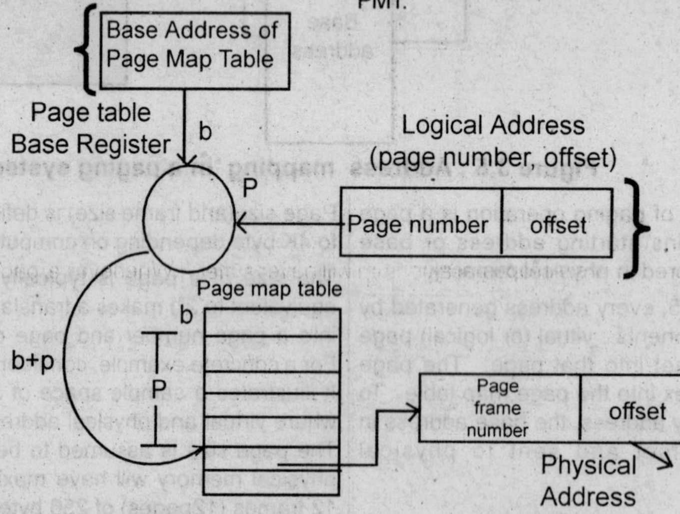

Figure 3.7 : Hardware for paging Operation

Problem with this approach is memory access time.Suppose we want to access a location L, we first index into PMT by counting a base address of PMT(b) offset by page number (p) for that location i.e b+p. This task requires a memory access which in turn gives a frame number, which is combined with page offset to produce actual address. With this scheme, two memory accesses are required (one for PMT, and one for the data). Thus, memory is slowed by a factor of two. Therefore, faster translation methods must be used. The standard solution to this problem is to store the complete page map table into an associative memory also called look-aside memory or content addressable memory.

Address Translation by Associative Memory : A set of associative registers are built of high speed memory. Unlike primary memory, associative memories can be searched by contents rather than by addresses. Each register consists of two parts : a key and a value.

If associative registers are presented with a question "Whether a particular page is stored in associative memory"? it is compared with all keys simultaneously. If the page is found in associative registers, its corresponding value is output. Although the hardware is quite expensive, the searching operation is very fast.

Associative memory works with page map table in the following manner:

Associative memory contains only a part of PMT. The page entries maintained in this PMT correspond to the most recently referenced pages only using the heuristic that a page referenced recently in the past is likely to be referenced again in the future.

Dynamic translation using associative memory proceeds as follows:

When a virtual address which contains page p and offset d is generated by a CPU, address translation mechanism first tries to find page p in the partial associative page map table which contains page number and their corresponding frame numbers.If the page number is found in associative registers, its frame number is immediately available which is combined with offset d to form the real address that corresponds to virtual address.This operation is shown in the figure 3.8.

If the target page number is not in the associative registers, a reference to PMT in memory is made when the frame number is obtained. We can use it to access memory. In addition, we add the page number and frame number to the associative registers, so that they can be found very quickly on the next reference.

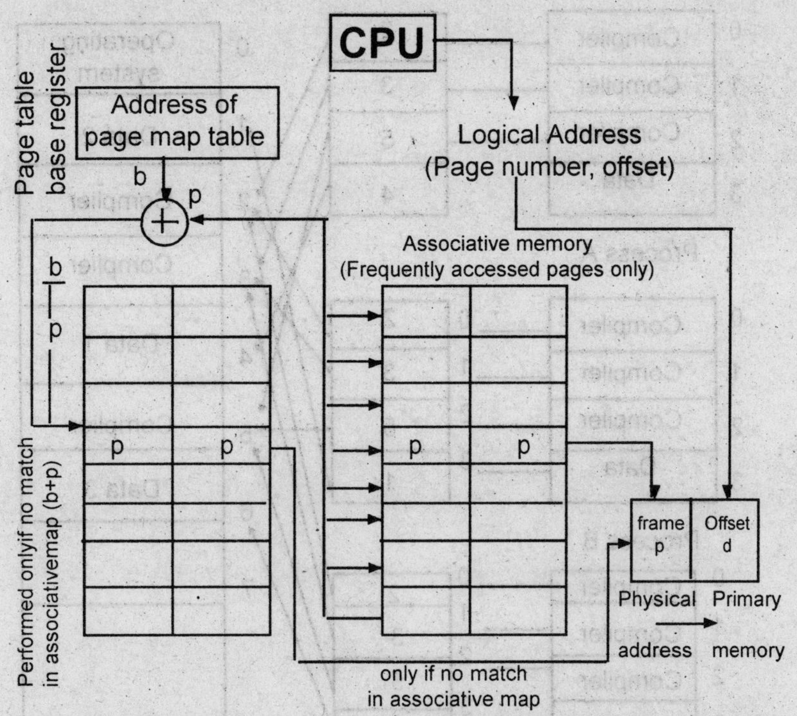

Figure 3.8 : Address translation through associative memory

The percentage of times that a page number is found in the associative registers is called hit ratio. A 70% hit ratio means that 70% of the time we find the desired page number in the associative registers. If it takes 60 nanosecond to search the associative registers and 740 nanosecond to access physical memory, then a total memory access time is 60+740=800 ns, when the target page is in associative registers. If the target page (60 nanosecond) is not found in associative registers, then we must first access primary memory (740 nanosecond) for page map table for getting the corresponding frame number ana then access the desired location in memory another (740 ns) for a total of 1480 ns.

The hit ration is clearly related to the number of associative registers. With 8 or 16 associative registers, a hit ratio of 80 percent ot 90 percent can be obtained.

Sharing and Protection in a Paging System : In a multiprogramming environment, where several users want to execute the same software, keeping a separate copy of the same for individual users will cause wastage of much of primary memory. For example, consider a time sharing system which supports 20 terminals (users), each of which run a compiler.If 50k bytes are required for compiler and 5 bytes for data storage for the total amount of memory required to support the 20 users will be

$$50K \times 20 + 5K \times 20 = 1100K$$

A little over 1 million byte. The obvious solution to this problem is to share those pages among all users.

Sharing must be controlled to prevent one process from modifying data that another process is reading. In today's systems that support sharing, programs consists of two separate parts, one for procedure code and another for data. To be sharable, the code must be non-self-modifiable or reentrant or pure code. The meaning of reentrant code is that it is read-only. There should not be any attempt on part of users to modify this code. Modifiable data or code cannot be shared. Since reentrant code never changes while execution, two or more processes, can execute the same code at the same time without having a personnel copy of it. But data for two different processes for the same code will of course be different. Each process has its own copy of registers and data storage to hold the data for its own execution.

Only one copy of complier needs to be kept in primary memory. The page map table of each process maps onto same physical copy of comiler but data pages are mapped into different frames. (figure 3.9)

Therefore to support only one copy of compiler (50K) plus 20 copies of the 5K of data space per user, the total memory requirement will be now just 150K instead of 1 mega byte, a significant saving.

To implement sharing each page is identified as a sharable or non-sharable. In this figure, pages 2,3 and 5 are identified as sharing. By having page map table entries of different processes pointing to the same page frame, then that page frame is shared among those processes. Sharing reduces the amount of primary storage needed for several users to run efficiently and make it possible for a given system to support more users.

Protection : Memory protection is usually done by protection bits associated with each page. These bits are usually kept in the page map table. One protection bit can define a page to be read/write or a read-only. Since reference to each page is passed through the page map

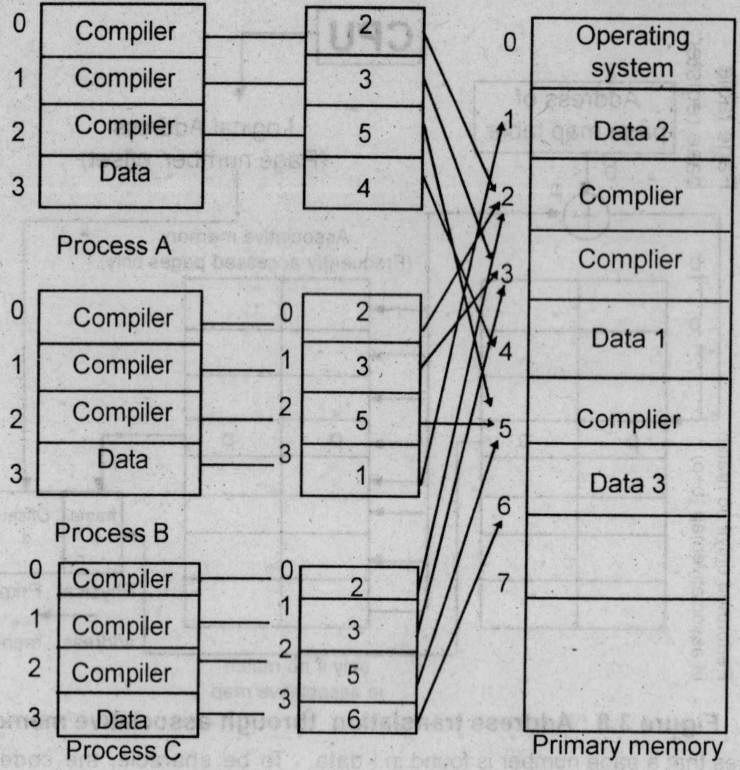

Figure 3.9 : Sharing in a Paging system

table to find the correct frame number, protection bits can be checked to verify that no write operation is being used to read-only page. An attempt to do this type of operation is easily trapped.

8. Segmentation : Segmentation is a memory management scheme which supports programmers view of memory. Programers never think of their programs as a linear array of words. Rather, they think of their programs as a collection of logically related entities, such as subroutines or procedures, functions, global or local data areas, stack etc.(figure 3.10)

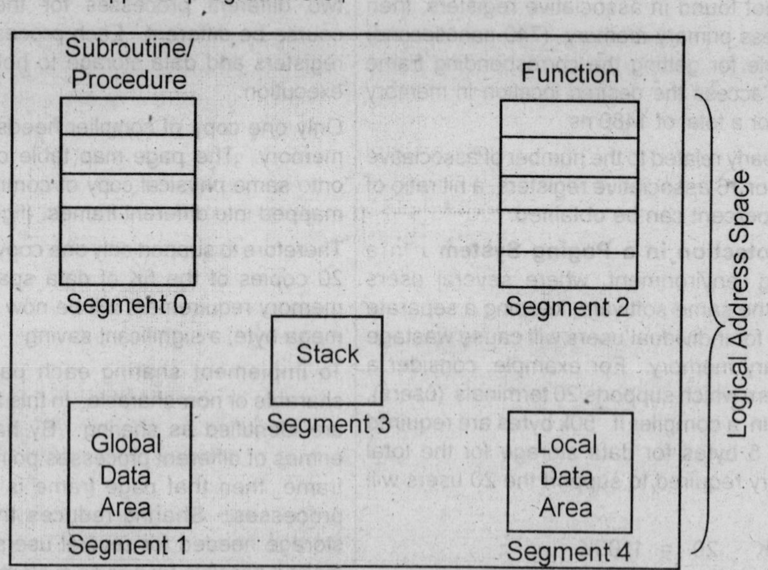

Figure 3.10 :Programmers' view of a Program

Segments are formed at program translation time by grouping together logically related entities.Formation of these segments vary from one compiler to another. A pascal compiler might create separate segments for (1) code of each procedure(2) global data (3) local data or variable (4) stack for calling procedure and storing its parameter.A Fortran compiler, for example, might create a separate segment for each common block. Arrays might be formed as separate segments.

To simplify implementation procedures, each segment in a program is numbered and referred to by a segment number rather than a segment name. In segmented systems, components belonging to a single segment reside in one contiguous area but different segments belonging to the same process occupy non-contiguous area of physical memory because each segment is individually relocated.

Address Mapping in a Segmented System : An important component of address mapping in a segmented system is a segment table. Its use is illustrated in the following figure 3.11.

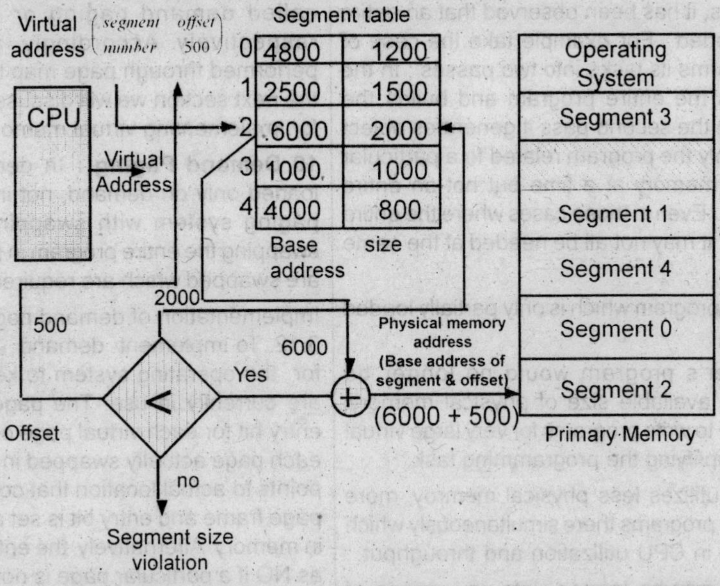

Figure 3.11: Virstual to Physical mapping in a segmented system

A virtual (logical) address consists of two parts: a segment number and an offset into that segment. The segment number provided in the virtual address is used as an index into the segment table. Each row of the segment table contains a starting address (base address) of segment and a size of the segment.The offset of the virtual address must be within (less than or equla to) the size of the segment. If the offset of virtual address is not within the range, it is trapped by the operating system otherwise the offset is added to the base address of the segment to produce physical address of the desired segment. As an example, consider the figure 3.10. There are 5 segments numbered from 0 to 4. The segment table (figure 3.11) has a separate entry for each segments having the starting address (base address) of segment in physical memory and the size of that segment.

For example, segment 1 is 1500 words long, starting at location 2500. Thus, a reference word at 500 location within segment 1 mapped is onto location 2500 (base address) + 500 = 3000 (actual physical address). A reference to segment 2 at location 300 will be mapped to 6000+300 =6300 etc. But a reference to a word 900 of segment 4 would cause segment size violation and it will be trapped by the operating system.

Fragmentation : Segmentation may then cause external fragmentation, when all blocks of free memory are too small to accommodate a segment. In this case, the process may simply have to wait until more memory (or at least a larger hole) becomes available, or compaction may be used to create a larger hole. Because segmentation is by its nature a dynamic relocation algorithm, we can compact memory whenever we want.

9. Virtual Memory : The common problem before programmers few years ago was how to fit large programs into the small memory. The solution developed was to break programs into small pieces called overlays. The first overlay (overlay 0) would be loaded and executed first, which in turn call another overlay. The overlays used to be on hard disk and swapped in and out of memory by operating system.

Althought the swapping was done by the operating system, the work of creating number of layers in a program was done by a programmer which was time consuming job. The virtual memory is a memory management technique which does splitting of a program into number pieces as well as swapping.

The basic idea behind virtual memory is that the combined size of the program, data and stack may exceed the amount of physical memory. The operating system keeps those parts of the program in the memory which are required during execution and the rest on the disk. For example, a 2MB program can run on a 640K RAM machine by carefully choosing 640K to keep in memory at each instant and swapping pieces of a program between disk and memory as needed.

The main objectives of various memory management strategies discussed so far, were to support multiprogramming, i.e to keep many processes in memory simultaneously. But all these schemes require the entire process to be in memory before the execution starts. Virtual memory is a memory management scheme that allows the execution of processes that may not be completeley in main memory. In other words, virtual memory allows execution of partially loaded processes.

As a consequence user programs can be larger than the physical memory.

Advantage of Virtual Memory : In virtual memory scheme, programmers get illusion of much larger memory than physical memory, therefore, they are not concerned about fitting their programs into samll physical memory. In most of the cases, it has been observed that an entire program is not needed. For example take the case of assembler. It performs its tasks into two passes ; In the first pass, it scans the entire program and builds the symbol table and in the second pass it generates object code. Therefore only the program related to a particular pass should be in memory at a time but not an entire assembler program. Even in those cases where the entire program is needed, it may not all be needed at the same time.

The advantage of a program which is only partially loaded in memory are:

1.The size of user's program would no longer be constrained by the available size of physical memory. Users would be able to write programs for very large virtual address space, simplifying the programming task.

2.Since each user utilizes less physical memroy, more users can keep their programs there simultaneously which will cause increase in CPU utilization and throughput.

3.Since a process may be loaded into a space of arbitrary size, which in turn reduces external fragmentation without the need to change the scheduled order of process execution. Moreover, the amount of space in use by a given process may be changing during its memory residence. As a result the operating system may speed up the execution of important processes by allocating more physical memory.

Virtual memeory can be implemented as an extension of paged or segmented memory management scheme, also called demand paging or demand segmentation respectively. Accordingly, address transaction is performed through page map table or segment-table. In the next section we will discuss demand paging scheme for implementing virtual memory management.

10. Demand Paging : In demand paging pages are loaded only on demand, not in advance. It is similar to paging system with swapping feature. Rather than swapping the entire program in memory, only those pages are swapped which are required currently by the system.

Implementation of demand paging is illustrated in figure 3.12. To implement demand paging, it is necessary for the operating system to keep track of which pages are currently in use. The page map table contains an entry bit for each virtual page of the related process. For each page actually swapped in memory, page map table points to actual location that contains the corresponding page frame and entry bit is set and marked as YES if it is in memory. Alternatively, the entry bit is reset and marked as NO if a particular page is not in memory. If a program during execution never accesses those pages which are marked as NO, there will be no problem and execution proceeds normally.

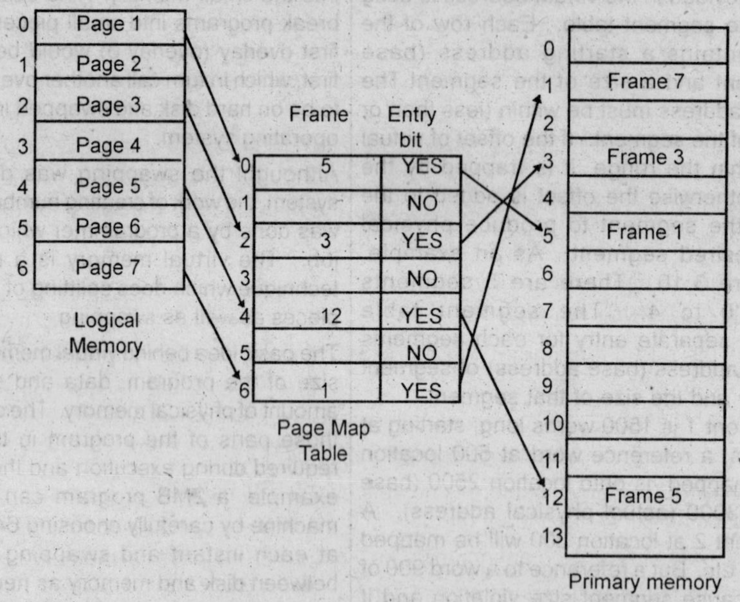

Figure 3.12 : Implementation Of Demand Paging Scheme

But what will happen if the program tries to access a page that was not swapped in memory ? In that case, page fault trap occurs. Page fault trap is the result of the operating system's failure to bring a valid part of the program into memory in an attempt to mininize swapping overhead and physical memory requirement. When the running program experiences a page fault, it must be suspended until the missing page is swapped in main memory. Since disk access time is usually several orders of magnitude longer than main memory cycle time, operating system usually schedules another processs during this period. Here is a list of steps operating system follows in handling a page fault:

1.If a process refers to a page which is not in the physical memory, then an internal table kept with a process control block is checked to verify whether a memory reference to a page was valid or invalid.

2. If the memory reference to a page was valid, but the page is missing, the process of bringing a page into the physical memory starts.

3. Free memory location is identified to bring a missing page.

4. By reaing a disk, the desired page is brought back into the free memory location.

5. Once the page is in the physical memory, the internal table kept with the process and page map table is updated to indicated that the page is now in memory.

6. Restart the instruction that was interrupted due to the missing page.

These steps are also illustrated through figure 3.13.

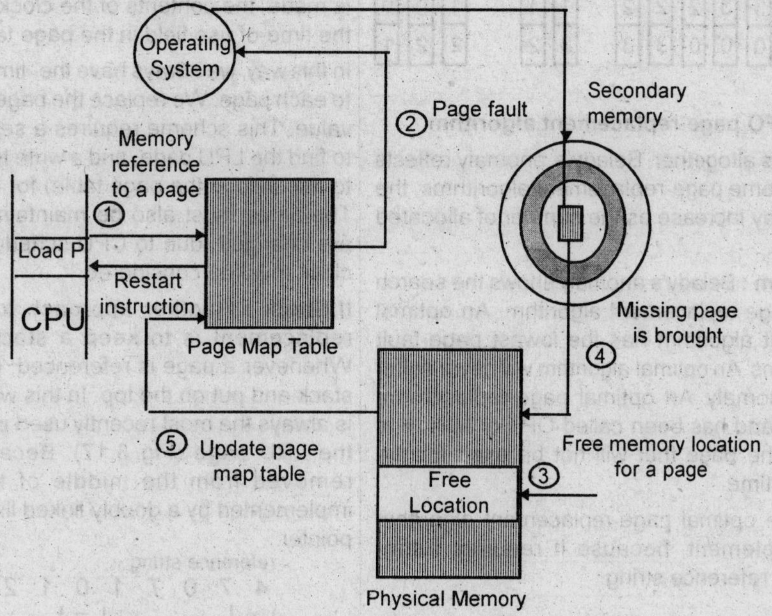

Figure 3.13 :Steps in handling a Page Fault

Whenever there is an interrupt due to any reason, the system saves the state (registers, stack program counter) of the interrupted process. After the condition is fulfilled, the interrupted process restarts from the same place where it was interrupted.

The hardware requirement to support demand paging is the same as the hardware for dmand paging:

- A page-map table with the ability to mark YES or NO for page entry.
- A hard disk to hold those pages not in memory.

11. Virtual Memory Management Policies.

The collection of only the part of a program into a physical memory requires the incorporation of certain policies into the virtual memory management scheme. These policies are classified as:

1. **Allocation policy:-** How much real memory to allocate to each active process ?
2. **Fetch policy :-** Which page to bring from secondary memory?
3. **Repalcement (Swap-out) policy:-** If there is no space for a new page in a physical memory, which page is to be selected for swapping out to secondary memory in order to mark room for a new page?
4. **Placement policy:-** Where to place an incoming new page into the physical memory?

Performance Of Demand Paging : Demand paging can have a significant effect on the performance of a computer system. To see why, let us compute the effective access time for a demand-paged memory.

Page-Replacement Algorithms : We evaluate an algorithm by running it on a particular string of memory references and computing the number of page faults. The string of memory references is called a reference string.

To illustrate the page-replacement algorithms, we shall use the reference string

7, 0, 1, 2, 0, 3, 0, 4, 2, 3, 0, 3, 2, 1, 2, 0, 1, 7, 0, 1

for a memory with three frames.

FIFO Algorithm : The simplest page-replacement algorithm is a FIFO algorithm. A FIFO replacement algorithm associates with each page the time when that page is chosen. It is not necessary to record the time when a page is brought in. We can create a FIFO queue to hold all pages in memory. We replace the page at the head of the queue. When a page is brought into memory, we insert it at the tail of the queue.

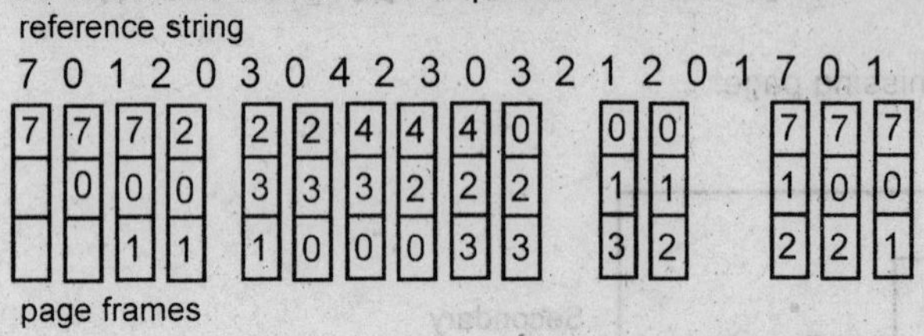

Fig. 3.14. FIFO page-replacement algorithm.

There are 15 faults altogether. Belady's anomaly reflects the fact that, for some page-replacement algorithms, the page-fault rate may increase as the number of allocated frames increases.

Optimal Algorithm : Belady's anomaly shows the search for an optimal page-replacement algorithm. An optimal page-replacement algorithm has the lowest page-fault rate of all algorithms. An optimal algorithm will never suffer from Belady's anomaly. An optimal page-replacement algorithm exists, and has been called OPT or MIN. It is simply Replace the page that will not be used for the longest period of time.

Unfortunately, the optimal page-replacement algorithm is difficult to implement, because it requires future knowledge of the reference string.

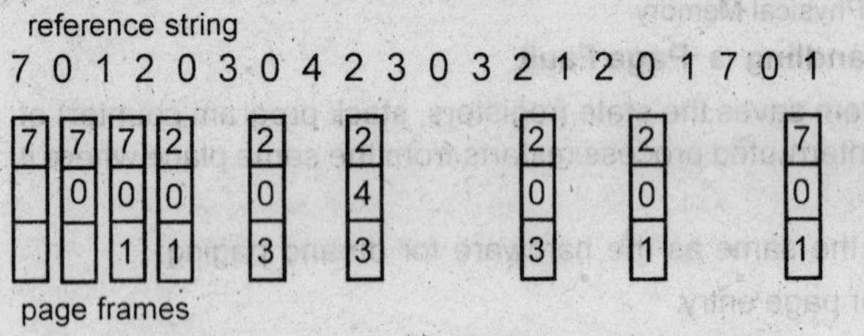

Fig. 3.15. Optimal page-replacement alogorithm

LRU Algorithm : We can replace the page that has not been used for the longest period of time. This approach is the least recently used (LRU) algorithm.

LRU replacement associates with each page the time of that page's last use. When a page must be replaced, LRU chooses that page that has not been used for the longest period of time. This strategy is the optimal page-replacement algorithm looking backward in time, rather than forward. The result of applying LRU replacement to out example reference string is shown in Fig.3.16.

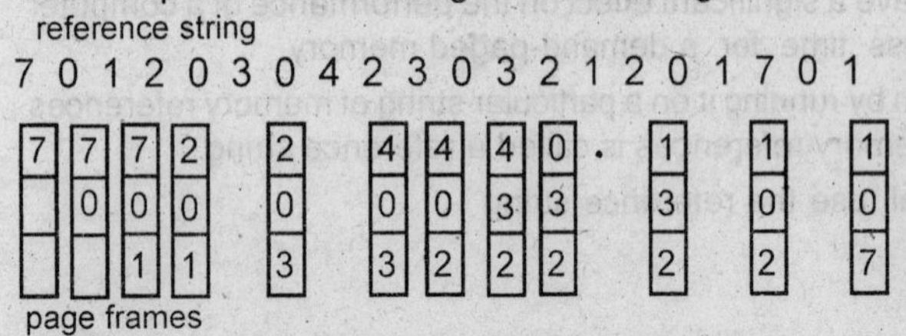

Fig.3.16. LRU page-replacement algorithm.

The LRU policy is often used as a page-replacement algorithm and is considered to be quite good. The major problem is how to implement LRU replacement. An LRU page-replacement algorithm may require substantial hardware assistance. The problem is to determine an order for the frames defined by the time of last use.

Two implementations are feasible :

I. Counters : In the simplest case, we associate with each page-table entry a time-of-use field, and add to the CPU a logical clock or counter. The clock is incremented for every memory reference. Whenever a reference to a page is made, the contents of the clock register are copied to the time-of use field in the page table for that page.

In this way, we always have the "time" of the last reference to each page. We replace the page with the smallest time value. This scheme requires a search of the page table to find the LRU page, and a write to memory (to the time-to-use field in the page table) for each memory access. The times must also be maintained when page tables are changed (due to CPU scheduling). Overflow of the clock must be considered.

II.Stack : Another approach to implementing LRU replacement is to keep a stack of page numbers. Whenever a page is referenced, it is removed from the stack and put on the top. In this way, the top of the sack is always the most recently used page and the bottom is the LRU page (Fig 3.17). Because entries must be removed from the middle of the stack, it is best implemented by a doubly linked list, with a head and tail pointer.

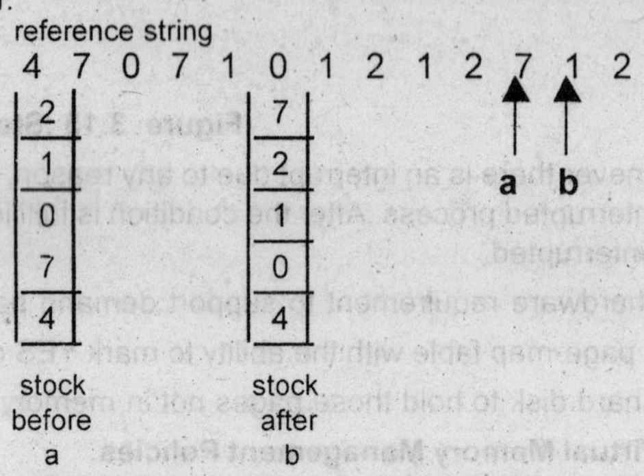

Fig. 3.17 Use of a stack to record the most recent page references.

12. Thrasing : If the number of frames allocated to a low-priority process falls below the minimum number required by the computer architecture, we must suspend that process' execution. If the process does not have this number of frames, it will very quickly page fault. At this point, it must replace some page. However, since all its pages are in active use, it must replace a page that will be needed again right away. Consequently, it very quickly faults again, and again. The process continues to fault, replacing pages for which it will then fault and bring back in right away.

This high paging activity is called thrashing. A process is thrasing if it is spending more time paging than executing.

SOLVED EXAMPLES

1. On a system using first-fit allocation, assume memory is allocated as specified in Fig. Before additional requests for 20K, 10K, and 5K (in that order) are received. At what starting address will each of the additional requests be allocated ?

Used	Hole	Used	Hole	Used	Hole	Used	Hole	Used	Hole	Used	Hole
10 K	10 K	20 K	30 K	10 K	5 K	30 K	20 K	10 K	15 K	20 K	20 K

Fig : Variable Partition memory allocation

Sol : First - fit searches for the first hole larger than or equal to 20K. The first hole is too small (10K). The second hole (30K) can accommodate the request. So the first request would be allocated at the start of the second hole, location 40K. This would reduce the size of the second hole to 10K. Memory would now look as follows.

Used	Hole	Used	Hole	Used	Hole	Used	Hole	Used	Hole	Used	Hole
10 K	10 K	40 K	10 K	10 K	5 K	30 K	20 K	10 K	15 K	20 K	20 K

The second request for 10K would be allocated into the first hole at starting location 10K, eliminating the first hole. Memory would now look as follows.

Used	Hole	Used	Hole	Used	Hole	Used	Hole	Used	Hole
60 K	10 K	10 K	5 K	30 K	20 K	10 K	15 K	20 K	20 K

For the final request, the first hole not smaller than 5K starts at location 60K.

2. On a system using best-fit allocation, assume memory is allocated as specified in Fig. Before additional requests for 20K, 10K and 5K (in that order) are received. At what starting address will each of the additional requests be allocated ?

Sol : Best fit searches for the smallest hole larger than or equal to 20K. The fourth hole is exactly 20K, making it is best fit. That hole starts at location 10K+10K+20K+30K+10K+5K+30K=115K. Memory would now look as follows.

Used	Hole	Used	Hole	Used	Hole	Used	Hole	Used	Hole
10 K	10 K	20 K	30 K	10 K	5 K	60 K	15 K	20 K	20 K

The second request for 10K would be allocated into the first hole at starting location 10K, eliminating the first hole. Memory would now look as follows.

Used	Hole	Used	Hole	Used	Hole	Used	Hole
40 K	30 K	10 K	5 K	60 K	15 K	20 K	20 K

The final request of 5K would be allocated to the hole of 5K that starts at location 60K+10K+10K=80K.

3. On a system using worst-fit allocation, assume memory is allocated as specified in Fig. Before additional requests for 20K, 10K and 5K (in that order) are received. At what starting address will each of the additional requests be allocated ?

Sol : Worst fit always allocates the largest hole, so the first request would be allocated to the second hole, 30K. The second hole starts at location 40K. Memory would now look as follows.

Used	Hole	Used	Hole	Used	Hole	Used	Hole	Used	Hole	Used	Hole
10 K	10 K	40 K	10 K	10 K	5 K	30 K	20 K	10 K	15 K	20 K	20 K

The fourth hole is now the largest and would be allocated to the second request. (The sixth hole is also 20K. It would be equally correct to derive a solution based on worst fit choosing the sixth hole instead of the fourth hole.) This hole starts at location 10K+10K+40K+10K+10K+5K+30K=115K. Memory would now look as follows.

Used	Hole	Used	Hole	Used	Hole	Used	Hole	Used	Hole	Used	Hole
10 K	10 K	40 K	10 K	10 K	5 K	40 K	10 K	10 K	15 K	20 K	20 K

Now the largest hole is the last one. The final request would be allocated starting at location 10K+10K+40K+10K+10K+5K+40K+10K+10K+15K+20K=180K.

4. On a system using next-fit allocation, assume memory is allocated as specified in Fig. Before additional requests for 20K, 10K and 5K (in that order) are received. At what starting address will each of the additional requests be allocated ?

Sol : Next fit searches for the first hole larger than or equal to 20K. The first hole is too small (10K). The second hole (30K) can accommodate the request. So the first request would be allocated at the start of the second hole, location 40K. This would reduce the size of the second hole to 10K. Memory would now look as follows.

Used	Hole	Used	Hole	Used	Hole	Used	Hole	Used	Hole	Used	Hole
10 K	10 K	40K	10 K	10 K	5 K	30 K	20 K	10 K	15 K	20 K	20 K

For the next request, next fit starts searching with the last allocated block. Starting the search with the second hole, the second hole is found to be able to accommodate the 10K request. Thus, the second request would be allocated starting at location 60K. Memory would now look as follows.

Used	Hole	Used	Hole	Used	Hole	Used	Hole	Used	Hole
10 K	10 K	60K	5 K	30 K	20 K	10 K	15 K	20 K	20 K

The search for a hole for the third request would start at location 80K; the hole at that location would be selected for the third request.

5. On a system with 1MB of memory using the buddy system, what is the first request that will fail in the following string of requests due to lack of available memory ? Requests : 50K, 150K, 90K, 130K, 70K, 80K, 120K, 180K, 60K.

Sol : Memory blocks will be adjusted for each request as follows :

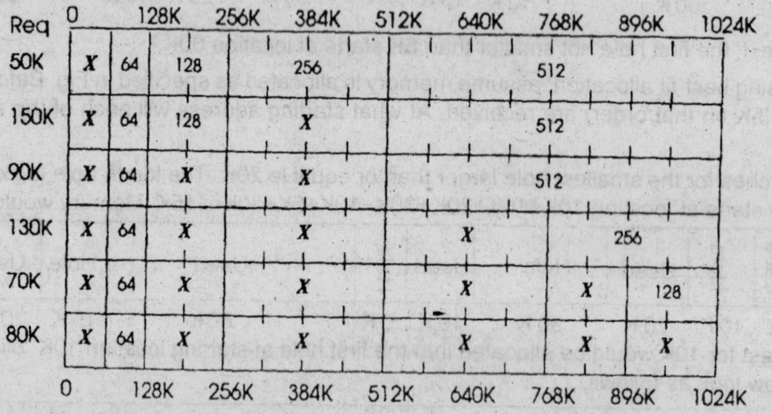

The next request for 120K could not be satisfied.

6. In the previous problem, at the time of the failed request, how much memory is wasted due to internal fragmentation and how much is wasted due to external fragmentation?

Sol : There is one 64K block that is unallocated but unusable. Thus, external fragmentation is 64K. Of the remaining 1 024K - 64K = 960K of allocated space, only 50K + 150K + 90K + 130K + 70K + 80K = 570K has been requested. Thus, 960K - 570K = 390K is wasted due to internal fragmentation.

7. Why are page sizes always a power of 2?

Sol : All addresses are binary and are divided into page or page frame number and offset. By making the page size a power of 2, the page number, the page frame number, and the offset can be determined by looking at particular bits in the address; no mathematical calculations are required. The physical address can be generated by concatenating the offset to the frame number.

8. On a simple paging system with 2^{24} bytes of physical memory, 256 pages of logical address space, and a page size of 2^{10} bytes, how many bits are in a logical address?

Sol : The logical address space contains $256 = 2^8$ pages of 2^{10} bytes, making the total logical address space $2^{10} \times 2^8 = 2^{18}$ bytes. An 18-bit address is required to cover a 2^{18} byte address space.

9. On a simple paging system with a page table containing 64 entries of 11 bits (including valid/invalid bit) each, and a page size of 512 bytes, how many bits are in a logical address?

Sol : The 6 most significant bits of the logical address would specify the page number, the 9 least significant bits would specify the page offset, making the total number of bits 15.

10. On a simple paging system with a page table containing 64 entries of 11 bits (including valid/invalid bit) each, and a page size of 512 bytes, what is the size of the logical address space?

Sol : In the previous problem, the logical address was determined to have 15 bits. A 15-bit address creates an address space of 32,768= 2^{15}.

11. On a simple paging system with a page table containing 64 entries of 11 bits (including valid/ invalid bit) each, and a page size of 512 bytes, how many bits are in a physical address?

Sol : The 10 most significant bits of the physical address would specify the page frame number, the 9 least significant bits would specify the page offset, making the total number of bits 19.

12. On a simple paging system with a page table containing 64 entries of 11 bits (including valid/invalid bit) each, and a page size of 512 bytes, what is the size of the physical address space?

Sol : A 19-bit address creates an address space of 2^{19}.

13. A system that uses a two-level page table has 2^{12} -byte pages and 32-bit virtual addresses. The first 8 bits of the address serve as the index into the first-level page table.

(a) How many bits specify the second-level index?
(b) How many entries are in a level-one page table?
(c) How many entries are in a level-two page table?
(d) How many pages are in the virtual address space?

Sol : a) 12 bits

Given a page size of 2^{12} bytes, 12 bits are needed to specify the offset. That leaves 32 - (12 + 8) = 12 bits.

b) 2^8 Eight bits can specify up to 2^8 entries.

c) 2^{12} Twelve bits can specify up to 2^{12} entries.

d) 2^{20} Twenty bits can address 2^{20} pages. Or, if you prefer, 2^8 level-one page table entires each access a level-two page table-with 2^{12} entries. $2^8 \times 2^{12} = 2^{20}$

14. A computer has a 32-bit virtual address space and 1024 byte pages. A page table entry takes 4 bytes. A multilevel page table is used because each table must be contained with a page. How many levels are required?

Sol : A page table can contain 1024/4 = 256 = 2^8 entries. Given the page size of 2^{10}, 32 - 10 = 22 bits specify the page number. Each level of page table can handle 8 of the 22 bits, so three levels would be required. Tables at two of the levels would have 2^8 entries; the table size at the other level would be only 2^6.

15. On system using simple segmentation, compute the physical address for each of the logical addresses, given the following segment table. If the address generates a segment fault, indicate so

Segment	Base	Length
0	330	124
1	876	211
2	111	99
3	498	302

(a) 0,99 (b) 2,78 (c) 1,265 (d) 3,222 (e) 0,111

Sol : a) 429

Offset 99 is less than the segment length of 124. Segment 0 begins at location 330, so offset 99 is at physical address 99 + 330 = 429

b) 177

Offset 78 is less than the segment length of 99. Segment 2 begins at location 111, so offset 78 is at physical address 78 + 111 = 189.

c) Fault

Offset 265 is greater than the segment length of 211. This address results in a segment fault.

d) 720.

Offset 222 is less than the segment length of 302. Segment 3 begins at location 498, so offset 222 is at physical address 222 +498=720.

e) 441

Offset 111 is less than the segment length of 124. Segment 0 begins at location 330. so offset 111 is at physical address 111 +330=441.

16. On a system using paging and segmentation, the virtual address space consists of up to 8 segments where each segment can be up to 2^{29} bytes long. The hardware pages each segment into 256-byte pages. How many bits in the virtual address specify the:

(a) Segment number? (b) Page number?

(c) Offset within page? (d) Entire virtual address?

Sol : a) 3

Since 8 =2^3,3 bits are needed to specify the segment number.

b) 21

With 256 = 2^8 byte pages, a 2^{29} -byte segment can have $2^{29}/2^8 = 2^{21}$ pages.

Thus, 21 bits are needed to specify the page number.

c) 8

To specify the offset into the 2^8-byte page, 8 bits are needed.

d) 32

3 + 21 + 8 = 32.

17. If an instruction takes time m if there is no page fault, and time n if there is a page fault, what is the effective instruction time if page faults occur once every i instructions?

Sol : For every i instruction, i - 1 instructions will take time m, and one instruction will take time n. The average instruction time is

$$\frac{(i-1)\times m+n}{i}=m+\frac{(n-m)}{i}$$

18. Give a reference string for the FIFO replacement algorithm that causes Belady's anomaly to occur when the number of frames increases from three to four.

Sol : 1, 2, 3, 4, 1, 2, 5, 1, 2, 3, 4, 5.

If three frames are allocated, pages will be swapped as follows.

Before	Page	Fault	After
-, -, -	1	*	1, -, -
1, -, -	2	*	2, 1, -
2, 1, -	3	*	3, 2, 1
3, 2, 1	4	*	4, 3, 2
4, 3, 2	1	*	1, 4, 3
1, 4, 3	2	*	2, 1, 4
2, 1, 4	5	*	5, 2, 1
5, 2, 1	1		5, 2, 1
5, 2, 1	2		5, 2, 1
5, 2, 1	3	*	3, 5, 2
3, 5, 2	4	*	4, 3, 5
4, 3, 5	5		4, 3, 5

The number of page faults for three frames is nine. If four frames are allocated, pages will be swapped as follows.The number of page faults for four frames is ten.

Before	Page	Fault	After
-, -, -, -	1	*	1, -, -, -
1, -, -, -	2	*	2, 1, -, -
2, 1, -, -	3	*	3, 2, 1, -
3, 2, 1, -	4	*	4, 3, 2, 1
4, 3, 2, 1	1		4, 3, 2, 1
4, 3, 2, 1	2		4, 3, 2, 1
4, 3, 2, 1	5	*	5, 4, 3, 2
5, 4, 3, 2	1	*	1, 5, 4, 3
1, 5, 4, 3	2	*	2, 1, 5, 4
2, 1, 5, 4	3	*	3, 2, 1, 5
3, 2, 1, 5	4	*	4, 3, 2, 1
4, 3, 2, 1	5	*	5, 4, 3, 2

19. Given a system with four page frames, the following table indicates page, load time, last reference time, dirty bit, and reference bit.

Page	Load Time	Last Reference	Dirty Bit	Reference Bit
0	167	374	1	1
1	321	321	0	0
2	254	306	1	0
3	154	331	0	1

a) Which page will FIFO replace ?

b) Which page will LRU repalce ?

c) Which page will NRU repalce ?

d) Which page will second chance replace ?

Sol : a) 3 ; FIFO selects the page loaded first, page 3 at time 154.

b) 2 ; LRU selects the page reference last, page 2 at time 306.

c) 1 ; NRU first looks for a clean page that was not recently referenced. Page 1 is the only such page

d) 2 ; Second chance first checks the oldest page, page 3. Since its reference bit is 1, it resets that bit to 0 and sets the entry's load time to the current time. The same thing happens to page 0, the next oldest page. The reference bit for page 2 is then checked, and since it is 0 it is the selected page.

20. Given references to the following pages by a program, 0,9,0,1,8, 1,8,7,8,7, 1,2,8,2,7, 8,2,3,8,3,

how many page faults will occur if the program has three page frames available to it and uses:

a) FIFO replacement?

b) LRU replacement?

c) Optimal replacement?

Sol : a) FIFO replacement

8 faults.

Before	Page	Fault	After
-, -, -	0	*	0, -, -
0, -, -	9	*	9, 0,
9, 0,	0		9, 0, -
9, 0, -	1	*	1, 9, 0
1, 9, 0	8	*	8, 1, 9
8, 1, 9	1		8, 1, 9
8, 1, 9	8		8, 1, 9
8, 1, 9	7	*	7, 8, 1
7, 8, 1	8		7, 8, 1
7, 8, 1	7		7, 8, 1
7, 8, 1	1		7, 8, 1
7, 8, 1	2	*	2, 7, 8
2, 7, 8	8		2, 7, 8
2, 7, 8	2		2, 7, 8
2, 7, 8	7		2, 7, 8
2, 7, 8	8		2, 7, 8
2, 7, 8	2		2, 7, 8
2, 7, 8	3	*	3, 2, 7
3, 2, 7	8	*	8, 3, 2
8, 3, 2	3		8, 3, 2

c) Optimal replacement

Before	Page	Fault	After
-, -, -	0	*	0, -, -
0, -, -	9	*	9, 0, -
9, 0, -	0		9, 0, -
9, 0, -	1	*	1, 9, 0
1, 9, 0	8	*	8, 1, 9
8, 1, 9	1		8, 1, 9
8, 1, 9	8		8, 1, 9
8, 1, 9	7	*	7, 8, 1
7, 8, 1	8		7, 8, 1
7, 8, 1	7		7, 8, 1
7, 8, 1	1		7, 8, 1
7, 8, 1	2	*	2, 7, 8
2, 7, 8	8		2, 7, 8
2, 7, 8	2		2, 7, 8
2, 7, 8	7		2, 7, 8
2, 7, 8	8		2, 7, 8
2, 7, 8	2		2, 7, 8
2, 7, 8	3	*	3, 2, 8
3, 2, 8	8		3, 2, 8
3, 2, 8	3		3, 2, 8

7 faults

21. You have a demand-paged system where intergers are stored in 4 bytes, pages are 256 bytes, LRU page replacement is used, and each process is allocated 3 frames. A process executes the following code.

```
int [ ] [ ] a = new int [200] [200];
int i = 0;
int j = 0;
while (i++<200) {
j = 0;
while ( j++<200)
a[ i ] [ j ] = 0;}
```

The code occupies page 0 and, since a location from page 0 is referenced each instruction, page 0 is always swapped-in. Variables i and j are stored in fast registers.
a) How many pages are needed for the array, assuming all elements of the array are stored in contiguous memory locations?
b) How many page faults will this program generate?
Sol :
a) The array has 200 x 200 = 40,000 elements, each requiring 4 bytes of storage. 160,000 bytes occupy 625 pages.
b) The program accesses the locations in the array in the order they are stored. Thus, it will swap-in the instruction page, and each of the 625 data pages once, for a total of 626 swaps.

22. On a system using paging and segmentation, references to swapped-in locations accessible through an entry in an associative table take 200 ns. If the main memory page table must be used, the reference takes 400 ns. References that result in page faults require 10 ms if the page to be replaced has been modified, 5 ms otherwise. If the page fault rate is 5%, the associative table hit rate is 65%, and 40% of replaced pages are "dirty," what is the effective access time? Assume the system is running only a single process, and the CPU is idle during page swaps.

Sol : The probability and time, converted to μs for the various types of accesses are as follows.

Type	Probability	Time (μs)
In, associative	95% x 65% = 0.6175	0.2
In, main memory	95% x. 35% = 0.3325	0.4
Out, clean	05% x 60% = 0.0300	5000
Out, dirty	05% x 40% = 0.0200	10000

Effective access time = 0.6175 x 0.2 + 0.3325 x 0.4 + 0.03 x 5000 + 0.02 x 10000

Effective access time = 350.2565

23. In a computer system where the 'best fit' algorithm is used for allocating 'jobs' to 'memory partitions', the following situation was encountered :

Partitions sizes in KB	4K	8K	20K	2K				
Jobs sizes in KB	2K	14K	3K	6K	6K	10K	20K	2K
Time for execution	4	10	2	1	4	1	8	6

When will the 20K job complete?

Sol : The 20 K partition will be occupied by the 14 K job which will require 10 units of time for execution. After its completion the 10K job will be occupying this 20 K partition. And on its completion 20 K will be taking up the partition.

Hence total time = 10 + 1 + 8 = 19 units

Other jobs will be occupying the other partitions.

OBJECTIVE QUESTIONS - I

1. The Optimal page replacement algorithm will select the page that **[GATE-2002]**
a) Has not been used for the longest time in the past.
b) Will not be used for the longest time in the feature.
c) Has been used least number of times.
d) Has been used most number of times.

2. Which of the following is / are advantages of virtual memory? **[GATE-1999]**
a) Faster access to memory on an average
b) Processes can be given protected address spaces.
c) Linker can assign addresses independent of where the program will be loaded in physical memory.
d) Programs than the physical memory size can be run.

3. Thrashing **[GATE-1997]**
a) Reduces page I/O
b) Decreases the degree of multi programming
c) Implies excessive page I/O
d) Improve the system performance.

4. Dirty bit for a page in a page table: **[GATE-1997]**
a) Helps avoid unnecessary writes on a paging device
b) Helps maintain LRU information
c) Allows only read on a page
d) None

5. Which of the following page replacement algorithms suffers from Belady's anamoly? **[GATE-1995]**
a) Optimal replacement b) LRU
c) FIFO d) both (a) and (c)

6. A memory page containing a heavily used variable that was initialized very early and is in constant use is removed when **[GATE-1994]**
a) LRU page replacement algorithm is used
b) FIFO page replacement algorithm is used
c) LRU page replacement algorithm is used
d) None

7. For effective swapping, binding of instructions and data should be done at
a) Load time b) Compile time
c) Execution time d) None

8. Which memory management scheme does not support multiprogramming
a) Single partition allocation
b) Multiple partition allocation
c) paging d) Segmentation

9. Which memory management technique causes no external fragmentation
a) Multiple partition with fixed partitions
b) Paging c) Segmentation d) None

10. Which technique is used to solve the problem of external fragmentation in multiple partition schemes?

a) Compaction b) Paging

11. Which memory management scheme, which allows the memory, occupied by a process to be non-contiguous.

a) Segmentation b) Paging

12. The page replacement algorithm which gives the lowest page fault rate is

a) LRU b) FIFO
c) Optimal page replacement d) None

13. Page fault occurs when

a) The page is corrupted by application software
b) The page is in main memory
c) The page is not in main memory
d) One tries to divide a number by 0

14. Overlay is

a) A part of an operating system
b) A single memory location
c) Overloading the system with many user files
d) A single contiguous memory that is used in olden days for running large programs by swapping.

15. The page replacement policy that some times leads to more page faults when the size of the memory is increased

a) FIFO b) LRU
c) No such policy exists d) None

16. Fragmentation is :

a) Dividing the secondary memory into equal sized fragments
b)Dividing the main memory into equal sized fragments
c) Fragments of memory words used in a page
d) Fragments of memory words unused in a page

17. In paged memory systems, if the page size is increased then the internal fragmentation generally

a) Becomes less b) Becomes more
c) Remains constant d) None

18. If the property of locality of reference is well pronounced in a program

a) The number of page faults will be more
b) The number of page faults will be less
c) The number of page faults will remain the same
d) Execution will be faster

19. The first-fit, best-fit and worst-fit algorithms can be used for

a) Contiguous allocation of memory
b) Linked allocation of memory
c) Indexed allocation of memory
d) All of the above

20. In a system that does not support swapping

a) The compiler normally binds symbolic addresses (variables) to relocate addresses
b) The compiler normally binds symbolic addresses to physical addresses
c) The loader binds relocatable addresses to physical addresses
d) Binding of symbolic addresses to physical addresses normally takes place during execution

21. Which of the following is TRUE?

a) Overlays are used to increase the size of physical memory
b) Overlays are used to increase the logical address space
c) When overlays are used the size of a process is not limited to the size of physical memory
d) Overlays are used whenever the physical address space is smaller than logical address space

22. In partitioned memory allocation scheme

a) The best fit algorithm is always better than the first fit algorithm
b) The first fit algorithm is always better than the best fit algorithm
c) The superiority of the first fit and best fit algorithms depend on the sequence of memory requests
d) None of these

23. The main function of shared memory is:

a) To use primary memory efficiently
b) To do interposes communication
c) To do intraprocess communication
d) None of the these.

24. Aging is

a) keeping track of cache contents
b) keeping track of what pages are currently residing in the memory
c) keeping track of how many times a given page is referenced
d) increasing the priority of jobs to ensure termination in a finite time

25. If there are 32 segments, each of size 1k byte, then the logical address should have

a) 13 bits b) 14 bits
c) 15 bits d) 16 bits

26. Which of the following page replacement algorithms suffer from Belady's anomaly?

a) Optimal replacement b) LRU
c) FIFO d) Both (a) and (c)

27. Which of the following statement(s) is FALSE?

a) Segmentation suffers from external fragmentation
b) Paging suffers from internal fragmentation
c) Segmented memory can be paged
d) Virtual memory is used only in multi-user systems

28. In virtual memory the page fault frequency is reduced when the

a) Page size is reduced
b) Processes are mostly I/O bound
c) Locality of reference is applicable to the process
d) Shortest remaining time scheduling is used

29. Page fault frequency in an operating system is reduced when the **[GRE-CS]**

a) Processes tend to be CPU-bound
b) Locality of reference is applicable to the process
c) Processes tend to be I/O –bound
d) Shortest remaining time scheduling is used
e) Size of pages is reduced

30. The main purpose of Translation Lookaside Buffers(LTBs) is to **[UGC NET-2000]**

a) reduce the main memory access time
b) increase the speed of CPU
c) reduce the requirement for main memory
d) increase the page size in paging

31. In the context of an Operating system Least Recently Used refers to **[Hyd.Univ-2000]**

a) The policy by which programs are allowed to perform low level input/output operations
b) A method by which deadlocks (if detected) are resolved
c) A policy which decides how storage space previously allocated should be reallocated
d) The policy of assigning device numbers to serial input/output terminals

ANSWERS

1. b	2. b	3. c	4. a	5. c
6. b	7. c	8. a	9. b	10. a
11. b	12. c	13. c	14. d	15. a
16. d	17. b	18. b,d	19. a	20. a,c
21. c	22. c	23. b	24. d	25. c
26. c	27. d	28. c	29. b	30. a
31. c				

OBJECTIVE QUESTIONS - II

1. Consider a virtual memory system with FIFO page replacement policy. For an arbitrary page access pattern, increasing the number of page frames in main memory will **[GATE2001]**

a) Always decrease the number of page faults.
b) Always increase the number of page faults.
c) Some times increase the number of page faults.
d) Never affect the number of page faults.

2. Consider a machine with 64-MB physical memory and a 32-bit virtual address space. If the page size is 4 KB, what is the approximate size of the page table? **[GATE-2001]**

a) 16 MB b) 8 MB
c) 2 MB d) 24 MB

3. Suppose the time to service a page fault is on the average 10 milliseconds, while a memory access takes 1 microsecond then a 99.99 % hit ratio results in average memory access time of **[GATE-2000]**

a) 1.9999 milliseconds b) 1 milliseconds
c) 9.999 milliseconds d) 1.9999 microseconds

4. If an instruction takes i microseconds and a page fault takes an additional j microseconds the effective instruction time if on the average a page fault occurs every k instructions is: **[GATE-1998]**

a) i + j / k b) i + j * k
c) (i + j) / k d) (i + j) * k

5. The overlay tree for a program is as shown below:What will be the size of the partition (in physical memory) required to load (and run) this program? **[GATE-1998]**

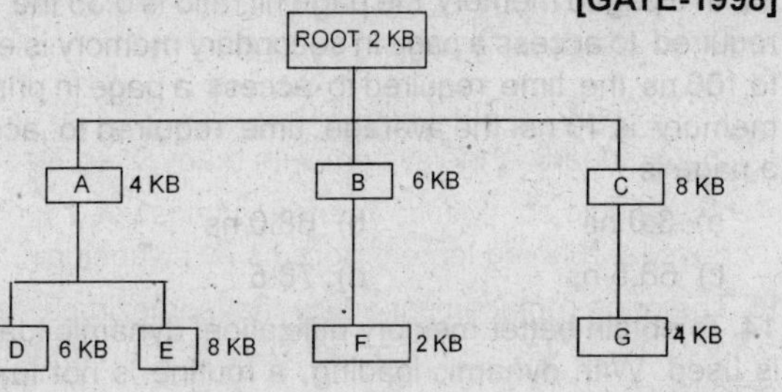

a) 8 KB b) 10 KB c) 12 KB d) 14 KB

6. A 1000 K byte memory is managed using variable partitions but to compaction. It currently has two partitions of sizes 200 K bytes and 260 K bytes respectively. The smallest allocation request in K bytes that could be denied is for **[GATE-1996]**

a) 151 b) 181 c) 231 d) 541

7. In a paged segment scheme of memory management, the segment table it self must have a page table because **[GATE-1995]**

a) The segment table is often too large to fit in one page
b) Each segment is spread over a number of pages
c) Segment tables point to page table and not to the physical locations of the segment
d) The processor description base registers points to a page table.

8. The address sequence generated by tracing a particular program executing, in a pure demand paging system with 100 records per page with 1 free main memory frame is recorded as follows what is the number of page faults?

0100, 0200, 0430, 0499, 0510, 0530, 0560, 0120, 0220, 0240, 0260, 0320, 0370. **[GATE-1995]**

a) 7 b) 8 c) 13 d) 10

9. In a virtual memory system the address space specified by the address lines of the CPU must be.... than the physical memory size and... than the secondary storage size. **[GATE-1995]**

a) Smaller, smaller b) Smaller, large
c) larger, smaller d) larger, larger

10. The memory management scheme which allows the execution of a process which is partially loaded into physical Memory is:

a) Virtual memory b) Segmentation
c) Paging d) None

11. The page replacement algorithm which replaces the page which will not be used for a longest Period of time is

a) LRU b) FIFO
c) Optimal page replacement d) None

12. Determine the number of page faults when references to pages occur in the following order: 1,2,4,5,2,1,2,4. Assume that the main memory can accommodate 3 pages and the main memory already has pages 1 and 2, with page 1 having been brought earlier than page2. (LRU algorithm is used)

a) 3 b) 4 c) 5 d) None

13. In a paged memory, the page hit ratio is 0.35 the time required to access a page in secondary memory is equal to 100 ns the time required to access a page in primary memory is 10 ns. the average time required to access a page is :

a) 3.0 ns b) 68.0 ns
c) 68.5 ns d) 78.5

14. To obtain better memory utilization, dynamic loading is used. With dynamic loading, a routine is not loaded until it is called for. For implementing dynamic loading:

a) Special support from h/w is essential
Special support from OS is essential
b) Special support from both h/w and OS are essential
c) User programs can implement dynamic loading with out any special support from the OS or the h/w

15. A process refers to 5 pages A,B,C,D and E in the following order A; B; C; D; A; B; E;A;B;C;D;E. if the page replacement algorithm is FIFO, the number of page transfer with an empty internal store of 3 frames is:

a) 7 b) 8 c) 9 d) 10

16. In a paged segmented scheme of memory management, the segment table itself must have a page table because

a) The segment table is often too large to fit in one page
b) Each segment is spread over a number of pages
c) Segment tables point to page tables and not to the physical location of the segment
d) The processor's description base register points to a page table

17. A computer installation has 1000k of main memory. The jobs arrive and finish in the following sequence

JOB1	requiring	200 k	arrives
JOB2	requiring	350 k	arrives
JOB3	requiring	300 k	arrives
JOB1	finishes		
JOB4	requiring	120 K	arrives
JOB5	requiring	150 k	arrives
JOB6	requiring	80 k	arrives

Among best-fit and first –fit, which performs better for this sequence:

a) First-fit b) Best-fit
c) Both perform the same d) None

18. Consider the following heap:

JOB1-50 FREE-150 JOB2 - 300 FREE-350 JOB3-600

The sequence of requests for blocks of sizes 300,25,125,50 can be satisfied if we use

a) Either first-fit or best –fit policy
b) First-fit but not best-fit
c) Best-fit but not first-fit d) None

19. An Operating system uses a LRU page replacement algorithm for managing memory. considered the following page reference string where each reference is made in one unit of time 1, 8,1,7,8,2, 7,2,1,8,3,8,2,1,3,1,7,1, 3, 7 which of the following number of page faults that are generated for this particular LRU case, assuming that the process has been allocated four page frames?

a) 2 b) 4 c) 5 d) 6

Question 21 and 22 refer to following information:

An operating system uses a working set storage management strategy. In order to analyze the performance of this operating system, some diagnostic measurements were made while a user process P1 was running. The results are shown in the following graph. **[GRE-CS]**

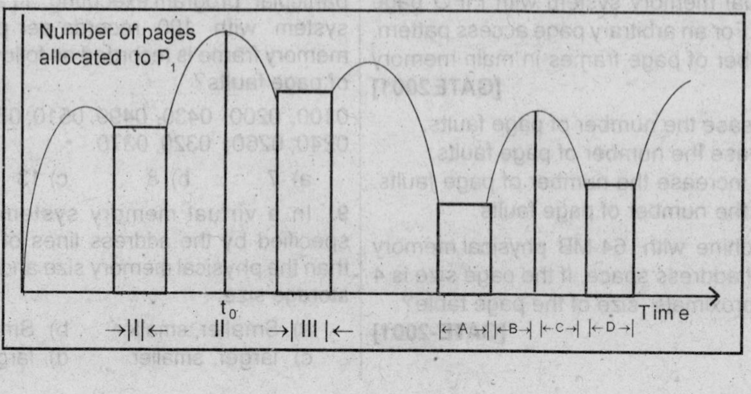

20. A working set for the process P for an optimal use of memory occurs during

a) time interval I. b) time interval II.

c) time interval III. d) time intervals I and III.

21. At which time interval is the page fault the greatest?

a) A b) B c) C d) D

22. An operating system uses the First-In-First-out technique for page replacement. If the number of main memory frames for a process is increased, then the number of page faults for that process **[UGC NET-1997]**

a) always decrease b) always increase

c) may increase d) will remain the same

23. Which of the following statement is FALSE? **[UGC NET-1997]**

a) The technique of storage compaction involves moving all occupied areas of storage to one end or the other of main memory

b) Compaction does not involve relocation of programs

c) Compaction is also known as garbage collection

d) The system must stop everything while it performs compaction

24. The memory management technique employed by DOS is **[UGC NET-1996]**

a) paging b) garbage collection

c) dynamic allocation d) none of these

25. In which of the storage placement strategies a program is placed in the smallest available hole in main memory **[UGC NET-1995]**

a) FIRST FIT b) BEST FIT

c) WORST FIT d) BUDDY

26. Which one of the following memory management schemes causes external fragmentation? **[UGC NET-1998]**

a) Simple paging b) Multiple fixed task

c) Demand paging d) Multiple variable task

27. Which of the following memory management schemes creates the minimum external fragmentation? **[UGC NET-1998]**

a) worst fit b) Best fit

c) First fit d) static

28. An LRU algorithm for page replacement is based on **[UGC NET-1999]**

a) Memory Reference string generated in the past

b) Memory Reference string forecast in future

c) FIFO algorithm

d) Does not use a reference string

29. A page fault **[UGC NET-1999]**

a) is an error specific page

b) occurs when a program accesses a page of memory

c) is an access to a page currently not in memory

d) is a interface to page belonging to another program

ANSWER KEY - II

1. c	**2.** c	**3.** d	**4.** a	**5.** d
6. d	**7.** a	**8.** a	**9.** c	**10.** a
11. c	**12.** b	**13.** c	**14.** c	**15.** c
16. b	**17.** a	**18.** b	**19.** d	**20.** d
21. b	**22.** c	**23.** b,c	**24.** c	**25.** b
26. d	**27.** d	**28.** a	**29.** c	

IV. FILE MANAGEMENT

- **File Concept and Directories**
- **Disk Organisation**
- **Disk Allocation Methods**
- **Disk Scheduling**
- **File protection**

1. File Concept : File management is one of the most visible services of an operating system. Files are commonly stored on physical devices such as floppy disk, hard disk, magnetic tape whose characteristics and organisation are different from each other.

While creating or manipulating files in any operating system environment users need not be aware of physical characteristics of devices. File mapping is done by file management (i.e. by operating system). The file management is supposed to hide all device specific aspects of file manipulation from users and provide them with an abstract of a simple, uniform space of named files. File often appears as a sequence of bits, bytes, lines or records whose meaning is defined by programmer or user. This is a general concept about files. Some systems support a single uniform set of file manipulation features for both file management and I /O devices management. This feature is also referred to as device independent I / O of simply device independence. It means that to users all I / O devices must appear as a set of files. For example a file may be printed by means of a copy command where the "destination file" is the "printer device".

Some operating systems are better at device independence than others. In Unix operating system for example a file system (e.g. a disk) can be mounted any where in the file tree, allowing any file to be accessible by its path name, wihtout regard to which device it is on, In DOS environment, on the other hand, the user must explicitly specify which device each file is on (except that one drive is default and may be omitted). Thus if the default device is drive c:, to run a program located on drive A : with input and output files on drive B:, one would have to type :

A : cprog < B : Input > B : output

There are three common file organisations which are illustrated in figure 4.1.

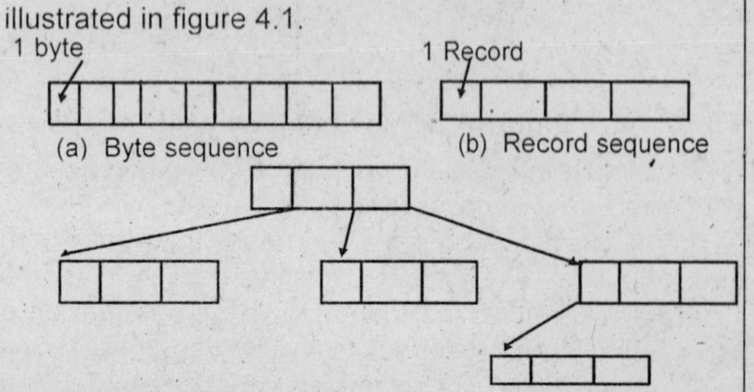

Figure 4.1: Three kinds of file organisations

The first file oraganization is a simple byte sequence (figure 4. 1(a)). The operating system does not impose any structure on the file organisation. This type of scheme has been adopted in unix operating system among others. Unix considers each file to be a sequence of 8 bit bytes. No interpretation of theses bits is made by the operating system. Each application program must incorporate its own code to interpret an input file into the appropriate structure. The advantage of this scheme is that there is a maximum flexibilitiy for user but the disadvantage is that there is no support from operating system side.

The second way is a sequence of fixed size records (figure 4.1(b)). Arbitrary records can be read or written, but records cannot be inserted or deleted in the middle of a file. CP/ M operating system supports this type of scheme.

The third file organization scheme (figure 4.1 (c)) is a free of disk blocks, each block holding n keyed records. Records can be searched by key value and new records can be inserted anywhere in the file structure. In case a block is full, for adding new records, the block A is broken into two other blocks, both of which are then added to the free blocks in their alphabetical order. This sequence is used on mainframe system where it is called ISAM (Indexed Sequential Access Method). Files are usually stored on disks. All disks I /O operations are in terms of blocks (physical records), and all blocks are of equal size. Its size is determined by the size of a sector.

It should be noticed that during allocation of a file on disk in terms of blocks, there is a wastage of space for the last block of a file. If each block is 512 bytes, then a file of 2660 bytes would be allocated five blocks, and the last 100 bytes will be wasted. The wasted bytes allocated to keep everything in units of blocks (instead of bytes) is internal fragmentation.

Some operating systems examine the creation time of a file before it will be executed. Then it locates the source file and checks if the source has been modified since binary code was made. If it has, it automatically recompiles the source. The **Make** utility serves the similar purpose in Unix operating system.

The operation available on files differs slightly from one operating system to another operating system. Some common operations viz. creating a file, reading a file, writing a file, deleting a file are available on every system.

The common responsibilities of the file management system include the following :

- Mapping of logical file address to physical disk address.
- Management of disk space and allocation and deallocation.
- Keeping track of all files in the system.
- Support for protection and sharing of files recovery. We will take up these issuses in the subsequent sections.

2. Directories : In many systems, directories are treated as files which keep track of all other files. A single flat directory can contain a list of all files in a system. Flat directories are those directories in which the root directory contains all system files and there is no other subdirectory. When hierarchical directories (i.e. a tree if directories) are used, the collection of all directory and subdirectory entries define the totality of system files. This is more natural way to organize files. This is illustrated in the figure 4.2.

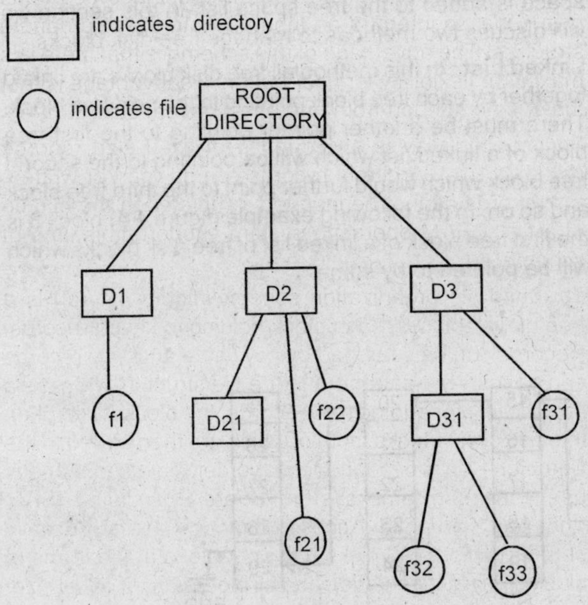

Figure 4.2 : Hierarchical directory

There are two basic methods to access a file organised in a hierarchical form. In the first method, each file is given an absolute pathname consisting of the path from the root directory to the files. For example, the path / d3 / d31/ f 32 (figure 4.2) means that the root directory contains a subdirectory d3, which in turn contains a subdirectory d31, which contains the file f 32. Absolute path name always start at the root directory and are unique. The other kind of method is the relative path name. This is used in conjunction with the concept of the working directory (also called the current directory).

A user can designate one directory as the current working directory in which case all path names not beginnig at the root directory are taken to the working directory. In Unix, all path names beginning with a slash are absolute, all others are relative to the user's current working directory. If the current working directory is / d3 / d31, then the file whose absolute path is / d3 / d31 / f 31 can be referenced simply as f 31.

One important question related to directories is where to keep them ? If we keep them in main memory, it will take lots of memory but searching of a file will be faster. On the other hand if we keep them on secondary storage device, searching will be slower. Most systems resolve this problem by keeping direcrtories on secondary storage system and by copying to main memory the actively used entries only. The identities of the activelly used entries are made known to the system when files are opened.

The most frequent operations on directories are searching, addition and removal of entries. Since directories are usually stored on disks, the efficiency of directory operations is, therefore primarily effected by the numbers of disk access required to complete a search, to append, or to remove an entry. Therefore, directories are organized in such a way that these operations should take less time.

Linear list Organization of directories is not very suitable for searching since it takes O(n) (order of n) comparisons to locate a given entry where n is the number of all entries in a directory. Hashing technique supports fast searching and appending operation but in the worst behaviour of hashing it is no better than that of linear list. Moreover,when a directory exceeds the limit that the related hash function can address, all of its entries may be hashed again. In that case directories can be organised in trees structure form.A balanced binary tree may be searched in O (log n) time. Other operations like appending and removing entries can also be performed in this time. An additional advantage of trees is that they provide sorted listing of directory entries which is required by many systems.

3. Disk Organisation : Information on the disk is referenced by a multi-part address, which includes the drive number, the surface and the track. The collection of tracks on all surfaces that are at the same distance is called a cylinder (figure 4.3)

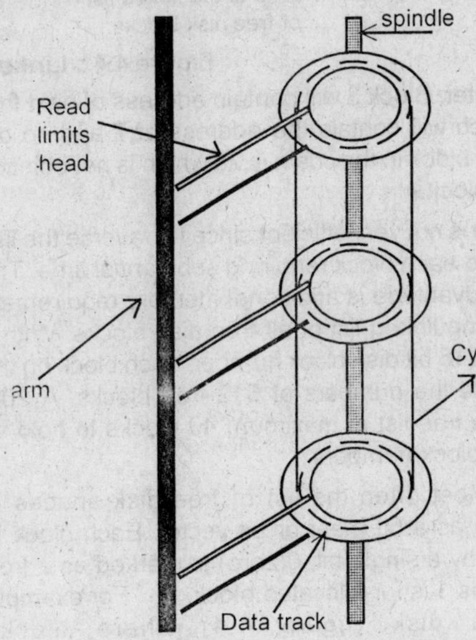

Figure 4.3 : Moving head disk

For example the disk shown in figure 4.3 has 3 platters and 6 recording surfaces (two on each platter). Therefore, each of its cylinder consists of 6 tracks (one track from each surface). The number of distinct tracks on a single surface determines the total number of cylinders on a given disk. The disk surface is divided into tracks which are further divided into sectors. Information is organized in the form of blocks within a track whose size is equivalent to a sector. Each sector can be separately read or written. The second approach to organize information on track is in the form of variable length. Records are separated by record gaps. Although this scheme is more flexible but is more difficult too. In either case information is read and written in blocks.

A sector is the smallest unit of information that can be read from or written to the disk. Depending upon the disk drive,sectors vary from 32 bytes. to 4096 bytes; there are 4 to 32 sectors per track and from 75 to 500 tracks per disk surface. To access a sector we must specify the surface, track and sector read/write heads. The read/write heads are moved to the correct track (seek time), electronically switched to the correct surface and then we wait (latency time) for the requested sector to rotate below the heads. On the average it equals one-half the disk revolution time, which is on the order of million records for the typical disk-rotational speeds.

For example, assuming that the disk is rotated 3000 times per minute, one revolution takes approximately 20 ms. The average rotational latency of such disks is usually quoted as 10 ms.

4. Disk Space Management Methods : Operating system maintains a list of free disk spaces to keep track of all disk blocks which are not being used by any file. Whenever a file has to be created, the list of free disk space is searched for and then allocated to the new file. The amount of space allocated to this file is then removed from the free space list. When a file is deleted, its disk space is added to the free space list. In this section we will discuss two methods to manage free disk blocks.

Linked List : In this method all free disk blocks are linked together by each free block pointing to the next free block. There must be another pointer pointing to the first free block of a linked list which will be pointing to the second free block which would further point to the third free block and so on. In the following example (figure 4.4) block 3 is the first free block of a linked list of free disk blocks which will be pointed to by some.

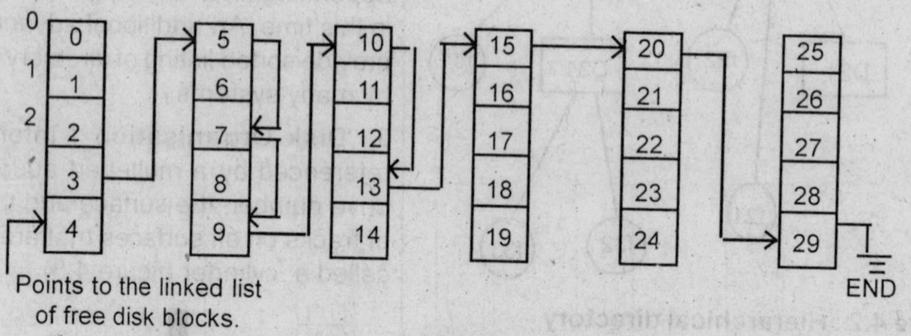

Figure 4.4 : Linked list of free disk blocks

starting pointer. Block 3 will contain address of next free block 5 which will contain the address of 7 and so on. The last free block in the chain is 28 which is not pointing to any free block.

This scheme is not very efficient since to traverse the list, we must read each block requiring substantial time. The second disadvantage is additional memory requirement for maintaining linked list of all free disk blocks. With a 1K block and 16 bit disk block number, each block on the free list holds the numbers of 512 free blocks. A 20M disk needs a free list of maximum 40 blocks to hold all 20,000 disk block numbers.

Bit Map : Most often the list of free disk spaces is implemented as a bit mapr or bit vector. Each block is represented by a single bit. 0(zero) is marked as a free block whereas 1 is for allocated block. For example consider a disk (figure 4.4) where blocks 3,5,7,9,10,13,15,20 and 28 are free and the rest of the blocks are allocated to a file. The bit map for free disk blocks would be 1 1 1 0 1 01 0 1 00 1 1 0 1 0 1 1 1 1 0 1 1 1 1 1 1 1 0 1. One advantage of this approach is that it is a simple method and efficient to find n consecutive free blocks on the disk. But the disadvantage is that extra disk space is required to store bit map. A disk with n blocks require a bit map of n bits. A 80 MB disk requires 80000 bits (1 BLOCK = 1 K byte) for the map. Compared to linked list the bit map requires less space since it uses 1 bit per block, versus 16 bits in the linked list model. Only if the disk is nearly full with the linked list it requires fewer blocks than the bit map. In case there is enough main memory to hold bit map, this method is better than linked list. If, however, only the block of memory can be spared for keeping track of free disk blocks, and the disk is nearly full, then the linked list may be preferable

5. Disk Allocation Methods : An important function of the file system is to manage space on the secondary storage, which includes keeping track of both disk blocks allocated to files and the free blocks available for allocation.

The main problems in allocating space to files are :

1. Effective utilization of disk space
2. Fast accessing of files

Management of disk blocks is a familiar problem that we have encountered and discussed in relation to main-memory management. But, secondary storage introdues two additional problems:

(I) Slows disk access time and

(II) Larger number of blocks to deal with

In spite of that, many considerations are similar to both environments, particularly, contiguous and non contiguous allocation of files. Three widely used allocation technques are contiguous, linked and indexed. The last two belongs to non contiguous allocation of files. Each method has its advantages and disadvantages.

While discussing those allocation strategies we will consider that a file is a sequence of blocks. All I /O operations on disk occur in term of blocks. The conversion from logical records to physical blocks is done by the software.

Contiguous Allocation : In contiguous allocation, files are assigned to contiguous areas of secondary storage. A user specifies in advance the size of the area needed to hold a file to be created. If the desired amount of contiguous space is not available, the file cannot be created.

One advantage of contiguous allocation is that all successive records of a file are normally physically adjacent to each other. This increases the accessing speed of records. It means that if records are scattered through the disk its accessing will be slower. Accessing a file which has been contiguously allocated is fairly easy. For sequential access the file system remembers the disk address of the last block and when necessary reads the next block. For direct access to block B of a file that start at location L, we can immediately access block L + B. Thus contiguous allocation supports both sequential and direct accessing.

The file directories in contiguous allocation systems are relatively straight forward to implement. For each file it is necessary merely to retain the starting address of the file and the file's size. If the size of the file is N blocks long, and starts at location L then it ocupies blocks L, L +1 .... L + N-1. The directory entry for each file indicates the address of the starting block and the size of the length for this file. This is illustrated in figure 4.5.

Directory

File	start	size
F1	2	3
F2	12	7
F3	22	6
F4	6	5

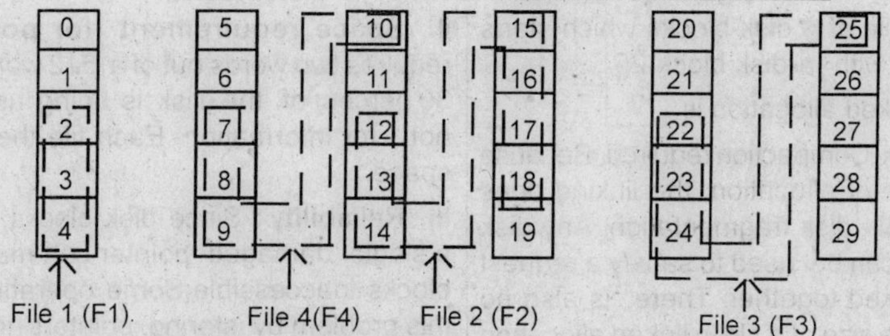

Figure 4.5: Contiguous allocation of files

The difficulty with contiguous allocation is how to find a space for a new file, if the implementation of a free space list is done through a bit map method then for the creation of a n-bytes long file, we need to find n 0 bits in a row. To understand the problem of contiguous storage allocation, let us consider disk space as a collection of free and used segments. Each segment is a contiguous set of disk blocks. To satisfy a request of a file of n free contiguous blocks, set of unused blocks are searched to determine which hole (an unallocated segment is called a hole) is best suited for allocation.This problem can be seen as a particular application of the dynamic storage allocation problem discussed in the previous chapter, which is how to satisfy a request of size N from a list of free holes.

Two most common strategies are :

(I) First - fit (II) Best - fit

First - fit : In this case as soon as the first hole (that is big enough) is encountered, searching is stopped and memory is allocated for creating a file. Searching can start either at the beginning of the set of holes or where the previous first - fit search ended.

Best - fit : In this case the entire list is searched for and the smallest hole, that is big enough, is allocated for creating a file.

Neither first-fit nor best-fit is clearly best in terms of storage utilization, but first-fit is generally faster.

These algorithms suffer from external fragmentations. As files are allocated and deleted the free disk space is brotken into little pieces. External fragmentations refer to the scattered groups of free blocks that are too small for allocation but which as a collection may represent a large percentage of disk size. As in the similar situations found when managing dynamically partitioned memory, this problem can be dealt with by occasional compaction of the disk. Depending upon the total amount of disk storage and the average file size, external fragmentation may be either a minor or a major problem.

The major problem with contiguous allocation is how to determine the space needed for a file. While this is not a problem when copying files, accurate estimation of file size is generally difficult and unreliable. For example, what may be reasonable basis for estimating the size of a new file having created by editor ? One way to handle this problem is to make some initial guessing and to allocate a suitable free area for the file in question. If the file turns out to be larger than guessed, its extension can be placed in some other disk area. This is often called file overfolw. Accessing of overflowed contiguous area is cumbersome,and it loses most of the appeal of contiguous allocation.

Non Contiguous Allocation : Because files do tend either to grow or shrink overtime and because users rarely know in advance how large their files will be contiguous. Storage allocation systems are being replaced by more dynamic non contiguous storage allocation systems. In this section we will discuss two schemes :

(I) Linked allocation (II) Indexed allocation

Linked allocation : Linked allocation is essentially a disk - based version of the linked list. With linked list allocation each fie is linked list of disk blocks. These disk blocks may be scattered through the disk. A few bytes of each disk block contains the address of the next block.

The directory contains a pointer to the first (and last) blocks of the file. This is shown in figure 4.6.

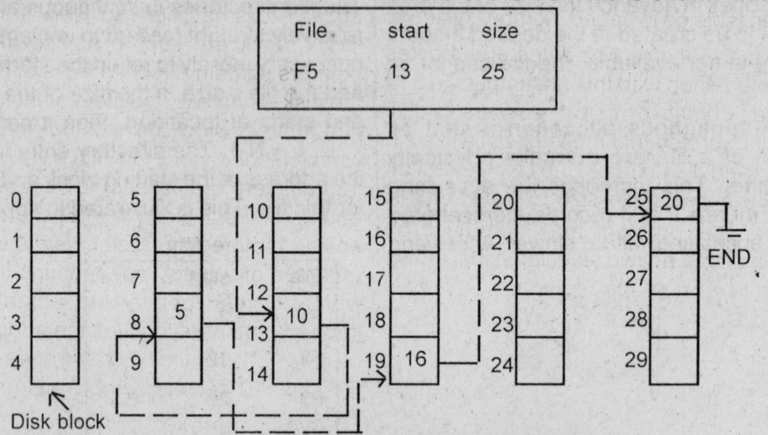

Figue 4. 6 : Linked allocation of disk blocks

In this figure there is a file of 4 disk blocks which starts from block 13 and ends with a disk block 25.

The advantage of linked allocation is :

(I) its simplicity (II) no disk Compaction required. Because of noncontiguous nature of allocation, the linking does not produce any external disk fragmentation. Any disk block on the free space can be used to satisfy a request ,since all blocks are linked together. There is also no need of declaration of the size of a file in linked allocation while it is created. A file can continue to grow as there are free blocks. Consequently, it is never necessary to have disk compaction.

The disadvantages of linked allocation are :

I. Slow direct accessing of any disk block: To find out the Nth block of a file, we must start at the beginning of that file and follow the pointer until we get to the Nth block.

II. Space requirement for pointers : If a pointer requires two words out of a 512 words block, then around 39 percent of the d isk is being used only for pointers, not for information. Each file therefore requires more space.

III. Reliability : Since disk blocks are linked by pointers, a single damaged pointer can make thousands of disk blocks inaccessible.Some operating systems address this problem by storing pointers in a dedicated file and making redundant copies of tit. The idea is to copy the list of pointers into the main memory and thus facilitate faster access of disk blocks. Redundacy of pointers files is employed for safer recovery.

Indexed allocation : One disadvantage with linked allocation method is that it does not support direct accessing since blocks are scattered all over the disk.

This problem is solved by indexed allocation by placing all of the pointers together into a index block. The principle of indexed allocation is illustrated in figure 4.7.

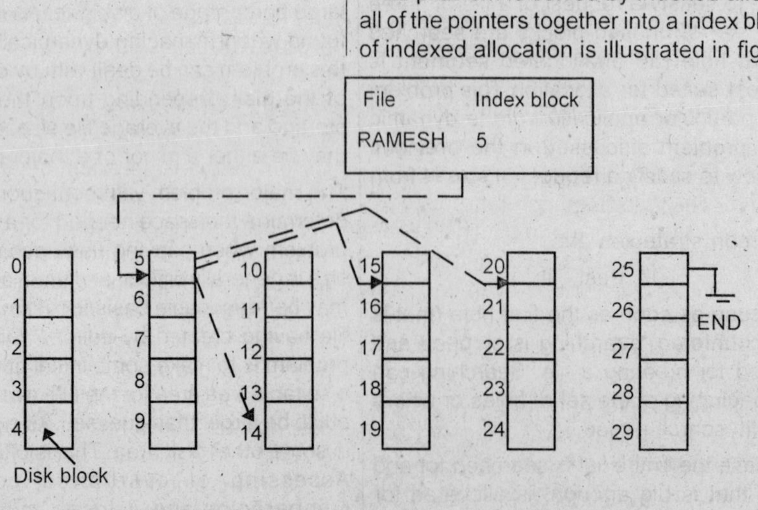

Figure 4.7 : Indexed allocation of disk space

In this scheme each file is provided with its own index block, which is an array of disk block pointers (addresses). The Kth entry in the index block points to the Kth disk block of the file. The directory contains the address of the index block. To read the Kth disk block the pointer in the Kth index block entry is used to find the desired block and then read. Although quite different in their implementations index blocks basically serve the same purpose as page map table do in paged memory systems.

Suppose disk blocks are of size 512 bytes and 4 bytes are allocated to pointers then with this assumption a 512 bytes index block can contain upto 128 pointers to disk blocks. This can be interpreted that the largest file in such a system cannot exceed 128 blocks of 512 bytes each. In most systems, the typically file size exceeds 64 KB (128 × 512 bytes). This problem may be solved by multiple level indexing such as two level indexing shown in figure 4.8.

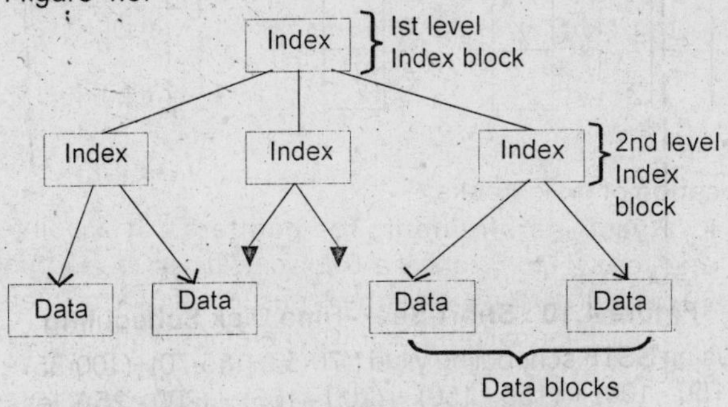

Figure 4. 8 : Tow-level indexing

In order to search for a disk block in this scheme, the operating system uses the first level index to find the second level index which in turn contain pointers to disk blocks. This approach could be continued to a third or fourth level, but two levels of indexes are generally sufficient. With the same assumption, this arrangement allows addressing of 128 × 128 second level index blocks, capable of supporting file sizes upto 8MB of file space. With the same assumption, the tripe-level indexing allows addressing of 1 GB file.

The advantage of indexed file allocation include the absence of external fragmentation and the efficiency of random accessing. Moreover, indexing can easily map around the bad disk blocks. Indexing of free space can be accomplished by means of the bit map.

The primary disadvantage of indexing is the number of disk accesses necessary to retrieve the address of the target block on disk. This overhead can be reduced by keeping some index blocks in memory. The second disadvantage is that indexed allocation requires lots of space for keeping pointers. The pointer overhead of the index block is generally worse than the pointer overhead of linked allocation. Most files are generally small. With linked allocation for a file of only one or two blocks, we only need a space for one pointer per block. With indexed allocation, an entire index block must be allocated even if one or two pointers will be non-nil.

This drawback raises the question of what should be the size of index block. Every file must have an index block, therefore, it should be smaller in size. But if the index block is too small, however, it will not be able to hold enough pointers for a large file and thus there should be a proper mechanism to deal with the situation. We have already discussed one mechanism of two level indexing to deal with the type of situation. An alternative used in unix system is to combine direct and indexed addressing in order to save space and to provide faster access to small files. In Unix first 15 pointers of the index block are kept in device directory. The first 10 pointers contain direct address of the 10 data blocks on the disk. The next 3 pointers point to indirect blocks. The first indirect block pointer that is the eleventh pointer, when needed, point to a first level index block containing 138 pointers (addresses) to dat ablocks.Then there is a double indirect block pointer that is the twelth pointer pointing to the first block of two level indexing structure (address of a block that contains the address fo blocks which contain pointers to actual data blocks). The thirteenth pointer contains the address of the first block of a three-level indexing structure. With this structure, files upto10 sectors can be addressed directly.

One level indexing is sufficient for files upto 137 setors,two level indexing is sufficient for upto 16,521 sectors and three-level indexing is sufficient for files on the order of 1 GB.

Indexed allocation is more complex. If the index block is already in memroy, then the access can be made directly. However, keeping the index block in memory requires considerable space. If the memory space is not available then we may have to read first the index block and then the desired data block. For a two level index, two index block reads might be needed. For a very large file accessing a block near the end of a file would require reading in all of the index blocks to follow the pointer chain before the data block could finally be read. Thus the performance of indexed allocation depends upon the index structure, the size of file and the position of block desired.

Some systems combine contiguous allocation with indexed allocation by using contiguous allocation for small files (upto 3 to 4 blocks) automtically switch to an indexed allocation if the file grows large. Since most files are small, and contiguous allocation is efficient for small files, average performance can be quite good.

6. Disk Scheduling : The performance of a computer system is to a large extent dependent upon how fast a disk request is serviced. Since most jobs nearly depend upon the disk for input and output purpose. In multiprogrammed environment, many processes may be generating request for reading and writing disk records because these processes often make requests faster than they can be serviced by disk systems, waiting lines or queues build up for each device. To service a request, a disk system requires that the head be moved to the desired track, then a wait for latency and finally the transfer of data. In the following section we will examine several scheduling algorithms.

First-Come-First-Served (FCFS) Scheduling : This is the simplest form of disk scheduling in which the first request to arrive is the first one serviced. FCFS has a fair policy in the sense that once a request has arrived, its place in the schedule is fixed irrespective of arrival of a higher priority request. FCFS is easy to program, however, it may not provide the best service. Consider, for example, disk queue with request involving tracks to read,

100, 200, 50, 150, 25, 155, 70 and 85

The starting track to read is 100 and the last one is 85.

If the head position of a disk system is initially at 50, it will first move from 50 to 100, then to 200, 50, 150,25, 155, 70 and 85 for a total head movement of 755 (100- 50) + (200-100) + (200-50) + (150-50) + (150 - 25) + (155-25)+ (155-70) + (85-70) tracks. The movement of a head for these tracks is illustrated in the following figure 4.9.

Movement of disk head :=

50 → 100 → 200 → 50 → 150 → 25 → 155 → 70 → 85

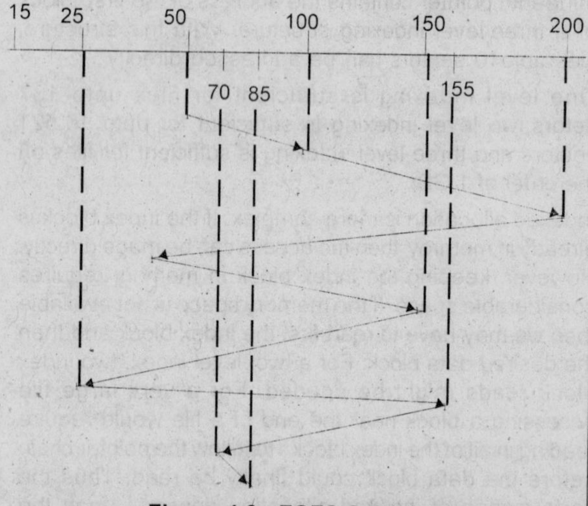

Figure 4.9 : FCFS Scheduling

One drawback with this scheduling approach as seen in the figure 4.9 is that there is a big swing from 150 to 25 back to 155. If the request or writing tracks 50 and 25 and serviced together either before or after the requests of 150 and 155, then the total head movement will be drastically reduced, thus increasing disk throughput.

FCFS is acceptable when the load on a disk is light. But as the load grows, FCFS tends to saturate the device and response time becomes longer.

Shortest- Seek - Time - First Scheduling : In Shortest-Seek-Time - First (SSTF) seheduling priority is given to those processes which need the shortest seek, even if these requests are not the first ones in the queue. It means that all request nearer to the current head positions are serviced together before moving head to distant tracks. Since seek time is generally proportional to the track difference between the requests this approach is implemented by moving the head to the closest track in the request queue.

Consider the previous example of request queue of 100, 200, 50, 150, 25, 155, 70 and 85 track, the nearest track for a service from an initial head position (50) is at track 70. Once we are at 70, the next closest track is 85. (the difference is of 15 tracks only).From here it can go to 100. Continuing this way, the request for track 155 is serviced next and then switch over to 200 and finally at 25 (figure 4.10).

Movement of disk head :

= 50 → 70 → 85 → 150 → 155 → 200 → 25

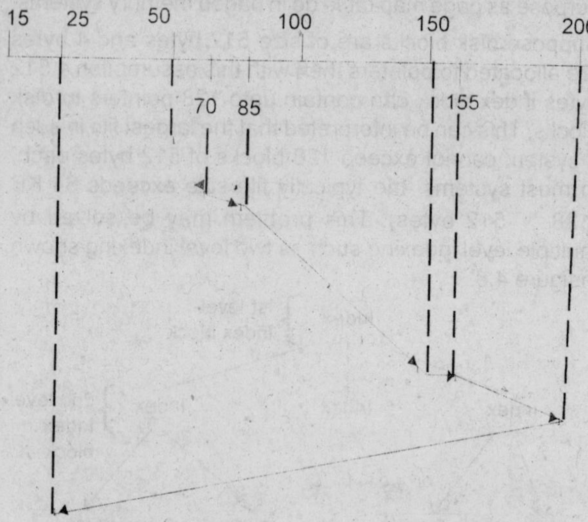

Figure 4.10 : Short-Seek-Time Disk Scheduling

Using SSTF scheduling yield ((70-50)+(85-70)+(100-85)+ (150 - 100) + (155 - 150) + (200 - 155) + (200 - 25)), less than a half of distance required for FCFS, which is a substantial improvement in disk throughput. SSTF is essentially a form of Shortest job First (SJF) Scheduling and like SJF it can create the shortage of requests. In real Systems, requests arrive randomly. Suppose thee are three requests 10,20 and 150. If read/write head is near 10 then 10 and 20 will be served and 150 will be kept waiting. If another request comes which is nearer 20,150 will have to wait further. Theoretically, a continued stream of comes which is near each other could arrive, causing the request for track 150 to wait for a very long period.

SSTF tends to discriminate sharply against certain requests. SSTF services requests for those tracks which are highly localized due to which the innermost and outermost tracks can receive poor service compared with the mid-range tracks.

Scan Scheduling : SCAN Scheduling Strategy was developed to overcome the discrimination as well as to recognize the dynamic nature of the request queue. The read/write head of a disk starts from one end and move towards the other end, services requests as it reaches each track until it reaches to another end of the disk. After reaching another end of disk, disk head reverses its path-direction while continuing with services whichnever comes on the way. This way disk head continuously oscillates from end to end.

Before applying Scan algorithm to the previous example of a request queue for 100,200,50, 150, 25, 155, 70 and 85 tracks, two informations must be known (i) the direction of head movement and (ii) the last position of the disk-head. Suppose disk head is moving from 0 and its last position was 45, then it would services 50, 70, 85, 100, 150, 155 and 200 as it moves in that direction (figure 4.11) . If any request come on its way it will be serviced immediately, while request arriving just behind the head will Movement of disk head :=

$$50 \rightarrow 70 \rightarrow 85 \rightarrow 100 \rightarrow 150 \rightarrow 155 \rightarrow 200 \rightarrow 25$$

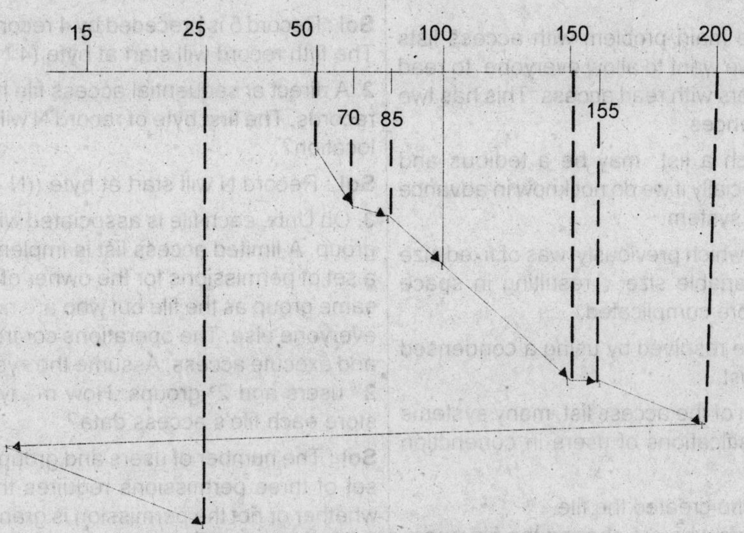

Figure 4.11 : SCAN disk scheduling

have to wait until disk head moves to the end of the disk, reverses direction and returns before being serviced. Scan has been the basis of most disk scheduling strategy actually implemented.

The Scan algorithm is sometimes called the "Elevator" algorithm, since it is similar to the behaviour of elevators algorithm , as they service requests to move from floor to floor in building . SCAN behaves very much like SSTF in terms if improved throughput but it eliminates much of the discrimination inherent in SSTF schemes. Because of the continuous scanning of disk from end to end to end, the outer tracks are visited less often than the mid-range tracks, but this is not as serious as the discrimination of SSTF.

7. File Protection : The protection of a file is mostly needed in multi-user environment where a file is shared among several users. On system which does not permit access to the files of other users, protection is not required. Protection mechanism must provide controlled access by restricting the types of files which can be made. Access is permitted of denied depending upon several factors, one of which is the type of access requested. Several different types of operations may be controlled. These operations are :

- Reading from the file.
- Writing on the file.
- Executing the file by loading into main memory.
- Writing new information at the end of a file.
- Deleting the file and releasing the space.

Many different protection mechanisms have been proposed. Each one has got advantages and disadvantages and must be selected which is suitable for its intended application. A large computer system requires different types of protection mechanisms compared to a small computer system used by a small group of people.

The protection can be supported with file itself but the more common scheme is to provide protection with path. Thus if a path name refers to a file in a directory, the user must be allowed access to both the directory and the file. In systems where files may have numerous path names, a given user may then have different access rights to a file depending upon the path name used.

In this subsection we will deal with several mechanisms to support protection.

Passwords : In this approach, password is associated with each file. Just as a password is required to access a computer system, access to each file will be also controlled by a password.

There are, however, several disadvantages to this scheme:

1. If we associate a sepaate password with each file, the number of passwords that need to be remembered are quite large, making the scheme impractical. If only one password is used for all the files, then once it is discovered all files are accessible. Some systems allow a user to associate a password with a subdirectory rather than an individual file, to deal with this problem.

2. Commonly, only one password is associated with each file. Thus protection is on an all-or-nothing basis. To provide protection on a more detailed level, multiple passwords are needed.

Access Lists : Another approach is to make access dependent on the identity of the user. Various users may need different types of access to a file or directory. An access list can be associated with each file and directory, specifying the user name and the types of access allowed for each user . When a user requests access to particular file, the operating system checks the access list associated with each file. If that user is listed for the requested access, the access is allowed. Otherwise, a protection violation occurs and the user job is denied access to the file.

Access Groups : The main problem with access lists can be their length. If we want to allow everyone to read a file, we must list all uers with read access. This has two undesirable consequences :

- Constructing such a list may be a tedious and unrewarding task, especially if we do not know in advance the list of users in the system.
- The directory entry which previously was of fixed size needs now to be of variable size, resulting in space management being more complicated.

These problems can be resolved by using a condensed version of the access list.

To condense the length of the access list, many systems recongnize three classifications of users in conenction with each file :

Owner - The user who created the file.
Group - A set of users who are sharing the file and need similar access.
Universe - All other users in the systems.

As an example consider a person, X, who is working on a new project. She has hired three gaduate students (A, B, C) to work on the project. The project program is kept in a file named project. The protection associated with this file is as follows :

- X should be able to invoke all operations on the file
- A,B and C should only be able to read, write, and execute the file. They sould not be allowed to delete the file.
- All other users should be able to execute the file . (X is interested in letting as many people as possible play the game in order to obtain appropriate feedback).

To achieve such a protection, a new group must be created with member A,B,C. The name of the group must be then associated with the file project, and the access-right must be set in accorance with the policy we have outlined above.

Note that for this scheme to work properly, group membership must be tightly controlled. This can be done in a number of different ways. For example, in the Unix system, groups can be created and modified only by the manager of the facility (or anyone who has the super password). Thus, this control is achieved through human interaction. In the VMS system, group can be created and controlled just as files are with appropriate control.

With this classification, only three fields are needed to define protection. Each field is often a collection of bits which either allow or prevent the access associated with that bit.

For example the Unix system defines three fields of three bits each: rwx, where r controls read access, w controls write access, and x controls execution. A separate field is kept for the file owner, for the owner's group and for all other users. In this scheme, nine bits per file are needed to record protection information.

SOLVED EXAMPLES

1. A sequential access file has fixed-size 15-byte records. Assuming the first record is record 1, the first byte of record 5 will be at what logical location?

Sol : Record 5 is preceded by 4 records, each of 15 bytes. The fifth record will start at byte (4 *15) + 1 =61.

2. A direct or sequential access file has fixed-size S-byte records. The first byte of record N will start at what logical location?

Sol : Record N will start at byte ((N - 1) * S) + 1.

3. On Unix, each file is associated with one user and one group. A limited access list is implemented by assigning a set of permissions for the owner of the file, those in the same group as the file but who are not the owner, and for everyone else. The operations controlled are read, write, and execute access. Assume the system may have up to 2^{16} users and 2^{8} groups. How many bits are needed to store each file's access data?

Sol : The number of users and groups is irrelevant. Each set of three permissions requires three bits to indicate whether or not the permission is granted. There are three sets of permissions. Nine bits total are needed per file.

4. On a capability-based system, each file is associated with a unique 16-bit number. For each file, each user may have the read or write capability. How many bytes are needed to store each user's access data?

Sol : Two bits are needed for each of 65,536 (2^{16}) files. Thus, 131,072 bits or 16,384 bytes would be needed.

5. On a system using contiguous allocation, compute the number of the physical block corresponding to the logical block given the file is stored starting at the indicated physical block (assume block numbers start with 1).

a)Starting physical block: 1000; logical block: 12
b) Starting physical block: 75; logical block: 2000
c) Starting physical block: 150; logical block: 25

Sol : a) If logical block 1 is stored in physical block 1000, 2 will be stored in 1001, 3 in 1002, and 12 in 101 1.

1000- 1 +12= 1011

b) 75 -1 + 2000 = 2074

c) 150 - 1 + 25 = 174

6. A file system uses 256-byte physical blocks. Each file has a directory entry giving the file name, location of the first block, length of file, and last block position. Assume the last physical block read and the directory entry are already in main memory. For the following, indicate how many physical blocks must be read to access the specified block (including the reading of the specified block) on a system using contiguous allocation.

a) Last block read: 100; block to be read: 600

b) Last block read: 500; block to be read: 200

c) Last block read: 20; block to be read: 21

d) Last block read: 21; block to be read: 20

Sol : The answer for all four scenarios is one. The location of the physical block to be read can be computed from the block number and the location of the first block.

7. Repeat the previous problem for a system using linked allocation.

Sol : a) Last block read: 100; block to be read: 600

500. Blocks 101 through 599 must be read to obtain the link to the next block.

b) Last block read: 500; block to be read: 200

200. To find block 200, the chain of blocks starting with block 1 must be read.

c) Last block read: 20; block to be read: 21

1. Block 20 contains a pointer to block 21 .

d) Last block read: 21; block to be read: 20

21. To find block 21, the chain of blocks starting with block 1 must be read.

8. Repeat the previous problem for a system using indexed allocation. Assume the directory entry contains the location of the first index block (not the first block of the file). Each index block contains pointers to 127 file blocks plus a pointer to the next index block. In addition to the last block read, assume the system has in main memory the index block that contains the pointer to the last block read. However, no other index blocks are in main memory.

Sol : a) Last block read: 100; block to be read: 600

5. The last block read is pointed to by index block 1, which will also be in memory. The pointer to block 600 is in the fifth index block. Index blocks 2 through 5 must be read before reading block 600.

b) Last block read: 500; block to be read: 200

3. Index block 4 is in memory. The pointer to block 200 is in the second index block. Index blocks 1 and 2 must be read before reading block 200.

c) Last block read: 20; block to be read: 21

1. The pointer to block 21 is in the first index block which is the same index block used by block 20.

d) Last block read: 21; block to be read: 20

1. The pointer to block 21 is in the first index block which is the same index block used by block 20.

9. Which allocation scheme would work best for a file system implemented on a device that can only be accessed sequentially, a tape drive, for instance.

Sol : Contiguous allocation. Once the start of the file was accessed, subsequent accesses would be relatively efficient. With linked or index allocation, the various physical blocks in which the file was stored could be scattered all over the device, making access very inefficient.

10. Consider a system in which a directory entry can store up to 16 disk block addresses. For files no larger than 16 blocks, the 16 addresses serve as the file's index table. For files larger than 16 blocks, the addresses point to indirect blocks which in turn point to 256 file blocks each. A block is 1024 bytes. How big can a file be?

Sol : The largest possible file will have $16 = 2^4$ indirect blocks. Since each indirect block can point to $256 = 2^8$ blocks, the largest file has $256 = 2^{8+4}$ blocks. Given a block size of $1024 = 2^{10}$, the maximum file size is

$2^{4+8+10} = 2^{22}$.

11. How many bits would be needed to store the free list under the following conditions ; if a bit map were used to implement the free list?

a) 16 bits per disk address; 500,000 blocks total; 200,000 free blocks

b) 16 bits per disk address; 500,000 blocks total; 0 free blocks

Sol : In either case, the number of bits per address and the number of free blocks are irrelevant. The status of each of the 500,000 blocksis stored in a bit. 500,000 bits are needed.

12. How many bits would be needed *to* store the-free list under the following conditions if an array *of* free blocks were used *to* implement the free list?

a) 16 bits per disk address; 500,000 blocks total; 200,000 free blocks

b) 16 bits per disk address; 500,000 blocks total; 0 free blocks

c) 16 bits per disk address; 500,000 blocks total; 1,000,000 free blocks

d) D bits per disk address; B blocks total; F free blocks

Sol : a) 200,000 addresses x 16 bits per address = 3,200,000 bits

b) 0 addresses x 16 bits per address = 0 bits

c) 500,000 addresses x 16 bits per address = 8,000,000 bits

d) D addresses x 16 bits per address = 16D bits,

13. (a) On a system with D bits per disk address, B blocks total,and F free blocks, under what conditions will the free list array use less space than the bit map?

(b) For D having the value *of* 32 bits, what is the fraction of the disk space that must be free? .

Sol : a) D x F <B

b) 32 x F < B, F < B/32

Less than one thirty-second of the blocks can be free.

14. On a system using a free hole array, and that has K physical blocks, what is the maximum length *of* a free hole array? The minimum length?

Sol : The maximum number of holes is created by having alternating free and used blocks. Thus, the maximum number of holes is (K + 1)/2 (where '/' represents integer division), The minimum number of holes is zero, which occurs when all of memory is in use.

15. On a computer system with a single 32-bit 100 MHz bus (1 0 ns cycle), the disk controller uses DMA to transfer data to/from memory at a rate of 40Mb per second. Assume the computer fetches and executes one 32-bit instruction every cycle when no cycles are stolen. By what percentage will the disk controller slow down instruction execution?

Sol : Given the bus is 32 bits wide, the controller can transfer 4 bytes 10,000,000 times per second or 4 bytes every 100ns. The controller will steal a cycle once every 10 instruction fetches for a slowdown of 10%.

16. A computer system that uses memory-mapped I/O has a 16-bit address space. Addresses with ones in the two most significant bits refer to devices. What is the maximum amount of memory that can be referenced in such a system?

Sol : At most 2^{14} memory locations can be accessed. All addresses 2^{14} and higher refer to I/O devices.

17. A disk file contains 32-byte fixed-size records. Physical I/O is in terms of 512-byte blocks that are stored in an operating system buffer. If a process sequentially reads through a file's records, what percentage of the read requests will result in I/O operations?

Sol : I/O will be in terms of blocks. Each 512-byte block input can hold sixteen 32-byte records. All but one in sixteen I/O requests can be satisfied by information already read into the buffer. 6.25% of I/O requests will result in I/O operations.

18. A system has two disk drives connected to one disk controller, a line printer connected to a parallel port, and 10 terminals connected to 10 serial lines controlled by a single controller. The disk controller can issue a command to one disk while another disk is processing a command (for example, both disks can be seeking at the same time), but it may have only one disk transferring I/O at a time. The serial controller can handle simultaneous I/O from all its connected devices. All the controllers use DMA. The transfer rates for the devices are: disk device, 1000 Kb/s; printer, 1 Kb/s; and terminal, 5 Kb/s. What is the maximum I/O transfer rate for this system?

Sol : Only one disk can be transferring at a time but all the terminals I/O can be handled simultaneously by the terminal controller. The maximum I/O rate is 1000 + 1 + 50 = 1051 Kb/s.

19. On a system using a disk cache, the mean access time is dependent on the mean cache access time, the mean disk access time, and the hit rate. For the following, what is the mean access time?

(a) Cache: 1 ms; disk: 100 ms; hit rate: 25%

(b) Cache: 1ms; disk: 100 ms; hit rate: 50%

Sol : (a) 100 x 0.75 + 1 x 0.25 = 75.25

(b) 100 x 0.50 + 1x 0.50 = 50.50

20. On a system using a disk cache, the mean access time is 41.2 ms, the mean cache access time is 2 ms, the mean disk access time is 100 ms, and the system has 8Mb of cache memory. For each doubling of the amount of memory, the miss rate is halved. How much memory must be added to reduce the mean access time to 20 ms? Assume the amount of memory may only increase by doubling (8Mb, 16Mb, 32Mb, 64Mb, etc.).

Sol : The hit ratio can be computed from the equation

$$2 \times (1 - m) + 100m = 41.2$$

$$2 - 2m + 100m = 41.2$$

$$39.2 = 98m$$

$$0.40 = m$$

Doubling memory to 16Mb would lower the miss rate to 0.20. 2 x 0.80 + 100 x 0.20 = 21.6

An additional doubling to 32Mb would be needed to lower the mean access time below 20 ms.

$$2 \times 0.90 + 100 \times 0.10 = 11.8$$

21. If video memory starts at address A0000000 (hex) for a monitor with 16-bit color for a resolution of 1024 x 768, where will video memory end?

Sol : A resolution of 1024 x 768 contains 786,432 pixels. Using two bytes per pixel, 1,572,864 bytes of video memory are required. A0000000(hex) + 1572864(dec) = A0180000(hex).

22. A disk has 8 sectors per track and spins at 600 rpm. It takes the controller 10 ms from the end of one I/O operation before it can issue a subsequent one. How long does it take to read all 8 sectors using the following interleaving systems?

(a) No interleaving

(b) Single interleaving

(c) Double interleaving

Sol : The disk makes 10 revolutions per second or one revolution in 100 ms. In 10 ms, less than one sector will have passed under the read/write head.

(a) The next sector cannot be read until the disk makes almost a complete revolution. It will require 8 revolutions to read all 8 sectors. At 100 ms per revolution, it will take 800ms.

(b) The next sector will spin under the read/write head almost as soon as the next I/O operation is issued. Two revolutions will be needed to read all 8 sectors, making the total read time 200 ms.

(c) A total of 2.75 revolutions will be needed to read all 8 sectors, for a total time of 275 ms.

23. On a disk with 1000 cylinders, numbers 0 to 999, compute the number of tracks the disk arm must move to satisfy all the requests in the disk queue. Assume the last request serviced was at track 345 and the head is moving toward track 0. The queue in FIFO order contains requests for the following tracks: 123, 874, 692, 475, 105, 376. Perform the computation for the following scheduling algorithms.

(a) FIFO (b) SSTF (c) SCAN

(d) LOOK (e) C-SCAN (f) C-LOOK

Sol : (a) 2013

The tracks traveled to will be 345,123, 874,692,475,105, and 376, making the total distance

222 + 751 + 182 + 217 + 370 + 271 = 2013.

(b) 1298

The tracks traveled to will be 345, 376, 475, 692, 874, 123, and 105, making the total distance 529 + 769 = 1298.

(c) 1219

The tracks traveled to will be 345,123,105,0,376,475,692, and 874, making the total distance 345 + 874 = 1219.

(d) 1009

The tracks traveled to will be 345,123,105,376,475,692, and 874, making the total distance 240 + 769 = 1009.

(e) 1967

The tracks traveled to will be 345, 123, 105, 0, 999, 874, 692, 475, and 376, making the total distance

345 + 999 +623 =1967.

(f) 1507

The tracks traveled to will be 345, 123, 105, 874,692,475, and 376, making the total distance

240 + 769 + 498 = 1507.

24. Which disk-scheduling algorithm would best optimize the performance of a RAM disk?

Sol : All the disk-scheduling algorithms would yield the same device performance. A RAM disk is a virtual device created from main memory. The order in which requests are handled has no effect on performance.

25. A file system with a one-level directory structure is implemented on a disk with disk block size of 4 K bytes. The disk is used as follows: **[GATE - 96]**

Disk-block 0 : File Allocation Table, consisting of one 8 bit entry per data block, representing the data block address of the next data block in the file;

Disk-block 1 : Directory, with one 32 bit entry per file:

Disk-block 2 : Data block 1;

Disk-block 3 : Data block 2; etc.

(a) What is the maximum possible number of files ?

(b) What is the maximum possible file size in blocks **?**

Sol : (a) As there must be at least one directory entry per file, Maximum number of files

$$= \frac{\text{Directory size}}{\text{Directory entry size}} = \frac{4\text{kB}}{32\text{b}} = 1024$$

(b) A single file can occupy as must as many data blocks as can be addressed by 8 bit FAT entries.

Number of distinct FAT entry representations $= 2^8 = 256$

As some x of these representations must be used to indicate end of file, unusual blocks, bad block, etc.

Maximum possible file size = 256 – x blocks.

For blocks of small x, or (256 – x) × 4 k Bytes.

26. Consider a disk with the 100 tracks numbered form 0 to 99 rotating at 3000 rpm. The number of sectors per track is 100. The time to move the head between two successive tracks is 0.2 millisecond.

(a) Consider a set of disk requests to read data from tracks 32, 7, 45, 5 and 10. Assuming that the elevator algorithm is used to schedule disk requests, and the head is initially at track 25 moving up (tocards larger track numbers), what is the total seek time for servicing the requests ?

(b) Consider an initial set of 100 arbitrary disk requests and assume that no new disk requests arrive while servicing these requests. If the head is initially at track 0 and the elevator algorithm is used to schedule disk requests, what is the worst case time to complete all the requests ? **[GATE- 2001]**

Sol : (a) The head is initially at 25th track of disk and it is going upwards. It will go up to 45th track and moves down and stops at 5th track.

So total head movement is

$(45-25)+(45-5)=20+40=60$ tracks

Time taken for movement between successive tracks = 0.2 ms.

Total time taken to process all requests

$= 60 \times 0.2 = 12\text{ms}.$

(b) In worst case, one of the request is for a sector on the 99th track. So total head movement is 99 tracks.

Time taken = 99 × 0.2 = 19.8 ms.

OBJECTIVE QUESTIONS - I

1. Formatting of a floppy disk refers to **[GATE-1998]**
   a) Arranging the data on the disk in contiguous fashion
   b) Writing the directory
   c) Erasing the system area
   d) Writing identification information on all tracks and sectors
2. Sector interleaving in disks is done by
   a) The disk manufacturer b) The disk controller card
   c) The operating system d) None of these
3. Disk scheduling involves deciding
   a) Which disk should be accessed next
   b) The order in which disk access requests must be serviced
   c) The physical location where files should be accessed in the disk
   d) None of these
4. The total time to prepare a disk drive mechanism for a block of data to be read from it is **[GRE-CS]**
   a) Seek time b) Latency
   c) Latency plus seek time d) Transmission time
5. A disk scheduling algorithm in an operating system causes the disk arm to move back and forth across the disk surface in order to service all requests in This path.This is a **[GRE-CS]**
   a) First come first served b) Scan
   c) Shortest seek time first d) None
6. Which one of the following file allocation strategies is used by UNIX operating system **[UGC NET-1998]**
   a) Contigeous allocation b) Linked allocation
   c) Indexed allocation d) Sequential allocation
7. The root directory of a disk should be placed:
   a) At a fixed address in main memory
   b) Any where on the disk
   c) At a fixed location on the system disk
   d) None
8. Disk I/O is done in terms of
   a) Blocks b) Sectors
   c) Bits d) Bytes

9. Direct access methods are not effectively supported by

a) Contiguous allocation b) Linked allocation
c) Indexed allocation d) None

10. Which allocation method incurs the greatest overhead for small files (i.e. files which occupy one block)

a) Contiguous allocation b) Linked allocation
c) Indexed allocation d) None

11. In which of the following directory systems is it possible to have multiple-paths for a file, starting from the root directory?

a) Single level directory b) Two level directory
c) Tree structured directory d) A cyclic graph directory

ANSWER KEY - I

1. d	2. c	3. b	4. c	5. b
6. c	7. c	8. a	9. b	10. c
11. d				

OBJECTIVE QUESTIONS - II

1. Which of the following disk-scheduling strategies is likely to give best throughput ? **[GATE-1999]**

a) Farthest cylinder next b) Nearest cylinder next
c) First come first served d) Elevator algorithm

2. The data transfer rate of a double-density floppy disk system is about **[GATE-1987]**

a) 5 k bits/sec b) 50 k bits/sec
c) 500 k bits/sec d) 5000 k bits/sec

3. Disk requests come to a disk drive for cylinders 10,22,20,2,40,6 and 38 in that order at a time when the disk drive is reading from cylinder 20. The seek time is 6 ms per cylinder. The total seek time if the disk arm scheduling algorithm is first-cum-first served is

a) 360 ms b) 850 ms c) 900 ms d) None

4. Consider the above question, if the scheduling algorithm is the closest cylinder next, then the total will be:

a) 360 ms b) 876 ms c) 850 ms d) 900 ms

Questions 5 through 8 refer to the following information:

In the operation of a certain disk drive mechanism, a disk formatted into 20 sectors and 10 tracks. The disk can be rotated either clockwise or counterclockwise. The times required to perform certain operations are as follows:

I. Rotate the disk through one revolution = 200 ms.
II. Move the disk head from the center to the rim =20ms.
III. Read and transmit one block of data = 0.3 ms.

Three files are stored on the disk:
File A: 2 blocks at sector 1, track 6.
File B: 5 blocks at sector 5, track 2.
File C: 1 block at sector 3, track 5.

5. The disk head is initially at sector 0, track 0. If all three files A,B, and C are to be read in the minimum amount of time, they should be read in the following order:

a) A,B,C b) A,C,B c) B,A,C d) C,A,B

6. The disk head is initially at sector 0, track 0. The files are read in the order C,B,A. The total time to read the files is

a) 143.9 ms. b) 116.4 ms. c) 114.0 ms. d) 102.6 ms.

7. The most nearly average latency time is

a) 1 ms. b) 7 ms.
c) 30 ms. d) 50 ms.

8. The most nearly average seek time is

a) 1 ms. b) 7 ms.
c) 30 ms. d) 50 ms.

A file system with a one level directory structure is implemented on a disk with block size of 4 bytes. the disk is used as following:

Disk block-0: File allocation table, consisting of one 8 bit entry per data block, representing the data block address of the next data blocks in file.

Disk block-1: Directory with, one 32 bit entry pr file

Disk block-2: Data block1

Disk block-3: Data block2, etc.

9. The maximum no. Of possible files is

a) 2^{12} files b) 2^{8} files
c) 2^{10} files d) None

10. The maximum possible file size in blocks in the above question is :

a) $2^{12} - 2$ blocks b) $2^{12} + 2$ blocks
c) $2^{12} - 1$ blocks d) $2^{12} + 1$ blocks

11. Consider six files : F_1, F_2, F_3, F_4, F_5 and F_6 with corresponding sizes 100, 200, 70, 40, 250 and 50 respectively. The files are to be stored on a sequential device (i.e., tape) in such a way as to optimize access time. In what order should the files be stored?

a) $F_6, F_5, F_4, F_3, F_2, F_1$
b) $F_1, F_2, F_3, F_4, F_5, F_6$
c) $F_5, F_2, F_1, F_3, F_6, F_4$
d) $F_4, F_6, F_3, F_1, F_2, F_5$

12. Consider six files F_1, F_2, F_3, F_4, F_5 and F_6 of corresponding sizes : 100, 200, 70, 40, 250 and 50 respectively.

If these files are stored in such a manner to optimise access time, what will be the approximate average access time of a record from one of the six files on the sequential device ?

a) 131 b) 198 c) 286 d) 433

ANSWER KEY - II

1. b	2. c	3. d	4. a	5. b
6. b	7. d	8. b	9. b	10. a
11. d	12. c			

V. PROTECTION AND SECURITY

- Introduction to Protection
- Goals of Protection
- Introduction to Security
- The Security Problems
- Encryption

1. Introduction

Protection mechanisms provide controlled access by limiting the types of file access that can be made by the various users. Protection must also be available to ensure that besides the files, memory segments, CPU, and other resources can be operated on by only those processes that have gained proper authorization from the operating system.

Protection is provided by a mechanism that controls the access of programs, processes, or users to the resources defined by a computer system. This mechanism must provide a means for specification of the controls to be imposed, together with a means of enforcement.

2. Goals of Protection

As computer systems have become more sophisticated and pervasive in their applications, the need to protect their integirty has also grown. Protection was originally conceived as an adjunct to multiprogramming operating systems, so that untrustworthy users might safely share a common logical name space, such as a directory of files, or share a common physical name space, such as a directory of files, or share a common physical name space, such as memory. Modern protection concepts have evolved to increase the reliability of any complex system that makes use of shared resources.

There are several reasons for providing protection. Most obvious is the need to prevent mischievous, intentional violation of an access restriction by a user. Of more general importance, however, is the need to ensure that each program component active in a system uses system resources only in ways consistent with the stated policies for the uses of these resources. This requirement is an absolute one for a reliable system.

Protection can improve reliability by detecting latent errors at the interfaces between component subsystems. Early detection of interface errors can often prevent contamination of a healthy subsystem by a subsystem that is malfunctioning. An unprotected resource cannot defend against use (or misuse) by an unauthorized or incompetent user. A protection-oriented system provides means to distinguish between authorized and unauthorized usage.

The role of protection in a computer system is to provide a mechanism for the enforcement of the policies governing resource use. These policies can be established in a variety of ways. Some are fixed in the design of the system, whereas others are formulated by the management of a system.Still others are defined by the individual users to protect their own files and programs. A protection system must have the flexibility to enforce a variety of policies that can be declared to it.

Policies for resource use may vary, depending on the applicatiion, and they may be subject to change over time. For these reasons, protection can no longer be considered solely as a matter of concern to the designer of an operating system. It should also be available as a tool for the applications programmer, so that resources created and supported by an applilcations subsystem can be guarded against misuse. In this chapter, we describe the protection mechanisms the operating system should provide, so that an application designer can use them in designing her own protection software.

3. Domain of Protection

A computer system is a collection of processes and objects. By objects, we mean both hardware objects (such as the CPU, memory segments, printers, disks and tape drives), and software objects (such as files, programs, and semaphores). Each object has a unique name that differentiates it from all other objects in the system, and each can be accessed only through well defined and meaningful operations. Objects are essentially abstract data types.

The operations that are possible may depend on the object. For example, a CPU can only be executed on. Memory segments can be read and written, whereas a card reader can only be read. Tape drives can be read, written, and rewound. Data files can be created, opened, read, written, closed, and deleted; program files can be read, written, executed, and deleted.

Obviously, a process should be allowed to access only those resources it has been authorized to access. Furthermore, at any time, it should be able to access only those resources that it currently requires to complete its task. This requirment, commonly referred to a the need-to-know principle, is useful in limiting the amount of damae a faulty process can cause in the system. For example, when process p invokes procedure A, the procedure should be allowed to access only its own variables and the formal parameters passed to it; it should not be able to access all the variables of process p. Similarly, consider the case where process p invokes a compiler to compile a particular file. The compiler should not be able to access any arbitrary files, but only a well-defined subset of files (such as the source file, listing file, and so on) related to the file to be compiled. Conversely, the compiler may have private files u sed for accounting or optimization purpose, which process p should not be able to access.

Domain Structure

To facilitate this scheme, we introduce the concept of a protection domain. A process operates within a protection domain, which specifies the resources that the process may access. Each domain defines a set of objects and the types of operations that may be invoked on each object. The ability to execute an operation on an object is an access right. A domain is a collection of access rights, each of which is an ordered pair < object-name, right-set>. For example, if domain D has the access right <file F,{ read, write}>, then a process executing in domain D can both read and write file F; it cannot, however, perform any other operation on that object.

Domains do not need to be disjoint; they may share access rights. For eample, in Figure 5.1, we have three domains: D_1, D_2, and D_3.The accss right <O_4, {print}>is shared by both D_2 and D_3, implying that a process executing in their one of these two domains can print object O_4.Note that a process must be executing in domain D_1 to read and write object O_1. On the other hand, only processes in domain D_3 may execute object O_1.

The association between a process and a domain may be either static (if the set of resources available to a process is fixed throughout the latter's lifetime) or dynamic. As might be exepected, the problems inherent in

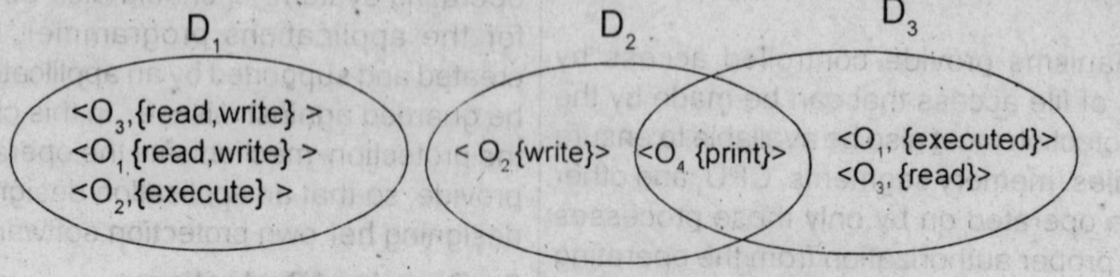

Figure 5.1 System with three protection domains.

establishing dynamic protection domains require more careful solution than do the simpler problems of the static case.

If the association between processes and domains is fixed, and we want to adhere to the need-to-know principle, then a mechanism must be available to change the content of a domain. A process may execute in two different phases. For example, it may need read access in one phase and write access in another. If a domain is static, we must define the domain to include both read and write access. However, this arrangement provides more rights than are needed in each of the two phases, since we have read access in the phase where we need only write access, and vice versa. Thus, the need-to-know principle is violated. We must allow the contents of a domain to be modified, so that it always reflects the minimum necessary access rights.

If the association is dynamic, a mechanism is available to allow a process to switch from one domain to another. We may also want to allow the content of a domain to be changes. If we cannot change the content of a domain, we can provide the same effect by creating a new domain with the changed content, and switching to that new domain when we want to changed the domain content.

We note that a domain can be realized in a variety of ways:

- Each user may be a domain. In this case, the set of objects that can be accessed depends on the identity of the user. Domain switching occurs when the user is changed -generally when one user logs out and another user logs in.
- Each process may be a domain. In this case, the set of objects that can be accessed depends on the identity of the process. Domain switching corresponds to one process sending a message to another process, and then waiting for a response.
- Each procedure may be a domain. In this case, the set of objects that can be accessed corresponds to the local variables defined within the procedure. Domain switching occurs when a procedure call is made.

4. Access Matrix

Our model of protection can be viewed abstractly as a matrix, called an access matrix. The rows of the access matrix represent domains, and the columns represent objects.Each entry in the matrix consists of a set of access rights. Because objects are defined explicitly by the column, we can omit the object name from the access right. The entry **access** (i, j) defines the set of operations that a process, executing in domain D_i can invoke on object O_j.

To illustrate these concepts, we consider the access matrix shown in Figure 5.2. There are four domains and four objects: three files (F_1, F_2, F_3), and one laster printer. When a process executes in domain D_1, it can

object / domain	F_1	F_2	F_3	printer
D_1	read		read	
D_2				print
D_3		read	execute	
D_4	read write		read write	

Figure 5.2 : Access matrix

read files F_1 and F_3. A process executing in domain D_4 has the same privileges as it does in Domain D_1, but in addition, it can also write onto files F_1 and F_3. Note that the laser printer can be accessed only by a process executing in domain D_2.

The access-matrix scheme provides us with the mechanism for specifying a variety of policies. The mechanism consists of implementing the access matrix and ensuring that the semantic properties we have outlined indeed hold. More specifically, we must ensure that a process executing in domain D_i can access only those objects specified in row i, and then only as followed by the access-matrix entries.

Policy decisions concerning protection can be implemented by the access matrix. The policy decisions involve which rights should be included in (i, j)th entry. We must also decide the domain in which each process executes. This last policy is usually decided by the operating system.

The users normally decide the contents of the access-matrix entries, When a user creates a new object O_j, the column O_j is added to the access matrix with the appropriate initialization entries, as dicted by the creator. The user may decide to enter some rights in some entries in column j and other rights in other entries, as needed.

The access matrix provides an appropriate mechanism for defining and implementing strict control for both the static and dynamic association between processes and domain. When we switch a process from one domain to another, we are executing an operation (switch) on an object (the domain). We can control domain switching by including domains among the objects of the access matrix. Similarly, when we change the content of the access matrix, we are performing an operation on an object: the access matrix. Again, we can control these changes by including the access matrix itself as an object. Actually, since each entry in the access matrix may be modified individually, we must consider each entry in the access matrix as an object to be protected.

domain \ Object	F_1	F_2	F_3	laser printer	D_1	D_2	D_3	D_4
D_1	read		read			switch		
D_2				print			switch	switch
D_3		read	execute					
D_4	read write		read write		switch			

Fig 5.3: Access matrix of Figure 5.2 with domains as objects

Now, we need to consider only the operations that are possible on these new objects (domains and the access matrix), and decide how we want processes to be able to execute these operations

Processes should be able to switch from one domain to another. Domain switching from domain D_i to domain D_j is allowed to occur if and only if the access right switch $\in$ **access** (i, j). Thus, in Figure 5.3, a process executing in domain D_2 can switch to domain D_3 or to domain D_4. A process in domain D_4 can switch to D_1, and one in domain D_1 can switch to domain D_2.

Allowing controlled change to the contents of the access-matrix entries requires three additional operations: **copy, owner, andcontrol.**

The ability to copy an access right from one domain (row) of the access matrix to another is denoted by an asterisk (*) appended to the access right. The copy right allows the copying of the access right only within the column (that is, for the object) for which the right is defined. For example, in Figure 5.4 (a), a process executing in domain D_2 can copy the read operation into any entry associated with file F_2. Hence, the access matrix of Figure 5.4 (a), a process executing in domain D_2 can copy the read operation into any entry associated with file F_2. Hence, the access matrix of Figure 5.4 (a) can be modified to the access matrix shown in Figure 5.4 (b).

domain \ object	F_1	F_2	F_3
D_1	execute		write*
D_2	execute	read*	execute
D_3	execute		

(a)

domain \ object	F_1	F_2	F_3
D_1	execute		write*
D_2	execute	read*	execute
D_3	execute	read	

(b)

Figure 5.4 Access Matrix with copy rights

There are two variants to this scheme.

1. A right is copied from **access** (i, j) to access (k, j); it is then removed from **access** (i, j); this action is a transfer of a right, rather than a copy.

2. Propagation of the copy right may be limited. That is, when the right R* is copied from **access** (i, j); to **access** (k, j), only the right R (not R *) is created. A process executing in domain D_k cannot further copy the right R.

A system may select only one of these three copy rights, or it may provide all three by identifying them as separate rights: copy, transfer, and limited copy.

The copy right allows a process to copy some rights from an entry in one column to another entry in the same column. We also need a mechanism to allow addition of new rights and removal of some rights. The owner right controls these operations. If **access** (i, j) includes the owner right, then a process executing in domain D_i can add and remove any right in any entry in column j. For example, in Figure 5.5 (a), domain D_1 is the owner of F_1, and thus can add and delete any valid right in column F_1. Similarly, domain D_2 is the owner of F_2 and F_3, and thus can add and remove any valid right within these two columns. Thus, the access matrix of Figure 5.5 (a) can be modified to the access matrix shown in Figure 5.5 (b).

(a)

object / domain	F_1	F_2	F_3
D_1	owner execute		write
D_2		read* owner	read* owner write*
D_3	execute		

(b)

object / domain	F_1	F_2	F_3
D_1	owner execute		
D_2		owner read* write*	read* owner write*
D_3		write	write

Figure 5.5 Access matrix with owner matrix

The copy and owner rights allow a process to change the entries in a column. A mechanism is also needed to change the entries in a row. The control right is applicable to only domain objects. If **access** (i, j) includes the control right, then a process executing in domain D_i can remove any access right from row j. For example, suppose that, in Figure 5.3, we include the control right in **access** (D_2, D_4). Then, a process executing in domain D_2 could modify domain D_4, as shown in Figure 5.6.

Although the copy and owner rights provide us with a mechanism to limit the propagation of access rights, they do not, however, provide us with the appropriate tools for preventing the propagation of information (that is,disclosure of information). The problem of guaranteeing that no information initially held in an object can migrate outside of its execution environment is called the confinement problem. This problem is in general unsolvable.

These operations on the domains and the access matrix are not in themselves particularly important. What is more important is that they illustrate the ability of the access-matrix model to allow the implementation and control of dynamic protection requirements. New objects and new domains can be created dynamically and included in the access-matrix model. However, we have shown only that the basic mechanism is here; the policy decisions concerning which domains are to have access to which objects in which ways must be made by the system desiners and users.

Object / domain	F_1	F_2	F_3	laser printer	D_1	D_2	D_3	D_4
D_1	read		read			switch		
D_2				print			switch	switch control
D_3		read	execute					
D_4	write		write		switch			

Figure 5.6 Modified access matrix of Figure 5.3 **Revocation of Access Rights**

Revocation of Access Rights

In a dynamic protection system, it may sometimes be necessary to revoke access rights to objects that are shared by different users. Various questions about revocation may arise:

- **Immediate versus delayed:** Does revocation occur immediately, or is it delayed? If revocation is delayed, can we find out when it will take place?
- **Selective versus general :** When an access rights to an object is revoked, does it affect all the users who have an access right to that object, or can we specify a select group of users whose access rights should be revoked?
- **Partial versus total :** Can a subset of the right associated with an object be revoked, or must we revoke all access rights for this object?

- **Temporary versus permanent :** Can access be revoked permanently (that is, the revoked access right will never again be available), or can access be revoked and later be obtained again?

With an access-list scheme, revocation is easy. The access list is searched for the access right(s)to be revoked, and they are deleted from the list. Revocation is immediate, and can be general or selective, total or partial, and permanent or temporary.

Capabilities, however, present a much more difficult revocation problem. Since the capabilities are distributed throughout the system, we must find them before we can revoke them. There are several different schemes for implementing revocation for capabilities, including the following:

- **Reacquisition.** Periodically, capabilities are deleted from each domain. If a process wants to use a capability, it may find that that capability has been deleted. The process may then try to reacquire the capability. If access has been revoked, the process will not be able to reacquire the capability.

- **Back-pointers.** A list of pointers is maintained with each object, pointing to all capabilities associated with that object. When revocation is required, we can follow these pointers, changing the capabilities as necessary. This scheme has been adopted in the MULTICS system. It is quite general, although it is a costly implementation.

- **Indirection.** The capabilities do not point to the objects directly, but instead point directly. Each capability points to a unique entry in a global table, which in turn points to the object. We implement revocation by searching the global table for the desired entry and deleting it. When an access is attempted, the capability is found to point to an illegal table entry. Table entries can be reused for other capabilities without difficulty, since both the capability and the table entry contain the unique name of the object. The object for a capability and its table entry must match. This scheme was adopted in the CAL system. It does not allow selective revocatiion.

- **Keys.** A key is a unique bit pattern that can be associated with each capability. This key is defined when the capability is created, and it can be neither modified nor inspected by the process owning that capability. A master key associated with each object can be defined or replaced with the **set-key** operation. When a capability is created, the current value of the master key is associated with the capability.When the capability is exercised, it key is compared to the master key. If the keys match, the operation is allowed to continue; otherwise, an exception condition is raised. Revocation replaces the master key with a new value by the **set-key** operation, invalidating all previous capabilities for this object.

5. Security

Protection, is strictly an internal problem : How do we provide controlled access to programs and data stored in a computer system? Security, on the other hand, requires not only an adequate protection system, but also consideration of the external environment within which the system operates. Internal protection is not useful if the operator's console is exposed to unauthorized personnel, or if files (stored, for example, on tapes and disks) can simply be removed from the computer system and taken to a system with no protection. These security poblems are essentially management, not operating-system, problems.

The information stored in the system (both data and code), as well as the physical resources of the computer system, need to be protected from unauthorized access, malicious destruction or alteration, and accidental introduction of inconsistency. In this chapter, we examine the ways in which information may be misused or intentionally made inconsistent. We then present machanisms to guard against this occurrence.

6. The Security Problem

Security violations (misuse) of the system can be categorized as being either intentional (malicious) or accidental. It is easier to protect against accidental misue than to protect against malicious misuse. Among the forms of malicious access are the following:

- Unauthorized reading of data (theft of information)
- Unauthorized modification of data
- Unauthorized destruction of data

Absolute protection of the system from malicious abuse is not possible, but the cost to the perpetrator can be made sufficiently high to deter most if not all attempts to access, without proper authority, the information residing in the system.

To protect the system, we must take security measures at two levels :

- **Physical :** The site or sites containing the computer systems must be physically secured against armed or surreptitious entry by intruders.

- **Human :** Users must be authorized carefully to reduce the chance of any such user giving access to an intruder in sex change for a bribe or other favors .

Security at both levels must be maintained to ensure operating system security. A weakness at a high level of security (physical or human) allows circumvention of strict low-level (operating sytem) security measures.

7. Authentication

A major security problem for operating systems is the authentication problem. The protection system depends on an ability to identify the programs and processes that are executing. This ability, in turn, eventually rests on our power to identify each user of the system. A user normally identifies himself.How do we determine if a user's identity is authentic? Generally, authentication is based on some combination of three sets of items: user possession (a key or card), users knowledge (a user identifier and password), and a user attribute (fingerprint, retina pattern, or signature).

The most common approach to authenticating a user identity is the use of user passwords. When the user identifies herself by user - id or account name, she is asked for a password. If the user-supplied password matches the password stored in the system, the system assumes that the user is legitimate.

Passwords can be either system-generated or user-selected. System-generated passwords may be difficult to remember, and thus may be commonly written down. User-selected passwords, however,are often easy to guess (the user's name or favorite car, for example). At some sites, admistrators occasionally check user passwords and notify the users if the password is too short or easy to guess. Some systems also age passwords, forcing users to change them at regular intervals (every month, for instance). This method is not foolproof either, since users may easily toggle between two passwords.

8. Program Threats

In an environment where a program written by one user may be used by another user, there is an opportunity for misuse, which may result in unexpected behavior. Below, we describe two common methods for achieving this.

Trojan Horse : Many systems have mechanisms for allowing programs written by users to be executed by other users. If these programs are executed in a domain that provides the access rights of the executing user, they may misuse these rights. Inside a text-editor program, for example, there may be code to search the file to be edited for certain keywords. If any are found, the entire file may be copied to a special area accessible to the creator of the text editor. A code segment that misuses its environmet is called a Trojan horse. The Trojan-horse problem is exacerbated by long search paths (such as are common on UNIX systems). The search path lists the set of directories to search when an ambiguous program name is given. The path is searched for a file of that name and the file is executed. All the directories in the search path must be secure, or a Trojan horse could be slipped into the user's path and executed accidentally.

For instance , consider the use of the " ." character in a search path. The " ." tells the shell to include the current directory in the search. Thus, if a user has " ." in her search path, has set her current directory to a friend's directory, and enters the name of a normal system command, the command may be executed from the friend's directory instead. The program would run within the user's domain, allowing the program to do anything that the user is allowed to do, including deleting the user's files for instance.

Trap Door : The designer of a program or system might leave a hole in the software that only he is capable of using. This type of security breach was shown in the movie " War Gmes." For instance, the code might check for a specific user identifier or password and circumvent normal security procedures. There have been cases of programmers being arrested for embezzling from banks by including rounding errors in their code, and having the occasional half-cent credited to their accounts. This account crediting can add up to quite a large amount of money, considering the number of transactions a bank executes !.

A clever trap door could be included in a compiler. The compiler could generate standard object code as well as a trap door, regardless of the source code being compiled. This activity is particularly nefarious, since a search of the source code of the program would not reveal any problems. Only the source code of the compiler would contain the information. Trap doors pose a difficult problem because, to detect them, we would have to analyze all the source code for all components of a system. Given that software systems may consists of millions of lines of code, this analysis is not done frequently.

System Threats : Most operating systems provide a means for processes to spawn other processes. In such an environment, it is possible to create a situation where operating - system resources and user files misused. The two most common methods for doing this are worms and viruses, which are discussed below.

Worms : A worm is a process that uses the spawn mechanism to clobber system performance. The worm spawns copies of itself, using up system resources and perhaps locking out system use by all other processes. On computer networks, worms are particularly potent, since they may reproduce themselves among systems and thus shut down the entire network. Such an event occurred in 1988 to UNIX systems on the worldwide Internet network, causing millions of dollars of lost system and programmer time.

Viruses : Another form of computer attack is a virus. Like worms, viruses are designed to spread into other programs and can wreak havoc in a system, including modifying or destroying files and causing system crashes and program malfunctions. Whereas a worm is structured as a complete, standalone program, a virus is a fragment of code embedded in a legitimate program. Viruses are a major problem for computer users, especially users of microcomputer systems.Multiuser computers generally are not prone to viruses because the executable programs are protected from writing by the operating system. Even if a virus does infect a program, its powers are limited because other aspects of the sytem are protected. Single-user systems have no such protections and, as a result a virus has free run.

Viruses usually spread by users downloading viral programs from public bulletin boards or exchanging floppy disks containing an infection.

The best protection against computer viruses is prevention, or the practice of safe computing. Purchasing unopened software from vendors and avoiding free or pirated copies form public sources or floppy-disk exchange is the safest route to preventing infection. However, even new copies of legitimate software applications are not immune to virus infection.

Another safeguard, while not preventing infection, does permit early detection. A user must begin by completely reformatting the hard disk, especially the boot sector, which is often targeted for viral attack. Only secure software is uploaded, and a checksum for each file is calculated. The checksum list must be then be kept free from unauthorized access.

Following any system reboot, a program can recompute the checksums and compare them to the original list; any differences serve as a warning of possible infection. Because they usually work among systems, wroms and viruses are generally considered to pose security, rather than protection, problems.

9. Threat Monitoring

The security of a system can be improved by two management techniques. One is threat monitoring. The system can check for suspicious patterns of activity in an attempt to detect a security violation. A common example of this scheme is a time-sharing system that counts the number of incorrect passwords given when a user is trying to log in. More than a few incorrect attempts may signal an attempt to guess a password.

Another common technique is an audit log. An audit log simply records the time, user, and type of all accesses to an object. After security has been violated, the audit log can be used to determine how and when the problem occurred and perhaps the amount of damage done. This information can be useful, both for recovery from the violation and, possibly, in the development of better security measures to prevent future problems.Unfortunately, logs can become large, and logging uses system resources which are them unavailable to the users.

Rather than log system activities, we can scan the system periodically for security holes. These scans can be done when the computer is relatively unused, and therefore have less effect than logging . Such a scan can check a variety of aspects of the system:

- Short or easy-to-guess passwords
- Unauthorized set-uid programs, if the system supports this mechanism.
- Unauthorized programs in system directories
- Unexpected long-running processes
- Improper directory protections on system data files, such as the password file, device drivers, or even the operating system kernel itself .
- Dangerous entries in the program search path (for example, Trojan horse
- Changes to system programs; we can detect unexpected changes by keeping a list of the checksum values of all system programs, and comparing this list against the current checksum values of the programs-checksums do not change unless the contents of the file have changed

Any problems found by a security scan can either be fixed automatically or be reported to managers of the system.

Encryption : The various provisions that an operating system may make for authorization may not offer sufficient protection for highly sensitive data. Moreover, as computer networks gain popularity, more sensitive (classified) information is being transmitted over channels where eavesdropping and message interception are possible. To keep such sensitive finformation secure, we need mechanisms to allow a user to protect data transferred over the network.

Encryption is one common method of protecting information transmitted over unreliable links. The basic mechanism works as follows :

1. The information (text) is encrypted (encoded) from its initial readable form (called clear text), to an internal form (called cipher text). This internal text form, although readable, does not make any sense.

2. The cipher text can be stored in a readable file, or transmitted over unprotected channels.

3. To make sense of the cipher text, the receiver must decrypt(decode) it back into clear text.

Even if the encrypted information is accessed by an unauthorized person, it will be useless unless it can be decoded. The main issue is the development if encryption schemes that are impossible (or at least exceedingly difficult) to break.

There is a variety of methods to accomplish this task. The most common ones provide a general encryption algorithm E, a general decryption algorithm D, and a secret key (or keys) to be supplied for each application. Let E_k and D_k denote the encryption and decryption algorithms, respectively, for a particular application with a key k. Then the encryption algorithm must satisfy the following properties for any massage m :

1. $D_k(E_k(m)) = m$

2. Both E_k and D_k can be computed efficiently.

3. The security of the system depends only on the secrecy of the key, and not on the secrecy of the algorithms E and D.

One such scheme, called the Data Encryption Standard, was recently adopted by the National Bureau of Standards. This scheme suffers from the key-distribution problem: Before communication can take place, the secret keys must be sent securely to both the sender and receiver. This task cannot be done effectively in a communication-network environment . A solution to this problem is to use a public key-encryption scheme. Each user has both a public and a private key, and two users can communicate knowing only each other's public key.

An algorithm based on this concept follows. This algorithm is believed to be almost unbreakable. The public encryption key is a pair (e,n); the private key is a pair (d,n), where e,d, and n are positive integers. Each message is represented as an integer between 0 and n-1. (A long message is broken into a series of smaller messages, each of which can be represented as such an integer.) The functions E and D are defined as

$$E(m) = m^e \ \textbf{mod}\ n = C,$$

$$D(C) = C^d \ \textbf{mod}\ n.$$

The main problem is choosing the encryption and decryption keys. The integer n is computed as the product of two large (100 or more digits) randomly chosen prime numbers p and q with $n = p \times q$

The value of d is chosen to be a large, randomly chosen integer relatively prime to $(p-1)\times(q-1)$. That is, d satisfies greatest common divisor $[d,(p-1)\times(q-1)] = 1$.

Finally, the integer e is computed from p,q, and d to be the multiplicative inverse of d modulo $(p-1)\times(q-1)$. That is, e satisfies $e \times d$ **mod** $(p-1)\times(q-1)=1$.

We should point out that, although n is publicly known, p and q are not. This condition is allowed because of the well-known fact that it is difficult to factor n. Consequently, the integers d and e cannot be guessed easily.

Let us illustrate this scheme with an example. Let p = 5 and q = 7 Then, n = 35 and $(p-1)\times(q-1)=24$. Since 11 is relatively prime to 24, we can choose d = 11; and since 11 x 11 **mod** 24 = 121 **mod** 24 = 1, e = 11. Suppose now that m = 3. Then ,

$C = m^e$ **mod** $n = 3^{11}$ **mod** $35 = 12$

and

C^d **mod** $n = 12^{11}$ **mod** $35 = 3 = m$

Thus,if we encode m using e,we can decode m using d.

SOLVED EXAMPLES

1. How many six - letter passwords can be constructed using lowercase letters and digits ?

Sol : There are 36 possibilities for each of six characters, making the total possibilities 36^6= 2, 176, 782, 336.

2. What is the minimum, average, and maximum time it would take to crack a six - digit password if one password can be checked every millisecond ?

Sol : The minimum time is 1ms, if the first guess is correct, only one password need be checked. On average half the passwords must be checked to find the correct one. There are 1,000,000 six-digit passwords. The average search would take 500,000 ms = 500s. If the password was not found until all the passwords were checked, the search would take 1000s.

3. How many eight-letter passwords can be constructed using digits, uppercase letters, and lowervase letters ?

Sol : 2.1834×10^{14}.

4. What is the average time in hours it would take to crack a six-letter (lowercase only) password if one password can be checked ever 500 μs ?

Sol : Between 21 and 22 h.

OBJECTIVE QUESTIONS

1. The most common systems security method is

   a) passwords b) key-card systems

   c) mantrap d) none of the above

2. Trojan-Horse programs

   a) are legitimate programs that allow unauthorized access

   b) are hacker programs that do not show up on the system

   c) really do not usually work

   d) usually are immediately discovered

3. Files can have

   a) Read access b) Copy access

   c) Write access d) All of the above

4. A public key encryption system

   a) allows anyone to decode the transmissions

   b) allows only the correct sender to decode the data

   c) allows only the correct receiver to decode the data

   d) does not encode the data before transmitting it

5. Link encryption

   a) is more secure than end-to-end encryption

   b) is less secure than end - to - end encryption

   c) cannot be used in a large network

   d) is used only to detect errors

6. Data encryption

   a) is most used by financial networks

   b) is most used by public billboard networks

   c) cannot be used by private installations

   d) is not necessary, since data cannot be intercepted

7. The most common security failure is

   a) insufficient technology used to prevent breaches

   b) too much emphasis on preventing physical access

   c) depending on passwords

   d) carelessness by users

ANSWER KEY

1. a	2. a	3. d	4. c	5. b
6. a	7. a			

TEST YOURSELF - I

1. Consider the methods used by processes P1 and P2 for accessing their critical sections whenever needed, as given below. The initial values of shared boolean variables S1 and S2 are randomly assigned.

Method used by PI	**Method used by P2**
While (S1 = = S2) ; Critical Section S1 = S2 ;	While (S1 ! = S2) ; Critical Section S2 = not (S1) ;

Which one of the following statements describes the properties achieved ?

a) Mutual exclusion but not progress
b) Progress but not mutual exclusion
c) Neither mutual exclusion nor progress
d) Both mutual exclusion and progress

2. A system uses FIFO policy for page replacement. It has 4 page frames with no pages loaded to begin with. The system first accesses 100 distinct pages in some order and then accesses the same 100 pages but now in the reverse order. How many page faults will occur ? **[GATE - 10]**

a) 196 b) 192 c) 197 d) 195

3. Which of the following statements are true ?

I. Shortest remaining time first scheduling may cause starvation

II. Preemptive scheduling may cause starvation

III. Round robin is better than FCFS in terms of response time **[GATE - 10]**

a) I only b) I and III only
c) II and III only d) I, II and III

4. The following program is to be tested for statement coverage :

```
begin
        if (a = = b )          { S1 ; exit ;}
        else if (c = = d)      { S2;}
                        else {S3; exit;}
        S4;
end
```

The test cases T1, T2, T3 and T4 given below are expressed in terms of the properties satisfied by the values of variables a, b, c and d. The exact values are not given.

T1 : a, b, c and d are all equal
T2 : a, b, c and d are all distinct
T3 : a = b and c ! = d
T4 : a ! = b and c = d

Which of the test suites given below ensures coverage of statements S1, S2, S3 and S4 ? **[GATE - 10]**

a) $T1.T2.T3$ b) $T2.T4$
c) $T3.T4$ d) $T1.T2.T4$

5. The following program consists of 3 concurrent processes and 3 binary semaphores. The semaphores are initialized as S0=1, S1=0, S2=0 .

Process P0	**ProcessP1**	**ProcessP2**
while (true) { wait(S0) ; print '0' release (S1); release(S2); }	wait(S1); Release(S0);	wait(S2); release(S0);

How many times will process P0 print '0' ? **[GATE - 10]**

a) At least twice b) Exactly twice
c) Exactly thrice d) Exactly once

6. A system has n resources $R_0 ...... R_{n-1}$ and k processes $P_0 ..... P_{k-1}$. The implementation of the resource request logic of each process P_i . is as follows:

```
if (i% 2 = = 0) {
    if (i<n) request R_i ;
    if (i+2<n) request R_{i+2} ;
}
else {
    if (i<n) request R_{n-i} ;
    if ( i+2<n) request R_{n-i-2} ;
}
```

In which one of the following situations is a deadlock possible ? **[GATE - 10]**

a) n = 40, k = 26 b) n = 21, k = 12
c) n = 20 , k = 10 d) n = 41, k = 19

7. In which one of the following page replacement policies, Belady's anomaly may occur ? **[GATE - 09]**

a) FIFO b) Optimal c) LRU d) MRU

8. The essential content(s) in each entry of a page table is / are **[GATE - 09]**

a) Virtual page number
b) Page frame number
c) Both virtual page number and page frame number
d) Access right information

9. Consider a system with 4 types of resources R1 (3 units), R2 (2 units), R3 (3 units), R4 (2 units). A non - preemptive resource allocation policy is used. At any given instance, a request is not entertained if it cannot be completely satisfied. Three processes P1, P2, P3 request the sources as follows if executed independently.

Process P1:	Process P2:	Process P3:
t=0: requests 2 units of R2	t=0: requests 2 units of R3	t=0: requests 1 unit of R4
t=1: requests 1 unit of R3	t=2: requests 1 unit of R4	t=2: requests 2 units of R1
t=3: requests 2 units of R1	t=4: requests 1 unit of R1	t=5: releases 2 units of R1
t=5: releases 1 unit of R2 and 1 unit of R1.	t=6: releases 1 unit of R3	t=7: requests 1 unit of R2
t=7: releases 1 unit of R3	t=8: Finishes	t=8: requests 1 unit of R3
t=8: requests 2 units of R4		t=9: Finishes
t=10: Finishes		

Which one of the following statements is TRUE if all three processes run concurrently starting at time t = 0 ? **[GATE - 09]**

a) All processes will finish without any deadlock

b) Only P1 and P2 will be in deadlock

c) Only P1 and P3 will be in a deadlock

d) All three processes will be in deadlock

10. Consider a disk system with 100 cylinders. The requests to access the cylinders occur in following sequence :

4, 34, 10, 7, 19, 73, 2, 15, 6, 20

Assuming that the head is currently at cylinder 50, what is the time taken to satisfy all requests if it takes 1ms to move from one cylinder to adjacent one and shortest seek time first policy is used ? **[GATE - 09]**

a) 95 ms b) 119 ms

c) 233 ms d) 276 ms

11. In the following process state transition diagram for a uniprocessor system, assume that there are always some processes in the ready state :

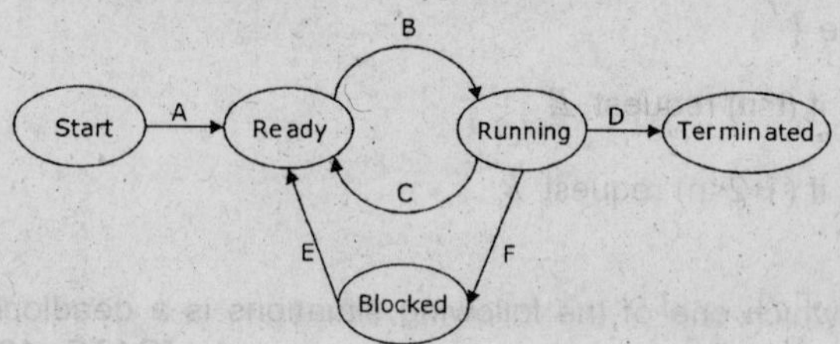

Now consider the following statements :

I. If a process makes a transition D, it would result in another process making transition A immediately.

II. A process P_2 in blocked state can make transition E while another process P_1 is in running state.

III. The OS uses preemptive scheduling.

IV. The OS uses non - preemptive scheduling.

Which of the above statements are TRUE ? **[GATE - 09]**

a) I and II b) I and III

c) II and III d) II and IV

12. The enter_CS( ) and Leave_CS( ) functions to implement critical section of a process are realized using test - and - set instruction as follows :

```
void enter_CS(x)
{
    while (test - and - set(x));
}
```

```
void leave_CS(x)
{
    x = 0 ;
}
```

In the above solution, x is a memory location associated with the CS and is initialized to 0. Now consider the following statements :

I. The above solution to CS problem is deadlock - free

II. The solution is starvation free.

III. The processes enter CS in FIFO order.

IV. More than one process can enter CS at the same time.

Which of the above statements is TRUE ? **[GATE - 09]**

a) I only b) I and II

c) II and III d) IV only

13. A multilevel page table is preferred in comparison to a single level page table for translating virtual address to physical address because **[GATE - 09]**

a) It reduces the memory access time to read or write a memory location

b) It helps to reduce the size of page table needed to implement the virtual address space of a process.

c) It is required by the translation lookaside buffer.

d) It helps to reduce the number of page faults in page replacement algorithms

14. The data blocks of a very large file in the Unix file system are allocated using **[GATE - 08]**

(a) contiguous allocation

(b) linked allocation

(c) indexed allocation

(d) an extension of indexed allocation

15. The P and V operations on counting semaphores, where is a counting semaphore, are defined as follows

```
P(s) : s = s-1;
    if s<0 then wait;
V(s): s=s+1;
    if s<=0 then wakeup a process
        waiting on s;
```

Assume that p_b and v_b, the wait and signal operations on binary semaphores are provided. Two binary semaphores x_b and y_b are used to implement the semaphore operations P(s) and V(s) as follows.

P (s) : $p_b(x_b)$;

s = s -1;

if (s < 0) {

$v_b(x_b)$;

$P_b(y_b)$;

}

else $V_b(x_b)$;

V (s) : $P_b(X_b)$;

s = s + 1 ;

if (s <= 0) $V_b(y_b)$;

$V_b(X_b)$;

The initial values of x_b the y_b are respectively. **[GATE - 08]**

(a) 0 and 0 (b) 0 and 1
(c) 1 and 0 (d) 1 and 1

16. Which of the following statements about synchronous and asynchronous I/O is NOT true? **[GATE - 08]**

(a) An ISR is invoked on completion of I/O in synchronous I/O but not in asynchronous I/O

(b) In both synchronous and asynchronous I/O, an ISR (Interrupt Service Routine) is invoked after completion of the I/O.

(c) A process making a synchronous I/O call waits until I/O is complete, but a process making an asynchronous I/O call does not wait for completion of the I/O

(d) In the case of synchronous I/O, the process waiting for the completion of I/O is woken up the by ISR that is invoked after the completion of I/O.

17. Which of the following is NOT true of deadlock prevention and deadlock avoidance schemes? **[GATE - 08]**

(a) In deadlock prevention, the request for resources is always granted if the resulting state is safe

(b) In deadlock avoidance, the request for resources is always granted if the resulting state is safe

(c) Deadlock avoidance is less restrictive than deadlock prevention

(d) Deadlock avoidance requires knowledge of resource requirements a priori.

18. A process executes the following code

for (i = 0; i<n; i++) for k ( );

The total number of child processes created is **[GATE - 08]**

(a) n (b) $2^n - 1$
(c) 2^n (d) $2^{n+1} - 1$

19. A processor uses 36 bit physical addresses and 32 bit virtual addresses, with a page frame size of 4 Kbytes. Each page table entry is of size 4 bytes. A three level page table is used for virtual-to-physical address translation, where the virtual address is sued as follows.

* bits 30-31 are used to index into the first level page table,

* bits 21-29 are used to index into the second level page table

* bits 12-20 are used to index into the third level page table, and

* bits 0-11 are used as offset within the page

The number of bits required for addressing the next level page table (or page frame) in the page table entry of the first, second and third level page tables are respectively. **[GATE - 08]**

(a) 20, 20 and 20 (b) 24, 24 and 24
(c) 24, 24 and 20 (d) 25, 25 and 24

20. The following is a code with two threads, producer and consumer, that can run in parallel. Further, S and Q are binary semaphores equipped with the standard P and V operations.

```
semaphore S=1; Q=0 ;
integer x ;
producer :              consumer :
while (true) do         While (true) do
P(S);                   P(Q) ;
x=produce( );           consume (x) ;
V(Q) ;                  V (S) ;
done                    done
```

Which of the following is TRUE about the program above ? **[GATE - 08]**

(a) The process can deadlock

(b) One of the threads can starve

(c) Some of the items produced by the producer may be lost

(d) Values generated and stored in 'x' by the producer will always be consumed before the producer can generate a new value.

21. An operating system implements a policy that requires a process to release all resources before making a request for another resource. Select the TRUE statement from the following : **[GATE - 08]**

(a) Both starvation and deadlock can occur

(b) Starvation can occur but deadlock cannot occur

(c) Starvation cannot occur but deadlock can occur

(d) Neither starvation nor deadlock can occur

22. If the time-slice used in the round-robin scheduling policy is more than the maximum time required to execute any process, then the policy will [GATE - 08]
(a) degenerate to shortest job first
(b) degenerate to priority scheduling
(c) degenerate to first come first serve
(d) none of the above

23. Match the following flag bits used in the context of virtual memory management on the left side with the different purposes on the right side of the table below. [GATE - 08]

Name of the bit		Purpose
1. Dirty	a	Page initialization
2. R/W	b.	Write-back policy
3. Reference	c.	Page protection
4. Valid	d.	Page replacement policy

(a) I-d, II-a, III-b, IV-c
(b) I-b, II-c, III-a, IV-d
(c) I-c, II-d, III-a, IV-b
(d) I-b, II-c, III-d, IV-a

24. Processes P1 and P2 use critical_flag in the following routine to achieve mutual exclusion. Assume that critical_flag is initialized to FALSE in the main program.

```
get_exclusive_access ( )
{   if (critical_flag = = FALSE)  {
        critical_flag = TRUE ;
        critical_region ( ) ;
        critical_flag = FALSE ;
    }
}
```

Consider the following statements
(i) It is possible for both P1 and P2 to acess critical_region concurrently.
(ii) This may lead to a deadlock.
Which of the following holds ? [GATE - 07]
(a) (i) is false and (ii) is true
(b) Both (i) and (ii) are false
(c) (i) is true and (ii) is false
(d) Both (i) and (ii) are true

25. Let a memory have four free blocks of sizes 4k, 8k, 20k, 2k. These blocks are allocated following the best-fit strategy. The allocation requests are stored in a queue as shown below.

Request No	J1	J2	J3	J4	J5	J6	J7	J8
Request sizes	2k	14k	3k	6k	6k	10k	7k	20k
Usage Time	1	40	2	8	4	1	8	6

The time at which the request for 17 will be completed will be [GATE - 07]
(a) 16 (b) 19
(c) 20 (d) 37

26. The address sequence generated by tracing a particular program executing in a pure demand paging system with 100 bytes per page is
0100, 0200, 0430, 0499, 0510, 0530, 0560, 0120, 0220, 0240, 0260, 0320, 0410.
Suppose that the memory can store only one page and if x is the address which causes a page fault then the bytes from addresses x to x + 99 are loaded on to the memory.
How many page faults will occur ? [GATE - 07]
(a) 0 (b) 4
(c) 7 (d) 8

27. Consider n jobs $J_1, J_2, \ldots\ldots, J_n$ such that job J_i has execution time t_i and a non-negative integer weight w_i. The weighted mean completion time of the jobs is defined to be

$$\frac{\sum_{i=1}^{n} w_i T_i}{\sum_{i=1}^{n} w_i}$$

where T_i is the completion time of job J_i. Assuming that there is only one processor available, in what order must the jobs be executed in order to minimize the weighted mean completion time of the jobs ? [GATE - 07]
(a) Non-decreasing order of t_i
(b) Non - increasing order of w_i
(c) Non - increasing order of $w_i t_i$
(d) Non - increasing order of w_i / t_i

28. Synchronization in the classical readers and writers problem can be achieved through use of semaphores. In the following incomplete code for readers-writers problem, two binary semaphores mutex and wrt are used to obtain synchronization

```
wait (wrt)
writing is performed
signal (wrt)
wait (mutex)
```

readcount = readcount + 1
if readcount = 1 then S1
S2
reading is performed
S3
readcount = readcount -1
if readcount = 0 then S4
Signal (mutex)

The values of S1, S2, S3, S4, (in that order) are [GATE - 07]

(a) signal (mutex), wait (wrt), signal (wrt), wait (mutex)

(b) signal (wrt), signal (mutex), wait (mutex), wait (wrt)

(c) wait (wrt), signal (mutex), wait (mutex), signal (wrt)

(d) signal (mutex), wait (mutex), signal (mutex), wait (mutex)

29. In a multi-user operating system on an average, 20 requests are made to use a particular resource per hour. The arrival of requests follows a Poisson distribution. The probability that either one, three or five requests are made in 45 minutes is given by [GATE - 07]

(a) $6.9 \times 10^6 \times e^{-20}$ (b) $1.02 \times 10^6 \times e^{-20}$

(c) $6.9 \times 10^3 \times e^{-20}$ (d) $1.02 \times 10^3 \times e^{-20}$

30. A demand paging system takes 100 time units to service a page fault and 300 time units to replace a dirty page. Memory access time is 1 time unit. The probability of a page fault is p. In case of a page fault, the probability of page being dirty is also p. It is observed that the average access time is 3 time units. Then the value of p is [GATE - 07]

(a) 0.194 (b) 0.233
(c) 0.514 (d) 0.981

31. The contents of the text file t1 txt containing four lines are as follows : [GATE - 07]

a1 b1
a2 b2
a3 b3
a4 b1

The contents of the text file t2 txt containing five lines are as follows :

a1 c1
a2 c2
a3 c3
a4 c3
a5 c4

Consider the following Bourne shell script :

```
awk - F ' ' ' {Print $1, $2} ' t1 . txt |
while read a b ; do
        awk - v aV = $ a - v bv =  $ b - F ' '
        aV  = = $1 ( print aV, bV, $2)  '  t2  .  txt
done
```

Which one of the following strings will NOT be present in the output generated when the above script in run ? (Note that the given strings may be substrings of a printed line.)

(a) "*b1 c1*" (b) "*b2 c3*"

(c) "*b1 c2*" (d) "*b1 c3*"

Statement for Linked Answer Questions 32 and 33 :

The head of a hard disk serves requests following the shortest seek time first (SSTF) policy. The head is initially positioned at truck number 180.

32. Which of the request sets will cause the head to change its direction after servicing every request assuming that the head does not change direction if there is a tie in SSTF and all the requests arrive before the servicing starts ? [GATE - 07]

(a) 11, 139, 170, 178, 181, 184, 201, 265

(b) 10, 138, 170, 178, 181, 185, 201, 265

(c) 10, 139, 169, 178, 181, 184, 201, 265

(d) 10, 138, 170, 178, 181, 185, 200, 265

33. What is the maximum cardinality of the request set, so that the head changes its direction after servicing every request if the total number of tracks are 2048 and the head can start from any track ? [GATE - 07]

(a) 9 (b) 10
(c) 11 (d) 12

34. Group 1 contains some CPU scheduling algorithms and Group 2 contains some applications. Match entries in Group 1 to entries in Group 2. [GATE - 07]

Group 1	Group 2
P. Gang Scheduling	1. Guaranteed Scheduling
Q. Rate Monotonic Scheduling	2. Real - time Scheduling
R. Fair Share Scheduling	3. Thread Scheduling

(a) P-3; Q-2; R-1 (b) P-1; Q-2; R-3
(c) P-2; Q-3; R -1 (d) P-1; Q-3; R-2

35. Consider the following statements about user level threads and kernel level threads. Which one of the following statements is FALSE? **[GATE - 07]**

(a) Context switch time is longer for kernel level threads than for user level threads.

(b) User level threads do not need any hardware support

(c) Related kernel level threads can be scheduled on different processors in a multiprocessor system

(d) Blocking one kernel level thread blocks all related threads.

36. An operating system uses Shortest Remaining Time first (SRT) process scheduling algorithm. Consider the arrival times and execution times for the following processes :

Process	Execution time	Arrival time
P1	20	0
P2	25	15
P3	10	30
P4	15	45

What is the total waiting time for process P2? **[GATE - 07]**

(a) 5 (b) 15 (c) 40 (d) 55

37. A virtual memory system uses first In First Out (FIFO) page replacement policy and allocates a fixed number of frames to a process. Consider the following statements

P : Increasing the number of page frames allocated to a process sometimes increases the rate fault rate.

Q : Some programs do not exhibit locality of reference.

Which one of the following is TRUE ? **[GATE - 07]**

(a) Both P and Q are true, and Q is the reason for P.

(b) Both P and Q are true, but Q is not the reason for P.

(c) P is false, but Q is true.

(d) Both P and Q are false.

38. A single processor system has three resource types X, Y and Z, which are shared by three processes. There are 5 units of each resource type. Consider the following scenario, where the column alloc denotes the number of units of each resource type allocated to each process, and the column request denotes the number of units of each resource type requested by a process in order to complete execution. Which of these processes will finish LAST ?

	alloc X Y Z	request X Y Z
P0	1 2 1	1 0 3
P1	2 0 1	0 1 2
P2	2 2 1	1 2 0

[GATE - 07]

(a) P0 (b) P1 (c) P2

(d) None of the above, since the system is in a deadlock

39. Two processes, P1 and P2, need to access a critical section of code. Consider the following synchronization construct used by the processes.

/* P1 */	/* P2 */
while (true){	while (true) {
wants1 = true ;	wants2 = true ;
while (wants2 = = true) ;	while (wants1 = = true) ;
/* Critical Section */	/* Critical Section */
wants 1 = false ;	wants 2 = false ;
}	}
/* Remainder section */	/*Remainder section */

Here, wants 1 and wants2 are shared variables, Which are initialized to false. Which one of the following statements is TRUE about the above construct ? **[GATE - 07]**

(a) It does not ensure mutual exclusion

(b) It does not ensure bounded waiting.

(c) It requires that processes enter the critical section in strict alternation.

(d) It does not prevent deadlocks, but ensures mutual exclusion.

Statement for Linked Answer Questions 40 and 41 :

A process has been allocated 3 page frames. Assume that none of the pages of the process are available in the memory initially. The process makes the following sequence of page references (reference string) : 1, 2, 1, 3, 7, 4, 5, 6, 3, 1

40. If optimal page replacement policy is used, how many page faults occur for the above reference string ? **[GATE - 07]**

(a) 7 (b) 8 (c) 9 (d) 10

41. **Least Recently Used (LRU) page replacement policy is a practical approximation to optimal page replacement. For the above reference string, how many more page faults occur with LRU than with the optimal page replacement policy? [GATE - 07]**

(a) 0 (b) 1 (c) 2 (d) 3

42. **Consider three CPU - intensive processes, which require 10, 20, and 30 time units and arrive at times 0, 2, and 6, respectively. How many context switches are needed if the operating system implements a shortes remaining time first scheduling algorithm ? Do not count the context switches at time zero and at the end. [GATE - 06]**

(A) 1 (B) 2

(C) 3 (D) 4

43. **The atomic fetch-and-set x, y instruction unconditionally sets the memory location x to 1 and fetches the old value of x in y without allowing any intervening access to the memory location x. Consider the following implementation of P and V functions on a binary semaphore S.**

```
void p (binary_semaphore *S) {
  unsigned y;
  unsigned * X = & (S - > value );}
    do {
      fetch - and - set x, y;
    } while (y);
}
void V (binary_semaphore * S) {
  {S - > value = 0;
}
```
[GATE - 06]

which one of the following is true ?

(A) The implementation may not work if context switching is disabled in P

(B) Instead of using fetch-and-set, a pair of normal load/store can be used

(C) The implementation of V is wrong

(D) The code does not implement a binary semaphore

44. **A CPU generates 32-bit virtual addresses. The page size is 4 KB. The processor has a translation look-aside buffer (TLB) which can hold a total of 128 page table entires and is 4 way set associative. The minimum size of the TLB tag is [GATE - 06]**

(A) 11 bits (B) 13 bits

(C) 15 bits (D) 20 bits

45. **A computer system supports 32-bit virtual address as well as 32-bit physical addresses. Since the virtual address space is of the same size as the physical address space, the operating system designers decide to get rid of the virtual memory entirely. Which one of the following is true ? [GATE - 06]**

(A) Efficient implementation of multi-user support is no longer possible

(B) The processor cache organization can be made more efficient now

(C) Hardware support for memory management is no longer needed

(D) CPU scheduling can be made more efficient now

46. **Consider three processes (process id 0, 1, 2, respectively) with compute time bursts 2, 4 and 8 time units. All processes arrive at time zero. Consider the longest remaining time first (LRTF) scheduling algorithm. In LRTF ties are broken by giving priority to the process with the lowest process id. The average turn around time is [GATE - 06]**

(A) 13 units (B) 14 units

(C) 15 units (D) 16 units

47. **Consider three processes, all arriving at time zero, with total execution time of 10, 20, and 30 units, respectively. Each process spends the first 20% of execution time doing I/O, the next 70% of time doing computation, and the last 10% of time doing I/O again. The operating system uses a shortest remaining compute time first scheduling algorithm and schedules a new process either when the running process gets blocked on I/O or when the running process finishes its compute burst. Assume that all I/O operations can be overlapped as much as possible. For what percentage of time does the CPU remain idle ?**

[GATE - 06]

(A) 0% (B) 10.6%

(C) 30.0% (D) 89.4%

48. Consider the following snapshot of a system running n processes. Process i is holding x_i instances of a resource R, for $1 \le i \le n$. Currently, all instances of R are occupied. Further, for all i, process i has placed a request for an additional y_i instances while holding the x_i instances it already has. There are exactly two processes p and q such that $y_p = y_q = 0$. Which one of the following can serve as a necessary condition to guarantee that the system is not approaching a deadlock ? [GATE - 06]

(A) $\min(x_p, x_q) < \max_{k \neq p,q} y_k$

(B) $x_p + x_q \ge \min_{k \neq p'q} y_k$

(C) $\max(x_{p'} x_q) > 1$

(D) $\min(x_{p'} x_q) > 1$

Statement for Linked Answer Questions 49 and 50 :

Barrier is a synchronization construct where a set of processes synchronizes globally i.e. each process in the set arrives at the barrier and waits for all others to arrive and then all processes leave the barrier. Let the number of processes in the set be three and S be a binary semaphore with the usual P and V functions. Consider the following C implementation of a barrier with line numbers shown on the left.

```
Void barrier (void)  {
1 :   P(S) ;
2 :   Process_arrived ++;
3 :   V(S);
4 :   while (process_arrived ! = 3);
5 :   P(S) ;
6 :   process_left ++ ;
7 :   if (process_left = = 3 )
8 :      process_arrived = 0 ;
9 :      process_left = 0;
10 :  }
11 :  V (S) ;
}
```

The variables process_arrived and process_left are shared among all processes and are initialized to zero. In a concurrent program all the three processes call the barrier function when they need to synchronize globally.

49. The above implementation of barrier is incorrect. Which one of the following is true ? [GATE - 06]

(A) The barrier implementation is wrong due to the use of binary semaphore S

(B) The barrier implementation may lead to a deadlock if two barrier invocations are used in immediate succession

(C) Lines 6 to 10 need not be inside a critical section

(D) The barrier implementation is correct if there are only two processes instead of three

50. Which one of the following rectifies the problem in the implementation ? [GATE - 06]

(A) Lines 6 to 10 are simply replaced by process_arrived --

(B) At the beginning of the barrier the first process to enter the barrier waits until process_arrived becomes zero before proceeding to execute P(S)

(C) Context switch is disabled at the beginning of the barrier and re-enabled at the end

(D) The variable process_left is made private instead of shared

51. In the working-set strategy, which of the following is done by the operating system to prevent thrashing ? [GATE - 06]

I) It initiates another process if there are enough extra frames

II) It selects a process to suspend if the sum of the sizes of the working - sets.

II) Exceeds the total number of available frames.

a) I only b) II only

c) Neither I nor II d) Both I and II

52. **The process state transition diagram of an operating system is as given below. Which of the following must be FALSE about the above operating system ? [GATE - 06]**

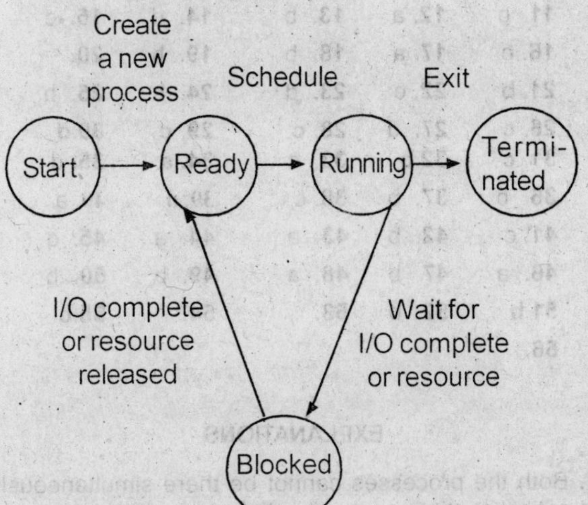

a) It is a mutiprogrammed operating system

b) It uses preemptive scheduling

c) It uses non-preemptive scheduling

d) It is a multi-user operating system

53. **The arrival time, priority, and durations of the CPU and I/O bursts for each of three processes P_1, P_2 and P_3 are given in the table below. Each process has a CPU burst followed by an I/O burst followed by another CPU burst. Assume that each process has its own I/O resource**

Process	Arrival time	Priority	Burst durations CPU, I/O, CPU)
P_1	0	2	1, 5, 3
P_2	2	3 (lowest)	3, 3, 1
P_3	3	1 (highest)	2, 3, 1

The multi-programmed operating system uses preemptive priority scheduling. What are the finish times of the processes P_1, P_2 and P_3 ? [GATE - 06]

a) 11, 15, 9 b) 10, 15, 9

c) 11, 16, 1 d) 12, 17, 11

54. **Consider the solution to the bounded buffer producer / consumer problem by using general semaphores S, F, and E. The semaphore S is the mutual exclusion semaphore initialized to 1. The semaphore F corresponds to the number of free slots in the buffer and is initialized to N. The semaphore E corresponds to the number of elements in the buffer and is initialized to 0.**

Producer Process	Consumer Process
Produce an item ;	Wait (E) ;
Wait (F) ;	Wait (S) ;
Wait (S) ;	Remove an item from the buffer ;
Append the item to the buffer ;	Signal (S) ;
Signal (S) ;	Signal (F) ;
Signal (E) ;	Consume the item ;

Which of the following interchange operations may result in a deadlock ?

(I) Interchanging Wait (F) and Wait (S) in the Producer process

(II) Interchanging Signal (S) and Signal (F) in the Consumer process [GATE - 06]

a) I only b) II only

c) Neither I nor II d) Both I and II

55. **For each of the four processes P_1, P_2, P_3 and P_4. The total size in kilobytes (KB) and the number of segments are given below .**

Process	Total size (in KB)	Number of segments
P1	195	4
P2	254	5
P3	45	3
P4	364	8

The page size is 1 KB. The size of an entry in the page table is 4 bytes. The size of an entry in the segment table is 8 bytes. The maximum size of a segment is 256 KB. The paging method for memory management uses two-level paging, and its storage overhead is P. The storage overhead for the segmentation method is S. The storage overhead for the segmentation and paging method is T. What is the relation among the overheads for the different methods of memory management in the concurrent execution of the above four processes ? [GATE - 06]

a) $P < S < T$ b) $S < P < T$

c) $S < T < P$ d) $T < S < P$

56. The wait and signal operations of a monitor are implemented using semaphores as follows. In the following.

x is a condition variable,

mutex is a semaphore initialized to 1,

x_sem is a semaphore initialized to 0,

x_count is the number of processes waiting on semaphore x_sem,

initially 0, next is a semaphore initialized to 0,

next_count is the number of processes waiting on semaphore next, initially 0.

The body of each procedure that is visible outside the monitor is replaced with the following : [GATE - 06]

```
P (mutex) ;
.
.
body of procedure
if (next_count > 0)
    V (next)
else
    V (mutex) ;
```

Each occurrence of x.wait is replaced with the following :

```
x_count = x_count + 1 ;
if (next_count > 0)
    V (next)
else
    V (mutex) ;
------------------------------------------E1 ;
x_count = x_count - 1 ;
```

Each occurrence of x.signal is replaced with the following :

```
if (x_count > 0)
{
    next_count = next _ count + 1 ;
    ------------------------------E2;
    P (next) ;
    next_count = next_count - 1 ;
}
```

For correct implementation of the monitor, statements E1 and E2 are, respectively,

a) $P(x\_sem), V(next)$ b) $V(next), P(x\_sem)$

c) $P(next), V(x\_sem)$ d) $P(x\_sem), V(x\_sem)$

ANSWER KEY

1. a	2. a	3. d	4. d	5. a
6. b	7. a	8. b	9. a	10.b
11. c	12. a	13. b	14. d	15. c
16. b	17. a	18. b	19. b	20.
21. b	22. c	23. d	24. b	25. b
26. c	27. d	28. c	29. c	30.d
31. c	32.b	33. c	34. a	35. d
36. b	37. b	38. c	39. d	40. a
41. c	42. b	43. a	44. a	45. c
46. a	47. b	48. a	49. b	50. b
51.b	52. b	53.	54.	55.c
56.				

EXPLANATIONS

1. Both the processes cannot be there simultaneously in critical section, so it will satisfy mutual exclusion.

2. Since we have 4 page frames, 1 to 100 pages will get 100 faults, But in reverse, there wont be any fault for page No's 100, 99, 98, 97, and for remaining 96 faults, Total = 100 + 96 = 196.

3. Any preemptive scheduling strategy may cause starvation, SRTF is also preemptive SJF. In FCFS late arriving short job have to wait for long time, but in RR, the time QUANTUM expiry will preempt the present job and give the response to the next job in Queue so all are TRUE.

5. Initially neither P_1 nor P_2 can go, only P_3 can go since It uses the semaphore so whose value is '1'

It prints '0' then, after releasing s_1, s_2, again either

P_1 or P_2 releases ' S_0 so that it will go to print '0' at least twice.

7. FIFO may have Belady's Anamoly

e.g. In FIFO for three frames the reference string

1, 2,3,4, 1, 2, 5, 1, 2, 3, 4, 5

there are 9 page faults and for four frames for the reference string there will be 10 faults.

8. Page frame number is essential content in each entry of the page table.

Alternately

The page table has the ability to mark on entry invalid through a valid invalid bit or special value of protection bit, which is often called page frame number.

9. At resource used

t =0 R2 - 2units R3-2 units R4 - 1units
t =2 R2 - 2units R3-3 units R4 - 2units R1-1units
t =3 R2 - 2units R3-3 units R4 - 2units <u>R1-4units</u>
t =4 R2 - 2units R3-3 units R4 - 2units <u>R1-5units</u>
t =5 R2 - 1units R3-3 units R4 - 2units R1-2units
t =6 R2 - 1units R3-2 units R4 - 2units R1-2units

t =7 R2 - 2units R3-1 units R4 - 2units R1-2units
t =8 R2 - 2units R3-1 units R4 - 3units R1-1units
t=9
The under line show that at that time the requirement of resource is less than availability but it does not cause dead lock. All process finish without any dead lock.

10. Requests : 4, 34,10, 7, 19, 73, 2, 15, 6, 20

Current Position of head is 50

Current	50	34	20	19	15	10	7	6	4	2
Next Access	34	20	19	15	10	7	6	4	2	73
Difference	16	14	1	4	5	3	1	2	2	71

Total Time = 16+14+1+4+5+3+1+2+2+71 = 119 ms

11. In premptive scheduling the process which allocated the CPU keeps it and release it only when it switching to waiting state. Also a process P_2 in blocked state can make transition E while other process P_1 in running.

Alternately

A process P_2 in blocked state can make transition E while another P_1 is in running state.

The OS uses preemptive Scheduling.

12. In given program the process test -and -set in critical section and in function leaves CS initialise memory location x to zero. So this process is dead lock free.

13. Most computer system support a large logical address space ($2^{32} to\ 2^{64}$). The page table become excessively large for example. Consider a system with a 32 bit logical address space. If page size $4K(2^{12})$ then page table may consist 1 million page. For this we use multilevels page table.

14 Index allocation cannot allocate very large file.

15. $P(I): P_6(x_1): S = S - 1:$

if $(S < 0)\{V_6(K_6): p_6(y_5):\}$

else $V_6(x_6)$: value of $x_6 = 1$ satisfy

V(S) : $P_6(x_6); P_6(x_6); S = S + 2$

if $(S <= 0)V_6(y_6): V_6(x_5)$

$Y_6 = 0$ satisfy

16. ISR is invoked after completion of I/O irrespective of the type of I/O

17. In deadlock avoidance, the request for resources is granted if only the state guarantees deadlock would never occur (safe state)

Deadlock prevention provides a set of methods that ensures that at least one of the necessary condition cannot hold -(mutual exclusion, hold & wait, no preemption & circular wait). These methods prevents deadlock by constraining how request can be made.

18. Number of child processes $2^n - 1$

21. If a process releases all resources before making a request for another resource, starvation may or may not occur but deadlock can't occur.

Deadlock: The implementation of a semaphore with a waiting queue may result in a situation where two or more processes are waiting indefinitely for an event that can be caused only by one of the waiting processes. When such a state is reached, processes are said to be deadlocked.

Starvation (Indefinite blocking). A situation in which processes have to wait indefinitely.

22. **Round Robin scheduling algorithm:** It is designed specially for time sharing systems. It is similar to FCFS scheduling, but preemption is added to switch between processes. A small unit of time, called a time quantum or time slice is defined.

To implement R scheduling, we keep the ready queue as a FIFO queue of processes. New processes are added to the tail of the ready queue. The CPU scheduler picks the first process from the ready queue, sets the times to interrupt after 1 time quantum and dispatches the process.

So, we can see that, if the time slice used in RR scheduling policy is more than the maximum time required to execute any process, then the policy will degenerate to first come first serve.

23.

Name of bit		Purpose
Dirty	→	Write-back policy
Read/write	→	Page protection
Reference	→	Page replacement policy
Valid	→	Page initialisation

24. Both process will not access critical region concurrently. So (i) is false and there is no chance of deadlocking. So (ii) is also false.

25. There are four free blocks of sizes 4k, 8k, 2k

Request No.	J1	J2		J3	J4	J5	J6	J7
Allocate block	2k	20k		4k	6k	8k	20k	10k
Remaining block	4k,8k,20k	4k,		k, 8k,6k	k,8k	1k,2k	1k,2k,3k	1k,2k,3k
Usage time	4	10		2	8	4	1	8

It will wait to free block used by J2. So it ait for 10 unit time, then J6 will complete in 1 unit time. So total time 11 unit is taken tc ess request of J7 and J7 takes unit time to complete the request.

Total time taken by J7 = 10 + 1 +

= 19 unit t

27. Which job have least execution time must completed first to get low completion time.

So w_i / t_i is high to low value.

28. By inspecting this code

If buffer is full, then gives wait (wrt)

So S1 = wait (wrt)

Similarly if buffer is empty, then gives signal (wrt) to writer

So Sy = signal (wrt)

31.

t_1	$t \times t$	t_2	$t \times t$
a_1	b_1	a_1	c_1
a_2	b_2	a_2	c_2
a_3	b_3	a_3	c_3
a_4	b_1	a_4	c_4
		a_5	c_5

So its output contain

$a_1 \quad b_1 \quad c_1$

$a_2 \quad b_2 \quad c_2$

$a_3 \quad b_2 \quad c_3$

$a_4 \quad b_1 \quad c_3$

Hence anser is (c)

32.

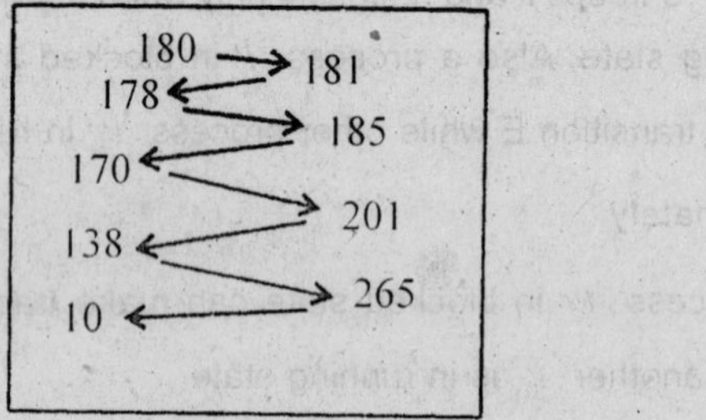

35. Blocking one kernel level thread doesn't block all related threads.

36. Gant chart of scheduling can be given as

P_1	P_2	P_3	P_2	P_4

0 20 30 40 55 70

Waiting time of process,

$P_2 = (20-15) + 10$

= 15

37. Both P and Q are true But Q is not the reason for P because page fault is concerned with pages present in frame not with pages nearly the page accessed recently.

38. In present situation free resources are 0.12, so

P_1 can be processed, after this resources will be 213 so

P_0 can be processed therefore P_2 will be last to be processed.

39. Since there is no way to synchronize two statement

wants 1 = true (in P_1) and wants 2 = true (in P_2).

So both of they may get executed one after other. In this situation there will be a dead lock.

But if only of the two is true other will not go beyond cuhile loop, hence ensure nutual exclusion.

40. 1 will occupy location 0.

3 will occupy location 6.

8 hashed to location 0 which is already occupied, so it will be hashed to one location next to it (0) i.e. to location 1.

Since 10 also clashes, so it will be hashed to location 2.

41. In case of unweighted, undirected graphs, BFS gives the most time efficient computation for shortest path. It is guaranteed to find first shortest path.

42. Context-Switches will be needed only at the time of switching over to some other process. The following table shows the time remaining for the three processes at different instants of time

Time\Processes →	I	II	III	Process continued
0	10	-	-	I
2	8	20	-	I
6	4	20	30	I
10	0	20	30	II
30	0	0	30	III
60	0	0	0	-

∴ Context-witches are needed only at times 10 and 30.

44. The number of bits required for various purposes in the 32 bit address is as follows.

To find the memory cell out of 4-way associative memory -2 bits are needed.

To find the page out of 128 page table entries -7 bits are needed.

To find out the required byte in the page of size 4 KB - 12 bits are needed.

Total bits consumed in this way = 21

TLB tag's minimum size $= 32 - 21 = 11$ bits

51. The working set of a program is the set of program pages which will be referenced in the next w instructions of the program, when w is defined constant.

The memory allocator uses this information to determine the allocation of a program. The program can be suspended due to shortage of memory and pages are moved to paging device.

52. From given operating system, we conclude that

(i) It is multi-programmed as scheduling of processes is done.

(ii) It is non-preemptive, as a process switches from the running state to the waiting state.

(iii) It is a multi-user operating system.

(iv) It is not pre-emptive as in the same process switches from running to ready state.

55. T = Storage overhead for segmentation and paging method.

S = Storage overhead for segmentation method.

P = Storage overhead for paging method.

∴ $S = 4\times8+5\times8+3\times8+8\times8 = 160$ bytes.

T = Stores the software program needed.

Thus, S < T, and

$P = 195\times4+254\times4+45\times4+364\times4$

= 343

$\therefore S < T < P$.

TEST YOURSELF - II

1. Using a larger block size in a fixed block size file system leads to [GATE - 03]

a) better disk throughput but poorer disk space utilization

b) better disk throughput and better disk space utilization

c) poorer disk throughput but better disk space utilization

d) poorer disk throughput and poorer disk space utilization

2. In a system with 32 bit virtual addresses and 1 KB page size, use of one-level page tables for virtual to physical address translation is not practical because of [GATE - 03]

a) the large amount of internal fragmentation

b) the large amount of external fragmentation

c) the large memory overhead in maintaining page tables

d) the large computation overhead in the translation process

3. Which of the following is NOT an advantage of using shared, dynamically linked libraries as opposed to using statically linked libraries ? [GATE - 03]

a) Smaller sizes of executable files.

b) Lesser overall page fault rate in the system

c) Faster program startup

d) Existing program need not be re-linked to take advantage of newer versions of libraries.

4. A uni-processor computer system only has two processes, both of which alternate 10 ms CPU bursts with 90 ms I/O bursts. Both the processes were created at nearly the same time. The I/O of both processes can proceed in parallel. Which of the following scheduling strategies will result in the least CPU utilization (over a long period of time) for this system ? [GATE - 03]

a) First come first served scheduling

b) Shortest remaining time first scheduling

c) Static priority scheduling with different priorities for the two processes.

d) Round robin scheduling with a time quantum of 5 ms

The following information pertains to Q.5-6

A processor uses 2-level page tables for virtual to physical address translation. Page tables for both levels are stored in the main memory. Virtual and physical addresses are both 32 bits wide. the memory is byte addressable. For virtual to physical address translation, the 10 most significant bits of the virtual address are used as index into the first level page table while the next 10 bits are used as index into the second level page table. The 12 least significant bits of the virtual address are used as offset within the page. Assume that the page table entries in both levels of page tables are 4 bytes wide. Further, the processor has a translatioin look-aside buffer (TLB), with a hit rate of 96%. The TLB caches recently used virtual page numbers and the corresponding physical page numbers. The processor also has a physically addressed cache with a hit rate of 90%. Main memory access time is 10 ns, cache access time is 1 ns, and TLB access time is also 1 ns.

5. Assuming that no page faults occur, the average time taken to access a virtual address is approximately (to the nearest 0.5 ns) [GATE - 03]

a) 1.5 ns b) 2 ns c) 3 ns d) 4 ns

6. Suppose a process has only the following pages in its virtual address space: two contiguous code pages starting at virtual address 0x00000000, two contiguous data pages starting at virtual address 0x00400000, and a stack page starting at virtual address 0xFFFFF000. The amount of memory required for storing the page tables of this process is [GATE - 03]

a) 8 KB b) 12 KB c) 16 KB d) 20 KB

7. Consider the following statements with respect to user-level threads and kernel-supported threads

i) Context switch is faster with kernel-supported threads

ii) For user-level threads, a system call can block the entire process

iii) Kernel-supported threads can be scheduled independently

iv) User-level threads are transparent to the kernal

Which of the above statements are true ? [GATE - 04]

a) ii), iii) and iv) only b) ii) and iii) only

c) i), and iii) only d) i) and ii) only

8. Consider an operating system capable of loading and executing a single sequential user process at a time. The disk head scheduling algorithm used is First Come First Served (FCFS). If FCFS is replaced by Shortest Seek Time First (SSTF), claimed by the vendor to give 50% better benchmark results, what is the expected improvement in the I/O performance of user programs ? [GATE - 04]

a) 50% b) 40% c) 25% d) 0%

9. Consider the following set of processes, with the arrival times and the CPU burst times given in milliseconds.

Process	Arrival Time	Burst Time
P1	0	5
P2	1	3
P3	2	3
P4	4	1

What is the average turnaround time for these processes with the preemptive shortest remaining processing time first (SRPT) algorithm? [GATE - 04]

a) 5.50 b) 5.75 c) 6.00 d) 6.25

10. Consider a system with a two-level paging scheme in which a regular memory access takes 150 nanoseconds, and servicing a page fault takes 8 milliseconds. An average instruction takes 100 nanoseconds of CPU time, and two memory accesses. The TLB hit ratio is 90%, and the page fault rate is one in every 10,000 instructions. What is the effective average instruction execution time? [GATE - 04]

a) 645 nanoseconds b) 1050 nanoseconds

c) 1215 nanoseconds d) 1230 nanoseconds

11. Consider two processes P_1 and P_2 accessing the shared variables X and Y protected by two binary semaphores S_x and S_y respectively, both initialized to 1. P and V denote the usual semaphore operators, where P decrements the semaphore value, and V increments the semaphore value. The pseudo-code of P_1 and P_2 is as follows :

P_1 :	P_2 :
while true do {	While true do {
L_1 :	L_3 : ....
L_2 :	L_4 : ....
X = X + 1 ;	Y = Y + 1 ;
Y = Y - 1 ;	X = Y - 1;
$V(S_x)$;	$V(S_y)$;
$V(S_y)$; }	$V(S_x)$; }

In order to avoid deadlock, the correct operators at L_1, L_2, L_3 and L_4 are respectively [GATE - 04]

a) $P(S_Y), P(S_X);\ P(S_X), P(S_Y)$

b) $P(S_X), P(S_Y);\ P(S_Y), P(S_X)$

c) $P(S_X), P(S_X);\ P(S_Y), P(S_Y)$

d) $P(S_X), P(S_Y);\ P(S_X), P(S_Y)$

12. A Unix-style i-node has 10 direct pointers and one single, one double and one triple indirect pointers. Disk block size is 1 Kbyte, disk block address is 32 bits, and 48-bit integers are used. What is the maximum possible file size ? [GATE - 04]

a) 2^{24} bytes b) 2^{32} bytes c) 2^{34} bytes d) 2^{48} bytes

13. Which of the following commands or sequence of commands will rename of file x to file y in a Unix system?

I. mv y, x II. mv x,y [GATE - 04]

III. cp y, x (rm x) IV. cp x, y (rm x)

a) II and III b) II and IV

c) I and III d) II only

14. A disk has 200 tracks (numbered 0 through 199). At a given time, it was servicing the request of reading data from track 120, and at the previous request, service was for track 90. The pending requests (in order of their arrival) are for track numbers.

30 70 115 130 110 80 20 25.

How many times will the head change its direction for the disk scheduling policies SSTF (Shortest Seek Time First) and FCFS (First Come First Serve) ? [GATE - 04]

a) 2 and 3 b) 3 and 3 c) 3 and 4 d) 4 and 4

15. In a certain operating system, deadlock prevention is attemped using the following scheme. Each process is assigned a unique timestamp, and is restarted with the same timestamp if killed. Let P_h be the process holding a resource R, P_r be a process requesting for the same resource R, and $T(P_h)$ and $T(P_r)$ be their timestamps respectively. The decision to wait or preempt one of the processes is based on the following algorithm.

```
if T(Pr) < T(Ph) then
    kill Pr
else
    wait
```

Which one of the following is TRUE ? [GATE - 04]

a) The scheme is deadlock-free, but not starvation-free

b) The scheme is not deadlock-free, but starvation-free

c) The scheme is neither deadlock-free nor starvation-free

d) The scheme is both deadlock-free and starvation-free

16. A process executes the following segment of code :

```
for(i=1; i<=n; i++)
fork ( );
```

[GATE - 04]

The number of new processes created is

a) n b) $\frac{n(n+1)}{2}$ c) $2^n - 1$ d) $3^n - 1$

17. The semaphore variables full, empty and mutex are initialized to 0, n and 1, respectively, Process P_1 repeatedly adds one item at a time to a buffer of size n, and process P_2 repeatedly removes one item at a time from the same buffer using the programs given below. In the programs, K, L, M and N are unspecified statements.

[GATE - 04]

```
P1:  while (1) {
     K; P(mutex);
     Add an item to the buffer;
     V(mutex); L;
     }
P2:  while (1) {
     M; P(mutex);
     Remove an item from the buffer;
     V(mutex); N;
     }
```

The statements K, L, M and N are respectively

a) P(full), V(empty), P(full), V(empty)

b) P(full), V(empty), P(empty), V(full)

c) P(empty), V(full), P(empty), V(full)

d) P(empty), V(full), P(full), V(empty)

18. In a virtual memory system, size of virtual address is 32-bit, size of physical address is 30-bit, page size is 4 Kbyte and size of each page table entry is 32-bit. The main memory is byte addressable. Which one of the following is the maximum number of bits that can be used for storing protection and other information in each page table entry ? [GATE - 04]

a) 2 b) 10 c) 12 d) 14

19. In a particular Unix OS, each data block is of size 1024 bytes, each node has 10 direct data block addresses and three additional addresses: one for single indirect block, one for double indirect block and one for triple indirect block. Also, each block can contain addresses for 128 blocks. Which one of the following is approximately the maximum size of a file in the file system ? [GATE - 04]

a) 512 MB b) 2 GB c) 8 GB d) 16 GB

20. What is the swap space in the disk used for ? [GATE - 05]

a) Saving temporary html pages
b) Saving process data
c) Storing the super-block
d) Storing device drivers

21. Increasing the RAM of a computer typically improves performance because [GATE - 05]

a) Virtual memory increases
b) Larger RAMs are faster
c) Fewer pages faults occur
d) Fewer segmentation faults occur

22. Suppose n processes, $P_1, ...P_n$ share m identical resource units, which can be reserved and released one at a time. The maximum resource requirement of process P_i is s_p where $s_i > 0$. Which one of the following is a sufficient condition for ensuring that deadlock does not occur ? [GATE - 05]

a) $\forall$ i, $s_i < m$ b) $\forall$ i, $s_i < n$

c) $\sum_{i=1}^{n} s_i < (m+n)$ d) $\sum_{i=1}^{n} s_i\ (m*n)$

23. Consider the following code fragment : [GATE - 05]

```
if (fork ( ) = = 0)
{a = a + 5; printf("%d,%d\n",a,and a);}
else {a=a-5;printf("%d,%d\n",a,&a);}
```

Let u, v be the values printed by the parent process, and x, y be the values printed by the child process. Which one of the following is TRUE ?

a) $u = x + 10$ and $v = y$ b) $u = x + 10$ and $v \neq y$

c) $u + 10 = x$ and $v = y$ d) $u + 10 = x$ and $v \neq y$

Statement for Linked Answer Questions 24 and 25

We are given 9 tasks $T_1, T_2, T_3........T_9$. The execution of each task requires one unit of time. We can execute one task at a time. Each task T_i has a profit P_i and a deadline d_i Profit P_i is earned if the task is completed before the end of the d_i^{th} unit of time. [GATE - 05]

Task	T_1	T_2	T_3	T_4	T_5	T_6	T_7	T_8	T_9
Profit	15	20	30	18	18	10	23	16	25
Deadline	7	2	5	3	4	5	2	7	3

24. Are all tasks completed in the schedule that gives maximum profit ?

a) All tasks are completed
b) T_1 and T_6 are left out
c) T_1 and T_8 are left out
d) T_4 and T_6 are left out

25. What is the maximum profit earned ?

a) 147 b) 165 c) 167 d) 175

26. A student wishes to create symbolic links in a computer system running Unix. The text files named "file 1" "file 2" and "file 3" exist in her current working directory, and the student has read and write permissions for all three files. Assume that file 1 contains information about her hobbies, file 2 contains information about her friends and file 3 contains information about her courses. The student executes the following sequence of commands from her current working directory.

ln-s file 1 file 2
ln-s file 1 file 3

Which of the following types of information would be lost from her file system ? [GATE - 05]

(I) Hobbies (II) Friends (III) Courses

a) (I) and (II) only b) (II) and (III) only
c) (II) only d) (I) and (III) only

27. The shell command

find-name passwd-print

is executed in/etc directory of a computer system running Unix. Which of the following shell commands will give the same information as the above command when executed in the same directory ? [GATE - 05]

a) *ls* passwd b) cat passwd
c) grep name passwd d) grep print passwd

28. A user level process in Unix traps the signal sent on a Ctrl-C input, and has a signal handling routine that saves appropriate files before terminating the process. When a Ctrl-C input is given to this process, what is the mode in which the signal handling routine executes ? [GATE - 05]

a) kernel mode b) superuser mode
c) privileged mode d) user mode

29. Given below is a program which when executed spawns two concurrent processes : Semaphore X : = 0;
/* Process now forks into concurrent processes P1 & P2 */

```
P1 : repeat forever      P2: repeat forever
V(X) ;                   P(X);
Compute ;                Compute ;
P (X) ;                  V(X) ;
```

Consider the following statements about processes P1 and P2 : [GATE - 05]

I : It is possible for process P1 to starve.
II : It is possible for process P2 to starve.

Which of the following holds ?

a) Both I and II are true. b) I is true but II is false.
c) II is true but I is false d) Both I and II are false.

30. Two concurrent process P1 and P2 use four shared resources R1, R2, R3 and R4 as shown below. [GATE - 05]

```
P1 :             P2:
Compute ;        Compute ;
Use R1;          Use R1;
Use R2;          Use R2;
Use R3;          Use R3;
Use R4;          Use R4;
```

Both processes are started at the same time, and each resource can be accessed by only one process at a time. The following scheduling constraints exist between the access of resources by the processes :

P2 must complete use of R1 before P1 gets access to R1.

P1 must complete use of R2 before P2 gets access to R2.

P2 must complete use of R3 before P1 gets access to R3.

P1 must complete use of R4 before P2 gets access to R4.

There are no other scheduling constraints between the processes. If only binary semaphores are used to enforce the above scheduling constraints, what is the minimum number of binary semaphores needed ?

a) 1 b) 2 c) 3 d) 4

31. We with to schedule three processes P1, P2 and P3 on a uniprocessor system. The priorities, CPU time requirements and arrival times of the processes are as shown below.

Process	Priority	CPU time required	Arrival time (hh:mm:ss)
P1	10(highest)	20 sec	00 : 00 : 05
P2	9	10 sec	00 : 00 : 03
P3	8 (lowest)	15 sec	00 : 00 : 00

We have a choice of preemptive or non-preemptive scheduling. In preemptive scheduling, a late-arriving higher priority process can preempt a currently running process with lower priority. In non-preemptive scheduling, a late-arriving higher priority process must wait for the currently executing process to complete before it can be scheduled on the processor.

What are the turnaround times (time from arrival till completion) of P2 using preemptive and non-preemptive scheduling respectively ? **[GATE - 05]**

a) 30 sec, 30 sec b) 30 sec, 10 sec

c) 42 sec, 42 sec d) 30 sec, 42 sec

32. Consider a 2-way set associative cache memory with 4 sets and total 8 cache blocks (0-7) and a main memory with 128 blocks (0-127). What memory blocks will be present in the cache after the following sequence of memory block references if LRU policy is used for cache block replacement. Assuming that initially the cache did not have any memory block from the current job ?

0 5 3 9 7 0 16 55 **[GATE - 05]**

a) 0 3 5 7 16 55 b) 0 3 5 7 9 16 55

c) 0 5 7 9 16 55 d) 3 5 7 9 16 55

33. Two shared resources R_1 and R_2 are used by processes P_1 and P_2. Each process has a certain priority for accessing each resource. Let T_{ij} denote the priority of P_i for accessing R_j. A process P_i can snatch a resource R_k from proces P_j if T_{ik} is greater than T_{jk}.

Given the following :

I. $T_{11} > T_{21}$ II. $T_{12} > T_{22}$

III. $T_{11} < T_{21}$ IV. $T_{12} < T_{22}$

Which of the following conditions ensures that P_1 and P_2 can never deadlock ? **[GATE - 05]**

a) (I) and (IV) b) (II) and (III)

c) (I) and (II) c) None of the above

34. In a computer system, four files of size 11050 bytes, 4990 bytes, 5170 bytes and 12640 bytes need to be stored. For storing these files on disk, we can use either 100 byte disk blocks or 200 byte disk blocks (but can't mix block sizes). For each block used to store a file, 4 bytes of bookkeeping information also needs to be stored on the disk. Thus, the total space used to store a file is the sum of the space taken to store the file and the space taken to store the bookkeeping information for the blocks allocated for storing the file. A disk block can store either bookkeeping information for a file or data from a file, but not both.

What is the total space required for storing the files using 100 byte disk blocks and 200 byte disk blocks respectively? **[GATE - 05]**

a) 35400 and 35800 bytes

b) 35800 and 35400 bytes

c) 35600 and 35400 bytes

d) 35400 and 35600 bytes

Statement for Linked Answer Questions 35 and 36

A disk has 8 equidistant tracks. The diameters of the innermost and outermost tracks are 1 cm and 8 cm respectively. The innermost track has a storage capacity of 10 MB. **[GATE - 05]**

35. What is the total amount of data that can be stored on the disk if it is used with a drive that rotates it with (i) Constant Linear Velocity (ii) constant Angular Velocity ?

a) (i) 80 MB (ii) 2040 MB

b) (i) 2040 MB (ii) 80 MB

c) (i) 80 MB (ii) 360 MB

d) (i) 360 MB (ii) 80 MB

36. If the disk has 20 sectors per track and is currently at the end of 5th sector of the inner-most track and the head can move at a speed of 10 meters/sec and it is rotating at constant angular velocity of 6000 RPM, how much time will it take to read 1 MB contiguous data starting from the sector 4 of the outer-most track ?

a) 13.5 ms b) 10 ms

c) 9.5 ms d) 20 ms

ANSWER KEY

1. d	2. a	3. d	4. a	5. b
6. b	7. a	8. d	9. a	10.
11. b	12. b	13. b	14. a	15. a
16. c	17. d	18. a	19. d	20. b
21. c	22. c	23. b	24.	25. d
26. b	27. a	28. a	29. d	30. b
31. d	32. c	33. c	34. c	35. a
36. c				

EXPLANATORY NOTES

2. Number of pages $= \frac{2^{32}}{2^{12}} = 2^{20}$

Hence, number of pages is much in number, thus, chances to internal fragmentation due to size of pages is larger, then overhead is reduced.

3. With dynamic linking, a stub is included in the image for each library routine reference. The stub is a small piece of code that indicates how to locate the appropriate memory-resident library routine, or how to load the library if the routine is not already present.

A library may be replaced by a new version and all programs that reference library will automatically use the new version. Without dynamic linking, all such programs would need to be relinked to gain access to the new library.

4. CPU utilization is always least when we take First come First Serve. CPU utilization is maximum when we take Round Robin Scheduling.

5. TLB hit ratio = 96%

TLB access time = 1 ns

Virtual page hit ratio = 90%

Main memory access time = 10 ns

For no page faults

Average time = hit ratio $\times$ cache access time + (1-hit ratio) $\times$ memory access time

$= 0.9 \times 1 + 0.1 \times 10$

$= 1.9$ ns

$\cong 2$ ns

9.

Process	Arrival time	Burst time
P_1	0	5
P_2	1	3
P_3	2	3
P_4	4	1

P_1	P_2	P_2	P_2	P_4	P_3	P_3	P_3	P_1	P_1	P_1	P_1

0 1 2 3 4 5 6 7 8 9 10 11 12

Gantt chart for scheduling.

Turn arround time for

$P_1 = 12$; $P_2 = 4 - 1 = 3$; $P_3 = 6$; $P_4 = 1$

Average turn around time

$= \frac{P_1 + P_2 + P_3 + P_4}{4} = \frac{12 + 3 + 6 + 1}{4} = \frac{22}{4} = 5.5$

11. First P_1 issue signal wait (x) at that time process P_2 issue wait (y) and when P_1 issue wait (y) the process P_2 issue wait (x).

So correct sequence is $P(S_x), P(S_y); P(S_x), P(S_Y)$

12. Disk block address = 32 bit

It can address 2^{32} location, i.e. 2^{32} byte

So, size of file is 2^{32}.

14.

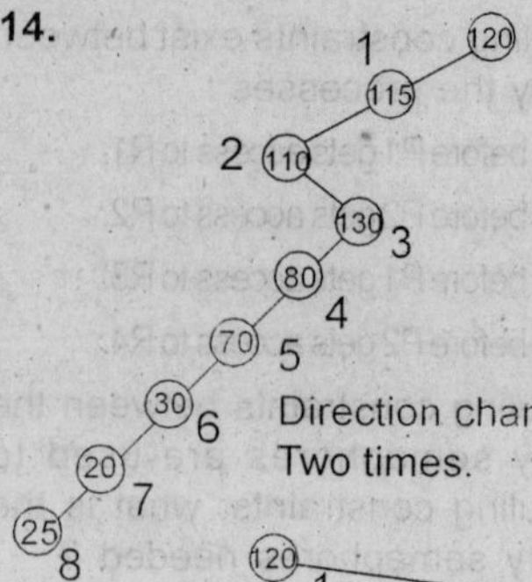

SSTF

Direction change only at points 2 and 3, Two times.

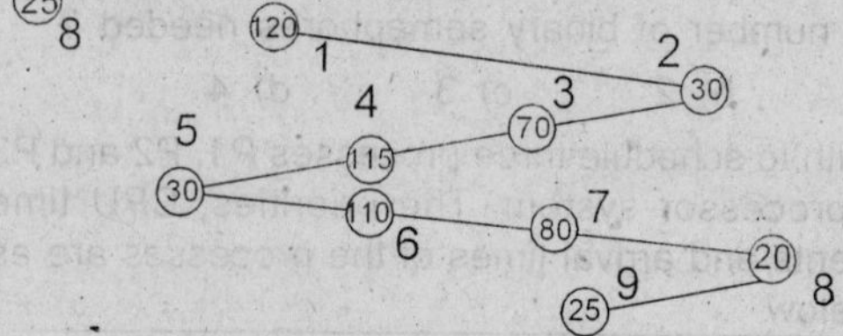

Direction changes at points 2, 5 and 8. Three times.

17. This is the algorithm of solution of consumer-producer process with the help of semaphore.

So, K=P (empty) P for wait

L = V(full) V for signal

M=P(full)

N=V(empty)

21. As size increase, page frame increase. If we apply any replacement rule which is not cause 'belady anamoly' is always reduce page fault.

23. Here, we follow concept of copy on write.

So, we have two different pages one for child and one for parent.

Because value of υ and y are different, so they read (assume value 10),

then child increases to '15' and parent decrement to '5'.

26. (*ln*-s) works as

it links first file with the second file in argument

it works just as windows shortcut

So, with first statement, we have

file 2 $\rightarrow$ file 1

So, contents of file 2 is removed and now it is pointing to file 2

By second command, we got

file 3 $\rightarrow$ file 2 $\rightarrow$ file 1

So, finally only contents we have are of file 1.

27.

find	name	passwd	print
↑	↑	↑	↑
for current directory	option for specify what we have to match here (-name means match name)	(name of) file	What to do when matched (Print on unix) (Only name is printed)

Same work is performed by *ls* password

28. Remember that only two signals that can't be catch by 440 processor

SIGKILL

SIGTOP

Sigkill is signal which is generated by Ctrl-C is passed.

29. Here P1 is always 'signal' first. So there is only way that P1 goes to starve is on P(X) wait (X) become waiting occurs only in wait (P) when 'X' have value zero waiting is done by P2 not P1. It may be possible for P1 that after signal V(X) context switch occur and P2 gets chance after exiling wait P(u) it is blocked in that case P1 is starve on condition P(X).

30. Consider here possible solution

X and Y = 0

P1	P2
Wait (X)	Use P1
Use R1	Signal (X)
Use R1	Wait (Y)
Signal (Y)	Use R2
Wait (X)	Use R3
Use R3	Signal (Y)
Use R4	Wait (Y)
Signal (Y)	Use R4

31. For non peemptive

total time = 45 - 3 (arrive time) = 42

15 sec	20 sec	10 sec

P_3 25 P_1 35 P_2 45

For preemptive

P3	P2	P1	P2

3 5 25 33

33-3=30

32. Notice here conflict occurs in placément of block 55.

Because 3 is come first and never reuse so in set 3 this is only victim for replacement.

33. Consider cases why option A and B are wrong.

What A says

$T_{11} > T_{21}$

$T_{12} > T_{22}$

P_1	P_2
required R_1	required R_1
required R_2	required R_2

Assume that P_1 is running and after a context switch P_2 goes to running that it snatch resource from P_1, because for R_2 it have high priority both process not required another resource. Hence deadlock.

34. Size of files are

	100 bytes block	Block need information	For	200 bytes needed	Block
11050	111	444-5	56	224	2
4990	50	200-2	25	100	1
5170	52	208-3	26	104	1
12640	127	508-6	64	256	2
35600	340	16		171	6

Please note here is that a block can be used to store either data or a single file or (book=keeping information for a single file).

In other words, a block can't be used to store both data and book-keeping and block can't be used to store data for two or more file or book keeping information for two or more file.

Unit - IX

Data Base Management Systems

UNIT - IX DATA BASE MANAGEMENT SYSTEMS

I. RELATIONAL MODEL

- Elements of Relational Model
- Characterstics of Relations
- Relational Constraints & Integrity Constraints
- Entity-Relationship Model
- Relational Algebra & Relational Calculas

1. Relational Model Concepts

The relational model represents the database as a collection of relations. Informally, each relation resembles a table of values or to some extent,a "flat" file of records. When a relation is thought of as a table of values, each row in the table represents a collection of related data values. In the relational model, each row in the table represents a fact that typically corresponds to real-world entity or relationship. The table name and column names are used to help in interpreting the meaning of the values in each row. All values in a column are of the same data type.

In the formal relational model terminology, a row is called a tuple, a column header is called an attribute, and the table is called a relation. The data type describing the types of values that can appear in each column is called a domain. We now define these terms domain, tuple, attribute, and relation-more precisely.

2. Domains, Attributes, Tuples, and Relations

A domain D is a set of atomic values. By atomic we mean that each value in the domain is indivisible as far as the relational model is concerned. A common method of specifying a domain is to specify a data type from which the data values forming the domain are drawn. It is also useful to specify a name for the domain, to help in interpreting its values. Some examples of domains follow

A relation schema R, denoted by R $(A_1, A_2, A_3, \ldots A_n)$, is made up of a relation name R and a list of attributes $A_1, A_2, \ldots A_n$. Each attribute A_i is the name of a role played by some domain D in the relation schema R. D is called the domain of A_i and is denoted by dom (A_i).

A relation schema is used to describe a relation; R is called the name of this relation. The degree of a relation is the number of attributes n of its relation schema.

An example of a relation schema for a relation of degree 7, which describes university students, is the following :

STUDENT (Name, SSN, HomePhone, Address, OfficePhone, Age, GPA)

Relation name — Attributes — Tuples

STUDENT	Name	SSN	HomePhone	Address	OfficePhone	Age	GPA
	Benjamin Bayer	305-61-2435	373-1616	2918B luebonnet lane	null	1 9	3.21
	Katherine Ashly	381-62-1245	375-4409	125 Kirby Road	null	18	2.89
	Dick Davidson	422-11-2320	null	3452 Elgin Road	749-1253	25	3.53
	Charles Cooper	489-22-1100	376-9821	265 Lark Lane	749-6492	28	3.93
	Barbara Benson	533-9-1238	839-8461	7384 Fontana Lane	null	19	3.25

Fig. 1.1 The attributes and tuples of a relation STUDENT

For this relation schema, STUDENT is the name of the relation, which has seven attributes. We can specify the following previously defined domains for some of the attributes of the STUDENT relation: dom(Name) = Names; dom(SSN) = Social_security_numbers ; dom (HomePhone) = Local_phone_numbers, dom (OfficalPhone) = Local_phone_numbers, and dom(GPA) = Grade_point_averages.

3. Relational Model Notation

We will use the following notation in our presentation:

- A relation schema R of degree n is denoted by R=(A1, A2,43 ,..........An)
- An n-tuple t in a relation r(R) is denoted by $t = < v_1, v_2, \ldots\ldots v_n >$, where v_i is the value corresponding to attribute A_i. The following notation refers to component values of tuples.
- Both $t[A_i]$ and $t.A_i$ refer to the value v_i in t for attributeA_i.
- Both $t[A_u, A_w, \ldots, A_z]$ and $t(A_u, A_w, \ldots\ldots A_z)$, where $A_u, A_w, \ldots, A_z$ is a list of attributes from R, refer to the subtuple of values $< V_u, V_w, \ldots\ldots V_z >$ from t corresponding to the attributes specified in the llist.
- The letters Q,R,S denote relation names
- The q,r,s denote relation states.
- The letter t,u,v, denote tuples.
- In general, the name of a relation schema such as STUDENT also indicates the current set of tuples in that relation--that current relation state--whereas STUDENT(Name,SSN,...) refers only to the relation schema.
- An attribute A can be qualified with the relation name R to which it belongs by using the dot notation R.A--for example, STUDENT. Name or STUDENT.Age. This is

because the same name may be used for two attributes in different relations. However, all attribute names in a particular relation must be distinct.

As an example, consider the tuple t=<'Barbara Benson', 533-69-1238', '839-8491', '7384 Fontana Lane', null, 19, 3.25> from the STUDENT relation in Figure 1.2 ; we have [Name]=<'Barbara Benson'>, and t[SSN,GPA, Age] = <'533-69-1238', 3.25, 19>.

4. Relational Constraints and Relational Database Schemas

Domain Constraints

Domain constraints specify that the value of each attribue A must be an atomic value from the domain dom (A).

The data types associated with domains typically include standard numeric data types for integers (such as short - integer, integer, long - integer) and real numbers (float and double precision float). Characters, fixed - length strings, and variable-length strings are also available, as are date, time, timestamp, and money data types. Other possible domains may be described by a subrange of values from a data type or as an enumerated data type where all possible values are explicitly listed.

Key Constraints and Constraints on Null

A relation is defined as a set of tuples. By definition, all elements of a set are distinct; hence all tuples in a relation must also be distinct. This means that no two tuples can have the same combination of values for all their attributes. Usually, there are other subsets of attributes of a relation schema R with the property that no two tuples in any relation state r of R should have the same combination of values for these attributes. Suppose that we denote one such subset of attributes by SK; then for any two distinct tuples t_1 and t_2 in a relation state r of R, we have the constraint that $t_1[SK] \neq t_2[SK]$

Any such set of attributes SK is called a superkey of the relation schema R. A superkey SK specifies a uniqueness constraint that no two distinct tuples in a state r of R can have the same value for SK. Every relation has at least one default superkey-the set of all its attributes.

A superkey can have redundant attributes, however, so a more useful concept is that of a key, which has no redundancy. A key K of a relation schema R is a superkey of R with the additional property that removing any attribute A from K leaves a set of attributes K' that is not a superkey of R. Hence, a key is a minimal superkey-that is, a superkey from which we cannot remove any attributes and still have the uniqueness constraint hold.

For example, consider the STUDENT relation of Figure 1.1. The attribute set {SSN} is a key of STUDENT because no two student tuples can have the same value for SSN. Any set of attributes that includes SSN-for example, {SSN, Name, Age} - is a superkey. However, the superkey {SSN, Name, Age} is not a key of STUDENT, because removing Name or Age or both from the set still leaves us with a superkey.

The value of a key attribute can be used to identify uniquely each tuple in the relation. For example, the SSN value 305-61-2435 identifies uniquely the tuple corresponding to Benjamin Bayer in the STUDENT relation. Notice that a set of attributes constituting a key is a property of the relation schema; it is a constraint that should hold on every relation state of the schema. A key is derermined from the meaning of the attributes, and the property is time-invariant; it must continue to hold when we insert new tuples in the relation. For example, we cannot and should not designate the Name attribute of the STUDENT relation in Figure 1.1 as a key, because there is no guarantee that two students with identical names will never exist.

In general, a relation schema may have more than one key. In this case, each of the keys is called a candidate key. For example, the CAR relation in Figure 1.2 has two candidate keys; License Number and Engine Serial Number. It is common to designate one of the candidate keys as the primary key of the relation. This is the candidate key whose values are used to identify tuples in the relation. We use the convention that the attributes that form the primary key of a relation schema are underlined, as shown in Figure 1.2. Notice that, when a relation schema has several candidate keys, the choice of one to

CAR	LicenseNumber	EngineSerial Number	Make	Model	Year
	Texas ABC-739	A69352	Ford	Mustang	96
	Florida TVP-347	B43696	Oldsmobile	cutlass	9
	New York MPO-22	X83554	Oldmobile	Delta	95
	California 432-TFY	C43742	Mercedes	190-D	93
	California RSK-629	Y82935	Toyota	Camry	98
	Texas RSK-629	U028365	Jaguar	XJS	98

Figure 1.5 The CAR relation with two candidate keys: License Number and Engine Serial Number.

become primary key is arbitrary; however, it is usually better to choose a primary key with a single attribute or a small number of attributes.

Another constraint on attributes specifies whether null values are or are not permitted. For exmaple, if every STUDENT tuple must have a valid, non-null value for the Name attribute, then Name of STUDENT is constrained to be NOT NULL.

5. Relational Databases and Relational Database Schemas

So far, we have discussed single relations and single relation schemas. A relational database usually contains many relations, with tuples in relations that are related in various ways. In this section we define a relational database and a relational database schema. A relational database schema S is a set of relation schemas

$S = \{R_1, R_2, \ldots\ldots R_m\}$ and a set of integrity constrainst IC. A relational database state DB of S is a set of relation states DB $= \{r_1, r_2, \ldots\ldots r_m\}$ such that each r_i is a state of R_i and such that the r_i relation states satisfy the integrity constraints specified in IC. Figure 1.3 shows a relational database schema.

EMPLOYEE

FNAME	MINIT	LNAME	SSN	BDATE	ADDRESS	SEX	SALARY	SUPERSSN	DNO

DEPARTMENT

DNAME	DNUMBER	MGRSSN	MGRSTARTDATE

DEPT_LOCATIONS

DNUMBER	DLOCATION

PROJECT

PNAME	PNUMBER	PLOCATION	DNUM

WORKS ON

ESSN	PNO	HOURS

DEPENDENT

ESSN	DEPENDENT_NAME	SEX	BDATE	RELATIONSHIP

Figure 1.3 Schema diagram for the COMPANY relational database schema; the primary keys are underlined.

Each relational DBMS must have a Data Definition Language (DDL) for defining a relational database schema. Current relational DBMSs are mostly using SQL for this purpose.

Integrity constraints are specified on a database schema and are expected to hold on every database state of that schema. In addition to domain and key constraints, two other types of constraints are considered part of the relational model: entity integrity and referential integrity.

6. Entity Integrity, Referential Integrity, and Foreign Keys

The entity integrity constraint states that no primary key value can be null. This is because the primary key value is used to identify individual tuples in a relation; having null values for the primary key implies that we cannot identify some tuples. For example, if two or more tuples had null for their primary keys, we might not be able to distinguish them.

Key constraints and entity integrity constraints are specified on individual relations. The referential integrity constraint is specified between two relations and is used to maintain the consisitency among tuples of the two relations. Informally, the referential integrity constraint states that a tuple in one relation that refers to another relation must refer to an existing tuple in that relation.

To define referential integrity more formally, we first define the concept of a forgeign key. The conditions for a foreign key, given below, specify a referential integrity constraint between the two relation schemas R_1 and R_2. A set of attributes FK in relation schema R_1 is a foreign key of R_1 that references relation R_2 if it satisfies the following two rules :

1. The attributes in FK have the same domain(s) as the primary key attributes PK of R_2; the attributes FK are said to reference or refer to the relation R_2.

2. A value of FK in a tuple t_1 of the current state $r_1(R_1)$ either occurs as a value of PK for some tuple t_2 in the current state $r_2(R_2)$ or is null. In the former case,we have $t_1[FK] = t_2[PK]$, and we say that the tuple t_1 references or refers to the tuple t_2. R_1 is called the referencing relation and R_2 is the referenced relation.

In a database of many relations, there are usually many referential integrity constraints. To specify these constraints, we must first have a clear understanding of the meaning or role that each set of attributes plays in the various relation schemas of the database. Referential integrity constraints typically arise from the relationships among the entities represented by the relation schemas.

7. Entity Relationship (E - R) Model

E-R model grew out of the excercise of using commercially available DBMS to model application database. Earlier DBMS were based on hierarchical and network approach. E-R is a generalisation of these models. Although it has some means of describing the physical database model, it is basically useful in the design of logical database model. This analysis is then used to organise data as a relation, normalising relations and finally obtaining a relational database model.

The entity-relationship model for data uses three features to describe data. These are :

1. **Entities** which specify distinct real-world items in an application.

2. **Relationships** which connect entities and represent meaningful dependencies between them.

3. **Attributes** which specify properties of entities and relationships.

We illustrate these terms with an example. A **vendor** supplying items to a company, for example, is an **entity.** The **item** he supplies is another entity. A vendor and an item are related in the sense that a vendor supplies an item. The act of supplying defines a relationship between a vendor and an item. An **entity set** is a collection of similar entities. We can thus define a **vendor set** and an **item set.** Each member of an entity set is described by some attributes. For example, a vendor may be described by the attributes

(vendor code, vendor name, address)

An item may be described by the attributes

(item code, item name)

Relationship also can be characterised by a number of attributes. We can think of the relationship as **supply** between vendor and item entities. The relationship supply can be described by the attributes: (order_no, date of supply).

Relationship Between Entity Sets

The relationship between entity sets may be many-to-many (M:N), one-to-many (1:M), many-to-one (M:1) or one-to-one (1-1).The1:1 relationship between entity sets E_1 and E_2 indicates that for each entity in either set there is at most one entity in the second set that is associated with it.The1:M relationship from entity set E_1 to E_2 indicates that for an occurrence of the entity from the set E_1, there could be zero, one, or more entities from the entity set E_2 associated with it. Each entity in E_2 is associated with at most one entity in the entity set E_1. In the M:N relationship between entity sets E_1 and E_2, there is no restriction as to the number of entities in one set associated with an entity in the other set. The database structure, employing the E-R model is usually shown pictorially using entity-relationship (E-R) diagram.

8. Relational Algebra

Relational algebra is a formal query language based on a set of five basic relational algebraic operations. These operations can be performed on relations. In the relational model, a query regarding data within the database is answered by constructing a new relation from the existing relations by performing a sequence of these basic operations. Two operations, namely, selection and projection are unary operations, i.e. they can be performed on one relation.

Selection Operation

Consider the following relation : **teach**

Name	Address	Course
Zohar	40B, East Behala	MIS
Biswanathan	16/2, Joka	MIS
Zohar	40B, East Behala	PDS
Ashim	B-2, IIMC Campus	DBMS

Example 1.1.

To answer the query: is the teacher with name Johar teaching the course PDS? We may write.

$$\sigma_{\text{Name = "Johar" course = "PDS"}}\text{(teach)}$$

Where σ (sigma) is the selection operation symbol.

$\sigma_F(R)$ is the set of those tuples of R components of each of which make the logical expression F true.

The logical expression F is formed with constants (string, integer, or real), names or positions of attributes of R, arithmetic comparison operators and logical operators as shown in Tables 1.2 and 1.3

Table 1.2 Arithmetic comparison operators

Operation symbol	Function
<	less than
≤	less than or equal to
=	equal to
≠	not equal to
>	greater than
≥	greater than or equal to

Table 1.3 Logical operators

Logical operation symbol	Function
˥	Not (unary)
I	OR (binary) {inclusive}
∧	AND (binary)

Projection Operation

With the help of this operation, any number of columns can be omitted from a table or columns of the table can be rearranged.

Example 1.2.

Consider the query : Print the names and addresses of teachers who are teaching the MIS course.

To answer this, first a selection operation is performed as follows.

$$\sigma_{\text{course}} = \text{"MIS" (teach)}$$

The result is the following new relation

Name	Address	Course
Zohar.	40B, East Behala	MIS
Biswanathan	16/2 Joka	MIS

If we want to get only names and addresses of those teachers, we may write

$$\pi_{1.2}\ (\sigma_{\text{course}} = \text{"MIS" (teach))}$$

where π is the projection operation symbol and the subscripts indicate the positions of the attributes (columns) of the operand (table.)

This operation when performed produces the following relation

Name	Address
Zohar	40B, East Behala
Biswanathan	16/2, Joka

The other three relational algebraic operations are **set union, set difference**, and **cartesian product.**

Set Union

Let n be the arity (number of columns) of the relations R and S. Union of R and S is denoted as R + S . It is the set of tuples, with arity n, which are either in R, or in S, or in both R and S.

Example 1.3.

R:

a	b	c
d	c	a
b	f	e

S:

b	c	a
c	d	f
b	f	e

R + S ≡

a	b	c
d	c	a
b	f	e
b	c	a
c	d	f

Set Difference

If is denoted as R - S It is the set of tuples which are in R but not in S.

Example 1.4.

R - S =

a	b	c
d	c	a

Cartesian Product

It is denoted as $R \times S$. If R and S have arity r and s, respectively, then $R \times S$ is the set of tuples with arity $r + s$, whose first r components are composed of the tuples in R and remaining s components are composed of the tuples in S.

Example 1.5.

R × S =

a	b	c	b	c	a
a	b	c	c	d	f
a	b	c	b	f	e
d	c	a	b	c	a
d	c	a	c	d	f
d	c	a	b	f	e
b	f	e	b	c	a
b	f	e	c	d	f
b	f	e	b	f	e

Additional Operations

Besides the five operations there are additional relational algebraic operations. They are **Natural join, Intersection,** Θ - join and Quotient operations.

Natural Join

It is denoted as $R \bowtie S$. It is applicable only when there is at least one attribute common in R and S.

To compute $R \bowtie S$, we will

1. Compute $R \times S$
2. Select those tuples from $R \times S$ whose components corresponding to the common attributes have the same values.
3. Project out the columns corresponding to the attributes which have been considered once.

Example 1.6.

A	B	C
b	c	e
f	g	h
a	d	d
a	c	e

R

B	C	D
d	d	f
g	h	k
g	h	i
c	m	k

S

A	B	C	D
f	g	h	k
f	g	h	i
a	d	d	f

$R \bowtie S$

Usage

Example 1.7.

Consider the following schema of a relational data base

Employee (emp_no, name, address)

Project (p_no, p_name)

Work_on (emp_no, p_no)

Part (part_no, part_name, qty_on_hand, size)

Use(emp_no, P_no, part_no, number)

Let us try to answer the following queries:

1. Print the names of the employees who are working on a project named "DBMS"

2. Print the names of the employees who are not working in any project

3. List the names of the employees and the projects with which the employees are associated with and for which they have used no part so far.

4. List the names of the projects such that every employee working on these projects has used a part "bolts" with size 6.

5. List the part numbers and names of the parts used in both the projects "DBMS" and "MIS"

6. List the names of the employees who are working in both the projects "DBMS" and "MIS"

Answers

1. π_{name} (employee $\bowtie$ (σ_{p_name} = "DBMS" (project)) $\bowtie$ work_on)

2. π_{name} (employee) - π_{name} (employee $\bowtie$ work_on

3. $\pi_{name', \; p\_name}$ ((work_on - $\pi_{1,2}$(use)) $\bowtie$ project $\bowtie$ employee)

4. A = $\pi_{emp\_no.\, p\_no}$ ($\sigma_{part\_name = \text{"bolts"} \; size = 6}$(part) $\bowtie$ use)

{(emp_no, proj_no) pairs related to the part "bolts" with size "6"}

B = work_on $\bowtie$ $\pi_{p\_no}(A)$

{(emp_no, p_no) pairs such that at least one employee who is associated with the project has used the part "bolts" with size "6" for the project)}

C = B - A

{(emp_no, p_no) pairs not related to "bolts"}

D= π_{p_no} (A) π_{p_no} (C)

{project nos of the projects such that every one working there has used "bolts")

N= π_{p_name} (project $\bowtie$ D)

5. $\pi_{part\_no,\, part\text{-}name}$ (part $\bowtie$ ($\pi_{part\text{-}no}$ (σ_{p_name}= "DBMS" (project) $\bowtie$ use) $\cap$ $\pi_{part\text{-}no}$ (σ_{p_name}= "MIS" (project) $\bowtie$ use))

6. A= π_{p_no} (σ_{p_name}= "MIS" I $_{p\_name}$= "DBMS" (project)

B= π_{e_no} (work-on)- π_{e_no} ((π_{e_no} (work_on) $\times$ A) - work_on)

C= π_{name} (employee $\bowtie$ B)

Intersection

It is denoted as $R \cap S$. It is the set of tuples which are both in R and S

Example 1.8.

b	f	e

$R \cap S$

Quotient

Let R and S be two relations with arity r and s respectively. The quotient operation which is denoted as ÷ can be performed on R and S, if

I. $r > s$ and

II. The last s attributes of R are the attributes of S arranged in the same order. R ÷ S is the set of tuples (with arity r-s) each of whose remaining s components in R form the entire relation S.

Example 1.9.

A	B	C	D
b	c	e	f
a	b	i	j
b	c	g	h
b	c	a	d
d	i	g	h
d	i	j	k
d	i	e	f

R

C	D
e	f
g	h

S

A	B
b	c
d	i

R ÷ S

Here $R \div S = \pi_{1,2}(R) - \pi_{1,2}(\pi_{1,2}(R) \times S - R)$

In general,

$R \div S = \pi_{1,2,\ldots(r-s)}(R) - \pi_{1,2,\ldots(r-s)}(\pi_{1,2,\ldots(r-s)}(R) \times S - R)$

If $\pi_{1,2,\ldots(r-s)}(R) \times S \subseteq R$ then R is said to be divisible by S and R ÷ S can be computed as $\pi_{1,2,\ldots(r-s)}(R)$.

Usage

Consider the following two relations

Name	address	flight_no	date
n_1	a_1	IC271	25.10.87
n_1	a_1	IC272	28.10.87
n_2	a_2	IC561	25.10.87
n_2	a_2	IC562	5.11.87
n_3	a_3	IC271	26.10.87
n_3	a_3	IC272	28.10.87
n_4	a_4	IC271	25.10.87
n_4	a_4	IC272	28.10.87
n_4	a_4	IC561	25.10.87
n_4	a_4	IC562	5.11.87

Booked_on

And

flight_no	date
IC271	25.10.87
IC272	28.10.87

S

Booked_on ÷ S gives names addresses of passengers who have booked on both the departures: flight IC271 on 25.10.87 and flight IC272 on 28.10.87

Example 1.10.

Consider the following two queries regarding data in the database of Example 1.7

1. List the names of the employees who are working in both the projects "DBMS" and "MIS"

2. List the names of the employees who are participating in every project and have used every part

Answer to these queries are given below by using quotient operator

1. π_{name} (employee ⋈ (work_on ÷ π_{p_no} (σ_{p_name} = " MIS" | p_name = " DBMS" (project)))

2. π_{name} (employee ⋈ (((work_on) ÷ π_{p_no} (project) ∩ ($\pi_{emp\_no,\ part\_no}$ (use)) ÷ π_{part_no} (part))))

Θ - join

It is denoted as R ⋈ S, where i and j are, respectively, a column of R and a column of S and Θ is one of the arithmetic comparison operation symbols.

R ⋈ S is the set of those tuples (with arity r + s) of R × S whose components corresponding to columns i and j satisfy the logical operator Θ

11. Relational Calculus

Relational calculus is an alternative to relational algebra. In contrast to the algebra, which is procedural, the calculus is nonprocedural, or declarative, in that it allows us to describe the set of answers without being explicit about how they should be computed. Relational calculus has had a big influence on the design of commercial query languages such as SQL and, especially, Query-by-Example (QBE).

The variant of the calculus that we present in detail is called the tuple relational calculus (TRC). In another variant, called the domain relational calculus (DRC), the variables range over field values. TRC has had more of an influence on SQL, while DRC has strongly influenced QBE.

We present a number of sample queries using the following schema.

Sailors(<u>sid: integer,</u> sname: string, rating: integer, age: real)

Boats(<u>bid: integer,</u> bname: string, color: string)

Reserves(<u>sid: integer, bid: integer, day: dates</u>)

The key fields are underlined, and the domain of each field is listed after the field name. Thus sid is the key for Sailors, bid is the key for Boats, and all three fields together form the key for Reserves. Fields in an instance of one of these relations will be referred to by name, or positionally, using the order in which they are listed above.

In several examples, we will use the instances S1 and S2 (of Sailors) and R1 (of Reserves) shown in Figures 1.18, 1.19, and 1.20, respectively.

Sid	Sname	Rating	Age
22	Dustin	7	45.0
31	Lubber	8	55.5
58	Rusty	10	35.0

Figure 1.18 An Instance S1 of Sailors

Sid	sname	rating	age
28	yuppy	9	35.0
31	lubber	8	55.5
44	guppy	5	35.0
58	rusty	10	35.0

Figure 1.19 An Instance S2 of Sailors

Sid	Bid	Day
22	101	10/10/96
58	103	11/12/96

Figure 1.20 An Instance R1 of Reserves.

Tuple Relational Calculus

A **tuple variable** is a variable that takes on tuples of a particular relational schema as values. That is, every value assigned to a given tuple variable has the same number and type of fields.

A tuple relational calculus query has the form {T | p (T)}, where T is a tuple variable and p(T) denotes a formula that describes T; we will shortly define formulas and queries rigorously. The result of this query is the set of all tuples t for which the formula p(T) evaluates to true with T = t. The language for writing formulas p(T) is thus at the heart of TRC and is essentially a simple subset of first-order logic. As a simple example, consider the following query.

1. Find all sailors with a rating above 7.

{S | S ∈ Sailors ∧ S. rating > 7 }

This query can be evaluated on an instance of the Sailors relation. The tuple variable S is instantiated successively to each tuple in this instance of Sailors, and the test S. rating > 7 is applied. The answer contains those instances of S that pass this test.

Syntax of TRC Queries

We now define these concepts formally, beginning with the notion of a formula. Let Rel be a relation name, R and S be tuple variables, a an attribute of R, and b an attribute of S. Let op denote an operator in the set {<, >, =, ≤, ≥, ≠}. An **atomic formula** is one of the following

- R ∈ Rel
- R.a op S.b
- R.a op constant, or constant op R.a

A **formula** is recursively defined to be one of the following, where p and q are themselves formulas, and p(R) denotes a formula in which the variable R appears:

- any atomic formula
- ¬p, p ∧ q, p ∨ q, or p ⇒ q
- ∃R(p(R)), where R is a tuple variable
- ∀ R(p(R)), where R is a tuple variable

In the last two clauses above, the **quantifiers** ∃ and ∀ are said to **bind** the variable R. A variable is said to be **free** in a formula or subformula (a formula contained in a larger formula) if the (sub) formula does not contain an occurrence of a quantifier that binds it.

R ∈ Rel gives R the type of tuples in Rel, and comparisons such as R.a op S.b and R.a op constant induce type restrictions on the field R.a. If a variable R does not appear in an atomic formula of the form R ∈ Rel (i.e., it appears only in atomic formulas that are comparisons), we will follow the convention that the type of R is a tuple whose fields include all (and only) fields of R that appear in the formula.

A **TRC query** is defined as an expression of the form { T | p(T)}, where T is the only free variable in the formula p.

Semantics of TRC Queries

What does a TRC query mean? More precisely, what is the set of answers for a given TRC query? The **answer** to a TRC query {T | p(T)}, as we noted earlier, is the set of all tuples t for which the formula p(T) evaluates to **true** with variable T assigned the tuple value t. To complete this definition, we must state which assignments of tuple values to the free variables in a formula make the formula evaluate to **true**.

A query is evaluated on a given instance of the database. Let each free variable in a formula F be bound to a tuple value. For the given assignment of tuples to variables, with respect to the given database instance, F evaluates to (or simply 'is') true if one of the following holds:

- F is an atomic formula R ∈ Rel, and R is assigned a tuple in the instance of relation Rel.
- F is a comparison R.a op S.b, R.a op constant, or constant op R.a, and the tuples assigned to R and S have field values R.a and S.b that make the comparison true.
- F is the form ¬p, and p is not **true**; or of the form p ∧ q, and both p and q are **true**; or of the form p ∨ q, and one of them is **true**, or of the form p ⇒ q and q is true whenever p is true.
- F is of the form ∃R(p(R)), and there is some assignment of tuples to the free variables in p(R), including the variable R, that makes the formula p(R) true.
- F is of the form ∀ R(p(R)), and there is some assignment of tuples to the free variables in p(R) that makes the formula p(R) true no matter what tuple is assigned to R.

Example of TRC Queries

We now illustrate the calculus through several examples. We will use parentheses as needed to make our formulas unambiguous. Often, a formula p(R) includes a condition R ∈ Rel, and the meaning of the phrases some tuple R and for all tuples R is intuitive. We will use the notation ∃ R ∈ Rel(p(R)) for ∃R(R ∈ Rel ∧ p(R)). Similarly, we use the notation ∀ R ∈ Rel (p(R)) for ∀ R(R ∈ Rel ⇒ p(R)).

2. Find the names and ages of sailors with a rating above 7.

{S/ ∃ S1 ∈ Sailors (S1.rating >7 ∧ S.name =S1.S name ∧ S . age =S1.age}

This query illustrates a useful convention: S is considered to be a tuple variable with exactly two fields, which are called name and age, because these are the only fields of S that are mentioned and S does not range over any of the relations in the query; that is, there is no subformula of the form S ∈ Relname. The result of this query is a relation with two fields, name and age. The atomic formulas S.name = S1.sname and S.age = S1.age give values to the fields of an answer tuple S.

3. Print the sailor name, boat id, and reservation date for each reservation.

{ P | ∃ R ∈ Reserves(∃ S ∈ Sailors (R.sid = S.sid ∧ P.bid= R.bid ∧ P.day = S.day ∧ P.sname = S.sname))}

For each Reserves tuple, we look for a tuple in Sailors with the same sid. Given a pair of such tuples, we construct an answer tuple P with fields sname, bid and day by copying the corresponding fields from these two tuples. This query illustrates how we can combine values from different relations in an answer tuple.

4. Find all sailors with a rating above 7 who have reserved boat 103.

{S/S ∈ Sailors ∧ S.rating > 7
∧ ∃ R ∈ Reserves(R.sid=S.sid ∧ R.bid = 103)}

This query can be read as follows: "Retrieve all sailor tuples with a rating greater than 7 for which there exists a tuple in Reserves, having the same value in the sid field, and with bid=103". That is, for each sailor tuple with rating greater than 7, we look for a tuple in Reserves that shows that this sailor has reserved boat 103.

5. Find all sailors with a rating above 7 who have reserved a red boat.

(S/S ∈ sailors ∧ S.rating >7
∧ ∃ R ∈ Reserves(R.sid = S.sid
∧ ∃ B ∈ Boats(B.bid = R.bid ∧ B.color = 'red'))}

This query can be read as follows: "Retrieve all sailor tuples S with a rating greater than 7 for which there exist tuples R in Reserves and B in Boats such that S.sid = R.sid, R.bid = B.bid and B. color = 'red' ". The three tuples S,R and B can be regarded as a 'proof-by-example' that sailor S has reserved a red boat. Another way to write this query, which corresponds more closely to this reading, is as follows:

{S\S ∈ Sailors(S . rating >7 ∧ ∃R ∈ Reserves ∃B ∈ Boats
(R.sid = S.sid ∧ B.bid = R.bid ∧ B.color = 'red')}

Find sailors who have reserved at least two boats.

{S\S∈Sailors ∧ ∃R1∈ Reserves ∃R2 ∈ Reserves
(S.sid = R1.sid ∧ R1.sid = R2.sid ∧ R1.bid ≠ R2.bid)}

Contrast this query with the algebra version and see how much simpler the calculus version is. In part, this difference is due to the cumbersome renaming of fields in the algebra version, but the calculus version really is simpler.

7. Find sailors who have reserved all boats.

{S\S∈Sailors ∧ ∀B∈Boats
(∃R∈Resrves(S.sid = R.sid ∧ R.bid=B.bid))}

This query was expressed using the division operator in relational algebra. Notice how easily it is expressed in the calculus. The calculus query directly reflects how we might express the query in English. "Find sailors S such that for all boats B, there is a Reserves tuple showing that sailor S has reserved boat B".

8. Find sailors who have reserved all red boats

{S\S ∈ Sailors ∧ ∀ B ∈ Boats
(B.color ≠'red' ⇒(∃ R ∈ Resrves(S.sid = R.sid ∧ R.bid=B.bid))) }

For each candidate (sailor), if a boat is red, the sailor must have reserved it.That is, for a candidate sailor, a boat being red must imply the sailor having reserved it. We can writed this query without using implication, by observing that an expression of the form p ⇒ q is logically equivalent to ¬p ∨ q.{S\S∈Sailors ∧ ∀B∈ Boats
(B.color≠'red' ∨ (∃R ∈ Resrves(S.sid =
R.sid ∧ R.bid=B.bid)))}

This query should be read as follows:"Find sailors S such that for all boats B, either the boat is not red or a Reserves tuple shows that sailorShas reserved boatB".

Domain Relational Calculus.

A **domain variable** is a variable that ranges over the values in the domain of some attribute (e.g., the variable can be assigned an integer if it appears in an attribute whose domain is the set of integers). A DRC query has the form $\{\langle x_1, x_2, \ldots, x_n\rangle \mid p\langle x_1, x_2, \ldots, x_n\rangle\}$, where each x_i is either a domain variable or a constant and $p\langle x_1, x_2 \ldots x_n\rangle$ denotes a **DRC formula** whose only free variables are the variables among the x_i, $1 \le i \le n$. The result of this query is the set of all tuples $\langle x_1, x_2, \ldots, x_n\rangle$ for which the formula evaluates to **true**.

A DRC formula is defined in a manner that is very similar to the definition of a TRC formula. The main difference is that the variables are now domain variables. Let op denote an operator in the set n {<, >, =, ≤, ≥, ≠} and let X and Y be domain variables. An atomic formula in DRC is one of the following:

- $\langle x_1, x_2 \ldots x_n\rangle$ ∈ Rel where Rel is a relation with n attributes; each x_i, $1 \le i \le n$ is either a variable or a constant.
- X op Y
- X op constant, or constant op x

A formula is recursively defined to be one of the following, where p and q are themselves formulas, and p (x) denotes a formula in which the variable X appears:

- Any atomic formula
- ¬p, p ∧ q, p ∨ q, p ⇒ q
- ∃ X(p(X)), where X is a domain variable
- ∀ X(p(X)), where X is a domain variable

Examples of DRC Queries

We now illustrate DRC through several examples. The reader is invited to compare these with the TRC versions.

9. Find all sailors with a rating above 7.

$\{\langle I, N, T, A\rangle / \langle I, N, T, A\rangle \in Sailors \wedge T > 7\}$

This differs from the TRC version in giving each attribute a (variable) name. The condition $\langle I, N, T, A\rangle \in$ Sailors ensures that the domain variables I, N, T and A are restricted to be fields of the same tuple. Thus in comparison with the TRC query, we can say T > 7 instead of R. rating > 7, but we must specify the tuple $\langle I, N, T, A\rangle$ in the result, rather than just R.

10. Find all sailors with a rating above 7 who have reserved boat 103.

$\{\langle I, N, T, A\rangle / \langle I, N, T, A\rangle \in \text{Sailors} \wedge T > 7$
$\wedge \exists\, Ir, Br, D\, (\langle Ir, Br, D\rangle \in Reserves \wedge Ir = I \wedge Br = 103)\}$

Note that we use the notation ∃Ir, Br, D (...) as a shorthand for ∃Ir (∃Br(∃D(....))). Very often, all the quantified variables appear in a single relation, as in this example. An even more compact notation in this case is ∃ ⟨Ir,Br,D⟩ ∈Reserves. With this notation, which we will use henceforth, the above query would be as follows:

$\langle I, N, T, A\rangle \mid \langle I, N, T, A\rangle \in \text{Sailors} \wedge T > 7$
$\wedge \exists \langle Ir, Br, D\rangle \in Reserves (Ir = I \wedge Br = 103)\}$

The comparison with the corresponding TRC formula should now be straight forward. This query can also be written as follows; notice the repetition of variable I and the use of the constant 103:

$\langle I,N,T,A\rangle \mid \langle I,N,T,A\rangle \in Sailors \wedge T > 7$

$\wedge \exists D \langle I,103,D\rangle \in Reserves)\}$

11. Find the names of all sailors with a rating above 7 who have reserved a red boat.

$$\{\langle N\rangle / \exists I,T,A(\langle I,N,T,A\rangle \in Sailors \wedge T > 7$$
$$\wedge \exists \langle Ir,Br,D\rangle \in Reserves(Ir = I$$
$$\wedge \exists \langle B,BN,'red'\rangle \in Boats(B = Br)))\}$$

Notice that only the sname field is retained in the answer and that only N is a free variable.

12. Find sailors who have reserved at least two boats.

$\langle I,N,T,A\rangle \mid \langle I,N,T,A\rangle \in Sailors \wedge$

$\exists.\ \langle I,Br1,D1\rangle \in Reserves\ \ \exists\langle I,Br2,D2\rangle \in Reserves$

$(Br1 \neq Br2)$

Notice how the repeated use of variable I ensures that the same sailor has reserved both the boats in question.

OBJECTIVE QUESTIONS - I

1. Which one is not a traditional set operator defined on relational algebra
   a) Union b) Intersection c) Difference d) Join
2. Consider the Join of a relation R with a relation S. If R has m tuples and S has ntuples ,then the maximum and minimum sizes of the Join respectively are: **[GATE-1999]**
   a) m+n and 0 b) mn and 0
   c) m+n and $|m-n|$ d) mn and m+n
3. Given two union compatible relations $R_1(A,B)$ and $R_2(C,D)$, what is the result of the operation
   $R_1 A=C\ AB=D\ R_2$? **[GATE-1998]**
   a) $R_1 \cup R_2$ b) $R_1 \times R_2$ c) $R_1 - R_2$ d) $R_1 \cap R_2$
4. An instance of a relation schema R(A,B,C) has distinct values for attribute A. can you conclude that A is a candidate key for R? **[GATE-1994]**
   a) True b) False
5. Give a relational algebra expression using only the minimum number of operators from ($\cup$,-) which is equivalent to $R \cap S$.
   [GATE-1994]
   a) True b) False
6. In airline reservation system, the entities are date, flight number, place of departure, destination, type of plane and seats available. The primary key is
   a) flight number b) flight number + destination
   c) flight number + place of departure
   d) flight number + date
7. Which of the following is/are record based logcal model(s)?
   a) Network model b) Object-Oriented Model
   c) Relational Model d) Entity – relationship model
8. The set of permitted values for each attribute is called its
   a) Attribute set b) Attribute range
   c) Domain d) Group
9. Choose the incorrect statements.
   a) In network model, data is represented by a collection of records,and relationship among data are represented by links.
   b) In hierarchical model, data and relationships among data are represented by records and links respectively
   c) In hierarchical model , the records are organized as acollection of arbitrary graphs
   d) In network model,the records are organized as a collection of trees.
10. Student and courses enrolled, is an example of
    a) One-to-one relationship b) One-to-many relationship
    c) Many-to-one relationship d) Many-to-many reation ship
11. Relations produced from an E-R model will always be in
    a) First normal form b) Second normal form
    c) Third normal form d) Fourth normal form
12. E-R modeling technique is a
    a) Top-down approach b) Bottom-up approach
    c) Left-right approach d) None of the above
13. Redundancy is dangerous as it is a potential threat to data
    a) Integrity b) Consistency
    c) Sufficiency d) None of the above
14. Who of the following is more concerned about the level of the DBMS?
    a) DBA b) The end user
    c) The systems programmer d) Client
15. An attribute of one table matching the primary key of another table, is called as
    a) Foreign key b) Secondary key
    c) Candidate key d) Composite key
16. A primary key, if combined with a foreign key creates
    a) Parent child relationship between the tables that connect them
    b) Many-to-many relationship between the tables that connect them
    c) Network model between the tables that connect them
    d) None of the above
17. Choose the correct statement
    a) Network models are complicated by physical keys,but the relational model is faster because it uses logical keys
    b) Network models are complicated by logical keys, but the relational model is faster because it uses physical keys
    c) Network models are complicated by logical keys, but the relational model is slower because it uses physical keys.
    d) Network models are complicated by physical keys, but the relational model is slower because it uses logical keys

18. The employee salary should not be greater than Rs.2000. This is
a) Integrity constraint b) Referential constraint
c) Over-defined constraint d) Feasible constraint

19. Manager's salary details are hidden from the employee. This is
a) Conceptual level data hiding
b) Physical level data hiding
c) External level data hiding
d) None of the above

20. A trigger is
a) A statement that enables to start any DBMS
b) A statement that is executed by the user when debugging an application program
c) A condition the system tests for the validity of the database user.
d) A statement that is executed automatically by the system as a side effect of a modification to the database.

The next two Questions are based on the following information.
Entity set TRANSACTION has the attributes transaction number, date, amount.
Entity set ACCOUNT has the attributes account number, customer name, balance.

21. Identify the discriminator of the weak entity.
a) Account number b) Transaction number
c) {Account number, date} d) date

22. Identify the primary key of the weak entity
a) Account number
b) { Account number,transaction number}
c) {Account number,date}
d) {Transaction number, date}

23. In a relational schema, each tuple is divided into fields called
a) Relations b) Domains
c) Queries d) All of the above

24. Generally speaking, for a weak entity set to be meaningful it must be part of a
a) One-to-One relationship
b) One-to-Many relationship
c) Many-to-Many relationship
d) None of the above

25. A top-to-bottom relationship among the items in a database is established by a
a) Hierarchical schema b) Network schema
c) Relational Schema d) All of the above

26. A logical schema
a) Is the entire database.
b) Is a standard way of organizing information into accessible parts
c) Describes how data is actually stored on disk
d) All of the above

27. A database management system
a) Allows simultaneous access to multiple files
b) Can do more than a record management system
c) Is a collection of programs for managing data in a single file.
d) Both (a) and (b)

28. A network schema
a) Restricts the structure to a one-to-many relationship
b) Permits many-to-many relationships
c) Stores data in tables d) None

29. What is (are) the serious problem(s) of file management systems?
a) Data redundancy b) Difficult to update
c) Program dependence d) All of the above

30. The data dictionary tells the DBMS
a) What files are in the database
b) What attributes are possessed by the data
c) What these files contain d) All of the above.

31. The level of data abstraction which describes how the data is actually stored is
a) Physical level b) Conceptual level
c) Storage level d) File level

32. The level of abstraction describes only part of the entire database is called
a) Physical level b) Conceptual level
c) View level d) Procedural level

33. The overall logical structure of a database can be expressed graphically by
a) Data Flow chart b) Flow chart
c) Directed Graph
d) Entity-Relationship diagram

34. A relational model which allows non-atomic domains is......
a) Nested relational data model
b) Non-atomic data model
c) Hierarchical data model d) None

35. In a relational database a referential integrity constraint is specified with the help of
a) Primary key b) Foreign key
c) Secondary key d) None

36. The data model in which data is represented by collection of records and relationship among data are represented by links, is called
a) Network data model b) Hierarchical data model
c) Relational data model d) None

37. In a network model, the presence of the following create certain problems
a) Cycles b) Duplication of data
c) Weak entities d) Many-to-many relation

38. Which one of the following is an example for network database
a) IDMS b) Oracle c) Ingress d) DBMS

39. A hierarchical data model is
a) Totally ordered b) Partially ordered
c) Not ordered d) None

40. Any relational calculus expression can
a) Be expressed using relational algebra
b) Not always be expressed using relational algebra
c) Be decomposed d) None

41. The operation of a relation x, produces Y such that Y contains only selected attributes of X. Such an operation is....
a) Projection b) Intersection
c) Union d) Difference

42. The following entities/attributes in a relational database should not have null values:
a) Keys b) Relations
c) Variables d) All of these

43. In context Hierarchical data model, which one of the following is false? **[UGC NET-2000]**
a) This data model provides a natural and straight forward method for implementing all types of relationship between record types,whether relationships are one-to-one, one -to - many or many-to-many
b) The relationship between two record types is expressed using links or pointers
c) A virtual record is a machanism to point to an occurrence of aphysical record.
d) In this model, when records to be modified are virtual records, detailed rules have to be specified to keep the database in aconsistent state

44. Identify the false statement? **[UGC NET-2000]**
a) Hierarchical model evolved from file based system
b) Hierarchical model is a superset of Network model
c) Network model is a superset of Hierarchical model
d) Relational model is based on the mathematical model of relation.

45. The ascending order of a data hierarchy is **[UGC NET-1997]**
a) bit-byte-record-field-file-database
b) byte-bit-field-record-file-database
c) byte-bit-record-file-field-database
d) bit-byte-field-record-file-database

46. Data management systems may be implemented as **[UGC NET-1997]**
a) system software b) application software
c) both (a) and (b) d) none of (a) and (b)

47. Which of the following is not a type of DBMS? **[UGC NET-1997]**
a) Network b) Relational
c) Hierarchical d) Modular

48. The database environment has all of the following components except **[UGC NET-1997]**
a) separate files b) users
c) database administrator d) querry language

49. Database management systems are intended to **[UGC NET-1997]**
a) eliminate data redundancy
b) establish relationship among records in different files
c) maintain data integrity
d) manage faster access to data bulk

50. Which of the following is not a type of DBMS? **[UGC NET-1996]**
a) Hierarchial b) Networ
c) Relational d) sequential

51. Which of the following is not a type of databases? **[UGC NET-1996]**
a) Network b) Hierarchial
c) Relational d) Modular

52. Which of the following is not one of the fundamental approaches to the design of DBMS software? **[UGC NET-1995]**
a) Hierarchical b) Circle
c) Relational d) Network

53. In a relationship database model a tripple is equivalent to a **[UGC NET-1998]**
a) data item b) file
c) field d) record

ANSWER KEY - I

1. d	**2.** b	**3.** d	**4.** b	**5.** a
6. d	**7.** a,c	**8.** c	**9.** c,d	**10.** d
11. c	**12.** a	**13.** a,b	**14.** a	**15.** a
16. a	**17.** a	**18.** a	**19.** c	**20.** d
21. b	**22.** b	**23.** b	**24.** b	**25.** a
26. b	**27.** d	**28.** b	**29.** d	**30.** d
31. a	**32.** c	**33.** d.	**34.** a	**35.** b
36. a	**37.** a	**38.** a	**39.** a	**40.** a
41. a	**42.** a	**43.** a	**44.** b	**45.** d
46. c	**47.** d	**48.** a	**49.** a	**50.** d
51. d	**52.** b	**53.** d		

OBJECTIVE QUESTIONS - II

1. With regard to the expressive power of the formal relational Query languages which of the following statements is TRUE? **[GATE-2002]**
a) Relational algebra is more powerful than relational calculus.
b) Relational algebra has the same power as relational calculus.
c) Relational algebra has the same power as safe relational calculus. d) None of these

2. Suppose the adjacency relation of vertices in a graph is represented in a table Adj(x,y). which of the following Queries can not be expressed by a relational algebra expression of constant length? **[GATE-2001]**
a) List all vertices adjacent to a given vertex
b) List all vertices which have self loops.
c) List all vertices which belong to cycles of less than three vertices.
d) List all vertices reachable from a given vertex.

3. Let r and s be two relations over the relation schemes R and S respectively, and let A be an attribute in R. Then the relational algebra expression **[GATE-2001]**

$\sigma_{A=a}(r \bowtie s)$ is always equal to

a) $\sigma_{A=a}(r)$ b) r c) $\sigma_{A=a}(r) \bowtie s$ d) None

4. Which of the following relational calculus expressions is not safe? **[GATE-2001]**

a) $\{t / \exists u \in R_1 (t[A] = u[A]) \wedge -\exists S \in R_2 (t[A] = s[A])\}$

b) $\{t / \neg(t \in R_1)\}$

c) $\{t / \exists u \in R_1 (t[A] = u[A]) \wedge \exists S \in R_2 (t[A] = s[A])\}$

d) $\{t / \forall u \in R_1 (u[A] = "x" \Rightarrow \exists S \in R_2 (t[A] \wedge s[A] = u[A]))\}$

5. Given the relations

employee(name, salary, depno), and

department (depno,depname,address)

Which of the following Queries cannot be expressed using the basic relational algebra operations ($\sigma, \pi, \times, \bowtie, \cup, \cap, -$)? **[GATE-2000]**

a) Department address of every employee

b) Employees whose name is the same as their department name.

c) The sum of all employee salaries.

d) All employees of a given department.

6. The relational algebra expression equivalent to the following tuple calculus expression **[GATE-1999]**

$\{t / t \in r(t[A] = 10 \wedge t[B] = 20)\}$ **is:**

a) $\sigma_{(A=10 \vee B=20)}(r)$ b) $\sigma_{(A=10)}(r) \cup \sigma_{(B=20)}(r)$

c) $\sigma_{(A=10)}(r) \cap \sigma_{(B=20)}(r)$ d) $\sigma_{(A=10)}(r) - \sigma_{(B=20)}(r)$

7. Which of the following Query transformations(i.e., replacing the L.H.S expression by the R.H.S expression) is incorrect? R_1 and R_2 are relations, C_1 and C_2 are selection conditions and A_1 and A_2 are attributes of R_1. **[GATE-1998]**

a) $\sigma_{C1}(\sigma_{C1}(R_1)) \rightarrow \sigma_{C2}(\sigma_{C2}(R_1))$

b) $\sigma_{C1}(\pi_{A1}(R_1)) \rightarrow \pi_{A1}(\sigma_{C1}(R_1))$

c) $\sigma_{C1}(R_1 \cup R_2) \rightarrow \sigma_{C1}(R_1) \cup \sigma_{C1}(R_2)$

d) $\pi_{A2}(\sigma_{C1}(R_1)) \rightarrow \sigma_{C1}(\pi_{A2}(R_1))$

8. Let R(a,b,c) and S (d,e,f) be two relations in which d is the foreign key of S that refers to the primary key of R. consider the following four operations R and S

I. Insert into R II. Insert into S

III. Delete from R IV. Delete from S

which of the following is TRUE about the referential Integrity constraint above? **[GATE-1997]**

a) None of I,II,III or IV can cause its violation

b) All of I, II, III and IV can cause its violation

c) Both I and IV can cause its violation

d) Both II and III can cause its violation

9. Match the following: **[GATE-1990]**

A. Secondary Index	I. Functional Dependency
B. Non-Procedural Query language	II. B-Tree
C. Closure of a set of Attributes	III. Domain Calculus
D. Natural- Join	IV. Relational Algebraic Operations

a) A-II,B-III,C-I,D-IV b) A-III,B-II,C-IV,D-I

c) A-III,B-I,C-II,D-IV d) A-II,B-I,D-IV,C-III

The next four questions are based on the following details. Consider the given schemes.

Branch _ scheme = (Branch _ name, assets, Branch _ city)

Customer _ scheme = (Customer _ name , street, Customer _city)

Deposit _ scheme = (Branch _ name , account _ number,Customer _ name, balance)

Borrow _ scheme = (Branch_name, loan _ number, Customer _ name, amount)

Client _ scheme = (Customer _ name, banker _ name)

10. Using relational algebra, the query that finds customers, who have a balance of over 1000 is

a) $\pi_{customer\_name}(\sigma_{balance>1000}(Deposit))$

b) $\sigma_{customer\_name}(\sigma_{balance>1000}(Deposit))$

c) $\pi_{customer\_name}(\sigma_{balance>1000}(Borrow))$

d) $(\sigma_{customer\_name}(\pi_{balance>1000}(Borrow))$

11. Which of the following queries finds the clients of banker Agassi and the city they live in?

a) $\pi_{client.\ customer\_name.customer\_city}$ $(\sigma_{client.customer\_name = customer.customer\_name}$ $(\sigma_{Banker\_name="Agassi"}(Client \times Customer))$

b) $\pi_{Customer\_name.Customer\_city}$ $(\sigma_{Banker\_name="Aggasi"}(Client \times Customer))$

c) $\pi_{Client.Customer\_name.Customer\_city}(\sigma_{Banker\_name="Aggasi"}$ $(\sigma_{Client.Customer\_name = Customer.\ customer\_name}$ $(Client \times Customer)))$

d) $\pi_{Customer\_name.Customer\_city}(\sigma_{Banker\_name="Aggasi"}$ $(Client \times Customer))$

12. Which of the following tuple relational calculus finds all the customers who have a loan amount more than 1200.

a){t(Customer_name) / t $\in$ borrow $\wedge$ t [amount >1200}

b) {t / t(Customer_name) $\in$ borrow $\wedge$ t[amount > 1200 }

c) {t / $\exists$ s $\in$ borrow(t[Customer_name] = s[Customer_name] $\wedge$ s[amount] > 1200))

d) {t / $\exists$ s $\in$ barrow(t[Customer_name] = s[Customer_name] $\wedge$ s[amount] > 1200))

13. Which of the following Domain relational calculus finds all customers who have a loan amount of over 1200?

a) {<c> I $\exists$ b, l, a (<b, l, c, a> $\in$ borrow V a > 1200)}

b) {<c> I $\exists$ b, l, a (<b, l, c, a> $\in$ borrow $\wedge$ a > 1200)}

c) {<c> I $\exists$ <b, l, c, a> $\in$ borrow $\wedge$ a > 1200)}

d) {<c> I <b, l,c,a> $\in$ borrow $\wedge$ a > 1200)}

14. Choose the correct statements.

a) Relational algebra and relational calculus are both procedural Query languages

b) Relational algebra and relational calculus are both non-proce-dural Query languages.

c) Relational algebra is a procedural Query language and relational calculus is a non-procedural Query language.

d) Relational algebra is a non-procedural Query language and relational calculus is a procedural Query language.

15. Choose the correct remark.

a) An alternate key is a candidate key, that is not a primary key

b) An alternate key is a primary key, that is not a candidate key

c) An alternate key is a candidate key ,that is also a primary key

d) None of the key

16. If P and Q are predicates and P is the relational algebra expression, then which of the following equivalence are valid?

a) $\sigma_P(\sigma_Q(e)) = \sigma_Q(\sigma_P(e))$

b) $\sigma_P(\sigma_Q(e)) = \sigma_{P \wedge Q}(e)$

c) $\sigma_Q(\sigma_P(e)) = \sigma_{P \wedge Q}(e)$

d) None of these

17. If D1,D2,....Dn are domains in a relational model then the relation is table which is a subset of ......

a) {D1,D2,....,Dn} b) D1 × D2 × ..... × Dn

c) D1 U D2 U.....Dn d) None

18. If E1 and E2 are relational algebra expressions, then which of the following is NOT a relational algebra expression.

a) E1 U E2 b) E1/E2 c) E1 – E2 d) E1 × E2

19. Match the following

A. Key	I. Combines tables
B. Join	II. Referential Integrity
C. Foreign Key	III.Attribute.
D. NULL	IV. Mimal set of attributes hat uniquely Identifies a row

a) A-IV,B-I,C-II,D-III b) A-II,B-III,C-IV,D-I

c) A-I, B-II,C-III,D-IV d) None

20. The PROJECT operator of relational algebra creates a new table that has always **[UGC NET-1998]**

a) more number of columns than the original table

b) more number of rows than the original table

c) same number of rows as the original table

d) same number of columns as the original table

21. The MIS database management system design is based on **[UGC NET-1998]**

a) a hierarchical model b) a relational model

c) a network model d) an object model

22. Referential integrity is directly related to **[UGC NET-1998]**

a) Foreign key b) Primary key

c) Candidate key d) Super key

23. A compound key **[UGC NET-1998]**

a) is made up of a several pieces of information

b) uniquely identifies an item in a list

c) is a combination of each unique key

d) both (a) and (b)

24. The data dictionary tells the DBMS **[UGC NET-1998]**

a) What files are in the database?

b) What attributes are possessed by the data?

c) What these files contain?

d) All of the above

25. Which of the following is not characteristic of a relational database model? **[UGC NET-1999]**

a) Tables b) Treelike structure

c) Complex logical relationships d) Records

26. A relational database management (RDBM) package manages data in more than one file at once. How does it organize these file ? As **[UGC NET-1999]**

a) tables b) relations

c) tuple d) both (a) and (b)

ANSWER KEY - II				
1. c	2. d	3. c	4. b	5. c
6. c	7. a	8. d	9. a	10. a
11. a	12. c	13. b	14. c	15. a
16. a,b,c	17. b	18. b	19. a	20. c
21. b	22. a	23. d	24. d	25. b
26. d				

II. STRUCTURED QUERY LANGUAGE

- Categories of SQL commands
- Data Definition Statements
- Data Manipulation Statements
- Data Control Statements

1. Categories of SQL Commands : SQL commands can be roughly divided into three major categories with regard to their functionality. Firstly, there are those used to create and maintain the data base structure. The second category includes those commands that manipulate the data in such structures and thirdly there are those that control the use of the database.

2. The Data Definition Statements : To construct and administer the database there are two major DDL statements - CREATE and DROP, which form the backone of many commands :

CREATE DATA BASE to create a database DROP DATA BASE to remove a database CREATE TABLE to create a table DROP TABLE to drop a table CREATE INDEX to create an index on a column DROP INDEX to drop an index CREATE VIEW to create a view DROP VIEW to drop a view.

There may be some additional ones, such as ALTER TABLE or MODIFY DATABASE, which are vendor specific.

3. The Data Manipulation Statements : To manipulate data in tables directly through views we use the four standard DML statements :

SELECT DELETE INSERT UPDATE

These statements are now universally accepted as is their functionality although the degree to which these commands support this functionality varies somewhat between products compare the functionality of different implementation of UPDATE for example.

4. Data Control : This deals with three issues.

(a) Recovery and Concurrency : Concurrency is concerned with the manner in which multiple users operate upon the data base.

Each individual user can either reflect the updates of a transaction by using the COMMIT or can cancel all the updates of a transaction by using ROLLBACK.

(b) Security : Security has two aspects to it.

The first is the VIEW mechanism. A view of a relation can be created which hides the sensitive information and defines only that part of a relation which should be visible. A user can then be allowed to access this view.

```
CREATE VIEW LOCAL AS
SELECT * FROM SUPPLIER
WHERE SUPPLIER. CITY = 'Delhi'
```

The above view reveals only the suppliers of Delhi.

The second is by using GRANT operation. This shall grant one or more access rights to perform the data manipulative operations on the relations.

(c) Integrity Constraints : Integrity constraints are enforced by the system. For example, one can specify that an attribute of a relation will not take on null values.

5. Data Definition : Data definition in SQL is via the **create** statement. The statement can be used to create a table, index, or view (i.e., a virtual table based on existing tables). To create a table, the create statement specifies the name of the table and the names and data types of each column of the table. Its format is :

create table < relation > (< attribute list >)

where the attribute list is specified as :

<attribute list> :: = < attribute name> (<data type>) [not null]

<attribute list>

The data types supported by SQL depends on the particular implementation. However the following data types are generally include: integer, decimal, real (i.e., floating point values), and character strings both of fixed size and varying length. A number of ranges of values for the integer data type are generally supported, for example, integer and smallint. The decimal value declaration requires the specification of the total number of decimal digits for the value and (optionally) the number of digits to the right of the decimal point. The number of fractional decimal digits is assumed to be zero if only the total number of digits is specified.

<data type> :: = <integer> I <smallint> I <char (n)> I <float>I <decimal (p [,q])>

In addition some implementations can support additional data types such as bit strings, graphical strings, logical, data, and time. Some DBMSs support the concept of date. One possible implementation of date could be as eight unsigned decimal digits representing the data in the yyyymmdd format. Here yyyy represents the year, mm represents the month and dd represents the day. Two dates can be compared to find the one that is larger and hence occuring later. The system ensures that only legal date values are inserted (19860536 for the date would be illegal) and functions are provided to perform operations such as adding a number of days to a date to come up with another date or subtracting a date from the current date to find the number of days, months, or years. Data constants are provided in either the format given above or as a character string in one of the following formats : mm/dd/yy; mm/dd/yyyy; dd-mmm-yy; dd-mmm-yyyy. In this text we represent a date constant as eight unsigned decimal digits in the format yyyymmdd.

The employee relation for the hotel database can be defined using the create table statement given below. Here, the Empl_No is specified to be **not null** to disallow this unique identifier from having a null value. SQL supports the concept of null values and, unless a column is declared with the not null option, it could be assigned a null value.

```
create table    EMPLOYEE
Empl_No         integer not null,
Name            char (25),
skill           char (20)
Pay_Rate        decimal (10,2)
```

The definition of an existing relation can be altered by using the alter statement. This statement allows a row column to be added to an existing relation. The existing tuples of the altered relation are logically considered to be assigned the null value for the added column. The physical alteration occurs to a tuple only during an update of the record.

```
alter table existing - table - name
add column - name data - type [.....]
alter table EMPLOYEE
add Phone _ Number decimal (10)
```

The create index statement allows the creation of an index for any already existing relation. The columns to be used in the generation of the index are also specified. The index is named and the ordering for each column used in the index can be specified as either ascending or descending. The cluster option could be specified to indicate that the records are to be placed in physical proximity to each other. The unique option specifies that only one record could exist at any time with given value for the column (s) specified in the statement to create the index. (Even though this is just an access aid and a wrong place to declare the primary key). Such columns, for instance, could form the primary key of the relation and hence duplicate tuples are not allowed. One case is the ORDER relation where the key is the combination of the attribute Bill#, Dish#. In the case of an existing relation, an attempt to create an index with the unique option will not succeed if the relation does not satisfy this uniqueness criterion. The syntax of the create index statement is shown below:

```
create [unique] index name-of-index
on existing-table-name
(column-name [ascending or descending]
[ column-name [order] .......] )
[cluster]
```

The following statement causes an index called empindex to be built on the columns Name and Pay_Rate. The entries in the index are ascending by Name value and descending by Pay_Rate. In this example there are no restrictions on the number of records with the same Name and Pay_Rate.

```
Create index expindex
on EMPLOYEE (Name asc, Pay_Rate desc):
```

An existing relation or index could be deleted from the data base by the drop SQL statement.

The syntax of the drop statement is as follows:

```
drop table existing - table - name;
drop tabel existing - index- name;
```

7. Data Manipulation

Date manipulation capabilities allows one to retrieve and modify contents of the data base. The most important of these is the SELECT operation which allows data to be retrieved from the data base.

The relation definitions that shall be used in this chapter is are given below.

There are parts which are supplied by supplieress contains the details about each supplier. Turnover for a supplier is in terms of lakhs of rupees. Information regarding supliers of specific parts is contained in SP whereas information about the parts themselves is contained in P.

S

S#	SNAME	SCITY	TURNOVER
10	CAUVERY	BANGALORE	50
11	NARMADA	BOMBAY	100
12	YAMUNA	DELHI	70
13	TAPI	BOMBAY	20

P

P#	WEIGHT	COLOUR	COST	SELLING PRICE
1	25	RED	10	30
2	30	BLUE	15	45
3	45	RED	20	45

SP

S#	P#	QTY
10	1	100
11	1	5
10	2	50
11	2	30
10	3	10
12	3	100
13	1	20

SELECT - The Basic Form

The Select statement specifies the method of selecting the tuples of the relation(s). The tuples processed are from one or more relations specified by the form clause of the select statment.

The basic form of SELECT is

```
SELECT< target list>
FROM < relation list>
[WHERE < predicate>]
```

SELECT lists the attributes to be selected

FROM relations from which information is to be used

WHERE condition. The rows that qualify are those for which the condition evaluates to ture.

Condition is a single predicate or a collection of predicates combinded using the Boolean operators AND, OR and NOT.

The column names following SELECT are to be retrieved from the relations specified in the FROM part. WHERE specifies the condition that tuples must satisfy in order to be part of the result.

Below we shall state first the retrieveal query in English and they specify its SQL equivalent.

Unqualified Retrieval

1. Get the part numbers of all the parts being supplied.

```
SELECT P#
FROM SP
```

P#
1
1
2
2
3
3
1

Part numbers getting repeated ? That's right, SELECT does not eliminate duplicate rows (unlike the porject operation of the relational algegra). In order to do that

2.Get the part numbers of all the parts being supplied with no duplicates.

```
SELECT DISTINCT P#
FORM SP
```

P#
1
2
3

If all the columns of the relaton are to be retrieved then one need n't list all of them.

A * can be specified after SELECT to indicate retrieval of the entire relation.

3. Get full details of all suppliers.

```
SELECT *
FROM S
```

S#	SNAME	SCITY	TURNOVER
10	CAUVERY	BANGALORE	50
11	NARMADA	BOMBAY	100
12	YAMUNA	DELHI	70
13	TAPI	BOMBAY	20

The ORDER BY clause

The result of a query can be ordered either in ascending (ASC) order or in descending (DESC) order.

4. Get the supplier numbers and turnover in descending order of turnover.

```
SELECT S#, TURNOVER
FROM S
ORDER BY TURNOVER DESC
```

S#	TURNOVER
11	100
12	70
10	50
13	20

Instead of a column name, the ordinal position of the column in the result can be used.

That is, the above query can be rewritten as

```
5. SELECT S#, TURNOVER
FROM S
ORDER BY 2 DESC
```

The format of the order clause is

ORDER BY (int/col [ASC/DESC],..........)

col - column name

int - ordinal position of the column in the result table

ASC/DESC - Ascending or descending

If there is more than one specification, then the left - to - right specification corresponds to major - to minor ordering. This is shown bleow.

6. Get the supplier number and part number in ascending order of supplier number and descending order for the part supplied for each supplier.

```
SELECT S#, P#, QTY
FROM SP
ORDER BY S#, P# DESC
```

S#	P#	QTY
10	3	10
10	2	50
10	1	100
11	2	30
11	1	5
12	3	100
13	1	20

Qualified retrieval

The expression following WHERE specifies the condition that must be satisfied, Below we consider a few examples.

7.Get the details of suppliers who operate from Bombay with turnover 50.

```
SELECT *
FROM S
WHERE CITY = 'BOMBAY'. AND TURNOVER > 50
```

S#	SNAME	SCITY	TURNOVER
11	NARMADA	BOMBAY	100

The above form is a conjundtion of comparison predicates. A comparison predicate is of the form

scalar - expr O scalar - expr

Where O is any of the six relational operators =, < >, <, >, <=, >= and a scalar experssion is an arithmetic expression with operators as +, -, *, / operands as col., function, constant

BETWEEN Predicate

8. Get the part numbers weighing between 25 and 35

```
SELECT P#, WEIGHT
FROM P
WHERE WEIGHT BETWEEN 25 AND 35
```

P#	WEIGHT
1	25
2	30

The use of BETWEEN gives the range within which the values must lie. If the value should lie outside a range then BETWEEN is to be preceded by NOT. For example,

```
SELECT  P#
FROM    P
WHERE   WEIGHT  NOT BETWEEN 25 AND 35
```

P#	WEIGHT
3	45

would retrieve all part numbers whose weight is less than 25 or greater than 35 as shown above.

LIKE Predicate

This predicate is used for pattern matching. A column of type char can be compared with a string constant. The use of the word LIKE doesn't look for exact match but a form of wild string match. A % or - can appear in the string constant where.

% stands for a sequence of n(>=0) characters

- stands for a single character

Examples

ADDRESS LIKE ' % Bangalore%' - ADDRESS should have Bangalore somewhere as a part of it if the match is to succeed.

9.Get the names and cities of suppliers whose name begin with C

```
SELECT  SNAME,   SCITY
FROM  S
WHERE  SNAME LIKE 'C %'
```

SNAME	SCITY
CAUVERY	BANGALORE

When the data is to be retrieved from more than one relation then both the relation names is specified in the FROM clause and the join condition in the WHERE part.

10. For each part supplied, get part number and names of all cities supplying the part.

```
SELECT  P#, CITY
FROM    SP, S
WHERE   SP.S# = S.S#
```

P#	SCITY
1	BANGALORE
1	BOMBAY
2	BANGALORE
2	BOMBAY
3	BANGALORE
3	DELHI
1	BOMBAY

How does, then, one specify a join on the same relation?

11. Get pairs of supplier numbers such that both operate from the same city.

```
SELECT   FIRST. S# , SECOND. S#
FROM     S FIRST, S SECOND
WHERE    FIRST.CITY = SECOND. CITY
AND      FIRST.S# < >  SECOND. S#
```

FIRST and SECOND are tuple variables, both ranging over . The last line eliminates a supplier getting compared with himself.

S#	S#
11	13
13	11

But, we see that suppliers with numbers 11 and 13 are getting compared twice. Can that be avoided ? How about < instead of < > ?

```
SELECT   FIRST.S#, SECOND. S#
FROM     S FIRST, S SECOND
WHERE    FIRST.CITY = SECOND. CITY
AND      FIRST.S# = SECOND. S#
```

Tests for NULL

An attribute can be tested for the presence or absence of null.

12. Get the supplier numbers whose turn over is null

```
SELECT  S#
FROM    S
WHERE   TURNOVER  IS  NULL
```

There is no tuple in the result of this query as in the sample Can the last line in the above query be replaced by

WHERE TURNOVER = NULL

Not really! It is incorrect as nothing (even NULL) is equal to NULL.

The format for specifying NULL is

col. ref IS [NOT] NULL

IN Predicate

This is to be used whenever you want to test whether an attribute value is one of a set of values. For example,

13. Get the part numbers that cost 20, 30 or 40 rupees.

```
SELECT  P. P# , SELLING PRICE
FROM    P
WHERE   SELLING PRICE  IN  (20, 40, 45)
```

P#	SELLING PRICE
2	45
3	45

It's a quicker way of specifying comparison.

The format of the predicate

scalar - expr [NOT] IN (atom list)

Subqueries

The expression follwing WHERE can be either a simple predicate as explained above or it can be a query itself! This part of the query following WHERE is called a Subquery.

A subquery, which in turn is a query, can have its own subquery and the process of specifying subqueries can contiune and infinitum! More practically, the process ends once the query has been fully expressed as a SQL statement.

Subqueries can appear when using the comparison predicate, the IN predicate and when quantifiers are used

Comparison Predicate

14. Get the supplier numbers of suppliers who are located in the same city as Tapi.

```
SELECT  S. S#, S NAME
FROMS
WHERE  S. CITY =
(SELECT S. CITY
FROMS
WHERE  SNAME = 'TAPI')
```

The inner select (subquery) retrieves the city of the supplier named Tapi. The outer select (the main one) then compares the city of each supplier in the supplier relation and picks up those where the comparison succeeds.

S#	SNAME
11	NARMADA
13	TAPI

Notice that the subquery appears after the comparison operator. The format of this form of expression

scalar - expr operator subquery

IN Predicate

In this form the subquery selects a set of values. The outer query checks whether the value of a specified attribute is in this set.

15. Get the names of suppliers who supply part 2

```
SELECT    S . SNAME
FROM      S
WHERE     S. S# IN
(SELECT   SP. S#
FROM      SP
WHERE     SP. P#= 2)
```

SNAME
CAUVERY
NARMADA

The above query can be equivalently expressed as

```
SELECT    S. S NAME
FROM      S
WHERE     2 IN
(SELECT   P#
FROM      SP
WHERE     S. S# = S#)
```

S# is unqualified and therefore, refers to SP. That is because every unqualifed attribute name is implicitly qualifed with the relation name from the nearest applicable FROM clause.

Quantified predicates

The two quantifiers that can be used are the ALL and ANY. Any stands for the existential quantifier and ALL for the universal quantifier.

Lets first look at ANY. It can be specified in a comparison predicate just following the comparison operator. That is,

scalar - expr O ANY subquery

The subquery is first evaluated to give a set of values. The above expression is ture if the scalar - expr is O comparable with any of the values that form the result of the subquery.

16. Get the part numbers for parts whose cost is less than the current maximum cost.

```
SELECT  P#, COST
FROM    P
WHERE   COST < ANY
(SELECT COST
FROM    P)
```

P#	COST
1	10
2	15

The inner select gets the cost of all the parts. In the outer select, a P# is selected if its cost is less than some element of the set selected in the earlier step.

17. Get the supplier names of suppliers who do not supply part 2.

```
SELECT  SNAME
FROM    S
WHERE   2 < > ALL
(SELECT P#
FROM    SP
WHERE   S# = S.S#)
```

SNAME
YAMUNA
TAPI

For each supplier, all the parts supplied by him are collected in the inner select. If none of them is equal to P2 then the condition evaluates to ture and supplier name forms part of the result.

Existence Test

This kind of an experssion is used when it is necessary to find out if the set of values retrieved by using a subquery contains an element or not.

18. Get the supplier names of supliers who supply at least one part.

```
SELECT  S NAME
FROMS
WHERE   EXISTS
(SELECT *
FROMSP
WHERE   SP. S#= S. S#)
```

For a given supplier, if the subquery selects at least one tuple then the condition (which follows WHERE) evaluates to true. Then, the name of the supplier is selected. IN our data base every supplier is supplying at least one part. So the names of all of them would be part of the result.

Functions

Some standard functions are defined in SQL and can be used when framing queries. There are five built - in functions. These are

COUNT	-	Number of values of a column
SUM	-	Sum of values in a column
AVG	-	Average of values in a column
MAX	-	Maximum of all the values in a column
MIN	-	Minimum of all the values in a column

If the function is followed by the wourd DISTINCT then unique values are used. On the other hand, if ALL follows the function then all the values are used for evaluating the function. ALL is the default.

COUNT (*) has a special meanig in that it counts the number of rows of a relation. COUNT in any other form must make use of DISTINCT. In other words, except when rows are counted, COUNT always returns the number of distinct values in a column.

19. Get the total number of suppliers

```
SELECT  COUNT (*)
FROMS
```

COUNT (*) counts the number of tuples of S and hence, the number of suppliers.

20. Get the total quantity of Part 2 that is supplied.

```
SELECT  SUM (SP. QTY)
FROMSP
WHERE   SP. P# = 2
```

The answer is one of

a) 25 b) 40 c) 80 d) 100

21. Get the part numbers whose cost is greater than the average cost.

```
SELECT  P#
FROM    P
WHERE   COST
(SELECT AVG (COST)
FROM    P)
```

22. Get the names of suppliers who supply from a city where there is atleast one more supplier.

```
SELECT  SNAME
FROM S  FIRST
WHERE   2
(SELECT COUNT (CITY)
FROM    S
WHERE   CITY = FIRST. CITY)
```

GROUP BY Feature

This feature allows one to partition the result into a number of groups such that all rows of the group have the same value in some specified column.

23. Get the part number and the total quantity.

```
SELECT  P#, SUM (QTY)
FROM    SP
GROUP   BY P#
```

P#	SUM(QTY)
1	125
2	80
3	110

GROUP BY groups together all the rows which have the same value for P#. The function SUM is then applied to each group. That is, the result consists of a part number along with the total quantity in which it is supplied.

Whenever GROUP BY is used then the phrase WHERE is to be replaced by HAVING.Is the same as WHERE except that the condition is now applicable to each group.

24. Get the part numbers for parts supplied by more than one supplier.

```
SELECT      P#
FROM        SP
GROUP BY    P#
HAVING      COUNT (*) > 1
```

Each group contains one or more tuples which have the same part number. COUNT (*) is applied to each such group.

The result before COUNT(*) is applies is-

P#
1
2
3

In this case all the part numbers will be selected.

8. Updating the Database

The contents of the database can be modified by inserting a new tuple, decleting an existing tuple or changing the values of attributes of one or more tuples.

Insert

The insertion facility allows new tuples to be inserted into given relations. Attributes which are not specified by the insertion statments are given null values. Consider.

1. Add a part with number 14, weight 10, coloured red, with the cost and selling price as 20 and 60 respectively.

```
INSERT INTO P :
< 14, 10, 'red', 20, 60 >
```

The tuple is inserted into P.

If all the fields are not known then a tuple can still be added. The attributes whose values are not sjpecified will have null value.

INSERT INTO P :

< 15, 'GREEN', >

The values for fileds the weight, cost and the selling price which are not specified are assumed to be null.

2. Let us assume that there is a relation called RED - PART with one column P#.

```
INSERT  INTO  RED - PART :
SELECT  P#
FROM    P
WHERE   COLOUR = 'red'
```

The various attributes of P having red colour are identified and inserted into the relation RED - PART.

Delete

The deletion facility removes specified tuples form the database. Consider

1. Delete supplier 13

```
DELETE   S
WHERE    S# = 13 .
```

Since S# is the primary key only one tuple will be deleted from S.

2. Delete all suppliers who supply from Bangalore

```
DELETE  S
WHERE   SCITY = 'BANGALORE'
```

Here, more than one supplier can get deleted.

3. Delete all the suppliesrs

```
DELETE  S
```

The definition of S exists but the relation is empty.

Update

When columns are to be modified SET clause is used. This clause specifies the update to be made to selected tuples.

1. Change the city of supplier 13 to Bangalore and increase the turnover by 20 Lakhs.

```
UPDATE  S
SET  CITY = ' BANGALORE'
TURNOVER = TURNOVER + 20
WHERE   S# = 13
```

2. Increase quantity by 10 for all supplies of red coloured parts.

```
UPDATE  SP
SET  QTY = QTY + 10
WHERE   P#  IN.
(SELECT P#
FROM    P
WHERE   COLOUR = 'RED')
```

9. Views

A view is a virtual table, that is one that does not actually exist. It is made up of a query on other tables in the database. It could include only certain columns or rows from a talbe or from many talbes. A view which restricts the user to certain rows is called a horizontal view and a vertical view restricts the user to certain columns. You are not restricted to purely horizonatal or vertical slices of data.

A view can be as complicated as you like, You can have grouped views where the query contains a GROUP BY clause. This makes the view a summary of the data in a table or tables.

If the list of column names is omitted the columns in the view take the same name as in the underlying tables. You must specify column names if the qurey includes calculated columns or two columns with the same name. There are several advantages to views, including :

- **Security** : Users can be given access to only the rows/ columns of tables that concern them. The table may contain data for thewhole firm but they only see data for their department.
- **Data integrity :** The WITH CHECK OPTION clause is used toenforce the query conditions on any updates to the view. If the view shows data for a particular office the user can only enter data for that office.
- **Simplicity :** Even if a view is a multi - table query, querying the view still looks like a single - table query.
- **Protection from change :** If the structure of the database changes, the user's view of the data can remain the same.

There are two disadvangages to views :

- **Performance :** A view may look like a single table but underneath the DBMS is usually still running multi - table queries. If the view is complex then even simple queries can take a long time.
- **Update restrictions :** Updating the data through a view may or may not be possible. If the view is complex the DBMS may decide it can't perform updates and make the view read - only.

The ISO standard specifies five conditions that a view must meet in order to allow updates :

- The view must not have a DISTINCT clause
- The view must only name one table in the FROM clause
- All columns must be real columns - no expressions, calculated columns or column functions.
- The WHERE clause must not contain a sub-query
- There must be no GROUP BY or HAVING clause

You will find that most dialects of SQL are not quite so restrictive. The underlying principle is that updates are allowed if the rows and columns of the view are traceable back to actual rows and columns in tables.

The format of view statement is as follows :

create view < view name > as query expression

A view is a relation (virtual rather than base) and can be used in query expressions, that is, queries can be written using the view as a realation. Views generally are not stored, since the data, in the base relations may change. The base relations on which a view is based are sometimes called the **existing relations**.

The definition of a view in a **create view** statement is stored in the system catalog. Having been defined it can be used as if the view really represented a real relation. However, such a virtual relation defined by a view is recomputed. Whenever a Query refers to it.

Example

(a) For reasons of confidentiality, not all users are permitted to see the Pay_Rate of anemployee. For such user the DBA can create a view, for example, EMP_VIEW defined as :

create view EMP_VIEW as
(**select** Empl_No, Name, Skill **from** EMPLOYEE)

(b) A view can be creatd for a subset of the tuples of a relation, as in this example. For assigning employees to particular jobs,the manager requires a list of the employees who have not been assigned to any jobs :

create view FREE **as**
(**select** Empl_No
from EMPLOYEE)
minus
(**select** Empl_No
from DUTY_ALLOCATION)

(c) The view in part (b) above can also be created using the following statments :

create view FREE **as**
(**select** Empl_No,
from EMPLOYEE)
where Empl_No **any**
(**select** Empl_No
from DUTY_ALLOCATION)

In the above examples, the names of the attributes in the views are implicitly taken from the base relation. The data types of the attribute of the view are inherited from the corresponding attributes in the base relation. We can, however, given new names to the attributes of the view. This is illustrated in the syntax of the create view statement given below :

create view VIEW_NAME
(Name1, Name 2,......)
as (select.......)

Here the attributes in the view are given as Name 1, Name 2,........ and these names are associated with the existing relation by order correspondence. The definition of a view is accomplished by means of a subquery involving a select statement as given in the syntax above. Since a view can be used in a select statment, a view can be defined on another existing view.

OBJECTIVE QUESTIONS - I

1. Which of the following is/are correct? **[GATE-1999]**

a) An SQL Query automatically eliminates duplicates

b) An SQL Query will not work if there are no indexes on the relations.

c) SQL permits attribute names to be repeated in the same relation d) None

2. Choose the incorrect Statement (s):

a) Procedural data manipulation language (DML) requires user to specify what data is needed and how to get it.

b) Procedural DML requires user to specify what data is needed without specifying how to get it.

c) Non-procedural DML requires the user to specify what data is needed with out specifying how to get it

3. A Query is

a) A miniature program that tells the DBMS what to do

b) Expressed using a Query language.

c) Both (a) and (b) d) None

4. Wild-card specifiers

a) Provide an easy way of finding groups of related files

b) Are only used when printing the contents of files

c) Can be used when writing a file

d) Allow several files to be read simultaneously.

5. The language used in application programs to request data from the DBMS is referred to as the

a) DML b) DDL c) Query language d) None

6. Which command is used to establish a link between two database files?

a) Join b) Update c) Set relation d) Browse

7. Tuple variables in SQL are defined in

a) From clause b) Select clause

c) Where clause d) All of these

8. Which one of these is an aggregate function in SQL?

a) AVG b) Select c) COUNT d) SUM

9. The strategy for processing a Query is improvedby

a) Query evaluation b) Query optimization

c) Decomposition d) None

10. ROLL-BACK of transactions is normally used to

a) Recover from the transaction failure

b) Update the transaction

c) Retrieve old records

d) Restore the record

11. In Query-By-Example , the Query is formed from the relations displayed by filling in the following.

a) SQL form b) Templates c) DML d) DDI

12. A language that enables users to access or manipulate data:

a) DML b) DDL c) SQL d) All of these

13. SQL is relationally.....

a) Complete language b) Incomplete language

c) Cannot handle certain relations

d) Sound language.

14. A command that lets you change one or more fields in arecord is **[UGC NET-1998]**

a) Insert b) Look-up

c) Modify d) Create

15. Which of the following is/are correct ?

a) An SQL query automatically eliminates duplicates.

b) An SQL query will not work if there are no indexes on the relations.

c) SQL permits attribute name to be repeated in the same relation.

d) None of the above.

ANSWER KEY - I

1. d	2. b	3. c	4. a	5. a
6. a	7. a	8. a	9. b	10. a
11. b	12. a	13. a	14. c	15. d

OBJECTIVE QUESTIONS - II

1. Consider a relation geq which represents "greater than or equal to ", that is,(x,y) $\in$ geq only if y >=x.

Create table geq(lb integer not null

ub integer not null

primary key lb

foreign key(ub) references geq on delete cascade)

Which of the following is possible if a tuple (x,y) is deleted? **[GATE-2001]**

a) A tuple (z,w) with z>y is deleted

b) A tuple (z,w) with z>x is deleted

c) A tuple (z,w) with w<x is deleted

d) The deletion of (x,y) is prohibited.

2. Given relations r(w,x) and s(y,z) ,the result of **[GATE-2000]**

select distinct w,x

from r,s

is guaranteed to be the same as r, provided:

a) r has no duplicates and s is non-empty

b) r and s have no duplicates

c) s has no duplicates and r is non-empty

d) r and s have the same number of tuples

3. In SQL , relations can contain null values, and comparisons with null values are treated as unknown suppose all comparisons with a null value are treated as false. Which of the following pairs is not equivalent?

[GATE-2000]

a) x = 5 not (not (x = 5))

b) x = 5 x > 4 and x < 6 , where x is an integer

c) x $\neq$ 5 not (x = 5)

d) None

4. The SQL expression

Select distinct T.branchname from branch T, branch S where T.assets>S.assets and S.branchcity="TENALI"

Finds the names of

a) All branches that have greater assets than some branch located in TENALI

b) All branches that have greater assets than all branches located in TENALI

c) The branch that has the greatest asset in TENALI

d) Any branch that has greater asset than any branch located in TENALI.

5. SQL is used for **[UGC NET-1997]**

a) data processing in batch mode

b) querry for relational databases

c) DTP work d) None of the above

6. The standard for data access query language is **[UGC NET-1995]**

a) SQL b) 4 GL c) ISO d) CASE

7. A high level language is a procedural language, but a .... proprietary non-procedural language that interacts with a DBMS to store, manipulate and retrieve data **[UGC NET-1995]**

a) C b) 4 GL

c) CODASYL d) None of the above

ANSWER KEY - II				
1. c	2. a	3. c	4. a	5. b
6. a	7. b			

III. NORMALIZATION

- **Introduction**
- **Functional Dependency**
- **Normal Forms**

1. Introduction

The basic objectives of normalization is to reduce redundancy which means that information is to be stored only once. Storing information several times leads to wastage of storage space and increase in the total size of the data stored. Relations are normalized so that when relations in a database are to be altered during the life time of the database, we do not lose information or introduce inconsistencies. The type of alterations normally needed for relations are :

1. Insertion of new data values to a relation. This should be possible without being forced to leave blank fields for some attributes.

2. Deletion of a tuple, namely, a row of a relation. This should be possible without losing vital information unknowingly.

3. Updating or changing a value of an attribute in a tuple. This should be possible without exhaustively searching all the tuples in the relation.

2. Functional Dependency

As the concept of dependency is very important, it is essential that we first understand it well and then proceed to the idea of normalization. There is no fool-proof algorithmic method of identifying dependency. We have to use our commonsense and judgement of specify dependencies.

Let X and Y be two attributes of a relation. Given the value of X, if there is only one value of Y corresponding to it, then Y is said to be functionally dependent on X. This is indicated by notation:

$$X \rightarrow Y$$

For example, given the value of item code, there is only one value of item name for it. Thus item name is functionally dependent on item code. This is shows as:

Item code $\rightarrow$ item name

Similarly in table 3.1 given an order number, the date of the order is known. Thus : order no. Order date

Functional dependency may also be based on a composite attribute. For example, if we write X, Z$\rightarrow$ Y it means that there is only one value of Y corresponding to given values of X, Z. In other words, Y is functionally dependent on the composite X, Z. In other words, Y is functionally dependent on the composite X, Z. In table 3.1 mentioned below, for example, Order no., and item code together determine Qty. and Price. Thus :

Order no., Item code $\rightarrow$ Qty., Price

As another example, consider the relation

Student (**Roll no.,** Name, Address, Dept., Year of study)

Order no.	Order date	Item code	Quantity	Price/unit
1456	260289	3687	52	50.40
1456	260289	4627	38	60.20
1456	260289	3214	20	17.50
1886	040389	4629	45	20.25
1886	040389	4627	30	60.20
1788	040489	4627	40	60.20

Table 3.1: Normalized Form of the Relation

In this relation, Name is functionally dependent on Roll no. In fact, given the value of Roll no., the values of all the other attributes can be uniquely determined. Name and Department are not functionally dependent because given the name of a student, one cannot find his department uniquely. This is due to the fact that there may be more than one student with the same name. Name in this case is not a key. Department and Year of study are not functionally dependent as Year of study pertains to a student whereas Department is an independent attribute. The functional dependency in this relation is shown in figure 3.1 are very useful in normalization.

Relation Key : Consider the relation of table 3.1. Given the Vendor code, the Vendor name and Address are uniquely determined. Thus Vendor code is the relation key. Given a relation, if the value of an attribute X uniquely determines the values of all other attributes in

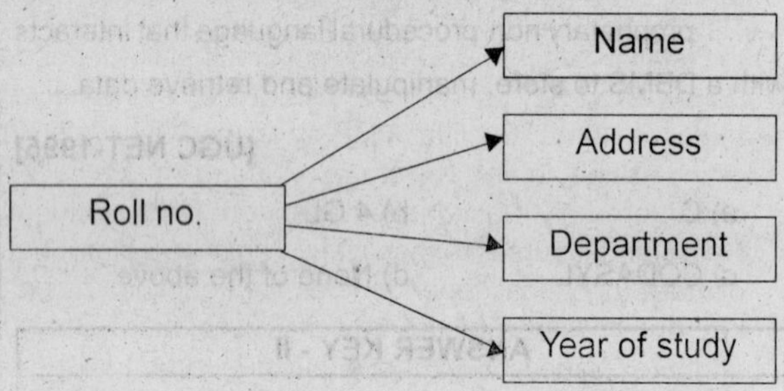

Figure 3.1: Dependency diagram for the relation "Student"

a row, then X is said to be the key of that relation. Sometimes more than one attribute is needed to uniquely determine other attributes in a relation row. In that case such a set of attributes is the key. In table 3.1, Order no. and Item code together form the key. In the relation "supplies""(Vendor code, Item code, Qty. supplied, Date of supply, Price/unit), Vendor code and Item code together form the key. This dependency is shown in the following diagram (figure 3.2).

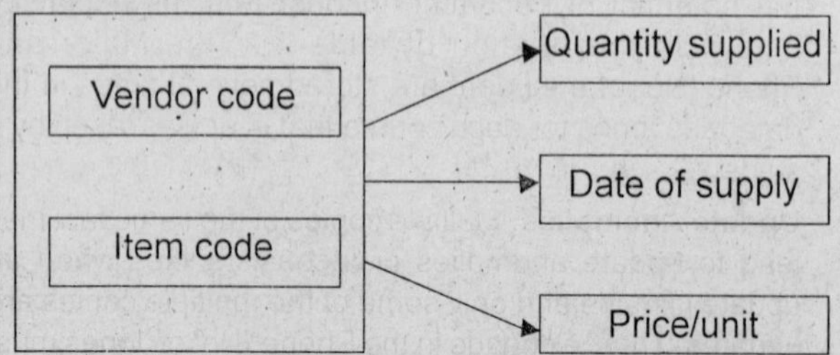

Figure 3.2 : Dependency diagram for the relation "Supplies"

Observe that in the figure the fact that Vendor code and Item code together form a composite key is clearly shown by enclosing them together in a rectangle.

3. Anomalies in a Database

Consider the following relation scheme pertaining to the information about a student maintained by a university:

STDINF(Name, Course, Phone_No, Major, Prof, Grade)

Table 3.2 shows some tuples of a relation on the relation scheme STDINF (Name, Course, Phone_No, Major, Prof, Grade). The functional dependencies amonng its attributes are shown in Figure 3.3. The key of the relation is Name Course and the relation has, in addition, the following functional dependencies {Name $\rightarrow$ Phone_No, Name $\rightarrow$ Major, Name Course $\rightarrow$ Grade, Course $\rightarrow$ Prof }.

Name	Course	Phone_No.	Major	Prof	Grade
Jones	353	237-4539	Comp Sci	Smith	A
Ng	329	427-7390	Chemistry	Turner	B
Jones	328	237-4539	Comp Sci	Clark	B
Martin	456	388-5183	Physics	James	A
Dulles	293	371-6259	Decision Sci	Cook	C
Duke	491	823-7293	Mathematics	Lamb	B
Duke	356	823-7293	Mathematics	Bond	in prog
Jones	492	237-4539	Comp Sci	Cross	in prog
Baxter	379	839-0827	English	Broes	C

Table 3.2 : Student Data Representation in Relation STDINF

Here the attribute Phone_No, Which is not in any key of the relation scheme STDINF, is not functionally dependent on the whole key but only one part of the key, namely, the attribute Name. Similarly, the attributes Major and Prof, which are not in any key of the relation scheme STDINF either, are fully functionally dependent on the attributes Name and Course, respectively. Thus the determinants of these functional dependencies are again not the entire key but only part of the key of the relation. Only the attribute Grade is fully functionally dependent on the key Name Course.

The relation Scheme STDINF can lead to several undesirable Problems.

Redundancy: The aim of the database system is to reduce redundancy,meaning that information is to be stored only once. Storing information several times leads to the waste of storage space and an increase in the total size of the data stored.

Updates to the database with such redundancies have the potential of becoming inconsistent, as explained below. In the relation of table 3.2, the Major and Phone_No. of a student are stored several times in the database: once for each course that is or was taken by a student.

Update Anomalies: Multiple copies of the same fact may lead to update anomalies or inconsistencies when an update is made and only some of the multiple copies are updated. Thus, a change in the Phone_No. of Jones must be made, for consistency, in all tuples pertaining to the student Jones. If one of the three tuples of Figure 3.3 is not changed to reflect the new Phone_No. of Jones, there will be an inconsistency in the data.

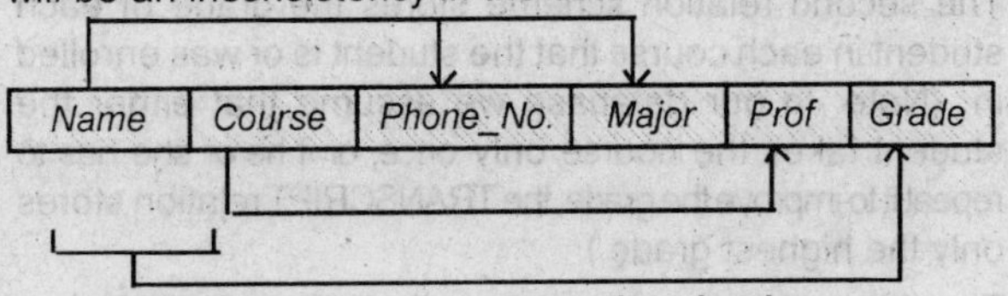

Figure 3.3: Function dependencies in STDINF

Insertion Anomalies : If this is the only relation in the database showing the association between a faculty member and the course he or she teaches, the fact that a given processor is teaching a given course cannot be entered in the database unless a student is registered in the course. Also, if another relation also establishes a relationship between a course and a professor who teaches that course the information stored in these relations has to be consistent.

Deletion Anomalies : If the only student registered in a given course discontinues the course, the information as to which professor is offering the course will be lost if this is the only relation in the database showing the association between a faculty member and the course she or he teaches. If another relation in the database also establishes the relationship between a course and a professor who teaches that course, the deletion of the last tuple in STDINF for a given course will not cause will not cause the information about the course's teacher to be lost.

The problems of database inconsistency and redundancy of data are similar to the problems that exist in the hierarchical and network models. These problems are addressed in the network model by the introduction of virtual fields and in the hierarchical model by the introduction of virtual records. In the relational model, the above problems can be remedied by decomposition. We define decomposition as follows:

Decomposition: The decomposition of a relation scheme $R = (A_1, A_2 .....A_n)$ is its replacement by a set of relation schemes $[R_1, R_2,....R_m]$, such that $R_i \leq R$ for $1 \leq i \leq m$ and $R_1 \cup R_2 \cup R_m = R$.

A relation scheme R can be decomposed into a collection of relation schemes $\{R_1, R_2, R_3.... R_m\}$ to eliminate some of the anomalies contained in the original relation R. Here the relation schemes R_i $(1 \leq i \leq m)$ are subsets of R and the intersection of $R_i \cap R_j$ for $i \neq j$ need not be empty. Furthermore, the union of Rj $(1 \leq i \leq m)$ is equal to R, i.e $R = R_1 R_2 .... R_m$.

The problems in the relation schemes STDINF can be resolved if we replace it with the following relation schemes:

STUDENT_INFO (Name, Phone_No, Major)

TRANSCRIPT (Name, Course, Grade)

TEACHER (Course, Prof)

The first relation scheme gives the phone number and the major of each student and such information will be stored only once for each student. Any change in the phone number will thus require a change in only one tuple of this relation.

The second relation scheme stores the grade of each student in each course that the student is or was enrolled in. (Note: In our database we assume that either the student takes the course only once, or if he or she has to repeat it to improve the grade, the TRANSCRIPT relation stores only the highest grade.)

The third relation scheme records the teacher of each course. One of the disadvantages of replacing the original relation scheme STDINF with the three relation schemes is that the retrieval of certain information requires a natural join operation to be performed. For instance, to find the majors of a student who obtained a grade of A in course 353 requires a join to be performed: (STUDENT_INFO $\bowtie$ TRANSCRIPT). The same information could be derived from the original relation STDINF by selection and projection.

When we replace the original scheme STDINF with the relation schemes STUDENT_INFO, TRANSCRIPT and TEACHER, the consistency and referential integrity constraints have to be enforced. The referential integrity enforcement implies that if a tuple in the relation TRANSCRIPT exists, such as (Jones, 353, in prog), a tuple must exist in STUDENT_INFO with name = Jones and futhermore, a tuple must exist in STUDENT - INFO with Course = 353. The attribute Name, which forms part of the key of the relation TRANSCRIPT, is a key of the relation STUDENT_INFO. Such an attribute (or a group of attributes), which establishes a relationship between specific tuples (of the same or two distinct relations), is called a foreign key. Notice that the attribute Course in relation TRANSCRIPT is also a foreign key. Notice that the attribute Course in relation TRANSCRIPT is also a foreign key, since it is a key of the relation TEACHER.

Note that the decomposition of STDINF into the relation schemes STUDENT (Name, Phone_No, Major, Grade) and COURSE (Course, Prof.) is a bad decomposition for the following reasons:

1. Redundancy and update anomaly, because the data for the attributes Phone_no and Major are repeated.

2. Loss of information, because we lose the fact that a student has a given grade in a particular course.

4. **Properties Of Normalized Relations:** Ideal relations after normalization should have following properties so that the problems mentioned above do not occur for relations in the (ideal) normalized form:

1 No data value should be duplicated in different rows unnecessarily.

2. A value must be specified (and required) for every attribute in a row.

3. Each relation should be self-contained. In other words, if a row from a relation is deleted, important information should not be accidentally lost.

4. When a row is added to a relation, other relations in the data base should not be affected.

5. A value of an attribute in a tuple may be changed independent of other tuples in the relation and other relations.

The idea of normalizing relations to higher and higher normal forms is to attain the goals of having a set of ideal relations meeting the above criteria.

5. **First Normalization:** The relation shown in table 3.1 is said to be in First Normal Form, abbreviated as 1NF. This form is also called a flat **file**. There are no composite attributes, and every attribute is single and describes one property.

Converting a relation to the 1NF form is the first essential step normalization. There are successive higher normal forms known as 2NF, 3NF, BCNF, 4NF and 5NF. Each form is an improvement over the earlier form. In other words, 2NF is an improvement on 1NF, 3NF is an improvement on 2NF, and so on. A higher normal form relation is a subset of lower normal form as shown in the following figure 3.4. The higher normalization steps are based on three important concepts:

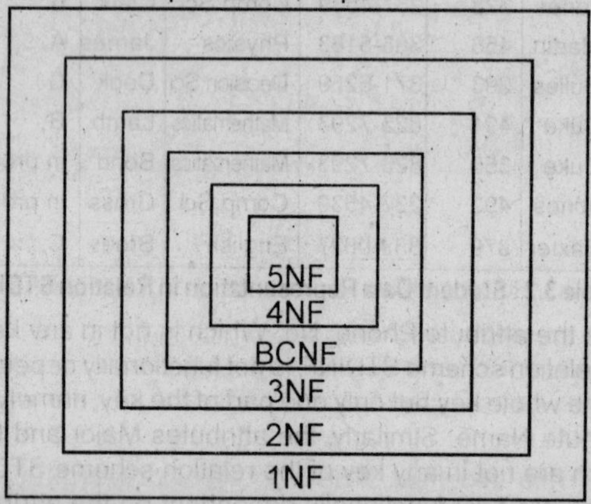

Figure 3.4: Illustration of successive normal forms of a relation

1. Dependency among attributes in a relation.

2. Identification of an attribute or a set of attributes as the key of a relation.

3. Multivalued dependency between attributes.

6. **Second Normal Form Relation:** We will now define a relation in the Second Normal Form (2NF). A relation is said to be in 2NF if it is in 1NF and non-key attributes are functionally dependent on the key attributes(s). Further, if the key has more than one attribute then no non-key attributes should be functionally dependent upon a part of the key attributes. Consider, for example, the relation given in table 1.1. This relation is in 1NF. The key is (Order no., Item code). The dependency diagram for attributes of this relation is shown in figure 3.5. The non-key attribute Price/Unit is functionally dependent on Item code which is part of the relation key. Also, the non-key attribute Order date is functionally dependent on Order no. which is a part of the relation key. Thus the relation is not in 2NF. It can be transformed to 2NF by splitting it into three relations as shown in table 3.3.

In table 3.3 the relation **Orders** has **Order no.** as the key. The relation **Order details** has the composite key **Order no.** and **Item code.** In both relations the non-key attributes are functionally dependent on the whole key. Observe that by transforming to 2NF relations the

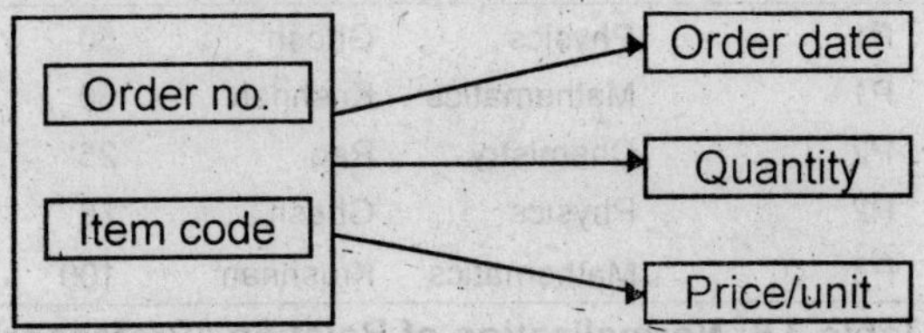

Figure 3.5: Dependency diagram for the relation given in table.

repetition of Order date (table 3.1) has been removed. Further, if an order for an item is cancelled, the price of an item is not lost. For example, If Order no. **1886** for Item code **4629** is cancelled in table 3.1, then the fourth row will be removed and the price of the item is lost. In table 3.3 only the fourth row of the table 3.3 (b) is omitted. The item price is not lost as it is available in table 3.3 (c). The data of the order is also not lost as it is in table 3.3 (a).

(a) Orders

Order no	Order date
1456	260289
1886	040389
1788	040489

(b) Orders Details

Order no	Item code	Qty.
1456	3687	52
1456	4627	38
1456	3214	20
1886	4629	45
1886	4627	30
1788	4627	40

(c) Prices

Item Code	Price/ unit
3687	50.40
4627	60.20
3214	17.50
4629	20.25

Table 3.3 : Splitting of Relation given in table 3.1 into 2NF Relations

These relations in 2NF form meet all the "ideal" conditions specified. Observe that the three relations obtained as self-contained. There is no duplication of data within a relation

7. Third Normal Form : A Third Normal Form normalization will be needed where all attributes in a relation tuple are not functionally dependent only on the key attribute. If two non-key attributes are functionally dependent, then there will be unnecessary duplication of data. Consider the relation given in table 3.4. Here, Roll no. is the key and all other attributes are

Roll no.	Name	Department	Year	Hostel name
1784	Raman	Physics	1	Ganga
1648	Krishnan	Chemistry	1	Ganga
1768	Gopalan	Mathematics	2	Kaveri
1848	Raja	Botany	2	Kaveri
1682	Manju	Geology	3	Krishna
1485	Singh	Zoology	4	Godavari

Table 3.4: A2NF Relation

Functionally dependent on it. Thus it is in 2NF. If it is known that in the college all first year students are accommodated in Ganga hostel, all second year students in Kaveri, all third year students in Krishna, and all fourth year students in Godavari, then the non-key attribute Hostel name is dependent on the non-key attribute Year. This dependency is shown in figure 3.6.

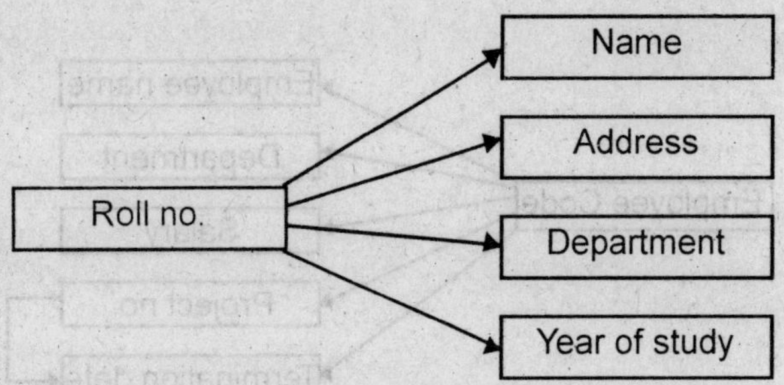

Figure 3.6: Dependency diagram for the relation

Observe that given the year of student, his hostel is known and vice versa. The dependency of hostel on year leads to duplication of data as is evident from table 3.4. If it is decided to ask all first year students to move to Kaveri hostel, and all second year students to Ganga hostel, this change should be made in many places in table 3.4. Also, when a student's year of study changes, his hostel change should also be noted in Table 3.4. This is undesirable. Table 3.4 is said to be in 3NF if it is in 2NF and no non-key attributes is functionally dependent on any other non-key attribute. Table 3.4 is thus not in 3NF. To transform it to 3NF, we should introduce another relation which includes the functionally related non-key attributes. This is shown in table 3.5.

Roll no. name	Name	Department	Year
1784	Raman	Physics	1
1648	Krishnan	Chemistry	1
1768	Gopalan	Mathematics	2
1848	Raja	Botany	2
1682	Manju	Geology	3
1485	Singh	Zoology	4

Year	Hostel
1	Ganga
2	Kaveri
3	Krishna
4	Godavari

Table 3.5: Conversion of table 4 into two 3NF relations

It should be stressed again that dependency between attributes is a semantic property and has to be stated in the problem specification. In this example the dependency between Year and Hostel is clearly stated. In case hostel allocated to students do not depend on their year in college, then the table 3.4 is already in 3NF.

Let us consider another example of a relation. The relation Employee is given below and its dependency diagram in figure 3.7.

Employee (**Employee code,** Employee name, Dept., Salary, Project no., Termination date of project).

As can be seen from the figure, the termination date of a project is dependent on the Project no. Thus this relation is not in 3NF. The 3NF relations are:

Employee (**Employee code**, employee name, salary, Project no.)

Project (**Project no.** Termination date)

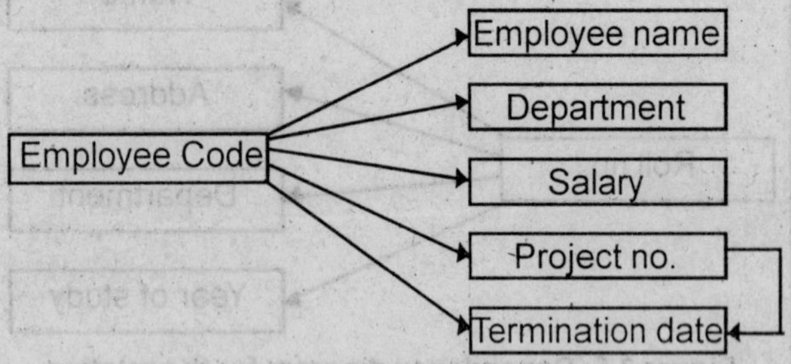

Figure 3.7: Dependency diagram of employee relation

8. Boyce- Codd Normal Form (BCNF): Assume that a relation has more than one possible key. Assume further that the composite keys have a common attribute. If an attribute of a composite key is dependent on an attribute of the other composite key, a normalization called BCNF is needed. Consider, as an example, the relation **Professor:**

Professor (**Professor code**, Dept, Head of Dept, parent time) It is assumed that

1. A professor can work in more than one department.
2. The percentage of the time he spends in each department is given.
3. Each department has only one Head of Department.

The relationship diagram for the above relation is given in figure 3.8. Table 3.6 gives the relation attributes. The two possible composite keys are professor code and Dept. or Professor code and Head of Dept. Observe that department as well as Head of Dept. are not non-key attributes. They are a part of a composite key.

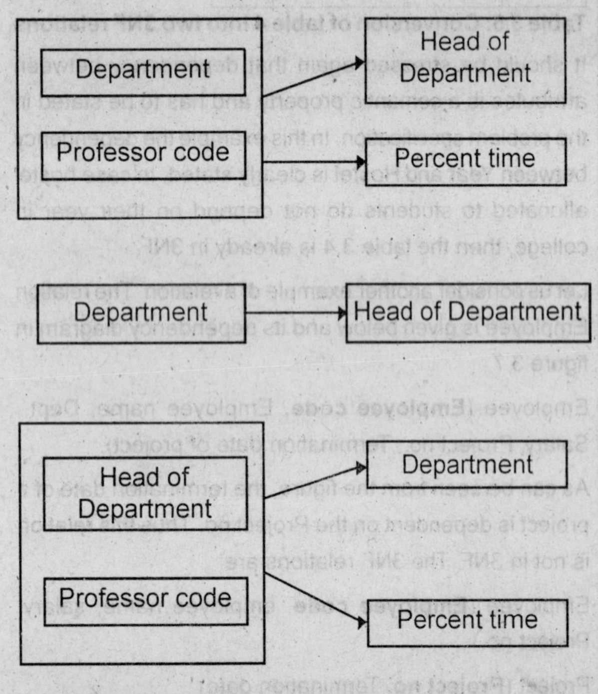

Figure 3.8: Dependency diagram of Professor relation

Professor Code	Department	Head of Dept.	Percent time
P1	Physics	Ghosh	50
P1	Mathematics	Krishnan	50
P2	Chemistry	Rao	25
P2	Physics	Ghosh	75
P3	Mathematics	Krushnan	100

Table 3.6: Normalization of Relation "Professor"

The relation given in table 3.6 is in 3NF. Observe, however, that the names of Dept. and Head of Dept. are duplicated. Further, if Professor P2 resigns, rows 3 and 4 are deleted. We lose the information that Rao is the Head of Department of Chemistry.

The normalization of the relation is done by creating a new relation for Dept. and Head of Dept. and deleting Head of Dept. from Professor relation. The normalized relations are shown in the following table 3.7.

(a)

Professor Code	Department	Percent time
P1	Physics	50
P1	Mathematics	50
P2	Chemistry	25
P2	Physics	75
P3	Mathematics	100

(b)

Department	Head of dept.
Physics	Ghosh
Mathematics	Krishnan
Chemistry	Rao

Table 3. 7: Normalization of Professor Relation in BCNF

and the dependency diagrams for these new relations in figure 3.8. The dependency diagram gives the important clue to this normalization step as is clear from figures 3.8 and 3.9.

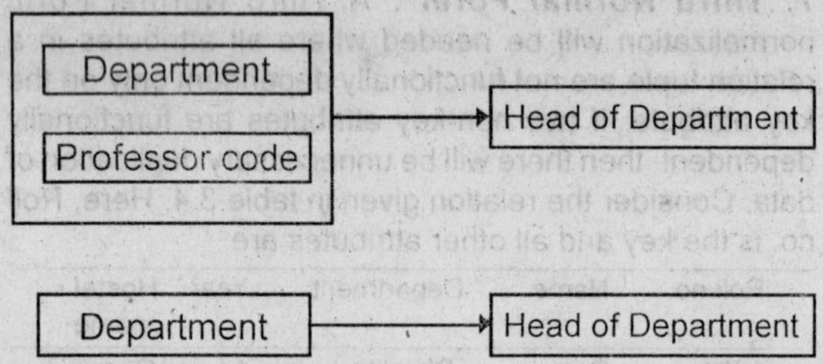

Figures 3.9: Dependency diagram of Professor relation

9. Fourth Normal Form (4 NF): Consider an unnormalized relation, CTX, containing information about courses, teachers and texts. The meaning of the given relation is that the indicated course can be taught by any of the indicated teachers, and uses all the indicated texts.

CTX	COURSE	TEACHER	TEXT
	Physics	Prof. Green	Basic Mechanics (BM)
		Prof. Brown	Principles of Optics (POO)
		Prof. Black	
	Maths	Prof. White	Modern Algebra
			Protective Geometry

Fig. 3.10 Sample tabulation of CTX (unnormalized)

Now let us convert this structure to an equivalent normalized form. Observe first that there are no FDS in the data at all (apart from trivial ones such as COURSE → COURSE).

CTX	COURSE	TEACHER	TEXT
	Physics	Prof. Green	Basic Mechanics
	Physics	Prof. Green	Principles of Optics
	Physics	Prof. Brown	Basic Mechanics
	Physics	Prof. Brown	Principles of Optics
	Physics	Prof. Black	Basic Mechanics
	Physics	Prof. Black	Principles of Optics
	Maths	Prof. White	Modern Algebra
	Maths	Prof. White	Protective Geometry

Fig. 3.11. Sample tabulation of CTX (normalized)

For a given course, all possible combinations of teacher and text appear – i.e., CTX satisfies the constraint if tuples < c, t_1, x_1 >, < c, t_2, x_2 > both appear, then tuples < c, t_1, x_2 >, < c, t_2, x_1 > both appear also.

This form contains a good deal of redundancy leading to problems over up date operations. CTX is in BCNF, since it is "all key" and there are no functional determinants.

The difficulties in relation CTX are caused due to the fact that the teachers and texts are independent of each other. The problems can be solved if CTX were replaced by its two projections CT (COURSE, TEACHER) and CX (COURSE, TEXT). (CT and CX are both "all key" and are thus both BCNF).

CT	COURSE	TEACHER	CX	COURSE	TEXT
	Physics	Prof. Green		Physics	Basic Mech.
	Physics	Prof. Brown		Physics	Principles of Optics
	Physics	Prof. Black		Maths	Modern Algebra
	Maths	Prof. White		Maths	Protective Geometry

This decomposition cannot be made on the basis of FDs. Instead, it is made on the basis of another type of dependency, the **Multivalued** dependency.

Multi-valued dependencies, (MVDs) are a generalisation of FDs. There are two MVDs in relation CTX :

CTX. COURSE → → CTX. TEACHER

CTX. COURSE → → CTX. TEXT.

(The MVD statement "R.A → → R.B" is read as "attribute R.B is multi-dependent on attribute R.A.", or, equivalently, "attribute R.A multi-determines attributes R.B")

TEACHER is not functionally dependent on COURSE. Same with TEXT and COURSE.

Definition of MVD. Given a relation R with attributes A, B, C, the multi-valued dependence

R.A → → R.B

Holds in R if and only if the set of B – values matching a given (A-value, C-value), pair in R depends only on the A-value and is independent of the C-value. As usual A, B and C may be composite.

This MVDs can exist only if the relation R has at least three attributes. It is common to express both statements as a single statement.

For example

COURSE → → TEACHER | TEXT

An FD is an MVD in which the "set" of department values actually consists of a single value. The two projections CT and CX are made on the basis of Fagin's theorem for replacement, according to which

- Relation R, with attributes A, B and C, can be non – loss decomposed into its two projections R1 (A, B) and R2 (A, C) if and only if the MVD A → → B/C holds in R.

Definition of 4 NF. A relation R is in fourth normal form (4NF) IF and only if, whenever there exists an MVD in R, say A → → B, then all attributes of R are also functionally dependent on A (i.e. A → X for all attributes X of R).

Relation CTX is not in 4NF, since it involves an MVD that is not an FD at all, let alone an FD in which the determinant is a candidate key. The two projections CT and CX are in 4NF, however.

Also,

- 4NF is strictly stronger than BCNF ; any 4NF relation is necessarily in BCNF.
- Any relation can be non loss-decomposed into an equivalent collection of 4NF relations.

SOLVED EXAMPLES

1. Consider the relational scheme of the relation SCHEDULE shown below. What is the highest normal form of this relation? What type of data anomalies does this relation have? Given an example of the type of data anomalies that this relation may experience.

SCHEDULE (Student-ID, Class - No, Student-Name, Student-Major, Class-Time, Building-Room, Instructor). Assume the following functional dependencies.

Student - ID → Student- Name.

Student - ID → Student- Major

Class-No → Class-Time

Class-No → Building-Room

Class-No → Instructor

Sol : There are some partial dependencies on the key, for example, class-No → Building-Room or student-ID → Student -Major. Therefore, the highest form of the relation is 1 NF.

2. Consider the relation scheme (Sales- Transaction-No, Item-No, Item-Price, Item-Quantity-Sold, Seller, Seller-District) and the FDs shown below. What is the key of this relation? Transform this relation to 3NF.

Sales -Transaction-No, Item -No → Item-Quantity-Sold

Item - No → Item-Price

Sales- Transaction-No → Seller.

Seller → Seller-District

The key of this relation is the composite attribute Sales

Transaction-No, Item-No. Infact, these two attributes determine all other attributes as indicated next.

a. Sales- Transaction-No, Item-No → Sales-Transaction-No (Reflexivity axiom)

b. Sales-Transaction-No, Item-No → Item-No (Reflexivity axiom)

c. Sales-Transaction-No, Item-No → Item-Quantity-Sold (given)

d. From sales- Transaction-No, Item-No → Item-No (Reflexivity axiom) and Item-No → Item- Price (given) we have that Sales Transaction-No, Item-No → Item-price (Transitivity axiom)

e. From Sales-Transaction-No, Item-No → Sales-Transaction- No (Reflexivity axiom) and Sales Transaction - No → Seller (given) we have that Sales- Transaction-No, Item-No → Seller (Transitivity axiom)

f. From Sales-Transaction-No, Item-no → Seller (Step e) and Seller → Seller-District we have that Sales-Transaction-No, Item-No → Seller-District (Transitivity axiom)

Sol : The highest form of this relation is 1 NF because there are partial dependencies on the primary key.

To transform it into 2NF, However, notice that we need to consider each of the partial dependencies separately. Therefore, instead of replacing the original relation with two, we need to replace it with three new relations. These relations are:

a. Item-Sold (Sales- Transaction-No, Item-No, Item - Quantity - Sold)

b. Item (Item - No, Item-Price)

c. Seller (Sales - Transaction-No, Seller, Region)

Relations (a) and (b) are in 3NF since the only nonprime attribute is fully dependent on the key. Relation (c) is not in 3NF because one of the nonprime attributes (Seller) functionally determines another nonprime attribute (Region). This relation can be transformed into 3NF. The new two relations are:

Seller-Transaction (Sales-transaction-No, Seller)

Seller-Region (Seller, Region)

3. The relation and all its dependencies are reproduced below.

SCHEDULE (Student-ID. Class No. Student-Name, Student Major, Class-Time, Building-Room, Instructor).

Assume the additional functional dependencies also hold:

Student - ID → Student - Name

Student - ID → Student - Major

Class-No → Class-Time

Class-No → Building - Room

Class - No → Instructor

Sol : In all cases where we need to transform a relation into 2NF. However, in this example, the solution is not as direct as in all previous cases. The reason for this is that the key is composite and each one of the prime attributes partially determines some other nonprime attributes. Therefore, instead of two relations we have to create at least three relations. These relations are:

STUDENT (Student -ID, Student-Name, Student-Major). Notice that all the attributes are dependent on the primary key student - ID.

CLASS (Class-No, Class-Time, Building-Room, Instructor).

Notice that all the attributes are dependent on the primary key Class-No.

Since the normalization process is reversible we should be able to recover the original relation, therefore, we need an extra relation to make this possible. In this case, we create a new relation STUDENT-CLASS (Student-ID, Class - No.)

4. Consider the relation r(X,Y,Z,W) and a set F=(Y↔W,XY→Z) where the symbol ↔ means that Y → W and W → Y simultaneously. What are the candidate keys of this relation? What is the highest normal form of this relation?

Sol : This relation has two candidate keys: XY and WX. XY is a candidate key because XY determines all other attributes as shown below.

a. XY → X (Reflexivity axiom)

b. XY → Y (Reflexivity axiom)

c. XY → Z (given)

d. XY→Y and Y→W then XY →W (Transitivity Axiom)

WX is a candidate key because WX determines all other attributes as shown below.

a. WX → W (Reflexivity)

b. WX → X (Reflexivity)

c. We already know that XY → Z (given). We also know that W → Y (given). Using the Pseudo transitivity axiom, we have that XW →Z.

d. WX →W (step a) and W→Y (given) therefore WX→Y (Transitivity axiom)

The prime attributes' are: X, Y and W. The only nonprime attribute is Z.

The relation is in 2NF because there are no partial dependencies of the non prime attribute on the key. In both cases we have that Z is fully dependent on both keys.

The relation is in 3NF because there is only one non prime attribute. Therefore, we cannot have one non prime attribute determining and the non prime.

The relation is not in BCNF because the determinants of either W → Y or Y → W are not keys of the relation. Therefore, the highest form of this relation is 3NF.

5. Show that every two-attribute relation is in BCNF. That is, if r(X,Y) then r(X,Y) is in BCNF.

Sol : Let us consider the following cases:

a. X is the sole key of the relation. In this case, the nontrivial dependency X → Y has X as a superkey since X ⊂ X.

b. Y is the sole key of the relation. In this case, the nontrivial dependency $Y \rightarrow X$ has Y as a superkey since $Y \subset Y$.

c. Both $X \rightarrow Y$ and $Y \rightarrow X$ hold simultaneously. Then whatever PK we consider for the relation we will have either X or Y as its determinant. Either one of the two possible cases has already been considered under (a) or (b).

6. Consider the relation r(A,B,C,D,E) and the set

$F = \{AB \rightarrow CE, E \rightarrow AB, C \rightarrow D\}$. What is the highest normal form of this relation?

Sol : Since r is a relation by definition the relation is already in 1 NF. To determine the normal form of the given relation, we need to determine all possible keys of the relation. The two possible keys of this relation are AB and E. Let us see why this is So. AB is a key

1. $AB \rightarrow A$ (Reflexivity axiom)
2. $AB \rightarrow B$ (Reflexivity axiom)
3. From $AB \rightarrow CE$ (given)we have that $AB \rightarrow C$ (Projectivity axiom)
4. From $AB \rightarrow CE$ (given) we have that $AB \rightarrow E$ (Projectivity axiom)
5. From $AB \rightarrow C$ (step 3) and $C \rightarrow D$ (given) we have that $AB \rightarrow D$ (Transitivity axiom)

Since AB determines all attributes of the relation AB is a key. E is a key

Since $E \rightarrow AB$ (given) and AB is a key then E is also a key because by application of the Transitivity axiom E can functionally determine all other attributes of the relation.

The prime attributes are: A, B and E.

The nonprime attributes are: C and D.

The relation is in 2NF because there are no partial dependencies on any of the key.

The relation is not in 3NF because there is a transitivity dependency on the keys. Notice that $C \rightarrow D$ and both attributes are nonprime.

7. Given the relation r(A,B,C) and the FDs $A \rightarrow B$ and $B \rightarrow C$, we know that r is not in 3NF because there is a transitivity dependency on the key A. Would this relation be in 3NF if $C \rightarrow B$ is true for this relation?

Sol : The relation is not in 3NF even if this functional dependency is true.

8. Consider the following relation REALTOR (Property-No, Date-Sold, License-No. Commission, Bonus). Assume also that the following dependencies hold in this database: Date-sold $\rightarrow$ Bonus, and License-No $\rightarrow$ Commission. If we assume that attributes Property-No and License-No form a composite key, what is the highest normal form of this relation?

Sol : There is a partial dependency on the key. Consider the FD: License-No $\rightarrow$ Commission. Therefore, the highest normal form is 1 NF.

9. Consider the relation r(A,B,C,D,E,F) and the set $F = \{A \rightarrow B, C \rightarrow DF, AC \rightarrow E, D \rightarrow F\}$, What is the key of the relation? What is the highest normal form of this relation? If it is not in 3NF find a decomposition that is lossless and dependency preserving.

Sol : The key of the relation is the composite attribute AC, As it stands this relation is only 1 NF. A 3NF decomposition of this relation is R_1 (A,C,E), R_2 (C,D), R_3 (D,F) and R_4 (A,B). This decomposition is lossless and preserves the dependencies.

10. Consider the relation r(A,B,C,D), the set $F= \{A \rightarrow B, C \rightarrow D\}$ and the decomposition R_1(AB) and R_2 (CD). Is this decomposition lossless? Does it preserve the dependencies of the set F?

Sol : The decomposition preserves the functional dependencies but is not lossless.

11. Consider the set of FDs = { Employee $\rightarrow$ Department, Employee $\rightarrow$ Title, Employee $\rightarrow$ Pay-Scale, Title $\rightarrow$ Pay-Scale}, and find at least two different loss less decompositions for the given relation.

Sol : R_1 (Employee, Department, Title) and R_2 (Title, Pay-Scale), Two other decompositions are: R_1 (Employee, Title) and R_2 (Employee, Department, Pay-Scale) or R_1 (Employee, Pay-Scale) and R_2 (Employee, Department, Title)

12. Find the highest normal form of the relation r(A, B, C, D) if the following FDs are satisfied by the relation

$F = \{AB \rightarrow D, AC \rightarrow BD, B \rightarrow C\}$

Sol : The keys of this relation are: AB and AC. The prime attributes are: {A,B,C}. Attribute D is the sole nonprime attribute. Attribute D is not partially dependent on any key. Therefore, the relation is in 2NF. The relation is also in 3NF because there is only one nonprime attribute; therefore, a transitive dependency between two different nonprime attributes cannot exist. The relation is not in BCNF because $B \rightarrow C$ and B is not a superkey of the relation.

13. Consider the relation for release RECORD (Title, Performer, Style, Price, Label, Producer) and the dependencies shown below. Indicate what is the highest normal form of this relation. Indicate a possible 3NF decomposition of this relation..

Title $\rightarrow$Label, Style Style $\rightarrow$Price Performer$\rightarrow$ Producer

Sol : The key of the relation is the composite attribute Title, Performer. As it stands the relation is in 1 NF because there is at least one partial dependency on the key: For example, Performer $\rightarrow$ Producer. A possible decomposition may be Record (Title, Performer), Genre (Title, Label, Style), Producer (Performer, Producer-Name), Cost (Style, Price).

14. Consider the following database relations containing the attributes

Book - id
Subject - Category - of - book
Name - of - author
Nationality - of - Author
with Book - id as the primary key.

(a) What is the highest normal form satisfied by this relation

(b) Suppose the attributes Book –title and Author – address are added to the relation, and the primary key is changed to {Name– of Author, Book – Title}, what will be the highest normal form satisfied by the relation? **[GATE-1998]**

Sol:

(a) Following diagrams shows the functional dependencies in the relation

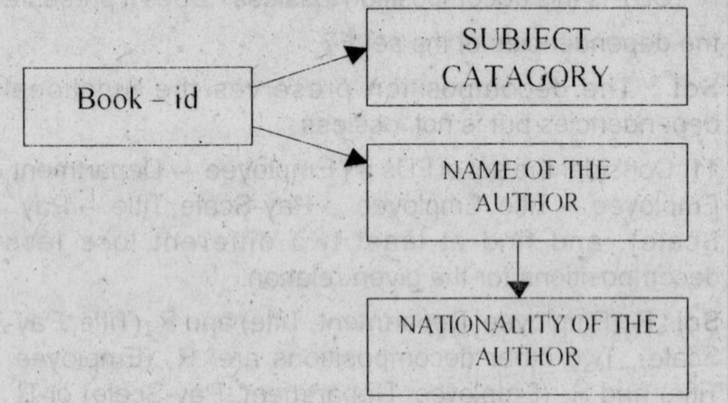

Nationality of the Author transitivily depends on Book – id, primary key. So it can't be in the 3NF but is in the 2 NF.

(b) Following diagrams shows the functional dependencies in the relation

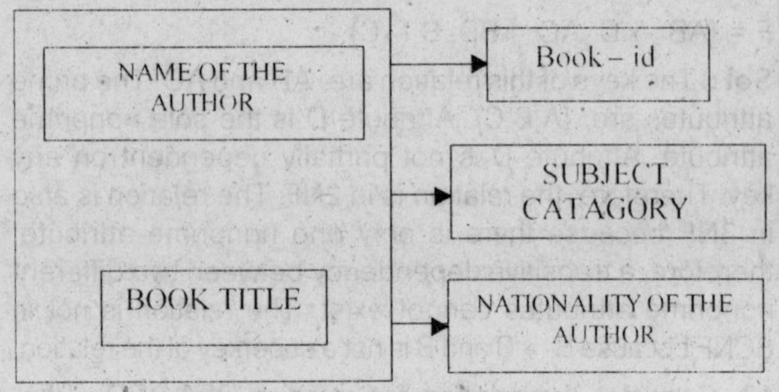

The above relation is in 3NF.

OBJECTIVE QUESTIONS - I

1. Let R=(A,B,C,D,E,F) be a relation scheme with the following dependencies: $C \to F, E \to A, EC \to D, A \to B$. Which of the following is a key for R? **[GATE-1999]**

a) CD b) EC c) AE d) AC

2. Consider the schema R = (S T U V) and the dependencies $S \to T$, $T \to U$, $U \to V$ and $V \to S$. Let R=(R1 and R2) be a decomposition such that $R1 \cap R2 = \phi$. The decomposition is : **[GATE-1999]**

a) Not in 2NF b) In 2NF but not in 3NF

c) In 3NF butnot in 2NF d) In both 2NF and 3 NF

3. Which normal form is considered adequate for normal relational database design? **[GATE-1998]**

a) 2NF b) 3NF c) 4NF d) 5NF

4. A functional dependency of the form $x \to y$ is trivial if

a) $y \subseteq x$ b) $y \subset x$ c) $x \subseteq y$ d) $x \subset y$ and $y \subset x$

5. If every non-key attribute is functionally dependent on the primary key, then the relation will be in

a) 1NF b) 2NF c) 3NF d) 4NF

6. Given the functional dependencies

$x \to w$; $x \to y$; $y \to z$ and $z \to pq$

which of the following does not hold good?

a) $x \to z$ b) $w \to z$ c) $x \to wy$ d) None

7. Third normal form is inadequate insituations where the relation

a) Has multiple candidate keys

b) Has candidate keys that arecomposite

c) Has overlapped candidate keys d) None.

8. If a relation scheme is in BCNF, then it is also in

a) 1NF b) 2NF c) 3NF d) None

9. Any Binary relation is in 3NF.

a) True b) False

10: Any Binary relation is in BCNF

a) True b) False

11. Normalization of database is used to **[AUCET-2000]**

a) Eliminate redundancy b) Improve security

c) Improve Efficiency d) Minimize errors

12. Functional dependencies are a generalization of ....

a) key dependencies b) Relation dependencies

c) Database dependencies d) None

13. Every Boyce-Codd Normal Form (BCNF) decompositon is....

a) Dependency preserving

b) Not dependency preserving

c) Need be dependency preserving

d) None

14. Which normal form is most desirable?

a) Domain-key normal form (DKNF)

b) Boyce-Codd normal form (BCNF)

c) Third normal form (3NF)

d) Fourth normal form (4NF)

15. The multi valued transitivity rule is

a) Sound b) Complete

c) Not sound but complete

d) Soundness does not apply

ANSWER KEY - I				
1. b	2. d	3. b	4. a	5. c
6. b	7. a,b,c	8. c	9. a	10. a
11. a	12. a	13. c	14. a	15. a

OBJECTIVE QUESTIONS - II

1. Relation R with an associated set of functional dependencies F, is decomposed into BCNF. The redundancy (arising out of functional dependencies) in the resulting set of relations is [GATE-2002]

a) zero

b) More than zero but lessthan that of an equivalent 3NF decomposition

c) Proportional to the size of F^+

d) Indeterminate

2. Relation R is decomposed using a set of functional dependencies F, and relation S is decomposed using another set of functional dependencies, G. one decomposition is definitely BCNF, the other is definitely 3NF, but it is not known which is which. To make a guaranteed identification, which one of the following tests should be used on the decompositions? (Assume that the closure of F and G are available). **[GATE-2002]**

a) Dependency-preservation

b) Lossless-Join

c) BCNF definition d) 3NF definition

3. From the following instance of a relation schema R(A,B,C), we can conclude that [GATE-2002]

A	B	C
1	1	1
1	1	0
2	3	2
2	3	2

a) A functionally determines B and B functionally determines C.

b) A functionally determines B and B does not functionally determine C.

c) B does not functionally determine C.

d) A does not functionally determine B and B does not functionally determine C.

4. Consider a schema R(A,B,C,D)and functional dependencies $A\rightarrow B$ and $C\rightarrow D$. Then the decomposition of R into R1(A B) and R2(C D) is [GATE-2001]

a) Dependency preserving and lossless Join.

b) Lossless Join but not dependency preserving

c) Dependency preserving but not lossless Join

d) Not dependency preserving and not lossless Join

5. R(A, B, C, D) is a relation. Which of the following does not have a lossless Join. Dependency preserving BCNF decomposition? [GATE-2001]

a) $A\rightarrow B, B\rightarrow CD$. b) $A\rightarrow B, B\rightarrow C, C\rightarrow d$

c) $AB\rightarrow C, C\rightarrow AD$ d) $A\rightarrow BCD$

6. Given the following relation instance [GATE-2000]

X	Y	Z
1	4	2
1	5	3
1	6	3
3	2	2

Which of the following functional dependencies are satisfied by the instance?

a) $xy\rightarrow z$ and $z\rightarrow y$ b) $yz\rightarrow x$ and $y\rightarrow z$

c) $yz\rightarrow x$ and $x\rightarrow z$ d) $xz\rightarrow y$ and $y\rightarrow x$.

7. For a database relation R(a,b,c,d), where the domains of a,b,c,d include only atomic values, only the following functional dependencies and those that can be inferred from them hold: $a\rightarrow c, b\rightarrow d$; The relation is:

[GATE-1997]

a) In First Normal form butnot in second normal form

b) In Second normal form butnot in third normal form

c) In Third normal form

d) None

8. Consider the following set of functional dependencies on the scheme (A,B,C).

$A\rightarrow BC$

$B\rightarrow C$

$A\rightarrow B$

$AB\rightarrow C$

The canonical cover for this set is

a) $A\rightarrow BC$ and $B\rightarrow C$ b) $A\rightarrow BC$ and $AB\rightarrow C$

c) $A\rightarrow BC$ and $A\rightarrow B$ d) $A\rightarrow B$ and $B\rightarrow C$

9. If a relation with a schema R is decomposed into tworelations R_1 and R_2 such that $(R_1\cup R_2)=R$, then which one of the following is to be satisfied for a lossless joint decomposition (→ indicates functional dependency)

[UGC NET-1998]

a) $R_1\cap R_2\rightarrow R_1$ b) $R_1\cap R_2\rightarrow R_2$

c) $R_1\cap R_2\rightarrow R_1$ or $R_1\cap R_2\rightarrow R_2$

d) $R_1\cap R_2\rightarrow R_1$ and $R_1\cap R_2\rightarrow R_2$

10. Given a relation R with attributes A, B, C, D denoted by R(A, B, C, D) with functional dependencies F ={A →B, A→C, A→D}, the decomposition of R into R_1 (A, B, D) with functional dependencies F_1 ={A→B, A→D} and R_2(B, C) with functional dependencies F_2={ }. The decomposition is **[UGC NET-2000]**

a) lossless and not dependency preserving

b) not lossless but dependency preserving

c) neither lossless nor dependency preserving

d) both lossless and dependency preserving

11. 'Every determinant should be a candidate key' is the definition of ..... normal form. **[UGC NET-1999]**

a) Project-Join NF b) Second NF

c) Boyce Codd NF d) Third NF

ANSWER KEY - II				
1. a	2. b	3. a	4. c	5. c
6. b	7. b	8. a	9. c	10. c
11. c				

IV. FILE ORGANISATION AND INDEXING TECHNIQUES

- File Organisation Techniques
- Indexing Techniques
- Search Trees, B-trees and B⁺- trees

1. File Organisation

The technique used to represent and store the recordson a file is called the file organisation. The four fundamental file organisation techniques that we will discuss are the following :

1. Sequential
2. Relative
3. Indexed sequential
4. Multi-key

There are two basic ways, that the file organisation techniques differ. First, the organisation determines the file's record sequencing, which is the physical ordering of the record in storage.

Second, the file organisation determines the set of operarions necessary to find particular records. Individual records are typically identified by having particular values in search-key fields. This data field may or may not have duplicate values in file, the field can be a group or elementary item. Some file organisation techniques provide rapid accessibility on a variety of search key; other techniques support direct access only on the value of a single one.

The organisation most apporpriate for a particular file is determined by the operational characteristics of the storage medium used an the nature of the operations to be performed on the data. The most important characteristic of a storage device that influences selection of a storage device once the appropriate file organisation technique has been determined) is whether the device allows direct access to particular record occurrences without accessing all physical prior record occurrences that are stored on the device,or allows only **sequential access** to record occurrences. Magnetic disks are examples of direct access storage devices (abbreviated DASD's); magnetic tapes are examples of sequential storage devices.

Sequential File Orgnaisation

The most basic way to organise the collection of records that from a file is to use sequential organisation. In a sequentially organised file records are written consecutively when the file is created and must be accessed consecutively when the file is later used for input (figure 4.1).

In a sequential file, records are maintained in the logical sequence of their primary key values. The processing of a sequential file is conceptually simple but inefficient for random access. However, if access to the file is strictly sequential, a sequential file is suitable. A sequential file could be stored on a sequential storage device such as a magnetic tape.

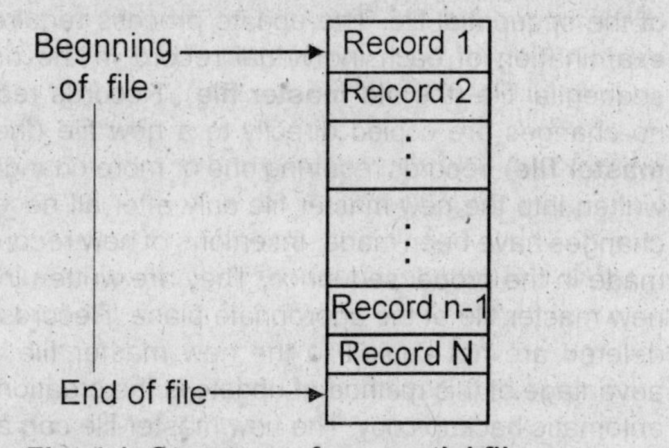

Fig. 4.1: Structure of sequential file

Search for a given record in a sequential file requires, on average, access to half the records in the file. Consider a system where the file is stored on a direct access device such as a disk. Suppose the key value is separated from the rest of the record and a pointer is used to indicate the location of the record. In such a system, the device may scan over the key values at rotation speeds and only read in the desired record. A binary or logarithmic search technique may also be used to search for a record. In this method, the cylinder on which the required record is stored is located by a series of decreasing head movements. The search, having been localised to a cylinder, may require the reading of half the tracks, on average,in the case where keys are embedded in the physical records, or require only a scan over the tracks in the case where keys are also stroed separately.

Updating usually requires the creation of a new file. To maintain file sequence, records are copied to the point where amendment is required. The changes are then made and copied into the new file. Following this, the remaining records in the original file are copied to the new file. This method of updating a sequential file creates an automatic backup copy. It permits updates of the type U_1 through U_4.

Addition can be handled in a manner similar to updating. Adding a record necessitates the shifting of all records from the appropriate point to the end of file to create space for the new record. Inversely, deletion of a record requires a compression of the file space, achieved by the shifting of records. Changes to an existing record may also require shifting if the record size expands or shrinks.

The basic advantage offered by a sequential file is the ease of access to the next record, the simplicity of organisation and the absence of auxiliary data structures. However, replies to simple queries are time consuming for large files. Updates, as seen above, usually require the creation of a new file. A single update is an expensive proposition if a new file must be created. To reduce the cost per update, all such requests are batched, sorted in the order of the sequential file, and then used to update the sequential file in a single pass. Such a file, containing the updates to be made to a sequential file, is sometimes referred to a **transaction file**.

In the batched mode of updating, a transaction file of update records is made and then sorted in the sequence of the sequential file. The update process requires the examination of each individual record in the original sequential file (the **old master file**). Records requiring no changes are copied directly to a new file (the **new master file**); records requiring one or more changes are written into the new master file only after all necessary changes have been made. Insertions of new records are made in the proper sequence. They are written into the new master file at the appropriate place. Records to be deleted are not copied to the new master file. A big advantage of this method of update is the creation of an automatic backup copy. The new master file can always be recreated by processing the old master file and the transaction file.

Figure 4.2: A file with empty spaces for record insertions

A possible method of reducing the creation of a new file at each update run is to create the original file with "holes" (space left for the addition of new records, as shown in the last figure). As such, if a block could hold K records, then at initial creation it is made to contain only $L * K$ records, where $0<L\leq 1$ is known as the loading factor. Additional space may also be earmarked for records that may "overflow" their blocks, e.g., if the record ri logically belongs to block Bi but the physical block Bi does not contain the requisite free space. This additional free space is known as the overflow area. A similar technique is employed in index-sequential files.

Index-Sequential File Organisation

The retrieval of a record from a sequential file, on average, requires access to half the records in the file, making such enquiries not only inefficient but very time consuming for large files. To improve the query response time of a sequential file, a type of indexing technique can be added.

An index is a set of y, address pairs, Indexing associates a set of objects to a set of orderable quantities, which are usually smaller in numer or their properties provide a mechanism for faster search. The purpose of indexing is to expedite the search process. Indexes created from a sequential (or sorted) set of primary keys are referred to as index sequential. Although the indices and the data blocks are held toghether physically, we distinguish between them logically. We shall use the term **index file** to dexribe the indexes and **data file** to refer to the data records. The index is usually small enough to be read into the processor memory.

A sequential (or sorted on primary keys) file that is indexed is called an **index sequential file**. The index provides for random access to records, while the sequential nature of the file provides easy access to the subsequent records as well as sequential processing. An additional feature of this file system is the **overflow area**. This feature provides additional space for record addition without necessitating the creation of a new file. Before starting discussion on index sequential file structure,, let us, discuss types of indexes.

2. Types of Indexes

The idea behind an index access structure is similar to that behind the indexes used commonly in textbooks. A textbook index lists important terms at the end of the book in alphabetic ordr. Along with each term, a list of page numbers where the term appears is given. We can search the index to find a list of **addresses** - page numbers in this case- and use these addresses to locate the term in the textbook by searching the specified pages. The alternative, if no other guidance is given, is to sift slowly through the whole textbook word by word to find term we are interested in, which corresponds to doing a linear search on a file. Of course, most books do have additional information, such as chapter and section titles, which can help us find a term without having to search through the whole book. However, the index is the only exact indication of where each term occurs in the book.

An index is usually defined on a single field of a file, called an **indexing field**. The index typically stores each value of the index field along with a list of pointers to all disk blocks that contain a record with that field value. The values in the index are ordered so that we can do a binary search on the index. The index file is much smaller than the data file, so searching the index using binary search is reasonably efficient. Multilevel indexing does away with the need for binary search at the expense of creating indexes to the index itself !

There are several types of indexes. A **primary index** is an index specified on the ordering key field of an ordered file of records. Recall that an ordering key field is used to physically order the file records on disk, and every record has a unique value for that field. If the ordering field is not a key field that is, several records in the file can have the same value for the ordering field another type of index, called a **clustering index**, can be used. Notice that a file can have at most one physical ordering field, so it can have at most one primary index or one clustering index, but not both. A third type of index, called a **secondary index**, can be specified on any nonordering field of a file. A file can have several secondary indexes in addition to its primary access method. In the next three subsections we discuss these three types of indexes.

Primary Indexes

A **primary index** is an ordered file whose records are of fixed length with two fields. The first field is of the same data types as the ordering key field of the data file and the second filed is a pointer to a disk block - a block address. The ordering key field is called the **primary key** of the data file. There is one index entry (or **index record**) in the index file for each block in the data file. Each index entry has the value of the primary key field for the first record in a block and a pointer to other block as its two field values. We will refer to the two field values of index entry i as K (i) , P (i).

To create a primary index on the ordered file shown in figure 4.3, we use the Name filed as the primary key, because that is the ordering key field of the file (assuming that each value of

	NAME	SSN	BIRTHDATE	JOB	SALARY	SEX
block 1	Aaron, Ed					
	Abbott, Diane					
	⋮					
	Acosta,Marc					
block 2	Adams, John					
	Adams, Robin					
	⋮					
	Akers, Jan					
block 3	Alexander, Ed					
	Alfred, Bob					
	⋮					
	Allen, sam					

Figure 4.3: Some blocks on an ordered (sequential) file of EMPLOYEE records with NAME as the ordering field

NAME is unique). Each entry in the index will have a NAME value and a pointer. The first three index entries would be :

< K (1) = (Aaron, Ed), P (1) = address of block 1 >

< K (2) = (Adams, John), P (1) = address of block 2 >

< K (3) = (Alexander, Ed), P (3) = address of block 3 >

Figure 4.4 illustrates this primary index. The total number of entries in the index will be the same as the number of disk blocks in the ordered data file. The first record in each block of the data file is called the **anchor record** of the block, or simply the **block anchor** (a scheme similar to the one described here can be used, with the last record in each block, rather than the first, as the block anchor. A primary index is an example of what is called a **nondense index** because it includes an entry for each disk block of the data file rather than for every record in the data file. A **dense index,** on the other, hand, contains an entry for every record in the file.

The index file for a primary index needs substantially fewer blocks than the data file for two reasons. First, there are fewer index entries than there are records in the data file because an entry exists for each whole block of the data file rather than for each record. Second, each index entry is typically smaller in size than a data record because it has only two fields, so more index entries than data records will fit in one block. A binary search on the index file will hence require fewer block accesses than a binary search on the data file.

A record whose primary key value is K will be in the block whose address is P (i), where $K i \le K \le (i+1)$. The ith block in the data file contains all such records because of the physical ordering of the file records on the primary key field, we do a binary search on the index file to find the appropriate index entry i, then retrieve the data file block whose address is P (i). Notice that the above formula would not be correct if the data file was ordered on a nonkey field that allows multiple records to have the same ordering field value. In that case the same index values as that in the block anchor could be repeated in the last records of the previous block.

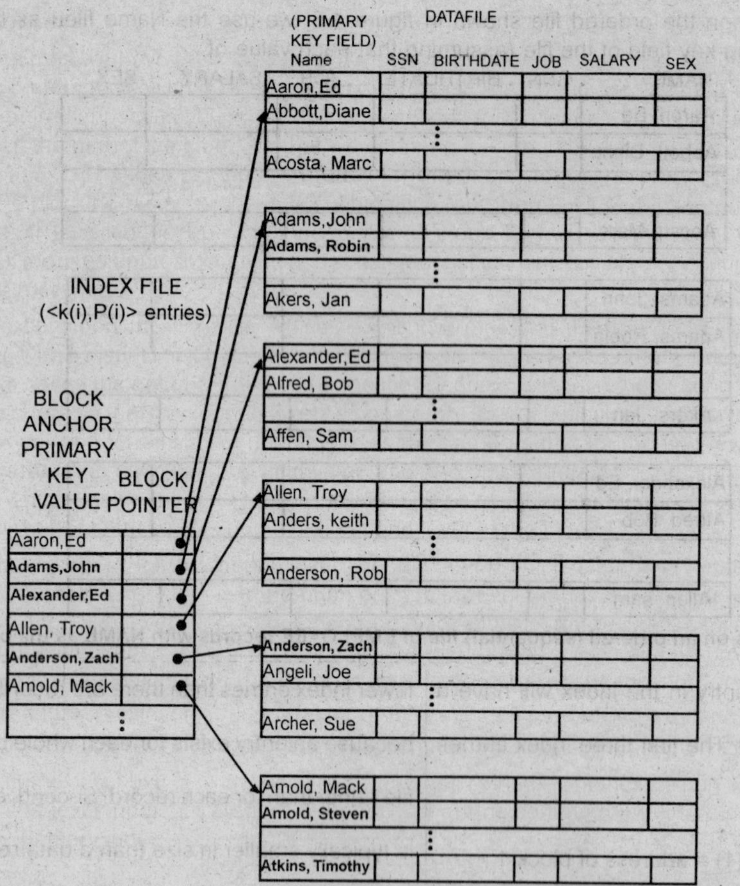

Figure 4.4 Primary index on the ordering key field of the files shown in the Figure 4.5

3. Clustering Indexes

If records of a file are physically ordered on a nonkey field--which does not have a distinct value for each record--that fields is called the clustering field. We can create a different type of index, called a clustering index, to speed up retreival of records that have the same value for the clustering field. This differs from a primay index, which requires that the ordering field of the data file have a distinct value for each record. A clustering index is also an ordered file with two fields; the first field is of the same type as the clustering field of the data file, and the second field is a block pointer. There is one entry in the clustering index for each distinct value of the clustering field, containing the value and a pointer to the first block in the data file that has a record with that value for its clustering field. Figure 4.5 shows an example.

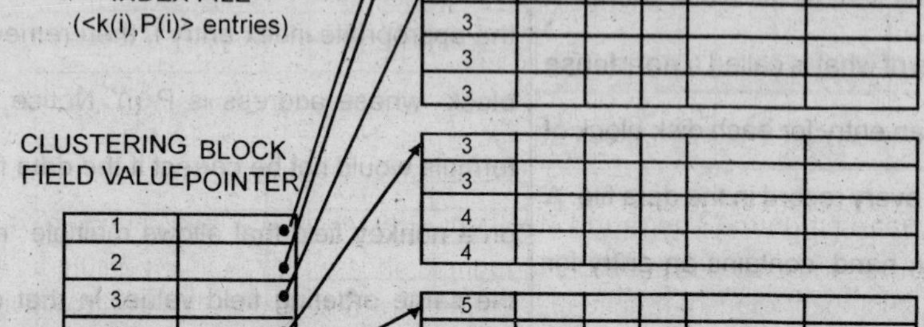

Figure 4.5 A clustering index on the DEPTNUMBER ordering nonkey field of an EMPLOYEE file

Notice that record insertion and deletion still cause problems, because the data records are physically ordered. To alleviate the problem of insertion, it is common to reserve a whole block (or a cluster of contiguous blocks) for each value of the clustering field; all records with that value are placed in the block (or a block cluster). This makes insertion and deletion relatively straight forward. Figure 4.6 shows this scheme.

A clustering index is another example of a nondense index, because it has an entry for every distinct value of the indexing field rather than for every record in the file.

4. Secondary Indexes

A secondary index is also an ordered file with two fields. The first field is of the same data type as some nonordering field of the data file that is an indexing field. The second field is either a block pointer or a record pointer. There can be many secondary indexes (and hence, indexing fields) for the same file.

We first consider a secondary index access structure on a key field that has a distinct value for every record. Such a field is sometimes called a secondary key. In this case there is one index entry for each record in the data file, which contains the value of the secondary key for the record and a pointer either to the block in which the record is stored or to the record itself. Hence, such an index is dense.

We again refer to the two field values of index entry i as <K(i), P(i)>. The entries are ordered by value of K(i), so we can perform a binary search. Because the records of the data file are not physically ordered by values of the secondary key field,we cannot use block anchors. That is why an index entry is created for each record in the data file, reather than for each block, as in the case of a primary index. Once the appropriate block is transferred to main memroy, a search for the desired record within the block can be carried out.

A secondary index usually needs more storage space and longer search time than does a primary index, because of its larger number of entries. However, the improvement in search time for an arbitrary record is much greater for a secondary index than for a primary index, since we would have to do a linear search on the data file if the secondary index did not exist. For a primary index, we could still use a binary search on the main file,even if the index did not exist.

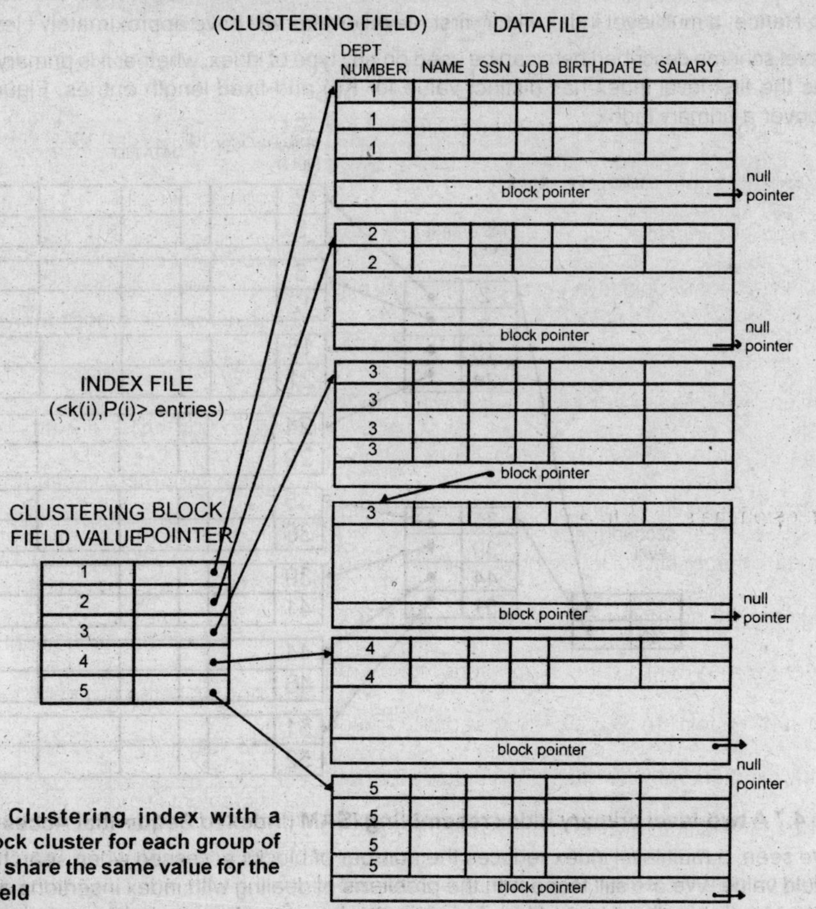

Figure 4.6 Clustering index with a separate block cluster for each group of records that share the same value for the clustering field

Notice that a secondary index provides a logical ordering on the records by the indexing field. If we access the records in order of the entries in the secondary index, we get them in order of the indexing field.

5. Multilevel Indexes.

The indexing schemes we have described thus far involve an ordered index file. A binary search is applied to the index to locate pointers to a disk block or to a record (or records) in the file having a specific index field value. A binary search requires apporximately ($\log_2 b_i$) block accesses for an index with b_i blocks, because each step of the algorithm reduces the part of the index file that we continue to search by a factor of 2. This is why we take the log function to the base 2. The idea behind a **multilevel index** is to reduce the part of the index that we continue to search by bfr_i, the blocking factor for the index, which is larger than 2. Hence, the search space is reduced much faster. The value bfr_i is called the **fan-out** of the multilevel index, and we will refer to it by the symbol, **fo**. Searching a multilevel index requires approximately ($\log_{fo} b_i$) block accesses, which is a smaller number than for binary search if the fan-out is larger than 2.

A multilevel index considers the index file, which we will now refer to as the **first** (or base) **level** of a multilevel index, as an ordered file with a distinct value for each K(i). Hence we can create a primary index for the first level; this index to the first level is called the **second level** of the multilevel index. Because the second level is a primary index, we can use block anchors so that the second level has one entry for each block of the first level.The blocking factor bfr_i for the second level--and for all subsequent levels--is the same as that for the first-level index, because all index entries are the same size; each has one field value and one block address. If the first level has r_1 entries, and the blocking factor--which is also the fan-out--for the index is $bfr_i=f_0$, then the first level needs $[r_1/fo]$ blocks, which is therefore the number of entries r_2 needed at the second level of the index.

We can repeat this process for the second level.The third level, which is a primary index for the second level, has an enty for each second-level block, so the number of third-level entries is $r_3= [(r_2/fo)]$. Notice that we require a second level only if the first level needs more than one block of disk storage, and, similarly, we require a third level only if the second level needs more than one block. We can repeat the preceding process until all the entries of some index level t fit in a single block. This block at the tth level is called the top index level. Each level reduces the number of entries at the previous level by a factor of fo--the index fan-out--so we can use formula $1 \leq (r_1/((fo)^t))$ to calculate t. Hence, a multilevel index with r_1 first - level entries will have approximately t levels, where $t = [(\log_{fo}(r_1))]$.

The multilevel scheme described here can be used on any type of index, whether it is primary, clustering, or secondray--as long as the first-level index has distinct value for K(i) and fixed-length entries. Figure 4.7 shows a multilevel index built over a primary index.

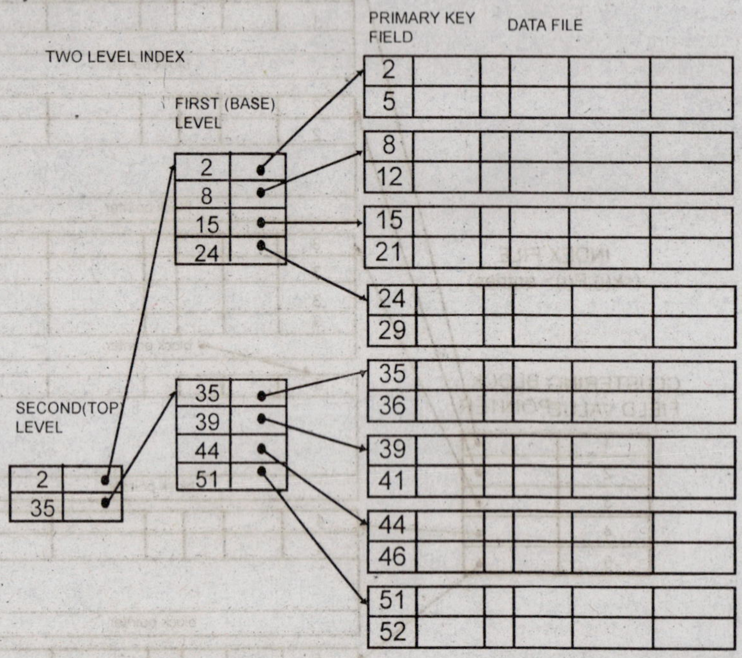

Figure 4.7 A two-level primary index resembling ISAM (Indexed Sequential Access Method) organization.

As we have seen, a multilevel index reduces the number of blocks accessed when searching for a record, given its indexing field value. We are still faced with the problems of dealing with index insertions and deletions, because all index levels are physically ordered files. To retain the benefits of using multilevel indexing while reducing index insertion and deletion problems, designers adopted a multilevel index that leaves some space in each of its blocks for inserting new entries. This is called a dynamic multilevel index and is oftern implemented by using data structures called B-trees and B^+ - trees.

6. Dynamic Multilevel Indexes Using B - Trees and B+ - Trees

B - trees and B^+ - trees are special cases of the well-known tree data structure. We introduce very briefly the terminology used in discussing tree data structures. A tree is formed of nodes. Each node in the tree, except for a special node called the root, has one parent node and several--zero or more--child nodes. The root node has no parent. A node that does not have any child nodes is called a leaf node; a nonleaf node is called an internal node. The level of a node is always one more than the level of its parent, with the level of the root node being zero. A subtree of a node consists of that node and all its descendant nodes--its child nodes, the child nodes of its child nodes, and so on. A precise recursive definition of a subtree is that it consists of a node n and the subtrees of all the child nodes of n, Figure 4.8 illustrates a tree data structure. In this figure the root node is A, and its child nodes are B,C,and D. Nodes E,J,C,,H,and K are leaf nodes.

Usually, we display a tree with the root node at the top, as shown in Figure 4.8. One way to implement a tree is to have as many pointers in each node as there are child nodes of that node. In some cases, a parent pointer is also stored in each node. In addition to pointers, a node usually contains some kind of stored information. When a multilevel index is implemental as a tree structure, this information includes the values of the file's indexing field that are used to guide the search for a particular record.

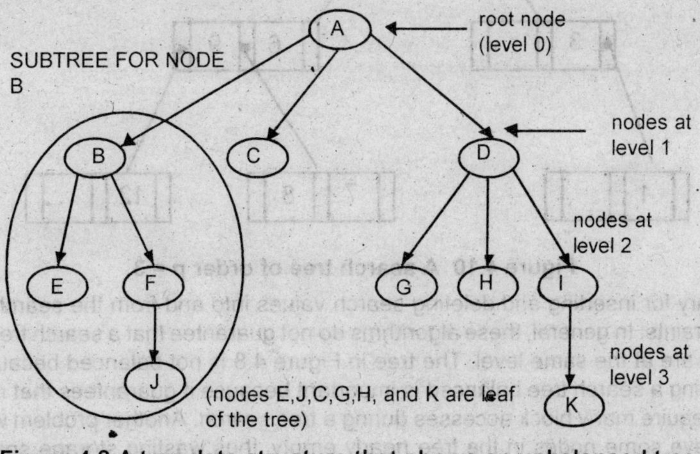

Figure 4.8 A tree data structure that shows an unbalanced tree.

7. Search Trees and B-Trees

A search tree is a special type of tree that is used to guide the search for a record, given the value of one of the record's fields. The multilevel indexes can be thought of as a variation of a search tree; each node in the multilevel index can have as many as fo pointers and fo key values,where fo is the index fan-out.The index field values in each node guide us to the next node, until we reach the data file block that contains the required records. By following a pointer, we restrict our search at each level to a subtree of the search tree and ignore all nodes not in this subtree.

Search Trees A search tree is slightly different from a multilevel index. A search tree of order p is a tree such that each node contains at most p - 1 search values and p pointers in the order $< P_1, K_1, P_2, K_2, \ldots, P_{q-1}, K_{q-1}, P_Q >$, Where $q \le p$; each P_i is a pointer to a child mode (or a null pinter); and each K_i is a search value from some ordered set If values. All search values are assumed to be unique. Figure 4.9 illstrates a node in a search tree. Two constraints must hold at all times on the search tree :

1. Within each node, $K_1 < K_2 < \ldots\ldots K_{q-1}$.
2. For all values X in the subtree pointed at by P_i we have $K_{i-1} < x < K_i$ for $1 < i < q$; $X < K_i$ for i = 1; and $K_{i-1} < X$ for i=q(see figure 4.13)

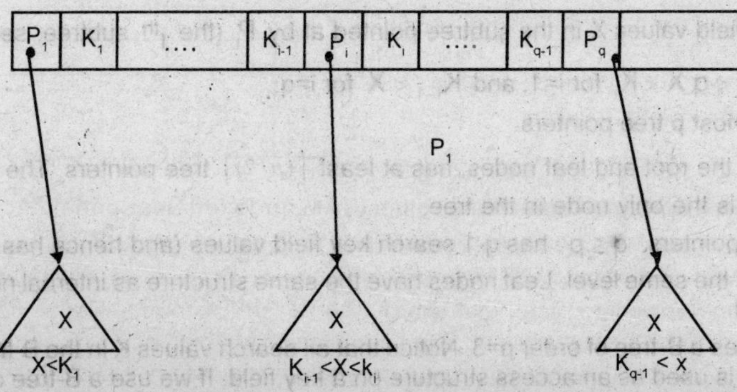

Figure 4.9 A node in a search tree with subtrees below it.

Whenever we search for a value X, we follow the appropriate pointer P_i according to the formulas in condition 2 above. Figure 4.10 illustratres a search tree of order p = 3 and integer search values. Notice that some of the pointers P_i in a node may be null pointers.

We can use a search tree as a machanism to search for records stroed in a disk file. The values in the tree can be the values of one of the fiels of the file, called the search field (which is the same as the index field if a multilevel index guides the search). Each key value in the tree is associated with a pointer to the record in the data file having that value. Alternatively, the pointer could be to the disk block containing that record. The search tree itself can be stored on disk by assigning each tree node to a disk block. When a new record is inserted, we must update the search tree by inserting an entry in the tree containing the search field value of the new record and a pointer to the new record.

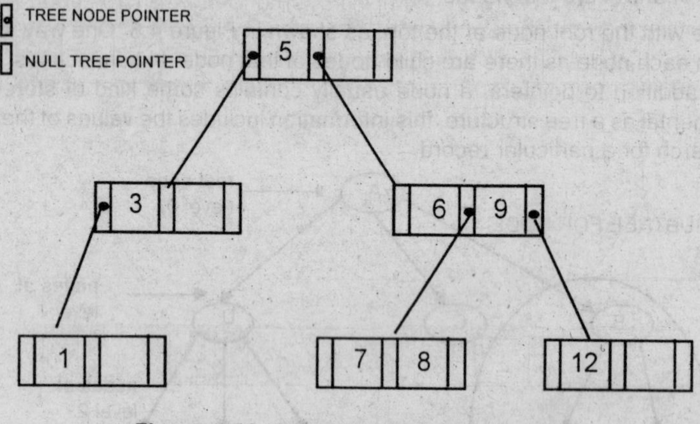

Figure 4.10 A search tree of order p = 3

Algorithms are necessary for inserting and deleting search values into and from the searchtree while mainatining the preceding two constraints. In general, these algorithms do not guarantee that a search tree is balanced, meaning that all of its leaf nodes are at the same level. The tree in Figure 4.8 is not balanced because it has leaf nodes at levels 1, 2, and 3. Keeping a search tree balanced is important because it guarantees that no nodes will be at very high levels and hence require many block accesses during a tree search. Another problem with search trees is that record deletion may leave some nodes in the tree nearly empty, thus wasting storage space and increasing the number of levels. The B-tree address both of these problems by specifying additional constraints on the search tree.

B-Trees. The B-tree has additional constraints that ensure that the tree is always balanced and that the space wasted by deletion, if any never becomes excessive. The algorithms for insertion and deletion, though,become more complex in order to maintain these constraints. Nonetheless, most insertions and deletions are simple processes; they become complicated only under special circumstances--namely, whenever we attempt an insertion into a node that is already full or a delection from a node that makes it less than half full. More formally, a B-tree of order p, when used as an access structure on a key field to search for records in a data file, can be defined as follows :

1. Each internal node in the B-tree (figure 4.11 a) is of the form $< P_1, < K_1, P_{r_1} >, P_2, < K_2, P_{r_2} >, \ldots\ldots, < K_{q-1}, P_{r_{q-1}} >, P_q >$ where $q \leq p$, Each P_i is a **tree pointer**--a pointer to another node in the B-tree. Each P_{r_i} is a **data pointer** -- a pointer to the record whose search key field value is equal to K_i (or to the data file block containing that record).
2. With in each node, $K_1 < K_2 < \ldots\ldots < K_{q-1}$
3. For all search key field values X in the subtree pointed at by P_i (the ith subtree, see Figure 4.15 a), we have: $K_{i-1} < X < K_i$ for $1 < i < q$; $X < K_i$ for i=1; and $K_{i-1} < X$ for i=q.
4. Each node has at most p tree pointers.
5. Each node, except the root and leaf nodes, has at least $\lceil (p/2) \rceil$ tree pointers. The root node has at least two tree pointers unless it is the only node in the tree.
6. A node with q tree pointers, $q \leq p$, has q-1 search key field values (and hence has q-1 data pointers).
7. All leaf nodes are at the same level. Leaf nodes have the same structure as internal nodes except that all of their tree pointer P_i are null.

Figure 4.11 (b) illustrates a B-tree of order p=3. Notice that all search values K in the B-tree are unique because we assumed that the tree is used as an access structure on a key field. If we use a B-tree on a nonkey field, we must change the definition of the file pointers Pr_i to point to a block--or cluster of blocks--that contain the pointers to the file records.

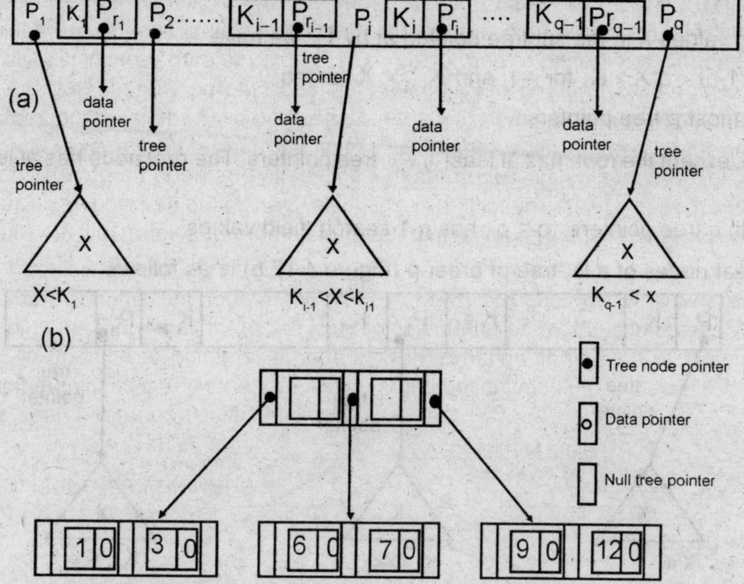

Figure 4.11 B-tree structures (a) A node in a B-tree with q-1 search values (b) A B-tree of order p=3. The values were inserted in the order 8,5,1,7,3,12,9,6

A B-tree starts with a single root node (which is also a leaf node) at level 0 (zero). Once the root node is full with p-1 search key values and we attempt to insert another entry in the tree, the root node splits into two nodes at level 1. Only the middle value is kept in the root node, and the rest of the values are split evenly between the other two nodes. When a nonroot node is full and a new entry is inserted into it, that node is split into two nodes at the same level, and the middle entry is moved to the parent node along with two pointers to the new split nodes. If the parent node is full, it is also split. Splitting can propagate all the way to the root node, creating a new level if the root is split.

We do not discuss algorithms for B-trees in detail here; rather, we outline search and insertion porcedures for B^+ - trees in the next section.

If deletion of a value causes a node to be less than half full, it is combined with its neighboring nodes, and this can also propagate all the way to the root. Hence, delection can reduce the number of tree levels. It has been shown by analysis and simulation that,after numerous random insertions and deletions on a B-tree, the nodes are approximately 69 percent full when the number of values in the tree stabilizes. This is alo true of B^+ trees. If this happens, node splitting and combining will occur only rerely, so insertion and deletion become quite efficient. If the number of values grows, the tree will expand without a problem--although splitting of nodes may occur, so some insertions will take more time

B-trees are sometimes used as primary file organizations. Inthis case, whole records are stored within the B-tree nodes rather than just the <searchkey, record pointer> entries. This works well for files with a relatively small number of records, and a small record size. Otherwise, the fan-out and the number of levels become too great to permit efficient access.

In summary, B-trees provide a multilevel access structure that is a balanced tree structure in which each node is at least half full. Each node in a B-tree of order p can have at most p-1 search values.

B⁺ - Tree

Most implmentations of a dynamic multilevel index use a variation of the B-tree data structure called a B^+-tree. In a B-tree, every value of the search field appears once at some level in the tree, along with a data pointer. In a B^+-tree, data pointers are stored only at the leaf nodes of the tree; hence, the structure of leaf nodes differs from the structure of internal nodes. The leaf nodes have an entry for every value of the search field, along with a data pointer to the record (or to the block that contains this record) if the search field is a key field. For a nonkey search field, the pointer points to a block containing pointers ot the data file records,creating an extra level of indirection.

The leaf nodes of the B^+-tree are usually linked together to porivde ordered acces on the search field to the records. These leaf nodes are similar to the first (base) level of an index. Internal nodes of the B^+-tree correspond to the other levels of a multilevel index. Some search field values from the leaf nodes are repeated in the internal nodes of the B^+-tree of order p (Figure 4.12a) is as follows:

1. Each internal node is of the form $< P_1, K_1, P_2, K_2, \ldots\ldots, P_{q-1}, K_{q-1}, P_q >$ where $q \leq p$, each P_i is a tree pointer

2. With in each node, $K_1 < K_2 < \ldots\ldots < K_{q-1}$
3. For all search field values X in the subtree pointed at by P_i we have:

   $K_{i-1} < X \le K_i$ for $1 < i < q$; $X < K_i$ for i=1; and $K_{i-1} < X$ for i=q.
4. Each node has at most p tree pointers.
5. Each internal node, except the root, has at least $q \le p$ tree pointers. The root node has at least two tree pointers if it is a internal node.
6. A internal node with q tree pointers, $q \le p$, has q-1 search field values.

The structure of the leaf nodes of a B⁺-tree of order p (Figure 4.12 b) is as follows:

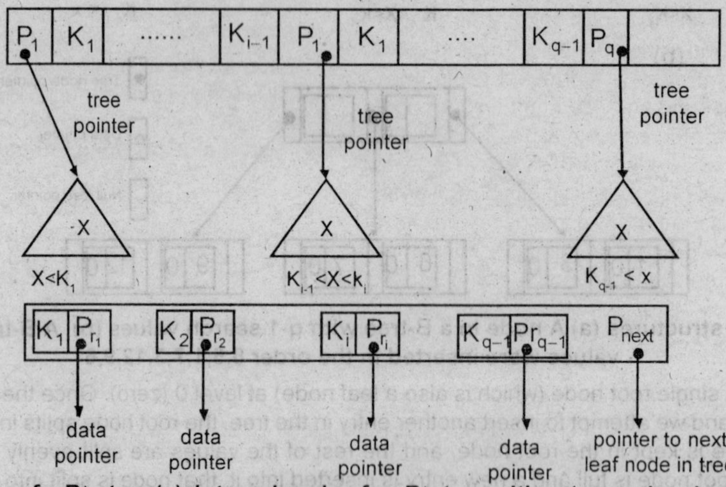

Figure 4.12 The nodes of a B⁺ - tree (a) Internal node of a B⁺- tree with q-1 search values. (b) Leaf node of a B⁺ - tree with q-1 search values and q-1 data pointers.

1. Each leaf node is of the form $<<K_1, P_{r_1}>, <K_2, P_{r_2}>, \ldots\ldots, <K_{q-1}, P_{r_{q-1}}>, P_{next}>$ where $q \le p$, each P_{r_i} is a data pointer, and P_{next} points to the next leaf node of the B⁺-tree.
2. Within each leaf node, $K_1 < K_2 < K_3 \ldots\ldots < K_{q-1}, q \le p$.
3. Each P_{r_i} is a data pointer that points to the record whose search field value is K_i or to a file block containing the record (or to a block of record pointers that point to records whose search field value is K_i if the search field is not a key.)
4. Each leaf node has at least $\lceil (p/2) \rceil$ values.
5. All leaf nodes are at the same level.

The pointers in internal nodes are tree pointers to blocks that are tree nodes, whereas the pointers in leaf nodes are data pointers to blocks--except for the P_{next} pointer, which is a tree pointer to the next leaf node. By starting at the leftmost leaf node, it is possible to traverse leaf nodes as a linked list, using the P_{next} pointers. This provides ordered access to the data records on the indexing field. A $P_{previous}$ pointer can also be included. For a B⁺-tree on a nonkey field, an extra level of indirection is needed, so the Pr pointers are block pointers to blocks that contain a set of record pointers to the actual records in the data file.

Because entries in the internal nodes of a B⁺-tree include search values and tree pointers without any data pointers, more entries canbe packed into an internal node of a B⁺-tree than for a similar B-tree. Thus, for the same block (node) size, the order p will be larger for the B⁺ tree tha for the B-tree. This can lead to fewer B⁺-tree levels, improving search time. Because the structures for internal and for leaf nodes of a B⁺-tree are different, the order p can be different. We will use p to denote the order for internal nodes and P_{leaf} to denote the order for leaf nodes, which we define as being the maximum number of data pointers in a leaf node.

Search, Inserion, and Deletion with B⁺-Trees.

Figure 4.13 illustrates insertion of records in a B⁺-tree of order $p = 3$ and $P_{leaf} = 2$. First, we observe that the root is the only node in the tree, so it is also a leaf node. As soon as more than one level is created, the tree is divided into internal nodes and leaf nodes.

Notice that every key value must exist at the leaf level, because all data pointers are at the leaf level. However, only some values exist in internal node to guide the search. Notice also that every value appearing in an internal node also appears as the rightmost value in the leaf level of the subtree pointed at by the tree pointer to the left of the value.

When a leaf node is full and a new entry is inserted there, the node overflows and must be split. The first $j = [((P_{leaf}+1)/2)]$ entries in the original node are kept there, and the remaining entries are moved to a new leafnode. The jth search value is replicated in the parent internal node, and an extra pointer to the new node is created in the parent. These must be inserted in the parent node in their correct sequence. If the parent internal node is full, the new value will cause it to overflow also, so it must be split. The entries in the internal node up to P_j--the jth tree pointer after inserting the new value and pointer, where $j = [((P+1)/2)]$ -- are kept, while the j^{th}search value is moved to the. Parent, not replicated. A new internal node will hold the entries from P_{j+1} to the end of the entries in the node. This splitting can propagate all the way up to create a new root node and hence a new level for the B^+-tree.

Figure 4.14 illustrates deletion from a B^+-tree. When an entry is deleted,it is always removed from the leaf level. If it happens to occur in an internal node, it must also be removed from there. In the latter case, the value to its left in the leaf node must replace it in the internal node, because that value is now the rightmost entry in the subtree. Deletion may cause **underflow** by reducing the number of entries in the leaf node directly to below the minimum required. In this case we try to find a sibling leaf node - a leaf node directly to the left or to the right of the node with underflow--and redistribuite the entries among the node and its sibling so that both are at least half full; otherwise,the node is merged with its suiblings and the number of leaf nodes is reduced. A common method is to try redistributing entries with left sibling; if this is not possible, an attempt to redistribute with the right sibling is made. If this is not possible either, the three nodes are merged into two leaf nodes.In such a case, underflow may propagate to internal nodes because one fewer tree pointer and search value are needed. This can propagate and reduce the tree levels.

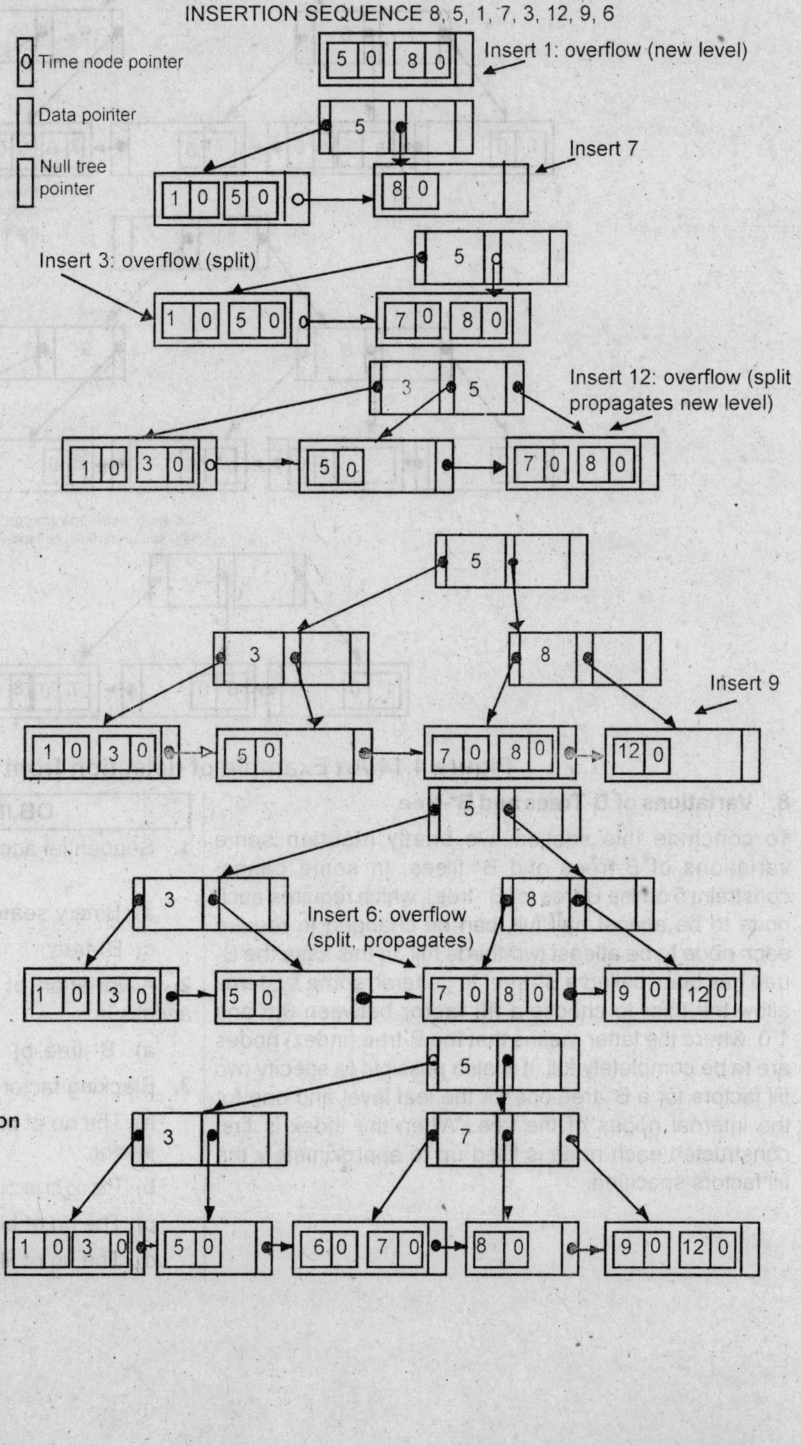

Figure 4.13 An example of insertion in a B^+ - tree with p=3 and P_{leaf}=2

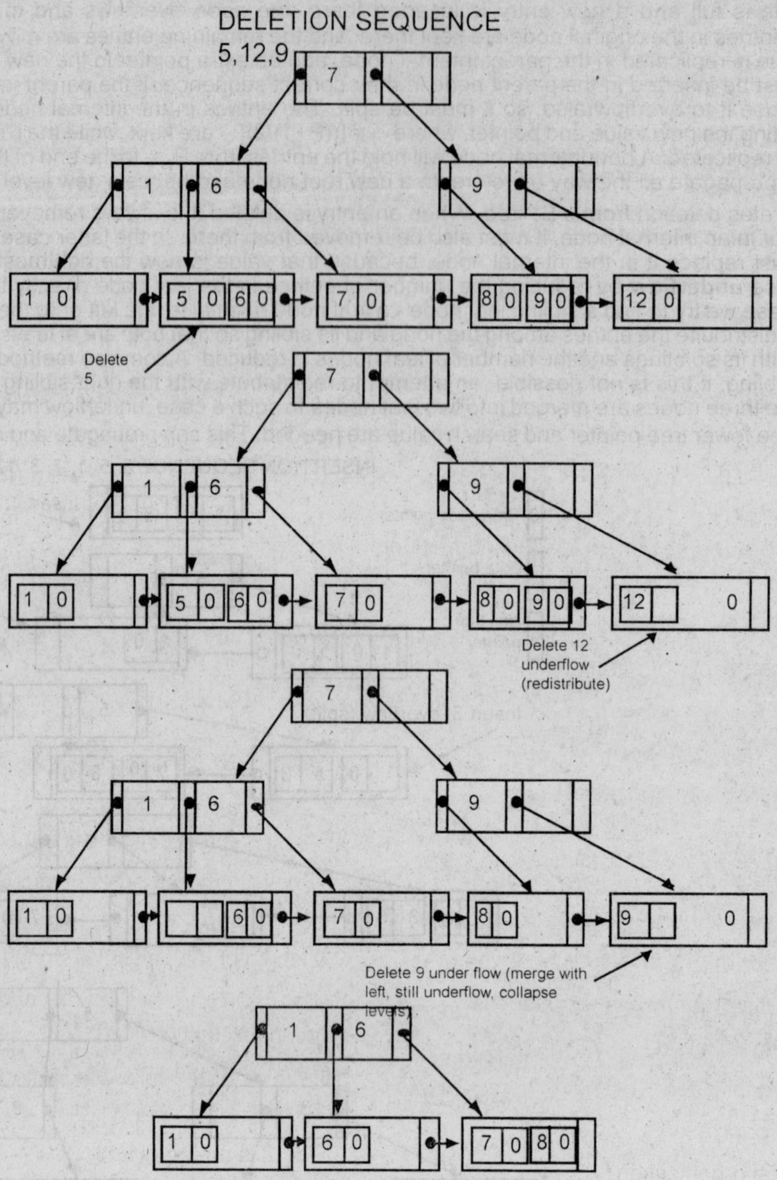

Figure 4.14 An Example of delection from a B⁺ - tree

8. Variations of B-Trees and B⁺-tree.

To conclude this section, we briefly mention some variations of B-trees and B⁺-trees. In some cases, constraint 5 on the B-tree (or B⁺-tree), which requires each node to be atleast half full, can be changed to require each node to be atleast two-thirds full. In this case the B-tree has been called a B*-tree. In general, some systems allow the user to choose a **fill factor** between 0.5 and 1.0, where the latter means that the B-tree (index) nodes are to be completely full. It is also possible to specify two fill factors for a B⁺-tree:one for the leaf level and one for the internal nodes of the tree. When the index is first constructed,each node is filled up to approximately the fill factors specified.

OBJECTIVE QUESTIONS - I

1. Sequential access to data is not possible in **[AUCET-2001]**

   a) Binary search tree b) Indexed sequential file

   c) B⁺-tree d) None

2. A variation of B-tree that allows rapid sequential access is: **[AUCET-2001]**

   a) B⁺-tree b) m-ary tree c) B'-tree d) AVL tree

3. Blocking factor refers to **[AUCET-2001]**

   a) The no.of blocks that can be accommodated in a sector.

   b) The no.of records that can be accommodated in a block.

   c) The no.of blocks into which the file is divided

   d) The no.of files that can fit into a single block

4. In file I/O [AUCET-2001]

a) Every logical read has a corresponding physical read

b) Every logical read corresponds to many physical reads.

c) Every physical read corresponds to many logical reads.

d) Every logical read corresponds to a single logical write.

5. A VSAM file contains the following area's[AUCET-2001]

a) Control intervals, control areas, sequence set, and index set

b) Data area and index block

c) Cylinder index, prime data area, and independent overflow area.

d) A hash table and a linked list of data items.

6. An indexing operation

a) Sorts a file using a single key

b) Sorts a file using two keys.

c) Establishes on index for a file

d) Both (b) and (c)

7. The index consists of

a) A list of keys b) Pointers to the master lest

c) Both (a) and (b) d) None

8. The simplest file organization technique is

a) Inverted organization

b) Multilist organization

c) Regular organization

d) Hierarchical organization

9. The file organization technique which provides a very fast direct access is

a) Inverted organization b) Multilist organization

c) Regular organization d) Hashed organization

10. The main disadvantage of Indexed sequential File is

a) It consumes lot of space

b) Performance degrades with the increase in the file size.

c) Accessing is very difficult d) None.

11. The maximum height of a B^+-tree of order n with k key values is

a) (n + k) / 2 b) log nk

c) log n/ 2 (k/2 + 1) d) log k/2n

12. Insertion of data/record into a B-tree may cause leaf and internal nodes to

a) Split b) be deleted

c) Maintain the same order d) Any of these

13. A multilevel indexing may lead to

a) Deadlock b) B-tree

c) Binary tree d) Transaction failure

14. The primary key indexing technique does not allow

a) Duplicate data in a field b) Multiple attributes

c) Set of relations d) Many-to-many relations

15. In a hierarchical sequential access method (HSAM) the following operation is not permitted:

a) Deletion b) Insertion

c) Indexing d) Sorting

16. Indexed sequential file is created using the following storage device

a) Direct access b) Random access

c) Sequential access d) Primary

17. Which of the following remarks about Trie indexing are TRUE?

a) It is efficient in dealing with strings of variable length

b) It is efficient if there are few number of data items

c) The no.of disk accesses can't execeed the length of the particular string that is searched.

d) It can handle insertions and deletions dynamically and efficiently

18. Which of the following remarks about Trie indexing are TRUE?

a) It is an m-ary tree

b) It is a search tree of order m

c) successful searches should terminate in leaf nodes

d) Unsuccessful searches should may terminate at any level of the tree structure.

19. Pick the correct statements.

a) Sequential file organization is suitable for batch processing.

b) Sequential file organization is suitable for interactive processing

c) Indexed sequential file organization supports both batch processing and interactive processing

d) Relative file can't be accessed sequentially.

20. For secondary key processing which of the following file organization is preferred?

a) Indexed sequential file organization

b) Two-way linked list

c) Inverted file organization

d) sequential file organization

ANSWER KEY - I				
1. a	2. a	3. b	4. a	5. c
6. c	7. c	8. c	9. d	10. b
11. c	12. a	13. b	14. a	15. b
16. a	17. a,c,d	18. a,c,d	19. a,c	20. c

OBJECTIVE QUESTIONS - II

1. In the index allocation scheme of blocks to a file the maximum possible size of the file depends on **[GATE-2002]**

a) The size of the blocks, and the size of the address of the blocks.

b) The number of blocks used for the index,and the size of the blocks.

c) The size of the blocks, the number of blocks used for the index,and the size of the address of the blocks.

d) None of these

2. A B^+-tree index is to be built on the Name attribute of the relation STUDENT. Assume that all student names are of length 8 bytes disk blocks are of size 512 bytes, and index pointers are of size 4 bytes. Given this scenario, what would be the best choice of the degree(i.e., the number of pointers per node) of the B^+-tree. **[GATE-2002]**

a) 16 b) 42 c) 43 d) 44

3. B^+-trees are preferred to binary trees in data bases because **[GATE-2000]**

a) Disk capacities are greaterthan memory capacities

b) Disk access is much slower than memory access

c) Disk data transfer rates are much lessthan memory data transfer rates.

d) Disks are more reliable than memory.

4. Which of the following is correct **[GATE-1999]**

a) Range Queries are faster on B^+ trees

b) B-trees are for storing data on disk and B^+- trees are for main memory

c) B-trees are for primary indexes and B^+-trees are for secondary indexes.

d) The height of a B^+-tree is independent of the number of records.

5. The file organization technique which includes an index on every secondary field is:

a) Inverted organization b) Multilist organization

c) Regular organization d) Hierarchical organization

6. Complete indexing is defined as an index existing for every distinct field combination in the indexed file. The no.of indexeds necessary to provide complete indexing for a file defined on 4 fields.

a) 6 b) 8 c) 10 d) 16

7. The redundant storage of search key values in B^+-trees is eliminated in

a) B-trees b) Indexed sequential File

c) AVL trees d) Binary trees.

8. Match the following

A. Storage structure	I. Two phase commit protocol
B. Hashing	II. B^+-trees
C. Blocking problem	III. Indexed sequential file organization
D. Cylinder Index	IV. Synonyms.

a) A-II,B-IV,C-I,D-III b) A-III,B-II,C-IV,D-I

c) A-I,B-II,C-III,D-IV d) A-IV,B-III,C-II,D-I

9. 6 files F1,F2,F3,F4,F5 and F6 have 100,200,50, 80,120,150 number of records respectively. In what order should they be stored so as to optimize access time. Assume each file is accessed with the same frequency.

a) F3,F4,F1,F5,F6,F2 b) F2,F6,F5,F1,F4,F3

c) F1,F2,F3,F4,F5,F6

d) Ordering is immaterial as all files are accessed with the same frequency

10. In context of file organisation, which of the following is false? **[UGC NET-2000]**

a) In a serial file organisation, no logical.sequence of records apply

b) In a sequential file organisition, logical sequence of records is same as physical sequence

c) In a direct file organisation, physical location of a record is given directly or indirectly by a hash address

d) In sequential file organisation, no logical sequence of records apply

11. Indexed Organization is good for **[UGC NET-1996]**

a) Low activity ratio files b) High activity ratio files

c) Both low and high activity ratio files

d) Neither low nor High activity ratio files

12. The file organization in which the physical location of a record is determined by a mathematical function that maps the key to address of the record is **[UGC NET-1998]**

a) hashed file organization

b) B tree file organization c) B^+ tree file organization

d) Sequential file organization

ANSWER KEY - II				
1. b	**2.** a	**3.** d	**4.** a	**5.** a
6. a	**7.** a	**8.** a	**9.** a	**10.** a
11. b	**12.** a			

V. TRANSACTIONS AND CONCURRENCY CONTROL TECHNIQUES

- **Transaction Processing Concepts**
- **Types of Failures**
- **Properties of Transactions**
- **Serializability**
- **Recoverability**
- **Concurrency Control Techniques**
- **Lock-Based Protocols**
- **Dealing with Deadlock and Starvation**
- **Time-Stamp Based Protocols**

Introduction to Transaction Processing

In this section we informally introduce the concepts of concurrent execution of transactions and recovery from transaction failures.

Single-User versus Multiuser Systems

One criterion for classifying a database system is according to the number of users who can use the system concurrently - that is, at the same time. A DBMS is single user if at most one user at a time can use the system, and it is multiuser if many users can use the system - and hence access the database - concurrently. Single - user DBMSs are mostly restricted to some microcomputer systems; most other DBMSs are multiuser. For example, an airline reservations system is used by hundreds of travel agents and reservation clerks concurrently. Systems in banks, insurance agencies, stock exchanges, supermarkets, and the like are also operated on by many users who submit transactions concurrently to the system.

Multiple users can access databases - and use computer systems - simultaneously because of the concept of multiprogramming, which allows the computer to execute multiple programs - or processes - at the same time.

If only a single central processing unit (CPU) exists, it can actually execute at most one process at a time.

However, multiprogramming operating systems execute some commands from one process, then suspend that process and execute some commands from the next process, and so on. A process is resumed at the point where it was suspended whenever it gets its turn to use the CPU again. Hence, concurrent execution of processes A and B executing concurrently in an interleaved fashion. Interleaving keeps the CPU busy when a process requires an input or output (I/O) operation, such as reading a block from disk. The CPU is switched to execute another process rather than remaining idle during I/O time. Interleaving also prevents a long process from delaying other processes.

If the computer system has multiple hardware processors (CPU's) parallel processing of multiple processes is possible, as illustrated by processes C and D in Figure 5.1. Most of the theory concerning concurrency control in databases is developed in terms of interleaved concurrency, so for the remainder of this chapter we assume this model. In a multiuser DBMS, the stored data items are the primary resources that may be accessed concurrently by interactive users or application programs,

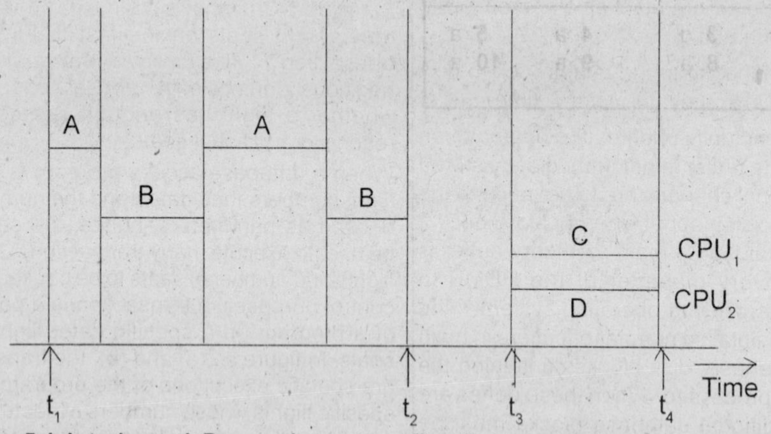

Fig. 5.1 Interleaved Processing versus Parallel processing of cocurrent transactions

which are constantly retrieving information from and modifying the database.

3. Transactions, Read and Write Operations, and DBMS Buffers

A transaction is a logical unit of database processing that includes one or more database access operations -- these can include insertion, deletion, modification, or retrieval operations.

The database operations that form a transaction can either be embedded within an application program or they can be specified interactively via a high-level query language such as SQL. One way of specifying the transaction boundaries is by specifying explicit begin transaction and end transaction statements in an application program; in this case, all database access operations between the two are considered as forming one transaction. A single application program may contain more than one transaction if it contains several transaction boundaries. If the database operations in a transaction do not update the database but only retrieve data,the transaction is called a read-only transaction.

The model of a database that is used to explain transaction processing concepts is much simplified. A database is basically represented as a collection of named data items. The size of a data item is called its granularity, and it can be a field of some record in the database, or it may be a large unit such as a record or even a whole disk block, but the concepts we discuss are independent of the data item granularity. Using this simplified database model, the basic database access operations that a transaction can include are as follows:

- **read_item(X) :** Reads a database item named X into a program variable. To simplify our notation, we assume that the program variable is also named X.
- **write_item (X):** Writes the value of program variable X into the database item named X.

The basic unit of data transfer from disk to main memory is one block. Executing a read_item (X) command includes the follwing steps:

1. Find the address of the disk block that contains item X.
2. Copy that disk block into a buffer in main memory (if that disk block is not already in some main memory buffer).
3. Copy item X from the buffer to the program variable named X.

Executing a write_item (X) command includes the following steps:

1. Find the address of the disk block that contains item X.
2. Copy that disk block into a buffer in main memory (if that disk block is not already in some main memory buffer).
3. Copy item X from the program variable named X into its correct location in the buffer.
4. Store the updated block from the buffer back to disk (either immediately or at some later point in time).

Step 4 is the one that actually updates the database on disk. In some cases the buffer is not immediately stored to disk, in case additional changes are to be made to the buffer. Usually, the decision about when to store back a modified disk block that is in a main memeory buffer is handled by the recovery manager of the DBMS in cooperation with the underlying operating system. The DBMS will generally maintain a number of buffers in main memory that hold database disk blocks containing the database items being processed. When these buffes are all occupied, and additional database blocks must be copied into memory, some buffer replacement policy is used to choose which of the current buffers is to be replaced. If the cohsen buffer has been modified, it must be written back o disk before it is reused.

A transaction includes read_item and write_item operations to access and update the database. Figure 5.2 shows examples of two very simple transactions. The read-set of a transaction is the set of all items that the transaction reads,and the write-set is the set

(a) T_1	(b) T_2
read_item(X);	read_item(X);
X:= X-N;	X:= X+M;
	write _ item (X);
	read_item(X);
write_item(X);	
	Y:=Y+N;
	write_item(Y);

Fig. 5.2 Two sample transactions. (a) Transaction T_1 (b) Transaction T_2

of all items that the transaction writes. For example, the read-set of T_1 in Figure 5.2 is {X,Y} and its write-set is also {X,Y}.

Concurrency control and recovery mechanisms are mainly concerned with the database access commands in a transaction. Transactions submitted by the various users may execute concurrently and may access and update the same database items. If this concurrent execution is uncontrolled, it may lead to problems, such as an inconsistent database.

4. Why Concurrency Control Is Needed

Several problems can occur when concurrent transactions execute in an uncontrolled manner.

We illustrate some of these problems by referrring to a much simplified airline reservations database in which a record is stored for each airline flight. Each record includes the number of reserved seats on that flight as a named data item, among other information. Figure 5.2(a) shows a transaction T_1 that transfers N reservations from one flight whose number of reserved seats is stored in the database item named X to another flight whose number of reserved seats is stored in the database item named Y. Figure 5.2(b) shows a simpler transaction T_2 that just reserves M seats on the first flight (X) referenced in transaction T_1. To simpligy our example, we do not show additional portions of the transactions, such as checking whether a flight has enough seats available before reserving additional seats.

When a database access program is written, it has the flight numbers,their dates,and the number of seats to be booked as parameters; hence, the same program can be usedto execute many transactions, each with different flights and number of seats to be booked. For concurrency control purposes, a transaction is a particular execution of a program on a specific date, flight, and number of seats. In figure 5.2(a) and (b), the transactions T_1 and T_2 are specific executions of the programs that refer to the specific flights whose numbers of seats are stored in data items X and Y in the database.We now discuss the types of problems we may encounter with these two transactions if they run concurrently.

The Lost Update Problem. This problem occurs when two transactions that access the same database items have their opeations interleaved in a way that makes the value of some database item incorrect. Suppose that transactions T_1 and T_2 are submitted at approximately the same time, and suppose that their operations are interleaved as shown in Figure 5.3(a); then the final value of item X is incorrect, because T_2 reads the value of X before T_1 changes it in the database, and hence the udated value resulting from T_1 is lost.

For example, if X = 80 at the start (originally there were 80 reservations on the flight), N=5 (T_1 transfers 5 seat reservations from the flight corresponding to X to the flight corresponding to Y), and M = 4 (T_2 reserves 4 seats on X), the final result should be X = 79; but in the interleaving of operations shown in Figure 5.3(a), it is X = 84 because the update inT_1 that removed the five seats from X was lost.

The Temporary Update (or Dirty Read) Problem. This problem occurs when one transaction updaes a database item and then the transaction fails for some reason. The updated item is accessed by another transaction before it is changed back to its original value. Figure 5.3(b) shows an example where T_1 updates item X and then fails before completion, so the system must change X back to its original value. Before it can do so, however, transaction T_2 reads the "temporary" value of X , which will not be recorded permanently in the database because of the failure of T_1. The value of item X that is read by T_2 is called dirty data, because it has been created by a transaction that has not completed and committed yet; hence, this problem is also known as the dirty read problem.

(a)

Time ↓	T_1	T_2
	read_item(X);	
	X:= X-N;	
		read_item(X);
		X:= X+M;
	write_item(X);	
	read_item(Y);	
		write_item(X); ⟵ Item X has an incorrect value because its update by T_1 is "lost" (overwritten)
	Y=Y+N;	
	write_item(Y);	

(b)

Time ↓	T_1	T_2
	read_item(X);	
	X:=X-N;	
	write_item(X);	
		read_item(X);
		X:=X+M;
		write_item(X);
	read_item(Y);	

TransactionT$_1$ fails and must change the value of X back to its old value; meanwhile T_2 has read the "temporary" incorrect value of X ⟶

Figure 5.3 Some probelms that occur when concurrent execution is uncontrolled.

(a) The lost update problem.

(b) The temporary update problem.

(c)

T_1	T_3
	sum:=0;
	read_item(A);
	sum:=sum+A
read_item(X);	
X:=X-N;	
write_item(X);	
	read_item(X);
	sum:=sum+X; ⟵ T_3 reads X after N is subtracted and reads Y before N isadded; a wrong summary is the result (off by N)
	read_item(Y);
read_item(Y);	sum:=sum+Y;
Y:=Y+N;	
write_item(Y);	

Figure 5.3 Some problems that occur when concurrent execution is uncontrolled.

(c) The incorrect summary problem.

The Incorrect Summary Problem.

If one transaction is calculating an aggregate summary function on a number of records while other transactions are updating some of these records,the aggregate function may calculate some values before they are updated and others after they are updated. For example, suppose that a transaction T_3 is calculating total number of reservations on all the flights; meanwhile, transaction T_1 is executing. If the interleaving of operations shown in Figure 5.3 (c) occurs, the result of T_3 will be off by an amount N because T_3 reads the value of X after N seats have been subtracted from it but reads the value of Y before those N seats have been added to it.

5. Why Recovery Is Needed

Whenever a transaction is submitted to a DBMS for execution, the system is responsible for making sure that either (1)all the operations in the transaction are completed successfully and their effect is recorded permanently in the database, or (2) the transaction has no effect what so ever on the database or on any other transactions. The DBMS must not permit some operations of a transaction T to be applied to the database while other operations of T are not. This may happen if a transaction fails after executing some of its operations but before executing all of them.

6. Types of Failures.

Failures are generally classified as transaction, system, and media failures. There are several possible reasons for a transaction to fail in the middle of execution:

1. A computer failure (system crash): A hardware, software, or network error occurs in the computer system during transaction execution. Hardware crashes are usually media failures -- for example, main memory failure.

2. A transaction or system error : Some operation in the transaction may cause it to fail, such as integer overflow or division by zero.Transaction failure may also occur because of erroneous parameter values or because of a logical programming error. In addition, the user may interrupt the transaction during its execution.

3. Local errors or exception conditions detected by the transaction : During transaction execution, certain conditions may occur that necessitate cancellation of the transaction. For example data for the transaction may not be found. Notice that an exception condition, such as insufficient account balance in a banking database, may cause a transaction, such as a fund withdrawal, to be canceled. This exception should be programmed in the transaction itself, and hence would not be considered a failure.

4. Concurrency control enforcement : The concurrency control method may decide to abort the transaction, to be restarted later, because it violates serializability or because several transactions are in a state of deadlock.

5. Disk failure : Some disk blocks may lose their data because of a read or write malfunction or because of a disk read/write head crash. This may happen during a read or a write operation of the transaction.

6.Physical problems and catastrophes : This refers to an endless list of problems that includes power or air-conditioning failure, fire, theft, sabotage,overwriting disks or tapes by mistake, and mounting of a wrong tape by the operator. Failures of types 1, 2, 3, and 4 are more common than those of types 5 or 6. When ever a failure of type 1 through 4 occurs, the system must keep sufficient information to recover from the failure. Disk failure or other catastrophic failures of type 5 or 6 do not happen frequently; if they do occur, recovery is a major task. The concept of transaction is fundamental to many techniques for concurrency control and recovery from failures.

7. Transaction and System Concepts.

Transaction States and Additional Operations

A transaction is an atomic unit of work that is either completed in its entirety or not done at all. For recovery purposes, the system needs to keep track of when the transaction starts, terminates, and commits or aborts (see below). Hence, the recovery manager keeps track of the following operations:

- **BEGIN_TRANSACTION:** This marks the beginning of transaction execution.
- **READ or WRITE :** These specify that read or write operations on the database items that are executed as part of a transaction.
- **END_TRANSACTION:** This specifies that READ and WRITE transaction operations have ended marks the end of transaction execution. However, at this point it may be necessary to check whether the changes introduced by the transaction can be permanently applied to the database (committed) or whether the transaction has to be aborted because it violates serializability or for some other reason.
- **COMMIT_TRANSACTION:** This signals a successful end of the transaction so that any changes (updates) executed by the transaction can be safely committed to the database and will not be undone.
- **ROLLBACK (or ABORT) :** This signals that the transaction has ended unsuccessfully, so that any changes or effects that the transaction may have applied to the database must be undone.

Figure 5.4 shows a state transition diagram that describes how a transaction moves through its execution states. A transaction goes into an active state immediately after it starts execution, where it can issue READ and WRITE operations. When the transaction ends, it moves to the partially committed state. At this point, some recovery protocols need to ensure that a system failure will not result in an inability to record the changes of the transaction permanently (usually by recording changes in the system log). Once this check is successful, the transaction is said to have reached its commit point and enters the committed state.

Once a transaction is committed, it has concluded its execution successfully and all its changes must be recorded permanently in the database.

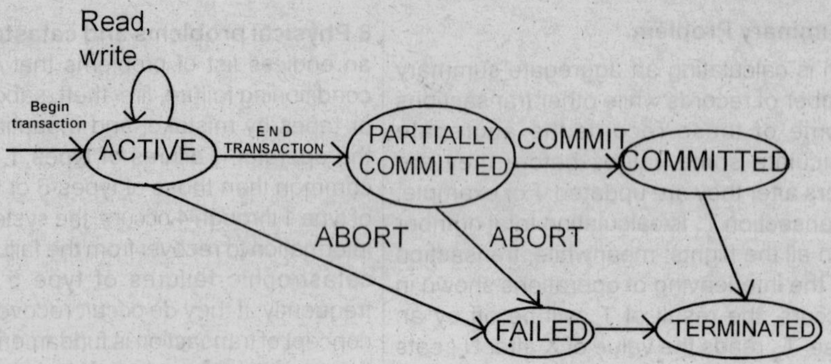

Fig. 5.4 State transition diagram illustrating the States for transaction execution

database at a given point in time. A consistent state of the database satisfies the constraints specified in the schema as well as any other constraints that should hold on the database is in a consistent state before executing the transaction, it will be in a consistent state after the complete execution of the transaction, assuming that no interference with other transactions occurs.

Isolation is enforced by the concurrency control sybsystem of the DBMS. If every transaction does not make its updates visible to other transactions until it is committed, one form of isolation is enforced that solves the temporary update problem and eliminates cascading rollbacks . There have been attempts to define the level of isolation of a transaction. A transaction is said to have level 0 (zero) isolation if it does not overwrite the dirty reads of higher - leverl transactions. A level 1 (one) isolation transaction has no lost updates; and level 2 isolation has no lost updates; and no dirty reads. Finally, level 3 isolation (also called true isolation) has, in addition to degree 2 properties, repeatable reads.

Finally, the durability property is the responsibility of the recovery subsystem of the DBMS.

9. Serializabilitv

The database systems must control concurrent execution of transactions, to ensure thatt he database state remains consistent. Before we examine how the database system can carry out this task, we must first understand which schedules will ensure consistency, and which schedules will not.

Since transactions are programs, it is computationally difficult to determine what are the exact operations that a transaction performs, and how operations of various transactions interact.For this reason, we shall not interpret the type of operations that a transaction can perform on a data item. Instead, we consider only two operations : read and write. We thus assume that, between a read (Q) instruction and a write (Q) instruction on a item Q, a transaction may perform an arbitrary sequence of operations on the copy of Q that is residing in the local buffer of the transaction. Thus, the only significant operations of a transaction, from a scheduling point of view, are its read and write instructions. We shall in the representation of schedule 3 that is shown in Figure 5.5

In this section,we diuscuss different forms of schedule equivalence; they lead to the notions of confict serializability and view serializability.

However, a transaction can go to the failed state if one of the checks fails or if the transaction is aborted during its active state. The transaction may then have to be rolled back to undo the effect of its WRITE operations on the database. The terminated state corresponds to the transaction leaving the system. The transaction information that is maintained in system tables while the transaction has been running is removed when the transaction terminates. Failed or aborted transactions may be restarted later -either automatically or after resubmitted by the user - as brand new transactions.

8. Desirable Properties of Transactions

Transactions should possesss several properties. These are often called the ACID properties, and they should be enforced by the concurrency control and recovery methods of the DBMS. The following are the ACID properties :

1. **Atomicity:** A transaction is an atomic unit of processing; it is either performed in its entirety or not performed at all.

2. **Consistency preservation:** A transaction is consistency preserving if its complete execution take (s) the database from one consistent state to another.

3. **Isolation :** A transaction should appear as though it is being executed in isolation from other transactions. That is, the execution of a transaction should not be interfered with by any other transactions executing concurrently.

4. **Durability or permanency :** The changes applied to the database by a committed transaction must persist in the database. These changes must not be lost because of any failure.

The atomicity property requires that we execute a transaction to completion.

It is the responsibility of the transaction revovery subsystem of a DBMS to ensure atomicity. If a transaction fails to complete for some reason, such as a system crash in the midst of transaction execution, the recovery technique must undo any effects of the transaction on the database.

The preservation of consistency is generally considered to be the responsibility of the programmers who write the database programs or of the DBMS module that enforces integrity constraints. Recall that a database state is a collection of all the stored data items (values) in the

T_1	T_2
read (A)	
write (A)	
	read (A)
	write (A)
read (B)	
write (B)	
	read (B)
	write (B)

Fig.5.5 Schedule 3 - Showing only the read and write instructructions

Conflict Serializability

Let us consider a schedule S in which there are two consecutive instructions I_i and I_j of transactions T_i and T_j ,respectively. $(i \neq j)$. If I_i and I_j refer to different data items, then we can swap I_i and I_j with out affecting the results of any instruction in the schedule. However, if I_i and I_j refer to the same data item Q, then the order of the two steps may matter. Since we are dealing with only read and write instructions, there are four cases that we need to consider:

1. I_i = read (Q), I_j = read (Q). The order of I_i and I_j does not matter, since the same value of Q is read by T_i and T_j regardless of the order.

2. I_i = read (Q), I_j = write (Q). If I_i comes before I_j, then T_i does not read the value of Q that is written by T_j in instrucion I_j. If I_j comes before I_i, then T_i reads the value of Q that is written by T_j. Thus, the order of I_i and I_j matters.

3. I_i =write(Q), I_j = read(Q). The order of I_i and I_j matters for reasons similar to those of the previous case.

4. I_i =write(Q), I_j = write(Q). Since both instructions are write operations, the order of these instructions does not affect either T_i or T_j.

However, the value obtained by the next read(Q) instruction of S is affected, since the result of only the latter of the two write instructions is preserved in the database. If there is no other write(Q) instruction after I_i and I_j in S, then the order of I_i and I_j directly affects the final value of Q in the database state that results from schedule S.

Thus, only in the case where both I_i and I_j are read instructions does the relative order of their execution not matter.

We say that I_i and I_j conflict if they are operations by different transactions on the same data item, and at least one of these instructions is a write operation.

View Serilizablity

In this section, we consider a form of equivalence that is less stringent than conflict equivalence, but that, like conflict equivalence, is based on only the read and write operations of transactions.

Consider two schedules S and S', where the same set of transactions participates in both schedules. The schedules S and S' are said to be view equivalent if the following three conditions are met:

1. For each data item Q, if transaction T_i reads the initial value of Q in schedule S, then transaction T_i must, in schedule S', also read the initial value of Q.

2. For each data item Q, if transaction T_i executes read(Q) in schedule S, and that value was produced by transaction T_j (if any) then transaction T_i must, in schedule S', also read the value of Q that was produced by transaction T_j.

3. For each data item Q, the transaction (if any) that performs the final write(Q) operation in schedule S must perform the final write(Q) operation in schedule S'.

Conditions 1 and 2 ensure that each transaction reads the same values in both schedules and, therefore, performs the same computation. Condition 3, coupled with conditions 1 and 2, ensures that both schedules result in the same final system state.

The concept of view equivalence leads to the concept of view serializability.

We say that a schedule S is view serializable if it is view equivalent to a serial schedule.

Every conflict-serializable schedule is view serializable, but there are view-serializable schedules that are not conflict serializable.

10. Recoverability

So far, we have studied what schedules are acceptable from the view point of consistency of the database, assuming implicitly that there are no transaction failures. We now address the effect of transaction failures during concurrent execution.

If a transaction T_i fails, for whatever reason, we need to undo the effect of this transaction to ensure the atomicity property of the transaction. In a system that allows concurrent execution, it is necessary also to ensure that any transaction T_j that is dependent on T_i (that is T_j has read data written by T_i) is also aborted.

To achieve this surely, we need to place restrictions on the type of schedules permitted in the system.

12. Locking Techniques for Concurrency Control

Some of the main techniques used to control concurrent execution of transactions are based on the concept of locking data items. A lock is a variable associated with a data item that describes the status of the item with respect to possible operations that can be applied to it.

Generally, there is one lock for each data item in the database. Locks are used as a means of synchronizing the access by concurrent transactions to the database items.

Lock-Based Protocols

One way to ensure serializability is to require that access to data items be done in a mutually exclusive manner, that is, while one transaction is accessing a data item, no other transaction can modify that data item. The most common method used to implement this requriement is to allow a transaction to access a data item only if it is currently holding a lock on that item.

Locks

There are various modes in which a data item may be locked. In this section, we restrict our attention to two modes:

1. Shared. If a transaction T_i has obtained a shared-mode lock (denoted by S) on item Q, then T_i can read, cannot write Q.

2. Exclusive. If a transaction T_i has obtained an exclusive - mode lock (denoted by X) on item Q, then T_i can both read and write Q.

We require that every transaction request a lock in an appropriate mode on data item Q, depending on the types of operations that it will perform on Q. The request is made to the concurrency-control manager. The transaction can proceed with the operation only after the concurrency-control manager grants the lock to the transaction.

Given a set of lock modes. We can define a compatibility function on them as follows. Let A and B represent arbitrary lock modes. Suppose that a transaction T_i requests a lock of mode A on item Q on which transaction T_j ($T_i \neq T_j$) currently holds a lock of mode B.

If transaction T_i can be granted a lock on Q immediately, in spite of the presence of the mode B lock, then we say mode A is compatible with mode B. Such a function can be represented conveniently by a matrix. The compatibility relation between the two modes of locking used in this section is given by the matrix **comp** of Figure 5.6. An element comp (A,B) of the matrix has the value true if and only if mode A is cpmpatible with mode B.

Note that shared mode is compatible with shared mode, but not with exclusive mode. At any time, several shared-mode locks can be held simultaneously (by different transactions) on a particular data item. A subsequent exclusive mode lock request has to wait until the currently held shared-mode locks are released.

A transaction requests a shared lock on data item Q by executing the lock-S(Q) instruction. Similarly, an exclusive lock is requested through the lock - X(Q) instruction.

A data item Q can be- unlocked via the unlock (Q) instruction.

To access a data item, transaction T_i must first lock that item. If the data item is already locked by another transaction in an imcompatible mode, the concurrency-control manager will not grant the lock until all imcompatible locks held by other transactions have been released. Thus, T_i is made to wait until all incompatible locks held by other transactions have been released..

Transaction T_i may unlock a data item that it had locked at some earlier point. Note that a transaction must hold a lock on a data item as long as it accesses that item. Moreover, for a transaction to unlock a data item immediately after its final access of that data item is not always desirable, since serializability may not be ensured.

	S	X
S	true	false
X	false	false

Fig.5.6 **Lock - Compatibility matrix comp**

As an illustration, consider the simplified banking system. Let A and B be two accounts that are accessed by transactions T_1 and T_2. Transaction T_1 transfers $50 from account B, to acount A and is defined as

T_1 :
```
lock -X(B);
read(B);
B:=B - 50 ;
write(B);
unlock(B);
lock -X(A);
read(A);
A:=A+50;
write (A);
unlock(A).
```

TransactionT_2 displays the total amont of money in accounts A and B -- that is, the sum A+B --- and is defined as

T_2 :
```
lock-S(A);
read(A);
unlock(A);
lock-S(B);
read (B);
unlock(B);
display(A+B)
```

Suppose that the values of accounts A and B are $100 and $200, respectively. If these two transactions are executed serially, either in the order T_1, T_2 or the order T_2,T_1, then transaction T_2 will dispaly the value $300. If, however, these transactions are executed concurrently, then schedule 1 as shown in Figure 5.7 is possible. In this case, transaction T_2 displays $250, which is incorrect. The reason for this mistake is that the transaction T_1 unlocked data item B too early,as a result of which T_2 saw an inconsistent state.

The schedule shows the actions executed by the transactions, as well as the points when the locks are granted by the concurrency-control manager. The transaction making a lock request cannot execute its next action until the lock is granted by the concurrency-control manager; hence, the lock must be granted in the interval of time between the lock-request operation and the following action of the transaction.Exactly when within this interval the lock is granted is not important; the lock can be safely assumed to be granted just before the following action of the transaction.

Suppose now that unlocking is delayed to the end of the transaction. Transaction T_3 correspons to T_1 with unlocking delayed, and is defined as

T_3 :
```
lock-X(B);
read(B);
B:=B - 50;
write(B);
lock - X(A);
read(A);
```

```
A:= A+50;
write(A);
unlock(B);
unlock(A);
```

Transaction T_4 corresponds to T_2 with unlocking delayed, and is defind as

```
T4 :   lock-S(A);
       read(A);
       lock-S(B);
       read(B);
       display(A+B);
       unlock(A);
       unlock(B);
```

T_1	T_2	Concurrecy-control manager
lock-X(B)		grant-X(B,T_1)
read(B)		
B := B - 50		
write(B)		
unlock(B)	lock-S(A)	
		grant-S(A,T_2)
	read(A)	
	unlock(A)	
	lock-S(B)	
		grant-S(B,T_2)
	read(B)	
	unlock(B)	
	display(A+B)	
lock-X(A)		
		grant-X(A,T_2)
read(A)		
A:= A + 50		
write(A)		
unlock(A)		

Fig. 5.7 Schedule 1

T_3	T_4
lock-X(B)	
read(B)	
B := B - 50	
write(B)	
	lock-S(A)
	read(A)
	lock-S(B)
lock-X(A)	

Fig. 5.8 Schedule 2

You should verify that the sequence of reads and writes in schedule 1, which lead to an incorrect total of $250 being displayed, is no longer possible with T_3 and T_4. Other schedules are possible. T_4 will not print out an inconsistent result in any of them;

Unfortunately, the use of locking can lead to an undesirable situation. Consider the partial schedule of Figure 5.8 for T_3 and T_4. Since T_3 is holding an exclusive-mode lock on B and T_4 is requesting a shared-mode lock on B, T_4 is waiting for T_3 to unlock B. Similarly,since T_4 is holding a shared-mode lock on A and T_3 is requesting an exclusive-mode lock on A, T_3 is waiting for T_4 to unlock A. Thus, we have arrived at a state where neither of these transactions can ever procced with tis normal execution. This situation is called deadlock. When deadlock occurs, the system must roll back one of the two transactions. Once a transaction has been rolled back, the data items that were locked by that transaction are unlocked. These data items are then available to the other transaction, which can continue with its execution.

If we do not use locking, or unlock data items as soon as possible after reading or writing them, we may get inconsistent states. On the other hand, if we do not unlock a data item before requesting a lock on another data item, deadlocks may occur. There are ways to avoid deadlock.However, in general , deadlocks are a necessary evil associated with locking,if we want to avoid inconsistent states. Deadlocks are definitely preferable to inconsistent states since they can be handled by rolling back of transactions, whereas inconsistent states may lead to real-world problems that cannot be handled by the database system.

Granting of Locks

When a transaction requests a lock on a data item in a particular mode, and no other transaction has a lock on the same data item in a conflicting mode, the lock can be granted. However,care must be taken to avoid the following scenario. Suppose a transaction T_2 has a shared-mode lock on a data item, and another transactionT_1 requests an exclusive-mode lock on the data item. Clearly, T_1 has to wait for T_2 to release the shared-mode lock meanwhile, a transactionT_3 may request a shared-mode lock on the same data item.

The lock request is compatible with the lick granted to T_2. so T_3 may be granted the shared-mode lock. At this point T_2 may release the lock, but still T_1 has to wait for T_3 to finish. But again, there may be a new transaction T_4 that requests a shard-mode lock on the data item, and is granted the lock before T_3 releases it.

In fact it is possible that there is a sequence of transactions that each requests a shared-mode lock on the data item,and each transaction releases the lock a short while after it is granted, but T_1 never gets the exclusive-mode lock on the data item.The transaction T_1 may never make progress, and is not is said to be starved.

We can avoid starvation of transactions by granting locks as follows. When a transaction T_i requests a lock on a data item Q in a particular mode M, the lock is granted provided that

1. There is no other transaction holding a lock on Q in a mode that conflicts with M.

2. There is no other transaction that is waiting for a lock on Q, and that made its lock request before T_i.

The Two-Phase Locking Protocol

One protocol that ensures serializability is the two - phase locking protocol. This protocol requires that each transaction issue lock and unlock requests in two phases:

1. Growing phase: A transaction may obtain locks, but may not release any lock.

2. Shrinking phase : A transaction may release locks, but may not obtain any new locks.

Initially, a transaction is in the growing phase. The transaction acquires locks as needed. Once the transaction releases a lock, it enters the shrinking phase, and it can issue no more lock requests.

For example, transactions T3 and T4 are two phase. On the other hand, transactions T1 and T2 are not two phase. Note that the unlock instructions do not need to appear at the end of the transaction. For example, in the case of transaction T3, we could move the unlock(B) instruction to just after the lock-X(A) instruction, and still retain the two phase locking property.

We can show that the two-phase locking protocol ensures conflict serializability. Consider any transaction. The point in the schedule where the transaction has obtained its final lock (the end of its growing phase) is called the lock point of the transaction. Now, transactions can be ordered according to their lock points this ordering is, in fact, a serializability ordering for the transactions.

Two-phase locking does not ensure freedom from deadlock. Observe that transactions T3 and T4 are two-phase, but in schedule 2(figure. 5.8), they are deadlocked.

13. Dealing with Deadlock and Starvation

Deadlock occurs when each transaction T in a set of two or more transactions is waiting for some item that is locked by some other transaction T′ in the set. Hence, each transaction in the set is on a waiting queue, waiting for one of the other transactions in the set to release the lock on an item. A simple example is shown in figure 5.9 (a), where the two transactions T_1' and T_2' are deadlocked in a partial schedule; T_1' is on the waiting queue for X, which is locked by T_2' while T_2' is on the waiting queue for Y, which is locked by T_1' .Meanwhile, neither T_1' nor T_2' nor any other transaction can access items X and Y.

Deadlock Prevention Protocols

One way to prevent deadlock is to use a deadlock prevention protocol. One deadlock prevention protocol, which is used in conservative two-phase locking, requires that every transaction lock all the items it needs in advance (which is generally not a parctical assumption) - if any of the items cannot be obtained, none of the items are locked. Rather, the transaction waits and then tries again to lock all the items it needs. This solution obviously further limits concurrency. A second protocol, which also limits concurrency,

(a)

Time ↓	T_1'	T_2'
	read_lock(Y);	
	read_item(Y)	
		read_lock(X);
		read_item(X);
	write_lock(X);	
		write_lock(Y);

b)

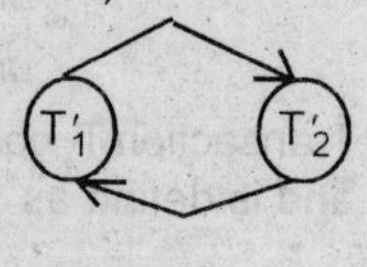

Fig. 5.9 Illustrating the deadlock Problem

(a) A Partial schedule of T_1' and T_2' that is in a state of deadlock (b) A wait-for graph for the Patial Schedule in (a)

involves ordering all the items in the database and making sure that a transaction that needs several items will lock them according to that order. This requires that the programmer (or the system) be aware of the chosen order of the items, which is also not practical in the database context.

A number of other deadlock prevention schemes have been proposed that make a decision about what to do with a trancsaction involved in a possible deadlock situation: Should it be blocked and made to wait or should it be aborted, or should the transaction preempt and abort another transaction ?

These techniques use the concept of transaction time stamp TS(T), which is a unique identifier assigned to each transaction. The timestamps are typically based on the order in which transactions are started; hence if transaction T_1 starts before transaction T_2, then $TS(T_1) < TS(T_2)$. Notice that the older transaction has the smaller timestamp value. Two schemes that prevent deadlock are called wait-die and wound-wait. Suppose that transaction T_i tries to lock an item X but is not able to because X is locked by some other transaction T_j with a conflicting lock. The rules followed by these schemes are as follows:

- **Wait-die:**If $TS(T_i) < TS(T_j)$, then (T_i older than T_j) T_i is allowed to wait; otherwise (T_i younger than T_j) abort T_i (T_i dies) and restrat it later with the same timestamp.

- **Wound-wait :** If $TS(T_i) < TS(T_j)$, then (T_i older than T_j) abort T_j (T_i wounds T_j) and restart it later with the same timestamp; otherwise (T_i younger than T_j) is allowed to wait.

In wait-die, an older transaction is allowed to wait on a younger transaction, whereas a younger transaction requesting an item held by an older transaction is aborted and restarted. The wound-wait approach does the opposite: A younger transaction is allowed to wait on an older one, whereas an older transaction requesting an item held by a younger transaction preempts the younger transaction by aborting it. Both schemes end up aborting the younger of the two transactions that may be involved in a deadlock. It can be shown that these two techniques are deadlock-free, since in wait-die, transactions only wait on younger transactions so no cycle is created. Similarly, in wound-wait, transactions only wait on older transactions so no cycle is created. However, both techniques many cause some transactions to be aborted and restarted needlessly, even though those transactions may never actually cause a deadlock.

Another group of protocols that prevent deadlock do not require timestamps. These include the no waiting (NW) and cautious waiting (CW) algorithms. In the no waiting algorithm, if a transaction is unable to obtain a lock, it is immediately aborted and then restarted after a certain time delay without checking whether a deadlock will actually occur or not.

Because this scheme can cause transactions to abort and restart needlessly, the cautious waiting algorithm was proposed to try to reduce the number of needless aborts/ restarts. Suppose that transaction T_i tries to lock an item X but is not able to do so because X is locked by some other transaction T_j with a conflicting lock. The cautious wiating rules are as follows.

Cautious waiting: if T_j is not blocked (not waiting for some other locked item,), then T_i is blocked and allowed to wait; otherwise abort T_i .

It can be shown that cautious waiting is deadlock-free, by considering the time b(T) at which each blocked transaction T was blocked. If the two transactions T_i and T_j above both become blocked, and T_i is waiting on T_j, then $b(T_i)<b(T_j)$, since T_i can only wait on T_j at a time when T_j is not blocked. Hence, the blocking times form a total ordering on all blocked transactions, so no cycle that causes deadlock can occur.

Deadlock Detection and Timeouts.

A second-more practical-approach to dealing with deadlock is deadlock detection, where the system checks if a state of deadlock actually exists. This solution is attractive if we know there will be little interference among the transactions - that is, if different transactions will rarely access the same items at the same time. This can happen if the transactions are short and each transaction locks only a few items, or if the transaction load is light. On the other hand, if transactions are long and each transaction uses many items, or if the transaction load is quite heavy, it may be advantageous to use a dead lock prevention scheme.

Another simple scheme to deal with deadlock is the use of timeouts. This method is practical because of its low overhead and simplicity. In this method, if a transaction waits for a period longer than a system-defined time out period, the system assumes that the transaction may be deadlocked and aborts it - regardless of whether a deadlock actually exists or not.

Starvation. Another problem that may occur when we use locking is starvation, which occurs when a transaction cannot proceed for an indefinite period of time while other transactions in the system continue normally.

This may occur if the waiting scheme for locked items is unfair, giving priority to some transactions over others.

One solution for starvation is to have a fair waiting scheme, such as using a first-come-first-serve queue; transactions are enabled to lock an item in the order in which they originally requested the lock. Another scheme allows some transactions to have priority over others but increases the priority of a transaction the longer it waits, until it eventually gets the highest priority and proceeds. Starvation can also occur because of victim selection if the algorithm selects the same transaction as victim repeatedly, thus causing it to abort and never finish execution. The algorithm can use hgiher priorities for transactions that have been aborted multiple times to avoid this problem. The wait-die and wound-wait schemes discussed above avoid starvation.

14. Timestamp-Based Protocols

In the locking protocols that we have described thus far, the order between every pair of conflicting transactions is determined at execution time by the first lock that they both request that involves incompatible modes. Another method for determining the serializability order is to select an ordering among transactions in advance. The most common method for doing so is to use a timestamp-ordering scheme.

Timestamps

With each transaction T_i in the system, we associate a unique, fixed timestamp, denoted by $TS(T_i)$. This timestamp is assigned by the database system before the transaction T_i starts execution. If a transaction T_i has been assigned timestamp TS (T_i) and a new transaction T_j enters the system, then $TS(T_i)<TS(T_j)$. There are two simple methods for implementing this scheme:

1. Use the value of the system clock as the timestamp; that is a transaction's timestamp is equal to the value of the clock when the transaction enters the system.

2. Use a logical counter thatis incremented after a new timestamp has been assigned; that is a transaction's timestamp is equal to the value of the counter when the transaction enters the system. The timestamps of the trnasactions determine the serializability order.

Thus, if $TS(T_i)<TS(T_j)$, then the system must ensure that the produced schedule is equivalent to a serial schedule in which transaction T_i appears before transaction T_j.

To implement this scheme, we associate with each data item Q two timestamp values:

- **W-timestamp (Q)** denotes the largest timestamp of any transaction that executed write (Q) successfully.
- **R-timestamp(Q)** denotes the largest timestamp of any transaction that executed read (Q) successfully.

These timestamps are updated whenever a new read (Q) or write (Q) instruction is executed.

The Timestamp-Ordering Protocol

The timestamp-ordering protocol ensures that any conflicting read and write operations are executed in timestamp order. This protocol operates as follows:

1. Suppose that transaction T_i issues read (Q).

a. If TS(T_i) < W-timestamp(Q), then T_i needs to read a value of Q that was already overwritten. Hence, the read operation is rejected, and T_i is rolled back.

b. If TS(T_i) $\geq$ W-timestamp(Q), then the read operation is executed, and R-timestamp (Q) is set to the maximum of R - timestamp(Q) and TS (T_i).

2. Suppose that transaction T_i issues write (Q).

a. If TS(T_i)< R-timestamp (Q), then the value of Q that T_i is producing was needed previously, and the system assumed that the value would never be produced. Hence, the write operation is rejected, and T_i is rolled back.

b. If TS(T_i)< W-timestamp(Q), then T_i is attempting to write an obsolete value of Q. Hence, this write operationis rejected, and T_i is rolled back.

c. Otherwise, the write operation is executed, and W-timestamp(Q) is set to TS (T_i).

A transaction T_i, that is rolled back by the concurrecy-control scheme as result of either a read or write operation being issued, is assigned a new timestamp and is restarted.

To illustrate this protocol, we consider transactions T_{14} and T_{15}. Transaction T_{14} displays the contents of accounts A and B, and is defined as

```
T14:    read(B);
        read (A);
        display(A+B).
```

Transaction T_{15} transfers $50 from account A to account B, and then displays the contents of both:

```
T15:    read(B);
        B:= B-50;
        write(B);
        read(A);
        A := A+50;
        write (A);
        display(A+B)
```

In presenting schedules under the timestamp protocol, we shall assume that a transaction is assigned a timestamp immediately before its first instruction. Thus, in schedule 3 of Figure 5.10, TS(T_{14})< TS(T_{15}), and the schedule is possible under the timestamp protocol.

We note that the preceding execution can also be produced by the two-phase locking protocol. There are, however, schedules that are possible under the two phase locking protocol, but are not possible under the timestamp protocol, and vice versa.

T_{14}	T_{15}
read(B)	
	read(B)
	B:=B-50
	write (B)
read(A)	
	read(A)
display(A+B)	A:=A+50
	write(A)
	display(A+B)

Fig. 5.10 Schedule 3

The timestamp-ordering protocol ensures conflict serializability. This assertion follows from the fact that conflicting operations are processed in timestamp order. The protocol ensures freedom from deadlock, since no transaction ever waits. The protocol can generate schedules that are not recoverable; however it can be extended to make the schedules cascadeless.

Validation-Based Protocols

In cases where the majority of transactions are read-only transactions, the rate of conflicts among transactions may below. Thus, many of these transactions, if executed without the supervision of a concurrency-control scheme, would nevertheless leave the system in a consistent state. A concurrency control scheme imposes overhead of code execution and possible delay of transactions. It may be desirable to use an alternative scheme that imposes less overhead. A difficulty in reducing the overhead is that we do not know in advance which transactions will be involved in a conflict. To gain that knowledge, we need a scheme for monitoring the system.

We assume that each transaction T_i executes in two or three different phases in its lifetime, depending on whether it is a read-only or an update transaction. The phases are, in order, as follows:

1. **Read phase :** During this phase, the execution of transaction T_i takes place. The values of the various data items are read. and are stored in variables local to T_i. All write operations are performed on temporary local variables, without updates of the actual database.

2. **Validation phase :** Transaction T_i performs a validation test to determine whether it can copy to the database the temporary local variables that hold the results of write operations without causing a violation of serializability.

3. **Write phase :** If transaction T_i succeeds validation (step2), then the actual updates are applied to the database. Otherwise, T_i is rolled back.

Each transaction must go through the three phases in the order shown. However, all three phases of concurrently executing transactions can be interleaved.

The read and write phases are self-explanatory. The only phase that needs further discussion is the validation phase.

To perform the validation test, we need to know when the various phases of transactions T_i took place. We shall, therefore, associate three different timestamps with transaction T_i :

1. **Start(** T_i **)**, the time when T_i started its execution.

2. **Validation(** T_i **)**, the time when T_i finished its read phase and started its validation phase.

3. **Finish(** T_i **)**, the time when T_i finished its write phase.

We determine the seializability order by the timestamp-ordering technique using the value of the timestamp validation (T_i). Thus, the value TS(T_i)=Validation(T_i) and if TS(T_j)< TS(T_k), then any producted schedule must be equivalent to a serial schedule in which transaction T_j appears before transaction T_k.

The reason we have chosen Validation(T_i) rather than start (T_i), as the timestamp of transaction T_i is that we can expect faster response time provided that conflict rates among transactions are indeed low.

The validation test for transaction T_j requires that, for all transactions T_i with TS(T_i) < TS(T_j), one of the following two conditions must hold:

1. Finish (T_i)< start(T_j). Since T_i completes its execution before T_j started the serializability order is indeed maintained.

2. The set of data items written by T_i does not interesect with the set of data items read by T_j and T_i completes its write phase before T_j starts its validation phase (Start (T_j)< Finish (T_j) < validation (T_j)).This condition ensures that the writes of T_i and T_j do not overlap. Since the writes of T_i the serializability order is indeed maintained.

As an illustration, consider again transactions T_{14} and T_{15}. Suppose that TS(T_{14}) < TS(T_{15}). Then, the validation phase succeeds in producing schedule 5, which is depicted in figure 5.11. Note that the writes to the actual variables are performed only after the validation phase of T_{15}. Thus T_{14} reads the old values of B and A and this schedule is serializable.

T_{14}	T_{15}
read(B)	
	read(B)
	B:=B-50
	read(A)
	A:=A+50
read(A)	
< validate >	
display(A+B)	
	< validate>
	write(B)
	write(A)

Fig. 5.11 Sehedule 5, a Schedule produced using validation

The validation scheme automatically guards against cascading rollbacks, since the actual writes take place only after the transaction issuing the write has committed.

15. Multiple Granularity

In the concurrency control schemes described thus far, we have used each individual data item as the unit on which synchronization is performed.

There are circumstances, however, where it would be advantageous to group several data items, and to treat them as one individual synchronization unit. For example, if a transaction T_i needs to access the entire database, and a locking protocol is used, then T_i must lock each item in the database. Clearly, executing these locks is time consuming. It would be better if T_i could issue a single lock request to lock the entire database. On the other hand, if transaction T_j needs to access only a few data items, it should not be required to lock the entire database, since otherwise concurrency is lost.

16. Insert and Delete Operations

Until now, we have restricted our attention to read and write operations. This restriction limits transactions to data items already in the database. Some transactions require not only access to existing data items, but also the ability to create new data items.

Others require the abiligy to delete data items. To examine how such transactions affect concurrency control, we introduce the following additional operations:

- **delete(Q)** deletes data item Q from the database.
- **insert(Q)** inserts a new data item Q into the database and assigns Q an initial value.

An attempt by a transaction T_i to perform a read (Q) operation after Q has been deleted results in a logical error in T_i. Likewise, an attempt by a transaction T_i to perform a read (Q) operation before Q has been inserted results in a logical error in T_i. It is also a logical error to attempt to delete a nonexistent data item.

OBJECTIVE QUESTIONS

1. The main purpose of locking is to have efficient **[UGC NET-1998]**

   a) Tunability b) Concurrency

   c) Security d) Stability

2. A locked file can be **[UGC NET-1999]**

   a) access by only one user

   b) modified by users with the correct password

   c) used to hide sensitive information

   d) both (b) and (c)

3. The Concept of transaction provides a mechanism for

   a) Describing logical units of database processing

   b) Describing the queries required for retrieving the date from database

   c) Both (a) and (b) d) None

4. Transaction processing systems are

   a) Systems with small database and hundreds of concurrent users that are executing database transactions.

   b) Systems with large database and hundreds of concurrent users that are executing database transactions

   c) Systems with large database and single user executing database transactions

   d) None

5. Examples of transaction systems

a) Banking b) Credit card processing

c) Stock markets d) All of the above

6. If the database operations in a transaction do not update the database but only retrieve data, the transaction is called

a) Read-only transaction b) Select transaction

c) Both a and b d) None

7. What is granularity

a) The size of a data base

b) The size of a record

c) The size of a data item d) None

8. What are the basic data base access operations that a transaction include

a) Read _ item (x) b) Write _ item (y)

c) Both (a) and (b) d) None

9. The problem that occurs when two transactions that access the same database items have their operations interleaved in a way that makes the value of same database item incorrect.

a) The lost modify problem b) The lost select problem

c) The lost update problem d) None

10. The problem that occurs when one transaction updates a database item and then the transaction fails for some reason.

a) Temporary modify problem

b) Temporary select problem

c) Dirty read problem d) None

11. The problem that occurs if one transaction is calculations an aggregate summary function on a number of records while of the transactions are updating some of these records

a) The incorrect select problem

b) The incorrect summary problem

c) The incorrect modify problem

d) None

12. Which of the following failure that can occur in transactions

a) Transaction failuresb) System failures

c) Media failures d) All the above

13. Which of the following are transaction states

a) BEGIN _ TRANSACTION and READ or WRITE

b) END _ TRANSACTION

c) COMMIT _ TRANSACTION and ROLLBACK

d) All of the above

14. Which of the following are desirable properties of Transactions

a) Atomicity and consistency preservation

b) Isolation and Durability or permanency

c) Both (a) and (b) d) None

15. A level 1 transaction has

a) No lost updates b) No dirty read problem

c) Both (a) and (b) d) None

16. A level 2 isolation has

a) No lost updates b) No dirty reads

c) Both (a) and (b) d) None

17. Which following property is the reponisibility of the recovery subsystem of the DBMS

a) Serializability b) Durability

c) Both (a) and (b) d) None

18. The protocols that are used to prevent multiple transactions from accessing the items concurrently

a) Concurrency control protocols

b) Locking protocols

c) Serializability d) None

19. What is a time stamp ?

a) It is a unique identifier for each transaction generated by the user

b) It is a unique identifier for each transaction generated by the system

c) It is one type of stamp d) None

20. Which of the following factor that affects cocurrency control

a) Granularity of the data items b) Queries

c) Both (a) and (b) d) None

21. A variable associated with a data item that describes the states of item w.r.t to possible operations that can be applied to it is

a) Time stamp b) Block

c) Lock d) None

22. Find out the modes in which a data item may be locked

a) Shared b) Exclusive

c) Both (a) and (b) d) None

23. Which is the state where neither of the transactions can never proceed with its normal execution

a) Lock state b) Block state

c) Dead lock state d) None

24. Which of the following phases presented in Two-Phase locking protocol

a) Growing Phase b) Shrinking Phase

c) Both (a) and (b) d) None

25. Which of the following Scheme prevent deadlock

a) Wait-die b) Wound - wait

c) Both (a) and (b) d) None

26. Which of the following is scheme to deal with deadlock

a) Timeouts b) Timeins

c) Both (a) and (b) d) None

27. The problem that occurs when a transaction can not proceed for an indefinite period of time while other transactions in the system continue normally is called

a) Confliction b) Congestion

c) Starvation d) None

28. Which of the following are time-stamp values that can be associated with each data item Q

a) W _ time stamp (Q) , S _ time stamp (Q)

b) W _ time stamp (Q), R _ time stamp (Q)

c) R _ time stamp (Q), S _ time stamp (Q)

d) None

29. The phase in which that each transaction executes

a) Read Phase b) Validation Phase

c) Write Phase d) All of the above

30. To perform the validation test, we need to know when the various phase of transactions T_i when we need to associate which of the following time stamp with transaction T_i.

a) Start (T_i) b) Validation (T_i)

c) Finish (T_i) d) All the above

31. A single individual synchronization unit that obtained by grouping several dataitems is called

a) Group item b) Granularity

c) Multiple Granularity d) None

32. Which of the following operations that make transactions affect concurrency control

a) Delete b) Insert

c) Both (a) and (b) d) None

33. If there exists a set of transactions such that every transaction in the set is waiting for another transaction in the set, then the system is called

a) Transaction failure b) Transaction Processing

c) Dead-lock d) Two-phase lock

34. To ensure that the system will never enter a deadlock, we can use

a) Deadlock prevention protocol

b) Boyce-codd Normal Form

c) Time-Stamped transaction

d) Validation protocol

35. Locking the data before the beginning of transaction execution in order to prevent deadlock may lead to

a) Program abortion b) Starvation

c) Delay d) Poor performance

36. A deadlock prevention scheme which avoid starvation is

a) Locking of data by transactions before execution

b) A deadlock prevention scheme using time stamps

c) Deadlock detection and prevention scheme.

d) All of these

37. A deadlock exists in the system if and only if the wait-for graph

a) Contains a Cycle

b) There is a path from first node to last node

c) There is no path from first node to last node

d) Contains a non-reachable node from the first node

38. A directed graph which represent the deadlock is called

a) Cyclic graph b) Deadlock graph

c) Wait-for graph d) Deadlock detection graph

39. In order to reduce the overhead imposed on transaction processing by the O.S., one can use

a) Transaction monitor b) Parallel processing

c) Concurrency in transactions

d) High-performance transactions

40. When the detection algorithm detects a deadlock, the recovery is normally accomplished by

a) Roll-back of transactions

b) Consistency checking c) Locking of data

d) Calling back the same transaction

41. The two phase commit protocol is resilient to

a) Deadlock b) Site and link failures

c) Dependencies d) Starving

42. The two phase commit protocol may introduce

a) Deadlock b) Starving

c) Blocking d) None of the above

ANSWER KEY				
1. b	**2.** d	**3.** a	**4.** b	**5.** d
6. a	**7.** c	**8.** c	**9.** c	**10.** c
11. b	**12.** d	**13.** d	**14.** c	**15.** a
16. c	**17.** b	**18.** b	**19.** b	**20.** a
21. c	**22.** c	**23.** c	**24.** c	**25.** c
26. a	**27.** c	**28.** b	**29.** d	**30.** d
31. c	**32.** c	**33.** c	**34.** a	**35.** b
36. b	**37.** a	**38.** c	**39.** a	**40.** a
41. b	**42.** c			

TEST YOURSELF - I

1. Two alternative packages A and B are available for processing a database having 10^k records. Package A requires $0.0001n^2$ time units and package B requires $10n\log_{10} n$ time units to process n records. What is the smallest value of k for which package B will be preferred over A? **[GATE - 10]**

a) 12 b) 10 c) 6 d) 5

2. Consider a B^+ – tree in which the maximum number of keys in a node is 5. What is the minimum number of keys in any non - root node? **[GATE - 10]**

a) 1 b) 2 c) 3 d) 4

3. A relational schema for a train reservation database is given below

Passenger (pid, pname, age)

Re servation (pid, cass, tid) **[GATE - 10]**

Table : Passenger

pid	'pname	age
0	'Sachin'	65
1	'Rahul'	66
2	'Sourav'	67
3	'Anil'	69

Table : Reservation

pid	class	tid
0	'AC'	8200
1	'AC'	8201
2	'SC'	8201
5	'AC'	8203
1	'SC'	8204
3	'AC'	8202

What pids are returned by the following SQL query for the above instance of the tables?

```
SELECT    pid
FROM      Reservation
WHERE     class = 'AC' AND
          EXISTS ( SELECT *
               FROM Passenger
               WHERE age > 65 AND
               Passenger.pid=Reservation.pid)
```

a) 1,0 b) 1,2 c) 1,3 (d) 1,5

4. Which of the following concurrency control protocols ensure both conflict serializability and freedom from deadlock? **[GATE - 10]**

I. 2 - phase locking

II. Time - stamp ordering

a) I only b) II only

c) Both I and II d) Neither I nor II

5. Consider the following schedule for transactions T1, T2 and T3 : **[GATE - 10]**

T1	T2	T3
Read(x)		
	Read(Y)	
		Read(Y)
	Write(Y)	
Write(x)		
		Write(x)
	Read(x)	
	Write(x)	

Which one of the schedules below is the correct serialization of the above?

a) $T1 \to T3 \to T2$ b) $T2 \to T1 \to T3$

c) $T2 \to T3 \to T1$ d) $T3 \to T1 \to T2$

6. The following functional dependencies hold for relations R(A, B, C) and S (B, D, E)

$B \to A$,

$A \to C$

The relation R contains 200 tuples and the relation S contains 100 tuples. What is the maximum number of tuples possible in the natural join R⋈S? **[GATE - 10]**

a) 100 b) 200 c) 300 d) 2000

7. Consider two transactions T_1 and T_2 and four schedules S_1, S_2, S_3, S_4 of T_1 and T_2 as given below:

$T_1 : R_1[x]W_1[x]W_1[y]$

$T_2 : R_2[x]R_2[y]W_2[y]$

$S_1 : R_1[x]R_2[x]R_2[y]W_1[x]W_1[y]W_2[y]$

$S_2 : R_1[x]R_2[x]R_2[y]W_1[x]W_2[y]W_1[y]$

$S_3 : R_1[x]W_1[x]R_2[x]W_1[y]R_2[y]W_2[y]$

$S_4 : R_2[x]R_2[y]R_1[x]W_1[x]W_1[y]W_2[y]$ **[GATE - 09]**

Which of the above schedules are conflict - serializable?

a) S_1 and S_2 b) S_2 and S_3

c) S_3 only d) S_4 only

8. The following key values are inserted into a B + - tree in which order of the internal nodes is 3, and that of the leaf nodes is 2, in the sequence given below. The order of internal nodes is the maximum number of tree pointers in each node, and the order of leaf nodes is the maximum number of data items that can be stored in it. The B+- tree is initially empty.

10, 3,6, 8, 4, 2, 1

The maximum number of times leaf nodes would get split up as a result of these insertions is **[GATE - 09]**

a) 2 b) 3 c) 4 d) 5

9. Let R and S be relational schemes such that R = {a,b,c} and S = {c}. Now consider the following queries on the database :

I. $\pi_{R-S}(r) - \pi_{R-S}(\pi_{R-S}(r) \times S - \pi_{R-S,S}(r))$

II. $\{t \mid t \in \pi_{R-S}(r) \wedge \forall u \in s(\exists v \in r(u = v[s] \wedge t = v[R-S]))\}$

III. $\{t \mid t \in \pi_{R-S}(r) \wedge \forall v \in r(\exists u \in s(u = v[s] \wedge t = v[R-S]))\}$

IV. Select R. a, R. b
From R, S
Where R.c = S.c

Which of the above queries are equivalent ?**[GATE - 09]**

a) I and II b) I and III
c) II and IV d) III and IV

COMMON DATA QUESTIONS : 10 & 11

Consider the following relational schema :

Suppliers (sid : integer, sname :string, city:string, street:string)

Parts(pid:integer, pname:string, color:string)

Catalog(sid:integer,pid:integer,cost : real)

10. Consider the following relational query on the above database:

```
SELECT      S.sname
FROM        Suppliers S
WHERE S.sid NOT IN(SELECT C.sid
                   FROM Catalog C
            WHERE C.pid NOT (SELECT P.pid
                             FROM Parts P
                             WHERE P. color < > 'blue'))
```

Assume that relations corresponding to the above schema are not empty. which one of the following is the correct interpretation of the above query ? **[GATE - 09]**

a) Find the names of all suppliers who have supplied a non - blue part

b) Find the names of all suppliers who have not supplied a non - blue part

c) Find the names of all suppliers who have supplied only blue parts

d) Find the names of all suppliers who have not supplied only blue parts

11. Assume that, in the suppliers relation above, each supplier and each street within a city has a unique name, and (sname, city) forms a candidate key. No other functional dependencies are implied other than those implied by primary and candidate keys. Which one of the following is TRUE about the above schema ? [GATE - 09]

a) The schema is in BCNF

b) The schema is in 3 NF but not in BCNF

c) The schema is in 2NF but not in 3NF

d) The schema is not in 2NF

12. Which of the following tuple relational calculus expression(s) is/are equivalent to $\forall t \in r(P(t))$?

I. $\neg\exists t \in r(P(t))$

II. $\exists t \notin r(P(t))$

III. $\neg\exists t \in r(\neg P(t))$

IV. $\exists t \notin r(\neg P(t))$

[GATE - 08]

(a) I only (b) II only
(c) III only (d) III and IV only

13. A clustering index is defined on the fields which are of type **[GATE - 08]**

(a) non-key and ordering
(b) non-key and non-ordering
(c) key and ordering
(d) key and non ordering

14. Let R and S be two relations with the following schema

R (<u>P</u>, <u>Q</u>, R1, R2, R3)
S (<u>P</u>, <u>Q</u>, S1, S2)

Where {P,Q} is the key for both schemas. Which of the following queries are equivalent ?

I. Π_P (R ⋈ S)

II. Π_P (R) ⋈ Π_P (S)

III. $\Pi_P(\Pi_{P,Q}(R) \cap \Pi_{P,Q}(S))$

IV. $\Pi_P(\Pi_{P,Q}(R) - (\Pi_{P,Q}(R) - (\Pi_{P,Q}(S)))$

[GATE - 08]

(a) Only I and II (b) Only I and III
(c) Only I,II and III (d)Only I,III and IV

15. Consider the following relational schemes for a library database:

Book (Title, Author, Catalog_no, Publisher, Year, Price)

Collection (Title, Author, Catalog_no)

With the following functional dependencies:

I. Title Author $\rightarrow$ Catalong_no

II. Catalog_no $\rightarrow$ Title Author Publisher year

III. Publisher Title Year $\rightarrow$ Price

Assume {Author, Title} is the key for both schemes. Which of the following statements is true? **[GATE - 08]**

(a) Both Book and Collection are in BCNF
(b) Both Book and Collection are in 3NF only
(c) Book is in 2NF and Collection is in 3 NF
(d) Both Book and Collection are in 2NF only

16. Consider a file of 16384 records. Each record is 32 bytes long and its key field is of size 6 bytes. The file is ordered on a non-key field, and the file organization is unspanned. The file is stored in a file system with block size 1024 bytes, and the size of a block pointer is 10 bytes. If the secondary index is built on the key field of the file, and a multi-level index scheme is used to store the secondary index, the number of first-level and second -level blocks in the multi-level index are respectively. **[GATE - 08]**

(a) 8 and 0 (b) 128 and 6
(c) 256 and 4 (d) 512 and 5

Statement for Linked Answer Questions 17 to 18

Consider the following ER diagram.

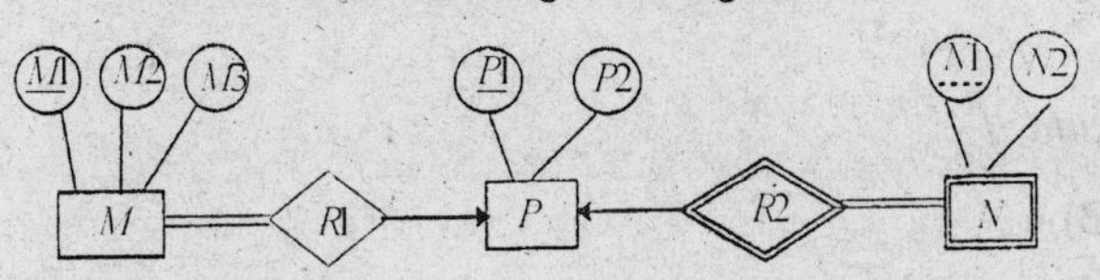

17. The minimum number of tables needed to represent M, N, P, R1, R2 is **[GATE - 08]**

(a) 2 (b) 3
(c) 4 (d) 5

18. Which of the following is a correct attribute set for one of the tables for the correct answer to the above question ? **[GATE - 08]**

(a) { M1, M2, M3, P1} (b) {M1, P1, N1, N2}
(c) {M1, P1, N1} (d) {M1, P1}

19. Let R (A, B ,C, D) be a relational schema with the following functional dependencies:
$A \to B$, $B \to C$,
$C \to D$ and $D \to B$. The decomposition of R into (A,B), (B,C), (B,D) **[GATE - 08]**

(a) gives a lossless join, and is dependency preserving
(b) gives a lossless join but is not dependencey preserving
(c) does not give a lossless join, but is dependencey preserving
(d) does not give a lossless join, and is not dependencey preserving

20. Let R(A,B,C,D,E,P,G) be a relational schema in which the following functional dependencies are known to hold : $AB \to CD$, $DE \to P$, $C \to E$, $P \to C$ and $B \to G$. The relational schema R is **[GATE - 08]**

(a) in BCNF
(b) in 3NF, but not in BCNF
(c) in 2NF, but not in 3NF
(d) not in 2NF

21. Consider the following three schedules of transactions T1, T2 and T3 [Notation: In the following NYO represents the action Y (R for read, W for write) performed by transaction N on object O.]

(S1) 2RA 2WA 3RC 2WB 3WA 3WC 1RA 1RB 1WA 1WB
(S2) 3RC 2RA 2WA 2WB 3WA 1RA 1RB 1WA 1WB 3WC
(S3) 2RA 3RC 3WA 2WA 2WB 3WC 1RA 1RB 1WA 1WB

Which of the following statements is TRUE ? **[GATE - 08]**

(a) S1, S2 and S3 are all conflict equivalent to each other
(b) No two of S1, S2 and S3 are conflict equivalent to each other
(c) S2 is conflict equivalent to S3, but not to S1
(d) S1 is conflict equivalent to S2, but not to S3

Common Data for Questions 22 and 23 :

Consider the following relational schema :

Student (<u>school-id, sch-roll-no,</u> sname, saddress)
School (<u>school-id</u>, sch-name, sch-address, sch-phone)
Enrolment (<u>school-id,sch-roll-no,</u> erollno, examname)
ExamResult (<u>erollno, examname</u>, marks)

22. What does the following SQL query output ?

```
SELECT   sch-name, COUNT(*)
FROM     SchoolC,Enrolment  E,  ExamResult R
WHERE    E.School-id = C.school-id
         AND
         E.examname = R.examname AND E.Erollno = R.erollno
         AND
         R.marks = 100 AND S.school-id IN      (SELECT School-id
                                                FROM student GROUP BY
                                                school-id HAVING COUNT (*) >200)
```

GROUP by school-id. **[GATE - 08]**

(a) for each school with more than 200 students appearing in exams, the name of the school and the number of 100s scored by its students.

(b) for each school with more than 200 students in it, the name of the school and the number of 100s scored by its students.

(c) for each school with more than 200 students in it, the name of the school and the number of its students scoring 100 in at least one exam..

(d) nothing; the query has a syntax error

23. Consider the following tuple relational calculus query **[GATE - 08]**

$\{t \mid \exists E \in$ Enrolment t=E.school-id $\wedge$

$\mid \{x \mid x \in$ Enrolment $\wedge$ x.school-id=t $\wedge$

$\exists B \in$ ExamResult

B.erollno=x.erollno $\wedge$ B.examname=x.examname

$\wedge$ B.marks>35}| $\div$

$\mid \{x \mid x \in$ Enrolment $\wedge$ x.school-id =t {| * 100>35 }

If a student needs to score more than 35 marks to pass an exam, what does the query return ?

(a) The empty set

(b) schools with more than 35% of its students enrolled in some exam or the other

(c) schools with a pass percentage above 35% over all exams taken together.

(d) schools with a pass percentage above 35% over each exam

24. Consider a selection of the form $\sigma_{A \le 100}(r)$, where r is a relation with 1000 tuples. Assume that the attribute values for A among the tuples are uniformly distributed in the interval $[0,500]$. Which one of the following options is the best estimate of the number of tuples returned by the given selection query? [GATE - 07]

(a) 50 (b) 100

(c) 150 (d) 200

25. Consider the following two transactions : T_1 *and* T_2.

T_1 : read (A) ;
read (B) ;
if A = 0 then $B \leftarrow B+1$;
write (B) ;

T_2 : read (B)
read (A)
if $B \neq 0$ then $A \leftarrow A-1$;
write (A) ;

Which of the following schemes, using shared and exclusive locks, satify the requirements for strict two phase locking for the above transactions ? [GATE - 07]

(a) S1 : lock S (A)
read (A) ;
lockS(B) ;
read(B) ;
if A = 0
then $B \leftarrow B+1$;
write (B) ;
commit ;
unlock (A) ;
unlock (B) ;

S2 : lock S (B)
read (B) ;
lockS(A) ;
read (A) ;
if $B \neq 0$
then $A \leftarrow A-1$;
write (A) ;
commit ;
unlock (B) ;
unlock (A) ;

(b) S1 : lock X (A)
read (A) ;
lockX(B) ;
read(B) ;
if A = 0
then $B \leftarrow B+1$;
write (B) ;
unlock (A) ;
commit ;
unlock (B) ;

S2 : lock X (B)
read (B) ;
lockX(A) ;
read (A) ;
if $B \neq 0$
then $A \rightarrow A-1$;
write (A) ;
unlock (A) ;
commit ;
unlock (B) ;

(c) S1 : lock S (A)
read (A) ;
lockX(B) ;
read(B) ;
if A = 0
then $B \leftarrow B+1$;
write (B) ;
unlock (A) ;
commit ;
unlock (B) ;

S2 : lock S (B)
read (B) ;
lockX(A) ;
read (A) ;
if $B \neq 0$
then $A \leftarrow A-1$;
write (A) ;
unlock (A) ;
commit ;
unlock (B) ;

(d) S1 : lock S (A)
read (A) ;
lockX(B) ;
read(B) ;
if A = 0
then $B \leftarrow B+1$;
write (B) ;
unlock (A) ;
unlock (B) ;
commit ;

S2 : lock S (B)
read (B) ;
lockX(A) ;
read (A) ;
if $B \neq 0$
then $A \leftarrow A-1$;
write (A) ;
unlock (A) ;
unlock (B) ;
commit ;

26. Consider the following implications relating to functional and multivalued dependencies given below, which may or may not be correct.

(i) If $A \to B$ and $A \to C$ then $A \to BC$

(ii) If $A \to B$ and $A \to C$ then $A \to BC$

(iii) If $A \to BC$ and $A \to B$ and $A \to C$

(iv) If $A \to BC$ and $A \to B$ and $A \to C$

Exactly how many of the above implications are valid ? **[GATE - 07]**

(a) 0 (b) 1 (c) 2 (d) 3

27. Consider the following relation schemas :

b-Schema = (b-name, b-city, assets)

a-Schema = (a-num, b-name, bal)

d-Schema = (c-name, a-number).

Let branch, account and depositor be respectively instances of the above schemas. Assume that account and depositor relations are much bigger than the branch relation.

Consider the following query :

$\Pi_{c-name}\left(\sigma_{b-city="Agra" \wedge bal<0}\left(branch \bowtie \left(account \bowtie depositer\right)\right)\right)$ **[GATE - 07]**

Which one of the following queries is the most efficient version of the above query ?

(a) $\Pi_{c-name}\left(\sigma_{bal<0}\left(\sigma_{b-city="Agra"}\ branch \bowtie account\right) \bowtie depositer\right)$

(b) $\Pi_{c-name}\left(\sigma_{b-city="Agra"}\ branch \bowtie \left(\sigma_{bal<0}\ account \bowtie depositer\right)\right)$

(c) $\Pi_{c-name}\left(\sigma_{b-city="Agra"}\ branch \bowtie \sigma_{b-city="Agra" \wedge bal<0}\ account \bowtie depositer\right)$

(d) $\Pi_{c-name}\left(\sigma_{b-city="Agra"}\ branch \bowtie \left(\sigma_{b-city="Agra" \wedge bal<0}\ account \bowtie depositer\right)\right)$

Statement for Linked Answer Questions 28 and 29 :

Consider the B tree in the adjoining figure, where each node has at most two keys and three links.

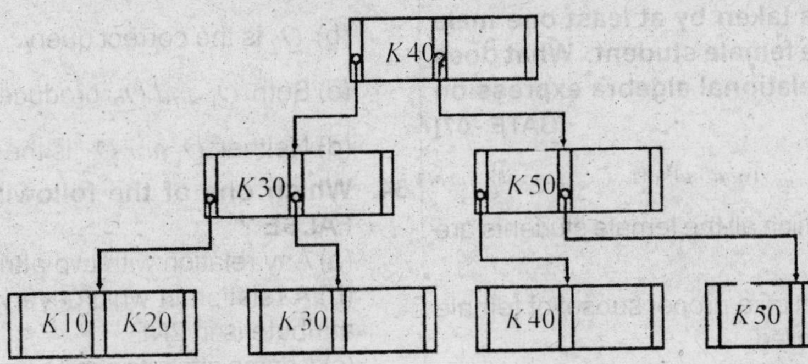

28. Keys K 15 and then K 25 are inserted into this tree in that order. Exactly how many of the following nodes (disregarding the links) will be present in the tree after the two insertions ? **[GATE - 07]**

| K30 | K50 | | K25 | K30 | | K20 | K25 | | K15 | K20 |

(a) 1 (b) 2 (c) 3 (d) 4

29. Now the key K 50 is deleted from the B^+ tree resulting after the two insertions made earlier.

Consider the following statements about the B^+ tree resulting after this deletion.

(i) The height of the tree remains the same.

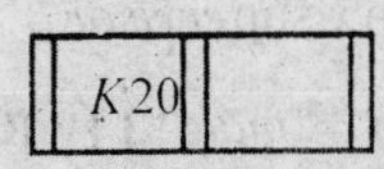

(ii) The node (disregarding the links) is present in the tree.

(iii) The root node remains unchanged (disregarding the links).

Which one of the following options is true ? **[GATE - 07]**

(a) Statements (i) and (ii) are true
(b) Statements (ii) and (iii) are true
(c) Statements (iii) and (i) are true
(d) All the statements are false

30. Which of the following is TRUE about formulae in Conjunctive Normal Form ? **[GATE - 07]**

(a) For any formula, there is a truth assignment for which at least half the clauses evaluate to true (b) For any formula, there is a truth assignment for a which all the clauses evaluate to true.

(c) There is a formula such that for each truth assignment at most one-fourth of the clauses evaluate to true

(d) None of the above.

31. Information about a collection of students is given by the relation studInfo (<u>studId</u>, name, sex). The relation enroll (<u>studId</u>, <u>courseId</u>) gives which student has enrolled for (or taken) what course (s). Assume that every course is taken by at least one male and at least one female student. What does the following relational algebra expression represent ? **[GATE - 07]**

$$\Pi_{courseId}\left(\left(\Pi_{studId}\left(\sigma_{sex="female"}(studInfo)\right) \times \Pi_{courseId}(enroll)\right) - enroll\right)$$

(a) Courses in which all the female students are enrolled.

(b) Courses in which a proper subset of female students are enrolled.

(c) Courses in which only male students are enrolled.

(d) None of the above

32. Consider the relation employee (<u>name</u>, sex, supervisorName (with name as the key. supervisor. Name gives the name of the supervisor of the employee under consideration. What does the following Tuple Relational Calculus query produce ?

{ e. name | employee (e) ^

$(\forall x)[\neg employee(x) v x.supervisor$

$Name \neq e.name\ v x.sex = "male"$] } **[GATE - 07]**

(a) Names of employees with a male supervisor.

(b) Names of employees with no immediate make subordinates.

(c) Names of employees with no immediate female subordinates.

(d) Names of employees with a female supervisor.

33. Consider the table employee (<u>empId</u>, name, department, salary) and the two queries Q_1, Q_2 below. Assuming that department 5 has more than one employee, and we want to find the employees who get higher salary than anyone in the department 5, which one of the statements is TRUE for any arbitrary employee table

Q_1 **: Select e.empId**
From employee e
Where not exists
(Select * from employee s Where S. department = "5" and s.salary > = e. salary)

Q_2 **: Select e.empId**
From employee e
Where e.salary > Any
(Select distinct salary From employee S Where S. department = "5") **[GATE - 07]**

(a) Q_1 is the correct query.

(b) Q_2 is the correct query

(c) Both Q_1 *and* Q_2 produce the same answer

(d) Neither Q_1 *nor* Q_2 is the correct query

34. Which one of the following statements is FALSE ? **[GATE - 07]**

(a) Any relation with two attributes is in BCNF.
(b) A relation in which every key has only one attribute is in 2NF
(c) A prime attribute can be transitively dependent on a key in a 3 NF relation.
(d) A prime attribute can be transitively dependent on a key in a BCNF relation.

35. The order of a leaf node in a B^+ - tree is the maximum number of (value, data record pointer) pairs it can hold. Given that the block size is 1 K bytes, data record pointer is 7 bytes long, the value field is 9 bytes long and a block pointer is 6 bytes long, what is the order of the leaf node? **[GATE - 07]**

(a) 63 (b) 64
(c) 67 (d) 68

36. Consider the following schedules involving two transactions. Which one of the following statements is TRUE :

$$S_1 : r_1(X); r_1(Y); r_2(X); r_2(Y); w_2(Y); w_1(X)$$

$$S_2 : r_1(X); r_2(X); r_2(Y); w_2(Y); r_1(Y); w_1(X)$$

[GATE - 07]

(a) Both S_1 *and* S_2 are conflict serializable

(b) S_1 is conflict serializable and S_2 is not conflict serializable

(c) S_1 is not conflict serializable and S_2 is conflict serializable

(d) Both S_1 *and* S_2 are not conflict serializable.

37. Consider the following log sequence of two transactions on a bank account, with initial balance 12000, that transfer 2000 to a mortgage payment and, then apply a 5% interest.

1. T1 start
2. T1 B old = 12000 new = 10000
3. T1 M old = 0 new = 2000
4. T1 commit
5. T2 start
6. T2 B old = 10000 new = 10500
7. T2 commit

Suppose the database system crashed just before log record 7 is written. When the system is restarted, which one statement is true of the recovery procedure ? [GATE - 06]

(A) We must redo log record 6 to set B to 10500

(B) We must undo log record 6 to set B to 10000 and then redo log records 2 and 3

(C) We need not redo log records 2 and 3 because transaction T1 has committed

(D) We can apply redo and undo operations in arbitrary order because they are idempotent

38. Consider the relation account (customer, balance) where customer is a primary key and there are no null values. We would like to rank customers according to decreasing balance. The customer with the largest balance gets rank 1. Ties are not broken but ranks are skipped: if exactly two customers have the largest balance they each get rank 1 and rank 2 is not assigned.

Query 1 :

Select A. customer, count (B. customer)

from account A, account B where A.

A. balance < = B. balance

group by A. customer

Query 2 :

select A.customer, 1+ count (B.customer)

From account A, account B

Where A.balance < B. balance

group by A.customer

Consider these statements about Query 1 and Query 2.

1. Query 1 will produce the same row set as Query 2 for some but not all databases

2. Both Query 1 and Query 2 are correct implementations of the specification

3. Query 1 is a correct implementation of the specification but Query 2 is not

4. Neither Query 1 nor Query 2 is a correct implementation of the specification

5. Assigning rank with a pure relational Query takes less time than scanning in decreasing balance order and assigning ranks using ODBC.

Which two of the above statements are correct ? [GATE - 06]

(A) 2 and 5 (B) 1 and 3

(C) 1 and 4 (D) 3 and 5

39. Consider the relation enrolled (student, course) in which (student, course) is the primary key, and the relation paid (student, amount) where student is the primary key. Assume no null values and no foreign keys or integrity constraints. Given the following four queries :

Query 1 : select student form enrolled where student in (select student form paid)

Query 2 : select student from paid where student in (select student from enrolled)

Query 3 : select E. student from enrolled E, paid P where E.student = P.student

Query 4 : select student from paid where exists

(select * from enrolled where enrolled.student = paid. student)

Which one of the following statements is correct ? [GATE - 06]

(A) All queries return identical row sets for any database

(B) Query 2 and Query 4 return identical row sets for all databases but there exist databases for which Query 1 and Query 2 return different row sets

(C) There exist databases for which Query 3 returns strictly fewer rows than Query 2

(D) There exist databases for which Query 4 will encounter an intergrity violation at runtime

40. Consider the relation enrolled (student, course) in which (student, course) is the primary key, and the relation paid (student, amount) where student is the primary key. Assume no null values and no foreign keys or integrity constraints. Assume that amounts 6000, 7000, 8000, 9000 and 10000 were each paid by 20 % of the students. Consider these query plans (Plan 1 on left, Plan 2 on right) to "list all courses taken by students who have paid more than x".

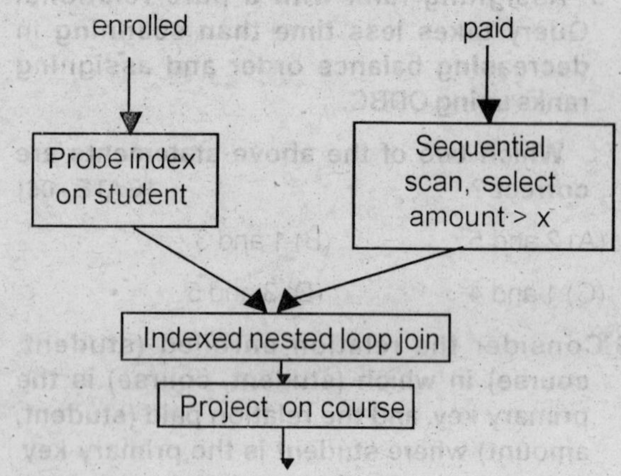

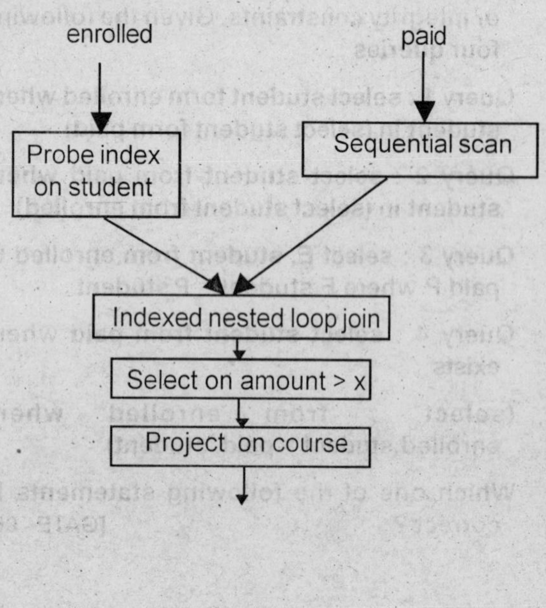

A disk seek takes 4 ms, disk data transfer bandwidth is 300 MB/s and checking a tuple to see if amount is greater than x takes $10\mu s$. Which of the following statements is correct ? [GATE - 06]

(A) Plan 1 and Plan 2 will not output identical row sets for all databases

(B) A course may be listed more than once in the output of Plan 1 for some databases

(C) For x = 5000, Plan 1 executes faster than Plan 2 for all databases

(D) For x = 9000, Plan 1 executes slower than plan 2 for all databases

41. The following functional dependencies are given :

$AB \to CD, AF \to D, DE \to F, C \to G,$

$F \to E, G \to A.$

Which one of the following options is false ? [GATE - 06]

(A) $\{CF\}^+ = \{ACDEFG\}$

(B) $\{BG\}^+ = \{ABCDG\}$

(C) $\{AF\}^+ = \{ACDEFG\}$

(D) $\{AB\}^+ = \{ABCDFG\}$

42. Consider the relations $r_1(P,Q,R)$ and $r_2(R,S,T)$ with primary keys P and R respectively. The relation r_1 contains 2000 tuples and r_2 contains 2500 tuples. The maximum size of the join $r_1 \bowtie r_2$ is [GATE - 06]

a) 2000 b) 2500

c) 4500 d) 5000

43. Which of the following relational query languages have the same expressive power?

(I) Relational algebra

(II) Tuple relational calculus restricted to safe expressions

(III) Domain relational calculus restricted to safe expressions [GATE - 06]

a) II and III only b) I and II only

c) I and III only d) I, II and III

44. Consider a relation R with five attributes V, W, X, Y, and Z. The following functional dependencies hold : $VY \to W, WX \to Z,$ and $ZY \to V$. Which of the following is a candidate key for R ? [GATE - 06]

a) VXZ b) VXY

c) VWXY d) VWXYZ

45.In a database file structure, the search key field is 9 bytes long, the block size is 512 bytes, a record pointer is 7 bytes and a block pointer is 6 bytes. The largest possible order of a non-leaf node in a B+ tree implementing this file structure is [GATE - 06]

a) 23 b) 24
c) 34 d) 44

Statement for linked Answer Questions 46 & 47

Consider a database with three relation instances shown below. The primary keys for the Drivers and Cars relation are did and cid respectively and the records are stored in ascending order of these primary keys as given in the tables. No indexing is available in the database.

D : Drivers relation

did	dname	rating	age
22	Karthikeyan	7	25
29	Salman	1	33
31	Boris	8	55
32	Amoldt	8	25
58	Schumacher	10	35
64	Sachin	7	35
71	Senna	10	16
74	Sachin	9	35
85	Rahul	3	25
95	Ralph	3	53

R : Reserves relation

did	cid	day
22	101	10/10/06
22	102	10/10/06
22	103	08/10/06
22	104	07/10/06
31	102	10/11/06
31	103	06/11/06
31	104	12/11/06
64	101	05/09/06
64	102	08/09/06
74	103	08/09/06

C : Cars relation

cid	cname	colour
101	Renault	blue
102	Renault	red
103	Ferrari	green
104	Jaguar	red

46.What is the output of the following SQL query?

```
select D.dname
from Drivers D
where D.did in (
        select R.did
        from Cars. C, Reserves R
        where R.cid = C.cid and C.colour
            = 'red'
        intersect
        select R.did
        from Cars C, Reserves R
        where R.cid = C.cid and C.colour
        = 'green'
        )
```

[GATE - 06]

a) Karthikeyan, Boris
b) Sachin, Salman
c) Karthikeyan, Boris, Sachin
d) Schumacher, Senna

47.Let n be the number of comparisons performed when the above SQL query is optimally executed. If linear search is used to locate a tuple in a relation using primary key, then n lies in the range. [GATE - 06]

a) $36-40$ b) $44-48$
c) $60-64$ d) $100-104$

ANSWER KEY

1. c	2. c	3. c	4. b	5. a
6. a	7. b	8. b	9. c	10. a
11. b	12. d	13. c	14. c	15. c
16. c	17. a	18. a	19.	20.
21.	22. b	23.c	24. d	25. c
26.c	27.b	28.a	29.a	30. b
31.b	32.b	33. b	34. d	35. b
36.c	37. c	38. c	39.a	40. c
41. c	42. a	43. d	44. b	45.
46. a	47. c			

EXPLANATORY NOTES

7.

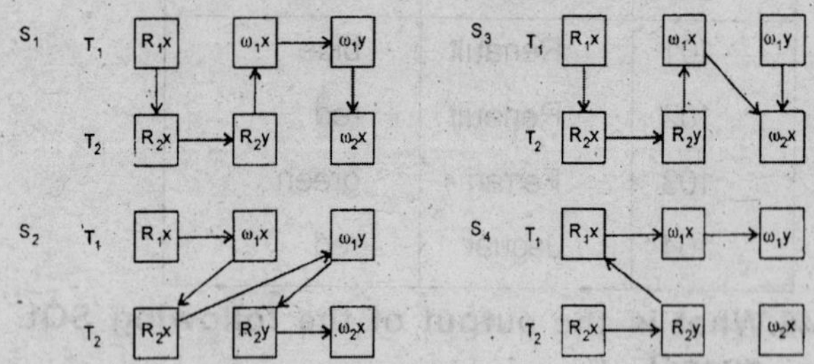

Clearly S_2 and S_3 are conflict = serializable.

8.

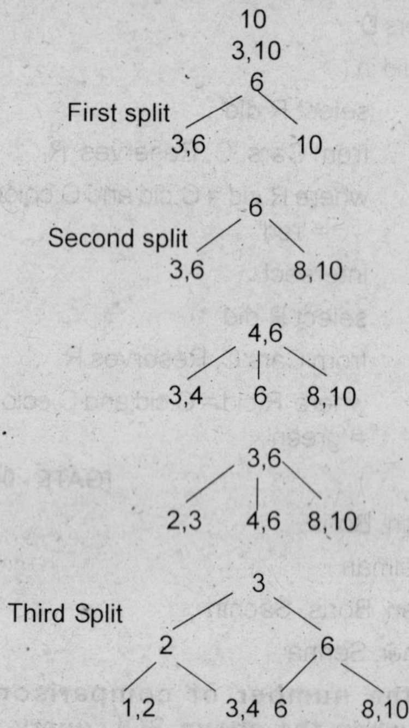

9. Query (II) and (IV) given equal data so they are equivalent

10. Here we have to select S. name from supplier S where S .sid (supplier Id) not in catalog select C.pid (product id) form parts where P. center< > 'blue' that means all names of all suppliers who have supplied a non blue part.

11. There is no non - prime to non - prime transitivity but prime to prime transitivity exists so it is in 3NF but not in BCNF.

12. $\forall t \in r(P(t))$ can also be expressed as $\neg\exists t \in r(\neg P(t))$ and $\exists t\, r(\neg P(t))$ of $\exists t \neq r(\neg P(t))$ according to equivalence rules.

13. Clustering index is unique and is created automatically when primary key is declared it is ordered.

14. I, II & III express following quering select P from the join (natural) of R and S, Which can be expressed as select P then joining (II) or selecting P from the union of R & S by first selecting R & Q (key values)

15. A table in the second normal from if all the non key fields fully dependent on whole key. A table is in third normal from if all the non key fields of the table are independent of all other non key field of table.

Alternately

If we have a function dependency of kind

$$\{A, B\} \rightarrow \{C, D\}$$

$$C \rightarrow B$$

Then the Tables are not in BCNF, so Book is not in BCNF

If a non key value is dependent on other non key, then table is not in 3NF, so Book is also not in 3NF and collection is in 3NF

16. Block size is given 1024 so first level and second level multiplication must be 2^{10}, so only (c) satisfy

$$2^8 \times 2^2 = 2^{10}$$

17. The diagram consists of three entity set M, P & N and relates through binary relation R1 & R2. So minimum 2 tables are needed to represent this diagram

Alternately

Relation R1 is one to many from P to M & R2 is one to many from P to N. But N is a weak entity. There is also total participation of tables in relations so we can merge M and P as one table and P & N as other table.

18. {M1, M2, M3, P1} shows correct attributes set as they are connected with binary Relation R1

Alternately

Since mergeing P & M into one result is non-normalised tables as P2 depends only one of the key values P1, so it has to be removed. This can be done without the loss of information as P1 & P2 are also in other table which has attributes P1, P2, N1 & N2 having primary key as P1.

So attribute set for tables are

{M1, M2, M3, P1 } and {P1, P2, N1, N2}

24. Given : $\sigma A \le 100(r)$

There are 100 tuples.

These are uniformly distributed, in the interval so [0, 500]

So for values of $A \le 100$. There will be 200 tuples returned by the given selected query

25. T_1 : A requires shared lock because it reads only

B requires exclusive lock because it perform read and write operation.

T_2 : similarly it performs shared lock on B and exclusive lock of A.

Before releasing exclusive lock, commit operation is necessary.

Hence answer is (c).

26. In DBMS (Ref.korth) we can see that

if $A \to\to B$ and $A \to\to C$ then $A \to BC$

and $A \to BC$ then $A \to\to B$ and $A \to\to BC$

Hence answer is (c)

27. b. Schema = (b-name, b-city, assets)

a. Schema = (a-num, b-name, bal)

d. Schema = (c-name, a-number)

$\Pi_{c-name}\left(\sigma_{b-city="Agra" \wedge bal<0}\left(branch \bowtie (account \bowtie depositor)\right)\right)$

Most efficient version of the above query.

$\Pi_{c-name}\left(\sigma_{b-city="Agra"}\ branch \bowtie \left(\sigma_{bal<0}\ account \bowtie depositor\right)\right)$

28.

Final tree is

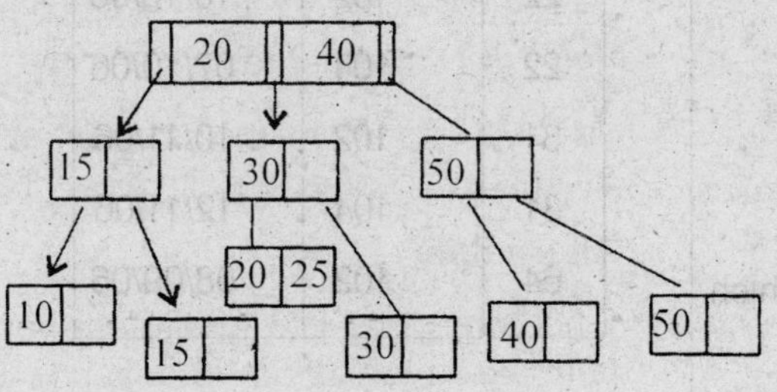

29. After deletion of K - 50, we get following B^+ tree

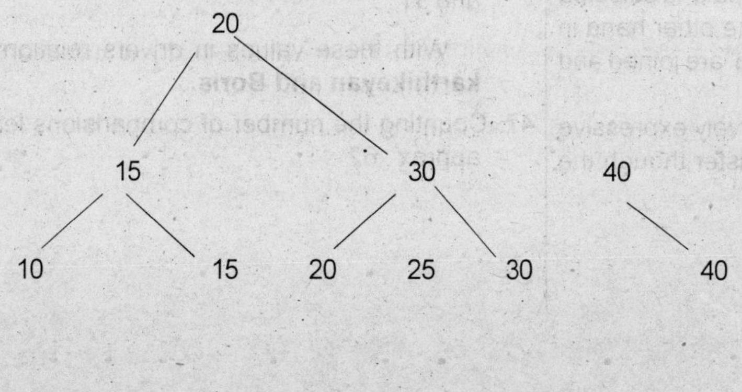

So (i) is true

(ii) is also true

(iii) is false because root not remain unchanged.

30. Since all clauses must be true for a fomula (in conjunctive Normal Form) to be true, so atleast for one truth assignment all clauses will evaluate to true.

33. Query Q_1 doesn't give correct result because it contain not exists clause which lead to wrong answer. Q_2 is correct.

34. In BCNF, a prime attribute cannot be transitively dependent on a key.

35. Let the order be n

So, $n \times 7 + (n-1)9 = 1024$

$\Rightarrow$ 16 n = 1033

$\Rightarrow$ $n = \left\lceil \frac{1033}{16} \right\rceil = 64$

36. Here S_1 is not conflict serializable whereas S_2 is conflict serializable.

37. Once a transaction commits, the changes made by it to the database get permanent and there is no effect of system failure over the data. Since T_2 is not allowed to commit, we must undo log record 6.

38. Lets take an example of the balances

1.	20	1
2.	21	2
3.	21	2
4.	21	2
5.	22	5

Query 1 will assign rank 4 to account (4) which is wrong.

39. All queries give one and the same output.

40. In Plan 1, first the records from 'paid' is selected and then they are joined. On the other hand in Plan 2, all the records from 'paid' are joined and then the records are checked.
As join is an iterative and relatively expressive process. Plan 1 will execute faster though the output will be the same.

41. The algorithm used is $(for\ \alpha^+)$

result = α,
While (changes to result) do
for each functional dependency $\beta \rightarrow \gamma$ in F
do begin
if $\beta \subseteq$ result, then
result = result $\cup \gamma$,
end

For(C), $\{AF\}^+ = \{AFDE\}$

42. $\bowtie$ is natural inner join operator.
It considers only pairs of tuples that have the same value on R.

44. $VY \rightarrow W$

$WX \rightarrow Z$

$ZY \rightarrow V$

$\because$ This key can be used to distinguish between tuples.

$\therefore$ Candidate Key is VXY

46. When R.cid = c.cid and c.colour = 'Green'

did	cid	day
22	103	08/10/06
31	103	06/11/06
74	103	08/09/06

Table - 1

When R.cid = c.cid and colour = 'Red;

did	cid	day
22	102	10/10/06
22	104	07/10/06
31	102	10/11/06
31	104	12/11/06
64	102	08/09/06

Table - 2

$\Rightarrow$ Table I $\cap$ Table 2 gives 2 values of did = 22 and 31.

$\therefore$ With these values in drivers relation gives **karthikeyan and Boris.**

47. Counting the number of comparisions leads to approx . 62

TEST YOURSELF - II

1. Which of the following scenarios may lead to an irrecoverable error in a database system ? **[GATE - 03]**

a) A transaction writes a data item it is read by an uncommitted transaction.

b) A transaction reads a data item after it is read by an uncommitted transaction.

c) A transaction reads a data item after it is written by a committed transaction.

d) A transaction reads a data item after it is written by an uncommitted transaction.

2. Consider the following SQL query **[GATE - 03]**

select distinct $a_1, a_2, \ldots\ldots\ldots, a_n$
from $r_1, r_2 \ldots\ldots r_m$
where P

For an arbitrary predicate P, this query is equivalent to which of the following relational algebra expressions?

a) $\Pi_{a_1,a_2,\ldots a_n} \sigma_P(r_1 \times r_2 \times \ldots\ldots \times r_m)$

b) $\Pi_{a_1,a_2,\ldots a_n} \sigma_P(r_1 \bowtie r_2) \bowtie \ldots\ldots \bowtie r_m)$

c) $\Pi_{a_1,a_2,\ldots a_n} \sigma_P(r_1 \cup r_2 \cup \ldots\ldots \cup r_m)$

d) $\Pi_{a_1,a_2,\ldots a_n} \sigma_P(r_1 \cap r_2 \cap \ldots\ldots \cap r_m)$

3. Consider the following functional dependencies in a database. **[GATE - 03]**

Date_of_Birth → Age   Age → Eligibility

Name → Roll_number   Roll_number → Name

Couse_number → Course_name

Course_number → Instructor

(Roll_number, Couse_number) → Grade

The relation (Roll_number, Name Date_of_birth, Age) is

a) in second normal form but not in third normal form

b) in third normal form but not in BCNF

c) in BCNF

d) none of the above.

4. Consider the set of relations shown below and the SQL query that follows.

Students : (Roll_number, Name, Date_of_birth)

Courses : (Course_number, Course_name, Instructor)

Grades : (Roll_number, Course_number, Grade)

select dinstinct Name
from Students, Courses, Grades
where Students. Roll_number = Grades. Roll_number
and Courses. Instructor = Korth
and Courses. Course_number = Grades.Course_number
and Grades.grade = A

Which of the following sets is computed by the above query ? **[GATE - 03]**

a) Name of students who have got an A grade in all courses taught by Korth.

b) Names of students who have got an A grade in all courses.

c) Names of Students who have got an A grade in at least one of the courses.

d) None of the above.

5. Consider three data items D1, D2 and D3 and the following execution schedule of transactions T1, T2 and T3. In the diagram, R(D) and W(D) denote the actions reading and writing the data item D respectively. **[GATE - 03]**

time ↓	T1	T2	T3
		R(D3);	
		R(D2);	
		W(D2) ;	
			R(D2);
			R(D3);
	R(D1);		
	W(D1);		
			W(D2);
			W(D3);
		R(D1);	
	R(D2);		
	W(D2);		
		W(D1);	

Which of the following statements is correct ?

a) The schedule is serializable as T2; T3; T1

b) The schedule is serializable as T2; T1; T3

c) The schedule is serializable as T3; T2; T1

d) The schedule is not serializable

6. Let R_1(A, B, C) and R_2 (D, E) be two relation schemas, where the primary keys are shown underlined, and let C be a foregin key in R_1 referring to R_2. Suppose there is no violation of the above referential integrity constraint in the corresponding relation instances r_1 and r_2. Which one of the following relational algebra expressions would necessarily produce an empty relation? **[GATE - 04]**

a) $\Pi_D(r_2) - \Pi_C(r_1)$ b) $\Pi_C(r_1) - \Pi_D(r_2)$

c) $\Pi_D(r_1 \bowtie_{C \neq D} r_2)$ d) $\Pi_C(r_1 \bowtie_{C \neq D} r_2)$

7. Consider the following relation schema pertaining to a students database: **[GATE - 04]**

Student (<u>rollno</u>, name, address)

Enroll (<u>rollno, courseno,</u> coursename)

where the primary keys are shown underlined. The number of tuples in the Student and Enroll tables are 120 and 8 respectively. What are the maximum and minimum number of the tuples that can be present in (Student * Enroll), where '*' denotes natural join ?

a) 8, 8   b) 120, 8   c) 960, 8   d) 960, 120

8. The relation scheme Student Performance (name, courseNo, rollNo, grade) has the following functional dependencies: **[GATE - 04]**

name, courseNo $\rightarrow$ grade

rollNo, courseNo $\rightarrow$ grade

name $\rightarrow$ rollNo

rollNo $\rightarrow$ name

The highest normal form of this relation scheme is

a) 2 NF b) 3 NF c) BCNF d) 4NF

9. Consider the relation Student (name, sex, marks) where the primary key is shown underlined, pertaining to students in a class that has at least one boy and one girl. What does the following relation algebra expression produce ? (Note : r is the rename operator). **[GATE - 04]**

$$\Pi_{name}(\sigma_{sex=female}(Student)) - \Pi_{name}(Student \bowtie_{(sex=female \wedge x=male \wedge marks \leq m)} \rho_{n,x,m}(Student))$$

a) names of girl students with the highest marks

b) names of girl students with more marks than some boy student

c) names of girl students with marks not less than some boy student

d) names of girl students with more marks than all the boy students

10. The order of an internal node in a B$^+$ tree index is the maximum number of children it can have. Suppose that a child pointer takes 6 bytes, the search field value takes 14 bytes, and the block size is 512 bytes. What is the order of the internal node ? **[GATE - 04]**

a) 24 b) 25 c) 26 d) 27

11. The employee information in the company is stored in the relation **[GATE - 04]**

Employee (<u>name</u>, sex, salary, deptName)

Consider the following SQL query

```
select deptName
    from Employee
    where sex = 'M'
group by deptName
having avg (salary) >
    (select avg (salary) from Employee)
```

It returns the names of the department in which

a) the average salary is more than the average salary in the company.

b) the average salary of male employees is more than the average salary of all male employees in the company.

c) the average salary of male employees is more than the average salary of employees in the same department

d) the average salary of male employees is more than the average salary in the company

12. Which level of locking provides the highest degree of concurrency in a relational database ? **[GATE - 04]**

a) Page b) Table c) Row

d) Page, table and row level locking allow the same degree of concurrency

13. Consider the following entity relationship diagram (ERD), where two entities E1 and E2 have a relation R of cardinality 1:m. **[GATE - 04]**

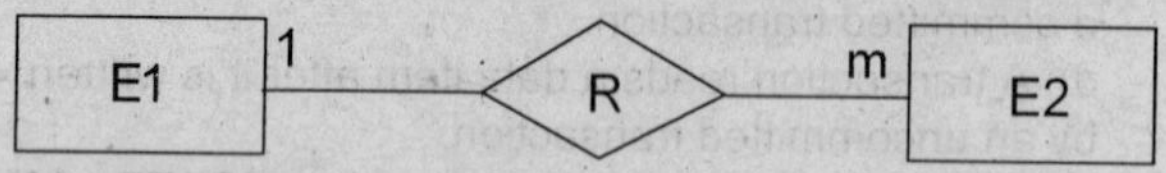

The attributes of E1 and A11, A12 and A13 where A11 is the key attribute. The attributes of E2 are A21, A22 and A23 where A21 is the key attribute and A23 is a multi-valued attribute. Relation R does not have any attribute. A relational database containing minimum number of tables with each table satisfying the requirements of the third normal form (3NF) is designed from the above ERD. The number of tables in the database is

a) 2 b) 3 c) 5 d) 4

14. A relational database contains two tables student and department in which student table has columns roll_no, name and dept_id and department table has columns dept_id and dept_name. The following insert statements were executed successfully to populate the empty tables:

```
Insert into department values (1, 'Mathematics')
Insert into department values (2, 'Physics')
Insert into student values (1, 'Navin', 1)
Insert into student values (2, 'Mukesh', 2)
Insert into student values (3, 'Gita', 1)
```

How many rows and columns will be retrieved by the following SQL statement ? **[GATE - 04]**

```
Select * from student, department
```

a) 0 row and 4 columns b) 3 rows and 4 columns

c) 3 rows and 5 columns d) 6 rows and 5 columns

15. A relation Empdtl is defined with attributes empcode (unique), name, street, city, state and pincode. For any pincode, there is only one city and state. Also, for any given street, city and state, there is just one pincode. In normalization terms, Empdtl is a relation in **[GATE - 04]**

a) 1NF only b) 2 NF and hence also in 1NF

c) 3 NF and hence also in 2 NF and 1 NF

d) BCNF and hence also in 3NF, 2NF and 1NF

16. A table T1 in a relational database has the following rows and columns: **[GATE - 04]**

roll no	marks
1	10
2	20
3	30
4	Null

The following sequence of SQL statements was successfully executed on table T1.

Update T1 set marks = marks + 5 Select avg (marks) from T1

What is the output of the select statement ?

a) 18.75 b) 20

c) 25 d) Null

17. Consider the following schedule S of transactions T1 and T2 : **[GATE - 04]**

T1	T2
Read (A)	
A = A - 10	
	Read (A)
	Temp = 0.2*A
	Write (A)
	Read (B)
Write (A)	
Read (B)	
B = B + 10	
Write (B)	
	B = B + Temp
	Write (B)

Which of the following is TRUE about the schedule S ?

a) S is serializable only as T1, T2

b) S is serializable only as T2, T1

c) S is serializable both as T1, T2 and T2, T1

d) S is serializable either as T1 or as T2

18. Consider two tables in a relational database with columns and rows as follows :

Table : Student

Roll_no	Name	Dept_id
1	ABC	1
2	DEF	1
3	GHI	2
4	JKL	3

Table : Department

Dept_id	dept_name
1	A
2	B
3	C

Roll_no is the primary key of the Student table, Dept_id is the primary key of the Department table and Student Dept_id is a foreign key from Department. Dept_id what will happen if we try to execute the following two SQL statements ? **[GATE - 04]**

i) update Student set Dept_id = Null where Roll_no = 1

ii) update Department set Dept_id = Null where Dept_id = 1

a) Both i) and ii) will fail

b) i) will fail but ii) will succeed

c) i) will succeed by ii) will fail

d) Both i) and ii) will succeed

19. Consider a table T in a relational database with a key field K. A B- tree of order p is used as an access structure on K, where p denotes the maximum number of tree pointers in a B-tree index node. Assume that K is 10 bytes long; disk block size is 512 bytes; each data pointer P_D is 8 bytes long and each block pointer P_B is 5 bytes long. In order for each B-tree node to fit in a single disk block, the maximum value of p is **[GATE - 04]**

a) 20 b) 22 c) 23 d) 32

20. Which one of the following is a key for prefering B^+ - trees to binary search trees for indexing database relations ? **[GATE - 05]**

a) Database relations have a large number of records.

b) Database relations are sorted on the primary key.

c) B^+ - trees require less memory than binary search trees

d) Data transfer from disks in blocks.

21. Which one of the following statements about normal forms is FALSE ? **[GATE - 05]**

a) BCNF is stricter than 3NF

b) Lossless, dependency preserving decomposition into 3NF is always possible.

c) Lossless, dependency preserving decomposition into BCNF is always possible.

d) Any relation with two attributes is in BCNF

22. Let r be a relation instance with schema $R = (A,B,C,D)$. We define $r_1 = \Pi_{A,B,C}(r)$ and $r_2 = \Pi_{A,D}(r)$. Let $s = r_1 * r_2$ where * denotes natural join. Given that the decomposition of r into r_1 and r_2 is lossy, which one of the following is TRUE ? **[GATE - 05]**

a) $S \subset r$ b) $r \cup s = r$ c) $r \subset s$ d) $r * s = s$

23. Let E_1 and E_2 be two entities in an E/R diagram with simple single-valued attributes. R_1 and R_2 are two relationships between E_1 and E_2, where R_1 is one-to-many and R_2 is many-to-many. R_1 and R_2 do not have any attributes of their own. What is the minimum number of tables required to represent this situation in the relational model ? **[GATE - 05]**

a) 2 b) 3 c) 4 d) 5

24. The following table has two attributes A and C where A is the primary key and C is the foreign key referencing A with on-delete cascade.

A	C
2	4
3	4
4	3
5	2
7	2
9	5
6	4

The set of all tuples that must be additionally deleted to preserve referential integrity when the tuple (2, 4) is deleted is : **[GATE - 05]**

a) (3, 4) and (6, 4) b) (5, 2) and (7, 2)

c) (5, 2) (7, 2) and (9, 5) d) (3, 4) (4, 3) and (6, 4)

25. The relation book (title, price) contains the titles and prices of different books. Assuming that no two books have the same price, what does the following SQL

```
select title                                   [GATE - 05]
from book as B
where (select count(*)
        from book as T
        where T.price > B.price) < 5
```

a) Titles of the four most expensive books
b) Title of the fifth most inexpensive book
c) Title of the fifth most expensive book
d) Titles of the five most expensive books

26. Consider a relation scheme R = {A, B, C, D, E, H} on which the following functional dependencies hold : $\{A \rightarrow B, BC \rightarrow D, E \rightarrow C, D \rightarrow A\}$. What are the candidate keys of R ? **[GATE - 05]**

a) AE, BE b) AE, BE, DE
c) AEH, BEH, BCH d) AEH, BEH, DEH

27. A table has fields F1, F2, F3, F4, F5 with the following functional dependencies

$F1 \rightarrow F3$ $F2 \rightarrow F4$ $(F1.F2) \rightarrow F5$

In terms of Normalization, this table is in **[GATE - 05]**

a) 1 NF 2) 2 NF 3) 3 NF 4) None of these

28. A B-Tree used as an index for a large database table has four levels including the root node. If a new key is inserted in this index, then the maximum number of nodes that could be newly created in the process are **[GATE - 05]**

a) 5 b) 4 c) 3 d) 2

29. Among the ACID properties of a transaction, the 'Durability' property requires that the changes made to the database by a successful transaction persist **[GATE - 05]**

a) Except in case of an Operating System crash
b) Except in case of a Disk crash
c) Except in case of a power failure
d) Always, even if there is a failure of any kind

30. A company maintains records of sales made by its salespersons and pays them commission based on each individual's total sales made in a year. This data is maintained in a table with following schema;
salesinfo = (salespersonid, totalsales, commission)
In a certain year; due to better business results, the company decides to further reward its salespersons by enhancing the commission paid to them as per the following formula
If commission < = 50000, enhance it by 2%
If 50000 < commission < = 100000, enhance it by 4%
If commission > 100000, enhance it by 6%
The IT staff has written three different SQL scripts to calculate enhancement for each slab, each of these scripts is to run as a separate transaction as follows :

```
T1   Update salesinfo
     Set commission = commission * 1.02
     Where commission < = 50000;
T2   Update salesinfo
     Set commission = commission * 1.04
     Where commission > 50000 and
     commission is < = 100000;
T3   Update salesinfo
     Set commission = commission * 1.06
     Where commission > 100000;
```

Which of the following options of running these transactions will update the commission of all salespersons correctly **[GATE - 05]**

a) Execute T1 followed by T2 followed by T3
b) Execute T2, followed by T3; T1 running concurrently throughout
c) Execute T3 followed by T2; T1 running concurrently throughout
d) Execute T3 followed by T2 followed by T1

31. A table 'student' with schema (roll, name, hostel, marks), and another table 'hobby' with schema (roll, hobbyname) contains records as shown below :

Table student				Table hobby	
Roll	Name	Hostel	Marks	Roll	hobbyname
1798	Manoj Rathod	7	95	1798	chess
2154	Soumic Banerjee	5	68	1798	music
2369	Gumma Reddy	7	86	2154	music
2581	Pradeep Pendse	6	92	2369	swimming
2643	Suhas Kulkarni	5	78	2581	cricket
2711	Nitin Kadam	8	72	2643	chess
2872	Kiran Vora	5	92	2643	hockey
2926	Manoj Kunkalikar	5	94	2711	volleyball
2959	Hemant Karkhanis	7	88	2872	football
3125	Rajesh Doshi	5	82	2926	cricket
				2959	photography
				3125	music
				3125	chess

The following SQL query is executed on the above tables:

```
select hostel
from student natural join hobby
where marks > = 75 and roll between 2000 and 3000;
```

Relatiions S and H with the same schema as those of these two tables respectively contain the same information as tuples. A new relation S' is obtained by the following relational algebra operation: **[GATE - 05]**

$S' = \pi_{hostel}((\sigma_{s.roll=H.roll}(\sigma_{marks>75 and roll>2000 and roll<3000}(S)) X (H))$

The difference between the number of rows output by the SQL statement and the number of tuples in S' is

a) 6 b) 4 c) 2 d) 0

32. In an inventory management system implemented at a trading corporation, there are several tables designed to hold all the information. Amongst these, the following two tables hold information on which items are supplied by which suppliers, and which warehouse keeps which items along with the stock-level of these items.

Supply = (supplierid, itemcode)
Iventory = (itemcode, warehouse, stocklevel)

For a specific information required by the management, following SQL query has been written

```
Select distinct STMP supplierid
From Supply as STMP
Where not unique (Select ITMP. supplier id
```

From Inventory, Supply as ITMP
Where STMP. supplierid = ITMP.supplierid
And ITMP.itemcode = Inventory.itemcode
And Inventory.warehouse = 'Nagpur');

For the warehouse at Nagpur, this query will find all suppliers who **[GATE - 05]**

a) do not supply any item b) supply exactly one item
c) supply one or more items
d) supply two or more items

33. In a schema with attributes A, B, C, D and E following set of functional dependencies are given **[GATE - 05]**

$A \rightarrow B$ $A \rightarrow C$ $CD \rightarrow E$ $B \rightarrow D$ $E \rightarrow A$

Which of the following functional dependencies is NOT implied by the above set

a) $CD \rightarrow AC$ b) $BD \rightarrow CD$
c) $BC \rightarrow CD$ d) $AC \rightarrow BC$

Statement for Linked Answer Questions 34 and 35

A database table T1 has 2000 records and occupies 80 disk blocks. Another table T2 has 400 records and occupies 20 disks blocks. These two tables have to joined as per a specified join condition that needs to be evaluated for every pair of records from these two tables. The memory buffer space available can hold exactly one block of records for T1 and one block of records for T2 simultaneously at any point in time. No index is available on either table. **[GATE - 05]**

34. If Nested-loop join algorithm is employed to perform the join, with the most appropriate choice of table to be used in outer loop, the number of block accesses required for reading the data are

a) 800000 b) 40080 c) 32020 d) 100

35. If, instead of Nested-loop join, Block nested-loop join is used, again with the most appropriate choice of table in the outer loop, the reduction in number of block accesses required for reading the data will be

a) 0 b) 30400 c) 38400 d) 798400

ANSWER KEY

1. d	2. a	3. a	4. c	5. c
6. b	7. c	8. a	9. d	10. c
11. d	12.	13. c	14. d	15. b
16. a	17. d	18. c	19. b	20. d
21. c	22. c	23. b	24. c	25. a
26. d	27. a	28. a	29. d	30. d
31. a	32. d	33. b	34. a	35. d

EXPLANATORY NOTES

3. Date-of-birth $\rightarrow$ Age
Age $\rightarrow$ Eligibility
Name $\rightarrow$ Roll-number, Roll Number $\rightarrow$ Name
Course-Number $\rightarrow$ Course-Name
Course - Name $\rightarrow$ Instructor
(Roll Number, Course - number) $\rightarrow$ Grade

From given relation,
Date-of-Birth $\rightarrow$ Eligibility (By transitively)
Course-Number $\rightarrow$ Instructor (By transitively)

So, it is not in 3rd Normal form because in third normal form relation, every non prime attribute is non-transitively and fully dependent on the every candidate key.

In 2nd Normal form, all non-prime attributes are fully functionally dependent on the relation keys.

5.

T1	T2	T3
	R(D3); R(D2); W(D2);	
		R(D2); W(D1);
R(D1); W(D1);		
R(D2); W(D2);	R(D1); W(D1);	W(D2); W(D3);

In transation T_3 R(D2) is conflict with original R(D2), because R(D2) is read W(D2). Again the transaction T_2, R(D1) is conflict with R(D1) in T_1 because it read (D1) as W(D1). In transaction T_1, R(D2) conflicts with R(D2) in T_2, because it is W(D2) in T_2. So, finally transaction is conflict serialization schedule as $T_3 ; T_2 ; T_1$

7. Maximum number of tuple is 120 × 8 = 960
Minimum number of tuple is minimum of these two, is 8.

8. From the given FD's it is clear that there is no repetition of attributes, hence it is in 1 NF. There is no partial dependency, hence it is in 2NF, and
name $\rightarrow$ roll no.
roll no $\rightarrow$ name
Since there is transitive dependency it is not in 3NF, so highest normal from of this relation is 2NF.

10. Child pointer takes b byte
Searh field = 14 byte
block size = 512 byte
The order of internal node is 9 + 14 + 3 = 26

11. Given relation is
Employee (<u>Name</u>, Sex, Salary, deptName)
SQL query is
Select deptName from Employee Where Sex = 'M'
group by deptName
having avg (Salary) > (Select avg (Salary) from Employee)
This will retrieve the department in which average salary of male employee is more than the average salary in the company.

14. After populating the table
Student

Roll no.	Name	Dept-id
1	Navin	1
2	Mukesh	2
3	Gita	1

Department

dept – id	dept name
1	Mathematics
2	Physics

Select * from student, department.

This query will execute natural join first and will retrieve all the row and column. Hence 6 row and 5 column will retrieve, which is same as cartesian product of these table.

15. empt {empcode, name, street, city, state, pincode}

Given functional dependency

Pincode $\rightarrow$ (city) (state)

(street) (city) (state) $\rightarrow$ Pincode.

This is in 1 NF and 2 NF but not in 3 NF because it contain transitive dependency.

16. After the updation, new table will be

roll no.	marks
1	10
2	20
3	30
4	5

$$\text{average} = \frac{15+25+30+5}{4} = \frac{75}{4} = 18.75$$

18. First query will successfully executed and after executing this statement attribute dept-id of student table become NULL which is foreign key from department's dept-id. So second statement will not execute, condition will not satisfy in this case, by any row.

20. Because data transfer take place in the form of block's frame disk, so any data structure which store as large as can, information in a single block is more efficient.

21. Note that B.C.N.F. is not dependency-preserving in all cases.

22. s have more tuples than required one

it have all tuples in r but some which are not is r also.

so, $r \subset s$

23. One table for E_1;

One table for E_2;

One table for R_2

* Remember that only one to many relationship can be represented with entity set from one's side, because each entity in this entity set can be associated with at most one entity of another set.

24. When (2, 4) is deleted any record that have '2' in the foreign key must have to be deleted.

Here, records are $< 5, 2 >, < 7, 2 >$

Then because of cascade any record that have 5 or 7 as foreign key must be deleted. There is no record that have 7 as foreign key (9, 5) is gone to delete. Now, any record that have 9 has an foreign key have to delete but there is no one.

26. When we consider

$A = \{A, B\}$, $B = \{B\}$ and $H = \{H\}$

$\{A, B, H\} = \{A, B, C, D, E, H\}$

$\{B, E, H\} = \{A, B, C, D, F, H\}$

$\{D, E, H\} = \{A, B, C, D, E, H\}$

and no one option can generate this result.

27. Primary key (candidate key) for this relation is [F1,F2] because

[F1, F2] column contains [F1, F2, F3, F4, F5]

so there is, functional dependency

F1 $\rightarrow$ F3

F2 $\rightarrow$ F4

It also have partial dependency so it is not in 2nd, 3rd NF

28. Consider here levels

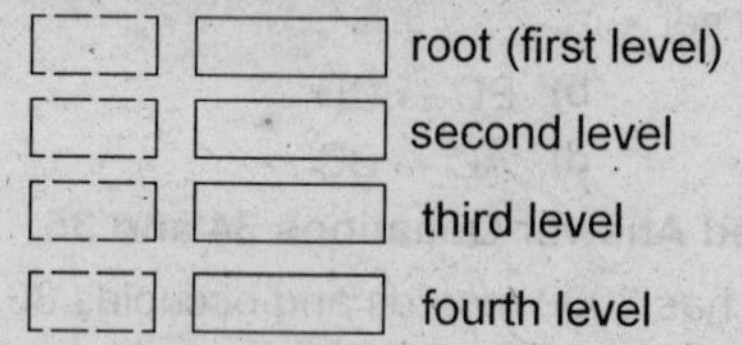

If insert takes place then new node can be inserted at each level shown by dashed box and in this process the new root can be created. Hence 5 is the answer.

30. For a person whose commission is 8000 than if modified by T2.

If T2 run first than T3, its commission will be modified twice by first update (T2), its commission will be, '12800'. Then after that if we appy T3, then its commissiion will be '12800' which is in range of T3 also.

31. By executing SQL query, we have table of the form.

Table Student				Table hobby
Roll	Name	Hostel	Marks	hobby name
1798	Manoj Rathod	7	95	chess
1798	Manoj Rathod	7	95	music
2154	Soumic Banerjee	5	68	music
*2369	Gumma Reddy	7	86	swimming
*2581	Pradeep Pendse	6	92	cricket
*2643	Suhas Kulkarni	5	78	hockey
*2872	Kiran Vora	5	92	football
*2711	Nitin Kadam	8	72	volleyball
*2926	Manoj Kunkalikar	5	94	cricket
*2959	Hemant Karkhanis	7	88	photography
*3125	Rajesh Doshi	5	82	music
*3125	Rajesh Doshi	5	82	chess
*2643	Suhas Kulkarni	5	78	chess

* line are related by final query.

{7, 6, 5, 5, 8, 5, 7, 5, 5, 5} - {7, 6, 5, 8} = 6

Main Point is R.A.

33. $CD \rightarrow A \; \{CD \rightarrow E; E \rightarrow A; E \rightarrow C\}$

$CD \rightarrow C$

$BC \rightarrow CD$

$B \rightarrow D;\; CD \rightarrow E, E \rightarrow A; A \rightarrow C$

$AC \rightarrow BC;\; A \rightarrow B; A \rightarrow C$

Unit - X
Computer Networks

UNIT - X COMPUTER NETWORKS

I. ISO/OSI STACK, SLIDING WINDOW PROTOCOL, TCP/UDP/IP

Introduction

The TCP/IP has been used since the 1960s as a method to connect large mainframe computers together to share information among the research community and the Department of Defense. Now TCP/IP is used to support the largest computer network, the Internet. Most manufacturers incorporate TCP/IP into their operating systems, allowing all types of computers to communicate with each other.

The ISO/OSI model breaks network communication down into seven layers. In contrast, the TCP/IP network model contains only five layers. These particular network models are also called the protocol stacks.

The Physical layer is responsible only for sending and receiving digital data, nothing else. Any errors that show up in the data at the Physical layer are handled by the next higher layer up in the protocol stack. In the TCP/IP network model, the next higher level up is referred to as the Network Interface layer. The Network Interface layer performs the same functions as the Data-Link layer in the ISO/OSI model. THe functions of the Physical layer and the Data-Link layer are determined by the type of LAN hardware in use, such as Ethernet, token ring, and so forth.

The next layer up, called the internet layer, performs the same function as the Network layer in the ISO/OSI model. The internet layer uses the Internet Protocol (IP) and provides what is called machine-to machine communication.

The Transport layer is the same in both the ISO/OSI and TCP/IP network stacks. There are two transport layer protocols in TCP/IP, the Transmission Control Protocol (TCP) and the User Datagram Protocol (UDP). The transport protocols provide application-to-application communication. TCP provides for connection-oriented reliable transport and UDP provides for connectionless unreliable transport.

The TCP/IP Application layer provides the same functionality as the Session, Presentation and Application layers in the ISO/OSI network model. The Application layer is where the communicating application programs running on the source computer and the destination computer reside. The packet of data that is transmitted from the source computer contains an informational header for each layer in the protocol stack (except layer1).

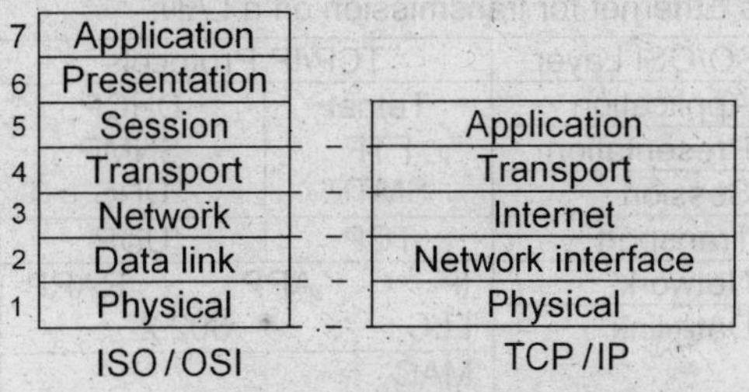

Fig. 1 : ISO/OSI and TCP/IP protocol stacks

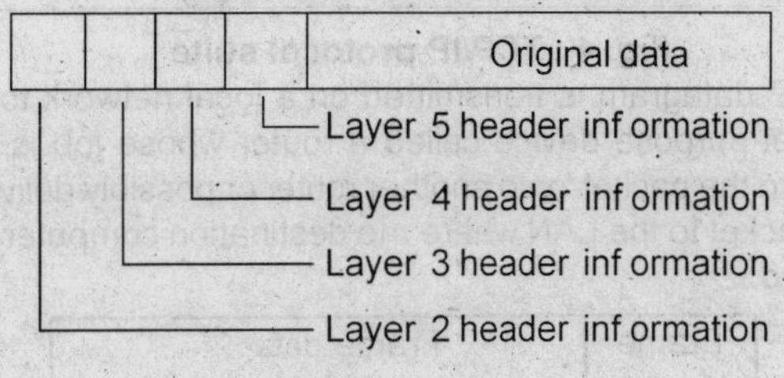

Fig. 2 : TCP/IP stack layering

When a hardware frame is sent out onto the network, it contains all of the information necessary to be forwarded and received properly by the destination computer without errors. At the destination computer, as the received message moves up the stack toward the application, identical copies of the packaged information are available at each layer, just as it was on the source computer.

RFCs

Official standards for the Internet are published as electonic documents called Request for Comments, or RFCs. The RFC series of documents on networking began in 1969 as part of the original Advanced Research Projects Agency wide-area networking project called ARPAnet. RFCs cover a wide range of topics in addition to Internet standards, from early discussion of new research concepts to status memos about the current state of the Internet.

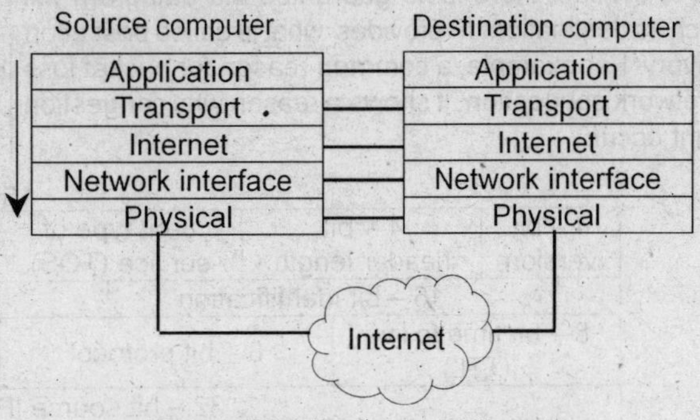

Fig. 3 : TCP/IP message layering

IP

The Internet Protocol (IP) is the base layer of the TCP/IP protocol suite. IP is used at the Internet layer in the TCP/IP stack. All TCP/IP data is packaged in units called IP datagrams. All other TCP/IP protocols, except for the address resolution protocols, are encapsulated inside of an IP datagram for delivery on the network. IP datagrams are eventually encapsulated inside of a hardware frame, such as Ethernet for transmission on a LAN.

ISO/OSI Layer	TCP/IP Protocols		
7. Application	Telnet		DHCP
6. Presentation	FTP		SNMP
5. Session	SMTP		DNS
4. Transport	TCP		UDP
3. Network	IP	ARP	RARP
2. Data link	LLC	802.2	
	MAC		
1. Physical	802.3 Ethernet 802.5 Token Ring		

Fig. 4 : TCP/IP protocol suite

The IP datagram is transmitted on a local network to a special purpose device called a router whose job is to forward the packet onto another router or possibly deliver the packet to the LAN where the destination computer is located.

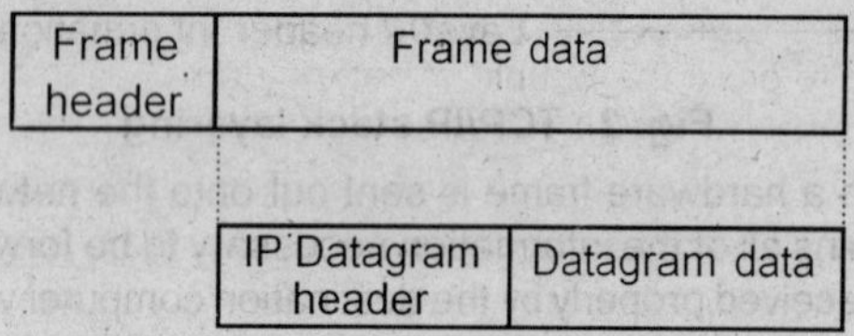

Fig. 5 : An IP diagram encapsulated into an Ethernet frame

There is no direct link between the source and destination computers on the internet. The source computer system sends packets out onto the Local Area Network (LAN) where they are forwarded onto their destination. When an IP datagram is sent, it is treated as an idependent entity on the network, with no relationship to any other datagrams. Datagrams sent to the same location may take different routes as network conditions change. This may cause the destination computer to receive the packets out of order if packets are delayed.

The Internet Protocol is considered to be unreliable due to the fact there is no guarantee the datagram will reach its destination. IP provides what is called best effort delivery. For example, a common reason for packet loss is network congestion. It shows a reason why congestion might occur.

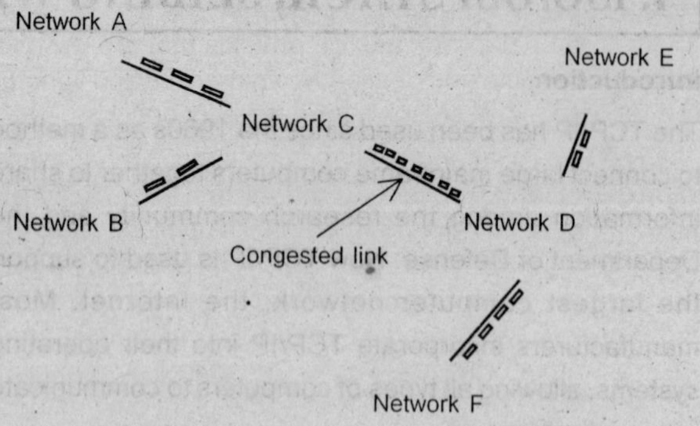

Fig. 6 : A source of network congestion

Consider the situation in which computers connected to network C are trying to communicate with computers in the E and F networks that are connected to network D. The link between networks C and D may not be able to handle all of the traffic, creating network congestion. On a congested network, some network traffic may simply be discarded to eliminate the congestion. Usually, when an IP datagram runs into trouble on the network, it is simply discarded. An error contained in an ICMP (Internet Control Message Protocol) message may or may not be returned to the sender.

Figure 7 illustrates an example of how an ICMP message is encapsulated in an IP datagram. Additional information about ICMP will be provided later in this chapter. Because of IP's best effort delivery, the highest layers of the Internet Protocol provide for transmission reliability (using TCP) if it is required

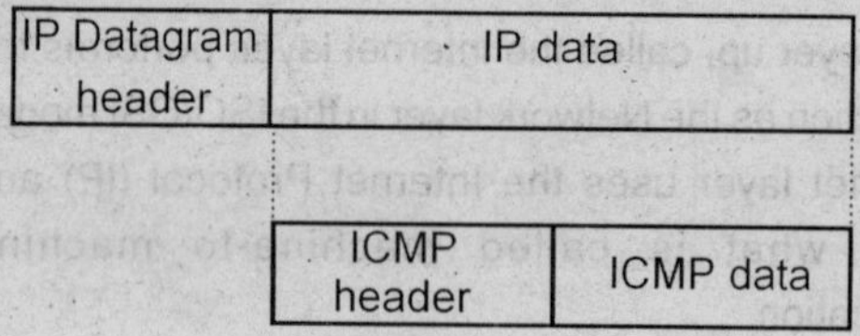

Fig. 7 : ICMP encapsulated in an IP diagram

The IP datagram header and data field containing the ICMP message. An IP datagram has a maximum size of 64 K, or 65536 bytes. This total includes the header and the data area. Basically, the IP datagram header contains

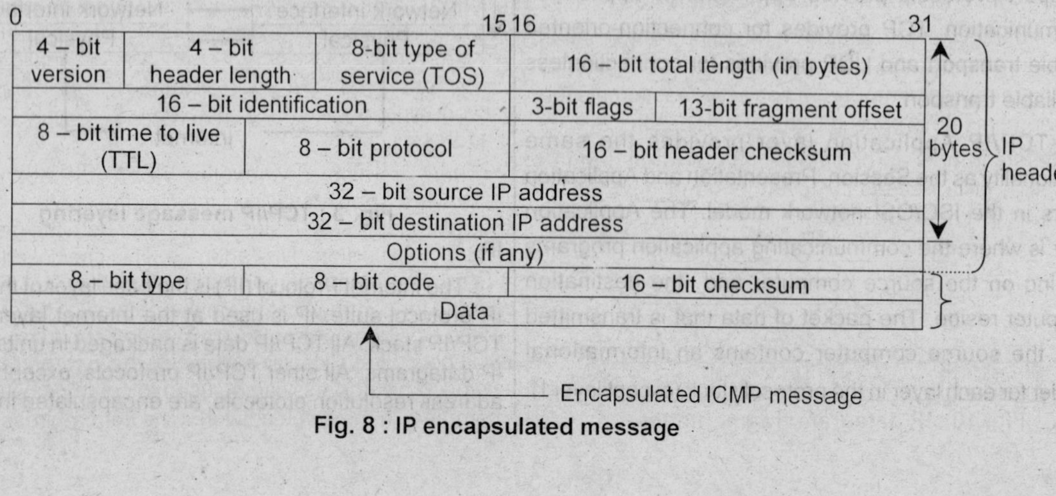

Fig. 8 : IP encapsulated message

data fields to identify the version of IP, the length of the IP header (usually 20 bytes), the type of service, the total length of the datagram, and other fields that are used to identify the source and destination addresses and various error detection, reassembly, and delivery options.

When an IP datagram is encapsulated in a hardware frame, it is done so according to the local network maximum transmission unit, or MTU. A typical value for an MTU is 1500 bytes, This MTU value specifies the maximum size of an IP packet that may be transmitted on that particular network. Some networks may set the MTU to a larger or smaller value. During transit , a packet generated with an MTU value of 1500 might encounter a network where the MTU is smaller. When this situation occurs, it is necessary to fragment the IP datagram to comply with the network's lower MTU value.

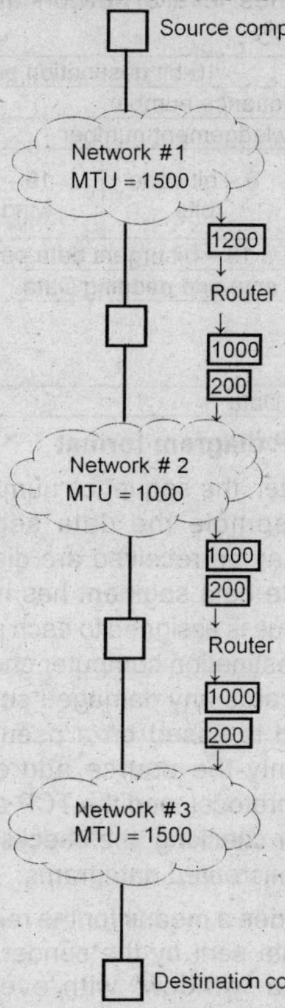

Fig. 9 : Networks with different MTUs

When a datagram is fragmented, additional information is coded in the fragment offset field in the IP header to allow the receiving computer to correctly reassemble the data segments data back into the original state.

To understand how an IP datagram is forwarded to the destination, it is necessary to understand how the IP datagrams are addressed.

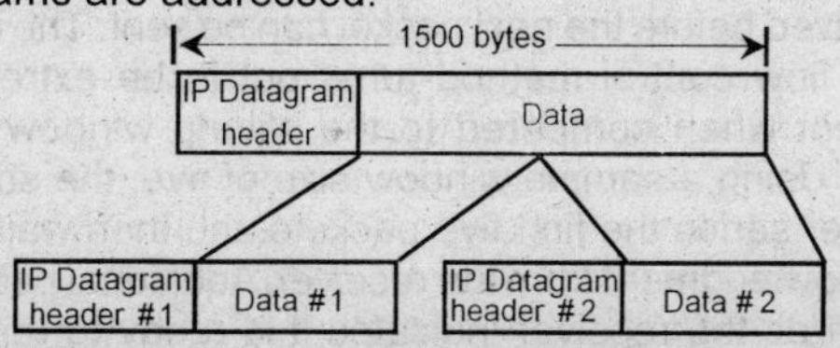

Fig. 10 : IP datagram fragmentation

IP ADDRESSES

IP datagrams are routed on the network using an IP address. The IP address consists of a 32-bit number, divided into four sections, each containing 8 bits. These four sections are called octets and may contain a decimal value in the range 0 to 255. The four values from each section are separated by periods in the IP address. This is called dotted decimal notation.

192.203.131.37

Network ID (24 bits) **host ID** (8 bits)

Class C network IP address

The IP address is assigned by software (statically or dynamically) and must be a unique address on the network. IP addresses thus differ from MAC addresses, which are fixed 48-bit addresses encoded into the hardware of every Ethernet controller, assigned by the manufacturer.

Address Classes

The Internet addresses are classified into five address classes. Each address class consists of a network ID and a host ID.

Address Properties

Class	31				24 23	16 15	87 0
(0-127) A	0	Network ID 128 possible				Host ID 16,777,216 possible	
(128-191) B	1	0	Network ID 16,384 possible			Host ID 65,536 possible	
(192-223)C	1	1	0	Network ID 2,097,152 possible			Host ID 256 possible
(224-239) D	1	1	1	0	Multicast address		
(240-247)E	1	1	1	1	0	Reserved	

Fig. 11 : IP address classes

The network ID in a class A adress contains a maximum number of 128 possible networks and over 16 million hosts. A class B address allocates more bits in the IP address to the network ID and less bits to the host ID, creating a possible 16,384 networks and 65,536 hosts. The class C address provides over two million networks, each with a possible 256 hosts. Class D addresses are reserved for multicast data and class E addresses are reserved.

Some of the IP addresses are reserved for special functions. For example, the class A address 127 is reserved for loopback testing, allowing a local method to test functionality of the TCP/IP software and applications. Other addresses are used as masks to identify each network type, such as 255.255.255.0 to identify a class C address.

There are also serveral other types of IP addresses that have a special use and meaning as well. The first special IP address contains all zeros for both the network and the host portion of the address, This special address is used on a computer during booting. It is not a vaild address for communicating on the network, The second adress type consisting of zeros in the host portion and a valid host address identifies a host on the current (default) network. This address can be used to communicate with a computer in the same network. The third address type, consisting of all ones in the network and host portion of the address, is called a limited broadcast. A limited broadcast message will be received by all computers on a local network. The fourth special address type contains a network ID in the network portion of the address and all ones in the host portion. This is called a directed broadcast. A directed broadcast is used to send a broadcast message to all computers on a remote network. The directed broadcast is usually done to obtain the physical address of a computer if only the IP address is known. This is known as address resolution.

One of the shortcomings in the design of IP version 4 is in the addressing scheme. Class A networks, while being able to support 16 million hosts rarely contain that many hosts. The same is true for class B addresses (though there are not as many wasted addresses, since calss B networks have a limit of 65,536 hosts). The popular class C networks have had such a high demand that there is now a shortage of these addresses. A solution to this problem was established by introducing a classless IP address that can take advantage of the unused addresses. This addressing is accomplished using the CIDR (Classless inter-Domain Routing) protocol

TCP

The Transmission Control Protocol defines a standard way that two computers can reliably communicate together over interconnected networks. Applications using TCP establish connections with each other, through the use of predefined ports or sockets.

A TCP connection is reliable, with error checking, acknowledgment for received packets and packet sequencing provided to guarantee the data arrives properly at its destination. Telnet and FTP are examples of TCP/IP applications that use TCP.

The TCP provides the communication link between the application program and IP. A set of function calls provides an application process a number of different options. For example, there are calls to open and close connections and to send and receive data on previously established connections.

The primary purpose of the TCP is to provide reliable communication between applications running on a source computer and a destination computer. To provide guaranteed delivery on top of IP requires attention to the following areas:

- Data transfer
- Reliability
- Flow control
- Multiplexing
- Connections
- Message precedence and security

TCP is used to transfer a stream of data in each direction between the applications by packaging some data into segments for transmission through the network. When a user needs to be sure that all the data has been sent to the destination computer, a special function called **PUSH** is defined. This function indicates that the data that has been sent should be pushed through to the receiving application. A PUSH causes the TCP to initiate delivery of all the data currently in the pipeline.

To provide reliability, the TCP must be able to recover from data that is lost, damaged, delivered out of order, or duplicated by the network. This reliability is accomplished by assigning sequence numbers to the data it is transmitted and requiring a positive acknowledgment (ACK) from the destination computer after the data has been received without error. If the ACK is not received within a certain timeout interval, tha data is automatically retransmitted by the source computer. The timeout interval is based on an estimate of the time it takes for a packet to be sent to the destination computer and then acknowledged. This value is also called the round trip estimation time. A good estimated value for the round trip time allows for the highest level of network utilization.

0	4	10	16 31
16-bit source port		16-bit destination port	
32 – bit sequence number			
32 – bit acknowledgement number			
4 – bit header length	6 – bit reserved	6 – bit code bits	16 – bit window
16 – bit checksum value		16 – bit urgent data pointer	
32 – bit options (if any) and padding Data			
Data			

Fig. 12 : TCP diagram format

At the destination computer, the sequence numbers are used to correctly reassemble the data segments. Duplicate packets that may be received are discarded. To ensure that a complete data segment has not been damaged, a checksum value is assigned to each segment that is transmitted. The destination computer checks the checksum value and discards any damaged segments. The TCP checksum field is based on a pseudo TCP header that contains only the source address, the destination address, the protocol, and the TCP segment length. In addition to error checking, the checksum field gives protection against misrouted datagrams.

TCP flow control provides a means for the receiver to govern the amount of data sent by the sender. This is achieved by resuming a "window" with every ACK indicating a range of acceptable sequence numbers beyond the last segment successfully received. The window indicates an allowed number of octets that the sender may transmit before receiving further permission to send more. As the data is acknowledged, the window can be slid over the data that must still be transmitted. This is known as sliding window flow control. It shows a simplistic example of two flow control methods. In the stop-and-go flow control requires that an acknowledgment be received before the next packet can be sent. The stop-and-go flow control method turns out to be extremely inefficient when compared to the sliding window flow control. Using a sample window size of five, the source computer sends the first five packets and then waits for an acknowledgment from the receiver. Additional packets are sent as the receiver indicates it is ready to receive them. In practice, window sizes are much larger and can accommodate large transfers of data very efficiently. One of the problems associated with the sliding window is called the silly window syndrome. The silly window syndrome occurs when the receiver repeatedly advertises a small window size due to what began as a temporary situation. The sender then transmits small segments to fit the small window size. Whereas less information is transmitted, some of the network bandwidth is unused and therefore wasted. When examining TCP packets using Lan Explorer, pay attention to the window size field. Try to determine how often the window size changes, if at all.

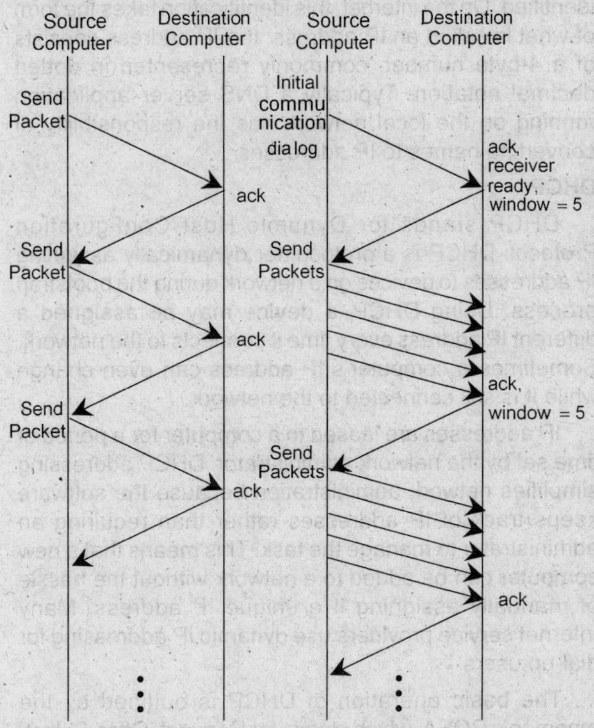

Fig. 13 : Stop - and - go versus sliding window flow control

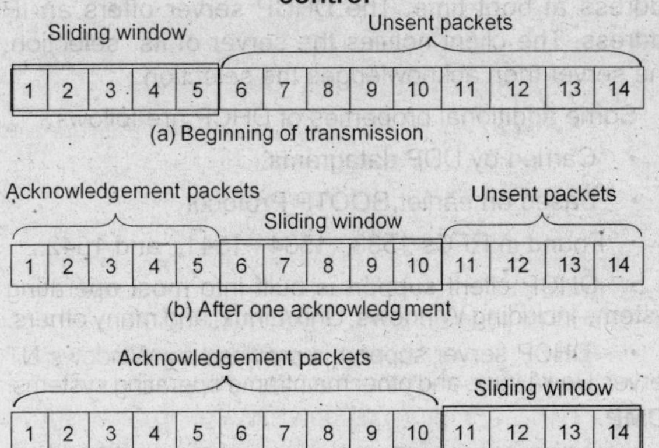

Fig. 14 : Sliding window flow control in operation .

Port numbers

All data communicated between two TCP applications take place through specific ports. A port is associated with the network socket created and used by the application. A pair of sockets uniquely identifies a connection.

So, a single computer may receive several TCP packets, with the port number associated with the TCP application stored in each packet. The port number is used to demultiplex the packet stream and forward the correct TCP packet to each application's input buffer. The binding of port numbers to application programs is handled independently by each operating system. However, commonly used network applications are assigned to fixed socket numbers (or port numbers) that are publicly known.The services then can be accessed through the well-known port numbers. In general, port numbers less than 1024 are reserved for well known services. Port numbers greater than or equal to 1024 but less than 65536 are available for general use by applications. The official list of well-known port numbers are published in RFC 762.

When two applications want to communicate, the TCP is used to establish a connection (initialize the status information on both sides). After the connection has been established, data can be exchanged as necessary. After the commmunication is complete, the connection can be closed. Since these connections are established between hosts using the unreliable Internet Protocol, a three-way hand-shake is used to avoid the error condition when a connection is closed before all of the data has been transmitted. Essentially, this handshaking consists of three messages sent back and forth between the hosts, with each successive message containing information based on the previously received message.

Lastly, the precedence and security of TCP communication may be specified. The TCP may have a requirement to indicate the security and precedence of the data or it may use default values when these special features are not needed.

UDP

The User Datagram Protocol is similar to TCP except it is connectionless. Data is transmitted with no acknowledgment of whether it is received or not. UDP is thus not as reliable as TCP. There are many applications that do not require the additional overhead and complexity of TCP handshaking, such as the Domain Name Service (DNS), the Dynamic Host Configuration Protocol (DHCP), and network games. For example, a DNS or DHCP request will use the UDP and will retransmit the request if a response is not obtained in a timely fashion. A multiplayer network game might use UDP because it is simple to implement, requires less overhead to manage than TCP, and also because the game may not be severely affected if a few packets are lost now and then.

0	16 31
16 –bit UDC source port	16 – bit UDP destination port
16 – bit message length	16 – bit UDP checksum
16 – bit checksum value	16 – bit urgent data pointer
Data	
⋮	
Data	

Fig. 15 : UDP diagram format

The UDP format is quite a bit simpler than TCP, with fields used to identify the port number of the source computer, the port number on the destination computer, the length of the message in bytes, and a checksum value used to insure that the data was received properly. The port numbers used in UDP are used to identify the specific application running on the source computer and the destination computer. Similar to TCP, port numbers less than 1024 are reserved (4 for well-known services). Port numbers greater than or equal to 1024 but less than 65536 are available for general use by applications. The complete list of wellknown port numbers is found in RFC 762.

The UDP checksum field is based on a pseudo header that will work under any conditions. The pseudo header contains the source address, the destination address, the protocol, and the UDP segment length. The use of the UDP checksum field is optional. It must contain zeros to indicate that no checksum is supplied. A good reason to use the checksum field is that it gives protection against misrouted datagrams.

ARP AND RARP

Before any packet can be transmitted, the hardware address of its destination computer must be known. This address is called a MAC address. MAC stands for media access control, and it takes the from of a 48-bit binary number that uniquely identifies one machine from every other. Every network interface card manufactured has a preassigned MAC address that it responds to.

The Address Resolution Protocol (ARP) uses a directed broadcast message to obtain the MAC address for a given IP address. For example, an ARP request may say, "What is the MAC address for 204.210.133.51?" The ARP reply may be "The MAC address is 00-60-97-2B-E6-0F."

0 — 8	8 — 16	16 — 31
16 – bit hardware type		16 – bit protocol type
8 – bit hardware length	8 – bit protocol length	16 – bit operation code
Send hardware address (first 4 bytes)		
Sender hardware address (last 2 bytes)		Sender IP address (first 2 bytes)
Sender IP address (last 2 bytes)		Target hardware address (first 2 bytes)
Target hardware address (last 4 bytes)		
Target IP address (all 4 bytes)		

Fig. 16 : ARP/RARP message format

Once the ARP reply has been received, it is possible for the source computer and destination computer to communicate directly.

The Reverse Address Resolution Protocol (RARP) preforms the opposite of ARP, providing the IP address for a specific MAC address. The message format for RARP is the same as the ARP message. RARP is usually performed on diskless computer workstations that do not have any other way of obtaining an IP address. A RARP server is used to perform the required translations.

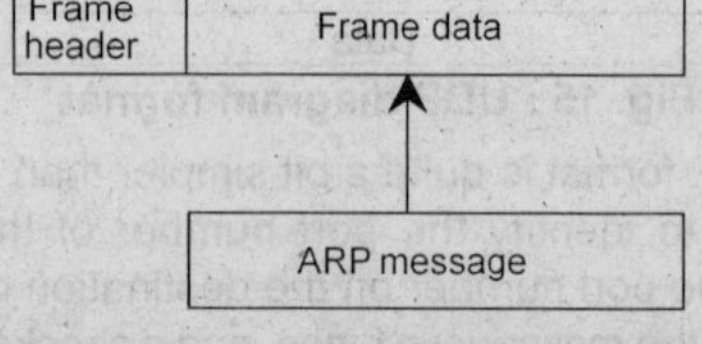

(a) ARP message in an Ethernet frame

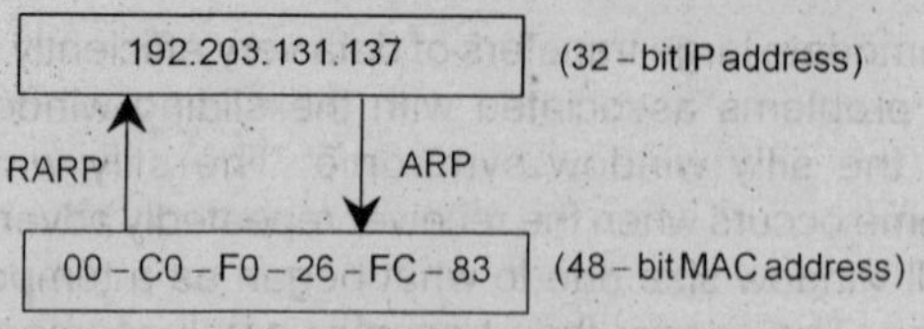

(b) Using ARP and RARP

Fig. 17

DNS

Every machine on a network must be uniquely identified. On the internet, this identification takes the form of what is called an IP address. the IP address consists of a 4-byte number, commonly represented in dotted decimal notation. Typically a DNS server application running on the local network has the responsibility of converting names to IP addresses.

DHCP

DHCP stands for Dynamic Host Configuration Protocol. DHCP is a protocol for dynamically assigning IP addresses to devices on a network during the bootstrap process. Using DHCP, a device may be assigned a different IP address every time it connects to the network. Sometimes a computer's IP address can even change while it is still connected to the network.

IP addresses are leased to a computer for a period of time set by the network administrator. DHCP addressing simplifies network administration because the software keeps track of IP addresses rather than requiring an administrator to manage the task. This means that a new computer can be added to a network without the hassle of manually assigning it a unique IP address. Many Internet service providers use dynamic IP addressing for dial-up users.

The basic operation of DHCP is outlined by the mnemonic ROSA, which stands for Request, Offer, Select, and Acknowledge. The DHCP client requests an IP address at boot time. The DHCP server offers an IP address. The client notifies the server of its selection. The server then acknowledges the selection.

Some additional properties of DHCP are follows:

- Carried by UDP datagrams.
- Based on earlier BOOTP Protocol
- Found in RFCs 1533, 1534, 1541, and 1542.
- DHCP client support is built into most operating systems including Windows, Unix/Linux, and many others.
- DHCP server support is provided by Windows NT Server, Unix/Linux, and other mainframe operating systems.

ICMP

Usually, when an IP datagram runs into trouble on the network, an error message is generated (which is sent back to the source computer) and then it is simply discarded. An error contained in an ICMP message may or may not be returned to the sender due to use of the UDP as the transport protocol. If an ICMP message must be discarded, no ICMP error message is generated. This is because of a specific rule in the implementation that prohibits error messages being generated about other error messages.

SMTP

The Simple Mail Transport Protocol (SMTP) is responsible for routing electronic mail on the internet using TCP and IP. The process usually requires connecting to a remote computer and transferring the e-mail message, but due to problems with the network or a remote computer, messages can be temporarily undeliverable. The e-mail server will try to deliver any messages by periodically trying to contact the remote destination. When the remote computer becomes available, the message is delivered using SMTP.

SNMP

Network managers responsible for monitoring and controlling the network hardware and software use the Simple Network Management Protocol. SNMP defines the format and meaning of messages exchanged by the manager and agents. The network manager uses SNMP to interrogate network devices(agents) such as routers, switches, and bridges in order to determine their status and also retrieve statistical information.

HTTP

HTTP, the Hyper Text Transport Proctocol, is used to transfer multimedia information over the internet, such as web pages, images, audio, and video. An HTTP client, such as a browser, sends a request to an HTTP server (also called an HTTP daemon), which services the request.

It is important to note that HTTP is carried via., TCP and not UDP. There is a great deal of back and forth between the client and server, with many acknowledgment messages insuring reliable communication.

PING

PING is a TCP/IP application that sends datagrams once every second in the hope of an echo response from the machine being PINGed. If the machine is connected and running a TCP/IP protocol stack, it should respond to the PING datagram with a datagram of its own. If PING encounters an error condition, an ICMP message is returned. PING displays the time of the return response in milliseconds or one of several error messages (request timed out, destination host unreachable, etc.,).

TRACERT

TRACERT (trace route) is a TCP/IP application that determines the path through the network to a destination entered by the user. Creative use of ICMP message is the basis for trace route application.

FTP

The File Transfer Protocol (FTP) allows a user to login to a remote computer and transfer files back and forth through simple commands. Many FTP sites allow you to login as an anonymous user, an open account with limited privileges, but still capable of file transfers.

Telnet

A Telnet session allows a user to establish a terminal emulation connection on a remote computer.

INTERFACES AND SERVICES

the function of each layer is to provide services to the layer above it.

The active elements in each layer are often called entities. An entity can be a software entity (such as a process), or a hardware entity (such as an intelligent I/O chip). Entities in the same layer on different machines are called peer entities. The entities in layer n implement a service used by layer n + 1. In this case layer n is called the service provider an layer n+1 is called the service user. Layer n may use the services of layer n - 1 in order to provide its service. It may offer several classes of service, for example, fast, expensive communication and slow, cheap commuication. Services are available at **SAPs (Service Access Points).** The layer n SAPs are the places where layer n + 1 can access the services offered. Each SAP has an address that uniquely identifies it. To make this point clearer, the SAPs in the telephone system are the sockets into which modular telephones can be plugged, and the SAP addresses are the telephone numbers of these sockets. To call someone, you must know the callee's SAP address. Similarly, in the postal system, the SAP addresses are street addresses and post office box numbers. To send a letter, you must know the addressee's SAP address.

In order for two layers to exchange information, there has to be an agreed upon set rules about the interface. At a typical interface, the layer n + 1 entity passes an **IDU (interface Data Unit)** to the layer n entity through the SAP as shown in **Figure 18.** The IDU consists of an **SDU (Service Data Unit)** and some control information. The SDU is the information passed across the network to the peer entity and then up to layer n + 1. The control information is needed to help the lower layer do its job (e.g., the number of bytes in the SDU) but is not part of the data itself.

In order to transfer the SDU, the layer n entity may have to fragment it into several pieces, each of which is given a header and sent as a separate **PDU (Protocol Data Unit)** such as a packet. The PDU headers are used by the peer entities to carry out their peer protocol. They

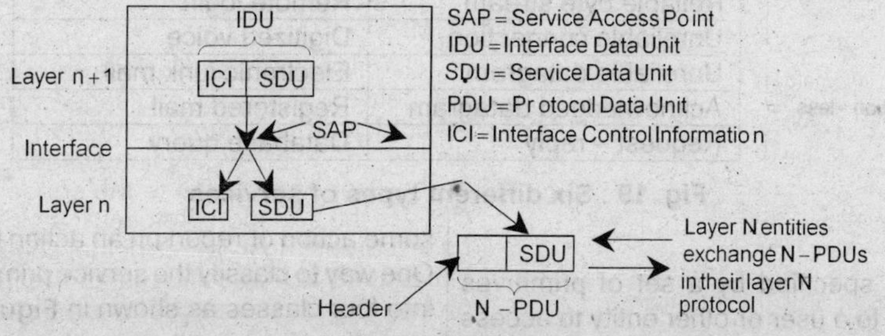

Fig. 18 : Relation between layers at an interface

identify which PDUs contain data and which contain control information, provide sequence numbers and counts, and so on.

Connnection-oriented and Connectionless Services

Layers can offer two different types of service to the layers above them: connection-oriented and connectionless. Here we see these two types and examine the differences between them.

Connection-oriented service is modeled after the telephone system. To talk to someone, you pick up the phone, dial the number , talk, and then hang up. Similarly, to use a connection-oriented network service the service user first establishes a connection, uses the connection, and then releases the connection. The essential aspect of a connection is that it acts like a tube: the sender pushes objects (bits) in at one end , and the receiver takes them out in the same order at the other end.

In contrast, **connectionless service** is modeled after the postal system. Each message (letter) carries the full destination address, and each one is routed through the system independent of all the others. Normally, when two messages are sent to the same destination, the first one sent will be the first one to arrive. However, it is possible that the first one sent can be delayed so that the second one arrives first. With a connection-oriented service this is impossible.

each service can be characterized by a **quality of service.** Some services are reliable in the sense that they never lose data. Usually, a reliable service is implemented by having the receiver acknowledge the receipt of each message, so the sender is sure that it arrived. The acknowledgement process introduces overhead and delays, which are often worth it but are sometimes undesirable.

A typical situation in which a reliable connection-oriented service is appropriate is file transfer. The owner of the file wants to be sure that all the bits arrive correctly and in the same order they were sent. Very few file transfer customers would prefer a service that occasionally scrambles or loses a few bits, even if it is much faster.

Reliable connection-oriented service has two minor variations: message sequences and byte streams. In the former, the message boundaries are preserved. when two 1-KB messages are sent, they arrive as two distinct 1-KB messages, never as one 2-K message. (Note : KB means kilobytes; kb means kilobits.) In the latter, the connection is simply a stream of bytes, with no message boundaries. When 2K bytes arrive at the receiver, there is no way to tell if they were sent as one 2-KB message, two 1-KB messages, or 2048 1-byte messages. If the pages of a book are sent over a network to a phototypesetter as separate messages, it might be important to preserve the message boundaries. On the other hand, with a terminal logging into a remote timesharing system, a byte stream from the terminal to the computer is all that is needed.

As mentioned above, for some applications, the delays introduced by acknowledgments are unacceptable. One such application is digitized voice traffic. It is preferable for telephone users to hear a bit of noise on the line or a garbled word from time to time than to introduce a delay to wait for acknowledgments. Similarly, when transmitting a video film, having a few pixels wrong is no problem, but having the film jerk along as the flow stops to correct errors is very irritating.

Not all applications require connections. For example,as electronic mail becomes more common, can electronic junk mail be far behind ? The electronic junk mail sender probably does not want to go to the trouble of setting up and later tearing down a connection just to send one item. Nor is 100 percent reliable delivery essential, especially if it costs more. All that is needed is a way to send a single message that has a high probability of arrival, but no guarantee. Unreliable (meaning not acknowledged) connectionless service is often called datagram sevice, in analogy with telegram service, which also does not provide an acknowledement back to the sender.

In other situations, the convenience of not having to establish a connection to send one short message is desired, but reliability is essential. the **acknowledged datagram service** can be provided for these applications. It is like sending a registered letter and requesting a return receipt. When the receipt comes back, the sender is absolutely sure that the letter was delivered to the intended party and not lost along the way.

Still another service is the **request-reply service.** In this service the sender transmits a single datagram containing a request; the reply contains the answer. For example , a query to the local library asking where uighur is spoken falls into this category. Request-reply is commomly used to implement communication in the client-server model: the client issues a request and the server responds to it.**Figure 19** summarizes the types of services discussed above.

	Service	Examples
Connection – oriented	Reliable message stream	Sequence of pages
	Reliable byte stream	Remote login
	Unreliable connection	Digitized voice
Connection – less	Unreliable datagram	Electronic junk mail
	Acknowledged datagram	Registered mail
	Request – reply	Database query

Fig. 19 : Six different types of services

Service Primitives

A service is formally specified by a set of primitives (operations) available to a user or other entity to access the service. these primitives tell the service to perform some action or report on an action taken by a peer entity. One way to classify the service primitives is to divide them into four classes as shown in **Figure 20** .

Primitive	Meaning
Request	An entity wants the service to do some work
Indication	An entity is to be informed about an event
Response	An entity wants to respond to an event
Confirm	The response to an earlier request has come back

Fig. 20 : Four classes of serivce primitives

To illustrate the uses of the primitives, consider how a connection is established and released. The initiating entity does a CONNECT request which results in a packet being sent. The receiver then gets a CONNECT indication announcing that an entity somewhere wants to set up a connection to it. The entity getting the CONNECT indication then uses the CONNECT response primitive to tell whether it wants to accept or reject the proposed connection. Either way, the entity issuing the initial CONNECT request finds out what happened via a CONNECT confirm primitive.

Primitives can have parameters, and most of them do. the parameters to a CONNECT request might specify the machine to connect to, the type of service desired, and the maximum message size to be used on the connection.

The parameters to a CONNECT indication might contain the caller's identity, the type of service desired, and the proposed maximum message size. If the called entity did not agree to the proposed maximum message size, it could make a counterproposal in its response primitive, which would be made available to the original caller in the confirm. The details of this negotiation are part of the protocol. For example, in the case of two conflicting proposals about maximum message size, the protocol might specify that the smaller value is always chosen.

As an aside on terminolgy, we will carefully avoid the terms "open a connection" and "close a connection" because to electrical engineers, an "open circuit" is one with a gap or break in it. Electricity can only flow over "closed circuits" computer scientists would never agree to having information flow over a closed circuit. To keep both camps pacified, we will use the terms "establish a connection" and "release a connection."

Services can be either confirmed or unconfirmed. In a confirmed service there is a request, an indication, a response, and a confirm. In an unconfirmed service, there is just a request and an indication CONNECT is always a confirmed service because the remote peer must agree to establish a connection. Data transfer, on the other hand, can be either confirmed or unconfirmed, depending on whether or not the sender needs an acknowledgement. both kinds of services are used in networks.

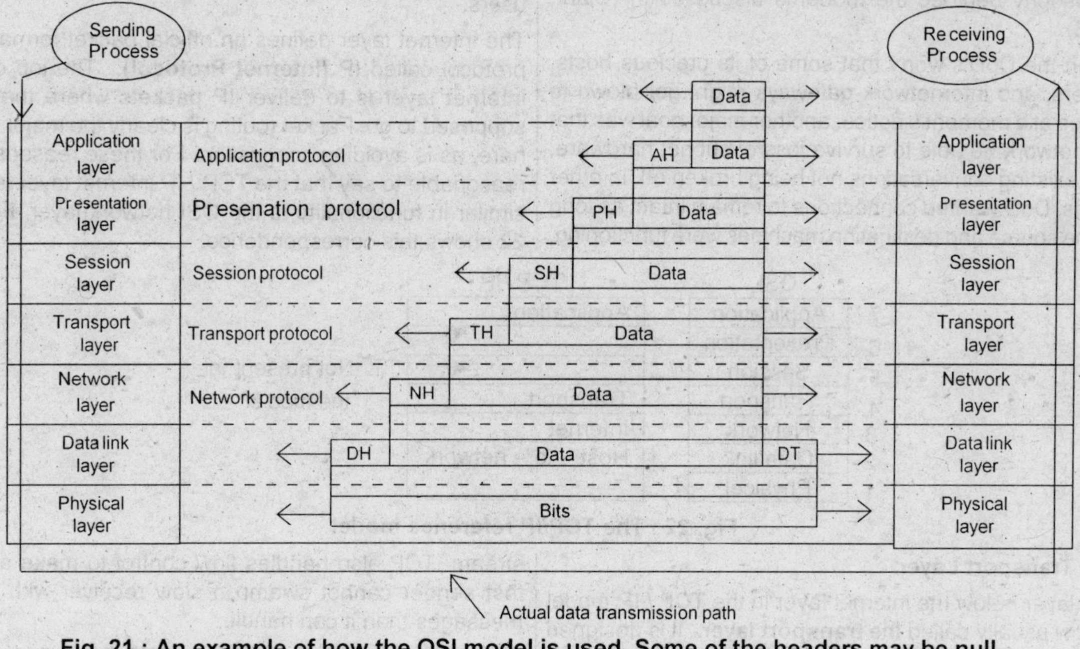

Fig. 21 : An example of how the OSI model is used. Some of the headers may be null

Data Transmission in the OSI Model

Figure 21 shows an example of how data can be transmitted using the OSI model. The sending process has some data it wants to send to the receiving process. It gives the data to the application layer, which then attaches the application header, AH (which may be null), to the front of it and gives the resulting item to the presentation layer. The presentaion layer may transfrom this item in various ways and possible add a header to the front, giving the result to the session layer. It is important to realize that the presentation layer is not aware of which portion of the data given to it by the application layer is AH, if any, and which is true user data.

This process is repeated until the data reach the physical layer, where they are actually transmitted to the receiving machine. On that machine the various headers are stripped off one by one as the message propagates up the layers until it finally arrives at the receiving process.

The key idea throughout is that although actual data transmission is vertical in **Figure 21**, each layer is programmed as though it were horizontal. When the sending transport layer, for example, gets a message from the session layer, it attaches a transport header and sends it to the receiving transport layer. From its point of view, the fact that it must actually hand the message to the network layer on its own machine is an unimportant technicality.

The TCP/IP Reference Model

Let us now turn from the OSI reference model to the reference model used in the grandparent of all computer networks, the ARPANET, and its successor, the worldwide Internet. Although we will give a brief history of the ARPANET later, it is useful to mention a few key aspects of it now. The ARPANET was a research network sponsored by the DoD (U.S. Department of Defense). It eventually connected hundreds of Universities and Government installations using leased telephone lines. When satellite and radio networks were added later, the existing protocols had trouble interworking with them, so a new reference architecture was needed. Thus the ability to connect multiple networks together in a seamless way was one of the major design goals from the very beginning. This architecture later became known as the **TCP/ IP reference Model,** after its two primary protocols. It was frist defined in (Cerf and Kahn, 1974). A later perspective is given (Leiner et al., 1985). The design philosophy behined the model is discussed in (Clark, 1988).

Given the DoD's worry that some of its precious hosts, routers, and internetwork gateways might get blown to pieces at a moment's notice, another major goal was that the network be able to survive loss of subnet hardware, with existing conversations not being broken off. In other words, DoD wanted connections to remain intact as long as the source and destination machines were functioning, even if some of the machines or transmission lines in between were suddenly put out of operation. Furthermore, a flexible architecture was needed, since applications with divergent requirements were envisioned, ranging from transferring files to real-time speech transmission.

The Internet Layer

All these requirements led to the choice of a packet-switching network based on a connectionless internetwork layer. this layer, called the **internet layer,** is the linchpin that holds the whole architecture together. Its job is to permit hosts to inject packets into any network and have them travel independently to the destination (potentially on a different network). They may even arrive in a different order than they were sent, in which case it is the job of higher layers to rearrange them, if in-order delivery is desired. Note that "internet" is used here in a generic sence, even though this layer is present in the Internet.

The analogy here is with the (snail) mail system. A person can drop a sequence of international letters into a mail box in one country, and with a little luck, most of them will be delivered to the correct address in the destination country. Probably the letters will travel through one or more international mail gateways along the way, but this is transparent to the users. Furthermore, that each country (i.e., each network) has its own stamps, preferred envelope sizes, and delivery rules is hidden from the users.

The internet layer defines an official packet format and protocol called **IP (Internet Protocol).** The job of the internet layer is to deliver IP packets where they are supposed to go. Packet routing is clearly the major issue here, as is avoiding congestion. For these reasons, it is reasonable to say that the TCP / IP internet layer is very similar in functionality to the OSI network layer. **Figure 22** shows this correspondence.

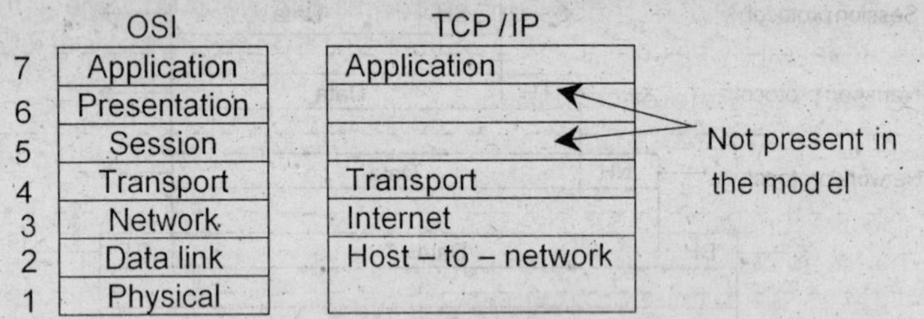

Fig. 22 : The TCP/IP reference model

The Transport Layer

The layer below the internet layer in the **TCP / IP** model is now usually called the **transport layer.** It is designed to allow peer entities on the source and destination hosts to carry on a conversation, the same as in the OSI transport layer. Two end-to-end protocols have been defined here. The first one, **TCP (Transmission Control Protocol)** is a reliable connection-oriented protocol that allows a byte stream originating on one machine to be delivered without error on any other machine in the Internet. It fragments the incoming byte stream into discrete messages and passes each one onto the internet layer. At the destination, the receiving TCP process reassembles the received messages into the output stream. TCP also handles flow control to make sure a fast sender cannot swamp a slow receiver with more messages than it can handle.

The second protocol in this layer, **UDP (User Datagram Protocol),** is an unreliable, connectionless protocol for applications that do not want TCP's sequencing or flow control and wish to provide their own. It is also widely used for one-shot, client-server type request-reply queries and applications in which prompt delivery is more important than accurate delivery, such as transmitting speech or video. The relation of IP,TCP, and UDP is shown in **Figure 23.** Since the model was developed, IP has been implemented on many other networks.

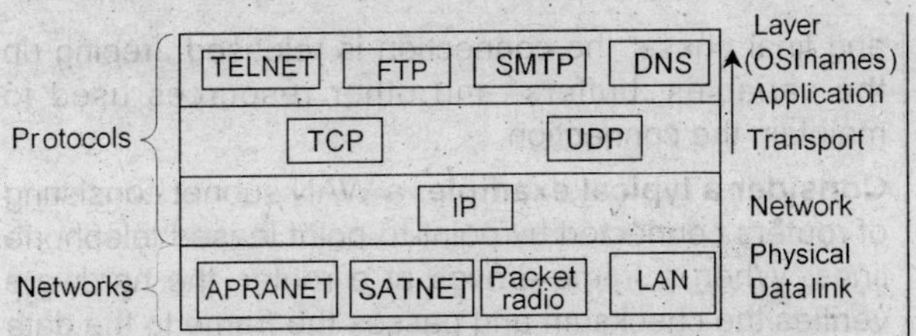

Fig. 23 : Protocols and networks in the TCP/IP model initially

The Application Layer

The TCP/IP model does not have session or presentation layers. No need for them was perceived, so they were not included. Experience with the OSI model has proven this view correct: they are of little use to most applications.

On top of the transport layer is the **application layer.** It contains all the higher-level protocols. The early ones included virtual terminal (TELNET), file transfer (FTP), and electronic mail (SMTP), as shown in **Figure 23.** The virtual terminal protocol allows a user on one machine to login to a distant machine and work there. The file transfer protocol provides a way to move data efficiently from one machine to another. Electronic mail was originally just a kind of file transfer, but later a specialized protocol was developed for it. Many other protocols have been added to these over the years, such as the Domain Name Service (DNS) for mapping host names onto their network addresses, NNTP, the protocol used for moving news articles around, and HTTP, the protocol used for fetching pages on the World Wide Web, and many others.

The Host-to-Network Layer

Below the Internet layer is a great void. The TCP/IP reference model does not really say much about what happens here, except to point out that the host has to connect to the network using some protocol so it can send IP packets over it. This protocol is not defined and varies from host to host and network to network. Books and papers about the TCP/IP model rarely discuss it.

A Comparison of the OSI and TCP Reference Models

The OSI and TCP/IP reference models have much in common. Both are based on the concept of a stack of independent protocols. Also, the functionality of the layers in roughly similar. For example, in both models the layers up through and including the transport layer are there to provide an end-to-end network-independent transport service to processes wishing to communicate. These layers form the transport provider. Again in both models, the layers above transport are application-oriented users of the transport service.

Despite these fundamental similarities, the two models also have many differences. in this section we will focus on the key differences between the two reference models. It is important to note that we are comparing the reference models here, not the corresponding protocol stacks.

Three concepts are central to the OSI model:

1. Services 2. Interfaces 3. Protocols

Probably the biggest contribution of the OSI model is to make the distinction between these three concepts explicit. Each layer performs some services for the layer above it. The service definition tells what the layer does, not how entities above it access it or how the layer works.

A layer's interface tells the processes above it how to access it. It specifies what the parameters are and what results to expect. It, too says nothing about how the layer works inside.

Finally, the peer protocols used in a layer are the layer's own business. It can use any protocols it wants to, as long as it gets the job done (i.e., provides the offered services). It can also change them without affecting software in higher layers.

These ideas fit very nicely with modern ideas about object-oriented programming. An object, like a layer, has a set of methods (operations) that processes.

DATA LINK LAYER DESIGN ISSUES

The data link layer has a number of specific functions to carry out. These functions include providing a well-defined service interface to the network layer, determining how the bits of the physical layer are grouped into frames, dealing with transmission errors, and regulating the flow of frames so that slow receivers are not swamped by fast senders. In the following sections we will examine each of these issues in trun.

Services Provided to the Network Layer

The function of the data link layer is to provide services to the network layer. The principal service is transferring data from the network layer on the source machine to the network layer on the destination machine. On the source machine there is an entity, call it a process, in the network layer that hands some bits to the data link layer for transmission to the destination. The job of the data link layer is to transmit the bits to the destination machine, so they can be handed over to the network layer there, as shown in **Figure 24(a).** The actual transmission follows the path of **Figure 24(b),** but it is easier to think in terms of two data link layer processes communicating using a data link protocol. For this reason, we will implicitly use the model of **Figure 24(a)** throughout this chapter.

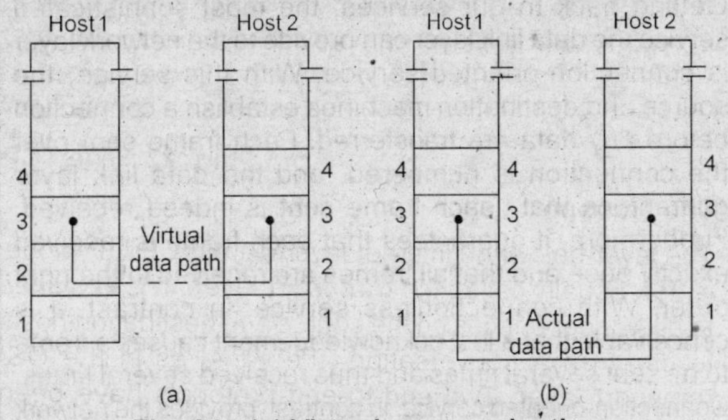

Fig. 24 : (a) Virtual communication (b) Actual communication

The data link layer can be designed to offer various services. The actual services offered can vary from system to system. Three reasonable possibilities that are commonly provided are

1. unacknowledged connectionless service.
2. Acknowledged connectionless service.
3. Acknowledged connection-oriented service.

Let us consider each of these in turn.

Unacknowledged connectionless service consists of having the source machine send independent frames to the destination machine without having the destination machine acknowledge them. No connection is established beforehand or released afterward. If a frame is lost due to noise on the line, no attempt is made to recover it in the data link layer. This class of service is appropriate when the error rate is very low so recovery is left to higher layers. It is also appropriate for real-time traffic, such as speech, in which late data are worse than bad data. Most LANs use unacknowledged connectionless service in the data link layer.

The next step up in terms of reliability is acknowledged connectionless service. When this service is offered, there are still no connections used, but each frame sent is individually acknowledged. In this way, the sender knows whether or not a frame has arrived safely. If it has not arrived within a specified time interval, it can be sent again. This service is useful over unreliable channels, such as wireless systems.

It is perhaps worth emphasizing that providing acknowledgements in the data link layer is just an optimization, never a requirement. The transport layer can always send a message and wait for it to be acknowledged. If the acknowledgement is not forthcoming before the timer goes off, the sender can just send the entire message again. The trouble with this strategy is that if the average message is broken up into, say, 10 frames, and 20 percent of all frames are lost, it may take a very long time for the message to get through. If individual frames are acknowledged and retransmitted, entire messages get through much faster. On reliable channels, such as fiber, the overhead of a heavyweight data link protocol may be unnecessary, but on wireless channels it is well worth the cost due to their inherent unreliability.

Getting back to our services, the most sophisticated service the data link layer can provide to the network layer is connection-oriented service. With this service, the source and destination machines establish a connection before any data are transferred. Each frame sent over the connection is numbered, and the data link layer guarantees that each frame sent is indeed received. Furthermore, it guarantees that each frame is received exactly once and that all frames are received in the right order. With connectionless service, in contrast, it is conceivable that a lost acknowledgement causes a frame to be sent several times and thus received several times. Connection-oriented service, in contrast, provides the network layer processes with the equivalent of a reliable bit stream.

When connection-oriented service is used, transfers have three distinct phases. In the first phase the connection is established by having both sides initialize variables and counters needed to keep track of which frames have been received and which ones have not. In the second phase, one or more frames are actually transmitted. In the third and final phase, the connection is released, freeing up the variables, buffers, and other resources used to maintain the connection.

Consider a typical example: a WAN subnet consisting of routers connected by point-to-point leased telephone lines. When a frame arrives at a router, the hardware verifies the checksum and passes the frame to the data link layer software (which might be embedded in a chip on the network adaptor board). The data link layer software checks to see if this is the frame expected, and if so, gives the packet contained in the payload field to the routing software. The routing software chooses the appropriate outgoing line and passes the packet back down to the data link layer software, which then transmits it. The flow over two routers is shown in **Figure 25**

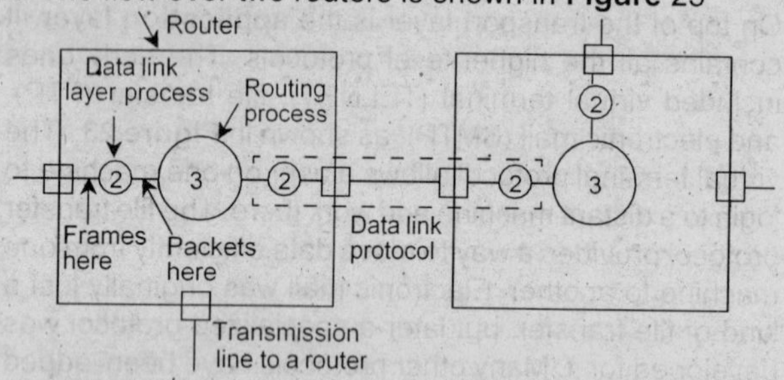

Fig. 25 : Placement of the data link protocol

The routing code frequently wants the job done right, that is, reliable, sequenced connections on each of the point-to-point lines. It does not want to be bothered too often with packets that got lost on the way. It is up to the data link protocol, shown in the dotted rectangle, to make unreliable communication lines look perfect, or at least, fairly good. This property is especially important for wireless links, which are inherently very unreliable. As an aside, although we have shown multiple copies of the data link layer software in each router. In fact, one copy handles all the lines, with different tables and data structures for each one.

Framing

In order to provide service to the network layer, the data link layer must use the service provided to it by the physical layer. What the physical layer does is accept a raw bit stream and attempt to deliver it to the destination. This bit stream is not guaranteed to be error free. The number of bits received may be less than, equal to, or more than the number of bits transmitted, and they have different values. It is up to the data link layer to detect, and if necessary, correct errors.

The usual approach is for the data link layer to break the bit stream up into discrete frames and compute the checksum for each frame. (Checksum algorithms will be discussed later in this chapter.) When a frame arrives at the destination, the checksum is recomputed. If the newly computed checksum is different from the one contained in the frame, the data link layer knows that an error has occurred and takes steps to deal with it (e.g., discarding the bad frame and sending back an error report).

Breaking the bit stream up into frames is more difficult than it at first appears. One way to achieve this framing is to insert time gaps between frames, much like the spaces between words in ordinary text. However, networks rarely make any guarantees about timing, so it is possible these gaps might be squeezed out, or other gaps might be inserted during transmission.

Since it is too risky to count on timing to mark the start and end of each frame, other methods have been devised. In this section we will look at four methods:

1. Character count.
2. Starting and ending characters, with character stuffing.
3. Starting and ending flags, with bit stuffing.
4. Physical layer coding violations.

The first framing method uses a field in the header to specify the number of characters in the frame. When the data link layer at the destination sees the character count, it knows how many characters follow, and hence where the end of the frame is. This technique is shown in Figure 26(a) for four frames of sizes 5, 5, 8, and 9 characters respectively.

The trouble with this algorithm is that the count can be garbled by a transmission error. For example, if the character count of 5 in the second frame of Figure (b) becomes a 7, the destination will get out of synchronization and will be unable to locate the start of the next frame. Even if the checksum is incorrect so the destination knows that the frame is bad, it still has no way of telling where the next frame starts. Sending a frame back to the source asking for a retransmission does not help either, since the destination does not know how many characters to skip over to get to the start of the retransmission. For this reason, the character count method is rarely used anymore.

The second framing method gets around the problem of resynchronization after an error by having each frame start with the ASCII character sequence DLE STX and end with the sequence DLE ETX. (DLE is Data Link Escape, STX is Start of TeXt, and ETX is End of TeXt.) In this way, if the destination ever loses track of the frame boundaries, all it has to do is look for DLE STX or DLE ETX characters to figure out where it is.

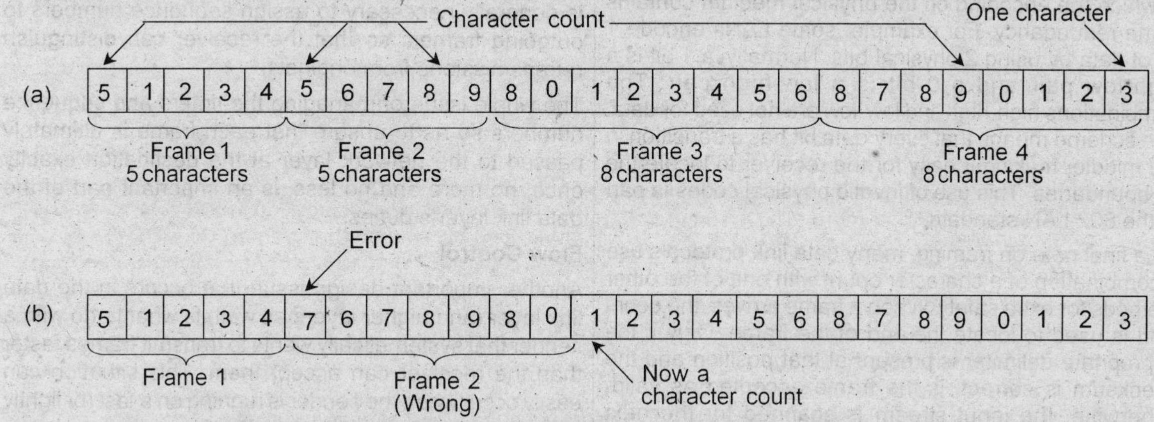

Fig. 26 : A character stream. (a) Without errors, (b) With one error

A serious problem occurs with this method when binary data, such as object programs or floating-point numbers, are being transmitted. It may easily happen that the characters for DLE STX or DLE ETX occur in the data, which will interfere with the framing. One way to solve this problem is to have the sender's data link layer insert an ASCII DLE character just before each "accidental" DLE character in the data. The data link layer on the receiving end removes the DLE before the data are given to the network layer. This technique is called character stuffing. Thus a framing DLE STX or DLE ETX can be distinguished from one in the data by the absence or presence of a single DLE. DLEs in the data are always doubled. Figure 27 gives an example data stream before stuffing, after stuffing, and after destuffing.

A major disadvantage of using this framing method is that it is closely tied to 8-bit characters in general and the ASCII character code in particular. As networks developed, the disadvantages of embedding the character code in the framing mechanism became more and more obvious so a new technique had to be developed to allow arbitrary sized characters.

The new technique allows data frames to contain an arbitrary number of bits and allows character codes with an arbitrary number of bits per character. It works like this. Each frame begins and ends with a special bit pattern, 01111110, called a flag byte. Whenever the sender's data link layer encounters five consecutive ones in the data, it automatically stuffs a 0 bit into the outgoing bit stream. This bit stuffing is analogous to character stuffing, in which a DLE is stuffed into the outgoing character stream before DLE in the data.

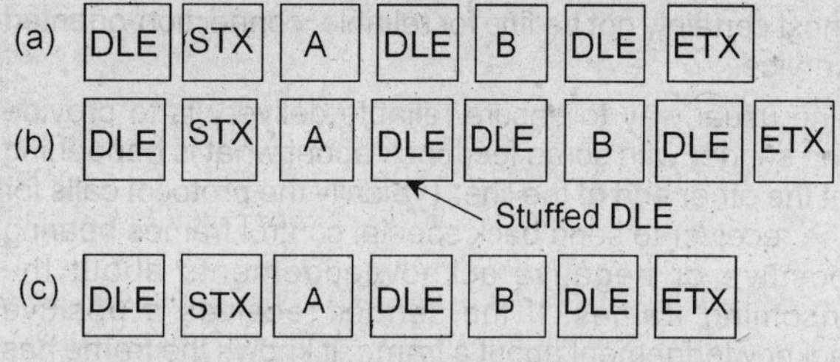

Fig. 27 : (a) Data sent by the network layer. (b) Data after being character stuffed by the data link layer. (c) Data passed to the network layer on the receiving side

When the receiver sees five consecutive incoming 1 bits, followed by a 0 bit, it automatically destuffs (i.e., deletes) the 0 bit. Just as character stuffing is completely transparent to the network layer in both computers, so is bit stuffing. If the user data contain the flag pattern, 01111110, this flag is transmitted as 011111010 but stored in the receiver's memory as 01111110. Figure 28 gives an example of bit stuffing.

(a) 0 1 1 0 1 1 1 1 1 1 1 1 1 1 1 1 1 1 1 1 0 0 1 0

(b) 0 1 1 0 1 1 1 1 1 0 1 1 1 1 1 0 1 1 1 1 1 0 1 0 0 1 0

Stuffed bits

(c) 0 1 1 0 1 1 1 1 1 1 1 1 1 1 1 1 1 1 1 1 0 0 1 0

Fig. 28 : Bit stuffing (a) The original data. (b) The data as they appear on the line (c) The data as they are stored in the receiver's memory after Restuffing.

With bit stuffing, the boundary between two frames can be unambiguously recognized by the flag pattern. Thus if the receiver loses track of where it is all it has to do is scan the input for flag sequences, since they can only occur at frame boundaries and never within the data.

The last method of framing is only applicable to networks in which the encoding on the physical medium contains some redundancy. For example, some LANs encode 1 bit of data by using 2 physical bits. Normally, a 1 bit is a high-low pair and a 0 bit is a low-high pair. The combinations high-high and low-low are not used for data. The scheme means that every data bit has a transition in the middle, making it easy for the receiver to locate the bit boundaries. This use of invalid physical codes is part of the 802 LAN standard.

As a final note on framing, many data link protocols use a combination of a character count with one of the other methods for extra safety. When a frame arrives, the count field is used to locate the end of the frame. Only if the appropriate delimiter is present at that position and the checksum is correct, is the frame accepted as valid. Otherwise, the input stream is scanned for the next delimiter.

Error Control

Having solved the problem of marking the start and end of each frame, we come to the next problem: how to make sure all frames are eventually delivered to the network layer at the destination, and in the proper order. Suppose that the sender just kept outputting frames without regard to whether they were arriving properly. This might be fine for unacknowledged connectionless service but would most certainly not be fine for reliable, connection-oriented service.

The usual way to ensure reliable delivery is to provide the sender with some feedback about what is happening at the other end of the line. Typically the protocol calls for the receiver to send back special control frames bearing positive or negative acknowledgements about the incoming frames. If the sender receives a positive acknowledgement about a frame, it knows the frame has arrived safely. On the other hand, a negative acknowledgement means that something has gone wrong, and the frame must be transmitted again.

An additional complication comes from the possibility that hardware troubles may cause a frame to vanish completely (e.g., in a noise burst). In this case, the receiver will not react at all, since it has no reason to react, It should be clear that a protocol in which the sender transmitted a frame and then waited for an acknowledgement, positive or negative, would hang forever if a frame were ever completely lost due to malfunctioning hardware.

This possibility is dealt with by introducing timers into the data link layer. When the sender transmits a frame, it generally also starts a timer. The timer is set to go off after an interval long enough for the frame to reach the destination, be processed there, and have the acknowledgement propagate back to the sender. Normally, the frame will be correctly received and the acknowledgement will get back before the timer runs out in which case it will be cancelled.

However, if either the frame or the acknowledgement is lost, the timer will go off, alerting the sender to a potential problem. The obvious solution is to just transmit the frame again. However, when frames may be transmitted multiple times there is a danger that the receiver will accept the same frame two or more times, and pass it to the network layer more than once. To prevent this from happening, it is generally necessary to assign sequence numbers to outgoing frames, so that the receiver can distinguish retransmissions from originals.

The whole issue of managing the timers and sequence numbers so as to ensure that each frame is ultimately passed to the network layer at the destination exactly once, no more and no less, is an important part of the data link layer's duties.

Flow Control

Another important design issue that occurs in the data link layer (and higher layers as well) is what to do with a sender that systematically wants to transmit frames faster than the receiver can accept them. This situation can easily occur when the sender is running on a fast (or lightly loaded) computer and the receiver is running on a slow (or heavily loaded) machine. The sender keeps pumping the frames out at a high rate until the receiver is completely swamped. Even if the transmission is error free, at a certain point the receiver will simply not be able to handle the frames as they arrive and will start to lose some. Clearly, some thing has to be done to prevent this situation.

The usual solution is to introduce flow control to throttle the sender into sending no faster than the receiver can handle traffic. This throttling generally requires some kind of a feedback mechanism, so the sender can be made aware of whether or not the receiver is able to keep up.

Various flow control schemes are known, but most of them use the same basic principle. The protocol contains well -defined rules about when a sender may transmit the next frame. These rules often prohibit frames from being sent until the receiver has granted permission, either implicitly or explicitly. For example, when a connection is setup, the receiver might say : "you may send me n frames now, but after they have been sent, do not send any more until I have told you to continue".

Services provided to the transport layer

The network layer provides services to the transport layer at the network layer/transport layer interface.

The network layer services have been designed with the following goals.

(1) The services should be independent of the subnet technology.

(2) The transport layer should be shielded from the number, type and topology of the subnets present.

(3) The network addresses made available to the transport layer should use a uniform numbering, plan, even across LANs and WANs.

In network layer connections should have the following properties

(1) Before sending data, a network layer process on the sending side must set up a connection to its peer on the receiving side.

(2) When a connection is established, the two processes can enter into a negotiation about the parameters, quality and cost of service provided.

(3) Communication is in both directions and packets are delivered in sequence.

(4) Flow control is provided automatically to prevent fast sender from slow receiver.

The main function of the network layer is routing packets from the source machine to the destination machine. The algorithms that choose the routes and the data structures that they use are a major area of network layer design.

The routing algorithm is that part of the network layer software responsible for deciding which output line an incoming packet should be transmitted on. Stability is one of the important goal in routing algorithm.

Routing algorithms can be divided into two groups

(1) Non-adaptive and (2) adaptive. Non-adaptive algorithms do not base their routing decisions on measurements or the estimates of current traffic or topology. This procedure is also called as 'Static routing'. Adaptive algorithms change their routing decisions to reflect changes in the topology and usually the traffic also.

Many routing algorithms are used in computer networks. Static algorithms include shortest path routing, flooding and flow based routing and dynamic algorithms like distance vector routing and link state routing. Most actual networks use one of there.

Transport layer

The transport layer is not just another layer. It is the heart of the whole protocol hierarchy. Its task is to provide reliable, cost effective data transport from the source machine to destination machine independent of the physical network(s) currently in use.

II. LAN TECHNOLOGIES

Ethernet

In this topic we will examine the details and operation of the popular LAN technology called Ethernet. Since its development by Xerox, Intel, and Digital Equipment Corporation in 1980. Ethernet has evolved from its initial 10-Mbps data rate, to Fast Ethernet (100 Mbps), and now Giga-bit Ethernet (1000 Mbps).

All of the properties, procedures, and definitions associated with Ethernet are contained in the IEEE Standard 802.3. This document includes such information as such

Standard	Technology
802.3	10Base5
802.3a	10Base2
802.3i	10BaseT
802.3j	10BaseFL
802.3u	100BaseT
802.3z	1000BaseX

- Flowcharts for transmitting and receiving a bit.
- Signal speed, noise, and other parameters for various media.
- Data encoding methods (Manchester, 4B5B, etc.) used by each technology.
- The method for computing the frame check sequence.
- Detailed discussion of collision detection.
- Autonegotiation using fast link pulses.
- Repeater operation.

Table shows some of the many different Ethernet technologies covered by the IEEE 802.3 Standard. We will examine each technology, beginning with the 10-Mbps systems. Since, each new technology must accomplish the same goal, reliable transmission of an Ethernet frame, our discussion begins there.

THE ETHERNET FRAME FORMAT

Figure 1 shows the format of an Ethernet 802.3 frame. A separate and practically identical frame format called Ethernet II has an 8-byte preamble of identical 10101010 patterns (no 10101011 SFD) and a 2-byte Type field in place of the Length field. Since the Data field of an 802.3 Ethernet frame is limited to 1500 bytes, a value larger than 1500 in the Type/Length field indicates an Ethernet II frame. Many companies, such as Xerox, DEC, and Novell, use Ethernet II frames in their networks.

All data is packaged into one or more frames for transmission over an Ethernet LAN. Let us look at each portion of the Ethernet frame.

Preamble (7 bytes)

The Preamble contains 7 bytes with the identical pattern 10101010. This alternating sequence of 1s and 0s are the first bytes transmitted. They are provided to help listening stations synchronize quickly on the new data stream.

SFD (1 byte)

The SFD (Start Frame Delimiter) is the 1 -byte pattern 10101011. Note that the LSB is high, whereas all the bytes of the Preamble had their LSB low.

Destination Address (6 bytes)

This is the destination MAC address of the station that is to receive the frame. Every Ethernet device manufactured (NIC, router) contains a unique 48-bit MAC address assigned by the manufacturer. An example of a MAC address is 00-C0-F0-27-64-E2. The first 24 bits are the manufacturer code (00-C0-F0 is Kingston). The last 24 bits are chosen by the manufacturer. The WINIPCFG utility can be used to determine the MAC address (called the adapter address).

A MAC address of FF-FF-FF-FF-FF-FF (all 1s) is reserved for use as a broadcast address.

Source Address (6 bytes)

This is the MAC address of the source station transmitting the frame.

Length (2 bytes)

This field indicates the number of bytes in the Data field.

Data (46 to 1500 bytes)

The Data area is where all data from the upper networking layers is carried.

FCS (4 bytes)

The FCS (frame check sequence) is a 32-bit CRC (cyclic redundancy check) value used to determine the validity of the received Ethernet frame. The FCS is found using a generator polynomial defined mathematically as

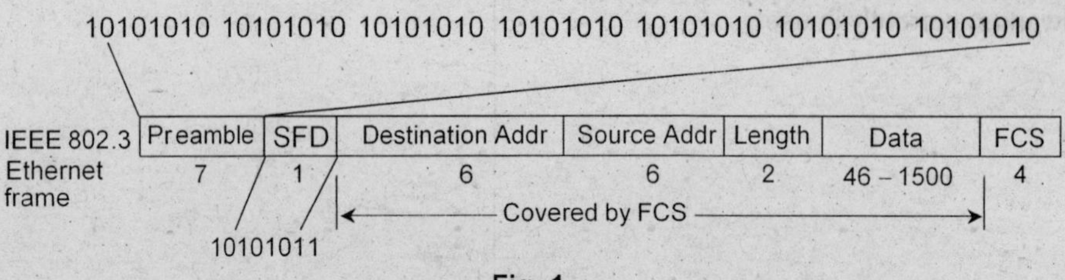

Fig. 1

$$G(X) = x^{32}+x^{26}+x^{23}+x^{22}+x^{16}+x^{12}+x^{11}+x^{10}+x^{8}+x^{7}+x^{5}+x^{4}+x^{2}+x+1$$

which is equivalent to the binary value 100000100110000010001110110111. The 1s in the pattern indicate where exclusive OR gates are used in a recirculating shift register circuit that is able to generate/check a valid stream of data of variable length. If even a single bit in the frame is received in error, the shift register will not generate the proper output and the error will be discovered.

All bits in the frame except those in the FCS are transmitted from LSB to MSB.

From the numbers provided in Figure 1, the minimum and maximum sizes of an Ethernet frame are 72 bytes and 1526 bytes, respectively. Many individuals refer to the smallest Ethernet frame as 64 bytes, by not including the 8 bytes of Preamble and SFD. The largest frame would then be 1518 bytes. For the purposes of the discussion in this exercise, the Preamble and SFD will be included in the length of the frame.

THE INTERFRAME GAP

The interframe gap is a self-imposed quiet time appended to the end of every frame. This idle time gives the network media a chance to stabilize, and other network components time to process the frame. Figure 2 shows a sequence of frames separated by the fixed-size interframe gap. For 10-Mbps Ethernet, the interframe gap is 9.6 microseconds. This corresponds to 96 bit times (device 9.6 microseconds by 100 nanoseconds/bit). Thus the 576 bits of a minimum length Ethernet frame are followed by 96 bit times of silence. Dividing 10 Mbps by 672 bits for each frame plus interframe gap gives a frame rate of 14,880 frames/second (minimum-sized frames), since each frame is followed by 96 bits of silence, there are a total of 14,880 times 96, or 1,428,480 bits of the 10 Mbps bandwidth (14.28%) lost due to the interframe gap. These calculations are shown.

When the frame size is larger, more of the available bandwidth is utilized. Note that as the frame size increases, the 96 bits of interframe gap becomes less significant. The actual frame lengths in a real network will be constantly fluctuating, as shown in Figure 2 so the bandwidth utilization will be constantly changing.

72 bytes/minimum Ethernet frame
x 8 bits/byte
576 bits/frame

+ 96 bits/interframe gap (IFG)
672 bits/frame and IFG

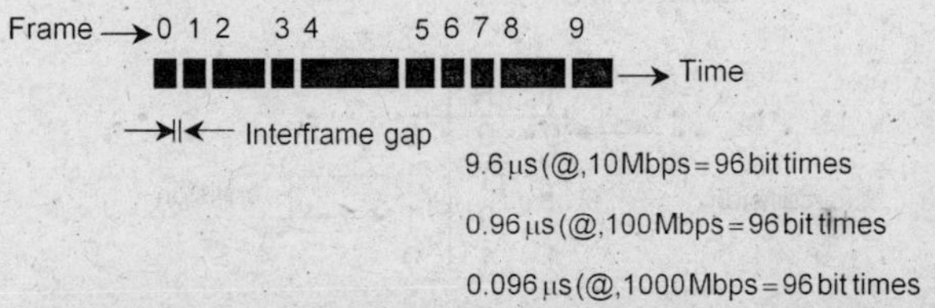

Fig. 2 : Interframe gap separates each Ethernet frame

10,000,000 bits / sec ond
÷ 672 bits / frame and IFG
14,880 frames / sec ond

14,880 frames / sec ond
× 96 bits / IFG
1,428,480 bits lost / sec ond

$$\frac{1,428,480}{10,000,000} \times 100\% = 14,28\% \text{ lost bandwidth}$$

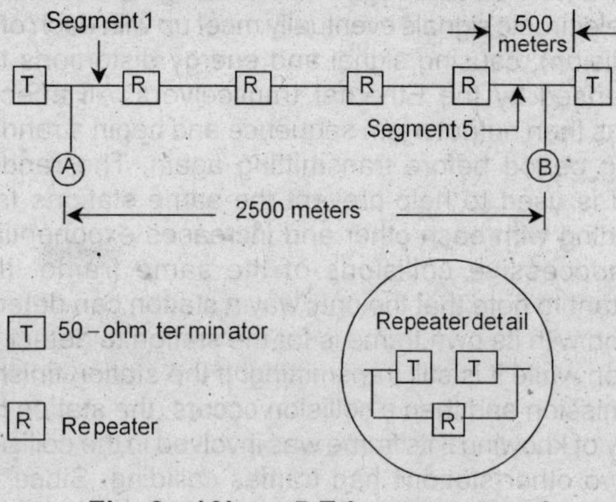

Fig. 3 : 10base5 Ethernet network

CSMA/CD

CSMA/CD stands for Carrier Sense Multiple Access with Collision Detection. This is the tech nique used to share access to the available bandwidth. To describe the basic operation of CSMA/CD, let us use the 10base5 network shown in Figure 3. This network consists of five 500 meter coaxial segments connected by four repeaters. This is the largest network possible using10base5 technology. Two stations, A and B are located at the farthest ends of the network. There may be many more stations connected to each segment, all competing for bandwidth using the same CSMA/CD method. Collectively, these stations operate in a collision domain. A collision domain is a portion of a LAN (or the entire LAN) where two or more stations transmitting at the same time will interfere with each other. For example, in Figure 3 the entire 2500 meter network is a single collision domain, due to the use of repeaters between each segment. In general, all ports on repeaters or hubs are in the same collision domain. Ports on switches and routers operate in their own individual collision domains.

Let us examine what occurs during good and bad frame transmissions in a collision domain.

No Collision During Frame

A station wishing to transmit a frame first listens to the coax, waiting for an idle period indicating no transmissions. Once the coax is quiet for a time equal to (or longer than) the interframe gap, the station begins transmitting the frame, one bit at a time. The electronic signal representing each bit travels at a limited speed within the thickwire coax, requiring 10.8 microseconds worst case to travel the 2500 meters from station A to station B or vice versa. This time is based on the speed coefficient of the cable (0.77 for thickwire coax), the length of the cable, and the speed of light (300,000.000 meters/second).

The signal is absorbed at the end of each coaxial segment by the terminating resistor. There are no reflections to cause a collision, and no other stations begins transmitting during the frame. One or more stations receive and process the frame.

A Collision Occurs During the Frame

At some point during the transmission of one station's frame, a new station begins transmitting its own frame. Their electronic signals eventually meet up wilh each other (a collision), causing signal and energy distortions that are sensed by the Ethernet transceivers. All affected stations then output a jam sequence and begin a random waiting period before transmitting again. The random period is used to help prevent the same stations from recolliding with each other and increases exponentially with successive collisions of the same frame. It is important to note that the only way a station can detect a collision wilh its own frame is for the station to detect the collision while it is still transmitting. If the station finishes transmission and then a collision occurs, the station has no way of knowing if its frame was involved in the collision, or if two other stations had frames colliding. Since the diameter of the 10base5 network from Figure 3 is limited to 2500 meters, the worst - case round trip time of a signal is 21.6 microseconds (not including the delays associated with the four repeaters). Now, let us consider why the round-trip time is important to collision detection. Figure 4(a) shows an example timeline of how station A detects a collision with a frame from station B. In Figure 4 station A has listened to the coax and found it idle and has begun transmitting its frame. Station B is also listening and finding the coax idle.

The signal representing station A's frame travels towards station B taking at least 10.8 micro seconds for the 2500 meter trip, ignoring the delays of the four repeaters. Now, just as the signal is about to reach station B; station B begins transmitting its own frame (recall that station B also found the coax free of transmission). An instant later the collision occurs, indicated by the time t_1 in Figure 4(b). A distorted signal begins travelling back to station A, requiring an additional 10.8 microseconds (ignoring repeater delays again) to travel another 2500 meters. At time t_2 in Figure 4(c) station A detects the collision. The total time is 21.6 microseconds for the round trip. With each bit requiring 100 nanoseconds, this corresponds to 216 bits (27 bytes) transmitted during the round trip. So an Ethernet frame must be at least 27 bytes long to detect a collision. But remember that we have ignored the delays associated with the four repeaters. The IEEE 802.3 standard limits repeater delays to 8 bit times This adds up to a total of 64 bits for the round trip, or another 8 bytes of length we must add to the Ethernet frame for collision detection. This gets us to 35 bytes for the minimum frame. Since an actual Ethernet frame is at least 72 bytes long, there is plenty of time for a round-trip collision signal to be detected.

Detecting Errors

Figure 5 shows a simplified circuit for detecting a collision by comparing the transmitted and received bit streams with each other. In other words, the Ethernet transceiver listens to itself as it is transmitting. If a zero signal was transmitted, the receiver should see a zero. The same is true for a one signal. If the signals do not match, it is most likely due to a collision or some other network malfunction. As indicated in Figure 5 an exclusive-OR gate is used to identify when the transmitted and received bit streams are different. The delay is required on the transmitter input of the exclusive-OR gate to compensate for the time required by the transmitter and receiver circuits.

Collision example

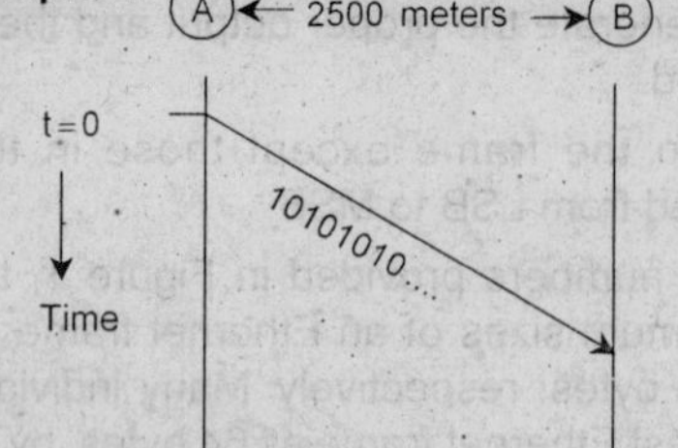

(a) Station A begins transmitting a frame

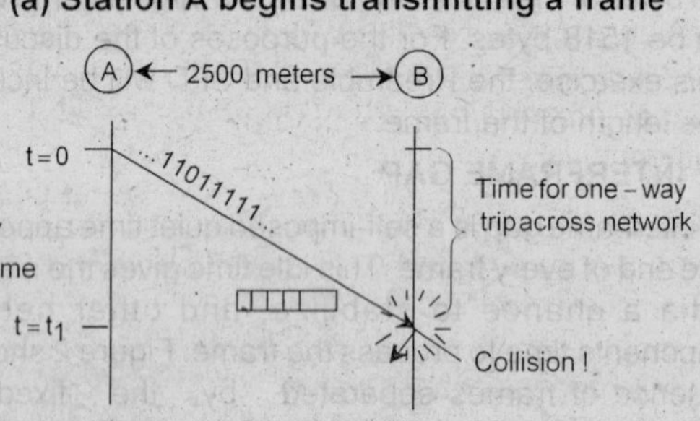

(b) Station B begins transmitting a frame an instant before station A's frame arrives

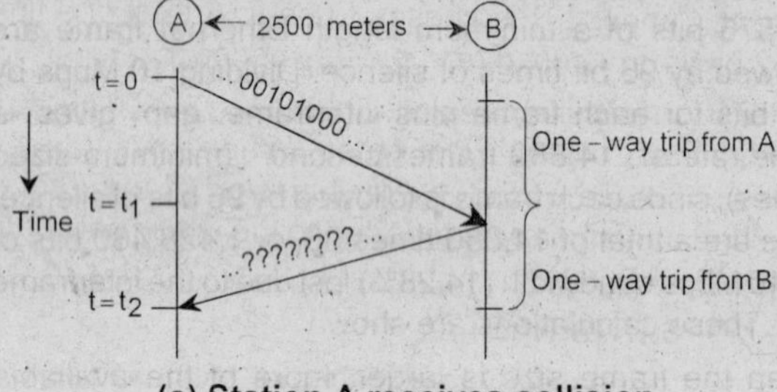

(c) Station A receives collision

Fig. 4

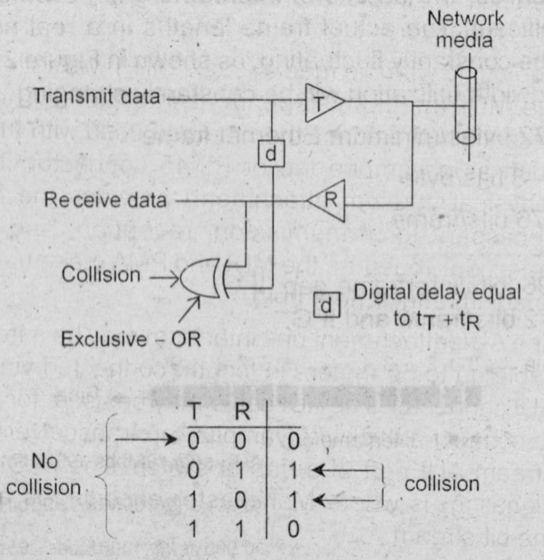

Fig. 5 : Hardware detection of collision

Figure 5 is based on information contained in the original Ethernet patent, which is titled MULTIPOINT DATA COMMUNICATION SYSTEM WITH COLLISION DETECTION.

Jam Sequence

The jam sequence generated by stations detecting a collision is a 32-bit pattern designed to propagate the collision throughout the network. The contents of the jam pattern may not represent a valid FCS for the current frame.

Random Waiting Period

Retransmission of a frame is attempted after a random waiting period that is a multiple of an Ethernet slot time, the time required to transmit 512 bits (a total of 51.2 microseconds for 10 Mbps Ethernet). The actual name of the algorithm is "truncated binary exponential backoff," due to the method used to increase the waiting period. A collision counter keeps track of successive collisions (collisions involving the same frame). The higher the number of collisions, the longer the waiting period. Mathematically, we have

$$0 \le R \le 2^K$$

where K equals minimum (N,10) and N is the number of successive collisions. R is a random integer chosen from the range 0 to 2^K. As K increases, the upper limit on the range increases exponentially (1, 2, 4, 8, 16, 32, 64, 128, 256, 512, 1024). The random integer R is thus chosen from an ever-increasing set of integers. As an example, after three collisions, R is chosen from the set of numbers 0 to 8. Suppose R is chosen to be 5. A total of 5 times 512 bit times is used as the delay period.

After 10 successive collisions, the waiting period becomes a maximum of 1024 slot times. When 16 successive collisions have occurred, transmission of the frame is aborted and an error is reported to the upper networking layer.

10-MBPS ETHERNET

The first three widely used Ethernet technologies were 10base5, 10base2, and 10baseT. Figure 6 shows the general 10-Mbps architecture of the Data-Link and Physical layers.

Beginning with the media (coax, UTP) we first encounter the MDI (medium dependent interface). The MDI is essentially the connection method used with the media, such as a vampire tap or RJ-45 connector. The PMA (physical medium attachment) provides the functions necessary for transmission, reception, and collision detection. Together, the MDI and PMA make up the MAU (medium attachment unit)

The AUI (attachment unit interface) may be a transceiver cable up to 50 meters in length, connected via a 15-pin AUI connector. The AUI cable may be used, for example, to connect a thickwire vampire-tap transceiver to the up stream AUI port of a hub or switch. The PLS (physical signaling) is where Manchester encoding is applied to the bit stream.

Let us examine the properties of each 10 Mbps technology.

10base5

1. Media: Thickwire coax
2. Propagation velocity. 0.77c
3. Impedance: 50 ohms
4. Connector: Vampire tap
5. Segment length: 500 meters
6. Maximum nodes/segment: 100
7. Node spacing: 2.5 meters
8. Topology: Bus

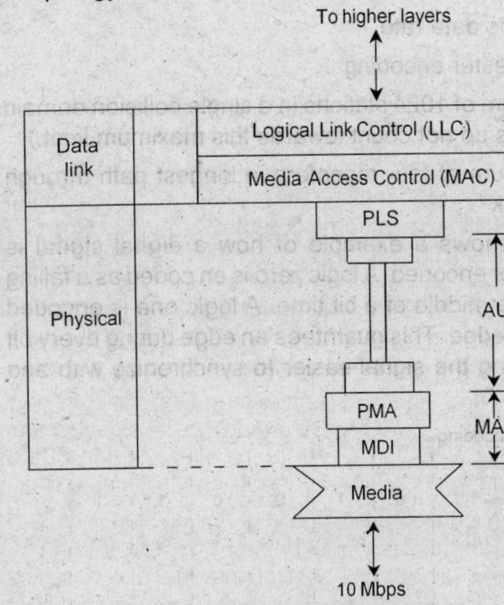

Fig. 6 : 10 - Mbps architecture

PLS	Physical Signaling
PMA	Physical Medium Attachment
AUI	Attachment Unit Interface
MAU	Medium Attachment Unit
MDI	Medium Dependent Interface

The 2.5 meter spacing requirement is designed to prevent signal distortions from each station from adding together in phase. The number of nodes allowed on a segment is limited by the electrical properties of the cable.

10base2

(1) Media: Thinwire coax
(2) Propagation velocily: 0.65 c
(3) Impedance: 50 ohms
(4) Connector: BNCT
(5) Segment length: 185 meters
(6) Maximum nodes/segment: 30
(7) Node spacing: 0.5 meters
(8) Topology: Bus

Here we can see that different coaxial cable affects the allowable cable length as well as the number of nodes/ segment.

10baseT

1. Media: Category 3, 4, or 5 UTP
2. Propagation velocity: 0.585 c
3. Impedance: 100 ohms

4. Connector: RJ-45
5. Segment length: 100 meters
6. Maximum nodes/segment: 2
7. Topology: Star

The number of nodes/segment is misleading, since the UTP cable requires a point-to-point connection, typically a NIC to a port on a hub, or a hub-to-hub connection.

All three technologies have some common properties. These include

- 10 Mbps data rate
- Manchester encoding
- Maximum of 1024 stations in a single collision domain (Repeaters do not count towards this maximum limit.)
- Maximum of four repeaters in longest path through the network

Figure 7 shows a example of how a digital signal is Manchester encoded. A logic zero is en coded as a falling edge in the middle of a bit time. A logic one is encoded as a rising edge. This guarantees an edge during every bit time, making the signal easier to synchronize with and decode.

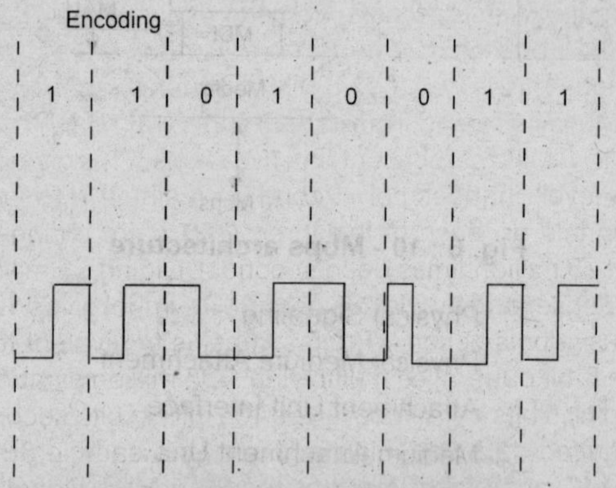

Fig. 7 : Manchester Encoding

The four repeater limit, chosen to maintain SMA/CD operation, is part of a larger set of restrictions, commonly referred to as the 5-4-3 rule, which following properties:

- 5 segments in the longest path
- 4 repeaters in the longest path
- 3 segments with nodes

again, these rules apply to a single collision domain. Figure 8 shows two ways of applying the 5-4-3 rule. In Figure 8(a) 10base5/10base2 network system is illustrated. Three of the five segments contain nodes. The longest path between (stations A and Z) is five segments, with four repeaters in between.

Figure 8(b) shows a 10 baseT network.The A and Z stations are separated by five UTP segments with four hubs in between. Each UTP cable from a station to a hub, or from a hub to another hub, is considered a segment and may be up to 100 meters in length.

Collisions are detected in the 10 baseT network whenever the transmit and receive pair in a single UTP cable are active at the same time. For example a frame being transmitted from port 3 on a collides with a frame coming into port 3. Even though the transmit and receive wire pairs are different, and there is no electrical collision (as we on coaxial cable), there is an attempt by two stations to use the network simultaneously which results in a collision. This indicates half - duplex generation, since any station may send or receive but not at the same time. New auto negotiating Ethernet transceivers are capable of operating in full - duplex.

Utilizing the 5-4-3 rule

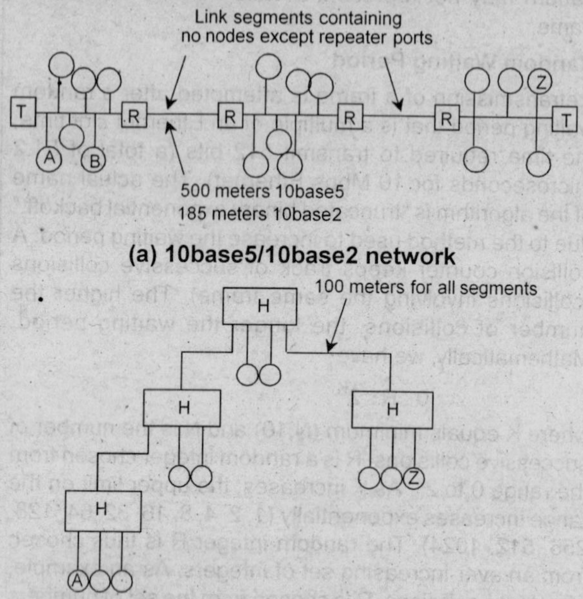

(a) 10base5/10base2 network

(b) 10baseT network

Fig.8

10 base F

Originally, the FOIRL (fiber optic inter-repeater link) specification was used to standardize Ethernet Communication over fiber. It allowed for a 1000-meter fiber between repeaters.

Eventually, the demand for fiber to the PC exceeded the FOIRL specifications and a new set of fiber specifications was designed called 10baseF, it contains three categories: 10baseFL, 10base FB, and 10base FP.

10baseFL

. Fiber Link specification
. Station-to-station, station-to-hub, and hub-to-hub connection.
. 2000-meter segment length
. Manchester encoding
. Replaces FOIRL, works with existing FOIRL hardware.
. Connector : ST

This specification is the most widely used. Figure 9 shows a portion of a 10baseFL network. Before fiber NICs became available, a 10baseFL to 10baseT (or 10base2) transceiver was used to interface fiber with the PC.

10baseFB

- Fiber backbone specification
- Hub-to-hub connection
- Synchronous operation
- 2000-meter segment length

The synchronous operation on the 10baseFB link is used to reduce the delays normally associated with Ethernet repeaters that cause the interframe gap to shrink as it propagates through multiple repeaters. This allows the segment distance to be extended without compromising the collision detection mechanism.

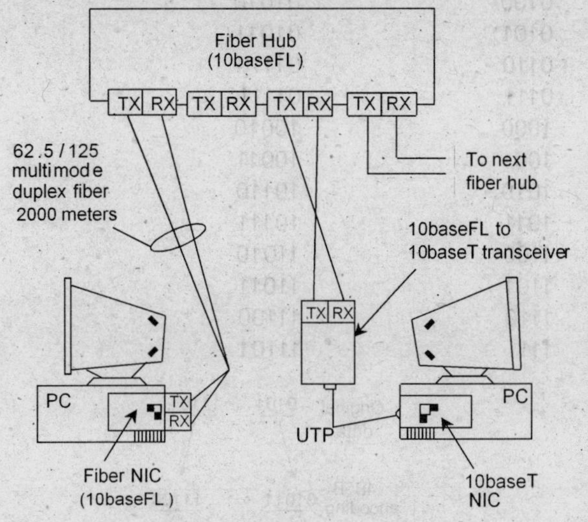

Fig. 9 : 10baseFL wiring

10baseFP

- Fiber passive specification
- Station-to-hub, hub-to-hub connection
- Hubs link up to 33 stations
- 500 meter segment length

This specification allows groups of 10baseFP computers to be passively connected via., a hub that optically shares signals transmitted from any station. This technology is well suited for low-power environments.

100-MBPS ETHERNET (FAST ETHERNET)

Demand for bandwidth quickly exceeded the capacity of 10 Mbps Ethernet. Even moving from hub-based 10baseT networks to switch-based networks only provided temporary relief. Increasing the data rate by a factor of 10 over 10 Mbps Ethernet, 100 Mbps Ethernet, or Fast Ethernet as it is commonly called, is implemented in several different ways, all collectively referred to as 100baseT technology. One disadvantage of Fast Ethernet is its smaller network diameter, typically one 10th that of 10 Mbps Ethernet, or around 200 meters. This reduction in network diameter is necessary to maintain the parameters of CSMA/CD at the faster data rate, since the signals still move at the same speed in the cable, but the frame times are shorter by a factor of 10.

Figure 10 shows the 100 Mbps Ethernet interface definition. Several new sublayers have been added, due to the requirements of 100-Mbps transmission. For example, the Manchester encoding used in 10-Mbps Ethernet is not well suited to high-frequency operation. Other data encoding and signaling techniques are used instead, using special bit patterns and multilevel signaling to transfer data in 4-bit chunks instead of one bit at a time (as in 10 Mbps Ethernet).

Let us take a look at each 100 Mbps technology

100baseT4

1. Data is exchanged over 3 pairs of Category 3 (or higher) UTP.
2. Data rate is 33.3 Mbps (25 M baud, 125 MHz) on each pair (total of 100 Mbps).
3. 8B6T encoding is used
4. Fourth pair is used for collision detection.
5. Half-duplex only.
6. Maximum segment length : 100 meters
7. Maximum network diameter : 205 meters
8. Maximum number of repeaters : 2
9. Connector : RJ-45

8B6T coding replaces 8-bit data values with 6 ternary codes that may have the values -, +, or 0. Table shows a small sample of the 256 code patterns used in 8B6T encoding. The patterns are chosen to provide good DC characteristics, error detection, and reduced high-frequency effects. Special patterns can also be used as markers or control codes.

A multilevel signaling scheme is used, which allows more than one bit of data to be encoded into a signal transition. This is why a 12.5 MHz frequency carries a 33.3 Mbps stream. Think about it this way: Each cycle of the 12.5 MHz carrier contains two levels. This gives 25 million level changes per second on a single UTP pair. The signals on each of the three UTP pairs change a total of 75 million times each second. Dividing 75 million levels per second by 6 levels per 8B6T symbol gives 12.5 million symbols/second. Each symbol is equivalent to a unique 8-bit pattern, so multiplying 12.5 million symbols/second by 8 bits per symbol gives 100 million bits/second, the required data rate. Figure 11 shows a sample 8B6T encoded waveform. Note that the 12.5 MHz signaling frequency is within the 16 MHz limit of Category 3 cable.

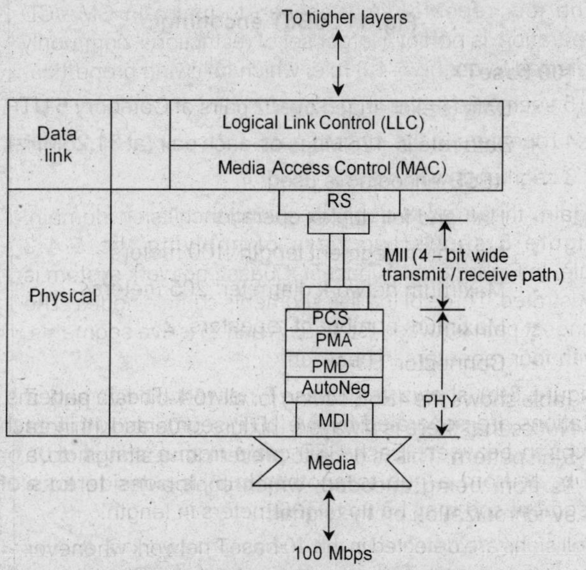

Fig. 10 : 100baseT architecture

RS	Reconcilation Sublayer
PCS	Physical Coding Sublayer
PMA	Physical Medium Attachment
PHY	Physical Layer Device
PMD	Physical Medium Dependent
MII	Media Independent Interface
MDI	Medium Dependent Interface

Hex Value	6T Code Group					
00	+	-	0	0	+	-
01	0	+	-	+	-	0
02	+	-	0	+	-	0
03	-	0	+	+	-	0
04	-	0	+	0	+	-
05	0	+	-	-	0	+
06	+	-	0	-	0	+
07	-	0	+	-	0	+
08	-	+	0	0	+	-
10	+	0	+	-	-	0
3F	+	0	-	+	0	-
5E	-	-	+	+	+	0
7F	0	0	+	-	-	+
80	+	-	+	0	0	-
C0	+	-	+	0	+	-
FF	+	0	-	+	0	0

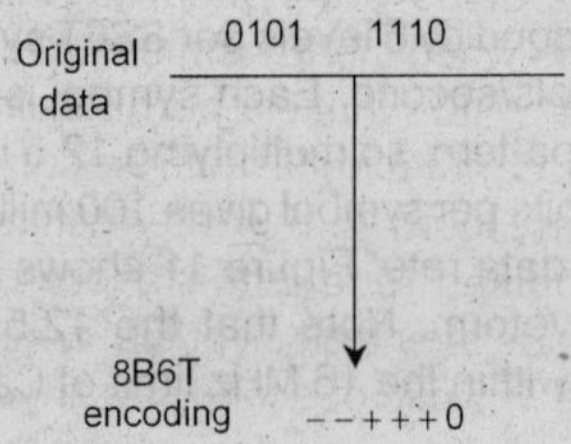

Fig. 11 : 8B6T encoding

100 baseTX

- Data is exchanged over 2 pairs of Category 5 UTP.
- Data rate is 125 Mbps on each pair (at 31.25MHz).
- 4B5B encoding is used
- Half and full duplex operation
- Maximum segment length: 100 meters
- Maximum network diameter: 205 meters
- Maximum number of repeaters: 2
- Connector: RJ-45

Table shows the 4B5B coding for all 16 4-bit data patterns. Notice that there is always a mixture of 0s and 1s in each 5-bit pattern. This is done to prevent long strings of 0s or 1s from being encoded, which contributes to loss of synchronization on the signal.

4-bit Data	5-bit Code
0000	11110
0001	01001
0010	10100
0011	10101
0100	01010
0101	01011
0110	01110
0111	01111
1000	10010
1001	10011
1010	10110
1011	10111
1100	11010
1101	11011
1110	11100
1111	11101

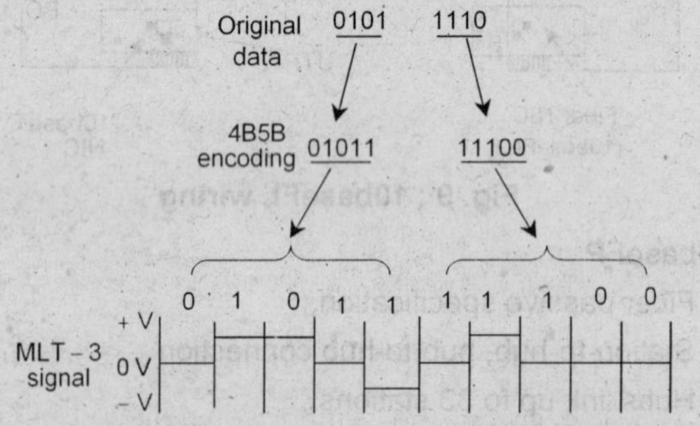

Fig. 12 : MLT-3 signaling for 4B5B encoded data.

Figure 12 shows how a three-level signal called MLT-3 (multiple level transition) is used to represent the 4B5B bit stream. Each 4-bit data value is replaced by its 5-bit 4B5B counterpart. Thus, the 100-Mbps data stream becomes a 125-Mbps data stream becomes a 125-Mbps 4B5B encoded data stream. Using MLT-3 allows the 125 Mbps 4B5B data stream to be carried using a signal rate of 31.25 MHz (31.25 MHz times 4 bits/cycle equals 125 Mbps).

Since the signaling frequency of 31.25 MHz is greater than the 16 MHz limit of Category 3 cable, a better cable, Category 5, is required. Category 5 cable has a frequency limit of 100 MHz

100base FX

- Single mode fiber limit: 10,000 meters
- Multimode fiber limit : 2000 meters (412 meters half duplex).
- 4B5B encoding is used.
- Connector: duplex SC, ST and FDDI MIC allowed

In this technology, the 4B5B encoded data is transmitted using NRZI (non-return to zero, invert on one). A 4B5B data rate of 125 Mbps is obtained using a 62.5 MHz carrier. Figure 13 illustrates a sample encoding and waveform. NRZI is well suited for fiber, due to its bi-level nature.

100baseT2

- Data is exchanged over 2 pairs of Category 3 (or higher) UTP.
- PAM 5×5 encoding is used.
- Connector: RJ-45

Sending 100 Mbps over only two pairs of UTP requires yet another encoding and signaling scheme. Here, two 5 level PAM (pulse amplitude modulation) signals are sent over the UTP pairs with a signalling rate of 12.5 MHz. Each cycle of the signal provides two PAM 5x5 level changes, so there are 25 million level changes per UTP pair. Each pair of PAM signals (called A and B) encode a different 4-bit pattern (along with other, special patterns for idle mode) using combinations of these levels: {+2, +1, 0, -1, -2}. So 25 million PAM 5x5 pairs times 4 bits/ pair gives 100 Mbps.

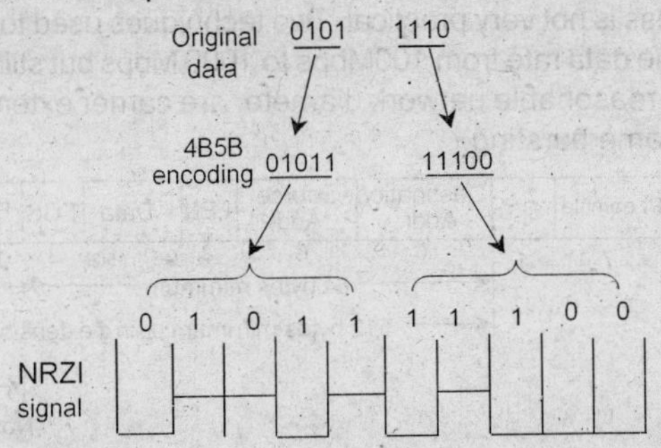

Fig. 13 : NRZI signaling for 4B5B encoded data.

Figure 14 shows the symbol constellations found in PAM5x5 encoding, as well as a sample pair of waveforms. Note that, when not transmitting data, the 100baseT2 link transmits an idle signal to maintain synchronization. During idle mode, the signals on A and B alternate between {+1, -1} and {+2,0,-2}.

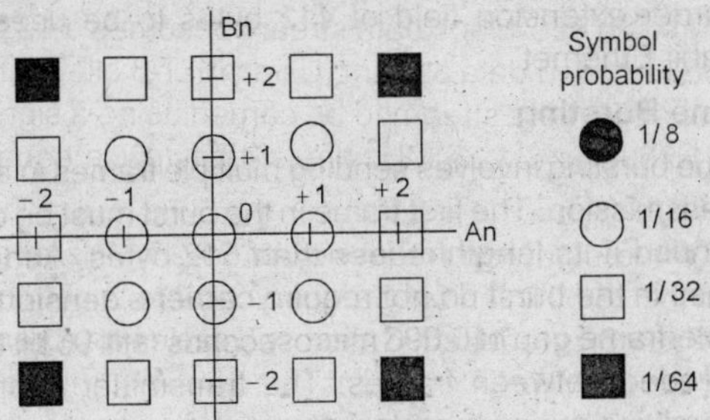

(a) Data mode

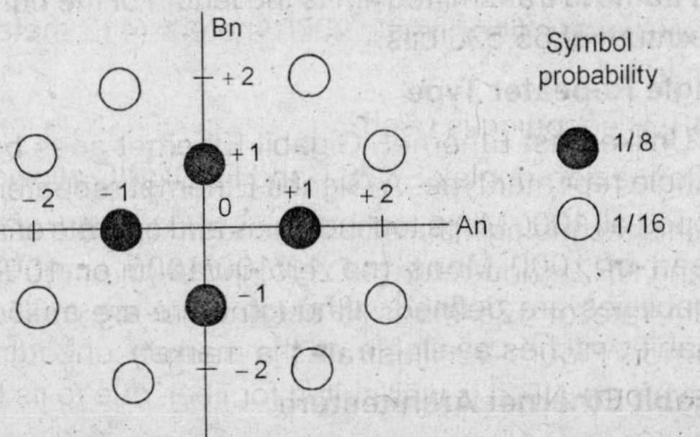

(b) Idle mode

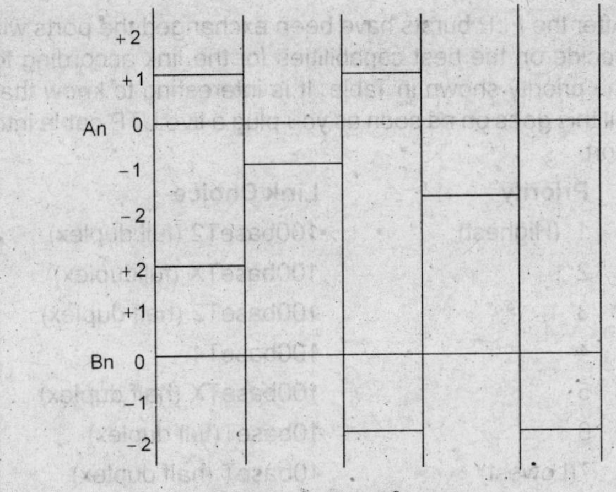

(c) Sample waveforms.

Fig. 14 : PAM 5 × 5 symbol constellations.

Fast Link Pulses

Beginning with 100baseT technology came the ability to perform autonegotiation between each end of a 100baseT connection. When the connection is established (plugging both ends of the UTP cabe into their respective ports), a series of fast link pulses (FLP) will be exchanged between the ports. The 33 pulses contain 17 clock pulses and 16 data pulses. The 16 data pulses form a 16-bit code indicating the capabilities of the port, such as communication mode (half-duplex, full-duplex) and speed (10,100,10/100). Originally, 10baseT NICs used a single normal link pulse (NLP) to perform a link integrity test. An indicator LED on the NIC showed the status of the link. NLP pulses are typically generated every 16 milliseconds when the transmitter is idle, as indicated in Figure 15(a). NICs that support fast link pulses send an FLP burst containing 2 milli seconds of pulses, as illustrated in Figure 15(b). Note that the even pulses are the data pulses (one when a pulse is present 62.5 microseconds after a clock pulse, zero if there is no data pulse). The 125 microsecond spacing between the clock pulses allows the entire burst to complete in 2 milliseconds.

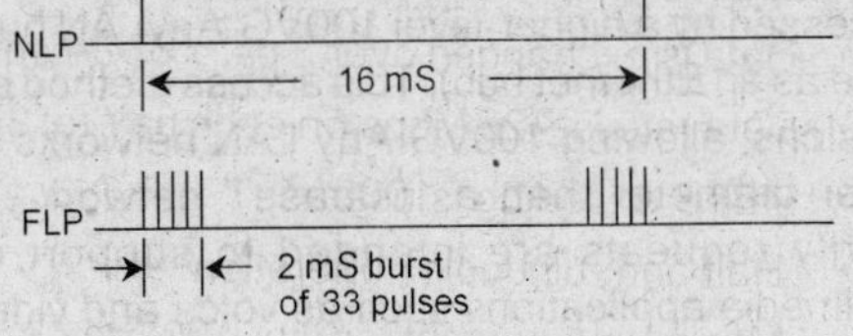

(a) General timing

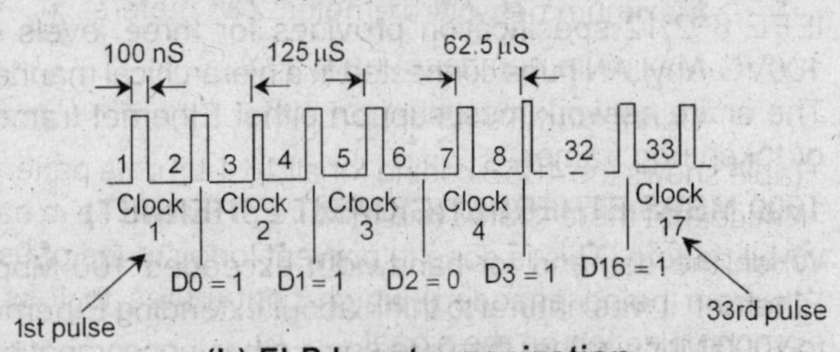

(b) FLP burst organization

Fig. 15 : Fast link pulse details

After the FLP bursts have been exchanged the ports will decide on the best capabilities for the link according to the priority shown in Table. It is interesting to know that all this goes on as soon as you plug a live UTP cable into port.

Priority	Link Choice
1. (Highest)	100baseT2 (full duplex)
2	100baseTX (fullduplex)
3	100baseT2 (half duplex)
4	100baseT4
5	100baseTX (half duplex)
6	10baseT(full duplex)
7(Lowest)	10baseT (half duplex)

Two Repeater Types

The original 10 Mbps Ethernet specified a single repeater type to propagate frames between segments. Fast Ethernet contains specifications for two types of repeaters. Class I and Class II. Class I repeaters are slower (140 bit times for its round trip delay) than Class II repeaters (92 bits times or less), but provide functions such as translation between the many different 100baseT technologies. Class II repeaters, although faster, only support a single technology.

Standard topologies for 100baseT networks are

- One Class I repeater. This provides a network diameter of 200 meters using copper cable. Stations may be 100 meters from the repeater.
- Two Class II repeaters connected via a 5 meter cable. This provides a diameter of 205 meters. Stations may be 100 meters from each repeater.

100VG-AnyLAN

Developed by Hewlett-Packard, the IEEE-802.12 standard 100 VG-Any LAN technology is a 100-Mbps LAN technology capable of handling both Ethernet and token ring frames. 100VG-AnyLAN uses domain-based priority access, an access method whereby stations are polled in a round-robin fashion. Each polled station may make a normal-priority or high-priority request, which is processed by a higher-level 100VG-AnyLAN hub (not the same as an Ethernet hub). This access method eliminates collisions, allowing 100VG-Any LAN networks to have a larger diameter than a 100baseT network. The high priority requests are intended to support real-time multimedia applications such as voice and video.

Figure 16 shows a sample 100VG-AnyLAN network. The IEEE 802.12 specification provides for three levels of 100VG-AnyLAN hubs connected in a hierarchical manner. The entire network must support either Ethernet frames or token ring frames.

1000 MBPS ETHERNET(GIGABIT ETHERNET)

When the demand for bandwidth exceeded 100-Mbps Ethernet, it was natural to think about extending Ethernet to 1000 Mbps, rather than use some other, noncompatible technology like ATM (asynchronous transfer mode) or FDDI (fiber distributed data interface)

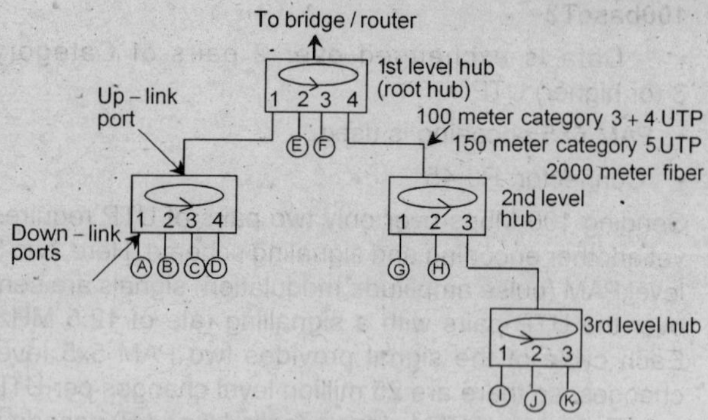

Fig. 16 : 100VG - Any Lan network

Extending the data rate, however, leads to a decrease in the network diameter (so that CSMA/CD is maintained). You might agree that a network diameter of 25 meters or less is not very practical. Two techniques used to increase the data rate from 100Mbps to 1000 Mbps but still maintain a reasonable network diameter are carrier extension and frame bursting.

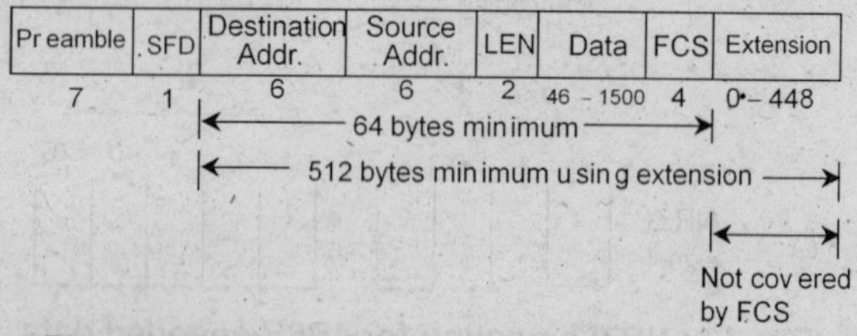

Fig. 17 : Ethernet frame with carrier extension

Carrier Extension

Figure 17 shows an Ethernet frame with a 0 to 448 byte carrier extension field. The carrier extension is used to maintain a minimum 512- byte Ethernet frame (not including the Preamble and SFD). So a 10/100 Mbps Ethernet frame containing only 100 bytes would require a carrier extension field of 412 bytes to be used over Gigabit Ethemet.

Frame Bursting

Frame bursting involves sending multiple frames in a burst of transmission. The first frame in the burst must be carrier extended if its length is less than 512 bytes. Additional frames in the burst do not require carrier extensions, but an interframe gap of 0.096 microseconds (still 96 bit times) is needed between frames. The transmitter continues transmitting during the interframe gap to maintain its hold on the network media. A burst timer started when the first frame is transmitted limits the length of the burst to a maximum of 65,536 bits.

Single Repeater Type

Unlike Fast Ethernet, Gigabit Ethernet goes back to a single repeater type. A Gigabit Ethernet repeater must support all 1000 Mbps technologies and operate at a fixed speed of 1000 Mbps (no 10/100/1000 or 100/1000 capabilities are defined, although there are multispeed Gigabit switches available on the market).

Gigabit Ethernet Architecture

Figure 18 shows the multilayer architectural model for Gigabit Ethernet. You may notice the similarities to 10

Mbps and 100 Mbps Ethernet. One no table difference is the new 8-bit wide transmit and receive paths.

In addition, you will notice that full duplex operation is available in every Gigabit technology, which was not the case for 10-Mbps Ethernet and Fast Ethernet. The MAC layer, described in IEE 802.3z, deals with issues such a half/full duplex operation, carrier extension, and frame bursting.

'Let us examine the different Gigabit Ethernet technologies.

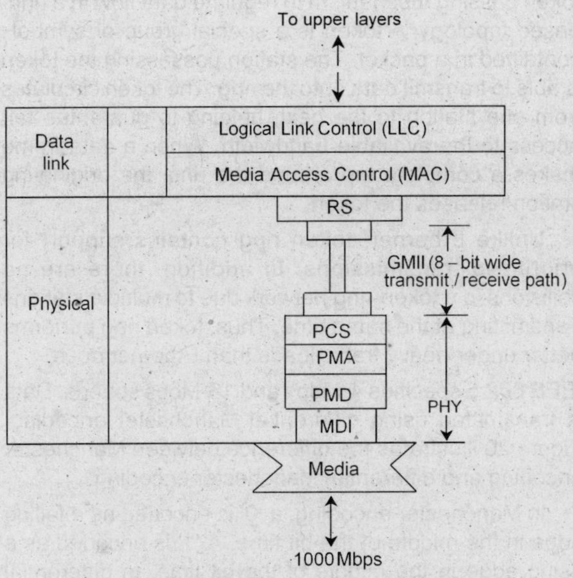

Fig. 18 : Gigabit Ethernet architecture

1000 base T

. IEEE standard 802.3ab

. Cable Type : 4 pairs of Category 5 (or higher) UTP

. Maximum cable length : 100 meters

RS	Reconciliation Sublayer
PCS	Physical Coding Sublayer
PMA	Physical Medium Attachment
PHY	Physical Layer Device
PMD	Physical Medium Dependent
GMII	Gigabit Media Independent Interface
MDI	Medium Dependent Interface

(a) Data rate: 1000 Mbps (2000 Mbps full duplex)

(b) PAM 5x5 encoding (c) Connector: RJ-45

The two-dimensional PAM5x5 encoding we were introduced to in l00baseT is extended here to a four-dimensional PAM5x5 system.

1000 baseCX

Short haul copper

Cable: shielded, balanced copper (twinax)

Maximum cable length: 25 meters

Data rate: 1000 Mbps (2000 Mbps full duplex) 8B10B encoding

Connector : 9-pin D or 8-pin Fibre Channel Type 2 (HSSC)

Transferring 8-bits of data reliably at Gigabit speeds requires another change in the method used to encode the data. The 8B10B coding method (originally developed for Fibre Channel) replaces 8-bit data values with 10-bit code words. The code words are chosen from groups such that the number of 0s and 1s transmitted is kept in balance. A signaling rate of 1.25 Gbps is required to encode the 1 Gbps data stream.

1000baseSX

1. Short wavelength laser
2. Wavelength: 770-860 nanometers
3. Data rate: 1000 Mbps (2000 Mbps full duplex)
4. 8B10B encoding
5. 62.5/125 multimode fiber length limit : 275 meters (half and full duplex)
6. 50/125 multimode fiber length limit: 550 meters (full-duplex), 316 meters (half duplex)
7. Connector : SC

1000 base LX

1. Long wavelength laser
2. Wavelength: 1270-1355 nanometers
3. Data rate : 1000 Mbps (2000 Mbps full duplex)
4. 8B10B encoding
5. Single-mode fiber length limit: 5000 meters (full duplex), 316 meters (half duplex)
6. Multimode fiber length limit: 550 meters (full duplex), 316 meters (half duplex)
7. Connector: SC

Table summarizes the 10-, 100-. and 1000 Mbps Ethernet copper and fiber technologies. Even with this large number of technologies, there is still one more to examine: wireless Ethernet.

WIRELESS ETHERNET

Wireless Ethernet, the use of Ethernet over radio frequency (RF) or Infrared (IR), is covered by the IEEE 802.11 Wireless LAN Standard. A wireless Ethernet network consists of one or more fixed stations (base stations) that service multiple mobile stations. Some implementation details are as follows:

- Frame formats for IEEE 802.3 (Ethernet) and IEEE 802.5 (token ring) remain the same.
- CSMA/CA utilized.
- 1-Mbps, and 2-Mbps operation. Faster speeds are becoming available.

CSMA/CA stands for Carrier Sense Multiple Access with Collision Avoidance. This method differs from CSMA/CD in that the wireless transceiver cannot listen to Fiber is the network for other transmissions while it is transmitting. Its transmitter simply drowns out any other signal that may be present. Instead, stations attempt to avoid collisions by using random backoff delays to delay transmission when the network is busy (when some other station is transmitting). A handshaking sequence is used between communicating stations (ready and acknowledge packets) to help maintain reliable delivery of messages over the air.

Types of Wireless LANs

There are two primary types of Wireless Ethernet LANs. They are RF-based and IR-based.

RF-based Signals can propagate through objects, such as walls, reducing security. The ISM band (industrial, scientific, medical) is used for transmission at the following frequencies:

Industrial: 902 to 928 MHz

Scientific: 2.40 GHz to 2.4835 GHz

Medical: 5.725 GHz to 5.850 GHz

Data is transmitted using the spread spectrum technologies frequency hopping and direct sequence. In frequency hopping, the transmitter hops from frequency to frequency, seemingly at random, transmitting a portion of each frame at each frequency. The receiver hops to the same frequencies, using the same pseudo-random sequence as the transmitter. A measure of security is added to the data, since it is difficult to eavesdrop on all the associated frequencies and reassemble the frame fragments.

The Direct Sequence method involves exclusive-ORing a pseudo-random bitstream with the data before transmission. The same pseudorandom bit sequence is used in the receiver to get the original data back.

IR-based Two types of IR wireless Ethernet are used: diffused IR and point to point IR. Diffused IR bounces signal off walls, ceilings, and floors. The data rate is limited by the multipath effect, whereby multiple signals radiate from a single transmission, each taking a different path to the receiving stations. This is illustrated in Figure 19

Point-to-point IR uses line of sight IR lasers and provides a faster data rate than diffused IR. It also works over larger distances (up to 1 mile).

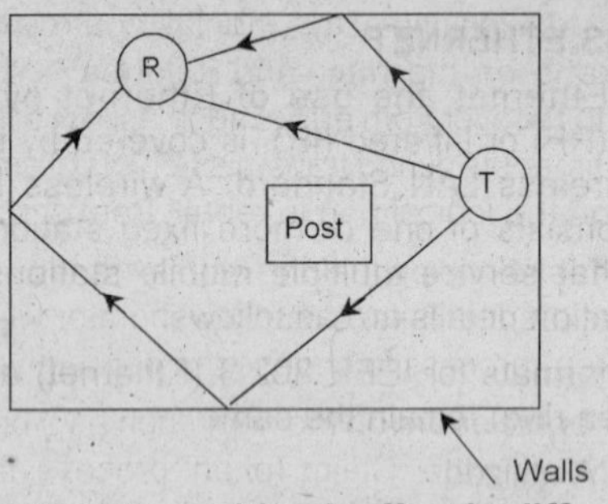

Fig. 19 : The multipath effect in diffused wireless Ethernet

TOKEN RING

In this, we take a survey of several additional networking technologies. Even though Ethernet is by far the most popular, each of the LAN technologies examined in this chapter have their own particular advantages and uses. In order, we will look at the following types of networks:

- Token ring
- Token bus
- FDDI
- Broadband LAN
- ARCnet

When you add Ethernet and wireless Ethernet LANs to the list, you have an impressive number of LAN technologies to choose from.

Currently, token ring is the second most popular networking technology after Ethernet. A token-ring network, specified by the IEEE 802.5 standard, uses a token passing mechanism to regulate data flow in a ring-based topology. A token is a special group of symbols contained in a packet. The station possessing the token is able to transmit data onto the ring. The token circulates from one station to the next, helping to guarantee fair access to the available bandwidth. When a data frame makes a complete trip around the ring, the originating station releases the token.

Unlike Ethernet, token ring contains support for prioritized transmissions. In addition, there are no collisions in a token-ring network due to multiple stations transmitting at the same time. Thus, token ring performs better under heavy traffic loads than Ethernet does.

IEEE 802.5 specifies 4-Mbps and 16-Mbps speeds. Data is transmitted using differential Manchester encoding. Figure 20 illustrates the difference between Manchester encoding and differential Manchester encoding.

In Manchester encoding, a '0' is encoded as a falling edge in the middle of the bit time. A '1' is encoded as a rising edge in the middle of the bit time. In differential Manchester encoding, there is always a transition in the middle of the bit time. A transition at the beginning of a bit interval indicates a '0.' No transition indicates a '1'. An advantage of differential Manchester encoding is that the polarity of the signal is independent of the bitstream. In other words, inverting a differential Manchester signal will result in the same received data, whereas inverting an ordinary Manchester encoded signal will invert all the data bits.

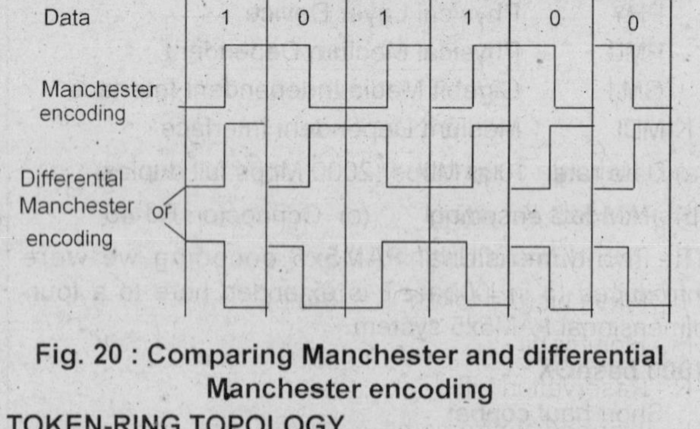

Fig. 20 : Comparing Manchester and differential Manchester encoding

TOKEN-RING TOPOLOGY

Though the logical topology of a token-ring network is a ring, the physical topology is a combination of star and ring. As shown in Figure 21 stations (or lobes) in a token-ring network connect to an MAU (multistation access unit). Multiple MAUs are connected serially to form a ring. Shielded twisted pair (STP) is typically used for wiring, although UTP and fiber arc also supported. The number of MAUs in the ring affects the maximum distance between MAUs. IEEE 802.5 limits the number of stations on any ring to 250 and the number of MAUs to 33. Stations can be added or removed without major disruption to the ring.

TOKEN-RING FRAME FORMATS

Figures 22 and 23 illustrate the formats of the token and information frame utilized by token ring. The token format in Figure 22 is a 24-bit frame generated by the active monitor (described in the following section). Notice that the Access Control field contains Priority and Reservation bits. The Priority bits indicate the current priority of the token and any associated information frames. The Reservation bits are used to reserve the next token at a specific priority. The Token bit is used to differentiate between Token frames and Information frames.

The Monitor bit helps identify a frame that is stuck circulating on the ring.

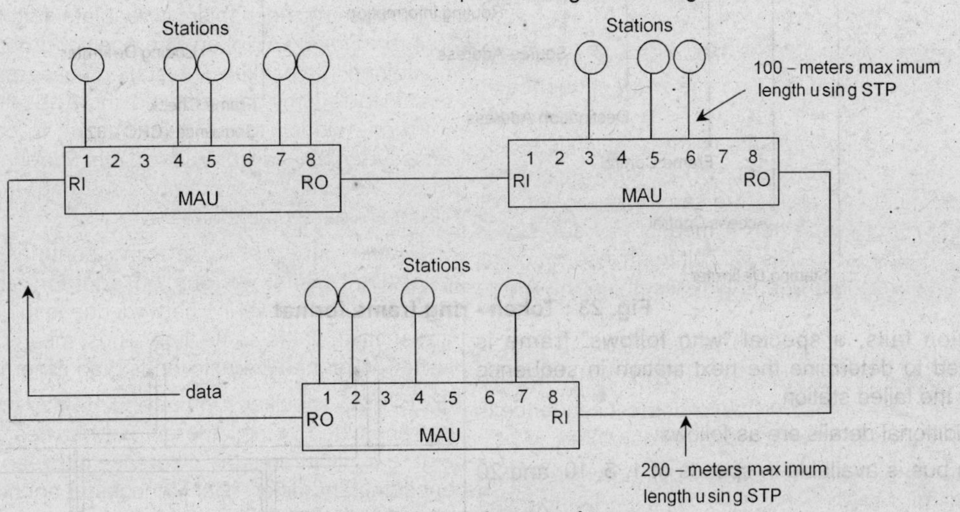

Fig. 21 : Token - ring topology

The J and K symbols are illegal differential Manchester codes (no transition in the middle of the bit time) used to identify the Starting and Ending delimiters.

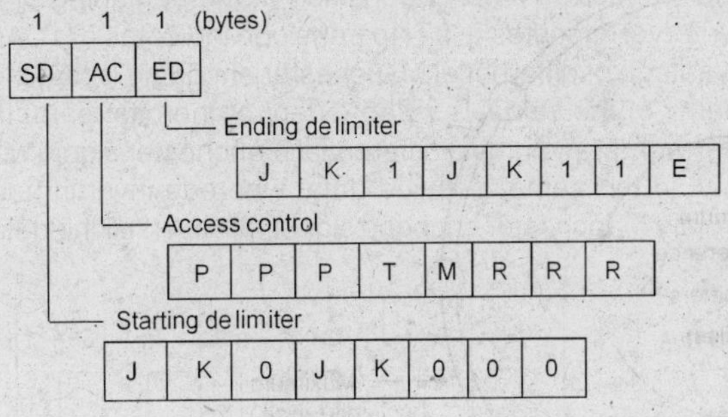

Key :

J - Nondata J symbol

K - Nondata K symbol

P - Priority bits (111 highest)

T - Token bit (0 = token, 1 = frame)

M - Monitor bit

R - Reservation bits

I - Intermediate frame bit

E - Error-detected bit

Fig. 22 : Token-ring token format

The Information frame diagramed in the below Figure 23 contains many of the same fields as the Ethernet frame. One significant difference is the length of the INFO field, which may be 4550 bytes in length in 4-Mbps token ring, or 18,200 bytes for 16-Mbps token ring. Compare this with Ethernet's 1500-byte limit (for all Ethernet speeds).

RING MANAGEMENT

One station on the ring acts as the active monitor and is responsible for ensuring that valid tokens circulate on the ring. In addition, the active monitor controls the timing of the tokens on the ring, and initiates a ring purge operation to recover from errors such as lost or corrupted tokens, or timeouts during transmission of a frame. During a ring purge the logical ring is rebuilt from scratch.

All other stations on the ring act as standby monitors. If the active monitor fails, a standby monitor will initiate a "Claim Token" process to choose the new active monitor.

Catastrophic errors on the ring, such as a broken ring or an error during the "Claim Token" process, initiate the beaconing procedure. Beacon frames are then circulated to determine the nature of the fault (the fault domain) and possibly isolate the cause of the failure.

TOKEN BUS

The IEEE 802.4 token-bus standard describes the mechanism by which multiple stations share access to a common bus. Figure 24 shows a sample token-bus network. Each station knows the address of the previous and next stations in the sequence of stations. Tokens are circulated in a logical ring, from station 10 to stations 8, 7, 5, and finally 3. Station 3 transmits back to station 10 to complete the logical ring. Note that the physical location of the station on the bus does not affect the sequence.

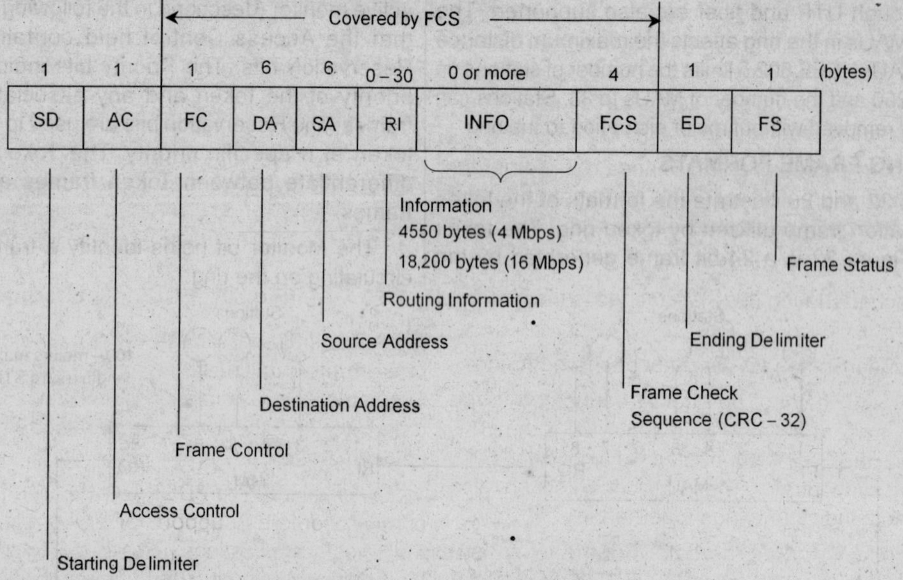

Fig. 23 : Token - ring frame format

If a station fails, a special "who follows" frame is transmitted to determine the next station in sequence following the failed station.

Some additional details are as follows:

- Token bus is available in speeds of 1, 5, 10, and 20 Mbps
- Data frames may contain over 8000 bytes of information
- Four levels of priority are available
- Operates over coax and fiber

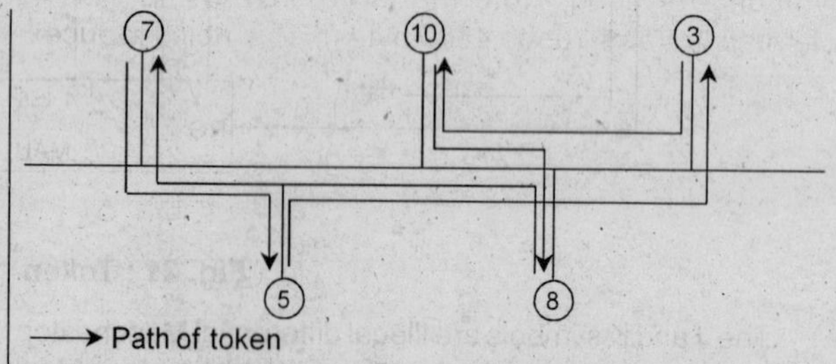

Fig. 24 : Token - bus operation

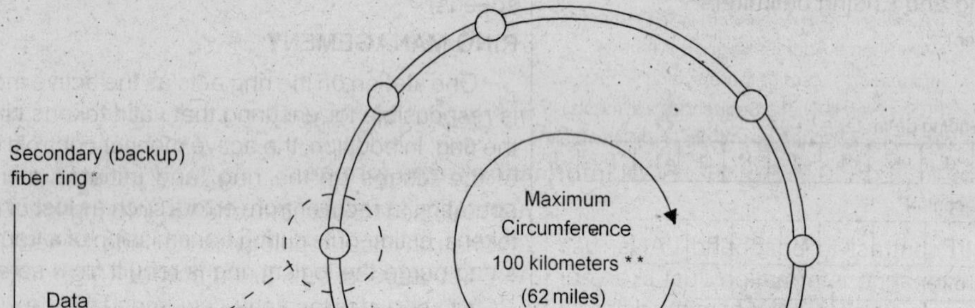

Fig. 25 : FDDI topology

FDDI

FDDI (fiber-distributed data interface) is a self-healing baseband network technology similar to token ring. As indicated in Figure 25 an FDDI network consists of nodes connected together with dual fiber rings. Both rings, called primary and secondary (or backup), operate at 100 Mbps using 4B/5B encoding. An error detected with the primary ring will cause an instant changeover to the secondary ring to prevent loss of communication.

Using 4B/5B encoding, 4 bits of data are encoded as a 5-bit symbol. Bits are transmitted over the fiber as transitions of light (off-to-on or on-to-off). LEDs operating at 1300 nanometers provide the light source for the fiber. 62.5/125 micrometer multimode fiber is typically used.

The large circumference of an FDDI ring (100 kilometers for dual ring, 200 kilometers for single ring) makes it ideal for use over a large geographical area.

FDDI nodes come in two types:

- Single Attached Station (SAS). Attaches to the primary ring only.
- Dual Attached Station (DAS). Attaches to both rings.

Note that an SAS may not be directly connected to the primary ring, since the entire ring would fail if the SAS failed. Instead, the SAS is connected via a modified DAS called a wiring concentrator. These concentrators are also used to connect individual rings together.

Figures 26 and 27 provide the details of the token and information frames used over the FDDI ring 4B/5B symbols are used. Note the large size of the FDDI information frame (up to 4500 bytes). This helps reduce overhead

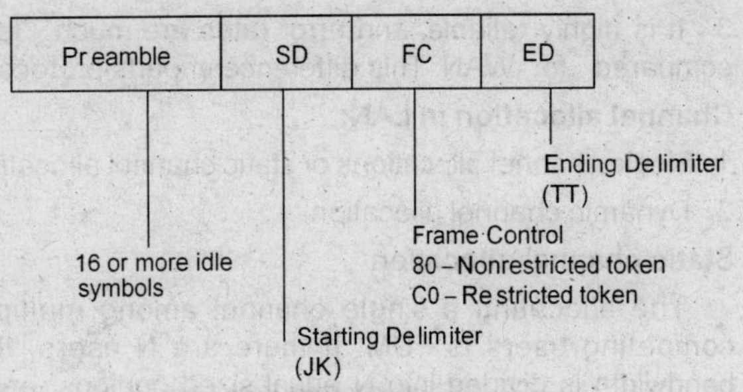

Fig. 26 : FDDI token format

when transmitting a large block of data that must be broken into multiple frames. The 4500-byte limit is a function of the maximum circumference of the ring, the data rate, and the allowable variation in clock speeds between nodes.

A newer variation of FDDI called FDDI-II(or hybrid ring control) contains support for voice, video, and other multimedia applications. This is accomplished through the addition of circuit-switching capability to the original packet-switching technique.

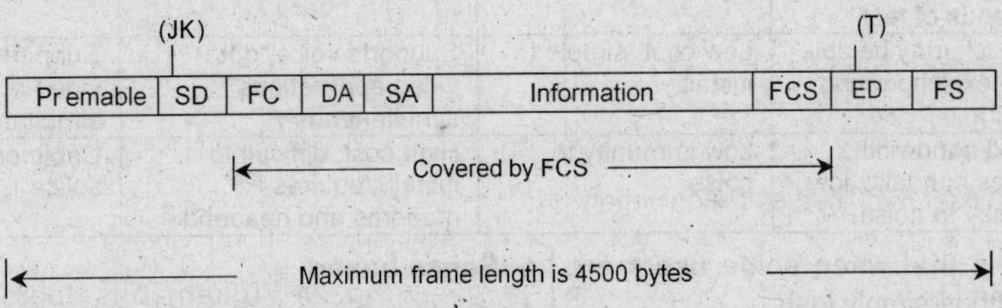

DA – Destination Address (4 or 12 symbols)
SA – Source Address (4 or 12 symbols)
Information – Zero or more symbols
FCS – Frame Check Sequence (8 symbols)
FS – Frame Status (3 or more R or S symbols)

Fig. 27 : FDDI information frame format

BROADBAND LANS

Broadband LANs, in contrast to baseband LANs, use multiple carriers to exchange information Just like your cable television allows you to select one of many different channels, broadband LAN technology divides the available bandwidth of the cable system into multiple channels (frequency division multiplexing). A head-end is used to maintain communication with each node on the broadband network. The broadband information signal is varied in both amplitude and phase to encode the digital data. One 6-MHz channel is capable of providing 27 Mbps using encoding technique called 64 QAM (Quadrature Amplitude Modulation).

The IEEE 802.7 Standard describes broadband LAN operation. Broadband LAN cable technology is becoming increasingly popular in the home computing market due to its high speed compared to ordinary telephone modems.

ARCnet

ARCnet (Attached Resource Computer Network) is a low-cost baseband network developed by the Datapoint Corporation in the late 1970s. Like token ring and FDDI, ARCnet uses tokens to exchange information. However, ARCnet physical topology is not ring based. Instead, ARCnet uses a bus architecture (high-impedance ARCnet) or a star topology (low-impedance ARCnet).

Nodes in an ARCnet network are addressed from 1 to 255 and organize themselves into a logical ring (not a physical ring). Address 0 is reserved for broadcasts. Nodes can be added or removed at will, their presence or absence automatically detected and accounted for communication speed is 2.5 Mbps (with a newer ARCnet Plus standard providing 20 Mbps). Coax, UTP, and fiber can all be used to connect ARCnet components. Both passive and active hubs are used to extend the size of the network.

LAN

The advantages of LAN

1. The superior price/performance ratio of a network of workstations is the attraction.

2. LAN designers have their own high bandwidth cable.

3. It is highly reliable, and error rates are much less compared to WAN. This difference imparts protocols.

Channel allocation in LAN:

1. Single channel allocations or static channel allocation
2. Dynamic channel allocation.

Static channel allocation

The allocating a single channel among multiple competing users is FDM. If there are N users, the bandwidth is divided into N equal sized portions, each user being assigned one portion. So, each user has his own private frequency band, there is no interference between users.

When the number of stations is large and continuously varying, or the traffic is burstly, FDM presents some problems. If more than N users want to communicate, some of them will be denied permission, for lack of bandwidth, even if some of the users who have been assigned a frequency band hardly ever transmit or receive anything.

Table : Technical characteristics of LANs

Characteristic	Transmission medium			
	Twisted pair wire	Baseband coaxial cable	Broadband coaxial cable	Fiber optic cable
Topology	Bus, star or ring	Bus or ring	Bus or ring	Bus, ring or star
Channels	Single channel	Single channel	Multi channel	Single, multi channel
Data rate	Normally up to 16 Mbps, with 100 and 155 Mbps obtainable at distance up to 100 meters	Normally 2 to 10 Mbps, up to 100 Mbps obtainable	Up to 400 Mbps	Up to Gbps
Maximum Nodes on net	Usually < 255	Usually < 1024	Several Thousand	Several Thousand
Geographical Coverage	In hundreds or thousands of feet	In miles	In tens of miles	In tens of miles
Major advantages	Low cost, may be able to use existing wiring	Low cost, simple to install	Supports voice, data video applications simultaneously	Supports voice, data, video applications simultaneously
Major disadvantages	Limited bandwidth, requires conduits, low immunity to noise	Low immunity to noise	High cost, difficult to install, requires RF modems and headend	Cable cost, difficult to splice

The basic problem is that when some users are quiescent, their bandwidth is simply lost.

Dynamic Channel Allocation: Five keys assumptions of this area arc:-

1. Station model: The model consists of N independent stations, each with a program or user that generates frames for transmission. Once a frame has been generated, the station is blocked and does nothing until the frame has been successfully transmitted:

2. Single Channel Assumption: A single channel is available for all communication. All stations can transmit on it and all can receive from it. All the stations are equivalent, although protocol software may assign priorities to them.

3. Collision Assumption: If two frames are transmitted simultaneously, they overlap in time and the resulting signal is garbled. This event is called a collision. All stations can detect collisions. A Collided frames must be transmitted again later. There are no errors other than those generated by collisions.

4. Time base

1. Continuous Time: Frame transmission can begin at any instant. There is no master clock dividing time into discrete intervals.

2. Slotted Time: Time is divided into discrete intervals. Frame transmissions always begin at the start of a slot. A slot may contain 0, 1 or more frames, corresponding to an idle slot a successful transmission or a collision, respectively.

5. Sense based

1. Carrier Sense: Stations can tell if the channel is in use before trying to use it. If the channel is sensed as busy, no station will attempt to use it until it goes idle.

2. No carrier Sense: Stations cannot sense the channel before trying to use it. They just go ahead and transmit. Only later can they determine whether or not the transmission was successful.

In some systems, when a collision occurs, stronger signals can still be detected and this is known as Capture effect.

A network can either have carrier sensing or not, but not both. LANs generally have carrier sense, but satellite networks do not (due to long propagation delay). Stations on carrier sense networks can terminate their transmission prematurely if they discover that it is colliding with another transmission.

ALOHA Protocol: ALOHA protocol systems were used to save the channel allocation problems in ground-based radio broadcasting and the basic idea is applicable to any system in which uncoordinated - users are competing for the use of a single shared channel.

Pure ALOHA and slotted ALOHA

The basic idea of an ALOHA system is: Let users transmit whenever they have data to be sent. There will be collisions and the colliding frames will be destroyed. If the frame was destroyed, the sender just waits a random amount of time and sends it again. Systems in which multiple users share a common channel in a way that can lead to conflicts are widely known as contention system. The frames are made with same length because the throughput of ALOHA system is maximized by having a uniform frame size rather than allowing variable length frames.

Whenever two frames try to occupy the channel at the same time, there will be a collision and both will be garbled. If the first bit of a new frame overlaps with just the last bit of a frame almost finished, both frames will be totally destroyed and both will have to be retransmitted later. The checksum cannot distinguish between a total loss and a near miss.

Let, the 'frame time' denote the amount of time needed to transmit the standard; fixed length frame (i.e., the frame length divided by the bit rate). In this, the infinite population of users generates new frames according to a Poisson distribution with mean S frames per frame time. If $S > 1$, the user community is generating frames at a higher rate than the channel can handle, and nearly every frame will suffer a collision. So throughput $0<S<1$. It is called pure ALOHA.

In slotted ALOHA, a terminal is not permitted to send whenever a carriage return is typed. Instead it is required to wait for the beginning of the next slot. Thus the continuous pure ALOHA is turned into a discrete one. Since the vulnerable period is now reduced in half, the probability of no other traffic during the same slot frame is e^{-G} which leads to $S=Ge^{-G}$

With slotted ALOHA, the best channel utilization that can be achieved is 1/e

LAN Protocols: A protocol is a set of rules for communicating among computers. These rules govern format, timing, sequencing, and error control. Protocol in which stations listen for a carrier and act accordingly are called carrier sense protocols

The first carrier sense protocol is 1-persistent CSMA (Carrier Sense Multiple Access). When a station has data to send, it first listens to the channel to see if any one else is transmitting.

If the channel is busy, the station waits until it become idle. When the station detects an idle channel, it transmits a frame. If a collision occurs, the station waits a random amount of time and starts all over again. The protocol is called 1-persistent because the station transmits with a probability of 1 whenever it finds the channel idle. A second carrier sense protocol is non-persistent CSMA. In this, a conscious attempt is made to be less greedy. Before sending, a station sense the channel. If no one else is sending, the station begins doing so itself. If the channel is already in use, the station does not continually sense it for the purpose of seizing it immediately upon detecting the end of the previous transmission. Intuitively, this algorithm should lead to be better channel utilization and longer delays than 1-persistent CSMA.

The last protocol is p-persistent CSMA. It applies lo slotted channels. When a station becomes ready to send, it senses the channel. If it is idle, it transmits with a probability p. With a probability q =1 - p, it defers until the next slot. If that slot is also idle, it either transmits or defers again, with probabilities p and q. This process is repeated until either the frame has been transmitted or another station has begun transmitting.

CSMA with collision detection: If two stations sense the channel to be idle and begin transmitting their frames, which are irretrievably garbled anyway, they should abruptly stop transmitting as soon as the collision is detected. Quickly terminating damaged frames saves time and bandwidth. This protocol is CSMA/CD.

In this, it allocate alternating contention and transmission periods, with idle periods occurring when all stations are quiet.

In this, collision detection is an analog process. The station's hardware must listen to the cable while it is transmitting. If what it reads back is different from what it is putting out, it knows a collision is occurring. The implication is that the signal encoding must allows collision to be detected. In this, Manchester encoding is used.

Collision - free protocols : The collision do not occur with CSMA/CD once a station has unambiguously seized the channel, they can still occur during the contention period. These collisions adversely affect the system performance, especially when the cable is long and the frames short.

The first collision-free protocol is basic bit-map method, each contention period consists of exactly N slots. If station O has a frame to send, it transmits a 1 bit during the first slot. No other station is allowed to transmit during this slot. Then station 1 get a chance to transfer the frame if it wants to transfer. In this, they begin transmitting in numerical order. Everyone agrees on who goes next, there will never be any collisions. If a station becomes ready just after its bit slot has passed by, it is out of luck and it should wait upto next chance.

Drawbacks

1. Asymmetry with respect to station number. Higher numbered stations get better service than lower numbered ones. High numbered stations have to wait on an average 0.5 N slots, while low numbered stations have to wait on an average 1.5 N slot.

2. Light lead a station must always wait for the current scan to be finished before it may transmit. The mean for all stations is N slots. The channel efficiency at low load is $d/(N + d)$.

Where N bits is the overhead per frame and d bit is the amount of data.

At high load, when all stations have something to send all the time, the N bit contention period is prorated over N frames. This yields an overhead of 1 bit per frame or, an efficiency of $d/(d + 1)$. These drawbacks are overcome in Broadcast Recognition with Alternating Priorities (BRAP).

BRAP (Broadcast Recognition with Alternating Priorities)

In this, a station insert a 1 bit into in slot, it begins transmission of in frame immediately. In addition, instead of starting the bit scan with station O each time, it is started with the station following the one that just transmitted.

The permission to send rotates among the stations in a round-robin fashion. Any station not wishing to send, lets its bit slot go idle.

The problem in BRAP is, the channel utilization is poor for lightly loaded. It is excellent for high load.

When no stations are ready, there are no data frames, and the N-bit headers just go on and on until some station insert a 1 bit in its bit slot. On the average, a station will have to wait N/2 bit slot before it may begin sending.

MLMA (The Multi-Level Multi-Access)

Protocol: The drawback of BRAP is overcomes in MLMA. In this method, a station announces that it want to send by broadcasting its address in a particular format. An address in this system consists of three decimal digits, each decimal digit represented by a group of 10 bits called a 'decade' for e.g., 472 is represented by setting bit 4 in the first decade, bit 7 in the middle decade and bit 2 in the last decade.

If only one station attempts to transmit during a frame slot, it uses the 30-bit header to announce itself and then it sends its frame. If two or more stations try to insert their addresses into the same header, the station behaves like: the first decade in every frame slot corresponds to the hundreds place in the station number. After the first decade is finished, stations that have not transmitted a bit must remain silent until all the stations that did set a bit have transmitted their data. Call the highest occupied bit function in the first decade x. In the second decade, all stations with x as their leading digit announce their ten's place. Call the highest occupied bit here y. In the third decade, all the stations whose addresses begin with by may set the bit corresponding to their last digits. These are atmost 10 of them.

The Channel efficiency is d/(d + 30) for minimum load for heavy load d/(d+ 1.11).

The limiting case of MLMA consist of having 1 bit at each level for N stations, the number of levels needed is log2 N rounded upward to an integer.

Binary countdown: A station writes in address into the header as a binary, that protocol is called binary countdown. In this, a high-order bit function that is 0 in its address has been overwritten with a 1, it gives up. After the winning station has transmitted in address, there is no information available telling how many other stations want to send. The channel efficiency is d/(d + ln N). The Channel efficiency in binary countdown is better than decimal MLMA when there are many bursty stations, but slightly less under a full load.

The binary countdown use parallel interface. They use virtual station numbers, with the virtual station numbers from 0 upto and including the successful station being circularly permuted after each transmission, in order to give higher priority to stations that have been silent unusually long.

Limited-contention protocol: Combination of best properties of the contention and collision-free protocols, arriving at a new protocol that used contention at low loads to provide low delay, but used a collision-free technique at high load to provide good channel efficiency. Such protocol is called limited-contention protocol.

The adaptive tree walk protocol: In this, all stations are organized in binary form. The first contention slot following a successful frame transmission, slot 0, all stations are permitted to try to acquire the channel. If one of them does so fine. If there is a collision, then during slot 1 only those stations falling under node B may compete. If one of them acquires the channel, the slot following the frame is reserved for those stations under node C. If, on the other hand, two or more stations under node B want to transmit, there will be a collision during slot 1, in which case it is node D's turn during slot 2.

If a collision occurs during slot 0, the entire tree is searched depth first, to locate all ready stations. Each bit slot is associated with some particular node in the tree. If a collision occurs, the search continuous recursively with the left and right children of that node. If a bit slot goes idle or if there is exactly one station that transmit into it, the searching of in node can stop, because all ready station have been located.

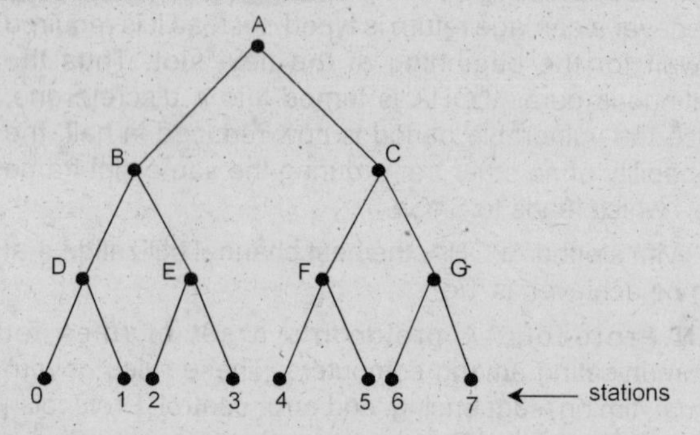

Fig. 28 : The tree for eight stations

LAN

IEEE (The Institute of Electrical and Electronic Engineers, USA) has produced several standards for LANs. These standards known, as IEEE 802, include CSMA/CD, token bus and token ring. The various standards differ at the physical layer and MAC sublayer, but are compatible at the data link layer. The standards are divided in parts, each published as a separate book.

Table: IEEE Series 802 committees

802.1	Architecture and Overview
802.2	Logical Link Control
802.3	CSMA/CD
802 3 μ	Fast Ethernet
802.4	Token Bus
802.5	Token-Ring
802.6	Metropolitan Area Networks
802.7	Broadband Technical Advisory Group
802.8	Fiber Optic Technical Advisory Group
802.9	Integrated Voice and Data Networks
802.10	Network Security
802.11	Wireless LANs
802.12	100VG Any LAN

The IEEE 802.1 standard gives an introduction to the set of standards and defines the interface primitivies (Architecture and Overview). 802.2 standard describes the upper part of the data link layer which uses the logical link control protocol. 802.3. to 802.5 describe the three LAN standard protocols, CSMA/CD token bus and token ring. Each standard cover physical and MAC layer protocol.

IEEE 802.3 and Ethernet: The IEEE 802.3 standard is for a 1-persistent CSMA/CD LAN. When a station wants to transmit, it listens to the cable. If the cable is busy, the station waits until it goes idle, otherwise it transmits immediately. If two or more stations simultaneously begin transmitting on an idle cable, they will collide. All colliding stations then terminate their transmission, waits a random time, and repeat the whole process all over again. This CSMA/CD protocols calls Ethernet. Generally, Ethernet refers to the cable. There are two types of cables, (1) Thick Ethernet (2) Thin ether net.

Thick Ethernet resembles a yellow garden host, with markings every 2.5 m to show where the taps go. Thin ethernet is smaller and more flexible and it uses industry standard BNC connectors to form T junctions, rather than using taps and it is much cheaper, but it can only run for short distances. In Ethernet network, the number of connections (taps) and their intervening distances are limiting factors. Repeaters generate the signal every 500 m or so. If, repeaters were not used, standing waves (additive signal reflections) would distort the signal and cause errors. Because collision detection depends partly on timing, only five 500 m segments and 4 repeaters can be placed in a series before the propagation delay becomes larger than the maximum allowed period for the detection of a collision otherwise the work stations farthest from the centre would be usable to determine whether a collision had occurred.

Table reveals the distance limitations of different kinds of LAN cable.

Manufactures of Ethernet products have made it possible to create Ethernet networks in star, branch and tree designs that overcome the basic limitations mentioned, upon LAN have 1000'8 of work stations on a complex Ethernet network.

The two types are compatible and can be connected in certain ways.

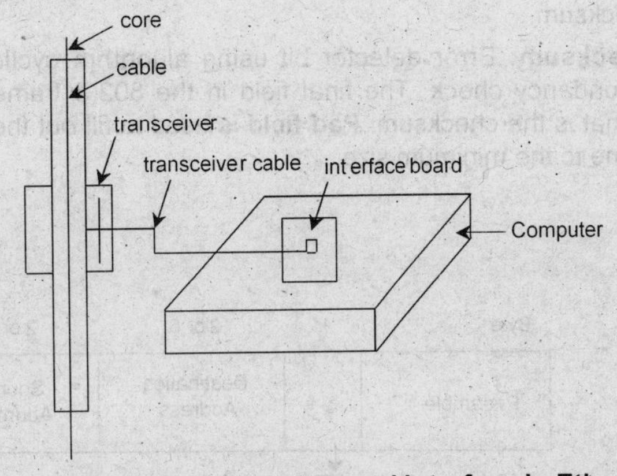

Function of the transceiver and interface in Ethernet

In the Ethernet configuration, transceiver clamped securely onto the cable so that its tap makes contact with the inner core. The transceiver contains the electronics that handle carrier detection and collision detection. When a collision is detected, the transceiver also puts a special invalid signal on the cable to insure that all other transceivers also realize that a collision has occurred.

A transceiver cable connects the transceiver to an interface board in the computer. It have five individual shielder twisted pairs. Two of the pairs are for data in and data out. Two more are for control signals in and out. The fifth pair, allows the computer to power the transceiver electronics.

Table : Network Distance Limitations

Network Adapter	Cable Type	Maximum	Minimum
Ethernet		607 ft.	20 in.
	Thin	164 ft.	8 ft
	Thick (drop cable)	1,640 ft.	8 ft
	Thick (backbones)	328 ft.	8 ft
Token – Ring	UTP	328 ft.	8ft
	STP	148 ft.	8ft.
ARCnet (passive hub)	UTP	393 ft.	depends on cable
ARCnet (active hub)		1,988 ft.	depends on cable

The transceiver cable terminates on an interface board inside the computer. The interface board contains a controller chip that transmit frames to and receives frames from, the transceiver. The controller is responsible for assembling the data into the proper frame format, as well as computing checksums on outgoing frames and verifying them on incoming frames. Some controller chips also manage a pool of buffers for incoming frames, a queue of buffers to be transmitted, DMA transfers with the host computers, 1 and other aspects of network management.

The maximum cable length is 500 m. If it is mere, multiple cables can be connected by repeaters. A repeater receives, amplifies and retransmits signal in both directions.

Cable network organization is a collection of separate segments connected by bridges (also called selective repeaters). Bridges examine each frame and only forward those that need to reach the other segment. Bridges have to know the locations of all stations, in order to know whether or not to copy a frame onto a given segment.

Start of frame: The frame is 10101011, to denote the start of the frame itself.

Source and destination address: It allows 2 or 6 bytes. The high-order bit of the destination address is a 0 for ordinary addresses and 1 for group addresses. Group address for multicast-single bit is reserved for broadcast.

Length of Data: These 2 bytes express the length of the data portion of the frame. **Data:** It is number of bytes are present in the data field, from 0 - 1500. If it is 0 means, data are collide. 802.3 states that valid Games must be atleast 64 bytes long, from destination address to checksum.

Checksum: Error detector bit using algorithm cyclic redundancy check. The final field in the 802.3 frame format is the checksum. **Pad field** is used to fill out the frame to the minimum size.

The pad field contains enough bytes of files to ensure that the frame has at least a certain overall size. If the data portion is large enough, the pad field does not appear in the frame (pad has zero length).

Channel efficiency:

$$\text{Channel efficiency} = \frac{1}{1 + 2BLe/cf}$$

f - frame length
B - network bandwidth
L cable length
C - speed of signal propagation
e - contentions slots/frame
p - mean time to transmit a frame
p - f/B

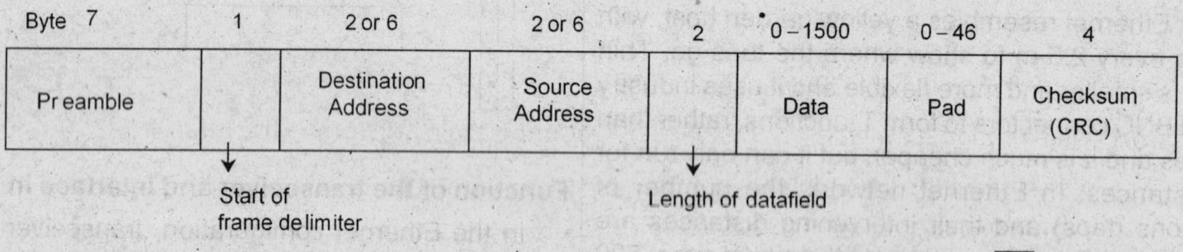

Preamble : 7 bytes - bit pattern is 10101010

It allow the receiver's clock to synchronize with the sender's. The Manchester encoding of this pattern produces a 10 MHz square wave for 5.6 μ secs.

So channel efficiency = $\dfrac{p}{p + 2\tau / A}$

where 2τ is the duration of each slot.

A - slot probability

The probability that some station acquires the channel in that slot.

Increasing network bandwidth or distance (the BL product), reduces efficiency for a given frame size.

IEEE Std 802.4: Token bus: In 802.3, frames do not have priorities (making them unsuited for real time systems) in which important frames should not be held up waiting for unimportant frames.

The token bus is a linear or tree-shaped cable onto which the stations are attached. The stations are organised into a ring with each station knowing the address of the station to its left and right.

802.4 - Protocol Frame Format :

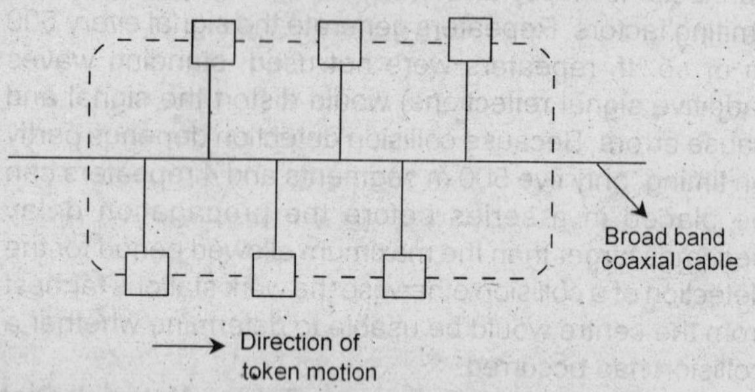

When the logical ring is initialized, the highest numbered station may send the first frame. After it is done, it passes permission to its immediate neighbour by sending the neighbour a special control frame called a 'token'. The token propagates around the logical ring, with only the token holder being permitted to transmit frames. Since only one station at a time holds the token, collisions do not occur.

When a station passes the token, it sends a token frame specifically addressed to its logical neighbour in the ring, irrespective of where that station is physically located on the cable. When stations are first powered on, they are not in the ring, the MAC protocol has provisions for adding to, and deleting from, the ring.

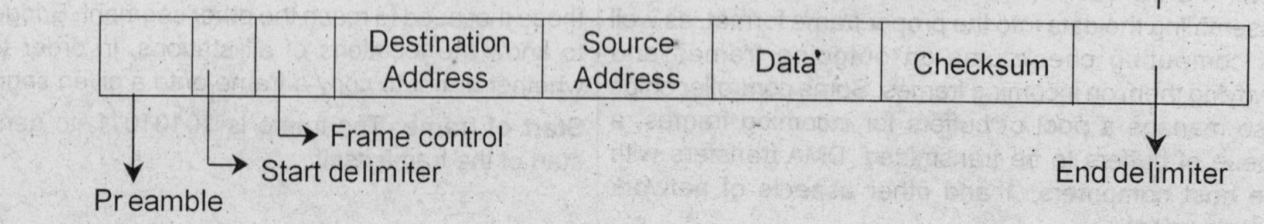

When the ring is initialized, stations are inserted into it, in order of station address, from highest to lowest. Token passing is done from high to low addresses. Each time a station acquires the token, it can transmit frames for a certain amount of time, then it must pass the token on. If a station has no data, it passes the token immediately upon receiving it.

The token bus defines four priority classes 0, 2, 4 and 6 for traffic, with 0 the lowest and 6 the highest. Each station internally divided into four substations, one at each priority level. So each data-inputs are checked for priority and routed to one of the four substations. So each substation maintains its own queue of frames to be transmitted.

Preamble is used to synchronize the receiver's clock.

Staffing and Ending Delimiters are used to mark the frame boundaries.

Frame Control is used to distinguish data frames from control frames.

Data frame	-	frame's priority
Control frames	-	specify the frame type
Data	-	size of data
Checksum	-	to detect transmission errors
Destination and source address	-	starting and ending address of data

Token bus Control Frames

Claim-token	-	Claim token during ring initialization
Solicit-successor	- 1 -	Allow stations to enter the ring
Solicit-successor	-2 -	Allow stations to enter the ring
Who-follows	-	Recover from lost token
Resolve-Contention	-	Used when multiple stations want to enter the ring.
Token	-	Pass the token
Set-successor	-	Allow station to leave the ring.

802.5 - Token Ring: In principle, the token is not a frame but simply a means by which each work station can recognize when its turn to transmit has arrived.

A Ring consists of a collection of ring interfaces connected by point-to-point lines. Each bit arriving at an interface is copied into a 1-bit buffer and then copied out onto the ring again. While in the buffer, the bit can be inspected and possibly modified before being written out. This copying step introduces a 1-bit delay at each interface.

In a ring, a special bit pattern called the token, circulates around the ring whenever all stations are idle. When a station wants to transmit a frame, it is required to seize the token and remove it from the ring before transmitting. Only one token is available. So only one station can transmit at a given instant. So collision is avoided. The delay has two components, the 1-bit delay introduced by each station, and the signal propagation delay. Ring interfaces have two operating modes: listen and transmit. In listen mode, the input bits are simply copied to output, with a delay of 1 bit time. In transmit mode, which is entered only after the token has been seized, the interface breaks the connection between input and output, entering its own data onto the ring. In 1 bit time, to switch from listen to transmit mode, the interface usually needs to buffer one or more frames.

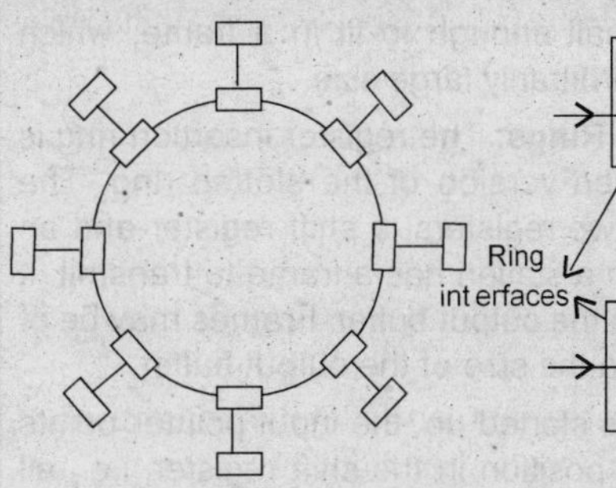

A ring network

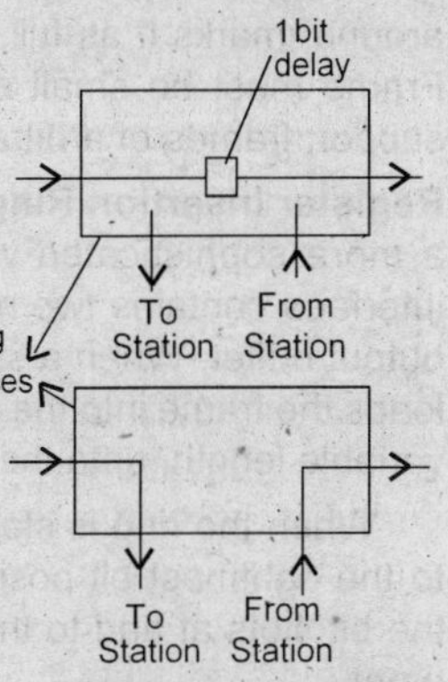

Transmit mode

The sending station can either save them, or discard them. The ring architecture puts no limit on the size of the frames. After a station has finished transmitting the last bit of its last frame, it must regenerate the token.

When traffic is light, the token will spend most of its time idly circulating around the ring. When the traffic is heavy, the stations are in queue.

Token Ring data frame (Length of each field in bytes)

Starting delimiter and ending delimiter-mark the beginning and ending of the frame.

Token Ring Protocol

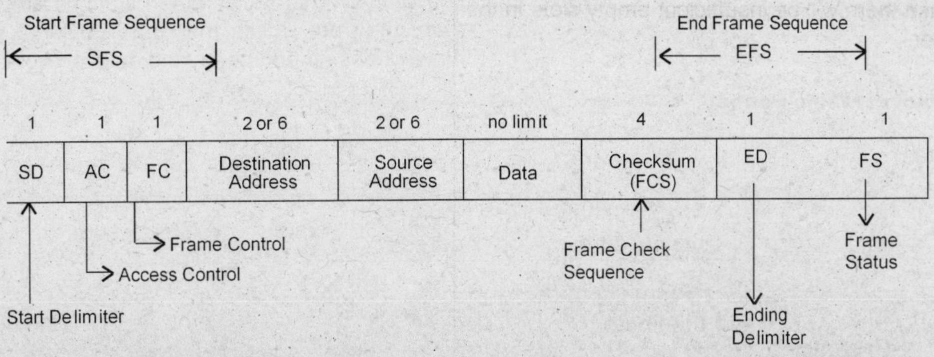

Frame Status byte is present in this protocol while in the earlier two protocols, they were not present.

It contains the A and C bits.

A = 0 and C = 0 : destination not present or not **powered up.**

A = 1 and C = 0 : destination present but frame not accepted.

A = 1 and C = 1 : destination present and frame copied.

This arrangement provides an automatic acknowledgment for each frame. If a frame is rejected but the station is present, the sender has the option of trying again. Frame status byte increases the reliability. Destination address, source address and checksum are similar to the previous protocol.

Slotted Rings: Slotted ring is slotted into a number of fixed size frames. Each frame contains a bit that tells whether it is full or empty. When a station wishes to transmit, it simply waits for an empty frame to come around, marks it as full, and puts its data in the frame. Frame must be small enough to fit in a frame, which support frames of arbitrarily large size.

Register Insertion Rings: The register insertion ring is a more sophisticated version of the slotted ring. The interface contains two registers, a shift register and an output buffer. When a station has a frame to transmit, it loads the frame into the output buffer. Frames may be of variable length, upto the size of the output buffer.

When the ring is started up, the input pointer points to the rightmost bit position in the shift register, i.e., all the bit slots at and to the left of where it is pointing are empty.

When a bit arrives from the ring, it is placed at the pointed to by the input pointer, and the pointer is moved 1 bit to the left. The address field of the frame has arrived, the interface can determine whether or not the frame is addressed to it. If the rest of the frame is diverted to the station, removing the frame from the ring, the input pointer is then reset to the extreme right.

An important property of ring architecture is the way it prevents a station from monopolizing the ring.

If the ring is idle, the shift register will be empty when an output frame has been finished, so the next output frame can be sent as soon as it can be loaded into the output buffer by the station. If the ring is busy, after sending a frame the station will not be allowed to send another one because there will be insufficient empty slots in the shift register.

III. SWITCHES

Introduction

In this topic we take a look at the routing protocols used to transport the higher-layer protocols between LANs (across a WAN). We will also explore the details of switching and make comparisons between hubs, switches, and routers.

Hubs Versus Switches

The essential difference between hubs and switches is that hubs broadcast frames received on one port to all other ports, whereas a switch will forward a received frame to a specific port. This is illustrated in Figure 1 in which a small network of six stations (A through F) are connected two different ways.

In Figure 1 (a), station A transmits a frame whose destination is station F. This does not matter to the 4-port hubs, which simply broadcast copies of the frame from station A to the other five stations. This amounts to a good deal of wasted bandwidth. Furthermore, all six stations operate in the same collision domain, making them compete for bandwidth.

Figure 1 (b) shows the same network with the 4-port hubs replaced by 4-port switches. Now, a frame transmitted from station A with a destination of station F is forwarded between the switches and sent directly to station F on port 4. Stations B, C, D, and E do not receive copies of the frame as they do in Figure 1(a). Thus, network traffic has been reduced.

The switches also partition the network into six separate collision domains. Now each station has unrestricted access to its own dedicated bandwidth (the speed of the switch port).

The switch is capable of specific forwarding because it learns what MAC addresses are associated with each port. Recall that every Ethernet frame contains a source MAC address and a destination MAC address. When a frame is received by a port on a switch, the switch will save a copy of the source MAC address and its associated port number in a special internal lookup table. Though we are storing the source MAC address, it is a destination MAC address to every other station in the network. Now, when a frame requires forwarding, the switch examines the destination MAC address stored in the frame, and looks for it in the lookup table. If the destination MAC address is found in the table, the frame is forwarded to the associated port. If the destination MAC address is not found, the frame is broadcast to all ports. Eventually the destination station will most likely respond with its own frame, and its port will be identified. Further broadcasts for that station will not be required.

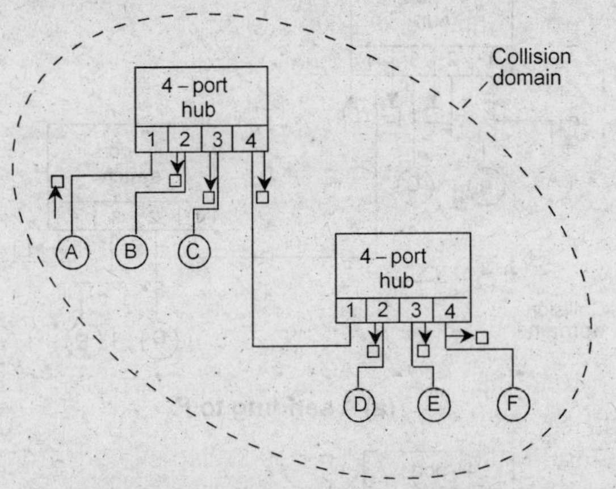

(a) Hubs broadcast on all ports

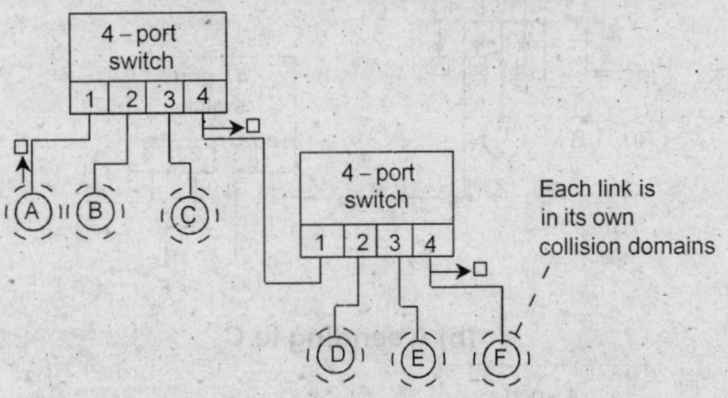

(b) Switches forward frames to specific ports.

Fig. 1 : Comparing a hub and a switch

Figure 2 shows the results obtained when a hub and switch are used together. Stations A, B and C are in one collision domain (competing for bandwidth). Stations D, E and F are in their own collision domains, each having full access to the available bandwidth. In the below Figure 2(a), station A transmits a frame destined for station F. The frame is broadcast by the hub and forwarded by the switch. Stations B and C must contend with the broadcast frame, waiting their turn for access. Neither station is allowed to transmit while the hub is broadcasting (or a collision will result). So, even though station A is sending a frame to station F. stations B and C are affected.

Figure 2(b) shows station A sending a frame to station C. The hub still broadcasts the frame, which affects station B but not stations D, E, or F.

In Figure 2(c), station F sends a frame to station A. The frame is forwarded by the switch and broadcast by the hub. Stations B and C are affected by F's frame; stations D and E are not.

Figure 2(d) shows- station F sending a frame to station D. Station E is unaffected and may transmit a frame to stations A, B or C without affecting the F to D transmission.

The switch contains the following components:

- Input port logic
- Output port logic
- Switching fabric
- Control logic

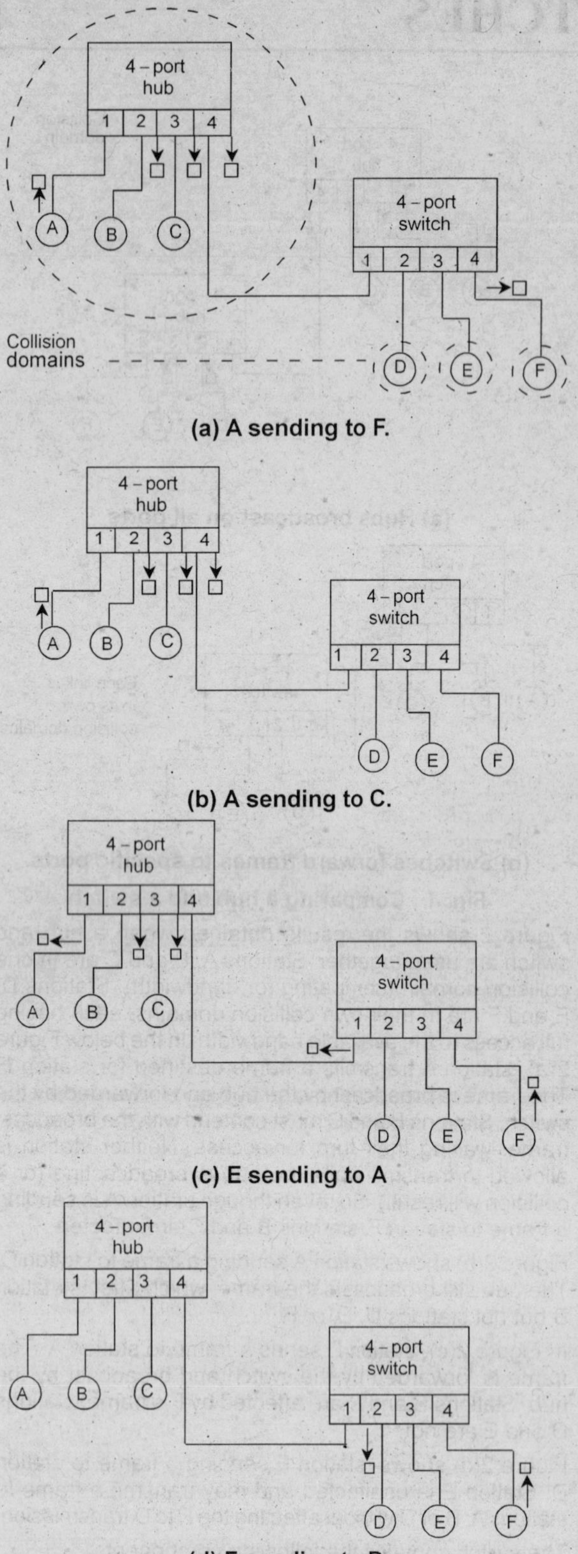

(a) A sending to F.

(b) A sending to C.

(c) E sending to A

(d) F sending to D.

Fig. 2 : Mixing hubs and switches

Input Port Logic

This section contains the Ethernet receiving logic and buffers for received frames. Buffering received frames lowers the rate of collisions and allows the switching fabric to be busy for short periods of time without losing data. If the frame buffer fills up, any new frame received by the port will trigger a collision. The random delay of CSMA/CD will then give the switch time to empty a portion of the input buffer before the station attempts retransmission.

Output Port Logic

Each output port contains an Ethernet transmitter and output frame buffer. Again, the buffer allows the switch fabric to service multiple output ports on a demand basis. For example, several frames may arrive simultaneously, with each frame directed to the same output port. The buffer is required to prevent the switching fabric from stalling. In addition, the input and output frame buffers allow different speeds between ports (port 1 operating at 10 Mbps and port 3 operating at 100 Mbps, for example). The buffers may be filled at one speed and emptied at another speed.

Switching Fabric

The switching fabric is responsible for directing the received frames from each input port to the appropriate output port. In addition, the switching fabric must also be able to handle a broadcast to all output ports. In general, there are two ways to build the switching fabric: crossbar switch and high-speed multiplexed bus. Both methods are shown in Figure 3.

The crossbar switch in Figure 3(a) is a two-dimensional set of data buses. Any combination of input-to-output connections is possible, even broadcasting. Each intersection of input wires and output wires in the crossbar switch contains an electronic switch that is either open or closed. A small amount of control information is required to configure the crossbar switch. Changing the control information changes the input-to-output connections.

The multiplexed bus in the below Figure 3(b) effectively makes one input-output connection at a time, with each input port getting its turn at using the bus. When many signals are multiplexed in this fashion, the data rate on the multiplexed bus must be much faster than the individual speeds of each port. For example, on a 4-port switch, with each port running at 100 Mbps, the multiplexed bus would need to operate at 400 Mbps. An 8-port switch would require an 800-Mbps bus. The speed requirement of this technique makes it unsuitable for switching at high speeds. This problem is overcome by the parallel nature of the crossbar switch.

Control Logic

The control logic must perform several chores, including

- Updating and searching the MAC address table
- Configuring the switching fabric
- Maintaining proper flow control through the switch fabric

Recall that the switch learns which ports are associated with specific stations by storing copies of the source MAC address from each received frame. The MAC address and port number are stored in a special high-speed memory called CAM (content addressable memory). The hard ware architecture of the CAM allows its internal memory to be searched for a desired data value, such as

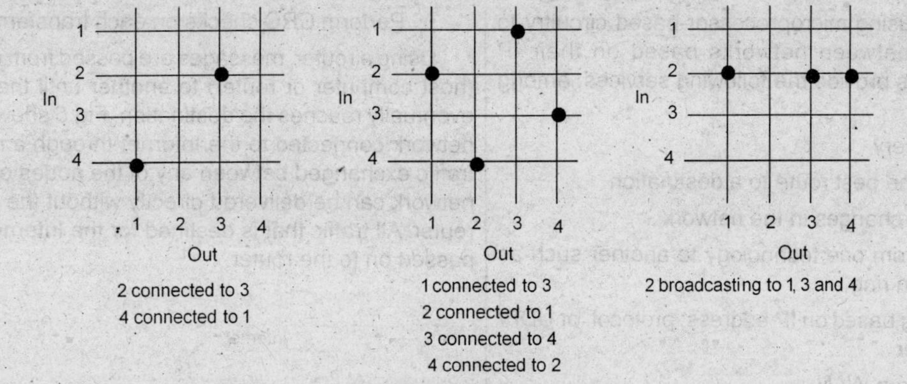

(a) Crossbar switch

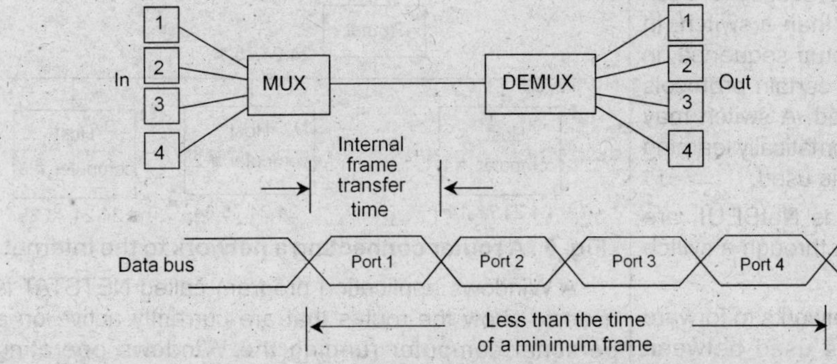

(b) Multiplexed bus (round - robin TDM)

Fig. 3 : Switching fabrics

a 48-bit MAC address, figure 4 shows a simple example of a CAM being searched for the MAC address 00 CO F0-3C-88-17. It is important to note that all of the MAC addresses being stored in the CAM are compared with the input value, at the same time.

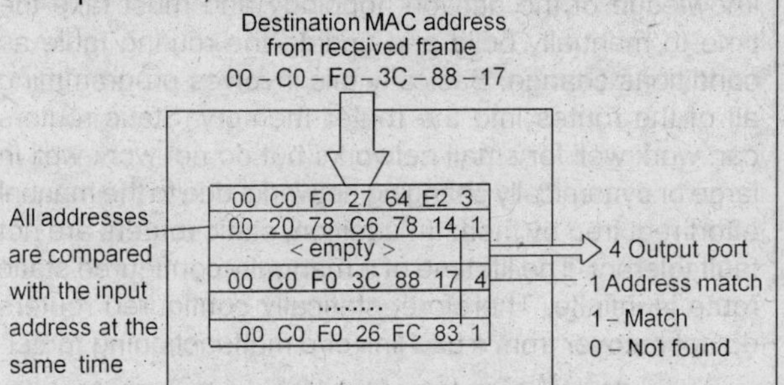

Fig. 4 : CAM organisation

The control logic uses the lookup results from the CAM to configure the switching fabric. In the event that an output port becomes unavailable due to congestion or some other problem, a flow control mechanism will prevent access to the port until it becomes available again.

Store-and-Forward Switching

Initially, switches handled frames using a technique called store-and-forward switching. In this technique, the entire frame is stored as it is received. If the FCS is valid, the destination MAC address is used to select an output port, and the frame is forwarded to the appropriate output port via the switching fabric. Since the entire frame is stored before any decisions are made, there is a delay between the time the frame is received and the time it begins transmission on the appropriate output port. This delay is called latency, and it varies depending on the length of the frame. The minimum latency is obtained with a minimum size frame. For 10 Mbps Ethernet, the minimum latency is 57.6 microseconds (576 bit times at 100 nanoseconds/bit, including Preamble). Maximum length frames have a latency just over 1.2 milliseconds.

Some applications, such as streaming audio and video, are sensitive to latency.

Cut-Through Switching

Cut-through switching reduces the latency of a switch tremendously. In this technique, as soon as the destination MAC address of an incoming frame is received, the forwarding process can begin (assuming there is a free output port and the switching fabric is available). This reduces the latency to just 11.2 microseconds, plus any additional time for internal switch operations. In addition, the latency of the cut-through method is fixed, since forwarding can always begin as soon as the destination MAC address is received. Un fortunately, errors can be propagated using the cut-through method, since there is no way to know if a frame being forwarded is good until it has been completely received. Cut-through switches will revert to store-and-forward when multiple errors occur while using the cut-through method.

Switches Versus Routers

Switches are considered Layer 2 (Data-Link) devices, using MAC addresses to forward frames to their proper destination. Routers, Layer 3 (Network) devices, are much more complex, using microprocessor-based circuitry to route packets between networks based on their IP address. Routers provide the following services, among others:

- Route discovery
- Selection of the best route to a destination
- Adaptation to changes in the network
- Translation from one technology to another such as Ethernet to token ring
- Packet filtering based on IP address, protocol, or UDP/TCP port number
- Connection to a WAN

Because of the additional processing required for each packet, a router has a higher latency than a switch. In addition, a router requires an initial setup sequence, in which the ports are programmed and certain protocols and characteristics enabled or disabled. A switch may be simply plugged into the network, automatically learning how to forward frames as the network is used.

Note that some protocols, such as NetBEUI, are nonroutable. These protocols will pass through a switch but not a router.

Finally, switches are used within networks to forward local traffic intelligently. Routers are used between networks, to route packets between networks in the most efficient manner. In the following sections, we will examine the many different routing protocols used to enable communication between LANs.

Routing protocols perform a different type of switching than discussed previously. Routing protocols operate at the Network layer (Layer 3) in the TCP/IP and OSI protocol stacks. Using a device called a router, various types of networks are connected together to form one logical network. The Internet is an example of one logical network. A router is a special purpose computer whose only responsibility is to move the data around on the network between source and destination computers. The individual networks can be made up of different types of LAN hardware and topologies, such as Ethernet, token-ring, or ATM networks, to name just a few. The router is able to move the data between the different network types and perform any necessary translations, such as between ASCII and EBCDIC and vice versa. On the Internet, the routing protocols are based on the Internet Protocol and use IP addresses. In general, each router must follow a few ground rules to allow it to process Network layer data:

- Communicate on a LAN just like any other station. For example, on an Ethernet network, a router communicates using CSMA/CD and monitors the media for their MAC address and any broadcast messages.
- Maintain tables with routing information for all reachable networks.
- Forward or block traffic based on the destination network address.
- Drop all frames to unknown destinations.
- Forward directed broadcasts (such as ARP) between networks.
- Perform CRC checks on each transferred packet.

Using a router, messages are passed from one device (host computer or router) to another until the message eventually reaches the destination. Fig.5 shows a typical network connected to the Internet through a router. Any traffic exchanged between any of the nodes on the local network can be delivered directly without the need for a router. All traffic that is destined for the Internet must be passed on to the router.

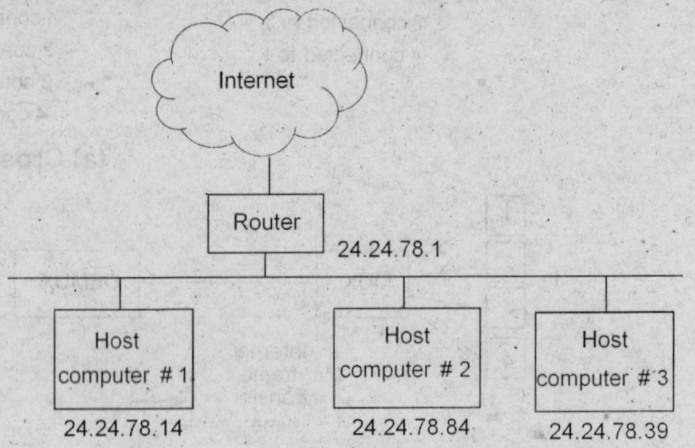

Fig. 5 : A router connecting a network to the Internet

A Windows application program called NETSTAT is used to show the routes that are currently active on a personal computer running the Windows operating system.

The routing table in a router can be created and maintained using two different methods: statically or dynamically. In a static router, a number of predefined routes are created and the router lacks the ability to discover new routes. In a router with statically configured routing tables, the network administrator needs a detailed knowledge of the network topology and must take the time to manually build and update the routing table as conditions change. Basically this involves programming all of the routes into the router memory. Static routers can work well for small networks but do not work well in large or dynamically changing networks due to the manual effort required by them. In addition, static routers are not fault tolerant. The lifetime of a manually configured static route is infinite. Therefore, statically configured routers do not recover from a bad link or a malfunctioning router.

In contrast, using dynamic routing, new routes can be discovered, and old routes can be updated as required. Dynamic routing consists of maintaining routing tables automatically using either periodic or on-demand messages through an ongoing communication between routers using the routing protocols. Except for their initial configuration, dynamic routers require little ongoing maintenance. Dynamic routing is fault tolerant. Dynamic routes learned from other routers have a finite lifetime. If a router or link goes down, the routers sense the change in the network topology through the expiration of the lifetime of the learned route in the routing table. This change can then be propagated to other routers so that all the routers on the network realize the new network topology. The router chooses the "best" path to send the data by implementing a distance-vector or link-state algorithm. In the distance-vector algorithm, each router in the network contains a partial view of the complete network topology. In the link-state algorithm, each router is aware of the entire network.

Before we can discuss how the routing algorithms and protocols operate, it is necessary to understand where a LAN fits into the logical network as a whole. Each connected network is part of a larger network called an autonomous system.

Autonomous Systems

Routing is based on the individual networks that are called autonomous systems, commonly abbreviated to AS. An autonomous system is a network or group of networks and routers controlled by a single administrative authority. The authority can be an institution, corporation, or any other type of organization. An autonomous system number is associated with each autonomous system. Each autonomous system has a single and clearly defined external routing policy. A new AS needs to be created if a network connect to more than one AS with a different routing policy. Different routing protocols are used when routing inside autonomous system and between them. The Interior Gateway Protocol (IGP) is used inside of autonomous systems. Exterior Gateway Protocols (EGP) exchange information between different autonomous systems. Figure 9 illustrates the how the different routing protocols are used to connect autonomous systems together.

Layer 3 Switching

A layer 3 switch is essentially a switch and a router combined into one package. Layer 3 switching has become popular due to the ever-increasing demand for bandwidth and services. Traditional routers have become bottlenecks in the campus and corporate LAN environments, due to their microprocessor-based operation and high latency. Layer 3 switches utilize ASIC (application specific integrated circuit) technology to implement the routing functions in hardware. This enables the layer 3 switch to perform router duties while forwarding frames significantly faster than an ordinary router. In fact, layer 3 switches are capable of forwarding millions of frames per second, compared to several hundred thousand frames per second for a router.

Replacing the campus or corporate routers with layer 3 switches, or adding layer 3 switching to a router-less network, has many benefits:

- Less expensive than routers
- Fewer network components to manage (via SNMP)
- Faster forwarding (close to wire speed, the speed of the frames on the wire)
- Helps provide QoS (quality of service) to the LAN environment
- Compatible with existing routing protocols (RIP, OSPF)
- Easier to configure than a router

Inside an ISP

Figure 6 shows an overhead view of the networking and telecommunications room at a small ISP (Internet service provider). Along the east wall are the incoming phone lines (200 pairs), modem bank (groups of sixty-four 56K-baud modems in a rack-mountable case), and the 44.7-Mbps T3 connection (to a higher-level ISP).

The west wall contains the routers and switches that make up the ISP topology and logical networks.

The center of the room contains the server farm, where all of the servers required for operation of the ISP reside. These include servers for DNS, DHCP, electronic mail. Web pages, and authentication. One machine is dedicated to monitoring the network (via SNMP) and another for performing backups.

The south wall contains server space for individual and corporate servers, which, along with the dial-up users, help generate income for the ISP.

Numerous UPS (uninterruptible power source) units provide 30 minutes of power in the event of a main power loss.

Figure 7 below illustrates the actual layout of the network. The T3 connection is the WAN connection to the higher-level ISP providing the actual Internet connection. Traffic in the T3 connection is filtered by the firewall. The I-router connects the individual subnetworks together and acts as the gateway to the Internet through the firewall. Employee computers (some of which have 100-Mbps switched service) communicate with their own file server or may tap into the server farm via the I-router. Individual and corporate servers share their own switch, as do the modems in the modem bank and the servers in the server farm. The F-router is used to lighten the load on the I-router for traffic moving between the server farm and the modem bank.

The network was designed in this fashion to allow the subnetworks to keep operating in the event that the main I-router goes down. Dial-up users can still check e-mail or work on their Web pages. Employees can continue to work as well, although without access to the Internet or the server farm.

Different Switching circuits are.

1. Circuit switch
2. Message switch
3. Packet switch

Circuit Switching: In this, each of the six rectangles represents a carrier switching office. In this, each office has three incoming lines and three outgoing lines. When a call passes through a switching office, a physical connection is established between the line on which the call came in and one of the output lines.

Once a call has been setup, a dedicated path between both ends exists and will continue to exist until the call is finished.

An important property of this is, the need to set up an end-to-end path before any data can be sent. The elapsed time between the end of dialing and the start of ringing

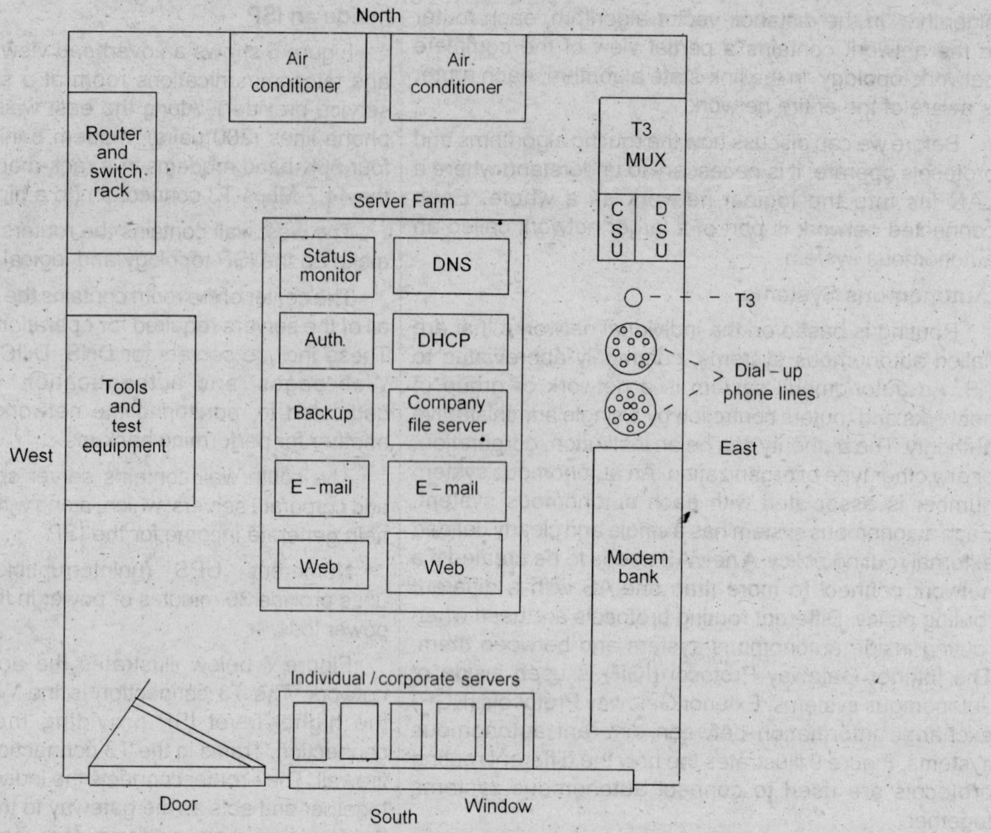

Fig. 6 : Overhead view of ISP network hardware

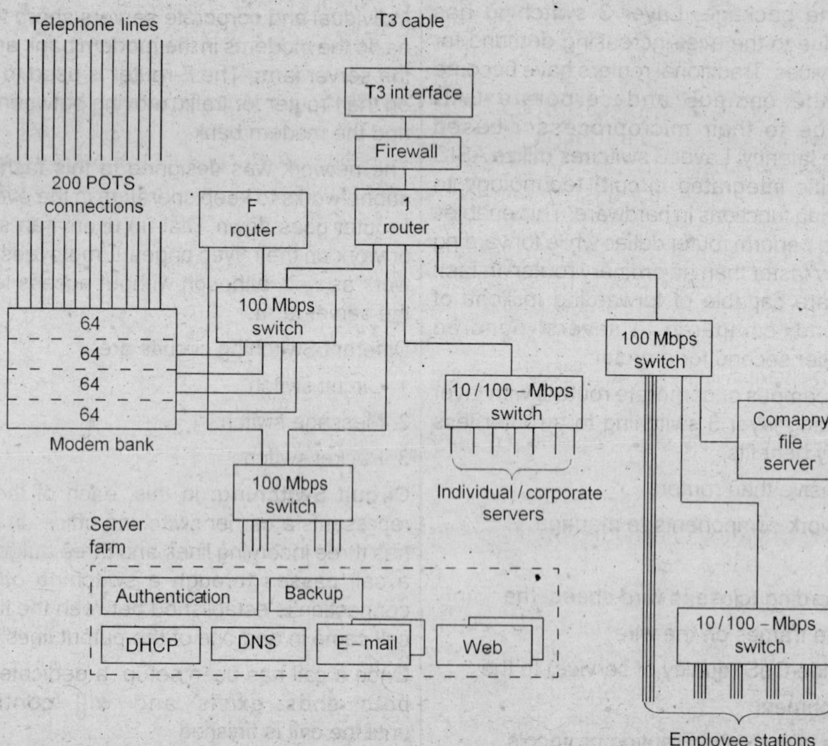

Fig. 7 : ISP network diagram

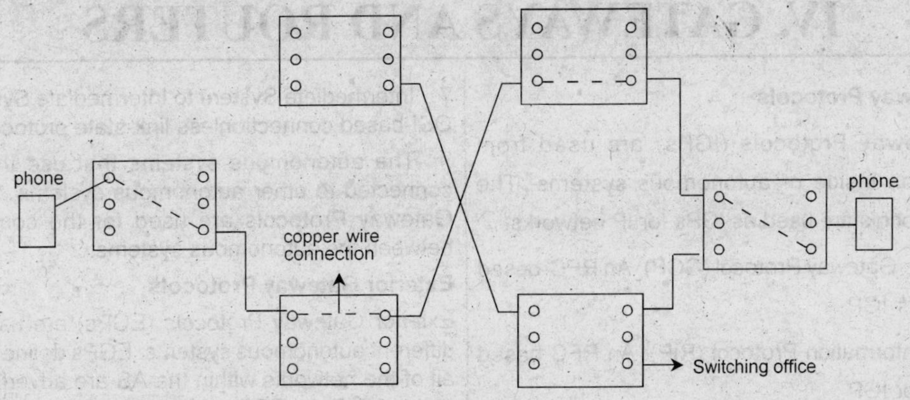

Fig. 8

can easily be 10 secs, more on long distance or international calls.

Packet switching: An alternative switching strategy is message switching. In this, no physical copper path is established in advance between sender and receiver. When the sender has a block of data to be sent, it is stored in the first switching office and then forwarded later, one hop at a time. Each block is received in its entirety, inspected for errors, and then retransmitted. This is called store and forward network.

With message switching, there is no limit on block size, which means that, IMPs must have disks to buffer long blocks. It also means that a single block may tie up are IMP-IMP line for many minutes, rendering message switching useless for interactive traffic.

This drawback is overcome in packet switching. In this networks place a tight upper limit on block size, allowing packets to be buffered in IMP main memory instead of a disk.

No user can monopolize any transmission line for more than a few tens of milliseconds; packet switching networks are well suited to handling interactive traffic. Another advantage is, the first multipacket message can be forwarded the second one has fully arrived, reducing delay and improving throughput. Computer networks are usually packet switched, occasionally circuit switched, but never message switched.

Difference between circuit switching and packet switching are:

1. Circuit switching statically reserves the required bandwidth in advance. In packet switching acquires and releases it as it is needed.

2. With circuit switching, any unused bandwidth on an allocated circuit is just wasted.

3. With packet switching it may be utilized by other packets from unrelated, sources going, to unrelated destinations, because circuits are never dedicated.

No circuits are dedicated, a sudden surge of input traffic may overwhelm an IMP, exceeding its storage capacity and causing it to lose packets.

When packet switching is used, it is straight forward for the IMPs to provide speed and code conversion. Also, they can provide error correction to some extent. In packet-switched networks, packets may be delivered in the wrong order to the destination. But reordering of packets never happen in circuit switch.

The charging algorithm also differs in the two methods. Packet carriers usually base their charge on both the numbers of bytes carried and the connect time. In this, transmission distance does not matter except perhaps internationally. With circuit switching, the charge is based on the distance and time only, not the traffic.

The main reason packet switching was invented is to get around the long call connection time present in the existing telephone system. A much more direct approach is to build a new telephone system, one in which calls are put through in milliseconds instead of seconds. It is called fast connect circuit switching.

Fast connect networks are a variation on circuit switching, there are variations on packet switching next one is time-division switching. In this, each IMP scans its input lines in strict rotation. Each packet is immediately retransmitted on the correct O/P line by using fixed-size packets and a rigid time synchronization, no buffer space is needed and the whole IMP can be reduced to a few chips.

IV. GATEWAYS AND ROUTERS

Interior Gateway Protocols

Interior Gateway Protocols (IGPs) are used from communication inside of autonomous systems. The following protocols are used as IGPs for IP networks:

1. Gateway to Gateway Protocol (GGP): An RFC-based distance-vector IGP.

2. Routing Information Protocol (RIP): An RFC based distance-vector IGP

3. Routing Information Protocol 2 (RIP-2): An enhanced version of RIP that includes support for Classless Inter-Domain Routing (CIDR)

4. Interior Gateway Routing Protocol (IGRP): A. distance-vector IGP developed by Cisco Systems. IGRP is capable of load balancing multiple network paths based on delay and bandwidth.

5. Extended Interior Gateway Routing Protocol (EIGRP): An enhanced distance vector IGP developed by Cisco Systems, Inc. EIGRP extends IGRP to support CIDR and provides several other enhancements.

6. Open Shortest Path First (OSPF): Link-state IP protocol that is primarily used within an autonomous system but can also be used as an EGP as well. OSPF includes authentication and has become the IP routing protocol of choice in large environments.

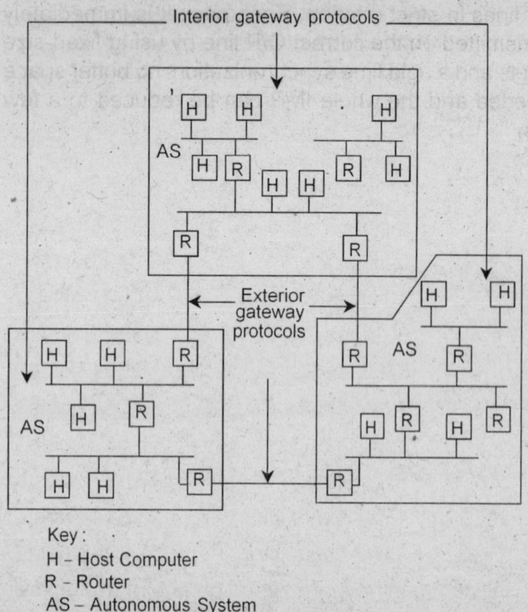

Fig. 9 : Interior versus Exterior Gateway Protocols

7. Intermediate System to Intermediate System (IS-IS): OSI-based connectionless link-state protocol.

The autonomous systems that use the IGPs are connected to other autonomous systems. The Exterior Gateway Protocols are used for the communication between the autonomous systems.

Exterior Gateway Protocols

Exterior Gateway Protocols (EGPs) are used between different autonomous systems. EGPs define the way that all of the networks within the AS are advertised outside of the AS. Each EGP advertises the "reachability" to the networks it can connect to Figure 10 shows what messages an EGP might broadcast to other EGPs. Note that the EGPs are independent of the IGPs used within the autonomous system. EGPs can facilitate the exchange of routes between autonomous systems that may use different IGPs. Each autonomous system contains two EGP routers for redundancy. If one EGP is down, the second must temporarily handle the additional load. In practice, the internal and external router may be located within the same box, performing both internal and external routing decisions simultaneously. For example, on a Cisco router, as many as 30 different routing protocols may be operating simultaneously. The combinations are made up as follows:

1. Up to 30-EGP routing processes
2. Up to 30 IGRP routing processes
3. Up to 30 OSPF routing processes
4. One RIP routing process
5. One IS-IS process
6. One BGP routing process

The following protocols are used for EGPs in IP networks:

Exterior Gateway Protocol (EGP): An RFC-based protocol that was developed for use between ASs on the Internet. EGP is no longer used on the Internet due to its lack of support for complex, multipath environments and Classless Inter-Doniain Routing (CIDR)

Border Gateway Protocol (BGP): An RFC-based protocol that is currently used between ASs on the Internet. BGP overcomes the weaknesses of EGP. For example, BGP is better at detecting routing loops than EGP.

Open Shortest Path First (OSPF): Link-state IP protocol. OSPF includes authentication and has become the IP routing protocol of choice in large environments. OSPF is also used as an IGP.

Classless Inter-Domain Routing

Classless Inter-Domain Routing (CIDR) was initially developed to recover many of the used addresses in class A and class B networks. The CIDR technique is supported by interior and exterior gateway protocols and is based on route aggregation. CIDR is also known as supernetting. CIDR is a new way of looking at IP address that eliminates the concept of classes (class A, class B and so on).

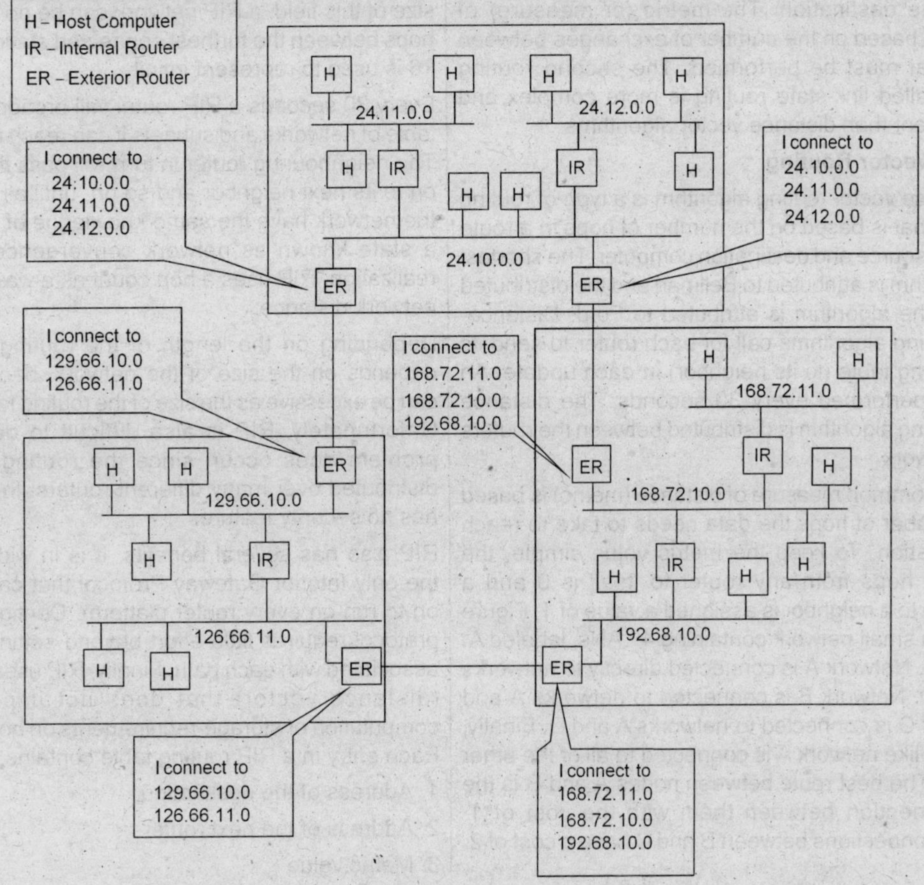

Fig. 10 : EGP messages

With CIDR, IP addresses and their subset masks are written as 4 octets, separated by periods, followed by a forward slash and a 2-digit number that represents the length of the subset mask. For example, the class B network 178.217.0.0. which is an illegal class C network address, is a legal class C supernet address, when it is represented in CIDR notation as 178.217.0.0/24. The /24 indicates that the subnet mask consists of 24-bits (counting from the left). Therefore, l78.217.0.0/24 represents the address 178.217.0.0 with a mask of 255.255.255.0.

CIDR makes it easy to aggregate or combine routes. Route aggregation is the process of using several different routes in such a way that a single route can be advertised, which minimizes the size of the routing tables maintained by a router. Table shows various CIDR prefixes with the number of class C equivalent addresses. Currently, big blocks of addresses are assigned to the large Internet service provides (ISPs), who then reallocate portions of their address blocks to their customers. The implementation of CIDR has been critical to the continued growth of the Internet; allowing more organizations and users to take advantage of this increasingly vital global networking and information resource.

CIDR Prefix	Number of Equivalent class C Addresses	Number of Host Addresses
/28	1/16 of one class C	16 hosts
/27	1/8 of one class C	32 hosts
/26	1/4 of one class C	64 hosts
/25	1/2 of one class C	128 hosts
/24	1 class C	256 hosts
/23	2classesCs	512 hosts
/22	4 class Cs	1,024 hosts
/21	8 class Cs	2,048 hosts
/20	16 class Cs	4,096 hosts
/19	32 class Cs	8,192 hosts
/18	64 class Cs	16,384 hosts
/17	128 classCs	32,768 hosts
/16	256 class Cs	65,536 hosts
/15	512 classCs	131,072 hosts
/14	1,024 class Cs	262,144 hosts
/13	2,048 class Cs	524,288 hosts

The routing algorithms are based on one of two different approaches to find the "best" or shortest path to the destination. The first method, called distance vector routing, is based on determining the fewest number of hops to the destination. The metric (or measure) of distance is based on the number of exchanges between routers that must be performed. The second routing method, called link-state routing is more complex and more efficient than distance vector algorithms.

Distance-Vector Routing

The distance vector routing algorithm is a type of routing algorithm that is based on the number of hops in a route between a source and destination computer. The shortest path algorithm is attributed to Bellman and the distributed nature of the algorithm is attributed to Ford. Distance-vector routing algorithms call for each router to send its entire routing table (to its neighbor) in each update. An update is performed every 30 seconds. The distance vector routing algorithm is distributed between the routers on the network.

The most common measure of distance (metric) is based on the number of hops the data needs to take to reach the destination. To keep the metric value simple, the number of hops from any router to itself is 0 and a connection to a neighbor is assigned a value of 1. Figure 11 shows a small network containing 4 LANs, labeled A, B, C and D. Network A is connected directly to networks B, C and D. Network B is connected to networks A and D. Network C is connected to networks A and D. Finally, network D, like network A is connected to all of the other networks. The best route between nodes A and B is the direct connection between them with the cost of 1. Similarly, connections between B and C have a cost of 2.

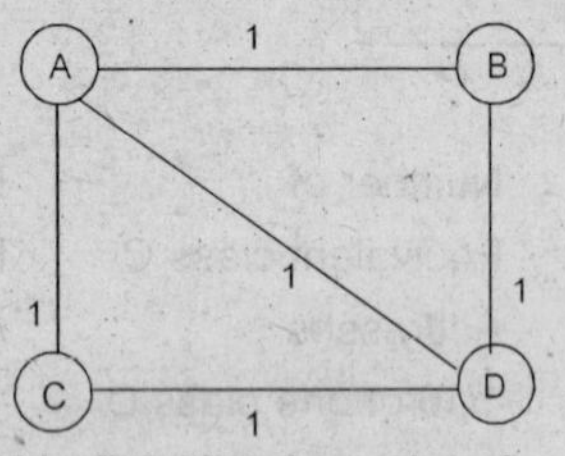

Fig. 11 : Routing based on hop

Although distance vector protocols are usually easy to implement, they have many weaknesses. For example, the distance-vector algorithm is susceptible to routing loops, cannot differentiate between the speeds of different links, cannot determine the lifetime of a single packet, and the time it takes to propagate all messages to every router in a large network is painfully slow.

Each router that implements the distance vector routing algorithm maintains a routing table that contains one entry for each router in the network. The routing table entry contains two fields: one is used to indicate the distance to the destination and the other one indicates the network interface adapter to use to get there. The distance vector routing algorithm was chosen as the original ARPA net routing algorithm, the predecessor to the modern Internet. A common distance-vector protocol is called the Routing Information Protocol, more commonly known as RIP.

RIP

The Routing Information Protocol (RIP) is a commonly used distance-vector routing protocol. RIP uses a 4-bit metric to count router hops to a destination. Due to the size of this field, a RIP network can be no larger than 15 hops between the furthest connected stations. The value 16 is used to represent infinity.

Every 30 seconds a RIP router will broadcast its routing table of networks and subnets it can reach to its neighbor. The neighbouring router in turn will pass the information on to its next neighbor and so on, until all routers within the network have the same knowledge of routing paths, a state known as network convergence, or network realization. RIP uses a hop count as a way to determine network distance.

Depending on the length of the routing table, which depends on the size of the network, bandwidth usage can be excessive as the size of the routing table increases. Unfortunately, RIP is also difficult to debug when a problem does occur, since the routing algorithm is distributed over many different routers. In addition, RIP has no security features.

RIP also has several benefits. It is in widespread use, the only Interior Gateway Protocol that can be counted on to run on every router platform. Configuring the RIP protocol requires little effort beyond setting up the cost associated with each path. Finally, RIP uses an algorithm (distance vector) that does not impose serious computation or storage-requirements on hosts or routers. Each entry in a RIP routing table contains,

1. Address of the destination
2. Address of the next router
3. Metric value
4. Recently Updated Flag
5. Various timers

RIP-2 maintain compatibility with RIP-1. RIP-2 provides for these additions routing features.

* Authentication
* Classless Inter-Domain Routing (CIDR)
* Next Hop
* Multicasting

0	8	16 31
Command	Version	(all zeros)
Address family		(all zeros)
IP Address		
(all zeros)		
(all zeros)		
Metric		

(a) RIP - 1

0	8	16 31
Command	Version	Routing Domain
Address family		Route tag
IP Address		
Subnet mask		
IP Address – next hop		
Metric		

(b) RIP - 2

Fig. 12 : Routing information Protocol Packets

Inter-Gateway Routing Protocol

The Inter-Gateway Routing Protocol (IGRP) is a Cisco-proprietary solution to many of the problems associated with the RIP protocol. In general, IGRP is characterized by the following properties:

1. IP transport
2. Updates broadcast every 90 seconds instead of 30
3. Hold down-protocol enhancement
4. New metrics Bandwidth Delay Load Reliability
5. Protection against loops
6. Multipath routing
7. Default route handling

IGRP is regarded as an Interior Gateway Protocol (IGP) but has also been used extensively as an Exterior Gateway Protocol for inter-domain routing. IGRP uses distance-vector routing technology. The methods used to implement IGRP are one of the reasons why Cisco routers are often used on the Internet. The proprietary nature of the IGRP protocol means that Cisco routers are the only routers that can use it.

The most important properties that set IGRP apart from RIP include less bandwidth consumed with broadcast updates, newer, more efficient methods to compensate for changing network conditions, additional metric parameters that can be used to allow a network administrator to provide additional measure of control as to how routes are chosen, increased prevention against routing loops, better handling of multiple paths, and the selection of a default route.

To prevent routing loops several new features are included. These features include

1. Hold down
2. Poison-reverse update
3. Split horizon

A hold down feature is used to prevent the condition where a route that has become unstable is used prematurely. A poison reverse update is used to eliminate routing loops by removing the routes from routing tables. Finally a split horizon is used to prevent information from being sent back on a link from the direction in which it was originally received.

The IGRP enhancements also include the addition of several new timer variables. These timers are,

1. Flush
2. Hold-time
3. Invalid
4. Update

The flush timer variable is used to control the lifetime of each entry in the routing table. The hold-time timer variable determines how long a route is kept in a hold down condition. The invalid timer variable determines how long a route remains valid in the absence of an update message. The update timer variable is used to determine how often update messages are distributed. The values specified for each of these timer variables allows for the most efficient use of the router processor and network bandwidth utilization.

Although IGRP has many advantages over RIP, a more advanced version called EIGRP provided many other capabilities.

Enhanced Inter-Gateway Routing Protocol

Cisco developed Enhanced IGRP in the early 1990s to improve the operating efficiency of IGRP. Improvements to IGRP include

A distributed update algorithm

MDS authentication

Protocol Independent Routing

Metric changes (not entire routing tables) exchanged every 90 seconds

CIDR support

Due to the proprietary nature of the EIGRP protocol, it is only supported by cisco routers.

Another set of routing protocols that offer additional capabilities are based on the link-state routing algorithm. It is simply a different method to select the "best" route to the destination.

Link-State Routing

In contrast to the distance vector algorithm, the link-state routing algorithm broadcasts information about the cost of reaching each of its neighbours to all over routers in the network. This allows the link-state algorithm to create a consistent view of the network at each router. The method used to compute the shortest distance is based on Dijkstra's algorithm, an open shortest path algorithm. The primary difference between distance vector routing and link-state routing is that a path with the least hops may not be chosen as the least cost route.

The implementation of the link-state routing algorithm is specified in the following steps:

1. Initialize a set called "S" to contain all of the routers in the network except for the source router.

2. Initialize a destination array to the metric value associated between each of the routers in set S and the source router if connection exists. If no connections exist, the distance is set to infinity.

3. Initialize values in a routing table array such that the metric distance value is assigned, otherwise the value is set to zero.

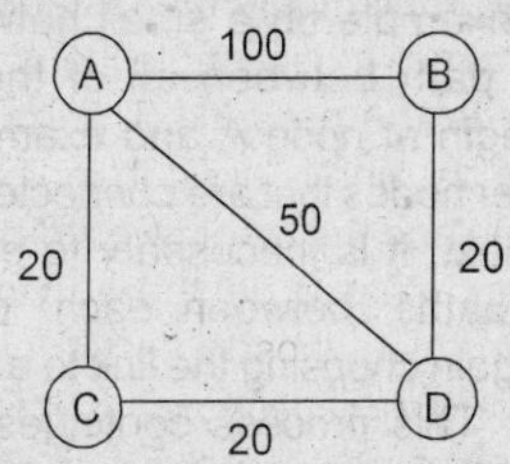

Fig. 13 : Routing based on weighted connections

4. While the set S it not empty

(a) Choose a router from set S such that the metric distance is minimum.

(b) If the minimum distance is infinity. No path exists to the router.

(c) Else

(i) Remove the router from set S.

(ii) For each router in set S.

(1) Compute a sum of the metric value.

(2) If the sum of the metric value is less than the current destination value.

(a) Store the sum metric value in the distance array.

(b) Identify this router as the "best" path in the routing table array.

In a small network, it may be easy to work through the algorithm. In a large network, it is much more difficult.

The link-state routing algorithm is not prone to routing loops or any of the other problems that are common to distance-vector routing algorithms. The problem with link-state algorithms lies in its complexity.

A link-state router has the following characteristics:

- Automatically discover any neighboring routers
- Shared information via., multicast transmissions
- Routing tables are built using first-hand information only
- Hello packets containing route information are forwarded to all routers
- Use metrics other than hop count
- Supports load balancing
- No assumption that routing information is accurate and has been received from a trusted source
- Compute the shortest path to every router in the network.

In link-state protocols, the router keeps a list of all routers that it knows in a table. In the routing table the distance to each network is specified.

Routers update their routing Tables periodically according to other router's routing tables and advertise their own routing tables to other routers. Each of the routes is calculated during every link-state modification. A special flooding protocol is used to propagate the changes in a link's status quickly throughout the network. To reduce the amount of multicast router traffic on the network, a spanning tree is used to create nonlooping paths between the routers in the network. Figure 14(a) illustrates a spanning tree example on a small network. To build a spanning tree path between all of the nodes in the network, we begin at node A and examine each of the paths to the other nodes that are connected to A, choosing the smallest. Next, it is, necessary to examine each of the remaining paths between each of the currently linked nodes, again choosing the link to a new node which is the smallest. This process continues until all nodes are in the tree. Figure 14(b) shows each of the individual steps required to build the spanning tree between each of the nodes in the small network. Figure 14(c) shows the complete spanning tree with the darkened lines indicating the resulting path.

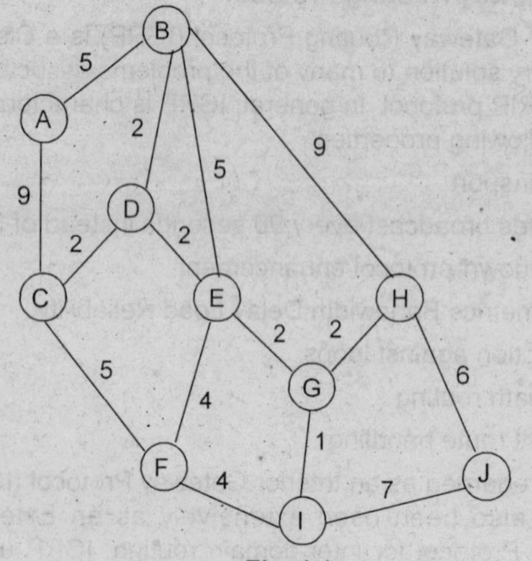

Fig. (a)

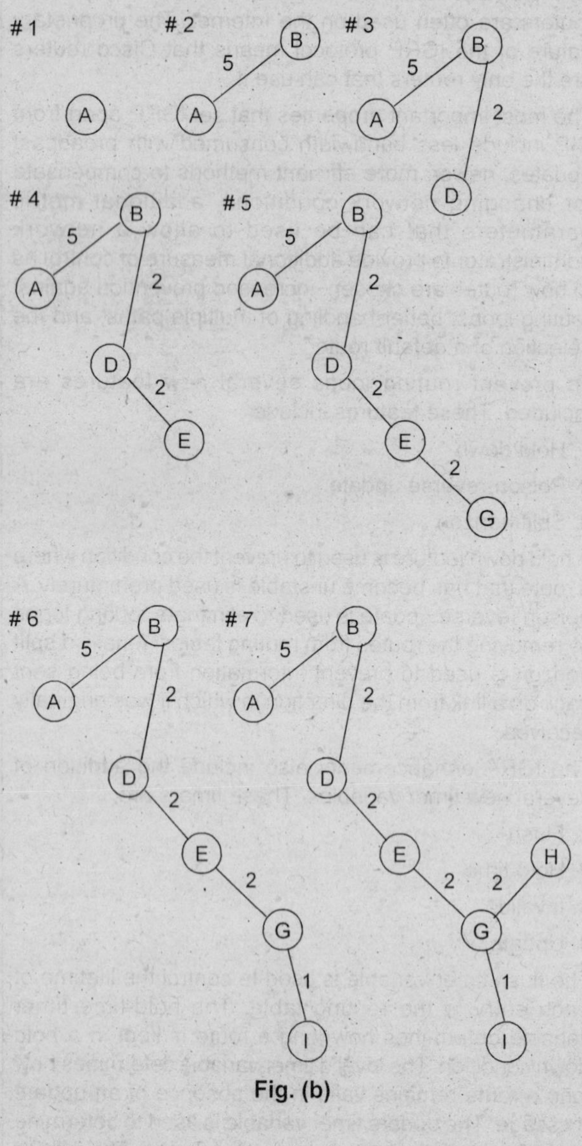

Fig. (b)

8

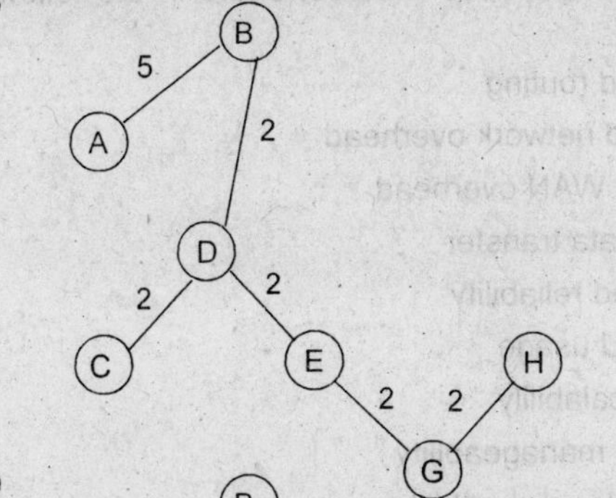

9

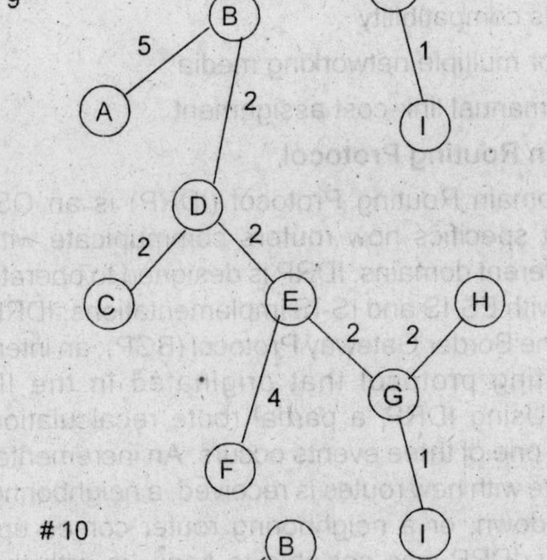

10

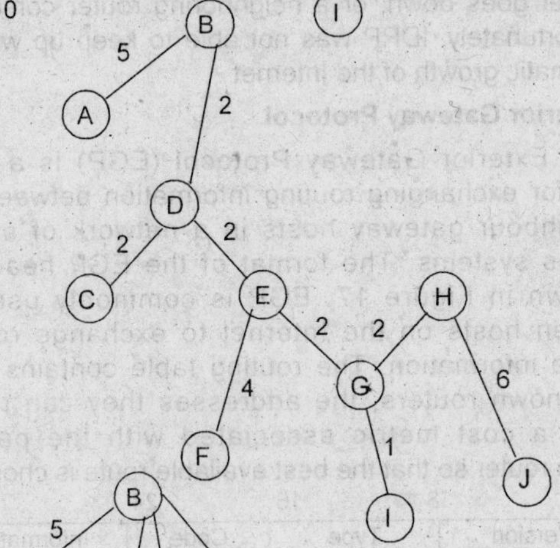

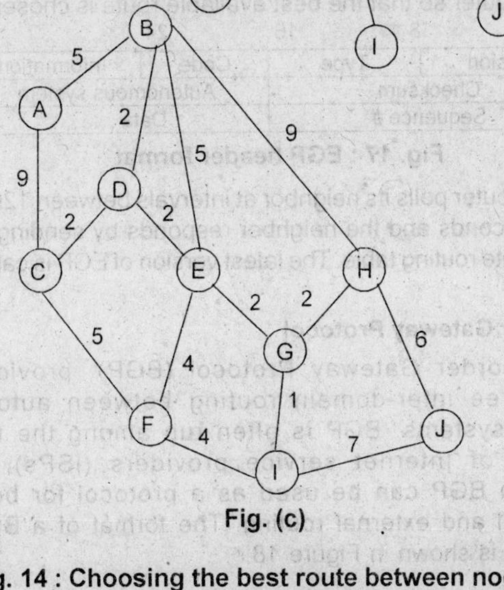

Fig. (c)

Fig. 14 : Choosing the best route between nodes using a spanning tree

There are six basic steps used to create the tree:

1. Let the distance D(n) = the set of the weighted links for a given path.
2. Let the cost C(i, j) = the cost between the two nodes i and j.
3. Initialize the set N = {1}.
4. For each node (n), not in N set D(n) = C(n, j), for all nodes j in N.
5. Find a new node (m) not in N for which the value of D(m) is the minimum.
6. Add m to set N.
7. Repeat steps 4 through 6 until all nodes are in set N.

More efficient methods called reverse-path forwarding, Steiner trees, and center-based trees are also used to deliver routing messages across a network.

Although link-state routing requires much more computation power it is more efficient. The advantages of link-state routing include.

(1) Fast, loopless convergence

(2) Support for multiple metrics

(3) Support for multiple paths

The link-state header is shown in Figure 15.

0	16	24 31
Age	Options	Type
Identification		
Advertising router		
Sequence number		
Checksum	Length	

Fig. 15 : Link - state header fields

One of the first types of routing protocols based on a link-state algorithm was end system (hosts) to intermediate systems (routers) (ES-IS) and intermediate system to intermediate system (IS-IS). ES-IS and IS-IS routing were originally used on the NSFnet when it was spun off from the ARPA net.

End System to intermediate System

End system to intermediate system (ES-IS) is an OSI protocol that defines how end systems (hosts) and intermediate systems (routers) learn about each other, a process known as configuration. Configuration happens before routing between end systems can occur. Once configured, data can be exchanged between the end systems and intermediate systems and vice versa. EI-IS is more of a discovery protocol than a routing protocol but is a necessary step in the routing process. Being able to communicate between end systems and an intermediate system is the first of two steps. Next, the data must be transferred between the intermediate systems until the data reaches the intermediate system that can communicate with the end system. This is where the intermediate system-to-intermediate system protocol is used.

Intermediate System to Intermediate System

Intermediate system to intermediate system (IS-IS) is an OSI link-state hierarchical routing protocol that, floods the network with link-state information to build a complete, consistent picture of network topology. To simplify router design and operation, IS-IS distinguishes between Level

1 and Level 2 ISs. Level 1 ISs communicate with other Level 1 ISs in the same area. Level 2 ISs route between Level 1 areas and form an intradomain routing backbone. Hierarchical routing simplifies backbone design because Level 1 ISs only need to know how to get to the nearest Level 2 IS. The backbone routing protocol also can change without impacting the intra-area routing protocol.

IS-IS uses a single required default metric with a maximum path value of 1,024. IS-IS uses three basic packet formats: IS-IS hello packets, link-state packets (LSPs), and sequence-number packets (SNPs). Each of the three IS-IS packets has a complex format with the following three different logical parts. The first part consists of an 8-byte fixed header shared by all three packet types. The second part is a packet-type-specific portion with a fixed format. The third part is also packet-type specific but of variable length. Figure 16 illustrates the logical format of IS-IS packets.

Integrated IS-IS

Integrated IS-IS is a routing protocol based on the OSI routing protocol, but it also supports IF and other protocols. Integrated IS-IS implementations send only one set of routing updates, making it more efficient than two separate implementations, formerly called the dual IS-IS routing method. Integrated IS-IS is a version of the OST IS-IS routing protocol that uses a single routing algorithm to support connection-oriented communication also.

The Netware Link Services Protocol is based on IS-IS routing.

0 — 8	8 — 16	16 — 24	24 — 31
Protocol Identifier	Header Length	Version	ID length
Packet type	Version	Reserved	Maximum Area address

Fig. 16 : IS - IS packet header

Netware Link Services Protocol

The Netware Link Services Protocol (NLSP) is a link-state routing protocol based on IS-IS. NLSP is the successor to the IPX protocol. Routers employing NLSP use incremental updates (as opposed to periodic updates) when exchanging network topology changes with their immediate neighbours.

As compared to RIP and SAP, NLSP provides improved routing, better efficiency, and scalability. In addition, NLSP-based routers are backward compatible with RIP-based routers. NLSP-based routers use a reliable delivery protocol, so delivery is guaranteed. Furthermore, NLSP facilitates improved routing decisions because NLSP-based routers store a complete map of the network, not just next-hop information such as RIP-based routers use. Routing information is transmitted only when the topology has changed, not every 30 seconds as RIP-based routers do, regardless of whether the topology has changed. Additionally, NLSP-based routers send service-information updates only when services change, not every 60 seconds as SAP does.

In terms of scalability, NLSP can support up to 127 hops. Recall that RIP supports only 15 hops. Compared to RIP, Netware Link Services Protocol offers in the following benefits:

(a) Improved routing
(b) Reduced network overhead
(c) Very low WAN overhead
(d) Faster data transfer
(e) Increased reliability
(f) Less CPU usage
(g) Better scalabilily
(h) Superior manageability
(i) Backwards compatibility
(j) Support for multiple networking media
(k) Optional manual link-cost assignment

Inter-Domain Routing Protocol

The Inter-Domain Routing Protocol (IDRP) is an OSI protocol that specifics how routers communicate with routers in different domains. IDRP is designed to operate seamlessly with ES-IS and IS-IS implementations. IDRP is based on the Border Gateway Protocol (BGP), an inter-domain routing protocol that originated in the IP community. Using IDRP, a partial route recalculation occurs when one of three events occurs: An incremental routing update with new routes is received, a neighboring router goes down, or a neighboring router comes up. Unfortunately, IDRP was not able to keep up with the dramatic growth of the Internet .

Exterior Gateway Protocol

The Exterior Gateway Protocol (EGP) is a proto col for exchanging routing information between two neighbour gateway hosts in a network of autono mous systems. The format of the EGP header is shown in Figure 17. EGP is commonly used be tween hosts on the Internet to exchange routing table information. The routing table contains a list of known routers, the addresses they can reach, and a cost metric associated with the path to each router so that the best available route is chosen.

0 — 8	8 — 16	16 — 24	24 — 31
Version	Type	Code	Information
Checksum		Autonomous system	
Sequence #		Data	

Fig. 17 : EGP header format

Each router polls its neighbor at intervals between 120 to 480 seconds and the neighbor responds by sending its complete routing table. The latest version of EGP is called EGP-2.

Border Gateway Protocol

The Border Gateway Protocol (BGP), provides loop-free inter-domain routing between autono mous systems. BGP is often run among the net works of Internet service providers (ISPs), al though BGP can be used as a protocol for both internal and external routing. The format of a BGP header is shown in Figure 18.

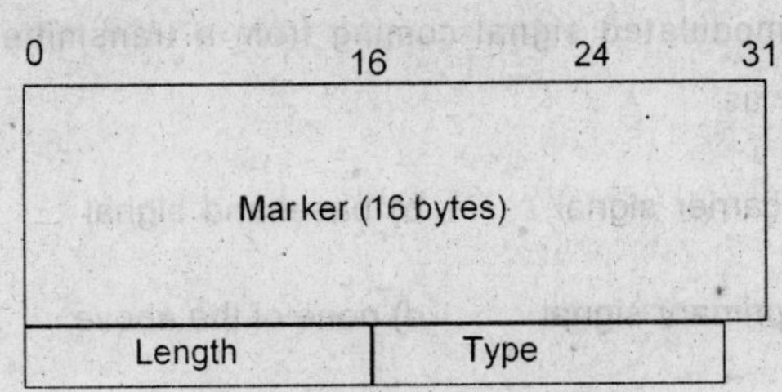

Fig. 18 : BGP header fields

Before BGP exchanges information with an external autonomous system it ensures that networks within the AS are reachable. This is done by a combination of internal BGP peering among routers within the AS and by redistributing BGP routing information to Interior Gateway Protocols (IGPs) that run within the AS, such as Interior Gateway Routing Protocol (IGRP), Intermediate System to Intermediate System (IS-IS), Routing Information Protocol (RDP), and Open Shortest Path First (OSPF).

BGP uses the Transmission Control Protocol (TCP) as its transport protocol. Any two routers that have opened a TCP connection to each other for the purpose of exchanging routing information are known as peers or neighbours. BGP peers initially exchange their full BGP routing tables. Thereafter, BGP peers send incremental updates only. BGP peers also exchange keep-alive messages (to ensure that the connection is up) and notification messages (in response to errors or special conditions).

When a BGP router receives updates that describe different paths to the same destination, it chooses the single best path for reaching that destination. Once chosen, BGP propagates the best path to its neighbours. The decision is based on the value of attributes (such as next hop, administrative weights, local preference, the origin of the route, and path length) that the update contain and other BGP-configurable factors.

Policy Routing

Policy-based routing refers to any type of routing that is based on factors other than the selection of the "shortest path". One of the primary uses of policy routing is used to accommodate acceptable use policies of the various interconnected networks. Other considerations are also incorporated, such as

1. Contract obligations

2. Quality of service (resource reservation)

3. Service provider selection

The BGP is used to support policy-based routing. Policy-based routing is a newer type of protocol that is even more complex to set up and manage.

Other types of routing algorithms are also possible, such as those based on routing IP information over ATM, like Multi-Protocol Label Switching and the Private Network-Network Interface.

OBJECTIVE QUESTIONS

1. Protocols are

a) agreements on how communications components and DTE's are to communicate

b) logical communication channels used for transferring data

c) physical communication channels used for transferring data

d) none of the above

2. The method of communication in which transmission takes place in both directions, but only in one direction at a time, is called

a) simplex b) four wire circuit

c) full duplex d) half duplex

3. Error detection at the data link level is achieved by

a) bit stuffing b) cyclic redundancy codes

c) Hamming codes d) equalization

4. Which of the following is a wrong example of a network layer?

a) Internet Protocol (IP)-ARPANET

b) X.25 Packet Level Protocol (PLP)-ISO

c) Source routing and domain naming-USENET

d) X.25 level 2-ISO

5. The topology with highest reliability is

a) bus topology b) star topology

c) ring topology d) mesh topology

6. baud means

a) the number of bits transmitted per unit time

b) the number of bytes transmitted per unit time

c) the rate at which the signal changes

d) none of the above

7. Start and stop bits are used in serial communication for

a) error detection b) error correction

c) synchronization

d) slowing down the communication

8. Unmodulated signal coming from a transmitter is known as

a) carrier signal b) baseband signal

c) primary signal d) none of the above.

9. Manchester code is a

a) Bi-polar code b) non return to zero code

c) polar code d) none of the above

10. Pick the incorrect statements.

a) Another name for primary/secondary protocol is master/slave

b) Peer to peer protocol provides equal status to all sites on the channel

c) Priority, non-priority types come under master/slave protocol

d) TDM is a primary/secondary non-priority system

11. Pick the correct statements.

a) A switched circuit is a dial-up circuit that may encounter blockage (busy signal)

b) Non switched leased line supports higher data volume and quality than switched lines

c) Non switched lines are expensive for high volume data

d) Switched circuit provides faster response time

12. Pick the incorrect statements in error retransmission used in continuous ARQ method.

a) Go-back-N method requires more storage at the receiving site

b) Selective Repeat involves complex login than Go-back-N

c) Go-back-N has better line utilization

d) Selective Repeat has better line utilization

13. In the carrier sense network if the prevailing condition is a 'channel busy', then which of the following are correct?

a) If the technique used is non-persistent then it results in randomised wait & sense

b) If the technique used is 1-persistent then the channel is continually sensed

c) If the technique used is p-persistent then randomised retransmission is done

d) If the method used is non-persistent then continuous sensing results

14. Which of the following are non-polling system?

a) TDMA b) Stop & Wait

c) Xon/Xoff d) Continuous ARQ

15. Pick the systems that can be used in both priority and non-priority modes.

a) TDM b) Register insertion

c) Carrier sense systems d) Token passing

16. How many characters per see (7 bits + 1 parity) can be transmitted over a 2400 bps line if the transfer is synchronous (1 start and 1 stop bit)?

a) 300 b) 240 c) 250 d) 275

17. How many characters per sec (7 bits + 1 parity) can be transmitted over a 2400 bps line if the transfer is asynchronous (1 start and 1 stop bit)?

a) 300 b) 240 c) 250 d) 275

18. Match the following

1. Session layer	(i) connects DCE into physical channel
2. Transport layer	(ii) provides end to end accountability
3. Application layer	(iii) provides organised means to exchange data between users. (like synchronization points)
4. MDI (Medium Dependent Interface)	iv) supports an end user process & performs required file transfer

a) 1-(iii), 2-(iv), 3-(ii), 4-(i)

b) 1-(iii), 2-(ii), 3-(iv), 4-(i)

c) 1-(ii), 2-(iv), 3-(i), 4-(iii)

d) 1-(iv), 2-(iii), 3-(ii), 4-(i)

19. The Hamming distance between 001111 and 010011 is

a) 1 b) 2 c) 3 d) 4

20. BSC is a

a) character oriented protocol b) bit-oriented protocol

c) full-duplex protocol d) half-duplex protocol

21. HDLC is

a) bit oriented b) code transparent

c) code dependent d) none of the above

22. Bit stuffing refers to

a) inserting a '0' in user data stream to differentiate it with a flag

b) inserting a '0' in flag stream to avoid ambiguity

c) appending a nibble to the flag sequence

d) appending a nibble to the user data stream

23. Choose the correct statement(s).

a) Baseband network uses analog technology

b) Baseband network are Time Division Multiplexed

c) Broadband network uses digital technology

d) In broadband network the carrier signals operate at lower frequency

24. In Ethernet CSMA/CD, the special bit sequence transmitted by media access management for collision handling is called as

a) preamble b) postamble

c) jam d) none of the above

25. Adaptive or dynamic directory used in packet routing changes

a) within each user session b) with each user session

c) at system generation time only d) both (a) & (b)

26. The method of network routing where every possible path between transmitting and receiving DTE is used is called

a) random routing b) packet flooding

c) directory routing d) message switching

27. Which one of the following network uses dynamic or adaptive routing?

a) TYMNET b) ARPANET

c) SNA (IBM's System Network Architecture)

d) None of the above

28. The number of cross point needed for 10 lines in a cross point switch which is full duplex in nature and there are no self connection is

a) 100 b) 45 c) 50 d) 90

29. A terminal multiplexer has six 1200 bps terminals and 'n' 300 bps terminals connected to it. The outgoing line is 9600 bps. What is the maximum value of n?

a) 4 b) 16 c) 8 d) 28

30. The difference between a multiplexer and a statistical multiplexer is

a) multiplexers use TDM (time division multiplexing), while statistical multiplexer uses FDM (frequency division multiplexing)

b) multiplexers often waste the output link capacity, while statistical multiplexers optimize its use

c) statistical multiplexers need buffers while multiplexers do not need buffers

d) multiplexers use the X.25 protocol, while statistical multiplexers use the ALOHA protocol.

31. A modem constellation diagram has data points at (0, 1) and (0, 2). What type of modulation does the modem use?

a) Phase modulation b) Amplitude modulation

c) Both (a) & (b) d) None of the above

32. Write the differential Manchester code for the given sketch

a) 111100101 b) 100010111

c) 101001111 d) 101001101

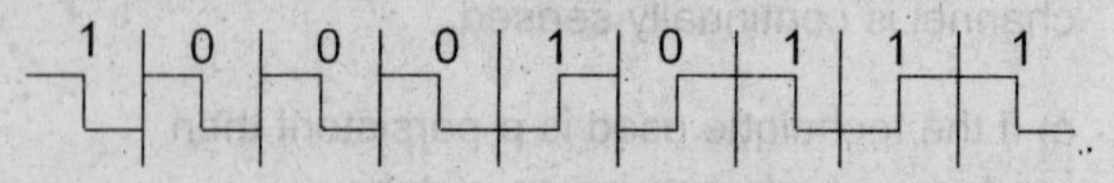

33. Maximum data rate of a channel for a noiseless 3-KHz binary channel is

a) 3000 bps b) 6000 bps

c) 1500 bps d) none of the above

34. The maximum data rate of a channel of 3000-Hz bandwidth and SNR 'of 30 dB is

a) 15,000 bps b) 60,000 bps

c) 30,000 bps d) 3,000 bps

35. In time division switches if each memory access takes 100 ns and one frame period is 125 μ s, then the maximum number of lines that can be supported is

a) 625 lines b) 1250 lines

c) 2300 lines d) 318 lines

36. If the bit string 0111101111101111110 is subjected to bit stuffing for the flag string 01111110, the output string is

a) 011110111110011111010

b) 01111011111011111100

c) 01111011111011111010

d) 0111101111101111110

37. End-to-end connectivity is provided from host-to-host in

a) the network layer b) the transport layer

c) the session layer

d) it is a combined functionality of the network and the data link layer.

38. Pick the correct statements.

a) In connection-oriented service, the destination address is to be specified only during the setup

b) Initial setup is not possible in connectionless service

c) Packet sequencing is not guaranteed in connection- oriented service

d) Initial setup is required for connectionless service

The next 3 questions are based on the following diagram representing shortest path adaptive routing.

FD	NN	OD
A		
B		
C		
E		
F		
G		

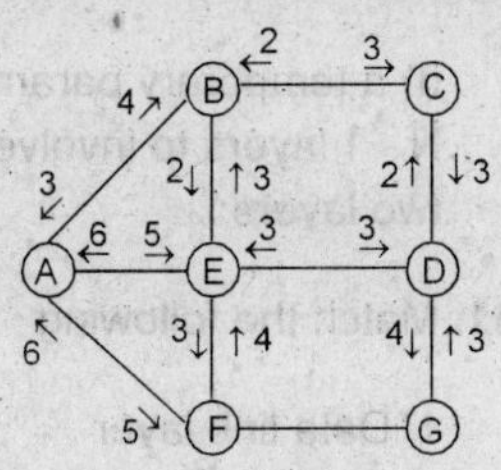

FD is the final destination; NN is the next node; OD is the overall delay.

In the above figure the routing table model is given for the node 'D'. Find out the following:

39. The entry for NN if FD is node B is

a) E b) C c) G d) none

40. The entry of OD for the above case is

a) 6 b) 14 c) 4 d) 3

41. The OD if FD is node A is

a) 6 b) 9 c) 13 d) 7

42. Pick the correct statements about flooding.

a) It is a type of isolated routing.

b) It is a method in which every incoming packet is sent out on every outgoing line except the one by which it arrived

c) Flooding does not always select the shortest path

d) Selective flooding is a type in which the packet are sent to those lines that are going approximately in the right direction.

43. The parameter which gives the probability of the transport layer itself spontaneously terminating a connection due to internal problems is called

a) protection b) resilience

c) option negotiation d) transfer failure

44. The___________measures the number of lost or garbled messages as a fraction of the total sent in the sampling period.

a) Residual Error rate

b) Transfer failure probability

c) Connection release failure probability

d) Connection establishment failure probability.

45. In session layer, during data transfer, the data stream responsible for the control purpose (i.e. control of the session layer itself) is

a) regular data b) typed data

c) capability data d) expedited data

The next three questions are based on Huffman's coding for the symbol A with probability 0.3, B with 0.15, C with 0.1, D with 0.25 and E with 0.2.

46. The minimum number of bits required to represent B is

a) 1 b) 2 c) 3 d) 4

47. The minimum number of bits required to represent all the symbols together is

a) 14 b) 11 c) 12 d) 15

48. The average code length of the given problem is

a) 2 b) 2.25 c) 2.45 d) 3

49. In cryptography the following uses transposition ciphers and the keyword is LAYER. Encrypt the following message. (Spaces are omitted during encryption)

WELCOME TO NETWORK SECURITY!

a) WMEKREETSILTWETCOOCYONRU!

b) EETSICOOCYWMEKRONRU! LTWET

c) LTWETONRU !WMEKRCOOCYEETSI

d) ONRU! COOCYLTWETEETSIWMEKR

50. Encrypt NEKEWNINRROGTTI using the above keyword in Transposition cipher method.

a) INTERWORKINGNET

b) INTERNETWORKING

c) WORKING INTERNET d) None

51. Assuming that for a given network layer implementation, connection establishment overhead is 100 bytes and disconnection overhead is 28 bytes. What would be the minimum size of a packet the transport layer needs to keep up, if it wishes to implement a datagram service above the network layer and needs to keep its overhead to a maximum of 12.5%. (Ignore transport layer overhead.)

a) 512 bytes b) 768 bytes

c) 1152 bytes d) 1024 bytes

52. Which of the following is not a standard RS-232C signal?

a) RTS b) CTS c) DSR d) VDR

53. A high speed communication equipment typically would not be needed for

a) E-mail b) transferring large volume of data

c) supporting communication between nodes in a LAN

d) all of the above

54. Which of the following ISO level is more closely related to the physical communications facilities?

a) Application b) Session

c) Network d) Data link

55. Which of the following is not a field in the Ethernet message packet?

a) Type b) Data c) Pin-code d) Address

56. The network topology that supports bidirectional links between each possible node is

a) ring b) star c) tree d) mesh

57. In a broad sense, .a railway track is an example of

a) simplex b) half-duplex

c) full-duplex d) all of the above.

58. The frequency range at which the land coaxial cables will be used is

a) 10^6 to 10^8 Hz b) 10^{10} to 10^{11} Hz

c) 10^3 to 10^4 Hz d) 10^{14} to 10^{15} Hz

59. If the data rate of ring is 20 Mbps, signal propagation speed is 200 m μ s, then the number of bits that can be placed on the channel of 200 km is

a) 2000 bits b) 20,000 bits

c) 1,000 bits d) none of the above

60. ICI (interface control information) is

a) used to transfer user data from layer to layer

b) used to exchange information by peer entities at different sites on the network to instruct an entity to perform a service function

c) a combination of service data unit (SDU) and protocol control information (PCI)

d) a temporary parameter passed between Nand N - 1 layers to involve service functions between two layers

61. Match the following

1. Data link layer	(i) the lowest layer whose function is to activate, deactivate & maintain the circuit between DTE and DCE
2. Physical layer	(ii) performs routing and communication
3. Presentation layer	(iii) detection and recovery from errors in the transmitted data
4. Network layer	(iv) Provides for the syntax of the data.

(a) 1-(iii), 2-(i), 3-(iv), 4-(ii)

(b) 1-(ii), 2-(i), 3-(iv), 4-(iii)

(c) 1-(iv), 2-(i), 3-(ii), 4-(iii)

(d) 1-(ii), 2-(i), 3-(iii), 4-(iv)

ANSWER KEY

1. a	**2.** d	**3.** b	**4.** d	**5.** d
6. c	**7.** c	**8.** b	**9.** b,c	**10.** c,d
11. a, b	**12.** a, c	**13.** a,b	**14.** a,c	**15.** c,d
16. a	**17.** b	**18.** b	**19.** c	**20.** a, d
21. a,b	**22.** a	**23.** b	**24.** c	**25.** a
26. b	**27.** b	**28.** b	**29.** c	**30.** c
31. b	**32.** b	**33.** b	**34.** c	**35.** a
36. a	**37.** b	**38.** a,b	**39.** b	**40.** c
41. d	**42.** a,b,d	**43.** b	**44.** a	**45.** c
46. c	**47.** c	**48.** b	**49.** b	**50.** b
51. d	**52.** d	**53.** a	**54.** d	**55.** c
56. d	**57.** b	**58.** a	**59.** b	**60.** d
61. a				

EXPLANATORY NOTES

9. In bipolar code the signal varies among three levels. In non-return to zero code the signal remains the same through-out the bit cell. In unipolar code, there will be no signal either below zero or above zero. In Manchester code, the signal level will not vary in the middle and is unipolar.

16. Start and stop bits are not needed in synchronous transfer of data. So, it is 2400/8 = 300.

22. Bit stuffing is required when there is a flag of bits to represent one of the incidents, like start of frame, end of frame, etc., If the same flag of bits appear in the data stream, a zero can be inserted. The receiver deletes this zero from the data stream.

28. As all lines are full-duplex and there are no self connections, only the cross points above the diagonal are needed. Hence formula for the number of cross points needed is n(n-l)/2

29. Since there are six 1200 bps terminals,

6 ×1200 + n × 300 = 9600. Solving, n = 8.

33. Maximum data rate = 2Hlogz V bps, where H is the bandwidth, V is the discrete levels.

Here H is 3 KHz and V is 2.

34. Maximum number of the bps = Hlogz (1 + SNR).

35. In time division switches 2nT = 1 frame period, where T is the memory access time.

46. The Huffman code for A will have 2 digits, B-3 digits, C- 3 digits, D-2 digits and E -2 digits. This can be obtained by constructing the binary tree corresponding to the given probabilities.

47. Refer to the explanation of the previous question.

48. Average code length is the sum of product of the length and probability of the occurrence of all the symbols. Here it is,

2 × 0.3 + 3 × 0. 15 + 3 × 0. 1 + 2 × 0. 25 + 2 × 0. 2 = 2.25.

TEST YOURSELF - I

1. One of the header fields in an IP datagram is the Time to Live (TTL) field. Which of the following statements best explains the need for this field ? [GATE - 10]

a) It can be used to prioritize packets
b) It can be used to reduce delays
c) It can be used to optimize throughput
d) It can be used to prevent packet looping

2. Which one of the following is not a client server application ? [GATE - 10]

a) Internet chat b) Web browsing
c) E - mail d) Ping

3. Suppose computers A and B have IP addresses 10.105.1.113 and 10.105.1.91 respectively and they both use the same net mask N. Which of the values of N given below should not be used if A and B should belong to the same network ? [GATE - 10]

a) 255.255.255.0 b) 255.255.255.128
c) 255.255.255.192 d) 255.255.255.224

Statements for Linked Answer Questions : 4 & 5

Consider a network with 6 routers R1 to R6 connected with links having weights as shown in the following diagram

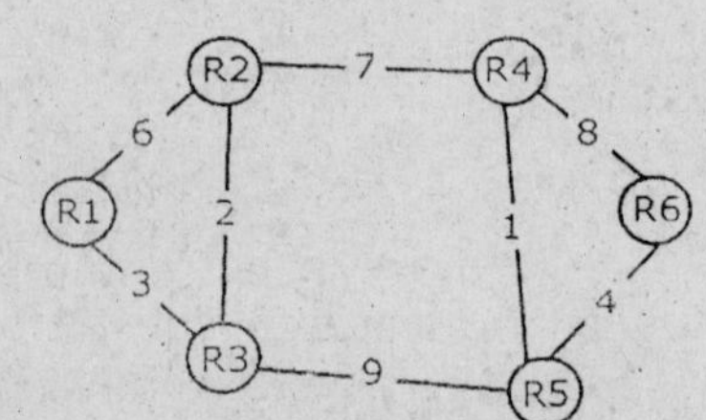

4. All the routers use the distance vector based routing algorithm to update their routing tables. Each router starts with its routing table initialized to contain an entry for each neighbour with the weight of the respective connecting link. After all the routing tables stabilize , how many links in the network will never be used for carrying any data ? [GATE - 10]

a) 4 b) 3 c) 2 d) 1

5. Suppose the weights of all unused links in the previous question are changed to 2 and the distance vector algorithm is used again until all routing tables stabilize. How many links will now remain unused ? [GATE - 10]

a) 0 b) 1 c) 2 d) 3

6. In the RSA public key cryptosystem, the private and public keys are (e,n) and (d, n) respectively, where n=p*q and p and q are large primes. Besides, n is public and p and q are private. Let M be an integer such that 0 <M<n and $\phi(n)=(p-1)(q-1)$. Now consider the following equations

I. $M' = M^e \bmod n$ II. ed ≡ 1 mod n

$M = (M')^d \bmod n$

III. $ed \equiv 1 \bmod \phi(n)$ IV. $M' = M^e \bmod \phi(n)$

$M = (M')^d \bmod \phi(n)$

Which of the above equations correctly represent RSA cryptosystem ? [GATE - 09]

a) I and II b) I and III c) II and IV d) III and IV

7. While opening a TCP connection, the initial sequence number is to be derived using a time - of - day (ToD) clock that keeps running even when the host is down. The low order 32 bits of the counter of the ToD clock is to be used for the initial sequence numbers. The clock counter increments once per millisecond. The maximum packet lifetime is given to be 64s.

Which one of the choices given below is closest to the minimum permissible rate at which sequence numbers used for packets of a connection can increase? [GATE - 09]

a) 0.015/s b) 0.064/s
c) 0.135/s d) 0.327/s

8. LetG(x) be the generator polynomial used for CRC checking. What is the condition that should be satisfied by G(x) to detect odd number of bits in error?[GATE - 09]

a) G(x) contains more than two terms
b) G(x) does not divide $1+x^k$, for any k not exceeding the frame length
c) 1 + x is a factor of G(x)
d) G(x) has an odd number of terms.

Statement for Linked Answer Questions : 9 & 10

Frames of 1000 bits are sent over a 10^6 bps duplex link between two hosts. The propagation time is 25 ms. Frames are to be transmitted into this link to maximally pack them in transit (within the link).

9. What is the minimum number of bits(I) that will be required to represent the sequence numbers distinctly? Assume that no time gap needs to be given between transmission of two frames. [GATE - 09]

a) I = 2 b) I = 3 c) I = 4 d) I = 5

10. Suppose that the sliding window protocol is used with the sender window size of 2^I, where I is the number of bits identified in the earlier part and acknowledgements are always piggy backed. After sending 2^I frames, what is the minimum time the sender will have to wait before starting transmission of the next frame ? (Identify the closest choice ignoring the frame processing time) [GATE - 09]

a) 16 ms b) 18 ms
c) 20 ms d) 22 ms

11. What is the maximum size of data that the application layer can pass on to the TCP layer below? **[GATE - 08]**
    (a) Any size
    (b) 2^{16} bytes-size of TCP header
    (c) 2^{16} bytes
    (d) 1500 bytes

12. Which of the following system calls results in the sending of SYN packets ? **[GATE - 08]**
    (a) socket (b) bind
    (c) listen (d) connect

13. In the slow start phase of the TCP congestion control algorithm, the size of the congestion window **[GATE - 08]**
    (a) does not increase
    (b) increase linearly
    (c) increases quadratically
    (d) increases exponentially

14. If a class B network on the Internet has a subnet mask of 255.255.248.0, what is the maximum number of hosts per subnet ? **[GATE - 08]**
    (a) 1022 (b) 1023
    (c) 2046 (d) 2047

15. A computer on a 10 Mbps network is regulated by a token bucket. The token bucket is filled at a rate of 2 Mbps. It is initially filled to capacity with 16 megabits. What is the maximum duration for which the computer can transmit at the full 10 Mbps ? **[GATE - 08]**
    (a) 1.6 seconds (b) 2 seconds
    (c) 5 seconds (d) 8seconds

16. A client process P needs to make a TCP connection to a server process S. Consider the following situation: the server process S executes a socket (), a bind () and a listen () system call in that order, following which it is preempted. Subsequently, the client process P executes a socket () system call followed by connect () system call to connect to the server process S. The server process has not executed any accept () system call. Which one of the following events could take place ? **[GATE - 08]**
    (a) connect ( ) system call returns successfully
    (b) connect ( ) system call blocks
    (c) connect ( ) system call returns an error
    (d) connect ( ) system call results in a core dump

17. How many bytes of data can be sent in 15 seconds over a serial link with baud rate of 9600 in asynchronous mode with odd parity and two stop bits in the frame? **[GATE - 08]**
    (a) 10,000 bytes (b) 12,000 bytes
    (c) 15,000 bytes (d) 27,000 bytes

18. Provide the best matching between the entries in the two columns given in the table below:

I. Proxy Server	a. Fire wall
II. KaZaA, DC++	b. Caching
III. SLIP	c. P2P
IV. DNS	d. PPP

[GATE - 08]
    (a) I-a, II-d, III-c, IV-b
    (b) I-b, II-d, III-c, IV-a
    (c) I-a, II-c, III-d, IV-b
    (d) I-b, II-c, III-d, IV-a

19. A 1 Mbps satellite link connects two ground stations. The altitude of the satellite is 36,504 km and speed of the signal is 3×10^8 m/s. What should be the packet size for a channel utilization of 25% for a satellite link using go-back-127 sliding window protocol? Assume that the acknowledgment packets are negligible in size and that there are no errors during communication. **[GATE - 08]**
    (a) 120 bytes (b) 60 bytes
    (c) 240 bytes (d) 90 bytes

20. The minimum frame size required for a CSMA/CD based computer network running at 1 Gbps on a 200m cable with a link speed of 2×10^8 m/s is **[GATE - 08]**
    (a) 125 bytes (b) 250 bytes
    (c) 500 bytes (d) None of the above

21. Data transmitted on a link uses the following 2D parity scheme for error detection: Each sequence of 28 bits is arranged in a 4 x 7 matrix(rows r_0 through r_3, and columns d_7 through d_1) and is padded with a column d_0 and row r_4 of partiy bits computed using the Even parity scheme. Each bit of column d_0 (respectively , row r_4) gives the parity of the corresponding row (respectively, column). These 40 bits are transmitted over the data link.

	d_7	d_6	d_5	d_4	d_3	d_2	d_1	d_0
r_0	0	1	0	1	0	0	1	1
r_1	1	1	0	0	1	1	1	0
r_2	0	0	0	1	0	1	0	0
r_3	0	1	1	0	1	0	1	0
r_4	1	1	0	0	0	1	1	0

The table shows data received by a receiver and has n corrupted bits. What is the minimum possible value of n? **[GATE - 08]**

(a) 1 (b) 2
(c) 3 (d) 4

22. Two popular routing algorithms are Distance Vector (DV) and Link State (LS) routing Which of the following are true ?

(S1) Count to infinity is a problem only with DV and not LS routing

(S2) in LS, the shortest path algorithm is run only at one node

(S3) in DV, the shortest path algorithm is run only at one node

(S4)DV requires lesser number of network messages than LS **[GATE - 08]**

(a) S1, S2 and S4 only
(b) S1,S3 and S4 only
(c) S2 and S3 only
(d) S1 and S4 only

23. Which of the following statements are TRUE ?

(S1) TCP handles both congestion and flow control

(S2) UDP handles congestion but not flow control

(S3) Fast retransmit deals with congestion but not flow control

(S4) Slow start mechanism deals with both congestion and flow control **[GATE - 08]**

(a) S1, S2 and S3 only
(b) S1and S3 only
(c) S3 and S4 only
(d) S1,S3 and S4 only

24. The three way handshake for TCP connection establishment is shown below

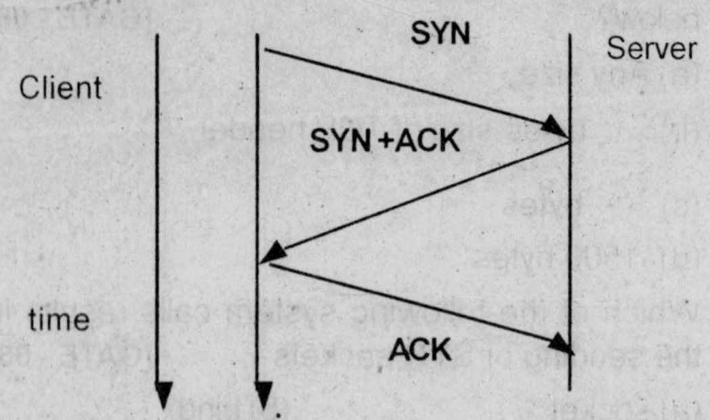

Which of the following statements are TRUE ?

(S1) Loss of SYN + ACK from the server will not establish a connection

(S2) Loss of ACK from the client cannot establish the connection

(S3) The server moves LISTEN $\rightarrow$ SYN_RCVD $\rightarrow$ SYN_SENT $\rightarrow$ ESTABLISHED in the state machine on no packet loss.

(S4) The server moves LISTEN $\rightarrow$ SYN_RCVD $\rightarrow$ ESTABLISHED in the state machine on no packet loss. **[GATE - 08]**

(a) S2 and S3 only
(b) S1and S4 only
(c) S1 and S3 only
(d) S2 and S4 only

25. The total number of keys required for a set of n individuals to be able to communicate with each other using secret key and public key cryptosystems, respectively are: **[GATE - 08]**

(a) n(n-1) and 2n

(b) 2n and $\frac{n(n-1)}{2}$

(c) $\frac{n(n-1)}{2}$ and 2n

(d) $\frac{n(n-1)}{2}$ and n

Statement for Linked Answer Questions 26 and 27 :

Host X has IP address 192.168.1.97 and is connected through two routers R1 and R2 to another host Y with IP address 192.168.1.80. Router R1 has IP addresses 192.168.1.135 and 192.168.1.110. R2 has IP addresses 192.168.1.67 and 192.168.1.155. The netmask used inthe network is 255.255.255.224.

26. Given the information above, how many distincy subnets are guaranteed to already exist in the network ? [GATE - 08]
(a) 1 (b) 2
(c) 3 (d) 6

27. Which IP address should X configure its gateway as ? [GATE - 08]
(a) 192.168.1.67
(b) 192.168.1.110
(c) 192.168.1.135
(d) 192.168.1.155

28. Consider the following statements about the timeout value used in TCP.

(i) The timeout value is set to the RTT (Round Trip Time) measured during TCP connection establishment for the entire duration of the connection.

(ii) Appropriate RTT estimation algorithm is used to set the timeout value of a TCP connection

(iii) Timeout value is set to twice the propagation delay from the sender to the receiver

Which of the following choices hold ? [GATE - 07]

(a) (i) is false, but (ii) and (iii) are true

(b) (i) and (iii) are false, but (ii) is true

(c) (i) and (ii) are false, but (iii) is true

(d) (i), (ii) and (iii) are false

29. Consider a TCP connection in a state where there are no outstanding ACKs. The sender sends two segments back to back. The sequence numbers of the first and second segments are 230 and 290 respectively. The first segment was lost, but the second segment was received correctly by the receiver. Let X be the amount of data carried in the first segment (in bytes), and Y be the ACK number sent by the receiver.

The values of X and Y (in that order) are [GATE - 07]

(a) 60 and 290
(b) 230 and 291
(c) 60 and 231
(d) 60 and 230

30. Consider the following two statements :

(i) A hash function (these are often used for computing digital signatures) is an injective function.

(ii) An encryption technique such as DES performs a permutation on the elements of its input alphabet.

Which one of the following options is valid for the above two statements ? [GATE - 07]

(a) Both are false
(b) Statement (i) is true and the other is false
(c) Statement (ii) is true and the other is false
(d) Both are true

31. The minimum positive integer p such that 3^p modulo 17 = 1 is [GATE - 07]

(a) 5 (b) 8
(c) 12 (d) 16

32. Exponentiation is a heavily used operation in public key cryptography. Which of the following options is the tightest upper bound on the number of multiplications required to compute b^n modulo m, $0 \le b, n \le m$? [GATE - 07]

(a) $O(\log n)$ (b) $O(\sqrt{n})$
(c) $O(n/\log n)$ (d) $O(n)$

33. A firewall is to be configured to allow hosts in a private network to freely open TCP connections and send packets on open connections. However, it will only allow external hosts to send packets on existing open TCP connections or connections that are being opened (by internal hosts) but not allow them to open TCP connections to hosts in the private network. To achieve this the minimum capability of the firewall should be that of [GATE - 07]

(a) A combinational circuit
(b) A finite automaton
(c) A pushdown automaton with one stack
(d) A pushdown automaton with two stacks

34. For the network given in the figure below, the routing tables of the four nodes A, E, D and G are shown. Suppose that F has estimated its delay to its neighbors, A, E, D and G as 8, 10, 12 and 6 msecs respectively and updates its routing table using distance vector routing technique.

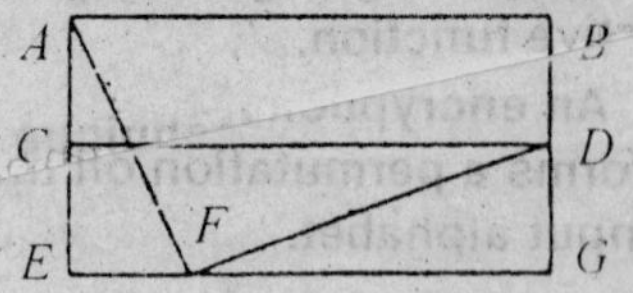

Routing Table of A		Routing Table of E		Routing Table of D		Routing Table of G	
A	0	A	24	A	20	A	21
B	40	B	27	B	8	B	24
C	14	C	7	C	30	C	22
D	17	D	20	D	0	D	19
E	21	E	0	E	14	E	22
F	9	F	11	F	7	F	10
G	24	G	22	G	22	G	0

Which one of the following options represents the updated routing table of F ? **[GATE - 07]**

(a)

A	8
B	20
C	17
D	12
E	10
F	0
G	6

(b)

A	21
B	8
C	7
D	19
E	14
F	0
G	22

(c)

A	8
B	20
C	17
D	12
E	10
F	16
G	6

(d)

A	8
B	8
C	7
D	12
E	10
F	0
G	6

35. In the waveform (a) given below, a bit stream is encoded by Manchester encoding scheme. The same bit stream is encoded in a different coding scheme in wave form (b) . The bit stream and the coding scheme are **[GATE - 07]**

(a) 1000010111 and Differential Manchester respectively

(b) 0111101000 and Differential Manchester respectively

(c) 1000010111 and Integral Manchester respectively

(d) 0111101000 and Integral Manchester respectively

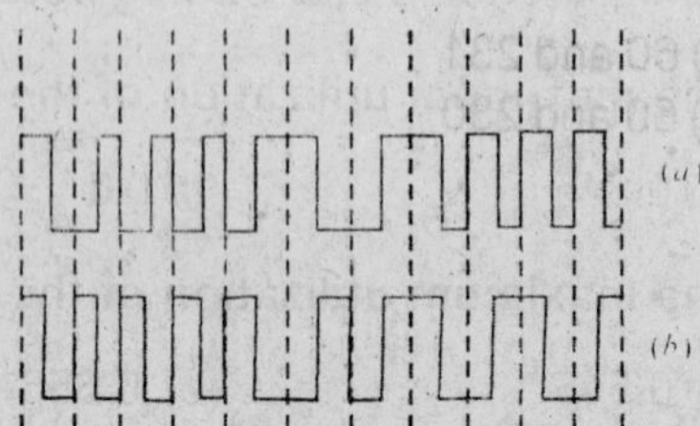

36. Let us consider a statistical time division multiplexing of packets. The number of sources is 10. In a time unit, a source transmits a packet of 1000 bits. The number of sources sending data for the first 20 time units is 6, 9, 3, 7, 2, 2, 2, 3, 4, 6, 1, 10, 7, 5, 8, 3, 6, 2, 9, 5 respectively. The output capacity of multiplexer is 5000 bits per time unit. Then the average number of backlogged of packets per time unit during the given period is [GATE - 07]

(a) 5 (b) 4.45

(c) 3.45 (d) 0

37. A group of 15 routers are interconnected in a centralized complete binary tree with a router at each tree node. Router i communicates with router j by sending a message to the root of the tree. The root then sends the message back down to router j. The mean number of hops per message, assuming all possible router pairs are equally likely is [GATE - 07]

(a) 3 (b) 4.26

(c) 4.53 (d) 5.26

38. A broadcast channel has 10 nodes and total capacity of 10 Mbps. It uses polling for medium access. Once a node finishes transmission, there is a polling delay of $80\mu s$ to poll the next node. Whenever a node is polled, it is allowed to transmit a maximum of 1000 bytes. The maximum throughput of the broadcast channel is [GATE - 07]

(a) 1 Mbps (b) 100 /11 Mbps

(c) 10 Mbps (d) 100 Mbps

39. Consider the following clauses :

(i) Not inherently suitable for client authentication.

(ii) Not a state sensitive protocol.

(iii) Must be operated with more than one server.

(iv) Suitable for structured message organization.

(v) May need two ports on the serve side for proper operation.

The option that has the maximum number of correct matches is [GATE - 07]

(a) IMAP - (i), FTP - (ii), HTTP - (iii), DNS - (iv), POP3 - (v)

(b) FTP - (i) , POP3 - (ii), SMTP - (iii), HTTP - (iv), IMAP - (v)

(c) POP3 - (i), SMTP - (ii), DNS - (iii), IMAP - (iv), HTTP - (v)

(d) SMTP - (i), HTTP - (ii) , IMAP - (iii), DNS - (iv), FTP - (v)

40. Your are given the following four bytes :

10100011	00110111	11101001	10101011

[GATE - 07]

Which of the following are substrings of the base 64 encoding of the above four bytes ?

(a) zdp (b) fpq

(c) qwA (d) oze

Common Data for Questions 41, 42 :

Consider a token ring topology with N stations (numbered 1 to N) running token ring protocol where the stations are equally spaced. When a station gets the token it is allowed to send one frame of fixed size. Ring latency is t_p , while the transmission time of a frame is t_t . All other latencies can be neglected.

41. The maximum utilization of the token ring when $t_t = 3ms, t_p = 5ms, N = 10$ is [GATE - 07]

(a) 0.545 (b) 0.6 (c) 0.857 (d) 0.961

42. The maximum utilization of the token ring when $t_t = 5ms, t_p = 3ms, N = 15$ is [GATE - 07]

(a) 0.545 (b) 0.655 (c) 0.9375 (d) 0.961

43. In Ethernet when Manchester encoding is used, the bit rate is [GATE - 07]

(a) Half the baud rate
(b) Twice the baud rate
(c) Same as the baud rate
(d) None of these

44. Which one of the following uses UDP as the transport protocol? [GATE - 07]

(a) HTTP (b) Telnet
(c) DNS (d) SMTP

45. There are n stations in a slotted LAN. Each station attempts to transmit with a probability p in each time slot. What is the probability that ONLY one station transmits in a given time slot ? [GATE - 07]

(a) $np(1-p)^{n-1}$ (b) $(1-p)^{n-1}$

(c) $p(1-p)^{n-1}$ (d) $1-(1-p)^{n-1}$

46. In a token ring network the transmission speed is 10^7 bps and the propagation speed is 200 metres / μs. The 1 - bit delay in this network is equivalent to : [GATE - 07]

(a) 500 metres of cable.
(b) 200 metres of cable.
(c) 20 metres of cable.
(d) 50 metres of cable.

47. The address of a class B host is to be split into subnets with a 6 - bit subnet number. What is the maximum number of subnets and the maximum number of hosts in each subnet ? [GATE - 07]

(a) 62 subnets and 262142 hosts.
(b) 64 subnets and 262142 hosts.
(c) 62 subnets and 1022 hosts.
(d) 64 subnets and 1024 hosts.

48. The message 11001001 is to be transmitted using the CRC polynomial x^3+1 to protect it from errors. The message that should be transmitted is : [GATE - 07]

(a) 11001001000 (b) 11001001011
(c) 11001010 (d) 110010010011

49. The distance between two stations M and N is L kilometres. All frames are K bits long. The propagation delay per kilometre is t seconds. Let R bits / second be the channel capacity. Assuming that processing delay is negligible, the minimum number of bits for the sequence number field in a frame for maximum utilization, when the sliding window protocol is used, is : [GATE - 07]

(a) $\left[\log_2 \frac{2LtR+2K}{K}\right]$

(b) $\left[\log_2 \frac{2LtR}{K}\right]$

(c) $\left[\log_2 \frac{2LtR+K}{K}\right]$

(d) $\left[\log_2 \frac{2LtR+K}{2K}\right]$

50. Match the following :

P. SMTP	**1. Application layer**
Q. BGP	**2. Transport layer**
R. TCP	**3. Data link layer**
S. PPP	**4. Network layer**
	5. Physical layer

[GATE - 07]

(a) $P-2, Q-1, R-3, S-5$
(b) $P-1, Q-4, R-2, S-3$
(c) $P-1, Q-4, R-2, S-5$
(d) $P-2, Q-4, R-1, S-3$

51. For which one of the following reasons does Internet Protocol (IP) use the time - to - live (TTL) field in the IP datagram header ? [GATE - 06]

(A) Ensure packets reach destination within that time
(B) Discard packets that reach later than that time
(C) Prevent packets from looping indefinitely
(D) Limit the time for which a packet gets queued in intermediate routers.

52. Station A uses 32 byte packets to transmit messages to Station B using a sliding window protocol. The round trip delay between A and B is 80 milliseconds and the bottleneck bandwidth on the path between A and B is 128 kbps. What is the optimal window size that A should use? [GATE - 06]

(A) 20 (B) 40 (C) 160 (D) 320

53. Two computers C1 and C2 are configured as follows. C1 has IP address 203. 197.2.53 and netmask 255. 255. 128.0. C2 has IP address 203.197.75.201 and netmask 255. 255. 192.0 Which one of the following statements is true ? [GATE - 06]

(A) C1 and C2 both assume they are on the same network

(B) C2 assumes C1 is on same network, but C1 assumes C2 is on a different network

(C) C1 assumes C2 is on same network, but C2 assumes C1 is on a different network

(D) C1 and C2 both assume they are on different networks.

54. Station A needs to send a message consisting of 9 packets to Station B using a sliding window (window size 3) and go-back-n error control strategy. All packets are ready and immediately available for transmission. If every 5th packet that A transmits gets lost (but no acks from B ever get lost), then what is the number of packets that A will transmit for sending the message to B ? [GATE - 06]

(A) 12 (B) 14 (C) 16 (D) 18

Statement for Linked Answer Questions 55 and 56 :

Consider the diagram shown below where a number of LANs are connected by (transparent) bridges. In order to avoid packets looping through circuits in the graph, the bridges organize themselves in a spanning tree. First, the root bridge is identified as the bridge with the least serial number. Next, the root sends out (one or more) data units to enable the setting up of the spanning tree of shortest paths from the root bridge to each bridge.

Each bridge identifies a port (the root port) through which it will forward frames to the root bridge. Port conflicts are always resolved in favour of the port with the lower index value. When there is a possibility of multiple bridges forwarding to the same LAN (but not through the root port), ties are broken as follows : bridges closest to the root get preference and between such bridges, the one with the lowest serial number is preferred.

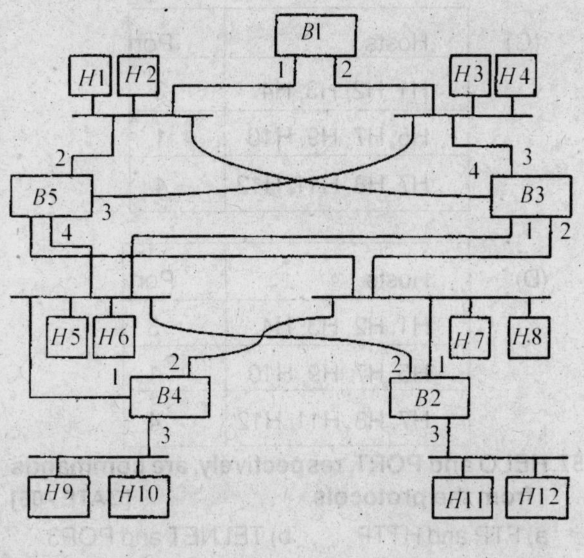

55. For the given connection of LANs by bridges, which one of the following choices represents the depth first traversal of the spanning tree of bridges ? [GATE - 06]

(A) B1, B5, B3, B4, B2

(B) B1, B3, B5, B2, B4

(C) B1, B5, B2, B3, B4

(D) B1, B3, B4, B5, B2

56. Consider the correct spanning tree for the previous question. Let host H1 send out a broadcast ping packet. Which of the following options represents the correct forwarding table on B3 ? [GATE - 06]

(A)

Hosts	Port
H1, H2, H3, H4	3
H5, H6, H9, H10	1
H7, H8, H11, H12	2

(B)

Hosts	Port
H1, H2	4
H3, H4	3
H5, H6	1
H7, H8, H9, H10	2
H11, H12	

(C)

Hosts	Port
H1, H2, H3, H4	3
H5, H7, H9, H10	1
H7, H8, H11, H12	4

(D)

Hosts	Port
H1, H2, H3, H4	3
H5, H7, H9, H10	1
H7, H8, H11, H12	4

57. HELO and PORT, respectively, are commands from the protocols [GATE - 06]

a) FTP and HTTP b) TELNET and POP3
c) HTTP and TELNET d) SMTP and FTP

58. Which of the following statements is TRUE? [GATE - 06]

a) Both Ethernet frame and IP packet include checksum fields

b) Ethernet frame includes a checksum field and IP packet includes a CRC field

c) Ethernet frame includes a CRC field and IP packet includes a checksum field

d) Both Ethernet frame and IP packet include CRC fields.

59. Which of the following statement(s) is TRUE?

(I) A hash function takes a message of arbitary length and generates a fixed length C.

(II) A hash function takes a message of fixed length and generates a code of variable length.

(III) A hash function may give the same hash value for distinct messages. [GATE - 06]

a) I only b) II and III only
c) I and III only d) II only

60. A router uses the following routing table :

Destination	Mask	Interface
144.16.0.0	255.255.0.0	eth0
144.16.64.0	255.255.224.0	eth1
144.16.68.0	255.255.255.0	eth2
144.16.68.64	255.255.255.224	eth3

A packet bearing a destination address 144.16.68.117 arrives at the router. On which interface will it be forwarded ? **[GATE - 06]**

a) eth 0 b) eth 1
c) eth 2 d) eth 3

61. Suppose that it takes 1 unit of time to transmit a packet (of fixed size) on a communication link. The link layer uses a window flow control protocol with a window size of N packets. Each packet causes an ack or a nak to be generated by the receiver, and ack / nak transmission times are negligible. Further, the round trip time on the link is equal to N units. Consider time $i > N$. If only acks have been received till time i (no naks), then the goodput evaluated at the transmitter at time i (in packets per unit time) is [GATE - 06]

a) $1 - N/i$ b) $i/(N+i)$
c) 1 d) $1-e^{(i/N)}$

62. In the 4B / 5B encoding scheme, every 4 bits of data are encoded in a 5-bit codeword. It is required that the codewords have at most 1 leading and at most 1 trailing zero. How many such codewords are possible? [GATE - 06]

a) 14 b) 16
c) 18 d) 20

63. A router has two full - duplex Ethernet interfaces each operating at 100 Mb/s. Ethernet frames are at least 84 bytes long (including the Preamble and the Inter-Packet-Gap). The maximum packet processing time at the router for wirespeed forwarding to be possible is (in micro-seconds) [GATE - 06]

a) 0.01 b) 3.36
c) 6.72 d) 8

64. A link of capacity 100 Mbps is carrying traffic from a number of sources. Each source generates an on-off traffic stream ; when the source is on, the rate of traffic is 10 Mbps, and when the source is off, the rate of traffic is zero. The duty cycle, which is the ratio of on-time to off-time, is 1 : 2. When there is no buffer at the link, the minimum number of sources that can be multiplexed on the link so that link capacity is not wasted and no data loss occurs is S_1. Assuming that all sources are synchronized and that the link is provided with a large buffer, the maximum number of sources that can be multiplexed

so that no data loss occurs is S_2. The values of S_1 *and* S_2 are, respectively, [GATE - 06]

a) 10 and 30 b) 12 and 25

c) 5 and 33 d) 15 and 22

65.On a wireless link, the probability of packet error is 0.2. A stop - and- wait protocol is used to transfer data across the link. The channel condition is assumed to be independent from transmission to transmission. What is the average number of transmission attempts required to transfer 100 packets ? [GATE - 06]

a) 100 b) 125

c) 150 d) 200

66.A program on machine X attempts to open a UDP connection to port 5376 on a machine Y, and a TCP connection to port 8632 on machine Z. However, there are no applications listening at the corresponding ports on Y and Z. An ICMP Port Unreachable error will be generated by [GATE - 06]

a) Y but not Z b) Z but not Y

c) Neither Y nor Z d) Both Y and Z

67.A subnetted Class B network has the following broadcase address : 144.16.95.255. Its subnet mask [GATE - 06]

a) is necessarily 255.255.224.0

b) is necessarily 255.255.240.0

c) is necessarily 255.255.248.0

d) could be any one of 255.255.224.0, 255.255.240.0, 255.255.248.0

ANSWER KEY

1. d	2. d	3. d	4. d	5. a
6. b	7. b	8. c	9. a	10. b
11. a	12. d	13. d	14. c	15. b
16. c	17.	18. d	19.	20. a
21. c	22. d	23. d	24. c	25.
26.	27.	28. c	29. d	30. d
31. b	32. c	33. d	34. a	35. c
36. c	37. d	38. a	39. d	40. a
41. c	42. a	43. b	44. c	45. a
46. c	47. c	48. d	49. b	50. b
51. c	52. b	53. d	54. a	55. c
56. a	57. d	58. c	59. c	60. c
61. c	62.	63.	64.	65.
66. a	67. d			

EXPLANATIONS

1. Whenever time to live field reaches '0' we discard the packet, so that we can prevent it from looping

3. Net mask is meant for giving the network address of a given IP address. When both IP - Addresses are in the same network, After performing the mask operation, same Id must be there for both, It will not come when we use 255.255.255.224

10. 105, 1. 113	10. 105. 1. 91
255.255.255.224	255.255.255.224
10. 105. 1. 106.	10. 105. 1. 64

Both are different, for all the others it will show the same Id.

Above Mask operation will be done in binary octants

4. In Distance vector, the Router will update its routing tables by exchanging the information from all its neighbours.

After All the routing tables stabilize the routing Table for 'R_1' will not have any entry to Router R_n., so that link will not be used.

So one link.

5. There will be no link unused since the link from R_1 *to* R_n is also filled will weight 2.

6. In RSA public key cryptography a public key provide for every one but they have their separate private key.

According to formula of public key crytography

I. $M' = M^e \bmod.\ M = (M')^d \bmod n$

IV. $M' = M^e \bmod \phi(n).\ M = (M')^d \bmod \phi(n)$

represent RSA cryptosystem.

7. Sequence number : Frame from a sending station are numbered sequentially we need to set a limit.

8. When the polynomial code method is employed the sender and receiver must agree upon a generator polynomial G(x) Both the high and low order bit of the generator must be 1.So 1 + x is a factor of G(x).

Alternately

For odd number of bits error, the error function will be having a factor of $(1 + x^2 + x^4 + ...)$ which can not be divided by the generator having (1+x) as a factor.

9. Pulse rate $= \dfrac{\textit{number of bits}}{\textit{time taken}} = \dfrac{1000}{25 \times 10^{-3}} = 4 \times 10^4$

Bit rate = pulse rate $\times 2^l$

Here, $2^l = 25$

$2^4 < 2^5 < 2^l$

So $l = 5$

Alternately

$10^3 ms..............10 bits$

$50''..............= \frac{1000}{50}\ 20\ bits$

Minimum number of bits required is 5

11. The application layer just takes the data from user, pass it on the TCP layers below. It is the work of other TCP layer to actually disintegrate the data into packets according to network parameters to transmit in the network and assemble different packets on receiving them and handling it to the application layer. So, application layer doesnot bother about size.
12. Connect result in sending SYN packets (TCP handshaking)
13. Congestion window increases exponentially
14. Starting address 255.255.248.0 & End address 255.255.255. 254
15. Let that duration is t sec.

    so 16 + 2t = 10 x t

    $\Rightarrow$ t = 2 sec
16. As there is no accept () at server side, the connect ( ) at client could not get anything that can respond to its requests, so error occurs.
18. Proxy Server $\rightarrow$ Caching

    KaZaA, DC ++ $\rightarrow$ P2P

    SLIP $\rightarrow$ PPP

    DNS $\rightarrow$ Firewall
20. Time = $\frac{distance}{speed}$

    $= \frac{200}{2\times10^8} m/s$

    $= 10^{-6} s = 1\mu s$

    1 Gbps x $1\mu s$

    $= 10^9 \times 10^{-6}$ bits

    $= 10^3$ bits

    =125 bytes
21.

	d_7	d_6	d_5	d_4	d_3	d_2	d_1	d_0
r_0	0	1	0	1	0	0	1	1
r_1	1	1	0	0	1	1	1	(0)
r_2	0	0	0	1	0	1	0	0
r_3	0	1	1	0	1	0	1	0
	1	1	(0)	0	0	(1)	1	0

Here two dimensional checking is done, i.e both horizontally and vertically.

If the total number of 1's is odd, parity bit is 1 and for even number of 1's , parity bit is 0.

So, above figure contains error at 3 places, i.e d_5, d_2 and r_1

24 Three way handshake for TCP connection establishment is

Client first sends connection request to server

Server replies with ACK, acknowledging client

After receiving acknowledgment, client and server establish a connection

29. When packet containing sequence number 230 is lost, then it will demand for this packet. So value of Y is 230. Difference between two consequtive packets is amount of data carried in first packet.

    Here 290 - 230 = 60

    X = 60

    Y = 230
31. 3^p modulo 17

    Check option

    In case of 5.3^5 modulo 17 = 243 mod 17 = 5

    In case of 8.3^8 modulo 7 = 1
34.

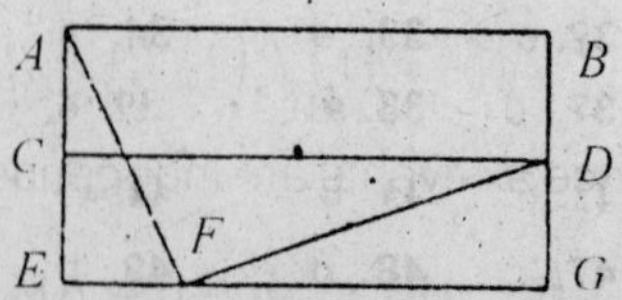

Given F has estimated its delay to its neighbours, A, E, D and G as 8, 10, 12 and 6 msec respectively and from routing table of E. E to C is 7 So from F to C is 10 + 7 =

17 and from routing table of D. D to B is 8. So F to B is F to D + D to B $\Rightarrow$ 12 + 8 = 20

So, updated routing table of F is

A	8
B	20
C	17
D	12
E	10
F	0
G	6

Hence answer is (a).

38. 10 nodes, 10 Mbps

Polling delay $80\ \mu s$

Maximum transmission is 1000 bytes

By inspecting we see that maximum throughput of broadcast channel is 1 Mbps.

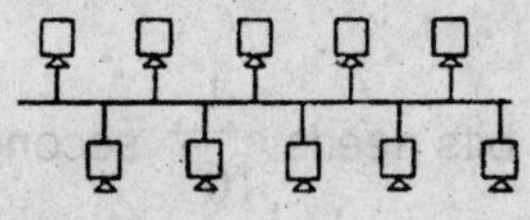

Hence answer is (a)

39. HTTP is non state sensitive protocol

FTP may need two parts on the reverse sides for proper operation

Hence answer is (d).

40.

Number	represented by
0 - 9	0 - q
10 - 35	0 - z
36 - 61	A - Z

Now took six bits from four bytes and convert it into base 4 digit we get Z as substring.

43. In Manchester encoding bit rate is twice the band rate.

44. DNS uses the UDP as the transport protocol.

45. Let 1 : event when station 1 transmits

2 : event when station 2 transmits

N : event when station N transmits

Given $P(i) = P \quad 1 \le i \le N \qquad P(\bar{i}) = (1-p) \quad (\le i \le N)$

Probability that only one station transmits is given as

$$P(1\ \bar{2}\ \bar{3} \ldots \bar{N} \cup \bar{1}\ 2\ \bar{3}\ \bar{4} \ldots \bar{N} \cup \bar{1}\ \bar{2}\ 3\ \bar{4}\ 5 \ldots \bar{N} \cup \ldots)$$

$$= P(1\ \bar{2}\ \bar{3} \ldots \bar{N}) + P(\bar{1}\ 2\ \bar{3}\ \bar{4} \ldots \bar{N}) + P(\bar{1}\ \bar{2}\ 3\ \bar{4}\ \bar{5} \ldots \bar{N}) + \ldots$$

Since all events are mutually exclusive, hence probability

$$= P(1)\,P(\bar{2})\,P(\bar{3}) \ldots P(\bar{N}) + P(\bar{1})\,P(2)\,P(\bar{3})\,P(\bar{4}) \ldots P(\bar{N})$$

$$+ P(\bar{1})\,P(\bar{2})\,P(3)\,P(\bar{4})\,P(\bar{5}) \ldots P(\bar{N}) + \ldots$$

Since all events are independent, hence probability

$$= p(1-p)(1-p) \ldots (1-p) + (1-p)\,p\,(1-p) \ldots (1-p) + (1-p)(1-p)\,p\,(1-p) \ldots (1-p) + \ldots$$

$$= n\ p\ (1-p)^{n-1}$$

46. 10^7 bits needs 1 second

1 bits needs $\frac{1}{10^7}$ second $= \frac{1}{10}\ \mu s$.

Since $1\ \mu s$ corresponds to 200 metre cable.

$\therefore\ \frac{1}{10}\ \mu s$ corresponds to 20 metre cable.

47. Maximum number of subnets =

$2^6 - 2 = 62$

Maximum number of host in each subnet

$= 2^{10} - 2 = 1022$

49. Total propagation delay = Lt Sec.

Roundtrip time =

$2 \times Lt = 2\ Lt$ sec.

Number of bits transmitted in roundtrip = 2 LtR bits

$$\text{Number of frames} = \frac{2LtR}{K}$$

Let the number of bit in sequent numbers be b.

$$\text{So,} \quad 2^b = \frac{2LtR}{K}$$

$$\Rightarrow \quad b = \left[\log_2 \frac{2LtR}{K}\right]$$

50. TCP protocol works on Transport layer.

SMTP protocol works on Application layer.

PPP Protocol works on data link layer.

BGP protocol works on physical layer.

52. Band width delay product = round trip delay × bottle neck bind width

$= 80 \times 128 \times 10^{-3} \times 1024\ bits$

$$= \frac{80 \times 128 \times 10^{-3} \times 1024}{8}\ bytes$$

$\therefore$ Optimal size window

$$= \frac{80 \times 128 \times 10^{-3} \times 1024}{8 \times 32} = 40$$

57. Helo and port commands are from the protocol SMTP and FTP.

60. eth0 and eth2 has a destination address which can prefer 144.16.68.117, but eth0 will be general and eth2 will be more specific.

61. The link layer uses a window from control protocol with a window size of N packets and round trip time is equal to N units.

Goodput = Packets / unit time

= N packets / N units = 1.

66. Because UDP generates the error that machine you are trying to reach is not responding unlike TCP.

67. Class B subnet mask begins with 255

Hence any of the three.

TEST YOURSELF - II

1. Which of the following assertions is FALSE about the Internet Protocol (IP) ? **[GATE - 03]**

a) It is possible for a computer to have multiple IP addresses

b) IP packets from the same source to the same destination can take different routes in the network

c) IP ensures that a packet is discarded if it is unable to reach its destination within a given number of hops

d) the packet source cannot set the route of an outgoing packets; the route is determined only by the routing tables in the routers on the way.

2. Which of the following functionalities must be implemented by a transport protocol over and above the network protocol ? **[GATE - 03]**

a) Recovery from packet losses

b)Detection of duplicate packets

c) Packet delivery in the correct order

d) End to end connectivity

3. The subnet mask for a particular network is 255.255.31.0. Which of the following pairs of IP addresses could belong to this network ? **[GATE - 03]**

a) 172.57.88.62 and 172.56.87.233

b) 10.35.28.2 and 10.35.29.4

c) 191.203.31.87 and 191.234.31.88

d) 128.8.129.43 and 128.8.161.55

4. A 2 km long broadcast LAN has 10^7 bps bandwidth and uses CSMA/CD. The signal travels along the wire at 2×10^8 m/s. What is the minimum packet size that can be used on this network ? **[GATE - 03]**

a) 50 bytes b) 100 bytes

c) 200 bytes d) None of the above

5. Host A is sending data to host B over a full duplex link. A and B are using the sliding window protocol for flow control. The send and receive window sizes are 5 packets each. Data packets (sent only from A to B) are all 1000 bytes long and the transmission time for such a packet is $50\,\mu s$. Acknowledgement packets (sent only from B to A) are very small and require negligible transmission time. The propagation delay over the link is $200\,\mu s$. What is the maximum achievable throughput in this communication ? **[GATE - 03]**

a) 7.69×10^6 bps b) 11.11×10^6 bps

c) 12.33×10^6 bps d) 15.00×10^6 bps

6. Choose the best matching between Group 1 and Group 2. **[GATE - 04]**

Group - 1	Group -2
P. Data link layer	1. Ensures reliable transport of data over a physical point-to-point link
Q. Network layer	2. Encodes/decodes data for physical transmission
R. Transport layer	3. Allows end-to-end communication between two processes
	4. Routes data from one network node to the next

a) P-1, Q-4, R-3 b) P-2, Q-4, R-1

c) P-2, Q-3, R-1 d) P-1, Q-3, R-2

7. Which of the following is NOT true with respect to a transparent bridge and a router ? **[GATE - 04]**

a) Both bridge and router selectively forward data packets

b) A bridge uses IP addresses while a router uses MAC addresses

c) A bridge builds up its routing table by inspecting incoming packets

d) A router can connect between a LAN and a WAN

8. A and B are the only two stations on an Ethernet. Each has a steady queue of frames to send. Both A and B attempt to transmit a frame, collide, and A wins the first backoff race. At the end of this successful transmission by A, both A and B attempt to transmit and collide. The probability that A wins the second backoff race is **[GATE - 04]**

a) 0.5 b) 0.625 c) 0.75 d) 1.0

9. The routing table of a router is shown below :

Destination	Subnet Mask	Interface
128.75.43.0	255.255.255.0	Eth0
128.75.43.0	255.255.255.128	Eth1
192.12.17.5	255.255.255.255.	Eth3
default		Eth2

On which interfaces will the router forward packets addressed to destinations 128.75.43.16 and 192.12.17.10 respectively? **[GATE - 04]**

a) Eth1 and Eth2 b) Eth0 and Eth2

c) Eth0 and Eth3 d) Eth1 and Eth3

The Following Information Pertains to Q. 10-11

Consider three IP networks A, B and C. Host H_A in network A sends messages each containing 180 bytes of application data to a host H_c in network C. The TCP layer prefixes a 20 byte header to the message. This passes through an intermediate network B. The maximum packet size, including 20 byte IP header, in each network is **[GATE - 04]**

A : 1000 bytes ; B : 100 bytes ; C : 1000 bytes

The network A and B are connected through a 1 Mbps link, while B and C are connected by a 512 Kbps link (bps=bits per second).

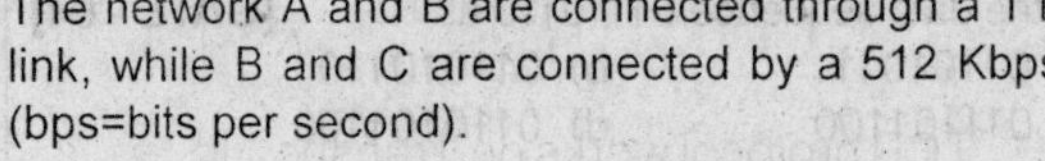

10. Assuming that the packets are correctly delivered, how many bytes, including headers, are delivered to the IP layer at the destination for one application message, in the best case ? Consider only data packets. **[GATE - 04]**

a) 200 b) 220 c) 240 d) 260

11. What is the rate at which application data is transferred to host H_c ? Ignore errors, acknowledgements, and other overheads. **[GATE - 04]**

a) 325.5 Kbps b) 354.5 Kbps

c) 409.6 Kbps d) 512.0 Kbps

12. Which one of the following statements is FALSE ? **[GATE - 04]**
a) Packet switching leads to better utilization of bandwidth resources than circuit switching.
b) Packet switching results in less variation in delay than circuit switching
c) Packet switching requires more per-packet processing than circuit switching
d) Packet switching can lead to reordering unlike in circuit switching

13. Which one of the following statements is FALSE ? **[GATE - 04]**
a) TCP guarantees a minimum communication rate
b) TCP ensures in-order delivery
c) TCP reacts to congestion by reducing sender window size
d) TCP employs retransmission to compensate for packet loss.

14. Which one of the following statement is FALSE ? **[GATE - 04]**
a) HTTP runs over TCP
b) HTTP describes the structure of web pages
c) HTTP allows information to be stored in a URL
d) HTTP can be used to test the validity of a hypertext link

15. A sender is employing public key cryptography to send a secret message to a receiver. Which one of the following statements is TRUE ? **[GATE - 04]**
a) Sender encrypts using receiver's public key
b) Sender encrypts using his own public key
c) Receiver decrypts using sender's public key
d) Receiver decrypts using his own public key

16. A subnet has been assigned a subnet mask of 255.255.255.192. What is the maximum number of hosts that can belong to this subnet? **[GATE - 04]**
a) 14 b) 30 c) 62 d) 126

17. A host is connected to a Department network which is part of a University network. The University network, in turn, is part of the Internet. The largest network in which the Ethernet address of the host is unique is **[GATE - 04]**
a) the subnet to which the host belongs
b) the Department network
c) the University network d) the Internet

18. In TCP, a unique sequence number is assigned to each **[GATE - 04]**
a) byte b) word c) segment d) message

19. In a data link protocol, the frame delimiter flag is given by 0111. Assuming that bit stuffing is employed, the transmitter sends the data sequence 01110110 as **[GATE - 04]**
a) 01101011 b) 011010110
c) 011101100 d) 0110101100

20. In a sliding window ARQ scheme, the transmitter's window size is N and the receiver's window size is M. The minimum number of distinct sequence numbers required to ensure correct operation of the ARQ scheme is **[GATE - 04]**
a) min (M,N) b) max (M,N) c) M + N d) MN

21. Consider a 10 Mbps token ring LAN with a ring latency of 400 μs. A host that needs to transmit seizes the token. Then it sends a frame of 1000 bytes, removes the frame after it has circulated all around the ring, and finally releases the token. This process is repeated for every frame. Assuming that only a single host wishes to transmit, the effective data rate is **[GATE - 04]**
a) 1 Mbps b) 2 Mbps c) 5 Mbps d) 6 Mbps

22. A 20 Kbps satellite link has a propagation delay of 400 ms. The transmitter employs the "go back n ARQ" scheme with n set to 10. Assuming that each frame is 100 bytes long, what is the maximum data rate possible? **[GATE - 04]**
a) 5 Kpbs b) 10 Kbps c) 15 Kbps d) 20 Kbps

23. Consider a parity check code with three data bits and four parity check bits. Three of the code words are 0101011, 1001101 and 1110001. Which of the following are also code words? **[GATE - 04]**
I. 0010111 II. 0110110 III. 1011010 IV. 0111010
a) I and III b) I, II and III
c) II and IV d) I, II, III and IV

24. Consider a simplified time slotted MAC protocol, where each host always has data to send and transmits with probability p=0.2 in every slot. There is no backoff and one frame can be transmitted in one slot. If more than one host transmits in the same slot, then the transmissions are unsuccessful due to collision. What is the maximum number of hosts which this protocol can support, if each host has to be provided a minimum throughput of 0.16 frames per time slot ? **[GATE - 04]**
a) 1 b) 2 c) 3 d) 4

25. In the TCP/IP protocol suite, which one of the following is NOT part of the IP header ? **[GATE - 04]**
a) Fragment Offset b) Source IP address
c) Destination IP address d) Destination port number

26. A TCP message consisting of 2100 bytes is passed to IP for delivery across two networks. The first network can carry a maximum payload of 1200 bytes per frame and the second network can carry a maximum payload of 400 bytes per frame, excluding network overhead. Assume that IP over head per packet is 20 bytes. What is the total IP overhead in the second network for this transmission? **[GATE - 04]**
a) 40 bytes b) 80 bytes c) 120 bytes d) 160 bytes

27. Suppose that the maximum transmit window size for a TCP connection is 12000 bytes. Each packet consists of 2000 bytes. At some point of time, the connection is in slow-start phase with a current transmit window of 4000 bytes. Subsequently, the transmitter receives two acknowledgement. Assume that no packets are lost and there are no time-outs. What is the maximum possible value of the current transmit window ? **[GATE - 04]**
a) 4000 bytes b) 8000 bytes
c) 10000 bytes d) 12000 bytes

28. Packets of the same session may be routed through different paths in **[GATE - 05]**
a) TCP, but not UDP b) TCP and UDP
c) UDP, but not TCP d) Neither TCP, nor UDP

29. The Address Resolution Protocol (ARP) is used for **[GATE - 05]**
a) Finding the IP address from the DNS
b) Finding the IP address of the default gateway
c) Finding the IP address that corresponds to a MAC address
d) Finding the MAC address that corresponds to an IP address

30. The maximum window size for data transmission using the selective reject protocol with n-bit frame sequence numbers is **[GATE - 05]**

a) 2^n b) 2^{n-1} c) $2^n - 1$ d) 2^{n-2}

31. In a network of LANs connected by bridges, packets are sent from one LAN to another through intermediate bridges. Since more than one path may exist between two LANs, packets may have to be routed through multiple bridges. Why is the spanning tree algorithm used for bridge-routing ? **[GATE - 05]**

a) For shortest path routing between LANs

b) For avoiding loops int he routing paths

c) For fault tolerance

d) For minimizing collisions

32. An organization has a class B network and wishes to form subnets for 64 departments. The subnet mask would be **[GATE - 05]**

a) 255.255.0.0 b) 255.255.64.0

c) 255.255.128.0 d) 255.255.252.0

33. In a packet switching network, packets are routed from source to destination along a single path having two intermediate nodes. If the message size is 24 bytes and each packet contains a header of 3 bytes, then the optimum packet size is **[GATE - 05]**

a) 4 b) 6 c) 7 d) 9

34. Suppose the round trip propagation delay for a 10 Mbps Ethernet having 48-bit jamming signal is 46.6 μs. The minimum frame size is: **[GATE - 05]**

a) 94 b) 416 c) 464 d) 512

35. Consider the three commands : PROMPT, HEAD and RCPT.

Which of the following options indicate a correct association of these commands with protocols where these are used ? **[GATE - 05]**

a) HTTP, SMTP, FTP b) FTP, HTTP, SMTP

c) HTTP, FTP, SMTP d) SMTP, HTTP, FTP

36. Traceroute reports a possible route that is taken by packets moving from some host A to some other host B. Which of the following options represents the technique used by traceroute to identify these hosts **[GATE - 05]**

a) By progressively querying routers about the next router on the path to B using ICMP packets, starting with the first router.

b) By requiring each router to append the address to the ICMP packet as it is forwarded to B. The list of all routers en-route to B is returned by B in an ICMP reply packet

c) By ensuring that an ICMP reply packet is returned to A by each router en-route to B, in the ascending order of their hop distance from A

d) By locally computing the shortest path from A to B

37. Which of the following statements is TRUE about CSMA/CD **[GATE - 05]**

a) IEEE 802.11 wireless LAN runs CSMA/CD protocol

b) Ethernet is not based on CSMA/CD protocol

c) CSMA/CD is not suitable for a high propagation delay network like satellite network

d) There is no contention in a CSMA/CD network

38. Which of the following statements is FALSE regarding a bridge **[GATE - 05]**

a) Bridge is a layer 2 device

b) Bridge reduces collision domain

c) Bridge is used to connect two or more LAN segments

d) Bridge reduces broadcast domain

39. Count to infinity is a problem associated with

a) link state routing protocol. **[GATE - 05]**

b) distance vector routing protocol

c) DNS while resolving host name

d) TCP for congestion control

40. A network with CSMA/CD protocol in the MAC layer is running at 1 Gbps over a 1 km cable with no repeaters. The signal speed in the cable is 2×10^8 m/sec. The minimum frame size for this network should be

a) 10000 bits b) 10000 bytes **[GATE - 05]**

c) 5000 bits d) 5000 bytes

41. A channel has a bit rate of 4 kbps and one-way propagation delay of 20 ms. The channel uses stop and wait protocol. The transmission time of the acknowledgement frame is negligible. To get a channel efficiency of at least 50%, the minimum frame size should be

a) 80 bytes b) 80 bits **[GATE - 05]**

c) 160 bytes d) 160 bits

42. On a TCP connection, current congestion window size is Congestiion Window = 4 KB. The window size advertised by the receiver is Advertise Window = 6 KB. The last byte sent by the sender is Last Byte Sent = 10240 and the last byte acknowledged by the receiver is Last Byte Acked = 8192. The current window size at the sender is **[GATE - 05]**

a) 2048 bytes b) 4096 bytes

c) 6144 bytes d) 8192 bytes

43. In a communication network, a packet of length L bits takes link L1 with a probability of p1 or link L2 with a probability of p2. Link L1 and L2 have bit error probability of b_1 and b_2 respectively. The probability that the packet will be received without error via either L1 or L2 is **[GATE - 05]**

a) $(1-b_1)^L p_1 + (1-b_2)^L p_2$ b) $\left[1-(b_1+b_2)^L\right] p_1 p_2$

c) $(1-b_1)^L (1-b_2)^L p_1 p_2$ d) $1-\left(b_1^L p_1 + b_2^L p_2\right)$

44. In a TDM medium access control bus LAN, each station is assigned one time slot per cycle for transmission. Assume that the length of each time slot is the time to transmit 100 bits plus the end-to-end propagation delay. Assume a propagation speed of 2×10^8 m/sec. The length of the LAN is 1 km with a bandwidth of 10 Mbps. The maximum number of stations that can be allowed in the LAN so that the throughput of each station can be 2/3 Mbps is [GATE - 05]

a) 3 b) 5 c) 10 d) 20

45. A company has a class C network address of 204.204.204.0. It wishes to have three subnets, one with 100 hosts and two with 50 hosts each. Which one of the following options represents a feasible set of subnet address/subnet mask pairs ? [GATE - 05]

a) 204.204.204.128/255.255.255.192
204.204.204.0/255.255.255.128
204.204.204.64/255.255.255.128

b) 204.204.204.0/255.255.255.192
204.204.204.192/255.255.255.128
204.204.204.64/255.255.255.128

c) 204.204.204.128/255.255.255.128
204.204.204.192/255.255.255.192
204.204.204.224/255.255.255.192

d) 204.204.204.128/255.255.255.128
204.204.204.64/255.255.255.192
204.204.204.0/255.255.255.192

46. Assume that "host1.mydomain.dom" has an IP address of 145.128.16.8. Which of the following options would be most appropriate as a subsequence of steps in performing the reverse lookup of 145.128.16.8 ? In the following options "NS" is an abbreviation of "nameserver"

a) Query a NS for the root domain and then NS for the "dom" domains. [GATE - 05]

b) Directly query a NS for "dom" and then a NS for "mydomain.dom" domains

c) Query a NS for in-addr.arpa and then a NS for 128.145.in-addr.arpa domains

d) Directly query a NS for 145.in-addr.arpa and then a NS for 128.145.in-addr.arpa domains.

47. Consider the following message M = 1010001101. The cyclic redundancy check (CRC) for this message using the divisor polynomial $x^5 + x^4 + x^2 + 1$ is : [GATE - 05]

a) 01110 b) 01011 c) 10101 d) 10110

48. Suppose that two parties A and B wish to setup a common secret key (D-H key) between themselves using the Diffie-Hellman key exchange technique. They agree on 7 as the modulus and 3 as the primitive root. Party A chooses 2 and party B chooses 5 as their respective secrets. Their D-H key is [GATE - 05]

a) 3 b) 4 c) 5 d) 6

ANSWER KEY

1. a	2. d	3. c	4. d	5. b
6. a	7. b	8. a	9. c	10. c
11. b	12. c	13. a	14. b	15. a
16. c	17. a	18. d	19. b	20. b
21. d	22. d	23.	24. b	25.
26. c	27. b	28. b	29. d	30. b
31. b	32. d	33. d	34. d	35. a
36. c	37. c	38. d	39. b	40. a
41. d	42. b	43. a	44. b	45. d
46.	47. c	48.		

EXPLANATORY NOTES

3. Subnect mask 255.255.31.0 shows class C network So, class C network. IP address is

191.203.31.87 and 191.234.31.88

4. Short message are called packet

length = 2km = 2000 m

$\lambda = 10^7$ bps

$\upsilon = 2 \times 10^8$ m/s

$s = \upsilon t$

$t = \dfrac{2 \times 10^3}{2 \times 10^8} = 10^{-5}$ sec = 10μ sec

Total number of packets $= 10^7 \times 10 \times 10^{-8} = 100$ bits

= 12 byte

5. Full Duplex

window size = 5 packets

Packet size = 1000 bytes

$= 2^{10}$ (approx.10bits)

Window size = 5000 bytes (13 bits approx.)

Transmission time = $5 \times 50 = 250\mu$ sec

Propagation delay = 200 μ sec

Total time = 250 + 200 μ sec = 450μ sec

Maximum achievable throughput $= \dfrac{5000}{450} \times 10^6$ bps

$= 11.11 \times 10^6$ bps

9. 128.75.43.16 lies in the range of Eth 0 and 192.12.17.10 lies in the range of Eth 3 so the correct answer is Eth 0 and Eth 3.

10. 1 Mbps 512Kbps

Network A — Network B — Network C

As 3 packets are formed.

Number of bytes delivered correctly including headers to the destination for one application message in best case

= 180 + 20 + 20 + 20 = 240 bytes

11. Ignore erros i,e. noise

B = 2 H (Nyquist theorem)= 354.5 kbps

13. TCP does not guarantees a minimum communication rate.

14. HTTP is hypertext transfer protocol does not contain any information about the web pages

15. If sender is employing public key cryptography to send a secret message to receiver the sender encrypts using receiver's public key.

16. The subnet mask is of class E, which can take value upto 255 . 255 . 255 . 255 . So number of host that belong to this subnet = 254 - 192 = 62.

18. In TCP sequence number is assigned to each packet which is message.

19. As bit stuffing is used and delimiter flag is given by 0111. so in stuffing it will put zero (0) after two consecutive 0's

20. Transmitted window size = N

Race window size = M

Minimum numbers required to execute correct operation of ARQ scheme is maximum of these two which is max (M, N).

21. Effective data transmission rate

$= (10 \text{ Mbps} - 4000 \times 1000)$

$= (10\text{Mbps} - 4\text{Mbps}) = 6 \text{ Mbps}$

28. Because routing is done at IP layer (Network layer), so the transport layer have no interface at network level.

30. Stalling networks assume here window size of 8, i.e., (0-7)

1. Station A under frame 0 through 6 to station B
2. Station B receive all seven frames, and ack with R R 7 (Receive Ready 7)
3. R R 7 is lost
4. A times out and retransmit frame 0.
5. B has already advanced its receive window to accept frames 7, 0, 1, 2, 3, 4 and 5 thus it assume that 7 has been lost and this is a new frame 0.

To overcome this problem, the maximum window size is 2^{n-1}

32. Since, we have to support 64 department

So our first

1 1 1 1 1 1 0 0 = 252

33. We consider optimum packet size which required minimum number of packet and does not waste any space.

Please consider here each packet header because they are all work on packet switching network.

3 bytes is header size;

9 - 3 = 6 is data size

24/6 = 4 are the required packet

For case $1 \rightarrow 24$ packets are required

For case $2 \rightarrow 8$ packets are required

For case $3 \rightarrow 6$ packets are required

34. Question have certain ambiguity. If we consider 46.4 μs total (round) propagation time, then aswer is (c).

If 46.4 μs total (round) propagation time with 48 bit jamming signal transmission time, then aswer is 512 i.e. (d).

Because now we have to subtract transmission time of 48 bit signal from Round propagation time to get actual round propagation time and this is 51.2 μs and hence answer is 512 bit.

37. Notice here that frame size depend on twice the largest propagation delay between any two stations in LAN. So, if propagation delay increase frame size increases and in this case a single station may transmit data most of time and increase waiting time for other station and it may be possible that station doesn't have that most of data in this case pudding have to increase so, channel utilizatiion decreases.

38. Whenever station sends broadcast, then it is received by all stations of LAN. It doesn't matter on which segment they are.

40. (round trip propagation) for covering 2000 m time taken is $\frac{2 \times 1000}{2 \times 10^8} = 1000 \times 10^{-8} = 10^{-5} \text{ sec} = 10 \ \mu \text{sec}$

bits transmit over 1 Gbps link with in 10 μs is 1 bit take 1 nsec = 1.10^{-9} . So 10,000 bits takes 10 μs .

41. $\frac{1}{1+2a}, \Rightarrow a = \frac{1}{2}$, where $a = \frac{\text{Propagation time}}{\text{Transmission time}}$

So transmission time should be 40 ms,

i.e. $\frac{160}{4000} = 40 \text{ ms}$

42. Notice here the point that sender window size is minimum for congestion window and window size advertise by receiver

So after getting acknowledge of 8192 byte sender still have 2048 byte outstanding without any (acknowledged) but sender can send up according to congestion window size.

Unit - XI
Compiler Design

UNIT - XI COMPILER DESIGN

I. INTRODUCTION TO COMPILERS

- **Compilers and Translators**
- **Structure of a compiler**
- **Compiler Writing Tools**

This lesson introduces the basic structure of a compiler. It is primarily a qualitative introduction to the different phases of compiler and the basic steps that go into its complete development.

1. Compilers and Translators

Translator: A translator is a program that takes as input a program written in one programming language(source language) and produces as output a program in another language(object language).

Compiler: If the source language is high-level language and the object language is a low-level language(assembly or machine), then such a translator is called a compiler.

Interpreter: An interpreter is a program that directly executes an intermediate code, where intermediate code is a simplified language which is transformed from a programming language.

Assembler: If the source language is assembly language and the target language is machine language, then the translator is called an assembler.

Preprocessor: Preprocessor is used for translators that take programs in one high-level language into equivalent programs in another high-level language.

Executing a program written in a high-level programming language is basically a two-step process, as illustrated in fig.1.1.

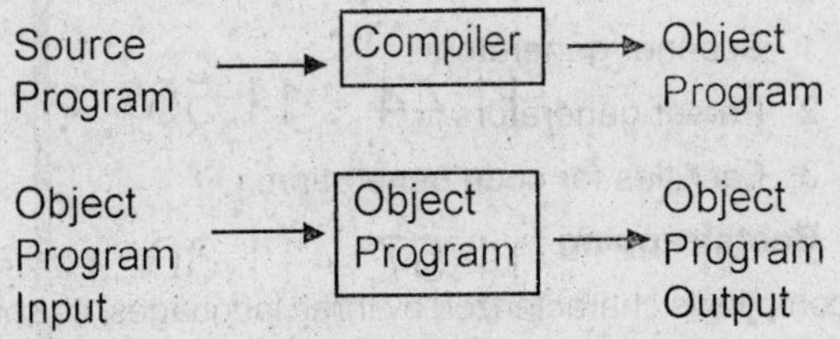

Fig. 1.1. Compilation and execution

The source program must first be compiled, that is, translated into the object program. Then the resulting object program is loaded into memory and executed.

2. The Structure of a compiler

A compiler takes as input a source program and produces as output an equivalent sequence of machine instructions. This process is so complex, and it is customary to partition the compilation process into a series of sub processes called phases, as shown in the figure 1.2

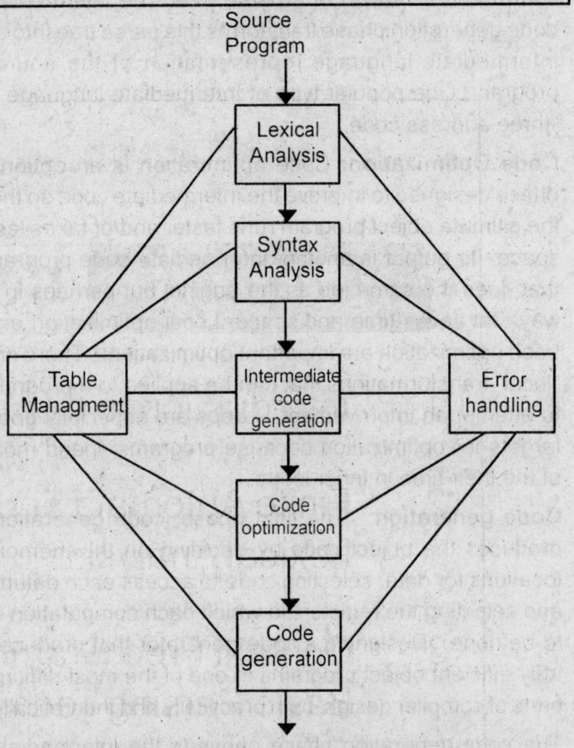

Fig. 1.2. Phases of a Compiler

Lexical Analyzer : The first phase, called the lexical analyzer, or scanner, separates characters of the source language into groups that logically belong together, these groups are called **tokens**. The usual tokens are keywords, identifiers, operator symbols and punctuation symbols. The output of the lexical analyzer is a stream of tokens, which is passed to the next phase, the syntax analyzer, or parser. The tokens in this stream can be represented by codes which we may regard as integers.

Syntax analyzer: The syntax analyzer groups tokens together into syntactic structures. For example, tokens might be grouped into a syntactic structures called an expression. Expressions might further be combined to form statements. Often the syntactic structure can be regarded as a tree whose leaves are the tokens. The interior nodes of the tree represent strings of tokens that logically belong together.

The parser has two functions. It checks that the tokens appearing in its input, which is the output of the lexical analyzer, occur in patterns that are permitted by the specification for the source language. It also imposes on the tokens a tree-like structure that is used by the subsequent phases of the compiler.

Intermediate code generator: The intermediate code generator uses the structure produced by the syntax analyzer to create a stream of simple instructions. Many styles of intermediate code are possible. One common style uses instructions with one operator and a small number of operands. The primary difference between intermediate code and assembly code is that the intermediate code need not specify the registers to be used for each operation.

On a logical level the output of the syntax analyzer is some representation of a parse tree. The intermediate code generation phase transforms this parse tree into an intermediate-language representation of the source program. One popular type of intermediate language is "three address code".

Code Optimization: Code optimization is an optional phase designed to improve the intermediate code so that the ultimate object program runs faster and/or takes less space. Its output is another intermediate code program that does the same job as the original but perhaps in a way that saves time and space. Local optimization and Loop optimization are important optimizations. There are "local" transformations that can be applied to a program to attempt an improvement. Loops are especially good targets for optimization because programs spend most of the their time in inner loops.

Code generation: The final phase, code generation, produces the object code by deciding on the memory locations for data, selecting code to access each datum, and selecting the registers in which each computation is to be done. Designing a code generator that produces truly efficient object programs in one of the most difficult parts of compiler design, both practically and theoretically.

The code-generation phase converts the intermediate code into a sequence of machine instructions. A simple-minded code generator might map the statement A:=B+C into the machine code sequence.

```
LOAD  B
ADD   C
STORE A
```

Table-management: The table-management or bookkeeping, portion of the compiler keeps track of the names used by the program and records essential information about each, such as type. The data structure used to record this information is called symbol table.

The information about data objects may be explicit, as a declarations, or implicit, as in the first letter of an identifier or in the context in which an identifier is used.

The information about data object is collected by the early phases of the compiler - lexical and syntactic analysis - and entered into the symbol table. The information collected about the data objects has a number of uses. The term **semantic analysis** is applied to the determination of type of intermediate results, the check that arguments are of types that are legal for an application of an operator, and the determination of the operation denoted by the operator.

Error handler: The error handler is invoked when a flaw in the source program is detected. It must warn the programmer by issuing a diagnostic, and adjust the information being passed from phase to phase so that each phase can proceed. It is desirable that compilation be completed on flawed programs, atleast through the syntax-analysis phase, so that as many errors as possible can be detected in one compilation.

Both the table management and error handling routines interact with all phases of the compiler.

Errors can be encountered by all the of the phases of a compiler.

Whenever a phase of the compiler discovers an error, it must report the error to the error handler, which issues an appropriate diagnostic message.

3. Compiler writing tools

A number of tools have been developed specifically to help construct compilers. These tools range from scanner and parser generators to complex systems variously called compiler-compilers, compiler-generators, or translator-writing systems, which produce a compiler from some form of specification of a source language and target language. The input specification for these systems may contain:

1. a description of the lexical and syntactic structure of the source language,

2. a description of what output is to be generated for each source language construct, and

3. a description of the target machine.

In many cases, the specification is merely a collection of programs fitted together into a framework by the compiler-compiler. While a number of useful compiler-compilers exist, they have limitations.

The principal aids provided by existing compiler-compilers are:

1. Scanner generators.
2. Parser generators.
3. Facilities for code generation.

4. Bootstrapping

A compiler is characterized by three languages: its source language, its object language, and the language in which it is written.

5. Cross compiler

A compiler which runs on one machine and produces object code for another machine is called as cross-compiler. Many minicomputer and microprocessor compilers are implemented in this way; they run on a bigger machine and produce object code for the smaller machine.

The below figure shows the representation of the statement position : = initial + rate * 60 after each phase of the compiler

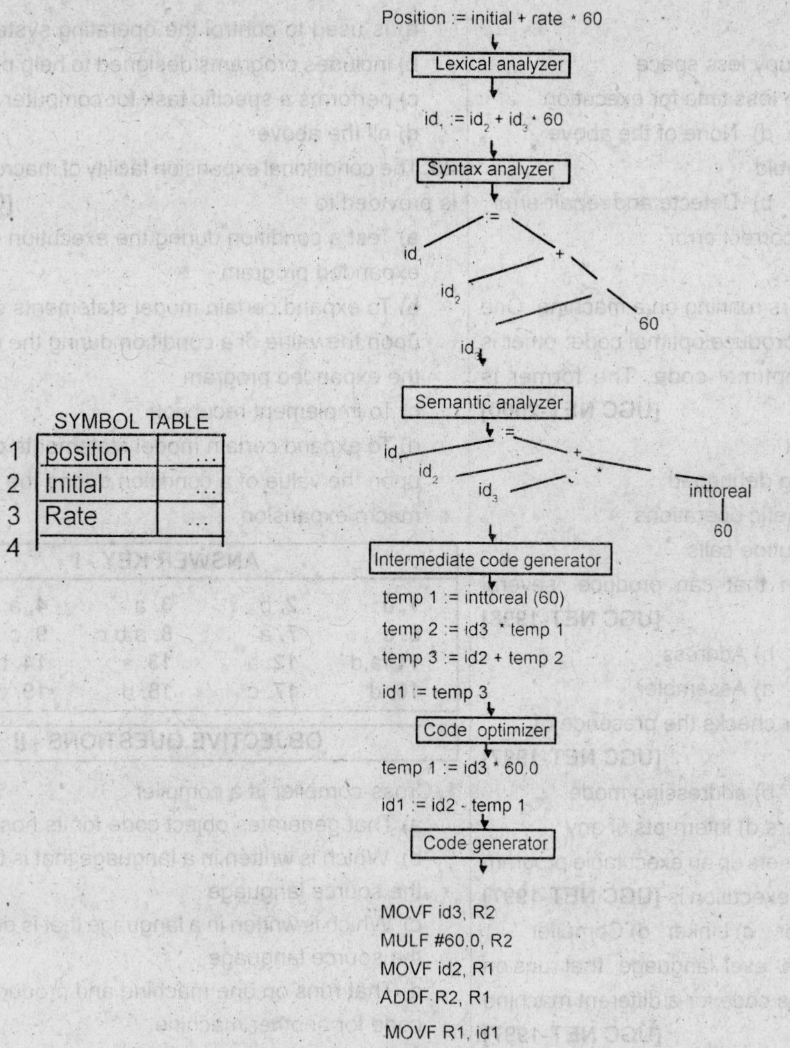

Fig. 1.3. Translation of a statement

OBJECTIVE QUESTIONS - I

1. An ideal compiler should

a) Be small in size

b) Produce object code that is smaller in size and executes faster

c) Take less time for compiling d) All of the above

2. An instruction in a programming language that is replaced by a sequence of instructions prior to assembly or compiling is known as

a) Procedure name b) Macro

c) Label d) Literal

3. Which of the following translator program converts assembly language program to object program

a) Assembler b) Compiler

c) Macro processor d) Linker

4. Which of the following modules does not incorporate initialization of values changed by the module?

a) Non reusable module

b) Serially reusable module

c) Reenterable module d) All of the above.

5. System programs such as Compilers are designed so that they are

a) Reenterable b) Non reusable

c) Serially usable d) Recursive

6. In what module, multiple instances of execution will yield the same result even if one instance has not terminated before the next one has begin?

a) Non reusable module b) Serially reusable

c) Reenterable module d) Recursive module

7. Which table is a permanent database that has entry for each terminal symbol

a) Terminal Table b) Literal Table

c) Identifier Table d) Reductions

8. The cost of developing a compiler

a) Is proportional to the complexity of the source language

b) Is proportional to the complexity of the architecture of the target machine.

c) Is proportional to the flexibility of the available instruction set.

9. An optimizing compiler

a) Is optimized to occupy less space

b) Is optimized to take less time for execution

c) Optimizes the code d) None of the above

10. An ideal compiler should

a) Detect error b) Detecte and repair error

c) Detect, repair and correct error

d) None of the above

11. There are two compilers running on a machine. One is very fast, but does not produce optimal code; other is slow but produces an optimal code. The former is preferred when **[UGC NET-2000]**

a) the program is short

b) the program is being debugged

c) there are few arithmetic operations

d) there are few subroutine calls

12. A type of instruction that can produce several lines of machine code is **[UGC NET-1998]**

a) Mnemonic b) Address

c) Macro d) Assembler

13. Pass-one in assembler checks the presence of **[UGC NET-1997]**

a) labels, if any b) addressing mode

c) hexadecimal numbers d) interrupts, if any

14. A system program that sets up an executable program in main memory ready for execution is **[UGC NET-1997]**

a) Assembler b) Loader c) Linker d) Compiler

15. A compiler for a high-level language that runs on one machine but produces code for a different machine is called **[UGC NET-1997]**

a) optimizing compiler b) multipass compiler

c) cross compiler d) interpreter

16. An 'assembler' in computer terminology means **[UGC NET-1996]**

a) A person who assemles a computer

b) A high level language

c) A programmer who makes a complete program by assembling smaller program segments

d) A program which converts assembly language program in machine code

17. Usage of subroutine over macros saves **[UGC NET-1996]**

a) Compilation time b) Execution time

c) Memory d) I/O time

18. The linker **[UGC NET-1999]**

a) is the same as the loader

b) is required to create a load module

c) is always used before programs are executed

d) uses object code as input

19. Application software **[UGC NET-1999]**

a) is used to control the operating system

b) includes programs designed to help programmers

c) performs a specific task for computer users

d) all the above

20. The conditional expansion facility of macro processors is provided to **[GATE-1997]**

a) Test a condition during the execution of the expanded program

b) To expand certain model statements depending upon the value of a condition during the execution of the expanded program

c) To implement recursion

d) To expand certain model statements depending upon the value of a condition during the process of macro expansion.

ANSWER KEY - I

1. d	**2.** b	**3.** a	**4.** a	**5.** a
6. c	**7.** a	**8.** a,b,c	**9.** c	**10.** c
11. a,d	**12.** a	**13.** a	**14.** b	**15.** c
16. d	**17.** c	**18.** d	**19.** d	**20.** d

OBJECTIVE QUESTIONS - II

1. Cross-compiler is a compiler

a) That generates object code for its host machine

b) Which is written in a language that is the same as the source language

c) Which is written in a language that is different from the source language

d) That runs on one machine and produces object code for another machine.

2. Which of the following class of statements usually produces no executable code when compiled?

a) Declaration statements b) Assignment statements

c) Input and output statements

d) Structural statements

3. Which of the following system program forgoes the production of object code to generate absolute machine code and load it into the physical main storage location from which it will be executed immediately upon completion of the assembly?

a) Two pass assembler b) Load-and-go assembler

c) Micro processor d) Compiler

4. A Compiler is

a) A program that places programs into memory and preparesthem for execution.

b) A program that automate the translation of assembly language into machine language.

c) A program that access a program written in a high-level language and produces an object program.

d) A program that appears to execute a source program as if it were machine language.

5. Which is a permanent database in the general model of compiler?

a) Literal table b) Identifier table
c) Terminal table d) Source code

6. Incremental-compiler is a compiler

a) Which is written in a language that is different from the source language.
b) That generates object code for its host machine.
c) Which is written in a language that is same as the source language.
d) That allows a modified portion of a program to be recompiled.

7. For which of the following reasons, an interpreter is preferred to a compiler?

a) It takes less time to execute
b) It is much helpful in the initial stages of program development.
c) Debugging can be faster and easier.
d) It needs less computer resources

8. For which of the following reasons, a compiler is preferable to an interpreter?

a) It can generate stand-alone programs that often take less time for execution.
b) It is much helpful in the initial stages of program development.
c) Debugging can be faster and easier.
d) If one changes a statement, only that statement needs recompilation.

9. A system program that combines the separately combined module of a program into a form suitable for execution **[UGC NET-1996]**

a) Assembler b) Cross-compiler
c) Linker d) Loader

ANSWER KEY - II				
1. d	**2.** d	**3.** b	**4.** c	**5.** c
6. d	**7.** b,c	**8.** a	**9.** d	

II. FINITE AUTOMATA AND LEXICAL ANALYSIS

- **Rolȩ of the Lexical Analyser**
- **Regular Expressions**
- **Finite Automata**
- **Issues in Lexical Analysis**

1. The Role of the lexical Analyzer

The lexical analyzer returns to the parser a representation for the token it has found. The representation is an integer code if the token is a simple construct such as a left parenthesis, comma, or colon. The representation is a pair consisting of an integer code and a pointer to a table if the token is a more complex element such as an identifier or constant. The integer code gives the token type; the pointer points to the value of that token.

2. The Need for Lexical Analysis

The purpose of splitting the analysis of the source program into two phases, lexical analysis and syntactic analysis, is to simplify the overall design of the compiler. It is easier to specify the structure of tokens than the syntactic structure of the source program. We can construct a more efficient recognizer for tokens than for syntactic structures. Other functions sometimes performed by a lexical analyzer are keeping track of line numbers, producing an output listing if necessary, stripping out white space, and deleting comments.

3. Input Buffering

The lexical analyzer scans the characters of the source program one at a time to discover tokens. Often, however, many characters beyond the next token may have to be examined before the next token itself can be determined. Therefore, it is desirable for the lexical analyzer to read its input from an input buffer.

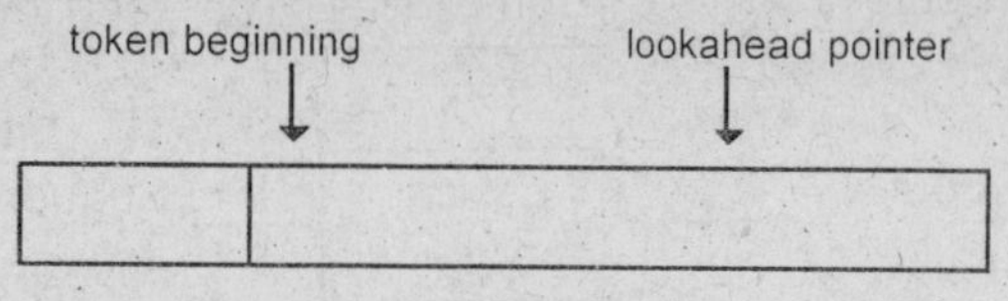

Fig. 2.1 Input Buffer

Fig. 2.1. shows a buffer divided into two halfs of, say 100 characters each. One pointer marks the beginning of the token being discovered. A lookahead pointer scans ahead of the beginning point, until the token is discovered. We view the position of each pointer as being between the character last read and the character next to be read. In practice, each buffering scheme adopts one convention; either a pointer is at the symbol last read or the symbol it is ready to read.

The distance which the lookahead pointer may have to travel past the actual token may be large. If the lookahead pointer travels beyond the buffer half in which it began, the other half must be loaded with the next characters from the source file.

4. Preliminary Scanning

There are certain processes that are best performed as characters are moved from the source file to the buffer. For example, we may delete comments, we can ignore blanks, we may collapse strings of several blanks to one blank. A count of lines may be kept by checking for new line characters. All these processes may be carried out with an extra buffer into which the source file is read and then copied, with modification, into the buffer of fig. 2.1. Alternatively, the condensation of the text may be carried out in place in the buffer of fig.2.1.

5. Regular Expressions

We shall now begin the discussion of the design of a program for generating lexical analyzers automatically. We first introduce a very useful notation, called regular expressions, suitable for describing tokens. We later show how regular expressions can be converted automatically into finite automata, which are just formal specifications for transition diagrams.

6. Finite Automata

A recognizer for a language L is a program that takes as input a string x and answers "yes" if x is a sentence of L and "no" otherwise. Clearly, the part of a lexical analyzer that identifies the presence of a token on the input is a recognizer for the language defining that token.

Suppose we have specified a language by a regular expression R, and we are given some string x. We want to know whether x is in the language L denoted by R. One way to attempt this test is to check that x can be decomposed into a sequence of substrings denoted by the primitive subexpressions in R.

Example : Suppose R is (a I b)* abb, the set of all strings ending in abb, and x is the string aabb. We see that $R = R_1R_2$ where

R_1 = (a I b)* and R_2 = abb. We can verify that a is an element of the language denoted by R_1 and the abb similarly matches R_2. In this way we show that aabb is in the language denoted by R.

7. Nondeterministic Automata

A better way to convert a regular expression to a recognizer is to construct a generalized transition diagram from the expression. This diagram is called a (nondeterministic) finite automaton. It cannot in general be easily simulated by a simple program, but a deterministic finite automaton can be simulated easily.

A nondeterministic finite automaton recognizing the language (a I b)* abb is shown in Figure.2.2.

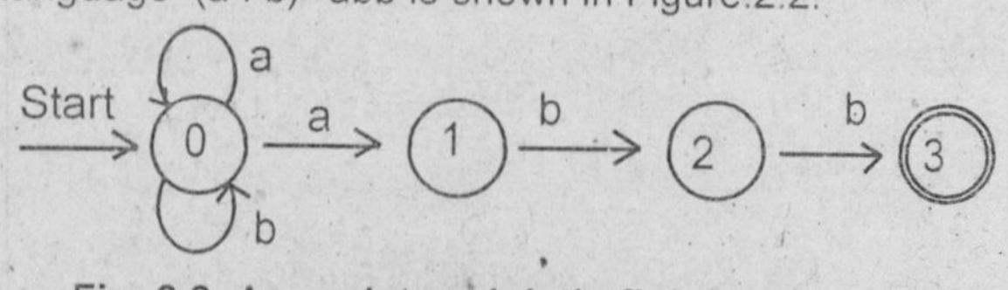

Fig. 2.2. A nondeterministic finite automaton

As can be seen, the nondeterministic finite automaton (hereafter abbreviated NDFA) is a labeled directed graph. The nodes are called states, and the labeled edges are called transitions. The NFA looks almost like a transition diagram, but edges can be labeled by ε as well as characters, and the same character can be label two or more transitions out of one state. One state is distinguished as the start state, and one or more states may be distinguished as final states.

The transitions of an NFA can be conveniently represented in tabular form by means of a transition table. The transition table for the NFA of Fig 2.2. is shown in Fig. 2.3. In the transition table, there is a row for each state and column for each admissible input symbol and ε , if necessary. The entry for row i and symbol a is the set of possible next states for state i on input a.

State	Input Symbol a	b
0	{0,1}	{0}
1	-	{2}
2	-	{3}

Fig. 2.3. Transition table

The NFA accepts an input string x if and only if there is a path from the start state to some accepting state such that the labels along that path spell out x.

The NFA of Fig will accept the input strings abb, aabb, babb, aaabb,....

The language defined by an NFA is the set of input strings it accepts. Note that the NFA of Fig.2.2. accepts (a I b)* abb.

Example : NFA in fig 2.4 recognizes a a* I b b*.

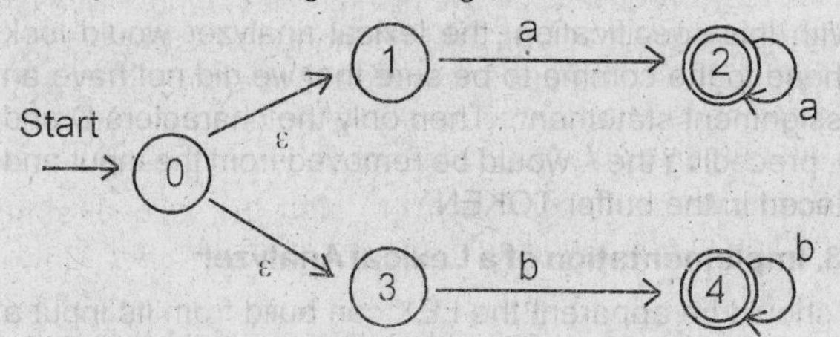

Fig. 2.4. NFA accepting aa* I bb*

8. Deterministic Automata.

We say a finite automaton is deterministic if

1. It has no transitions on input ε

2. For each state s and input symbol a, there is at most one edge labeled a leaving s.

Example: In fig 2.5 we see a deterministic finite automaton (DFA) accepting the language (a I b)* abb, which is the same language as that accepted by the NFA of Fig.2.2.

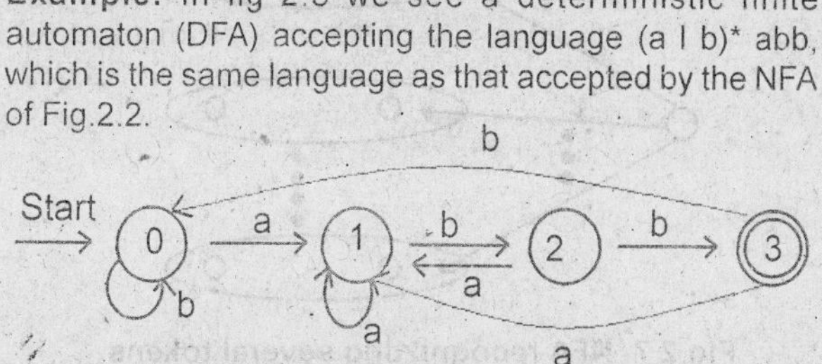

Fig. 2.5 DFA accepting (a|b)*abb

Since there is at most one transition out of any state on any input symbol, a DFA is easier to simulate by a program than is an NFA.

9. A Language for Specifying Lexical Analyzers

We shall now study the question of how one builds a lexical analyzer in the form of a list of regular expressions. The discussion centers around the design of an existing tool, called LEX , for automatically generating lexical analyzers. If the compiler writer does not have a tool like LEX, he may still follow a mental process similar to the method by which LEX constructs a lexical analysis program from its input.

A LEX source program is a specification of a lexical analyzer, consisting of a set of regular expressions together with an action for each regular expression. The action is a piece of code which is to be executed whenever a token specified by the corresponding regular expression is recognized. Typically, an action will pass an indication of the token found to the parser, perhaps with side effects such as making an entry in the symbol table. The output of LEX is a lexical analyzer program constructed from the LEX source specification.

We can view LEX as a compiler for a stylized language, a language that is very good for writing lexical analyzers and some other tasks of a text processing nature. Unlike most programming languages, a source program for LEX does not supply all the details of the intended computation. Rather, LEX itself supplies with its output a program that simulates a finite automaton. This program takes a transition table as data. The transition table is that portion of LEX's output that stems directly from LEX's input. The situation is depicted in Fig. 2.6. where the lexical analyzer L is the transition table plus the program to simulate an arbitrary finite automaton expressed as a transition table. Only L is to be included in the compiler being built.

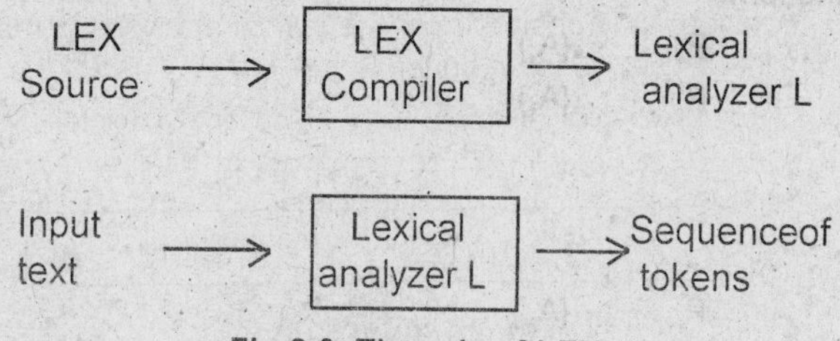

Fig.2.6. The role of LEX

10. Auxiliary Definitions

A LEX source program consists of two parts, a sequence of auxiliary definitions followed by a sequence of translation rules. The auxiliary definitions are statements of the form

$$D_1 = R_1$$
$$D_2 = R_2$$
$$\vdots$$
$$D_n = R_n$$

Where each D_i is a distinct name, and each R_i is regular expression whose symbols are chosen from S U { $D_1, D_2, \ldots, D_{i-1}$}, i.e., characters or previously defined names. The D's are shorthand names for regular expressions. S is our input symbol alphabet, e.g., the ASCII or EBCDIC character sets.To avoid confusion, lower-case strings are used as names for the D_i's, the regular expression names.

Example: We can define the class of identifiers for a typical programming language with the following sequence of auxiliary definitions.

Letter = A I B I ... I Z

digit = 0 I 1I ... I 9

identifier = letter(letter / digit)*

For a more elaborate example of the use of auxiliary definitions, let us consider the definition of FORTRAN arithmetic constants. Having had the definition of digit above, we can define

Integer = digit $^+$

sign = + /. - / ε

signed-integer = sign integer

Then we can define a decimal real number to be

decimal = signed-integer . integer

I signed . integer

I sign . integer

Next, an exponential form real number is a decimal or signed integer followed by E and a signed-integer. That is

exponential=(decimal I signed-integer) E signed-integer

Finally, a real number is defined by

real = decimal I exponential

11. Translation Rules

The translation rules of a LEX program are statements of the form

P_1 $\{A_1\}$

P_2 $\{A_2\}$

.

.

.

P_m $\{A_m\}$

Where each P_i is a regular expression called a pattern, over the alphabet consisting of Σ and the auxiliary definition names. The patterns describe the form of the tokens.

Each A_i, is a program fragment describing what action the lexical analyzer should take when token P_i is found. The A_i 's are written in a conventional programming language; however, rather then using any particular language here, we shall use our informal "pseudo-language" which should be clear to everyone.

To create the lexical analyzer L, each of the A_i 's must be compiled into machine code, just like any other program written in the language of the A_i 's.

The lexical analyzer L created by LEX behaves in the following manner. L reads its input, on e character at a time, until it has found the longest prefix of the input which matches one of the regular expressions, P_i.Once L has found that the prefix, L removes it from the input and places it in a buffer called TOKEN. (Actually, TOKEN may be pair of pointers to the beginning and end of the matched string in the input buffer itself). L then executes the action A_i. After completing A_i, L returns control to the parser. When requested to, L repeats this series of actions on the remaining input.

It is possible, however, that none of the regular expressions denoting the tokens matches any prefix of the input. In that case, an error has occurred, and L presumably transfers control to some error handling routine. It is also possible that two or more patterns match the same longest prefix of the remaining input. It that is the case, L will break the tie in favor of that token which came first in the list of translation rules.

12. The Lookahead Operator

Let us also mention at this point that certain programming languages require a "lookahead" feature in order to specify their lexical analyzers correctly.

A popular example from FORTRAN is the statement.

DO 10 I = 1.25

where we cannot tell until we see the decimal point that this is not a DO statement, but rather an assignment to an identifier DO 10 I. We shall use / as an operator in regular expressions indicating that the token, if found, ends at the point of the /. Then a specification for the keyword DO could be

DO / (letter/digit)* = (letter/digit)*

With this specification, the lexical analyzer would look ahead to the comma to be sure that we did not have an assignment statement. Then only the characters D and O, preceding the /, would be removed from the input and placed in the buffer TOKEN.

13. Implementation of a Lexical Analyzer

It should be apparent the LEX can build from its input a lexical analyzer that behaves roughly like a finite automaton. The idea is to construct a nondeterministic finite automaton N for each token pattern P in the translation rules, using Algorithm 'Constructing an NFA from a regular expression' and then link these NFA's together with a new start state as shown in Fig. 2.7.

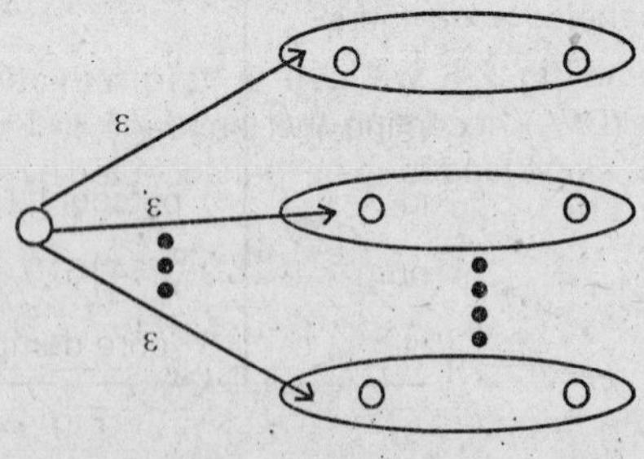

Fig.2.7. NFA recognizting several tokens simultaneously

Next we convert this NFA to a DFA, using the subset construction, Algorithm.

14. The Role of the Lexical Analyzer

The lexical analyzer is the first phase of a compiler. Its main task is to read the input characters and produce as output a sequence of tokens that the parser uses for syntax analysis. This interaction, summarized schematically in Fig. 2.8, is commonly implemented by making the lexical analyzer be a subroutine or a coroutine of the parser. Upon receiving a "get next token" command from the parser, the lexical analyzer reads input characters until it can identify the next token.

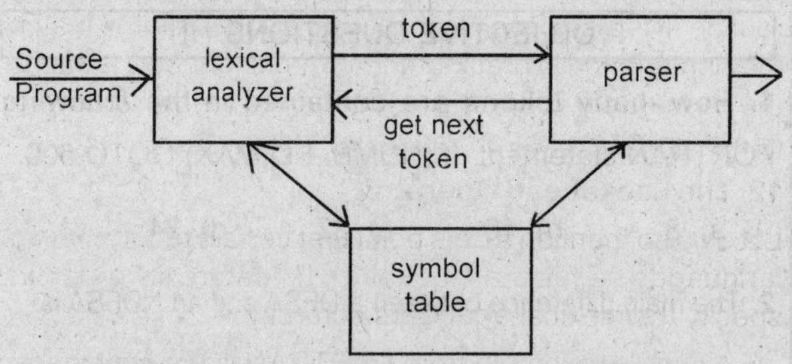

Fig. 2.8. Interaction of lexical analyzer with parser.

Since the lexical analyzer is the part of the compiler that reads the source text, it may also perform certain secondary tasks at the user interface. One such task is stripping out from the source program comments and white space in the form of blank, tab, and newline characters. Another is correlating error messages from the compiler with the source program.

Sometimes, lexical analyzers are divided into a cascade of two phases, the first called "scanning," and the second "lexical analysis." The scanner is responsible for doing simple tasks, while the lexical analyzer proper does the more complex operations. For example, a Fortran compiler might use a scanner to eliminate blanks from the input.

15. Issues in Lexical Analysis

There are several reasons for separating the analysis phase of compiling into lexical analysis and parsing.

1. Simpler design is perhaps the most important consideration. The separation of lexical analysis from syntax analysis often allows us to simplify one or the other of these phases. For example, a parser embodying the conventions for comments and white space is significantly more complex than one that can assume comments and white space have already been removed by a lexical analyzer.

If we are designing a new language, separating the lexical and Syntactic convention can lead to a cleaner overall language design

2. Compiler efficiency is improved. A separate lexical analyzer allows us to construct a specialized and potentially more efficient processor for the task. A large amount of time is spent reading the source program and partitioning it into tokens. Specialized buffering techniques for reading input characters and processing tokens can significantly speed up the performance of a compiler.

3. Compiler portability is enhanced. Input alphabet peculiarities and other device-specific anomalies can be restricted to the lexical analyzer. The representation of special or non-standard symbols, such as ↑ in Pascal, can be isolated in the lexical analyzer. Specialized tools have been designed to help automate the construction of lexical analyzers and parsers when they are separated.

16. Tokens, Patterns, Lexemes

When talking about lexical analysis, we use the terms "token," "pattern," and "lexeme" with specific meanings. Examples of their use are shown in Fig. 2.9. In general, there is a set of strings in the input for which the same token is produced as output.

This set of strings is described by a rule called a pattern associated with the token. The pattern is said to match each string in the set. A lexeme is a sequence of characters in the source program that is matched by the pattern for a token.

For example, in the Pascal statement

const pi = 3.1416;

the substring pi is a lexeme for the token "identifier,"

TOKEN	SAMPLE LEXEME	INFORMAL DESCRIPTION OF PATTERN
const	const	const
if	if	if
relation	<,<=,=,<>,>,>=	< or <= or = or <> or >= or >
id	pi, count, D2	letter followed by letters and digits
num	3.1416, 0, 6.02E23	any numeric constant
litera l	"core dumped"	any characters between "and" except".

Fig. 2.9 Examples of tokens.

A popular example that illustrates the potential difficulty of recognizing tokens is the DO statement of Fortran. In the statement DO 5 I = 1.25
we cannot tell until we have seen the decimalpoint that DO is not a keyword, but rather part of the identifier DO5I. On the other hand, in the statement DO 5 I = 1.25
We have seven tokens, corresponding to the keyword DO, the statement label 5, the identifier I, the operator =, the constant 1, the comma, and the constant 25. Here, we cannot be sure until we have seen the comma that DO is a keyword. To alleviate this uncertainty, Fortran 77 allows an optional comma between the label and index of the DO statement. The use of this comma is encouraged becuase it helps make the DO statement clearer and more readable.

OBJECTIVE QUESTIONS - I

1. Which of the following is used for grouping of characters into tokens (in a compiler)?

a) Parser b) Code optimizer

c) Code generator d) Scanner

2. Whether a given pattern constitutes a token or not

a) Depends on the source language

b) Depends on the target language

c) Depends on the compiler

d) None of the above comments is true

3. Which of the following statements is FALSE?

a) A sentence of a grammar should be derivable from the start state

b) A sentence of a grammar should be frontier of a derivation tree, in which the root nod has the start state as the label.

c) A sentence of a grammar is a sentential form without any terminals.

d) Two different grammars may generate the same language

4. The output of a lexical analyzer is

a) Machine code b) Intermediate code

c) A stream of tokens d) A parse tree

5. In an incompletely specified automata

a) No edge should be labeled epsilon

b) From any given state, there can't any token leading to two different states

c) Some states have no transition on some tokens

d) Start state may not be there

6. Two finite state machines are said to be equivalent if

a) They have the same number of states

b) They have the same number of edges

c) They have same number of states and edges

d) They recognize the same set of tokens

7. If two finite state machines are equivalent, they should have the same number of

a) States b) Edges

c) States and edges d) None of the above

ANSWER KEY - I

1. d	2. a	3. c	4. c	5. c
6. d	7. d			

OBJECTIVE QUESTIONS - II

1. How many tokens are contained in the following FORTRAN statement: If (NUMB. EQ.MAX) GOTO 500

a) 8 b) 10 c) 22 d) 24

2. The main difference between a DFSA and an NDFSA is

a) In DFSA, ε transition may be present

b) In NDFSA, ε transitions may be present

c) In DFSA, from any given state, there can't be any alphabet leading to two different states

d) In NDFSA, from any given state, there can't be any alphabet leading to two different states

3. If two finite states machine M and N are isomorphic, then

a) M can be transformed to N, merely relations re-labelling its states

b) M can be transformed to N, merely relations re-labeling its edges

c) M can be transformed to N, merely re-labelling its states and edges

d) None of the above

4. In some programming language, an identifier is permitted to be a letter followed by any number of letters or digits. If L and D denote the set of letters and digits respectively, which of the following expressions defines an identifier?

a) $(L \cup D)^+$ b) $L(L \cup D)^*$ c) $(L\,D)^*$ d) $L(L\,D)^*$

5. The number of tokens in the following C statement

Printf("i = %d, &i = %x", i, &i); is **[GATE-2000]**

a) 3 b) 26 c) 10 d) 21

6. The number of tokens in the Fortran statement DO 10 I=1.25 Is **[GATE-1999]**

a) 3 b) 4 c) 5 d) None

7. Which of the following strings can definitely be said to be tokens without looking at the next input character while compiling a Pascal program? **[GATE-1995]**

I. Begin II. Program III. <>

a) I b) II

c) III d) All the above

ANSWER KEY - II				
1. a	**2.** c	**3.** a	**4.** b	**5.** c
6. c	**7.** c			

III. CONTEXT FREE GRAMMARS AND BASIC PARSING TECHNIQUES

- Context-Free Grammar
- Parse Trees
- The Role of Parser
- Shift Reduce Parsing
- Operator-Precedence Parsing
- Top-Down Parsing
- Recursive-Descent Parsing
- Predictive Parsers
- Automatic Construction of Efficient Parsers

1. Context-Free Grammar

A grammar involves four quantities: **terminals, non terminals, a start symbol,** and **productions**.

The basic symbols of which strings in the language are composed we shall call terminals. The word "token" is a synonym for "terminal" when we are talking about programming languages. Nonterminals are special symbols. The term "syntactic variable" is synonym for "nonterminal". One nonterminal is selected as the start symbol. The productions (rules) define the ways in which the syntactic categories may be built up from one another and from the terminals. Each production consists of a nonterminal, followed by an arrow, followed by a string of nonterminals and terminals.

2. Notational conventions

1. The following symbols are usually nonterminals:

I. Lower-case names such as expression, statement, operator, etc.;

II. italic capital letters near the beginning of the alphabet;

III. the letter S, which, when it appears is usually the start symbol.

2. The following symbols are usually terminals:

I. single lower-case letters a,b,c,....;

II. operator symbols such as +,-,etc.;

III. punctuation symbols such as parentheses, comma, etc.;

IV. the digits 0,1,......,9;

V. boldface strings such as id or if

3. Capital symbols near the end of the alphabet, chiefly X,Y,Z, represent grammar symbols, that is, either nonterminals or terminals.

4. Small letters near the end of the alphabet, chiefly u,v,....z represent strings of terminals.

5. Lower-case Greek letters, α, β, γ, for example, represent strings of grammar symbols.

6. If $A \rightarrow \alpha_1$, $A \rightarrow \alpha_2$,.......... $A \rightarrow \alpha_k$ are all productions with A on the left (we call them A-productions), we may write $A \rightarrow \alpha_1 | \alpha_2 | ......| \alpha_k$. We call the alternates for A.

7. Unless otherwise stated, the left side of the first production is the start symbol.

3. Derivation And Parse Trees

If $A \rightarrow \gamma$ is a production and α and β are arbitrary strings of grammar symbols, then $\alpha A \beta \overset{\text{derives}}{==>} \alpha \gamma \beta$.

If $\alpha_1 ==> \alpha_2 ==> .... ==> \alpha_n$, then we say α_1 derives α_n

The symbol ==>means "derives in one step."

The symbol $\overset{*}{==>}$ means "derives in zero or more steps".

The symbol $\overset{+}{==>}$ means "derives in one or more steps"

Language generated by grammar G is the set of strings of terminals which are derived from start symbol S using the productions in G. that is,

$$L(G) = \{ w / S \overset{+}{=>} w \text{ and w is string of terminals} \}$$

The string w is called a sentence of G. If $S \overset{*}{==>} \alpha$, where α may contain nonterminals, then we say α is a sentential form of G.

Example : 3.1. Consider the grammar

$$E \rightarrow E + E | E * E | (E) | - E / id$$

Consider the following derivation using this grammar

$$E => -E => -(E) = -(E + E) => -(id + E) => -(id + id)$$

The string $-(id + id)$ is a sentence of this grammar and the strings. E, -E, -(E),...-(id+id) appearing in this derivation are all sentential forms of this grammar.

We can write $E => -(id + id)$. Note that the language of this grammar is the set of all arithme- tic expressions involving the binary operators + and *, the unary operator -, and the operand id.

4. Leftmost derivation and Rightmost derivations

Leftmost derivations are those in which the leftmost nonterminal is replaced at each step of derivation and similarly Rightmost derivations are those in which the rightmost nonterminal is replaced at each step of derivation.

Parse Trees : We can create a graphical representation for derivations that filters out the choice regarding replacement order. This representation is called the parse tree, and it has the important purpose of making explicit the hierarchical syntactic structure of sentences that is implied by the grammar.

Characteristics of a Parse tree

1. A Parse tree consists of a finite set of labeled nodes connected by edges.

2. The labels of the nodes are symbols of terminals and nonterminals.

3. Each interior node of the parse tree is labeled by some nonterminal A, and the children of the node are labeled, from left to right, by the symbols in the right side of the production by which this A was replaced in the derivation.

4. The leaves of the parse tree are labeled by nonterminals or terminals and, read from left to right, they constitute a sentential form, called the yield of the tree.

5. Ambiguity

Definition A grammar that produces more than one parse tree for some sentence is said to be ambiguous. In other words, an ambiguous grammar is one that produces more than one leftmost or more than one rightmost derivation for some sentence.

Example 3.2: Let us again consider the grammar in example 3.1, for which the sentence id + id * id has the two distinct leftmost derivations. They are

E =>	E + E	E =>	E * E
=>	Id + E	=>	E + E * E
=>	Id + E * E	=>	id + E * E
=>	Id + id * E	=>	id + id * E
=>	Id + id * id	=>	id + id * id

With the two corresponding parse trees shown inFig.3.1

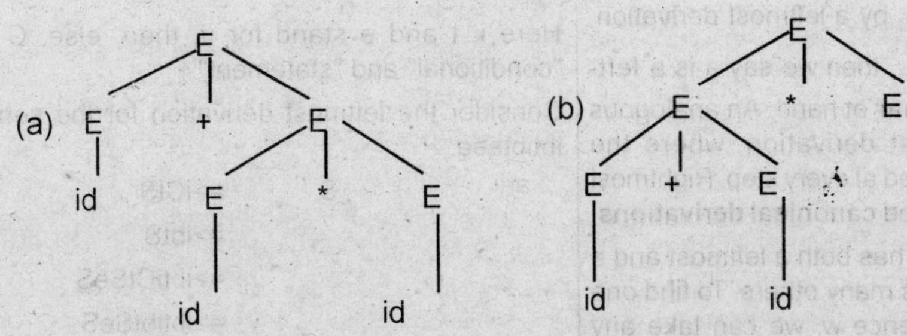

Fig. 3.1. Two parse trees for id + id * id

Suppose we wish to give the operators the following precedence in decreasing order:

- (unary minus)

↑

* /

+ -

and we wish ↑ to be right-associative and the other binary operators to be left-associative, then Fig.3.1(b) would not be a valid parse tree for id + id * id according to these rules because there + appears to have higher precedence than *.

Example 3.3: Let us consider the grammar

$S \rightarrow aS, \quad S \rightarrow a.$

The general derivation for this grammar is

$S \Rightarrow a^{n-1}S$ (applying $S \rightarrow aS$, (n-1) times)

$\Rightarrow a^n$ (applying $S \rightarrow a$ once)

This derivation is only one leftmost derivation of a^n. Thus this grammar is unambiguous.

6. Parser

A parser for grammar G is a program that takes as input string w and produces as output either a parse tree for w, if w is a sentence of G, or an error message indicating that w is not a sentence of G.

The Role of the Parser

In our compiler mode, the parser obtains a string of tokens from the lexical analyzer, as shown in Fig 3.2, and verifies that the string can be generated by the grammar for the source language. We except the parser to report any syntax errors in an intelligible fashion. It should also recover from commonly occuring errors so that it can continue processing the remainder of its input.

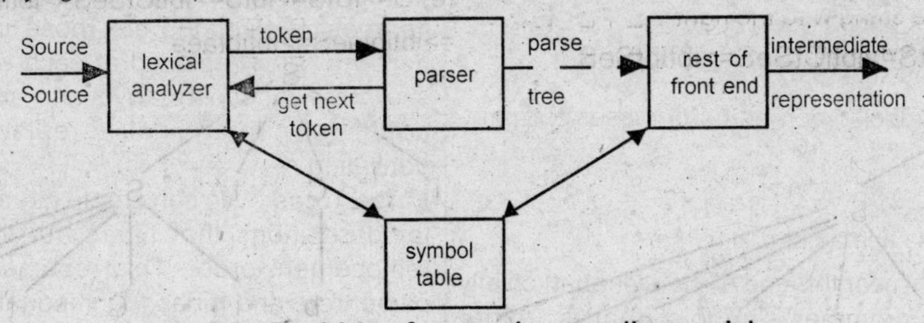

Fig. 3.2. Position of parser in compiler model

This lesson discusses the operation of two basic types of parsers for context-free grammars.

1. bottom-up 2. top-down.

As indicated by their name, bottom-up parsers build parse trees from the bottom (leaves) to the top(root), while top-down parsers start with the root and work down to the leaves. In both cases the input to the parser is being scanned from left to right, one symbol at a time. The bottom-up parsing method is called "shift-reduce" parsing and the top-down parsing method is called "recursive descent" parsing.

An operator-precedence parser is one kind of shift reduce parser and predictive parser is one kind of recursive descent parser.

7. Representation of Parse Tree

In this lesson we shall treat the output of a parser as representation of a parse tree for the input, if the input is syntactically well formed. There are two basis types of representations we shall consider

1. implicit 2. explicit.

The sequence of productions used in some derivation is an example of an implicit representation A linked list structure for the parse tree is an explicit representation.

If $\alpha => \beta$ by a step in which the leftmost nonterminal in α is replaced, we write $\alpha \underset{lm}{=>} \beta$.

Every leftmost step, using our Notational conventions, has the form $wA\gamma \underset{lm}{=>} w\delta\gamma$ in which w consists of terminals only. If α derives β by a leftmost derivation, we write a $\alpha \underset{lm}{\overset{*}{=>}} \beta$. If $S \underset{lm}{\overset{*}{=>}} \alpha$, then we say a is a **left-sentential form** of the **grammar** at hand. An analogous definition holds for rightmost derivation, where the rightmost nonterminal is replaced at every step. Rightmost derivations are sometimes called **canonical derivations**.

Every sentence of a language has both a leftmost and a rightmost derivation, as well as many others. To find one leftmost derivation for a sentence w, we can take any derivation for w and construct from it the corresponding parse tree T. From T we can then construct a leftmost derivation by traversing the **tree top-down**. We begin with the start symbol S, which corresponds to the root of T. We then construct the leftmost derivation

$$S = \alpha_1 \underset{lm}{=>} \alpha_2 \underset{lm}{=>} \ldots\ldots \underset{lm}{=>} \alpha_n = w$$

corresponding to T, one step at a time.

If we were to construct a parse tree in preorder, then the order in which the nodes are created corresponds to the order in which the productions are applied in a leftmost derivation.

Example 3.4 consider the grammar

1. $S \rightarrow iCtS$ 2. $S \rightarrow iCtSeS$ 3. $S \rightarrow a$
4. $C \rightarrow b$

Here i, t and e stand for if, then, else, C and S for "conditional" and "statement".

Consider the leftmost derivation for the sentence w = ibtibtaea.

S =>iCtS
=>ibtS
=>ibtiCtSeS
=>ibtibtSeS
=>ibtibtaeS
=>ibtibtaea

The portions of the parse tree corresponding to the steps of this derivation are shown in Fig.3.3.

(a) S=>iCtS

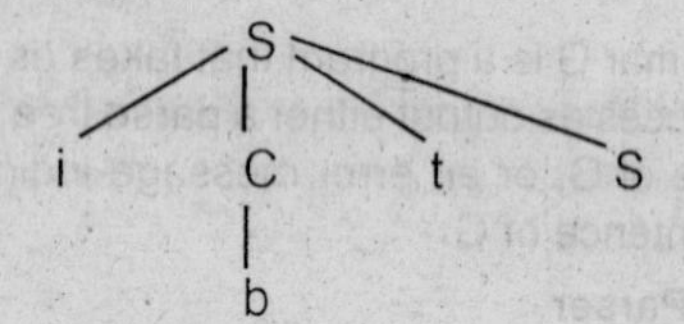

(b) S=>iCtS=>ibtS

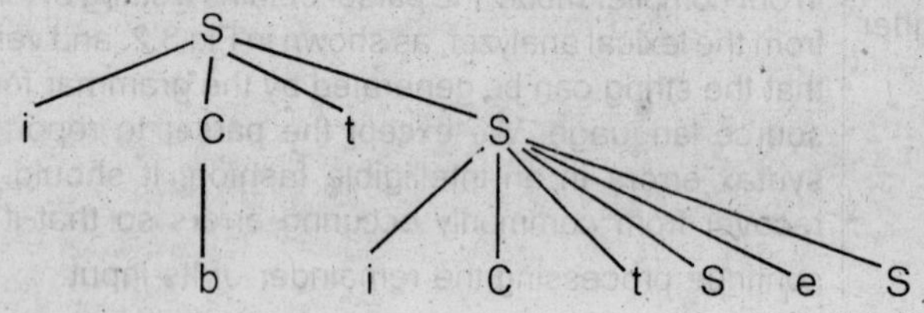

c) S => iCtS=>ibtS=>ibtiCiSeS

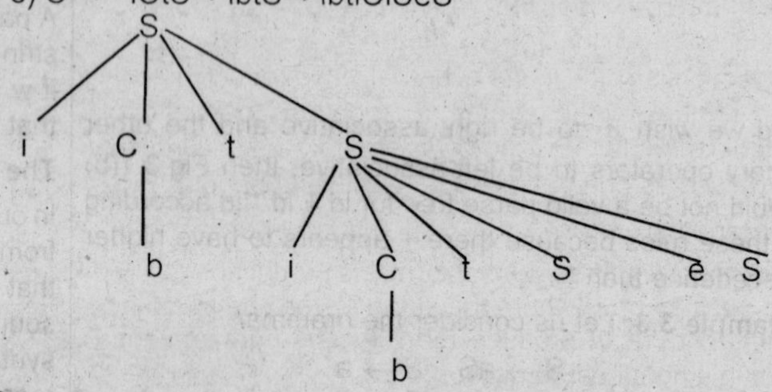

(d) S=>iCtS => ibtS=>ibtiCtSeS=>ibtibtSeS

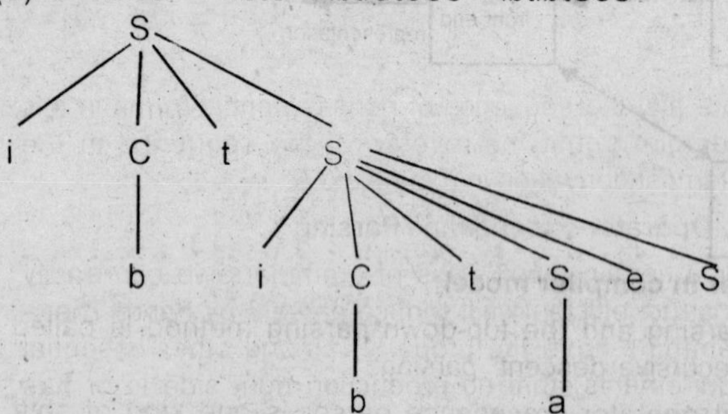

(e) S=>iCtS=>ibtS=>ibtiCtSeS=>ibtibtSeS =>ibtibtaeS=>ibtibtaea

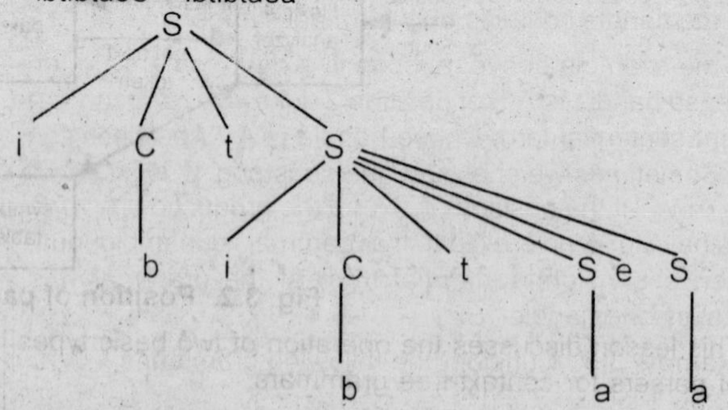

Fig. 3.3. Constructing a leftmost derivation

It should now be clear that a leftmost (or rightmost) derivation is equivalent to a parse tree, in that we can easily convert a leftmost (or rightmost) derivation to a parse tree, or a parse tree to a leftmost (or rightmost) derivation.

8. Shift-Reduce Parsing

In this section we discuss a bottom-up style of parsing called shift-reduce parsing. This parsing method is bottom-up because it attempts to construct a parse tree for an input string beginning at the leaves (the bottom)

and working up towards the root (the top). We can think of this process as one of "reducing" a string w to the start symbol of a grammar. At each step a string matching the right side of a production is replaced by the symbol on the left.

For example, consider the grammar

(1) $S \rightarrow aAcBe$ (2) $A \rightarrow Ab$ (3) $A \rightarrow b$
(4) $B \rightarrow d$

and the string abbcde. We want to reduce this string to S. We scan abbcde looking for substrings that match the right side of some production. The substrings b and d qualify.

Let us choose the leftmost b and replace it by A, the left side of the production $A \rightarrow b$.

We obtain the string aAbcde. We now find that Ab, b, and d each match the right side of some production. Suppose this time we choose to replace the substring Ab by A, the left side of the production $A \rightarrow Ab$. We now obtain aAbcde. Then replacing d by B, the left side of the production $B \rightarrow d$, we obtain aAcBe. We can now replace this entire string by S.

Each replacement of the right side of a production by the left side in the process above is called a reduction. Thus, by a sequence of four reductions we were able to reduce abbcde to S. These reductions, in fact, traced out a right most derivation in reverse.

Informally, a substring which is the right side of a production such that replacement of that substring by the production left side leads eventually to a reduction to the start symbol, by the reverse of a rightmost derivation, is called a "handle." The process of bottom-up parsing may be viewed as one of finding and reducing handles.

9. Handles

A handle of right-sentential form γ is production $A \rightarrow \beta$ and position of γ where the string β may be found and replaced by A to produce the previous right-sentential form in a rightmost derivation of γ.

That is, if $S \overset{*}{\underset{rm}{\Rightarrow}} \alpha Aw \underset{rm}{\Rightarrow} \alpha\beta w$ then $A \rightarrow \beta$ in position following a is a handle of $\alpha\beta w$. The string w to the right of the handle contains only terminal symbols.

In the example above, abbcde is a right-sentential form whose handle is A-b at position 2. Likewise, aAbcde is a right-sentential form whose handle is A - Ab at position 2. Sometimes we shall say "the substring β is a handle of $\alpha\beta w$" if the position of β and the production $A \rightarrow \beta$ we have in mind are clear. If a grammar is unambiguous, then every right-sentential form of the grammar has exactly one handle.

Example 3.5. Consider the following grammar

(1) $E \rightarrow E+E$, (2) $E \rightarrow E*E$, (3) $E \rightarrow (E)$, (4) $E \rightarrow id$

and consider the rightmost derivation

E => E + E
=> E + E * E
=> E + E * id3
=> E + id2 * id3
=> id1 + id2 * id3

id_1 is a handle of the right-sentential form $id_1 + id_2 * id_3$ because id is the right side of the production $E \rightarrow id$, and replacing id_1 by E produces the previous right-sentential form $E + id_2 * id_3$. Note that the string appearing to the right of a handle contains only terminal symbols.

10. Handle Pruning

A rightmost derivation in reverse, often called a canonical reduction sequence, is obtained by "handle pruning". That is, we start with a string of terminals w which we wish to parse. If w is sentence of the grammar at hand, then $w = \gamma_n$, where γ_n is the nth right-sentential form of some as yet unknown rightmost derivation.

$$S = \gamma_0 \Rightarrow \gamma_1 \Rightarrow \ldots\ldots \Rightarrow \gamma_{n-1} \Rightarrow \gamma_n = w$$

To reconstruct this derivation in reverse order, we locate the handle, β_n in γ_n and replace β_n by the left side of some production $A_n \rightarrow \beta_n$ to obtain the (n-1)st right-sentential form γ_{n-1}.

We then repeat this process. This is, we locate the handle β_{n-1} in γ_{n-1} and reduce this handle to obtain the right-sentential form γ_{n-2} consisting only of the start symbol S, then we halt and announce successful completion of parsing. The reverse of the sequence of productions used in the reductions is rightmost derivation for the input string.

Example 3.6. Consider the grammar

1. $E \rightarrow E+E$, 2. $E \rightarrow E*E$, 3. $E \rightarrow (E)$, 4. $E \rightarrow id$

and the input string $id_1 + id_2 * id_3$. The following sequence of reduction reduces $id_1 + id_2 * id_3$ to the start symbol E:

Right-sentential form	Handle	Reducing production
$id_1 + id_2 * id_3$	id_1	$E \rightarrow id$
$E + id_2 * id_3$	id_2	$E \rightarrow id$
$E + E * id_3$	id_3	$E \rightarrow id$
$E + E * E$	$E * E$	$E \rightarrow E * E$
$E + E$	$E + E$	$E \rightarrow E + E$
E		

Note that the sequence of right-sentential forms in this example is just the reverse of the sequence in the rightmost derivation in Example 3.5.

11. Operator-Precedence Parsing

For a certain small class of grammars we can easily construct efficient shift reduce parsers by hand. These grammars have the property (among other essential requirements) that no production right side is or has two adjacent nonterminals.

A grammar in which the right side production has two adjacent nonterminals is called Operator grammar.

Example 3.7. The following grammar for expressions

$$E \rightarrow EAE \mid (E) \mid -E \mid id$$

$$A \rightarrow + \mid - \mid * \mid / \mid \uparrow$$

is not an operator grammar, because the right side EAE has two (in fact three) consecutive nonterminals. However, if we substitute for A each of its alternates, we obtain the following operator grammar:

$$E \to E + E|E - E|E * E|E / E|E \uparrow E|(E)|-E|id$$

In operator-precedence parsing, we use three disjoint precedence relations, <. , = and . > , between certain pairs of terminals. These precedence relations guide the selection of handles.

If a<. b, a "yields precedence to" b

If a = b, a "has the same precedence as" b

If a .> b, a "takes precedence over" b.

There are two common ways of determining what precedence relation should hold between a pair of terminals. The first method we discuss is intuitive and is based on the traditional notions of associativity and precedence of operators.

For example, if * is to have higher precedence then +, we make + < . * and * . > +.

The second method of selecting operator-precedence relations is to first construct an unambiguous grammar for the language, a grammar which reflects the correct associativity and precedence in its parse trees. Having obtained an unambiguous grammar, there is a mechanical method for constructing operator-precedence relations from it.

12. Operator-Precedence Grammar

We shall now define the term "operator-precedence grammar", show how to compute its precedence relations, and explain the details of shift reduce parsing using precedence relations.

To begin, let G be an free operator grammar (i.e., no right side is and no right side has a pair of adjacent nonterminals). For example consider the grammar.

1. $S \to iCtS$ 2. $S \to iCtSeS$ 3. $s \to a$ 4. $C \to b$ -(*)

For each two terminal symbols a and b, we say:

I. a = b if there is a right side of a production of the form $\alpha a\beta b\gamma$, where β is either ε or a single nonterminal. That is a = b if a appears immediately to the left of b in a right side, or if they appear separated by one nonterminal. For example, in grammar (*) the production S - i CtSeS implies that $i = t$ and $t = e$.

II. a <. b if for some nonterminal A there is a right side of the form $\alpha aA\beta$ and $A \overset{+}{=>} \gamma b\delta$ where γ is either ε or a single nonterminal. That is, a.b if a nonterminal A appears immediately to the right of a and derives a string in which b is the first terminal symbol. For example, in grammar (*), we have i immediately to the left of C in

$S \to iCtS$, and $C \overset{+}{=>} b$, , so i <. b.

The derivation is one step, and $\gamma = \varepsilon$ here. Also define $ <. b if there is a derivation $S \overset{+}{=>} \gamma b\delta$ and γ is ε or a single nonterminal.

III. a.> b if for some nonterminal A there is a right side of the form α Abβ , and $A \overset{+}{=>} \gamma a\delta$, where δ is either ε or a single nonterminal. That is, a.b if a nonterminal appearing immediately to the left of b derives a string whose last terminal is a. For example, in grammar (*). $S \to iCtS$ and $C \overset{+}{=>} b$ imply b .> t. Also, define a.> $ if $S \overset{+}{=>} \gamma a\delta$ and δ is either ε or a single nonterminal.

Definition : An **operator-precedence grammar** is an free operator grammar in which the precedence relations <., = and .> constructed as above are disjoint. That is, for any pair of terminals a and b, never more than one of the relations a <. b, a = b , and a.>b is true.

Example 3.8. Let us consider the operator grammar in which there are only the operators + and *.

$$E \to E + E|E * E|(E)|id \qquad \text{--(*)}$$

This grammar is not an operator-precedence grammar because two precedence relations hold between certain pairs of terminals. We shall show that two precedence relations hold between + and +.

In rule (iii) defining the .> relation, we could let the right side α Abβ be E + E. That is, $\alpha = \varepsilon$, A = E, b = +, and $\beta = E$. Then the derivation $A \overset{+}{=>} \gamma a\delta$ could be the one-step derivation E => E+E, where $\gamma = E$, a = + and $\delta = E$. We have thus shown the relation + . > +.

We can similarly apply rule (ii), the definition <.. Let right side α a Ab β be E+E again; here a is + and A is E. The derivation $E \overset{+}{=>} E + E$ shows that E can derive strings whose first terminal is +. Thus, +.+. Since both + < . + and + . > +, grammar (*) is not operator-precedence grammar.

Now, we can transform grammar (*) into an equivalent grammar that is both operator-precedence and unambiguous. It is:

$$E \to E + T|T$$
$$T \to T * F|F \qquad \text{---(**)}$$
$$F \to (E)\,|\,id$$

Hence E, T, and F stand for expression, term, and factor.

Let us now develop the algorithm necessary to construct precedence relations. If we observe the methodology used in Example 3.8, we notice we first constructed two sets LEADING(A) and TRAILING(A) for each nonterminal A, definedby:

LEADING (A) = $\left\{ a | A \overset{+}{=>} \gamma a\delta, \text{ where } \gamma \text{ is } \varepsilon \text{ or a single nonterminal} \right\}$

TRAILING(A) = $\left\{ a | A \overset{+}{=>} \gamma a\delta, \text{ where } \delta \text{ is } \varepsilon \text{ or a single nonterminal} \right\}$

We determined by relation by looking in each side for adjacent symbols...aA.... Then a was related by . to each b is LEADING(A). The **TRAILING** sets similarly helped compute the relations.

Let us therefore begin with an algorithm to compute LEADING. The computation is based on two rules.

1. a is in LEADING(A) if there is a production of the form $A \rightarrow \gamma a\delta$, where γ is ε or a single nonterminal.

2. if a is in LEADING(B), and there is a production of the form A, $A \rightarrow B\alpha$, then a is in LEADING(A).

Our computational method will be to construct a Boolean array L[A,a], where L[A,a] is to be set to true if and only if terminal a is in LEADING(A), for nonterminal A. We initialize L[A,a] to be true if and only if it so follows from rule (1) above. We keep a stack of pairs (B,a) such that L[B,a] has been set to true, but the pair (B,a) has not been used in rule (2) to find new pairs (A,a) such that a is inLEADING(A).

Algorithm 3.1 Computation of LEADING.

Input: A context-free grammar G.

Output: The Boolean array L[A,a] in which the entry for indices A and a is true if and only if a is in LEADING(A).

Method: We need a procedure INSTALL(A,a), which sets L[A,a] to true, and if it was not previously true, places (A,a) on STACK, a stack of nonterminal-terminal pairs that have not yet been used in rule (2) above to find new pairs.

```
    procedure INSTALL(A,a);
        If not L[A,a] then
            begin
                L[A,a]:=true;
                push (A,a) onto STACK
    end
```

The main procedure is given in Fig.3.4

```
        begin /* initialize L according to rule (1) */
(1) for each nonterminal A and terminal a do L(A,a) :=false;
(2) for each production of form A → a or A → Baα do
    INSTALL (A, a);
    /* main loop considers each (B, a) on STACK to add
new pairs (A, a) by rule (2) */
(3) while STACK not empty do
    begin
        pop top pair (B, a) from STACK;
        for each production of the form A → Bα do
            INSTALL(A, a)
            end
    end
```

Fig. 3.4. computing LEADING

The algorithm for calculating TRAILING is analogous to Algorithm 3.1. The complete algorithm for finding precedence relations can now be stated.

Algorithm.3.2. Computing operator-precedence relations.

Input: An operator grammar G.

Output: the relations <. , $\doteq$, and . > for G.

Method:

1. Compute LEADING(A) and TRAILING(A) for each nonterminal A.

2. Execute the program of Fig. 3.5 examining each position of the right side of each production.

for each production $A \rightarrow X_1\ X_2\ \ldots\ldots\ X_n$ do
for i:=1 to n-1 do
begin
if X_i and X_{i+1} are both terminals then set $X_i \doteq X_{i+1}$;
if $i \leq n-2$ and X_i and X_{i+2} are terminals
and X_{i+1} is a nonterminal then set $X_i \doteq X_{i+2}$
if X_i is a terminal and X_{i+1} is a nonterminal
then for all a in LEADING (X_{i+1}) do set $X_i <\cdot a$;
if X_i is a nonterminal and X_{i+1} is terminal then
for all a in TRAILING (X_i) do set $a \cdot> X_{i+1}$
end

3. Set $ <. a for all a in LEADING(S) and set b > $ for all b in TRAINING(S), where S is the start symbol of G.

Fig.3.5. Calculation of operator-precedence relations

13. Top-Down Parsing

We first discuss a general from of top-down parsing that may involve backtracking, that is, making repeated scans of the input. We then discuss a special case, called recursive-descent parsing, which eliminates the need for backtracking over the input.

Top-down parsing can be viewed as an attempt to find a leftmost derivation for an input string. Equivalently, it can be viewed as attempting to construct a parse tree for the input starting from the root and creating the nodes of the parse tree in preorder.

For example, consider the grammar

$S \rightarrow cAd \quad A \rightarrow ab|a$

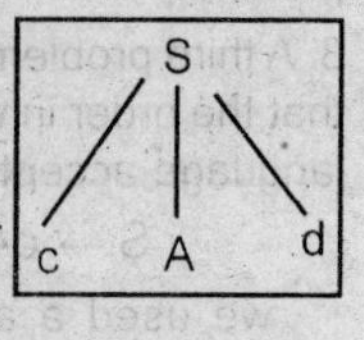

and the input w = cad. To construct a parse tree for this sentence top-down, we initially create a tree consisting of a single node labeled S. An input pointer points to c, the first symbol of w. We then use the first production for S to expand the tree and obtain.

The leftmost leaf, labeled c, matches the first symbol of w, so we now advance the input pointer to a, the second symbol of w, and consider the next leaf, labeled A. We can then expand A using the first alternate for A to obtain the tree

We now have a match for the second input symbol.

We now consider d, the third input symbol, and the next leaf, labeled b. Since b does not match d, we report failure and go back to A to see whether there is another alternate for A that we have not tried but which might produce a match.

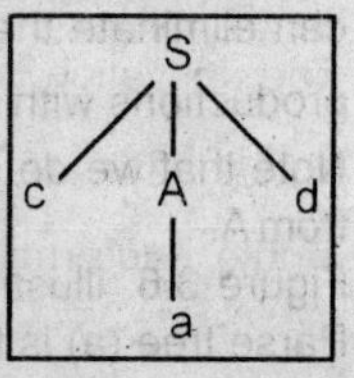

In going back to A we must reset the input pointer to position 2, the position it had when we first came to A. We now

try the second alternate for A to obtain the tree. The leaf a matches the second symbol of w and the leaf d matches the third symbol. Since we have now produced a parse tree for w, we halt and announce successful completion of parsing.

An easy way to implement such a parser is to create a procedure for each nonterminal. If grammars derive an infinite number of strings, recursive procedures are essential. There are several difficulties with top-down parsing as just presented.

1. The first concerns left-recursion. A grammar G is said to be left recursive if it has a nonterminal A such that there is derivation $A \overset{+}{\Rightarrow} A\alpha$ for some α. A left-recursive grammar can cause a top-down parser to go into an infinite loop. That is, when we try to expand A, we may eventually find overselves again trying to expand A without having consumed any input. This cycling will surely occur on an erroneous input string, and it may also occur on legal inputs, depending on the order in which the alternates for A are tried. Therefore, to use top-down parsing, we must eliminate all left-recursion from the grammar.

2. A second problem concerns backtracking. If we make a sequence of erroneous expansions and subsequently discover a mismatch, we may have to undo the semantic effects of making these erroneous expansions. For example, entries made in the symbol table might have to be removed. Since undoing semantic actions requires a substantial overhead, it is reasonable to consider top-down parsers that avoid backtracking. The recursive desent and predictive parsers are types of top-down parsers that avoid backtracking.

3. A third problem with top-down backtracking parsers is that the order in which alternates are tried can affect the language accepted. For example, if, in grammar

$$S \rightarrow cAd \quad A \rightarrow ab|a$$

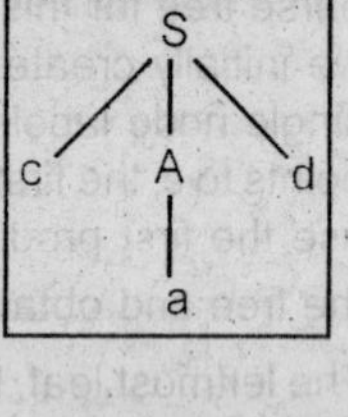

we used a and then ab as the order of the alternates for A, we could fail to accept cabd. That is, with parse tree and ca already matched, the failure of the next input symbol, b, to match, would imply that the alternate cAd for S was wrong, leading the rejection of cabd.

4. Another problem is that when failure is reported, we have very little idea where the error actually occurred. In the form given here, a top-down parser with backtrack simply returns failure no matter that the error is.

14. Elimination of Left-recursion

If we have the left-recursive pair of productions $A \rightarrow A\alpha|\beta$ where b does not begin with an A, then we can eliminate the left-recursion by replacing this pair of productions with $A \rightarrow \beta A' \quad A' \rightarrow \alpha A'|\varepsilon$

Note that we do not change the set of strings derivable from A.

Figure 3.6. illustrates the nature of this transformation. Parse tree (a) is the original grammar, (b) in the new.

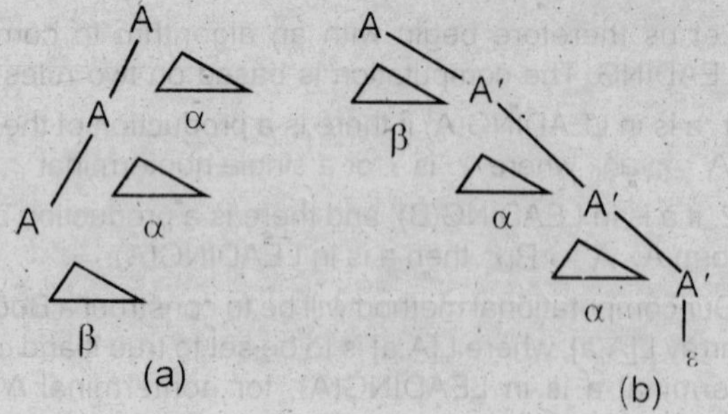

Fig.3.6. Equivalent parse trees

Example 3.9. Consider the following grammar

$$E \rightarrow E + T|T$$

$$T \rightarrow T * F|F \quad ..(*)$$

$$F \rightarrow (E)|id$$

Eliminating the immediate left-recursion (productions of the form $A \rightarrow A\alpha$), we obtain

$$E \rightarrow TE'$$

$$E' \rightarrow +TE'|\varepsilon$$

$$T \rightarrow FT'|\varepsilon \quad \_\_(**)$$

$$T' \rightarrow *FT'$$

$$F \rightarrow (E)|id$$

In general, to eliminate immediate left-recursion among all A-productions we group the A productions as

$$A \rightarrow A\alpha_1|A\alpha_2|........|A\alpha_m|\beta_1|\beta_2|.....|\beta_n$$

Where no β_i begins with an A. Then we replace the A-productions by $A \rightarrow \beta_1 A'|\beta_2 A'|.......|\beta_n A'$

$$A' \rightarrow \alpha_1 A'|\alpha_2 A'|......|\alpha_m A'|\varepsilon$$

This process will eliminate all immediate left-recursion (provided no is), but it will eliminate left-recursion involving derivations of two or more steps. For example, consider

$$S \rightarrow Aa|b$$

$$A \rightarrow Ac|Sd|e$$

The nonterminal S is left-recursive because

S ==> Aa ==> Sda.

15. Recursive-Descent Parsing

In many practical cases a top-down parser needs no backtrack. In order that no backtracking be required, we must know, given the current input symbol a and the nonterminal A to be expanded, which one of the alternates of production

$$A \rightarrow \alpha_1 | \alpha_2 | .....| \alpha_n$$

is the unique alternate that derives a string beginning with a. That is, the proper alternate is detectable by looking at only the first symbol it derives.

A parser that uses a set of recursive procedures to recognize its input with no backtracking is called a recursive- descent parser. The recursive procedures can be quite easy to write and fairly efficient if written in a language that implements procedure calls efficiently.

To avoid the necessity of a recursive language, we shall also consider a tabular implementation of recursive descent, called predictive parsing, where a stack is maintained by the parser, rather than by the language in which the parser is written.

16. Left-Factoring

Often the grammar one writes down is not suitable for recursive-descent parsing, even if there is no left-recursion. For example, if we have the two productions

Statement - - > if condition then statement else statement

| if condition then statement

we could not, on seeing input symbol if, tell which to choose to expand statement.

A useful method for manipulating grammars into a form suitable for recursive-descent parsing is left-factoring, the process of factoring out the common prefixes of alternates.

If $A --> \alpha\beta|\alpha\gamma$ are two A-productions, and the input begins with a nonempty string derived from α ,we do not know whether to expand A to $\alpha\beta$ or to $\alpha\gamma$. We may defer the decision by expanding A to $\alpha A'$. Then, after seeing the input derived from a , we expand A' to β or to γ . That is left-factored, the original productions become

$$A \rightarrow \alpha A',\ A' \rightarrow \beta / \gamma$$

Example : 3.10. Consider again the grammar

$S --> iCtS \mid iCtSeS \mid a$

$C --> b$

Which abstracts the dangling else problem. Left-factored, this grammar becomes:

$S --> iCtSS \mid a$

$S' --> eS$

$C --> b$

Thus we may expand S to iCtS S' on input i, and wait until iCtS has been seen to decide whether to expand S' to eS or to ε

17. Predictive Parsers

A predictive parser is an efficient way of implementing recursive-descent parsing by handling the stack of activation records explicitly. We can picture a predictive parser as in Fig. 3.7.

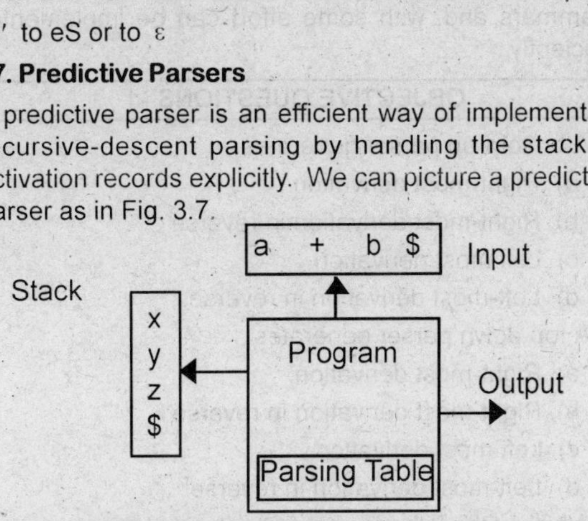

Fig.3.7. Model of a predictive parser

The predictive parser has an input, a stack, a parsing table, and an output. The input contains the string to be parsed, followed by $, the right endmarker. The stack contains a sequence of grammar symbols, preceded by $, the bottom-of - stack marker. Initially, the stack contains the start symbol of the grammar preceded by $. The parsing table is a two dimensional array M [A,a], where A is a nonterminal, and a is a terminal or the symbol $.

18. FIRST and FOLLOW

Now let us consider how to fill in the entries of a predictive parsing table. We need two functions associated with a grammar G. These functions, FIRST and FOLLOW, will indicate the proper entries in the table for G, if such a parsing table G exists. If α is any string of grammar symbols, let FIRST( ) be the set of terminals that begin strings derived from α . $\alpha => \varepsilon$ then is also in FIRST (α) .

Define FOLLOW(A), for nonterminal A, to be the set of terminals a that can appear immediately to the right of A in some sentential form, that is, $S \overset{*}{=>} \alpha A \alpha \beta$ for some α and β . If A can be the rightmost symbol in some sentential forms, then we add $ to FOLLOW(A).

To compute FIRST(X) for all grammar symbols X, apply the following rules until no more terminals or ε can be added to any FIRST set.

1. If X is terminal, then FIRST(X) is {X}.
2. If X is nonterminal and X -- > a α is a production, then add a to FIRST(X). If $X ---> \varepsilon$ is a production, then add ε to FIRST(X).
3. If X -- > $Y_1 Y_2 ..... Y_K$ is a production, then all i such that all of $Y_1, Y_2, ... Y_{i-1}$ are nonterminals and FIRST(Y_j) contains for j = 1,2,....,i-1 (i.e. $Y_1 Y_2 .... Y_{j-1} => \varepsilon$), add every non- ε symbol in FIRST (Y_i) to FIRST(X). If ε is in FIRST (Y_j) for all j=1,2,...k, then add ε to FIRST(X).

Now, we can compute FIRST for any string $X_1\ X_2\ ....\ X_n$ as follows:

1.Add to FIRST $(X_1 X_2 ..... X_n)$ all the non - ε symbols of FIRST(X1).

2. Also add the non - ε symbols of FIRST (X_2) if ε is in FIRST (X_1) , the non ε symbols of FIRST (X_3) if ε is in both FIRST (X_1) and FIRST (X_2) , and so on.

3. Finally, add ε to FIRST $(X_1 X_2 ..... X_n)$ if, for all i, FIRST (X_i) contains ε , or if n = 0.

To compute FOLLOW(A) for all nonterminals A, apply the following rules until nothing can be added to any FOLLOW set.

1. $ is in FOLLOWS(S), where S is the start symbol.
2. If there is a production $A --> \alpha B \beta,\ \beta \neq \varepsilon$ then everything in FIRST (β) but ε is in FOLLOW(B).
3. If there is a production $A --> \alpha B$ or a production $A --> \alpha B \beta$, where FIRST (β) contains ε (i.e., $\beta => \varepsilon$), then everything in FOLLOW(A) is in FOLLOW(B).

Example: 3.11. Consider the grammar

$E \rightarrow TE'$

$E \rightarrow TE' / \varepsilon$

$T \rightarrow FT'$

$T' \rightarrow *FT' | \varepsilon$

$F \rightarrow (E) | id \cdot$

Then :

FIRST(E) = FIRST(T) = FIRST(F) ={(,id}.

FIRST(E') = { +, ε }

FIRST(T') = { *, ε }

FOLLOW(E) = FOLLOW(E') = {),$ }

FOLLOW(T) = FOLLOW(T') = { +,),$ }

FOLLOW(F) = { +,*,),$ }

For example, id and left parenthesis are added to FIRST(F) by rule (2) in the definition of FIRST. Then by rule (3) with i=1, the production $T \text{- - >} FT'$ implies that id and left parenthesis are in FIRST(T) as well.

To compute FOLLOW sets, we put $ in FOLLOW(E) by rule (1). By rule (2) applied to production F - - (E), the right parenthesis is also in FOLLOW(E). By rule (3) applied to production $E - TE'$, $ and right parenthesis are in FOLLOW (E'). since $E' => \varepsilon$, they are also in FOLLOW(T). For a last example of how the FOLLOW rules are applied the production $E - TE'$ implies by rule (2) that everything other than ε in FIRST(E) must be placed in FOLLOW(T). We have already seen that $ is in FOLLOW(T).

A grammar whose parsing table has no multiply-defined entries is said to be LL(1). It can be shown that algorithm, produces a parsing table for every LL(1) grammar, and that this table parses all and only the sentences of the grammar.

It can also be shown that a grammar G is LL(1) iff whenever $A \rightarrow \alpha | \beta$ are two distinct productions of G the following conditions hold:

1. For no terminal a do and can derive strings beginning with
2. Atmost one of and can derive the empty string.
3. If $\beta => \varepsilon$, then does not derive any strings beginning with a terminal in FOLLOW(A).

Clearly, the grammar of example 3.11 for arithmetic expressions is LL(1).

19. Automatic Construction Of Efficient Parsers

This lesson section how to construct efficient bottom-up parsers for a large class of context-free grammars. These parsers are called LR parsers, because they scan the input from left-to-right and construct a rightmost derivation in reverse. LR parsers are attractive for a variety of reasons.

1. LR parsers can be constructed to recognize virtually all programming language constructs for which context-free grammars can be written.

2. The LR parsing method is more general than operator precedence or any of the other common shift-reduce techniques. LR parsing also dominates the common forms of top-down parsing without backtrack.

3. LR parsers can detect syntactic errors as soon as it is possible to do so on a left-to-right scan of the input.

Logically, an LR parser consists of two parts, a driver routine and a parsing table.

The driver routine is the same for all LR parsers; only the parsing table changes from one parser to another. The schematic form of an LR parser is shown in the fig.3.8.

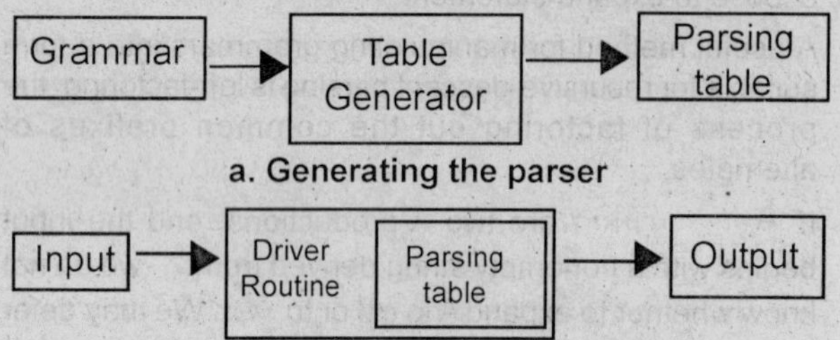

b. Operation of the Parser

Fig. 3.8. Generating an LR Parser

There are many different parsing tables that can be used an LR parser for a given grammar. Some parsing tables may detect errors sooner that others, but they all accept the same sentences, exactly the sentences generated by the grammar.

1. Simple LR (SLR) : This method is easiest to implement Unfortunately, it may fail to produce a table for certain grammars on which the other method succeed.

2. Canoncial LR : This is the most powerful and will work on a very large class of grammars. Unfortunately this method can be very expensive to implement.

3. Lookahead LR (LALR) : This is intermediate in power between the SLR and the canonical LR methods. This method will work on most programming language grammars and, with some effort can be implemented efficiently.

OBJECTIVE QUESTIONS - I

1. A bottom-up parser generates
   a) Right-most derivation
   b) Right-most derivation in reverse
   c) Left most derivation
   d) Left-most derivation in reverse
2. A top-down parser generates
   a) Right-most derivation
   b) Right-most derivation in reverse
   c) Left most derivation
   d) Left-most derivation in reverse
3. Which of the following actions an operator-precedence parser may take to recover from an error?
   a) Insert symbols onto the stack
   b) Delete symbols from the stack
   c) Insert or delete symbols from the input
   d) All of the above

4. Recursive descent parsing belongs to the class of:

a) Predictive parsing b) Top-down parsing

c) Bottom-up parsing d) None of the above.

5. Which of the following parsers is the most powerful?

a) Operator-precedence b) Canonical LR

c) LALR d) SLR

6. The bottom-up parsing method is also called

a) Recursive descent parsing

b) Shift reduce parsing

c) Predicative parsing d) None of the above.

7. Handle pruning is technique used to obtain

a) Canonical derivation sequence

b) Canonical reduction sequence

c) Both (a) and (b) d) None of the above.

8. The top-down parsing method is also called

a) Recursive descent parsing

b) Shift reducing parsing

c) Operator precedence parsing

d) None of the above.

9. The parsing technique that avoids backtracking is

a) Top-down parsing b) Recursive-descent parsing

c) Predicative parsing d) Both (b) and (c).

10. Left factoring is the process of factoring out the common

a) Prefixes of alternatesb) Suffixes of alternates

c) Both (a) and (b) d) None of the above.

11. The value of k, in LR(k) cannot be

a) 0 b) 1 c) 2 d) None

12. A grammar will be meaningless

a) If terminal set and non-terminal set are not disjoint.

b) If the left hand side of a production is a single terminal.

c) If the left hand side of a production has no non-terminal

d) If the left hand side of a production has more than two non-terminals

13. Which of the following grammar are not phase-structured?

a) Regular b) Context-free

c) Context-sensitive d) None of the above.

14. Which of the following is most general phase-structured grammar?

a) Regular b) Context-free

c) Context-sensitive d) None

15. CFG can be recognized by a

a) Push-down automata

b) 2-way linear bounded automata

c) Finite state automata d) None of the above

16. Choose the correct statements.

a) Sentence of a grammar is a sentential form without any terminals

b) A sentence of a grammar should be derivable from the start state

c) A sentence of a grammar should be frontier of a derivation tree, in which the root node has the start state as the label.

d) All of the above

17. A grammar

a) Can have a non-terminal A that can't derive any string of terminals

b) Can have a non-terminal A that can be present in any sentential form

c) Can have ε as the only symbol on the left hand side of a production

d) None of the above

18. Choose the correct statement

a) Language corresponding to a given grammar, is the set of all strings that can be generated by the given grammar

b) A given language is ambiguous if no unambiguous grammar exists for it

c) Two different grammars may generate the same language.

d) None of the above

19. A syntax tree

a) Is another name for a parse tree

b) Is a condensed form of parse tree

c) Should not have keywords as levels

d) None of the above

20. A parser with the valid prefix property is advantageous because

a) It detects error as soon as possible

b) It detects errors as and when they occur

c) It limits the amount of erroneous output passed to the next phase

d) All of the above

21. To recover from an error, the operator precedence parser may

a) Insert symbols onto the stack

b) Insert symbols onto the input

c) Delete symbols from the stack

d) Delete symbols from the input

22. Shift-reduce parsers are

a) Top-down parsers b) Bottom-up parsers

c) May be top-down or bottom up

d) None of the above

23. Recursive descent parsing is an example of

a) Top-down parsing b) Bottom-up parsing

c) Predictive parsing d) None of the above

24. In operator precedence parsing, precedence relations are defined

a) For all pair of non-terminals

b) For all pair of terminals

c) To delimit the handle

d) Only for a certain pair of terminals

25. LR stands for

a) Left to right b) Let to right reduction

c) Right to left

d) Left to right-most derivation in reverse

26. LR parsers are attractive because

a) It can be constructed to recognize CFG corresponding to almost all programming constructs

b) It does not backtrack

c) It detects error as and when they occur d) All of the above

27. Choose the correct statements.

a) There are CFG's that are not LR

b) An ambiguous grammar can never be LR

c) An ambiguous grammar can be LR d) Any CFG has to be LR

28. YACC builds up

a) SLR parsing table

b) Canonical LR parsing table

c) LALR parsing table d) None of the above.

29. LR(k) grammar

a) Can only examine a maximum of k input symbols

b) Can be used to identify handles

c) Can be used to identify the production associated with a handle

d) Covers the LL(k) class

30. The set of all viable prefixes of right sentential from of a given grammar

a) Can be recognized by a finite state machine

b) Cannot be recognized by a finite state machine

c) Can be used to control an LR(k) parser d) None of the above

31. Merging states with a common core may produce ..., conflicts and does not produce... conflicts in an LALR parser

a) Reduce-reduce; shift-reduce

b) Shift-reduce; reduce-reduce

c) Shift-reduce; shift-reduce d) None of the above

32. Which of the following features cannot be captured by CFG?

a) Syntax of if-then-else statements

b) Syntax of recursive procedures

c) Whether a variable is declared before its use

d) Matching nested parenthesis

33. Which of the following derivations does a top-down parser use while parsing an input string? The input is assumed to be scanned in left to right order. **[GATE-2000]**

a) Leftmost derivation

b) Leftmost derivation traced out in reverse

c) Right most derivation

d) Right most derivation traced out in reverse

34. Which of the following is the most powerful parsing method? **[GATE-1999]**

a) LL(1) b) Canonical LR c) SLR d) LALR

35. A grammar that is both left and right recursive for add non-terminal, is **[GATE-1999]**

a) Ambiguous b) Unambiguous

c) Information is not sufficient to decide whether it is ambiguous or unambiguous. d) None

36. Syntax analysis while compiling does the following: **[UGC NET-1997]**

a) Identify various tokens

b) Identify various type of statements

c) Checks the correctness of statements

d) Generates intermediate code

37. Which of the following passing methods handle left recursive grammars? **[UGC NET-1996]**

a) Top-down passing b) Bottom-up passing

c) Both top-down and Bottom-up passing

d) None of the above

38. Which of the following tool is known as 'parser generator'? **[UGC NET-1995]**

a) LEX b) YAAC c) TEX d) EMACS

39. Which of the following statements is wrong? **[UGC NET-1998]**

a) No left recursive grammar can be LL(1)

b) No LL(1) grammar can be ambiguous

c) Every LR(1) grammar is an LL(1) grammar

d) Every grammar can be converted into an equivalent operator grammar

ANSWER KEY - I

1. b	**2.** c	**3.** d	**4.** b	**5.** b
6. b	**7.** b	**8.** a	**9.** d	**10.** a
11. d	**12.** a,b,c	**13.** d	**14.** c	**15.** a
16. b,c	**17.** a,b	**18.** a,b,c	**19.** b,c	
20. a,c	**21.** a,b,c,d	**22.** b	**23.** a	**24.** c,d
25. d	**26.** a,b	**27.** a,b	**28.** c	
29. a,b,c,d	**30.** a,b	**31.** a	**32.** c	**33.** a
34. b	**35.** c	**36.** c	**37.** a	**38.** a
39. d				

OBJECTIVE QUESTIONS - II

1. Consider the following left-associative operators, in decreasing order of precedence:

- subtraction (highest precedence)

* multiplication

$ exponentiation (lowest precedence)

What is the result of the following expression?

3 – 2 * 4 $ 1 * 2 $ 3

a) –61 b) 64 c) 512 d) 4096

2. Consider the following parse tree:

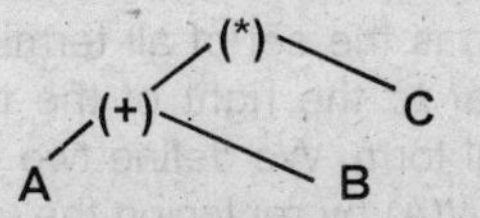

Which of the following expressions is represented by the parse tree?

a) (A+B).C b) A + * BC c) A + B.C d) A*C + B

3. Which of the following grammar is LR (1)?

a) A→aAa
A→bAb
A→a
A→b

b) A→ aAa
A→ aAb
A→ c

c) A→ A + A
A→ a

d) both (a) and (b).

4. Consider the left-recursive grammar:

S→Aa | b

A→Ac | Sd

When the left-recursion is removed, the grammar will become equivalent to the grammar:

a) S→b A′
A′ →c | da

b) A→ad | bd | cA
S→Aa | b

c) S→Aa | b
A→Ac | Aad | bd

d) S→Aa | b
A→bd A′
A′→c A′

5. In operator precedence parsing precedence relations are defined

a) Only for a certain pair of terminals and to delimit the handle

b) For all pairs of terminals

c) For all pairs of non-terminals

d) Both (b) and (c)

6. The important factor that justifies the use of a stack in shift-reduce parsing is

a) The handle will always appear on top of stack and never inside.

b) The handle will always appear inside the stack and never on top.

c) The handle will never appear in stack. d) None of the above.

7. The function(s) of the syntax phase is

a) To recognize the major constructs of the language and to call the appropriate action routines that will generate the intermediate form or matrix for these constructs.

b) To built a literal table and an identifier table.

c) To built a uniform symbol table.

d) To parse the source program into the basic elements or tokens of the language.

8. The action of parsing the source program into the proper syntactic classes is known as

a) Syntax analysis b) Lexical analysis

c) Interpretation analysis d) General syntax analysis

9. In a Context-sensitive grammar

a) ε can't be the right hand side of any production

b) Number of grammar symbols on the left hand side of a production can't be grater than the number of grammar symbols on the right hand side.

c) Number of grammar symbols on the lest hand side of a production can't be greater than the number of terminals on the right hand side

d) Number of grammar symbols on the lest hand side of a production can't be greater than the number of nonterminals on the right hand side

10. In a context-free grammar

a) ε can't be the right hand side of any production

b) Terminal symbols can't be present in theleft hand side of any production

c) The number of grammar symbols in the left hand side is not grater than the number of grammar symbols in the right hand side

d) All of the above

11. If W is a string of terminals and A, B are two non-terminals, then which of the following are right-linear grammars?

a) A → BW b) A→ BW/ W

c) A → WB / W d) None

12. If a is a terminal and S, A, B are three non-terminals, then which of the following are regular grammars?

a) S → ε
A → as| b

b) A → abB | aB

c) A → Ba | Bab

d) A → aB | a
B → bA | b

13. A given grammar is said to be ambiguous if

a) Two or more productions have the same non-terminal on the left hand side

b) A derivation tree has more than one associated sentence

c) There is a sentence with more than one derivation tree corresponding to it

d) Brackets are not present in the grammar

14. The grammar E→ E + E → E*E → a, is

a) Ambiguous b) Unambiguous

c) Ambiguous or not depends on the given sentence

d) None of the above

15. Consider the grammar

S → ABSc / Abc

BA→AB

Bb→ bb

Ab → ab

Aa → aa

Which of the following sentences can be derived by this grammar?

a) abc b) aab c) abcc d) abbc

16. For parsing arithmetic expressions,involving (,), + and -, the precedence relation between - and -, will be <, if

a) We are talking of unary minus

b) Minus is right associative

c) Minus is left associative

d) A (comes in between

17. Choose the correct statements.

a) LL (k) grammar has to be CFG

b) LL (k) grammar has to be unambiguous

c) There are LL (k) grammars that are not CFG

d) LL (k) grammars cannot have left recursive non-terminals

18. Consider an ε-free CFG. If for every pair of productions A → u and A → v

a) If FIRST(u) ∩ FIRST(v) is empty then the CFG has to be LL(1)

b) If the CFG is LL(1) then FIRST(u) ∩ FIRST(v) has to be empty

c) If FIRST(u) ∩ FIRST(v) is empty then the CFG cannot be LL(1)

d) None of the above.

Note :The following THREE questions are based on the following

$FIRST_k$ (x), where x is a string, is the set of all leading terminal strings of length k or less, in the strings derivable from x.

$FOLLOW_k$ (A), where A is a non-terminal, is the set of all derivable terminal strings of length k or less, that can follow A in some left-most sentential form.

19. Consider the grammar

E → T**E′**

E′ →+ T**E′** | ε

T′ → *F **T′** | ε

F → (E) | id

$FIRST_1$ (E) will be same as that of

a) $FIRST_1$ (T) b) $FIRST_1$ (F)

c) $FIRST_1$ (**T′**) d) All of the above

20. $FOLLOW_1$ (F) is

a) {+, *,), $} b) {+,), $}

c) {*,), $} d) {+, (,), *}

21. Which of the following remarks logically follows

a) $FIRST_k$ (ε) = {ε}

b) If $FOLLOW_k$ (A) contains ε, then A is the start symbol.

c) If A → w, is a production in the given grammar G, then $FIRST_k$ (A) contains $FIRST_k$ (w).

d) If A →w, is a production in the given grammar G, then $FIRST_k$ (w) contains $FIRST_k$ (A).

22. For a CFG, FOLLOW (A) is the set of all terminals that can immediately appear to the right of the non-terminal A in some sentential form. We define two sets LFOLLOW(A) and RFOLLOW(A) by replacing the word sentential by "Left most sentential" and " Right most sentential" respectively in the definition of FOLLOW(A). Choose the correct statement(s).

a) FOLLOW (A) and LFOLLOW(A) may be different

b) FOLLOW (A) and RFOLLOW(A) are always the same

c) All the three are same d) All the three are different

23. Which of the following statements is FALSE? **[GATE-2001]**

a) An unambiguous grammar has the same leftmost and rightmost derivation

b) An LL(1) parser is a top-down parser

c) LALR is more powerful than SLR

d) An ambiguous grammar can never be LR(k) for any k

24. Given the following expression grammar:

E → E ∗ F | F + E | F

F → F – F | id

Which of the following is TRUE? **[GATE-2000]**

a) ∗ has higher precedence than +

b) – has higher precedence than ∗

c) + and – have same precedence

d) + has higher precedence than ∗

25. Which of the following statements is TRUE? **[GATE-1998]**

a) SLR parser is more powerful than LALR.

b) LALR parser is more powerful than Canonical LR parser

c) Canonical LR parser is more powerful than LALR parser

d) The parsers SLR, Canonical CR, and LALR have the same power

ANSWER KEY - II

1. d	**2.** a	**3.** b	**4.** d	**5.** a
6. a	**7.** a	**8.** b	**9.** a,b,c,d	**10.** b
11. c	**12.** d	**13.** c	**14.** a	**15.** a
16. a,b	**17.** a,b,d	**18.** a,b	**19.** a,b	**20.** a
21. a,b,c	**22.** a,b	**23.** a	**24.** b	**25.** c

IV. SYNTAX DIRECTED TRANSLATION

- **Syntax Directed Translation Schemes**
- **Implementation of Syntax Directed Translators**
- **Intermediate Code**

1. Introduction

The syntax directed translation scheme allows subroutines or semantic actions to be attached to the productions of a context free grammer. These subroutines generate intermediate code when called at appropriate time by a parse for that grammer. The syntax-directed translation scheme is useful because it enables the complier designer to express the generation of intermediate code directly in terms of the syntactic structure of the source language. Unfortunately, there is no widely-accepted formalism, other than a general programming language, to describe the full range of semantic actions needed to produce intermediate code for typical programming languages.

2. Syntax-Directed Translation Schemes

First we must decide what intermediate code we should generate for each programming language construct. Then we must implement an algorithm for generating this code.

Semantic Actions : A syntax - directed translation scheme is merely a context-free grammer in which a program fragment called an output action (or sometimes a semantic action or semantic rule) is associated with each production.

Example: Suppose output action a is associated with production $A \rightarrow XYZ$. The action a is executed whenever the syntax analyzer recognizes in its input a substring w which has a derivation of the form $A \Rightarrow XYZ \Rightarrow w$. In a bottom-up parser the action is taken when X Y Z is reduced to A. In a top-down parser the action is taken when A, X, Y, or Z is expanded, whichever is appropriate.

The output action may involve the Computation of values for variables belonging to the compiler, the generation of intermediate code, the printing of an error diagnostic, or the placement of some value in a table, for example. The values computed by action α quite frequently are associated with the parse tree nodes corresponding to the instance of A to which α is reduced.

A value associated with a grammer symbol is called a translation of that symbol. The translation may be a structure consisting of fields of various types. The rules for computing the value of a translation can be as involved as we wish.

Suppose we have the production and semantic action

$$E \rightarrow E^{(1)} + E^{(2)} \{ E.VAL := E^{(1)}.VAL + E^{(2)}.VAL \}.$$

The semantic action is enclosed in braces, and it appears after the production. Here the semantic action is a formula which states that the translation E.VAL associated with the E on the left side of the production is determined by adding together the translation associated with the E's on the right side of the production. Note that the terminal symbol + in the production is translated into its usual meaning by the semantic rule. This translation is suitable not for a compiler, but for a "desk calculator" program that actually evaluates expressions rather than generating code for them. In most Compilers we need an action that generated code to perform the addition.

The action above is typical of many of the semantic actions we shall use. It defines the value of the translation of the nonterminal on the left side of the production as a function of the translations of the nonterminals on the right side. Such a translation is called a synthesized translation.

Consider the following production and action

$$A \rightarrow XYZ \{ Y.VAL := 2 * A.VAL \}$$

Here the translation of a nonterminal on the right side of the production is defined in terms of a translation of the non-terminal on the left. Such a translation is called an inherited translation.

3. Implementation of Syntax-Directed Translators

A Syntax-directed translation scheme is a Convenient description of what we would like done. The output defined is independent of the kind of parser used to construct the parse tree or the kind of mechanism used to compute the translations. Thus a Syntax- directed translation scheme provides a method for describing an input-output mapping, and that description is independent of any implementation. Another Convenience of this approach is that it is easy to modify. New productions and semantic actions can often be added without disturbing the existing translations being computed.

After writing a syntax directed translation scheme we have to convert it into a program that implements the input-output mapping described. Some type of LR1 parser would be quite adequate for most schemes.

To Compute the translation at a node. A associated with a production.

$A \rightarrow XYZ$, we need only the values of the translations associated with nodes labeled X, Y and Z. These nodes will be roots of subtrees in the forest representing the partially constructed parse tree. The nodes X, Y and Z will become children of node A after reduction by $A \rightarrow XYZ$. Once the reduction has acquired we do not need the translation of X, Y and Z any longer.

One way to implement a syntax-directed translator is to use extra fields in the parser stack entries corresponding to the grammer symbols. These extra fields hold the values of the corresponding translations. Let us suppose the stack is implemented by a pair of arrays STATE and VAL as shown below.

	STATE	VAL
Top →	Z	Z.VAL
	Y	Y.VAL
	X	X.VAL

Each STATE entry is a pointer (or index) to the LR(1) parsing table. (Note that the grammer symbol is implicit in the state and need not be present explicitly). It is convenient, however, to refer to the state by the unique grammer symbol which it covers when placed on the parsing stack. If the ith STATE symbol is E, then VAL[i] will hold the value of the translation E.VAL associated with the parse tree node corresponding to this E.

Top is a pointer to the Current of p of the stack. We assume semantic routines are executed just before each reduction. Before XYZ is reduced to A, the value of the translation of Z is in VAL[TOP], that of Y in VAL[Top+1] and that of X in VAL[Top+2]. After the reduction, Top is incremented by 2 and the value of A. VAL appears in VAL[Top].

Example : We shall now give an example of how a syntax-directed translation scheme can be used to specify a "desk calculator" program and how that translation scheme can be implemented by a bottom-up parser that invokes program fragments to compute the semantic actions. We Consider this desk Calculator is to evaluate arithmetic expressions involving integer operands and the operators +, -, $*$ and /. To design the translator we first write the grammer to describe the input. We use the nonterminals 'S' (for complete sentence), 'E' (for expressions) and I (for integer). The productions are,

$$S \rightarrow E\$$$
$$E \rightarrow E + E$$
$$E \rightarrow E - E$$
$$E \rightarrow E * E$$
$$E \rightarrow E / E$$
$$E \rightarrow (E)$$
$$I \rightarrow Id$$
$$I \rightarrow d,$$

where we assume the usual precedence levels and associativity for the operators +, -, * and /. The terminals are $, +, -, *, /, (,) and d, which we assume stands for any of the digits 0 to 9.

We must now add the semantic actions to the production with each of the nonterminals E and I we associate one integer valued translation called E.VAL and I.VAL, respectively, which denotes the numerical value of the expression or integer represented by a node of the parse tree labeled E or I. With the terminal d we associate the translation LEXVAL, which we assume is the second component of the pair (d, LEXVAL) returned by the lexical analyzer when a token of type digit is found.

One possible set of semantic action for the desk calculator grammer is shown below.

Production	Semantic Action
1. $S \rightarrow E\$$	{Print E.VAL} .
2. $E \rightarrow (E)^{(1)} + E^{(2)}$	$\{E.VAL := E^{(1)}VAL + E^{(2)}.VAL\}$
3. $E \rightarrow (E)^{(1)} + E^{(2)}$	$\{E.VAL := E^{(1)}VAL - E^{(2)}.VAL\}$
4. $E \rightarrow (E)^{(1)} * E^{(2)}$	$\{E.VAL := E^{(1)}VAL * E^{(2)}.VAL\}$
5. $E \rightarrow (E)^{(1)} / E^{(2)}$	$\{E.VAL := E^{(1)}VAL / E^{(2)}.VAL\}$
6. $E \rightarrow \left((E)^{(1)}\right)$	$\{E.VAL := E^{(1)}VAL\}$
7. $E \rightarrow I$	$\{E.VAL := I.VAL\}$
8. $I \rightarrow I^{(1)}d$	$\{I.VAL := 10 * I^{(1)}.VAL + LEXVAL\}$
9. $I \rightarrow d$	$\{I.VAL := LEXVAL\}$

To implement the semantic action we cause the parse 1 execute the following program fragments just before making the appropriate reduction.

Production	Program Fragment
1. $S \rightarrow E\$$	Print VAL [Top]
2. $E \rightarrow E + E$	VAL[Top]: = VAL[Top]+VAL[Top-2]
3. $E \rightarrow E - E$	VAL[Top]: = -VAL[Top] +VAL[Top-2]
4. $E \rightarrow E * E$	VAL[Top]: = VAL[Top] $*$ VAL [Top-2]
5. $E \rightarrow E / E$	VAL[Top]: = VAL[Top-2] / VAL[Top]
6. $E \rightarrow (E)$	VAL[Top]: = VAL [Top-1]
7. $E \rightarrow I$	none
8. $I \rightarrow Id$	VAL[Top[: = 10 $*$ VAL[Top]+LEXVAL
9. $I \rightarrow d$	VAL[Top]: = LEXVAL

For the input expression

24*7-47$, the prcgram is to produce the value 121. The following fig. Shows the sequence of moves made by the parser on input 24 * 7-47$.

Input	State	Value	Production Used
24*7-47$	-	-	
4*7-47$	2	-	
4*7-47$	I	2	$I \rightarrow d$
*7-47$	I4	2-	
*7-47$	I	24	$I \rightarrow Id$
*7-47$	E	24	$E \rightarrow I$
7-47$	E*	24 -	
-47$	E* 7	24 - -	
-47$	E*I	24 - 7	$I \rightarrow d$
-47$	E*E	24 - 7	$E \rightarrow I$
-47$	E	168	$E \rightarrow E * E$
47$	E -I	168 -	
7$	E-4	168 - -	
7$	E-I	168-4	$I \rightarrow d$
$	E-I 7	168-4-	
$	E-	168-47	$I \rightarrow Id$
$	E-E	168-47	$E \rightarrow I$
$	E	121	$E \rightarrow E - E$
-	E$	121	
-	S	121	$S \rightarrow E\$$

The program fragment for the production rule $S \to E\$$ is Print VAL[Top]. VAL[Top] is 121. So the value 121 is printed.

4. Intermediate Code

In many Compilers the source is translated into a language which is intermediate in complexity between a high level programming language and machine code. Therefore such a language is called intermediate code or intermediate text.

Five kinds of intermediate codes often used in compilers are

i) Postfix Notation ii) Syntax trees
iii) Three address code iv) Quadruples
v) Triples.

5. Postfix Notation

The ordinary (infix) way of writing the sum of a and b is with the operator in the middle: a+b. The postfix notation for the same expression places the operator at the right end as ab+. In general if e_1 and e_2 are any postfix expressions, and θ is any binary operator the result of applying θ to the values denoted by e_1 and e_2 is indicated in postfix notation by $e_1e_2\theta$. No parentheses are needed in postfix notation because the position and arity of the operators permits only one way to decode a postfix expression.

Example:

i) For the expression $(a+b)*C$, the postfix notation is ab + c∗.

ii) For a/(b-c), the postfix expression is abc-/

iii) For $(a+b)*(C-d)$,then the postfix notation is ab + cd−∗

The syntax directed translation scheme for a simple grammer is given below.

Production	Semantic Action
1) $E \to E^{(1)}\ op\ E^{(2)}$	$E.CODE := E^{(1)}.CODE \| E^{(2)}$ .CODE || 'op
2) $E \to (E^{(1)})$	$E.CODE := E^{(1)}.CODE$
3) $E \to id$	E.CODE:=id

In the first rule(1) E.CODE is a string valued translation which is the concatenation of the two translations $E^{(1)}.CODE, E^{(2)}.CODE$,and the symbol op, which is an operator symbol.

In the second rule(2) the translation of a parenthesized expression is the same as that for the unparenthesized expression. In the third rule, the translation of any identifier is the identifier itself.

The program fragment corresponding to the above semantic actions are given below.

Production	Program Fragment
$E \to E^{(1)} op E^{(2)}$	{Print op}
$E \to E^{(1)}$	{ }
$E \to id$	{Print id}

6. Syntax trees

A syntax tree is a tree in which each leaf represents an operand and each interior node an operator.

Example: i) For the expression $a*(b+c)/d$ the syntax tree is given below.

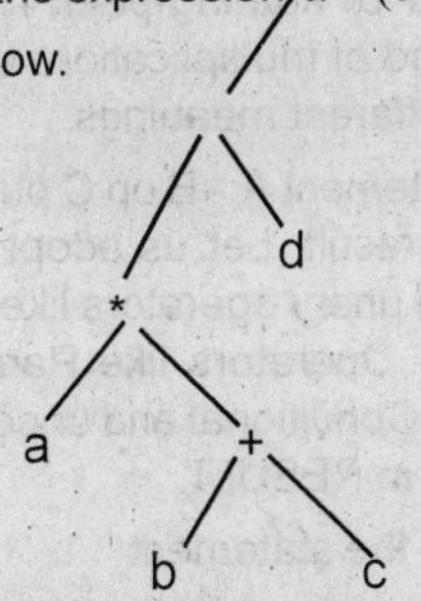

Syntax directed translation scheme for syntax trees:

Production	Semantic Action
1) $E \to E^{(1)} op E^{(2)}$	$\{E.VAL: NODE(op. E^{(1)}.VAL. E^{(2)}.VAL\}$
2) $E \to (E^{(1)})$	$\{E.VAL := E^{(1)}.VAL\}$
3) $E \to -E^{(1)}$	$\{E.VAL := UNARY(-, E^{(1)}.VAL\}$
4) $E \to id$	$\{E.VAL := LEAF(id)\}$

The function NODE(op,LEFT,RIGHT) takes three arguments. The first is the name of the operator the second and third are pointers to roots of subtrees. The function (creates a new node labeled by the first argument and makes the second and third arguments' the left and right children of the new node, returning a pointer to the created node. The function UNARY (op, CHILD) (Creates a new node labeled op and makes CHILD its child. Again, a pointer to the created node is returned. The function LEAF(ID) creates a new node labeled by ID and returns a pointer to that node. This node receives no children.

7. Three address code

This intermediate code is preferred in many compilers, especially those doing extensive code optimization, since it allows the intermediate code to be rearranged in a convenient manner.

The general form of a three-address code is, A:=B op C, where A, B and C are either programmer defined names, constants or compiler generated temporary names. op stands for any operator. The reason for the name "Three address code" is that each statement usually contains three address 2 for the operand and one for the result.

The three-address statement is an abstract form of intermediate code, in an actual compiler these statements can be implemented by either of the following way.

- Quadruples.
- Triples
- Indirect Triples.

8. Quadruples

We may use a record structure with four fields, which we shall call op, ARG1, ARG2 and RESULT. This representation of three- address statements is known as quadruples. Op Contains an internal code for the operator. There are actually two levels of specificity which the op field may have. We may, as we translate to intermediate code, determine whether a binary operator like * means fixed or floating-point multiplication, or perhaps another kind of multiplication. Using a different operator code for different meanings.

A three-address statement A:=B op C puts B in ARG1, C in ARG2, and A in result. Let us adopt the convention that statements with unary operators like A:= -B or A:=B do not use ARG2. Operators like Param use neither ARG2 nor RESULT. Conditional and unconditional jumps put the target Label in RESULT.

Example: Consider the statement

A : = -B * (C+D).

The translated three-address statement for the above expression is,

T_1 : = -B;

T_2 : = C+D

T_3 : = $T_1 * T_2$

A : = T_3

These statements are represented by quadruples as shown below.

	OP	ARG1	ARG2	RESULT
(0)	Uminus	B	-	T_1
(1)	+	C	D	T_2
(2)	*	T_1	T_2	T_3
(3)	: =	T_3	-	A

The contents of fields ARG1, ARG2 & RESULT are normally pointers to the symbol table entries for the names represented by these fields.

9. Triples

To avoid entering temporary names into the symbol table, one can allow the statement computing a temporary value to represent that value. If we do so, three address statements are representable by a structure with only three fields op, ARG1 and ARG2, where ARG1 and ARG2, the arguments of op, are either pointers to the symbol table (for programmer defined names or constants) or pointers into the structure itself (for temporary values). Since three fields are used, this intermediate code format is known as triples.

We use parenthesized numbers to represent pointers into the triple structure, while symbol-table pointers are represented by the names themselves. In practice, the information needed to interpret the different kinds of entries in the ARG1 and ARG2 fields can be encoded into the op field or some additional field.

Example: Consider the expression

A: = -B * (C+D) and the three address code for this is,

T_1: = -B

T_2 : = C+D

T_3 : = $T_1 * T_2$

A : = T_3 .

The triple implementation for this is shown below:

	OP	ARG1	ARG2
(0)	Uminus	B	-
(1)	+	C	-
(2)	*	(0)	(1)
(3)	:=	A	(2)

Consider the ternary operation A[I]:=B.

For this the triple representation is;

	OP	ARG1	ARG2
(0)	[ ] =	A	I
(1)	-	B	-

Consider the expression A:= B[I] and the triple representation for this is :

	OP	ARG1	ARG2
(0)	=[ ]	B	I
(1)	: =	(0)	A

10. Indirect Triples

Another implementation of three address code which has been considered is that of listing pointers to triples, rather than listing the triples themselves. This implementation is naturally called indirect triples.

Example: Let us use on array STATEMENT to list pointers to triples in the desired order. Then the three-address statement of the expression A:= - B * (C+D) might be represented as follows:

	STATEMENT
(0)	(14)
(1)	(15)
(2)	(16)
(3)	(17)

	OP	ARG1	ARG2
(14)	Uminus	B	-
(15)	+	C	D
(16)	*	(14)	(15)
(17)	:=	A	(16)

OBJECTIVE QUESTIONS - I

1. Consider a syntax directed translation scheme. If the value of an attribute of a node is a function of the attributes of its children, then the attribute is called a:

a) Canonical attribute b) Synthesized attribute

c) Inherited attribute d) A parse tree

2. Synthesized attribute can be easily simulated by a

a) LL grammar b) Ambiguous grammar

c) LR grammar d) None of the above

3. For which of the following situations, inherited attribute is a natural choice

a) Evaluation of arithmetic expressions

b) Keeping track of variable declaration

c) Checking for the correct use of L-values and R-values

d) All of the above.

4. The graph depicting the interdependencies of the attributes of different nodes in a parse tree is called

a) Flow graph b) Dependency graph

c) Karnaugh's graph d) Steffi graph

5. Syntax directed translation scheme is desirable because

a) It is based on the syntax

b) Its description is independent of any implementation

c) It is easy to modify

d) Only (a) and (c) are correct.

6. A basic block can be analyzed by

a) A DAG b) Flow graph

c) A graph with cycles d) None of the above.

7. The graph that shows the basic blocks and their successor relationship is called

a) DAG b) Flow graph

c) Control graph d) Hamilton graph.

ANSWER KEY - I

1. b	**2.** c	**3.** b,c	**4.** b	**5.** a,b,c
6. a	**7.** b			

OBJECTIVE QUESTIONS - II

1. A shift-reduce parser carries out the actions specified within braces immediately after reducing with the corresponding rules of grammar

S→AS {print "1"}
S→ AB {print "2"}
A→a {print "3"}
B→ bC {print "4"}
B→ dB {print "5"}
C→c {print "6"}

This syntax directed translation scheme translates a language whose terminal symbols are a, b, c and d into another language whose terminal symbols are 1, 2, 3, 4, 5 and 6. What is the translation of "aaadbc"?

a) 333546 b) 654211 c) 333645211 d) 645233311

2. A shift reduce parser carries out the actions specified with in braces immediately after reducing with the corresponding rule of grammar

S → xxW {pirnt "1"}
S → y {print "2"}
W → Sz {print "3"}

What is the translation of xxxxyzz using the syntax directed translation scheme described by the above rules? **[GATE-1995]**

a) 23131 b) 11233 c) 11231 d) 33211

3. Three address code involves

a) Exactly 3 address b) At most 3 address

c) No unary operators d) None of the above

4. Which of the following can not be used as an intermediate code form?

a) Post fix notation b) Three address codes

c) Syntax trees d) Quadruples

5. The DAG shown in the figure given below represents:

a) (a-b ∗ c) ∗ (b ∗ c+d) b) a - b ∗ c + d

c) (a – d) ∗ (b ∗ c) d) (a – d) + (b ∗ c)

6. Type checking is normally done during **[GATE-1998]**

a) lexical analysis b) Syntax analysis

c) Syntax directed translation d) Code optimization

7. Three address codes can be implemented by

a) Indirect triples b) Direct triples

c) Quadruples d) None of the above

ANSWER KEY - II

1. c	**2.** a	**3.** b	**4.** d	**5.** a
6. c	**7.** b,c			

V. RUNTIME ENVIRONMENTS

- Introduction
- Source Language Issues
- Storage Organisation
- Storage Allocation Strategies
- Scope of variables
- Parameter Passing Techniques
- Symbol Table Implementation

1. Source Language Issues

For specificity, suppose that a program is made up of procedures, as in Pascal. This section distinguishes between the source text of a procedure and its activations at run time.

2. Procedures

A procedure definition is a declaration that, in its simplest form, associates an identifier with a statement. The identifier is the procedure name, and the statement is the procedure body.

For example, the Pascal code in Fig. 5.1 contains the definition of the procedure named readarray on lines 3-7; the body of the procedure is on lines 5-7. Procedures that return values are called functions in many languages; however, it is convenient to refer them as procedures. A complete program will also be treated as a procedure.

```
1.  program sort (input, output);
2.  var a : array [0..10] of integer;
3.  procedure readarray;
4.  var i : integer;
5.  begin
6.  for i := 1 to 9 do read (a[i])
7.  end ;
8.  function partition (y, z: integer) : integer;
9.  var l, j, x, v: integer;
10. begin ...
11. end;
12. procedure quicksort (m, n: integer);
13. var i : integer;
14. begin
15. if ( n > m ) then begin
16. i := partition (m,n);
17. quicksort (m, i - 1)
18. quicksort ( i + 1, n)
19. end
20. end;
21. begin
22. a [ 0 ] := -9999; a [ 10 ] := 9999;
23. readarray;
24. quicksort ( 1,9 )
25. end.
```

Fig. 5.1. A Pascal program for reading and sorting integers.

When a procedure name appears within an executable statement, we say that the procedure is called at that point. The basic idea is that a procedure call executes the procedure body. The main program in lines 21-25 of Fig. 5.1 calls the procedure readarray at line 23 and then calls quicksort at line 24. Note that procedure calls can also occur within expressions, as on line 16.

Some of the identifiers appearing in a procedure definition are special, and are called formal parameters (or just formals) of the procedure. (C calls them "formal arguments" and Fortran calls them "dummy arguments"). The identifiers m and n on line 12 are formal parameters of quicksort. Arguments, known as actual parameters (or actuals) may be passed to a called procedure; they are substituted for the formals in the body. Line 18 of Fig. 5.1 is a call of quicksort with actual parameters i + 1 and n.

3. Activation Trees

We make the following assumptions about the flow of control among procedures during the execution of a program:

1. Control flows sequentially; that is, the execution of a program consists of a sequence of steps, with control being at some specific point in the program at each step.

2. Each execution of a procedure starts at the beginning of the procedure body and eventually returns control to the point immediately following the place where the procedure was called. This means the flow of control between procedures can be depicted using trees, as we shall soon see.

Each execution of a procedure body is referred to as an activation of the procedure. The lifetime of an activation of a procedure p is the sequence of steps between the first and last steps in the execution of the procedure body, including time spent executing procedures called by p, the procedures called by them, and so on. In general, the term "lifetime" refers to a consecutive sequence of steps during the execution of a program.

In languages like Pascal, each time control enters a procedure q from procedure p, it eventually returns to p (in the absence of a fatal error). More precisely, each time control flows from an activation of a procedure p to an activation of a procedure q, it returns to the same activation of p.

If a and b are procedure activations, then their lifetimes are either non-overlapping or are nested. That is, if b is entered before a is left, then control must leave b before it leaves a.

This nested property of activation lifetimes can be illustrated by inserting two print statements in each procedure, one before the first statement of the procedure body and the other after the last. The first statement prints enter followed by the name of the procedure and the values of the actual parameters; the last statement prints leave followed by the same information. One execution of the program in Fig. 5.1 with these print statements produced the output shown in Fig. 5.2. The lifetime of the activation quicksort (1,9) is the sequence of steps executed between printing enter quicksort (1,9) and printing leave quicksort (1,9). In Fig. 5.2, it is assumed that the value returned by partition (1,9) is 4.

A procedure is recursive if a new activation can begin before an earlier activation of the same procedure has ended. Figure 5.2 shows that control enters the activation of quicksort (1,9), from line 24, early in the execution of the program but leaves this activation almost at the end. In the meantime, there are several other activations of quicksort, so quicksort is recursive.

A recursive procedure p need not call itself directly; p may call another procedure q, which may then call p through some sequence of procedure calls. We can use a tree, called an activation tree, to depict the way control enters and leaves activations. In an activation tree,

Execution begins..........
Enter readarray
Leave readarray
Enter quicksort (1,9)
Enter partition (1,9)
Leave partition (1,9)
Enter quicksort (1,3)

Leave quicksort (1,3)
Enter quicksort (5,9)

Leave quicksort (5,9)
Leave quicksort (1,9)
Execution terminated.

Fig. 5.2 Output suggesting activations of procedures in Fig. 5.1.

1. each node represents an activation of procedure,
2. The root represents the activation of the main program,
3. The node for a is the parent of the node for b if and only if control flows from activation a to b, and
4. The node for a is to left of the node for b if and only if the lifetime of a occurs before the lifetime of b

Since each node represents a unique activation, and vice versa, it is convenient to talk of control begin at a node when it is in the activation represented by node.

Example 5.1. The activation tree in Fig. 5.3 is constructed from the output in Fig 5.2 To save space, only the first letter of each procedure is shown. The root of the activation tree is for the entire program sort. During the execution of sort, there is an activation of readarray, represented by the first child of the root, with label r. The next activation , represented by the second child of the root, is for quicksort, with actuals 1 and 9. During this activation, the calls of the partition and quicksort on lines 16-18 of Fig , 5.1 lead to the activations p(1,9) , q(1,3), and q(5,9).Note that the activations q(1,3)and q(5,9) are recursive, and that they begin and end before q(1,9)ends.

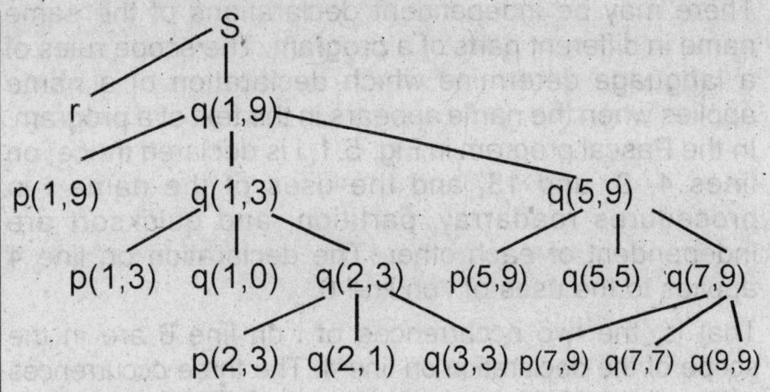

Fig. 5.3 An activation tree corresponding to the output in Fig.5.2.

4. Controls Stacks

The flow of control in a program corresponds to a depth-first traversal of the activation tree that starts at the root, visits a node before its children, and recursively visits children at each node in a left-to-right order. The output in Fig. 5.2 can therefore be reconstructed by traversing the activation tree in Fig. 5.3, printing enter when the node for an activation is reached for the first time and printing leave after the entire subtree of the node has been visited during the traversal.

We can use a stack, called a control stack to keep track of live procedure activations.

The idea is to push the node for an activation onto the control stack as the activation begins and to pop the node when the activation ends. Then the contents of the control stack are related to paths to the root of the activation tree. When node n is at the top of the control stack, the stack contains the nodes along the path from n to the root.

Example 5.2 : Figure 5.4 shows nodes from the activation tree of Fig. 5.3 that have been reached when control enters the activation represented by q(2,3). Activations with labels r, p(1,9), p(1,3), and q(1,0) have executed to completion, so the figure contains dashed lines to their nodes. The solid lines mark the path from q(2,3) to the root.

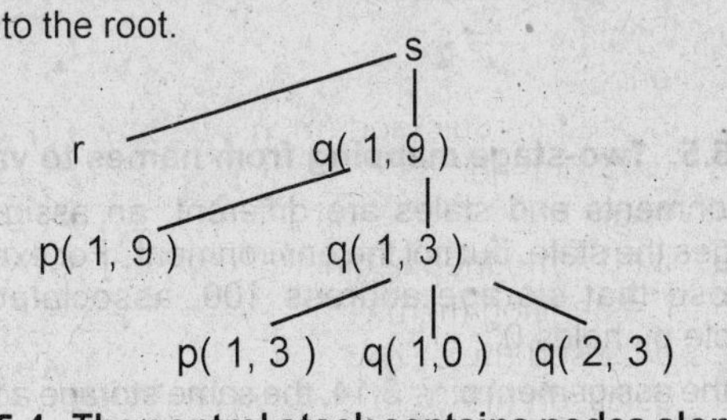

Fig. 5.4. The control stack contains nodes along a path to the root

At this point the control stack contains the following nodes along this path to the root S, q(1,9), q(1,3), q(2,3) and no other nodes.

5. The Scope of a Declaration

A declaration in a language is a syntactic construct that associates information with a name. Declarations may be explicit, as in the Pascal fragment

```
var i : integer ;
```

or they may be implicit. For example, any variable name starting with I is assumed to denote an integer in a Fortran program unless otherwise declared.

There may be independent declarations of the same name in different parts of a program. The scope rules of a language determine which declaration of a name applies when the name appears in the text of a program. In the Pascal program in Fig. 5.1, i is declared thrice, on lines 4, 9, and 13, and the uses of the name i in procedures readarray, partition, and quicksort are independent of each other. The declaration on line 4 applies to the uses of i on line 6.

That is, the two occurrences of i on line 6 are in the scope of the declaration on line 4. The three occurrences of i on lines 16-18 are in the scope of the declaration of i on line 13.

The portion of the program to which a declaration applies is called the scope of that declaration. An occurrence of a name in a procedure is said to be local to the procedure if it is in the scope of a declaration within the procedure; otherwise, the occurrence is said to be nonlocal. The distinction between local and nonlocal names carries over to any syntactic construct that can have declarations within it.

While scope is a property of the declaration of a name, it is sometimes convenient to use the abbreviation "the scope of name x" for "the scope of the declaration of name x that applies to this occurrence of x". In this sense, the scope of i on line 17 in Fig. 5.1 is the body of quicksort.

6. Bindings of Names

Even if each name is declared once in a program, the same name may denote different data objects at run time. The informal term "data object" corresponds to a storage location that can hold values.

In programming language semantics, the term environment refers to a function that maps a name to a storage location, and the term state refers to a function that maps a storage location to the value held there, as in Fig. 5.5. using the terms l-value and r-value, an environment maps a name to an l-value, and a state maps the l-value to an r-value.

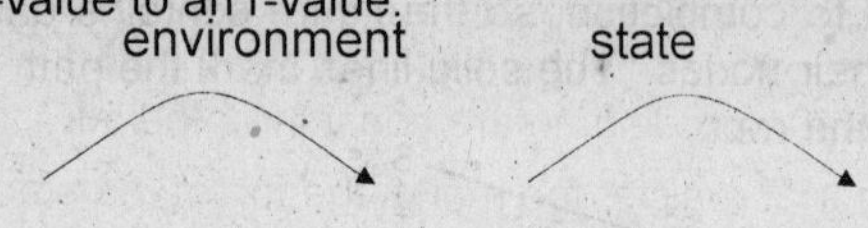

Fig. 5.5. Two-stage mapping from names to values.

Environments and states are different; an assignment changes the state, but not the environment. For example, suppose that storage address 100, associated with variable pi, holds 0.

After the assignment pi : = 3.14, the same storage address is associated with pi, but the value held there is 3.14.

When an environment associates storage location s with a name x, we say that x is bound to s; the association itself is referred to as a binding of x. The term storage "location" is to be taken figuratively. If x is not of a basic type, the storage s for x may be a collection of memory words.

A binding is the dynamic counterpart of a declaration, as shown in Fig. 5.6. As we have seen, more than one activation of a recursive procedure can be alive at the same time. In Pascal, a local variable name in a procedure is bound to a different storage location in each activation of a procedure.

STATIC NOTION	DYNAMIC COUNTERPART
definition of a procedure	activations of the procedure
declaration of a name	bindings of the name
scope of a declaration	lifetime of a binding

Fig. 5.6. Corresponding static and dynamic notions.

7. Storage Organization

The organization of run-time storage in this section can be used for languages such as Fortran, Pascal, and C.

Subdivision of Run-Time Memory

Suppose that the compiler obtains a block of storage from the operating system for the compiled program to run in. From the discussion in the last section, this run-time storage might be subdivided to hold:

1. The generated target code,
2. Data objects, and
3. A counterpart of the control stack to keep track of procedure activations.

The size of the generated target code is fixed at compile time, so the compiler can place it in a statically determined area, perhaps in the low end of memory. Similarly, the size of some of the data objects may also be known at compile time, and these too can be placed in a statically determined area, as in Fig. 5.7. One reason for statically allocating as many data objects as possible is that the addresses of these objects can be compiled into the target code.

All data objects in Fortran can be allocated statically.Implementations of languages like Pascal and C use extensions of the control stack to manage activations of procedures. When a call occurs, execution of an activation is interrupted and information about the status of the machine, such as the value of the program counter and machine registers, is saved on the stack.

When control returns from the call, this activation can be restarted after restoring the values of relevant registers and setting the program counter to the point immediately after the call. Data objects whose life-times are contained n that of an activation can be allocated on the stack, along with other information associated with the activation.

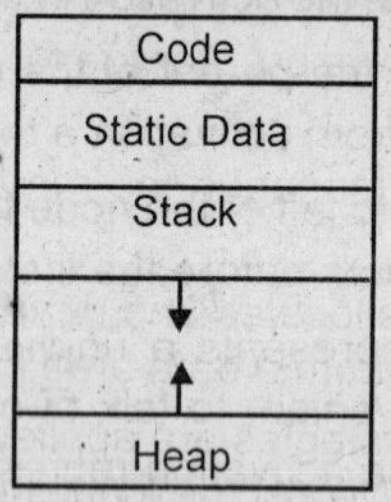

Fig. 5.7. Typical subdivision of run-time memory into code and data areas.

A separate area of run-time memory, called a heap, holds all other information. Pascal allows data to be allocated under program control, the storage for such data is taken from the heap. Implementations of languages in which the lifetimes of activations cannot be represented by an activation tree might use the heap to keep information about activations. The controlled way in which data is allocated and deallocated on a stack makes it cheaper to place data on the stack than on the heap.

The sizes of the stack and the heap can change as the program executes, so we show these at opposite ends of memory in Fig. 5.7, where they can grow toward each other as needed. Pascal and C need both a run-time stack and heap, but not all languages do.

By convention, stacks grow down. That is, the "top" of the stack is drawn towards the bottom of the page. Since memory addresses increase as we go down a page, "downwards-growing" means toward higher addresses. If top marks the top of the stack, offsets from the top of the stack can be computed by subtracting the offset from top. On many machines this computation can be done efficiently by keeping the value of top in a register. Stack addresses can then be represented as offsets from top.

8. Activation Records

Information needed by a single execution of a procedure is managed using a contiguous block of storage called an activation record or frame, consisting of the collection of fields shown in Fig. 5.8. Not all languages, nor all compilers use all of these fields; often registers can take the place of one or more of them. For languages like Pascal and C, it is customary to push the activation record of a procedure on the run-time stack when the procedure is called and to pop the activation record off the stack when control returns to the caller.

returned values
actual parameters
option control link
option access link
saved machine status
local data
temporaries

Fig. 5.8. A general activation record.

9. Storage-Allocation Strategies

A different storage-allocation strategy is used in each of the three data areas in the organization of Fig. 5.7.

1. Static allocation lays out storage for all data objects at compile time.
2. Stack allocation manages the run-time storage as a stack.
3. Heap allocation allocates and deallocates storage as needed at run time from a data area known as a heap.

These allocation strategies are applied in this section to activation records. We also describe how the target code of a procedure accesses the storage bound to a local name.

10. Static Allocation

In static allocation, names are bound to storage as the program is compiled, so there is no need for a run-time support package. Since the bindings do not change at run time, every time a procedure is activated, its names are bound to the same storage locations. This property allows the values of local names to be retained across activations of a procedure. That is, when control returns to procedure, the values of the locals are the same as they were when control left the last time.

From the type of name, the compiler determines the amount of storage to set aside for that name, the address of this storage consists of an offset from an end of the activation record for the procedure. The compiler must eventually decide where the activation records go, relative to the target code and to one another. Once this decision is made, the position of each activation record, and hence of the storage for each name in the record is fixed. At compile time we can therefore fill in the addresses at which the target code can find the data it operates on. Similarly, the addresses at which information is to be saved when a procedure call occurs are also known at compile time.

However, some limitations go along with using static allocation alone.

1. The size of a data object and constraints on its position in memory must be known at compile time.
2. Recursive procedures are restricted, because all activations of a procedure use the same bindings for local names.
3. Data structures cannot be created dynamically, since there is no mechanism for storage allocation at run time.

Fortran was designed to permit static storage allocation.

11. Stack Allocation

Stack allocation is based on the idea of a control stack; storage is organized as a stack, and activation records are pushed and popped as activations begin and end, respectively. Storage for the locals in each call of a procedure is contained in the activation record for that call. Thus locals are bound to fresh storage in each activation, because a new activation record is pushed onto the stack when a call is made. Furthermore, the values of locals are deleted when the activation ends; that is, the values are lost because the storage for locals disappears when the activation record is popped.

We first describe a form of stack allocation in which the sizes of all activation records are known at compile time. Situations in which incomplete information about sizes is available at compile time are considered below.

Suppose that register top marks the top of the stack. At run time, an activation record can be allocated and deallocated by incrementing and decrementing top, respectively, by the size of the record. If procedure q has an activation record of size a, then top is incremented by a just before the target code of q is executed. When control returns from q, top is decremented by a.

12. Dangling References

Whenever storage can be deallocated, the problem of dangling references arises. A dangling reference occurs when there is a reference to storage that has been deallocated. It is a logical error to use dangling references, since the value of deallocated storage is undefined according to the semantics of most languages. Worse, since that storage may later be allocated to another datum, mysterious bugs can appear in programs with dangling references.

13. Heap Allocation

The stack allocation strategy discussed above cannot be used if either of the following is possible.

1. The values of local names must be retained when an activation ends.

2. A called activation outlives the caller. This possibility cannot occur for those languages where activation trees correctly depict the flow of control between procedures.

In each of the above cases, the deallocation of activation records need not occur in a last-in first-out fashion, so storage cannot be organized as a stack.

Heap allocation parcels out pieces of contiguous storage, as needed for activation records or other objects. Pieces may be deallocated in any order, so over time the heap will consist of alternate areas that are free and is use.

14. Dynamic Scope

Under dynamic scope, a new activation inherits the existing bindings of non-local names to storage. A non local name a in the called activation refers to the same storage that it did in the calling activation. New bindings are set up for local names of the called procedure; the names refer to storage in the new activation record.

The program in Fig. 5.9 illustrates dynamic scope. Procedure show on lines 3-4 writes the value of non-local r. Under lexical scope in Pascal, the non local r is in the scope of the declaration on line 2, so the output of the program is

0.250 0.250

0.250 0.250

However, under dynamic scope, the output is

0.250 0.125

0.250 0.125

When show is called on lines 10-11 in the main program, 0.250 is written

```
1. program dynamic (input, output) ;
2. var r : real ;
3. procedure show;
4. begin write ( r : 5:3) end ;
5. procedure small ;
6. var r : real ;
7. begin r : = 0.125 ; show end ;
8. begin
9. r : = 0.25;
10. show; small; writeln;
11. show; small; writeln
12. end.
```

Fig. 5.9. The output depends on whether lexical or dynamic scope is used.

because the variable r local to the main program is used. However, when show is called on line 7 from within small, 0.125 is written because the variable r local to small is used.

15. Parameter Passing

When one procedure calls another, the usual method of communication between them is through nonlocal names and through parameters of the called procedure. Both nonlocals and parameters are used by the procedure in Fig.5.10 to exchange the values of a [i] and a [j]. Here the array a is nonlocal to the procedure exchange, and i and j are parameters.

```
Procedure exchange ( i, j: integer);
var x : integer;
begin
x : = a [ i ] ; a [ i ] : = a [ j] ; a [j ] : = x
end
```

Fig. 5.10. The Pascal procedure swap with nonlocals and parameters.

Several common methods for associating actual and formal parameters are discussed in this section. They are: call-by-value, call-by-reference, copy-restore, call-by-name, and macro expansion. It is important to know the parameter passing method a language (or compiler) uses, because the result of a program can depend on the method used.

Why so many methods? The different methods arise from differing interpretations of what an expression represents. In an assignment like expression a [j] represents a value, while a [i]

a [i] : = a [j]

represents a storage location into which the value of a [j] is placed. The decision of whether to use the location or the value represented by an expression is determined by whether the expression appears on the left or right, respectively, of the assignment symbol, the term l-value refers to the storage represented by an expression, and r-value to the contained value in the storage. The prefixes l- and r- come from "left" and "right" side of an assignment.

Differences between parameter-passing methods are based primarily on whether an actual parameter represents an r-value, an l-value, or the text of the actual parameter itself.

Call-by-Value

This is, in a sense, the simplest possible method of passing parameters. The actual parameters are evaluated and their r-values are passed to the called procedure. Call-by-value is used in C, and Pascal parameters are usually passed this way. All the programs so far in this chapter have relied on this method for passing parameters. Call-by-value can be implemented as follows.

1. A formal parameter is treated just like a local name, so the storage for the formals is in the activation record of the called procedure.

2. The caller evaluates the actual parameters and places their r-values in the storage for the formals.

A distinguishing feature of call-by-value is that operations on the formal parameters do not affect values in the activation record of the caller. If the keyword var on line 3 in Fig.5.11 is left out, Pascal will pass x and y by value to the procedure swap. The call swap (a,b) on line 12 then leaves the values of a and b undisturbed. Under call-by-value, the effect of the call swap (a,b) is equivalent to the sequence of steps.

```
x : = a
y : = b
temp : = x
x : = y
y : = temp
```

where x, y, and temp are local to swap. Although these assignments change the values of the locals x, y, and temp, the changes are lost when control returns from the call and the activation record for swap is deallocated. The call therefore has no effect on the activation record of the caller.

```
1. program reference (input, output) ;
2. var a, b : integer ;
3. procedure swap (var x, y ; integer ) ;
4. var temp : integer ;
5. begin
6. temp : = x ;
7. x : = y ;
8. y : = temp
9. end ;
10. begin
11. a : = 1 ; b : = 2 ;
12. swap (a, b) ;
13. write1n ('a = ', a) ; write1n('b = ', b)
14 end.
```

Fig. 5.11. Pascal program with procedure swap.

A procedure called by value can affect its caller either through nonlocal names (see exchange in Fig.5.10) or through pointers that are explicitly passed as values. In the C program in Fig. 5.12 , x and y are declared on line 2 to be pointers to integers; the & operator in the call swap (&a, &b) on line 8 results in pointers to a and b being passed to swap.

The output of this program is a is now 2, b is now 1

```
1. swap (x,y)
2. int * x, * y ;
3. {   int temp ;
4. temp = * x; * x = * y ; * y = temp ;
5. }
6. main ( )
7. {   int a = 1, b = 2 ;
8. swap ( &a, & b );
9. printf ("a is now %d, b is now %d\n", a,b);
10. }
```

Fig. 5.12. C program using pointers in a procedure called by value.

The use of pointers in this example suggests how a compiler using call-by-reference would exchange values.

Call-by-Reference

When parameters are passed by reference (also known as call-by-address or call-by-location), the caller passes to the called procedure a pointer to the storage address of each actual parameter.

1. If an actual parameter is a name or an expression having an l-value, then that l-value itself is passed.

2. However, if the actual parameter is an expression, like a + b or 2, that has no l-value, then the expression is evaluated in a new location, and the address of that location is passed.

A reference to a formal parameter in the called procedure becomes, in the target code, an indirect reference through the pointer passed to the called procedure.

Example 5.3. Consider the procedure swap in Fig. 5.11. A call to swap with actual parameters i and a [i], that is, swap (i, a [i]), would have the same effect as the following sequence of steps.

1. Copy the address (l-value) of i and a [i] into the activation record of the called procedure, say into locations arg1 and arg2 corresponding to x and y, respectively.

2. Set temp to the contents of the location pointed to by arg1 (i.e., set temp equal to I, where I is the initial value of i). This step corresponds to temp : = x on line 6 in the definition of swap.

3. Set the contents of the location pointed to by arg1 to the value of the location pointed to by arg2; that is, i : = $a[I_0]$. This step corresponds to x : = y on line 7 in swap.

4. Set the contents of the location pointed by arg2 equal to the value of temp; that is, set $a[I_0] := i$. This step corresponds to y : = temp

Call-by-reference is used by a number of languages; var parameters in Pascal are passed in this way. Arrays are usually passed by reference.

Copy-Restore

A hybrid between call-by-value and call-by-reference is copy-restore linkage, also known as copy-in copy-out, or value-result).

1. Before control flows to the called procedure, the actual prameters are evaluated. The r-values of the actuals are passed to the called procedure as in call-by-value. In addition, however, the l-values of those actual parameters having l-values are determined before the call.

2. When control returns, the current r-values of the formal parameters are copied back into the l-values of the actuals, using the l-values computed before the call. Only actuals having l-values are copied, of course.

The first step "copies in " the values of the actuals into the activation record of the called procedure (into the storage for the formals). The second step "copies out" the final values of the formals into the activation record of the caller (into l-values computed from the actuals before the call).

Note that swap (i, a [i]) works correctly using copy-restore, since the location of a [i] is computed and preserved by the calling program before the call. Thus, the final value of formal parameter y, which will be the initial value of i, is copied into the correct location, even though the location of a [i] is changed by the call (because the value of i changes).

Copy-restore is used by some Fortran implementations. However, others use call-by-reference. Differences between the two can show up if the called procedure has more than one way of accessing a location in the activation record of the caller. The activation set up by the call unsafe (a) on line 6 of Fig. 5.13 can access a as a nonlocal and through formal x. Under call-by-reference, the assignments to both x and a immediately affect a, so the final value of a is 0. Under copy-restore, however, the value 1 of actual a is copied into formal x. The final value 2 of x is copied out into the l-value of a just before control returns, so the final value of a is 2.

```
1. program copyout (input, output) ;
2. var a : integer ;
3. procedure unsafe (var x : integer);
4. begin x : = 2; a : = 0 end ;
5. begin
6. a : = 1; unsafe (a) ; writeln (a)
7. end.
```

Fig. 5.13. The output changes if call-by-reference is changed to copy-restore.

Call-by-Name

Call-by-name is traditionally defined by the copy-rule of Algol, which is:

1. The procedure is treated as if it were a macro; that is, its body is substituted for the call in the caller, with the actual parameters literally substituted for the formals. Such a literal substitution is called macro-expansion or in-line expansion.

2. The local names of the called procedure are kept distinct from the names of the calling procedure. We can think of each local of the called procedure being systematically renamed into a distinct new name before the macro-expansion is done.

3. The actual parameters are surrounded by parentheses if necessary to preserve their integrity.

Example 5.4. The call swap (i, a [i]) from Example 5.3 would be implemented as though it were

```
temp : = i
    i : = a [ i ]
a [ i ] : = temp
```

Thus, under call-by-name, swap sets i to a [i], as expected, but has the unexpected result of setting a[a[I_0]] - rather than a [I_0] - to I_0, where I_0 is the initial value of i. This phenomenon occurs because the location of x in the assignment x : = temp of swap is not evaluated until needed, by which time the value of i has already changed. A correctly working version of swap apparently cannot be written if call-by-name is used.

Although call-by-name is primarily of theoretical interest, the conceptually related technique of in-line expansion has been suggested for reducing the running time of a program. There is a certain cost associated with setting up an activation of a procedure - space is allocated for the activation record, the machine status is saved, links are set up, and then control is transferred. When a procedure body is small, the code devoted to the calling sequences may outweigh the code in the procedure body. It may therefore be more efficient to use in-line expansion of the body into the code for the caller, even if the size of the program grows a little. In the next example, in-line expansion is applied to a procedure called by value.

Example 5.5. Suppose that the function f in the assignment

$$X := f(A) + f(B)$$

is called by value. Here the actual parameters A and B are expressions. Substituting expressions A and B for each occurrence of the formal parameter in the body of f leads to call-by-name; recall a [i] in the last example.

Fresh temporary variables can be used to force the evaluation of the actual parameters before execution of the procedure body;

$$t_1 := A;$$
$$t_2 := B;$$
$$t_3 := f(t_1);$$
$$t_4 := f(t_2);$$
$$x := t_3 + t_4;$$

Now in-line expansion will replace all occurrences of the formal by t_1 and t_2 when the first and second calls, respectively, are expaned.

16. Symbol Tables

Before starting the execution of our program, the computer must know, what are all the identifiers we are using in our program, its type its form, its location in memory and other related information. These information are collected by the compiler and entered or placed into a data structure called a symbol table.

Each entry in the symbol table is a pair of the form (name, information). The compiler trace our program and place name and information about each identifier only once. The name of the identifier is new, it is entered into the table. The information about that name is entered in the table during lexical and Syntactic analysis.The information collected in the symbol table is used during several stages (like semantic analysis, code generation etc) in the compilation process. It is also used in error detection and correction process.

17. The contents of a symbol table

Symbol table is merely a table with two fields a name field and an information field. But by using this symbol table, we need to be able to,

i) determine whether a given name is in the table,

ii) add a new name to the table,

iii) Access the information associated with a given name,

iv) Add new information for a given name,

v) Delete a name or a group of names from the table.

The names in the symbol table denote object of various sorts. There may be separate table for variable name, labels, procedure and function names, constant, field names (for structured variable).etc. The space required to store name and information and the format is vary from one type of name to other.

So, we are using different table for each kind of name. Even there may be a separate table for reserve word.

18. Data structures for symbol tables

During Compilation the symbol table is searched every time an identifier is encountered. Data are added if a new name or new information about an existing name is discovered. Thus, in designing a symbol table mechanism, we would like scheme that allows us to add new entries and find existing entries in a table efficiently.

The three symbol table mechanisms which we are going to discuss here are,

- Linear Array/Linear lists.
- Trees
- Hash tables

We shall evaluate each scheme on the basis of the time required to add 'n' entries and 'm' inquiries

19. Linear Lists

It is the simplest scheme to implement, but its performance becomes poor when m and n get large. The conceptually simplest and easiest-to-implement data structure for a symbol table is the linear list of records given as follows.

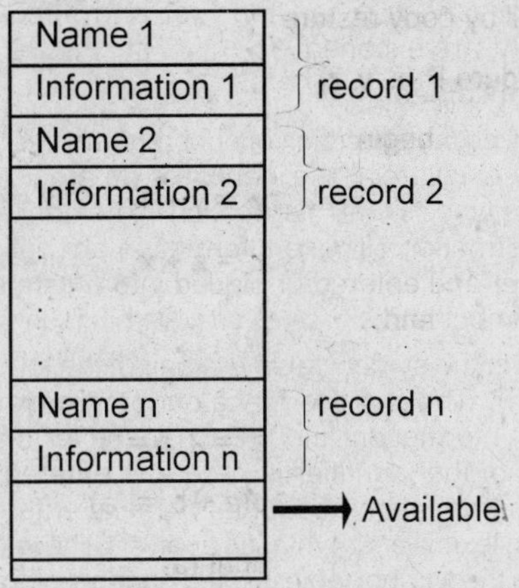

We use a single array or several arrays to store names and their-associated information. New names are added to the list in the order in which they are encountered. By using linear search procedure (ii) Start from the beginning. Search each record, until we reach the last record) we search the array to retrieve a particular information.

When the name is located, the associated information can be found in the words following next. Otherwise if we reach the position marked by pointer AVAILABLE without finding the name, we may use an undefined symbol.

Whenever we insert a new name, we must scan down the list to be sure it is not already there. If it is, we have another fault-a multiply defined name.

If the symbol table contains n names work necessary to insert a new name is proportion to n. To find data about a name we shall, as the average search n/2 names, so the cost of an inquiry is also proportional to n. To insert n names and m inquiries the total work is cn(m+n). Where c-is a constant representing the time necessary for a few machine operations.

Advantage of List organization

The minimum possible space is taken. In a simple non optimizing compiler, the space taken by the symbol table may consume most of the space used for the compiler's data.

20. Self-Organizing List

At the cost of a little extra space, we can use a trick that will often save a substantial fraction of the time spent in searching the symbol table. We add a LINK field to each record and according to LINK field direction we search for a particular record. The following figure shows the structure of self-organizing list.

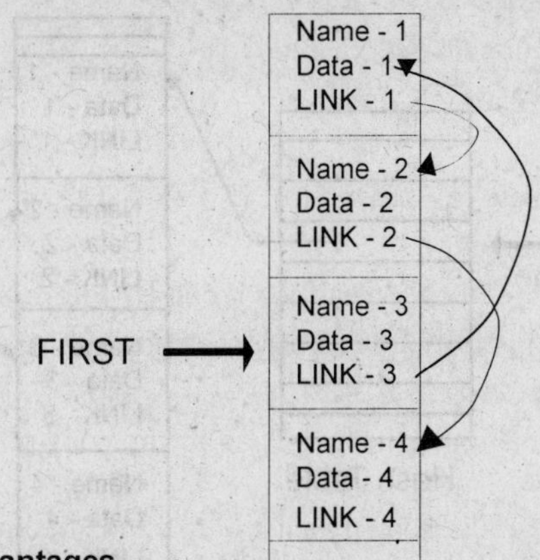

Advantages

- The names referenced frequently will tend to be at the front of the list. Where they will be found quickly.
- If a small set of names is heavily used in a small section of the program, these names will rise to the top while that section is being processed by the compiler.

21. Search tree

A more efficient approach to symbol table organization is to add two link fields, LEFT and RIGHT to each record. These fields are used to link the records into a binary tree.

The property of this tree is,

- All names (Name j) accessable from Name i by using the LEFT i is always lesser than Name i, ie; Name j < Name i
- All names (Name k) accessible from Name i is by using the RIGHT i is always greater than NAME i. Ie; Name k > Name i.

If we want to search a particular name from Name P then we have to only follow Left Pointer if Name < Name P otherwise ie; If Name P < Name then we have to only follow Right pointer If Name and Name P equal means ie; Name = Name P then we have found what we are looking for

22. Hash Tables

The basic hashing scheme is illustrated in the following table Here. We use two tables, a hash table and a storage table. The hash table consists of k words, numbered 0,1,2,......, k-1.

These words are pointers into the storage table to the heads of k separate linked list. Each record in the symbol table appears an exactly one of these lines.

To determine whether Name is in the symbol table, we apply to NAME a hash function h such that h(NAME) is an integer between 0 and k-1. It is on the list numbered h(NAME) that the record for NAME belongs.

To enquire about NAME we compute h(NAME) and search that list only. To enter name into the symbol table, we create a record for it at the first available place in the storage and link that record to the beginning of the $h(NAME)^{th}$ list.

If we use hash table we must give a great attention to the question of how to design the hash function h such that,

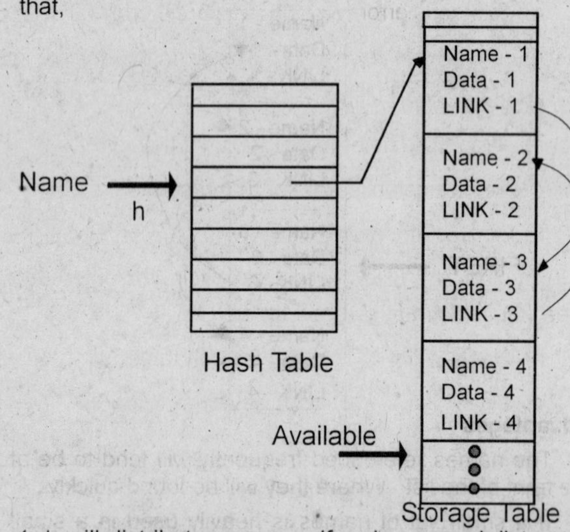

1. h will distribute names uniformly among k lists, and

2. h is easy to compute for names consisting of strings of characters.

SOLVED EXAMPLES

1. Consider the following program in pseudo-pascal syntax. What is printed by the program if parameter a in procedure test 1 is passed as

I. Call-by-reference parameter?

II. Call-by-value-result parameter?

```
Program Example (input, output);
var b: integer;
procedure test 2;
begin b := 10 end
procedure test 1 (a : integer);
begin
a := 5;
writeln ('point 1 : 'a, b)
test 2;
writeln ('point : ' a, b);
end;
begin (*Example*)
b := 3; test 1(b);
writeln ('point 3 : 'b);
end;
```

[GATE-1996]

Sol: I. call-by-reference

Point 1 : 5 5

Point 2 : 10 10

Point 3 : 10

II. call-by-value

Point 1 : 5 3

Point 2 : 5 10

Point 3 : 5

2. What will be the output of the following program assuming that parameter passing is

I. call by value, II. call by reference

III. call by copy restore

```
procedure P (x, y, z)
        begin
                y := y + 1;
                z := z + x;
        end;
begin
                a := 2; b:= 3;
                p(a + b, a, a)
                print (a)
end.
```

[GATE-1999]

Sol : I. In the case of call by value, the values of a and b in the main program don't change upon calling procedure p. So 2 is printed.

II. In the case of call by reference, only the second and third arguments to P('a' and 'a') – can be called by reference. So, variables 'y' and 'z' in procedure P point to the same location as 'a'. So, on executing y = y + 1, the value of a becomes 3. Then on executing z = z + x, the value of a becomes 3 + 5 = 8. So, 8 get printed.

III. The final output is same as call by reference. So, 8 gets printed.

3. Consider the following program in pseudo - Pascal syntax.

```
program main ;
    var x : integer ;
procedure Q (z : integer) ;
    begin
        z := z + x ;
        writeln (z)
    end ;
procedure P (y : integer) ;
    var x : integer ;
    begin
        x := y + 2 ;
        Q (x) ;
        writeln (x)
    end ;
begin
        x := 5 ;
        P (x) ;
        Q (x) ;
        writeln (x)
        end.
```

What is the output of the program, when

(a) The parameter passing mechanism is call-by-value and the scope rule is static scoping ?

(b) The parameter passing mechanism is call-by-reference and the scope rule is dynamic scoping ?

[GATE- 2000]

Sol :

(a) 12
7
10
5

(b) 14
14
10
10

OBJECTIVE QUESTIONS

1. Which of the following techniques is used to replace runtime computations by complie time computations?

a) Inavriant computation b) Peehole optimization
c) Constant foloding d) Code hosting.

2. Which of the following symbols table implementation is based on the property of locality of reference?

a) Hash table b) Search table
c) Self-organizing list d) Linear list.

3. Minimum Hamming Distance method is used for correction of

a) Syntactic errors b) Semantic errors
c) Algorithmic errors d) Transcription errors.

4. Semantic errors can be detected

a) At compile time only b) At run time only
c) Both at compile time and at run time.
d) None of the above.

5. Which of the following symbol table implementation has the minimum access time?

a) Hash table b) Search table
c) Linear list d) self-organizing list

6. Symbol table can be used for

a) Checking type compatibility
b) Suppressing duplication of error messages
c) Storage allocation d) None of the above

7. Access time of the symbol table will be logarithmic, if it is implemented by

a) Linear list b) Search tree
c) Hash table d) Self-organization list

8. Heap allocation is required for languages **[GATE-1997]**

a) That support recursion b) That support dynamic data structures

c) That use dynamic scope rules d) None

9. The identification of common sub-expression and replacement of run time computations by compile-time computations is

a) Local optimization b) Loop optimization

c) Constant folding d) Data flow analysis.

ANSWER KEY				
1. c	**2.** c	**3.** a	**4.** c	**5.** a
6. a,b,c	**7.** b	**8.** b	**9.** c	

VI. CODE GENERATION

- **Introduction**
- **Issues In Design of Code Generator**

1. Introduction

The final phase in our compiler model is the code genrator. It takes as input an intermediate represenation of the source program and produces as output an equivalent target program, as indicated in Fig. 6.1. The code generation techniques presented in this chapter can be used whether or not an optimization phase occurs before code genration, as in some so-called "optimizing" compilers. Such a phase tries to transform the intermediate code into a form from which more efficient target code can be produced.

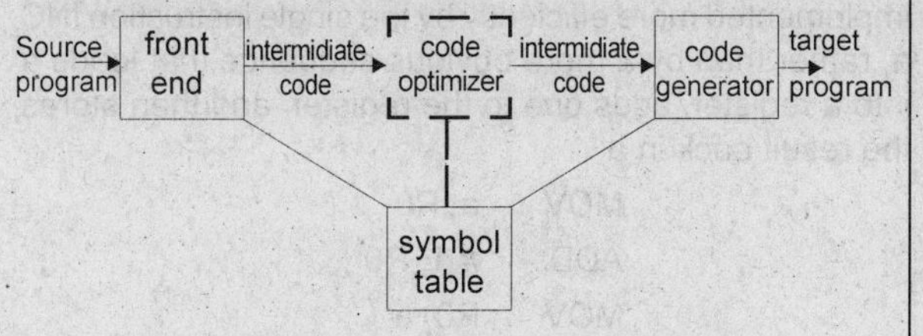

Fig. 6.1 Position of code generator.

The requirements traditionally imposed on a code generator are severe. The output code must be correct and of high quality, meaning that it should make effective use of the resources of the target machine. Moreover, the code generator itself should run efficiently.

Mathematically, the problem of generating optimal code is undecidable. In practice, we must be content with heuristic techniques that generate good, but not necessarily optimal, code. The choice of heuristics is important, in that a carefully designed code generation algorithm can easily produce code that is several times faster than that produced by a hastily conceived algorithm.

2. Issues In The Design Of A Code Generator

While the details are dependent on the target language and the operating system, issues such as memory management, instruction selection, register allocation, and evaluation order are inherent in almost all code-generation problems. In this section, we shall examine the generic issues in the design of code generators.

3. Input to the Code Generator

The input to the code generator consists of the intermediate representation of the source program produced by the front end, together with information in the symbol table that is used to determine the run-time addresses of the date objects denoted by the names in the intermediate representation.

There are several choices for the intermediate language, including: linear representations such as postfix notation, three-address representations such as quadruples, virtual machine representations such as stack machine code, and graphical representations such as syntax trees and dags. Although the algorithms in this chapter are couched in terms of three-address code, trees, and dags, many of the techniques also apply to the other intermediate representations.

We assume that prior to code generation the front end has scanned, parsed, and translated the source program into a reasonably detailed intermediate representation, so the values of names appearing in the intermediate language can be represented by quantities that the target machine can directly manipulate (bits, integers, reals, pointers, etc.). We also assume that the necessary type checking has taken place, so type-conversion operators have been inserted wherever necessary and obvious semantic errors (e.g., attempting to index an array by a floating-point number) have already been detected. The code generation phase can therefore proceed on the assumption that its input is free of errors. In some compilers, this kind of semantic checking is done together with code generation.

4. Target Programs

The output of the code generator is the target program. Like the intermediate code, this output may take on a variety of forms: absolute machine language, relocatable machine language, or assembly language.

Producing an absolute machine-language program as output has the advantage that it can be placed in a fixed location in memory and immediately executed. A small program can be compiled and executed quickly.

Producing a relocatable machine-language program (object module) as output allows subprograms to be compiled separately. A set of relocatable object modules can be linked together and loaded for execution by a linking loader. Although we must pay the added expense of linking and loading if we produce relocatable object modules, we gain a great deal of flexibility in being able to compile subroutines separately and to call other previously compiled programs from an object module. If the target machine does not handle relocation automatically, the compiler must provide explicit relocation information to the loader to link the separately compiled program segments.

Producing an assembly-language program as output makes the process of code generation somewhat easier. We can generate symbolic instructions and use the macro facilities of the assembler to help generate code. The price paid is the assembly step after code generation. Because producing assembly code does not duplicate the entire task of the assembler, this choice is another reasonable alternative, especially for a machine with a small memory, where a compiler must use several passes. In this chapter, we use assembly code as the target language for readability. However, we should emphasize that as long as addresses can be calculated from the offsets and other information stored in the symbol table, the code generator can produce relocatable or absolute addresses for names just as easily as symbolic addresses.

5. Memory Management

Mapping names in the source program to addresses of data objects in run-time memory is done cooperatively by the front end and the code generator.We assumed that a name in a three-address statement refers to a symbol-table entry for the name. Symbol-table entries were created as the declarations in a procedure were examined. The type in a declaration determines the width, i.e., the amount of storage, needed for the declared name. From the symbol-table information, a relative address can be determined for the name in a data area for the procedure.

If machine code is being generated, labels in three-address statements have to be converted to addresses of instructions. This process is analogous to the "backpatching" technique . Suppose that labels refer to quadruple numbers in a quadruple array. As we scan each quadruple in turn we can deduce the location of the first machine instruction generated for that quadruple, simply by maintaining a count of the number of words used for the instructions generated so far. This count can be kept in the quadruple array (in an extra field), so if a reference such as j: goto i is encountered, and i is less than j, the current quadruple number, we may simply generate a jump instruction with the target address equal to the machine location of the first instruction in the code for quadruple i. If, however, the jump is forward, so i exceeds j, we must store on a list for quadruple j. Then, when we process quadruple i, we fill in the proper machine location for all instructions that are forward jumps to i.

6. Instruction Selection

The nature of the instruction set of the target machine determines the difficulty of instruction selection. The uniformity and completeness of the instruction set are important factors. If the target machine does not support each data type in a uniform manner, then each exception to the general rule requires special handing.

Instruction speeds and machine idioms are other important factors. If we do not care about the efficiency of the target program, instruction selection is straightforward. For each type of three-address statement, we can design a code skeleton that outline the target code to be generated for that construct. For example, every three-address statement of the form x:=y+z, where x,y, and z are statically allocated, can be translated into the code sequence.

```
MOV y, R0    /* load y into register R0 */
ADD z, R0    /* add z to R0 */
MOV R0, x    /* store R0 into x */
```

Unfortunately, this kind of statement-by-statement code generation often produces poor code. For example, the sequence of statements

```
a : = b + c
d : = a + e
```

would be translated into

```
MOV b, R0
ADD c, R0
MOV R0, a
MOV a, R0
ADD e, R0
MOV R0, d
```

Here the fourth statement is redundant, and so is the third if a is not subsequently used.

The quality of the generated code is determined by its speed and size. A target machine with a rich instruction set may provide several ways of implementing a given operation. Since the cost differences between different implementations may be significant, a naïve translation of the intermediate code may lead to correct, but unacceptably inefficient target code. For example, if the target machine has an "increment" instruction (INC), then the three-address statement a : = a + 1 may be implemented more efficiently by the single instruction INC a, rather than by a more obvious sequence that loads a into a register, adds one to the register, and then stores the result back in a:

```
MOV    a, R0
ADD    # 1, R0
MOV    R0, a
```

Instruction speeds are needed to design good code sequences but, unfortunately, accurate timing information is often difficult to obtain. Deciding which machine-code sequence is best for a given three-address construct may also require knowledge about the context in which that construct appears.

7. Register Allocation

Instructions involving register operands are usually shorter and faster than those involving operands in memory. Therefore, efficient utilization of registers is particularly important in generating good code. The use of registers is often subdivided into two subproblems:

1. During registers allocation, we select the set of variables that will reside in registers at a point in the program.

2. During a subsequent register assignment phase, we pick the specific register that a variable will reside in.

Finding an optimal assignment of registers to variables is difficult, even with single-register values. Mathematically, the problem is NP-complete. The problem is further complicated because the hardware and/or the operating system of the target machine may require that certain register-usage conventions be observed.

Certain machines require register - pairs (an even and next odd-numbered register) for some operands and results. For example, in the IBM System/370 machines integer multiplication and integer division involve register pairs. The multiplication instruction is of the form

```
M    x, y
```

Where x, the multiplicated, is the even register of an even/odd register pair. The multiplicand value is taken from the odd register of the pair. The multiplier y is a single register. The product occupies the entire even/odd register pair.

The division instruction is of the form

D x, y

Where the 64-bit dividend occupies an even/odd register pair whose even register is x; y represents the divisor. After division, the even register holds the remainder and the odd register the quotient.

Now, consider the two three-address code sequences in Fir. 6.2(a) and (b), in which the only difference is the operator in the second statement. The shortest assembly-code sequences for (a) and (b) are given in Fig. 6.3.

Ri stands for register i, (SRDA R0, 32 shifts the dividend into R1 and clears R0 so all bits equal its sign bit.) L, ST, and A stand for load, store and add, respectively. Note that the optimal choice for the register into which

(a)	(b)
t := a + b	t := a + b
t := t * c	t := t * c
t := t / d	t := t / d

Fig. 6.2. Two three-address code sequence.

(a)		(b)	
L	R1, a	L	R0, a
A	R1, b	A	R0, b
M	R0, c	A	R0, c
D	R0, d	SRDA	R0, 32
St	R1, t	D	R0, d
		ST	R1, t

Fig. 6.3. Optimal machine-code sequences.

a is to be loaded depends on what will ultimately happen t.

8. Choice of Evaluation Order

The order in which computations are performed can affect the efficiency of the target code. Some computation orders require fewer registers to hold intermediate results than others, as we shall see. Picking a best order is another difficult, NP-complete problem. Initially, we shall avoid the problem by generating code for the three-address statements in the order in which they have been produced by the intermediate code generator.

OBJECTIVE QUESTIONS

1. Running time of the program depends on

a) The way the registers are used
b) The order in which computations are performed
c) The way the addressing modes are used
d) The usage of machine idioms

2. The process of assigning load address to the various parts of the program and adjusting the code and data in the program to reflect the assigned addresses is called **[GATE-2001]**

a) Assembly b) Parsing
c) Relocation d) Symbol resolution

3. A language L allows declaration of arrays whose sizes are not known during compilation. It is required to make efficient use of memory. Which one of the following is TRUE? **[GATE-1997]**

a) A compiler using static memory allocation can be written for L

b) A compiler cannot be written for L; an interrupt must be used.

c) A compiler using dynamic memory allocation can be written for L

d) None

4. A linker is given object modules for a set of programs that were compiled separately. What information need not be include in an object module? **[GATE-1995]**

a) Object code b) Relocation bits

c) Names and locations of all external symbols defined in the object module

d) Absolute address of internal symbols

ANSWER KEY

1. a,b,c,d	2. c	3. c	4. d

PRACTICE QUESTIONS

FUNDAMENTALS OF LANGUAGE TRANSLATORS

1. The capcity of a laguage to compile itself is called as
   (a)Cross-compiler
   (b) Compiler
   (c) bootstrapping
   (d) all the above

2. A compiler which runs on one machine and products target code for another machine is called as
   (a) Crisp-compiler (b) compact compiler
   (c) overlap compiler (d) cross-compiler

3. Error handler is attached to
   (a) All the phases of the compiler
   (b) few phases of the compiler
   (c) No phases
   (d) the symbol table

4. Intermediate code generation and code generation phases of a compiler gives
   (a) The same phases of compiler
   (b) few pahases of the compiler
   (c) ro result
   (d) the symbol table

5. The number of phases in the compiler are
   (a) 6 (b) 5
   (c) any number (d) 7

6. The special character that identifies the end of a starting is called as
   (a) end pointer (b) sentinel
   (c) lexeme d) all

7. The regular expression a* denotes
   (a) set of all strings of one or more number of a's
   (ɔ) set of all strings of zero or more number of a's
   (c) set of all strings of two consecutive a's
   d) all the above

8. The notation r? is a shorthand for
   (a) $r / \in$ (b) r/r
   (c) $r*$ (d) rr

9. A Lex specification consists of ______________parts
   (a) 2 (b) 3
   (c) 4 (d) 5

10. Which of them is called as the lexical analyzer generator
    (a) parser-generator
    b)compiler construction tool
    (c) lex comiler
    d) compiler-compiler

11. $\in$-Closure(s) is
    (a) set of DFA states reachable from state s on $\in$-transitions alone.·
    (b) set of DFA states reachable from state s on all transitions.
    (c) set of NFA states reachable from state s on $\in$-transitions alone
    d) set of DFA states reachable from other states to state s on $\in$-transitions.

12. The NFA for $\in$ is
    (a) $\xrightarrow{Start} i \xrightarrow{\in} f$
    b) $\xrightarrow{Start} i \xrightarrow{\phi} f$
    (c) $\xrightarrow{Start} i \xrightarrow{\in} f$
    d) None

13. Regular definitions are of the form, where d is the distinct name and *r* is a regular expression
    (a) $d -> dr$ (b) $r -> d$
    (c) $d -> r$ (d) all

14. The unary postfix operator + means
    (a)one or more instances
    (b) Zero or more instances
    (c) both a and b
    (d) none

15. The sequence of characters in the program that is matched by the pattern for a token is known as
    (a) token string
    (b) lexeme
    (c) sentine l
    (d) identifier

16. The (r)/(s) is a regular expression denoting
    (a) (r) (s)
    (b) $L(r) \cap L(s)$
    (c) $(L(r))^* \cap (L(s))^*$
    d) $L(r) \cup L(s)$

17. Which of the following are language processors
(a) assemblers (b) compilers
(c) interpeters (d) all of the above

18. In analyzing the compilation of a program, the term lexical analysis is associate with
(a) recognition of basic syntactic constructs through reductions
(b) recognition of basic elements and creation of uniform symbols
(c) creation of more optimal matrix.
(d) use of macro processor to procedure more optimal assembly code

19. A translator is best described as
(a) application software
(b) system software
(c) a hardware component
(d) None

20. Which of the following is not a phase of compilation process
(a) booting (b) optimizing
(c) parsing (d) scanning

21. A compiler is a program that
(a) Places programs into memory and prepares them for execution.
(b) Automates the translation of assembly language into machine language.
(c) Accept a program written in a high level language and produces an object program
(d) Appears to execute a source program as if it were machine language

22. Compilers can diagnose
(a) Grammatical errors only
(b) Logical errors only
(c) Grammatical as well as logical errors
(d) Neither grammatical not logical errors

23. Compilers are generally written by
(a) Computer manufacturer
(b) Computer users
(c) Professional programmers
(d) System analyst

24. Lexical analysis helps in
(a) Simpler design of compiler
(b) Improving effieciency of compiler
(c) Enhancing compiler portability
(d) All the above

25. The action of passing the source program into the proper syntatic classes is known as
(a) Syntax analysis
(b) Lexical analysis
(c) Interpretation analysis
(d) All the above

26. In analyzing the compilation of program, the term "lexical analysis" is associated with
(a) Recognition of basic syntactic constructs through reduction
(b) Recognition of basic elements creation uniform symbols
(c) Creation of more optimal matrix
(d) Use of macro processor to produce more optimal assembly code

27. The task of the lexical analysis phase is
(a) To parse the source program into the basic elements or tokens of the language.
(b) To build a literal table and an identifier table
(c) To build an uniform symbol table.
(d) All of these

28. Which is the permanent database in the general model of a complier
(a) Literal table
(b) Identifier
(c) Terminal table
(d) Reductions

29. Which of the following strings can definitely be said to be token without looking at the next input character while compiling a *C* program.
(I) main (II) while (III) if (VI) ++
(a) I (b) IV
(c) Both I & IV (d) None

30. Which one of the regular expressions over { 0, 1 } denotes the set of all strings not containg 100 as a sub string.
(a) $0^*(1+0)^*$ (b) 0^*1010^*
(c) 0^*1^*00 (d) $0^*(10+1)^*$

31. An ideal compiler should
(a) Be smaller in size and take less time for compiling
(b) Be written in a high level language
(c) Produce object code that is smaller in size and executes faster
(d) All the above

ANSWER KEY

1. (c)	2. (d)	3. (a)	4. (b)	5. (a)
6. (b)	7. (b)	8. (a)	9. (b)	10. (c)
11. (c)	12. (a)	13. (c)	14. (a)	15. (b)
16. (d)	17. (d)	18. (b)	19. (b)	20. (a)
21. (c)	22. (a)	23. (c)	24. (d)	25. (d)
26. (b)	27. (d)	28. (c)	29. (b)	30. (b)
31. (d)				

PARSERS AND SYNTAX DIRECTED TRANSLATION1.

1. Left factoring is the process of factoring out the common

(a) prefixed of alternates
(b) suffixes of alternates
(c) predicative parsing
(d) none

2. The graph depicting the inter-dependencies of the attribute of idfferent nodesin a parse tree is called

(a) Flow graph
(b) Steffi graph
(c) Karnaugh's graph
(d) Dependency graph

3. The function of the syntax phase is

(a) To recognize the major constructs of the language and to call the appropriate action routing that will generate the intermediate form or matrix for these constructs
(b) To build a literal table and a identifier table
(c) To build a uniform symbol table
(d) To parse the source program into the basic elements or tokens of the language

4. To value of K, in LR(K) cannot be

(a) 0 (b) 1
(c) 2 (d) None of the above

5. The most powerful parsing method is

(a) LL(1) (b) Canonical LR (c)
SLR (d) LALR

6. A shift reduce parser carries out the action specified within braces immediately after reducing with the corresponding rule of grammer.

$S \rightarrow xx\, W \{print "1"\}$

$S \rightarrow Y \{print "2"\}$

$W \rightarrow Sz \{print "3"\}$

what is translation of "xxxxyzz" using the syntax directed scheme described by the above rules?

(a) 23131 (b) 11233 (c)
11231 (d) 33211

7. A grammer can have

(a) A non-terminal A that can't derive any string of terminals
(b) A non-terminal A that can be present in any sentential form
(c) ε as the only symbol on the left hand side of a production
d) both (a) and (b)

8. A top down parser generates

(a) Left-most derivation
(b) Right-most derivation
(c) Right most derivation is reverse
(d) Left-most derivation in reverse

9. A bottom up parser generates

(a) Left-most derivation
(b) Rigt-most derivation
(c) Right most derivation in reverse
(d) Left-most derivation in reverse

10. A given grammer is said to be ambiguous if

(a) Two or more production have the same non-terminal to the left hand side
(b) a derivation tree has more than one associated sentence
(c) there is sentence wiht more than one derivation three corresponding to it (d)
brackets are not present in the grammer

11. The grammer $E\ a\ E + E / E * E / a$, is

(a) ambiguous
(b) unambiguous
(c) ambiguous or not depend on the given sentences
(d) None of these

12. Which of the following sentences can be derived by the grammer

$S \rightarrow ABCc / Abc$

$BA \rightarrow AB$

$Ab \rightarrow ab$

$Aa \rightarrow aa$

(a) abc (b) aab
(c) abcc (d) abbc

13. In a syntax dierected translation scheme, if the value of an attribute of a node is a function of the value of the attributes of the children, then it is called a

(a)synthesized
(b) inherited attribute
(c) both a and b
(d) nul! attribute

14. Inherited attributed is a natural choice in
(a) keeping track of variable declaration
(b) checking for the correct use of the L values and R-values
(c) both and b
(d) none

15. The advantage of panic of error recovery is that
(a) it is simple of implemented
(b) it never gets into an infinite loop
(c) both a and b
(d) none

16. The recover from an error, the operate - precedence parser may
(a) insert symbols onto the stack and onto the input
(b) delete symbols from the stack
(c) delete symbols from the input
(d) all of these

17. The most general phase-structured grammer is
(a) regular
(b) context-free
(c) context sensitve
(d) done

18. In a context-sensitive grammer, number of grammer symbols on the left hand side of a production cannot be greater than the number of
(a) grammar symbols on the right hand side
(b) terminal on the right hand side
(c) non terminal on the right hand side
(d) all of these

19. In a contex free grammer
(a) ε cannot be the right hand side of any production
(b) terminal symbols can't be present in the left hand side of any production
(c) the number of grammer symbols in the left hand side is not greater than the number if grammer symbols in the right hand side
(d) all of these

20. The main difference between a DFSA and NDFSA is
(a) in DFSA, e transition may be present
(b) in NDFSA, e transition may be present
(c) in DFSA, from any given state, there can't be any alphabet leading to two different state
(d) In NDFSA, from any given state, there can't be any alphabet leading to two different state.

21. Two finite state machine are said to be equivalent of they
(a) have the same number of state
(b) have the same number of edges
(c) have the same number of states and edges
(d) recognize the same number of states and edges

22. Shift-reduce parser are
(a) top-down parser
(b) bottom up parser
(c) predictive parser
(d) none of these

23. Recursive descent parsing in an example of
(a) top-down parser
(b) bottom-up parser
(c) predictive parser
(d) none of thes

24. In operator precedence parsing, precedence relations are defined
(a) for all pairs of non-terminals
(b) for all paris of terminals
(c) to delimit the handle
(d) non of these

26. LR parsers are attractive because;
(a) it can be constructed to recognize CFG corresponding to almost all programming constructs.
(b) it does not backtrack
(c) both a and b
(d) none

27. LL(K) grammar
(a) has to be CFG
(b) has to be unambiguos
(c) cannot have left recursive non-terminals
(d) all of these

Consider the grammar for the questions 28 through 30

$$E \rightarrow E / X \;\; X$$

$$X \rightarrow T - X \;\; X * T \;\; T$$

$$T \rightarrow T + F \;\; F$$

$$F \rightarrow (E) \;\; id$$

28. This grammar is
(a) unambiguos (b) ambiguos
(c) context free (d) both a and c

29. The above grammar is used to generate all valid arithmetic expressions in a hypothetical language in which
(a) / associates from the left
(b) $*$ associates from the left

(c) + associates from the left
(d) all

30. The above grammar is used to generate all valid arithmetic expressions in a hypothetical language in which
(a) + has the highest precedence
(b) $*$ has the highest precedence
(c) - has the highest precedence
(d) / has the highest precedence

31. In the grammar

$E \rightarrow TE'$

$E' \rightarrow TE' / \in$

$T \rightarrow FT'$

$T' \rightarrow *FT' / \in$

$F \rightarrow (E) / id$

FIRST(E) is the sam as that of
(a) FIRST (T)
(b) FIRST (F)
(c) both a and b
(d) none of the above

32. In the grammar of question 31, FOLLOW (F) is
(a) $\{+,*,),\$\}$ (b) $\{+,),\$\}$
(c) $\{*,),\$\}$ (d) $\{+,(,),*\}$

33. The grammar $S \rightarrow AB, B \rightarrow ab, A \rightarrow aa, A \rightarrow a, B \rightarrow b$ is
(a) ambiguous
(b) unambiguous
(c) maybe ambious or may not be ambiguous
(d) none

34. Which of the following suffices to convert an arbitrary CFG to an LL(1) grammar?
(a) removing left recursion alone
(b) factoring the grammar alone
(c) removing left recursion and factoring the grammar
(d) none of the above.

35. Assume that the SLR parser of a grammar G has n1 states and the LALR parser for G has n2 states. The relationship between n1 and n2 is
(a) n1 is necessarily less than n2
(b) n1 is necessarily equal than n2
(c) n1 is necessarily greater than n2
(d) none of the above.

36. In a bottom-up evaluation of a syntax directed definition, inherited attributes can
(a) always be evaluated
(b) be evaluated only if the definition in L-attributed
(c) be evaluated only if the definition h a s synthesized attributes
(d) never be evaluated

37. The regular expression $0*(10*)*$ denotes the same set as
(a) $(1*0*)*1*$ (b) $0+(0+10)*$
(c) $(0+1)*10(0+1)*$ (d) none of the above

38. Consider the grammar shown below

$S \rightarrow iEtSS' | a$

$S' \rightarrow eS | \in$

$E \rightarrow b$

In the predictive parse table M, of this grammar, the entries $M[S',e]$ and $M[S',\$]$ respectively are
(a) $\{S' \rightarrow eS\}$ & $\{S' \rightarrow \in\}$
(b) $\{S' \rightarrow eS\}$ & $\{\}$
(c) $\{S' \rightarrow \in\}$ & $\{S' \rightarrow \in\}$
(d) $\{S' \rightarrow S, S' \rightarrow \in\}$ & $\{S' \rightarrow \in\}$

39. Consider the grammar shown below:

$S \rightarrow CC$

$C' \rightarrow Cc | d$

(a) LL(1)
(b)SLR (1) but not LL (1)
(c) LALR (1) but not SLR (1)
(d) LR (1) but not LALR (1)

40. Consider the translation scheme shown below

$S \rightarrow TR$

$R \rightarrow +T\{print('+');\}R / \in$

$T \rightarrow num(print(num.val);\}$

Here numb is a token that represents an integer and num.val represents the corresponding integer value. For an input string '9+5+2', this translation scheme will print
(a) 9 + 5 + 2 (b) 95 + 2 +
(c) 952 + (d) ++952

41. Consider the syntax directed definiton shown below

$S \rightarrow id := E \quad \{gen(id.place = E.place;);\}$

$E \rightarrow E1 + E2 \quad \{t = newtemp();$

$Gen(t = E1.place + E2.place;);$

$E.place = t\}$

$Eid \quad \{E.place = id.place;\}$

Here gen is a function that generates the output code and newtemp is a function that returns the name of a new temporary variable on every call. Assume that it's are the temporary variable names generated by newtemp. For the statement "X:=Y+Z", the three-address sequence generated by this definition is

(a) $X = Y + Z$

(b) $t1 = Y + Z; X = t1$

(c) $t1 = y; \; t2 = t1 + z; \; X = 2t$

(d) $t1 = Y; t2 = Z; t3 = t1 + t2; \; X = t3$

42. Grammar $S \rightarrow aAb, A \rightarrow aAb \mid a$ is in

(a) LR (1) but not is LR (0)
b) both LR(o) and LR(1)
(c) LR (0) but not in LR (1)
d) neither in LR(o) and LR (1)

43. Synatx directed translator is a combination of

(a) lexical analyzer and an code generator
(b) syntax analyzer and code generator
(c) syntax analyzer and intermediate code generator
(d) lexical analyzer and intermediate code generator

44. The grammar used to specify the syntax of a language is

(a) context sensitive grammar
(b) type grammar
(c) regular grammar
(d) context free grammar

45. The grammar havig more than one parse tree is said to be

(a) consistent grammar
(b) ambiguous grammar
(c) irregular grammar
(d) none

46. The postfix notation for the expression (9-5)+2 is

(a) 95+2- (b) 952 - +
(c) 295+ - (d) 95 - 2 +

47. Pick the error recovery strategy from the four

(a) left recursion (b) panic mode
(c) left factoring (d) none

48. How many look ahead symbols can be used in LL (1) grammar

(a) 2 (b) more than 1 (c) only 1 (d) no lookahead symbol

49. The first L and second L in LL (1) stands for, respectively as

(a) leftmost derivation and left to right
(b) left right and leftmost derivation
(c) left most derivation in reverse order and left most deriation
(d) none

50. How many actions can be performed by a shift reduce parser

(a) 1 (b) 4
(c) 2 (d) no actions

51. L and R in LR(K) parsing stands respectively for

(a) left to right scanning and right most derivation in reverse
(b) left to right scanning
(c) left derivation and right derivation
(d) all tha above

52. The set of items which include the initial item $S' -> .S$ is called as

(a) non kernel items
(b) kernel items
(c) viable prefixes
(d) none

53. The attribute whose value at a node in a parse tree is defined in terms of attributes at the parent or siblings of that node is called as

(a) synthesized
(b) inherited attribute
(c) attribute grammar
(d) none

54. DAG identifies

(a) an operator in common
(b) an expression in common
(c) node which has more than one parent
(d) all the above

55. Pick the method to eliminate ambiguity in the grammar
(a) elimination of left recursion
(b) left factoring
(c) both a and b
(d) none

56. For what values of X and a, the predictive parsing gives successful completion, when X is the symbol on top of stack and a is the current input symbol
(a) $X = a = \$$ (b) $X = a \neq \$$
(c) $X = a = \in$ (d) $X = a = \phi$

57. What is FIRST (E) for the given grammar, $E->TE',\ T->FT',\ F->(E)/id$
(a) $\{\in, (, id\}$ (b) $\{(, id\}$
(c) $\{), id\}$ (d) $\{()\}$

58. Substring that matches the right side of a production is called as
(a) sentinels
(b) handle
(c) reducing production
(d) all the above

59. The production in a grammar G with a dot at some position of the right side is called as
(a) dot production (b) itmes
(c) kernel (d) none

60. Syntax analysis is also calleda as
(a) hierarchical analysis
(b) parsing
(c) both a and b
(d) neither a nor b

61. In LALR parsing LALR stands for
(a) Left derivation and Left recursive
(b) parsing
(c) both a and b
(d) neither a nor b

62. Which is the best parsing method
(a) LALR
(b) Canonical LR
(c) shift reduce
(d) Predictive

ANSWER KEY

1. (a)	2. (d)	3. (a)	4. (d)	5. (d)
6. (a)	7. (d)	8. (a)	9.(c)	10.(c)
11.(a)	12.(a)	13.(a)	14.(c)	15.(c)
16.(d)	17.(c)	18.(d)	19.(b)	20.(c)
21.(d)	22.(b)	23.(a)	24.(c)	25(c)
26.(c)	27.(d)	28.(d)	29.(d)	30.(a)
31.(c)	32.(a)	33.(a)	34.(c)	35.(b)
36.(c)	37.(d)	38.(d)	39.(d)	40.(b)
41.(c)	42.(a)	43.(c)	44.(d)	45.(b)
46.(d)	47.(b)	48.(c)	49.(b)	50.(b)
51.(a)	52.(b)	53.(b)	54.(b)	55.(c)
56.(a)	57.(b)	58.(b)	59.(b)	60.(c)
61.(b)	62.(a)			

TYPE CHECKING AND RUN-TIME ENVIRONMENT

1. Semantic errors can be detected
(a)at comiple time only
(b) at run time only
(c) both at compile time and at run time
(d) none

2. A language which can guarantee that the programs it accepts will execute without type errors is said to be
(a) strongly-typed
(b) weakly typed
(c) loosely typed
(d) none

3. Heap allocation is required for languages
(a) that support recursion
(b) that support dynamic data structures
(c) that use dynamic scope rules
(d) none

4. Type checking is normally done during
(a) lexical analysis
(b) syntax analysis
(c) syntax-directed translation
(d) code optimization

5. A language *L* allows declaratio of arrays whose sizes are not known during compilation. It is required to make effiecient use of memory. Which one of the following is true?
(a) A compiler using static memory allocation can be writtern for L
(b) a compiler cannot be written for L, an interpreter should be used
(c) a compiler using dynamic memory allocation can be writtern for L
(d) none

6. Symbol table can be used for
(a) checking type compatibility
(b) suppressing duplication of error message
(c) storage allocation
d) all of these

7. The best way to compare the different implementations of symbol table is to compare the time required to
(a) add a new name
(b) make an inquiry
(c) add a new name and make an inquiry
c) none

8. Which of the following symbolo table implementain is based on the propery of locality of reference
(a) linear list
(b) search tree
(c) hash table
(d) self organizing llist

9. Access time of the symbol table will be logarthmic if it is implemented by
(a) linear list
(b) search tree
(c) hash table
(d) self organizing list

10. Type checking done by the compiler is called as
(a) Dynamic checking
(b) Compilation
(c) Debugging
(d) Static checking

11. The position of type checker is in between
(a) parser and intermediate code generator
(b) parser and scanner
(c) semantic analyzer and intermediate code generator
(d) code optimizer and code generator

12. Type checker checks for
(a) incompatible operand
(b) flow of control
(c) uniqueness of the object
(d) all the above

13. In the production E - > E1 mod E2, if E1.type=integer and E2.type=real, what is E.type
(a) integer (b) real
(c) either integer or real (d) error

14. Coercions are
(a) the explicit type conversions (b) the implicit type conversions
(c) both a and b (d) none

15. Give the semantic rule for production E - > num.num
(a) E.type : = integer
(b) E.type : = real
(c) E.type : = lookup (id.entry)
(d) none

16. Execution of a procedure is called as
(a) transaction (b) lifetime
(c) activation (d) none

17. The one which keeps track of all live procedure activaions is
(a) activation tree
(b) control stack
(c) activation record (d) all

18. Informaiton needed by the exection of a procedure is present in a block of stroage called as
(a) relative address (b) frame
(c) both a and b (d) none

19. The difference between the addresses of the counter position and the data object is called as
(a) relative address
(b) absolute address
(c) effective address
(d) offset

20. The function of a caller program is
(a) to evaluate the actuals
(b) to save register values
(c) to initialize the local data
(d) all the above

21. In what storage strategy activation record is maintained even after the execution of a preocedure is completed
(a) stack allocation
(b) heap allocation
(c) static allocation
(d) dynamic allocation

22. Which is not the data structure for symbol table
(a) trees (b) list
(c) hash table (d) None

23. Which method is used in allocation variable sized blocks
(a) first fit method (b) FIFO
(c) LIFO (d) none

24. The problem of unused space in memory is called as
(a) padding (b) fragmentation
(c) a and b (d) a or b

25. The technique used in implict deallocation is
(a) relocation (b) overlays
(c) reference counts (d) a and c

26. A sequence which stores the state of the machine so that the calling procedure can continue exection is called as
(a) return sequence (b) caller
(c) call sequence (d) callee

27. A symbol that has different meanings depending on its content is said to be
(a) polymorphic (b) inherited
(c) extended (d) all

28. Which procedure when called allows the execution of statements in its body with arguments of different types
(a) open procedure
(b) macro
(c) polymorphic procedure
(d) all

29. Information about the types represented by vairables is formalized by defining a mapping from type vairables to type expressions called as
(a) mapping (b)unification
(c) substitution (d) type checking

30. The limitations in static allocation is
(a) size of the data should be known at compile time
(b) data structures cannot be created dynamically
(c) recursive procedures are not allowed
(d) all the above

ANSWER KEY

1.(c)	2.(a)	3.(b)	4.(c)	5.(c)	6.(d)
7.(c)	8.(c)	9.(b)	10.(d)	11.(a)	12.(d)
13.(d)	14.(b)	15.(b)	16.(c)	17.(b)	18.(c)
19.(d)	20.(d)	21.(b)	22.(d)	23.(a)	24.(c)
25.(c)	26.(b)	27.(d)	28.(c)	29.(c)	30.(d)

INTERMEDIATE CODE GENERATION AND CODE OPTIMIZATION

1. In analyzing the compilation of a program, the term "machine independent optimization" is associated with
   (a) recognition of basic syntactic construction though reduction
   (b) recognition of basic elements and creation of uniform symbols
   (c) creation of more optimal matrix
   (d) use of macro processor to produce more optimal assembly code
2. The identification of common sub-expression and replacement of run-time computaions by compile time computations is
   (a) local optimzation
   (b) loop optimization
   (c) constant folding
   (d) data flow analysis
3. Peephole optimization is form of
   (a) loop optimization
   (b) local optimization
   (c) constant
   (d) data flow analysis
4. Replacement of an expensive operation by a cheaper one is termed as
   (a) code motion
   (b) reduction strength
   (c) loop-invariant computation
   (d) none of these
5. Local and loop optimization inturn provide motication for
   (a) data flow analysis
   (b) constant folding
   (c) peephone optimization
   (d) DFA and constant folding
6. The optimization that avoids a test at every iteration is
   (a) loop unrolling
   (b) loop ramming
   (c) both a and b
   d) none
7. The method which merges the bodies of two loops is
   (a) loop unrolling
   (b) loop ramming
   (c) constant folding
   d) none
8. Loop is a collection of nodes that
   (a) is strongly connected
   (b) us loosely connected and has a unique entry
   (c) is strongly connected and has a unique entry
   (d) none
9. A basic block can be analyzed by
   (a) a DAG
   (b) flow graph
   (c) a graph with cycle
   (d) none
10. The graph that shows the basic blocks and their successor relationship is called
    (a) DAG
    (b) flow graph
    (c) control
    (d) Hamiltonain graph
11. Generation of intermediate code based on a abstract machine model is useful in compilers because
    (a) it makes implemetaion of lexical analysis and syntax analysis easier
    (b) syntax-directed translations can be written for intermediate code generation
    (c) it enhances the portability of the front end of the compiler
    (d) it is not possible to generate code for real machine directly from high level language programs.
13. The optimization technique which is typically applied on loops is
    (a) removal of invariant computation
    (b) peephole optimization
    (c) constant folding
    (d) all of these
14. Reduction is strength means
    (a) replacing tun-time computation by compile time computation
    (b) removing loop invariant computations
    (c) removing common sub-expressions
    (d) replacing a costly operation by a relatively cheaper one
15. Concept which can be used in identifying loops is
    (a) dominators
    (b) reducible graphs
    (c) depth first ordering
    (d) all of these
16. The peephole optimization
    (a) is applied to a small part of the code
    (b) can be used to optimize intermediate code
    (c) can be applied to a portion of the code that is not ambiguous
    (d) all of these
17. In compilers the source code is translated into a language which is intermediate in between a high level programming language and machine code is called as
    (a) intermediate code
    (b) assemly code
    (c) mnemonic
    (d) all the above

18. For the given expression $(a+b)*c$ the corresponding postfix notation is.
(a) $+abc*$ (b) $ab+*c$
(c) $abc+*$ (d) $ab+c*$

19. For the evaluation of postfix expression which data structure is used
(a) array (b) hash table
(c) queue (d) LIFO list

20. In optimizing compilers the intermediate code is generated by using
(a) abstract syntax tree (b)
syntax directed tree
(c) syntax translation tree
(d) none

21. The three-address code is statement usually contains three addresses in which
(a) two for the operands and one for the result
(b) three for the operands
(c) three for the operators
(d) none

22. The record structure with four fields which we call "op, ARGI, ARG2 and RESULT", this representation of three-address statement is known as
(a) quadruples
(b) triples
(c) indirect triples (d)
none

23. In intermediate code implementation of three-address code is that of listing pointers to triple, rather than listing the triples themselves is called
(a) triples (b)
indirect triples
(c) quadruples
(d) all the above

24. A syntax directed translation scheme deals with
(a) regular grammar (b)
context free grammar
(c) context sensitive grammar (d) all the above

25. The translation of a nonterminal on the right side of the production is defined in terms of a translation of the non terminal on the left, such a translation is called as
(a) inherited translation (b)
syntax directed translation
(c) synthesized translation (d)
none

26. In compilers to produce a better object language program than the most obvious for a given source program is called as
(a) code optimization
(b) source optimization
(c) reduce code (d)
redundant code

27. Code-optimization techniques are applied
(a) before code generation
(b) after code generation
(c) both a and b
(d) none

28. In optimization technique the meaning of "90-10" rule is
(a) that 90% of the time is spent in 10% of the code (b) that 10% of the time is spent in 90% of the code
(c) that 90% to 10% is the time taken to 100% of the code
(d) none of the above.

29. Code optimization applicable to
(a) local optimization only
(b) loop optimizations only
(c) data flow analysis only
(d) all the above

30. In loop optimization techniques a loop whose body is rarely executed is called.
(a) dead code
(b) blank code
(c) redundant code
(d) blank stripper

31. When there are two or more induction variables in a loop we have an opportunity to get rid of all but one, and we call this process as
(a) loop optimization
(b) induction variable elimination
(c) loop reduction
(d) none of the above

32. up-chaining is useful for
(a) determing whether a particular definition is used anywhere or not
(b) constant folding
(c) checking whether a variable is used, without prior assignment
(d) all of these

33. Which sorting technique we are using for evaluation of interior nodes of the DAG in any order is
(a) a quick sort (b) a selection sort
(c) a bubble sort (d) a topological sort

34. To improve the code generation from a DAG which law is used
(a) associative law
(b) distribute law
(c) PCNF
(d) commutative law

35. Application of DAGs are
(a) automatically detect common subexpression
(b) determine which identifiers have their values in the block for which a leaf is created at some time.
(c) we can determine which statements compute values which could be used outside the block
(d) all the above

36. The loop which does not contain no other loops is called as
(a) Natural loop
(b) Inner loop
(c) acyclic graph
(d) none

ANSWER KEY

1.(c)	2.(c)	3.(b)	4.(b)	5.(a)	6.(a)
7.(b)	8.(c)	9.(a)	10.(b)	11.(a)	12.(d)
13.(a)	14.(d)	15.(d)	16.(d)	17.(a)	18.(d)
19.(d)	20.(a)	21.(a)	22.(a)	23.(b)	24.(b)
25.(a)	26.(a)	27.(c)	28.(a)	29.(d)	30.(d)
31.(b)	32.(d)	33.(d)	34.(a)	35.(d)	36.(b)

TEST YOURSELF - I

1. Which data structure in a compiler is used for managing information about variables and their attributes? **[GATE - 10]**

a) Abstract syntax tree b) Symbol table

c) Semantic stack d) Parse table

2. Which languages necessarily need heap allocation in the runtime environment ? **[GATE - 10]**

a) Those that support recursion

b) Those that use dynamic scoping

c) Those that allow dynamic data structures

d) Those that use global variables

3. The grammar $S \rightarrow aSa \mid bS \mid c$ is

a) LL(1) but not LR(1) b) LR(1)but not LR(1)

c) Both LL(1) and LR(1) d) Neither LL(1)nor LR(1)

4. Match all items in Group 1 with correct options from those given in Group 2. **[GATE - 09]**

Group 1	Group 2
P. Regular expression	1. Syntax analysis
Q. Pushdown automata	2. Code generation
R. Dataflow analysis	3. Lexical analysis
S. Register allocation	4. Code optimization

a) P-4, Q-1, R-2, S-3 b) P-3, Q-1, R-4, S-2

c) P-3, Q-4, R-1, S-2 d) P-2, Q-1, R-4, S-3

5. Which of the following statements are TRUE ?

I. There exist parsing algorithms for some programming languages whose complexities are less than $\theta(n^3)$.

II. A programming language which allows recursion can be implemented with static storage allocation.

III. No L - attributed definition can be evaluated in the framework of bottom-up parsing. **[GATE - 09]**

IV. Code improving transformations can be performed at both source language and intermediate code level.

a) I and II b) I and IV

c) III and IV d) I, III and IV

6. Which of the following describes a handle (as applicable to LR-parsing) appropriately ? **[GATE - 08]**

(a) It is the position in a sentential form where the next shift or reduce operation will occur

(b) It is a non-terminal whose production will be used for reduction in the next step

(c) It is a production that may be used for reduction in a future step along with a position in the sentential form where the next shift or reduce operation will occur

(d) It is the production p that will be used for reduction in the next step along with a position in the sentential form where the right hand side of the production may be found.

7. Some code optimizations are carried out on the intermediate code because **[GATE - 08]**

(a) they enhance the portability of the compiler to other target processors

(b) program analysis is more accurate on intermediate code than on machine code

(c) the information from dataflow analysis cannot otherwise be used for optimization

(d) the information from the front end cannot otherwise be used for optimization

8. Which of the following are true?

I. A programming language which does not permit global variables of any kind and has no nesting of procedures/functions, but permits recursion can be implemented with static storage allocation.

II. Multi-level access link (or display) arrangement is needed to arrange activation records only if the programming language being implemented has nesting of procedures/functions.

III. Recursion in programming languages cannot be implemented with dynamic storage allocation.

IV. Nesting of procedures/functions and recursion require a dynamic heap allocation scheme and cannot be implemented with a stack-based allocation scheme for activation records.

V. Programming languages which permit a function to return a function as its result cannot be implemented with a stack-based storage allocation scheme for activation records.

[GATE - 08]

(a) II and V only (b) I, III and IV only

(c) I, II and V only (d) I, III and V only

9. AnLALR(1) parser for a grammar G can have shift-reduce (S-R) conflicts if and only if **[GATE - 08]**

(a) the SLR(1) parser G has S-R conflicts

(b) the LR(1) parser for G has S-R conflicts

(c) the LR(0) parser for G has S-R conflicts

(d) The LALR(1) parser for G has reduce-reduce conflicts

10. Early binding refers to a binding performed at compile time and late binding refers to a binding performed at execution time. Consider the following statements :

(i) Static scope facilitates $\underline{w1}$ bindings.

(ii) Dynamic scope requires $\underline{w2}$ bindings.

(iii) Early bindings $\underline{w3}$ execution efficiency.

(iv) Late bindings $\underline{w4}$ execution efficiency.

The right choices of $w1, w2, w3$ *and* $w4$ (in that order) are **[GATE - 07]**

(a) Early, late, decrease, increase
(b) Late, early, increase, decrease
(c) Late, early, decrease, increase
(d) Early, late, increase, decrease.

11. Which one of the following is a top-down parser? **[GATE - 07]**

(a) Recursive descent parser
(b) Operator precedence parser
(c) An LR(k) parser
(d) An LALR(k) parser

12. Consider the following two statements :
P : Every regular grammar is LL (1)
Q : Every regular set has a LR (1) grammar
Which of the following is TRUE? [GATE - 07]

(a) Both P and Q are true
(b) P is true and Q is false
(c) P is false and Q is true
(d) Both P and Q are false

13. Consider the following grammar.

$S \rightarrow S * E$

$S \rightarrow E$

$E \rightarrow F + E$

$E \rightarrow F$

$F \rightarrow id$

Consider the following LR(0) items corresponding to the grammar above.

(i) $S \rightarrow S * .E$

(II) $E \rightarrow F. + E$

(III) $E \rightarrow F + .E$

Given the items above, which two of them will appear in the same set in the canonical sets - of - items for the grammar ? **[GATE - 06]**

(A) (i) and (ii) (B) (ii) and (iii)
(C) (i) and (iii) (D) None of these

14. Consider the following translation scheme.

$S \rightarrow ER$

$R \rightarrow *E$ **(print** ('*'); $R \mid \varepsilon$

$E \rightarrow F + E\{print('+');\mid F$

$F \rightarrow (S) \mid id\{print(id.value);\}$

Here id is a taken that represents an integer and id. value represents the corresponding integer value. For an input '2*3+4', this translation scheme prints **[GATE - 06]**

(A) 2*3+4 (B) 2*+34
(C) 2 3*4+ (D) 2 3 4+*

15. Consider the following C code segment.

```
for (i = 0; i < N; i++)
  for (j = 0; j < N; j++)
    if (i % 2)
      {x+ = (4 * j + 5 * i);
       y+ = (7 + 4 * j);
      }
    }
}
```

Which one of the following is false? **[GATE - 06]**

(A) The code contains loop-in variant computation
(B) There is scope of common sub-expression elimination in this code
(C) There is scope of strength reduction in this code
(D) There is scope of dead code elimination in this code

ANSWER KEY

1. b	2. c	3. c	4. b	5. c
6. d	7. a	8. d	9. b	10. a
11. a	12. b	13. c	14. d	15.b

EXPLANATIONS

3. Grammar S = aSa | bS | C
First (S) = { a, b, c }
Follow $(S) = \{a, \$\}$

LL(1)

	a	b	c	$
S	$S \to aSa$	$S \to bS$	$S \to c$	

No multiple entries see this is LL(1) Grammar.

LR(1) : $S \to asa|bs|C$

1. $S \to asa$
2. $S \to bs$
3. $S \to c$

$I_0 : S' \to .S.\$$

$S' \to .asa.\$$

$S \to .bs.\$$

$S \to .c.\&$

$I_1 : S \to s..\$$

$I_2 : S \to a.sa.\$$

$S \to .asa.a$

$S \to .bs.a$

$S \to .c.a$

$I_3 : S \to b.s.\$$

$S \to .asa.\$$

$S \to .bs.\$$

$S \to .c.\$$

$I_4 : S \to c..\$$

$I_5 : S \to as.a.\$$

$I_6 : S \to a.sa.a$

$S \to .asa.a$

$S \to .bs.a$

$S \to .c.a$

$I_7 : S \to b.s.a$

$S \to .asa.a$

$S \to .bs.a$

$I_9 : S \to bs..\$$

$I_{10} : S \to b.s.\$$

$: S \to .asa.\$$

$: S \to .bs.\$$

$: S \to .c.\$$

$I_{11} : S \to asa..\$$

$I_{12} : S \to as.a.a$

$I_{13} : S \to bs..a$

$I_{14} : S \to bs..\$$

$I_{15} : S \to asa..a$

$S \to .c.a$

$I_8 : S \to c..a$

	a	b	c	$	S
0	S_2	S_3	S_4		1
1					
2	S_6	S_7	S_8		5
3	S_2	S_{10}	S_4		9
4				r_3	
5	S_{11}				
6	S_6	S_7	S_8		12
7	S_6	S_7	S_8		13
8	R_3				
9				r_2	
10	S_2	S_3	S_4		14
11				r_1	
12	S_{15}				
13	r_2				
14				r_2	
15	r_1				

No multiple entries so thus LR(1) also.

4. Regular expression denote structure of data specially text string - lexical analyser break input text into logical unit such as identifiers.
Push down automater - Syntax analysis. Study with the help of syntax.
Data flow analysis - Code optimization
Register allocation - Lexical analysis

5. It is not necessary there exists parsing algo for some programming language whose complexities are less than $\theta(n^3)$

A programming language which allow recursion can be implemented with static storage allocation.

6. According to definition, handle is a production rule $A \to \beta$ and a position in the sentential form where string β may be found and replaced to produce the previous right sentential form in a rightmost derivation of sentential form.

7. Code optimizations are carried out on the intermediate code so that same source code can be converted to machine language of the target machine depending upon the back end tools. It also increases the probability of the compiler, i. e. compiler can generate machine code of different machine codes.

9. An LALR(1) parser for a grammer G can have shift reduce S-R conflict if and only if the LR(1) parser for G has S-R conflicts.

11. Recursive descent parser is a kind of top - down parser.

15. The code an be converted to the following in order to reduce the number of cycles needed for execution

for $(i=0, i<N, i++)$

if $(i\%2)$

for $(j=0; j<N, j++)$

$\{x+=(4*j+5*c);$

$\{y+=(7+4*j);\}$

$\therefore$ The only false statement is B since no sub expression can be eliminated.

TEST YOURSELF - II

1. Which of the following suffices to convert an arbitrary CFG to an LL(1) grammar ? **[GATE - 03]**

a) Removing left recursion alone

b) Factoring the grammar alone

c) Removing left recursion and factoring the grammar

d) None of the above

2. Assume that the SLR parser for a grammar G has n_1 states and the LALR parser for G has n_2 states. The relationship between n_1 and n_2 is **[GATE - 03]**

a) n_1 is necessarily less than n_2

b) n_1 is necessarily equal to n_2

c) n_1 is necessarily greater than n_2

d) none of the above

3. In a bottom-up evaluation of a syntax directed definition, inherited attributes can **[GATE - 03]**

a) always be evaluated

b) be evaluated only if the definition is L-attributed

c) be evaluated only if the definition has synthesized attributes

d) never be evaluated

4. Which of the following statements is FALSE ?

a) In statically typed languages, each varible in a program has a fixed type **[GATE - 03]**

b) In Un-typed languages, values do not have any types

c) In dynamically typed languages, varibles have no types

d) In all statically typed languages, each varible in a program is associated with values of only a single type during the execution of the program

5. Consider the grammar shown below **[GATE - 03]**

$S \rightarrow iEtSS' | a$

$S' \rightarrow eS | \epsilon$

$E \rightarrow b$

In the predictive parse table, M, of this grammar, the entries $M[S', e]$ and $M[S', \$]$ respectively are

a) $\{S' \rightarrow eS\}$ and $\{S' \rightarrow \epsilon\}$

b) $\{S' \rightarrow eS\}$ and $\{\}$

c) $\{S' \rightarrow \epsilon\}$ and $\{S' \rightarrow \epsilon\}$

d) $\{S' \rightarrow eS, S' \rightarrow \epsilon\}$ and $\{S' \rightarrow \epsilon\}$

6. Consider the grammar shown below. **[GATE - 03]**

$S \rightarrow CC$

$C \rightarrow cC | d$

This grammar is

a) LL(1) b) SLR(1) but not LL(1)

c) LALR (1) but not SLR (1)

d) LR (1) but not LALR (1)

7. Consider the translation scheme shown below.

$S \rightarrow TR$ **[GATE - 03]**

$R \rightarrow +T \{print('+');\} R \backslash \epsilon$

$T \rightarrow$ num {print (num.val) ;}

Here num is a token that represents an integer and num.val represents the corresponding integer value. For an input string '9+5+2', this translation scheme will print

a) 9 + 5 + 2 b) 95 + 2 +

c) 952 + + d) + + 952

8. Consider the syntax directed definition shown below.

$S \rightarrow Id := E$ {gen(id.place = E.place);} **[GATE - 03]**

$E \rightarrow E_1 + E_2$ {t = newtemp ( );

gen (t = E_1.place + E_2.place;);

E.place = t ;}

$E \rightarrow id$ {E.place = id.place;}

Here, gen is a function that generates the output code, and newtemp is a function that returns the name of a new temporary variable on every call. Assume that t_i's are the temporary variable names generated by newtemp. For the statement 'X := Y + Z', the 3-address code sequence generated by this definition is

a) X = Y + Z b) $t_1 = Y + Z; X = t_1$

c) $t_1 = Y; t_2 = t_1 + Z; X = t_2$

d) $t_1 = Y; t_2 = Z; t_3 = t_1 + t_2; X = t_3$

9. Which of the following grammar rules violate the requirements of an operator grammar ? **[GATE - 04]**

P, Q, R are nonterminals, and r, s, t are terminals.

i) $P \rightarrow QR$ ii) $P \rightarrow QsR$

iii) $P \rightarrow \varepsilon$ iv) $P \rightarrow QtRr$

a) i) only b) i) and iii) only

c) ii) and iii) only d) iii) and iv) only

10. Consider a program P that consists of two source modules M_1 and M_2 contained in two different files. If M_1 contains a reference to a function defined in M_2 the reference will be resolved at **[GATE - 04]**

a) Edit-time b) Complie-time

c) Link-time d) Load-time

11. Consider the grammar rule $E \rightarrow E_1 - E_2$ for arithmetic expressions. The code generated is targeted to a CPU having a single user register. The subtraction operation requires the first operand to be in the register. If E_1 and E_2 do not have any common subexpression, in order to get the shortest possible code **[GATE - 04]**

a) E_1 should be evaluated first

b) E_2 should be evaluated first

c) Evaluation of E_1 and E_2 should necessarily be interleaved

d) Order of evaluation of E_1 and E_2 is of no consequence

12. Consider the grammar with the following translation rules and E as the start symbol. **[GATE - 04]**

$E \rightarrow E_1 \# T$ {E. value = E_1. value * T. Value}

$\quad | \; T$ {E.value = T. value}

$T \rightarrow T_1 \& F$ {T.value = T_1.value + F.value}

$\quad | \; F$ {T.value = F.value}

$F \rightarrow num$ {F.value = num. value}

Compute E. Value for the root of the parse tree for the expression : 2 # 3 & 5 # 6 & 4.

a) 200 b) 180 c) 160 d) 40

13. The grammar $A \rightarrow AA \,|\, (A) \,|\, \epsilon$ is not suitable for predictive-parsing because the grammar is **[GATE - 05]**

a) ambiguous b) left-recursive

c) right-recursive d) an operator grammar

14. Consider the following two problems on undirected graphs

α: Given G(V, E), does G have an independent set of size $|V| - 4$?

β: Given G(V, E), does G have an independent set of size 5 ?

Which one of the following is TRUE ? **[GATE - 05]**

a) α is in P and β is NP-complete

b) α is NP-complete and β is in P

c) Both α and β are NP-complete

d) Both α and β are in P

15. Consider the grammar

$$E \rightarrow E + n \,|\, E \times n \,|\, n$$

For a sentence $n + n \times n$, the handles in the right-sentential form of the reduction are **[GATE - 05]**

a) n, E + n and E + n × n b) n, E + n and E + E × n

c) n, n + n and n + n × n d) n, E + n and E × n

16. Consider the grammar **[GATE - 05]**

$S \rightarrow (S) \mid a$

Let the number of states in SLR (1), LR (1) and LALR (1) parsers for the grammar be n_1, n_2 and n_3 respectively. The following relationship holds good

a) $n_1 < n_2 < n_3$ b) $n_1 = n_3 < n_2$

c) $n_1 = n_2 = n_3$ d) $n_1 \geq n_3 \geq n_2$

17. Consider line number 3 of the following C-program.

```
int main ( ) {              /* Line 1*/
int I, N;                   /* Line 2*/
fro (I = 0, I < N, I + +);  /* Line 3*/
}
```

Identify the compiler's response about this line while coreating the object-module **[GATE - 05]**

a) No compilation error b) Only a lexical error

c) Only syntactic errors

d) Both lexical and syntactic errors

Statement for Linked Answer Questions 18 and 19

Consider the following expression grammar. The semantic rules for expression calculation are stated next to each grammar production. **[GATE - 05]**

$E \rightarrow$ number	E.val = number. val
|E'+'E	$E^{(1)}.val = E^{(2)}.val + E^{(3)}.val$
|E'×'E	$E^{(1)}.val = E^{(2)}.val \times E^{(3)}.val$

18. The above grammar and the semantic rules are fed to a yacc tool (which is an LALR (1) parser generator) for parsing and evaluating arithmetic expressions. Which one of the following is TRUE about the action of yacc for the given grammar ?

a) It detects recursion and eliminates recursion

b) It detects reduce-reduce conflict, and resolves

c) It detects shift-reduce conflict, and resolves the conflict in favour of a shift over a reduce action

d) It detects shift-reduce conflict, and resolves the conflict in favor of a reduce over a shift action.

19. Assume the conflicts in Part (a) of this question are resolved and an LALR (1) parser is generated for parsing arithmetic expressions as per the given grammar. Consider an expression $3 \times 2 + 1$. What precedence and associativity properties does the generated parser realize ?

a) Equal precedence and left associativity; expression is evaluated to 7.

b) Equal precedence and right associativity; expression is evaluated to 9.

c) Precedence of ' × ' is higher than that of '+', and both operators are left associative; expression is evaluated to 7.

d) Precedence of '+' is higher than that of ' × ', and both operators are left associative; expression is evaluated to 9.

ANSWER KEY				
1. c	**2.** b	**3.** c	**4.** a	**5.** a
6. d	**7.** d	**8.** c	**9.** a	**10.** c
11. d	**12.** c	**13.** b	**14.** c	**15.** d
16. b	**17.** c	**18.** c	**19.** a	

Unit - XII
Software Engineering

UNIT - XII SOFTWARE ENGINEERIG

XII. SOFTWARE ENGINEERING

Information Gathering

The specific methods analysts follow for collecting data about the requirements are **"FACT FINDING TECHNIQUES"**. The fact finding techniques include the interview, questionnaire, record review and observation. More than one of these methods are cascaded for a comprehensive gathering information.

On the other hand, fact finding techniques offers analysts an opportunity to go behind the scenes in an organization to learn the inside story to discover how things actually work in a new area of information.

Interview

The respondents for the interview are people currently operating with the system or the proposed users of the system. Analysts use interviews to collect information for individuals or from groups. This method is especially helpful for gathering information from individuals who do not communicate effectively in writing or who do not find time to fill in questionnaires.

Interviews can be categorized as Structured and Un-Structured. A structured interview has uniform wording of questions, objective evaluation, and short duration interviews. Structured interview can be used both as open response and closed response format. An unstructured interview has flexible wording of questions, is not time stipulated and interpretation of results may be lengthy but extraneous information can be gathered.

The success of the interview depends on the skill of the interviewer and the preparation behind the interview.

Questionnaire

The information in this technique is gathered from large group of people. The use of standardized question formats can yield more reliable data and the wide distribution ensures greater anonymity for respondents, leading to honest information. But the method does not allow the analyst to observe the reactions or expression of respondent and the response may be limited. There are two kinds of questionnaires, open-ended and close-ended questionnaires. The high cost of developing and distributing questionnaires is a major disadvantage.

Record Review

Analysts examine information that has been recorded about the system and users. This method can be performed as the study at the beginning as an introduction or later in the study as a basis for comparison with actual operations.

The records include written policy manuals, regulations, and standard operating procedures used by most organizations as a guide. They do not show what activities are actually occurring, where the decision making power lies, or how tasks are performed, but they can help analysts understand the system by familiarizing them with what operations must be supported and with formal relations within the organization.

Observation

Observation allows analysts to gain information that they cannot obtain by any other fact-finding method. Through this information; analyst can obtain first hand information about how activities are carried out. Experienced observers know what to look for and how to assess the significance of what they observe.

REQUIREMENT AND FEASIBILITY ANALYSIS

Requirements analysis involves studying the current business system to find out how it works and where improvements should be made. Systems studies result in an evaluation of how current methods are working and whether adjustments are necessary or possible. A requirement is a feature that must be included in a new system. It may include a way of capturing or processing data, producing information; controlling a business activity or supporting management. The determination of requirements thus entails studying the existing system and collecting details about it to find out what these requirements are.

During the requirement analysis phase of the traditional approach the focus is on determining user needs, studying the application area in-depth, assessing the strengths and weakness of the present system and reporting results to the management. How thoroughly the system is studied depends on the situation. To determine if the present system needs just a few adjustments instead of a complete overhaul, the system should be studied in-depth. If management is determined to replace the current system the analyst would probably waste time by studying it thoroughly. In this case it is better to spend time in determining exactly what management desires in the new system.

While studying the present system, the systems analyst must analyze how this system uses hardware, software, and human resources to convert the data of the organization into information for end users. He should also analyze how these resources are used to accomplish the activity of input, processing, output, storage and control.

Activities in requirements determination

The three major activities involved are

1. **Requirements anticipation.**

Foreseeing system characteristics based on previous experience, may cause the analyst to investigate areas and issues that could otherwise be overlooked. If may also introduce bias. Experience from previous studies may lead to investigation of areas that would otherwise go unnoticed by an inexperienced analyst. Having the background to know what to ask or which aspects to investigate can be a substantial benefit to the organization. The features they investigate for the current system, questions they arise, or methods employed may be based on familiarity.

2. Requirements investigation.

Study and documentation of the current system, using fact-finding techniques, data flow analysis and decision analysis. Using a variety of tools and skills, analysts study the current system and document its features for further analysis. This activity is at the heart of system analysis.

3. Requirements specification.

Analysis of data describing the system to determine how well it is performing, want requirements must be met and strategies for fulfilling them. This activity has three Interrelated parts :

- **Analysis of factual data :** The data collected during fact finding study and included in data flow and decision analysis documentation are examined to determine how well the system is performing and whether it will meet the organization's demands.
- **Identification of essential requirements :** Features that must be included in a new system, ranging from operational details to performance criteria are specified.
- **Selection of requirement fulfillment strategies :** The methods that will be used to achieve the stated requirements are selected. These form the basis for system design which follows requirement specification.

All the three activities are important and must be performed with accuracy.

Basic Requirements : Analysts structure their investigation by seeking answers to these four major questions :

What is the basic business process ?

What data are used or produced during that process?

What are the limits imposed by time and the volume of work ?

What performance controls are used ?

Understand the process : Begin with the basics. Analysts must raise questions that when answered, will provide a background of fundamental details about the system and describe it.

Identify data used and information produced : Analysts next need to find out what data are used to perform each activity. Most business transactions also produce information that is useful to managers when they evaluate employee, business, and systems performance and that may be useful in another context to both the manager and the analyst.

Determine process timing and volume : The frequency of business activities varies greatly. Analysts should learn how often the activity is repeated. The easiest way to get this information is to identify the reason for the activity. Timing alone does not determine the importance of an activity but it does affect the way analysts evaluate certain steps in carrying out the performance. The volume of items to be handled may increase the amount of time needed to complete the activity.

Identify controls : In business situations that are well controlled either by management or process monitoring, determining whether an activity has been performed properly may be no problem. But during the analysis stage the analysts must examine control methods: Are there specific performance standards ? How are mistakes caught ? How are errors handled ? Are the errors excessive ? Weak or missing controls are an important discovery in any systems investigation.

User Transaction Requirements

Transaction level system capture, process and store data for a reason. For example sales orders from customers are processed so that specified items can be shipped. This simple procedure is applied to every order that is received.

User Decision Requirements

Decisions unlike the transaction activities may not follow a specific procedure. Routines are not as clear-cut and controls may be very vague. Decisions are made by integrating information in such a way that managers can know what actions to take. Decision systems may focus on the past, the present or the future. Some may support recurring decisions, while others are unique and do not recur. They may use data that originate inside the firm such as through transaction processing or outside. In some other cases transaction data are processed to provide new information for decision making. The analyst investigating decision systems must be aware of supporting transaction systems and that effective decision systems require suitable transaction processing procedures to be in place first.

Organization Wide Requirements

Departments depend on each other to provide services, manufacture products and satisfy customers. The work that one department does affect other departments. When analysts study systems for one department they also access the implications for other departments that interact with the system being investigated. Systems involve work that takes place across several departments. Systems that operate separately on the other hand depend on the data provided from the other departments. It is the analysts responsibility to identify dependencies between departments and determine how a systems project may affect them.

Data Flow Diagrams

An information moves through software, it is modified by a series of transformations.

A 'Data Flow Diagram' is a graphical technique that depicts information flow and the transforms that are applied as data move from input to output. The DFD is also known as a 'data flow graph' or a 'bubble chart'.

DFDs may be partitioned into levels that represent increasing information flow and functional detail.

A level 0 DFD also called as 'a fundamental system' or a 'context model' represents the entire software element as a single bubble with input and output data indicated by incoming and outgoing arrows.

Additional processes (bubbles) and information flow paths are represented as the level 0 DFD is partitioned to reveal more detail.

A level 1 DFD might contain 5/6 bubbles with interconnecting arrows. Each of the processes represented at level 1 are sub functions of the overall system.

Developing DFD

System analyst must first study the current system, that is the actual activities and processes that occur called - the study of physical system.

The physical system is then translated into a logical system or description that focuses on data and processes.

DFD Notation :

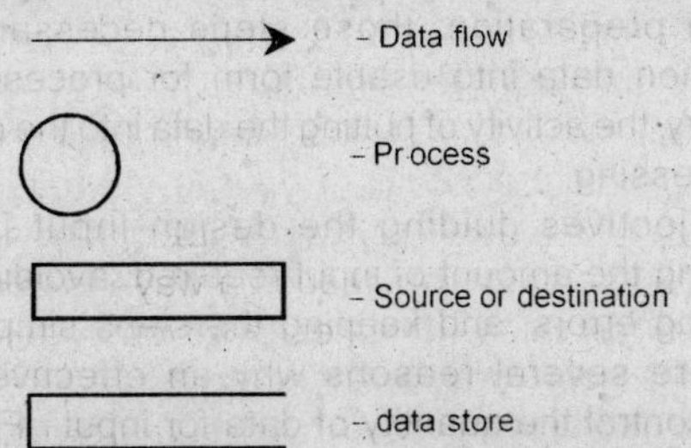

Each component in a DFD is labelled with a descriptive name. Process names are further identified with a number that will be used for identification purposes.

Sample DFD

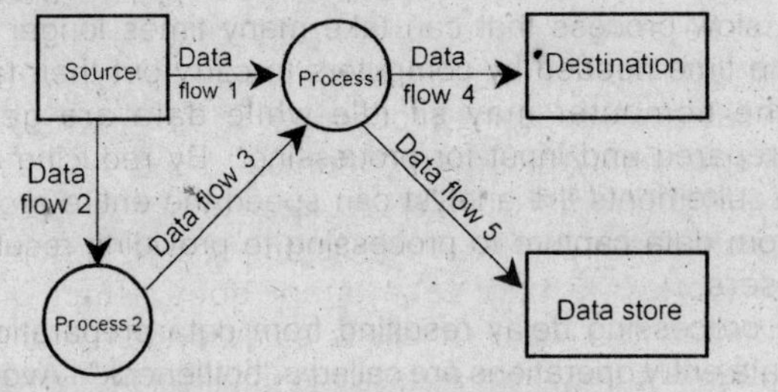

Sequence of Logical and Physical System

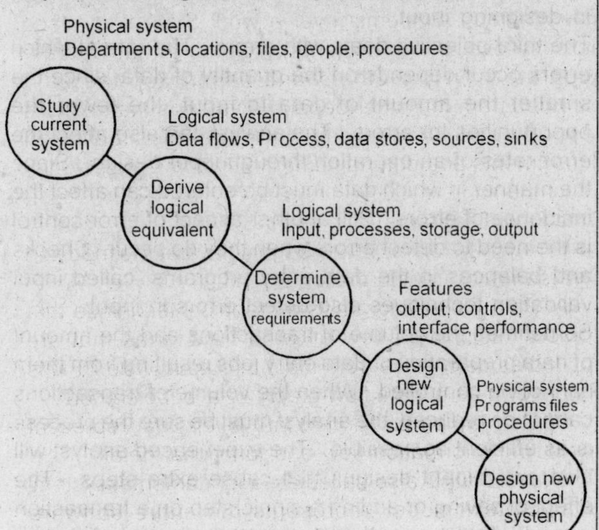

Drawing physical data flow diagrams

Step 1 : Drawing a context diagram

2 : Developing a process chart

3 : Developing the first-level DFD

4 : Exploding processes into greater detail

5 : Deriving the logical view

Logical Level 1

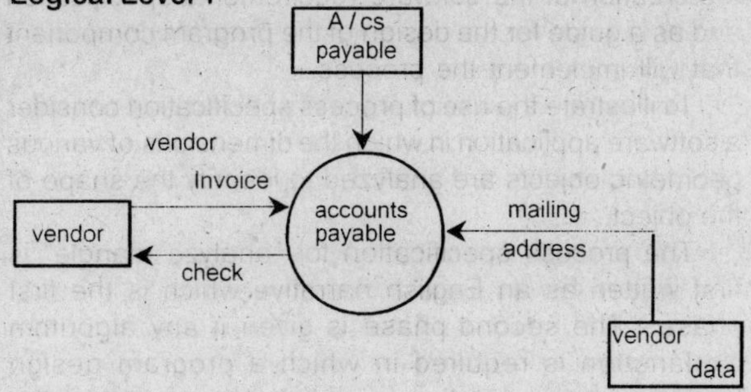

Program Structure Chart

Example

Context data flow diagram

Accounts payable system

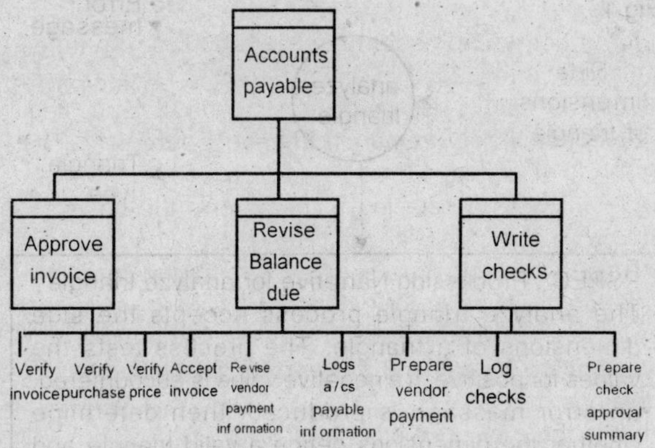

Physical Level

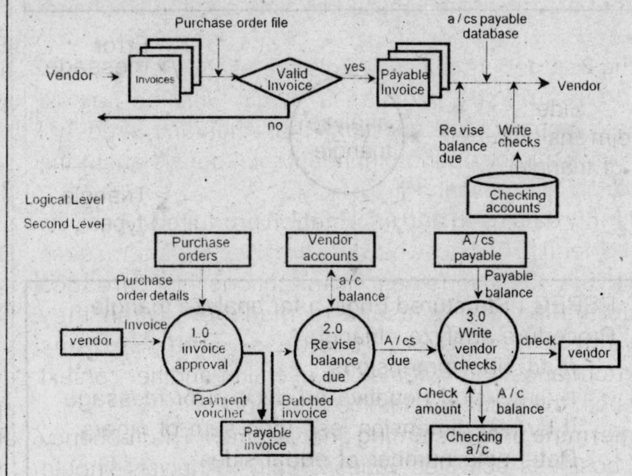

The A/cs payable example illustrates the physical and logical level DFD. The levels can be further exploded in greater detail for every process.

Process Specifications

The process specification is used to describe all the flow model processes that appear at the final level of refinement. The content of the process specification can include narrative text, a program design language (PDL) description of the process algorithm, mathematical equations, tables, diagrams or charts. By providing a process specification to accompany with each bubble in the flow model the software engineer creates a "mini-spec" that can serve as the first step in the creation of the software requirements specification and as a guide for the design of the program component that will implement the process.

To illustrate the use of process specification consider a software application in which the dimensions of various geometric objects are analyzed to identify the shape of the object.

The process specification for "analyze triangle" is first written as an English narrative which is the first phase. The second phase is given if any algorithm explanation is required in which a program design language representation is included as a part of the process specification. In most cases the PDL version is postponed until the design stage commences.

PROCESS SPECIFICATION ILLUSTRATION FIG.

PROCESS SPECIFICATION

Fig 1 :

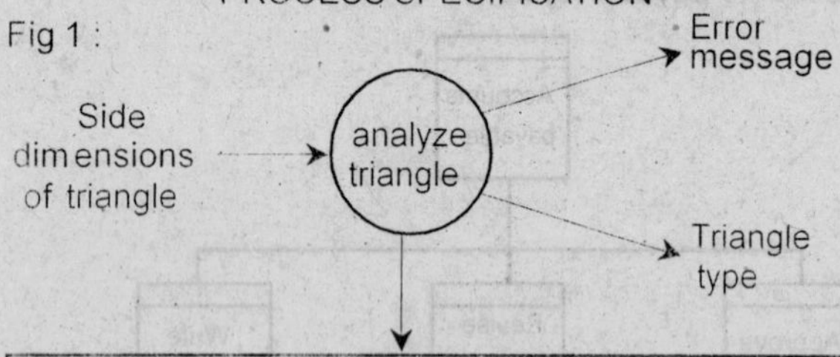

PSPEC : Processing Narrative for analyze triangle
The analyze triangle process accepts the side dimensions of a triangle. The process tests the values for positive. If a negative value is encountered, an error message is produced, then determine whether the dimensions define a valid triangle and type - equilateral, isosceles or scalene, type is output

Fig 2 :

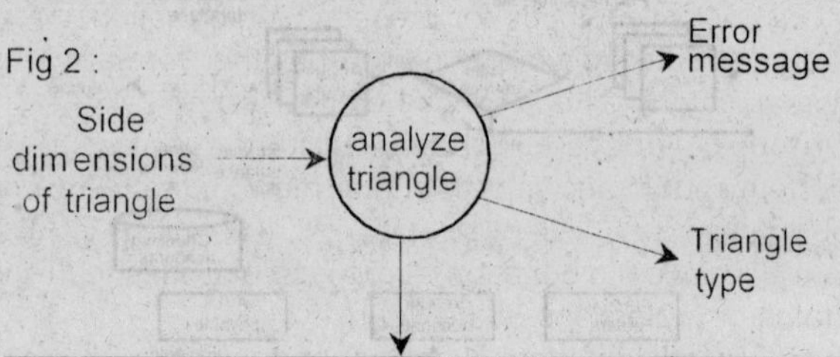


```
PSPEC : Structured English for analyze triangle.
Procedure analyze - triangle
    read side dimensions
    If dimension negative produce error message
    If largest dimension less than sum of others
    Determine number of equal sides
    If three sides equal then type - equilateral
    If two sides equal then type - isosceles
    If no sides equal then type - scalene
    output triangle type
   end
  else
    endif
endproc.
```

I/O DESIGN

INPUT DESIGN

Concerns guiding input design

The input design is the link that ties the information system into the world of its users. Some aspects of the design vary, depending on whether the system is batch oriented or online. The objectives of input design consists of developing specifications and procedures for data preparation, those steps necessary to put transaction data into usable form for processing and data entry, the activity of putting the data into the computer for processing.

Five objectives guiding the design input focus on controlling the amount of input required, avoiding delay, controlling errors, and keeping the steps simple.

There are several reasons why an effective design should control the quantity of data for input. First, data preparation and data entry operations depends on people. Because labour costs are high, the cost of preparing and entering data is also high. Reducing data requirements can lower costs by reducing labor expense. Second the input phase of computing can be a slow process that can take many times longer than the time needed by computers to carry out their tasks. The computer may sit idle while data are getting prepared and input for processing. By reducing input requirements the analyst can speed the entire process from data capture to processing to providing results to users.

A processing delay resulting from data preparation or data entry operations are called a "bottleneck". Avoiding bottlenecks should always be one objective of the analyst in designing input.

The third objective deals with errors. The rate at which errors occur depends on the quantity of data, since the smaller the amount of data to input, the fewer the opportunities for errors. The analyst can also affect the error rates of an operation through input design. Since the manner in which data must be entered can affect the incidence of errors. Still another aspect of error control is the need to detect errors when they do occur. Checks and balances in the data entry programs, called input validation techniques also detect errors in input.

Sometimes the volume of transactions and the amount of data preparation or data entry jobs resulting from them cannot be controlled. When the volume of transactions cannot be reduced, the analyst must be sure the process is as efficient as possible. The experienced analyst will also avoid input designs that cause extra steps. The effect of saving or adding a single step on a transaction is multiplied many times over the course of a day. Example feeding checks in a bank process.

The best advice to analyst is to achieve all of the objectives mentioned in simplest manner. It is possible to include so many controls or errors that people will have difficulty in using the system. The best-designed system fits the people who will use it in a way that is comfortable for them while providing the error control methods. Simplicity works and that is accepted by the users.

Capturing Data for Input : Data capture guidelines - the analyst should start by capturing only those times that must be actually input. There are two types of data that must be input when processing transactions :

1. **Variable data :** Data items that change for each transaction handled or decision made.
2. **Identification data :** The element of data that uniquely identifies the item being processed.

What no to enter is equally important. Input procedures should not require the entry of

1. Constant data.
2. Details that the system can retrieve.
3. Details that the system can calculate.

Design of Source Document : The source document is the form on which data are initially captured, or recorded.

Layout - Organizes the document by placing important information where it will be noticed and establishing the appropriate sequence of items. Most people fill in documents from left to right and top to bottom and the source document layout should be designed for use in the same way. A carefully designed for usually goes unnoticed even as it is easy to use and follow. The analyst must consider how the form will be completed with the spacing in between lines. A well designed source document is easily completed.

Captions - on source documents tell the user what data to provide and where they should be entered. The caption should be brief but easily understood with standard terms that all persons using the form should know. Abbreviations should be avoided.

Coding methods - in which conditions, ideas, words or relationships are expressed by a code are developed to reduce input, control errors and speed the entire process. A code is a brief number, title or symbol used instead of more lengthy or ambiguous descriptions. The types of coding methods available are :

1. **Classification codes** - place separate entities such as events, people or objects into distinct groups, called classes. A code is used to distinguish one class from another. Classification codes vastly simplify the input process because only a single digit code is required.
2. **Function codes** - state the activities or work to be performed without spelling out all of the details in narrative statements. Analysts use this type of code frequently in transaction data to tell the system how to process the data. Example for addition of records A, deletion D, and modification M.
3. **Sequence codes** - are either numbers or letter assigned in series. They tell the order in which events have occurred. They are used for identification purpose too.
4. **Significant-digit subset codes** - provides wealth of information. Assign numbers in sequence starting with first and moving through the last and adding a prefix identifying a different entry, example a sales record code may be made up of product class + item number + vendor number.
5. **Mnemonic codes** - use letters and symbols from the product to describe it in a way that communicates visually. Example CS-Computer Science. The data and transaction coding reduces the volume of data for input and simplifies the process thus lessening the likelihood of errors.

Methods of data capture : The method in which the data are captured from transaction markedly affects the analyst's ability to meet the objectives. There are four different methods to consider :

- Source document capture using keypunching.
- Key to storage devices.
- Optical character recognition.
- Direct entry through intelligent terminals.

Input Validation : The general term given to methods aimed at detecting errors in input is called "input validation".

The main categories of the method are concerned with

Checking the transaction
- Batch controls.

Checking the transaction data
- Sequence test.

Changing the transaction data
- Existence test.
- Limit and range test.
- Combination test.
- Duplicate processing.

Modifying the transaction data.
- Automatic correction.
- check digits.

Output Design : Computer output is to be identified before applied. When analyst designs computer output.

- Identify the specific output that is needed to meet the information requirements.
- Select methods for presenting information.
- Create document, report, or other formats that contain information produced by the system.

Output Objectives : The output form an information system should accomplish one or more of the following objectives :

1. Convey information about past activities, current status or projections of the future.
2. Signal important events, opportunities, problems, or warnings.
3. Trigger an action.
4. Confirm an action.

Types of Output : Whether the output is a formatted report or a simple listing of the contents of a file, a computer process will produce the output. System output may be :

1. Report. 2. Document. 3 Message.

Output contents originate from these sources :

1. Retrieval from a data store.
2. Transmission from a process or system activity.
3. Directly from an input source.

How to Present Information :

(a) Tabular format : The tabular format should be used under the following conditions :

- Details dominate and few narrative comments or explanations are needed.
- Details are presented in discrete categories.
- Each category must be labelled.

• totals must be drawn or comparisons made between components.

Certain information in a tabular format is more important and should be more visible than other information. This will vary by application but in general the following items should stand out :

- Exceptions to the normal exceptions.
- Major categories or groups of activities or entities.
- Summaries of major categories or activities.
- Unique identification information.
- Time dependent entities.

The most frequent complaint with tabular representation is too many details, too little information that is drawing with data but starving for information.

(b) Graphic format : Management presentations have been enhanced by graphics and visuals. It is an area in computer information systems that is exploding largely because low cost but powerful computer software is available that will use data from existing data bases and produce high quality charts and diagrams. There is a chance that presentation suffer because the graphics will be poorly used.

Types of business graphics : Business graphics makes use of various types of charts and maps, Pie-charts, Area charts, Curve charts, Step and Bar charts are the commonly familiar ones used in the arena. Business graphics are like road signs, they must make their point quickly or the reader will lose interest.

When to use graphics ?

Graphics are used for several reasons : to improve the effectiveness of output reporting for the targeted recipients, to manage information volume, and to suit personal preferences.

Using icons : Icons are pictorial representations of entities described by data. Icons are commonly used in computer interfaces. Icons are to be selected so that they will be immediately recognized and understood by anticipated users.

Color presentation : Color should enhance not replace or hinder good output design when used with presentations. The analyst must take care to maintain consistent color usage throughout the output and reports for a system.

But neither color nor graphics can enhance or compensate for poor design.

Designing Visual Display Output

Layout of display screens : Each display page is commonly called a screen or panel. Its layout will ease or impede use. Designing a layout begins by verifying the characteristics of the display screen. These include:

- Physical dimensions of the screen.
- Number of rows and columns of data that can be displayed.
- Degree of resolution.
- Number of colors available.
- Methods of highlighting.
- Methods of intensity control.

Design of files and use of auxiliary storage devices :

Basic file terminology :

1. Data item :

- Individual items of data
- Called as fields or items
- Example: date, cheque no., etc., in a cheque
- Identified by name and has specific values
- Can comprise subitems/subfields

2. Record

- Complete set of related data pertaining to an entry 2 types :

(i) Fixed length record : Number and size of each data item in the record are constant.

Advantage - always same size. Therefore, system need not determine length.

Disadvantage : does not make any efficient use of allocated space.

(ii) variable length record - Varying record size.

3. Record key

- Purpose - Identification
- Analyst selects a unique data item.
- Record key/key attribute/key.
- Already part of record.
- Example :- Cheque number in a cheque.

4. Entity

- Any person, place or event of interest to organization.
- About which data are captured, stored or processed.
- Example: Patients and tests are entities in hospitals.

5. File

- Collection of related records.
- Number of records determines the file size.

6. Database

- Integrated collection of data stored in different types of applications
- Accessible by multiple applications.
- Inter-relation of data is derived from data relationships, not their storage locations.
- Records for different entities are stored.
- Do not eliminate files in an information system.

Data structure diagrams

- Useful tool for developing border framework of data

Purpose : Graphical tool that shows logical data structure requirements on an information system application.

- Serves four purposes :

(i) Verify information requirements.

(ii) Describe data associated with entities.

(iii) Show relation between entities

(iv) Communicate the data requirements to file designer or database administrator.

- They organize the data.

Notation

- Entities represented by rectangles with entity name on top and a list of attributes.
- Each entity is identified by a key attribute.
- Key attribute is the first data item listed.

Use in file design

- To use data structure diagram, analyst should address the following about entity being described.

(i) What data items will uniquely identify an occurrence of the entity ?

(ii) By what means will the information about the entity be accessed ?

(iii) What other data items describe attributes of the entity ?

Types of files

1. Master file

- Collection of records about an important aspect of an organization's activities.

- May contain data describing current status of specific events.

- Two types

(i) That reflects current status of entities.

(ii) That reflects history of events affecting a particular entity.

- Should be accurate and up-to-date.

- They are kept up to date through the use of transaction file.

- They are permanent and exist as long as system exists.

2. Transaction file

- Temporary file.

- Two purposes.

(i) Accumulating data about events as they occur.

(ii) Updating master files to reflect the results of current transactions.

- 'Transaction' refers to any business event that affects the organization and about which data are captured.

- Data important to the organization are collected about each event and accumulated into transaction file.

- Used in updating master files.

- They are temporary, i.e., at a point of time, they are destroyed.

3. Table file

- Special type of master file.

- Contain reference data used in processing transactions, updating master files, or producing output.

- Used to store data that otherwise would be included in master files.

- Advantage

(i) Conserves storage space.

(ii) Ease program maintenance.

4. Report file

- Temporary files,

- Used when printing time is not available for all the reports.

- Computer writes report to a file on magnetic disk or tape, where it remains until being printed, which is called spooling.

- Reports files can be used with many I/O devices like graphic plotters, microfilm and microfiche units.

5. Other files

(i) Backup file :

- copy of master, transaction or table file
- ensure duplicate available.

(ii) Archival file :

- copies for long-term storage or data.
- Stored away from computer center.

Methods of file organization :

- Determines how records are stored, located or retrieved.

1. Sequential organization

- Records are stored one after other without concern for actual value in actual records.

- First record is placed at the beginning of the file, second record next, and so on.

- Order never changes.

- Records are stored by positions, and not address.

- Reading sequential files :

- Reading always starts at the beginning, one record at a time.

- Example, if a particular record is tenth in a file, the system starts at the first one, and reads one at a time until tenth is reached. If cannot go directly to the tenth.

- Based on nature of design, this can be on advantage or disadvantage.

- Searching for records :

- Sequential files do not use physical record keys.

- The 'search key' is used to retrieve records.

- Search process is continued till a match is found/eof.

- Only way to process sequential files.

- Evaluation of sequential files :

- In sequential organization, records are stored and read one after the other.

- Advantageous : When every record in a large file is to be accessed.

- Disadvantageous : When a specific record is to be accessed, in a very large file.

2. Direct access organization

- Keyed files

- Record with specific key value and particular storage location.

- All records are stored by key at addresses rather than by position.

- If 10% of the file is to be accessed, direct access.

- If > 40%, sequential organization.

- **direct addressing**

- Uses physical record key.

- Key serves as address and goes directly to the assigned location.

- Records are retrieved quickly.

- requirements for direct addressing

- Using record key as storage address is called 'direct addressing'.

- Direct addressing should have data set with the following characteristics:

(i) Few open gaps in key values are wanted.

(ii) There is a storage address for each actual or possible key value and therefore no duplicate key values.

- hash addressing

- Used when direct addressing is not possible, but direct access is necessary.

- Hashing is the process of deriving a storage address from a record key.

- An algorithm is used to change a key value into another that serves as a storage address.

- Types of hashing algorithms :

Any number of hashing algorithms can be developed.

Example, conversion of social security number into a value corresponding to a storage address.

The social security number consists of three major sections separated by dashes or blanks.

The algorithm consists of the following steps :

Step 1 : Strip off the first three digits of the Social security number.

Step 2 : divide the new key by a prime number.

Step 3 : modular division is used.

Other hashing algorithms other than division by prime are :

(i) Folding:- Split the key into pieces and process them further.

(ii) Extraction:- Select specific digits from the key and process them with the remaining digits.

(iii) Squaring:- Multiply the number by itself and apply other hashing methods.

- requirements for hashing algorithms:

(i) repeatable

(ii) even distribution

(iii) minimizes synonyms

3. Indexed organization

- Accessing through index.

- The basic form includes a record key and the storage address for a record.

- Takes less time to search with index.

- **Characteristics of an index**

- 'Index' is a separate file from the master file.

- Each record in an index has two parts :

(i) Record key (ii) Storage address

- To find a record, the index is first searched and the corresponding storage address is noted.

- This method uses a sequential scan of the index, followed by direct address to the appropriate record.

- **Indexed non-sequential organization**

- Index contains record keys for each record and the corresponding storage location.

- There is one entry in the index for every record in the master file.

- **Indexed sequential organization :**

- Creates pseudosequential file.

- Groups of records are stored in blocks with a capacity for a specified amount of data.

- It is a file of separate, full or partially full blocks, each in sequential order.

- Adjacent blocks are not in ascending order.

Searching for records

- Only one index entry is necessary for each block.

- Index entry contains the highest or lowest key value for only record.

- Highest entry contains the highest or lowest key value for any record.

- Highest key value is used in indexed sequential access method (ISAM).

- The index is first accessed.

- Index key is compared with search key.

- Search continues sequentially till the value of the search key is lower than the highest key value on that track.

- When keys match, processing starts.

- **Advantage** - Saves search time.

File maintenance under ISAM

- As records are added to indexed sequential files, the system inserts them in such a way as to preserve the sequential nature of each block.

- As each record is added, the block is automatically resorted by the ISAM software.

- As records are added, other are in effect moved down to make space for them to maintain the sequence.

- Separate overflow area is maintained when the track to which a record is to be added is already full.

Steps needed to find a record when it is needed for processing :

(i) The program searches the index to find proper block.

(ii) The block is scanned on record at a time and record keys are checked to find one with the correct value.

(iii) When record key is not found until end of file, overflow area is searched.

- changing and deleting also use same search procedure.

- Changes are made and replaced in its original position.

- Deletion denies access to that record. The space occupied by the deleted record is not reused.

- Indexes occupy extra space but saves time.

- Allows entry into middle of file and continue processing sequentially.

- ISAM is neither a true sequential nor a true direct access method.

- Additional feature of ISAM: is the flexibility to process the file in either way (sequentially or randomly).

Magnetic tape :

Auxillary file storage method.

- Storing data on magnetic tape :

- data are stored along the length of the tape at a specified density measured in bytes per inch (bpi).

- One byte stores one character.

- Recording density of a system is measured by the number of bpi.

- reading and writing:

- Data are read from or written to the tape in blocks or physical records.

- Data are read or written when the tape is being moved forward by the tape drive past the read-write heads.

- The tape stops and is restarted when the system tells the tape drive to read the physical record.

- To allow space for the tape drive to stop and restart without losing data, inter-block gaps are used to separate each block.

- An inter-block gap is a length of blank tape measuring 0.3 inch.

- space determination :

- Analyst must calculate the amount of space needed to store records when determining storage required during system design.

- **Blocking :**

- To save space and to speed up proceessing, inter-block gaps are reduced through blocking.

- Blocking occurs when the system stores multiple logical records within a single physical record or block.

- The number of records in t he block is called blocking factor.

- Maximum length of any block is the number of bytes the buffer can hold.

- **Time determination:**

- Analyst should know the time required to process the file.

- Following are the formulae to determine tape processing speeds:

(i) Tape speed
(in inches per second)

$$= \frac{\text{transfer rate (per second)}}{\text{recording density (bpi)}}$$

(ii) Time per block of data

$$= \frac{\text{bytes per block of data}}{\text{transfer rate (bytes per second)}}$$

(iii) Inter-block gap time

$$= \frac{\text{size of inter-block gap}}{\text{Tape speed}}$$

- **Sequential file processing:**

- Most efficient way to process master and transaction file is to have them both in the same sequence.

- Transaction file should be started into the same order as the master file in any processing.

- File updating:

- To update a master file, the following are the three possible transactions:

(i) Adding new records.
(ii) Deleting existing ones
(iii) Changing contents of existing records.

- Updating creates a new master file with the result of update activities.

- Previous master file is not destroyed.

- Not possible to update in a sequential master file without a copy of the master file on a different reel or tape.

- Error handling:

- During file processing, error checks are made continually.

- The following types of errors can occur:

(i) Attempting to add records to a master file when the record is already stored in the file and duplicate keys are not allowed.

(ii) Attempting to update records that are not stored in the file.

- Whenever errors are found, they should be printed out as error report.

Magnetic disk:

- Fomatting of disk permits organized data storage.

- disk layout and design :

- data are written or or read from, only when it is rotated by the disk drive.

- Rotation ranges from 200 to several thousand revolutions per minute.

- Read-write heads electronically sense data from the disk surface.

- Data are stored on surface in one of the tracks.

- Each track is identified by a number and a surface number.

- Read-write heads are mounted on access arm and all heads move together.

- The heads on each surface can reference one track on that surface.

- The tracks that can be reference at one time from one specific position are refererred to as 'cylinder'.

- To access a specific record, the cylinder and surface where the record is located must be known.

- Track/cylinder addressing:

Index point:

- Signals the start of a track.
- Considered as one large storage bucket.
- All tracks on the disk are synchronized to the index point

Home address:

During intialization, the system writes the home address on the track.

- Home address consists of four separate pieces of information:

(i) Flag byte: Indicates whether the track is a primary or alternate track.

(ii) Cylinder/read-write head number: identify a specific track on the disk.

(iii) A hashing algorithm.

(iv) Error control bytes.

- Track descriptor record:

- Describes how data on remaining part of the track are organized.

- Nine elements are included: flag byte, cylinder number, read-write head number, record number, key length, data length, check bytes, data.

- General formats:

- Track/sector addressing
- Disk storage capacity
- Time determination.

Back-up and recovery of files

- Information systems are not immune to problems that lead to loss of data.

- Therefore, Preservation of master and transaction files are important.

- Potential causes of data loss:

- As soon as the system is installed
- After specific intervals, parts and components may fall.
- Magnetic tape and magnetic disk wear out with use.
- Improper processing of data or operator error.
- Software failures.

- Natural disasters, such as fire, flood or earthquake.
- Sudden power fluctuations.

• Back-up methods:

(i) Generation of master files:

- Highly effective method.
- Grandparent-parent-child technique.
- This technique expands two generation of master files.

(ii) Master file dumps :

- Process of duplicating a file to create another copy of it.
- A system program is run that reads data from master file and writes a copy to a back-up disk.
- Provides additional copy.
- Original version is unaltered.
- Only original version is used for further processing.
- If master file is damaged, the copy is processed and updated.

(iii) Record image copies.

PROCESS LIFE CYCLE

System development refers to the process of examining a business situation with the intent of improving it though better procedures and methods. System design is the process of planning a new business system or one to replace or complement an existing system.

System development life cycle

The system development life cycle can be thought of as a set of activities that analysts, designers and users carry out to develop and implement an information system. System development life cycle method consists of the following activities:

1. Preliminary investigation.
2. Requirements analysis or system analysis.
3. Design of system.
4. Development of software.
5. System testing.
6. Implementation and maintenance.

Preliminary investigation

It is undertaken when users come across a problem or opportunity and submit a formal request for a new system to the MIS department. This activity consists of three parts:

(a) Request clarification.

(b) Feasibility study.

(c) Request approval.

The request which are submitted to the MIS department are not clearly stated. Before any systems investigations can be considered the system request must be examined to determine precisely what the originator wants. The analysts tries to determine whether the system requested is feasible or not. Aspects of technical, economic and operational feasibility study. The third part of investigation relates to approval of the request. Not all requested systems are deliverable or feasible. Based on the observations of the analyst the management decides which system should be taken up for development.

Requirements analysis

The needs of the users are studied after the preliminary investigation process. Details are gathered. Analysts study the present system to identify its problems and shortcomings and identify the features the new system should include to satisfy the new or changed user application environment.

Design of the system

The user requirements that arose by analyzing the applications environment are incorporated in the new systems design. The design of the information system produces the details that state how a system will meet the requirements defined. The analyst designs various reports or outputs, data entry procedures, inputs, files and database. These detailed design specifications are then passed detailed design specifications are then passed on to the programming staff so that software development can begin.

Acquisition and development of software

After the system design details are resolved resource needs such as specific type of hardware, software and services are determined. Choices are made as to which products to buy are lease from which vendors. Software developers may install purchased software or write new custom designed programs. The choice depends on many factors such as time, cost and availability of programmers if the software is to be developed in-house. During this phase the analyst also works with users to develop worthwhile documentation for software including various procedure manuals.

Systems testing

Before the information system can be used, it must be tested. System testing is done experimentally to ensure that the software does not fail. It will run according to the specifications and in a way users expect. Special test data are input for processing and results are examined. It is found satisfactory it is eventually tested with actual data from the current system.

Implementation and maintenance

After the system is found to be fit it is implemented with the actual data. Hardware is installed and users are then trained on the new system and eventually work on it is carried out independently. The results of the development efforts are reviewed to ensure that the new system satisfies user requirements. After implementation, the system is maintained. It is modified to adopt to changing users and business needs so that the system can be very useful to the organization as long as possible.

The system development life cycle should be viewed as a continuous iterative process that recycles through each stage for many applications. Even when a system is fully specified, designed, purchased and running it is continuously being enhanced or maintained. Enhancement and maintenance may require returning to one of the earlier stages of system development life cycle.

Achieving systems development objectives

There are many reasons why organizations fail to achieve their systems development objectives. Although we cannot catalog all the reasons a few representative samples are given below.

- Lack of senior management support for and involvement in information systems.
- Shifting user needs.
- Development of strategic systems.
- New technologies.
- Lack of standard project management and system development methodologies.
- Overworked or undertrained development staff.
- Resistance to change.
- Lack of user participation.
- Inadequate testing and user training.

Approaches to system development

The system development approaches used within an organization are as follows:

1. The activities are performed in a sequence. In this approach an activity is undertaken only after the prior step is fully completed.

2. Prototyping approach

A prototype is a usable system or system component that is built quickly at a lesser cost and with the intention of modifying or replacing it by a full scale and fully operational system.

3. End-user development approach

It is end user not the computer professional who is responsible for system development activities.

4. Top-down approach

It assumes high degree of top management involvement in the planning process and focuses on organizational goals, objectives and strategies.

5. Bottom-up approach

It starts with the development of life stream systems which is now the essential base for day today business systems.

6. Systematic approach for development in small organizations :

Many small systems use the systematic approach which consists of the following steps:

Identify requirements.

Locate, evaluate, and secure software and then hardware.

Implement the system.

PLANNING THE PROJECT

Purpose: To describe an organization wide direction for a firm's information system. This includes identifying key elements on which application depends are built. It may also include describing the interrelation of these elements and possibly documenting the current status of the organizations's information needs or outlining future plans.

Three widely used approaches to planning projects are:

1. Business system planning.
2. Computer architecture strategic planning method.
3. Critical success factors method.

Business system planning

The method focuses on identification of the data necessary to run on organization.

Computer architecture strategic planning method

It links an organization's current capabilities with its future needs. It emphasizes on development of a strong technical infrastructure, or foundation to support the applications.

Critical success factors method

It seeks to identify areas that are key to an organization's survival and ensure that these key elements are incorporated into the project.

Project planning requires a "vision" - a view of the impact the project has on an organization's long term corporate success, strategically and operationally. A vision is not a plan. It is not sufficiently well defined to be a plan but rather describes the directions for an organization and a conceptualization of the impact information systems have. In other words the vision describes whether a firm wished to go but not what means it will take to get there. The means of the firm's objectives will be developed during the formulation of the firm's business strategy.

Having a vision is essential but it must also be communicated, if the organization's personnel and other resources are to be successfully mobilized to build and maintain a viable information system plan for the project that is taken up.

MANAGING THE PROJECT

Basic Concepts

Software Project Management begins with the basic management cycle steps of setting the goal, planning, organization, scheduling, staffing, co-ordinating, and control to achieve the goal. The steps become operational after the project proposals accepted by the customer and requirement analysis is carried out resulting in an RDD confirmed by the customer. Boehm defines the project management cycle in four phases namely Inception, Elaboration, Construction and Transition.

SPM is built on four basic tools: people, product, process and project.

People are those who are directly or indirectly related with the system, including Users, Developers, Stakeholders, Vendors, and Partners. The focus is on the environment, communication, productivity, skills, motivation and team management, PM-CMM

SPM is a team effort, where every member has a role, responsibility and commitment to the task. The team composition is a function of numerous parameters:

Team composition = f (SRS complexity, Skill requirement, Development strategy, implementation strategy, Available resource).

Seldom Software Management fails for technical reasons. The failure incidence or customer dissatisfaction is more often a result of how the team is built and how effectively it performs to achieve different goals, such as quality, cost, time and delivery. The common team problems are many, but prominent among them are ineffective leadership, Lack of co-operation between team members, lack of participation, and unintentional or inadvertent entry of user requirement by members.

The solution to these problems is to build effective teams. An effective team is that one whose members form a cohesive group, demonstrating physical and emotional togetherness and communication without barriers. When large projects are taken up more teams would come up, coordination between teams becomes critically important. Teams must jell well to bring out a successful close to the project.

Product is the deliverables, scope, form and format, key functions and objectives, often as a solution in one-off case. It mainly targets on the cost, feasibility, quality and risk.

The product is the basic concept of the software which is the solution for the user proposal. They are described by four factors :

1. Context : The arena in business which has problems and which need solutions within the given domain, and with due regard to constraints and limitations prevailing in the environment.

2. Objective : The system objectives can be defined in terms of information, analysis, reports, etc; as output benefit of users to solve business problems.

3. Functions : Software system processes and functions, such as data sourcing and processing, information processing and analysis, transaction processing, applications processing and integrating the applications to achieve software product goals.

4. Performance : This stipulates non-functional requirements of speed quality and features such as ease of use, adaptability, achieving certain measurable objectives such as cost reduction, reduction in cycle time, improving management effectiveness.

Such a definition makes the product bounded, unambiguous and understandable. It paves way to analyze, model and design the software product and aids further to evolve a development and implementation methodology. A clear software product definition and its details help in business modeling and requirement engineering leading to a document called the RDD-Requirement Definition and Description.

Process includes the method through which product will be delivered - process models, choice of technology, processes of development. The main activities includes a syllogism of scope definition, analysis, estimation, feasibility, risk management and quality assurance and tasks such as monitoring and tracking, correcting and reporting.

The issue is how these steps will be executed to develop the product. There are several models to execute these steps on different strategies. The process models are Linear sequential, prototyping, Rapid application development, Incremental development, Spiral model, Component-based development, Concurrent development. The model is selected according to the application, incorrect choice will have an adverse effect on cost, quality and performance. A process is composed of activities and the most general activities include :

1. Communication - activities for interaction and information exchange.

2. Planning - software product scope determination, effort estimation, cost and resource evaluation, base 4 line product software definition.

3. Risk - risk identification, analysis, exposure, classification and resolution.

4. Engineering - requirement engineering, business modeling, system modeling, work break down structure, key deliverables including technologies.

5. Mechanism of delivery - implementation and deployment, change management, acceptance process, release and support for a stipulated period.

6. Closure - feedback of stages, reviews, demos, traceability of deliverables, acceptance and certification and building project experience database for learning.

Project details about the definition, scope, business goal, economics, resource budgets, configuration, clear start and closure. It mainly focuses on the deliverables, cost control, time management, meeting deadline and fulfilling commitments to customer satisfaction.

A project is an activity that has a definite start and a definite end, requires multiple resources and is composed of uncertainties. The operating functions comprises of scope, time, cost and quality. Project characteristics as in the case of the modern software development approach are contiguous integration, early risk resolution, component based development, user involvement and architecture first and design later.

PROJECT MANAGEMENT

They key to successful project management is composed of nine operational areas. The project manager is responsible for managing these knowledge ares and using appropriate tools and techniques to meet project goals gracefully.

The elaboration of knowledge areas :

Scope : Defines the requirements that when delivered will be the successful completion of the project.

Tools - Prototyping, modeling, structuring, partitioning and packaging.

Time : Estimating the time for project completion, for each activity in the project, activity scheduling, and monitoring activity progress.

Tools - Gantt charts, PERT networks and Critical path method.

Cost : Estimating the effort and resources required, costing them for budgeting and controlling then throughout the project duration. It also includes control of cost of outsourcing and cost of procurement items. It is maintained to keep in track with the profit margin.

Tools - Activity costing, Product costing, Budgeting and Estimation analysis tool for cost control.

Quality : Producing the quality of all deliverables as expected by the customer and stakeholders.

Tools - testing tools for design and coding, alpha testing and beta testing.

Human resource - HR means managing people and their skills effectively as estimated to deliver quality products within cost and time.

Tools - Resource estimation, FPA and COCOMO for size and effort estimation.

Communications : Capturing all aspects of the data on a project and its activities, processing the data and distributing it to the concerned people so as to respond effectively to meet the commitments.

Tools - mails and messages, reviews and reports and meetings.

Risk : Building monitoring and mitigation plans for controlling risks.

Tolls - elimination, prevention, avoidance and mitigation.

Procurement : Identifying procurement requirements such as development and engineering services, hardware and software as a support to the project.

Tools - vendor analysis, technology analysis, configuration management and sourcing.

Integration : Integrating various tasks in the scope to product project deliverable as per the expectations of the customer within the cost and time budgets.

Tools - Process models; life cycle models change management interfacing and connectivity tools.

DESIGN : System decomposing and modeling provides a system view in modular form using functional components.

System design is an engineering representation of customer and user requirement. Such a design can be traced back to the requirement and can be accessed for quality specification of good design.

The design process is an iterative process ensuring that the Requirements Determination Document (RDD) and Systems Requirement software (SRS) are translated into design to meet data, function and behavioural requirements. The goal of the design process is to implement all explicit and implicit requirements stated in the RDD and SRS. Design implementation should be such that it is easy to code, test and maintain subsequently.

The design process should produce a design that is easy to understand and should provide an implementation perspective for all viewers and users of the system.

An efficient and effective design process produces a design with the following characteristics :

- Modular
- Connectivity through interface
- Clarity
- Use of components

The design phase deals with

- Input-output design.
- Data and data structure.
- Data model
- Functional modules.
- Components
- Interfaces.
- System components.

The design architecture is built through four different models.

Structural model - shows the organization of components in a hierarchical manner.

Frame work model - helps to use design patterns tested and readily available.

Dynamic model - shows behavioral aspects of the architecture as a response to induced change.

Process model - models a business process that the system can accommodate.

To achieve the operational efficiency the design architecture is portioned. A suitably partitioned architecture has a number of benefits.

Easy to test

Easy to maintain

Fewer side effects

Easy to extend

There are two types of portioning.

1. Horizontal - separates major functions from each other. The portioned structure is built through control modules and operational modules. The operational module achieves the function, the control module executes operations and communicates with the next control module to move the system further.

2. Vertical - puts the work flow driven by decisions in the functional module in a top-down direction. Decision-making for control of operation is at the top while decision-making operations are put at bottom.

Effective modular design reduces system complexity, offers better understanding of the system and its component behaviour. It is easy to maintain and can be implemented after ,understanding customer environment. One can evoke a modular development and implementation strategy through parallel development of different parts of the system.

Effective modular design is built with three properties:

1. Functional independence - achieved by carving out a module that is responsible for any unique function and which does not need to interact with other parts of the system.

2. Cohesion - measures functional strength within the module. It measures how closely different steps in a module perform with least interaction with other parts of the system, subsystem, or modules.

3. Coupling - is a measure of degree of complexity of interconnection among modules. The connectivity between modules is built through interface. Interface is a channel to receive input and to send an output for further processing.

Architecture and non-functional requirements :

There is a close correlation between architecture and non-functional requirements of the system. The non-functional requirements of the system are :

- Performance
- Security
- Safety
- Availability
- Maintainability

DESIGN SPECIFICATION DOCUMENT : The purpose of design specification documentation is to provide a central reference document for software system design. The documentation covers :

Scope of design, data design specification, System design, Architecture design. Design specification, Design testing, Design constraints.

CODING : The design must be translated into a machine readable form. The code generation step performs this task. If design is performed in a detailed manner, code generation can be accomplished mechanistically.

TOOLS in coding : Tools extend the capability of the systems analyst in three ways :

- The potential to improve the productivity
- Facilitate more effective prodedures.
- Improve system quality.

The benefits of using computer assisted tools are decreased elapsed time, automate tedious tasks, ensure consistent procedures and capture system data.

Code generators

Code generators automate the preparation of computer software. They incorporate methods that allow the conversion of system specification into executable source code.

Code generation is not yet perfected. Thus best generators would produce approximately 75 percent of the source code for an application. The rest must be done manually. Becuase CASE tools are general purpose tools that is not limited to any specific application area. The challenge of fully automating the software generation process is susstantial.

The greatest benefits accrue when code generators are integrated with a central information repository. Such a combination achieves the objective of creating reusable computer code. When specifications change code can be generated by feeding details from the data dictionary through the code generator. The dictionary contents can be reused to prepar executable code. The most visible advantage of the feature is reduction of time to preprare an operational program. Code generation also ensures a consistent and standard program structure and reduction of errors thus improving software quality.

- The reliance on structured methods.
- Absence of standard levels of methodology support.
- Conflicting use of diagrams.
- Diagrams not used.
- Limited function and scope.
- Human tasks remain critical.

RE-ENGINEERING

Code restructuring : The most common type of re-engineering is code restructuring. Some legacy systems have a relatively solid program architecture, but individual modules were coded in a way that makes them difficult to understand, test and maintain. In such cases the code within the suspect modules can be restructured.

To accomplish this activity source code is analyzed using a restructuring tool. Violations of structured programming constructs are noted and code is then restricted. The resultant restructured code is then reviewed and tested to ensure that no anomalies have been introduced. Internal code documentation is updated.

Code restructuring is performed to yield a design that produces the same function but with higher quality than the original program. In general, code restructuring techniques model program logic using Boolean algebra and then apply a series of transformation rules that yield restructured logic.

TESTING and IMPLEMENTATION : A successful test is one which finds an error. Testing is a process which is done with the explicit intention to make a program fail. The 50% of budget for program development is spent on the testing process. It is an expensive but critical process. The person performing the testing can be analyst, programmer or a specially trained person in software testing.

Verification and validation - It is performed by executing the program in a simulated environment.

Validation is a process of using software in a live environment in order to find errors. The process may continue for several months.

Testing strategies : A test case is a set of data that the system will process as normal input. For example the test cases for inventory: quantity ordered exceed or equal or less than the quantity on hand. The two general strategies are :

1. **Code testing :** It examines the logic of the program. Every path through the program is tested. A path is defined as 'specific combinations of conditions that is handles by the program'.

The limitations of the testing include :

Virtual impossibility to do exhaustive testing of this nature

Financial considerations and time limitations alone preclude.

Performed in its entirety, it does not guarantee against software failures.

It does not indicate whether the code meets the specification.

Whether all aspects are implemented.

It does not check the range of data the program will accept.

2. **Specification testing :** Analyst examines the specifications and tests the program as for what the program should do and how it should perform under different conditions and with different test cases. The assumption is if the program meets the specifications it will not fall. A program is taken as a Black Box which is not concerned about whether every instruction or path through the program is tested.

Management of test practices : There are two levels by which testing is carried out,

1. **Unit testing :** It focuses firsts on the modules (independent of one another). The testing personnel detect errors in coding and logic that are within the module alone. Errors resulting from the interaction between modules are initially avoided. The test cases needed for unit testing should exercise each condition and option. Test cases are needed to test the range of values expected both valid and invalid data.

The two ways unit testing is performed are bottom-up testing and top-down testing.

Bottom-up testing starts with the smallest and lowest level modules proceeding one module at a time. A short program called the driver program executes the module and provides the necessary data. The module is performed in a way it will be executed in an actual manner. After the closure of first level testing moves to the next level after individual testing linked with the previously examined modules.

Top-down testing starts with the upper level modules. A stub program is normally written as a module shell that can be called by the upper level module and when reached properly will return a message to the calling module which indicates proper interaction. No attempts are made to test the correctness of the lower level modules.

The best strategy is to adopt top-down and bottom-up testing - lower level modules are unit tested and integrated in a top down testing program.

2. System testing : If tests the compatibility of individual modules. For example the different specifications for data length, type, name, etc The system may not report on error but the output may show unexpected results. System testing must also verify that adequate file sizes are available.

Special systems testing :

There are six essential tests to be performed under the special tests for systems.

1. Peak load testing : There are critical times in many systems, particularly online systems. Example - teller machines in banking system. It is used for testing in real time situations.

2. Storage testing : The analyst specify a capacity for the system when it is designed and constructed. Capacities are measured in terms of the number of records the file can handle or a file can contain. Storage testing often requires entering data until the capacity is reached. Comparing the actual and claimed capacities will verify the accuracy of the documentation and it allows for the judgement of actual capacity. Users find out too late that claims made during installation are not true. There is not enough storage capacity for transaction and master file records.

3. Performance time testing : Testing is conducted prior to implementation. Determination of how long it take to receive a response to an enquiry, make a back up copy of a file, etc. It also includes test runs to time indexing or resorting of large files.

A system that runs well with only a handful of test transaction may be unacceptably slow when fully loaded.

4. Recovery testing : By creating a failure or data loss event where users are forced to reload and recover from a backup copy is termed as recovery testing. Analyst can determine whether recovery procedures are adequate.

5. Procedure testing : Documentation and run manuals tells the user how to perform certain functions and are tested quite easily by asking the user to follow them exactly through a series of events. Example when to depress the enter key, paper-out light of printer, etc; this type of testing not only shows where they are needed but also where they are wrong.

6. Human factor testing

Finding answers to questions about how people will react to the system in ways not anticipated. It is the responsibility of the analyst to anticipate questions that will arise in the minds of the users as they interact with the system.

Test data that is used for testing can be designed in two ways.

1. Live test data 2. Artificial test data

Live test data are actually extracted from the organization files. It has to be keyed in by using some special personnel for the specific task. Live test data will not test all the combinations or formats that can enter the system.

Artificial test data are created solely for test purposes. They are generated to test all combinations or formats. An independent team of testers formulates a testing plan using the system specification.

Testing libraries

A testing library is a set of data developed to thoroughly test a system of programs. It is stored in machine readable form usually on a magnetic disk. The library is not just for initial testing, as the system evolves programs are modified and maintained so they must be retested. So library is to be maintained throughout the life of the system.

PROJECT MAINTENANCE: CORE FUNCTIONS

Scope refers to entire work in the project, and related tasks and activities. The scope defines the project in terms of goals and boundaries, deliverables, different knowledge and technology inputs going into the project. The scope description is given by SOW-scope of work agreed by users, stake holders and customers.

The processes involved in scope management are:

1. Initiation - a process to commit the organization to start the project and come out with a project document giving broad view of the project.

2. Planning - a process of creating a document SOW detailing the tasks and major delivarables. It is then used to plan the scope management.

3. Defining - a process which breaks SOW and scope plan into smaller work units with clear definition of start - end and its output preferably measurable as a milestone. Such a break down statement is called as Work Breakdown Structure - WBS representing tasks.

4. Verification - this is a key process where all concerned and affected people verify SOW and WBS and come to agreement on delivarables of the project, and processes which will produce them.

5. Change Control - It is a formal process to control the changes post verification and agreement to ensure the changes brought in are unavoidable and cannot be deferred to the next version of the product. Its inclusion though it disturbs the SOW and WBS is necessary to keep the project on the path of success.

The SOW generally has the following information:

- Project justification and motivation factors.
- Description of current scenraio.
- Summary of deliverables.
- High level WBS.
- Measures of successful project.

The characteristic features of a WBS are :

- Statement of manageable, meaningful tasks having a lucid goal.
- Monitoring a task units in terms of cost and time.
- Achievement of task goal to continue the project further.

Personnel and development management

The project teams are segregated into different concepts as follows :

1. Chief programmer team

The team members involved are the chief programers, back up programmer(s) and a programming librarian. The characteristics of the team include the chief programmer formulates the design tasks and writes code for all critical modules in the system, performs integration and testing of teams code, serves as primary contact for the user and the other teams.

2. Specialist team

The team contains a chief programmer, backup programmer (s), an administrator, a tool smith, test specialist, documentation editor, a programming librarian and other specialist. It contains a common core of team members and some for administrative support who handles budgets, facilities and make outsourcing arrangements.

3. Leader less team

The team contains the basic constituents or essence of a team without a leader depending upon the project. Team members are assigned leadership roles and work is carried as a whole.

Structured walkthrough

It is a planned review of the system by persons involved in the development effort. The structured walkthrough or peer review has its participants as the colleagues at the same level in the organization.

Purpose - to identify areas where improvement can be done in the system life cycle.

The review session does not result in the correction of errors or changes in specifications.

The emphasis is constantly on review not on the repair. It is a team process. The team may contain 7 - 9 people.

- The individual who formulated the specification or program code.
- Moderator - one who leads the review.
- Scribe/recorder - to capture details from the meeting.

The maintenance issues are also addressed during walkthrough, which enforces coding standards, modularity and documentation. The reviews normally exceed 90 mins.

Requirements review : RR or specification review is aimed at examing the functions, activities, and processes the new system will handle. The inconsistencies among requirements and the new system are uncovered by the walkthrough to be dealt in future. The documentation description includes the requirements, key components and the key persons involved. The review also uses DFDs, decision tables or data dictionary to assist in describing and understanding the processes, data captured and used, and storage requirements.

Design review : It focuses on design specifications for meeting previously identified system requirements. The purpose of it is the participants find discrepancies between the design and requirements they will point them out and discuss them. The walkthrough does not merely focus on the redesign of the code.

Code review : The conduct to examine the program code developed in a system along with its documentation. The participants access the execution efficiency, the use of standard data names and modules and program errors. The check list can be used for noting problems and their severity.

Testing review : The least common use of software is for the analysis of testing programs. The participants do not actually examine the output or search for the errors rather they focus on the testing strategy that will be used and assess its suitability for detecting errors.

Approved authority : If the team doesn't approve the specifications, design, or code problems identified by the team members it must be remedied before further development. The major changes may be at the worst case the team requests a final review before acceptance, the less changes might be the team that accepts the verbal or written notification and no additional review. Some organizations require unanimous approval of authority of all team members before a project is approved while others establish strict voting technique.

OBJECTIVE QUESTIONS - I

1. The set of activities involved in carrying out an information system is called

a) software development cycle
b) system development life cycle
c) software design cycle
d) system development activities

2. The first step in SDLC is

a) requirements analysis
b) system analysis
c) preliminary investigation
d) system design

3. The prototype built initial version is called

a) parallel system b) cut-over system
c) phase-in-system d) pilot system

4. The document prepared after system analysis is

a) system requirement, specification
b) requirements determination document
c) testing manual d) analysis document

5. A description of the data being used is
   a) data base b) data dictionary
   c) datafile d) data description

6. The modular design property that describes strength within the module is
   a) functional independence
   b) functional dependence
   c) cohesion d) coupling

7. The complexity of interconnection among modules is described by
   a) cohesion b) temporal cohesion
   c) logical cohesion d) coupling

8. Non-execution based testing is done through
   a) walkthrough b) report review
   c) observation d) questionnaire

9. What aspect is not tested in unit testing ?
   a) Data structure b) Boundary value conditions
   c) Specific value d) Flow of data

10. White box testing deals with
   a) software sensitivity b) flow and control structure
   c) validity d) behaviour

11. Black box testing addresses
   a) structure values b) structure definition
   c) validity d) behaviour

12. Testing done on development platform is
   a) Beta Test b) Eta Test
   c) Alpha Test d) Gamma Test

13. Test done on___ environment is beta test.
   a) simulated b) real-time
   c) host d) developed

14. Testing for abnormal situations is done through
   a) recovery testing b) security testing
   c) performance testing d) stress testing

15. A set of instructions designed to discover an error is
   a) Test plan b) Test cases
   c) Test examples d) Test exemplars

16. A document that consists of different test cases for different objectives and attributes is called
   a) Test case b) Test plan
   c) Test Management d) Test document

17. Design phase in SDLC includes
   a) user interface design
   b) project report design
   c) test case design
   d) quality design

18. The input design should focus mainly on
   a) user friendly design b) graphics usage
   c) data d) innovation

19. CASE Tools are used mainly for
   a) ease b) revision
   c) reusability d) research

20. Reusable code - focus in programming is called
   a) content programming
   b) component programming
   c) coding d) code usage

21. Estimation of resources, cost and schedule depends on
   a) project design b) application
   c) planning d) project size

22. Project risk can be measured by
   a) degree of uncertainty
   b) project plan
   c) design d) management

23. Process specification is used to describe all
   a) data b) design structure
   c) flow model structure d) data bases

24. Data dictionary should surely contain
   a) hame b) alias
   c) description d) miscellaneous

25. Measures of reliability is given by
   a) Mean time between success
   b) Mean reliable time
   c) Mean time between failure
   d) Mean time

26. A map for instituting software quality assurance is
   a) software plan b) SQA plan
   c) software preview d) SQA review

27. A top-down design strategy - a macroscopic statement of function is called
   a) modularity b) abstraction
   c) architecture d) step-wise refinement

28. Program structure is also called as
   a) structure chart b) control hierarchy
   c) control design d) program design

29. The logical relationship among individual elements of data is
   a) data structure b) data representation
   c) data system d) data view

30. Structured English is otherwise called as
   a) Pseudo-design
   b) Program design language
   c) Code-language d) Program-language

31. Change implementation checking is confirmed through
   a) software configuration audit
   b) auditing
   c) accountability d) audition

32. Status accounting is otherwise called as
   a) status reporting b) audit reporting
   c) configuration status reporting
   d) status review

33. In data dictionary, a notation for representing content is called
   a) content description b) content information
   c) data content d) data description

34. Prototyping paradigm can be called as
   a) use-case design b) object-oriented design
   c) structuring d) throwaway prototyping

35. RAD is
a) Rapid aid design
b) Rapid aided development
c) Rapid application development
d) Rapid application design

36. SRS includes information and ____ description
a) detail b) functional
c) design d) development

37. SRS is prepared at the culmination of the
a) requirements determination
b) analysis
c) design d) development

38. The number of occurrences of objects in a given relationship is
a) modality b) numeric constant
c) cardinality d) correlation constant

39. Errors occurring at one location and propagate through a system is called
a) parallel errors b) phase - in errors
c) ripple effect d) cascading

40. The primary activity during data design is
a) logical representation
b) physical representation
c) preliminary investigation
d) feasibility study

41. When a segment of a DFD exhibits data flow it is called
a) transform flow b) incoming flow
c) information flow d) overall flow

42. The path in which data flow is depicted is called
a) flow path b) action path
c) data path d) information path

43. A set of design steps that allows a DFD with transform flow characteristics is
a) transaction flow b) information map
c) design map d) transform mapping

44. Modules in DFD at various levels are
a) split b) imploded
c) exploded d) exported

45. The design of internal program interface is called
a) intermodular interface design
b) information design
c) design structure d) development structure

46. One characteristic that serves to distinguish real-time system is
a) real-time operation b) real-time environment
c) interrupt handling d) information exchange

47. Basis-path testing falls under
a) system testing b) white box testing
c) black box testing d) unit testing

48. Quantitative measure of the logical complexity of a program is called
a) cyclomatic complexity b) mccabe number
c) PERT chart d) Pie chart

49. Testing that exercise logical conditional in a program module is
a) general testing b) control testing
c) condition testing d) cyclomatic testing

50. ___ refers to set of activities that ensure that software correctly implements a specific function.
a) Validation b) Verification
c) Testing d) Design-maintenance

ANSWER KEY - I

1. b	2. c	3. d	4. a	5. b
6. c	7. d	8. a	9. d	10. b
11. a	12. c	13. c	14. d	15. b
16. b	17. a	18. a	19. c	20. b
21. d	22. a	23. c	24. a	25. c
26. b	27. d	28. b	29. a	30. b
31. a	32. c	33. a	34. d	35. c
36. b	37. b	38. c	39. c	40. a
41. a	42. a	43. d	44. c	45. a
46. c	47. b	48. a	49. c	50. b

OBJECTIVE TYPE QUESTIONS - II

1. Software engineering primarily aims on
a. reliable software b. cost effective software
c. reliable and cost effective software
d. none of the above

2. A good specification should be
a. unambiguous b. distinctly specific
c. functional d. none of the above

3. Which of the following is a tool in design phase ?
a. Abstraction b. Refinement
c. Information hiding d. none of the above

4. Information hiding is to hide from user details
a. that are relevant to him
b. that are not relevant to him
c. that may be maliciously handled by him
d. that are confidential

5. In object oriented design of software, which of the following is not true ?
a. Objects inherit the properties of the class
b. Classes are defined based on the attributes of objects.
c. An object can belong to two classes.
d. Classes are always different.

6. Design phase includes
a. data, architectural and procedural designs only
b. architectural, procedural and interface designs only
c. data, architectural and interface designs only
d. data, architectural, interface and procedural designs.

7. To completely write program in FORTRAN and rewrite the 1% code in assembly language, if a project team needs 13 days, the team consists of
a. 13 programmers b. 10 programmers
c. 8 programmers d. 100/13 programmers

8. If 99% of the program is written in FORTRAN and the remaining 1% in assembly language, the percentage increase in the programming time compared to writing the entire program in FORTRAN and rewriting the 1% in assembly language is

a. 10 b. 5 c. 13 d. 8

9. If the entire program is written in FORTRAN, the percentage increase in the execution time compared to writing the entire program in FORTRAN and rewriting the 1% in assembly language is

a. 0.9 b. 8 c. 0.8 d. 9

10. If 99% of the program is written in FORTRAN and the remaining 1% in assembly language, the percentage increase in the execution time, compared to writing the entire program in FORTRAN and rewriting the 1% in assembly language is

a. 0.9 b. 1 c. 0.1 d. 0

11. If a weightage of 3 is given to the programmers effort and a weightage of 2 is given to the execution time, then coding 99% in FORTRAN and the 1% in assembly language performs better than coding in FORTRAN completely and rewriting the 1% in assembly language by a factor of about

a. 1.5 b. 1.2 c. 1.1

d. it does not perform better

12. Data structure suitable for the application is discussed in

a. data design b. architectural design
c. procedural design d. interface design

13. Design phase will ususally be

a. top-down b. bottom-up
c. random d. centre fringing

14. Assertions are conditions which are true at the point of execution

a. always b. sometimes
c. many times d. no time

15. Assuming the existence of a start and end nodes for a program graph (PG), the total number of paths is equivalent to . . . . . . set of test data required to test software.

a. minimum b. maximum
c. optimum d. supremum

16. In a PG, if x is a if - then - else node and the number of paths from if part to end node is y and for else part is z with the number of paths form start node to end node is x, the total number of paths through x is

a. $xy + z$ b. $xz + y$ c. $x + y + z$ d. $x(y + z)$

17. If x is case statement in a PG with n cases instead of an if- then-else statement in the previous question with each case leading to only one path to end node, total number of paths through x is

a. $x + n$ b. x^n c. $x \log(n)$ d. xn

18. Structured programming codes include

a. sequencing b. alteration
c. iteration d. multiple exit from loops

19. Which of the following is a desirable property of a module ?

a. Independency b. Low choesiveness
c. High coupling d. Multi functional

20. Which of the following types of maintenance takes the maximum chunk of the total maintenance effort in a typical commercial application environment ?

a. Adaptive maintenance
b. Corrective maintenance
c. Preventive maintenance
d. Perfect maintenance

21. An important aspect in coding is

a. readability b. productivity
c. to use as small memory space as possible
d. brevity

22. One way to improve readability in coding is to

a. avoid go to statements
b. name variables and functions according to their use
c. modularize the program
d. none of the above

23. The data flow model of an application mainly shows

a. The underlying data and the relationship among them
b. Processing requirements and the flow of data
c. decision and control information
d. communication network structure

24. According to Brooks, if n is the number of programmers in a project team then the number of communication paths is

a. $n(n-1)/2$ b. $n \log n$
c. n d. $n(n+1)/2$

25. The extent to which the software can continue to operate correctly despite the introduction of invalid input is called as

a. reliability b. robustness
c. fault-tolerance d. portability

26. If the number of conditions in a decision table is n, the maximum number of rules (columns) possible is

a. n b. $2n$ c. 2^n d. $\log_2 n$

27. Which of the following statements is not true ?

a. Content coupling in a module is desirable
b. Logical cohesion in a module is desirable
c. Stamp coupling is preferred over functional coupling
d. None of the above

28. Configuration management is not concerned with

a. controlling changes to the source code
b. choice of hardware configuration for an application
c. controlling documentation changes
d. maintaining versions of software

29. The railway reservation system currently operational in India can be classified as a

a. batch processing system

b. real-time system

c. on-line system d. expert system

30. According to PUTNAM, project effort is inversely proportional to the fourth power of development time. Doubling the development schedule for a 100 PM would reduce the project effort to

a. 50 PM b. 7.16 PM c. 25 PM d. 6.25 PM

31. Data Flow Diagram, Regular Expression and Transition Table can be combined to provide .... for functional specification of systems software.

a. decision table b. Finite state automata

c. event table d. none of the above

32. Which of the following is an assertion ?

a. P is true, P and Q are true and K or not (Q) is true implies k is true.

b. P is true, P and Q are true and K or not (Q) is true implies K is false.

c. P is true, P and Q are false and K or not (Q) is true implies K is true

d. P is true, P and Q are true and K or not (K) is true implies K is true.

33. A program P calls two subprograms P1 and P2. P1 can fail 50% times and P2 40% times. P can fail

a. 50% b. 60% c. 10% d. 70%

34. The probability of success of two modules in unit testing is 0.9 each. The Probability of success of integration testing is 0.9. The joint probability of success is

a. 0.9 b. 0.81 c. 0.729 d. 0.1

35. Which of the following graph theoretic concept will be useful in software testing ?

a. Cyclomatic number b. Hamiltonian circuit

c. Eulerian cycle d. None of the above

36. In a decision table if there are 3 variables and 3 rules, implies

a. specification may not be complete

b. design could be faulty

c. coding will be incorrect

d. all of the above

37. Which of the following software engineering concept does ADA language support ?

a. Abstraction b. Generic

c. Information hiding d. None of the above

38. In unit testing of a module, it is found for a set of test data, at the maximum 90% of the code alone were tested with the probability of success 0.9. The reliability of the module is

a. atleast greater than 0.9

b. equal to 0.9 c. atmost 0.81

d. atleast 1/0.81

39. Which of the following testing methods is normally used as the acceptance test for software system ?

a. Regression testing b. Integration testing

c. Unit testing d. Functional testing

40. A computer program can often be a very satisfactory .... of a physical system such as road traffic conditions.

a. solution b. replacement

c. simulation d. model

41. for (i = 0, s = 0; i < n; i ++) s += a [i] ;

The symbolic execution with n = 3 at i = 2, s is

a. a0 + a1 + a2 + a3 b. a0 + a1+ a2

c. a0 + a1 d. a0 + a1 + a3

42. On an average, the programmer months is given by $3.6 \times (KDSI)^{1.2}$. If so, a project requiring one thousand source instructions will require.

a. 3.6 PM b. 0.36 PM

c. 0.0036 PM d. 7.23 PM

43. Considering a program graph (PG) with statement as vertices and control as edges, which of the following is not true for any program graph.

a. PG is always acyclic

b. PG is always a directed graph

c. there won't be any self loops

d. PG is always a connected graph

44. Consider the following code for finding the factorial of a given positive integer.

```
IFACT = 1
DO 100 I = 2,N,2
100 IFACT  = IFACT*I* (I-1)
```

For which values of N the above FORTRAN code won't work

a. N is even b. N is odd

c. N is perfect d. N mod 3=0

45. For the above code, using symbolic execution, after the iteration with N=5, IFACT is

a. 1 * 1 * 2 * 3 * 4 * 5 b. 1 * 2 * 3 * 4 * 5

c. 1 * 1 * 2 * 3 * 4 d. 1 * 2 * 3 * 4

46. Which of the following is not an assertion ?

a. P is true, P and Q are true and K or not (Q) is true implies K is true.

b. P is true, P and Q are true and K or not (Q) is true implies K is true.

c. P is true, P and Q are false and K or Q is true implies K is true.

d. P is true, P and Q are true and K or not (K) is true implies K is true.

47. Given a source code with 10 operators including 6 unique operators, and 6 operands including 2 unique operands. The program volume is

a. 48 b. 120 c. 720 d. insufficient

48. For the above data, the length of the program is

a. 80 b. 48 c. 320 d. 46

49. To increase reliability, fault tolerance is included in the system in the form of multiple modules. If the problem can be solved by 5 different modules, each with probability of success 0.7, the probability that it can be solved even if 4 modules fail is approximately.

a. 0.3 b. 0.03 c. 0.49 d. 0.05

50. In object-oriented design of software, objects have
a. attributes and name only
b. operations and name only
c. attributes, name and operations
d. none of the above

ANSWER KEY - II

1. c	2. a,b,c	3. a,b,c	4. c	5. c
6. d	7. c	8. b	9. c	10. d
11. d	12. a	13. a	14. a	15. a
16. d	17. d	18. a,b,c	19. a	20. d
21. a	22. a,b,c	23. b	24. a	25. b
26. c	27. a,b,c	28. b	29. c	30. d
31. b	32. a	33. d	34. c	35. a
36. a	37. a,b,c	38. c	39. d	40. c
41. b	42. a	43. a	44. b	45. c
46. d	47. a	48. d	49. b	50. c

EXPLANATORY NOTES - II

7. Writing the whole program in FORTRAN takes 100 man days. Rewriting the 1% code takes 4 man-days. So, altogether 104 man-days. If it is completed in 13 days, 104/13 = 8 men should be involved.

8. The first case takes 99+10 = 109 man-days. The second case takes 100+4 = 104 man-days. The required percentage is (109-104) x 100/100 = 5

9. Let the first case take 100 units of time to execute. The second case will take 99+(1/5) units of time as coding the 1% in assembly language will take 1/5 units of time. Hence the required percentage is 0.8 x 100/100 = 0.8.

10. In both the cases, the final program will have the same 99% of the code in FORTRAN and the remaining 1% in assembly language. Hence the execution time will remain the same.

11. The first case has a measure of $((3\times109)+2\times(99+1/5))/5$ and the second $((3\times104)+2\times(99+1/5))/5$ Lower the measure, preferable it is.

16. As there are x paths from root to node X which is a if - then - else statement, it can take either true case or false. Any how, the total number of paths will be $x(y+z)$.

30. Doubling the developing time reduces the project effort by a factor of $2^4 = 16$. So, the project effort will be $100/16 = 6.25$ PM

32. P is true, P and Q are true implies Q is also true. Now K or not (Q) is true implies K is true because atleast one should be true. Since not (Q) is false, K must be true.

33. $1-(1-0.5)(1-0.4) = 0.7$

38. Since only 90% code alone is tested and that too with 90% of success, the reliability will be atmost 0.81. It may increase or decrease if some more set of test data is given for testing the rest of the code. However, now the reliability is 0.81 only.

42. Here KDSI (Kilo Delivered Source Instructions) is 1. So, $3.6\times1^{1.2}$ PM.

PRACTICE QUESTIONS

1. In a software project, COCOMO (Constructive Cost Model) is used to estimate **[GATE - 04]**

a) effort and duration based on the size of the software

b) size and duration based on the effort of the software

c) effort and cost based on the duration of the software

d) size, effort and duration based on the cost of the software

2. The diagram that helps in understanding and representing user requirements for a software project using UML (Unified Modeling Language) is **[GATE - 04]**

a) Entity Relationship Diagram

b) Deployment Diagram

c) Data Flow Diagram d) Use Case Diagram

3. A software organization has been assessed at SEI CMM Level 4. Which of the following does the organization need to practice beside Process Change Management and Technology Change Management in order to achieve Level 5 ? **[GATE - 04]**

a) Defect Detection b) Defect Prevention

c) Defect Isolation d) Defect Propagation

4. A software configuration management tool helps in **[GATE - 04]**

a) keeping track of the schedule based on the milestones reached

b) maintaining different versions of the configurable items

c) managing manpower distribution by changing the project structure

d) all of the above

5. A software project involves execution of 5 tasks, T1, T2, T3 , T4 and T5 of duration 10, 15, 18, 30, 40 days respectively. T2 and T4 can start only after T1 completes. T3 can start after T2 completes. T5 can start only after both T3 and T4 complete. What is the slack time of the task T3 in days ? **[GATE - 04]**

a) 0 b) 3 c) 18 d) 30

6. Consider the following program module : **[GATE - 04]**

```
int module1 (int x, int y)
while (x!=y) {
if (x>y)
x=x-y;
else y = y - x; }
return x ; }
```

What is Cyclomatic complexity of the above module ?

a) 1 b) 2 c) 3 d) 4

7. Assume that the delivered lines of code L of a software is related to the effort E in person months and duration t in calendar months by the relation $LP*(E/B)^{1/3}*t^{1/3}$, where P and B are two constants for the software process and skills factor. For a software project, the effort was estimated to be 20 person months and the duration was estimated to be 8 months. However, the customer asked the project team to complete the software project in 4 months. What would be the required effort in person months ? [GATE - 04]

a) 10 b) 40 c) 160 d) 320

8. A software was tested using the error seeding strategy in which 20 errors were seeded in the code. When the code was tested using the complete test suite, 16 of the seeded errors were detected. The same test suite also detected 200 non-seeded errors. What is the estimated number of undetected errors in the code after this testing? [GATE - 04]

a) 4 b) 50 c) 200 d) 250

9. What is the availability of a software with the following reliability figures ? [GATE - 04]

Mean Time Between Failure (MTBF) = 25 days

Mean Time To Repair (MTTR) = 6 hours

a) 1% b) 24% c) 99% d) 99.009%

10. The Function Point (FP) calculated for a software project are often used to obtain an estimate of Lines of Code (LOC) required for that project. Which of the following statements is FALSE in this context. [GATE - 05]

a) The relationship between FP and LOC depends on the programming language used to implement the software.

b) LOC requirement for an assembly language implementation will be more for a given FP value, than LOC for implementation in COBOL.

c) On an average, one LOC of C++ provides approximately 1.6 times the functionality of a single LOC of FORTRAN

d) FP and LOC are not related to each other

11. Consider the entites 'hotel room', and 'person' with a many to many relationship 'lodging' as shown below fig:

If we wish to store information about the rent payment to be made by person(s) occupying different hotel rooms, then this information should appear as an attribute of

a) Person b) Hotel Room [GATE - 05]

c) Lodging d) None of these

12. The availability of a complex software is 90%. Its Mean Time Between Failure (MTBF) is 200 days. Because of the critical nature of the usage, the organization deploying the software further enhanced it to obtain an availability of 95%. In the process, the Mean Time to Repair (MTTR) increased by 5 days. [GATE - 05]

What is the MTBF of the enhanced software

a) 205 days b) 300 days

c) 500 days d) 700 days

13. To carry out white box testing of a program, its flow chart representation is obtained as shown in the figure below : [GATE - 05]

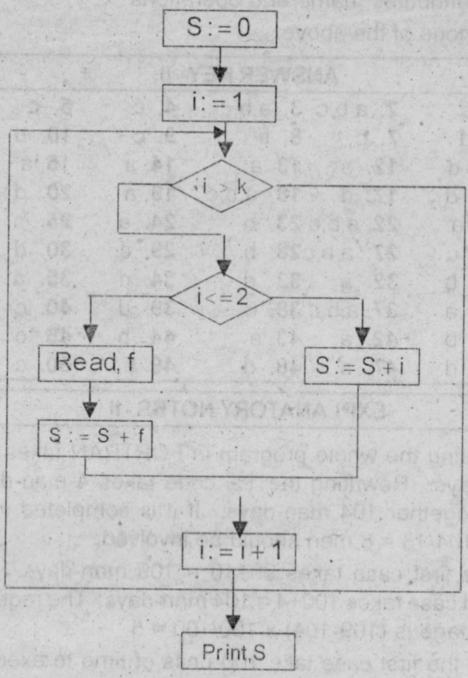

For basis path based testing of this program, its cyclomatic complexity is

a) 5 b) 4 c) 3 d) 2

14. In a data flow diagram, the segment shown below is identified as having transaction flow characteristics, with P2 identified as the transaction center

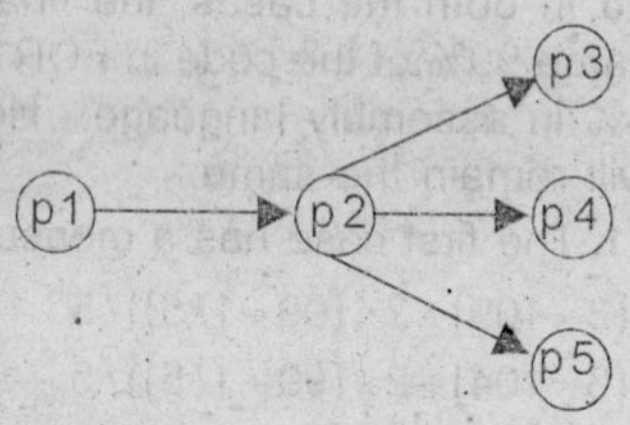

A first level architectural design of this segment will result in a set of process modules with an associated invocation sequence. The most appropriate architecture is [GATE - 05]

a) p1 invokes p2, p2 invokes either p3, or p4, or p5

b) p2 invokes p1, and then invokes p3, or p4, or p5

c) A new module Tc is defined to control the transaction flow. This module Tc first invokes p1 and then invokes p2. P2 in turn invokes p3, or p4, or p5

d) A new module Tc is defined to control the transactiion flow. The module Tc invokes p2. P2 invokes p1, and then invokes p3, or p4, or p5

ANSWER KEY				
1. c	2. a	3. b	4. b	5. b
6. c	7. d	8.	9. b	10. c
11. c	12. c	13. c	14. c	

PRACTICE QUESTIONS

SET - I

1. The goal of software engineering is
   (a) Quality (b)
   Productivity
   (c) Cost effective
   (d) All
2. According to IEEE ________ can be defined as "systematic approach to the development , operation maintenance and retirement of software.
   (a) software
   (b) software engineering
   (c) program
   (d) none
3. ______ is an important characteristic of high quality software component.
   (a) reusability
   (b) inheritance
   (c) quality
   (d) reliability
4. ______ software is a collection of programs written to service other programs.
   (a) real time software
   (b) embedded software
   (c) system software
   (d) none
5. ______ is an example for system software
   (a) compilers
   (b) editors
   (c) file management utilities
   (d) all
6. _______ software that monitors/analyzers/controls real world events as they occur.
   (a) real time software
   (b) embedded software
   (c) system software
   (d) none
7. Embedded software resides in ______
   (a) RAM (b) ROM
   (c) Harddisk (d) none
8. ________ model can be called as "meta model" in the development
   (a) prototype (b) RAD
   (b) life cycle (d) incremental
9. _______ measure accomodates applications in which algorithmic complexity is high.
   (a) feature points (b) function point (c) LOC
   (d) none
10. _______ is the formula for finding productivity for function oriented metrics.
    (a) FP/PM (b) defects / FP (c) $ / FP
    (d) DOC/FP
11. Example for the direct measures of the software engineering process is ______
    (a) cost (b) effort
    (c) KLOC (d) all
12. During the software engineering maintenance phase focuses on _________.
    (a) what (b) how
    (c) change (d) none
13. ______ is developed and controlled to ensure that complete information about the system and software will be available for later use
    (a) documentation
    (b) reviews
    (c) change control
    (d) none
14. ________ and model paradigm for software engineering is currently the most realistic approach to the development for large scale systems and software.
    (a) life cycle
    (b) prototype
    (c) UGT
    (d) spiral
15. ______ model takes risks into consideration, in the software engineering paradigms.
    (a) waterfal
    (b) spiral
    (c) RAD
    (d) incremental
16. _______ process translates requirements into a representation of software.
    (a) analysis
    (b) design
    (b) coding (d)
    planning
17. _______ are the glue that holds the methods and tools together and they enable rational and timely development of computer software in the software engineering.
    (a) tools
    (b) methods
    (c) procedures
    (d) none
18. ________ are the examples of descriptive languages for artificial neural networks.
    (a) APL (b) LISP
    (c) prolog (d) all of the above
19. ________ is an examples for object oriented languages.
    (a) C++ (b) C
    (c) pascal (d) COBOL
20. ______ language implies a program by specifying the desired result rather than specifying action required to achieve the result.
    (a) non procedural
    (b) procedural
    (c) high level
    (d) machine level

21. Multiprogramming and multiuser systems are introduced at ________ era.
(a) first ear (b)
second area
(c) third era
(d) fourth era.

22.________ softwares are the examples of the personnel computer software.
(a) word processing
(b) spreadsheets
(c) business applications
(d) all

FILL IN THE BLANKS

1. ________ is an important characteristics of high quality software component.

2. ______ and ______ are the two basic forms of computer software.

3. ________ and ________ are important factors in determining the nature of a software application.

4. _______ refers to the meaning and form of incoming and outgoing information.

5._______ algorithms are used by artificial intelligence to solve complex problems that are not amenable to computation.

6. Acronym for CASE is ___

7. ________ is a valid model of the development, ______ model of the development, if the system can be fully specified at the beginning.

8. If the requirements are uncertain, ________ model can be used to define the requirements completely.

9. ________ is the aim of feasibility study.

10. Formulae for the productivity and quality according to size oriented metrics________.

11. _______ is the probability that an attack of a specific type will occur with in a given time.

12. ______ is the probability that the attack of a specific type wi!' be repelled.

13. ______ can be measured throughout the software engineering process and after the software has been released to the customers and users.

14. The effort required to learn, operate, prepare input and interpret output of a program is known as________.

15. The extent to which access to software or data by unauthorized persons can be controlled known as ___________.

16. ________ is the measure for maintainability.

17. ________ is the measure for correctness.

18. ________ metric accommodates applications in which algorithmic complexity is high.

19. ________ metrics are indirect measures of software.

20. ________ is the measure for finding the length of a program.

21. ________ software metrics are direct measures of software.

22. ________ is the measure for the effort of a program.

23. According to generic view of software engineering phase focuses on ________.

24. According to generic view of software engineering focuses on ________.

25. ________ is a toy implementation of a system, usually exibiting limited functional capabilities, low availability and inefficient performance.

26. Function points are derived using an _______ relationship based on countable measures of software's information domain and assessments of software quality.

ANSWER KEY

1. (d) 2. (b) 3. (a) 4. (c) 5. (d)
6. (a) 7. (b) 8. (a) 9. (a) 10. (a)
11. (d) 12. (c) 13. (a) 14. (d) 15. (b)
16. (b) 17. (c) 18. (d) 19. (a) 20.a)
21. (b) 22. (d)

Fill In The Blanks

1. Reusability
2. Non machie executable components, machine executable components
3. information content, determinacy
4. information content
5. non numerical algorithm
6. computer aided software engineering
7. classic lifecycle model
8. prototype
9. to determine product is technically and financially feasible
10. P = KLOC/pm, Q = defects/KLOC
11. Threat 12. Security
13. Quality
14. usability 15. integrity
16. MTTC
17. defects per KLOC 18. Feature point 19. Function point
20. KLOC 21. Size oriented 22. person month
23. What to do 24. how to do
25. prototype
26. emperical.

SET - II

1. During planning, the project manager estimates _______ paramets.
(a) size
(b) Effort
(c) Duration
(d) Cost
(e) All of the above
(f) None

2. The following is the bridge between computer system engineering and software design ________
(a) Requirement analysis
(b) Planning
(c) Scheduling
(d) Estimation

3. _________ is the example for software specification language
(a) RSL
(b) IORL
(c) GYPSY (d) All of the above
(e) None

4. Preliminary user manual refers software as _______
(a) White box (b) Black box
(c) Green box (d) Red box

5. ________ describes function, performance, constraints, interfaces, and reliability
(a) S/W planning
(b) Analysis
(c) Software scope
(d) Project complexity

6. _______ take a "devide and conquer" approach to software project estimation
(a) Decomposition techniques
(b) Empirical models
(c) Heuristic techniques
(d) Autimated estimated tools

7. Effort is measured in __________
(a) Dollars (b) Person months
(c) kLOC (d) Defects

8. _______ models consist of one or more empirically derived equation that predict effort, project duration or other project data
(a) Analysis (b) Data
(c) Resource (d) None

9. _________ is an example of a static single - variable model
(a) PUtnam (b) Cocomo
(c) LOC (d) FP

10. _________ is an example of a dynamic - multi variable model
(a) PUtnam (b) Cocomo
(c) LOC (d) FP

11. _________ Cocomo computes software development effort as a function of program size expressed in estimated lines of code
(a) Basic (b) Intermediate (c) Advanced (d) None

12. According to Boehm, the complexities of organic mode, semi detached mode, embedded mode are in the ratio of _________
(a) 1 : 2 : 4 (b) 4 : 2 : 1 (c) 1 : 3 : 9
(d) 2 : 4 : 6

13. According to putnam estimation model, the technology constant Ck = 2000 refers to _________
(a) Poor S/W development
(b) Good S/W development
(c) Excellent environment
(d) None

14. Automated estimated tools B4L, WICOMO, BEC plan are based on _________
(a) Cocomo (b) Putnam (c) LOC
(d) FP

15. Automated estimated tool SLIM is based on _____
(a) COCOMO (b) PUTNAM
(c) LOC (d) FP

16. _________ risks identify potential budgetary, schedule, personnel, resource, customet and requirements problems and their impact on the S/W project
(a) Project (b) Program (c) Technical
(d) Business

17. ________ attempts to rate each risk in likelihood and the consequences of the problems associated with the risk
(a) Risk projection
(b) Risk assessment
(c) Risk management
(d) Risk identification

18. ________ performs or coordinates each of the tasks associated with software requirements analysis
(a) Programmer
(b) Project manager
(c) System analyst
(d) Designer

19. ________ represents the individual data and control items that comprise some larger item of information
(a) Information flow
(b) Information content
(c) Information structure
(d) All of the above

20. ________ should be conducted regardless of S/W engineering paradigm that is applied
(a) Analysis (b) Design
(c) Coding (d) Testing

21. _________ is produced at the culmination of the analysis task
(a) Documentation
(b) Prototype
(c) SRS
(d) Solution

22. Software requirement specification as a _________ betweeen customer and developer
(a) Legal contract
(b) 'Signed off"
(c) a & b
(d) None

FILL IN THE BLANKS

1. Lack of __________ are primary causes of schedule slippage, cost overrans, poor quality, high maintenance for software.

2. __________ is the element which is used as the estimated variable in the LOC & FP estimation.

3 Formula for finding expected value for LOC or FP.

4. ________is an example for statistic single variable model.

5.__________ model is an example for dynamic multivariable model.

6 __________ are the three important classes of software products according to Boehm.

7. Effort distribution for staffing-level estimation can be computed by using __________ model.

8. B4L, WIMCO are automated tools that are based on __________ method.

9. ________ is an automated costing system based on Reyleigh Nordon curve for the software life cycle.

10. ESTAMICS uses a __________ estimation method enhanced to accomodate a variety of project and.

11. The legal document between customer and developer is __________ when the project planning is done.

12. __________ is conducted for customer and developer assessment of the product.

13.________ can be drafted for cases in which a prototype has not been developed.

14. ________ chart representation of a project schedule is helpful in planning the utilization of resources in the scheduling.

15. "80 percent of the overall project risk can be accomodated for by only 20 percent of identified risks" known as __________.

16. __________ means every milestone is critical to meeting the project deadline in scheduling.

17. ________ is the triple used in the risks analysis.

18. At the time of referent point what type of decision will be taken by the Analyst __________

19. ________ techniques takes a divide and conquer approach to software project estimation.

20. ________ and __________ are the two basic activities of S/W project planning.

21. Examples of Decomposition techniques are __________

22. Examples of problem based Estimation techniques are __________

23. Example of process based Estimation technique is ________.

24. __________ is multiplied in Basic COCOMO equation to obtain intermediate COCOMO equation.

25. Formula for the putnam effort estimation is __________

26. Evaluation and solution synthesis step of analysis tasks focuses on __________

27. __________ is done to check and identify a successful software implementation.

ANSWER KEY

1. (e) 2. (a) 3. (d) 4. (b)
5. (c) 6. (a) 7. (b) 8. (c) 9. (b)
10. 11. (a) 12. (c) 13. (a) 14.
(a) 15. (b) 16. (a) 17. (a) 18. (c)
19. (b) 20. (a) 21. (c) 22. (c)

Fill In The Blanks

1. Project planning
2. Size of a program
3. $\left[E = \dfrac{a + u m + b}{6}\right]$
4. COCOMO
5. putnam
6. Organic, Semidetached, Embedded
7. Putnam estimation
8. COCOMO
9. SLIM
10. function point
11. Software requirements specification
12. Prototy pe
13. Preliminary user is manual
14. GANTT Chart
15. Pareto 80/20 rule
16. Critical path
17. [ri, li, xi]
18. either to stop the project or continue the project
19. Decomposition
20. Software scope, Estimating resources
21. LOC & FP
22. LOC & FP
23. Effort estimation
24. Effort adjustment factor
25. $\left[K = \dfrac{L^3}{C_k^3} t_d^4\right]$
26. "What but not now"
27. Validation criteria

SET - III

1. Objects manifest themselves in one of the ways___________
   (a) External entities
   (b) Things
   (c) Occurences
   (d) Events
   (e) Roles
   (f) All of the above
2. Objects name is determined by understanding _______
   (a) Nouns (b) Verbs
   (c) Adverbs
   (d) Adjectives
3. Operations of an object can be identified by using ___________
   (a) Verbs
   (b) Nouns
   (c) Adverbs
   (d) None
4. _______ changes the attributes values that are contained within an object
   (a) Class
   (b) Object
   (c) Operation
   (d) None
5. _________ are three categories of operations generally used in the analysis
   (a) Manipulate
   (b) Computation
   (c) Monitor (d) All of the above
6. Reusability is the property of _________
   (a) Encapsulation
   (b) Abstraction
   (c) Inheritance
   (d) Polymorphism
7. _________ is the combination of data and logic that represents real world entity
   (a) Class
   (b) Object
   (c) Instance
   (d) None
8. _______ diagram enables the analyst to represent information hierarchy in a compact manner
   (a) Warnier diagram
   (b) Entity diagram
   (c) Assembly line diagram
   (d) None
9. According to DSSD approach, the _________ is an entity diagram depicts a producer or consumer of information
   (a) Rectangle
   (b) Circle
   (c) Square (d) None
10. According to Warnier, information hierarchy is represented using ___________
    (a) Sequence
    (b) Selection
    (c) Repetition
    (d) All of the above
11. DFD starts at ___________ level
    (a) 0th level (b) 1st level
    (c) 2nd level (d) nth level
12. ___________development evolved out of work conducted on information domain analysis and its relationship to program and system design
    (a) Jackson system
    (b) Data Structure system
    (c) Data Structure oriented
    (d) Structured
13. Actions that affect each entity are ordered by time and represented with _________diagrams
    (a) Warnier
    (b) Assembly line
    (c) Jackson
    (d) System specification
14. _________ diagram is developed by beginning with the last numbered information flow and working backward showing the processing that derives the proceding numbered information item.
    (a) Warnier (b) Assembly line
    (c) Jackson (d) System specification
15. _________ symbol indicates a conditional occurance (selection) of an information item in the Warnier diagrams.
    (a) { (brace)
    (b) ( ) (paranthesis)
    (c) + (exclusive - or)
    (d) none
16. The corner stone notation for data modling is the _________ diagram
    (a) Warnier - orr
    (b) E - R diagram
    (c) Jackson diagram
    (d) All of the above
17. A mechanism for communication must be established between objects known as _______
    (a) Message
    (b) Information
    (c) Parameters
    (d) Arguments

FILL IN THE BLANKS

1. ___________ is the process of extracting the needs of a system and what the system must do to satisfy the user's requirements.
2. _________ is a combination of data and logic that represents some real world entity.
3. _________ is a set of objects that share a common structure and common behavior.
4. A single object is simply an instance of a _______
5. The data internal to an object are often called _________ of an object and functions supported by an object are called its ___________
6. __________ consists of focusing on the essential, inherent aspects of an entity and ignoring its acidental properties.

7. __________ consists of separating the external aspects of an object, which are accessible to other objects, from the internal implementation details of an object, which are hidden from other objects.

8. ____________ is the property, that allows objects to be built from other objects.

9. The same operation may behave differently on different classes is known as ____________

10. According to Coad & Yourdan "object-oriented" can be defined as__________

11. A file or database can provide support for objects having a longer life time, longer than the duration of the process for which they were created, this characteristic is called ____________

12. ________ is the relationship between a class and one or more refined versions of it.

13. ________ is the "a-part-of" relationship in which objects representing the components are associated with an object representing the entire assembly.

14. Objects are determined by understanding ________

15. Inter object communication can be done by passing ____________.

16. According to OOAD, a subject is nothing more than ________ or ________ to more detail in the analysis model.

17. According to Data modeling, Data object encapsulates data only, there is no reference with in a data object to ________ that act on the data.

18. ____________ focuses on solely on data, representing a data network that exists for a given system.

19. Attributes of a data object represents elementary data items, they should contain no__________

20. ________ analysis methods representing software requirements by focusing on data structure rather than data flow.

21. According to Warnier information hierarchy is represented using ________, ________ and ____________

22. DSSD was developed by__________

23. According to DSSD, ____________ representation is used to differentiate levels of information hierarchy.

24. ____________ development focuses on information domain analysis and its relationship to program and system design

25. According to structured analysis, ________ represents list of all data items appearing in a Data flow diagram.

26. According to data flow diagrams, "context diagram" represents____________ level of DFD

27. According to DSSD approach, __________diagram provides a mechanism for coupling information and the process that are applied to it.

ANSWER KEY

1. (f) 2. (a) 3. (a) 4. (c) 5. (d) 6. (c) 7. (b) 8. (a) 9. (b) 10. (d) 11. (a) 12. (a) 13. (c) 14. (b) 15. (c) 16. (b) 17. (a)

Fill In The Blanks

1. Analysis
2. Object
3. Class
4. Class
5. attributes, methods
6. Abstraction
7. Encapsulation
8. Inheritance
9. polymorphism
10. objects + classification + inheritance + communication
11. object persistence
12. Generalization
13. Aggregation
14. noun / noun clause
15. messages
16. reference, pointer
17. operations
18. data modeling
19. internal structure
20. Data structure oriented analysis
21. sequence, selection, repetition
22. Warnier-Orr
23.
24. Jackson system development
25. Data dictionary.
26. 0th level
27. assembly line diagram

SET - IIII

1. _______ sits at the technical kernel of the software engineering process is applied regardless of the development paradigm is applied.
(a) analysis
(b) design
(c) coding (d) testing
(e) maintenance

2. _______ is the way, can be accurately translated a customer's requirements into a finished software product or system.
(a) Design (b) coding
(c) testing
(d) none

3. _______ design concerned with the transformation of requirements into dat and software architecture.
(a) Architectural
(b) Data
(c) Preliminary
(d) Procedural
(e) None

4. _______ design estabishes the layout and interaction mechanisms for human-machine interaction.
(a) interface
(b) Data
(c) preliminary
(d) none

5. Throughout the design process_______ of the evolving design is assessed with a series of format technical reviews or walk throughs
(a) integrity (b) quality
(c) reliability
(d) maintainability

6. A design should load to _______ that reduce the complexity of corrections between modules and with the external environment.
(a) hierarchical (b) modular
(c) interfaces (d) repeatable

7. _______ abstraction is a named sequence of instructions that has a specific and limited function
(a) procedural
(b) data
(c) control (d) none

8. _______ is an early top-down design strategy proposed by Niklous with
(a) abstraction
(b) refinement
(c) modularity
(d) none

9. As the number of modules grows effort associated with _______ the modules also grows.
(a) cost (b) staff
(c) interfacing (d) documentation

10._______ is a representation of the logical relationship among individual elements of data
(a) control hierarchy
(b) S/W architecture
(c) refinement
(d) data structure

11. _______ is the principle suggests that modules be characterised by design decisions that hides from all others
(a) abstraction
(b) information hiding
(c) modularity
(d) None

12._______are the characteristics of the module.
(a) Time history incorporation
(b) activation mechanisms
(c) pattern of control
(d) all of the above
(e) none

13._______ modules maintain an entry pointer that allows the module to restart at the point of interruprion.
(a) Sequential (b) incremental
(c) parallel (d) None

14. Words such as "initialize", "cleanup" represents _______ binding.
(a) Logical (b) temporal
(c) procedural (d) functional

15. If the predicate of the sentence doesn't contain a single specific object following the verb, the module is probably _______ bound
(a) Logical (b) coincidental
(c) communicational
(d) functional

16. _______ coupling occurs when one module makes use of data or control information maintained within the boundary of another module.
(a) Data (b) stamp
(c) common (d) content

17. Information enters the system along paths that transform external data into an internal form and are identified as _______
(a) incoming flow
(b) outgoing flow
(c) Transaction
(d) None

18. _______ is a component of the real world that is mapped into the software domain
(a) class (b) object
(c) instance (d) entity

FILL IN THE BLANKS

1._______ design transforms the information domain model created during analysis into the data structures that will be required to implement the software.

2. _______ design defines the relationship among major structural components of the program.

3. _______ design transforms the structural components into a procedural description of the software.

4. A design should exhibit _______ organization that makes intelligent use of control among components of the software.

5.A design should be _______ because the software should be logically partitioned into components that perform specific functions and sub functions.

6. According to M.A. Jackson S/W design concepts provide necessary framework for__________
7. At the high level of abstraction, solution stated in ________ and at the lowels level of abstraction, solution is stated in __________
8. As the number of modules grows, the effort associated with___________ the modules also grows.
9. Software procedure focuses on the ________ details of each module individual
10. ________ represents the organization of program components and implies a hierarchy of control.
11.________ module performs a single task with in a software procedure, requiring little interaction with procedures being performed in other parts of a program.
12.Words such as "initialize", cleanup" imply ______ cohension.
13.Words relating to time such as "first", "next", "then", "when", "short" etc. imply the module probably has ________ or ________ cohesion.
14.If the sentence has to be a compound sentence, contains comma or contains more than one verb, the module is probably performing more than one function, it has_________ or _________cohesion.
15. ___________ is a measure of interconnection among modules in a software structure.
16.Two modules are ___________ coupled, if they communicate via an elementary data item which is passed between the two modules.
17. Two modules are __________ coupled, if they communicate via composite data item.
18. Two modules __________ coupled, if they share some global data areas.
19. A module having__________ cohension and __________ coupling is said to functionally independent.
20. ________ is a representation of the logical relationship among individual elements of data.
21.The principle of _________ suggests that modules be characterised by design decisions that hide from all others.
22.___________ module that is referenced and executed without apparent interruption by the application.
23. _______ module that is referenced and executed without apparent interruption by the application
24. A module is said to have__________ cohesion if it performs a set of tasks that are related to each other very loosly
25. A module is said to be __________ cohesive, if all the elements of the module performs similar operations.
26. _____________ coupling exists between two modules, if their code is shared between two modules.
27. __________ coupling exists between if data from one module is used to direct the order of execution of instructions in another.
28. ______________ description of an object that establishes the interface of an object by defining each message that the object can receive and the related operation that the object performs when it receives the message.
29. According to design heuristics, attempt to minimize structures with high __________, strive for __________ as depth increases.
30. To achieve low coupling the number of ________ between modules should be minimized
31. __________ and ___________ are the different types of help facilities available in HCID.
32. __________ and __________ are the two important characteristics of system response time according to HCID.
33. The length and complexity of the written specification of the system and its interfaces provide an indication of amount of _______ required.
34. The number of commands specified and the average of arguments per command provide an indication of ________ and _________ the system.

ANSWER KEY

1. (b) 2. (a) 3. (c) 4. (a) 5. (b)
6. (c) 7. (a) 8. (b) 9. (c) 10. (d)
11. (b) 12. (d) 13. (b) 14. (b) 15. (a)
16. (d) 17. (a) 18. (b)

Fill in the Blanks

1. data design
2. architectural
3. procedural design
4. hierarchial
5. modular
6. getting program right
7. more procedure orientation, direct implementation.
8. interfacing
9. processing
10. control hierarchy
11. cohesive
12. Temporal
13. Sequential, temporal
14. Sequential, communications
15. coupling
16. data
17. common
18. common
19.
20. data structure
21. information hiding
22. sequential
23. incremental

24. coincidental
25. logically
26. content
27. control
28. protocol
29. fanout, fanin
30. interfaces
31. integrated, addon
32. length, variability
33. learning
34. interaction time, overall efficiency

SET - V

1. Software requirements are the foundation from which ________ is measured
(a) Quality (b) Reliability
(c) Correctness
(d) Productivity

2. _______ are the examples for the software quality factors that can be directly measured
(a) Errors
(b) kLOC
(c) Unit time
(d) All of the above

3. _______ are the examples for the software quality factors that can be measured indirectly.
(a) Reliability
(b) Usability
(c) Maintainability
(d) Quality
(e) All of the above

4. According to McCall, the categorization of factors that affect software quality are _______
(a) Product operations (b) Product transition
(c) Product revision
(d) a,b,c (e) none

5. A walkthrough sessions should be limited to
(a) 2 hours (b) 4 hours
(c) 6 hours (d) none

6. _______ is the quantitative measure of software quality.
(a) Correctness
(b) Efficiency
(c) Usability
(d) Cyclomatic complexity

7. __________ is an incorrect intermediate state that may have been entered during program execution.
(a) Failure (b) Fault
(c) Error
(d) None

8. ________ testing method enables the test care designer to derive a logical complexity measure of a procedural design and use this measure as a guide for defining a set of execution paths.
(a) Basis path
(b) Condition
(c) Data flow
(d) None

9. _________ engineering can extract design information from source code, but the abstraction level, the completeness of the documentation the degree to which tools and human analyst work to gather and the directionality of the process are highly variable.
(a) Reverse engineering
(b) Reengineering
(c) System engineering (d) Software engineering

10. __________ combines the analysis and design extaction features of reverse engineering with a restructuring capability for program data, architectural and logic.
(a) Reverse engineering
(b) Reeingineering
(c) Software engineerng
(d) All of the above

11. Reengineering also called as_________
(a) Renovation
(b) Reclamation
(c) Revocation
(d) a, b (e) b, c

12. According to Vassey and Weber program complexity and programming style were the most significant factors in the occurance of ________ maintenance.
(a) Corrective
(b) Adaptive
(c) Perfective
(d) Preventive

13. _________ maintenance is characterised by reverse engineering and re-engineering techniques.
(a) Corrective (b) Adaptive
(c) Preventive (d) Perfective

14. ________ has one overriding objective "to find and correct the cause of a software error ".
(a) Testing (b) Debugging
(c) Designing
(d) Maintenance

15. ________ is a fairly common debugging approach that can be used successfully in small programs.
(a) Brute force
(d) Back tracking
(c) Cause elimination
(d) Program slicing

16. _________ tests are designed to confront programs with abnormal situations
(a) Recovery
(b) Security
(c) Stress
(d) Performance

17. _________ testing attempts to uncover data combinations with in valier input classes that may cause instability or improper processing
(a) System (b) Stress
(c) Recovery
(d) Security

18. ________ testing is actually a series of different test whose primary purpose is to fully exercise the computer based system
(a) White box
(b) Black box
(c) Integration
(d) System

19. _________ test is conducted at one or more customer sites by the end user of the software
(a) Validation
(b) Alpha
(c) Beta
(d) Gamma

20.________ test is conducted at the developer's site by a custormer
(a) Alphe
(b) Beta
(c) Gamma
(d) None

21.The number of possible tests _________ geometrically as the level of resting of program complexity increases
(a) Decreases
(b) Grows
(c) Same
(d) None

22.______ testing method can be applied to a detailed procedural design or to source code
(a) Basis path
(b) Equivalence partitioning
(c) Boundary value analysis
(d) None

23.________ cannot show the absence of defects, it cannot only show that S/W defects are present
(a) Coding (b)
Testing
(c) Debugging
(d) Maintenance

24.________ is an "umbrella activity" that is applied throughout the software engineering process
(a) Quality (b)
Testing
(c) Software quality assurance
(d) Designing

FILL IN THE BLANKS

1. ________ is a planned and systematic pattern of actions that are required to ensure quality in software.
2. Software requirements are the foundation from which __________ is measured.
3. Hewlett-packard has developed a set of software quality factors has been the acronyms__________
4. The central activity that accomplishes quality assesment is the _______
5.__________ is the stylized meeting conducted by techincal staff with the sole purpose of uncovering quality problems.
6. Strategy with a series of test care design methods that help ensure effective ________ error detection.
7. ________ and ________ for software quality assurance, provide procedures for the collection and dissemination of SQA information.
8. According ti IEEE standard 982.1-1988 , the airforce uses information obtained from data and architectural design to derive a design structure quality index that ranges from _________
9. According to IEEE 982.1-1988 suggests ______ that provides an indication of the stability of the S/W product.
10. SMI can be computed by using ________
11. According to Stead's software science, the length of a program can be found by calculating the total number of ________in the program.
12. ________ is the quantitative measure of software quality.
13._______ is a dynamic system characteristic which is the function of the number of software failures.
14. If the ______ of the software is increases, it reduces the program execution speed and increases the amount of store required by the program.
15. The goal of _______ is to design a series of test cases that have high likelihood of finding errors.
16. _________ is a manifestation of an error.
17. __________ is an incorrect intermediate state thay may have been entered during program execution.
18. ___________ is a triplet [I, S, O] here I is the data input to the system, S is the state is the system at which the data is input, and O is the expected output of the system.
19. Testing is a process of executing a program with the intent of finding on ________
20. Knowing the specified function that a product has been designed to perform, tests can be conducted that demonstrate each function is fully operational while at the same time searching for errors in functions is called ________
21. Knowing the internal working of a product, tests can be conducted to ensure that all internal operations are performed according to specifications and all internal components have been adequately exercised is called ______
22. Black bos testing alludes to tests that are conducted at the __________
23. White box testing of software is predicted to close examination of ________ detail.
24. White box testing, sometimes called ________ testing
25. Basis path testing is a_________testing technique proposed by JOM Mc Cabe.
26. The basis path testing method enables the test care designer to derive a ____________ measure of a procedural design.
27. Cyclomatic complexity, V(a) for a graph G, is defined as ___________
28. The number of regions of a flow graph correspond to the _________
29. ________ matrix is nothing more than a tabular representation of a flow graph.

30. _________ possible tests are required, for a boolean expression with n variables.

31. _______ testing technique guarantees the detection of branch and relational operations errors in a condition provided that all boolean variables and relational operators in the condition occur only once and have uncommon variables.

32. _______ testing method selects the test paths of a program according to the locations of the definitions and uses of different variables in a program.

33. _________ testing technique loads to a selection of test cases that exercise boundary values.

34. ________ testing involves delivering a system to a number of potential customers who agree to use the sytem and to report problems to the system developers.

35. _______ testing is a system test that forces the software to fail in a variety of ways and verifies that recovery is properly performed.

36. _______ testing attempts to verify that protection mechanisms built into a system will protect it from improper penetration.

37. _______ tests are designed to confront programs with abnormal situations or to handle a specified load.

38. _______ is the process of isolating and correcting the causes of known errors.

39. ________ is the debugging approach method, in which the program is loaded with print statements to print the intermediate values with the hope that some of the printed values will help to identify the statement in error.

40. The process of modifying a program after it has been delivered and is in use is known as ______

41. The process that includes the diagnosis and correction of one or more errors is called ________ maintenance.

42. An activity that modifies S/W to properly interface with a changing environment is called _______ maintenance.

43. Acronym for "SEI - CMM" ________

44. ________ uses design information to alter or reconstitute the existing system in an effort to improve its overall quality.

45. __________ is not testing, but always occurs as a consequence of testing.

ANSWER KEY

1. (a)	2. (d)	3. (e)	4. (d)	5. (a)
6. (d)	7. (b)	8. (a)	9. (a)	10. (b)
11. (d)	12. (a)	13. (c)	14. (b)	15. (b)
16. (c)	17. (c)	18. (d)	19. (c)	20. (a)
21. (b)	22. (a)	23. (b)	24.(c)	

Fill in the Blanks

1. Software quality assurance
2. Quality 3. FURPS
4. formal technical review 5. FIR
6. S/W testing 7. Record keeping, Reporting
8. 0 to 1 9. Software maturity index

10. $\left[SMI = \left[\frac{M_T - (F_a + F_c + F_d)}{M_T} \right] \right]$

11. [operation & operands]
12. Cyclomatic complexity 13. Reliability 14. Reliability 15. S/W testing 16. Failure 17. Falut
18. Test care 19. Error
20. black box testing 21. white box testing 22. software interface 23. Procedural
24. glass box 25. White box 26. logical complexity

27. $[G(u) = E - N + 2/P + 1]$

28. cyclomatic complexity

29. Graph matrix 30. $\lfloor 2^n \rfloor$

31. BRO
32. Equivalence partitioning
33. Boundary value analysis
34. Beta 35. Recovery
36. Security 37. Stress
38. debugging 39. Brute force
40. maintenace
41. corrective
42. adaptive
43. software engineering institute group capability maturity model
44. Re-engineering
45. debugging

TEST YOURSELF

1. The cyclomatic complexity of each of the modules A and B shown below is 10. What is the cyclomatic complexity of the sequential integration shown on the right hand side ? **[GATE - 10]**

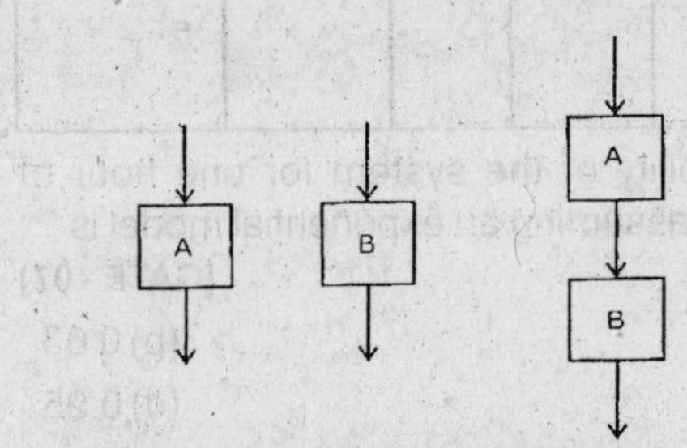

a) 19 b) 21 c) 20 d) 10

2. What is the appropriate pairing of items in the two columns listing various activities encountered in a software life cycle ? **[GATE - 10]**

P. Requirements Capture	1. Module Development and Integration
Q. Design	2. Domain Analysis
R. Implementation	3. Structural and Behavioral Modeling
S. Maintenance	4. Performance Tuning

a) P-3, Q-2, R-4,S-1 b) P-2, Q-3, R-1,S-4
c) P-3, Q-2, R-1,S-4 d) P-2, Q-3, R-4,S-1

3. The coupling between different modules of a software is categorized as follows :

I. Content coupling II. Common coupling
III. Control coupling IV. Stamp coupling
V. Data coupling

Coupling between modules can be ranked in the order of strongest (least desirable) to weakest (most desirable) as follows : **[GATE - 09]**

a) I-II-III-IV-V b) V-IV-III-II-I
c) I-III-V-II-IV d) IV-II-V-III-I

4. Which of the following statements are TRUE ?

I. The context diagram should depict the. system as a single bubble.

II. External entities should be identified clearly at all levels of DFDs.

III. Control information should not be represented in a DFD.

IV. A data store can be connected either to another data store or to an external entity. **[GATE - 09]**

a) II and III b) II and III
c) I and III d) I,II and III

5. Consider the following statements about the cyclomatic complexity of the control flow graph of a program module. Which of these are TRUE?

I. The cyclomatic complexity of a module is equal to the maximum number of linearly independent circuits in the graph.

II. The cyclomatic complexity of a module is the number of decisions in the module plus one, where a decision is effectively any conditional statement in the module.

III. The cyclomatic complexity can also be used as a number of linearly independent paths that should be tested during path coverage testing. **[GATE - 09]**

a) I and II b) II and III
c) I and III d) I, II and III

6. Find if the following statements in the context of software testing are TRUE or FALSE.

(S1) Statement coverage cannot guarantee execution of loops in a program under test.

(S2) Use of independent path testing criterion guarantees execution of each loop in a program under test more than once. **[GATE - 08]**

(a) True, True (b) True, False
(c) False,True (d) False,False

7. Which of the following are NOT considered when computing function points for a software project?

(01) External inputs and outputs

(02) Programming language to be used for the implementation

(03) User interactions

(04) External interfaces

(05) Number of programmers in the software project

(06) Files used by the system **[GATE - 08]**

(a) O2, 03 (b) O1, O5
(c) O4, O6 (d) 02, 05

8. A software project plan has identified ten tasks with each having dependencies as given in the following table:

Task	Depends On
T1	-
T2	T1
T3	T1
T4	T1
T5	T2
T6	T3
T7	T3, T4
T8	T4
T9	T5, T7,T8
T10	T6,T9

Answer the following questions :

(Q1) What is the maximum number of tasks that can be done concurrently ?

(Q2) What is the minimum time required to complete the project, assuming that each task requires one time unit and there is no restriction on the number of tasks that can be done in parallel? **[GATE - 08]**

(a) 5,5 (b) 4,4
(c) 4,5 (d) 5,4

9. A software engineer is required to implement two sets of algorithms for a single set of matrix operations in an object oriented programming language; the two sets of algorithms are to provide precisions of 10^{-3} and 10^{-6}, respectively. She decides to implement two classes, Low Precision Matrix and High Precision Matrix, providing precisions 10^{-3} and 10^{-6} respectively. Which one of the following is the best alternative for the implementation ?

(S1) The two classes should be kept independent

(S2) Low Precision Matrix should be derived from High Precision Matrix

(S3) High Precisoin Matrix should be derived from Low Precision Matrix

(S4) One class should be derived from the other; the hierarchy is immaterial **[GATE - 08]**

(a) S1 (b) S2
(c) S3 (d) S4

10. Which of the following requirement specifications can be validated ?

(S1) If the system fails during any operation, there should not be any loss of data

(S2) The system must provide reasonable performance even under maximum load conditions

(S3) The software executable must be deployable under Ms windows 95, 2000 and XP.

(S4) User interface windows must fit on a standard monitors' screen **[GATE - 08]**

(a) S4 and S3 (b) S4 and S2
(c) S3 and S1 (d) S2 and S1

11. In the Spiral model of software development, the primary determinant in selecting activities in each iteration is **[GATE - 07]**

(a) iteration sive

(b) Cost

(c) Adopted process such as Rational Unified Process of Extreme Programming

(d) Risk.

12. The following table shows the time between failures for a software system.

Error Number	1	2	3	4	5
Time since last failure (hours)	6	4	8	5	6

The reliability of the system for one hour of operation assuming an exponential model is **[GATE - 07]**

(a) 0.45 (b) 0.63
(c) 0.84 (d) 0.95

13. Given the following algorithm for shorting an array X and N numbers :

```
SUBROUTINE SHORT (X, N)
    IF (N < 2)
        RETURN
FOR ( i = 2 ) TO N INCREMENT BY 1)
)
    FOR ( j = 1 ) TO i
INCREMENT BY 1)
        IF ( X [i] > X ( [j] )
            CONTINUE
        TEMP = X [i]
        X [i] = X [j]
        X [j] = TEMP
    END FOR
END SUBROUTINE
```

A good approximation of Halstead's estimated program length is **[GATE - 07]**

(a) 20 (b) 50
(c) 80 (d) 110

14. In the simplified flowchart given below, the shaded boxes represent code that is executed during a test case.

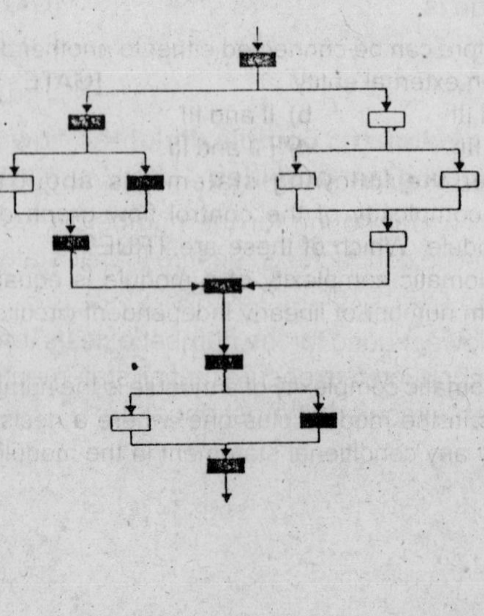

The branch coverage is **[GATE - 07]**

(a) 3 / 4 (b) 2 / 3

(c) 1 / 2 (d) 3 / 8

15. Consider the CPM activity chart where an arc connecting two milestones is labeled with a task identifier and the time taken in days. For example in order to go from A to B, task T1 takes 180 days. A dashed line depicts an additional dependency that is equivalent to a zero time task.

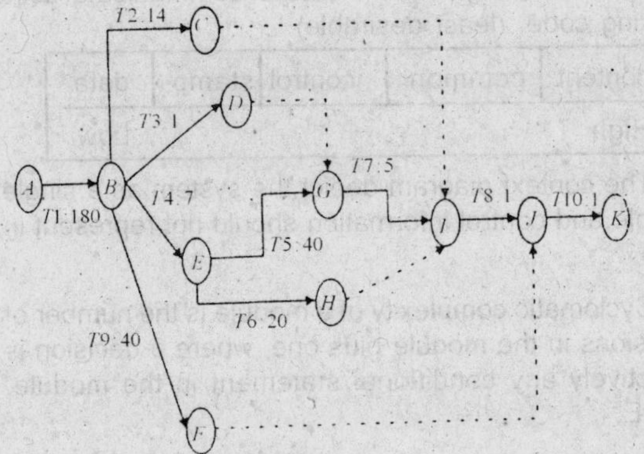

The set of activities that lie on the critical path are **[GATE - 07]**

(a) T1, T2, T8, T10

(b) T1, T3, T8, T10

(c) T1, T2, T3, T4, T5, T6, T7, T8, T9, T10

(d) T1, T4, T5, T7, T8, T10

16. Consider the following pseudo - code :

```
IF ( A > B ) AND ( C > D ) ) THEN
    A = A + 1
    B = B + 1
ENDIF
```

The cyclomatic complexity of the pseudo-code is **[GATE - 07]**

(a) 2 (b) 3

(c) 4 (d) 5

17. The cyclomatic complexity of the flow graph of a program provides **[GATE - 06]**

a) an upper bound for the number of tests that must be conducted to ensure that all statements have been executed at most once.

b) a lower bound for the number of tests that must be conducted to ensure that all statements have been executed at most once.

c) an upper bound for the number of tests that must be conducted to ensure that all statements have been executed at least once.

d) a lower bound for the number of tests that must be conducted to ensure that all statements have been executed at least once.

18. With respect to software testing, consider a flow graph G with one connected component. Let E be the number of edges, N be the number of nodes, and P be the number of predicate nodes of G. Consider the following four expressions : [GATE - 06]

(I) $E - N + P$ **(II)** $E - N + 2$

(III) $P + 2$ **(IV)** $P + 1$

The cyclomatic complexity of G is given by

a) I or III b) II or III

c) II or IV d) I or IV

19. A software program consists of two modules M_1 and M_2 that can fail independently, but never simultaneously. The program is considered to have failed if any of these modules fails. Both the modules are 'repairable' and so the program starts working again as soon as the repair is done. Assume that the mean time to failure (MTTF) of M_1 is T_1 with a mean time to repair (MTTR) of R_1. The MTTF of M_2 is T_2 with an MTTR of R_2. What is the availability of the overall program given that the failure and repair times are all exponentially distributed random variables ? [GATE - 06]

a) $\frac{T_1 T_2}{T_1 R_1 + T_2 R_2}$ b) $\frac{R_1 R_2}{T_1 R_1 + T_2 R_2}$

c) $\frac{T_1 T_2}{T_1 T_2 + T_1 R_1 + T_2 R_2}$ d) $\frac{T_1 T_2}{T_1 T_2 + T_1 R_2 + T_2 R_1}$

20. Consider the following structure chart diagram. The boxes have function names embedded in them, while the variables are indicated along the arcs.

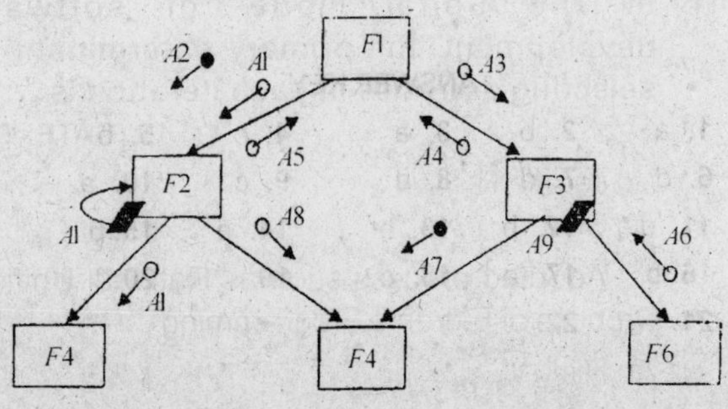

Given below are a set of statements relevant to the above diagram.

I) F3 and F6 can be in the same module.

II) F4 and F6 can be in the same module.

III) A4 is both an output and a control variable.

IV) It is incorrect to pass A1 as data and use it as a control variable.

Which combination of these statements is TRUE ? [GATE - 06]

a) III and IV b) I and IV

c) II and IV d) I, II and Iv

Statement for linked Answer Questions 21 & 22

A software project has four phases P1, P2, P3 and P4. Of these phases, P1 is the first one and needs to be completed before any other phase can commence. Phases P2 and P3 can be executed in parallel. Phase P4 cannot commence until both P2 and P3 are completed. The optimistic, most likely, and pessimistic estimates of the phase completion times in days, for P1, P2, P3 and P4 are, respectively, (11, 15, 25), (7, 8, 15), (8, 9, 22), and (3, 8, 19).

21. The critical path for the above project and the slack of P2 are, respectively. [GATE - 06]

a) $P1-P2-P4$, 1 day. b) $P1-P3-P4$, 1 day

c) $P1-P3-P4$, 2 days d) $P1-P2-P4$, 2 days

22. The costs (in Rupees per day) of crashing the expected phase completion times for the four phases, respectively, are 100, 2000, 50, and 1000. Assume that the expected phase completion times of the phases cannot be crashed below their respective most likely completion times. The minimum and the maximum amounts (in Rupees) that can be spent on crashing so that ALL paths are critical are, respectively. [GATE - 06]

a) 100 and 1000 b) 100 and 1200

c) 150 and 1200 d) 200 and 2000

ANSWER KEY

1. a	2. b	3. a	4. c	5. b
6. d	7. d	8. d	9. c	10. a
11. d	12. b	13. b	14. a	15. b
16. b	17. a	18. c	19.	20.
21.	22.			

EXPLANATORY NOTES

3. Data coupling : Two modules communicate using elementary data types (most desirable)

Stamp Coupling. : Two modules communicate using composite data items (i. e. Array)

Control coupling : Two modules communicate using data items which are used to control the order of execution of one of the module by another.

Common coupling : Two modules communicate using share global data items.

Content Coupling : Two modules communicate using sharing code. (least desirable)

content	common	control	stamp	data
High				Low

4. The context diagram depict the system as a single bubble and control information should not represent in DFD.

5. Cyclomatic complexity of a module is the number of decisions in the module plus one, where a decision is effectively any conditional statement in the module. TRUE.

Cyclomatic Complexity can also be used as a number of linearly independent paths that should be tested during path converage testing. TRUE

6. Both the statements are false

7. While computing function points, project planner estimates inputs, outputs, inquiries, files and external interfaces for the software. Thus programming language to be used for implementation and files used by the system all not considered.

8. T2, T3, T4 depends on T1

T5 depends on T2

T6 depends on T3

T7 depends on T3, T4

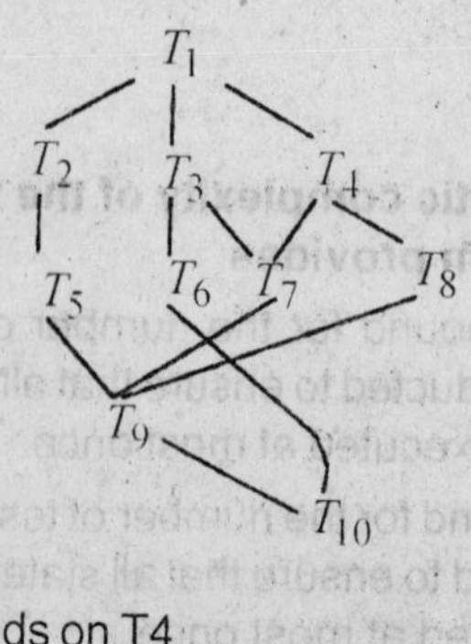

T8 depends on T4

T9 depends on T5, T7, T8

T10 depends on T6, T9

So maximum number of tasks that can be done concurrently is 4, i.e T5,T6,T7,T8

Total time taken to complete the project is 4 units.

10. Validation is the process of evaluating software at the end of the software development to ensure compliance with the software requirements.

    For validation of software, we must have clearly specified conditions which are not given in S1 and S2.

11. In the spiral model of software development, the primary determinant in selecting activities in each interation is risk.

16. Cyclomatic complexity = predicate node + 1

    there are two predicate nodes.

    one is A > B and other is C > D

    Since so, cyclomatic complexity =

    2 + 1 = 3

Unit - XIII
Web Technologies

UNIT - XIII WEB TECHNOLOGIES

XIII. WEB TECHNOLOGIES

I. JAVA SERVER PAGES

Introducing Java Server Pages

1. Integrating applications with web services
2. The Java 2 Enterprise Edition
3. Comparisons with other technologies
4. A First JSP and what's happening under the covers
5. The future development of JSP

The success of the HTTP protocol, in its simplicity, scalability, is obvious.

The current push is to put more of business processes on the web, whether in business-to-consumer (B2C) or business-to-business (B2B) market (BEA recently forecast a $billion market for B2C, and $trillion market for B2B applications). These applications show up some drawbacks in conventional website design, using CGI for instance. The main problems facing developers seeking to develop web applicaions are:

- Scalability - a successful site will have more users than the biggest corporation intranet. And the number of users is rising all the time, so your solutions have to scale.
- Integration of backend data and business logic - the web is just another way to conduct business and so it should be able to use the same middletier and data-access code.
- Manageability - sites just keep getting bigger, and we need some way to manage the content and its interaction with business systems. Object oriented programming techniques have proven their worth in client-server systems and they should help us in web application development.
- Personalization - adding a personal touch to the page isn't just cosmetic, its essential keeping our customer, knowing their preferences, allowing them to configure the information they see, remembering past purchases or frequent search terms are all important in providing feedback and interaction from what is otherwise a fairly one-sided conversation.

The Java 2 Enterprise Edition

J2EE consists of the following APIs, ordered roughly by where in a three-tier app they are used:

- Java server Pages
- Servlets
- Java support for XML
- Enterprise Java Beans (EJBs)
- Java Messaging
- Java Transaction Support
- JavaMail
- Java Naming and Directory Interface (JNDI)
- JDBC
- Java support for CORBA

The J2EE specification meets the needs of web applications because it provides:

- Rich interaction with a web Server.
- Built-in support for sessions, both in servlets and EJBs.
- The use of EJBs to mirror the user interaction with data by providing automatic session and transaction support to EJBs operating in the EJB server.
- Entity EJBs to represent data as an object.
- Seamless integration with the Java data access APIs.
- Flexible template-based output using JSP and XML.

Java Server Pages

The Java Server Pages 1.1 specification provides web developers with a framework to create dynamic content on the server using HTML and XML templates and Java code, which is secure, fast and independent of server platform.

The JSP 1.1 specification extends the Java Servlet API, both form part of the Java 2 Platform Enterprise Edition (J2EE), JSP-specific information can be downloaded from Sun's JSP website at http://java.sun.com/products/jsp/.

How does a JSP do its stuff?

We create a simple HTML form, simple JSP.html, which allows a user to type a number into a form in their web browser and submit the form to the server. The server will respond with "Hello, world!" repeated that number of times followed by a standard footer at the bottom of the page.

The contents of Simple JSP.html are:

```
<HTML>
<HEAD>
<TITLE>Simple JSP Example</TITLE>
</HEAD>
<BODY>
<P> How many times? </p>
<FORM METHOD="GET"ACTION="SimpleJSP.jsp">
<INPUT TYPE="TEXT"SIZE="2"NAME="numtimes">
<INPUT TYPE = "SUBMIT">
</FORM>
</BODY>
</HTML>
```

When viewed in a Web browser, it looks like this:

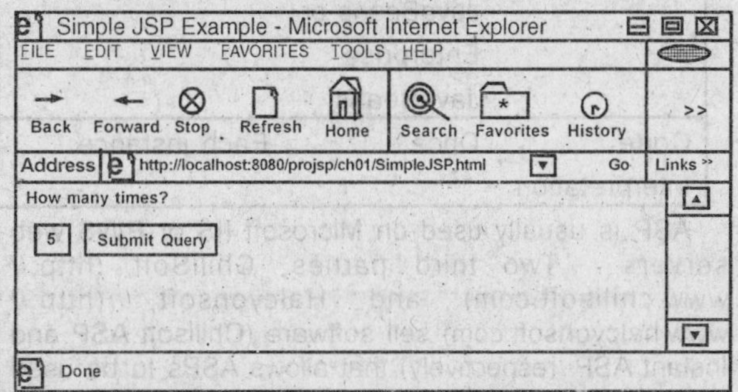

When the user clicks the Submit Query button, the browser sends a request to the web server for Simple

JSP.jsp and includes with its request the value entered by the user. The server interprets the JSP code and creates a response HTML code, which is sent back to the browser.

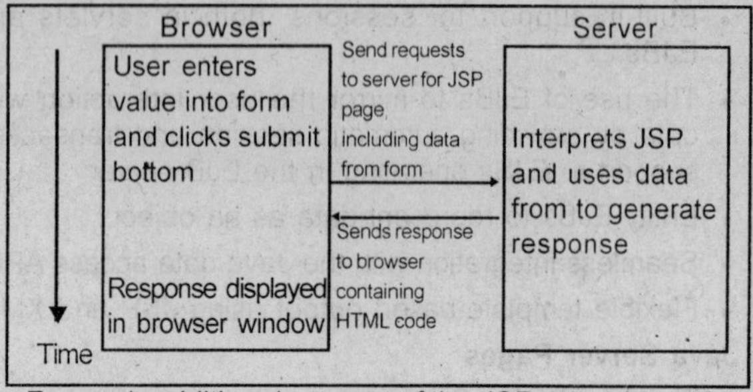

For each additional request of the JSP page thereafter, there is no delay because the request goes to the servlet byte code already in memory.

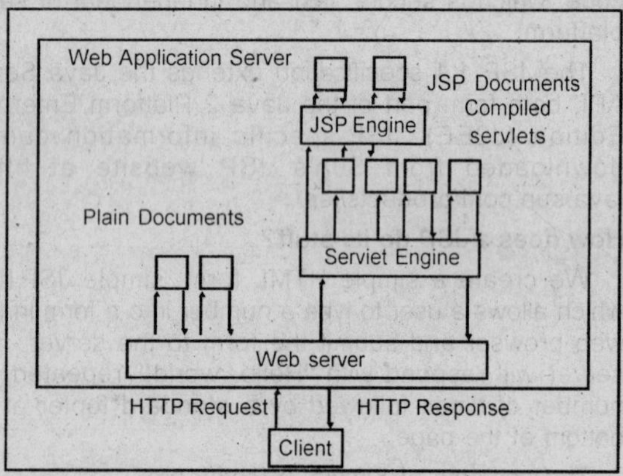

ASP

The main competing technology for JSPs in Active Server Pages (ASP) from microsoft. The technologies are quite similar in the way they support the creation of dynamic web pages, using HTML templates, scripting code and components for business logic. Both also interface to enterprise application frame works, J2EE and Microsoft DNA respectively.

	JSP	ASP
Platforms	All major web Platforms	Microsoft only (see below)
Base Language	Java	Jscript or VBScript
Components	JSP Tags, JavaBeans or Enterprise JavaBeans	Com/DCom
Code Interpretation	Once	Each instance

ASP is usually used on Microsoft IIS or PWS web servers. Two third parties, ChiliSoft (http://www.chilisoft.com) and Halcyonsoft, (http://www.halcyonsoft.com) sell software (Chilisoft ASP and Instant ASP, respectively) that allows ASPs to be used with other platforms. The main problem these companies face is in porting the COM components to the new platform - Halcyonsoft convert them to JavaBeans, or employ a JavaBeans - ActiveX bridge.

JSPs score over ASP in that:

- JSPs are interpreted only once, to Java byte-code, and re-interpreted only when the file is modified.
- JSPs run on all the main web servers
- JSPs provide better facilities for separation of page code and template data by means of JavaBeans, Enterprise Java Beans and custom tag libraries.

But don't be complacent, ASP+ is in the works, Client-side scripting (JavaScript, JScript, VBScript)

Using client-side scripting languages such as JavaScript, JScript, and VBScript brings into play the issue of varying interpretations by different browser versions, a user can also decide to disable scripting altogether. Since JSP runs on the server as such, the browser is not an issue (unless, of course, the JSP produces HTML code which contains client side scripting code).

Servlets

The JSP specification is built on top of the Java Servlet API. Servlets provide the ability to build dynamic content for web sites using the Java language and are supported on all major web server platforms.

While anything that can be done with a JSP can be done with a servlet, JSPs provide a much cleaner separation of presentation from logic, and are simpler to write.

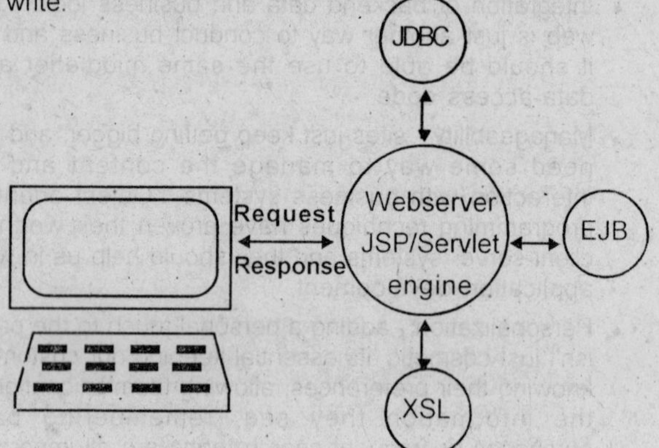

JSP technology is continuing to develop rapidly to include a new features.

- Better support for application localization
- Proper support for including JSP pages, without forcing flushing of buffers
- Application event support
- Improved debugging and tool support
- Improved XML support
- Improved tag library support, with standard tags
- Improved JSP authoring support
- Better support for component composition
- Enabling of WebDAV and WAP requests

JSP SESSIONS

The Hyper Text Transfer protocol (HTTP) is a stateless protocol.

HTTP is the network protocol used to deliver resources on the web to the client, whether they are HTML files, image files, dynamically generated query results, the output of a servlet or JSP, or any CGI scripts.

HTTP is a communication protocol that allows a server and client to understand each other, assuming an underlying TCP/IP connection. A good analogy is two people talking on the telephone, where the conversation is like HTTP and the telephone connection is like TCP/IP. HTTP uses the client-server model; an HTTP client connects to the server over TCP/IP and sends a request message to the HTTP server. The server then returns a response message, usually containing the resource that was requested. After delivering the response, the server can close the connection, as the server's work is done-the conversation is over, the question has been answered. It treats the next request as Independent from any previous request, and serves the requested resource, HTTP is therefore a stateless protocol.

Techniques for maintaining user Information across pages

The idea of maintaining state between different requests is known as session tracking. JSP and servlets may be new, but session tracking is not. Here are some of the techniques used in web applications.

HTTP Information

This included trapping HTTP headers, like the e-mail header and the IP information. The techniques quickly became obsolete with the sprouting of proxy servers and privacy issues.

Hidden Fields

This method is still popular and uses the HTML HIDDEN tag to embed information in each page.

<INPUT TYPE = HIDDEN NAME = "MYKEY" VALUE = "MYVALUE">

Extended Path info and URL-rewriting

This involves appending identification Information to the end of every link on the page and uses the PATH-INFO header to extract it. For example a normal link like:

< a href = "mypage jsp"> Next </a>

would be modified to:

< a href = "/mypag.jsp? login = james & item=book1"> Next </a>

When the link is clicked the login and item name-value pairs are passed to the resource. Not only does this suffer the same drawbacks as hidden fields - identification information used by the server is clearly visible to the client - but there is also a physical limit in bytes on the length of a URL, dictated by the browser and server.

The HTTP1.1 specification RFC 2616 is available from http://www.w3.org and http://www. ietf.org. It doesn't place a limit on the size of the URI (Uniform Resource Identifier) but explicitly states that the server should be capable of handling any size URIs if they support HTTP GET operations. This value is server specific and the server can send back an error code,414 (Requested URI too long), if it is exceeded.

COOKIES

Cookies are small pieces of information deposited on the client by the server. A server, when returning an HTTP resource to a client, may also send a cookie containing some state information, which the client will then store. Included in that state information is a description of the range of URLs for which that state is valid. Any Future HTTP requests made by the client, which fall in that range of URLs, will include a transmission of thye current value of the cookie from the client back to the server.

Cookies were originally specified by Netscape (http://home. netscape.com/newsref/std/ cookie_spec.html), but are now part of an internet standard (RFC 2109, the HTTP state management mechanism).

A cookie is introduced to the client by including a set-cookie HTTP header as part of an HTTP response:

Set-cookie: NAME = VALUE; expires = DATE; path = PATH; domain = DOMAIN_NAME; secure

Where the NAME = VALUE is a name-value pair containing the information and expires = DATE specifies a data string that defines the valid lifetime of that cookie. Consider the earlier example where a client requests a resource on the server and receives this in the response:

Set cookie : CUSTOMER = JAMES_M:

path = /;

expires=Wednesday. 09-NOV-99 23:12:40 GMT

This is stored in the browser in one or two ways depending on its age.

- If the cookie has a well-defined age, it is either written to a permanent storage (that is, appended to the cookies.txt file in Netscape Navigator or stored as an individual text file in the Temporary Internet files directory for Internet Explorer)
- If the cookie is not assigned a particular age, it is stored in memory while the browser program is running. Once the browser is closed the cookie is lost. This type of cookie tracks a user session.

In both cases when the client requests any other resource on the server, a part of the request includes an HTTP header containing details of the cookie, which can be parsed by the server:

Cookie: CUSTOMER = JAMES_M

The Servlet API has a javax. servlet.http. Cookie class that allows server applications to set cookies from JSP in order to maintain state:

Cookie uid = new cookie ("uid", 112kkjk222kjj989c");

response.add cookie (uid);

There are two drawbacks to cookies:

- Users can choose not to accept them. This is really related to privacy issues. With the increased use of cookies to store and pass information between groups and associated vendors, people are becoming more sensitive to cookies and declining to accept them.
- The browser may not work with cookies. Almost all current web browsers, however, accept cookies.

Cookies are harmless and can greatly improve the user experience when used appropriately. People who don't accept cookies should be treated as the special case, not vice-versa.

It is important to remember that cookies have a size limitation in terms of the total bytes per cookies (usually 4 kb), a number of cookies per domain that can be stored in a browser (no more than 20 cookies from a particular domain).

Working with cookies and Java

Developing applications that use cookies is simple in Java because of the object-based abstraction that the javax.servlet.http.Cookie class provides.

Here are the public methods in the javax.servlet.http.Cookie class that can be used from servlets or JSPs.

METHOD	DESCRIPTION
object clone ( )	Oversides the standard java.lang.Object.clone ( ) method to return a copy of this cookie.
void setComment (String purpose)	Set a comment that describes a cookie's purpose.
String getComment ( )	Returns the comment, or null if there is no comment.
void setDomain (String pattern)	Set the domain, the URLs for which this cookie shoul be presented to the server as part of any request.
string getDomain( )	Returns the domain name for this cookie.
void setMaxAge (int expiry)	Set the maximum age of the cookie in seconds. A default return value of-1 indicates that the cookie will last untill the browser shutdowns.
int getMaxAge( )	Returns the maximum age of the cookie in seconds.
String getName( )	Returns the name of the cookie.
Void setPath (Stringuri)	Set a path on the server to which the client should return the cookie.
String getPath( )	Returns the path on the server to which the browser returns this cookie.
Viod setSecure (boolean flag)	Indicates to the browser whether the cookie should only be sent using a secure protocol, such as HTTPs or SSL.
boolean getSecure ( )	Returns true if the browser only sends the cookie over a secure protocol, or false if the browser can use a standard protocol.
void setValue (string new Value)	Set a value for the cookie.
String getValue ( )	Returns the value of the cookie.
Void setVersion (int V)	Set the version of the cookie protocol.
Int getVersion ( )	Returns the version of the protocol with which the cookie complies.

Sessions

A session can be defined as a series of related interactions between a single client and the server, which take place over a period of time. In terms of objects, a session can be thought of as an object, resident and exposed to applications on the server (servlets and JSPs), that can be used for storing and retrieving objects.

Once a session object is created on the server a unique identifier called session ID is associated with it and this ID is the only information that is given to the client (by one of the mechanisms we saw in the last section)

Every time the client accesses a resource on the server, the client provides the session ID that it was assigned. Using this session ID, servlets can obtain the associated session object and retrieve any stored information. The simplest analogy would be to imagine a big hash table of hash tables living on the server with strings (session IDs) as keys. The key is mapped to a hash table that is created for a client, as shown in the following figure.

Key	Value	
	Key	Value
Session ID for client 1	"Key 1"	Object 1
	"Key 2"	Object 2
	"Key 3"	Object 3
Session ID for client 2		
Session ID for client 3		

The **Figure** illustrates the life cycle of a session.

A session lifecycle is fairly simple. It is created on the server as a result of a client request and assigned a

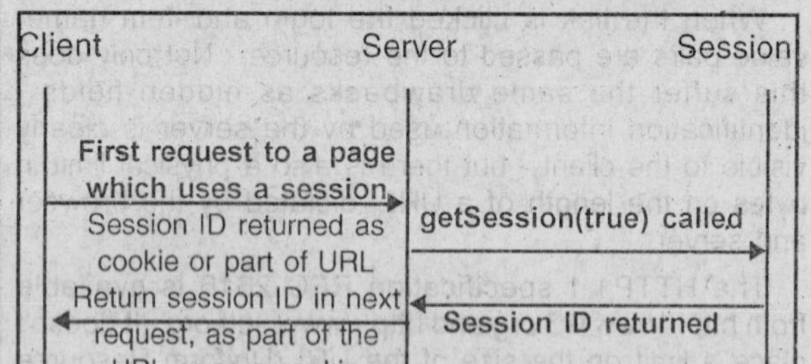

unique session ID and this ID is then passed to the client (as a cookie or a hidden variable). The session itself, is however considered "new" until the client returns the session ID to the server (through a cookie or as a part of the requested URL) indicating that a session has been established. This associates the client with that particular session object. If the client doesn't support cookies (or rejects them) and URL rewriting has not been enabled on the server, no session ID will be returned to the server and the session will continue to be treated as new. A session exists on the server until it is invalidated or the server is stopped.

Contexts

JSP is designed to be portable between different servers. It therefore needs to abstract information held by the web server from the JSPs and Java objects involved in creating the response. The JSP specification relies on contexts to provide that abstraction and bridge the gap between server-specific implementations and portable JSP code.

A context provides an invisible container for resources and an interface for them to communicate with their environment. For example a servlet executes in a context (Javax.servlet.ServletContext). Everything that the servlet needs to know about this server can be extracted from this context and everything the server wants to communicate to the servlet goes through the context. By using interfaces to define contexts, the vendor has flexibility in implementing the interface, while the developer is assured that the contract represented by the context between the server and the resource will always be fulfilled.

A rule of thumb is that everything in JSP is associated with a context, and every context has a scope.

There's another context interface for JSPs, Javax.servlet.jsp.PageContext to which objects can be bound with different scopes.

Consider the following tag:

```
<jsp: useBean id = "catalog"
                    scope = "page I request I
session I application".
                    Class = "projsp.Catalog"/>
```

What happens when this tag is complied for the different values of the scope can be seen below. An object instance of the com.wrox.projsp.Catalog class is created end bound according to the details below:

Page Scope

Page scope is intrpreted as PageContext, Page_scope. The object is bound to the javax.servlet.jsp.PageContext.

This is relatively simple - the object reference is discarded upon completion of the current Servlet.service ( ) invocation. When generating the servlet, the servlet engine creates an object in the service( ) method which follows the usual object scope convention in Java. The object exists for every client request to the resource.

Request Scope

Request scope is interpreted as PageContext.REQUEST_SCOPE and the object bound to the javax.servlet.jsp.PageContext.

The object reference is available as long as the HttpRequest object exists, even if the request is passed/ chained to different pages. The underlying, generated servlet relies on binding the object to the HttpServletRequest using the setAttribute (String key, Object value) method in the HttpServletRequest. This is transparent to the user. The object is distinct for every client request.

Session Scope

Session scope is interpreted as PageContext.SESSION_SCOPE and the object bound to the javax.servlet.jsp.PageContext.

In earlier implementations of the servlet API, a session always ran under a separate session context. (Though deprecated for security reasons, an implementation for the HttpSessionContext interface is still available on most servers. It allows us to get a list of sessions on a particular session by its ID -and that's not really that secure). These days, the only session ID we can get at is the one that's part of the request from the client, passed to the PageContext and made available to the JSP.

Application scope

Application scope is interpreted as PageContext. Application_scope and the object bound to the javax servlet.ServletContext.

This is the most persistent scope. The generated servlet relies on binding the object to the ServletContext using the SetAttribute (String key, Object value) method in the ServletContext. This is not unique to individual clients and, consequently, all clients access the same object.

The Item class

This bean abstracts an item in the vendor catalog. Each item has the following attributes:

- String itemId, the ID of the item in the catalog
- Float price, the price of the item.
- String description, the textual detail about the item.
- Boolean available, true if it is available in stock.
- Int quantity_in_stock, the quantity instock.

The order class

This bean abstracts the concept of a client order. Every such order has an assigned ID, a user, and some items in it.

- String itemIds [ ], which holds an array of item Ids in an order
- Long order ID, the id for that order
- User user, the user placing the order
- Boolean processed, denotes whether the order has been completed or not.

The user class

This bean represents the user on the system. Every user in the application must be unique and thus has a unique user id describing the user (like the SSN or e-mail address). Also, for security purposes, we need to capture the IP address for the machine from which the user is logging on. In effect this means that our user bean has two attributes an id and an IP address:

- String userId, the users ID
- String IPAddr, the IP address from where the user is logging on

The login Bean

This bean is fairly simple. Its purpose in life is to take two attributes (a username and a password) and authenticate them

- String username, the username
- String password, the password

This class also has one simple method, User authenticate ( ), which returns to user.

```
package com.wrox.projsp.ch04;
public class login{
private string username = " ";
private string password = " " ;
public login ( ) { }
public void setUsername (string username)
{this.username = username;
}
public void getpassword (string password)
{this.password = password;
}
public User authenticate ( )
{if   username   equals   ("duke")   &&
password.equals ("jsp") ) {
return new User ("duke");
} else if (username.equals ("wrox") &&
password.equals ("JSP") ) {
return new user ("wrox");
return null;
}
}
}
```

ERROR HANDLING WITH JSP

Types or Errors and Exceptions

There are two primary types of exceptions you can run into when processing with JSP pages:

- Translation time errors
- Run time exceptions

We'll look at these in more detail shortly. In short, if a translation error occurs during page compilation, this results in an internal Servlet Error (error code 500). An exception, on the other hand, occurs once the page is compiled and the servlet is running. Exceptions can be dealt with within the JSP, using the standard Java exception-handling mechanism, but if they are not then it is possible, using the special error-handling JSP directives and JSP error pages, to ensure a more pleasurable user experience instead of showing a stack dump of where the exception was thrown.

Handling Run Time Exceptions

Like Java programs, JSP files may catch exceptions that occur during execution:

```
<%
try {
URL url = new URL (param);
{ } catch (MalformedURLException e) {
response.sendRedirect (BAD-URL-PAGE);
}
%>
```

try-catch blocks are perfectly legal Java source to include in a JSP script. If, however, an exception is thrown during the handling of a client request and is not caught, the user will again see a screen similar that shown at the beginning of the chapter.

If exception handling is new to us, we should see either the "Handling Errors with Exceptions" trail in the online Sun tutorial of

http://java.sun.com/docs/books/tutorial/essential'/exceptions/index.html, or read up on exception handling in Beginning Java 2-JDK 1.3 Edition by Ivor Horton, from Wrox Press, ISBN1-861-003-66-8.

With the JSP file already converted into a servlet, the default action when an exception is thrown is to display a stack dump. However, the JSP page directive provides the facility to override this default behavior and have the exception handling redirected to a new JSP file:

<%@page.......errorPage="/error.html"%>

If the errorPage attribute is specified within the page directive, the relative URL specified is displayed when an uncaught Throwable object is encountered - any Throw - able object, not just an Error or an Exception. The relative URL can be a fixed HTML file or another JSP file.

If an exception occurs, the output buffer is cleared before the error page is displayed. If, however, the autoFlush attribute of the page directive is set to "true", then the dispatching of an uncaught exception could result in an error (HTTP error code 500-Internal Server Error) if the ServletResponse has already received some output.

JSP-Specific Exception Classes

The JSP 1.1 specification added two **java.lang.Exception** subclasses into the mix;the Figure below shows the class hierarchy for these classes. If you aren't writing JSP tag extensions or creating a JSP runtime container, you'll probably never need to use them.

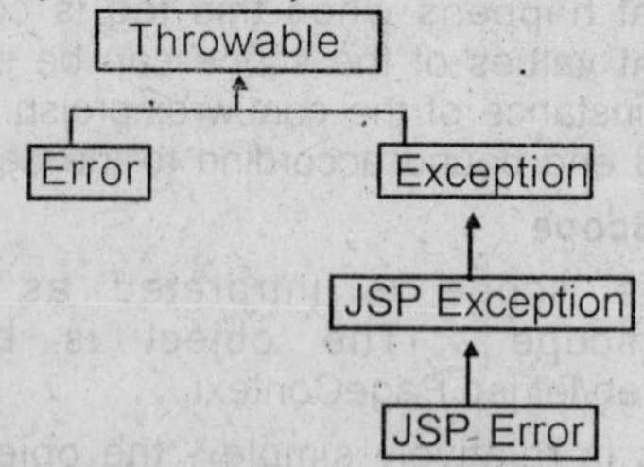

JSPException

The **JSPException** class represents a generic exception for the JSP packages. It can be found in the **Javax.servlet.jsp** package.

JSPError

The JSPError class in the **javax.servlet.jsp** package is the only predefined subclass of **JSPE xception.** It is meant to represent an unrecoverable error, as opposed to an exception. Why it doesn't subclass **java.lang.Error** is unknown, but it is a subclass of **Exception.**

JSP ARCHITECTURE

This chapter will examine a variety of ways to architect a system with JavaServer Pages, Servlets, and JavaBeans.

Redirecting and Forwarding

Redirecting and forwarding requests in JSPs and Servlets takes place often, and it is important to understand the subtle difference between these two mechanisms even though they achieve the same goal (that is, a client asks for a resource on the server and a different resource is served to it):

- When a Servlet or JSP resource chooses to redirect the client (using a **response, sendRedirect (url))** the request object does not reach the second resource directly since the underlying implementation is an HTTP redirect. The server sends an HTTP 302 message back to the client telling it that the resource has moved to another URL, and that the client should access it there. The bottom line is that the lifecycle of the initial request object that was accessed in the first JSP terminates with the end of the service method in the first JSP, or with the reply from the server.
- In a forward mechanism the request object is forwarded to the second resource, thus maintaining any object bindings to the request and its state, without a round trip to the client on the network.

  JSPs and Servlets can utilize the forwarding mechanism to delegate tasks amongst themselves, in the process of separating dynamic and static content.

Architectures

Two basic ways of using the JSP technology are:

- The first method is referred to here as the **page-centric** (or **Client-server)** approach. This approach involves request invocations being made directly to JSP page.
- In the second method, the **dispatcher** (or N-tier) approach, a basic Servlet or JSP acts as a mediator or controller, delegating requests to JSP pages and JavaBeans.

The "Page -Centric" Approach

we will look at the **page-view** and **page-view with bean** architectures.

Applications built using a client-server approach have been around for some time; they consist of one or more application programs running on client machines, and connecting to a server-based application to work. (A good example would be a powerBuilder or Oracle Forms based system.) CGIs and pre-Servlet applications were generally based on this simple 2-tier model, and with the introduction of servlets, 2-tier applications could also be created in Java.

The advantage of such an approach is that it is simple to program, and allows the page author to generate dynamic content easily, based upon the request and th e state of the resources.

However this architecture does not scale up well for a large number of simultaneous clients since there would be a significant amount of request processing to be performed, and each request must establish or share a potentially scarce/expensive connection to the resource in question. (A good example would be JDBC connections in servlets or JSPs and the need for connection pools.)

The RequestDispatcher Apl

These architectures rely on a Servlet's dispatching a request to another resource. Prior to the Servlet API 2.1, dispatching of requests to another dynamic resource was cumbersome, and limited because it was not directly supported in the specification.

The Servlet API 2.1 introduces the RequestDispatcher, accessible via the ServletContext. The Request Dispatcher implementation can dispatch a request to either a static resource, such as an HTML page, or another dynamic resource, such as a Servlet or JSP. A request can be either 'forwarded' or 'included', and state can be made available to the dispatched resource.

The include ( ) method allows the programmatic equivalent of Server-Side includes (SSI), while the forward ( ) method transfers control to the designated dynamic resource. A Servlet that invokes RequestDispatcher. Include ( ) retains overall responsibility for the service request, handling the response to the client. In this case, the Servlet has the opportunity to do any sort of clean up after all the nested invocations have been completed.

Once the Servlet dispatches to another dynamic resource with an invocation to Request Dispatcher.forward ( ), though, the Servlet has completed its work. An example of the latter type of forward dispatch occurs in both the Mediator-View and Mediator Composite View architectures.

Mediator-Composite View

There is content being generated dynamically, and the template text that surrounds this data is also being constantly changed. If we simply type our header and footer text directly into our JSP source, then we will need to modify that source each time we want to change the template text, and our modified source will need to be re-translated in order to service another request, causing this run time latency, and this scenario will repeat itself each and every time the text undergoes even slight modification.

A portion of the template text from our JSP source to another file, replacing it with the JSP include action:

<jsp: include page = "header.html"flush = "true"/>

Unlike its counterpart the include directive (signified with <%@), which includes text at translation time, the JSP include action substitutes the designated page into the JSP source at run time. Not only have we made our solution more modular, we have provided a way to change template text on the fly without the need for recompiling our JSP source. Of course, there is a run time cost for this processing, but it is far less expensive than a full source compile. Such flexibility is useful, especially since the JSP specification does not include much coverage of pre-compilation.

Case Study: Publishing Data to the Web

This chapter is useful if we want a simple and practical introduction to publishing data on the web using a database and JSP.

We can access the latest version at http://www.magiccookie.com/photography/ photodb/, and download the complete source code from http://www.magiccookie.com/software/.

Overview

The PhotoDB system consists of:

- A user interface rendered as HTML
- Some Java source code and JSP files
- Meta data stored in a relational database
- Image files

The purpose of the Java and JSP source code is to map between the HTML interface and the data stored in the database and on disk. The basic architecture looks like the following:

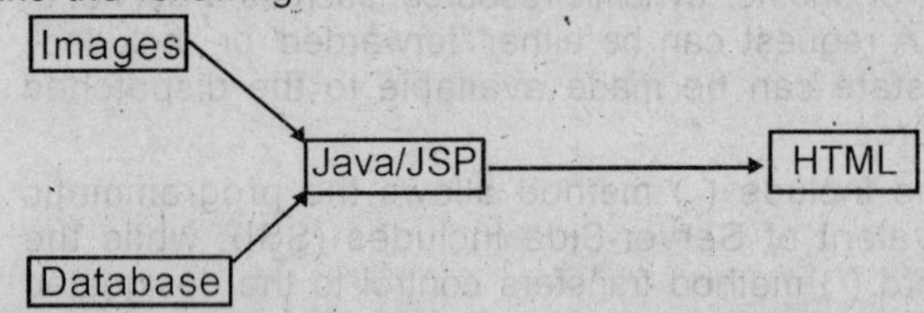

User Interface

Perhaps the best way to understand PhotoDB is to start by accessing some of the web pages that it generates. The full PhotoDB is accessible from our web site at

http://www.magiccookie.com/photography/photodb.

We use the term **photograph** to refer to the original film slide or negative, and the term image to refer to an electronic replica of a photograph. Thus, there is only ever a single specific photograph, but there may be multiple **images** of the photograph. For instance, we make small and large scans of a photograph and store the scans as two different image files.

PhotoDB provides two ways to organize photographs: by location and by category. Each photograph belongs to a single location, which corresponds to the physical place at which the original photograph was taken, for example, Yellowstone National Park. Every photograph must belong to one and only one location. In contrast, categories allow Images to belong to a list of keyword-based categories. A photograph may belong to any number of categories. For example, a photograph might be of an insect and also of a flower, and therefore it would belong to both the insect and flower categories.

Viewing a Single Image

Clicking on a thumbnail of a photograph displays a larger version of the photograph. This detailed view also includes the photograph's description, notes, and links to the location and categories to which the photograph belongs.

Embedded Images

There are images interspersed among our web pages. For example, our home page contains several images. Clicking on any thumbnail image displays a larger version of the photograph along with details about the photograph.

Thumbnails can be coded by hand, that is, the link and image tags are coded directly in the HTML. Thumbnails can also be automatically generated using server-side includes and a servlet included with photoDB.

Index Pages

Photographs can be accessed using an index consisting of the names of locations, categories, or photographs. Clicking on the name of the item displays images belonging to the item. The names of the items are displayed in a narrow frame on the left side of the browser window, and small 150 x 100 pixel images are displayed in a larger frame on the right side of the browser window. Clicking on one of the small images brings up a larger version of the image, along with details about the photograph and image.

Gallery Pages

A gallery of locations or categories shows one image from each location or category. Clicking on the image brings up a gallery of all the photographs belonging to the slected item. This provides a convenient way to showcase the best photographs from each location or category.

Custom Index Page

PhotoDB also allows the web site maintainer to use an HTML file to display the images from a location. Before generating an index page for a location, PhotoDB looks in the directory containing the index page's image files for a file named index.html index.jhtml, or index.jsp. If an index file is found, then photoDB sends a redirect to the index file rather than automatically generating an index.

Editing Data

PhotoDB includes a set of pages for maintaining the database. Access to these pages requires a valid PhotoDB login and for further security, login is limited to a restricted set of hosts. These pages allow the webmaster to add, delete, and modify data in the database. When logged in, PhotoDB provides access to:

- pages for editing, creating, and deleting data
- photographs and locations that are designated as private
- a list of all the rolls of film

Marking items as private allows us to catalog photographs that are not intended for public access.

Presentation Layer

JSPs provide a bridge between pure Java code and the character data that make up a web page. Outputting HTML directly from pure Java code, such as Java Servlet, can be tedious and may be inflexible. A better approach is to separate most or all of the logic of the application into business objects and provide a separate presentation layer. Some approaches to this problem are to output XML data that are formatted using XSL stylesheets or to define JSP tag libraries so that the amount of Java code embedded in the presentation layer is minimized. PhotoDB was developed before XML/XSL became readily available and was originally developed based on the JSP v0.9 specification. Hence, PhotoDB does not use XML or tag libraries.

Data Storage

PhotoDB has two generic types of data: image data and descriptive meta data. The images are binary data, typically around 10Kb for small thumbnail images and 30-150Kb for larger images. The descriptive meta data include information such as the name of each photograph, where and when the photograph was taken, etc. These two types of data have different characteristics that affect their storage and retrieval.

The descriptive meta data in PhotoDB is amenable to representation in a relational database because:

- the data is structured.
- There are many small pieces of data.
- It is all character or integer data.
- The data need to be searched, retrieved, and recombined in various ways.

An object database could have provided an alternative to a relational database. Using an object database would simplify much of the database access code since there would have been no need for a separate data access layer. However, relational databases, including free databases such as PostgreSQL and my SQL, are more widely available and accessible, using SQL, is more standardized.

Images Files

The standard and compact format for display of photographic color images on the Internet is JPEG. Hence, all the image files are stored as JPEGs on our website. In principle, it should be possible to store images in any arbitrary format, though PhotoDB currently includes code to extract image attributes only from JPEG images.

Images must be stored digitally for access via PhotoDB. We do all my photography using film, either slide or negative. For more details see http:// www. magiccookie.com/photography.

JPEG is a poor storage format, it does not retain all the original information in an image. This makes JPEG a poor format for storage of original scans or digital images. A better image format for permanent storage would be TIFF or PNG, which retain image data and are highly portable. Automated tools can then be used to convert the high-quality images into lower-quality JPEG images.

Database

This section describes the database tables used in PhotoDB.

The objects manipulated by PhotoDB that must be stored in persistent storage correspond to photographs, folders, categories, and rolls of film. Each of these objects is mapped to a single database table and a corresponding Java Class. Additional tables are used to provide many-to-many mappings between objects and to support generating unique object IDS.

Table	Description
folder	Represents folders.
category	Represents categories.
photo	Represents photographs.
roll	Represents a roll of film.
photo_category	Associates photographs with categories.
object_id	Keeps track of the last ID assigned to a row in the database. Used to generate unique objects IDs.

The folder, category, photo, and roll tables correspond to persistent objects. These tables include the columns id and ts, which are used by the Java database mapping layer to identify individual objects and to support optimistic row locking (optimistic locking is explained in the section on data access).

Database Portability

SQL is supposed to be relatively standard. However, each vendor implements a slightly different version of SQL. Even things as simple as varchar and text types may differ between vendors. The solution I adopted in PhotoDB was two-fold:

- Code to the lowest-common denominator.
- Use a macro preprocessor to generate the SQL statements that create the initial database tables and objects.

The generic initialization scripts are in the directory src/sql. Two files are provided: create sql.m4 contains the data definition statements for constructing the database tables and indexes. The file sample-sql.m4 contains statements that insert data for a small sample database and photographs included with photoDB. Scripts for a desired target RDBMS are automatically generated from these files using the bin/mksql script. This script Uses Perl and GNU's m4 macro preprocessor to merge the macros defined Pin src/sql/ include.m4 with the SQL macro files to generate scripts for different RDBMSs. The resulting scripts are written to the sql directory.

Any text in create- sql.m4 that is is all caps in a macro defined either in the same file or in the include.m4 file. Several of the macros map to an underlying data type supported by the database. For instance, the TIMESTAMP type maps to a numeric type at least 32 bits long, but uses a larger value, such as bigint or int8, if possible. The following table contains descriptions of some of the macros (some of the other macros are used only within the macro definition file).

Object-Relational Mapping

When accessing a relational database from an object-oriented language one typically encounters what is known as the 'object/relational impedance mismatch'. This so-called impedance mismatch arises because a relational database accesses data as rows in tables. Data is accessed using a declarative query language which itself generates tables of data. However, Java is an object-oriented language, and represents data using a static inheritance graph of types combined with runtime instantiation of objects, a very different method of representing information than a RDBMS.

A programmer accessing a relational database using an object-oriented language may end up thinking and programming in two different paradigms. This can lead to code in which two essentially incompatible methods of representing information are interwoven. The programmer may, instead, use an object relational mapping layer. Essentially, what the mapping layer does is convert data stored in a relational database into data stored in objects and vice versa. The mapping layer can also insulate the application from the database so that the application logic and database schema can be modified some-what independently.

The impedance mismatch is reduced or eliminated if you use an object-oriented database. These are databases that let an application store and mainpulate objects instead of relations. Unfortunately, there is less industry standardization on how to access object databases than for relational databases. Each vendor may provide its own API and/or query language. The market for such databases is also currently much smaller than the market for relational databases and it may be harder to find a suitable object database.

Optimistic Locking

Database object uses optimistic locking when updating objects, which is both efficient and portable since it avoids the need for rowlevel locking in the DBMS. When a row is loaded into memory, the value of the row's ts column is kept with the data for the row. When updated data are written to the row, a check is made to ensure that the row's ts column has maintained its old value, and in the same SQL update statement, the row's ts column is incremented. If the value of the ts column is differed, then no data would be written and a com.magiccookie.db.ConcurrentUpdateException would be thrown. The entire compare/update/increment operation is contained in a single SQL update statement, and so is guaranteed to execute atomically.

Transactions

PhotoDB uses database transactions if they are supported by RDBMS. PhotoDB uses objects of type com.magiccookie.database. Transaction (part of my utility classes mentioned earlier). to manage transactions. In the event that there is an error while storing data to the database PhotoDB will rollback the transaction.

```
Transaction trans = getDatabase ( ).
openTransaction ( );
Try {
        trans.begin ( );
        ...
        trans.commit( );
        } finally {
        trans.end( );
        }
```

As shown in the above listing, a transaction is opened using the begin method of class transaction. If the operation completes successfully then the commit method is called. The end method of the transaction class is always called, both when the transaction was committed and when there is an error. If the end method is called and the transaction was not committed, then the transaction is rolled back, otherwise the method does nothing.

Persistent Objects

The class com.magiccookie.database. DatabaseObject provides an interface to a row in a database table. This class provides methods to automatically load column values from a row, to store an object in a row and to delete a row. Objects are identified using a unique object ID. A new object ID is automatically generated for each new row. The table object_id holds the value of the last reserved object ID. (A regular table is used to ensure portability across RDBMSs.)

The com.magiccookie.photodb.PhotoObject class extends the generic Database Object class and adds a few methods that are used by all objects stored in the database. Photo object is not meant to be used directly, rather it is meant to be subclassed with representations of specific database tables. The classes Category, Folder, Photo, and Roll correspond to tables in the database and all extend PhotoObject. These classes are very similar and simple, they do not implement any logic, they just provide a convenient way to represent the data stored in the database.

PhotoObject provides several constructors. The default constructor is required by the database mapping layer to allow it to create a database object via a factory method. The other constructors are provided to construct a PhotoObject based on whether it represents an entirely new object with no corresponding entry in the database, an existing row in the database (lookup can be done by name or by ID), data retrieved using a prior SQL query and converted into a Hashtable. Since the persistent objects have a name attribute, PhotoObject also provides name accessor functions using the methods getName and setName. PhotoObject also provides a couple of static methods, exists and lookup By Name, for testing for and retrieving an object by its name.

ASP
Questions and Answers

What is Active Server pages?

Microsoft R Active Server pages (ASP) is the server-side execution environment in Microsoft internet information Server (IIS) 3.0 that enables us to run ActiveX™scripts and ActiveX server components on the server. By combining scripts and components, developers can create dynamic content and powerful Web-based applications easily.

What is dynamic content?

Web pages that are customized for each user on the fly, based upon their actions or requests. For example, new visitors to our site can be shown a different welcome page than returning users see, or pages in an online catalog can be queries to a database so customers always see the most current information and availability.

What do we need to run Active Server Pages?

The Active Server Pages feature or IIS 3.0 requires Microsoft Windows NTR Server 4.0 running IIS 2.0 or Windows NTR Workstation 4.0 running Peer Web Services.

The ASP feature does not require Service Pack 2 for Windows NTR 4.0, although it is recommended that all Windows NT systems receive the Service Pack 2 updates. The ASP feature contains a subset of the updates found in Service Pack 2.

Installation of ASP will upgrade IIS version 2.0 to version 3.0. The other IIS 3.0 features Index Server 1.1, Microsoft NetShow ™, FrontPageR 97 Server Extensions, and Crystal Reports - add significant functionality to IIS, but are not required to take advantage of ASP.

What ActiveX Server Components are supported?

Active Server Pages allows organizations to extend the power of scripting on the server with ActiveX server components. These components can be created using Microsoft Visual Basic, Visual C++, Java, and other languages.

What scripting languages does Active Server Pages support?

Active Server Pages provides native support for both Microsoft JScript and VBScript. ActiveX scripting plug-ins are available for REXX, PERL, and Python.

What browsers does Active Server Pages support?

Active Server Pages can work with any Web browser. The output of an ASP file is plain HTML, the content of which can be customized for the capabilities of the client.

On which platforms does Active Server Pages run?

Active Server Pages will run on Microsoft Windows NT Server 4.0, Windows NT workstation 4.0 with Peer Web Services, and Microsoft Windows R 95 with Personal Web Server. Windows NT 3.51 is NOT supported. Windows NT 4.0 on MIPS is also not supported by ASP.

Is Active Server Pages secure?

Yes. Active Server Pages is a component of internet information server, and thus uses Windows NT Security. ASP files can be easily restricted to just certain users through secure Windows NT authentication, basic Web authentication, or client-side certificates. For additional security, all client-to-server communications can be secured with Secure Sockets Layer (SSL).

What data sources can our Web application integrates with?

An Active Server pages application can integrate with any ODBC -compliant databases including Microsoft SQL Server™ Oracle, Sybase, Informix, and DB2 databases. Any OLE 2 application, such as Lotus Notes or Microsoft Excel, can also be scripted to access or process information. We can also write components to access online data feeds and legacy mainframes.

How does Active Server pages compare to ISAPI applicatons?

ISAPI applications require all of the programming and layout to be contained in a .dll file written In C++. ISAPI applications are thus more difficult to create and maintain. With ASP files, an HTML writer can script an external component and format the output. ASP separates the layout and design from the business logic.

Cache Frequently-Used Data in the Application or Session Objects.

The ASP Application and Session objects provide convenient containers for caching data in memory. We can assign data to both Application and Session objects, and this data will remain in memory between HTTP calls. Session data is stored per user, while Application data is shared between all users.

Do not Cache Database Connections in the application or Session Objects.

Caching ADO Connections is usually a bad strategy. If one Connection object is stored in the Application object and used on all pages, then all pages will contend for use of this connection. If the Connection object is stored in the ASP Session object, then a database connection will be created for every user. This defeats the benefits of connection pooling and puts unnecessarily high stress on both the Web server and the database.

Encapsulate Code in COM objects

If we have a lot of VBScript or JScript, we can often improve performance by moving the code to a compiled COM object. Compiled code typically runs faster than interpreted code. Compiled COM objects can access other COM objects through "early binding", a more efficient means of invoking COM object methods than the "late binding" employed by script.

Acquire Resources Late, Release Early

Here is a short tip for us. In general, it is best to acquire resources late and release them early. This goes for COM objects as well as file handles and other resources.

ADO Coonections and recordsets are the prime candidates for this optimization. When we are done using a recordset, say after painting a table with its data, release it immediately, rather than waiting until the end of the page. Setting our VBScript variable to Nothing is a best practice. Do not let the recordset simply fall out of scope. Also, release any related Command or Connection objects.. (Do not forget to call Close( )) on recordsets or Connections before setting them = Nothing). This shortens the time span in which the database must juggle resources for us, and releases the database connection to the connection pool as quickly as possible.

Avoid Redimensioning Arrays

Try to avoid Redim arrays. As far as performance is concerned, if we have a machine that is constrained by physical memory size, it is much better to set the initial dimension of the array to its worst-case scenario-or to set the dimension to its optimal case and redim as necessary. This does not mean that we should just go out and allocate a couple of megabytes of memory if we know we are not going to need it.

Use Response Buffering

We can buffer a whole page worth of output by turning on "response buffering". This minimizes the amount of writes to the browser and thus improves overall performances. Each write has a lot of overhead (both in IIS and in the amount of data sent down the wire), so the fewer the writes there are, the better. TCP/IP works much more efficiently when it can send a few large blocks of data than when it has to send many blocks because of its slow start and Nagling algorithms (used to minimize network congestion).

Use Response. IsClientConnected Before Embarking on Long Trips

If the user gets impatient, he or she may abandon our ASP page before we even start executing their request. If he clicks Referesh or moves to a different page on our server, we will have a new request sitting at the end of the ASP request queue and a disconnected request sitting in the middle of the queue. Often this happens when our server is under high load (so it has a long request queue, with correspondingly high response times) and this only makes the situation worse. There is no point executing an ASP page (especially a slow, heavyweight ASP page) if the user is no longer connected. We can check for this condition by using the Response. IsClient Connected property. If it returns False, we should call Response. End and abandon the rest of the page.

Take Advantage of our Browser's Validation Abilities

Modern browsers have advanced support for features such as XML, DHTML, Java applets, and the Remote Data Service. Take advantage of these features whenever we can. All of these technologies can save round trips to the Web server by performing clientside validation as well as data caching. If we are running a smart browser, the browser is capable of doing some validation for us (for example, checking that a credit card has a valid checksum before executing POST). Again, take advantage of this whenever we can. By cutting down on client-server round trips, we will reduce the stress on the web server and cut down network traffic (though the initial page sent to the browser is likely to be larger), as well as any back-end resources that the server accesses. Furthermore, the user will not have to fetch new pages as often, improving the experience. This does not relieve us of the need to do server-side validation we should always do server-side validation as well. This protects against bad data coming from the client for some reason, such as hacking, or browsers that do not run our client-side validation routines.

Use Server. Transfer Instead of Response.Redirect Whenever Possible

Response.Redirect tells the browser to request a different page. This function is often used to redirect the user to a log on or error page. Since a redirect forces a new page request, the result is that the browser has to make two roundtrips to the Web server, and the Web server has to handle an extra request. IIS 5.0 introduces a new function, Server.Transfer, which transfers execution to a different ASP page on the same server. This avoids the extra browser-to. Web-server round trip, resulting in better overall system performance, as well as better response time for the user. Check our New Directions in Redirection, which talks about Server. Transfer and Server. Execute.

Avoid Using Server Variables

Accessing server variables causes our Web site to make a special request to the server and collect all server variables, not just the one that we requested. This is akin to needing to retrieve a specific item in a folder that we have in that musty attic of ours. When we want that one item, we have to go to the attic to get the folder first, before we can access the item. This is the same thing that happens when we request a server variable the performance hit occurs the first time we request a server variable. Subsequent requests for other server variables do not cause performance hits.

Never access the Request object unqualified (for example, Request ("Data")). For items not in Request.Cookies, Request.Form, Request.QueryString, or Request.ClientCertificate, there is an implicit call to Request.Server Variables. The Request. Server Variables collection is much slower than the other collections.

QUESTIONS

1. Name some of the built-in ASP objects.
2. What can we do using GLOBAL.ASA?
3. Choose (a) or (b). The contents collection of the application object contains
   (a) Application-level scoped variables and objects added to the current application only through scripts
   (b) Application-level scoped variables and objects added to the current application through scripts and through <OBJECT> tag
4. SetComplete, SetAbort methods are for what object and what will they do?
5. Code after Response.End will be processed by the web server. (True/False)
6. What is the error in this piece of code?

```
<HTML>
<% Response.Buffer = True %>
<HEAD>
<TITLE> One Long Script </TITLE>
</HEAD>
</HTML>
```

7. Can we create an instance of the built in objects using CreateObject method?
8. Set ScriptTimeout property of Server Object to 10 secs. Consider ASPScriptTimeout in IIS metabase is 90 secs. After how many secs. the script will time out?
9. What do we understand by HTMLEncode, URLEncode?
10. Chose the applicable.
    Session.SessionId is
    - Read only value
    - Stored as a cookie on the client machine
    - Stored as a cookie on the server machine
    - Cookie is reinitialized at every Session.Timeout
    - Cookie is reinitialized during only either the client browser restarts or Web server restarts.
11. What is SafeArray?
12. Which preprocessing directive is used in HTML as well as in ASP files?
13. Which object has PICS property?
14. Name some of the installable components that come with IIS 4.0.
15. What is Static-Object?
16. What is the difference between GET and POST methods?
17. Application_OnStart is called
    (a) Only once per application
    (b) Once per client
    (c) On both Event
    (d) None of the above
18. We can have list of named string values that can be used often throughout the site is in
    (a) Content Linking Component
    (b) File Access Component
    (c) Myinfo Component
    (d) Browser capabilities Component

Answers

1. Built-in ASP objects are
   Application,
   Object Context,

Request,

Response,

Server,

Session.

2. The Global.ASA file is where we can declare objects, variables and event handlers for the OnStart and OnEnd event procedures for the Application and Session objects, specially that have session or application scope.
3. (a)
4. SetComplete and SetAbort Methods are for Object Context Object.

   SetAbort - Aborts the transaction as a whole.

   SetComplete - Signals the successful completion of a transaction.
5. False
6. <% Response.Buffer = True %> statement should come before <HTML> tag.
7. We can't create built-in objects using CreateObject method.
8. After 90 secs. The scriptTimeout should be greater than ASPScripTimeout in IIS metabase. Otherwise it always will take ASPScriptTimeout as ScriptTimeout. If we wants, we can change ASPScriptTimeout in IIS metabase.
9. HTMLEncode - if we want to display actual HTML code or ASP script we should encrypt using HTMLEncode method. So browser won't interpret the code.

   URLEncode - Encodes a string that can be sent over the address line as a query string.
10. Session.SessionId is
    - Read only value
    - Stored as a cookie on the client machine
    - Cookie is reinitialized during only either the client browser restarts or Web server restarts.
11. A Safe Array is not an array at all. It is a special type of structure used internally to maintain information held in its array portion. The dimensions and upper bounds values are available only from C/C++ elements of the structure. We cannot manipulate these values through script.
12. #include
13. Response Object
14. Installable Components with IIS 4.0 are

    ActiveX data Objects 1.5

    Ad Rotator Component

    Browser capabilities Component

    Collaboration Data Objects for Windows NT Server

    Content Linking Component

    Counters Component

    File Access Component

    Mynfo Component

    Page Counter Component

    Permission Checker Component
15. Static-Object is a object which is added to the application using <OBJECT> tag
16. I) GET can be used to retrieve any document, POST cannot.

    II) Both GET and POST can be used to pass data to the object indicated by the URL. When GET is used for this purpose, the data is included in the URL as the argument string; when POST is used the data is passed to the server in the body of the request message.
17. (a)
18. (c)

J2EE

The Java 2 Enterprise Edition is a specification that aims to provide a standard for Java application servers. This is a good thing for several reasons.

- No investment in a single proprietary platform.
- Best-of-breed inplementation. we can select the implementation that best suits our needs and provides our application with the best scalability and performance.
- Price competition between different vendors ensures good pricing.
- The broad knowledge base that comes with an established standard makes it easier to find good developers, good education, and everything that comes with it.

Overview

The Java 2 Enterprise Edition is a specification consisting of many parts. J2EE is made up of a lot of different technologies, and a specification, for how these work together. The main parts of the J2EE runtime environment are:

- **Application components.** There are four mandatory types of application components that a J2EE compliant product must support.
- Application clients. These are fat clients written as normal Java applications. They normally access the application server using Remote Method Invocation (RMI).
- Applets. These are graphical clients that normally execute within a web browser.
- Servlets and Java Server Pages.
- Enterprise Java Beans. These are component that execute in a container in the application server that manages them and handles must of the "plumbing" for them. These are more thoroughly discussed in a later section.
- **Containers.** Every kind of component lives inside a container that provides the components with runtime services. There is one type of container for every type of component. The most interesting containers for us are the web container that the servlets and Java Server Pages live in, and the EJB container that manages the Enterprise Java Beans.
- **Resource manager drivers.** A resource manager driver is a driver that provides some kind of connectivity to an external component. The only type is supported today is driver that implements any of the J2EE

standard services, like Java Messaging Service (JMS), Java DataBase Connectivity (JDBC), or Java Mail.

- **Database.** Databases are accessible via JDBC and need to accessible from all component types except for applets.

The J2EE platform also includes some standard services that must be provided by a J2EE compliant product. They include the following mandatory services:

- HTTP. The J2EE server must avail it services via the HTTP protocol through support for servlets and JSP pages.
- HTTP. A J2EE server must also be able to layer HTTP over SSL to make connections secure.
- Java Transaction API (JTA). A J2EE server must imp[lement transaction support.
- JavaIDL. J2EE clients are required to be able to connect to CORBA objects using the JavaIDL ORB.
- JDBC
- Java Naming and Directory Interfaces (JNDI). Naming and directory services are used server wide and have to be present.
- Java Mail and Java Beans Activation Frame-work. The J2EE platform Includes services for sending email over the internet. To support this the Java Beans Activation Frame-work is also needed.

The following services, however, are optional:

- **RMI-IIOP.** RMI-IIOP support is not yet required by a J2EE implementation, but it will become so in future versions. However, a J2EE application is already required to use the RMI-IIOP APIs when accesing Enterprise Java Beans via RMI.
- **Java Message Service (**JMS) allows enterprise messaging. A J2EE server does not yet have to implement this service, but it will be required in later versions.

Enterprise Applications

The J2EE specification introduces the concept of Enterprise applications. An Enterprise application contains J2EE modules, which could be web applications, EJBs, application clients (normal Java applications talking to a J2EE server), and applets. A J2EE Enterprise archive is packaged as a Java Archive file with aear filename extension.

Enterprise Java Beans

Enterprise Java Beans (EJB) is a technology specification from sun, which standardizes how to build server side components for distributed applications.

The components

The components, the actual beans, are large-grained objects representing either some functionality (like a shopping cart) or some data (like a customer bean that holds information about a customer). There is a big difference between these kinds of components, and because of this EJBs exist in two basic types: Session beans and Entity beans. The main differences are outlined in this table.

EJB Container

Enterprise Java Beans run within a container, which controls the beans and provides them with system services. The services a container offeres are:

- Transactions. The EJB specification provides transactional semantics for componets, allowing the component developer to specify transactional properties of components.
- Security
- Remote access (RMI)
- Life cycle management
- Database connection pooling. To provide efficient access to databases it is better to make clients share, database connections, this means a new database connection doesn't have to be opened for every client request. Instead there is a pool of a number of connections, and when a client wants a database connection it gets an already opened connection to use so there is no start-up time. When the client doesn't need the connection any more it's returned to the pool, so that it can be refused. The EJB server handles this pooling.

JMS

The Java Message Server (JMS) is a standardized interface that enables Java programs to create, send, receive and read messages using any enterprise messaging system that supports JMS. Just as JDBC provides an abstraction for database access so that your code will not be database dependent, JMS gives you the possibility to write code that will not be dependent on the underlying messaging system, be it IBM MQSeries, Java Message Queue, BEA Weblogic, Orion or any JMS compliant messaging system.

Servlets/JSP

Servlets and Java Server Pages are an important part of J2EE and are described extensively in this book.

The underlying transaction Service in a J2EE product is Vendor-specific, and does not affect application code. Not all J2EE prduct vendors support distributed transactions.

- Notsupported: Never participate in a transaction.
- Supports: If invoked from inside a transaction, take part in it; if not, do not start a new transaction.
- RequiresNew: Every method invocation on the component starts a new transaction.
- Mandatory: If invoked with an active transaction, participate in it if not, throw javax.transaction. Transaction Required Exception.
- Never: If invoked with an active transaction, throw java.rmi.Remote-Exception.
- Required: If invoked inside a transaction, take part it; otherwise start a new transaction.
- Transaction support for on EJB is set to one of the following.

XML and XSL

In the current J2EE version (1.2) XML support is not mandated, but it will become mandated in the future. (Extensible. Stylesheet Language). XSL is two things:

- A language for transforming XML documents.
- An XML vocabulary for specifying formatting semantics.

The first part is referred to as XSLT (XSL Transformations, http://www.w3.org/TR/xslt). It can be used to transform XML documents into HTML. The Second part is similar to Cascading Style Sheets (CSS), and is used for providing formatting.

For more information about XSL, see http://ww.w3.org/Style/XSL/

.NET SYSTEMS

1. ASP.NET versus ASP.

The key differences between ASP.NET and ASP are:

ASP.NET is much more event-driven, with the event handlers running on the sever.

ASP.NET separates code from HTML.

The code in ASP.NET is compiled, not interpreted.

Configuration and deployment are greatly simplified.

Type of bean	Purpose	Client access	Persistence	Transactions
Session bean	Perform work	One client	Normally transient	Usually initiates transactions
Entity bean	Represent persistent object	May have many clients	Persistents	Can take part in transactions

Use CMP if you need	Use BMP if you need
Performance Simplicity Database portability	Full control

There are many other minor differences, but these four are the key changes, and they change everything. The event-driven model in ASP.NET is very powerful. The separation of HTML from code, and the fact that the code is compiled rather than interpreted, allows for the creation of larger, easier to scale, easier to maintain web sites. The configuration and deployment simplifications make working with ASP.NET web sites, both large and small, much easier.

2. Visual Studio.NET

Since all the ASP.NET source files are plain text, we can develop all our web applications using our favorite text editor. Visual Studio.NET, the integrated development environment that is being released in conjunction. With the .NET framework, offers many advantages and productivity geins.

These include:

Visual development of web pages

Drag-and-drop web form design

IntelliSense and automatic code completion

Integrated debugging

Automated build and complie

Integration with the Visual SourceSafe source control program

Fully-integrated, dynamic help.

3. Accessing Data with ADO.NET

ADO.NET provides a rich set of objects to manage database interaction.

ADO.NET looks, at first glance, very similar to ADO, its predecessor technology. The key difference, however, is that ADO.NET is modeled around a disconnected data architecture. Database connections are "expensive" (that is, they use a lot of resources), and it is difficult to have thousands (or tens of thousands) of simultaneous continuous connections. A disconnected architecture, on the other hand, is resourcefrugal. Connections are used only briefly. Of course, ADO.NET does connect to the database to retrieve data, and connects again to update data when we have made changes. When not updating data to or from the database, the connection is broken. Most web applications spend most of their time simply reading through data and displaying it, and ADO.NET provides a disconnected subset of the data for our use while reading and displaying. As we might imagine, disconnected datasets can have scale and performance problems of their own. There is overhead in creating and tearing down connections, and if we drop the connection each time we fill the database, and then must reestablish it each tme we update the data, we will find that performance begins to degrade quickly. This problem is mitigated by the use of connection pooling. While it looks to our application like we are creating and destroying connections, we are actually borrowing and returning connections from a pool that ADO.NET manages on our behalf.

4. Bug Database Design

Imagine that we have been asked to create a tool to manage bugs for a large development effort. We will be supporting three developers who will work in C# and VB.NET, along with a user interface designer and a few quality control engineers. We would like our design to be reasonably flexible so that we can reapply our bug tracking application to future projects.

Our first decision is that we will create a web application. This has the great advantage that all the participants will be able to access the application from their home computers. Since the developers work off-site, this is almost a necessity. We will, of course, develop our web application in ASP.NET.

We imagine that there will be a web page for entering bugs, as well as a page for reviewing and editing bugs. To support this, we will need to design a relational database, and for a number of reasons beyond the scope of this book, we decide to create that database using SQL Server.

We begin by thinking about the kinds of information we want to capture in the database, and how that information will be used. We want to allow any user of the system to create a bug report. We also want eretain users (e.g., developers and QA) to update the bug reports. Developers will want to be able to record progress in fixing a bug, or to mark a bug fixed. QA will want to check the fix and either close the bug or reopen it for further investigation.

The original reporter of the bug will want to find out who is working on the bug, and track progress. One requirement imposed early in the design process is that the bug database ought to provide an "audit trail". If the bug is modified we want to be able to say who modified it and when they did so. In fact, we want to be able to track all the changes to the bug, so that we can generate a report like the excerpt shown in following Example.

Example. Excerpt from a bug report

Bug 101-System crashes on login

1-Reporter: Osborn

Date: 1/1/2004 Original bug filed

Description: When I login I crash.

Status: Open

Owner: QA

2-Modified by: Smith

Date: 1/2/2004 Changed Status, Owner

Action: Confirmed bug.

Status: Assigned

Owner: Hurwitz

To track this information we need to know the date and time of each modification, as well as who made the modification and what they did. There will probably be other information we want to capture as well, though this may become more obvious as we build the application.

One way to meet these requirements is to create two tables to represent each Bug. Each record in the Bugs table will represent a single bug, but we need an additional table to keep track of the revisions.

A Bug record will have a BugID and will include the information that is constant for the bug throughout its history. A BugHistory record will have the information specific to each revision.

The bug database design includes three significant tables: Bugs, BugHistory, and Peopl. Bugs and BugHistory work together to track the progress of a bug. For any given bug, a single record is created in the Bugs table, and a record is created in BugHistory each time the bug is revised in any way. The People table tracks the developers, QA, and other personnel who might be referred to in a Bug report.

5. HTTP-POST

In POST requests, the name/value pairs are also unencoded, but instead of being appended to the URL,they are sent as part of the request message.

QUESTIONS AND ANSWERS

1. What is .NET?

That's difficult to sum up in a sentence. According to Microsoft, .NET is a "revolutionary new platform, built on open internet protocols and standards, with tools and services that meld computing and communications in new ways". A more practical definition would be that .NET is a new environment for developing and running software applications, featuring ease of development of web-based services, rich standard run-time services available to components written in a variety of programming languages, and inter-language and inter-machine interoperability.

Note that when the term ".NET" is used in this FAQ it refers only to the new .NET runtime and associated technologies. This is sometimes called the ".NET Framework". This FAQ does NOT cover any of the various other existing and new products/technologies that Microsoft are attaching the .NET name to (e.g. SQL Server.NET).

2. Does .NET only apply to people building web sites?

No. If we write any Windows software, .NET may offer a viable alternative (or addition) to the current practice.

3. What platforms does the .NET Framework run on?

The runtime supports Windows XP, Windows 2000, NT4 SP6a and Windows ME/98. Windows 95 is not supported. Some parts of the framework do not work on all platforms for example, ASP.NET is only supported on Windows XP and Windows 2000. Windows 98/ME cannot be used for development.

IIS is not supported on Windows XP Home Edition, and so cannot be used to host ASP.NET. However, the ASP.NET web Matrix web server **does** run on XP Home.

The mono project is attempting to implement the .NET Franework on Linux.

4. What languages does the .NET Framework support?

MS provides compilers for C#, C++, VB and JScript. .NET compilers are under development for languages such as COBOL, Eiffel, Perl, Smalltalk and Python.

5. What is the CLR?

CLR = Common Language Runtime. The CLR is a set of standard resource that (in theory) any .NET program can take advantage of regardless of prgramming language. Some of the CLR rescurces are

- Object-oriented programming model (inheritance, polymorphism, exception handling, garbage collection)
- Security model
- Type system
- All .NET base classes
- Many .NET Framework classes
- Development, debugging, and profiling tools
- Execution and code management
- IL-to-native translators and optimizers

What this means is that in the .NET world, different programming languages will be more equal in capability than they have ever been before, although clearly not all languages will support all CLR services.

6. What is the CTS?

CTS = Common Type System. This is the range of types that the .NET runtime understands, and therefore that .NET applications can use. However note that not all .NET languages will support all the types in the CTS. The CTS is a superset of the CLS.

7. What is the CLS?

CLS = Common Language Specificaiton. This is a subset of the CTS which all .NET languages are expected to support. The idea is that any program which uses CLS-compliant types can interoperate with any .NET program written in any language.

In theory this allows very tight interop between different .NET languages - for example allowing a C# class to inherit from a VB class.

8. What is IL?

IL = Intermediate Language. Also known as MSIL (Microsoft Intermediate Language) or CIL (Common Intermediate Language). All .NET source code (of any language) is compiled to IL. The IL is then converted to machine code at the point where the software is installed, or at run-time by a Just-in-Time (JIT) compiler.

9. What is C#?

C# is a new language designed by Microsoft to work with the .NET Framework. In their "Introduction to C#" whitepaper, Microsoft describe C# as follows:

"C# is a simple, modern, object oriented, and type-safe programming language derived from C and C++. C# (pronounced C sharp) is firmly planted in the C and C++ family tree of languages, and will immediately be familiar to C and C++ programmers. C# aims to combine the high productivity of Visual Basic and the raw power of C++."

If one is C++ programmer, he might like to check out his C# FAQ.

10. What does 'managed' means in the .NET context?

The .NET Framework provides several core run-time services to the programs that run within it - for example exception handling and security. For these services to work, the code must provide a minimum level of information to the runtime. Such code is called managed code. All C# and Visual Basic.NET code is managed by default. VS7 C++ code is not managed by default but the compiler can produce managed code by specifying a command-line switch (/com+).

Managed data: This is data that is allocated and de-allocated by the .NET runtime's garbage collector. C# and VB.NET data is always managed. VS7 C++ data is unmanaged by default, even when using the /com+ switch, but it can be marked as managed using the_gc keyword.

Managed classes: This is usually referred to in the context of Managed Extensions (ME) for C++. When using ME C++, a class can be marked with the _gc keyword. As the name suggests, this means that the memory for instances of the class is managed by the garbage collector, but it also means more than that. The class becomes a fully paid-up member of the .NET community with the benefits and restrictions that brings. An example of a benefit is proper interop with classes written in other languages - for example, a managed C++ class can inherit from a VB class. An example of a restriction is that a managed class can only inherit from one base class.

11. What is reflection?

All .NET compilers produce metadata about the types defined in the modules they produce. This metadata is packaged along with the module (modules in turn are packaged together in assemblies), and can be accessed by a mechanism called reflection. The System.Reflection namespace contains classes that can be used to interrogate the types for a module/assembly.

Using reflection to access .NET metadata is very similar to using ITypeLib/ITypeInfo to access type library data in COM, and it is used for similar purposes - e.g. determining data type size for marshaling data across context/process/machine boundaries.

Reflection can also be used to dynamically invoke methods (see System.Type.InvokeMember), or even create types dynamically at run-time (see System.Reflection.Emit.TypeBuilder).

12. What is an assembly?

An assembly is sometimes described as a logical .EXE or .DLL, and can be an application (with a main entry point) or a library. An assembly consists of one or more files (dlls, exes, html files etc), and represents a group of resources, type definitions, and implementations of those types. An assembly may also contain references to other assemblies. These resources, types and reference are described in a block of data called a manifest. The manifest is part of the assembly, thus making the assembly self describing.

An important aspect of assemblies is that they are part of the identity of a type. The identity of a type is the assembly that houses it combined with the type name. This means, for example, that if assembly A exports a type called T, and assembly B exports a type called T, the .NET runtime sees these as two completely different types. Furthermore, we should not confused between assemblies and namespaces - namespaces are merely a hierarchical way of organising type names. To the runtime, type names are type names, regardless of whether namespaces are used to organise the names. It's the assembly plus the typename (regardless of whether the type name belongs to a namespace) that uniquely identifies a type to the runtime.

Assemblies are also important in .NET with respect to security - many of the security restrictions are enforced at the assembly boundary.

13. How can we produce an assembly?

The simplest way to produce an assembly is directly from a .NET compiler. For example, the following C# program:

```
public class CTest
{
public CTest( )
{
    System.Console.WriteLine("Hello from
    CTest").
}
```

Can be compiled into a library assembly (dll) like this:

csc/t:library ctest.cs

The contents of the assembly can be viewed by running the "IL Disassembler" tool that comes with the .NET SDK.

Alternatively we can compile our source into **modules,** and then combine the modules into an assembly using the assembly linker (al.exe). For the C# compiler, the/ target:module switch is used to generate a module instead of an assembly.

14. Write the difference between a private assembly and a shared assembly.

- **Location and visibility:** A private assembly is normally used by a single application, and is stored in the application's directory, or a subdirectory beneath. A shared assembly is normally stored in the global assembly cache, which is a repository of assemblies maintained by the .NET runtime. Shared assemblies are usually libraries of code which many applications will find useful, e.g. the .NET Framework classes.
- **Versioning:** The runtime enforces versioning constraints only on shared assemblies, not on private assemblies.

15. How do assemblies find each other?

Assemblies find each other by searching directory paths. There are several factors which can affect the path (such as the Application Domain host, and application configuration files), but for private assemblies the search path is normally the application's directory and its sub-directories. For shared assemblies, the search path is normally same as the private assembly path plus the shared assembly cache.

16. How does assembly versioning work?

Each assembly has a version number called the compatibility version. Also each reference to an assembly (from another assembly) includes both the name and version of the referenced assembly.

The version number has four numeric parts. Assemblies with either or the first two parts different are normally viewed as incompatible. If the first two parts are the same, but the third is different, the assemblies are deemed as 'maybe compatible'. If only the fourth part is different, the assemblies are deemed compatible. However, this is just the default guideline - it is the version policy that decides to what extent these rules are enforced. The version policy can be specified via the application configuration file.

Remember: Versioning is only applied to shared assemblies, not private assemblies.

17. What is an Application Domain?

An application domain can be thought of as a lightweight process. Multiple Application Domains can exist inside a Win32 process. The primary purpose of the Application Domain is to isolate an application from other applications.

Win32 processes provide isolation by having distinct memory address spaces. This is effective, but it is expensive and doesn't scale well. The .NET runtime enforces Application Domain isolation by keeping control over the use of memory - all memory in the Application Domain is managed by the .NET runtime, so the runtime can ensure the Application Domains do not access each other's memory.

18. How does an Application Domain get created?

Application Domains are usually created by hosts. Examples of hosts are the Windows shell, ASP.NET and IE. A .NET application is run from the command line, the host is the Shell. The shell creates a new Application Domain for every application.

Application Domains can also be explicitly created by .NET applications. Here is a C# sample which creates an Application Domain, creates an instance of an object inside it, and then executes one of the object's methods.

19. Explain the differences between Serverside and Client-side code.

Server side code basically gets executed on the server (for example on a webserver) per request/call basis, while client side code gets executed and rendered on the client side (for example web browser as a platform) per response basis.

20. What type of code (server or client) is found in a Code-Behind class?

In the Code-behind class the server side code resides, and it generates the responses to the client appropriately while it gets called or requested.

21. Should validation (did the user enter a real date) occur server-side or client-side? Why?

That depends. In the up browsers (like IE 5.0 and up and Netscape 6.0) this would help if it gets validated on the client side, because it reduces number of round trips between client and server. But for the down level browsers (IE 4.0 and below and Netscape 5.0 and below) it has to be on server side. Reason being the validation code requires some scripting on client side.

22. What does the "Enable ViewState" property do? Why would we want it on or off?

EnableViewState stores the current state of the page and the objects in it like text boxes, buttons, tables etc. So this helps not losing the state between the roundtrips between client and server. But this is a very expensive on browser. It delays rendering on the browser, so we should enable it only for the important fields/objects.

23. Write the difference between Server.Transfer and Response.Redirect.

Server.Transfer transfers the current context of the page to the next page and also avoids double roundtrips. Whereas Response.Redirect could only pass querystring and also requires roundtrip.

24. Can we give an example of when it would be appropriate to use a web service as opposed to a non-serviced .NET component?

Webservice is one of main component in Service Oriented Architecture. We can use webservices when our clients and servers are running on different networks and also different platforms. This provides a loosely coupled system. And also if the client is behind the firewall it would be easy to use webservice since it runs on port 80 (by default) instead of having some thing else in SOA apps.

25. Let's say we have an existing application written using Visual Studio 6 (VB 6, InterDev 6) and this application utilizes Windows 2000 COM+ transaction services. How would we approach migrating this application to .NET?

We have to use System.EnterpriseServices namespace and also COMinterop the existing application.

26. Explain the difference between an ADO.NET Dataset and an ADO Recordset.

ADO.NET DataSet is a mini RDBMS based on XML, where as RecordSet is collection of rows. DataSet is independent of connection and communicates to the database through DataAdapter, so it could be attached to any well defined collections like hashtable, dictionary, tables, array lists etc. practically. And also it can be bound to DataGrid etc. controls straightaway. RecordSet on the otherhand is tightly coupled to Database System.

27. Give an example of what might be best suited to place in the Application_Start and Session_Start subroutines.

In the Application_Start event we could store the data, which is used throughout the lifetime of an application for example application name, where as Session_Start could be used to store the information, which is required for that session of the application say for example user id or user name.

28. If we are developing an application that must accomodate multiple security levels though secure login and my ASP.NET web application is spanned across three webservers (using round-robbin load balancing) what would be the best approach to maintain login in state for the users?

State server is to be used or the state in the database is to be stored. This can be easily done through simple setting change in the web.config.

```
<sessionState mode="inProc"
state ConnectionString="tcpip=127.0.0.1:42424"
sqlConnectionString="data
source = 127.0.0.1;userid=sa;password="
    cookieless="false"timeout="30" />
```

in the above one instead of mode='inProc', you specify stateserver or sqlserver.

29. What are ASP.NET Web Forms? How is this technology different than what is available though ASP (1.0-3.0)?

ASP.NET webforms are analogous to Windows Forms which are available to most VB developers. A webform is essentially a core container in a Page. An empty webform is nothing but a HTML Form tag(control) running at server and posting form to itself by default, but we can change it to post it to something else. This is a container, and we can place the web controls, user controls and HTML Controls in that one and interact with user on a postback basis.

30. How does VB.NET/C# achieve polymorphism?

Polymorphism is achieved through virtual, overloaded, overridden methods in C# and VB.NET.

31. How would we implement inheritance using VB.NET/C#?

Inheritance is extending the properties, behaviour, methods to child classes from super classes. VB.NET and C# provide single inheritance, means the subclasses can be derived from only one parent unlike C++, where true multiple inheritance is possible. As an alternate to implement multiple inheritance, we could do the same to implement interfaces to the parent classes and implement the same interfaces to derive the child classes.

32. What is other name for shared assembly?

. Global Assembly

33. How do we create and use shared assemblies?

Using GACUtil.exe

34. How do we use thread in our .NET application?

Derive our class from Thread Class

35. What are the main differences between Web Server controls and HTML Server controls?

HTML Server controls are the nothing but some html tags with runat=server attribute in that tag. Web Server controls have richer event handling. HTML Server Controls need to be nested within a form tag and so on.

36. Write the difference between ref and out parameters.

Out parameter need not be initialized, where as ref parameter needs to be. This clarifies that ref is both I/O parameter and out is only Output parameter as name suggests.

37. How web services in ASP.NET is consumed?

In order to consume .NET web services, we need to create a proxy (if we are not using VS.NET, use wsdl.exe) then compile that proxy class and add reference to our webservice and start using it in our class (ASP.NET Application). If we are using VS.NET, then we could straightaway add web reference on our project to that webservice and start using it.

38. To test web service we must create a windows application or web application to consume this service (True or False).

This is tricky. If the web service doesn't provide a test form (hiding it, through web.config by disallowing HTTP GET method) then we need to create a web application or windows app to test it.

39. When on the internet whether we would look for web services?

The question may be where on the internet would we look for web services. Then we should go to UDDI Server and find.

40. Does .NET supports Pessimistic record locking or Optimistic record locking or both?

It supports both.

41. What is Catch API?

We are not aware of any Catch API. If it is Cache then look for Cache class and if it is exception handling (try catch throw block) look for it.

42. How can we handle concurrency errors?

Through .NET Exception handling.

43. In .NET Exception handling model, what are the benefits of over return code?

Return code returns only the code and we need to implement a message related to that. If we use exception handling it throws the complete exception including code (if any), message and stack trace. So the debugging would be a lot easier.

We could return only one value from a function unless we pass 'out' parameters. Where are if we use Exception Handling model, we can show a lot of information and also do gracefull terminate the execution of the application and we have a chance to close all opened resources in the finally block and also dispose the objects used.

44. What is the difference between Managed OLEDB and Managed SQLServer provider?

The managed OLEDB Provider is a generic provider and Managed Sqlserver Provider (SqlClient) is optimised for SqlServer, so it has SqlServer specific calls internally.

45. What are the parts of .NET assembly and explain?

Assemblies can be made up of one or more files. Each file can be either a module or another assembly.

The contents of assembly could include:

Manifest, Type MetaData, IL Code and Resources.

But one of the file must contain a manifest.

46. What is AutoPostBack?

This property is used to specify whether the state of the control is posted back to the server when clicked, changes, list.

47. When and why do we use AutoGenerateColumn property?

Gets or sets a value that indicates whether BoundColumn objects are automatically created and displayed in the DataGrid control for each field in the data source.

This property is used to automatically create a BoundColumn object for each field in the data source. Each field is then rendered as a column in the DataGrid control in the order that the fields appear in the data source.

48. How can View State be used?

The current property settings of an ASP.NET page and those of any ASP.NET server controls contained within the page. ASP.NET can detect when a form is requested for the first time versus when the form is posted (sent to the server), which allows us to program accordingly.

49. Write the difference between Managed code and Unmanaged Code.

When we compile our Visual Studio.NET source code, the compiler gives out MSIL (Microsoft Intermediate Language). MSIL code converts into native code at runtime by a Just In Time (JIT) Compiler. MSIL code should supply all the metadata necessary for the Common Language runtime (CLR) to provide services such as Garbage Collection, Memory Management, Cross-Language Intergration, Code Access Security etc. All code that is based on MSIL executes as Managed Code.

Unmanaged code is the one that was created without taking the requirements of the CLR into account. This code executes in the runtime environment with minimal services from runtime. For example, the above COM object that we created using VB 6 executes as unmanaged code.

50. Write the differences between DataGrid, DataList, and Repeater.

A Brief Overview of the DataGrid Web Control:

The DataGrid Web control was designed to display data in an HTML <table>. Each row in the DataGrid's DataSource is displayed as a row in the HTML <table>. The DataGrid has an AutoGenerateColumns property, which can be set to either True or False. If AutoGenerateColumns is set to True (the default), each field in the DataSource is displayed as a column in the resulting HTML<table>. If, however, AutoGenerateColumns is set to False, then the developer must explicitly specify what columns should appear.

One downside of a DataGrid is its rather "blocky" display. That is, the DataGrid displays each DataSource record as a row in an HTML <table>, and each field as a column in said table. While the DataGrid allows for the developer to customize the output for a particular column through the <asp:TemplateColumn>, it still is restrictive in that each DataSource record occupies precisely one HTML <table> row. (For an example of customizing a DataGrid's column using templates, see Combining Two DataSource Fields Into One DataGrid Column.)

Despite its limitations on specifying the display of the DataGrid's data, the DataGrid is the most commonly used data Web control due to its impressive feature list. For example, with minimal code and effort, the DataGrid can provide sorting, paging, and in-place editing of its data. (For more on the DataGrid Web control, as well as thorough coverage of its features, be sure to read An Extensive Examination of the DataGrid Web Control.)

A Brief Overview of the DataList Web Control:

The DataList Web control is useful for displaying data that can be highly customized in its layut. By default, the DataList displays its data in an HTML <table>. However, unlike the DataGrid, with the DataList we can specify via the RepeatColumns how many DataSource records should appear per HTML <table> row. For example, the following code and live demo illustrates having five DataSource records displayed per HTML <table> row.

```
<asp: DataList runat = "Server".
                    id="dlPopularFAQs"
          ...
              RepeatColumns="5">
<ItemTemplate>
<b>Question:</b><br/>
<%# DataBinder.Eval(Container.DataItem,
"Description")%>
<p>
<b>Views:</b><br/>
<%#DataBinder.Eval(container.DataItem,
"ViewCount","{0:#,###}") %>
</ItemTemplate>
</asp:DataList>
```

As the code example above shows, DataLists are comprised of a number of templates. Templates can contain a mix of HTML syntax and databinding expressions, as shown in the above ItemTemplate.

Databinding expressions are ones delimited by <%# and %>, and contain code that's executed when the DataList Item's DataBind( ) method is called. The ItemTemplate specifies the template used for each record in the DataSource. The DataList contains other templates, which are listed below:

AlternatigItemTemplate - If specified, each alternating Item in DataSource uses this template instead of the itemTemplate.

EditItemTemplate - the template used when a DataList row is in "edit mode". HeaderTemplate - the template for the DataList's header item. Precisely one header Item is created for the DataList item if the DataList's ShowHeader property is set to True. FooterTemplate - the template for the Datalist's footer item. Precisely one footer item is created for the DataList item if the DataList's ShowFooter property is set to True. SeparatorTemplate -this template, if specified, is applied after each DataListitem has been added.

The DataList is capable of performing sorting, paging, and editing of its data, but to do so requires quite a bit more programming than to accomplish these tasks with the DataGrid. There fore, if we knwo we will be needing these functionalities, it is likely best to proceed with the DataGrid. If, however, we will not need these funcitonalities, but do necessarily need the extra control over the formatting, consider a DataList.

A Brief Overview of the Repeater Control:

The Repeater contro, unlike the DataGrid and DataList, is not derived from the WebControl class, and therefore does not have the stylistic properties that are common to all Web controls. (These stylistic properties that are common to all Web controls include Font, ForeColor, BackColor, BorderStyle, and so on.)

The Repeater, like the DataList, only supports templates; however, the Repeater has only a subset of the DataList's template options. Specifically, the following templates can be used with the Repeater:

AlternatingitemTemplate, ItemTemplate, HeaderTemplate, FooterTemplate, and SeparatorTemplate

The Repeater control provides the maximum amount of flexibility over the HTML produced. Whereas the DataGrid wraps the DataSource contents in an HTML <table>, and the Datalist wraps the contents in either an HTML <table> or <SPan> tags depending on the DataList's RepeatLayout property), the Repeater adds absolutely no HTML content other than what we explicitly specify in the templates.

Therefore, the Repeater is a good data Web control choice if we want to display data in, say, an unordered list. As the following code and live demo show, displaying database data in an unordered list using a Repeater is relatively straightforward. All we have to do is add the <ul> tag to the HeaderTemplate, the </ul> tag to the FooterTemplate, and an <li> tag along with the DataSource field we want to display to the ItemTemplate:

```
<asp:Repeater runat="server"
id="rptULforFAQs">
  <HeaderTemplate><ul>
  </HeaderTemplate>
<itemTemplate> <li><%# DataBinder.Eval
(Container.DataItem, "Description")%></li>
</ItemTemplate>
  <FooterTemplate></ul>
  </FooterTemplate>
</asp:Repeater>
```

The Repeater is a good choice if we need to display our data in a format different than an HTML <table>. Unfortunately, the Repeater does not provide any builtin means for editing, sorting, or paging of data. However, these mechanisms could be added programmatically, but would result in a lot of code and effort.

51. What is garbage collection?

Garbage collection is a system whereby a run-time component takes responsibility for managing the lifetime of objects and the heap memory that they occupy. This concept is not new to .NET -Java and many other languages/runtimes have used garbage collection for some time.

52. Why doesn't the .NET runtime offer deterministic destruction?

because of the garbage collection algorithm, the .NET garbage collector works by periodically running through a list of all the objects that are currently being referenced by an application. All the objects that it doesn't find during this search are ready to be destroyed and the memory reclaimed. The implication of this algorithm is that the runtime doesn't get notified Immediately when the final reference on an object goes away - it only finds out during the next sweep of the heap.

Furthermore, this type of algorithm works best by performing the garbage collection sweep as rarely as possible. Normally heap exhaustion is the trigger for a collection sweep.

53. Is the lack of deterministic destruction in .NET a problem?

It's certainly an issue that affects component design. If we have objects that maintain expensive or scarce resources (e.g. database locks), we need to provide some way for the client to tell the object to release the resource when it is done. Microsoft recommend that we provide a method called Dispose( ) for this purpose. However, this causes problems for distributed objects - in a distributed system who calls the Dispose( ) method? some form of reference-counting or ownership-management mechanism is needed to handle distributed objects - unfortunately the runtime offers no help with this.

54. Does non-deterministic destruction affect the usage of COM objects from managed code?

Yes, When using a COM object from managed code, we are effectively relying on the garbage collector to call the final release on our object. If our COM object holds onto an expensive resource which is only cleanedup after the final release, we may need to provide a new interface on our object which supports an explicit Dispose( ) method.

55. What is serialization?

Serialization is the process of converting an object into a stream of bytes. Deserialization is the opposite process of creating an object from a stream of bytes. Serialization/Deserialization is mostly used to transport objects (e.g. during remoting), or to persist objects (e.g. to a file or database).

56. Does the .NET Framework have in-built support for serialization?

There are two separate mechanisms provided by the .NET class library - XmlSerializer and SoapFormatter/ BinaryFormatter. Microsoft uses XmlSerializer for Web Services, and uses SoapFormatter/BinaryFormatter for remoting. Both are available for use in our own code.

57. Why is XmlSerializer so slow?

There is a once-per-process-per-type overhead with XmlSerializer. So the first time we serialize or deserialize an object of a given type in an application, there is a significant delay. This normally doesn't matter, but it may mean, for example, that XmlSerializer is a poor choice for loading configuration settings during startup of a GUI application.

58. What are attributes?

There are at least two types of .NET attribute. The first type we refer to as a metadata attribute - it allows some data to be attached to a class or method. This data becomes part of the metadata for the class, and (like other class metadata) can be accessed via reflection. An example of a metadata attribute is (serializable), which can be attached to a class and means that instances of the class can be serialized.

[serializable] public class CTest { }

The other type of attribute is a context attribute. context attributes use a similar syntax to metadata attributes but they are fundamentally different. Context attributes provide and interception mechanism whereby Instance activation and method calls can be pre-and/or post-processed. If we come across Kielth Brown's universal delegator we will be familiar with this idea.

59. What is Code Access Security (CAS)?

CAS is the part of the .NET security model that determines whetner or not a piece of code is allowed to run, and what resources it can use when it is running. For example, it is CAS that will prevent a .NET web applet from formatting our hard disk.

60. How does CAS work?

The CAS security policy revolves around two key concepts - code groups and permissions. Each .NET assembly is a member of a particular **code group,** and each code group is granted the permissions specified in a **named permission set.**

For example, using the default security policy, a control downloaded from a web site belongs to the 'Zone - Internet' code group, which adheres to the permissions defined by the 'Internet' named permission set. (Naturally the 'Internet' named permission set represents a very restrictive range of permissions.)

61. Who defines the CAS code groups?

Microsoft defines some default ones, but we can modify these and even create our own. To see the code groups defined on our system, run 'caspol -lg' from the commandline. On our system it looks like this:

Level=Machine

Code Groups:

1. All code: Nothing

1.1.Zone - MyComputer: FullTrust

1.1.1. Honor Skip Verification requests: SkipVerification

1.2. Zone - Intranet: LocalIntranet

1.3. Zone - Internet: Internet

1.4. Zone - Untrusted: Nothing

1.5. Zone - Trusted: Internet

1.6. StrongName - 002400000480000094000000060200000024000052534131000400000300000 0CFCB3291AA715FE99D40D49040336F9056D7886FED46775BC7BB5430BA4444FEF8348EBD06

F962F39776AE4DC3B7B04A7FE6F49F25F740423EBF2C0B89698D8D08AC48D69CED0FC8F83B465E08

07AC11EC1DCC7D054E807A43336DDE408A5393A48556123272CEEEE72F1660B71927D38561AABF5C

AC1DF1734633C602F8F2D5: Everything.

Note the hierarchy of code groups - the top of the hierarchy is the most general (All code), which is then sub-divided into several groups, each of which in turn can be subdivided. It may be noted that (somewhat counter-intuitively) a sub-group can be associated with a more permissive permission set than its parent.

62. Can source code be reverse-engineered from IL?

Yes, it is often relatively straightforward to regenerate high-level source (e.g. C#) from IL.

63. Is DCOM dead?

Pretty much, for .NET developers. The .NET Framework has a new remoting model which is not based on DCOM. Of course DCOM will still be used in interop scenarios.

64. Is MTS/COM+ dead?

No. The approach for the first .NET release is to provide access to the existing COM+ services (through an interop layer) rather than replace the services with native .NET ones. Various tools and attributes are provided to try to make this as painless as possible. The PDC release of the .NET SDK includes interop support for core services (JIT activation, transactions) but not some of the higher level services (e.g. COM+ Events, Queued components).

65. Is ATL redundant in the .NET world?

Yes, if we are writing applications that live inside the .NET framework. Of course many developers may wish to continue using ATL to write C++ COM components that live outside the framework, but if we are inside we will almost certainly want to use C#. Raw C++ (and therefore ATL which is based on it) doesn't have much of a place in the .NET world-it's just too near the metal and provides too much flexibility for the runtime to be able to manage it.

66. How does .NET remoting work?

.NET remoting involves sending messages along channels. Two of the standard channels are HTTP and TCP. TCP is intended for LANs only -HTTP can be used for LANs or WANs (internet).

Support is provided for multiple message serialization formats. Examples are SOAP (XML - based) and binary.

By default, the HTTP channel uses SOAP and the TCP channel uses binary. But either channel can use either serializaiton format.

There are a number of styles of remote access:

- **SingleCall:** Each incoming request from a client is serviced by a new object. The object is thrown away when the request has finished. This (essentially stateless) model can be made stateful in the ASP.NET environment by using the ASP.NET state service to store application or session state.
- **Singleton:** All incoming requests from clients are processed by a single server object.
- **Client-activated object:** This is the old stateful (D)COM model whereby the client receives a reference to the remote object and holds that reference (thus keeping the remote object alive) until it is finished with it.

Distributed garbage collection of objects is managed by a system called 'leased based lifetime'. Each object has a lease time, and when that time expires the object is disconnected from the .NET runtime remoting infrastructure. Objects have a default renew time - the lease is renewed when a successful call is made from the client to the object. The client can also explicitly renew the lease.

67. Is DOM supported?

Yes Let us have the example XML document:

```
<PEOPLE>
   <PERSON>Fred</PERSON>
   <PERSON>Bill</PERSON>.
</PEOPLE>
```

This document can be parsed as follows:

```
XmlDocument doc = new XmlDocument( );
doc.Load("test.xml");
XmlNode root = doc.DocumentElement;
foreach(XmlNode personElement in
   root.ChildNodes)
Console. WriteLine
(personElement.FirstChild.Value.ToString( ));
```

The output is:

Fred

Bill

68. Is SAX supported?

No. Instead, a new XmlReader/XmlWriter API is offered. Like SAX it is stream-based but it uses a 'pull' model rather than SAX's 'push' model. Here's an example:

```
XmlTextReader reader = new XmlTextReader
                                    ("test.xml");
while(reader.Read( ))
{
   if(reader.NodeType==
XmlNodeType.Element && reader.Name==
"PERSON")
   {
   reader.Read( ); //Skip to the child text
   Console.WriteLine(reader.Value);
   }
}
```

69. Is XPath supported?

Yes, via the XPath XXX classes:

```
XPathDocument xpdoc
         =new XpathDocument("test.xml");
XPathNavigator nav = xpdoc.CreateNavigator( );
XPathExpression expr = nav.Compile
         ("descendant::PEOPLE/PERSON");
XPathNodeiterator iterator = nav.Select(expr);
while (iterator.MoveNext())
         Console.WriteLine(iterator.Current);
```

70. Is multi-threading supported?

Yes, there is extensive support for multithreading. New threads can be spawned, and there is a system-provided threadpool which applications can use.

71. Is there built-in support for tracing/logging?

Yes, in the System.Diagnostics namespace. There are two main classes that deal with tracing - Debug and Trace. They both work in a similar way - the difference is that tracing from the Debug class only works in builds that have the DEBUG symbol defined, whereas tracing from the Trace class only works in builds that have the TRACE symbol defined. Tpically this means that we should use System.Diagnostics.Trace.WriteLine for tracing that we want to work in debug and release builds, and System.Diagnostics.Debug.Writeline for tracing that we want to work only in debug builds.

EXERCISE

1. Fill in the missing word or words.

1. Before a Visual Basic .NET application can be run it must be compiled into a language called ....
2. The primary components of the .NET Framework are the .NET Framework Class Library and the ....
3. When we add a control to a form, we start by finding the control in the ....
4. An object is derived from a ....
5. For business applications, the two data types that we'll use the most for numeric data are the integer and the .... data types.
6. To override the order of precedence in an arithmetic expression, we can use ....
7. If dNumber has a value of 1200.333, the shared method that follows returns a value of ....
8. The difference between a Function and a Sub procedure is that a function always .... a value.
9. If we want a Sub procedure to update a variable that's in the calling program, the variable must be passed by ....
10. To cast data implicitly, we use .... statements.

2. Choose the correct answer.

11. The Common Language Runtime provides services for

(A) executing any application developed with Visual Basic

(B) executing any application developed with a Visual Studio language.

(C) executing any application developed with a .NET language

(D) executing any application

12. We can

(A) open a project or solution, but we can only close a solution.

(B) open a project or a solution, and we can close a project or a solution.

(C) only open a project, but we can close either a project or a solution.

(D) only open a solution, but we can close either a project or a solution.

13. When we double-click on a button named btnProcess in the Form Designer, Visual Basic starts a procedure named

(A) btnProcess.Click

(B) btnProcess_Click

(C) btnProcess.DoubleClick

(D) btnProcess_DoubleClick

14. A syntax error is identified in

(A) the Code Editor window

(B) the Task List window

(C) the Run-Time dialog box

(D) the Code Editor window and the Task List window.

15. What is the value of this arithmetic expression?

(250-200 / 200*100

(A) 0.25 (B) 25

(C) 25000 (D) None of the above

16. What is the value of sCityState after these statements are executed?

scitystate = "Milwaykee"

scitystate & = ","

scitystate = ","

scitystate = "wisconsin"

(A) Milwaukee, Wisconsin

(B) Milwaukee, Wisconsi

(C) MilwaukeeWisconsin

(D) Wisconsin

17. If dNumber has a value of 12345, which of the following functions will format the data as: 12,345.00?

(A) FormatCurrency(dNumber)

(B) FormatCurrency(dNumber,2)

(C) FormatNumeber (dNumber)

(D) None of the above

18. We can't, if we do numeric validation and numeric range checking as parts of a compound condition then

(A) the range test will cause a run-time error if the data isn't numeric

(B) the Visual Basic syntax doesn't provide for compound validation tests

(C) the range checking has to be done before the numeric validation

(D) None of the above

19. Assuming the standard prefixes for variable names, which of the following could cause a casting exception?

(A) sValue=txtValue.Text

(B) dValue=Math.Round(dValue,2)

(C) dValue=iValue (D) iValue=iValue/2

For question 20, please refer to the Function statement that follows:

Private Function Interest(ByVal years As Integer,_

ByVal InterestRate As Decimal,_

ByVal Principle As Decimal) As Decimal

20. Assuming the standard prefixes for variables, which of the following is a valid call of this function?

(A) dInterest = Interest (iYrs, dIntRate,dPrin)

(B) dInterest = Interest(dintRate, dPrin,iYrs)

(C) dInterest = Interest(iyears,iInterestRate, dPrinciple)

(D) dInterest = Interest(dInterestRate, dPrinciple, iYears)

HTML

WHAT IS AN HTML FILE?

An HTML file is what a Web browser uses to generate a web page. At its heart, the World Wide Web is nothing but a vast collection of HTML files residing on the hard drives of computers spread throughout the world and a transport **protocol** for transferring these files from computer to computer. These HTML files in turn, are simply text files - files that can be easily read and understood by human eyes.

Web browsers such as Netscape Navigator or Microsoft Internet Explorer, interpret HTML files in order to display Web pages. This is the main function of a Web browser. Whenever we use a browser to view a page on the World Wide Web, the browser has converted that Web page from an HTML file.

INTRODUCTION TO HTML TAGS

HTML files are different from other text files because they include special codes called HTML tags. The HTML tags are surrounded by the two angle bracket characters (< and >). Tags are used to denote the various elements in a HTML document. Tags are usually paired (e.g., <h1> and </h1>) to start and end the tag instruction. The end tag is similar to the start tag except a slash (/) precedes the text within the brackets.

HTML is not case sensitive.

e.g., <title> is equivalent to <TITLE>. Not all tags are supported by all World Web browsers.

Tags define

- headings
- paragraphs
- lists

- character styles
- links

Minimal HTML Document

Every HTML, document should contain standard HTMlLtags. Each document consists of head and body text. The head contains the title, and the body contains the actual text that is made up of paragraphs, lists and other elements. Browsers expect specific information because they are programmed to HTML and SGML specifications.

Required elements are shown in the below example:

```
<html>
<head>
<TITLE> A simple HTML Example </TITLE>
</head>
<body>
<h1> HTML is easy to learn </h1>
<P> welcome to the world of HTML.
This is the first paragraph. While short, it is still a paragraph ! </P>.
<P> And this is the second paragraph
</P>
</body>
</html>
```

FUNCTIONAL MARKUP TAG

HTML

This element our browser that the file contains HTML-coded information. The file extension .html also Indicates this an HTML document and must be used. (If we are restricted to 8.3 filenames. (e.g., LeeHome. html, use only for our extension.)

HEAD

The head element identifies the first part of HTML-coded document that contains the title. The title is shown as part of our browser's window.

TITLE

The title element contains our document title and identifies its content in a global context. The title is displayed somewhere on the browser window (usually at the top), but not within the next area. The title is also what is displayed on someone's hotlist or book mark list, so choose something descriptive unique and relatively short.

BODY

The second and largest-part of our HTML document is the body, which contains the content our document (displayed within the next area of our browser window).

```
<html>
<head>
<title> My first HTML Document </title>
</head>
<body>
<h1> Welcome to HTML </h1>
</body>
</html>
```

TEXT MARKUP TAG

Headings

HTML has six levels of headings, numbered 1 through 6 with 1 being the most prominent. Headings are displayed larger and/or bolder fonts than normal body text. The first heading in each document should be tagged <H1>. The syntax of the heading element is:

```
<Hy> Text of heading </Hy>
```

where y is a number between 1 and 6 specifying the level of the heading.

Paragraphs

In the example shown in the minimal HTML document section, the first paragraph is coded as

```
<P> Welcome to the world of HTML.
This is the first paragraph
While short it is still a paragraph !
</P>
```

In the source file there is a line break between the sentences. A Web browser ignores this line break and starts a new paragraph only when it encounters another <P> tag.

NOTE: The </P> closing tag can be omitted. This is because browsers understand that when they encounter a <P> tag, it implies that there is an end to the previous paragraph. Using the <P> and </P> as a paragraph container means that we can center a paragraph by including the ALIGN = alignment attribute in our source file.

```
<PALIGN = CENTER>
This is a centered paragraph.
(See the formatted version below)
</P>
```

OUTPUT

This is a centered paragraph. (see the formatted version below.)

Lists

HTML supports unnumbered, numbered and definition lists.

Unnumbered Lists

Below is a sample of three-item list:

```
<UL>
<LI> apples
<LI> bananas
<LI> grape fruit
</UL>
```

The output is:

- apples
- bananas
- grape fruit

Numbered Lists

A numbered list (also called an ordered list) is identical to an unnumbered list, except it uses <O1> instead of <UL>. The items are tagged using the same <LI> tag.

The following HTML code:

```
<OL>
<LI> oranges
<LI> peaches
<LI> grapes
</OL>
```

produces this formatted output:

1. oranges
2. peaches
3. grapes

Definition Lists

A definition list is a set of definitions, references or cross indexes. The list contained within a pair of definition list tags: <DL> and </DL>.

Within the pair of listings, a definition has two parts.

- Definition Term (<DT>)
- Definition Description (<DD>)

The following is an example of a definition list:

```
<DL>
<DT> NCSA
<DD>NCSA, the National Center for Supercomputing Applications, is located on the campus of the University of Ilinois at Urbana-champaign.
<DT> Cornell Theory Center
<DD> CTC is located on the campus of Cornell University, in Itchca, New York.
</DL>
```

The output looks like:

NCSA

NCSA, the National Center for Super computing Applications, is located on the campus of the University of Ilinois at urbana champaign.

Cornell Theory Center

CTC is located on the campus of Cornell University in Itchca, New York.

The <DT> and <DD> entries can contain multiple paragraphs (indicated by <P> paragraph tags), lists or other definition information.

The COMPACT attribute can be used routinely in case our definition terms are very short. If, for example, we are showing some computer options, the options may fit on the same line as the start of the definition.

```
<DL COMPACT>
<DT>-I
<DD> invokes NCSA Mosaic for Microsoft Windows using the initialization file defined in the path.
<DT> -k
<DD> invokes NCSA Mosaic for Microsoft Windows in kiosk mode.
</DL>
```

The output looks like:

-i invokes NCSA Mosaic for Microsoft Windows using the initialization file defined in the path.

-k invokes NCS mosaic for Microsoft Windows in kiosk mode.

Nested Lists

Lists can be nested. We can also have a number of paragraphs, each containing a nested list, in a single list item.

Here is a sample nested list:

```
<UL>
<LI> A few New England states:
<UL>
<LI>Vermont
<LI>New Hampshire
<LI> Maine
</UL>
<LI> Two Midwestern states:
<UL>
<LI> Michigan
<LI> Indiana
</UL>
</UL>
```

The nested list is displayed as

A few New England states:

- Vermont
- New Hampshire
- Maine

Two Midwestern states:

- Michigan
- Indiana

Preformatted Text

Use the <PRF> tag (which stands for "preformatted") to generate text in a fixed width font. This tag also make spaces, new lines, and tabs significant (multiple spaces are displayed as Multiple spaces, and lines break in the same locations as in the source HTML file). This is useful for program listings, among other things. For example, the follwing lines:

```
<HTML>
        Dave's Desk <BR>
        <P>
        Dear Reader, <BR>
        <P>
Thank you for connecting to my Web server.
        <PRE>
                                    Yours sincerely,
                                         Dave Taylor
        </PRE>
        </HTML>
```

display as:

```
        Dave's Desk
        Dear Reader,
Thank you for connecting to my Web server.
                                    Yours sincerely,
                                         Dave Taylor
```

The <PRE> tag can be used with an optional WIDTH attribute that specifies the maximum number of characters for a line. WIDTH also signals your browser to choose an appropriate font and indentation for the text.

Extended Quotations

Use the <BLOCKQUOTE> tag to include lengthy quotations in a separate block on the screen. Most browsers generally change the margins for the quotation to separate it from surrounding text.

<BLOCKQUOTE>

<P> Omit needless words. </P>

<P> vigorous writing is concise. A sentence should contain no unnecessary words, a paragraph no unnecessary sentences, for the same reason that a drawing should have no unnecessary lines and a machine to a unnecessary parts. </P>

-- William Strunk, Jr., 1918

</BLOCKQUOTE>

The result is:

Omit needless words.

Vigorous writing is concise. A sentence should contain no unnecessary words, a paragraph no unnecessary sentences, for the same reason that a drawing should have no unnecessary lines and a machine no unnecssary parts.

- William Strunk, Jr.,1918

Address

The <ADDRESS> tag is generally used to specify the author of a document, a way to contact the author (e.g., an email address), and a revision date. It is usually the last item in a file.

For example, the last line of the online version of this guide is:

<ADDRESS>

A Beginner's Guide to HTML / NCSA / pubs@ncsa.uiuc.edu./rvd April 96

<ADDRESS>

The result is:

A Beginner's Guide to HTML / NCSA / pubs@ncsa.uiuc.edu./rvd April 96

Forced Line Breaks / Postal Addresses

The <BR> tag forces a line break with no extra (white) space between lines. Using <P> elements for short lines of text such as postal addresses results in unwanted additional white space. For example, with <BR>:

National Center for Supercomputing

Applications

605 East Springfield Avenue

Champaign, Ilinois 61820-5518

The output is:

National Center for Supercomputing Applications

Applications

605 East Springfield Avenue

Champaign, Ilinois 61820-5518

The output is:

National center for supercomputing

Applications

605 East Springfield Avenue

Champaign, Ilinois 61820-5518

Horizontal Rules

The <HR> tag produces a horizontal line to the width of the browser window. A horizontal rule is useful to separate sections of our document. For example, many people add a rule at the end of their text and before the <address> information.

Character Formatting

HTML has two types of styles for individual words or sentences: logical and physical. Logical styles tag text according to its meaning. While physical styles indicate the specific appearance of a section.

Logical Styles

<DFN>

for a word being defined. Typically displayed in italics.

E.g., NCSA Mosaic is a World Wide Web browser.

<EM>

for emphasis. Typically displayed in italics.

E.g., Consultants cannot reset our password unless we call the help line.

<CITE>

for titles of books, films, etc. Typically displayed in italics.

E.g., A Beginner's Guide to HTML.

<CODE>

for computer code. Displayed in a fixed width font.

E.g., The <stdio.h> header file

<KBD>

for user keyboard entry. Typically displayed in plain fixed-width font.

E.g., Enter password to change your password.

<SAMP>

for a sequence of literal characters.

Displayed in a fixed-width font.

E.g., segmentation fault: core dumped.

<STRONG>

for strong emphasis. Typically displayed in bold.

E.g., NOTE: Always check your links.

<VAR>

for a variable. Typically displayed in italics.

E.g., rm filename deletes the file.

Physical styles

<B> bold text

<I> italic text

<TT> typewriter text, e.g., fixed-width font.

LINKING HTML PAGES

Linking

The chief power of HTML comes from its ability to link text and /or an image to another document or section of a document. A browser highlights the identified text or image with color and/or underlines to indicate that it is a hypertext link (often shortened to hyperlink or link).

HTML's single hypertext - related tag is <A>, which stands for anchor. To include an anchor in our document.

1. start the anchor with <A (include a space after the A)
2. specify the document we're linking to by entering the parameter HREF = "filename" followed by a closing right angle bracket (>).
3. enter the text that will serve as the hypertext link in the current document
4. enter the ending anchor tag: </A> (no space is needed before the end anchor tag)

Here is a sample hypertext reference in a file called US.html:

<A HREF = "MaineStates.html"> Maine</A>

This entry makes the word Maine the hyperlink to the document MaineStates.html, which is in the same directory as the first document.

Relative pathnames versus Absolute pathnames

We can link to documents in other directories by specifying the relative path from the current document to the linked document. For example, a line to a file NYStates.html located in the subdirectory Atlantic States would be:

<A HREF = "AtlanticStates/NYStates.html">

New York </A>

These are called relative links because we are specifying the path to the linked file relative to the location of the current file. We can also use the absolute pathname (the complete URL) of the file, but relative links are more efficient in accessing a server.

If we were in the NYStates.html file and were referring to the original document US.html, our link would look like this:

<A HREF="../US.html"> United States </A>

In general, we should use relative links because:

1. It's easier to move a group of documents to another location (because the relative path names will still be valid)
2. It's more efficient connecting to the server
3. there is less to type

However, use absolute pathnames when linking to documents that are not directly related. For example, consider a group of documents that comprise a user manual. Links within this group should be relative links. Links to other documents (perhaps a reference to related software) should use full path names. This way if we move the user manual to a different directory, none of the links would have to be updated.

Links to specific sections

Anchors can also be used to move a reader to a particular section in a document (either the same or a different document) rather than to the top, which is the default. This type of an anchor is commonly called a named anchor because to create the links, we insert HTML names within the document.

This material is a good example of using named anchors in one document. The material is constructed as one document to make printing easier, But as one (long) document, it can be time-consuming to move through when all we really want to know about is one bit of information about HTML.

Internal hyperlinks are used to create a "table of contents" at the top of this document. These hyperlinks move us front one location in the document to another location in the same document. (Go to the top of this document and then click on the links to specific sections hyperlink in the table of contents.)

The information is presented first because understanding that helps us understand linking within one document.

Links between sections of Different Documents

Suppose we want to set a link from document A (document A.html) to a specific section in another document (maineStates.html).

Enter the HTML coding for a link to a named anchor:

document A.html:

<a href="MaineStates.html#ANE">

Acadia National Park </a>

Think of the characters after the hash (#) mark as a tab within the MaineStates.html file. This tab tells our browser what should be displayed at the top of the window when the link is activated. In other words, the first line in our browser window should be the Acadia National Park heading.

Next, create the named anchor (in this example "ANP") in MaineStates.html:

<H2> <A NAME = "ANP"> Acadia National Park </A> </H2>

With both of these elements in place, we can bring a reader directly to the Acadia reference in MaineStates.html.

Links to specific sections within the current document

The technique is the same except the file name is omitted.

For example, to link to the ANP achor from within MaineStates, enter:

...... More information about <A HREF = "#ANP"> Acadia National Park </A) is available elsewhere in this document.

Be sure to include the <A NAME ⇒ tag at the place in our document where we want the link to jump to (<H2><A NAME = "ANP"> Acadia National Park </a></H2>).

Named anchors are particularly useful when we think readers will print a document in its entirely or when we have a lot of short information we want to place online in the file:

Mailto

The format is:

<A HREF = "mailto:emailinfo@host"> Name

</A>

For example,

<A HREF = "mailto:pubs@ncsa.uiuc.edu">

NCSA Publications Group </A>

To create a mail window that is already configured to open a mail window for the NCSA Publications Group alias.

INCLUDING IMAGES

Inline Images

Most Web browsers can display inline images (that is, images next to text) that are in X Bitmap (XBM), GIF or JPEG format. Other image formats are being incorporated into Web browsers (e.g., the Portable Network Graphic (PNG) format). Each image takes time to process and slows down the initial display of a document. Carefully select the images and the number of images in a document.

To include an inline image, enter:

<IMG SRC = ImageName>

where ImageName is the URL of the image file.

The syntax for <IMG SRC> URLs is identical to that used in an anchor HREF. If the image file is a GIF file, then the filename part of ImageName must end with. gif. Filenames of X Bitmap images must end with .xbm: JPEG image files must end with .jpg or .jpeg: and Portable Network Graphic files must end with .png.

Image Size Attributes

The two other attributes on <IMG> tags to browser the size of the images it is downloading with the text. The HEIGHT and WIDTH attributes let the browser set aside the appropriate space (in pixels) for the images as if downloads the rest of the file. (Get the pixel size from our image-processing software. Such as Adobe Photoshop.)

For example, to include a self portrait image in a file along with the portrait's dimensions, enter:

<IMG SRC = SelfPortrait.gif HEIGHT = 100 WIDTH = 85>

Aligning Images

We have some flexibility when displaying images. We can have images separated from text and aligned to the left or right or centered. Or we can have an image aligned with text. Try several possibilities to see how the information looks best.

Aligning Text with an Image

By default, the bottom of an image is aligned with the following text, as shown in this paragraph. We can align images to the top or center or a paragraph using the ALIGN attributes TOP and CENTER.

This text is aligned with the top of the image (<IMG SRC = "BarHotlist.gif". ALIGN = TOP>). Notice how the browser aligns only one line and then jumps to the bottom of the image for the rest of the text.

And this text is centered on the image (<IMG SRC = "BarHotlist.gif" ALIGN = CENTER>)

Again, only one line of text is centered: the rest is below the image.

Images Without Text

To display an image without any associated text (e.g., our organisation's logo), make it a separate paragraph. Use the paragraph ALIGN attribute to center the image or adjust it to the right side of the window as shown below:

<P ALIGN = CENTER>

<IMG SRC = "BarHotlist.gif">

</P>

which results in:

The image is centered: this paragraph starts below it and left justified.

Alternate Text Images

Some World Wide Web browsers - primarily those that run on VT 100 terminals - cannot display images. Some users turn off image loading even if their software can display images (especially if they are using a modem or have a slow connection).

The ALT attribute lets you specifuy text to be displayed instead of an image. For example:

<IMG SRC = "UpArrow.gif" ALT = "Up">

where UpArrow.gif is the picture of an upward pointing arrow. With graphicscapable viewers that have image-loading turned on; we see the up arrow graphic. Wigh a VT 100 browser or if image. Loading is turned off., the word up is shown in the window.

Background Graphics

Newer versions of Web browsers can load an image and use it as a background when displaying a page. Some people like background images and some don't. In general, if we want to include a background, make sure the text can be read easily when displayed on top of the image.

Background images can be a texture (linen finished paper, for example) or an image of an object (a logo possibly). We create the background image as we do any image.

However, we only have to create a small piece of the image. Using a feature called tiling, a browser takes the image and repeats it across and down to fill the browser window. In sum, we generate one image,and the browser replicates it enough times to fill our window. This action is automatic when we use the background tag shown below.

The tag to include a background image is included in the <BODY> statement as an attribute:

<BODY BACKGROUND = "filename.gif">

Background color

By default browsers display text in black on a gray background. However, we can change both elements if we want. Some HTML authors select a background color and coordinate it with a change in the color of the text.

Always preview changes like this to make sure the pages are readable. (For example, many people find red text on a black background difficult to read!)

We change the color of text, links, visited links, and active links using attributes of the <BODY> tag.

For example, enter:

<BODY BGCOLOR = "# 000000" TEXT = "# FFFFFF" LINK = "# 9690CC">

This creates a window with a black background (BGCOLOR) , white text (TEXT) and silvery hyperlinks (LINK).

The six-digit number and letter combinations represent colors by giving their RGB (red, green, blue) value. The six digits are actually three two-digit numbers in sequence representing the amount of red, green, or blue as a hexadecimal value in the range 00-FF. For example, 000000 is black (no color at all), FF0000 is bright red, and FFFFFF is white (fully saturated with all three colors). These number and letter combinations are cryptie. Fortunately, an online resource is available to help the track down the combinations that map to specific colors and there is software available for us to do this on the workstation.

ColorPro Web Server

(http://www.biola.edu/egi-bin/colorpro/)

WebColor-downloadable software site

(http://www.macworld.com/egi-bin/

software.pl/Telecominternet

/Software.767.html)

External Images, Sounds and Animations If we want to have an image open as a separate document when a user activates a link on either a word or a smaller, inline version of the image included in the document. This is called an external image. and it is useful if we do not wish to slow down the loading of the main document with large inline images.

To include a reference to an external image enter:

<A HREF = "Myimage.gif"> link anchor </A>

We can also use a smaller image as a link to a larger image. Enter:

<A HREF = "Largerimage.gif">

<IMG SRC = "Smallimage.gif"></A>

The reader sees the Smallimage.gif image and clicks on it to open the LargerImage.gif file.

Use the same syntax for links to external animations and sounds. The only difference is the file.

Extension of the linked file. For example.

<A HREF = "AdamsRib.mov">link anchor </A> specifies a link to a Quick time movie.

Most UNIX workstations, for instance, cannot view Quick time movies.

Example:

```
<HTML>
<HEAD>
<TITLE> The Dark Box </TITLE>
</HEAD><BODY>
<IMG SRC = "start-thrust-bottom.gif">
<H1> Welcome to the Dark Box <? H1>
We've never seen anything quite as terrifying as
<1> The Dark Box <?I>
Dare you continue? <A HREF = "black box2.html">
Yes </A> no.
<BODY></HTML>
```

Tables

Tables are found in most of the Web pages. Any business site has to include tables in its documents. It may be product list, services, inventory status or accounts statements, tables are present everywhere. The Table tag helps incorporate tables in Web pages. The format of the Table statement is shown below.

```
<TABLE> → Table begins
    <B><CAPTION "table title">
<TR> → First row (heading)
    <TH> First column heading
    <TH> Second column heading
    <TH> Third column heading
</TR>
<TR> → Second row (data)
    <TD> First column data
    <TD> Second column data
    <TD> Third column data
</TR>
    So on
</TABLE> → Table ends
```

The tags are self-explanatory to a certain extent. The illustration is shown below:

Code Reading

```
<!doctype html public "-// W3C // DTD. HTML 4.0 // EN">
<html lang = "EN">
<head>
<title > Devaki Departmental Shop < /title >
< /head >
<body >
<centre> <h2 > Devaki Departmental Shop < /h2 > < /centre>
<table >
    <caption>
            <b > provision </b>
< /caption>
< tr>
    < th > product < /th >
    <th > Category < /th >
    <th > packing < /th >
    <th > price < /th >
< /tr >
< tr>
    < td > Soap < /td >
    < td > Detergent < / td >
    < td > Box < / td >
    <td > 600 < / td >
< / tr >
< tr >
    < td > Fruits < /td >
    < td > Ooty Apples < /td >
    < td > 1 doz. Packets < /td >
    < td > 200 < / td >
< /tr >
< tr >
    < td > Processed Food < /td >
    <td > Kraft cheese < /td >
```

```
        < td > Tins < /td >
        < td > 150 < /td >
    < /tr >
    < /table >
    < /body > < /html >
```

Output of the table Feature

Devaki Departmental Shop

Provision

Product	Category	Packing	Price
Soap	Detergent	Box	600
Fruits	Ooty Apples	1 doz. packed	200
Processed Food	Kraft cheese	Tins	150

This is a borderless table. Though this table is quite readable, the appearance can be improved to a great extent by placing it inside a border. When the border attribute is used with the table tape, it covers the table inside a border.

The following table illustrates border table.

< Table Border >

There are many options that are available with the Table and Border feature.

Border = n willincrease the width of the border to the n size

Cellspacing=n will fragment the table with the n size

< Table border = 3 cellspacing = 15 > will have border of size 3 and fragment the table with 15 pixels spacing.

Cellpadding = n will keep the standard size for the cells.

If cellpadding = 15 the cell space will be adjusted according to the size of the table data to give breathing space for the table.

Cellspacing determines the space between the cells. Cellpadding determines the space within the cells.

<TD ALIGN > indicates how the table content has to be aligned. Standard LEFT, RIGHT, CENTRE table alignments can be used.

FRAMES AND FORMS

Frames: Building frames using multiple frames for real-life use, defining/restricting the user's intervention on frames and cleaning frames.

Forms: Creating form scrolling options and the beautification of forms.

Using Frames: For a regular surfer, frame is not an alien word. Many web sites operate with frames, which makes surfing very convenient. A frame divides a browser window into parts. Usually, a web page occupies the entire browser window. By using frames, it can be divided into two (or more) and different pages can be viewed in the same window.

Though viewing multiple pages in a single window does not make any practical sense, it is very useful. One frame can be kept for more options and the other frame for documents. Two (or more) different documents can be shown in a single window for comparing or contrasting information.

< Frame > is a tag that is used to create and manipulate frames. The first part of this session describes the syntax usage of the tag.

The following code explains the usage of th e frame tag.

```
< HTML >
< HEAD > < TITLE > Frame Page < /Title > < /Head >
< Frameset cols = "50% or 80% " >
< Frame SCR
        = "E : \ module 1 \ HTML \ frame-page 1.html" >
< Frame SCR
        = "E : \ module 1 \ HTML \ frame-page 2. html" >
< /Frameset >
< /HTML >
Frame-page 1. HTML
< HTML >
< Head > < Title > First frame < / Title > < / Head >
< Body >
<HI> First frame </HI>
<Body>
</HTML>
Frame - Page 2.html
<HTML>
<Head> <Title> second frame
</Title> </Head>
<Body>
        <H1 > Second frame < /H1 >
< Body >
< / HTML >
```

The code listing of three titles are given here - frame-head.html, frame-page1. html, frame-page2.html.

The first code frame-head.html creates the frameset. It can be called a head document or master document. This divides the window into two parts.

<FRAMESET COLS ="50%, 50%" >

This specifies that a frameset should be created column-wise. COLS stands for columns. It is supplied with a number which states that the column is divided equally. COLS "75%, 25% " will be an easy guess.

This statement is followed by the Frame SRC tag. FRAME SRC indicates the source for these two Web pages. The contents of the two documents are also given though they do not have new tags. An important aspect is that the FRAME-HEAD.HTML does not have a <Body>. It is not required as it is only a master file which does not perforn any task other than activating the other two Web pages that actually contain the body.

The output is

First page	Second page

Frameset builds a frameset and indicates their properties. viz., COLS, Rows are the two properties of frameset. COLS builds the frames vertically and rows does it horizontally.

The size of the column can be given in two styles. One is an explicit percentage style. Another is dividing the frames into pieces. E.g., COLS = "1*5" will divided the window into 2. The first one will get one sixth of the space and the second the remaining.

Like images, frames are also speciality elements that are confined to selective browsers.

```
<Frameset cols="50%, 50%">
<Frame SRC
    = "E:\Module 1\HTML \ frame-page 1.html">
<Frame SRC
    ="E:\Module 1 \HTML \ frame-page2.html">
<No frames> Your browser does not support frames view the documents separately.
< A HREF
    ="E:\Module 1 \HTML\frame-page1.html">
First Document </A><P>
<A HREF
    ="E:\Module1\HTML\frame-page2.html">
Second Document </A><P>
</No frames>
</Frameset>
```

Nested frames

Here we have two frames or more inside one frame. The source code is

```
<FRAMESET ROWS = "75%, 25%">
<FRAMESET COLS = "50%, 50%">
<FRAME SRC = "E:\Module 1\HTML\1.html">
</FRAMESET>
<FRAME SRC = "E:\Module 1\HTML\2.html">
</FRAMESET>
<FRAME SRC = "E:\Module 1\HTML\3.html">
</FRAMESET>
```

The main window is divided into two rows. One is 75% and the other is 25%.

The 75% portion is further divided into two equal portions. 1.html is loaded onto the first column; 2.html is loaded onto the second column and the last one, 3.html is loaded onto the bottom-most frame (i.e.,row).

The output is

<table>
<tr><td>First Frame</td><td>Second Frame</td></tr>
<tr><td colspan="2">One more Frame</td></tr>
</table>

MAIN.HTML is the master document. This document creates two frames (as shown in the output) and loads the two documents into the frames.

ARCHILE.HTML contains the product categories traded at OSD.

LIST.HTML as the name implies, shows the list of item under each category at the bottom of the window and provided hyperlinks.

The only factor that needs attention is the Name = "archive" attribute. This is very important as this attribute sets a name for this document. The entire document is given a reference named "archive".

<NOFRAMES> does not perform any crucial task here.

The list archive.html is as follows:

```
<! DOCTYPE HTML PUBLIC "-//W3C//DTD HTML
4.0 FRAMESET//EN">
<HTML LANG = "EN">
<HEAD>
<TITLE> A sample demonstration of frames
</TITLE>
</HEAD>
<BODY>
<H1 > One stop shop </H 1> <BR>
<H2><A NAME
        ="a"> Groceries</A></H2>
<UL>
<LI>Rice
<LI> Wheat
<LI> Pulses
<LI>Oil
<LI> Sugar
<LI> Salt
</UL>
<H2> <A NAME
    ="b"> Fruits and vegetables </A></H2>
<LI> ............ content...................
<H2> <A NAME
    ="c">Cakes and confectioner </A></H2>
<LI> ............. content ..............
<H2><A NAME
        ="d"> Beverages </A></H2>
<LI> ........... contents .................
</Body>
</HTML>
```

It lists all the items of all the product categories. One aspect to be noted here is naming the strategic locations. Before listing the items, the beginning is named as. a, b, c, ... and so.

```
<FRAME SRC = "archive.html" NAME="archive">
```

Inside this document, each location is named a, b, c and so on.

The Groceries list begins at "a" in the archive

The Fruits and vegetables list begins at "b" in the archive.

In HTML terminology, this can be written as

'archive.html#a'- Groceries

'archive.html#b'- Fruits and vegetables

```
<!DOCTYPE HTML PUBLIC
    "-//W3C//DTD HTML 4.0 FRAmESET//EN">
<HTML LANG = "EN">
<HEAD>
    <TITLE> Your Guide to Good shopping </TITLE>
</HEAD>
    <BODY>
        <CENTER><BIG>
```

```
<A HREF = "E:\ Module1 \HTML\archive.html#a"
  TARGET = "archive">
Groceries </A>
<A HREF = "E:\ Module1 \HTML\archive.html#b"
  TARGET = "archive">
Fruits and vegetables </A>
<A HREF = "E:\ Module1 \HTML\archive.html#c"
  TARGET = "archive">
Cakes and confectionery </A>
<A HREF = "E:\ Module \HTML\archive.html#d"
  TARGET = "archive">
Beverages </A>
</Big></Centre>
</Body>
</HTML>
```

The output of this code is

One Stop Shop

Groceries

- Rice
- Wheat
- Pulses
- Oil
- Sugar
- Salt
- Groceries
- Fruits
- Vegetables
- Cakes
- Confectionery
- Beverages

Frame Options

The FRAMES tag comes with certain options. These options are available for formatting the frames in a better manner.

The size of the frame can be specified through MARGINHEIGHT and MARGINWIDTH.

Usually, the surfer has the freedom to manually change the size of the frame. When the page is involved, it appears in the specified size. This manual resizing is allowed for the convenience of the surfer. It becomes necessary especially when the browser window is resized and viewed in a portion of the screen. But the web author can prevent manual resizing by including the NORESIZE attribute.

The frame options are

MARGINHEIGHT- Sets the frame's margin height.

MARGINWIDTH- Sets the frame's margin width.

NORESIZE- Prevents the surfer from resizing the margin.

SCROLLING- Scroll bar options

SCROLLING="No"- Switches off the scroll bar

SCROLLING="Yes"- Keeps the bar on, even when it is not needed.

SCROLLING = "Auto"- Switches on the bar only when needed. It is the default option.

Forms

Many companies have order forms for taking online orders for this products.

A form typically consists of the following components;

1. A title to indicate what this form is all about.
2. User prompt, like 'Name:', 'Mail-id:', etc.
3. A control for the surfer to provide an answer. A column bar is displayed next to the prompt for the surfer to supply data.
4. Buttons for forms. Usually, all forms have two standard buttons - submit and clear. A submit button prompts the page to accept the data supplied and act upon it. A clear button refers to erasing all the typed data and starting a fresh.
5. Though creating forms involves some understanding of input tags of HTML, they are as easy as any other HTML feature, when understood.

A simple form is explained here with a simple example.

```
<HTML LANG = "EN">
<HEAD>
    <TITLE> Forms and More </TITLE>
</HEAD>
<BODY>
    <H1><CENTER> A sample form
    </CENTRE></H1>
Want to create an e-mail account?
<FORM ACTION
="http.//www.hotmail.com"METHOD="GET">
<INPUT TYPE="Submit"VALUE="YES">
</FORM>
</BODY>
</HTML>
```

Creating a form

Creating a form is explained here with a simple example. The <FORM> tag, like FRAMESET, invokes a form. It is accompanied by a mandatory ACTION attribute. The action refers to the task to be performed upon submitting the form data. According to the above example, hotmail.com gets invoked when the form is submitted.

The METHOD attribute indicates the nature of accepting the input.

<INPUT TYPE = "Submit" VALUE = "Yes"> indicates a data to be accepted in a button form. Submit is a button. The text to be shown on the button is "yes" on the following **Figure.**

A sample form
Want to create an e-mail account?
[Yes] [No]

A typical form structure is shown below:

```
<FORM>
    First input
    Second input
    Third input and so on ...
```

```
</FORM>
```

All the input structures are embedded within the FORM tag.

The METHOD attributes has two options GET and POST.GET accepts each data individually, and POST considers the entire data set as a single block.

INPUT TYPE = "text" accepts the data as a text. Name, address, e-mail address, etc., are text type data. TYPE = "Submit" and TYPE = "RESET" both create buttons. "Submit" submits the data to the ACTION attribute. Reset clears the form elements and allows the surfer to type the data all over again.

When a data is entered, it has to be stored in a place. Name = "XXX" is a repository for the data. When the surfer keys the data in, it gets stored under the given label.

The label is supplied through NAME attributes. Enter your mail address: <INPUT TYPE = "Text" NAME = "mail-id> accepts the mail address from the surfer and stores its mailID.

<TEXTAREA NEM = "XXX" > accepts textual information in the form of a block text. It is mainly used for general feedbacks, comments, etc.

Code Reading.

```
<HTML>
<HEAD>
<TITLE> SHOP </TITLE>
</HEAD>
<BODY>
        <H 1> SHOP </H 1>
        <H 3> Everything under one
        category </H3>
        Send us your comments <P>
<FORM ACTION
        ="mail to:feedback@oss. com"
METHOD= "Post" ENCTYPE = "text/plain">
Name:< INPUT TYPE = "Text" NAME
                    = "name"><P>
Mail-id:<INPUT TYPE="Text" NAME
                    ="MAIL-ID"><P>
Favourite Item:<INPUT TYPE = "Text"
NAME = "favourite"><P>
Kindly fill in your feedback about our site
<P>
<TEXTAREA NAME = "Comments" ROWS = "4"COLS="20">
</TEXTAREA><BR><BR>
<INPUT TYPE = "Submit" VALUE="Submit">
<INPUT TYPE = "Reset" VALUE="Clear">
</FORM>
</Body>
</HTML>
```

A typical code that makes use of form elements is listed here.

```
<FORM ACTION
        ="mail to:feedback@oss.com"
METHOD="post"ENCTYPE="Text/Plain">
```

Initiates a form, specifies an action,and mentions the mode of data input (method).

```
<TEXTAREA NAME ="comments" Rows = "4" Cols="20">
```

Creates a text entry box with 4 rows and 20 columns (scroll bar if necessary). Free flow text is accepted and stored in a name "comments".

```
<INPUT TYPE="Submit" VALUE = "Submit">
<INPUT TYPE="Reset" VALUE = "Clear">
```

Two buttons are created, **Submit** for submitting the form, the **Clear** for erasing the contents of the form.

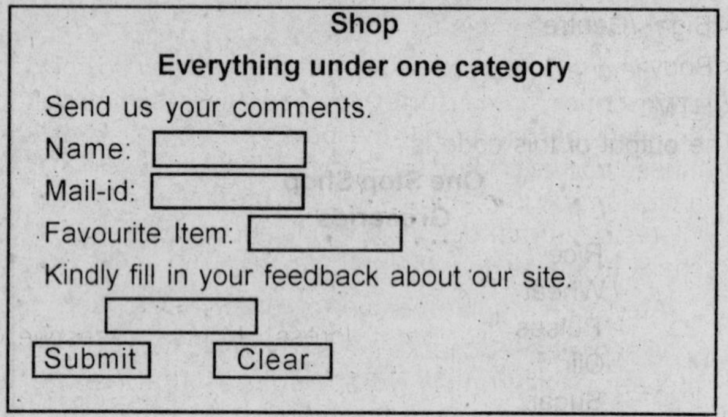

A typical output of the document is shown here. The TEXTAREA tag is interesting to view. The user can type the comments. If the comments exceed the base size, they get scrolled.

The Submit and Clear buttons are also shown here. They are the output of INPUT TYPE ="Submit" and INPUT TYPE = "Reset" tags. When a user type in the required data and hits the submit button, the data gets pushed to the ACTION attribute.

Scrolling Options

A surfer can be made to select one option from a pre-defined list which can be given in the form of a scroll list. It is quite useful if a surfer has to select an option from a list of choices. It would be a better idea to list those choices and let the surfer select one from the list. For example, the city can be selected from a list of cities.

The format of the command is shown below:

```
<SELECT NAME = 'data storage'COLS
                    ='No. of columns'>
<Option> Option 1
<Option> Option 2
<Option> Option 3
</SELECT>
```

Code Reading

```
<SELECT NAME ="favourite" COLS=3>
<Option> Chocolates
<Option> Cakes
<Option> Beverages
<Option> Others
</SELECT>
```

A sample code is given here. The SELECT tag begins the options list. OPTION lists all the options for the surfer to choose from </SELECT> ends the list.

COLS mentions the number of columns. If COLS = "1" is given, it creates a pull-down menu.

There are many options that are possible with the type attribute. The first three options, TEXT SUBMIT and RESET have been explained.

Your Membership-ID:< INPUT TYPE = "PASS WORD" NAME = "mem-id"><P>

PASSWORD is another option that hides the characters while typing. Remember that it only prevents anybody from sneaking above your shoulders and getting to know the password. But when it gets mailed, it sends it as plain data. You will have to employ different techniques to protect it before sending it on air.

Your Membership-ID: xxxxxxxx

The check box shows an option with a tiny box next to each option. If a surfer (or user) wants that option enabled, he can check the box, i.e., he can keep the mouse pointer on that box and click on it. This will initiate a tick mark in the box, which means the surfer is interested in that option. For instance, you can have a check box next to the following option.

Are you a regular member of OSS?

When a surfer checks the box that is provided next to the above option, it would indicate that he/she is interested in becoming the member and that you can provide further online information and mail it to him.

The format of the check box option will be:

Are you a regular member of OSS < INPUT TYPE = "CHECKBOX" NAME = "mem"> <P>

This is how the output will look.

Are you a regular member of OSS

Radio invokes radio buttons, which are different from check boxes in the sense that check boxes allow the user to "check" or ignore the option. Radio buttons show two or more options. The surfer can select only one option.

Do you own more than one credit card? Yes/No.

Here the surfer needs to click only one option. So, he is allowed to click only one button.

Do you own more than one credit card?

Yes<INPUT TYPE = "RADIO" NAME="Yes">

No<INPUT TYPE = "RADIO" NAME="No">

The output is shown below:

Do you own more than one credit card?

Yes ☐ No ☐

Intermediate HTML (in detail)

This chapter extends our knowledge of the basic tags of HTML by showing us how to work with forms, lists, and tables. Using forms we can create Web pages that interact with the visitors to our Web site. This chapter also shows us how to use the HTML tags needed to formatlists of information. After reading this chapter we'll even be able to create Web pages with advanced layouts only by using HTML tables.

Working with HTML forms

To create a genuinely interactive Web site, we need to use HTML forms. HTML forms enable us to gather and respond to the information provided by the visitors to ours Web site. Using forms, we can create such things as check boxes, radio buttons and text areas.

ASP

A mastery of forms is crucial to effective Active Server Pages Programming.

One of the main functions of ASP scripts is to respond to the information entered in HTML forms. Therefore, we should think of forms as the primary user interface to our ASP applications.

To create HTML forms, we use the <FORM> tag. This tag works as a container tag enclosing other form elements and specifying, through its attributes, what should be done with the information gathered within the form. Here's an example of a very simple form:

```
<HTML>
<HEAD>
<TITLE> Simple Form </TITLE>
</HEAD>
<BODY>
<FORM>
<INPUT>
</FORM>
</BODY>
</HTML>
```

When this HTML file is displayed in a Web browser, a single three-dimensional box appears. We can type whatever text we want into this box. However, as it stands, nothing happens after we enter the text. This form is too simple to be useful. The first problem with this example is that the form doesn't know when we're finished entering information. To rectify this problem, we need to add another form element: a submit button. Here's the same form, with the addition of a submit button.

```
<HTML>
<HEAD>
<TITLE> Simple Form </TITLE>
</HEAD>
<BODY>
<FORM>
<INPUT>
<INPUT TYPE = SUBMIT VALUE = "Submit me!">
</FORM>
</BOD>
</HTML>
```

When we display this modified HTML file, a button appears with the words Submit Me! By clicking the Submit Me! Button, we indicate that we're finished entering information in the text box. This lets the browser know that it must now do something with the information.

Currently, however, the form has no idea what to do with the information that it has gathered. We have to tell the form how to handle the infomration, by using the ACTION attribute of the <FORM> tag. The ACTION attribute determines what action the form will take with the form information. The following example shows how this attribute is used:

```
<HTML>
<HEAD>
<TITLE> Simple Form </TITLE>
</HEAD>
<BODY>
<FORM> ACTION
        = "MAILTO: billg@microsoft.com">
<INPUT>
<INPUT TYPE = SUBMIT VALUE = "Submit me!">
</FORM>
</BODY>
</HTML>
```

When we fill out and submit this form, the ACTION attribute tells the form that the information should be sent immediately to Bill Gates at Microsoft. We can have form information sent to any internet address by using this syntax:

```
<FORM ACTION
    ="MAIL TO:name@domain.com">
```

However, we'll rarely need to send form information to an e-mail address. In most cases we want the form information to be sent back to our web site be processed. To sent form information to an Active Server Page, we use the following HTML:

```
<HTML>
<HEAD>
<TITLE> Simple Form </TITLE>
</HEAD>
<BODY>
<FORM> "/Somedirectory
        /mypage.asp" METHOD = "POST">
<INPUT>
<INPUT TYPE = SUBMIT VALUE = "Submit me!">
</FORM>
</BODY>
</HTML>
```

In this example, the attributes ACTION and METHOD specify that the form information should be posted to an Active Server Page named mypage.asp to be processed. The ACTION attribute gives the path to the Active Server Page. The METHOD attribute indicates how the form infomration should be sent. In this case the form informant that's entered in the text box is posted to the Active Server Page.

In almost all cases, our <FORM> tag will have the same general characteristics as in this example. We'll specify an Active Server Page that will process the form by using the ACTION attribute, and we'll specify that the form information should be posted by using the METHOD attribute.

One final addition needs to be made to our form example to make it useful. When we include form elements such as text boxes, we need to give each form element a name. We do this with the NAME attribute of the <INPUT> tag. This example has two different text boxes named text 1 and text 2:

```
<HTML>
<HEAD>
<TITLE> Simple Form </TITLE> </HEAD>
<BODY>
<FORM ACTION="/Somedirectory
        /mypage.asp" METHOD = "POST">
<INPUT NAME = "text 1">
<INPUT NAME = "text 2">
<INPUT TYPE = SUBMIT VALUE
        = "Submit me!">
</FORM>
</BODY>
</HTML>
```

In this example information entered into the text box named text [will be associated with the name text] when it's sent to the Web site for the processing. This will distinguish the text entered into this first text box from any information entered into any other form element.

Form information is sent to the destination specified in the ACTION attribute in name and value pairs. Separated by an ampersand (&) character. For example, if we enter our first name in the first text box and our last name in the second text box, the information sent to be processed would look like this:

text. 1=our first name & text 2 = our last name. Actually, prior to being sent the form information will be URL-encoded. When text is URL = encoded, certain characters are replaced with others.

For example, the space character is replaced with a plus sign (+). So, if we enter the text, this is text box 1 in the first text box and the text. This is text box 2 in the second text box, the following text would be sent for processing:

text 1 = This + is + text box 1 & text 2
= This + is + text box 2

Mixing Forms with other HTML Tags

This feature is useful for creating labels for our form elements.

```
<HTML>
<HEAD>
<TITLE> Simple Form </TITLE>
</HEAD>
<BODY>
<FORM ACTION="/Somedirectory
        /mypage.asp" METHOD = "POST">
<BR> <B> First Name:</B><INPUT NAME
        = "first name">
<BR> <B> Last Name:</B><INPUT NAME
        = "Last name">
<INPUT TYPE = SUBMIT VALUE
        = "Submit me!">
</FORM>
</BODY>
</HTML>
```

In this example, the text box named first name is labelled (in bold) with the text First Name:

Sadly, one thing that we can't do is use HTML tags to alter the text that's displayed in form elements. For example, we can't use HTML to make the text Submit Me! on the submit button appear bold. Also, we can't control the appearance of the text that's entered into a text box.

NOTE

Although we can't use standard HTML tags to control the appearance of text with form elements, we can use cascading style sheets to do this with Netscape Navigator 4.0 and Internet Explorer 4.0.

We can place our <FORM> tags any where in the body of an HTML document. Also and we'll often find this necessary - you can have multiple forms in the same HTML document. This is useful when we need to invoke different Active Server Pages scripts to process different Form information.

Just make sure that our forms do overlap or contain one another. For example the following HTML code is illegal.

```
<HTML>
<HEAD>
<TITLE> Bad Form </TITLE>
</HEAD>
<BODY>
<FORM ACTION="/Somedirectory
            /mypage.asp" METHOD = "POST">
<INPUT NAME = "text box 1">
<INPUT NAME = "text box 2">
<BR> <INPUT TYPE = SUBMIT VALUE
                     = "Submit me!">
<FORM ACTION="/Some other directory
            /my other page.asp">
< METHOD = "POST">
<INPUT NAME = "another text box 1">
<INPUT NAME = "another text box 2">
<BR> <INPUT TYPE = SUBMIT VALUE
                 = "Submit me Also!">
</FORM>
</BODY>
</HTML>
```

This HTML code won't work because on <FORM> tag is embedded in another. If we ever find ourself in a situation where one of our forms doesn't work as it should, check to make sure that we haven't accidentally overlappe our <FORM> tags or forgotten to close one of our forms.

The most basic form element is a text box. We can create a text box with nothing more than the tag <INPUT> with no attributes. However, to create a useful text box, we need to include the Name attribute. The next example includes two text boxes named text 1 and text2.

```
<HTML>
<HEAD>
<TITLE> Simple Form </TITLE>
</HEAD>
<BODY>
<FORM ACTION="/Somedirectory
            /mypage.asp" METHOD = "POST">
<INPUT NAME = "text box 1" TYPE="text">
<INPUT NAME = "text box 2">
<INPUT TYPE = SUBMIT VALUE
                     = "Submit me!">
</FORM>
</BODY>
</HTML>
```

Notice that the first text box has its TYPE attribute set to "text". We can include this attribute for a text box if we prefer, but it really has no effect; the <INPUT> tag defaults to TYPE = "text" anyway.

We can also use the VALUE attribute of the <INPUT> tag to specify the default text that will appear in the text box before the user makes any changes.

For example, suppose we have a text box in which users can enter the name of their county. If we expect more of our users to originate from the United States, we can make the default country, USA by using the tag <INPUT> NAME="country" VALUE = "USA">

Another useful attribute of the <INPUT> tag is the SIZE attribute. Using the SIZE attribute we can control how wide a text box will appear. The value we provide to the SIZE attribute is the number of characters that we want the text box to display.

Form Buttons

Three types of buttons are typically used in HTML forms, we've already been introduced to the first type of button, the submit button. When we click the submit button the form is processed by whatever program is referred to in the Action attribute of the <FORM> tag. The syntax of the submit button is as follows:

```
<INPUT TYPE = SUBMIT VALUE = "DO IT!">
```

The VALUE attribute determines the text that appears on the button.

The might button has almost exactly the same effect as the submit button. This type of button however, appears as an image of the typical ugly rectangle. Following is an example of how to use an image button:

```
<HTML>
<HEAD>
<TITLE> Simple Form </TITLE>
</HEAD>
<BODY>
<FORM ACTION="/Somedirectory
            /mypage.asp" METHOD = "POST">
<INPUT NAME = "text box 1">
<INPUT NAME = "text box 2">
<BR> <INPUT TYPE = "IMAGE SRC=
            "my image.gif" BORDER=0>
</FORM>
</BODY>
</HTML>
```

As in the case of image used for links, we should set the BORDER attribute to zero to hide the border that appears around images on browsers such as Netscape Navigator. In some cases, the extra image border can be distracting. We can use any of the other attributes of the <IMG> tag with image buttons as well.

NOTE

An interesting aspect of the image button is that when clicked it not only submits the form date, it also submits the exact coordinates of where the image was clicked. For example, if you click dead center on an image button that has a width and height of five pixels. The x, y coordinates of 3, 3 are passed with the rest of the form information.

This feature of form buttons is useful if we need to combine the function of a form with an image map. A normal image map doesn't allow to pass the contents of form fields.

The final type of button is the reset button. When a user of a form clicks the reset button, all the form field are returned to their initial state. For example, text boxes with no default value will be cleared; those with there VALUE attributes set will be reinitialized to the specialized value. As in the case of the submit button we specify the text that appears on a rest button by supplying a VALUE attribute, as in the next example:

```
<HTML>
<HEAD>
<TITLE> Simple Form </TITLE>
</HEAD>
<BODY>
<FORM ACTION="/Somedirectory
            mypage.asp" METHOD = "POST">
<INPUT NAME = "text box 1">
<INPUT NAME = "text box 2" value="my default">
<BR><INPUT TYPE = SUBMIT VALUE
                        = "Submit me!">
<BR><INPUT TYPE = RESET VALUE
                        = "clear me!">
</FORM>
</BODY>
</HTML>
```

In this example, when a user of the form clicks the button labelled clear me! all the form fields are cleared back to their initial state.

To use multiple buttons, we need to supply the NAME attribute for the buttons.

Example

```
<HTML>
<HEAD>
<TITLE> Simple Form </TITLE>
</HEAD>
<BODY>
<FORM ACTION="/Somedirectory
            /mypage.asp" METHOD = "POST">
<BR><INPUT name = "YES" TYPE
                        = SUBMIT VALUE = "yes!">
<BR><INPUT name = "NO" TYPE
                        = SUBMIT VALUE = "no!">
</FORM>
</BODY>
</HTML>
```

Here, clicking either yes! or no! will invoke the same form action. However, when the form is processed, the user's vote can be determined. Just to reiterate, in this section we have learned how to use the three types of buttons most commonly used in HTML forms:

Submit buttons: Used to submit the contents of a form to the server for processing or to e-mail a form to a particular e-mail address.

Image buttons: Used for the same purpose as a submit button, but contains an image instead of text.

Reset buttons: Reinitializes all the elements of a form to their initial state.

Password Boxes

Using standard text boxes, we could create an HTML form that demands the user's name and password. To protect the user's password, we need to use a password box. A password box works like a text box except that when information is entered, it's hidden. Here's an example:

```
<HTML>
<HEAD>
<TITLE> Simple Form </TITLE>
</HEAD>
<BODY>
<FORM ACTION="/Somedirectory
            /mypage.asp" METHOD = "POST">
Please enter your name and password:
<BR>Name:<INPUT name= "username">
<BR>password:<INPUT name=
            "password:TYPE=PASSWORD">
<INPUT TYPE=SUBMIT VALUE="continue">
</FORM>
</BODY>
</HTML>
```

NOTE

It's important to realize that next entered in a password box, though hidden, is not encrypted when it's submitted. In theory, this means that someone could steal the text entered into a password box off the wires as it moves across the interment to our Web site. For most applications, this shouldn't cause great concern.

However, we can encrypt the information that's submitted in an HTML form to prevent even this every unlikely event from happening.

When the preceding HTML file is displayed in a Web browser, we enter text into the password box in the same way as we would with a normal text box. However, all of the text entered will be masked (typically with asterisks). We can use both the SIZE and MAXLENGTH attributes with password boxes to control the screen width of the box and the maximum number of characters that can be entered into it.

Check Boxes

Check boxes are useful in two types of situations. In the simplest case, we can use a check box to pose a simple true or false question to a user, as in the following example:

```
<HTML>
<HEAD>
<TITLE> Simple Form </TITLE>
</HEAD>
<BODY>
<FORM ACTION="/Somedirectory
        /mypage.asp" METHOD = "POST">
<BR>Do you like this web site?
<BR><INPUT NAME = "Like"TYPE=
        CHECKBOX VALUE = "yes">
<BR><INPUT TYPE = SUBMIT VALUE
                = "Submit Me!">
</FORM>
</BODY>
</HTML>
```

When this HTML code is displayed in a Web browser, an empty box appears beneath the question, Do you like this Web site? If the box is checked when he submit button is clicked, the value of the VALUE attribute is also submitted. In this case, the name and value pair of Like and Yes is submitted.

If the check box isn't checked when the user clicks the submit button, on the other hand, absolutely nothing is submitted. Not even the name of the check box is submitted. We can't specify the "off" value of a check box.

We can use multiple check boxes with the same name to gather information. For example, suppose we want to find out how the visitors of our Web site discovered the site, and we want to allow for the possibility that some users discovered our site in more than one way, we could use the following HTML code:

```
<HTML>
<HEAD>
<TITLE> Simple Form </TITLE>
</HEAD>
<BODY>
<FORM ACTION="/Somedirectory
        /mypage.asp" METHOD = "POST">
<BR>How did you find out about this web site?
<BR>Magazine: <INPUT NAME
    ="discover" TYPE=CHECKBOX
            VALUE="Magazine">
<BR>"Search Engine" <INPUT NAME
    ="discover" TYPE=CHECKBOX
            VALUE="search">
<BR>Friend: <INPUT NAME
    ="discover" TYPE=CHECKBOX
            VALUE="Friend">
<BR><INPUT TYPE = SUBMIT VALUE
                = "submit me!">
</FORM>
</BODY>
</HTML>
```

Notice in this example that all three check boxes have the same name. When a user clicks the submit button, the value of each check box that has been selected will be submitted. In this case, every value will be associated with the name discover.

NOTE

Don't confuse multiple check boxes with radio buttons. Unlike radio buttons, more than one check box can be selected at the same time, even if the check boxes have the same name.

Radio Buttons

Radio buttons have an old name, but we most likely have encountered them in both Web pages and normal computer application. Using radio buttons, we can present choice of multiple values unlike when using check boxes. However, the user can select only one radio button at a time.

For example, suppose we want to determine the gender of visitors to our Web site. We could use radio buttons to allow visitors to specify whether they are male or female, since no visitor can be both. Here's an example of how you could do this:

```
<HTML>
<HEAD>
<TITLE> Radio </TITLE>
</HEAD>
<BODY>
<FORM ACTION="/Somedirectory
        /mypage.asp" METHOD = "POST">
<BR>Please indicate your sex.
<BR>Male : < INPUT NAME = "sex"
TYPE=RADIOVALUE = "male">
<BR>Female : < INPUT NAME = "sex"
TYPE=RADIOVALUE = "female">
<BR> <INPUT TYPE = SUBMIT VALUE
                = "submit me!">
</FORM>
</BODY>
</HTML>
```

Notice that both radio buttons are given the same name. When the user of this Web page hits the submit button, the value of only one radio button will be sent. The value sent is specified in the VALUE attribute of the radio button.

As in the case of check boxes, we can use the CHECKED attribute to specify the radio button that will be selected when the Web page is first loaded. We can also have as many radio buttons as we like in a form. If we want to have multiple groups of radio buttons, simply give the radio buttons in each group different names.

Creating Drop-Down List Boxes

As an alternative to check boxes and radio buttons, we can create drop-down list boxes to display a menu of alternatives. A drop-down list box shows only one selection at a time. The view the other possible selections, we click the arrow attached to the box. One advantage to using a drop-down list box is that it occupies less space on our Web page than either radio buttons or check boxes.

We create a drop-down list box using the <SELECT> and <OPTION> tags, like this

```
<HTML>
<HEAD>
<TITLE> Drop-down List Box </TITLE>
</HEAD>
<BODY>
<FORM ACTION="/Somedirectory
          /mypage.asp" METHOD = "POST">
<BR>Please indicate your sex.
<BR><SELECT NAME = "sex">
<OPTION VALUE = "Is Male"> Male
<OPTION VALUE = "Is Female"> Female
</SELECT>
<INPUT TYPE = SUBMIT VALUE
                         = "submit me!">
</FORM>
</BODY>
</HTML>
```

In this example, the NAME attribute of the <SELECT> tag provides a name for our dropdown list box. Each of the <OPTION> tags presents a possible selection. The VALUE attribute of the <OPTION> tag specifies the value that will be sent if that option is selected when the form is submitted. Finally, the next that appears after the <OPTION> tag determines how that option appears in the drop-down list box.

The list box in this example has the same function as a group of radio buttons. We are forced to select only one option at a time. For example, if we select Female from the list box, the only value sent when the form is submitted would be is female. This value would be associated with the NAME of the list box sex. We can never select more than a single value.

By default, the following first <OPTION> tag will appear as the initial value of the drop-down list box. We can override this default by using the SELECTED attribute of the <OPTION> tag. In the following example, Female is selected when the list box appears, even though Male appears before it in the list of options:

```
<HTML>
<HEAD>
<TITLE> Scrolling-List Box </TITLE>
</HEAD>
<BODY>
<FORM ACTION="/Somedirectory
          /mypage.asp" METHOD = "POST">
<BR>Please indicate your sex.
<BR><SELECT NAME = "sex">
<OPTION VALUE = "Is Male"> Male
<OPTION VALUE = "Is Female"SELECTED>
    Female
</SELECT>
<BR> <INPUT TYPE = SUBMIT VALUE ="submit Me!">
</FORM>
</BODY>
</HTML>
```

One situation when specifying a different default value is useful when we have an alphabetical list of items, but we want to choose the default value from within the list. For example, suppose we have a list of countries, but we want the default country to be the United States. In that case, we can use the SELECTED attribute to specify that United States is the default value, even though this country name appears near the end of the alphabetical list of country names.

Creating Scrolling List Boxes

Yet another alternative to a group of radio buttons or check boxes is a scrolling list box. Using scrolling list boxes, we can create a list of alternatives from which the user can make one or more selections. To create a scrolling list box, we use the same tags used to create a drop-down list box, but with different attributes. Here's an example:

```
<HTML>
<HEAD>
<TITLE> Scrolling-List Box </TITLE>
</HEAD>
<BODY>
<FORM ACTION="/Somedirectory/
          mypage.asp" METHOD = "POST">
<BR>Please indicate your sex.
<BR><SELECT NAME = "sex"SIZE=2>
<OPTION VALUE = "Is Male"> Male
<OPTION VALUE = "Is Female"> Female
</SELECT>
<INPUT TYPE = SUBMIT VALUE ="submit Me!">
</FORM>
</BODY>
</HTML>
```

The SIZE attribute of the <SELECT> tag converts a drop-down list box into a scrolling list box. The SIZE attribute specifies how many options should appear on screen at a time.

When using scrolling list box instead of drop-down list boxes, we lose the advantage of saving screen space. However, scrolling list boxes have one significant advantage: Using a scrolling list box, we can allow for the selection of more than one option. To do this, we use the MULTIPLE attribute of the <SELECT> tag:

```
<HTML>
<HEAD>
<TITLE> Simple Form </TITLE>
</HEAD>
<BODY>
<FORM ACTION="/Somedirectory/
```

```
            mypage.asp" METHOD = "POST">
<BR>How did you find out about this website?
<BR><SELECT NAME = "discover"SIZE>
                        = 3 MULTIPLE>
<OPTION VALUE = "Magazine"> Magazine
<OPTION VALUE = "Search">search Engine
<OPTION VALUE ="Friend"> Friend
</SELECT>
<BR> <INPUT TYPE = SUBMIT VALUE ="submit Me!">
</FORM>
</BODY>
</HTML>
```

When this HTML file is displayed in a browser, a normal scrolling list box appears. However, if we use the ctrl key on our keyboard while clicking the list box, we can choose more than one alternative. For example, we can select the options Magazine and Friend at the same time (we can also use the shift key to select a contiguous group of alternatives).

NOTE

Many visitors to a Web site won't know how to make multiple selections. Furthermore, the method for selecting more than one option varies according to the type of computer a person is using. For example, one must use the command key rather than the ctrl key. Consider using a group of check boxes instead when we want to allow a user to make multiple selections.

Creating Text Areas

With the form elements described so far, the user can't enter more than one line of text. The <TEXTAREA> form tag creates a text area where you can give the user more room for freedom of expression. Use this tag whenever you want someone to enter a paragraph of text, as in the following example.

```
<HTML>
<HEAD>
<TITLE> TextArea </TITLE>
</HEAD>
<BODY>
<FORM ACTION="/Somedirectory
            mypage.asp" METHOD = "POST">
<BR>Please enter your comments below:
<BR><TEXTAREA NAME = "comments"
COLS=40 ROWS = 20></TEXTAREA>
</FORM>
</BODY>
</HTML>
```

We should notice immediately that, in this example, the <INPUT> tag is not used to create the textarea. We create textareas using the <TEXTAREA> tag, like all other form elements, must always appear within the <FORM> tag.

NOTE

Some browsers allow us to use the following syntax.

<INPUT TYPE = TextArea COLS = 40 ROWS =20>. If we want our Web pages to be compatible with the maximum number of Web browsers, use the <TEXTAREA> tag and not the TextArea attribute.

To specify the width and height of a text area, you use the COLS and ROWS attributes. The COLS attribute specifies the width of the textarea in columns; the ROWS attributes specifies the height in rows. Note that both of these attributes are measured in average character widths; because textareas don't use fixed-pitch fonts, we may not get exactly 50 characters in a textarea with 50 columns.

We can't provide the COLS and ROWS attributes with a percentage value. This is unfortunate, because it results in textareas looking different on monitors with smaller or larger screen resolutions. This is a shortcoming of HTML.

Also, textareas have no MAXLENGTH attribute. There's no way to prevent someone from entering massive amounts of text into a textarea, we have no control over this.

Notice that the <TEXTAREA> tag is a container tag. If we want text to appear in the textarea when we first load the Web page, place the text within the <TEXTAERA> tag, like this.

```
<HTML>
<HEAD>
<TITLE> TextArea </TITLE>
</HEAD>
<BODY>
<FORM ACTION="/Somedirectory
            /mypage.asp" METHOD = "POST">
<BR>Please enter your comments below:
<BR><TEXTAREA NAME = "comment"
COLS=40 ROWS = 20>
I am the default text!
</TEXTAREA>
</FORM>
</BODY>
</HTML>
```

When this HTML file is interpreted by a Web browser, the text I am the default text! appears in the textarea. Be aware that we can only enter text in textareas. Any HTML tags within a <TEXTAREA> tag will simply be ignored.

The official HTML specification is silent on how text should wrap within a textarea. When we get to the right edge while typing in a textarea with internet Explorer, the text automatically wraps to a new line. When we do the same thing with Netscape Navigator, on the other hand, the textarea keeps scrolling to the right.

The attribute can accept three values: OFF, PHYSICAL, and VIRTUAL. By default, the value of the WRAP attribute is OFF, and text doesn't wrap to a new line. When WRAP = PHYSICAL, on the other hand, text is wrapped to a new line when the contents of the text area are submitted, carriage returns are added wherever the text wraps. If we want the textarea to wrap, but we don't want the extra carriage returns submitted with the form contents, use when WRAP = VIRTUAL.

When we create textareas, we almost include the WRAP = VIRTUAL attribute. When text in a textarea doesn't wrap, this confuses the person entering information. By using this attribute, can guarantee that text will wrap in both the Microsoft and Netscape browsers.

Also, typically, we don't want to deal with the extra carriage returns when interpreting the contents of a textarea using WRAP = VIRTUAL leaves those out.

Creating Hidden Fields

Using hidden fields, the entry in the form will never be displayed on screen. This information will be included, however, when the form is submitted.

Hidden fields are extremely valuable to the Active Server Page programmer. We'll probably use them quite often to pass hidden information page. For the moment, untill we learn how to use Active server pages, we'll have to take the usefulness of hidden fields on faith.

We can use hidden fields to create session variables that don't depend on cookies. There are both advantages and disadvantages in doing this. Here's an example of hidden field in a form.

```
<HTML>
<HEAD>
<TITLE> Hidden Field </TITLE>
</HEAD>
<BODY>
<FORM ACTION="/Somedirectory
              /mypage.asp" METHOD = "POST">
<INPUT NAME = "secret" TYPE = HIDDEN
    VALUE="you cannot see me!"
<INPUT TYPE = SUBMIT VALUE
                        = "submit me!">
</FORM>
</BODY>
</HTML>
```

When this HTML file is displayed in a Web browser, the only thing you see on the screen is the submit button. The hidden field, names secret, doesn't appear. However, when the form button is clicked, the value you cannot see me! is submitted as part of the contents of the form.

CAUTION

We should never place genuinely private information in hidden fields. A viewer of a Web page can see the value of a hidden field by using the source command in a Web browser.

Hidden fields are hidden from view, but they aren't hidden from the eyes of a knowledgeable user.

Creating File Upload Buttons

Suppose we want to create a Web site where people can advertise houses for sale. It would be nice if people could upload pictures of their homes. Or suppose we want to create a Web site devoted to short stories. It would be nice if people could upload their stories, for example, in Microsoft Word format.

In theory, we can not do this with the TYPE = FILE attribute of the <INPUT> tag. Using this attribute, we can create a file upload button on a form. When users of our Web site click this button, they can select a file to upload from their local hard drive. Following is an example of how this can be done.

```
<HTML>
<HEAD>
<TITLE> File upload </TITLE>
<HEAD>
</BODY>
</HTML>
```

When this HIML file is displyaed in Netscape Navigator (version 3 or greater) a normal looking text box appears next to a Browse button. The user can type the name of the file directly into the text box, or use the Browse button to select a file dialog box. When the form is submitted, the file the user selected is also submitted.

Notice the addition of the ENCTYPE attribute to the <FORM> tag in this example. The ENCTYPE attribute specifies the type of encoding to apply to the form information, when it's submitted. Normally, from information is URL- encoded (spaces are replaced with + and so on).

However, this is a bad encoding format to use with information that isn't text. To transfer files such as images efficiently, we should use ENCTYPE = "multipart/form-data".

The actual file upload button is created using this tag:

```
<INPUT NAME = "Picture" TYPE =FILE
    ACCEPT = "Image /*">
```

Because the value of the TYPE attribute is FILE, the browser should create a file upload button. The ACCEPT attribute restricts the files that will appear in the file dialog box to a particular file type. In this case, the file type is restricted to image files. We can specify any list of MIME types as the value of the ACCEPT attribute.

NOTE

MIME stands for multipurpose internet mail extension. MIME was originally developed as a mean for specifying the file types of e-mail attachments. Browsers use MIME types to associate files with appropriate programs.

Some examples of NAME types are image/gif for GIF images, image/jpeg for JPEG images, application/x-msexcel for Microsoft Excel spreadsheets, and application /Ms-Word for Microsoft Word documents.

To view the MIME types that our computer supports, choose view/File Types in Windows Explorer.

Potentially, file upload buttons could be very useful. Regrettably however, current browsers support file upload buttons only partially -or not at all. This is true even though file upload buttons are part of the official HTML 3.2 specification.

For example, Netscape Navigator (versions 3.0 and 4.0) ignores the ACCEPT attribute. Internet Explorer version 3.0 is even worse. It completely fails to interpret file upload buttons, and displays normal text boxes, instead. Internet Explorer version 4.0, however, does recognize the file upload button. Until file upload buttons are supported by more browsers, use them with caution.

XML SYNTAX

Markup syntax

XML markup describes and provides structure to the content of an XML file or data packet, more generally known as an **XML entity.** This markup is comprised of tags that delimit different sections of content, provide references to special symbols and text macros, or convey special instructions to the application software and comments to the document's editors.

Nearly all XML markup is case-sensitive, including element tag names and attribute values; i.e.,

Book ≠ BOOK ≠ book ≠ bOok

Characters

For worldwide use, characters are not limited to the 7-bit ASCII character set. XML specifies characters that are the numbers defined in the 16-bit + Unicode 2.1 character set.

Legal XML characters include three ASCII C0 control characters, all normal ASCII display characters and almost all other Unicode character values (shown in hexadecimal):

Character values (hexadecimal)	Description
09	Horizontal tab (HT)
0A	Line-feed (LF)
0D	Carriage-return (CR)
20 ... 7E	ASCII display characters
80 ... D7FF	Unicode characters
E000 ... F8FF	"Private Use Area"
F900 ... FFFD	CJK (Chinese - Japanese -Korean) Compatibility Ideographs
10000 ... 10FFFF	Surrogate equivalents and "High Private Use Area"

Names

Structures used in XML are almost always named. All XML **names** must begin with a letter, underscrore (_) or colon (:); and are continued using valid **name characters.** Valid name characters include the preceding, plus digits, hyphens (-), or fullstops (periods). In actual practice, the colon character should not be used, except as a namespace delimiter. Letters are not limited to ASCII characters, so that non-English-speaking users of XML may use their own language for markup.

One restriction on name is that they may not begin with the string "xmℓ", "XML", or any string which would match any variation of these three characters, in that sequence ("xMℓ", or "Xmℓ", for example)

The following are legal names:

Book

BOOK

Wrox : Book_Catalog

AΓΔ

Conseil_Europeen_Pour_la_Recherche_Nucleaire_(CERN)

XML names are case-sensitive. The first two names are not equivalent. The third is an example of a name that uses the proposed namespace delimiter (the colon). The last two examples are reminders that Greek and French are just as acceptable as English for XML names.

Document parts

a **well-formed** XML document is compries of three parts:

- An optional **prolog.**
- The **body** of the document, consisting of one or more **elements,** in the form of a hierarchical tree that may also contain **character data.**
- An optional "miscellaneous" epilog, comprised of comments, processing instructions (PIs) and/or white space that follows the end of the element tree.

Document

Prolog

<?xmℓ version = "1.0"?>

<!-comments and processing instructions allowed here - >

<! DOCTYPE BOOKS SYSTEM

"http: //myserver.com/books.dtd">

<! - comments and processing instructions allowed here - >

Epilog

<! - comments and processing instructions allowed here - >

Elements

Elements are the basic building blocks of an XML markup. They contain other elements, character data, character references, entity references, PIs, comments, and/or CDATA sections - these are collectively known as **element content.** All XML data (except for comments, PIs and white space) must be contained within elements.

Elements are delimited using tags, which consist of the **element type name** (a string literal) enclosed within a pair of angle brackets ("< >"). Every element must be delimited with a **start-tag** and an **end-tag.** The only exception to this rule is an element that has no content, known as an **empty element,** which may be represented with either a start-tag/end-tag pair, or using an abbreviated hybrid form the **empty-element tag.**

Start-tags

The beginning delimiter of an element is called the **start-tag.** Start-tags are comprised of an **element type name,** enclosed within angle brackets. We can also think of start-tags as "opening" an element as we would open a file or communications link.

Example:

<Book>

<BOOK>

<Wrox : Book_Catalog>

<AΓΔ>

End-tags

The ending delimiter of an element is called the **end-tag.** End-tags are comprised of a forward slash ("/") followed by an element type name, enclosed within angle brackets. Every end-tag must match a corresponding start-tag and x is considered as "closing" the element that was opened by the start-tag.

Example:

</Book>

</BOOK>

</Wrox : Book_Catalog>

</A ΓΔ >

A complete element with start-tags and end-tags would look like:

<some-tag> content goes here </some-tag>

Empty-Element Tags

Empty element may not have any content. XML specifies that an empty element may also be represented in an abbreviated form that is a hybrid of the start-tags and end-tags.

Empty element tags are comprised of an element type name followed b a forward slash ("/"), enclosed within angle brackets.

<point/>

Document Element

Well formed XML documents are defined as being in the form of a simple **hierarchical** tree, each document having one, and only one root node, called the **document entity** or **document root.** This node may contain PIs and/or comments and will always contain a sub-tree of elements, the root of which is called the **document element.** This element is the parent of all other elements in the tree and it may not be contained in an other element.

The beginning of the maximum document tree for any XML data object is:

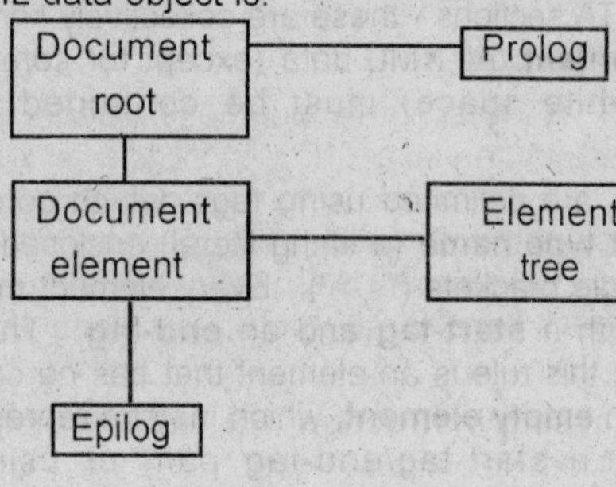

The beginning of the document tree that is implied by the last "textfile" example is:

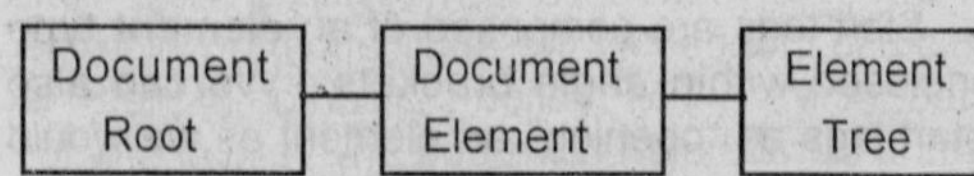

Any well-formed XML document must be comprised of elements that form a single hierarchical tree, with a singular root node, called the "document root". This contains a secondary tree of elements, also with a singular root node, called the "document element".

Child Elements

All other elements in an XML document are descendants ("children") of the document element. In the text file example earlier, the document element is the <textfile> element, and the <line> and <EOF> elements are its children.

Element Nesting

Element must be 'properly nested'. Consider how this book might have been transported to you. After printing this book and 23 others like it were packed into a box. Two boxes were packed into a carton and cartons were loaded on to a truck and delivered to a bookstore.

Improper Nesting

Consider the following

<name> Joe Lee <address> Park Lane </name> Any town </address>

The 'tree' implied by this example is:

name .... {name $\cap$ address}.... address

Since tags are not properly nested, a simple hierarchical tree of elements cannot be constructed. XML prohibits overlapping tags. If an improperly nested tag is encountered, the XML parser must report a "not well-formed" error and will quit processing the document and report a "fatal error".

String Literals

String literals are used for the values of attributes, Internal entities and external identifiers. All string literals in XML are enclosed by a pair of delimiters, using either apostrophe (') or quotation mark (") characters.

The one restriction upon these literals is that the character used for the delimiters may not appear within the literal - If an apostrophe appears in the literal, the quotation mark delimiter must be used and vice versa. If both delimiter characters must appear in the literal, the one being used within the string must be escaped, using an appropriate entity reference (' or &quot).

The following are legal string literals:

"string"

"string"

'... his friend ' s cow said "moo"'

Character Data is any text that is not markup and is the text content of an element or an attribute value.

The ampersand (&) and less than (<) **characters** are markup delimiters and thus may never appear in their literal form within character data. If these characters are needed within character data, they must be escaped using the & or < entity references.

Attributes

If elements are the "nouns" of XML, then attributes are its "adjectives".

There is some information about an element that we wish to attach to it, as opposed to the information that is contained within the element. This can be done using attributes, each of which is comprised of a name-value pair, using one of these two formats:

attribute_name ="attribute_value"

attribute_name='attribute_value'

Attribute values must be delimited strings that may contain entity references, character references and/or text characters. Only one instance of an attribute name is allowed within the same start-tag or empty-element tag. Both start-tags and empty-element tags may include attributes within the tag.

Special Attributes

Two special attributes have been defined by XML 1.0 recommendation: xml : space and xml : lang.

These attributes may be used by the XML document author to pass a signal to the XML application (browser, for e.g.). Both of these use the XML Namespace syntax, that is a namespace prefix ("xml"), followed by a colon (:) and then the attribute name ("space" of "lang").

These attribute values are not case-sensitive. The value of the xml: lang attribute is not just applied to the element including the attribute; it also applies to all child elements and attributes of that element, xml: lang attribute will be an essential feature of almost all international XML documents.

White Space

White space is an important linguistic concept for both human and computer languages. Only four characters are treated as white space in XML data:

Character Values (in hexadecimal)	Description
09	Horizontal tab (HT)
0A	Line-feed (LF)
0D	Carriage-return (CR)
20	ASCII space character

There is no expansion of tabs to more than one character, so each of these is simply treated as a single character.

XML is rule for handling white space is very simple: all white space characters within the content are preservd by the parser and passed unmodified to the application, while white space within element tags and attribute values may be removed.

End-of-Line Handling

XML data objects will often be stored in discrete computer files that are divided into "lines" of text. Two of the four XML white space characters are the standard ASCII end-of-line control characters. There are three common combinations of these two characters that are used to indicate end-of-line : CR-LF, LF only and CR only.

Character and Entity References

XML provides two simple methods of representing characters that don't exist in the ASCII character set:

Character references

Entity references

Character References

Character references are typically used as a substitute for the literal form of a character in places where the direct proccessing of the literal form would result in a violation of the well-formedness constraints of XML syntax.

Character references represent a displayable character and are comprised of a decimal or hexadecimal number, preceded by "&#x" string literals respectively and followed by a semi-colon (:) character:

&# NNNNN;

&#X XXXX;

The "NNNNN" and "XXXX" strings above may be one or more digits that correspond to any unicode character value that is allowed by XML.

Entity References

Entity references allow the insertion of any string literal into element content or attribute values, as well as providing mnemonic alternatives to character references.

Entity references are a legal XML name, preceded by an ampersand (&) and followed by a semi-colon (:) character.

&name;

Five entities are defined as an integral part of XML and are used as escape sequences for the XML markup delimiter characters:

Entity	Usage
&	Always used to escape the & character (except within CDATA sections)
<	Always used to escape the <character (except with CDATA sections)
>	May be used to escape the > character -within a CDATA section, the entity must be used if the > follows "]]" string literal.
'	May be used to escape the 'character in string literals.
"	May be used to escape the ' character in string literals.

Processing Instructions (PIs)

Since XML is a descriptive markup language, it does not presume to try to explain how to handle an element or its contents. This is a powerful advantage, that it provides presentation flexibility and OS-and application independence. However, there are times when it is desirable to pass hints to an application along with the document.

The **Processing Instruction (PI)** is the mechanism that XML provides for this purpose.

PIs use a variation of XML element syntax:

<? target ..... instruction....?>

The PI target is required and must be a valid XML name that is used to identify the application to which the PI is directed. The .... instruction...... portion of the PI is merely a string literal that may include any valid character string, except the "?>" string literal (the PI is ending delimiter.)

The other example of a nearly ubiquitous PI is the one used to anociate a style sheet with an XML data object:

<?xml - style sheet .....?>

This PI was not included in the XML 1.0 recommendation.

Comments

It can be useful to insert notes or **comments**, into a document.

The basic syntax of an XML comment is:

<! -- ... comment text .....-->

Where the "... comment text ...." portion can be any character string that does not include the "--" string literal. Further more, the "...." portion cannot end with a hyphen ("-") since this could cause misinterpretation of the closing delimiter.

Comments are not part of a documents character data ! Within a comments section, entities are not expanded, nor is any markup interpreted.

Consider some examples.

<tag > .... some content .... </tag> <! -- this is a legal comment -->

<! -- ======== 'Beginning of some more comments ======-->

<! - -

This is a comment containing unexpanded <tag> and &entity - reference; is ..... that is continued on another line ....

- - >

<! --===== End of comments ===-- >

The following are illegal comments.

<tag> .... some content .... </tag <! -- this is an illegal comment -- >>

<! - - this is an illegal comment - - because of the double hyphen within -->

CDATA Sections

CDATA sections are a method of including text that cotains characters that would otherwise be recognised as markup. This feature is primarily useful to authors who wish to include examples of XML markup in their documents.

CDATA sections may occur any where that character data may occur, but they **cannot be nested.** The only markup string that is recognised within a CDATA section is its end delimiter ("]]>") - less-than characters (<) and ampersands (&) may occur in their literal form; the need not be escaped.

The basic syntax of a CDATA section is:

<![CDATA[.....]]>

where the "...." portion can be any character string that does not include the "]]" string literal.

Prolog

An XML document starts with a prolog. The prolog is used to signal the beginning of XML data, describe its character encoding method and provide some other configuration hints to the XML parser and application.

The prolog consists of an optional XML declaration followed by zero or more comments, PIs and white space characters, followed in turn by an optional document type declaration. Since all these parts are effectively optional, this entails that the prolog may be omitted and the document can still be well-formed.

XML Declaration

All XML documents should begin with a single **XML declaration.**

<? xml version = "1.0"?>

Example of complete syntax of XML declaration

<?xml version = "1.0" encoding = "UTF-8" standalone = "yes"?>

Character Encodings

Unicode was originally designed as a simple set of 16-bit scalar values, since this would include most of the special symbols or characters needed for any modern language.

Unicode has added the **surrogate block** mechanism to allow another 1,048,576 additional characters. Although unicode identifies characters using scalar values (numbers), there are different acceptable ways of storing those numbers.

The **UTF-8** encoding treats 7-bit ASCII normally, but requires 2 to 5 bytes for everything else. "UTP" means "Unicode Transformation Format".

The **UTF-16** encoding is the simplest, storing 16-bit characters.

The **UTF-32** encoding has the advantage of programming simplicity - all Unicode characters can be encoded using this method. Such an encoding doubles the storage space required by each individual character.

Surrogate Blocks

Surrogate blocks are Unicode's method of expanding its character number space beyond that provided by 16-bit number's. Two blocks of 1,024 characters have been reserved (hex D800 to DBFF and DC00 to DFFF) to allow for 1,048,576 additional characters. A 16-bit value from the first block is the high order part of the 32-bit number, the second is the low-order part. This allows a program that doesn't understand these combined characters simply show one or two place holders.

Surrogate block character pairs are **not** legal XML characters.

Byte-Order Marks (BOMs)

Unicode data can use a BOM as a signature to specify the encoding that is being used. For example:

Bytes	Encoding form
00 00 FE FF	UTF-32 UCS-4, big-endian
FF FE 00 00	UTF-32 UCS-4, little-endian
FE FF 00 ##	UTF-16 big-endian
FF FE ## 00	UTF-16 little-endian
EF BE BF	UTF-8

These values are not legal within XML data, they immediately precede the valid XML data in a transmitted data object and can provide valuable encoding information to the XML parser or application. An XML parser must be able to handle BOMs prior to XML data, since support of both UTF-8 and UTF-16 encodings is a requirement.

Document Type Declaration

This contains the internal subset of, and/or refers to the external subset of, the Document Type Declaration (DTD).

All valid XML documents must include this declaration, but simple well-formed documents do not require its inclusion as long as they do not contain any entity references. If additional character entities are used, in a well formed but not valid document that will be parsed with a non-validating parser, they may be declared in an internal subset of a standalone XML data object, without requiring the use of an external DTD or scheme.

External Subset References

There are two forms of Document Type Declaration that refer to the external subset of the DTD:

<! DOCTYPE root_element_name SYSTEM "system_identifier">

<! DOCTYPE root_element_name PUBLIC "public_identifier", "system_identifier">.

The root_element_name is the point-of-attachment for the DTD. All children in the XML Document's element tree will thus inherit the DTD's declarations from this root node.

Both the SYSTEM identifierc in the above example) and the PUBLIC identifier values are URIs (Uniform Resource Identifiers). At the time of writing this, a URI can be considered equivalent to a URL for all practical purposes.

The PUBLIC identifier is used to refer a DTD that is catalogued and referenced using some method that is not defined in XML, but is the result of a standard within an organisation, or an agreement between parties exchanging the XML data.

An XML parser may try to generate URI using the PUBLIC identifier. If it is unable to do so, the SYSTEM identifier URI must be used.

Since the commonly used fragment identifier (# character) is not actually part of the URI, the parser may signal an error if a fragment identifier is encountered in a SYSTEM identifier.

An example of a reference to an external DTD:

<?xml version= "1.0" standalone = "no"?>

<! DOCTYPE Catalog SYSTEM

"http://www.wrox.com/DTDS/Pubcatalog.dtd">

<Catalog>

....

</Catalog>

Internal subset declarations

If no external DTD is available, an XML data object can still use entity reference, provided that they are declared in the internal subset of the DTD, which is done using an expanded form of the <! DOCTYPE ....> declaration.

Internal subset declarations are delimited within the <! DOCTYPE ....> declaration using brackets "[......]" and that the <! ENTITY ....> declarations are used to define the expansion of entity references.

Epilog

The XML epilog include comments, PIs and/or white space. Since XML does not define any end-of document indicator, most applications will use the end-tag of the document element for this purpose. Thus, a network link is likely to be closed after encountering the root element end-tag, without any handling of an epilog.

Summary of XML Syntax

The XML tag forms:

Syntax	Component
<tagname>	Element start-tag
<tagname attribute ="value">	Start-tag with an attribute
</tagname>	End-tag
<tagname/>	Empty-element-tag
<tagname attri ="value 1" attr2 ="value 2"/>	Empty-element tag with two attributes.

Special XML instructions and declarations

Syntax	Component
<?xml version = "1.0" encoding = "UTF-8" standalone = "yes"?>	The XML declaration
<? name string?>	Processing Instruction (PI), for the XML application.
<!-- string -->	Comment, for the human writer / editor
<! [CDATA [string ....]]>	Unparsed character Data (CDATA) section
<!DOCTYPE string ...>	Document Type Declaration (not the "DTD")

XML entity references :

Syntax	Component
&# decimal; Example: ®	Reference to a character number, using its decimal value
&# XHEX; Example: &# xAF4	Reference to a character number, using its hexadecimal value
&ref; macro)	Reference to a predefined XML entity (such as a text

Data Modeling and XML

The three aspects of the design process:

- Information Modeling: Which is about understanding the structure and meaning of the information carried in documents.
- Document Design: Which is about translating your information model into a set of rules (or schema) for creating actual documents.
- Schema Notations: Techniques for recording your document design so it is accessible both to processing software and to human users.

Information Modeling

The first rule of Information modeling is to focus on the "real world", not the technology. An Information model is a description of the information used in an organization, independent of any IT systems.

There are two objectives in an information modeling exercise:

- To achieve absolutely precise definitions.
- To communicate effectively with users.

Static and Dynamic Models

There are two main types of information model: static models and **dynamic** models.

Static models concentrate on describing the permissible states of the system.

Dynamic models concentrate on describing what happens to information.

In general, static models are immediately relevant to the design of a database, where information is stored for long periods of time to serve a variety of purposes; while dynamic models are most directly relevant to the design of messages, which have only transient existence and a very specific purpose.

The Static information Model

To define a static information model, we will take the following steps.

- Step 1 - Identify things, name them and define them.
- Step 2 - Organize things into a class hierarchy.
- Step 3 - Define relationships cardinality and constraints.
- Step 4 - Add properties to flushout the details of values associated with the objects.

Dynamic Modeling: What Happens to the Data?

If we are going to use XML for representing the messages that flow around the system, we also need to understand what happens to the data, where it comes from, and where it goes next.

There are several approaches you can take to dynamic modeling. They are

- Process and work flow models
- Data flow models
- Object models
- Object life histories
- Use cases
- Object Interaction diagrams

Two Roles for XML

In data flow models there are two kinds of data in the system: data stores, which hold long-term persistent data for reference purposes and message flows, which move transient information from one subsystem (or person, or data store) to another.

XML is useful for both kinds of data, but the design considerations are rather different.

XML for Messages

Using XML for the messages in the system poses server design problems than using it for persistent data. This is because each message is generally fairly self-contained.

The message might be an EDI-style message sent between organizations to represent a transaction such as a hotel reservation. Or it might be data passed when one subsystem calls another, for example in a bank the fraud detection system might ask the operational backing system for a transaction history on a particular account, and get the answer back in the form of an XML message. It might also be a message intended for display to people, for example when we use our browser to get a weather forecast for a particular holiday resort. In this case the answer to our enquiry might come back from the back-end database system as an XML message, for the front-end system (or the browser) to turn into a visible page by rendering it using an appropriate style sheet. It might also be a message that originates from a person, such as a sales enquiry created using a firm on our web site.

XML for Persistent Data

When we use XML for persistent data, it is the static model that is important.

How Big is a Document?

XML is not well suited for direct access; to access any part of such a large document with current technology, it will be necessary to parse the whole thing, which may take hours.

At the opposite extreme, having a very large number of tiny documents is usually not ideal either, because it makes it impossible to use the rich structure of XML to represent the relationships in the information model.

Sometimes, where business objects have a significant size and internal complexity, there is a natural mapping of one XML document to one business object. For example, in a human resources system it may be natural to use one XML document for each employee's record, or in a medical records system there might be one XML document for each patient episode.

When we use XML documents to hold persistent data, finding information will always be a two-part operation; first find the right document, then find the facts we are interested in. The tools and techniques available for the two parts of this operation are quite different, so the choice of what information to put in each document is of critical importance.

How many Document types should we have?

Where we conceptually have several different document types in our model, there are several ways we can represent them in XML.

- We can have a different DTD for each one. If the different document types have parts in common, we can define external parameter entities which are shared between the diffeent DTDs.
- We can use the same DTD for all of the document types, but using a different top-level element tag for each section. There is nothing in a DTD that says which element should be the document element, so it is quite possible to define several types in the same DTD. Components used in a multiple document types can then easily be referenced in each one.

Representing Object Types

Generally an object type in our information model will translate into an element type in the XML structure.

Representing Relationships

Some of the relationships in our model can be represented by using nested elements in the XML document structure. The obvious ones to represent this way are the "contains" relationships, for example <Footnote> elements can be nested within a <character> element, and <hotel> elements can be nested within a <resort> element.

An element can only be contained by one parent element, so the other relationships in the model must be handled in some other way.

There are several ways of representing a link from one element to another in XML.

Example: In a Holiday itinerary for a party travelling together, there are optimal events and excursions, but we recognize that not all the members of the party will always take the same options. For example, Mum and Dad might want to go to the casino while the kids go off to the zoo. In an itinerary for a holiday, we'll represent this by giving each traveller an ID value, and against excursion we'll include an IDREFs field that lists the travellers taking part something like this,

```
<itinerary>
    <traveler id = "t01">
    <name> Mrs. Mary Higgins </name>
</traveler>
    <traveler id = "t02">
    <name> Mr. John Higgins </name>
</traveler>
    <traveler id = "t03">
    <name> Mr. Rory Higgins </name>
    <age> 12 </age>
</traveler>
    <traveler id = "t03">
    <name> Kylie Higgins </name>
    <age> 9 </age>
</traveler>
    <excursion participants = "t01 t02">
    <venue> Casino </venue>
    <date> 2000-06-15 </date>
    </excursion>
    <excursion participants = "t03 t04">
    <venue> 700 </venue>
    <date> 2000-06-15 </date>
    </excursion>
</itinerary>
```

Elements or Attributes

Here is an example whre the properties of a book are represented as XML attributes:

```
<book
    author = "Nelson Mandela"
    title = "King walk to Freedom"
    publisher = "Abacus"
    isbn = "0-349-10653-3" />
```

Here is the same information, this time representing the properties as child elements

```
<book>
    <author> Nelson Mandela </author>
    <title> Long Walk to Freedom </title>
    <publisher> Abacus </publisher>
    <isbn> 0-349--10653-3 </isbn>
</book>
```

Encoding property values

Whether we represent the properties of an object using elements or attributes, we have to make a decision on how to encode their values.

Some of the common situations encountered are listed here.

Quantities, such as Height, Width and Weight

The first question is whether to standardize the unit of measure (all lengths in meters, for example) or to allow different units of measure. If the unit is included as part of the data, there is the option of making it part of the value (height = "1.86 m") or making it a separate item (<height units = "m"> 1.86 </height>). Finally, the decimal number format must be defined, for example which symbols are used for the decimal point and thousands separator.

Yes/No values

An obvious representation, in an English speaking context, is to use the strings yes and no or y and n if space is at a premium. If attributes are used, an alternative is to use the SGML convention, also encountered in HTML. For example border = "border" means yes, and an absent border attribute means no.

Other Small Value Sets

The basic decision is whether to use full names or codes, such as W or Withdrawn, UK or United Kingdom. Apart from space considerations, the main benefit of using an expanded name is that it avoids misunderstanding.

Dates and Times

Many would advocate standardizing on the ISO 8601 date format (YYYY-MM-DD) to avoid ambiguity. ISO 8601 also allows times to be represented, including the time zone.

There are actually many standards available for particular types of property. The international Organization for Standardization (ISO) have published a good many, among them the following:

Property Names

When we use elements to represent properties in the information model, and to a lesser extent also when we use attributes, the question arises as to what names to use.. There are two choices. We can name either the data type, or its role in the parent object, example, the date of a document's creation can be named either date-created or date; the address of a customer can be named either address or billing-address.

Binary Data

Not all properties of objects can be conveniently represented as character strings: multimedia data, in particular is binary in nature to represent binary data.

There are two main approaches: the binary objects can be **internal** or **external.** Internal means they are represented as part of the XML data stream, external means they are in a separate file.

For internal storage, most people use base 64 encoding. This is a technique of encoding each sequence of binary digits as an ASCII character. It works well with XML, because the ASCII charaters used do not clash with the markup sequences (such as "<" and "]]") that have a special meaning in XML.

For external storage, the "pure XML" way of doing it is to use external unparsed entities and notations. For example, to include a reference to the image file picture.gif. We would write in the DTD (typically in the internal subset):

<! NOTATION gif SYSTEM "gifeditor. exe">

<! ENTITY picturel SYSTEM "picturel.gif" NDATA gif>

Using XML entities

Entities in XML are supposed to be part of the physical document structure, not part of its logical structure. An entiry reference is supposed to mean exactly the same thing as if the entity were expanded in line.

So its natural to regard entities as part of the physical design stage - the XML design equivalent of the physical design stage in relational database design.

There are however some logical uses for entities. The main advantage of extracting something from the body of document into an external entity is that it can then be updated separately from the rest of the document. This makes it useful for text or other document components that have a different update cycle or update authority from the main body. This control aspect is far more important than any space saving obtained by sharing common content.

The Role of a Schema

The concept of a schema has been present in both the database and the document worlds. The formal role of the schema is to define the set of all possible valid documents or to look at it more negatively, to define what constraints beyond those of XML itself, documents must meet for them to be meaningful.

In the XML standard, being "valid" means something quite specific. Informally, it means that a document conforms to the rules with DTD. Many constraints that cannot be expressed in a DTD. And in an XML schema either-like a rule that a sequence number at the start of a message must be one greater than the sequence number at the start of the previous message.

Why do we need Schemas?

The purpose of a schema is to define and describe a class of XML documents by using there (markup) constructs to constrain and document the meaning, usage and relationships of their constituent parts: datatypes, elements and their content, attributes and their content entities and their contents and notations. Schema constructs may also provide for the specification of additional information such as default values. Schemas are intended to document their own message, meaning, usage and Function through a common documentation vocabulary. Thus XML schemas Structures can be used to define, describe and catalogue XML vocabularies for classes of XML documents.

The Schema as a Set of Constraints

One purpose of a schema is to define the difference between a valid document and an invalid one. As far as possible the rules should be expressed in such a way that software can decide whether a document is valid or not.

The Schema as an Explanation

The second purpose of a schema is to explain to document the interpretation and usage of the constructs provided, so that the sender and the recipient share a common understanding of the meaning of the message.

The both the document and database traditions, this role of a schema has tended to play second fiddle, though potentially it is the more important. Part of the problem is that the schema doesn't usually reach the person who needs to know, namely the person entering the data on the screen. This is why many systems suffer from what has been called **semantic drift**,a tendency for the user community over time to change the way in which they use the system and the meanings they attach to the data fields, eventhough the software structure remains unchanged.

The fact that end users won't look at the schema directly means there is a need for the schema to be application readable: for example, it should be possible to extract explanations of data fields and display them as advisory text on data entry screens. This concept of a schema was fashinable in the 1970s and 1980s under the title of **a data dictionary;** another term that emphasizes its role in defining the meaning of names in the systems vocabulary though it eventually dropped out of favour through software companies whose data dictionary was little more than an internal systems catalog.

What Constraints can you Express in a DTD?

As a constraint language, DTDs are very limited. They provide some control over which elements can be rested within each other, but say nothing about the text contained within the elements. They offer slightly more control over attributes, but even this is very limited, for example there is no way of saying that an attribute must be numeric. This inevitably means that much of the real validation has to be done in the application.

Can you Rely on DTD Validation?

Even within the limits of the rules we can define in a DTD, there is the odd ability in XML for a document to set its own agenda: it is the document itself that decides whether it is going to reference a DTD or not, which DTD it is going to reference, and whether it is going to override any of the declarations in the DTD in its private internal subset. Even if the application knows it is using a validating parser, the constraints it is enforcing are those that the application wishes to impose.

If the application is going to avoid doing validation itself, it needs to be sure that the input document conforms to the DTD called itinerary. dtd.

Writing DTDs

The syntax of DTDs does not conform to the XML syntax and can be difficult to read and even harder to write, betraying its late 1960s SGML legacy with every choice of delimiter. A DTD as an explanation of the meaning of elements and attributes is about as useful as explaining a business process by showing compiled Javabyte codes: there are even places in the syntax where comments are not allowed.

DTDs and Namespaces

Namespaces are supposed to allow to mix elements from more than one information model in the same document, for example to use chemical formulae as well as geographical coding in a document about the spread of pollution. The intent is that we can choose our own prefixes for names without changing the meaning, for example we could use <GEO:LI> to refer Lithuania and <CHEM:LI> For Lithium. There are two problems with this. Firstly our document can only refer to one external DTD. Secondly the remaining of elements by adding prefixes is not recognized by the DTD.

Deriving a DTD from the informaiton Model

Although the modeling capabilities of the DTD are limited, a great deal can be achieved by exploiting the very flexible mechanism of parameter entities. Although on proper parameter entities simply allow text to be shared between diffeent definitions in the DTD in practice they can be used as definitions in their own right and a great deal of the out of writing effective DTDs depends on creative use of parameter entities.

Namespaces and schemas

We have seen the basic rules of wellformed XML as well as DTDS - the ones provided in the W3C 1.0 Recommendation.

The problem is that everyone has the ability to create their own tags. We can imagine cases of people using the same element names to mean different things; example, element name 'monitior'. If we have a DTD for computer peripherals then monitor may refer to the screen, while in a music studio speakers are called monitors. Even if meanings were the same the possible content for the elements may chage between definitions. With these possibilities of different usage, there is a need to distinguish the particular use of the element, especially if we mix different vocabularies in one XML document. To solve this problem, there is a specification called XML Namespaces, from the W3C, which allows us to define the context of an element within a namespace.

Furthermore, there are likely to be situations where we need to combine XML documents from different sources, which conform to different DTDs.

Namespaces help XML vocabulary designers to break complex problems into smaller pieces and mix multiple vocabularies as needed to fully describe a problem in a single XML document. Schemas permit vocabulary designers to create a more precise definition of the vocabulary than is possible with DTD and do so using XML syntax.

Namespaces and Schemas allow XML designers and programmers to do the following:

- Better organise the vocabularies surrounding a complex problem.
- Provide a way to retain strong typing of data when converting it to and from XML.
- Describe vocabularies with more precision and flexibility than DTDs permit.
- "Read" vocabulary rules in XML, permitting access to vocabulary definitions without increasing parser complexity.

XML Namespaces reached W3C Recommendation status on Jan 14, 1999. Schemas are working their way through the standards process, but a Recommendation is expected soon.

Mixing Vocabularies

After building a site that exposes our book catalogs as XML written to PubCatalog.dtd vocabulary we might decide to sell the books online. This means we need to be able to take orders for the books in the catalog. So we need a DTD that covers ordering of books.

We might start adding to the PubCatalog.dtd file, because the two areas address different parts of the same problem domain : sharing data about books. They can be seen as different problem domains because the one addresses the whole of the catalog details, while the other addresses sale of items from the catalog. While there is some overlap in the information conveyed by the two topics, if we tried to use one DTD to cover both acreas, we might end up with one very large confusing or DTD.

There is a better solution, merging data that conforms to separate catalog and order DTDs in a single document using namespaces.

Segmenting a Problem

When we write a large program we break the overall problem into smaller pieces. Modules classes, components, packages and functions are some of the constructs programming languages offer for this purpose. Designing a vocabulary can be seen as a similar problem to writing a program. We need ways to segment large problem into multiple vocabularies.

Reuse

In our PubCatalog.dtd we made use of the book element. We have talked about marking up our data in a way that describes contents of the element. But when we consider taking online orders for books, we are likely to use the same element name again, when referring to the book that a customer wants to order. The book element in the order might be child of an order element, whereas it is a child of catalog in the PubCatalog.dtd.

When describing real world concepts, we will continually find that common constructs keep appearing. after all, complex creations are build from simple building blocks - color, shape, price and dimensions, for example and simple things don't go undefined fo iong, so there will be many instances of element names that already have definitions and content models.

If someone else already created a DTD that uses these elements, our task will be made easier by borrowing from proven DTDs. This is the concept of **reuse.** Reuse saves time, effort and money.

Ambiguity and Name Collision

If we are reusing useful definitions from another designer's DTD or combining segmented DTDs to create a document describing a composite problem, we risk the problems of **AMBIGUITY** and **name collision** if the documents we are using feature elements of the same name. For example, Books. There may be several DTDs that declare a book element, at least for publishers and printers, retailers and libraries. A single usage of the name Book in a document needs resolution to match it with the proper Book element declaration. In our example, Book is a name common to both catalogs and orders.

A document marked up using the PubCatalog.dtd may include the following use of the element <Book>:

```
<Title> Professional XML </Title>
<Abstract> Compendium book containing
        everything we need to learn
        to use XML in our
        programming solutions
        today. </Abstract>
<ReeSubjCategories>
<Category> XML </Category>
<Category> Programming </Category>
<Category> Internet</Category>
</RecSubjCategories>
</Book>
```

If we are reading an XML document that includes data from both of the vocabularies we would not know which definition it refers to.

The problem becomes acute when we use instances of a name drawn from multiple DTDs. Example, when we talk about lighting, we want to draw on pre-existing DTDs for traffic lights and street lights. The respective designers choose the word <Light> as an element name. Had they known the eventual use of their DTDs, they might have chosen <TrafficSignal> and <StreetLamp>. Now the documents have ambiguous Light elements.

The declarations for the two uses of <Light> are very different. The first declaration covers traffic signals and has an enumeration for its color attribute. This is important because there are only three valid colors for traffic signals. An application can be expected to do some error checking based on the value of this attribute

```
<! ELEMENT Light EMPTY>
<! ATTLIST Light color (red | yello | green)
                        #REQUIRED>
```

The second declaration has no such restrictions on its color attribute's value. Indeed, lamps are often

The practical pros and cons of each approach.

	Advantages	Disadvantages
XML Attributes	DTD can constrain the values; useful when there is a small set of allowed values such as "yes" and "no".	Simple string values only
		No support for metadata (or "attributes of attributes")
	DTD can define a default value.	
		Unordered
	ID and IDREF validation.	
	Lower space overhead (makes a difference when you are sending gigabytes of data over the network)	
	White space normalization is available for certain data types (such as NMTOKENS), which saves the application some parsing effort.	
	Easier to process using DOM and SAX interfaces.	
	Access to unparsed external entities, such as binary data.	
Child Elements	Support arbitrarily complex values and repeating values.	Slightly higher space usage
	Ordered	
	support "attributes of attributes".	More complex programming
	Extensible when the data model changes.	

chosen on the basis of cost, not color, although the color is still specified :

```
<! ELEMENT Light EMPTY>
<! ATTLIST Light color CDATA #REQUIRED>
```

Consider the following XML document written by an application that mixes the two DTDs :

```
<Inventory>
<Light color = "red"/>
..............................
<Light color="white"/>
..............................
<Inventory>
```

Namespaces

XML **namespaces** are the solution to the problems of ambiguity and name collisions. A namespace is " a collection of names, identified by URI reference, which are used in XML documents as element types and attribute names".

A DTD can be a namespace. In this case, the URI could be the address of the DTD on your server, for example:

http://www.wrox.com/xmldtds/PubCatalog.dtd.

If this case the namespace refers to the names used in the Pubcatalog.dtd.

In the namespace is a DTD or schema; the definitions we use must remain consistent with the structure and syntax specified therein.

So, in order to use the namespaces effectively in a document that combines elements from different sources, we need two parts.

- A reference to the URI that defines the use of the element.
- An alias that we can use to identify which namespace our elements is taken from, this takes the form of a prefix for the element (for eg. <catalog: Book> where catalog is the alias for the ambiguous Book element.

Declaring a Namespace

We need a namespace declaration to recognize it. The Namespace Recommendation gives us xmlns. The value of the attribute is the URI that uniquely defines the namespace in use. This URI is often a URL pointing to a DTD, but it doesn't have to be.

Example of namespace declarations:

```
xmlns="http://www.wrox.com/bookdefs/book.dtd"
xmlns=urn: wrox-publishing-orderdefs"
```

We can provide an alias we could use throughout a document that would refer to the declaration. We do this by appending a colon and your alias to the xmlns attribute.

Thus, the examples above become:

```
xmlns: catalog="http://www.wrox.com/book defs/PubCatalog.dtd"
xmlns: order="urn: wrox-publishing-sales-order defs"
```

Here the prefix catalog will refer to elements from the PubCatalog.dtd, while order will refer to elements declared in the order.dtd. After these declarations appear, we can just use book to refer to the first namespace declaration, and order to refer to the other one (without the URI).

Qualified Names

We have to tie a namespace to a specific name we want to use. This is done through the use of **qualified names.** We create a qualified name by taking the alias, known as a **namespace prefix,** and back it on to the beginning of the name. Going back to the question of including a Book element in both catalog and ordering DTDs, assume that we declare a catalog namespace with the prefix catalog like.

```
xmlns: catalog = "http://www.wrox.com/
book defs/PubCatalog.dtd"
```

We can now use the prefix catalog to make it clear which namespace the element came from. So,

Stanard	Scope	Description
ISO 2955	Metric measurements (systeme Internationale)	Method for representing SI and other units in information systems with limited character sets (without using Greek letters).
ISO 3166	Country codes	Specification of codes for the representation of names of countries; a set of two letter and three letter codes. The two letter codes are most familiar as internet domain names such as de for Germany.
ISO 4217	Currency codes	Codes for national currencies.
ISO 5218	Human Series	Codes to represent male and female (surely this must be the shortest international standared ever published?)
ISO 6093	Numeric values	Three presentations of numerical values, represented in character strings in machine readable form and recognizable by humans
ISO 6709	Location of points	A Format and representation for unique identification of points on, above, or below the earth's surface by longitude, latitude and altitude.
ISO 8601	Dates and Times	Expressing of dates, including calender dates, ordinal dates, week numbers and times in numeric form, with punctuation to avoid ambiguity.

<Catalog:Book>

Here's an example that mixes some namespaces:

<catalog:Book order: ISBN = "1-861005 - 11 - 0">

The element <Book> is drawn from the first namespace, while the attribute ISBN is drawn from the order namespace.

Scope

Namespace declarations have scope in the same way that variable declaration do in programming lanaguages. Namespaces can be included within a later section of the document. A namespace declaration therefore applies to the element in which the declaration appears, as well as children of that element even if it is not explicitly specified in the element. A name can refer to a namespace only if it is used within the scope of the namespace declaration.

We will also need to mix namespaces where elements would otherwise inherit the scope of a namespace, so there are two ways in which scope can be declared, **default** and **qualified.**

Default

To add a prefix to every name is tiresome. By introducing the concept of name scope to our tool set, we can dispense with a lot of prefixes. If we define a default namespace, all unqualified names within the scope of the declaration are presumed to belong to that default. So, if we declare a default namespace in the root element, it is treated as the default namespace for the whole document.

Here's how we might use this to embed some HTML within an XML document marked up according to a DTD designed for book content, called BookContent.dtd:

<Chapter xmlns = "http: //

www.wrox.com/bookdefs/BookContent.dtd">

<Title number = "7"> Namespaces and Schemas </Title>

<Author> I.M. Named </Author>

<Content>

<Paragraph>

Let's have a table:

<table xmlns = "http: //www.w3.org/TR/REC/REC-html 40">

<tr>

<td>A tisket </td> <td> A tasket </td>

</tr>

<tr>

<td> one fish </td><td> Two fish </td>

</tr>

</table>

</Paragraph>

<Paragraph> This is a very short paragraph

</Paragraph>

</Content>

</Chapter>

The elements <Title>, <Author>, <Content>, and <Paragraph> and the attribute number come from the default namespace defined in the <Chapter> element. Within the chapter element, we can see the table element and its children - tr and td. These belong to the HTML namespace declared in the table element.

The scope of the HTML namespace declaration in this eg. ends when the table element closes. The second occurrence of Paragraph does not come from the HTML namespace.

Qualified

Sometimes, we will want to sprinkle names from foreign namespaces through a document. Rather than declaring namespaces all over the place, we can make use of qualified names. Declare the namespaces we will need at the beginning of the document and then qualify them at the point of use.

<Measurements xmlns = "um:mydecs-sclence - measurements"

xmlns:units="urn:mydecs-science-unitsofmeasure">

xmlns:prop="urn:mydecs-Science-things

measusred">

<OutsideAirunits: units - "Fahrenheit">

86</OutsideAir>

<FuelTank>

<Prop: Volume units:units="liters">120</prop: Volume>

<Prop: Temperature units:units="Celsius">20</prop: Temperature>

</FuelTank>

</Measurements>

Schemas

A DTD is actually a type of schema. However, when people in the XML community refer to schemas they often mean a replacement for DTDs written in XML syntax, a term that is used in this chapter.

We can refer to schemas as **metadata,** or data about data. They also explain the relationships between certain types of data. Like a DTD, a schema is a description of the components and rules of an XML vocabulary. Schemas refine DTDs by permitting more precision in expressing some concepts in the vocabulary. They use a wholly different syntax than DTDs. The permit us to borrow from other schemas, there by solving the validation problem. They offer data typing of elements and attributes. Schemas are a better answer to the problem of specifying vocabularies..

The Problem with DTDs

DTDs have a few shortcomings:

- they are difficult to write and understand
- programmatic processing of their metadata is difficult
- they are not extensible
- they do not provide support for namespaces
- there is no support for datatypes
- there is no support for inheritance

The Problems in detail

DTDs are difficult to write and understand

DTDs use a syntax other than XML, namely Extended Backus Naur Form (EBNF)and many people find it difficult to read and use. The proposed XML schemas, however, actually use XML to describe the languages they define, removing the difficulty to learning EBNF before learning to read and write them.

Programmatic Processing of Metadata is Difficult

The use of EBNF also makes the automated processing of metadata in DTDs difficult.

DTDs are not extensibles and do not provide support for Namespaces

DTDs do not support datatypes.

DTDs do not support inheritance

An Abundance of Help Creating Schemas

Metadata - data about data is about describing information.

The academic community - and some of the XML - related metadata proposals - tend toward the ambitions end of this spectrum. One eg. is **Resource Description Framework** (RDF), a W3C backed effort for describing resource so that they may be discovered automatically. Because of the desire for an XML - based schema language to replace and extend DTDs, a number of proposals were put forward. These include:

- XML - Data
- Document Content Description (DCD)
- Schema for Object - Oriented XML (SOX)
- Document Definition Markup Language (DDML previously known as X Schema)

None of these have directly received formal work backed by the W3C, however each has been considered in the W3C work on XML schemas. In this section we will look at some XML-based metadata proposals.

Resource Description Frame work

RDF allows a designer to describe objects add properties to define and describe them, and also to make complicated statements about the objects, such as statements about relationships between resources. Its proposed uses include sitemaps, content ratings, stream channel definitions, search engine data collection (web crawling), digital library collections, and distributed authoring. The specifications come in 2 sections:

- Model and syntax
- RDF Schemas

The basic RDF model is a full Recommendation (22nd Feb 1999). It covers the descriptive data model that can be expressed in XML, as well as other syntaxes. RDF schemas are a proposd Recommendation (3rd March 1999) covering an XML vocabulary for expressing RDF data models. RDF drawn on the experince of developing the Platform for internet Content Selection (PICS), a scheme for defining web content and implementing rating systems, and also draws on earlier academic work in metadata.

Schemas developed with RDF can define not only names and structure, but can also make assertions, such as relationships about the things under discussions. RDF can be complicated, but it offers tremendous expressive power and depth, that its complexity is required for it to be so descriptive.

RDF is oriented around three concepts:

resources, properties, and **statements.**

Resources

Resources can almost be anything - any tangible entity in a conceptual domain that can be referred to by a URI, from an entire website to a single element in an HTML or XML page. It could even include something that is not available on the web, such as a printed book.

Resources are typed: a class system is used to define categories from which specific resource instances are drawn. Class inheritance is supported, so a designer can specify levels of definition ranging from highly general to narrowly specific. Here are 2 simple class definitions,the first defines a general Rocket class, and the 2nd defines that class through inheritance into a chemical Rocket class. The rdfs and rdf namespaces are part of the RDF Recommendation:

```
<rdfs: Class rdf : ID ="Rocket">
<rdfs : subClassOf
rdf : resource = "http://www.w3.org/
TR/WD-rds-Schema # Resource"/>
</rdfs : Class>
<rdfs : Class rdf : ID ="ChemicalRocket">
rdfs : ClassOf rdf : resource = "# Rocket"/>
</rdfs :. Class>
```

Properties

Resources are said to have properties that define and describe them. Constraints are placed on properties to give them shape. These constraints limit the types of values that can be assigned to a property and the range of literal values from the type that can be chosen. Let's give our chemical rocket some fuel:

```
<rdfs : Class rdf : ID="Fuels">
<rdfs : subCalssof rdf : resource
="http://www.w3.org/TR/WD-rdf-schema# Resource"/>
</rdfs: Class>
<rdf : Property ID = "fuel">
<rdfs : range rdf : resource ="# Fuels"/>
<rdfs : domain rdf : resource="#ChemicalRocket"/>
</rdf : Propety>
```

The fuel property is typed as being of the Fuels class, and the property can take on values from this range. For this, we have to make a class declaration similar to the ones above somewhere else in our schema to define this, perhaps providing literal values for the rocket fuels we wish to discuss. The fuel property applies to the chemical Rocket class, its domain.

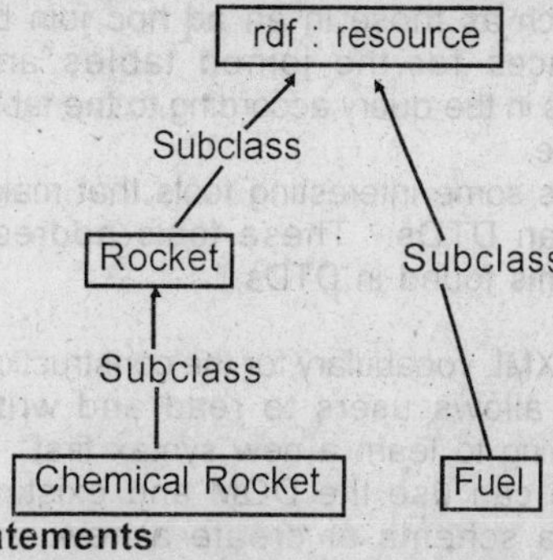

Statements

Once names and structures are defined through

resources and proprties, statements about conceptual domain can be made. This is done by composing triplets of subject resources, property predicates, and value objects. The values can be literals for specific statements, or resources for powerful and sweeping statements about entire classes. Let's make a simple statement about a particular rocket in a document conforming to our RDF schema. First, you begin by declaring an instance of the Chemical Rocket class and giving it a name.

```
<? xml version = "1.0" ?>
<rdf : RDF xmlns = "http://www.w3.org / 1999/02/22-rdf-syntax-ns#"
<ChemicalRocket ID="Moonship" xmlns = "urn : my-rdf-rocket-schema"/>
<rdf : Description about = "Moonship">
<fuel> hydrogen </fuel>
</rdf : Description>
</rdf : RDF>
```

XML Data

XML data aims for a more modest scope than RDF. This proposal was submitted to the W3C by Arbor Text, Datachannel, Inso and Microsoft, and is clearly focussed toward automated documents and processing, but is still more ambitious than DTDs.

XML Data makes a distinction between syntactic and conceptual schemas. While both use the same language, they provide different ways for us to think about the data we are marking up.

A syntactic model is a set of rules describing how to write documents using markup, as such DTDs are an eg. of syntactic schemas. In an XML document marked up according to our catalog DTD, a <Book> element can legally contain <Title>, <Abstract>, <Recsubjcategories> and <Price> elements. A syntactic XML Data schema of this would represent similar constriants on the structure of the vocabulary.

Conceptual models describe relationships between concepts or objects and as such they are ideal for modeling relational data bases. We could use an XML Data Schema to suggest the relationships that books have titles and prices, in a manner separate from the syntax of an XML document. In this sense XML Data was intended to broaden XML's reach to readily encompass information from relational databases. The principle relationships captured by the keys in a relational database can be captured formally in an XML Data Schema. With namespaces, we can capture ad hoc relationships such as those in an ad hoc join by declaring namespaces for the joined tables and qualifying the columns in the query according to the table from which they come.

XML Data provides some interesting tools that make it more powerful than DTDs. These tools address several of the problems found in DTDs.

Written in XML

XML data uses an XML vocabulary for the construction of schemas, which allows users to read and write schemas without having to learn a new syntax first. It also means that we can use the DOM and existing parsers to peruse a schema or create a new one dynamically.

Continuing the conceptual schema metaphor we could dynamically create a schema for an ad hoc SQL query based on the query itself. The recipient of the data would have both data and formal structure and would never know that this was dynamically generated.

Date Typing

XML Data adds strong typing of elements and attributes, thereby answering one of our prime objections to DTDs. These may be basic types defined by the datatypes namespace, or complex, user-defined types provided in a schema provided by the designer.

Constraints on Allowed Values

XML Data allows constraints on the range of values for elements and attributes to be defined, such as minimum and maximum. This can be extremely helpful in a lot of situations where you are validating XML documents. You could use constraints to prevent people from spending any money if their account has no funds, or prevent them from inputting values that were not valid.

Inheritance of Types

An interesting reuse mechanism is XML Data's support for inheritance of types. This lets us evolve and extend elements as we describe the entities in the problem we are trying to solve with XML. We can write some generally expressive super types, then refine them into more specific classes of elements by adding members to or replacing members of the supertype declaration.

Open and Closed Content Models

Another powerful feature of XML Data is the notion of **open** and **closed content models.** A classical DTD is a closed model.

If a schema is open, documents conforming to it may include other information not declared in the DTD. The parts that conform to the schema must obey the rules laid down in the schema, but we can insert other items without restriction from the current schema. These items may be defined in another schema or may be completely unconstrained. We might insert ad hoc values.

Open model documents are the way we can mix namespaces. We can embed a chuk of information conforming to one schema right in the middle of a document conforming to another. Individual elements may be explicitly declared to have open or closed content models. This is done through the content attribute. The default value for this attribute is open. Here is an example:

```
<element Type id="Person"
                    content="closed">
<element Type="#name"/>
<element type="#address" />
</element type >
<!- - This document fragment is invalid due to the added Telephone element - ->
<Person>
        <Name>John Doe </Name>
< Address >123 Anywhere Street Blasted Rock, NV</Address>
<Telephone>555-1212</Telephone>
</Person>
```

Had the content attribute in the above example been given the value open, the fragment would have been valid.

Expanded ID and IDREF constructs

XML Data extend ID and IDREF constructs with relations. In a **relation,** one element acts as a key or index into another element's content. This is directly applicable to the primary and foreign keys of relational databases. It is also particularly useful in bilingual documents. Two are of particular interest: aliases and correlatives.

An **alias** is used to define an equivalent element, so in our example. We may have <Book> in the English document and want to translate the tags to <Livre> in the equivalent French elements.

Other times we will want to suggest that two tags describe identical things, this is done using a **correlative.**

Document content Description

The Document Content Description (DCD) proposal followed on the heels of the XML Data proposal. It was submitted by IBM, Microsoft and Textuality. It is an RDF vocabulary designed for the purpose of declaring XML vocabularies.

W3C Work on XML Schemas

The W3C XML schema working Group has a two part working draft for XML schemas dated 17 Dec. 1999. Particular features and syntax are subject to change in later versions. These schemas answer our main objections to DTDs. They are written in XML syntax, they permit the use of multiple namespaces, and they provide for strong typing of content. They are moreover, a superset of the capabilities of XML 1.0 DTDs. Their expressive power is greater than DCD, but it is far less abstract than RDF. In short, this is a promising metadata effort.

The working Draft is divided into two sections: **structures and datatypes.**

XML Schema Part 1: Structures, deals with the description and declaration of elements and attributes. The material provided therein allows an XML designer to specify complex element structure and set constraints on the permitted values of the content of those elements. This part of the specification can be found at **http://www.w3.org/TR/xmlschema-1/.**

XML Schema Part 2: Datatypes, sets forth a standard set of content datatypes as well as the rules for generating new types from them. This part of the specification can be found at **http://www.w3.org/TR/xmlschema-2/.**

DTD vs. XML Schema : A contrast

We shall consider a very simple DTD and its translation into XML Schema form. Consider the following DTD for naming a person:

```
< ! ELEMENT Name        (Honorific ?, First,
                         MI?, Last, Suffix?)>
< ! ELEMENT Honorific   (#PCDATA) >
< ! ELEMENT First       (#PCDATA) >
< ! ELEMENT MI          (#PCDATA) >
< ! ELEMENT Last        (#PCDATA) >
< ! ELEMENT Suffix      (#PCDATA) >
```

We must minimally have first and last names, but we may optionally have a middle initial, honorific (Mr., Ms., Dr., etc.) and a suffix (Jr., III, etc.). Here is what it looks like in a schema:

```
<Schema ...>
  <element name="Name">
      <type>
  <element name="Honorific">
      type="string" minOccurs = "0"
MaxOccurs="1"/>
  <element name="First"type="string"/>
  <element name="MI"
      <type="string" minOccurs = "0"
MaxOccurs="1"/>
  <element name="Last"type="string"/>
  <element name="suffix"
      type = "string" minOccurs = "0"
maxOccurs="1"/>
      </type>
  </ element>
</ schema>
```

The schema form is somewhat longer but we specify a bit more information.

Structures

As XMl Schemas are written in XML syntax, structures refer to the XML constructs that we can use to define our mark up. XML Schemas are actually just another application of XML and as such could have a schema to describe itself.

So the structures section of the specification is the part where the elements and attributes for defining schemas are set out. Content models explicitly specify the allowable internal structure of elements. Structures are the heart of XML Schemas.

Writing Schemas

A schema consists of a **preamble** and zero or more definitions and declarations.

Preanble

The preanble is found within the root element, schema. This must include atleast three pieces of information in attributes.

- targetNS, which is the namespace and URI of the schema we are using
- version to specify the version of this schema
- xmlns which provides the namespace for the XML Schemas specification
- optionally, final Default and/or exact-Default, to provide defaults for two types of extension.

It may also include export, import and include constructs. A sample schema showing the preamble:

```
<? xml version="1.0"?>
<schema targetNS
    ="http:// myserver/myschema.xsd"
  version="1.0"
xmlns
  ="http://www.w3.org/1999/XMLSchema">
...
</schema>
```

Here, our hypothetical schema is residing on myserver and is called myschema.xsd.

Simple Type Definitions

A **simple type definition** is used to constrain information that does not include elements. It consists of a name and a specification that is either a reference to another type definition or consists of a series of facts.

A free-standing simple type definition is found in a datatype element:

```
  <datatype name = "smallint"
              source="integer"/>
  <minExclusive value="0"/>
  <maxExclusive value="10"/>
  </ datatype>
```

Complex Type Definitions

These are essential constructions in XML schemas. Without them, we would be unable to compose nontrivial content models for elements. The <type> element encloses a complex type definition. Nested within it, we have declarations for elements and attributes, or references to model groups. For example:

```
  <type name = "someContent">
```

<element .../>
<attribute ../>
< / type>

Attributes and Attribute Groups

Attribute declarations consist of an <attribute> element, which must minimally include a name attribute. The <attribute> element also has optional cardinality attributes, minOccurs and maxOccurs, which are used to indicate whether the attribute must appear. A type attribute specifies the datatype of the attribute, such as string or integer. An attribute declaration may also have default and fixed attributes. These function much like the IMPLIED and FIXED keywords in DTDs. The value of the fixed attribute is the value the attribute must always have. The value of the default attribute is the value which is assumed if the attribute does not explicitly appear in an element within an XML document. Example:

```
< attribute name="simpleAttr"/>
< attribute name="sequenceNo"
        type="integer"default="0"/>
```

we will often encounter a group of related attributes that are applied to multiple element declarations in a schema. XML schema structures accommodate this with the idea of attribute groups. This is a named collection of attribute declarations:

```
<attributeGroup name="troopParameters">
<attribute name ="SerialNum"type="String"/>
<attribute name ="rank"type = "string"/>
</attributeGroup>
<type name = "officerParms">
<attributeGroup ref="troopParameters"/>
</type>
```

Here we've declared the troop parameters attribute group, then used it within the officer Parms type definition.

Content Models

XML Schemes provide us with mechanisms for describing content models with a lot more accuracy than DTDs. These use complex type definitions and a new structure, the <group> element, to build the internal contents of an element declaration. We need another attribute for type elements, the content attribute. The content attribute tell us what elements can be contained.

Content attribute value	Meaning
Unconstrained	Content of any kind
empty	Empty element
mixed	Elements and character data

We need some content operators-termed **compositors** in the schema draft-to show how content may be composed. These compositors are the value of the order attribute of a <group> element. This new element gives us a way to provide ordered bodies of elements in a declaration. The compositors are shown in the following table:

Compositor keyword	Meaning	DTD equivalent
seq	Elements must follow in exact order	/(comma)
choice	Exactly one of the model elements appears	I(pipe)

Element Declarations

We can see how XML is used to make the syntax of schemas an XML application, where we had to use the <! ELEMENT syntax to declare a <Book> element in a DTD we now put element declarations inside an XML element, so we use:

```
<element name="Book"/>
```

Simple elements are composed of a reference to a data type and a series of attribute declarations or a reference to an attribute group. This is analogous to a DTD declaration where the element contains only PCDATA, except that the content is strongly typed. For example.

```
<element name="ZIP" type="string"/>
<element name="windspeed"
            type="float"/>
```

These would correspond to:

```
<! ELEMENT ZIP ## PCDATA>
<! ELEMENT Windspeed ## PCDATA>
```

There would be no notion of the string and floating point numeric types from the DTD declarations. When we want to define an element with structure, we replace the daa type reference with a content model.

Some examples. and their DTD equivalents:

```
<element name = "ListOfNames">
<type>
<group order="seq">
<element type="CustomerName"/>
<element type="SalesName"/>
<element type="ProductName"/>
</group>
</type>
</element>
<! ELEMENT (CustomerName, SalesName, ProductName)>
<element name="PickOne">
<type order="choice">
<group order="choice">
<element type="ColumnOne"/>
<element type="ColumnTwo"/>
</group>
</type>
</element>
<! ELEMENT PickOne
            (ColumnOne/ColumnTwo)>
```

To be able to specify multiple occurences of element content, we use minoccurs and maxoccurs attributes on the element references.

Model Groups

Some other schema constructs give us the ability to compose building blocks of definitions and declarations. We can have a **model group** within a particular type, to which we can then give a name. This construct enables us to build complex content models as we can refer to a **named model group** to build some part of an element content model for reuse in types and element declarations by putting a name to a model group, thereby allowing us to reference it elsewhere. Here are some samples:

```
<type minOccurs="1" maxOccurs="2">
<group order="seq">
<element type = "A"/>
<element type="B"/>
</group>
<group       order="choice"       minOccurs="3"
```

```
maxOccurs="7">
  <element type="C"/>
  <element type="D"/>
  </group>
  </type>
```

In this model, every document will start with a sequence of AB. This will occur at least once. We can choose between C and D and make the choice three to seven times. Finally we bring all our elements back one last time in any order. The following would be a legal document fragment conforming to this content model.

```
< A / > < B / > < A / > < B / > <!-sequence-->
< C / > < C / > < D / > < C / > < !--choice-->
```

Wildcards

XML schemas provide the any element that allows us to introduce a wildcard into a schema at any particular point. Schemas provide for departures from the written schema in any of the following four ways:

- Any well-formed XML element construction
- Any well-formed element construction so long as it is in any namespace other than the one in which the wildcard appears
- Any well-formed element construction, provided it is from a specific namespace.
- Any well-formed element construction provided it is from the current namespace.

Wildcards may also be, used in conjunction with attributes, in which case we can use any attribute element. Example for each of the four cases:

```
<any/>
<any namespace="##other"/>
<any namespace
=http://www.myserver.com/Other Schema/>
<any namespace="## targetNamespace"/>
```

The following is an example using a wildcard in conjunction with attributes within an element declaration.

```
<element name="someElement!">
<type>
<any Attribute namespace
=http://www.w3.org/1999?XML Schema/>
<element name = "someNum" type = "integer"/>
</ type>
</ element>
```

Here we have declared an element that has a single child element. <someNum>, and may have any attribute declared in the W3C schema for XML schemas.

Deriving Type Definitions

When we use the source attribute on a type, we are in effect deriving a new type from an existing one. We can extend a type or restrict it.

Derivation

A new type extends another when it adds additional content to its source type. In this case, all the content declared in the source type will appear in the derived type. For example, we extend a PersonName type declaration by adding an honorific element to the existing content.

```
<type name = "PersonName">
<element name = "FirstName"type="string"/>
<element name = "MI"type="string"/>
<element name = "Lastname" type="string"/>
</type>
<type name = "FormalPersonName" source =
"PersonName"derivedBy="extension">
<element name="honorific"type="string"/>
</type>
```

However, if we wish to restrict a type when we derive a new type from it, we can give the derived by attributing the value restriction and add a <restrictions> element:

```
<type name = "ShortName"source=
"PersonName"derivedBy="restriction">
<restrictions>
<element name="MI" maxOccurs="0"/>
<restrictions>
</type>
```

Composition

We can combine schemas and namespaces together to allow users to build document instances from multiple schemas. Schemas allow designers to use other schemas in building their own schema document. This is termed composition.

Import

We can import parts of another schema to use in ours provided the namespace of the other schema is referenced in an <import> element.

This element has the namespace attribute whose value is a URI for the schema we want to use. We may also provide a schema location attribute to point to the schema file desired. Once we have imported a namespace, we can use some construction from other schema within our schema:

```
<schema name="SomeOtherSchema.xsd"
xmlns:other = "http://www.OtherOrg. org/Schemas/
Useful. xsd">
<import namespace = "http://www.other Org. org/
SomeUsefulSchema"
Schemalocation = "http://www.otherOrg. org/
schemas/useful.xsd"/>
...
<element ref = "other:stuff" name="someName"/>
</schema>
```

When a construct is imported into a shcema, it remains an external resource. We are composing a new schema in effect, by linking in parts of another schema rather than including them whole in the new schema.

Inclusion

Inclusion is specified with the <include> element. This appears in a schema after the <import> element and before the <export> element, if any. The <include> element is an empty element with the required attribute schemaLocation, whose value is a URI to the included schema. When this element appears in a schema, then schema is understood to consist of its declared types as well as all the types declared in the included schema provided several criteria are met: The URI must resolve to another schema, and the schema thus designated must have a targetNamespace attribute identical to the containing schema's targetNamespace attribute value.

Annotating Schemas

No body of computing definitions or code is complete without a mechanism for providing additional comments or processing information. Schemas provide for this with the <annotation> element. This element may contain <info> elements, which consist of character data intended for human consumption, or <appinfo> elements, which do the same for schema processors. Either element may have an infoSource attribute that provides a URI reference to further information.

```
<element name="HardToRemember">
<annotation>
<info>
I want to remember the following about this element declaration ....
</info>
</annotation>
...
</element>
```

Datatypes

The addition of datatypes to XML schemas will be a great asset to programmers using XML for data in this applications. This support for datatypes includes the ability to check the validity of a value in a document as well as aiding an appropriate conversion from text to the native type when processing an XML document. So, we need to capture the data types of the information we mark up if we are going to use XML documents as the basis for integrating programs and systems.

Schema data types are said to have a set of distinct values called their **value space.** This is the abstract collection of values the type can take on.

XML Schemas : Datatypes is all about specifying value spaces, then listing the constraining properties of the type. It provides a set of primitive data types, and then elaborates a mechanism for generating new types derived from those primitives. The draft includes a number of generated types of wide utility, but schema designers are welcome to generate their own types intended for application specific use.

Some properties termed **facets,** are provided to specify datatypes. Facets refine the value space to give us the permissible values for the new type. Facets are either **fundamental or constraining.** Fundamental facets define some fundamental property of the datatype. Constraining facets place restrictions on the value space but do not define its nature. Strings, for example, have length.

Primitive Types

Primitive datatypes are those that are not defined in terms of other types. They are axiomatic. Here are the primitive types introduced by XML schemas:

Schema Primitive Type	Definition
String	Finite sequence of ISO 10646 or Unicode characters such as "this is a string"
boolean	The set (true, false)
float	Standard mathematical concept of real numbrs, corresponding to a single precision 32 bit floating point type.
double	Standard mathematical concept of real numbers, corresponding to a double precision 64 bit floating point type: doubles consist of a decimal mantissa, followed optionally by the letter E and an integer exponent, for example, 6.02E23
decimal	Standard mathematical concept of a real numeric type: it covers a smaller range than double, and consists of a sequence of digits separated by a period, such as 9.06.
time-Instant	The combination of date and time to define a specific instant in time, encoded as a string, 2000-01-01108 : 12 : 00.000 represents 8 : 12 on 1 Jan 2000, expressed as YYYY-MM-DDThh : mm : SS. SSS, but can be immediately followed by a Z, to specify that the time is a Co-ordinated Universal Time.
time-Duration	A combination of date and time to define a period, interval or duration of time. For example, one month is represented as P0Y1M0DT0H0M0S, where the lexical pattern is PnYnMnDTnHnMnS and can be preceeded by a + or - . The representation may be truncated on the right when the finer time intervals are not needed, for example P2Y3M for 2 years and three months. Note that the number, precedes the character representing the interval. Seconds may be expressed by a number: including a decimal to represent fractional seconds. A minus sign preceding the lexical representation indicates a negative duration.
recurring-Instant	An instant of time that recurs with some regular frequency, such as, everyday, represented by substituting a dash for any period not provided in the lexical pattern for time instant. For example an instant that occurs at 08 : 00 everyday would be expressed as _ _ _ T08 : 00 : 00.000.
binary	Arbitrarily long bodies of binary data.
uri	URI reference

Generated and User Defined Types

A generated datatype builds from an existing type. The type on wihch it builds is the **basetype.** XML Schemas specify some generated types that are broadly useful. These are shown in the following table:

Generated type	Base type	Meaning
Language	String	Natural language identifiers: a token that meets the Language ID production in XML, for example "en"
NMTOKEN	NMTOKENS	XML 1.0 NMTOKEN
NMTOKENS	String	XML 1.0 NMTOKENS
Name	NMTOKEN	XML 1.0 name
Qname	Name	XML 1.0 qualified name
NCNAME	Name	XML 1.0 "non-colonized"name
ID	NCName	XML 1.0 attribute type ID
IDREF	IDREFS	XML 1.0 attribute type IDREF
ENTITY	ENTITIES	XML 1.0 ENTITY
ENTITIES	string	XML 1.0 ENTITIES
NOTATION	NCName	XML 1.0 NOTATION
integer	decimal	standard mathematical concept of a discrete numeric type
non-negative-integer	integer	Standard mathematical concept of a negative integers
positive integer	integer	Standard mathematical concept of positive integers
non-positive-integer	integer	Standard mathematical concept of non-negative integer, or zero.
negative-integer	integer	Standard mathematical concept of a strictly negative integers.
date	recurring-Instant	Standard concept of a day, this is, an Interval beginning at midnight and lasting 24 hours.
time	recurring-Instant	Same as the left truncated representation for time instant. hh : mm : ss.sss

XML Data-Reduced

XML Data-Reduced (XML-DR) is a subset of the full XML Data proposal. In terms of how much that subset covers, it provides roughly the same functionality as the Document content Description specification containing those constructs needed to perform the tasks of a DTD. It also provides a few extensions to the capabilities DTDs offer. It is implemented as a **technology preview** in the XML parser that ships with internet Explorer 5.0. It is also supported in some commercial tools, notably DTD/ Schema editors such as Extensibility's XML Authority.

Schema Support

Conceptually XML Data-Reduced is similar to the core construct of XML schemas, even though the syntax is slightly different. The more complicated constructs, such as types, are not reproduced, but everything we need to define a vocabulary in XML is here, using similar datatypes. Here are the elements specified in XML Data - Reduced and their XML schemas equivalents:

Note carefully the case of the names as there are subtle differences between XML schemas and XML-DR schemas:

XML Schemas Construct	XML-DR Construct
schema	Schema
element	ElementType
elementRef	element
attribute	AttributeType
none	attribute
datatype	datatype
none	description
ModelGroup,group	group

Schemas

The Schema element in XML-DR is quite similar to the schema element in XML Schemas. This element performs the following functions:

- Contains element and attribute declarations
- Names the schema
- Declares namespaces used in the schema

Unlike XML Schemas, schemas in XML-DR do not use a preamble containing import, export, and include elements. Instead, they use namespace declarations. Every XML-DR schema must declare the XML Data and Microsoft datatypes namespaces. If a certain naming convention is observed, external content from another namespace may be used and validated in a schema. Here is a sample schema omitting the content:

```
<Schema name = "ShortSchema.xml" xmls="um : schemas - microsoft - com : xml - data".

xmlns : dt = "urn : schemas - microsoft - com : datatypes">
...<!--Declarations here -->
</Schema>
```

Elements and Attributes

Elements and attributes are declared in ElementType and AttributeType elements, respectively.

```
<elementType name= "myElement"/>
```

The <ElementType> element has five important attributes.

Element Type Attribute	Meaning
name	Name of the element
content	Describes the content that may be contained by the element : empty, textOnly (PCDATA only), eltOnly (element content only), mixed (PCDATA and elements)
dt : type	Denotes the type of the element. This attribute corresponds to the <datatype> element in XML Schemas. Valid values are taken from the XML Data Types Preview implementation.
model	Open or closed content model
order	Basic ordering of child elements; one (one chosen from a list of elements), seq (a specified sequence of elements), many (specified elements may appear or not appear, in any order)

Elements can contain one of four types of content described in the value of the content attribute of the <ElementType> element:

- no content : empty
- text only : textOnly
- subelements only : eltOnly
- a mix of text and sub elements : mixed

We can use the <element> and <attribute> elements to constrain the content of the declared element. These elements declare the child elements and attributes that may be applied to an element.

The <element> element can take three attributes:

Attribute	Description
type	Corresponds to the value of the name attribute of the <Element Type> defined in the schema.
minOccurs	Minimum number of times the reference element type can occur on the element, takes the values 0 where the minimum value is zero, and the element is optional or 1 where the element must occur at least once (default is 1)
max Occurs	Maximum number of times the element can occur on the element, takes the values 1 where it can occur once at the most, or where the occurrences are unlimited (default is 1)

The <attribute> element also can take three attributes:

Attribute	Description
default	Default value for the attribute, overrides any default provided in <AttributeType> element it referes to.
type	Corresponds to the vaue of the name attribute of the <Attribute Type> element defined in this schema.
required	Indicates whether the attribute must be present on the element, takes the value yes if it is required. Not needed if specified in the <Attribute Type> element.

Consider some simple element declaration and their DTD equivalents. First we have a parent element called <Fex>, which can contain a child element called <Tex>.

```
<ElementType name = "Fex" content = "mixed" order="many">
<element type="Tex"/>
</ElementType>
```

In a DTD, the declaration we have just seen would be:

```
<! ELEMENT Fex (# PCDATA 1 Tex)>
```

Next, we have a <Person> element, which has the child elements <FirstName>, <MI> and <LastName>:

```
<! ELEMENT Person (FirstName, MI, LastName)>
```

Using XML-DR this could look as follows:

```
<ElementType name="Person"content = "eltOnly" order = "seq">
<element type ="FirstName"/>
<element type ="MI"/>
<element type ="LastName"/>
</ElementType>
```

To declare attributes, we need the <AttributeType> element, which has one required attribute and four optional attributes. We associate an attribute with an element through the inclusion of an <attribuite> element as a child of the <ElementType> used to declare the element, as we did with elements. The type attribute of the <attribute> element refers to the <AttributeType> that declares the attribute.

<AttributeType> attributes	Description
name	Required : name of the attribute
default	Default value of the attribute; must be consistent with dt : type
dt : type	Data type of the attribute defined
dt : values	A list of possible values when dt : type is enumeration
required	Indicates whether the attribute must appear on all instances of the element; true or false.

We can add a required age attribute to the <Person> element we declared previously in a DTD.

```
<! ELEMENT Person (FirstName, MI, LastName)>.
<! ATLIST Person age CDATA#REQUIRED>
```

while in XML-DR this becomes:

```
<AttributeType name = "age" required = "yes"/>
<ElementType name="Person" content= "eltOnly"order="seq">
<attribute type'"age">
<element type="FirstName"/>
<element type="MI"/>
<element type="LastName"/>
</ElementType>
```

Groups

We saw the <group> element used with <ElementType> element declarations. The values permitted on the attributes of this element differ from those under XML Schemas. It is not flexible in its cardinality support. The all enumerated type for order ing is also not supported. <Group> takes the maxOccurs, minOccurs and order attributes.

Comments

XML-DR provides the description element for inline documentation.

<description> This is how you use the description element </description>

We can use the description element for capturing schema-related comments, such as usage notes that could be read by a particular tool.

Data Types

XML-DR makes use of the datatypes technology preview in IE 5.0 to provide strong typing for elements and attributes. This is in addition to the dt : type attribute used with <AttributeType> elements. This element, <datatype>, takes the single attribute dt : type to indicate the type of the parent element.

<ElementType name="Age".>

<datatype dt : type="int"/>

</ElementType>

Additional types supported are:

Data type	Meaning
bin.base 64	MIME style base64 encoded binary
bin.hex	Hexadecimal digits representing octets
boolean	0 (false) or 1 (true)
char	Single character
date	ISO 8601 date without the time.
dateTime	ISO 8601 date time, with optional time and fractional seconds to nanosecond resolution
DateTime.tz	dateTime, but with optional time zone added
fixed.14.4	Numeric type with no more than 14 digits to the left of the decimal point and no more than four to the right
float	Real number with optional sign, fractional digits, and exponent.
int	Integral numeric type
number	General numeric type with no limit on digits; may have optional sign, fractional digits, and exponent
time	ISO 8601 formt time group
time.tz	As for time, but with time zone added
i1	One byte signed integer
i2	Two byte signed integer
i4	Four byte signed integer
r4	Four byte real number
r8	Eight byte real number
ui 1	Single byte unsigned integer
ui 2	Two byte unsigned integer
Ui 4	Four byte unsigned integer
uri	URI string
uuid	Hexadecimal digits denoting octets and composing a COM style UUID; hypens are optional and ignored if present.

Namespaces and Schema Support in MSXML

MSXML supports XML-DR and name spaces through extensions to the XML DOM.

Namespaces

Support for namespaces in MSXML is quite robust. When accessing a node in the DOM that is namespace-qualified, we can use the basename, or prefix, and namespaceURI properties to obtain namespace-related information.

Property	Meaning
basename	Returns a string whose value is the unqualified basename of the node
prefix	Returns a string with the namespace prefix
namespaceURI	Returns a string with the name space URI corresponding to the nodes namespace prefix

If we want to create a namespace qualified node using the extended Microsoft XML DOM, we cannot use createElement ( ) to create it directly. Instead, we must use the document object's createNode ( ) method. It takes an enumerated value to indicate the type of node to create, the qualified name, and the associated URI:

```
qualifiedNode = doc.createNode
    (1,"pub : Book", "urn ; myschemas-pub");
```

Alternately, we can create a default namespace by using createElement ( ) followed by setting the xmlns attribute :

```
qualifiedNode = doc.createElement("Book");
qualifiedNode.setAttribute ("xmlns", "urn: myschemes-pub");
```

However in the latter case we lose the ability to qualify other names using the declared namespace.

A partial list of the node type enumerations supported by MSXML:

Enumeration	Value
NODE_ELEMENT	1
NODE_ATTRIBUTE	2
NODE_ENTITY-REFERENCE	5
NODE_ENTITY	6
NODE_NOTATION	12

The attributes collection of a node supports the getQualifiedItem ( ) method, allowing us to search for attributes by their qualified name.

To search for an attribute pub : isbn, we will make the following call:

```
FoundAttr = nodeAttrs.getqualifiedItem ("isbn", "pub");
```

Validation

MSXML will validate a document against a DTD like any other validation parser. If a schema is provided with the prefix x-schema, MSXML assumes that the name that follows the colon is an XML-DR style schema and will attempt to load it and use it to validate the document.

```
<Book xmlns="x-schema:
            Pubcatalog.xml">
<! - - Pubcatalog is assumed to be a schema -- >
```

If we have validated our document against an XML-DR schema, we have the opportunity to gain access to the schema node that defines a particular node in our document. This is done with the definition property. If called with a node corresponding to an element or attribute, we will receive a node containing the applicable < ElementType> or <AttributeType> element:

```
SchemaNode=node.definition:
```

Data types

The data type support in MSXML is separate but complementary to the XML-DR preview. Even if we do not use schemas. We can use strongly typed elements and attributes in well formed XML. To do this, we must declare the Microsoft data type namespace in our document:

```
<MyRootElement xmlns:dt = "urn:schemas microsoft-com: data types">
```

Variations on the Book Catalog

We can apply namespace and schema information to our book catalog example. We should break up the publishing catalog domain into two separate schemas, one reflecting a namespace dealing with authors, the other catalog information. We can provide strong typing on some of our attributes and elements. Since XML schemas is in a state of flux as it nears Recommendation status, we will use the XML-DR version of schemas as implemented in MSXML.

Segmentation

If we need to write a schema for marking up the actual content of individual books, we may want to include author information as well. We split the Book Catalog DTD into two separate schemas: **Catalog** and **Author.**When we want to create a catalog document, we can declare a default namespace for the Catalog schema and then use a qualified namespace to bring in the Author schema.

MetaData Discovery

Creating a schema in XML syntax is useful to programmers, in that we give them a little bit of support for writing programs that manipulate book catalog documents marked up according to our schema. We take the DTD into schema form. Once it is in XML syntax, programmers can use the same parser that they use with XML document instances to discover the meaning behind the metadata. If you are unfamiliar with our schema, you could investigate individual elements with the <definition> element.

The Catalog is split into three sections:

- Publisher information about the publisher
- Threads that contain descriptive information
- Books containing information about the books

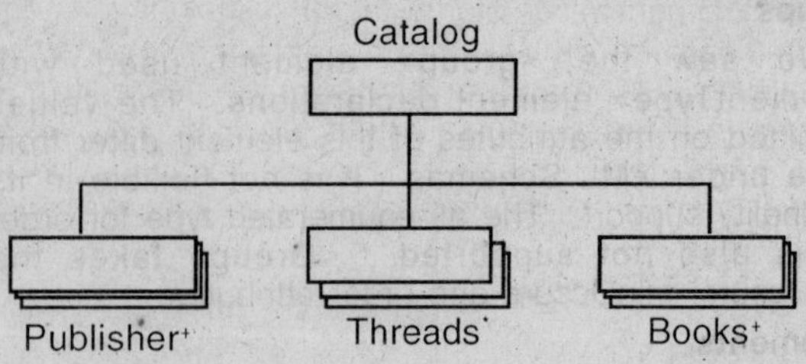

Author Schema

The first thing to do is to cut out the declaration of the <Author> element and everything subordinate to it and create a new schema file, authors.xml. The top of the file should declare conformance to XML 1.0, name the schema, and declare the XML-DR and datatypes namespaces:

```
<?xml version="1.0"?>
<Scherm name ="authors.xml"
xmlns="urn: schemas - microsoft -com:
            xml - data"
xmls: dt="urn:schemas - microsoft - com:
            datatypes">
```

The default namespace is XML-DR and the datatypes namespace will be aliased with the prefix dt. The Author element is our starting place. It cantains only element content consisting of the name-related, <Biographical>, and <Portraint> elements in sequence:

```
< ! ELEMENT Author (FirstName, MI, LastName, Biographical, Portrait)>
< ! ATTLIST Author authorCiteID # REQUIRED >
```

in XML-DR, this becomes:

```
< AttributeType name = "authorCiteID"
dt: type = "ID" required="yes"/>
<ElementType name = "Author" content =
"eltOnly" order="seq">
< attribute type="authorCiteID"/>
<element type="FirstName"/>
<element type="MI" minOccurs="0"
      maxOccurs="1"/>
<element type="LastName"/>
<element type="Biographical"/>
<element type="="Portrait"/>
</ElementType>
```

Catalog Schema

Now we recreate the catalog data in XML. We call this schema pubCatalog.xml. This will borrow from the author schema when we need to contain author details. Here's the opening information.

```
<?xml version="1.0"?>
<Schema name="PubCatalog.xml"
xmlns="urn: schemas - microsoft - com:xml- data" .
xmlns:dt = "urn: schemas - microsoft - com: datatypes"
xmlns : "athr="x-schema:. authors.xml"0>
```

Note how we've added a namespace declaration for our newly created author Schema authors.xml with the alias prefix athr.

Let's dig right in: we start with the <Catalog> element, which contains other elements. This contains the <Publisher>, <Thread>, and <Book> elements, just as we had in the earlier catalog.dtd, each of which may occur many times.

```
<ElementType name="Catalog" content="eltOnly" order="seq">
<element type="Publisher" minOccurs="1" maxOccurs="."/>
<element type="Thread" minOccurs="0" maxOccurs="."/>
<element type="Book" minOccurs="1" maxOccurs="."/>
</ElementType>
```

Next we need to declare the isbn attribute which will be used within both the <Publisher> and <Book> element that we have just declared.

```
<AttributeType name = "isbn" required = "yes"/>
```

Thread

<Thread> was used to specify the category area of the book. If we look above the bar code on the back of the book, we can see three different threads are used to categorize the book.

Schema concordance

It would be nice to have a simple listing of all the elements in a shema and their contents. That is, for each element declaration, we'd like a list of the permissible child elements and attributes used by that element. That way, we'd be able to gauge the impact of changing any particular element or attribute. Since XML-DR schema use XML syntax, we can use MSXML and some JavaScript to produce this utility.

Finding the element

We know that a schema document starts with a root <schema> element. Its child elements will be <ElementType> and <AttributeType> elements. The elements are declared with <ElementType> elements; and each such element contains a list of the elements and attributes it contains. This simplifies our task somewhat. All we have to do is walk the list of the <Schema> element's child nodes and process each <ElementType> element we find.

Simple Object Access Protocol

Simple Object Access Protocol (SOAP) does not fit in the family of B2B Vocabularies. It is a protocol. It cannot only exchange documents between servers in a special wrapper called an envelope but it can also exchange special messages that contain, for example, method calls to be executed on a remote server.

SOAP is based on the concept of the remote procedure call (RPC). The basic idea is that a program on one computer that is connected by a network can issue a procedure call to a program on another computer. The remote computer, be it down the hall or on another continent, can accept the call and hand it to a local program. The local program may issue a return value or some other response, then send back the return value or whatever to the computer that issued the call in the first place. SOAP is ideal for trading information rapidly, especially small amounts of information or information that can change quickly. Some data suited for SOAP includes time of day, weather reports, stock quotes, commodities prices, postal codes, word or phrase translations, we name it. There are no restrictions on the size of document we can send in a SOAP message.

Microsoft. Distributed Component Object Model (DCOM), the object Management Group's Unix-based Common Object Request Broker Architecture (COBRA), Internet Inter-ORB Protocol (IIOP), and Sun's Remote Method Invocation (RMI) are all similar technologies with the same basic thing in mind, but that require infrastructure (software licenses usually) on both local and remote systems. SOAP is different in that it relies on two ubiquitous technologies - XML and HTTP – to accomplish the same thing, thus reducing overhead in time, money and harsle. Any system connected to the internet is already prepared more or less to handle HTTP and XML, making SOAP relatively easy to implement.

The main technology employed for SOAP transport now is HTTP, but SOAP specification does not mandate HTTP.

The SOAP Envelope Element

The SOAP Envelope is the container that holdes the content of the SOAP message. The first or root element of this or any other SOAP document must always be an Envelope element.

The SOAP Envelope Namespace

The root element in this example contains an XML namespace declaration (Xmln). The conventional namespace prefix is SOAP-ENV. We can use a prefix of our choice, but should use a prefix. In some earlier XML examples, we used the default name space, that is, with no prefix, but we recommend using prefixes with SOAP.

The namespace is http://Schemas. xmlsoap.org/ soap/envelope. This identifier is unlikely to change. According to the XML Namespaces specification, you can use a URI or a URN as a unique namespace identifier. This is permissible but is not always the case: The namespace ID is simply supposed to distinguish the namespace as unique. The URI may point at a physical resource, such as a file that contains a schema but this is not a requirement.

The SOAP Header Element

The Header element is optional, but if we do include it, it must be the first child element after the Envelope element it must come right after Envelope, that is, with no elements in between.

The Header element is defined in the SOAP schema, but the child elements of Header are not. A qualifying namespace URI is required for the elements that follow Header, we are using a Uniform Resource Name (URN) Syntax with a prefix of program. The URN syntax is simple.

urn:eddys.stockomatic:program .xsd.

The urn prefix, followed by a colon, can be either upper case or lower case. The next phrase is the namespace identifier (NID). Legal characters for NIDs can consist of A - Z, a - z, 0 - 9, and hypen (-). An NID provides a unique namespace name.

After the second colon comes the namespace specific string (NSS). Legal characters for an NSS are also A - Z, a - z, 0 - 9, plus the characters ( ) +, . : = @ ; $ - ! * dot & column but the four characters % / ? # are reserved and should not be used in an NSS. Spaces are also a +no-no. encode a reserved or forbidden character by using the hexadecimal notation. For example, we could encode % 20 for a space or % 3 F for a question mark (?)

The Must Understand Attribute

The must understand attribute is defined in SOAP's Schema (notice the SOAP-ENV Namespace prefix). It can take a Boolean value, either 0 or 1. If its value is 1, the server on the receiving end must be able to understand the information contained in the Header element. If it does not recognize the information, the message must fail. If must understand is absent, the meaning must understand="o" (zero) is implied.

The actor Attribute

Another attribute is the actor attribute, which names a recipient for the information in the Header element. The Header element, if present, does not need to be consumed by every SOAP server (hop) along a message's path. All we care about is that the first destination server receives the content of Header not the SOAP intermediary server that are just passing the document along.

If the value of actor is http://Schemas.xmlsoap.org/soap/actor/next, only the very first SOAP application server
that processes the message should process the Header element.

The SOAP Body Element

The Body element like Header must be a direct child of Envelope, but it need not be the first immediate child if Header is present. If a Header element is not present, the Body element must be the first immediate child of Envelope. In other words, Body is first if Header isn't. The Body element is what carries the payload of the SOAP message. Child elements of Body are called body entries.

The GetQuote element declares the namespace. Using again the URN syntax as in urn: eddys stockomatic : stocks.xsd.

Response Message

The whole big deal behind SOAP is that after we have issued a method call or some other instruction in a message, we should be able to expect a message in return that will satisfy your request. This is not unlike order. Request and order Response in xCBL or cXML. Here is a response message that you could very well expect from the first example:

```
<SOAP - ENV : Envelope
xmlns : Soap ="http://schemas.xml
soap. org/soap/envelope">
<SOAP - ENV : Body>
<stocks: get quote
xmlns : stocks="urn:eddys - stockomatic: stocks.xsd">
<Stocks:Price stocks: Currency="USD"
63.97</Stocks: Price>
</Stocks : GetQuote>
</SOAP - ENV : Body>
</SOAP - ENV : Envelope>
```

The SOAP Fault Element

SOAP provides the optional Fault element for reporting error and status information. The Fault element may appear in a SOAP response if there is an error in the preceding request message. It must appear as a body entry, only once, if used at all. There are four possible child elements for Fault following is an example of a response with a Fault element as a body entry:

```
<SOAP - ENV : Envelope
xmlns : Soap ="http://schemas.xml
soap. org/soap/envelope">
<SOAP - ENV : Body>
<SOAP - ENV : Fault>
<SOAP - ENV : Fault code> server
</SOAP - ENV: Fault code>
</SOAP - ENV : Fault string> Error. Try again.
</SOAP - ENV : Fault string>
</SOAP - ENV: Fault>
</SOAP - ENV:Body>
</SOAP - ENV:Envelope>
```

If Fault appears in a SOAP response message, it must also contain the fault code and fault string elements. The fault actor and detail eidments are necessary under other circumstances: fault actor must be present if one of the network hops or forwarding servers fails, and detail spells out the something went wrong within the body element of the request.

SOAP and HTTP

Add an HTTP header to our SOAP request message, as if we were going to transport it. The HTTP method POST sends data to a fictitious program at / SOAP / stockomatic on the host or server test b2b. org. the content type (also called a MIME or media type) is plain text in XML format with a standard Latin-1 (Western European) character set, that is, ISO -8859-1

```
POST/SOAP/Stockomatic HTTP/1.1
Host : test b2b.org
Content-Type:textUxml;charset = "ISO-8859-1"
SOAP Action:http://test b2b.org/SOAP/ intent #
GetQuote
```

It accepts the method call GetQuote with two parameters, one for the stock symbol, another, time, specifying a date and time.

The HTTP Response

Content-Type is an example of an entity header. An entity header provides meta information about entity bodies-entity bodies consist of the document or pay load that is part of the HTTP message of which the SOAP message is an example. The HTTP version 1.1 Specification provides for extension headers. Extension headers are entity headers that go beyond standard headers to allow, under certain conditions, additional information about a message. SOAP Action is an example of such an extension header.

The SOAP Action Header

The SOAP Action header is mandatory in any HTTP SOAP request. It expects as a field value a URI that identifies the intent of the associated message. The URI may point to a resolvable or even human rendable document, or it may just serve as an identifier, not unlike a namespace URI. The preceding example shows a simple URI with no actual document associated with it.

Whatever the case, the header field can help servers or firewalls figure out if they want anything to do with the request. A server can parse the value of the SOAP Action header to filter messages and see if the message is from a trusted partner or a renegode.

The HTTP Response

Following is an example of an HTTP response consistent with the previous HTTP request:

```
HTTP/1.1 200 ok
Date: Fri, 01 Sep 2000 20: 01 : 58 GMT
Content Type: textUxml; charset = "ISO 8859-1"
Content - Length: 277
<SOAP - ENV : Envelope
xmlns : soap="http://Schemas.xmlsoap.
org/soap/envelope">
<SOAP - ENV : Body>
```

```
<Stocks: get quote
xmlns: stocks="urn : eddys
stockomatic:stocks.xsd">
<stocks:price stocks:currency="USD">
63.97</stocks:price>
</stocks : get quote>
</SOAP-ENV: Body>
</SOAP-ENV : Envelope>
```

SOAP, HTTP, and Firewalls

TCP/IP relies on ports to make connections between hosts. For IP version 4, these ports are identified with integers in the range 0 - 65535. Often a firewall will block ports from receiving packets. Nonetheless, the default HTTP port (port 80) that SOAP uses must be left unblocked if the system is to function as a public SOAP server. This reliance on trusted protocols like HTTP makes SOAP versatile, and, in addition, an attractive choice because no new software is needed to enable SOAP messaging, unlike CORBA, DCOM, or RMI.

SOAP Encoding

SOAP provides its own encoding scheme based on XML Schema. Standard SOAP encoding is provided as a simple system for describing SOAP content. If you want to use SOAP's encoding, the Envelope element must also contain a SOAP - ENV : encoding style attribute with a value of http://schemas.xml soap .org/soap/encoding/.

The SOAP encoding scheme provides an XML. Schema-based grammar for describing the data in a SOAP message. Because it supports XML schema, such as simple types (data types), complex types, and so on. One handy way to pass values to a SOAP server is with compound types in the form of structs and arrays. Familiar terms to a programmer. For example,

```
<ord:order>
<Date> 2001-01-02</Date>
<Item type="title"> SOAP Essentials </Item>
<Quantity>18</Quantity>
<Comments></Comments>
<Shipping method class = "1 st"> USPS
            </shipping method>
</ord:order>
```

A Working SOAP Example

A working SOAP example that uses HTML and scripting is availble online at the address http://soap. develop. com/xslt write/client.html.

This example was created by Don Box of Develop *r* Mentor. With if we can either add or subtract numbers by sending values and getting responses via SOAP. The example page is shown in above **Figure** and it is deceptively simple.

Just enter numbers in the first and second boxes, and then click either the add or subtract button. When we click on the Add button, a script on the page sends the values to the script at http://soap.develop.com/xsl/ twice/calculator.xslt in the following SOAP envelope:

```
<env : Envelope
xmlns:ev
="http://Schemas.xml soap.org/ Soap/envelope/"
xmlns:enc = "http://
Schemas.xml soap.org/Soap/encoding/"
xml ns: 1 cal="Uuid:84124454-FF27
     -4C41-8F21-dff5f2aa.24Id">
<env:Body>
<1 calc:Add>
<a>g</a><b>4</b>
</1 calc:Add>
</env : Body>
</env : Envelope>
```

An SMTP Example For SOAP

Simon Feel has put together an SMTP end point that allows you to send a SOAP message to an email address and get a SOAP response message in reply. This is part of Fell's work "Simon's SOAP Server Services for COM", which is composed of SOAP-to-COM server bindings that can invoke methods. (COM stands for Microsoft's Component Object Model).

If we would like to check it out, just send a plain text email message to the address soaplist_ener@zaks. Demon. co.uk with the following SOAP request in the body of the email message:

```
<SOAP - ENV : Envelope
xmlns : SOAP - ENV
="http://schemas.xml soap.org/soap/envelope/"
xmlns : SOAP-ENC
="http://schemas.xml soap.org/ Soap/encoding/"
xmlns : Xsd="htp:/lwww.w3org/1999? XMLSchema"
SOAP-ENV: Body>
 <m: doubler xmlns : m = "http://SimonFell.com/calc">
<nums SOAP - ENC:array Type = "xsd : int [1]">
<number> 10</number>
</nums>
</m : doubler>
</SOAP - ENV :Body>
</SOAP - ENV : Envelope>
```

SOAP and MIME

Even though it is not part of the original W3C SOAP 1.1 note, an additional note, "SOAP Messages with Attachments", was submitted to the W3C in December 2000. This note outlines the use of MIME headers in SOAP. This will allow for the transmittal of attach ments, which assumes also that we could send multipart documents (multipart/related) in a SOAP message. This capability is a logical addition to SOAP, and I imagine it will be incorporated into a future release of SOAP.

SOAP Programming Resources

Microsoft has already built a SOAP toolkit into its visual studio and it is likely that other interactive development environments (IDEs) will follow. I like phalanx systems little SOAP demo written by Mutt Long (www. phallanxsys.com/Soap/Soaphyproxy.asp). 'We can get a stock quote, time from the U.S. Naval observatory or word and phrase translations on the fly. James Snell's site is also a good starting point (ww. Soap - wrc.com/ Webservices/).

Apache SOAP Calculator Example

In order to get the Apache SOAP examples to work on our local machine, we have to install the Apache Tomcat server on it.

After Tomcat is installed and our class path is properly set, we can start the Tomcat Server with this command:

Jakarta-tomacatbin/tomcat start

SOAP and the Future

SOAP is the basis for a new W3C activity called the XML Protocol or XP. XP is based on SOAP 1.1, but we will be surprised if XP doesn't take SOAP by the horns. The goal for XP is, like SOAP, to encapsulate peer-to-peer messages in XML, with perhaps a broader swathe. The W3C, for example.

Microsoft appears to have gone to the bank on SOAP 1.1, having recreated Biz Talk as a SOAP implementation, as if it were a fully approved W3C recommendation. Biz Talk, it seems to me, is the most highly developed implementation of SOAP available.

Putting It All Together

Why XML?

XML is not a panacea to all the ailments of information processing.

If you are using computers to strore and share information, XML should be the basis for the structure of that information.

Unparsed entities - that is, including nonXML data in an XML document such as JPEG or GIF image - have never worked satisfactorily, at least not with a browser. Supporting specifications, such as X Link, have been tiresomely slow incoming. One of the best thing about XML is that it is a wide-open frontier land. It is not owned by a big, clutching, secretive, power-hungry behemoth of a corporation. It is managed by a consortium (the W3C,), but it's there for any one to claim and use.

XML vocabularies and protocols

A number of the front-running XML vocabularies and protocols for B2B; ebXML, xCBL,cXML, SOAP, and BizTalk are covered. There are others, too, such as RosettaNet for the IT, semiconductor, and electronic components world, Extensible Business Reporting Language (XBRL) for exchanging financial and other business reports and data, and BizTalk's Xlang.

ebXML

ebXML will win big. Because of its association with OASIS and UN/CEFACT, it's being developed on relatively neutral ground, UN/CEFACT'S overt awareness of the global needs of small-and medium-sized businesses.

ebXML is broad in its scope. It will define not only a messaging protocol, not unlike SOAP, but also core business components and processes, a registry, and collaborative partner arrangements as well. An integrated security strategy is not yet spelled out.

xCBL

Commerce one has done a good job with xCBL. With nearly 600 elements, we can create a large range of XML B2B documents. Auction documents are a strength, especially when compared to its competition, as are its security measures. If we are interacting with any Commerce One infrastructure on our supply chain, then xCBL is a natural choice.

Its support of the SOX and XDR schema languages narrows its scope somewhat, but, of course,xCBL

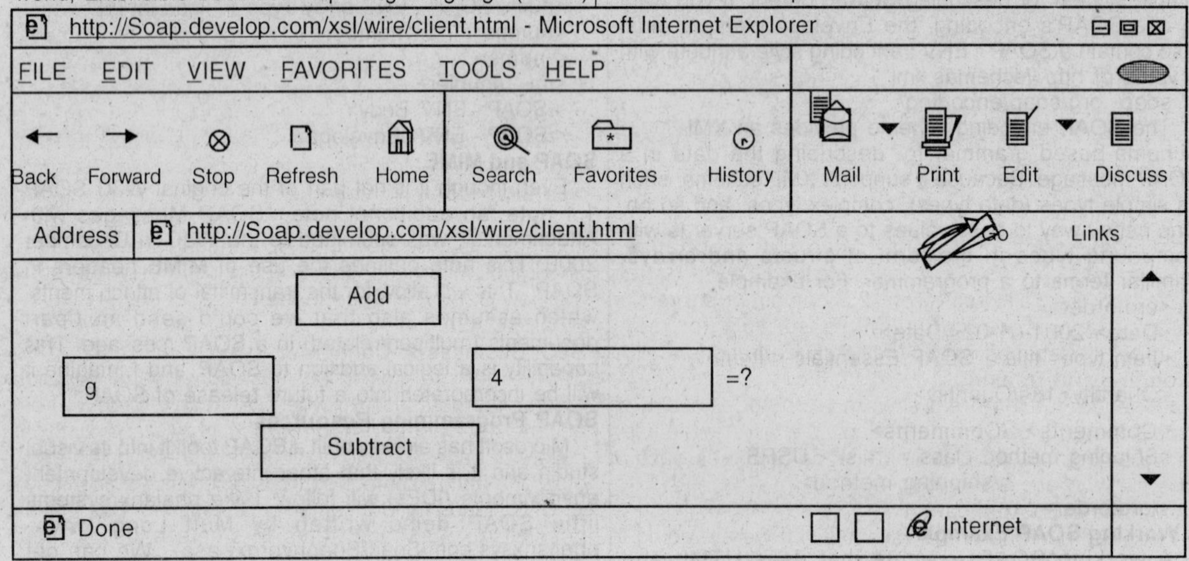

Fig: Develop Mentor SOAP example

If you clicked the subtract button, the element 1 calc: Add would be replaced by 1 calc:subtract. The namespace for 1 calc is a Universally identifier (UUID), a 128-bit value that is guaranteed to be unique in a Distribute Computing Environment (DCE). The Web page is updated with the sum of 13 (or dividend of 5) by passing back a value with a SOAP response message:

```
<?xml version = "1.0" encoding = "UTF - 16"?>
<env : Envelope xmlns : env = "http://Schemas.
xmlsoap.org/soap/envelope/">
<env : Body>
<1 calc : Add Response
xmlns : 1 calc = "Uuid : 84124454 - FF 27 - 4C41 - 8F21/ -dfF5f2aa241d">
<result> 13 < / result>
</1 calc : Add Response>
< / env : Body>
</env : Envelope>
```

If you clicked the subtract button, the element

1 calc : Add response would be replaced by

1 calc : subtract Response and the value in result would be 5.

documents can be validated against plain, old DTDs, as any legal, valid XML must.

cXML

Ariba's cXML, like xCBL, is mature and has, had a few years to spread itself around. Its innovative catalog management and punchout site facilities make it stand out among its peers. If we plan to use Ariba products anywhere along the line, cXML will be important to us. Like xCBL, Ariba does not use W3C namespaces.

Rosetta Net

Rosetta Net is a very strong offering, especially for those in the semiconductor or electronic component business. Many big players are behind it including intel, IBM, Cisco, Lucent Technologies, Motorola, NEC, and Toshiba. They have developed an innovative way to document and support business process with their Partner Interface Process (called PIP for short). Rosetta Net's specifications and supporting documents are well thought out and of high quality.

SOAP

SOAP provides a neat, orderly way to wrap up XML documents and method calls and send them zipping all over the Internet. It is not an XML vocabulay perse, but it is an important protocol for enveloping and transporting data.

SOAP is an important protocol for enveloping B2B documents, but XML protocol (XP) is hot on its heels. It stands a chance of overtaking SOAP. The ebXML messaging envelope is remarkably similar to the SOAP envelope.

Biz Talk

The Biz Talk protocol is Microsoft's enhanced implementation of SOAP. It relies on the SOAP. Biz Talk relies on Biz Talk framework 2.0 compliant servers in order to do its work. Biz Talk is suitable for many purposes, including heavy-duty B2B operations.

Logging and Tracking

One useful feature we want to include in a B2B application is logging and tracking. We will want to keep a log or logs of what our programs do and keep track of the files we are managing.

A Minimal B2B Application

We can use one of the vocabulary. If we are going to need to interoperate with a variety of vendors or partners, on the other hand, perhaps we are working in an area of business that does not have its own mark up language. Nothing is stopping us from forming a consortium and creating a suitable mark up language.

Create B2B documents in XML, with an ordinary text editor or with a specialized XML editor.

Choose a naming scheme for our documents, before moving a files, if privacy and security are important to us.

Once our documents are created, we are ready to transport them to or from our trading partners, over the internet.

In many instances, a reply message is necessary to confirm that a message has been sent successfully and arrived safely.

The person who receives the documents must take responsibility for them, gleaning such information as is necessary to enact business.

Logging, tracking, and storing files. We should also keep logs of your activities, as demonstrated in Get Order Log.

For example, we might store copies of documents sent after this model:

C : \AcmeAlliedInc \ sent \ 2001-01-01\2001-01-01-7788-order.xml

or :

\Usr\AcmeAlliedInc\sent\2001-01-01-\2001-01-01-7788-order.xml

Likewise we can store copies of documents received using this model:

C:\AcmeAliedInc\received\2001-01-01-2001-01-01-6375-order.xml

or :

\Usr\AcmeAlliedInc/received/2001-01-01-/2001-01-01-6375-order.xml

BUSINESS-to-BUSINESS (B2B)

What is Business-to-Business?

B2B means businesses doing business with each other, across the internet and using virtually instant communication to make their enterprises not only increase sales but run more smoothly, quickly and cheaply.

It means exchanging documents, such as orders, invoices, price lists, scientific papers. B2B is about legal arguments signed digitally by parties on different continents, executing them in minutes rather than weeks. It's about increasing the velocity and volume of commerce between remote system running on incompatible platforms, now bridged via a mutually agreed-upon B2B interchange.

B2B is about sharing ideas, as fast as the internet can carry them. More efficiently, more accurately, and for less expensively.

One of the most technologies relating to B2B commerce is Extensible Markup Language, or XML. XML undergrids most, not all, the technology that will support the new B2B commerce model.

B2C,C2C,A2A

B2C: Business-to-Consumer

This refers to a direct business-consumer connection via the internet, such as when we order a product from an internet-based vendor, which markets directly to buyers or traditional role of wholesaler, hawking an away of goods from disparate sources, such as Amazon.com.

C2C stands for Consumer-to Consumer. This makes possible to purchase items in single units or small quantities person-to person such as through an online auction house - @ Bay.com. Yahoo.com. These sites lets individuals purchase from individuals somewhat efficiently.

B2C and C2C are important components in the internet marketplace, and they sometimes even store space, such as at Icpenney.com.

A2A stands for application-to application. The application can communicate directly with another application, usually across a network. This process is automatic and requires no human intervention. Eventhough it can also require human interaction.

EDI: Electronic Data Interchange or EDI, is a format for electronic trade and commerce that has evolved into national and international standards (ANSI, X12, ISO 9735 and UN(EDIFACT). A number of major corporations have adopted EDI, but it has not made its way down to smaller companies because of cost and other barriers. The simplicity and minimal cost of Hyper Text Markup Language (HTML) opened the doors for the small operator to get on the web, so B2B will make it possible for just about anybody to automate and quicker the pace of business transactions.

The XML Foundation

History of Markup Languages

Proof readers marks have been around for centuries. These marks are a shorthand system somewhat standard way for editors to communicate changes to manuscripts that are being prepared for publication.

These handwritten marks suggest changes to authors or helps typesetters know how to format a printed page. They provide information about information, which is an underlying concept behind markup languages.

Generalized Markup Language

Mainframe computers came into common use in large businesses in the 1960s, computerized typesetting systems were also developed around this time. The Graphic Communications Association (GCA) created a standard for embedding formatting and typesetting information in text files. The standard was called Gen Code. A little later, IBM created something similar called Generalized Markup Language, or GML.

In 1969, the year the original Internet came into being, IBM was putting together a system for publishing legal information. The various computer systems in the company could not talk to each other because of differing command languages and file formats.

Three researchers at IBM, Charles Goldfarb, Ed Mosher, and Ray Lorie, solved this problem with GML, a markup system that allowed output coming form different sources to be marked up and processed uniformly.

Other markup system have emerged since that time, such as Troff and Rich Text Format (RTF).

Troff is a unix system designed for typesetting plain text files....

RTF is a kind of markup language that produces a file format allowing word processors like Microsoft Word or Corel Word Perfect to store typographical information.

Standard Generalized Markup Language

the American National Standards Institute (ANSI) in the late 1970s invited Charles Goldfarb and others produced the 1980 draft of standard Generalized Markup language, or SGML. By 1986, SGML grew into an international standard (ISO 8879 : 1986).

Tim Berners-Lee and HTML

In March 1989, Tim Berners-Lee was working for CERN, the European particle physics Laboratory. He circulated a paper he had written entitled "Information Management : A Proposal". This document explored a method for scientists to share papers using hypertext and the internet with that little paper, the seeds of the World Wide Web were planted.

Berners-Less and his associates created an SGML application or document type called Hyper Text Markup Language, that is HTML. HTML put an application of SGML into the hands of common folk, providing a set of comparatively simple tags for marking up text for presentation and display in early browsers.

Here is a simple example :

```
<html>
<head>
<title>HTML</title>
</head>
<body>
<p> we are likely familiar with what an HTML document
looks like.  Here is a simple example.  </p>
</body>
</html>
```

Berners-Lee founded the World Wide Web consortium (W3C) in 1994 to keep tabs on Web-related de facto standards, such as HTML.

After HTML version 2.0 was released, a struggle ensued over what tags and attributes to include in the next release of HTML. HTML 3.0 should look like that XML was born, born it seems of a need to get beyond the limitations of HTML, to tap into the might of SGML and to address the future needs of the Web.

Where did XML come from ?.

Activity on XML began in the summer of 1996. Originally it was called the "Web SGML" activity at W3C. Jon Bosak of Sun Microsystems the Father of XML - Jean Paoli of Microsoft, Tim Bray, and James clark, who coined the term XML invented a language that was like its predecessors SGML and HTML, but different in some important respects.

Three consequential differences between SGML and HTML

- XML does not lock its users into predetermined elements or tags.
- XML left behind a great deal of SGML's complexity, making it easier to create document types.
- XML allows us to easily use tags to describe not just documents but data, making it easier to parse and exchange data between different applications on a variety of platforms.

XML is free.

XML is structured. Structure is inherent in XML as element in XML have parent-child relationships. This allows us to structure textbased data in a clear, logical way.

XML is the basis for a file format. XML structure is orderly and predictable and because its data content can be clearly and easily labeled, XML documents can potentially be shared beetween entirely different computer and software systems.

XML is nonproprietary. It does work within proprietary software, such as the Microsoft Internet Explore or Netscape Navigator browsers.

XML is platform independent. We can develop XML on any platform that can handle a plain text file, which covers all of them.

XML provides a framework - the rules and the tools. for creating own Markup Language. We can use these rules and conventions to design, name and organize XML elements, attributes, and entities.

HTML is limited to a set of predefined elements.

- Body for the body of a Web page
- p for a paragraph
- hr for a horizontal rule or line

XML does not offer ready-made predefined elements. We must define them our self or use someone else's vocabulary or collection of XML elements. Some examples of XML vocabularies for B2B are the following:

- Electronic Business XML (ebXML) is jointly sponsored by OASIS and UN/CEFACT
- Rosetta Net makes possible the exchange of business documents and other business-to business exchages
- Commerce XML (cXML) is an XML vocabulary developed by Ariba for online catalog transactions, among other things

The XML Declaration

The first line of the XML document is called the XML declaration.

It looks like this:

<? xml version="1.0" encoding = "iso-8895-1"?>

This declaration tells the processing software that this is an XML document. The version attribute says that the document that follows conforms to the version 1.0 recommendation 08 XML.

The optimal encoding attribute indicates the character set or encoding used in the document. This document uses the ISO-8859-1 or Latin-1 character setXML 1.0 supports character set registered by the internet Assigned Numbers Authority (IANA), but XML processors are not required to support all the character sets registered by IANA. Some of the most common are UTF-8, UTF-16 and ISO-8859-1 through ISO-8859-9.

The stand alone attribute state explicitly the presence or absence of external declarations, such as those in a document type definition (DTD).

The XML declaration is an example of a processing instruction that process the XML document; they begin with<? and end with>? it is always a good idea to include an XML declaration because it is a precise indicator, both to the human eye and to the software reading the file, that the document is supposed to be XML.

The Root Element in XML

An XML document must have a root element. In XML, though, the root element can be whatever we want. In the case if this example, the root is the order element.

The root element must always be the first element in an XML document. The root is like a container that holds all the other elements in the doument, the root element is also called the document element.

Child Elements

The root element order is also a parent element. It has five child element in this example, namely Data, Item, Quantity, Comment, and Shipping Method. Each of these child elements is a sibling of the others. DTDs and XML schema are used to validate or constrain XML documents so that elements may be presented.

Well Formed XML

XML must be well formed. This means that it follows certain rules enforced by a complying XML processor. Four important rules of well-formedness are as follows:

Each element must have a start-tag and an end-tag.

Empty elements must terminate propely.

XML is case sensitive, so start-tags and end-tags must match.

In XML, attribute values must always be enclosed in either single (') or double (") quotes.

Valid XML

We can validate XML documents against a special document that defines the elements, attributes, and so forth contained in an XML document. In order for a document to be valid, it must conform to an associated DTD and also be well formed. Validating an XML document against a schema requires a validating XML parser.

The Document Type Declaration

In the following example, a document type declaration, highlighted in bold, is added to order. xml. If present, this declaration must appear before any elements in an XML document. The XML and document type declarations make up what is called the prolog.

```
<?xml version="1.0" encoding="180-8859-1"?>
<1 DOCTYPE Order SYSTEM "Order. dtd">
<Order partner = "06-853-2535">
<DATE>2001-03-05</Date>
<Item type="ISBN">/>0471404012</Item>
<Quantity>22</Quantity>
<Comments> None. </Comments>
<Shipping Method class = "4th"> USPS
</ShippingMethod>
</order>
```

Content Models

The content model determines the sequence and number of occurences of these elements. Generally speaking, when an element is constrained in a DTD, it can contain three things: (1) parsed character data (#PCDATA) , (2) other ordered elements, or (3) mixed content. Mixed content consists of parsed character data and other elements in no particular order.

In its own syntax, the model states the element content order:

There will be one and only one occurrence of the Date element. Date can contain parsed character data (# PCDATA). The Date element is expected to contain a date based on the ISO 8601 format, but this cannot be enforced by the # PCDATA type alone.

Following the order declaration is an attribute declaration. The ATTLIST keyword is preceded by <! as are other declarations. It then names, the element it is providing an attribute for, namely, order. Then it names the attribute partner, followed by the CDATA keyword. The expected value is a nine-digit DUNS number, identifying the business entity that is making the order. The # REQUIRED keyword means that the partner attribute is required, but it does not have a default value. Attribute values can occur once in a start-tag or an empty element tag, but never in an end tag.

The Date element will be followed immediately by one or more occurences of Item. The t operator called an occurrence operator, indicates that an element can appear one or more times. other occurrence operators are for zero or more times and? For zero or one occurrence. The absence of an operator in a content model means that the element can occur exactly once.

Follwing this is a declaration for an attribute called type. The attribute type has an enumerated list of two possible values.

Next come declarations for two more elements, quantity and comments, both containing # PCDATA. Quantity should be a positive integer. Comments contain, well comments. This element can appear zero or more times (X). Shipping Method contains # PCDATA as well.

Entities

Another markup declaration in the example DTD, defines entities. An entity in XML is like an abbreviation. With an entity name contained in an ampersand (&) and semicolon (;), it can represent a general entity, a built-in entity, or a character reference. DTDs can also contain parameter entities. Parameter entities work only in DTDs.

The following declaration in a DTD would define a general entity:

<! ENTITY nt" Note: An entity is like an abbreviation.">.

Built-in Entities

XML 1.0 comes with several built-in entities that represent special, markup-related characters so that they can be present in parsed character data without a parser chewing them up and spitting them out.

The built-in entities are listed in below table.

Table XML 1.0 Built-in Entities

ENTITY	DESCRIPTION
<	Less than or left angle bracket (<)
>	Greater than or right angle bracket (>)
&	Ampersand (&)
'	Apostrophe or single quote (')
"	Double quote (")

Character References

XML 1.0 supports ISO/IEC 10646 or the implementation of 10646 called Unicode. If you have a Unicode complaint browser, you can display these Unicode characters in XML documents with character references.

For example, the Latin capital letter A is represented with A, the registered trademark sign is &#XOOAE; R and Greek small letter alpha is α (α)

XML Schema

The XML DTD syntax trickled down from SGML and was part of the original XML 1.0 recommendation. XML schema, another W3C - sponsored initiative, picks up where the DTD left off as means to define and describe valid XML documents and data- XML schema uses XML syntax whereas DTDs have their own syntax.

One of XML schema's strengths is that it offers much stronger datatyping than the DTD offers XML 1.0 has somewhat limited datatypes.

The XML Schema Root Element

An XML Schema begins with its root element schema. Inside the start-tag for this element is a declaration for the schema language namespace (http://www.w3.org/ 2000/08/XMLSchema), plus declarations for the target namespace (http://testb2b.org/order) and its perfix (order).

Complex and Simple Types

A complex type in XML schema contains elements and attributes. A simple type, on the other hand, contains only character data. In this schema, the name of the first complex type is the element order, it contains five elements that must appear in the instance document in the same sequence given in the schema, namely, in the order Date, item, Quantity, Comments, and shipping Method.

The Date Datatype

The Date element has a simple type of date. An element of this type must have content that conforms to the ISO 8601 standard for date formats-for example, 2001-01-01 (CCYV-MM-DD).

In a DTD, we can't constrain element data other than with #PCDATA or with an element or mixed content model, but XML schema allows us to constrain element content based on a variety of datatypes. It even allows us to define our own datatypes.

Below table shows a sampling data types that come standard with XML schema.

Table A Sample of XML Schema Datatypes

DATATYPE	DESCRIPTION	EXAMPLES
Int	Integer in the range -214748647 to 214748648	-13,7,144,32655
float	32-bit floating point number	2064, 12.63,INF, Nan
date	Time instant with a duration of one day. In the form CCYY-MM-DD C=century, Y= Year. M=Month, D=Day	2001-08-19 2004-01-01
boolean	Boolean value	True, false
Uri reference	Absolute or relative URI	Http:/lwww.wley.com/ index.html(absoulte)or widex.html (relative)

The Item and Shipping Method Elements

Both the Item and shipping Method element definitions carry the maxoccurs attribute with a value of unbounded.

The min occurs and max-occurs attributes in XML schema a dictate how often elements appear. By default (or if omitted), min occurs is assumed to equal 1 and max occurs is likewise assumed to equal 1, so an element so constrained appears exactly once.

In a DTD, the declaration would look like this

```
<! Element Order (Item)>
```

In XML schema, you would achieve the same result by declaring order in this way

```
<complexType name="Order">
<sequence>
<element name = "Item"minOccurs="1"
maxOccurs="unbounded">
</sequence>
</complexType>
```

If the attributes min Occurs and max Occurs are omitted, so that the default value (1) is assumed for each:

```
<complexType name="Order">
<sequence>
<element name="Item">
</sequence>
</complexType>
```

For zero or more (*) occurances of item, the declaration in a DTD would be:

```
<!Element Order (Item*)>
```

In XML Schema, the same declaration would be

```
<complexType name="Order">
<sequence>
<element name="Item"minOccurs="0"
maxOccurs="unbounded">
</sequence>
</complexType>
```

Finally, to declare zero or exactly one (?) occurrence of item, in a DTD, declare order this way.

```
<!Element Order (Item?)>
```

Converting this to schema would make the declaration look like this:

```
<complexType name="Order">
<sequence>
<element name="Item"minOccurs="0"
maxOccurs="1 ">
</sequence>
</complexType>
```

Attributes in XML Schema

DUNS is a type that restricts the string type to a special pattern. The value attribute of the parttern element uses a regualr expression to form a nine-digit number.

Attributes and Anonymous Types

Simple types in XML schema cannot attributes, but you can associate an attribute with a simple type by using an anonymous complex type. Follwing the starttags for Item and shipping Method we can see the complex type element without a name attribute. This occurs because it is anonymous or in essence, nameless.

The enumeration element (a facet) lists the precise legal values for the attributes. that is ISBN and title for type (or Type Item Type), and Express, 1 st, Parcel, and 4th for class (or type mail class).

The Quantity and Comments Elements

The Quantity element has a simple type of positive integer. This is another type that is built into XML schema.

The possible range for this type is 1 through infinity. The comments element is a string that can occur zero or more times.

Validating against XML Schema

Finally, a few lines inserted in root element start - tag will let the schema processor know where our schema doucment is located, with the schema location attribute that belong to the XML schema instance namespace.

The value of schema location is first the namespace identifier associated with the target schema and then, after a space, the URI for the location of the resource, that is the actual file containing the schema against which the XML document is to be validated.

Transport

Transport-related technologies are used regularly in B2B implementations or are integral parts of them. For eg. Multipurpose Internet Message Extensions (MIME) is an essential protocol for B2B systems. Every B2B Implementations uses Hypertext Transfer protocol (HTTP). Simple Mail Transfer Protocol (SMT) and File Transfer Protocol (FTP) are other means for transporting files across the internet.

The programs that move data from one place to another or are used to encapsulate or describe data that is being moved. These protocols are TCP/IP, HTTP, FTP, RFE822, SMTP, POP3, IMAP4, and MIME).

Transmission Control Protocol and internet Protocol :

The Undubelly of the Internet lies in a pair of layered programs. Transmission Control Protocol (TCP) and internet Protocol (IP), called TCP/IP for short. These programs transport billions of packets of data all over the Internet every single day.

User Datagram Protocol (UDP) is another example but by and large. TCP/IP is the dominant networking protocol used. It was created to connect a vast array of computers over great distances many of which had in compatible network protocols.

TCP and IP work together, but each has a different purpose. TCP is the higher level, top layer, and IP is the low-level, bottom layer. To put it simply, IP moves the data and TCP referes.

How IP Works

IP (Internet Protocol), transmits huge amount of data all over the world at any given second on any given day, IP doesn't send everything out in one big chunk, but it breaks data into fragments, or packets, and then sends them out toward a destination identified by an addressing scheme. After the packets arrive IP with the help of TCP reassembles them in the correct order at the destination.

IP could route its packets around the damaged areas of the network until it found their destination.

IP also does not keep a connection open until it finishes sending all the data. Packets of data sent by IP are send to be unreliable. If data is undeliverable over a certain period of time, it just self destructs as phone and fibre optic lines, is down or out of commission, IP can find alternative paths to get the data to where it needs to go.

When IP gives the data it sends a header. The header provides information about the data, such as the IP version number, the length of the packet, and source and destination addresses. These addresses are known as IP addresses.

The IP Address

An IP address, is sometimes called a dotted decimal. It is a part of the IP header. An IP address is a unique identifier that is composed of four bytes or octets (32 bits). For human consumption, each octet is converted into a decimal number in the range 0 - 255 and separated by a dot or period.

How TCP Works

TCP works on the layer just above IP. When TCP opens a connection across the internet, unlike IP, it keeps that connection open until it has completed its work. As TCP fires packets of data through the IP layer and then lacross a network connection, the receiving TCP process sends an acknowledgment for each chunk of data. IP, the layer below TCP, can explode packets into 64bit fragments, reassembling them at the terminal end. TCP similarly reassembles the larger packets.

Each packet is assigned a sequence number so that these blocks of data can be reassembled in their proper sequence at the receiving end. A check sum is also added, so that TCP can check to see if the data it receives is sound and correct. TCP was designed to provide a range of addressable points, allowing what is known as mutliplexing. These points are called ports. Each port is numbered, and we can have up to 65,535 ports on a machine. A socket is a port that connects applications or operating systems to TCP/IP and allows them to talk to each other TCP connections are made between sockets at each end. TCP keeps the connection alive until, with the help of IP, it finishes its job.

Hypertext Transfer Protocol

Tim Berners. Lee invent the systematic use of the internet that we call the world wide web or the web for short. The web was built on the foundation of HTML and Hypertext Transfer Protocol (HTTP). Web browsers help us skip cumbersome steps such as moving documents separately with one program and then using a local application to read them in retrieving data in the form of documents.

One of the initial programs or clients available to use on the internet was file Transfer protocol, or FTP. FTP allows us to transfer files, individually or in groups, between remote computers.

With HTML we can define a document that could be readable across the internet, and with HTTP, we can transport and HTML document to a local client such as web browser, thus avoiding a couple of steps or obstacles.

- We no need to download or upload the file. We could just view a temporary copy of it in a user agent such as a browser.
- We no need to open the file with another local program. We could read the file with the browser we used to retrieve it.
- HTML provides a means to share not only text but also binary graphics.

How HTTP Works

In 1996 Berners.Lee and other described HTTP version 1.0 in an internet Engineering Task Force (IETF) document called Request for Comment (RFC) 1945.

HTTP is yet another application, like FTP that operates in the layer that sits a top TCP/IP and whose simple commands retrieves resources (documents and other kinds of sites or entities) over the internet. HTTP requires that a computer that stores web documents must be connected to the internet and have a web server installed on it to handle the HTTP requests.

A web server listens to Port 80, the default HTTP port, waiting patienetly for requests from a client for documents and other resources. From another computer, perhaps a remote one, a client such as a web browser sends a request to a server with a given IP address. The server then retrieves the requested source and sends it to the requester. Once it arrives, the client usually renders or otherwise processes the resource. We can also push resources with HTTP.

This is an example of a response from the HTTP server on test b2b. org

HTTP/1.1 200 OK

Date: Mon, 05 Mar 2001 13: 1 0:24 GMT
Server: Apache/1.3.4 (unix) Front Page/4.0. 4.3pHp/ 3.0.14
Last-Modified: Mon, 05 Mar 2001 12:45:05 GMT
Etag:"1 0b0b4-BC-39F7 e 116"
Accept-Ranges:bytes
Content-Length :421
Connection: close
Content Type:text/plain
<?xml version="1.0"encoding="iso-8859-"?>
<!DOCTYPE order SYSTEM order.dtd">
<order:order order: partner "06-853-2535" xmln s: order=" http: Iltestb2 b. org/ ord e r" >
<order:Date>2001-03-05</order:Date>
<order:Item order:type="ISBN">0471404012
</Order Item>
<order: Quantity> 22 </order: Quantity>
<order: Comments> None < Order:comments>
<Order: Shipping Method order :Class="4th" > USPS</Shipping Method>
</order : order>

Java and XML

Java and XML are a prety good match. While XML is a neutral language that describes data that can be handled by a variety of applications. Java applications complement XML's portability as well. The Java/XML combination has not achieved perfection, but it is about the best thing going for business and partners that operate networks that connect different platforms using different operating systems.

Compiling and Running Java

First get Java on our computer. We may already have a Java Runtime Environment (JRE) on our computer.
A JRE includes a Java Virtual Machine (JVM), Which is the engine that actually executes Java programs. A JVM can be installed on just about any platform, be it windows, unix, Linux, Mac os, or whatever, so that a compiled Java program will run on any platform.
We might have a JRE on our system, without knowing it because many times a JRE gets installed as a part of another program. A simple test to see if we have Java on our system is to go to a command line and type.

Java - Version

If we have java on our system, we will get a message back stating the version number. If Java is not on our system or in the proper path. We will get hiccup saying that the command was not recognized.

Compiling and Running a Java Program

Assuming that the JDK installed successfully, we can now compile and run Java programs from the command line. Before that, we must set that path and the calsspath variables on your system so that we can run the Java executables. we'll cover the path first.

The Path Variable

On windows, it is likely that we installed the JDK at c:1jdkl 3. Under that directory will be the bin directory, which holds executables such as java c.exe and java.exe. our command interpreter must be able to "see" where these executables are in order to run them. It does so through the path variable. We can set the path in a variety of ways.

The Class path Variable

The classpath variable informs the system where to find class (Java executable) and Java Archive (JAR) files. A class file essentially is a bytecode file or the result of compiling a Java program. The JVM interprets these bytecodes when we run a Java program. A JAR file is a collection of class files bundled into a ZIP file.

The default classpath is the current directory, represented by a single period (.). Sometimes the classpath gets messed up in one way or another so that you get the error Exception in thread "main" java.lang. No class Def Found Error: Program Name when we try to run a program.

Compiling a Java Program

With the JDK installed and our path and classpath variables properly set, we can now compile a Java program. In an archive file on the companion web site for this book (www. Wiley. com/compbooks/filzgrald/ testb2b.zip), we will find, among others, a program called Get. Java. After we download this archive file to a local system and extract Get. java, we can compile this program by typing the following on a command line:

javaGet.java

This will produce the file Get. Class, a byte code file that is interpretable by a JVM. If we get an error, it is one of four things; (1) the JDK is not installed properly or installed at all; (2) the path variable is not set correctly; (3) the classpath variable is not set correctly; or (4) any combination of the first three.

Retrieving order. xml with Get.java

The Java program Get.java requests a document with HTTP. This program assumes that we have an internet connection at execution time. Provided that the path and classpath variables are in good shape, extract Get. Class from the program archive to our current working directory and just type in the following at a command line:

java GetB2B/order.xml

What's Going on in Get.java?

After a few comments at the beginning (11 comments out a single line, / * * /can comment out multiple lines), the program imports two packages. These packages are composed of groups of classes and/or interfaces (types). Classes from these packages will be used later in the program. Classes contain definitions for variables and methods (functions or routines), among their things. The socket class is part of the java.net. *package, while the Input stream class is part of java. util. * An asterisk (*) is a convenience that lets we import all the classes that are part of a given package.

The user-defined public (accessible) class Get encapsulates the entire Java program, and this class name must be the same name as the program file. The main method is where the program begins its execution. The try block defines a chunt of code that may generate errors (exceptions). The catch block that follows try will handle the output of any excePtions that get generated.

The if statement tests to see if Get has an argument, throwing an exception if it does not. The socket class creates a new socket, connecting it to a port (80) at a specified last address (tests 2b-org). Input stream creates a means to get input from the socket and printwriter, a means for output to the socket. Using print in, the program submits the HTTP GET method to the socket, then pushes the request out the hatch (flush ( )).

The input from the socket is read into a buffer 1024 bytes at a time, then written to the output stream (standard output or the screen) until there is nothing left to read (-1). The socket is closed for good measure, and the main block ends. Without, we hope, catch having to throw any exceptions.

Retrieving order.xml with Get order.Java

The following simple Java program, Get order. Java, requests the document order. Xml via the Java URL class and then saves the document in the same local directory where the program was invoked. Like the last program, this one assumes that we have a live connection with the internet at execution time. Extract Get order. Class from the archive to our current working directory and just type in the following at a command line.

Java Get Order.

What's Happening in Get Order. java?

New input and output streams are assigned null values. A new random number object is created (rn) as well as a date format (today) that conforms to ISO 8601, a standard for the representation of dates and times (yyyy -MM-dd).

A new object of type URL is created with the address of http://testb2b.org/B2B/ order.xml. The input stream the program created earlier (input) is hooked up with the stream from the URL, which amounts to the contents of the file order. xml. The output stream is assigned a file output stream for the unique file ufn. The date is followed by a random number and then ends with the original filename, order.xml.

Retrieving Any Document with Get Any.java

Perhaps we would like to retrieve other documents besides order.xml. With a few alterations to the program, we can specify a URL for a document on the command line at runtime. The new program Get Any. Java allows we to do just that (available from the program archive). At the command line, type:

java Get Anyhttp://test.
b2b.org/B2B/summ.xml

What's Happening in Get Any.java?

Unlike Get Order.java, Get Any.java accepts a URL as a commanding argument (args (0)). It extracts the filename from the URL, breaking it down to a filename part (such as order) plus the extension part (such as xml) and placing the name in a string variable (fn). If it can't get a filename from, for example, a URL like http:/ / test b2b.org/, it uses a default filename (any txt). From this filename (fn), it creates a unique filename (ufn), and like Getorder.java, writes the file to the current local directory.

Putting a Document on a Server with Put.java

The following Java program is similar to the Get.java program, but rather than using the GET method it uses PUT. In order for it to work, the PUT method must be enabled on the server, presenting potential security risks, which can be dealt with in different ways, depending on our server.

From the program archive, extract Put class and ack.xml and place them in our local working directory. Then at a command line, type:

java.put-2001-02-28-8558-ack.xml

what's Happening in Put.java?

The Put.java program uses the HTTP method PUT to place a copy of the local file order. xml on http://local hostB2B. It first tests to see if we have included an argument on the command line to use as the remote filename. If not, it uses a default filename (new.xml) for the file we will put on the host. Then it sets up a URL connection for the new file and sets the method to PUT. The local file order.xml is read and this file is buffered. Some HTTP header information is then printed on standard output (our screen), demonstrating that the operation was successful.

W3C's Winie

In addition to this program, W3C's Winie product (down loadable from http://jigsaw.w3.org/winie) offers several programs that use the PUT HTTP methods, either from a command line or with a nice user interface.

Web DAV

Web based Distributed Authoring and versioning (Web DAV) deserves a mention here. It is a collection of extensions to HTTP that allows us to edit and manage resources, such as XML files, on remote web servers. Some of the features of web DAV include the ability to lock files so others can't overwrite them, the use of properties to identify authors, dates, and so forth, and the ability to collect documents and manipulate them as groups. Popular tools, such as Macromedia Dreamweaver and Adobe Go Live, have recently added Web DAV support. Web DAV features may prove to be useful tools in B2B solutions.

HTTP Methods

Object-oriented programming languages also used the term method to refer to what are called routines or procedures in traditional programming. HTTP has a number of methods other than GET or PUT, and HTTP version 1.1 offers a few more than HTTP 1.0. These methods are listed in Table below.

Table: HTTP Methods

METHOD	VERSION	DESCRIPTION
GET	1.0,1.1	Requests a resource such as a document
HEAD	1.0,1.1	Requests only that header be returned
POST	1.0,1.1	Submit data to a server, such as from an HTML form
PUT	1.0,1.1	Places a resource on the server rather tha retrieves it
DELETE	1.0,1.1	Deletes a named resource
LINK	1.0	Establishes links between resources
UNLINK	1.0	Removes links between resources
OPTIONS	1.1	Tests a server for capability but does not request a resource
TRACE	1.1	Traces what is being received at the destinatio
CONNECT	1.1	Is reserved for TCP-based or SSL tunneling

In HTTP 1.1, we have much more control over caching. A server can restrict what can or cannot be cached. Further, the expiration model has been improved so that a server can assign a heuristic expiration time and calculate that age of items in the cache in a more sophisticated way. The version 1.0 model makes it necessary to close a network connection after each and every transaction. Persistent connections will hold those connections open long enough to conduct a number of transactions in one session, reducing network overhead.

File Transfer Protocol

File Transfer Protocol, or FTP, is standard Internet program for moving files over the Net. It's easy to use, and we can create scripts to move the files automatically or at a certain given time, such as with scheduled Tasks on windows, or at on unix.

We get started by simply connecting to a host using the anonymous FTP convention. Normally an Internet host is set up with a special, protected directory where we can upload and download files, even though we do not have an account on that particular machine. To get started, type the following line at a command prompt:

ftp test b2b.org

ETP Scripts

We can place FTP commands in a batch file in windows. The first catch is that we have to store the FTP commands in a separate file. The example that follows is *for* windows only. On the windows platform, in an editor that can save a plain-text file, such as Notepad.

OBJECTIVE QUESTIONS

1. How many types of JDBC drivers are present?
(a) One (b) Two
(c) Three (d) None of the above

2. Which of the following statements is true?
(a) A URL instance represents a link for accessing or communicating with the resource at the location.
(b) A URL connection instance represents the location of a resource.
(c) A URL instance represents the location of a resource.
(d) None of the above

3. What are the two types of assemblies available in.NET?
(a) Private and Public (b) Public and Shared
(c) Private and Shared (d) Local and Global

4. What is a Postaway web form in ASP.NET?
(a) It is a web form with AutoPostback set to false.
(b) It is web form with no runat = server, but with method and action attributes.
(c) It is a web *form* with runat = server and method and action attributes.
(d) None of the above

5. What is the syntax to explicitly kill a user's session?
(a) session. Close (b) session. Discard
(c) session.Abandon (d) session.End

6. Which of the following statements is true?
(a) Response. Write ( ) allows us to write formatted output
(b) Response.Output. Write ( ) allows us to write formatted output
(c) Both (a) and (b) are true (d) None of the above

7. What data type does the Range Validator Control Support?
(a) Integer (b) String
(c) Date (d) None of the above
(e) All the above

8. What type of code is found in a Code-Behind Class?
(a) Client Side Code (b) Server-Side Code
(c) Both (a) and (b) (d) None of the above

9. Which of the following statements is true?
(a) Server.Transfer is used to redirect the user to another page or site.
(b) Response.Redirect is used to post a form to another page.
(c) Server.Transfer is used to post a form to another page.
(d) None of the above

10. Which of the following statements is true?
(a) There is concept of cursor types in a Dataset.
(b) Assemblies are the building blocks of .NETframework.
(c) Both (a) and (b) are true. (d) None

11. Which of the following statements is true?
(a) Inline code is code written in a separate file and referenced by the .aspx page.
(b) Code-behind Code is code written along side the html in a page.
(c) The .Fill ( ) method is used to load our generated dataset with data.
(d) All the above are true. (e) None

12. Which of the following statements is true above Repeater Control?
(a) Data in Repeater Control can be edited.
(b) Item template is used to provide alternating color scheme in Repeater Control.
(c) Both (a) and (b) are true.
(d) None of the above is true.

13. Which of the following properties are common in every Validation Control?
(a) Control to validate property (b) Text property
(c)Both (a) and (b) are true (d) None

14. Which of the following statements about web service is true?
(a) Web service can only be written in .NET.
(b) SOAP is the preferred protocol to call a Web service.
(c) WSDL stand for web services description language.
(d) Both (a) and (b) (e) Both (b) and (c)
(f) All the above are true

15. Can we implement an interface in a JSP?
(a) Yse (b) No

16. Choose the Valid statement
(a) Request.getRequestDispatcher (path): In order to create it we need to give the absolute path of the resource.
(b) Context.getRequestDispatcher (path): In order to create it we need to give relative path of the resource.
(c) RequestgetRequestDispatcher (path): In order to create it we need to give relative path of the resource. (d) None of the above.

17. Dam stands for
(a) Dynamic Object Model (b) Document Object Model
(c) Document Object Memory (d) None of the above.

18. Essentially how many related types of XML documents are there?
(a) One (b) Two
(c) Three (e) None of the above

19. What does DSSSL stand for?
(a) Document Style Semantics and Specification Language.
(b) Design Style Semantics and Specification Language.
(c) Development Style Semantics and Specification Language.
(d) None of the above.

20. Which of the following statements is true about HTTP?
(a) It is a generic stateless object oriented protocol.
(b) an application level. protocol for the www.
(c) HTTP supports dynamic data representation through Client-Server negotiation.
(d) (a) and (b) only (e) (b) and (c) only
(f) All above are ture

21. Assume our ASP.NET application manages order entry data by using a Dataset object named OrderEntry. The OrderEntry object includes two data table objects named OrderNames and OrderDetails. A foreign Key Constraint object named order detailskey is defined between the two data table objects. When we attempt to delete a row in order Names while there are related rows in Order Details, and an exception is generated. What is the most likely cause of the problem?
(a) The current value of the OrderDetails.KeyDeleteRule is Rule.Cascade.
(b) The current value of the OrderDetails.KeyDeleteRule is Rule.SetNull.
(c) The current value of the OrderDetails.KeyDeleteRule is Rule.SetDefault.
(d) The current value of the OrderDetails.KeyDeleteRule is Rule.Null.

22. Suppose we are a member of a team of developers creating several ASP.NET applications for our company. We want to create a reusable toolbar that will be used in each of the applications. The toolbar will be displayed at the top of each page viewed by the user. The content of tool bar will vary depending on options each user selects when creating profile. If we want to be able to add the toolbar to the ASP.NET toolbox for each of the developers on our team what should we do?
(a) Create a new web Control Library project. Create a toolbar within a web custom control.
(b) Add a new web user control to our ASP.NET project. Create the toolbar within the web user control.
(c) Add a new web form our ASP.NET project and design the tool bar within the web form.
(d) Add a new component class to project. Design the toolbar within the designer of the component class.

23. Suppose we are creating an ASP.NET application. The application will be deployed on our company's intranet. Suppose our company uses Microsoft Windows authentication. If we want the application to run in the security context of the user, What should we do?
(a) Add the following element to the authentication section of the web.config file.
(b) Add the following element to the system.web section of the web.config file.
(c) Use the configuration Manager for our project to designate the user's security context.
(d) Write code in the Application AuthenticationRequest event handler to configure the application of run.

24. Which ASP.NET authentication mode is best suited for identifying and authorizing users who belong to corporate network?
(a) Custom authentication (b) Passport authentication
(c) Form authentication
(d) Windows integrated authentication.

25. Each ASP.NET Page is parsed and compiled into a class that extends
(a) System.Web.UI.Webcontrols
(b) System.Web.UI.Htmlcontrols
(c) System.Web.UI.Page (d) None of the above

26. What is MVC?
(a) Model View Controller (b) Model Vertical Controller
(c) Metal View Controller (d) None of the above

27. What does stayInsync denote?
(a) Status of content record
(b) Creates a duplicate record set
(c) Indication that a hierarchical record set should stay In contact with the data source
(d) None of the above

28. Which of the following represent value of TYPE which is an attribute of <INPUT> tag?
(a) CHECKBOX (b) RADIO
(c) SUBMIT (d) (a) and (b) only
(e) (b) and (c) only (f) all the above
(g) None of the above

29. Which of the following represents parts of a URL?
(a) Protocol (b) host name
(c) directory and filename (d) None of the above
(e) All the above

30. Which of the following indicate attribute of <IMG> tag?
(a) SRC (b) ALIGN (c) ALT
(d) (a) and (b) only (e) (b) and (c) only
(f) None of the above (g) All the above

31. Which of the following represents attribute of <EMBED> tag?
(a) SRC (b) WIDTH (c) HEIGHT.
(d) (a) and (b) only (e) (b) and (c) only
(f) None of the above (g) All the above

32. If the document contains
<META HTTP.EQUIV = "Expires" CONTENT ="Sat, 06 Jan 1990 00:00:01 GMT">
<META HTTP.EQUIV = "From" CONTENT = "nick@htmlib.com">
<META HTTP.EQUIV= "Reply-to" CONTENT ="Stephen@ htmlib.com"
then the HTTP response header word be
(a) Error
(b) Expires: Sat 06 Jan 1990 00:00:01 GMT
From: nick@htmlib.com
Reply-to: stephen@htmlib.com
(c) Expires: Sat 06 Jan 1990 00:00:01 GMT
From: nick@htmlib.com
Reply-to: stephen@htmlib.com
(d) None of the above

33. Default value for COLSPAN is
(a) 0 (b) 1
(c) NULL (d) None of the above

34. Default value for ALIGN in (<TH >) is:
(a) RIGHT (b) LEFT
(c) Center (d) None of the above

35. Methods of combo Object is
(a) Click ( ) (b) Focus ( )
(c) Clear ( ) (d) (a) and (b) only
(e) (b) and (c) only (f) All the above

36. Which of the following statements is true?
(a) Node type property is used to get node's value
(b) Node value property is to get node's type
(c) Both (a) and (b) (d) None of the above

37. Which of the following represent ways of style?
(a) Embedded style sheets
(b) External style sheets
(c) Inline styles (d) Both (a) and (b)
(e) Both (b) and (c) (f) None of the above
(g) All the above

38. Which of the following values below can be set to DATAFORMATS of <PARAM> tag?
(a) HTML (b) PLAINTEXT
(c) TEXT (d) Both (a) and (b)
(e) Both (b) and (c) (f) All the above

39. <LEGEND> element is used to
(a) Group elements in a form together in a box
(b) Add caption to the box
(c) Lable elements in the form (d) None

40. Which of the following represent attributes of <ISINDEX>?
(a) ACTION (b) CLASS
(c) CHALLENGE (d) Both (a) and (b)
(e) Both (b) and (c) (f) All the above
(g) None of the above

41. What is the default value of ROWSPAN in <TD> tag?
(a) 0 (b) 1 (c) NULL (d) None

42. Which of the following represent standard format for web servers and browsers?
(a) BMP (b) GIF (c) JPEG
(d) Both (a) and (b) (d) Both (b) and (c)
(f) All the above (g) None of the above

43. The document below <?xml version="1.0"?>
<TITLE>
A Noisy Noise Annoys An Oyster
<HEADING>
<!TITLE>

A study of shellfish and Audio Disturbances </HEADING>

(a) Is well formed but not valid
(b) Neither valid nor well formed
(c) Valid as well as well formed (d) None

44. Position attribute of style property can hold which of these values?
(a) absolute (b) rlative
(c) Both (a) and (b) (d) None of the above

45. Which of the following about <BUTTON> is ture?
(a) Attribute <LANG> denote scripting language used for the tag.
(b) Attribute <LANGUAGE> denote base language used for the tag.
(c) Attribute <LANGUAGE> denote scripting language used for the tag.
(d) None of the kbove

46. There is attribute for <NOLAYER> attribute:
(a) True (b) False

47. Default value of volume of <BGSOUND> is
(a) 0 (b) 1 (c) 10,000 (d) None

48. Which of the following are values of CLEAR attribute of <BR> element?
(a) ALL (b) LEFT
(c) RIGHT (d) Both (a) and (b)
(e) Both (b) and (c) (f) All the above

49. Hyperlinks and hypertexts were created in
(a) 1930's (b) 1960's
(c) 1980's (d) 1990's

50. Which of the following is default value of SHAPE attribute of <A> (Anchor)?
(a) CIRC (c) RECT
(b) POLYGON (d) None of the above

ANSWERS

1. (d) There are 4 types of JDBC drivers. They are
   - JDBC-ODBC Bridge Driver
   - Native API Partly Java Driver
   - Network Protocol Driver
   - JDBC Net Pure Java Driver
2. (c) A URL instance represents the location of a resource, and a URL connection instance represents a link for accessing or communicating with the resource at the location.
3. (c) 2 types of assemblies available in .NET are private and shared.
4. (b) 5. (c)
6. (b) Response.Output.Write( ) allows us to write formatted output.
7. (e) Range Validator Control Support Integer, string and Data datatype.
8. (b) Server-Side Code is found in Code Behind class.
9. (c) Server.Transfer is used to post a form to another page. Response.Redirect is used to redirect the user to another Page or site.
10. (b) There is no concept of cursor types in a dataset and Assemblies are the building blocks of .NET framework.
11. (c) Inline code written along side the html in a page and Code-behind is Code written in a separate file and referenced by the .aspx page. The .Fill ( ) method is a method on DataAdapter control that is used to load our generated dataset with data.
12. (d) Data in Repeater Control cannot be edited. It just reads the information from its data source. Alternating item Template is used to provide alternating color scheme in a Repeater control and item Template is used to display data in a Repeater control.
13. (c) 14. (e) 15. (b)
16. (c) Request.getRequestDispatcher (path): In order to create it we need to give the relative path of the resource.
    Context.getRequestDispatcher (path): In order to create it we need to give the absolute path of the resource.
17. (b) XML's real strength for the web is how it interacts with **Document Object Model** (DOM) an interface that defines the mechanisms for accessing data in a document. Using DOM, programmers can script dnamic content in a standardized way.
18. (b) There are essentially 2 related types of XML documents **well-formed and valid.** A well - formed XML document conforms to the general rules of XML syntax.Valid XML documents are documents that conform to a specific Dcocument Type Definition (DTD).
19. (a). DSSSL (Document Style Sematics and Specification Language). an ISO (International Standards Organisation) standard that's popular with high-end publishers using SGML.
20. (f) 21. (d) 22. (a) 23. (b)
24. (d) 25. (c)
26. (a) A Model View Controller (MVC) Model architectural design pattern that was in a J2EE container with servlets and JSPS that uses TopLink to access data without EJBS.
27. (c)
28. (f) Possible values of TYPE area: CHECKBOX, HIDDEN, IMAGE, RADIO, RESET, SUBMIT, TEXT
29. (e) 30. (g)
31. (g) SRC attribute is the path name or URL to the media file, WIDTH and HEIGHT represent dimensions of the embedded object's bounding box.
32. (c)
33. (b) COLSPAN attribute can appear in any table cell (<TH> or <TD>). It specifies how many columns of the table this cell should span. Defoult value for COLSPAN is 1.
34. (c) Default value of ALIGN - Center.
35. (f) Click ( ) → simulates a click on the combo field
    Clear ( ) → clear the contents of the combo field
    Focus ( ) → gives focus to the combo field
36. (c)
37. (g) Embedded style sheets → styles are specified in the <STYLE> element.
    External style sheets → useful when we want all the pages on our web site to adhere to the same style.
    Inline styles → STYLE attribute of the tags we want to style is used, instead of the <STYLE> element.
38. (f) 39. (b) 40. (d)

TEST YOURSELF

1. Consider the HTML table definition given below :

```
<table border =1>
        <tr> <td rowspan = 2> ab < /td>
                <td colspan=2> cd</td>
        </tr>
        <tr> <td> ef </td>
                <td rowspan=2> gh </td>
        </tr>
        <tr> <td colspan =2> ik </td>
        </tr>
</table>
```

The number of rows in each column and the number of columns in each row are : **[GATE - 09]**

a) $\langle 2,2,3\rangle$ and $\langle 2,3,2\rangle$ b) $\langle 2,2,3\rangle$ and $\langle 2,2,3\rangle$

c) $\langle 2,3,2\rangle$ and $\langle 2,3,2\rangle$ d) $\langle 2,3,2\rangle$ and $\langle 2,2,3\rangle$

2. Which of the following is TRUE only of XML but NOT HTML ? **[GATE - 08]**

(a) It is derived from SGML

(b) It describes content and layout

(c) It allows user defined tags

(d) It is restricted only to be used with web browsers

3. Given below are some HTML lines.

```
< img src = " pict.jpg" width = "256" height = " 256" border = "0"
usemap = " #map" / >
< map name = "map" >
<area shape =" poly"
   coords = "50, 50, 50, 100, 100, 100, 75, 75, 100, 50"href = "f1. html " >
< area shape = " circle" coords = " 100, 75, 5 "href = " .. / cgi - bin / f2 . p1 ? v1 = ask abc's age " >
< area shape = " default " href = " fd. html ">
< / map >
```

With reference to the HTML lines given above, consider the following statements.

(i) Clicking on the point < 80, 75> does not have any effect.

(ii) The web browser can identify the area applicable to the mouse-click within the image and the subsequent action to be taken without additional responses from the web server.

(iii) The dots in the cgi-bin URL will be resolved by the web browser before it is sent to the web server

(iv) The "fd.html" request when sent to the web server will result in a GET request.

Exactly how many of the statements given above are correct ? **[GATE - 07]**

(a) 0 (b) 1

(c) 2 (d) 3

4. Consider the XML document fragment given below :

```
<Book>
<Title> GATE 2K7 Example </Title>
<Content>
    One of many lines
< / Content>
<TOC>
One of many content entries
</TOC>
</Book>
```

Consider the XPath expression : * [not (self) : : TOC]

What would be the result of the given XPath expression when the current node is Book ? **[GATE - 07]**

(a) The Title and Content elements

(b) The Content and TOC elements

(c) The Title and TOC elements

(d) The Title Content and TOC elements

5. Consider the following XML DTD describing course information in a university :

```
<!ELEMENT Univ (Course+, Prof+) >
<!ELEMENT Course (Title, Eval*)>
<!ATTLIST Course Number ID #REQUIRED
 Instructor IDREF #IMPLIED>
<!ELEMENT Title (#PCDATA) >
<!ELEMENT Eval (#PCDATA) >
<!ATTLIST Eval Score CDATA#REQUIRED>
<!ELEMENT Prof EMPTY>
<!ATTLIST Prof Name ID #REQUIRED Teaches
 IDREF #IMPLIED>
```

What is returned by the following XQuery ?

```
let $as := / / @Score
for $c in /Univ / Course [Eval]
let $cs := $c / Eval? @Score
where min ($cs) > avg ($as)
return $c
```

[GATE - 06]

a) The professor with the lowest course evaluation

b) Professors who have all their course evaluations above the university average

c) The course with the lowest evaluation

d) Courses with all evaluations above the university average.

ANSWER KEY

1. c **2.** c **3.** d **4.** d **5.** d

EXPLANATORY NOTES

1. Here two <td> command used in first<tr> command and three <td>command used in second <tr> command, so required rows in each column is <2, 3, 2>

Alternately

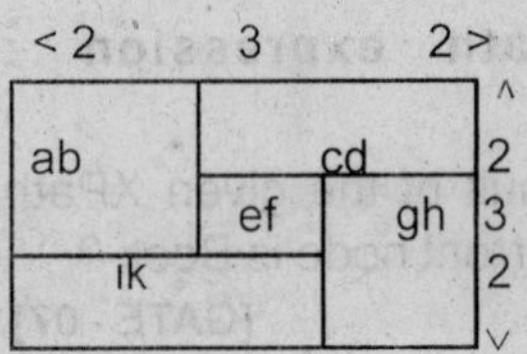

2. XML → Extensible mark up language

HTML → Hypertext mark up language

Only XML allows user defined tags.

Unit - XIV
Verbal Ability

UNIT - XIV VERBAL ABILITY

1. SPOTTING ERRORS

In this type of test each question contains a sentence which is divided into three parts (or four parts) labelled as ABC (or ABCD) or 1, 2, 3 (or 1,2,3,4). The error if any will be in one of the three parts (or four parts) of the sentence. The number or the letter of that part is the answer. If there is no error, the answer is 4 (or 5) or D (or E) which stands for 'No Error'.

Example:

1. The Chief Minister (A)/ as well as the ministers (B)/ have offered to resign (C)/ No Error (D)

Answer : (C) has offered to resign.

Explanation : In this sentence, the error lies in the part (C) The rule is: "When words added to a singular subject by 'as well as', with, etc, the verb should be in the singular". 'The Chief Minister' is the singular subject and it should be followed by a singular verb. Hence, 'Has' should be used instead of 'have'.

STUDY MATERIAL

RULES FOR SPOTTING ERRORS

Remember the following rules which will help you in spotting errors.

RULES OF NOUNS

Rule – 1 : *Scenery, information, poetry, furniture, language, advice, business, audience, mischief, bread, news, innings, Mathematics, Economics, Physics and Politics should be used in the singular form. Neither these should be preceded by an indefinite article nor pluralised.*

Incorrect : The audience are requested to be in its seats.
Correct : The audience is requested to be in its seats.

Incorrect : The Sceneries of Kashmir are beautiful.
Correct : The scenery of Kashmir is beautiful.

Rule – 2 : *Cattle, poultry, people, gentry, peasantry, artillery, infantry. clergy, police, company, swine, alphabet and prey should be always used as plural. They are themselves plural and should not be made plural by adding 's'.*

Incorrect : The cattle is grazing in the field.
Correct : The cattle are grazing in the field.

Incorrect : The people of our village is rich.
Correct : The people of our village are rich.

Rule – 3 : *Some nouns like scissors, spectacles, tongs, trousers, measles, remains, proceedings, tidings, odds, annals, auspises assets, credentials, alms, riches, tactics, congratulations, thanks, binoculars, arrears, etc. are always used in the plural and take a plural verb.*

Incorrect : Where is my spectacle ?
Correct : Where are my spectacles ?

Incorrect : Riches has wings.
Correct : Riches have wings.

Rule – 4 : *Some nouns like deer, sheep, fish, yolk (of Oxen), brace (of birds), dozen, score (twenty), hundred, thousand, etc. have the singular and the plural alike.*

Incorrect : The sheeps are grazing in the field.
Correct : The sheep are grazing in the field.

RULES OF AGREEMENT OF SUBJECT & PREDICATE

Agreement is the grammatical relationship existing between the subject and its verb.

Rule – 5 : *The general rule isthe verb should agree with its subject in number and person.*

Incorrect : They is farmers.
Correct : They are farmers.

Incorrect : I is a teacher.
Correct : I am a teacher.

Rule – 6 : *A verb should not be made agree with a noun near it instead of its proper subject.*

Incorrect : The quality of mangoes are not good.
Correct : The quality of mangoes is not good.

Rule - 7 : *When two or more singular subjects are joined by 'and', the verb should be in the plural.*

Incorrect : Gold and silver is precious metals.
Correct : Gold and silver are precious metals.

Rule - 8 : *When the singular subjects connected by 'and' suggest one idea and refer to the same person or thing, the verb should be in the singular.*

Incorrect : Time and tide wait for no man.
Correct : Time and tide waits for no man.

Incorrect : Bread and butter are his only food.
Correct : Bread and butter is his only food.

Rule - 9 : *When words are joined to a singular subject by 'as well as' 'besides', 'like', 'in addition to', 'with', 'together with', 'and not', etc., the verb should be in the singular.*

Incorrect : The Chief minister, with his colleagues have offered to resign.

Correct : The Chief Minister, with his Colleagues has offered to resign.

Rule - 10 : *When two or more singular subjects are connected by 'or', 'nor', 'either....or', 'neither...nor', the verb should be in the singular.*

Incorrect : Either Ravi or Raju have done this.
Correct : Either Ravi ro Raju has done this.

Rule - 11 : *Either, neither, each, one of, everyone, many a should be follwed by a singular verb.*

Incorrect : Neither of these two applicants were qualified.
Correct : Neither of these two applicants was qualified.

Incorrect : One of my teachers live next door to me.
Correct : One of my teachers lives next door to me.

Rule - 12 : *When the plural noun is a proper name for some single object or some collective unit, it must be followed by a singular verb.*

Incorrect : Gulliver's Travels were written by Swift.
Correct : Gulliver's Travels was written by Swift.

Incorrect : The United States have a big navy.
Correct : The United States has a big navy.

Rule - 13 : *When a plural noun expresses some specific quantity or amount considered as a whole, the verb should be in the singular.*

Incorrect : Fifty thousand rupees are a large sum.
Correct : Fifty thousand rupees is a large sum.

Rule - 14 : *A collective noun takes a singular verb when the collection is thought of as one whole and a plural verb when the individuals of which it is composed are thought of.*

Incorrect : The committee have submitted its report.
Correct : The committee has submitted its report.

Incorrect : The jury was divided in its opinion.
Correct : The jury were divided in their opinions.

Rule - 15 : *When a noun is used as an adjective in a compound word, it is always in the singular, even though it is preceded by a plural numeral.*

Incorrect : He gave me a ten rupees note.
Correct : He gave me a ten rupee note.

RULES OF PRONOUNS

Rule - 16 : *A pronoun must agree with its antecedent in person, number and gender.*

Incorrect : Each of these boys has done their homework.
Correct : Each of these boy has done his homework.

Rule - 17 : *When two or more singular nouns are joined by 'and, the pronoun used for them must be in plural.*

Incorrect : Both Raju and Ravi have done his homework.
Correct : Both Raju and Ravi have done their homework.

Rule - 18 : *When two or more singular nouns joined by 'and' refer to the same person or thing, the pronoun used for them must be with in the singular.*

Incorrect : The collector and District Magistrate is not negligent in their duty.

Correct : The collector and District Magistrate is not negligent in his duty.

Rule - 19 : *When two or more singular nouns joined by 'or' or 'nor', either..or, neither...nor, the pronoun used for them should be in the singular.*

Incorrect : Neither Ravi nor Raju has done their home-work.

Correct : Neither Ravi nor Raju has done his home-work.

Rule - 20 : *When two or more singular pronouns of different persons come together, the pronoun of second person singular (you) comes first, the pronoun of the first person singular (I) comes last and the pronoun of the third person singular (he) comes in between.*

Incorrect : I, You and he must work together.
Correct : You, he and I must work together.

Rule - 21 : *when two or more plural pronouns of different persons come together first person plural (we) comes first, then second person plural (you) and last of all third person plural (they).*

Incorrect : You, they and we must work together.
Correct : We, you and they must work together.

Rule - 22 : *When a pronoun is used as the complement of the verb 'to be', it should be in the nominative case.*

Incorrect : If I were him, I would not do it.
Correct : If I were he, I would not do it.

Rule - 23 : *When the pronoun is used as the object of a verb or of a preposition, it should be in the objective case.*

Incorrect : Let You and I do it.

Correct : Let You and me do it.

Incorrect : These presents are for you and I.

Correct : These presents are for you and me.

Rule - 24 : *Emphatic pronouns can not stand alone as subjects.*

Incorrect : Himself did it.

Correct : He himself did it.

Rule - 25 : *The Indefinite pronoun 'one' should be used throughout if used at all.*

Incorrect : One must not boast of his own success.

Correct : One must not boast of one's own success.

Rule - 26 : *'Either' or 'neither' is used only in speaking of two persons or things; 'any', 'no one' and 'none' is used in speaking of more than two.*

Incorrect : Any one of these two roads leads to the railway station.

Correct : Either of these two roads leads to the railway station.

Incorrect : Neither of these three boys did their homework.

Correct : No one of these three boys did their homework.

Rule - 27 : *'Each other' is used in speaking of two persons or things; 'one another' is used in speaking of more than two.*

Incorrect : The two brothers loved one another.

Correct : The two brothers loved each other.

Rule - 28 : *A noun or pronoun goveming a gerund should be put in the possessive case.*

Incorrect : Please excuse me being late.

Correct : Please excuse my being late.

Rule - 29 : *A Relative Pronoun should agree with its antecedent in number and person.*

Incorrect : The boys, who was lazy were punished.

Correct : The boys, who were lazy were punished.

Incorrect : I, who is your king will lead you.

Correct : I, who am your king will lead you.

Rule - 30 : *The Relative Pronoun 'as' or 'that' should be used after 'such' or same.*

Incorrect : He is such a man who I honour.

Correct : He is such a man as I honour.

Incorrect : This is the same man who came yesterday.

Correct : This is the same man that came yesterday.

RULES OF ADJECTIVES

Rule - 31 : *In comparing two things, the comparative should be used. The Superlative should not be used.*

Incorrect : Which is the best of the two?

Correct : Which is the better of the two?

Rule - 32 : *When two qualities of the same person or thing are compared, the comparative in-er is not used. 'More' is used for this purpose.*

Incorrect : He is wiser than brave.

Correct : He is more wise than brave.

Rule - 33 : *When comparison is made by means of a comparative, the thing compared should be excluded from the class of things with which it is compared by using 'other' or some such word.*

Incorrect : He is cleverer than any boy in the class.

Correct : He is cleverer than any other boy in the class.

Rule - 34 : *When comparison is made by means of a superlative, the thing compared should include the class of things with which it is compared.*

Incorrect : He is the strongest of all other men.

Correct : He is the strongest of all men.

Rule - 35 : *When two persons or things are compared, it is important that the same parts of things should be compared.*

Incorrect : The population of Bombay is greater than Delhi.

Correct : The population of Bombay is greater than that of Delhi.

Rule - 36 : *Double comparatives and superlatives should not be used.*

Incorrect : He is the most cleverest boy in the class.

Correct : He is the cleverest boy in the class.

Incorrect : He is more wiser than his brother.

Correct : He is wiser than his brother.

Rule - 37 : *The comparative adjectives superior, inferior, senior, junior, prior, anterior, posterior, prefer, etc. should be followed by 'to' instead of 'than'.*

Incorrect : He is senior than me.

Correct : He is senior to **me.**

Rule - 38 : *Adjectives like 'unique', ideal, perfect, complete, universal, entire, extreme, chief, full, square and round, which do not admit different degrees of comparison should not be compared.*

Incorrect : It is the most unique thing.

Correct : It is a unique thing.

Rule - 39 : *All the adjectives which refer to the same noun should be in the same degree of comparison.*

Incorrect **:** He is the wisest and honest worker in the office.

Correct **:** He is the wisest and most honest worker in the office.

Rule - 40 : *'Elder' and 'eldest' should be used for persons only, strictly speaking, they are used for the members of the same family only. 'Older' and 'oldest' are used for both persons and things.*

Incorrect **:** He is my older brother.

Correct **:** He is my elder brother.

RULES OF VERBS

Rule - 41 : *A single verb should be made to serve two subjects, only when the form of verb is same for both the subjects.*

Incorrect **:** I am eighteen years old and my sister sixteen.

Correct **:** I am eighteen years old and my sister is sixteen.

Rule - 42 : *Two auxiliaries can be used with one principal verb, only when the form of the principal verb is appropriate to both the auxiliaries.*

Incorrect **:** He never has, and never will take such strong measures.

Correct **:** He never has taken, and never will take such strong measures.

Rule - 43 : *When there is only one auxiliary to two principal verbs it should be correctly associated with the both.*

Incorrect **:** Ten candidates have passed, one failed.

Correct **:** Ten candidates have passed, one has failed.

Rule - 44 : *A past tense in the main clause should be followed by a past tense in the subordinate clause.*

Incorrect **:** He succeeded because he works hard.

Correct **:** He succeeded because he worked hard.

Rule - 45 : *A past tense in main clause may be followed by a present tense in the subordinate clause when the subordinate clause expresses a universal truth.*

Incorrect **:** Our teacher said that the earth moved round the sun.

Correct **:** Our teacher said that the earth moves round the sun.

Rule - 46 : *When the subordinate clause comes after 'lest', the auxiliary verb 'should' must be used, whatever be the tense of the verb in the main clause.*

Incorrect **:** We start early lest we shall miss the train.

Correct **:** We start early lest we should miss the train.

Rule - 47 : *An adverb or adverbial phrase should not be placed between 'to' and the verbal part of the infinitive. (This is called the split infinitive).*

Incorrect **:** I hoped to immediately reply to your letter.

Correct **:** I hoped to reply immediately to your letter.

Rule - 48 : *An infinitive should be in the present tense unless it represents an action prior to that of the governing verb.*

Incorrect **:** I should have liked to have gone there.

Correct **:** I should have liked to go there.

Incorrect **:** He seems to enjoy his stay in Madras.

Correct **;** He seems to have enjoyed his stay in Madras.

Rule - 49 : *A participle should not be unrelated or falsely related. Since it is an Adjective, it must qualify a Noun or Pronoun.*

Incorrect **:** Sitting on the gate, a scorpion strung him.

Correct **:** Sitting on the gate, he was strung by a scorpion. (or)
While he was sitting on the gate, a scorpion strung him.

Rule - 50 : *The auxiliary verbs like will, would, shall, should, can, could, may, might, must and expressions like 'would rather', would sooner, rather than, sooner than, had better etc. should be followed by the infinitive without 'to'.*

Incorrect **:** He might went there.

Correct **:** He might go there.

Rule - 51 : *The form of the verb that follows a preposition must be the gerund.*

Incorrect **:** They were punished for come late.

Correct **:** They were punished for coming late.

Rule - 52 : *The future indefinite tense is not used in the clauses of time, place and condition. Here the present Indefinitive Tense is used.*

Incorrect **:** I shall wait you till you will finish your work.

Correct **:** I shall wait you, till you finish your work.

Rule - 53 : *The present perfect tense is not used with the adverbs of past time like yesterday, in 1990 etc; Here Past Indefinite Tense is used.*

Incorrect **:** I have bought a cycle yesterday.

Correct **:** I bought a cycle yesterday.

Rule - 54 : *The past perfect tense is used to represent the earlier of the two past actions.*

Incorrect : When I reached the station, the train already left.

Correct : When I reached the station, the train had already left.

Rule - 55 : *The present perfect continuous tense is used for an action that began in the past time and still going at the time of speaking. It is used with adverbials of time introduced by 'since', 'for' and 'how long'.*

Incorrect : How long are you working in this office ?

Correct : How long have you been working in this office ?

RULES OF ADVERBS

Rule - 56 : *Two negatives destroy each other. Hence two negatives should not be used in the same sentence unless we make an affirmation.*

Incorrect : I have not got none.

Correct : I have not got any.

Rule - 57 : *'At present' means 'at the present time', 'presently' means 'shortly'. These should not be confused.*

Incorrect : Nothing more can be done presently.

Correct : Nothing more can be done at present.

Incorrect : He will come back at present.

Correct : He will come back presently.

Rule - 58 : *'Hard' means 'diligently', strenuously', 'Hardly' means 'scarcely at all'. These two adverbial forms of 'hard' must not be confused.*

Incorrect : He tried hardly to win the race.

Correct : He tried hard to win the race.

Incorrect : She has eaten hard anything today.

Correct : She has eaten hardly anything today.

Rule - 59 : *'Much' is used before past participles and Adjectives or Adverbs of comparative degree.*

'Very' is used before the present participles and adjectives and adverbs of positive degree.

Incorrect : The news is much surprising.

Correct : The news is very surprising.

Incorrect : I was very surprised at hearing the news.

Correct : I was much surprised at hearing the news.

RULES OF CONJUNCTIONS

Rule - 60 : *Two conjunctions should not be used in the same sentence.*

Incorrect : Although she was tired, but she still went on working.

Correct : Although she was tired, she still went on working.

Rule - 61 : *'Both' should be followed by 'and'. It should be used in the positive sense. In the negative sense, 'neither'nor' should be used in place of 'both'.*

Incorrect : Both Ravi as well as Raja were present there.

Correct : Both Ravi and Raja were present there.

Rule - 62 : *'Either......or', 'neither......nor', 'both.....and, 'not only.....but also' should be followed by the same parts of speech.*

Incorrect : He not only lost his ticket, but also his luggage.

Correct : He lost not only his ticket but also his luggage.

Rule - 63 : *'Neither' should be followed by 'nor'; 'either' should be followed by 'or'. Both these should not be confused.*

Incorrect : He washed neither his hands or his face.

Correct : He washed neither his hands nor his face.

Rule - 64 : *'No sooner' should be followed by 'than', not by 'but' or 'then'.*

Incorrect : No sooner do I finish this book then I shall begin another.

Correct : No sooner do I finish the book, than I shall begin another.

Rule - 65 : *'Hardly' and 'scarcely' should be followed by 'when' or 'before', but not by 'than' or 'that'.*

Incorrect : Hardly did I reach the station, than the train left it.

Correct : Hardly did I reach the station when the train left it.

Rule - 66 : *'That' should not be used before a sentence in Direct Speech and before Interrogative adverbs and pronouns in the Indirect Speech.*

Incorrect : He said that "I shall go there".

Correct : He said, "I shall go there".

Incorrect : He asked me that who I was.

Correct : He asked me who I was.

PRACTICE TEST - 1

SPOTTING ERRORS

***DIRECTIONS :* In each of the following questions, find out which part of the sentence has an error. If there is no mistake, the answer is 'No error'.**

1. Dr. Addison has been trying (a) / to develop a medicine for diabetes (b) / for the last ten years. (c) / No error (d)
2. As the party was going (a) / we noticed Mr. Roy fall back (b) / in his chair, gasping for breath (c) / No error (d)
3. In this age of despair and despondency, (a) / only our rich cultural heritage (b) / can help us escape to the menacing doom. (c) / No error (d)
4. Rekha was trying for admission (a) / in the Engineering College (b) / even though her parents wanted her to take up medicine. (c) / No error (d)
5. All the players (a) except sam / (b) were present on the playground / (c) No error (d)
6. It proves (a) / year after year (b) / that when it comes into judging new writing we are on slippery ground (c) / No error (d)
7. This kind of behaviour (a) / has not been seen here (b) / since the last four years. (c) / No error (d)
8. He is in the habit (a) / of going our (b) / on a morning walk regularly. (c) / No error (d)
9. The child freed himself (a) / from his father's hold (b) / and ran on the road (c) / No error (d)
10. Without a moment's hesitation (a) / the Principal agreed to write (b) / a foreword to my book (c) / No error (d)
11. When the teachers are on strike (a) / and a notice to this effect is pasted on the college gate (b) / there is no sense to go there. (c) / No error (d)
12. The gain reputation among your clients, (a) / you ought to finish (b) / every assignment you undertake (c) / within the stipulated deadline. (d) / No error (e)
13. Children often (a) / quarrel on (b) / petty issues. (c) / No error (d)
14. He says that (a) / he has a car (b) / beside a scooter (c) / No error (d)
15. Mahatma Gandhi laboured (a) / to the good (b) / of humanity (c) / No error (d)
16. The blind old man was (a) / Knocked down by a car (b) / Upon crossing the road (c) / No error (d)
17. In a report issued by Indian Statistical Institute, (a) / the Iron and Steel Industry is investing more than any other (b) / Indian industry in fighting pollution (c) / No error (d)
18. I found it difficult (a) / to remove (b) / The chewing gum that stuck with my cap. (c) / No error (d)
19. My answer (a) / is different (b) / than yours (c) / No error (d)
20. Of all the men (a) / known to me none is (b) / less inclined that he is to think ill to others (c) / No error (d)
21. The management did not (a) / select him as he was (b) / not eligible to the post (c) / No error (d)
22. Both the form and substance (a) / of what he says (b) / about his life and work (c) / testify for a great reserve of inner energy (d) / No error (e)
23. I am vexed (a)/ at him about what (b) / he has done (c) / No error (d)
24. Beside chocolates (a) / they also bought (b) / many toys for the child (c) / No error (d)
25. Young children should not sit close (a) / to the TV set (b) / as it affects on their eyes (c) / No error (d)

EXPLANATIONS

1. (c) Replace 'for' by 'since'
2. (a) Add 'on' after 'going'
3. (d) Replace 'to' by 'from'
4. (b) Replace 'in' by 'to'
5. (d) Replace 'on' by 'in'
6. (c) Replace 'into' by 'to'
7. (c) Replace 'since' by 'for'
8. (c) Replace 'on' by 'for'
9. (c) Replace 'on' by 'across'
10. (c) Replace 'to' by 'for'
11. (c) Replace 'to go' by 'in going'
12. (d) Replace 'within' by 'by'
13. (b) Replace 'on' by 'over'
14. (c) Replace 'beside' by 'besides'
    'Beside' means 'by the side of',
    'besides' means 'apart from'.
15. (b) Replace 'to' by 'for'
16. (c) Replace 'upon' by 'while'.
17. (d) Replace 'in' by 'on'
18. (d) 'Stuck' takes the preposition 'to', not 'with'.
19. (c) The correct usage is 'different from'
20. (d) The correct form is 'to think ill of others'
21. (c) The correct form is 'eligible for ...'
22. (d) Replace 'for' by 'to'
23. (b) Replace 'at' by 'with'
24. (a) Replace 'beside' by 'besides'
25. (c) Remove 'on'

PRACTICE TEST - 2
SPOTTING ERRORS

***DIRECTIONS* : In each of the following questions, find out which part of the sentence has an error. The error may be idiomatic or grammatical. If the sentence is correct, the answer is 'No error'.**

1. It was apparent for everyone present (a) / that if the patient did not receive (b) / medical attention fast he would die (c) / No error (d)
2. He knows very well (a) / what is expected from him (b) / but he is not able to fulfil all the expectations (c) / No error (d)
3. My brother (a) / has ordered (b) / for a new book (c) / No error (d)
4. That Brutus, who was his trusted friend (a) / had attacked on him (b) / and caused heart break to Julius Caesar (c) / No error (d)
5. Rana Pratap was (a) / blind with (b) / one eye (c) / No error (d)
6. The doctor (a) / attended to the patient (b) / very quietly (c) / No error (d)
7. I was shocked to hear (a) / that his father (b) / died of an accident (c) / No error (d)
8. I must start (a) / at dawn (b) / to reach the station in time (c) / No error (d)
9. The widely publicised manifesto (a) / of the new party (b) / is not much different than ours (c) / No error (d)
10. I was taken with surprise (a) / when I saw (b) / the glamorous Appu Ghar (c) / No error (d)
11. Man needs (a) / security and leisure (b) / of free thinking (c) / No error (d)
12. This watch (a) / is superior and (b) / more expensive than that (c) / No error (d)
13. The venue of examination (a) / is one mile further up (b) / the hill (c) / No error (d)
14. The doctor referred (a) / the patient for the O.P.D. (b) / without examining him (c) / No error
15. A man who always (a) / connives on the faults (b) / of his children is their worst enemy (c) / No error (d)
16. I do not understand (a) / why he is (b) / So angry at me (c) / No error (d)
17. I am hearing (a) / a lot about (b) / the problem of AIDS these days (c) / No error (d)
18. Because of his innocence (a) / he cannot (b) / distinguish a cheat for an honest person (c) / No error (d)
19. I will put on (a) / a note in this regard (b) / for your consideration and necessary decision (c) / No error (d)
20. None could (a) / dare to encroach (b) / on his rights (c) / No error (d)
21. The father brought the sweets (a) / and distributed then (b) / between his five children (c) / No error (d)
22. Raman developed the (a) / habit for sleeping late (b) / when he was staying in the hostel (c) / No error (d)
23. It is the duty of every right thinking citizen (a) / to try to make the whole world (b) / a happier place to live (c) / No error (d)
24. The top-ranking candidates (a) / will be appointed in (b) / senior jobs in good companies (c) / No error (d)
25. My niece has been (a) / married with (b) / the richest man of the town (c) / No error (d)

EXPLANATIONS

1. (a) Replace 'for' by 'to'
2. (b) Replace 'from' by 'of'
3. (c) Remove 'for'
4. (b) Remove 'on'
   The verb 'attack' does not take a preposition
5. (b) Replace 'with' by 'in'
6. (b) Replace 'to' by 'upon'
7. (c) The correct usage is 'died in an accident' and 'died of malaria'
8. (b) Replace 'at' by ' by'
9. (c) The correct usage is 'different from ....'
10. (a) Replace 'with' by ' by
11. (c) Replace 'of ' by 'for'
12. (b) Add 'to' after 'superior'.
    Adjectives ending in-ior take 'to' after them.
13. (a) Replace 'of ' by 'for'
14. (b) Replace 'for' by 'to'
    The verb 'refer' takes preposition 'to'
15. (b) Replace 'on' by 'at'.
    'Connive' takes the preposition 'at'
16. (c) The correct form is 'angry with.....'
17. (b) Replace 'about' by 'of '
18. (d) Replace 'for' by 'from'
19. (a) Replace 'on' by 'up'
20. (c) Replace 'on' by 'upon'. the verb 'encroach' takes 'upon' after it.
21. (c) Replace 'between' by 'among'. 'Between' is used for two persons and 'among' for more than two persons.
22. (b) Replace 'for' by 'of'
23. (c) Add 'in' after 'live'
24. (b) Replace 'in' by 'to'
25. (b) Replace 'with' by 'to'
    The verb 'marry' takes preposition 'to'

PRACTICE TEST - 3
SPOTTING ERRORS

DIRECTIONS : **In the following questions, some of the sentences have errors and some have none. Find out which part of a sentence has an error. If there is no error, mark your answer as 'No error'.**

1. He, you and I (a) / will manage (b) / this problem together (c) / No error (d)
2. Was it him (a) / who got injured (b) / in an accident this morning (c) / No error (d)
3. I suppose (a) / you have heard the latest news (b) / John has married that girl (c) / No error (d).
4. As soon as he (a) / saw his mother (b) / be ran to her and embraced (c) / No error (d)
5. Due to me being a newcomer (a) / I was unable to get a house (b) / suitable for my wife and me (c)/ No error (d)
6. The audience (a) / are requested (b) / to be in its seats (c) / No error (d)
7. The scientist must follow (a) / his hunches and his data (b) / wherever it may lead (c) / No error (d)
8. He is the general (a) / whom the reporters agree (b) / was most popular with the troops (c) / No error (d)
9. There is no question (a) / of my failing (b) in the examination (c) / No error (d)
10. The interviewer asked the candidate (a) / to what district (b) / he came from (c) / No error (d)
11. The master did not know (a) / who of the servants (b) / broke the glass (c) / No error (d)
12. The ruling party stood (a) / for implementation of the Bill (b) / and was ready to stake their political existence (c) / No error (d)
13. This is the boy (a) / whom I think had won (b) / the gold medal in the Dance competition (c) / No error (d)
14. Wherever they go (a) / Indians easily adapt to (b)/ local circumstances (c) / No error (d)
15. He is (a) / taller (b) / than me (c) / No error (d)
16. It is not easy for anyone to command (a) / respect from both one's friends as well as critics (b) / as Dr. Johnson did for his integrity and honesty (c) / No error (d)
17. Mahatma Gandhi taught us (a) / that one should respect (b) / the religions of others (c) / as much as his own (d) / No error (e)
18. The number of vehicles (a) / plying on this road (b) / is more than on the main road (c) / No error (d)
19. Being a destitute (a) / I admitted him (b) / to an old people's home (c) / No error (d)
20. This is not the (a) / sense which (b) / concerns us here (c) / No error (d)
21. One should make (a) / his best efforts (b) / if one wishes to achieve (c) / success in this organisation (d) / No error (e)
22. May I (a) / know who you want (b) / to see please (c) / No error (d)
23. Our is the only (a) / country in the world (b) / than can boast of (c) / unity in diversity (d) / No error (e)
24. Last summer he went (a) / to his uncle's village (b) / and enjoyed very much (c) / No error (d)
25. If I were him, (a) / I would have taught (b) / those cheats a lesson (c) / No error (d)

EXPLANATIONS

1. (a) The correct form is 'You, he and I'
2. (a) Replace 'him' by 'he'
3. (a) Add 'that' after 'suppose'
4. (d) Add 'her' after 'embraced'
5. (a) Replace 'me' by 'my'
6. (c) Replace 'its' by 'their'
7. (c) Replace 'it' by 'they'
8. (b) Replace 'whom' by 'who'
9. (b) Replace 'my' by 'me'
10. (b) Replace 'to what' by 'which'
11. (b) Replace 'who' by 'which'
12. (c) Replace 'their' by 'its'
13. (b) Replace 'whom' by 'who'
14. (b) Add 'themselves'after 'adapt'
15. (c) Replace 'me' by 'I'
16. (b) Add 'one's' before 'critics'
17. (c) Replace 'his' by 'one's'
18. (c) Add 'that' after 'than'
19. (a) Add 'He' before 'being'
20. (b) Replace 'which' by 'that'
21. (b) Replace 'his' by 'one's'
22. (b) Replace 'who' by 'whom'
23. (b) Replace 'our' by 'ours'
24. (c) Add 'himself' after 'enjoyed'
25. (a) Replace 'him' by 'he'

PRACTICE TEST - 4

SPOTTING ERRORS

DIRECTIONS : **In the following questions, some of the sentences have errors and some have none. Find out which part of a sentence has an error. If there is no error, mark your answer as 'No error'.**

1. Please put away (a) / the candle (b) / before you leave (c) / No error (d)
2. All the doctors were (a) / puzzled on the (b) / strange symptoms reported by the patient (c) / No error (d)
3. Most of the Uttar Pradesh (a) / is an extremely fertile plain (b) / between which the river Ganga flows (c) / No error (d)
4. He has been working on (a) / the problem from a long time (b) / but is still not able to solve it (c)/ No error (d)
5. At the last day (a) / of your holidays (b) / you have started thinking about how to finish your work (c) / No error (d)
6. India is in no way (a) / inferior than the U.S.A. (b) / in the fertility of soil and in the richness of mineral resources (c) / No error (d)
7. The students were (a) / awaiting for (b) / the arrival of the chief guest (c) / No error (d)
8. I could not convince them (a) / because they persisted to suggest (b) / that I was lying (c) / No error (d)
9. It was apparent for everyone present (a) / that if the patient (b) / did not receive medical attention fast he would die (c) / No error (d)
10. The detective says that (a) / there is no chance for (b) / finding the person who wrote these letters (c) / No error (d)
11. Vishal is (a) / one year junior (b) / than Madan in our office (c) / No error (d).
12. They walked (a) / besides each other (b) / in silence (c) / No error (d)
13. Our Mathematics teacher (a) / often emphasises on (b) / the ned for a lot of practice (c) / No error (d)
14. In urban society (a) / the social circle is limited with the family (b) / but in the villages it encompasses the entire village (c) / No error (d)
15. Being most loquacious (a) / among her brothers and sisters (b) / she related a good many tales in each breath (c) / No error (d)
16. The committee appointed to investigate into (a)/ the said murder (b) / will convene next week to further discuss their strategy (c) / No error (d)
17. Kamal's habit of (a) / delaying his work (b) / put his colleagues to lot of trouble (c) / No error (d)
18. No effort has been made at all (a) / to cash on (b)/ the refurbished image of Indian tennis (c) / No error (d)
19. After opening the door (a) / we entered into the room (b) / next to the kitchen (c) / No error (d)
20. As the meeting was about to end (a) / he insisted to ask (b) / several questions (c) / No error (d)
21. The watchman was kind enough (a) / to inform us about the conspiracy (b) / but declined to name the person behind it (c) / No error (d)
22. The captain and his wife (a) / were invited (b)/ for the cultural function at my home (c) / No error (d)
23. The engineer came out (a) / to a novel solution (b) / which may even reduce daily energy consumption (c) / No error (d)
24. On a holiday (a) / Madhu prefers reading (b) / than going out visiting friends (c) / No error (d)
25. People who are averse with (a) / hard work (b) / generally do not succeed in life (c) / No error (d)

EXPLANATIONS

1. (a) The correct usage is 'put out the candle'
2. (b) Replace 'on'by 'at'
3. (c) Replace 'between' by 'through'
4. (e) No error
5. (a) Replace 'at' by 'on'
6. (b) Replace 'than' by 'to'
7. (b) Remove 'for'
8. (b) Replace 'to suggest' by 'in suggesting'
9. (a) Replace 'for' by 'to'
10. (b) Replace 'for' by 'of'
11. (c) Replace 'than' by 'to'
    Adjective ending in - ior take 'to'.
12. (b) Replace 'besides' by 'beside'
13. (b) Remove 'on'
14. (b) Replace 'with' by 'to'
15. (b) Replace 'among' by 'of '
16. (a) Remove 'into'
17. (d) Replace 'to' by 'into'
18. (b) The correct form is 'to cash in on'
19. (b) Remove 'into'
    'Enter' does not take any preposition
20. (b) The correct form is 'he insisted on asking'
21. (b) Replace 'about' by 'of '
22. (c) Replace 'for' by 'to'
23. (b) Replace 'to' by 'with'
24. (c) Replace 'than' by 'to'
    The verb 'prefer' takes preposition 'to'
25. (a) Replace 'with' by 'to'

2. FILL IN THE BLANKS

In this type of test each question consists of a sentence with a blank. You have to fill the blank by choosing the correct one from among the four options given below the question.

***Example – 1* (*Correct* Preposition)**

1. We are confident ________ winning the race.

A) in B) for C) of D) with

Answer : (C) of

Explanation : In this question the correct preposition to fill in the blank is 'of'. Hence the answer is (C).

***Example – 2* (*Correct* Form of Verb)**

1. Everyone wants to _____ for one's achievements.

A) admire B) be admiring C) being admired D) be admired

Answer : (D) be admired

Explanation : In this question the correct form of the verb to be filled in the blank is 'be admired'. Hence the correct answer is (D).

***Example – 3* (*Correct* Word)**

1. The traveller slept under the _______ shade of the banyan tree.

A) cold B) cool C) cooling D) dark

Answer : (B) cool

Explanation : The correct word for filling the blank is 'cool'. Hence the correct answer is (B).

How to prepare for this test :

As the questions in the fill in the blanks test requires choice of (1) Correct Preposition (2) Correct form of the Verb (3) Correct Word you have to prepare for the test accordingly.

Here we are giving study material and Practice Tests for the following types of Fill in the blanks.

1. Correct Form of the Verb 2. Correct Prepositions 3. Correct Articles 4. Confusing Words

STUDY MATERIAL

I. CORRECT FORM OF VERB

In this type of Test, a sentence is given with a blank with four or five responses (answers). You have to select correct form of verb from the given responses.

Example:

1. He __________to office everyday.

a) goes b) gone

c) went d) going

Answer : (A) Here, the action expressed by the verb is habitual. So the correct form is 'goes'.

FORMS OF VERBS

There are five forms of verbs in English. They are:

	Be forms	Write forms
1. Infinitive Form	Be,	Write
2. Present Form	am, is, are,	write, writes
3. Past Form	was, were,	wrote
4. Past Participle Form	been	written
5. Present Participle Form	being	writing

Use of infinitive form :

1. The Infinitive form is the basic form of the verb. Except for 'be', it is same as the present form of the verb without 's'.
2. The Infinitive form of the verb is used after the auxiliary verbs like do, did, does, can, could, shall, should, will, would, may, might, must, ought to, used to, etc.

Examples :

1. I shall do it tomorrow.
2. He may come tomorrow.
3. He must go there.

3. The Infinitive form of the verb is used after the expressions like 'would rather', would sooner', rather than, sooner than, had better, etc.

Examples :

1. You had better tell him.
2. I would do rather than suffer.

Use of Present Form

1. Except for 'be', the present form of the verb is same as infinitive form.
2. When the subject is in third person singular s or es is added to the present form.

Examples :

1. He writes a letter everyday.
2. She goes to temple everyday.
3. The sun rises in the east.

3. Only present form of the verb is used with Adverbs or Adverb Phrases such as 'often, usually, regularly, daily, generally, always, never, occationally, every day, every week, on Sundays, twice a week, etc'.

Examples :

1. I always play tennis.
2. He goes to school every day.

4. The present form of verb is used with habitual or regular actions and general truths.

Examples :

1. He goes to school regularly.
2. He drinks milk every morning.
3. The sun rises in the east.
4. The earth moves round the sun.

5. The present form of the verb is used in proverbs.

Example :

Slow and steady wins the race.

Use of Past Form

1. The past form of the verb is generally formed by adding d, ed to the Regular verb.

Example :

call - called.

2. The past forms of the irregular verbs are formed differently. So you should learn them by heart. List of common Irregular verbs of given at the end of this topic. Get them by heart.

Example :

Write - wrote, see - saw.

3. The past form of the verb is used with the adverbs of past time like yesterday, in 1990, last year, last sunday, etc.

Examples :

1. I met him yesterday.
2. India got independence in 1947.

4. The past of the verb is used with expressions like 'It is high time', 'It is time', 'I wish', 'would rather', 'as though', 'suppose', etc.

Examples :

1. It is high time that we left.
2. It is time we went.
3. I wish I were rich.
4. He acts as if he were a king.
5. I would rather you stayed at home.

Use of past participle form

1. The past participle form of the verb is same as the past form for regular verbs. It is formed by adding - 'd' or 'ed' to the infinitive.
2. The past participle form of the Irregular verb is formed differently. So you should learn them by heart.

Example :

write – wrote – written

3. The past participle form of the verb is used with 'have', 'has', 'had' when they are used as auxiliaries.

Example :

I have written that letter.

4. The past participle form of the verb is used in passive voice. (It is generally recognised by the presence of 'by'. and 'is, am, are, was, were, been, being').

Examples :

1. The rat was killed by the cat.
2. The letter has been written by me.

5. The past participle form of the verb is used in present, past and future perfect tenses (have/has/had + past participle)

Example :

1. He has written a letter just now.
2. He had gone out, when he came here.
3. I shall have done it, by the time you come.

Use of Present Participle

1. The present participle form of the verb is formed by adding -ing to the infinitive form.

2. The present participle form is used in present, past, future continuous tenses. (am/is/are/was/were/will be/ shall be + ing form)

   Examples :

   1. He is writing a letter now. (Present continuous tenses)
   2. He was writing a letter yesterday at 9 a.m. (Past continuous tense)
   3. He will be writing a letter tomorrow at 9 a.m. (Future continuous tense)

3. The present participle form is used in perfect continuous tenses. (has/have/had/shall have/will have/ + been + ing form)

   Examples :

   1. He has been learning English since 1990. (Present Perfect Continuous Tense)
   2. He had been learning English for four years by 1994. (Past Perfect Continuous Tense)
   3. He will have been learning English for ten years by 2000. (Future perfect continuous tense)

Some other Short-cut Rules for using *Correct* Form of the Verb

1. If one part of the sentence is in future, then the other part will be present.

   Examples :

   1. The match will start.
   2. When the rain stops.

2. If one part of the sentence is in simple past and 'when' is used in the sentence, the other part will be in the past perfect.

   Example :

   When I reached the station the train had already left.

3. If one part of the sentence is in simple past tense and 'while' is used in the sentence, the part will be in the past continuous tense.

   Example :

   It started raining while we were playing cricket.

4. A past tense in the main clause should be followed by a past tense in the subordinate clause.

   Example :

   He succeeded because he worked hard.

5. A past tense in the main clause may be followed by a Present Tense in the subordinate clause when the subordinate clause expresses a universal truth.

   Example :

   Our teacher said that the earth moves round the sun

TENSES

There are 12 tenses in English. They are :

1. **Present Indefinite Tense** or Simple Present Tense

   Structure : (Subject) + Present form

   Examples :

   1. I write a letter everyday.
   2. He writes a letter every day.

2. **Past Indefinite Tense** or Simple Past tense

   Structure : (Subject) + Past form

   Examples :

   1. I wrote a letter yesterday
   2. He wrote a letter yesterday.

3. **Future Indefinite Tense** or Simple Future Tense

   Structure : (Subject) + Shall/will + Infinitive

   Examples :

   1. I shall write a letter tomorrow.
   2. He will write a letter tomorrow.

4. **Present continuous Tense**

   Structure : (Subject) + am/is/are + Present participle

   Examples :

   1. I am writing a letter now.
   2. You are writing a letter now.
   3. He is writing a letter now.

5. **Past continuous Tense**

   Structure : (Subject) + was/were+Present Participle

   Examples :

   1. I was writing a letter yesterday at 9 a.m.
   2. They were writing letters yesterday at 9 a.m.

6. **Future Continuous Tense**

   Structure :(Subject) + shall/will+be+ Present Participle

   Examples :

   1. I shall be writing a letter tomorrow at 9 a.m.
   2. He will be writing a letter tomorrow at 9 a.m.

7. **Present Perfect Tense**

   Structure: (Subject) + have/has+Past Participle

   Examples :

   1. I have written a letter just now.
   2. He has written a letter just now.

8. **Past Perfect Tense**

   Structure : (Subject) + had + Past Participle

   Examples :

   1. I had written a letter yesterday by 10 a.m.
   2. He had written a letter yesterday by 10 a.m.

9. **Future perfect tense**

   Structure :(Subject) + shall/will + have + Past Participle

Examples :

1. I shall have written a letter by 10 a.m. tomorrow.
2. He will have written a letter by 10 a.m. tomorrow.

10. Present Perfect Continuous Tense

Structure :(Subject) + have/has + been + Present Participle

Examples :

1. I have been learning English since 1990.
2. He has been learning English since 1990.

11. Past Perfect Continuous Tense

Structure :(Subject) +had+been+Present Participle

Example :

I had been learning English for four years by 1994.

12. Future Perfect Continuous Tense

Structure : (Subject) + shall/will+have+been + Present Participle

Examples :

1. I shall have been learning English for Ten years by 2000.
2. He will have been learning English for ten years by 2000.

IMPORTANT USES OF TENSES

1. The **Simple Present Tense** is used to express habitual or regular actions and universal truths.

Examples :

1. I go to school every day.
2. The sun rises in the east.
3. The earth goes round the sun.

2. The **Simple Present Tense** is used in the present or Future Clauses of time and condition.

Examples :

1. If you work hard, you will pass the examination.
2. I shall meet him when he comes here.

3. The **Present continuous tense** is used to show an action that is taking place at the time of speaking.

Example :

He is reading a book now.

4. The **Present Perfect Tense** is used with 'Just (now)' to express a recently completed action.

Example :

He has just gone out.

5. The **Present Perfect Tense** is used for past actions whose time is not given and not definite.

Examples :

1. I have already read the book.
2. She has not arrived at.

6. The **Present Perfect Tense** is used to express an action that began in the past and still continuing.

Examples :

1. I have lived here since 1975.
2. I have known him for five years.

7. The **Present Perfect Tense** is never used with the Adverbs of Past time. (Eg. Yesterday, on Monday last, in 1990 etc.) Here Simple past tense is used.

Example :

He went to Guntur Yesterday.

8. The **Present Perfect Continuous** tense is used for an action which began in the past and is still continuing at the present.

Example :

He has been learning English for five years.

9. The **Simple Past tense** is used for an action that took place in the past.

Example :

I met him yesterday.

10. The **Simple past tense** is used in the since-clause showing past action.

Example :

I have not played cricket since I left the college.

11. The **Simple past tense** is used after the expressions like 'It is high time', 'It is time', 'I wish', 'would rather', 'as if', 'as though', 'Suppose' etc.

Examples :

1. I wish I were rich.
2. He acts as if he were a king.
3. It is time we went.

12. The **Past continuous tense** is used for an action which was going on at some point in the past.

Examples :

1. At seven in the morning I was reading the news paper.
2. I was writing, when he called upon me.

13. The **Past continuous tense** is used to show two or more actions that were going on at some time in the past.

Example :

While Hari was doing his homework, his sister was reading here lessons.

14. The **Past perfect Tense** is used for an action completed before a certain point in the past.

Example :

I had done my homework by 8 0' clock yesterday.

15. The **past perfect tense** is used for the earlier of the two actions happened in the past.

Example :

When I reached the station, the train had already left.

16. The **Past Perfect continuous Tense** is used for an action that began before a certain point in the past and continued upto that point and beyond.

Examples :

1. He had been learning English for four years by 1996.
2. When I joined the school in 1990, Mr. Hari had already been teaching there for five years.

17. The **simple future tense** is used to show a future action.

Example :

I shall go to Guntur tomorrow.

18. The **Future continuous tense** is used for an action that will be going on at some point in the future.

Example :

I shall be going to office tomorrow at 9 a.m.

19. The **Future perfect Tense** is used to show an action that will be completed before some point in the Future.

Example :

I shall have completed my homework by 9 a.m. tomorrow.

20. The **Future perfect continuous Tense** is used for an action that will be going on upto some point in the Future.

Example :

He will have been learning English for five years by next year.

FORMS OF IRREGULAR VERBS

Preset Form	Past Form	Past Participle Form
Abide	Abode	Abode
Am, is	was	Been
Be, Are	were	Been
Arise	Arose	Arisen
Awake	Awoke	Awaked
Awaken	Awakened	Awakened
Bear	Bore	Borne
Beat	Beat	Beaten
Begin	Began	Begun
Behold	Beheld	Beheld
Bend	Bent	Bent
Bereave	Bereaved	Bereaved
Bind	Bound	Bound
Bid	Bid	Bid
Bite	Bit	Bitten
Bleed	Bled	Bled
Blow	Blew	Blown
Break	Broke	Broken
Breed	Bred	Bred
Bring	Brought	Brought
Build	Built	Built
Burn	Burned, Burnt	Burned, Burnt
Burst	Burst	Burst
Buy	Bought	Bought
Catch	Caught	Caught
Chide	Chid	Chid, Chidden
Choose	Chose	Chosen
Cling	Clung	Clung
Come	Came	Come
Creep	Crept	Crept
Dare	Dared	Dared
Deal	Dealt	Dealt
Dig	Dug	Dug

Preset Form	Past Form	Past Participle Form
Do	Did	Done
Draw	Drew	Drawn
Dream	Dreamt	Dreamt
Drink	Drank	Drunk
Drive	Drove	Driven
Dwell	Dwelt	Dwelt
Eat	Ate	Eaten
Emberrass	Emberrassed	Emberrassed
Fall	Fell	Fallen
Feed	Fed	Fed
Feel	Felt	Felt
Fight	Fought	Fought
Find	Found	Found
Flee	Fled	Fled
Fling	Flung	Flung
Fly	Flew	Flown
Forecast	Forecast	Forecast
Forget	Forgot	Forgotten
Forgive	Forgave	Forgiven
Forsake	Forsook	Forsaken
Freeze	Froze	Frozen
Get	Got	Got
Give	Gave	Given
Grind	Ground	Ground
Hang	Hanged	Hanged
Hang	Hung	Hung
Harass	Harassed	Harassed
Have	Had	Had
Hear	Heared	Heared
Hide	Hid	Hidden
Hit	Hit	Hit

Preset Form	Past Form	Past Participle Form
Hold	Held	Held
Hurt	Hurt	Hurt
Keep	Kept	Kept
Kneel	Knelt	Knelt
Knit	Knit	Knit
Know	Knew	Known
Lay	Laid	Laid
Lead	Led	Led
Learn	Learnt, Learned	Learned
Leap	Leaped, Leapt	Leaped
Leave	Left	Left
Lend	Lent	Lent
Let	Let	Let
Lie	Lay	Lain
Light	Lit	Lit
Lose	Lost	Lost
Mean	Meant	Meant
Meet	Met	Met
Mislead	Misled	Misled
Mistake	Mistook	Mistaken
Pay	Paid	Paid
Put	Put	Put
Read	Read	Read
Ride	Rode	Ridden
Ring	Rang	Rung
Rise	Rose	Risen
Saw	Sawed	Sawn
Say	Said	Said
See	Saw	Seen
Seek	Sought	Sought
Sell	Sold	Sold
Set	Set	Set
Shake	Shook	Shaken
Shed	Shed	Shed
Shine	Shone	Shone
Show	Showed	Shown
Shrink	Shrank	Shrunk
Shut	Shut	Shut
Sing	Sang	Sung
Sink	Sank	Sunk
Sit	Sat	Sat
Slay	Slew	Slain
Sleep	Slept	Slept

Preset Form	Past Form	Past Participle Form
Speak	Spoke	Spoken
Speed	Sped	Sped
Spell	Spelt	Spelt
Spin	Spun	Spun
Spit	Spat	Spat
Split	Split	Split
Spread	Spread	Spread
Spring	Sprang	Sprung
Stand	Stood	Stood
Steal	Stole	Stolen
Stick	Stuck	Stuck
Sting	Stung	Stung
Stink	Stank	Stunk
Strew	Strewed	Strewn
Stride	Strode	Stridden
Strive	Strove	Striven
Swear	Swore	Sworn
Swell	Swelled	Swollen
Sweep	Swept	Swept
Swim	Swam	Swum
Swing	Swung	Swung
Take	Took	Taken
Teach	Taught	Taught
Tear	Tore	Torn
Tell	Told	Told
Think	Thought	Thought
Thrive	Throve, Thrived	Thriven
Throw	Threw	Thrown
Thrust	Thrust	Thrust
Tread	Trod	Trodden
Wake	Woke	Woken
Waken	Wakened	Wakened
Wear	Wore	Worn
Weave	Wove	Woven
Wed	Wed, Wedded	Wed, Wedded
Weep	Wept	Wept
Wet	Wet	Wet
Win	Won	Won
Wind	Wound	Wound
Work	Worked, Wrought	Worked, Wrought
Write	Wrote	Written
Wring	Wrung	Wrung

PRACTICE TEST - 1

FILL IN THE BLANKS

DIRECTIONS : **Fill in the blanks with correct form of verb.**

1. She _____unconscious since four O' clock.
   a) is b) was c) has been d) had been
2. Honesty is _____on her face.
   a) written b) writing c) write d) wrote
3. It_____since early morning.
   a) is raining b) has been raining
   c) was raining d) has rained
4. He_____out five minutes ago.
   a) had gone b) has gone c) went d) go
5. The lesson_____in the class by the teacher yesterday.
   a) was taught b) is taught
   c) taught d) has taught
6. He has_____his case to my satisfaction.
   a) proved b) prove
   c) proving d) none of these
7. The baby_____every morning.
   a) cry b) cries c) crying d) cried
8. I have_____of a plan.
   a) think b) thinking c) thought d) thinks
9. We_____our breakfast half an hour before.
   a) finished b) have finished
   c) had finished d) finish
10. I_____the Prime Minister Yesterday.
    a) see b) saw c) seen d) have seen
11. He wondered if ......I......by so strict rules.
    a) could abide b) can abide
    c) abide d) will abide
12. He was surprised to know that I......with my friend.
    a) break relations
    b) will break relations
    c) have broken relations
    d) had broken relations
13. He said that change.........law of life.
    a) is the b) was the
    c) has been the d) will be the
14. He helps more than Ravi.........
    a) has helped b) will help
    c) is helping d) helps
15. I..............that he told a lie.
    a) think b) will think
    c) had thought d) will have thought
16. When I distributed sweets he......away.
    a) was running
    b) has been running
    c) had run
    d) ran.
17. I..............my pen yesterday.
    a) lost b) have lost
    c) had lost d) would lose
18. He went to the exhibition but I..............
    a) had not
    b) have not
    c) did not
    d) would not
19. I..............him in the restaurant last night.
    a) meet b) have met
    c) an meeting d) met
20. I..............him since he met in kenya.
    a) knew
    b) had known
    c) have known
    d) would have known
21. I.........to my college daily.
    a) go b) will go
    c) am going d) have been going
22. When I reached there, he.........gone to Mumbai.
    a) had b) has
    c) was d) would have
23. I will not stay if you..............ask me.
    a) do not b) will not
    c) shall not d) should not
24. I wish I..............a bird.
    a) were b) was
    c) am d) had
25. If you had informed me, I.........cautious.
    a) will be b) would be
    c) had been d) would have been

ANSWERS

1. c	2. a	3. b	4. c	5. a
6. a	7. b	8. c	9. a	10. b
11. a	12. d	13. a	14. d	15. a
16. c	17. a	18. c	19. d	20. c
21. a	22. a	23. a	24. a	25. d

II. PREPOSITIONS

A preposition is word placed before a Noun or Pronoun to show its relation to some other word in the sentence.

Example :

The book is on the table.

In the preposition Test, a sentence is given as a question with a blank in it. You have to choose the correct preposition from the given alternatives.

Example :

1. The snake was killed .... him.

A) by B) with C) about D) on

Answer : (A) Here 'by' is the correct preposition. So the correct answer is (A).

STUDY MATERIAL

USE OF PREPOSITIONS : IMPORTANT RULES

1. Beside, Besides : 'Beside' means 'by the side of' and 'besides' means 'in addition to'.

Examples :

1. He stood beside me. (by the side of)
2. He has a pen besides pencil. (in addition to)

2. Between, among : 'Between' is used with two persons or things. 'Among' is used with more than two persons or things.

Examples :

1. I stood between Hari and Rama.
2. The children quarrelled among themselves.

3. By, with : 'By' is used with the doer of the action. 'with' is used with the instrument of the action.

Example :

The dog was killed by him with a stick.

4. In, at : 'In' is used with the names of large towns and countries; 'at' is used with names of small towns and villages.

Example :

He lives at Tandur in Andhra Pradesh.

5. On, in, at, by : While speaking of time, 'at' indicates an exact point of time, 'on' a more general point of time and in a period of time.

Examples :

1. I shall be there at 4 P.M. on Friday.
2. It is very hot in the day but quite cold at night.

'By' is used to show the latest time at which an action will be finished. So it is usually used with the Future tenses.

Example :

I hope to finish the work by Monday.

6. In, Into : 'In' is used in speaking of things at rest. 'Into' is used in speaking of things in motion.

Examples :

1. He is in bed.
2. The cat fell into the well.

7. On, upon : 'On' is used in speaking of things at rest, 'upon' used in speaking of things in motion.

Examples :

1. He sat on a chair.
2. The cat jumped upon the table.

8. In, within : With reference to time, 'in' means at the end of a certain period whereas 'within' means before the end of a certain period.

Examples :

1. The loan will be repaid in a year. (at the end of one year)
2. The loan will be repaired within a year. (before the end of one year)

9. Till, To : Till is used of time; 'to' is used of place.

Examples :

1. He slept till six 0' clock.
2. He walked to the end of the street.

10. Since, for : 'Since' is with point of time; 'For' is used with period of time.

Examples :

1. He has lived here for three years.
2. He has been working in this office since 1975.

11. Certain Verbs, nouns, adjectives and participles are always followed by particular prepositions.

Examples :

1. He is fond of sweets.
2. He prefers tea to coffee.

WORDS FOLLOWED BY APPROPRIATE PREPOSITIONS

In ordinary usage 'by' is used for a person and 'with' for a thing. The mango was cut by Mridula with a knife. 'To' is generally used for purpose. In this way a verb may take more than one preposition for different purposes. But certain verbs take only one preposition. Only that preposition is used with that verb. Following is a list of words together with the appropriate prepositions.

Abide by (a person)
Abide with (a decision etc.)
Abide at/in (a place/house)
Abound in (have abundance)
Abound with (are infested with dacoits, mosquitoes etc.)
Absent from (a place)
Absolve from (a promise/sin)
Absorbed in (work, thoughts)
Abstain from (duty)
Abstract from (a statement etc.)
Abundance of (things)
Accede to (a demand/request)
Acceptable to (a person)
Access to (an authority)
Accommodate to (circumstances)
Accommodate with (a person)
Accompany to (a place)
Accompany by (a person)
Accordance with (a thing)
According to (a person/principle)
Account to (a person)
Account for (an act)
Account of (a crime)
Accuse of (something)
Accuse by (a person/body)
Accustomed to (behaviour etc.)
Acquainted with (a person)
Acquitted of (a charge/offence)
Adapt to (conditions)
Add to
Adequate to (demand)
Adequate for (purpose)
Addicted to (drugs/smoking/drinks)
Adhere to (a principle/decision)
Appoint to (a post)
Adjacent to (a place/building)
Adjourn to (date) for (reason)
Admit to (an institution/place)

Admit of (allow)
Affiliate to (a university)
Afflict with (a disease/sorrow)
Afraid of
Agree to (a proposal etc.)
Agree with (a person)
Agree upon (an action/subject)
Aim at (a goal)
Alight from (a plane/train etc.)
Alliance with (person)
Alliance for (purpose)
Allotted to (assigned)
Allude to (refer indirectly)
Amateur in (imperfect in)
Amazed at (surprised)
Amuse at (an outside thing)
Amuse with (one's own action/something)
Anger at (action) towards (person)
Annoyed at (an action)
Annoyed with (a person)
Answer to (a person)
Answer for (an action)
Anxious to (do a thing)
Anxious about/for (worried about)
Apologise to (something)
Appeal to (a person)
Appeal against (a decision)
Appear from
Appetite for (hunger for)
Apply to (a person)
Apply for (a thing)
Brood over (an idea, an event)
Approve of
Aptitude for (a thing)
Argue with (person), for against about (a thing)
Arbitrate between (parties)
Arbitrate in (a dispute)
Arm against (a danger)

Arm with (a weapon)
Arrive at (a place) in (a vehicle)
Ask about (thing) after (person)
Aspire to (a position)
Assure of
Astonish at
Atone to (a person)
Atone for (an act)
Attached to (a thing, a person)
Attached with (an instrument)
Attend to (duty/speech)
Attend upon (serve a person)
Attribute to
Authority over (a person)
Authority on (a subject)
Avail to (an opportunity)
Avenge on
Averse to
Aware of
Back-see verbal phrases
Backward in (something)
Bank on (depend on)
Bargain for
Bear-see verbal phrase
Beg of/from (a person)
Beg for (something)
Believe in (a philosophy/action)
Beneficial to
Bent on (doing something)
Bestow on (a person)
Beware of (a thing or a person)
Blind to (a fault)
Blow out (ruin tire/gadget)
Blow over (anger/storm) up (explode)
Blush at
Boast of
Born of (someone)
Bound for (a destination)
Break-see verbal phrases

Bring-see verbal phrases
Build of
Busy with (some work)
Call-See verbal phrases
Care for (others' action/opinion)
Care of, Take (ones own thing)
Carry-see verbal phrases
Cater to
Cause for (anxiety)
Cause of (a trouble)
Caution against
Certain of
Charge with (an offence)
Claim on (a person/thing)
Clear to (mind) in (argument)
Come-see verbal phrases
Comment on (something)
Commence on (a day/date)
Commit to (a person) for (trial)
Compare to (of different classes)
Compare with (of the same class)
Compete for (something)
Compete with (a person)
Complain of (a pain etc.)
Complain against (injustice)
Comply with (a request)
Conceal from (a person)
Concede to (a person)
Conceive of (a thing)
Concerned about (progress) for (safety)
Concur with (a person) in (a view)
Condemn to (death) for (crime)
Confer with (a person)
Confer on (a situation)
Confide in (a person)
Confidence in (a person)
Confident of (something)
Confine to (limited to a place/subject)
Congratulate on (an action)
Conscious of (something)
Consent to (a proposal etc.)
Consist of
Consistent with

Conspire against
Contend with
Contrary to
Contrast with
Contribute to (a fund)
Contribute towards (a project)
Contrary to (an expectation)
Control over
Controversy with (a person)
Controversy on/about (a matter)
Conversant with (have knowledge of)
Converse with
Convicted of (punished for)
Convince of
Correspond with (a person) about (a thing)
Count on (rely/depend)
Crave for (intense desire)
Cure of (action over disease)
Cure for (thing that cures)
Cut-see verbal phrases
Deaf to (request etc.)
Deal in (something for business)
Deal with (a person)
Decide upon (plan/matter)
Defence against
Deficient in
Delight with (a person)
Delight in (a thing)
Delivery from
Depend on (a person/one's faculty)
Deprive of
Derive from
Desirous of (an action)
Despair of
Destined for (a good thing)
Destined to (an evil thing)
Deviate from (an act)
Devoid of
Devote to
Die of (a disease)
Differ with (a person)
Differ from (one thing from another)
Differ on (an issue/thinking)

Diffident about
Disagree on (an issue)
Disagree with (a person)
Disgrace to (an institution/a family)
Disgusted at (something)
Disgusted with (a person/a course of life)
Dispense with
Dispose of
disqualified for (a post)
Dispute with (a person)
Dispute on (a thing)
Dissuad from
Distinct from
Dinstinguish from/between
Divert from
Divide between (two)
Divide among (more)
Do-see verbal phrases
Dream of/about
Drive to/from (a place) at (an idea, a conclusion)
Due to
Dwell on (a matter) at (a place)
Eager for (a desire/an opportunity)
Eager to (do a thing)
Efficient in (something)
Elaborate on
Eligible for (a post)
Eliminate from
Embark on (plan/project)
Emerge from
Encroach on
Endow with (wealth/qualities)
Engage in (a work or discussion)
Engaged to (married to)
Engross in (something)
Enrage at (an action)
Enrage with (a person)
Enter on (a plan)
Enter into (an agreement)
Enticed into (an evil thing)
Entitled to (something/opinion)
Entrusted with (a work)

Entrusted to (a person)
Envious of (a person/a thing)
Equal to
Escape from (a place)
Essential for/to
Esteem for (a person)
Estranged from (people)
Evidence of (in favour of)
Evidence against
Excel in
Exchange with (a person)
Exchange for (a thing)
Exclude from
Excuse for (a fault)
Excused from (allowed not to)
Explain to
Expend on
Experience in
Export to (a place)
Exposure to
Exult at/in (a success)
Faced with (a problem/trouble)
Fair to (a person)
Faint with (hunoer etc.)
Faith in (a person or ideology)
Fall-see verbal phases
Familiar with (a person, a thing)
Fancy to (a person/doing some thing)
Fascination for
Favourable to
Favour with
Fawn on (some one for favour)
Feed on (something)
Feel for (sympathise with)
Feel like (have a sensation of)
Figure out (solve)
Figure up (to total)
Fill with (something/ideas)
Fit for (a post/job)
Fit in (adjust according to.)
Flirt with (someone)
Fly into (anger-rage etc.)
Focus on (a matter/person)

Fond of (have liking)
Foreign to (an idea/ideal)
Forgetful of
Free to (allowed to do)
Frown at (a person)
Frown on (an evil)
Full of (something/ideas)
Furnish with (equip with)
Get-see verbal phrases
Gifted with
Give-see verbal phrases
Glad of (something)
Glad to (do something)
Glance at (to look at)
Glance over (examine rapidly)
Glance through (see hurriedly)
Glory in (something/victory)
Go-see verbal phrases
Grapple with (a problem)
Grasp at (something)
Grateful to/for (person/thing)
Grieve about (something)
Grieve for (a person)
Grow up (be mature)
Grumble at/about (something)
Guarantee for
Guard against (an evil person)
Guilty of (an offence)
Hand-see verbal phrases
Hang-see verbal phrases
Hanker after (something)
Happen in (a place) at (a time) to (a person)
Hard on (a person)
Hear about/of (a matter)
Hear from (receive news from)
Heir to (property/assets)
Hide from (a person)
Hinder from (a thing) by (person)
Hit on or upon (devise a plan)
Hold-see verbal phrases
Hope for (something)
Hopeful of (some result)
Hostile to (someone)

Hunger for (something)
Identical with (similar to)
Indentify with (a person/achievement)
Identity of (a person)
Ignorant of (something)
Impart to (a person)
Impatient of (a restriction)
Impatient to (do something)
Implicit in (what is said/written)
Impose on (a person)
Impress on (a person)
Impudent to (a senior person)
Incline to
Include in
Inconsiderate of (others'..)
Incorporate in (include in)
Indebted to
Indifferent to
Indignant at
Indulge in
Infect with
Infer from
Inferior to
Inform of
Infuse with
Inherit from
Injurious to
Inquir into (an accident/complaint)
Inquire of (ask about)
Insensible to
Insist on/upon
Intent on
Interested in
Interfere with
Interrupt in
Intervene in
Introduce to
Intrude on/upon
Invest in (something)
Invite to (a place/function)
Involve in
Irrelevant to
Jealous of (a person/a thing)

Jeer at
Jest at
Join in (participate)
Join to (connect with)
Jump to (a conclusion)
Keen about (a person/thing)
Keen on (doing something)
Keep-see verbal phrases
Kill with (a thing) for (a purpose)
Knack for
Knock-see verbal phrases
Know about
Lament for
Lapse into
Last for (have a life of - time)
Laugh at
Lay-see verbal phrases
Lead to
Lean against (a wall) on (a staff) to (a principle)
Levy (tax etc.) on (people)
Live at (a place)
Live in (a city/country/house)
Live on (edibles)
Live by (a profession)
Live for (desire for)Long for
Look-see verbal phrases
Lost to (a person)
Lost in (absorbed in)
Love for (something-music etc)
Loyal to (a person/country)
make-see verbal phrases
Manage with (to do with)
Mania for
Married to
Marvel at
Match for - no (equal to) with
Meddle with
Meditate on/upon
Meet with
Migrate from-to
Mistaken in
Mix up (confuse)

Mix with (be friendly with)
Mourn for
Moved by (a good act) with (pity)
Murmur against
Muse upon
Name after (give name of)
Necessary/Necessity for
Negligent in
Obedient to
Object to (have objection to)
Obliged to (a person) for (a thing)
Oblivious of/to
Occur to (a person)
Officiate for (a person)
Open to (persons/suggestion)
Operate on
Opposed to
Originate in (started in-Place)
Overwhelmed by (a person/a scene/ a calamity)
Overwhelmed with (a feeling)
Owing to (because of)
Part with (to relinquish)
Part from (be separated from)
Partake of
Partial to
Participate in
Pass-see verbal phrases
Patience/Patient with (a person)
Pay for
Peculiar to
Persist in
Pertain to
Pertinent to
Petition for
Pick-see verbal phrases
Pine away (because of grief)
Pine for (desire for)
Plead with (a person) for (a cause)
Plentiful in
Plot against (a person) for (a thing)
Point at (a person)
Ponder on

Popular with (people)
Prefer to
Preference for
Prejudice against
Prepare for
Preside over
Prevail on
Prevent from
Pride-in-take
Proceed with
Proficient in
Prohibit from
Protect from
Protest against
Proud of (a person, achievement)
Provide for (something)
Provide to (someone)
Pull-see verbal phrases
Puzzled over (Something)
Quake with (fear, cold etc.)
Quality for (a post)
Quarrel with/over (a person/a matter)
Quest for
Quick to (understand, deny etc.)
Raid on
Reach for
React to (an action)
Ready for (something)
Reason with (a person) about (a subject) for (an opinion)
Rebel against
Reckon on
Recommend to/for (person/thing)
Reconcile to/with (something/person
Recover from
Reduce to
Refer to(a person/something)
Reflect on
Refrain from (doing something)
Regard for (consideration for)
Rejoice at (other's in (one's own)
Relate to (a person/something)
Relation with

Relieve of (a feeling/a post)
Rely on
Remedy for
Remind of (a happening)
Reply to
Reprimand for
Reproach for
Request for
Rescue from
Resolve on
Resort to
Respect for
Respond to
Responsible to (a person) for (something)
Response from
Restrain from
Retire from/to
Return to/from
Revolt against
Reward with (a thing)
Reward for (a service)
Rich in (have abundance)
Rob of (valuables)
Rude to
Rule over
Run-see verbal phrases
Save from (a danger)
Search for (something)
See-see verbal phrases
Seek for
Send for (call someone)
Sensitive to (criticism or action)
Sentence to
Serious about
Set-see verbal phrases
Share in
Shield from
Shop for (something)
Short of (have scarce)
Show off (make a display of)

Shudder at
Side with (a person)
Sick of (tired of)
Similar to
Sink beneath/under
Smack of
Smile at
Sneer at
Speak of (mention)
Speak on (a topic)
Speak to (a person)
Specialize in
Stamp out (eradicate)
Stand-see verbal phrases
Stare at (a person)
Stay up (remain awake)
Stay at (a place)
Stay with (a person)
Stick to (a principle/job)
Stoop to
Strange to (not known to)
Strip of
Strive for (a great achievement)
Struggle against)
Submit to
Substitute for
Succeed/successful in (something)
Succeed to (a position)
Sue for
Suffer from
Suggest to (a person)
Suitable for
Superior to
Supply to (a person) with (thing)
Supremacy over
Sure of
Susceptible to
Suspected of
Suspicious of/about
Sympathy for (a reason)
Take-see verbal phrases

Talk about/of (discuss)
Talk over (discuss to decide)
Talk to (a person)
Tamper with
Taste for (something)
Testify to
Thank for
Thankful to (person) for (thing)
think about (consider)
Think of (remember)
Threaten with (an action)
Thrive on
Throw at (throw to hit)
Throw away (discard)
Tired of
Tolerance of (others opinion)
Tolerance for (things/people)
Touch at (stop briefly-ship etc.)
Touch on/upon (speak/write briefly on a topic)
Transform into
Triumph over
True to (ideals etc.)
Turn-see verbal phrases
Unaware of
Uncalled for (inappropriate)
Understanding with (a person)
Used to (in the habit of)
Victim of (fraud/circumstances)
Void of
Vouch for
Wanting in (lack something)
Wary of (cautious)
Wedded to (stick to)
Withdraw from
Worthy of
Yearn for
Yield to
Zeal for
Zeast for

Note : 'Un' and 'in' are prefixes for making negative. The same preposition is used for un /in words as for the original ones e.g. unsuitable for.

PRACTICE TEST - 2
PREPOSITIONS

DIRECTION : **Fill in the blanks with appropriate prepositions from the choice given.**

1. Man is entirely different .... other animals.
A) to B) from
C) on D) with

2. We are accountable .... God.
A) to B) from
C) of D) on

3. He is not ashamed .... his conduct.
A) at B) of
C) from D) on

4. Hard work is key .... success.
A) of B) from
C) to D) at

5. He died .... heart attack.
A) of B) from
C) for D) on

6. He is ignorant .... facts.
A) from B) of
C) about D) on

7. His master dispensed .... his services.
A) with B) of
C) on D) from

8. Who is knocking .... the door.
A) on B) of
C) at D) about

9. My friend invited me .... dinner.
A) to B) for
C) with D) from

10. His parents are disgusted .... him.
A) about B) with
C) from D) on

11. The village was destroyed .... fire.
A) by B) with
C) on D) from

12. Do not quarrel .... your neighbours.
A) at B) with
C) on D) of

13. Our parliament is composed .... two houses.
A) of B) from
C) with D) on

14. He has been ill .... three days.
A) since B) for
C) of D) from

15. He has no control .... his children.
A) over B) on
C) about D) at

16. He goes to college .... foot.
A) by B) with
C) on D) at

17. He is blind .... one eye.
A) in B) on
C) with D) to

18. I was alarmed .... the news.
A) by B) at
C) with D) about

19. He was born .... rich parents.
A) of B) to
C) off D) for

20. Health is preferable .... wealth.
A) than B) from
C) to D) with

21. The young cricketer batted .... a long time.
A) for B) at
C) to D) in

22. The young man has been endowed .... many qualities.
A) for B) in
C) with D) at

23. The road is .... repair.
A) by B) in
C) at D) under

24. You must abide .... the rules.
A) in B) by
C) with D) to

25. The two brothers differ .... each other in appearance.
A) from B) with
C) in D) at

ANSWERS

1. b	2. a	3. b	4. c	5. a
6. b	7. a	8. c	9. a	10. b
11. a	12. b	13. a	14. b	15. a
16. c	17. a	18. b	19. a	20. c
21. a	22. c	23. d	24. b	25. a

PRACTICE TEST - 3
PREPOSITIONS

***IRECTIONS* : Fill in the blanks with appropriate prepositions choosing from the given choices.**

1. He looked............another long street.
   a) at b) on c) by d) to
2. I can venture............point out the fact.
   a) at b) on c) to d) by
3. I wanted you to come .... the library.
   a) to b) at c) on d) for
4. She can turn round me any time.
   a) at b) to
   c) in d) between
5. A judge can not diseriminate .... either of the two parties.
   a) between b) at
   c) from d) in
6. You are not bound........... take any quantity.
   a) at b) in
   c) to d) from
7. I can expect nothing better .... him.
   a) at b) from
   c) on d) in
8. I wondered .... her behaviour.
   a) at b) from
   c) on d) in
9. There was laughter .... her expense.
   a) at b) to
   c) on d) in
10. The work must be finished .... the end of the week.
    a) to b) by c) in d) at
11. The thief entered the house .... the owner's absence.
    a) on b) at
    c) in d) under
12. They ceased work .... sun set.
    a) on b) at
    c) in d) from
13. He pointed .... the tree
    a) at b) in
    c) to d) from
14. The regulations apply .... all.
    a) to b) for
    c) at d) in
15. We hope .... the best.
    a) to b) for
    c) at d) in
16. Whom are you looking ....
    a) for b) to
    c) at d) in
17. I am proud ....that inheritance.
    a) to b) of
    c) at d) from
18. People flocked .... him.
    a) for b) to
    c) at d) on
19. If consists .... hundreds of separate scenes.
    a) of b) in
    c) at d) under
20. Send .... a doctor.
    a) for b) from
    c) to d) of
21. You must call .... him today.
    a) from b) on
    c) at d) against
22. I am driving .... this point.
    a) to b) at
    c) on d) against
23. The police defend people .... harm.
    a) at b) from
    c) against d) for
24. You can dispense .... his services.
    a) with b) on
    c) at d) against.
25. Do not laugh......... him.
    a) to b) at
    c) on d) against.

ANSWERS

1. d	2. c	3. a	4. b	5. a
6. c	7. b	8. a	9. a	10. b
11. c	12. b	13. c	14. a	15. b
16. a	17. b	18. b	19. a	20. a
21. b	22. a	23. c	24. a	25. b

III. ARTICLES

The Adjectives 'a', 'an' and 'the' are called Articles. 'A' or 'an' is called the Indefinite Article and 'the' is called the Definite Article.

Example :

This is an apple. There is a book on the table.

In Articles Test, a sentence is given as a question with a blank in it. You have to insert the correct 'article' by choosing from the given alternation.

Example :

1. This is __________ interesting book.

A) a B) an C) the D) No article

Answer : (B) an. Here 'interesting' begins with a vowel sound. So 'an' is the correct answer.

STUDY MATERIAL

USE OF ARTICLES - IMPORTANT RULES

1. 'An' is used before the word beginning with a vowel sound.

Example :

An umbrella, an ant, an apple, an heir, an honest man, an hotel, an historical novel, an M.L.A., an M.P.

2. 'A' is used before the word beginning with a consonant sound.

Example :

a year, a women, a box, a European, a University, a Union, a hero.

3. The Definite Article 'the' is used :

1) before the names of rivers, mountains, oceans.

Example :

The Ganga, the Godavari, the Tungabadra, the Pacific, the Indian Ocean, the Antarctic Ocean, the Himalayas, the Vindhyas.

2) before the name of religious books.

Example :

the Ramayan, the bible, the quaran

3) Before the names of things of unique nature.

Example :

the sun, the moon, the earth, the sky.

4) Before an adjective in the Superlative Degree.

Example :

the cleverest boy, the cleverest girl, the best boy, the best girl.

5) Before a noun which is qualified by who, which, where, why clauses.

Example :

She is the girl who stood first.

He is the boy who jumped yesterday.

6. Before a proper noun when it is used as a common noun.

Example :

Kalidas is the Shakespear of India.

7. Before an adjective when the noun is understood

Example :

the poor

4. No Article is used before proper Nouns, Material Nouns, and Abstract Nouns.

Examples :

1. Rama is a clever boy.
2. Sita is a bad girl.
3. Delhi is the capital of India.
4. Hyderabad is the capital of Andhra Pradesh.
5. Gold is a precious metal.
6. Platinum is a valuable metal.
7. Honesty is the best policy.

PRACTICE TEST - 4
ARTICLES

DIRECTIONS : **Each of the following sentences has a blank space. You have to insert correct article by choosing from the given choices. If no article is required, your choice is (D) (No article)**

1. .... elephant is a wild animal.
   a) a b) an c) the d) No article
2. Ravi got .... appointment.
   a) a b) an c) the d) No article
3. It is .... heroic deed.
   a) a b) an c) the d) No article
4. His father is .... M.P.
   a) a b) an c) the d) No article
5. We must help......... poor.
   a) a b) an c) the d) No article
6. The cow is .... useful animal.
   a) a b) an c) the d) No article
7. He is .... heir to the throne.
   a) a b) an c) the d) No article
8. .... Ganga is a holy river.
   a) a b) an c) the d) No article
9. Let us discuss .... matter seriously.
   a) a b) an c) the d) No article
10. He returned in......... hour.
    a) a b) an c) the d) No article
11. His father is .... university lecturer.
    a) a b) an c) the d) No article
12. I met him .... year ago.
    a) a b) an c) the d) No article
13. He is .... European.
    a) a b) an c) the d) No article
14. Hindi is .... easy language.
    a) a b) an c) the d) No article
15. .... sun rises in the east.
    a) a b) an c) the d) No article
16. Honest men speak .... truth.
    a) a b) an c) the d) No article
17. Sanskrit is .... sacred language.
    a) a b) an c) the d) No article
18. He is .... honourable person.
    a) a b) an c) the d) No article
19. She went out with .... umbrella.
    a) a b) an c) the d) No article
20. Man is .... only animal that uses fire.
    a) a b) an c) the d) No article
21. .... union is strength.
    a) a b) an c) the d) No article
22. .... Duke of wellington was a greatman.
    a) a b) an c) the d) No article
23. .... gold is a precious metal.
    a) a b) an c) the d) No article
24. .... Peacock is a national bird of India.
    a) a b) an c) the d) No article
25. .... Prejudice is the child of ignorance.
    a) a b) an c) the d) No article

ANSWERS

1. b	2. b	3. b	4. b	5. c
6. b	7. b	8. c	9. c	10. b
11. a	12. a	13. a	14. b	15. c
16. c	17. a	18. b	19. b	20. c
21. d	22. c	23. d	24. d	25. d

IV. SENTENCE COMPLETION

In this type of test each question consists of a sentence with a blank in it. You have to fill the blank by choosing the most effective word from the given words to make the sentence meaning fully complete.

Examples **:**

1. His companions prevailed upon him not to .... to violence.
   a) refer b) resort c) provoke d) prone.
   Answer : (b) resort.
2. He .... his shoes till they shone.
   a) brushed b) scrubbed c) polished d) wiped
   Answer : (c) Polished.
3. The .... is working on wood.
   a) artefact b) artiste c) artist d) artisan
   Answer : (d) artisan.
4. The prolonged illness left him totally ....
   a) enervated b) invalid c) indisposed d) healthy
   Answer : (a) enervated.
5. He is .... dancer.
   a) a skilled b) an adept c) an adapt d) an adopt
   Answer : (b) an adept

How to prepare for this test :

As this test is mainly on vocabulary, you have to build up strong vocabulary by reading english news papers with eye on improving vocabulary. While reading the news paper you have to learn the meaning and the usage of difficult words and remember the sentences in which they occured. You have also to learn the meanings and usage of different words of the same origin (Paronyms) and same pronounciation (Homonyms).

Here is a list of such Homonyms and paronyms.

STUDY MATERIAL

Some words have some what similar pronounciations but have different spellings and, of course, different meanings. They are known as Homonyms. Some words have similar roots but different meanings. They are called Paronyms. There are many words which get confused in usage. The following list explains all these words.

Abject (miserable) He is living in abject poverty.

Object (purpose) My object was to change his opinion.

Accede (agree) The principal acceded to our request.

Exceed (go beyond) The cost exceeded the estimate.

Accept (take) She accepted my present proposal.

Except (leaving out) All were present except Sundaram.

Access (approach) I have no access to any Minister.

Excess (more than necessary) Excess of tea affects digestion.

Accessary (Helper in crime) He was an accessary to the murder of his wife committed by his mother.

Accessory (extra useful things) Lamp and pump are the accessories of a bicycle.

Accident (that happens by chance) He died in an accident. His first class was just an accident.

Incident (an event) There are many incidents of terrorism in Punjab.

Adapt (make suitable) Adapt yourself to changing conditions.

Adept (expert) He is an adept on photography.

Adopt (take in one's family as son/daughter) I have adopted my sister's son.
(take) We are adopting Western culture.

Addition (putting more) Addition of lemon made it tasteful.

Edition (number or form of a publication) The first edition of the book was sold in no time.

Admission (being admitted) His admission to the college was cancelled.

Admittance (letting in) He was allowed admittance to the private museum with great difficulty.

Affect (to act-generally harmfully) Smoking affects our health.

Effect (Verb-accomplish) We effected our plans.
(Noun-result) The effect of this medicine is good.

Aggregate (total) The aggregate of my marks is 360.

Aggravate (to worsen) Rise in population aggravates poverty.

Allusion (reference) He made allusions to Upanishads.

Illusion (false notion) She has an illusion that she is the most beautiful girl in the college.

Altar (sacred place of worship) Revolutionaries offered their lives at the altar of the nation.

Alter (change) The plan has been altered.

Alternate (Perform by turns) English lectures are on alternate days.

Alternative (choice) There is no alternative to character.

Amiable (lovable) He is an amiable person liked by all.

Amicable (friendly way) We had an amicable settlement.

Apathy (indifference) The rich have an apathy towards the poor.

Antipathy (hostility) His antipathy towards his wife spoils the children.

Apposite (proper) He made an apposite remark.

Opposite (contrary) Black is the opposite of white.

Appreciable (that can be felt) There is no appreciable difference between the two sets.

Appreciative (feeling of liking) He had an appreciative audience.

Apprehensive (full of fear) She is apprehensive of his actions.

Comprehensive (in details-easily understood) He has written a comprehensive article on AIDS.

Ark (a boat) Ark is an old word for boat.

Arc (a part of a circle) Divide the circle into three arcs.

Arch (a curved structure supporting weight above) We passed under the arch of the bridge.

Artist (who practises fine arts) Lata is a great cine artist.

Artiste (a performer in singing/dancing) The opera had twenty artistes.

Artisan (a craftsman) Village artisans have lost their work due to big mills.

Artistic (have aesthetic qualities) He has an artistic taste.

Artful (cunning) He is an artful rogue of Bombay.

Artificial (Opposed to natural or genuine) People keep artificial plastic flowers in drawing rooms.

Ascetic (one leading an austere life) He has left worldly pleasures and has become an ascetic.

Aesthetic (Having a love for beauty and art) You can't enjoy moonlight with him. He has no aesthetic taste.

Assay (test fineness/purity) He assayed the ornaments.

Essay (a piece of composition) You have written a fine essay.

Assent (concurrence) The President gave his assent to the bill.

Ascent (climbing up) The ascent to this mountain is difficult.

Avert (ward off) A big accident was averted because of the presence of mind of the driver.

Invert (to turn upside down) He inverted the bottle to make it empty.

Aught (anything) For aught they know about his misery.

Ought (should) We ought to help the poor

Bale (big packing) They received 100 bales of cotton.

Bail (security) He was released on bail.

Bare (uncovered) I went to the temple barefooted.

Bear (tolerate) She bears all difficulties.

Beer (a mild drink) Many people take beer not wine.

Beach (sea shore) People go to the beach for a sea bath.

Beech (a tree) Beech trees grow generally on mountains.

Beneficial (useful) Exercise is beneficial to health.

Beneficient (kind and good) Ashok was a beneficient king.

Beside (by the side of) His hut is beside the canal.

Besides (in addition to) Besides my elder son the younger one too is a doctor.

Berth (a sleeping seat in train) He got reservation for a berth.

Birth (coming to life) He took birth on the 3rd of February.

Born (take birth) He was born in August.

Borne (Past Perfect of bear) She has borne many difficulties.

Borrow (take loan) Do not borrow money.

Burrow (hole in earth) Serpents live in burrows.

Bough (branch of tree) Birds perch on boughs.

Bow (bend in respect) Let us bow to the great saints.

Brake (Lever to stop wheel) Brakes of my bicycle are good.

Break (make apart) Do not break the glass.

Bridal (pertaining to bride) We enjoyed the bridal feast.

Bridle (control) Bridle your tongue.
(reins of borse) It is difficult to put bridle on a horse.

Calendar (chart of days-years etc.) The calendar shows that 21st is a Saturday.

Calender (giving finish to cloth) After manufacture cloth is sent for calendering.

Cannon (a big gun) wars were fought with cannons in the past.

Canon (principles, rules) We we are taught canons of Taxation.

Canvas (a rough cloth) We wear canvas shoes in sports.

Canvass (solicit votes) I went for canvassing during elections.

Capital (seat of administration) Madras is the Capital of Tamil Nadu.

Capitol (Building of US Parliament) The Congress meets in the Capitol.

Capture (arrest, conquer) The dacoits were captured.

Captivate (attract) Her charms captivated his heart.

Career (profession) Nehru falled in his legal career.

Carrier (one who carries) Some patients treated in the USA are carriers of AIDS.

Cast (throw) Do not cast pearls before swine.
(give vote) She did not cast her vote in the election.
(group of actors/actresses) This film has a good cast.

Caste (narrow class of society) There are many castes even among backward classes.

Casual (occasional) We enjoy 15 casual leaves.

Causal (relating to cause) There is no causal relation between superstitions and diseases.

Cease (discontinue) He has ceased to be Manager.

Seize (catch) Seize the opportunity when it comes.

Siege (being surrounded) Golden Temple lay siege to the army.

Ceiling (inner roof) The house being old cement falls from the ceiling.

Sealing (mark with a stamp) The sealing of the parcel was done carefully.

Cell (underground room) He kept secret documents in a cell.

Sell (give for a price) We sell only good books.

Censer (a pot to burn incense) Fragrance came out of the censer before the idol.

Censor (an official body to check or to check) News were censored during Emergency.

Censure (disapprove, criticise) The Prime Minister censured the attitude of the opposition.

Ceremonious (formal) His letter was ceremonious having no warmth.

Ceremonial (connected with ceremony) Dasehra is a ceremonial festival.

Cession (yielding) Pakistan was formed by the cession of Punjab, Bengal and Sindh by us.

Session (period of working of a Committee, College etc.) The Congress session lasted for three days.

Check (prevent) It is difficult to check corruption.

Cheque (order of payment on a bank) He gave me a bearer cheque.

Childlike (like a child) J.P. had childlike simplicity.

Childish (immature) Even on becoming a Minister his conduct remained childish.

Cite (quote) He cited many examples from the Bible.

Site (local position) The engineer went to the site of construction.

Sight (view, scene) He has lost his eye-sight.

Coarse (rough) It is a coarse cloth.

Course (way) His course of life is smooth.
(syllabus of a subject) The Physics course is difficult.
(successive dishes served in a hotel) It was a rich dinner of ten courses.

Collision (clashing) 40 persons died in a bus and train collision.

Collusion (secret agreement for an evil purpose) Some government officers were found in collusion with smugglers.

Coma (unconsciousness) He remained in diabetic coma before his death.

Comma (a mark of punctuation) Comma is not used at many places in sentences now.

Comity (friendly recognition of laws of other nations) India has a say in the comity of nations.

Committee (body of persons) he is not a member of the Working Committee.

Complement (that completes) Practicals are the complement of theory.

Compliment (expression of regard) Convey my compliments to your better half.

Complacent (over-self satisfied) We are complacent about our achievements.

Complaisant (obliging) Nirupama has a complaisant nature. She helps her classmates.

Comprehensive (detailed) Thakkar Commission gave a comprehensive report ot the government on the death of Mrs. Indira Gandhi.

Comprehensible (that can be understood fully) This book is comprehensible only to a specialist.

Confidant (one entrusted with secret) Mr. Antulay was the confidant of Mrs. Gandhi.

Confident (sure) I am confident of my success.

Confidential (secret) Confidential files are kept by the officers.

Confirm (ratify-make permanent) He was confirmed after two years.

Conform (comply with) The goods do not conform to the samples shown.

Congenial (agreeable) One feels at home in congenial surroundings.

Congenital (from or before birth) His blindness is congenital.

Conscientious (honest, scrupulous) A conscientious person suffers a lot in this corrupt world.

Consensus (general agreement of opinion) There is a consensus against terrorism.

Considerable (much, sufficient) We have spent a considerable amount on this project.

Considerate (having regard for others) Sheela is very considerate. She helps her friends.

Contagious (spreading by contact) Tuberculosis is a contagious disease.

Contiguous (neighbouring) China has occupied some Indian territory contiguous to her border.

Infectious (water or atmosphere borne) Typhoid is an infectious disease.

Contemptible (deserving contempt) The workers went on strike as the Manager's conduct was contemptible.

Contemptuous (expressing contempt) He dismissed him with a contemptuous gesture.

Continual (frequent) He was continually disturbed in his speech.

Continuous (unceasing) he studied continuously to get a position.

Convenience (comfortably) you may do it at your convenience.

Conveyance (means of travelling) He had to go by bus as he had no conveyance of his own.

Core (innermost part) She loves the child from the core of her heart.

Corps (division of army) He is a member of National Cadet Corps.

Corpse (dead body) The corpse was taken for post mortem.

Corporal (physical) Corporal punishment is banned in schools.

(NCO in army) He is a corporal in the 5th Regiment.

Corporeal (of body) Food and drinks are our corporeal needs. Prayer is a spiritual one.

Corporate (shared by members of a group of persons) We do not lead a corporate life in India.

Council (assembly of elected or nominated members) He is a member of Legislative Council.

Counsel (advice, suggestions) You may keep your counsel to yourself. I am well advised.

Credible (which can be believed) That people can burn men alive is not credible.

Credulous (believe as real) Indians are credulous people believing in the promises of politicians.

Creditable (praiseworthy) Dr. Khurana's achievement in genetic science is creditable.

Dairy (place where cream, butter, milk are produced) Amul Dairy of Gujarat supplies milk to Bombay.

Diary (a personal record of daily events) Diary for the new year will be available in December.

Dam (barrier for water) Many dams were built after independence for irrigation.

Damn (condemn) He is a damned person.

Decease (death) People collected to pray for the soul of the deceased.

Disease (illness) Tuberculosis is a curable disease.

Decry (cry down) The muslim Women's Bill was decried even by Muslim intelligentsia.

Descry (see dimly) We could descry a bird in the sky.

Deduce (infer, conclude) Different people deduce different conclusions from the P.M.'s speech.

Deduct (take away-amout) Rs. 30.00 have been deducted from my salary for P.F. account.

Decent (looking nice) He wears a decent dress.

Descent (coming down) This road from the mountain has a steep descent.

Dissent (disagreement) The dissent of the opposition was overlooked by the government.

Deference (respect) He shows deference to the old.

Difference (dissimilarity) There is a great difference of opinion between the husband and the wife.

Defy (resist, refuse to obey) They defied the orders of the Magistrate.

Deify (worship as a God) some people have deified Dr. Ambedkar.

Delivery (distributing postal articles - giving birth to child) The delivery of letter was late. The delivery of the child was smooth.

Deliverance (making free) Deliverance from poverty should be the aim of the UNO

Dependent (relying on) A sincere man is dependent on none.

Dependant (one who depends on others) He has four dependants to support.

Depositary (a person trustee of a deposit) He is trusted depositary of the secret documents.

Depository (a store house) Strong room is the depository of people's valuables in a bank.

Deprecate (disapprove) The President deprecated the indiscipline in the country.

Depreciate (to under value) the rupee was depreciated too much during the last decade.

Desert (waste sandy land) Sahara is the biggest desert in the world.
(betray) He deserted his friend.

Dessert (Sweet of fruit dish served after dinner) We had ice-cream for dessert in the dinner.

Desist (remain away) It becomes difficult to desist from an evil path.

Resist (oppose) She could not resist his approaches and was ruined.

Destination (the place to reach, goal) Her destination was Cochin.

Destiny (fate) We decide our destiny ourselves.

Device (a plan/implements) The USSR and the USA have developed new nuclear devices.

Devise (to find a plan or method) Japan has devised a new device to be used in robots.

Dew (drops of moisture in the morning) The petals were covered with dew drops.

Due (what one deserves) Give the devil his due.

Differ (disagree) Shanmukham and his wife differ in their views.

Defer (Postpone) The matter was deferred to the next meeting

Disinterested (Unselfish) Subhash rendered disinterested service to the nation.

Uninterested (lacking in interest) He is uninterested in social matters.

Disposal (get rid of, sale) I purchased these motors at a military disposal.

Disposition (arrangement) The disposition of furniture in his room is beautiful.

Divers (several) The problem was solved by divers opinions not by one.

Diverse (different) Their likings are as diverse as Bombay and an interior village.

Dose (quantity taken at a time) The doctor gave me four doses of the medicine.

Doze (Sleep lightly) The teacher asked him to sleep as he was dozing in the class.

Draught (Quantity of liquid drunk at a time) He could not get even a draught of water in the desert.

Drought (want of rain) There was drought in a portion of Andrha Pradesh for three years.

Dual (double) There is a dual control over food supply in India.

Duel (a fight between two persons with pistols in the past to settle a point of honour) The Noble defeated in a duel had to surrender.

Duet (a song sung by two persons) Indian pictures are full of duets of Lata and Rafi.

Dye (colour/to colour) I got my sari dyed in red.

Die (expire) He died of blood pressure.

Economic (adj.) The economic condition of women is not satisfactory.

Economical (frugal, thrifty, inexpensive) Costly fans are ultimately more economical.

Economy (control and management of money) The economy of backward countries depends upon foreign help.

Economics (science of consumption, production etc.) Many students offer Economics at M.A.

Effective (resulting in great effect) The medicine was very effective.

Efficacious (sure to have desired effect) The methods adopted to check population are not efficacious. (Generally used of medicines and treatment)

Egoist (one who believes in systematic selfishness) He is an egoist and won't help you unless he is benefitted by helping you.

Egotist (one who talks too much about oneself-self conceited) Indian politicians and bureaucrats are great egotists.

Elemental (connected with natural elements) It is still beyond human ability to control elemental problems like flood and storms.

Elementary (introductory) Elementary education is free and compulsory in the USA.

Elicit (draw out ideas or opinions of others) The police elicited many secrets about terrorists from the culprit.

Illicit (illegal) He has illicit relations with a girl working under him. Illicit distillation of wine is a common feature of villages in Saurashtra.

Eligible (having required qualifications) One is eligible for an officer's post only after he is 21.

Illegible (which cannot be read easily) Gandhi's handwriting was illegible.

Elusive (baffling, not clear) His interrogation was not of much use as he gave elusive answers.

Illusive (deceptive, false) The concept of ghosts is quite illusive.

Emerge (come out suddenly) A black figure emerged out of the darkness.

Immerge (to immerse oneself) She immerged herself in the river.

Eminent (distinguished) Sathyamurti was an eminent leader of the South.

Imminent (approaching) We are facing imminent danger of further partition of the country.

Envelop (verb-to surround) The hotel was enveloped in fire.

Envelope (Noun-a cover) He sent the letter in an envelope.

Enviable (qualities arousing envy) Dr. Raman was an enviable scientist.

Envious (feeling envy) He is envious of the progress of even his wife.

Eruption (bursting out) More than 2,000 persons died in a volcanic eruption in North-West Cameroon on 25th August, 1986.

Irruption (invasion, abrupt increase) Population irruption has become a problem in third world countries.

Errant (erring, wandering) There were many knights-errant in the Middle Ages. She has an errant husband.

Arrant (notorious downright) He is an arrant liar. I can't believe him.

Errand (N. Short journey to bring message, goods etc.) She has appointed an errand boy to run her errands.

Erratic (irregular, not predicable)Her erratic nature created problems for the son and the daughter-in-law.

Erroneous (mistaken, incorrect) We have an erroneous impression about the culture of the West.

Especial (pre-eminent, exceptional) It was an especial case that a law student rose to be the Chief Minister.

Special (of a particular kind, peculiar, not general) Special favour was shown to the candidate by a member of the interview committee.

Estimate (approximate valuation) An estimate was submitted by the engineer about the cost of construction.

Estimation (esteem, respect, judgement) Bismil was held in high estimation by all revolutionaries.

Exceptional (unusal, rare) Bhagat Singh had exceptional courage in him.

Exceptionable (objectionable) He was not allowed to speak further by the speaker in the assembly as he passed exceptionable remarks.

Exhausting (which tires) A labourer's work is an exhausting one.

Exhaustive (comprehensive, detailed) he has prepared an exhaustive report on water pollution.

Expedient (advisable though against principle) It was quite expedient on her part not to work under such a corrupt man.

Expeditious (speedy, quick) The bureaucracy is far from taking an expeditious action even in the most urgent cases.

Exposure (exposed to danger-air, light, cold etc.) Over exposure of the film has spoiled the photograph.

Exposition (commentary, explanation) Shankaracharya's exposition of the Upanishads is remarkable.

Facility (ease, aids to do things easily) Many facilities have been provided to the backward people.

Felicity (intense happiness) Only a self-satisfied person can enjoy felicity.

Fain (gladly) I would fain have your company.

Feign (pretend) He feigned illness to take leave.

Faint (become unconscious) He fainted and fell.

Feint (sham attack) We made a feint against the enemy as a strategy to attack from the other side.

Fair (A gathering for fun) We went to the Dasehra fair.
(right) It is a fair deal.
(white complexioned) She is a fair girl.

Fare (money paid for a journey) What is the railway fare from Madras to Bangalore?

Fatal (deadly) He received a fatal wound and died.

Fatalist (one who believes in fate) Many fatalists starve as they don't work.

Fateful (important-good or bad) World War was a fateful event in the World.

Feat (surprising trick, deed of valour) It was a feat for Mrs. Gandhi to come to power again. The rope walker showed many feats.

Feet (plural of foot) Wear socks on your feet.

Festal (of a feast, keeping holiday) Marriage season is a festal season in India.

Festive (joyous, jovial) Diwali is a festive occasion.

Fir (a tree) Fir trees are found in the North.

Fur (soft hair of animals) She wore a fur hat.

Flagrant (scandalous) He showed a flagrant violation of moral conduct.

Fragrant (sweet smelling) The flowers made the atmosphere fragrant.

Forceful (Full of force or impetus) He made a forceful appeal to the people.

Forcible (by force or compulsion) Police made a forcible entry to the den of dacoits.

Forego (go before) for applied grammar study the foregone chapters. It is a foregone fact.

Forgo (go without) A good man forgoes his pleasurers to help the poor.

Formerly (Previously) He was formerly a lecture. Then he became Principal.

Fromally (connected with outward show, not sincere) He met him formally.

Foul (dirty, unfair) He adopted foul means to become rich. Foul smell.

Fowl (old word for a bird) One can see many species of fowls in a zoo.

Funeral (pertaining to ceremonies after death) The body was put on a funeral pyre.

Funereal (gloomy, dismal) He had funereal looks on the day of his dismissal.

Gage (old word for security or guarantee) He offered a gold bangle as gage for the money borrowed.

Gauge (a standard measure) The distance between the rails of metre gauge trains is one metre.

Gaze (look fixedly) He gazed at the girl.

Gait (manner of walking-used for tender women) Her gait was graceful

Gate (outer door) Temples and mosques have big gates to enter.

Gaol (old spellings of Jail) Many juveniles remain in Indian gaols without trail for years.

Goal (aim) Many youngmen have no goal in their lives.

Gamble (play game for high stakes or money) Yudhishthira gambled away his whole kingdom.

Gambol (skip about in play, frisk) Thy young deer gambolled in the park.

Gild (apply thin cover of gold) Now-a-days it is advisable to go out wearing gilded ornaments not real gold ones.

Guild (society of persons for mutual help in profession or welfare work) Artisans organised themselves in guilds in Europe in the past.

Graceful (having grace of beauty, movement or expression) Ashok Kumar remained a graceful character even in his old age.

Gracious (showing kindness, courtesy) Jamshedji was a gracious industrialist

Hail (frozen rain drops) Hailstorm has ruined the crop.
(greet, give a welcome cry) Churchill was hailed as the saviour of England in the II World War.

Hale (strong and healthy) My father is hale and hearty even in his seventies.

Hair (thread like growth on skin) Comb your hair.

Hare (a fast running small animal) Hare is larger than a rabbit.

Heir (one having legal right to inherit property) Ram was heir to the throne of Ayodhya.

Hart (adult male of deer) Hart is male deer known as stag too.

Heart (blood distributing organ) He died of heart failure. Heart represents emotions, mind intellect.

Healthful (which promotes health, wholesome) Fruits are healthful.

Healthy (enjoying good health) Because he is quite healthy he will be selected in the army.

Historic (notable or memorable in history) 1857 was a historic year in India.

Historical (belonging to history) Partition of India is a historical fact.

Hoard (to store for undue profit Sometimes certain things are not available as they are hoarded by traders.

Horde (a group of people) A horde of people attacked the police post.

Honorary (holding office without payment) He was elected honorary Secretary of the club.

Honourable (won by of honour) Subramanyam was an honourable person

Human (pertaining to mankind) It is in human nature to envy others.

Humane (tender, kind-hearted) He is a humane officer. He helps the poor.

Imaginary (unreal) He has imaginary fear.

Imaginative (having the faculty of imagination) Poets and Artists are imaginative people.

Impassive (without expression on the face) He remained impassive even when he was sentenced to death.

Impulsive (acting on impulse) Rash people have to pay for their impulsive nature.

Imperial (connected with empire) Imperial rule in India ended in 1947.

Imperious (dominating) She behaves in an imperious manner in the family.

Incite (rouse, stir up) Insult incites resentment.

Insight (Understanding, power of seeing through mind) Scientists have great insight in the working of nature.

Inconstant (changing) He failed as he is inconstant in his studies.

Inconsistent (contradictory, not in harmony) His actions are inconsistent with what he preaches.

Indite (compose a poem/speech) He indited a poem full of emotions.

Indict (to accuse) He was indicted on a charge of terrorising people.

Industrious (laborious) Being industrious Mr. Patel has created a national market for Nirma.

Industrial (Pertaining to industries) Bangalore is known for industrial growth.

Ingenious (clever and skilful) He is an ingenious carpenter.

Ingenuous (frank, simple) Jaiprakash Narain was an ingenuous person liked by all.

Intelligent (wise) He is so intelligent that he topped in the examination.

Intelligible (which can be easily followed) His lecture is not intelligible to the weak students.

Intensive (deep) Intensive study is required to go to the depth of a matter.

Extensive (wide) Extensive study sometimes does not help in an examination.

Jealous (having envy and ill will) he is jealous of his friend's position.

Zealous (enthusiastic) She is zealous about her research work.

Judicious (wise) Be judicious in choosing a life partner.

Judicial (relating to judge) A judicial enquiry was ordered in the Ahmedabad riots.

Knotty (full of complications) It is a knotty problem.

Naughty (mischievous) She is a naughty girl.

Later (afterwards in time) I reached there. Later my sister too came.

Latter (the second one) Susamma and Sunitha came to my house. The latter gave me a present.

Lightening (decreasing) He appealed for lightening the burden on landless labourers.

Lightning (flash in the cloud) Four people were killed by the lightning during rains.

Look (see attentively) The student looked at the black board.

See (simply perceive) I saw many men in the street.

Loose (relaxed) The screw was loose.

Lose (to have no more) Do not lose your books and friends

Loudly (in a loud voice) He spoke so loudly in the library that all were disturbed.

Aloud (in audible voice) She spoke aloud so that all in the hall could hear her.

Lovable (worthy of love) He has many lovable qualities.

Lovely (Charming) The baby has lovely eyes.

Luxuriant (rich in growth) The fields are luxuriant with green grass after the rains.

Luxurious (full of luxury) The rich lead a luxurious life.

Main (chief) Terrorism was the main issue in the conference.

Mane (long hair on an animal's neck) My horse has a thick mane.

Maize (Indian corn) Maize is the main crop in many areas in India.

Maze (labyrinth) The poorman's case was lost in the maze of red tapism.

Mantel (shelf over a fire place) He placed a portrait of Vivekanand on the mantel piece.

Mantle (a cloak) Burqa worn by Muslim ladies is like a mantle.

Mental (of mind) He does too much of mental work.

Marshal (a high army rank) There is no Marshal in Indian army.

Martial (Pertaining to war) Marathas were a martial race in the past.

Medal (a small flat piece of metal given as an award etc.) The General won many medals for his bravery.

Meddle (interfere) It is wise not to meddle in the affairs of others.

Memorable (worth remembering) 1857 is a memorable year in the history of India.

Memorial (something built in the memory of) India Gate, in New Delhi is the War memorial for the men who died in the I World War.

Metal (a mineral substance like iron, gold etc.) Gold is a precious metal.

Mettle (quality of endurance and courage) Indian soldiers showed their mettle in Indo-Pak War.

Meter (a measuring apparatus) Electric meter records the consumption of electricity.

Metre (Unit of length) This cloth is five metre in length. (unit of rhythm in a poem) This poem is in tetrametre.

Miner (one who works in a mine) The life of miners in Bihar is miserable.

Minor (below the age of 18) A guardian has been appointed for him as he is a minor. (of less importance) Over population is not a minor problem in India.

Moat (a ditch surrounding a fort) There is a moat round the Red Fort of Delhi.

Mote (a particle of dust) Sometimes even a mote makes us blind.

Momentary (lasting for a short time) Although the flash of the atom bomb was momentary it destroyed the whole city of Hiroshima.

Momentous (of great importance) The election of Mr. Gorbachev as head of the communist party of Russia was a mementous step towards rejuvenation of the party.

Necessaries (thing necessary to lead life) Food, cloth and residence are the necessaries of life.

Necessities (urgent needs) A car is a necessity for a good physician.

Negligent (one who is careless) He is negligent in his studies.

Negligible (an unimportant thing or matter that may be disregarded) The age factor is quite negligible in their case as they have a love affair.

Oar (a rod for rowing) The boat sank as the oar slipped from the hand of the boatsman.

Ore (mineral from which metal is extracted) Copper is extracted from the ore.

Obscure (not well-known) Ramulu is an obscure writer. No criticism is available on his writings.

Vague (not clear) He gave vague replies in the interview and was not selected.

Official (adj. of office) He went to the secretariat for some offical work.

Officious (eager or ready to help) People generally seek advice from him as he is very officious.

Opportunity (a chance/ or occasion) We should always avail the opportunities.

Opportunism (to adapt to circumstances to advance ones interests without regard to principles) Opportunism has become the creed of modern politicians.

Ordinance (a law enacted by the government outside the Parliament or Assembly) The president issued a number of ordinances before the budget session.

Ordnance (all military weapons together and servicing) He serves in the Central Ordnance Depot.

Pair (two together) Give me a pair of socks.

Pare (to trim, to divide into two parts) Pare your nails.

Patrol (to go round to keep a watch) The police has been patrolling the riot torn area.

Petrol (Liquid used to move automobiles) There is no petrol in my car.

Peal (loud sound or continuous sound) He burst into a peal of laughter.

Peel (to remove skin) She peeled the bananas.

Persecute (to harass) The Black are persecuted in South Africa.

Prosecute (to pursue) I am prosecuting my studies.
(to bring to court by police) He was prosecuted on a charge of theft.

Physic (medicine) No physic can cure him.

Physique (body structure) He has a sound physique.

Plain (simple) it is the plain truth about him. She is not a plain girl.

Plane (surface) This room is at a higher plane.

Plan (scheme) he draw another plan for his business.

Popular (well-known good) He is a popular figure in the city.

Populous (thickly populated) Calcutta is the most populous city in India.

Practicable (which can be turned into action) His ideas are not practicable.

Practical (connected with reality, useful) He has practical experience of automobiles.

Practice (N) His practice is better than that of other doctors.

Practise (V) Practise what you preach.

Pray (offer prayer) Pray for mercy.

Prey (animal, bird hunted for food) The hawk ate its prey.
(greatly troubled) He fell a prey to anxiety.

Precede (go before) One precedes two, three follows it.

Proceed (continue) He proceeded with the project.

Prefer (choose rather) I would prefer coffee to tea.

Proffer (old usage for offer of intangible things) Mr. Srinivasan proffered me his friendship.

Prescribe (to advise a medicine) The doctor prescribed him a long treatment.
(set down for studies) this book is prescribed by the university.

Proscribe (to forbid or prohibit) British Government prescribed many books on patriotic themes in India.

Primary (basic, chief, of the beginning) The child was admitted to the primary school. The issue is of primary importance.

Primitive (of the earliest times) The primitive man lived in the jungles with other animals.

Principal (chief, main, head) The Principal has not yet come to the college. This is the Principal clause.

Principle (basic truth, guiding rule for behaviour) He believes in certain moral principles. The principles of Economics change according to socio-economic set-up.

Prophecy (Noun-prediction forecast) The astrologer made a thrilling prophecy about the death of a leader.

Prophesy (Verb-to predict) None can prophesy about the future of the world under the shadow of atom bombs.

Proportional (determined by proportion) Some communities want proportional representation in government jobs.

Proportionate (corresponding in degree or amount) The salary was not proportionate to the labour he did.

Provident (having foresight, providing for future needs) He proved provident in running the organisation/family.

Providential (by divine providence) he had a providential escape as he was not in the car which was fired at.

Prudent (cautious of his interest, having sound judgement in practical matters)
He is very prudent and will rush the family to great success.

Prudential (relating to actions of prudence) His prudential schemes will save the business even in this recession.

Raise (lift up) She raised the child from the ground.

Raze (to destroy totally) The building was razed to ground by bulldozers.

Rain (water from the clouds) Heavy rain has resulted in floods in many areas.

Rein (bridle, control) the reins of the government have gone loose in India.

Reign (rule) There was a reign of terror in South Africa.

Rap (sound made on the door by a knocker) She rapped the door loudly to awake the people.

Wrap (to enclose) The box was wrapped in a piece of cloth.

Reduce (make less) The remuneration was reduced by Rs.5/-.

Deduce (to infer by reasoning) Newton deduced that all bodies are attracted towards the centre of gravitation.

Regretful (full of regret-a person) He was regretful of his wrong actions.

Regrettable (some action/thing causing regret) It is regrettable that you have not left smoking.

Rest (peace, comfort) As he was tired he took rest for an hour.

Wrest (snatch away) He wrested her necklace and ran away.

Right (proper, correct) He is right in consuming only the things made in India.

Rite (ceremonial things) There are many rites among Hindus. The last one is funeral rite.

Write (compose) Please write a letter to my mother.

Roll (form in cylindrical shape) She rolled the paper.

Role (part) She plays an important role in the drama.

Rout (flight) We routed the enemies.

Route (way) He went to Kanyakumari enroute Madras.

Sanguinary (eager for bloodshed) the terrorists have a sanguinary nature. They don't hesitate in killing any one.

Sanguine (hopeful, cheerful) His sanguine nature has made him popular among his friends.

Sensual (having weakness for sex and food) Let him not mix among girls. He is a sensual boy.

Sensuous (appreciative of beauty) Keats was a sensuous poet.

Serge (a worsted cloth) He has a beautiful suit of serge.

Surge (swelling motion of water) The ship tossed on the surging waves of the sea.
(verb) The water surged out of the crack in the dam.

Servility (too much submitting behaviour) Many bureaucrats lead a life of servility under politicians.

Servitude (almost slavish life) The life of bonded labourers is nothing less than servitude

Several (an indefinite but small number) Several friends came to see the patient.

Numerous (consisting of many persons or things) Museum of Hyderabad is a collection of numerous ancient articles. He takes a week to see them.

Shear (clip the wool of sheep) Hex is not a barber but shears sheep.

Sheer (utter, absolute) The working of nationalised banks is an example of sheer incompetence.

Social (pertaining to society) Running free classes for the poor is a social activity.

Sociable (having a tendency to mix with others) The doctor has a good practice as he is very sociable.

Sole (bottom surface of shoes) The sole of this shoe is of hard rubber.
(the only, exclusive) M/s. Steelsons ar the sole distributors of Tata Steel in this region.

Soul (spirit) The soul is immortal. It is a painting without soul.

Sooth (old word for truth) Soothsayers in ancient Rome predicted many things correctly.

Soothe (to calm, assuage) Religious songs soothe disturbed minds.

Sore (giving physical pain) He has a sore throat

Soar (to fly high) The glider soared high in the sky. Soaring prices disturb the family budget. A poet soars in the heights of imagination.

Sour (having sharp acid taste) Green guavas are generally sour. No one would talk to you if you keep your mood sour.

Spacious (having enough space) He had four spacious rooms in his house.

Specious (seeming good, logical etc. Without really being so) His specious arguments convince people at least for the time being.

Spiritual (connected with soul) We must try to have not only material but spiritual progress too.

spirituous (having distilled alcohol) Most of the alcoholic drinks are spirituous not fermental ones.

Stationary (not movable) Earth is not stationary as was believed in the past.

Stationery (writing material) I bought this pen from a stationery shop.

Statue (an image in stone etc.) There is the statue of Mahatma Gandhi near India Gate in New Delhi.

Statute (a law of Rules and Regulations of an organisation, Industry etc.) The students are admitted according to the statutes of the University. All the laws passed by the Parliament are noted in the statute book.

Stimulant (something that stimulates) Coffee is a common stimulant taken in the South.

Stimulus (that rouses to action, incentive) A man requires stimulus to do something important.

Story (a tale) My grandma told me the story of 'Sleeping Beauty'.

Storey (floor of a building) He lives on the third storey.

Straight (not curved or circular) A triangle has three straight lines. Parliament Street is a straight road between Parliament House and Connaught Place.

Strait (narrow passage, narow water between two seas) Palk strait joins Bay of Bengal with Arab Sea.

Suit (a set of clothes or an action in a law court) She has a beautiful suit of 'Kurta and Salwar'. He has filed a suit against his landlord. (V. agree) European culture does not suit India.

Suite (a set of rooms) He stayed in a two roomed suite in the hotel.

Superficial (shallow) His knowledge of Malayalam is rather superficial. He can't go deep.

Suprerfluous (more than needed) The stock in his shop is superfluous. Most of it will remain after Diwali shopping.

Superstitious (believing in unscientific things out of ignorance) He won't go as a cat has crossed the road. He is very superstitious.

Supecilious (arogant, haughty) He has no friends due to his supercilious nature.

Symmetry (similarity of form or arrangement, proportion) The four sides of Taj mahal have a perfect symmetry.

Cemetry (graveyard) Her grave is in the centre of the cemetry where here brothers were burried.

Team (a group specially of players, working together) Indian Cricket team won the rubber. There is no team work in the Public Sector organisations.

Teem (full of, abound in) The tree is teeming with flowers.

Temperament (mental disposition) He has an even temperament.

Temperance (sobriety, moderation) Temperance is necessary to maintain good health and sweet relations.

Temporal (of this world not spiritual) People make efforts generally for temporal gains.

Temporary (not permanent) He has been offered a temporary job.

Tenor (general course or tendency) His tenor of life was peaceful.

Tenure (the act, right or period holding a post/office or property) The underground drainage was completed during his tenure as Mayor.

Toe (forepart of foot, socks, shoes) He kicked the ball with his toe.

Tow (to pull or drag with a rope) We had to tow the car behind a truck.

Tolerable (a thing that can be tolerated) The food at the railway station was just tolerable.

Tolerant (a person who tolerates) He won't mind you. He has a tolerant nature.

Transient (which lasts or stays for a short time) It is a transient sorrow. He will be normal soon.

Transitory (which by its nature passes or ends sooner or late) Life is transitory. We are passing through a transitory period.

Umpire (a referee) The Pakistani Cricket Team made a complaint about the umpire.

Empire (Dominion) British Empire was the biggest one before II World War

Unity (oneness in spirit, aims, feelings etc. of diverse individuals) National unity is necessary to be strong in the world.

Union (state of being united into a single organisation for a common purpose) The union of bank clerks is the most powerful one in India.

Unision (harmony in music) They sang in unison with the musical instruments.

Vacation (long holidays) They went on a historical tour during summer vacation.

Vocation (any trade, profession, occupation) He has taken automobile servicing as his vocation.

Avocation (something done for pleasure, as hobby) Dancing and painting are her regular avocations.

Vain (useless) A vain attempt was made on the life of the Chief Priest of Golden Temple.

Vein (blood vessel taking blood to the heart) He was given dextrose through vein.

Van (a closed truck or wagon or railway car) The household things were carried in a van on his transfer.

Vane (weather cock) Many churches have a vane over the dome.

Wane (to decrease after reaching the peak) His fame has been waning rapidly.

Vale (poetic word for valley) The vale and the fields are green.

Wail (to express grief by loud cries) A group of women came wailing to the house of the deceased.

Veracity (truthfulness) The veracity of his statement is yet to be ascertained.

Voracity (greed, eagerness, insatiable desire) His voracity for reading keeps him locked in the library.

Verbal (by means of words) The writer drew a verbal image.

Verbose (using more words than needed) He writes in a verbose style.

Venal (characterised by corruption and bribery) it is common to see venal bureaucrats making venal bargains.

Venial (pardonable) Although they have crossed a certain limit yet being sincere in love it is a venial affair.

Virtual (in effect) His Secretary is the virtual proprietor of the firm.

Virtuous (Possessing moral virtues, chaste) She is a virtuous girl and won't take up such an evil profession.

Waive (forgo) The age limit was waived for backward classes.

Wave (to move to and fro) He waved the handkerchief to give him a send off. They waved the flags.

Weather (general condition or atmosphere at a certain time) The weather was cool in the morning.

Whether (if) I wonder whether he is coming tomorrow.

Climate (permanent atmospheric conditions) Tropical climate would suit him.

(prevailing condition) The political climate of the country is not satisfactory.

Willing (ready to do) He is willing to serve you.

Wilful (deliberate) Wilful negligence is not pardonable.

Wither (fade away) The flower has withered.

Whither (old word for where) Whither go the old morals and virtues?

Wreak (to inflict) Terrorists have wreaked vengeance upon the Sinhalese army.

Wreck (destruction) The ship was wrecked to pieces during the storm.

Yarn (any fibre spun for cloth) Nylon yarn is mixed with cotton yarn for gay colours.

Yearn (to be filled with longing or desire) She yearns to see her sons married before she dies.

Yoke (mark of bondage) We are no more under the yoke of slavery.

Yolk (the yellow portion of an egg) Yolk is the most nourishing part of an egg.

PRACTICE TEST - 5
SENTENCE COMPLETION

DIRECTIONS : **Each of the following sentences has a blank space and four words are given after the sentence. Select whichever word you consider most appropriate for the blank space and indicate your choice.**

1. The young man lost his way in the forest and found that he had become a/an ....
   a) enemy b) adversary c) decoy d) quarry
2. The firm ...... its workers well.
   a) treated b) regarded c) behaved d) cared
3. If negotiations are to prove fruitful, there must not only be sincerity on each side, but there must also be ..... in the sincerity of the other side.
   a) certainty b) substance c) faith d) belief
4. Jail officials ..... that the plot for the escape may have been hatched by local criminals.
   a) doubt b) suspect c) hold d) believe
5. Known as devout and serious person, she also has a ..... sense of humour.
   a) good b) better c) quick d) beautiful
6. Judicial decrees may not change the heart, but they can ..... the heartless.
   a) subdue b) regulate c) transform d) disarm
7. Untouchability is still in vogue and atrocities are being ..... against poor people.
   a) perpetuated b) transmitted
   c) executed d) perpetrated
8. The ..... is working on wood.
   a) artefact b) artiste c) artist d) artisan
9. The .... arts are those that pertain to writing or drawing.
   a) cartographic b) graphic
   c) elocutionery d) theatrical
10. I tried to give him some good advice but he ..... to listen.
    a) prevented b) avoided c) refused d) denied
11. Due to stringent financial conditions of the firm, several workers have been declared ....
    a) extra b) abundant
    c) redundant d) superfluous
12. One of the time-tested ways of remembering a series of items is known as a .... device.
    a) intellectual b) schematic
    c) mnemonic d) ingenious
13. When the bus was at fall speed, its brakes failed and an accident was ......
    a) inevitable b) infallibie
    c) essential d) undeniable
14. Barring last minute surprise he is .... to become the President of the Committee.
    a) designed b) poised c) elected d) endorsed
15. He sold his property because he was under a lot of ....
    a) account b) debt c) loan d) credit
16. My boss ....my request to increase my allowance.
    a) refused b) denied c) resisted d) withheld
17. The accused .... having made any statement.
    a) declined b) refused c) denied d) rejected
18. The problems suggested by style as a sign and index of personality may be ..... from many points of view.
    a) approached b) encountered
    c) confronted d) upheld
19. The Committee's .... to establish a 'Home For The Destitute' has been rejected.
    a) argument b) decision c) view d) proposal
20. Some people in Punjab .... from voting at the last elections.
    a) refused b) refrained c) denied d) abstained
21. He is a person of sound character and .... disposition.
    a) amicable b) amiable c) morous d) beneficent
22. To break the stalemate over the controversial issue, the Prime Minister held discussions today with four other leaders to .... a consensus.
    a) emerge b) win c) evolve d) capture
23. Walking at three o' clock, I heard the .... of thunder.
    a) clank b) crackle c) rumble d) ripple
24. Pratibha had a old and couldn't go to the party, so I bought her a cake to make up for her ....
    a) dissillusion b) disappointment
    c) disgust d) depression
25. What's so special about it is that the author has tried to .... all the worth while information in this book into a few pages.
    a) compact b) abstract c) contract d) condense

ANSWERS

1. d	2. a	3. c	4. b	5. a
6. a	7. d	8. d	9. b	10. c
11. d	12. c	13. a	14. b	15. b
16. a	17. c	18. a	19. d	20. b
21. b	22. c	23. c	24. b	25. d

PRACTICE TEST - 6
SENTENCE COMPLETION

DIRECTIONS : **Each of the following sentences has a blank space and four words are given after the sentence. Select whichever word you consider most appropriate for the blank space and indicate your choice.**

1. The boy made good .... in his studies.
   a) progress b) development
   c) growth d) advancement
2. It was .... that a mind so pure and searching could miss the truth.
   a) likely b) unlikely
   c) possibly d) scarcely
3. The punch made the boxer .... with pain.
   a) grumble b) gape
   c) wince d) fumble
4. He tried to .... himself against a horde of ruffians.
   a) save b) collect
   c) defend d) support
5. It is equally important to .... one's goal, as it is to set one.
   a) obtain b) realise c) establish d) prove
6. The poor fellow can't even afford daily food, ... alone buying a house.
   a) cast b) forget c) let d) leave
7. it is a rugged, restless and uncertain existence that the .... lead.
   a) travellers b) nomads c) sailors d) tourists
8. There is a .... possibility of an understanding between India and Pakistan on the nuclear issue.
   a) high b) distinct c) sound d) clear
9. The higher class students should be careful as regards discipline as the youngsters tend to .... them.
   a) mimic b) ape c) simulate d) emulate
10. The prolonged illness left him totally ....
    a) enervated b) invalid
    c) indisposed d) healthy
11. The .... arguments put forth for not disclosing the facts did not impress anybody.
    a) convincing b) specious
    c) intemperate d) spurious
12. Tension is likely to .... in the Kashmir valley now.
    a) decline b) drop c) ease d) fall
13. The degrees were awarded in the annual ....
    a) conference b) convention
    c) convolution d) convocation
14. He .... all his energies to bear upon the task.
    a) brought b) devoted
    c) put d) organised
15. A cup of water was enough to .... his thirst.
    a) satisfy b) appease
    c) quench d) extinguish
16. The .... of the Minister's statement cannot be verified by people who have no access to official records.
    a) validity b) veracity
    c) ambiguity d) verbosity
17. The lost child was .... to his parents.
    a) restored b) sent
    c) dispatched d) returned
18. If something is beyond the .... of human knowledge man can know nothing about it.
    a) edge b) view c) end d) boundary
19. The liquid seems .... but I am sure that it will prove poisonous.
    a) innocuous b) insane
    c) incongruent d) fragile
20. Handle with care, it is ....
    a) agile b) volatile
    c) ductile d) Inflammable
21. The annual .... of industrial products has risen enormously in recent years.
    a) outlay b) outcome
    c) output d) outbreak
22. Prospective candidates have to prove that they have sufficient money to cover their .... and tuition.
    a) boarding b) lodging
    c) sustenance d) maintenance
23. We had a ....of warm weather in March.
    a) period b) phase
    c) time d) spell
24. He is .... dancer.
    a) a skilled b) an adept
    c) an adapt d) an adopt
25. Her parents will never give their .... to such an unsuitable match.,
    a) willingness b) agreement
    c) consent d) acquiescence

ANSWERS

1. a	2. b	3. c	4. c	5. b
6. d	7. b	8. b	9. d	10. a
11. b	12. c	13. d	14. a	15. c
16. b	17. a	18. d	19. a	20. d
21. c	22. a	23. d	24. a	25. c

PRACTICE TEST - 7
SENTENCE COMPLETION

DIRECTIONS : **Each of the following sentences has a blank space and four words are given after the sentence. Select whichever word you consider most appropriate for the blank space and indicate your choice.**

1. The bill in the Parliament was ...... by fifty five votes.
   a) accepted b) voted c) carried d) opposed
2. The principal and staff have made .... efforts to enable the students to attend college on the days of the bus strike.
   a) integrated b) deliberate
   c) concerted d) systematic
3. We .... the family members after expressing our grief at the tragedy.
   a) condoled b) satisfied
   c) mourned d) consoled
4. The defending champion .... to victory in just 30 minutes.
   a) led b) rushed c) reached d) cruised
5. Inspite of our best efforts, we failed to .... any new facts from him.
   a) elicit b) evoke c) eject d) enlist
6. A number of advances in medicine would have been .... sooner if free enquiry had been common.
   a) secured b) persisted
   c) achieved d) propagated
7. His fired speeches .... the people to cry for justice and achieve it by their united efforts.
   a) prompted b) inflamed
   c) instigated d) inspired
8. This book is a useful .... to our library.
   a) arrival b) discovery c) asset d) addition
9. The interior of the concert hall is .... to the eve.
   a) delicious b) veritable c) visual d) hopeless
10. No one likes to be forced into .... situations.
    a) humble b) insulting
    c) humiliating d) infuriating
11. He admired precision in everything, but it never hampered his quick ....
    a) finalisation b) dealing c) action d) decision
12. My friend took his first pay to the bank to .... it in his account.
    a) deposit b) deficit c) debit d) demote
13. When Satish..... that he was guilty, he was pardoned.
    a) claimed b) boasted c) confessed d) confided
14. Anita .... me of a girl I used to know.
    a) remembers b) recalls c) recollects d) reminds
15. The steam engine was .... by James Watt.
    a) discovered b) manufactured
    c) created d) invented
16. His most striking .... is the enthusiasm which he brings to everything he does.
    a) character b) factor c) characteristic d) attitude
17. Cholera is mainly .... through contaminated food and drinks.
    a) conducted b) transferred
    c) circulated d) transmitted
18. Authority .... when it is not supported by the moral purity of its user.
    a) waits b)crumbles c) empowers d) prevails
19. When the courtier had advanced to the highest position attainable, his friends felt jealous about his having reached the ....
    a) vigil b) precipice c) threshold d) pinnacle
20. He was not willing to accept the ....
    a) wrong b) blame c) fault d) sorry
21. The Government's economic policy includes certain projects for .... the living conditions of the poor.
    a) bettering b) harmonising
    c) manipulating d) doing away with
22. The reward is a .... of her service to mankind.
    a) momento b) memorial c) recognition d) witness
23. They have .... up the deficit:
    a) covered b) filled c) made d) pulled
24. A restaurant that presents spick and span look will naturally .... more customers.
    a) accomodate b) invite c) avoid d) attract
25. Though they did not speak to each other much, there was a .... understanding between them.
    a) placid b) contractual c) tacit d) verbal

ANSWERS

1. c	2. c	3. d	4. d	5. a
6. c	7. d	8. d	9. b	10. c
11. d	12. a	13. c	14. d	15. d
16. c	17. d	18. b	19. d	20. b
21. a	22. c	23. c	24. d	25. c

3. SYNONYMS

Synonyms are words having similar meanings. In the synonyms Test, One word is given as the question accompanied by four or five responses. You have to choose the correct word from the given responses which best expresses the meaning of the given word.

Example :

1. IMPECCABLE

A) Faultless B) Fussy C) precise D) Having integrity

Answer : (A) 'Impeccable' means 'faultless'. So the correct answer is (A).

STUDY MATERIAL

Learn the following synonyms so that you can get good marks in your examination.

Word	Synonyms
Abandon	Leave, forsake
Abbreviate	Abridge, Shorten
Ability	Skill, Talent
Abnormal	Irregular, Unusual, Unnatural
Abode	Dwelling, Residence, lodging
Abortive	Fruitless, Futile
Abolish	Finish, Annul, Nullify
Abstain	Refrain, Desist
Absurd	Silly, Ridiculous, Irrational
Abundant	Plentiful, Ample, Copious
Accelerate	Quicken, Hasten
Accept	Agree, Accord, Consent, Assent
Accomplish	Achieve, Perform
Accord	Harmony, Agreement
Accuse	Blame, Charge, Indict
Actual	Real, Genuine
Advantage	Profit, Gain
Adoration	Worship, Admiration Reverence
Adversary	Opponent
Adversity	Calamity, Misfortune, Distress
Advice	Counsel, Instruction
Advocate	Plead, Pleader
Affiction	Distress, Sorrow
Affront	Insult, Outrage, Indignity
Aggravate	Heighten, Intensify
Aggression	Attack, Invasion, Assault
Aid	Help, Succour, Relief
Aim	Purpose, End
Air	Wind
Alacrity	Alertness, Promptitude, Readiness
Alarm	Danger
Allay	Calm, Soothe, Asuae
Allow	Permit, Grant
Alteration	Change, Modification.
Ameliorate	Improve, Better
Amiable	Lovable , Charming
Amusement	Recreation ,Diversion ,Entertainment
Annual	yeariy
Answer	Reply ,Response
Antagonistic	Hostile ,Adverse,Inimical
Antipathy	Dislike , Hostility
Antithesis	Contrast, Opposition
Ascend	Rise, Soar,Climb
Astonish	Wonder,Surprise,Amaze
Audacious	Bold,Daring
Augment	Increase Enlarge
Authentic	Genuine,True
Autocrat	Despot, Tyrant
Avaricious	Greedy, Covetous
Aversion	Dislike, Antipathy
Awkward	Clumsy, Ungainly
Baby	Child, Infant
Bad	Worthless, Evil, Foul, Rotten
Barbarous	Uncivilized, Wild, Inhuman
Base	Low, Mean, Ignoble
Beast	Animal, Brute
Beautiful	Charming, Lovely, comely, Fair
Begin	Start, Initiate, Commence
Beneficent	Generous, Liberal, Kindly
Bias	Prejudice, Partiality

Blend - Mix, Mingle
Bliss - Happiness, Felicity, Joy
Boisterous - Noisy, Stormy, Clamorous
Bondage - Slavery, Servitude, Subjection
Brave - Courageous, Bold, Daring
Brawl - Fight, Quarrel, Squabble
Breach - Rift, Violence
Brevity - shortness, Conciseness
Brief - Concise, Short, Abridged
Bright - Shining, Glistening, Sparkling, Luminous
Brisk - Vigorous, Animated, Sprightly
Brutal - Savage, Cruel
Build - Make, Construct, Erect
Burden - Load
Burlesque - Farce, Caricature
Callous - Hard, Cruel, Indifferent
Calumny - Defamation, Slander
Cause - Reason, Base, Source
Censure - Blame, Condemn, Criticize, Chide, Upbraid
Charm - Magic, Fascinate, Hypnotize, Bewitch, Enchant
Check - Restrain, Stop, Curb
Cheerful - Happy, Gay, Pleased
Circumscribe - Restrict, Limit, Confine
Cite - Mention, Quote
Clamour - Outcry, Din, Uproar
Clandestine - Secret, Stealthy Surreptitious
Clemency - Mercy, Pity
Coalition - Alliance, Combination
Coerce - Compel, Force
Colossal - Huge, Enormous, Gigantic
Command - Order, Bid
Commodious - Roomy, Spacious
Companion - Friend, Associate
Colleague - Pal, Comrade, Churn
Compassion - Pity, Sympathy
Compatible - Consistent, Harmonious, Congruous
Compel - Coerce, Pressurize, Force
Competition - Rivalry
Complete - Whole, Entire, Full
Comprehend - Understand, Grasp, Follow
Conciliate - Reconcile, Appease, Pacify
Concise - Short, Brief, Abridged
Condense - Compress, Abridge, Thicken, Harden
Confer - Grant, Bestow
Confess - Admit, Avow
Confuse - Confound, Perplex
Conscientious - Scrupulous, Incorruptible
Conscious - Aware, Percipient
Consider - Think, Brood, Weigh, Ponder
Conspicuous - Noticeable, Manifest
Constancy - Steadliness, Steadiastness
Contemplate - Meditate, Muse
Contemptible - Despicable, Mean
Contemptuous - Scornful, Disdainful, Hateful
Contrary - Opposite, Counter
Convert - Change, Transform
Cordial - Warm, Friendly
Correct - Right, Amend, Accurate
Corrupt - Debased, Tainted, Depraved
Costly - Dear Expensive
Counterfeit - False, Forged, Sham
Couple - Pair, Brace
Courteous - Polite, Civil
Covert - Hidden, Secret
Crafty - Cunning, Sly
Credible - Believable, Probable
Creditable - Praiseworthy, Honourable
Cruel - Savage, Fierce, Hard
Cursory - Rapid, Superficial
Damage - Injury, Loss, Harm
Danger - Risk, Peril, Hazard
Death - Lifeless, Decease, Demise
Deadly - Fatal, Destructive, Lethal
Dearth - Scarcity, Lack, Want
Decay - Decline, Wither, Fade
Decide - Settle, Determine
Decline - Refuse
Decorate - Adorn, Embellish, Ornament
Decry - Disparage, Denounce
Deference - Respect, Reverence
Deficient - Lacking, Inadequate, Wanting
Deformity - Malformation, Disfigurement
Degeneration - Deterioration, Demoralisation
Deleterious - Harmful, Injurious, Noxious
Delicate - Tender, Subtle
Delight - Pleasure, Joy
Delusion - Illusion
Deny - Refuse
Deplorable - Regrettable, Lamentable
Desecrate - Pollute, Profane

Despair - Wish, Want, Need
Desolate - Lonely, Deserted
Despis - Dejection, Despondency
Despicable - Contemptible, Hateful, Disdainful
Despise - Scorn, Disdain, Hate
Destitute - Wanting, Indigent, Lacking
Devil - Demon, Satan, Fiend
Dexterity - Skill, Adroitness
Dexterous - Skiful, Clever
Diffidence - Shyness, Modesty
Diligence - Industry, Perseverance
Dim - Faint, Dull, Misty
Disclose - Reveal, Divulge, uncover
Disease - Illness
Disgrace - Dishonour, Discredit, Shame
Disinterested - Fair, Impartial, Unbiased
Dislike - Hate, Aversion, Depressing
Dismal - Gloomy, Cheerless, Depressing
Dispute - Quarrel, Controversy
Disreputable - Disgraceful, Discreditable, Infamous
Disseminate - Spread, Scatter, Propagate
Disseminate - Spread, Scatter, Propagate
Distinct - Clear, Explicit
Distress - Discomfort, Trouble
Diverse - Different, Variable
Divide - Part, Separate
Divine - Godlike, Heavenly, Celestial
Divine - Predict, Guess
Dizzy - Giddy, Dazed, Unsteady
Dwindle - Decrease, Shrink, Diminish
Earnest - Serious, Ardent
Eat - Dine, Feed, Devour
Eccentric - Odd, Singular, Peculiar
Efficient - Able, Competent, Capable
Effort - Attempt, Endeavour
Elegant - Refined, Tasteful, Careful
Eligible - Qualified, Suitable
Emancipate - Free, Liberate, Deliver, Release
Embellish - Adorn, Beautify
Emphasize - Stress - Accentuate
Endure - Suffer, Tolerate
Enemy - Foe, Adversary
Energy - Force, Vigour, Power
Enormous - Huge, Immense
Enrage - Infuriate, Incense
Error - Mistake, Blunder
Eternal - timeless, Perpetual
Ever - Always, Perpetually
Exceptional - Unusual, Rare
Exquisite - Fine, Refined
Extensive - Large, Broad, Wide
Extraordinary - Exception, Uncommon, Abnormal
Extravagance - Lavishness, Prodigality
Exult - Rejoice, Jubilate
Fable - Story, Myth, Fiction
Fabricate - Forge, Construct
Fabulous - Fictitious, Mythical
Fair - Beautiful, Handsome, Pretty
Fallacious - Deceptive, False
False - Mendacious, Spurious, Untrue
Famous - Celebrated, Renowned
Fanciful - Imaginative, Fantastic Visionary, Queer
Farcical - Ludicrous, Laughable
Fastidious - Particular, Squeamish, Finical
Fecund - Productive, Fertile
Feeble - Weak, Infirm
Felicity - Bliss, Joy, Happiness
Ferocious - Fierce, Savage, Barbarous
Fictitious - False, Untrue
Fiendish - Devilish, Diabolical
Final - Last, Ultimate
Flashy - Showy, Goddy
Flimsy - Slight, Trivial, Thin
Foolish - Silly, Stupid
Forest - Jungle, Woods
Frailty - Weakness, Failing, Foible
Frank - Candid, Open
Fraud - Deceit, Trickery
Funeral - Dismal, Gloomy
Gaiety - Joyousness, Hilarity, Joility
Garrulous - Talkative Loquacious
General - Universal, Common
Gentle - Tender, Mild, Kind
Ghastly - Spectral, Horrible
Gigantic - Colossal, Huge, Stupendous
Giggle - Laugh, Titter
Gorgeous - Splendid, Superb Magnificent
Grandeur - Splendour, Majesty
Graphic - Vivid, Pictorial
Grasping - Avaricious, Greedy

Gratification - Enjoyment, Satisfaction, Indulgence
Gratutious - Voluntary, Spontaneous, Unwarranted
Grave - Sober - Serious
Grief - Sorrow, Pain, Agony
Grievous - Painful, Hurtful
Grisly - Hideos, Frightful
Grotesque - Odd, Bizzare, Unnatural
Guard - Defend, Protect
Guerdon - Reward, Prize
Guile - Fraud, Trickery
Habit - Custom, Practice
Hamper - Impede, Hinder
Hasty - Rash, Hurried, Impatient
Haughty - Arrogant, Proud
Hazardous - Dangerous, Perilous
Headstrong - Obstinate, Stubborn
Hear - Listen, Hark
Heave - Life, Raise
Heterogenious - Dissimilar
Hideous - Repulsive, ugly
Hinder - Obstruct, Prevent
Hindrance - Impediment, Obstruction
Holy - Sacred, Godly, Pious
Homogeneous - Uniform, Similar
Humane - Kind, Benevolent
Humble - Meek, Modest
Humility - Humbleness, Modesty
Idle - Lazy, Indolent
Ignominy - Shame, Disgrace
Illegal - Unlawful, Lawless
Immaterial - Incorporeal, Unimportant
Imminent - Impending
Imperious - Authoritative, Dictatorial
Implacable - Unrelenting, Unappeasable
Imposter - Cheat, Charlatan
Improvident - Prodigal, Careless
Impudent - Impertinent, Insolent
Inadvertent - Careless, Negligent, Unintentional
Inanimate - Lifeless, Dead
Incessant - Unceasing, Continue
Include - Comprise, Contain
Increase - Enlarge, Augment
Indignant - Angry, Wrathful
Indiscretion - Rashness, Error
Ingenious - Inventive, Clever
Ingenuity - Inventiveness, Skill
Injure - Harm, Hurt, Wound
Intelligent - Clever, Nimble
Intemperate - Uncontrolled, Unrestrained
Intention - Purpose, Motive
Irresolute - Undecided, Wavering
Irritation - Vexation, Annoyance
Jealous - Envious
Jeer - Mock, Sneer
Jolly - Jovial, Merry
Joy - Delight, Pleasure
Jubilant - Exultant, Joyful
Judicious - Discrete, Prudent
Just - Fair, Impartial, Lawful
Juvenile - Adolescent, Minor, Youngster
Keen - Sharp, Eager, Ardent
Keeper - Custodian, Warden
Keepsake - Memento, Souvenir
Kind - Benevolent, Benign
King - Monarch, Ruler
Knavery - Fraud, Villainy
Knack - Art, Dexterity
Laborious - Industrious, Assiduous
Labour - Work, Toil
Laconic - Brief, Concise, short
Lament - Mourn, Grieve
Latent - Dormant, Hidden
Lax - Loose, Negligent, Careless
Laxity - Slackness, Looseness
Lazy - Idle, Slow, Sluggish
Leap - Jump, Hop, Spring
Learning - Knowledge, Scholarship, Erudition
Lecherous - Lustful, Lacivious
Legend - Myth, Tradition
Legendary - Mythical, Traditional
Lenient - Mild, forbearing, Gentle, Tolerant
Lethal - Deadly, Fatal
Lethargy - Apathy, Sluggishness
Liberal - Generous, Tolerant
Liberty - Freedom, Independence
Lie - Falsehood, Untruth, Guile
Likeness - Similarity, Resemblance
Lively - Animated, Active
Loathe - Detest, Abhor
Loquacious - Talkative, Garrulous

Loth - Unwilling, Reluctant
Loyal - Faithful, Devoted
Lucky - Fortunate, Auspicious
Lucrative - Profitable, Remunerative
Luxuriant - Abundant, Profuse
Mad - Insane, Lunatic
Magnanimous - Generous, Largehearted
Magnificent - Splendid, Grand, Stately
Magnify - Exaggerate, Glorify
Malady - Illness, Ailment, Disease
Malice - Spite, Ill-Will
Malign - Defame, Vilify
Manifest - Exhibit, Show
Manifest - (Adj) - Obvious, Evident
Marvel - Miracle, Wonder
Marvellous - Wonderful, Astonishing
Massive - Heavy , Huge
Meagre - Small, Thin
Mean - Low, Abject
Meal - Food, (Dinner, Supper, Lunch)
Melancholy - Gloom Sadness
Mendacious - False, Untruthful
Mercenary - Hired, Venal
Merciful - Pitiful, Compassionate
Might - Strength, Power
Miserly - Stingy, Closefisted, Niggardly
Misery - Sorrow, Distress, Suffering, Tribulation
Mock - Jeer, Ridicule
Mockery - Ridicule, Derision
Modest - Bashful, Unassuming
Morbid - Sick, Diseased
Morose - Gloomy, Melancholic
Mourn - Lament, Grieve, Bewail
Mournful - Sorrowful, Sad
Myth - Fable, Legend
Naive - Artless, Simple
Nasty - Filthy, Dirty
Neat - Clean, Tidy, Orderly
Need - Necessity, Requirement
Negligent - Careless, Heedless
Nice - Pleasant, Agreeable
Niggardly - Miserly
Noble - Lofty, Illustrious
Nonplus - Puzzle, Confound
Norm - Standard, Principle
Notable - Remarkable, Memorable
Notorious - Infamous, Disreputable
Novice - Beginner, Learner
Nuptials - Wedding, Marriage
Obdurate - Stubborn, Obstinate
Obscene - Indecent, Filthy
Observe - Watch, Remark
Obsolete - Antiquated, Out of date
Obstacle - Hindrances, Obstruction
Obstinate - Stubborn, Obdurate
Obtrude - Intrude, Encroach
Obvious - Clear, Evident
Occurrence - Event, Happening
Odd - Strange, Peculiar
Onerous - Heavy, Burdensome
Opportune - Timely, Seasonable
Opportunity - Occasion, Opening, Chance
Opporbrium - Criticism, Censure
Option - Choice, Discretion
Opulence - Wealth, Riches
Ornament - Adornment, Jewel
Ostensibly - Apparently, Seemingly
Ostentation - Show, Display
Palatable - Tasty, Delicious
Paradise - Heaven
Pardon - Forgive, Excuse
Pathetic - Touching, Moving
Paucity - Dearth, Shortage
Pay - Salary, Wage
Pensive - Thoughtful Contemplative
Perennial - Perpetual, Permanent
Permit - Allow, Grant
Persuade - Induce, Urge
Pious - Holy, Devout, Godly, Religious
Pithy - Short, Terse
Plentiful - Abundant, Copious
Pompous - Inflated, Ostentatious
Ponderous - Heavy, Weighty
Portray - Draw, sketch
Praise (V) - Appalaud, Eulogize
Praise (N) - Applause, Eulogy
Precarious - Risky, Uncertain
Precise - Exact, Accurate
Premature - Hasty, Untimely
Pretence - Pretext, Excuse
Pretty - Attractive, Comely, Beautiful
Prodigal - Extravagant, Improvident

Word	Meaning
Proficient	Adept, Expert
Profuse	Lavish, Extravagant
Progeny	Offspring, Descendant
Project	Plan, Scheme
Propagate	Breed, Circulate
Proprietor	Owner, Master
Proscribe	Ban, Banish
Protect	Defend, Shield
Proverb	Adage, Saying
Prudent	Discrete, Wise, Sagacious
Quack	Charlatan, Mountebank
Quaint	Odd, Singular
Quake	Shake, Quiver, Tremble
Quantum	Amount, Share
Queer	Strange, Eccentric
Quell	Suppress, subdue
Questionable	Disputable, Doubtful
Quest	Search, Pursuit
Quit	Leave, Depart
Quiver	Tremble, shiver
Radiant	Bright, Brilliant
Rare	Scarce, Infrequent
Rebellion	Revolt, Mutiny
Rebuff	Rebuke, Reprimand
Recover	Regain, Retrieve
Refined	Polished, Elegant
Rejoice	Exult, Delight
Relate	Tell, Narrate
Remember	Recollect, Recall
Remorse	Regret, Repentance
Remote	Far, Distant
Renown	Fame, Reputation
Reserved	Reticent
Residue	Remainder, Remnant
Resistance	Opposition, Hindrance
Respect	Esteem, Regard
Restraint	Restriction, Curb
Revenge	Vengeance, Retribution
Reverence	Veneration, Honour, Respect
Ridicule	Jeer, Deride
Ridiculous	Absurd, Laughable
Righteous	Just, Godly
Rigid	Stiff, Unyielding
Rigour	Rigidity, Austerity
Rude	Impolite, Discourteous
Ruin	Destruction, Downfall

Word	Meaning
Ruinous	Destructive, Injurious
Sacred	Holy, Consecrated
Sacrilege	Impiety, Profanity
Sad	Dejected, Gloomy
Salubrious	Healthy, Wholesome
Sane	Sensible, Rational
Savage	Wild, barbarous, Cruel
Scandal	Infamy, Slander
Scanty	Slender, Meagre
Scold	Chide, Rubuke, Reprove, Reprimand, Censure
Secret	Hidden, Mystery
Sense	Meaning, Import
Serious	Grave, Earnest
Serpent	Snake
Severe	Stern, Strict
Shapely	Graceful, Elegant
Shrewd	Sagacious, Penetrating
Shy	Bashful, Coy
Significant	Expressive Important
Silent	Quiet
Simple	Plain, Natural
Slander	Defamation, Calumny
Slender	Scant, Slight
Sloth	Indolence, Lethargy
Sly	Crafty, Cunning
Solictious	Apprehensive, Anxious
Solitary	Single, Lonely
Sordid	Mean, Dirty
Sombre	Gloomy, Dark
Special	Particular, Distinctive
Specimen	Sample, Model
Splendid	Magnificent, Grand
Spurious	Sham, False
Squander	Waste, Dissipate
Stagnant	Still, Motionless
Static	Fixed, Quiescent
Stationary	Motionless, Fixed
Sterile	Barren, Unproductive
Stiff	Rigid, Stern
Stricture	Criticism, Censure
Strive	Try, Endeavour
Stubborn	Obstinate, Obdurate
Submission	Surrender, Obedience
Subservient	Subject, Subordinate
Sufficient	Enough, Ample, Adequate
Superfluous	Unnecessary, Needless

Word		Synonyms
Superficial	-	shallow, Cursory
Surmise	-	Conjecture, Guess
Surplus	-	Excess, Glut
Sympathy	-	Compassion, Fellow feeling
Synonymous	-	Equivalent, Identical
Tacit	-	Silent, Unspoken, Implied
Taciturn	-	Silent, Reserved
Talkative	-	Garrulous, Loquatious
Tame	-	Gentle, Mild
Teach	-	Instruct, Educate
Tedious	-	Tiresome, Wearisome, Monotonous
Teeming	-	Replete, Abounding
Temperate	-	Moderate
Temporal	-	worldly, secular
Tender	-	Soft, Delicate
Terrestrial	-	Earthly, Mundane
Thankful	-	Grateful, Obliged
Theatrical	-	Dramatic, Showy
Thin	-	Slim, slender, Lean
Thoughtful	-	Contemplative, Pensive
Thrifty	-	Economical, Provident, Frugal
Thrive	-	Prosper, Flourish
Timid	-	Shy, Diffident
Tortuous	-	Devious, Circuitous
Tough	-	Hard, Strong
Tradition	-	Custom, Convention
Tragic	-	Sorrowful, Distressing
Transient	-	Fleeting, Transitory, Temporary
Transparent	-	Clear, Lucid
Trouble	-	Worry
Truism	-	Platitude, Commonplace
Trusty	-	Trustworthy , Reliable
Tyranny	-	Despotism,Autocracy
Tyrant	-	Despot,Autocrat
Ubiquitous	-	Omnipresent
Ugly	-	Unsightly,Repulsive
Ultimate	-	Final,Last
Umbrage	-	Offence
Underhand	-	Dishonest,Clandestine
Undergo	-	Suffer,Endure
Unique	-	Unequalled, Incomparable
Urge	-	Press, Incite
Useful	-	Advantageous, Serviceable, Beneficial
Usual	-	Common, Ordinary
Vacant	-	Empty, Void
Vacation	-	Holiday
Vacillate	-	Waver, Flater, fluctuate
Vagrant	-	Vagabond, Roving
Vague	-	Indistinct, Obscure, Blurred
Vain	-	Fruitless, Conceited
Valiant	-	Brave, Gallant, Daring
Vanquish	-	Conquer, Defeat
Vehemence	-	Passion, Force
Venerable	-	Reverend, Respectable
Venom	-	Poison
Vertical	-	Perpendicular
Vexation	-	Annoyance, Irritation
Vigilant	-	Watchful, Wary
Vigilance	-	Watchfulness, Wariness
Vindicate	-	Defend, Justify
Vindictive	-	Revengeful, Vengeful, Rancorous
Vow	-	Promise, Swear, Pledge
Vulgar	-	Coarse, Crude, Indecent
Wane	-	Decrease, Ebb, Fade
Warlike	-	Martial, Soldierly
Wax	-	Increase, Grow
Weary	-	Tired
Wealthy	-	Rich, Affluent
Wholesome	-	Healthy, Sound
Wicked	-	Evil, Vicious, Depraved
Wild	-	Savage, Rough
Winsome	-	Charming, Attractive
Withstand	-	Resist, Oppose
Wizard	-	Magician
Woeful	-	Sad, Sorrowful
Wonderful	-	Astonishing, Amazing
Workable	-	Practicable
Wrathful	-	Furious, Indignant
Wretched	-	Unfortunate, Miserable
Wreck	-	Ruin, Destroy
Yearn	-	Crave, Hanker
Yield	-	Surrender, Give, Submit, Produce
Yielding	-	Submissive, Supple
Yoke	-	Oppression, Bondage
Zeal	-	Passion, Fervour, Enthusiasm
Zealot	-	Bigot, Fanatic
Zealous	-	Fervent, Enthusiastic
Zenith	-	Top, Summit
Zest	-	Gusto, Relish, Enthusiasm

PRACTICE TEST - 1
SYNONYMS

DIRECTIONS : **In each of the following questions, out of the given alternatives, choose the one which is the nearest in meaning to the given word.**

1. JUVENILE
   a) Hidden b) Young c) Humorous d) Tedious
2. VENTURE
   a) Entrust b) Risk
   c) Travel d) Adventure
3. IMPERTINENT
   a) Impudent b) Sceptical
   c) Thoughtless d) Irritable
4. IMPEDIMENT
   a) Passage b) Obstruction
   c) Fateful d) Destruction
5. LURID
   a) Sensational b) Old
   c) Obscene d) Pale
6. MASTERPIECE
   a) Sample b) Wonder
   c) Miracle d) Model
7. ALERT
   a) Watchful b) Intelligent
   c) Observant d) Energetic
8. ARTICULATE
   a) Distinct b) Aggressive
   c) Wayword d) Confused
9. VEER
   a) Shift b) Dominate
   c) Delegate d) Concentrate
10. SUPERCILIOUS
    a) Indifferent b) Annoyed
    c) Haughty d) Angry
11. RECALL
    a) Speak b) Receive
    c) Face d) Remember
12. JEOPARDY
    a) Magic b) Adventure
    c) Enmity d) Danger
13. ECONOMISE
    a) Accumulate b) Minimise
    c) Save d) Reduce
14. INDICT
    a) Accuse b) Condemn
    c) Allege d) Reprimand
15. SUPERANNUATED
    a) Accepted b) Senile
    c) Retired d) Experienced
16. ENDORSEMENT
    a) Reprimand b) Censure
    c) Commendation d) Reproach
17. STAMINA
    a) Endurance b) Intensity
    c) Eagerness d) Energy
18. BRASH
    a) Invective b) Rude
    c) Abusive d) Superficial
19. DECRY
    a) Deviate b) Despoil
    c) Demure d) Disparage
20. MOLLIFY
    a) Chastise b) Testify
    c) Rebuke d) Appease
21. DEMIT
    a) Promote b) Judge
    c) Appoint d) Resign
22. BERSERK
    a) Clever b) Morose
    c) Cheerful d) Wild
23. GROTESQUE
    a) Fantastic b) Exaggerated
    c) Colourful d) Pompous
24. COUNSEL
    a) Correct b) Publish
    c) Oppose d) Advise
25. SUCCULENT
    a) Fluid b) Sucking
    c) Juicy d) Sweet

ANSWERS

1. b	2. b	3. a	4. b	5. a
6. d	7. a	8. a	9. a	10. c
11. d	12. d	13. c	14. a	15. c
16. c	17. d	18. b	19. d	20. d
21. d	22. d	23. a	24. d	25. c

PRACTICE TEST - 2
SYNONYMS

DIRECTIONS : **In each of the following questions, out of the given alternatives, choose the one which is the nearest in meaning to the given word.**

1. NUMINOUS
   a) Dull b) Awful c) Bright d) Dark
2. STUPID
   a) silly b) Disobedient
   c) Insane d) Incapable
3. BUCOLIC
   a) Helpless b) Intoxicated
   c) Shameless d) Rustic
4. GROGGY
   a) Unsteady b) Sensible
   c) Sensitive d) Pensive
5. VIE
   a) Hesitate b) Compete
   c) Refuse d) Shirk
6. CONSOLE
   a) Sympathise b) Share
   c) Pacify d) Assist
7. GENESIS
   a) Relevant b) Beginning
   c) Style d) Movement
8. INDOLENCE
   a) Leniency b) Relaxation
   c) Laziness d) Stagnation
9. ACCENTUATED
   a) Projected b) Exhibited
   c) Sharpened d) Mitigated
10. DISRUPT
    a) Deny b) Refuse c) Break d) Join
11. SUMPTUOUS
    a) Lavish b) Fancy
    c) Meagre d) Irritable
12. LATENT
    a) Prompt b) Lethargic
    c) Apparent d) concealed
13. TESTIFY
    a) Witness b) Describe
    c) Recognise d) Appreciate
14. ASCEND
    a) Leap b) Mount c) Deviate d) Grow
15. OSTRACIZE
    a) Cut off b) Expedite
    c) Welcome d) Include
16. FORLORN
    a) Uncared b) Funny
    c) Fearful d) Fortunate
17. RILE
    a) Drag b) Agree
    c) Please d) Annoy
18. LAZE
    a) Abdicate b) Rest
    c) Waste d) Wander
19. UGLY
    a) Fearful b) Hateful
    c) Evilsome d) Repulsive
20. EFFETE
    a) Affected b) Injured
    c) Feeble d) Sick
21. SLOPPY
    a) Drunk b) Careless
    c) Tired d) Untrustworthy
22. HARBINGER
    a) Pilot b) Forerunner
    c) Steward d) Messenger
23. SULTRY
    a) Unpleasant b) Ill-smelling
    c) Impure d) Hot & oppressive
24. HEDGE
    a) Curse b) Injure
    c) Evade d) Harm
25. WAN
    a) Tired b) Strong
    c) Cheerful d) Unhappy

ANSWERS

1. b	2. a	3. d	4. a	5. b
6. c	7. b	8. c	9. a	10. c
11. a	12. d	13. a	14. b	15. a
16. d	17. d	18. b	19. b	20. c
21. a	22. d	23. d	24. c	25. a

PRACTICE TEST - 3
SYNONYMS

DIRECTIONS : **In each of the following questions, out of the given alternatives, choose the one which best epressess the meaning of the italicised bold part of the sentence.**

1. The **bounties** of nature are being exploited by man.
   a) gifts b) products
   c) rules d) ecological balances
2. There is no **efficacious** remedy to unemployment problem in a country which has no ideological convictions.
   a) suitable b) possible
   c) effective d) proper
3. A bone got stuck in his **gullet.**
   a) stomach b) molars c) chest d) throat
4. India is still a **virgin** land exploring its tourism potential.
   a) undone b) unexplored
   c) uncouth d) unexploited
5. The need of the hour is to initiate the **renaissance** of moral conscience and halt the rampant corruption, for it could even hasten the end of our civilisation.
   a) rise b) introduction
   c) revival d) significance
6. I have told him **many** times not to do that.
   a) several b) unlimited
   c) endless d) numberless e) infinite
7. Everyone tries to avoid his company because he is in the habit of using too many **jibes.**
   a) jokes b) abuses
   c) criticism d) taunts
8. The story is too fantastic to be **credible.**
   a) praiseworthy b) readable
   c) believable d) false
9. A government rocked by frequent scandals is an **effete** set of people with no credibility.
   a) innocent b) stupid
   c) excited d) exhausted
10. She does not get her work done punctually and properly because she is always **dawdling.**
    a) confused b) playing
    c) loitering d) chit-chatting
11. The little **coy** bride was admired by everyone.
    a) beautiful b) shy c) talented d) sweet
12. The small boy was able to give a **graphic** description of the thief.
    a) vague b) broad
    c) vivid d) drawing e) picture.
13. **Sporadic** rise in his temperature has caused us much worry.
    a) Frequent b) Irksome
    c) Irregular d) Scattered
14. Ant is believed to be the most **industrious** creature.
    a) sensible b) diligent
    c) successful d) punctual
15. The tour was cancelled on account of **incessant** rain.
    a) constant b) heavy
    c) intermittent d) unexpected
16. I **disdain** those who tell lies.
    a) condemn b) scorn
    c) hate d) despise
17. The **benevolence** of the God of Rain has seen a very successful monsoon this year too.
    a) vision b) morbidity
    c) kindness d) ill-will
18. The country's resources should be **judiciously** used.
    a) sparingly b) reasonably
    c) adequately d) immensely
19. Methods of spreading the family planning message have been as **fecund** as they have been abortive.
    a) fruitful b) failing c) peculiar d) false
20. He was **dismissed** from service because they found him dishonest.
    a) prevented b) removed
    c) retired d) stopped
21. If you **waver** at the last minute, you shall face failure.
    a) resolve b) decide c) settle d) hesitate
22. Reading of poetry is not **congenial** to his taste.
    a) suited b) beneficial
    c) helpful d) preferable
23. Gandhiji always advocated the cause of **indigenous** industries.
    a) foreign b) big c) cottage d) native
24. Kamini could not **recollect** the incident that had happened in her childhood.
    a) remind b) recall
    c) memorise d) assemble
25. He was warned at the **outset** of his career.
    a) end b) beginning
    c) middle d) entrance

ANSWERS

1. a	2. c	3. d	4. b	5. c
6. a	7. d	8. c	9. d	10. c
11. b	12. c	13. c	14. b	15. a
16. c	17. c	18. b	19. a	20. b
21. d	22. a	23. d	24. b	25. b

PRACTICE TEST - 4
SYNONYMS

DIRECTIONS : **In each of the following questions, out of the given alternatives, choose the one which best epresses the meaning of the italicised bold part of the sentence.**

1. What is the **ultimate** goal of life ?
   a) fixed b) final c) desired d) immediate
2. Everyone who has worked for him **hammers** home that point because they feel that it is widely unappreciated.
   a) hints b) stresses c) strikes d) directs
3. We must **eradicate** corruption.
   a) minimise b) control c) condemn d) uproot
4. During his stay in Canada, he had quite a few **frustrating** experiences.
   a) disappointing b) disenchanting c) humiliating d) repulsive
5. I tried to persuade him to change his mind, but he remained **adamant.**
   a) foolish b) aggressive c) stubborn d) defensive
6. The weavers have to do **monotonous** work.
   a) repititive b) exhausting c) irksome d) autonomous
7. There was always an **imminent** danger of the falling of that damaged roof.
   a) constant b) impending c) remote d) favourable
8. He is very **discreet** while deciding such matters.
   a) obstinate b) cunning c) prudent d) trustworthy
9. Science has **revealed** the mysteries of nature.to man.
   a) released b) disclosed c) opened d) cleared
10. It is a **scandal** that the murderer was declared innocent.
    a) rumour b) silly notion c) talk d) disgraceful action
11. Ever since the death of his mother, my friend is utterly **forlorn** and wretched.
    a) frustrated b) lonely c) fearful d) gloomy
12. The **rancous** shouts off the audience forced the management to call off the show.
    a) unpleasant b) loud c) harsh d) harmful
13. He was **admonished** for the delay.
    a) dismissed b) admired c) censured d) advised
14. All the streets looked **just** the same.
    a) decisively b) normally c) exactly d) simply
15. At the Hardy house there was great **consternation** when Aunt Gertrude saw Joe helping Frank up the stairs.
    a) surprise b) panic c) gaity d) anxiety
16. He is a **candid** politician.
    a) frank b) faithful c) fearless d) soft spoken
17. The purpose of his speech was to **lampoon** the leading politicians of the day.
    a) praise b) expose c) ridicule d) defame
18. **Sagacity** increases with age.
    a) Wisdom b) Love c) Maturity d) Efficiency
19. She has an **insatiable** love for music.
    a) undesirable b) irreconcilable c) unchanging d) unsatisfiable
20. I was shocked to see my friend lying at the **edge** of the road.
    a) side b) bank c) corner d) border
21. After his graduation, he had to pass through a period of **privation.**
    a) hardship b) uncertainty c) prosperity d) privacy
22. Now the **fury** of the demonstrators turned against the machines.
    a) asperity b) passion c) rage d) acrimony
23. His commands were so **peremptory** that we felt humiliated.
    a) specious b) poignant c) dictatorial d) rough
24. He would have been his close **associate** had he not been disloyal.
    a) employee b) competitor c) colleague d) executive
25. Every body laughed at the puerility of his statement.
    a) inaccuracy b) childishness c) insincerity d) falsehood

ANSWERS

1. b	2. b	3. d	4. a	5. c
6. a	7. b	8. c	9. b	10. d
11. b	12. c	13. c	14. c	15. b
16. a	17. a	18. a	19. d	20. a
21. a	22. c	23. c	24. c	25. b

4. ANTONYMS

Antonyms are words having opposite meanings. In the Antonyms Test each question consists of a single word or Key word with four alternatives given below it, labelled as A B C and D.

You have to choose the correct alternative which is opposite in meaning to the key word.

***Example*:**

1. **REPEL**

A) Red B) Concentrate C) Continue D) attract

Answer (D) Attract. The opposite word of 'repel' is 'attract'. So the correct answer is (D).

STUDY MATERIAL

Learn the following antonyms so that you can get good marks in the examination.

Word	Antonym
Abbreviate	Lengthen, Exapnd
Abhor	Like, Approve
Abiding	Fleeting, Passive
Above	Below, Beneath
Abstract	Concrete
Absurd	Rational, Sane
Accelerate	Retard
Accept	Reject, Refuse
Accord	Discord, Disagreement
Acquit	Condemn, Sentence
Active	Passive
Admire	Dislike
Adopt	Reject
Advance	Retreat
Adventurous	Timid
Adversity	Prosperity
Affinity	antipathy
Affirm	Deny
Affluence	Poverty
Aggravate	Alleviate, Lessen
Agree	Differ, Disagree
Allow	Forbid
Always	Never
Ample	Scanty
Analysis	Synthesis
Ancestor	Descendant
Angry	Pleased
Ancient	Modern
Appreciate	Depreciate
Aristocrat	Commoner
Armed	Unarmed
Arrival	Departure
Arrive	Depart
Arrogant	Humble
Ascent	Descent
Asleep	Awake
Asset	Liability
Attack	Defend
Attract	Repel
Authentic	Spurious
Awake	Asleep
Awkward	Graceful
Barbarous	Civilized, Refined
Barren	Fertile
Base	Noble, Exalted
Beautiful	Ugly
Before	After
Beginning	End
Belief	Doubt
Beneficent	Maleficent
Benediction	Malediction
Benevolent	Malevolent
Big	Small
Bless	Misery, Infelicity
Blunt	Sharp
Boisterous	Quiet, Clam
Bloody	Timd, Cowardly
Bonafides	Malafides
Boom	Slump
Brave	Coward
Bravery	Cowardice
Bright	Dim, Dull
Broad	Narrow
Brutal	Humane, Kindly
Callous	Soft, Tender
Care	Neglect
Carnal	Spiritual
Censure	Praise, Commend
Chaste	Impure, Unchaste
Cheap	Dear, Costly
Cheerful	Gloomy, Depressed
Clean	Dirty
Clergy	Laity
Clever	Stupid
Cold	Hot
Comic	Tragic, Serious
Comfortable	Uncomfortable
Common	Rare
Comply	Refuse
Compress	Expand

Compulsory - Optional
Concise - Diffuse, Lengthy
Concord - Discord
Condemn - Approve, Acquit
Condense - Lengthen, Expand
Confess - Deny
Confident - Diffident
Confidence - Distrust, Doubt
Conquest - Defeat
Contemptible - Likable, Worthy
Convex - Concave
Covert - Overt
Convenience - Inconvenience
Cordial - Cold, Frigid
Costly - Cheap
Create - Destroy
Credit - Debit
Crude - Refined
Cruel - Kind, Gentle
Cursory - Through
Dark - Light
Dawn - Dusk
Dear - Hateful, Cheap
Dearth - Plenty
Debit - Credit
Debtor - Creditor
Decrease - Increase
Deduction - Induction
Deep - Shallow
Defendant - Plaintiff
Defence - Offence
Defensive - Offensive
Deficit - Surplus
Definite - Indefinite
Delight - Displeasure, Sorrow
Deflate - Infalte
Deliberate - Unintentional, Accidental
Democracy - Autocracy
Despair - Hope
Destruction - Construction

Devious - Straight
Difficult - Easy
Diffident - Confident
Diligent - Lazy, Idle
Dilute - Concentrated
Diminish - Increase
Discourage - Encourage
Distant - Near
Dwarf - Giant
Dwindle - Grow, Increase
Early - Late
Earthly - Celestial, Heavenly
Ebb - Flow
Economy - Extravagance, Prodigality
Efflux - Influx
Egoism - Altruism
Egoistic - Altruistic
Eldest - Youngest
Elevation - Depression
Emancipate - Enslave
Emigrant - Immigrant
Encourage - Discourage
End - Beginning
Energetic - Weak, Nerveless
Enrich - Impoverish
Enmity - Friendship
Enthrone - Dethrone
Enthusiasm - Indifference
Eulogy - Dispraise, condemnation
Evil - Good
Ever - Never
Exceptional - Commonplace, Ordinary
Exit - Entrance
Exotic - Indigenous
Expand - Contract
Explicit - Implicit
Exterior - Interior
External - Internal

Extraordinary - Commonplace, Ordinary
Extravagant - Thrifty, Economical Frugal
Exult - Grieve
Fabulous - Actual, Real
Failure - Success
Fair - Foul
Faithful - Unfaithful
Fall - Rise
Fallacious - True, Actual
False - True
Famous - Obscrure, Unknown
Fanciful - Practical
Fact - Fiction
Fast - slow, Loose
Fat - thin
Felicity - Sorrow, Misfortune
Ferocious - Mild Gentle
Few - Many
Fickle - Constant
Fictitious - Real, True, Genuine
Fierce - Mild, Gentle
Final - Initial
Fine - Coarse
Finite - Infinite
Flattery - Detraction
Flexible - Stiff
Flimsy - Substantial
Folly - Wisdom
For - Against
Foreign - Native
Forget - Remember
Fortunate - Unfortunate
Frank - Reserve
Frequent - Rare
Fresh - Stale, Faded
Frindly - Hostile
Frugal - Wasteful, Extravagant
Full - Empty

Futility - Utility
Gaiety - Mourning, Dullness
Gain - Loss
Gather - Scatter
Generosity - Stinginess
Generous - Mean
Gentle - Rough
Genuine - Spurious, Fictitious
Giant - Dwarf
Gloomy - Gay
Glut - Scarcity
Good - Bad, Wicked
Gradual - abrupt, Sudden
Grant - Withhold
Grave - Gay
Great - small
Growth - Decline
Guilty - Innocent
Happy - Sad
Happiness - Sorrow, Sadness
Hard - Soft
Harsh - Gentle
Hasty - Leisurely
Hate - Love
Haughty - Meek, Humble
Heavy - Light
Help - Hinder
Hero - Villain
Heterogenous - Homogeneous
High - Low
Hit - Miss
Honour - shame
Hope - Despair
Host - Guest
Humane - Cruel
Humble - Proud
Humility - Arrogance
Hurt - Heal
Hypocrisy - Sincerity
Ideal - Actual
Idle - Busy
Imagination - Reality
Immanent - Transcendent
Immortal - Mortal
Import - Export
Improvident - Provident, Economical
Impure - Pure
Increase - Decrease
Individual - General
Industrious - Indolent, Slothful, Lazy
Intenor - Superior
Inhale - Exhale
Input - Output
Insert - Extract
Intrinsic - Extrinisic
Intricate - Simple
Jovial - Morose, Gloomy
Jolly - Morose, Gloomy
Joyful - Sad, Depressed
Joy - Sadness, Sorrow
Joint - Separate
Junior - Senior
Justice - Injustice, Partiality
Justify - Condemn
Keen - Indifferent, Dull, Blunt
Kernel - Shell
Kill - Resurrect
Kind - Cruel
Kindle - Extinguish
King - Subject
Knowing - Ignorant
Knowledge - Ignorance
Lack - Plenty
Laconic - Lengthy, Verbose
Landlord - Tenant
Languid - Vigorous, Energetic
Languor - Vigour, Activity
Lascivious - Chaste, Moral, Virtuous
Licentious - Chaste, Moral, Virtuous
Lewd - Chaste, Moral, Virtuous
Late - Earlier
Latter - Former
Latitude - Longitude
Laxity - Firmness
Lead - Follow
Lend - Borrow
Lenient - Rigorous, Severe
Let - Hire
Liberate - Enslave
Liberty - Slavery
Light - Heavy
Likeness - Difference
Liquid - Solid
Literal - Figurative
Loose - Tight, Concise
Loss - Gain, Profit
Loquacious (talkative) - Taciturn (Silent)
Lose - Find
Love - Hate
Loyal - Treacherous
Luxury - Asceticism
Mad - Sane
Madness - Sanity
Make - Mar
Malice - Goodwill
Malign - Praise, Commend
Major - Minor
Manifest - Conceal, Suppress
Manifest - Puzzling, Obscure
Many - Few
Marriage - Celibacy
Masculine - Feminine
Massive - Light, small
Master - Servant
Material - Spiritual
Maximum - Minimum

Meagre - Plentiful, Large
Melancholy - Gaiety
Memory - Oblivion
Microcosm - Macrocosm
Mighty - Weak
Mild - Stern, Severe
Miser - Spend thrift
Miserly - Generous, Prodigal
Moderate - Extreme, Excessive
Monogamy - Polygamy
Monotony - Variety, Diversity
Monotheism - Polytheism
Morbid - Healthy
Motion - Rest
Naive - Sophisticated
Narrow - Broad
Native - Foreign
Natural - Artificial
Nebulous - Distinct, Clear
Neat - Untidy
Negative - Positive
Neutral - Partial
Niggard - Spendthrift
Noble - Ignoble
Normal - Abnormal
Notorious - Reputable
Noxious - Wholesome
Numerous - Few
Obedient - Disobedient
Obscurity - Publicity
Obstinate - Obedient, Yielding
Obverse - Reverse
Odd - Even
Offensive - Defensive, Pleasing, Polite
Often - Seldom
.Ominous - Auspicious
Omission - Commission
Onerous - Light, Easy
Open - Shut

Opposition - Support, Cooperation
Optimistic - Pessimistic
Oral - Written
Oriental - Occidental
Original - Duplicate
Orthodox - Heterodox
Ostensible - Actual
Outward - Inward, Internal
Part (N) - Whole
Part (V) - Join, Unite
Partial - Fair, Impartial
Partly - Wholly
Particular - General
Passion - Coolness
Pathetic - Joyous, Cheery
Peace - War
Permanent - Passing, Fleeting, Temporary
Permit - Prohibit
Persuade - Dissuade
Philanthropist- Misanthropist
Physical - Spiritual
Pleasure - Pain
Plenty - Scarcity
Poor - Rich, Wealthy
Practice - Theory
Pragmatic - Idealistic
Praise - Condemn, Blame
Precede - Succeed
Precious - Cheap, Worthless
Present - Absent
Pride - Humility
Profit - Loss
Progress - Regress, Retrogression
Progressive - Reactionary, Retrograde
Promote - Demote
Prospect - Retrospect

Prosperity - Adversity, Proverty
Public - Private
Punish - Reward
Queer - Normal
Queel - Agitate, Arouse
Quench - Ignite, Excite
Question - Answer
Quick - slow, Tardy
Quiet - Noise, Uproar
Rapid - Slow
Rare - Common, Ordinary
Rash - Steady, Cautious, Careful
Real - False, Imaginary
Recede - Advance
Receive - Give, Bestow
Rectify - Falsify
Reduce - Lengthen, Strengthen
Reject - Accept, Admit
Relative - Absolute
Religious - Secular, Unholy
Relish - Dislike
Reluctant - Ready, Willing
Remember - Forget
Remote - Near
Remitter - Receiver, Recipient
Repugnant - Agreeable, Pleasing
Repulsive - Attractive
Reputable - Worthless
Reserved - Sociable
Resisit - Submit
Retail - Wholesale
Return - Depart
Reward - Punish
Rich - Poor, Needy
Right - Wrong
Rigid - Flexible
Ripe - Raw
Rise - Fall

Rough - Smooth
Rude - Polite
Sacred - Secular, Profane, Unholy
Safe - Risky, Dangerous
Saint - Sinner
Sane - Insane
Satisfaction - Dissatisfaction
Savage - Civilized, Tame
Scanty - Plentiful
Secret - Open, Public
Sensible - Senseless
Separate - Join
Servant - Master
Service - Disservice
Service - Masterly, Lordly
Sharp - Blunt, Dull
Shy - Bold, Impudent
Sick - Healthy
Silence - Noise, Clamour
Simple - Complex, Complicated
Sin - Virtue
Sink - Swim, Float
Slender - Stout, Adequate
Smile - Frown
Smooth - Rough, Rugged
Sober - Drunk, Rash, Excited
Soft - Hard, Strubborn
Solid - Liquid
Sorrow - Joy, Happiness
Special - Ordinary
Speed - Slowness
Spiritual - Material
Splendid - Squalid
Starboard - Port, Larboard
Stationary - Moving
Straight - Crooked, Curved
Strait - Broad
Strange - Familiar
Strict - Lax

Subjective - Objective
Sublime - Ridiculous
Succeed - Rail
Superior - Inferior
Sure - Doubtful
Surplus - Deficit
Sweet - Bitter, Sour
Swift - Slow
Sympathy - Antipathy
Synonymous - antonymous
Tall - Short
Tame - Savage, Wild
Tedious - Agreeable, Lively
Temperate - Intemperate, Immoderate
Temporal - Spiritual, Eternal
Temporary - Permanent
Tender - Hard, Strong
Tense - Relaxed
Terrestrial - Celestial
Thankful - Thankless
Thesis - Antithesis
Thick - Thin
Thrifty - Prodigal, Extravagant
Thrive - Decline
Tight - Loose, Slack
Timid - Plucky, Bold
Top - Bottom
Tortuous - Straight, Honest
Tough - Flexible
Tragic - Comic
Tranquil - Disturbed, Agitated
Transient - Lasting, Durable
Transparent - Opaque
Triumph - Defeat
Trivial - Important
True - False
Tutor - Pupil
Teacher - Pupil, Student
Ugly - Beautiful, Attractive

Uniform - Variable, Inconstant
Unite - Separate
Upper - Lower
Upright - Dishonest
Urbance - Rude, Uncultured
Useful - Useless, Harmful
Usual - Rare, Unusual
Vacant - Occupied
Vague - Clear, Definite
Vain - Modest
Vanity - Modesty
Variable - Constant
Vanguard - Rearguard
Verbal - Written
Vertical - Horizontal
Victory - Defeat
Virtue - Vice
Voluntary - Compulsory
Vulgar - Refined, Cultured
Wake - Sleep
Want - Abundance
Warmth - Coolness
Wax - Wane
Wealth - Poverty
Weighty - Light, Unimportant
Wide - Narrow
Win - Lose
Wise - Foolish
Woeful - Cheerful
Work - Idlenes
Work - Rest
Wrong - Right
Yield - Resist, Fight
Yoke - Freedom, Liberty
Youth - Age
Youthful - Aged
Zeal - Apathy, Indifference
Zealous - Cold, Indifferent
Zenith - Nadir

PRACTICE TEST - 1
ANTONYMS

DIRECTIONS : **In each of the following questions, choose the word opposite in meaning to the given word.**

1. ARTIFICIAL
   a) Solid b) Truthful c) Authentic d) Natural
2. DEBACLE
   a) Regain b) Progress
   c) Rise d) Movement
3. ARCHAIC
   a) Modern b) Broken
   c) Old b) Dilapidated
4. CAPITULATE
   a) Conquer b) Venerate
   c) Destory d) Surrender
5. MUNDANE
   a) Extraordinary b) Superb
   c) Heavenly d) Excellent
6. RESERVED
   a) Likeable b) Talkative
   c) Popular d) Companionable
7. SUB ROSA
   a) Fresh b) Intoxicating
   c) Open d) Repulsive
8. CULPRIT
   a) Jury b) Witness
   c) Accused d) Victim
9. INVETERATE
   a) Stupid b) Uneducated
   c) Ignorant d) Inexperienced
10. ZANY
    a) Cautious b) Calculating
    c) Cunning d) Sane
11. DISDAIN
    a) Depreciate b) Admiration
    c) Penitence d) Contempt
12. OBSOLETE
    a) Conducive b) Rare
    c) Useless d) Recent
13. INDOLENT
    a) Desirable b) Adequate
    c) Energetic d) Consistent
14. COSSET
    a) Neglect b) Divert
    c) Pamper d) Pinch
15. DOLEFUL
    a) Poor b) Happy
    c) Rich d) Steady
16. MORTAL
    a) Eternal b) Spiritual
    c) Immortal d) Divine
17. MASTICATE
    a) Gobble b) Conceal
    c) Chew d) Review
18. HAWK
    a) Conciliation b) Dove
    c) Pacifist d) Pigeon
19. ANFRACTUOUS
    a) Devious b) Straight
    c) Remote d) Attractive
20. INSANITY
    a) Sanity b) Normality
    c) Lucidity d) Sobriety
21. IMMEDIATE
    a) Delayed b) Gradual
    c) Leisurely d) Slow
22. QUIESCENT
    a) Troublesome b) Weak
    c) Indifferent d) UnHappy
23. CHURLISH
    a) Accomodating b) Polite
    c) Helpful d) Unconcerned
24. HARMONY
    a) Hatred b) Friction
    c) Discord d) Enmity
25. GROTESQUE
    a) Imaginary b) Familiar
    c) Hateful d) Natural

ANSWERS

1. d	2. c	3. a	4. a	5. c
6. b	7. c	8. d	9. d	10. d
11. b	12. d	13. c	14. a	15. b
16. c	17. a	18. b	19. b	20. a
21. d	22. a	23. b	24. c	25. d

PRACTICE TEST - 2
ANTONYMS

DIRECTIONS: **In each of the following questions, choose the word opposite in meaning to the given word.**

1. CHIDE
   a) Praise b) Fear
   c) Criticise d) Flatter
2. UNCOUTH
   a) Crude b) Awkward
   c) Courteous d) Refined
3. BLEMISH
   a) Preserve b) Purify
   c) Defect d) Tarnish
4. RELIGIOUS
   a) Atheistic b) Immoral
   c) Secular d) Sinful
5. REPRISAL
   a) Relief b) Forgiveness
   c) Exemption d) Relaxation
6. DISCREPANCY
   a) Variance b) Inappropriate
   c) Consistency d) Inconsistency
7. UNDERHAND
   a) Cruel b) Quiet
   c) Secret d) Open
8. NEAT
   a) Sloppy b) Fragrant
   c) Spruce d) Prodigal
9. TERSE
   a) Concise b) Detailed
   c) Expressive d) Descriptive
10. ILLUSTRIOUS
    a) uneducated b) Uncivilised
    c) Unintelligent d) Unknown
11. PASTEL
    a) Urban b) Delicate
    c) Bright d) Sweet
12. LETHAL
    a) Safe b) Playful
    c) Virulent d) Forgiving
13. ENCOURAGE
    a) Warn b) Discourage
    c) Dampen d) Disapprove
14. JUSTIFY
    a) Accuse b) Infuriate
    c) Absolve d) Vindicate
15. ERRATIC
    a) Reliable b) Right
    c) Punctual d) Free
16. CRITICISE
    a) Judge b) Appreciate
    c) Flatter d) Analyse
17. MONOLOGUE
    a) Dialogue b) Prologue
    c) Epilogue d) Catalogue
18. CATHOLIC
    a) Rigid b) Orthodox
    c) Strict d) Narrow-minded
19. INNOCENCE
    a) Crime b) Mischief
    c) Guilt d) Sin
20. DIABOLIC
    a) Patient b) Generous
    c) Kind d) Simple
21. FRUGAL
    a) Gaudy b) Generous
    c) Extravagant d) Charitable
22. FRIEND
    a) Rival b) Acquaintance
    c) Foe d) Competitor
23. DIVERGE
    a) Repulse b) Converge
    c) Resurge d) Constrict
24. ALLURE
    a) Repulse b) Develop
    c) Entice d) Decoy
25. CRASS
    a) Gross b) Refined
    c) Coarse d) Dense

ANSWERS

1. a	2. c	3. b	4. c	5. a
6. c	7. d	8. a	9. b	10. d
11. c	12. a	13. b	14. a	15. a
16. b	17. a	18. d	19. c	20. c
21. c	22. c	23. b	24. a	25. b

PRACTICE TEST - 3
ANTONYMS

***DIRECTIONS :* In each of the following sentences, a word has been printed in bold. Out of the given alternatives, choose the one which is closest to the opposite in meaning of the italicised word.**

1. I thought about her a lot during the **following** months.
   a) receding b) preceding
   c) proceeding d) succeeding
2. This is the **cardinal** point of the issue.
   a) minor b) vital
   c) debatable d) insignificant
3. She is indeed **human.**
   a) universal b) devilish
   c) terrestrial d) divine
4. It was universally characterised as a **progressive** measure.
   a) abhorrent b) retrograde
   c) obstructive d) regressive
5. People know him for his **vanity.**
   a) humanity b) honesty
   c) courtesy d) modesty
6. Let us not **aggravate** the sufferings of the poor.
   a) abbreviate b) alleviate
   c) advocate d) appreciate
7. It was a mystery as to where the young girl had acquired such a **cynical** attitude.
   a) mature b) naive c) eccentric d) crazy
8. Avogadro's **hypothesis** has proved significant till the present day.
   a) fact b) theory
   c) conclusion d) experiment
9. This shop keeps only **genuin** articles.
   a) imported b) spurious
   c) antique d) duplicate
10. He has been **commended** for all that he did.
    a) dismissed b) condemned
    c) censured d) rejected
11. The drug will have **pernicious** effect on your health.
    a) beneficial b) prolonged
    c) ruinous d) pornographic
12. It was a **unanimous** decision.
    a) uncertain b) partial
    c) discordant d) divergent
13. Phoolan Devi had to **surrender** under the most pressing circumstances.
    a) release b) claim c) plunder d) attack
14. Geetanjali is loved by all for her **meehness.**
    a) harshness b) pride c) anger d) passion
15. It was a **voluntary** gesture.
    a) compulsory b) violent
    c) deliberate d) valuable
16. There has always been a feeling of **rancour** between the two families.
    a) rivalry b) competition
    c) friendliness d) suspicion
17. The government is taking measures to **augment** the country's food supply.
    a) prohibit b) decrease
    c) surpass d) compensate
18. He stood gazing at the **serene** expanse of the sea.
    a) scenic b) clear c) tranquil d) ruffled
19. The authorities inflicted harsh **penalty** on him for indulging in forgery.
    a) commendation b) default
    c) concession d) reward
20. He is always **hungry** for wealth.
    a) ravenous b) famished
    c) satiated d) greedy
21. Machine-civilisation has made human life **artificial**
    a) genuine b) natural
    c) true d) authentic
22. The students **assembled** in the meeting hall.
    a) removed b) diffused
    c) dispersed d) eliminated
23. His **boorish** ways amused everyone.
    a) refined b) funny
    c) eccentric d) friendly
24. How can you cut an apple with this **blunt** knife ?
    a) pointed b) sharpened c) polished d) filed
25. He has suspended his secretary on a **flimsy** ground.
    a) vigorous b) strong c) sound d) salutory

ANSWERS

1. b	2. d	3. d	4. d	5. d
6. b	7. b	8. a	9. b	10. c
11. a	12. c	13. b	14. b	15. a
16. c	17. b	18. d	19. d	20. c
21. b	22. c	23. a	24. b	25. b

PRACTICE TEST - 4
ANTONYMS

DIRECTIONS : **In each of the following sentences, a word has been printed in bold. Out of the given alternatives, choose the one which is closest to the opposite in meaning of the italicised word.**

1. We must realise the **futility** of wars.
   a) urgency b) usefulness
   c) value d) importance
2. The scheme proved harmful for the people, the middle income group in **particular.**
   a) common b) usual c) specific d) general
3. His punctuality and regularity **propitiales** everyone with whom he deals.
   a) depresses b) excites
   c) enrages d) appeases
4. Most surfaces are made **glossy** by polishing
   a) rough b) crude c) dull d) ugly
5. It is **obligatory** for a common citizen to follow the rules
   a) optional b) superfluous
   c) necessary d) advisable
6. Every detail of the painting is **prominent** enough to be seen.
   a) negligible b) insignificant
   c) trifling d) inconspieuous
7. A faithful officer is always **vigilant** towards his duties.
   a) innocent b) ignorant
   c) irresponsible d) careless
8. The **gross** domestic product (GDP) is an essential component of economy.
   a) rough b) coarse
   c) refined d) estimated
9. It was really a **gracious** occasion for me.
   a) benign b) infructuous
   c) churlish d) wasteful
10. The man at the gate had a **forbidding** appearance.
    a) lenient b) handsome
    c) tranquil d) mild
11. Terrorism is a serious **hazard** to the country.
    a) chance b) peril c) safety d) problem
12. **Nourishing** food is a necessity both for a pregnant mother and a baby.
    a) Unhygienic b) Poor
    c) Undercooked d) Heavy
13. There is a **dearth** of milk in Delhi nowadays.
    a) extravagance b) scarcity
    c) abundance d) sufficiency
14. Hydra is biologically believed to be **immortal.**
    a) undying b) perishable c) ancient d) eternal
15. **Crestfallen** he returned as he had never faced such humiliation in the whole of his life.
    a) Disturbed b) Vainglorious
    c) Triumphant d) Indignant
16. A feeling of brotherhood should be **propagated** amongst the masses.
    a) disseminated b) suppressed
    c) dissipated d) crushed
17. She is beautiful as well as **frivolous.**
    a) indecent b) serious c) insane d) rude
18. The problem of dowry in our country has assumed **gargantuan** proportions.
    a) negligible b) bearable c) minute d) minimal
19. The boy found himself in a **pathetic** situation.
    a) comical b) marvellous
    c) common d) surprising
20. Gandhi did not welcome the idea of being a mute spectator to the **atrocious** behaviour of the whites towards the non-whites in Africa.
    a) gracious b) noble c) civilised d) pleasing
21. A **friendly** dog met us at the farmgate.
    a) hostile b) quiet
    c) understanding d) helpful
22. The scene was indeed **captivating.**
    a) hateful b) disgusting c) repulsive d) obscene
23. That police officer is known to be **humane** in his approach.
    a) unsympathetic b) uncultured
    c) uncompromising d) uncivilised
24. A crowd **gathered** quickly when the police came.
    a) collected b) dispersed
    c) spread d) disappeared
25. Matter **expands** on heating.
    a) shrinks b) reduces
    c) diminishes d) contracts

ANSWERS

1. b	2. d	3. c	4. c	5. a
6. d	7. d	8. c	9. c	10. b
11. c	12. a	13. c	14. b	15. c
16. b	17. b	18. c	19. a	20. b
21. a	22. c	23. a	24. b	25. d

5. SENTENCE PARTS RESHUFLING

In this type of test the first and the last parts of a sentence are numbered 1 and 6. The rest of the sentence is split into four parts and named P, Q, R and S.

The four parts are not given in their proper order. You have to read the sentence and find out which of four combinations A), B), C) and D) is correct. That is your answer.

Example :

Q. 1 1. I have not come.

P. even if it means some humiliation.

Q. but the boy must learn

R. to complain, he said.

S. to be honest.

6. and admit he broke our window pane.

A) SPRQ B) QSPR C) RQSP D) PRQS

Answer : (c) RQSP.

PRACTICE TEST
SENTENCE PARTS RESHUFLING

DIRECTIONS : **In questions 1-10 the first and the last parts of the sentences are numbered 1 & 6. The rest of the sentence is split into four parts and named P, Q, R and S. These four parts are not given in their proper order. Read the sentence and find out which of the four combinations is correct. Then find the correct answer and indicate it.**

1. 1. I have not come

P. even if it means some humiliation

Q. but the boy must learn.

R. to complain, he said,

S. to be honest

6. and admit he broke our window pane.

A) SPRQ B) QSPR

C) RQSP D) PRQS

2. 1. India has been a land

P. but in the sense that learning has always been very highly valued.

Q. not indeed in the sense that education has been universal

R. and the learned man has been held in higher esteem

S. of learning throughout the ages.

6. than the warrior or the administrator.

A) PQSR B) RSQP

C) SQPR D) RQPS

3. 1. It was obvious

P. made by him

Q. submitted at the meeting

R. from the comments

S. on the draft proposals

6. that he was not satisfied with them.

A) QRSP B) SQRP

C) RPSQ D) PSRQ

4. 1. A sub-inspector of police.

P. disturbing a public meeting

Q. justice to his uniform.

R. rounded up a group of 28 miscreants

S. bent on doing

6. and marched them to the local rail in Shimoga.

A) PRSQ B) SQRP

C) QRPS D) RPSQ

5. 1. Fair play demands that

P. which may tend to give a poor impression

Q. before any comments are made

R. the person concerned should be asked

S. about a person

6. to give his side of the story

A) RQPS B) QPSR

C) PSRQ D) SRQP

6\. 1. I wasn't sure.

P. to hear his voice

Q. in my heart of hearts

R. through that window once more,

S. what I really wished

6\. or never to see him there again

A) QPRS B) SQPR

C) PRSQ D) RSQP

7\. 1. Yet so few of us have

P. Walls of daily routine

Q. the time or the means

R. that enclose our lives

S. to break through the narrow

6\. to get to know this land.

A) SRQP B) RPSQ

C) QSPR D) PQRS

8\. 1. Religion has been used

P. both as a weapon of isolation

Q. to dull awareness

R about real problems

S. and as morphia

6\. like education, health and employment.

A) PQRS B) PSQR

C) QPSR D) RPQS

9\. 1. When all the members were seated,

P. during the previous year

Q. the secretary came to the dais

R. and read out the report

S. detailing the activities of the organisation

A) PQRS B) SPQR

C) RSPQ D) QRSP

10\. 1. Inspite of an unprecedented boom in the market.

P. and had to remain content

Q. the paper-dealer could not

R. push up his sales

S. with a volume of sale lower than

6\. what he had sold the previous year

A) QRPS B) SPRQ

C) PRQS D) RPSQ

11\. 1. The district collector

P. The next of kin

Q. Would be provided to

R. announced that financial assistance

S. of the deceased

6\. From the Chief Minister's Relief Fund

A) PSRQ B) QSPR

C) RQPS D) SQRP

12\. 1. The Arabs live in tents

P. so that they can move from one oasis to another

Q. and taken down very easily and quickly

R. seeking grass and water

S. than can be put up

6\. For their sheep, goats, camels and horses

A) PRSQ B) QSPR

C) SQPR D) SRPQ

13\. 1. In the coming session,

P. of our computer management services

Q. to make them

R. more useful, efficient and cost-effective

S. we are undertaking a thorough review

6\. for your needs

A) PQRS B) QRSP

C) RSPQ D) SPQR

14\. 1. Though the Government claims

P. it has failed to arrest

Q. the rate of inflation is down

R. or the decrease

S. the rise in prices

6\. in the per capita income

A) PQRS B) PQSR

C) PSQR D) QPSR

15\. 1. The greatest tragedy of Bhopal

P. We in India

Q. is that ever after the Bhopal disaster

R. have not realized

S. the necessity

6\. the town planning

A) PRQS B) PRSQ

C) QPRS D) RSQP

16\. 1. If the parachutes had failed to open

P. the whole of his couch would

Q. he could have pulled a liver to provide an emergency ejection

R. and special parachutes would have unfolded

S. then have beeen fired from the cabin

6\. to bring him down

A) PSRQ B) QPSR C) RSQP D) SQPR

17. 1. The next stop
P. lying in the centre
Q. of a bay
R. was a port
S. that our ship halted at
6. in South Wales
A) PQRS B) RSQP
C) SRPQ D) SRQP

18. 1. Ashoka was successful
P. by the cruelty and horrors of war
Q. he was so disgusted
R. in his military operations
S. and alone among conquerors
6. that he renounced it
A) PSQR B) QPRS
C) RSQP D) SQPR

19. 1. Some remarks
P. put the police on the right scent
Q. by a woman
R. and they discovered
S. casually dropped
6. the whole gang of brigands
A) QSPR B) RPSQ
C) RQSP D) SQPR

20. 1. The advantage of travelling
P. customs and manners, is that
Q. we learn to appreciate the good points of others
R. in different places and having
S. the opportunities of observing people's
6. and our prejudice against them disappear
A) PQRS B) QRSP
C) RSPQ D) SPQR

21. 1. Sugar
P. and our average consumption of sugar
Q. provides something like
R. itself is rich in calories
S. one-fifth of our
6. total calorie intake
A) PQRS B) QRPS
C) RPQS D) RPSQ

22. 1. At the end of the lecture
P. for sparing their valuable time
Q. the speaker thanked the audience
R. and listening to him
S. in coming to the meeting
6. patiently
A) PSRQ B) QPSR
C) RSPQ D) SRQP

23. 1. Charles Goodyear once accidentally
P. and found that the resulting rubber
Q. dropped a mixture of natural rubber
R. was harder, stronger, more elastic
S. and sulphur over a hot stove
6. and non-sticky
A) PSQR B) QSPR
C) RSQP D) SQPR

24. 1. The Minister of State for power
P. laying emphasis
Q. in conservation of electricity in industries
R. has written to his counterparts in State Governments
S. on bringing about improvement
6. by introduction of energy efficient equipment.
A) QPSR B) RPSQ
C) SPQR D) SQPR

25. 1. In reply to a question
P. that securing extradition
Q. operating from the U.K. soil remained
R. of anti-India elements
S. the spokesman said
6. New Delhi's 'first priority.'
A) PRQS B) QSPR
C) RQSP D) SPRQ

ANSWERS

1. c	2. c	3. b	4. b	5. b
6. b	7. c	8. b	9. d	10. a
11. c	12. c	13. d	14. d	15. c
16. b	17. c	18. c	19. d	20. c
21. c	22. b	23. b	24. b	25. d

10. SENTENCE RESHUFLING

In this type of test the first and the last sentences of a passage are numbered as 1 and 6. The rest of the passage is split into four parts and named as P, Q, R and S. These four parts are not given in their proper order. You have to read the passage and find out which of the four combinations is correct and indicate it as your answer.

Example-1 :

Q.1. 1. You live either in a village or a town of India.

P. Many villages and towns form a tehsil or a taluk.

Q. There are also some areas in our country called Union Territories.

R. Many tahsils or talukas form a district and many districts form a state.

S. These, together with all the states of our Country make India.

6. India is our home land.

A) PQRS B) PRQS C) RPQS D) QPRS

Answer : (B) PRQS

This test may be given in the following way also.

DIRECTIONS : *Rearrange the following five sentences A, B, C, D, E in the proper sequence so as to form a meaningful paragraph. On the basis of the arranged sequence answer the questions based on it.*

A) Modern Science has produced some good drugs for these ailments.

B) Heart ailments are very common these days.

C) So I always prefer taking milk in my meals as it contains no fat.

D) But this can also be prevented by taking fatless food.

E) That is why some people call this era as the age of heart troubles.

Q.1. Which of the following is the fourth Sentence?

A) A B) B C) C D) D E) E

Q.2. Which of the following should be the third sentence?

A) A B) B C) C D) D E) E

Q.3. Which of the following should be the first sentence?

A) A B) B C) C D) D E) E

Q.4. Which of the following should be the second sentence?

A) A B) B C) C D) D E) E

Q.5. Which of the following should be the last sentence?

A) A B) B C) C D) D E) E

Answers :

1. d 2. a 3. b 4. c 5. c

PRACTICE TEST
SENTENCE PARTS RESHUFLING

DIRECTIONS : **In questions 1-25 the 1st and the last sentences of the passages are numbered 1 & 6. The rest of the passage is split into four parts and named P, Q, R, S. These four parts are not given in their proper order. Read the passage and find out which of the four combinations is correct. Then find the correct answer out of A), B) C) and D) :**

1. 1. There was a time Egypt faced economic crisis.
   P. Cotton is the chief export commodity of Egypt.
   Q. Foreign trade depends on cultivation of cotton on large scale.
   R. It became necessary for Egypt to boost cotton crops.
   S. Only by means of increasing foreign trade Egypt could survive.
   6. Egypt was able to sustain itself by its cotton produce.
   a) PQRS b) QRPS
   c) RPSQ d) SRQP

2. 1. While crossing a busy road we should obey the policeman on duty.
   P. We should always cross the road at the zebra crossing.
   Q. We must lock to the signal lights and cross the road only when the road is clear.
   R. If there are no signal lights at the crossing, we should look to the right, then to the left and again to the right before crossing the road.
   S. If the road is not clear we should wait.
   6. We should never run while crossing a road.
   a) PQRS b) PSRQ
   c) QRPS d) RQSP

3. 1. A black-haired, young woman came tripping along.
   P. She was leading a young man wearing a hat.
   Q. The woman swept it off and tossed it in the air.
   R. The child jumped upto catch the hat.
   S. The young man tossed his head to shake the hat back.
   6. Both disappeared from view.
   a) PSQR b) RPSQ
   c) QRPS d) SQRP

4. 1. "You know my wife, Madhavi, always urged me to give up smoking.
   P. I really gave it up.
   Q. And so when I went to jail I said to myself I really must give it up, if for no other reason than of being self-reliant.
   R. When I emerged from jail, I wanted to tell her of my great triumph!
   S. But when I met her, there she was with a pcket of cigarettes.
   6. Poor girl!
   a) PSRQ b) SPQR
   c) QPRS d) RSPQ

5. 1. Religion is not a matter of mere dogmatic conformity.
   P. It is not merely going through the ritual prescribed to us.
   Q. It is not a question of ceremonial piety.
   R. Unless that kind of transformation occurs, you are not an authentically religious man.
   S. It is the remaking of your ownself, the transformation of your nature.
   6. A man of that character is free from fear, free from hatred.
   a) SPRQ b) QPSR
   c) PSRQ d) SPQR

6. 1. Proverbs contain homely but universal truths.
   P. They point out the incongruities of situations in life.
   Q. Naturally, therefore, they are translatable from one language to another.
   R. Therefore, their appeal is direct.
   S. Many of them had their birth in folk literature.
   6. They are everyman's philosophy.
   a) QPRS b) QSPR
   c) RPSQ d) SPQR

7. 1. Proverbs contain homely but universal truths.
   P. They point out the incongruities of situations in life.
   Q. Naturally, therefore, they are translatable from one language to another.
   R. Therefore, their appeal is direct.

S. Many of them had their birth in folk literature.

6. They are everyman's philosophy.

a) QPRS b) QSPR
c) RPSQ d) SPQR

8. 1. In ancient Indian history the city of Ujjain was quite famous.

P. Here lived at one time the poet Kalidas.

Q. He was a famous learned astronomer.

R. And here also worked and visited Rajah Jaysingh of Jaipur.

S. It was always renowned as a seat of learning.

6. So one can see what a great love all who care for India must feel for the ancientry of Ujjain.

a) PSRQ b) SPRQ
c) QSRP d) SRPQ

9. 1. An Ameer was on a voyage with one of his slaves.

P. The captain of the ship threatened the slave that he would order him to be thrown into the sea.

Q. The slave got frightened and stopped crying.

R. He requested the captain to quieten the slave.

S. The slave had never been to sea before and began to cry in fear.

6. The Ameer was surprised at the sudden change in his slave.

a) QRPS b) RSPQ
c) QPSR d) RSPQ

10. 1. Gopal worked as a labourer at the building site.

P. But Gopal made a bid and he got the box.

Q. There was no key to it and it seemed unless but Gopal took it home.

R. Once while returning from his work Gopal stopped at an suction sale.

S. No one seemed to want a rough old box when it was put up for sale.

6. He unscrewed the lid and found a valuable collection of old silver in it.

a) QRPS b) SRPQ
c) RPQS d) RSPQ

11. 1. Egypt is located in North-east of Africa.

P. The land should be productive.

Q. It is on the coast of Mediterranean Sea.

R. The important thing is not just the area.

S. Its area is almost four times that of England.

6. Egypt is mostly desert or semi-desert.

a) PQSR b) QSRP
c) RPQS d) SRPQ

12. 1. For making a film, first producer and a director are appointed.

P. Scenes are shot according to convenience and not in the same order in which they appear in the film.

Q. The story is converted into a film script.

R. These separate scenes are joined together in the proper order by the film editor.

S. A story is selected.

6. Finally sound is added to the scenes.

a) QRPS b) SPRQ
c) QPSR d) SQPR

13. 1. There are many obstacles in our way to enlightment.

P. Instruction from a true Guru alone is the remedy.

Q. This Maya is the cause of our ignorance.

R. It is ignorance that binds the self to the material world.

S. Illusion or Maya is the one chief flaw of human mind.

6. From ignorance to understanding, then, and to salvation is possible.

a) QRPS b) RPSQ
c) PSQR d) SQRP

14. 1. Ingratitude stings strongest where relationship is closest.

P. Expectation turns innocent relationship into commerce

Q. Human relationship is adulterated with sly commerce.

R. In commerce, of course, give and take in understood.

S. Most relationships are founded on mutual expectations.

6. From any warm and healthy human relationship expectation of returns has to be weeded out.

a) SRQP b) RPSQ
c) QSPR d) PQRS

15. 1. The mail is first collected from different letter boxes.

P. From there it is sent to the head post office.

Q. It is then sorted out at the sorting office.

R. The mail is again sorted out at the head office by the concerned beat postman.

S. The sorted mail is sent to the zonal post office.

6. Finally it is delivered to us.

a) QSPR b) RSQP
c) QPRS d) RPQS

16. 1. Gandhi's first political fast was made soon after his return from Africa.

P. He had also received help from this man's sister.

Q. This was when the poor labourers of the cotton mills of Ahmedabad were on strike.

R. He was a friend of the largest millowner.

S. Gandhi had made the strikers promise to remain on strike until the owners agreed to accept the decision of an arbitrator.

6. He did not fast against the mill owners, but in order to strengthen the determination of the strikers.

a) PQSR b) QSRP
c) RPQS d) SRPQ

17. 1. Kabir knew that Ramananda got up very early in the morning and went down on the steps of the 'ghat' to bathe in the waters of the sacred Ganges.

P. As Ramananda came down the steps before daybreak for his usual bath, he trod on the sleeping man.

Q. Kabir at once jumped up and threw himself at the feet of the preacher.

R. "Ram, Ram", he exclaimed in astonishment.

S. One dark night, Kabir went to the 'ghat' and lay down on one of the river steps.

6. He said, "You have given me the mantra, 'Ram, Ram' I have become your disciple".

a) PQSR b) PRQS
c) RQPS d) SPRQ

18. 1. Smoke oozed up between the planks.

P. Passengers were told to be ready to quit the ship.

Q. The rising gale fanned the smouldering fire.

R. Every one now knew there was a fire on board.

S. Flames broke out here and there.

6. Most people bore the shock bravely.

a) QPSR b) QSRP
c) RSPQ d) SRQP

19. 1. The houses in the Indus Valley were built of baked bricks.

P. This staircase sometimes continued upwards on to the roof.

Q. Access to the upstairs rooms was by a narrow stone staircase at the back of the house.

R. The drains were incorporated in the walls.

S. The houses had bathrooms and water closets, rubbish chutes and excellent drainage systems.

6. They led outside into covered sewers which ran down the side of the streets.

a) PSQR b) QPSR
c) QRPS d) SPQR

20. 1. A transformation of consciousness is now beginning to express itself in the field of theoretical architecture.

P. In the still theoretical structure an attempt is being made to create a house that is "a domestication of an ecosystem."

Q. What is happening in the architecture is a shift from the international style of the post - industrial era to a symbolic structure.

R. Since architecture is the collective unconscious made visible, the architect does not himself always understand the full cultural implications of his own work.

S. The new form is not a celebration of power over new materials, but a celebration of cooperation with ecosystem.

6. The relationship between culture and nature is changed, for the architect grows a house like a garden.

a) PRSQ b) QPSR
c) RQSP d) SQRP

ANSWERS

1. d	2. c	3. a	4. c	5. b
6. d	7. d	8. b	9. d	10. d
11. b	12. d	13. d	14. b	15. c
16. a	17. d	18. d	19. b	20. b

7. COMPREHENSION

In this type of test a passage is given followed by five or four questions each with four responses (answers). You have to read the passage and answer the questions by choosing the correct answer from among the four responses (answers).

Points of remember:

As this examination is time bound and you have to answer all the questions with in limitted time, follow the following suggestions or method to save time.

1. First read the questions instead of the passage.
2. After reading the questions, read the passage carefully and wherever you find the exact words or the answer of the questions then mark the right choice.
3. If there is any doubt, read the passage again and answer the questions.
4. Sometimes the heading or theme of the passage is also asked. You will find the heading or the theme of the passage generally either in the first or in the last sentence.

PRACTICE TEST - 1
COMPREHENSION

***DIRECTIONS* : Read the following passages carefully and answer the questions that follow choosing the correct alternative from the given choices.**

PASSAGE-1

As a young child, you were not afraid of making mistakes. You were willing to try anything and if you weren't very good at it at first you were like the kitten that missed its first fifty mice : You got a little smarter and a little quicker with every try until finally you mastered ice-skating, sowing or making soup. In fact, if you learned to ice-skate, the first thing your body worked out was how to fall down on the ice without getting hurt: how to relax and tuck your legs under you or roll or slide so that you just wouldn't crack your head or break a leg, so that you could come up laughing every time, anxious to try again. 'Failure' was nothing to be ashamed of or avoided. In fact, you welcomed it because you instinctively know that you couldn't learn anything unless you were willing to fail at it first. Thus, as a child you were a natural experimenter, accepting that nobody knows how to throw a ball, swim in deep water, ride a bicycle, or anything else until they give it a try, make an error and give it another try. If children were made in such a way that they were afraid to try new things because they feared failure, they would never get out of their cribs. Likewise, adults who fear failure simply vegetate. Fear of failure becomes fear of success for those who never try anything new.

1. The passage is mainly about
   A) Children getting frightened to make mistakes
   B) Learning ice-skating, sewing and soup-making
   C) Adults who are failures
   D) Learning through failures
2. According to the writer, a child learns anything after initial failures because
   A) he is not ashamed of failures
   B) his world is very small
   C) he is a natural experimenter
   D) no one finds fault with him.
3. "Adults who fear failure simply vegetate" means
   A) Fearful adults are like vegetables
   B) Failed adults are only fit to grow vegetables
   C) Adults who fear failure live an uneventful life
   D) Adults who fear are vegetarians
4. In the sentence... so that you could come up laughing every time anxious to try again the word that can replace anxious is
   A) nervous B) longing
   C) eager D) wishing
5. In the sentence... you welcomed it because you instinctively knew that you couldn't learn....... the expression that comes closest in meaning to instinctively is

A) all of a sudden
B) on an inborn impulse
C) through good reasoning
D) understanding and knowing

PASSAGE-2

Aside from the construction of tombs, mosques and other buildings the Adilshahi rulers paid great attention to the construction of waterworks: Indeed it became imperative to devise arrangements of a long-term nature that would supply water to the city of Bijapur. These schemes were so ingenious it is hard to believe that what we know today as a city with a perennial water problem could have been so lavishly provided with water, not just for essential purposes, but also to fill the tanks, fountains and artificially constructed channels of the residences of the rulers and their nobles and for the countless other public buildings.

6. The writer says that the Adilshahi rulers.
   A) built more tombs and mosques than waterworks
   B) built tombs, mosques and other buildings only
   C) built tombs, mosques, other buildings as well as waterworks
   D) did a lot to construct huge dams

7. The writer is of the opinion that making arrangements for the supply of water to the city of Bijapur was
   A) matter of high priority task
   B) not a very urgent task
   C) not thought of for a long time
   D) not feasible during those days

8. "These schemes" refers to
   A) palace intrigues
   B) water supply arrangements
   C) irrigation facilities
   D) construction of dams.

9. The word "Ingenious" means
   A) faulty B) scientific
   C) primitive D) skilful

10. During the Adishahi days the city of Bijapur
   A) suffered from water shortage
   B) had just enough water for essential purposes
   C) had enough supply of water for all purposes
   D) had plenty of water from time to time

PASSAGE-3

There is no good reason why people should tire more easily as they grow older. Constant fatigue is not normal at any age. I believe that it is caused largely by dietary deficiencies. People who eat natural foods are not likely to feel tired. An abundance of B vitamins is most important in outwitting fatigue because they contain 'thiamin', necessary for boundless energy. In other words B vitamins provide the spark which releases the energy from the foods we eat. One type of fatigue in older people may be caused by inadequate amounts of protein foods. Proteins, as we know, contain the amino acids which manufacture the enzymes necessary for the production of energy. It is amazing how many older people still eat only one or two small portions of good protein foods daily. This is in spite of the fact that more than any other foods, proteins are necessary for muscles, glands and energy production. But it is equally important to relax and lead, wherever possible, a tension-free life. The wise ones have learned by experience that there are two secrets for living and working without strain, without, overdoing, without knowing the meaning of fatigue. These two secrets are : a balanced vitamin-and-mineral-rich diet, and the art of relaxation.

11. Fatigue is generally caused by
   A) Old age
   B) Deficiency in food
   C) Lack of vitamins in food
   D) Unbalanced diet

12. B-vitamins are a source of help in overcoming fatigue because they
   A) contain a spark
   B) provide a supplement to the over-refined foods
   C) contain thiamin
   D) break up the food

13. Older people eat very little good protein food because
   A) they are ignorant about scientific matters
   B) they eat very little anyway
   C) they believe vitamin-rich food is sufficient
   D) they are generally not aware of the value of protein food

14. The function of enzymes is to
   A) produce energy B) build up muscle
   C) relieve tension D) make food nutritious.

15. The main idea discussed in the passage is
   A) balanced diet B) food and relaxation
   C) fatigue D) nutrition

PASSAGE-4

The cockroach has learnt more about survival than any other creature. He can live anywhere, from the middle of the Sahara desert to the kitchens in freezing Labrador. The cockroach's ability to survive comes from his ability to eat anything. His tastes in food include everything from flower buds to shoes. He eats his own cast-off skin and if there is nothing else, dines on the eggs of his own species. He can live for a month without any food or water. Nearly all cockroaches have wings - even household species can fly when other means of escape fail - but they mainly depend on six, powerful, short legs. What is most remarkable about the cockroach are his two antennae. They help him feel his way in the dark. They contain his sense of smell, detect food and water and when he is interested in a lady cockroach, he uses them to stroke her affectionately.

16. The cockroach can survive
    A) in Sahara desert
    B) in hot as well as freezing cold temperatures
    C) in the kitchens having refrigerators
    D) everywhere except damp places
17. A cockroach saves itself mainly by
    A) flying away with its wings
    B) running away with its quick legs
    C) seeing the enemy through its antennae
    D) eating up the enemy
18. A cockroach feels its way in the dark with the help of
    A) its wings B) its powerful legs
    C) its antennae D) its eyes
19. If a cockroach cannot get anything to eat, he even eats.
    A) flower buds
    B) the eggs of the cockroach
    C) shoes
    D) cast-off skin
20. A male cockroach woos the lady cockroach by
    A) waving his antennae affectionately
    B) stroking her with his front legs
    C) dancing around her
    D) stroking her with his antennae

PASSAGE-5

It is difficult to reconcile the ideas of different schools of thought on the question of education. Some people maintain that pupils at school should concentrate on a narrow range of subjects which will benefit them directly in their subsequent careers. Others contend that they should study a wide range of subjects so that they have not only the specialised knowledge necessary for their chosen careers but also sound general knowledge about the world they will have to work and live in. Supporters of the first theory state that the greatest contributions to civilisation are made by those who are most expert in their trade or profession. Those on the other side say that, unless they have a broad general education, the experts will be too narrow in their outlook to have sympathy with their fellows or a proper sense of responsibility towards humanity as a whole.

21. Broad general knowledge is necessary because
    A) it broadens one's outlook
    B) it teaches us about different things
    C) without it no one would get a job
    D) specialisation is incomplete without it.
22. Supporters of the first theory say that
    A) experts have contributed most to progress in the modern world
    B) experts have done nothing to help mankind
    C) People with general knowledge are more useful than experts
    D) people with general knowledge have contributed to civilisation
23. 'Schools of thought' can be explained as
    A) groups of people who are schooled to think
    B) groups of people whose job is to think
    C) groups of people having the same ideas but with different perception on a particular subject
    D) groups of people who study in a particular school thoughtfully
24. The idea of the first school of thought in the passage is that
    A) studetns should study a few subjects that will help them in their profession
    B) students should study all the subjects they want to
    C) students should not undertake any specialised work
    D) students should concentrate on studies
25. According to the second school of thought education will not be very effective if pupils
    A) have nothing but general knowledge
    B) ignore the study of fine arts
    C) do not have a wide general education
    D) have inadequate knowledge of their own work

ANSWERS

1. d	2. c	3. c	4. c	5. b
6. c	7. a	8. b	9. b	10. c
11. c	12. c	13. d	14. a	15. c
16. b	17. b	18. c	19. b	20. d
21. a	22. a	23. c	24. a	25. c

PRACTICE TEST-2

COMPREHENSION

PASSAGE-1

A pioneering scheme has been started recently in Southampton on England's south coast to educate motorists who have been convicted of drunken driving.

The penalty for drunken driving might be the loss of a driving licence and a heavy fine. But under the new scheme, convicted drivers do not pay the fine. Instead they have to attend eight training sessions - one a week - organised by the local authority probation service. Designed to demonstrate the damage alcohol can do, the scheme was devised by senior probation officer John Cook. He said about a quarter of the people who came to him had a drink problem, but had not realised how much they were drinking. One way of getting the message across was to make the drivers pour out their usual ration of alcohol - and then measure it.

Almost everyone pours out not a single measure but a double at least - an example of how easy it is to have more than "just one drink" and to encourage other people to do the same. The instructors on the course are giving clinical evidence of the effects of alcohol on the body and brain. The sober truth is that drink badly affects driving skills, although the drinker might like to believe otherwise.

1. The Southampton scheme requires convicted drivers
   A) to pay a heavy fine
   B) to attend eight driving sessions - one a week
   C) to undergo a probation service
   D) to surrender their driving licence
2. John Cook devised the scheme
   A) as a demonstration technique for driving
   B) to demonstrate the harmful effects of alcohol
   C) to show that South-ampton was concerned about drivers
   D) to prove that alcohol does influence driving
3. The problem with a quarter of the people who went to John Cook was that they
   A) did not want to stop drinking
   B) were unaware of the fact that they could get drunk
   C) would not admit that they had a drinking problem
   D) did not know-how much they were drinking
4. Most drivers start off with at least
   A) a double measure
   B) a single measure
   C) a little less than a single measure
   D) two doubles
5. The truth is that alcohol
   A) does not effect the body but only the brain
   B) affects only the brain
   C) affects the body and the brain
   D) has no effect on the body or the brain

PASSAGE-2

The train was on time. Mrs. Tickoo picked up her suitcase and got into a first-class compartment. There was a woman with a baby in the compartment. Mrs. Tickoo put her suitcase under the seat and sat near the window. Just then, a boy and a girl rushed up. The girl cried, "We nearly missed saying goodbye to you!" The boy said, "Our car broke down fifty yards from the station - and we ran all the way!" "Where is your mother?" asked Mrs. Tickoo, "She can't run like us", said the boy. At that moment, the guard blew his whistle and waved his green flag. "Thank you very much for seeing me off", said Mrs. Tickoo, "but you shouldn't have run all the way. Say goodbye to your mother for me" "Goodbye!" cried the children, as the train steamed out of the station.

6. The children had come to the station.
   A) to meet Mrs. Tickoo
   B) to say goodbye to Mrs. Tickoo
   C) to receive Mrs. Tickoo
   D) to give a gift to Mrs. Tickoo
7. The children were delayed because
   A) they had to walk their way to the station
   B) the train had arrived late
   C) they had started late from home
   D) their car had developed some trouble.
8. The children's mother could not reach the station because
   A) the car had broken down
   B) she was held up at home
   C) she couldn't run fast enough
   D) she didn't want to meet Mrs. Tickoo

9. "The train steamed out of the station" means
   A) the train produced a lot of steam
   B) the train moved out of the station
   C) the train arrived at the station
   D) the train produced steam but did not move.
10. From this passage, we gather that the children
   A) were fond of Mrs. Tickoo
   B) did not care much for Mrs. Tickoo
   C) hated Mrs. Tickoo a lot
   D) were happy that Mrs. Tickoo was leaving

PASSAGE-3

Nehru's was a many sided personality. He enjoyed reading and writing books as much as he enjoyed fighting political and social evils or resisting tyranny. In him, the scientist and the humanist were held in perfect balance. While he kept looking at special problems from a scientific standpoint, he never forgot that we should nourish the total man. As a scientist, he refused to believe in a benevolent power interested in men's affairs; but, as a self-proclaimed non-believer, he loved affirming his faith in life and the beauty of nature. Children he adored. Unlike Wordsworth, he did not see them trailing clouds of glory from their recent sojourn in heaven. He saw them as blossoms of promise and renewal, the only hope for mankind.

11. A 'many'- side personality' means
   A) a complex personality
   B) a secretive person
   C) a person having varied interests
   D) a capable person
12. Nehru enjoyed
   A) reading and writing books
   B) fighting political and social evils
   C) resisting tyranny
   D) doing all the above and much more
13. Which of the following statements reflects Nehru's point of view?
   A) Humanism is more important than science
   B) Science is supreme and humanism is subordinate to it
   C) Science and humanism are equally important
   D) There is no common ground between science and humanism
14. In this passage, 'a benevolent power interested in men's affairs' means
   A) a supernatural power or God
   B) beauty of nature
   C) the spirit of science
   D) the total man
15. Nehru thought that children
   A) were trailing clouds of glory
   B) held promise for a better future
   C) were like flowers to be loved and admired
   D) held no hope for mankind.

PASSAGE-4

When my wife, daughter and I were travelling in a rented car through Nottingham, England, on our way to a cousin's home in the northern suburbs, I became thoroughly confused by the maze of circular and one-way streets. Finally, we spotted a policeman and pulled up to ask directions. We were surprised when he removed his tall helmet, opened the back door and slid in beside our daughter. "It'll be much easier to show you", he explained.

Off we went through a series of left and right turns, traffic lights and even a short jaunt up a one-way lane the wrong way. Twenty minutes later we arrived at a main intersection. "Now you'll be all right", announced our guide. "Just turn here, stay on the road for eight or nine kilometres, and you're there". With that, he opened the door, got out, replaced his helmet and strode briskly off.

16. The tone of narration in the paragraph is
   A) sombre B) ironic
   C) comic D) satiric
17. The writer was confused by
   A) a network of circuclar roads
   B) wrong direction by traffic cops
   C) confusing road signals
   D) road map that he had with him
18. The word 'suburbs' means
   A) a remote village
   B) the city centre
   C) residential area outside a town
   D) an industrial town
19. The expression "our guide" in the paragraph means
   A) policeman B) cousin
   C) passerby D) professional guide
20. The response of the policeman can be said to be
   A) rude B) greedy
   C) imposing D) helpful

PASSAGE-5

He always remembered where he had come from and when he had some real money and moved into a comfortable house, his mother and father went with him. Later, when he grew rich and bought Graceland, he sent for his grandmother and widowed aunt. He also hired cousins and friends who needed work. He said "Sir" and 'maam' to his seniors, loved singing spirituals and when he was conscripted he served two years in the US Army with out a complaint or special favour. "When they saw me working in the kitchen and doing guard duty", Elvis Presley once said, "they realised well he's just like us".

21. Which of the following adjectives sums up Elvis as a person?
    A) humble and human
    B) cold and stiff
    C) mean and petty
    D) rude and arrogant

22. The word 'conscription' means
    A) working in the kitchen
    B) doing guard duty
    C) compulsory service in the army
    D) singing spiritual songs

23. Elvis must have come from
    A) a poor background
    B) an aristocratic background
    C) a rich background
    D) an uppder-middle-class background.

24. After getting rich, Elvis Presley
    A) turned cool towards his friends and relatives
    B) didn't do much for his friends and relatives
    C) helped his friends and relatives
    D) scorned his friends and relatives

25. The pronoun 'he' in this passage refers to
    A) US Army Officer
    B) Graceland
    C) Elvis cousin
    D) Elvis himself

ANSWERS

1. c	2. b	3. d	4. a	5. c
6. b	7. d	8. c	9. b	10. a
11. c	12. d	13. c	14. a	15. b
16. c	17. a	18. c	19. a	20. d
21. a	22. c	23. a	24. c	25. d

Unit - XV
Numerical Computation

UNIT - XV NUMERICAL COMPUTATION

1. AVERAGE

Average is defined as the sum of the items divided by the total number of items.

$$\text{Average} = \frac{\text{Sum of all items}}{\text{Number of items}}$$

SOLVED EXAMPLES

EXAMPLE-1

The ages of four brothers in a family is 22, 20, 18 and 16 respectively. Find their average age :

Sol. Average age $= \frac{22 + 20 + 18 + 16}{4} = \frac{76}{4} = 19$

EXAMPLE-2

Find the average of all prime numbers between 30 and 50.

Sol. There are five prime numbers between 30 & 50. They are 31, 37, 41, 43 and 47.

$\therefore$ Required average $= \frac{31 + 37 + 41 + 43 + 47}{5}$

$= \frac{199}{5} = 39.8$

EXAMPLE-3

Find the average of a class of 39 students is 15 years. If the age of the teacher be included, then the average increases by 3 months. Find the age of the teacher.

Sol. Total age of 39 persons = (39 × 15) years = 585 years.

Average age of 40 persons = 15 years 3 months $= \frac{61}{4}$ years.

Total age of 40 persons $= \left[\frac{61}{4} \times 40\right]$ years = 610 years.

$\therefore$ Age of the teacher = (610 − 585) years = 25 years.

EXAMPLE-4

The average weight of 10 oarsmen in a boat is increased by 1.8 kg. When one of the crew, who weighs 53 kg is replaced by a new man. Find the weight of the new man.

Sol. Total weight increased = (1.8×10) Kg = 18 Kg.

$\therefore$ Weight of the new man = (53 + 18) Kg = 71 Kg.

EXAMPLE-5

A batsman makes a score of 87 runs in the 17th inning and thus increases his average by 3. Find his average after 17th inning.

Sol. Let the average after 17th inning = x.

Then, average after 16th inning = (x − 3)

$\therefore$ 16(x - 3) + 87 = 17x

(or) x = (87 - 48) = 39

EXAMPLE-6

There were 35 students in a hostel. Due to the admission of 7 new students, the expenses of the mess were increased by Rs.42 per day while the average expenditure per head diminished by Re.1. What was the original expenditure of the mess ?

Sol. Let the original average expenditure be Rs. x then,

42(x−1) − 35x = 42

$\Rightarrow$ 42x − 42 − 35x = 42

$\Rightarrow$ 42x−35x = 42 + 42

$\Rightarrow$ 7x = 84 $\Rightarrow$ x = 12

$\therefore$ Original expenditure = Rs. (35 × 12) = Rs.420

EXAMPLE-7

The average salary per head of all workers in a factory is Rs. 95. The average salary of 15 officers is Rs. 525 and the average salary per head of rest is Rs. 85. Find the total number of workers in the workshop.

Sol. Let 'x' be the total number of workers.

The average salary of all workers i.e., x workers in the factory = Rs. 95

The total salary of all workers = Rs. 95 × x = 95x

The average salary of 15 officers = Rs. 525

The total salary of 15 officers = Rs. 525 × 15

The average salary of the rest of workers = Rs. 85

95x = (525 × 15) + (x − 15) 85

= 525 × 15 + 85x − (85 × 15)

= 7875 + 85x − 1275

95x − 85x = 7875 − 1275

10x = 6600

$x = \frac{6600}{10} = 660$

$\therefore$ Total workers = 660

PRACTICE TEST
AVERAGE

1. Average temperature of first 4 days of a weak is 38.6°C and that of the last 4 days is 40.3°C. If the average temperature of the week be 39.1°C, the temperature on 4th day is :
   a) 39.8°C b) 41.9°C c) 36.7°C d) 38.6°C

2. The mean temperature of Monday to Wednesday was 37°C and of Tuesday to Thursday was 34°C. If the temperature on Thursday was 4/5th that of Monday, what was the temperature on Thursday ?
   a) 34°C b) 35.5°C c) 36°C d) 36.5°C

3. The average age of 5 members of a committee is the same as it was 3 years ago, because an old member has been replaced by a new member. The difference between the ages of old and new member is ?
   a) 8 years b) 2 years c) 4 years d) 15 years

4. Ten years ago, the average age of a family of 4 members was 24 years. Two children having been born, the average age of the family is same today. What is the present age of the youngest child if they differ in age by 2 years ?
   a) 1 year b) 2 years c) 3 years d) 5 years

5. The average age of 11 players of a cricket team is decreased by 2 months when two of them aged 17 years and 20 years are replaced by two reserves. The average age of the reserves is :
   a) 18 years 3 months b) 17 years 1 month
   c) 17 years 7 months d) 17 years 11 months

6. The average age of a husband and a wife was 23 years when they were married 5 years ago. The average of the husband, the wife and a child, who was born during the interval, is 20 years now. How old is the child now ?
   a) less than 1 year b) 1 year
   c) 3 years d) 4 years

7. There were 35 students in a hostel. If the number of students increased by 7, the expenses of the mess were increased by Rs. 42 per day, while the average expenditure per head diminished by Re. 1. The original expenditure of the mess was :
   a) Rs. 400 b) Rs. 420
   c) Rs. 432 d) Rs. 442

8. A batsman has a certain average runs for 11 innings. In the 12th inning he made a score of 90 runs and thereby decreases his average by 5. His average after 12th inning is :
   a) 127 b) 150 c) 145 d) 217

9. The average score of a cricketer for 10 matches is 43.9 runs. If the average for the first six matches is 53, the average for the last 4 matches is :
   a) 30.25 b) 29.75 c) 31. d) 17.15

10. The average score of a cricketer in 2 matches is 27 and that in 3 others is 32. Then his average score in 5 matches is :
   a) 25 b) 29.5 c) 30 d) 11.8

11. The average weight of A, B, C is 45 kg. If the average weight of A and B be 40 kg and that of B and C be 43 Kg, then the weight of B is.
   a) 17 kg b) 20 Kg c) 26 kg d) 31 kg

12. A man covers a journey from a village P to another village Q at 5km/hr. and returns from Q to P at 3 km/hr. His average speed during the whole journey is.
   a) 4 km/hr. b) 3.75 km/hr.
   c) 3.5 km/hr. d) 4.5 km/hr.

13. With an average speed of 40 km/hr. a train reaches its destination in time. If it goes with an average speed of 35 km/hr, it is late by 15 minutes. The total journey is.
   a) 40 km. b) 70 km. c) 30 km. d) 80 km.

14. The average weight of three men A, B and C is 84 kg. Another man D joins the group and the average now becomes 80 kg. If another man E, whose weight is 3 kg more thao that of D, replaces A, then the average weight of B, C, D and E becomes 79 kg. The weight of A is.
   a) 70 kg. b) 72 kg.
   c) 75 kg. d) 80 kg.

15. The average weight of a class of 40 students is 40 kg. If the weight of the teacher be included, the average weight increases by 500 gms. The weight of the teacher is.
   a) 40.5 kg. b) 60 kg.
   c) 62 kg. c) 60.5 kg.

16. The average age of three boys 15 years. If their ages are in the ratio 3 : 5 : 7, the age of the youngest boy is.

a) 21 years b) 18 years
c) 15 years d) 9 years

17. The average age of A, B, C, D five years ago was 45 years. By including X, the present average age of all the five is 49 years. The present age of X is.

a) 64 years b) 48 years
c) 45 years d) 40 years

18. The average salary of 20 workers in an office is Rs. 1900 per month. If the Manager's salary is added, the average salary becomes Rs. 2000 p.m. What is the Manager's salary ?

a) Rs. 24,000 b) Rs. 25,200
c) Rs. 45,600 d) None of these

19. If a, b, c, d, e are five consecutive odd numbers, their average is

a) 5 (a+4) b) $\frac{abcde}{5}$
c) 5 (a+b+c+d+e) d) none of these

20. The average of first five multiples of 3 is

a) 3 b) 9
c) 12 d) 15

ANSWERS

1. b	2. c	3. d	4. c	5. c
6. d	7. b	8. c	9. a	10. c
11. d	12. b	13. b	14. c	15. d
16. d	17. c	18. d	19. d	20. b

EXPLANATIONS

1. **Ans : (b)**

Temperature on 4th day

$= (38.6 \times 4 + 40.3 \times 4 - 39.1 \times 7)^0C = 41.9^0C.$

2. **Ans : (c)**

Sum of temperatures on (M+T+W) $= (37 \times 3)^0$

$= 111^0$ ......(i)

Sum of temperatures on (T+M+h) $= (34 \times 3)^0$

$= 102^0.$

$\therefore$ Sum of temperatures on $\left(T + W + \frac{4}{5}M\right) = 102^0.$

......(ii)

$\therefore$ (i) – (ii) gives :

$^1/_5$th of temperature on Monday = (111 – 102)

$= 9^0.$

(or) Temperature on Monday $= 45^0.$

$\therefore$ Temperature on Thursday $= \frac{4}{5} \times 45^0$

$= 36^0C.$

3. **Ans : (d)**

During these 3 years, the total age would have increased (5 × 3) = 15 years.

But, it remains the same as it was 3 years ago.

$\therefore$ Difference in ages of the old and the new member = 15 years.

4. **Ans : (c)**

Total age now = (24 × 6) years = 144 years.

Total age of 4 members 10 years ago = (24 × 4) years

= 96 years.

Total age of these 4 members now = (96 + 40) years

= 136 years

Sum of the ages of 2 children now = (144 – 136) years

= 8 years

$\therefore$ $x + x + 2 = 8$ or $x = 3.$

5. **Ans : (c)**

Total decrease = (11 × 2) months = 1 year 10 months.

$\therefore$ Total age of reserves = (17+20) years – (1 year 10 months)

= 35 years 2 months.

$\therefore$ Average age of reserves = 17 years 7 months.

6. **Ans : (d)**

Total present age of husband and wife = (23×2+5×2)

= 56 years.

Total age of husband, wife and child, now = (20×3)

= 60 years.

Age of the child = (60–56) years = 4 years.

7. **Ans : (b)**

Let the original expenditure per head be Rs. x. Then,

35x + 42 = (x–1) × 42 or x = 12.

$\therefore$ Original expenditure = Rs. (35×12) = Rs. 420.

8. Ans : (c)

Let the average for 11 innings be x.

Then, $\frac{11x + 90}{12} = (x-5)$ or $x = 150.$

$\therefore$ Average for 12 innings = (150 – 5)

= 145.

9. Ans : (a)

Total of last 4 matches = $(43.9 \times 10 - 53 \times 6)$

= 121.

Average of these matches = (121 ÷ 4)

= 30.25.

10. Ans : (c)

Total score in 5 matches = $(2 \times 27 + 3 \times 32)$

= 150.

Average score = (150 ÷ 5)

= 30.

11. Ans : (d)

Total weight of A, B, C = (45×3) kg

= 135 kg.

Total weight of A and B = (40×2) kg

= 80 Kg.

Total weight of B and C = (43×2) kg

= 86 kg.

Total weight of (A + 2B + C) = (80 + 86) kg

= 166 kg.

$\therefore$ Weight of B = (166 – 135) kg

= 31 kg.

12. Ans : (b)

Average speed = $\frac{2xy}{(x+y)}$ km/hr.

$= \left(\frac{2 \times 5 \times 3}{5+3}\right)$ km/hr.

= 3.75 km/hr.

13. Ans : (b)

Let the required distance be x km.

Then, $\frac{x}{35} - \frac{x}{40} = \frac{1}{4}$

(or) $\frac{x}{280} = \frac{1}{4}$

(or) $x = \frac{280}{4}$ = 70 km.

14. Ans : (c)

Weight of D = $(80 \times 4 - 84 \times 3)$ kg

= 68 kg.

Weight of E = (68+3) kg

= 71 kg.

(B + C + D + E) weight = (79×4) kg

= 316 kg.

$\therefore$ (B+C) weight = [316 – (268 + 71) kg.

= 177 kg.

Hence, A weight = [(84 × 3) – 177} kg

= 75 kg.

15. Ans : (d)

Total weight of 40 students = (40×40) = 1600 kg.

Total weight of 40 persons = (41×40.5) kg.

= 1660.5 kg.

Weight of the teacher = (1660.5 – 1600) kg

= 60.5kg.

16. Ans : (d)

Let their ages be 3x, 5x & 7x years.

Then, $3x + 5x + 7x = 3 \times 15$ (or) $15x = 45$ (or) $x = 3$.

$\therefore$ Age of the youngest one = 9 years.

17. Ans : (c)

Total age of A, B, C, D five years ago = (45×4) years

= 180 years.

Total present age of A, B, C, D and X = (49×5) years

= 245 years.

Present age of A, B, C and D = $(180 + 5 \times 4)$ years

= 200 years.

Present age of X = 45 years.

18. Ans : (d)

Total monthly salary of 21 personals = (21×2000)

= Rs. 42,000.

Total monthly salary of 20 personals = (20×1900)

= Rs. 38,000.

Monthly salary of the Manager = Rs. 4000.

Annual salary of the Manager = Rs. 48,000.

19. Ans : (d)

Average = $\frac{a + a + 2 + a + 4 + a + 6 + a + 8}{5}$ = (a+4).

20. Ans : (b)

Average = $\frac{3(1 + 2 + 3 + 4 + 5)}{5}$

$= \left(\frac{3 \times 15}{5}\right) = 9.$

2. PERCENTAGE

A per cent is a fraction with a denominator of 100. Instead of writing the denominator, we write the symbol %.

To convert a per cent into fraction, write original number as numerator and 100 as denominator. To convert a per cent into decimal fraction, put the decimal point to the left of the original number.

eg. $5\% = \frac{5}{100} = \frac{1}{20} = 0.05$

$25\% = \frac{25}{100} = \frac{1}{4} = 0.25$

$50\% = \frac{50}{100} = \frac{1}{2} = 0.5$

By a certain percent we mean that many hundredths. Thus, x percent means x hundredths, written as x%.

To express x% as a fraction : we have $x\% = \frac{x}{100}$.

Thus, $15\% = \frac{15}{100} = \frac{3}{20}$; $24\% = \frac{24}{100} = \frac{6}{25}$ etc.

SOLVED EXAMPLES

To express $\frac{a}{b}$ as a percent : we have

$\frac{a}{b} = \left(\frac{a}{b} \times 100\right)\%$

thus, $\frac{3}{4} = \left(\frac{3}{4} \times 100\right)\% = 75\%$,

$\frac{6}{5} = \left(\frac{6}{5} \times 100\right)\% = 120\%$ etc.

$0.6 = \frac{6}{10} = \frac{3}{5} = \left(\frac{3}{5} \times 100\right)\% = 60\%$.

EXAMPLE PROBLEMS

EXAMPLE-1

Express each of the following as a fraction :

(i) 56% (ii) 0.08%

Sol. i) $56\% = \frac{56}{100} = \frac{14}{25}$ ii) $0.08\% = \frac{8}{10000} = \frac{1}{1250}$

EXAMPLE-2

Express each of the following as a decimal :

(i) 28% (ii) 0.04%

Sol. (i) $28\% = \frac{28}{100}$

$= 0.28$

(ii) $0.04\% = \frac{0.04}{100}$

$= .0004$

EXAMPLE-3

Express each of the following as rate percent :

(i) $\frac{23}{36}$, (ii) .004

Sol. (i) $\frac{23}{36} = \left(\frac{23}{36} \times 100\right)\% = \left(\frac{575}{9}\right)\% = 63\frac{8}{9}\%$

(ii) $.004 = \frac{4}{1000} = \left(\frac{4}{1000} \times 100\right)\% = .4\%$

EXAMPLE-4

(i) What percent of 72 is 6 ?

(ii) What percent of 6.5 litres is 130 ml ?

Sol. (i) Required percentage $= \left(\frac{6}{72} \times 100\right)\% = 8\frac{1}{3}\%$.

(ii) Required percentage

$= \left[\frac{130}{6.5 \times 1000} \times 100\right]\% = 2\%$.

Two Important Rules (short cut Methods) :

(i) If A is R% more than B, then :

B is less than A by $\left\{\frac{R}{(100+R)} \times 100\right\}\%$

(ii) If A is R% less than B, then B is more than A by

$\left\{\frac{R}{(100-R)} \times 100\right\}\%$

EXAMPLE-5

If A's income is 33% more than that of B, then how much percent is B's income less than that of A ?

Sol. Required percentage = $\left\{\frac{33}{(100+33)} \times 100\right\}\% = 24.8\%$

EXAMPLE-6

If A's height is 24% less than that of B, then how much percent is B's height more than that of A ?

Sol. Required percentage = $\left\{\frac{24}{(100-24)} \times 100\right\}\%$

$= \left\{\frac{24}{70} \times 100\right\}\% = 31.57\%$

Two Important Rules (Short cut Method) :

(i) If the price of a commodity increases by R% then reduction in consumption, not to increase the expenditure is :

$$\left\{\frac{R}{100+R} \times 100\right\}\%$$

(ii) If the price of a commodity decreased by R%, then the increase in consumption, not to decrease the expenditure is :

$$\left\{\frac{R}{(100-R)} \times 100\right\}\%$$

EXAMPLE-7a

If the price of tea is increased by 20%, find by how much percent must a householder reduce her consumption of tea so as not to increase the expenditure ?

Sol. Reduction in consumption

$= \left\{\frac{R}{(100+R)} \times 100\right\}\%$

$= \left\{\frac{20}{120} \times 100\right\}\% = 16\frac{2}{3}\%$

EXAMPLE-7b

If the price of sugar decreases by 20% find how much percent must a house holder increase her consumption of sugar so as not to decrease the expenditure ?

Sol. Increase in consumption

$= \left(\frac{R}{(100-R)} \times 100\right)\% = \frac{20}{80} \times 100 = 25\%$

Results on population (Formulae) :

Let the population of a town be p now and suppose it increases at the rate of R% per annum, then :

(i) population after 'n' years = $p\left[1 + \frac{R}{100}\right]^n$

(ii) population 'n' years ago = $\frac{p}{\left(1 + \frac{R}{100}\right)^n}$

EXAMPLE-8

The population of a town is 176400. If it increases at the rate of 5% per annum, what will be its population 2 years hence ?

What was it 2 years ago ?

Sol. Population after 2 years

$= 176400 \times \left[1 + \frac{5}{100}\right]^2$

$= 176400 \times \frac{21}{20} \times \frac{21}{20} = 194481$

Population 2 years ago = $\frac{176400}{\left(1 + \frac{5}{100}\right)^2}$

$= \left(176400 \times \frac{20}{21} \times \frac{20}{21}\right)$

$= 160000.$

Results on Depreciation Formulae :

Let the present value of a machine be p. suppose it depreciates at the rate of R% per annum then :

(i) Value of the machine after 'n' years

$= p\left[1 - \frac{R}{100}\right]^n$

(ii) Value of the machine 'n' years ago

$= \frac{p}{\left[1 - \frac{R}{100}\right]^n}$

EXAMPLE-9

The value of a machine depreciates at the rate of 10% per annum. If its present value is Rs.162000, what will be its worth after 2 years ? What was the value of the machine 2 years ago ?

Sol. Value of the machine after 2 years.

$= \text{Rs.}\left[162000 \times \left[1 - \frac{10}{100}\right]^2\right]$

$= \text{Rs.}\left[162000 \times \frac{9}{10} \times \frac{9}{10}\right] = \text{Rs. } 1312200$

Value of the machine 2 years ago.

$$= \text{Rs.} \left\{ \frac{162000}{\left\{1 - \frac{10}{100}\right\}^2} \right\}$$

$$= \text{Rs.} \left[162000 \times \frac{10}{9} \times \frac{10}{9} \right]$$

= Rs.200000.

EXAMPLE-10

Two numbers are less than a third number by 30% and 37% respectively. How much percent is the second number less than the first ?

Sol. Let the third number be 100.

Then, first number = 70 and Second number = 63

$\therefore$ Second number is less than the first by

$$= \left[\frac{70 - 63}{70} \times 100 \right] \% = 10\%$$

EXAMPLE-11

1 Litre of water is added to 5 litres of a 20% solution of alcohol in water. Find the strength of alcohol in the new solution.

Sol. Alcohol in 5 Litres = (20% of 5 Litres)

$$= \frac{20}{100} \times 5 = 1 \text{ Litre.}$$

Alcohol in 6 litres of new mix = 1 litre.

$$\therefore \text{ percentage of alcohol} = \left[\frac{1}{6} \times 100 \right] \%$$

$$= 16\frac{2}{3}\%$$

EXAMPLE-12

From the salary of an officer, 10% is deducted as house rent, 15% of the rest he spends on children's education and 10% of the balance, he spends on clothes. Then, he is left with Rs.4131. Find his total salary.

Sol. Let his total salary be Rs. x Then,

90 % of 85% of 90% of x = 4131

$$\therefore x = \frac{4131 \times 100 \times 100 \times 100}{90 \times 85 \times 90}$$

= 6000

EXAMPLE-13

Due to a reduction of $6\frac{1}{4}$ % in the prices of sugar, a man is able to buy 1 Kg more for Rs. 120. Find the original and reduced rate of sugar.

Sol. Let Original rate be Rs. x per kg.

Reduced Rate

$$= \text{Rs.} \left[100 - \frac{25}{4} \right] \cdot \frac{1}{100} \cdot x = \frac{15x}{16} \text{ per kg.}$$

$$\therefore \frac{120}{\frac{15x}{16}} - \frac{120}{x} = 1$$

$$\text{(or) } \frac{128}{x} - \frac{120}{x} = 1 \text{ (or) } x = 8.$$

$\therefore$ Original rate = Rs. 8 per kg.

$$\text{Reduced rate} = \text{Rs.} \left[\frac{15}{16} \times 8 \right] \text{ per kg}$$

= Rs.7.50 per kg.

EXAMPLE-14

If A earns $33\frac{1}{3}$ % more than B, then B earns less than A by : (a) $16\frac{2}{3}$ %, (b) 20 % (c) 25% (d) $33\frac{1}{3}$ %

Sol. Required percentage

$$= \left[\frac{\left(\frac{100}{3}\right)}{\left[100 + \frac{100}{3}\right]} \times 100 \right] \%$$

$$= \left(\frac{100}{400} \times 100 \right) \%$$

= 25% Answer is C.

EXAMPLE-15

Find the missing figures :

(i) (...... ? ....) % of 25 = 2.125

(ii) 0.25 % of (..... ? ......) = 0.04

Sol. (i) Let x % of 25 = 2.125 then,

$$\frac{x}{100} \times 25 = 2.125 \text{ (or) } x$$

= (2.125 × 4) = 8.5

(ii) Let 0.25 % of x = 0.04

$$\text{Then } \frac{0.25x}{100} = 0.04 \text{ (or) } x = \frac{.04 \times 100}{0.25} = 16$$

EXAMPLE-16

Which is greatest in $16\frac{2}{3}$%, $\frac{2}{15}$ and 0.17 ?

Sol. $16\frac{2}{3}\% = \left(\frac{50}{3} \times \frac{1}{100} \right) = \frac{1}{6} = 0.166$

$\frac{2}{15} = 0.133$ & the third number is 0.17

clearly 0.17 is the greatest.

PRACTICE TEST
PERCENTAGE

1. .02 = (...?...)%

a) 20 b) 2 c) .02 d) .2

2. Which is greatest in $33\frac{1}{3}\%$, $\frac{4}{15}$ and .35 ?

a) $33\frac{1}{3}\%$ b) $\frac{4}{15}$

c) .35 d) can not be compared

3. If the side of a square is increased by 30%, then its area is increased by

a) 30% b) 9% c) 60% d) 69%

4. By how much percent is four fifth of 70 lesser than five-seventh of 122 ?

a) 42% b) 30% c) 24% d) 36%

5. If 70% of the students in a school are boys and the number of girls be 504, the number of boys is :

a) 1176 b) 1008 c) 1208 d) 3024

6. A man spends Rs. 1750 per month and saves $12\frac{1}{2}\%$ of his income. His monthly income is :

a) Rs. 2000 b) Rs. 2135 c) Rs. 2200 d) Rs. 1968.75

7. An ore contains 12% copper. How many kilograms of the ore are required to get 69 kg of copper ?

a) 828 kg. b) 575 kg.

c) $1736\frac{3}{23}$ kg d) 424 kg.

8. 96% of the population of a village is 23040. The total population of the village is :

a) 32256 b) 24000

c) 24936 d) 25640

9. After deducting a commission of 5% a T.V. set costs Rs. 7600. Its grass value is :

a) Rs. 7980 b) Rs. 8000

c) Rs. 8200 d) None of these

10. A fruit seller had some apples. He sells 40% and still he has 420 apples. Originally, he had :

a) 588 apples b) 600 apples

c) 700 apples d) 672 apples

11. In an examination, 65% of the total examinees passed. If the number of failures is 420, the total number of examinees is :

a) 693 b) 567 c) 1000 d) 1200

12. In an examination, it is required to get 36% of maximum marks to pass. A student got 113 marks and declared failed by 85 marks. The maximum marks are :

a) 500 b) 550 c) 640 d) 1008

13. In an examination, there were 2000 candidates, out of which 900 candidates were boys and rest are girls. If 32% of the boys and 38% of the girls passed, then the total percentage of failed candidates is :

a) 35.3% b) 70% c) 64.7% d) 68.5%

14. In an examination 42% students failed in Hindi and 52% failed in English. If 17% failed in both the subjects, the percentage who passed in both the subjects is :

a) 23% b) 40% c) 34% d) 27%

15. A man received 10% increase in his salary. His new salary is Rs. 506. His original salary was.

a) Rs. 455.40 b) Rs. 440

c) Rs. 460 d) Rs. 456

16. A man's wages were decreased by 50%. Again the reduced wares were increased by 50%. He has a loss of :

a) 0% b) 2.5% c) .25% d) 25%

17. From the salary of an officer 10% is deducted as house rent, 15% of the rest he spends on children's education and 10% of the balance, he spends on clothes. After this expenditure, he is left with :

a) Rs. 2200 b) Rs. 2000

c) Rs. 2100 d) Rs. 2040

18. Two numbers are less than a third number by 30% and 37% respectively. How much percent is the second number less than the first ?

a) 7% b) 10% c) 4% d) 3%

19. 30 quintals is what percent of 2 metric tonnes :

a) 15% b) 1.5% c) 150% d) 30%

20. x% of y is y% of (?)

a) x b) $\frac{y}{100}$ c) $\frac{x}{100}$ d) 100 x

21. The price of rice has increased by 60%. In order to restore to the original price, the new price must be reduced by :

a) 33 1/3% b) 37 1/2%

c) 40% d) 45%

22. 1 litre of water is added to 5 litres of a 20% solution of alcohol in water. The strength of alcohol is now:
a) 12 1/2% b) 16 1/2% c) 24% d) 16%

23. On decreasing the price of T.V. sets by 30% its sale is increased by 20%. What is the effect on the revenue received by the shopkeeper ?
a) 10% increase b) 10% decrease
c) 16% increase d) 16% decrease

24. The population of a town is 8000. It increases by 10% during first year and by 20% during second year. The population after 2 years will be :
a) 10400 b) 10560
c) 10620 d) None of these

25. The population of a town increases by 5% annually. If it is 15435 now, its population 2 years ago was:
a) 14000 b) 15000
c) 13700 d) 14800

ANSWERS

1. b	2. c	3. d	4. b	5. a
6. a	7. b	8. b	9. b	10. c
11. d	12. b	13. c	14. a	15. c
16. d	17. b	18. b	19. c	20. a
21. b	22. b	23. d	24. b	25. a

EXPLANATIONS

1. **Ans : (b)**
$02 = \frac{2}{100} = 2\%.$

2. **Ans : (c)**
$\frac{4}{15} = \left(\frac{4}{15}\times 100\right)\% = 26\frac{2}{3}\%;$
$\therefore\ .35 = \left(\frac{35}{100}\times 100\right)\% = 35\%.$
Hence, .35 is greatest.

3. **Ans : (d)**
Let the side of the square be 100 metres.
Area of the square = (100 × 100) sq. metres
= 10000 sq.m.
Increased length of side = 130 metres.
New area = (130 × 130) sq. metres = 16900 sq. m.
$\therefore$ Increase in area $= \left(\frac{6900}{10000}\times 100\right) = 69\%.$

4. **Ans : (b)**
$\left(\frac{4}{5}\text{of } 70\right) = 56$ and $\left(\frac{5}{7}\times 112\right) = 80.$
So, it is less by 24.
$\therefore$ Less % $= \left(\frac{24}{80}\times 100\right)\% = 30\%$

5. **Ans : (a)**
30% of total number of students = 504.
$\therefore$ Total number of students $= \frac{504\times 100}{30}$
= 1680.
So, number of boys = 70% of 1680 = 1176.

6. **Ans : (a)**
$87\frac{1}{2}\%$ of x = 1750.
$\therefore\ x = \frac{1750\times 100\times 2}{175}$ = Rs. 2000.

7. **Ans : (b)**
12 : 69 : : 100 : x
$\therefore\ x = \frac{69\times 100}{12}$ = 575 kg.

8. **Ans : (b)**
Let the total population be x. Then, 96% of x = 23040. or $x = \left(\frac{23040\times 100}{96}\right) = 24000.$

9. **Ans : (b)**
95% of x = 7600.
$\therefore\ x = \left(\frac{7600\times 100}{95}\right) = 8000.$

10. **Ans : (c)**
Let 60% of x = 420.
Then, $\frac{60}{100}\times x = 420$
(or) $x = \frac{420\times 100}{60} = 700$

11. Ans : (d)

35% of x = 420

$\therefore \frac{35}{100} \times x = 420$ (or) $x = \frac{420 \times 100}{35} = 1200.$

12. Ans : (b)

Let maximum marks be x.

Then, 36% of x = (113 + 85).

$\therefore x = \left(\frac{198 \times 100}{36}\right) = 550$

13. Ans : (c)

Candidates failed

= (68% of 900) + (62% of 1100)

$= \left(\frac{68}{100} \times 900\right) + \left(\frac{62}{100} \times 1100\right)$

= (612 + 682) = 1294.

Percentage of failed candidates

$= \left(\frac{1294}{2000} \times 100\right) = 64.7\%.$

14. Ans : (a)

Failed in Hindi only = (42–17) = 25%.

Failed in English only = (52–17) = 35%.

Failed in one or both the subjects = (25+35+17)% = 77%.

∴ Students passed = (100 – 77)% = 23%.

15. Ans : (c)

110% of original salary = Rs. 506.

$\therefore \frac{110}{100} \times x = 506$ or $x = \left(\frac{506 \times 100}{110}\right) =$ Rs. 460.

16. Ans : (d)

Let the wages at initial stage be Rs. 100.

Wages after reduction = Rs. 50.

New wages after increase.

= (150% of Rs. 50) = Rs. $\left(\frac{150}{100} \times 50\right)$ = Rs. 75.

$\therefore$ Loss $= \left(\frac{25}{100} \times 100\right)\% = 25\%.$

17. Ans : (b)

[90% of 85% of 90% of X] = 1377.

OR $\left(\frac{90}{100} \times \frac{85}{100} \times \frac{90}{100} \times x\right) = 1377.$

OR $\frac{1377x}{2000} = 1377$ or $x = 2000.$

18. Ans : (b)

Let the third number be 100.

Then, 1st number = 70 and second number = 63.

$\therefore$ Second is less than the first by $\left(\frac{7}{70} \times 100\right)\% = 10\%.$

19. Ans : (c)

Required percentage $= \left(\frac{30}{20} \times 100\right) = 150\%.$

20. Ans : (a)

Let (x%) of y = (y%) of z

or $\frac{xy}{100} = \frac{yz}{100}$ or $z = \frac{100xy}{100y} = x.$

21. Ans : (b)

Clearly, the cost Rs. 160 must be reduced to Rs. 100.

$\therefore$ Reduction % $= \left(\frac{60}{160} \times 100\right)\% = 37\ 1/2\%.$

22. Ans : (b)

Alcohol in 6 litres = (20% of 5 litre) = 1 litre.

$\therefore$ Alcohol percent $= \left(\frac{1}{6} \times 100\right)\% = 16\ 2/3\%.$

23. Ans : (d)

Let there be 100 sets sold for Rs. 100 each.

Then, revenue received = Rs. (10000).

New price of each set = Rs. 70 and

sets sold now are 120.

∴ Revenue received now = Rs. (70 × 120) = Rs. 8400.

Hence, decrease in revenue $= \left(\frac{1600}{10000} \times 100\right)\% = 16\%.$

24. Ans : (b)

Population after 2 years

$= 8000 \times \left(1 + \frac{10}{100}\right)\left(1 + \frac{20}{100}\right).$

$= \left(8000 \times \frac{11}{10} \times \frac{6}{5}\right) = 10560.$

25. Ans : (a)

Population 2 years ago.

$= \frac{15435}{\left(1 + \frac{5}{100}\right)^2}$

$= \left(15435 \times \frac{20}{21} \times \frac{20}{21}\right) = 14000.$

3. RATIO & PROPORTION

The relationship between two quantitites which tells how many times one is if the other is called the ratio between two quantities.

Suppose A's salary is Rs. 20,000 and B's salary is Rs. 15,000. The ratio between A's salary and B's salary is

$$20000 : 15000 \text{ (or) } 4 : 3 \text{ or } \frac{4}{3}$$

The equality of two ratios is called proportion 4 : 3 : : 8 : 6

DEFINITIONS - FORMULAE - SOLVED EXAMPLES

Ratio : The ratio of two quantities in the same units is a fraction that one quantity is of the other. The ratio a : b represents a fraction $\frac{a}{b}$. The first term of a ratio is called antecedent while the second term is known as consequent.

Rule : The multiplication or division of each term of a ratio by a same non-zero number does not effect the ratio.

Thus, $4 : 5 = 8 : 10 = 12 : 15 = 16 : 20 = \frac{4}{5} : 1$ etc.

Proportion : The quality of two ratios is called proportion. If a : b = c : d, we write,

a : b : : c: d and we say that a, b, c, d are in proportion.

In a propotion, the first and fourth terms are known as extremes, while second and third terms are known as means.

We have,

product of means = product of extremes.

Fourth proportional : If a : b = c : d, then d is called the fourth proportional to a, b, c.

Third proportional : The third proportional to a, b is the fourth proportional to a, b, b.

Mean proportional : Mean proportional between a and b is $\sqrt{ab}$.

Comparison of Ratios : We say that (a : b) > (c : d) if $\frac{a}{b} > \frac{c}{d}$

Compounded Ratio : The compounded ratio of the ratios (a : b), (c : d), (e : f) is (ace) : (bdf)

Some more Definitions :

(i) $a^2 : b^2$ is called the 'duplicate ratio of a : b'

(ii) $\sqrt{a} : \sqrt{b}$ is called the sub-duplicate ratio of a:b

(iii) $a^3 : b^3$ is called the triplicate ratio of a : b

(iv) $a^{\frac{1}{3}} : b^{\frac{1}{3}}$ is called the sub-triplicate ratio of a : b

(v) If $\frac{a}{b} = \frac{c}{d}$ then $\frac{a+b}{a-b} = \frac{c+d}{c-d}$

(componendo & dividendo)

Variation : We say that x is directly proportional to y if x = ky for some constant k and we write,

$x \propto y$. Also we say that x is inversly proportional to y. If $x = \frac{k}{y}$ for some constant k and we write, $x \propto \frac{1}{y}$.

EXAMPLE PROBLEMS

EXAMPLE-1

If a : b = 5 : 9 and b : c = 4 : 7, find a : b : c and a : c.

Sol. $a : b = 5 : 9,\ b : c = 4 : 7 = 4 \times \frac{9}{4} : 7 \times \frac{9}{4} = 9 : \frac{63}{4}$

$\therefore a : b : c = 5 : 9 : \frac{63}{4}$

$= 20 : 36 : 63$

Also $\frac{a}{c} = \frac{a}{b} \times \frac{b}{c} = \frac{5}{9} \times \frac{4}{7} = \frac{20}{63}$

So a : c = 20 : 63.

EXAMPLE-2

Find

(i) Fourth proportional to 4, 9, 12 ;

(ii) third proportional to 16 and 36 ;

(iii) mean proportion between 0.08 and 0.18.

Sol. (i) Let 4 : 9 : : 12 : x

then, $4x = 9 \times 12$ or $x = \left[\frac{9 \times 12}{4}\right] = 27.$

(ii) Let 16 : 36 : : 36 : x

then 16 : 36 : : 36 : x

then, $16x = 36 \times 36$ or

$x = \frac{36 \times 36}{16} = 81.$

(iii) Mean proportion between 0.08 and 0.18

$$= \sqrt{0.08 \times 0.18} = \sqrt{\frac{144}{100 \times 100}} = \frac{12}{100} = 0.12$$

EXAMPLE-3

Find

(i) the duplicate ratio of 3 : 7 ;

(ii) the triplicate ratio of 2 : 5 ;

(iii) the sub – duplicate ratio of 25 : 16 ;

(iv) the sub – triplicate ratio of $(125)^{1/3} : (64)^{1/3}$;

(v) the compounded ratio of (2 : 5), (3 : 4) and (4 : 9)

Sol. (i) Duplicate ratio of 3 : 7 is $3^2 : 7^2 = 9 : 49$.

(ii) Triplicate ratio of 2 : 5 is $2^3 : 5^3 = 8:125$

(iii) The sub -duplicate ratio of 25:16

$$= \sqrt{25} : \sqrt{16} = 5 : 4$$

(iv) The sub-triplicate ratio of 125 : 64

$$= (125)^{\frac{1}{3}} : (64)^{\frac{1}{3}} = 5 : 4$$

(v) compounded ratio of (2 : 5) (3 : 4) and

$$(4 : 9) = \left(\frac{2}{5} \times \frac{3}{4} \times \frac{4}{9}\right) = \frac{2}{15} \text{ i.e., } 2 : 15$$

EXAMPLE-4

If x : y = 3 : 4, find (4x+5y) : (5x-2y).

Sol. Given $\frac{x}{y} = \frac{3}{4}$

$$\therefore \frac{4x + 5y}{5x - 2y} = \frac{4\left(\frac{x}{y}\right) + 5}{5\left(\frac{x}{y}\right) - 2} \text{ (Dividing Nr \& Dr by y)}$$

$$= \frac{4 \times \frac{3}{4} + 5}{5 \times \frac{3}{4} - 2} = \frac{8 \times 4}{7} = \frac{32}{7}.$$

EXAMPLE-5

Divide Rs. 672 in the ratio 5 : 3.

Sol. Sum of the terms of the ratio = (5 + 3) = 8.

First part = Rs. $\left[672 \times \frac{5}{8}\right]$

= Rs.420

Second part = Rs. $\left[672 \times \frac{3}{8}\right]$

= Rs.252

EXAMPLE-6

A bag contains 50 paise, 25 paise and 10 paise coins in the ratio 5 : 9 : 4, amounting to Rs. 206. Find the number of coins of each type.

Sol. Ratio of values $= \frac{5}{2} : \frac{9}{4} : \frac{4}{10} = 50 : 48 : 8$

Value of 50 paise coins = Rs. $\left[206 \times \frac{50}{103}\right]$ = Rs.100

Value of 25 paise coins = Rs. $\left[206 \times \frac{45}{103}\right]$ = Rs. 90.

Value of 10 paise coins = Rs. $\left[206 \times \frac{8}{103}\right]$ = Rs. 16

∴ Number of 50 paise coins = $(100 \times 2) = 200$

Number of 25 paise coins = $(90 \times 4) = 360$

Number of 10 paise coins = $(16 \times 10) = 160$

EXAMPLE-7

Which one of 2 : 3 ; 4 : 5, 7 : 9 and 11: 13 is the largest ratio ?

Sol. $2 : 3 = \frac{2}{3} = 0.666$; $4 : 5 = \frac{4}{5} = 0.8$;

$7 : 9 = \frac{7}{9} = 0.777$; $11 : 13 = \frac{11}{13} = 0.846$

clearly, $0.846 > 0.8 > 0.777 = 0.666$

Hence, 11 : 13 is the largest ratio.

EXAMPLE-8

A mixture contains alcohol and water in the ratio 4 : 3. If 5 litres of water is added to the mixture, the ratio becomes 4 : 5. Find the quantity of alcohol in the given mixture.

Sol. Let the quantity of alcohol & water be 4x & 3x respectively.

Then, $\frac{4x}{3x+5} = \frac{4}{5} \Leftrightarrow 20x = 12x + 20 \Leftrightarrow x = 2.5$

∴ Alcohol in given mixture = 4x litres = 10 litres.

EXAMPLE-9

Suppose x varies inversely as square of y and when y = 3, then x = 4. Find x, when y = 6.

Sol. Let $x = \frac{k}{y^2}$, where k is a constant. When

y = 3, we have x = 4.

$\therefore 4 = \frac{k}{9}$ or $k = 36$ $\therefore x = \frac{36}{y^2}$.

Now, when $y = 6 \Rightarrow x = \frac{36}{36} = 1$.

PRACTICE TEST
RATIO & PROPORTION

1. Rs. 53 is divided among three persons A, B and C in such a way that B gets Rs. 7 more than what B gets and B gets Rs. 8 more than what C gets. Then the ratio of their shares is :
   a) 16 : 9 : 18 b) 25 : 18 : 10
   c) 18 : 25 : 10 d) 15 : 8 : 30
2. A sum of Rs. 1300 is divided between A, B, C and D. What is A's share, if :
   $$\frac{\text{A's share}}{\text{B's share}} = \frac{\text{B's share}}{\text{C's share}} = \frac{\text{C's share}}{\text{D's share}} = \frac{2}{3}$$
   a) Rs. 140 b) Rs. 160
   c) Rs. 240 d) Rs. 320
3. A bag contains Rs. 102 in the form of rupee, 50 paise and 10 paise coins in the ratio of 3 : 4 : 10. The number of 10 paise coins is
   a) 60 b) 80 c) 170 d) 340
4. A sum of money is divided among A, B and C so that to each rupee A gets, B gets 65 paise and C gets 35 paise. If C's share is Rs. 28 the sum is :
   a) Rs. 120 b) Rs. 140
   c) Rs. 160 d) Rs. 180
5. The ratio between two numbers is 3 : 4 and their L.C.M. is 180. The first number is :
   a) 60 b) 45 c) 15 d) 20
6. The L.C.M. and H.C.F of two numbers x and y is l and h respectively. Then
   a) l : h = x : y b) x : h = l : y
   c) x : l : y : h d) x : h = y : l
7. A train travels at 72 kmph and another at 30 metres per second. What is the ratio of the speeds of the first and the second train ?
   a) 2 : 3 b) 3 : 2 c) 4 : 3 d) 3 : 4
8. A man has some hens and cows. If the number of heads be 48 and number of feet equals 140, the number of hens will be :
   a) 26 b) 22 c) 23 d) 24
9. The ratio of the number of boys and girls in a school of 720 students is 7 : 5. How many more girls should be admitted to make the ratio 1 : 1 ?
   a) 90 b) 120
   c) 220 d) 240
10. The average age of 3 girls is 20 years and their ages are in the proportion 3 : 5 : 7. The age of youngest girl is :
    a) 4 years
    b) 6 years and 8 months
    c) 8 years and 3 months
    d) 12 years
11. A boy 1.4 metres tall casts a shadow 1.2 metres long at the same time when a building casts a shadow 5.4 metres long. The height of the building is :
    a) 4.63 m (nearly) b) 3.21 m (nearly)
    c) 6.3 m d) 5.6 m
12. The speeds of three cars in the ratio 2 : 3 : 4. The ratio between the times taken by these cars to travel the same distance is :
    a) 2 : 3 : 4 b) 4 : 3 : 2
    c) 4 : 3 : 6 d) 6 : 4 : 3
13. Two whole numbers whose sum is 64, cannot be in the ratio :
    a) 5 : 3 b) 7 : 1
    c) 3 : 4 d) 9 : 7
14. The income of A and B are in the ratio 3 : 2 and their expenditures in the ratio 5 : 3. If each saves Rs. 1000, A's income is :
    a) Rs. 3000 b) Rs. 4000
    c) Rs. 6000 d) Rs. 9000
15. The ratio of the first and second class fares between two stations is 4 : 1 and that of the number of passengers travelling by first and second class is 1 : 40. If Rs. 1100 is collected as fare, the amount collected from the first class passengers is :
    a) Rs. 275 b) Rs. 315
    c) Rs. 137.50 d) Rs. 100
16. The prices of a scooter and a moped are in the ratio of 9 : 5. If a scooter costs Rs. 4200 more than in a moped, the price of a moped is :
    a) Rs. 5200 b) Rs. 5250
    c) Rs. 5050 d) Rs. 5000
17. A father's age was 5 times his son's age 5 years ago and will be three times the son's age after 2 years. The ratio of their present ages is :
    a) 5 : 2 b) 5 : 3
    c) 10 : 3 d) 11 : 5

18. A mixture contains milk and water in the ratio 5 : 1. On adding 5 litres of water in the ratio of milk to water becomes 5 : 2. The quantity of milk in the mixture is :

a) 16 litres b) 25 litres
c) 32.5 litres d) 22.75 litres

19. The ratio of milk and water in 85kg of adulterated milk is 27 : 7. The amount of water which must be added to make the ratio 3 : 1 is :

a) 5 kg b) 6.5 kg
c) 7.25 kg d) 8 kg

20. In what ratio must 25% alcohol be mixed with 50% alcohol to get a mixture of 40% alcohol strength?

a) 1 : 2 b) 2 : 1
c) 2 : 3 b) 3 : 2

ANSWERS

1. b	2. b	3. c	4. c	5. b
6. b	7. c	8. a	9. b	10. d
11. c	12. d	13. c	14. c	15. d
16. b	17. c	18. b	19. a	20. c

EXPLANATIONS

1. **Ans : (b)**

Suppose C gets Rs. x.

Then B gets Rs. (x + 8) and A gets Rs. (x + 15)

$\therefore x + x + 8 + x + 15 = 53$ or $x = 10$

Thus,

A gets Rs. 25, B gets Rs. 18 and C gets Rs. 10.

$\therefore A : B : C = 25 : 18 : 10$

2. **Ans : (b)**

$A : B = 2 : 3 = 4 : 6; \; B : C = 2 : 3 = 6 : 9$

and $C : D = 2 : 3 = 9 : \frac{27}{2}$

$\therefore A : B : C : D = 4 : 6 : 9 : \frac{27}{2} = 8 : 12 : 18 : 27$

Hence A's share = Rs. $\left(1300 \times \frac{8}{65}\right)$ = Rs. 160.

3. **Ans : (c)**

The Ratio of the values of 3 rupee coins, 4 fifty paise coins and 10 ten paise coins is

$$3 : 4 \times \frac{1}{2} : 10 \times \frac{1}{10} \text{ i.e., } 3 : 2 : 1$$

Value of 10 paise coins = Rs. $\left(102 \times \frac{1}{6}\right)$ = Rs. 17

Number of 10–paise coins = $(17 \times 10) = 170$

4. **Ans : (c)**

$A : B : C = 100 : 65 : 35$ or $20 : 13 : 7$

If C gets Rs. 7, total sum = Rs. 40

If C gets Rs. 28, total sum = Rs. $\left(\frac{40}{7} \times 28\right)$

= Rs. 160

5. **Ans : (b)**

If the numbers are 3x & 4x, then their L.C.M. is 12x.

$\therefore 12x = 180$ or $x = 15$.

So, the first number is 45.

6. **Ans : (b)**

Since the product of H.C.F. and L.C.M. of two numbers is equal to the product of numbers, we have

$$xy = lh \text{ or } \frac{x}{h} = \frac{l}{y} \text{ or } x : h = l : y$$

7. **Ans : (c)**

Speed of 1st train = $\left(72 \times \frac{5}{18}\right)$ m/sec

= 20 m/sec.

Speed of 2nd train = 15 m/sec

Ratio of their speeds = 20 : 15

= 4 : 3

8. **Ans : (a)**

Let there be x hens and y cows. Then,

$x + y = 48$ and $2x + 4y = 140$

Solving these equations, we get x = 26

9. **Ans : (b)**

Number of boys = $\left(720 \times \frac{7}{12}\right) = 420$

Number of girls = (720 – 420)

= 300

So, to make the ratio 1 : 1, the number of more girls to be admitted = (420 – 300)

= 120

10. Ans : (d)

Age of youngest girl $= \left(60 \times \frac{3}{15}\right)$

$= 12$ years

11. Ans : (c)

Let $\frac{1.2}{5.4} = \frac{1.4}{x}$

Then, $x = \frac{1.4 \times 5.4}{1.2}$

$= 6.3$ metres

12. Ans : (d)

Times taken are in the reverse ratio of speeds.

$\therefore$ Ratio of times taken $= \frac{1}{2} : \frac{1}{3} : \frac{1}{4}$ or $6 : 4 : 3$

13. Ans : (c)

The sum of two parts of the ratio must be a factor of 64.

But $(3 + 4) = 7$ is not a factor of 64.

So, 64 can not be divided in the ratio 3 : 4 to get whole numbers.

14. Ans : (c)

Let their incomes be 3x, 2x and expenditure 5y, 3y

Then, $3x - 5y = 1000$ and $2x - 3y$

$= 1000$

Solving these equations, we get $x = 2000$

$\therefore$ A's income = Rs. 6000

15. Ans : (d)

Ratio of amounts collected from 1st & 2nd class

$= (4 \times 1 : 1 \times 40)$

$= (1 : 10)$

$\therefore$ Amount collected from 1st class passengers

$=$ Rs. $\left(1100 \times \frac{1}{11}\right)$

$=$ Rs. 100

16. Ans : (b)

Let the prices of a scooter & a moped be Rs.9x and Rs. 5x respectively. Then,

$(9x - 5x) = 4200$ or $x = 1050$

$\therefore$ Price of a moped = Rs. (5×1050)

$=$ Rs. 5250

17. Ans : (c)

Let the present ages of father and son be 'x' and 'y' years respectively. Then,

$x - 5 = 5(y - 5)$ and $x + 2$

$= 3(y + 2)$

or $5y - x = 20$ and $x - 3y = 4$.

Solving these equations,

we get $x = 40$ and $y = 12$

$\therefore$ Ratio of their present ages = 40 : 12

(or) 10 : 3

18. Ans : (b)

Let the mixture contain milk 5x litres and water x litres.

Then, $\frac{5x}{x+5} = \frac{5}{2}$ or $x = 5$

$\therefore$ Quantity of milk in the given mixture = 25 litres.

19. Ans : (a)

Quantity of milk $= \left(85 \times \frac{27}{34}\right)$

$= \frac{135}{2}$ kg

$= 67.5$ kg

Quantity of water $= \left(85 - \frac{135}{2}\right)$

$= \frac{35}{2}$ kg

$= 17.5$ kg

Let x kg. of water be added to make the ratio 3 : 1.

Then, $\frac{67.5}{17.5 + x} = \frac{3}{1}$

or) $x = 5$ kg.

20. Ans : (c)

By the rule of allegation, we have

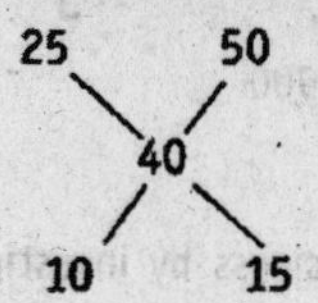

Required ratio is 10 : 15 or 2: 3

4. PARTNERSHIP

When two or more than two persons run a business jointly, they are called partners and the deal is known as partnership.

DEFINITIONS - FORMULAE - SOLVED EXAMPLES

RATIO OF DIVISION OF GAINS

i) When investments of all the partners are for the same time; the gain or loss is distributed among the partners in th EXAMPLE PROBLEMS

ii) When investments are for different time, then equivalent capitals are calculated for a unit of time by taking (capital × number of units of time). Now gain or loss is divided in the ratio of these capitals.

WORKING & SLEEPING PARTNERS

A partner who manages the business is known as a working partner and the one who simply invests the money is a sleeping partner.

EXAMPLE PROBLEMS

EXAMPLE-1

Manoj and Vinod started a business by investing Rs.120000 and Rs.135000 respectively. Find the share of each, out of an annual profit of Rs. 35700

Sol. Ratio of their shares = 120000 : 135000

$= 8 : 9$

$\therefore$ Manoj's share = Rs. $\left[35700 \times \frac{8}{17}\right]$

= Rs.16800

Vinod's share = Rs. $\left[35700 \times \frac{9}{17}\right]$

= Rs.18900

EXAMPLE-2

Sanjeev started a business by investing Rs. 36000. After 3 months Rajeev joined him by investing Rs. 36000. Out of an annual profit of Rs.37100, find the share of each.

Sol. Ratio of their capitals = $36000 \times 12 : 36000 \times 9$

$= 4 : 3$

Sanjeev's share = Rs. $\left[37100 \times \frac{4}{7}\right]$

= Rs.21200

Rajeev's share = Rs. $\left[37100 \times \frac{3}{7}\right]$

= Rs.15900.

EXAMPLE-3

A, B and C start a business each investing Rs.20000. After 5 months A withdraw Rs.5000, B withdraw Rs.4000 and C invests Rs. 6000 more. At the end of the year, a total profit of Rs. 69900 was recorded. Find the share of each.

Sol. Ratio of the capitals of A, B and C.

$= 20000 \times 5 + 15000 \times 7 : 20000 \times 5 + 16000 \times 7 : 20000 \times 5 + 26000 \times 7$

$= 205000 : 212000 : 282000$

$= 205 : 212 : 282$

$\therefore$ A's share = Rs. $\left[69900 \times \frac{205}{699}\right]$

= Rs.20500

B's share = Rs. $\left[69900 \times \frac{212}{699}\right]$

= Rs.21200

C's share = Rs. $\left[69900 \times \frac{282}{699}\right]$

= Rs.28200

EXAMPLE-4

A started a business with Rs. 21000 and is joined afterwards by B with Rs. 36000. After how many months did B join, if the profits at the end of the year are divided equally?

Sol. Suppose B joined after x months.

Then B's money was invested for (12–x) months.

$\therefore 21000 \times 12 = 36000 \times (12 - x)$

$\Rightarrow 21 \times 12 = 36 \times (12 - x)$

$\Rightarrow 36 \times 12 - 36x = 21 \times 12$

$\Rightarrow 36x = 36 \times 12 - 21 \times 12$

$\Rightarrow 36x = 180$

$x = 5$

So, B joined after 5 months.

PRACTICE TEST

PARTNERSHIP

1. A is a working and B a sleeping partner in a business. A puts in Rs. 12000 and B Rs. 2000. A receives 10% of the profits for managing the rest being divided in proportion to their capitals. Out of a total profit of Rs. 9600, the money received by A is:

   a) Rs. 3240 b) Rs. 4200
   c) Rs. 3600 d) Rs. 4500

2. A and B invest in a business in the ratio 3 : 2. Five percent of the total profit goes to charity. If A's share of profit is Rs. 855, the total profit is :

   a) Rs.1425 b) Rs.1500
   c) Rs.1537.50 d) Rs.1576

3. A, B and C enter into a partnership and their capitals are in the proportion of $\frac{1}{3}:\frac{1}{4}:\frac{1}{5}$. A withdraws half his capital at the end of 4 months. Out of a total annual profit of Rs. 847. A's share is :

   a) Rs. 315 b) Rs. 252
   c) Rs. 280 d) Rs. 412

4. A, B, C and D enter into a partnership. A subscribe (1/4) of the capital, B(1/5) of the capital, C(1/6) of the capital and the remainder is subscribed by D. If the total profit is Rs.180000, the share of D is

   a) Rs.45000 b) Rs.69000
   c) Rs.57000 d) Rs.36000

5. A, B and C together start a business. A invests 3 times as much as B invests and B invest two -third of what C invests. Then, the ratio of capitals of A, B and C is :

   a) 3 : 9 : 2 b) 6 : 10 : 15
   c) 5 : 3 : 2 d) 6 : 2 : 3

6. A, B and C enter into partnership. A invests some money at the beginning ; B invests double the amount after 6 months and C invests thrice the amount after 8 months. If the annual profit be Rs. 18000, C's share is :

   a) Rs. 5750 b) Rs. 7200 c) Rs. 6000 d) Rs. 7500

7. A starts a business with Rs. 30000 and 4 months later B joins. If at the end of the year the profits are divided by A and B in the proportion of 9 : 4, what was B's capital ?

   a) Rs.20000 b) Rs.25000
   c) Rs.20700 d) Rs.18000

8. A and B invest Rs. 60000 and Rs.80000 in a business. A receives Rs. 100 per month out of profit for running the business and the rest of the profit is divided in the ratio of investments. If A receives Rs. 3900 totally, then B receives :

   a) Rs. 3200 b) Rs. 2700
   c) Rs. 3600 d) Rs. 2925

9. A' s share in a partnership is Rs.7500 more than B's but A's capital is invested for 8 months while B's for 12 months. If their annual shares of profit be equal, then A's capital is:

   a) Rs.25600 b) Rs.27500
   c) Rs.22500 d) Rs.30000

10. In a partnership A invests (1/6) of the capital for (1/6) of the time B invests 1/3 of the capital for (1/3) of the time and C, the rest of the capital for the whole time. Out of a profit of Rs.4600, B's share is :

    a) Rs. 800 b) Rs. 1000 c) Rs. 650 d) Rs. 960

11. A and B enter into partnership. A invests Rs.16000 for 8 months and B remains in the business for 4 months. Out of a total profit B claims (2/7) of the profit. How much money did B contribute ?

    a) Rs. 13600 b) Rs. 12800
    c) Rs. 10500 d) Rs. 11900

12. A, B and C invest Rs. 3512420; Rs. 4222180 and Rs. 4065400 in a partnership. At the end of second year, A withdraws Rs. 11 lakhs and C invests Rs. 8 lakhs more. What would be the share of C in the profit of Rs. 1053000 at the end of third year?

    a) Rs. 379966.2 b) Rs. 283117.8
    c) Rs. 389886 d) None of these

13. A, B and C invest Rs. 2000, Rs. 3000 and Rs. 4000 in a business. After one year A removed his money. B and C continued the business for one more year. If the net profit after 2 years be Rs. 3200, then A's share in the profit is :

    a) Rs. 1000 b) Rs. 600
    c) Rs. 800 d) Rs. 400

14. Suresh invested Rs. 12000 in a shop and Dinesh joined him after 4 months by investing Rs. 7000. If the net profit after one year be Rs. 13300, Dinesh's share in the profit is :

    a) Rs. 9576 b) Rs. 4900 c) Rs. 8400 d) Rs. 3724

15. Dilip, Ram and Amar started a shop by investing Rs. 2700, Rs. 8100 and Rs. 7200 respectively. At the end of one year, the profit was distributed. If Ram's share was Rs. 3600, what was their total profit?

a) Rs. 10800 b) Rs. 11600
c) Rs. 8000 d) None of these

16. A and B enter into a partnership investing Rs. 12000 and Rs. 16000 respectively. After 8 months, C also joins the business with a capital of Rs. 15000. The share of C in a profit of Rs. 45600 after 2 years will be

a) Rs. 12000 b) Rs. 14400
c) Rs. 19200 d) Rs. 21200

17. Four milkmen rented a pasture. A grazed 18 cows for 4 months ; B 25 cows for 2 months ; C 28 cows for 5 months and D 21 cows for 3 months. If A's share of rent is Rs. 360, the total rent of the field is

a) Rs. 1500 b) Rs. 1625 c) Rs. 1650 d) Rs. 1600

18. A, B, C enter into a partnership with shares in the ratio $\frac{7}{2}:\frac{4}{3}:\frac{6}{5}$. After 4 months, A increases his share by 50%. If the total profit at the end of one year be Rs. 21600, then B's share in the profit is

a) Rs. 2100 b) Rs. 4000 c) Rs. 3600 d) Rs. 2400

19. Jagmohan, Rooplal and Pandey ji rented a video casset for one week at a rent of Rs. 350. If they use it for 6 hours, 10 hours and 12 hours respectively, what is the rent to be paid by Pandey ji ?

a) Rs. 75 b) Rs. 125
c) Rs. 35 d) Rs. 150

20. A, B and C enter into a partnership by making investments in the ratio 3 : 5 : 7. After a year C invests another Rs. 337600 while A withdraws Rs. 45600. The ratio of investments then changes to 24 : : 9 : 167. How much did A invest initially ?

a) Rs. 45600
b) Rs. 96000
c) Rs. 141600
d) None of these

ANSWERS

1. b	2. b	3. c	4. b	5. d
6. c	7. a	8. c	9. c	10. a
11. b	12. c	13. d	14. d	15. c
16. a	17. b	18. b	19. d	20. c

EXPLANATIONS

1. Ans : (b)

Ratio of capitals = 12000 : 20000 = 3 : 5

A's salary = Rs. $\left(9600 \times \frac{10}{100}\right)$

= Rs. 960

Balance profit = Rs. (9600 − 960)

= Rs. 8640

Out of this profit, A's share = Rs. $\left(8640 \times \frac{3}{8}\right)$

= Rs. 3240

Total money received by A = Rs. (3240 + 960)

= Rs. 4200.

2. Ans : (b)

Let the total profit be Rs. 100

After paying to charity, A's share = Rs. $\left(95 \times \frac{3}{5}\right)$

= Rs. 57

$\therefore \frac{\text{A's actual share}}{\text{actual total profit}} = \frac{57}{100}$

i.e., Total profit

= Rs. $\left(\frac{855 \times 100}{57}\right)$

= Rs. 1500.

3. Ans : (c)

Ratio of capitals in the beginning = $\frac{1}{3}:\frac{1}{4}:\frac{1}{5}$ or 20 : 15 : 12.

Ratio of investments for the whole year

= (20 × 4 + 10 × 8) : (15 × 12) : (12 × 12)

= 40 : 45 : 36.

$\therefore$ A's share = Rs. $\left(\frac{847 \times 40}{121}\right)$

= Rs. 280

4. **Ans : (b)**

Let the total capital be Rs. x.

Then, the ratio of their capitals

$$\frac{x}{4}:\frac{x}{5}:\frac{x}{6}:\left\{1-\left(\frac{1}{4}+\frac{1}{5}+\frac{1}{6}\right)\right\}x$$

i.e., $\frac{x}{4}:\frac{x}{5}:\frac{x}{6}:\frac{23x}{60}$ (or) 15 : 12 : 10 : 23.

$\therefore$ D's share of profit = Rs. $\left(180000\times\frac{23}{60}\right)$

= Rs. 69000

5. **Ans : (d)**

Suppose C invests Rs. x.

Then, B invests Rs. $\frac{2}{3}x$ and A invests Rs. 2x.

$\therefore$ Ratio of investments of A, B and C

$= 2x : \frac{2}{3}x : x$ (or) 6 : 2 : 3

6. **Ans : (c)**

Ratio of investments of A, B and C

$= (12\times x : 6\times 2x : 4\times 3x)$

= 1 : 1 : 1

$\therefore$ C's share of profit = Rs. 6000

7. **Ans : (a)**

$$\frac{30000\times 12}{x\times 8}=\frac{9}{4}$$

(or) $x=\left(\frac{30000\times 12\times 4}{9\times 8}\right)$

= Rs. 20000.

8. **Ans : (c)**

Clearly, Rs. $(3900-1200)$= Rs. 2700 is received by A as a share of profit divided in the ratio of investments.

$\therefore\ \frac{2700}{\text{B's share}}=\frac{60000}{80000}=\frac{3}{4}$(or)

B's share = Rs. $\left(\frac{2700\times 4}{3}\right)$

= Rs. 3600.

9. **Ans : (c)**

Let A's capital be Rs. x.

Then, 8x = (x − 7500) × 12 (or)

4x = 7500 × 12 (or)

x = Rs. 22500

10. **Ans : (a)**

Let the total capital be Rs. x and the business carried for 12 months. Then, the ratio of the investments of A, B, C is

$$\left(\frac{x}{6}\times 2\right):\left(\frac{x}{3}\times 4\right):\left\{x-\left(\frac{x}{6}+\frac{x}{3}\right)\right\}\times 12 \text{ (or)}$$

$\frac{1}{3}:\frac{4}{3}:6$ (or) 1 : 4 : 18

$\therefore$ B's share of profit = Rs. $\left(\frac{4600\times 4}{25}\right)$

= Rs. 800

11. **Ans : (b)**

A's share = $\frac{5}{7}$ of total profit.

$\therefore$ Ratio of shares of profits of A and B is $\frac{5}{7}:\frac{2}{7}$

i.e., 5 : 2

So, $\frac{16000\times 8}{x\times 4}=\frac{5}{2}$

(or) $x=\frac{16000\times 8\times 2}{4\times 5}$

= Rs. 12800

$\therefore$ B contributed Rs. 12800.

12. **Ans : (c)**

Ratio of investments = (3512420 × 2 + 2412420) : (4222180 × 3) : (4065400 × 2 + 4665400)

= 9437260 : 12666540 : 12996200

Sum of ratios = 35100000

$\therefore$ C's share = Rs. $\left(\frac{12996200}{35100000}\times 1053000\right)$

= Rs. 389886

13. **Ans : (d)**

Ratio of investments = (2000 × 1) : (3000 × 2) : (4000 × 2)

= 1 : 3 : 4

$\therefore$ A's share = Rs. $\left(3200\times\frac{1}{8}\right)$

= Rs. 400

14. **Ans : (d)**

Ratio of investments = (12000 × 12) : (7000 × 8)

= 18 : 7

Dinesh's share in the profit = Rs. $\left(13300 \times \frac{7}{25}\right)$

= Rs. 3724

15. Ans : (c)

Ratio of investments = 2700 : 8100 : 7200

= 3 : 9 : 8

Let the total profit be Rs. x.

Then, Ram's share = Rs. $\left(x \times \frac{9}{20}\right)$

$\therefore \frac{9x}{20} = 3600$

(or) $x = \frac{20 \times 3600}{9}$

= 8000

16. Ans : (a)

Ratio of investments

= (12000 × 24) : (16000 × 24) : (15000 × 16)

= 6 : 8 : 5

Share of C in the profit = Rs. $\left(45600 \times \frac{5}{19}\right)$

= Rs. 12000

17. Ans : (b)

Ratio of rents

= (18 × 4) : (25 × 2) : (28 × 5) : (21 × 3)

= 72 : 50 : 140 : 63.

If total rent is Rs. x, then A's share = Rs. $\left(\frac{x \times 72}{325}\right)$

$\therefore \frac{72x}{325} = 360$

(or) $x = \frac{325 \times 360}{72} = 1625$

18. Ans : (b)

The given ratio is $\frac{7}{2} : \frac{4}{3} : \frac{6}{5}$

(or) 105 : 40 : 36.

Let them initially invest Rs. 105, Rs. 40 and Rs. 36 respectively,

Ratio of investments = [105 × 4 + (150% of 105) × 8]:

(40 × 12) : (36 × 12)

= 1680 : 480 : 432

= 35 : 10 : 9

$\therefore$ B's share in the profit = Rs. $\left(21600 \times \frac{10}{54}\right)$

= Rs. 4000.

19. Ans : (d)

Ratio of rents = 6 : 10 : 12 or 3 : 5 : 6.

Pandey's share of rent = Rs. $\left(350 \times \frac{6}{14}\right)$

= Rs. 150

20. Ans : (c)

Let the initial investments made by A, B, C be 3x, 5x and 7x respectively. Then,

(3x – 45600) : 5x : (7x + 337600) : 24 : 59 : 167

$$\therefore \frac{3x - 45600}{24} = \frac{5x}{59} = \frac{7x + 337600}{167}$$

This gives, x = 47200.

Hence, the initial investment made by A = Rs. 141600

5. CHAIN RULE

Chain rule is the rule of three. It is a method of finding the fourth proportional when the other three are given.

DEFINITIONS - SOLVED EXAMPLES

DIRECT PROPORTION

Two quantities are said to be directly proportional if on the increase (or decrease) of the one, the other increases (or decreases) to the same extent.

EXAMPLE-1

The cost of articles is directly proportional to the number of articles.

(More articles, More cost) & (Less articles, less cost).

EXAMPLE-2

The work done is directly proportional to the number of men working at it.

(more men, more work) & (less men, less work)

INDIRECT PROPORTION

Two quantities are said to be indirectly proportional if on the increase of the one, the other decreases to the same extent and vice versa.

EXAMPLE-1

Time taken to cover a distance is inversely proportional to the speed of the car. (More speed, less is the time taken to cover a distance)

EXAMPLE-2

Time taken to finish a work is inversely proportional to the number of persons working at it.

(More persons, less is the time taken to finish a job)

Remark : *In solving questions on chain rule, we make repeated use of finding the fourth proportional. We compare every item with the term to be found out.*

EXAMPLE PROBLEMS

EXAMPLE-1

If 15 toys cost Rs. 234, what do 35 toys cost ?

Sol. More toys, More cost (Direct proportion) Let the required cost be x Rs.

then $15 : 35 : : 234 : x$ (or) $15 \times x = 35 \times 234$

$\therefore x = \left[\dfrac{35 \times 234}{15}\right] = 546.$

$\therefore$ Cost of 35 toys = Rs. 546

EXAMPLE-2

If 36 men can do a certain piece of work in 25 days, in how many days will 15 men do it ?

Sol. Less men, more days (Indirect proportion). Let the required number of days be x. So, inverse ratio of men is equal to ratio of time taken.

$\therefore 15 : 36 : : 25 : x$ (or) $15 \times x = 36 \times 25$

$\therefore x = \left[\dfrac{36 \times 25}{15}\right] = 60.$

Hence the required number of days = 60.

EXAMPLE-3

If 20 men can build a wall 56 metres long in 6 days, what length of a similar wall can be built by 35 men in 3 days ?

Sol. Since the length is to be found out, we compare each item with the length as shown below. More men, more length built (Direct)

Less days, Less length built (Direct)

Less days, less length built (Direct)

$$\left.\begin{array}{ll}\text{Men} & 20 : 35 \\ \text{Days} & 6 : 3\end{array}\right\} : : 56 : x$$

$\therefore 20 \times 6 \times x = 35 \times 3 \times 56$ (or)

$$x = \frac{35 \times 3 \times 56}{20 \times 6} = 49$$

Hence, the required length = 49 metres.

EXAMPLE-4

If 15 men working 9 hours a day can reap a field in 16 days, in how many days will 18 men reap the field, working 8 hours a day ?

Sol. Since the number of days is to be found out, we compare each item with the number of days. More men, less days (Indirect)

Less working hours, more days (Indirect)

$$\left.\begin{array}{ll}\text{Men} & 18 : 15 \\ \text{Working hours} & 8 : 9\end{array}\right\} : : 16 : x$$

$\therefore 18 \times 8 \times x = 15 \times 9 \times 16$ (or)

$$x = \frac{15 \times 9 \times 16}{18 \times 8} = 15$$

∴ Required number of days = 15

EXAMPLE-5

If 8 men working 9 hours a day can build a wall 18 m long, 2 m wide and 12 m high in 10 days, how many men will be required to build a wall 32 m long, 5 m wide and 9 m high, working 6 hours a day in 8 days?

Sol. We shall compare each item with the number of men.

Less working hours, more men (Indirect)
More length, more men (Direct)
More breadth, more men (Direct)
Less hight, less men (Direct)
Less days, more men (indirect)

$$\left.\begin{array}{ll} \text{working hours} & 6:9 \\ \text{length} & 18:32 \\ \text{Breadth} & 2:5 \\ \text{Height} & 12:9 \\ \text{Days} & 8:10 \end{array}\right\} :: 8 : x$$

$$\therefore\ 6 \times 18 \times 2 \times 12 \times 8 \times x = 9 \times 32 \times 5 \times 9 \times 10 \times 8$$

$$\therefore\ x = \left\{\frac{9 \times 32 \times 5 \times 9 \times 10 \times 8}{6 \times 18 \times 2 \times 12 \times 8}\right\} = 50$$

∴ Required number of men = 50

EXAMPLE-6

If 9 engines consume 24 metric tonnes of coal, when each is working 8 hours a day ; how much coal will be required for 8 engines, each running 13 hours a day, it being given that 3 engines of the former type consume as much as 4 engines of latter type.

Sol. We shall compare each item with the quantity of coal.

Less engines, less coal consumed (Direct)

More working hours, more coal consumed (Direct)

If 3 engines of former type consume 1 unit, then 1 engine will consume $\frac{1}{3}$ unit.

If 4 engines of latter type consume 1 unit, then 1 engine will consume $\frac{1}{4}$ unit.

Less rate of consumption, less coal consumed (direct)

$$\left.\begin{array}{ll} \therefore \text{Number of Engines} & 9:8 \\ \text{Working hours} & 8:13 \\ \text{Rate of consumption} & \frac{1}{3}:\frac{1}{4} \end{array}\right\} :: 24 : x$$

$$9 \times 8 \times \frac{1}{3} \times x = 8 \times 13 \times \frac{1}{4} \times 24 \text{ (or)}$$

$$24x = 48 \times 13 \text{ (or) } x = \frac{48 \times 13}{24} = 26$$

∴ Required consumption of coal = 26 metric tonnes.

EXAMPLE-7

A contract is to be completed in 46 days and 117 men were set to work, each working 8 hours a day. After 33 days $\frac{4}{7}$ of the work is completed. How many additional men may be employed so that the work may be completed in time, each man now working 9 hours a day?

Sol. Remaining work = $\left[1 - \frac{4}{7}\right] = \frac{3}{7}$.

Remaining period = (46 – 33) days = 13 days

Less work, less men (Direct)
Less days, more men (Indirect)
More Hrs, Day, Less men (Indirect)

$$\left.\begin{array}{ll} \therefore \text{Work} & \frac{4}{7}:\frac{3}{7} \\ \text{Days} & 13:33 \\ \text{Hrs./day} & 9:8 \end{array}\right\} :: 117 : x$$

$$\therefore\ \frac{4}{7} \times 13 \times 9 \times x = \frac{3}{7} \times 33 \times 8 \times 117 \text{ (or)}$$

$$x = \left\{\frac{3 \times 33 \times 8 \times 117}{4 \times 13 \times 9}\right\} = 198$$

∴ Additional men to be employed = (198 –117) = 81

EXAMPLE-8

A garrison of 3300 men had provisions for 32 days, **when given at the rate of 850 gms per head. At the end of 7 days, a reinforcement arrives and it was found that the provisions will last 17 days more, when given at the rate of 825 gms per head. What is the strength of the reinforcement ?**

Sol. *The problem becomes :*

3300 men taking 850 gms per head have provisions **for (32 – 7) or 25 days. How many men taking 825** grams each have provisions for 17 days ?

Less ratio per head, more men (Indirect)

Less days, more men (Indirect)

$$\left.\begin{array}{ll} \text{Ratio} & 825 : 850 \\ \text{Days} & 17 : 25 \end{array}\right\} :: 3300 : x$$

$$\therefore\ 825 \times 17 \times x = 850 \times 25 \times 3300 \text{ (or)}$$

$$x = \frac{850 \times 25 \times 3300}{825 \times 17} = 5000$$

∴ Strength of reinforcement

= (5000 – 3300) = 1700

PRACTICE TEST

CHAIN RULE

1. The price of 357 mangoes is Rs. 1517.25, what will be the approximate price of 49 dozens of such mangoes ?

   a) Rs. 3000 b) Rs. 3500 c) Rs. 4000 d) Rs. 2500

2. If 10 men can reap a field in 8 days, then 8 men will reap the same field in :

   a) 6 days b) 4 days c) 10 days d) 20 days

3. 14 pumps of equal capacity can fill a tank in 6 days. If the tank has to be filled in 4 days, the number of extra pumps needed is :

   a) 7 b) 14 c) 21 d) 28

4. If 18 persons can build a wall 140 m long in 42 days, the number of days that 30 persons will take to complete a similar wall 10m long is:

   a) 180 b) 21 c) 18 d) 24

5. If 18 binders bind 900 books in 10 days, how many binders will be required to bind 660 books in 12 days?

   a) 22 b) 14 c) 13 d) 11

6. If 6 men working 8 hours a day earn Rs.840 per week, then 9 men working 6 hours a day will earn how much per week ?

   a) Rs. 840 b) Rs. 945
   c) Rs. 1620 d) Rs. 1680

7. If 21 cows eat that much as 15 buffaloes, how many cows will eat that much as 105 buffaloes ?

   a) 147 b) 75 c) 63 d) None of these

8. A garrison had provision for a certain number of days. After 10 days $\frac{1}{5}$ th of the men desert and it is found that the provisions will now last just as long as before. How long was that ?

   a) 15 days b) 25 days
   c) 35 days d) 50 days

9. If 9 men working $7\frac{1}{2}$ hours a day can finish a piece of work in 20 days, then how many days will be taken by 12 men, working 6 hours a day to finish the work, it being given that 2 men of latter type work as much as 3 men of the former type?

   a) $9\frac{1}{2}$ b) 11
   c) $12\frac{1}{2}$ d) 13

10. A contractor employed 30 men to do a piece of work in 38 days. After 25 days, he employed 5 men more and the work was finished one day earlier. How many he would have been behind, if he had not employed additional men ?

    a) 1 b) $1\frac{1}{4}$ c) $1\frac{3}{4}$ d) $1\frac{1}{2}$

ANSWERS

1. d	2. c	3. a	4. c	5. d
6. b	7. a	8. d	9. c	10. a

EXPLANATIONS

1. Ans : (d)

1. (d) More mangoes, more cost (Direct)

$357 : 49 \times 12 :: 1517.25 : x$

$\therefore x \times 357 = 49 \times 12 \times 1517.25$

$\therefore x = \frac{49 \times 12 \times 1517.25}{357}$

$= 2499$

$\therefore$ Approximate cost = Rs.2500

2. Ans : (c)

Less men, more days (Indirect) $8 : 10 :: 8 : x$ (or)

$8x = 10 \times 8$ (or) $x = \left[\frac{10 \times 8}{8}\right] = 10$

$\therefore$ Required number of days = 10.

3. Ans : (a)

Less days, more pumps (Indirect)

$4 : 6 :: 14 : x$ (or) $4x = 6 \times 14$ (or)

$x = \left[\frac{6 \times 14}{4}\right] = 21.$

$\therefore$ Extra pumps needed = (21–14) = 7

4. Ans : (c)

Less length, less days (Direct)

More persons, less days (Indirect)

$$\left.\begin{array}{ll}\text{Length} & 140 : 100 \\ \text{persons} & 30 : 18\end{array}\right\} :: 42 : x$$

$\therefore 140 \times 30 \times x = 100 \times 18 \times 42$

$$\text{or } x = \left[\frac{100 \times 18 \times 42}{140 \times 30}\right]$$

$= 18$

$\therefore$ Required number of days = 18

5. Ans : (d)

Less books, less binders (Direct)

More days, less binders (Indirect)

$$\left.\begin{matrix}\text{Books} & 900 : 660 \\ \text{Days} & 12 : 10\end{matrix}\right\} :: 18 : x$$

$\therefore 900 \times 12 \times x = 660 \times 10 \times 18$ (or)

$$x = \left[\frac{660 \times 10 \times 18}{900 \times 12}\right]$$

$= 11$

$\therefore$ Required number of binders = 11

6. Ans : (b)

More men, more earning (Direct)

Less hrs/day, less earning (Direct)

$$\left.\begin{matrix}\text{Men} & 6 : 9 \\ \text{Hrs./Day} & 8 : 6\end{matrix}\right\} :: 840 : x$$

$\therefore 6 \times 8 \times x = 9 \times 6 \times 840$ (or)

$$x = \frac{9 \times 6 \times 840}{6 \times 8}$$

$= 945$

$\therefore$ Required earning = Rs.945

7. Ans : (a)

15 buffaloes = 21 cows

$$105 \text{ buffaloes} = \left\{\frac{21}{15} \times 105\right\}$$

cows = 147 cows

8. Ans : (d)

Initially, let there be x men having food for y days.

After 10 days,

x men had food for (y–10) days.

i.e., $\left(x - \frac{x}{5}\right)$ or $\frac{4x}{5}$ men had food for y days.

$$\therefore x(y - 10) = \left(\frac{4x}{5}\right) y$$

(or) $xy - 50x = 0$

(or) $x(y - 50) = 0$

So, $y = 50$ $[\because x \neq 0]$

9. Ans : (c)

More men, less days (Indirect)

Less Hrs/day, more days (Indirect)

Less speed, more days (Indirect)

$$\left.\begin{matrix}\text{Men} & 12 : 9 \\ \text{Hrs./day} & 6 : \frac{15}{2} \\ \text{Speed} & \frac{1}{2} : \frac{1}{3}\end{matrix}\right\} :: 20 : x$$

$$\therefore 12 \times 6 \times \frac{1}{2} \times x = 9 \times \frac{15}{2} \times \frac{1}{3} \times 20$$

$$\text{(or) } x = \frac{9 \times \frac{1.5}{2} \times \frac{1}{3} \times 20}{12 \times 6 \times \frac{1}{2}}$$

$$= \frac{\frac{45}{2} \times 20}{\frac{72}{2}}$$

$$= \frac{45}{72} \times 20$$

$$= \frac{25}{2} = 12\frac{1}{2}.$$

$\therefore$ Required number of days = $12\frac{1}{2}$

10. Ans : (a)

After 25 days, 35 men complete the work in 12 days.

35 men can finish the remaining work in 12 days.

$\therefore$ 30 men can finish it in $\left[\frac{12 \times 35}{30}\right]$

= 14 days i.e., 1 day behind.

6. TIME & WORK

In solving problems on time and work, it is necessary to find the work done in a unit time which is generally an hour or a day.

DEFINITIONS - FORMULAE - SOLVED EXAMPLES

GENERAL RULES

i) If A can do a piece of work in n days, then A's 1 day's work = $\frac{1}{n}$

ii) If A's 1 day's work = $\frac{1}{n}$, then A can finish the work in n days.

iii) If A is thrice as good a work man as B, then:
Ratio of work done by A and B = 3 : 1
Ratio of times taken by A & B to finish a work = 1 : 3

EXAMPLE PROBLEMS

EXAMPLE-1

A can do a piece of work in 10 days which B alone can do in 12 days. In how many days will they finish the work, both working together ?

Sol. A's 1 day's work $= \frac{1}{10}$

B's 1 day's work $= \frac{1}{12}$

(A + B)'s 1 day's work $= \left[\frac{1}{10} + \frac{1}{12}\right]$

$= \frac{11}{60}.$

$\therefore$ Both will finish the work in $\frac{60}{11} = 5\frac{5}{11}$ days

EXAMPLE-2

Two persons A and B working together can dig a trench in 8 hours while A alone can dig it in 12 hours. In how many hours B alone can dig such a trench?

Sol. (A + B)'s 1 hour's work $= \frac{1}{8}$

A's 1 hour's work $= \frac{1}{12}$

$\therefore$ B's 1 hour's work $= \left[\frac{1}{8} - \frac{1}{12}\right]$

$= \frac{1}{24}$

Hence B alone can dig the trench in 24 hours.

EXAMPLE-3

A and B can do a piece of work in 12 days, B and C can do it in 15 days; A and C can do it in 20 days. In how many days will A, B and C finish it, working all together? Also, find the number of days taken by each to finish it working alone.

Sol. (A + B)'s 1 day's work $= \frac{1}{12}$;

(B + C)'s 1 day's work $= \frac{1}{15}$ and

(A + C)'s 1 day's work $= \frac{1}{20}$.

Adding, we get = 2(A + B + C)'s 1 day's work.

$= \left[\frac{1}{12} + \frac{1}{15} + \frac{1}{20}\right]$

$= \frac{1}{5}$

$\therefore$ (A + B + C)'s one day's work $= \frac{1}{10}$.

Thus A, B and C together can finish the work in 10 days.

Now A's 1 day's work

= [(A + B + C)'s 1 day's work]

− [(B + C)'s 1 day's work]

$= \left[\frac{1}{10} - \frac{1}{15}\right] = \frac{1}{30}.$

$\therefore$ A alone can finish the work in 30 days.

Similarly B's 1 day's work $= \left[\frac{1}{10} - \frac{1}{20}\right]$

$= \frac{1}{20}.$

$\therefore$ B alone can finish the work in 20 days.

And C's 1 day's work $= \left[\frac{1}{10} - \frac{1}{12}\right]$

$= \frac{1}{60}$

$\therefore$ C alone can finish the work in 60 days.

EXAMPLE-4

3 men can complete a piece of work in 6 days. Two days after they started the work, 3 more men joined them. How many days will they take to complete the remaining work ?

Sol. Work done by 3 men in 2 days = $\frac{1}{6} \times 2$

$= \frac{1}{3}$

Remaining work = $\left[1 - \frac{1}{3}\right]$

$= \frac{2}{3}$

Now, 3 men's 1 day's work = $\frac{1}{6}$

1 man's 1 day's work = $\frac{1}{18}$

6 men's 1 day's work = $\left[6 \times \frac{1}{18}\right]$

$= \frac{1}{3}$

Now, $\frac{1}{3}$ work is done by them in 1 day

$\because \frac{2}{3}$ work is done by them in $\left[3 \times \frac{2}{3}\right] = 2$ days.

EXAMPLE-5

2 men and 3 boys can do a piece of work in 10 days while 3 men and 2 boys can do the same work in 8 days. In how many days can 2 men and 1 boy do the work?

Sol. Let 1 man's 1 day's work = x & boy's 1 day's work = y

then, $2x + 3y = \frac{1}{10}$ .... (1)

and $3x + 2y = \frac{1}{8}$ .... (2)

Solving. Multiplying (1) with 3 and (2) with 2 we get

$$6x + 9y = \frac{3}{10}$$

$$6x + 4y = \frac{1}{4}$$

$$5y = \frac{3}{10} - \frac{1}{4}$$

$$\Rightarrow 5y = \frac{1}{20}$$

$$\Rightarrow y = \frac{1}{100}$$

Substituting y's value in (1) and we get

$$2x + 3 \times \frac{1}{100} = \frac{1}{10}$$

$$200x + 3 = 10$$

$$200x = 7$$

$$x = \frac{7}{200}$$

$\therefore$ (2 men + 1 boys) 1 day's work

$$= \left[2 \times \frac{7}{200} + 1 \times \frac{1}{100}\right]$$

$$= \frac{16}{200} = \frac{2}{25}$$

So, 2 men & 1 boy together can finish the work in $\frac{25}{2} = 12\frac{1}{2}$ days.

EXAMPLE-6

A is twice as good a workman as B and together they finish a piece of work in 18 days. In how many days will A alone finish the work ?

Sol. (A's 1 day's work) : (B's 1 day's work) = 2 :1

(A + B)'s 1 day's work = $\frac{1}{18}$

Divide f(1,18) in the ratio 2 :1

$\therefore$ A's 1 day's work = $\left[\frac{1}{18} \times \frac{2}{3}\right]$

$= \frac{1}{27}$

Hence A alone can finish the work in 27 days.

EXAMPLE-7

A and B undertake to do a piece of work for Rs. 600. A alone can do it in 6 days while B alone can do it in 8 days with the help of C, they finish it in 3 days. Find the share of each.

Sol. C's 1 day's work = $\frac{1}{3} - \left[\frac{1}{6} + \frac{1}{8}\right]$

$= \frac{1}{24}$

$\therefore$ A : B : C = Ratio of their 1 day's work

$= \frac{1}{6} : \frac{1}{8} : \frac{1}{24}$

= 4 : 3 : 1

$\therefore$ A's share = Rs. $\left[600 \times \frac{4}{8}\right]$

= Rs.300;

B's share = Rs. $\left[600 \times \frac{3}{8}\right]$

= Rs.225

C's share = Rs. [600 – (300 + 225)]

= Rs.75.

PIPES AND CISTERNS

A pipe connected with a cistern is called an inlet pipe or an outlet pipe according as it fills it or empties it respectively.

Filled part of a tank :

= (sum of work done by inlets)

– (sum of work done by outlets)

EXAMPLE PROBLEMS

EXAMPLE-1

When a cistern is in thorough repair it is filled by the inlet in 3 1/2 hours. But, due to a leakage in the bottom, it takes half an hour longer. When the cistern is full, how long would the leak takes to empty the cistern?

Sol. Work done by inlet only in 1 hour = $\frac{1}{(7/2)}$

$= \frac{2}{7}$.

Time taken to fill the cistern (with leak) = 4 hours.

Work done by inlet and leak in 1 hour = (1/4).

Work done by leak only in 1 hour = $\left(\frac{2}{7} - \frac{1}{4}\right)$

$= \frac{1}{28}$.

Time taken by leak to empty the cistern = 28 hours.

EXAMPLE-2

A cistern is filled by pipe A in 10 hours and the full cistern can be leaked out by an exhaust pipe B in 12 hours. If both the pipes are opened, in what time the cistern is full?

Sol. The part of the cistern filled up in 1 hour, when both the pipes are opened = $\left(\frac{1}{10} - \frac{1}{12}\right) = \frac{1}{60}$.

∴ The cistern shall be full in 60 hours.

EXAMPLE-3

Two pipes can fill cistern in 14 and 16 hours respectively. The pipes are opened simultaneously and it is found that due to leakage in the bottom 32 minutes extra are taken for the cistern to be filled up. When the cistern is full, in what time will the leak empty it ?

Sol. Work done by the two pipes in 1 hour.

$= \left(\frac{1}{14} + \frac{1}{16}\right)$

$= \frac{15}{112}$

Time taken by these pipes (without leak) to fill the cistern = $\frac{112}{15}$ hours

= (7 hours 28 min.)

Due to leakage, times taken

= (7 hours 28 min.) + 32 min.

= 8 hours.

Work done by 2 pipes and leak in 1 hour = (1/8).

Work done by leak only in 1 hour = $\left(\frac{15}{112} - \frac{1}{8}\right)$

$= \frac{1}{112}$.

∴ Time taken by the leak to empty the full cistern = 112 hrs.

EXAMPLE-4

Pipe A can fill a tank in 6 hours and pipe B can fill it in 4 hours. If they opened on alternate hours and if pipe A is opened first, in how many hours, the tank shall be full?

Sol. A's work in 1 hour = (1/6)

B's work in 1 hour = (1/4)

(A + B)'s 2 hours work when opened alternately

$= \left(\frac{1}{6} + \frac{1}{4}\right)$

$= \frac{5}{12}$

(A + B)'s 4 hours work when opened alternately

$= \left(\frac{5}{12} + \frac{5}{12}\right)$

$= \frac{10}{12} = \frac{5}{6}$

[Approx, 12/5 i.e., 2 pairs will be taken]

Remaining part = $\left(1 - \frac{5}{6}\right)$

$= \frac{1}{6}$

Now, this is A's turn and (1/6) part of tank will be filled by A in 1 hour.

So, total time taken = (4+1) = 5 hours.

PRACTICE TEST
TIME & WORK

1. A and B together can complete a piece of work in 35 days while A alone can complete the same work in 60 days. B alone will be able to complete the same work in :
   a) 42 days b) 72 days
   c) 84 days d) 96 days
2. A can do a piece of work in 15 days and B alone can do it in 10 days. B works at it for 5 days and then leaves. A alone can finish the remaining work in :
   a) $6\frac{1}{2}$ days b) $7\frac{1}{2}$ days
   c) 8 days d) 9 days
3. A and B can do a piece of work in 45 days and 40 days respectively. They began to do the work together but A leaves after some days and then B completed the remaining work in 23 days. The number of days after which A left the work was :
   a) 6 b) 8
   c) 9 d) 12
4. If Ramesh, Suresh and Harish can do a piece of work in 15 days, 10 days and 6 days respectively, how long will they take to do it, if all the three work at it together ?
   a) 3 days b) $3\frac{1}{2}$ days
   c) $3\frac{9}{20}$ days d) $3\frac{3}{20}$ days
5. A and B can do a piece of work in 72 days, B and C can do it in 120 days; A and C can do it in 90 days. In what time can A alone do it?
   a) 150 days b) 120 days
   c) 100 days d) 80 days
6. 12 men can complete work in 18 days. Six days after they started working. 4 men joined them. How many days will all of them take to finish the remaining work ?
   a) 9 b) 10 c) 12 d) 15
7. A father can do a job as fast as his two sons working together. If one son does the job in 3 hours and the other in 6 hours, how many hours does it take the father to do the job ?
   a) 1 b) 2 c) 3 d) 4
8. If 10 men or 18 boys can do a piece of work in 15 days, then 25 men and 15 boys together will do twice the work in :
   a) $4\frac{1}{2}$ days b) 9 days
   c) 8 days d) 36 days
9. A certain number of men complete a piece of work in 60 days. If there were 8 men more, the work could be finished in 10 days less. How many men were originally there ?
   a) 30 b) 32 c) 36 d) 40
10. A does half as much work as B and C does half as much work as A and B together. If C alone can finish the work in 40 days, then together all finish the work in :
    a) $13\frac{1}{3}$ days b) 15 days
    c) 20 days d) 30 days
11. A cistern can be filled by pipes A and B in 12 minutes and 16 minutes respectively. When full, the tank can be emptied by a third pipe C in 8 minutes only. If all the taps be turned on at the same time, the cirstern will be full in :
    a) 20 min. b) 24 min. c) 36 min. d) 48 min.
12. A pipe can fill a tank in x hours and another can empty it in y hours. They can together fill it in $(y > x)$:
    a) (x–y) hours b) (y–x) hours
    c) $\frac{xy}{(x-y)}$ hours d) $\frac{xy}{(y-x)}$ hours
13. Pipe A can fill an empty tank in 5 hours while pipe B can empty the full tank in 6 hours. If both are opened in the empty tank it will be full in :
    a) 30 hours b) 11 hours
    c) 5 1/2 hours d) 1 hour
14. A pipe can fill a cistern in 9 hours. Due to a leak in the bottom the cistern is just full in 10 hours. When the cistern is full, the leak can empty it in :
    a) 1 hour b) 19 hours
    c) 1 1/2 hours d) 90 hours
15. Pipes A and B can fill a tank in 10 hours and 15 hours respectively. Both together can fill it in :
    a) 12 1/2 hours b) 5 hours
    c) 6 hours d) None of these

16. A tap can fill a tank in 16 minutes and another can empty it in 8 minutes. If the tank is already half full and both the tanks are opened together, the tank will be :
a) filled in 12 min. b) emptied in 12 min.
c) filled in 8 min. d) emptied in 8 min.

17. A cistern has two taps which fill it in 12 min. and 15 min. respectively. There is also a waste pipe in the cistern. When all the pipes are opened the empty cistern is full in 20 min. How long will the waste pipe take to empty a full cistern ?
a) 8 min b) 10 min
c) 12 min d) 16 min.

18. A tank can be filled by one top in 20 min. and by another in 25 min. Both the taps are kept open for 5 min. and then the second is turned off. In how many minutes more is the tank completely filled ?
a) 6 min b) 17 1/2 min.
c) 11 min. d) 12 min.

19. Two pipes A and B can fill a cistern in 24 minutes and 32 minutes respectively., If both the pipes are opened together, then after how much time B should be closed so that the tank is full in 18 minutes ?
a) 10 min. b) 12 min.
c) 8 min. d) 6 min.

20. A leak in the bottom of a tank can empty the full tank in 6 hours. An inlet pipe fills water at the rate of 4 litres per minute. When the tank is full, the inlet is opened and due to the leak the tank is empty in 8 hours. The capacity of the tank is :
a) 5760 litres b) 6970 Litres
c) 5846 litres d) 5260 litres

21. A pipe can fill a cistern in 12 minutes and another pipe in 15 minutes, but a third pipe can empty it in 6 minutes. The first two pipes are kept open for 5 minutes in the beginning and then the third pipe is also opened. In what time is the cistern emptied?
a) 30 min. b) 37 1/2 min.
c) 45 min. d) 33 min.

22. Two pipes A and B can fill a tank in 15 hours and 20 hours respectively while a third pipe C can empty the full tank in 25 hours. All the three pipes are opened in the beginning. After 10 hours. C is closed. Find, in how much time will the tank be full?
a) 12 hours b) 13 1/2 hours
c) 16 hours d) 18 hours

23. One tap can fill a cistern in 2 hours and another tap can empty the cistern in 3 hours. How long will they take to fill the cistern if both the taps are opened ?
a) 5 hours b) 6 hours c) 7 hours d) 8 hours

24. If two pipes function simultaneously, the reservoir will be filled in 12 hours. One pipe fills the reservoir 10 hours faster than the other. How many hours it takes the second pipe to fill the reservoir?
a) 35 hours b) 25 hours c) 30 hours d) 28 hours

ANSWERS

1. c	2. b	3. c	4. a	5. b
6. a	7. b	8. b	9. d	10. a
11. d	12. d	13. a	14. d	15. c
16. d	17. b	18. c	19. c	20. a
21. c	22. a	23. b	24. c	

EXPLANATIONS

1. **Ans : (c)**

$(A+B)$'s 1 day's work $= \frac{1}{35}$ & A's 1 day's work $= \frac{1}{60}$

$\therefore$ B's 1 day's work $= \left[\frac{1}{35} - \frac{1}{60}\right] = \frac{5}{420} = \frac{1}{84}$

Hence B alone will finish the work in 84 days.

2. **Ans : (b)**

B's 5 days work $= \left[\frac{1}{10} \times 5\right] = \frac{1}{2}$

Remaining work $= \left[1 - \frac{1}{2}\right] = \frac{1}{2}$

A can do $\frac{1}{2}$ work in $7\frac{1}{2}$ days.

3. **Ans : (c)**

$(A+B)$'s 1 days work $= \left[\frac{1}{45} + \frac{1}{40}\right] = \frac{17}{360}$

work done by B in 23 days $= \left[\frac{1}{40} \times 23\right] = \frac{23}{40}$

Now, $\frac{17}{360}$ work was done by (A+B) in 1 day.

$\frac{17}{40}$ work was done by (A+B) in $\left[1 \times \frac{360}{17} \times \frac{17}{40}\right] = 9$ days.

$\therefore$ A left after 9 days.

4. Ans : (a)

(Ramesh + Suresh + Harish)'s 1 day's work

$$= \left[\frac{1}{15} + \frac{1}{10} + \frac{1}{6}\right] = \frac{10}{30} = \frac{1}{3}$$

So, all the three will finish the work in 3 days.

5. Ans : (b)

(A + B)'s 1 day's work $= \frac{1}{72}$; (B + C)'s 1 day's work $= \frac{1}{120}$

(A + C)'s 1 day's work $= \frac{1}{90}$

Adding 2 (A + B + C)'s 1 day's work

$$= \left[\frac{1}{72} + \frac{1}{120} + \frac{1}{90}\right] = \frac{12}{360} = \frac{1}{30}$$

$\therefore$ (A + B + C)'s 1 day's work $= \frac{1}{60}$

A's 1 day's work $= \left[\frac{1}{60} - \frac{1}{120}\right] = \frac{1}{120}$

$\therefore$ A alone can finish the work in 120 days.

6. Ans : (a)

(12 × 18) men can complete the work in 1 day.

$\therefore$ 1 man's 1 day's work $= \frac{1}{216}$

12 men's 6 day's work $= \left[\frac{1}{18} \times 6\right] = \frac{1}{3}$

Remaining work $= \left[1 - \frac{1}{3}\right] = \frac{2}{3}$

16 men's 1 day's work $= \frac{16}{216} = \frac{2}{27}$.

$\frac{2}{27}$ work is done by them in 1 day.

$\frac{2}{3}$ work is done by them in $\left[\frac{27}{2} \times \frac{2}{3}\right] = 9$ days

7. Ans : (b)

Father's 1 hour's work $= \left[\frac{1}{3} + \frac{1}{6}\right] = \frac{1}{2}$

$\therefore$ Time taken by father to complete the work = 2 hours.

8. Ans : (b)

10 men = 18 boys $\Leftrightarrow$ 1 man $= \frac{18}{10}$ boys

$\therefore$ 25 men + 15 boys $= \left[25 \times \frac{18}{10} + 15\right]$ boys

= 60 boys.

Now, more work, more days

More boys, less days.

$$\left.\begin{array}{ll} \text{work} & 1:2 \\ \text{Boys} & 60:18 \end{array}\right\} :: 15 : x$$

$\therefore 1 \times 60 \times x = 2 \times 18 \times 15$ (or) $x = \frac{2 \times 18 \times 15}{60}$

= 9 days.

9. Ans : (d)

Originally, let there be x men. More men, less days.

$\therefore (x + 8) : x :: 60 : 50$

So, $\frac{x+8}{x} = \frac{60}{50}$ or $x = 40$

Hence there were 40 men originally

10. Ans : (a)

C alone can finish the work in 40 days.

$\therefore$ (A + B) can do it in 20 days.

$\therefore$ (A + B)'s 1 day's work $= \frac{1}{20}$

A's 1 day's work : B's 1 day's work $= \frac{1}{2} : 1 = 1:2$

$\therefore$ A's 1 days work $= \left[\frac{1}{20} \times \frac{1}{3}\right] = \frac{1}{60}$

[Divide $\frac{1}{20}$ in the ratio 1 : 2]

B's 1 day's work $= \left[\frac{1}{20} \times \frac{2}{3}\right] = \frac{1}{30}$

(A + B + C)'s 1 day's work $= \left[\frac{1}{60} + \frac{1}{30} + \frac{1}{40}\right]$

$$= \frac{9}{120} = \frac{3}{40}$$

$\therefore$ All the 3 together will finish it in $\frac{40}{3} = 13\frac{1}{3}$ days.

11. Ans : (d)

Net filling in 1 minute $= \left(\frac{1}{12} + \frac{1}{16} - \frac{1}{8}\right) = \frac{1}{48}$.

$\therefore$ The time taken to fill the cistern = 48 minutes.

12. Ans : (d)

Net filling in 1 hour $= \left(\frac{1}{x} - \frac{1}{y}\right) = \left(\frac{y-x}{xy}\right)$.

$\therefore$ Time taken to fill the tank $= \frac{xy}{(y-x)}$ hours.

13. Ans : (a)

Net filling in 1 hour $= \left(\frac{1}{5} - \frac{1}{6}\right) = \frac{1}{30}$.

$\therefore$ The time taken to fill the tank = 30 hours.

14. Ans : (d)

Work done by the leak only in 1 hour = $\left(\frac{1}{9}-\frac{1}{10}\right)=\frac{1}{90}$.

So the leak can empty the full cistern in 90 hours.

15. Ans : (b)

Work done by the pipes in 1 hour = $\left(\frac{1}{10}+\frac{1}{15}\right)=\frac{1}{6}$.

So, both the pipes will fill it in 6 hours.

16. Ans : (d)

Clearly, the rate of waste pipe is more, so the tank will be emptied, if both the pipes are opened.

Net work done by both the pipes in 1 minute.

$= \left(\frac{1}{16}-\frac{1}{8}\right)=-\frac{1}{16}$ = [–ve sign means emptying]

So, both the pipes together will empty the full tank in 16 min. So, they will empty the half-full tank in 8 min.

17. Ans : (b)

Work done by waste pipe in 1 min. = $\frac{1}{20}-\left(\frac{1}{12}+\frac{1}{15}\right)$

$= \frac{-1}{10}$ [–ve sign means emptying]

So the waste pipe will empty the full cistern in 10 min.

18. Ans : (c)

Part filled in 5 min. by both the pipes = $5\left(\frac{1}{20}+\frac{1}{25}\right)$

$= \frac{9}{20}$

Remaining part = $\left(1-\frac{9}{20}\right)=\frac{11}{20}$.

Now, (1/20) part is filled by first tap in 1 minute.

$\therefore \frac{11}{20}$ part will be filled by it in $\left(\frac{20\times 11}{20}\right)$ min. = 11 min.

19. Ans : (c)

Let B be closed after x min.

Part filled by both in x min. = $x\left(\frac{1}{24}+\frac{1}{32}\right)=\frac{7x}{96}$.

Part filled by A in (18–x) min. = $\frac{18-x}{24}$.

$\therefore \frac{7x}{96}+\frac{18-x}{24}=1$ (or) $x = 8$ min.

20. Ans : (a)

Work done by inlet in 1 hour = $\left(\frac{1}{6}-\frac{1}{8}\right)=\frac{1}{24}$

Work done by inlet in 4 min. = $\left(\frac{1}{24\times 60}\right)=\frac{1}{1440}$.

$\therefore$ Volume of $\frac{1}{1440}$ part = 4 litres.

Hence, the volume of the tank = (1440 × 4) = 5760 litres

21. Ans : (c)

Cistern filled in 5 minutes = $5\left(\frac{1}{12}+\frac{1}{15}\right)=\frac{3}{4}$.

Net work done by 3 pipes in 1 min. = $\left(\frac{1}{12}+\frac{1}{15}-\frac{1}{6}\right)$

$= -\frac{1}{60}$

Thus, (1/60) part is emptied in 1 min.

$\therefore \frac{3}{4}$ part will be emptied in $\left(60\times\frac{3}{4}\right)$ = 45 minutes.

22. Ans : (a)

Tank filled in 10 hours = $10\left(\frac{1}{15}+\frac{1}{20}-\frac{1}{25}\right)=\frac{23}{30}$.

Remaining part = $\left(1-\frac{23}{30}\right)=\frac{7}{30}$.

Net work done by A and B in 1 hour = $\left(\frac{1}{15}+\frac{1}{20}\right)$

$= \frac{7}{60}$.

Now, $\frac{7}{60}$ part is filled in 1 hour.

$\therefore \frac{7}{30}$ part will be filled in $\left(\frac{60}{7}\times\frac{7}{30}\right)$ = 2 hours.

Hence, the tank will be full in 12 hours.

23. Ans : (b)

Net filling in 1 hour = $\left(\frac{1}{2}-\frac{1}{3}\right)=\frac{1}{6}$.

$\therefore$ The cistern will be full in 6 hours.

24. Ans : (c)

Let the reservoir be filled by the first pipe in x hours. Then, it will be filled by the second pipe in (x+10) hours.

$\therefore \frac{1}{x}+\frac{1}{x+10}=\frac{1}{12}$ (or)

$12(x+10-x) = x(x+10)$ (or) $x^2 - 14x - 120 = 0$

(or) $(x-20)(x+6) = 0$ (or) $x = 20$.

So, the second pipe takes 30 hours to fill the reservoir.

7. TIME AND DISTANCE

Suppose a train travels for 5 hours and covers a distance of 150 km. It travels at a rate of 30 km per hour. Here 30 is the speed of the train, 5 hours is the time of travel and 500 km is the distance.

FORMULAE - SOLVED EXAMPLES

FORMULAE

i) Speed = $\left[\frac{\text{Distance}}{\text{time}}\right]$

Time = $\left[\frac{\text{distance}}{\text{speed}}\right]$

ii) Distance = [Speed × Time]

iii) 1 km/hour = $\frac{5}{18}$ m/sec

iv) 1 m/sec = $\frac{18}{5}$ km/hr

v) If the ratio of the speeds of A and B is a : b, then the ratio of the times taken by them to cover the same distance is $\frac{1}{a} : \frac{1}{b}$ or b : a

(vi) Suppose a man covers a certain distance at 'x' kmph and an equal distance at 'y' kmph. Then the average speed during the whole journey is $\left[\frac{2xy}{x+y}\right]$ kmph.

EXAMPLE PROBLEMS

EXAMPLE-1

A scooterist covers a certain distance at 36 kmph. How many metres does he cover in 2 minutes?

Sol. Speed = 36 kmph

$= \left[36 \times \frac{5}{18}\right]$ m/sec

= 10m/sec

∴ Distance covered in 2 min

= (10 × 2 × 60 m)

= 1200 m

EXAMPLE-2

If a man runs at 3 metres per second, how many kilometres does he run in 1 hour 40 minutes.

Sol. Speed of the man = $\left[3 \times \frac{18}{5}\right]$ km/hr

$= \frac{54}{5}$ km/hr

Distance covered in $\frac{5}{3}$ hours

$= \left[\frac{54}{5} \times \frac{5}{3}\right]$ km

= 18km

EXAMPLE-3

There are two towns A and B. Anil goes from A to B at 40 kmph and comes back to the starting point at 60 kmph. What is his average speed during the whole journey ?

Sol. Average speed = $\left[\frac{2xy}{x+y}\right]$ kmph

$= \frac{(2 \times 40 \times 60)}{40 + 60}$ kmph

= 48 kmph.

EXAMPLE-4

If a man walks at the rate of 5 kmph, he misses a train by only 7 minutes. However, if he walks at the rate of 6 kmph, he reaches the station 5 minutes before the arrival of the train. Find the distance covered by him to reach the station.

Sol. Let the required distance be x km.

Difference in the times taken at two speeds

$= 12 \text{ min} = \frac{1}{5}$ hr

$\therefore \frac{x}{5} - \frac{x}{6} = \frac{1}{5}$ (or) 6x − 5x = 6

(or) x = 6 km

Hence, the required distance is 6 km.

EXAMPLE-5

A man travels for 5 hrs 15 min. If he covers the first half of the journey at 60 kmph and the rest at 45 kmph. Find the total distance travelled by him.

Sol. Let the total distance be x km. Then,

$$\frac{(x/2)}{60} + \frac{(x/2)}{45} = \frac{21}{4}$$

(or) $\frac{x}{120} + \frac{x}{90} = \frac{21}{4}$

(or) $3x + 4x = 21 \times 90$ (or) $x = 270$

∴ Required distance = 270 km.

EXAMPLE-6

A and B are two stations 330 km apart. A train starts from A at 8 a.m. and travels towards B at 60 kmph. Another train starts from B at 9 a.m. and travels towards A at 75 kmph. At what time do they meet ?

Sol. Suppose they meet x hrs after 8 a.m. Then,

(Distance moved by first in x hrs] + [Distance moved by second in (x – 1) hrs = 330

∴ $60x + 75(x - 1) = 330$ (or)

$60x + 75x = 330 + 75$

$135x = 405 \quad x = 3$

So, they meet at 11 a.m.

EXAMPLE-7

A goods train leaves a station at a certain time and at a fixed speed. After 6 hours, an express train leaves the same station and moves in the same direction at a uniform speed of 90 kmph. This train catches up the goods train in 4 hours. Find the speed of the goods train.

Sol. Let the speed of the goods train be x kmph.

Distance covered by goods train in 10 hours

= Distance covered by the express train in 4 hours.

∴ $10x = 4 \times 90$ (or) $x = 36$.

∴ Speed of goods train = 36 kmph.

EXAMPLE-8

Walking $\frac{5}{6}$ of its usual speed, a train is 10 minutes too late. Find its usual time to cover the journey.

Sol. New speed = $\frac{5}{6}$ of the usual speed.

∴ New time taken = $\frac{6}{5}$ of the usual time.

∴ $\left[\frac{6}{5}\text{ of the usual time}\right]$ – (usual time)= 10 min.

∴ $\frac{1}{5}$ of the usual time = 10 min. (or)

usual time = 50 min.

EXAMPLE-9

Two bicyclists cover the same place at 15 kmph and 16 kmph respectively. Find the distance travelled by each, if one takes 16 minutes longer than the other.

Sol. Let the required distance be x km. Then,

$$\frac{x}{15} - \frac{x}{16} = \frac{16}{60} \text{ (or)}$$

$$\Rightarrow \frac{16x - 15x}{15 \times 16} = \frac{16}{60}$$

$$\Rightarrow 60(16x - 15x) = 16 \times 16 \times 15$$

$$x = \frac{16 \times 16 \times 15}{60} = 64.$$

Hence, the required distance = 64 km.

EXAMPLE-10

A thief is spotted by a police man from a distance of 200 m. When the police man starts the chase, the thief also starts running. If the speed of the thief be 10 km/hr and that of policeman 12 km/hr, how far the thief will have run before he is overtaken ?

Sol. Relative speed of the policeman = 2 kmph

Time taken by policeman to cover 200 m

$= \left[\frac{200}{1000} \times \frac{1}{2}\right]$ hrs.

$= \frac{1}{10}$ hrs.

In $\frac{1}{10}$ hrs, the thief covers a distance of $\left[10 \times \frac{1}{10}\right]$ km

= 1 km.

EXAMPLE-11

A man cycles from A to B, a distance of 21 km in 1 hour 40 min. The road from A is level for 13 km and then it is uphill to B. The man's average speed on level is 15 km/hr. Find his average uphill speed.

Sol. Let the average uphill speed be x km/hr.

Then, $\frac{13}{15} + \frac{8}{x} = \frac{5}{3}$

$$\Rightarrow \frac{8}{x} = \frac{5}{3} - \frac{13}{15} \Rightarrow \frac{8}{x} = \frac{12}{15}$$

$$\Rightarrow \frac{8}{x} = \frac{4}{5}\ x = \frac{5 \times 8}{4} = 10$$

Average uphill speed = 10 km/hr.

PRACTICE TEST

TIME AND DISTANCE

1. A car can finish a certain Journey in 10 hrs at a speed of 48 kmph. In order to cover the same distance in 8 hrs, the speed of the car must be increased by:
a) 6 km/hr b) 7.5 km/hr c) 12 km/hr d) 5 km/hr

2. A car takes 6 hours to cover a journey at a speed of 45 kmph. At what speed must it travel in order to complete the journey in 5 hours ?
a) 55 km/hr b) 54 km/hr c) 53 km/hr d) 52 km/hr

3. Ravi runs at 15.6 kmph. How many metres does he run in 2 minutes ?
a) 260 m b) 312 m c) 520 m d) 1040 m

4. A walks at 4 kmph and 4 hours after his start, B cycles after him at 10 kmph. How far from the start does B catch up with A ?
a) 16.7 km b) 18.6 km c) 21.5 km d) 26.7 km

5. A train starts from station X at the rate of 60 kmph and reaches station Y in 45 minutes. If the speed is reduced by 6 kmph, how much more time will the train take to return from station 'Y' to station 'X' ?
a) 5 min b) $7\frac{1}{2}$ min c) 6 min d) 4 min

6. A motorist covers a distance of 39 km in 45 minutes by moving at a speed of x kmph for the first 15 minutes, then moving at double the speed for the next 20 minutes and then again moving at his original speed for the rest of the journey. Then, x is equal to:
a) 31.2 b) 36 c) 40 d) 52

7. A man drives 150 km from A to B in 3 hours 20 min. and returns to A in 4 hours 10 min. Then average speed from A to B exceeds the average speed for the entire trip by :
a) 5 km/hr b) 4.5 km/hr
c) 4 km/hr d) 2.5 km/hr

8. A car completes a certain journey in 8 hours. It covers half the distance at 40 kmph and the rest at 60 kmph. The total distance of the journey is :
a) 350 km b) 384 km c) 400 km d) 420 km

9. A bullock cart has to cover a distance of 80 km in 10 hours. If it covers half of the journey in $\frac{3}{5}$th time, what should be its speed to cover the remaining distance in the time left ?
a) 8 km/hr b) 6.4 km/hr c) 10 km/hr d) 20 km/hr

10. Two buses travel to a place at 45 kmph and 60 kmph respectively. If the second bus takes $5\frac{1}{2}$ hours less than the first for the same journey, the length of journey is :
a) 900 km b) 945 km c) 1350 km d) 990 km

11. A train 100 metres long completely crosses a bridge 300 metres long in 30 seconds. How many seconds will it take to pass a telegraph pole ?
a) 10 seconds b) $7\frac{1}{2}$ seconds
c) 9 seconds d) 3 seconds

12. A train travelling at 36 kmph. completely crosses another train having half its length and travelling in the opposite direction at 54 kmph., in 12 seconds. If it also passes a railway platform in $1\frac{1}{2}$ minutes, the length of the platform is :
a) 560 metres b) 620 metres
c) 700 metres d) 750 metres

13. Two trains travelling in the same direction at 40 kmph, and 22 kmph. completely pass one another in 1 minute. If the length of the first train is 125 metres the length of the second train is :
a) 125 metres b) 150 metres
c) 175 metres d) 200 metres

14. If a train overtakes two persons who are walking in the same direction in which the train is going at the rate of 2 kmph, and 4 kmph, and passes them completely in 9 and 10 seconds respectively. The length of the train is :
a) 45 metres b) 50 metres
c) 54 metres d) 72 metres

15. A train 100 metres in length passes a milestone in 10 seconds and another train of the same length travelling in opposite direction in 8 seconds. The speed of the second train is:
a) 36 kmph b) 48 kmph
c) 54 kmph d) 60 kmph

16. A train 50 metres long passes a platform 100 metres long in 10 seconds. What is the speed of the train in metres per second ?
a) 150 b) 50 c) 10 d) 15

17. A train 280 metres long is moving at 60 km per hour. The time taken by the train to cross a platform 220 metres long is

a) 20 seconds b) 25 seconds
c) 30 seconds d) 35 seconds

18. A train 270 metres long is moving at a speed of 25km per hour. it will cross a man coming from the opposite direction at a speed of 2 km per hour in

a) 36 seconds b) 32 seconds
c) 28 seconds d) 24 seconds

19. Two stations A and B are 110 Kms apart on a straight line. One train starts from A at 7 A.M. and travels towards B at 20 km per hour speed. Another train starts from B at 8 A.M. and travels towards A at 25 km per hour speed. At what time will they meet ?

a) 9 A.M. b) 10 A.M.
c) 11 A.M. d) None of these

20. A train 100 metres long moving at a speed of 50 km per hour crosses a train 120 metres long coming from opposite direction in 6 seconds. What is the speed of the second train.

a) 132 kmph
b) 82 kmph
c) 60 Kmph
d) 50 kmph

ANSWERS

1. c	2. b	3. c	4. d	5. a
6. b	7. a	8. b	9. c	10. d
11. b	12. c	13. c	14. b	15. c
16. d	17. c	18. a	19. b	20. b

EXPLANATIONS

1. **Ans : (c)**

Time = 10 hours, speed = 48 kmph

$\therefore$ Distance = (Speed × time)

$= (48 \times 10)$ km = 480 km

2. **Ans : (b)**

Time = 6 hrs, speed = 45 kmph

$\therefore$ Distance = (45×6) Km = 270 Km

New time taken = 5 hrs.

$\therefore$ New speed = $\left[\frac{270}{5}\right]$ km/hr = 54 km/hr

3. **Ans : (a)**

Speed = $\left[15.6 \times \frac{5}{18}\right]$ m/sec = $\left(\frac{13}{3}\right)$ m/sec

Distance covered = (Speed × time)

$= \left[\frac{13}{3} \times 2 \times 60\right]$ m = 520 m

4. **Ans : (d)**

Suppose after x km from the start B catches up with A. Then, the distance in the time taken by A to cover x km and that taken by B to cover x km in 4 hours.

$\therefore \frac{x}{4} - \frac{x}{10} = 4$ (or) x = 26.7 km.

5. **Ans : (a)**

Distance xy = $\left[60 \times \frac{45}{60}\right]$ km = 45 Km/hr.

New speed = (60 – 6) kmph = 54 kmph

Time taken = $\left(\frac{45}{54} \times 60\right)$ min = 50 min

$\therefore$ Extra time taken = (50 – 45) min = 5min.

6. **Ans : (b)**

$x \times \frac{15}{60} + 2x \times \frac{20}{60} + x \times \frac{10}{60} = 39 \Rightarrow \frac{x}{4} + \frac{2x}{3} + \frac{x}{6} = 39$

$\therefore 3x + 8x + 2x = 468$ (or) $x = 36$

7. **Ans : (a)**

Average speed from A to B = $\left[150 \times \frac{3}{10}\right]$ km/hr

= 45 km/ph

Average speed from B to A = $\left[150 \times \frac{6}{25}\right]$ km/hr

= 36 km/hr.

$\therefore$ Average speed for the entire trip

$= \left[\frac{2 \times 45 \times 36}{45 + 36}\right]$ km/hr

$\therefore$ Required difference = 5km/hr

8. **Ans : (b)**

Let the total journey be x km, then

$\frac{(x/2)}{40} + \frac{(x/2)}{60} = 8$ (or) $\frac{x}{80} + \frac{x}{120} = 8$ (or)

$3x + 2x = 240 \times 8$ (or) $x = 384$

9. **Ans : (c)**

Distance covered = $\left[\frac{1}{2}\times 80\right]$ km = 40 km

Time elapsed = $\left[\frac{3}{5}\times 10\right]$ hrs = 6 hrs.

Distance left = 40 km, Time left = 4 hrs.

$\therefore$ Speed required = $\left[\frac{40}{4}\right]$ km/hr = 10 km/hr

10. **Ans : (d)**

Let the length of journey be x km

then, $\frac{x}{45}-\frac{x}{60}=\frac{11}{2}$ (or) x = 990 km.

11. **Ans : (b)**

Sum of lengths of train and bridge = 400 metres.

Speed of train = $\left(\frac{400}{30}\right)$m/sec = $\left(\frac{40}{3}\right)$m/sec.

Time taken to pass a telegraph pole

$$= \left(100\times\frac{3}{40}\right)\text{sec.} = 7\frac{1}{2}\text{ seconds.}$$

12. **Ans : (c)**

Relative speed = (36+ 54) kmph = 90 kmph.

$$= \left(90\times\frac{5}{18}\right)\text{metres / sec} = 25\text{ metres/sec.}$$

Let 'l' be the length of the slower train now,

$$\frac{l+\frac{l}{2}}{25} = 12 \text{ (or) } l = 200\text{m.}$$

Also, speed of slower train = $\left(36\times\frac{5}{18}\right)$m/sec

= 10 m/sec.

Now, if x be the length of platform, then $\frac{x+200}{10}$ = 90 or x = 700 metres.

13. **Ans : (c)**

Relative speed of faster train = (40 – 22) Kmph.

$$= \left(18\times\frac{5}{18}\right)\text{m/sec.} = 5\text{m/sec.}$$

Now, $\frac{125+l}{5}$ = 60 (or) l = (300 – 125) = 175 metres.

14. **Ans : (b)**

Let the speed of the train be x kmph.

then, $\frac{(x-2)\times 1000}{60\times 60}\times 9=\frac{(x-4)\times 1000}{60\times 60}\times 10$

(or) $\frac{5x-10}{2}=\frac{25x-100}{9}$ (or) x = 22 kmph.

$\therefore$ length of train = $\left(\frac{5x-10}{2}\right)$metres = 50 metres.

15. **Ans : (c)**

Speed of first train = $\left(\frac{100}{10}\right)$ m/sec. = 10 m/sec.

Let the speed of second train be x m/sec.

Then. $\frac{200}{x+10}$ = 8 (or) x = 15 m/sec.

$$= \left(15\times\frac{18}{5}\right)\text{Kmph} = 54\text{ kmph.}$$

16. **Ans : (d)**

The train covers 150 metres in 10 seconds. So, its speed is 15 metres per second.

17. **Ans : (c)**

Speed of the train = 60 Km/hr = $\left(16\times\frac{5}{18}\right)$ m/sec.

$= \frac{50}{3}$ m/sec.

Time taken by the train to cover (220 + 280) metres.

$$= \left(500\div\frac{50}{3}\right)\text{sec} = \left(500\times\frac{3}{50}\right)\text{sec} = 30\text{ sec.}$$

18. **Ans : (a)**

Speed of the train relative to man = 27 Km/hr.

$$= \left(27\times\frac{2}{18}\right)\text{metres/sec} = \left(\frac{15}{2}\right)\text{metres/sec.}$$

Required time = $\left(270\times\frac{2}{15}\right)$sec = 36 seconds.

19. **Ans : (b)**

Suppose they meet x hours after 7 A.M. Distance travelled by the train from A in x hours = (20x) Km. Distance travelled by the train from B in (x –1) hours.

= 25 (x – 1) Km.

$\therefore$ 20x + 25 (x – 1) = 110 (or) x = 3 hours.

So the trains meet at 10 A.M.

20. **Ans : (b)**

Let the speed of the required train be x kmph. Time taken by the trains in passing each other

$$= \frac{\left(\frac{1}{10}+\frac{12}{100}\right)}{(50+x)}\text{ hrs} = \left[\frac{22}{100(50+x)}\times 60\times 60\right]\text{sec.}$$

$$\therefore \frac{22\times 60\times 60}{100(50+x)} = 6 \text{ (or) } x = 82\text{ Km/hr.}$$

8. SIMPLE INTEREST

If the interest on a sum borrowed for a certain period is its reckoned uniformly it is called simple interest

DEFINITIONS - FORMULAE - SOLVED EXAMPLES

GENERAL CONCEPTS

Principal or sum : The money borrowed or lent out for a certain period is called the principal or the sum.

Interest : Extra money paid for using other's money is called interest.

Simple interest : If the interest on a sum borrowed for a certain period is reckoned uniformly, then it is called simple interest.

FORMULAE

Let Principal = P, Rate = R% per annum and Time = T years, then,

i) $S.I. = \left[\frac{P \times R \times T}{100}\right]$

ii) $P = \left[\frac{100 \times S.I.}{R \times T}\right]$

$R = \left[\frac{100 \times S.I.}{P \times T}\right]$ and

$T = \left[\frac{100 \times S.I.}{P \times R}\right]$

EXAMPLE PROBLEMS

EXAMPLE-1

Find

i) S.I. on Rs. 68000 at $16\frac{2}{3}$ % per annum for 9 months.

ii) S.I. on Rs. 6250 at 14% per annum for 146 days.

iii) S.I. on Rs. 3000 at 18% per annum for the period from 4th Feb. 1995 to 18th April 1995.

Sol. i) P = 68000

$R = \frac{50}{3}$ % p.a. &

$T = \frac{9}{12}$ year $= \frac{3}{4}$ year.

$\therefore S.I. = \left[\frac{P \times R \times T}{100}\right]$

$= Rs. \left[68000 \times \frac{50}{3} \times \frac{3}{4} \times \frac{1}{100}\right]$

= Rs.85000

ii) P = Rs.6250

R = 14% p.a. and

$T = \left[\frac{146}{365}\right]$ year

$= \frac{2}{5}$ year.

$\therefore$ S.I. Rs. $\left[6250 \times 14 \times \frac{1}{100}\right]$ = Rs. 350

iii) Time = (24 + 31 + 18) days

= 73 days

$= \frac{1}{5}$ year

P = Rs. 3000 and R = 18 % p.a.

$\therefore$ S.I.= Rs. $\left[3000 \times 18 \times \frac{1}{5} \times \frac{1}{100}\right]$

= Rs.108

Remark : The day on which money is deposited is not counted while the day on which money is withdrawn is counted.

EXAMPLE-2

A sum at simple interest at $13\frac{1}{2}$% per annum amounts to Rs. 2502.50 after 4 years. Find the sum.

Sol. Let sum be x, then.

$S.I. = \left[x \times \frac{27}{2} \times 4 \times \frac{1}{100}\right]$

$= \frac{27x}{50}$

$\therefore$ Amount $= \left[x + \frac{27x}{50}\right] = \frac{77x}{50}$

$\therefore \frac{77x}{50} = 2502.50$ (or) $x = \frac{2502.50 \times 50}{77}$

= 1625

Hence, sum = Rs.1625.

EXAMPLE-3

A certain sum of money amounts to Rs.1008 in 2 years and to Rs. 1164 in $3\frac{1}{2}$ years. Find the sum and the rate of interest.

Sol. S.I. for $1\frac{1}{2}$ years = Rs.[1164 – 1008]

= Rs.156.

S.I. for 2 years = Rs. $\left[156 \times \frac{2}{3} \times 2\right]$

= Rs.208.

$\therefore$ Principal = Rs. (1008 – 208)

= Rs.800

Now, P = 800, T = 2 and S.I. = 208

$$\therefore \text{Rate} = \left[\frac{100 \times S.I.}{P \times T}\right]$$

$$= \left[\frac{100 \times 208}{800 \times 2}\right]\%$$

= 13%.

EXAMPLE-4

At what rate percent per annum will a sum of money double in 8 years ?

Sol. Let principal = P.

Then, S.I. = P and Time = 8 years.

$$\therefore \text{Rate} = \left[\frac{100 \times P}{P \times 8}\right]\%$$

= 12.5 % per annum.

EXAMPLE-5

A sum was put at simple interest at a certain rate for 3 years. Had it been put at 2% higher rate, it would have fetched Rs.360 more. Find the sum.

Sol. Let, sum = P and original rate = R, then,

$$\frac{P \times (R+2) \times 3}{100} - \frac{P \times R \times 3}{100} = 360$$

(or) 3PR + 6P – 3PR = 36000

(or) 6p = 36000

$\therefore$ P = 6000

Hence sum = Rs.6000

EXAMPLE-6

Simple interest on a certain sum is $\frac{16}{25}$ of the sum. Find the rate percent and time, if both are numerically equal.

Sol. Let sum = x.

Then, S.I. = $\frac{16x}{25}$

Let rate = R% and time = R years.

$$\therefore \frac{x \times R \times R}{100} = \frac{16x}{25}$$

$$\text{(or) } R^2 = \frac{1600}{25}$$

$$R = \frac{40}{5} = 8$$

$\therefore$ Rate = 8% and Time = 8 years.

EXAMPLE-7

A man borrowed Rs.24000 from two money lenders. For one loan, he paid 15% per annum and for the other 18% per annum. At the end of one year, he paid Rs.4050. How much did he borrow at each rate ?

Sol. Let the sum at 15% be Rs. x and that at 18% be Rs. (24000 – x).

$$\therefore \frac{x \times 15 \times 1}{100} + \frac{(24000 - x) \times 18 \times 1}{100} = 4050$$

(or) 15x + 432000 – 18x = 405000

(or) x = 9000

$\therefore$ Money borrowed at 18% = Rs.15000

EXAMPLE-8

What annual instalment will discharge a debt of Rs.1092 due in 3 years at 12% simple interest ?

Sol. Let each instalment be Rs. x then,

$$\left[x + \frac{x \times 12 \times 1}{100}\right] + \left[x + \frac{x \times 12 \times 2}{100}\right] + x = 1092$$

$$= \frac{100x + 12x}{100} + \frac{100x + 24x}{100} + x$$

$$= \frac{112x}{100} + \frac{124x}{100} + x$$

= 1092 (or)

$$= \frac{28x}{25} + \frac{31x}{25} + x$$

= 1092 (or)

= 28x + 31x + 25x

= 1092 × 25

= 84x = 1092 × 25 (or)

$$x = \left[\frac{1092 \times 25}{84}\right] = 325$$

$\therefore$ Each instalment = Rs.325.

PRACTICE TEST
SIMPLE INTEREST

1. At the rate of 6% p.a. simple interest, a sum of Rs.2500 will earn how much interest by the end of 5 years ?

   a) Rs. 150 b) Rs. 700 c) Rs. 750 d) Rs.3250

2. A person borrowed Rs. 500 at the rate of 5% per annum S.I. What amount will he pay to clear the debt after 4 years ?

   a) Rs. 200 b) Rs. 600 c) Rs. 550 d) Rs. 700

3. Avinash borrowed Rs. 5000 from Sanjay at simple interest. After 3 years, Sanjay got Rs.300 more than what he had given to Avinash. What was the rate of interest per annum ?

   a) 2 % b) 5 % c) 8 % d) 10 %

4. Rs. 800 amounts to Rs. 920 in 3 years at simple interest. If the interest rate is increased by 3%, it would amount to how much ?

   a) Rs. 1056 b) Rs. 992
   c) Rs. 1112 d) Rs. 1182

5. The interest on a certain deposit at 4.5 % p.a. is Rs. 202.50 in one year. How much will the additional interest in one year be on the same deposit at 5% p.a. ?

   a) Rs. 22.5 b) Rs. 20.25
   c) Rs. 225 d) Rs. 427.50

6. If a sum of money doubles itself in 8 years at simple interest, the rate percent per annum is:

   a) 11.5 b) 12 c) 12.5 d) 13

7. The simple interest accured on a sum of money at the end of four years is $\frac{1}{5}$th of its principal. What is the rate of interest per annum?

   a) 4% b) 5%
   c) 6 % d) Data Inadequate

8. How long will it take a sum of money invested at 5% p.a. S.I. to increase its value by 40% ?

   a) 5 years b) 6 years c) 7 years d) 8 years.

9. A monthly instalment of Rs. 180 is required to be paid for repayment of an interest free loan in 40 months. If it is decided to pay it in 30 months, how much will be the monthly instalment in rupees ?

   a) 60 b) 198 c) 240 d) 330

10. The rates of simple interest in two banks A and B are in the ratio 5 : 4. A person wants to deposit his total savings in two banks in such a way that he received equal half yearly interest from both. He should deposit the savings in banks A and B in the ratio :

    a) 2 : 5 b) 4 : 5 c) 5 : 2 d) 5 : 4

11. The simple interest at x% for x years will be Rs. x on a sum of :

    a) Rs. x b) Rs. 100x
    c) Rs. $\left(\frac{100}{x}\right)$ d) Rs. $\left(\frac{100}{x^2}\right)$

12. A sum of Rs. 10 is lent to be returned in 11 monthly instalments of Re.1 each, interest being simple. The rate of interest is :

    a) 10% b) 11%
    c) $9\frac{1}{11}$% d) $21\frac{9}{11}$%

13. What annual payment will discharge a debt of Rs. 580 due in 5 years, the rate being 8% per annum?

    a) Rs. 166.40 b) Rs. 65.60
    c) Rs. 100 d) Rs. 120

14. A sum of Rs. 2 is lent to be paid back in 3 equal monthly instalments of Re 1 each. The rate percent is :

    a) 100% b) 150% c) 200% d) 400%

15. A man invested $\frac{1}{3}$ of his capital at 7%; $\frac{1}{4}$ at 8% and the remainder at 10%. If his annual income is Rs. 561; the capital is :

    a) Rs. 5400 b) Rs. 6600
    c) Rs. 7200 d) Rs. 6000

16. A lent Rs. 600 to B for 2 years and Rs. 150 to C for 4 years and received altogether from both Rs. 90 as simple interest. The rate of interest is :

    a) 12% b) 4%
    c) 5% d) 10%

17. A certain sum of money at simple interest amounts to Rs. 1260 in 2 years and to Rs. 1350 in 5 years. The rate percent per annum is :

    a) 2.5% b) 3.75%
    c) 5% d) 7.5%

18. At a certain rate of simple interest Rs. 800 in 3 years amount to Rs. 920. If the rate of interest be increased by 3%, what will be the amount after 3 years ?

a) Rs. 1056 b) Rs. 1112
c) Rs. 1182 d) Rs. 992

19. Interest on a certain sum of money for $2\frac{1}{3}$ years at $3\frac{1}{4}$ % per annum is Rs. 210. The sum is

a) Rs. 2800
b) Rs. 2400
c) Rs. 1580
d) None of these

20. The simple interest on a sum of money will be Rs. 600 after 10 years .If the principal is trebled after 5 years, what will be the total interest at the end of the tenth year ?

a) Rs.600 b) Rs.900
c) Rs.1200 d) Data inadequate

ANSWERS

1. c	2. b	3. a	4. b	5. a
6. c	7. b	8. d	9. c	10. b
11. c	12. d	13. c	14. d	15. b
16. c	17. a	18. d	19. b	20. c

EXPLANATIONS

1. **Ans : (c)**

$$S.I. = Rs. \left[2500 \times 6 \times \frac{5}{100}\right]$$
$$= Rs.750$$

2. **Ans : (b)**

$$Amount = Rs. \left[500 + \frac{500 \times 5 \times 4}{100}\right]$$
$$= Rs.600$$

3. **Ans : (a)**

$$Rate = \left[\frac{100 \times 300}{5000 \times 3}\right] \%$$
$$= 2\%$$

4. **Ans : (b)**

Principal = Rs.800

S.I. = Rs.120

Time = 3 years.

$$\therefore Rate = \left[\frac{100 \times 120}{800 \times 3}\right] \% = 5\%$$

New rate = 8%

Principal = Rs.800

Time = 3 years.

$$S.I = Rs. \left[\frac{800 \times 8 \times 3}{100}\right]$$
$$= Rs.192$$

$$\therefore \text{New amount} = Rs.\ (800 + 192)$$
$$= Rs.992$$

5. **Ans : (a)**

$$Sum = \left[\frac{100 \times 202.50}{4.5 \times 1}\right] = Rs.4500$$

$$\text{Additional interest} = Rs. \left[\frac{4500 \times 5 \times 1}{100} - 202.50\right]$$
$$= Rs.22.50$$

6. **Ans : (c)**

Let sum = x, then, S.I. = x

$$Rate = \left[\frac{100 \times x}{x \times 8}\right] \%$$
$$= 12\frac{1}{2} \%$$

7. **Ans : (b)**

Let sum = x

Then S.I. = $\frac{x}{5}$

Time = 4 years.

$$\therefore Rate = \left[100 \times \frac{x}{5} \times \frac{1}{x \times 4}\right] \% = 5\%$$

8. **Ans : (d)**

Let the sum be x, then amount = $\frac{140x}{100} = \frac{7x}{5}$

$$\therefore S.I. = \left[\frac{7x}{5} - x\right]$$
$$= \frac{2x}{5}$$

Thus, sum = x, S.I. = $\frac{2x}{5}$ & Rate = 5%

$$\therefore Time = \left[100 \times \frac{2x}{5} \times \frac{1}{x \times 5}\right]$$
$$= 8 \text{ years.}$$

9. **Ans : (c)**

180 × 40 = x × 30 or x = 240

10. Ans : (b)

Let the savings be X and Y and the rates of Simple interest be 5x and 4x respectively.

Then $x \times 5x \times \frac{1}{2} \times \frac{1}{100} = Y \times 4x \times \frac{1}{2} \times \frac{1}{100}$ (or)

$\frac{X}{Y} = \frac{4}{5}$. i.e., X : Y = 4 : 5.

11. Ans : (c)

Principal = Rs. $\left(\frac{100 \times x}{x \times x}\right)$ = Rs.$\left(\frac{100}{x}\right)$

12. Ans : (d)

Rs. 10 + S.I. on Rs. 10 for 11 months

= Rs. 11 + S.I. on Re. 1 for 55 months

or Rs. 10 + S.I. on Re 1 for 110 months

= Rs. 11 + S.I. on Re 1 for 55 months

or S.I. on Re 1 for 55 months = Re 1.

$\therefore$ Rate = $\left(\frac{100 \times 1}{1 \times 55} \times 12\right)$%

$= 21\frac{9}{11}$%

13. Ans : (c)

Let the annual instalment be Rs. x.

Then, $\left[x + \left(\frac{x \times 4 \times 8}{100}\right)\right] + \left[x + \left(\frac{x \times 3 \times 8}{100}\right)\right]$

$+ \left[x + \left(\frac{x \times 2 \times 8}{100}\right)\right] + \left[x + \left(\frac{x \times 1 \times 8}{100}\right)\right] + x = 580$

(or) $\frac{33x}{25} + \frac{31x}{25} + \frac{29x}{25} + \frac{27x}{25} + x = 580$

(or) x = 100

14. Ans : (d)

Rs 2 + (S.I. on Rs. 2 for 3 months)

= Rs. 3 + (S.I. on Re. 1 for 3 months)

or Rs. 2 + (S.I. on Re. 1 for 6 months)

= Rs 3 + S.I. on Re. 1 for 3 months.

S.I. on Re 1 for 3 months = Re 1.

Rate% = $\left(\frac{100 \times 1 \times 12}{3}\right) = 400\%$

15. Ans : (b)

Let the capital be Rs. x. Then,

$\left(\frac{x}{3} \times \frac{7}{100} \times 1\right) + \left(\frac{x}{4} \times \frac{8}{100} \times 1\right) + \left(\frac{5x}{12} \times \frac{10}{100} \times 1\right)$

= 561 (or) x = 6600

16. Ans : (c)

$\left(\frac{600 \times x \times 2}{100}\right) + \left(\frac{150 \times x \times 4}{100}\right) = 90$

(or) x = 5%

17. Ans : (a)

S.I. for 3 years = Rs. (1350 - 1260)

= Rs. 90.

S.I. for 2 years = Rs {(90/3) × 2}

= Rs. 60.

$\therefore$ Principal = Rs (1260 - 60)

= Rs. 1200.

Hence, rate = $\left(\frac{100 \times 60}{1200 \times 2}\right)$% = 2.5%

18. Ans : (d)

Rate = $\left(\frac{120 \times 100}{800 \times 3}\right) = 5\%$

New rate = (5 + 3)% = 8%

S.I. at the new rate = $\left(\frac{800 \times 3 \times 8}{100}\right)$

= Rs. 192.

$\therefore$ Amount at the new rate = Rs. (800 + 192)

= Rs. 992.

19. Ans : (b)

Sum = Rs $\left(210 \times 100 \times \frac{3}{7} \times \frac{4}{15}\right)$ = Rs. 2400.

20. Ans : (c)

Let the sum be Rs x.

S.I. = Rs. 600

Time = 10 Years.

$\therefore$ Rate = $\left(\frac{600 \times 100}{x \times 10}\right)$%

$= \left(\frac{6000}{x}\right)$% per annum.

S.I. for 1st five years = Rs. $\left(\frac{x \times 5 \times 6000}{100 \times x}\right)$

= Rs. 300

And, S.I. for last 5 years = Rs. $\left(\frac{3x \times 5 \times 6000}{100 \times x}\right)$

= Rs. 900

Hence, total interest at the end of 10th year = Rs. 1200

9. COMPOUND INTEREST

Sometimes it so happens that the borrower and the lender agree to fix up a certain unit of time, say yearly or half-yearly or quarterly to settle the previous account.

In such cases, the amount after first unit of time becomes the principal for the second unit, the amount after second unit becomes the principal for the third unit and so on.

After a specified period, the difference between the amount and the money borrowed is called the compound interest (abbreviated as C.I.) for that period.

FORMULAE - SOLVED EXAMPLES

FORMULAE

Let Principal = P,

Rate = R% per annum Time = n years.

i) When interest is compounded Annually :

$$\text{Amount} = P\left[1+\frac{R}{100}\right]^n$$

ii) When interest is compounded Half-yearly :

$$\text{Amount} = P\left[1+\frac{(R/2)}{100}\right]^{2n}$$

iii) When Interest is compounded quarterly :

$$\text{Amount} = P\left[1+\frac{(R/4)}{100}\right]^{4n}$$

iv) When interest is compounded Annually but time is in fraction, say $3\frac{2}{5}$ years.

$$\text{Amount} = P\left[1+\frac{R}{100}\right]^3 \times \left[1 \times \frac{\frac{2}{5}R}{100}\right]$$

v) When rates are different for different years, Say R_1%, R_2%, R_3% for 1st, 2nd and 3rd year respectively.

Then, Amount

$$= P\left[1+\frac{R_1}{100}\right]\left[1+\frac{R_2}{100}\right]\left[1+\frac{R_3}{100}\right]$$

vi) Present worth of Rs. x due n years hence is given by:

$$\text{Present worth} = \frac{x}{\left(1+\frac{R}{100}\right)^n}$$

EXAMPLE PROBLEMS

EXAMPLE-1

Find compound interest on Rs. 6250 at 16% per annum for 2 years, compounded annually.

Sol. $\text{Amount} = \text{Rs.}\left[6250 \times \left[1+\frac{16}{100}\right]^2\right]$

$$= \text{Rs.}\left[6250 \times \frac{116}{100} \times \frac{116}{100}\right]$$

$$= \text{Rs.}\left[6250 \times \frac{29}{25} \times \frac{29}{25}\right]$$

= Rs.8410

∴ C.I. = Rs.(8410 − 6250)

= Rs.2160

EXAMPLE-2

Find the compound interest on Rs.5000 at 12% per annum for 1 year compounded half-yearly.

Sol. Principal = Rs.5000, Rate = 6% per half year

Time = 1 year = 2 half years.

$$\therefore \text{Amount} = \text{Rs.}\left[5000 \times \left[1+\frac{6}{100}\right]^2\right]$$

$$= \text{Rs.}\left[5000 \times \frac{106}{100} \times \frac{106}{100}\right]$$

$$= \text{Rs.}\left[5000 \times \frac{53}{50} \times \frac{53}{50}\right]$$

= Rs.5618

∴ C.I. = Rs. [5618 − 5000]

= Rs.618.

EXAMPLE-3

Find compound interest on Rs.16000 at 20% per annum for 9 months, compounded quarterly.

Sol. Principal = Rs. 16000; Time = 9 months = 3 quarter;

Rate = 20% per annum = 5% per quarter.

$$\therefore \text{Amount} = \text{Rs.}\left[16000 \times \left[1+\frac{5}{100}\right]^3\right]$$

$$= \text{Rs.}\left[16000 \times \frac{21}{20} \times \frac{21}{20} \times \frac{21}{20}\right] = \text{Rs.}18522$$

$\therefore$ C.I. = Rs.[18522 − 16000]

= Rs.2522.

EXAMPLE-4

The difference between the compound interest and simple interest on a certain sum at 10% per annum for 2 years is Rs. 631. Find the sum.

Sol. Let the sum be Rs. x then, C.I.= $x\left[1+\frac{10}{100}\right]^2 - x$

$$\text{S.I.} = \left[\frac{x \times 10 \times 2}{100}\right] = \frac{x}{5}$$

$$\Rightarrow x \times \frac{11}{10} \times \frac{11}{10}$$

$$\Rightarrow x = \frac{121x - 100x}{100} = \frac{21x}{100}$$

$$\therefore (\text{C.I}) - (\text{S.I.}) = \left(\frac{21x}{100} - \frac{x}{5}\right)$$

$$= \frac{21x - 20x}{100} = \frac{x}{100}$$

$$\therefore \frac{x}{100} = 631 \Leftrightarrow x = 63100$$

Hence the sum is Rs. 63100

EXAMPLE-5

If the compound interest on a certain sum for 2 years at 12% per annum is Rs.1590, what would be the simple interest?

Sol. Let the sum be Rs. x then, $x\left[1+\frac{12}{100}\right]^2 - x = 1590$

$$x\left[\frac{112}{100}\right]^2 - x = 1590$$

$$= x \times \frac{28}{25} \times \frac{28}{25} - x = 1590$$

$$\text{(or)}\quad x\left(\frac{28 \times 28 - 625}{625}\right) = 1590$$

$$\Rightarrow x\left(\frac{784 - 625}{625}\right) = 1590$$

$$\Rightarrow \frac{159x}{625} = 1590$$

$$\Rightarrow 159\,x = 1590 \times 625$$

$$\therefore x = \frac{1590 \times 625}{159} = 6250$$

Sum = 6250 Rs.

$$\therefore \text{S.I.} = \text{Rs.}\left[\frac{6250 \times 12 \times 2}{100}\right]$$

= Rs.1500

EXAMPLE-6

A sum of money amounts to Rs.6690 after 3 years and to Rs. 10035 after 6 years on compound interest. Find the sum.

Sol. Let the sum be Rs. P then,

$$P\left[1+\frac{R}{100}\right]^3 = 6690 \quad \text{.... (i)}$$

$$\text{and } P\left[1+\frac{R}{100}\right]^6 = 10035 \quad \text{.... (ii)}$$

on dividing we get

$$\left[1+\frac{R}{100}\right]^3 = \frac{10035}{6690} = \frac{3}{2}.$$

Substituting this value in (i) we get : $P \times \frac{3}{2}$

$$= 6690 \text{ (or) } P = 6690 \times \frac{2}{3} = 4460$$

Hence, the sum is Rs.4460

EXAMPLE-7

A sum of money doubles itself at compound interest in 15 years. In how many years will it become eight times ?

Sol. $P\left[1+\frac{R}{100}\right]^{15} = 2P \Leftrightarrow \left[1+\frac{R}{100}\right]^{15} = 2 \quad ...(i)$

$$\text{Let } P\left[1+\frac{R}{100}\right]^n = 8P$$

$$\Rightarrow \left[1+\frac{R}{100}\right]^n = 8 = 2^3 = \left\{1+\left(\frac{R}{100}\right)\right\}^{15}$$

$$\Rightarrow \left[1+\frac{R}{100}\right]^n = \left[1+\frac{R}{100}\right]^{45}$$

$\Rightarrow n = 45$

Thus, the required time = 45 years.

EXAMPLE-8

A certain sum amounts to Rs.7350 in 2 years and to Rs.8575 in 3 years. Find the sum and rate per cent.

Sol. S.I. on Rs.7350 for 1 year = Rs.(8575−7350)

= Rs.1225

$$\therefore \text{Rate} = \left[\frac{100 \times 1225}{7350 \times 1}\right]\% = 16\frac{2}{3}\%$$

Let the sum be Rs. x. then $x\left[1+\frac{50}{3 \times 100}\right]^2 = 7350$

$$\Leftrightarrow x \times \frac{7}{6} \times \frac{7}{6} = 7350$$

$$\Leftrightarrow x = \left[7350 \times \frac{36}{49}\right] = 5400$$

$\therefore$ Sum = Rs. 5400

PRACTICE TEST

COMPOUND INTEREST

1. The amount of Rs.7500 at compound interest at 4% per annum for 2 years is :

a) Rs. 7800 b) Rs. 8100

c) Rs. 8112 d) Rs. 8082

2. The compound interest on a certain sum at 5% per annum for 2 years is Rs.328. The simple interest for that sum at the same rate and for the same period will be :

a) Rs. 320 b) Rs. 322

c) Rs. 325 d) Rs. 326

3. The compound interest on Rs.5600 for $1\frac{1}{2}$ years at 10% per annum, compounded annually is :

a) Rs. 882.70 b) Rs. 873.50

c) Rs. 868 d) Rs. 840

4. What is the principal amount which earns Rs.132 as compound interest for the second year at 10% per annum ?

a) Rs. 1000 b) Rs. 1200

c) Rs. 1320 d) Rs. 1188

5. To find out the total compound interest accured on a sum of money after 5 years, which of the following informations given in the statements A and B is / are sufficient ?

A : the rate of interest was 6% per annum.

B : the total simple interest on the same amount after 5 years at the same rate will be Rs.600

a) Only A is sufficient

b) Either A or B is sufficient

c) Both A & B together are needed

d) Only B is sufficient

e) Both A and B are not sufficient

6. A tree increases annually by $\frac{1}{8}$ th of its height. By how much will it increase after 2 years, if it stands today 64 cm high ?

a) 72 cm b) 74 cm

c) 75 cm d) 81 cm

7. The compound interest on a sum for 2 years is Rs.832 and the simple interest on the same sum for the same period is Rs. 800. The difference between the compound interest and the simple interest for 3 years will be :

a) Rs. 48 b) Rs. 66.56

c) Rs. 98.56 d) None of these

8. The difference between simple interest and compound interest on Rs.1200 for one year at 10% per annum, reckoned half-yearly is :

a) Nil b) Rs. 13.20 c) Rs. 8.80 d) Rs. 3

9. A sum of money invested at compound interest amounts to Rs. 800 in 3 years and Rs.840 in 4 years. What is the rate of interest per annum ?

a) 2% b) 4% c) 5% d) 10%

10. A sum of money becomes 8 times of itself in 3 years at compound interest. The rate of interest is:

a) 100% b) 8% c) 1% d) Data inadequate

11. The compound interest of Rs. 20480 at $6\frac{1}{2}$% per annum for 2 years 73 days is :

a) Rs. 3000 b) Rs. 3131

c) Rs. 2929 d) Rs. 3636

12. The compound interest on a certain sum for 2 years is Rs. 208 and the simple interest is Rs. 200. The sum is :

a) Rs. 750 b) Rs. 1050

c) Rs. 1250 d) Rs. 1650

13. The difference between the compound interest and the simple interest on a certain sum at 5% per annum for 2 years is Rs. 1.50. The sum is

a) Rs. 600 b) Rs. 500 c) Rs. 400 d) Rs. 300

14. Simple interest on a sum at 4% per annum for 2 years is Rs. 80. The compound interest on the same sum for the same period is :

a) Rs. 1081.60 b) Rs. 81.60

c) Rs. 160 d) None of these

15. The difference between simple interest and compound interest on a sum for 2 years at 8%, when the interest is compounded annually is Rs. 16. If the interest were compounded half yearly, the difference in two years would be nearly.

a) Rs: 16 b) Rs. 16.80 c) Rs. 21.85 d) Rs. 24.64

16. At what rate of compound interest percent per annum will a sum of Rs. 1200 become Rs. 1348.32 in 2 years?

a) 7% b) 7.5% c) 6% d) 6.5%

17. The difference in compound interest and simple interest for two years on a sum of money is Rs. 160. If the simple interest for 2 years be Rs. 2880, the rate percent is

a) $5\frac{5}{9}$% b) $12\frac{1}{2}$% c) $11\frac{1}{9}$% d) 9%

18. A sum of Rs. 12000 deposited at compound interest becomes double after 5 years. After 20 years it will become.
    a) 120000 b) 192000
    c) 124000 d) 96000
19. The compound interest on a sum for 2 years is Rs. 832 and the simple interest on the same sum for the same period is Rs. 800. The difference between the compound and simple interests for 3 years will be
    a) 98.56
    b) Rs. 66.56
    c) Rs. 48
    d) None of these
20. A certain sum of money at compound interest becomes Rs. 3472.87 in 3 years and Rs. 3307.50 in 2 years. The rate of interest is
    a) 6% b) 5.5%
    c) 5% d) 6.75%

ANSWERS

1. c	2. a	3. c	4. b	5. c
6. d	7. c	8. d	9. c	10. a
11. c	12. c	13. a	14. b	15. d
16. c	17. c	18. b	19. a	20. c

EXPLANATIONS

1. **Ans : (c)**

$$\text{Amount} = \text{Rs.}\left[7500 \times \left(1+\frac{4}{100}\right)^2\right]$$

$$= \text{Rs.}\left[7500 \times \left(\frac{104}{100}\times\frac{104}{100}\right)\right]$$

$$= \text{Rs.}\left[7500 \times \frac{26}{25}\times\frac{26}{25}\right]$$

$$= \text{Rs. } 8112$$

2. **Ans : (a)**

Let the sum be Rs. x. Then,

$$\text{C.I.} = x\left[1+\frac{5}{100}\right]^2 - x$$

$$= \left[\frac{441x}{400} - x\right]$$

$$= \frac{41x}{400}$$

$$\therefore \frac{41x}{400} = 328$$

$$\text{(or) } x = \frac{328 \times 400}{41}$$

$$= 3200$$

$$\therefore \text{S.I.} = \text{Rs.}\left[\frac{3200 \times 5 \times 2}{100}\right]$$

$$= \text{Rs.}320$$

3. **Ans : (c)**

$$\text{Amount} = \text{Rs}\left[5600 \times \left[1 \times \frac{10}{100}\right] \times \left[1 + \frac{\frac{1}{2}\times 10}{100}\right]\right]$$

$$= \text{Rs.}\left[5600 \times \frac{11}{10} \times \frac{21}{20}\right]$$

$$= \text{Rs.}6468$$

$$\therefore \text{C.I} = \text{Rs. } (6468 - 5600)$$

$$= \text{Rs.}868$$

4. **Ans : (b)**

Let the principal at the end of first year be Rs.x

$$\text{Then, } \frac{x \times 10 \times 1}{100} = 132$$

$$\text{(or) } x = 1320$$

Now, let the original principal be Rs. P. then, amount

$$\text{after 1 year} = P + \frac{P \times 10 \times 1}{100}$$

$$= \frac{11p}{10}$$

$$\therefore \frac{11p}{10} = 1320$$

$$\text{(or) } P = \frac{1320 \times 10}{11} = \text{Rs.}1200$$

5. **Ans : (c)**

Using A and B both, we may calculate

$$\text{Sum} = \left[\frac{100 \times 600}{6 \times 5}\right]$$

$$= \text{Rs.}2000$$

So, the compound interest may be obtained. Thus, both A and B together are needed.

6. **Ans : (d)**

$$\text{Increase \%} = \left[\frac{1}{8} \times 100\right]\% = 12\frac{1}{2}\%$$

Height after 2 years $= \left[64 \times \left[1+\frac{25}{2\times100}\right]^2\right]$ cm

$= \left[64 \times \frac{9}{8} \times \frac{9}{8}\right]$ cm

$= 81$ cm

7. Ans : (c)

Diff. in C.I. & S.I. for 2 years = Rs.32

S.I. for 1 year = Rs.400

$\therefore$ S.I. on Rs. 400 for 1 year = Rs.32

$\therefore$ Rate $= \left[\frac{100\times32}{400\times1}\right]$ %

$= 8\%$

Hence diff. in C.I. and S.I for 3rd year

= S.I. on Rs. 832

$=$ Rs. $\left[832 \times \frac{8}{100} \times 1\right]$

= Rs.66.56

Total difference = Rs. (32 + 66.56)

= Rs.98.56

8. Ans : (d)

S.I. = Rs. $\left[\frac{1200\times10\times1}{100}\right]$

= Rs.120

C.I. = Rs. $\left[1200 \times \left[1+\frac{5}{100}\right]^2 - 1200\right]$

= Rs.123.

$\therefore$ (C.I.) – (S.I.) = Rs.(123 – 120)

= Rs.3

9. Ans : (c)

Interest on Rs. 800 for 1 year = Rs.(840–800)

= Rs.40

$\therefore$ Rate $= \left[\frac{100\times40}{800\times1}\right]$ %

$= 5\%$

10. Ans : (a)

$P\left[1+\frac{R}{100}\right]^3 = 8P$

(or) $\left[1+\frac{R}{100}\right]^3 = 8 = 2^3$

$\therefore 1+\frac{R}{100} = 2$

(or) $\frac{R}{100} = 1$

(or) R = 100 %.

11. Ans : (c)

C.I. = Rs. $\left[20480 \times \left(1+\frac{25}{4\times100}\right)^2 \left(1+\frac{1}{5}\times\frac{25}{4\times100}\right) - 20480\right]$

$=$ Rs. $\left[\left(20480 \times \frac{17}{16} \times \frac{17}{16} \times \frac{81}{80}\right) - 20480\right]$

= Rs. 2929

12. Ans : (c)

Clearly, Rs.8 is S.I. on Rs.100 for 1 year

$\therefore$ Rate $= \left(\frac{8\times100}{100\times1}\right)$

$= 8\%$

And, sum $= \frac{100\times200}{8\times2}$

= Rs. 1250

13. Ans : (a)

Let the sum be Rs. 100. Then,

S.I. = Rs. $\left(\frac{100\times5\times2}{100}\right)$

= Rs. 10

C.I. = Rs. $\left[\left\{100 \times \left(1+\frac{5}{100}\right)^2\right\} - 100\right]$

$=$ Rs. $\frac{41}{4}$

Difference between C.I. & S.I. $= \left(\frac{41}{4} - 10\right)$

= Re. 25

= 25 : 1.50 : : 100 : x

$\therefore x = \left(\frac{1.50\times100}{.25}\right)$

= Rs. 600

14. Ans : (b)

Principal – Rs. $\left(\frac{100\times80}{4\times2}\right)$ = Rs. 1000.

$\therefore$ C.I. = Rs. $\left[\left\{1000 \times \left(1+\frac{4}{100}\right)^2\right\} - 1000\right]$

= Rs. 81.60.

15. Ans : (d)

For first year, S.I. = C.I.

Thus, Rs. 16 is the S.I. on S.I. for 1 year, which at 8% is thus

Rs. 200. i.e., S.I. on the principal for 1 year is Rs. 200.

$\therefore$ Principal = Rs. $\left(\frac{100 \times 200}{8 \times 1}\right)$

= Rs. 2500

Amount for 2 years, compounded half yearly

= Rs. $\left[2500 \times \left(1 + \frac{4}{100}\right)^4\right]$

= Rs. 2924.64.

$\therefore$ C.I. = Rs. 424.64.

Also, S.I. = Rs. $\left(\frac{2500 \times 8 \times 2}{100}\right)$

= Rs. 400

Hence, [(CI.) - (S.I.)] = Rs (424.64 - 400)

= Rs. 24.64

16. Ans : (c)

Let the rate be x% per annum.

Then, $1200 \times \left(1 + \frac{x}{100}\right)^2 = 1348.32$ (or)

$= \left(1 + \frac{x}{100}\right)^2$

$= \frac{1348.32}{1200}$

$= 1.1236 = (1.06)^2$

$\therefore \left(1 + \frac{x}{100}\right) = 1.06$

(or) $\frac{x}{100} = 0.06$

(or) x = 6%

17. Ans : (c)

S.I. for 1 year = Rs. 1440

$\therefore$ S.I. on Rs. 1440 for 1 year Rs. 160.

Hence, rate percent = $\left(\frac{100 \times 160}{1440 \times 1}\right)$

$= 11\frac{1}{9}\%$

18. Ans : (b)

$2P = P\left(1 + \frac{R}{100}\right)^5$

(or) $\left(1 + \frac{R}{100}\right)^5 = 2$

$\therefore \left(1 + \frac{R}{100}\right)^{20} = 2^4 = 16$

Thus, $P\left(1 + \frac{R}{100}\right)^{20} = 16P$

= Rs. (12000 × 16)

= Rs. 192000

19. Ans : (a)

S.I. for first year = Rs. 400.

S.I. on Rs. 400 for 1 year = Rs. 32.

$\therefore$ Rate = $\left(\frac{100 \times 32}{400 \times 1}\right)$

= 8%.

Hence, the difference for 3rd year is S.I. on Rs. 832

= Rs. $\left(832 \times \frac{8}{100}\right)$

= Rs. 66.56

$\therefore$ Total difference = Rs. (32 + 66.56)

= Rs. 98.56.

20. Ans : (c)

Let p be the principal and R% per annum be the rate.

Then, $P\left(1 + \frac{R}{100}\right)^3 = 3472.87$ .... (i)

and $P\left(1 + \frac{R}{100}\right)^2 = 3307.50$ .... (ii)

on dividing (i) and (ii), we get : $\left(1 + \frac{R}{100}\right)$

$= \frac{3472.87}{3307.50}$

= 1.05

$\therefore \frac{R}{100} = 0.5$ (or) R = 5%.

10. AGES

SOLVED EXAMPLES

EXAMPLE PROBLEMS

EXAMPLE-1

The ratio of the ages of Mona and Sona is 4 : 5. Twelve years hence, their ages will be in the ratio of 5 : 6. What will be Sona's age after 6 years ?

Sol. Let their present ages be 4x & 6x.

Then, $\frac{4x + 12}{5x + 12} = \frac{5}{6}$

$\Rightarrow (4x + 12)\,6 = 5\,(5x + 12) \Rightarrow 24x + 72$

$= 25x + 60 \Rightarrow x = 12.$

EXAMPLE-2

Ramu was 4 times as old as his son 8 years ago. After 8 years, Ramu will be twice as old as his son. What are their present ages ?

Sol. Let son's age 8 years ago be x years.

Then Ramu's age at that time = 4x years.

Son's age after 8 years = (x + 8) + 8 = (x+16) yrs.

Ramu's age after 8 years = (4x + 8) + 8 = (4x+16) yrs.

$\therefore$ 2(x + 16) = 4x + 16 (or) x = 8

$\therefore$ Son's present age = (x + 8) = 16 years.

Ramu's present age = (4x + 8) = 40 years.

EXAMPLE-3

The sum of the ages of Aruna and her mother is 49 years. Also, 7 years ago, the mother's age was 4 times Aruna's age. Find the present age of Aruna's mother.

Sol. Let Aruna's age 7 years ago be x.

Mother's age 7 years ago = 4x.

$\therefore$ (x + 7) + (4x + 7) = 49 (or) x = 7

$\therefore$ Mother's present age = (4x + 7)

= 35 years.

EXAMPLE-4

The ages of A and B differ by 16 years. If 6 years ago, the elder one be 3 times as old as the younger one. Find their present ages.

Sol. Let A's age = x &

B's age = (x + 16)

3 (x – 6) = (x + 16 – 6) (or) x = 14

$\therefore$ A's Age = 14 years and

B's age = 30 years.

EXAMPLE-5

The sum of the ages of a mother and a dauther is 50 years. Also, 5 years ago, the mother's age was 7 times the age of the daughter. The present ages of the mother and the daughter ?

Sol. Let the daugher's present age be x years.

Then, mother's present age = (50 – x) years.

Now, 7 (x – 5) = (50 – x – 5) or x = 10

So, their present ages are 40 yrs and 10 yrs.

EXAMPLE-6

The sum of the ages of a son and father is 56 years. After four years, the age of the father will be three times that of the son. Find their ages ?

Sol. Let the present ages of son and father be x years and (56 – x) years respectively.

Then ,(56 – x + 4) = 3 (x + 4)

(or) 4x = 48 (or) x = 12

So, their ages are 12 years, 44 years respectively.

EXAMPLE-7

Rajan's age is 3 times that of Ashok. In 12 years, Rajan's age will be double the age of Ashok. Rajan's present age is ?

Sol. Let Ashok's age be x years.

Then, Rajan's age = 3x years

$\therefore$ 2 (x + 12) = 93x + 12)

(or) x = 12

Hence, Rajan's present age is 36 years.

EXAMPLE-8

The age of Arvind's father is 4 times his age. If 5 years ago, father's age was 7 times of the age of his son at that time, what is Arvind's father's present age ?

Sol. Let Arvind's age be x years.

Then, his father's age = 4x years.

$\therefore$ (4x – 5) = 7 (x – 5)

(or) 3x = 30

(or) x = 10

PRACTICE TEST

AGES

1. The ratio of the ages of Meena and Meera is 4 : 3. The sum of their ages is 28 years. The ratio of their ages after 8 years will be :
   a) 4 : 3 b) 12 : 11 c) 7 : 4 d) 6 : 5
2. The ratio of the ages of Swati and Varun is 2:5. After 8 years, their ages will be in the ratio of 1 : 2. The difference in their present ages (in years) is :
   a) 24 b) 26 c) 29 d) 32
3. One year ago, Mrs. Promila was four times as old as her daughter Swati. Six years hence, Mrs. Promila's age will exceed her daughter's age by 9 years. The ratio of the present ages of Promila and her daughter is:
   a) 9 : 2 b) 11 : 3 c) 12 : 5 d) 13 : 4
4. If 6 years are substracted from the present age of Gulzar and the remainder is divided by 18, then the present age of his grandson Anup is obtained. If Anup is 2 years younger to Mahesh whose age is 5 years, then what is the present age of Gulzar ?
   a) 96 years b) 84 years
   c) 60 years d) 48 years.
5. Sneh's age is $\frac{1}{6}$ th of her father's age. Sneh's father's age will be twice of Vimal's age after 10 years. If Vimal's eighth birthday was celebrated two years before, then what is Sneh's present age?
   a) 24 years b) 30 years
   c) $6\frac{2}{3}$ years d) None of these
6. Jayesh is as much younger to Anil as he is older to Prashant. If the sum of the ages of Anil and Prasant is 48 years, what is the age of Jayesh ?
   a) 20 years b) 30 years
   c) 24 years d)cannot be determined
7. The ages of two persons differ by 20 years. If 5 years ago, the elder one be 5 times as old as the younger one, their present ages (in years) are respectively.
   a) 30, 10 b) 25, 5
   c) 29, 9 d) 50, 30
8. The ratio of the father's age of his son's age is 7 : 3. The product of their ages is 756. The ratio of their ages after 6 years will be :
   a) 5 : 2 b) 2 : 1
   c) 11 : 7 d) 13 : 9
9. Average age of A and B is 24 years and average age of B, C and D is 22 years. The sum of the ages of A, B, C and D is :
   a) 90 years b) 96 years
   c) 114 years d) Data inadequate
10. The average age of an adult class is 40 years. Twelve new students with an average age of 32 years join the class, thereby decreasing the average of the class by 4 years. The original strength of the class was :
    a) 10 b) 11 c) 12 d) 15
11. One year ago a father was four times as old as his son. In 6 years time his age exceeds twice his son's age by 9 years. Ratio of their ages is :
    a) 13 : 4 b) 12 : 5
    c) 11 : 3 d) 9 : 2
12. The age of Arvind's father is 4 times his age. If 5 year's ago, father's age was 7times of the age of his son at that time, what is Arvind's father's present age?
    a) 84 years b) 70 years
    c) 40 years d) 35 years
13. 10 years ago Chandravati's mother was 4 time older than her daughter. After 10 years, the mother will be twice older than the daughter. The present age of Chandravati is :
    a) 5 years b) 10 years c) 20 years d) 30 years
14. Ratio of Ashok's age to Pradeep's age is equal to 4 : 3. Ashok will be 26 years old after 6 years, How old is Pradeep now ?
    a) $19\frac{1}{2}$ years b) 21 years
    c) 12 years d) 15 years
15. The ratio of the father's age to the son's age is 4 : 1. The product of their ages is 196. The ratio of their ages after 5 years will be.
    a) 3 : 1 b) 10 : 3 c) 11 : 4 d) 14 : 5
16. The ages of A and B are in the ratio 2 : 5. After 8 years their ages will be in the ratio 1 : 2. The difference of their ages is
    a) 20 years b) 24 yeras
    c) 26 years d) 29 years
17. Deepak is 4 times as old as his son. Four years hence the sum of their ages will be 43 years. How old is Deepak's son now ?
    a) 5 years b) 7 years c) 8 years d) 10 years

18. The ratio of Mona's age to the age of her mother is 3 : 11. The difference of their ages is 24 years. The ratio of their ages after 3 years will be
    a) 1 : 3 b) 2 : 3
    c) 3 : 5 d) None of these

19. Three years ago the average age of A and B was 18 years. With C joining them now, the average becomes 22 years. How old is C now ?
    a) 27 years b) 30 years
    c) 28 years d) 24 years

20. Kamla got married 6 years ago. Today her age is $1\frac{1}{4}$ times her age at the time of marriage. His son's age is (1/10) times her age. The age of her son is
    a) 2 years b) 3 years c) 4 years d) 5 years

ANSWERS

1. d	2. a	3. d	4. c	5. d
6. c	7. a	8. b	9. d	10. c
11. c	12. c	13. c	14. d	15. c
16. b	17. b	18. a	19. d	20. b

EXPLANATIONS

1. Ans : (d)

Let Meena's age = 4x & Meera's age = 3x

Then, $4x + 3x = 28 \Leftrightarrow x = 4$

$\therefore$ Meena's age = 16 years &

Meera's age = 12years.

2. Ans : (a)

Let Swati's age = 2x & Varun's age = 5x.

$\therefore \frac{2x+8}{5x+8} = \frac{1}{2} \Leftrightarrow 2(2x+8) = (5x+8)$

$\Leftrightarrow x=8$

$\therefore$ Swati's age = 16 years &

Varun's age = 40 years.

Difference of their ages = 24 years.

3. Ans : (d)

Let Swati's age 1 year ago = x

Then, Promila's age 1 year ago = 4x

$(4x + 6) - (x + 6) = 9$

(or) $x = 3$

Present age of Promila = (12 + 1) years

= 13 years.

$\therefore$ Ratio of their ages = 13 : 4

4. Ans : (c)

Anup's age = (5 − 2) years

= 3 years.

Let Gulzar's age be x years.

then, $\frac{x-6}{18} = 3 \Leftrightarrow x - 6 = 54$

$\Leftrightarrow x = 60$

5. Ans : (d)

Vimal's age after 10 years = (8+2+10) years

= 20 years.

Sneh's father's age after 10 years = 40 years.

Sneh's father's present age = 30 years

Sneh's age = $\left(\frac{1}{6} \times 30\right)$ years

= 5 years.

6. Ans : (c)

A − J = J − P (or) A + P

= 2J.

So, 2J = 48

(or) J = 24.

7. Ans : (a)

Let their ages be x and (x + 20) years

$5(x-5) = (x + 20 - 5)$

(or) $4x = 40$

(or) $x = 10$

$\therefore$ Their present ages are 30 years & 10 years.

8. Ans : (b)

Let their ages be 7x and 3x years.

Then, $7x \times 3x = 756 \Leftrightarrow x^2 = 36 \Leftrightarrow x = 6$

$\therefore$ Their present ages are 42 years & 18 years.

Ratio of their ages after 6 years = 48 : 24 = 2 : 1

9. Ans : (d)

$(A + B) = (24 \times 2) = 48$ & $B + C + D$

$= (22 \times 3) = 66$

with this data, we cannot find A+B+C+D. So the data is inadequate.

10. Ans : (c)

Let, Original strength = x.

$\therefore \dfrac{40x + 12 \times 32}{x + 12} = 36$

(or) x = 12.

11. Ans : (c)

Let the present ages of father & son be x & y respectively.

Then, $(x - 1) = 4(y - 1)$

(or) $4y - x = 3$ ...... (i)

And, $(x + 6) - 2(y + 6) = 9$

(or) $-2y + x = 15$ ...... (ii)

Solving (i) & (ii)

We get, x = 33 and y = 9

$\therefore$ Ratio of their ages = 33 : 9 (or) 11 : 3.

12. Ans : (c)

Let Arvind's age be x years.

Then, his father's age = 4x years.

$\therefore (4x - 5) = 7(x - 5)$

(or) $3x = 30$

(or) $x = 10$

Hence, Arvind's father's age is 40 yeras.

13. Ans : (c)

Let the present ages of Chandravati and her mother be x years and y years respectively.

Then, $(y - 10) = 4(x - 10)$

(or) $4x - y = 30$ ...... (i)

And, $(y + 10) = 2(x + 10)$

(or) $y - 2x = 10$ ..... (ii)

Solving (i) & (ii) we get,

Chandravati's present age = 20 years

$\therefore x = 20$

14. Ans : (d)

Ashok's present age = (26 – 6) years

= 20 years.

$\therefore$ pradeep's present age = $\left(\dfrac{3}{4} \times 20\right)$ years

= 15 years.

15. Ans : (c)

Let their ages be 4x and x respectively.

Then, $4x^2 = 196$

(or) $x^2 = 49$ (or) $x = 7$

$\therefore$ Their present ages are 28 years & 7 years.

Ratio of their ages after 5 years = 33 : 12

(or) 11 : 4.

16. Ans : (b)

Let their ages be 2x & 5x years.

Then, $\dfrac{2x + 8}{5x + 8} = \dfrac{1}{2}$

(or) $x = 8$

$\therefore$ Their ages are 16 years and 40 years.

Hence, the difference of their ages = 24 years.

17. Ans : (b)

Let the son's age be x years.

Then, $(x + 4) + (4x + 4) = 43$

(or) $5x = 35$

(or) $x = 7$

18. Ans : (a)

Let the ages of Mona and her mother be 3x and 11x years respectively.

Then $(11x - 3x) = 24$

(or) $x = 3$

So, their present ages are 9 years and 33 years.

Ratio of their ages after 3 years 12 : 36 (or) 1 : 3.

19. Ans : (d)

(A + B)'s total present age = (36 + 6)

= 42 years

(A + B + C)'s total present age = (22 × 3)

= 66 yeras.

$\therefore$ C's age = (66 – 42) years

= 24 years.

20. Ans : (b)

Let son's age be x.

Then, Kamla's age = 10x years.

Kamla's age at the time of marriage = (10x – 6) years

$\therefore 10x = \dfrac{5}{4}(10x - 6)$ (or)

$= 40x$

$= 50x - 30$

(or) $x = 3$

11. NUMBER SYSTEM

All numbers are written by using the symbols 0, 1, 2, 3, 4, 5, 6, 7, 8, 9 called digits. A number contains digits.

Even Number : Any number ending with 0, 2, 4, 6, 8 is always divisible by 2. Hence they are called even numbers.

Odd number : Any number ending with 1, 3, 5, 7, 9 are not divisible by 2. Hence they are called odd numbers.

Local value : The local value of a number is the face value of that number. For example, the local value of 5 in 1520 is five.

Place value : The place value of number is the value of the place it occupies times its local value. For example, the place value of 5 in 1520 is 100 times 5, i.e., 500.

Prime Numbers : The numbers which are divisible by either 1 or themselves are called prime numbers.

Eg : 2, 3, 5, 7, 11, 13, 17, 19, 23, 29, 31, 37, 41, 43, 47, 53, 59, 61, 67, 71, 73, 79, 83, 89, 97, 101 etc.

Composite Numbers : The numbers which are not prime that is, which are divisible by atleast one natural number other than 1 are called composite numbers.

Eg : 4, 6, 8, 9, 10, 12, 14, 16, 18, 20, 21 etc.

The number 1 is neither prime number nor a composite number.

SOLVED EXAMPLES

EXAMPLE-1

If 30% of a certain number is 12.6, then find the number.

Sol. Let the required number be x.

Then $\frac{30}{100} x = 12.6$

(or) $x = \left[\frac{12.6 \times 100}{30}\right]$

$= 42$

$\therefore$ Required number = 42.

EXAMPLE-2

A number consists of two digits. The sum of the digit is 9. If 63 is substracted from the number, its digits are inter changed. Find the number.

Sol. Let ten's digit = x, Then, unit's digit = (9–x).

$10x + (9 - x) - 63 = 10(9 - x) + x$

(or) $9x - 54 = 90 - 9x$

(or) $18x = 144$

(or) $x = 8$

$\therefore$ Ten's digit = 8 and unit's digit = 1.

Hence, the required number is 81.

EXAMPLE-3

A fraction becomes $\frac{2}{3}$ when 1 is added to both, its numerator and denominator. And, it becomes $\frac{1}{2}$ when 1 is subtracted from both the numerator and denominator. Find the fraction.

Sol. Let the required fraction be $\frac{x}{y}$

Then, $\frac{x+1}{y+1} = \frac{2}{3}$

$\Rightarrow 3x - 2y = -1$ ...... (i)

And $\frac{x-1}{y-1} = \frac{1}{2}$

$\Rightarrow 2x - y = 1$ ...... (ii)

Solving (i) & (ii),

We get x = 3,

y = 5.

$\therefore$ Required fraction $= \frac{3}{5}$.

PRACTICE TEST
NUMBER SYSTEM

1. Three – fourth's of a number exceeds its one-third by 60. The number is :
a) 108 b) 144 c) 184 d) None of these
2. A number exceeds its four seventh by 18. What is the number ?
a) 36 b) 49 c) 63 d) None of these
3. The product of two numbers is 120 and the sum of their squares is 289. The sum of the numbers is :
a) 20 b) 23 c) 169 d) None of these
4. The difference of two numbers is 11 and $\frac{1}{5}$ th of their sum is 9. The numbers are :
a) 31, 20 b) 30, 19 c) 29, 18 d) 28, 17
5. The sum of two consecutive odd numbers in a set of three consecutive odd numbers is 5 more than the third number. What is the second of these numbers ?
a) 5 b) 7 c) 9 d) 11
6. The difference between a two - digit number and the number obtained by interchanging the digits is 27. What is the difference between the digits of the number ?
a) 3 b) 6 c) 9 d) 5
7. The sum of the numerator and denominator of a fraction is 11. If 1 is added to the numerator and 2 is subtracted from the denominator, it becomes $\frac{2}{3}$. The fraction is :
a) $\frac{5}{6}$ b) $\frac{6}{5}$ c) $\frac{3}{8}$ d) $\frac{8}{3}$
8. The sum of digits of a two digit number is 9 less than the number. Which of the following digits is at unit place of the number ?
a) 1 b) 2
c) 4 d) Data inadequate
9. The sum of two numbers is twice their difference. If one of the numbers is 10, the other one is :
a) $3\frac{1}{3}$ b) $-3\frac{1}{3}$ c) $4\frac{1}{4}$ d) 30
10. If one-seventh of a number exceeds is eleventh part by 100, then the number is :
a) 770 b) 1100
c) 1925 d) 1900
11. The ratio between two numbers is 2 : 3. If the consequent of the ratio is 24, the antecedent is :
a) 36 b) 16 c) $\frac{48}{5}$ d) $\frac{72}{5}$
12. The difference between two numbers is 5 and the difference between their squares is 65. The numbers are :
a) 15, 10 b) 14, 9 c) 12, 7 d) 9, 4
13. The product of two positive numbers is 112 and one of the numbers is larger than the other by 6. The larger number is :
a) 16 b) 14 c) 7 d) 8
14. Of the two numbers, 4 times the first is equal to 5 times the other and the sum of 3 times the first and 5 times the second is 105. The first number is :
a) 10 b) 12 c) 13 d) 15
15. Of the two numbers, 4 times the smaller one is less than 3 times the larger one by 5. But, the sum of the numbers is larger than 6 times their difference by 6. The larger number is :
a) 59 b) 69 c) 79 d) 89
16. We are given a fraction. If 3 is added to its denominator, it becomes $\frac{1}{3}$ and if 4 be added to its numerator, it becomes (3/4). The given fraction is:
a) $\frac{4}{9}$ b) $\frac{3}{20}$ c) $\frac{7}{24}$ d) $\frac{5}{12}$
17. Of the three numbers, the sum of first two is 45; the sum of the second and the third is 55 and the sum of the third and thrice the first is 90. The third number is :
a) 20 b) 25 c) 30 d) 35
18. If 10 be added to four times a certain number, the result is 5 less than 5 times the number. The number is :
a) 35 b) 25 c) 20 d) 15
19. The sum of three consecutive odd numbers is 21. The middle one is :
a) 11 b) 9 c) 7 d) 5
20. One number is greater by 2 than thrice the another number; and 6 times the smaller number exceeds the greater by 1. The smaller number is :
a) 6 b) 5 c) 3 d) 1

ANSWERS

1. b	2. d	3. b	4. d	5. c
6. a	7. c	8. d	9. d	10. c
11. b	12. d	13. d	14. d	15. a
16. d	17. c	18. d	19. c	20. d

EXPLANATIONS

1. **Ans : (b)**
$\frac{3}{4}x - \frac{1}{3}x = 6 \Leftrightarrow 9x - 4x = 720 \Rightarrow x = 144.$

2. **Ans : (d)**
$x - \frac{4x}{7} = 18 \Rightarrow \frac{3x}{7} = 18 \Leftrightarrow x = \left(\frac{18 \times 7}{3}\right) = 42$

3. **Ans : (b)**
Let the numbers be x and y. Then, $xy = 120$ and $x^2 + y^2 = 289$
$(x + y)^2 = ((x^2 + y^2) + 2xy = (289 + 240) = 529$
$\therefore x + y = \sqrt{529} = 23$

4. **Ans : (d)**
Let the numbers be x and y. Then, $x-y = 11$ .....(1)
and $\frac{1}{5}(x + y) = 9$ (or) $x + y = 45$ ..... (2)
Solving (1) & (2) we get, $x = 28; y = 17$

5. **Ans : (c)**
Let the numbers be x, x+2, x+4
$(x + x + 2) - (x + 4) = 5 \Leftrightarrow x = 7$
$\therefore$ Second number = 9.

6. **Ans : (a)**
Let ten's digit be x and unit's digit be y. Then, $(10x + y) - (10y + x) = 27 = 9(x - y) = 27 \Leftrightarrow x - y = 3.$

7. **Ans : (c)**
Let the fraction be $\frac{x}{y}$.
Then, $x + y = 11$ .... (1)
$\frac{x+1}{y-2} = \frac{2}{3} \Rightarrow 3(x + 1)$
$= 2(y - 2)$ (or) $3x - 2y = 7$ .... (2)
Solving (1) & (2) we get $x = 3$ & $y = 8$
$\therefore$ Fraction is $\frac{3}{8}$.

8. **Ans : (d)**
Let the tens digit be x and the unit's digit be y.
$(10x + y) - (x + y) = 9$ (or) $x = 1$
From this data, we can not find y. The unit's digit.
$\therefore$ Data is inadequate.

9. **Ans : (d)**
Let the other number be x. Then $10+x = 2(x-10) \Rightarrow x = 30$

10. **Ans : (c)**
Let the number be x. Then,
$\frac{x}{7} - \frac{x}{11} = 100 \Leftrightarrow 11x - 7x = 7700 \Leftrightarrow x = 1925$

11. **Ans : (b)**
If consequent is 3, antecedent = 2
If consequent is 24, antecedent = $\left(\frac{2}{3} \times 24\right)$
$= 16$

12. **Ans : (d)**
Let the numbers be x and y.
Then, $x - y = 5$ and $x^2 - y^2 = 65$
On dividing, we get $x + y = 13$ Solving $x - y = 5$ & $x + y = 13$ we get, $x = 9$ & $y = 4$

13. **Ans : (d)**
Let the numbers be x and $(x - 6)$. Then, $x(x - 6) = 112$ (or) $x^2 - 6x - 112 = 0$ (or) $(x - 14)(x + 8) = 0$
$\therefore x = 14$ (neglecting the negative value, $x = -8$).

14. **Ans : (d)**
Let the numbers be x and y.
Then, $4x = 5y$ and $3x + 5y = 105$
$\therefore 4x - 5y = 0$ and $3x + 5y = 105$
Adding, we get $7x = 105$ (or) $x = 15$.

15. **Ans : (a)**
Let the numbers be x and y.
Then, $3x - 4y = 5$ and $(x + y) - 6(x - y) = 6$
(or) $3x - 4y = 5$ and $-5x + 7y = 6$
On solving these equations, we get $x = 59$.

16. **Ans : (d)**
Let the given fraction be (x/y).
Then, $\frac{x}{y+3} = \frac{1}{3}$ and $\frac{x+4}{y} = \frac{3}{4}$ (or) $3x - y = 3$
and $3y - 4x = 16$
On solving, we get $x = 5$ and $y = 12$
Hence, the required fraction is $\frac{5}{12}$

17. **Ans : (c)**
Let the numbers be x, $45 - x$ and y.
Then, $45 - x + y = 55$ and $y + 3x = 90$
This gives $-x + y = 10$ and $3x + y = 90$
On solving, we get $y = 30$.

18. **Ans : (d)**
$4x + 10 = 5x - 5$ or $x = 15$.

19. **Ans : (c)**
Let the numbers be $(2x - 1)$, $(2x + 1)$, $(2x + 3)$.
Then, $(2x-1+2x+1+2x+3) = 21$ (or) $6x = 18$ (or) $x = 3$.
$\therefore$ Middle one is $= (2x + 1) = 7$

20. **Ans : (d)**
Let the larger number be x and the smaller one be y.
Then, $x - 3y = 2$ and $6y - x = 1$
On solving, we get $y = 1$.

Unit - XVI
Numerical Reasoning

UNIT - XVI NUMERCIAL REASONING

1. ANALOGIES

Analogy means similarity i.e., having some relationship between the two given things. The purpose of this test is to test one's ability to reason and how far a candidate is able to compare the relationship that exists between two objects things or diagrams or pictures.

For Example : Sun : Moon : : Day : Night

This abbreviated form conveys the idea that day is related to night in the same way as the sun is related to the moon.

STUDY MATERIAL

KINDS OF RELATIONSHIP

Working with verbal relationships is a kind of reasoning similar to mathematical thinking. There are many possibilities in establishing a relationship. Some of the more common ones are briefly illustrated here.

1. Symbol relationship. Examples :
   Nation : Flag
   Rank : Insignia
2. Part to whole relationship. Examples :
   Ship : Fleet
   Star : Constellation
3. Antonyms relationship. Examples :
   Slavery : Freedom
   Few : Umpteen
4. Synonyms relationship. Examples :
   Clean : Tidy
   Neat : Clear
5. Grammatical relationship. Examples :
   I : Mine (first person, nominative case; first person possessive case)
   Wolf : Vulpine (noun : adjective)
   Have : Had (present tense : past tense)
6. Degree of intensity relationship. Examples :
   Ecstasy : Joy
   Hot Warm
7. Time sequence relationship. Examples :
   January : December
   Sunday : Monday
8. Association relationship. Examples :
   Alchemist : Gold
   Ink : Writing
9. Type and characteristics relationship. Examples :
   Cow : Herbivorous
   Tiger : Carnivorous
10. Sex relationship. Consider the examples :
    Cow : Bull
    Girl : Boy
11. Class - species relation. Consider the examples :
    Mammal : Whale
    Reptile : Snake
    Woman : Girl
12. Action to object and object to action relationship consider the examples :
    Kick : Football (throws)
    Brake : Car (stops)
    Saw : Wood (cuts)
13. Tool and object relationship. Consider the examples:
    Pencil : Paper
    Saw : Wood
14. Worker and tool relationship. Consider the example :
    Carpenter : Saw
    Writer : Pen
    Blacksmith : Hammer
15. Producer and product relationship. Consider the given example :
    ARCHITECT : BUILDING
    a) Producer : Distributor
    b) Consumer : Goods
    c) Civil Engineer : Dam
    d) Demand : Supply
    Answer is C, because the relationship is produce, and product.
16. Cause and effect relationship. Consider the following example :
    GERM : DISEASE
    a) Carelessness : Accident b) Brake : Car
    c) Insect : Grasshopper d) Dog : Bark

Germ is the cause of disease or disease is the effect of a germ. Similarly carelessness is the cause of accident. So choice (a) is correct.

Rules for Analogy : In dealing with an analogy question, an analysis must be made of the established relationship between the pair of words or alphabets. The following steps may be kept in mind:

1. For clarification always return to the question pair.
2. Eliminate answers that do not appear to be the best answer.
3. Refine and expand the relationship in the question pair.
4. If more than one pair of words seem to fit, return to the question pair.
5. Look among the given alternatives for the pair of words with a similar relationship.
6. Analyse the relationship between the question pair.

Note : Always remember that parts of speech must be the same in the related words / phrases.

EXAMPLES

SLAVERY : FREEDOM : :

a) Disease : Health b) Work : Play
c) In : Out d) War : Peace
e) Enemy : Friend

Analysis :

Opposite	:	All are opposities (return to the question pair)
Negative	:	Eliminate (b) and (c) since they may be either negative or positive depending on the situation (return to the question pair)
Involve people	:	Eliminate (a) since it involves all living things and not just people (return to question pair)
A state or condition in which people find themselves.	:	Eliminate (e) since these are people themselves and not just conditions or states of being.

The best answer must be (d)

Format of Analogy Questions : Usually analogy questions are asked in any of the four forms given below :

1. CONDONE : OFFENCE : :
   a) Mitigate : Remittance
   b) Punish : Criminal
   c) Overlook : Aberration
   d) Mistake : Judgement
2. Condone is to offence as :
   a) mitigate is to penitence
   b) punish is to criminal
   c) overlook is to aberration
   d) mistake is to judgement
3. Condone is to offence as ..... is to aberration.
   a) mitigate b) punish
   c) overlook d) judgement
4. Sometimes questions are asked in a set of groups/ pairs of words or alphabets.

   For example : Cap : shirt : pyjamas : dress

   a) paper : pencil : pen : book
   b) white : blue : red : green
   c) pedical : sepals : petals : flowers
   d) love : service : humility : morals

   The first three words, cap, shirt, pyjamas are the parts of a dress; the first 3 words of a group; pedical, petals, sepals, are the parts of a flower. So the answer is (c).

ALPHABET ANALOGIES

Introduction : There exists certain types of relationship between the alphabets even if they are jumbled together. Observing the pattern of alphabets in a set of letters, the candidate should try to establish the same pattern or relationship among the required questions.

For examples :

KILM : MILK : : EAKM ?

a) KAME b) MAKE
c) EMKA d) MEKA

The relationship between KILM and MILK is that the first letter becomes the last letter, while IL remains constant in both the cases. If we interchange the position of first and last letter in EAKM; it will read MAKE. So (b) is correct.

Rules of alphabet analogies : Some of the common factors are denoted over here :

1. The letters in each group may have the same number of intervening letters of the alphabet left out.
2. The letters may be consonants followed by vowels or vice-versa.
3. The letters may be in a pattern which reverses their position in the alphabetical order.

   In any case keep the alphabet and its respective order in your mind.

PRACTICE TEST - 1

ANALOGIES

DIRECTIONS (Qs. 1 to 15) **: The following questions each consists of words that have a certain relationship to each other, followed by four lettered pairs of words (a), (b), (c), (d). Select the lettered pair of words that has the same relationship as the original pair of words :**

1. LACKADAISICAL : ENERVATING : :
   a)Listless : weakening b) listless : energetic
   c) wishful : thinking d) flowery : hot
2. HYPOCRISY : HONESTY : :
   a) pugnacity : pacifism
   b) duplicity : deceit
   c) belligerence : inhumanity
   d) authenticity : quality
3. SCHOLARLY : ERUDITE : :
   a) teacher : lawyer b) teacher : headmaster
   c) reader : hermit d) ignorant : illiterate
4. NECROMANCY : GHOSTS : :
   a) fortune-telling : gypsies
   b) magic : legerdemain
   c) romance : stories
   d) sorcery : spirits
5. PORTENTOUS : OMINOUS : :
   a) foreshadowing : threatening
   b) magnificent : treacherous
   c) significant : pertinent
   d) good : evil
6. PROJECTILE : TRAJECTORY : :
   a) satellite : orbit
   b) bullet : weapon
   c) project : tragedy
   d) rejection : renunciation
7. PEDAGOGUE : LEARNING : :
   a) teaching : books
   b) professor : erudition
   c) plato : pedant
   d) schoolmaster : ABCs
8. INSULT : INVULNERABLE :
   a) success : capable
   b) poverty : miserable
   c) purchase : refundable
   d) assault : impregnable
9. ARCHAEOLOGIST : ANTIQUITY :
   a)lower : horticulture
   b) ichthyologist : marine life
   c) theology : minister
   d) Bible : psalms
10. YELL : UTTER : :
    a) scream : deafen
    b) shout : call
    c) chaild : infant
    d) bend : break
11. THROW : BALL : :
    a) shoot : trigger
    b) pat : dog
    c) mew : cat
    d) boil : shell
12. DEPRESSION : UNEMPLOYMENT : :
    a) legislation : lobbying
    b) emaciation : debilitation
    c) capital : interest
    d) deterioration : rust
13. ENERGY : DISSIPATE : :
    a) battery : recharge
    b) atom : split
    c) food : heat
    d) money : squander
14. DEBATE : SOLILOQUY : :
    a) crowd : mob
    b) fight : defend
    c) group : hermit
    d) Hamlet : Macbeth
15. DEER : VENSION : :
    a) pig : ham
    b) sheep : mutton
    c) lamb : veal
    d) duck : roast

ANSWERS

1. a	2. a	3. d	4. d	5. a
6. a	7. b	8. d	9. b	10. b
11. b	12. d	13. d	14. c	15. b

EXPLANATIONS

1. **(a)** Lackadaisical and listless are synonyms; enervating and weakening are synonyms.
2. **(a)** Hypocrisy is the opposite of honesty; pugnacity, the opposite of pacifism.
3. **(d)** Scholarly and erudite are synonyms; ignorant and illiterate are synonyms.
4. **(d)** Necromancy involves foretelling the future by comunicating with the dead or with ghosts; sorcery deals with spirits and uses witchcraft, usually for a harmful or evil purpose.
5. **(a)** Something that is portentous is foreshadowing solemn events; that which is ominous is threatening.
6. **(a)** A trajectory is the path of a projectile; an orbit is the path of a satellite.
7. **(b)** A pedagogue encourages learning; a professor encourages erudition.
8. **(d)** An insult cannot hurt one who is invulnerable; an assault cannot hurt a city which is impregnable.
9. **(b)** An archaeologist is a specialist in the study of antiquity; an ichthyologist is a specialist in the study of marine life.
10. **(b)** Yell is the extreme of utter; shout is the extreme of call.
11. **(b)** You can throw a ball, you can pat a dog.
12. **(d)** Unemployment can be caused by a depression; rust can be caused by a deterioration.
13. **(d)** Energy can be wasted or dissipated just as money can be wasted or squandered.
14. **(c)** A debate is a formal argument involving more than one person; a soliloquy is a dramatic speech given by an individual expressing his interior state. Similarly a group is a social unit composed of more than one person, a hermit is an anti-social individual.
15. **(b)** Venison is deer meat; mutton is sheep meat.

PRACTICE TEST - 2
ANALOGIES

DIRECTIONS (Qs. 1 to 15) **: The following questions each consists of words that have a certain relationship to each other, followed by four lettered pairs of words (a), (b), (c), (d). Select the lettered pair of words that has the same relationship as the original pair of words :**

1. CORRUGATED : STRIPED : :
   a) wall : board
   b) paint : crayon
   c) box : zebra
   d) pit : dot
2. BIOGRAPHY : AUTOBIOGRAPHY : :
   a) rickshaw : autorickshaw
   b) mobile : automobile
   c) testimony : confession
   d) dead : living
3. BASEMENT : ATTIC : :
   a) nadir : zenith
   b) apex : pinnacle
   c) zenith : apex
   d) zenith : root
4. ASCETIC : PHILANDERING : :
   a) continent : wandering
   b) continent : immoral
   c) sour : sad
   d) monastery : castle
5. AGILE : ACROBAT : :
   a) plant : fruit
   b) eloquent : orator
   c)grease : mechanic
   d) iron : heavy
6. POET : ECLOGUE ::
   a) philosopher is to nature
   b) dramatist is to scenery
   c) teacher is to truth
   d) seamstress is to gown
7. DELETE : DELETERIOUS : :
   a) mystery is to mysterious
   b) erase is to erasing
   c) expunge is to harmful
   d) expand is to expandable

8. CAUCASIAN : SAXON : :
a) lamp is to a store
b) hammer is to a nail
c) furniture is to a chair
d) carriage is to a horse

9. COURT : JUSTICE : :
a) doctor is to sickness
b) chief is to boss
c) mechanist is to product
d) auditor is to accuracy

10. BACTERIA : ILLNESS : :
a) knife is to laceration
b) medicine is to germs
c) calcium is to bone
d) fire is to explosion

11. BUOY : CHANNEL : :
a) red light is to street
b) conscience is to man
c) road map is to travel
d) white line is to highway

12. AMULET : EVIL : :
a) fire is to house
b) eggs are to rotten
c) police is to law
d) fort is to attack

13. AGGRESSOR : DEFENDER : :
a) a dictator is to foe
b) a dictator is to despot
c) a plaintiff is to defendant
d) a hero is to accused

14. ASYLUM : REFUGEE : :
a) destination is to traveller
b) peace is to war
c) accident is to injury
d) flight is to escape

15. APOGEE : PERIGEE : :
a) apposite is to composite
b) inappropriate is to apposite
c) paradoxical is to incredible
d) dog is to pedigree

ANSWERS

1. d	2. c	3. a	4. b	5. b
6. d	7. c	8. c	9. d	10. a
11. d	12. d	13. c	14. a	15. b

EXPLANATIONS

1. **(d)** Something striped may be corrugated just as a dot may be pitted.
2. **(c)** An autobiography is written by the principal character; a confession is made by the individual who has committed the act that is being discussed.
3. **(a)** Nadir (lowest point and zenith (highest point) can be compared-to basement and attic respectively.
4. **(b)** An ascetic, leads a restrained or continent existence. The philandering individual is immoral.
5. **(b)** An acrobat must be agile and an orator must be eloquent.
6. **(d)** An eclogue, a pastoral poem, is written by poet; a gown is made by a seamstress.
7. **(c)** Delete and expunge are synonyms like deleterious and harmful are synonyms.
8. **(c)** A chair is a piece of furniture; a Saxon is a member of Caucasian race.
9. **(d)** A court's duty is to provide justice; an auditor's duty is to provide accuracy in financial matters.
10. **(a)** Bacteria can lead to illness : a knife can cause a laceration.
11. **(d)** A buoy indicates the proper channel or path for a ship to follow, the white line indicates the highway for a motorist to follow.
12. **(d)** An amulet is a charm designed to repel evil, a fort is a stronghold designed to repel attack.
13. **(c)** If an aggressor brings his dispute to court, he is plaintiff and the defender becomes the defendant.
14. **(a)** A refugee seeks asylum, a traveller seeks destination.
15. **(b)** Apogee and perigee are opposites; so are inappropriate and apposite.

PRACTICE TEST - 3
ANALOGIES

DIRECTIONS (Qs. 1 to 15) : **In the following sets of analogies one word is missing. Select that word from the four alternatives which exhibits the same analogy as established among the three question words.**

1. ? : Temporary : : Ephemeral : Transient
   a) Emanating b) Evanescent
   c) Permanent d) Stationary
2. Callow : : Maturity : ? : Fruition
   a) Young b) Old c) Incipient d) Senility
3. Inoculation : ? : : Exposure : Toughening
   a) Immunity b) Punctuality
   c) Vulnerability d) Contagious
4. Perch : Freshwater : : ? : Saltwater
   a) Fish b) Seawater c) Crocodile d) Cod
5. Truculent : Agitator : : Pacifier : ?
   a) Powerful b) Amenable c) Subversive d) Feeble
6. ? : Drama : : Choreographer : Dance
   a) Composer b) Conductor c) Director d) Producer
7. Grain : Gram : : ? : Litre
   a) Quart b) Meter c) Millilitre d) 1000 gram
8. Errata : ? : : Flaws : Jewels
   a) Manuscripts b) Books c) Literature d) Prints
9. Ethmoidal : Mouth : : ? : Eyes
   a) Abdomen b) Ear c) Maxilla d) Nose
10. Halcyon : ? : : Placid : Serene
    a) Calm b) Irritated c) Harmful d) Peaceful
11. Sartorial : Tailor : : ? : Dancer
    a) Musician b) Thespian
    c) Terpsichorean d) Singer
12. Hirsute : Hairy : : Emaciated : ?
    a) Cadaverous b) Bald c) Unkempt d) Short hair
13. ? : Entrepreneur : : Knowledge : Scholar
    a) Loss b) Profit c) Charity d) Deficit
14. Bow : Rifle : : Hourglass : ?
    a) Clock tower b) Bullet
    c) Diameter d) Chronometer
15. Congenial : Animosity : : ? : Hatred
    a) Friendly b) Hostile c) Brotherly d) Courteous

ANSWERS

1. b	2. c	3. a	4. d	5. b
6. c	7. a	8. b	9. c	10. d
11. c	12. a	13. b	14. d	15. c

EXPLANATIONS

1. **(b)** Ephemeral and transient are synonyms; so is in the case of temporary and evanescent.
2. **(c)** A callow individual is immature while maturity refers to fulfilment or fruition. The immature or callow person can be compared with something that is beginning or incipient which best fits here giving a relationship of antonyms.
3. **(a)** As exposure to a hardship produces a toughening against it, inoculation induces a mild form of disease in order to produce immunity.
4. **(d)** A perch is a freshwater fish. Opposite to it, a cod is a saltwater fish.
5. **(b)** An agitator may be truculent or belligerent while a pacifier is usually amenable and tractable.
6. **(c)** A choreographer plans the movements of the dancer in a dance; similarly a director plans the movements of the actors in a drama.
7. **(a)** A grain is unit of weight in the English system; a gram is a unit of weight in the metric system. Likewise a quart is a unit in the English system and a litre in the metric system.
8. **(b)** Flaws are defects in jewels; errata are errors in books.
9. **(c)** Ethmoidal pertains to mouth orbit; Maxilla pertains to eyes orbit.
10. **(d)** Placid means seren; halcyon means peaceful.
11. **(c)** Sartorial is an adjective that pertains to tailor or clothes; similarly terpsichorean pertains to dancing or dancer.
12. **(a)** A person who is hirsute is hairy. A person who is emaciated is cadaverous. The relationship is matching synonyms.
13. **(b)** As knowledge is sought by a scholar profit is sought by an entrepreneur.
14. **(d)** Bow is the primitive projectile weapon while a rifle is the modern one; similarly hourglass is a primitive tool for measuring time and chronometer is the modern device for telling time.
15. **(c)** A congenial person does not exhibit animosity; similarly a brotherly person is devoid of hatred.

PRACTICE TEST - 4
ANALOGIES

DIRECTIONS (Qs. 1 to 15) **: In each of the following questions, there is a certain relation between two given words on one side of : : and one word is given on another side of : : while another word is to be found from the given alternatives, having the same relation with this word as the words of the given pair bear. Choose the correct alternative.**

1. Quail : Partridges : : Yak : ?
   a) Cows b) Deer c) Oxen d) Antelopes
2. Tarlatan : Muslin : : Poplin : ?
   a) Rayon b) Silk c) Wool d) Cotton
3. Poodle : Dog : : Moose : ?
   a) Duck b) Donkey c) Fowl d) Deer
4. Matricide : Mother : : Homicide : ?
   a) Human b) Children c) Father d) Apes
5. Horde : Nomads : : Cavalcade : ?
   a) Merchants b) Riders
   c) Salesmen d) Players
6. Scrap : Food : : Lees : ?
   a) Bread b) Tea c) Wine d) Rice
7. Porcupine : Rodent : : Mildew : ?
   a) Fungus b) Germ
   c) Insect d) pathogen
8. Microphone : Loud : : Microscope : ?
   a) Increase b) Investigate
   c) Examine d) magnify
9. Engineer : Map : : Bricklayer : ?
   a) Design b) Template
   c) Mould d) Cement
10. Teheran : Iran : : Beijing : ?
    a) China b) Japan c) Turkey d) Malaysia
11. Pigeon : Peace : : White Flag : ?
    a) Friendship b) Victory
    c) Surrender d) War
12. Glucose : Carbohydrate : : Soyabean : ?
    a) Proteins b) Vitamins
    c) Minerals d) Legumes
13. Conscience : Wrong : : Police : ?
    a) Thief b) Law
    c) Discipline d) Crime
14. Sarat Chandra : West Bengal : : K.M. Munshi : ?
    a) Bihar b) Gujarat
    c) Maharashtra d) Haryana
15. Doctor : Diagnosis : : Judge : ?
    a) Court b) Punishment
    c) Lawyer d) Judgement

ANSWERS

1. c	2. b	3. d	4. a	5. b
6. c	7. a	8. d	9. b	10. a
11. c	12. a	13. d	14. b	15. d

EXPLANATIONS

1. **(c)** First belongs to the family of second.
2. **(b)** Tarlatan is a variety of muslin and poplin is a variety of silk.
3. **(d)** Poodle is a breed of dog & moose is a breed of deer.
4. **(a)** First implies killing the second.
5. **(b)** First is a collective moving group of the second.
6. **(c)** First is the left over of the second.
7. **(a)** Procupine is a rodent and mildew is a fungus.
8. **(d)** A microphone makes the sound louder and a microscope magnifies an object.
9. **(b)** Second gives the pattern to be followed by the first.
10. **(a)** Teheran is the capital of Iran and Beijing is the capital of China.
11. **(c)** Pigeon is a symbol of peace and white flag is a symbol of surrender.
12. **(a)** Glucose is a source of carbohydrates and soyabean is a source of proteins.
13. **(d)** First prevents the second.
14. **(b)** First is a famous novelist of second state.
15. **(d)** The function of a doctor is to diagnise a disease and that of a judge is to give judgement.

PRACTICE TEST - 5
ANALOGIES

DIRECTIONS : **In each of the following questions, there is a certain relation between two given words on one side of : : and one word is given on another side of : : while another word is to be found from the given alternatives, having the same relation with this word as the words of the given pair bear. Choose the correct alternative.**

1. Flower : Butterfly : : Dirt : ?
   a) Rats b) Fly c) Bugs d) Sweeper
2. Giant : Dwarf : : Genius : ?
   a) Wicked b) Gentle c) Idiot d) Tiny
3. Tree : Forest : : Grass : ?
   a) Lawn b) Field c) Garden d) Farm
4. Conference : Chairman : : Newspaper : ?
   a) Reporter b) Distributor
   c) Printer d) Editor
5. Darwin : Evolution : : Archimedes : ?
   a) Friction b) Lubrication
   c) Buoyancy d) Liquids
6. Earth : Planet : : Moon : ?
   a) Sun b) Satellite
   c) solar system d) Asteroid
7. Typhoid : Bacteria : : ? : Virus
   a) Bacillus b) Parasite
   c) Protozoa d) Influenza
8. Melt : Liquid : : Freeze : ?
   a) Ice b) Condense
   c) Solid d) Crystal
9. Muslims : Mosque : : Parsi : ?
   a) Golden Temple b) Madina
   c) Fire Temple d) Gurudwara
10. Teeth : Pyorrhoea : : ? : Myopia
    a) Eye b) Cataract
    c) Trachoma d) Eczema
11. Scribble : Write : : ? : Speak
    a) Walk b) Play
    c) Speach d) Stammer
12. Flaws : Metals : : Errata : ?
    a) Manuscripts b) Books
    c) Speech d) Charter
13. Drizzle : Downpour : : Breeze : ?
    a) Earthquake b) Storm
    c) Floor d) Cyclone
14. Fast : Hunger : : Race : ?
    a) Food b) Appetite
    c) Fatigue d) Weakness
15. Gambling : Casino : : ? : Theatre
    a) Acting b) Club
    c) Bar d) Gymn

ANSWERS

1. b	2. c	3. a	4. d	5. c
6. b	7. d	8. c	9. c	10. a
11. d	12. b	13. d	14. c	15. a

EXPLANATIONS

1. **(b)** First attracts the second.
2. **(c)** The given words are opposites of each other.
3. **(a)** A forest consists of trees and a lawn is made up of grass.
4. **(d)** Chairman is the highest authority in a conference. Similarly, editor is the highest authority in a newspaper agency.
5. **(c)** Darwin gave the Theory of Evolution. Likewise. Archimedes gave the principle of Buoyancy.
6. **(b)** Moon is a satellite and earth is a planet.
7. **(d)** First is the disease caused by the second.
8. **(c)** First is the process of formation of the second.
9. **(c)** Muslims worship in a mosque and Parsis worship in a fire temple.
10. **(a)** Second is a disease of the first.
11. **(d)** First is an improper form of the second.
12. **(b)** Errata comprises of errors in books. Similarly, flaws are the defects in metals.
13. **(d)** Second is more intense than the first.
14. **(c)** First causes the second.
15. **(a)** Second is the place for performing the first.

PRACTICE TEST - 6
ANALOGIES

DIRECTIONS **: In each of the following questions, there is a certain relation between two given words on one side of : : and one word is given on another side of : : while another word is to be found from the given alternatives, having the same relation with this word as the words of the given pair bear. Choose the correct alternative.**

1. Water : Convection : : Space : ?
   a) Conduction b) Transference
   c) Vacuum d) Radiation
2. Oxygen : Burn : : Carbon dioxide : ?
   a) Isolate b) Foam
   c) Extinguish d) Explode
3. Stick : ? : : Grain : Stock
   a) Heap b) Bundle
   c) Collection d) String
4. Planet : Orbit : : Projectile : ?
   a) Trajectory b) Track
   c) Milky way d) Path
5. Carpenter : Wood : : ? : Leather
   a) Tailor b) Cobbler c) Hammer d) Weaver
6. Mirage : Illusion : : Genuine : ?
   a) Image b) Transpiration
   c) Reflection d) Authentic
7. Waves : Sand dunes : : Oceans : ?
   a) Sea b) Dust c) Deserts d) Ripples
8. Dungeon : Confinement : : Asylum : ?
   a) Refuge b) Mercy c) Truancy d) Remorse
9. Colt : Stallion : : Piglet : ?
   a) Doe b) Stag c) Leopard d) Pig
10. Drill : Bore : : Sieve : ?
    a) Thresh b) Sift c) Pry d) Rinse
11. Tile : Mosaic : : Knot : ?
    a) Embroidery b) Abacus
    c) Macrame d) Easle
12. Country : President : : State : ?
    a) Governor b) Minister
    c) Chief Minister d) Citizen
13. Car : Garage : : Aeroplane ?
    a) Airport b) Depot c) Hangar d) Harbour
14. Bread : Yeast : : Curd : ?
    a) Fungi b) Bacteria c) Germs d) Virus
15. Quartz : Radio : : Gypsum : ?
    a) Glass b) Porcelain
    c) Cement d) Powder

ANSWERS

1. d	2. c	3. b	4. a	5. b
6. d	7. c	8. a	9. d	10. b
11. c	12. a	13. c	14. b	15. c

EXPLANATIONS

1. **(d)** Second is the mode of transference of heat by the first.
2. **(c)** Oxygen helps in burning while carbon dioxide extinguishes fire.
3. **(b)** Second is a collection of the first.
4. **(a)** Second is the path traced by the first.
5. **(b)** Second is the raw material used by the first.
6. **(d)** The given words are opposites of each other.
7. **(c)** If oceans were deserts, waves would be sand dunes.
8. **(a)** A prisoner is confined within the dungeon, and an unsheltered person takes refuge within the asylum.
9. **(d)** First is a young one of the second.
10. **(b)** Second denotes the function performed by the first.
11. **(c)** Just as tiles in a mosaic make a pattern, so too the knots in a piece of macrame make a pattern.
12. **(a)** President and Governor are the nominal heads of the country and the state respectively.
13. **(c)** First is temporarily parked in the second.
14. **(b)** First is produced by the action of second.
15. **(c)** First is used to make the second.

2. CLASSIFICATION

In classification items are grouped together on the basis of similarity and the one to be picked out is that which does not belong to that same group. For example among shoe, socks, shirt, trousers, coat, all the materials of wear. But except, shoes remaining form group of clothing wear. Since shoes is different, hence the answer is shoes.

STUDY MATERIAL

The classification is generally divided into two types.

1. Alphabet classification
2. Word classification

1. **Alphabet classification :** In this type a Jumble of letters form a set or group and we have to find out that which does not belong to the same group for example :

SELF, CHILD, FOOL, WOOD

In the above example the answer is WOOD. Because if you suffix ish to the above words all the three words except 'WOOD' make sense. Since there is no word 'WOODISH'. We conclude it as the odd man out.

2. **Word classification :** Questions are asked among names, things, places, verbs, nouns, sexes, races and any matter pertaining to verbal classification.

MODEL - 1

In this type of questions, certain pairs of words are given out of which the words in all the pairs barring one, bear a certain common relationship. The candidate is required to derive this relationship and choose the pair in which the words are differently related, as the answer.

Here are a few examples :

Directions : **Choose the odd pair of words.**

1. a) China : Beijing
   b) Russia : Moscow
   c) Japan : Singapore
   d) Spain : Madrid

Ans. (c) : In all others except c the second one is the capital of first one.

2. a) Sprinkle : Pour
   b) Happiness : Merriment
   c) Mist : Fog
   d) Sad : Unhappy

Ans. (d) : In others second one is lighter intensity of the first.

3. a) Volume : Litre
   b) Time : Seconds
   c) Length : Metre
   d) Pressure : Barometer

Ans. (d) : Second is the unit to measure the first.

4. a) Cow : Calf
   b) Dog : Bitch
   c) Tortoise : Turtle
   d) Lion : Cub

Ans. (b) : In all others second is the younger one of the first.

5. a) Blacksmith : Anvil
   b) Carpenter : Saw
   c) Barber : Scissors
   d) Goldsmith : Ornaments

Ans. (d) : Second is the tool used by the first.

6. a) Painter : Gallery
   b) Actor : Stage
   c) Mason : Wall
   d) Farmer : Field

Ans. (c) : Second is the place where the first works.

7. Eyes : Tears
   a) Sea : Water
   b) Volcano : Lava
   c) Heart : Artery
   d) Hunger : Bread

Ans. (b)

8. Knowledge : Ignorance
   a) Cure : Health
   b) Conceal : Hide
   c) Breath : Suffocation
   d) Construction : War

Ans. (c)

MODEL - 2

In this type of questions four numbers are given out which three are alike in some manner while one is different and this number is to be chosen as the answer. The examples given below will make it more clear.

***Directions* : Choose the number which is different from others in the group.**

1. a) 48 b) 12 c) 36 d) 59

Ans. (d) : In others the units digit is twice the ten's digit.

2. a) 8314 b) 2709 c) 1315 d) 2518

Ans. (a) : In others the sum of the first three digits is placed as fourth digit.

3. a) 381 b) 552 c) 729 d) 903

Ans. (b): Remaining are odd numbers.

4. a) 751 b) 734 c) 981 d) 853

Ans. (a): In others the difference of first and last digits is placed at middle.

5. a) 21 b) 36 c) 49 d) 56

Ans. (b) : Remaining are multiply of 7.

6. a) 8 b) 64 c) 125 d) 28

Ans. (d) : Remaining are cubes.

MODEL - 3

In this type of questions, four groups of letters are given. Three of them are similar to each other in some manner while one is different and this is to be chosen by the candidate.

***Directions* : Choose the group of letters which is different from others.**

1. a) DXCLQZ
   b) PFZUBM
   c) XGKNTY
   d) NWMBHJ

Ans. (b) : This is the only group containing a vowel.

2. a) AUgPZ
   b) MXiDV
   c) KFeCO
   d) YGLhT

Ans. (d) : In all other groups the smaller letter is the middle one.

3. a) FCGDE
   b) TRQPS
   c) KJHMF
   d) KHGJI

Ans. (c) : Other groups consists of consecutive letters though not in order.

4. a) DkUZ
   b) LPuB
   c) FoMY
   d) UXeN

Ans. (a) : In all other groups the small letter is a vowel.

5. a) CFIL
   b) PSVX
   c) JMPS
   d) ORUX

Ans. (b) : In all other groups each letter moves 3 steps forward to obtain the next letter.

6. a) BDGK b) JLCS c) NPSW d) MORU

Ans. (d) : The gap between the letter is 1, 2, 3 and so on.

7. a) CZHK b) MLAG c) XUBU d) SENO

Ans. (c) : This is the only group in which one letter has been repeated.

8. a) POCG b) KLIZ c) BUDX d) FQMV

Ans. (d) : This group doesn't contain any vowel.

9. a) ACEG b) JLNP c) SUWY d) BDFG

Ans. (d) : Other groups consists of alternate letters.

10. a) BCD b) KMN c) QRS d) GHI

Ans. (b) : Other groups consists of three consecutive letters.

PRACTICE TEST - 1
CLASSIFICATION

DIRECTIONS : **In each of the following questions, there are four choices (a), (b), (c), (d). Three of them are alike and form a group. One is different and does not belong to the group. Find the odd man out.**

1. a) Cyclone b) Doldrums c) Anticyclone d) Storms
2. a) Kerry b) Kentucky c) Utah d) Alabama
3. a) Andhra Pradesh b) Chandigarh d) Madhya Pradesh d) Uttar Pradesh
4. a) Chess b) Cricket c) Hockey d) Football
5. a) Clove hitch b) Cat's cradle c) Granny d) Reef
6. a) Vixen b) Hen c) Peacock d) Filly
7. a) Sofa b) Chair c) Table d) Settee
8. a) Oil b) Coal c) Lignite d) Coke
9. a) Roof b) Wall c) Window d) House
10. a) Good b) Bad c) Nice d) Better
11. a) Black b) Blacksmith c) Red d) Green
12. a) Man b) Animal c) Plant d) Soil
13. a) Man b) Wife c) Husband d) Sister
14. a) India b) China c) Japan d) Asia
15. a) West Bengal b) Bihar c) Manipur d) Pondicherry
16. a) Gold b) Silver c) Metal d) Iron
17. a) Hydrogen b) Mercury c) Oxygen d) Ozone
18. a) Earth b) Moon c) Mercury d) Venus
19. a) Circle b) Rectangle c) Triangle d) Square
20. a) Rice b) Wheat c) Peanut d) Gram
21. a) Sugarcane b) Tobacco c) Tea d) Rice
22. a) May b) June c) Summer d) July
23. a) Sunday b) Week c) Monday d) Friday
24. a) Come b) Go c) Eat d) Campa Cola
25. a) Clouds b) Thunder c) Lightning d) Water

ANSWERS

1. b	2. a	3. b	4. a	5. b
6. c	7. d	8. a	9. d	10. b
11. b	12. d	13. a	14. d	15. d
16. c	17. b	18. b	19. a	20. a
21. d	22. c	23. b	24. d	25. d

EXPLANATIONS

1. **(b)** Others represent the types of moving winds; Doldrums is a place.
2. **(a)** Kerry is in Ireland, others are in USA.
3. **(b)** Others are States.
4. **(a)** Chess is an indoor game while others form a group of outdoor games.
5. **(b)** Cat's cradle is a game, others are knots.
6. **(c)** All except peacock are female.
7. **(d)** Others belong to the group of furniture.
8. **(a)** Oil is the liquid form of fuel while others are solid form of fuel.
9. **(d)** Others are parts of a house.
10. **(b)** Others belong to group of greatness, while bad does not.
11. **(b)** Others form a group of colours.
12. **(d)** Others represent living beings while soil is non living.
13. **(a)** Others represent some relationship.
14. **(d)** Others are countries while Asia is a continent.
15. **(d)** All are states while Pondicherry is a Union Territory.
16. **(c)** Others are specific metals but metal itself is a general term.
17. **(b)** Others are gases while mercury is a liquid.
18. **(b)** Others are planets while the moon is a satellite.
19. **(a)** Others represent formation of straight line figures.
20. **(a)** Rice is kharif crop while others belong to rabi crop.
21. **(d)** Rice is a cereal food-grain while all others are cash crops.
22. **(c)** Summer is a season, others are months.
23. **(b)** Others are days .
24. **(d)** Others are verbs; Campa Cola is a noun.
25. **(d)** Others are associated with a single phenomenon of rain.

PRACTICE TEST - 2
CLASSIFICATION

DIRECTIONS : In each of the following questions there are four choices (a) (b) (c) (d). Three of them are alike and form a group. One is different and does not belong to the group. Find the oddman out.

1. a) Dice b) Counter c) Chess-man d) Domino
2. a) Unfed b) Monmouth c) Sponsor d) Indefinite
3. a) TAB b) POT c) DEAR d) LIFER
4. a) EKMA b) DINIA c) RUMBA d) NAJPA
5. a) ECLAP b) YSORT c) PHYPA d) QREAT
6. a) EVFU b) CXDW c) AZBZ d) GTHS
7. a) MMPQ b) NNRS c) QQXY d) PQVW
8. a) CRDT b) APBQ c) EUFV d) GWHX
9. a) SNPqR b) STUm c) IjKM d) VwxZ
10. a) SRT b) PQO c) KJL d) VUW
11. a) BAT b) CAT c) DOT d) EAT
12. a) JOT b) OUT c) FED d) DIN
13. a) POP b) TOP c) LOT d) GET
14. a) TURNIP b) PEAR c) LUPIN d) APPLE
15. a) SIR b) MAN c) TON d) HOG
16. a) INDIA b) ICELAND c) EIRE d) FRANCE
17. a) CHEESE b) LARD c) BUTTER d) SAFFRON
18. a) CIVIC b) DRUID c) LEVEL d) FRETFUL
19. a) BEAR b) DEAR c) FEAR d) EAR
20. a) IRAN b) IRAQ c) INDIA d) ITAWA
21. a) ABC b) EfD c) eHK d) LMn
22. a) BCd b) CDe c) DeF d) EFg
23. a) AbBC b) BccD c) CddE d) KuMu
24. a) ToWer b) PoWeR c) MoWer d) LOWeR
25. a) abCD b) klMN c) EFgh d) MnOP

ANSWERS

1. a	2. d	3. c	4. a	5. d
6. c	7. d	8. a	9. d	10. b
11. c	12. b	13. a	14. d	15. c
16. d	17. b	18. d	19. d	20. c
21. a	22. c	23. a	24. d	25. d

EXPLANATIONS

1. **(a)** The only possible way to find odd man out is that DICE is plural. All the others are singular.
2. **(d)** All the other words contain three consecutive letters in reverse order (unFED, mONMouth, sPONsor) whereas INDEFINITE contains three consecutive letters in correct order - in DEFinte.
3. **(c)** Other words when read backward given meaningful words.
4. **(a)** Others are anagrams of country names e.g., (b) INDIA (c) BURMA (d) JAPAN
5. **(d)** Others are anagrams of words : ECLAP = PLACE, YSORT = STORY; PHYPA = HAPPY.
6. **(c)** In the other alternatives, first and the third letters are consecutive in the alphabet and second and fourth are also consecutive though the fourth comes earlier than the third in the alphabet.
7. **(d)** In the others first two letters are the same.
8. **(a)** In all the other groups third letter is the next one to the first in the alphabet and the fourth is next to the second in the alphabet. For example in (b) 'A' is the first letter and 'B' is the third letter, 'P' is the second 'Q' is the fourth.
9. **(d)** Inthis group there are two small letters whereas in the others ther is only one small letter.
10. **(b)** In the other groups the first letter which precedes the other two in the alphabet comes in the middle but in this it comes in the end.
11. **(c)** Others end with AT.
12. **(b)** Middle letter in each word is a vowel. First consonant is same number of letter before the vowel as the second letter after the vowel e.g., in DIN first consonant D is five letters before I and second consonant N is five letters after I.
13. **(a)** Only word having the same consonant twice. Middle letter in all words is vowel.

14. **(d)** All the others begin and end with a consonant while APPLE begins and ends with vowel.

15. **(c)** In the other groups two consonants are consecutive letters in the alphabet.

16. **(d)** Others contain three vowels while FRANCE contain only two vowels.

17. **(b)** Other groups contain double letters.

18. **(d)** Others begin and end with the same letter.

19. **(d)** Others start with a consonant.

20. **(c)** Others form a group by having one consonant between two vowels I and A.

21. **(a)** In the others one letter is small, others are capital.

22. **(c)** Middle letter is small while others are capital.

23. **(a)** Others are having two small letters.

24. **(d)** Other words contain vowels in small letters while in LOWER one vowel is in capital letter.

25. **(d)** In other groups two letters are capital and two small in a series while in MnOp alternate letters are small and capital.

PRACTICE TEST - 3
CLASSIFICATION

***DIRECTIONS :* In each of the following questions, four words have been given out of which three are alike in some manner and the fourth one is different. Choose out the odd one.**

1. a) Gorges b) Bars c) Canyons d) Meanders
2. a) Magnalium b) Germanium c) Duralumin d) Bronze
3. a) Hydrazine b) Warfarin c) Malathion d) Agrosan
4. a) Canoe b) Igloo c) Yacht d) Dinghy
5. a) Bishop b) Knight c) Pawn d) Jockey
6. a) Pituitary b) Pancreas d) Thalamus d) Adrenal
7. a) Granite b) Lignite c) Peat d) Anthracite
8. a) Galileo b) Copernicus c) Columbus d) Bhaskara
9. a) Cricket b) Baseball c) Football d) Billiards
10. a) Rigveda b) Yajurveda c) Atharvanaveda d) Ayurveda
11. a) Epicentre b) Seismology c) Focus d) Crater
12. a) Breathing b) Swimming c) Dancing d) Playing
13. a) Thyroxine b) Adrenalin c) Iodine d) Insulin
14. a) Brook b) Bridge c) Steam d) Canal
15. a) Henri Becquerel b) Roentgen c) Madam Curie d) Einstein
16. a) Dinosaur b) Sphinx c) Unicorn d) Mermaid
17. a) Mackenzie b) Yenisei c) Michigan d) Danube
18. a) Tellurium b) Francium c) Strontium d) Scandium
19. a) Beaver b) Alpaca c) Walrus d) Koala
20. a) Lamprey b) Salmon c) Rhea d) Trout
21. a) Sugarcane b) Coffee c) Tobacco d) Rice
22. a) Leprosy b) Influenza c) Typhoid d) Dysentry
23. a) Buffalo b) Stag c) Camel d) Rhinoceros
24. a) Almirah b) Rack c) Safe d) Cupboard
25. a) Advise b) Counsel c) Suggest d) Direct

ANSWERS

1. b	2. b	3. a	4. b	5. d
6. c	7. a	8. c	9. d	10. d
11. d	12. a	13. c	14. b	15. d
16. a	17. c	18. a	19. c	20. c
21. d	22. d	23. c	24. b	25. d

EXPLANATIONS

1. **(b)** All except bars are structures formed by rivers while bars are formed by sea.
2. **(b)** All except Germanium are alloys while Germanium is a metal.
3. **(a)** All except Hydrazine are pesticides while it is a rocket fuel.
4. **(b)** All except Igloo are types of boats while Igloo is a house at polars.
5. **(d)** Jockey is a professional horse rider, others are chessmen.
6. **(c)** Others are Harmone secreting glands.
7. **(a)** Others are types of coal while Granite is a rock.
8. **(c)** Columbus is an explorer while others are astronomers.
9. **(d)** Others are outdoor games.
10. **(d)** Others are holy scriptures
11. **(d)** Others are terms associated with earthquakes.
12. **(a)** Breathing is the only natural action.
13. **(c)** Others are harmones.
14. **(b)** Others contain water.
15. **(d)** Others are scientists related to radioactivity.
16. **(a)** Others are imaginary creatures.
17. **(c)** Others are rivers while Michigan is a lake.
18. **(a)** Others are metals while Tellurium is a non metal.
19. **(c)** All except Walrus are fur bearing animals.
20. **(c)** All except Rhea are kinds of fishes.
21. **(d)** All except Rice are cash crops, while rice is a food crop.
22. **(b)** All except influenza are caused by bacteria, while influenza is caused by virus.
23. **(c)** All except Camel have horns.
24. **(b)** All except Rack can be closed.
25. **(d)** All except Direct are synonyms.

PRACTICE TEST - 4
CLASSIFICATION

***DIRECTIONS* : In each of the following questions four words have been given out of which three are alike in same manner and the fourth one is different. Choose out the odd one.**

1. a) Archaeology b) Ecology c) Epipgraphy d) Palaeontology
2. a) Duece b) Pitch c) Crease d) Stump
3. a) Lymphocytes b) Plasma c) Fibrinogen d) Pepsinogen
4. a) Tricycle b) Tricolour c) Trifle d) Trident
5. a) Pellagra b) Beri beri c) Scurvy d) Anaemia
6. a) Doe b) Bitch c) Sorceress d) Drone
7. a) Virgo b) Pisces c) Sagittarius d) Orion
8. a) Admiral b) Colonel c) Captain d) Commodore
9. a) Morarji Desai b) Jawaharlal Nehru c) Charan Singh d) S.Radhakrishnan
10. a) Physics b) Chemistry c) Geography d) Zoology
11. a) Spectacles b) Goggles c) Binoculars d) Microphone
12. a) Snore b) Slumber c) Yawn d) Doze
13. a) Jordan b) Bhutan c) Turkey d) Norway
14. a) Raddish b) Carrot c) Turnip d) Spinach
15. a) Kathakali b) Disco c) Garba d) Bhangra
16. a) House b) Cottage c) School d) Palace
17. a) Hydrogen b) Oxygen c) Iodine d) Nitrogen
18. a) Coriander b) Jasmine c) Lotus d) Lily
19. a) Swimming b) Diving c) Driving d) Sailing
20. a) Actor b) Artist c) Musician d) Poet
21. a) Engineer b) Architect c) Mechanic d) Mason
22. a) Shoulder b) Foot c) Elbow d) Arm
23. a) Treachery b) Fraud c) Deceit d) Morbid
24. a) Assassinate b) Kill c) Kidnap d) Stab
25. a) Mustard b) Rapeseed c) Cashewnut d) Sesame

ANSWERS

1. b	2. a	3. d	4. c	5. d
6. d	7. d	8. b	9. d	10. c
11. d	12. c	13. c	14. d	15. b
16. c	17. c	18. a	19. c	20. d
21. c	22. b	23. d	24. c	25. c

EXPLANATIONS

1. (b) All except Ecology provide evidences of history.
2. (a) All except Deuce are terms associated with cricket.
3. (e) All except Pepsinogen are constituents of blood, while Pepsinogen is an enzyme.
4. (c) In all except Trifle, tri indicates 'three'.
5. (d) All except Goitre are caused due to deficiency of vitamins ,while goitre is caused due to deficiency of iodine.
6. (d) All except Drone and females.
7. (d) All except Orion are zodiac signs, While Orion is a constellation.
8. (b) All except Colonel are ranks of the navy.
9. (d) All except S. Radha Krishnan were the Prime Ministers of India.
10. (c) All except Geography are science subjects.
11. (d) All except Microphone are related to eyes.
12. (c) All except Yawn are actions in sleep
13. (c) All except Turkey are ruled by kings.
14. (d) All except Spinach are forms of roots.
15. (b) All except Disco are folk dances.
16. (c) All except school are dwelling places.
17. (c) All except Iodine are gases, while iodine is a liquid.
18. (a) All except Coriander are flowers.
19. (c) All except Driving are activities performed in water.
20. (d) All except Poet perform on stage.
21. (c) All except Mechanic help in building a house.
22. (b) Others are parts of hand.
23. (d) Others are synonyms of DECEIT.
24. (c) Others are actions of killing.
25. (c) Others can be used for extracting oil.

PRACTICE TEST -5
CLASSIFICATION

***DIRECTIONS* : Choose the odd pair of words.**

1. a) Mulder : Proteins
   b) Curie : Radium
   c) Becquerel : Radioactivity
   d) Einstein : Television
2. a) Sheep : Bleat
   b) Horse : Neigh
   c) Ass : Grunt
   d) Lion : Roar
3. a) Profit : Loss
   b) Wise : Foolish
   c) Virtue : Vice
   d) Seduce : Attract
4. a) Deer : Flesh b) Crane : Fish
   c) Crow : Carrion d) Mongoose : Snake
5. a) Saw : Wood b) Pen : Paper
   c) Author : Book d) Chalk : Blackboard
6. a) Beautician : Parlour b) Chemist : Medicine
   c) Lawyer : Court d) Painter : Gallery
7. a) Car : Road b) Ship : Sea
   c) Rocket : Space d) Aeroplane : Pilot
8. a) Iran : Asia b) Canberra : Australia
   c) Noway : Europe d) Algeria : Africa
9. a) Jupiter : Planet b) Musician : Artist
   c) Merchant : Business d) Maize : Cereal
10. a) Flurry : Blizzard b) Moisten : Drench
    c) Prick : Stab d) Scrub : Polish
11. a) Principal : School b) Soldier : Barrack
    c) Artist : Troupe d) Singer : Chorus
12. a) Petrol : Car b) Ink : Pen
    c) Garbage : Dustbin d) Lead : Pencil
13. a) Door : Bang b) Piano : Play
    c) Rain : Patter d) Drum : Beat
14. a) Chandragupta : Mauryan
    b) Babar : Mughal
    c) Kanishka : Kushan
    d) Mahavira : Jainism
15. a) Army : General b) Team : Captain
    c) Meeting : Chairman d) Crache : Infant
16. a) Beans : Pulses b) Rice : Cereals
    c) Tea : Beverages d) Legumes : Nodules
17. a) Tongue : Taste b) Eye : Blind
    c) Ear : Deaf d) Leg : Lame

18. a) Donald : Comedy
    b) Holmes : Suspense
    c) Premchand: Novel
    d) Robinson: Adventure
19. a) Twigs : Nest   b) Wood : Furniture
    c) Pitcher : Pottery   d) Gold : Ornaments
20. a) Husband : Wife   b) Lion : Fox
    c) Dog : Cat   d) King : Minister
21. a) Shopkeeper : Customer
    b) Doctor : Patient
    c) Lawyer : Client
    d) Clerk : file
22. a) Mercury : Sun
    b) Moon : Earth
    c) Star : Galaxy
    d) Wheel : Axle
23. a) Crime : Punishment
    b) Judgement : Advocacy
    c) Enterprise : Success
    d) Exercise : Health
24. a) Wine : Grapes   b) Paper : Pulp
    c) Shoes : Leather   d) Wheat : Crop
25. a) Man : Child   b) Dog : Puppy
    c) Horse : Mare   d) Pig : Farrow

ANSWERS

1. d	2. c	3. d	4. a	5. c
6. b	7. d	8. b	9. c	10. a
11. a	12. c	13. b	14. d	15. d
16. d	17. a	18. c	19. c	20. d
21. d	22. c	23. b	24. d	25. c

EXPLANATIONS

1. **(d)** In all other groups, first is the name of a scientist who discovered the second.
2. **(c)** In all other groups, second is the sound made by the first.
3. **(d)** All other groups consist of pairs of antonyms.
4. **(a)** In all other groups, second is the food on which the first feeds.
5. **(c)** In all other groups, first is the tool which works over the second.
6. **(b)** In all other groups, second is the place where the first works.
7. **(d)** In all other groups, first is the means of transport on the medium denoted by the second.
8. **(b)** In all other groups, second is the continent to which the country denoted by the first belongs.
9. **(c)** In all other groups, second is the class to which the first belongs.
10. **(a)** In all other groups, second is higher intesity than the first.
11. **(a)** In all other groups, second higher intensity than the first.
12. **(c)** In all other groups, second requires the first to function.
13. **(b)** In all other groups, second is the noise made by the first.
14. **(d)** In all other groups, second is the name of the dynasty founded by the first.
15. **(d)** In all other groups, first is the head of the second.
16. **(d)** In all other groups, second is the class of plants to which the first belongs.
17. **(a)** In all other groups, second is the one suffering from a defect of the first.
18. **(c)** In all other groups, first is a character of the type of movie denoted by the second.
19. **(c)** In all other groups, first is the material used to make the second.
20. **(d)** In all other groups, the first is masculine while the second is feminine.
21. **(d)** In all other groups, second is the person for whom that first works to earn money.
22. **(c)** In all other groups, first moves about the second.
23. **(b)** In all other groups, second is the result of the first.
24. **(d)** In all other groups, second is the substance which is used to prepare the first.
25. **(c)** In all other groups, second is the young one of the first.

PRACTICE TEST - 6
CLASSIFICATION

***DIRECTIONS* : Choose the odd pair of words.**

1. a) Solder : Tin
   b) Haematite : Iron
   c) Bauxite : Aluminium
   d) Malachite : Copper
2. a) Whale : Mammal
   b) Salamander : Insect
   c) Snake : Reptile
   d) Frog : Amphibian
3. a) Onomatology : Names
   b) Nidology : Nests
   c) Phycology : Algae
   d) Concology : Shells
4. a) Scalpel : Surgeon
   b) Chisel : Soldier
   c) Awl : Cobbler
   d) Knife : Chef
5. a) Tree : Branch
   b) Hand : Finger
   c) Table : Chair
   d) Room : Floor
6. a) Student : Scholar b) Paddy : Rice
   c) Soldier : Warrior d) Politician : Leader
7. a) Needle : Prick b) Gun : Fire
   c) Auger : Bore d) Chisel : Carve
8. a) Captain : Team
   b) Boss : Gang
   c) Prime Minister : Cabinet
   d) Artist : Troupe
9. a) Class : Students
   b) Sentence : Words
   c) Tree : Forest
   d) Hour : Minutes
10. a) Newspaper : Editor
    b) Film : Director
    c) Stamps : Philatelist
    d) Book : Author
11. a) Aphid : Paper
    b) Moth : Wool
    c) Termite : Wood
    d) Locust : Plant
12. a) Cockroach : Antennae
    b) Lizard : Flagella
    c) Hydra : Tentacles
    d) Plasmodium : Cilia
13. a) Malaria : Protozoa
    b) Yeast : Fungi
    c) Typhoid : Bacteria
    d) Polio : Virus
14. a) Pyroheliometer : Radiation
    b) Calorimeter : Heat
    c) Planimeter : Area
    d) Hygrometer : Pressure
15. a) Sahara : Africa b) Thar : India
    c) Kalahari : America c) Gobi : Mongolia
16. a) Pelican:Reptile b) Gnu:Antelope
    c) Elk : Deer d) Shark : Fish
17. a) Avesta : Parsi b) Torah : Jew
    c) Tripitaka : Buddhist d) Temple : Hindu
18. a) Taiwan : Taipei b) China : Mongolia
    c) Iran : Tehran d) Japan : Tokyo
19. a) Chaff : Wheat b) Grit : Pulses
    c) Grain : Crop d) Drugs : Wine
20. a) Wrist : Band b) Neck : Tie
    c) Waist : Belt d) Shoe : Laces
21. a) Broom : Sweep b) Spoon : Feed
    c) Nut : Crack d) Soap : Bathe
22. a) Proteins : Marasmus b) Sodium : Rickets
    c) Iodine : Goitre d) Iron : Anaemia
23. a) Apple : Jam b) Lemon : Citrus
    c) Orange : Squash d) Tomatto : Pury
24. a) Cat : Paw b) Man : Leg
    c) Horse : Hoof d) Lizard : Pad
25. a) Cow : Fodder b) Crow : Carrion
    c) Poultry : Farm d) Vulture : Prey

ANSWERS

1. a	2. b	3. d	4. b	5. c
6. b	7. a	8. d	9. c	10. c
11. a	12. b	13. b	14. d	15. c
16. a	17. d	18. b	19. c	20. d
21. c	22. b	23. b	24. d	25. c

EXPLANATIONS

1. **(a)** In all other groups, first is the name of a metal of which the second is an ore. On the other hand, Solder is an alloy.
2. **(b)** In all other groups, second is the class of animals to which the first belongs.
3. **(d)** In all other groups, first is the study of second.
4. **(b)** In all other groups, first is a tool used by the second.
5. **(c)** In all other groups, second is a part of the first.
6. **(b)** In all other groups, the first, when becomes an expert, is given the name of second.
7. **(a)** In all other groups, second is the action of the first.
8. **(d)** In all other groups, first is the head of the second.
9. **(c)** In all other groups, second is unit of the first.
10. **(c)** In all other groups, first is prepared by the second.
11. **(a)** In all other groups, first is the insect which damages the second.
12. **(b)** In all other groups, second is the organ for movement of the first.
13. **(b)** In all other groups, first is the disease caused by the second.
14. **(d)** In all other groups, first is the instrument used to measure the second.
15. **(c)** In all other groups, first is the name of a desert which is situated in the country denoted by the second.
16. **(a)** In all other groups, first is a type of the second.
17. **(d)** In all other groups, first is a religious book of the second.
18. **(b)** In all other groups, second is the capital of the first.
19. **(c)** In all other groups, first is the waste obtained from the second.
20. **(d)** In all other groups, first is the body part over which the second is worn.
21. **(c)** In all other groups, second is the purpose for which the first is used.
22. **(b)** In all other groups, second is the disease caused by the deficiency of first.
23. **(b)** In all other groups, second is the form in which the first is preserved.
24. **(d)** In all other groups, second is the name given to the foot of the first.
25. **(c)** In all other groups, second is the food over which the first feeds.

PRACTICE TEST - 7
CLASSIFICATION

DIRECTIONS : **In each of the following questions, there are four numbers out of these, three are alike in a certain way but the rest one is different. Choose the one which is different from the rest three.**

1. a) 7851 b) 6432 c) 5789 d) 1325
2. a) 27 b) 125 c) 343 d) 1321
3. a) 1532 b) 8749 c) 4268 d) 5846
4. a) 5698 b) 4321 c) 7963 d) 4232
5. a) 7487 b) 5963 c) 8218 d) 6596
6. a) 263 b) 111 c) 242 d) 551
7. a) 3740 b) 4635 c) 5869 d) 7946
8. a) 325 b) 236 c) 178 d) 639
9. a) 43 b) 53 c) 63 d) 73
10. a) 7359 b) 1593 c) 9175 d) 3781
11. a) 45 b) 99 c) 109 d) 126
12. a) 5698 b) 7894 c) 9865 d) 8793
13. a) 3759 b) 2936 c) 6927 d) 4836
14. a) 2384 b) 1592 c) 3756 d) 3639
15. a).385 b) 572 c) 671 d) 427
16. a) 324 b) 244 c) 136 d) 352
17. a) 15 b) 21 c) 24 d) 28
18. a) 9 b) 16 c) 23 d) 25
19. a) 10 b) 26 c) 24 d) 21
20. a) 7 b) 12 c) 13 d) 17

ANSWERS

1. b	2. d	3. c	4. d	5. b
6. d	7. a	8. b	9. c	10. d
11. c	12. d	13. b	14. d	15. d
16. a	17. d	18. c	19. d	20. b

EXPLANATIONS

1. **(b)** Each of the numbers except 6432, is an odd number.
2. **(d)** All other numbers are cubes of odd numbers.
3. **(c)** In all other numbers, the last digit is one more than the first digit.
4. **(d)** This is the only number in which a digit has been repeated.
5. **(b)** In all other numbers, the first and the last digits are the same.
6. **(d)** In all other numbers, the middle digit is equal to the product of other two digits.
7. **(a)** In all other numbers, the sum of the first and the last digits is equal to the sum of the other two.
8. **(b)** In all other numbers, the last digit is the sum of the first two.
9. **(c)** Each of the numbers except 63, is a prime number.
10. **(d)** All other numbers consists of odd digits.
11. **(c)** Each of the numbers except 109, is divisible by 9.
12. **(d)** Sum of the digits in each other number is 28.
13. **(b)** In all others, the sum of second and last digits is twice the sum of first and third digits.
14. **(d)** In all other numbers, the last digit is two times the first.
15. **(d)** In all other numbers, the middle digit is sum of the other two.
16. **(a)** Sum of the digits in each other number is 10.
17. **(d)** Each of the numbers except 28, is divisible by 3.
18. **(c)** Each of the numbers except 23, is a perfect square.
19. **(d)** Each of the numbers except 21, is an even number.
20. **(b)** Each of the numbers except 12, is a prime number.

PRACTICE TEST - 8

CLASSIFICATION

***DIRECTIONS* : Choose the group of letters which is different from others.**

1. a) CDE b) JKL c) PQS d) TUV
2. a) ABD b) FGI c) LMO d) STU
3. a) WVU b) NML c) HGF d) DBA
4. a) ACE b) PRT c) UWY d) MNO
5. a) ECA b) JHF c) OMK d) UWY
6. a) BEH b) DGJ c) KNQ d) MPR
7. a) AOT b) CPA c) REB d) TIW
8. a) DAH b) IFM c) ROV d) QNT
9. a) BAC b) POQ c) STV d) YXZ
10. a) PQO b) AZY c) TWS d) VBU
11. a) VXZ b) ACF c) PRU d) GIL
12. a) GDA b) OLI c) VSP d) WYZ
13. a) NPM b) IJL c) QSZ d) BHK
14. a) AEC b) PTR c) FJH d) KPM
15. a) BDH b) IKP c) QSW d) TVZ
16. a) APoQ b) DXeM c) SFiK d) OWjB
17. a) ACeG b) BfGH c) PZyE d) XTuW
18. a) MEWGN b) PBQTX c) DRYSN d) CGLKV
19. a) EDCBA b) PONML c) UTSRQ d) YXWVZ
20. a) ACFJO b) DFIMR c) GILPU d) EGIKM
21. a) ACEGH b) PRTVX c) KMOQS d) FHJLN
22. a) FKEBQ b) ADXAT c) FTZEM d) APDQL
23. a) VYAKB b) MYGHZ c) LMVOX d) FSYLD
24. a) ECBFD b) LQPOM c) WSVTU d) ROQNP
25. a) EMGLK b) BHJFD c) WUSQY d) NOSUX
26. a) BCDEI b) PQRSW c) LMNOS d) HIKLO
27. a) ADGJM b) PSVYB c) HKNQT d) SVXAD
28. a) TREAT b) LATER c) TABLE d) RATES
29. a) JOEHNP b) YUVWON c) TVWORU d) SFXPMG
30. a) LAHMQW b) HUTMCX c) CLOVIK d) IXMLBC
31. a) MOTXYZ b) GKRVWX c) PSBEFG d) ORNODF
32. a) STUTTER b) RESURRECT c) SURRENDER d) SUCCEED

33. a) QePFoLA b) OrDFkV c) TuMBiN d) XZaWoB
34. a) CBUV b) YXTU c) RQOP d) IJSO
35. a) LNMO b) SUTV c) EUFV d) CRDT
36. a) BXCW b) OTPS c) GQHR d) JNKM
37. a) BVPS b) OLMZ c) RPSR d) XTCD
38. a) RNJH b) SOKG c) QMIE d) MIEA
39. a) QLJS b) GVTI c) KRPM d) ONMP
40. a) PEAR b) TORE c) REAP d) TEAR

ANSWERS

1. c	2. d	3. d	4. d	5. d
6. d	7. b	8. d	9. c	10. b
11. a	12. d	13. b	14. d	15. b
16. d	17. b	18. a	19. d	20. d
21. a	22. b	23. d	24. b	25. d
26. d	27. d	28. c	29. d	30. c
31. d	32. d	33. b	34. d	35. d
36. c	37. c	38. a	39. d	40. b

EXPLANATIONS

1. **(c)** In all others, the three letters are consecutive.
2. **(d)** In all others, the first two letters are consecutive and third letter is obtained by skipping one letter from the second.
3. **(d)** In all others, the three letters are consecutive but in reverse order.
4. **(d)** All other groups contain alternate letters from left to right.
5. **(d)** All other groups contain alternate letters from right to left.
6. **(d)** In all others, there is a gap of two letters between first and second as well as between second and third.
7. **(b)** In all other groups, the middle letter is a vowel.
8. **(d)** In all other groups, first letter is 3 letters ahead of the second and third letter is 4 letters ahead of the first.
9. **(c)** In all other groups, the second, first and third letters are in alpnabetical order.
10. **(b)** In all other groups, the third and first letters are in alphabetical order.
11. **(a)** In all other groups, the first two letters are alternate and third letter is 3 letters ahead of the second.
12. **(d)** In all other groups, the second and first letters are three steps ahead of third and second letter respectively.
13. **(b)** No other group contains a vowel.
14. **(d)** In all other groups, first, third and second letters are alternate.
15. **(b)** In all other groups, the first two letters are alternate and third letter is 4 steps a head of the second.
16. **(d)** All other groups contain third letter which is small and a vowel.
17. **(b)** In all other groups, the third letter is small.
18. **(a)** Only this group contains a vowel.
19. **(d)** In all other groups, the letters are in reverse alphabetical order.
20. **(d)** In all other groups, there is a gap of 1 letter between first and second, 2 letters between second and third, 3 letters between third and fourth and 4 letters between fourth and fifth.
21. **(a)** In all other groups, the letters are alternate.
22. **(b)** No letter is repeated in any other group.
23. **(d)** In all other groups, two consecutive letters appear in a random order.
24. **(b)** All other groups contain consecutive letters though not in order.
25. **(d)** All other groups contain alternate letters, though not in order.
26. **(d)** In all other groups, the first four letters are consecutive and there is a gap of 3 letters between last two letters.
27. **(d)** In all other groups, there is a gap of two letters between any two consecutive letters.
28. **(c)** All other groups contain the letters T, E, A, R.
29. **(d)** All other groups contain three consecutive letters, though not in order.
30. **(c)** All other groups, contain only one vowel.
31. **(d)** In all other groups, the last three letters are consecutive.
32. **(d)** In all other groups, one letter is repeated three times.
33. **(b)** In all other groups, the small letters are vowels.
34. **(d)** In all other groups, first two letters are in backward sequence and next two in forward sequence.
35. **(d)** In all other groups, first and third letters and second and fourth letters are in alphabetical sequence.
36. **(c)** In all other groups, first and third letters are consecutive and second and fourth letters are in reverse alphabetical order.
37. **(c)** No letter is repeated in any other group.
38. **(a)** In all other groups, there is a gap of three letters between two consecutive letters.
39. **(d)** In all other groups, first and third letters are moved two steps forward to obtain fourth and second letters respectively.
40. **(b)** All other groups contain E, A and R.

3. SERIES

Remember this letter and number Relationship which is very useful in solving problems on series.

Ascending order	1	2	3	4	5	6	7	8	9	10	11	12	13	14	15	16	17	18	19	20	21	22	23	24	25	26
Alphabet Order	A	B	C	D	E	F	G	H	I	J	K	L	M	N	O	P	Q	R	S	T	U	V	W	X	Y	Z
Descending Order	26	25	24	23	22	21	20	19	18	17	16	15	14	13	12	11	10	9	8	7	6	5	4	3	2	1

STUDY MATERIAL

LETTER SERIES

Letter series contains only letters of the alphabet placed in specific pattern.

Example : U B I P ?

Answer : W - If you carefully study the pattern that is followed here, you will find that every seventh letter is taken alphabetically. After U the letters VWXYZA are omitted, and 'B' is taken. After B, CDEFGH are omitted, and 'I' is taken. After I, JKLMNO are omitted and 'P' is taken to follow the same pattern, naturally after P, QRSTUV should be omitted and 'W' will be the next letter in the series.

It may, at present, appear rather difficult for you to hit upon the pattern followed. However, with some practice, you will be able to find the pattern quite quickly. Here is a technique you will find useful in solving questions of this nature.

When you come across a group of questions of the 'letter series' type, quickly write down the alphabet and number the letters as above. A few examples will show you how this is helpful. Write the alphabet down on a piece of paper. Number it as shown above, draw the lines, and keep it in front of you.

Example :

a) JLNPRT ? b) CFILOR ? c) BFJRV ?

Answers :

a) V. Look at J, N, P, R and T on the numbered alphabet. It is easy to see that the pattern is to skip one letter.

b) U. Look at C, F, I, L, O and R on the numbered alphabet. Its easy to see that the pattern is to skip two letters.

c) Z skip 3 letters.

Example :

a) A C F J O ?

b) A D H M S ?

c) D F I M R ?

Answers :

a. U. Look at letters A, C, F, J and O on the numbered alphabet. You will see the pattern : A (skip) C (skip 2) F (skip 3) J (skip 4) O. Therefore the next 'skip' should be 5 letters, bringing you to the answer, 'U'.

b. Z. The skipping pattern is 2, 3, 4, 5 and 6 letters.

c. X. The skipping pattern is 1, 2, 3, 4 and 5 letters.

In this example, letters are skipped in increasing order.

NUMBER SERIES

In these tests, the series consists of numbers (digits). Number series are generally formed by :

A. Addition of figures given in the row.
B. Subtraction of figures from one another.
C. Division of figures by one another.
D. Multiplication of figures.
E. Logical transposition of figures.
F. Increase/decrease of numbers in a specific pattern.

A careful examination of the example given here will familiarise you with the type of questions that you are likely to face :

Example :

Each of the following series follows a regular pattern. Write down the number which will complete the sequence and replace the question mark ?

a. 5 10 15 25 40 ?
b. 3 12 48 192 ?
c. 48 24 72 36 108 ?

Answer :

a. 65. Each term is the additional of the two previous numbers.
b. 768. Multiply each term by 4 to get the next number.
c. 54. Divide by 2 and multiply by 3 alternately.

PRACTICE TEST - 1
SERIES

DIRECTIONS **: Supply the right letters for question (?) marks in the following questions :**

01. A E J P T ?
a) U b) Y c) R d) W
02. Z W S P L I E ?
a) D b) F c) G d) B
03. A C B E C G D ??
a) MN b) LM c) IE d) GH
04. AM BN EI FJ CO DP GK ------
a) PQ b) QR c) QT d) HL
05. AI BJ CK ------
a) LM b) GH c) QR d) DL
06. XD WC ---------------- YA
a) XY b) CD c) OP d) ZB
07. CD GH KL OP ST ------
a) RS b) QR c) GH d) WX
08. BC FG JK --------------- RS VW
a) LM b) OP c) QR d) NO
09. AY BZ CW ---------------
a) EF b) GH c) MN d) DX
10. AD EH IL --------------- QT UX
a) AY b) VB c) CW d) MP
11. A C F J O ?
a) P b) Q c) U d) V
12. A D E H I L ??
a) MP b) MN c) MO d) MQ
13. A Z B Y C X D ?
a) E b) W c) R d) E
14. CD HI MN ?
a) QS b) RS c) OP d) PQ
15. AU BY CZ ?
a) BC b) KL c) MN d) DX
16. A G L P S ?
a) X b) Y c) W d) U
17. R U X A D ?
a) E b) F c) G d) I
18. C F I L O ?
a) P b) Q c) R d) S
19. E G J L O Q ?
a) S b) R c) U d) T
20. A I P V A E ?
a) F b) G c) H d) I
21. X W V U T S ?
a) R b) T c) E d) X
22. W T P M I F B ?
a) W b) P c) F d) Y
23. R T P R N P ?
a) Q b) L c) S d) F
24. B F K Q ?
a) R b) S c) W d) X
25. B E I N T ?
a) R b) S c) U d) A
26. S P L G ?
a) R b) S c) U d) A
27. T S Q N J ?
a) S b) E c) K d) L
28. U B I P W ?
a) V b) F c) X d) D
29. HV GT FR EP DN ??
a) KL b) LM c) MN d) CI
30. J E Z U P K ?
a) F b) M c) O d) P
31. I M Q U Y C ?
a) C b) D c) F d) G
32. C E I K O Q ?
a) R b) S c) T d) U
33. Z X V T R P ?
a) Q b) R c) S d) N
34. X U S P N K ?
a) I b) L c) M d) Q
35. T Q N K H ?
a) I b) J c) S d) E
36. D I L Q T Y B G ?
a) H b) I c) O d) J
37. X U S P N K I ?
a) J b) K c) M d) F
38. D F I M R ?
a) S b) U c) M d) X
39. B D G I L N ?
a) O b) Q c) S d) U
40. D I L Q T Y B G ?
a) H b) J c) R d) I

ANSWERS

1. b	2. d	3. c	4. d	5. d
6. d	7. d	8. d	9. d	10. d
11. c	12. a	13. b	14. b	15. d
16. d	17. c	18. c	19. d	20. c
21. a	22. d	23. b	24. d	25. d
26. d	27. b	28. d	29. d	30. a
31. d	32. d	33. d	34. a	35. d
36. d	37. d	38. d	39. b	40. b

EXPLANATIONS

1. **(b)** From A to J there are 10 letters (both inclusive) 3 of which are written in the question and between A to J seven letters are skipped. Based on this between P to (?) there should be 10 letters, out of which 2 are written (P & T) skipping only 7 letters. Hence counting from P to (?), based on A-J pattern, 8 letters are missing and therefore, after P, Y should come to complete the pattern in series.
2. **(d)** Skip 2 and 3 steps backwards in the alphabetic order alternately.
3. **(c)** There are 2 alternate series, one from A and other from alternate letter C.
4. **(d)** Vertically 1st and 4th letter of the row are taken skipping two letters in between.
   To fully understand the basis of these questions, make the 4-letter block of the alphabet as follows and the answers will be quite clear without any explanation :
   A B C D
   E F G H
   I J K L
   M N O P
   Q R S T
   U V W X
   Y Z - -
5. **(d)** Vertically 1st and 3rd letter of the row is taken, skipping one in between.
6. **(d)** Initial and last letter taken from downwards to upwards in the 4-letter block.
7. **(d)** Last pair of letters are taken from the 4-letter block.
8. **(d)** Middle pair of letters are taken from the 4-letter block.
9. **(d)** Initial and final letters of vertical rows are taken as 4 letter rows of alphabets as made for Q.
10. **(d)** Alphabet arranged in a block of 4 letter-rows A-D, E-H, I-L, M-P, Q-T, U-X and YZ. The series comprises of initial and final letter of horizontal rows arranged as above.
11. **(c)** Letters following alphabetic order and letter in between skipped in ascending order, increasing each time.
    A **(B)** C (DE) F (GHI) J (KLMN) O (PQRST) U
    1 2 3 4 5
12. **(a)** Alphabets are grouped in rows of 4 (ABCD, EFGH, ...) and an initial and final letter of the row in horizontal position is taken.
13. **(b)** Letter sets are formed by taking a letter from forward order (A-Z) and one letter from backward order (Z-A).
14. **(b)** After a set of 2 consecutive letters, 3 letters are skipped . .
    CD (EFG) HI (JKL) MN (OPQ) RS
    3 3 3
15. **(d)** Alphabet blocked into a 4-letter row (4 horizontally in the row) and 6 (vertically in the row, except row 2 and 3 which have 7 letters). Vertically A row ends in U, B row ends in Y, C row ends in Z and obviously the last row starting from D will end in X.
16. **(d)** Letters follow alphabetic order. Skipping is done in decreasing order.
    A (BCDEF) G (HIJK) L (MNO) P (QR) S (T) U
    5 4 3 2 1
17. **(c)** Two intervening letters are skipped in alphabetic order : R (ST) U (VW) X (YZ) A (BC) D (EF) G
    As explained earlier, circular continuity is maintained i.e. after Z again ABCD order follow.
18. **(c)** 2 consecutive letters are skipped after each letter.
19. **(d)** First 1 letter and 2 letters skipped alternatively.
20. **(c)** Letters maintain alphabetic order and number of letters skipped decrease by one each time. First 7 letters are skipped then 6, 5, 4, 3, and so on.
21. **(a)** Letters are in alphabetic order from backward.
22. **(d)** Skip 2 and 3 steps back in alphabetic order alternately.
    W (VU) T (SRQ) P (ON) M (LKJ) I (HG) F (EDC) B (AZ) Y
23. **(b)** Move 2 letters forward and 4 letters backwards alternately.
24. **(d)** Letters go up by 3,4,5,7,6
    B (CDE) F (GHIJ) K (LMNOP) Q (RSTUVW) X
    3 4 5 6
25. **(d)** Letters to be skipped at the rate of 2,3,4,5,6
    B(CD)E(FGH) I(JKLM)N(OPQRS)T(UVWXYZ)A
    2 3 4 5 6
    Circular continuation follows: After Z again A starts.

26. **(d)** The letters go back by 2, 3, 4, 5 and 6 in that order.
S (RQ) P (ONM) L (KJIH) G (FEDCB)
2 3 4 5

27. **(b)** Backwards the interval increases by one each time. TS (R) Q (PO) N (MLK) J (IHGF) E

28. **(d)** Every 7th letter alphabetically.

29. **(d)** 2 alternating series, one going backwards from H and the other every 2nd letter forwards from V.

30. **(a)** Every 5th letter backward from J.

31. **(d)** 3 letters are skipped.

32. **(d)** In a alphabetic order, letter skipped in the order of 1 and 3 alternatively.
C (D) E (FGH) I (J) K (LMN) O (P) Q (RST) U.

33. **(d)** Backward order of letter is taken, skipping one letter alternatively - Z (Y) X (W) V (U) T (S) R (Q) P (O) N.

34. **(a)** Backward order is taken. First 2 letters and 1 letter are skipped alternatively - X (WV) U (T) S (RQ) P (O) N (ML) K (J) I.

35. **(d)** Backward order is followed skipping 2 letters alternatively. T(SR)Q (PO) N (ML) K (JI) H (GF) E.

36. **(d)** After D, 4 letters are skipped and I follows after which 2 letters are skipped. Trend flows alternatively.
D (EFGH) I (JK) L (MNOP) Q (RS) T (UVWX) Y (ZA) B (CDEF) G (HI) J

37. **(d)** Before X,2 letters are skipped and U follows, after which 1 letter is skipped and S follows. In this manner after every letter 2 and 1 letter alternatively are skipped.

38. **(d)** Letters skipped in the order as 1,2,3,4,5
D (E) F (GH) I (JKL) M (NOPQ) R (STUVW)
1 2 3 4 5

39. **(b)** Letters skipping 1 and 2 alternatively.
B (C) D (EF) G (H) I (JK) L (M) N (OP) Q

40. **(b)** First 4 and 2 letters are skipped alternatively.
D (EFGH) I (JK) J (MNOP) Q (RS) T (UVWX) Y (ZA) B (CDEF) G (HI) J.

PRACTICE TEST - 2
SERIES

***DIRECTIONS* : In each of the following questions, various terms of a letter series are given with one term missing as shown by (?) Choose the mising term out of the given alternatives.**

01. DKY FJW HIU JHS ?
a) KGR b) LFQ c) KFR d) LGQ

02. ACD GHI ? UVWXY
a) LMNO b) MNO c) NOPQ d) MNOP

03. E A C G C E I E G ?
a) JHI b) JGI c) KGI d) KHI

04. BL FP IT ? RB VF
a) MW b) NX c) OY d) MX

05. BXJ ETL HPN KLP ?
a) NHR b) MHQ c) MIP d) NIR

06. CDF DEG EFH ?
a) FGI b) FGH c) FHG d) EFI

07. EJOT DHLP CFIL ?
a) BDFH b) BHLM c) DEIJ d) DGKL

08. CE GI KM OQ ?
a) TW b) TV c) SU d) RT

09. BD GI LN QS ?
a) TV b) UW c) WX d) WY

10. OTE PUF QVG RWH ?
a) SYJ b) TXI c) SXJ d) SXI

11. HUA GTZ FSY ERX ?
a) DWQ b) DQW c) WDQ d) WQD

12. C, B, A, E, D, Z, G, F, (.....)
a) X b) Y c) V d) H

13. Y, V, S, P, M, (.....)
a) N b) K c) L d) J

14. R, M, (.....) F, D, (.....)
a) C, B b) J, H c) B, H d) I, C

15. BS, DT, FU, (.....)
a) XZ b) YA c) HV d) WC

16. CMW, HRB, (.....), RBL, WGQ, BLV
a) MWG b) LVF c) LWG d) MXG

17. DFE, JIH, MLN, (.....), VUT
a) OQP b) PSR c) PRQ d) RSP

18. D-4, F-6, H-8, J-10 (.....), (.....)
a) K-12, M-13 b) L-12, M-14
c) L-12, N-14 d) K-12, M-14

19. HS, JQ, LO, NM, (.....)
a) PK b) RH c) PL d) TG

20. QPO, SRQ, UTS, WVU, (.....)
a) XVZ b) ZYA c) YXW d) VWX

21. AZXBVTCR (.....), (.....)
a) P, D b) E, O c) Q, E d) O, Q

22. P3C, R5F, T81, V12L
a) Y170 b) X17M c) X170 d) X160

23. ABP, CDQ, EFR (.....)
a) GHS b) GHT c) HGS d) GHR

24. Y W U S Q (.....), (.....)
a) N, J b) M, L c) J, R d) O, M

25. G, H, J, M, (.....), V
a) T b) S c) R d) Q

26. (.....), AYW, GEC, MKI, SQO
a) ZXW b) BZW c) USQ d) MAY

27. C, G, L, R (.....)
a) Y b) S c) U d) Z

28. AB DEF HIJK ? STUVWX
a) MNOPQ b) LMNOP
c) LMNO d) QRSTU

29. DCXW FEVU HGTS ?
a) LKPO b) ABYZ
c) JIRQ d) LMRS

30. DF GJ KM NQ RT ?
a) UW b) YZ c) XZ d) UX

31. DEF HIJ MNO ?
a) STU b) RST c) RTV d) SRQ

32. AZ BY CX ?
a) EF b) GH c) IJ d) DW

33. AYD BVF DRH ? KGL
a) FMI b) GMJ c) HLK d) GLJ

34. ? SIY OEU KAQ GWM CRI
a) WNC b) WNB c) VNC d) WMC

35. DEB IJG NOL ? XYV
a) RSP b) STP c) RSQ d) STQ

36. EJO TYD INS XCH ?
a) NRW b) MRW c) MSX d) NSX

37. PRT ? BDF HJL NPR
a) UWY b) VYA c) XZB d) VXZ

38. CX FU IR ? OL RI
a) LO b) MN c) NO d) OP

39. ND IY DT YO TJ ?
a) MP b) NQ c) OF d) OE

40. AK EO IS ? QA UE
a) LV b) MW c) NX d) LW

ANSWERS

1. d	2. d	3. c	4. b	5. a
6. a	7. a	8. c	9. c	10. d
11. b	12. b	13. d	14. d	15. c
16. a	17. d	18. c	19. a	20. c
21. a	22. c	23. a	24. d	25. d
26. c	27. a	28. a	29. c	30. d
31. a	32. d	33. b	34. d	35. d
36. b	37. d	38. a	39. d	40. d

EXPLANATIONS

1. **(d)** The first letters of all the terms are alternate. The second letter of each term is moved one step backward to obtain the second letter of the successive term. The third letter of each term is moved two steps backward to obtain the third letter of the subsequent term.

2. **(d)** The first term consists of one letter, the second term consists of two letters and the third term consists of three letters. So, the required term would consist of four letters. Also, there is a gap of one letter between the letter of the first term and first letter of the second term and a gap of two letters between the last letter of the second term and first letter of the third term. So, the first letter of the required term would be four steps ahead of the last letter of the third term.

3. **(c)** The first letters of the terms are alternate, and similarly the second and the third letters.

4. **(b)** Each letter of a term of the series is four steps ahead of the corresponding letter of the preceding term.

5. **(a)** The first, second and third letters of each term are moved three steps forward, four steps backward and two steps forward respectively to obtain the corresponding letters of the successive term .

6. **(a)** The first letters of the terms are consecutive, and so are the second letters and the third letters.
7. **(a)** The letters of each term are moved backward one, two, three and four steps to obtain the corresponding letters of the next term.
8. **(c)** The letters of each term are alternate and also the last letter of each term and the first letter of the next term are alternate.
9. **(c)** Each term of the series consists of two alternate letters and there is a gap of two letters between the last letter of each term and the first letter of the next term.
10. **(d)** The first letters of the terms are in alphabetical order, and so are the second and third letters.
11. **(b)** All the letters of each term are moved one step backward to obtain the corresponding letters of the next term.
12. **(b)** Third, sixth and ninth letters of the series are in reverse alphabetical order.
13. **(d)** There is a gap of two letters between any two consecutive terms of the series.
14. **(d)** Letters are in reverse order in which from the last 0,1,2,3 and 4 letters are missing between two consecutive letters.
15. **(c)** The first letters of the terms are alternate while the second letters are in alphabetical order.
16. **(a)** All the letters of each term are moved five steps forward to obtain the corresponding letters of the next term.

17. **(d)** 18. **(c)** 19. **(a)** 20. **(c)**
21. **(a)** 22. **(c)** 23. **(a)** 24. **(d)**
25. **(d)** 26. **(c)**

27. **(a)** There is a gap of three letters between the first and the second term, four letters between the second and the third term; and five letters between the third and the fourth term. So, there should be a gap of six letters between the fourth term and the missing term.
28. **(a)** The number of letters in the terms goes on increasing by 1 at each step, Each term consists of letters in alphabetical order. The last letter of each term and the first letter of the next term are alternate.
29. **(c)** First two letters of each term are in reverse order. Similarly third and fourth letters are also in reverse order. Besides, the second letter of each term is the letter next to the first letter of the preceding term.
30. **(d)** There is a gap of one letter between both the letters of first term, a gap of two letters between both the letters of second term and again a gap of one and two letters between the letters of third and fourth terms respectively. Besides, the last letter of each term and the first letter of next term are in alphabetical order.
31. **(a)** The letters in each term are consecutive. There is a gap of one letter between the last letter of the first term and the first letter of the second term and a gap of two letters between the last letter of the second term and the first letter of third term. So, there would be gap of three letters between the last letter of the third term and the first letter of the fourth term.
32. **(d)** The first letter of each term is moved one step forward and the second letter is moved one step backward to obtain the corresponding letters of the next term.
33. **(b)** The first letters of the first, second, third and fourth terms are moved one, two, three and four steps forward respectively to obtain the first letter of the successive terms. The second letters of the first, second, third and fourth terms are moved three, four, five and six steps backward respectively to obtain the second letters of the successive terms. The last letters of the terms are alternate.
34. **(d)** The letters in each term are moved four steps backward to obtain the letters of the next term.
35. **(d)** The letters in each term are moved five steps forward to obtain the letters of the next term.
36. **(b)** There is a gap of four letters between the first and second; and second and third letters of each term, and also between the last letter of a term and the first letter of the next term.
37. **(d)** The letters in each term are alternate. Also, each letter in a term is moved six steps forward to obtain the next term.
38. **(a)** The first letter of each term is moved three steps forward and the second letter is moved three steps backward to obtain the corresponding letters of the next term.
39. **(d)** The first and second letters of each term are five steps ahead of the correponding letters of the successive term.
40. **(b)** The first and second letters of each term are skipped four steps forward to obtain the next term.

PRACTICE TEST - 3
SERIES

***DIRECTIONS* : Figure given in each question follows a set pattern. Determine the pattern and find the correct answer :**

01. 12 32 72 152 ? 632
a) 515 b) 613 c) 815 d) 312

02. 3 8 22 63 185 ?
a) 285 b) 295 c) 310 d) 550

03. 18 30 48 72 96 ?
a) 106 b) 115 c) 120 d) 162

04. 15 20 30 ?
a) 35 b) 40 c) 45 d) 50

05. 6 12 21 33 ?
a) 38 b) 40 c) 45 d) 48

06. 4 9 20 43 90 ?
a) 180 b) 182 c) 179 d) 185

07. 5 9 16 29 54 103 ?
a) 102 b) 94 c) 103 d) 200

08. 5 8 12 17 23 ? 38
a) 26 b) 28 c) 30 d) 29

09. 4 5 7 ? 19
a) 56 b) 66 c) 17 d) 11

10. 4 10 22 46 ??
a) 56 b) 66 c) 76 d) 94

11. 9 15 23 33 ?
a) 44 b) 36 c) 38 d) 45

12. 12 8 14 6 16 ?
a) 18 b) 32 c) 05 d) 04

13. 9 6 16 10 30 18 ? 34
a) 36 b) 60 c) 58 d) 60

14. 68 81 96 ? 132
a) 105 b) 110 c) 130 d) 113

15. 2 5 9 ? 20 27
a) 48 b) 12 c) 14 d) 24

16. 30 23 17 12 ? 5
a) 6 b) 7 c) 8 d) 9

17. 10 18 34 ?? 130 258
a) 32 b) 60 c) 68 d) 66

18. 18 10 6 4 3 ?
a) 8 b) 4 c) 3.5 d) 2.5

19. 6 11 18 27 38 ? 66
a) 58 b) 57 c) 51 d) 59

20. 1 8 27 ?
a) 37 b) 47 c) 57 d) 64

21. 5, 10, 13, 26, 29, 58, 61, ...
a) 122 b) 125 c) 128 d) 64

22. 2, 3, 5, 7, 11, ..., 17
a) 12 b) 13 c) 14 d) 15

23. 5, 11, 17, 25, 33, 43, ...
a) 49 b) 51 c) 52 d) 53

24. 2, 9, 28, 65, 126 .....
a) 137 b) 223 c) 217 d) 199

25. 3, 7, 15, 31, 63
a) 92 b) 127 c) 115 d) 131

26. 4, 9, 13, 22, 35, ...
a) 57 b) 70 c) 63 d) 75

27. 18, 36, 54, 72, 90 .....
a) 106 b) 98 c) 108 d) 112

28. 2, 4, 7, 11, 16, ...
a) 18 b) 20 c) 22 d) 25

29. 0, 2, 6, ..., 20, 30, 42
a) 8 b) 10 c) 12 d) 14

30. 1, 8, 27, 64, 125, 216 .....
a) 354 b) 343 c) 392 d) 245

31. 1, 2, 3, 6, 9, 18, ..., 54
a) 18 b) 36 c) 81 d) 27

32. 5, 9, 6, 11, 7, ...
a) 13 b) 15 c) 17 d) 19

33. 128, 110, 90, ..., 44
a) 56 b) 68 c) 70 d) 72

34. 11, 13, 17, 19, 23, 29, 31, 37, 41, ...
a) 43 b) 47 c) 51 d) 53

35. 11, 13, 17, 19, 23, 25, ...
a) 25 b) 27 c) 29 d) 31

36. 2, 6, 11, 17, ..., 32
a) 22 b) 23 c) 24 d) 28

37. 97, 86, 73, 58, 45, ...
a) 34 b) 54 c) 55 d) 56

38. 66, 36, 18, ...
a) 9 b) 3 c) 6 d) 8

39. 6, 24, 12, ..., 18, 8, 24
a) 4 b) 8 c) 16 d) 6

40. 1, 3, 4, 8, 15, 27, ...
a) 37 b) 44 c) 50 d) 55

ANSWERS

1. d	2. d	3. d	4. d	5. d
6. d	7. d	8. c	9. d	10. d
11. d	12. d	13. c	14. d	15. c
16. c	17. d	18. d	19. c	20. d
21. a	22. b	23. d	24. c	25. b
26. a	27. c	28. c	29. c	30. b
31. d	32. a	33. b	34. a	35. c
36. c	37. a	38. d	39. c	40. c

EXPLANATIONS

1. **(d)** The scheme is : 12 + 4 = 16 x 2 = 32
32 + 4 = 36 x 2 = 72
72 + 4 = 76 x 2 = 152
and so on.

2. **(d)** The scheme is : 3 x 3 = 9 - 1 = 8
8 x 3 = 24 - 2 = 22
22 x 3 = 66 - 3 = 63 and so on

3. **(d)** The scheme is 18 x 2 = 36 - 6 = 30
30 x 2 = 60 - 12 = 48
48 x 2 = 96 - 24 = 72
and so on

4. **(d)** Number increasing in order of 5,10,20, 30 and so on
5. **(d)** Numbers increases @ 6,9,12,15 and so on.
6. **(d)** Double each figure and add 1,2,3,4 and so on. e.g. 4 doubled = 8 + 1 = 9, then 9 doubled = 18 add 2 = 20 and so on.
7. **(d)** Multiply each number by 2 and reduce 1,2,3,4 and so on from each figure. e.g. 5 x 2 = 10 – 1=9
8. **(c)** Starting from first figure, increase each time by 3,4,5 and so on.
9. **(d)** Each number is twice the previous one and minus 3.
10. **(d)** Each number is twice the previous one plus 2.
11. **(d)** The series increases by 6, then 8,10 and finally 12,
12. **(d)** Two alternating series increasing in twos from 12 and reducing in twos from 8.
13. **(c)** Two alternate series, in each of which figures are doubled and result reduced by 2 to form the next number of the series.
14. **(d)** Series increases by steps 13,15,17,19
15. **(c)** Interval increases by 1 each time.
16. **(c)** Interval reduces by 1 each time.
17. **(d)** Double each figure and subtract 2 each time.
18. **(d)** Each figure is obtained by adding 2 to the previous one and then dividing by 2.
19. **(c)** Square numbers 2,3,4,5 respectively adding 2 each time.
20. **(d)** Cube numbers 1,2,3,4 respectively.
21. **(a)** The numebrs are alternately multiplied by 2 and increased by 3. Thus,
5 x 2=10, 10+3 = 13, 13 x 2 = 26, 26 + 3 = 29 and so on.
∴ Missing number = (61 x 2) = 122
22. **(b)** Clearly, the given series consists of prime numbers starting from 2. The prime number after 11 is 13. So, 13 is the missing number.
23. **(d)** The sequence is + 6, +6, +8, +8, + 10 .....
So missing number = 43 + 10 = 53.
24. **(c)** The sequence is $1^3+1, 2^3+1, 3^3+1, 4^3+1, 5^3+1$ ...
∴ Missing number = $6^3 + 1 = 216 + 1 = 217$
25. **(b)** Each number in the series is the preceding number multiplied by 2 and then increased by 1.
Thus, (3 x 2) + 1 = 7, (7 x 2) + 1 = 15, (15 x 2) + 1 = 31 and so on.
∴ Missing number = (63 x 2) + 1 = 126 +1= 127.
26. **(a)** The sum of two consecutive numbers of the series gives the next number.
Thus, 4 + 9 = 13, 9 + 13 = 22, 13 + 22 = 25 and so on.
∴ Missing number = 22 + 35 = 57
27. **(c)** The sequence is 18 x 1, 18 x 2, 18 x 3, 18 x 4, 18 x5, ...
∴ Missing number = 18 x 6 = 108.
28. **(c)** The difference between consecutive numbers increases by 1.
Thus, the sequence is +2, +3, +4, +5,....
∴ Missing number = 16 + 6 = 22 .
29. **(c)** The sequence is +2, +4, ...,....., + 10, + 12.
Clearly, missing number = 6 + 6 = 12.
30. **(b)** The numbers are $1^3, 2^3, 3^3, 4^3, 5^3, 6^3$.
∴ Missing number = 7^3 = 343.
31. **(d)** The numbers are alternately multiplied by 2 and 3/2.
Thus, 1 x 2 = 2, 2 x 3/2 = 3, 3 x 2 = 6, 6 x 3/2 = 9 and so on.
∴ Missing number = 18 x 3/2 = 9 x 3 = 27.
32. **(a)** The given sequence is a combination of two series -5,6,7 and 9,11, (.....) Obviously, the pattern followed in the second series is +2.
So, missing term = 11 + 2 = 13.
33. **(b)** The sequence is –18,–20,.....,....
∴ Missing number = 90 – 22 = 68 .
34. **(a)** The series consists of prime numbers.
∴ The missing number is the next prime number, which is 43.
35. **(c)** The sequence is +2,+4,+2,+4, +2,....
∴ Missing number = 25 + 4 = 29 .

36. (c) The sequence is +4,+5,+6 .....

$\therefore$ Missing number = 17 + 7 = 24.

37. (a) The sequence is −11,−13,−15,−13,...

$\therefore$ Missing number = 45 − 11 = 34.

38. (d) Each number in the series is the product of the digits of the preceding number.

Thus, 6 x 6 = 36, 3 x 6 = 18 and so on.

$\therefore$ Missing number = 1 x 8 = 8

39. (c) The given sequence is a combination of two series 6,12,18,24 and 24, (...),8.

Clearly, the first series consists of consecutive multiples of 6 and the second series consists of multiples of 8. Thus, the missing number is the multiple of 8, lying between 8 and 24, which is 16.

$\therefore$ Missing term = $\frac{1}{24} \times \frac{3}{2} = \frac{1}{16}$

40. (c) The sum of any three consecutive terms of the series gives the next term.

Thus, 1 + 3 + 4 = 8; 3 + 4+8=15, 4+8+15= 27 and so on.

PRACTICE TEST - 4
SERIES

***DIRECTIONS* : Figure given in each question follows a set pattern. Determine the pattern and find the correct answer :**

01. 13, 31, 63, 127, 255, ...
a) 513 b) 511 c) 517 d) 523

02. 2, 6, 12, 20, 30, 42, 56, ...
a) 60 b) 64 c) 70 d) 72

03. 8, 24, 12, 36, 18, 54, ...
a) 37 b) 68 c) 72 d) 108

04. 165, 195, 255, 285, 345, ...
a) 375 b) 420 c) 435 d) 390

05. 2, 4, 12, 48, 240, ...
a) 960 b) 1080 c) 1440 d) 1920

06. 10, 5, 13, 10, 16, 20, 19, ...
a) 22 b) 23 c) 38 d) 40

07. 1, 2, 4, 8, 16, 32, 64, ..., 256
a) 148 b) 128 c) 154 d) 164

08. 71, 76, 69, 74, 67, 72, ...
a) 65 b) 76 c) 77 d) 80

09. 2, 5, 9, 19, 37, ...
a) 76 b) 75 c) 74 d) 72

10. 3, 5, 6, 10, 9, 15, 12, A, B
a) A = 18, B = 20 b) A = 15, B = 20
c) A = 15, B = 18 d) A = 20, B = 15

11. 9, 12, 11, 14, 13, ..., 15
a) 12 b) 16 c) 10 d) 17

12. 4, 9, 5, 12, 7, 15, 8, ..., 10
a) 25 b) 18 c) 21 d) 19

13. 50, 49, 46, 42, 34, ...
a) 32 b) 25 c) 21 d) 19

14. 1, 5, 11, 19, 29, ...
a) 45 b) 39 c) 41 d) 47

15. 2, 6, 3, 4, 20, 5, 6, ..., 7
a) 25 b) 30 c) 42 d) 28

16. 2, 5, 11, 23, 47, ...
a) 49 b) 52 c) 95 d) 106

17. 3, 15, 35, ..., 99, 143
a) 48 b) 63 c) 80 d) 95

18. 64, 32, 16, 8, ...
a) 0 b) 1 c) 2 d) 4

19. 1, 9, 17, 33, 49, 73, ...
a) 97 b) 98 c) 99 d) 100

20. 3, 10, 20, 33, 49, 68, ...
a) 75 b) 85 c) 90 d) 91

21. 3, 4, 10, 32, 136, 685, 4116
a) 136 b) 10 c) 4116 d) 32

22. 125, 126, 124, 127, 123, 129, ...
a) 123 b) 124 c) 126 d) 129

23. 112, 114, 120, 124, 132, 142, 154
a) 114 b) 120 c) 124 d) 132

24. 325, 259, 202, 160, 127, 105, 94, ...
a) 94 b) 127 c) 105 d) 202

25. 105, 85, 60, 30,
a) −5 b) 85 c) −10 d) 45

26. 196, 169, 144, 121, 100, 80, 64, ...
a) 169 b) 114 c) 121 d) 100

27. 8, 14, 26, 48, 98, 194, 386
a) 194 b) 98 c) 14 d) 48

28. 8, 13, 21, 32, 47, 63, 83, ...

a) 21 b) 13 c) 32 d) 47

29. 380, 188, 92, 48, 20, 8, 2

a) 188 b) 48 c) 92 d) 20

30. 1, 3, 10, 21, 64, 129, 356, 777, ...

a) 10 b) 21 c) 356 d) 129

31. 6, 12, 48, 100, 384, 768, 3072

a) 768 b) 384 c) 100 d) 48

32. 10, 26, 74, 218, 654, 1946, 5834, ...

a) 26 b) 74 c) 218 d) 654

33. 3, 7, 15, 39, 63, 127, 255, 511, ...

a) 39 b) 15 c) 7 d) 63

34. 445, 221, 109, 46, 25, 11, 4, ...

a) 221 b) 109 c) 46 d) 25

35. 8, 27, 125, 343, 1331, ...

a) 1331 b) 343 c) 125 d) None of these

36. 10, 14, 28, 32, 64, 68, 132, ...

a) 32 b) 68 c) 64 d) 132

37. 6, 13, 18, 25, 30, 37, 40, ...

a) 25 b) 30 c) 37 d) 40

38. 56, 72, 90, 11, 132, 150, ...

a) 72 b) 90 c) 110 d) 150

39. 8, 13, 21, 32, 47, 63, 83, ...

a) 13 b) 47 c) 63 d) 32

40. 25, 36, 49, 81, 121, 169, 225, ...

a) 36 b) 49 c) 121 d) 169

ANSWERS

1. b	2. d	3. a	4. c	5. c
6. d	7. b	8. a	9. b	10. d
11. b	12. b	13. b	14. c	15. c
16. c	17. b	18. d	19. a	20. c
21. d	22. d	23. b	24. d	25. a
26. d	27. d	28. c	29. b	30. c
31. c	32. d	33. a	34. c	35. d
36. a	37. d	38. d	39. b	40. a

EXPLANATIONS

1. **(b)** Each number is double the preceding one plus 1.

So, the next number is (255 x 2) + 1 = 511.

2. **(d)** The sequence is 1 x 2, 2 x 3, 3 x 4, 4 x 5, 5 x 6, 6 x 7, 7 x 8.

∴ Next number = 8 x 9 = 72.

3. **(a)** The terms are alternately multiplied by 3 and divided by 2.

Thus 8 x 3 = 24, 24 ÷ 2 = 12, 12 x 3 = 36, 36 ÷ 2 = 18 and so on.

∴ Missing term = 54÷2 = 27.

4. **(c)** Each number is 15 multiplied by a prime number i.e. 15 x 11, 15 x 13, 15 x 17, 15 x 19, 15 x 23.

∴ Missing term = 15 x 29 = 435.

5. **(c)** The terms are respectively multiplied by 2,3,4 and 5 to obtain the successive terms. Clearly, missing term = 240 x 6 = 1440.

6. **(d)** There are two series : 10,13,16,19 and 5,10,20, (....)

∴ Missing number = 40.

7. **(b)** Each term is multiplied by 2 to obtain the next term.

8. **(a)** Alternately 5 is added and 7 is subtracted.

Thus, 71 + 5 = 76, 76−7 = 69, 69 + 5 = 74, 74 - 7 = 67 and so on.

∴ Missing number = 72 − 7 = 65.

9. **(b)** The second number is one more than twice the first; third number is one less than twice the second, fourth number is one more than twice the third; fifth number is one less than twice the fourth.

Thus, the sixth number is one more than twice the fifth.

So, the missing number is 75.

10. **(d)** The given sequence is a combination of two series – 3,6,9,12, B and 5,10,15,A

Clearly, A = 20 and B = 15

11. **(b)** Alternately we add 3 and subtract 1.

Thus, 9 + 3, = 12, 12 − 1 = 11, 11 + 3 = 14 and so on.

12. **(b)** The given sequence is a combination of two series : 4,5,7,8,10 and 9,12,15, (...)

Clearly, the missing term is 18.

13. **(b)** The sequence is -1, -3, -5, -7.....

∴ Missing number = 34 - 9 = 25.

14. **(c)** The sequence is +4,+6,+8,+10,...

   ∴ Missing number = 29 + 12 = 41.

15. **(c)** The arrangement in the series is as under :

   2 x 3 = 6; 4 x 5 = 20 and so on.

   ∴ Missing number = 6 x 7 = 42.

16. **(c)** The sequence is +3,+6,+12, + 24,.......

   ∴ Missing number = 47 + 48 = 95.

17. **(b)** The terms of the series are 2^2 - 1, 4^2 - 1, 6^2 -1, ...,10^2 - 1, 12^2-1.

   ∴ Missing number = $8^2 - 1 = 64 - 1 = 63$.

18. **(d)** Each number is half of the preceding number.

   ∴ Missing number = $\frac{1}{2} \times 8 = 4$

19. **(a)** The sequence is +8,+8,+16,+16,+24,...

   ∴ Missing number = 73+24 = 97.

20. **(c)** The sequence is +7,+10,+13,+16,+19....

   ∴ Missing number = 68+22 = 90.

21. **(d)** The sequence is as follows :

   2nd term = (1st term + 1) x 1

   3rd term= (2nd term + 1) x 2

   4th term = (3rd term +1) x 3 and so on.

   So, 32 is wrong, the correct term being 33.

22. **(d)** The sequence is +1,–2,+3,–4,+5.

   So, 129 is wrong. The correct term should be 128.

23. **(b)** The sequence is +2,+4,+6,+8,+10,+12.

   So, 120 is wrong.

24. **(d)** The sequence is –66,–55, –44, –33, –22, –11.

   So, 202 is wrong. The correct term is 204.

25. **(a)** The sequence is -20,–25, –30, –35.

   So –5 is the correct answer.

26. **(d)** The sequence is $(14)^2$,$(13)^2$, $(12)^2$,$(11)^2$, $(10)^2$, $(9)^2$, $(8)^2$.

   So, 80 is wrong, since $9^2 = 81$.

27. **(d)** Each term, except 48 in the series is less than twice the preceding term by 2. The correct term is 50.

28. **(e)** The sequence is +5,+8,+11,+14,+17,+20.

   However, 47 breaks this sequence. The correct term is 46.

29. **(b)** Each term, except 48 in the series is four more than two times the next term. The correct term is 44.

30. **(c)** Suppose the terms of the series are A,B,C,D,... then, the sequence is Ax2 + 1, Bx3 + 1, Cx2+1, Dx3 + 1 and so on.

   ∴ 356 is wrong.

31. **(c)** Each term at even place in the series, except 100 is obtained by multiplying the preceding term by 2.

   ∴ 100 is wrong, the correct term being 96.

32. **(d)** Each term is four less than, the preceding number multiplied by 3.

   Thus, 26 = (10x 3 - 4) ; 74 = (26x3-4) & so on.

   ∴ 654 is wrong, the correct term being 650

33. **(a)** Each number in the series is multiplied by 2 and the result increased by 1 to obtain the next number.

   ∴ 39 is wrong, the correct term being 31.

34. **(c)** 3 is subtracted from each number and the result is divided by 2 to obtain the next number of the series. Thus, $221 = \frac{(445 - 3)}{2}$;

   $109 = \frac{(221 - 3)}{2}$ and so on.

   ∴ 46 is wrong, the correct term being 53.

35. **(d)** The numbers are cubes of prime numbers i.e. $2^3,3^3,5^3,7^3,11^3$. Clearly, none is wrong.

36. **(a)** Alternately, the numbers are increased by four and doubled to get the next number. Thus, 10 + 4 = 14; 14 x 2 = 28; 28 + 4 = 32; 32 x 2 = 64 and so on. So 132 is wrong. It must be (68 x 2) i.e. 136.

37. **(d)** The sequence is +7,+5,+7,+5,+7,+5.

   Clearly, 40 is wrong, the correct term being 42.

38. **(d)** The numbers are 7 x 8, 8 x 9, 9 x 10, 10 x 11, 11 x 12, 12 x 13. So, 150 is wrong.

39. **(b)** The sequence is +5,+8,+11,+14,+17,+20. So, the number 47 is wrong. The corect term is 46.

40. **(a)** The correct sequence is $5^2,7^2,9^2,11^2,13^2,15^2$

4. DISTANCE AND DIRECTION

In this test, the questions consist of a sort of direction puzzle, A successive follow-up of directions is formulated and the candidate is required to ascertain the final direction or the distance between two points. The test is meant to judge the candidate's ability to trace and follow correctly and sense the direction correctly.

STUDY MATERIAL

The adjoining figure shows the four main direction (North N, South S, East E, West W) and the four cardinal (North East NE, North West NW, South East SE, South West SW) to help the candidates know the directions.

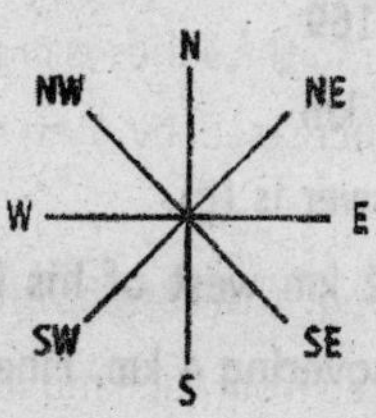

1. John left for his office in his car. He drove 15 km towards north and then 10 km towards west. He then tuned to the south and covered 5 km. Further, he turned to the east and moved 8 km. Finally, he turned right and drove 10 km. How far and in which direction is he from his starting point ?

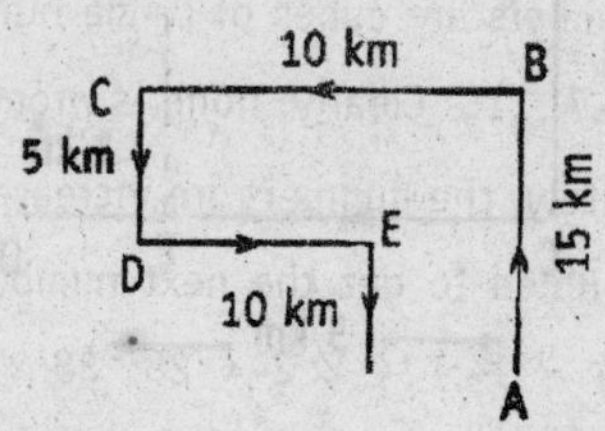

a) 2 km West

b) 5 km East

c) 3 km North

d) 6 km South

Sol. Clearly, John drove 15 km from A to B northwards and then 10 km from B to C towards west. He then moves 5 km southwards from C to D and 8 km eastwards upto E. Finally, he turned right and moved 10 km upto F.

∴ A and F lie in the same straight line and F lies to the west of A.

So, John's distance from the starting point A

= AF = (BC – DE) = (10 – 8) km = 2 km

Hence, the answer is (a).

2. Divya moved a distance of 75 meters towards the north. She then turned to the left and walking for about 25 meters, turned left again and walked 80 meters. Finally, she turned to the right at an angle of 45°. In which direction was she moving finally?

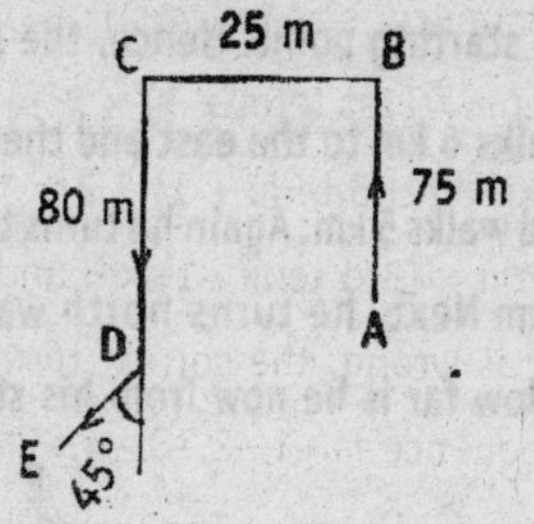

a) North – East

b) North – West

c) South

d) South – West

Sol. Divya started from A, moved 75m upto B, turned left and walked 25 m upto C m upto D. Turning to the right at an angle of 45°, she was finally moving in the direction DE i.e., South – West. Hence, the answer is (d).

3. Krishna faces towards north. Turning to his right, he walks 25 meters. He then turns to his left and walks 30 meters. Next, he moves 25 meters to his right. He then turns to his right again and walks 55 meters. Finally, he turns to the right and moves 40 meters. In which direction is he now from his starting point ?

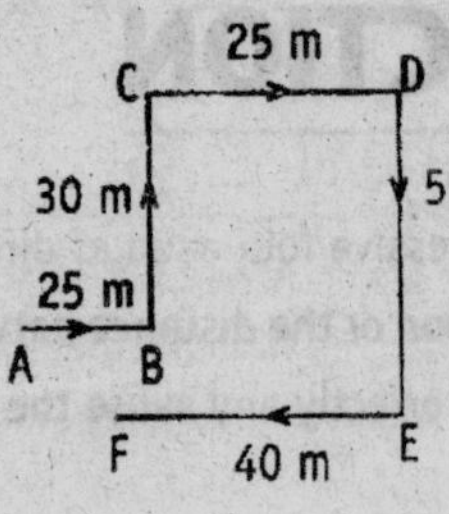

a) South – West

b) South

c) North – West

d) South – East

Sol. Krishna turns towards right from north direction. So, he walks 25m towards east upto B, turns left and moves 30 m upto C, turns right and goes 25m upto D. At D, he turns to right towards the south and walks 55 m upto E. Next, he again turns to right and walks 40 m upto F, which is his final position. F is to the south – east of A. So, he is to the south – east from his starting point. Hence, the answer is (d).

4. A man walks 6 km to the east and then turns to the south and walks 5 km. Again he turns to the east and walks 6 km Next, he turns north wards and walks 10 km. How far is he now from his starting point?

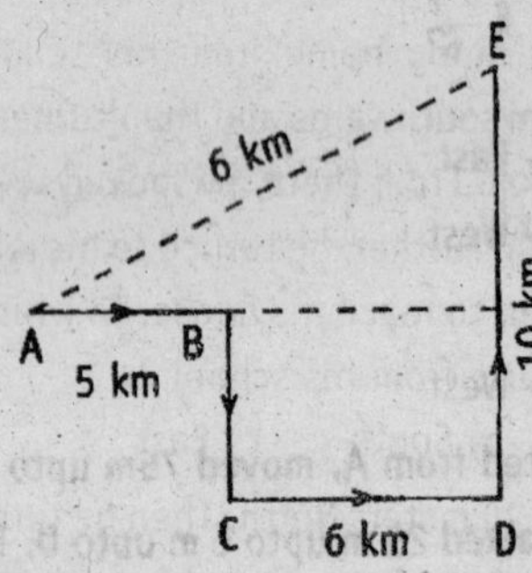

a) 5 km

b) 12 km

c) 13 km

d) 17 km

Sol. A man starts from A and walks 6 km east upto B, turns southwards and moves 5 km upto C. At C, he turns to the east and walks 6 km upto D. He then tuns northwards and walks 10 km upto E.
Now, draw BO E

AO = (AB + BO)

= (6 + 6) km

= 12 km

OE = (DE – OD)

= (DE – BC)

= (10 – 5) km

= 5 km

∴ Man's distance from the starting point A

$= AE = \sqrt{AO^2 + OE^2}$

$= \sqrt{(12)^2 + (5)^2}$

$= \sqrt{169}$

= 13 km

Hence, the answer is (c).

5. Arjun walked 2 km west of his house and then turned south covering 4 km. Finally, he moved 3 km towards east and then again 1 km west. How far is he from his initial position?

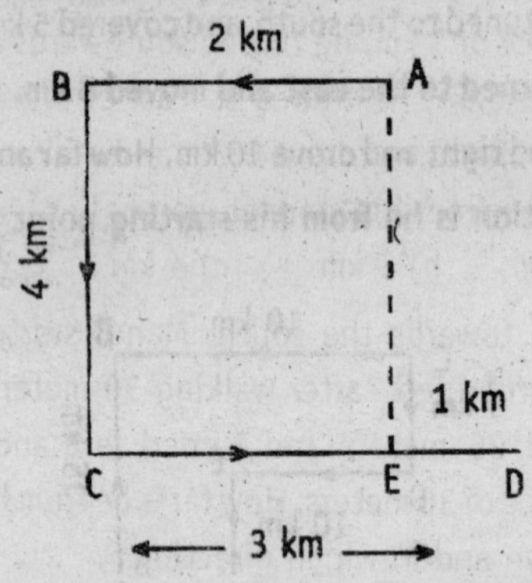

a) 2 km

b) 4 km

c) 9 km

d) 10 km

Sol. Arjun starts from his house at A, moves 2 km west upto B, then 4 km to the south upto C, 3 km east upto D and finally 1 km west upto E. Thus, his distance from the initial position A = AE = BC = 4 km.
Hence, the answer is (b).

PRACTICE TEST - 1
DISTANCE AND DIRECTION

1. Jonny starts walking towards East. After moving a distance of 20 meters, he turns to his left and walks 15 meters. Then, he turns to his right and moves 25 meters. Next, he turns to his right again and walks 15 meters. How far in meters is he from his starting point?
   a) 0 b) 25 c) 40 d) 45
2. Starting from a point, P, Sachin walked 20 meters towards South. He turned left and walked 30 meters. He then turned left and walked 20 meters. He again turned left and walked 40 meters and reached a point Q. How far and in which direction is the point Q from the point P ?
   a) 20 meters west b) 10 meters east
   c) 10 meters west d) 10 meters north
3. Ramesh walks northwards. After a while, he turns to his right and a little further to his left, Finally, after walking a distance of one kilometers, he turns to his left again. In which direction is he moving now?
   a) North b) South c) East d) West
4. Jatin started walking towards East. After moving a distance of 1 km, he turned southwards and walked 5 km. Again, he turned to East and walked 2 km. Finally, he turns to the North and walked 9 km. How far is he from his starting point?
   a) 7 km b) 5 km c) 4 km d) 3 km
5. Facing towards the south, Manoj started walking and turned left after walking 30 meters. He then walked 25 meters and turned left and moved a distance of 30 meters. How far is he from his starting position and in which direction ?
   a) At the starting point b) 25 meters west
   c) 25 meters East d) 30 meters East
6. Raju travelled from a point X straight to Y at a distance of 80 meters. He turned right and walked 50 meters, then again turned right and walked 70 meters. Finally, he turned right and walked 50 meters. How far is he from the starting point?
   a) 20 meters b) 50 meters
   c) 70 meters d) 10 meters
7. Lokesh went 15 kms to the west from my house, then turned left and walked 20 kms. He then turned east and walked 25 kms and finally turning left covered 20 kms. How far was he from my house?
   a) 5 kms b) 10 kms c) 40 kms d) 80 kms
8. Vali travelled 7 km eastward, then turned left and travelled 5 km. Next, he turned left and travelled 7 km. How far was he from the starting point?
   a) 19 km b) 12 km c) 7 km d) 5 km
9. From his house, Laal went 15 kms to the North. Then he turned west and covered 10 kms. Then, he turned South and covered 5 kms. Finally, turning to East, he covered 10 kms. In which direction is he from his house ?
   a) East b) West c) North d) South
10. Going 50m to the south of her house, Radha turns left and goes another 20m. Then, turning to the North, she goes 30m and then starts walking to her house. In which direction is she walking now ?
    a) North - West b) West
    c) North d) East
11. The door of Arun house faces the East, from the backside of the house, he walks straight 50 meters, then turns to the right and walks 50 meters again. Finally, he turns towards left and stops after walking 25 meters. Now, Arun is in which direction from the starting point ?
    a) South - East b) North - East
    c) South - West d) North - West
12. Vinod was going home from his school. He first moved 3 km southwards and then turned to his left, moving 2 km. From there, he moved to his left and walked 3 km. Further, he turned to his left again and moved 1 km, to reach his house. In which direction was his house from his school ?
    a) North b) South c) East d) West
13. Dilip walks 20 meters North. Then he turns right and walks 30 meters. Now, he turns right and walks 35 meters. Now, turning left, he walks 15 meters. Again, he turns left and moves 15 meters. Finally, turning left he again walks 15 meters. In which direction and how far is he from his original position?
    a) 15 meters East b) 45 meters East
    c) 15 meters West d) 45 meters West
14. Rita who is facing South turns to her left and walks 15 meters, then she turns to her left and walks 7 meters, then facing West she walks 15 meters. How far is she from her original position ?
    a) 22 meters b) 37 meters
    c) 44 meters d) 7 meters

15. X walks 10 meters in front and 10 meters to the right. Then every time turning to his left, he walks, 5,15 and 15 meters respectively. How far is he now from his starting point ?
   a) 10 meters b) 20 meters
   c) 5 meters d) 15 meters

16. Deeru starts walking straight towards east. After walking 75 meters, he turns to the left and walks 25 meters straight, Again he turns to the left, walks a distance of 40 meters straight, again he turns to the left and walks a distance of 25 meters. How far is he from the starting point ?
   a) 140 meters b) 50 meters
   c) 115 meters d) 35 meters

17. Kishore walks 10 kilometers towards North. From there, he walks 6 kilometers towards South. Then, he walks 3 kilometers towards East. How far and in which direction is he with reference to his starting point ?
   a) 7 kilometers East
   b) 5 kilometers West
   c) 5 kilometers North – East
   d) 7 kilometers West

18. Ravi leaves for his office from his house. He walks towards East. After moving a distance of 20 m, he turns towards South and walks 10 m. Then he walks 35 m towards the west and further 5 m towards the north. He then turns towards east and walks 15m. What is the straight distance in meters between his initial and final positions ?
   a) 0 b) 5
   c) 10 d) None of these

19. Giri walks 20 meters towards North. He then turns left and walks 40 meters. He again turns left and walks 20 meters. Further, he moves 20 meters after turning to the right. How far is he from his original position ?
   a) 20 meters b) 30 meters
   c) 50 meters d) 60 meters

20. Ajay started walking towards North. After walking 30 meters he turned left and walked 40 meters. He then turned left and walked 30 meters. He again turned left and walked 50 meters. How far is he from his original position ?
   a) 50 meters b) 40 meters
   c) 30 meters d) 10 meters

21. Ganesh walks from his house towards West. After walking a distance of 30 meters, he turned towards right and walked 20 meters. He then turned left and moving a distance of 10 meters, turned to his left again and walked 40 meters. He now turns to the left and walks 5 meters. Finally he turns to his left. In which direction is he a walking now ?
   a) North b) South
   c) East d) South – West

22. Sarala leaves from her home. She first walks 30 meters in North – West direction and then 30 meters in South – West direction. Next, she walks 30 meters in South – East direction. Finally, she turns towards her house. In which direction is she moving ?
   a) North – East b) North – West
   c) South – East d) South – West

23. Sanjay walks 10 meters towards the south. Turning to the left, he walks 20 meters and then moves to his right. After moving a distance of 20 meters, he turns to the right and walks 20 meters. Finally, he turns to the right and moves a distance of 10 meters. How far and in which direction is he from the starting point?
   a) 10 meters North b) 20 meters South
   c) 20 meters North d) 10 meters South

24. Kumar goes 30 meters North, then turns right and walks 40 meters, then again turns right and walks 20 meters, then again turns right and walks 40 meters. How many meters is he from his original position ?
   a) 0 b) 10 c) 20 d) 40

25. Vipul walks 30 meters towards South. Then, turning to his right, he walks, 30 meters. Then, turning to his left, he walks 20 meters. Again, he turns to his left and walks 30 meters. How far is he from his initial position ?
   a) 30 meters b) 20 meters
   c) 80 meters d) None

ANSWERS

1. d	2. c	3. d	4. b	5. c
6. d	7. b	8. d	9. c	10. a
11. d	12. d	13. b	14. d	15. c
16. d	17. c	18. b	19. d	20. d
21. a	22. a	23. b	24. b	25. d

EXPLANATIONS

1. **(d)** The movements of Jonny are as shown in figure.

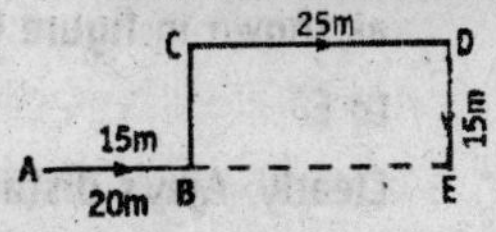

∴ Jonny's distance from the starting point A.
= AE = (AB + BE) = (AB + CD) = (20 + 25) m
= 45m

2. **(c)** The movements of Sachin are as shown in figure.

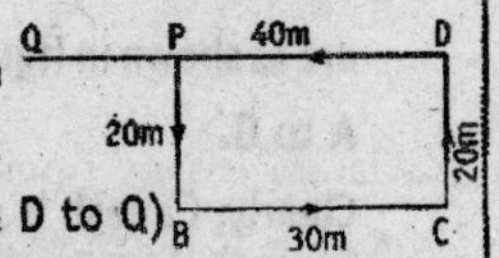

(P to B, B to C, C to D & D to Q)
Clearly, Sachin's distance from P
= PQ = (DQ – PD) = (DQ – BC) = (40 – 30) m = 10 m.
Also, Q is to the west of P.
∴ Q is 10 m west of P.

3. **(d)** The movements of Ramesh are as shown in figure.

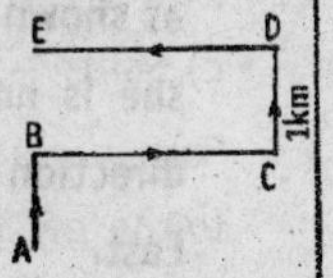

Clearly, he is now walking in the direction DE i.e. west.

4. **(b)** The movements of Jatin are as shown in figure.

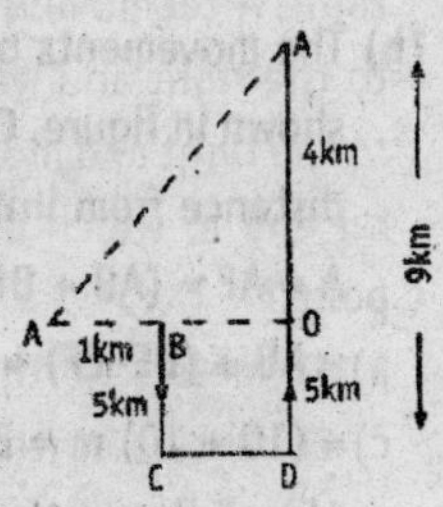

(A to B, B to C, C to D and D to E)
Draw BO ⊥ DE.
Clearly, BO = CD = 2 km.
AO = (AB + BO) = 3 km.
OE = (DE – DO) = (DE – BC)
= (9 – 5) km=4 km.
∴ Joseph's distance from starting point A.

$$= AE = \sqrt{AO^2 + OE^2} = \sqrt{(3)^2 + (4)^2}$$

$$= \sqrt{9 + 16} = 5 \text{ km}.$$

5. **(c)** The movements of Manoj are as shown in figure.

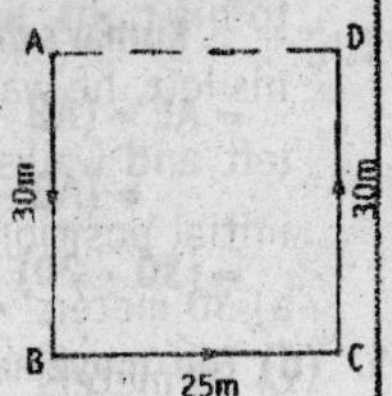

∴ Manoj's distance from starting point
A = AD = BC = 25 m.
Also D is to the East of A.
∴ Manoj is 25 m to the East from his starting point.

6. **(d)** The movements of Raju are as shown in figure.

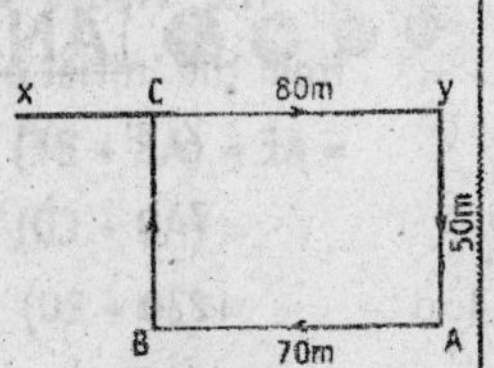

(X to Y, Y to A, A to B and B to C).
∴ Raju's distance from the starting point
= XC = (XY – YC)
= (XY – BA) = (80 – 70) m = 10m.

7. **(b)** The movements of Lokesh are as shown in figure.

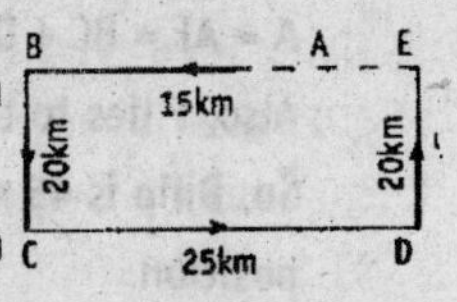

∴ Lokesh's distance from his house at A.
= AE = (BE – AB) = (CD – AB)
= (25 – 15) km = 10 km

8. **(d)** The movements of Vali are as shown in figure.

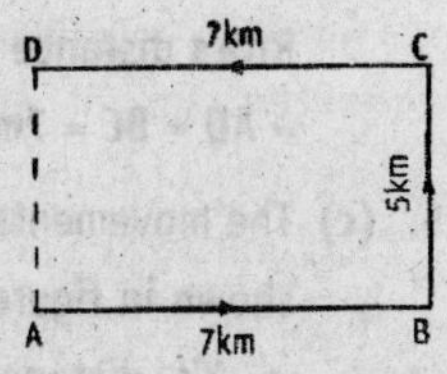

∴ Vali's distance from the starting point
A = AD = BC = 5 km.

9. **(c)** The movements of Laal are as shown in figure.

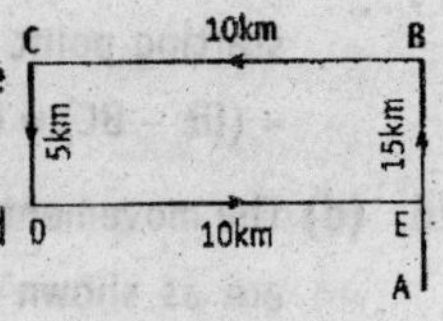

(A to B, B to C, C to D and D to E)
Clearly, his final position is E which is the North of his house at A.

10. **(a)** The movements of Radha are as shown in figure.

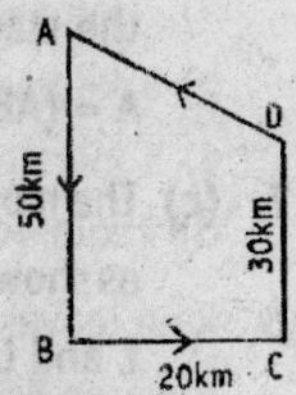

Thus, she is now moving in the direction DA i.e. North – West.

11. **(d)** The movements of Arun are as shown in figure. Since his house faces towards East and he walks from the backside of his house, he will start walking towards West. Now Arun's final position is D which is to the North – west of his starting point A.

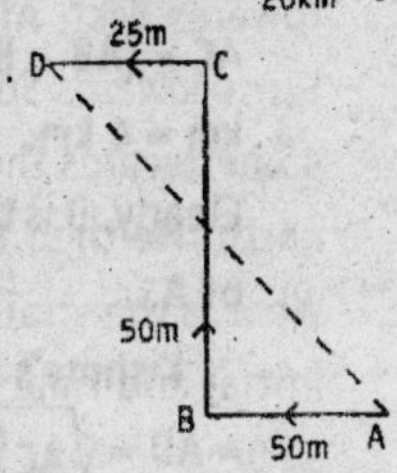

12. **(d)** The movements of Vinod are as shown in figure. Clearly, Vinod's house is at A & his school is at E.

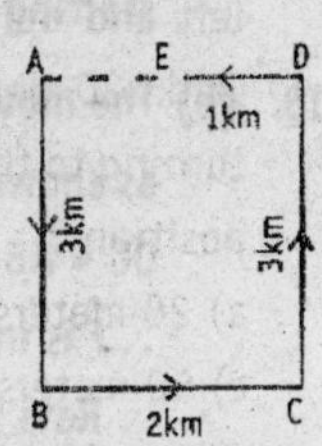

A is to the West of E.
So Vinod's house is to the West of his school.

13. **(b)** The movements of Dilip are as shown in figure. (A to B, B to C, C to D, D to E and E to F).

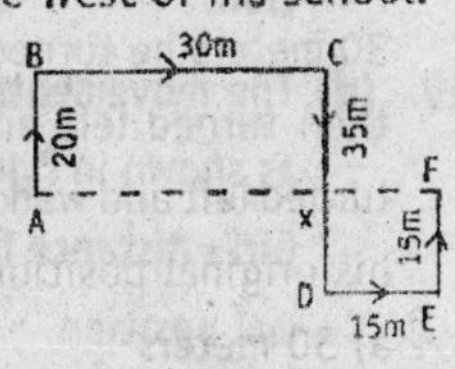

Join AF, meeting CD at X.

Dilip's distance from his original position.

A = AF = BC + DE = 45 m.

Also, F lies to the East of A.

So, Dilip is 45 metres to the east of his initial position.

14. **(d)** The movements of Rita are as shown in figure.

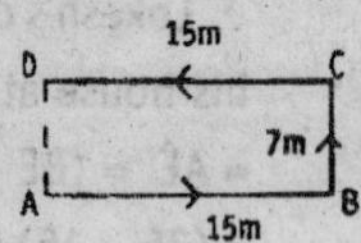

Rita's distance from the starting point A = AD = BC = 7m.

15. **(c)** The movements of X are as shown in figure.

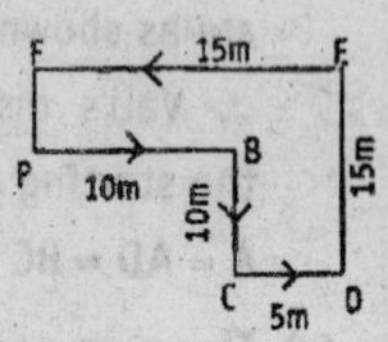

∴ X's distance from his starting point P = PE
= (DE – BC) = (15 – 10) m = 5 m.

16. **(d)** The movements of Deeru are as shown in figure.

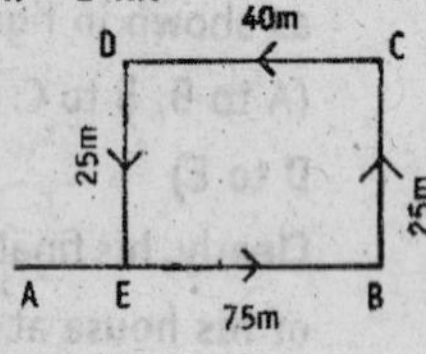

Clearly, EB = DC = 40 m.

∴ Deepak's distance from the starting point A = (AB – EB) = (75 – 40) m = 35 m.

17. **(c)** The movements of Kishore are as shown in figure (A to B, B to C and C to D)

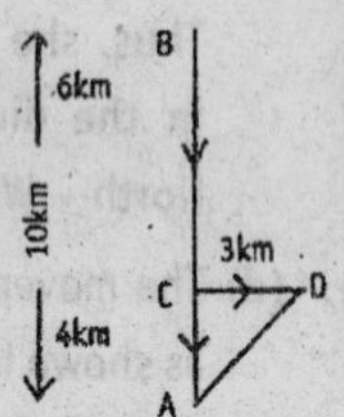

AC = (AB – BC) km = (10 – 6) km = 4 km.

Clearly, D is to the North – East of A.

∴ Kishore's distance from starting point A.

$= AD = \sqrt{AC^2 + CD^2} = \sqrt{(4)^2 + (3)^2}$

$= \sqrt{16+9} = \sqrt{25} = 5$ km.

So, Kishore is 5 km to the North – East from his starting point.

18. **(b)** The movements of Ravi are as shown in figure.

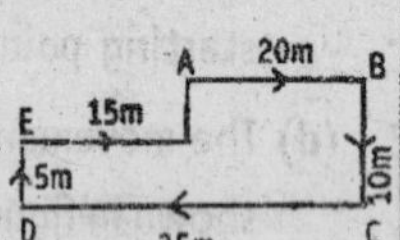

Clearly, DC = AB + EF.

∴ F is in a line with A. Also, AF=(BC – DE)=5 m.

∴ Ravi is 5 metres away from his initial position.

19. **(d)** The movements of Giri are as shown in figure.

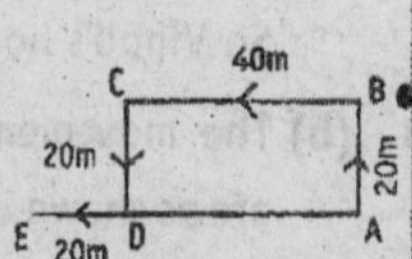

Clearly, Giri's distance from his initial position.

A = AE = (AD + DE) = (BC + DE) = 60 m.

20. **(d)** The movements of Ajay are as shown in figure from A to E.

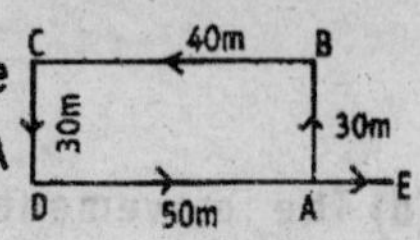

Clearly, Ajay's distance from his original position.

= AE = (DE – DA) = (DE – BC) = 10 m.

21. **(a)** The movements of Ganesh are as shown in figure from A to G.

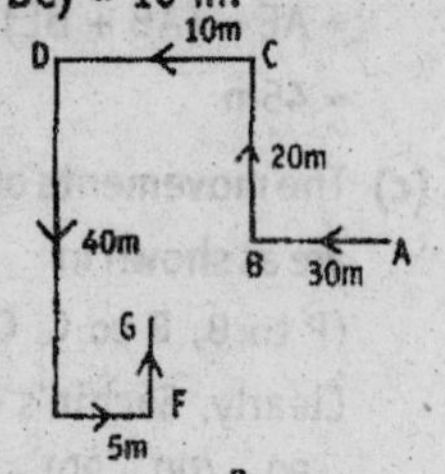

Clearly, Ganesh is walking in the direction FB i.e. North.

22. **(a)** The movements of Sarala are as shown in figure.

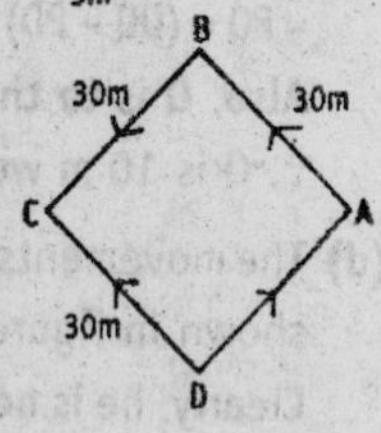

Clearly, she is now moving in the direction DA i.e., North – East.

23. **(b)** The movements of Sanjay are as shown in figure.

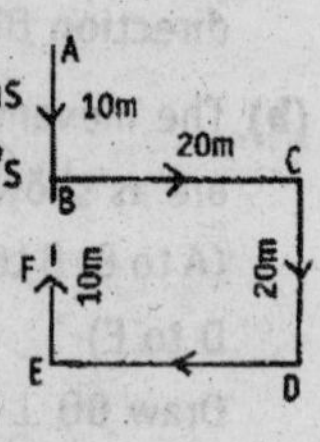

Clearly, Sanjay's distance from initial position.

A = AF = (AB + BF)
= AB + (BE–EF) = AB + (CD–EF)
= (10 + 10) m = 20 m.

Also, F lies to the South of A.

So, Sanjay is 20 metres to the South from his starting point.

24. **(b)** The movements of Kumar are as shown in figure. (A to B, B to C, C to D, D to E).

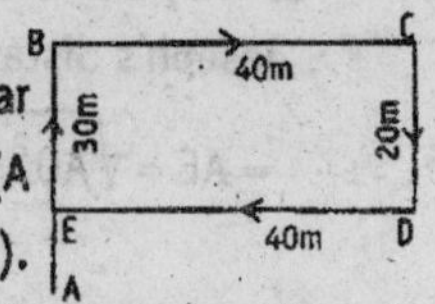

∴ Kumar's distance from his original position A.

= AE = (AB – BE)
= (AB – CD)
= (30 – 20) m = 10 m.

25. **(d)** The movements of Vipul are as shown in figure.

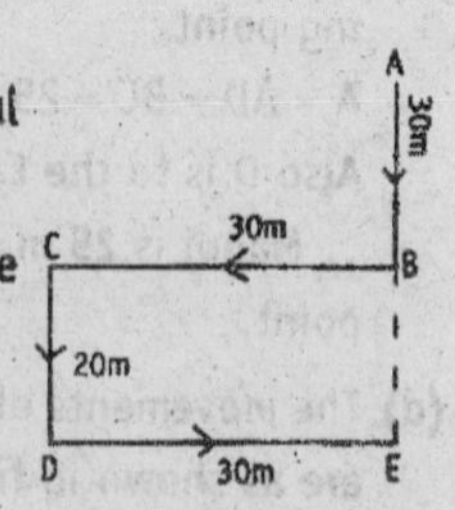

∴ The Vipul's distance from the initial position

= AE = (AB + BE)
= (AB + CD)
= (30 + 20)
= 50 m.

5. TIME SEQUENCE

This concept of 'Time Sequence' is aimed, at testing the ability of the candidate in scheduling and calculating the time point in a given arrangement or sequence of days, hours, minutes etc. Like most of the chapters of 'Analytical Ability' the case in solving these type of problems lies in the ability of connecting the links among the informative points of the given data logically.

There are three basic divisions in time sequence, They are : clocks, calenders and arrivals & departures (schedules).

Here, we have provided all varieties of possible models based on time sequence for exercise with arrivals & departures, along with seperate practice tests for clocks and calenders.

STUDY MATERIAL

1. CLOCKS

The questions under this head covers various questioning patterns based on clocks and hours and minute hands. Mainly they were concentrated on the arrangement of hands of the clock, angles subtended by them, hours and minutes sequencing etc.

2. CALENDARS

In order to fairly go through calenders, the student is required to have through knowledge on dates, days and months, rarely regarding leap years etc. The important point that is to be remembered in this type is that the month February consists of 28 days in an ordinary year and 29 days in a leap year, leap year is that which when divided by 4 leaves the remainder zero.

Examples :

1. If the seventh day of a month is three days earlier than Friday, what day will it be on the nineteenth day of the month ?

   a) Sunday b) Monday c) Wednesday d) Friday

Sol. As mentioned, the seventh day of the month is three days earlier than Friday, which is Tuesday.
So, the fourteenth day is also Tuesday and thus, the nineteenth day is Sunday.
Hence, the answer is (a)

2. If it was Saturday on 17th December, 1982 what will be the day on 22nd December, 1984 ?

   a) Monday b) Tuesday c) Wednesday d) Sunday

Sol. Clearly, every day repeats itself on the seventh day. Now, 17th Dec. 1982– 17th Dec. 1983 is a period of 365 days. Dividing by 7, we get 52 weeks and one day. Thus, the 365th day will be the same as the first day i.e., 16th Dec. 1983 is also Saturday.
Now, 16th Dec, 1983 – 16th Dec, 1984 is a period of 366 days (because 1984, being a leap year, has 29 days in February). Thus, as shown above, 14th Dec. 1984 will be the same as 16th Dec. 1983 i.e, Saturday. So, 21st Dec. 1984 is also saturday and thus, 22nd Dec. 1984 is a Sunday. Hence, the answer is (d)

3. ARRIVAL & DEPARTURE

The problems in this type, tests the candidates presence of mind in sequencing the scheduled timining of trains, buses etc according to their arriving and departing timings. In this exercise you can have a glance of all other possible varieties of problems related to time sequence.

Examples :

1. Sekhar remembers that his brother's birthday is after fifteenth but before eighteenth of February whereas his sister Kiran remembers that her brother's birthday is after sixteenth but before nineteenth of February. On which day in February is Sekhar's brother's birthday?

   a) 16th b) 17th c) 18th d) 19th

Sol. According to Sekhar, the brother's birthday is on one of the days among 16th and 17th February.
According to Kiran, the brother's birthday is on one of the days among 17th and 18th, February.
Clearly, Sekhar's brother's birthday is on the day common to both the above groups i.e., 17th February.
Hence the answer is (b).

2. A train for Shirdi leaves every thirty minutes from a railway station. An enquiry clerk told a passenger that the train had already left ten minutes ago and the next train will leave at 9.35 a.m. At what time did the enquiry clerk give this information to the passenger ?

   a) 9.10 a.m. b) 8.55 a.m. c) 9.08 pm. d) 9.15 a.m.

Sol. The next train will leave at 9.35 a.m. This means that the previous train had left at 9.05 a.m. But it happened ten minutes before the clerk gave the information to the passenger.
Thus, the enquiry clerk gave the information at 9.15 a.m.
Hence, the answer is (d).

PRACTICE TEST - 1

TIME SEQUENCE

1. An alarm clock gives sirens as follows : Once each at 12.30, 1.30, 2.30, 3.30,......etc., as also once at 1.00, twice at 2.00, thrice at 3.00 ....etc. In a span of 24 hours, how many sirens would it give ?

   a) 180 b) 170

   c) 156 d) 200

2. What is the angle between two hands of the clock at 2.15 p.m. ?

   a) 7.5° b) 27.5°

   c) 12.5° d) 22. 5°

3. A clock gains 8 minutes per two hours. It is set right at 10.10 a.m. When the clock shows 4.34 p.m. on the same day, what should be the correct time ?

   a) 4.24 p.m. b) 4.34 p.m.

   c) 4.10 p.m. d) 4.44p.m.

4. A clock gains 15 minutes every two hours. If it is set right at 3.00 P.M. today, what time would it show at 3.00 p.m. tomorrow ?

   a) 10.00 p.m. b) 6.00 p.m.

   c) 7.00 p.m. d) 9.00 p.m.

5. A clock loses 10 minutes every hour. It is set right at 8.00 a.m. today. What would be the correct time when the clock shows 8.00 p.m. today ?

   a) 10.00 p.m b) 8.30 p.m.

   c) 10.30 p.m. d) 6.30 p.m.

6. Two clocks gain 5 minutes and 10 minutes every hour respectively. Both the clocks are set right at 3.00 p.m. today. What would be the difference in the timings shown by the two clocks when the correct time is 3.00 p.m. tomorrow ?

   a) 1 hour

   b) 2 hours

   c) 3 hours

   d) 4 hours

***DIRECTIONS (7–10) :* Read the following information carefully and answer the questions given below it :**

i) Eight doctors P, Q, R, S, T, U, V and W visit a charitable dispensary everyday except on a holiday i.e., Monday.

ii) Each doctor visits for one hour from Tuesday to Sunday except Saturday. The timings are 9 a.m. to 1 p.m. and 2 p.m. to 6 p.m.; 1p.m. to 2 p.m. is lunch break.

iii) On Saturday, it is open only in the morning i.e. 9 a.m. to 1 p.m. and each doctor visits for only half an hour .

iv) No other doctor visits the dispensary before doctor Q and after doctor U.

v) Doctor W comes immediately after lunch break and is followed by R.

vi) S comes in the same order as P in the afternoon session.

7. Doctor P visits in between which of the following pairs of doctors?

   a) S and V b) U and W

   c) R and W d) R and U

8. At What time the visit of doctor R is over on Sunday?

   a) 1 p.m. b) 3 p.m.

   c) 4 p.m. d) 5 p.m.

9. At what time the visit of Doctor T would be over on Saturday ?

   a) 10 a.m.

   b) 11 a.m.

   c) Either 10 a.m. or 11 a.m.

   d) Data inadequate

10. If the lunch break and subsequent visiting hours are reduced by 15 minutes, at what time Doctor U is expected to attend the dispensary ?

    a) 3.15 p.m.

    b) 4 p.m.

    c) 4.15 p.m.

    d) 4.45 p.m.

ANSWERS

1. a	2. d	3. c	4. d	5. c
6. b	7. d	8. c	9. c	10. b

EXPLANATIONS

1. **Ans : [a]**

In a span of 12 hours, total sirens given out by the alarm clock = { (1 + 2 + 3 + ... + 1b) + (1 + 1 + 1 + 1 + ...12 times)

$$= \frac{12\times 3}{2} + 12$$

$$= 78 + 12 = 90.$$

Hence, in the span of 24 hours, the number of sirens would be 180.

2. **Ans : [d]**

See the diagrammatic representation below :

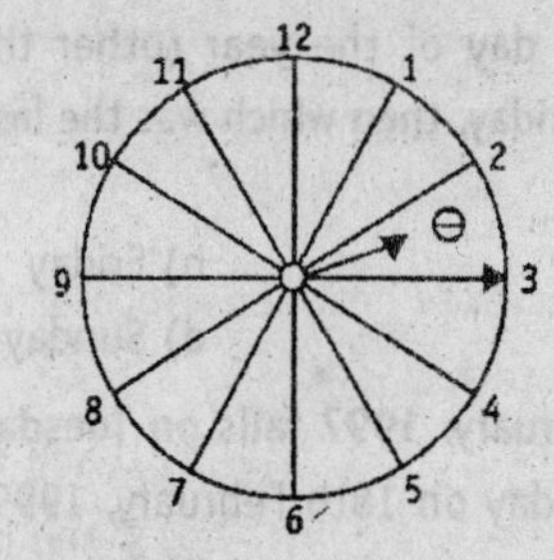

When the minute hand rotates through 90° (one-fourth of 360°) the hour hand would rotate through 7.5° (one – fourth of 30°).

Thus, $\theta = 30° - 7.5° = 22.5°$

3. **Ans : [c]**

Since the clock gains 8 minutes per 2 hours, it goes ahead by 128 minutes when the correct time has gone ahead by 120 minutes. Now, the time gap between 4.34 p.m. & 10.10 a.m. (on the same day is 384 minutes. Thus, when the clock has gone ahead by 384 minutes, the correct time passed must be equivalent to a period of $384 \times 120 \div 128$) = 360 minutes (or 6 hours). Thus, the correct time must be 4.10 p.m.

4. **Ans : [d]**

The clock gains 15 minutes every two hours. Thus in a span of 24 hours, it would gain $(15 \times 24 \div b)$ = 180 minutes; equivalent to 6 hours. Thus, instead of 3.00 p.m.; it would show 9.00 p.m.

5. **Ans : [c]**

When the correct time of 60 minutes is complete, the clock shows only 50 minutes. Thus, we have,

$$50 \text{ minutes} \equiv 60 \text{ minutes}$$

$$(12 \times 60) \text{ minutes} \equiv ?$$

$$\therefore ? = \frac{(12 \times 60) \times 60}{50}$$

$$= 870 \text{ minutes}$$

$$= 14\frac{1}{2} \text{ hours.}$$

Thus, the correct time would be 2.5 hours ahead of 8.00 p.m. Hence, the answer is 10.30 p.m.

6. **Ans : [b]**

In every one hour, the difference in the timings of the two clocks would be five minutes. Thus, in a span of 24 hours, the difference should be 5×24 = 120 minutes or 2 hour.

***QUESTIONS (7-10)* : We first form the Sequence of visit using (iv), (v) and (vi)**

From (iv), we know that Q visits first and U visits last.

From (v), we know that w visits first after break and is followed by R.

From (vi), we know that P visits after break.

Thus, the sequence of visit after break becomes WRPU.

Also, S has the same position in morning session as P in afternoon session. So, Sequence of visit before break is Q, T/V, S, V/T.

7. **Ans : [d]**

Clearly, P visits between R and U.

8. **Ans : [c]**

The time of visit of W is 2 p.m, that of doctor R is 3 p.m. to 4 p.m. So, the visit of doctor R is over at 4 p.m.

9. **Ans : [c]**

clearly, T visits either second or fourth. So, the time of visit on Saturday will be either 9.30 a.m. or 10.30 a.m. Thus, T's visit will be over at either 10 a.m. or 11a.m.

10. **Ans : [b]**

Clearly, as mentioned, lunch break will be over and doctor W will visit at 1.45 p.m., doctor R will visit at 2.30 p.m, doctor P will visit at 3.15 p.m. and U will visit at 4 p.m.

PRACTICE TEST - 2

TIME SEQUENCE

1. An application was received by inward clerk in the afternoon of a week day and next day he forwarded it to the table of the senior clerk, who was on leave that day. The senior clerk next day evening put up the application to the desk officer. Desk officer studied the application and disposed off the matter on the same day i.e., Friday. Which day was the application received by the inward clerk ?
   a) Monday b) Tuesday
   c) Wednesday d) Earlier week's Saturday
2. If the day before yesterday was Thursday, when will Sunday be ?
   a) Today b) Two days after today
   c) Tomorrow d) Day after tomorrow
3. If the day before yesterday was Saturday, what day will fall on the day after tomorrow ?
   a) Friday b) Thursday
   c) Wednesday d) Tuesday
4. Menaka went to the movies nine days ago. She goes to the movies only on Thursday. What day of the week is today ?
   a) Thursday b) Saturday
   c) Sunday d) Tuesday
5. If the third day of a month is Monday, which of the following will be the fifth day from 21st of the month?
   a) Monday b) Tuesday
   c) Wednesday d) Thursday
6. 1.12.91 is the first Sunday. Which is the fourth Tuesday of December 91 ?
   a) 17.12.91 b) 24.12.91
   c) 26.12.91 d) 31.12.91
7. If Thursday was the day after the day before yesterday five days ago, what is the least number of days ago when Sunday was three days before the day after tomorrow ?
   a) Two b) Three
   c) Four d) Five
8. If the 25th of August in a year is Thursday, the number of Mondays in that month is
   a) 3 b) 4
   c) 5 d) 6
9. If 1st October is Sunday, then 1st November will be
   a) Monday b) Tuesday
   c) Wednesday d) Thursday
10. If 3rd December, 1990 is Sunday, what day is 3rd January, 1991 ?
    a) Tuesday b) Wednesday
    c) Thursday d) Friday
11. If February 1, 1996 is Wednesday, what day is March 3, 1996 ?
    a) Monday b) Sunday c) Saturday d) Friday
12. If the first day of the year (other than the leap year) was Friday, then which was the last day of that year ?
    a) Monday b) Friday
    c) Saturday d) Sunday
13. If 18th February, 1997 falls on Tuesday then what will be the day on 18th February, 1999 ?
    a) Monday b) Tuesday
    c) Thursday d) Friday
14. How many days will there be from 26th January, 1996 to 15th May, 1996 (both days included) ?
    a) 110 b) 111 c) 112 d) 113
15. Which two months in a year have the same calendar?
    a) June, October b) April, November
    c) April, July d) October, December
16. Between two book ends in your study are displayed your five favourite puzzle books. If you decide to arrange the five books in every possible combination and moved just one book every minute, how long would it take you ?
    a) 1 hr b) 2 hrs
    c) 3 hrs d) 4 hrs
17. If August 18, 1995 falls on Friday, on which of the following dates would the first Friday of January 1995 fall ?
    a) 4th b) 5th c) 6th d) 7th
18. If yesterday of yesterday was Wednesday, which of the following days would represent tomorrow's yesterday ?
    a) Wednesday b) Thursday
    c) Friday d) Saturday

19. The rainfall on June 15 was less than that on June 13 but more than that on June 21. The rainfall on June 17 was more than that on June 21 and June 15 but less than that on June 12. If the rainfall on June 12 was more than that on June 13, on which of the following days was the rainfall the maximum?

a) June 12 b) June 15 c) June 17 d) June 13

20. If tomorrow of yesterday of yesterday falls on Saturday, which day would be yesterday of tomorrow of yesterday ?

a) Friday b) Sunday c) Saturday d) Thursday

21. If August 15, 1995 falls on Tuesday, on which of the following days would January 26, 1996 fall ?

a) Sunday b) Tuesday c) Thursday d) Friday

22. If May 5, 1996 falls on a Sunday, on which of the following days would May 5, 1997 fall ?

a) Saturday b) Sunday c) Monday d) Thursday

23. If January 08, 1995 falls on Sunday, which day would be four days before the October 17, 1995?

a) Thursday b) Saturday
c) Wednesday d) Friday

24. If December 31, 1995 falls on Sunday, on which of the following days must December 31, 1994 have fallen ?

a) Sunday b) Monday
c) Thursday d) Saturday

ANSWERS

1. c	2. c	3. c	4. b	5. c
6. b	7. a	8. c	9. c	10. b
11. c	12. b	13. c	14. b	15. c
16. b	17. c	18. c	19. a	20. c
21. d	22. c	23. d	24. d	

EXPLANATIONS

1. **Ans : [c]**

Desk officer received the application on Friday. Clearly, the application was forwarded to the table of the senior clerk on Thursday. So, the application was received by the inward clerk on Wednesday.

2. **Ans : [c]**

If day before yesterday was Thursday, So today is Saturday.

∴ Tomorrow will be Sunday.

3. **Ans : [c]**

If day before yesterday was Saturday, so today is Monday. Thus, tomorrow will be Tuesday and day after tomorrow will be Wednesday.

4. **Ans : [b]**

Clearly, nine days ago, it was Thursday.

∴ Today is Saturday.

5. **Ans : [c]**

The 3 rd day is Monday. So, the 10th and 7th days are also Mondays. Thus, the 21st day is Friday.

∴ The fifth day from the 21st will be Wednesday.

6. **Ans : [b]**

1.12.91 is the first sunday of December 91.

So, 3.12. 91 is the first Tuesday of the month.

Clearly, 10.12.91, 17.12.91, 24.12.91 and 31. 12.91 are also Tuesdays.

So, 24. 12. 91 is the fourth Tuesday.

7. **Ans : [a]**

Day after the day before yesterday is yesterday.

Now, five days ago, yesterday was Thursday.

So, five days ago, it was Friday.

∴ Today is Wednesday.

Now, three days before the day after tomorrow is yesterday.

Now, it is on Monday that we say 'Yesterday was Sunday'.

8. **Ans : [c]**

25th August is a Thursday.

So, 22nd August is a Monday.

So, Mondays fall on 1st, 8th, 15th, 22nd and 29th of August.

Thus, there are five Mondays.

9. **Ans : [c]**

Clearly 1st, 8th, 15th, 22nd, and 29th October are Sundays.

So, 31st October is Tuesday.

∴ 1st November will be Wednesday.

10. **Ans : [b]**

Clearly, 3rd, 10th, 17th, 24th and 31st December 1990 are Sundays.

So, 1st January 1991 is Monday and 3rd January 1991 is Wednesday.

11. **Ans : [c]**
1996 is a leap year and so February has 29 days. Now, 1st, 8th, 15th, 22nd and 29th February are Wednesdays. So, 1st March is Thursday and 3rd March is Saturday.

12. **Ans : [b]**
If the year is not a leap year, then the last day of the year is the same as the first day.

13. **Ans : [c]**
18th February, 1997 was Tuesday.
So, 18th February, 1998 was Wednesday.
$\therefore$ 18th February, 1999 will be Thursday.

14. **Ans : [b]**
Number of days = (6 + 29 + 31 + 30 + 15) = 111.
Note : 1988 is a leap year. So, number of days in February = 29

15. **Ans : [c]**
Two months will have the same calendar if the period between them is divisible by 7. Now,
1) June + July + Aug. + Sep. = 30 + 31 + 31 + 30 = 122 (not divisible by 7)
2) Apr. + May + June + July + Aug. + Sep. + Oct.
= 30 + 31 + 30 + 31 + 31 + 30 + 31
= 213 (not divisible by 7)
3) Apr. + May + June = 30 + 31 + 30 = 91 (divisible by 7)
4) October + November = 31 + 30 = 61 (not divisible by 7)

16. **Ans : [b]**
Clearly, number of ways of arranging 5 books = 5 ! = $5 \times 4 \times 3 \times 2 \times 1 = 120$.
So, the total time taken = 120 minutes = 2hrs.

17. **Ans : [c]**
August 18, 1995 falls on Friday. Now, calculate the number of days between 1st January to August 18. They are (31 + 28 + 31+30 + 31 + 30 + 31+18) = 230. Now, (230 – 7) = 32 weeks and remainder as 6. Thus, 1st January must fall on Sunday (counting from Friday, 6 days backwards). Hence, the first Friday must fall on 6th.

18. **Ans : [c]**
Since yesterday of yesterday was Wednesday, today must be Friday. Now, since "tomorrow's" yesterday "implies" "today", it must be Friday.

19. **Ans : [a]**
As per the information given. June 12 would register maximum rainfall whereas June 21, the least one.

20. **Ans : [c]**
"Tomorrow of yesterday of yesterday falls on Saturday" implies "Yesterday was Saturday". Similarly, "Yesterday of tomorrow of Yesterday" also implies "Yesterday". Thus, Saturday is the answer.

21. **Ans : [d]**
Count the number of days as follows : 1) August : 31 – 15 = 16, 2) September = 30 3) October = 31, 4) November = 30, 5) December = 31 and 6) January = 26.
$\therefore$ Total number of days = 16 + 30 + 31 + 30 + 31 + 26 = 164. Now, 164 – 7 $\Rightarrow$ 23 weeks and 3 days (remainder). Thus, January 26, 1996 would fall 3 days ahead of that of August 15, 1995. Hence, from Tuesday, we get Friday.

22. **Ans : [c]**
In one normal year (of 365 days), the same date would fall 1 day ahead in the next year. Thus, May 05, 1995 would fall on Monday. (Note that in case of leap year, the no. of days would go ahead by 2, instead 1).

23. **Ans : [d]**
January 08, 1995 falls on Sunday. Counting the days till October 17, 1995, we get, 1) January = 31 – 08 = 23, 2) February = 28, (3) March = 31, 4) April = 30, 5) May = 31, 6) June = 30, 7) July =31, 8) August =31, 9) September = 30 and 10) October = 17. Thus, total no. of days = (23 + 28 + 31 + 30 + 31 + 30 + 31 +31 + 30 + 17 = 282 ÷ 7 $\Rightarrow$ 40 weeks with remainder as 2. Thus, October 17, 1995 would fall 2 days ahead of Sunday, viz., on Tuesday. Hence, the day which is four days before October 17, 1995 would be Friday.

24. **Ans : [d]**
December 31, 1995 falls on Sunday, the same date of the immediate preceding year falls one day behind (in a normal year of 365 days) December 31, 1994 must have fallen on Saturday. (Note that, in case of December 31, 1993, it must be a Thursday, i.e., two days behind Saturday, since 1994 is a leap year, consisting 366 days)

PRACTICE TEST - 3

TIME SEQUENCE

1. A doctor gives Seeta five tablets asking her to take each one after half an hour. How long will she take to finish all the tablets ?
a) 1.5 hrs. b) 2 hrs. c) 2.5 hrs. d) 3 hrs.

2. Kumar remembers that his brother Dilip's birthday falls after 20th May but before 28th May, while Geeta remembers that Dilip's birthday falls before 22nd May but after 12th May. On what date Dilip's birthday falls ?
a) 20th May b) 21st May
c) 22nd May d) Cannot be determined

3. Geeta remembers that her father's birthday was certainly after eighth but before thirteenth of December. Her sister Asha remembers that their father's birthday was definitely after ninth but before fourteenth of December. On which date of December was their father's birthday ?
a) 10th b) 11th
c) 12th d) Data inadequate

4. Arun leaves his house at 20 minutes to seven in the morning, reaches Kumar's house in 25 minutes, they finish their breakfast in another 15 minutes and leave for their office which takes another 35 minutes. At what time do they leave Kumar's house to reach their office ?
a) 7.40 a.m. b) 7.20 a.m.
c) 7.45 a.m. d) 8.15 a.m.

5. Jay left home for the bus stop 15 minutes earlier than usual. It takes 10 minutes to reach the stop. He reached the stop at 8.40 a.m. What time does he usually leave home for the bus stop ?
a) 8.30 a.m. b) 8.45 p.m. c) 8.55 a.m. d) None

6. Reaching the place of meeting on Tuesday 15 minutes before 08.30 hours, Amar found himself half an hour earlier than the man who was 40 minutes late. What was the scheduled time of the meeting ?
a) 8.00 hrs. b) 8.05 hrs. c) 8.15 hrs. d) 8.45 hrs.

7. The priest told the devotee, "The temple bell is rung at regular intervals of 45 minutes. The last bell was rung five minutes ago. The next bell is due to be rung at 7.45 a.m." At what time did the priest give this information to the devotee ?
a) 7.40 a.m. b) 7.05 a.m.
c) 7.00 a.m. d) 6.55 a.m.

8. The train for Hyderabad leaves every two and a half hours from Vijayawada Railway Station. An announcement was made at the station that the train for Hyderabad had left 40 minutes ago and the next train will leave at 18.00 hrs. At what time was the announcement made?
a) 15.30 hrs. b) 17.10 hrs. c) 16.00 hrs. d) None

9. There are twenty people working in an office. The first group of five works between 8.00 A.M. and 2.00 P.M. The second group of ten works between 10.00 A.M. and 4.00 P.M. And the third group of five works between 12 noon and 6.00 P.M. There are three computers in the office which all the employees frequently use. During which of the following hours the computers are likely to be used most?
a) 10.00 A.M. – 12 noon b) 12 noon – 2.00 P.M.
c) 1.00 P.M. – 3.00 P.M. d) 2.00 P.M. – 4.00 P.M.

10. A climbs 30 feet at the beginning of each hour and rests for a while when he slips back 20 feet before he again starts climbing in the beginning of the next hour. If he begins his ascent at 8.00 a.m., at what time will he first touch a flag at 120 feet from the ground?
a) 4 p.m. b) 5 p.m. c) 6 p.m. d) None

***DIRECTIONS* (11–12) : Study the following information carefully and answer the questions given below it :**

I) Komal is available at home from 12 noon to 4 p.m. on Tuesday, Thursday and Sunday.

II) His younger brother Nitish is available at home on Monday, Thursday, Friday and Sunday between 10 a.m. to 2. p.m.

III) The eldest brother Ravi is available between 9 a.m to 12 noon on Monday, Wednesday and Thursday and 2 p.m. to 4 p.m. on Friday, Saturday and Sunday.

11. At a time, on which day of a week all the three brothers are available at home?
a) None b) Sunday
c) Thursday d) Cannot be determined

12. For how many days only one brother is available at a particular time in a week?
a) One b) Two c) Three d) Four

ANSWERS

1. b	2. b	3. d	4. b	5. d
6. b	7. b	8. d	9. b	10. c
11. a	12. d			

EXPLANATIONS

1. **Ans : [b]**
Seeta takes a tablet.
half hour passes S takes 2nd Tablet
half hour passes S takes 3rd Tablet
half hour passes S takes 4th Tablet
half hour passes S takes 5th Tablet
Total time elapsed = 2 hours

2. **Ans : [b]**
According to Kumar, Dilip's birthday falls on one of the days among 21st, 22nd, 23rd, 24th, 25th, 26th and 27th May.
According to Geeta, Dilip's birthday falls on one of the days among 13th, 14th, 15th, 16th, 17th, 18th, 19th, 20th and 21st May.
The day common to both the groups is 21st May.
Dilip's birthday falls on 21st May.

3. **Ans : [d]**
According to Geeta, the father's birthday falls on one of the days among 9 th, 10th, 11th and 12th December. According to Asha, the father's birthday falls on one of the days among 10th, 11th, 12th and 13th December.
The days common to both the groups are 10th, 11th and 12th December. So, the father's birthday falls on any one of these days.

4. **Ans : [b]**
Arun leaves his house at 6.40 a.m.
He reaches kumar's house in 25 minutes i.e., at 7.05 a.m.
Both leave for office 15 minutes after 7.05 a.m. i.e., at 7.20 a.m.

5. **Ans : [d]**
Clearly, Jay left home 10 minutes before 8.40 a.m. i.e., at 8.30 a.m. But it was 15 minutes earlier than usual. So, he usually left for the stop at 8.45 a.m.

6. **Ans : [b]**
Amar reached the place at 08.15 hours.
Clearly, the man who was 40 minutes late would reach the place at 08.45 hours.
So, the scheduled time of the meeting was 08.05 hours.

7. **Ans : [b]**
Clearly, the last bell rang 45 minutes before 7.45 a.m. i.e., at 7.00 a.m. But it happened five minutes before the priest gave the information to the devotee. So, the information was given at 7.05 a.m.

8. **Ans : [d]**
Clearly, the last train left two and a half hours before 18.00 hours i.e. at 15. 30 hours. But this happened 40 minutes before the announcement was made. So, the announcement was made at 16.10 hours.

9. **Ans : [b]**
Clearly, the computers would be used most when all the three groups are working simultaneously and this happens during the period 12 noon to 2 p.m.

10. **Ans : [c]**
Clearly, the monkey climbs 10 feet in one hour. So, it will climb upto a height of 90 feet in 9 hours i.e., at 5.00 p.m. It will then ascend a height of 30 feet in the next hour to touch the peak at 6.00 p.m.

QUESTIONS 11- 12 :

We prepare a table as under :

	Mon	Tue	Wed	Thu	Fri	Sat	Sun
9 a.m. to 10 a.m.	R		R	R			
10 a.m. to 12 noon	N, R		R	N, R	N		N
12 noon to 2 p.m.	N	K		K,N	N		K,N
2 p.m. to 4 p.m.		K		K	R	R	K,R

11. **Ans : [a]**
Clearly, all the three brothers are not available at the same time on any day of the week.

12. **Ans : [d]**
Clearly, one brother is available at a particular time on all seven days of the week.

6. SEATING ARRANGEMENTS

The ability in arranging the seating pattern of the given number of persons in the required form of placement, is tested through the concept of 'Seating Arrangements'.

STUDY MATERIAL

These Questions are based on set of informative conditions. The possible arranging patterns regularly questioned in various examinations are rows, columns, matrix forms, circles, squares, pentagons, hexagons, and rarely octagons etc.

Candidates are required to arrange the objects in the required pattern on the basis of the given conditions. The information given in the question is presented in distorted form to create confusion and to test candidate's ability to analyse the information step by step in order to answer the question correctly.

Examples :

i) Read the information carefully and answer the questions based on it.

Five persons are sitting in a row. One of the two persons at the extreme ends is intelligent and other one is fair. A fat person is sitting to the right of a weak person. A tall person is to the left of the fair person and the weak person is sitting between the intelligent and the fat person.

1. Tall person is at which place counting from right ?

   a) First b) Second
   c) Third d) Fourth

2. Person to the left of weak person possesses which of the following characteristics ?

   a) Intelligent b) fat
   c) Fair d) Tall

3. Which of the following persons is sitting at the centre?

   a) Intelligent b) Fat c) Fair d) Weak

Sol. First information given in the question that one of the two persons at the extreme ends is intelligent and other one is fair suggest two figures as shown in fig. (a) and (b).

Fig. (a)

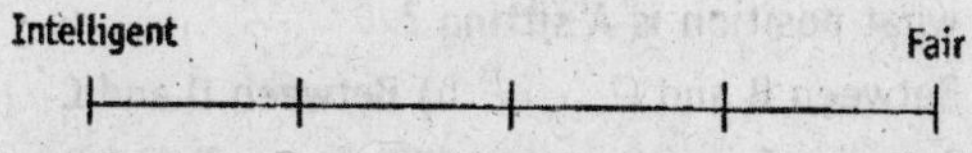

Fig. (b)

Information that a tall person is sitting to the left of fair person rules out the possibility of fig (a) as no person in fig (a) can sit to the left of fair person. Therefore, only fig (b) shows the correct positions of intelligent and fair persons.

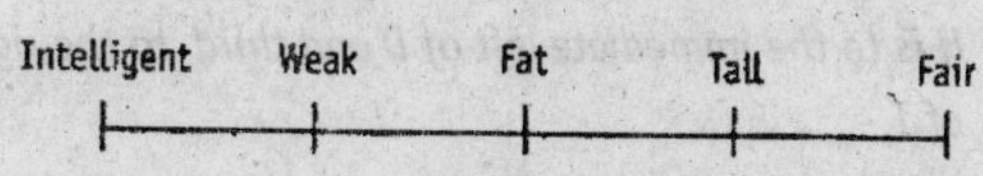

Fig. (c)

Now rest of the information regarding the position of other persons can easily be inserted. The final ranking of their sitting arrangement is as shown in fig. (c).

Sol :

1. Fat person is sitting at the centre. So the answer is (b).
2. Tall person is at second place counting from right. So our answer is option (b).
3. Person left of weak person possesses intelligence. Hence our answer is option (a).

ii) Four men A, B, C, and D and four women W, X, Y and Z are sitting round a table facing each other.

(i) No two men or women are sitting together.
(ii) W is to the right of B.
(iii) Y is facing X and is to the left of A.
(iv) C is to the right of Z.

1. **Who are the two persons sitting adjacent to D ?**

   a) W and Y b) X and W c) X and Z d) W and Z

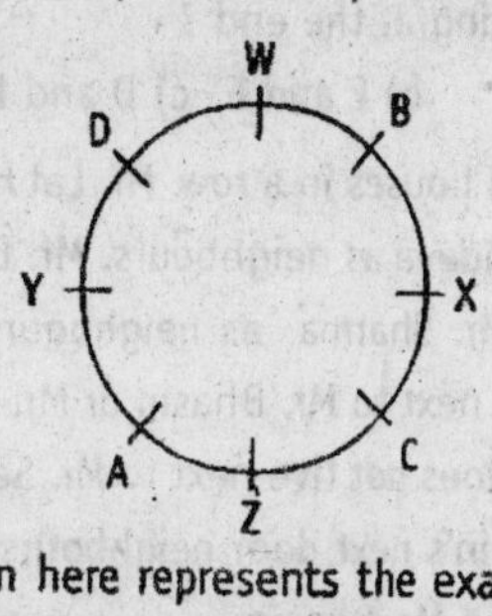

Sol. Figure given here represents the exact position of all the eight persons. The sitting arrangement fulfills all the conditions given in the question. We observe from here that D is sitting between W and Y. Hence our answer is (a).

PRACTICE TEST - 1

SEATING ARRANGEMENTS

***DIRECTIONS (1–5)* : Study the following information carefully and answer the questions given below it :**

i) Eleven students A, B, C, D, E, F, G, H, I, J and K are sitting in a row of the class facing the teacher.

ii) D, who is to the immediate left of F, is second to the right of C.

iii) A, is second to the right of E, who is at one of the ends.

iv) J is the immediate neighbour of A and B and third to the left of G.

v) H is to the immediate left of D and third to the right of I.

1. Who is sitting in the middle of the row ?
   a) C b) I c) B d) G
2. Which of the following groups of friends is sitting to the right of G ?
   a) IBJA b) ICHDF
   c) CHDF d) CHDE
3. In the above sitting arrangement, which of the following statements is superfluous ?
   a) i b) ii
   c) iii d) None is superfluous
4. Which of the following statements is true in the context of the above sitting arrangements ?
   a) There are three students sitting between D and G.
   b) G and C are neighbours sitting to immediate right of H.
   c) B is sitting between J and I.
   d) K is sitting between A and J.
5. If E and D, C and B, A and H &K and F interchange their positions, which of the following pairs of students is sitting at the end ?
   a) D and E b) E and F c) D and K d) K and F
6. There are six houses in a row. Mr. Lal has Mr. Bhasin and Mr. Sachdeva as neighbours. Mr. Bhatia has Mr. Gupta and Mr. Sharma as neighbours. Mr. Gupta's house is not next to Mr. Bhasin or Mr. Sachdeva and Mr. Sharma does not live next to Mr. Sachdeva. Who are Mr. Bhasin's next door neighbours?
   a) Mr. Lal and Mr. Bhasin
   b) Mr. Lal and Mr. Sachdeva
   c) Mr. Sharma and Mr. Lal
   d) Only Mr. Lal

***DIRECTIONS (7–11)* : Read the following information carefully and answer the questions based on it.**

i) Eight rooms P, Q, R, S, T, U, V and W are located adjacent to one another in two different rows, four on each row.

ii) Doors of the rooms of one row open to the doors of the rooms of another row.

iii) No room of consecutive letter is either adjacent or opposite to each other.

iv) P and Q are the rooms located at the end of two different rows.

v) R is just left of T and is located in the same row in which P is located.

vi) U is just left of Q.

7. Which room is just opposite to P ?
   a) S b) U
   c) W d) Can't be determined
8. Which room is diagonally opposite to S ?
   a) Q b) W
   c) V d) Can't be determined.
9. Which room is just opposite to R ?
   a) U b) W
   c) Q d) Can't be determined
10. Which room/(s) is/are adjacent to W ?
    a) R b) U–S
    c) Q–U d) Can't be determined
11. Which of the following rooms is not located at either ends ?
    a) Q b) V c) S d) W

***DIRECTIONS (12–15)* : Read the following information carefully and answer the questions given below it.**

Five friends A, B, C, D and E are sitting on a bench.

i) A is sitting next to B.

ii) C is sitting next to D.

iii) D is not sitting with E.

iv) E is on the left end of the bench.

v) C is on second position from right.

vi) A is on the right side of B and to the right side of E.

vii) A and C are sitting together.

12. At what position is A sitting ?
    a) Between B and C b) Between D and C
    c) Between E and D d) Between C and E

13. **Who is sitting at the centre ?**
a) A b) B c) C d) D

14. What is the position of B ?
a)Second from right b) Centre
c) Extreme left d) Second from left

15. What is the position of D ?
a) Extreme left b) Extreme right
c) Third from left d) Second from left

***DIRECTIONS* (16–20) : Read the following information and answer the questions based on it.**

Ten students A, B, C, D, E, F, G, H, I and J are sitting in a row facing west.

i) B and F are not sitting on either of the edges.
ii) G is sitting to the left of D and H is sitting to the right of J.
iii) There are four persons between E and A.
iv) I is to the north of B and F is to the south of D.
v) J is in between A and D and G is in between E and F.
vi) There are two persons between H and C.

16. **Who is sitting at the seventh place counting from the left ?**
a) H b) C
c) J d) Either H or C

17. **Who among the following is definitely sitting at one of the ends ?**
a) C b) H
c) E d) Can't be determined

18. **Who are the immediate neighbour of I ?**
a) BC
b) BH
c) AH
d) Can't be determined

19. **Who is sitting second left of D ?**
a) G b) F c) E d) J

20. **If G and A interchange their positions, then who become the immediate neighbours of E ?**
a) G and F b) F only
c) A only d) J and H

ANSWERS

1. b	2. c	3. d	4. c	5. c
6. c	7. a	8. c	9. a	10. b
11. d	12. a	13. d	14. a	15. b
16. d	17. c	18. a	19. a	20. c

EXPLANATIONS

(1–5) : Eleven students occupy the position in a row as shown in the figure.

1. **Ans : [b]**
It is clear from the figure drawn on the basis of the information that student I is sitting in the middle of the row.

2. **Ans : [c]**
CHDF are sitting to the right of G.

3. **Ans : [d]**
All informations are required to know the position of all students in the row.

4. **Ans : [c]**
B is sitting between J and I.

5. **Ans : [c]**
D and K will be sitting at the two ends if the positions as given in the question are interchanged.

6. **Ans : [c]**
We see from the fig. that Mr. Bhasin's next door neighbours are Mr. Lal and Mr. Sharma.

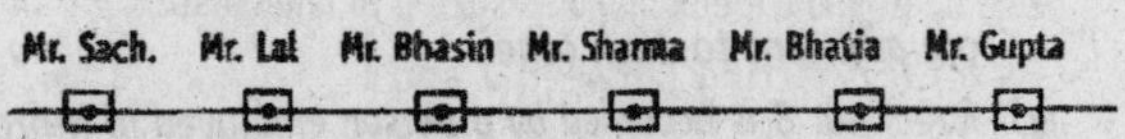

7. **Ans : [a]**
Room S is just opposite to room P.

8. **Ans : [c]**
Room V is diagonally opposite to room S.

9. **Ans : [a]**
Room U is opposite to room R.

10. **Ans : [b]**
Rooms U and S are just adjacent to room W.

11. **Ans : [d]**
Room W is not located at either ends.

(12–16) : Fig. shows the correct position of five persons sitting in a row.

12. **Ans : [a]**
A is sitting between B and C.

E B A C D

13. **Ans : [d]**
A is sitting at the centre of the row.

14. **Ans : [a]**
B is sitting second from the left.

15. **Ans : [b]**
D is sitting at the extreme right of the bench.

16. **Ans : [d]**
Since the exact position of H and C is not known. However, they will occupy either seventh or tenth position from the left end.

17. **Ans : [c]**
E is sitting at the left end.

18. **Ans : [a]**
One of the immediate neighbours of I will either be H or C.

19. **Ans : [a]**
G is sitting to the second left of D.

20. **Ans : [c]**
If positions of G and A are interchanged, A will become the immediate neighbour of E.

PRACTICE TEST - 2
SEATING ARRANGEMENTS

DIRECTIONS (1–6) : Read the following information carefully and answer the questions based on it.

i) Ten rooms numbered 1, 2, 3, 4, 5, 6, 7, 8, 9 and 10 are situated adjacent to one another in two different rows. Doors of the rooms of one row open to the doors of the rooms of another row.

ii) Persons living in different rooms are engaged in different professions M, N, O, P, Q, R, S, T, U and V but not in this order.

iii) Rooms with odd numbers are situated opposite to the rooms with even numbers. And no two rooms with even numbers and no two rooms with odd numbers are adjacent to each other.

iv) Room no. 8 is occupied by a person employed in N profession and his room is to the extreme left of one row.

v) Room no. 4 and 6 are not in the row of room no. 8. And any of the rooms numbered 4, 6, 2, is not in the middle of the row.

vi) Person engaged in profession T lives in the room no. 5 and his room is not situated in the row in which room no. 6 is situated.

(vii) person related to profession O has odd number room and resides in the row in which person engaged in profession N resides but both are not adjacent to each other.

viii) Q's and R's rooms are adjacent to room no. 6. However, R's room is not adjacent to room no. 4

(ix) M's and V's rooms are in the same row.

x) P's room is even numbered and is diagonally opposite to room no. 1

xi) Room no. 6 is second left from its own row.

xii) No room has consecutive number on them and S's room is neither opposite to Q's room nor just adjacent to it.

1. What is the number of room of the person related to profession O ?
a) 10 b) 7
c) 2 d) Can't be determined

2. What is the no. of room in which person engaged in Q resides ?
a) 7 b) 8
c) 9 d) Can't be determined

3. In which company, person who lives just opposite to room no. 5, is employed ?
a) V b) M
c) R d) Can't be determined

4. Which room is opposite to room no. 2 ?
a) 4 b) 9
c) 1 d) None of these

5. Which of the following options represents the correct numbers of rooms of any row ?
a) 2, 8, 10, 9, 5 b) 1, 3, 6, 4, 9
c) 6, 4, 9, 8, 2 d) 1, 3, 9, 5, 6

6. Which of the following statements is not correct ?
a) Persons related with V and O are in the same row.
b) Person related with M lives either in room no. 6 or 4.
c) Q's room is adjacent to V's room.
d) All correct

DIRECTIONS (7–10) : Study the following information to answer the given questions.

i) Eight persons E, F, G, H, I, J, K and L are seated around a square table two on each side.

ii) There are three lady members and they are not seated next to each other.

iii) J is between L and F.

iv) G is between I and F.

v) H, a lady member, is second to the left of J.

vi) F, a male member, is seated opposite to E, a lady member.

vii) There is a lady member between F and I.

7. Who among the following are the three lady members ?
   a) E, H and J b) E, G and J,
   c) G, H and J d) None of these

8. Which of the following is true about J ?
   a) J is a male member
   b) J is a female member
   c) Sex of J cannot be determined
   d) Position of J cannot be determined

9. Who among the following is seated between E and H ?
   a) I b) J
   c) F d) None

10. Who among the following is to the immediate left of F ?
   a) G b) J
   c) I d) Can't be determined

DIRECTIONS (11–13) : Read the information below and answer the questions that follow.

Four ladies A, B, C & D and four gentlemen E, F, G & H are sitting in a circle round a table facing each other.

i) No two ladies or two gentlemen are sitting side by side.

ii) C, who is sitting between G and E, is facing D.

iii) F is between D and A and is facing G.

iv) H is to the right of B.

11. Who is sitting to the left of A ?
   a) E b) F c) G d) H

12. E is facing whom ?
   a) F b) B c) G d) H

13. Who are immediate neighbours of B ?
   a) G and H b) E and F
   c) E and H d) F and H

DIRECTIONS (14–18) : Read the following information and answer the questions :

i) A, B, C, D, E, F, G and H are sitting in a row facing north.

ii) A is fourth to the right of E.

iii) H is fourth to the left of D.

iv) C and F, which are not at the ends, are neighbours of B and E respectively.

v) H is next to the left of A and A is the neighbour of B.

14. What is the position of F ?
   a) Next to the right of E b) Next to the right of G
   c) Sixth to the right of D d) Between G and H

15. Which of the following statements is not true ?
   a) G is the neighbour of H and F
   b) B is next to the right of A
   c) E is at left end
   d) D is next to the right of B

16. Who is/are the neighbours of D ?
   a) F alone b) C alone
   c) B and C d) Can't be determined

17. Which of the following statements is not true ?
   a) H is second to the right of F.
   b) E is fourth to the left of A.
   c) D is fourth to the right of H.
   d) None of these

18. Who are sitting at the ends ?
   a) E and C b) F and D
   c) G and B d) None of these

DIRECTIONS (19–20) : Read the information carefully and answer the following questions.

Eight friends A, B, C, D, E, F, G and H are sitting in a circle facing the centre. B is sitting between G and D, H is third to the left of B and second to the right of A, C is sitting between A & G and B & E are not sitting opposite to each other.

19. Who is third to the left of D ?
   a) F b) E
   c) A d) Can't be determined

20. Which of the following statements is not correct ?
   a) D and A are sitting opposite to each other
   b) C is third to the right of D
   c) E is sitting between F and D
   d) A is sitting between C and F

ANSWERS

1. b	2. c	3. d	4. d	5. b
6. a	7. d	8. a	9. d	10. b
11. b	12. d	13. a	14. a	15. d
16. b	17. d	18. d	19. a	20. c

EXPLANATIONS

(1–6) : On the basis of the information given in the questions, we have the arrangement of rooms and numbers on it as per fig. But room number 4 and 6 may have been occupied by the person with the profession either M and V.

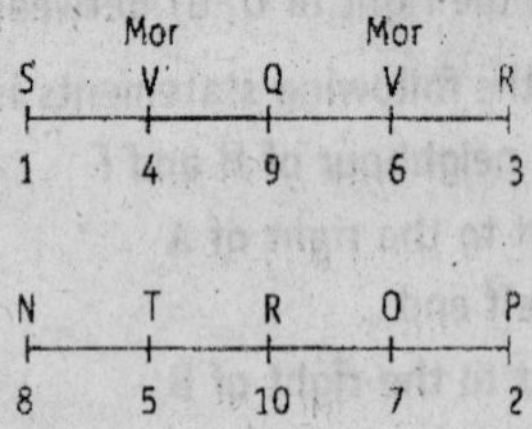

1. **Ans : [b]**
Room number 7 is occupied by the person related with profession N.

2. **Ans : [c]**
Room number 9 is occupied by the person related with profession Q.

3. **Ans : [d]**
We cannot determine it because it can be either M or V.

4. **Ans : [d]**
Room number 3 is opposite to room number 2.

5. **Ans : [b]**
Alternative (b) –1, 3, 6, 4, 9 represents the correct order.

6. **Ans : [a]**
Persons related with profession V and U are in the different rows.

7. **Ans : [d]**
It is clear from the figure that the three lady members are H, E, G, since none of the options shows this combination, hence option (d) is our answer.

8. **Ans : [a]**
It is clear that J is a male member.

9. **Ans : [d]**
K is sitting between E and H.

10. **Ans : [b]**
J is sitting immediately left to F.

(11–13) : On the basis of information given in the question the sitting arrangement of the persons will be as given in the figure.

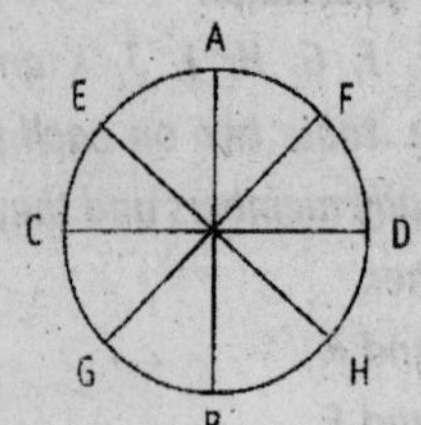

11. **Ans : [b]**
From the Fig. it is clear that F is sitting to the left of A.

12. **Ans : [d]**
E is facing H.

13. **Ans : [a]**
G and H are immediate neighbours of B.

(14 –18) : The order of sitting of all the eight persons which satisfies all the conditions will be as per figure.

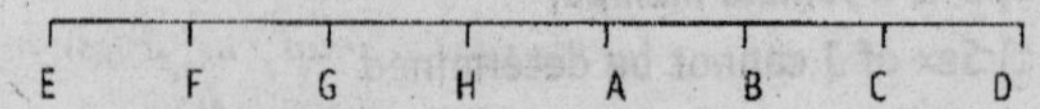

14. **Ans : [a]**
It is clear from the above order that F is next to the right of E.

15. **Ans : [d]**
Statement D is wrong because C is next to the right of B.

16. **Ans : [b]**
Because D is to the extreme right hence has only one neighbour C.

17. **Ans : [d]**
All the statements are correct.

18. **Ans : [d]**
E and D are sitting at the extreme ends.

(19–20) : Figure Shows the correct position of persons sitting around the circular table.

19. **Ans : [a]**
F is third to the left of D.

20. **Ans : [c]**
E is sitting between H and D. Hence, alternative (c) is not correct.

7. SITUATION ANALYSIS

In questions of these types a set of statements providing information on various aspects relating to different variables is given. On the basis of this information we have to establish the correlation between different variables. This establishment can best be demonstrated through examples as follows.

STUDY MATERIAL

Example :

Directions (*Qs. 1 to 5*) Five professors residing at five different cities, teach five different subjects. From the clues, answer the questions that follow :

I. Mr. A does not stay at either Bangalore or Lucknow. He teaches Philosophy.

II. Mr. B does not stay at either Hyderabad or Lucknow. He teaches Maths.

III. Mr. D stays at Jaipur.

IV. Mr. E does not stay either at Bangalore or at Delhi. He teaches Geography.

V. Mr. C does not takes History and he says at Delhi.

1. Who stays at Bangalore ?
   a) Mr. A b) Mr. B
   c) Mr. C d) Mr. D
2. Which of the following subjects does Mr. C teach?
   a) Philosophy b) Maths
   c) History d) Economics
3. At which of the following places does Mr. E teach?
   a) Bangalore
   b) Hyderabad
   c) Lucknow
   d) Delhi
4. Which of the following subjects does Mr. D teach?
   a) Maths
   b) Philosophy
   c) Economics
   d) History
5. Which of the following combinations is wrong?
   a) Mr. A – Hyderabad
   b) Mr. B – Geography
   c) Mr. C – Delhi
   d) Mr. D – History

Procedure in solving Matrix :

Step (1) : In examples of this type, first locate different variables. Here, the variables are :

1. **Number of men :** Five (given) Mr. A, Mr. B, Mr. C, Mr. D, Mr. E.
2. **Number of subjects :** Five (given) Philosophy, History, Maths, Economics and Geography.
3. **Number of cities :** Five (given) - Bangalore, Lucknow. Hyderabad, Delhi & Jaipur.

Prepare a table (twin - Matrix) in the following form:

Step (2)

Section I

	Phil.	Hist.	Maths	Eco.	Geo.
Mr. A					
Mr. B					
Mr. C					
Mr. D					
Mr. E					

Section II

	Bang.	Luck.	Hyd.	Delhi	Jaipur
Mr. A					
Mr. B					
Mr. C					
Mr. D					
Mr. E					

Observe that there are five rows and ten columns (five columns for section I and five columns for section II) in this case.

Step (3) : Now, read the first statement. "Mr. A does not stay at either Bangalore or Lucknow. He teaches Philosophy." The first part of the statement gives two negative facts. Therefore, make a cross mark (x) in Mr. A's row in the columns below "Bangalore" and Lucknow" in Section II. The second part of the statement gives a positive fact. Therefore, make a tick mark (✔) in Mr. A's row and in the column below "Philosophy" in Section I.

The table will now look as follows :

Section I

	Phil.	Hist.	Maths	Eco.	Geo.
Mr. A	✔				
Mr. B					
Mr. C					
Mr. D					
Mr. E					

Section II

	Bang.	Luck.	Hyd.	Delhi	Jaipur
Mr. A	x	x			
Mr. B					
Mr. C					
Mr. D					
Mr. E					

Step (4) : Now, since Mr. A teaches Philosophy, he will not be teaching any other subject. So make four crosses in Mr. A's row after the tick mark (in Section I). Similarly, since it is Mr. A who teaches Philosophy, no one else will be teaching that subject. So make four crosses in the column "Philosophy" below the tick mark (in Section I).

This leads us to the logic that, by the principle of mutual exclusion, whenever a tick is made anywhere in Section I or in Section II, it will automatically generate crosses in its entire row and column. Also, as corollary, whenever we find four crosses in any row or column, the fifth place must be filled in with a tick. This reasoning is to be remembered and upto to practice in all the steps that follow.

The position now looks like this :

Section I

	Phil.	Hist.	Maths	Eco.	Geo.
Mr. A	✔	x	x	x	x
Mr. B	x				
Mr. C	x				
Mr. D	x				
Mr. E	x				

Section II

	Bang.	Luck.	Hyd.	Delhi	Jaipur
Mr. A	x	x			
Mr. B					
Mr. C					
Mr. D					
Mr. E					

Step (5) : The second statement gives rise to two crosses in Mr. B's row in the columns below "Hyderabad" and "Lucknow" (in Section II) and tick mark in Mr. B's row under "Maths" column. Completing the crosses in Mr. B's row and "Maths" column (in section I) brings the position to

Section I

	Phil.	Hist.	Maths	Eco.	Geo.
Mr. A	✔	x	x	x	x
Mr. B	x	x	✔	x	x
Mr. C	x		x		
Mr. D	x		x		
Mr. E	x		x		

Section II

	Bang.	Luck.	Hyd.	Delhi	Jaipur
Mr. A	x	x			
Mr. B		x	x		
Mr. C					
Mr. D					
Mr. E					

Step (6) : The third statement, "Mr. D" stays at Jaipur" generates a tick (✔) mark in Mr. D's row under the column of "Jaipur" in section (II). Completing the 'x' marks in that row (and column) under section (II) the picture looks as follows :

Section I

	Phil.	Hist.	Maths	Eco.	Geo.
Mr. A	✔	x	x	x	x
Mr. B	x	x	✔	x	x
Mr. C	x		x		
Mr. D	x		x		
Mr. E	x		x		

Section II

	Bang.	Luck.	Hyd.	Delhi	Jaipur
Mr. A	x	x			x
Mr. B		x	x		x
Mr. C					x
Mr. D	x	x	x	x	✔
Mr. E					x

Step (7) : The fourth statement, "Mr. E does not stay either at Bangalore or at Delhi. He teaches Geography." Generates a tick (✔) mark in Mr. E's row and the column under "Geography in section (I) and two crosses in Mr. E's row and the columns under Bangalore and Delhi in section (II). Further, it will generate crosses in the row and the column containing the tick (✔) mark under "Geography" of section (I).The picture then looks as follows :

Section I

	Phil.	Hist.	Maths	Eco.	Geo.
Mr. A	✔	x	x	x	x
Mr. B	x	x	✔	x	x
Mr. C	x		x		x
Mr. D	x		x		x
Mr. E	x	x	x	x	✔

Section II

	Bang.	Luck.	Hyd.	Delhi	Jaipur
Mr. A	x	x			x
Mr. B		x	x		x
Mr. C					x
Mr. D	x	x	x	x	✔
Mr. E	x			x	x

Step (8) : The fifth statement "Mr. C does not take History and he stays at Delhi" generates a tick (✔) mark in Mr. C's row and the column under "Delhi" is section (II) and a cross under the column "History" in section (I). It will also generate cross marks in the row and column of the tick (✔) mark under "Delhi" [in Section (II)]. The picture then looks as follows :

Section I

	Phil.	Hist.	Maths	Eco.	Geo.
Mr. A	✔	x	x	x	x
Mr. B	x	x	✔	x	x
Mr. C	x	x	x		x
Mr. D	x		x		x
Mr. E	x	x	x	x	✔

Section II

	Bang.	Luck.	Hyd.	Delhi	Jaipur
Mr. A	x	x		x	x
Mr. B		x	x	x	x
Mr. C	x	x	x	✔	x
Mr. D	x	x	x	x	✔
Mr. E	x			x	x

Step (9) : Now consider the position at step (8) above. In section (I), the four cross marks in C's row would generate tick marks (✔) under the column **"Economics"**. This (✔) mark would generate a cross mark under "Economics" coloumn against D. This cross mark generates a tick mark (✔) under column "History" against D. **In section (II),** the four cross marks against Mr. A generate a tick mark (✔) under the column "Hyderabad" this tick mark under "Hyderabad" generates a cross mark against Mr. E this cross mark generates a tick mark (✔) against Mr. E under the column "Lucknow". The last position in Mr. B's row under column "Bangalore" then gets filled with a tick (✔) mark. The completed table then looks as below :

Section I

	Phil.	Hist.	Maths	Eco.	Geo.
Mr. A	✔	x	x	x	x
Mr. B	x	x	✔	x	x
Mr. C	x	x	x	✔	x
Mr. D	x	✔	x	x	x
Mr. E	x	x	x	x	✔

Section II

	Bang.	Luck.	Hyd.	Delhi	Jaipur
Mr. A	x	x	✔	x	x
Mr. B	✔	x	x	x	x
Mr. C	x	x	x	✔	x
Mr. D	x	x	x	x	✔
Mr. E	x	✔	x	x	x

The questions can now be answered as follows :

1. (b) Mr. B stays at Bangalore.
2. (d) Mr. C teaches Economics.
3. (c) Mr. E stays at Lucknow.
4. (d) Mr. D teaches History.
5. (b) Mr. B Geography is a wrong combination

PRACTICE TEST - 1
SITUATION ANALYSIS

DIRECTIONS (Qs. 1 to 5) : Read the following information carefully. Four different states (S, T, U & V) of a country are the chief producers of five different types of foodgrains (Rice, Wheat, Jowar, Maize, Bajra) These states also have four major industries (Steel, Machine - tools, Cement and Petroleum).

A. The state producing maize has petroleum industry.

B. The state "S" produces wheat only and has cement industry.

C. The state "T" produces neither maize nor bajra.

D. The petroleum industry is in State "U" which is a major producer of rice.

E. The Bajra producing State has machine - tool industry.

1. Which foodgrain does the State T produce ?
   a) Rice   b) Wheat
   c) Jowar   d) Bajra
2. Which state is the chief producer of two different types of food grains ?
   a) S   b) U
   c) T   d) V
3. Which state has the steel industry ?
   a) T   b) U
   c) V   d) S
4. Which of the following combinations is correct ?
   a) Jowar - Machine - Tool
   b) State S - Petroleum
   c) Wheat - Steel
   d) State V - Bajra
5. Which of the following combinations is false?
   a) State S - Wheat
   b) Steel - State T
   c) Rice - Petroleum
   d) State V - Cement

DIRECTIONS (Qs. 6 to 10) : Read the following information carefully and answer the questions given below it.

Pratap, Meena, Uday, Sonia and Vinay live in a Co-operative Housing Society. Three of them occupy positions of secretary, chairman and treasurer and the other two are ordinary members. These 5 persons are engaged in five different professions such as Scientist, Teacher; Accountant, Engineer and Doctor.

A. Pratap is scientist and works as Treasurer in society.

B. Neither of the ladies holds the position of secretary.

C. Uday is a teacher and holds the position of the Chairman.

D. The Secretary is neither Accountant nor Doctor.

E. Neither Vinay or Sonia is Doctor.

6. Who are the ordinary members ?
   a) Meena & Sonia
   b) Meena & Pratap
   c) Vinay & Sonia
   d) Uday & Vinay
7. What is the profession of Vinay ?
   a) Accountant   b) Scientist
   c) Teacher   d) Engineer
8. Who works as the Secretary ?
   a) Meena   b) Uday
   c) Pratap   d) Vinay
9. What is the profession of Meena ?
   a) Engineer   b) Teacher
   c) Doctor   d) Scientist
10. Which of the following combinations is right?
   a) Scientist - Uday
   b) Accountant - Sonia
   c) Chairman - Meena
   d) Treasurer - Teacher

DIRECTIONS (Qs. 11 to 15) : In a Research Institute, certain projects have to be carried out jointly by a pair of a boy and a girl. There are 5 boys (A, B, C, D, and E) and 5 girls (F. G. H. I and J) who are selected for the projects. The four teachers K, L, M, and N will be guiding them. Each teacher should guide atleast one pair of students. Based on the following constraints about the formation of pairs, answer the questions that follow :

A. A can go only with I and teacher K is the guide for them.

B. H cannot form pair with B or C and she doesn't want teacher M or L as the guide either.

C. B cannot form pair with any other girl except G and he wants the guide as L only.

D. E cannot go either with H or with J.

E. C does not want either M or L as the guide.

F. Teachers K & L are to be the guides only for one pair each.

11. Which of the following teachers is the guide for two pairs ?

a) L b) M

c) N d) K

12. With whom the girl F forms the pair ?

a) C b) D

c) A d) E

13. Teacher M is the guide for which of the following pairs of Students ?

a) E and F b) A and I

c) C and J d) B and G

14. Who among the following forms the pair with J ?

a) b b) C

c) A d) E

15. Which of the following combinations is correct ?

a) B - G - k b) C - J - L

c) A - H - K d) D - H - N

ANSWERS

1. c	2. b	3. a	4. d	5. d
6. a	7. d	8. d	9. c	10. d
11. c	12. d	13. a	14. b	15. d

EXPLANATIONS

Questions 1 to 5 :

By the twin matrix procedure, we can prepare following table :

	Food grains					Major Industry			
State	Rice	Wht.	Jr.	Mxe.	Baj.	Sti.	M/ctool	Cem.	Petro.
S	x	✔	x	x	x	x	x	✔	x
T	x	x	✔	x	x	✔	x	x	x
U	✔	x	x	✔	x	x	x	x	✔
V	x	x	x	x	✔	x	✔	x	x

1. **(c)** The state "T" produces jowar.
2. **(b)** The state "U" is a chief producer of two different types of food grains (Viz Rice and Maize).
3. **(a)** The state "T" has steel industry.
4. **(d)** The combination "State V - Bajra" is correct.
5. **(d)** The combination "State V - Cement" is false.

Questions 6 to 10 :

By the twin matrix procedure, we can prepare the following table :

	Position				Professions				
	Set.	chair.	Trea	Ordi.	Sci	Teac.	A/c.	Engg	Dr.
Pratap	x	x	✔	x	✔	x	x	x	x
Meena	x	x	x	✔	x	x	x	x	✔
Uday	x	✔	x	x	x	✔	x	x	x
Sonia	x	x	x	✔	x	x	✔	x	x
Vinay	✔	x	x	x	x	x	x	✔	x

6. **(a)** Meena & Sonia are the ordinary members.
7. **(d)** Vinay is an Engineer.
8. **(d)** Vinay works as the secretary.
9. **(c)** Meena is a Doctor.
10. **(d)** "Accountant - Sonia" is the right combination

Questions 11 to 15 :

By the twin matrix procedure, we can prepare following table :

	Girls					Treachers			
Boys	F	G	H	I	J	K	L	M	N
A	x	x	x	✔	x	✔	x	x	x
B	x	✔	x	x	x	x	✔	x	x
C	x	x	x	x	✔	x	x	x	✔
D	x	x	✔	x	x	x	x	x	✔
E	✔	x	x	x	x	x	x	✔	x

11. **(c)**
12. **(d)**
13. **(a)**
14. **(b)**
15. **(d)**

PRACTICE TEST - 2
SITUATION ANALYSIS

DIRECTIONS (Qs. 1 to 5) : Read the following information carefully and answer the questions given below it.

Five artists (Amit, Beena, Chander, Dolly and Rakesh) have five different types of performances (Actor, Writer, Musician, Singer and Director) and are engaged in three different media (T.V., films and stage).

A. Rakesh directs T.V. serials only.
B. Amit is neither musician nor writer and he works neither for T.V. nor for films.
C. Dolly sings only for films.
D. The male musician works for films only.
E. The female writer neither writes for T.V. nor for films.

1. Which of the following pairs works for stage ?
a) Amit & Dolly b) Amit & Rakesh
c) Beena & Amit d) Beena & Dolly

2. Which media engages only one artist ?
a) T.V. b) Film
c) Stage d) Both TV & film

3. Which of the following is the profession of Amit ?
a) Writer b) Actor
c) Singer d) Musician

4. Which of the following is the musician ?
a) Rakesh b) Amit
c) Dolly d) Chander

5. Which of the following is a wrong combination ?
a) Beena - Writer b) Films - Chander
c) Rakesh - T.V. d) Musician - Dolly

DIRECTIONS (Qs. 6 to 10) : Read the following information carefully and answer the questions given below it.

A University offers five specialisation disciplines (Agriculture, Marketing, Finance, Systems and personnel) for Post - Graduate studies in Management Sciences. Five students (Abhijit, Vijay, Saira, Deepak and Meeta) opt for different specialisations while studying from four different institutes (A, B, C & D).

A. Institute D doesn't provide facility to study Agriculture and Systems Management.
B. Only Abhijit and Deepak have taken Marketing as the specialisation and they are studying in different Institutes.
C. Both the lady students are studying in Institute D.
D. Vijay is the only student who has taken Finance. He is studying in Institute B.
E. Deepak does not study in Institute A.

6. In which of the following institutes is Abhijit studying ?
a) A b) B c) C d) D

7. In which of the following institutes is Deepak studying ?
a) A b) B c) C d) D

8. Which discipline (has n', havn't) not been opted by any student ?
a) Agriculture only
b) Systems only
c) Personnel only
d) Both Agriculture and Systems

9. Which of the following combinations is right?
a) Deepak - Finance
b) Meena - Personnel
c) Marketing - Institute D
d) Vijay - Institute C

10. Which are the specialisations opted by two students each ?
a) Marketing only
b) Personnel only
c) Finance only
d) Both Marketing & Personnel.

DIRECTIONS (Qs. 11 to 15) : Read the following information carefully and answer the questions given below it.

A. In a college having three disciplines (Arts, Science and Commerce). Competitions are to be held for five items (Drama, Music, Debate, Sports and General knowledge Quiz). Four students are selected as the incharge of these five competitions. They are Harish, Smita, Leena and Jatin.
B. The science students is incharge of G.K. Quiz only
C. Harish an arts student, is incharge of Music competition.
D. Leena, a commerce student, is incharge of debate competition.
E. The Drama competition is neither headed by Harish nor Jatin, while Arts students heads both Music & Sports competitions.
F. There are two students of commerce discipline, one of whom heads Drama competition.

11. Who is the incharge of two competitions ?
a) Smita b) Harish
c) Leena d) Jatin

12. Jatin is the in-charge of which of the following competitions ?
    a) Drama b) Music c) Sports d) G.K. Quiz
13. Which two competitions are headed by lady students ?
    a) Music and Drama b) Music and Debate
    c) Sports and Debate d) Drama and Debate
14. Which of the following combinations is right?
    a) Harish - Sports - Science
    b) Music - Arts - Leena
    c) Smita - Commerce - Drama
    d) Commerce - Drama - Leena
15. Which of the following combinations is wrong?
    a) Science - Sports
    b) Harish - Sports
    c) Drama - Smita
    d) Jatin - Science

ANSWERS

1. c	2. a	3. b	4. d	5. d
6. a	7. c	8. d	9. b	10. d
11. b	12. d	13. d	14. c	15. a

EXPLANATIONS

Questions (1 to 5) :

By the twin matrix procedure, we can prepare following table :

Artists	Acto	Writer	Musn	Singer	Director	T.V.	Film	Stage
Amit	✔	x	x	x	x	x	x	✔
Beena	x	✔	x	x	x	x	x	✔
Chander	x	x	✔	x	x	x	✔	x
Dolly	x	x	x	✔	x	x	✔	x
Rakesh	x	x	x	x	✔	✔	x	x

1. **(c)** The pair "Beena & Amit" works for the stage.
2. **(a)** T.V. engages only one artist.
3. **(b)** Amit perfoms as an Actor.
4. **(d)** Chander is the musician.
5. **(d)** The combination "Musician - Dolly" is a wrong combination.

Questions 6 to 10 :

By the twin matrix procedure, we can prepare following table :

	Agri	Mktg.	Fin	Syst.	Perso.	A	B	C	D
Abhijit	x	✔	x	x	x	✔	x	x	x
Vijay	x	x	✔	x	x	x	✔	x	x
Saira	x	x	x	x	✔	x	x	x	✔
Deepak	x	✔	x	x	x	x	x	✔	x
Meeta	x	x	x	x	✔	x	x	x	✔

6. **(a)** A
7. **(c)** C
8. **(d)** Both Agriculture & Systems
9. **(b)** The combination "Meeta Personnel" is right
10. **(d)** Both Marketing & Personnel.

Questions (11 to 15) :

By the twin matrix procedure, we can prepare following table :

Competition	Harish	Smiths	Leena	Jatin	Arts	Sci.	Com.
Drama	x	✔	x	x	x	x	✔
Music	✔	x	x	x	✔	x	x
Debate	x	x	✔	x	x	x	✔
Sports	✔	x	x	x	✔	x	x
GK Quiz	x	x	x	✔	x	✔	x

11. **(b)** Harish is the incharge of two competitions.
12. **(d)** Jatin is the incharge of General Knowledge Quiz competition.
13. **(d)** Drama and debate are the two competitions headed by lady students (Smita and Leena).
14. **(c)** The combination "Smita - Commerce - Drama" is right.
15. **(a)** The combination "Science - Sports" is wrong.

Unit - XVII

Previous Gate Papers

UNIT - XVII PREVIOUS GATE PAPERS

GATE - 2010

COMPUTER SCIENCE & INFORMATION TECHNOLOGY

Duration : Three Hours **Maximum Marks : 100**

Reading the following instructions carefully

1. This are a total of 65 questions carrying 100 marks.
2. Questions 1 through 25 are 1 - mark questions, question 26 through 55 are 2 - mark questions.
3. Questions 48 and 51(2 pairs) common data questions and question pairs (Q.52 and Q.53) and (Q.54 and Q.55) are linked answer questions. The answer to the second question of the above pair depends on the answer to the first question of the pair. If the first question in the linked pair is wrongly answered or is un-attempted, then the answer to the second question in the pair will not be evaluated.
4. Questions 56 - 65 belong to general aptitude(GA). Questions 56-60 will carry 1-mark each, and questions 61-65 will carry 2-marks each. The GA questions will begin on a fresh page.
5. Wrong answers will carry NEGATIVE marks. For Q.1 to Q.25 and Q.56 - Q.60, 1/3 mark will be deducted for each wrong answer. For Q.26 to Q.51, and Q.61 - Q.65, 2/3 mark will be deducted for each wrong answer. The question pairs (Q.52,Q.53) and (Q.54, Q.55) are questions with linked answers. There will be negative marks only for wrong answer to the first question of the linked answer question pair i.e. for Q.52 and Q.54, 2/3 mark will be deducted for each wrong answer. There is no negative marking for Q.53 and Q.55

Q. 1 -25 CARRY ONE MARK EACH

1. Let G = (V, E) be a graph. Define $\xi(G)=\sum_d i_d \times d$, where i_d is the number of vertices of degree d in G. If S and T are two different trees with $\xi(S)=\xi(T)$, then

a) $|S|=2|T|$ b) $|S|=|T|-1$

c) $|S|=|T|$ d) $|S|=|T|+1$

2. Newton - Raphson method is used to compute a root of the equation $x^2-13=0$ with 3.5 as the initial value. The approximation after one iteration is

a) 3.676 b) 3.575 c) 3.607 d) 3.667

3. What is the possible number of reflexive relations on a set of 5 elements ?

a) 2^{10} b) 2^{15} c) 2^{20} d) 2^{25}

4. Consider the set $S=\{1,\omega,\omega^2\}$, where ω and ω^2 are cube roots of unity. If * denotes the multiplication operation, the structure (S, *) forms

a) A group (b) A ring

c) An integral domain d) A field

5. What is the value of $\lim_{n\to\infty}\left(1-\frac{1}{n}\right)^{2n}$?

a) 0 b) e^{-2} c) $e^{-1/2}$ d) 1

6. The minterm expansion of $f(P,Q,R)=PQ+Q\overline{R}+P\overline{R}$ is

a) $m_2+m_4+m_6+m_7$ b) $m_0+m_1+m_3+m_5$

c) $m_0+m_1+m_6+m_7$ d) $m_2+m_3+m_4+m_5$

7. A main memory unit with a capacity of 4 megabytes is built using $1M\times 1-bit$ DRAM chips. Each DRAM chip has 1K rows of cells with 1K cells in each row. The time taken for a single refresh operation is 100 nanoseconds. The time required to perform one refresh operation on all the cells in the memory unit is

a) 100 nanoseconds

b) $100*2^{10}$ nanoseconds

c) $100*2^{20}$ nanoseconds

d) $3200*2^{20}$ nanoseconds

8. P is a 16 - bit signed integer. The 2's complement representation of P is $(F87B)_{16}$. The 2's complement representation of 8*P is

a) $(C3D8)_{16}$ b) $(187B)_{16}$

c) $(F878)_{16}$ d) $(987B)_{16}$

9. The Boolean expression for the output f of the multiplexer shown below is

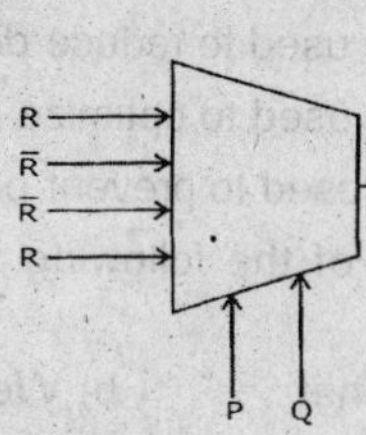

a) $\overline{P\oplus Q\oplus R}$

b) $P\oplus Q\oplus R$

c) $P+Q+R$

d) $\overline{P+Q+R}$

10. In a binary tree with n nodes, every node has an odd number of descendants. Every node is considered to be its own descendant. What is the number of nodes in the tree that have exactly one child ?

a) 0 b) 1 c) (n-1)/2 d) n - 1

11. What does the following program print ?

```
# include <stdio.h>
void f(int *p, int *g) {
    p = q ;
    *p=2;
}
int i =0, j = 1;
```

```
int main( ) {
        f ( &i, &j);
        print f ("%d %d\n", i, j );
        return 0;
}
```

a) 2 2 b) 2 1 c) 0 1 d) 0 2

12. Two alternative packages A and B are available for processing a database having 10^k records. Package A requires $0.0001n^2$ time units and package B requires $10n\log_{10} n$ time units to process n records. What is the smallest value of k for which package B will be preferred over A ?

a) 12 b) 10 c) 6 d) 5

13. Which data structure in a compiler is used for managing information about variables and their attributes?

a) Abstract syntax tree b) Symbol table
c) Semantic stack d) Parse table

14. Which languages necessarily need heap allocation in the runtime environment ?

a) Those that support recursion
b) Those that use dynamic scoping
c) Those that allow dynamic data structures
d) Those that use global variables

15. One of the header fields in an IP datagram is the Time to Live (TTL) field. Which of the following statements best explains the need for this field ?

a) It can be used to prioritize packets
b) It can be used to reduce delays
c) It can be used to optimize throughput
d) It can be used to prevent packet looping

16. Which one of the following is not a client server application ?

a) Internet chat b) Web browsing
c) E - mail d) Ping

17. Let L1 be a recursive language. Let L2 and L3 be languages that are recursively enumerable but not recursive. Which of the following statements is not necessarily true ?

a) L2 - L1 is recursively enumerable
b) L1 - L3 is recursively enumerable
c) $L2 \cap L1$ is recursively enumerable
d) $L2 \cup L1$ is recursively enumerable

18. Consider a B^+ – tree in which the maximum number of keys in a node is 5. What is the minimum number of keys in any non - root node ?

a) 1 b) 2 c) 3 d) 4

19. A relational schema for a train reservation database is given below

Passenger (pid, pname, age)
Re servation (pid, cass, tid)

Table : Passenger

pid	'pname	age
0	'Sachin'	65
1	'Rahul'	66
2	'Sourav'	67
3	'Anil'	69

Table : Reservation

pid	class	tid
0	'AC'	8200
1	'AC'	8201
2	'SC'	8201
5	'AC'	8203
1	'SC'	8204
3	'AC'	8202

What pids are returned by the following SQL query for the above instance of the tables ?

```
SELECT   pid
FROM     Reservation
WHERE    class = 'AC' AND
         EXISTS ( SELECT *
             FROM Passenger
             WHERE age > 65 AND
             Passenger.pid=Reservation.pid)
```

a) 1,0 b) 1,2 c) 1,3 (d) 1,5

20. Which of the following concurrency control protocols ensure both conflict serializability and freedom from deadlock ?

I. 2 - phase locking
II. Time - stamp ordering

a) I only b) II only
c) Both I and II d) Neither I nor II

21. The cyclomatic complexity of each of the modules A and B shown below is 10. What is the cyclomatic complexity of the sequential integration shown on the right hand side ?

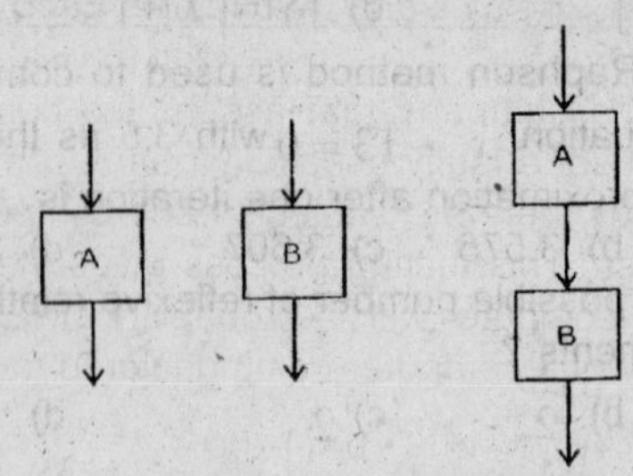

a) 19 b) 21 c) 20 d) 10

22. What is the appropriate pairing of items in the two columns listing various activities encountered in a software life cycle ?

P. Requirements Capture 1. Module Development and Integration
Q. Design 2. Domain Analysis

R. Implementation	3. Structural and Behavioral Modeling
S. Maintenance	4. Performance Tuning

a) P-3, Q-2, R-4, S-1 b) P-2, Q-3, R-1, S-4

c) P-3, Q-2, R-1, S-4 d) P-2, Q-3, R-4, S-1

23. Consider the methods used by processes P1 and P2 for accessing their critical sections whenever needed, as given below. The initial values of shared boolean variables S1 and S2 are randomly assigned.

Method used by PI	Method used by P2
While (S1 = = S2) ;	While (S1 ! = S2);
Critical Section	Critical Section
S1 = S2 ;	S2 = not (S1) ;

Which one of the following statements describes the properties achieved ?

a) Mutual exclusion but not progress

b) Progress but not mutual exclusion

c) Neither mutual exclusion nor progress

d) Both mutual exclusion and progress

24. A system uses FIFO policy for page replacement. It has 4 page frames with no pages loaded to begin with. The system first accesses 100 distinct pages in some order and then accesses the same 100 pages but now in the reverse order. How many page faults will occur ?

a) 196 b) 192 c) 197 d) 195

25. Which of the following statements are true ?

I. Shortest remaining time first scheduling may cause starvation

II. Preemptive scheduling may cause starvation

III. Round robin is better than FCFS in terms of response time

a) I only b) I and III only

c) II and III only d) I, II and III

Q. No. 26 - 51 Carry Two Marks Each

26. Consider a company that assembles computers. The probability of a faulty assembly of any computer is *p*. The company therefore subjects each computer to a testing process. This testing process gives the correct result for any computer with a probability of q. What is the probability of a computer being declared faulty ?

a) pq + (1 - p)(1 - q) b) (1-q)p

c) (1 - p) q d) pq

27. What is the probability that divisor of 10^{99} is a multiple of 10^{96} ?

a) 1/625 b) 4/625 c) 12/625 d) 16/625

28. The degree sequence of a simple graph is the sequence of the degrees of the nodes in the graph in decreasing order. Which of the following sequences can not be the degree sequence of any graph ?

I. 7, 6, 5, 4, 4, 3, 2, 1 II. 6, 6, 6, 6, 3, 3, 2, 2

III. 7, 6, 6, 4, 4, 3, 2, 2 IV. 8, 7, 7, 6, 4, 2, 1, 1

a) I and II b) III and IV

c) IV only d) II and IV

29. Consider the following matrix

$$A = \begin{bmatrix} 2 & 3 \\ x & y \end{bmatrix}$$

If the eigenvalues of A are 4 and 8, then

a) $x = 4, y = 10$ b) $x = 5, y = 8$

c) $x = -3, y = 9$ d) $x = -4, y = 10$

30. Suppose the predicate F(x, y, t) is used to represent the statement that person x can fool person y at time t. Which one of the statements below expresses best the meaning of the formula $\forall x \exists y \exists t(\neg F(x,y,t))$?

a) Everyone can fool some person at some time

b) No one can fool everyone all the time

c) Everyone cannot fool some person all the time

d) No one can fool some person at some time

31. What is the Boolean expression for the output f of the combinational logic circuit of NOR gates given below?

a) $\overline{Q+R}$

b) $\overline{P+Q}$

c) $\overline{P+R}$

d) $\overline{P+Q+R}$

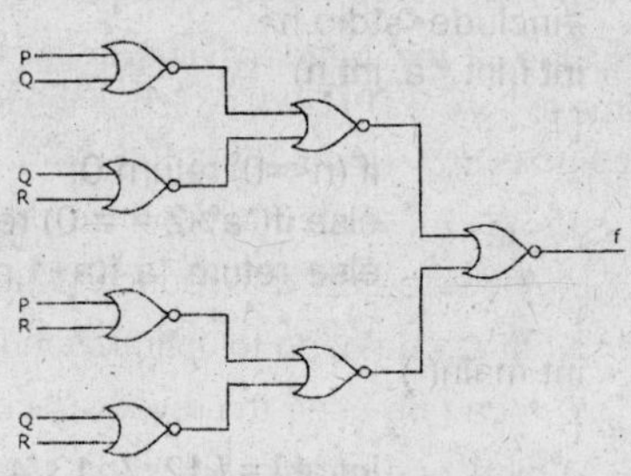

32. In the sequential circuit shown below, if the initial value of the output Q_1Q_0 is 00, what are the next four values of Q_1Q_0 ?

a) 11,10,01,00

b) 10,11,01,00

c) 10,00,01,11

d) 11,10,00,01

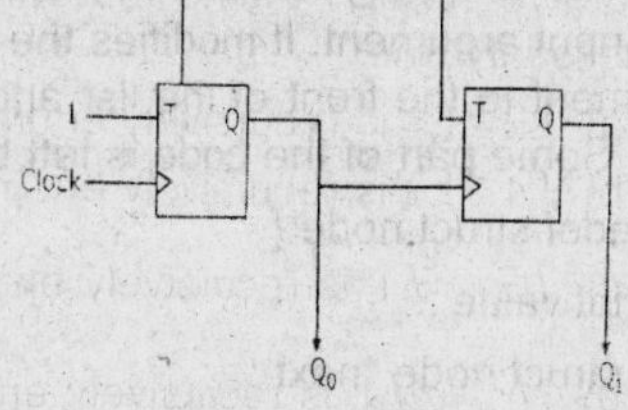

33. A 5 - stage pipelined processor has Instruction Fetch(IF), Instruction Decode(ID), Operand Fetch(OF), Perform Operation (PO) and Write Operand(WO) stages. The IF, ID, OF and WO stages take 1 clock cycle each for any instruction. The PO stage takes 1 clock cycle for

ADD and SUB instructions, 3 clock cycles for MUL instruction, and 6 clock cycles for DIV instruction respectively. Operand forwarding is used in the pipeline. What is the number of clock cycles needed to execute the following sequence of instructions ?

Instruction	Meaning of Instruction
$I_0: MUL R_2, R_0, R_1$	$R_2 \leftarrow R_0 * R_1$
$I_1: DIV\ R_5, R_3, R_4$	$R_5 \leftarrow R_3 / R_4$
$I_2: ADD\ R_2, R_5, R_2$	$R_2 \leftarrow R_5 + R_2$
$I_3: SUB\ R_5, R_2, R_6$	$R_5 \leftarrow R_2 - R_6$

a) 13 b) 15 c) 17 d) 19

34. The weight of a sequence $a_0, a_1, \ldots, a_{n-1}$ of real numbers is defined as $a_0 + a_1/2 + \ldots + a_{n-1}/2^{n-1}$. A subsequence of a sequence is obtained by deleting some elements from the sequence, keeping the order of the remaining elements the same. Let X denote the maximum possible weight of a subsequence of $a_0, a_1, \ldots, a_{n-1}$. Then X is equal to

a) $\max(Y, a_0 + Y)$ b) $\max(Y, a_0 + Y/2)$

c) $\max(Y, a_0 + 2Y)$ d) $a_0 + Y/2$

35. What is the value printed by the following C program?

```
#include<stdio.h>
int f(int *a, int n)
{
        if (n<=0) return 0;
        else if(*a%2 = = 0) return *a+f(a+1,n-1);
        else return *a-f(a+1,n-1);
}
int main( )
{
        int a[ ] = { 12, 7, 13, 4, 11, 6 } ;
        printf("%d", f(a, 6));
        return 0;
}
```

a) -9 b) 5 c) 15 d) 19

36. The following C function takes a simply - linked list as input argument. It modifies the list by moving the last element to the front of the list and returns the modified list. Some part of the code is left blank.

```
typedef struct node {
    int value ;
    struct node *next ;
} Node ;
Node * move_to_front(Node * head) {
    Node *p , *q;
    if (( head = = NULL: || (head->next= = NULL)) return
                                              head ;
    q = NULL: p = head ;
    while (p-> next ! =NULL) {
                q = p;
                p = p -> next ;
    }
    ____________________
    return head ;
}
```

Choose the correct alternative to replace the blank line.

a) q =NULL;p->next=head; head=p;

b) q->next=NULL;head=p; p->next=head;

c) head=p; p->next=q; q->next=NULL;

d) q->next =NULL;p->next=head;head=p;

37. The program below uses six temporary variables a,b,c,d,e,f.

```
a = 1
b = 10
c = 20
d = a + b
e = c + d
f = c + e
b = c + e
e = b + f
d = 5 + e
return d + f
```

Assuming that all operations take their operands from registers, what is the minimum number of registers needed to execute this program without spilling ?

a) 2 b) 3 c) 4 d) 6

38. The grammar S $\rightarrow$ aSa | bS | c is

a) LL(1) but not LR(1) b) LR(1)but not LR(1)

c) Both LL(1) and LR(1) d) Neither LL(1)nor LR(1)

39. Let $L = \{w \in (0+1)^* |$w has even number of 1S $\}$, i.e. L is the set of all bit strings with even number of 1s. Which one of the regular expressions below represents L?

a) $(0^*10^*1)^*$ b) $0^*(10^*10^*)^*$

c) $0^*(10^*1^*)^*0^*$ d) $0^*1(10^*1)^*10^*$

40. Consider the languages $L1 = \{0^i1^j | i \neq j\}$. $L2 = \{0^i1^j | i = j\}$. $L3 = \{0^i1^j | i = 2j+1\}$ $L4 = \{0^i1^j | i \neq 2j\}$. Which one of the following statements is true ?

a) Only L2 is context free

b) Only L2 and L3 are context free

c) Only L1 and L2 are context free

d) All are context free

41. Let w be any string of length n in { 0, 1 }*. Let L be the set of all substrings of w. What is the minimum number of states in a non - deterministic finite automaton that accepts L?

a) n - 1 b) n c) n+1 d) 2^{n-1}

42. Consider the following schedule for transactions T1, T2 and T3 :

T1	T2	T3
Read(x)		
	Read(Y)	
		Read(Y)
	Write(Y)	
Write(x)		
		Write(x)
	Read(x)	
	Write(x)	

Which one of the schedules below is the correct serialization of the above ?

a) $T1 \rightarrow T3 \rightarrow T2$ b) $T2 \rightarrow T1 \rightarrow T3$

c) $T2 \rightarrow T3 \rightarrow T1$ d) $T3 \rightarrow T1 \rightarrow T2$

43. The following functional dependencies hold for relations R(A, B, C) and S (B, D, E)

$B \rightarrow A$,

$A \rightarrow C$

The relation R contains 200 tuples and the relation S contains 100 tuples. What is the maximum number of tuples possible in the natural join R⋈S ?

a) 100 b) 200 c) 300 d) 2000

44. The following program is to be tested for statement coverage :

```
begin
    if (a = = b )          { S1 ; exit ;}
    else if (c = = d)      { S2;}
         else {S3; exit;}
    S4;
end
```

The test cases T1, T2, T3 and T4 given below are expressed in terms of the properties satisfied by the values of variables a, b, c and d. The exact values are not given.

T1 : a, b, c and d are all equal

T2 : a, b, c and d are all distinct

T3 : a = b and c ! = d

T4 : a ! = b and c = d

Which of the test suites given below ensures coverage of statements S1, S2, S3 and S4 ?

a) $T1, T2, T3$ b) $T2, T4$

c) $T3, T4$ d) $T1, T2, T4$

45. The following program consists of 3 concurrent processes and 3 binary semaphores. The semaphores are initialized as S0=1, S1=0, S2=0

Process P0	ProcessP1	ProcessP2
while (true) {	wait(S1);	wait(S2);
wait(S0) ;	Release(S0);	release(S0);
print '0'		
release (S1);		
release(S2);		
}		

How many times will process P0 print '0' ?

a) At least twice b) Exactly twice

c) Exactly thrice d) Exactly once

46. A system has n resources $R_0, \ldots, R_{n-1}$ and k processes $P_0, \ldots, P_{k-1}$. The implementation of the resource request logic of each process P_i is as follows:

```
if (i% 2 = = 0) {
    if (i<n) request R_i ;
    if (i+2<n) request R_{i+2} ;
}
else {
    if (i<n) request R_{n-i} ;
    if ( i+2<n) request R_{n-i-2} ;
}
```

In which one of the following situations is a deadlock possible ?

a) n = 40, k = 26 b) n = 21, k = 12

c) n = 20 , k = 10 d) n = 41, k = 19

47. Suppose computers A and B have IP addresses 10.105.1.113 and 10.105.1.91 respectively and they both use the same net mask N. Which of the values of N given below should not be used if A and B should belong to the same network ?

a) 255.255.255.0 b) 255.255.255.128

c) 255.255.255.192 d) 255.255.255.224

Common Data Questions Q.48 & 49

A computer system has an L1 cache, an L2 cache, and a main memory unit connected as shown below. The block size in L1 cache is 4 words. The block size in L2 cache is 16 words. The memory access times are 2 nanoseconds. 20 nanoseconds and 200 nanoseconds for L1 cache, L2 cache and main memory unit respectively.

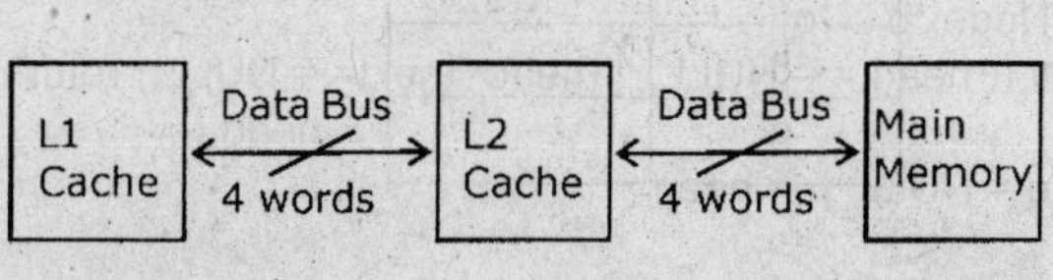

48. When there is a miss in L1 cache and a hit in L2 cache, a block is transferred from L2 cache to L1 cache. What is the time taken for this transfer ?

a) 2 nanoseconds b) 20 nanoseconds
c) 22 nanoseconds d) 88 nanoseconds

49. When there is a miss in both L1 cache and L2 cache, first a block is transferred from main memory to L2 cache, and then a block is transferred from L2 cache to L1 cache. What is the total time taken for these transfers ?

a) 222 nanoseconds b) 888 nanoseconds
c) 902 nanoseconds d) 968 nanoseconds

Common Data Questions : 50 & 51

Consider a complete undirected graph with vertex set { 0, 1, 2, 3, 4}. Entry W_{ij} in

$$W = \begin{pmatrix} 0 & 1 & 8 & 1 & 4 \\ 1 & 0 & 12 & 4 & 9 \\ 8 & 12 & 0 & 7 & 3 \\ 1 & 4 & 7 & 0 & 2 \\ 4 & 9 & 3 & 2 & 0 \end{pmatrix}$$

50. What is the minimum possible weight of a spanning tree T in this graph such that vertex 0 is a leaf node in the tree T ?

a) 7 b) 8 c) 9 d) 10

51. What is the minimum possible weight of a path P from vertex 1 to vertex 2 in this graph such that P contains at most 3 edges ?

a) 7 b) 8 c) 9 d) 10

Linked Answer Questions : Q.52 to Q55 Carry Two marks Each Statement for Linked Answer Questions : 52 53

A hash table of length 10 uses open addressing with hash function h(k)=k mod 10, and linear probing. After insertir g 6 values into an empty hash table, the table is as shown below

0	
1	
2	42
3	23
4	34
5	52
6	46
7	33
8	
9	

52. Which one of the following choices gives a possible order in which the key values could have been inserted in the table ?

a) 46,42, 34, 52, 23, 33
b) 34, 42, 23, 52, 33, 46
c) 46, 34, 42, 23, 52, 33
d) 42, 46, 33, 23,34, 52

53. How many different insertion sequences of the key values using the same hash function and linear probing will result in the hash table shown above ?

a) 10 b) 20 c) 30 d) 40

Statements for Linked Answer Questions : 54 & 55

Consider a network with 6 routers R1 to R6 connected with links having weights as shown in the following diagram

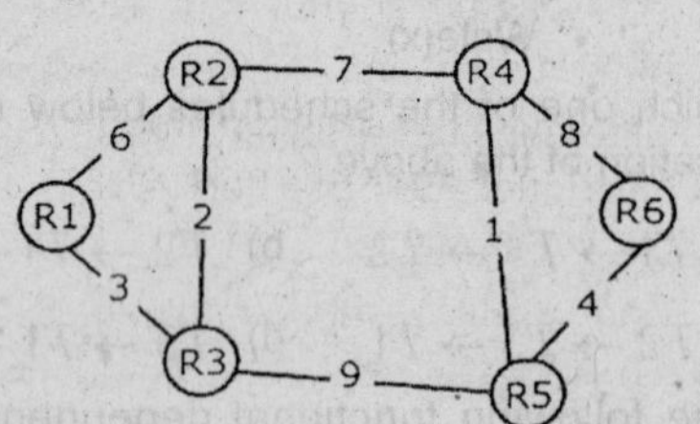

54. All the routers use the distance vector based routing algorithm to update their routing tables. Each router starts with its routing table initialized to contain an entry for each neighbour with the weight of the respective connecting link. After all the routing tables stabilize , how many links in the network will never be used for carrying any data ?

a) 4 b) 3 c) 2 d) 1

55. Suppose the weights of all unused links in the previous question are changed to 2 and the distance vector algorithm is used again until all routing tables stabilize. How many links will now remain unused ?

a) 0 b) 1 c) 2 d) 3

Q. No. 56 - 60 Carry One Mark Each

56. Choose the most appropriate word from the options given below to the complete the following sentence :

His rather casual remarks on politics _______ his lack of seriousness about the subject.

a) masked b) belied
c) betrayed d) suppressed

57. Which of the following options is closest in meaning to the word Circuitous.

a) cyclic b) indirect
c) confusing d) crooked

58. Choose the most appropriate word from the options given below to complete the following sentence :
If we manage to ____ our natural resources, we would leave a better planet for our children.

a) uphold b) restrain c) cherish d) conserve

59. 25 persons are in a room. 15 of them play hockey, 17 of them play football and 10 of them play both hockey and football. Then the number of persons playing neither hockey nor football is :

a) 2 b) 17 c) 13 d) 3

60. The question below consists of a pair of related words followed by four pairs of words. Select the pair that best expresses the relation in the original pair.

Unemployed : Worker

a) fallow : land b) unaware : sleeper

c) wit : jester d) renovated : house

Q.No. 61- 65 Carry Two Marks Each

61. If 137 + 276 = 435 how much is 731 + 672 ?

a) 534 b) 1403 c) 1623 d) 1513

62. Hari (H), Gita(G), Irfan(I) and Saira(S) are siblings (i.e. brothers and sisters). All were born on 1st january. The age difference between any two successive siblings (that is born one after another) is less than 3 years. Given the following facts :

i. Hari's age + Gita' sage>Irfan'sage+Saira's age

ii. The age difference between Gita and Saira is 1 year. However Gita is not the oldest and Saira is not the youngest.

iii. There are no twins.

In what order were they born (oldest first) ?

a) HSIG b) SGHI c) IGSH d) IHSG

63. 5 Skilled workers can build a wall in 20 days : 8 semi-skilled workers can build a wall in 25 days 10 unskilled workers can build a wall in 30 days. If a team has 2 skilled, 6 semi - skilled and 5 unskilled workers, how long will it take to build the wall ?

a) 20 b) 18 c) 16 d)15

64. Modern warfare has changed from large scale clashes of armies to suppression of civilian populations. Chemical agents that do their work silently appear to be suited to such warfare ; and regretfully, there exist people in military establishments who think that chemical agents are useful tools for their cause.

Which of the following statements best sums up the meaning of the above passage :

a) Modern warfare has resulted in civil strife

b) Chemical agents are useful in modern warfare

c) Use of chemical agents in warfare would be undesirable.

d) People in military establishments like to use chemical agents in war

65. Given digits 2, 2, 3, 3, 4, 4, 4, 4 how many distinct 4 digit numbers greater than 3000 can be formed

a) 50 b) 51 c) 52 d) 54

ANSWER KEY

1. c	2. c	3. c	4. a	5. b
6. a	7. d	8. a	9. b	10. a
11. d	12. c	13. b	14. c	15. d
16. d	17. b	18. c	19. c	20. b
21. a	22. b	23. a	24. a	25. d
26. a	27. a	28. d	29. d	30. d
31. a	32. a	33. b	34. a	35. c
36. d	37.	38. c	39. b	40. d
41. c	42. a	43. a	44. d	45. a
46. b	47. d	48. c	49. a	50. d
51. b	52. c	53.	54. d	55. a
56. c	57. b	58. d	59. d	60. a,c
61. c	62. b	63. c	64. d	65. b

EXPLANATIONS

2. $X_{n+1} = X_n - \dfrac{(x_n)}{f'(x_n)}$

$n = 0$

$x_1 = x_0 - \dfrac{f(x_0)}{f'(x_0)}$

given $x_0 = 3.5,\ f(x) = x^2 - 13$

$f(3.5) = (3.5)^2 - 13 = -0.75$

f '(x) = 2x

$f'(3.5) = 2 \times 3.5 = 7$

So after one iteration

$x_1 = x_0 - \dfrac{f(x_0)}{f_1(x_0)} = 3.5 - \dfrac{(-0.75)}{7} = 3.607$

3. Total number of reflexive relations are

$= 2^{n^2-n} = 2^{25-5} = 2^{20}$

4. A group satisfies four properties

1. Closure 2. Associative 3. Identity 4. Inverse

So let $S = \{(1, \omega, \omega^2), *\}$

1. Closure

$1 * \omega = \omega \in S$

$\omega * \omega^2 = 1 \in S$

$\omega^2 * 1 = \omega^2 \in S$

closed under set s

2. Associative

$a * (b * c) = (a * b) * c$

$1 * (\omega * \omega^2) \quad (1 * \omega) * \omega^2$

$= 1 * \omega^3 \qquad = \omega * \omega^2$

$= \omega^3 \qquad = \omega^3$

i. e., associative property hold

3. Identity

$a * e = a \in G,\ \exists e \in G$

$$\left.\begin{array}{c} 1 * \omega = \omega \in G \\ 1 * \omega^2 = \omega^2 \in G \\ 1 * 1 = 1 \in G \end{array}\right\} Identity\ hold$$

4. Inverse

$x * a = e, x, a, e \in G$

$$\left.\begin{array}{l} 1 * 1 = 1 \\ \omega * \omega^2 = 1 \\ \omega^2 * \omega = 1 \end{array}\right] \text{hold so given set is a Group}$$

5. $\lim_{n\to\infty}\left(1-\frac{1}{n}\right)^{2n} = \left(\lim_{n\to\infty}\left(1-\frac{1}{n}\right)^{-n}\right)^{-2} = e^{-2}$

6.

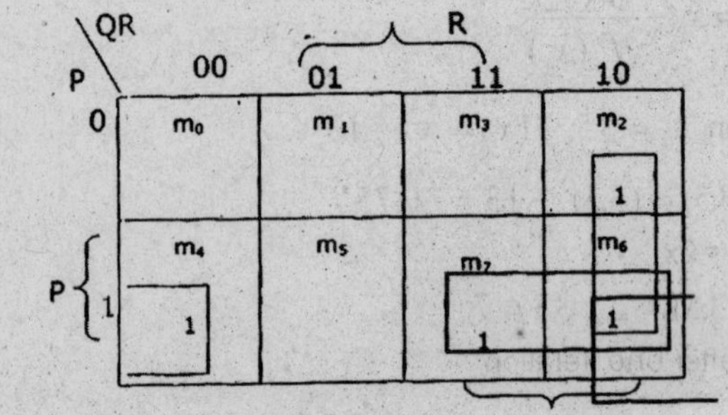

Minterms are 1 at $(m_2 + m_4 + m_6 + m_7)$

$(m_2 m_6)$ Map is for $Q\overline{R}$

$(m_7 m_6)$ Map is for PQ

$(m_4 m_6)$ Map is for $p\overline{R}$

8. $P = (F87B)_{16}$ is -1111 1000 0111 1011 = - 1925

P * 8 = 1925 $\times$ 8 = -15400.

2's complement is 11000011 1101 1000

It is equivalent to $(C3D8)_{16}$

9.

P	Q	F
0	0	R
0	1	$\overline{R}$
1	0	$\overline{R}$
1	1	R

So is

$p'Q'R + p'QR' + pQ'R' + PQR$

$p'(Q'R + QR') + p(Q'R' + QR)$

$p'(Q \oplus R) + p(Q \odot R)$

$= P \oplus Q \oplus R$

15. Whenever time to live field reaches '0' we discard the packet, so that we can prevent it from looping.

17. Union of two recursive enumerable is not R. E and also the Intersection, $L_2 - L_1$ is r.E is True but $L_1 - L_2$ need not to be recursive enumerable

23. Both the processes cannot be there simultaneously in critical section, so it will satisfy mutual exclusion.

24. Since we have 4 page frames, 1 to 100 pages will get 100 faults, But in reverse, there wont be any fault for page No's 100, 99, 98, 97, and for remaining 96 faults, Total = 100 + 96 = 196.

25. Any preemptive scheduling strategy may cause starvation, SRTF is also preemptive SJF. In FCFS late arriving short job have to wait for long time, but in RR, the time QUANTUM expiry will preempt the present job and give the response to the next job in Queue so all are TRUE.

26. Given, the probability of faulty assembly of the computers = p

$\therefore$ The probability of non faulty assembly of the computers = 1 - p

The probability of correct result for any computer in testing process = q

$\therefore$ The probability of correct result for any computer in testing process = 1-q

The probability of a computer being declared faulty = probability of being faulty and probability of being declared or probability of not being faulty and probability of declaring incorrectly. $= pq + (1-p)(1-q)$

27. The divisors of 10^{96} which are multiples of 10^{99} are

$10^{96}, 2\times10^{96}, 4\times10^{96}, 5\times10^{96}, 8\times10^{96}, 10\times10^{96},$

$20\times10^{96}, 25\times10^{96}, 40\times10^{96}, 50\times10^{96}, 100\times10^{96},$

$125\times1096, 200\times1096, 250\times1096,$

$500\times1096, 1000\times10^{96}$

That is a total of 16

The total number of divisors of

$10^{99} = 2^{99}\times5^{99}(99+1)(99+1) = 10000$

The required probability = 16/10000=1/625

29. $A = \begin{bmatrix} 2 & 3 \\ x & 4 \end{bmatrix}$

$$[A - \lambda I] = 0$$

$$\begin{bmatrix} 2-\lambda & 3 \\ x & y-\lambda \end{bmatrix} = 0$$

$$(2-\lambda)(y-\lambda) - 3x = 0$$

given eigen values : $\lambda = 4,8$

if we use $\lambda = 4$ then equation , - 2 (y-4)-3x= 0 ....(i)

if we use $\lambda = 8$ then equation, -6(y-8)-3x=0 ....(ii)

Solving equations (i) & (ii), we get, x = -4, y = 10

31. $\left(\left(\overline{P+Q}\right)+\left(\overline{Q+R}\right)\right)+\left(\left(\overline{P+R}\right)+\left(\overline{Q+R}\right)\right)$

So it can be written as

$$\left(\left(\overline{P+Q}\right)+\left(\overline{Q+R}\right)\right) \cdot \left(\left(\overline{P+R}\right)+\left(\overline{Q+R}\right)\right)$$

$$[\because (A+B)' = A'B']$$

$$(P'Q' + Q'R').(P'R' + Q'R')$$

$$P'Q'P'R' + P'Q'Q'R' + Q'R'P'R' + Q'R'Q'R'$$

$$P'Q'R' + P'Q'R' + Q'P'R' + Q'R' [\because xx = x]$$

$$P'Q'R' + Q'R'$$

$$Q'R'[P'+1] \Rightarrow Q'R' \Rightarrow \left(\overline{Q+R}\right)$$

32. Initially $\begin{matrix} Q_1 & Q_0 \\ 0 & 0 \end{matrix}$

Here T is T - flip flop

If T = 0 the next state is equal to present state

When T = 1 the next state is compliment of present state.

For both flip flops the T input is '1' (constant) the O|P 'Q_0' is given as a clock to the next T - flip flop.

Initially both will be complimented it will become 11,

Next second Q_1 remains 1 since Q_0 became '0' when T i/p is 1, so when clock is not present it remains in same state, ...

So it will go for sequence 11, 10, 01, 00

33. Clocks.

	1	2	3	4	5	6	7	8	9	10	11	12	13	14	15	16
I_0	IF	ID	OF	PO	PO	PO	WO									
I_1		IF	ID	OF	Stall	Stall	PO	PO	PO	PO	PO	PO	WO			
I_2			IF	ID	OF		-	-	-	-	-	-	PO	WO		
I_3				IF	ID	OF	-	-	-	-	-	-	-	PO	WO	

So total 15 cycles required

35.

12	7	13	4	11	6

A = 100 102 104 106 108 110

$$f(a,6) \rightarrow 12 + f(a+1, n+1)$$

$$\downarrow$$

$$7 - f(a+1, n-1) \rightarrow 13 - f(a+1, n-1)$$

$$\downarrow$$

$$4 + f(a+1)n+1)$$

$$\downarrow$$

$$11 - f(a+1, n-1)$$

$$(12 + (7 - (13 - (4 + (11 - (6+0)))))) \downarrow$$

= 15 $\qquad 6 + f(a+1, 0)$

$$\downarrow$$

$$0$$

38. Grammar S = aSa | bS | C

First (S) = { a, b, c }

Follow $(S) = \{a, \$\}$

LL(1)

	a	b	c	$
S	$S \rightarrow aSa$	$S \rightarrow bS$	$S \rightarrow c$	

No multiple entries see this is LL(1) Grammar.

LR(1) : $S \rightarrow asa|bs|C$

1. $S \rightarrow asa$
2. $S \rightarrow bs$
3. $S \rightarrow c$

$I_0 : S' \rightarrow .S, \$$

$S' \rightarrow .asa, \$$

$S \rightarrow .bs, \$$

$S \rightarrow .c, \&$

$I_1 : S \rightarrow s., \$$

$I_2 : S \rightarrow a.sa, \$$

$S \rightarrow .asa, a$

$S \rightarrow .bs, a$

$S \rightarrow .c, a$

$I_3 : S \rightarrow b.s, \$$

$S \rightarrow .asa, \$$

$I_9 : S \rightarrow bs., \$$

$I_{10} : S \rightarrow b.s, \$$

$: S \rightarrow .asa, \$$

$: S \rightarrow .bs, \$$

$: S \rightarrow .c, \$$

$I_{11} : S \rightarrow asa., \$$

$I_{12} : S \rightarrow as.a, a$

$S \rightarrow .bs,\$$ $\quad I_{13}: S \rightarrow bs..a$

$S \rightarrow .c,\$$ $\quad I_{14}: S \rightarrow bs..\$$

$I_4: S \rightarrow c..\$$ $\quad I_{15}: S \rightarrow asa..a$

$I_5: S \rightarrow as.a,\$$

$I_6: S \rightarrow a.sa,a$

$S \rightarrow .asa,a$

$S \rightarrow .bs,a$

$S \rightarrow .c,a$

$I_7: S \rightarrow b.s,a$

$S \rightarrow .asa,a$

$S \rightarrow .bs,a$

$S \rightarrow .c,a$

$I_8: S \rightarrow c.,a$

	a	b	c	$	S
0	S_2	S_3	S_4		1
1					
2	S_6	S_7	S_8		5
3	S_2	S_{10}	S_4		9
4				r_3	
5	S_{11}				
6	S_6	S_7	S_8		12
7	S_6	S_7	S_8		13
8	R_3				
9				r_2	
10	S_2	S_3	S_4		14
11				r_1	
12	S_{15}				
13	r_2				
14				r_2	
15	r_1				

No multiple entries so thus LR(1) also.

39. Every string should contain even number of '1' s. min string are ε, 0, 11, 101, 1010 ...... we get those string from the expression 0*(10* 10*)*

In option C and D Min. string is not ε. In option A we don't get, 0,00,000 ....

40. All the given languages can be recognized by the push - Down Automation

41. Ex : Let w = 01

Substring are 0, 1, 01

NFA is 

If requires Min 3,

So, If string length is 'n' It required n + 1 states.

45. Initially neither P_1 nor P_2 can go, only P_0 can go since It uses the semaphore so whose value is '1'

It prints '0' then, after releasing s_1, s_2, again either P_1 *or* P_2 releases ' S_0 so that it will go to print '0' at least twice

47. Net mask is meant for giving the network address of a given IP address. When both IP - Addresses are in the same network, After performing the mask operation, same Id must be there for both, It will not come when we use 255.255.255.224

10. 105, 1. 113	10. 105. 1. 91
255.255.255.224	255.255.255.224
10. 105. 1. 106.	10. 105. 1. 64

Both are different, for all the others it will show the same Id.

Above Mask operation will be done in binary octants

48. A block to access in L_2 cache requires 20 nanoseconds, and 2 sec to place in L_1 - cache, so is total 22 nanoseconds

49. When there is a miss in both L_1, L_2, It has to access in Main which takes 200 nanoseans, and 20 nanoseconds to transfer to L_2 cache and 2 for L_1 cache, Total = 200 + 20 + 2 = 222

52. 46, 34, 42, 23, 52, 33

hash function h(k) = k mod 10

sol, 46, h(46) =46 mod10=6.(index)

so, 46 will be stored on index 6

Similarly, 34

h(34) = 34 mod 10 = 4

$42 \rightarrow 2$

$23 \rightarrow 3$

$52 \rightarrow 2$ (location is occupied)

so it will be stored next at next rempty location

$33 \rightarrow 3$ (location is occupied)

so it will be stored next at next rempty location

54. In Distance vector, the Router will update its routing tables by exchanging the information from all its neighbours.
After All the routing tables stabilize the routing Table for 'R_1' will not have any entry to Router R_n., so that link will not be used.
So one link.

55. There will be no link unused since the link from R_1 *to* R_n is also filled will weight 2.

56. The key words in the statement are 'casual remarks' and 'lack of seriousness'. The blank should be filled with a word meaning 'showed' or 'revealed'. Hence, 'betrayed' is the correct answer.

57. Circuitous means round about or not direct. Therefore the closest in meaning will be indirect

58. The clue in this sentence is 'If we manage to _____ our natural resources' and 'better planet'. This implies that the blank should be filled by a word which means 'preserve' or 'keep for long time'. Therefore the word 'conserve' is the right answer.

59. Representing the given information in the Venn diagram, we have

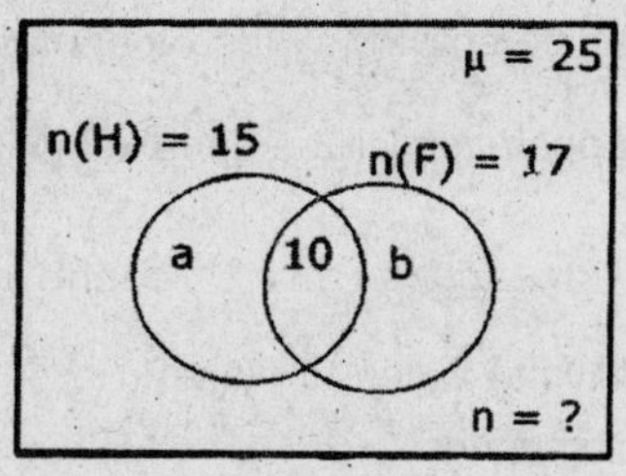

Let, the number of people who play only hockey = a
The number of people who play only football = b
Now, a = n(H) - 10 = 15 - 10 = 5
b = n(F) - 10 = 17 - 10 = 7
Clearly, a + b + 10 + n = 25

$\Rightarrow$ n = 25 - 7 - 5 - 10 $\Rightarrow$ n = 3.

$\therefore$ The number of people who play neither Hockey nor football is 3

60. A worker who is inactive or not working is termed as unemployed, similarly land which is inactive or not in use is called fallow.

61. Given,
137 + 276 = 435
Adding units digits i. e. 7 + 6 = 13, but given as 5, which is 13 - 8 and also 1 is carry forwarded to the tens place.

i. e.,

```
  +1
   7   6
   3   7
  --------
      05
```

Here, 7+3+1=11 i. e., 11 - 8 = 3 and 1 is carry forwarded to hundreds place

```
  +1   +1
   1    7    6
   2    3    7
  --------------
        3    5
```

Now, the sum of digits in hundred's place is
1 + 1 + 2 = 4
i. e.,

```
   1    7    6
   2    3    7
  --------------
   4    3    5
```

Using the same logic, we have

```
        +1
  +1     7    3    1
         6    7    2
  ---------------------
   1     6    2    3
  ---------------------
```

Sum of units digits 1 + 2 = 3, sum of tens digit = 3 + 7 =10 i. e. 10- 2 and +1 carry forward and Sum of hundreds digits = 1+7+6=14 i.e., 14 - 8 = 6 and one carry forward.

62. i) H + G > I + S
ii) | G - S | =1

Meaning G & S will be next to each other in the order. So the option A is ruled out.

G not oldest
S not youngest

iii) No twins.
Going by the options, we will try to solve the equation,
Taking an example with youngest aged 1, we can try to solve the equation,
and correct the age (started with ages 4, 3, 2, 1) to suit condition (i) and (ii) which gives 5, 4, 3, 1

S	4	5 I + 4	Generalizing, we can take their ages in terms of I's age,
G	3	4 I + 3	In this case, H + G > I + S
H	2	3 I + 2	Since 2 I + 5 > 2 I + 4
I	1	1 I	

I 4 In this order, G is always less than I and H is always less than S.
G 3 So G < I and H < S
S 2 Implies G + H < I + S, all values are positive
H 1 Defies condition i) Hence incorrect.

I 4 In this order H < I, G<S
H 3 Hence H + G < I + S
S 2 Defies Condition i)
G 1 Hence incorrect.

63. Among the answer choices, the three options B, C and D can be inferred from the passage. But the main essence of the passage is that chemical agents are being used my military establishments in warfare which is not desirable. Therefore option C is the statement which best sums up the meaning of the passage.

64. Given,

5 skilled workers can build a wall in 20 days i. e., 1 skilled worker can build the same wall in 100 days

$\therefore$ The capacity of ach skilled worker is $\frac{1}{100}$

8 semi - skilled workers can build a wall in 25 days i.e., 1 semi - skilled worker can build the same wall in 200 days

$\therefore$ the capacity of each semi - skilled worker is $\frac{1}{200}$

Similarly, the capacity of 1 unskilled worker is $\frac{1}{300}$

Now, the capacity of 2 skilled +6 semi - skilled +5 unskilled workers is $2\left(\frac{1}{100}\right)+6\left(\frac{1}{200}\right)+\frac{5}{300}$

$$=\frac{2}{100}+\frac{3}{100}+\frac{5}{100}=\frac{20}{300}=\frac{1}{15}$$

$\therefore$ The required numbers of days is 15

65. The given digits are 2, 2, 3, 3, 3, 4, 4, 4, 4, we have to find the numbers that are greater than 3000

$\therefore$ The first digit can be 3 or 4 but not 2.

Now, let us fix the first, second and third digits as 3, 2, 2 and then the fourth place can be filled in 3 ways.

i. e.,

3	2	2	2 or 3 or 4 (3 ways)

$\therefore$ the number of ways is 3.

Similarly, we fix first, third and fourth places as 3, 2 and 2 respectively, so the second place can be filled in 3 ways again

i. e.,

3	2 or 3 or 4	2	2

The number of ways is 3.

Now, we fix first, second and fourth places just as previous cases and we obtain the same result.

$\therefore$ The number of ways is 3 so; the total number of ways is 9.

Similarly this can do by fixing the numbers as 3 and 4 (instead of 2) and thereby we obtain the 9 ways in each case.

$\therefore$ The number of numbers greater than 3000 starting with 3 is 27

Similarly by taking 4 as the first digit and applying the same process, we get 27 numbers

$\therefore$ The total number of numbers that are greater than 3000 is 27 + 27 = 54.

But, 3222 and 4222 is not possible as there are only two 2's (given), 3333 is also not possible as there are only three 3's (given)

$\therefore$ The total number of numbers that are greater than 3000 is 54 - 3 = 51.

GATE - 2009

COMPUTER SCIENCE & INFORMATION TECHNOLOGY

Duration : Three Hours **Maximum Marks : 100**

Reading the following instructions carefully

1. Questions must be answered on Optical Response Sheet (ORS) by darkening the appropriate bubble (marked A, B, C, D) using HB pencil against the question number on the left hand side of the ORS. Each questions has only one correct answer. In case you wish to change an answer, erase the old answer completely. More than one answer bubbled against a question will be treated as an incorrect response.

2. There are a total of 60 questions carrying 100 marks. Questions 1 through 20 are 1 - mark questions, question 21 through 60 are 2-mark questions.

3. Questions 51 through 56(3 Pairs) are common data questions and question pairs(57, 58) and 959, 60) are linked answer questions. The answer to the second question of the above 2 pairs depends on the answer to the first question of the pair. If the first question in the linked pair is wrongly answered or is un - attempted, then the answer to the second question in the pair will not be evaluated.

6. Wrong answers will carry NEGATIVE marks. For Q.1 to Q.20, 1/3 mark will be deducated for each wrong answer. For Q.21 to Q.56, 2/3 mark will be deducated for each wrong answer. The question pairs(Q.57, Q.58), and (Q.29, Q.60) are questions with linked answers. There will be negative marks only for wrong answer to the first question of the linked answer question pair i. e. for Q.57 and Q.59, 2/3 mark will be deduced for each wrong answer. There is no negative marking for Q.58 and Q.60

Q. 1 - 20 CARRY ONE MARK EACH

1. Which one of the following in NOT necessarily a property of a Group ?
   a) Commutativity
   b) Associativity
   c) Existence of inverse for every element
   d) Existence of identity

2. What is the chromatic number of an n - vertex simple connected graph which does not contain any odd length cycle ? Assume $n \geq 2$.
   a) 2 b) 3 c) n -1 d) n

3. Which one of the following is TRUE for any simple connected undirected graph with more than 2 vertices?
   a) No two vertices have the same degree
   b) At least two vertices have the same degree
   c) At least three vertices have the same degree
   d) All vertices have the same degree.

4. Consider the binary relation R = { (x, y), (x,z), (z,x), (z,y)} on the set { x, y, z}. Which one of the following is TRUE?
   a) R is symmetric but NOT antisymmetric
   b) R is NOT symmetric but antisymmetric
   c) R is both symmetric and antisymmetric
   d) R is neither symmetric nor antisymmetric

5. $(1217)_8$ is equivalent to
   a) $(1217)_{16}$ b) $(028F)_{16}$
   c) $(2297)_{10}$ d) $(0B17)_{16}$

6. What is the minimum number of gates required to implement the Boolean function (AB+C) if we have to use only 2 - input NOR gates ?
   a) 2 b) 3 c) 4 d) 5

7. How many 32 K x 1 RAM chips are needed to provide a memory capacity of 256K - bytes ?
   a) 8 b) 32 c) 64 d) 128

8. A CPU generally handles an interrupt by executing an interrupt service routine
   a) As soon as an interrupt is raised
   b) By checking the interrupt register at the end of fetch cycle
   c) By checking the interrupt register after finishing the executing of the current instruction.
   d) By checking the interrupt register at fixed time intervals.

9. In which one of the following page replacement policies, Belady's anomaly may occur ?
   a) FIFO b) Optimal c) LRU d) MRU

10. The essential content(s) in each entry of page table is / are
    a) Virtual page number
    b) Page frame number
    c) Both virtual page number and page frame number
    d) Access right information

11. What is the number of swaps required to sort n elements using selection sort, in the worst case ?
    a) $\theta(n)$ b) $\theta(n \log n)$
    c) $\theta(n^2)$ d) $\theta(n^2 \log n)$

12. S $\rightarrow$ aSa | bSb | a | b ; The language generated by the above grammar over the alphabet { a, b } is the set of
    a) All palindromes
    b) All odd length palindromes
    c) Strings that begin and end with the same symbol
    d) All even length palindromes

13. Which of the following statement(s) is / are correct regarding Bellman - Ford shortest path algorithm ?
    P. Always finds a negative weighted cycle, if one exists.
    Q. Finds whether any negative weighted cycle is reachable from the source.
    a) P only b) Q only
    c) both P and Q d) Neither P nor Q

14. Let π_i be a problem that belongs to the class NP. Then which one of the following is TRUE ?

a) There is no polynomial time algorithm for π_i

b) If π_i can be solved deterministically in polynomial time, then P = NP

c) If π_i is NP - hard, then it is NP - complete

d) π_i may be undecidable

15. Which one of the following languages over the alphabet { 0, 1 } is described by the regular expression : (0 + 1)* 0(0+1)*0(0+1)* ?

a) The set of all strings containing the substring 00

b) The set of all strings containing at most two 0's

c) The set of all strings containing at least two 0's

d) The set of all strings that begin and end with either 0 or 1

16. Which one of the following is FALSE ?

a) There is unique minimal DFA for every regular language

b) Every NFA can be converted to an equivalent PDA

c) Complement of every context - free language is recursive

d) Every nondeterministic PDA can be converted to an equivalent deterministic PDA

17. Match all items in Group 1 with correct options from those given in Group 2.

Group 1	Group 2
P. Regular expression	1. Syntax analysis
Q. Pushdown automata	2. Code generation
R. Dataflow analysis	3. Lexical analysis
S. Register allocation	4. Code optimization

a) P-4, Q-1, R-2, S-3 b) P-3, Q-1, R-4, S-2

c) P-3, Q-4, R-1, S-2 d) P-2, Q-1, R-4, S-3

18. Consider the program below :

```
# include <stdio.h>
int fun (int n, int * f_p) {
        int t, f ;
        if ( n < = 1 ) {
                *f_p=1;
                return 1;
        }
        t = fun (n-1, f_p);
        f = t + *f _ p ;
        * f _ p = t ;
        return f ;
}
int main( ) {
        int x = 15 ;
        printf("%d\n", fun (5, &x));
        return 0;
}
```

The value printed is

a) 6 b) 8 c) 14 d) 15

19. The coupling between different modules of a software is categorized as follows :

I. Content coupling II. Common coupling

III. Control coupling IV. Stamp coupling

V. Data coupling

Coupling between modules can be ranked in the order of strongest (least desirable) to weakest (most desirable) as follows :

a) I-II-III-IV-V b) V-IV-III-II-I

c) I-III-V-II-IV d) IV-II-V-III-I

20. Consider the HTML table definition given below :

<table border =1>

<tr> <td rowspan = 2> ab < /td>

<td colspan=2> cd</td>

</tr>

<tr> <td> ef </td>

<td rowspan=2> gh </td>

</tr>

<tr> <td colspan =2> ik </td>

</tr>

</table>

The number of rows in each column and the number of columns in each row are :

a) $\langle 2,2,3 \rangle$ and $\langle 2,3,2 \rangle$ b) $\langle 2,2,3 \rangle$ and $\langle 2,2,3 \rangle$

c) $\langle 2,3,2 \rangle$ and $\langle 2,3,2 \rangle$ d) $\langle 2,3,2 \rangle$ and $\langle 2,2,3 \rangle$

Q.No. 21 - 56 Carry Two Marks Each

21. An unbalanced dice (with 6 faces, numbered from 1 to 6) is thrown. The probability that the face value is odd is 90% of the probability that the face value is even. The probability of getting any even numbered face is the same.

If the probability that the face is even given that it is greater than 3 is 0.75, which one of the following options is closest to the probability that the face value exceeds 3?

a) 0.453 b) 0.468 c) 0.485 d) 0.492

22. For the composition table of a cyclic group shown below

*	a	b	c	d
a	a	b	c	d
b	b	a	d	c
c	c	d	b	a
d	d	c	a	b

Which one of the following choices is correct ?

a) a, b are generators b) b, c are generators

c) c,d are generators d) d,a are generators

23. Which one of the following is the most appropriate logical formula to represent the statement ? "Gold and silver ornaments are precious ".

The following notations are used :

G(x) : x is a gold ornament

S(x) : x is a silver ornament

P(x) : x is precious

a) $\forall X((P(X) \rightarrow (G(X) \wedge S(X)))$

b) $\forall X((G(X) \wedge S(X)) \rightarrow P(X))$

c) $\exists X((G(X) \wedge S(X)) \rightarrow P(X))$

d) $\forall X((G(X) \vee S(X)) \rightarrow P(X))$

24. The binary operation $\Box$ is defined as follows

P	Q	P $\Box$ Q
T	T	T
T	F	T
F	T	F
F	F	T

Which one of the following is equivalent to $P \vee Q$?

a) $\neg Q \Box \neg P$ b) $P \Box \neg Q$

c) $\neg P \Box Q$ d) $\neg P \Box \neg Q$

25. $\int_0^{\pi/4} (1-\tan x)/(1+\tan x)dx$ evaluates to

a) 0 b) 1 c) ln 2 d) ½ ln 2

26. Consider the following well - formed formulae :

I. $\neg\forall x(P(x))$ II. $\neg\exists x(P(x))$

III. $\neg\exists x(\neg P(x))$ IV. $\neg\exists x(\neg P(x))$

Which of the above are equivalent ?

a) I and III b) I and IV

c) II and III d) II and IV

27. Given the following state table of an FSM with two states A and B, one input and one output :

Present State A	Present State B	Input	Next State A	Next State B	output
0	0	0	0	0	1
0	1	0	1	0	0
1	0	0	0	1	0
1	1	0	1	0	0
0	0	1	0	1	0
0	1	1	0	0	1
1	0	1	0	1	1
1	1	1	0	0	1

If the initial state is A = 0, B=0, what is the minimum length of an input string which will take the machine to the state A =0, B=1 with Output =1?

a) 3 b) 4 c) 5 d) 6

28. Consider a 4 stage pipeline processor. The number or cycles needed by the four instructions I1, I2, I3, I4 in stages S1, S2, S3, S4 is shown below :

	S1	S2	S3	S4
I1	2	1	1	1
I2	1	3	2	2
I3	2	1	1	3
I4	1	2	2	2

What is the number of cycles needed to execute the following loop ?

For (i = 1 to 2) { I1; I2; I3; I4; }

a) 16 b) 23 c) 28 d) 30

29. Consider a 4-way set associative cache (initially empty) with total 16 cache blocks. The main memory consists of 256 blocks and the request for memroy blocks is in the following order :

0, 255, 1, 4, 3, 8, 133, 159, 216, 129, 63, 8, 48, 32, 73, 92, 155.

Which one of the following memory block will NOT be in cache if LRU replacement policy is used ?

a) 3 b) 8 c)129 d) 216

30. Consider a system with 4 types of resources R1 (3 units), R2 (2 units), R3 (3 units), R4 (2 units). A non - preemptive resource allocation policy is used. At any given instance, a request is not entertained if it cannot be completely satisfied. Three processes P1, P2, P3 request the sources as follows if executed independently.

Process P1:	Process P2:	Process P3:
t=0: requests 2 units of R2	t=0: requests 2 units of R3	t=0: requests 1 unit of R4
t=1: requests 1 unit of R3	t=2: requests 1 unit of R4	t=2: requests 2 units of R1
t=3: requests 2 units of R1	t=4: requests 1 unit of R1	t=5: releases 2 units of R1
t=5: releases 1 unit of R2 and 1 unit of R1.	t=6: releases 1 unit of R3	t=7: requests 1 unit of R2
t=7: releases 1 unit of R3	t=8: Finishes	t=8: requests 1 unit of R3
t=8: requests 2 units of R4		t=9: Finishes
t=10: Finishes		

Which one of the following statements is TRUE if all three processes run concurrently starting at time t = 0 ?

a) All processes will finish without any deadlock

b) Only P1 and P2 will be in deadlock

c) Only P1 and P3 will be in a deadlock

d) All three processes will be in deadlock

31. Consider a disk system with 100 cylinders. The requests to access the cylinders occur in following sequence :

4, 34, 10, 7, 19, 73, 2, 15, 6, 20

Assuming that the head is currently at cylinder 50, what is the time taken to satisfy all requests if it takes 1ms to move from one cylinder to adjacent one and shortest seek time first policy is used ?

a) 95 ms b) 119 ms
c) 233 ms d) 276 ms

32. In the following process state transition diagram for a uniprocessor system, assume that there are always some processes in the ready state :

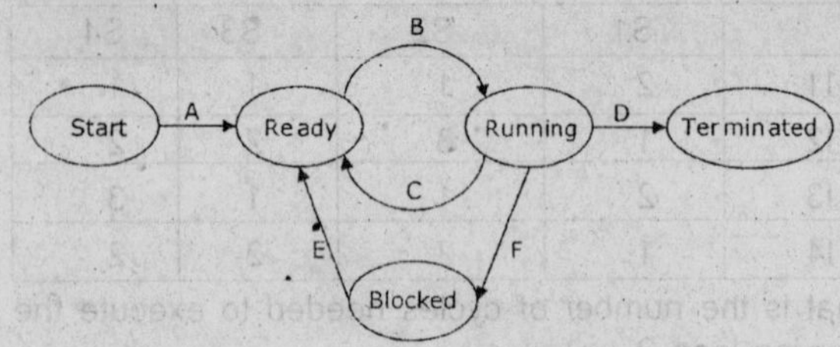

Now consider the following statements :

I. If a process makes a transition D, it would result in another process making transition A immediately.

II. A process P_2 in blocked state can make transition E while another process P_1 is in running state.

III. The OS uses preemptive scheduling.

IV. The OS uses non - preemptive scheduling.

Which of the above statements are TRUE ?

a) I and II b) I and III
c) II and III d) II and IV

33. The enter_SC( ) and Leave_CS( ) functions to implement critical section of a process are realized using test - and - set instruction as follows :

```
void enter_CS(x)
{
    while (test - and - set(x));
}
void leave_CS(x)
{
    x = 0 ;
}
```

In the above solution, x is a memory location associated with the CS and is initialized to 0. Now consider the following statements :

I. The above solution to CS problem is deadlock - free

II. The solution is starvation free.

III. The processes enter CS in FIFO order.

IV. More than one process can enter CS at the same time.

Which of the above statements is TRUE ?

a) I only b) I and II
c) II and III d) IV only

34. A multilevel page table is preferred in comparison to a single level page table for translating virtual address to physical address because

a) It reduces the memory access time to read or write a memory location

b) It helps to reduce the size of page table needed to implement the virtual address space of a process.

c) It is required by the translation lookaside buffer.

d) It helps to reduce the number of page faults in page replacement algorithms

35. The running time of an algorithm is represented by the following recurrence relation :

$$T(n) = \begin{cases} n & n \le 3 \\ T\left(\frac{n}{3}\right) + cn & otherwise \end{cases}$$

Which one of the following represents the time complexity of the algorithm ?

a) $\theta(n)$ b) $\theta(n \log n)$
c) $\theta(n^2)$ d) $\theta(n^2 \log n)$

36. The keys 12, 18, 13, 2, 3, 23, 5 and 15 are inserted into an initially empty hash table of length 10 using open addressing with hash function h(k) = k mod 10 and linear probing. What is the resultant hash table ?

	(A)	(B)	(C)	(D)
0				
1				
2	2	12	12	12,2
3	23	13	13	13,3,23
4			2	
5	15	5	3	5,15
6			23	
7			5	
8	18	18	18	18
9			15	

37. What is the maximum height of any AVL - tree with 7 nodes? Assume that the height of a tree with a single node is 0 .

a) 2 b) 3 c) 4 d) 5

38. Consider the following graph :

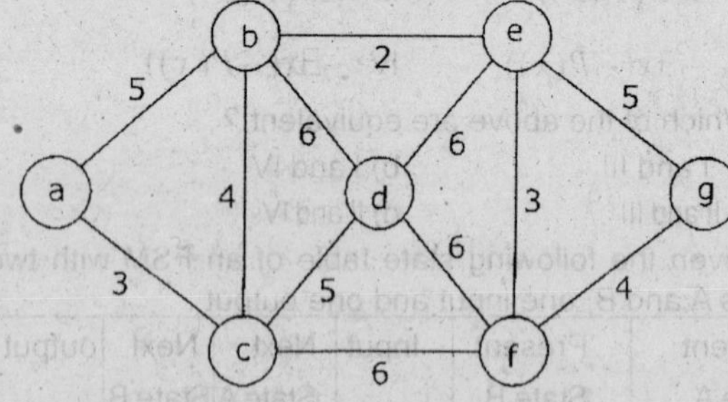

Which one of the following is NOT the sequence of edges added to the minimum spanning tree using Kruskal's algorithm ?

a) (b,e) (e, f) (a, c), (b, c), (f,g) (c, d)

b) (b, e) (e, f) (a, c) (f,g) (b, c) (c,d)

c) (b, e) (a,c) (e, f) (b, c) (f,g) (c,d)

d) (b,e) (e, f) (b,c) (a,c) (f,g) (c,d)

39. In quick sort, for sorting n elements, the $(n/4)^{th}$ smallest element is selected as pivot using an O(n) time algorithm. What is the worst case time complexity of the quick sort ?

a) $\Theta(n)$ b) $\Theta(n \log n)$

c) $\Theta(n^2)$ d) $\Theta(n^2 \log n)$

40. Let $L = L_1 \cap L_2$, where L_1 and L_2 are languages as defined below :

$$L_1 = \{a^m b^m c\, a^n b^m \mid m, n \geq 0\}$$

$$L_2 = \{a^i b^j c^k \mid i, j, k \geq 0\}$$

Then L is

a) Not recursive

b) Regular

c) Context free but not regular

d) Recursively enumerable but not context free.

41.

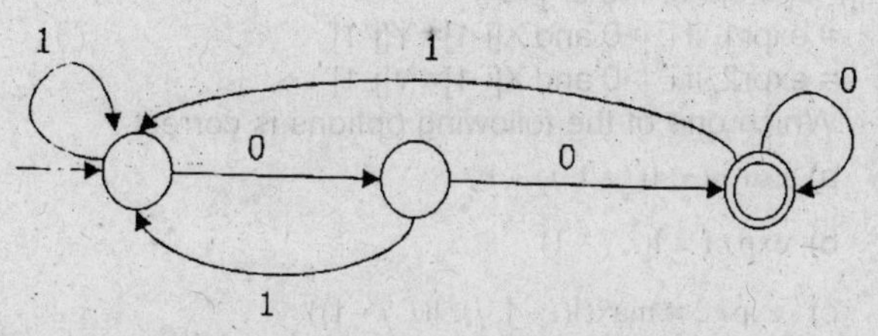

The above DFA accepts the set of all strings over {0, 1} that

a) begin either with 0 or 1

b) end with 0

c) end with 00

d) contain the substring 00

42. Which of the following statements are TRUE ?

I. There exist parsing algorithms for some programming languages whose complexities are less than $\Theta(n^3)$.

II. A programming language which allows recursion can be implemented with static storage allocation.

III. No L - attributed definition can be evaluated in the framework of bottom-up parsing.

IV. Code improving transformations can be performed at both source language and intermediate code level.

a) I and II b) I and IV

c) III and IV d) I, III and IV

43. Consider two transactions T_1 and T_2 and four schedules S_1, S_2, S_3, S_4 of T_1 and T_2 as given below:

$T_1 : R_1[x]W_1[x]W_1[y]$

$T_2 : R_2[x]R_2[y]W_2[y]$

$S_1 : R_1[x]R_2[y]R_2[y]W_1[x]W_1[y]W_2[y]$

$S_2 : R_1[x]R_2[y]R_2[y]W_1[x]W_2[y]W_1[y]$

$S_3 : R_1[x]W_1[y]R_2[x]W_1[y]R_2[y]W_2[y]$

$S_4 : R_2[x]R_2[y]R_1[x]W_1[x]W_1[y]W_2[y]$

Which of the above schedules are conflict - serializable?

a) S_1 *and* S_2 b) S_2 *and* S_3

c) S_3 only d) S_4 only

44. The following key values are inserted into a B + - tree in which order of the internal nodes is 3, and that of the leaf nodes is 2, in the sequence given below. The order of internal nodes is the maximum number of tree pointers in each node, and the order of leaf nodes is the maximum number of data items that can be stored in it. The B+- tree is initially empty.

10, 3,6, 8, 4, 2, 1

The maximum number of times leaf nodes would get split up as a result of these insertions is

a) 2 b) 3 c) 4 d) 5

45. Let R and S be relational schemes such that R = {a,b,c} and S = {c}. Now consider the following queries on the database :

I. $\pi_{R-S}(r) - \pi_{R-S}(\pi_{R-S}(r) \times S - \pi_{R-S,S}(r))$

II. $\{t \mid t \in \pi_{R-S}(r) \wedge \forall u \in s(\exists v \in r(u = v[s] \wedge t = v[R-S]))\}$

III. $\{t \mid t \in \pi_{R-S}(r) \wedge \forall v \in r(\exists u \in s(u = v[s] \wedge t = v[R-S]))\}$

IV. Select R. a, R. b

From R, S

Where R.c = S.c

Which of the above queries are equivalent ?

a) I and II b) I and III

c) II and IV d) III and IV

46. In the RSA public key cryptosystem, the private and public keys are (e,n) and (d, n) respectively, where n=p*q and p and q are large primes. Besides, n is public and p and q are private. Let M be an integer such that 0 <M<n and $\phi(n) = (p-1)(q-1)$. Now consider the following equations.

I. $M' = M^e \bmod n$

$M = (M')^d \bmod n$

II. ed = 1 mod n

III. $ed \equiv 1 \bmod \phi(n)$

IV. $M' = M^e \bmod \phi(n)$

$M = (M')^d \bmod \phi(n)$

Which of the above equations correctly represent RSA cryptosystem ?

a) I and II b) I and III c) II and IV d) III and IV

47. While opening a TCP connection, the initial sequence number is to be derived using a time - of - day (ToD) clock that keeps running even when the host is down. The low order 32 bits of the counter of the ToD clock is to be used for the initial sequence numbers. The clock counter increments once per millisecond. The maximum packet lifetime is given to be 64s.

Which one of the choices given below is closest to the minimum permissible rate at which sequence numbers used for packets of a connection can increase ?

a) 0.015/s b) 0.064/s
c) 0.135/s d) 0.327/s

48. LetG(x) be the generator polynomial used for CRC checking. What is the condition that should be satisfied by G(x) to detect odd number of bits in error ?

a) G(x) contains more than two terms
b) G(x) does not divide $1+x^k$, for any k not exceeding the frame length
c) 1 + x is a factor of G(x)
d) G(x) has an odd number of terms.

49. Which of the following statements are TRUE ?

I. The context diagram should depict the. system as a single bubble.

II. External entities should be identified clearly at all levels of DFDs.

III. Control information should not be represented in a DFD.

IV. A data store can be connected either to another data store or to an external entity.

a) II and III b) II and III
c) I and III d) I,II and III

50. Consider the following statements about the cyclomatic complexity of the control flow graph of a program module. Which of these are TRUE ?

I. The cyclomatic complexity of a module is equal to the maximum number of linearly independent circuits in the graph.

II. The cyclomatic complexity of a module is the number of decisions in the module plus one, where a decision is effectively any conditional statement in the module.

III. The cyclomatic complexity can also be used as a number of linearly independent paths that should be tested during path coverage testing.

a) I and II b) II and III
c) I and III d) I, II and III

Common Data Questions 51 & 52

A hard disk has 63 sectors per track, 10 platters each with 2 recording surfaces and 1000 cylinders. The address of a sector is given as a triple $\langle c,h,s \rangle$, where c is the cylinder number, h is the surface number and s is the sector number. Thus, the 0th sector is addressed as $\langle 0,0,0 \rangle$, the 1st sector as $\langle 0,0,1 \rangle$, and so on

51. The address < 400, 16, 29> corre4sponds tp sector number :

a) 505035 b) 505036
c) 505037 d) 505038

52. The address of the 1039th sector is

a) $\langle 0,15,31 \rangle$ b) $\langle 0,16,30 \rangle$

c) $\langle 0,16,31 \rangle$ d) $\langle 0,17,31 \rangle$

Common Data Questions 53 & 54

A sub-sequence of a given sequence is just the given sequence with some elements (possibly none or all) left out. We are given two sequences X[m] and Y[n] of lengths m and n, respectively, with indexes of X and Y starting from 0.

53. We wish to find the length of the longest common sub - sequence (LCS) of X[m] and Y[n] as l(m,n), where an incomplete recursive definition for the function l(i,j) to compute the length of the LCS of X[m] and Y[n] is given below :

l(i,j)=0, if either i=0 or j=0
= expr1, if i, j>0 and X[i-1]= Y[j-1]
= expr2, if i, j>0 and X[i-1]= Y[j-1]

Which one of the following options is correct ?

a) $\exp r1 \equiv l(i-1,j)+1$

b) $\exp r1 \equiv l(i,j-1)$

c) $\exp r2 \equiv \max(l(i-1,j), l(i,j-1))$

d) $\exp r2 \equiv \max(l(i-1,j-1), l(i,j))$

54. The values of l(i, j) could be obtained by dynamic programming based on the correct recursive definition of l(i, j) of the form given above, using an array L[M,N], where M=m+1 and N=n+1, such that L[i,j] =l(i,j) . Which one of the following statements would be TRUE regarding the dynamic programming solution for the recursive definition of 1 (i, j) ?

a) All elements L should be initialized to 0 for the values of i(i, j) to be properly computed
b) The values of l(i,j) may be computed in a row major order or column major order of L(M,N)
c) The values of l(i,j) cannot be computed in either row major order or column major order of L(M,N)
d) L[p,q] needs to be computed beforeL[r,s] if either p<r or q<s.

COMMON DATA QUESTIONS : 55 & 56

Consider the following relational schema :

Suppliers (sid : integer, sname :string, city:string, street:string)

Parts(pid:integer, pname:string, color:string)

Catalog(sid:integer,pid:integer,cost : real)

55. Consider the following relational query on the above database:

SELECT S.sname

```
FROM         Suppliers S
WHERE S.sid NOT IN(SELECT C.sid
                   FROM Catalog C
       WHERE C.pid NOT (SELECT P.pid
                        FROM Parts P
                        WHERE P. color < > 'blue'))
```

Assume that relations corresponding to the above schema are not empty. which one of the following is the correct interpretation of the above query ?

a) Find the names of all suppliers who have supplied a non - blue part

b) Find the names of all suppliers who have not supplied a non - blue part

c) Find the names of all suppliers who have supplied only blue parts

d) Find the names of all suppliers who have not supplied only blue parts

56. Assume that, in the suppliers relation above, each supplier and each street within a city has a unique name, and (sname, city) forms a candidate key. No other functional dependencies are implied other than those implied by primary and candidate keys. Which one of the following is TRUE about the above schema ?

a) The schema is in BCNF

b) The schema is in 3 NF but not in BCNF

c) The schema is in 2NF but not in 3NF

d) The schema is not in 2NF

Linked Answer Questions : Q.57 to Q.60 Carry Two Marks Each

Statement for Linked Answer Questions : 57 & 58

Frames of 1000 bits are sent over a 10^6 bps duplex link between two hosts. The propagation time is 25 ms. Frames are to be transmitted into this link to maximally pack them in transit (within the link).

57. What is the minimum number of bits(l) that will be required to represent the sequence numbers distinctly? Assume that no time gap needs to be given between transmission of two frames.

a) l = 2   b) l = 3   c) l = 4   d) l = 5

58. Suppose that the sliding window protocol is used with the sender window size of 2^l, where l is the number of bits identified in the earlier part and acknowledgements are always piggy backed. After sending 2^l frames, what is the minimum time the sender will have to wait before starting transmission of the next frame ? (Identify the closest choice ignoring the frame processing time).

a) 16 ms   b) 18 ms

c) 20 ms   d) 22 ms

STATEMENT for Linked Answer Questions : 59 & 60

Consider a binary max - heap implemented using an array.

59. Which one of the following array represents a binary max - heap ?

a) { 25, 12, 16, 13, 10, 8, 14 }

b) { 25, 14, 13, 16, 10, 8, 12 }

c) { 25, 14, 16, 13, 10, 8, 12 }

d) { 25, 14, 12, 13, 10, 8 16 }

60. What is the content of the array after two delete operations on the correct answer to the previous question?

a) { 14, 13, 12, 10, 8 } b) { 14, 12, 13, 8, 10 }

c) { 14, 13, 8, 12, 10 } d) { 14, 13, 12, 8, 10 }

ANSWER KEY

1. a	2. a	3. b	4. d	5. b
6. d	7. c	8. c	9. a	10. b
11. a	12. b	13. b	14. c	15. c
16. d	17. b	18. b	19. a	20. c
21. b	22. c	23. d	24. b	25. d
26. b	27. a	28. b	29. d	30. a
31. b	32. c	33. a	34. b	35. a
36. c	37. b	38. d	39. b	40. c
41. c	42. c	43. b	44. b	45. c
46. b	47. b	48. c	49. c	50. b
51. c	52. c	53. c	54. b	55. a
56. b	57. a	58. b	59. c	60. d

EXPLANATIONS

1. Group is an algebraic system having property of closed under operation, associative, contain identity element and exist inverse of every element.

Alternately

A group(G, *) is a monoid with identity e that the additional property that for every element $a \varepsilon$ G there exist an element a' ε G such that a * a' $= a' \times a = e$. Thus group is set together with binary operation *. It follows property

(i) Associativity : $(a*b)*c = a*(b*c)$

(ii) Inverse for every element a*e = e *a for every a ε G

(iii) Existence of identity : a*a' =a' *a

2. Chromatic Number of n - vertex simple connected undirected graph which does not contain any odd length cycle will be 2(n > = 2).

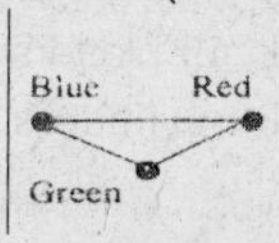

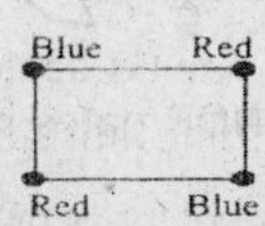

(For n =3 Odd Length Cycle) (For n =4 Even length Cycle)

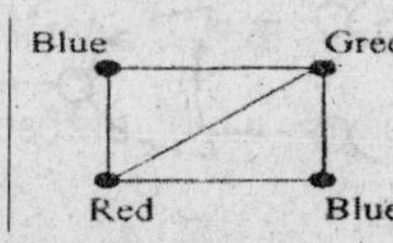

(For n=4 Odd Length Cycle)

3. At least 2 vertices have the same degree.

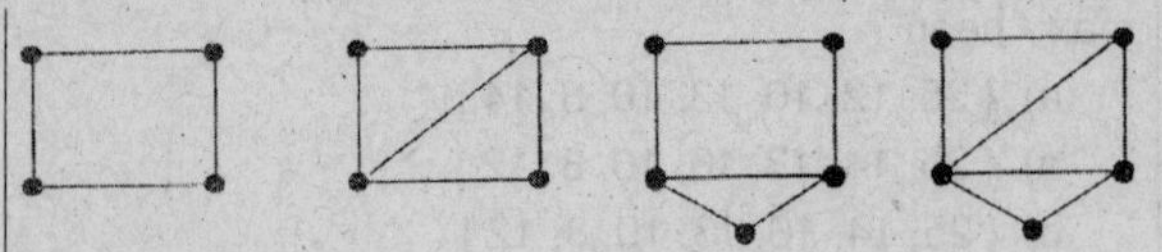

Alternately

For simple corrected graph with more than two vertices at least 2 vertices have the same degree

$n = 3$ 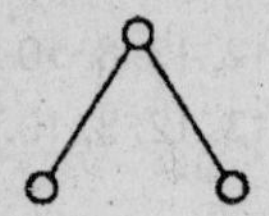 $n = 4$ 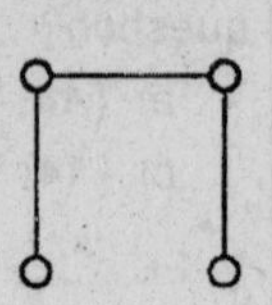

4. $R = \{(x,y),(x,z),(z,x),(z,y)\}$

If (a, b) $\in$ R then $(b,a) \notin R$ to be symmetric but (x,y) $\in R$ and $(y,x) \notin R$

So R is NOT Symmetric

If $(a,b) \in R$ then $(b,a) \in R$ to be Antisymmetric but $(x,z) \in R$ and $(z,x) \in R$

So R is NOT Antisymmetric

5. $(1217)_8 = 001\ 010\ 001\ 111$

$0010\ 10000\ 1111 = (028F)_{16}$

Alternately

Given digit in octal 1217

Convert them in binary $\overline{001}\,\overline{010}\,\overline{001}\,\overline{111}$

Now convert binary to hexa $\dfrac{\overline{001010001111}}{2 \quad 8 \quad F} = (028F)_{16}$

6. For implementing (AB+C) very first we implement AB. We know that for AND operation 3 NOR GATE required

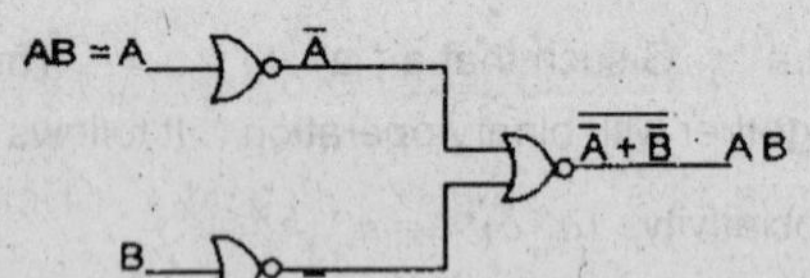

Now for (AB+C)

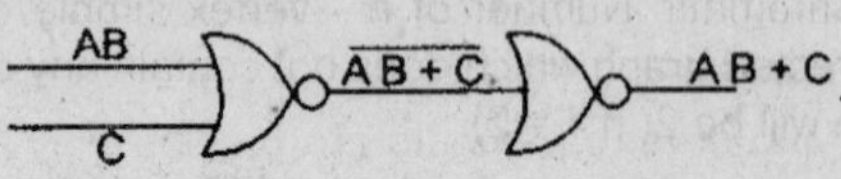

Hence five NOR gates required.

Alternately

7. Number of chips required $= \dfrac{256 * 1024 * 8}{32 * 1024 * 1} = 64$ chips

8. A CPU generally handles an interrupt by executing ISR by checking the interrupt register after finishing the execution of the current instruction.

9. FIFO may have Belady's Anamoly

e.g. In FIFO for three frames as the reference string 1, 2,3,4, 1, 2, 5, 1, 2, 3, 4, 5

there are 9 page faults and foı four frames for the reference string there will be 10 faults.

10. Page frame number is essential content in each entry of the page table.

Alternately

The page table has the ability to mark on entry invalid through a valid invalid bit or special value of protection bit, which is often called page frame number.

11. If there are n elements then is worst case, total swaps will be (n-1) is selection sort. So the number of swaps is $\theta(n)$.

Alternately

The selection sort is similar to bubble sort except it does not swap elements with every move. The sorting algorithm first finds the smallest element in the list and then puts it in to place in single swap. So n swap required for n elements.

12. S -> aSa | bSb | a |b

will generate all odd length palindromes

e.g. S = > aSa => abSba => abbSbba => abbabba.

13. In Bellman Ford shortest path algorithm , we choose a node first and then we find any negative weighted cycle is reachable from source.

14. If π_A is NP - hard, then it is NP - complete

Alternately

π_A be a problem (given) in class NP. we say that π_A is NP complete if the following statements are true about L.

(1) π_A is in NP

(2) For every π_A' in NP there is polynomial time reduction of π_A' to π_A.

15. R. E = (0+1)*0(0+1)*0(0+1)*

Accepting Language will be

L = { 00, 000, 100, 001, 010, 0000, 0001, 1000,1001,0100, 1100, 0010, 0011, 0110, 0101, 1010, ...}

At least two zeros.

16. (a) There is a unique minimal DFA for every regular language. TRUE
(b) Every NFA can be converted to an equivalent PDA. TRUE
(c) Complement of every context -free language is recursive. TRUE
Because context -free language is not closed under complementation
(d) Every nondeterministic PDA can be converted to an equivalent deterministic PDA. FALSE . Because ww^R can not be accepted by deterministic PDA but nondeterministic PDA can accept it. So there does not exist any equivalent deterministic PDA.

17. Regular expression denote structure of data specially text string - lexical analyser break input text into logical unit such as identifiers.
Push down automater - Syntax analysis. Study with the help of syntax.
Data flow analysis - Code optimization
Register allocation - Lexical analysis

18. Stack Positions :

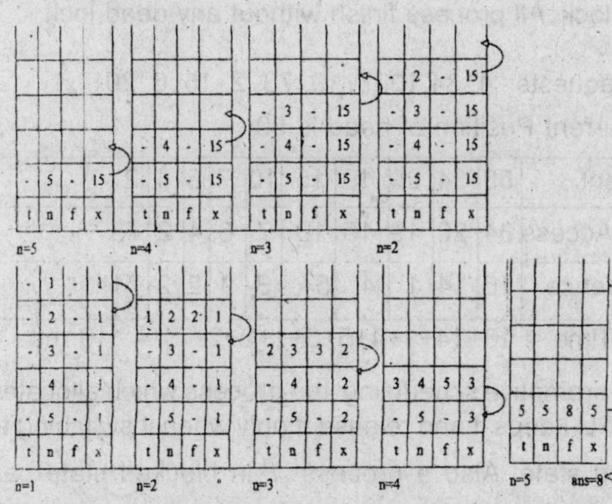

19. Data coupling : Two modules communicate using elementary data types (most desirable)

Stamp Coupling : Two modules communicate using composite data items (i. e. Array)

Control coupling : Two modules communicate using data items which are used to control the order of execution of one of the module by another.

Common coupling : Two modules communicate using share global data items.

Content Coupling : Two modules communicate using sharing code. (least desirable)

content	common	control	stamp	data

High Low

20. Here two <td> command used in first<tr> command and three <td>command used in second <tr> command, so required rows in each column is <2, 3, 2>

Alternately

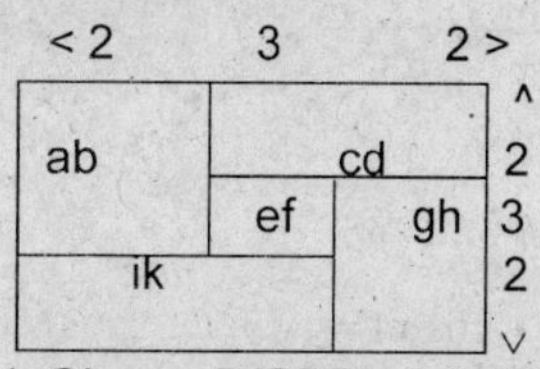

21. Given : P(ODD)=0.9*P(EVEN) # P(2)=P(4) = P(6) # P(EVEN | Exceeds 3)=0.75
Now as we know
P(ODD)=1-P(EVEN)
or0.9*P(EVEN)=1-P(EVEN)
{ Given P(ODD)=0.9*P(EVEN) }
or (PEVEN)=1/1.9
Because of independent events
$P(2) = P(4) = P(6) = P(2 \cup 4 \cup 6)/3 = P(EVEN)/3 = 1(3*1.9)$
Now P(EVEN|Exceeds 3)=P(EVEN $\cap$ Exceeds3)/ P(Exceeds3)
P(Exceeds3) =P(EVEN Exceeds 3) /P(EVEN Exceeds3)
P(Exceeds3) = $P(4 \cup 6)/0.75$
P(Exceeds3) = (P(4)+P(6))/0.75
P(Exceeds3) = (1/(3*1.9)+1/(1.9*3))/0.75
P(Exceeds3) = 0.468

22.

*	a	b	c	d
a	a	b	c	d
b	b	a	d	c
c	c	d	b	a
d	d	c	a	b

Hence c, d is generator

Alternately
Every element in Cyclic group can be represented in the power of any generator.
From given table
a*a=a b*b=a

$c*c = c^2 = b \quad c^3 = c*c^2 = c*b = d \quad c^4 = c*c^3 = c*d = a$

$d*d = d^2 = b \quad d^3 = d*d^2 = d*b = c \quad d^4 = d*d^3 = d*c = a$

23. Here given that gold and silver ornament are precious.
We use
gold and silver gold $\vee$ silver
$\forall((G(x) \vee S(x))$
gold and silver $\rightarrow$ precious
$\forall x((G(x) \vee S(x)) \rightarrow P(x)$

Alternately
"Gold and silver ornaments are precious"
Gold ornaments are alone precious.
Silver ornaments are also alone precious.
Gold and silver ornaments are also precious
$\forall x((G(x) \vee S(x))) \rightarrow P(x)$

24.

P	P	P□Q	$P \vee Q$	P□¬Q
T	T	T	T	T
T	F	T	T	T
F	T	F	T	T
F	F	T	F	F

25. $\frac{1-\tan x}{1+\tan x}dx = \frac{\cos x - \sin x}{\cos x + \sin x}dx$

Let $\cos x + \sin x = t$

$\therefore (-\sin x + \cos x)dx = dt$

So $(1/t)dt = \ln t = \ln(\sin x + \cos x) =$

$$\ln\left(\sin\frac{\pi}{4} + \cos\frac{\pi}{4}\right) = \ln\left(\frac{1}{\sqrt{2}} + \frac{1}{\sqrt{2}}\right) = \frac{1}{2}\ln 2$$

26. The universal qualification of P(x) is denoted $\forall x P(x)$ The symbol $\forall$ is called the universal quantifier. The existential qualification of predicate P(x) is statement, there exist a value of x for which P(x) is true. The existential qualification of P(x) is denoted by $\exists(x)\ P(x)$. The symbol $\exists$ is called the existential quantier.

so $\neg \forall x P(x) = \exists x(\neg P(x))$

$\neg$ sign is for not

27.

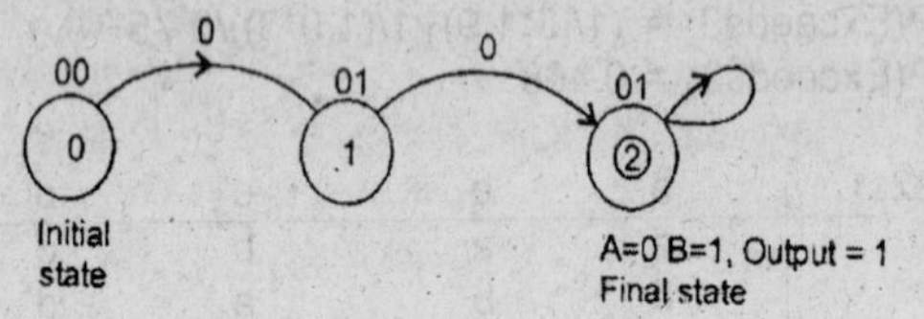

Hence 3 stakes required

Alternately

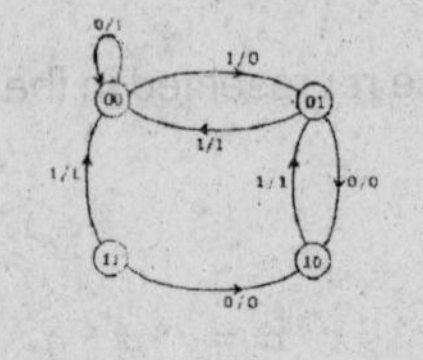

So from state 00 to 01 with output 1 required minimum 3 length of string i.e. 101

28.

	1	2	3	4	5	6	7	8	9	10	11	12	13	14	15	16	17	18	19	20	21	22	23
S_1	I_1		I_2	I_3		I_4	I_1		I_2	I_3		I_4											
S_2			I_1	I_2			I_3	I_4		I_1	I_2			I_3	I_4								
S_3				I_1			I_2		I_3	I_4		I_1		I_2		I_3	I_4						
S_4					I_1				I_2		I_3			I_4		I_1	I_2		I_3			I_4	

29.

Set 0	Ø48
	4 32
	8
	216 92
Set 1	1
	133
	129
	73

Set 2	
Set 3	255 155
	3
	159
	63

30. At refuse used
t =0 R2 - 2units R3-2 units R4 - 1units
t =2 R2 - 2units R3-3 units R4 - 2units R1-1units
t =3 R2 - 2units R3-3 units R4 - 2units R1-4units
t =4 R2 - 2units R3-3 units R4 - 2units R1-5units
t =5 R2 - 1units R3-3 units R4 - 2units R1-2units
t =6 R2 - 1units R3-2 units R4 - 2units R1-2units
t =7 R2 - 2units R3-1 units R4 - 2units R1-2units
t =8 R2 - 2units R3-1 units R4 - 3units R1-1units
t=9
The under line show that at that time the requirement of resource is less than availability but it does not cause dead lock. All process finish without any dead lock.

31. Requests : 4, 34,10, 7, 19, 73, 2, 15, 6, 20

Current Position of head is 50

Current	50	34	20	19	15	10	7	6	4	2
Next Access	34	20	19	15	10	7	6	4	2	73
Difference	16	14	1	4	5	3	1	2	2	71

Total Time = 16+14+1+4+5+3+1+2+2+71 = 119 ms

32. In premptive scheduling the process which allocated the CPU keeps it and release it only when it switching to waiting state. Also a process P_2 in blocked state can make transition E while other process P_1 in running.

Alternately

A process P_2 in blocked state can make transition E while another P_1 is in running state.

The OS uses preemptive Scheduling.

33. In given program the process test -and -set in critical section and in function leaves CS initialise memory location x to zero. So this process is dead lock free.

34. Most computer system support a large logical address space $(2^{32}\ to\ 2^{64})$. The page table become excessively large for example. Consider a system with a 32 bit logical address space. If page size $4K(2^{12})$ then page table may consist 1 million page. For this we use multilevels page table.

35. Applying Mater's Theorem

$$cn > n^{\ln\frac{1}{3}}$$

$cn > n^0$

Now Checking af(n/b) $\leq$ kf(n) for some k<1

1*cn/3 $\leq$ kcn

This is true for k > 1/3

so the solution is $\theta(n)$.

36. In linear probing, if there is already an element then the new element is fed into the next blank value field in the table.

37.

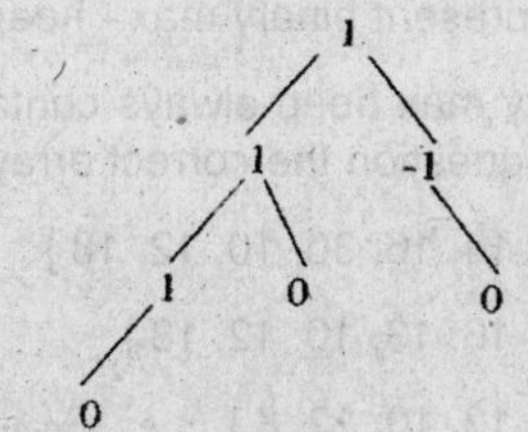

Maximum Height = 3

38. In Kruskal's Algorithm we choose on edge of G which has smallest weight among the edges of G.

So (b,e)=2,(e,f)=3,(b,c)=4, (a,c)=8, (f,g)=4 (c,d)=5

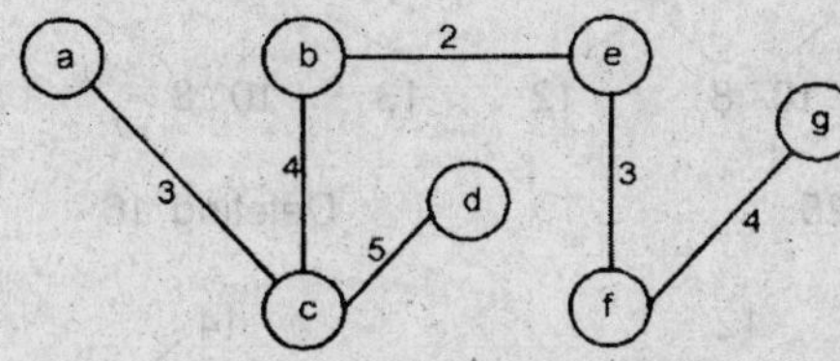

Alternately

Order the edges in non - decreasing order and pick edge one by one until all the nodes are completed with no edge making cycle on addition.

39. Average complexity of quick sort is $\theta(n \log n)$.

40. The inter section of two regular language is regular

Given $L_1 = \{a^m b^m c a^n b^n \mid m,n \geq 0\}$

If n = 1 $\{a.b.c.a.b\}$ not regular

m = 1

If $n/1\ a^m.b^m c\ a^n b^n$ is regular

Alternately

41. Let A be the starting state

A to A	input 1 0 output 0·0
B to A	input 1 0 output 0 0
C to A	input 1, 0 output 0 0

So end with 0 0

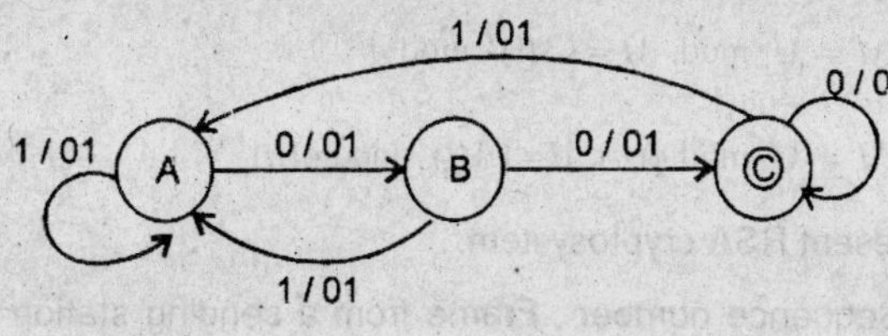

Alternately

R. E. = (0 + 1)*00

42. It is not necessary there exists parsing algo for some programming language whose complexities are less than $\theta(n^3)$

A programming language which allow recursion can be implemented with static storage allocation.

43.

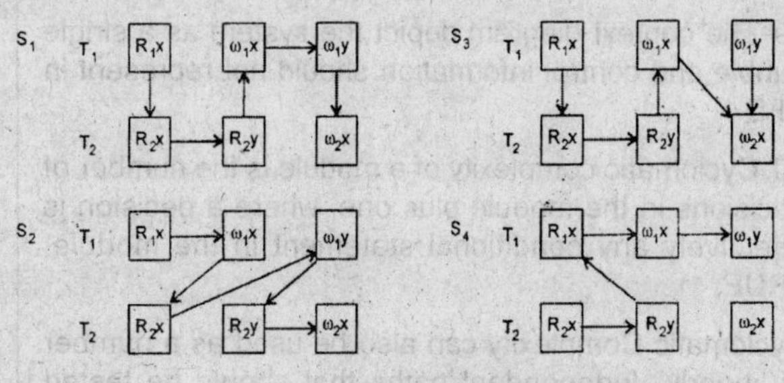

Clearly S_2 *and* S_3 are conflict = serializable.

44.

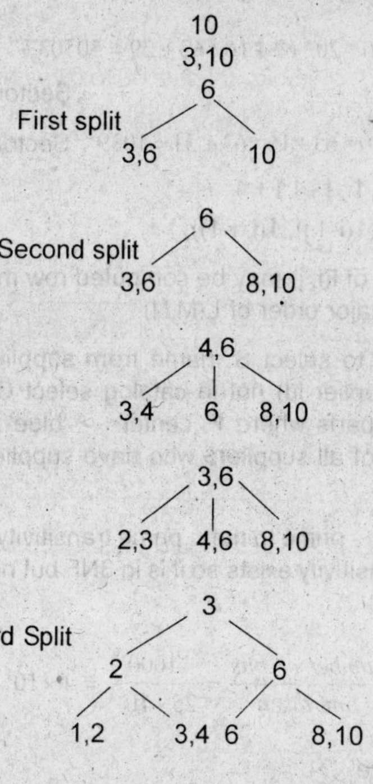

45. Query (II) and (IV) given equal data so they are equivalent

46. In RSA public key cryptography a public key provide for every one but they have their separate private key.

According to formula of public key crytography

I. $M' = M^e \text{ mod}, M = (M')^d \text{ mod } n$

IV. $M' = M^e \text{ mod } \phi(n), M = (M')^d \text{ mod } \phi(n)$

represent RSA cryptosystem.

47. Sequence number : Frame from a sending station are numbered sequentially we need to set a limit.

48. When the polynomial code method is employed the sender and receiver must agree upon a generator polynomial G(x) Both the high and low order bit of the generator must be 1.So 1 + x is a factor of G(x).

Alternately

For odd number of bits error, the error function will be having a factor of $(1 + x^2 + x^4 + ...)$ which can not be divided by the generator having (1+x) as a factor.

49. The context diagram depict the system as a single bubble and control information should not represent in DFD.

50. Cyclomatic complexity of a module is the number of decisions in the module plus one, where a decision is effectively any conditional statement in the module. TRUE.

Cyclomatic Complexity can also be used as a number of linearly independent paths that should be tested during path converage testing. TRUE

51. $\langle 400,16,29 \rangle = 400 * 20 * 63 + 16 * 63 + 29 = 505037^{th}$ Sector

52. $\langle 0,16,31 \rangle = 0 * 20 * 63 + 16 * 63 + 31 = 1039^{th}$ Sector

53. expr 1 $\equiv$ 1 (i - 1 , j - 1) + 1

expr 2 $\equiv$ max (1(i-1,j), 1(i, j-1))

54. Here the value of I(i, j) may be computed row major order or column major order of L(M,N).

55. Here we have to select S. name from supplier S where S .sid (supplier Id) not in catalog select C.pid (product id) form parts where P. center< > 'blue' that means all names of all suppliers who have supplied a non blue part.

56. There is no non - prime to non - prime transitivity but prime to prime transitivity exists so it is in 3NF but not in BCNF.

57. Pulse rate $= \dfrac{number\ of\ bits}{time taken} = \dfrac{1000}{25 \times 10^{-3}} = 4 \times 10^4$

Bit rate = pulse rate $\times 2^l$

Here, $2^l = 25$

$2^4 < 2^5 < 2^l$

So $l = 5$

Alternately

$10^3\, ms$..............$10 bits$

$50''$...............$= \dfrac{1000}{50}\ 20\ bits$

Minimum number of bits required is 5

59. In binary max heap, the difference between two level mode or child of a single mode is atleast differ by one so the correct array with represent binary max - heap.

60. The deletion from binary max heap always contain greater element . In above question the correct array

{ 25, 14, 16, 30, 10, 12, 18 }

{ 14, 16, 13, 10, 12, 18 }

after one deletion { 14, 13, 10, 12, 8 }

after two deletions

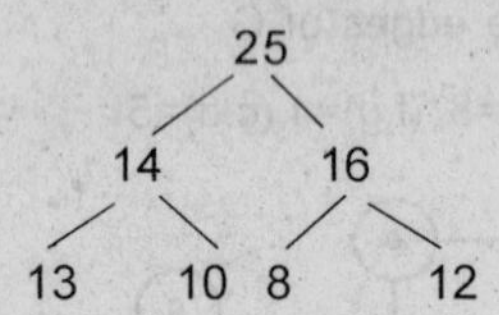

After Heapifying

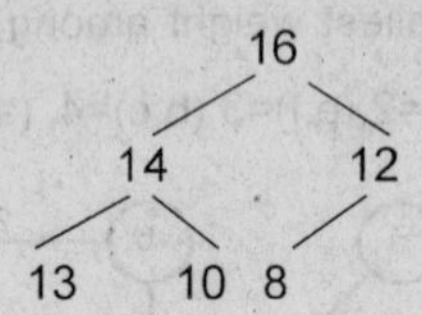

Deleting 25

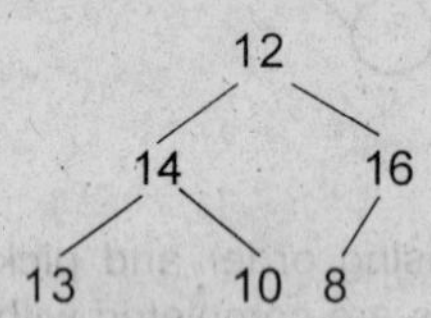

Deleting 16

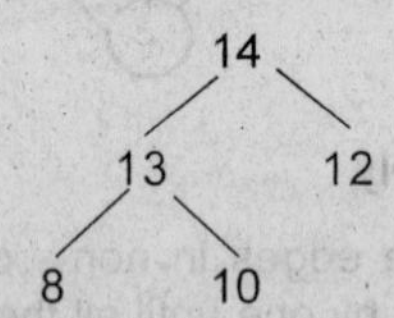

So ans { 14, 13, 12, 8, 10 }

GATE - 2008

CS : COMPUTER SCIENCE AND ENGINEERING

Read the following instructions carefully

1. Using HB pencil, darken the appropriate bubble under each digit of your registration number and the letters corresponding to your paper code.

2. All the questions in this question paper are of objective type.

3. Questions must be answered on **O**bjective **R**esponse **S**heet **(ORS)** by darkening the appropriate bubble (marked A, B, C, D) using HB pencil against the question number on the left hand side of the ORS. **Each question has only one correct answer.** In case you wish to change an answer, erase the old answer completely. More than one answer bubbled against a question will be treated as a wrong answer.

4. Questions 1 through 20 are 1 - mark questions and questions 21 through 85 are 2 - mark questions.

5. Questions 71 through 73 is one set of common data questions, questions 74 and 75 is another pair of common data questions. The question Pairs (76, 77), (78, 79), (80, 81), (82, 83) and (84, 85) are questions with linked answers. The answer to the second question of the above pairs will depend on the answer to the first question of the pair. If the first question in the linked pair is wrongly answered or is un - attempted, then the answer to the second question in the pair will not be evaluated.

6. Un - attempted questions will carry zero marks.

7. **NEGATIVE MARKING :** For Q.1 to Q. 20, **0.25** mark will be deducted for each wrong answer. For Q.21 to Q.75, **0.5** mark will be deducted for each wrong answer. For the pairs of questions with linked answers, there will be negative marks only for wrong answer to the first question, i. e; for Q. 76, Q.78, Q.80, Q.82 and Q.84, **0.5** mark will be deducted for each wrong answer. There is no negative marking for Q.77, Q.79, Q81, Q.83 and Q.85.

8. Calculator **without data connectivity** is allowed in the examination hall.

9. Charts, graph sheets and tables are NOT allowed in the examination hall.

10. Rough work can be done on the question paper itself. Additional blank pages are given at the end of the question paper for rough work.

Q.1 to Q. 20 Carry one mark each

1. $\lim_{x\to\infty} \dfrac{x-\sin x}{x+\cos x}$ equals

(a) 1 (b) -1 (c) ∞ (d) $-\infty$

2. If P, Q, R are subsets of the universal set U, then $(P\cap Q\cap R)\cup(P^c\cap Q\cap R)\cup Q^c\cup R^c$ is

(a) $Q^c\cup R^c$ (b) $P\cup Q^c\cup R^c$ (c) $P^c\cup Q^c\cup R^c$ (d) U

3. The following system of equations

$$x_1+x_2+2x_3=1$$

$$x_1+2x_2+3x_3=2$$

$$x_1+4x_2+\alpha x_3=4$$

has a unique solution. The only possible value (s) for α is/are

(a) 0 (b) either 0 or 1

(c) one of 0, 1 or -1 (d) any real number

4. In the IEEE floating point representation the hexadecimal value 0x00000000 corresponds to

(a) the normalized value 2^{-127} (b) the normalized value 2^{-126}

(c) the normalized value +0 (d) the special value +0

5. In the Karnaugh map shown below, X denotes a don't care term. What is the minimal form of the function represented by the Karnaugh map ?

cd \ ab	00	01	11	10
00	1	1		1
01	X			
11	X			
10	1	1		X

(a) $\bar{b}.\bar{d}+\bar{a}.\bar{d}$

(b) $\bar{a}.\bar{b}+\bar{b}.\bar{d}+\bar{a}.b.\bar{d}$

(c) $\bar{b}.\bar{d}+\bar{a}.b.\bar{d}$

(d) $\bar{a}.\bar{b}+\bar{b}.\bar{d}+\bar{a}.\bar{d}$

6. Let **r** denote number system radix. The only value(s) of **r** that satisfy the equation $\sqrt{121_r}=11_r$ is/are

(a) decimal 10 (b) decimal 11

(c) decimal 10 and 11 (d) any value >2

7. The most efficient algorithm for finding the number of connected components in an undirected graph on n vertices and m edges has time complexity

(a) Θ(n) (b) Θ(m) (c) Θ(m+n) (d) Θ(mn)

8. Given f_1, f_3 and f in canonical sum of products form (in decimal) for the circuit

$f_1=\Sigma m(4,5,6,7,8)$

$f_3=\Sigma m(1,6,15)$

$f=\Sigma m(1,6,8,15)$

then f_2 is

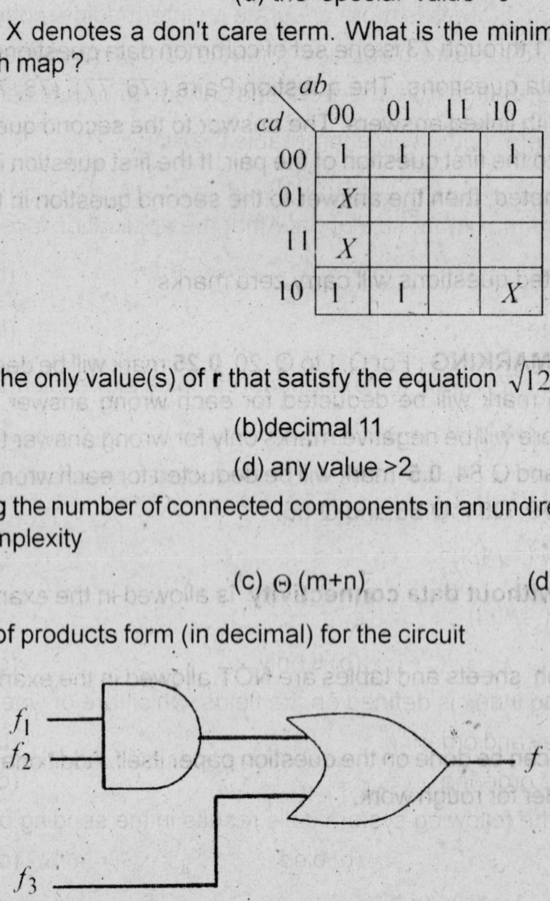

(a) $\Sigma m(4.6)$ (b) $\Sigma m(4.8)$ (c) $\Sigma m(6.8)$ (d) $\Sigma m(4,6,8)$

9. Which of the following is true for the language { a^p |p is a prime} ?

(a) It is not accepted by a Turning Machine (b) Itis regular but not context -free

(c) It is context-free but not regular

(d) It is neither regular nor context-free, but accepted by a Turning machine.

10. Which of the following are decidable ?

I. Whether the intersection of two regular languages is infinite

II. Whether a given context-free language is regular

III. Whether two push-down automata accept the same language

IV. Whether a given grammar is context-free

(a) I and II (b) I and IV (c) II and III (d) II and IV

11. Which of the following describes a handle (as applicable to LR-parsing) appropriately ?

(a) It is the position in a sentential form where the next shift or reduce operation will occur

(b) It is a non-terminal whose production will be used fro reduction in the next step

(c) It is a production that may be used for reduction in a future step along with a position in the sentential form where the next shift or reduce operation will occur

(d) It is the production p that will be used for reduction in the next step along with a position in the sentential form where the right hand side of the production may be found.

12. Some code optimizations are carried out on the intermediate code because

(a) they enhance the portability of the compiler to other target processors

(b) program analysis is more accurate on intermediate code than on machine code

(c) the information from dataflow analysis cannot otherwise be used for optimization

(d) the information from the front end cannot otherwise be used for optimization

13. If L and $\overline{L}$ are recursively enumerable then L is

(a) regular (b) context-free (c) context-sensitive (d) recursive

14. What is the maximum size of data that the application layer can pass on to the TCP layer below?

(a) Any size (b) 2^{16} bytes-size of TCP header

(c) 2^{16} bytes (d) 1500 bytes

15. Which of the following tuple relational calculus expression(s) is/are equivalent to $\forall t \in r(P(t))$?

I. $\neg\exists t \in r(P(t))$

II. $\exists t \notin r(P(t))$

III. $\neg\exists t \in r(\neg P(t))$

IV. $\exists t \notin r(\neg P(t))$

(a) I only (b) II only (c) III only (d) III and IV only

16. A clustering index is defined on the fields which are of type

(a) non-key and ordering (b) non-key and non-ordering

(c) key and ordering (d) key and non ordering

17. Which of the following system calls results in the sending of SYN packets ?

(a) socket (b) bind (c) listen (d) connect

18. Which combination of the integer variables x, y and z makes the variable a get the value 4 in the following expression ?

$$a = (x > y)?((x > z)?x : z) : ((y > z)?y : z)$$

(a) x=3, y=4, z=2 (b) x=6, y=5, z=3

(c) x=6, y=3, z=5 (d) x=5, y=4, z=5

19. The Breadth First Search algorithm has been implemented using the queue data structure, One possible order of visiting the nodes of the following graph is

(a) MNOPQR

(b) NQMPOR

(c) QMNPRO

(d) QMNPOR

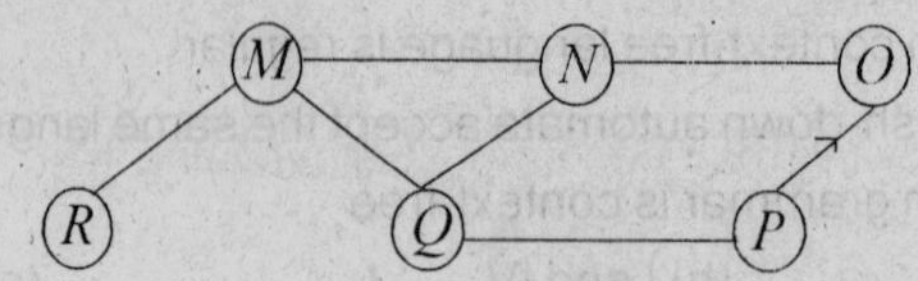

20. The data blocks of a very large file in the Unix file system are allocated using

(a) contiguous allocation (b) linked allocation

(c) indexed allocation (d) an extension of indexed allocation

Q.21 to Q.75 two marks each:

21. The minimum number of equal length subintervals needed to approximate $\int_1^2 xe^x$ dx to an accuracy of at least $\frac{1}{3} \times 10^{-6}$ using the trapezoidal rule is

(a) 1000 e (b) 1000 (c) 100 e (d) 100

22. The Newton-Raphson iteration $x_{n+1} = \frac{1}{2}\left(x_n + \frac{R}{x_n}\right)$ can be used to compute the

(a) square of R (b) reciprocal of R

(c) square root of R (d) logarithm of R

23. Which of the following statements is true for every planer graph on n vertices ?

(a) The graph is connected (b) The graph is Eulerian

(c) The graph has a vertex-cover of size at most $\frac{3n}{4}$

(d) The graph has a independent set of size at least $\frac{n}{3}$

24. Let $P = \sum_{\substack{1 \le i \le 2k \\ i \, odd}} i$ and $Q = \sum_{\substack{1 \le i \le 2k \\ i \, even}} i$, where k is a positive integer. Then

(a) P = Q-K (b) P=Q+K (c) P=Q (d) P=Q+2K

25. A point on a curve is said to be an extremum if it is a local minimum or a local maximum. The number of distinct extrema for the curve $3x^4 - 16x^3 + 24x^2 + 37$ is

(a) 0 (b) 1 (c) 2 (d) 3

26. If P,Q,R are Boolean variables, then

Simplifies to $\left(P+\overline{Q}\right)\left(P.\overline{Q}+P.R\right)\left(\overline{P}.\overline{R}+\overline{Q}\right)$

(a) $P.\overline{Q}$ (b) $P.\overline{R}$ (c) $P.\overline{Q}+R$ (d) $P.\overline{R}+Q$

27. Aishwarya studies either computer science or mathematics everyday. If she studies computer science on a day, then the probability that she studies mathematics the next day is 0.6. If she studies mathematics on a day, then the probability that she studies computer science the next day is 0.4. Given that Aishwarya studies computer science on Monday, what is the probability that she studies computer science on Wednesday ?

(a) 0.24 (b) 0.36 (c) 0.4 (d) 0.6

28. How many of the following matrices have an eigenvalue 1 ?

$\begin{bmatrix}1 & 0\\0 & 1\end{bmatrix}, \begin{bmatrix}0 & 1\\0 & 0\end{bmatrix}, \begin{bmatrix}1 & -1\\1 & 1\end{bmatrix}$ and $\begin{bmatrix}-1 & 0\\1 & -1\end{bmatrix}$

(a) one (b) two (c) three (d) four

29. Let **X** be a random variable following normal distribution with mean +1 and variance 4. Let **Y** be another normal variable with mean - 1 and variance unknown. If $P(X \leq -1) = P(P \geq 2)$. the standard deviation of **Y** is

(a) 3 (b) 2 (c) $\sqrt{2}$ (d) 1

30. Let fsa and pda be two predicates such that fsa(x) means x is a finite state automaton, and pda(y) means that y is a pushdown automaton. Let equivalent be another predicate such that equivalent (a,b) means a and b are equivalent. Which of the following first order logic statements represents the following Each finite state automaton has an equivalent pushdown automaton.

(a) $(\forall x\ fsa(x)) \Rightarrow (\exists y\ pda(y) \wedge$ equivalent (x,y))

(b) $\sim \forall y (\exists x\ fsa(x)) \Rightarrow pda(y) \wedge$ equivalent (x,y))

(c) $\forall x \exists y\ (fsa(x)) \wedge pda(y) \wedge$ equivalent (x,y))

(d) $\forall x \exists y\ (fsa(y)) \wedge pda(x) \wedge$ equivalent (x,y))

31. P and Q are two propositions. Which of the following logical expressions are equivalent ?

I $P \vee \sim Q$

II. $\sim (\sim P \wedge Q)$

III. $(P \wedge Q) \vee (P \wedge \sim Q) \vee (\sim P \wedge \sim Q)$ IV. $(P \wedge Q) \vee (P \wedge \sim Q) \vee (\sim P \wedge Q)$

(a) Only I and II (b) Only I,II and III

(c) Only I,II and IV (d) All of I, II, III and IV

32. For a magnetic disk with concentric circular tracks, the seek latency is not linearly proportional to the seek distance due to

(a) non-uniform distribution of requests (b) arm starting and stopping inertia

(c) higher capacity of tracks on the periphery of the platter

(d) use of unfair arm scheduling policies

33. Which of the following is/are true of the auto-increment addressing mode?

I. It is useful in creating self-relocating code

II. If it is included in an Instruction Set Architecture, then an additional ALU is required for effective address calculation

III. The amount of increment depends on the size of the data item accessed

(a) I only (b) II only (c) III only (d) II and III only

34. Which of the following must be true for the RFE(Return from Exception) instruction on a general purpose processor ?

I. It must be a trap instruction

II. It must be a privileged instruction

III. An exception cannot be allowed to occur during execution of an RFE instruction

(a) I only (b) II only (c) I and II (d) I, II and III only

35. For inclusion to hold between two cache levels L1 and L2 in a multi-level cache hierarchy, which of the following are necessarily ?

I. L1 must be a write-through cache

II. L2 must be a write-through cache

III. The associativity of L2 must be greater than that of L1

IV. The L2 cache must be at least as large as the L1 cache

(a) IV only (b) I and IV only (c) I,III and IV only (d) I,II,III and IV

36. Which of the following are NOT true in a pipelined processor ?

I. Bypassing can handle all RAW hazards.

II. Register renaming can eliminate all register carried WAR hazards

III. Control hazard penalties can be eliminated by dynamic branch prediction.

(a) I and II only (b) I and III only (c) II and III only (d) I,II and III

37. The use of multiple register windows with overlap causes a reduction in the number of memory accesses for

I. function locals and parameters

II. register saves and restores

III. instruction fetches.

(a) I only (b) II only (c) III only (d) I,II and III

38. In an instruction execution pipeline, the earliest that the data TLB (Translation Lookside Buffer) can be accessed is

(a) before effective address calculation has started

(b) before effective address calculation

(c) after effective address calculation has completed

(d) after data cache lookup has completed.

39. Consider the following functions:

$f(n) = 2^n$

$g(n) = n!$

$h(n) = n^{\log n}$

Which of the following statements about the asymptotic behaviour of f(n), g(n) and h(n) is true?

(a) f(n)=O(g(n)); g(n)= O(h(n)) (b) f(n)= Ω (g(n)); g(n)=0(h(n))

(c) g(n)= O(f(n)); h(n)= O(f(n)) (d) h(n)=O(f(n)); g(n)= Ω (f(n))

40. The minimum number of comparisons required to determine if an integer appears more than $\frac{n}{2}$ times in a sorted array of n integers is

(a) $\Theta(n)$ (b) $\Theta(\log n)$ (c) $\Theta(\log^* n)$ (d) $\Theta(1)$

41. A B -tree of order 4 is built from scratch by 10 successive insertions. What is the maximum number of node sphitting operations that may take place?

(a) 3 (b) 4 (c) 5 (d) 6

42. G is a graph on n vertices and 2n-2 edges. The edges of G can be partitioned into two edge-disjoint spannint trees. Which of the following is NOT true for G ?

(a) For every subset of k vertices, the induced subgraph has at most 2k-2 edges

(b) The minimum cut in G has at least two edges

(c) There are two edge-disjoint paths between every pair of vertices

(d) There are two vertex--disjoint paths between every pair of vertices

43. Consider the Quicksort algorithm.Suppose there is a procedure for finding a pivot element which splits the list into two sub-lists each of which contains at least one-fifth of the elements. Let T(n) be the number of comparisons required to sort n elements. Then

(a) $T(n) \leq 2T\left(\frac{n}{5}\right) + n$ (b) $T(n) \leq T\left(\frac{n}{5}\right) + T\left(\frac{4n}{5}\right) + n$

(c) $T(n) \leq 2T\left(\frac{4n}{5}\right) + n$ (d) $T(n) \leq 2T\left(\frac{n}{2}\right) + n$

44. The subset-sum problem is defined as follows: Given a set S of n positive integers and a positive integer W, determine whether there is a subset of S whose elements sum to W.

An algorithm Q solves this problem in O(nW) time. Which of the following statements is false?

(a) Q solves the subset-sum problem in polynomial time when the input is encode in unary

(b) Q solves the subset-sum problem in polynomial time when the input is encode in binary

(c) The subset sum problem belongs to the class NP

(d) The subset sum problem is NP-hard

45. Dijkstra's single source shortest path algorithm when run from vertex a in the above graph, computes the correct shortest path distance to

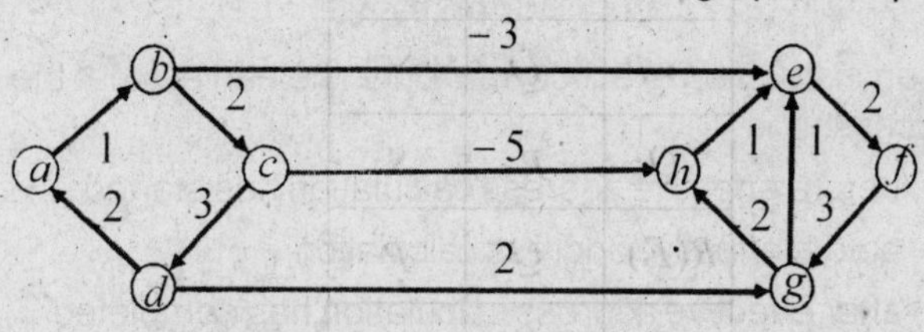

(a) only vertex a

(b) only vertices a,e,f,g,h

(c) only vertices a,b,c,d

(d) all the vertices.

46. You are given the postorder traversal, P, of a binary search tree on the n elements 1,2...n. You have to determine the unique binary search tree that has P as its postorder traversal. What is the time complexity of the most efficient algorithm for doing this ?

(a) Θ (log n) (b) Θ (n)

(c) Θ (nlog n) (d) none of the above, as the tree cannot be uniquely determined.

47. We have a binary heap on n elements and wish to insert n more elements (not necessarily one after another) into this heap. The total time required for this is

(a) Θ (log n) (b) Θ (n) (c) Θ (nlog n) (d) Θ (n^2)

48. Which of the following statements is false ?

(a) Every NFA can be converted to an equivalent DFA

(b) Every non-deterministic Turing machine can be converted to an equivalent deterministic Turing machine.

(c) Every regular language is also a context-free language

(d) Every subset of a recusively enumerable set is recursive.

49. Given below are two finite state automata ($\rightarrow$ indicates the start state and F indicates a final state)

Y :

	a	b
$\rightarrow 1$	1	2
$2(F)$	2	1

Z :

	a	b
$\rightarrow 1$	2	2
$2(f)$	1	1

Which of the following represents the product automaton Z x Y ?

(a)

	a	b
$\rightarrow P$	S	R
Q	R	S
$R(F)$	Q	P
S	Q	P

(b)

	a	b
$\rightarrow P$	S	Q
Q	R	S
$R(F)$	Q	P
S	P	Q

(c)

	a	b
$\rightarrow P$	Q	S
Q	R	S
$R(F)$	Q	P
S	Q	P

(d)

	a	b
$\rightarrow P$	S	Q
Q	S	R
$R(F)$	Q	P
A	Q	P

50. Which of the following statements are true ?

I. Every left-recursive grammar can be converted to a right-recursive grammar and vice-versa.

II. All ε -productions can be removed from any context-free grammar by suitable transformations.

III. The language generated by a context-free grammar all of whose productions are of the form $X \rightarrow w$ or $X \rightarrow w$ Y (where, w is a string of terminals and Y is a non-terminal), is always regular

IV. The derivation trees of strings generated by a context-free grammar in Chomsky Normal from are always binary trees.

(a) I, II, III and IV
(b) II, III and IV only
(c) I, III and IV only
(d) I,II and IV only

51. Match the following

(E) Checking that identifiers are declared before their use	(P) L= $\{a^n b^m c^n d^m \mid n \geq 1, m \geq 1\}$
(F) Number of formal parameters in the declaration of a function agrees with the number of actual parameters in a use of that function	(Q) $X \rightarrow XbX \mid XcX \mid dXf \mid g$
(G) Arithmetic expressions with matched pairs of parentheses	(R) = $L = \{wcw \mid w \in (a \mid b)^*\}$
(H) Palindromes	(S) $X \rightarrow bXb \mid cXc \mid \varepsilon$

(a) E-P, F-R, G-Q, H-S (b) E-R, F-P, G-S, H-Q

(c) E-R, F-P, G-Q, H-S (d) E-P, F-R, G-S, H-Q

52. Match the following NFAs with the regular expressions they correspond to

(P)

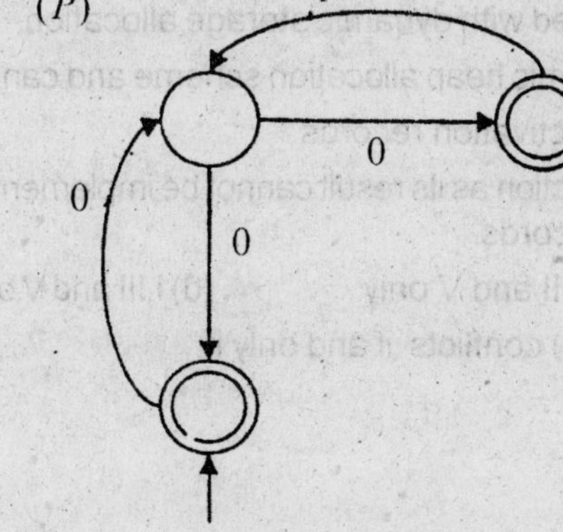

(Q)

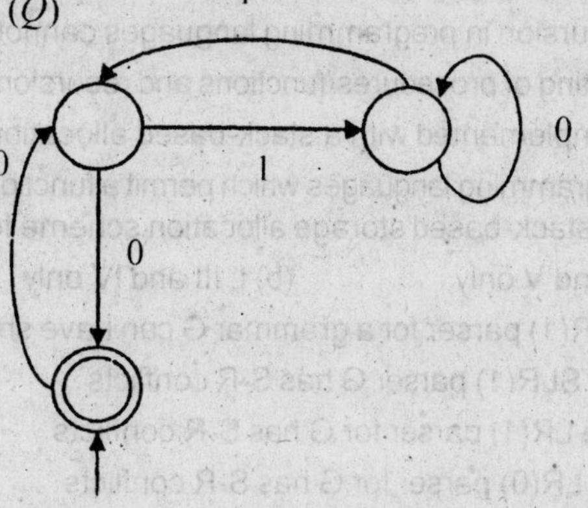

(R)

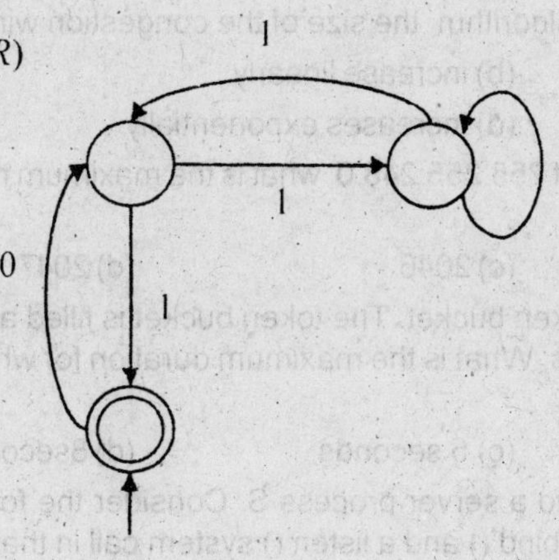

(S)

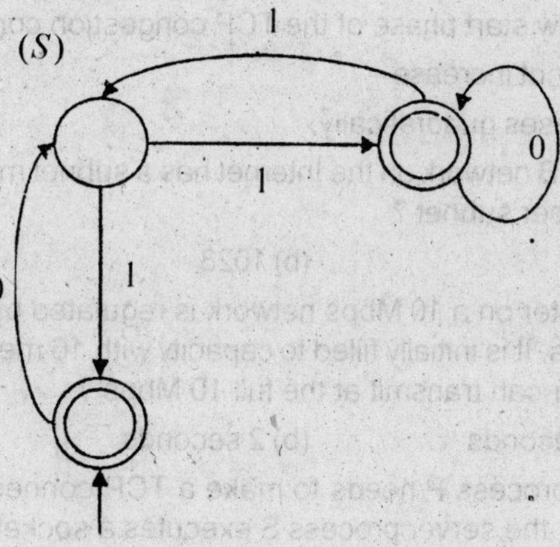

1. ε + 0 (01*1+00)*01*
2. ε + 0 (10*1+00)*0
3. ε + 0 (10*1+10) *1
4. ε + 0 (10*1+10) *10*

(a) P-2, Q-1, R-3, S-4 (b) P-1, Q-3, R-2, S-4

(c) P-3, Q-2, R-3, S-4 (d) P-3, Q-2, R-1, S-4

53. Which of the following are regular sets ?

I. $\{a^n b^{2m} \mid n \geq 0, m \geq 0\}$

II. $\{a^n b^m \mid n = 2m\}$

III. $\{a^n b^m \mid n \neq m\}$

IV. $\{xcy \mid x, y \in \{a,b\}^*\}$

(a) I and IV only (b) I and III only (c) I only (d) IV only

54. Which of the following are true?

I. A programming language which does not permit global variables of any kind and has no nesting of procedures/functions, but permits recursion can be implemented with static storage allocation.

II. Multi-level access link (or display) arrangement is needed to arrange activation records only if the programming language being implemented has nesting of procedures/functions.

III. Recursion in programming languages cannot be implemented with dynamic storage allocation.

IV. Nesting of procedures/functions and recursion require a dynamic heap allocation scheme and cannot be implemented with a stack-based allocation scheme for activation records.

V. Programming languages which permit a function to return a function as its result cannot be implemented with a stack-based storage allocation scheme for activation records.

(a) II and V only (b) I, III and IV only (c) I, II and V only (d) I, III and V only

55. N LALR(1) parser for a grammar G can have shift-reduce (S-R) conflicts if and only if

(a) the SLR(1) parser G has S-R conflicts

(b) the LR(1) parser for G has S-R conflicts

(c) the LR(0) parser for G has S-R conflicts

(d) The LALR(1) parser for G has reduce-reduce conflicts

56. In the slow start phase of the TCP congestion control algorithm, the size of the congestion window

(a) does not increase (b) increase linearly

(c) increases quadratically (d) increases exponentially

57. If a class B network on the Internet has a subnet mask of 255.255.248.0, what is the maximum number of hosts per subnet ?

(a) 1022 (b) 1023 (c) 2046 (d) 2047

58. A computer on a 10 Mbps network is regulated by a token bucket. The token bucket is filled at a rate of 2 Mbps. It is initially filled to capacity with 16 megabits. What is the maximum duration for which the computer can transmit at the full 10 Mbps ?

(a) 1.6 seconds (b) 2 seconds (c) 5 seconds (d) 8seconds

59. A client process P needs to make a TCP connection to a server process S. Consider the following situation: the server process S executes a socket (), a bind () and a listen () system call in that order, following which it is preempted. Subsequently, the client process P executes a socket () system call followed by connect () system call to connect to the server process S. The server process has not executed any accept ( ) system call. Which one of the following events could take place ?

(a) connect ( ) system call returns successfully

(b) connect ( ) system call blocks

(c) connect ( ) system call returns an error

(d) connect ( ) system call results in a core dump

60. What is printed by the following C program ?

```
int f (int x, int *py, int **ppz)
{
int y, z;
**ppz + =1; z = *ppz;
*py + =2 ; y = -py
x + = 3;
return x + y + z;
}
```

```
void main ( )
{
int c, *b, **a;
c =4 ; b = &c; a = &b
printf( "%d", f(c,b,a);
}
```

(a) 18 (b) 19 (c) 21 (d) 22

61. Choose the correct option to fill ?1 and ?2 so that the program below prints an input string in reverse order. Assume that the input string is terminated by a newline character.

```
Void reverse (void) {
 Int c;
 If (?1) reverse () ;
 ?2
}
main ( ) {
 printf ( "Enter text") ; printf ("\n");
 reverse ( ) ; printf ("\n");
}
```

(a) ?1 is (getchar( ) != '\n')
 ?2 is getchar (c);

(b) ?1 is (c = getchar( )) != '\n')
 ?2 is getchar (c);

(c) ?1 is (c! = '\n')
 ?2 is putchar (c);

(d) ?1 is (c = getchar( )) != '\n')
 ?2 is putchar (c);

62. The following C function takes a singly-linked list of integers as a parameter and rearranges the elements of the list. The function is called with the list containing the integers 1,2,3,4,5,6, 7 in the given order. What will be the contents of the list after the function completes execution ?

```
struct node {
  int value ;
  struct node *next;
};
void rearrange (struct node *list) {
  struct node *p, *q;
  int temp ;
  if (! list || ! list -> next) return ;
  p = list ; q = list -> next;
  while (q) {
  temp = p -> value; p - > value = q -> value;
```

```
        q → value = temp;  p = q → next ;
        q = p ? p -> next : 0 ;
      }
  }
```

(a) 1,2,3,4,5,6,7 (b) 2,1,4,3,6,5,7 (c) 1,3,2,5,4,7,6 (d)2,3,4,5,6,7,1

63. The P and V operations on counting semaphores, where is a counting semaphore, are defined as follows

```
P(s) : s = s-1;
 if s<0 then wait;
V(s): s=s+1;
if s<=0 then wakeup a process waiting on s;
```

Assume that p_b and v_b the wait and signal operations on binary semaphores are provided. Two binary semaphores x_b and y_b are used to implement the semaphore operations P(s) and V(s) as follows.

P (s) : $p_b(x_b)$;

  s = s -1;

  if (s < 0) {

   $v_b(x_b)$;

   $P_b(x_b)$;

 }

 else $V_b(x_b)$;

V (s) : $P_b(X_b)$;

  s = s + 1 ;

  if (s <= 0) $V_b(y_b)$;

  $V_b(X_b)$;

The initial values of x_b the y_b are respectively.

(a) 0 and 0 (b) 0 and 1 (c) 1 and 0 (d) 1 and 1

64. Which of the following statements about synchronous and asynchronous I/O is NOT true ?

(a) An ISR is invoked on completion of I/O in synchronous I/O but not in asynchronous I/O

(b) In both synchronous and asynchronous I/O, an ISR (Interrupt Service Routine) is invoked after completion of the I/O.

(c) A process making a synchronous I/O call waits until I/O is complete, but a process making an asynchronous I/O call does not wait for completion of the I/O

(d) In the case of synchronous I/O, the process waiting for the completion of I/O is woken up the by ISR that is invoked after the completion of I/O.

65. Which of the following is NOT true of deadlock prevention and deadlock avoidance schemes ?

(a) In deadlock prevention, the request for resources is always granted if the resulting state is safe

(b) In deadlock avoidance, the request for resources is always granted if the resulting state is safe

(c) Deadlock avoidance is less restrictive than deadlock prevention

(d) Deadlock avoidance requires knowledge of resource requirements a priori.

66. A process executes the following code

for (i = 0; i<n; i++) for k ( );

The total number of child processes created is

(a) n (b) $2^n - 1$ (c) 2^n (d) $2^{n+1} - 1$

67. A processor uses 36 bit physical addresses and 32 bit virtual addresses, with a page frame size of 4 Kbytes. Each page table entry is of size 4 bytes. A three level page table is used for virtual-to-physical address translation, where the virtual address is sued as follows.

* bits 30-31 are used to index into the first level page table
* bits 12-29 are used to index into the second level page table
* bits 12-20 are used to index into the third level page table, and
* bits 0-11 are used as offset within the page

The number of bits required for addressing the next level page table (or page frame) in the page table entry of the first, second and third level page tables are respectively.

(a) 20, 20 and 20 (b) 24, 24 and 24 (c) 24, 24 and 20 (d) 25, 25 and 24

68. Let R and S be two relations with the following schema

R (P, Q, R1, R2, R3)

S (P, Q, S1, S2)

Where {P,Q} is the key for both schemas. Which of the following queries are equivalent ?

I. Π_p (R ⋈ S)

II. Π_p (R) ⋈ Π_p (S)

III. $\Pi_p(\Pi_{p,Q}(R) \cap \Pi_{p,Q}(S))$

IV. $\Pi_p(\Pi_{P,Q}(R) - (\Pi_{p,Q}(R) - (\Pi_{p,Q}(S))$

(a) Only I and II (b) Only I and III

(c) Only I,II and III (d)Only I,III and IV

69. Consider the following relational schemes for a library database:

Book (Title, Author, Catalog_no, Publisher, Year, Price)

Collection (Title, Author, Catalog_no)

With the following functional dependencies:

I. Title Author $\rightarrow$ Catalong_no

II. Catalog_no $\rightarrow$ Title Author Publisher year

III. Publisher Title Year $\rightarrow$ Price

Assume {Author, Title} is the key for both schemes. Which of the following statements is true?

(a) Both Book and Collection are in BCNF

(b) Both Book and Collection are in 3NF only

(c) Book is in 2NF and Collection is in 3 NF

(d) Both Book and Collection are in 2NF only

70. Consider a file of 16384 records. Each record is 32 bytes long and its key field is of size 6 bytes. The file is ordered on a non-key field, and the file organization is unspanned. The file is stored in a file system with block size 1024 bytes, and the size of a block pointer is 10 bytes. If the secondary index is built on the key field of the file, and a multi-level index scheme is used to store the secondary index, the number of first-level and second -level blocks in the multi-level index are respectively.

(a) 8 and 0 (b) 128 and 6 (c) 256 and 4 (d) 512 and 5

Common Data Questions :

Common Data for Questions 71,72 and 73 :

Consider a machine with a 2-way set associative data cache of size 64K bytes and block size 16 bytes. The cache is managed using 32 bit virtual addresses and the page size is 4 Kbytes. A program to be run on this machine begins as follows :

```
double Arr [1024] [1024];
int i, j ;
/* Initialize array ARR to 0.0*/
for ( i = 0; i < 1024; i++)
for ( j = 0; j < 1024; j++)
ARR [i ] [j] = 0.0 ;
```

The size of double is 8 bytes. Array ARR is located in memory starting at the beginning of virtual page 0xFF000 and stored in row major order. The cache is initially empty and no pre-fetching is done. The only data memory references made by the program are those to array ARR.

71. The total size of the tags i the cache directory is

(a) 32 Kbits (b) 34 Kbits (c) 64 Kbits (d) 68 Kbits

72. Which of the following array elements has the same cache index as ARR [0] [0]?

(a) ARR [0] [4] (b) ARR [4] [0] (c) ARR [0] [5] (d) ARR [5] [0]

73. The cache hit ratio for this initialization loop is

(a) 0% (b) 25% (c) 50% (d) 75%

Common Data for Questions 74 and 75 :

Consider the following C functions.

```
int f1 ( int n)
{
 if (n == 0 | | n == 1)
  return n;
 else
 return (2*f1 (n-1) + 3* f1 (n-2));
}
int f2 (int n)
{
 int i;
 int X[N], Y[N], Z[N];
 X[0] = Y[0] = Z[0] = 0;
 X[1] =1; Y[1] =2; Z[1] = 3;
 for ( i = 2; i <= n; i++) {
  X [i] = Y [i - 1] + z [i - 2];
  Y [i ] = 2 * X [ i ] ;
  Z [i ] = 3 * X [ i ] ;
 }
return X[n];
}
```

74. The running time of f1 (n) and f2(n) are

(a) Θ (n) and Θ (n) (b) Θ (2^n) and Θ (n)

(c) Θ (n) and Θ (2^n) (d) Θ (2^n) and Θ (2^n)

75. f1(8) and f2(8) return the values

(a) 1661 and 1640 (b) 59 and 59

(c) 1640 and 1640 (d) 1640 and 1661

Linked Answer Questions Q.76 to Q.85 carry two marks each

Statement for Linked Answer Questions 76 to 77

Delayed branching can help in the handling of control hazards

76. For all delayed conditional branch instructions, irrespective of whether the condition evaluates to true or false.

(a) the instruction following the conditional branch instruction in memory is executed

(b) the first instruction in the fall through path is executed

(c) the first instruction in the taken path is executed

(d) the branch takes longer to execute than any other instruction.

77. The following code is to run on a pipelined processor with one branch delay slot

11: ADD R2 $\leftarrow$ R7 + R8

12: SUB R4 $\leftarrow$ R5 - R6

13: ADD R1 $\leftarrow$ R2 + R3

14: STORE Memory [R4] $\leftarrow$ ==0

BRANCH to Label if R1==0

Which of the instructions I1, I2, I3 or I4 can legitimately occupy the delay slot without any other program modification?

(a) I1 (b) I2 (c) I3 (d) I4

Statement for Linked Answer Questions 78 to 79

Let x_n denote the number of binary strings of length n that contain no consecutive 0s.

78. Which of the following recurrences does x_n satisfy ?

(a) $x_n = 2x_{n-1}$ (b) $x_n = x_{\lceil \frac{n}{2} \rceil} + 1$

(c) $x_n = x_{\lceil \frac{n}{2} \rceil} + n$ (d) $x_n = x_{n-1} + x_{n-2}$

79. The value of x_5 is

(a) 5 (b) 7 (c) 8 (d) 16

Statement for Linked Answer Questions 80 to 81

The subset-sum problem is defined as follows. Given a set of n positive integers, $S = \{a_1, a_2, a_3 ... a_n\}$ and a positive integer W, is there a subset of S whose elements sum to W ? A dynamic program for solving this problem uses a 2-dimensional Boolean array X, with n rows and W+1 columns. X[i, j], $1 \le i \le n, 0 \le j \le W$, is TRUE if and only if there is a subset of $\{a_1, a_2, ... a_i\}$ whose elements sum to j.

80. Which of the following is valid for $2 \le i \le n$ and $a_i \le j \le W$?

(a) $X[i, j] = X[i-1, j] \vee X[i, j-a_i]$ (b) $X[i, j] = X[i-1, j] \vee X[i-1, j-a_i]$

(c) $X[i, j] = X[i-1, j] \wedge X[i, j-a_i]$ (b) $X[i, j] = X[i-1, j] \wedge X[i-1, j-a_i]$

81. Which entry of the array X, if TRUE, implies that there is a subset whose elements sum to W?

(a) X [1, W] (b) X [n,0] (c) X[n, W] (d) X[n-1, n]

Statement for Linked Answer Questions 82 to 83

Consider the following ER diagram.

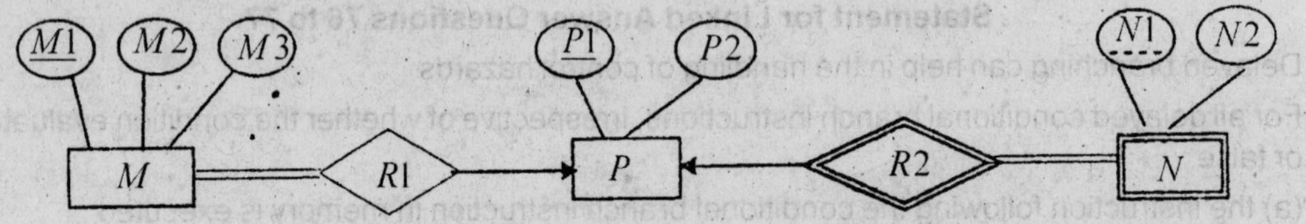

82. The minimum number of tables needed to represent M, N, P, R1, R2 is

(a) 2 (b) 3 (c) 4 (d) 5

83. Which of the following is a correct attribute set for one of the tables for the correct answer to the above question ?

(a) { M1, M2, M3, P1} (b) {M1, P1, N1, N2} (c) {M1, P1, N1} (d) {M1, P1}

Statement for Linked Answer Questions 84 to 85

Consider the following C program that attempts to locate an element x in an array Y[] using binary search. The program is erroneous.

```
1.  f (int Y[10], int x ) {
2.    int i, j, k;
3.  i = 0;  j = 9;
4.  do {
5.       k = (i + j) /2
6.    if ( Y[k] < x) i = k; else j = k;
7.   } while (( Y [k]  !=x ) && ( i < j) ) ;
8.  if ( Y[k] == x) printf ("x is in the array ");
9.  else printf ("x is not in the array ") :
10. }
```

84. On which of the following contents of Y and X does the program fail ?

(a) Y is [1 2 3 4 5 6 7 8 9 10] and x < 10

(b) Y is [1 3 5 7 9 11 13 15 17 19] and x < 1

(c) Y is [2 2 2 2 2 2 2 2 2 2] and x >2

(d) Y is [2 4 6 8 10 12 14 16 18 20] and 2 < x < 20 and x is even.

85. The correction needed in the program to make it work properly is

(a) change line 6 to : if (Y[k] < x) i = k + 1; else j = k - 1;

(b) change line 6 to : if (Y[k] < x) i = k - 1; else j = k + 1;

(c) change line 6 to : if (Y[k] <= x) i = k; else j = k ;

(d) change line 7 to : } while (Y[k] = = x) && (i<j));

ANSWERS

1. (a)	2. (d)	3. (d)	4. (d)	5. (a)	6. (d)	7. (c)	8. (c)	9. (d)	10.(b)
11.(d)	12.(a)	13. (d)	14.(a)	15.(d)	16. (c)	17.(d)	18. (a)	19.(c)	20.(d)
21. (a)	22. (c)	23. (a)	24.(a)	25. (b)	26.(a)	27.(c)	28. (a)	29.(a)	30.(a)
31. (b)	32. (b)	33. (c)	34. (d)	35. (b)	36. (a)	37.(b)	38. (b)	39.(d)	40.(a)
41. (a)	42. (d)	43. (b)	44. (b)	45. (d)	46. (d)	47.(a)	48. (d)	49.(b)	50.(d)
51. (c)	52. (c)	53. (a)	54. (d)	55. (b)	56. (d)	57.(c)	58. (b)	59.(c)	60.(d)
61. (d)	62. (b)	63. (c)	64. (b)	65. (a)	66. (b)	67.(b)	68. (c)	69.(c)	70.(c)
71. (c)	72. (b)	73. (b)	74. (b)	75. (c)	76. (d)	77.(d)	78. (d)	79.(c)	80.(c)
81. (c)	82. (a)	83. (a)	84. (c)	85. (a)					

GATE - 2008

IT : INFORMATION TECHNOLOGY

Read the following instructions carefully

1. Using HB pencil, darken the appropriate bubble under each digit of your registration number and the letters corresponding to your paper code.

2. All the questions in this question paper are of objective type.

3. Questions must be answered on **O**bjective **R**esponse **S**heet **(ORS)** by darkening the appropriate bubble (marked A, B, C, D) using HB pencil against the question number on the left hand side of the ORS. **Each question has only one correct answer.** In case you wish to change an answer, erase the old answer completely. More than one answer bubbled against a question will be treated as a wrong answer.

4. Questions 1 through 20 are 1 - mark questions and questions 21 through 85 are 2 - mark questions.

5. Questions 71 through 73 is one set of common data questions, questions 74 and 75 is another pair of common data questions. The question Pairs (76, 77), (78, 79), (80, 81), (82, 83) and (84, 85) are questions with linked answers. The answer to the second question of the above pairs will depend on the answer to the first question of the pair. If the first question in the linked pair is wrongly answered or is un - attempted, then the answer to the second question in the pair will not be evaluated.

6. Un - attempted questions will carry zero marks.

7. **NEGATIVE MARKING :** For Q.1 to Q. 20, **0.25** mark will be deducted for each wrong answer. For Q.21 to Q.75, **0.5** mark will be deducted for each wrong answer. For the pairs of questions with linked answers, there will be negative marks only for wrong answer to the first question, i. e; for Q. 76, Q.78, Q.80, Q.82 and Q.84, **0.5** mark will be deducted for each wrong answer. There is no negative marking for Q.77, Q.79, Q81, Q.83 and Q.85.

8. Calculator **without data connectivity** is allowed in the examination hall.

9. Charts, graph sheets and tables are NOT allowed in the examination hall.

10. Rough work can be done on the question paper itself. Additional blank pages are given at the end of the question paper for rough work.

Q.1 to Q. 20 Carry one mark each

1. A set of Boolean connectives is functionally complete if all Boolean functions can be synthesized using those. Which of the following sets of connectives is NOT functionally complete ?

   (a) EX-NOR (b) implication, negation (c) OR,negation (d)NAND

2. A Sample space has two events A and B such that probabilities $P(A \cap B) = 1/2, P(\overline{A}) = 1/3, P(\overline{B}) = 1/3$

   What is $P(A \cup B)$?

   (a) 11/12 (b) 10/12 (c) 9/12 (d) 8/12

3. What is the chromatic number of the following graph ?

   (a) 2

   (b) 3

   (c) 4

   (d) 5

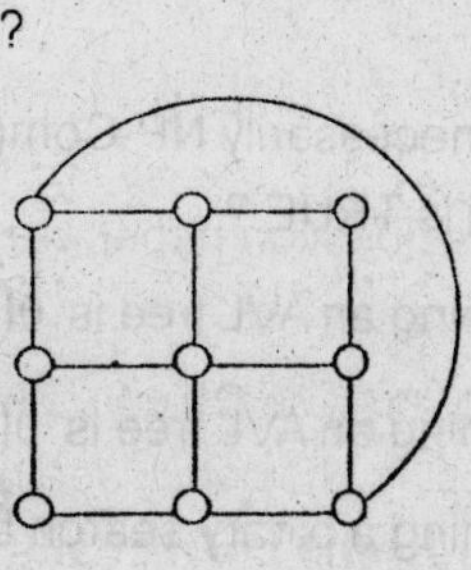

4. What is the size of the smallest MIS(Maximal Independent Set) of a chain of nine nodes ?

   (a) 5 (b) 4 (c)3 (d) 2

5. Which of the following regular expressions describes the language over {0,1} consisting of strings that contain exactly two 1's ?

   (a) (0+1) * 11(0+1)* (b) 0*110*

   (c) 0*10*10* (d)(0+1)*1(0+1)*(0+1)*

6. Let N be an NFA with n states and let M be the minimized DFA with m states recognizing the same language. Which of the following in NECESSARILY true ?

   (a) $m \le 2^n$ (b) $n \le m$ (c) M has one accept state (d) $m = 2^n$

7. The following bit pattern represents a floating point number in JEEE 754 single precision format

   1 10000011 10100000000000000000000

   The value of the number in decimal form is

   (a) -10 (b) -13 (c) -26 (d) None of the above

8. Consider the following Boolean function of four variables

   $f(A,B,C,D) = \Sigma(2,3,6,7,8,9,10,11,12,13)$

   The function is

   (a) independent of one variable (b) independent of two variables

   (c) independent of three variables (d) independent on all the variables

9. What Boolean function does the circuit below realise ?

   (a) $xz + \overline{x}\overline{z}$

   (b) $x\overline{z} + \overline{x}z$

   (c) $\overline{xy} + yz$

   (d) $xy + yz$

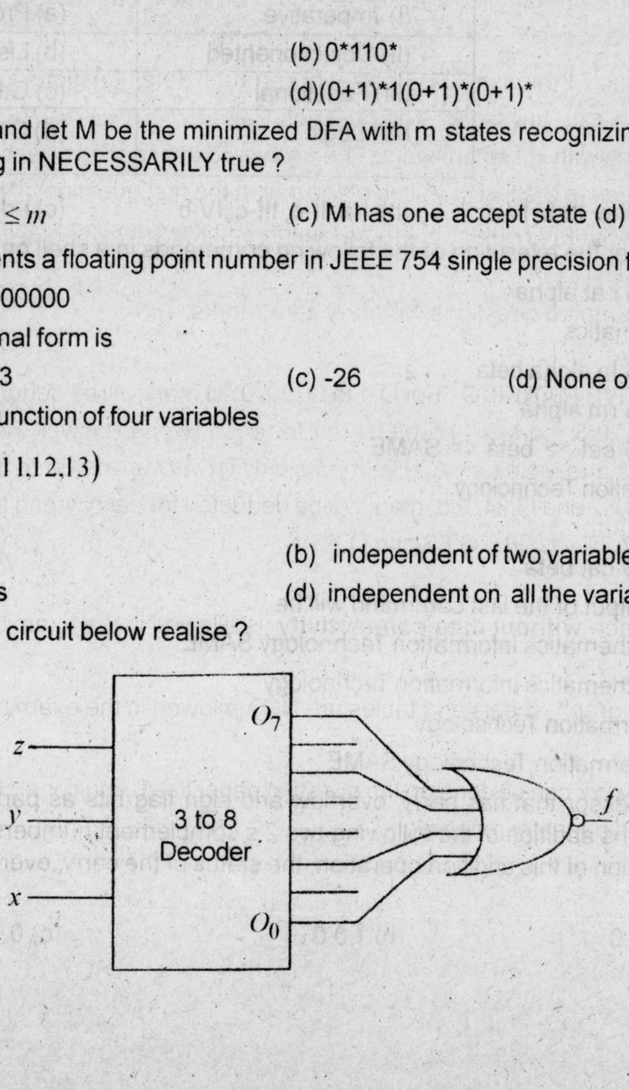

10. Arrange the following functions in increasing asymptotic order.

A. $n^{1/3}$ B. e^{n} C. $n^{7/4}$ D. $n \log^{9} n$

E. 1.0000001^{n}

(a) A,D,C,E,B (b) D,A,,C,E,B (c) A,C,D, E,B (d) A,C,D,B,E

11. For problems X and Y, Y is NP-complete and X reduces to Y in polynomial time. Which of the following is TRUE ?

(a) If X can be solved in polynomial time, then so can Y

(b) X is NP-complete

(c) X is NP-hard

(d) X is in NP, but not necessarily NP-Complete

12. Which of the following is TRUE ?

(a) The cost of searching an AVL tree is $\theta(\log n)$ but that of a binary search tree is 0(n)

(b) The cost of searching an AVL tree is $\theta(\log n)$ but that of a complete binary tree is $\theta(n \log n)$

(c) The cost of searching a binary search tree is O(log n) but that of an AVL tree is $\theta(n)$

(d) The cost of searching an AVL tree is $\theta(n \log n)$ but that of a binary search tree is O(n).

13. Match the programming paradigms and languages given in the following table.

Paradigms	Languages
(I) Imperative	(a) Prolog
(II) Object oriented	(b) Lisp
(III) Functional	(c) C,Fortran 77, Pascal
(IV) Logic	(d) C++, Smalltalk, Java

(a) I-c, II-d, III-b, IV-a (b) I-a, II-d, III-c, IV-b (c) I-d, II-c, III-b, IV-a (d) I-c, II-d, III-a, IV-b

14. Consider the execution of the following commands in a shell on a Linux operating system.

```
BASH$ cat alpha
Mathematics
BASH$ In alpha beta
BASH$ rm alpha
BASH$ cat >> beta << SAME
Information Technology
SAME
BASH$ cat beta
```

The output of the last command will be :

(a) Mathematics Information Technology SAME

(b) Mathematics Information Technology

(c) information Technology

(d) Information Technology SAME

15. A processor that has carry, overflow and sign flag bits as part of its program status word (PSW) performs addition of the following two 2's complement numbers 01001101 and 11101001. After the execution of this addition operation, the status of the carry, over-flow and sign flags, respectively will be:

(a) 1,1,0 (b) 1,0,0 (c) 0, 1, 0 (d) 1, 0, 1

16. A paging scheme uses a Translation Look-aside Buffer (TLB). A TLB-access takes 10 ns and a main memory access takes 50 ns. What is the effective access time (in ns) if the TLB hit ratio is 90% and there is no page-fault ?

(a) 54 (b) 60 (c) 65 (d) 75

17. Find if the following statements in the context of software testing are TRUE or FALSE.

(S1) Statement coverage cannot guarantee execution of loops in a program under test.

(S2) Use of independent path testing criterion guarantees execution of each loop in a program under test more than once.

(a) True, True (b) True, False (c) False, True (d) False, False

18. How many bytes of data can be sent in 15 seconds over a serial link with baud rate of 9600 in asynchronous mode with odd parity and two stop bits in the frame?

(a) 10,000 bytes (b) 12,000 bytes (c) 15,000 bytes (d) 27,000 bytes

19. Which of the following is TRUE only of XML but NOT HTML ?

(a) It is derived from SGML

(b) It describes content and layout

(c) It allows user defined tags

(d) It is restricted only to be used with web browsers

20. Provide the best matching between the entries in the two columns given in the table below:

I.	Proxy Server	a.	Fire wall
II.	KaZaA, DC++	b.	Caching
III.	SLIP	c.	P2P
IV.	DNS	d.	PPP

(a) I-a, II-d, III-c, IV-b (b) I-b, II-d, III-c, IV-a

(c) I-a, II-c, III-d, IV-b (d) I-b, II-c, III-d, IV-a

Q.21 to Q.75 carry two marks each.

21. Which of the following first order formulae is logically valid ? Here $\alpha(x)$ is a first order formula with x as a free variable, and β is a first order formula with no free variable.

(a) $[\beta \rightarrow (\exists x, \alpha(x))] \rightarrow [\forall x, \beta \rightarrow \alpha(x)]$ (b) $[\exists x, \beta \rightarrow \alpha(x)] \rightarrow [\beta \rightarrow [\forall x, \alpha(x))]$

(c) $[(\exists x, \alpha(x)] \rightarrow \beta] \rightarrow [\forall x, \alpha(x) \rightarrow \beta]$ (d) $[\forall x, \alpha(x)) \rightarrow \beta] \rightarrow [\forall x, \alpha(x)) \rightarrow \beta]$

22. Which of the following is the negation of $[\forall x, \alpha \rightarrow (\exists y, \beta \rightarrow [\forall u, \exists v, \gamma)))]$

(a) $[\exists x, \alpha \rightarrow (\forall y, \beta \rightarrow (\exists u, \forall v, \gamma)))]$ (b) $[\exists x, \alpha \rightarrow (\forall y, \beta \rightarrow (\exists u, \forall v, \neg\gamma)))]$

(c) $[\forall x, \neg\alpha \rightarrow (\exists y, \neg\beta \rightarrow (\forall u, \exists v - \neg\gamma)))]$ (d) $[\exists x, \alpha \wedge (\forall y, \beta \wedge (\exists u, \forall v, \neg\gamma)))]$

23. What is the probability that in a randomly chosen group of r people at least three people have the same birthday?

(a) $1 - \dfrac{365.364...(365 - r + 1)}{365^r}$

(b) $\dfrac{365.364...(365 - r + 1)}{365^r} + rC_2.365.\dfrac{364.363...(364 - (r - 2) + 1)}{365^{r-2}}$

(c) $1-\dfrac{365.364...(365-r+1)}{365^r}-{}^rC_2.365.\dfrac{364.363...(364-(r-2)+1)}{365^{r-2}}$

(d) $\dfrac{365.364...(365-r+1)}{365^r}$

24. The exponent of 11 in the prime factorization of 300! is

(a) 27 (b) 28 (c) 29 (d) 30

25. In how many ways can b blue balls and r red balls be distributed in n distinct boxes ?

(a) $\dfrac{(n+b-1)!(n+r-1)!}{(n-1)!b!(n-1)!r!}$ (b) $\dfrac{(n+(b+r)-1)!}{(n-1)!(n-1)!(b+r)!}$

(c) $\dfrac{n!}{b!r!}$ (d) $\dfrac{(n+(b+r)-1)!}{n!(b+r-1)!}$

26. Consider the field C of complex numbers with addition and multiplication. Which of the following form(s) a subfield of C with addition and multiplication?

(S1) the set of real numbers

(S2) {a+ib) | a and b are rational numbers }

(S3) {a + ib} | $(a^2+b^2)\leq 1$}

(S4) {ia | a is real }

(a) only SI (b) S1 and S3 (c) S2 and S3 (d) S1 and S2

27. G is a simple undirected graph. Some vertices of G are of odd degree. Add a node v to G and make it adjacent to each odd degree vertex of G. The resultant graph is sure to be

(a) regular (b) complete (c) Hamiltonian (d) Euler

28. Consider the following Hasse diagrams

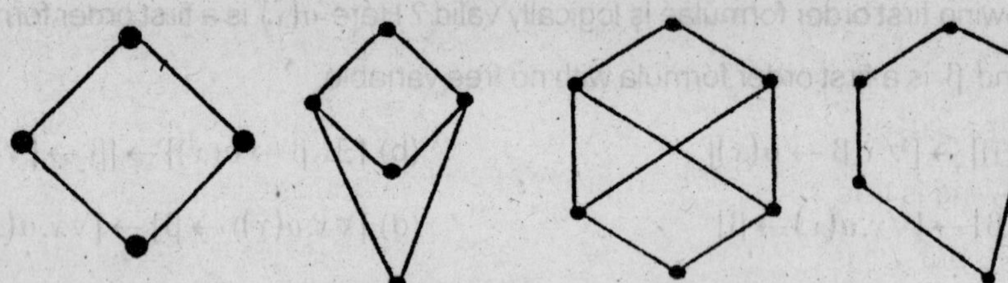

(i) (ii) (iii) (iv)

Which of the above represent a lattice ?

(a) (i) and (iv) only (b) (ii) and (iii) only

(c) (iii) only (d) (i),(ii) and (iv) only

29. If M is a square matrix with a zero determinant, which of the following assertion (s) is (are) correct ?

(S1) each row of M can be represented as a linear combination of the other rows

(S2) Each column of M can be represented as a linear combination of the other columns

(S3) MX=0 has a nontrivial solution

(S4) M has an inverse

(a) S3 and S2 (b) S1 and S4 (c) S1 and S3 (d) S1,S2 and S3

30. Consider the function $f(x) = x^2 - 2x - 1$. Suppose an execution of the Newton-Raphson method to find a zero of $f(x)$ starts with an approximation $x_0 = 2$ of x. What is the value of x_2, the approximation of x that algorithm produces after two iterations, rounded to three decimal places.

(a) 2.417 (b) 2.419 (c) 2.423 (d) 2.425

31. If $f(x)$ is defined as follows, what is the minimum value of $f(x)$ for $x \in (0,2)$?

$$f(x) = \begin{cases} \frac{25}{8x} & when\, x \le \frac{3}{2} \\ x + \frac{1}{x} & otherwise \end{cases}$$

(a) 2 (b) $2\frac{1}{12}$ (c) $2\frac{1}{6}$ (d) $2\frac{1}{2}$

32. If the final states and non-final states in the DFA below are interchanged, then which of the following languages over the alphabet {a,b} will be accepted by the new DFA ?

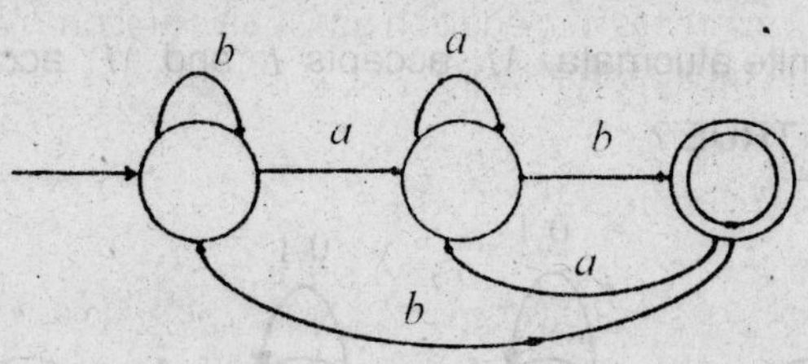

(a) Set of all strings that do not end with ab

(b) Set of all strings that begin with either an a or a b

(c) Set of all strings that do not contain the substring ab

(d) The set described by the regular expression b*aa*(ba)*b*

33. Consider the following languages

$L_1 = \{a^i b^j c^k \mid i = j, k \ge 1\}$

$L_2 = \{a^i b^j \mid j = 2i, i \ge 0\}$

Which of the following is true ?

(a) L_1 is not a CFL but L_2 is

(b) $L_1 \cap L_2 = \Phi$ and L_1 is non-regular

(c) $L_1 \cup L_2$ is not a CFL but L_2 is

(d) There is a 4-state PDA that accepts L_1, but there is no DPDA that accepts L_2

34. Consider a CFG with the following productions

$S \to AA / B$

$A \to 0A \mid A0 \mid 1$

$B \to 0B00 \mid 1$

S is the start symbol, A and B are non-terminals and 0 and 1 are the terminals. The language generated by this grammar is

(a) $\{0^n 10^{2n} \mid n \geq 1\}$

(b) $\{0^i 10^j 10^k \mid i, j, k \geq 0\} \cup \{0^n 10^{2n} \mid n \geq 1\}$

(c) $\{0^i 10^j \mid i, j \geq 0\} \cup \{0^n 10^{2n} \mid n \geq 1\}$

(d) The set of all strings over {0, 1} containing at least two 0's

35. Which of the following languages is (are) non-regular ?

$L_1 = \{0^m 1^n \mid 0 \leq m \leq n \leq 10000\}$

$L_2 = \{w \mid w$ reads the same forward and backward$\}$

$L_3 = \{w \in \{0,1\}^* \mid$ w contains an even number of 0's and an even number of 1's$\}$

(a) L_2 and L_3 only
(b) L_1 and L_2 only
(c) L_3 only
(d) L_2 only

36. Consider the following two finite atuomata. M_1 accepts L_1 and M_2 accepts L_2

Which one of the following is TRUE ?

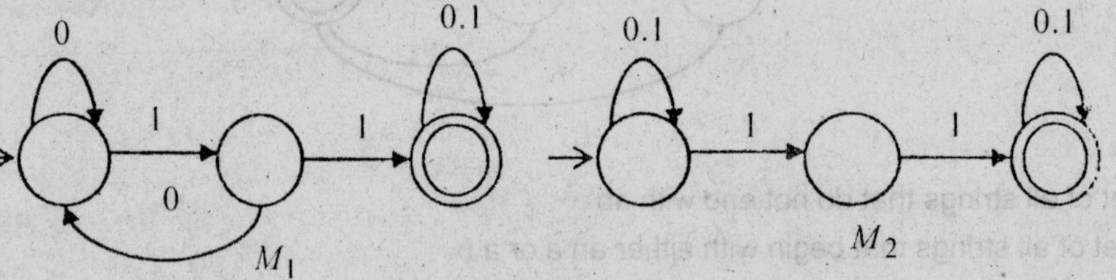

(a) $L_1 = L_2$
(b) $L_1 \subset L_2$
(c) $L_1 \cap \overline{L}_2 = \Phi$
(d) $L_1 \cup L_2 \neq L_1$

37. Consider the following state diagram and its realization by a JK flip flop

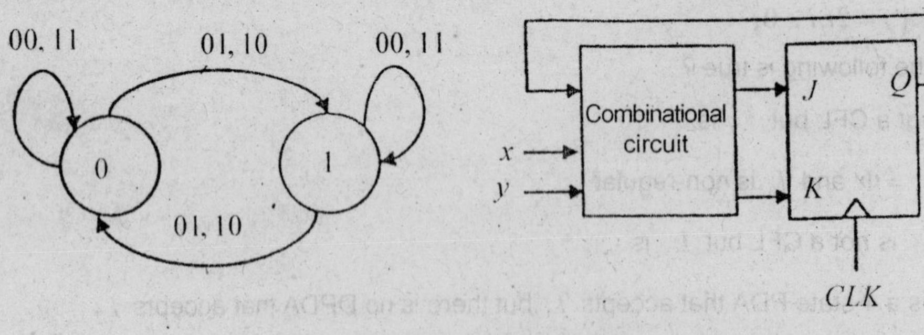

The combinational circuit generates J and K in terms of x,y and Q

The Boolean expressions for J and K are:

(a) $\overline{x \oplus y}$ and $\overline{x \oplus y}$
(b) $\overline{x \oplus y}$ and $x \oplus y$
(c) $x \oplus y$ and $\overline{x \oplus y}$
(d) $x \oplus y$ and $x \oplus y$

38. Assume that EA=(X)+ is the effective address equal to the contents of location X. with X incremented by one word length after the effective address is calculated; EA=-(X) is the effective address equal to the contents of locations X, with X decremented by one word length before the effective address is calculated; EA=(X)- is the effective address equal to the contents of location X with X decremented by one word length after the effective address is calculated. The format of the instruction is (opcode, source, destination), which means (destination $\leftarrow$ source op destination) Using X as a stack pointer, which of the following instructions can pop the top two elements from the stack, perform the addition operation and push the result back to the stack.

(a) ADD (X)-, (X) (b) ADD (X), (X)-
(c) ADD-(X), (X)+ (d) ADD - (X), (X)

39. Consider a CPU where all the instructions require 7 clock cycles to complete execution. There are 140 instructions in the instruction set. It is found that 125 control signals are needed to be generated by the control unit. While designing the horizontal microprogrammed control unit, single address field format is used for branch control logic. What is the minimum size of the control work and control address register ?

(a) 125, 7 (b) 125,10 (c) 135, 9 (d) 135, 10

40. A nonpipelined single cycle processor operating at 100 MHz is converted into a synchronous pipelined processor with five stages requiring 2.5 nsec, 1.5 nsec, 2nsec, 1.5 nsec and 2.5 nesec, respectively. The delay of the latches is 0.5 nsec. The speedup of the pipeline processor for a large number of instructions is

(a) 4.5 (b) 4.0 (c) 3.33 (d) 3.0

41. Assume that a main memory with only 4 pages, each of 16 bytes, is initially empty. The CPU generates the following sequence of virtual addresses and uses the Least Recently Used (LRU) page replacement policy.

0,4,8,20,24,36,44,12,68,72,80,84,28,32,88,92

How many page faults does this sequence cause ? What are the page numbers of the pages present in the main memory at the end of the sequence ?

(a) 6 and 1,2,3,4 (b) 7 and 1,2,4,5
(c) 8 and 1,2,4,5 (d) 9 and 1,2,3,5

42. The two numbers given below are multiplied using the Booth's algorithm.

Multiplicand : 0101 1010 1110 1110
Multiplier : 0111 0111 1011 1101

How many additions/Substraction are required for the multiplication of the above two numbers?

(a) 6 (b) 8 (c) 10 (d) 12

43. If we use Radix Sort to sort n integers in the range $\left(n^{k/2}, n^{k}\right)$, for some k>0 which is independent of n, the time taken would be

(a) $\theta(n)$ (b) $\theta(kn)$ (c) $\theta(n\log n)$ (d) $\theta\left(n^{2}\right)$

44. When $n = 2^{2k}$ for some $k \geq 0$ the recurrence relation

$$T(n) = \sqrt{2}\,T(n/2) + \sqrt{n},\; T(1) = 1$$

evaluate to

(a) $\sqrt{n}\left(\log n + 1\right)$ (b) $\sqrt{n}\log n$ (c) $\sqrt{n}\log\sqrt{n}$ (d) $n\log\sqrt{n}$

45. For the undirected, weighted graph given below, which of the following sequences of edges represents a correct execution of Prim's algorithm to construct a Minimum Spanning Tree?

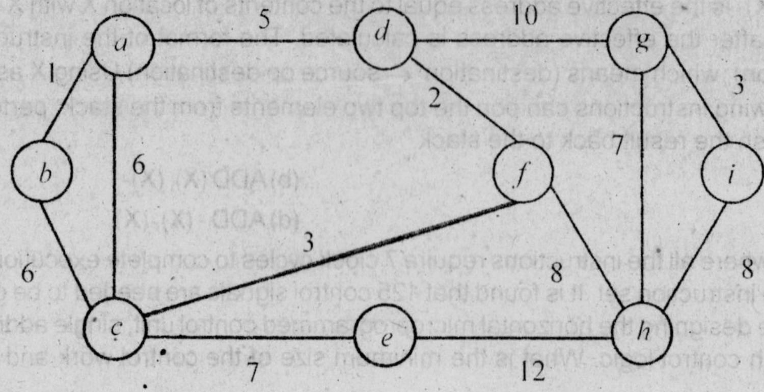

(a) (a,b), (d,f), (f,c), (g,i), (d,a), (g,h), (c,e) , (f,h)

(b) (c,e), (c,f), (f,d), (d,a), (a,b),(g,h), (h,f), (g,i)

(c) (d,f), (f,c), (d,a), (a,b), (c,e), (f,h), (g,h),(g,i)

(d) (h,g), (g,i), (h,f) , (f,c), (f,d), (d,a), (a,b), (c,e)

46. The following three are known to be the preorder, inorder and postorder sequences of a binary tree. But it is not known which is which

I. MBCAFHPYK

II. KAMCBYPFH

III. MABCKYFPH

Pick the true statement from the following

(a) I and II are preorder and inorder sequences, respectively

(b) I and III are preorder and postorder sequences, respectively

(c) II is the inorder sequence, but nothing more can be said about the other two sequences

(d) II and III are the preorder and inorder sequences, respectively.

47. Consider the following sequence of nodes for the undirected graph given below

I. a b e f d g c

II. a b e f c g d

III. a d g e b c f

IV. a d b c g e f

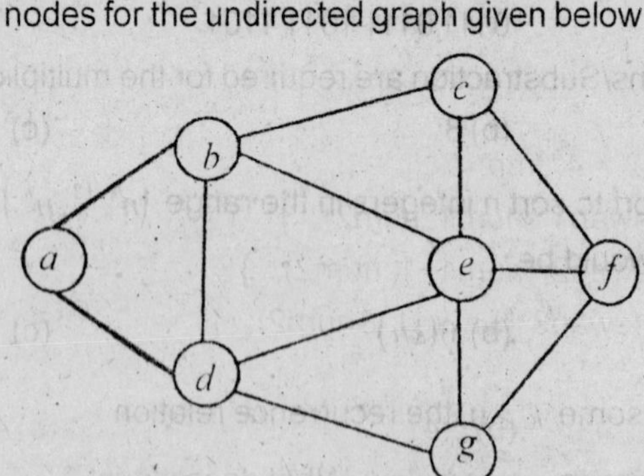

A depth First Search (DFS) is started at node a. The nodes are listed in the order they are first visited. Which all of the above is (are) possible output(s) ?

(a) I and III only
(b) II and III only
(c) II,IIII and IV only
(d) I,II and III only

48. Consider a hash table of size 11 that uses open addressing with linear probing. Let $h(k) = k$ mod 11 be the hash function used. A sequence of records with keys.

43 36 92 87 11 4 71 13 14

is inserted into an initially empty hash table, the bins of which are indexed from zero to ten. What is the index of the bin into which the last record is inserted ?

(a) 3 (b) 4 (c) 6 (d) 7

49. What is the output printed by the following C code ?

```
# include <stdio.h>
int main ( )
{
  char a[6] = "world";
  int i, j ;
  for (i = 0, j=5; i<j;  a[i++] =a[j--])
  printf ("%s\n", a);
 }
```

(a) dlrow (b) Null string (c) dlrld (d) worow

50. Consider the C program below. What does it print ?

```
# include <stdio.h>
# define swapl (a,b) tmp=a; a=b; b=tmp
void swap2 ( int a , int b)
{
    int tmp;
    tmp=a; a=b; b=tmp;
}
void swap3 (int *a, int *b)
{
  int tmp;
  tamp = *a;  *a=*b;  *b=tmp;
}
int main ( )
{
 int num1 =5, num2 = 4, tmp;
 if (num1<num2) {swap1  (num1, num2);  }
 if (num1<num2) {swap2  (num1+1, num2);  }
 if (num1>=num2) {swap3  (&num1, &num2);  }
}
```

(a) 5,5 (b) 5,4 (c) 4,5 (d) 4,4

51. Consider the C program given below. What does it print ?

```
# include <stdio.h>
int main ( )
{
int i, s;
  int a[8] = {1, 2, 3, 4, 5, 6, 7, 8};
  for (i=0;  i<3;  i++) {
```

```
    a[i] = a[i] + 1;
    i++;
  }
  i-- ;
  for (j=7; j>4, j--) {
  int i=j/2;
  a[i] = a[i] -1;
  }
  printf ("%d, %d", i, a[i]);
}
```

(a) 2,3 (b) 2,4 (c) 3,2 (d) 3,3

52. A C program is given below:

```
# include <stdio.h>
int main ( )
{
  int i, j;
  char a[2] [3] = {'a', 'b', 'c'}, {'d', 'e', 'f'} };
  char b[3] [2];
  char *p = *b;
  for (i=0; i<2; i++0 {
      for ( i = 0; j < 3 ; j ++ ) {
      * (p + 2 *j +1) = a [i] [j];
      }
  }
}
```

What should be the contents of the array b at the end of the program ?

(a) $\begin{matrix} a & b \\ c & d \\ e & f \end{matrix}$ (b) $\begin{matrix} a & d \\ b & e \\ c & f \end{matrix}$ (c) $\begin{matrix} a & c \\ e & b \\ d & f \end{matrix}$ (d) $\begin{matrix} a & e \\ d & c \\ b & f \end{matrix}$

53. The following is a code with two threads, producer and consumer, that can run in parallel. Further, S and Q are binary semaphores equipped with he standard P and V operations.

```
semaphore S=1; Q=0 ;
integer x ;
producer :              consumer :
while (true) do         While (true) do
P(S);                   P(Q) ;
x=produce( );           consume (x) ;
V(Q) ;                  V (S) ;
done                    done
```

Which of the following is TRUE about the program above ?

(a) The process can deadlock

(b) One of the threads can starve

(c) Some of the items produced by the producer may be lost

(d) Values generated and stored in 'x' by the producer will always be consumed before the producer can generate a new value.

54. An operating system implements a policy that requires a process to release all resources before making a request for another resource. Select the TRUE statement from the following :

(a) Both starvation and deadlock can occur

(b) Starvation can occur but deadlock cannot occur

(c) Starvation cannot occur but deadlock can occur

(d) Neither starvation nor deadlock can occur

55. If the time-slice used in the round-robin scheduling policy is more than the maximum time required to execute any process, then the policy will

(a) degenerate to shortest job first

(b) degenerate to priority scheduling

(c) degenerate to first come first serve

(d) none of these

56. Match the following flag bits used in the context of virtual memory management on the left side with the different purposes on the right side of the table below.

Name of the bit	Purpose
1. Dirty	a. Page initialization
2. R/W	b. Write-back policy
3. Reference	c. Page protection
4. Valid	d. Page replacement policy

(a) I-d, II-a, III-b, IV-c (b) I-b, II-c, III-a, IV-d

(c) I-c, II-d, III-a, IV-b (d) I-b, II-c, III-d, IV-a

57. Which of the following are NOT considered when computing function points for a software project?

(01) External inputs and outputs

(02) Programming language to be used for the implementation

(03) User interactions

(04) External interfaces

(05) Number of programmers in the software project

(06) Files used by the system

(a) O2, 03 (b) O1, O5 (c) O4, O6 (d) 02, 05

58. A software project plan has identified ten tasks with each having dependencies as given in the following table:

Task	Depends On
T1	-
T2	T1
T3	T1
T4	T1
T5	T2
T6	T3
T7	T3, T4
T8	T4
T9	T5, T7,T8
T10	T6,T9

Answer the following questions :

(Q1) What is the maximum number of tasks that can be done concurrently ?

(Q2) What is the minimum time required to complete the project, assuming that each task requires one time unit and there is no restriction on the number of tasks that can be done in parallel ?

(a) 5,5 (b) 4,4 (c) 4,5 (d) 5,4

59. A software engineer is required to implement two sets of algorithms for a single set of matrix operations in an object oriented programming language; the two sets of algorithms are to provide precisions of 10^{-3} and 10^{-6}, respectively. She decides to implement two classes, Low Precision Matrix and High Precision Matrix, providing precisions 10^{-3} and 10^{-6} respectively. Which one of the following is the best alternative for the implementation ?

(S1) The two classes should be kept independent

(S2) Low Precision Matrix should be derived from High Precision Matrix

(S3) High Precisoin Matrix should be derived from Low Precision Matrix

(S4) One class should be derived from the other; the hierarchy is immaterial

(a) S1 (b) S2 (c) S3 (d) S4

60. Which of the following requirement specifications can be validated ?

(S1) If the system fails during any operation, there should not be any loss of data

(S2) The system must provide reasonable performance even under maximum load conditions

(S3) The software executable must be deployable under Ms windows 95, 2000 and XP.

(S4) User interface windows must fit on a standard monitors' screen

(a) S4 and S3 (b) S4 and S2 (c) S3 and S1 (d) S2 and S1

61. Let R (A, B ,C, D) be a relational schema with the following functional dependencies:

$A \rightarrow B$, $B \rightarrow C$,

$C \rightarrow D$ and $D \rightarrow B$. The decomposition of R into (A,B), (B,C), (B,D)

(a) gives a lossless join, and is dependency preserving

(b) gives a lossless join but is not dependencey preserving

(c) does not give a lossless join, but is dependencey preserving

(d) does not give a lossless join, and is not dependencey preserving

62. Let R(A,B,C,D,E,P,G) be a relational schema in which the following functional dependencies are known to hold : $AB \rightarrow CD$, $DE \rightarrow P$, $C \rightarrow E$, $P \rightarrow C$ and $B \rightarrow G$. The relational schema R is

(a) in BCNF (b) in 3NF, but not in BCNF (c) in 2NF, but not in 3NF (d) not in 2NF

63. Consider the following three schedules of transactions T1, T2 and T3 [Notation: In the following NYO represents the action Y (R for read, W for write) performed by transaction N on object O.]

(S1) 2RA 2WA 3RC 2WB 3WA 3WC 1RA 1RB 1WA 1WB

(S2) 3RC 2RA 2WA 2WB 3WA 1RA 1RB 1WA 1WB 3WC

(S3) 2RA 3RC 3WA 2WA 2WB 3WC 1RA 1RB 1WA 1WB

Which of the following statements is TRUE ?

(a) S1, S2 and S3 are all conflict equivalent to each other

(b) No two of S1, S2 and S3 are conflict equivalent to each other

(c) S2 is conflict equivalent to S3, but not to S1

(d) S1 is conflict equivalent to S2, but not to S3

64. A 1 Mbps satellite link connects two ground stations. The altitude of the satellite is 36,504 km and speed of the signal is 3 x 10^8 m/s. What should be the packet size for a channel utilization of 25% for a satellite link using go-back-127 sliding window protocol? Assume that the acknowledgment packets are negligible in size and that there are no errors during communication.

(a) 120 bytes (b) 60 bytes (c) 240 bytes (d) 90 bytes

65. The minimum frame size required for a CSMA/CD based computer network running at 1 Gbps on a 200m cable with a link speed of 2×10^8 m/s is

(a) 125 bytes (b) 250 bytes (c) 500 bytes (d) None of these

66. Data transmitted on a link uses the following 2D parity scheme for error detection: Each sequence of 28 bits is arranged in a 4 x 7 matrix(rows r_0 through r_3, and columns d_7 through d_1) and is padded with a column d_0 and row r_4 of partiy bits computed using the Even parity scheme. Each bit of column d_0 (respectively., row r_4) gives the parity of the corresponding row (respectively, column). These 40 bits are transmitted over the data link.

	d_7	d_6	d_5	d_4	d_3	d_2	d_1	d_0
r_0	0	1	0	1	0	0	1	1
r_1	1	1	0	0	1	1	1	0
r_2	0	0	0	1	0	1	0	0
r_3	0	1	1	0	1	0	1	0
r_4	1	1	0	0	0	1	1	0

The table shows data received by a receiver and has n corrupted bits. What is the minimum possible value of n?

(a) 1 (b) 2 (c) 3 (d) 4

67. Two popular routing algorithms are Distance Vector (DV) and Link State (LS) routing Which of the following are true ?

(S1) Count to infinity is a problem only with DV and not LS routing

(S2) in LS, the shortest path algorithm is run only at one node

(S3) in DV, the shortest path algorithm is run only at one node

(S4)DV requires lesser number of network messages than LS

(a) S1, S2 and S4 only (b) S1,S3 and S4 only

(c) S2 and S3 only (d) S1 and S4 only

68. Which of the following statements are TRUE ?

(S1) TCP handles both congestion and flow control

(S2) UDP handles congestion but not flow control

(S3) Fast retransmit deals with congestion but not flow control

(S4) Slow start mechanism deals with both congestion and flow control

(a) S1, S2 and S3 only (b) S1and S3 only

(c) S3 and S4 only (d) S1,S3 and S4 only

69. The three way handshake for TCP connection establishment is shown below

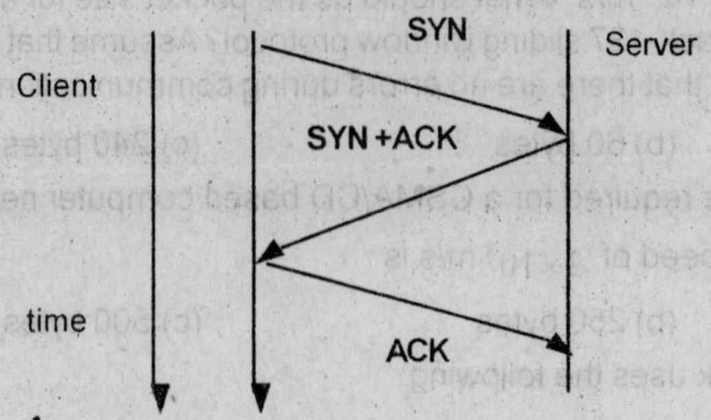

Which of the following statements are TRUE ?

(S1) Loss of SYN + ACK from the server will not establish a connection

(S2) Loss of ACK from the client cannot establish the connection

(S3) The server moves LISTEN $\rightarrow$ SYN_RCVD $\rightarrow$ SYN_SENT $\rightarrow$ ESTABLISHED in the state machine on no packet loss.

(S4) The server moves LISTEN $\rightarrow$ SYN_RCVD $\rightarrow$ ESTABLISHED in the state machine on no packet loss.

(a) S2 and S3 only (b) S1and S4 only

(c) S1 and S3 only (d) S2 and S4 only

70. The total number of keys required for a set of n individuals to be able to communicate with each other using secret key and public key cryptosystems, respectively are:

(a) n(n-1) and 2n (b) 2n and $\frac{n(n-1)}{2}$

(c) $\frac{n(n-1)}{2}$ and 2n (d) $\frac{n(n-1)}{2}$ and n

Common Data Questions

Common Data for Questions 71, 72 and 73 :

A Binary Search Tree (BST) stores values in the range 37 to 573. Consider the following sequence of keys.

1. 81, 537, 102, 439, 285, 376, 305

II. 52, 97, 121, 195, 242, 381, 472

III. 142, 248, 520, 386, 345, 270, 307

IV. 550, 149, 507, 395, 463, 402, 270

71. Suppose the BST has been unsuccessfully searched for key 273. Which all of the above sequences list nodes in the order in which we could have encountered them in the search ?

(a) II and III only (b) I and III only (c) III and IV only (d) III only

72. Which of the following statements is TRUE ?

(a) I, II and IV are inorder sequences of three different BSTs

(b) I is a preorder sequence of some BST with 439 as the root

(c) II is an inorder sequence of some BST where 121 is the root and 52 is a leaf

(d) IV is a postorder sequence of some BST with 149 as the root

73. How many distinct BSTs can be constructed with 3 distinct keys ?

(a) 4 (b) 5 (c) 6 (d) 9

Common Data for Questions 74 and 75 :

Consider the following relational schema :

Student (school-id, sch-roll-no, sname, saddress)

School (school-id, sch-name, sch-address, sch-phone)

Enrolment (school-id,sch-roll-no, erollno, examname)

ExamResult (erollno, examname, marks)

74. What does the following SQL query output ?

```
SELECT   sch-name, COUNT(*)
FROM     SchoolC, Enrolment E, ExamResult R
WHERE    E.School-id = C.school-id
AND
  E.examname = R.examname AND E.Erollno = R.erollno
AND
  R.marks = 100 AND S.school-id IN (SELECT School-id FROM student GROUP BY
  school-id HAVING COUNT (*) >200) GROUP by school-id.
```

(a) for each school with more than 200 students appearing in exams, the name of the school and the number of 100s scored by its students.

(b) for each school with more than 200 students in it, the name of the school and the number of 100s scored by its students.

(c) for each school with more than 200 students in it, the name of the school and the number of its students scoring 100 in at least one exam..

(d) nothing; the query has a syntax error

75. Consider the following tuple relational calculus query

$\{t \mid \exists E \in$ Enrolment t=E.school-id $\wedge$

$|\{x \mid x \in$ Enrolment $\wedge$ x.school-id=t $\wedge$

$(\exists B \in$ ExamResult B.erollno=x.erollno $\wedge$ B.examname=x.examname $\wedge$ B.marks>35$\}| \div$

$|\{x \mid x \in$ Enrolment $\wedge$ x.school-id =t $|$ * 100>35 $\}$

If a student needs to score more than 35 marks to pass an exam, what does the query return ?

(a) The empty set

(b) schools with more than 35% of its students enrolled in some exam or the other

(c) schools with a pass percentage above 35% over all exams taken together.

(d) schools with a pass percentage above 35% over each exam

Linked Answer Questions : Q.76 to Q.85 carry two marks each.

Statement for Linked Answer Questions 76 and 77 :

A binary tree with n>1 nodes has n_1, n_2 and n_3 nodes of degree one, two and three respectively. The degree of a node is defined as the number of its neighbours.

76. n_3 can be expressed as :

(a) $n_1 + n_2 - 1$ (b) $n_1 - 2$ (c) $\left[\frac{n_1 + n_2}{2}\right]$ (d) $n_2 - 1$

77. Starting with the above tree, while there remains a node v of degree two in the tree, add an edge between the two neighbours of v and then remove v from the tree. How many edges will remain at the end of the process ?

(a) $2 * n_1 - 3$ (b) $n_2 + 2 * n_1 - 2$ (c) $n_3 - n_2$ (d) $n_2 + n_1 - 2$

Statement for Linked Answer Questions 78 and 79:

A CFG G is given with the following productions where S is the start symbol, A is a non-terminal and a and b are terminals.

$S \to aS \mid A$

$A \to aAb \mid bAa \mid \in$

78. Which of the following strings is generated by the grammar above ?

(a) aabbaba (b) aabaaba (c) abababb (d) aabbaab

79. For the correct answer in Q.78 how many steps are required to derive the string and how many parse trees are there ?

(a) 6 and 1 (b) 6 and 2 (c) 7 and 2 (d) 4 and 2

Statement for Linked Answer Questions 80 and 81:

Consider a computer with a 4-ways set-associative mapped cache of the following characteristics: a total of 1 MB of main memory, a word size of 1 byte, a block size of 128 words and a cache size of 8 KB.

80. The number of bits inthe TAG, SET and WORD fields, respectively are :

(a) 7, 6, 7 (b) 8, 5, 7 (c) 8, 6,6 (d) 9,4,7

81. While accessing the memory location 0C795H by the CPU, the contents of the TAG field of the corresponding cache line is

(a) 000011000 (b) 110001111

(c) 00011000 (d) 110010101

Statement for Linked Answer Questions 82 and 83 :

Consider the code fragment written in C below :

```
Void f (int n )
{
 if (n<=f) {
 printf("%d", n);
 }
 else {
 f (n/2);
 printf("%d", n%2);
  }
}
```

82. What does f(173) print ?

(a) 010110101 (b) 010101101 (c) 10110101 (d) 10101101

83. Which of the following implementations will produce the same output for f(173) as the one from Q.82?

P1:
```
Void f (int n)
{
if (n/2) {
f (n/2);
}
printf("%d", n%2);
 }
```

P2:
```
Void f (int n)
{
if (n<= 1) {
printf ("%d", n);
}
else {
printf("%d", n%2);
f (n/2);
}
 }
```

(a) Both P1 and P2 (b) P2 only
(c) P1 only (d) Neither P1 nor P2

Statement for Linked Answer Questions 84 and 85 :

Host X has IP address 192.168.1.97 and is connected through two routers R1 and R2 to another host Y with IP address 192.168.1.80. Router R1 has IP addresses 192.168.1.135 and 192.168.1.110. R2 has IP addresses 192.168.1.67 and 192.168.1.155. The netmask used inthe network is 255.255.255.224.

84. Given the information above, how many distincy subnets are guaranteed to already exist in the network ?
(a) 1 (b) 2 (c) 3 (d) 6

85. Which IP address should X configure its gateway as ?
(a) 192.168.1.67 (b)192.168.1.110
(c) 192.168.1.135 (d) 192.168.1.155

ANSWERS

1. (b)	2. (b)	3. (b)	4. (c)	5. (c)	6. (b)	7.(d)	8.(a)	9.(b)	10.(c)
11.(b)	12.(a)	13. (c)	14. (b)	15.(b)	16. (c)	17. (d)	18. (*)	19.(c)	20.(d)
21. (*)	22. (*)	23. (c)	24. (a)	25. (a)	26. (d)	27.(d)	28.(a)	29.(d)	30.(a)
31. (a)	32. (a)	33. (b)	34. (c)	35. (*)	36. (a)	37.(*)	38.(*)	39.(*)	40.(*)
41. (b)	42. (b)	43. (c)	44. (a)	45. (c)	46. (d)	47.(b)	48.(d)	49.(b)	50.(c)
51. (a)	52. (b)	53. (*)	54. (b)	55. (c)	56. (d)	57.(d)	58.(d	59.(c)	60.(a)
61. (*)	62. (*)	63. (*)	64. (*)	65. (a)	66. (c)	67.(d)	68.(d)	69.(c)	70.(*)
71. (*)	72. (c)	73. (b)	74.(b)	75. (c)	76. (b)	77.(d)	78.(d)	79.(b)	80.(*)
81. (*)	82. (d)	83. (c)	84.(*)	85. (*)					

GATE - 2007

CS : Computer Science and Engineering

Q. 1 to Q. 20 Carry one mark each

1. **Consider the following two statements about the function a** $f(x)\left|S\right|$ $P.\ f(x)$ **is continuous for all real values of x**

   $Q.\ f(x)$ **is differentiable for all real values of x**

   Which of the following is TRUE?

   (a) P is true and Q is false
   (b) P is false and Q is true
   (c) Both P and Q are true
   (d) Both P and Q are false

2. **Let S be a set of n elements. The number of ordered pairs in the largest and the smallest equivalence relations on S are**

   (a) n and n
   (b) n^2 *and* n
   (c) n^2 *and* 0
   (d) n *and* 1

3. **What is the maximum number of different Boolean functions involving n Boolean variables ?**

   (a) n^2
   (b) 2^n
   (c) 2^{2^n}
   (d) 2^{n^2}

4. **Let G be the non-planar graph with the minimum possible number of edges. Then G has**

   (a) 9 edges and 5 vertices
   (b) 9 edges and 6 vertices
   (c) 10 edges and 5 vertices
   (d) 10 edges and 6 vertices

5. **Consider the DAG with** $V = \{1,2,3,4,5,6\}$**, shown below**

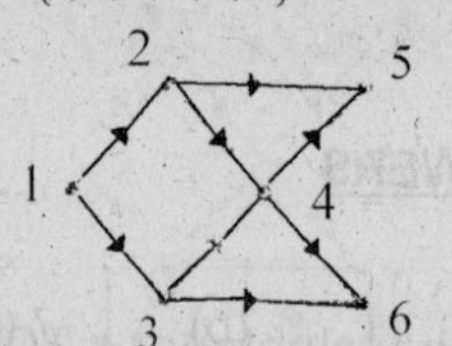

   Which of the following is NOT a topological ordering ?

   (a) 1 2 3 4 5 6
   (b) 1 3 2 4 5 6
   (c) 1 3 2 4 6 5
   (d) 3 2 4 1 6 5

6. **Which of the following problems is undecidable?**

   (a) Membership problem for CFGs.
   (b) Ambiguity problem for CFGs.
   (c) Finiteness problem for FSAs
   (d) Equivalence problem for FSAs

7. **Which of the following is TRUE?**

   (a) Every subset of a regular set is regular
   (b) Every finite subset of a non-regular set is regular
   (c) The union of two non-regular sets is not regular
   (d) Infinite union of finite sets is regular

8. **How many 3-to-8 line decoders with an enable input are needed to construct a 6-to-64 line decoder without using any other logic gates ?**

(a) 7 (b) 8 (c) 9 (d) 10

9. **Consider the following Boolean function of four variables**

$$f(w,x,y,z) = \sum(1,3,4,6,9,11,12,14)$$

The function is

(a) independent of one variable (b) independent of two variables

(c) independent of three variables (d) dependent on all the variables

10. **Consider a 4-way set associative cache consisting of 128 lines with a line size of 64 words. The CPU generates a 20-bit address of a word in main memory. The number of bits in the TAG LINE and WORD fields are respectively**

(a) 9, 6, 5 (b) 7, 7, 6

(c) 7, 5, 8 (d) 9, 5, 6

11. **Consider a disk pack with 16 surfaces, 128 tracks per surface and 256 sectors per track 512 bytes of data are stored in a bit serial manner in a sector. The capacity of the disk pack and the number of bits required to specify a particular sector in the disk are respectively**

(a) 256 Mbyte, 19 bits (b) 256 Mbyte, 28 bits

(c) 512 Mbyte, 20 bits (d) 64 Gbyte, 28 bits

12. **The height of a binary tree is the maximum number of edges in any root to leaf path. The maximum number of nodes in a binary tree of height h is**

(a) 2^h (b) $2^{h-1} - 1$ (c) $2^{h+1} - 1$ (d) 2^{h+1}

13. **The maximum number of binary trees that can be formed with three unlabeled nodes is**

(a) 1 (b) 5 (c) 4 (d) 3

14. **Which of the following sorting algorithms has the lowest worst - case complexity?**

(a) Merge sort (b) Bubble sort

(c) Quick sort (d) Selection sort

15. **Consider the following segment of C - code**

```
int J , n ;
    j = 1 ;
while ( j < = n )
    j = j * 2 ;
```

The number of comparisons made in the execution of the loop for any $n > 0$ is

(a) $\lceil \log_2 n \rceil + 1$ (b) n (c) $\lceil \log_2 n \rceil$ (d) $\lfloor \log_2 n \rfloor + 1$

16. **Group 1 contains some CPU scheduling algorithms and Group 2 contains some applications. Match entries in Group 1 to entries in Group 2.**

Group 1	Group 2
P. Gang Scheduling	1. Guaranteed Scheduling
Q. Rate Monotonic Scheduling	2. Real - time Scheduling
R. Fair Share Scheduling	3. Thread Scheduling

(a) P-3; Q-2; R-1 (b) P-1; Q-2; R-3 (c) P-2; Q-3; R -1 (d) P-1; Q-3; R-2

17. Consider the following statements about user level threads and kernel level threads. Which one of the following statements is FALSE?

(a) Context switch time is longer for kernel level threads than for user level threads.

(b) User level threads do not need any hardware support

(c) Related kernel level threads can be scheduled on different processors in a multiprocessor system

(d) Blocking one kernel level thread blocks all related threads.

18. Which one of the following is a top-down parser?

(a) Recursive descent parser (b) Operator precedence parser

(c) An LR(k) parser (d) An LALR(k) parser

19. In Ethernet when Manchester encoding is used, the bit rate is

(a) Half the baud rate (b) Twice the baud rate

(c) Same as the baud rate (d) None of these

20. Which one of the following uses UDP as the transport protocol?

(a) HTTP (b) Telnet (c) DNS (d) SMTP

Q.21 to Q.75 carry two marks each.

21. How many different non-isomorphic Abelian groups of order 4 are there?

(a) 2 (b) 3 (c) 4 (d) 5

22. Let Graph(x) be a predicate which denotes that x is a graph. Let Connected(x) be a predicate which denotes that x is connected. Which of the following first order logic sentences DOES NOT represent the statement; " Not every graph is connected"?

(a) $\neg\forall \dot{S}\,(Graph(x) \Rightarrow Connected(x))$

(b) $\exists \dot{S}\,(Graph(x) \wedge \neg\, Connected(x))$

(c) $\neg\forall \dot{S}\,\bar{A}(\neg Graph(x) \vee Connected(x))$

(d) $\forall \dot{S}\,(Graph(x) \Rightarrow \neg Connected(x))$

23. Which of the following graphs has an Eulerian circuit?

(a) Any k-regular graph where k is an even number

(b) A complete graph on 90 vertices

(c) The complement of a cycle on 25 vertices

(d) None of the above

24. Suppose we uniformly and randomly select a permutation from the 20! permulations of $1,2,3......,20$. What is the probability that 2 appears at an earlier position than any other even number in the selected permutation?

(a) $\frac{1}{2}$ (b) $\frac{1}{10}$ (c) $\frac{9!}{20!}$ (d) None of these

25. Let A be a 4×4 matrix with eigenvalues -5, -2, 1, 4. Which of the following is an eigenvalue of $\begin{bmatrix} A & I \\ I & A \end{bmatrix}$, where I is the 4×4 identity matrix?

(a) - 5 (b) - 7 (c) 2 (d) 1

26. Consider the set $S=\{a,b,c,d\}$. Consider the following 4 partitions π_1,π_2,π_3,π_4 on $S:\pi_1=\{\overline{abcd}\},\ \pi_2\ \{\overline{ab},\overline{cd}\},\pi_3=\{\overline{abc},\overline{d}\},\pi_4=\{\overline{a},\overline{b},\overline{c},\overline{d}\}$. Let $\prec$ be the partial order on the set of partitions $S'=\{\pi_1,\pi_2,\pi_3,\pi_4\}$ defined as follows : $\pi_\eta \prec \pi_{\overline{O}}$ if and only if π_η refines $\pi_{\overline{O}}$. The poset digram for $(S',\prec)$ is

(a)

(b)

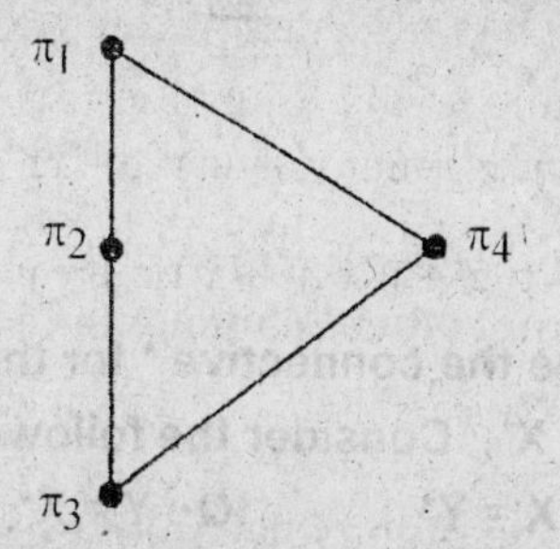

(c)

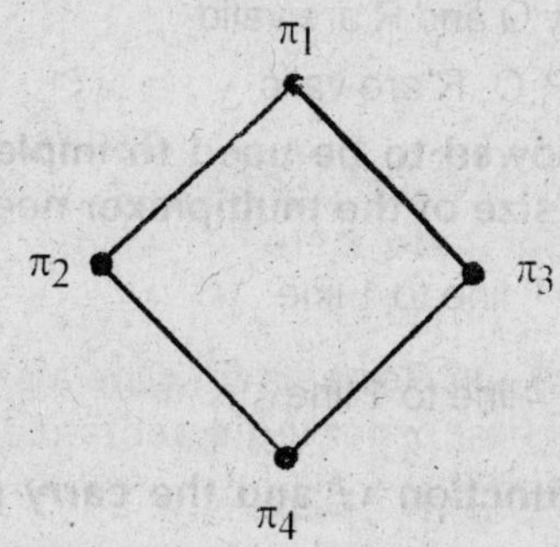

(d)

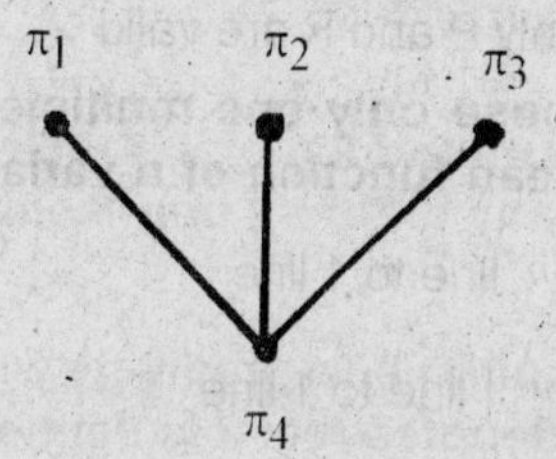

27. Consider the set of (column) vectors defined by $X=\{X\in R^3 \mid x_1+x_2+x_3=0$: where $x^T=[x_1,x_2,x_3]^T\}$. Which of the following is TRUE?

(a) $\{[1,-1,0]^T,[1,0,-1]^T\}$ is a basis for the subspace X.

(b) $\{[1,-1,0]^T,[1,0,-1]^T\}$ is a linearly independent set, but it does not span X and therefore is not a basis of X

(c) X is not a subspace of R^3

(d) None of the above

28. Consider the series $x_{n+1}=\dfrac{x_n}{2}+\dfrac{9}{8x_n}, x_0=0.5$ obtained from the Newton - Raphson method. The series converges to

(a) 1.5 (b) $\sqrt{2}$ (c) 1.6 (d) 1.4

29. A minimum state deterministic finite automaton accepting the language $L=\{w \mid w\in\{0,1\}^*$, number of 0s and 1s in w are divisible by 3 and 5, respectively } has

(a) 15 states (b) 11 states (c) 10 states (d) 9 states

30. The language $L=\{0^i 21^i \mid i\geq 0\}$ over the alphabet $\{0,1,2\}$ is

(a) not recursive

(b) is recursive and is a deterministic CFL

(c) is a regular language

(d) is not a deterministic CFL but a CFL

31. Which of the following languages is regular?

(a) $\{ww^R \mid w \in \{0,1\}^+\}$ (b) $\{ww^R x \mid x, w \in \{0,1\}^+\}$

(c) $\{wxw^R \mid x, w \in \{0,1\}^+\}$ (d) $\{xww^R \mid x, w \in \{0,1\}^+\}$

32. Let $f(w,x,y,z) = \sum(0,4,5,7,8,9,13,15)$. Which of the following expressions are NOT equivalent to f ?

(a) $x'y'z' + w'xy' + wy'z + xz$ (b) $w'y'z' + wx'y' + xz$

(c) $w'y'z' + wx'y' + xyz + xy'z$ (d) $x'y'z' + wx'y' + w'y$

33. Define the connective * for the Boolean variable X and Y as : X * Y = XY + XL'

Let = X*. Consider the following expression P, Q and R.

P : X = Y* **Q: Y = X*** **R : X* Y* = 1**

Which of the following is TRUE ?

(a) Only P and Q are valid (b) Only Q and R are valid

(c) Only P and R are valid (d) All P, Q, R are valid

34. Suppose only one multiplexer and one inverter are allowed to be used to implement any Boolean function of n variables. What is the minimum size of the multiplexer needed ?

(a) 2^n line to 1 line (b) 2^{n+1} line to 1 line

(c) 2^{n-1} line to 1 line (d) 2^{n-2} line to 1 line

35. In a look - ahead carry generator, the carry generate function G_i and the carry propagate function P_i for inputs, A_i and B_i are given by :

$$P_i = A_i \oplus B_i \text{ and } G_i = A_i B_i$$

The expressions for the sum bit S_i and the carry bit C_{i+1} of the look - ahead carry adder are given by :

$$S_i = P_i \oplus C_i \text{ and } C_{i+1} = G_i + P_i C_i \text{, where } C_0 \text{ is the input carry.}$$

Consider a two-level logic implementation of the look-ahead carry generator. Assume that all P_i and G_i are available for the carry generator circuit and that the AND and OR gates can have any number of inputs. The number of AND gates and OR gates needed to implement the look-ahead carry generator for a 4-bit adder with S_3, S_2, S_1, S_0 and C_4 as its outputs are respectively

(a) 6, 3 (b) 10, 4 (c) 6, 4 (d) 10, 5

36. The control signal functions of a 4-bit binary counter are given below (where X is "don't care") :

Clear	Clock	Load	Count	Function
0	X	X	X	Clear to 0
0	X	0	0	No change
0	↑	1	X	Load input
0	↑	0	1	Count next

The counter is connected as follows :

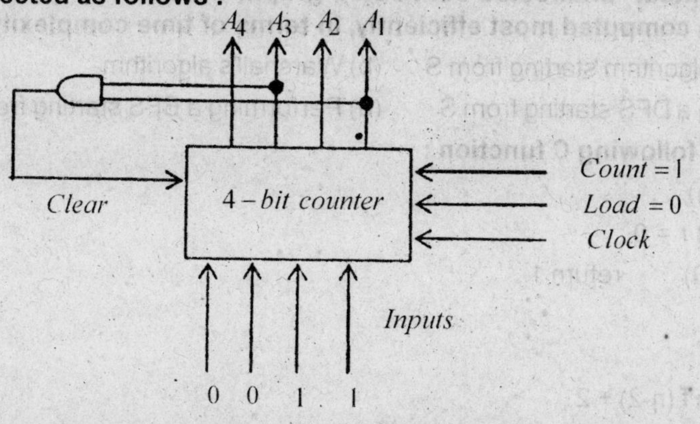

Assume that the counter and gate delays are negligible. If the counter starts at 0, then it cycles through the following sequence

(a) 0, 3, 4 (b) 0, 3, 4, 5 (c) 0, 1, 2, 3, 4 (d) 0, 1, 2, 3, 4, 5

37. Consider a pipelined processor with the following four stages :

IF : Instruction Fetch

ID : Instruction Decode and Operand Fetch

EX : Execute

WB : Write Back

The IF, ID and WB stages take one clock cycle each to complete the operation. The number of clock cycles for the EX stage depends on the instruction. The ADD and SUB instructions need 1 clock cycle and the MUL instruction need 3 clock cycles in the EX stage. Operand forwarding is used in the pipelined processor. What is the number of clock cycles taken to complete the following sequence of instructions?

ADD R2, R1, R0 $R2 \leftarrow R1 + R0$

MUL R4, R3, R2 $R4 \leftarrow R3 * R2$

SUB R6, R5, R4 $R6 \leftarrow R5 - R4$

(a) 7 (b) 8 (c) 10 (d) 14

38. The following postfix expression with single digit operands is evaluated using a stack

$$8\,2\,3\,\hat{}\,/\,2\,3\,*\,+\,5\,1\,*\,-$$

Note that ^ is the exponentiation operator. The top two elements of the stack after the first * is evaluated are

(a) 6, 1 (b) 5, 7 (c) 3, 2 (d) 1, 5

39. The inorder and preorder traversal of a binary tree are

d b e a f c g and a b d e c f g respectively.

The postorder traversal of the binary tree is

(a) d e b f g c a (b) e d b g f c a (c) e d b f g c a (d) d e f g b c a

40. Consider a hash table of size seven, with starting index zero, and a hash function (3x + 4) mod 7. Assuming the hash table is initially empty, which of the following is the contents of the table when the sequence 1, 3, 8, 10 is inserted into the table using closed hashing? Note that denotes an empty location in the table.

(a) 8, -, -, -, -, -, 10 (b) 1, 8, 10, -, -, -, 3

(c) 1, -, -, -, -, -, 3 (d) 1, 10, 8, -, -, -, 3

41. In an unweighted, undirected connected graph, the shortest path from a mode S to every other node is computed most efficiently, in terms of time complexity, by

(a) Dijkstra's algorithm starting from S (b) Warshall's algorithm

(c) Performing a DFS starting from S (d) Performing a BFS starting from S

42. Consider the following C function :

```
int f (int n)
{static int r = 0 ;
if (n < = 0)      return 1 ;
if (n > 3)
    (r = n ;
    return f (n-2) + 2 ;
    }
return f (n-1) + r ;
}
```

What is the value of f (5) ?

(a) 5 (b) 7 (c) 9 (d) 18

43. A complete n-ary tree is a tree in which each node has n children or no children. Let I be the number of internal nodes and L be the number of leaves in a complete n-ary tree. If L=41, and I = 10, what is the value of n

(a) 3 (b) 4 (c) 5 (d) 6

44. In the following C function, let $n \geq m$.

```
int gcd (n, m)
{
    if (n % m = = 0) return m ;
    n = n % m ;
    return gcd (m, n ) ;
}
```

How many recursive calls are made by this function?

(a) $\Theta(\log_2 n)$ (b) $\Theta(n)$ (c) $\Theta(\log_2 \log_2 n)$ (d) $\Theta(\sqrt{n})$

45. What is the time complexity of the following recursive function :

```
int DoSomething (int n) {
if (n < = 2)
else return | :
    return (Do something (floor (sqrt (n) ) + n) ;
```

(a) $\Theta(n^2)$ (b) $\Theta(n \log_2 n)$ (c) $\Theta(\log_2 n)$ (d) $\Theta(\log_2 \log_2 n)$

46. Consider the following C program segment shere Cell Node represents a node in a binary tree

```
struct CellNode {
struct cellNode * leftchild ;
int element ;
struct CellNode * right child ;
} ;
int GetValue (struct CellNode * ptr) {
int value = 0 ;
```

```
if (ptr ! = NULL) {
  if ( ( ptr - > leftchild = = NULL) &&
            (ptr -> rightchild = = NULL ) )
  value = 1 ;
else
  value = value + GetValue (ptr -> leftchild)
            + GetValue (ptr -> rightchild) ;
}
return (value)
}
```

The value returned by Get Value when a pointer to the root of a binary tree is passed as its argument is

(a) the number of nodes in the tree
(b) the number of internal nodes in the tree
(c) the number of leaf nodes in the tree
(d) the height of the tree

47. Consider the process of inserting an element into a Max Heap, where the Max Heap is represented by an array. Suppose we perform a binary search on the path from the new leaf to the root to find the position for the newly inserted element, the number of comparisons performed is

(a) $\Theta(\log_2 n)$ (b) $\Theta(\log_2 \log_2 n)$ (c) Θn (d) $\Theta(n \log_2 n)$

48. Which of the following is TRUE about formulae in Conjunctive Normal Form ?

(a) For any formula, there is a truth assignment for which at least half the clauses evaluate to true
(b) For any formula, there is a truth assignment for a which all the clauses evaluate to true.
(c) There is a formula such that for each truth assignment at most one-fourth of the clauses evaluate to true
(d) None of the above.

49. Let w be the minimum weight among all edge weights in an undirected connected graph. LEt e be a specific edge of weight w. Which of the following is FALSE ?

(a) There is a minimum spanning tree containing e.
(b) If e is not in a minimum spanning tree T, then in the cycle formed by adding e to T, all edges have the same weight.
(c) Every minimum spanning tree has an edge of weight w
(d) e is present in every minimum spanning tree

50. An array of n numbers is given, where n is an even number. The maximum as well as the minimum of these n numbers needs to be determined. Which of the following is TRUE about the number of comparisons needed ?

(a) At least 2n-c comparisons, for some constant c, are needed.
(b) At most 1.5 n-2 comparisons are needed.
(c) At least $n \log_2 n$ comparisons are needed
(d) None of the above

51. Consider the following C code segment :

```
int         IsPrime (n)
    {
            int i, n ;
            for ( i = 2 ; i < = sqrt (n) ; i++ )
            uf*b% u = = 0 )
```

```
        if ( n% i = = 0)
          {printf ('Not Prime \ n" } ; return 0 ;)
      return 1 ;
      }
```

Let T(n) denote the number of times the for loop is executed by the program on input n. Which of the following is TRUE ?

(a) $T(n) = O(\sqrt{n})$ and $T(n)=\Omega(\sqrt{n})$

(b) $T(n) = O(\sqrt{n})$ and $T(n)=\Omega(1)$

(c) $T(n) = O(n)$ and $T(n)=\Omega(\sqrt{n})$

(d) None of these

52. Consider the grammer with non-terminals $N = \{S, C, S_1\}$, terminals $T = \{a, b, i, t, e\}$ with S as the start symbol, and the following set of rules.

$$S \to iCtSS_1 \mid a$$

$$S_1 \to eS \mid \varepsilon$$

$$C \to b$$

The grammar is NOT LL(1) because :

(a) It is left recursive
(b) it is right recursive
(c) it is ambiguous
(d) it is not context - free

53. Consider the following two statements :

P : Every regular grammar is LL (1)

Q : Every regular set has a LR (1) grammar

Which of the following is TRUE ?

(a) Both P and Q are true
(b) P is true and Q is false
(c) P is false and Q is true
(d) Both P and Q are false

54. In a simplified computer the instructions are :

$OP\ R_j, R_i$	-	**Performs $R_j\ OP\ P_i$ and stores the result in register R_i**
$OP\ m, R_i$	-	**Performs val $OP\ R_i$ and stores the result in R_i. val denotes the content of memory location m.**
$MOV\ m, R_i$	-	**Moves the content of memory location m to register R_i**
$MOV\ R_i, m$	-	**Moves the content of register R_i to memory location m.**

The computer has only two registers, and OP is either ADD or SUB. Consider the following basic block :

$$t_1 = a+b$$

$$t_2 = c+d$$

$$t_3 = e-t_2$$

$$t_4 = t_1-t_3$$

Assume that all operands are initially in memory. The final value of the computation should

be in memory. What is the minimum number of MOV instructions in the code generated for this basic block ?

(a) 2 (b) 3 (c) 5 (d) 6

55. An operating system uses Shortest Remaining Time first (SRT) process scheduling algorithm. Consider the arrival times and execution times for the following processes :

Process	Execution time	Arrival time
P1	20	0
P2	25	15
P3	10	30
P4	15	45

What is the total waiting time for process P2?

(a) 5 (b) 15 (c) 40 (d) 55

56. A virtual memory system uses first In First Out (FIFO) page replacement policy and allocates a fixed number of frames to a process. Consider the following statements :

P : Increasing the number of page frames allocated to a process sometimes increases the rate fault rate.

Q : Some programs do not exhibit locality of reference.

Which one of the following is TRUE ?

(a) Both P and Q are true, and Q is the reason for P.

(b) Both P and Q are true, but Q is not the reason for P.

(c) P is false, but Q is true.

(d) Both P and Q are false.

57. A single processor system has three resource types X, Y and Z, which are shared by three processes. There are 5 units of each resource type. Consider the following scenario, where the column alloc denotes the number of units of each resource type allocated to each process, and the column request denotes the number of units of each resource type requested by a process in order to complete execution. Which of these processes will finish LAST ?

	alloc	request
	X Y Z	X Y Z
P0	1 2 1	1 0 3
P1	2 0 1	0 1 2
P2	2 2 1	1 2 0

(a) P0 (b) P1 (c) P2

(d) None of the above, since the system is in a deadlock

58. Two processes, P1 and P2, need to access a critical section of code. Consider the following synchronization construct used by the processes.

```
/*   P1          */
while (true)       {
   wants1 = true ;
   while (wants2 = = true) ;
   / * Critical
      Section * /
   wants 1 = false ;
}
/*  Remainder section */
```

```
/*          P2      */
while (true) {
wants2  = true ;
     while (wants1 = = true) ;
     /*         Critical
                Section */
wants 2 = false ;
}
/*  Remainder section * /
```

Here, wants 1 and wants2 are shared variables, Which are initialized to false. Which one of the following statements is TRUE about the above construct ?

(a) It does not ensure mutual exclusion

(b) It does not ensure bounded waiting.

(c) It requires that processes enter the critical section in strict alternation.

(d) It does not prevent deadlocks, but ensures mutual exclusion.

59. Information about a collection of students is given by the relation studInfo (studId, name, sex). The relation enroll (studId, courseId) gives which student has enrolled for (or taken) what course (s). Assume that every course is taken by at least one male and at least one female student. What does the following relational algebra expression represent ?

$$\Pi_{courseId}\left(\left(\Pi_{studId}\left(\sigma_{sex="female"}(studInfo)\right) \times \Pi_{courseId}(enroll)\right) - enroll\right)$$

(a) Courses in which all the female students are enrolled.

(b) Courses in which a proper subset of female students are enrolled.

(c) Courses in which only male students are enrolled.

(d) None of the above

60. Consider the relation employee (name, sex, supervisorName) with name as the key. supervisor. Name gives the name of the supervisor of the employee under consideration. What does the following Tuple Relational Calculus query produce ?

{ e. name | employee (e) ^

$$(\forall x)[\neg employee(x) \vee x.supervisorName \neq e.name \vee x.sex = "male"]\}$$

(a) Names of employees with a male supervisor.

(b) Names of employees with no immediate make subordinates.

(c) Names of employees with no immediate female subordinates.

(d) Names of employees with a female supervisor.

61. Consider the table employee (empId, name, department, salary) and the two queries Q_1, Q_2 below. Assuming that department 5 has more than one employee, and we want to find the employees who get higher salary than anyone in the department 5, which one of the statements is TRUE for any arbitrary employee table

Q_1 : **Select e.empId**

From employee e

Where not exists

(Select * from employees Where S. department = "5" and s.salary > = e. salary)

Q_2 : **Select e.empId**

From employee e

Where e.salary > Any

(Select distinct salary From employee s Where S. department = "5")

(a) Q_1 is the correct query.

(b) Q_2 is the correct query

(c) Both Q_1 and Q_2 produce the same answer

(d) Neither Q_1 nor Q_2 is the correct query

62. Which one of the following statements is FALSE ?

(a) Any relation with two attributes is in BCNF.

(b) A relation in which every key has only one attribute is in 2NF

(c) A prime attribute can be transitively dependent on a key in a 3 NF relation.

(d) A prime attribute can be transitively dependent on a key in a BCNF relation.

63. The order of a leaf node in a B^+ - tree is the maximum number of (value, data record pointer) pairs it can hold. Given that the block size is 1 K bytes, data record pointer is 7 bytes long, the value field is 9 bytes long and a block pointer is 6 bytes long, what is the order of the leaf node?

(a) 63 (b) 64 (c) 67 (d) 68

64. Consider the following schedules involving two transactions. Which one of the following statements is TRUE :

$$S_1 : r_1(X); r_1(Y); r_2(X); r_2(Y); w_2(Y); w_1(X)$$

$$S_2 : r_1(X); r_2(X); r_2(Y); w_2(Y); r_1(Y); w_1(X)$$

(a) Both S_1 *and* S_2 are conflict serializable

(b) S_1 is conflict serializable and S_2 is not conflict serializable

(c) S_1 is not conflict serializable and S_2 is conflict serializable

(d) Both S_1 *and* S_2 are not conflict serializable.

65. There are n stations in a slotted LAN. Each station attempts to transmit with a probability p in each time slot. What is the probability that ONLY one station transmits in a given time slot ?

(a) $np(1-p)^{n-1}$ (b) $(1-p)^{n-1}$

(c) $p(1-p)^{n-1}$ (d) $1-(1-p)^{n-1}$

66. In a token ring network the transmission speed is 10^7 bps and the propagation speed is 200 metres / μs. The 1 - bit delay in this network is equivalent to :

(a) 500 metres of cable. (b) 200 metres of cable.

(c) 20 metres of cable. (d) 50 metres of cable.

67. The address of a class B host is to be split into subnets with a 6 - bit subnet number. What is the maximum number of subnets and the maximum number of hosts in each subnet ?

(a) 62 subnets and 262142 hosts. (b) 64 subnets and 262142 hosts.

(c) 62 subnets and 1022 hosts. (d) 64 subnets and 1024 hosts.

68. The message 11001001 is to be transmitted using the CRC polynomial x^3+1 to protect it from errors. The message that should be transmitted is :

(a) 11001001000 (b) 11001001011

(c) 11001010 (d) 110010010011

69. The distance between two stations M and N is Lkilometres. All frames are K bits long. The propagation delay per kilometre is t seconds. Let R bits / second be the channel capacity. Assuming that processing delay is negligible, the minimum number of bits for the sequence number field in a frame for maximum utilization, when the sliding window protocol is used, is:

(a) $\left[\log_2 \frac{2LtR+2K}{K}\right]$ (b) $\left[\log_2 \frac{2LtR}{K}\right]$

(c) $\left[\log_2 \frac{2LtR+K}{K}\right]$ (d) $\left[\log_2 \frac{2LtR+K}{2K}\right]$

70. Match the following :

P. SMTP	**1. Application layer**
Q. BGP	**2. Transport layer**
R. TCP	**3. Data link layer**
S. PPP	**4. Network layer**
	5. Physical layer

(a) $P-2, Q-1, R-3, S-5$ (b) $P-1, Q-4, R-2, S-3$

(c) $P-1, Q-4, R-2, S-5$ (d) $P-2, Q-4, R-1, S-3$

Common Data Questions

Common data for questions 71, 72, 73 :

Consider the following program segment. Here R1, R2, and R3 are the general purpose registers.

	Instruction	Operation	Instruction size (no.of words)
	MOV R1, (3000)	$R1 \leftarrow M[3000]$	2
LOOP:	MOV R2, (R3)	$R2 \leftarrow M[R3]$	1
	ADD R2, R1	$R2 \leftarrow R1 + R2$	1
	MOV (R3), R2	$M[R3] \leftarrow R2$	1
	INC R3	$R3 \leftarrow R3 + 1$	1
	DEC R1	$R1 \leftarrow R1 - 1$	1
	BNZ LOOP	Branch on not zero	2
	HALT	Stop	1

Assume that the content of memory location 3000 is 10 and the content of the register R3 is 2000. The content of each of the memory locations from 2000 to 2010 is 100. The program is loaded from the memory location 100. All the numbers are in decimal.

71. Assume that the memory is word addressable. The number of memory references for accessing the data in executing the program completely is

(a) 10 (b) 11 (c) 20 (d) 21

72. Assume that the memory is word addressable. After the execution of this program, the content of memory location 2010 is

(a) 100 (b) 101 (c) 102 (d) 110

73. Assume that the memory is byte addressable and the word size is 32 bits. If an interrupt occurs during the execution of the instruction "INC R3", what return address will be pushed on to the stack?

(a) 1005 (b) 1020 (c) 1024 (d) 1040

Common Data for Questions 74, 75 :

Consider the following Finite State Automation :

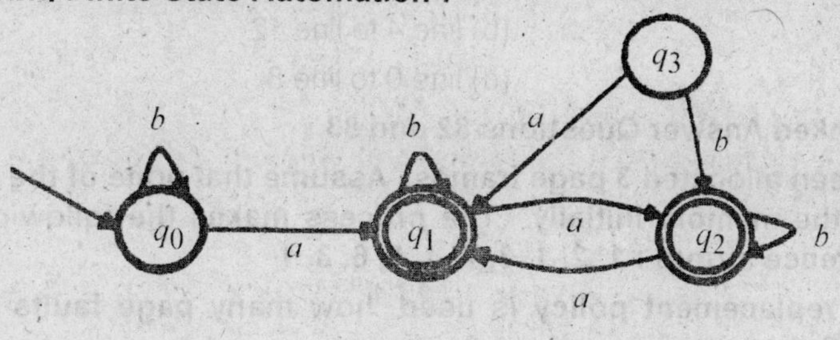

74. The language accepted by this automaton is given by the regular expression

(a) $b^*ab^*ab^*ab^*$ (b) $(a+b)^*$

(c) $b^*a(a+b)^*$ (d) $b^*ab^*ab^*$

75. The minimum state automaton equivalent to the above FSA has the following number of states

(a) 1 (b) 2 (c) 3 (d) 4

Linked answer Questions : Q.76 to Q.85 carry two marks each.

Statement for Linked Answer Questions 76 and 77 :

Suppose the letters a,b,c,d,e,f have probabilities $\frac{1}{2}, \frac{1}{4}, \frac{1}{8}, \frac{1}{16}, \frac{1}{32}, \frac{1}{32}$**, respectively**

76. Which of the following is the Huffnan code for the letter a,b,c,d,e,f ?

(a) 0, 10, 110, 1110, 11110, 11111 (b) 11, 10, 011, 010, 001, 000

(c) 11, 10, 01, 001, 0001, 0000 (d) 110, 100, 010, 000, 001, 111

77. What is the average length of the correct answer to Q.76 ?

(a) 3 (b) 2.1875 (c) 2.25 (d) 1.9375

Statement for Linked Answwer Questions 78 to 79 :

Consider the CFG with {S, A, B} as the non-terminal alphabet, {a, b} as the terminal alphabet, S as the start symbol and the following set of production rules :

$S \rightarrow aB$ $S \rightarrow bA$

$B \rightarrow b$ $A \rightarrow a$

$B \rightarrow bS$ $A \rightarrow aS$

$B \rightarrow aBB$ $A \rightarrow bAA$

78. Which of the following strings is generated by the grammar ?

(a) aaaabb (b) aabbbb (c) aabbab (d) abbbba

79. For the correct answer string to Q.78, how many derivation trees are there ?

(a) 1 (b) 2 (c) 3 (d) 4

Statement for Linked Answer Questions 80 and 81: Consider a machine with a byte addressable main memory of 2^{16} bytes. Assume that a direct mapped data chache consisting of 32 line of 64 bytes each is used in the system. A 50×50 two dimesional array of bytes is stored in the main memory starting from memory location 1100H. Assume that the data chache is initially empty. The complete arrary is accussed twice. Assume that the contents of the data chache do not change in between the two accesses.

80. How many data cache misses will occur in total ?

(a) 48 (b) 50 (c) 56 (d) 59

81. Which of the following lines of the data cache will be replaced by new blocks in accessing the array for the second time?

(a) line 4 to line 11 (b) line 4 to line 12

(c) line 0 to line 7 (d) line 0 to line 8

Statement for Linked Answer Questions 82 and 83 :

A process has been allocated 3 page frames. Assume that none of the pages of the process are available in the memory initially. The process makes the following sequence of page references (reference string) : 1, 2, 1, 3, 7, 4, 5, 6, 3, 1

82. If optimal page replacement policy is used, how many page faults occur for the above reference string ?

(a) 7 (b) 8 (c) 9 (d) 10

83. Least Recently Used (LRU) page replacement policy is a practical approximation to optimal page replacement. For the above reference string, how many more page faults occur with LRU than with the optimal page replacement policy?

(a) 0 (b) 1 (c) 2 (d) 3

Statement for Linked Answer Questions 84 and 85 :

Suppose that a robot is placed on the cartesian plane. At each step it is allowed to move either one unit up or one unit right, i.e., if it is at (i,j) then it can move to either $(i+1, j)$ *or* $(i, j+1)$.

84. How many distinct paths are there for the robot to reach the point (10, 10) starting from the initial position (0, 0) ?

(a) $\binom{20}{10}$ (b) 2^{20}

(c) 2^{10} (d) None of the above

85. Suppose that the robot is not allowed to traverse the line segment from $(4,4)$ to $(5,4)$. With this constraint, how many distinct paths are there for the robot to reach $(10,10)$ starting from $(0,0)$?

(a) 2^9 (b) 2^{19} (c) $\binom{8}{4} \times \binom{11}{5}$ (d) $\binom{20}{10} - \binom{8}{4} \times \binom{11}{5}$

ANSWERS

1. (a)	2. (b)	3. (c)	4. (b)	5. (d)	6. (b)	7. (b)	8. (b)	9. (b)	10. (b)
11. (a)	12. (c)	13. (b)	14. (a)	15. (d)	16. (a)	17. (d)	18. (a)	19. (b)	20. (c)
21. (b)	22. (d)	23. (a)	24. (d)	25. (d)	26. (c)	27. (b)	28. (a)	29. (a)	30. (b)
31. (c)	32. (d)	33. (d)	34. (c)	35. (c)	36. (d)	37. (b)	38. (a)	39. (a)	40. (b)
41. (d)	42. (d)	43. (c)	44. (c)	45. (d)	46. (c)	47. (a)	48. (b)	49. (d)	50. (d)
51. (b)	52. (c)	53. (b)	54. (b)	55. (b)	56. (b)	57. (c)	58. (d)	59. (b)	60. (b)
61. (b)	62. (d)	63. (b)	64. (c)	65. (a)	66. (c)	67. (c)	68. (d)	69. (b)	70. (b)
71. (d)	72. (a)	73. (a)	74. (c)	75. (b)	76. (a)	77. (d)	78. (c)	79. (b)	80. (c)
81. (c)	82. (a)	83. (c)	84. (a)	85. (d)					

GATE - 2007

IT : Information Technology

Q.1 to Q. 20 Carry one mark each

1. **Suppose there are two coins. The first coin gives heads with probability 5/8 when tossed, while the second coin gives heads with probability 1/4. One of the two coins is picked up at random with equal probability and tossed. What is the probability of obtaining heads ?**

(a) $7/8$ (b) $1/2$

(c) $7/16$ (d) $5/32$

2. **Let A be the matrix $\begin{bmatrix} 3 & 1 \\ 1 & 2 \end{bmatrix}$. What is the maximum value of $x^T Ax$ where the maximum is taken over all x that are the unit eigenvectors of A ?**

(a) 5 (b) $\frac{5+\sqrt{5}}{2}$

(c) 3 (d) $\frac{5-\sqrt{5}}{2}$

3. **Consider a weighted undirected graph with positive edge weights and let uv be an edge in the graph. It is known that the shortest path from the source vertex s to u has weight 53 and the shortest path from s to v has weight 65. Which one of the following statements is always true?**

(a) weight $(u,v) < 12$ (b) weight $(u,v) \le 12$

(c) weight $(u,v) > 12$ (d) weight $(u,v) \ge 12$

4. **In the Spiral model of software development, the primary determinant in selecting activities in each iteration is**

(a) iteration sive

(b) Cost

(c) Adopted process such as Rational Unified Process of Extreme Programming

(d) Risk.

5. **Which of the following systems is a most likely candidate example of a pipe and filter architecture ?**

(a) Expert system (b) DB repository

(c) Aircraft flight controller (d) Signal processing

6. **A processor takes 12 cycles to complete an instruction 1. The corresponding pipelined processor uses 6 stages with the execution times of 3, 2, 5, 4, 6 and 2 cycles respectively. What is the asymptotic speedup assuming that a very large number of instructions are to be executed ?**

(a) 1.83 (b) 2

(c) 3 (d) 6

7. **Which of the following input sequences for a cross-coupled R-S flip-flop realized with two NAND gates may lead to an oscillation ?**

(a) 11, 00 (b) 01, 10

(c) 10, 01 (d) 00, 11

8. The following circuit implements a two-input AND gate using two 2-1 multiplexers.

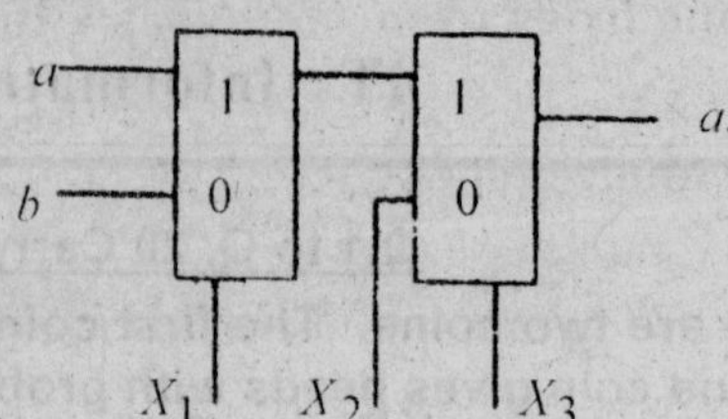

What are the values of X1, X2, X3 ?

(a) $X1 = b,\ X2 = 0, X3 = a$

(b) $X1 = b,\ X2 = 1, X3 = b$

(c) $X1 = a,\ X2 = b, X3 = 1$

(d) $X1 = a,\ X2 = 0, X3 = b$

9. Consider an ambiguous grammar G and its disambiguated version D. Let the language recognized by the two grammars be denoted by L(G) and L(D) respectively. Which one of the following is true ?

(a) $L(D) \subset L(G)$

(b) $L(D) \supset L(G)$

(c) $L(D) = L(G)$

(d) $L(D)$ is empty

10. Processes P1 and P2 use critical_flag in the following routine to achieve mutual exclusion. Assume that critical_flag is initialized to FALSE in the main program.

```
get_exclusive_access ( )
{       if (critical_flag = = FALSE)  {
            critical_flag = TRUE ;
            critical_region ( ) ;
            critical_flag = FALSE ;
        }
}
```

Consider the following statements

(i) It is possible for both P1 and P2 to acess critical_region concurrently.

(ii) This may lead to a deadlock.

Which of the following holds ?

(a) (i) is false and (ii) is true

(b) Both (i) and (ii) are false

(c) (i) is true and (ii) is false

(d) Both (i) and (ii) are true

11. Let a memory have four free blocks of sizes 4k, 8k, 20k, 2k. These blocks are allocated following the best-fit strategy. The allocation requests are stored in a queue as shown below.

Request No	J1	J2	J3	J4	J5	J6	J7	J8
Request sizes	2k	14k	3k	6k	6k	10k	7k	20k
Usage Time	1	40	2	8	4	1	8	6

The time at which the request for 17 will be completed will be

(a) 16

(b) 19

(c) 20

(d) 37

12. The address sequence generated by tracing a particular program executing in a pure demand paging system with 100 bytes per page is

0100, 0200, 0430, 0499, 0510, 0530, 0560, 0120, 0220, 0240, 0260, 0320, 0410.

Suppose that the memory can store only one page and if x is the address which causes a page fault then the bytes from addresses x to x + 99 are loaded on to the memory.

How many page faults will occur ?

(a) 0 (b) 4 (c) 7 (d) 8

13. Consider the following statements about the timeout value used in TCP.

(i) The timeout value is set to the RTT (Round Trip Time) measured during TCP connection establishment for the entire duration of the connection.

(ii) Appropriate RTT estimation algorithm is used to set the timeout value of a TCP connection

(iii) Timeout value is set to twice the propagation delay from the sender to the receiver

Which of the following choices hold ?

(a) (i) is false, but (ii) and (iii) are true

(b) (i) and (iii) are false, but (ii) is true

(c) (i) and (ii) are false, but (iii) is true

(d) (i), (ii) and (iii) are false

14. Consider a TCP connection in a state where there are no outstanding ACKs. The sender sends two segments back to back. The sequence numbers of the first and second segments are 230 and 290 respectively. The first segment was lost, but the second segment was received correctly by the receiver. Let X be the amount of data carried in the first segment (in bytes), and Y be the ACK number sent by the receiver.

The values of X and Y (in that order) are

(a) 60 and 290 (b) 230 and 291

(c) 60 and 231 (d) 60 and 230

15. Consider the following two statements :

(i) A hash function (these are often used for computing digital signatures) is an injective function.

(ii) An encryption technique such as DES performs a permutation on the elements of its input alphabet.

Which one of the following options is valid for the above two statements ?

(a) Both are false (b) Statement (i) is true and the other is false

(c) Statement (ii) is true and the other is false (d) Both are true

16. The minimum positive integer p such that 3^p modulo 17 = 1 is

(a) 5 (b) 8 (c) 12 (d) 16

17. Exponentiation is a heavily used operation in public key cryptography. Which of the following options is the tightest upper bound on the number of multiplications required to compute b^n modulo m, $0 \le b, n \le m$?

(a) $O(\log n)$ (b) $O(\sqrt{n})$

(c) $O(n / \log n)$ (d) $O(n)$

18. A firewall is to be configured to allow hosts in a private network to freely open TCP connections and send packets on open connections. However, it will only allow external hosts to send packets on existing open TCP connections or connections that are being opened (by internal

hosts) but not allow them to open TCP connections to hosts in the private network. To achieve this the minimum capability of the firewall should be that of

(a) A combinational circuit
(b) A finite automaton
(c) A pushdown automaton with one stack
(d) A pushdown automaton with two stacks

19. Given below are some HTML lines.

< img src = " pict.jpg" width = "256" height = " 256" border = "0"

usemap = " #map" / >

< map name = "map" >

<area shape =" poly"

coords = "50, 50, 50, 100, 100, 100, 75, 75, 100, 50"

href = "f1. html " >

< area shape = " circle" coords = " 100, 75, 5 "

href = " .. / cgi - bin / f2 . p1 ? v1 = ask abc's age " >

< area shape = " default " href = " fd. html ">

< / map >

With reference to the HTML lines given above, consider the following statements.

(i) Clicking on the point < 80, 75> does not have any effect.

(ii) The web browser can identify the area applicable to the mouse-click within the image and the subsequent action to be taken without additional responses from the web server.

(iii) The dots in the cgi-bin URL will be resolved by the web browser before it is sent to the web server

(iv) The "fd.html" request when sent to the web server will result in a GET request.

Exactly how many of the statements given above are correct ?

(a) 0
(b) 1
(c) 2
(d) 3

20. Consider the XML document fragment given below :

<Book>

<Title> GATE 2K7 Example </Title>

<Content>

One of many lines

< / Content>

<TOC>

One of many content entries

</TOC>

</Book>

Consider the XPath expression : * [not (self) : : TOC]

What would be the result of the given XPath expression when the current node is Book ?

(a) The Title and Content elements
(b) The Content and TOC elements
(c) The Title and TOC elements
(d) The Title Content and TOC elements

Q.21 to Q.75 carry two marks each.

21. Which one of these first - order logic formulae is valid ?

(a) $\forall x(P(x) \Rightarrow Q(x)) \Rightarrow ((\forall x P(x)) \Rightarrow (\forall x Q(x)))$

(b) $\exists x(P(x) \vee Q(x)) \Rightarrow ((\exists x P(x)) \Rightarrow (\exists x Q(x)))$

(c) $\exists x(P(x) \wedge Q(x)) \Leftrightarrow ((\exists x P(x)) \wedge (\exists x Q(x)))$

(d) $\forall x \exists y p(x,y) \Rightarrow \exists y \forall x\, p(x,y)$

22. The trapezoidal method is used to evaluate the numerical value of $\int_0^1 e^x\,dx$. Consider the following values for the step size h.

(i) 10^{-2} (ii) 10^{-3} (iii) 10^{-4} (iv) 10^{-5}

For which of these values of the step size h, is the computed value guaranteed to be correct to seven decimal places. Assume that there are no round - off errors in the computation.

(a) (iv) only
(b) (iii) and (iv) only
(c) (ii), (iii) and (iv) only
(d) (i), (ii), (iii) and (iv)

23. A partial order P is defined on the set of natural numbers as follows. Here x/y denotes integer division.

(i) $(0,0) \in P$.

(ii) $(a,b) \in P$ **if and only if** $a\%10 \le b\%10$ **and** $(a/10, b/10) \in P$.

Consider the following ordered pairs :

(i) $(101,22)$ **(ii)** $(22,101)$ **(iii)** $(145,265)$ **(iv)** $(0,153)$

Which of these ordered pairs of natural numbers are contained in P ?

(a) (i) and (iii)
(b) (ii) and (iv)
(c) (i) and (iv)
(d) (iii) and (iv)

24. A depth-first search is performed on a directed acyclic graph. Let $d[u]$ denote the time at which vertex u is visited for the first time and $f[u]$ the time at which the dfs call to the vertex u terminates. Which of the following statements is always true for all edges (u, v) in the graph?

(a) $d[u] < d[v]$
(b) $d[u] < f[v]$
(c) $f[u] < f[v]$
(d) $f[u] > f[v]$

25. What is the largest integer m such that every simple connected graph with n vertices and n edges contains at least m different spanning trees ?

(a) 1 (b) 2 (c) 3 (d) n

26. **Consider n jobs $J_1, J_2, ... J_n$ such that job J_i has execution time t_i and a non-negative integer weight w_i. The weighted mean completion time of the jobs is defined to be**

$$\frac{\sum_{i=1}^{n} w_i T_i}{\sum_{i=1}^{n} w_i}$$

where T_i is the completion time of job J_i. Assuming that there is only one processor available, in what order must the jobs be executed in order to minimize the weighted mean completion time of the jobs ?

(a) Non-decreasing order of t_i (b) Non - increasing order of w_i

(c) Non - increasing order of $w_i t_i$ (d) Non - increasing order of w_i / t_i

27. **The function f is defined as follows :**

```
int f (int n) {
if (n < = 1 ) return 1 ;
else if (n % 2 = = 0 ) return f (n/2) ;
else return f (3n - 1) ;
```

Assuming that arbitrarily large integers can be passed as a parameter to the function, consider the following statements.

(i) The function f terminates for finitely many different values of $n \geq 1$.

(ii) The function f terminates for infinitely many different values of $n \geq 1$

(iii) The function f does not terminate for finitely many different values of $n \geq 1$.

(iv) The function f does not terminate for infinitely many different values of $n \geq 1$.

Which one of the following options is true of the above

(a) (i) and (iii) (b) (i) and (iv)

(c) (ii) and (iii) (d) (ii) and (iv)

28. **Consider a hash function that distributes keys uniformly. The hash table size is 20. After hashing of how many keys will the probability that any new key hashed collides with an existing one exceed 0.5.**

(a) 5 (b) 6 (c) 7 (d) 10

29. **When searching for the key value 60 in a binary search tree, nodes containing the key values 10, 20, 40, 50, 70, 80, 90 are traversed, not necessarily in the order given. How many different orders are possible in which these key values can occur on the search path from the root to the node containing the value 60?**

(a) 35 (b) 64 (c) 128 (d) 5040

30. **Suppose you are given an implementation of a queue of integers. The operations that can be performed on the queue are :**

(i) is Empty (Q) - returns true if the queue is empty, false otherwise.

(ii) delete (Q) - deletes the element at the front of the queue and returns its value.

(iii) insert (Q, i) - inserts the integer i at the rear of the queue.

Consider the following function :

```
void f (queue Q)  {
int i ;
if (! is Empty (Q) ) {
  i = delete (Q) ;
  f (Q) ;
  insert (Q, i) ;
  }
}
```

What operation is performed by the above function f ?

(a) Leaves the queue Q unchanged

(b) Reverses the order of the elements in the queue Q

(c) Deletes the element at the front of the queue Q and inserts it at the rear keeping the other elements in the same order

(d) Empties the queue Q

31. Consider the C program given below :

```
# include <stdio.h>
int main ( )     {
  int  sum = 0, maxsum = 0 ; i ,  n  =  6 ;
  int a ( ) = { 2, -2, -1, 3, 4, 2 } ;
  for ( i = 0 ; i < n ; i + + )              {
     if ( i = = 0 | | a [i] < 0  | | a [i] < a [ i -1] )       {
        if (sum > maxsum) maxsum  =  sum ;
        sum = ( a [i]  >  0 )  ? a [i]  :  0  ;
     }
     else sum + = a [i] ;
     if (sum > maxsum)  maxsum = sum ;
     print f (11 %d\n" , maxsum) ;
  }
```

What is the value printed out when this program is executed?

(a) 9 (b) 8

(c) 7 (d) 6

32. Consider the following C program :

```
# include  <stdio . h >
# define EOF - 1
Void push (int) ;  / * push the argument on the stack * /
int pop (void)    ;  /* pop the top of the stack */
void flagError ( ) ;
int main ( )
{   int c, m, n, r ;
    while ( ( c = getchar ( ) )   !  =  EOF)
                    {if    (isdigit  (c) )
                            push (c) ;
```

```
        else if ( ( c = = '+' ) | | ( c = = '*' ) )
        {  m = pop ( ) ;
           n = pop ( ) ;
           r = ( c == '+' ) ? n + m : n * m ;
           push (r) ;
        }
        else if (c ! = ' ' )
           flagError ( ) ;
    }
    print f ( " % c " , pop ( ) ) ;
}
```

What is the output of the program for the following input ?

5 2 * 3 3 2 + * +

(a) 15 (b) 25 (c) 30 (d) 150

33. Consider the program below in a hypothetical language which allows global variable and a choice of call by reference or call by value methods of parameter passing.

```
int i ;
program main ( )
{  int j = 60 ;
   i = 50 ;
   call f ( i, j ) ;
   print i, j ;
}
procedure f (x , y )
{  i = 100 ;
   x = 10 ;
   y = y + i ;
}
```

Which one of the following options represents the correct output of the program for the two parameter passing mechanisms ?

(a) Call by value : i = 70, j = 10 ; Call by reference : i = 60 , j = 70
(b) Call by value : i = 50, j = 60 ; Call by reference : i = 50 , j = 70
(c) Call by value : i = 10, j = 70 ; Call by reference : i = 100 , j = 60
(d) Call by value : i = 100, j = 60 ; Call by reference : i = 10 , j = 70

34. Consider the program below in a hypothetical programming language which allows global variables and a choice of static or dynamic scoping.

```
int i ;
Program main ( )
{     i = 10 ;
      call f ( ) ;
```

```
}
procedure f ( )
{       int i = 20 ;
        call g ( ) ;
}
procedure g ( )
{       print i ;
}
```

Let x be the value printed under static scoping and y be the value printed under dynamic scoping. Then, x and y are

(a) $x = 10,\ y = 20$ (b) $x = 20,\ y = 10$

(c) $x = 10,\ y = 10$ (d) $x = 20,\ y = 20$

35. Early binding refers to a binding performed at compile time and late binding refers to a binding performed at execution time. Consider the following statements :

(i) Static scope facilitates $\underline{w1}$ bindings.

(ii) Dynamic scope requires $\underline{w2}$ bindings.

(iii) Early bindings $\underline{w3}$ execution efficiency.

(iv) Late bindings $\underline{w4}$ execution efficiency.

The right choices of $w1, w2, w3$ *and* $w4$ (in that order) are

(a) Early, late, decrease, increase (b) Late, early, increase, decrease

(c) Late, early, decrease, increase (d) Early, late, increase, decrease.

36. The floating point unit of a processor using a design D takes 2t cycles compared to t cycles taken by the fixed point unit. There are two more design suggestions D1 and D2. D1 uses 30% more cycles for fixed point unit but 30% less cycles for floating point unit as compared to design D. D2 uses 40% less cycles for fixed point unit but 10% more cycles for floating point unit as compared to design D. For a given program which has 80% fixed point operations and 20% floating point operations, which of the following ordering reflects the relative performances of three designs ?

($Di > Dj$ denotes that Di is faster than Dj)

(a) $D1 > D > D2$ (b) $D2 > D > D1$

(c) $D > D2 > D1$ (d) $D > D1 > D2$

37. Consider a Direct Mapped Cache with 8 cache blocks (numbered 0 - 7). If the memory block requests are in the following order

3, 5, 2, 8, 0, 63, 9, 16, 20, 17, 25, 18, 30, 24, 2, 63, 5, 82, 17, 24.

Which of the following memory blocks will not be in the cache at the end of the sequence?

(a) 3 (b) 18 (c) 20 (d) 30

38. The following expression was to be realized using 2 - input AND and OR gates. However,

during the fabrication all 2 - input AND gates were mistakenly substituted by 2 - input NAND gates.

$$(a.b).c+(\overline{a}.c).d+(b.c).d+a.d$$

What is the function finally realized ?

(a) 1 (b) $\overline{a}+\overline{b}+\overline{c}+\overline{d}$

(c) $\overline{a}+b+\overline{c}+\overline{d}$ (d) $\overline{a}+\overline{b}+c+\overline{d}$

39. Data forwarding techniques can be used to speed up the operation in presence of data dependencies. Consider the following replacements of LHS with RHS.

(i) $R1 \rightarrow Loc$, $Loc \rightarrow R2$ = $R1 \rightarrow R2$, $R1 \rightarrow Loc$

(ii) $R1 \rightarrow Loc$, $Loc \rightarrow R2$ = $R1 \rightarrow R2$

(iii) $R1 \rightarrow Loc$, $R2 \rightarrow Loc$ = $R1 \rightarrow Loc$

(iv) $R1 \rightarrow Loc$, $R2 \rightarrow Loc$ = $R1 \rightarrow Loc$

In which of the following options, will the result of executing the RHS be the same as executing the LHS irrespective of the instructions that follow ?

(a) (i) and (iii) (b) (i) and (iv)

(c) (ii) and (iii) (d) (ii) and (iv)

40. What is the final value stored in the linear feedback shift register if the input is 101101 ?

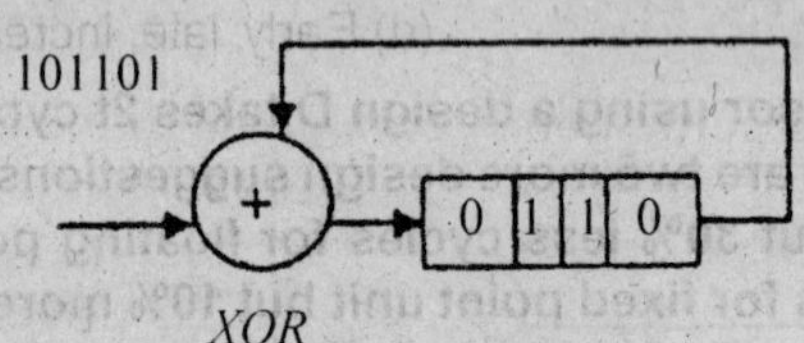

(a) 0110 (b) 1011

(c) 1101 (d) 1111

41. Following table indicates the latencies of operations between the instruction producing the result and instruction using the result.

Instruction production the result	Instruction using the result	Latency
A L U Operation	ALU Operation	2
ALU Operation	Store	2
Load	ALU Operation	1
Load	Store	0

Consider the following code segment.

```
Load R1,   Loc 1 ;       Load R1 from memory location Loc 1
Load R2,   Loc 2 ;       Load R2 from memory location Loc 2
Add R1, R2, R1 ;         Add R1 and R2 and save result in R1
Dec R2 ;                 Decrement R2
```

Dec R1 ;	Decrement R1
Mpy R1, R2, R3 ;	Multiply R1 and R2 and save result in R3
Store R3, Loc 3 ;	Store R3 in memory location Loc 3

What is the number of cycles needed to execute the above code segment assuming each instruction takes one cycle to execute ?

[a] [b]
[c] [d]

42. $(C012.25)_H - (10.111001110.101)_B =$

(a) $(135103.412)_O$ (b) $(564411.412)_O$

(c) $(564411.205)_O$ (d) $(135103.205)_O$

43. An error correcting code has the following code words : 00000000, 00001111, 01010101, 10101010, 11110000. What is the maximum number of bit errors that can be corrected ?

(a) 0 (b) 1 (c) 2 (d) 3

44. A hard disk system has the following parameters :

Number of tracks = 500

Number of sectors / track = 100

Number of bytes / sector = 500

Time taken by the head to move from one track to adjacent track = 1 ms

Rotation speed = 600 rpm.

What is the average time taken for tranferring 250 bytes from the disk ?

(a) 300.5 ms (b) 255.5 ms

(c) 255 ms (d) 300 ms

45. The line T in the following figure is permanently connected to the ground.

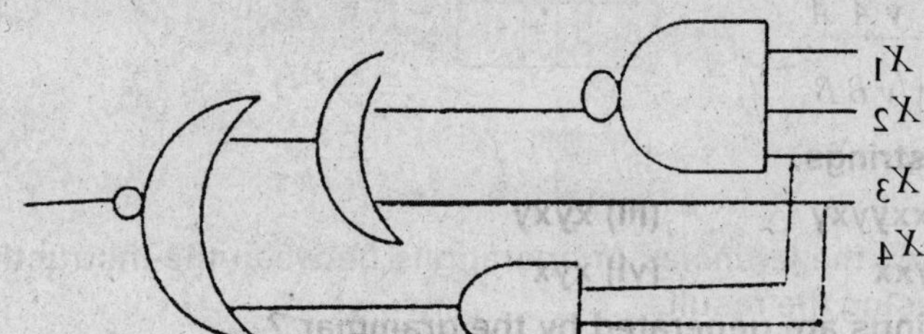

Which of the following inputs (X1 X2 X3 X4) will detect the fault ?

(a) 0000 (b) 0111 (c) 1111 (d) None of these

46. The two grammars given below generate a language over the alphabet {x, y, z}

G1 : $S \rightarrow x \mid z \mid xS \mid zS \mid yB$

$B \rightarrow y \mid z \mid yB \mid zB$

G2 : $S \rightarrow y \mid z \mid yS \mid zS \mid xB$

$B \rightarrow y \mid yS$

Which one of the following choices describes the properties satisfied by the strings in these languages ?

(a) G1 : No y appears before any x

G2 : Every x is followed by at least one y

(b) G1 : No y appears before any x

G2 : No x appears before any y

(c) G1 : No y appears after any x

G2 : Every x is followed by at least one y

(d) G1 : No y appears after any x

G2 : Every y is followed by at least one x

47. Consider the following DFA in which s_0 is the start state and s_1, s_3 are the final states.

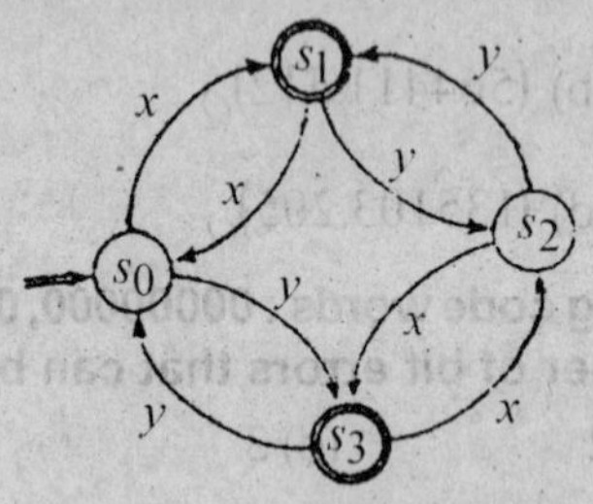

What language does this DFA recognize ?

(a) All strings of x and y

(b) All strings of x and y which have either even number of x and even number of y of odd number or x and odd number of y

(c) All strings of x and y which have equal number of x and y

(d) All strings of x and y with either even number of x and odd number of y or odd number of x and even number of y

48. Consider the grammar given below

$$S \rightarrow xB \mid yA$$
$$A \rightarrow x \mid xS \mid yAA$$
$$B \rightarrow y \mid yS \mid yBB$$

Consider the following strings.

(i) xxyyx	**(ii) xxyyxy**	**(iii) xyxy**
(iv) yxxy	**(v) yxx**	**(vi) xyx**

Which of the above strings are generated by the grammar ?

(a) (i) , (ii) and (iii) (b) (ii) , (v) and (vi)

(c) (ii), (iii) and (iv) (d) (i), (iii), and (iv)

49. Consider the following grammars. Names representing terminals have been specified in capital letters.

G_1	**stmnt**	$\rightarrow$	**WHILE (expr) stmnt**
	stmnt	$\rightarrow$	**OTHER**
	expr	$\rightarrow$	**ID**
G_2 :	**stmnt**	$\rightarrow$	**WHILE (expr) stmnt**
	stmnt	$\rightarrow$	**OTHER**

expr → expr + expr

expr → expr * expr

expr → ID

Which one of the following statements is true ?

(a) G_1 is context - free but not regular and G_2 is regular

(b) G_2 is context-free but not regular and G_1 is regular

(c) Both G_1 *and* G_2 are regular

(d) Both G_1 *and* G_2 are context-free but neither of them is regular

50. Consider the following finite automata P and Q over the alphabet { a, b, c}. The start states are indicated by a double arrow and final states are indicated by a double circle. Let the languages recognized by them be denoted by L(P) and L(Q) respectively.

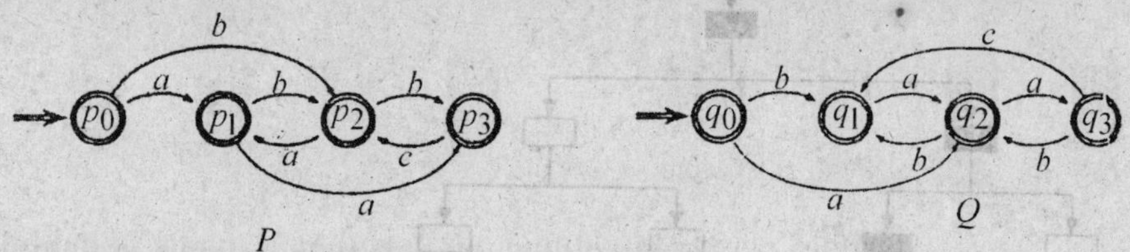

The automation which recognizes the language $L(P) \cap L(Q)$ is

(a) 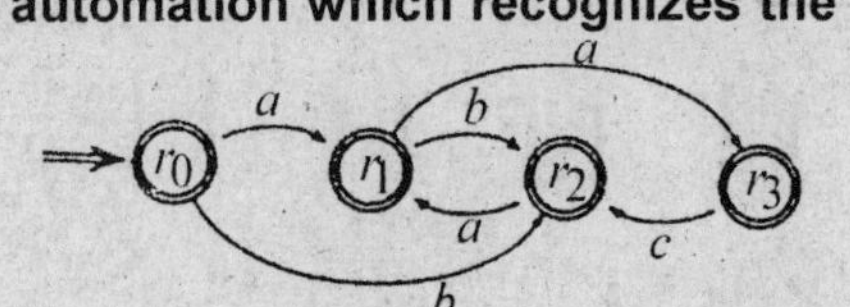

(b) 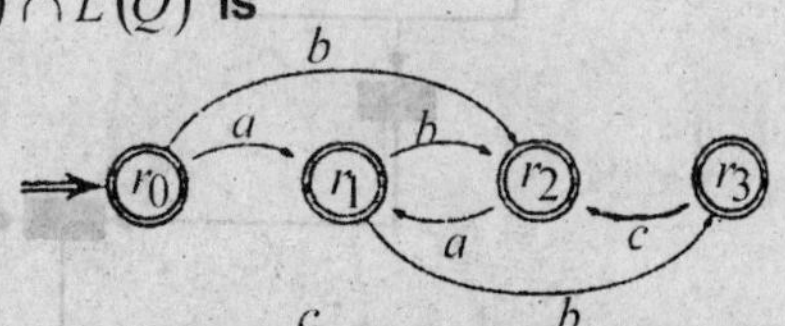

(c) 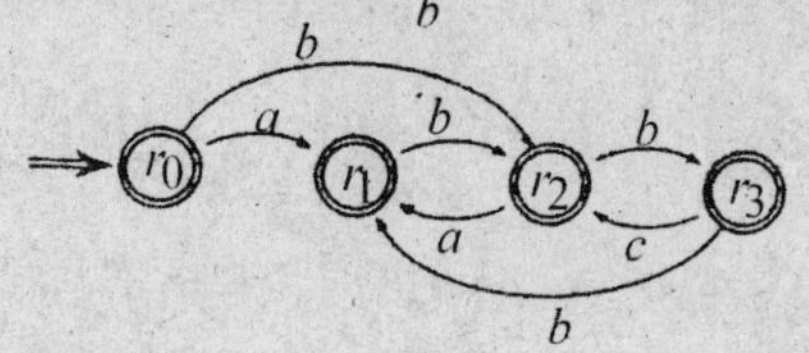

(d) 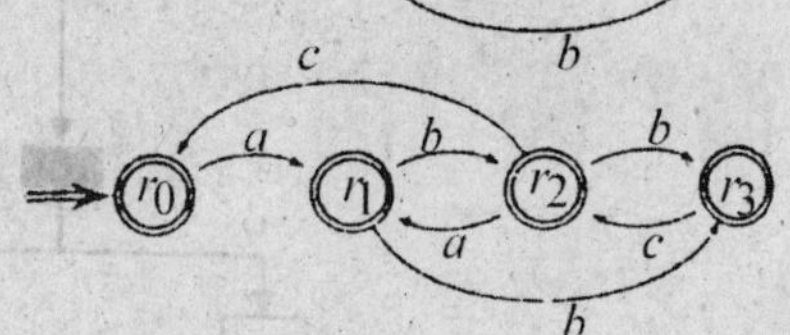

51. The following table shows the time between failures for a software system.

Error	1	2	3	4	5
Number Time since last failure (hours)	6	4	8	5	6

The reliability of the system for one hour of operation assuming an exponential model is

(a) 0.45 (b) 0.63

(c) 0.84 (d) 0.95

52. Given the following algorithm for shorting an array X and N numbers :

```
SUBROUTINE SHORT (X, N)
    IF (N < 2)
        RETURN
FOR ( i = 2 ) TO N INCREMENT BY 1 )
        FOR ( j = 1 ) TO i INCREMENT BY 1)
```

```
            IF ( X [i] > X ( [j] )
                 CONTINUE
            TEMP = X [i]
            X [i] = X [j]
            X [j] = TEMP
      END   FOR
END SUBROUTINE
```

A good approximation of Halstead's estimated program length is

(a) 20 (b) 50

(c) 80 (d) 110

53. In the simplified flowchart given below, the shaded boxes represent code that is executed during a test case.

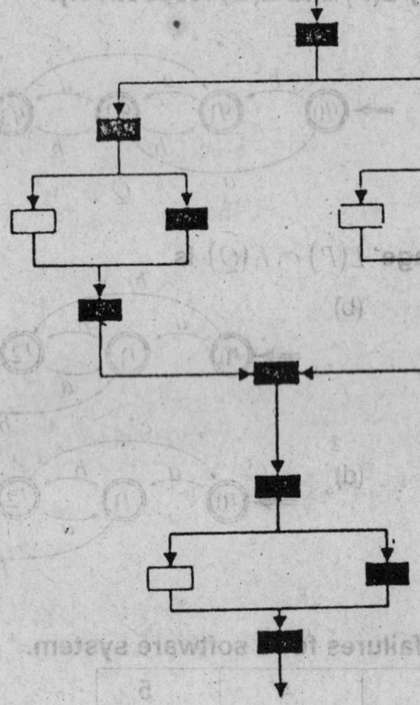

The branch coverage is

(a) 3 / 4 (b) 2 / 3

(c) 1 / 2 (d) 3 / 8

54. Consider the CPM activity chart where an arc connecting two milestones is labeled with a task identifier and the time taken in days. For example in order to go from A to B, task T1 takes 180 days. A dashed line depicts an additional dependency that is equivalent to a zero time task.

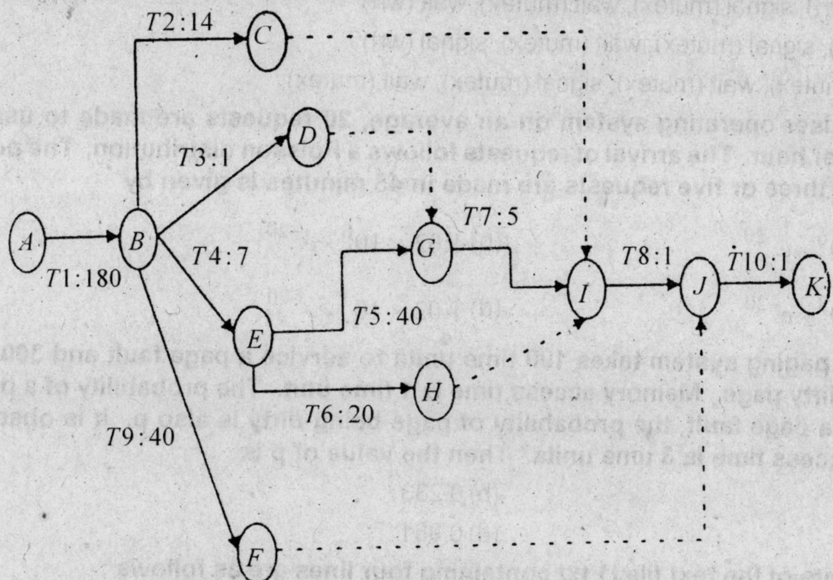

The set of activities that lie on the critical path are

(a) T1, T2, T8, T10 (b) T1, T3, T8, T10

(c) T1, T2, T3, T4, T5, T6, T7, T8, T9, T10 (d) T1, T4, T5, T7, T8, T10

55. Consider the following pseudo - code :

```
IF  ( A > B ) AND ( C > D )  )   THEN
    A  =  A + 1
    B  =  B + 1
ENDIF
```

The cyclomatic complexity of the pseudo-code is

(a) 2 (b) 3

(c) 4 (d) 5

56. Synchronization in the classical readers and writers problem can be achieved through use of semaphores. In the following incomplete code for readers-writers problem, two binary semaphores mutex and wrt are used to obtain synchronization

```
wait (wrt)
writing is performed
signal (wrt)
wait (mutex)
  readcount = readcount + 1
  if readcount = 1 then S1
S2
reading is performed
S3
  readcount = readcount -1
  if readcount = 0 then S4
Signal (mutex)
```

The values of S1, S2, S3, S4, (in that order) are

(a) signal (mutex), wait (wrt), signal (wrt), wait (mutex)

(b) signal (wrt), signal (mutex), wait (mutex), wait (wrt)

(c) wait (wrt), signal (mutex), wait (mutex), signal (wrt)

(d) signal (mutex), wait (mutex), signal (mutex), wait (mutex)

57. In a multi-user operating system on an average, 20 requests are made to use a particular resource per hour. The arrival of requests follows a Poisson distribution. The probability that either one, three or five requests are made in 45 minutes is given by

(a) $6.9 \times 10^6 \times e^{-20}$ (b) $1.02 \times 10^6 \times e^{-20}$

(c) $6.9 \times 10^3 \times e^{-20}$ (d) $1.02 \times 10^3 \times e^{-20}$

58. A demand paging system takes 100 time units to service a page fault and 300 time units to replace a dirty page. Memory access time is 1 time unit. The probability of a page fault is p. In case of a page fault, the probability of page being dirty is also p. It is observed that the average access time is 3 time units. Then the value of p is

(a) 0.194 (b) 0.233

(c) 0.514 (d) 0.981

59. The contents of the text file t1 txt containing four lines are as follows :

```
a1    b1
a2    b2
a3    b3
a4    b1
```

The contents of the text files t2 txt containing five lines are as follows :

```
a1    c1
a2    c2
a3    c3
a4    c3
a5    c4
```

Consider the following Bourne shell script :

```
awk - F ' ' {Print $1, $2} ' t1 . txt |
while read a b ; do
     awk - v aV = $ a - v bv  =  $ b - F ' '
     aV  = = $1 ( print aV, bV, $2)  '  t2  .  txt
done
```

Which one of the following strings will NOT be present in the output generated when the above script in run ? (Note that the given strings may be substrings of a printed line.)

(a) "$b1\ c1$" (b) "$b2\ c3$"

(c) "$b1\ c2$" (d) "$b1\ c3$"

60. For the network given in the figure below, the routing tables of the four nodes A, E, D and G are shown. Suppose that F has estimated its delay to its neighbors, A, E, D and G as 8, 10, 12 and 6 msecs respectively and updates its routing table using distance vector routing technique.

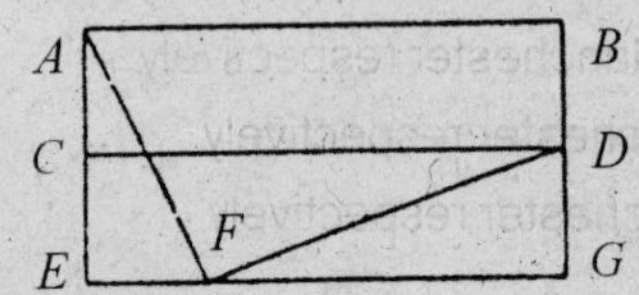

Routing Table of A	
A	0
B	40
C	14
D	17
E	21
F	9
G	24

Routing Table of E	
A	24
B	27
C	7
D	20
E	0
F	11
G	22

Routing Table of D	
A	20
B	8
C	30
D	0
E	14
F	7
G	22

Routing Table of G	
A	21
B	24
C	22
D	19
E	22
F	10
G	0

Which of the following options represents the updated routing table of F ?

(a)

A	28
B	20
C	17
D	12
E	10
F	0
G	6

(b)

A	21
B	8
C	7
D	19
E	14
F	0
G	22

(c)

A	8
B	20
C	17
D	12
E	10
F	16
G	6

(d)

A	8
B	8
C	7
D	12
E	10
F	0
G	6

61. In the waveform (a) given below, a bit stream is encoded by Manchester encoding scheme. The same bit stream is encoded in a different coding scheme in wave form (b) . The bit stream and the coding scheme are

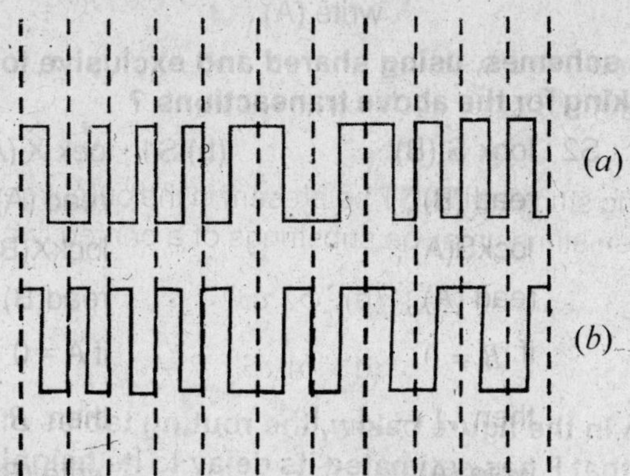

(a) 1000010111 and Differential Manchester respectively

(b) 0111101000 and Differential Manchester respectively

(c) 1000010111 and Integral Manchester respectively

(d) 0111101000 and Integral Manchester respectively

62. Let us consider a statistical time division multiplexing of packets. The number of sources is 10. In a time unit, a source transmits a packet of 1000 bits. The number of sources sending data for the first 20 time units is 6, 9, 3, 7, 2, 2, 2, 3, 4, 6, 1, 10, 7, 5, 8, 3, 6, 2, 9, 5 respectively. The output capacity of multiplexer is 5000 bits per time unit. Then the average number of backlogged of packets per time unit during the given period is

(a) 5 (b) 4.45

(c) 3.45 (d) 0

63. A group of 15 routers are interconnected in a centralized complete binary tree with a router at each tree node. Router i communicates with router j by sending a message to the root of the tree. The root then sends the message back down to router j. The mean number of hops per message, assuming all possible router pairs are equally likely is

(a) 3 (b) 4.26

(c) 4.53 (d) 5.26

64. A broadcast channel has 10 nodes and total capacity of 10 Mbps. It uses polling for medium access. Once a node finishes transmission, there is a polling delay of $80\mu s$ to poll the next node. Whenever a node is polled, it is allowed to transmit a maximum of 1000 bytes. The maximum throughput of the broadcast channel is

(a) 1 Mbps (b) 100 /11 Mbps

(c) 10 Mbps (d) 100 Mbps

65. Consider a selection of the form $\sigma_{A\leq 100}(r)$, where r is a relation with 1000 tuples. Assume that the attribute values for A among the tuples are uniformly distributed in the interval $[0,500]$. Which one of the following options is the best estimate of the number of tuples returned by the given selection query?

(a) 50 (b) 100

(c) 150 (d) 200

66. Consider the following two transactions : T_1 *and* T_2.

T_1 : read (A) ;	T_2 : read (B)
read (B) ;	read (A)
if A = 0 then $B \leftarrow B+1$;	if $B \neq 0$ then $A \leftarrow A-1$;
write (B) ;	write (A) ;

Which of the following schemes, using shared and exclusive locks, satify the requirements for strict two phase locking for the above transactions ?

(a)

S1 : lock S (A)	S2 : lock S (B)
read (A) ;	read (B) ;
lockS(B) ;	lockS(A) ;
read(B) ;	read (A) ;
if A = 0	if $B \neq 0$
then $B \leftarrow B+1$;	then $A \leftarrow A-1$;
write (B) ;	write (A) ;
commit ;	commit ;
unlock (A) ;	unlock (B) ;
unlock (B) ;	unlock (A) ;

(b)

S1 : lock X (A)	S2 : lock X (B)
read (A) ;	read (B) ;
lockX(B) ;	lockX(A) ;
read(B) ;	read (A) ;
if A = 0	if $B \neq 0$
then $B \leftarrow B+1$;	then $A \rightarrow A-1$;
write (B) ;	write (A) ;
unlock (A) ;	unlock (A) ;
commit ;	commit ;
unlock (B) ;	unlock (B) ;

(c) S1 : lock S (A) S2 : lock S (B)

(d) S1 : lock S (A) S2 : lock S (B)

read (A) ;	read (B) ;	read (A) ;	read (B) ;
lockX(B) ;	lockX(A) ;	lockX(B) ;	lockX(A) ;
read(B) ;	read (A) ;	read(B) ;	read (A) ;
if A = 0	if $B \neq 0$	if A = 0	if $B \neq 0$
then $B \leftarrow B+1;$	then $A \leftarrow A-1;$	then $B \leftarrow B+1;$	then $A \leftarrow A-1;$
write (B) ;	write (A) ;	write (B) ;	write (A) ;
unlock (A) ;	unlock (A) ;	unlock (A) ;	unlock (A) ;
commit ;	commit ;	unlock (B) ;	unlock (B) ;
unlock (B) ;	unlock (B) ;	commit ;	commit ;

67. Consider the following implications relating to functional and multivalued dependencies given below, which may or may not be correct.

(i) If $A \to B$ and $A \to C$ then $A \to BC$

(ii) If $A \to B$ and $A \twoheadrightarrow C$ then $A \to BC$

(iii) If $A \to BC$ and $A \to B$ then $A \to C$

(iv) If $A \to BC$ and $A \to B$ then $A \to C$

Exactly how many of the above implications are valid ?

(a) 0 (b) 1 (c) 2 (d) 3

68. Consider the following relation schemas :

b-Schema = (b-name, b-city, assets)

a-Schema = (a-num, b-name, bal)

d-Schema = (c-name, a-number).

Let branch, account and depositor be respectively instances of the above schemas. Assume that account and depositor relations are much bigger than the branch relation.

Consider the following query :

$$\Pi_{c-name}\left(\sigma_{b-city="Agra" \wedge bal<0}\left(branch \bowtie (account \bowtie depositer)\right)\right)$$

Which one of the following queries is the most efficient version of the above query ?

(a) $\Pi_{c-name}\left(\sigma_{bal<0}\left(\sigma_{b-city="Agra"}\, branch \bowtie account\right) \bowtie depositer\right)$

(b) $\Pi_{c-name}\left(\sigma_{b-city="Agra"}\, branch \bowtie \left(\sigma_{bal<0}\, account \bowtie depositer\right)\right)$

(c) $\Pi_{c-name}\left(\sigma_{b-city="Agra"}\, branch \bowtie \sigma_{b-city="Agra" \wedge bal<0}\, account \bowtie depositer\right)$

(d) $\Pi_{c-name}\left(\sigma_{b-city="Agra"}\, branch \bowtie \left(\sigma_{b-city="Agra" \wedge bal<0}\, account \bowtie depositer\right)\right)$

69. Consider the following clauses :

(i) Not inherently suitable for client authentication.

(ii) Not a state sensitive protocol.

(iii) Must be operated with more than one server.

(iv) Suitable for structured message organization.

(v) May need two ports on the serve side for proper operation.

The option that has the maximum number of correct matches is

(a) IMAP - (i), FTP - (ii), HTTP - (iii), DNS - (iv), POP3 - (v)

(b) FTP - (i), POP3 - (ii), SMTP - (iii), HTTP - (iv), IMAP - (v)

(c) POP3 - (i), SMTP - (ii), DNS - (iii), IMAP - (iv), HTTP - (v)

(d) SMTP - (i), HTTP - (ii), IMAP - (iii), DNS - (iv), FTP - (v)

70. Your are given the following four bytes :

10100011	00110111	11101001	10101011

Which of the following are substrings of the base 64 encoding of the above four bytes ?

(a) zdp (b) fpq

(c) qwA (d) oze

COMMON DATA QUESTIONS

Common data for questions 71, 72, 73 :

Consider the regular expression $R = (a+b)^*(aa+bb)(a+b)^*$

71. Which of the following non-deterministic finite automata recognizes the language defined by the regular expression R ? Edges labeled λ denote transitions on the empty string.

(a)

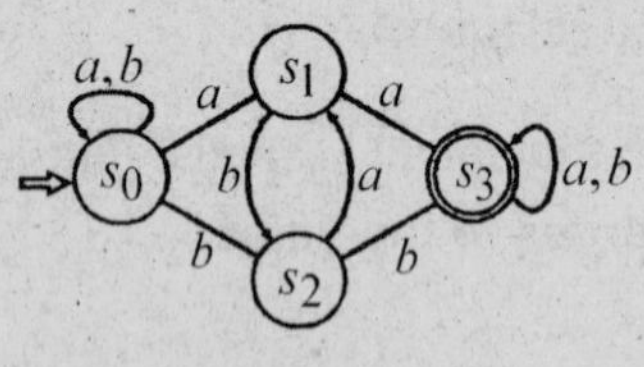

(b)

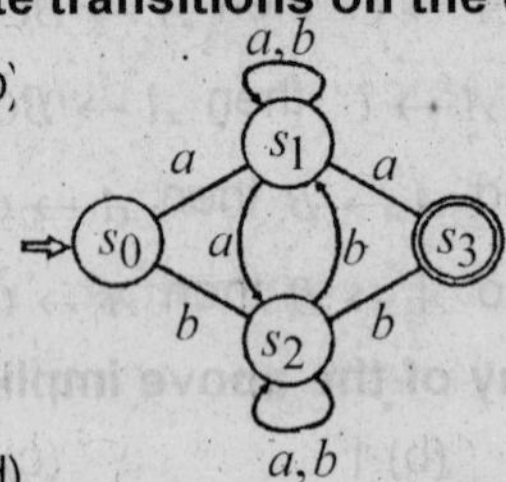

(c)

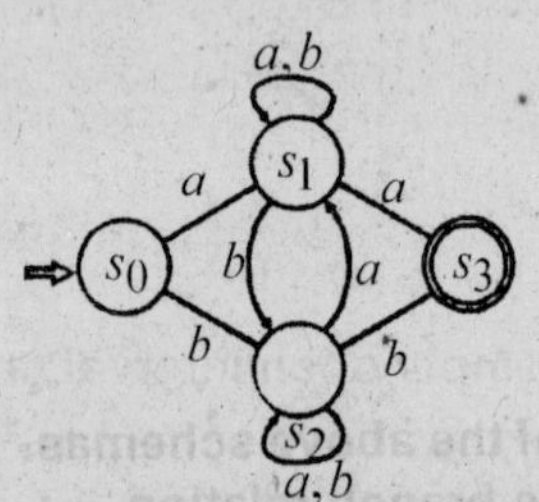

(d)

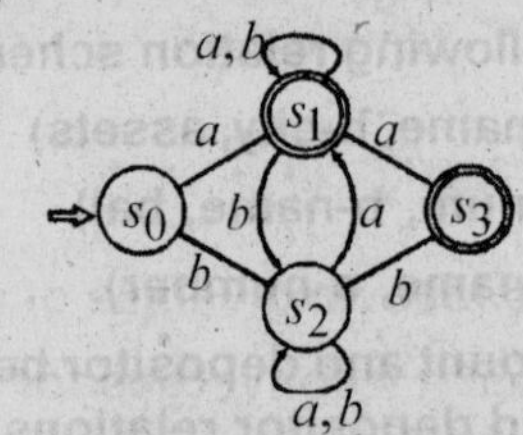

72. Which deterministic finite automaton accepts the language represented by the regular expression R ?

(a)

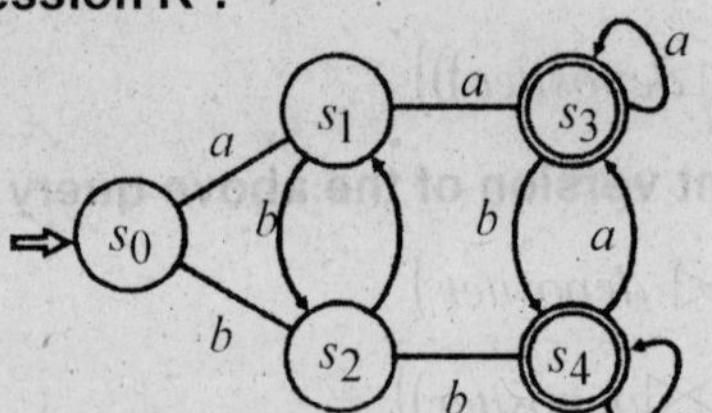

(b)

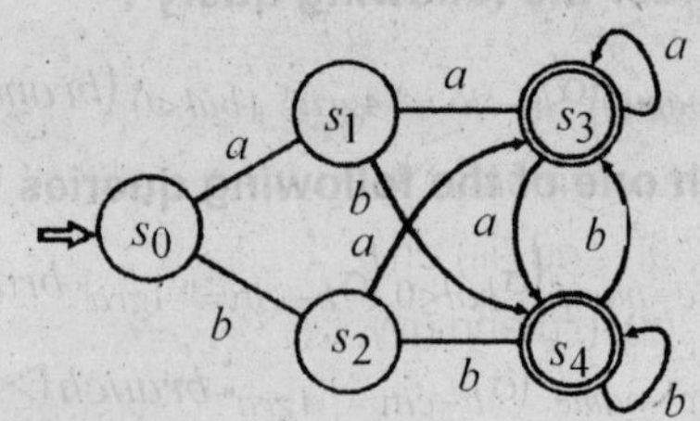

(c)

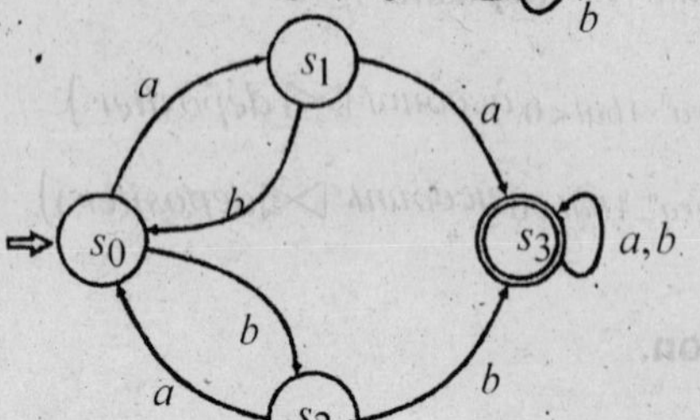

(d)

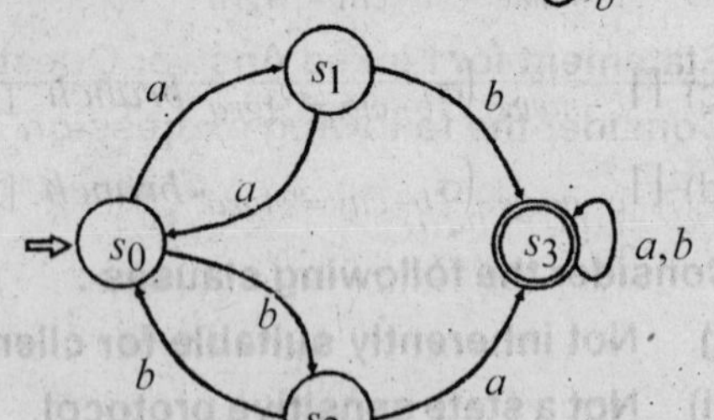

73. Which one of the regular expressions given below defines the same language as defined by the regular expression R ?

(a) $\left(a(ba)^{*}+b(ab)^{*}\right)(a+b)^{+}$

(b) $\left(a(ba)^{*}+b(ab)^{*}\right)^{*}(a+b)^{*}$

(c) $\left(a(ba)^{*}(a+bb)+b(ab)^{*}(b+aa)\right)(a+b)^{*}$

(d) $\left(a(ba)^{*}(a+bb)+b(ab)^{*}(b+aa)\right)(a+b)^{+}$

Common Data for Questions 74, 75 :

Consider a token ring topology with N stations (numbered 1 to N) running token ring protocol where the stations are equally spaced. When a station gets the token it is allowed to send one frame of fixed size. Ring latency is t_p , while the transmission time of a frame is t_t . All other latencies can be neglected.

74. The maximum utilization of the token ring when $t_t = 3\,ms,\ t_p = 5\,ms,\ N = 10$ is

(a) 0.545 (b) 0.6 (c) 0.857 (d) 0.961

75. The maximum utilization of the token ring when $t_t = 5\,ms,\ t_p = 3\,ms,\ N = 15$ is

(a) 0.545 (b) 0.655 (c) 0.9375 (d) 0.961

Linked answer Questions : Q.76 to Q.85 carry two marks each.

Statement for Linked Answer Questions 76 and 77 :

Consider the sequence $\{x_n\}_{n\geq 0}$ defined by the recurrence relation $x_{n+1} = cx_n^2 - 2,$ where c >0

76. Suppose there exists a non - empty open interval (a, b) such that for all x_0 satisfying $a < x_0 < b$, the sequence converges to a limit. The sequence converges to the value

(a) $\frac{1+\sqrt{1+8c}}{2c}$ (b) $\frac{1-\sqrt{1+8c}}{2c}$

(c) 2 (d) $\frac{2}{2c-1}$

77. For which of the following values of c, does there exist a non-empty open interval (a, b) such that the sequence x_n converges for all x_0 satisfying $a < x_0 < b$?

(i) 0.25 (ii) 0.35 (iii) 0.45 (iv) 0.5

(a) (i) only
(b) (i) and (ii) only
(c) (i), (ii) and (iii) only
(d) (i), (ii), (iii) and (iv)

Statement for Linked Answer Questions 78 to 79 :

Consider the following expression

$a\bar{d} + \overline{ac} + b\bar{c}d$

78. Which of the following Karnaugh Maps correctly represents the expression ?

(a)

	$\overline{cd}$	$\bar{c}d$	cd	$c\bar{d}$
$\overline{ab}$	×	×		
$\bar{a}b$	×	×		
ab	×	×		×
$a\bar{b}$	×			×

(b)

	$\overline{cd}$	$\bar{c}d$	cd	$c\bar{d}$
$\overline{ab}$	×	×		
$\bar{a}b$	×			
ab	×	×		×
$a\bar{b}$	×	×		×

(c)

	$\overline{cd}$	$\bar{c}d$	cd	$c\bar{d}$
$\overline{ab}$	×	×		
$\bar{a}b$	×	×		×
ab	×			×
$a\bar{b}$	×			×

(d)

	$\overline{cd}$	$\bar{c}d$	cd	$c\bar{d}$
$\overline{ab}$	×	×		
$\bar{a}b$	×	×		
ab	×	×		×
$a\bar{b}$	×		×	×

79. Which of the following expressions does not correspond to the Karnaugh Map obtained in Q 78 ?

(a) $c\bar{d} + a\bar{d} + ab\bar{c} + \bar{a}\bar{c}d$

(b) $\bar{a}\bar{c} + \bar{c}\bar{d} + a\bar{d} + ab\bar{c}d$

(c) $\bar{a}\bar{c} + a\bar{d} + ab\bar{c} + \bar{c}\bar{d}$

(d) $\bar{b}\bar{c}\bar{d} + ac\bar{d} + \bar{a}\bar{c} + ab\bar{c}$

<u>Statement for Linked Answer Questions 80 and 81</u>

Let $P_1, P_2, \ldots\ldots, P_n$ be n points in the xy - plane such that no three of them are collinear. For every pair of points P_i and P_j, let L_{ij} be the line passing through them. Let L_{ab} be the line with the steepest gradient amongst all $n(n-1)/2$ lines.

80. Which one of the following properties should necessarily be satisfied ?

(a) P_a *and* P_b are adjacent to each other with respect to their x - coordinate

(b) Either P_a *or* P_b has the largest or the smallest y - coordinate among all the points

(c) The difference between x - coordinatef P_a *and* P_b is minimum

(d) None of the above

81. The time complexity of the best algorithm for finding P_a *and* P_b is

(a) $\theta(n)$

(b) $\theta(n \log n)$

(c) $\theta(n \log^2 n)$

(d) $\theta(n^2)$

Statement for Linked Answer Questions 82 and 83 :

The head of a hard disk serves requests following the shortest seek time first (SSTF) policy. The head is initially positioned at truck number 180.

82. Which of the request sets will cause the head to change its direction after servicing every request assuming that the head does not change direction if there is a tie in SSTF and all the requests arrive before the servicing starts ?

(a) 11, 139, 170, 178, 181, 184, 201, 265

(b) 10, 138, 170, 178, 181, 185, 201, 265

(c) 10, 139, 169, 178, 181, 184, 201, 265

(d) 10, 138, 170, 178, 181, 185, 200, 265

83. What is the maximum cardinality of the request set, so that the head changes its direction after servicing every request if the total number of tracks are 2048 and the head can start from any track ?

(a) 9 (b) 10 (c) 11 (d) 12

Statement for Linked Answer Questions 84 and 85 :

Consider the B tree in the adjoining figure, where each node has at most two keys and three links.

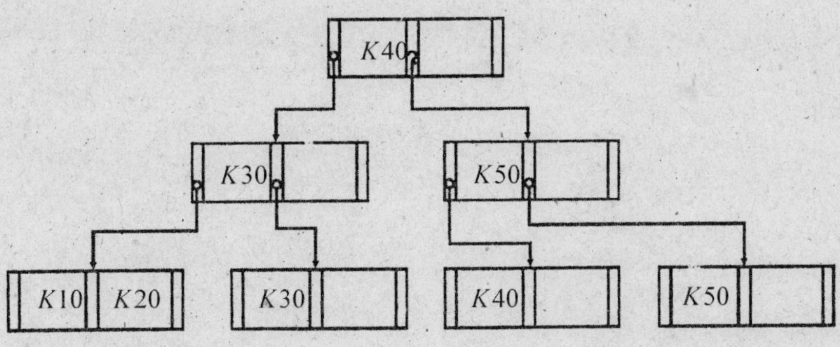

84. Keys K 15 and then K 25 are inserted into this tree in that order. Exactly how many of the following nodes (disregarding the links) will be present in the tree after the two insertions ?

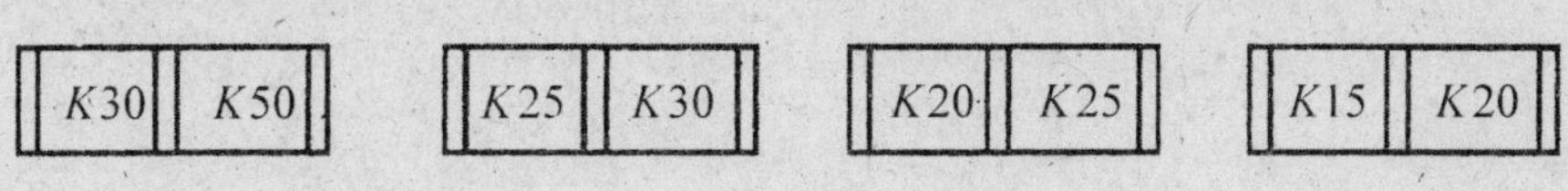

(a) 1 (b) 2 (c) 3 (d) 4

85. Now the key K 50 is deleted from the B^+ tree resulting after the two insertions made earlier.

Consider the following statements about the B^+ tree resulting after this deletion.

(i) The height of the tree remains the same.

(ii) The node  **(disregarding the links) is present in the tree.**

(iii) The root node remains unchanged (disregarding the links).

Which one of the following options is true ?

(a) Statements (i) and (ii) are true (b) Statements (ii) and (iii) are true

(c) Statements (iii) and (i) are true (d) All the statements are false

ANSWERS

1. (c)	2. (b)	3.(d)	4. (d)	5. (d)	6. (a)	7. (b)	8.(a)	9. (a)	10. (b)
11.(b)	12.(c)	13.(c)	14.(d)	15. (d)	16.(b)	17.(c)	18.(d)	19.(d)	20. (d)
21.(a)	22.(d)	23.(d)	24.(a)	25. (d)	26.(d)	27.(c)	28.(d)	29.(c)	30. (b)
31.(c)	32.(b)	33.(d)	34.(c)	35. (a)	36.(b)	37.(b)	38.(b)	39.(b)	40. (c)
41.(b)	42.(a)	43.(d)	44.(b)	45. (d)	46.(a)	47.(d)	48.(d)	49.(b)	50. (b)
51.(b)	52.(b)	53.(a)	54.(b)	55. (b)	56.(c)	57.(c)	58.(d)	59.(c)	60. (a)
61.(c)	62.(c)	63.(d)	64.(a)	65. (d)	66.(c)	67.(c)	68.(b)	69.(d)	70. (a)
71.(b)	72.(c)	73.(c)	74.(c)	75. (a)	76.(d)	77.(c)	78.(a)	79.(a)	80. (b)
81.(a)	82.(b)	83.(c)	84.(a)	85. (a)					

GATE - 2006
COMPUTER SCIENCE AND ENGINEERING

Duration : Three Hours **Maximum Marks : 150**

Read the following instructions carefully

1. This question paper contains objective questions. Q.1-20 carry one mark each and Q.20-75 carry two marks each. Q. 75-85 carry two marks each.
2. Answer all the questions.
3. Questions must be answred on the special machine gradable Objective Response Sheet (ORS) by darkening the appropriate bubble (marked A, B, C, D) using HB pencil against the question number on the left hand side of the ORS. Each question has only one correct answer. In case you wish to change an answer, erase the old answer completely using a good soft eraser.
4. There will be NEGATIVE marking. In Q.1 to Q.20, 0.25 mark will be deducted for each wrong answer and in Q.20 to Q.75, 0.5 mark will be deducted for a wrong answer. Marks for correct answer to part "b" of Q. 76 to Q.85 will be given only if the answer to the corresponding part "a" is correct. However, there is no negative marking for part "b" of Q. 75 to Q.85. More than one answer bubbled against will be deemed as an incorrect response.
5. Write your registration number, name and name of the centre at the specified loacations on the right half of the **ORS.**
6. Using HB pencil, darken the appropriate bubble under each digit of your registration number and the letter corresponding to your paper code.
7. Calculator is allowed in the examination hall.
8. Charts, graph sheets or tables are not allowed.

Q. 1 - 20 CARRY ONE MARK EACH

1. Consider the polynomial $p(x) = a_0 + a_1x + a_2x^2 + a_3x^3$, where $a_i \neq 0, \forall i$. The minimum number of multiplications needed to evaluate p on an input x is

(b) 3 (b) 4

(c) 6 (d) 9

2. Let, X,Y,Z be sets of sizes x,y and z respectively. Let $W = X \times Y$ and E be the set of all subsets of W. The number of functions from Z to E is

(A) z (B) $z \times 2^{xy}$

(C) z (D) 2^{xyz}

3. The set { 1, 2, 3, 5, 7, 8, 9 } under multiplication modulo 10 is not a group. Given below are four plausible reasons. Which one of them is false ?

(A) It is not closed

(B) 2 does not have an inverse

(C) 3 does not have an inverse

(D) 8 does not have an inverse

4. A relation R is defined on ordered pairs of integers as follows

$(x, y) R (u, v)$ *if* $x<u$ **and** $y>v$. **Then** R **is**

(A) Neither a Partial Order nor an Equivalent Relation

(B) A partial Order but not a Total Order

(C) A Total Order

(D) An Equivalence Relation

5. For which one of the following reasons does Internet Protocol (IP) use the time - to - live (TTL) field in the IP datagram header ?

(A) Ensure packets reach destination within that time

(B) Discard packets that reach later than that time

(C) Prevent packets from looping indefinitely

(D) Limit the time for which a packet gets queued in intermediate routers.

6. Consider three CPU - intensive processes, which require 10, 20, and 30 time units and arrive at times 0, 2, and 6, respectively. How many context switches are needed if the operating system implements a shortes remaining time first scheduling algorithm ? Do not count the context switches at time zero and at the end.

(A) 1 (B) 2

(C) 3 (D) 4

7. Consider the following grammar.

$S \to S * E$

$S \to E$

$E \to F + E$

$E \to F$

$F \to id$

Consider the following LR(0) items corresponding to the grammar above.

(i) $S \to S * .E$

(II) $E \to F. + E$

(III) $E \to F + .E$

Given the items above, which two of them will appear in the same set in the canonical sets - of - items for the grammar ?

(A) (i) and (ii) (B) (ii) and (iii)

(C) (i) and (iii) (D) None of these

8. You are given a free running clock with a duty cycle of 50% and a digital waveform f which changes only at the negative edge of the clock. Which one of the following circuits (using clocked D flip - flops) will delay the phase of f by $180°$?

(A)

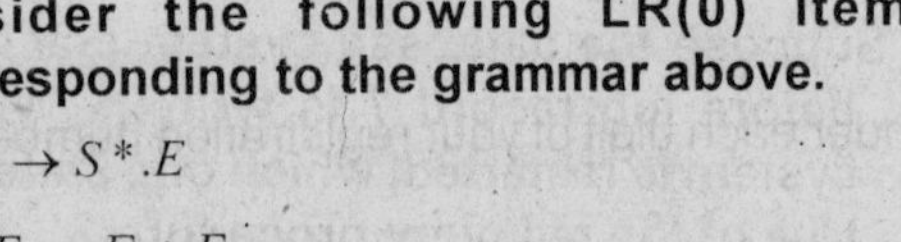

(B)

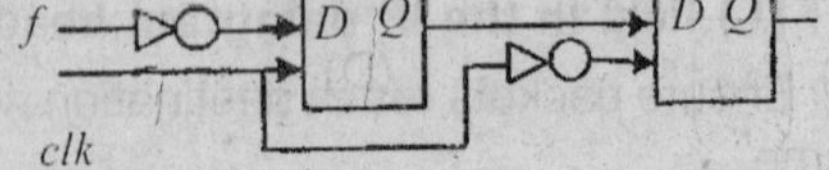

(C)

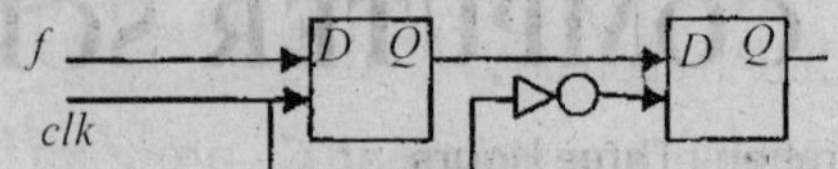

(D)

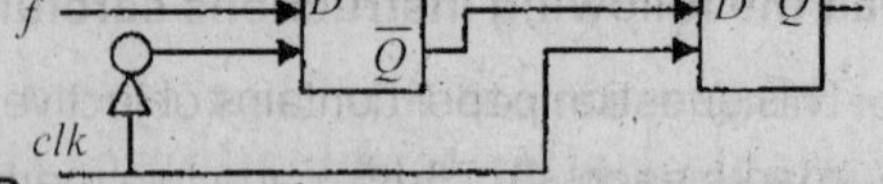

9. A CPU has 24 - bit instructions. A program starts at address 300 (in decimal). Which one of the following is a legal program counter (all values in decimal) ?

(A) 400 (B) 500

(C) 600 (D) 700

10. In a binary max heap containing n numbers, the smallest element can be found in time

(A) $\theta(n)$ (B) $\theta(\log n)$

(C) $\theta(\log \log n)$ (D) $\theta(1)$

11. Consider a weighted complete graph G on the vertex set $\{V_1, V_2, ..., V_n\}$ such that the weight of the edge (V_i, V_j) is $2|i-j|$. The weight of a minimum spanning tree of G is

(A) $n-1$ (B) $2n-2$

(C) $\binom{n}{2}$ (D) n^2

12. To implement Dijkstra's shortest path algorithm on unweighted graphs so that it runs in linear time, then data structure to be used is

(A) Queue (B) Stack

(C) Heap (D) B-Tree

13. A scheme for storing binary trees in an array X is as follows. Indexing of X starts at 1 instead of 0. The roots is stored at X[1]. For a node stored at X[i], the left child, if any, is stored in x[2i] and the right child, if any, in $X[2i+1]$. To be able to store any binary tree on n vertices, the minimum size of X should be

(A) $\log_2 n$ (B) n (C) $2n+1$ (D) $2^n - 1$

14. Which one of the following in place sorting algorithms needs the minimum number of swaps ?

(A) Quick sort (B) Insertion sort

(C) Selection sort (D) Heap sort

15. Consider the following C - program fragment in which i, j, and n are integer variables.

for $(i = n, j = 0; i > 0; i /= 2, j += i);$

Let val(j) = denote the value stored in the variable j after termination of the for loop. Which one of the following is true?

(A) $val(j) = \theta(\log n)$ (B) $val(j) = \theta(\sqrt{n})$

(C) $val(j) = \theta(n)$ (D) $val(j) = \theta(n \log n)$

16 Let S be an NP-complete problem Q and R be two other problems not known to be in NP. Q is polynomial-time reducible to S and S is polynominal-time reducible to R. Which one of the following statements is true ?

(A) R is NP-complete (B) R is NP-hard
(C) Q is NP-complete (D) Q is NP-hard

17. An element in an array X is called a leader if it is greater than all elements to the right of it in X. The best algorithm to find all leaders in an array

(A) Solves it in linear time using a left to right pass of the array

(B) Solves in linear time using a right to left pass of the array

(C) Solves it using divide and conquer in time $\theta(n \log n)$

(D) Solves it in time $\theta(n^2)$

18. We are given a set $X = \{x_1, \ldots\ldots, x_n\}$ **where** $x_i = 2^i$ **. A sample** $S \subseteq X$ **is drawn by selecting each** x_i **independently with probability** $p_i = \frac{1}{2}$**. The expected value of the smallest number in sample S is**

(A) $\frac{1}{n}$ (B) 2 (C) $\sqrt{n}$ D) n

19. Let $L_1 = \{0^{n+m} 1^n 0^m \mid n, m \geq 0\}$,

$L_2 = \{0^{n+m} 1^{n+m} 0^m \mid n, m \geq 0\}$ **and**

$L_3 = \{0^{n+m} 1^{n+m} 0^{n+m} \mid n, m \geq 0\}$, **Which of these languages are NOT context free ?**

(A) L_1 only (B) L_3 only

(C) L_1 and L_2 (D) L_2 and L_3

20. Consider the following log sequence of two transactions on a bank account, with initial balance 12000, that transfer 2000 to a mortgage payment and, then apply a 5% interest.

1. T1 start

2. T1 B old = 12000 new = 10000

3. T1 M old = 0 new = 2000

4. T1 commit

5. T2 start

6. T2 B old = 10000 new = 10500

7. T2 commit

Suppose the database system crashed just before log record 7 is written. When the system is restarted, which one statement is true of the recovery procedure ?

(A) We must redo log record 6 to set B to 10500

(B) We must undo log record 6 to set B to 10000 and then redo log records 2 and 3

(C) We need not redo log records 2 and 3 because transaction T1 has committed

(D) We can apply redo and undo operations in arbitrary order because they are idempotent

Q.21 to Q.75 carry Two Marks each.

21. For each element in a set of size 2n, an unbiased coin is tossed. The 2n coin tosses are independent. An element is chosen if the corresponding coin toss were head. The probability that exactly n elements are chosen is

(A) $\frac{\binom{2n}{n}}{4^n}$ (B) $\frac{\binom{2n}{n}}{2^n}$

(C) $\frac{1}{\binom{2n}{n}}$ (D) $\frac{1}{2}$

22. Let E,F and G be finite sets.

Let $X=(E\cap F)-(F\cap G)$ **and** $Y=(E-(E\cap G))-(E-F)$. **Which one of the following is true ?**

(A) $X\subset Y$

(B) $X\supset Y$

(C) $X=Y$

(D) $X-Y\neq\varnothing$ and $Y-X\neq\varnothing$

23. F is an $n\times n$ **real matrix, b is an** $n\times 1$ **real vector. Suppose there are two** $n\times 1$ **vectors, u and v such that** $u\neq v$, **and** $Fu=b, Fv=b$. **Which one of the following statements is false ?**

(A) Determinant of F is zero

(B) There are an infinite number of solutions to $Fx=b$

(C) There is an $x\neq 0$ such that $Fx=0$

(D) F must have two identical rows

24. Given a set of elements $N=\{1,2,.........,n\}$ **and two arbitrary subsets** $A\subseteq N$ and $B\subseteq N$, **how many of the n! permutations** π **from N to N satisfy min** $(\pi(A))=\min(\pi(B))$, **where** $\min(S)$ **is the smallest integer in the set of integers** S, **and** $\pi(S)$ **is the set of integers obtained by applying permutation** π **to each element of S ?**

(A) $(n-|A\cup B|)|A||B|$

(B) $(|A|^2+|B|^2)n^2$

(C) $n!\dfrac{|A\cap B|}{|A\cup B|}$ (D) $\dfrac{|A\cap B|^2}{\binom{n}{|A\cup B|}}$

25. Let $S=\{1,2,3,.....,m\}\ m>3$. **Let** $X_1,......,X_n$ **be subsets of S each of size 3. Define a function f from S to the set of natural numbers as,** $f(i)$ **is the number of sets** X_j **that contain the element i. That is** $f(i)=|\{j\mid i\in X_j\}|$

Then $\sum_{i=1}^{m} f(i)$ is

(A) 3m (B) 3n

(C) 2m+1 (D) 2n+1

26. Which one of the first order predicate calculus statements given below correctly expresses the following English statement ?

Tigers and lions attack if they are hungry or threatened.

(A) $\forall x$ [(tiger (x) $\wedge$ lion (x)) $\rightarrow$ {(hungery (x) $\vee$ threatened (x)) $\rightarrow$ attacks (x) }]

(B) $\forall x$ [(tiger (x) $\vee$ lion (x)) $\rightarrow$ {(hungry (x) $\vee$ threatened (x)) $\wedge$ attacks (x) }]

(C) $\forall x$ [(tiger (x) $\vee$ lion (x)) $\rightarrow$ {(attacks (x) $\rightarrow$ (hungry (x) $\vee$ threatened (x))}]

(D) $\forall x$ [tiger (x) $\vee$ lion (x)) $\rightarrow$ {(hungry (x) $\vee$ threatened (x)) $\rightarrow$ attacks (x) }].

27. Consider the following propositional statements

$P1:((A\wedge B)\rightarrow C))\equiv((A\rightarrow C)\wedge(B\rightarrow C))$

$P2:((A\vee B)\rightarrow C))\equiv((A\rightarrow C)\vee(B\rightarrow C))$

Which one of the following is true ?

(A) P_1 is a tautology, but not $P2$

(B) $P2$ is a tautology but not P_1

(C) P_1 and $P2$ are both tautologies

(D) Both P_1 and $P2$ are not tautologies.

28. A logical binary relation ⊙, is defined as follows

A	B	A⊙B
True	True	True
True	False	True
False	True	False
False	False	True

Let ~ be the unary negation (NOT) operator, with higher precedence, then ⊙. Which one of the following is equivalent to $A\wedge B$**?**

(A) $(\sim A \odot B)$

(B) $\sim(A \odot \sim B)$

(C) $\sim(\sim A \odot \sim B)$

(D) $\sim(\sim A \odot B)$

29. If s is a string over $(0+1)^*$, then let $n_0(s)$ denote the number of $0's$ in s and $n_1(s)$ the number of $1's$ in s. Which one of the following languages is not regular ?

(A) $L = \{s \in (0+1)^* \mid n_0(s)$ is a 3-digit prime$\}$

(B) $L = \{s \in (0+1)^* \mid$ for every prefixs' of s, $|n_0(s') - n_1(s')| \leq 2\}$

(C) $L = \{s \in (0+1)^* \mid . \mid n_0(s) - n_1(s) \leq 4\}$

D) $L = \{s \in (0+1)^* \mid n_0(s) \bmod 7 = n_1(s) \bmod 5 = 0\}$

30. For $s \in (0+1)^*$ let d(s) denote the decimal value of s(e.g. d (101) = 5).

Let $L = \{s \in (0+1)^* \mid d(s) \bmod 5 = 2$ *and* $\bmod 7 \neq 4\}$

Which one of the following statements is true?

(a) L is recursively enumerable, but not recursive

(b) L is recursive, but not context-free

(c) L is context-free, but not regular

(d) L is regular

31. Let $SHAM_3$ be the problem of finding a Hamiltonian cycle in a graph $G = (V, E)$ with $|V|$ divisible by 3 and $DHAM_3$ be the problem of determining if a Hamiltonian cycle exists in such graphs. Which one of the following is true ?

(A) Both $DHAM_3$ and $SHAM_3$ are NP-hard

(B) $SHAM_3$ is NP-hard, but $DHAM_3$ is not

(C) $DHAM_3$ is NP-hard, but $SHAM_3$ is not

(D) Neither $DHAM_3$ nor $SHAM_3$ is NP-hard

32. Consider the following statements about the context-free grammar,

$$G = \{S \to SS, S \to ab, S \to ba, S \to \epsilon\}$$

I. G is ambiguous

II. G produces all strings with equal number of a's and b's

III. G can be accepted by a deterministic PDA.

Which combination below expresses all the true statements about G ?

(A) I only

(B) I and III only

(C) II and III only

(D) I, II, and III

33. Let L_1 be regular language, L_2 be a deterministic context-free language and L_3 a recursively enumerable, but not recursive, language. Which one of the following statements is false ?

(A) $L_1 \cap L_2$ is a deterministic CFL

(B) $L_3 \cap L_1$ is recursive

(C) $L_1 \cup L_2$ is context free

(D) $L_1 \cap L_2 \cap L_3$ is recursively enumerable

34. Consider the regular language $L = (111 + 11111)^*$. The minimum number of states in any DFA accepting this languages is

(A) 3 (B) 5 (C) 8 (D) 9

35.

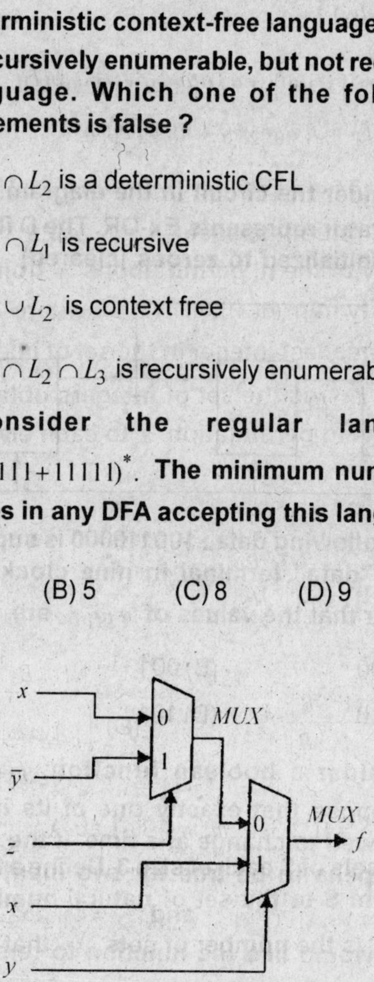

Consider the circuit above. Which one of the following options correctly represents f(x,y,z)?

(A) $x\bar{z}+xy+\bar{y}z$ (B) $x\bar{z}+xy+\overline{yz}$

(C) $xz+xy+\overline{yz}$ (D) $xz+x\bar{y}+\bar{y}z$

36. Given two three bit numbers $a_2 a_1 a_0$ and $b_2 b_1 b_0$ and c, the carry in, the function that represents the carry generate function when these two numbers are added is

(A) $a_2b_2+a_2a_1b_1+a_2a_1a_0b_0+a_2a_0b_1b_0$ $+a_1b_2b_1+a_1a_0b_2b_0+a_0b_2b_1b_0$

(B) $a_2b_2+a_2b_1b_0+a_2a_1b_1b_0+a_1a_0b_2b_1+a_1a_0b_2$ $+a_1a_0b_2b_0+a_2a_0b_1b_0$

(C) $a_2+b_2+(a_2\oplus b_2)(a_1+b_1+(a_1\oplus b_1)$ $(a_0+b_0))$

(D) $a_2b_2+\overline{a_2}a_1b_1+\overline{a_2 a_1}a_0b_0+\overline{a_2}a_0\overline{b_1}b_0$ $a_1\overline{b_2}b_1+\overline{a_1}a_0\overline{b_2}b_0+a_0\overline{b_2 b_1}b_0$

37. Consider the circuit in the diagram. The $\oplus$ operator represents Ex-OR. The D flip-flops are initialized to zeroes (cleared).

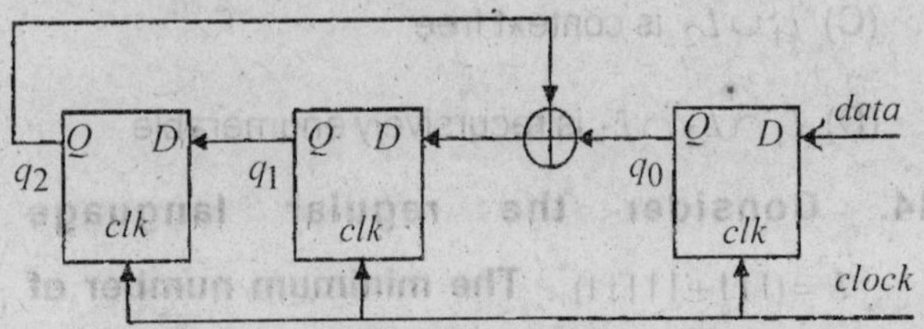

The following data : 100110000 is supplied to the "data" terminal in nine clock cycles. After that the values of $q_2q_1q_0$ are

(A) 000 (B) 001

(C) 010 (D) 101

38. Consider a boolean function $f(w,x,y,z)$. Suppose that exactly one of its inputs is allowed to change at a time. If the function happens to be true for two input vectors $i_1 = \langle w_1, x_1, y_1, z_1 \rangle$ and $i_2 = \langle w_2, x_2, y_2, z_2 \rangle$, we would like the function to remain true as the input changes from i_1 to i_2 (i_1 and i_2 differ in exactly one bit position), without becoming false momentarily. Let $f(w,x,y,z)=\sum(5,7,11,12,13,15)$. Which of the following cube covers of f will ensure that the required property is satisfied ?

(A) $\overline{w}xz, wx\overline{y}, x\overline{y}z, xyz, wyz$

(B) $wxy, \overline{w}xz, wyz$

(C) $wx\overline{yz}, xz, w\overline{x}yz$

(D) $wzy, wyz, wxz, \overline{w}xz, x\overline{y}z, xyz$

39. We consider the addition of two 2's complement numbers $b_{n-1}b_{n-2}\ldots b_0$ and $a_{n-1}a_{n-2}\ldots a_0$. A binary adder for adding unsigned binary numbers is used to add the two numbers. The sum is denoted by $c_{n-1}c_{n-2}\ldots c_0$ and the carry-out by c_{out}. Which one of the following options correctly identifies the overflow condition ?

(A) $c_{out}\overline{(a_{n-1}\oplus b_{n-1})}$

(B) $a_{n-1}b_{n-1}\overline{c_{n-1}}+\overline{a_{n-1}b_{n-1}}c_{n-1}$

(C) $c_{out}\oplus c_{n-1}$

(D) $a_{n-1}\oplus b_{n-1}\oplus c_{n-1}$

40. Consider numbers represented in 4-bit gray code/ Let $h_3h_2h_1h_0$ be the gray code representation of a number n and let $g_3g_2g_1g_0$ be the gray code of (n+1) (modulo 16) value of the number. Which one of the following functions is correct ?

(A) $g_0(h_3h_2h_1h_0)=\sum(1,2,3,6,10,13,14,15)$

(B) $g_1(h_3h_2h_1h_0)=\sum(4,9,10,11,12,13,14,15)$

(C) $g_2(h_3h_2h_1h_0)=\sum(2,4,5,6,7,12,13,15)$

(D) $g_3(h_3h_2h_1h_0)=\sum(0,1,6,7,10,11,12,13)$

41. A CPU has a cache with block size 64 bytes. The main memory has k banks, each bank being c bytes wide. Consecutive c-byte chunks are mapped on consecutive banks with warp-around. All the k banks can be

accessed in parallel, but two accesses to the same bank must be serialized. A cache block access may involve multiple iterations of parallel bank accesses depending on the amount of data obtained by accessing all the k banks in parallel. Each iteration requires decoding the bank numbers to be accessed in parallel and this takes k/2 ns. The latency of one bank access is 80 ns. If c = 2 and k = 24, then latency of retrieving a cache block starting at address zero from main memory is

(A) 92 ns (B) 104 ns

(C) 172 ns (D) 184 ns

42. A CPU has five-stage pipeline and runs at 1 GHz frequency. instruction fetch happens in the first stage of the pipeline. A conditional branch instruction computes the target address and evaluates the condition in the third stage of the pipeline. The processor stops fetching new instructions following a conditional branch until the branch outcome is known. A program executes 10^9 instructions out of which 20% are conditional branches. If each instruction takes one cycle to complete on average, then total execution time of the program is

(A) 1.0 second (B) 1.2 seconds

(C) 1.4 seconds (D) 1.6 seconds

43. Consider a new instruction named branch-on-bit-set (mnemonic bbs). The instruction "bbs reg, pos, label" jumps to label if bit in position pos of register operand reg is one. A register is 32 bits wide and the bits are numbered 0 to 31, bit in position 0 being the least significant. Consider the following emulation of this instruction on a processor that does not have bbs implemented.

$temp \leftarrow reg\ and\ mask$

Branch to label if temp is non-zero

The variable temp is a temporary register. For correct emulation, the variable mask must be generated by

(A) $mask \leftarrow 0x1 << pos$

(B) $mask \leftarrow 0x\,ffffffff >> pos$

(C) $mask \leftarrow pos$

(D) $mask \leftarrow 0xf$

44. Station A uses 32 byte packets to transmit messages to Station B using a sliding window protocol. The round trip delay between A and B is 80 milliseconds and the bottleneck bandwidth on the path between A and B is 128 kbps. What is the optimal window size that A should use ?

(A) 20 (B) 40 (C) 160 (D) 320

45. Two computers C1 and C2 are configured as follows. C1 has IP address 203. 197.2.53 and netmask 255. 255. 128.0. C2 has IP address 203.197.75.201 and netmask 255. 255. 192.0 Which one of the following statements is true ?

(A) C1 and C2 both assume they are on the same network

(B) C2 assumes C1 is on same network, but C1 assumes C2 is on a different network

(C) C1 assumes C2 is on same network, but C2 assumes C1 is on a different network

(D) C1 and C2 both assume they are on different networks.

46. Station A needs to send a message consisting of 9 packets to Station B using a sliding window (window size 3) and go-back-n error control strategy. All packets are ready and immediately available for transmission. If every 5th packet that A transmits gets lost (but no acks from B ever get lost), then what is the number of packets that A will transmit for sending the message to B ?

(A) 12 (B) 14 (C) 16 (D) 18

47. Consider the following graph :

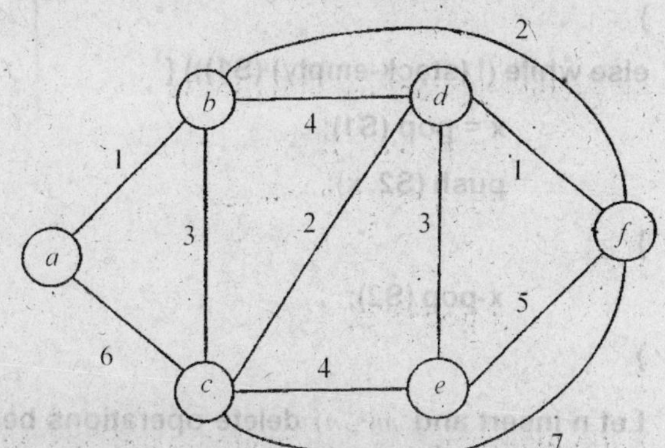

Which one of the following cannot be the sequence of edges added, in that order, to a minimum spanning tree using Kruskal's algorithm ?

(A) $(a-b),(d-f),(b-f),(d-c),(d-e)$

(B) $(a-b),(d-f),(d-c),(b-f),(d-e)$

(C) $(d-f),(a-b),(d-c),(b-f),(d-e)$

(D) $(d-f),(a-b),(b-f),(d-e),(d-c)$

48. Let T be a depth first search tree in a undirected graph G. Vertices u and v are leaves of this tree T. The degrees of both u and v in G are at least 2. Which one of the following statements is true ?

(A) There must exist a vertex w adjacent to both u and v in G

(B) There must exist a vertex w whose removal disconnects u and v in G

(C) There must exist a cycle in G containing u and v

(D) There must exist a cycle in G containing u and all its neighbours in G

49. An Implementation of a queue Q, using two stacks S1 and S2, is given below

```
Void insert (Q, X) {
        Push (S1, x);
}
void delete (Q) {
        if (stack-empty (S2)) then {
                if (stack-empty (S1)) then {
        print (Q is empty ");
        return;
}
else while (! (stack-empty) (S1))) {
        x = pop (S1);
        push (S2, x);
}
        x-pop (S2);
}
```

Let n insert and $m(\leq n)$ delete operations be performed in an arbitrary order on an empty queue Q. Let x and y be the number of push and pop operations performed respectively in the process. Which one of the following is true for all m and n ?

(A) $n+m \leq x < 2n$ and $2m \leq n+m$

(B) $n+m \leq x < 2n$ and $2m \leq y \leq 2n$

(C) $2m \leq x < 2n$ and $2m \leq y \leq n+m$

(D) $2m \leq x < 2n$ and $2m \leq y \leq 2n$

50. A set X can be represented by an array x[n] as follows

$$x[i] = \begin{cases} 1 & if\ i \in X \\ 0 & otherwise \end{cases}$$

Consider the following algorithm is which x,y, and z are boolean arrays of size n:

algorithm zzz (x[], y[], z[] {

int i;

for ($i = 0; i < n; ++i$)

$$z[i] = (x[i] \wedge \sim y[i]) \vee (\sim x[i] \wedge y[i])$$

}

The set Z computed by the algorithm is

(A) $(X \cup Y)$ (B) $(X \cap Y)$

(C) $(X-Y)\cap(Y-X)$ (D) $(X-Y)\cup(Y-X)$

51. Consider the following recurrence

$$T(n) = 2T(\lceil\sqrt{n}\rceil) + 1, T(1) = 1$$

Which one of the following is true ?

(A) $T(n) = \theta(\log \log n)$ (B) $T(n) = \theta(\log n)$

(C) $T(n) = \theta(\sqrt{n})$. (D) $T(n) = \theta(n)$

52. The median of n elements can be found in $O(n)$ time. Which one of the following is correct about the complexity of quick sort, in which median is selected as pivot ?

(A) $\theta(n)$ (B) $\theta(n \log n)$

(C) $\theta(n^2)$ (D) $\theta(n^3)$

53. Consider the following C-function in which a[n] and b[m] are two sorted integer arrays and $c[n+m]$ be another integer array.

```
void xyz (int a [ ], int b [ ], int c[ ] ) {
int i, j, k;
i = j = k = 0;
while ((i < n) && (j < m))
    if (a[i] < b[j]) c[k++] = a[i++];
    else c[k++] = b[j++];
}
```

Which of the following condition (s) hold(s) after the termination of the while loop.

(i) $j < m, k = n + j - 1$, **and** $a[n-1] < b[j]$ *if* $i = n$

(ii) $i < n, k = m + i - 1$, **and** $b[m-1] \le a[i]$ *if* $j = m$

(A) only (i)

(B) only (ii)

(C) either (i) or (ii) but not both

(D) neither (i) nor (ii)

54. Given two arrays of numbers $a_1, ..., a_n$ and $b_1, ..., b_n$ where each number is 0 or 1, the fastest algorithm to find the largest span (i,j) such that $a_i + a_{i+1} + ... + a_j = b_i + b_{i+1} + ... + b_j$, or report that there is no such span,

(A) Takes $O(3^n)$ and $\Omega(2^n)$ time if hashing is permitted

(B) Takes $O(n^3)$ and $\Omega(n^{2.5})$ time in the key comparison model

(C) Takes $\Theta(n)$ time and space

(D) Takes $O(\sqrt{n})$ time only if the sum of the 2n elements is an even number

55. Consider these two functions and two statements S1 and S2 about them.

```
int work 1 (int * a, int i, int j )
{
    int x = a[i + 2];
    a[j] = x + 1;
    return a[i + 2] - 3;
}
```

```
int work 2 (int * a, int i, int j )
{
    int t1 = i + 2;
    int t2 = a[t1];
    a[j] = t2 + 1;
    return t2 - 3;
}
```

S1: The transformation from work 1 to work 2 is valid, i.e., for any program state and input arguments, work 2 will compute the same output and hvae the same effect on program state as work 1.

S2 : All the transformations applied to work 1 to get work 2 will always improve the performance (i.e. reduce CPU time) of work 2 compared to work 1

(A) S1 is false and S2 is false

(B) S1 is false and S2 is true

(C) S1 is true and S2 is false

(D) S1 is true and S2 is true

56. Consider the following code written in a pass-by-reference language like FORTRAN and these statements about the code.

```
        Subroutine swap (ix, iy)
        it = ix
L1:     ix = iy
L2:     iy = it
    end
    ia = 3
    ib = 8
    call swap (ia, ib +5)
    print * , ia, ib
    end
```

S1: The compiler will generate code to allocate a temporary nameless cell, initialize it to 13, and pass the address of the cell to swap

S2: On execution the code will generate a runtime error on line L1

S3: On execution the code will generate a runtime error on line L2

S4: The program will print 13 and 8

S5 : The program will print 13 and -2

Exactly the following set of statement (s) is correct :

(A) S1 and S2 (B) S1 and S4

(C) S3 (D) S1 and S5

57.Consider this C code to swap two integers and these five statements : the code

```
void swap (int* px, int* py){
*px = *px - *py;
*py = *px + *py;
*px = *py - *px;
```

S1 : will generate a compilation error

S2 : may generate a segmentation fault at runtime depending on the arguments passed

S3 : Correctly implements the swap procedure for all input pointers referring to integers stored in memory locations accessible to the process

S4 : implements the swap procedure correctly for some but not all valid input pointers

S5 : may add or subtract integers and pointers

(A) S1 (B) S2 and S3

(C) S2 and S4 (D) S2 and S5

58. Consider the following grammar

$S \to FR$

$R \to *S \mid \varepsilon$

$F \to id$

In the predictive parser table, M, of the grammar the entries $M[S, id]$ and $M[R, \$]$ respectively

(A) $(S \to FR)$ and $(R \to \varepsilon)$

(B) $(S \to FR)$ and $\{\}$

(C) $(S \to FR)$ and $\{R \to S*\}$

(D) $\{F \to id\}$ and $\{R \to \varepsilon\}$

59.Consider the following translation scheme.

$S \to ER$

$R \to *E$ **{print** ('*'); $R \mid \varepsilon$

$E \to F + E\{print('+');\mid F$

$F \to (S) \mid id\{print(id.value);\}$

Here id is a taken that represents an integer and id. value represents the corresponding integer value. For an input '2*3+4'. **this translation scheme prints**

(A) 2*3+4 (B) 2*+34

(C) 2 3*4+ (D) 2 3 4+*

60. Consider the following C code segment.

```
for (i = 0; i < N; i++)
    for (j = 0; j < N; j++)
      if (i % 2)
        {x+ = (4*j+5*i);
        y+ = (7+4*j);
        }
    }
}
```

Which one of the following is false ?

(A) The code contains loop-in variant computation

(B) There is scope of common sub-expression elimination in this code

(C) There is scope of strength reduction in this code

(D) There is scope of dead code elimination in this code

61. The atmic fetch-and-set x, y instruction unconditionally sets the memory location x

to 1 and fetches the old value of x in y without allowing any intervening access to the memory location x. Consider the following implementation of P and V functions on a binary semaphore S.

```
void p (binary_semaphore *S) {
  unsigned y;
  unsigned * X = & (S - > value );}
    do {
      fetch - and - set x, y;
    } while (y);
}
void V (binary_semaphore * S) {
    {S - > value = 0;
}
```

which one of the following is ture ?

(A) The implementation may not work if context switching is disabled in P

(B) Instead of using fetch-and-set, a pair of normal load/store can be used

(C) The implementation of V is wrong

(D) The code does not implement a binary semaphore

62. A CPU generates 32-bit virtual addresses. The page size is 4 KB. The processor has a translation look-aside buffer (TLB) which can hold a total of 128 page table entires and is 4 way set associative. The minimum size of the TLB tag is

(A) 11 bits (B) 13 bits

(C) 15 bits (D) 20 bits

63. A computer system supports 32-bit virtual address as well as 32-bit physical addresses. Since the virtual address space is of the same size as the physical address space, the operating system designers decide to get rid of the virtual memory entirely. Which one of the following is true ?

(A) Efficient implementation of multi-user support is no longer possible

(B) The processor cache organization can be made more efficient now

(C) Hardware support for memory management is no longer needed

(D) CPU scheduling can be made more efficient now

64. Consider three processes (process id 0, 1, 2, respectively) with compute time bursts 2, 4 and 8 time units. All processes arrive at time zero. Consider the longest remaining time first (LRTF) scheduling algorithm. In LRTF ties are broken by giving priority to the process with the lowest process id. The average turn around time is

(A) 13 units (B) 14 units

(C) 15 units (D) 16 units

65. Consider three processes, all arriving at time zero, with total execution time of 10, 20, and 30 units, respectively. Each process spends the first 20% of execution time doing I/O, the next 70% of time doing computation, and the last 10% of time doing I/O again. The operating system uses a shortest remaining compute time first scheduling algorithm and schedules a new process either when the running process gets blocked on I/O or when the running process finishes its compute burst. Assume that all I/O operations can be overlapped as much as possible. For what percentage of time does the CPU remain idle ?

(A) 0% (B) 10.6%

(C) 30.0% (D) 89.4%

66. Consider the following snapshot of a system running n processes. Process i is holding x_i instances of a resource R, for $1 \le i \le n$. Currently, all instances of R are occupied. Further, for all i, process i has placed a request for an additional y_i instances while holding the x_i instances it already has. There are exactly two processes p and q such that $y_p = y_q = 0$. Which one of the following can serve as a necessary condition to guarantee that the system is not approaching a deadlock ?

(A) $\min(x_p, x_q) < \max_{k \ne p.q} y_k$

(B) $x_p + x_q \ge \min_{k \ne p'q} y_k$

(C) $\max(x_p, x_q) > 1$

(D) $\min(x_p, x_q) > 1$

67. Consider the relation account (customer, balance) where customer is a primary key and there are no null values. We would like to rank customers according to decreasing balance. The customer with the largest balance gets rank 1. Ties are not broken but ranks are skipped: if exactly two customers have the largest balance they each get rank 1 and rank 2 is not assigned.

Query 1 :

```
Select A. customer, count (B. customer)
from account A, account B
where A. balance < = B. balance
group by A. customer
```

Query 2 :

```
select A.customer, 1+ count (B.customer)
From account A, account B
Where A.balance < B. balance
group by A.customer
```

Consider these statements about Query 1 and Query 2.

1. Query 1 will produce the same row set as query 2 for some but not all databases

2. Both Query 1 and Query 2 are correct implementations of the specification

3. Query 1 is a correct implementation of the specification but Query 2 is not

4. Neither Query 1 nor Query 2 is a correct implementation of the specification

5. Assigning rank with a pure relational Query takes less time than scanning in decreasing balance order and assigning ranks using ODBC.

Which two of the above statements are correct ?

(A) 2 and 5 (B) 1 and 3

(C) 1 and 4 (D) 3 and 5

68. Consider the relation enrolled (student, course) in which (student, course) is the primary key, and the relation paid (student, amount) where student is the primary key. ;Assume no null values and no foreign keys or integrity constraints. Given the following four queries :

Query 1 : select student form enrolled where student in (select student form paid)

Query 2 : select student from paid where student in (select student from enrolled)

Query 3 : select E. student from enrolled E, paid P where E.student = P.student

Query 4 : select student from paid where exists

(select * from enrolled where enrolled.student = paid. student)

Which one of the following statements is correct ?

(A) All queries return identical row sets for any database

(B) Query 2 and Query 4 return identical row sets for all databases but there exist databases for which Query 1 and Query 2 return different row sets

(C) There exist databases for which Query 3 returns strictly fewer rows than Query 2

(D) There exist databases for which Query 4 will encounter an intergrity violation at runtime

69. Consider the relation enrolled (student, course) in which (student, course) is the primary key, and the relation paid (student, amount) where student is the primary key. Assume no null values and no foreign keys or integrity constraints. Assume that amounts 6000, 7000, 8000, 9000 and 10000 were each paid by 20 % of the students. Consider these query plans (Plan 1 on left, Plan 2 on right) to "list all courses taken by students who have paid more than x".

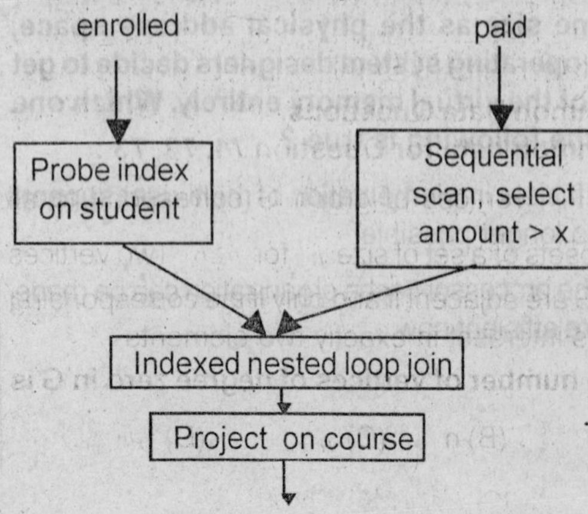

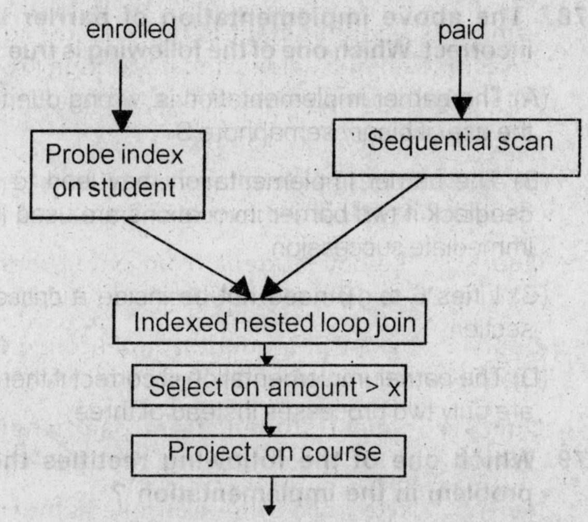

A disk seek takes 4 ms, disk data transfer bandwidth is 300 MB/s and checking a tuple to see if amount is greater than x takes $10\mu s$. Which of the following statements is correct ?

(A) Plan 1 and Plan 2 will not output identical row sets for all databases

(B) A course may be listed more than once in the output of Plan 1 for some databases

(C) For x = 5000, Plan 1 executes faster than Plan 2 for all databases

(D) For x = 9000, Plan 1 executes slower than plan 2 for all databases

70. The following functional dependencies are given :

$AB \to CD, AF \to D, DE \to F, C \to G,$
$F \to E, G \to A.$

Which one of the following options is false ?

(A) $\{CF\}^+ = \{ACDEFG\}$

(B) $\{BG\}^+ = \{ABCDG\}$

(C) $\{AF\}^+ = \{ACDEFG\}$

(D) $\{AB\}^+ = \{ABCDFG\}$

Common Data Questions

Common data for Question 71, 72, 73 :

The 2^n vertices of graph G correspond to all subsets of a set of size n, for $n \geq 6$. Two vertices of G are adjacent if and only if the corresponding sets intersect in exactly two elements

71. The number of vertices of degree zero in G is

(A) 1 (B) n (C) $n+1$ (D) 2^n

72. The maximum degree of a vertex in G is

(A) $\binom{n/2}{2} 2^{n/2}$ (B) 2^{n-2}

(C) $2^{n-3} \times 3$ (D) 2^{n-1}

73. The number of connected components in G is

(A) n (B) $n+2$

(C) $2^{n/2}$ (D) $\frac{2^n}{n}$

Common data for questions 74, 75 :

Consider two cache organizations : The first one is 32 KB 2-way set associative with 32-byte block size. The second one is of the same size but direct mapped. The size of an address 32 bits in both cases. A 2- to-1 multiplexer has latency of 0.6 ns while a k-bit comparator has a latency to k/10ns. The hit latency of the set associative organization is h_1 while that of the direct mapped one is h_2.

74. The value of h_1 is

(A) 2.4 ns (B) 2.3 ns

(C) 1.8 ns (D) 1.7 ns

75. The value of h_2 is

(A) 2.4 ns (B) 2.3 ns

(C) 1.8 ns (D) 1.7 ns

Linked answer Questions : Q. 76 to Q.85 carry two marks each

Statement for Linked Answer Questions 76 and 77 :

A 3 -ary max heap is like a binary max heap, but instead of 2 children, nodes have 3 children. A 3-ary heap can be represented by an array as follows : The root is stored in the first location, $a[0]$, nodes in the next level, from left to right, is stored from $a[1]$ to $a[3]$. The nodes from the second level of the tree from left to right are stored from $a[4]$ location onward. An item x can be inserted into a 3-ary heap containing n items by placing x in the location $a[n]$ and pushing it up the tree to satisfy the heap property.

76. Which one of the following is a valid sequence of elements in an array representing 3-ary max heap ?

(A) 1, 3, 5, 6, 8, 9 (B) 9, 6, 3, 1, 8, 5

(C) 9, 3, 6, 8, 5, 1 (D) 9, 5, 6, 8, 3, 1

77. Suppose the elements 7, 2, 10, and 4 are insetrted, in that order, into the valid 3-ary max heap found in the above question, Q.76. Which one of the following is the sequence of items in the array representing the resultant heap ?

(A) 10, 7, 9, 8, 3, 1, 5, 2, 6, 4

(B) 10, 9, 8, 7, 6, 5, 4, 3, 2, 1

(C) 10, 9, 4, 5, 7, 6, 8, 2, 1, 3

(D) 10, 8, 6, 9, 7, 2, 3, 4, 1, 5

Statement for Linked Answer Questions 78 and 79 :

Barrier is a synchronization construct where a set of processes synchronizes globally i.e. each process in the set arrives at the barrier and waits for all others to arrive and then all processes leave the barrier. Let the number of processes in the set be three and S be a binary semaphore with the usual P and V functions. Consider the following C implementation of a barrier with line numbers shown on the left.

```
Void barrier (void) {
1 :    P(S) ;
2 :    Process_arrived ++;
3 :    V(S);
4 :    while (process_arrived ! = 3);
5 :    P(S) ;
6 :    process_left ++ ;
7 :    if (process_left = = 3 )
8 :        process_arrived = 0 ;
9 :        process_left = 0;
10 :   }
11 :   V (S) ;
}
```

The variables process_arrived and process_left are shared among all processes and are initialized to zero. In a concurrent program all the three processes call the barrier function when they need to synchronize globally.

78. The above implementation of barrier is incorrect. Which one of the following is true ?

(A) The barrier implementation is wrong due to the use of binary semaphore S

(B) The barrier implementation may lead to a deadlock if two barrier invocations are used in immediate succession

(C) Lines 6 to 10 need not be inside a critical section

(D) The barrier implementation is correct if there are only two processes instead of three

79. Which one of the following rectifies the problem in the implementation ?

(A) Lines 6 to 10 are simply replaced by process_arrived --

(B) At the beginning of the barrier the first process to enter the barrier waits until process_arrived becomes zero before proceeding to execute P(S)

(C) Context switch is disabled at the beginning of the barrier and re-enabled at the end

(D) The variable process_left is made private instead of shared

Statement for Linked Answer Questions 80 and 81 :

A CPU has a 32 KB direct mapped cache with 128-byte block size. Suppose A is a two dimensional array of size 512×512 with elements that occupy 8-bytes each. Consider the following two C code segments, P1 and P2

P1 : for $(i = 0; i < 512; i++)$ {
   for $(j = 0; j < 512; j++)$ {
      $x += A[i][j];$
   }
}

P2 : for $(i = 0; i < 512; i++)$ {
   for $(j = 0; j < 512; j++)$ {
      $\{x += A[j][i];\}$
   }
}

P1 and P2 are executed independently with the same initial state, namely, the array A is not in the cache and i,j, x are in registers. Let the number of cache misses experienced by P1 be M_1 and that for P2 be M_2.

80. The value of M_1 is

(A) 0 (B) 2048 (C) 16384 (D) 262144

81. The value of the ratio M_1/M_2 is

(A) 0 (B) $\frac{1}{16}$ (C) $\frac{1}{8}$ (D) 16

Statement for Linked Answer Questions 82 and 83 :

Consider the diagram shown below where a number of LANs are connected by (transparent) bridges. In order to avoid packets looping through circuits in the graph, the bridges organize themselves in a spanning tree. First, the root bridge is identified as the bridge with the least serial number. Next, the root sends out (one or more) data units to enable the setting up of the spanning tree of shortest paths from the root bridge to each bridge.

Each bridge identifies a port (the root port) through which it will forward frames to the root bridge. Port conflicts are always resolved in favour of the port with the lower index value. When there is a possibility of multiple bridges forwarding to the same LAN (but not through the root port), ties are broken as follows : bridges closest to the root get preference and between such bridges, the one with the lowest serial number is preferred.

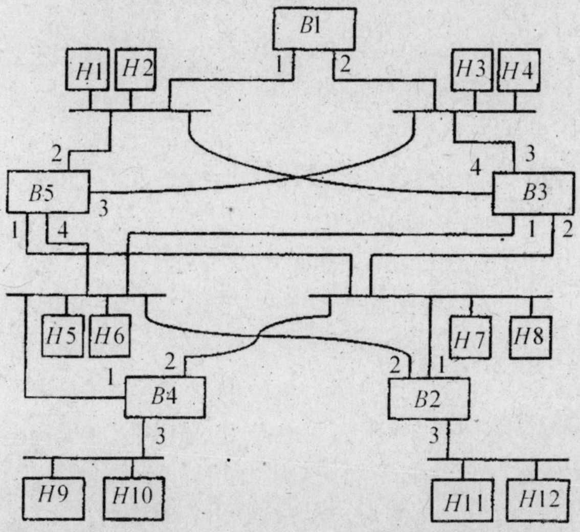

82. For the given connection of LANs by bridges, which one of the following choices represents the depth first traversal of the spanning tree of bridges ?

(A) B1, B5, B3, B4, B2

(B) B1, B3, B5, B2, B4

(C) B1, B5, B2, B3, B4

(D) B1, B3, B4, B5, B2

83. Consider the correct spanning tree for the previous question. Let host H1 send out a broadcast ping packet. Which of the following options represents the correct forwarding table on B3 ?

(A)

Hosts	Port
H1, H2, H3, H4	3
H5, H6, H9, H10	1
H7, H8, H11, H12	2

(B)

Hosts	Port
H1, H2	4
H3, H4	3
H5, H6	1
H7, H8, H9, H10	
H11, H12	2

(C)

Hosts	Port
H1, H2, H3, H4	3
H5, H7, H9, H10	1
H7, H8, H11, H12	4

(D)

Hosts	Port
H1, H2, H3, H4	3
H5, H7, H9, H10	1
H7, H8, H11, H12	4

Statement for linked answers questions 84 and 85 :

84. Which one of the following grammars generates the language $L = \{a^i b^j \mid i \neq j\}$?

(A) $S \to AC \mid CB$

$C \to aCb \mid a \mid b$

$A \rightarrow aA \mid \in$

$B \rightarrow Bb \mid \in$

(B) $S \rightarrow aS \mid Sb \mid a \mid b$.

(C) $S \rightarrow AC\ CB$

$C \rightarrow aCb \mid \in$

$A \rightarrow a\ A \mid \in$

$B \rightarrow Bb \mid \in$

(D) $S \rightarrow AC \mid CB$

$C \rightarrow aCb \mid \in$

$A \rightarrow aA \mid a$

$B \rightarrow b\ B \mid b$

85. In the correct grammar above, what is the length of the derivation (number of steps starting from S) to generate the string $a^l b^m$ with $l \neq m$?

(A) max (l, m)+2 (B) $l+m+2$

(C) $l+m+3$ (D) $\max(l, m)+3$

ANSWER KEY

1. (A)	2. (D)	3. (C)	4. (A)	5. (C)
6. (B)	7. (C)	8. (C)	9. (C)	10. (B)
11. (B)	12. (C)	13. (D)	14. (C)	15. (C)
16. (C)	17. (C)	18. (B)	19. (D)	20. (C)
21. (A)	22. (C)	23. (D)	24. (D)	25. (B)
26. (D)	27. (D)	28. (D)	29. (B)	30. (B)
31. (A)	32. (D)	33. (D)	34. (D)	35. (A)
36. (A)	37. (C)	38. (A)	39. (*)	40. (B)
41. (D)	42. (B)	43. (A)	44. (B)	45. (D)
46. (A)	47. (D)	48. (D)	49. (A)	50. (D)
51. (A)	52. (B)	53. (C)	54. (C)	55. (C)
56. (A)	57. (B)	58. (A)	59. (D)	60. (B)
61. (A)	62. (A)	63. (C)	64. (A)	65. (B)
66. (A)	67. (C)	68. (A)	69. (C)	70. (C)
71. (C)	72. (A)	73. (B)	74. (A)	75. (D)
76. (D)	77. (A)	78. (B)	79. (B)	80. (C)
81 (B)	82 (C)	83. (A)	84. (D)	85. (D)

GATE - 2006
INFORMATION TECHNOLOGY

Duration : Three Hours **Maximum Marks : 150**

Read the following instructions carefully

1. This question paper contains objective questions. Q.1-20 carry one mark each and Q.21-75 carry two marks each. Q. 81-85 carry two marks each.
2. Answer all the questions.
3. Questions must be answered on the special machine gradable Objective Response Sheet (ORS) by darkening the appropriate bubble (marked A, B, C, D) against the question number on the left hand side of the ORS, using HB pencil. Each question has only one correct answer. In case you wish to change an answer, erase the old answer completely using a good soft eraser.
4. There will be NEGATIVE marking. In Q.1 to Q.20, 0.25 mark will be deducted for each wrong answer and in Q.21 to Q.75, 0.5 mark will be deducted for each wrong answer. In Q.75 to Q 85, For the part "a", 0.5 mark will be deducted for a wrong anwer . Marks for correct answer to part "b" of Q. 75 to Q.85 will be given only, if the answer to the corresponding part "a" is correct. However, there is no negative marking for part "b" of Q. 75 to Q.85. More than one answer bubbled against a question will be deemed as an incorrect response.
5. Write your registration number, name and name of the centre at the specified loacations on the right half of the **ORS.**
6. Using HB pencil, darken the appropriate bubble under each digit of your registration number and the letters corresponding to your paper code.
7. Calculator is allowed in the examination hall.
8. Charts, graph sheets or tables are not allowed.
9. Use the blank Paper given at the end of the question paper for rough work.

Q. 1 - 20 CARRY ONE MARK EACH

1. In a certain town, the probability that it will rain in the afternoon is known to be 0.6. Moreover, meteorological data indicates that if the temperature at noon is less than or equal to $25°C$, the probability that it will rain in the afternoon is 0.4. The temperature at noon is equally likely to be above $25°C$, ot at/below $25°C$. What is the probability that it will rain in the afternoon on a day when the temperature at noon is above $25°C$?

a) 0.4 b) 0.6

c) 0.8 d) 0.9

2. For the set N of natural numbers and a binary operation $f : N \times N \rightarrow N$, an element $z \in N$ is called an identity for f, if $f(a,z) = a = f(z,a)$, for all $a \in N$. Which of the following binary operations have a identity?

(I) $f(x,y) = x + y - 3$

(II) $f(x,y) = max(x,y)$

(III) $f(x,y) = x^y$

a) I and II only b) II and III only

c) I and III only d) None of these

3. In the automaton below, s is the start state and t is the only final state.

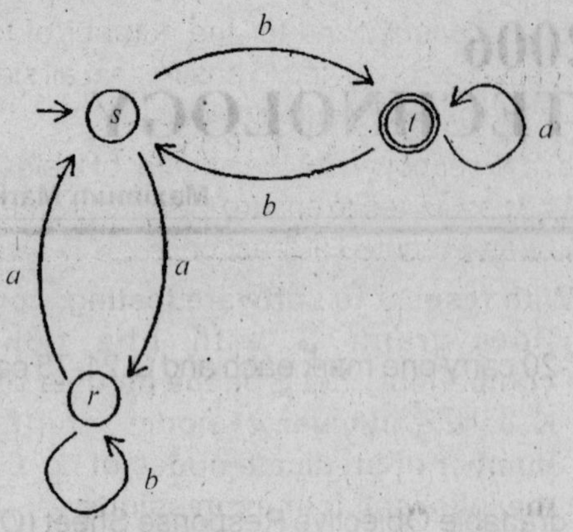

Consider the strings $u = abbaba$, $v = bab$, and $w = aabb$. Which of the following statements is true ?

a) The automaton accepts u *and* v but not w

b) The automaton accepts each of $u, v,$ *and* w

c) The automation rejects each of $u, v,$ *and* w

d) The automaton accepts u but rejects v *and* w

4. **In the context-free grammar below, S is the start symbol, a *and* b are terminals and ε denotes the empty string**

$$S \to aSa \,|bSb|\, a \,|\, b \,|\, \varepsilon$$

Which of the following strings is NOT generated by the grammar ?

a) aaaa b) baba

c) abba d) babaaabab

5. **Which regular expression best describes the language accepted by the non-deterministic automaton below ?**

a) $(a+b)*a(a+b)b$ b) $(abb)*$

c) $(a+b)*a(a+b)*b(a+b)*$

d) $(a+b)*$

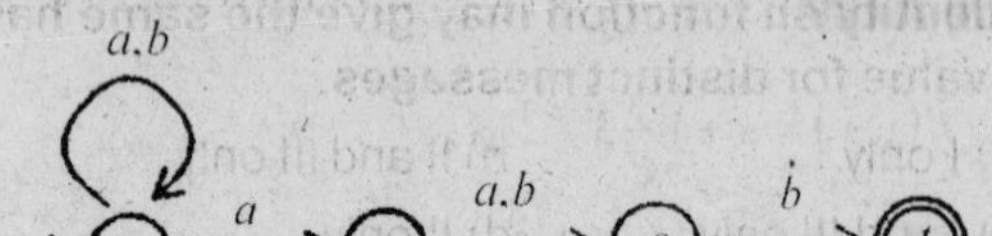

6. **Given a boolean function $f(x_1, x_2, ..., x_n)$, which of the following equations is NOT true**

a) $f(x_1, x_2, ..., x_n) = x_1' f(x_1, x_2, ... x_n) + x_1 f(x_1, x_2, ..., x_n)$

b) $f(x_1, x_2, ..., x_n) = x_2 f(x_1, x_2, ... x_n) + x_2' f(x_1, x_2, ..., x_n)$

c) $f(x_1, x_2, ..., x_n) = x_n' f(x_1, x_2, ..., 0) + x_n f(x_1, x_2, ..., 1)$

d) $f(x_1, x_2, ..., x_n) = f(0, x_2, ..., x_n) + f(1, x_2, ..., x_n)$

7. **The addition of 4-bit, two's complement, binary numbers 1101 and 0100 results in**

a) 0001 and an overflow

b) 1001 and no overflow

c) 0001 and no overflow

d) 1001 and an overflow

8. **Which of the following DMA transfer modes and interrupt handling mechanisms will enable the highest I/O band-width ?**

a) Transparent DMA and Polling interrupts

b) Cycle-stealing and Vectrored interrupts

c) Block transfer and Vectored interrupts

d) Block transfer and Polling interrupts

9. **In a binary tree, the number of internal nodes of degree 1 is 5, and the number of internal nodes of degree 2 is 10. The number of leaf nodes in the binary tree is**

a) 10 b) 11

c) 12 d) 15

10. **A problem in NP is NP - complete if**

a) it can be reduced to the 3 - SAT problem in polynomial time

b) the 3-SAT problem can be reduced to it in polynomial time

c) it can be reduced to any other problem in NP in polynomial time

d) some problem in NP can be reduced to it in polynomial time

11. **If all the edge weights of an undirected graph are positive, then any subset of edges that connects all the vertices and has minimum total weight is a**

a) Hamiltonian cycle b) grid

c) hypercube d) tree

12. **In the working-set strategy, which of the following is done by the operating system to prevent thrashing ?**

I) It initiates another process if there are enough extra frames

II) It selects a process to suspend if the sum of the sizes of the working - sets.

II) Exceeds the total number of available frames.

a) I only b) II only

c) Neither I nor II d) Both I and II

13. The process state transition diagram of an operating system is as given below. Which of the following must be FALSE about the above operating system ?

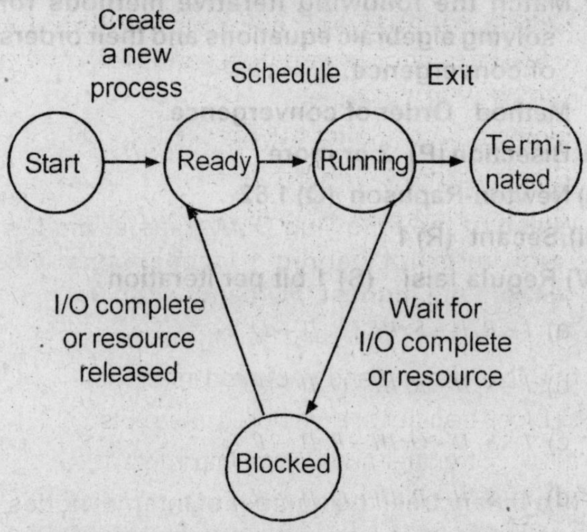

a) It is a mutiprogrammed operating system

b) It uses preemptive scheduling

c) It uses non-preemptive scheduling

d) It is a multi-user operating system

14. Consider the relations $r_1(P,Q,R)$ and $r_2(R,S,T)$ with primary keys P and R respectively. The relation r_1 contains 2000 tuples and r_2 contains 2500 tuples. The maximum size of the join $r_1 \bowtie r_2$ is

a) 2000 b) 2500

c) 4500 d) 5000

15. Which of the following relational query languages have the same expressive power?

(I) Relational algebra

(II) Tuple relational calculus restricted to safe expressions

(III) Domain relational calculus restricted to safe expressions

a) II and III only b) I and II only

c) I and III only d) I, II and III

16. The cyclomatic complexity of the flow graph of a program provides

a) an upper bound for the number of tests that must be conducted to ensure that all statements have been executed at most once.

b) a lower bound for the number of tests that must be conducted to ensure that all statements have been executed at most once.

c) an upper bound for the number of tests that must be conducted to ensure that all statements have been executed at least once.

d) a lower bound for the number of tests that must be conducted to ensure that all statements have been executed at least once.

17. With respect to software testing, consider a flow graph G with one connected component. Let E be the number of edges, N be the number of nodes, and P be the number of predicate nodes of G. Consider the following four expressions :

(I) $E-N+P$ **(II)** $E-N+2$

(III) $P+2$ **(IV)** $P+1$

The cyclomatic complexity of G is given by

a) I or III b) II or III

c) II or IV d) I or IV

18. HELO and PORT, respectively, are commands from the protocols

a) FTP and HTTP b) TELNET and POP3

c) HTTP and TELNET d) SMTP and FTP

19. Which of the following statements is TRUE?

a) Both Ethernet frame and IP packet include checksum fields

b) Ethernet frame includes a checksum field and IP packet includes a CRC field

c) Ethernet frame includes a CRC field and IP packet includes a checksum field

d) Both Ethernet frame and IP packet include CRC fields.

20. Which of the following statement(s) is TRUE?

(I) A hash function takes a message of arbitary length and generates a fixed length code.

(II) A hash function takes a message of fixed length and generates a code of variable length.

(III) A hash function may give the same hash value for distinct messages.

a) I only b) II and III only

c) I and III only d) II only

Q. 21 - 75 CARRY TWO MARKS EACH

21. Consider the following first order logic formula in which R is a binary relation symbol.

$\forall x \forall y (Rx,y) \Rightarrow R(y,x)$

The formula is

a) satisfiable and valid

b) satisfiable and so is its negation

c) unsatisfiable but its negation is valid

d) satisfiable but its negation is unsatisfiable

22. When a coin is tossed, the probability of getting a Head is $p, 0 < p < 1$. Let N be the random variable denoting the number of tosses till the first Head appears, including the toss where the Head appears. Assuming that successive tosses are independent, the expected value of N is

a) $\frac{1}{P}$ b) $\frac{1}{(1-P)}$

c) $\frac{1}{P^2}$ d) $\frac{1}{(1-P)^2}$

23. Let P, Q and R be sets let Δ denote the symmetric difference operator defined as $P \Delta Q = (P \cup Q) - (P \cap Q)$. Using Venn diagrams, determine which of the following is/are TRUE?

(I) $P \Delta (Q \cap R) = (P \Delta Q) \cap (P \Delta R)$

(II) $P \cap (Q \cap R) = (P \cap Q) \Delta (P \cap R)$

a) I only b) II only

c) Neither I nor II d) Both I and II

24. What is the cardinality of the set of integers X defined below ?

$X = \{n \mid 1 \le n \le 123, n \text{ is not divisible by either } 2, 3 \text{ or } 5\}$

a) 28 b) 33 c) 37 d) 44

25. Consider the undirected graph G defined as follows. The vertices of G are bit strings of length n. We have an edge between vertex u and vertex v if and only if u and v differ in exactly one bit position (in other words, v can be obtained from u by flipping a single bit). The ratio of the chromatic number of G to the diameter of G is

a) $\frac{1}{2^{n-1}}$ b) $\frac{1}{n}$ c) $\frac{2}{n}$ d) $\frac{3}{n}$

26. What are the eigenvalues of the matrix P given below

$$P = \begin{pmatrix} a & 1 & 0 \\ 1 & a & 1 \\ 0 & 1 & a \end{pmatrix}$$

a) $a, a-\sqrt{2}, a+\sqrt{2}$ b) a, a, a

c) $0, a, 2a$ d) $-a, 2a, 2a$

27. Match the following iterative methods for solving algebraic equations and their orders of convergence.

Method Order of convergence

(I) Bisection (P) 2 or more

(II) Newton-Raphson (Q) 1.62

(III) Secant (R) 1

(IV) Regula falsi (S) 1 bit per iteration

a) $I-R, II-S, III-P, IV-Q$

b) $I-S, II-R, III-Q, IV-P$

c) $I-S, II-Q, III-R, IV-P$

d) $I-S, II-P, III-Q, IV-R$

28. The following definite integral evaluates to

$$\int_{-\infty}^{0} e^{-\left(\frac{x^2}{20}\right)} dx$$

a) $\frac{1}{2}$ b) $\pi\sqrt{10}$ c) $\sqrt{10}$ d) π

29. Consider the regular grammar below

$S \to bS \mid aA \mid \varepsilon$

$A \to aS \mid bA$

The Myhill - Nerode equivalence classes for the language generated by the grammar are

a) $\{w \in (a+b)^* \mid \#_a(w) \text{ is even}\}$ and $\{w \in (a+b)^* \mid \#_a(w) \text{ is odd}\}$

b) $\{w \in (a+b)^* \mid \#_b(w) \text{ is even}\}$ and $\{w \in (a+b)^* \mid \#_b(w) \text{ is odd}\}$

c) $\{w \in (a+b)^* \mid \#_a(w) = \#_b(w) \text{ and }$ $w \in (a+b)^* \mid \#_a(w) \ne \#_b(w)\}$

d) $\{\varepsilon\}, \{wa \mid w \in (a+b)^* \text{ and } \{wb \mid w \in (a+b)^*\}$

30. Which of the following statements about regular languages is NOT true ?

a) Every language has a regular superset

b) Every language has a regular subset

c) Every subset of a regular language is regular

d) Every subset of a finite language is regular.

31. Which of the following languages is accepted by a non-deterministic pushdown automaton (PDA) but NOT by a deterministic PDA ?

a) $\{a^n b^n c^n \mid n \geq 0\}$

b) $\{a^l b^m c^n \mid l \neq m \text{ or } m \neq n\}$

c) $\{a^n b^n \mid n \geq 0\}$

d) $\{a^m b^n \mid m, n \geq 0\}$

32. Let L be a context-free language and M a regular language. Then the language $L \cap M$ is

a) always regular

b) never regular

c) always a deterministic context - free language

d) always a context-free language

33. Consider the pushdown automaton (PDA) below which runs over the input alphabet (a, b, c). It has the stack alphabet $\{Z_0, X\}$ where Z_0 is the bottom-of-stack marker. The set of states of the PDA is $\{s, t, u, f\}$ where s is the start state and f is the final state. The PDA accepts by final state. The transitions of the PDA given below are depicted in a standard manner. For example, the transition $(s, b, X) \rightarrow (t, XZ_0)$ means that if the PDA is in state s and the symbol on the top of the stack is X, then it can read b from the input and move to state t after popping the top of stack and pushing the symbols Z_0 and X (in that order) on the stack.

$(s, a, Z_0) \rightarrow (s, XXZ_0)$

$(s, \varepsilon, Z_0) \rightarrow (f, \varepsilon)$

$(s, a, X) \rightarrow (s, XXX)$

$(s, b, X) \rightarrow (t, \varepsilon)$

$(t, b, X) \rightarrow (t, \varepsilon)$

$(t, c, X) \rightarrow (u, \varepsilon)$

$(u, c, X) \rightarrow (u, \varepsilon)$

$(u, \varepsilon, Z_0) \rightarrow (f, \varepsilon)$

The language accepted by the PDA is

a) $\{a^l b^m c^n \mid l = m = n\}$

b) $\{a^l b^m c^n \mid l = m\}$

c) $\{a^l b^m c^n \mid 2l = m + n\}$

d) $\{a^l b^m c^n \mid m = n\}$

34. In the context-free grammar below, S is the start symbol, a and b are terminals, and ε denotes the empty string.

$S \rightarrow aSAb \mid \varepsilon$

$A \rightarrow bA \mid \varepsilon$

The grammar generates the language

a) $((a+b)^* b)^*$

b) $\{a^m b^n \mid m \leq n\}$

c) $\{a^m b^n \mid m = n\}$

d) $a^* b^*$

35. The boolean function for a combinational circuit with four inputs is represented by the following Karnaugh map.

RS \ PQ	00	01	11	10
00	1	0	0	1
01	0	0	1	1
11	1	1	1	0
10	1	0	0	1

Which of the product terms given below is an essential prime implicant of the function ?

a) QRS

b) PQS

c) PQ'S'

d) Q'S'

36. The majority function is a Boolean function $f(x, y, z)$ that takes the value 1 whenever a majority of the variables x, y, z and 1. In the circuit diagram for the majority function shown below, the logic gates for the boxes labeled P and Q are, respectively,

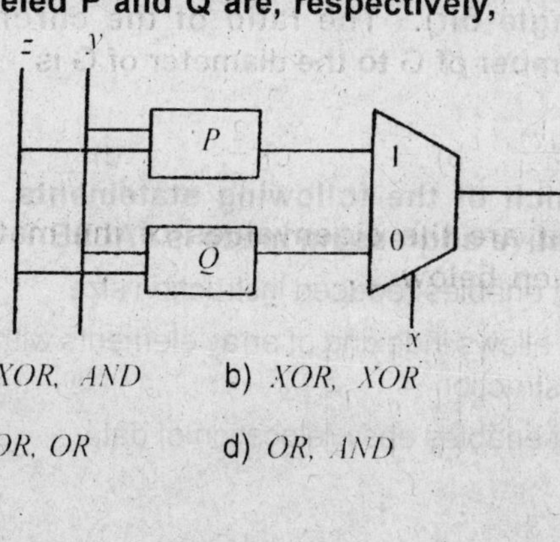

a) XOR, AND

b) XOR, XOR

c) OR, OR

d) OR, AND

37. For a state machine with the following state diagram the expression for the next state S^+ in terms of the current state S and the input variables x and y is

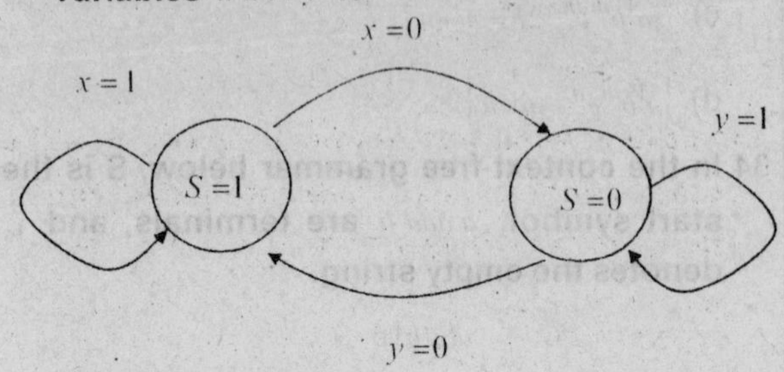

a) $S^+ = S'.y' + S.x$

b) $S^+ = S.x.y' + S'.y.x'$

c) $S^+ = x.y'$

d) $S^+ = S'.y + S.x'$

38. When multiplicand Y is multiplied by multiplier $X = x_{n-1} x_{n-2} \ldots x_0$ using bit-pair recoding in Booth's algorithm, partial products are generated according to the following table.

Row	x_{i+1}	x_i	x_{i-1}	Partial Product
1	0	0	0	0
2	0	0	1	Y
3	0	1	0	Y
4	0	1	1	2Y
5	1	0	0	?
6	1	0	1	-Y
7	1	1	0	-Y
8	1	1	1	?

The partial products for rows 5 and 8 are

a) $2Y$ and Y

b) $-2Y$ and $2Y$

c) $-2Y$ and 0

d) 0 and Y

39. Which of the following statements about relative addressing mode is FALSE ?

a) It enables reduced instruction size

b) It allows indexing of array elements with same instruction

c) It enables easy relocation of data

d) It enables faster address calculations than absolute addressing

40. The memory locations 1000, 1001 and 1020 have data values 18, 1 and 16 respectively before the following program is executed.

```
MOV1     Rs, 1          ; Move immediate
LOAD     Rd,1000(Rs)    ; Load from memory
ADD1     Rd, 1000       ; Add immediate
STOREI   0(Rd), 20      ; Store immdiate
```

Which of the statements below is TRUE after the program is executed ?

a) Memory location 1000 has value 20

b) Memory location 1020 has value 20

c) Memory location 1021 has value 20

d) Memory location 1001 has value 20

41. The data path shown in the figure computes the number of 1s in the 32 - bit input word corresponding to an unsigned even integer stored in the shift register.

The unsigned counter, initially zero, is incremented if the most significant bit of the shift register is 1.

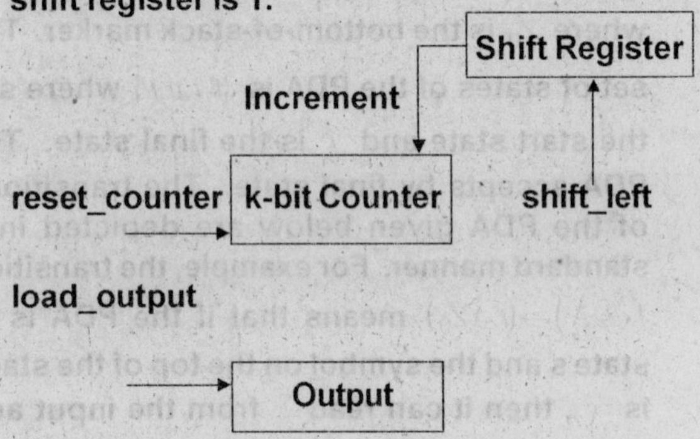

The microprogram for the control is shown in the table below with missing control words for microinstructions $I_1, I_2 \ldots I_n$.

Microinstruction	reset_counter	shift_left	load_output
BEGIN	1	0	0
I_1	?	?	?
:	:	:	:
I_n	?	?	?
END	0	0	1

The counter width (k), the number of missing microinstructions (n), and the control word for microinstructions $I_1 I_2 \ldots I_n$ are, respectively.

a) 32, 5 010 b) 5, 32 010

c) 5, 31 011 d) 5, 31 010

42.A cache line is 64 bytes. The main memory has latency 32ns and bandwidth 1G. Bytes/s. The time required to fetch the entire cache line from the main memory is

a) 32 ns b) 64 ns

c) 96 ns d) 128 ns

43.A computer system has a level - 1 instruction cache (1-cache), a level-1 data cache (D-cache) and a level-2 cache (L2-cache) with the following specifications :

	Capacity	Mapping Method	Block size
1 - cache	4K words	Direct mapping	4 words
D-cache	4K words	2-way set-associative mapping	4 words
L2-cache	64K words	4-way set-associative mapping	16 words

The length of the physical address of a word in the main memory is 30 bits. The capacity of the tag memory in the I-cache, D-cache and L2-cache is, respectively,

a) $1K \times 18-bit, 1K \times 19-bit, 4K \times 16-bit$ b) $1K \times 16-bit, 1K \times 19-bit, 4K \times 18-bit$

c) $1K \times 16-bit, 512 \times 18-bit, 1K \times 16-bit$ d) $1K \times 18-bit, 512 \times 18-bit, 1K \times 18-bit$

44.Which of the following sequences of array elements forms a heap ?

a) $\{23,17,14,6,13,10,1,12,7,5\}$

b) $\{23,17,14,6,13,10,1,5,7,12\}$

c) $\{23,17,14,7,13,10,1,5,6,12\}$

d) $\{23,17,14,7,13,10,1,12,5,7\}$

45.Suppose that we have numbers between 1 and 100 in a binary search tree and want to search for the number 55. Which of the following sequencies CANNOT be the sequence of nodes examined ?

a) $\{10,75,64,43,60,57,55\}$

b) $\{90,12,68,34,62,45,55\}$

c) $\{9,85,47,68,43,57,55\}$

d) $\{79,14,72,56,16,53,55\}$

46.Which of the following is the correct decomposition of the directed graph given below into its strongly connected components?

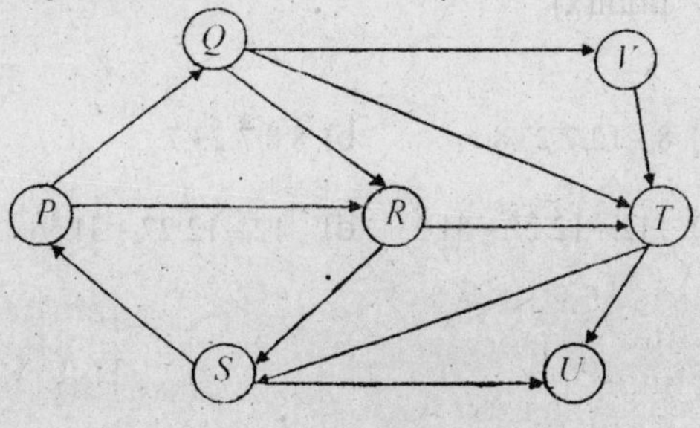

a) $\{P,Q,R,S\},\{T\},\{U\},\{V\}$

b) $\{P,Q,R,S,T,V\},\{U\}$

c) $\{P,Q,S,T,V\},\{R\},\{U\}$

d) $\{P,Q,R,S,T,U,V\}$

47.Consider the depth-first search of an undirected graph with 3 vertices P, Q, and R. Let discovery time $d(u)$ represent the time instant when the vertex u is first visited, and finish time $f(u)$ represent the time instant when the vertex u is last visited. Given that

$d(P)=5$ units $f(P)=12$ units

$d(Q)=6$ units $f(Q)=10$ units

$d(R)=14$ units $f(R)=18$ units

Which one of the following statements is TRUE about the graph

a) There is only one connected component

b) There are two connected components, and P and R are connected.

c) There are two connected components, and Q and R are connected

d) There are two connected components, and P and Q are connected.

48.The characters a to h have the set of frequencies based on the first 8 Fibonacci numbers as follows

$a:1, b:1, c:2, d:3, e:5, f:8, g:13, h:21$

A Huffman code is used to represent the characters. What is the sequence of characters corresponding to the following code ?

110111100111010

a) fdheg b) ecgdf

c) dchfg d) fehdg

49. Which one of the choices given below would be printed when the following program is executed ?

```
# include <stdio.h>
struct test {
   int i ;
   char * c ;
} st [] = {5, "become", 4, "better", 6, "jungle", 8,
"ancestor", 7,
"brother"} ;
main ( )
{ struct test *p=st ;
p+=1 ;
   ++p -> c ;
printf("%s," , p++->c) ;
printf("%c,", *++p->c);
printf("%d,", p[0].i);
   printf("%s\n", p--->c) ;
```

a) jungle, n, 8, ncestor

b) etter, u, 6, ungle

c) cetter, k, 6, jungle

d) etter, u, 8, ncestor

50. Which one of the choices given below would be printed when the following program is executed?

```
# include <stdio.h>
void swap (int *x, int *y)
{
      static int *temp;
      temp = x ;
      x = y ;
      y = temp ;
}
void printab ()
{
      static int i, a = -3, b = -6;
      i = 0 ;
      while (i<=4)
      {
      if ((i++)  %2 ==1) continue ;
      a = a+i ;
      b = b+i ;
}
swap (&a, &b) ;
printf("a =%d, b=%d\n", a,b);
}
main ( )
{
      printab ( ) ;
      printab ( ) ;
}
```

a) $a=0,\ b=3$ b) $a=3,\ b=0$

$a=0,\ b=3$ $a=12,\ b=9$

c) $a=3,\ b=6$ d) $a=6,\ b=3$

$a=3,\ b=6$ $a=15,\ b=12$

51. Which one of the choices given below would be printed when the following program is executed ?

```
# include <stdio.h>
int al[] = {6, 7, 8,18, 34, 67} ;
int a2[] = {23, 56, 28, 29};
int a3[] = {-12, 27, -31};
int *x[] = {a1, a2, a3 };
void print ( int *a[] )
{
   pintf("%d,", a[0] [2] ) ;
   printf("%d,", *a[2] ) ;
   printf("%d,", *++a[0] ) ;
   printf("%d,", *(++a) [0] ) ;
   printf("%d\n",  a[-1] [+1] ) ;
}
main ()
{
   print(x) ;
}
```

a) 8, –12, 7, 23, 8 b) 8, 8, 7, 23, 7

c) –12, –12, 27, –31, 23 d) –12, –12, 27, –31, 56

52. The following function computes the value of $\binom{m}{n}$ correctely for all legal values m and n $(m \geq 1, n \geq 0 \text{ and } m > n)$

```
int func (int m, int n)
{
    if (E) return 1 ;
    else return (func (m-1, n) + func(m-1, n-1)) ;
}
```

In the above function, which of the following is the correct expression for E ?

a) $(n == 0) || (m == 1)$

b) $(n == 0) \&\& (m == 1)$

c) $(n == 0) || (m == n)$

d) $(n == 0) \&\& (m == n)$

53. Match the following concepts and their best possible desctiptions.

Concept	Description
(i) overloading	(a) allows to define a class to have properties of another class.
(ii) friend	(b) defining a set of similar functions
(iii) constructor	(c) used in dereferencing
(iv) protected	(d) used to give a non-member function access to the private parts of an object
(v) this	(e) a function which is automatically called when an object is created.
(vi) inheritance	(f) allows a derived class to have access to the private parts of the base class.
	(g) a pointer to the object associated with the current function.
	(h) used to obtain object persistence

a) $i-b, ii-d, iii-e, iv-f, v-g, vi-a$

b) $i-c, ii-a, iii-e, iv-d, v-h, vi-f$

c) $i-c, ii-f, iii-h, iv-a, v-g, vi-d$

d) $i-b, ii-e, iii-c, iv-f, v-g, vi-s$

54. The arrival time, priority, and durations of the CPU and I/O bursts for each of three processes P_1, P_2 and P_3 are given in the table below. Each process has a CPU burst followed by an I/O burst followed by another CPU burst. Assume that each process has its own I/O resource

Process	Arrival time	Priority	Burst durations CPU, I/O, CPU)
P_1	0	2	1, 5, 3
P_2	2	3 (lowest)	3, 3, 1
P_3	3	1 (highest)	2, 3, 1

The multi-programmed operating system uses preemptive priority scheduling. What are the finish times of the processes P_1, P_2 and P_3 ?

a) 11, 15, 9 b) 10, 15, 9

c) 11, 16, 10 d) 12, 17, 11

55. Consider the solution to the bounded buffer producer / consumer problem by using general semaphores S, F, and E. The semaphore S is the mutual exclusion semaphore initialized to 1. The semaphore F corresponds to the number of free slots in the buffer and is initialized to N. The semaphore E corresponds to the number of elements in the buffer and is initialized to 0.

Producer Process	Consumer Process
Produce an item ;	Wait (E) ;
Wait (F) ;	Wait (S) ;
Wait (S) ;	Remove an item from the buffer ;
Append the item to the buffer ;	Signal (S) ;
Signal (S) ;	Signal (F) ;
Signal (E) ;	Consume the item ;

Which of the following interchange operations may result in a deadlock ?

(I) Interchanging Wait (F) and Wait (S) in the Product process

(II) Interchanging Signal (S) and Signal (F) in the Consumer process

a) I only b) II only

c) Neither I nor II d) Both I and II

56. For each of the four processes P_1, P_2, P_3 and P_4. The total size in kilobytes (KB) and the

number of segments are given below .

Process	Total size (in KB)	Number of segments
P1	195	4
P2	254	5
P3	45	3
P4	364	8

The page size is 1 KB. The size of an entry in the page table is 4 bytes. The size of an entry in the segment table is 8 bytes. The maximum size of a segment is 256 KB. The paging method for memory management uses two-level paging, and its storage overhead is P. The storage overhead for the segmentation method is S. The storage overhead for the segmentation and paging method is T. What is the relation among the overheads for the different methods of memory management in the concurrent execution of the above four processes ?

a) $P < S < T$ b) $S < P < T$

c) $S < T < P$ d) $T < S < P$

57. The wait and signal operations of a monitor are implemented using semaphores as follows. In the following.

x is a condition variable,

mutex is a semaphore initialized to 1,

x_sem is a semaphore initialized to 0,

x_count is the number of processes waiting on semaphore x_sem,

initially 0, next is a semaphore initialized to 0,

next_count is the number of processes waiting on semaphore next, initially 0.

The body of each procedure that is visible outside the monitor is replaced with the following :

```
    P (mutex) ;
        .
        .
    body of procedure
    if (next_count > 0)
        V (next)
    else
        V (mutex) ;
```

Each occurrence of x.wait is replaced with the following :

```
    x_count = x_count + 1 ;
    if (next_count > 0)
        V (next)
    else
        V (mutex) ;
------------------------------------------------E1 ;
x_count = x_count - 1 ;
```

Each occurrence of x.signal is replaced with the following :

```
        if (x_count > 0)
    {
        next_count = next _ count + 1 ;
        ________________________________E2;
        P (next) ;
        next_count = next_count - 1 ;
    }
```

For correct implementation of the monitor, statements E1 and E2 are, respectively,

a) $P(x\_sem), V(next)$ b) $V(next), P(x\_sem)$

c) $P(next), V(x\_sem)$ d) $P(x\_sem), V(x\_sem)$

58. A software program consists of two modules M_1 *and* M_2 that can fail independently, but never simultaneously. The program is considered to have failed if any of these modules fails. Both the modules are 'repairable' and so the program starts working again as soon as the repair is done. Assume that the mean time to failure (MTTF) of M_1 *and* T_1 with a mean time to repair (MTTR) of R_1. The MTTF of M_2 *is* T_2 with an MTTR of R_2. What is the availability of the overall program given that the failure and repair times are all exponentially distributed random variables ?

a) $\frac{T_1 T_2}{T_1 R_1 + T_2 R_2}$ b) $\frac{R_1 R_2}{T_1 R_1 + T_2 R_2}$

c) $\frac{T_1 T_2}{T_1 T_2 + T_1 R_1 + T_2 R_2}$ d) $\frac{T_1 T_2}{T_1 T_2 + T_1 R_2 + T_2 R_1}$

59. Consider the following structure chart diagram. The boxes have function names embedded in them, while the variables are indicated along the arcs.

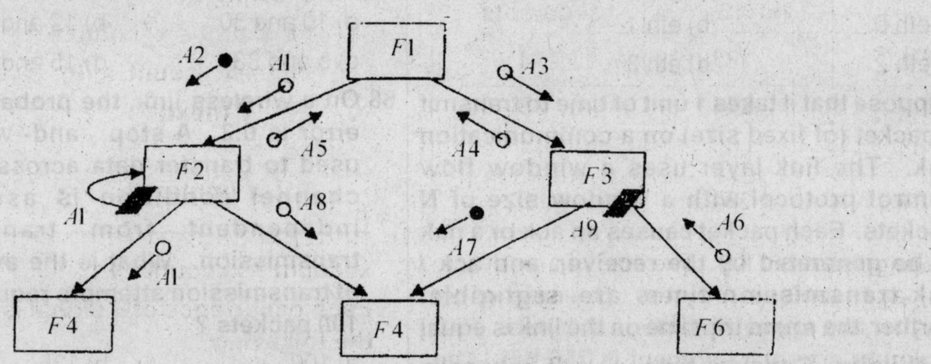

Given below are a set of statements relevant to the above diagram.

I) F3 and F6 can be in the same module.

II) F4 and F6 can be in the same module.

III) A4 is both an output and a control variable.

IV) It is incorrect to pass A1 as data and use it as a control variable.

Which combination of these statements is TRUE ?

a) III and IV b) I and IV c) II and IV d) I, II and Iv

60. Consider a relation R with five attributes V, W, X, Y, and Z. The following functional dependencies hold : $VY \to W$, $WX \to Z$, and $ZY \to V$. Which of the following is a candidate key for R ?

a) VXZ b) VXY

c) VWXY d) VWXYZ

61. In a database file structure, the search key field is 9 bytes long, the block size is 512 bytes, a record pointer is 7 bytes and a block pointer is 6 bytes. The largest possible order of a non-leaf node in a B+ tree implementing this file structure is

a) 23 b) 24

c) 34 d) 44

62. Consider the following XML DTD describing course information in a university :

<!ELEMENT Univ (Course+, Prof+) >

<!ELEMENT Course (Title, Eval*)>

<!ATTLIST Course Number ID #REQUIRED Instructor IDREF #IMPLIED>

<!ELEMENT Title (#PCDATA) >

<!ELEMENT Eval (#PCDATA) >

<!ATTLIST Eval Score CDATA#REQUIRED>

<!ELEMENT Prof EMPTY>

<!ATTLIST Prof Name ID #REQUIRED Teaches IDREF #IMPLIED>

What is returned by the following XQuery ?

```
let $as := / / @Score
for $c in /Univ / Course [Eval]
let $cs := $c / Eval? @Score
where min ($cs) > avg ($as)
return $c
```

a) The professor with the lowest course evaluation

b) Professors who have all their course evaluations above the university average

c) The course with the lowest evaluation

d) Courses with all evaluations above the university average.

63. A router uses the following routing table :

Destination	Mask	Interface
144.16.0.0	255.255.0.0	eth0
144.16.64.0	255.255.224.0	eth1
144.16.68.0	255.255.255.0	eth2
144.16.68.64	255.255.255.224	eth3

A packet bearing a destination address 144.16.68.117 arrives at the router. On which interface will it be forwarded ?

a) eth 0 b) eth 1

c) eth 2 d) eth 3

64.Suppose that it takes 1 unit of time to transmit a packet (of fixed size) on a communication link. The link layer uses a window flow control protocol with a window size of N packets. Each packet causes an ack or a nak to be generated by the receiver, and ack / nak transmission times are negligible. Further, the round trip time on the link is equal to N units. Consider time $i > N$. If only acks have been received till time i (no naks), then the goodput evaluated at the transmitter at time i (in packets per unit time) is

a) $1 - N/i$ b) $i/(N+i)$

c) 1 d) $1 - e^{(i/N)}$

65.In the 4B / 5B encoding scheme, every 4 bits of data are encoded in a 5-bit codeword. It is required that the codewords have at most 1 leading and at most 1 trailing zero. How many such codewords are possible ?

a) 14 b) 16

c) 18 d) 20

66.A router has two full - duplex Ethernet interfaces each operating at 100 Mb/s. Ethernet frames are at least 84 bytes long (including the Preamble and the Inter-Packet-Gap). The maximum packet processing time at the router for wirespeed forwarding to be possible is (in micro-seconds)

a) 0.01 b) 3.36

c) 6.72 d) 8

67.A link of capacity 100 Mbps is carrying traffic from a number of sources. Each source generates an on-off traffic stream ; when the source is on, the rate of traffic is 10 Mbps, and when the source is off, the rate of traffic is zero. The duty cycle, which is the ratio of on-time to off-time, is 1 : 2. When there is no buffer at the link, the minimum number of sources that can be multiplexed on the link so that link capacity is not wasted and no data loss occurs is S_1. Assuming that all sources are synchronized and that the link is provided with a large buffer, the maximum number of sources that can be multiplexed so that no data loss occurs is S_2. The values of S_1 *and* S_2 are, respectively,

a) 10 and 30 b) 12 and 25

c) 5 and 33 d) 15 and 22

68.On a wireless link, the probability of packet error is 0.2. A stop - and- wait protocol is used to transfer data across the link. The channel condition is assumed to be independent from transmission to transmission. What is the average number of transmission attempts required to transfer 100 packets ?

a) 100 b) 125

c) 150 d) 200

69.A program on machine X attempts to open a UDP connection to port 5376 on a machine Y, and a TCP connection to port 8632 on machine Z. However, there are no applications listening at the corresponding ports on Y and Z. An ICMP Port Unreachable error will be generated by

a) Y but not Z b) Z but not Y

c) Neither Y nor Z d) Both Y and Z

70.A subnetted Class B network has the following broadcase address : 144.16.95.255. Its subnet mask

a) is necessarily 255.255.224.0

b) is necessarily 255.255.240.0

c) is necessarily 255.255.248.0

d) could be any one of 255.255.224.0, 255.255.240.0, 255.255.248.0

<u>Common Data for Questions 71,72,73</u>

An array X of n distinct integers is interpreted as a complete binary tree. The index of the first element of the array is 0.

71.The index of the parent of element $X[i], i \neq 0$, **is**

a) $\lfloor i/2 \rfloor$ b) $\lceil (i-1)/2 \rceil$

c) $\lceil i/2 \rceil$ d) $\lceil i/2 \rceil - 1$

72.If only the root node does not satisfy the heap property, the algorithm to convert the complete binary tree into a heap has the best asympototic time complexity of

a) $O(n)$ b) $O(\log n)$

c) $O(n \log n)$ d) $O(n \log \log n)$

73.If the root node is at level 0, the level of element $X[i], i \neq 0$, is

a) $\lfloor log_2 i \rfloor$ b) $\lceil log_2 (i+1) \rceil$

c) $\lfloor log_2 (i+1) \rfloor$ d) $\lceil log_2 i \rceil$

Common Data for Questions 74,75

Consider the following program module :

```
void swap (float*A1, float * A2)
{
    float temp ;
    if (*A1 = = *A2) return ;
    temp = *A1 ;
    *A1 = *A2 ;
    *A2 = temp ;
    return ;
}
```

74.The program volume for the above module using Halstead's method is

a) 60 b) 63

c) 66 d) 69

75.The program effort for the above module using Halstead's method is

a) 315 b) 330

c) 393 d) 403

Linked Answer Questions : Q.76-85 carry two marks each

Statement for linked Answer Questions 76 & 77:

Consider the following set of linear equations :

$x + y/2 = 9$

$3x + y = 10$

76.The value of the Frobenius norm for the above system of equations is

a) 0.5 b) 0.75

c) 1.55 d) 2.0

77.What can be said about the Gauss-Siedel iterative method for solving the above set of linear equations ?

a) It will converge

b) It will diverse

c) It will neither converge nor diverse

d) It is not applicable

Statement for linked Answer Questions 78 & 79:

A pipelined processor uses a 4-stage instruction pipeline with the following stages : Instruction fetch (IF), Instruction decode (ID), Execute (EX) and Writeback (WB). The arithmetic operations as well as the load and store operations are carried out in the EX stage. The sequence of instructions corresponding to the statement $X = (S - R*(P+Q))/T$ is given below. The values of variables P, Q, R, S and T are available in the registers R0, R1, R2, R3 and R4 respectively, before the execution of the instruction sequence.

ADD	R5, R0, R1	;	$R5 \rightarrow R0 + R1$
MUL	R6, R2, R5	;	$R6 \rightarrow R2 * R5$
SUB	R5, R3, R6	;	$R5 \rightarrow R3 - R6$
DIV	R6, R5, R4	;	$R6 \rightarrow R5/R4$
STORE	R6; X	;	$X \leftarrow R6$

78.The number of Read-After-Write (RAW) dependencies, Write-After-Read (WAR) dependencies, and Write-After-Write (WAW) dependencies in the sequence of instructions are, respectively,

a) 2, 2, 4 b) 3, 2, 3

c) 4, 2, 2 d) 3, 3, 2

79.The IF, ID and WB stages take 1 clock cycle each. The EX stage takes 1 clock each for the ADD, SUB and STORE operations, and 3 clock cycles each for MUL and DIV operations. Operand forwarding from the EX stage to the ID stage is used. The number of clock cycles required to complete the sequence of instructions is

a) 10 b) 12

c) 14 d) 16

Statement for linked Answer Questions 80 & 81:

Let L be a regular language : Consider the constructions on L below :

(I) repeat (L) = $\{ww \mid w \in L\}$

(II) prefix (L) = $\{u \mid \exists v : uv \in L\}$

(III) suffix (L) = $\{v \mid \exists u : uv \in L\}$

(IV) half (L) = $\{u \mid \exists v : |v| = |u| \text{ and } uv \in L\}$

80. Which of the constructions could lead to a non-regular language ?

a) Both I and IV b) Only I

c) Only IV d) Both II and III

81. Which choice of L is best suited to support your answer above ?

a) $(a+b)^*$ b) $\{\varepsilon, a, ab, bab\}$

c) $(ab)^*$ d) $\{a^n b^n \mid n \geq 0\}$

Statement for linked Answer Questions 82 & 83

A software project has four phases P1, P2, P3 and P4. Of these phase , P1 is the first one and needs to be completed before any other phase can commence. Phases P2 and P3 can be executed in parallel. Phase P4 cannot commence until bith P2 and P3 are completed. The optimistic, most likely, and pessimistic estimates of the phase completion times in days, for P1, P2, P3 and P4 are, respectively, (11, 15, 25), (7, 8, 15), (8, 9, 22), and (3, 8, 19).

82. The critical path for the above project and the slack of P2 are, respectively.

a) $P1-P2-P4$, 1 day b) $P1-P3-P4$, 1 day

c) $P1-P3-P4$, 2 days d) $P1-P2-P4$, 2 days

83. The costs (in Rupees per day) of crashing the expected phase completion times for the four phases, respectively, are 100, 2000, 50, and 1000. Assume that the expected phase completion times of the phases cannot be crashed below their respective most likely completion times. The minimum and the maximum amounts (in Rupees) that can be spent on crashing so that ALL paths are critical are, respectively.

a) 100 and 1000 b) 100 and 1200

c) 150 and 1200 d) 200 and 2000

Statement for linked Answer Questions 84 & 85

Consider a database with three relation instances shown below. The primary keys for the Drivers and Cars relation are did and cid respectively and the records are stored in ascending order of these primary keys as given in the tables. No indexing is available in the database.

D : Drivers relation

did	dname	rating	age
22	Karthikeyan	7	25
29	Salman	1	33
31	Boris	8	55
32	Amoldt	8	25
58	Schumacher	10	35
64	Sachin	7	35
71	Senna	10	16
74	Sachin	9	35
85	Rahul	3	25
95	Ralph	3	53

R : Reserves relation

did	cid	day
22	101	10/10/06
22	102	10/10/06
22	103	08/10/06
22	104	07/10/06
31	102	10/11/06
31	103	06/11/06
31	104	12/11/06
64	101	05/09/06
64	102	08/09/06
74	103	08/09/06

C : Cars relation

cid	cname	colour
101	Renault	blue
102	Renault	red
103	Ferrari	green
104	Jaguar	red

84. What is the output of the following SQL query?

```
select D.dname
from Drivers D
where D.did in (
        select R.did
```

from Cars, C, Reverses R
where R.cid = C.cid and C.colour = 'red'
intersect
select R.did
from Cars C, Reverses R
where R.cid = C.cid and C.colour = 'green'
)

a) Karthikeyan, Boris

b) Sachin, Salman

c) Karthikeyan, Boris, Sachin

d) Schumacher, Senna

85. Let n be the number of comparisons performed when the above SQL query is optimally executed. If linear search is used to locate a tuple in a relations using primary key, then n lies in the range.

a) $36-40$ b) $44-48$

c) $60-64$ d) $100-104$

ANSWERS

1.(c)	2. (a)	3. (d)	4.(b)	5. (a)
6.(c)	7. (c)	8. (d)	9.(b)	10.(c)
11.(b)	12.(b)	13.(b)	14.(a)	15.(d)
16.(a)	17.(c)	18.(d)	19.(c)	20.(c)
21.(a)	22.(a)	23.(b)	24.(b)	25.(*)
26.(a)	27.(d)	28.(*)	29.(c)	30.(a)
31.(b)	32.(d)	33.(b)	34.(b)	35.(d)
36.(c)	37.(a)	38.(*)	39.(b)	40.(d)
41.(*)	42.(c)	43.(d)	44.(c)	45.(c)
46.(c)	47.(d)	48.(*)	49.(a)	50.(d)
51.(a)	52.(c)	53.(a)	54.(*)	55.(*)
56.(c)	57.(*)	58.(*)	59.(*)	60.(b)
61.(*)	62.(d)	63.(c)	64.(c)	65.(*)
66.(*)	67.(*)	68.(*)	69.(a)	70.(d)
71.(a)	72.(c)	73.(a)	74.(*)	75.(*)
76.(a)	77.(d)	78.(c)	79.(b)	80.(c)
81.(d)	82.(*)	83.(*)	84.(a)	85.(c)

GATE - 2005
COMPUTER SCIENCE & ENGINEERING

Duration : Three Hours | **Maximum Marks : 150**

Reading the following instructions carefully

1. The question paper will be fully objective type for a total of 150 marks divided into three groups.

Group I : Question numbers 1 to 30 (30 Questions) will carry **ONE** mark each.

Group II : Question numbers 31 to 80 (50 Questions) will carry **TWO** marks each

Group III : Question numbers 81a to 85b (10 Questions) will carry **TWO** marks each. Each number in this series (81, 82, 83, 84, 85) will have two sub-questions (a & b). The answer to Part 'b' will be linked to the correct answer to Part 'a'.

2. Answer all the questions. Each question will have four choices for the answer. Only one choice is correct.

3. Wrong answers carry 25% NEGATIVE marks in Q.1 to Q.80 and Q.81a, 82a, 83a, 84a and 85a. Marks for the correct answers to Q.81b, 82b, 83b, 84b and 85b will be given only if the answer to the corresponding Part 'a' is correct. However Q81b, 82b, 83b, 84b and 85b will not carry any NEGATIVE marks.

Q. 1 - 30 CARRY ONE MARK EACH

1. What does the following C-statement declare ?
int (* f) (int *);

a) A function that takes an integer pointer as argument and returns an integer

b) A function that takes an integer as argument and returns an integer pointer

c) A pointer to a function that takes an integer pointer as argument and returns an integer

d) A function that takes an integer pointer as argument and returns a function pointer.

2. An abstract Data Type (ADT) is

a) same as an abstract class

b) a data type that cannot be instantiated

c) a data type for which only the operations defined on it can be used, but none else

d) all of the above

3. A common property of logic programming languages and functional languages is

a) both are procedural languages

b) both are based on λ - calculus

c) both are declarative

d) both use Horn-clauses

4. Which one of the following are essential features of an object-oriented programming language ?

(i) Abstraction and encapsulation

(ii) Strictly typedness

(iii) Type safe property coupled with subtype rule

(iv) Polymorphism in the presence of inheritance

a) (i) and (ii) only   b) (i) nad (iv) only

c) (i), (ii) and (iv) only   d) (i), (iii) and (iv) only

5. A program P reads in 500 integers in the range (0, 100) representing the scores of 500 students. It then prints the frequency of each score above 50. What would be the best way for P to store the frequencies ?

a) An array of 50 numbers

b) An array of 100 numbers

c) An array of 500 numbers

d) A dynamically allocated of 550 numbers

6. An undirected graph G has n nodes. Its adjacency matrix is given by an n x n square matrix whose

(i) diagonal elements are 0's and

(ii) non-diagonal elements are 1's. Which one of the following is TRUE ?

a) Graph G has no minimum spanning tree (MST)

b) Graph G has a unique MST of cost n-1

c) Graph G has multiple distinct MSTs, each of cost n-1

d) Graph G has multiple spanning trees of different costs.

7. The time complexity of computing the transitive closure of a binary relation on a set of n elements is known to be

a) $O(n)$ b) $O(n \log n)$ c) $O\left(n^{3/2}\right)$ d) $O\left(n^3\right)$

8. Let A, B and C be non empty sets and let

$X = (A-B)-C$ and $Y = (A-C)-(B-C)$

Which one of the following is TRUE ?

a) $X = Y$ b) $X \subset Y$

c) $Y \subset X$ d) none of these

9. The following is the Hasse diagram of the poset $[\{a, b, c, d, e\}, \le]$

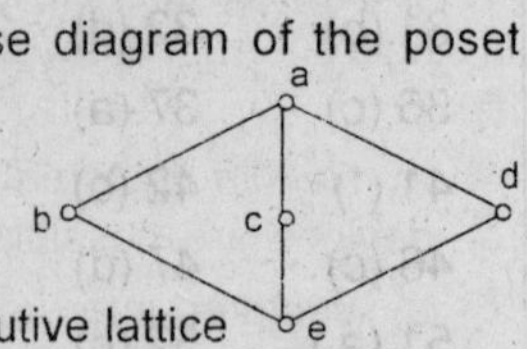

The Poset is

a) not a lattice

b) a lattice but not a distributive lattice

c) a distributive lattice but not a Boolean algebra

d) a Boolean algebra

10. Let G be a simple connected planar graph with 13 vertices and 19 edges. Then, the number of faces in the planar embedding of the graph is

a) 6   b) 8   c) 9   d) 13

11. Let G be a simple graph will 20 vertices and 100 edges. The size of the minimum vertex cover of G is 8. Then, the size of the maximum indepedent set of G is

a) 12   b) 8

c) Less than 8   d) More than 12

12. Let f(x) be the continuous probability density function of a random variable X. The probability that a $a < X \le b$

a) $f(b-a)$ b) $f(b)-f(a)$ c) $\int_a^b f(x)\,dx$ d) $\int_a^b xf(x)\,dx$

13. The set $\{1,2,4,7,8,11,13,14\}$ is a group under multiplication modulo 15. The inverses of 4 and 7 are respectively

a) 3 and 13 b) 2 and 11
c) 4 and 13 d) 8 and 14

14. The grammar $A \to AA|(A)|\epsilon$ is not suitable for predictive-parsing because the grammar is

a) ambiguous b) left-recursive
c) right-recursive d) an operator grammar

15. Consider the following circuit.

Which one of the following is TRUE ?

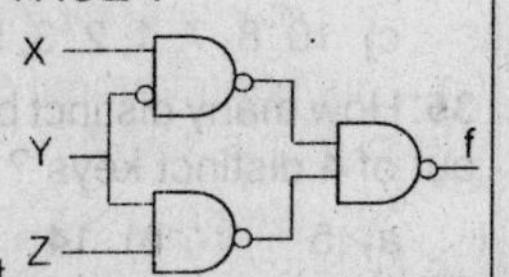

a) f is independent of X
b) f is independent of Y
c) f is independent of Z
d) None of X,Y,Z is redundant

16. The range of integers that can be represented by an n bit 2's complement number system is

a) -2^{n-1} to $(2^{n-1}-1)$ b) $-(2^{n-1}-1)$ to $(2^{n-1}-1)$
c) -2^{n-1} to 2^{n-1} d) $-(2^{n-1}+1)$ to $(2^{n-1}-1)$

17. The hexadecimal representation of 657_8 is

a) 1AF b) D78 c) D71 d) 32F

18. The switching expression corresponding to $f(A,B,C,D) = \Sigma(1,4,5,9,11,12)$

a) $BC'D' + A'C'D + AB'D$ b) $ABC' + ACD + B'C'D$
c) $ACD' + A'BC' + AC'D'$ d) $A'BD + ACD' + BCD'$

19. Which one of the following is TRUE for a CPU having a single interrupt request line and a single interrupt grant line ?

a) Neither vectored interrupt nor multiple interrupt devices are possible
b) Vectored interrupts are not possible but multiple interrupting devices are possible.
c) Vectored interrupts and multiple interrupting devices are both possible
d) Vectored interrupt is possible but multiple interrupting devices are not possible

20. Normally user programs are prevented from handling I/O directly by I/O instructions in them. For CPUs having explicit I/O instructions, such I/O protection is ensured by having the I/O instructions priviliged. In a CPU with memory mapped I/O, there is no explicit I/O instruction. Which one of the following is TRUE for a CPU with memory mapped I/O ?

a) I/O protection is ensured by operationg system routine (s)
b) I/O protection is ensured by a hardware trap
c) I/O protection is ensured during system configuration
d) I/O protection is not possible

21. What is the swap space in the disk used for ?

a) Saving temporary html pages
b) Saving process data
c) Storing the super-block
d) Storing device drivers

22. Increasing the RAM of a computer typically improves performance because

a) Virtual memory increases
b) Larger RAMs are faster
c) Fewer pages faults occur
d) Fewer segmentation faults occur

23. Packets of the same session may be routed through different paths in

a) TCP, but not UDP b) TCP and UDP
c) UDP, but not TCP d) Neither TCP, nor UDP

24. The Address Resolution Protocol (ARP) is used for

a) Finding the IP address from the DNS
b) Finding the IP address of the default gateway
c) Finding the IP address that corresponds to a MAC address
d) Finding the MAC address that corresponds to an IP address

25. The maximum window size for data transmission using the selective reject protocol with n-bit frame sequence numbers is

a) 2^n b) 2^{n-1} c) 2^n-1 d) 2^{n-2}

26. In a network of LANs connected by bridges, packets are sent from one LAN to another through intermediate bridges. Since more than one path may exist between two LANs, packets may have to be routed through multiple bridges. Why is the spanning tree algorithm used for bridge-routing ?

a) For shortest path routing between LANs
b) For avoiding loops int he routing paths
c) For fault tolerance
d) For minimizing collisions

27. An organization has a class B network and wishes to form subnets for 64 departments. The subnet mask would be

a) 255.255.0.0 b) 255.255.64.0
c) 255.255.128.0 d) 255.255.252.0

28. Which one of the following is a key for prefering B^+-trees to binary search trees for indexing database relations ?

a) Database relations have a large number of records.
b) Database relations are sorted on the primary key.

c) B^+ - trees require less memory than binary search trees

d) Data transfer from disks is in blocks.

29. Which one of the following statements about normal forms is FALSE ?

a) BCNF is stricter than 3NF

b) Lossless, dependency preserving decomposition into 3NF is always possible.

c) Lossless, dependency preserving decomposition into BCNF is always possible.

d) Any relation with two attributes is in BCNF

30. Let r be a relation instance with schema $R = (A,B,C,D)$. We define $r_1 = \Pi_{A,B,C}(r)$ and $r_2 = \Pi_{A,D}(r)$ Let $s = r_1 * r_2$ where * denotes natural join. Given that the decomposition of r into r_1 and r_2 is lossy, which one of the following is TRUE ?

a) $S \subset r$ b) $r \cup s = r$ c) $r \subset s$ d) $r * s = s$

Q. 31 - Q. 80 CARRY TWO MARKS EACH

31. Consider the following C-program

```
Void foo (int n, int sum) {
    int k = 0, j = 0;
    if (n ==0) return;
    k = n% 10 ; j = n/10;
    sum = sum + k;
    foo ( j, sum);
    printf ("%d," k);
    }
    int main ( ) {
    int a = 2048, sum = 0;
    foo (a, sum);
    printf ("% d\n", sum);
    }
```

What does the above program print ?

a) 8,4,0,2,14 b) 8,4,0,2,0

c) 2,0,4,8,14 d) 2,0,4,8,0

32. Consider the following C-program

```
double foo (double) ; /*Line 1*/
    int main ( ) {
    double da,db;
     // input da
    db = foo (da);
    }
double foo (double a) {
     return a;
    }
```

The above code compiled without any error or warning. If Line 1 is deleted, the above code will show

a) no compile warning or error

b) some compiler-warnings not leading to unintended results

c) some compiler-warnings due to type-mismatch eventually leading to unintended results

d) Compiler errors

33. Postorder traversal of a given binary search tree, T produces the following sequence of keys

10, 9, 23, 22, 27, 25, 15, 50, 95, 60, 40, 29

Which one of the following sequences of keys can be the result of an inorder traversal of the tree T ?

a) 9,10, 15, 22, 23, 25, 27, 29, 40, 50, 60, 95

b) 9, 10, 15, 22, 40, 50, 60, 95, 23, 25, 27, 29

c) 29, 15, 9, 10, 25, 22, 23, 27, 40, 60, 50, 95

d) 95, 50, 60, 40, 27, 23, 22, 25, 10, 9, 15, 29

34. A Priority-Queue is implemented as a Max-Heap. Initially, it has 5 elements. The level-order traversal of the heap is given below :

10, 8, 5, 3, 2

Two new elements '1' and '7' are inserted in the heap in that order. The level order traversal of the heap after the insertion of the elements is

a) 10, 8, 7, 5, 3, 2, 1 b) 10, 8, 7, 2, 3, 1, 5

c) 10, 8, 7, 1, 2, 3, 5 d) 10, 8, 7, 3, 2, 1, 5

35. How many distinct binary search trees can be created out of 4 distinct keys ?

a) 5 b) 14 c) 24 d) 42

36. In a complete k-ary tree, every internal node has exactly k children. The number of leaves in such a tree with n internal nodes is

a) nk b) $(n-1)k+1$ c) $n(k-1)+1$ d) $n(k-1)$

37. Suppose $T(n) = 2T(n/2) + n$, $T(0) = T(1) = 1$

Which one of the following is FALSE ?

a) $T(n) = O(n^2)$ b) $T(n) = \theta(n \log n)$

c) $T(n) = \Omega(n^2)$ d) $T(n) = O(n \log n)$

38. Let G (V,E) an undirected graph with positive edge weights. Dijkstra's single source shortest path algorithm can be implemented using the binary heap data structure with time complexity :

a) $O(|V|^2)$ b) $O(|E| + |V| \log |V|)$

c) $O(|V| \log |V|)$ d) $O((|E| + |V|) \log |V|)$

39. Suppose there are log n sorted lists of n log n elements each. The time complexity of producing a sorted list of all these elements is :

(Hint : Use a heap data structure)

a) $O(n \log \log n)$ b) $\theta(n \log n)$

c) $\Omega(n \log n)$ d) $\Omega(n^{3/2})$

40. Let P, Q and R be three atomic propositional assertions. Let X denote $(P \vee Q) \rightarrow R$ and Y denote $(P \rightarrow R) \vee (Q \rightarrow R)$. Which one of the following is a tautology ?

a) $X \equiv Y$ b) $X \rightarrow Y$ c) $Y \rightarrow X$ d) $\neg Y \rightarrow X$

41. What is the first order predicate calculus statement equivalent to the following ?

Every teacher is liked by some student

a) $\forall(x)\left[\text{teacher}(x) \rightarrow \exists(y)\left[\text{student}(y) \rightarrow \text{likes}(y,x)\right]\right]$

b) $\forall(x)\left[\text{teacher}(x) \rightarrow \exists(y)\left[\text{student}(y) \wedge \text{likes}(y,x)\right]\right]$

c) $\exists(y)\forall(x)\left[\text{teacher}(x) \rightarrow \left[\text{student}(y) \wedge \text{likes}(y,x)\right]\right]$

d) $\forall(x)\left[\text{teacher}(x) \wedge \exists(y)\left[\text{student}(y) \rightarrow \text{likes}(y,x)\right]\right]$

42. Let R and S be any two equivalence relations on a non-empty set A. Which one of the following statements is TRUE ?

a) $R \cup S$, $R \cap S$ are both equivalence relations

b) $R \cup S$ is an equivalence relation

c) $R \cap S$ is an equivalence relation

d) Neither $R \cup S$ nor $R \cap S$ is an equivalence relation

43. Let $f: B \rightarrow C$ and $g: A \rightarrow B$ be two functions and let $h = f o g$,. Given that h is an onto function. Which one of the following is TRUE ?

a) f and g should both be onto functions

b) f should be onto but g need not be onto

c) g should be onto but f need not be onto

d) both f and g need not be onto

44. What is the minimum number of ordered pairs of non-negative numbers that should be chosen to ensure that there are two pairs (a,b) and (c,d) in the chosen set such that $a \equiv c \bmod 3$ and $b \equiv d \bmod 5$

a) 4 b) 6 c) 16 d) 24

45. Consider three decision problems P_1, P_2 and P_3. It is known that P_1 is decidable and P_2 is undecidable. Which one of the following is TRUE ?

a) P_3 is decidable if P_1 reducible to P_3

b) P_3 is undecidable if P_3 is reducible to P_2

c) P_3 is undecidable if P_2 is ruducible to P_3

d) P_3 is decidable if P_3 reducible to P_2. complement

46. Consider the set H of all 3×3 matrices of the type

$$\begin{bmatrix} a & f & e \\ 0 & b & d \\ 0 & 0 & c \end{bmatrix}$$

Where a,b,c,d,e and f are real numbers and $abc \neq 0$. Under the matrix multiplication operation, the set H is

a) a group b) a monoid but not a group

c) a semigroup but not a monoid

d) neither a group nor a semigroup

47. Which one of the following graphs is NOT planar ?

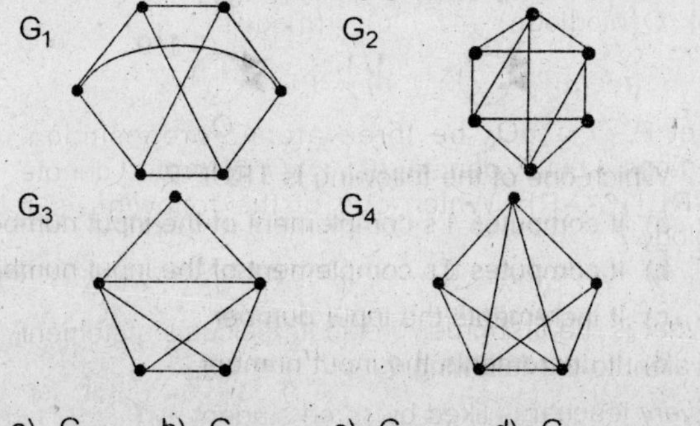

a) G_1 b) G_2 c) G_3 d) G_4

48. Consider the following system of equations in three real variables x_1, x_2 and x_3

$$2x_1 - x_2 + 3x_3 = 1$$
$$3x_1 - 2x_2 + 5x_3 = 2$$
$$-x_1 + 4x_2 + x_3 = 3$$

This system of equations has

a) no solution b) a unique solution

c) more than one but a finite number of solutions

d) an infinite number of solutions

49. What are the eigen values of the following 2×2 matrix? $\begin{bmatrix} 2 & -1 \\ -4 & 5 \end{bmatrix}$

a) -1 and 1 b) 1 and 6

c) 2 and 5 d) 4 and -1

50. Let $G(x) = 1/(1-x)^2 = \sum_{i=0}^{\infty} g(i)x^i$ where $|x| < 1$. What is g(i) ?

a) 1 b) i + 1 c) 2i d) 2^i

51. Box P has 2 red balls and 3 blue balls and box Q has 3 red balls and 1 blue ball. A ball is selected as follows :

(i) select a box

(ii) choose a ball from the selected box such that each ball in the box is equally likely to be chosen. The probabilities of selecting boxes P and Q are $\frac{1}{3}$ and $\frac{2}{3}$, respectively. Given that a bll selected in the above process is a red ball, the probability that it came from the box P is

a) $\frac{4}{19}$ b) $\frac{5}{19}$ c) $\frac{2}{9}$ d) $\frac{19}{30}$

52. A random bit string of length n is constructed by tossing a fair coin n times and setting a bit to 0 or 1 depending on outcomes head and tail, respectively. The probability that two such randomly generated strings are not identical is

a) $\frac{1}{2^n}$ b) $1-\frac{1}{n}$ c) $\frac{1}{n!}$ d) $1-\frac{1}{2^n}$

53. Consider the machine M :

The language recognized by M is :

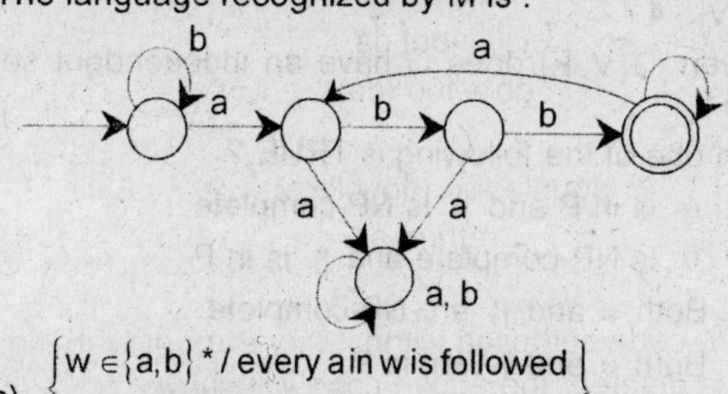

a) $\left\{ w \in \{a,b\}^* / \text{every a in w is followed by exactly two b's} \right\}$

b) $\left\{ w \in \{a,b\}^* \text{ every a in w is followed by atleast two b's} \right\}$

c) $\{w \in \{a,b\}^* \text{ w contains the substing 'abb'}\}$

d) $\{w \in \{a,b\}^* \text{ w does not contain 'aa' as a substring}\}$

54. Let N_f and N_p denote the classes of languages accepted by non-deterministic finite automata and non-deterministic push-down automata, respectively. Let D_f and D_p denote the classes of languages accepted by deterministic finite automata and deterministic push-down autpmata, respecively. Which one of the following is TRUE ?

a) $D_f \subset N_f$ and $D_p \subset N_p$ b) $D_f \subset N_f$ and $D_p = N_p$

c) $D_f = N_f$ and $D_p = N_p$ d) $D_f = N_f$ and $D_p \subset N_p$

55. Consider the languages

$L_1 = \{a^n b^n c^m \mid n, m > 0\}$ and $L_2 = \{a^n b^m c^m \mid n, m > 0\}$

Which one of the following statements is FALSE ?

a) $L_1 \cap L_2$ is a context-free language

b) $L_1 \cup L_2$ is a context-free language

c) L_1 and L_2 are context-free language

d) $L_1 \cap L_2$ is a context sensitive language

56. Let L_1 be a recursive language, and let L_2 be a recursively enumerable but not a recursive language. Which one of the following is TRUE ?

a) $\bar{L}_1$ is recursive and $\bar{L}_2$ is recursively enumerable

b) $\bar{L}_1$ is recursive and $\bar{L}_2$ is not recursively enumerable

c) $\bar{L}_1$ and $\bar{L}_2$ are recursively enumerable

d) $\bar{L}_1$ is recursively enumerable and $\bar{L}_2$ is recursive.

57. Consider the languages

$L_1 = \{ww^R \mid w \in \{0, 1\}^*\}$

$L_2 = \{w\#w^R \mid w \in \{0, 1\}^*\}$, where # is a special symbol

$L_3 = \{ww \mid w \in \{0, 1\}^*\}$

Which of the following is TRUE ?

a) L_1 is a deterministic CFL

b) L_2 is a deterministic CFL

c) L_3 is a CFL, but not a deterministic CFL

d) L_3 is a deterministic CFL

58. Consider the following two problems on undirected graphs

α: Given $G(V, E)$, does G have an independent set of size $|V| - 4$?

β: Given $G(V, E)$, does G have an independent set of size 5 ?

Which one of the following is TRUE ?

a) α is in P and β is NP-complete

b) α is NP-complete and β is in P

c) Both α and β are NP-complete

d) Both α and β are in P

59. Consider the grammar

$E \rightarrow E + n \mid E \times n \mid n$

For a sentence $n + n \times n$, the handles in the right-sentential form of the reduction are

a) n, E + n and E + n × n b) n, E + n and E + E × n

c) n, n + n and n + n × n d) n, E + n and E × n

60. Consider the grammar

$S \rightarrow (S) \mid a$

Let the number of states in SLR (1), LR (1) and LALR (1) parsers for the grammar be n_1, n_2 and n_3 respectively. The following relationship holds good

a) $n_1 < n_2 < n_3$ b) $n_1 = n_3 < n_2$

c) $n_1 = n_2 = n_3$ d) $n_1 \geq n_3 \geq n_2$

61. Consider line number 3 of the following C-program.

```
int main ( ) {                /* Line 1*/
int I, N;                     /* Line 2*/
fro (I = 0, I < N, I + +);    /* Line 3*/
}
```

Identify the compiler's response about this line while coreating the object-module

a) No compilation error b) Only a lexical error

c) Only syntactic errors

d) Both lexical and syntactic errors

62. Consider the following circuit involving a positive edge triggered DFF.

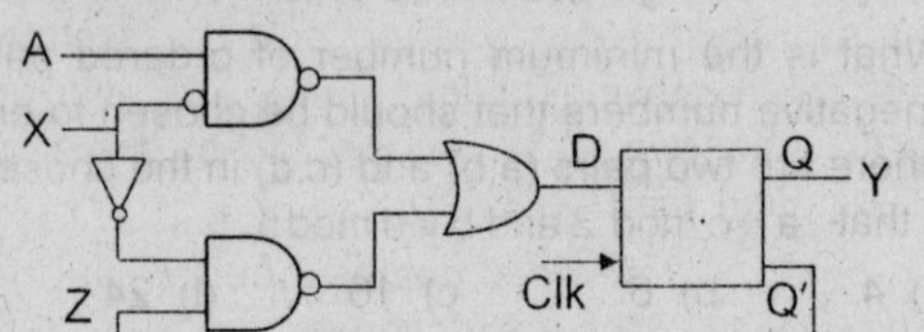

Consider the following timing diagram. Let A_i represent the logic level on the line A in the i-th clock period.

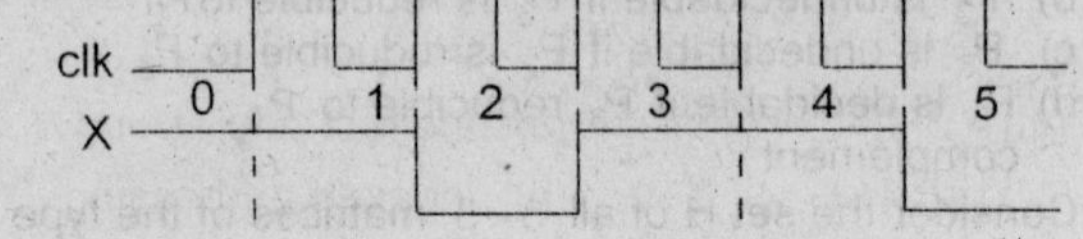

Let A' represent the complement of A. The correct output sequence on Y over the clock periods 1 through 5 is

a) $A_0\ A_1\ A'_1\ A_3\ A_4$ b) $A_0\ A_1\ A'_2\ A_3\ A_4$

c) $A_1\ A_2\ A'_2\ A_3\ A_4$ d) $A_1\ A'_2\ A_3\ A_4\ A'_5$

63. The following diagram represents a finite state machine which takes as input a binary number from the least significant bit.

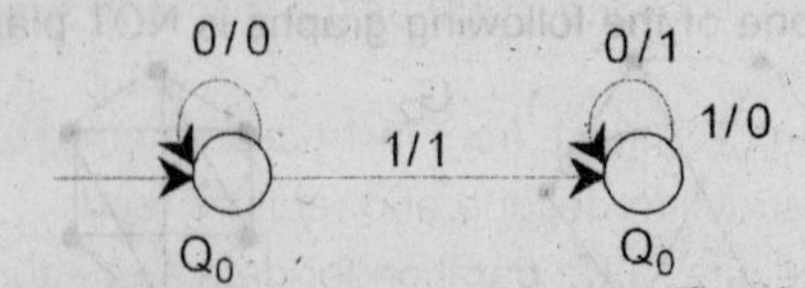

Which one of the following is TRUE ?

a) It computes 1's complement of the input number

b) It computes 2's complement of the input number

c) It increments the input number

d) It decrements the input number

64. Consider the following circuit

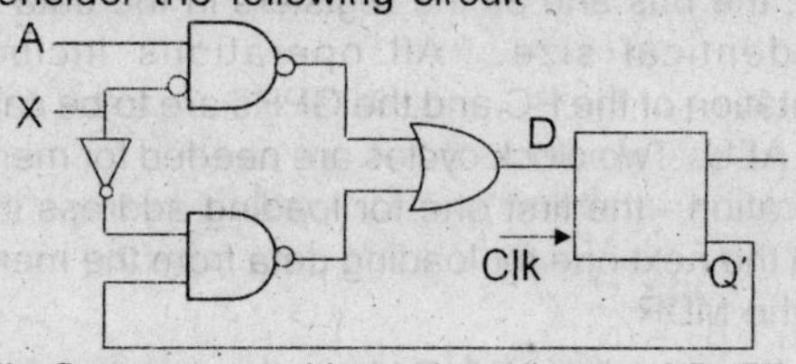

The flip-flops are positive edge triggered D FFs. Each state is designated as a two bit string Q_0Q_1. Let the initial state be 00. The state transition sequence is :

a) $00 \to 11 \to 01$ b) $00 \to 11$

c) $00 \to 10 \to 01 \to 11$ d) $00 \to 11 \to 01 \to 10$

65. Consider a three word machine instruction

ADD A[R0], @ B

The first operand (destination) "A[R0]" uses indexed addressing mode with R0 as the index register. The second operand (source) "@ B" uses indirect addressing mode. A and B are memory addrsses residing at the second and the third words, respectively. The first word of the instruction specifies the opcode, the index register designation and the source and destination addressing modes. During execution of ADD instruction, the two operands are added and stored in the destination (first operand).

The number of memory cycles needed during the execution cycle of the instruction is

a) 3 b) 4 c) 5 d) 6

66. Match each of the high level language statements given on the left hand side with the most natural addressing mode from those listed on the right hand side.

(1) A[I] = B[J] ; (a) Indirect addressing

(2) While [*A++]; (b) Indexed addressing

(3) int temp = *x; (c) Autoincrement

a) (1, c), (2, b), (3, a) b) (1, a), (2, c), (3, b)

c) (1, b), (2, c), (3, a) d) (1, a), (2, b), (3, c)

67. Consider a direct mapped cache of size 32 KB with block size 32 bytes. The CPU generates 32 bit addresses. The number of bits needed for cache indexing and the number of tag bits are respectively.

a) 10, 17 b) 10, 22 c) 15, 17 d) 5, 17

68. A 5 stage pipelined CPU has the following sequence of stages :

IF - Instruction fetch from instruction memory,

RD - Instruction decode and register read,

EX - Execute:ALU operation for data and address computation,

MA - Data memory access - for write access, the register read at RD stage is used,

WB - Register write back.

Consider the following sequence of instructions :

I_1: L R0, 1oc1; R0 <= M[1oc1]

I_2: A R0, R0 ; R0 <= R0 + R0

I_3: A R2, R0; R2 <= R2 - R0

Let each stage take one clock cycle.

What is the number of clock cycles taken to complete the above sequence of instructions starting from the fetch of I_1 ?

a) 8 b) 10 c) 12 d) 15

69. A device with data transfer rate 10 KB/sec is connected to a CPU. Data is transferred byte-wise. Let the interrupt overhead be $4\ \mu$sec. The byte transfer time between the device interface register and CPU or memory is negligible. What is the minimum performance gain of operating the device under interrupt mode over operating it under program controlled mode?

a) 15 b) 25 c) 35 d) 45

70. Consider a disk drive with the following specifications:

16 surfaces, 512 tracks/surface, 512 sectors/track, 1KB/sector, rotation speed 3000 rpm. The disk is operated in cycle stealing mode whereby whenever one byte word is ready it is sent to memory; similarly, for writing, the disk interface reads a 4 byte word from the memory in each DMA cycle. Memory cycle time is 40 nsec. The maximum percentage of time that the CPU gets blocked during DMA operation is

a) 10 b) 25 c) 40 d) 50

71. Suppose n processes, P_1,......P_n share m identical resource units, which can be reserved and released one at a time. The maximum resource requirement of process P_i is s_p where $s_i > 0$. Which one of the following is a sufficient condition for ensuring that deadlock does not occur ?

a) $\forall i, s_i < m$ b) $\forall i, s_i < n$

c) $\sum_{i=1}^{n} s_i < (m+n)$ d) $\sum_{i=1}^{n} s_i\ (m * n)$

72. Consider the following code fragment :

if (fork ( ) = = 0)

{a = a + 5; printf("%d,%d\n", a, &a);}

else {a=a–5; printf("%d, %d\n", a,&a);}

Let u, v be the values printed by the parent process, and x, y be the values printed by the child process. Which one of the following is TRUE ?

a) $u = x + 10$ and $v = y$ b) $u = x + 10$ and $v \neq y$

c) $u + 10 = x$ and $v = y$ d) $u + 10 = x$ and $v \neq y$

73. In a packet switching network, packets are routed from source to destination along a single path having two intermediate nodes. If the message size is 24 bytes and each packet contains a header of 3 bytes, then the optimum packet size is

a) 4 b) 6 c) 7 d) 9

74. Suppose the round trip propagation delay for a 10 Mbps Ethernet having 48-bit jamming signal is 46.6 μs .The minimum frame size is:

a) 94 b) 416 c) 464 d) 512

75. Let E_1 and E_2 be two entities in an E/R diagram with simple single-valued attributes. R_1 and R_2 are two relationships between E_1 and E_2, where R_1 is one-to-many and R_2 is many-to-many. R_1 and R_2 do not have any attributes of their own. What is the minimum number of tables required to represent this situation in the relational model ?

a) 2 b) 3 c) 4 d) 5

76. The following table has two attributes A and C where A is the primary key and C is the foreign key referencing A with on-delete cascade.

A	C
2	4
3	4
4	3
5	2
7	2
9	5
6	4

The set of all tuples that must be additionally deleted to preserve referential integrity when the tuple (2, 4) is deleted is :

a) (3, 4) and (6, 4) b) (5, 2) and (7, 2)

c) (5, 2) (7, 2) and (9, 5) d) (3, 4) (4, 3) and (6, 4)

77. The relation book (title, price) contains the titles and prices of different books. Assuming that no two books have the same price, what does the following SQL

```
select title
from book as B
where (select count(*)
        from book as T
        where T.price > B.price) < 5
```

a) Titles of the four most expensive books

b) Title of the fifth most inexpensive book

c) Title of the fifth most expensive book

d) Titles of the five most expensive books

78. Consider a relation scheme R = {A, B, C, D, E, H} on which the following functional dependencies hold : $\{A \rightarrow B, BC \rightarrow D, E \rightarrow C, D \rightarrow A\}$. What are the candidate keys of R ?

a) AE, BE b) AE, BE, DE

c) AEH, BEH, BCH d) AEH, BEH, DEH

COMMON DATA FOR QUESTIONS 79,80

Consider the following data path of a CPU

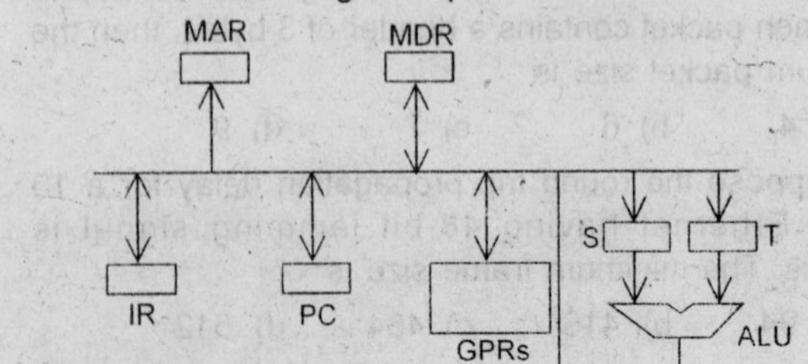

The ALU, the bus and all the registers in the data path are of identical size. All operations including incrementation of the PC and the GPRs are to be carried out in the ALU. Two clock cycles are needed for memory read operation - the first one for loading address in the MAR and the next one for loading data from the memory bus into the MDR.

79. The instruction "add R0, R1" has the register transfer interpretation R0 < = R0 + R1. The minimum number of clock cycles needed for execution cycle of this instruction is

a) 2 b) 3 c) 4 d) 5

80. The instruction "call Rn, sub" is a two word instruction. Assuming that PC is incremented during the fetch cycle of the first word of the instruction, its register transfer interpretation is :

Rn < = PC + 1 ;

Pc < = M[PC];

The minimum number of CPU clock cycles needed during the execution cycle of this instruction is :

a) 2 b) 3 c) 4 d) 5

LINKED ANSWER QUESTIONS : Q.81A - Q.85B
CARRY TWO MARKS EACH

Statement for Linked Answer Questions 81a and 81b

Consider the following C - function :

```
double foo (int n) {
    int i;
    double sum;
    if (n==0) return 1.0;
    else {
        Sum = 0.0;
        for (i =0; i < n; i++)
            sum + = foo (i) ;
        return sum;
    }}
```

81a. The space complexity of the above function is

a) O(1) b) O(n) c) O(n!) d) $O(n^n)$

81b. Suppose we modify the above function foo ( ) and store the values of foo (i), 0 < = i < n, as and when they are computed. With this modification, the time complexity for function foo ( ) is significantly reduced. The space complexity of the modified function would be :

a) O(1) b) O(n) c) $O(n^2)$ d) O(n!)

Statement for Linked Answer Questions 82a and 82b

Let s and t be two vertices in a undirected graph G + (V, E) having distinct positive edge weights. Let (X, Y) be a partition of V such that $s \in X$ and $t \in Y$. Consider the edge e having the minimum weight amongst all those edges that have one vertex in X and one vertex in Y.

82a. The edge e must difinitely belong to :

a) the minimum weighted spanning tree of G

b) the weighted shortest path from s to t

c) each path from s to t

d) the weighted longest path from s to t

82b. Let the weight of an edge e denote the congestion on that edge. The congestion on a path is defined to be the maximum of the congestions on the edges of the path. We wish to find the path from s to t having minimum congestion. Which one of the following paths is always such a path of minimum congestion ?

a) a path from s to t in the minimum weighted spanning tree

b) a weighted shortest path from s to t

c) an Euler walk from s to t

d) a Hamiltonian path from s to t

Statement for Linked Answer Questions 83a and 83b

Consider the following expression grammar. The semantic rules for expression calculation are stated next to each grammar production.

$E \to$ number	E.val = number. val
$\mid E'+'E$	$E^{(1)}.\text{val} = E^{(2)}.\text{val} + E^{(3)}.\text{val}$
$\mid E'\times'E$	$E^{(1)}.\text{val} = E^{(2)}.\text{val} \times E^{(3)}.\text{val}$

83a. The above grammar and the semantic rules are fed to a yacc tool (which is an LALR (1) parser generator) for parsing and evaluating arithmetic expressions. Which one of the following is TRUE about the action of yacc for the given grammar ?

a) It detects recursion and eliminates recursion

b) It detects reduce-reduce conflict, and resolves

c) It detects shift-reduce conflict, and resolves the conflict in favour of a shift over a reduce action

d) It detects shift-reduce conflict, and resolves the conflict in favor of a reduce over a shift action.

83 b. Assume the conflicts in Part (a) of this question are resolved and an LALR (1) parser is generated for parsing arithmetic expressions as per the given grammar. Consider an expression $3 \times 2 + 1$. What precedence and associativity properties does the generated parser realize ?

a) Equal precedence and left associativity; expression is evaluated to 7.

b) Equal precedence and right associativity; expression is evaluated to 9.

c) Precedence of '×' is higher than that of '+', and both operators are left associative; expression is evaluated to 7.

d) Precedence of '+' is higher than that of '×', and both operators are left associative; expression is evaluated to 9.

Statement for Linked Answer Questions 84a and 84b

We are given 9 tasks $T_1, T_2, T_3 \ldots\ldots T_9$. The execution of each task requires one unit of time. We can execute one task at a time. Each task T_i has a profit P_i and a deadline d_i Profit P_i is earned if the task is completed before the end of the d_i^{th} unit of time.

Task	T_1	T_2	T_3	T_4	T_5	T_6	T_7	T_8	T_9
Profit	15	20	30	18	18	10	23	16	25
Deadline	7	2	5	3	4	5	2	7	3

84a. Are all tasks completed in the schedule that gives maximum profit ?

a) All tasks are completed

b) T_1 and T_6 are left out

c) T_1 and T_8 are left out

d) T_4 and T_6 are left out

84b. What is the maximum profit earned ?

a) 147 b) 165 c) 167 d) 175

Statement for Linked Answer Questions 85a and 85b

Consider the following floating point format.

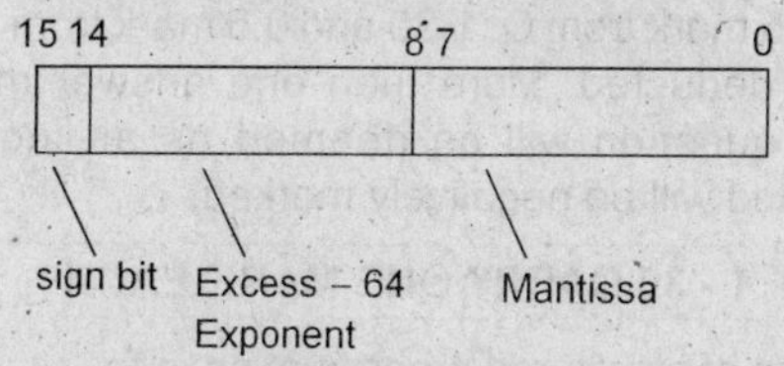

Mantissa is a pure fraction in sign-magnitude form.

85a. The decimal number 0.239×2^{13} has the following hexadecimal representation (without normalizatiion and rounding off).

a) 0D 24 b) 0D 4D c) 4D 0D d) 4D 3D

85b. The normalized representation for the above format is specified as follows.

The mantissa has an implicit 1 preceding the binary (radix) point. Assume that only 0's are padded in while shifting a field.

The normalized representation of the above number $\left(0.239 \times 2^{13}\right)$ is :

a) 0A 20 b) 11 34 c) 4D D0 d) 4A E8

ANSWER KEY

1. c	2. c	3. c	4. b	5. a
6. c	7. d	8. a	9. b	10. a
11. a	12. c	13. c	14. b	15. a
16. a	17. a	18. a	19. b	20. a
21. b	22. c	23. b	24. d	25. b
26. b	27. d	28. d	29. c	30. c
31. d	32. d	33. a	34. d	35. b
36. c	37. b	38. b	39. a	40. b
41. b	42. c	43. b	44.	45. a
46. b	47. a	48.	49.	50.
51.	52.	53. b	54. d	55. a
56. b	57. b	58. c	59. d	60. b
61. c	62. a	63. b	64. d	65. c
66. c	67. a	68. a	69. b	70. b
71. c	72. b	73. d	74. c, d	75. b
76. c	77. a	78. d	79. a	80. d
81a. c	81b. b	82a. a	82b. b	83a. c
83b. a	84a. d	84b. a	85a. d	85b. d

GATE - 2004
COMPUTER SCIENCE & ENGINEERING

Duration : Three Hours **Maximum Marks : 150**

Reading the following instructions carefully

1. This question paper contains 90 objective questions. Q. 1-30 carry one mark each and Q. 31-90 carry two marks each.

2. Answer all the questions

3. Questions must be answered on the special machine gradable Objective Response Sheet (ORS) by darkening the appropriate bubble (marked A, B, C, D) using HB pencil against the question number on the left hand side of the ORS. Each question has only one correct answer. In case you wish to change an answer, erase the old answer completely using a good soft eraser.

4. There will be NEGATIVE marking. For each wrong answer 0.25 mark from Q. 1-30 and 0.5 mark from Q. 31-90 will be deducted. More than one answer marked against a question will be deemed as an incorrect response and will be negatively marked.

Q. 1 - 30 CARRY ONE MARK EACH

1. The goal of structured programming is to

a) have well indented programs
b) be able to infer the flow of control from the compiled code
c) be able to infer the flow of control from the program text
d) avoid the use of GOTO statements

2. Consider the following C function

```
void swap (int a, int b)
{ int temp ;
temp = a ;
a = b ;
b = temp ;
}
```

In order to exchange the values of two variables x and y.

a) call swap (x, y) b) call swap (&x, &y)
c) swap (x, y) cannot be used as it does not return any value
d) swap (x, y) cannot be used as the parameters are passed by value

3. A single array A [1..MAXSIZE] is used to implement two stacks. The two stacks grow from opposite ends of the array. Variables top 1 and top 2 (top 1< top 2) point to the location of the topmost element in each of the stacks. If the space is to be used efficiently, the condition for "stack full" is

a) (top1 = MAXSIZE/2) and (top2 = MAXSIZE/2+1)
b) top1 + top2 = MAXSIZE
c) (top1 = MAXSIZE/2) or (top2 = MAXSIZE)
d) top1 = top2 -1

4. The following numbers are inserted into an empty binary search tree in the given order : 10, 1, 3, 5, 15, 12, 16. What is the height of the binary search tree (the height is the maximum distance of a leaf node from the root) ?

a) 2 b) 3 c) 4 d) 6

5. The best data structure to check whether an arithmetic expression has balanced parentheses is a

a) queue b) stack c) tree d) list

6. Level order traversal of a rooted tree can be done by starting from the root and performing

a) preorder traversal b) inorder traversal
c) depth first search d) breadth first search

7. Given the following input (4322, 1334, 1471, 9679, 1989, 6171, 6173, 4199) and the hash function x mod 10, which of the following statements are true ?

i) 9679, 1989, 4199 hash to the same value
ii) 1471, 6171 hash to the same value
iii) All elements hash to the same value
iv) Each element hashes to a different value

a) i) only b) ii) only
c) i) and ii) only d) iii) or iv) only

8. Which of the following grammar rules violate the requirements of an operator grammar ?

P, Q, R are nonterminals, and r, s, t are terminals.

i) $P \to QR$ ii) $P \to QsR$
iii) $P \to \varepsilon$ iv) $P \to QtRr$

a) i) only b) i) and iii) only
c) ii) and iii) only d) iii) and iv) only

9. Consider a program P that consists of two source modules M_1 and M_2 contained in two different files. If M_1 contains a reference to a function defined in M_2 the reference will be resolved at

a) Edit-time b) Complie-time
c) Link-time d) Load-time

10. Consider the grammar rule $E \to E_1 - E_2$ for arithmetic expressions. The code generated is targeted to a CPU having a single user register. The subtraction operation requires the first operand to be in the register. If E_1 and E_2 do not have any common subexpression, in order to get the shortest possible code

a) E_1 should be evaluated first
b) E_2 should be evaluated first
c) Evaluation of E_1 and E_2 should necessarily be interleaved
d) Order of evaluation of E_1 and E_2 is of no consequence

11. Consider the following statements with respect to user-level threads and kernel-supported threads

i) Context switch is faster with kernel-supported threads
ii) For user-level threads, a system call can block the entire process

iii) Kernel-supported threads can be scheduled independently

iv) User-level threads are transparent to the kernal

Which of the above statements are true ?

a) ii), iii) and iv) only b) ii) and iii) only

c) i), and iii) only d) i) and ii) only

12. Consider an operating system capable of loading and executing a single sequential user process at a time. The disk head scheduling algorithm used is First Come First Served (FCFS). If FCFS is replaced by Shortest Seek Time First (SSTF), claimed by the vendor to give 50% better benchmark results, what is the expected improvement in the I/O performance of user programs ?

a) 50% b) 40% c) 25% d) 0%

13. Let R_1(A, B, C) and R_2 (D, E) be two relation schemas, where the primary keys are shown underlined, and let C be a foregin key in R_1 referring to R_2. Suppose there is no violation of the above referential integrity constraint in the corresponding relation instances r_1 and r_2. Which one of the following relational algebra expressions would necessarily produce an empty relation?

a) $\Pi_D(r_2) - \Pi_C(r_1)$ b) $\Pi_C(r_1) - \Pi_D(r_2)$

c) $\Pi_D(r_1 \bowtie_{C \neq D} r_2)$ d) $\Pi_C(r_1 \bowtie_{C \neq D} r_2)$

14. Consider the following relation schema pertaining to a students database:

Student (<u>rollno</u>, name, address)

Enroll (<u>rollno, courseno,</u> coursename)

where the primary keys are shown underlined. The number of tuples in the Student and Enroll tables are 120 and 8 respectively. What are the maximum and minimum number of the tuples that can be present in (Student * Enroll), where '*' denotes natural join ?

a) 8, 8 b) 120, 8 c) 960, 8 d) 960, 120

15. Choose the best matching between Group 1 and Group 2.

Group - 1	Group -2
P. Data link layer	1. Ensures reliable transport of data over a physical point-to-point link
Q. Network layer	2. Encodes/decodes data for physical transmission
R. Transport layer	3. Allows end-to-end communication between two processes
	4. Routes data from one network node to the next

a) P-1, Q-4, R-3 b) P-2, Q-4, R-1

c) P-2, Q-3, R-1 d) P-1, Q-3, R-2

16. Which of the following is NOT true with respect to a transparent bridge and a router ?

a) Both bridge and router selectively forward data packets

b) A bridge uses IP addresses while a router uses MAC addresses

c) A bridge builds up its routing table by inspecting incoming packets

d) A router can connect between a LAN and a WAN

17. The Boolean function $x'y' + xy + x'y$ is equivalent to

a) $x' + y'$ b) x+y c) $x + y'$ d) $x' + y$

18. In an SR latch made by cross-coupling two NAND gates, if both S and R inputs are set to 0, then it will result in

a) Q = 0, Q' = 1 b) Q = 1, Q' = 0

c) Q =1, Q' = 1 d) Indeterminate states

19. If 73_x (in base-x number system) is equal to 54_y (in base-y number system), the possible values of x and y are

a) 8, 16 b) 10, 12 c) 9, 13 d) 8, 11

20. Which of the following addressing modes are suitable for program relocation at run time ?

i) Absolute addressing ii) Based addressing

iii) Relative addressing iv) Indirect addressing

a) i) and iv) b) i) and ii)

c) ii) and iii) d) i), ii) and iv)

21. The minimum number of page frames that must be allocated to a running process in a virtual memory environment is determined by

a) the instruction set architecture b) page size

c) physical memory size

d) number of processes in memory

22. How many 8-bit characters can be transmitted per second over a 9600 baud serial communication link using asynchronous mode of transmission with one start bit, eight data bits, two stop bits, and one parity bit ?

a) 600 b) 800 c) 876 d) 1200

23. Identify the correct translation into logical notation of the following assertion.

Some boys in the class are taller than all the girls

Note : taller (x,y) is true if x is taller than y.

a) $(\exists x)(boy(x) \rightarrow (\forall y)(girl(y) \wedge taller(x,y)))$

b) $(\exists x)(boy(x) \wedge (\forall y)(girl(y) \wedge taller(x,y)))$

c) $(\exists x)(boy(x) \rightarrow (\forall y)(girl(y) \rightarrow taller(x,y)))$

d) $(\exists x)(boy(x) \wedge (\forall y)(girl(y) \wedge taller(x,y)))$

24. Consider the binary relation :

$S = \{(x,y) \mid y = x+1 \text{ and } x, y \in \{0,1,2,........\}\}$

The reflexive transitive closure of S is

a) $\{(x,y) \mid y > x \text{ and } x, y \in \{0,1,2,.......\}\}$

b) $\{(x,y) \mid y \geq x \text{ and } x, y \in \{0,1,2,......\}\}$

c) $\{(x,y) \mid y < x \text{ and } x, y \in \{0,1,2,.......\}\}$

d) $\{(x,y) \mid y \leq x \text{ and } x, y \in \{0,1,2,.......\}\}$

25. If a fair coin is tossed four times. What is the probability that two heads and two tails will result ?

a) 3/8 b) 1/2 c) 5/8 d) 3/4

26. The number of different n × n symmetric matrices with each element being either 0 or 1 is : (Note power (2, x) is same as 2^x)

a) power (2, n) b) power (2, n^2)

c) power (2, (n^2+n)/2) d) power(2, (n^2-n)/2)

27. Let A, B, C, D be n × n matrices, each with non-zero determinant. If ABCD = I then B^{-1} is

a) $D^{-1}C^{-1}A^{-1}$ b) CDA

c) ADC d) Does not necessarily exist

28. What is the result of evaluating the following two expressions using three-digit floating point arithmetic with rounding ?

(113. + -111.) + 7.51

113. + (-111. + 7.51)

a) 9.51 and 10.0 respectively

b) 10.0 and 9.51 respectively

c) 9.51 and 9.51 respectively

d) 10.0 and 10.0 respectively

29. The tightest lower bound on the number of comparisons, in the worst case, for comparison-based sorting is of the order of

a) n b) n^2 c) nlogn d) $n\log^2 n$

30. The problems 3-SAT and 2-SAT are

a) both in P b) both NP-complete

c) NP-complete and in P respectively

d) undecidable and NP-complete respectively

Q. 31 - 90 CARRY TWO MARKS EACH

31. Consider the following C function :

```
int f (int n)
{ static int i = 1 ;
    if (n>=5) return n ;
    n = n+i ;
    i ++;
    return f (n) ; }
```

The value returned by f(1) is

a) 5 b) 6 c) 7 d) 8

32. Consider the following program fragment for reversing the digits in a given integer to obtain a new integer. Let $n=d_1 d_2 \ldots d_m$

```
int n, rev ;
rev = 0 ;
while (n>0) {
    rev = rev * 10 + n % 10 ;
    n = n / 10 ;
}
```

The loop invariant condition at the end of the ith iteration is :

a) $n=d_1 d_2 \ldots\ldots d_{m-1}$ and $rev=d_m d_{m-1} \ldots\ldots d_{m-i+1}$

b) $n=d_{m-i+1} \ldots d_{m-1} d_m$ or $rev=d_{m-i} \ldots d_2 d_1$

c) $n \neq rev$

d) $n=d_1 d_2 \ldots d_m$ or $rev = d_m \ldots d_2 d_1$

33. Consider the following C program segment :

```
char p [20] ;
char * s = "string" ;
int length = strlen (s) ;
for (i = 0 ; i < length ; i ++)
    p [ i ] = s [length - i] ;
printf ("%s", p) ;
```

The output of the program is

a) gnirts b) string

c) gnirt d) no output is printed

34. It is desired to design an object-oriented employee record system for a company. Each employee has a name, unique id and salary. Employees belong to different categories and their salary is determined by their category. The functions get Name, getId and compute Salary are required. Given the class hierarchy below, possible locations for these functions are :

i) getId is implemented in the supercalss

ii) getId is implemented in the subclass

iii) getName is an abstract function in the superclass

iv) getName is implemented in the superclass

v) getName is implemented in the subclass

vi) getSalary is an abstract function in the superclass

vii) getSalary is implemented in the superclass

viii) getSalary is implemented in the subclass

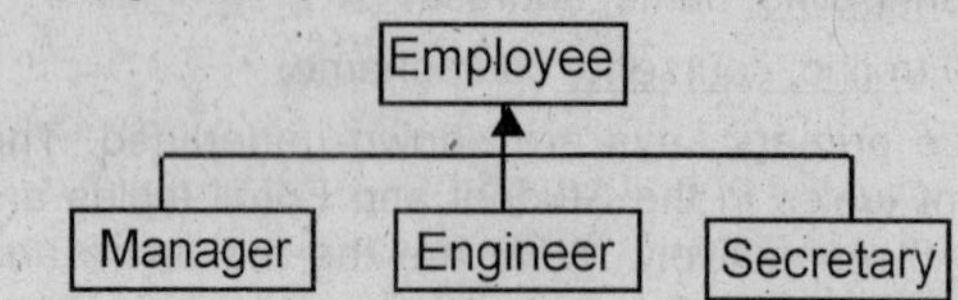

Choose the best design

a) i), iv), vi), viii) b) i), iv), vii)

c) i), iii), v), vi), viii) d) ii), v), viii)

35. Consider the label sequences obtained by the following pairs of traversals on a labeled binary tree. Which of these pairs identify a tree uniquely ?

i) preorder and postorder

ii) inorder and postorder

iii) preorder and inorder

iv) level order and postorder

a) i) only b) ii), iii) c) iii) only d) iv) only

36. A circularly linked list is used to represent a Queue. A single variable p is used to access the Queue. To which node should p point such that both the operations enQueue and deQueue can be performed in constant time ?

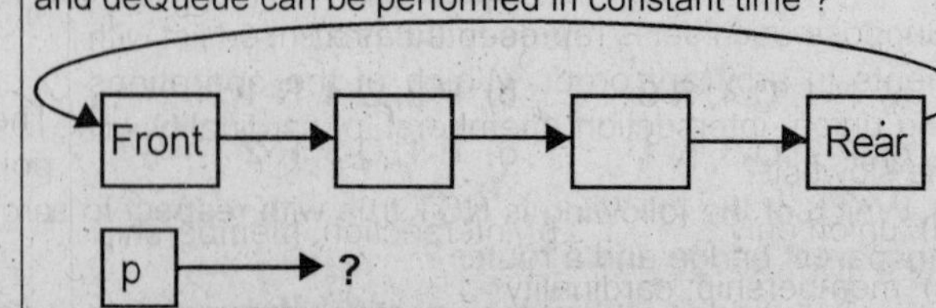

a) rear node b) front node

c) not possible with a single pointer

d) node next to front

37. The elements 32, 15, 20, 30, 12, 25, 16 are inserted one by one in the given order into a maxHeap. The resultant maxHeap is

a)

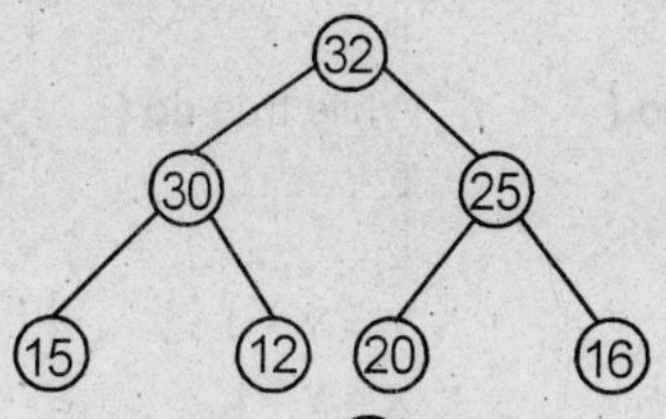

b)

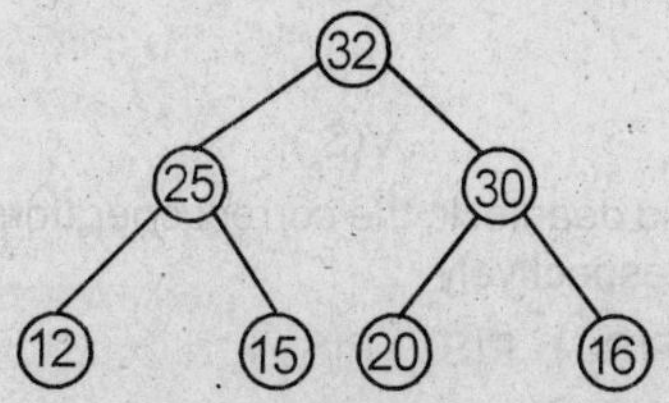

c)

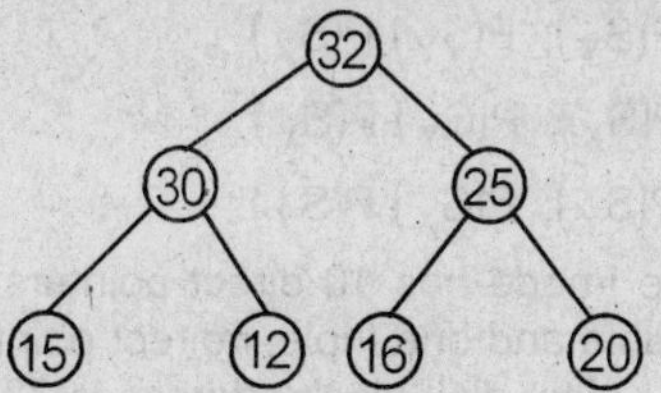

d)

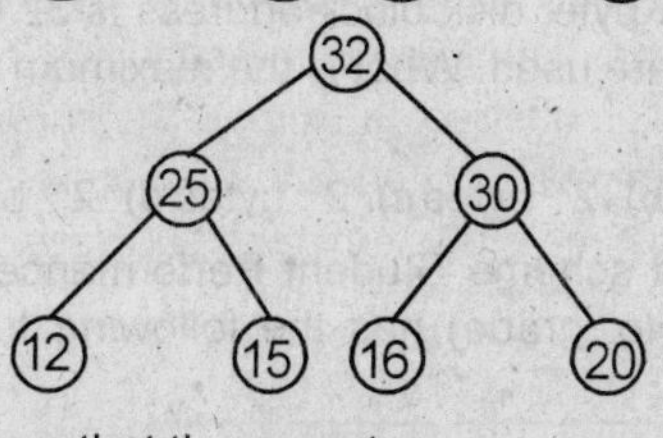

38. Assume that the operators +, -, × are left associative and $\wedge$ is right associative. The order of precedence (from highest to lowest) is $\wedge$, ×, +, -. The postfix expression corresponding to the infix expression $a+b\times c-d\hat{}e\hat{}f$ is

a) abc×+def^^− b) abc×+de^f^−

c) $ab+c\times d-e\hat{}f\hat{}$ d) $-+a\times bc\hat{}\hat{}def$

39. Two matrices M_1 and M_2 are to be stored in arrays A and B respectively. Each array can be stored either in row-major or column-major order in contiguous memory locations. The time complexity of an algorithm to compute $M_1 \times M_2$ will be

a) best if A is in row-major, and B is in column-major order

b) best if both are in row-major order

c) best if both are in column-major order

d) independent of the storage scheme

40. Suppose each set is represented as a linked list with elements in arbitrary order. Which of the operations among union, intersection, membership, cardinality will be the slowest ?

a) union only b) intersection, membership

c) membership, cardinality

d) union, intersection

41. Consider the following C program

```
main ( )
{ int x, y, m, n ;
   scanf ("%d %d", &x, &y) ;
   / * Assume x>0 and y>0 * /
   m = x ; n = y ;
   while (m!=n)
    { if (m > n)
        m = m - n ;
      else
        n = n - m ;
    }
   printf("%d", n);
   }
```

The program computes

a) x $\div$ y using repeated subtraction

b) x mod y using repeated subtraction

c) the greatest common divisor of x and y

d) the least common multiple of x and y

42. What does the following algorithm approximate ? (Assume m > 1, $\in$>0).

```
x = m ;
y = 1 ;
while (x - y > ∈)
  { x = (x + y) / 2 ;
    y = m / x ;
  }
print (x) ;
```

a) log m b) m^2 c) $m^{1/2}$ d) $m^{1/3}$

43. Consider the following C program segment

```
struct CellNode {
  struct CellNode *leftChild ;
  int element ;
  struct CellNode *rightChild ;
  } ;
int DoSomething (struct CellNode *ptr)
{
  int value = 0 ;
  if (ptr ! = NULL)
   { if (prt -> leftChild != NULL)
       value = 1 + DoSomething (ptr-> leftChild) ;
     if (ptr-> rightChild ! = NULL)
Value = max (value, 1 + DoSomething (ptr - > rightChild));
   }
   return (value) ;
}
```

The value returned by the function DoSomething when a pointer to the root of a non-empty tree is passed as argument is

a) The number of leaf nodes in the tree

b) The number of nodes in the tree

c) The number of internal nodes in the tree

d) The height of the tree

44. Suppose we run Dijkstra's single source shortest-path algorithm on the following edge-weighted directed graph with vertex P as the source.

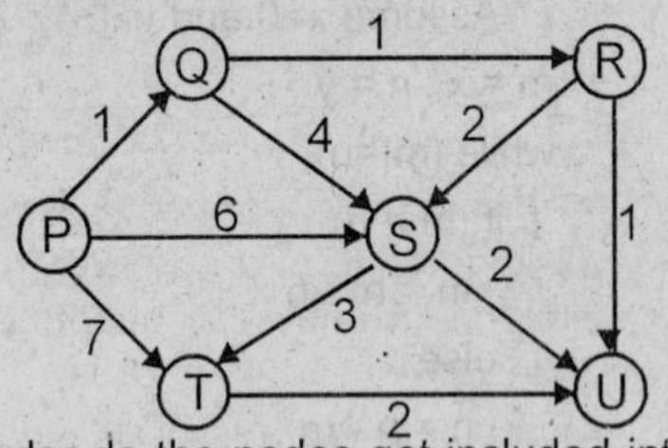

In what order do the nodes get included into the set of vertices for which the shortest path distances are finalized?

a) P, Q, R, S, T, U b) P, Q, R, U, S, T

c) P, Q, R, U, T, S d) P, Q, T, R, U, S

45. Consider the grammar with the following translation rules and E as the start symbol.

$E \rightarrow E_1 \# T$	{E. value = E_1. value * T. Value}
| T	{E.value = T. value}
$T \rightarrow T_1 \& F$	{T.value = T_1.value + F.value}
| F	{T.value = F.value}
$F \rightarrow num$	{F.value = num. value}

Compute E. Value for the root of the parse tree for the expression : 2 # 3 & 5 # 6 & 4.

a) 200 b) 180 c) 160 d) 40

46. Consider the following set of processes, with the arrival times and the CPU burst times given in milliseconds.

Process	Arrival Time	Burst Time
P1	0	5
P2	1	3
P3	2	3
P4	4	1

What is the average turnaround time for these processes with the preemptive shortest remaining processing time first (SRPT) algorithm?

a) 5.50 b) 5.75 c) 6.00 d) 6.25

47. Consider a system with a two-level paging scheme in which a regular memory access takes 150 nanoseconds, and servicing a page fault takes 8 milliseconds. An average instruction takes 100 nanoseconds of CPU time, and two memory accesses. The TLB hit ratio is 90%, and the page fault rate is one in every 10,000 instructions. What is the effective average instruction execution time?

a) 645 nanoseconds b) 1050 nanoseconds

c) 1215 nanoseconds d) 1230 nanoseconds

48. Consider two processes P_1 and P_2 accessing the shared variables X and Y protected by two binary semaphores S_x and S_y respectively, both initialized to 1. P and V denote the usual semaphore operators, where P decrements the semaphore value, and V increments the semaphore value. The pseudo-code of P_1 and P_2 is as follows :

P_1 :	P_2 :
while true do {	While true do {
L_1 :	L_3 : ....
L_2 :	L_4 : ....
X = X + 1 ;	Y = Y + 1 ;
Y = Y - 1 ;	X = Y - 1;
V(S_x) ;	V(S_y);
V(S_y); }	V(S_x); }

In order to avoid deadlock, the correct operators at L_1, L_2, L_3 and L_4 are respectively

a) $P(S_Y), P(S_X); P(S_X), P(S_Y)$

b) $P(S_X), P(S_Y); P(S_Y), P(S_X)$

c) $P(S_X), P(S_X); P(S_Y), P(S_Y)$

d) $P(S_X), P(S_Y); P(S_X), P(S_Y)$

49. A Unix-style i-node has 10 direct pointers and one single, one double and one triple indirect pointers. Disk block size is 1 Kbyte, disk block address is 32 bits, and 48-bit integers are used. What is the maximum possible file size ?

a) 2^{24} bytes b) 2^{32} bytes c) 2^{34} bytes d) 2^{48} bytes

50. The relation scheme Student Performance (name, courseNo, rollNo, grade) has the following functional dependencies:

name, courseNo $\rightarrow$ grade

rollNo, courseNo $\rightarrow$ grade

name $\rightarrow$ rollNo

rollNo $\rightarrow$ name

The highest normal form of this relation scheme is

a) 2 NF b) 3 NF c) BCNF d) 4NF

51. Consider the relation Student (name, sex, marks) where the primary key is shown underlined, pertaining to students in a class that has at least one boy and one girl. What does the following relation algebra expression produce ? (Note : r is the rename operator).

$$\Pi_{name}(\sigma_{sex=female}(Student)) - \Pi_{name}(Student \bowtie_{(sex=female \wedge x=male \wedge marks \leq m)} \rho_{n,x,m}(Student))$$

a) names of girl students with the highest marks

b) names of girl students with more marks than some boy student

c) names of girl students with marks not less than some boy student

d) names of girl students with more marks than all the boy students

52. The order of an internal node in a B^+ tree index is the maximum number of children it can have. Suppose that a child pointer takes 6 bytes, the search field value takes 14 bytes, and the block size is 512 bytes. What is the order of the internal node ?

a) 24 b) 25 c) 26 d) 27

53. The employee information in the company is stored in the relation

Employee (<u>name</u>, sex, salary, deptName)

Consider the following SQL query

```
select deptName
    from Employee
   where sex = 'M'
group by deptName
having avg (salary) >
     (select avg (salary) from Employee)
```

It returns the names of the department in which

a) the average salary is more than the average salary in the company.

b) the average salary of male employees is more than the average salary of all male employees in the company.

c) the average salary of male employees is more than the average salary of employees in the same department

d) the average salary of male employees is more than the average salary in the company

54. A and B are the only two stations on an Ethernet. Each has a steady queue of frames to send. Both A and B attempt to transmit a frame, collide, and A wins the first backoff race. At the end of this successful transmission by A, both A and B attempt to transmit and collide. The probability that A wins the second backoff race is

a) 0.5 b) 0.625 c) 0.75 d) 1.0

55. The routing table of a router is shown below :

Destination	Subnet Mask	Interface
128.75.43.0	255.255.255.0	Eth0
128.75.43.0	255.255.255.128	Eth1
192.12.17.5	255.255.255.255.	Eth3
default		Eth2

On which interfaces will the router forward packets addressed to destinations 128.75.43.16 and 192.12.17.10 respectively?

a) Eth1 and Eth2 b) Eth0 and Eth2

c) Eth0 and Eth3 d) Eth1 and Eth3

The Following Information Pertains to Q. 56-57

Consider three IP networks A, B and C. Host H_A in network A sends messages each containing 180 bytes of application data to a host H_c in network C. The TCP layer prefixes a 20 byte header to the message. This passes through an intermediate network B. The maximum packet size, including 20 byte IP header, in each network is

A : 1000 bytes ; B : 100 bytes ; C : 1000 bytes

The network A and B are connected through a 1 Mbps link, while B and C are connected by a 512 Kbps link (bps=bits per second).

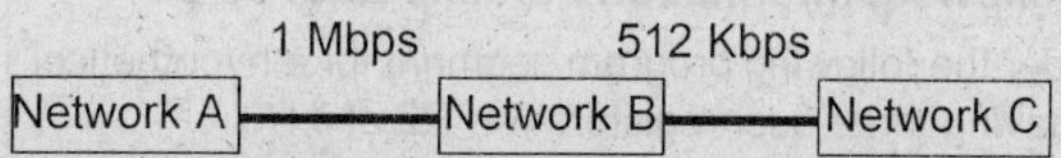

56. Assuming that the packets are correctly delivered, how many bytes, including headers, are delivered to the IP layer at the destination for one application message, in the best case ? Consider only data packets.

a) 200 b) 220 c) 240 d) 260

57. What is the rate at which application data is transferred to host H_c ? Ignore errors, acknowledgements, and other overheads.

a) 325.5 Kbps b) 354.5 Kbps

c) 409.6 Kbps d) 512.0 Kbps

58. A circuit outputs a digit in the form of 4 bits. 0 is represented by 0000, 1 by 0001, ..., 9 by 1001. A combinational circuit is to be designed which takes these 4 bits as input and outputs 1 if the digit ≥ 5, and 0 otherwise. If only AND, OR and NOT gates may be used, what is the minimum number of gates required ?

a) 2 b) 3 c) 4 d) 5

59. Which are the essential prime implicants of the following Boolean function ?

$$f(a,b,c) = a'c + ac' + b'c$$

a) $a'c$ and ac' b) $a'c$ and $b'c$

c) $a'c$ only d) ac' and bc'

60. Consider a multiplexer with X and Y as data inputs and Z as control input. Z=0 selects input X, and Z=1 selects input Y. What are the connections required to realize the 2-variable Boolean function f=T+R, without using any additional hardware ?

a) R to X, 1 to Y, T to Z b) T to X, R to Y, T to Z

c) T to X, R to Y, 0 to Z d) R to X, 0 to Y, T to Z

61. Consider the partial implementation of a 2-bit counter using T flip-flops following the sequence 0-2-3-1-0, as shown below.

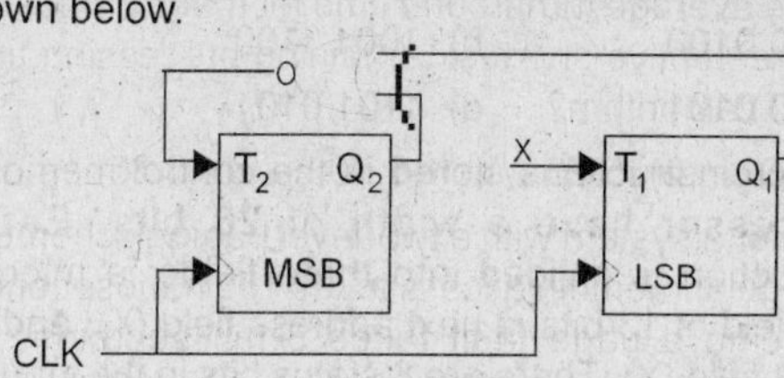

To complete the circuit, the input X should be

a) Q_2' b) $Q_2 + Q_1$

c) $(Q_1 \oplus Q_2)'$ d) $Q_1 \oplus Q_2$

62. A 4-bit carry lookahead adder, which adds two 4-bit numbers, is designed using AND, OR, NOT, NAND, NOR gates only. Assuming that all the inputs are available in both complemented and uncomplemented forms and the delay of each gate is one time unit, what is the overall propagation delay of the adder ? Assume that the carry network has been implemented using two-level AND-OR logic.

a) 4 time units b) 6 time units
c) 10 time units d) 12 time units

The Following Information Pertains to Q. 63-64

Consider the following program segment for a hypothetical CPU having three user registers R1, R2 and R3.

Instruction	Operation	Instruction Size (in words)
MOV R1, 5000	; R1 ← Memory [5000]	2
MOV R2, (R1)	; R2 ← Memory [(R1)]	1
ADD R2, R3	; R2 ← R2+R3	1
MOV 6000, R2	; Memeory [6000] ← R2	2
HALT	; Machine halts	1

63. Consider that the memory is byte addressable with size 32 bits, and the program has been loaded starting from memory location 1000 (decimal). If an interrupt occurs while the CPU has been halted after executing the HALT instruction, the return address (in decimal) saved in the stack will be

a) 1007 b) 1020 c) 1024 d) 1028

64. Let the clock cycles required for various operations be as follows :

Register to/ from memory transfer : 3 clock cycles

ADD with both operands in register : 1 clock cycle

Instruction fetch and decode : 2 clock cycles per word

The total number of clock cycles required to execute the program is

a) 29 b) 24 c) 23 d) 20

65. Consider a small two-way-set-associative cache memory, consisting of four blocks. For choosing the block to be replaced, use the least recently used (LRU) scheme. The number of cache misses for the following sequence of block addresses is 8, 12, 0, 12, 8

a) 2 b) 3 c) 4 d) 5

66. Let A = 1111 1010 and B = 0000 1010 be two 8-bit 2's complement numbers. Their product in 2's complement is

a) 1100 0100 b) 1001 1100
c) 1010 0101 d) 1101 0101

67. The microinstructions stored in the control memory of a processor have a width of 26 bits. Each microinstruction is divided into three fields: a micro-operation field of 13 bits, a next address field (X), and a MUX select field (Y). There are 8 status bits in the inputs of the MUX.

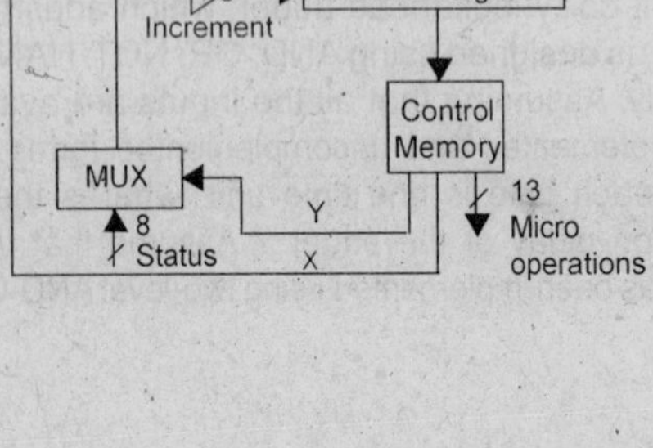

How many bits are there in the X and Y fields, and what is the size of the control memory in number of words ?

a) 10, 3, 1024 b) 8, 5, 256
c) 5, 8, 2048 d) 10, 3, 512

68. A hard disk with a transfer rate of 10 Mbytes/second is constantly transferring data to memory using DMA. The processor runs at 600 MHz, and takes 300 and 900 clock cycles to initiate and complete DMA transfer respectively. If the size of the transfer is 20 Kbytes, what is the precentage of processor time consumed for the transfer operation ?

a) 5.0% b) 1.0% c) 0.5% d) 0.1%

69. A 4-stage pipeline has the stage delays as 150, 120, 160 and 140 nanoseconds respectively. Registers that are used between the stages have a delay of 5 nanoseconds each. Assuming constant clocking rate, the total time taken to process 1000 data items on this pipeline will be

a) 120.4 microseconds b) 160.5 microseconds
c) 165.5 microseconds d) 590.0 microseconds

70. The following propositional statement is

$(P \rightarrow (Q \vee R)) \rightarrow ((P \wedge Q) \rightarrow R)$

a) satisfiable but not valid b) valid
c) a contradiction d) none of the above

71. How many solutions does the following system of linear equations have ?

$-x + 5y = -1$

$x - y = 2$

$x + 3y = 3$

a) infinitely many b) two distinct solutions
c) unique d) none

72. The following is the incomplete operation table of a 4-element group.

The last row of the table is

*	e	a	b	c
e	e	a	b	c
a	a	b	c	e
b				
c				

a) c a e b b) c b a e c) c b e a d) c e a b

73. The inclusion of which of the following sets into

$S = \{\{1,2\}, \{1,2,3\}, \{1,3,5\}, \{1,2,4\}, \{1,2,3,4,5\}\}$

is necessary and sufficient to make S a complete lattice under the partial order defined by set containment ?

a) $\{1\}$ b) $\{1\}, \{2,3\}$
c) $\{1\}, \{1,3\}$ d) $\{1\}, \{1,3\}, \{1,2,3,4\}, \{1,2,3,5\}$

74. An examination paper has 150 multiple-choice questions of one mark each, with each question having four choices. Each incorrect answer fetches -0.25 mark. Suppose 1000 students choose all their answers randomly with uniform probablity. The sum total of the expected marks obtained by all these students is

a) 0 b) 2550 c) 7525 d) 9375

75. Mala has a colouring book in which each English letter is drawn two times. She wants to paint each of these 52 prints with one of k colours, such that the colour-pairs used to colour any two letters are different. Both prints of a letter can also be coloured with the same colour. What is the minimum value of k that satisfies this requirement?

a) 9 b) 8 c) 7 d) 6

76. In an $M \times N$ matrix such that all non-zero entries are covered in a rows and b columns. Then the maximum number of non-zero entries, such that no two are on the same row or column, is

a) $\le a+b$ b) $\le \max\{a,b\}$

c) $\le \min\{M-a, N-b\}$ d) $\le \min\{a,b\}$

77. The minium number of colours required to colour the following graph, such that no two adjacent vertices are assigned the same colour, is

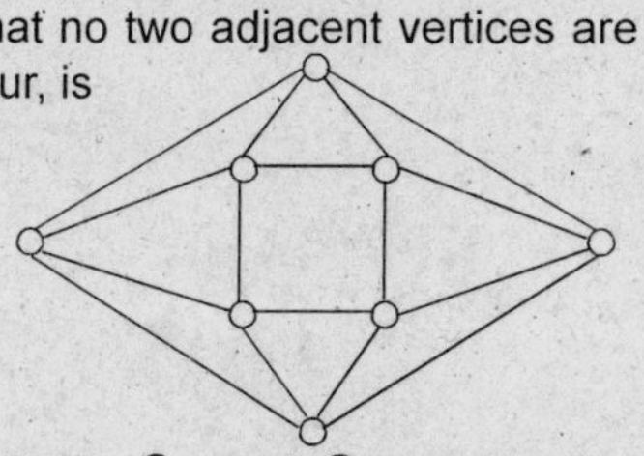

a) 2

b) 3

c) 4

d) 5

78. Two n bit binary strings, S_1 and S_2 are chosen randomly with uniform probability. The probability that the Hamming distance between these strings (the number of bit positions where the two strings differ) is equal to d is

a) ${}^{n}C_d/2^n$ b) ${}^{n}C_d/2^d$

c) $d/2^n$ d) $1/2^d$

79. How many graphs on n labeled vertices exist which have at least $(n^2-3n)/2$ edges ?

a) ${}^{(n^2-n)/2}C_{(n^2-3n)/2}$ b) $\sum_{k=0}^{(n^2-3n)/2} {}^{(n^2-n)}C_k$

c) ${}^{(n^2-n)/2}C_n$ d) $\sum_{k=0}^{n} {}^{(n^2-n)/2}C_k$

80. A point is randomly selected with uniform probability in the X-Y plane within the rectangle with corners at $(0,0)$, $(1,0)$, $(1,2)$ and $(0,2)$. If p is the length of the position vector of the point, the expected value of p^2 is

a) 2/3 b) 1 c) 4/3 d) 5/3

81. Let $G_1 = (V, E_1)$ and $G_2 = (V, E_2)$ be connected graphs on the same vertex set V with more than two vertices. If $G_1 \cap G_2 = (V, E_1 \cap E_2)$ is not a connected graph, then the graph $G_1 \cup G_2 = (V, E_1 \cup E_2)$

a) cannot have a cut vertex

b) must have a cycle

c) must have a cut-edge (bridge)

d) has chromatic number strictly greater than those of G_1 and G_2

82. Let A [1,....... n] be an array storing a bit (1 or 0) at each location, and $f(m)$ is a function whose time complexity is $\theta(m)$. Consider the following program fragment written in a C like language:

```
counter = 0 ;
for (i=1; i<=n ; i++)
{ if (A [i] ==1) counter ++
  else {f (counter) ; counter =0 ; }
}
```

The complexity of this program fragment is

a) $\Omega(n^2)$ b) $\Omega(n \log n)$ and $O(n^2)$

c) $\theta(n)$ d) $o(n)$

83. The time complexity of the following C function is (assume n>0)

```
int recursive (int n) {
  if (n == 1)
    return (1) ;
  else
    return (recursive (n-1) + recursive (n-1)) ;
}
```

a) $O(n)$ b) $O(n \log n)$ c) $O(n^2)$ d) $O(2^n)$

84. The recurrence equation

$T(1) = 1$

$T(n) = 2T(n-1) + n, n \ge 2$

evaluates to

a) $2^{n+1} - n - 2$ b) $2^n - n$

c) $2^{n+1} - 2n - 2$ d) $2^n + n$

85. A program takes as input a balanced binary search tree with n leaf nodes and computes the value of a function g(x) for each node x. If the cost of computing g(x) is min{number of leaf-nodes in left-subtree of x, number of leaf-nodes in right-subtree of x} then the worst-case time complexity of the program is

a) $\Theta(n)$ b) $\Theta(n \log n)$ c) $\Theta(n^2)$ d) $\Theta(n^2 \log n)$

86. The following finite state machine accepts all those binary strings in which the number of 1's and 0's are respectively

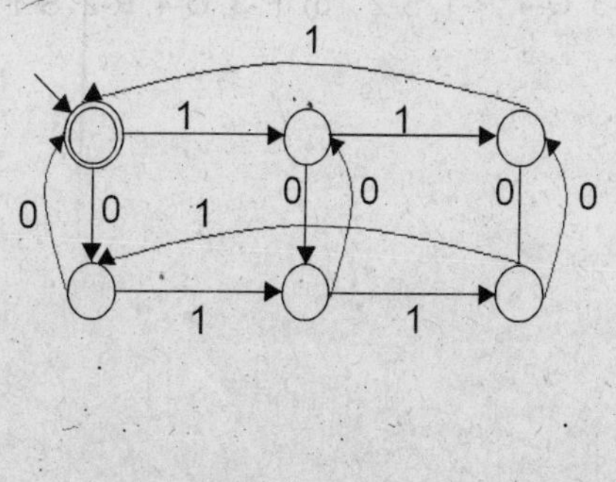

a) divisible by 3 and 2 b) odd and even
c) even and odd d) divisible by 2 and 3

87. The language $\{a^m b^n c^{m+n} \mid m,n \geq 1\}$ is
a) regular b) context free but not regular
c) context sensitive but not context free
d) type-0 but not context sensitive

88. Consider the following grammar G :

$$S \rightarrow bS \mid aA \mid b$$
$$A \rightarrow bA \mid aB$$
$$B \rightarrow bB \mid aS \mid a$$

Let $N_a(w)$ and $N_b(w)$ denote the number of a's and b's in a string w respectively. The language $L(G) \subseteq \{a,b\}^+$ generated by G is
a) $\{w \mid N_a(w) > 3N_b(w)\}$
b) $\{w \mid N_b(w) > 3N_a(w)\}$
c) $\{w \mid N_a(w) = 3k, k \in \{0,1,2,........\}\}$
d) $\{w \mid N_b(w) = 3k, k \in \{0,1,2,........\}\}$

89. L_1 is a recursively enumerable language over Σ. An algorithm A effectively enumerates its words as $w_1, w_2, w_3.........$ Define another language L_2 over $\Sigma \cup \{\#\}$ as $\{w_i \# w_j : w_i, w_j \in L_1,\ i < j\}$. Here # is a new symbol. Consider the following assertions.

S_1 : L_1 is recursive implies L_2 is recursive

S_2 : L_2 is recursive implies L_1 is recursive

Which of the following statements is true ?
a) Both S_1 and S_2 are true
b) S_1 is true but S_2 is not necessarily true
c) S_2 is true but S_1 is not necessarily true
d) Neither is necessarily true

90. Choose the best matching between the programming styles in Group 1 and their characteristics in Group 2.

Group - 1	Group - 2
P. Functional	1. Command-based, procedural
Q. Logic	2. Imperative, abstract data types
R. Object-oriented	3. Side-effect free, declarative, expression evaluation
S. Imperative	4. Declarative, Clausal representation, theorem proving

a) P-2, Q-3, R-4, S-1 b) P-4, Q-3, R-2, S-1
c) P-3. Q-4, R-1, S-2 d) P-3, Q-4, R-2, S-1

ANSWER KEY

1. c	2. d	3. d	4. b	5. b
6. d	7. c	8. a	9. c	10. d
11. a	12. d	13. b	14. c	15.a
16. b	17. d	18. d	19. d	20. c
21. d	22. b	23. d	24. b	25. a
26. d	27. b	28. a	29. b	30. c
31. c	32. a	33. a	34. a	35. b
36. c	37. a	38. a	39. a	40. d
41. c	42. c	43. d	44. b	45. c
46. a	47. -	48. b	49. b	50. a
51. d	52. c	53. d	54. a	55. c
56. c	57. b	58. c	59. a	60. a
61. d	62. a	63. a	64. b	65. b
66. a	67. a	68. d	69. b	70. a
71. c	72. d	73. a	74. d	75. a
76. a	77. c	78. a	79. b	80. d
81. c	82. c	83. d	84. a	85. a
86. a	87. b	88. c	89. b	90. d

GATE - 2003
COMPUTER SCIENCE & ENGINEERING

Duration : Three Hours **Maximum Marks : 150**

Reading the following instructions carefully

1. This question paper contains 90 objective quetstions. Q.1-30 carry one mark each and Q.31-90 carry two marks each.

2. Answer all the questions.

3. Questions must be answered on the special machine gradable Objective Response Sheet (ORS) by darkening the appropriate bubble (marked A, B, C, D) using HB pencil against the question number on the left hand side of the ORS. Each question has only one correct answer. In case you wish to change an answer, erase the old answer completely using a good soft eraser.

4. There will be NEGATIVE marking. For each wrong answer 0.25 mark from Q.1-30 and 0.5 mark from Q.31-90 will be deducted. More than one answer marked against a aquestion will be deemed as an incorrect response and will be negatively marked.

Q. 1 - 30 CARRY ONE MARK EACH

1. Consider the following C function.

```
float f (float x, int y)    {
    float p, s; int i ;
    for (s = 1, p = 1, i = 1; i < y ; i++) {
        P * = x / i;
        S + = p ;
    }
    return s ;
}
```

For large values of y, the return value of the function f best approximates

a) x^y b) e^x c) $\ln(1+x)$ d) X^x

2. Assume the following C variable declaration

int * A [10], B[10] [10];

Of the following expressions

I. A [2] II. A [2] [3] III. B [1] IV. B [2] [3]

Which will not give compile-time errors if used as left hand sides of assignments statements in a C program?

a) I, II and IV only b) II, III and IV only

c) II and IV only d) IV only

3. Let P(E) denote the probability of the event E. Given P(A) = 1, P(B) = ½, the values of P(A/B) and P(B/A) respectively are

a) ¼, ½ b) ½, ¼ c) ½, 1 d) 1, ½

4. Let A be a sequence of 8 distinct integers sorted in ascending order. How many distinct pairs of sequences, B and C are there such that (i) each is sorted in ascending order, (ii) B has 5 and C has 3 elements, and (iii) the result of merging B and C gives A ?

a) 2 b) 30 c) 56 d) 256

5. n couples are invited to a party with the condition that every husband should be accompanied by his wife. However, a wife need not be accompanied by her husband. The number of different gatherings possible at the party is

a) $\binom{2n}{n}*2^n$ b) $\frac{(2n)!}{2^n}$ c) 3^n d) $\binom{2n}{n}$

6. Let T(n) be the number of different binary search trees on n distinct elements. Then

$$T(n) = \sum_{k=1}^{n} T(k-1)T(x), \text{ where x is}$$

a) n - k + 1 b) n - k

c) n - k - 1 d) n - k -2

7. Consider the set Σ^* of all strings over the alphabet $\Sigma = \{0,1\}$. Σ^* with the concatenation operator for strings

a) does not form a group

b) forms a non-commutative group

c) does not have a right identity element

d) forms a group if the empty string is removed from Σ^*

8. Let G be an arbitrary graph with n nodes and k components. If a vertex is removed from G, the number of components in the resultant graph must necessarily lie between

a) k and n b) k - 1 and k + 1

c) k -1 and n - 1 d) k + 1 and n - k

9. Assuming all numbers are in 2's complement representation, which of the following numbers is divisible by 11111011 ?

a) 11100111 b) 11100100

c) 11010111 d) 11011011

10. For a pipelined CPU with a single ALU, consider the following situations.

I. The (J+1)'st instructiion uses the result of the (J)'th instruction as an operand

II. The executiion of a conditional jump instruction

III. The (J)'th and (J + 1)'st instructions require the ALU at the same time.

Which of the above can cause a hazard ?

a) I and II only b) II and III only

c) III only d) All the three

11. Consider an array multiplier for multiplying two n bit numbers. If each GATE in the circuit has a unit delay, the total delay of the multiplier is :

a) $\Theta(1)$ b) $\Theta(\log n)$ c) $\Theta(n)$ d) $\Theta(n^2)$

12. Ram and Shyam have been asked to show that a certain problem Π is NP-complete. Ram shows a polynomial time reduction from the 3-SAT problem to Π, and Shyam shows a polynomial time reduction from Π to 3-SAT. Which of the following can be inferred from these reductions ?

a) Π is NP-hard but not NP-complete

b) Π is NP, but is not NP-complete

c) Π is NP-complete

d) Π is neither NP-hard, nor in NP

13. No body knows yet if P = NP. Consider the language L defined as follows. $L = \begin{cases} (0+1)^* \text{ if } P = NP \\ \phi \text{ otherwise} \end{cases}$

Which of the following statements is true ?

a) L is recursive

b) L is recursively enumerable but not recursive

c) L is not recursively enumerable

d) Whether L is recursive or not will be known after we find out if P = NP

14. The regular expression $0^*(10^*)^*$ denotes the same set as

a) $(1^*0)^*1^*$ b) $0+(0+10)^*$

c) $(0+1)^*10(0+1)^*$ d) none of the above

15. If the strings of a language L can be effectively enumerated in lexicographic (i.e., alphabetic) order, which of the following statements is true ?

a) L is necessarily finite

b) L is regular but not necessarily finite

c) L is context free but not necessarily regular

d) L is recursive but necessarily context free

16. Which of the following suffices to convert an arbitrary CFG to an LL(1) grammar ?

a) Removing left recursion alone

b) Factoring the grammar alone

c) Removing left recursion and factoring the grammar

d) None of the above

17. Assume that the SLR parser for a grammar G has n_1 states and the LALR parser for G has n_2 states. The relationship between n_1 and n_2 is

a) n_1 is necessarily less than n_2

b) n_1 is necessarily equal to n_2

c) n_1 is necessarily greater than n_2

d) none of the above

18. In a bottom-up evaluation of a syntax directed definition, inherited attributes can

a) always be evaluated

b) be evaluated only if the definition is L-attributed

c) be evaluated only if the definition has synthesized attributes

d) never be evaluated

19. Suppose the numbers 7, 5, 1, 8, 3, 6, 0, 9, 4, 2 are inserted in that order into an initially empty binary search tree. The binary search tree uses the usual ordering on natural numbers. What is the in-order traversal sequence of the resultant tree ?

a) 7 5 1 0 3 2 4 6 8 9 b) 0 2 4 3 1 6 5 9 8 7

c) 0 1 2 3 4 5 6 7 8 9 d) 9 8 6 4 2 3 0 1 5 7

20. Consider the following three claims

I. $(n+k)^m = \Theta(n^m)$, where k and m are constants

II. $2^{n+1} = O(2^n)$ III. $2^{2n} = O(2^n)$

Which of these claims are correct ?

a) I and II b) I and III

c) II and III d) I, II and III

21. Consider the following graph

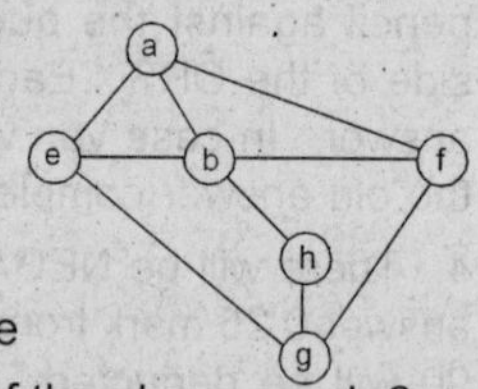

Among the following sequences

I. a b e g h f

II. a b f e h g

III. a b f h g e IV. a f g h b e

Whcih are depth first traversals of the above graph ?

a) I, II and IV only b) I and IV only

c) II, III and IV only d) I, III and IV only

22. The usual $\Theta(n^2)$ implementation of Insertion Sort to sort an array uses linear search to identify the position where an element is to be inserted into the already sorted part of the array. If, instead, we use binary search to identify the position, the worst case running time will

a) remain $\Theta(n^2)$ b) become $\Theta(n(\log n)^2)$

c) beocme $\Theta(n \log n)$ d) become $\Theta(n)$

23. In a heap n elements with the smallest element at the root, the 7th smallest element can be found in time

a) $\Theta(n\log n)$ b) $\Theta(n)$ c) $\Theta(\log n)$ d) $\Theta(1)$

24. Which of the following statements is FALSE ?

a) In statically typed languages, each varible in a program has a fixed type

b) In Un-typed languages, values do not have any types

c) In dynamically typed languages, varibles have no types

d) In all statically typed languages, each varible in a program is associated with values of only a single type during the execution of the program

25. Using a larger block size in a fixed block size file system leads to

a) better disk throughput but poorer disk space utilization

b) better disk throughput and better disk space utilization

c) poorer disk throughput but better disk space utilization

d) poorer disk throughput and poorer disk space utilization

26. In a system with 32 bit virtual addresses and 1 KB page size, use of one-level page tables for virtual to physical address translation is not practical because of

a) the large amount of internal fragmentation

b) the large amount of external fragmentation

c) the large memory overhead in maintaining page tables

d) the large computation overhead in the translation process

27. Which of the following assertions is FALSE about the Internet Protocol (IP) ?

a) It is possible for a computer to have multiple IP addresses

b) IP packets from the same source to the same destination can take different routes in the network

c) IP ensures that a packet is discarded if it is unable to reach its destination within a given number of hops

d) the packet source cannot set the route of an outgoing packets; the route is determined only by the routing tables in the routers on the way.

28. Which of the following functionalities must be implemented by a transport protocol over and above the network protocol ?

a) Recovery from packet losses

b) Detection of duplicate packets

c) Packet delivery in the correct order

d) End to end connectivity

29. Which of the following scenarios may lead to an irrecoverable error in a database system ?

a) A transaction writes a data item it is read by an uncommitted transaction.

b) A transaction reads a data item after it is read by an uncommitted transaction.

c) A transaction reads a data item after it is written by a committed transaction.

d) A transaction reads a data item after it is written by an uncommitted transaction.

30. Consider the following SQL query

select distinct $a_1, a_2, ..., a_n$
from $r_1, r_2 ......r_m$
where P

For an arbitrary predicate P, this query is equivalent to which of the following relational algebra expressions?

a) $\Pi_{a_1,a_2,...a_n} \sigma_P(r_1 \times r_2 \times ..... \times r_m)$

b) $\Pi_{a_1,a_2,...a_n} \sigma_P(r_1 \bowtie r_2) \bowtie ...... \bowtie r_m)$

c) $\Pi_{a_1,a_2,...a_n} \sigma_P(r_1 \cup r_2 \cup ...... \cup r_m)$

d) $\Pi_{a_1,a_2,...a_n} \sigma_P(r_1 \cap r_2 \cap ...... \cap r_m)$

Q. 31 - 90 CARRY TWO MARKS EACH

31. Let $(S, \leq)$ be a partial order with two minimal elements a and b, and a maximum element c. Let $P: S \rightarrow$ {True, False} be a predicate defined on S. Suppose that P(a) = True, P(b) = False, and $P(x) \Rightarrow P(y)$ for all $x, y \in S$ satisfying $x \leq y$, where $\Rightarrow$ stands for logical implication. Which of the following statements CANNOT be TRUE ?

a) P (x) = True for all $x \in S$ such that $x \neq b$

b) P (x) = False for all $x \in S$ such that $x \neq a$ and $x \neq c$

c) P (x) = False for all $x \in S$ such that $b \leq x$ and $x \neq c$

d) p(x) = False for all $x \in S$ such that $a \leq x$ and $b \leq x$

32. Which of the following is a valid first order formula ? (Here α and β are first order formulae with x as their only free variable)

a) $((\forall x)[\alpha] \Rightarrow (\forall x)[\beta]) \Rightarrow (\forall x)[\alpha \Rightarrow \beta]$

b) $(\forall x)[\alpha] \Rightarrow (\exists x)[\alpha \wedge \beta]$

c) $((\forall x[\alpha \wedge \beta] \Rightarrow (\exists x)[\alpha]) \Rightarrow (\forall x)[\alpha]$

d) $(\forall x)[\alpha \Rightarrow \beta] \Rightarrow ((\forall x)[\alpha] \Rightarrow (\forall x)[\beta])$

33. Consider the followng formula α and its two interpretations I_1 and I_2

$\alpha : (\forall x)[Px \Leftrightarrow (\forall y)[Qxy \Leftrightarrow \neg Qxy]] \Rightarrow (\forall x)[\neg Px]$

I_1 : Domain : the set of natural numbers

$P_x \equiv$ ' x is a prime number '

$Qxy \equiv$ ' y divides x '

I_2: same as I_1 except that $P_x \equiv$ 'x is a composite number'.

Which of the following statements is TRUE ?

a) I_1 satisfies α, I_2 does not

b) I_2 satisfies α, I_1 does not

c) Neither I_1 nor I_2 satisfies α

d) Both I_1 and I_2 sastisfy α

34. m identical balls are to be placed in n distinct bags. You are given that $m \geq kn$, where k is a natural number ≥ 1. In how many ways can the balls be placed in the bags if each bag must contain atleast k balls ?

a) $\binom{m-k}{n-1}$ b) $\binom{m-kn+n-1}{n-1}$

c) $\binom{m-1}{n-k}$ d) $\binom{m-kn+n+k-2}{n-k}$

35. Consider the following recurrence relation

$T(1) = 1,\ T(n+1) = T(n) + \lfloor \sqrt{(n+1)} \rfloor$ for all $n \geq 1$

The value of $T(m^2)$ for $m \geq 1$ is

a) $\frac{m}{6}(21m - 39) + 4$ b) $\frac{m}{6}(4m^2 - 3m + 5)$

c) $\frac{m}{2}(3m^{2.5} - 11m + 20) - 5$

d) $\frac{m}{6}(5m^3 - 34m^2 + 137m - 104) + \frac{5}{6}$

36. How many perfect matchings are there in a complete graph of 6 vertices ?

a) 15 b) 24 c) 30 d) 60

37. Let $f: A \rightarrow B$ be an injective (one-to-one) function. Define $g: 2^A \rightarrow 2^B$ as : $g(C) = \{f(x); x \in C\}$, for all subsets C of A. Define $h: 2^B \rightarrow 2^A$ as: $h(D) = \{x / x \in A, f(x) \in D\}$, for all subsets D of B. Which of the following statements is always TRUE ?

a) $g(h(D)) \subseteq D$ b) $g(h(D)) \supseteq D$

c) $g(h(D)) \cap D = \phi$ d) $g(h(D)) \cap (B - D) = \phi$

38. Consider the set {a, b, c} with binary operators + and x defined as follows :

+	a	b	c
a	b	a	c
b	a	b	c
c	a	c	b

×	a	b	c
a	a	b	c
b	b	c	a
c	c	c	b

For example, $a + c = c$, $c + a = a$, $c \times b = c$ and $b \times c = a$

Given the following set of equations :

$$(a \times x) + (a \times y) = c,\ (b \times x) + (c \times y) = c$$

the number of solution(s) (i.e., pair(s) (x, y) that satisfy the equation) is

a) 0 b) 1 c) 2 d) 3

39. Let $\Sigma = \{a, b, c, d, e\}$ be an alphabet. We define an encoding scheme as follows :

$$g(a) = 3, g(b) = 5, g(c) = 7, g(d) = 9, g(e) = 11$$

Let p_i denote the i-th prime number $(p_1 = 2)$

For a non-empty string $S = a_1 \ldots\ldots a_n$, where each $a_i \in \Sigma$, define $f(s) = \Pi_{i=1}^{n} P_i^{g(a_i)}$. For a non-empty sequence $< S_1, \ldots\ldots S_n >$ of strings from Σ^+, define $h(< S_1 \ldots\ldots S_n >) = \Pi_{i=1}^{n} P_i^{f(s_i)}$

Which of the following numbers is the encoding, h, of a non-empty sequence of strings ?

a) $2^7 3^7 5^7$ b) $2^8 3^8 5^8$
c) $2^9 3^9 5^9$ d) $2^{10} 5^{10} 7^{10}$

40. A graph $G = (V, E)$ satisfies $|E| \le 3|V| - 6$. The min-degree of G is defined as $\min_{v \in V} \{degree(V)\}$. Therefore, min-degree of G cannot be

a) 3 b) 4 c) 5 d) 6

41. Consider the following system of linear equations

$$\begin{bmatrix} 2 & 1 & -4 \\ 4 & 3 & -12 \\ 1 & 2 & -8 \end{bmatrix} \begin{bmatrix} x \\ y \\ z \end{bmatrix} = \begin{bmatrix} \alpha \\ 5 \\ 7 \end{bmatrix}$$

Notice that the second and the third columns of the coefficient matrix are linearly dependent. For how many values of α, does this system of equations have infinitely many solutions ?

a) 0 b) 1 c) 2 d) infinitely many

42. A piecewise linear function f(x) is plotted using thick solid lines in the figure below (the plot is drawn to scale)

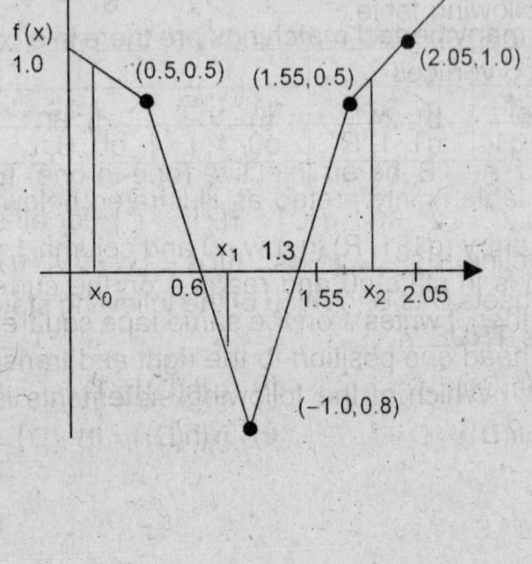

If we use the Newton-Raphson method to find the roots of f(x) = 0 using x_0, x_1 and x_2 respectively as initial guesses, the roots obtained would be

a) 1.3, 0.6 and 0.6 b) 0.6, 0.6 and 1.3
c) 1.3, 1.3, and 0.6 d) 1.3, 0.6 and 1.3

43. The following is a scheme for floating point number representation using 16 bits

Bit Position	15	14 ... 9	8 ... 0
	s	e	m
	Sign	Exponent	Mantissa

Let s, e and m be the numbers represented in binary in the sign, exponent, and mantissa fields respectively. then the floating point number represented is :

$$\begin{cases} (-1)^2 \left(1 + m \times 2^{-9}\right) 2^{e-31}, \text{ if the exponent} \ne 111111 \\ 0, \text{ otherwise} \end{cases}$$

What is the maximum difference between two successive real numbers representable in this system ?

a) 2^{-40} b) 2^{-9} c) 2^{22} d) 2^{31}

44. A 1-input, 2-output synchronous sequential circuit behaves as follows :

Let z_k, n_k denote the number of 0's and 1's respectively in initial k bits of the input $(z_k + n_k = k)$. The circuit outputs 00 until one of the following conditions holds.

- $z_k - n_k = 2$. In this case, the output at the k-th and all subsequent clock ticks is 10.
- $n_k - z_k = 2$. In this case, the output at the k-th and all subsequent clock ticks is 01.

what is the minimum number of states required in the state transition graph of the above circuit ?

a) 5 b) 6 c) 7 d) 8

45. The literal count of a boolean expression is the sum of the number of times each literal appears in the expression. For example, the literal count of $(xy + xz')$ is 4. What are the minimum possible literal counts of the product-of-sum and sum-of-product representations respectively of the function given by the following Karnaugh map ? Here, X denotes "don't care"

a) (11, 9)
b) (9, 13)
c) (9, 10)
d) (11, 11)

xy \ zw	00	01	11	10
00	X	1	0	1
01	0	1	X	0
11	1	X	X	0
10	X	0	0	X

46. Consider the ALU shown below.

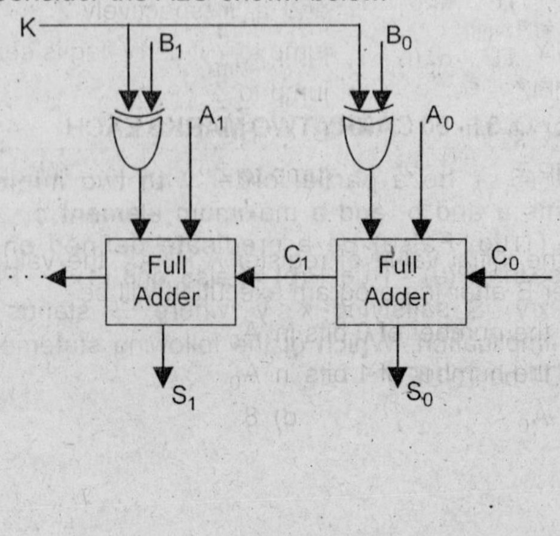

If the operands are in 2's complement representation, which of the following operations can be performed by suitable setting the control lines K and C_0 only (+ and - denote addition and subtraction respectively) ?

a) A+B, and A-B, but not A+1

b) A+B, and A+1, but not A-B

c) A+B, but not A-B or A+1

d) A+B, and A-B, and A+1

47. Consider the following circuit composed of XOR gates and non-inverting buffers.

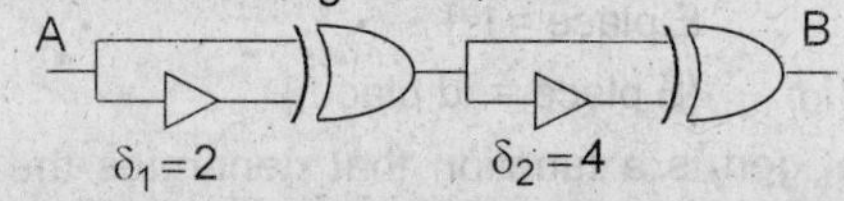

The non-inverting buffers have delays $\delta_1 = 2$ns and $\delta_2 = 4$ns as shown in the figure. Both XOR gates and all wires have zero delay. Assume that all gate inputs, outputs and wires are stable at logic level 0 at time 0. If the following waveform is applied at input A, how many transaction(s) (change of logic levels) occur(s) at B during the interval from 0 to 10 ns ?

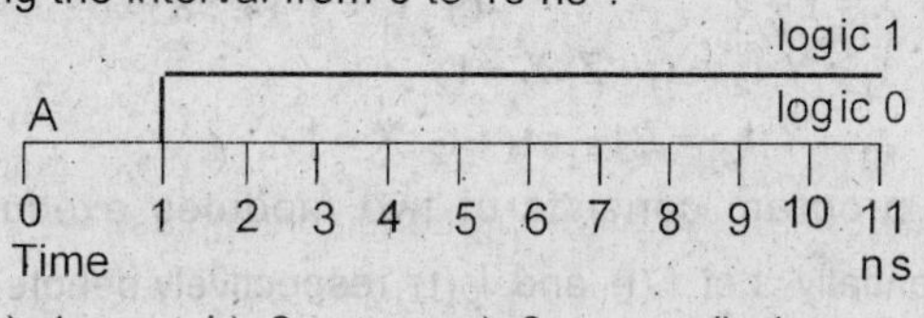

a) 1 b) 2 c) 3 d) 4

The following information pertains to Q.48-49

Consider the following assembly language program for a hypothetical processor. A, B and C are 8 bit registers. The meaning of various instructions are shown as comments.

```
   MOV B, #0    ;  B ← 0
   MOV C, #8    ;  C ← 8
Z  CMP, C #0    ;  Compare C with 0
   JZ X         ;  jump to X if zero flag is set
   SUB C, #1    ;  C ← C-1
   RRC A, #1    ;  right rotate A through carrry
                   by one bit. Thus
                ;  if the initial values of A and the
                   carry flag are a7....a0 and
                ;  c0 respectively, their values
                   after the execution of this
                ;  instruction will be c0a7....a1
                   and a0 respectively.
   JC Y         ;  jump to Y if carry flag is set
   JMP Z        ;  jump to Z
y: ADD B, #1    ;  B ← B+1
   JMP Z        ;  jump to Z
X:
```

48. If the initial value of register A is A_0 the value of register B after the program execution will be

a) the number of 0 bits in A_0

b) the number of 1 bits in A_0

c) A_0 d) 8

49. Which of the following instructions when inserted at location X will ensure that the value of register A after program execution is the same as its initial value ?

a) RRC A, #1

b) NOP ; no operation

c) LRC A, #1 ; left rotate A through carry flag by one bit

d) ADD A, #1

50. Consider the following deterministic finite state automation M.

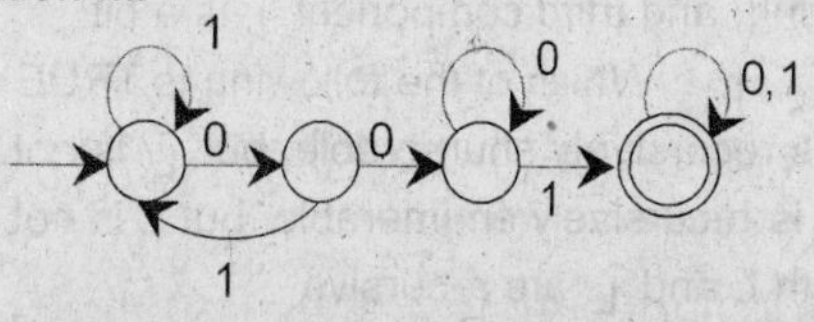

Let S denote the set of seven bit binary strings in which the first, the fourth, and the last bits are 1. The number of strings in S that are accepted by M is

a) 1 b) 5 c) 7 d) 8

51. Let $G=(\{S\},\{a,b\},R,S)$ be a context free grammar where the rule set R is $S \to aSb \mid SS \mid \in$

a) G is not ambiguous

b) There exist $x, y \in L(G)$ such that $xy \notin L(G)$

c) There is a deterministic pushdown automation that accepts L(G)

d) We can find a deterministic finite state automation that accepts L(G)

52. Consider two languages L_1 and L_2, each on the alphabet Σ. Let $f: \Sigma^* \to \Sigma^*$ be a polynomial time computable bijection such that $(\forall x)[x \in L_1 \text{ iff } f(x) \in L_2]$. Further, let f^{-1} be also polynomial time computable.

Which of the following CANNOT be true ?

a) $L_1 \in P$ and L_2 is finite

b) $L_1 \in NP$ and $L_2 \in P$

c) L_1 is undecidable and L_2 is decidable

d) L_1 is rescursively enumerable and L_2 is recursive.

53. A single tape turing Machine M has two states q0 and q1, of which q0 is the starting state. The tape alphabet of M is {0, 1, B} and its input alphabet is {0, 1}. The symbol B is the blank symbol used to indicate end of an input string. the transition function of M is described in the following table.

	0	1	B
q0	q1, 1, R	q1, 1, R	Halt
q1	q1, 1, R	q0, 1, L	q0, B, L

The table is interpreted as illustrated below.

The entry (q1, 1, R) in row q0 and column 1 signifies that if M is in state q0 and reads 1 on the current tape square, then it writes 1 on the same tape square, moves its tape head one position to the right and transitions to state q1. Which of the following statements is TRUE about M ?

a) M does not halt on any string in $(0+1)^+$

b) M does not halt on any string in $(00+1)^*$

c) M halts on all strings ending in a 0

d) M halts on all strings ending in a 1

54. Define languages L_0 and L_1 as follows :

$L_0 = \{< M, w, 0 > \backslash M \text{ halts on } w\}$

$L_1 = \{< M, w, 1 > \backslash M \text{ does not halt on } w\}$

Here <M, w, i> is a triple t, whose first component, M, is an encoding of a Turing Machine, second component, w, is a string, and third component, i, is a bit.

Let $L = L_0 \cup L_1$. Which of the following is TRUE ?

a) L is recursively enumerable, but $\overline{L}$ is not.

b) $\overline{L}$ is recursively enumerable, but L is not

c) Both L and $\overline{L}$ are recursive

d) Neither L nor $\overline{L}$ is recursively enumerable

55. Consider the NFA M shown below.

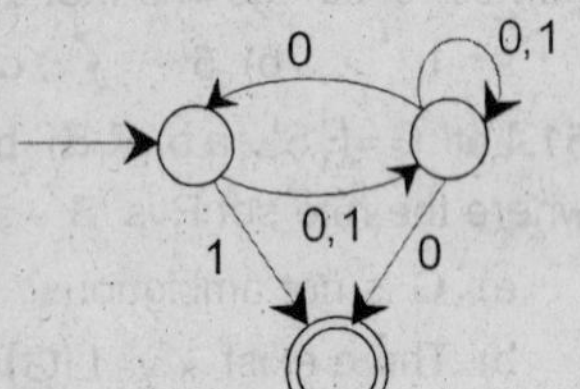

Let the language accepted by M be L. Let L_1 be the languae accepted by the NFA M_1, obtained by changing the accepting state of M to a non-accepting state and by changing the non-accepting states of M to accepting states. Which of the following statments is TRUE ?

a) $L_1 = \{0, 1\}^* - L$ b) $L_1 = \{0, 1\}^*$

c) $L_1 \subseteq L$ d) $L_1 = L$

56. Consider the grammar shown below

$S \to iEtSS'|a$

$S' \to eS|\in$

$E \to b$

In the predictive parse table, M, of this grammar, the entries $M[S', e]$ and $M[S', \$]$ respectively are

a) $\{S' \to eS\}$ and $\{S' \to \in\}$

b) $\{S' \to eS\}$ and $\{\ \}$

c) $\{S' \to \in\}$ and $\{S' \to \in\}$

d) $\{S' \to eS, S' \to \in\}$ and $\{S' \to \in\}$

57. Consider the grammar shown below.

$S \to CC$

$C \to cC\ |d$

This grammar is

a) LL(1) b) SLR(1) but not LL(1)

c) LALR (1) but not SLR (1)

d) LR (1) but not LALR (1)

58. Consider the translation scheme shown below.

$S \to TR$

$R \to +T\{\text{print}('+');\}R \backslash \in$

$T \to$ num {print (num.val) ;}

Here num is a token that represents an integer and num.val represents the corresponding integer value. For an input string '9+5+2', this translation scheme will print

a) 9 + 5 + 2 b) 95 + 2 +

c) 952 + + d) + + 952

59. Consider the syntax directed definition shown below.

$S \to Id := E\{\text{gen(id.place} = \text{E.place);}\}$

$E \to E_1 + E_2$ {t = newtemp ( );

gen (t = E_1.place + E_2.place;);

E.place = t ;}

$E \to id$ {E.place = id.place;}

Here, gen is a function that generates the output code, and newtemp is a function that returns the name of a new temporary variable on every call. Assume that t_i's are the temporary variable names generated by newtemp.

For the statement '$X := Y + Z$', the 3-address code sequence generated by this definition is

a) $X = Y + Z$ b) $t_1 = Y + Z;\ X = t_1$

c) $t_1 = Y; t_2 = t_1 + Z;\ X = t_2$

d) $t_1 = Y; t_2 = Z; t_3 = t_1 + t_2; X = t_3$

60. A program consists of two modules executed sequentially. Let $f_1(t)$ and $f_2(t)$ respectively denote the probability density functions of time taken to execute the two modules. The probability density function of the overall time taken to execute the program is given by

a) $f_1(t) + f_2(t)$ b) $\int_0^t f_1(x) f_2(x)\,dx$

c) $\int_0^t f_1(x) f_2(t-x)\,dx$ d) $\max\{f_1(t), f_2(t)\}$

The following information pertains to Q.61-62

In a permutation $a_1.....a_n$ of n distinct integers, an inversion is a pair (a_i, a_j) such that $i < j$ and $a_i > a_j$

61. If all permutations are equally likely, what is the expected number of inversions in a randomly chosen permutation of 1.....n ?

a) $n(n-1)/2$ b) $n(n-1)/4$

c) $n(n+1)/4$ d) $2n\lfloor \log_2 n \rfloor$

62. What would be the worst case time complexity of the Insertion Sort algorithm, if the inputs are restricted to permutations of 1........n with atmost n inversions ?

a) $\Theta(n^2)$ b) $\Theta(n\log)$ c) $\Theta(n^{1.5})$ d) $\Theta(n)$

63. A data structure is required for storing a set of integers such that each of the following operations can be done in O(log n) time, where n is the number of elements in the set.

I Deletion of the smallest element

II Insertion of an element if it is not already present in the set.

Which of the following data structures can be used for this purpose ?

a) A heap can be used but not a balanced binary search tree

b) A balanced binary search tree can be used but not a heap

c) Both balanced binary search tree and heap can be used

d) Neither balanced binary search tree nor heap can be used

64. Let S be a stack of size $n \geq 1$. Starting with the empty stack, suppose we push the first n natural numbers in sequence, and then perform n pop operations. Assume that Push and Pop operations take X seconds each, and Y seconds elapse between the end of one such stack operation and the start of the next operation. For $m \geq 1$, definie the stack-life of m as the time elapsed from the end of Push(m) to the start of the pop operation that removes m from S. The average stack-life of an element of this stack is

a) $n(X+Y)$ b) 3Y+2X

c) n(X+Y) - X d) Y + 2X

65. Consider the following 2-3-4 tree (i.e. B-tree with a minimum degree of two) in which each data item is a letter. The usual alphabetical ordering of letters is used in constructing the tree.

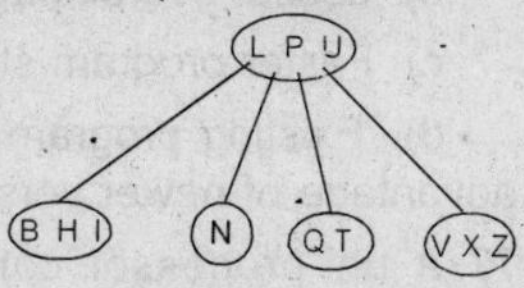

What is the result of inserting g in the above tree ?

a)

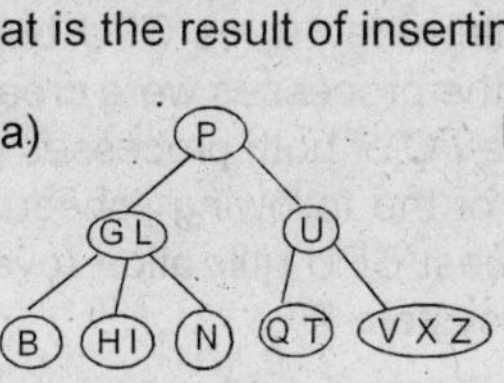

b)

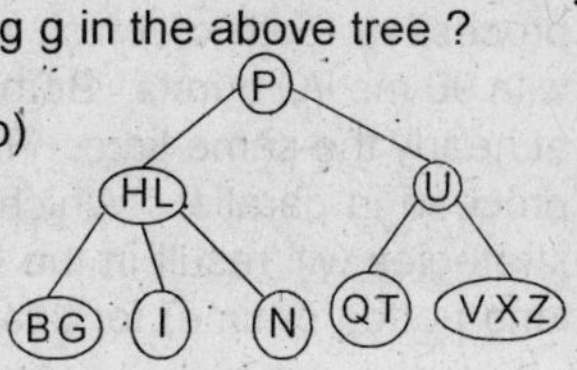

c)

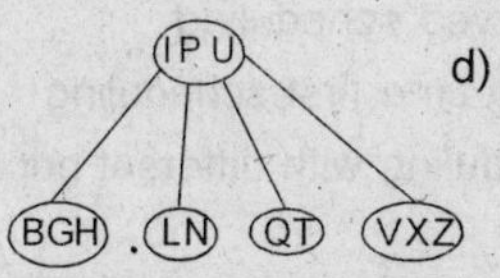

d) None of the above

66. The cube root of a natural number n is defined as the largest natural number m such that $m^3 \leq n$. The complexity of computing the cube root of n (n is represented in binary notation) is

a) O(n) but not $O(n^{0.5})$

b) $O(n^{0.5})$ but not $O((\log n)^k)$ for any constant $k > 0$

c) $O((\log n)^k)$ for some constant k>0, but not $O((\log \log n)^m)$ for any constant $m > 0$

d) $O((\log \log n)^k)$ for some constant k > 0.5 ; but not $O((\log \log n)^{0.5})$

67. Let G = (V, E) be an undirected graph with a subgraph $G_1 = (V_1, E_1)$. Weights are assigned to edges of G as follows.

$$w(e) = \begin{cases} 0 \text{ if } e \in E_1 \\ 1 \text{ otherwise} \end{cases}$$

A single source shortest path algorithm is executed on the weighted graph (V, E, w) with an arbitrary vertex v_1 of V_1 as the source. Which of the following can always be inferred from the path costs computed ?

a) The number of edges in the shortest paths from v_1 to all vertices of G

b) G_1 is connected.

c) V_1 forms a clique in G d) G_1 is a tree.

68. What is the weight of a minimum spanning tree of the following graph ?

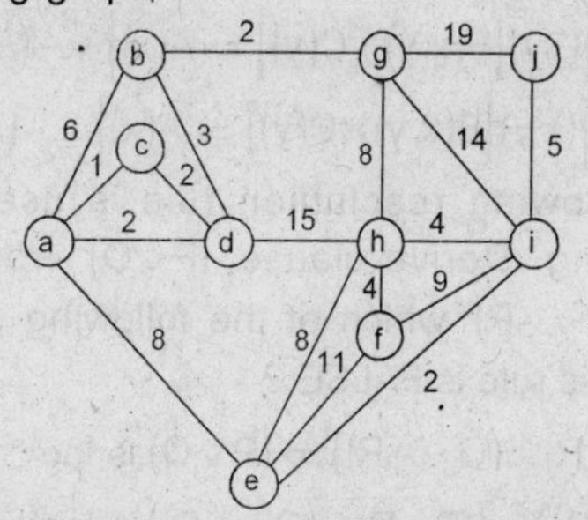

a) 29 b) 31 c) 38 d) 41

69. The following are the starting and ending times of activities A, B, C, D, E, F, G and H respectively in chronological order. "$a_s b_s c_s d_s c_s e_s f_s b_e d_e g_s e_e f_e h_s g_e h_e$". Here, x_s denotes the starting time and x_e denotes the ending time of activity X. We need to schedule the activities in a set of rooms available to us. An activity can be scheduled in a room only if the room is reserved for the activity for its entire duration. What is the minimum number of rooms required ?

a) 3 b) 4 c) 5 d) 6

70. Let G=(V, E) be a directed graph with n vertices. A path from v_i to v_j in G is a sequence of vertices $(v_i, v_{i+1}, \ldots, v_j)$ such that $(v_k, v_{k+1}) \in E$ for all k in i through j - 1. A simple path is a path in which no vertex appears more than once.

Let A be an $n \times n$ array initialized as follows.

$$A[j, k] = \begin{cases} 1 \text{ if } (j, k) \in E \\ 0 \text{ otherwise} \end{cases}$$

Consider the following algorithm

```
for i = 1 to n
   for j = 1 to n
      for k = 1 to n
         A[j, k] = max (A[j, k], A[j, i] + A[i, k]);
```

which of the following statements is necessarily TRUE for all j and k after termination of the above algorithm ?

a) $A[j,k] \le n$

b) If $A[j,j] \ge n-1$, then G has a Hamiltonian cycle

c) If there exists a path from j to k, A [j, k] contains the longest path length from j to k

d) If there exists a path from j to k, every simple path from j to k contains at most A [j, k] edges.

71. Consider the following logic program P

$$A(x) \leftarrow B(x,y), C(y)$$
$$\leftarrow B(x,x)$$

which of the following first order sentences is equivalent to P ?

a) $(\forall x)\left[(\exists y)[B(x,y) \wedge C(y)] \Rightarrow A(x)\right] \wedge \neg(\exists x)[B(x,x)]$

b) $(\forall x)\left[(\forall y)[B(x,y) \wedge C(y)] \Rightarrow A(x)\right] \wedge \neg(\exists x)[B(x,x)]$

c) $(\forall x)\left[(\exists y)[B(x,y) \wedge C(y)] \Rightarrow A(x)\right] \vee \neg(\exists x)[B(x,x)]$

d) $(\forall x)\left[(\forall y)[B(x,y) \wedge C(y)] \Rightarrow A(x)\right] \vee \neg(\exists x)[B(x,x)]$

72. The following resolutiion rule is used in logic programming. Derive clause $(P \vee Q)$ from clauses $(P \vee R)$, $(Q \vee \neg R)$ which of the following statements related to this rule is FALSE ?

a) $((P \vee R) \wedge (Q \vee \neg R)) \Rightarrow (P \vee Q)$ is logically valid

b) $(P \vee Q) \Rightarrow ((P \vee R) \wedge (Q \vee \neg R))$ is logically valid

c) $(P \vee Q)$ is satisfiable if and only if $(P \vee R) \wedge (Q \vee \neg R)$ is satisfiable

d) $(P \vee Q) \Rightarrow \text{FALSE}$ if and only if both P and Q are unsatisfiable.

The following information pertains to Q.73-74

The following program fragment is written in a programming language that allows global variables and does not allow nested declarations of functions.

```
global int i = 100, j = 5;
void P(X) {
  int i = 10 ;
  Print (x + 10) ;
  i = 200 ;
  j = 20 ;
  Print (X) ;
}
main ( ) { P(i+j) ; }
```

73. If the programming language uses static scoping and call by need parameter passing mechanism, the values printed by the above program are

a) 115, 220 b) 25, 220

c) 25, 1 d) 115, 105

74. If the programming language uses dynamic scoping and call by name parameter passing mechanism, the values printed by the above program are

a) 115, 220 b) 25, 220

c) 25, 15 d) 115, 105

75. Consider the following class definitions in a hypothetical Object Oriented language that supports inheritance and uses dynamic binding. The language should not be assumed to be either Java or C++, though the syntax is similar.

```
Class P {                    Class Q subclass of P {
  Void f(int i) {                Void f(int i)   {
    print (i) ;                    print (2*i) ;
  }                              }
}                            }
```

Now consider the folowing program fragment :

```
PX = new Q ( ) ;
Qy = new Q ( ) ;
Pz = new Q ( ) ;
```

x. f(1); ((P) Y) denotes a typecast of y to P. the output produced by executing the above program fragment will be

a) 121 b) 211 c) 212 d) 222

76. Which of the following is NOT an advantage of using shared, dynamically linked libraries as opposed to using statically linked libraries ?

a) Smaller sizes of executable files.

b) Lesser overall page fault rate in the system

c) Faster program startup

d) Existing program need not be re-linked to take advantage of newer versions of libraries.

77. A uni-processor computer system only has two processes, both of which alternate 10 ms CPU bursts with 90 ms I/O bursts. Both the processes were created at nearly the same time. The I/O of both processes can proceed in parallel. Which of the following scheduling strategies will result in the least CPU utilization (over a long period of time) for this system ?

a) First come first served scheduling

b) Shortest remaining time first scheduling

c) Static priority scheduling with different priorities for the two processes.

d) Round robin scheduling with a time quantum of 5 ms

The following information pertains to Q.78-79

A processor uses 2-level page tables for virtual to physical address translation. Page tables for both levels are stored in the main memory. Virtual and physical addresses are both 32 bits wide. the memory is byte addressable. For virtual to physical address translation, the 10 most significant bits of the virtual address are used as index into the first level page table while the next 10 bits are used as index into the second level page table. The 12 least significant bits of the virtual address are used as offset within the page. Assume that the page table entries in both levels of page tables are 4 bytes wide. Further, the processor has a translatioin look-aside buffer (TLB), with a hit rate of 96%. The TLB caches recently used virtual page numbers and the corresponding physical page numbers.

The processor also has a physically addressed cache with a hit rate of 90%. Main memory access time is 10 ns, cache access time is 1 ns, and TLB access time is also 1 ns.

78. Assuming that no page faults occur, the average time taken to access a virtual address is approximately (to the nearest 0.5 ns)

a) 1.5 ns b) 2 ns c) 3 ns d) 4 ns

79. Suppose a process has only the following pages in its virtual address space: two contiguous code pages starting at virtual address 0x00000000, two contiguous data pages starting at virtual address 0x00400000, and a stack page starting at virtual address 0xFFFFF000. The amount of memory required for storing the page tables of this process is

a) 8 KB b) 12 KB c) 16 KB d) 20 KB

The following information pertains to Q.80-81

Suppose we want to synchronize two concurrent processes P and Q using binary semaphores S and T. The code for the processes P and Q is shown below.

```
Process P :              Process Q :
While (1)  {             While (1)  {
W:                       Y:
    print'0' ;               print '1' ;
    print '0' ;              print '1' ;
X:                       Z:
}                        }
```

Synchronization statements can be inserted only at points W, X, Y and Z

80. Which of the following will always lead to an output starting with '001100110011' ?

a) P(S) at W, V(S) at X, P(T) at Y, V(T) at Z, S and T initially 1

b) P(S) at W, V(T) at X, P(T) at Y, V(S) at Z, S initially 1, and T initially 0

c) P(S) at W, V(T) at X, P(T) at Y, V(S) at Z, S and T initially 1

d) P(S) at W, V(S) at X, P(T) at Y, V(T) at Z, S initially 1, and T initially 0

81. Which of the following will ensure that the output string never contains a substring of the form 01^n0 or 10^n1 where n is odd ?

a) P(S) at W, V(S) at X, P(T) at Y, V(T) at Z, S and T initially 1

b) P(S) at W, V(T) at X, P(T) at Y, V(S) at Z, S and T initially 1

c) P(S) at W, V(S) at X, P(S) at Y, V(S) at Z, S initially 1

d) V(S) at W, V(T) at X, P(S) at Y, P(T) at Z, S and T initially 1

82. The subnet mask for a particular network is 255.255.31.0. Which of the following pairs of IP addresses could belong to this network ?

a) 172.57.88.62 and 172.56.87.233

b) 10.35.28.2 and 10.35.29.4

c) 191.203.31.87 and 191.234.31.88

d) 128.8.129.43 and 128.8.161.55

83. A 2 km long broadcast LAN has 10^7 bps bandwidth and uses CSMA/CD. The signal travels along the wire at 2×10^8 m/s. What is the minimum packet size that can be used on this network ?

a) 50 bytes b) 100 bytes

c) 200 bytes d) None of the above

84. Host A is sending data to host B over a full duplex link. A and B are using the sliding window protocol for flow control. The send and receive window sizes are 5 packets each. Data packets (sent only from A to B) are all 1000 bytes long and the transmission time for such a packet is $50\ \mu s$. Acknowledgement packets (sent only from B to A) are very small and require negligible transmission time. The propagation delay over the link is $200\ \mu s$. What is the maximum achievable throughput in this communication ?

a) 7.69×10^6 bps b) 11.11×10^6 bps

c) 12.33×10^6 bps d) 15.00×10^6 bps

85. Consider the following functional dependencies in a database.

Date_of_Birth → Age Age → Eligibility

Name → Roll_number Roll_number → Name

Couse_number → Course_name

Course_number → Instructor

(Roll_number, Couse_number) → Grade

The relation (Roll_number, Name Date_of_birth, Age) is

a) in second normal form but not in third normal form

b) in third normal form but not in BCNF

c) in BCNF

d) none of the above.

86. Consider the set of relations shown below and the SQL query that follows.

Students : (Roll_number, Name, Date_of_birth)

Courses : (Course_number, Course_name, Instructor)

Grades : (Roll_number, Course_number, Grade)

select dinstinct **Name**
from **Students, Courses, Grades**
where **Students. Roll_number = Grades. Roll_number**
and **Courses. Instructor = Korth**
and **Courses. Course_number = Grades.Course_number**
and **Grades.grade = A**

Which of the following sets is computed by the above query ?

a) Name of students who have got an A grade in all courses taught by Korth.

b) Names of students who have got an A grade in all courses.

c) Names of Students who have got an A grade in at least one of the courses.

d) None of the above.

87. Consider three data items D1, D2 and D3 and the following execution schedule of transactions T1, T2 and T3. In the diagram, R(D) and W(D) denote the actions reading and writing the data item D respectively.

T1	T2	T3
	R(D3);	
	R(D2);	
	W(D2) ;	
		R(D2);
		R(D3);
R(D1);		
W(D1);		
		W(D2);
		W(D3);
	R(D1);	
R(D2);		
W(D2);		
	W(D1);	

(time ↓)

Which of the following statements is correct ?

a) The schedule is serializable as T2; T3; T1

b) The schedule is serializable as T2; T1; T3

c) The schedule is serializable as T3; T2; T1

d) The schedule is not serializable

88. In the following C program fragment, j, k, n and Two log_n are integer variables, and A is an array of integers. The variable n is initialized to an integer ≥ 3, and Two Log_n is initialized to the value of $2 * \lceil \log_2(n) \rceil$

```
for (k = 3; k < =n ; k++)
    A [k] = 0 ;
for ( k = 2; k<= Two Log_n ; k++)
    for ( j = k+1; j <=n ; j++)
        A [ j ] = A [ j ]  | | ( j%k ) ;
for ( j = 3; j <=n; j++)
    if (! A [ j ] ) printf ("%d", j);
```

The set of numbers printed by this program fragment is

a) $\{n/m \leq n, (\exists i)\ [m = i!]\}$ b) $\{m/m \leq n, (\exists i)\ [m = i^2]\}$

c) $\{m/m \leq n, \text{m is prime}\}$ d) $\{\ \}$

89. Consider the c program shown below.

```
# include <stdio.h>
# define print(x) printf("%d, x)
int x;
Void Q (int z) {
    z + = x; print (z) ;
}
void P (int*y)    {
    int x = *y+2 ;
    Q(x);  *y = x - 1 ;
    print (x) ;
}
main (void)    {
    x = 5 ;
    p (&x) ;
    print (x) ;
}
```

The output of this program is

a) 12 7 6 b) 22 12 11 c) 14 6 6 d) 7 6 6

90. Consider the function f defined below.

```
Struct item {
    int data ;
    struct item * next ;
}
int f(struct item*p)    {
    return ((p==NULL | | (p - > next ==NULL)    | |
        (( p->data<=p->next->data) &&
        f(p->next) ) ) ;
```

For a given linked list p, the function f returns 1 if and only if

a) the list is empty or has exactly one element

b) the elements in the list are sorted in non-decreasing order of data value

c) the elements in the list are sorted in non-increasing order of data value,

d) not all elements in the list have the same data value.

ANSWER KEY

1. c	2. d	3. d	4. d	5. a
6. a	7. d	8. d	9. a	10. d
11. d	12. c	13. d	14. d	15. a
16. c	17. b	18. c	19. b	20. a
21. d	22. c	23. c	24. a	25. d
26. a	27. a	28. d	29. d	30. a
31. c	32. d	33. a	34. b	35. b
36. a	37. a	38. d	39. b	40. c
41. b	42. b	43. c	44. c	45. a
46.	47. c	48. a	49. a	50. a
51. d	52. c	53. d	54. a	55. c
56. a	57. d	58. d	59. c	60. c
61. b	62. b	63. b	64. d	65. c
66. c	67. b	68. b	69. b	70. d
71. c	72. b	73. d	74. a	75. d
76. d	77. a	78. b	79. b	80. a
81. c	82. c	83. d	84. b	85. a
86. c	87. c	88. b	89. a	90. d

GATE - 2002

COMPUTER SCIENCE & ENGINEERING

Duration : Three Hours | **Maximum Marks : 150**

1. This question consists of TWENTY FIVE sub-questions, 1.1 to 1.25 of ONE mark each.

1.1 The rank of the matrix $\begin{bmatrix} 1 & 1 \\ 0 & 0 \end{bmatrix}$ is

a) 4 b) 2 c) 1 d) 0

1.2 The trapezoidal rule for integration gives exact result when the integrand is a polynomial of degree

a) 0 but not 1 b) 1 but not 0 c) 0 or 1 d) 2

1.3 The solution to the recurrence equation

$$T(2^K) = 3T(2^{K-1}) + 1,\ T(1) = 1 \text{ is.}$$

a) 2^K b) $(3^{k+1} - 1)/2$ c) $3^{\log_2 k}$ d) $2^{\log_3 k}$

1.4 The minimum number of colours required to colour the vertices of a cycle with n nodes in such a way that no two adjacent nodes have the same colour is :

a) 2 b) 3 c) 4 d) $n - 2\lfloor n/2 \rfloor + 2$

1.5 In the worst case, the number of comparisons needed to search a singly linked list of length n for given element is

a) $\log_2 n$ b) n/2 c) $\log_2 n - 1$ d) n

1.6 Which of the following is TRUE?

a) The set of all rational negative numbers forms a group under multiplication.

b) The set of all non-singular matrices forms a group under multiplication.

c) The set of all matrices forms a group under multiplication.

d) Both (b) and (c) are true.

1.7 The language accepted by a Pushdown Automaton in which the stack is limited to 10 items is best described as

a) Context Free b) Regular

c) Deterministic Context Free d) Recursive

1.8 "If X then Y unless Z" is represented by which of the following formulas in propositional logic?("$\neg$" is negation "$\wedge$" is conjunction, and "$\rightarrow$" is implication)

a) $(X \wedge \neg Z) \rightarrow Y$ b) $(X \wedge Y) \rightarrow \neg Z$

c) $(X \rightarrow (Y \wedge \neg Z)$ d) $(X \rightarrow Y(\wedge \neg Z)$

1.9 A device employing INTR line for device interrupt puts the CALL instruction on the data bus while

a) $\overline{\text{INTA}}$ is active b) HOLD is active

c) READY is active d) None

1.10 In 8085 which of the following modifies the program counter?

a) Only PCHL instruction b) Only ADD instructions

c) Only JMP and CALL instructions d) All instructions

1.11 In serial data transmission, every byte of data is padded with a '0' in the beginning and one or two '1's at end of byte because

a) Receiver is to be synchronized for byte reception

b) Receiver recovers lost '0's and '1's from these padded bits

c) Padded bits are useful in parity computation

d) None of the above

1.12 Minimum sum of product expression for f (w,x,y,z) shown in karnaugh-map below is

a) $xz + y'z$

b) $xz' + zx'$

c) $x'y + zx'$

d) None

YZ \ WX	00	01	11	10
00	0	1	1	0
01	X	0	0	1
11	X	0	0	1
10	0	1	1	X

1.13 Which of the following is not a form of memory?

a) instruction cache b) instruction register

c) instruction opcode d) translation lookaside buffer

1.14 The decimal value 0.25

a) Is equivalent to the binary value 0.1

b) Is equivalent to the binary value 0.01

c) Is equivalent to the binary value 0.00111....

d) Cannot be represented precisely in binary

1.15 The 2's complement representation of the decimal value-15 is

a) 1111 b) 11111 c) 111111 d) 10001

1.16 Sign extension is a step in

a) floating point multiplication

b) signed 16 bit integer addition

c) arithmetic left shift

d) converting a signed integer from one size to another

1.17 In the C language

a) At most one activation record exists between the current activation record and the activation record for the main.

b) The number of activation records between the current activation record and the activation record for the main depends on the actual function calling sequence

c) The visibility of global variables depends on the actual function calling sequence.

d) Recursion requires the activation record for the recursive function to be saved on a different stack before the recursive function can be called.

1.18 The results returned by functions under value-result and reference parameter passing coventions

a) Do not differ

b) Differ in the presence of loops

c) Differ in all cases

d) May differ in the presence of exceptions

1.19 Relation R with an associated set of functional dependencies, F, is decomposed into BCNF. The redundancy (arising out of functional dependencies) in the resulting set of relations is

a) Zero

b) More than zero but less than that of an equivalent 3NF decomposition

c) Proportional to the size of F^+ d) Indeterminate

1.20 With regard to the expressive power of the formal relational query languages, which of the following statements is TRUE?

a) Relational algebra is more powerful than relational calculus

b) Relational algebra has the same power as relational calculus

c) Relational algebra has the same power as safe relational calculus

d) None of the above

1.21 In 2's complement addition, overflow

a) Is flagged whenever there is carry from sign bit addition

b) Cannot occur when a positive value is added to a negative value

c) Is flagged when the carries from sign bit and previous bit match

d) None of the above

1.22 Which of the following scheduling algorithms is non-preemptive?

a) Round Robin b) First-In-First- Out

c) Multilevel Queue Scheduling

d) Multilevel Queue Scheduling with Feedback

1.23 The optimal page replacement algorithm will select the page that

a) Has not been used for the longest time in the past.

b) Will not be used for the longest time in the future.

c) Has been used least number of times.

d) Has been used most number of times.

1.24 In the absolute addressing mode

a) The operand is inside the instruction

b) The address of the operand is inside the instruction

c) The register containing the address of the operand is specified inside the instruction

d) The location of the operand is implicit

1.25 Maximum number of edges in a n-node undirected graph without self loops is

a) n^2 b) n(n-1)/2 c) n-1 d) (n+1)(n)/2

CS2. This question consists of TWENTY FIVE sub - questions, 2.1 to 2.25, of TWO marks each.

2.1 Consider the following logic circuit whose inputs are functions f_1, f_2, f_3, and output is f.

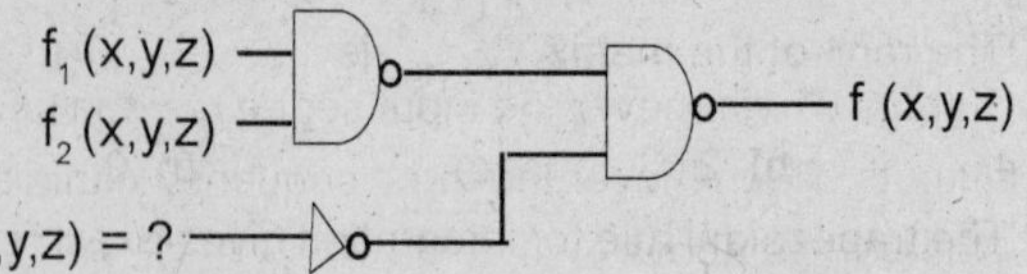

Given that

$f_1(x,y,z) = \Sigma(0,1,3,5)$

$f_2(x,y,z) = \Sigma(6,7)$ *and*

$f(x,y,z) = \Sigma(1, 4, 5)$

f_3 is

a) $\Sigma(1,4,5)$ b) $\Sigma(6,7)$ c) $\Sigma(0,1,3,5)$ d) None

2.2 Consider the following multiplexor where I0, I1,I2,I3 are four data input lines selected by two address line combinations A1A0 = 00,01,10,11 respectively and f is the output of the multiplexor. EN is the Enable input.

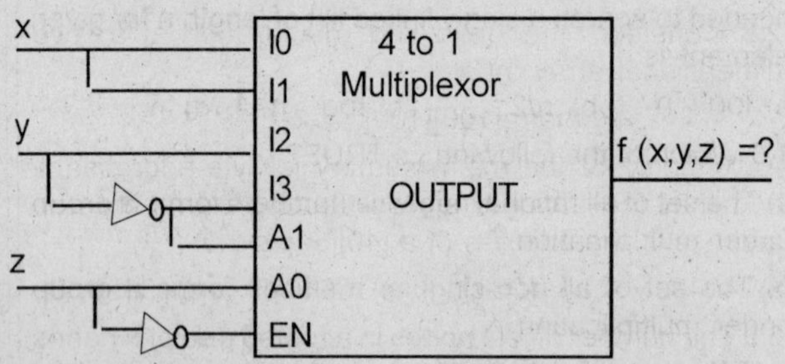

The function f(x,y,z) implemented by the above circuit is

a) x y z′ b) xy+z c) x+y d) None

2.3 Let f(A,B) = **A**′ + B. Simplified expression for functio f(f(x+y,y),z) is

a) x′ +z b) xyz c) X Y′ +z d) None

2.4 What are the states of the Auxillary Carry (AC) and Carry Flag (CY) after executing the following 8085 program?

```
MVI H, 5DH
MVI L, 6BH
MOV A, H
ADD L
```

a) AC = 0 and CY = 0 b) AC = 1 and CY =1

c) AC = 1 and CY =0 d) AC = 0 and CY =0

2.5 The Finite state machine described by the following state diagram with A as starting state, where an arc label is x/y and x stands for 1-bit input and y stands for 2-bit output

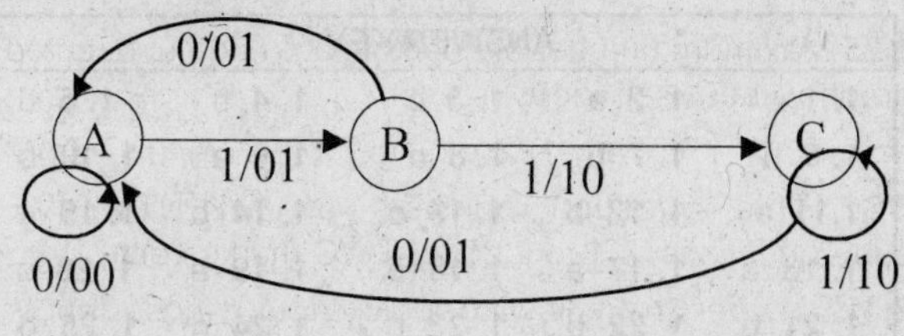

a) Outputs the sum of the present and the previous bits of the input.

b) Outputs 01 whenever the input sequence contains 11

c) Outputs 00 whenever the input sequence contains10

d) None of the above

2.6 The performance of a pipelined processor suffers if

a) The pipeline stages have different delays

b) Consecutive instructions are dependent on each other

c) The pipeline stages share hardware resources

d) All of the above

2.7 Horizontal microprogramming

a) Does not require use of signal decoders

b) Results in larger sized microinstructions than vertical microprogramming

c) Uses one bit for each control signal

d) All of the above

2.8 Consider the following declaration of a two-dimensional array inC:

char a[100][100];

Assuming that the main memory is byte-addressable and that the array is stored starting from memory address 0, the address of a [40][50] is

a) 4040 b) 4050 c) 5040 d) 5050

2.9 The number of leaf nodes in a rooted tree of n nodes, with each node having 0 or 3 childern is :

a) n/2 b) (n-1)/3 c) (n-1)/2 d) (2n+1)/3

2.10 Consider the following algorithm for searching for a given number x in an unsorted array A[1..n] having n distinct values :

I. Choose an i uniformly at random from 1...n,

II. If A [i] = x then Stop else Goto 1;

Assuming that x is present in A, what is the expected number of comparisons made by the algorithm before it terminates?

a) n b) n-1 c) 2n d) n/2

2.11 The running time of the following algorithm

Procedure A(n)

If n<=2 return(1) else return $\left(A\left(\left\lceil \sqrt{n} \right\rceil\right)\right)$

is best described by

a) O(n) b) O(log n) c) O(log log n) d) O(1)

2.12 A weight- balanced tree is a binary tree in which for each node, the number of nodes in the left subtree is at least half and at most twice the number of nodes in the right subtree. The maximum possible height (number of nodes on the path from the root to the furthest leaf) of such a tree on n nodes is best described by which of the following?

a) $\log_2 n$ b) $\log_{4/3} n$ c) $\log_3 n$ d) $\log_{3/2} n$

2.13 The smallest finite automaton which accepts the language { x | length of x is divisible by 3} has

a) 2 states b) 3states c) 4 states d) 5 states

2.14 Which of the following is TRUE?

a) The complement of a recursive language is recursive.

b) The complement of a recursively enumerable language is recursively enumerable

c) The complement of a recursive language is either recursive or recursively enumerable

d) The complement of a context-free language is context-free.

2.15 The Newton-Raphson iteration

$X_{n+1} = (X_n / 2) + 3/(2X_n)$ can be used to solve the equation

a) $X^2 = 3$ b) $X^3 = 3$ c) $X^2 = 2$ d) $X^3 = 2$

2.16 Four fair coins are tossed simultaneously. The probability that at least one head and one tail turn up is

a) 1/16 b) 1/8 c) 7/8 d) 15/16

2.17 The binary relation $S = \phi$ (empty set) on set A = {1,2,3 } is

a) Neither reflexive nor symmetric

b) Symmetric and reflexive

c) Transitive and reflexive d) Transitive and symmetric

2.18 The C language is

a) A context free language

b) A context sensitive language

c) A regular language

d) Parsable fully only by a Turing machine

2.19 To evaluate an expression without any embedded function calls

a) One stack is enough

b) Two stacks are needed

c) As many stacks as the height of the expression tree are needed

d) A Turing machine is needed in the general case

2.20 Dynamic linking can cause security concerns because

a) Security is dynamic

b) The path for searching dynamic libraries is not known till runtime

c) Linking is insecure

d) Cryptographic procedures are not available for dynamic linking

2.21 Which combination of the following features will suffice to characterize an OS as a multiprogrammed OS?

I. More than one program may be loaded into main memory at the same time for execution.

II. If a program waits for certain events such as I/O, another program is immediately scheduled for execution.

III. If the execution of a program terminates, another program is immediately scheduled for execution.

a) I b) I and II c) I and III d) I,II, and III

2.22 In the index allocation scheme of blocks to a file, the maximum possible size of the file depends on

a) The size of the blocks, and the size of the address of the blocks.

b) The number of blocks used for the index, and the size of the blocks

c) The size of the blocks, the number of blocks used for the index, and the size of the address of he blocks.

d) None of the above

2.23 A B^+ -tree index is to be built on the Name attribute of the relation STUDENT. Assume that all student names are of length 8 bytes, disk blocks are of size 512 bytes, and index pointers are of size 4 bytes. Given this scenario, what would be the best choice of the degree (i .e. the number of pointers per node) of the B^+-tree ?

a) 16 b) 42 c) 43 d) 44

2.24 Relation R is decomposed using a set of functional dependencies, F and relation S is decomposed using another set of functional dependencies, G. One decomposition is definitely BCNF, the other is definitely 3NF, but it is not known which is which. To make a guaranteed identification, which one of the following tests should be used on the decompositions ?(Assume that the closures of F and G are available)

a) Dependency- preservation b) Lossless- join

c) BCNF definition d) 3NF definition

2.25 From the following instance of a relation schema R (A,B,C), we can conclude that:

A	B	C
1	1	1
1	1	0
2	3	2
2	3	2

a) A functionally determines B and B functionally determines C

b) A functionally determines B and B does not functionally determine C

c) B does not functionally determine C

d) A does not functionally determine B and B does not functionally determine C

ANSWER KEY

1. 1 c	1. 2 a	1. 3 b	1. 4 b	1. 5 d
1. 6 b	1. 7 b	1. 8 d	1. 9 a	1. 10 b
1.11 a	1. 12 b	1. 13 c	1. 14 b	1. 15 d
1. 16 a	1. 17 a	1. 18 b	1. 19 a	1. 20 c
1. 21 b	1. 22 b	1. 23 b	1. 24 d	1. 25 b
2. 1 a	2. 2 a	2. 3 d	2. 4 c	2. 5 a
2. 6 d	2. 7 c	2. 8 b	2. 9 d	2. 10 a
2.11 a	2. 12 a	2. 13 c	2. 14 a	2. 15 a
2. 16 c	2. 17 d	2. 18 a	2. 19 a	2. 20 d
2. 21 d	2. 22 b	2. 23 a	2. 24 b	2. 25 a

GATE - 2001
COMPUTER SCIENCE & ENGINEERING

Duration : Three Hours **Maximum Marks : 150**

1. This question consist of 25 sub-questions (1.1 –1.25) of ONE mark each.

1.1. Consider the following statements:

S1 : The sum of two singular n × n matrices may be non-singular

S2 : The sum of two n × n non-singular matrices may be singular

Which of the following statements is CORRECT ?

a) S1 and S2 are both true b) S1 is true, S2 is false

c) S1 is false, S2 is true d) S1 and S2 are both false

1.2. Consider the following relations:

R1 (a, b) iff (a + b) is even over the set of integers

R2 (a, b) iff (a + b) is odd over the set of integers

R3 (a, b) iff a.b > 0 over the set of non-zero rational numbers

R4 (a, b) iff $|a - b| \leq 2$ over the set of natural numbers

Which of the following statement is CORRECT?

a) R1 and R2 are equivalence relations , R3 and R4 are not

b) R1 and R3 are equivalence relations , R2 and R4 are not

c) R1 and R4 are equivalence relations , R2 and R3 are not

d) R1, R2, R3 and R4 are all equivalence relations

1.3. Consider two well-formed formulas in propositional logic

F1: $P \Rightarrow \neg P$ F2: $(P \Rightarrow \neg P) \vee (\neg P \Rightarrow P)$

Which of the following statements is CORRECT?

a) F1 is satisfiable, F2 is valid

b) F1 is unsatisfiable, F2 is satisfiable

c) F1 is unsatisfiable, F2 is valid

d) F1 and F2 are both satisfiable

1.4. Consider the following two statements:

S1: $\{0^{2n} \mid n \geq 1\}$ is a regular language

S2: $\{0^m 1^n 0^{m+n} \mid m \geq 1 \text{ and } n \geq 1\}$ is a regular language

Which of the following statements is correct?

a) Only S1 is correct b) Only S2 is correct

c) Both S1 and S2 are correct

d) None of S1 and S2 is correct

1.5. Which of the following statements is TRUE?

a) If a language is context free it can always be accepted by a deterministic push-down automata.

b) The union of two context free languages is context free

c) The intersection of two context free languages is context free

d) The complement of a context free language is context free

1.6. Given an arbitrary non-deterministic finite automaton (NFA) with N states, the maximum number of states in an equivalent minimized DFA is at least

a) N^2 b) 2^N c) 2N d) N!

1.7. More than one word are put in one cache block to

a) Exploit the temporal locality of reference in a program

b) Exploit the spatial locality of reference in a program

c) Reduce the miss penalty d) None of the above

1.8. Which of the following statements is FALSE?

a) Virtual memory implements the translation of a program's address space into physical memory address space

b) Virtual memory allows each program to exceed the size of the primary memory

c) Virtual memory increases the degree of multiprogramming

d) Virtual memory reduces the context switching overhead

1.9. A slow memory can be connected to 8085 by using

a) INTR b) $\overline{\text{RESET IN}}$ c) HOLD d) READY

1.10. Suppose a processor does not have any stack pointer register. Which of the following statements is TRUE?

a) It can not have subroutine call instruction

b) It can have subroutine call instruction, but no nested subroutine calls

c) Nested subroutine calls are possible, but interrupts are not

d) All sequences of subroutine calls and also interrupts are possible

1.11. Given the following Karnaugh map, which one of the following represents the minimal Sum-Of-Products of the map?

a) $xy + y'z$

b) $wx'y' + xy + xz$

c) $w'x + y'z + xy$

d) $xz + y$

YZ \ WX	00	01	11	10
00	0	X	0	X
01	X	1	X	1
11	0	X	1	0
10	0	1	X	0

1.12. A processor needs software interrupt to

a) Test the interrupt system of the processor

b) Implement co-routines

c) Obtain system services which need execution of privileged instructions

d) Return from subroutine

1.13. A CPU has two modes – privileged and non-privileged. In order to change the mode from privileged to non-privileged

a) A hardware interrupt is needed

b) A software interrupt is needed

c) A privileged instruction (which does not generate an interrupt) is needed.

d) A non-privileged instruction (which does not generate an interrupt) is needed

1.14. Randomized quick sort is an extension of quick sort where the pivot is chosen randomly. What is the worst case complexity of sorting n numbers using randomized quicksort?

a) O(n) b) O(n log n) c) $O(n^2)$ d) O(n!)

1.15. Consider an array representation of an n element binary heap where the elements are stored from index 1 to index n of the array. For the element stored at index i of the array ($i \leq n$), the index of the parent is

a) i – 1 b) $\lfloor i/2 \rfloor$ c) $\lceil i/2 \rceil$ d) $(i+1)/2$

1.16. Let $f(n) = n^2 \log n$ and $g(n) = n(\log n)^{10}$ be two positive functions of n. Which of the following statements is correct?

a) f(n) = O(g(n)) and g(n) ≠ O(f(n))

b) g(n) = O(f(n)) and f(n) ≠ O(g(n))

c) f(n) ≠ O(g(n)) and g(n) ≠ O(f(n))

d) f(n) = O(g(n)) and g(n) = O(f(n))

1.17. The process of assigning load address to the various parts of the program and adjusting the code and data in the program to reflect the assigned addresses is called

a) Assembly b) Parsing

c) Relocation d) Symbol resolution

1.18. Which of the following statements is FALSE?

a) An unambiguous grammar has the same leftmost and rightmost derivation

b) An LL(1) parser is a top-down parser

c) LALR is more powerful than SLR

d) An ambiguous grammar can never be LR(k) for any k

1.19. Consider a set of n tasks with known runtimes r_1, r_2, ... r_n to be run on a uniprocessor machine . Which of the following processor scheduling algorithms will result in the maximum throughput?

a) Round-Robin b) Shortest-Job-First

c) Highest-Response-Ratio-Next

d) First-Come-First-Serve

1.20. Where does the swap space reside?

a) RAM b) Disk c) ROM d) On-chip-cache

1.21. Consider a virtual memory system with FIFO page replacement policy. For an arbitrary page access pattern, increasing the number of page frames in main memory will

a) Always decreases the number of page faults

b) Always increases the number of page faults

c) Sometimes increase the number of page faults

d) Never affect the number of page faults

1.22. Which of the following requires a device driver?

a) Register b) Cache c) Main memory d) Disk

1.23. Consider a schema R(A, B, C, D) and functional dependencies $A \rightarrow B$ and $C \rightarrow D$. Then the decomposition of R into R_1(AB) and R_2(CD) is

a) Dependency preserving and lossless join

b) Lossless join but not dependency preserving

c) Dependency preserving but not lossless join

d) Not dependency preserving and not lossless join

1.24. Suppose the adjacency relation of vertices in a graph is represented in a table Adj(X. Y). Which of the following queries cannot be expressed by a relational algebra expression of constant length?

a) List all vertices adjacent to a given vertex

b) List all vertices which have self loops

c) List all vertices which belong to cycles of less than three vertices

d) List all vertices reachable from a given vertex

1.25. Let r and s be two relations over the relation schemes R and S respectively, and let A be an attribute in R. Then the relational algebra expression $\sigma_{A=a}(r \bowtie s)$is always equal to

a) $\sigma_{A=a}(r)$ b) r c) $\sigma_{A=a}(r) \bowtie s$ d) None

2. This question consists of 25 sub-questions (2.2 – 2.25) of TWO marks each.

2.1. How many 4 digit even numbers have all 4 digits distinct?

a) 2240 b) 2296 c) 2620 d) 4536

2.2. Consider the following statements:

S1: There exist infinite sets A, B, C such that $A \cap (B \cup C)$ is finite.

S2: There exist two irrational numbers x and y such that (x + y) is rational

Which of the following is true about S1 and S2?

a) Only S1 is correct b) Only S2 is correct

c) Both S1 and S2 are correct

d) None of S1 and S2 are correct

2.3. Let $f: A \rightarrow B$ be a function, and let E and F be subsets of A. Consider the following statements about images.

S1: $f(E \cup F) = f(E) \cup f(F)$ S2: $f(E \cap F) = f(E) \cap f(F)$

Which of the following is true about S1 and S2?

a) Only S1 is correct b) Only S2 is correct

c) Both S1 and S2 are correct

d) None of S1 and S2 is correct

2.4. Seven (distinct) car accidents occurred in a week. What is the probability that they all occurred on the same day?

a) $1/7^7$ b) $1/7^6$ c) $1/2^7$ d) $7/2^7$

2.5. Consider a DFA over Σ ={a, b} accepting all strings which have number of a's divisible by 6 and number of b's divisible by 8. What is the minimum number of states that the DFA will have?

a) 8 b) 14 c) 15 d) 48

2.6. Consider the following languages:

L1 = { ww | w ∈ {a, b}*}

L2 = { ww^R | w ∈ {a, b}*, w^R is the reverse of w}

L3 = { 0^{2i} | i is an integer}

L4 = { 0^{i^2} | i is an integer}

Which of the languages are regular?

a) Only L1 and L2 b) Only L2, L3 and L4

c) Only L3 and L4 d) Only L3

2.7. Consider the following problem X.

Given a Turing machine M over the input alphabet Σ, any state q of M and a word w $\in \Sigma^*$, does the computation of M on w visit the state q?

Which of the following statements about X is CORRECT?

a) X is decidable

b) X is undecidable but partially decidable

c) X is undecidable and not even partially decidable

d) X is not a decision problem

2.8. Consider the following circuit with initial state $Q_0 = Q_1 = 0$.

The D Flip-Flops are positive edge triggered and have set up times 20 nanosecond and hold times 0.

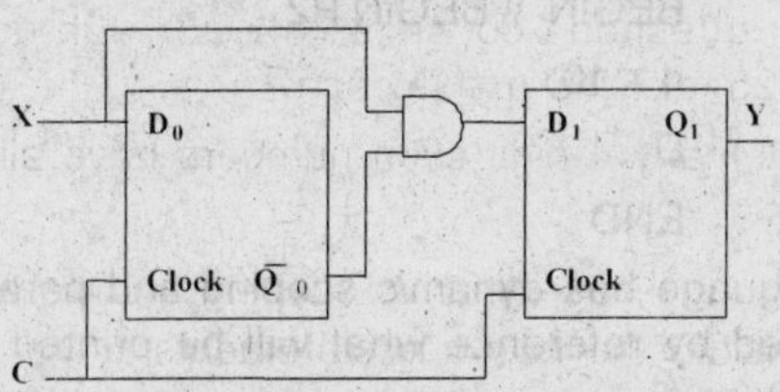

Consider the following timing diagrams of X and C; the clock period of C ≥ 40 nanosecond. Which one is the correct plot of Y?

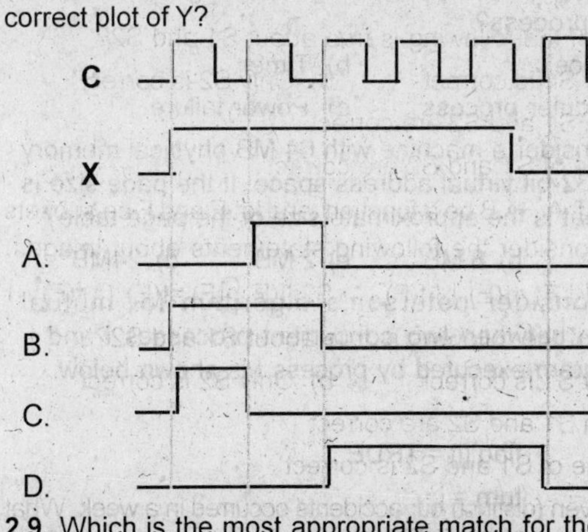

2.9. Which is the most appropriate match for the items in the first column with the items in the second column?

X. Indirect addressing I. Array implementation

Y. Indexed addressing II. Writing relocatable code

Z. Base register addressing III. Passing array as parameter

a) (X, III), (Y, I), (Z, II) b) (X, II), (Y, III), (Z, I)

c) (X, III), (Y, II), (Z, I) d) (X, I), (Y, III), (Z, II)

2.10. The 2's complement representation of $(-539)_{10}$ in hexadecimal is

a) ABE b) DBC c) DE5 d) 9E7

2.11. Consider the circuit shown below. The output of a 2:1 Mux is given by the function $(ac'+bc)$

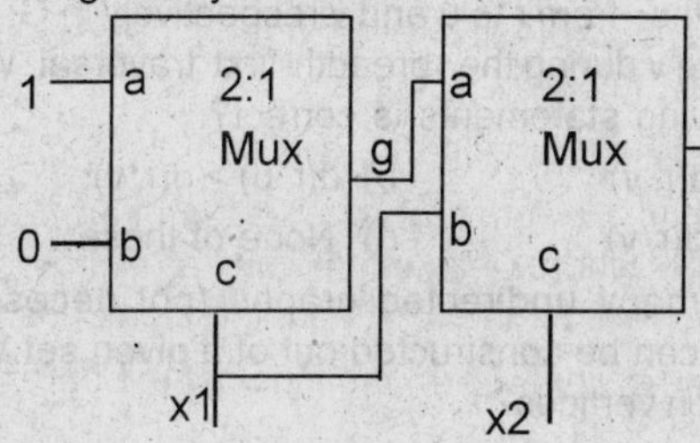

Which of the following is TRUE?

a) $f = x1' + x2$ b) $f = x1'x2 + x1x2'$

c) $f = x1x2 + x1'x2'$ d) $f = x1 + x2'$

2.12. Consider the circuit given below with initial state $Q_0 = 1$, $Q_1 = Q_2 = 0$. The state of the circuit is given by the value $4Q_2 + 2Q_1 + Q_0$.

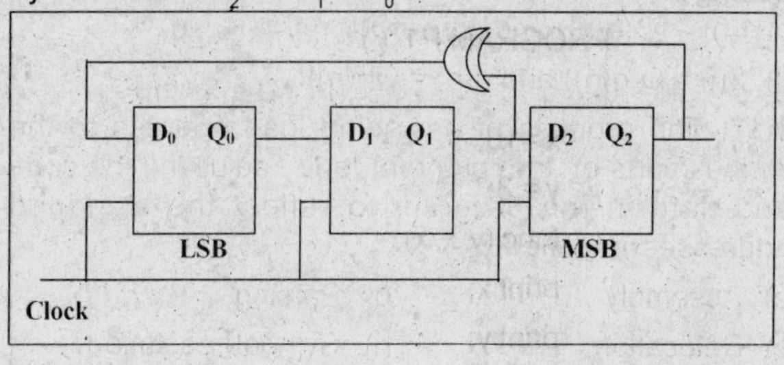

Which of the following is the correct state sequence of the circuit?

a) 1, 3, 4, 6, 7, 5, 2 b) 1, 2, 5, 3, 7, 6, 4

c) 1, 2, 7, 3, 5, 6, 4 d) 1, 6, 5, 7, 2, 3, 4

2.13. Consider the following datapath of a simple non-pipelined CPU. The registers A, B, A_1, A_2, MDR, the bus and the ALU are 8-bit wide. SP and MAR are 16-bit registers. The MUX is of size 8 x (2:1) and the DEMUX is of size 8 × (1:2). Each memory operation takes 2 CPU clock cycles and uses MAR (Memory Address Register) and MDR (Memory Data Register). SP can be decremented locally

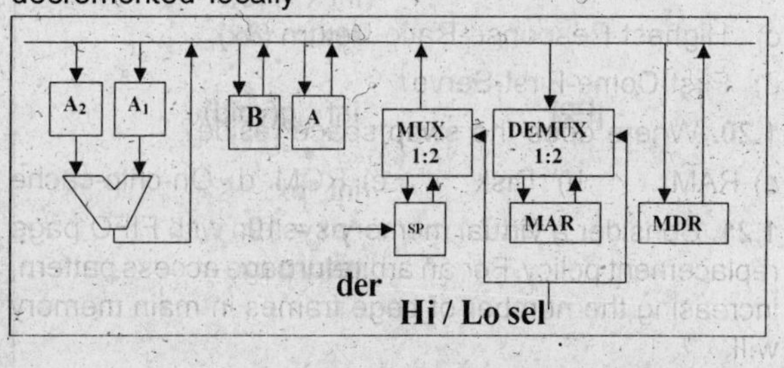

The CPU instruction "push r", where r = A or B, has the specification

$M[SP] \leftarrow r$

$SP \leftarrow SP - 1$

How many CPU clock cycles are needed to execute the "push r" instruction?

a) 2 b) 3 c) 4 d) 5

2.14. Consider an undirected unweighted graph G. Let a breadth-first traversal of G be done starting from a node r. Let d(r,u) and d(r, v) be the lengths of the shortest paths from r to u and v respectively in G. If u is visited before v during the breadth-first traversal. Which of the following statements is correct?

a) d(r, u) < d(r, v) b) d(r, u) > d(r, v)

c) d(r, u) $\leq$ d(r, v) d) None of these

2.15. How many undirected graphs (not necessarily connected) can be constructed out of a given set V={v_1, v_2, ... v_n} of n vertices?

a) n(n-1)/2 b) 2^n c) n! d) $2^{n(n-1)/2}$

2.16. What is the minimum number of stacks of size n required to implement a queue of size n?

a) One b) Two c) Three d) Four

2.17. What is printed by the print statements in thr program P1 assuming call by reference parameter passing?

```
PROGRAM P1 ( )
{
    x =10;
    y = 3;
    funcl(y, x, x);
    print x;
    print y;
}
funcl(x, y, z)
{
    y =y + 4;
    z = x + y + z;
}
```

a) 10, 3 b) 31, 3 c) 27, 7 d) None

2.18. Consider the following three C functions.

```
[P1]        int * g (void)
            {
                int x = 10;
                return (&x);
            }
[P2]        int * g(void)
            {
                int *px;
                *px = 10;
                return px;
            }
[P3]        int * g(void)
            {
                int *px;
    px=(int*)malloc(sizeof(int));
                *px =10;
                return px;
            }
```

Which of the above three functions are likely to cause problems with pointers?

a) Only P3 b) Only P1 and P3

c) Only P1 and P2 d) P1, P2 and P3

2.19. Consider the following program.

```
PROGRAM P2
    VAR n : int ;
    Procedure W(var x : int )
    BEGIN
    x = x + 1;
    Print x;
    END
    Procedure D
    BEGIN
    VAR n : int;
    n = 3
    W(n);
    END
    BEGIN \\ BEGIN P2
    n = 10;
    D;
    END
```

If the language has dynamic scoping and parameters are passed by reference what will be printed by the program?

a) 10 b) 11 c) 3 d) None

2.20. Which of the following does not interrupt a running process?

a) A device b) Timer

c) Scheduler process d) Power failure

2.21. Consider a machine with 64 MB physical memory and a 32-bit virtual address space. If the page size is 4KB. What is the approximate size of the page table?

a) 16MB b) 8 MB c) 2 MB d) 24MB

2.22. Consider peterson's algorithm for mutual exclusion between two concurrent processes i and j. The program executed by process i is shown below.

```
REPEAT
    flag [i] = TRUE;
    turn = j;
    WHILE (P) no-op;
Enter critical section. Perform actions,THEN exit critical section
```

Flag [i] = FALSE;

Perform other non-critical section actions.

Until FALSE;

For the program to guarantee mutual exclusion, the predicate P in the while loop should be

a) flag [j] = true and turn = i

b) flag [j] = true and turn = j

c) flag [i] = true and turn = j

d) flag [i] = true and turn = i

2.23. R(A, B, C, D) is a relation. Which of the following does not have a lossless-join. Dependency preserving BCNF decomposition?

a) $A \to B \quad B \to CD$ b) $A \to B, B \to C, C \to D$

c) $AB \to C, C \to AD$ d) $A \to BCD$

2.24. Which of the following relational calculus expressions is not safe?

a) $\{t \mid \exists u \in R_1 (t[A] = u[A]) \wedge \neg \exists s \in R_2 (t[A] = s[A])\}$

b) $\{t \mid \forall u \in R_1 (u[A] = \text{"x"} \Rightarrow \exists s \in R_2 (t[A] \wedge s[A] = u[A]))\}$

c) $\{t \mid \neg (t \in R_1)\}$

d) $\{t \mid \exists u \in R_1 (t[A] = u[A]) \wedge \exists s \in R_2 (t[A] = s[A])\}$

2.25. Consider a relation geq which represents "greater than or equal to", that is, (x, y) ∈ geq only y ≥ x.

```
create table geq
(lb integer not null
ub integer not null
primary key lb
foreign key (ub) references geq on delete cascade)
```

Which of the following is possible if a tuple (x, y) is deleted?

a) A tuple (z, w) with z > y is deleted

b) A tuple (z, w) with z > x is deleted

c) A tuple (z, w) with w < x is deleted

d) The detection of(x, y) is prohibited

ANSWER KEY

1. 1 a	1. 2 b	1. 3 c	1. 4 a	1. 5 b
1. 6 c	1. 7 b	1. 8 d	1. 9 d	1. 10 d
1.11 a	1. 12 c	1. 13 d	1. 14 c	1. 15 b
1. 16 b	1. 17 c	1. 18 a	1. 19 d	1. 20 b
1. 21 c	1. 22 d	1. 23 c	1. 24 d	1. 25 c
2. 1 b	2. 2 c	2. 3 a	2. 4 b	2. 5 d
2. 6 d	2. 7 b	2. 8 a	2. 9 a	2. 10 c
2. 11 c	2. 12 b	2. 13 b	2. 14 c	2. 15 d
2. 16 b	2. 17 b	2. 18 c	2. 19 d	2. 20 c
2. 21 c	2. 22 b	2. 23 c	2. 24 c	2. 25 c

GATE - 2000

COMPUTER SCIENCE & ENGINEERING

Duration : Three Hours **Maximum Marks : 150**

1. This question consists of 23 multiple choice questions, each carrying one mark.

1.1. The minimum number of cards to be dealt from an arbitrary shuffled deck of 52 cards to guarantee that three cards are from same suit is

a) 3 b) 8 c) 9 d) 12

1.2. An n × n array v is defined as follows

v [i, j] = i – j for all i, j, $1 \le i \le n$, $1 \le j \le n$,

The sum of the elements of the array v, is

a) 0 b) n – 1 c) $n^2 - 3n + 2$ d) $n^2 (n + 1) / 2$

1.3. The determinant of the matrix $\begin{bmatrix} 2 & 0 & 0 & 0 \\ 8 & 1 & 7 & 2 \\ 2 & 0 & 2 & 0 \\ 9 & 0 & 6 & 1 \end{bmatrix}$ is

a) 4 b) 0 c) 15 d) 20

1.4. Let S and T be languages over Σ ={a, b} represented by the regular expressions (a + b*)* and (a + b)*, respectively. Which of the following is TRUE?

a) $S \subset T$ b) $T \subset S$ c) S = T d) $S \cap T = \phi$

1.5. Let L denote the language generated by the grammar $S \rightarrow 0S0 \mid 00$. Which of the following is TRUE?

a) $L = 0^+$

b) L is regular but not 0^+

c) L is context free but not regular

d) L is not context free

1.6. The number 43 in 2's complement representation is

a) 01010101 b) 11010101

c) 00101011 d) 10101011

1.7. To put the 8085 microprocessor in the wait state

a) Lower the HOLD input

b) Lower the READY input

c) Raise the HOLD input

d) Raise the READY input

1.8. Comparing the time T1 taken for a single instruction on a pipelined CPU with time T2 taken on a non-pipelined but identical CPU, we can say that

a) T1 $\le$ T2 b) T1 $\ge$ T2 c) T1 < T2

d) T1 is T2 plus the time taken for one instruction fetch cycle

1.9. The 8085 microprocessor responds to the presence of an interrupt

a) As soon as the TRAP pin becomes 'high'

b) By checking the TRAP pin for 'high' status at the end of each instruction fetch

c) By checking the TRAP pin for 'high' status at the end of the execution of each instruction

d) By checking the TRAP pin for 'high status at regular intervals

1.10. The most appropriate matching for the following pairs

X. Indirect addressing I. Loops

Y. Immediate addressing II. Pointers

Z. Autodecrement addressing II. Constants

Is

a) X – III, Y – II, Z – I b) X – I, Y – III, Z – II

c) X – II, Y – III, Z – I d) X – III, Y – I, Z – II

1.11. The following C declarations

```
struct node {
    int i;
    float j;
};
struct node *s[10];
```

define s to be

a) An array, each element of which is a pointer to a structure of type node

b) A structure of 2 fields, each field being a pointer to an array of 10 elements

c) A structure of 3 fields, an integer, a float, and an array of 10 elements

d) An array, each element of which is a structure of type node

1.12. The most appropriate matching for the following pairs

X: m = malloc(5); m = NULL; I. using dangling pointers

Y: free(n); n $\rightarrow$ value = 5; II. using unintilized pointers

Z: char *p; *p = 'a'; III. lost memory

Is

a) X – I, Y – III, Z – II b) X – II, Y – I, Z – III

c) X – III, Y – II, Z – I d) X – III, Y – I, Z – II

1.13. The most appropriate matching for the following pairs

X : depth-first search I. heap

Y : breadth-first search II. queue

Z : sorting III. stack

Is

a) X – I, Y – II, Z – III b) X – III, Y – I, Z – II

c) X – III, Y – II, Z – I d) X – II, Y – III, Z – I

1.14. Consider the following nested representation of binary trees: (X Y Z) indicates Y and Z are the left and right subtrees, respectively, of node X. Note that Y and Z may be NULL, or further nested. Which of the following represents a valid binary tree

a) (1 2 (4 5 6 7)) b) (1 ((2 3 4) 5 6) 7)

c) (1 (2 3 4) (5 6 7)) d) (1 (2 3 NULL) (4 5))

1.15. Let s be a sorted array of n integers. Let t(n) denote the time taken for the most efficient algorithm to determine if there are two elements with sum less than 1000 in s. Which of the following statements is TRUE?

a) t(n) is O(1) b) $n \le t(n) \le n \log_2 n$

c) $n \log_2 n \le t(n) < \binom{n}{2}$ d) $t(n) = \binom{n}{2}$

1.16. Aliasing in the context of programming languages refers to

a) Multiple variables having the same memory location

b) Multiple variables having the same value

c) Multiple variables having the same identifier

d) Multiple uses of the same variable

1.17. Consider the following C declaration

```
struct {
    short s[5];
  union {
    float y;
    long z;
    }u;
  } t;
```

Assume that objects of the type short, float and long occupy 2 bytes, 4 bytes and 8 bytes, respectively. The memory requirement for variable t, ignoring alignment considerations, is

a) 22 bytes b) 14 bytes

c) 18 bytes d) 10 bytes

1.18. The number of tokens in the following C statement

Printf("i = %d, &i = %x", i, &i); is

a) 3 b) 26 c) 10 d) 21

1.19. Which of the following derivations does a top-down parser use while parsing an input string? The input is assumed to be scanned in left to right order.

a) Leftmost derivation

b) Leftmost derivation traced out in reverse

c) Right most derivation

d) Right most derivation traced out in reverse

1.20. Which of the following need not necessarily be saved on a context switch between processes?

a) Genaral purpose registers b) Translation lookaside buffer

c) Program counter d) All of the above

1.21. Let m[0] .... m [4] be mutexes (binary semaphores) and P[0]...P [4] be processes. Suppose each process P[i] executes the following:

wait(m [i]); wait (m [(i+1)mod 4]);

....

release (m [i]); release (m [(i+1)mod 4]);

This could cause

a) Thrashing b) Deadlock

c) Starvation, but not deadlock d) None of the above

1.22. B^+-trees are preferred to binary trees in databases because

a) Disk capacities are greater than memory capacities

b) Disk access is much slower than memory access

c) Disk data transfer rates are much less than memory data transfer rates

d) Disks are more reliable than memory

1.23. Given the relations

Employee (name, salary, deptno), and

Department (deptno, deptname, address)

Which of the following queries cannot be expressed using the basic relational algebra operations

$(\sigma, \pi, \times, \bowtie, \cup, \cap, -)$?

a) Department address of every employee

b) Employees whose name is the same as their department name

c) The sum of all employee salaries

d) All employees of a given department

2. This question consists of 26 multiple choice questions, eachcarrying two marks.

2.1. X, Y and Z are closed intervals of unit length on the real line. The overlap of X and Y is half a unit. The overlap of Y and Z is also half a unit. Let the overlap of X and Z be k units. Which of the following is TRUE?

a) k must be 1 b) k must be 0

c) k can take any value between 0 and 1

d) None of the above

2.2. E_1 and E_2 are events in a probability space satisfying the following constraints

- $Pr(E_1)=Pr(E_2)$
- $Pr(E_1 \cup E_2) = 1$
- E_1 and E_2 are independent

The value of $Pr(E_1)$, the probability of the event E_1, is

a) 0 b) 1/4 c) 1/2 d) 1

2.3. Let $S = \sum_{i=3}^{100} i \log_2 i$ and $T = \int_2^{100} x \log_2 x\, dx$

Which is the following statements is TRUE?

a) S > T b) S = T

c) S < T and 2S > T d) $2S \le T$

2.4. A polynomial p(x) satisfies the following

p (1) = p (3) = p (5) = 1

p (2) = p (4) = -1

The minimum degree of such a polynomial is

a) 1   b) 2   c) 3   d) 4

2.5. A relation R is defined on the set of integers as xRy iff (x + y) is even. Which of the following statements is TRUE?

a) R is not an equivalence relation

b) R is an equivalence relation having 1 equivalence class

c) R is an equivalence relation having 2 equivalence classes

d) R is an equivalence relation having 3 equivalence classes

2.6. Let P(S) denote the power set of S. which of the following is always TRUE?

a) P(P(S)) =P(S)   b) $P(S) \cap P(P(S)) = \{\phi\}$

c) $P(S) \cap S = P(S)$ d) $S \notin P(S)$

2.7. Let a, b, c ,d be propositions. Assume that the equivalences $a \leftrightarrow (b \vee \neg b)$ and $b \leftrightarrow c$ hold. Then the truth value of the formula $(a \wedge b) \rightarrow ((a \wedge c) \vee d)$ is always

a) True   b) False

c) Same as the truth value of b

d) Same as the truth value of d

2.8. What can be said about a regular language L over {a} whose minimal finite state automaton has two states?

a) L must be $\{a^n \mid n \text{ is odd}\}$

b) L must be $\{a^n \mid n \text{ is even}\}$

c) L must be $\{a^n \mid n \geq 0\}$

d) Either L must be $\{a^n \mid n \text{ is odd}\}$, or L must be $\{a^n \mid n \text{ is even}\}$

2.9. Consider the following decision problems:

P(1) Does a given finite state machine accept a given string

P(2) Does a given context free grammar generate an infinite number of strings.

Which of the following statements is TRUE?

a) Both (P1) and (P2) are decidable

b) Neither (P1) nor (P2) are decidable

c) Only (P1) is decidable

d) Only (P2) is decidable

2.10. The simultaneous equations on the Boolean variables x, y,z and w,

$$x + y + z = 1$$

$$xy = 0$$

$$xz + w = 1$$

$$xy + \overline{z}\,\overline{w} = 1$$

Have the following solution for x, y, z and w , respectively

a) 0 1 0 0   b) 1 1 0 1   c) 1 0 1 1   d) 1 0 0 0

2.11. Which function does NOT implement the Karnaugh map given below?

WZ→ / xy ↓	00	01	11	10
00	0	X	0	0
01	0	X	1	1
11	1	1	1	1
10	0	X	0	0

a) (w + x)y   b) xy + yw

c) $(w + x)(\overline{w} + y)(\overline{x} + y)$ d) None of the above

2.12. The following arrangement of master-slave flip flops

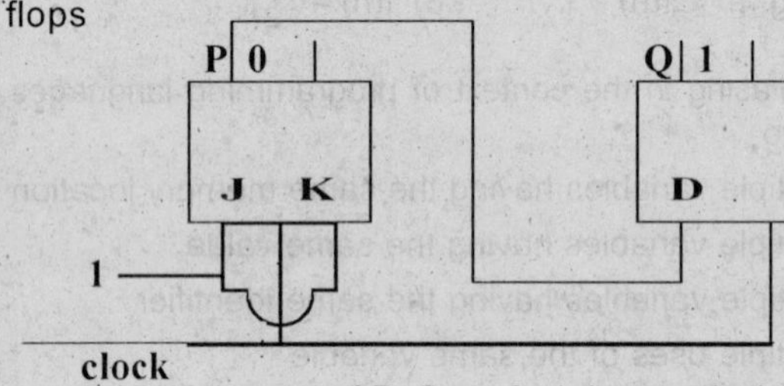

has the initial state of P, Q as 0, 1 (respectively). After three clock cycles the output state P, Q is (respectively),

a) 1, 0   b) 1,1   c) 0, 0   d) 0,1

2.13. A graphics card has on board memory of 1MB. Which of the following modes can the card not support?

a) 1600 × 400 resolution with 256 colours on a 17 inch monitor

b) 1600 × 400 resolution with 16 million colours on a 14 inch monitor

c) 800 × 400 resolution with 16 million colours on a 17 inch monitor

d) 800 × 800 resolution with 256 colours on a 14 inch monitor

2.14. Consider the values $A = 2.0 \times 10^{30}$, $B = -2.0 \times 10^{30}$, C = 1.0, and the sequence

X := A + B

Y := A + C

X := X + C

Y := Y + B

executed on a computer where floating point numbers are represented with 32 bits. The values for X and Y will be

a) X =1.0, Y =1.0   b) X =1.0, Y =0.0

c) X =0.0, Y =1.0   d) X =0.0, Y =0.0

2.15. Suppose you are given an array s[1..n] and a procedure reverse(s, i, j) which reverse the order of elements in s between positions i and j (both inclusive). What does the following sequence do, where $1 \leq k < n$.

```
reverse (s, 1, k);
reverse (s, k+1, n);
reverse (s, 1, n);
```

a) Rotates s left by k positions b) Leaves s unchanged
c) Reverse all elements of s d) None of the above

2.16. Let LASTPOST, LASTIN and LASTPRE denotes the last vertex visited in a postorder, inorder and preorder traversal respectively, of a complete binary tree. Which of the following is always TRUE?
a) LASTIN = LASTPOST b) LASTIN = LASTPRE
c) LASTPRE = LASTPOST d) None of the above

2.17. Consider the following functions

$f(n) = 3n^{\sqrt{n}}$ $g(n) = 2^{\sqrt{n}\log_2 n}$ $h(n) = n!$

which of the following is TRUE?
a) h(n) is O(f(n)) b) h(n) is O(g(n))
c) g(n) is not O(f(n)) d) f(n) is O(g(n))

2.18. Let G be an undirected connected graph with distinct edge weights. Let e_{max} be the edge with maximum weight and e_{min} the edge with minimum weight. Which of the following statements is FALSE ?
a) Every minimum spanning tree of G must contain e_{min}.
b) If e_{max} is in a minimum spanning tree, then its removal must disconnect G
c) No minimum spanning tree contains e_{max}.
d) G has a unique minimum spanning tree.

2.19. Let G be an undirected graph. Consider a depth-first traversal of G, and let T be the resulting depth-first search tree. Let u be a vertex in G and let v be the first new (unvisited) vertex visited after visiting u in the traversal. Which of the following statements is always TRUE?
a) {u, v} must be edge in G, and u is a descendant of v in T.
b) {u, v} must be edge in G, and v is a descendant of u in T.
c) If {u,v} is not an edge in G, and u is a leaf in T.
d) If {u,v} is an edge in G,then u and v must have the same parent in T.

2.20. The value of j at the end of the execution of the following C program

```
int incr(integer i)
{
static integer count = 0;
count = count + i;
return (count);
}
main()
{
int i, j;
for (i=0; i<=4; i++)
j = incr(i);
}
```

is
a) 10 b) 4 c) 6 d) 7

2.21. Given the following expression grammar:

$E \rightarrow E * F \mid F + E \mid F$

$F \rightarrow F - F \mid id$

Which of the following is TRUE?
a) * has higher precedence than +
b) – has higher precedence than *
c) + and – have same precedence
d) + has higher precedence than *

2.22. Suppose the time to service a page fault is on the average 10 milliseconds, while a memory access takes 1 microsecond. Then a 99% hit ratio results in average memory access time of
a) 1.9999 milliseconds b) 1 millisecond
c) 9.999 micro seconds d) 1.9999 microseconds

2.23. Which of the following is NOT a valid deadlock prevention scheme?
a) Release all resources before requesting a new resource
b) Number the resources uniquely and never request a lower numbered resource than the last one requested
c) Never request a resource after releasing any resource
d) Request and be allocated all required resources before execution

2.24. Given the following relation instance

X	Y	Z
1	4	2
1	5	3
1	6	3
3	2	2

Which of the following functional dependencies are satisfied by the instance?
a) $XY \rightarrow Z$ and $Z \rightarrow Y$ b) $YZ \rightarrow X$ and $Y \rightarrow Z$
c) $YZ \rightarrow X$ and $X \rightarrow Z$ d) $XZ \rightarrow Y$ and $Y \rightarrow X$

2.25. Given relations r(w, x) and s(y, z) , the result of

```
select distinct w, x
from r, s
```

is guaranteed to be the same as r, provided:
a) r has no duplicates and s is non-empty
b) r and s have no duplicates
c) s has no duplicates and r is non-empty
d) r and s have the same number of tuples

2.26. In SQL, relations can contain null values, and comparisions with null values are treated as unknown. Suppose all comparisions with a null value are treated as false. Which of the following pairs is not equivalent?
a) x = 5 not(not(x=5))
b) x = 5 x > 4 and x < 6, where x is an integer
c) $x \neq 5$ not (x = 5)
d) None of the above.

ANSWER KEY

1. 1 c	1. 2 a	1. 3 a	1. 4 c	1. 5 b
1. 6 c	1. 7 b	1. 8 b	1. 9 c	1. 10 c
1.11 a	1. 12 d	1. 13 c	1. 14 c	1. 15 a
1. 16 a	1. 17 c	1. 18 c	1. 19 a	1. 20 b
1. 21 b	1. 22 d	1. 23 c	1. 24	1. 25
2. 1 c	2. 2 d	2. 3 a	2. 4 d	2. 5 c
2. 6 b	2. 7 a	2. 8 d	2. 9 a	2. 10 c
2. 11 d	2. 12 a	2. 13 b	2. 14 b	2. 15 a
2. 16 b	2. 17 d	2. 18 c	2. 19 c	2. 20 a
2. 21 b	2. 22 d	2. 23 c	2. 24 b	2. 25 a
2. 26 c				

GATE - 1999
COMPUTER SCIENCE & ENGINEERING

Duration : Three Hours **Maximum Marks : 150**

1.This question consists of 25 multiple choice questions each carrying one mark.

1.1. Suppose that the expectation of a random variable X is 5. Which of the following statements is TRUE?

a) There is a sample point at which X has the value 5.

b) There is a sample point at which X has value greater than 5.

c) There is a sample point at which X has value greater than or equal to 5.

d) None of the above.

1.2. The number of binary relations on a set with n elements Is:

a) n^2 b) 2^n c) $_{2}n^{2}$ d) None

1.3. The number of binary strings of n zeroes and k ones such that no two ones are adjacent is

a) ${}^{n-1}C_k$ b) ${}^{n}C_k$ c) ${}^{n}C_{k+1}$ d) None

1.4. Consider the regular expression (0+1) (0+1) ..... n times. The minimum state finite automaton that recognizes the languages represented by this regular expression contains.

a) n states b) n + 1 states c) n + 2 states d) None

1.5. Context free languages are closed under:

a) Union, intersection b) Union, Kleene closure

c) Intersection, complement

d) Complement, Kleene closure

1.6 Let L_D is the set of all languages accepted by a PDA by final state and L_E the set of all languages accepted by empty stack. Which of the following is TRUE?

a) $L_D = L_E$ b) $L_D \supset L_E$ c) $L_D \subset L_E$ d) None

1.7. Which of the following expression is not equivalent to $\bar{x}$?

a) x NAND x b) x NOR x c) x NAND 1 d) x NOR 1

1.8. Which of the following implements the Karnaugh map shown below?

AB \ CD	00	01	11	10
00	0	0	1	0
01	x	x	1	x
11	0	1	1	0
10	0	1	1	0

a) $\bar{A}B + CD$ b) $D(C + A)$

c) $AD + \bar{A}B$ d) None

1.9. Listed below are some operating system abstractions (in the left column) and the hardware components or matching of pairs is correct?

A. Thread	I. Interrupt
B. Virtual address space	II. Memory
C. File system	III. CPU
D. Single	IV. Disk

a) A – II, B – IV, C – III, D – I

b) A – I, B – II, C – III, D – IV

c) A – III, B – II, C – IV, D – I

d) A – IV, B – I, C – II, D – III

1.10. Which of the following disk scheduling strategies is likely to give the best throughput?

a) Farthest cylinder next b) Nearest cylinder next

c) First come first served d) Elevator algorithm

1.11. System calls are usually invoked by using

a) A software interrupt b) Polling

c) An indirect jump d) A privileged instruction

1.12. A sorting technique is called stable if:

a) It takes O(n log n) time

b) It maintains the relative order of occurrence of non-distinct elements

c) It uses divide and conquer paradigm

d) It takes O(n) space

1.13. Suppose we want to arrange the n numbers stored in an array such that all negative values occur before all positive ones. Minimum number of exchanges required in the worst case is:

a) n-1 b) n c) n + 1 d) None

1.14. If one uses straight two-way merge sort algorithm to sort the following elements in ascending order

20, 47, 15, 8, 9, 4, 40, 30, 12, 17

then the order of these elements after the second pass of the algorithm is:

a) 8, 9, 15, 20, 47, 4, 12, 17, 30, 40

b) 8, 15, 20, 47, 4, 9, 30, 40, 12, 17

c) 15, 20, 47, 4, 8, 9, 12, 30, 40, 17

d) 4, 8, 9, 15, 20, 47, 12, 17, 30, 40

1.15. The number of articulation points of the following graph is:

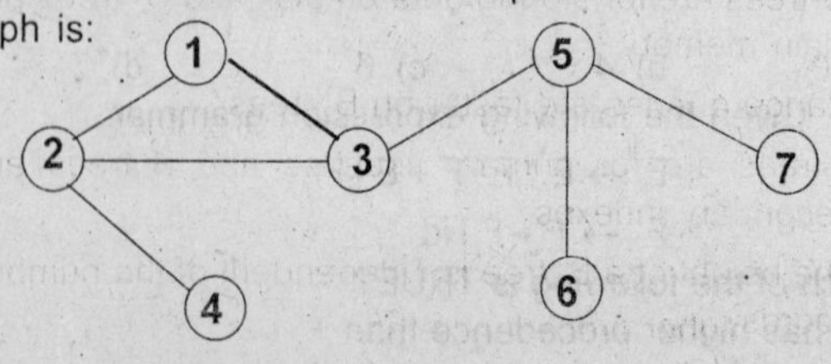

a) 0 b) 1 c) 2 d) 3

1.16. If n is a power of 2, then the minimum number of multipli- cations needed to compute a^n is

a) $\log_2 n$ b) $\sqrt{n}$ c) n – 1 d) n

1.17. Which of the following is the most powerful parsing method?

a) LL(1) b) Canonical LR c) SLR d) LALR

1.18. Consider the join of a relation R with a relation S. If R has m tuples and S has n tuples, then the maximum and minimum sizes of the join respectively are:

a) m + n and 0 b) mn and 0

c) m+n and |m-n| d) mn and m+n

1.19. The relational algebra expression equivalent to the following tuple calculus expression

{t / t ∈ r ∧ (t[A] = 10 ∧ t [B] = 20)} is :

a) $\sigma_{(A=10 \vee B=20)}(r)$ b) $\sigma_{(A=10)}(r) \cup \sigma_{(B=20)}(r)$

c) $\sigma_{(A=10)}(r) \cap \sigma_{(B=20)}(r)$ d) $\sigma_{(A=10)}(r) - \sigma_{(B=20)}(r)$

1.20. Booth's coding in 8 bits for the decimal number –57 is:

a) 0 – 1 0 0 + 1 0 0 0 b) 0 – 1 0 0 + 1 0 0 – 1

c) 0 – 1 + 1 0 0 – 1 0 + 1 d) 0 0 – 1 0 + 1 0 0 – 1

1.21. The maximum gate delay for any output to appear in an array multiplier for multiplying two n bit number is:

a) $O(n^2)$ b) O(n) c) O(log n) d) O(1)

1.22. The main memory of a computer has 2c.m. blocks while the cache has 2c blocks. If the cache uses the set associative mapping scheme with 2 blocks per set, then block k of the main memory maps to the set:

a) (k mod m) of the cache b) (k mod c) of the cache

c) (k mod 2c) of the cache d) (k mod 2cm) of the cache

1.23. The Newton-Raphson method is to be used to find the root of the equation f (x)=0 where x_0 is the initial approximation and **f′** is the derivative of f. The method converges:

a) Always b) Only if f is a polynomial

c) Only if $f(x_0) < 0$ d) None

1.24. Let R=(A, B, C, D, E, F) be a relation scheme with the following dependencies:

C → F, E → A, EC → D, A → B.

Which of the following is a key for R?

a) CD b) EC c) AE d) AC

1.25. Which of the following is CORRECT?

a) B-trees are for storing data on disk and B^+ trees are for main memory.

b) Range queries are faster on B^+ trees.

c) B-trees are for primary indexes and B^+ trees are for secondary indexes

d) The height of a B^+ tree is independent of the number of records.

2. This question consists of 25 multiple choice questions. Each carrying 2 marks

2.1. Consider two events E_1 and E_2 such that probability of E_1, $Pr[E_1] = 1/2$. Probability of E_2, $Pr[E_2] = 1/3$, and probability of E_1 and E_2, $Pr[E_1 \text{ and } E_2] = 1/5$. Which of the following statements is/are TRUE?

a) $Pr[E_1 \text{ or } E_2]$ is 2/3

b) Events E_1 and E_2 are independent

c) Events E_1 and E_2 are not independent

d) $Pr[E_1 / E_2] = 4/5$

2.2. Two girls have picked 10 roses, 15 sunflowers and 14 daffodils. What is the number of ways they can divide the flowers amongest themselves?

a) 1638 b) 2100 c) 2640 d) None

2.3. Let L be a set with a relation R which is transitive, anti-symmetric and reflexive and for any two elements a, b ∈ L let the least upper bound lub(a, b) and the greatest lower bound glb(a, b) exist. Which of the following is/are TRUE?

a) L is a poset b) L is a boolean algebra

c) L is a lattice d) None

2.4. If L1 is context free language and L2 is a regular language which of the following is/are FALSE?

a) L1 – L 2 is not context free

b) L1 ∩ L 2is context free

c) ~L1 is context free

d) ~ L 2is context free

2.5. Given the programming constructs

I. assignment

II. for loops where the loop parameter can not be changed with in the loop

III. if-then-else

IV. forward goto

V. arbitrary goto

VI. non-recursive procedure call

VII. recursive procedure/function call

VIII. repeat loop

Which constructs will you not include in a programming language such that it should be possible to program the terminates (i.e. halting) function in the same programming language

a) (II), (III), (IV) b) (V), (VII), (VIII)

c) (VI), (VII), (VIII) d) (III), (VII), (VIII)

2.6. For the schedule given below, which of the following is correct:

1. Read A	
2.	Read B
3. Write A	
4.	Read A
5.	Write A
6.	Write B
7. Read B	
8. Write B	

a) This schedule is serialisable and can occur in a scheme using 2PL protocal

b) This schedule is serialisable but can not occur in a scheme using 2PL protocal

c) This schedule is not serialisable but can occur in a scheme using 2PL protocal

d) This schedule is not serialisable and can not occur in a scheme using 2PL protocal

2.7. Consider the schema R=(S T U V) and the dependencies $S \rightarrow T$, $T \rightarrow U$, $U \rightarrow V$ and $V \rightarrow S$. Let R=(R1 and R2) be a decomposition such that $R1 \cap R2 = \emptyset$. The decomposition is:

a) Not in 2NF b) In 2NF but not in 3NF

c) In 3NF but not in 2NF d) In both 2NF and 3NF

2.8. Consider the circuit shown below. In a certain steady state, the line Y is at '1'. What are the possible values of A, B and C in this state?

a) A=0, B=0, C=1

b) A=0, B=1, C=1

c) A=1, B=0, C=1

d) A=1, B=1, C=1

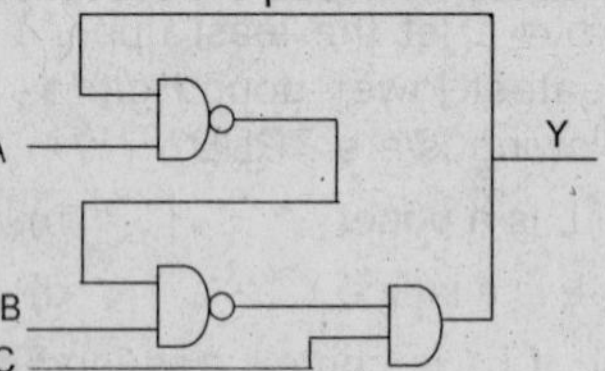

2.9. Which of the following sets of component(s) is/are sufficient to implement any arbitrary boolean function?

a) XOR gates, NOT gates b) 2 to 1 multiplexors

c) AND gates, XOR gates

d) Three-input gates that output (A.B) + C for the inputs A, B, and C

2.10. A multi-user, multi-processing operating system cannot be implemented on hardware that does not support:

a) Address translation b) DMA for disk transfer

c) At least two modes of CPU execution (privileged and non-privileged).

d) Demand paging

2.11. Which of the following is/are advantages of virtual memory?

a) Faster access to memory on an average

b) Processes can be given protected address spaces.

c) Linker can assign addresses independent of where the program will be loaded in physical memory

d) Programs larger than the physical memory size can be run

2.12. Which of the following actions is/are typically not performed by the operating system when switching context from process A to process B?

a) Saving current register values and restoring saved register values for process B

b) Changing address translation tables

c) Swapping out the memory image of process A to the disk

d) Invalidating the translation look-aside buffer

2.13. Consider the following program in a language that has dynamic scoping.

```
VAR x : real;
Procedure show :
  BEGIN print (x);END;
  Procedure small;
  VAR x : real;
BEGIN x : = 0.125; show; END;
BEGIN x : = 0.25;
  show ; small
END.
```

Then the output of the program is :

a) 0.125 0.125 b) 0.25 0.25

c) 0.25 0.125 d) 0.125 0.25

2.14. The number of tokens in the Fortran statement DO 10 I=1.25 Is

a) 3 b) 4 c) 5 d) None

2.15. A grammar that is both left and right recursive for non-terminal, is

a) Ambiguous b) Unambiguous

c) Information is not sufficient to decide whether it is ambiguous or unambiguous. d) None

2.16. The number of full and half-adders required to addition 16-bit numbers is:

a) 8 half-adders, 8 full adders

b) 1half-adder, 15 full adders

c) 16 half-adders, 0 full adders

d) 4 half-adders, 12 full adders

2.17. Zero has two representations in :

a) Sign magnitude b) 1's complement

c) 2's complement d) None

2.18. RAID configurations of disks are used to provide:

a) Fault-tolerance b) High speed

c) High data density d) None

2.19. Arrange the following configurations for CPU in decreasing order of operating speeds:Hardware control, Vertical micro-programming, horizontal micro-programming

a) Hardwired control, Vertical micro-programming, Horizontal micro-programming

b) Hardwired control, Horizontal micro-programming, Vertical micro-programming.

c) Horizontal micro-programming, Vertical micro-programming, Hardwired control,

d) Vertical micro-programming, Horizontal micro-programming, Hardwired control.

2.20. The minimum number of record movements required to merge five files A (with 10 records), B (with 20 record), C(with 15 records), D (with 5 records) and E(with 25 records) is:

a) 165 b) 90 c) 75 d) 65

2.21. If $T_1 = O(1)$, give the correct matching for the following pairs:

(M) $T_n = T_{n-1} + n$	(U) $T_n = O(n)$
(N) $T_n = T_{n/2} + n$	(V) $T_n = O(n \log n)$
(O) $T_n = T_{n/2} + n \log n$	(W) $T_n = O(n^2)$
(P) $T_n = T_{n-1} + \log n$	(X) $T_n = O(\log^2 n)$

a) M – W, N – V, O – U, P – X

b) M – W, N – U, O – V, P – V

c) M – V, N – W, O – X, P – U

d) M – W, N – U, O – V, P – X

2.22. The main difference(s) between a CISC and a RISC processor is/are that a RISC processor typically:

a) has fewer instructions

b) has fewer addressing modes

c) has more registers

d) is easier to implement using hard-wired control logic

2.23. A certain processor supports only the immediate and the direct addressing modes. Which of the following programming language features cannot be implemented on this processor?

a) Pointers b) Arrays c) Records

d) Recursive procedures with local variable

2.24. Consider the following C function definition

```
int trial (int a, int b, int c)
{
    if ((a >= b)&&(c < b)) return b;
    else if (a >= b) return trail (a, b, c);
    else return trail (b,a,c);
}
```

The function Trail:

a) finds the maximum of a, b and c

b) finds the minimum of a, b and c

c) finds the middle of a, b and c d) None

2.25. Which of the following is/are correct?

a) An SQL query automatically eliminates duplicates

b) An SQL query will not work if there are no indexes on the relations

c) SQL permits attribute names to be repeated in the same relation d) None

ANSWER KEY

1. 1 a	1. 2 c	1. 3 d	1. 4 c	1. 5 b
1. 6 a	1. 7 d	1. 8 b	1. 9 c	1. 10 b
1.11 a	1. 12 b	1. 13 d	1. 14 b	1. 15 d
1. 16 a	1. 17 b	1. 18 b	1. 19 c	1. 20 b
1. 21 d	1. 22 b	1. 23 d	1. 24 b	1. 25 b
2. 1 c	2. 2 c	2. 3 a,c	2. 4 a,c	2. 5 b
2. 6 d	2. 7 d	2. 8 a,c,d	2. 9 b,c	
2. 10 a,b,c		2. 11 b,d	2. 12 c,d	2. 13 c
2. 14 c	2. 15 c	2. 16 b	2. 17 a	2. 18 a
2. 19 b	2. 20 d	2. 21 b	2. 22 a,b,c,d	
2. 23 a,b,c,d		2. 24 c	2. 25 d	

GATE - 1998

COMPUTER SCIENCE & ENGINEERING

Duration : Three Hours **Maximum Marks : 150**

1.This question consists of 35 multiple choice questions, each carrying one mark.

1.1. A die is rolled three times. The probability that exactly one odd number turns up among the three outcomes is

a) $\frac{1}{6}$ b) $\frac{3}{8}$ c) $\frac{1}{8}$ d) $\frac{1}{2}$

1.2. Consider the following set of equations.

$$x + 2y = 5$$
$$4x + 8y = 12$$
$$3x + 6y + 3z = 15$$

This set

a) Has unique solution b) Has no solutions

c) Has finite number solutions

d) Has infinite number of solutions

1.3. Which of the following statements applies to the bisection method used for finding roots of functions:

a) Converges within a few iterations

b) Guaranteed to work for all continuous functions

c) Is faster than the Newton-Raphson method

d) Requires that there be no error in determining the sign of the function

1.4. Consider the function $y = |x|$ in the interval [-1, 1]. In this interval, the function is

a) Continuous and differentiable

b) Continuous but not differentiable

c) Differentiable but not continuous

d) Neither continuous not differentiable

1.5. What is the converse of the following assertion?

I stay only if you go

a) I stay if you go b) If I stay then you go

c) If you do not go then I do not stay

d) If I do not stay then you go

1.6. Suppose A is a finite set with n elements. The number of elements in the Largest equivalence relation of A is

a) n b) n^2 c) 1 d) n+1

1.7. Let R_1 and R_2 be two equivalence relations on a set. Consider the following assertions:

I. $R_1 \cup R_2$ is an equivalence relation

II. $R_1 \cap R_2$ is an equivalence relation

Which of the following is correct

a) both assertions are true

b) assertion I is true but assertion II is not true

c) assertion II is true but assertion I is not true

d) neither I nor II is true

1.8. The number of functions from an m element set to an n element set is

a) m + n b) m^n c) n^m d) m * n

1.9. If the regular set A is represented by A =(01 + 1)* and the regular set 'B' is represented by B =((01)* 1*)*, which of the following is TRUE?

a) $A \subset B$ b) $B \subset A$

c) A and B are incomparable d) A = B

1.10. Which of the following set can be recognized by a Deterministic Finite-state Automaton?

a) The numbers 1, 2, 4, 8, ..., 2^n, ... Written in binary

b) The numbers 1, 2, 4, 8, ..., 2^n, ... Written in unary

c) The set of binary strings in which the number of zeros is the same as the number of ones

d) The set {1, 101, 11011, 1110111, ...}

1.11. Regarding the power of recognition of languages, whichof the following statements is FALSE?

a) The non-deterministic finite state automata are equivalent to deterministic finite-state automata

b) Non-deterministic Push-down automata are equivalent to deterministic Push-down automata

c) Non-deterministic Turing machines are equivalent to deterministic Push-down automata

d) Multi-tape Turing machines are equivalent to Single-tape Turing machines

1.12. The string 1101 does not belong to the set represented by

a) 110*(0 + 1) b) 1(0 + 1)*101

c) (10)*(01)*(00 + 11)* d) (00 + (11)* 0)*

1.13. What happens when a bit-string is XORed with itself n-times as shown:

$[B \oplus (B \oplus (B \oplus (B \ldots$ n times]

a) Complements when n is even

b) Complements when n is odd

c) Divides by 2^n always

d) Remains unchanged when n is even

1.14. A multiplexor with a 4 bit data select input is a

a) 4 : 1 multiplexor b) 2 : 1 multiplexor

c) 16 : 1 multiplexor d) 8 : 1 multiplexor

1.15. The threshold level for logic 1 in the TTL family is

a) Any voltage above 2.5 V

b) Any voltage between 0.8 V and 5.0 V

c) Any voltage below 5.0 V

d) Any voltage below V_{cc} but above 2.8 V

1.16. In serial communication employing 8 data bits, a parity bit and 2 stop bits, the minimum band rate required to sustain a transfer rate of 300 characters per second is

a) 2400 band b) 19200 band
c) 4800 band d) 1200 band

1.17. The octal representation of an integer is 342_8. If this were to be treated as an eight bit integer in an 8085 based computer, its decimal equivalent is

a) 226 b) – 98 c) 76 d) – 30

1.18. Which of the following devices should get higher priority in assigning interrupts?

a) Hard disk b) Printer
c) Keyboard d) Floppy disk

1.19. Which of the following addressing modes permits relocation without any change whatsever in the code?

a) Indirect addressing b) Indexed addressing
c) Base register addressing d) PC relative addressing

1.20. Which of the following is TRUE?

a) Unless enabled, a CPU will not be able to process interrupts

b) Loop instructions cannot be interrupted till they complete

c) A processor checks for interrupts before executing a new instruction

d) Only level triggered interrupts are possible on microprocessors

1.21. Which one of the following algorithm design techniques is used in finding all pairs of shortest distances in a graph?

a) Dynamic programming b) Backtracking
c) Greedy d) Divide and Conquer

1.22. Give the correct matching for the following pairs:

A. O (log n) I. Selection
B. O (n) II. Insertion sort
C. O (n log n) III. Binary search
D. O (n^2) IV. Merge sort

a) A – R, B – P, C – Q, D – S
b) A – R, B – P, C – S, D – Q
c) A – P, B – R, C – S, D – Q
d) A – P, B – S, C – R, D – Q

1.23. How many substrings of different lengths (non-zero) can be formed from a character string of length n?

a) n b) n^2 c) 2^n d) n(n+1)/2

1.24. Which of the following statements is FALSE?

a) A tree with n nodes has (n – 1) edges.

b) A labeled rooted binary tree can be uniquely constructed given its postorder and preorder traversal results.

c) A complete binary tree with n internal nodes has (n + 1) leaves

d) The maximum number of nodes in a binary tree of length h is (2^{h+1} – 1)

1.25. In a resident –OS computer, which of the following systems software must reside in the main memory under all situations?

a) Assembler b) Linker
c) Loader d) Compiler

1.26. Which of the following statements is TRUE?

a) SLR parser is more powerful than LALR.

b) LALR parser is more powerful than Canonical LR parser

c) Canonical LR parser is more powerful than LALR parser

d) The parsers SLR, Canonical CR, and LALR have the same power

1.27. Type checking is normally done during

a) lexical analysis b) Syntax analysis
c) Syntax directed translation d) Code optimization

1.28. A linker reads four modules whose lengths are 200, 800, 600 and 500 words, respectively. If they are loaded in that order, what are the relocation constants?

a) 0, 200, 500, 600 b) 0, 200, 1000, 1600
c) 200, 500, 600, 800 d) 200, 700, 1300, 2100

1.29. Which of the following is an example of a spooled device?

a) The terminal used to enter the input data for the C program being executed

b) An output device used to print the output of a number of jobs

c) The secondary memory device in a virtual storage system

d) The swapping area on a disk used by the swapper

1.30. When the result of a computation depends on the speed of the processes involved there is said to be

a) Cycle stating b) Race condition
c) A time lock d) A deadlock

1.31. A counting semaphore was initialized to 10. Then 6P(wait) operations and 4V(signal) operations were completed on this semaphore. The resulting value of the semaphore is

a) 0 b) 8 c) 10 d) 12

1.32. A computer has six tape drives, with n processes competing for them. Each process may need two drives. What is the maximum value of n for the system to be deadlock free?

a) 6 b) 5 c) 4 d) 3

1.33. Given two union compatible relations R_1(A, B) and R_2(C, D), what is the result of the operation R_1 A = C A B = D R_2?

a) $R_1 \cup R_2$ b) $R_1 \times R_2$ c) $R_1 - R_2$ d) $R_1 \cap R_2$

1.34. Which normal form is considered adequate for normal relational database design?

a) 2 NF b) 5 NF c) 4 NF d) 3 NF

1.35. There are five records in a database.

Name	Age	Occupation	Category
Rama	27	CON	A
Abdul	22	ENG	A
Jebiffer	28	DOC	B
Maya	32	SER	D
Dev	24	MUS	C

There is an index file associated with this and it contains the values 1, 3, 2, 5 and 4. Which one of the fields is the index built from?

a) Age b) Name
c) Occupation d) Category

2. This question consist of 20 multiple choice questions, each carrying two marks.

2.1. The rank of the matrix given below is :

$$\begin{bmatrix} 1 & 4 & 8 & 7 \\ 0 & 0 & 3 & 0 \\ 4 & 2 & 3 & 1 \\ 3 & 12 & 24 & 2 \end{bmatrix}$$

a) 3 b) 1 c) 2 d) 4

2.2. Consider the following determinant $\Delta = \begin{vmatrix} 1 & a & bc \\ 1 & b & ca \\ 1 & c & ab \end{vmatrix}$

Which of the following is a factor of Δ ?

a) a + b b) a – b c) a + b + c d) abc

2.3. The binary relation R={(1, 1), (2, 1), (2, 2), (2, 3), (2, 4), (3, 1), (3, 2), (3, 3), (3, 4)} on the set A ={1, 2, 3, 4} is

a) Reflexive, symmetric and transitive
b) Neither reflexive, nor irreflexive but transitive
c) Irreflexive, symmetric and transitive
d) Irreflexive and anti-symmetric

2.4. In a room containing 28 people, there are 18 people who speak English, 15 people who speak Hindi and 22 people who speak Kannada, 9 persons speak both English and Hindi, 11 persons speak both Hindi and Kannada whereas 13 persons speak both Kannada and English. How many people speak all the three languages?

a) 9 b) 8 c) 7 d) 6

2.5. Let L be the set of all binary strings whose last two symbols are the same. The number of states in the minimum state deterministic finite state automaton accepting L is

a) 2 b) 5 c) 8 d) 3

2.6. Which of the following statements is FALSE?

a) Every finite subset of a non-regular set is regular
b) Every subset of a regular set is regular
c) Every finite subset of a regular set is regular
d) The intersection of two regular sets is regular

2.7. The function represented by the Karnaugh map given below is:

A \ BC	00	01	10	11
0	1	0	0	1
1	1	0	0	1

a) A . B b) AB + BC + CA
c) $\overline{B \oplus C}$ d) A . BC

2.8. Which of the following operations in commitative but not assosiative

a) AND b) OR c) NAND d) EXOR

2.9. Formatting of a floppy disk refers to

a) Arranging the data on the disk in contiguous fashion
b) Writing the directory
c) Erasing the system area
d) Writing identification information on all tracks and sectors.

2.10. The address space of 8086 CPU is

a) one Megabyte b) 256 Kilobytes
c) 1 K Megabyte d) 64 Kilobytes

2.11. A complete n-ary tree is one in which every node has 0 or n sons. If x is the number of internal nodes of a complete n-ary tree, the number of leaves in it is given by

a) x(n-1) + 1 b) xn – 1
c) xn + 1 d) x(n +1)

2.12. What value would the following function return for the input x = 95?

```
function fun(x : integer ) : integer ;
BEGIN
  IF x > 100 THEN fun : = x- 10
  ELSE fun : = fun(fun(x + 11))
END;
```

a) 89 b) 90 c) 91 d) 92

2.13. What is the result of the following program?

```
PROGRAM side-effect;
VAR X, result : integer;
Function f(var x : integer ): integer;
BEGIN
    x : = x + 1; F :=x
END
BEGIN
    x := 5
    result := f(x) * f(x);
    writeln (result);
END
```

a) 5 b) 25 c) 36 d) 42

2.14. Let A be a two dimensional array declared as follows:

A : array [1...10] [1...15] of integer;

Assuming that each integer takes one memory location the array is stored in row-major order and the first element of the array is stored at location 100, what is the address of the element A[i][j]?

a) 15i + j + 84 b) 15j + i + 84

c) 10i + j + 89 d) 10j + i + 89

2.15. Faster access to non-local variables is achieved using an array of pointers to activation records, called a

a) stack b) heap

c) display d) activation tree

2.16. The overlay tree for a program is as shown below:

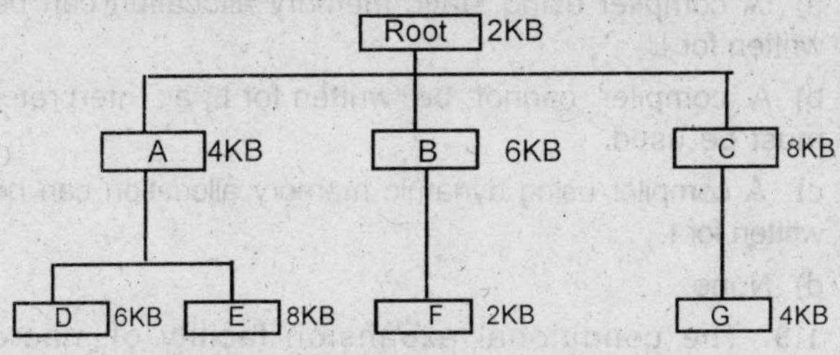

What will be the size of the partition (in physical memory) required to load(and run) this program?

a) 12KB b) 14KB c) 10KB d) 8KB

2.17. Consider n processes sharing the CPU in a round-robin fashion. Assuming that each process switch takes s seconds, what must be the quantum size q such that the overhead resulting from process switching is minimized but, at the same time each process is guaranteed to get is turn at the CPU at least every t seconds?

a) $q \le \frac{t - ns}{n - 1}$ b) $q \ge \frac{t - ns}{n - 1}$

c) $q \le \frac{t - ns}{n + 1}$ d) $q \ge \frac{t - ns}{n + 1}$

2.18. If an instruction takes i microseconds and a page fault takes an additional j microseconds, the effective instruction time if on the average a page fault occurs every k instruction is:

a) $i + \frac{j}{k}$ b) $i + j * k$ c) $\frac{i + j}{k}$ d) $(i + j) * k$

2.19. Which of the following query transformations (i.e. replacing the l.h.s expression by the r.h.s expression) is incorrect ? R_1 and R_2 are relations, C_1, C_2 are selection conditions and A_1, A_2 are attributes of R_1.

a) $\sigma_{c_1}(\sigma_{c_1}(R_1)) \rightarrow \sigma_{c_2}(\sigma_{c_2}(R_1))$

b) $\sigma_{c_1}(\pi_{A_1}(R_1)) \rightarrow \pi_{A_1}(\sigma_{c_1}(R_1))$

c) $\sigma_{c_1}(R_1 \cup R_2) \rightarrow \sigma_{c_1}(R_1) \cup \sigma_{c_1}(R_2)$

d) $\pi_{A_2}(\sigma_{c_1}(R_1)) \rightarrow \sigma_{c_1}(\pi_{A_2}(R_1))$

2.20. Suppose the domain set of an attribute consists of signed four digit numbers. What is the percentage of reduction in storage space of this attribute if it is stored as an integer rather than in character form?

a) 80% b) 20% c) 60% d) 40%

ANSWER KEY

1. 1 c	1. 2 b	1. 3 b	1. 4 b	1. 5 c
1. 6 b	1. 7 c	1. 8 c	1. 9 d	1. 10 a
1.11 b	1. 12 c	1. 13 d	1. 14 c	1. 15 d
1. 16 c	1. 17 b	1. 18 c	1. 19 c	1. 20 a
1. 21 a	1. 22 b	1. 23 c	1. 24 c	1. 25 c
1. 26 c	1. 27 c	1. 28 b	1. 29 b	1. 30 b
1. 31 b	1. 32 b	1. 33 d	1. 34 d	1. 35 c
2. 1 a	2. 2 b	2. 3 b	2. 4 d	2. 5 b
2. 6 a	2. 7 c	2. 8 c	2. 9 d	2. 10 a
2. 11 a	2. 12 c	2. 13 c	2. 14 b	2. 15 d
2. 16 b	2. 17 b	2. 18 a	2. 19 a	2. 20 c

GATE - 1997
COMPUTER SCIENCE & ENGINEERING

Duration : Three Hours **Maximum Marks : 150**

1. This question contains 10 subparts each carrying 1 mark. Each subpart contains a multiple choice question.

1.1. The probability that it will rain today is 0.5. The probability that it will be rain tomorrow is 0.6. The probability that it will rain either today or tomorrow is 0.7. What is the probability that it will rain today and tomorrow?

a) 0.3 b) 0.25 c) 0.35 d) 0.4

1.2. The Newton-Raphson method is used to find the root of the equation $x^2 - 2 = 0$. If the iterations are started from −1, the iterations will

a) Converge to −1 b) converge to $\sqrt{2}$

c) Converge to - $\sqrt{2}$ d) not converge

1.3. The determinant of the matrix $\begin{bmatrix} 6 & -8 & 1 & 1 \\ 0 & 2 & 4 & 6 \\ 0 & 0 & 4 & 8 \\ 0 & 0 & 0 & -1 \end{bmatrix}$ **is**

a) 11 b) − 48 c) 0 d) − 24

1.4. The concatenation of two lists is to be performed in O(1) time. Which of the following implementations of a list should be used?

a) singly linked list b) doubly linked list

c) circular doubly linked list

d) array implementation of list.

1.5. The correct matching for the following pairs is

A. All pairs shortest paths	I. Greedy
B. Quick sort	II. Depth-First Search
C. Minimum weight spanning tree	III. Dynamic programming
D. Connected Components	IV. Divide and Conquer

a) A-II, B-IV, C-I, D-III b) A-III, B-IV, C-I, D-II

c) A-III, B-IV, C-II, D-I d) A-IV, B-I, C-II, D-III

1.6. In the following grammar

X :: = X ⊕ Y / Y

Y :: = Z * Y/Z

Z :: = id

Which of the following is TRUE?

a) '⊕' is left associative while ' * ' is right associative

b) Both '⊕' and '*' are left associative

c) '⊕' is right associative while ' * ' is left associative

d) None

1.7. Which of the following is essential for converting an infix expression to the postfix form efficiently?

a) An operator stack b) An operand stack

c) An operand stack and an operator stack

d) A parse tree

1.8. A language L allows declaration of arrays whose sizes are not known during compilation. It is required to make efficient use of memory. Which one of the following is TRUE?

a) A compiler using static memory allocation can be written for L

b) A compiler cannot be written for L; an interpreter must be used.

c) A compiler using dynamic memory allocation can be written for L

d) None

1.9. The conditional expansion facility of macro processors is provided to

a) Test a condition during the execution of the expanded program

b) To expand certain model statements depending upon the value of a condition during the execution of the expanded program

c) To implement recursion

d) To expand certain model statements depending upon the value of a condition during the process of macro expansion.

1.10. Heap allocation is required for languages

a) That support recursion

b) That support dynamic data structures

c) That use dynamic scope rules d) None

2. The question contains 5 subparts, each carrying 1 mark. Each subpart contain a multiple choice question.

2.1. Let * be defined as $x * y = \bar{x} + y$. Let $z = x * y$. Value of $z * x$ is

a) $\bar{x} + y$ b) x c) 0 d) 1

2.2. RST 7.5 interrupt in 8085 microprocessor executes service routine from interrupt vector location

a) 0000H b) 0075H c) 003CH d) 0034H

2.3. Purpose of a start bit in RS 232 serial communication protocol is

a) To synchronise receiver for receiving every byte

b) To synchronise receiver for receiving a sequence of bytes

c) A parity bit

d) To synchronise receiver for receiving the last byte

2.4. The correct matching for the following pairs is

A. DMA I/O — I. High speed RAM

B. Cache — II. Disk

C. Interrupt I/O — III. Printer

D. Condition Code Register — IV. ALU

a) A-IV, B-III, C-I, D-II b) A-II, B-I, C-III, D-IV

c) A-IV, B-III, C-II, D-I d) A-II, B-III, C-IV, D-I

2.5. An N-bit carry lookhead adder, where N is a multiple of 4, employs ICs 74181 (4 bit ALU) and 74182(4 bit carry lookhead generator).The minimum addition time using the best architecture for this adder is

a) Proportional to N b) Proportional to log N

c) A constant d) None

3. This question contains 10 subparts, each carrying 1 mark. Each sub part contains a multiple choice question.

3.1. Let (Z, $*$) be an algebraic structure, where Z is the set of integers and the operation $*$ is defined by $n * m$ = maximum (n, m). Which of the following statements is true for (Z, $*$)?

a) (Z, $*$) is a monoid

b) (Z, $*$) is an Abelian group

c) (Z, $*$) is a group d) None

3.2. Which of the following propositions is a TAUTOLOGY?

a) $(p \vee q) \rightarrow p$ b) $p \vee (q \rightarrow p)$

c) $p \vee (p \rightarrow q)$ d) $p \rightarrow (p \rightarrow q)$

3.3. In the lattice defined by the Hasse diagram given in below fig. ,how many complements does the element 'e' have?

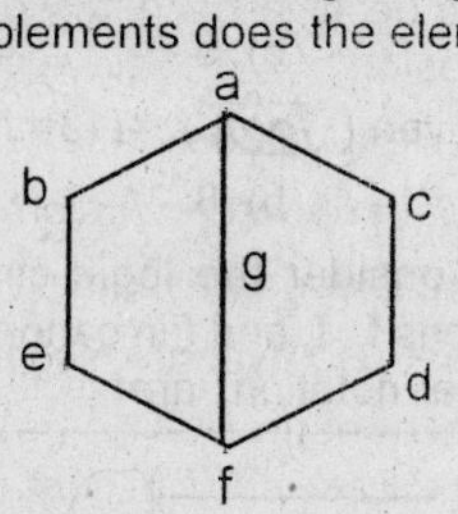

a) 2

b) 3

c) 0

d) 1

3.4. Given Σ ={a, b}, which one of the following sets is not countable?

a) Set of all strings over Σ

b) Set of all languages over Σ

c) Set of all regular languages over Σ

d) Set of all languages over Σ accepted by Turing machines

3.5. Locality of reference implies that the page reference being made by a process

a) Will always be to the page used in the previous page reference

b) is likely to be to one of the pages used in the last few page references

c) Will always be to one of the pages existing in memory

d) Will always lead to a page fault

3.6. The correct matching for the following pairs

A. Disk scheduling — I. Round robin

B. Batch processing — II. SCAN

C. Time sharing — III. LIFO

D. Interrupt processing — IV. FIFO

a) A-III, B-IV, C-II and D-I b) A-IV, B-III, C-II and D-I

c) A-II, B-IV, CI and D-III d) A-III, B-IV, C-I and D-II

3.7. I/O redirection

a) Implies changing the name of a file

b) Can be employed to use an existing file as input for a program.

c) Implies connection 2 programs through a pipe

d) None of the above

3.8. When an interrupt occurs, an operating system

a) Ignores the interrupt

b) Always changes state of interrupted process after processing the interrupt

c) Always resumes execution of interrupted process after processing the interrupt

d) May change state of interrupted process to 'blocked' and schedule another process

3.9. Thrashing

a) Reduces page I/O

b) Decreases the degree of multiprogramming

c) Implies excessive page I/O

d) Improve the system performance

3.10. Dirty bit for a page in a page table

a) Helps avoid unnecessary writes on a paging device

b) Helps maintain LRU information

c) Allows only read on a page d) None

4. This question contain 10 subparts each carrying 2 marks. Each sub part contains a multiple choice question.

4.1. What is the maximum value of the function $f(x) = 2x^2 - 2x + 6$ in the interval [0, 2]?

a) 6 b) 10 c) 12 d) 5.5

4.2. Let A =(a_{ij}) be an n-rowed square matrix and I_{12} be the matrix obtained by interchanging the first and second rows of the n-rowed Identity matrix. Then AI_{12} is such that its first.

a) Row is the same as its second row

b) Row is the same as the second row of A

c) Column is the same as the second column of A

d) Row is all zero.

4.3. Using the forward Eular method to solve y''(t) = f(t), y(0) = 0, y''(0) = 0 with a step size of h, we obtain the following values of y in the first iterations:

a) 0, hf(0), h(f(0) + f(h)) and h(f(0) + f(h) + f(2h))

b) 0, 0, h^2 f(0) and $2h^2$f(0) + f(h)

c) 0, 0, h^2 f(0) and $3h^2$f(0).

d) 0, 0, hf(0) + h^2f(0)and hf(0) + h^2f(0) + hf(h)

4.4. A polynomial p(x) is such that p(0) =5, p(1) =4, p(2) = 9 and p(3) =20. The minimum degree it can have is

a) 1 b) 2 c) 3 d) 4

4.5. A binary search tree contains the values 1, 2, 3, 4, 5, 6, 7, 8.The tree is traversed in pre-order and the values are printed out. Which of the following sequences is a valid output?

a) 5 3 1 2 4 7 8 6 b) 5 3 1 2 6 4 8 7

c) 5 3 2 4 1 6 7 8 d) 5 3 1 2 4 7 6 8

4.6. Let T(n) be the function defined by T(1) = 1, $T(n) = 2T\lfloor n/2 \rfloor + \sqrt{n}$ for $n \geq 2$. Which of the following statements is TRUE?

a) $T(n)=O(\sqrt{n})$ b) T(n)=O(n)

c) T(n)= O(log n) d) None

4.7. A priority queue Q is used to implement a stack S that stores characters. PUSH (C) is implemented as INSERT (Q, C, K) where K is an appropriate integer key chosen by the implementation. POP is implemented as DELETEMIN (Q). For a sequence of operations, the keys chosen are in

a) Non-increasing order b) Non-decreasing order

c) Strictly increasing order d) Strictly decreasing order

4.8. Given the following Pascal-like program segment

```
Procedure A;
        x, y : integer;
        Procedure B;
            x, z :real;
        S1
END B;
Procedure C;
        i : integer;
        S2
END C;
END A;
```

The variables accessible in S1 and S2 are

a) x of A, y, x of B and z in S1 and x of B, y and i in S2

b) x of B, y and z in S1 and x of B, i and z in S2

c) x of B, z and y in S1 and x of A, i and y in S2

d) None

4.9. The expression (a ∗ b) ∗ c op.....where 'op' is one of '+', '∗' and '↑'(exponentiation)can be evaluated on a CPU with a single register without storing the value of(a ∗ b) if

a) 'op' is '+' or '∗' b) 'op' is '↑' or '∗'

c) 'op' is '↑' or '+'

d) Not possible to evaluate without storing

4.10. The trapezoidal method to numerically obtain $\int_a^b f(x)dx$ has an error E bounded by $\frac{b-a}{12}h^2$ max $f''(x)$ $x \in [a, b]$ where h is the width of the trapezoids. The minimum number of trapezoids guaranteed to ensure $E \leq 10^{-4}$ in computing in 7 using $f = \frac{1}{x}$ is

a) 60 b) 100 c) 600 d) 10,000

5. This question contains 5 subparts, each carrying 2 marks. Each subpart contains a multiple choice question.

5.1. Let $f(x, y, z) = \bar{x} + \bar{y}x + xz$ be a switching function. which one of the following is valid?

a) $\bar{y}x$ is a prime implicant of f

b) xz is a minterm of f

c) xz is an implicant of f

d) y is a prime implicant of f

5.2. Contents of A register after the execution of the following 8085 microprocessor program is

```
MVI A, 55H
MVI C, 25H
ADD C
DAA
```

a) 7AH b) 80H c) 50H d) 22H

5.3. A micro instruction is to be designed to specify

A. none or one of the three micro operations of one kind

B. none or upto six micro-operations of another kind

The minimum number of bits in the micro-instruction is

a) 9 b) 5 c) 8 d) None

5.4. Given $(\sqrt{224})_r = (13)_r$, the value of the radix r is:

a) 10 b) 8 c) 5 d) 6

5.5. Consider the logic circuit shown below fig. The functions f_1, f_2 and f(in canonical sum of products form in decimal notation) are:

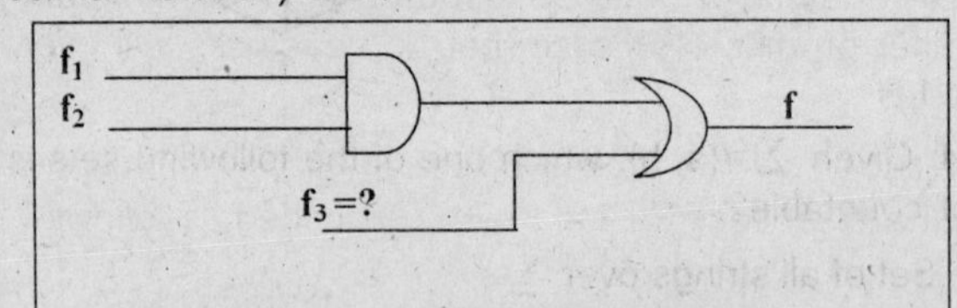

$f1(w, x, y, z) = \sum 8, 9, 10$

$f2(w, x, y, z) = \sum 7, 8, 12, 13, 14, 15$

$f(w, x, y, z) = \sum 8, 9$

The function f 3 is

a) $\sum 9, 10$ b) $\sum 9$

c) $\sum 1, 8, 9$ d) $\sum 8, 10, 15$

6. This question contains 10 subparts, each carrying 2 marks. Each subpart contains a multiple choice question.

6.1. A partial order $\leq$ is defined on the set $S = \{x, a_1, a_2, \ldots a_n, y\}$ as $x \leq a_i$ for all i and $a_i \leq y$ for all i, where $n \geq 1$. The number of total orders on the set S which contain the partial order $\leq$ is

a) n! b) n + 2 c) n d) 1

6.2. Let G be a graph 100 vertices numbered 1 to 100. Two vertices i and j are adjacent if $|i - j| = 8$ or $|i - j| = 12$. The number of connected components in G is

a) 8 b) 4 c) 12 d) 25

6.3. The number of equivalence relations of the set {1, 2, 3, 4} Is

a) 15 b) 16 c) 24 d) 4

6.4. Which one of the following regular expressions over {0, 1} denotes the set of all strings not containing 100 as a sub string?

a) $0^*(1^+ 0)^*$ b) 0^*1010^*

c) 0^*1^*01 d) $0^*(10 + 1)^*$

6.5. Which of the following is not decidable?

a) Given a Turing machine M, a string s and an integer k, M accepts s with in k steps

b) Equivalence of two given Turing machines

c) Language accepted by a given finite state machine is non empty

d) Language generated by a context free grammar is non empty

6.6. Which of the following languages over {a, b, c} is accepted by a deterministic pushdown automata?

a) $\{w \subset w^R \mid w \in \{a, b\}^*\}$ b) $\{ww^R \mid w \in \{a, b, c\}^*\}$

c) $\{a^n b^n c^n \mid n \geq 0\}$

d) {w | w is a palindrome over {a, b, c}}

6.7. An operating system contains 3 user processes each requiring 2 units of resource R. The minimum number of units of R such that no deadlocks will ever arise is

a) 3 b) 5 c) 4 d) 6

6.8. Each process P_i, i = 1 ...... 9 is coded as follows

```
repeat
    P(mutex)
    {critical section}
    v(mutex)
forever
```

The code for P_{10} is identical except that it uses v (mutex) in place of P (mutex). What is the largest number of processes that can be inside the critical section at any moment

a) 1 b) 2 c) 3 d) None

6.9. For a database relation R(a, b, c, d), where the domains of a, b, c, d include only atomic values, only the following functional dependencies and those that can be inferred from them hold:

$$a \rightarrow c \quad b \rightarrow d$$

The relation is

a) in first normal form but not in second normal form

b) in second normal form but not in third normal form

c) in third normal form d) None

6.10. Let R(a, b, c) and S(d, e, f) be two relations in which d is the foreign key of S that refers to the primary key of R. Consider the following four operations R and S

A. Insert into R B. Insert into S

C. Delete from R D. Delete form S

Which of the following is true about the referential integrity constraint above?

a) None of (A), (B), (C) or (D) can cause its violation

b) All of (A), (B), (C) and (D) can cause its violation

c) Both (A) and (D) can cause its violation

d) Both (B) and (C) can cause its violation

ANSWER KEY

1. 1 d	1. 2 c	1. 3 b	1. 4 c	1. 5 b
1. 6 a	1. 7 a	1. 8 c	1. 9 d	1. 10 b
2.1 b	2. 2 c	2. 3 a	2. 4 b	2. 5 b
3. 1 d	3. 2 c	3. 3 b	3. 4 b	3. 5 b
3. 6 c	3. 7 b	3. 8 d	3. 9 c	3. 10 a
4. 1 b	4. 2 c	4. 3 d	4. 4 b	4. 5 d
4. 6 b	4. 7 d	4. 8 c	4. 9 a	4. 10 c
5. 1 a	5. 2 b	5. 3 c	5. 4 c	5. 5 b
6. 1 a	6. 2 b	6. 3 a	6. 4 b	6. 5 ?
6. 6 a	6. 7 c	6. 8 d	6. 9 a	6. 10 d

GATE - 1996
COMPUTER SCIENCE & ENGINEERING

Duration : Three Hours **Maximum Marks : 150**

1. Write in your answer book the correct or the most appropriate answer to the following multiple choice questions by writing the corresponding letter a, b, c and d against the subquestion number. (25 x 1=25)

1.1. Let A and B be sets and let A^c and B^c denote the complements of the sets A and B. The set $(A-B) \cup (B-A) \cup (A \cap B)$ is equal to

a) $A \cup B$ b) $A^c \cup B^c$ c) $A \cap B$ d) $A^c \cap B^c$

1.2. Let X ={2, 3, 6, 12, 24}, let $\leq$ be the partial order defined by $X \leq Y$ if x divides y. Number of edges as in the Hasse diagram of $(X, \leq)$ is

a) 3 b) 4 c) 9 d) None

1.3. Suppose X and Y are sets and |X| and |Y| are their respective cardinalities, It is given that there are exactly 97 functions from X to Y. From this one can conclude that

a) |X| =1, |Y| =97 b) |X| =97, |Y| =1

c) |X| =97, |Y| =97 d) None of the above

1.4. Which of the following statements is FALSE?

a) The set of rational numbers is an abelian group under addtion

b) The set of integer in an abelian group under addition

c) The set of rational numbers form an abelian group under multiplication

d) The set of real numbers excluding zero is an abelian group under multiplication

1.5. Two dice are thrown simultaneously. The probability that at least one of them will have 6 facing up is.

a) $\frac{1}{36}$ b) $\frac{1}{3}$ c) $\frac{25}{36}$ d) $\frac{11}{36}$

1.6. The formula used to compute an approximation for the second derivative of a function f at a point x_o is.

a) $\frac{f(x_0+h)+f(x_0-h)}{2}$ b) $\frac{f(x_0+h)-f(x_0-h)}{2h}$

c) $\frac{f(x_0+h)+2f(x_0)+f(x_0-h)}{h^2}$

d) $\frac{f(x_0+h)-2f(x_0)+f(x_0-h)}{h^2}$

1.7. Let Ax =b be a system of linear equations where A is an m $\times$ n matrix and b is a m x 1 column vector and X is a n $\times$ 1 column vector of unknowns. Which of the following is FALSE?

a) The system has a solution if and only if, both A and the augmented matrix [A b] have the same rank

b) If m < n and b is the zero vector, then the system has infinitely many solutions.

c) If m = n and b is a non zero vector, then the system has a unique solution.

d) The system will have only a trivial solution when m = n, b is the zero vector and rank(A) = n.

1.8. Two of the following four regular expressions are equivalent which two?(ε is the empty string).

I. $(00)^*(\varepsilon+0)$
II. $(00)^*$
III. 0^*
IV. $0(00)^*$

a) I and II b) II and III
c) I and III d) III and IV

1.9. Which of the following statements is FALSE?

a) The Halting problem of Turing machines is undecidable.

b) Determining whether a context free grammar is ambiguous is undecidable.

c) Given two arbitrary context free grammars G_1 and G_2 it is undecidable whether $L(G_1) = L(G_2)$.

d) Given two regular grammars G_1 and G_2 it is undecidable whether $L(G_1) = L(G_2)$

1.10. Let $L \subseteq \Sigma^*$ where Σ ={a, b}; Which of the following is TRUE?

a) L ={x | x has an equal number of a's and b's} is regular

b) L ={a^nb^n | n $\geq$ 1} is regular

c) L ={x | x has more a's than b's} is regular

d) L ={ a^mb^n | m $\geq$, n $\geq$ 1} is regular

1.11. Which of the following is FALSE?

a) 100 n log = $\left(\frac{n \log n}{100}\right)$

b) $\sqrt{\log n}$ = 0(log log n)

c) If 0 < x < y then $n^x = 0(n^y)$

d) 2n $\neq$ o(nk)

1.12. Consider the following statements.

I. First-in-first- out types of computations are efficiently supported by STACKS

II. Implementing LISTS on linked lists is more efficient than implementing LISTS on an array for almost all the basic LIST operations.

III. Implementing QUEUES on a circular array is more efficient than implementing QUEUES on a linear array with two indices.

IV. Last-in-first-out type of computations are efficiently supported by QUEUES.

a) II and III are true b) I and II are true

c) III and IV are true d) II and IV are true

1.13. An advantage of chained hash table (external hashing) over the open addressing scheme is

a) Worst case complexity of search operations is less

b) Space used is less

c) Deletion is easier d) None of the above

1.14. In the balanced binary tree given below, how many nodes will become unbalanced when a node is inserted as a child of the node "g"?

a) 1

b) 3

c) 7

d) 8

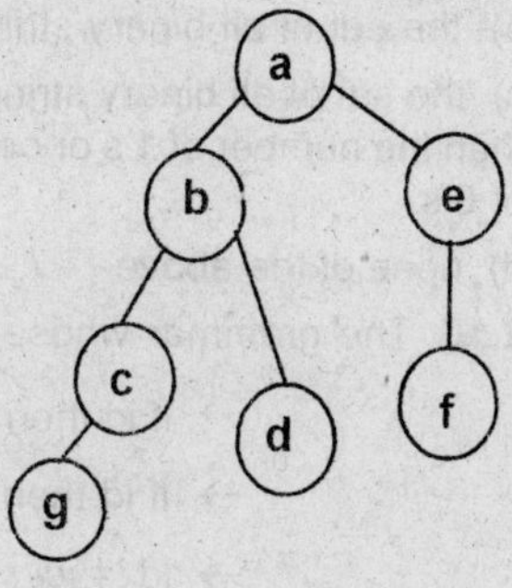

1.15. Which of the following sequences denotes the post order traversal sequence of the tree of question 1.14?

a) f e g c d b a b) g c b d a f e

c) g c d b f e a d) f e d g e b a

1.16. Relative mode of addressing is most relevant to writing.

a) coroutines

b) position-independent code

c) sharable code d) interrupt handlers

1.17. The pass numbers for each of the following activities

I. object code generation

II. literals added to literal table

III. listing printed

IV. address resolution of local symbols that occur in a two pass assembler respectively are

a) I, II, I, II b) II, I, II, I

c) II, I, I, II d) I, II, II, II

1.18. The process state transition diagram is shown below is representative of

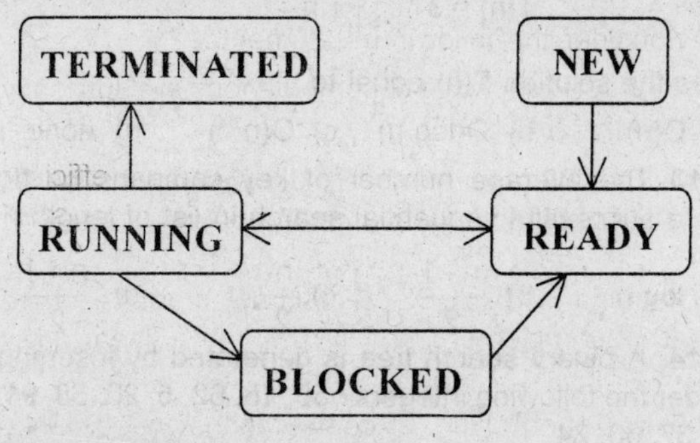

a) a batch operating system

b) an operating system with a preemptive scheduler

c) an operating system with a non-preemptive scheduler

d) a uni-programmed operating system

1.19. A critical section is a program segment.

a) Which should run in a certain specified amount of time

b) Which avoids deadlocks

c) Where shared resources are accessed

d) Which must be enclosed by a pair of semaphore operations, P and V

1.20. Which of the following is an example of spooled device?

a) a line printer used to print the output of a number of jobs

b) a terminal used to enter input data to a running program

c) a secondary storage device in a virtual memory system

d) a graphic display device

1.21. A ROM is used to store the table for multiplication of two 8-bit unsigned integers. The size of ROM required is

a) 256×16 b) $64K \times 8$

c) $4K \times 16$ d) $64K \times 16$

1.22. Number of machine cycles required for RET instruction in 8085 microprocessor is.

a) 1 b) 2 c) 3 d) 5

1.23. Booth's algorithm for integer multiplication gives worst performance when the multiplier pattern is.

a) 101010....10 b) 100000......0001

c) 111111.....11 d) 011111......1110

1.24. For the daisy chain scheme of connecting I/O devices, which of the following statements is TRUE?

a) It gives non-uniform priority to various devices

b) It gives uniform priority to all devices.

c) It is only useful for connecting slow devices to a processor device.

d) It requires a separate interrupt pin on the processor for each device

1.25. Consider the following floating point number representation.

31 — 24	23 — 0
Exponent	Mantissa

The exponent is in 2's complement representation and mantissa is in the sign magnitude representation. The range of the magnitude of the normalised numbers in this representation is

a) 0 to 1 b) 0.5 to 1

c) 2^{-23} to 0.5 d) 0.5 to $(1-2^{-23})$

2. Write in your answer book the correct or the most appropriate answer to the following multiple choice questions by writing the corresponding letter A, B, C or D against the subquestion number. (25 x 2 = 50)

2.1. Let R denote the set of real numbers. Let f: $R \times R \rightarrow R \times R$ be a bijective function defined by $f(x, y) = (x + y, x - y)$. The inverse function of f is given by

a) $f(x,y)=\left(\frac{1}{x+y},\frac{1}{x-y}\right)$ b) $f(x,y)=(x-y.\ x+y)$

c) $f(x,y)=\left(\frac{x+y}{2},\frac{x-y}{2}\right)$ d) $f(x.y)=[2(p.y).\ 2(x+y)]$

2.2. Let R be non-empty relation on a collection of sets defined by A R B if and only if A $\cap$ B = ϕ Then, (pick the TRUE statement)

a) R is reflexive and transitive

b) R is symmetric and not transitive

c) R is an equivalence relation

d) R is not reflexive and not symmetric

2.3. Which of the following is FALSE? Read $\wedge$ as AND, $\vee$ as OR, ~ as NOT, $\rightarrow$ as one way implication and $\leftrightarrow$ as two way implication

a) $((x \rightarrow y) \wedge x) \rightarrow y$

b) $((\sim x \rightarrow y) \wedge (\sim x \wedge \sim y)) \rightarrow x$

c) $(x \rightarrow (x \vee \psi))$ d) $((x \vee y) \leftrightarrow (\sim x \rightarrow \sim y))$

2.4. Which one of the following is FALSE?

a) The set of all bijective functions on a finite set forms a group under function composition

b) The set {1, 2, ..., p-1} forms a group under multiplication mod p where p is a prime number.

c) The set of all strings over a finite alphabet forms a group under concatenation

d) A subsets $\neq \phi$ of G is a sub group of the group <G, *> If and only if for any pair of elements a, b $\in$ s, a *b^{-1} $\in$ S.

2.5. Newton-Raphson iteration formula for finding $\sqrt[3]{c}$, where c > 0 is,

a) $X_{n+1}=\frac{2x^3_n+\sqrt[3]{c}}{3x^2_n}$ b) $X_{n+1}=\frac{2x^3_n-\sqrt[3]{c}}{3x^2_n}$

c) $X_{n+1}=\frac{2x^3_n+c}{3x^2_n}$ d) $X_{n+1}=\frac{2x^3_n-c}{3x^2_n}$

2.6. The matrices $\begin{pmatrix}\cos\theta & -\sin\theta\\ \sin\theta & \cos\theta\end{pmatrix}$ and $\begin{pmatrix}a & 0\\ 0 & b\end{pmatrix}$ commute under multiplication

a) if a = b or $\theta = n\pi$, n is an integer

b) always c) never d) if a cos $\theta \neq$ b sin θ

2.7. The probability that the top and bottom cards of a randomly shuffled deck are both aces is

a) $\frac{4}{52}\times\frac{4}{52}$ b) $\frac{4}{52}\times\frac{3}{52}$ c) $\frac{4}{52}\times\frac{3}{51}$ d) $\frac{4}{52}\times\frac{4}{51}$

2.8. If L_1 and L_2 are context free languages and R a regular set, one of the languages below is not necessarily a context free language. Which one?

a) $L_1 L_2$ b) $L_1 \cap L_2$ c) $L_1 \cap R$ d) $L_1 \cup L_2$

2.9. Define for a context free language L $\leq$ {0; 1} init (L) ={u / uv $\in$ L for some v in {0, 1}} (in other words, init (L) is the set of prefixes of L)

Let L= {w/w is nonempty and has on equal number of 0's and 1's} Then init(L) is

a) the set of all binary strings with unequal number of 0's and 1's

b) the set of all binary strings including the null string

c) the set of all binary strings with exactly one more 0's than the number of 1's or one or more 1 than the number of 0's.

d) none of the above

2.10. The grammar whose productions are

→ if id then <stmt>

→ if id then <stmt> else <stmt>

→ id := id

is ambiguous because

a) the sentence

if a then if b then c : = d

b) the left most and right most derivations of the sentence

if a then if b then c : = d

give rise two different parse trees

c) the sentence

if a then if b then c : = d else c : = f

has more than two parse trees

d) the sentence

if a then if b then c : = d else c : = f

has two parse trees.

2.11. The minimum number of interchanges needed to convert the array 89, 19, 40, 17, 12, 10, 2, 5, 7, 11, 6, 9, 70 into a heap with maximum element at the root is

a) 0 b) 1 c) 2 d) 3

2.12. The recurrence relation

$T(1) = 2$

$T(n) = 3T\left(\frac{n}{4}\right) + n$

Has the solution T(n) equal to

a) O(n) b) O(log n) c) $O(n^{3/4})$ d) none

2.13. The average number of key comparisons done on a successful sequential search in list of length n is

a) log n b) $\frac{n-1}{2}$ c) $\frac{n}{2}$ d) $\frac{n+1}{2}$

2.14. A binary search tree is generated by inserting in order the following integers : 50, 15, 62, 5, 20, 58, 91, 3, 8, 37, 60, 24

The number of nodes in the left subtree and right subtree of the root respectively is

a) (4, 7) b) (7, 4) c) (8, 3) d) (3, 8)

2.15. Quick sort is run on two inputs shown below to sort in ascending order

I. 1, 2, 3, ... n

II. n, n-1, n-2, ...2, 1

Let C_1 and C_2 be the number of comparisons made for the inputs (I) and (II) respectively. Then,

a) $C_1 < C_2$ b) $C_1 > C_2$ c) $C_1 = C_2$

d) we cannot say anything for arbitrary n.

2.16. Which of the following macros can put a macro assembler into an infinite loop?

```
I. MACRO     M1, X
 IF EQ, X:   if X = 0 then ....
 M1 X + 1
 ENDC
 IF NE, X:   if X '0 then ....
 WORD X :    address (X) is stored here
 ENDC
 ENDM
II.MACRO M2, X
 IF EQ, X
 M2 X
 ENDC
 IF NE, X
 WORD X + 1
 ENDC
 ENDM
```

a) II only b) I only

c) both I and II d) none

2.17. The correct matching for the following pairs is

A. Activation record	I. Linking loader
B. Location counter	II. Garbage collection
C. Reference counts	III. Subroutine call
D. Address relocation	IV. Assembler

a) A-III, B-IV, C-I, D-II b) A-IV, B-III, C-I, D-II

c) A-IV, B-III, C-II, D-I d) A-III, B-IV, C-II, D-I

2.18. A 1000 Kbyte memory is managed using variable partitions but to compaction. It currently has two partitions of sizes 200 Kbytes and 260 Kbytes respectively. The smallest allocation request in Kbytes that could be denied is for

a) 151 b) 181 c) 231 d) 541

2.19. A solution to the Dining Philosophers Problem which avoids deadlock is

a) ensure that all philosophers pick up the left fork before the right fork

b) ensure that all philosophers pick up the right fork before the left fork

c) ensure that one particular philosopher picks up the left fork before the right fork, and that all other philosophers pick up the right fork before the left fork

d) none of the above.

2.20. Four jobs to be executed on a single processor system arrive at time 0^+ in the order A, B, C, D. Their burst CPU time requirements are 4, 1, 8, 1 time units respectively. The completion time of A under round robin scheduling with time slice of one time unit is

a) 10 b) 4 c) 8 d) 9

2.21. Consider the circuit is shown below. Which has a four bit binary number $b_3\ b_2\ b_1\ b_0$ as input and a five bit binary number $d_4\ d_3\ d_2\ d_1\ d_0$ as output

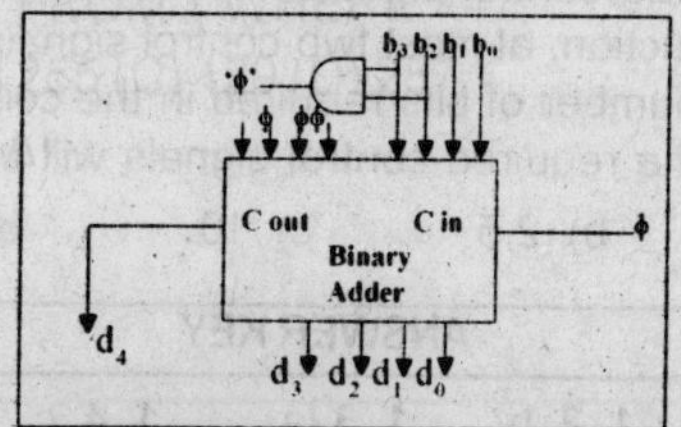

a) Binary to Hex conversion

b) Binary to BCD conversion

c) Binary to grey code conversion

d) Binary to radix-12 conversion

2.22. Consider the circuit shown below f implements

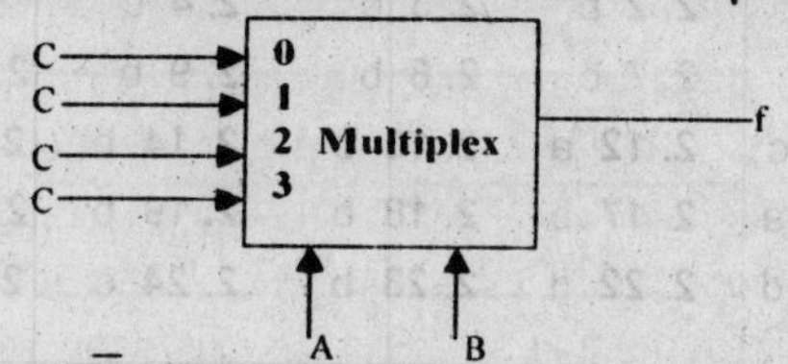

a) $\overline{A}\overline{B}C + \overline{A}B\overline{C} + ABC$ b) A + B + C

c) $A \oplus B \oplus C$ d) AB + BC + CA

2.23. Consider the following state table is shown below for a sequential machine. The number of states in the minimized machine will be

a) 4

b) 3

c) 2

d) 1

Present state \ Input	0	1
A	D0	B1
B	A0	C1
C	A0	B1
D	A1	C1

Next state output

2.24. What is the equivalent boolean expression in product-of-sums form for the Karnaugh map given in below

CD \ AB	00	01	11	20
00		1	1	
01	1			1
11	1			1
10		1	1	

a) $B\overline{D}+\overline{B}D$

b) $(B+\overline{C}+D)(\overline{B}+C+\overline{D})$

c) $(B+\overline{D})(\overline{B}+D)$

d) $(B+\overline{D})(\overline{B}+D)$

2.25. A micro program control unit is required to generate a total of 25 control signals. Assume that during any micro instruction, at most two control signals are active. Minimum number of bits required in the control word to generate the required control signals will be

a) 2 b) 2.5 c) 10 d) 12

ANSWER KEY

1. 1 a	1. 2 b	1. 3 a	1. 4 c	1. 5 d
1. 6 d	1. 7 c	1. 8 c	1. 9 d	1. 10 d
1.11 b	1. 12 a	1. 13 c	1. 14 b	1. 15 c
1. 16 b	1. 17 b	1. 18 b	1. 19 c	1. 20 a
1. 21 d	1. 22 d	1. 23 d	1. 24 a	1. 25 d
2. 1 c	2. 2 b	2. 3 d	2. 4 c	2. 5 c
2. 6 a	2. 7 c	2. 8 b	2. 9 b	2. 10 d
2. 11 c	2. 12 a	2. 13 b	2. 14 b	2. 15 c
2. 16 a	2. 17 b	2. 18 b	2. 19 b	2. 20 d
2. 21 d	2. 22 d	2. 23 b	2. 24 c	2. 25 c

GATE - 1995
COMPUTER SCIENCE & ENGINEERING

Duration : Three Hours **Maximum Marks : 150**

1. This question has 25 parts. Each part carries 1 Mark choose the correct alternative for each part (25 x1=25)

1.1. A single instruction to clear the lower four bits of the accumulator in 8085 assembly language is:

a) XRI OFH b) ANI FOH

c) XRI FOH d) ANI OFH

1.2. Which of the following statements is TRUE?

a) ROM is a Read / Write memory

b) PC points to the last instruction that was executed

c) Stack works on the principle of the LIFO

d) All instructions affect the flags

1.3. In a vectored interrupt

a) The branch address is assigned to a fixed location in memory

b) The interrupting source supplies the branch information to the processor the through an interrupt vector

c) The branch address is obtained from a register in the processor

d) None of the above

1.4. In the following Pascal program segment, what is the value of X after the execution of the program segment

X := -10; Y :=20;

If X > Y then if X < 0 then X :=abs(X) else X :=2 * X;

a) 10 b) –20 c) –10 d) None

1.5. Merge sort uses

a) Divide and conquer strategy

b) Backtracking approach

c) Heuristic search d) Greedy approach

1.6. The principle of locality justifies the use of

a) Interrupts b) DMA

c) Polling d) Cache memory

1.7. In a paged segmented scheme of memory management, the segment table itself must have a page table because

a) The segment table is often too large to fit in one page

b) Each segment is spread over a number of pages

c) Segment tables point to page table and not to the physical locations of the segment

d) The processor's description base register points to a page table

1.8. Which of the following page replacement algorithms suffers from Belady's anamoly?

a) Optimal replacement b) LRU

c) FIFO d) Both (a) and (c)

1.9. In some programming languages, an identifier is permitted to be a letter followed by any number of letters or digits. If L and D denotes the sets of letters and digits respectively, which of the following expressions defines an identifier?

a) $(L \cup D)^+$ b) $L(L \cup D)^*$

c) $(L . D)^*$ d) $L . (L . D)^*$

1.10. Consider a grammar with the following productions

$S \rightarrow a\ \alpha\ b \mid b\ \alpha\ c \mid aB$

$S \rightarrow \alpha\ S \mid b$

$S \rightarrow \alpha\ bb \mid ab$

$B \rightarrow bdb \mid b$

The above grammar is

a) Context free b) Regular

c) Context sensitive d) LR(k)

1.11. What are x and y in the following macro definition?

```
Macro    Add  x, y
         Load  y
         Mul   x
         Store y
         End Macro
```

a) Variables b) Identifiers

c) Actual parameters d) Formal parameters

1.12. What is the distance of the following code?

000000, 010101, 000111, 011001, 111111

a) 2 b) 3 c) 4 d) 1

1.13. Which of the following strings can definitely be said to be tokens without looking at the next input character while compiling a Pascal program?

I. Begin II. Program III. <>

a) I b) II c) III d) All the above

1.14. A linker is given object modules for a set of programs that were compiled separately. What information need not be include in an object module?

a) Object code b) Relocation bits

c) Names and locations of all external symbols defined in the object module

d) Absolute address of internal symbols

1.15. Which scheduling policy is most suitable for a time-shared operating system?

a) Shortest job first b) Round Robin

c) First Come First Serve d) Elevator

1.16. For merging two sorted lists of sizes m and n into a sorted list of size m + n, we require comparisons of

a) O(m) b) O(n)

c) O(m + n) d) O(log m + log n)

1.17. A binary tree T has n leaf nodes. The number of nodes of degree 2 in T is

a) $\log_2 n$ b) n-1 c) n d) 2^n

1.18. The probability that a number selected at random between 100 and 999 (both inclusive) will not contain the digit 7 is

a) 16 / 25 b) $(9/10)^3$ c) 27 / 75 d) 18 / 25

1.19. Let R be a symmetric and transitive relation on a set A. Then

a) R is reflexive and hence an equivalence relation

b) R is reflexive and hence a partial order

c) R is not reflexive and hence not an equivalence relation

d) None of the above

1.20. The number of elements in the power set P(S) of the set S ={ (ϕ), 1, (2, 3) } is

a) 2 b) 4 c) 8 d) None

1.21. In the interval [0, π] the equation x = cos x has

a) No solution b) Exactly one solution

c) Exactly two solutions

d) An infinite number of solutions

1.22. If at every point of a certain curve, the slope of the tangent equals - 2x / y the curve is

a) A straight line b) A parabola

c) An circle d) An ellipse

1.23. The value of k for which $4x^2 - 8xy + ky^2 = 0$ does not represent a pair of straight lines (both passing through the origin) is

a) 0 b) 2 c) 9 d) 3

1.24. The rank of the following (n + 1) x (n + 1) matrix, where a is a real number is

$$\begin{bmatrix} 1 & a & a^2 & \cdots & a^n \\ 1 & a & a^2 & \cdots & a^n \\ \cdot & & & & \cdot \\ \cdot & & & & \cdot \\ 1 & a & a^2 & \cdots & a^n \end{bmatrix}$$

a) 1 b) 2 c) n d) None

1.25. The minimum number of edges in a connected cyclic graph on n vertices is

a) n-1 b) n c) n + 1 d) None

2. This Question has 25 parts, Each part carries 2 Marks. Choose the correct alternative for each part (25 x 2=50)

2.1. A sequence of two instructions that multiplies the contents of the DE register pair by 2 and stores the result in the HL register pair (in 8085 assembly language) is

a) XCHG and DAD B b) XTHL and DAD H

c) PCHL and DAD D d) XCHG and DAD H

2.2. The capacity of a memory unit is defined by the number of words multiplied by the number of bits / word. How many separate address and data lines are needed for a memory of 4K x 16?

a) 10 address, 16 data lines

b) 11 address, 8 data lines

c) 12 address, 16 data lines

d) 12 address, 12 data lines

2.3. Assume that X and Y are nonzero positive integers. What does the following Pascal program segment do?

```
while X <> Y do
if X > Y then
X :=X – Y;
else
Y :=Y – X;
write (X);
```

a) Computes the LCM of two numbers

b) Divides the larger number by the smaller number

c) Computes the GCD of two numbers

d) None of the above

2.4. What is the value of X printed by the following program?

```
Program COMPUTE (input, output)
Var X : integer;
procedure FIND (X: real)
BEGIN
X: = sqrt(X);
END;
BEGIN
X : = 2
FIND (X)
writeln (X)
END
```

a) 2 b) $\sqrt{2}$

c) Run-time error d) None of these

2.5. What values of A, B, C and D satisfy the following simultaneous boolean equations?

$\overline{A} + AB = 0,\ AB = AC,\ AB + A\overline{C} + CD = \overline{CD}$

a) A = 1, B = 0, C = 0, D = 1

b) A = 1, B = 1, C = 0, D = 0

c) A = 1, B = 0, C = 1, D = 1

d) A = 1, B = 0, C = 0, D = 0

2.6. The sequence .... is an optimal non preemptive scheduling sequence for the following jobs which leaves the CPU idle for ....unit(s) of time

Job	Arrival time	Burst Time
1	0.0	9
2	0.6	5
3	1.0	1

a) {3, 2, 1}, 1 b) {2, 13}, 0
c) {3, 2, 1}, 0 d) {1, 2, 3}, 5

2.7. The address sequence generated by tracing a particular program executing in a pure demand paging system with 100 records per page with 1 free main memory frame is recorded as follows. What is the number of page faults?

0100, 0200, 0430, 0499, 0510, 0530, 0560, 0120, 0220, 0240, 0260, 0320, 0370

a) 13 b) 8 c) 7 d) 10

2.8. If the cube roots of unity are 1, ω and ω^2, then the roots of the following equation are $(x-1)^3 + 8 = 0$

a) $-1, 1+2\omega, 1+2\omega^2$ b) $1, 1-2\omega, 1-2\omega^2$
c) $-1, 1-2\omega, 1-2\omega^2$ d) $-1, -1+2\omega, -1+2\omega^2$

2.9. A language with string manipulation facilities uses the following operations

head (s) : first character of a string
tail (s) : all but the first character of a string
Concat (s1, s2) : s1 s2

For the string acbc what will be the output of

Concat (head(s), head(tail (tail(s))))

a) ac b) bc c) ab d) cc

2.10. A shift reduce parser carries out the actions specified with in braces immediately after reducing with the corresponding rule of grammar

S → xxW {pirnt "1"}
S → y {print "2"}
W → Sz {print "3"}

What is the translation of xxxxyzz using the syntax directed translation scheme described by the above rules?

a) 23131 b) 11233 c) 11231 d) 33211

2.11. A variant record in Pascal is defined by

```
type varirec = record
               number : integer;
               case(var1, var2) of
               var1 :(x, y :integer);
               var2 : (p, q :real)
               END
               END
```

Suppose an array of 100 such records was declared on a machine which use 4 bytes for an integer and 8 bytes for a real. How much space would the compiler have to reserve for the array?

a) 2800 b) 2400 c) 2000 d) 1200

2.12. The number of 1's in the binary representation of (3*4096 +15*256 + 5*16 +3) are

a) 8 b) 9 c) 10 d) 12

2.13. A unit vector perpendicular to both the vectors a =2i – 2j + k and b = i + j –2k is

a) $1/\sqrt{3}\,(i+j+k)$ b) $1/3(i+j-k)$
c) $1/3(i-j-k)$ d) $1/\sqrt{3}\,(i+j-k)$

2.14. A bag contains 10 white balls and 15 black balls. Two balls are drawn in succession. The probability that one of them in black and the other is white is

a) 2/3 b) 4/5 c) 1/2 d) 1/2

2.15. The iteration formula to find the square root of a positive real number b using the Newton Raphson method is

a) $x_{k+1} = 3(x_k + b)/2x_k$ b) $x_{k+1} = (x_{k2} + b)/2x_k$
c) $x_{k+1} = x_k - 2x_k/(x^2{}_k + b)$ d) None of these

2.16. In a virtual memory system the address space specified by the address lines of the CPU must be .... than the physical memory size and ..... than the secondary storage size.

a) smaller, smaller b) smaller, larger
c) larger, smaller d) larger, larger

2.17. Let A be the set of all nonsingular matrices over real number and let* be the matrix multiplication operator. Then

a) A is closed under * but <A, *> is not a semi group
b) <A, *> is a semi group but not a monoid
c) <A, *> is a monoid but not a group
d) <A, *> is a group but not an abelian group

2.18. The solution of differential equation y" + 3y' + 2y =0 is of the form

a) $C_1e^x + C_2e^{2x}$ b) $C_1e^{-x} + C_2e^{3x}$
c) $C_1e^{-x} + C_2e^{-2x}$ d) $C_1e^{-2x} + C_22^{-x}$

2.19. If the proposition $\neg p \Rightarrow v$ is true, then the truth value of the proposition $\neg p \vee (p \Rightarrow q)$, where $\neg$ is negation, 'V' is inclusive or and $\Rightarrow$ is implication, is

a) true b) multiple-valued
c) false d) cannot be determined

2.20. Which of the following definitions below generates the same language as L, where L={$x^n y^n$ such that n > = 1}?

I. E → xEy | xy
II. xy | (x^+ x yy^+)
III. x^+y^+

a) I only b) I and II
c) II and III d) II only

2.21. The Postfix expression for the infix expression

A + B*(C + D)/F + D* E is

a) AB+CD+*F/D+E* b) ABCD+*F/DE*++
c) A*B+CD/F*DE++ d) A+*BCD/F*DE++

2.22. Which of the following statements is TRUE?

I. As the number of entries in a hash table increases, the number of collisions increases

II. Recursive programs are efficient

III. The worst case complexity for Quick sort is $O(n^2)$

IV. Binary search using a linear linked list is efficient

a) I and II b) II and III

c) I and IV d) I and III

2.23. A finite state machine with the following state table has a single input x and a single out put z.

Present state	Next state, z X = 1	X = 0
A	D, 0	B, 0
B	B, 1	C, 1
C	B, 0	D, 1
D	B, 1	C, 0

If the initial state is unknown, then the shortest input sequence to reach the final state C is

a) 01 b) 10 c) 101 d) 110

2.24. Let $\Sigma = \{0,1\}$, $L = \Sigma^*$ and $R = \{0^n 1^n$ such that $n > 0\}$ then the languages $L \cup R$ and R are respectively.

a) Regular, regular

b) Not Regular, Regular

c) Regular, Not Regular

d) Not Regular, Not Regular

2.25. A computer system has a 4K word cache organised in block-set-associative manner with 4 blocks per set, 64 words per block. The number of bits in the SET and WORD fields of the main memory address format is

a) 15, 40 b) 6, 4

c) 7, 2 d) 4, 6

ANSWER KEY

1. 1 b	1. 2 c	1. 3 b	1. 4 c	1. 5 a
1. 6 d	1. 7 a	1. 8 c	1. 9 b	1. 10 c
1.11 d	1. 12 a	1. 13 c	1. 14 d	1. 15 b
1. 16 c	1. 17 b	1. 18 d	1. 19 d	1. 20 c
1. 21 b	1. 22 d	1. 23 c	1. 24 a	1. 25 b
2. 1 d	2. 2 c	2. 3 c	2. 4 a	2. 5 a
2. 6 a	2. 7 c	2. 8 c	2. 9 c	2. 10 a
2. 11 c	2. 12 c	2. 13 a	2. 14 c	2. 15 ?
2. 16 c	2. 17 d	2. 18 c	2. 19 d	2. 20 a
2. 21 b	2. 22 d	2. 23 b	2. 24 ?	2. 25 d

GATE - 2011
COMPUTER SCIENCE & ENGINEERING

Duration : Three Hours **Maximum Marks : 150**

Q. No. 1 – 25 Carry One Mark Each

1. The simplified SOP (Sum of Product) form of the Boolean expression $\left(P+\overline{Q}+\overline{R}\right).\left(P+\overline{Q}+R\right).\left(P+Q+\overline{R}\right)$ **is**

(A) $\left(\overline{PQ}+\overline{R}\right)$ (B) $\left(P+\overline{QR}\right)$

(C) $\left(\overline{PQ}+R\right)$ (D) $\left(PQ+R\right)$

Answer: (B)

Exp: -

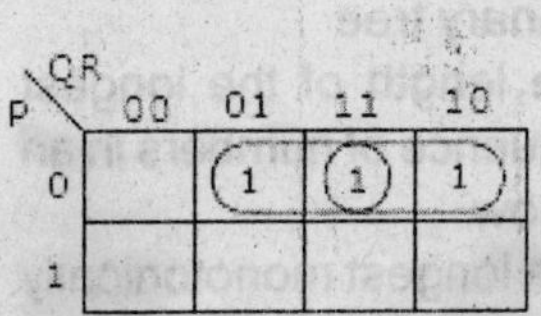

$f=\left(P+\overline{R}\right)\left(P+\overline{Q}\right)$

$=P+\overline{QR}$

Alternate method

$$\left(P+\overline{Q}+\overline{R}\right).\left(P+\overline{Q}+R\right).\left(P+Q+\overline{R}\right)$$

$$=\overline{\overline{\left(P+\overline{Q}+\overline{R}\right).\left(P+\overline{Q}+R\right).\left(P+Q+\overline{R}\right)}}$$

$$=\overline{\overline{P}QR+\overline{P}Q\overline{R}+\overline{P}\,\overline{Q}R}=\overline{\overline{P}Q\left(R+\overline{R}\right)+\overline{P}\,\overline{Q}R}$$

$$=\overline{\overline{P}Q+\overline{P}\,\overline{Q}R}=\overline{\overline{P}\left(Q+\overline{Q}R\right)}$$

$$=\overline{\overline{P}(Q+R)}=P+\overline{Q}\,\overline{R}$$

2. Which one of the following circuits is NOT equivalent to a 2-input XNOR (exclusive NOR) gate?

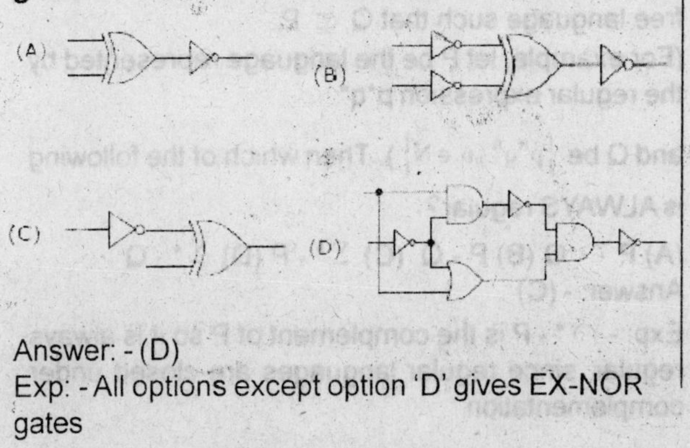

Answer: - (D)

Exp: - All options except option 'D' gives EX-NOR gates

3. The minimum number of D flip-flops needed to design a mod-258 counter is

(A) 9 (B) 8 (C) 512 (D) 258

Answer: - (A)

Exp: - $2^n>258\Rightarrow n=9$

4. A thread is usually defined as a 'light weight process' because an operating system (OS) maintains smaller data structures for a thread than for a process. In relation to this, which of the followings is TRUE?

(A) On per-thread basis, the OS maintains only CPU register state

(B) The OS does not maintain a separate stack for each thread

(C) On per-thread basis, the OS does not maintain virtual memory state

(D) On per thread basis, the OS maintains only scheduling and accounting information

Answer :- (A)

5. K4 and Q3 are graphs with the following structures

Which one of the following statements is TRUE in relation to these graphs?

K4 Q3

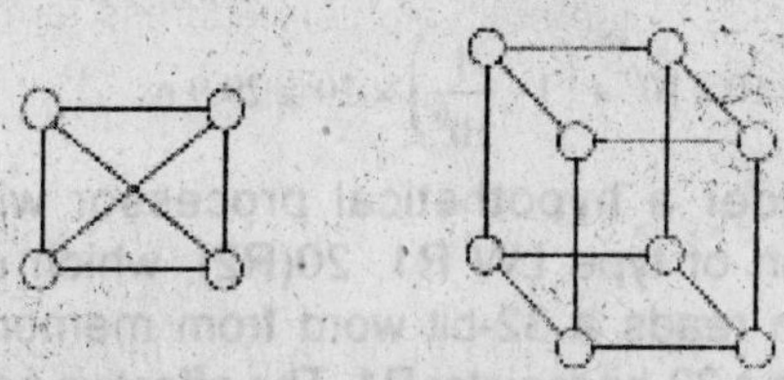

(A) K4 is planar while Q3 is not

(B) Both K4 and Q3 are planar

(C) Q3 is planar while K4 is not

(D) Neither K4 not Q3 is planar

Answer: - (B)

Exp: -

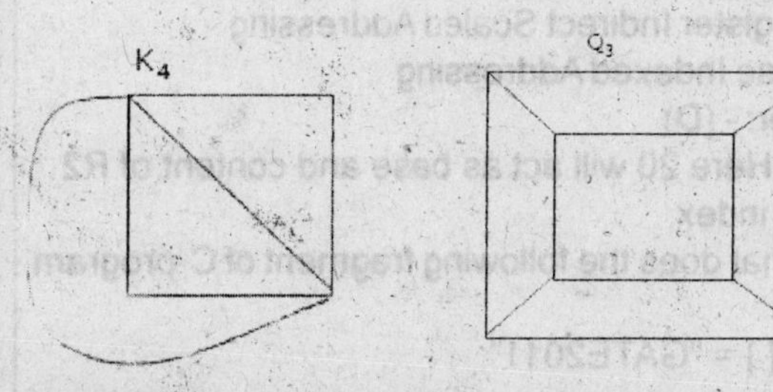

∴ Both K_4 and Q_3 are planar

6. If the difference between the expectation of the square of random variable $\left(E\left[X^2\right]\right)$ and the square of the expectation of the random variable $\left(E\left[X^2\right]\right)$ is denoted by R then
(A) R = 0 (B) R < 0 (C) R > 0 (D) R > 0
Answer: - (C)

7. The lexical analysis for a modern computer language such as Java needs the power of which one of the following machine models in a necessary and sufficient sense?
(A) Finite state automata
(B) Deterministic pushdown automata
(C) Non-Deterministic pushdown automata
(D) Turing machine
Answer: - (A)
Exp: - Lexical Analysis is implemented by finite automata

8. Let the page fault service time be 10ms in a computer with average memory access time being 20ns. If one page fault is generated for every 10^6 memory accesses, what is the effective access time for the memory?
(A) 21 ns (B) 30 ns (C) 23 ns (D) 35 ns
Answer: - (B)
Exp: - P = page fault rate
EA = p × page fault service time
+(1- p) × Memory access time

$$= \frac{1}{10^6} \times 10 \times 10^6 + \left(1 - \frac{1}{10^6}\right) \times 20 \cong 29.9 \text{ ns}$$

9. Consider a hypothetical processor with an instruction of type LW R1, 20(R2), which during execution reads a 32-bit word from memory and stores it in a 32-bit register R1. The effective address of the memory location is obtained by the addition of constant 20 and the contents of register R2. Which of the following best reflects the addressing mode implemented by this instruction for the operand in memory?
(A) Immediate Addressing
(B) Register Addressing
(C) Register Indirect Scaled Addressing
(D) Base Indexed Addressing
Answer: - (D)
Exp: - Here 20 will act as base and content of R2 will be index

10. What does the following fragment of C-program print?

```
char c[ ] = "GATE2011";
char *p =c;
printf ("%s", p+p [3] – p [1]);
```

(A) GATE2011 (B) E2011 (C) 2011 (D) 011
Answer: - (C)

11. A max-heap is a heap where the value of each parent is greater than or equal to the value of its children. Which of the following is a max-heap?

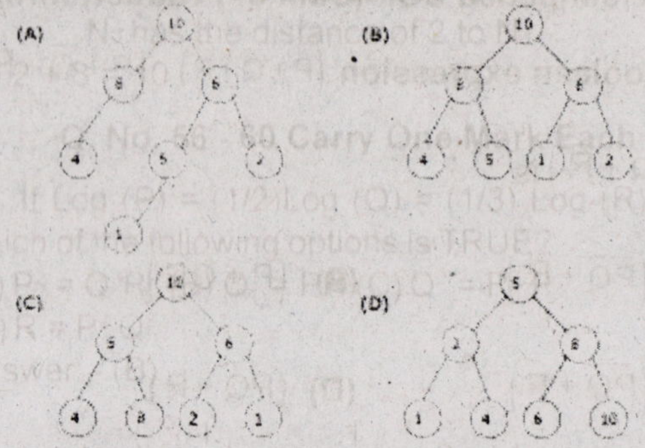

Answer: - (B)
Exp: - Heap is a complete binary tree

12. An algorithm to find the length of the longest monotonically increasing sequence of numbers in an array A [0 : n - 1] is given below.
Let L_i denote the length of the longest monotonically increasing sequence starting at index i in the array
Initialize L_{n-1} =1
For all i such that $0 \le i \le n-2$

$$L_i = \begin{cases} 1 + L_{i+1} & \text{if } A[i] < A[i+1] \\ 1 & \text{Otherwise} \end{cases}$$

Finally the length of the longest monotonically increasing sequence is Max $(L_0, L_1, \ldots, L_{n-1})$. Which of the following statements is TRUE?
(A) The algorithm uses dynamic programming paradigm
(B) The algorithm has a linear complexity and uses branch and bound paradigm
(C) The algorithm has a non-linear polynomial complexity and uses branch and bound paradigm
(D) The algorithm uses divide and conquer paradigm.
Answer: - (A)

13. Let P be a regular language and Q be a context free language such that $Q \subseteq P$.
(For example, let P be the language represented by the regular expression p*q* and Q be $\{p^n q^n \mid n \in N\}$). Then which of the following is ALWAYS regular?
(A) $P \cap Q$ (B) P - Q (C) Σ^* - P (D) Σ^* - Q
Answer: - (C)
Exp: - $\cap$ * - P is the complement of P so it is always regular, since regular languages are closed under complementation

14. In a compiler, keywords of a language are recognized during
(A) parsing of the program (B) the code generation
(C) the lexical analysis of the program
(D) dataflow analysis
Answer: - (C)
Exp: - Any identifier is also a token so it is recognized in lexical Analysis

15. A layer-4 firewall (a device that can look at all protocol headers up to the transport layer) CANNOT
(A) block entire HTTP traffic during 9:00PM and 5:00AM
(B) block all ICMP traffic
(C) stop incoming traffic from a specific IP address but allow outgoing traffic to the same IP address
(D) block TCP traffic from a specific user on a multi-user system during 9:00PM and 5:00AM
Answer: - (A)
Exp: - Since it is a layer 4 firewall it cannot block application layer protocol like HTTP.

16. If two fair coins are flipped and at least one of the outcomes is known to be a head, what is the probability that both outcomes are heads?
(A) 1/3 (B) 1/4 (C) 1/2 (D) 2/3
Answer: - (A)
Exp: - Sample space = {HH, HT, TH}

Required probability = $\frac{1}{3}$

17. Consider different activities related to email.
m1: Send an email from a mail client to a mail server
m2: Download an email from mailbox server to a mail client
m3: Checking email in a web browser
Which is the application level protocol used in each activity?

(A) m1:HTTP m2:SMTP m3:POP
(B) m1:SMTP m2:FTP m3:HTTP
(C) m1: SMTP m2: POP m3: HTTP
(D) m1: POP m2: SMTP m3:IMAP

Answer: - (C)
Exp: - Sending an email will be done through user agent and message transfer agent by SMTP, downloading an email from mail box is done through POP, checking email in a web browser is done through HTTP

18. A company needs to develop a strategy for software product development for which it has a choice of two programming languages L1 and L2. The number of lines of code (LOC) developed using L2 is estimated to be twice the LOC developed with L1. the product will have to be maintained for five years. Various parameters for the company are given in the table below.

Parameter	Language L1	Language L2
Man years needed for development	LOC / 10000	LOC / 10000
Development Cost per year	Rs. 10,00,000	Rs. 7,50,000
Maintenance time	5 years	5 years
Cost of maintenance per year	Rs. 1,00,000	Rs. 50,000

Total cost of the project includes cost of development and maintenance. What is the LOC for L1 for which the cost of the project using L1 is equal to the cost of the project using L2?
(A) 4000 (B) 5000 (C) 4333 (D) 4667
Answer: - (B)
Exp: - LOC $L_1 = x$

$L_2 = 2x$

Total cost of project

$$\frac{X}{10000} \times 1000000 + 5 \times 100000 = \frac{2X}{10000} \times 750000 + 50000 \times 5$$

100 × + 500000 – 250000

$$\Rightarrow 50x = 500000 - 250000$$

$$\therefore X = \frac{250000}{50} \Rightarrow X = 5000$$

19. Let the time taken to switch between user and kernel modes of execution be t_1 while the time taken to switch between two processes be t_2. Which of the following is TRUE?
(A) $t_1 > t_2$
(B) $t_1 = t_2$
(C) $t_1 < t_2$
(D) Nothing can be said about the relation between t_1 and t_2
Answer: - (C)
Exp: - Process switching also involves mode changing.

20. A company needs to develop digital signal processing software for one of its newest inventions. The software is expected to have 40000 lines of code. The company needs to determine the effort in person-months needed to develop this software using the basic COCOMO model. The multiplicative factor for this model is given as 2.8 for the software development on embedded systems, while the exponentiation factor is given as 1.20. What is the estimated effort in person months?
(A) 234.25 (B) 932.50 (C) 287.80 (D) 122.40
Answer: - (A)

Exp: - Effort person per month

$= \alpha\ (\text{KDSI})^{B}$

KDSI = Kilo LOC

$= 2.8 \times (40)^{1.20}$

$= 2.8 \times 83.6511$

= 234.22 person per month

21. Which of the following pairs have DIFFERENT expressive power?

(A) Deterministic finite automata (DFA) and Non-deterministic finite automata (NFA)

(B) Deterministic push down automata (DPDA) and Non-deterministic push down automata (NPDA)

(C) Deterministic single-tape Turing machine and Non-deterministic single tape Turing machine

(D) Single-tape Turing machine and multi-tape Turing machine

Answer: - (B)

Exp: - NPDA is more powerful than DPDA.

Hence answer is (B)

22. HTML (Hyper Text Markup Language) has language elements which permit certain actions other than describing the structure of the web document. Which one of the following actions is NOT supported by pure HTML (without any server or client side scripting) pages?

(A) Embed web objects from different sites into the same page

(B) Refresh the page automatically after a specified interval

(C) Automatically redirect to another page upon download

(D) Display the client time as part of the page

Answer: - (D)

23. Which of the following is NOT desired in a good Software Requirement Specifications (SRS) document?

(A) Functional Requirements

(B) Non Functional Requirements

(C) Goals of Implementation

(D) Algorithms for Software Implementation

Answer: - (D)

24. A computer handles several interrupt sources of which the following are relevant for this question.

Interrupt from CPU temperature sensor

Interrupt from Mouse

Interrupt from Keyboard

Interrupt from Hard Disk

(A) Interrupt from Hard Disk

(B) Interrupt from Mouse

(C) Interrupt from Keyboard

(D) Interrupt from CPU temp sensor

Answer: - (D)

25. Consider a relational table with a single record for each registered student with the following attributes.

1. Registration_Number: Unique registration number for each registered student
2. UID: Unique Identity number, unique at the national level for each citizen
3. BankAccount_Number: Unique account number at the bank. A student can have multiple accounts or joint accounts. This attributes stores the primary account number
4. Name: Name of the Student
5. Hostel_Room: Room number of the hostel

Which of the following options is INCORECT?

(A) BankAccount_Number is a candidate key

(B) Registration_Number can be a primary key

(C) UID is a candidate key if all students are from the same country

(D) If S is a superkey such that $S \cap UID$ is NULL then $S \cup UID$ is also a superkey

Answer: - (A)

Exp: - In case two students hold joint account then BankAccount_Num will not uniquely determine other attributes.

Q.No.26 - 51 Carry Two Marks Each

26. Which of the given options provides the increasing order of asymptotic complexity of functions f1, f2, f3 and f4?

$f_1(n) = 2^n; f_2(n) = n^{3/2}; f_3(n) = n\log_2 n; \quad f_4(n) = n^{\log_2 n}$

(A) f3, f2, f4, f1 (B) f3, f2, f1, f4

(C) f2, f3, f1, f4 (D) f2, f3, f4, f1

Answer: - (A)

Let n = 1024

$f_1(n) = 2^{1024}$

$f_2(n) = 2^{15}$

$f_3(n) = 10 \times 2^{10}$

$f_4(n) = 1024^{10} = 2^{100}$

$\therefore f_3\ f_2\ f_4\ f_1$ is the rquired increasing order

27. Four matrices M_1, M_2, M_3 and M_4 are dimensions p × q, q × r, r × s and s × t respectively can be multiplied in several ways with different number of total scalar multiplications. For example, when multiplied as $((M_1 \times M_2) \times (M_3 \times M_4))$ the total number of scalar multiplications is pqr + rst + prt. When multiplied as $(((M_1 \times M_2) \times M_3) \times M_4)$, the total number of scalar multiplications is pqr + prs + pst.

If p=10, q=100, r=20, s=5 and t=80, then the minimum number of scalar multiplications needed is

(A) 248000 (B) 44000 (C) 19000 (D) 25000

Answer: - (C)

Exp: - Multiply as $(M_1 \times (M_2 \times M_3)) \times M_4$

The total number of scalar multiplication is

= qrs pqs pst

= 10000 5000 4000 19000

28. Consider a relational table r with sufficient number of records, having attributes

$A_1, A_2, \ldots, A_n$ and let $1 \le p \le n$. Two queries Q1 and Q2 are given below.

Q1 : $\pi_{A1 \ldots An}(\sigma A_{p=c}(r))$ where c is a const

Q2 : $\pi_{A1 \ldots An}(\sigma_{c_1 \le A_p \le c_2}(r))$ where c_1 and c_2 are constants

The database can be configured to do ordered indexing on A_p or hashing on A_p.

Which of the following statements is TRUE?

(A) Ordered indexing will always outperform hashing for both queries

(B) Hashing will always outperform ordered indexing for both queries

(C) Hashing will outperform ordered indexing on Q1, but not on Q2

Answer: - (C)

29. Consider the matrix as given below.

$$\begin{bmatrix} 1 & 2 & 3 \\ 0 & 4 & 7 \\ 0 & 0 & 3 \end{bmatrix}$$

Which one of the following provides the CORRECT values of eigenvalues of the matrix?

(A) 1,4,3 (B) 3,7,3 (C) 7,3,2 (D) 1,2,3

Answer: - (A)

Exp: - Given matrix is upper triangular matrix and its diagonal elements are its eigen values = 1, 4, 3

30. Consider an instruction pipeline with four stages (S1, S2, S3 and S4) each with combinational circuit only. The pipeline registers are required between each stage and at the end of the last stage. Delays for the stages and for the pipeline registers are as given in the figure.

What is the approximate speed up of the pipeline in steady state under ideal conditions when compared to the corresponding non-pipeline implementation?

(A) 4.0 (B) 2.5 (C) 1.1 (D) 3.0

Answer: - (B)

Exp:- $\dfrac{(5+6+11+8)}{(11+1)} = \dfrac{30}{12} = 2.5$

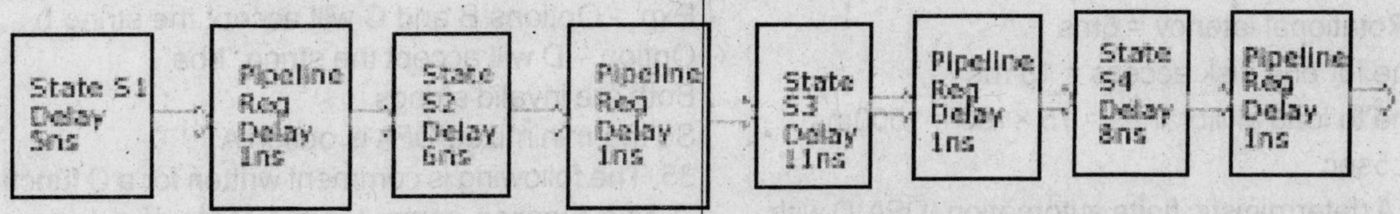

31. Definition of a language L with alphabet {a} is given as following

$L = \{a^{nk} \mid k > 0$, and n is a positive integer constant$\}$

What is the minimum number of states needed in a DFA to recognize L?

(A) k+1 (B) n+1 (C) 2^{n+1} (D) 2^{k+1}

Answer: - (B)

Exp: - Let n = 3 and k = 1

(n + 1) states

32. An 8KB direct mapped write-back cache is organized as multiple blocks, each of size 32-bytes. The processor generates 32-bit addresses. The cache controller maintains the tag information for each cache block comprising of the following.

1 Valid bit

1 Modified bit

As many bits as the minimum needed to identify the memory block mapped in the cache.

What is the total size of memory needed at the cache controller to store metadata (tags) for the cache?

(A) 4864 bits (B) 6144 bits (C) 6656 bits (D) 5376 bits

Answer: - (D)

Exp:-

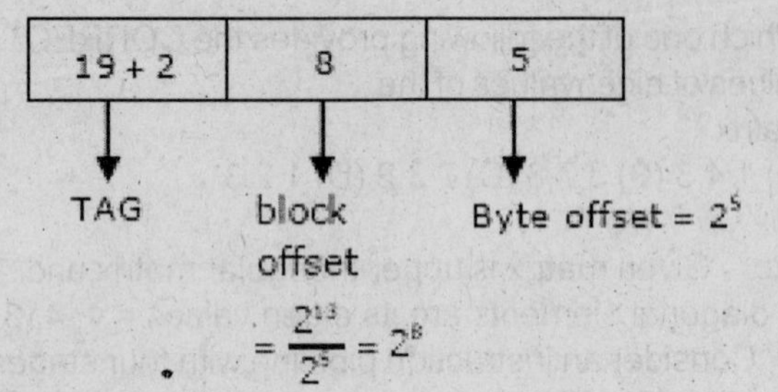

Required answer = 256 × (19 + 2) = 5376 bits

33. An application loads 100 libraries at startup. Loading each library requires exactly one disk access. The seek time of the disk to a random location is given as 10ms. Rotational speed of disk is 6000rpm. If all 100 libraries are loaded from random locations on the disk, how long does it take to load all libraries? (The time to transfer data from the disk block once the head has been positioned at the start of the block may be neglected)

(A) 0.50s (B) 1.50s (C) 1.25s (D) 1.00s

Answer: - (B)

Exp: - 6000 rotations ________ 60 sec

1 rotation ________ 10 ms

$\therefore$ Rotational latency = 5ms

Time for one disk access = 15 ms

Time to load all libraries = 15 ×100 = 1500ms

= 1.5sec

34. A deterministic finite automation (DFA)D with alphabet $\Sigma = \{a,b\}$ is given below

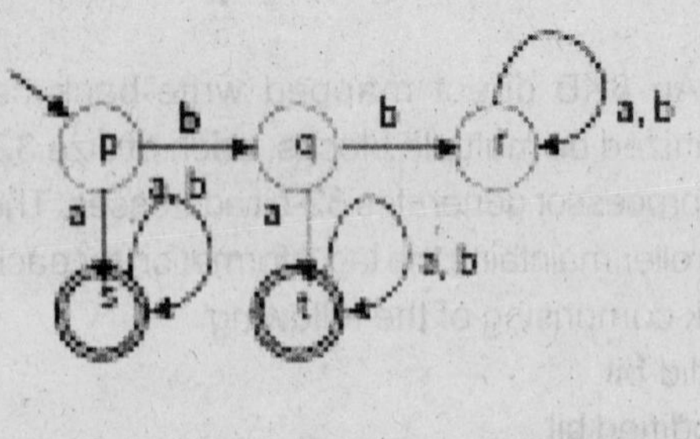

Which of the following finite state machines is a valid minimal DFA which accepts the same language as D?

(A)

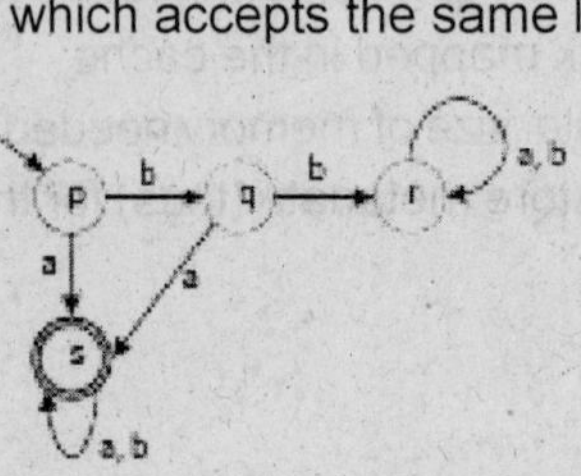

(B)

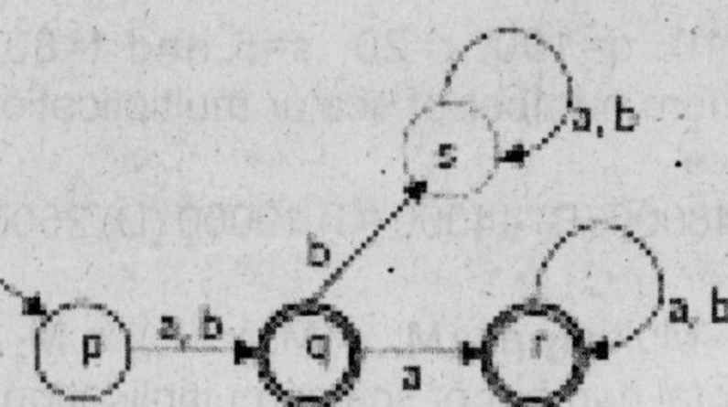

(C)

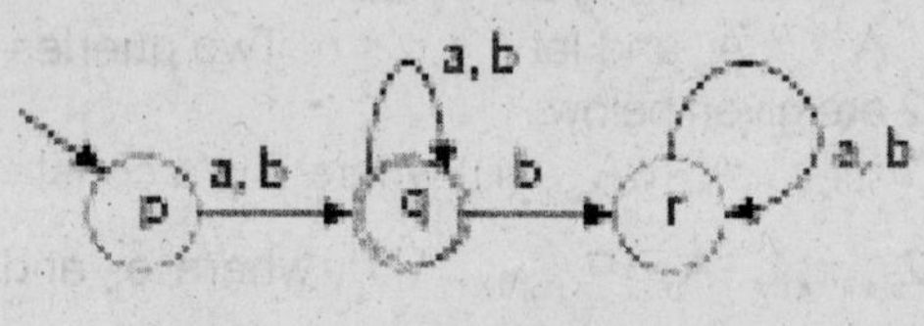

(D)

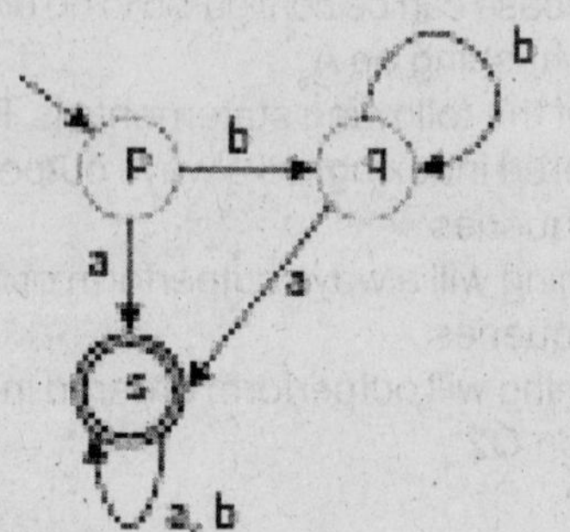

Answer: - (A)

Exp: - Options B and C will accept the string b

Option – D will accept the string “bba”

Both are invalid strings.

So the minimized DFA is option A

35. The following is comment written for a C function

/* This function computes the roots of a quadratic equation

a.x^2+b.x+c=0. The function stores two real roots in *root1 and *root2 and returns the status of validity of roots. It handles four different kinds of cases.

(i) When coefficient a is zero irrespective of discriminant

(ii) When discriminant is positive

(iii) When discrimanant is zero

(iv) When discrimanant is negative

Only in cases (ii) and (iii) , the stored roots are valid. Otherwise 0 is stored in the roots. the function returns 0 when the roots are valid and -1 otherwise.

The functin also ensures root1>=root2.

int get_QuadRoots (flloat a, float b, float c, float *root1, float *root2);

* /

A software test engineer is assigned the job of doing black box testing. He comes up with the following test cases, many of which are redundant.

Which one of the following options provide the set of non-redundant tests using equivalence class partitioning approach from input perspective for black box testing?

(A) T1,T2,T3,T6 (B) T1,T3,T4,T5 (C) T2,T4,T5,T6

(D) T2,T3,T4,T5

Answer: - (C)

Exp: - T_1 and T_2 checking same condition a = 0

Test Case	Input set			Expected Output set		
	a	b	C	Root1	Root2	Return Value
T1	0.0	0.0	7.0	0.0	0.0	-1
T2	0.0	1.0	3.0	0.0	0.0	-1
T3	1.0	2.0	1.0	-1.0	-1.0	0
T4	4.0	-12.0	9.0	1.5	1.5	0
T5	1.0	-2.0	-3.0	3.0	-1.0	0
T6	1.0	1.0	4.0	0.0	0.0	-1

hence, any one of T_1 and T_2 is redundant.

T_3, T_4: in both case discriminant (D) = b^2 — - 4ac = 0.

Hence any one of it is redundant.

T_5: D>0

T_6: D<0

36. Database table by name Loan_Records is given below.

What is the output of the following SQL query?

SELECT count(*)

Borrower	Bank_Manager	Loan_Amount
Ramesh	Sunderajan	10000.00
Suresh	Ramgopal	5000.00
Mahesh	Sunderajan	7000.00

FROM(

(SELECT Borrower. Bank_Manager FROM Loan_Records) AS S

NATURAL JOIN

(SELECT Bank_Manager, Loan_Amount FROM Loan_Records) AS T

);

(A) 3 (B) 9 (C) 5 (D) 6

Answer: - (C)

Exp: S

S

Borrower	Bank _ Manager
Ramesh	Sunderajan
Suresh	Ramgqpal
Mahesh	Sunderjan

T

Bank _ Manager	Loan _ Amount
Sunderajan	10000.00
Ramgopal	5000.00
Sunderjan	7000.00

After executing the given query, the output would be

Borrower	Bank_Manager	Load_Amount
Ramesh	Sunderajan	10000.00
Ramesh	Sunderajan	7000.00
Suresh	Ramgopal	5000.00
Mahesh	Sunderajan	10000.00
Mahesh	Sunderajan	7000.00

37. Consider two binary operators '↑' and '↓' with the precedence of operator ↓ being lower than that of the operator ↑. Operator - is right associative while operator ↓, is left associative. Which one of the following represents the parse tree for expression $(7\downarrow 3\uparrow 4\uparrow 3\downarrow 2)$?

(A)

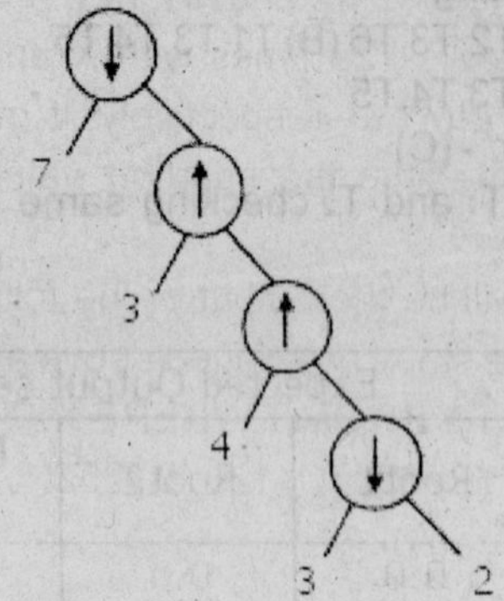

(B)

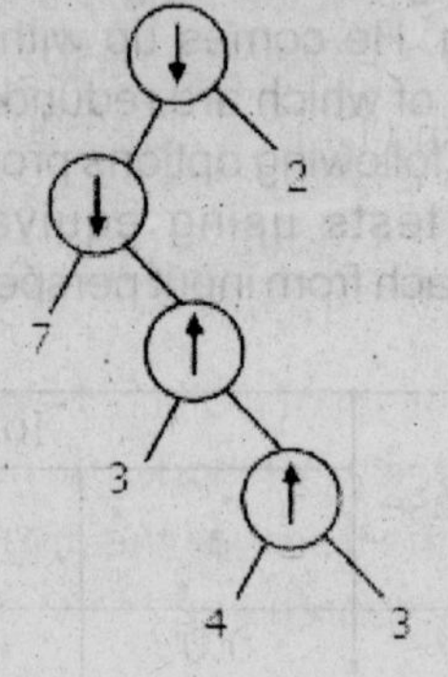

(C)

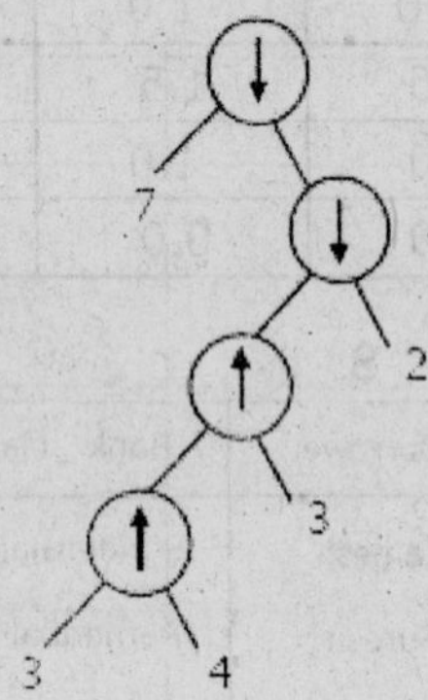

(D)

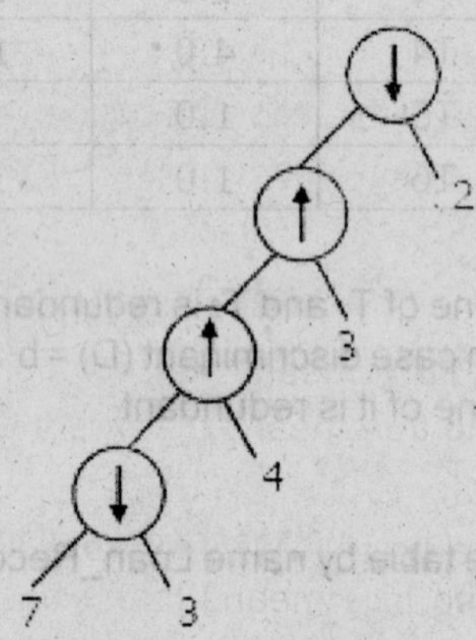

Answer: - (B)

Exp: - $7 \downarrow 3 \uparrow 4 \uparrow 3 \downarrow 2$

$\Rightarrow 7 \downarrow 3 \uparrow (4 \uparrow 3) \downarrow 2$ as - is right associative

$\Rightarrow 7 \downarrow (3 \uparrow (4 \uparrow 3)) \downarrow 2$

$\Rightarrow (7 \downarrow (3 \uparrow (4 \uparrow 3))) \downarrow 2$ as $\downarrow$ is left associative

38. Consider the languages L1, L2 and L3 as given below

L1 = $\{0^p 1^q \mid p, q \in N\}$

L2 = $\{0^p 1^q \mid p, q \in N \text{ and } p = q\}$ and

L3 = $\{0^p 1^q 0^r \mid p, q, r \in N \text{ and } p = q = r\}$

Which of the following statements is NOT TRUE?

(A) Push Down Automata (PDA) can be used to recognize L1 and L2

(B) L1 is a regular language

(C) All the three languages are context free

(D) Turing machines can be used to recognize all the languages

Answer: - (C)

Exp: - L1: regular language

L2: context free language

L3: context sensitive language

39. On a non-pipelined sequential processor, a program segment, which is a part of the interrupt service routine, is given to transfer 500 bytes from an I/O device to memory.

Initialize the address register

Initialize the count to 500

LOOP: Load a byte from device

Store in memory at address given by address register

Increment the address register

Decrement the count

If count ! = 0 go to LOOP

Assume that each statement in this program is equivalent to a machine instruction which takes one clock cycle to execute if it is a non-load/store instruction. The load-store instructions take two clock cycles to execute.

The designer of the system also has an alternate approach of using the DMA controller to implement the same transfer. The DMA controller requires 20 clock cycles for initialization and other overheads. Each DMA transfer cycle takes two clock cycles to transfer one byte of data from the device to the memory.

What is the approximate speedup when the DMA controller based design is used in place of the interrupt driven program based input-output?

(A) 3.4 (B) 4.4 (C) 5.1 (D) 6.7

Answer: - (A)

Exp: - No. of clock cycles required by using load-store approach = 2 + 500 × 7 = 3502

and that of by using DMA = 20 + 500 × 2 = 1020

Required speed up $= \frac{3502}{1020} = 3.4$

40. We are given a set of n distinct elements and an unlabeled binary tree with n nodes. In how many ways can we populate the tree with the given set so that it becomes a binary search tree?

(A) 0 (B) 1 (C) n! (D) $\frac{1}{n+1} \cdot {}^{2n}C_n$

Answer: - (D)

41. Which one of the following options is CORRECT given three positive integers x, y and z, and a predicate

$P(x) = \neg(x=1) \wedge \forall y(\exists z(x = y*z) \Rightarrow (y=x) \vee (y=1))$

(A) P(x) being true means that X is a prime number

(B) P(x) being true means that x is a number other than 1

(C) P(x) is always true irrespective of the value of x

(D) P(x) being true means that x has exactly two factors other than 1 and x

Answer: - (A)

42. Given i = $\sqrt{-1}$, what will be the evaluation of the definite integral $\int_0^{\pi/2} \frac{\cos x + i\sin x}{\cos x - i\sin x} dx$?

(A) 0 (B) 2 (C) –i (D) i

Answer: - (D)

Exp:- $\int_0^{\pi/2} \frac{e^{ix}}{e^{-ix}} dx = \int_0^{\pi/2} e^{2ix} dx$

$= \left(\frac{e^{2ix}}{2i}\right)_0^{\pi/2} = \frac{1}{2i}\left[e^{i\pi} - 1\right] = \frac{1}{2i}[\cos \pi + i\sin \pi - 1]$

$= \frac{1}{2i}[-1+0-1] = \frac{-2}{2i} = \frac{-1}{i} \times \frac{i}{i} = \frac{-i}{-1} = i$

43. Consider a database table T containing two columns X and Y each of type integer.

After the creation of the table, one record (X= 1, Y=l) is inserted in the table.

Let MX and MY denote the respective maximum values of X and Y among all records in the table at any point in time. Using MX and MY, new records are inserted in the table 128 times with X and Y values being MX+1, 2*MY+1 respectively. It may be noted that each time after the insertion, values of MX and MY change.

What will be the output of the following SQL query after the steps mentioned above are carried out?

SELECT Y FROM T WHERE X=7;

(A) 127 (B) 255 (C) 129 (D) 257

Answer: - (A)

Exp:

X	Y
1	1
2	3
3	7
4	15
5	31
6	63
7	127

44. Consider a finite sequence of random values $x = [x_1, x_2, ... x_n]$. Let μx be the mean and σ_x be the standard deviation of X. Let another finite sequence Y of equal length be derived from this as $y_i = a * x_i + b$, where a and b are positive constants. Let μ_y be the mean and σ_y be the standard deviation of this sequence. Which one of the following statements is INCORRECT?

(A) Index position of mode of X in X is the same as the index position of mode of Y in Y.

(B) Index position of median of X in X is the same as the index position of median of Y in Y.

(C) $\mu_y = a\mu_x + b$

(D) $\sigma_y = a\sigma_x + b$

Answer: - (D)

45. A deck of 5 cards (each carrying a distinct number from 1 to 5) is shuffled thoroughly. Two cards are then removed one at a time from the deck. What is the probability that the two cards are selected with the number on the first card being one higher than the number on the second card?

(A) 1/5 (B) 4/25 (C) 1/4 (D) 2/5

Answer :- (A)

Exp :- (2,1), (3,2), (4,3), (5,4)

Required probability $= \frac{4}{5 \times 4} = \frac{4}{20} = \frac{1}{5}$

46. Consider the following table of arrival time and burst time for three processes P0, P1 and P2.

Process	Arrival time	Burst Time
P0	0 ms	9 ms
P1	1 ms	4ms
P2	2 ms	9ms

The pre-emptive shortest job first scheduling algorithm is used. Scheduling is carried out only at arrival or completion of processes. What is the average waiting time for the three processes?

(A) 5.0 ms (B) 4.33 ms (C) 6.33 ms (D) 7.33 ms

Answer: - (A)

Average waiting time $= \frac{4+11}{3} = 5$ ms

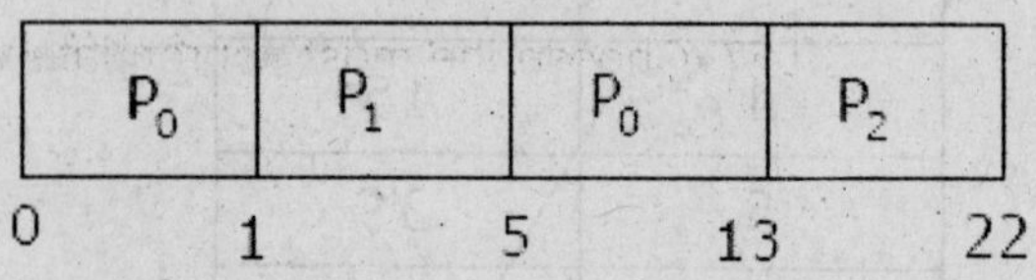

47. Consider evaluating the following expression tree on a machine with load-store architecture in which memory can be accessed only through load and store instructions. The variables a, b, c, d and e are initially stored in memory. The binary operators used in this expression tree can be evaluated by the machine only when the operands are in registers. The instructions produce result only in a register. If no intermediate results can be stored in memory, what is the minimum number of registers needed to evaluate this expression?

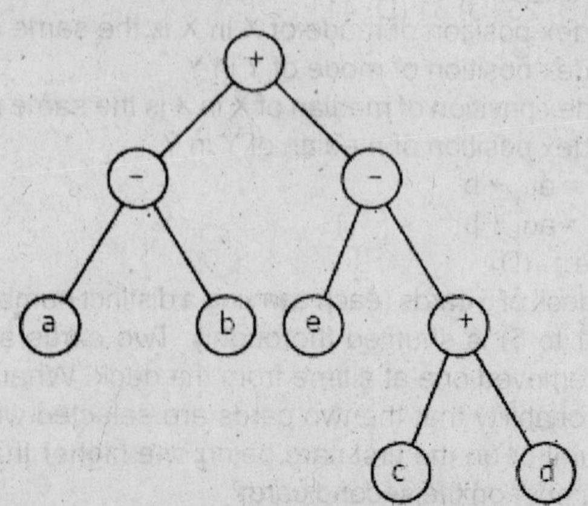

Exp:-

(A) 2 (B) 9 (C) 5 (D) 3

Answer: - (D)

Exp: - Load R_1, a ; $R_1 \leftarrow M[a]$

Load R_2, b ; $R_2 \leftarrow M[b]$

Sub R_1, R_2 ; $R_1 \leftarrow R_1 - R_2$

Load R_2, c ; $R_2 \leftarrow M[c]$

Load R_3, d ; $R_3 \leftarrow M[d]$

Add R_2, R_3 ; $R_2 \leftarrow R_2 + R_3$

Load R_3, e ; $R_3 \leftarrow M[e]$

Sub R_3, R_2 ; $R_3 \leftarrow R_3 - R_2$

Add R_1, R_3 ; $R_1 \leftarrow R_1 + R_3$

Total 3 Registers are required minimum

Common Data Questions: 48 & 49

Consider the following recursive C function that takes two arguments

```
unsigned int foo (unsigned int n; unsigned int r) {
if (n > 0) return (n%r) + foo (n / r, r));
else return 0;
}
```

48. What is the return value of the function foo when it is called as foo (513, 2)?

(A) 9 (B) 8 (C) 5 (D) 2

Answer: - (D)

Exp: -

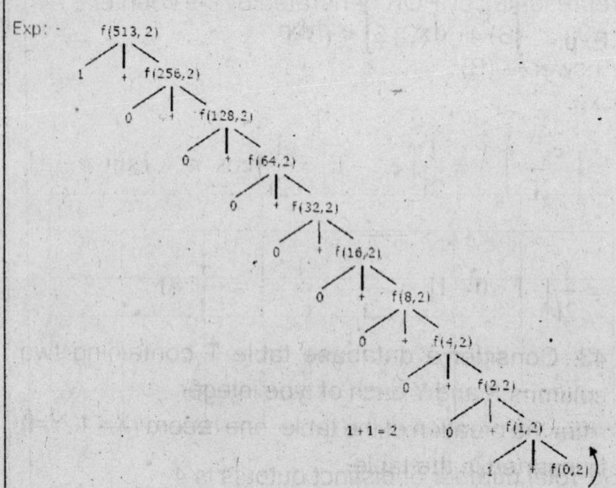

49. What is the return value of the function foo when it is called as foo (345, 10) ?
(A) 345 (B) 12 (C) 5 (D) 3

Exp: -

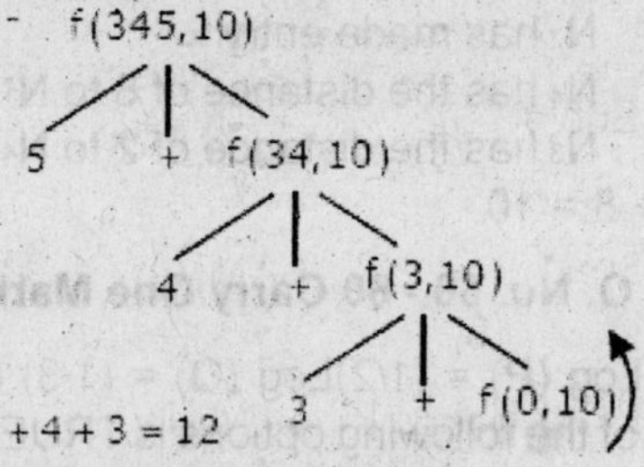

Common Data Questions: 50 & 51

Consider the following circuit involving three D-type flip-flops used in a certain type of counter configuration.

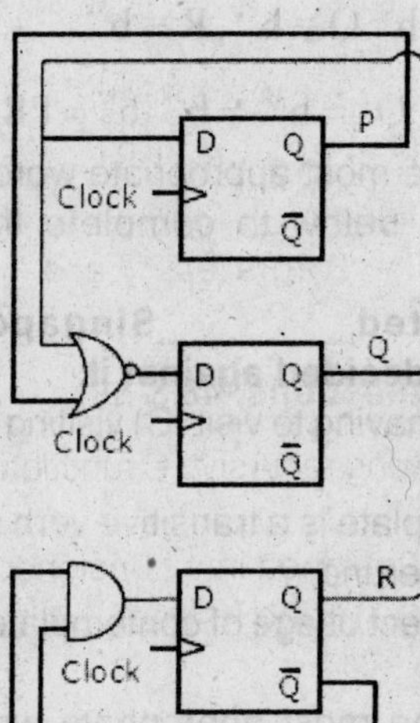

50. If all the flip-flops were reset to 0 at power on, what is the total number of distinct outputs (states) represented by PQR generated by the counter?
(A) 3 (B) 4 (C) 5 (D) 6
Answer : - (B)
Exp:

CLOCK	Inputs			Outputs		
	$D_1 = R$	$D_2 = \overline{(P+R)}$	$D_3 = Q\overline{R}$	P	Q	R
1	0	1	0	0	1	0
2	0	1	1	0	1	1
3	1	0	0	1	0	0
4	0	0	0	0	0	0

So Total number of distinct outputs is 4

51. If at some instance prior to the occurrence of the clock edge, P. Q and R have a value 0, 1 and 0 respectively, what shall be the value of PQR after the clock edge?
(A) 000 (B) 001 (C) 010 (D) 011
Answer: - (D)
Exp: - From the Table Shown in the explanation of question 50, if first state is 010 next State is 011

Linked Answer Questions: Q.52 to Q.55 Carry Two Marks Each

Statement for Linked Answer Questions: 52 & 53

An undirected graph G(V,E) contains n (n > 2) nodes named $v_1, v_2, \ldots v_n$. Two nodes v_i, v_j are connected if and only if 0 < |i - j| £ 2. Each edge (v_i, v_j) is assigned a weight i + j. A sample graph with n = 4 is shown below

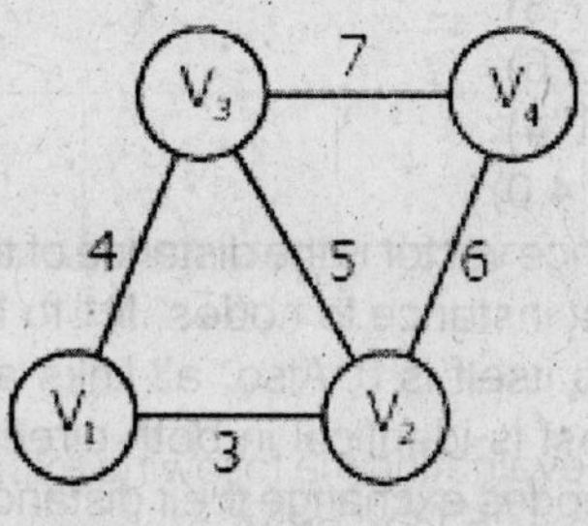

52. What will be the cost of the minimum spanning tree (MST) of such a graph with n nodes?
(A) $\frac{1}{12}\left(11n^2 - 5n\right)$ (B) $n^2 - n + 1$
(C) 6n – 11 (D) 2n + 1
Answer: - (B)

53. The length of the path from v_5 to v_6 in the MST of previous question with n = 10 is
(A) 11 (B) 25 (C) 31 (D) 41
Answer: - (C)
Exp : -

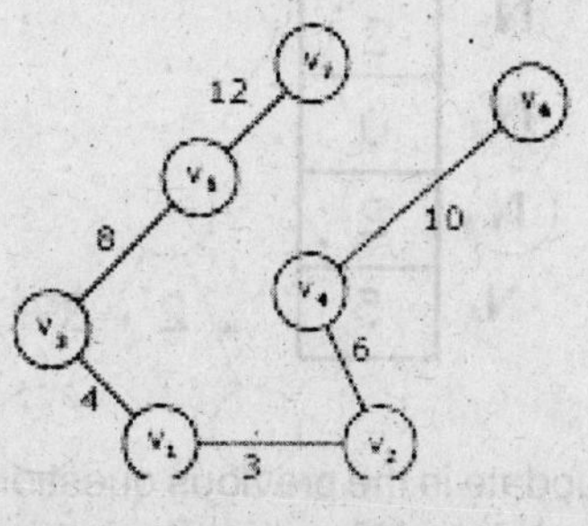

12 + 8 + 4 + 3 + 6 + 10 = 31

Statement for Linked Answer Questions: 54 & 55

Consider a network with five nodes, N1 to N5, as shown below

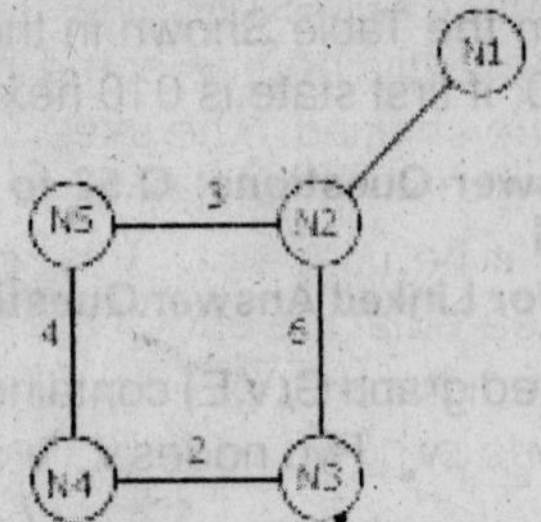

The net work uses a Distance Vector Routing protocol. Once the routes have stabilized, the distance vectors at different nodes are as following

N1: (0,1,7,8,4)
N2 :(1,0,6,7,3)
N3: (7,6,0,2,6)
N4 :(8,7,2,0,4)
N5 :(4,3,6, 4,0)

Each distance vector is the distance of the best known path at that instance to nodes, N1 to N5, where the distance to itself is 0. Also, all links are symmetric and the cost is identical in both directions. In each round, all nodes exchange their distance vectors with their respective neighbors. Then all nodes update their distance vectors. In between two rounds, any change in cost of a link will cause the two incident nodes to change only that entry in their distance vectors

54. The cost of link N2-N3 reduces to 2 in (both directions). After the next round of updates, what will be the new distance vector at node, N3?
(A) (3. 2, 0, 2, 5) (B) (3, 2, 0, 2, 6)
(C) (7, 2, 0, 2, 5) (D) (7, 2, 0, 2, 6)
Answer: - (A)
Exp: -

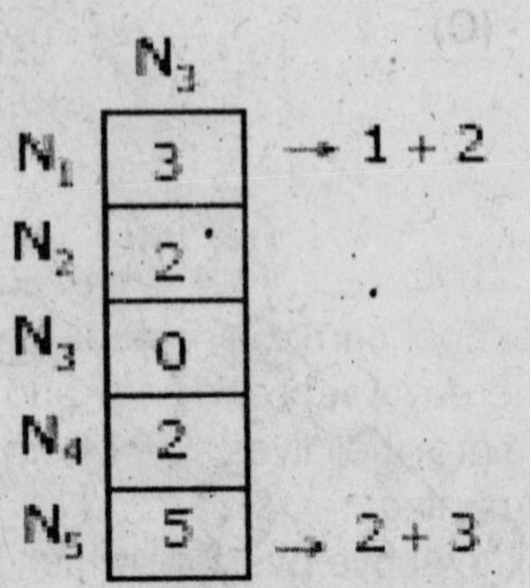

55. After the update in the previous question, the link N1-N2 goes down. N2 will reflect this change immediately in its distance vector as cost, ∞ . After the NEXT ROUND of update, what will be the cost to N1 in the distance vector of N3?
(A) 3 (B) 9 (C) 10 (D) ∞
Answer: - (C)
Exp: - N_3 has neighbors N_2 and N_4
N_2 has made entry ∞
N_4 has the distance of 8 to N_1
N_3 has the distance of 2 to N_4
So 2 + 8 = 10

Q. No. 56 - 60 Carry One Mark Each

56. If Log (P) = (1/2)Log (Q) = (1/3) Log (R), then which of the following options is TRUE?
(A) $P^2 = Q^3R^2$ (B) $Q^2 = PR$ (C) $Q^2 = R^3 P$
(D) $R = P^2 Q^2$
Answer: - (B)

Exp : - $\log P = \frac{1}{2}\log Q = \frac{1}{3}\log(R) = k$

$$\therefore P = b^k, Q = b^{2k}, R = b^{3k}$$

$$\text{Now, } Q^2 = b^{4k} = b^{3k}\, b^k = PR$$

57. Choose the most appropriate word(s) from the options given below to complete the following sentence.
I contemplated_________Singapore for my vacation but decided against it.
(A) To visit (B) having to visit (C) visiting (D) for a visit
Answer: - (C)
Exp: - Contemplate is a transitive verb and hence is followed by a gerund.
Hence the correct usage of contemplate is verb+ ing form.

58. Choose the most appropriate word from the options given below to complete the following sentence.
If you are trying to make a strong impression on your audience, you cannot do so by being understated, tentative or____________.
(A) Hyperbolic (B) Restrained
(C) Argumentative (D) Indifferent
Answer: - (B)
Exp: - The tone of the sentence clearly indicates a word that is similar to understated is needed for the blank. Alternatively, the word should be antonym of strong (fail to make strong impression). Therefore, the best choice is restrained which means controlled/reserved/timid.

59. Choose the word from the options given below that is most nearly opposite in meaning to the given word: **Amalgamate**
(A) Merge (B) Split (C) Collect (D) Separate
Answer: - (B)

Exp: - Amalgamate means combine or unite to form one organization or structure. So the best option here is split. Separate on the other hand, although a close synonym, it is too general to be the best antonym in the given question while Merge is the synonym; Collect is not related.

60. Which of the following options is the closest in the meaning to the word below:

Inexplicable

(A) Incomprehensible (B) Indelible
(C) Inextricable (D) Infallible

Answer: - (A)

Exp: - Inexplicable means not explicable; that cannot be explained, understood, or accounted for. So the best synonym here is incomprehensible.

Q. No. 61 - 65 Carry Two Marks Each

61. P, Q, R and S are four types of dangerous microbes recently found in a human habitat. The area of each circle with its diameter printed in brackets represents the growth of a single microbe surviving human immunity system within 24 hours of entering the body. The danger to human beings varies proportionately with the toxicity, potency and growth attributed to a microbe shown in the figure below

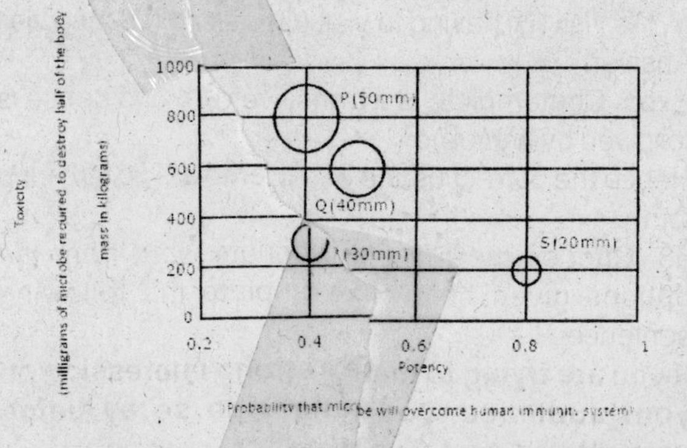

A pharmaceutical company is contemplating the development of a vaccine against the most dangerous microbe. Which microbe should the company target in its first attempt?

(A) P (B) Q (C) R (D) S

Answer: - (D)

Exp: - By observation of the table, we can say S

	P	Q	R	S
Requirement	800	600	300	200
Potency	0.4	0.5	0.4	0.8

62. The variable cost (V) of manufacturing a product varies according to the equation V = 4q, where q is the quantity produced. The fixed cost (F) of production of same product reduces with q according to the equation F = 100/q. How many units should be produced to minimize the total cost (V + F)?

(A) 5 (B) 4 (C) 7 (D) 6

Answer: (A)

Exp: - Checking with all options in formula: (4q + 100/q) i.e. (V + F). Option A gives the minimum cost.

63. A transporter receives the same number of orders each day. Currently, he has some pending orders (backlog) to be shipped. If he uses 7 trucks, then at the end of the 4th day he can clear all the orders. Alternatively, if he uses only 3 trucks, then all the orders are cleared at the end of the 10th day. What is the minimum number of trucks required so that there will be no pending order at the end of the 5th day?

(A) 4 (B) 5 (C) 6 (D) 7

Answer: - (C)

Exp: - Let each truck carry 100 units.

2800 = 4n + e n = normal

3000 = 10n + e e = excess/pending

$$\therefore n = \frac{100}{3}, \quad e = \frac{8000}{3}$$

$$5\text{days} \Rightarrow 500x = \frac{5.100}{3} + \frac{8000}{3}$$

$$\Rightarrow 500x = \frac{8500}{3} \Rightarrow x > 5$$

Minimum possible = 6

64. A container originally contains 10 litres of pure spirit. From this container 1 litre of spirit is replaced with 1 litre of water. Subsequently, 1 litre of the mixture is again replaced with 1 litre of water and this process is repeated one more time. How much spirit is now left in the container?

(A) 7.58 litres (B) 7.84 litres (C) 7 litres (D) 7.29 litres

Answer: - (D)

Exp: - $10\left(\frac{10-1}{10}\right)^3 = 10\left(\frac{9}{10}\right)^3 = \frac{729}{1000}$

$$\therefore \frac{729}{1000} \times 10 = 7.29 \text{ litres}$$

65. Few school curricula include a unit on how to deal with bereavement and grief, and yet all students at some point in their lives suffer from losses through death and parting.

Based on the above passage which topic would not be included in a unit on bereavement?

(A) how to write a letter of condolence

(B) what emotional stages are passed through in the healing process

(C) what the leading causes of death are
(D) how to give support to a grieving friend
Answer: - (C)
Exp: - The given passage clearly deals with how to deal with bereavement and grief and so after the tragedy occurs and not about precautions. Therefore, irrespective of the causes of death, a school student rarely gets into details of causes—which is beyond the scope of the context. Rest all are important in dealing with grief.